The Elements

Name	Atomic Number	Symbol	Atomic Mass[a]	Name	Atomic Number	Symbol	Atomic Mass[a]	Name	Atomic Number	Symbol	Atomic Mass[a]
Actinium	89	Ac	(227)[c]	Hassium	108	Hs	(265)[c]	Promethium	61	Pm	(145)[c]
Aluminum	13	Al	26.981 539[5]	Helium	2	He	4.002 602[2][d,e]	Protactinium	91	Pa	231.035 88[2][g]
Americium	95	Am	(243)[c]	Holmium	67	Ho	164.930 32[3]	Radium	88	Ra	(226)[c]
Antimony	51	Sb	121.757[3]	Hydrogen	1	H	1.007 94[7][d,e]	Radon	86	Rn	(222)[c]
Argon	18	Ar	39.948[1][d,e]	Indium	49	In	114.818[3]	Rhenium	75	Re	186.207[1]
Arsenic	33	As	74.921 59[2]	Iodine	53	I	126.904 47[3]	Rhodium	45	Rh	102.905 50[3]
Astatine	85	At	(210)[c]	Iridium	77	Ir	192.22[3]	Rubidium	37	Rb	85.4678[3][d]
Barium	56	Ba	137.327[7]	Iron	26	Fe	55.847[3]	Ruthenium	44	Ru	101.07[2][d]
Berkelium	97	Bk	(247)[c]	Krypton	36	Kr	83.80[1][d,f]	Rutherfordium	104	Rf	(261)[c]
Beryllium	4	Be	9.012 182[3]	Lanthanum	57	La	138.9055[2][d]	Samarium	62	Sm	150.36[3][d]
Bismuth	83	Bi	208.980 37[3]	Lawrencium	103	Lr	(262)[c]	Scandium	21	Sc	44.955 910[9]
Boron	5	B	10.811[5][d,e,f]	Lead	82	Pb	207.2[1][d,e]	Seaborgium	106	Sg	(263)[c]
Bromine	35	Br	79.904[1]	Lithium	3	Li	6.941[2][d,e,f]	Selenium	34	Se	78.96[3]
Cadmium	48	Cd	112.411[8][d]	Lutetium	71	Lu	174.967[1][d]	Silicon	14	Si	28.0855[3][e]
Calcium	20	Ca	40.078[4][d]	Magnesium	12	Mg	24.3050[6]	Silver	47	Ag	107.8682[2][d]
Californium	98	Cf	(251)[c]	Manganese	25	Mn	54.938 05[1]	Sodium	11	Na	22.989 768[6]
Carbon	6	C	12.011[1][d,e]	Meitnerium	109	Mt	(266)[c]	Strontium	38	Sr	87.62[1][d,e]
Cerium	58	Ce	140.115[4][d]	Mendelevium	101	Md	(258)[c]	Sulfur	16	S	32.066[6][d,e]
Cesium	55	Cs	132.905 43[5]	Mercury	80	Hg	200.59[3]	Tantalum	73	Ta	180.9479[1]
Chlorine	17	Cl	35.4527[9][f]	Molybdenum	42	Mo	95.94[1][d]	Technetium	43	Tc	(98)[c]
Chromium	24	Cr	51.9961[6]	Neodymium	60	Nd	144.24[3][d]	Tellurium	52	Te	127.60[3][d]
Cobalt	27	Co	58.933 20[1]	Neon	10	Ne	20.1797[6][d,f]	Terbium	65	Tb	158.925 34[3]
Copper	29	Cu	63.546[3][e]	Neptunium	93	Np	(237)[c]	Thallium	81	Tl	204.3833[2]
Curium	96	Cm	(247)[c]	Nickel	28	Ni	58.6934[2]	Thorium	90	Th	232.0381[1][d,g]
Dysprosium	66	Dy	162.50[3][d]	Nielsbohrium	107	Ns	(262)[c]	Thulium	69	Tm	168.934 21[3]
Einsteinium	99	Es	(252)[c]	Niobium	41	Nb	92.906 38[2]	Tin	50	Sn	118.710[7][d]
Erbium	68	Er	167.26[3][d]	Nitrogen	7	N	14.006 74[7][d,e]	Titanium	22	Ti	47.88[3]
Europium	63	Eu	151.965[9][d]	Nobelium	102	No	(259)[c]	Tungsten	74	W	183.84[1]
Fermium	100	Fm	(257)[c]	Osmium	76	Os	190.23[3][d]	Uranium	92	U	238.0289[1][d,f,g]
Fluorine	9	F	18.998 4032[9]	Oxygen	8	O	15.9994[3][d,e]	Vanadium	23	V	50.9415[1]
Francium	87	Fr	(223)[c]	Palladium	46	Pd	106.42[1][d]	Xenon	54	Xe	131.29[2][d,f]
Gadolinium	64	Gd	157.25[3][d]	Phosphorus	15	P	30.973 762[4]	Ytterbium	70	Yb	173.04[3][d]
Gallium	31	Ga	69.723[1]	Platinum	78	Pt	195.08[3]	Yttrium	39	Y	88.905 85[2]
Germanium	32	Ge	72.61[2]	Plutonium	94	Pu	(244)[c]	Zinc	30	Zn	65.39[2]
Gold	79	Au	196.966 54[3]	Polonium	84	Po	(209)[c]	Zirconium	40	Zr	91.224[2][d]
Hafnium	72	Hf	178.49[2]	Potassium	19	K	39.0983[1][d]				
Hahnium	105	Ha	(262)[c]	Praseodymium	59	Pr	140.907 65[3]				

[a] Atomic masses are 1991 IUPAC values based on carbon-12 = 12 u.

[b] Numbers in square brackets are the uncertainties in the last digits, for example, 26.981 539[5] means 26.981 539±0.000 005.

[c] Radioactive element that does not have a typical isotopic composition on earth. The mass given in parentheses is the mass of the most stable isotope.

[d] The difference between the isotopic composition of samples from different places may be greater than the uncertainty given in the table.

[e] The range in isotopic composition of different samples limits the precision of the atomic mass.

[f] Different isotopic compositions may be found in commercial material as a result of the separation of isotopes.

[g] Although the element has no stable nuclides (a stable nuclide has a half-life greater than 3×10^{10} years), some nuclides have long enough half-lives that the element has a typical isotopic composition on earth.

(Adapted from J. Phys. Chem. Ref. Data 1993, 22, 1571–1584, by permission of the publisher.)

SECOND EDITION

GENERAL CHEMISTRY

JEAN B. UMLAND

University of Houston—Downtown

JON M. BELLAMA

University of Maryland

WEST PUBLISHING COMPANY

Minneapolis/Saint Paul · New York · Los Angeles · San Francisco

PRODUCTION CREDITS

Copy Editor Betty Duncan   Design Diane Beasley   Layout and Illustrations TECHarts
Indexer William Ragsdale
Composition Monotype Composition Company, Inc.; TECHarts
Production, Prepress, Printing, and Binding by West Publishing Company.

This text contains descriptions, photographs, and other depictions of chemical reactions performed in a laboratory setting. These depictions are for reference only and should not be used as a basis for attempting similar chemical experiments. By this notice, the purchaser or reader of this book shall indemnify and hold harmless West Publishing Company and the text authors from any liability or damages incurred as the result of attempting chemical experiments described in this book.

The opinions expressed in the Guest Essays are those of the essayists and do not necessarily reflect the policy or opinions of West Publishing Company or the Authors of this textbook.

WEST'S COMMITMENT TO THE ENVIRONMENT

In 1906, West Publishing Company began recycling materials left over from the production of books. This began a tradition of efficient and responsible use of resources. Today, 100% of our legal bound volumes are printed on acid-free, recycled paper consisting of 50% new paper pulp and 50% paper that has undergone a de-inking process. We also use vegetable-based inks to print all of our books. West recycles nearly 27,700,000 pounds of scrap paper annually—the equivalent of 229,300 trees. Since the 1960s, West has devised ways to capture and recycle waste inks, solvents, oils, and vapors created in the printing process. We also recycle plastics of all kinds, wood, glass, corrugated cardboard, and batteries, and have eliminated the use of polystyrene book packaging. We at West are proud of the longevity and the scope of our commitment to the environment.

West pocket parts and advance sheets are printed on recyclable paper and can be collected and recycled with newspapers. Staples do not have to be removed. Bound volumes can be recycled after removing the cover.

PHOTO CREDITS

vi, top Gay Bumgarner/Tony Stone; **vi, bottom** Superstock; **vii, top** Bernard Asset/Photo Researchers; **vii, center** Stuart Westmorland/Tony Stone; **vii, bottom** Donald Johnson/Tony Stone; **viii, top** Tony Stone; **viii, bottom** Paul Conklin/Photo Edit; **ix, top** Mark Harwood/Tony Stone; **ix, center** Courtesy of Bufftech; **ix, bottom** Dr. Jeremy Burgess/Science Photo Library/Photo Researchers, Inc.; **x, top** John Neubauer/Photo Edit; **x, bottom** James Randklev/Tony Stone; **xi, top** Peter Aprahamian/Science Photo Library/Photo Researchers, Inc.; **xi, bottom** Gary Ladd/Photo Researchers, Inc.; **xii, top** Courtesy of Sue Monroe; **xii, bottom** Rhoda Sidney/Photo Edit; **xiii, top** Joel Gordon; **xiii, bottom** Dick Baker/ Third Coast Stock Source; **xiv, top** Tony Stone; **xiv, bottom** E. R. Degginger; **xv, top** George Holton/ Photo Researchers, Inc.; **xv, bottom** Jim Cambon/Tony Stone. *Photo credits continued following the Index.*

British Library Cataloguing-in-Publication Data. A catalogue record for this book is available from the British Library.

Printed with Printwise
Environmentally Advanced Water Washable Ink

Library of Congress Cataloging-in-Publication Data

Umland, Jean B.
  General chemistry / Jean B. Umland, Jon M. Bellama.—2nd ed.
    p. cm.
  Includes index,
    ISBN 0-314-06353-6 (hard : alk. paper)   ISBN 0-314-08924 (inst edition)
    1. Chemistry. I. Bellama, Jon M., 1938–   . II. Title.
QD33.U44 1996
540—dc20   95-41712
                                                                           CIP

To my father, Don Blanchard, who introduced me to publishing by hiring me to proof the galley of an engineering handbook with him when I was about 10 years old.

J.B.U.

To my parents, Edey J. and Jane I. Bellama, who first created an environment of encouragement of intellectual pursuits, and to my wife, Elaine E. Bellama, who with her love has created an atmosphere in which my love of knowledge has come to fruition.

J.M.B.

The first edition of this text benefited from the coherence and uniformity of presentation that is characteristic of a single-author text. To further enrich Jean Umland's achievement in the first edition of the book, we welcome to this edition a new author, Professor Jon M. Bellama of the University of Maryland–College Park. Dr. Bellama has 30 years of experience teaching general chemistry to majors and nonmajors. He brings substantial experience as a chemical educator that complements the strength of his co-author. Dr. Bellama's background in inorganic, organometallic, and environmental chemistry has contributed to the creation of this edition of the text, particularly with respect to the inclusion of inorganic and organic topics as an integral part of the various chapters, the greater emphasis on the chemistry of the elements in chapters usually covered in the first semester, and the addition of topics of current environmental interest.

JEAN B. UMLAND is a graduate of Swarthmore College and obtained her Ph.D. from the University of Wisconsin–Madison. Although her degree is in organic chemistry, she took all the course work required for a doctorate in physical chemistry, and her thesis research could equally well have been classified as physical chemistry. She began her teaching career at Mount Holyoke College, and also taught at Union College, Cranford, New Jersey, a two-year community college, before moving to the University of Houston–Downtown in 1975. She has worked in industry for American Cyanamid Company and Exxon Research and Engineering. Dr. Umland has just completed her second term on the First-Term General Chemistry Committee of the Examinations Institute of the American Chemical Society Division of Chemical Education. She served on the Scientific Advisory Committee for the new Welch Chemistry Hall at the Houston Museum of Natural Science, which opened in the Fall of 1993. The first edition of this book was used as a reference by the museum staff who prepared the exhibits, and was also used to train docents. In 1994, Dr. Umland received the Teaching Award at the University of Houston–Downtown and an Enron Teaching Excellence Award.

JON M. BELLAMA is a graduate of Allegheny College, Meadville, Pennsylvania, and obtained his Ph.D. in inorganic chemistry from the University of Pennsylvania. During his tenure at the University of Maryland–College Park he has served as director of the general chemistry program and as University of Maryland site director for the Institute of Chemical Education for the training both of pre-college chemistry and science teachers. Dr. Bellama serves as a faculty consultant to the Advanced Placement Program in chemistry of the Educational Testing Service, and is known for his demonstration lectures and for incorporating cooperative learning techniques into chemistry courses. He has received a variety of teaching awards including the 1991 Maryland Commission of Higher Education statewide award for innovation in teaching. Dr. Bellama was a Senator J.W. Fulbright scholar in Czechoslovakia in 1991–92 and has written or edited several research monographs on carbon-functional organosilicon chemistry, metals and organometals in the aquatic environment, and two dimensional nuclear magnetic resonance spectroscopy.

CONTENTS IN BRIEF

CONTENTS

The first chapter concerns chemicals and their reactions. The use of macroscopic, microscopic, and symbolic representations (pp. 17, 23), the periodic table (pp. 10–13, 25–27), and reasoning by analogy (pp. 17–18, 25–27) are introduced. The material on the scientific method is now in a separate section (Section 1.4).

In this text, students are not just told what to do; they are shown why they should do it (pp. 45–47). Careful attention is paid to details that are obvious to instructors but not always to beginning students, such as units in temperature conversion formulas (p. 55).

The new Related Topic on green chemistry and Guest Essay by a well-known green chemist show how chemistry is being applied to the solution of world problems.

Guest essays show chemistry as a human endeavor. From them, a student can see that many different kinds of people use chemistry in a variety of interesting and satisfying jobs.

Gases can be postponed until just before liquids and solids if you prefer. However, the discussion of the difference between reactions at constant pressure and reactions at constant volume in Chapter 6 will then have to be postponed until after gases have been covered.

Although enthalpy changes are the main focus of Chapter 6, the concepts of entropy and rate are introduced qualitatively as determinants of whether reactions and physical changes take place. These concepts are used repeatedly in the rest of the book.

6 CHEMICAL THERMODYNAMICS: THERMOCHEMISTRY 186

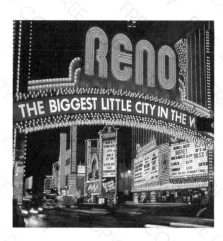

Sections that you may prefer to skip are placed at the ends of chapters whenever possible. Later chapters do *not* assume knowledge of the material in these sections. (Sections 8.8, 10.8, and 10.9 are examples.)

7 ATOMIC STRUCTURE 217

Section 8.7 has been rewritten to include the descriptive chemistry of the elements whose properties vary systematically down groups and across periods and a brief introduction to transition metals. (Transition metals are also included in the discussion of reactivity in Section 4.5.)

Chapter 9 now contains an introduction to polymers and the material on bond energies formerly in Chapter 17. Note the realistic example of the utility of bond energy calculations (pp. 326–27).

Because users of the first edition differed on when, or even if, they wished to teach the balancing of complicated redox equations, and because this is a skill with little conceptual content such as using exponential numbers, the balancing of complicated redox equations has been moved to Appendix F. Chapter 11 now includes additional sections on oxidation by oxygen and metallurgy.

Nothing in later chapters is based on Section 12.11.

CONTENTS

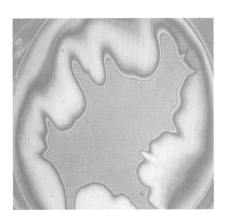

The unique organization of the end-of-chapter problems gives students practice classifying problems, answering multiple-choice questions, and combining material. "Applications" problems add interest and supply information concerning the uses of chemistry.

Section 14.6 can be postponed. On the other hand, some sections and problems from Chapters 21 and 24 can be used earlier if you wish to include more descriptive chemistry sooner.

About 15% of the problems in the text are new. New problems emphasize a microscopic view (for example, 3.30, 3.53, 15.105), material published since the first edition (2.131, 4.126), organic and biochemistry (3.106, 11.74, and many in Chapter 23), and how to organize material (1.104, 2.96). The 85% of the problems that were also in the first edition provide a variety of student-tested questions and answers.

16 MORE ABOUT EQUILIBRIA   601

As much of the material on complexes from Chapter 24 (Sections 24.2–24.10) as you wish to include can be done with Section 16.4. Section 16.8 can be omitted with no loss of continuity. See Appendix H for a discussion of the classical qual sheme.

17 CHEMICAL THERMODYNAMICS REVISITED: A CLOSER LOOK AT ENTHALPY, ENTROPY, AND EQUILIBRIUM   640

Formulas for calculating $\Delta H°_{rxn}$, $\Delta S°_{rxn}$, and $\Delta G°_{rxn}$ have been changed as a result of my own (JBU) experience. This type of problem was the *only* one that my students did not do better when using the first edition than when using other texts.

A discussion of the mechanisms of reactions with equilibria before the rate-determining step has been added to Chapter 18, and the section on the relationship between rate constants and equilibrium constants has been deleted. Chapter 18 can now easily be moved so that kinetics precedes equilibrium if you wish. Nothing in later chapters is based on Section 18.14.

Chapter 21 now contains all material on the nonsystematic chemistry of nonmetals. (The halogens and noble gases are discussed in Sections 4.5 and 8.7.) Repetition of material in different contexts and the amount of descriptive inorganic chemistry have been reduced to make room for more organic and biochemistry.

21 A CLOSER LOOK AT INORGANIC CHEMISTRY: NONMETALS AND SEMIMETALS AND THEIR COMPOUNDS 807

22 A CLOSER LOOK AT ORGANIC CHEMISTRY 857

23 POLYMERS: SYNTHETIC AND NATURAL   905

24 A CLOSER LOOK AT INORGANIC CHEMISTRY: TRANSITION METALS AND COMPLEXES   936

APPENDIXES

In Chapter 22, organic chemistry is shown as a part of general chemistry. Enantiomers are introduced here instead of with coordination compounds because of their importance in modern organic chemistry and because the organic examples of enantiomers are simpler.

Structure determination has been deleted from Chapter 22, as it is now covered in connection with proteins in the new chapter on polymers, Chapter 23. Both synthetic polymers, including silicones, and biopolymers are discussed in Chapter 23.

The last chapter now concerns transition metals and complexes. It concludes with a section on alloys and a Related Topic on superconductors. (Metallurgy is now in Section 11.9, and Group IA and IIA metals and aluminum in Section 8.7.)

Appendix I contains answers to all the in-chapter problems and to all of the end-of-chapter problems with blue numbers. Unlike answer appendixes in many other texts, verbal and pictorial answers (not just numerical answers) are included.

 **INDEX**   I-1

This book has been written to help you learn chemistry. Chemistry is one of the most useful subjects you will ever study. Of course, we're prejudiced, but we also think that it is one of the most interesting, and hope that you will come to think so too.

Chemistry is useful because it is central to the other sciences and to engineering as well as to life itself. In addition, your study of chemistry should help you to develop a logical approach to solving problems that is applicable to all kinds of problems, not just chemical problems.

As you begin, take a few minutes to get acquainted with this book. On the endpapers inside the front and back covers are tables of information that you will find useful, including a listing of "Where to find it" that gives the numbers of the pages where important information may be found. As usual, the table of contents is at the front and the index at the back. Each chapter begins with a chapter outline and a paragraph or two telling about the subject of the chapter. Then there are brief discussions of why the material in the chapter is important to you, and what you already know about the subject. You will probably be surprised to learn how much you do already know about chemistry from your everyday life, even if this is your first formal chemistry course. The main part of the chapter helps you to build on what you already know. The early sections usually include examples and discussions of chemical properties and behavior that will help you understand the theories presented later in the chapter.

Throughout the chapters, you will find *drawings, photographs, and graphs* that illustrate the material being discussed in the text. Study these and their captions carefully because they relate directly to the concepts you are learning. You may see some of the reactions again in lab.

Marginal notes are used throughout the book to help you make connections. The *green logos* signal learning hints, often a reminder of related material or a main idea. The *yellow logos* mark safety hints, frequently concerning something you may encounter in the lab. *Red logos* remind you to check your work. They are usually associated with the *"reasonableness checks"* that end many of the Sample Problems. Marginal notes without logos are used to present related information without interrupting the flow of the discussion.

Besides building on what you know, another main theme of this text is the need for a thoughtful, logical approach to problem-solving. The book provides you with lots of assistance in developing your problem-solving skills. In each chapter, many *Sample Problems* demonstrate the use of each important concept or skill, using step-by-step explanations; *Practice Problems* let you check your understanding as you go along.

The *Summary* at the end of each chapter resembles a glossary in narrative form and gives the terms a meaning in context. Use it for a general review of the chapter content.

The *Additional Practice Problems* are similar to the Practice Problems in the chapter. Try to work them without referring to the book. If you get stuck, look at the end of a problem for the section number (in parentheses) to which the problem is related; then restudy that section.

The *Stop and Test Yourself* questions are multiple-choice questions covering the basic skills and concepts in the chapter. Check your answers in Appendix I before going on to the higher-level problems that follow.

The *Putting Things Together* problems require you to combine skills from more than one section of the chapter and from different chapters.

The *Applications* problems deal with real-world applications of chemistry. These problems challenge you to demonstrate your mastery of chemical concepts and problem-solving skills.

Answers to the blue-numbered problems are available in Appendix I. Other appendices help you learn chemical nomenclature, review mathematics, and give you data you will need to solve problems. Solutions to the problems with blue numbers are in the *Student Solutions Manual*. Answers and solutions to the problems with black numbers may be available to you if your instructor wants you to have them.

Advice on Studying

1. Find other students who are in your class to study with.
2. Schedule a regular time to study each day. Don't wait until the week before an exam to begin studying. The athletes on your favorite team practice regularly; they do *not* stay up all night the night before the big game. In chemistry, explanations are usually based on concepts and skills learned previously; don't let yourself get behind.
3. Do the assignments before coming to class. Read slowly and carefully. Try to understand, not memorize. (However, some memorization is necessary; you cannot learn to think without knowing something to think about. Use flash cards for memorization.) Make sure you understand the worked-out examples, and do the in-chapter problems and check your answers to them in Appendix I as you go along. (The answers are in the appendix so that you don't see them too soon accidentally, and to help you resist the temptation to peek.) Write down any questions you have so you don't forget to ask them when you get to class.
4. After each class, review what you have just learned before beginning the next assignment.
5. Pay careful attention to vocabulary. An important part of a first course in any subject is learning the meaning of the terms used in the field (which, unfortunately, do not always mean the same thing they do in everyday life or in other fields). In this book, new terms are in **boldface type** and are defined (in *italics*) and explained as simply and accurately as possible. Most are also included in the summary at the end of the chapter. If you do not remember the meaning of a term when you meet it again later, use the index to find the definition. The index also includes references to other places where the term is used, which may be helpful in reviewing. If you come across unfamiliar words, look up their meanings in a dictionary.
6. If the amount of material to be learned seems overwhelming, use the first chapter of the *Study Guide* to learn how to organize it. Different people have different learning styles so a number of different methods of organization are described using examples from the first chapters of this book. Choose the ones that suit *you*.
7. If you are having difficulty, *use* the help that's available—your instructor's and teaching assistant's office hours, for example. Many colleges and universities also have computer programs, videodisks, and other materials to assist you, as well as services such as reading and math labs. In addition to other software that your instructor may make available to you, there is an outstanding and relatively inexpensive interactive CD-ROM available from the publisher. If the *Discover Chemistry CD-ROM* is not in your bookstore, you can order it through

the bookstore or directly from the publisher. There is more information on the back of your text.

In writing this book, we have done our best to make chemistry both understandable and interesting. Please let us know where we have succeeded and where we have failed so that we can do better in the third edition. Good luck!

Jean B. Umland
Department of Natural Sciences
University of Houston–Downtown
One Main Street
Houston, TX 77002

Jon M. Bellama
Department of Chemistry
  and Biochemistry
University of Maryland
College Park, MD 20742-2021

Audience

This book is intended for a full-year general chemistry course for pre-engineering students, pre-health profession students, and science majors, including chemistry majors. It assumes that students know how to solve simple algebraic equations; however, no previous study of chemistry or physics is assumed.

In our experience, most students are studying general chemistry for two reasons: (1) to learn the chemistry needed for their other courses and future professions; (2) to learn how to think scientifically. The second goal is often unconscious but is no less important than the first. These goals have influenced our overall approach, as well as many particular aspects of this book.

Approach

For students to learn to think scientifically, they must first have something to think about; therefore, in this text, descriptive chemistry and theory are integrated. Reactions, which we believe are the heart of chemistry, are introduced in Chapter 1, and are the subject of much of Chapters 4, 8, 11, and 21. Throughout the book marginal notes and footnotes, colored photographs, Related Topic boxes, and Applications problems supply a wealth of facts about chemicals and chemical reactions. Both common organic compounds and important inorganic compounds are used as examples, and applications include astronomy, biology, biochemistry, environmental chemistry, geology, and industrial chemistry, as well as everyday life. In the last four chapters, where the descriptive chemistry of both inorganic and organic compounds is the main focus, we have tried to pull together the material from earlier chapters as much as possible. Students only master material by meeting it several times in different contexts.

Readability

To make the material as easy as possible to understand, definitions and explanations are written in familiar words: "make easier" rather than "facilitate," for instance. Also, more detail is often provided than in other texts. For example, most texts simply list rules about the number of significant figures; this text explains where the rules come from. We find that students who understand the rules are more willing and better able to follow them. We think that one of students' major problems in understanding chemistry is their lack of information that authors and instructors assume is common knowledge. In the writing and in the presentation of problem-solving, we are careful not to make assumptions or to leave out or combine steps. Just as a stranger needs more detailed directions than someone who is familiar with a town, beginning students need more details to understand chemical concepts than people trained in the field.

Changes in the Second Edition

The second edition retains all the characteristic features of the first edition—such as the chapter openers, clear explanations, interesting problems, and Guest Essays—

that were so enthusiastically received by instructors and students alike. After reading the detailed comments provided by the reviewers of the first edition, many of whom had also taught from it, and considering suggestions sent by both student and faculty users, we went over the entire book correcting the seemingly inevitable mistakes, fine-tuning some parts and completely rewriting others that students or instructors told us were not clear, changing the position of some topics, and updating where necessary. We also had as a goal both shortening the book by one hundred pages and adding material on biopolymers requested by many instructors. (Unfortunately, while everybody agrees that general chemistry texts cover too much material and should be shorter, everybody does *not* agree on what should be left out.)

The major changes that were made are noted in the Table of Contents of the *Instructor's Annotated Edition*. Briefly, more descriptive chemistry is included in the chapters usually covered in the first semester. The section on chemical properties and the periodic table in Chapter 8 has been expanded, an introduction to polymers is included in Chapter 9, and metallurgy and oxygen as an oxidizing agent have been moved to Chapter 11. A chapter on polymers including proteins, carbohydrates, and nucleic acids has been added (Chapter 23). The chapters on descriptive inorganic chemistry at the end of the book, which many instructors seem not to have time for, have been significantly shortened.

There are many new problems throughout the book. A major portion of these problems are Applications problems based on information that has appeared in the literature since the problems in the first edition were written. Many are nonmathematical "concept" problems that involve either a microscopic-level drawing or a verbal answer.

The art and photos in the second edition received special attention. Drawings of molecular models have been enhanced for a more three-dimensional look. We significantly increased the number of figures that combine a macroscopic view with a microscopic view, usually by pairing a photo and a drawing. Users of the first edition praised the effectiveness of the figures and tables that compared different representations of chemical structure such as Figure 10.1 (p. 338).

In this edition, we have taken every opportunity to emphasize the importance of visualizing molecules by doing it ourselves and teaching students how to do it (see, for example, problems 15.103–15.105 on pages 594 and 595).

The first edition's focus on the positive aspects of chemistry has been increased by the addition of Related Topic boxes on risk and on green chemistry and a Guest Essay by a prominent green chemist.

Organization

Chapter Organization. Each chapter starts with a preview of its contents and ends with a summary. At the beginning of each chapter is a "sales-pitch" based on the answers to two questions that any student might reasonably pose. The first answer tells the student why the chapter is part of the study of general chemistry and why he or she should be interested in learning what is in the chapter. The answer to the second question reminds the student what he or she already knows about the subject from everyday experience and, in later chapters, from previous study. Next some relevant experimental observations are introduced and students are led to generalizations about the data. Theory to explain the generalizations is developed last. Our intention was to involve students in the discovery process and let them experience the "Aha!" feeling; we think that this is the best way for them to learn how science works and why scientists enjoy their work. Students who have used the text have responded to this approach and have commented favorably.

Text Organization. Most general chemistry texts cover measurement and significant figures in the first chapter. This book begins with an introduction to **chemical reactions,** so that the first lectures—where students form their impression of the subject—can include more interesting demonstrations. Early chapters provide plenty of material for laboratory work involving chemical reactions as well as for quantitative experiments.

The concepts of both **enthalpy** and **entropy** are introduced early (Chapter 6). Entropy is as important, if not more important, than enthalpy in determining whether a change will take place and should, we feel, have equal time. However, because experimental measurement of enthalpy changes is much simpler than experimental measurement of entropy changes, enthalpy is treated quantitatively in Chapter 6, whereas a quantitative treatment of entropy is postponed until Chapter 17.

The idea of rate as a factor in whether a change will take place is also introduced in Chapter 6 and is used again and again. The quantitative treatment of **kinetics** follows thermodynamics because, just as shooting at a moving target is harder than shooting at a stationary one, kinetics is more difficult than thermodynamics. (If you prefer kinetics earlier, it can be done before Chapter 14.)

In some texts, **gases** are covered just before liquids and solids. But since the material on gases forms the basis for the atomic theory, it is presented here before atomic theory. As solids can't very well precede chemical bonding, gases and solids are separated. If you prefer gases just before liquids and solids, Chapter 5 can easily be postponed. Early treatment of gases also makes this topic more meaningful to students who do lab experiments with gases early in the semester.

We have tried to write an organic chapter that will give students an idea of what modern organic chemistry is about and how the organic chemistry of carbon fits into the rest of chemistry. This approach seems preferable to trying to cram the reactions of all the major classes of compounds into one chapter as is often done. One has only to glance at the abstracts of the papers from the Organic Division of any meeting to see the importance of chirality in contemporary organic chemistry. Therefore, stereoisomerism comes early in Chapter 22. *If models are used,* stereochemistry is very concrete and our students do not find it difficult.

Industrial representatives claim that the primary topic that is missing from college chemistry courses is a treatment of polymers. Certainly, the things that non-chemistry majors or even non-science students will read about in the popular press will frequently involve synthetic polymers, and the general pervasiveness of polymers in everyday life makes an introduction of this topic essential. In addition, students are usually very interested in biological polymers (proteins, carbohydrates, and nucleic acids, in other words, the chemistry of their bodies). Chapter 23 provides a concise introduction to the properties of polymers and a coherent overview of some basic biochemistry.

Throughout the writing of this text, many reviewers mentioned topics that they felt were valuable but optional. These **optional topics** are usually located at the ends of chapters so that they may be easily assigned for reading only, or omitted entirely. See, for example, the section on the calculation of atomic and ionic radii and Avogadro's number in Chapter 12, or the section on colloids in Chapter 13. Later chapters do not assume knowledge of material in these optional sections.

Problems and Problem-Solving

In-Chapter Problems. Many worked-out *Sample Problems* are presented as examples throughout the chapters. Some teach students to use pictures and graphs—two things experienced scientists usually do but students do not. In the Sample Problems, we have tried to teach students to use a thoughtful and logical approach to problem solving rather than simply memorizing procedures or using dimensional analysis as a substitute for thought. Dimensional analysis is stressed as a way to check answers, and the importance of checking work is emphasized by the "reasonableness checks" that conclude many Sample Problems.

An example has not been included for every kind of problem, however. Where we feel that the explanation in the text is adequate, we ask the student to answer questions without a Sample Problem. Students need to realize that they are supposed to be learning to solve problems, not just to follow an example. The concentration of Sample Problems decreases toward the end of the book; by the time they reach the final chapters, students should need worked-out examples only for completely new processes.

In the chapters, Sample Problems are usually followed by one or more **Practice Problems**. These give students a chance to work problems similar to the examples and to reinforce their understanding of skills and concepts. We have included more problems than usual within the chapter text because in-chapter problems encourage students to be active, not passive, readers. Students who work through the in-chapter problems will gain a much better understanding of concepts and skills than those who simply read the text. The Practice Problems are generally paired with the end-of-chapter problems. Answers to all of the in-chapter problems appear in Appendix I, and solutions are available in the *Student Solutions Manual.*

End-of-Chapter Problems. The end-of-chapter problems begin with **Additional Practice Problems**. These are drill-and-practice problems similar in difficulty and covering the same skills as the in-chapter problems. These Additional Practice Problems are numbered in continuous sequence with the in-chapter Practice Problems. Approximately 85% of them are followed by a parenthetical notation indicating the section of the chapter to which the problem refers. Not all of these problems are classified, however; classifying problems is an important skill that students need to develop. The section or sections to which each unclassified problem relates are given in the *Instructor's Resource Manual* to aid you in assigning a good selection of problems.

The next group of problems is called **Stop and Test Yourself.** It is a multiple choice self-test covering the basic skills and concepts from the chapter so that students can quickly check their readiness to go on to the higher-level problems that follow. Also, skill in taking this type of test is important to many students (for admission to medical school, for example), and they need practice. Answers appear in Appendix I along with the number of the section to be re-studied if the student answers the question incorrectly. The answers are explained in the *Student Solutions Manual.* The self-tests are in the book, not in the *Study Guide,* because the students who most need the self-tests are, in our experience, those least likely to use ancillaries.

The self-test is followed by **Putting Things Together** problems. Many students can do all of the individual operations, but have trouble putting them together. The Putting Things Together problems help students become better problem-solvers by requiring them to combine material from different sections of the current chapter or material from the current chapter with material from earlier chapters.

Last come **Applications** problems in which students must apply chemical skills and concepts to real-world problems. These problems appear last, after students have had a lot of practice with "generic" problems.

The end-of-chapter problems range from simple to fairly challenging so that the level of the course can be tailored to suit different audiences. Many more problems are provided than any student will have time to do so that problem assignments can be varied from year to year. Solutions to many end-of-chapter problems (those with blue numbers) and all the in-chapter Practice Problems are given in the *Student Solutions Manual,* for the benefit of instructors who like their students to have solutions available. (For the same problems presented in the *Student Solutions Manual,* the answers only are provided in Appendix I.) Answers and solutions to the remaining end-of-chapter problems are given in the *Instructor's Solutions Manual* and therefore are available only to students whose instructors want them to have them.

Special Features

Guest Essays. Guest Essays between the chapters give students perspective on the importance of chemistry to life after college. In these, people of various backgrounds and ages (many not much older than most students) reflect on their experiences in general chemistry, and describe the role of chemistry in their work. As students study chemistry throughout the year, these essays encourage them to believe that they can succeed and remind them that an understanding of chemistry can lead to many opportunities and be of value in many different careers.

Accuracy. Research has shown that even competent scientists retain wrong information and inefficient ways of doing things if they learn them first. Material simplified for beginning students should be correct; those who go on in science should not have to unlearn anything. Therefore, in addition to the usual reviews by teachers of general chemistry, many chapters have also been reviewed by chemists who are specialists in the area concerned, and by colleagues in related fields such as astronomy, biology, geology, and physics. We have tried very hard to avoid "common textbook errors" and to use the best available data. The placement of La, Ac, Lu, and Lr in the periodic table is an example. Where the treatment is different from that in other texts, references to the literature are given in the *Instructor's Annotated Edition.*

Instructor's Annotated Edition. We have prepared a special edition of the text for the use of instructors. The *Instructor's Annotated Edition* contains marginal comments in magenta to signal the location of related material, to explain why a particular approach is used, to cite references to the chemical literature, and so forth. Most of these marginal notes to instructors are derived from "dialogues" that were carried on between the reviewers and ourselves through many drafts of the text. In the second edition, these marginal notes include references to the *Discover Chemistry CD-ROM,* which are identified by a disk logo. Instructors will find these references helpful if they use portions of the CD-ROM to enrich their lectures, or if their students are using the CD-ROM as a learning aid.

 ## SUPPLEMENTS

An extensive package of supplements has been created to support both the instructor and the student. It includes:

For Students

Discover Chemistry CD-ROM is an interactive encounter with the content of general chemistry. Students interact with a variety of graphical models and experimental setups, changing parameters, entering values, or predicting outcomes, and getting clear feedback at every stage. *Discover Chemistry CD-ROM* includes video of actual experiments and many animations to involve students in visualizing and understanding at the molecular level. It coordinates with and complements the Umland/Bellama text, but is not a scrolling textbook.

Study Guide. The *Study Guide* has been written by Kenneth J. Hughes of University of Wisconsin–Oshkosh. Each chapter includes a review of the text material followed by sample problems. Chapters conclude with diagnostic self-test questions and a variety of Practice Problems. Outline solutions are provided for all Practice Problems. An introductory chapter on reading and learning scientific material has been written with specific reference to the textbook; it provides practical strategies for learning chemistry, with particular emphasis on vocabulary.

Student Notetaking Guide provides copies of key text figures and tables on which students can take notes in class without the distraction of having to copy from the chalkboard or a projected transparency.

Student Solutions Manual. Prepared by Jean Umland, Juliette Bryson (Las Positas College), and Byron Christmas (University of Houston–Downtown), the *Student Solutions Manual* provides complete solutions to all of the in-chapter Practice Problems and to the end-of-chapter problems identified by blue numbers.

Essential Math for Chemistry by David Ball of Cleveland State University is a student workbook written to help students learn or review the math skills required for success in general chemistry. Numerous examples and exercises give students extensive opportunity for practice.

Chemistry and the Allied Health Career: A Book of Readings by Michael R. Slabaugh (Weber State University). This reader contains approximately 50 articles from a variety of general and scientific journals—all related to general and/or allied health chemistry. Each article has an introduction explaining its relevance to the introductory course and questions to provoke discussion.

For Instructors

Instructor's Solutions Manual. Written by Jean Umland, Juliette Bryson, and Byron Christmas, the *Instructor's Solutions Manual* contains answers and solutions to all problems for which answers are not provided in Appendix I.

Instructor's Resource Manual. Written by Jean Umland and Mike Bellama, this supplement provides a variety of information, including suggestions for adapting the text to courses of different organizations, lengths, levels, and emphasis; section number references for all unclassified problems; additional text sections (on normality, for example) for reproduction as handouts; suggested classroom demonstrations; and other resources.

Test Bank. Prepared by Jon M. Bellama and Cleta Kay Hanebuth of the University of South Alabama, the test bank includes over 2400 questions in multiple-choice format. The test bank is available in hard copy and on disk with a computerized test generator that allows instructors to modify, write, and display test questions. The testing program has outstanding graphics capability and has a full range of chemical symbols. Both IBM and Macintosh versions are availalbe.

Transparencies. Transparencies of about 175 of the most important illustrations from the text are available in full color.

Lecture Presentation Software. This PowerPoint software by James Hardy of the University of Akron provides more than 1200 colorful electronic transparencies for use in lecture. Clearly labeled images are grouped into 24 topical units and include 33 animations. There is extensive coverage of organic and biochemistry. Both Macintosh and Windows versions are available. Images may be edited using PowerPoint 4.0.

For the Laboratory

Experiments in General Chemistry, Second Edition. This laboratory manual by Steven Murov of Modesto Junior College provides 35 tested and clearly presented experiments covering a wide range of topics. The presentation often uses an "inquiry" approach, and emphasizes connections between text topics and lab experiences. Safe laboratory practices are emphasized, and the use of hazardous substances is limited to the minimum amount required for reliable results. Available with or without electronic lab worksheets customized for use with LabSystant software (see below).

Qualitative Inorganic Analysis. This combined text and lab manual by William T. Scroggins of Chabot College introduces qualitative analysis techniques and their underlying equilibrium principles and provides a wide range of experiments.

Multimedia/Electronic Supplements

Discover Chemistry CD-ROM. (See the list of supplements for students, above.)

Videodisc. *General Chemistry* videodisc contains demonstrations and experiments that are difficult, expensive, or dangerous to perform in class. Most of the experiment footage is original, developed specifically for this text. Much of the artwork from the text is also included, with animation when motion is inherent to the concept.

Tutorial Video Tapes. Instructional tapes on *Graphing Scientific Information* and *Using a Scientific Calculator* have been created for this text by Dr. Michael Clay of the College of San Mateo.

Tutorial Software. Concentrated Chemical Concepts by Trinity Software provides students with well-structured exercises for the range of topics covered by the text. Available for IBM compatible and Macintosh computers.

LabSystant by Trinity Software allows instructors to create electronic worksheets for quantitative labs. In addition a Checker program helps students find and correct errors in lab data and calculations, and prints out lab results. A disc available with the Murov lab manual provides electronic lab worksheets matching those printed in the lab manual.

Inorganic Qualitative Analysis by Trinity Software is a *PC* simulation of the laboratory study of seven inorganic cations and four anions. In the simulation, students choose to explore or attempt to solve more than 1000 unknowns at five different levels of difficulty. The program generates unknowns randomly and displays a video image of the sample both before and after water has been added. Students then choose from a number of operations as they attempt to identify the unknown.

Diatomic: Molecular Mechanics and Motion by Trinity Software illustrates the translational, rotational, and vibrational motion of diatomic molecules. It's an ideal vehicle for classroom lectures and demonstrations, lab assignments or independent study. Equipment required: MS/DOS and compatibles or Macintosh.

 ## ACKNOWLEDGMENTS

This project would not have been possible without the support, encouragement and contributions of our editor, Richard Mixter. The guidance and insight he has contributed have been invaluable in shaping this book from early ideas to its ultimate form. Keith Dodson, the developmental editor, has also been important in the success of the project, particularly by developing much of the supplement package and working with accuracy checkers. Finally, we are grateful to the publishing team in Eagan, Minnesota, especially to Tom Modl and Deborah Meyer for their outstanding work in the production of this complex book, and to Ellen Stanton, Mary Steiner, Cyndi Eller, and Lisa Lysne for their insight and effort in marketing.

We also wish to thank a number of colleagues for their help. At the University of Houston–Downtown we would like to thank: all of the members of the Department of Natural Sciences, who provided a variety of support during the writing and revision of this text; physicists Bowen Loftin and Peter Hoffmann-Pinther, who reviewed chapters with a sizeable physics component, such as Chapter 7 on Atomic Structure; chemists Larry Spears, who reviewed Electrochemistry, and Jim Driy, who read various chapters and consulted on problem solutions; biologists Al Avenoso, John Capeheart, and Ruth Sherman and geologist Glen Merrill, who answered innumerable questions pertaining to their disciplines; Barbara Bartholomew of the Writing Lab, who helped in the early stages of writing; Helen Allen of the Reading Lab, who wrote the chapter on reading science for the Study Guide; and Eugenia Adams, Laura Olejnik, Henri Achee and Anita Garza from the library staff, who helped locate hard-to-find information and processed numerous interlibrary loan requests. Thanks also to Dr. Barbara Sanborn of the Department of Biochemistry and Molecular Biology of the University of Texas–Houston Medical School for her review of the biochemistry sections of the new chapter on polymers.

At the University of Maryland, we are grateful to: Alfred E. Boyd Jr., James E. Huheey, Samuel O. Grim, Rinaldo E. Poli, Bryan W. Eichhorn, and Robert S. Pilato for many interesting and stimulating conversations about chemistry and teaching; and Prof. Bruce B. Jarvis, department chairman, for his cooperation and assistance.

In addition, several friends from industry very kindly reviewed chapters in their areas of expertise: Paul Deisler, chemical engineer, who was, at the time, vice-president of Health, Safety and Environment for Shell (kinetics), and his wife, Ellen, who is also a chemical engineer (Chapters 1–4); and Jim Lamb, president of IMP, Inc. (nuclear chemistry). Thanks also are due to numerous students who critiqued the text, and especially to Rose Depugh, who made many constructive and insightful suggestions at the beginning.

An important group of people has been helpful in checking the end-of-chapter problems and solutions for accuracy and clarity. Thanks to G. Myron Arcand, Byron Christmas, Jeff Korcz, Gerard L'Heureux, Paul O'Brien, Amer E. Villaruz and particularly to Juliette Bryson and Keith Dodson.

We especially appreciate the participation of the guest essayists and their commitment to giving students a perspective on the diverse careers that can emerge from an educational background in chemistry. Jill Torbett deserves special thanks for her work in coordinating these essays.

Last, but certainly not least, we express our sincere gratitude to the reviewers for this edition who offered their various perspectives and insights on how to make this edition of the text work even better for students: Janet S. Anderson, Union College; Jeffrey Appling, Clemson University; David W. Ball, Cleveland State University; Londa Borer, California State University; Juliette A. Bryson, Las Positas College; Mapi Cuevas, Santa Fe Community College; R. Scott Daniels, Acadia University; Norman Duffy, Kent State University; Gary Edvenson, Moorhead State University; Clark L. Fields, University of Northern Colorado; Michael Golde, University of Pittsburgh; Nancy Gordon, University of Southern Maine; Leland Harris, University of Arizona; Lisa Hibbard, Spellman College; Michael Johnson, New Mexico State University; Lance Lund, Anoka-Ramsey Community College; John Luoma, Cleveland State University; Chris McGowan, Tennessee Technological University; D. F. Nachman, Phoenix College; Earl F. Pearson, Western Kentucky University; Joseph Sarnesky, Fairfield University; Jesse Spencer, Valdosta State University; Herman Stein, Bronx Community College; Albert Thompson, Spellman College; Byron J. Wilson, Brigham Young University.

Special thanks to Jeff Appling, Julie Bryson, Michael Golde, and Michael Johnson—users of the first edition—who read *all* of the manuscript for the second edition.

Thanks also to the reviewers of the first edition: David Adams, Bradford College; Hugh Akers, Lamar University; Peter Baine, California State University–Long

Beach; Richard Baker, El Reno Junior College; Carl Bishop, Clemson University; Joan Bouillon, Mount San Antonio College; Arthur Breyer, Beaver College; Robert J. Bydalek, University of Minnesota–Duluth; Donald Campbell, University of Wisconsin–Eau Claire; Roy D. Caton, University of New Mexico; Gregory R. Choppin, Florida State University; Neil R. Coley, Chabot College; Corinna L. Czekaj, Oklahoma State University; Geoffrey Davies, Northeastern University; Michael Davis, University of Texas–El Paso; Daniel R. Decious, California State University–Sacramento; John M. DeKorte, Northern Arizona University; Robert Desiderato, University of North Texas; Jerry A. Driscoll, University of Utah; Julie Ellefson-Kuehn, Harper College; Gordon J. Ewing, New Mexico State University; Steven L. Fedder, Santa Clara University; Kenneth R. Fountain, Northeast Missouri State University; Donna G. Friedman, St. Louis Community College; John K. Garland, Washington State University; Robert Gayhart, Bradley University; Keith Harper, University of North Texas; S. J. Hawkes, Oregon State University; H. Fred Henneike, Georgia State University; James C. Hill, California State University–Sacramento; Andrew Holder, University of Missouri–Kansas City; Jeffrey A. Hurlbut, Metropolitan State College of Denver; Earl Huyser, University of Kansas; Paul B. Kelter, University of Wisconsin–Oshkosh; Leslie N. Kinsland, University of Southwestern Louisiana; James M. Landry, Loyola Marymount University; Anne G. Lenhert, Kansas State University; Gerard A. L'Heureux, Holyoke Community College; William Litchman, University of New Mexico; Roger V. Lloyd, Memphis State University; Richard A. Lungstrom, American River College; Joel Mague, Tulane University; R. Bruce Martin, University of Virginia; Paul J. Ogren, Richmond College; Richard S. Perkins, University of Southwestern Louisiana; Richard L. Petersen, Memphis State University; Robert C. Pfaff, University of Nebraska at Omaha; J. Robert Pipal, Alfred University; Dennis Regan, Green River Community College; Nancy C. Reitz, American River College; Don Roach, Miami-Dade Community College; Richard J. Ruch, Kent State University; Charles W. J. Scaife, Union College; Martha W. Sellers, Northern Virginia Community College; David B. Shaw, Madison Area Technical College; B. R. Siebring, University of Wisconsin–Milwaukee; Ernest F. Silversmith, Morgan State University; Robert W. Smith, University of Nebraska at Omaha; Theodore W. Sottery, University of Southern Maine; Mabel-Ruth Stephanic, Oklahoma State University; Mary E. Thompson, College of St. Catherine; Wayne Tikkanen, California State University–Los Angeles; Donald Titus, Temple University; Carl Trindle, University of Virginia.

We welcome your contributions to this book. Your comments and suggestions about any part of the text or its supplements will be appreciated. If you should find any errors, we would be grateful if you would let us know so that we can correct them.

Jean B. Umland
Department of Natural Sciences
University of Houston–Downtown
One Main Street
Houston, TX 77002

Jon M. Bellama
Department of Chemistry
  and Biochemistry
University of Maryland
College Park, MD 20742-2021

INTRODUCTION

Many chemical reactions are taking place all the time in all living things.

Chemistry is *the science that concerns the composition, structure, and properties of matter and the changes that matter undergoes.* **Matter** is *anything that occupies space.* Water, gold, rocks, buildings, plants, and people are matter. Ideas and actions are not matter. All the matter that you are likely to meet in your everyday life is made of chemicals. The photographs above and on the next page show chemicals undergoing changes!

This chapter is an introduction to many subjects that will be discussed in later chapters. It will provide you with a basic chemical vocabulary and with some of the skills you will need in your study of chemistry. As you read it, please keep in mind that you can use things without understanding them. People all over the world used chemical processes such as making glass and obtaining metals from ores for thousands of years before the science of chemistry existed. It is impossible to explain everything at once. There is no perfect order of topics. Be patient. (If you can't wait to learn more about a topic, use the index to find out where it is treated in more detail.) When science answers one question, it usually raises others. No person ever understands everything.

■ Why is a study of general chemistry a part of your education?

A knowledge of chemistry is needed in every modern science from astronomy to zoology. All of the materials engineers and technologists use are either made or modified by chemical reactions.

Many things in American life that you take for granted are modern inventions. In 1900 few automobiles or telephones existed. There were no synthetic fibers

Fireworks are chemicals.

except rayon, no plastics except celluloid. Radios, airplanes, television, handheld calculators, and computers did not exist. Chemistry has played an important part in the creation of thousands of new products. These products make life longer, healthier, easier, and more comfortable. People in the United States live an average of 30 years longer now than they did in 1900. This increase in life span is mainly due to the discovery of vaccines and drugs that prevent or cure diseases such as polio and pneumonia. Also, a better understanding of body chemistry has helped control heart disease and high blood pressure. The world's farms produced twice as much food in 1990 as in 1960. Greater food production has resulted largely from improved fertilizers and pesticides. Synthetic fibers and plastics are products of the twentieth-century chemical industry.

The chemical industry provides jobs for over a million people in the United States alone and accounts for more than 10% of the gross domestic product. It consistently has the best safety record of any group of industries. In addition, more chemicals are exported than imported, although overall the United States imports much more than it exports.

The same changes that have produced so many good effects have also produced some bad ones. Some of these could have been predicted. For example, mass production of automobiles put carriage makers out of business. (Many jobs, however, were created in the auto industry.) Some harmful effects couldn't have been predicted. The problem of atmospheric pollution by car exhausts is an example. However, whether problems are predictable or not, chemistry will be important to their solution and in supplying society's needs in the future. For example, chemistry will play an important part in cleaning up the environment. It will provide new materials with improved properties to substitute for scarce resources.

Another reason to study chemistry has to do with your role as a decision maker. Every one of us is a consumer. You may not think that what one person does makes much difference, but it does because individual decisions tend to be typical. As a result of the huge number of people living today, small effects are multiplied and become significant. For example, every American household throws away an average of 0.3 of an ounce (9 grams) of dangerous waste (dead batteries, cosmetics, insecticides, drain cleaners, leftover paint, used antifreeze, motor oil, and so forth) *per day*. For a city of 100,000 households, this adds up to about 1 ton per day. If a chemical plant threw away a ton of a poisonous compound, that action would be classified as a major industrial spill.

Your decisions as a voter will also be important. Many questions that will have to be decided have a chemical component. What to do about nuclear power is one example. A study of chemistry will provide you with a good foundation for making intelligent decisions. After graduation remember that, in a world of change, you'll have to continue learning as long as you live.

— What do you already know about chemistry?

Everything is a chemical or a mixture of chemicals: Air, water, food, plants, animals, people, and rocks are all chemicals. Chemical reactions are part of everything you do from breathing and eating to thinking. Everyone has had a large amount of chemical experience before going to college. Every chapter in this book begins by pointing out what you already know about the subject of the chapter.

1.1 OBSERVATIONS AND CONCLUSIONS

Chemistry, like all sciences, is based on observations. ■Figures 1.1 and 1.2 show a substance undergoing changes. What observations lead to the conclusion that the object in Figure 1.1 has undergone a change? First, the object shown in Figure 1.1(a) is a cloudy white block. In Figure 1.1(b), it is not as high and is more rounded. It is sitting in the middle of a thin irregularly shaped layer of clear, colorless material. If you could really observe the change instead of looking at a picture of it, you might touch the object and feel that it is hard and cold. You might stick your finger in the thin layer of clear, colorless material [as shown in Figure 1.1(c)] and find

(a)

(b)

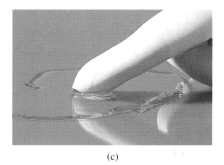
(c)

▪ FIGURE 1.1 What observations lead to the conclusion that the object has changed?

that you could put your finger through it. You could spread it around and feel that it is slippery and cool but not cold like the object. You might smell both the object and the thin layer of clear, colorless material and note that neither had any odor—that is, you might use senses other than sight to observe.

You would interpret your observations as meaning that the cloudy white object was a solid; it had a shape of its own and felt hard. The thin layer was a liquid; you could put your finger through it and push it around. Liquids are fluid—that is, they flow and take the shape of the bottom of their container. You would conclude from your observations that a change had taken place; solid had melted to liquid.

It is important to distinguish between observations and what you think about them. If observations are carefully made, you would make the same observations if you were to repeat the experiment. You might change your mind about what you thought about them because of something you had learned. Other people in other places or at other times would also make the same observations if they did the experiment. But they might explain their observations differently than you. For example, scientists today usually interpret their observations in terms of atoms and molecules. Before the seventeenth century, people usually accounted for their observations in terms of religion.

PRACTICE PROBLEM

1.1 (Answers for Practice Problems are in Appendix I. Detailed solutions are in the Student Solutions Manual.) What observation leads you to the conclusion that the pictures in Figures 1.1(b) and (c) were taken after the picture in Figure 1.1(a)?

◑ Use the material discussed in the preceding paragraphs to figure out the answers to Practice Problems.

1.2 PHYSICAL AND CHEMICAL CHANGES

You have probably guessed that the object pictured in Figure 1.1(a) was an ice cube. From the picture, you cannot make enough observations to be sure. The object might have been a cube of some liquid other than water frozen in an ice cube tray. You must be careful not to jump to conclusions based on too few observations.

An ice cube is the substance, water, frozen into a solid form. All parts of Figure 1.1 show the same substance, water. *A change* like this *that doesn't involve changing any substances into any other substances* is called a **physical change.** Another familiar example of a physical change is boiling water. When water boils, liquid water changes to steam. Steam is the substance, water, in the form of a gas. To a scientist, the term *gas* refers to one of the three states of matter: solid, liquid, and gas. Gases fill their containers and take the shape of the container. They're often invisible like air (see Figure 1.2).

▪ FIGURE 1.2 Steam is invisible. (Look very close to the top of the spout.) The "steam" that you can see is really tiny drops of liquid water that have condensed when the steam flows into room temperature air.

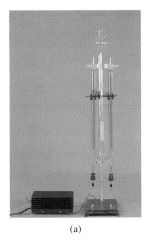
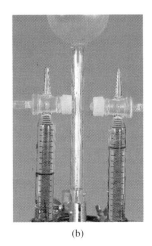
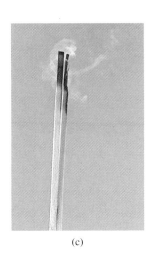

(a) (b) (c) (d)

■ FIGURE 1.3   The electrolysis of water. (a) At the beginning of the experiment, both gas burets are filled with water. (b) As electricity passes through the water, gases collect above the water. The volume of the gas in the left gas buret is twice the volume of the gas in the right gas buret. (c) and (d) A glowing splint inserted in a sample of the gas from the right gas buret bursts into flame.

The Fahrenheit temperature is given first in this chapter because it is the one most familiar to American students. After the Celsius scale is introduced in Chapter 2, it will be used throughout the book.

◑ Studying chemistry is like studying a foreign language. At the beginning, you must devote a lot of effort to learning the vocabulary.

Now let's consider the change pictured in ■Figure 1.3. In Figure 1.3(b), you can see boundaries between two phases in each of the gas burets. In chemistry, a **phase** is *a sample of matter that is uniform in composition and physical state and separated from other phases by definite boundaries.* The upper phases in the two burets are gases; the lower phase is a liquid.

If a test tube is held over the stopcock at the top of either buret and the stopcock is opened, a hissing noise is heard. The liquid level in the buret rises. Gas has flowed out of the buret into the test tube. Touching a burning splint to the mouth of a test tube containing a sample of the gas from the left buret produces a loud noise. This is a test for hydrogen gas. Inserting a glowing splint into a test tube containing a sample of gas from the right buret makes the glowing splint burst into flame [see Figures 1.3(c) and (d)]. This is a test for oxygen gas. The liquid, which originally filled the electrolysis apparatus, could be identified as water by suitable tests. [For example, water freezes at 32 °F (0 °C) and boils at 212 °F (100 °C)].

In Figure 1.3, the substance water has changed into different substances: hydrogen gas and oxygen gas. The change shown in Figure 1.3 is an example of a **chemical change** or **chemical reaction.** *In a chemical change, one or more substances are changed into other substances.*

A reaction in which a substance breaks down into simpler substances is called **decomposition.** The substance is said to **decompose.** Electricity has decomposed water into hydrogen and oxygen.

Substances are identified by their properties. *Properties that involve substances changing into other substances* are called **chemical properties.** The fact that water is decomposed to hydrogen gas and oxygen gas by electricity is a chemical property of water. *Properties that do not involve substances changing into other substances* are called **physical properties.** The facts that water freezes at 32 °F (0 °C) and boils at 212 °F (100 °C) are physical properties of water.

Definitions are very important. For clear communication, both the writer or speaker and the reader or listener must understand a word to mean the same thing. The word *substance* has been used often in this section. To a chemist, substance has a very exact meaning. A **substance** is *a distinct type of matter. All samples of a substance have the same properties; a substance has the same properties through-out the whole sample.* Pure water is a familiar example of a substance. All samples of pure water have the same melting and boiling points. Seawater is not a substance; it contains both salt (and other dissolved substances) and water. A sample of seawater from the Dead Sea tastes saltier and has a lower freezing point and a higher boiling point than a sample of seawater from the Atlantic Ocean. There is no difference between a substance obtained from natural sources and another sample of the same substance made in a chemical plant. Vitamin C is vitamin C whether it comes from rose hips or from a factory. To succeed in chemistry, you must be sure you understand the definitions of the words in dark type, especially ones that have different or less exact meanings in everyday life.

PRACTICE PROBLEMS

1.2 Identify the different phases present in a glass of cola with ice cubes in it.

1.3 Which of the following changes are physical changes and which are chemical changes? (a) Rubbing alcohol evaporates from your skin, (b) a seed grows into a flower, (c) aluminum metal is rolled into foil, (d) copper metal is drawn into wire, and (e) a candle burns.

1.4 Aspirin is a substance. Will expensive brand-name aspirin cure your headache any better than cheaper aspirin that is sold in a bottle without a brand name? Explain your answer.

1.3 ELEMENTS, COMPOUNDS, AND MIXTURES

There are two kinds of substances, elements and compounds. **Elements** *cannot be decomposed (broken down into simpler substances) by chemical reactions.* Hydrogen gas and oxygen gas are elements. So are many other familiar substances such as carbon, aluminum, copper, silver, and gold. One hundred and eleven elements are known; their names* are listed alphabetically in the table inside the front cover.

The "lead" in your pencil is actually a form of carbon called graphite mixed with clay.

Compounds can be decomposed into simpler substances by chemical reactions. For example, water is a compound; it is decomposed into hydrogen gas and oxygen gas by electricity. (Remember Figure 1.3.) Other familiar compounds are salt, sugar, and carbon dioxide. **Compounds** are *substances composed of two or more elements. No matter when or where a sample of a compound is found, the compound always has the same physical and chemical properties. A compound always contains the same percent (by weight) of each element.* All samples of pure water are 11.19% (by weight) hydrogen and 88.81% (by weight) oxygen. This summary of many observations is called the **law of constant composition** or the **law of definite proportions.** (In science, a **law** is *a summary of many observations.* Laws are often expressed by mathematical equations, although this particular one is not.)

Just as the properties of water are far different from the properties of hydrogen and oxygen, the properties of compounds are usually far different from the properties of the elements of which the compounds are composed.

Nonstoichiometric compounds and compounds of different isotopic composition are discussed later.

Law of Constant Composition

In a pure compound, the percent (by weight) of each element is always the same.

A reaction in which substances combine to form more complex substances is called a **combination reaction.** For example, in the test for hydrogen gas (burning splint touched to mouth of test tube produces pop or bang), hydrogen gas combines with oxygen gas from the air to form water. All compounds are combinations of two or more of the 111 known elements. An almost infinite number of compounds

---

*The person who discovers a new element has the right to name it. Many names were chosen because they described the element. For example, the name "chromium" is derived from the Greek word *chroma,* which means "color." Most compounds of chromium are colored. Other names were chosen to honor a person or place. Einsteinium is named after Albert Einstein and curium after Marie and Pierre Curie. Americium honors the Americas and europium, Europe.

is possible, and over 13 million are known. Almost 800,000 new compounds are being discovered each year. However, the discovery of a new element is an event that happens only every few years.

Many familiar materials, such as air and seawater, are neither elements nor compounds but mixtures. A **mixture** is *a sample of matter composed of two or more substances. Mixtures have varying compositions.* For example, air may contain from 0 to 5% by weight of water vapor. (Water in the gaseous state is called water vapor.) Seawater from the Atlantic Ocean contains about 3.5% salt; seawater from the Dead Sea contains about 30% salt. *The substances making up a mixture* (such as salt and water) are called **components** of the mixture.

Another difference between compounds and mixtures is that mixtures can be separated by physical changes. For example, salt is often obtained from seawater by letting seawater stand in the sun. Heat from the sun turns the water into water vapor. The water vapor escapes into the air, and the salt is left behind. ▪Figure 1.4 shows salt produced from seawater. Although compounds can be decomposed to simpler compounds or elements by chemical reactions, they can't be separated into simpler substances by physical changes.

Clear seawater is an example of a **homogeneous mixture.** *Homogeneous mixtures have only one phase; they have the same properties throughout a sample, although the properties of different samples may be different. Homogeneous mixtures* are called **solutions.**

Heterogeneous mixtures have *more than one phase. They do not have the same properties throughout a sample.* Bits of the phases can be seen either with the eye or with a microscope. The phases can be in the same or different physical states. Chocolate chip cookies and ice water are familiar examples of heterogeneous mixtures. In chocolate chip cookies, both phases (chocolate chips and cookie) are in the same physical state, the solid state. In ice water, one phase (ice) is a solid, and the other phase (water) is a liquid. Although the chocolate is broken up into chips and the ice is in cubes, chocolate and ice are each considered to be a single phase.

▪Figure 1.5 summarizes the information in this section.

▪ FIGURE 1.4 Mounds of salt up to 200 feet high produced by the evaporation of seawater on the island of Bonaire off the coast of Venezuela. Much of this salt is shipped to the United States where it is used on roads to prevent icing in winter. What can you conclude about the climate of Bonaire from this photograph?

PRACTICE PROBLEM

1.5   Classify each of the following as solution, heterogeneous mixture, compound, or element: (a) chlorine, (b) gelatin with fruit in it, (c) sodium, (d) sugar, and (e) white wine.

▪ FIGURE 1.5 Classification of materials.

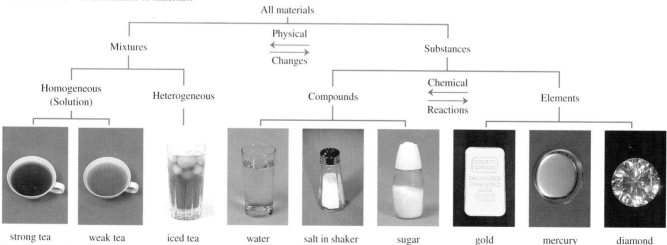

Chemistry from the Fifth Century B.C. to the Sixteenth Century A.D.

From the days of the ancient Greeks to the present, people have observed the world around them and tried to explain it. Alchemy, which many historians believe provided the basis for the science of chemistry, developed at about the same time in China, India, and Greece. Alchemists thought that five elements—air, water, earth, fire, and space—made up all matter and that matter functioned by means of opposing forces such as hot and cold and wet and dry.

Alchemy reached Europe when the Muslims conquered Spain in the eighth century. At first, European alchemists were chiefly interested in changing metals such as copper and lead into silver and gold. Then they switched to trying to find a substance that would make people live forever and finally to searching for superior medicines. Alchemists invented many useful pieces of laboratory apparatus and procedures and discovered a number of important substances.

The last great alchemist was Paracelsus who was born in 1493, the year after Columbus first explored America. Paracelsus was a German doctor who established the role of chemistry in medicine.

He took the name *Paracelsus* because it means "above and beyond Celsus," a famous Roman doctor who lived in the first century A.D.

Paracelsus attended a school in Austria that trained boys as overseers and analysts for mining operations and then studied at a number of European universities. He did not accept the usual view of his time that the stars and the planets control all parts of the human body. He proposed that the common miners' disease of silicosis resulted from breathing metal vapors rather than being a punishment for sin dispensed by mountain spirits. Paracelsus wrote a clinical description of syphilis, which he treated with mercury compounds taken internally, and was the first person to connect goiter (a swelling of the thyroid gland) with minerals in drinking water and to observe that, if given in small doses, "what makes a man ill also cures him." Paracelsus prepared and used a number of new drugs, and his work and ideas formed the basis for much of modern medicine including psychiatry.

The alchemists' dreams of turning other metals into gold were doomed to failure.

1.4 THE SCIENTIFIC METHOD

In science, a **theory** is *an explanation of a law or a series of observations.* Theories are invented by human beings to explain observations. They help people remember the vast number of scientific facts that have been discovered by making sense out of them. As soon as a new theory has been suggested, scientists start thinking of experiments to test it. If the results of these new experiments support the theory, it is accepted. If they do not, it must be corrected or a new theory invented. The corrected or new theory in turn leads to more experiments.

Science progresses by cycles of suggested theories and tests by experiment. A theory can be proved wrong, but it can never be proved right. There is always a chance that the results of some new experiment will not agree with the results predicted by the theory. The overturning of an important theory opens new frontiers in science. For example, before the sixteenth century, people thought that Earth was the center of the universe. Then Copernicus explained observations of the planets by suggesting that Earth and the other planets revolve around the sun. This was the beginning of the scientific revolution.

An explanation that has not been tested sufficiently is often referred to as a **hypothesis.** *A hypothesis that is supported by the results of experiments becomes a theory.*

1.5 ATOMS

▪ FIGURE 1.6 *Left photos:* When a plastic rod is rubbed with wool, electrons are transferred from the wool to the rod. The rod becomes negatively charged and attracts uncharged objects such as feathers. Charged objects also attract objects that have the opposite charge. "Static cling" results from charges caused by rubbing. *Right photos:* In an electroscope, two thin strips of gold leaf hang from a metal rod inside a glass container. When the top of the metal rod is touched with a charged object, such as the plastic rod, both gold leaves acquire the same charge and repel each other. The first recorded observation of an electric charge was made in Greece about 600 B.C. The word *electricity* comes from the Greek word for amber, which gets a negative charge when rubbed with wool.

Many observations made by chemists are explained by picturing *elements* as being *made up of very small particles called* **atoms.** The word "atom" is derived from the Greek word *átomos,* which means "things that can't be cut or divided." The idea that everything is made of small particles was suggested by Greek philosophers in the fifth century B.C. (However, this was a minority view in ancient Greece.)

The modern atomic theory was suggested by an English schoolteacher, John Dalton, in 1803–1807. Dalton saw that the existence of atoms would explain the law of constant composition (Section 1.3) and other observations that had been made about chemical reactions. However, the scientific establishment of the day was unwilling to approve the new idea, and more than 50 years passed before the atomic theory was generally accepted. Even as late as 1900, the author of one general chemistry text refused to mention Dalton's theory. The main ideas of Dalton's theory are still used today, although the discovery of new facts has made a few small changes necessary.

According to the atomic theory suggested by Dalton,

1. All matter is composed of atoms. An atom is the smallest particle of an element that takes part in chemical reactions.
2. All atoms of a given element are alike—that is, all atoms of gold are the same. Atoms of different elements are different—for example, an atom of copper is lighter than an atom of gold.
3. Compounds are combinations of atoms of more than one element; in a given compound, the relative number of each type of atom is always the same. For example, in water there are always two hydrogen atoms for each oxygen atom.
4. Atoms cannot be created or destroyed. Atoms of one element cannot be changed into atoms of another element by chemical reactions.

Around 1900 several different observations were made that were explained by picturing an atom as composed of a nucleus and one or more electrons. The **nucleus** *is at the center of the atom and is small, dense, and positively charged.* ▪Figure 1.6 explains the term *charged.* The positive charge on the nucleus is a result of the presence of protons in the nucleus. **Protons** are *small particles with a unit positive charge. The number of protons in the nucleus* is called the **atomic number** of the element. The atomic number of an element distinguishes atoms of one element from

atoms of another element and shows which element an atom is. An element is a substance composed of atoms that all have the same atomic number (same number of protons). The atomic numbers of all the elements are shown in the table of the elements inside the front cover.

The nucleus is surrounded by rapidly moving, negatively charged particles called electrons. An **electron** is *an extremely small particle (much smaller than an atom) with a unit negative charge.* Atoms are **electrically neutral**—that is, they *do not have a net electric charge.* In an atom, the positive charge of the protons in the nucleus is exactly balanced by the negative charge of the electrons outside the nucleus. The number of electrons outside the nucleus is equal to the number of protons in the nucleus (the atomic number of the element). All atoms of an element have the same number of electrons around the nucleus; atoms of different elements have different numbers of electrons. The number of electrons around the nucleus determines the properties of an atom.

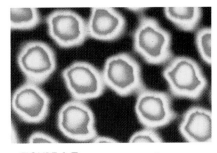

■ FIGURE 1.7 Scanning tunneling micrograph of the surface atoms in a crystal of silicon.

The electric current that lights your lights and runs your TV is a flow of electrons through a metal wire.

PRACTICE PROBLEMS

1.6 Use the table inside the front cover to find the atomic number of (a) aluminum, (b) carbon, (c) neon, (d) silver, and (e) uranium.

1.7 In an atom of each of the elements in Practice Problem 1.6, how many protons are there in the nucleus, what is the charge on the nucleus, and how many electrons are there outside the nucleus?

Beginning in 1956, many pictures of atoms have been taken with a variety of instruments, most recently with devices called scanning probe microscopes. ■Figure 1.7 shows a recent picture of atoms.

1.6 USE OF MODELS IN SCIENCE

Scientists often use models to help them understand their observations and predict the results of experiments. The models you are familiar with are smaller than the real thing. For example, model cars are tiny compared to real cars. However, the models chemists use are much larger than real atoms because real atoms are so very small. The models of atoms in ■Figure 1.8 are 65 million times as large as real atoms. Model cars look like real cars, but chemists do not believe that real atoms look like the models in Figure 1.8. However, these models are very useful. For example, they can be used to make clear the difference between mixtures and compounds. ■Figure 1.9(a) shows beakers containing mixtures of models of hydrogen atoms and oxygen atoms. Obviously, the beaker on the left contains a higher percentage of hydrogen atoms and a lower percentage of oxygen atoms than the beaker on the right. Figure 1.9(b) shows beakers containing models of water. Although the beaker on the left contains more models of water than the beaker on the right, both beakers contain the same percentage of hydrogen atoms (and the same percentage of oxygen atoms). Of the atoms in both beakers, 67% are hydrogen and 33% are oxygen.

The models scientists use don't have to be real physical models. They can be imaginary models that exist only in the scientist's mind. Models can also be mathematical. An important use of computers in chemistry is testing models to see how well they fit observations and predict results of future experiments. Computer models can also be used to predict the results of experiments that are too dangerous or too expensive to be carried out or that are impossible (such as testing the effect

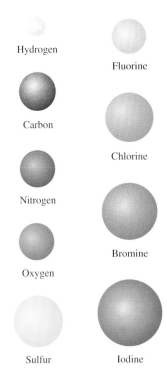

Hydrogen

Fluorine

Carbon

Chlorine

Nitrogen

Bromine

Oxygen

Sulfur   Iodine

■ FIGURE 1.8 Some common atoms. The drawings are 65 million times as large as real atoms. The easiest way to identify atoms in pictures of models (or in the models themselves) is by color.

(a)

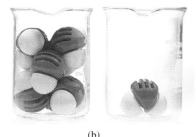

(b)

- FIGURE 1.9 (a) Mixtures of models of hydrogen atoms (white) and oxygen atoms (red) can have different proportions of hydrogen atoms and oxygen atoms. (b) Although the beaker on the left contains more models of the compound water, a combination of two hydrogen atoms and one oxygen atom, than the beaker on the right, the ratio of the number of hydrogen atoms to the number of oxygen atoms is 2 : 1 in both beakers.

For the evidence in favor of the placement of Lu, Lr, La, and Ac shown in this book, see Jensen, W. B. J. Chem. Educ. 1982, 59, 634–636. The names and symbols used for elements 104–109 are those adopted by the American Chemical Society. The IUPAC names and symbols for these elements are given in Section 8.8, where the IUPAC numbering system for the groups of the periodic table is discussed.

The periodic chart on your classroom wall may show La and Ac at the bottom of Group IIIB, Ce and Th at the beginning of the lanthanide and actinide series, and the old, temporary three letter symbols for elements 104–109. Elements 110 and 111 were discovered in 1994 and have not yet been named or assigned symbols.

- FIGURE 1.10 The periodic table.

of changes in Earth's atmosphere on the climate). Computer simulation is now an important part of research in chemistry.

1.7 SYMBOLS

In writing, *atoms are represented by* **symbols.** The symbols for all the elements are given in the table of the elements inside the front cover. Symbols always begin with a capital letter. This capital letter is often the first letter of the name of the element; for example, the symbol for hydrogen is H. However, because the names of more than one element often begin with the same letter, a second letter is frequently included in the symbol. This second letter is *always* a lowercase letter. For example, the symbol for helium is He, and the symbol for aluminum is Al. Sometimes the symbol for an element comes from the name in a language other than English; the symbol for gold is Au, the first two letters of the Latin word for gold, *aurum.*

PRACTICE PROBLEMS

1.8 Use the table inside the front cover to find the symbols for (a) iron, (b) argon, (c) phosphorus, (d) lead, and (e) uranium.

1.9 Use the table inside the front cover to find the name of each of the following elements: (a) Zn, (b) Se, (c) Pu, (d) Xe, and (e) Ba.

1.8 THE PERIODIC TABLE

One of the most useful organizers of chemical information is the **periodic table.** The form of the periodic table that is preferred by most American chemists is shown in -Figure 1.10. The numbers immediately above the symbols for the elements in

Period

	IA	IIA											IIIA	IVA	VA	VIA	VIIA	0
1							1 H											2 He
2	3 Li	4 Be											5 B	6 C	7 N	8 O	9 F	10 Ne
3	11 Na	12 Mg	IIIB	IVB	VB	VIB	VIIB		VIII		IB	IIB	13 Al	14 Si	15 P	16 S	17 Cl	18 Ar
4	19 K	20 Ca	21 Sc	22 Ti	23 V	24 Cr	25 Mn	26 Fe	27 Co	28 Ni	29 Cu	30 Zn	31 Ga	32 Ge	33 As	34 Se	35 Br	36 Kr
5	37 Rb	38 Sr	39 Y	40 Zr	41 Nb	42 Mo	43 Tc	44 Ru	45 Rh	46 Pd	47 Ag	48 Cd	49 In	50 Sn	51 Sb	52 Te	53 I	54 Xe
6	55 Cs	56 Ba *	71 Lu	72 Hf	73 Ta	74 W	75 Re	76 Os	77 Ir	78 Pt	79 Au	80 Hg	81 Tl	82 Pb	83 Bi	84 Po	85 At	86 Rn
7	87 Fr	88 Ra ◆	103 Lr	104 Rf	105 Ha	106 Sg	107 Ns	108 Hs	109 Mt	110	111							

■ Metal ▨ Semimetal
▨ Nonmetal

✱ Lanthanide series

57 La	58 Ce	59 Pr	60 Nd	61 Pm	62 Sm	63 Eu	64 Gd	65 Tb	66 Dy	67 Ho	68 Er	69 Tm	70 Yb

◆ Actinide series

89 Ac	90 Th	91 Pa	92 U	93 Np	94 Pu	95 Am	96 Cm	97 Bk	98 Cf	99 Es	100 Fm	101 Md	102 No

TABLE 1.1	Properties of Group IA Elements				
Symbol	Melting Point, °F	°C	Boiling Point, °F	°C	Reactivity Toward Water
Li	358	181	2419	1326	Reacts slowly with water at room temperature.
Na	208	98	1621	883	Reacts rapidly with water at room temperature.
K	147	64	1393	756	Reacts with water faster than Na does.
Rb	102	39	1270	688	Reacts violently with water.
Cs	84	29	1274	690	Reacts explosively with water.
Fr[a]	—	—	—	—	—

[a]Nobody has ever had enough francium at one time to observe its physical and chemical properties.

■ FIGURE 1.11 Sodium metal is soft enough to be cut with a knife. It looks like silver when freshly cut, but rapidly tarnishes in air.

Figure 1.10 are atomic numbers. In the modern periodic table, the elements are arranged in rows in order of increasing atomic number. The table is constructed so that the elements in a vertical column have similar properties. The *vertical columns in the periodic table* are numbered and are called **groups** or **families.** The *elements in the column labeled 0*—He, Ne, Ar, Kr, Xe, and Rn—are called **noble gases** because they are all unreactive gases. Some of them do not take part in any chemical reactions; none of them take part in very many.

The *elements in the column labeled IA*—Li, Na, K, Rb, and Cs—are all soft metals that look like silver when freshly cut (see ■Figure 1.11). They are called the **alkali metals.** The alkali metals are all very **reactive**—that is, they *take part in reactions with many elements and compounds, and their reactions are very rapid* (■Figure 1.12).

Some physical and chemical properties of the elements in Group IA are given in Table 1.1. ■Figure 1.13 shows the reactions of some of the elements in Group IA with water. As you can see from Table 1.1 and Figure 1.13, the properties of the elements in a group, although similar, are not exactly alike. For the Group IA metals, the properties gradually change going from the top to the bottom of the group. Melting and boiling points decrease from top to bottom of the group but reactivity increases.

◐ Fire and explosion may result from the reaction of sodium with water and will result from the reaction of potassium, rubidium, and cesium with water.

■ FIGURE 1.12 Sodium metal reacts vigorously with many elements and compounds. Sodium must be protected from air and moisture by storage under an unreactive liquid *(left)* or in a sealed can. The reactions with chlorine *(middle)* and water *(right)* are typical.

- FIGURE 1.13 *Top:* Lithium and water. *Bottom:* Potassium and water. As you can see, the reactivity of the alkali metals increases from top to bottom of the group. The metals of Group IA are so reactive that they are used mostly in the form of compounds.

The electronegativity of hydrogen is similar to the electronegativity of carbon rather than to the electronegativity of either lithium or fluorine, and hydrogen resembles carbon by forming an enormous number of covalent compounds.

Some chemists do not consider the elements of Group IIB to be transition elements.

- FIGURE 1.14 Some familiar metals. The cylinder at the lower right is iron. How many of the other metals can you identify?

The *elements in Group IIA* are called the **alkaline earth metals.** Those in *Group VIIA* are called the **halogens.**

PRACTICE PROBLEMS

1.10   Predict how francium will react with water.

1.11   Answer the following questions for these elements (numbers in parentheses are atomic numbers to help you find the elements in the periodic table): Ba (56), Cl (17), K (19), N (7), and Ne (10). (a) Which is a noble gas? (b) Which is an alkali metal? (c) Which is an alkaline earth metal? (d) Which is a halogen? (e) Which is in Group VA? (f) Which would you expect to be similar to Br (35) in its chemical properties?

The *horizontal rows in the periodic table* are called **periods.** The elements hydrogen (H) and helium (He) are in the first period; the elements lithium (Li), beryllium (Be), boron (B), carbon (C), nitrogen (N), oxygen (O), fluorine (F), and neon (Ne) are in the second period.

PRACTICE PROBLEM

1.12   List the symbols and names of the elements of the third period (in order of increasing atomic number).

Hydrogen (H) does not really belong in any group in the periodic table. Hydrogen is like the Group IA elements in some ways, like the Group IVA elements in some, and like the Group VIIA in some. Therefore, hydrogen is placed in the middle of the first period in the periodic table shown in Figure 1.10.

The *two rows at the bottom,* the **lanthanide series** (atomic numbers 57–70) and the **actinide series** (89–102), are placed there to save space. If these elements were put in their places between elements 56 and 71 and 88 and 103, the periodic table would be too wide to fit easily on a page.

PRACTICE PROBLEM

1.13   In what period is the element with atomic number 94?

The *elements in Groups IIIB–IIB* are called **transition elements,** and those in the *lanthanide and actinide series* are called **inner transition elements.** *All the other elements* are called **main group elements** (or sometimes representative elements or nontransition elements).

All of the *elements that are in gray spaces* are **metals.** As you can see, most of the elements are metals. Metals are shiny if their surfaces are clean; all metals except mercury, which is a liquid, are solids under ordinary conditions. Metals usually conduct heat and electricity well and can be rolled or hammered into sheets and pulled into wires (-Figure 1.14). Their chemical properties vary tremendously. Gold and platinum are used in jewelry because they do not react with water or oxygen in the air; rubidium not only reacts violently with water but begins to burn if it is exposed to air.

Elements that do *not* have the properties of metals are called **nonmetals.** In Figure 1.10, the *nonmetals are in purple spaces.* Both the physical and the chemical properties of nonmetals vary greatly. Some nonmetals (hydrogen, nitrogen, oxygen, fluorine, chlorine, and the noble gases) are gases under ordinary conditions; others are solids (carbon, phosphorus, sulfur, and iodine). Only bromine is a liquid (see ▬Figure 1.15). Nitrogen gas is quite unreactive, but one form of phosphorus catches fire when exposed to air and must be stored under water.

Elements that have some metallic properties and some nonmetallic properties are called **semimetals.** *Semimetals are in blue spaces in the periodic table.* As you can see, semimetals form a border between the metals and the nonmetals. Semimetals are all solids and look rather like metals (see ▬Figure 1.16). Notice how the elements change from metals to semimetals to nonmetals from left to right across periods. Also observe how the elements change from nonmetals to semimetals to metals from top to bottom of Groups IVA–VIA.

▬ FIGURE 1.15   Some nonmetals. Clockwise starting at the lower left: sulfur, phosphorus (yellowish white solid), bromine (liquid and gas), and carbon. Phosphorus must be stored underwater because it catches fire if exposed to air.

PRACTICE PROBLEMS

1.14   Answer the following questions for these elements (numbers in parentheses are atomic numbers to help you find the elements in the periodic table): antimony (51), argon (18), bromine (35), calcium (20), molybdenum (42), and uranium (92). (a) Which is or are metals? (b) Which is or are semimetals? (c) Which is or are nonmetals?

1.15   Of the elements in Practice Problem 1.14, (a) which are main group elements, and (b) which are transition elements?

▬ FIGURE 1.16   Semimetals look rather like metals. Germanium *(left)* and silicon *(right)* are used to make semiconducting devices such as computer chips.

 1.9   MOLECULES AND IONS

Molecules

When atoms combine to form compounds, some of their electrons arrange around the nuclei in a way different from the arrangement in the atoms. *When atoms of nonmetals combine to form compounds, molecules result. A molecule of a compound is composed of two or more atoms of at least two different elements.* For example, water is a compound of hydrogen and oxygen. Both hydrogen and oxygen are nonmetals, and the units of water are water molecules. A **molecule** is the *smallest unit of an element or compound that has the chemical properties of the element or compound. A molecule does not have a net electric charge*—that is, a molecule is electrically neutral. In a water molecule, two atoms of hydrogen and one atom of oxygen are joined together [▬Figure 1.17(a)]. A water molecule is the smallest unit into which water can be divided without chemical change. *Compounds that have molecules as units* are called **molecular compounds.**

A molecule of an element consists of one or more atoms of the same element. For example, a single atom of helium is a molecule of helium. A molecule of hydrogen gas is made up of two atoms of hydrogen joined together [see Figure

 The plural of nucleus is **nuclei.**

 Not all substances have molecules as units.

Models in the text are computer-generated. Hydrogen attached to oxygen is smaller than hydrogen attached to chlorine, which is smaller than hydrogen attached to hydrogen, because the electron density around hydrogen depends on the electronegativity of the atom to which hydrogen is bonded.

(a) Water   (b) Hydrogen   (c) Hydrogen chloride

▬ FIGURE 1.17   Models of molecules. Which molecules are diatomic?

1.17(b)]. *Molecules made up of two atoms* are called **diatomic molecules.** Besides hydrogen, other common elements that have diatomic molecules as their units under ordinary conditions are nitrogen, oxygen, fluorine, chlorine, bromine, and iodine. Molecules of some compounds are also diatomic, but the two atoms are different. Hydrogen chloride gas [see Figure 1.17(c)] is an example of a compound with diatomic molecules.

Formulas are used to *represent molecules. Formulas consist of the symbols for each of the elements making up a compound followed by a* **subscript** *showing how many of that kind of atom there are in a molecule of the compound. If no subscript is written after a symbol, the number 1 is assumed.* For example, the formula for a molecule of water is H_2O showing that one molecule of water is a combination of two atoms of hydrogen and one atom of oxygen:

$$H_2O$$

2 hydrogen atoms   1 oxygen atom
(The 1 is assumed when no number is shown.)

The formula for a hydrogen molecule is H_2, and the formula for a molecule of hydrogen chloride is HCl (see Figure 1.17). In formulas, the element that is to the left in the periodic table is usually written on the left side of the formula. If two elements are in the same group, the one that is lower in the column in the periodic table is usually written first. The formula SO_3, which is used in Sample Problem 1.1, is an example.

Compounds of the elements in Groups IVA and VA with hydrogen such as CH_4 (methane) and NH_3 (ammonia) are exceptions to the rule that the symbol for the element that is to the left in the periodic table is written on the left in the formula. You will simply have to memorize these exceptions.

SAMPLE PROBLEM

1.1   How many atoms of each element are there in (a) one molecule of SO_3? (b) Two molecules of SO_3?

Solution   (a) The formula is SO_3:

$$SO_3$$

1 sulfur atom (The 1 is not written.)   3 oxygen atoms

(b) In one molecule of SO_3 there are 1 S atom and 3 O atoms. In two molecules of SO_3, there will be twice as many sulfur and oxygen atoms, or 2 S atoms and 6 O atoms.

PRACTICE PROBLEMS

1.16   How many atoms of each element are there in (a) one molecule of CO_2? (b) One molecule of NH_3? (c) One molecule of H_2SO_4? (d) Three molecules of HCl?

1.17   (a) A molecule of nitrous oxide is composed of two nitrogen atoms and one oxygen atom. What is the formula for nitrous oxide? (b) How many nitrogen atoms are needed to make two molecules of nitrous oxide? (c) A molecule of sulfur dioxide is composed of one sulfur atom and two oxygen atoms. What is the formula for sulfur dioxide?

1.18   Write formulas for molecules of (a) nitrogen and (b) chlorine.

1.19   Which of the following formulas represent elements, and which represent compounds? (a) Br_2 (b) HF (c) CS_2 (d) S_8 (e) KI

Ions

Ions are *charged particles formed by the transfer of electrons from one element or combination of elements to another element or combination of elements.* When atoms of reactive metals such as the elements in Group IA and the elements toward the bottom of Group IIA in the periodic table combine with atoms of nonmetals to form compounds, enough electrons are transferred to give each atom the same number of electrons as the noble gas with the closest atomic number. Because the negative charge of the electrons in an atom equals the positive charge on the nucleus, the loss of electrons leaves an ion with a positive charge. Let's take the formation of a calcium ion from a calcium atom as an example. The atomic number of calcium is 20, and a calcium atom has 20 electrons outside its nucleus.

The noble gas that has an atomic number closest to 20, the atomic number of calcium, is argon with atomic number 18. Therefore, to get the same number of electrons as an argon atom, a calcium atom must lose two electrons. Because atoms are electrically neutral, loss of two electrons leaves a calcium ion with a $2+$ charge; the charge of an ion is written as a right superscript:

$$\underset{\text{calcium atom}}{Ca} \xrightarrow{\text{lose two electrons}} \underset{\text{calcium ion}}{Ca^{2+}}$$

Notice that in the formula for an ion, the charge is written with the number on the left followed by the sign of the charge. *Positively charged ions,* such as Ca^{2+}, are called **cations.** Metals usually form cations.

Nonmetals usually gain electrons when they form ions. Let's consider the fluorine atom, atomic number 9, as an example. The noble gas that has an atomic number closest to 9 is neon with atomic number 10. To get the same number of electrons as a neon atom, a fluorine atom must gain one electron. Because atoms are electrically neutral, a gain of one electron gives a fluoride ion a $1-$ charge. The number 1 is usually not written in the superscript:

$$\underset{\text{fluorine atom}}{F} \xrightarrow{\text{gain one electron}} \underset{\text{fluoride ion}}{F^-}$$

Negatively charged ions such as F^- are called **anions.** Nonmetals usually form anions.

Although a fluoride ion has the same number of electrons as a neon atom, a fluoride ion is different than a neon atom. Fluorine has an atomic number of 9, and a fluoride ion has nine protons in its nucleus; the atomic number of neon is 10, and a neon atom has ten protons in its nucleus. The number of protons in the nucleus determines the identity of an element. In addition, a fluoride ion has a $1-$ charge; a neon atom has no net charge.

Remember that molecules are *not* charged.

Joining the sides of the periodic table so that the atomic numbers are continuous shows that the noble gas closest to calcium, Ca, is argon, Ar.

To remember that cations are positively charged, note that the top of the letter *t* in the word *cation* looks like a + sign. Like molecules, ions are represented by formulas.

SAMPLE PROBLEM

1.2   Write formulas for (a) a lithium ion and (b) an oxide ion.

Solution   (a) The atomic number of lithium is 3; a lithium atom has three electrons. From the periodic table, the noble gas with the closest atomic number is helium, atomic number 2. Therefore, a lithium atom must lose one electron to become a lithium ion. A lithium ion has a $1+$ charge, and its formula is Li^+.

If this reasoning is not clear to you, make a drawing:

Lithium atom loses 1 electron.

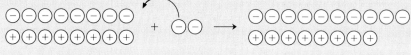

Lithium atom, Li,   gives a   lithium ion, Li$^+$,   +   1 electron, e$^-$

As you can see, the lithium ion has one more positive charge than it has negative charges.

(b) The atomic number of oxygen is 8; an oxygen atom has eight electrons. The noble gas with the closest atomic number is neon, atomic number 10. An oxygen atom with eight electrons must gain two electrons to have ten electrons like a neon atom. Therefore, an oxide ion has a $2-$ charge, and its formula is O^{2-}. Again, if this reasoning is not clear, make a drawing:

Oxygen atom, O,   gains 2 electrons and gives an   oxide ion, O^{2-}

As you can see, the oxide ion has two more negative charges than it has positive charges.

PRACTICE PROBLEMS

1.20   Classify each of the following as either a molecule or an ion: (a) NO_2^-, (b) NO_2, (c) NH_3, and (d) NH_4^+.

1.21   Classify each of the following ions as either an anion or a cation: (a) K^+, (b) F^-, and (c) H^+.

1.22   Write formulas for (a) a sulfide ion and (b) an aluminum ion.

In compounds of reactive metals with nonmetals, there are no molecules. The units of the compound are positive and negative ions. *Compounds composed of ions* are called **ionic compounds.** Compounds do not have net electric charges—that is, compounds are electrically neutral. Because compounds do not have net electric charges, *in compounds, ions must be combined in such a way that the sum of the positive charges equals the sum of the negative charges. Charges must cancel so that the net charge is zero.* For example, the compound sodium chloride, which is composed of sodium ions, Na^+, and chloride ions, Cl^-, must have one Na^+ for each Cl^-. The formula for the compound sodium chloride is NaCl. Note that cations are always written to the left of anions in formulas and named first. The name of a cation is the same as the name of the element from which the cation is formed. The ending of the name of the element from which a monatomic anion is formed is changed to -ide.

SAMPLE PROBLEM

1.3   (a) The units of the compound calcium bromide are calcium ions, Ca^{2+}, and bromide ions, Br$^-$. What is the formula for calcium bromide? (b) The units of the

You probably know from experience that if you touch an object that has built up a charge, such as a person who has walked across a carpet in a dry room, you get a shock. You certainly know that you do not get a shock when you touch spilled salt. Therefore, you know that ionic compounds such as salt have no net charge.

compound aluminum oxide are aluminum ions, Al^{3+}, and oxide ions, O^{2-}. What is the formula for aluminum oxide?

Solution (a) Compounds must not have a net electric charge. Therefore, in the compound calcium bromide, for each calcium ion with a $2+$ charge, there must be two bromide ions each with a $1-$ charge:

$$1(2+) + 2(1-) = 0$$

The formula for calcium bromide is $CaBr_2$.

(b) Compounds must not have a net electric charge. Therefore, in the compound aluminum oxide, there must be three oxide ions for every two aluminum ions:

$$2(3+) + 3(2-) = 0$$

The formula for aluminum oxide is Al_2O_3.

PRACTICE PROBLEMS

1.23 The units of the compound lithium oxide are lithium ions, Li^+, and oxide ions, O^{2-}. What is the formula for lithium oxide?

1.24 The units of the compound aluminum fluoride are aluminum ions, Al^{3+}, and fluoride ions, F^-. What is the formula for aluminum fluoride?

1.25 What is the formula for (a) calcium oxide? (b) Aluminum sulfide?

1.26 Based on the types of elements combined (metal and nonmetal or nonmetal and nonmetal), tell whether the units of each of the following compounds would be molecules or ions: (a) MgO, (b) CS_2, (c) HF, (d) KI, and (e) Mg_3N_2.

Sodium ions are a little smaller than sodium atoms; chloride ions are about twice as large as chlorine atoms. Both sodium ions and chloride ions, however, are so small that a grain of salt contains three billion billion sodium ions and an equal number of chloride ions. ▪Figure 1.18 shows part of a grain of salt magnified about 35 million times.

In solid sodium chloride, each sodium ion is surrounded by six chloride ions, and each chloride ion is surrounded by six sodium ions. The ions are held in place by the *attraction between positive and negative charges,* which is referred to as an **electrostatic force.** Because the attraction between positive and negative charges is strong, solids that are composed of ions such as sodium chloride are hard and have high melting points. (The melting point of sodium chloride is 801 °C or 1474 °F.) The formula for sodium chloride, NaCl, gives the simplest ratio of sodium ions and chloride ions, that is, one sodium ion for each chloride ion.

Polyatomic Ions

Besides simple **monatomic ions** *(formed from one atom)* such as Na^+ and Cl^-, a number of polyatomic ions also exist. A **polyatomic ion** is *a charged particle formed from more than one atom.* The formulas of some common polyatomic ions are given in Table 1.2. (At this stage, don't worry about why the atoms in the common polyatomic ions combine in the ratios they do.) You should memorize the names and formulas, including charge, of the common polyatomic ions in Table 1.2. From these names and formulas, you should be able to figure out the formulas of a number of other polyatomic ions. Sample Problem 1.4 shows the type of reasoning involved.

◗ A unit of $CaBr_2$ has 1 Ca^{2+} and 2 Br^-, not 1 Ca and 1 Br_2 or 1 Ca^{2+} and 1 Br_2^{2-}.

▪ FIGURE 1.18 Chemists think on three different levels: macroscopic, microscopic, and symbolic. The photo of a crystal of common table salt is a macroscopic view. The model of a part of the crystal is a microscopic view, and the formula NaCl is a symbolic representation. In the model, the gray spheres represent sodium ions, Na^+, and the green spheres represent chloride ions, Cl^-. Note that chloride ions are larger than sodium ions. The magnification in this figure is only half the magnification in Figure 1.8. Chloride ions are really almost twice as big as chlorine atoms.

TABLE 1.2

Common Polyatomic Ions[a]

Formula	Name
$C_2H_3O_2^-$	Acetate ion
NH_4^+	Ammonium ion
CO_3^{2-}	Carbonate ion
CN^-	Cyanide ion
HCO_3^-	Hydrogen carbonate ion (bicarbonate ion)
OH^-	Hydroxide ion
NO_3^-	Nitrate ion
ClO_4^-	Perchlorate ion
PO_4^{3-}	Phosphate ion
SiO_4^{4-}	Silicate ion
SO_4^{2-}	Sulfate ion
SO_3^{2-}	Sulfite ion

[a]A more complete list of anions is given in Appendix A.

SAMPLE PROBLEM

1.4 Write formulas for (a) a hydrogen sulfate ion and (b) a nitrite ion.

Solution (a) From Table 1.2, the formula for the carbonate ion is CO_3^{2-}, and the formula for the hydrogen carbonate ion is HCO_3^-. The hydrogen carbonate ion has a hydrogen and one less unit of negative charge than the carbonate ion.

Also from Table 1.2, the formula for the sulfate ion is SO_4^{2-}. Adding a hydrogen and decreasing the negative charge by one unit gives HSO_4^- for the formula of the hydrogen sulfate ion.

(b) The formula for the sulfate ion is SO_4^{2-}, and the formula for the sulfite ion is SO_3^{2-}. The sulfite ion has one less oxygen and the same charge as the sulfate ion. The formula for the nitrate ion is NO_3^-. Removing one oxygen gives NO_2^- for the formula of the nitrite ion.

PRACTICE PROBLEMS

1.27 The formula for the oxalate ion is $C_2O_4^{2-}$. What is the formula for the hydrogen oxalate ion?

1.28 The formula for the chlorate ion is ClO_3^-. What is the formula for the chlorite ion?

○ Compounds that contain polyatomic ions are ionic.

When formulas for compounds containing polyatomic ions are written, parentheses must be used around the polyatomic ion if there is more than one of them. Sample Problem 1.5 shows how to write and interpret the formula for a compound that contains polyatomic ions.

SAMPLE PROBLEM

1.5 (a) Write the formula for calcium nitrate. (b) In one unit of calcium nitrate, how many calcium atoms are there? How many nitrogen atoms? How many oxygen atoms?

Solution (a) A calcium atom has 20 electrons. To get 18 electrons like the nearest noble gas, argon, a calcium ion must lose 2 electrons. The formula for a calcium ion is Ca^{2+}. From Table 1.2, the formula for a nitrate ion is NO_3^-. The compound calcium nitrate must have zero net electric charge. Therefore, for each calcium ion there must be two nitrate ions, and the formula for calcium nitrate is $Ca(NO_3)_2$.

(b) The parentheses around the nitrate ion in the formula for calcium nitrate mean that there are two nitrate ions in each unit of calcium nitrate. In each nitrate ion, there is 1 N and 3 O. Therefore, in one unit of $Ca(NO_3)_2$, there is 1 Ca, 2(1 N) = 2 N, and 2(3 O) = 6 O.

PRACTICE PROBLEM

1.29 (a) Write the formula for ammonium sulfide. (b) In one unit of ammonium sulfide, how many nitrogen atoms are there? Hydrogen atoms? Sulfur atoms?

1. Find the charge on the cation and anion.
   Combine the cation and anion in the simplest way that will result in no net charge.
2. Enclose polyatomic ions in parentheses if there is more than one of them.

> ▶ To solve problems, always try to break down the problem into simpler problems that you know how to do. Divide and conquer works in chemistry too!

1.10 NAMING INORGANIC COMPOUNDS

Inorganic compounds are *compounds of all elements except carbon*. A few simple carbon compounds such as carbon dioxide are also usually classified as inorganic. All **organic compounds** *contain carbon; most also contain hydrogen*. Before 1828 people thought that organic compounds could come only from plants and animals, and inorganic compounds from nonliving sources. In 1828 the German chemist, Friedrich Wöhler, showed that this idea was wrong. Wöhler made urea, a typical organic compound present in urine, by heating a water solution of the inorganic compound ammonium cyanate until all of the water had evaporated. However, classification of compounds into organic and inorganic is still convenient.

Of the over 13 million compounds known, about 91% contain carbon.

The term aqueous *is introduced in Section 1.11.*

PRACTICE PROBLEM

1.30 Classify each of the following compounds as organic or inorganic:
(a) C_2H_2, (b) HCl, (c) $CaCl_2$, (d) H_2O, and (e) C_3H_6O.

Some compounds that have been known for a long time have common names. For example, H_2O is called water, and NH_3 is called ammonia. Common names are often not related to the chemical formula of the compound and must be memorized.

Fortunately, most compounds are named systematically. Systematic names are based on formulas. The system commonly used was developed by a committee of the International Union of Pure and Applied Chemistry (IUPAC).* The version used in the United States was published by the American Chemical Society (ACS).† This system is built on endings such as -ide, -ate, and -ite. The ending -ide usually means that the compound is a combination of two elements.

> ▶ Although you must be careful not to mistake memorization for understanding, memorization is sometimes necessary.

Compounds of Metals and Nonmetals

If one of the two elements in a compound is a metal and the other a nonmetal, the name of the metal is given first. The name of the nonmetal is given second, and the ending of its name is changed to -ide. For example, the compound of sodium and chlorine is called sodium chloride, and the compound of magnesium and oxygen is called magnesium oxide. Unfortunately, the method of changing the nonmetal

---

*The IUPAC is an association of organizations, each of which represents the chemists of a nation. IUPAC works toward cooperation between chemists of different countries, organizes a large variety of international meetings, and establishes international standards for physical constants and units as well as for the names of elements and compounds.
†The ACS, which has a membership of more than 149,000, is the member organization of IUPAC for the United States.

TABLE 1.3

Names and Formulas of Common Monatomic Anions

Formula	Name
Br^-	Bromide ion
Cl^-	Chloride ion
F^-	Fluoride ion
H^-	Hydride ion
I^-	Iodide ion
N^{3-}	Nitride ion
O^{2-}	Oxide ion
S^{2-}	Sulfide ion

name to the -ide ending is not always the same, and the names of the common monatomic anions, which are given in Table 1.3, must be memorized.

SAMPLE PROBLEM

1.6   What is the compound of calcium and fluorine called?

Solution   The compound is a combination of two elements. Calcium is a metal and fluorine is a nonmetal. The name of the metal should be given first. From Table 1.3, the name of the monatomic anion from fluorine is fluoride. The name of the compound is calcium fluoride.

PRACTICE PROBLEM

1.31   What is each of the following called? (a) The compound of potassium and iodine (b) The compound of bromine and magnesium

The endings -ate and -ite show that an anion is polyatomic (see Table 1.2). For example, SO_4^{2-} is called sulfate ion, and SO_3^{2-} is called sulfite ion. However, the names of a few polyatomic ions end in -ide; the most important are the hydroxide ion, OH^-, and the cyanide ion, CN^-. In naming compounds that contain polyatomic ions, the name of the cation is given first followed by the name of the anion. For example, $BaSO_4$ is called barium sulfate, and NH_4CN is called ammonium cyanide.

The metals in Groups IA and IIA and aluminum each form only one compound with a given anion. For example, there is only one kind of sodium chloride (NaCl), one kind of magnesium oxide (MgO), and one kind of aluminum sulfide (Al_2S_3). Many of the other metals form more than one compound with some anions. For example, there are two kinds of iron chloride, $FeCl_2$ and $FeCl_3$, and two kinds of tin oxide, SnO and SnO_2. Roman numerals in parentheses after the name of the metal are used to show which compound is meant. The Roman numerals show the charge on the metal ion (or the charge that the metal would have if it were a cation). For example, $FeCl_2$ is called iron(II) chloride because the iron ion has a 2+ charge, and $FeCl_3$ is called iron(III) chloride because the iron ion has a 3+ charge. In SnO, tin has a 2+ charge, and the name of this compound is tin(II) oxide. In SnO_2, tin has a 4+ charge, and the name of this compound is tin(IV) oxide. There is no space between the name of the metal and the opening parenthesis before the Roman numeral; there is a space between the closing parenthesis and the name of the negative ion.

Metals such as iron and tin do *not* lose enough electrons to get the same number of electrons as the noble gas with the closest atomic number when they form cations. They would have to lose too many electrons to do so. For example, iron (atomic number 26) would have to lose eight electrons to have the same number of electrons (18) as argon, and tin (atomic number 50) would have to lose 14 electrons to have the same number as krypton (atomic number 36). Unlike charges attract each other (remember Figure 1.6); the higher the positive charge on a cation, the harder it is to remove negatively charged electrons from the cation. Cations with charges greater than 4+ are not formed in ordinary chemical reactions. Sample Problem 1.7 shows how to figure out the charges on metal ions from the formulas of compounds so that the compounds can be named.

⬤ Note that you cannot tell how many cations and anions are found in an ionic compound simply from the name.

⬤ The Roman numeral or Stock system of nomenclature is explained more fully in Section 11.3.

1.7   Copper forms two oxides, Cu_2O and CuO. What is the systematic name of each of the oxides of copper?

Solution   Let's do Cu_2O first. From Table 1.3, the charge on an oxide ion is $2-$. Because Cu_2O is a compound, its net charge must be zero. For the net charge to be zero, the total charge of the two copper ions must be $2+$. Therefore, the charge on each copper ion is $1+$ and the name of the compound is copper(I) oxide.

For the net charge on CuO to be zero, the charge on the copper ion must be $2+$. The name of this oxide of copper is copper(II) oxide.

Tin(II) oxide, SnO (left) and tin(IV) oxide, SnO_2 (right). Tin(IV) oxide is used to polish glass and metals and to make glazes and enamels opaque. It is also the main source of tin metal.

1.32   Write formulas for the following compounds: (a) vanadium(II) chloride, (b) vanadium(II) oxide, and (c) vanadium(V) oxide.

1.33   Name the following compounds: (a) CrS, (b) Cr_2S_3, (c) TiO, and (d) TiO_2.

Compounds of Two Nonmetals

Compounds of two nonmetals are named similarly to compounds of a metal and a nonmetal. The symbol for the element that is to the left in the periodic table is written first in the formula, and this element is named first in the name. The other element is named as if it were an anion. For example, the formula for the compound of hydrogen and chlorine is written HCl, and the compound is called hydrogen chloride. If both elements are in the same group (vertical column) in the periodic table, the lower one is named first, and the higher one is named as if it were an anion. Thus, the formula for the compound of bromine and chlorine is written BrCl, and the compound is called bromine chloride.

If two nonmetals form more than one compound with each other, Greek prefixes are used to show which compound is meant. Table 1.4 shows some common Greek prefixes. The prefix mono- is often omitted from before the name of the first element. If there is no prefix before the name of the first element, mono- is understood. Thus, CO_2 is called carbon dioxide, SO_2 is called sulfur dioxide, and SO_3 is called sulfur trioxide.

The final *a* or *o* of a prefix is sometimes omitted before vowels. For example, N_2O_4 is called dinitrogen tetroxide, ~~not dinitrogen tetraoxide~~; CO is called carbon monoxide, ~~not carbon monooxide~~. Why is the system not always systematic? Chemists, like most people, prefer doing things the way they are used to. The IUPAC has been forced to accept a number of established customs.

The rules for naming inorganic compounds that we have learned are summarized in ▪Figure 1.19. We will learn additional common names and other rules for naming compounds when we need them.

TABLE 1.4

Common Greek Prefixes

Prefix	Number
Mono-	1
Di-	2
Tri-	3
Tetra-	4
Penta-	5
Hexa-	6
Hepta-	7
Octa-	8
Nona-	9
Deca-	10

▶ Other prefixes are not usually omitted. However, H_2S is commonly called hydrogen sulfide not dihydrogen sulfide.

▶ A brief summary of the current U.S. version of the rules for naming inorganic compounds is given in Appendix A.

To avoid confusing students when discussing what is wrong, the wrong expression is shown "crossed out."

1.34   Write the formula for (a) disulfur dichloride and (b) phosphorus tribromide.

1.35   Name the following compounds: (a) N_2O and (b) P_2S_5.

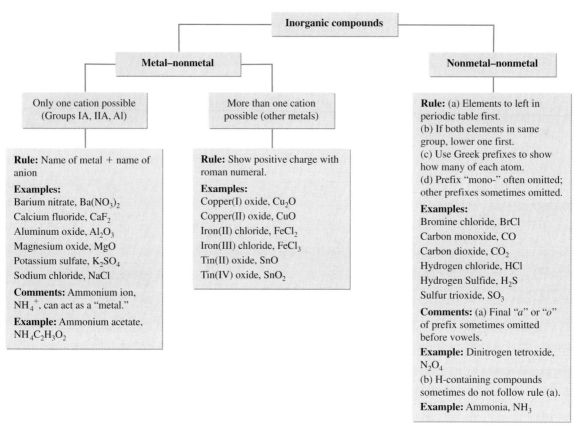

■ FIGURE 1.19 Naming inorganic compounds.

CHEMICAL EQUATIONS

Chemical equations summarize information about chemical reactions. To write a chemical equation, you must first identify the substances that are present before and after reaction. Then write a word equation. For example, the word equation for the decomposition of water to hydrogen gas and oxygen gas by electricity is

$$\text{water} \xrightarrow{\text{electricity}} \text{hydrogen gas} + \text{oxygen gas}$$

(As you gain experience, you will probably be able to skip this step and proceed directly to the second step.)

The substance or substances on the left side of the arrow, such as water, are called **reactants.** Reactants are *substances that are present before reaction takes place.* The substance or substances on the right side of the arrow, such as hydrogen gas and oxygen gas, are called **products.** Products are *substances that are formed in a reaction.* The arrow means "react to form." The plus sign means "and." Any special conditions needed to make the reaction take place in a useful period of time are written above or below the arrow. In the example, water does not decompose unless electricity is passed through it.

The second step in writing a chemical equation is to write the formulas for the reactants and products. If you have written a word equation, as you should at the beginning, write the formulas under the names. Elements are represented by their symbols except for elements such as hydrogen and oxygen that exist as molecules containing more than one atom:

$$\text{water} \xrightarrow{\text{electricity}} \text{hydrogen gas} + \text{oxygen gas}$$

$$\text{H}_2\text{O} \xrightarrow{\text{electricity}} \text{H}_2 + \text{O}_2 \qquad (1.1)$$

The third step is to **balance,** that is, to *make the expression show the same number of atoms of each kind on both sides of the arrow.* Models show why equations must be balanced:

1 water molecule   1 hydrogen molecule   1 oxygen atom

From one molecule of water, one molecule of hydrogen gas can be formed. But a molecule of oxygen gas, O_2, cannot be formed because there is only one oxygen atom in one molecule of water. If two molecules of water decompose, a molecule of oxygen gas can be formed:

2 water molecules   2 hydrogen molecules   1 oxygen molecule

Two molecules of hydrogen gas are formed from two molecules of water.

An equation is balanced by writing **coefficients** in front of the formulas. If no coefficient is shown, the coefficient is assumed to be the number 1. One way of balancing the expression for the decomposition of water (reaction 1.1) is to write a 2 in front of the formula for water so that two oxygens are shown on the left as well as on the right:

$$2\text{H}_2\text{O} \xrightarrow{\text{electricity}} \text{H}_2 + \text{O}_2$$

Now there are four hydrogen atoms on the left and only two on the right. Writing a 2 in front of the formula for hydrogen makes the number of each kind of atom on the left the same as the number of the same kind of atom on the right:

$$2\text{H}_2\text{O} \xrightarrow{\text{electricity}} 2\text{H}_2 + \text{O}_2$$

The expression is now balanced and is an equation for the reaction. The equation shows the same number of each kind of atom on the product side as on the reactant side: four atoms of hydrogen and two atoms of oxygen on each side.

An equation must be balanced by changing the coefficients in front of the formulas. It is wrong to change the formulas. Formulas are facts found by experiments; people cannot change them. For example, the expression

$$\text{H}_2\text{O}_2 \xrightarrow{\text{electricity}} \text{H}_2 + \text{O}_2$$

is a balanced equation because it shows the same number of each kind of atom on both sides. However, it is *not* the equation for the decomposition of water. The formula for water is H_2O not H_2O_2; H_2O_2 is the formula for a different compound of hydrogen and oxygen called hydrogen peroxide. Adding or removing a reactant or product to make an equation balance is also wrong because what the reactants and products are is an experimental fact.

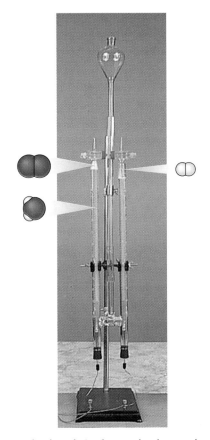

For the electrolysis of water, the photograph above is a macroscopic view. The models in the text show a microscopic view, and the equation is a symbolic representation.

Hydrogen peroxide is used as a rocket propellant. Aqueous solutions of hydrogen peroxide are used to bleach textiles and wood pulp and to kill germs.

Equations should be as simple as possible; that is, they should show the smallest possible number of units that can react. If the coefficients have a common denominator, an equation should be simplified by dividing by the common denominator. For example, the equation

$$4H_2O \xrightarrow{\text{electricity}} 4H_2 + 2O_2$$

should be simplified to

$$2H_2O \xrightarrow{\text{electricity}} 2H_2 + O_2$$

by dividing by 2.

An expression, such as (1.1), that shows the formulas of reactants and products but is not balanced is *not* an equation. *An expression is not an equation unless it is balanced; an equation must show the same number of each kind of atom on both sides of the arrow.*

The symbols (s) for a solid, (l) for a liquid, and (g) for a gas can be used after symbols and formulas to add more information to an equation:

$$2H_2O(l) \xrightarrow{\text{electricity}} 2H_2(g) + O_2(g)$$

The symbol (aq) is used for a solution of a substance in water. The letters *a* and *q* are the first two letters of *aqua*, the Latin word for water. *Water solutions* are usually referred to as **aqueous solutions.**

Biological reactions take place in aqueous solution. Many geological processes involve aqueous solutions as do a number of industrially important reactions and most of the reactions carried out in general chemistry lab.

A more systematic method of balancing equations by inspection is given in Chapter 3.

Summary of Steps for Writing a Chemical Equation

1. Identify reactants and products and write a word equation.
2. Write symbols for elements (formulas for elements existing as polyatomic molecules) and formulas for compounds.
3. Balance by changing coefficients in front of symbols and formulas. Do not change formulas or add or remove substances.
4. Check to be sure the same number of each kind of atom is shown on both sides. If coefficients have a common divisor, simplify.
5. Add symbols showing whether substances are solids, liquids, gases, or in aqueous solution.

SAMPLE PROBLEM

1.8 Write a chemical equation for each of the following reactions: (a) In sunlight, hydrogen gas reacts explosively with chlorine gas. Hydrogen chloride gas is formed. (b) When an aqueous solution of sodium chloride is added to an aqueous solution of lead(II) nitrate, solid lead(II) chloride and an aqueous solution of sodium nitrate are formed.

Solution (a) **Step 1.** The reactants are hydrogen gas and chlorine gas. Hydrogen chloride gas is the product. Sunlight is required for reaction to take place. The word equation is

$$\text{hydrogen gas} + \text{chlorine gas} \xrightarrow{\text{sunlight}} \text{hydrogen chloride gas}$$

Step 2. $H_2 \quad + \quad Cl_2 \xrightarrow{\text{sunlight}} \quad HCl$

Step 3. $H_2 \quad + \quad Cl_2 \xrightarrow{\text{sunlight}} \quad 2HCl$

Step 4. There are 2 H and 2 Cl on both the reactant and the product side.

Step 5. $H_2(g) \quad + \quad Cl_2(g) \xrightarrow{\text{sunlight}} \quad 2HCl(g)$

(b) **Step 1.** The reactants are aqueous solutions of sodium chloride and lead(II) nitrate; the products are solid lead(II) chloride and an aqueous solution of sodium nitrate. The word equation is

aqueous sodium chloride	+	aqueous lead(II) nitrate	$\longrightarrow$	solid lead(II) chloride	+	aqueous sodium nitrate

Step 2. $NaCl$ + $Pb(NO_3)_2$ $\longrightarrow$ $PbCl_2$ + $NaNO_3$

Step 3. $2NaCl$ + $Pb(NO_3)_2$ $\longrightarrow$ $PbCl_2$ + $2NaNO_3$

Step 4. There are 2 Na, 2 Cl, 1 Pb, 2 N, and 6 O on each side. (Because the nitrate ions are not changed in the reaction, balance could be checked by observing that there are 2 Na, 2 Cl, 1 Pb, and 2 NO_3 on each side.)

Step 5. $2NaCl(aq)$ + $Pb(NO_3)_2(aq)$ $\longrightarrow$ $PbCl_2(s)$ + $2NaNO_3(aq)$

Aqueous solutions of sodium chloride (left) and lead(II) nitrate (middle) are clear and colorless. When mixed (right), they yield lead(II) chloride, an insoluble white solid.

PRACTICE PROBLEMS

1.36   Given the equation

$$2AgNO_3(aq) + H_2S(g) \longrightarrow Ag_2S(s) + 2HNO_3(aq)$$

(a) List the formulas of the reactants. (b) List the formulas of the products. (c) How many atoms of each element appear on each side of the equation? (d) Which substances are in aqueous solution? (e) Which substance is a solid?

1.37   Which of the following expressions is a chemical equation? Explain your answer: (a) $CO(g) + H_2O(g) \longrightarrow CO_2(g) + H_2(g)$; (b) $HgO(s) \longrightarrow$ $Hg(l) + O_2(g)$; and (c) $Zn(s) + HCl(aq) \longrightarrow ZnCl_2(aq) + H_2(g)$.

1.38   Balance each of the following expressions: (a) $Mg(s) + O_2(g) \longrightarrow$ $MgO(s)$; (b) $N_2(g) + H_2(g) \longrightarrow NH_3(g)$; (c) $H_2SO_4(aq) + NaOH(aq) \longrightarrow$ $Na_2SO_4(aq) + H_2O(l)$.

1.39   State the information given in the following equation in words:

$$2NaCl(aq) + 2H_2O(l) \xrightarrow{\text{electrolysis}} H_2(g) + Cl_2(g) + 2NaOH(aq)$$

1.40   Write an equation for the reaction of hot copper metal with oxygen gas to form copper(II) oxide. Copper(II) oxide is a solid.

⊙ Don't forget to check your work.

1.12   PREDICTING REACTIONS

With the help of the periodic table, many chemical changes can be predicted from a few observations. For example, from an observation that sodium metal reacts with water to form a solution of sodium hydroxide and hydrogen gas:

$$2Na(s) + 2H_2O(l) \longrightarrow 2NaOH(aq) + H_2(g)$$

We can predict that the other members of Group IA will also react with water to give a solution of the metal hydroxide and hydrogen gas. In other words, we can predict that the following reactions will take place:

$$2Li(s) + 2H_2O(l) \longrightarrow 2LiOH(aq) + H_2(g)$$

$$2K(s) + 2H_2O(l) \longrightarrow 2KOH(aq) + H_2(g)$$

$$2Rb(s) + 2H_2O(l) \longrightarrow 2RbOH(aq) + H_2(g)$$

$$2Cs(s) + 2H_2O(l) \longrightarrow 2CsOH(aq) + H_2(g)$$

⊙ Remember that the reactions of some Group IA metals with water are shown in Figures 1.12 and 1.13.

The Oldest Reaction

The first reaction known to be carried out by humans was combustion (burning). Combustion is the rapid reaction of materials with oxygen, which normally comes from air. Both heat and light are usually given off during combustion. Energy, such as a lighted match, must commonly be supplied to start reaction. Once started, enough heat is ordinarily given off so that a material continues to burn.

Evidence for the use of fire is found with primitive tools and human bones from at least half a million years ago. Fire was used for cooking, which makes food safer to eat, for protection from hungry animals, and probably also for heating. Fires were used to drive game, to clear forests of underbrush so that game could be seen better, and to clear fields to increase the yield of food from farming. Repeated burning of plants, which changes the composition of the soil and leads to erosion by water and wind, probably led to the formation of grasslands and prairies, the pinewoods of the southeastern United States, and the teak forests of Burma—perhaps humans' first significant influence on the earth.

People probably first used fire for light about 40,000 years ago. They first learned to make fire by friction around 7000 B.C.

The ancient Greeks classified fire as one of the four elements—air, earth, fire, and water—that make up all matter both living and nonliving. They used fire to obtain iron, copper, and tin from their ores; to make bronze by alloying copper with tin; to make pottery; in war; and as signals.

The amount of energy released by combustion has continually increased as the population of the world has increased and each person has used more energy. Today combustion is still often used for

cooking, heating, lighting, and bringing about chemical reactions. Fire is used in war, and slash-and-burn agriculture is widely used in developing countries. In addition, combustion of gasoline powers cars, combustion of jet fuel propels airplanes, and combustion of rocket fuel powers rockets. Burning of coal and natural gas in power plants is used to generate electricity; in fact, combustion supplies more than 90% of the world's energy at the present time.

All of these reactions have been carried out many times in many places by many different people and always found to take place as predicted.

Practice Problems are continued at the end of the chapter under the heading Additional Practice Problems. To emphasize the importance of the in-chapter Practice Problems, the Additional Practice Problems at the end of the chapter continue the sequence of numbering begun in the chapter.

PRACTICE PROBLEM

1.41 The reaction,

$$2KNO_3(s) \xrightarrow{\text{heat}} 2KNO_2(s) + O_2(g)$$

can be used to make oxygen gas in the laboratory. Write equations for the reactions that will probably take place when the following compounds are heated:
(a) $LiNO_3(s)$ and (b) $NaNO_3(s)$.

Predictions based on the periodic table are correct often enough to be very useful. However, to be sure that reactions will really take place as expected, predictions must be tested by experiment in the laboratory. For example, the products of the reactions of potassium, rubidium, and cesium with oxygen are not similar to the products of the reaction of either lithium or sodium with oxygen:

$$4Li(s) + O_2(g) \longrightarrow 2Li_2O(s)$$

$$2Na(s) + O_2(g) \longrightarrow Na_2O_2(s)$$

$$K(s) + O_2(g) \longrightarrow KO_2(s)$$

$$Rb(s) + O_2(g) \longrightarrow RbO_2(s)$$

$$Cs(s) + O_2(g) \longrightarrow CsO_2(s)$$

Unexpected results like this are what make life interesting to a chemist. On obtaining an unexpected result like this, a chemist would first repeat the experiment to make sure that the result was correct. He or she would then try to figure out an explanation and publish the work so that other chemists could use it. If the results were important to other chemists, the other chemists would probably check the results by experiment before using them.

SUMMARY

Summaries are glossaries in narrative form. If at a later time you need to find the meaning of any terms introduced in this chapter, use the index at the back of the book (blue-edged pages) to find the number of the page where the term is defined.

Chemistry concerns the composition, structure, and properties of matter and the changes that matter undergoes. **Matter** is anything that occupies space; all matter is made of chemicals. A **substance** is a distinct kind of matter. All samples of a substance have the same properties; a substance has the same properties throughout the whole sample. Properties that do not involve substances changing into other substances are called **physical properties.** Properties that involve substances changing into other substances are called **chemical properties.** The changes are called **chemical changes** or **chemical reactions.** Substances that react rapidly with many other substances are said to be **reactive.** Reactions in which substances combine to form more complex substances are called **combination reactions** and reactions in which substances break down **(decompose)** into simpler substances are called **decomposition reactions.**

There are two kinds of substances, elements and compounds. **Elements** cannot be broken down into simpler substances by chemical reaction. **Compounds** are substances composed of two or more of the 111 known elements. Compounds can be broken down into simpler substances by decomposition reactions. Over 13 million compounds are known.

Although compounds can be decomposed into simpler substances by chemical reactions, they cannot be separated into simpler substances by physical changes. Mixtures have varying compositions and can be separated by physical changes. Substances making up mixtures are called **components. Homogeneous mixtures** have only one phase and are called **solutions.**

A **phase** is a sample of matter that is uniform in composition and physical state and is separated from other phases by definite boundaries. The term **aqueous** refers to a water solution. Mixtures that do not have the same properties throughout are called **heterogeneous mixtures.**

Chemistry, like all sciences, is based on observations. In science, a **law** is a summary of many observations and a **theory** is an explanation of a law. As soon as a new theory has been proposed, scientists start thinking of experiments to test it. Science progresses by cycles of proposed theories and tests by experiment. A theory can be proved wrong but it can never be proved right beyond any shadow of a doubt. The overturning of an important theory opens new frontiers in science. Scientists must be careful not to jump to conclusions based on too few or wrong observations; they must be alert to observations that do not fit their theories.

The **law of constant composition** states that the percent by weight of each element in a compound is always the same. This law is explained by the atomic theory. According to **Dalton's atomic theory** everything is made of small particles called **atoms.** All atoms of an element are alike, and atoms of different elements are different. Atoms of one element cannot be changed into atoms of another element by chemical reactions. In chemical reactions, atoms are combined in new ways; the relative number of each kind of atom in a compound is always the same.

According to modern atomic theory, atoms are pictured as being made up of a small, dense, positively charged **nucleus** that contains protons. **Protons** are small particles with a unit positive charge. The number of protons in the nucleus determines the **atomic number** of the element. All atoms of an element have the same atomic number. The nucleus is surrounded by rapidly moving **electrons,** which are very small particles with a unit negative charge. Atoms are **electrically neutral**—that is, they have no net charge; therefore, the number of electrons outside the nucleus must be equal to the number of protons in the nucleus (and to the atomic number).

The **periodic table** is one of the most useful organizers of chemical information. It can be used to predict many chemical and physical properties from a few observations. In the modern periodic table, the elements are arranged in rows in order of increasing atomic number. The table is constructed so that elements in a vertical column have similar properties; the vertical columns are called **groups** or **families.** The groups are numbered. Some of the groups also have names: Group IA, **alkali metals;** Group IIA, **alkaline earth metals;** Group VIIA, **halogens;** Group 0, **noble gases.** The properties of the elements in a group usually change gradually going from top to bottom of the group. In Groups IVA–VIA, properties change from nonmetallic to metallic. Horizontal rows in the periodic table are called **periods;** properties change from metallic to nonmetallic from left to right across periods. The elements in Groups IIIB–IIB and those in the lanthanide and actinide series are called **transition elements.** All other elements are called **main group elements.** In the periodic table inside the front cover, elements in gray spaces are **metals,** elements in blue spaces are **semimetals,** and elements in purple spaces are **nonmetals.**

When atoms combine to form compounds, some of their electrons arrange around the nuclei in a way different from the arrangement in the atoms. When atoms of nonmetals combine to form compounds, molecules result. A **molecule** is the smallest unit of an element or compound that has the chemical properties of the element or compound; a molecule does not have a net electric charge. Molecules made up of two atoms are called **diatomic molecules.** Compounds that have molecules as units are referred to as **molecular compounds.**

Ions are charged particles formed by transfer of electrons. When atoms of the metals of Group IA and the reactive metals toward the bottom of Group IIA combine with atoms of nonmetals to form compounds, enough electrons are transferred to give each atom the same number of electrons as the noble gas with the closest atomic number. Positively charged ions are called **cations,** and negatively charged ions are called **anions. Monatomic ions** are formed from a single atom; **polyatomic ions** are formed from more than one atom. Ions combine to form compounds so that charges cancel and the compounds are electrically neutral. Compounds composed of ions are called **ionic compounds.** In crystals of ionic compounds, oppositely charged ions are held together by **electrostatic force,** the attraction between opposite charges.

Inorganic compounds are compounds of all elements except carbon. Some simple carbon compounds are also usually classified as inorganic. All **organic compounds** contain carbon; most also contain hydrogen. Most compounds are named according to the IUPAC system. In naming inorganic compounds according to the IUPAC system, Roman numerals are used to distinguish between different compounds of a metal with the same anion. Prefixes are used to distinguish between different compounds of the same nonmetals.

Chemical equations summarize information about chemical reactions. In equations, compounds and molecules of elements are represented by **formulas,** and atoms of elements are represented by **symbols. Reactants,** substances present before reaction, are written on the left. **Products,** substances formed in a reaction, are written on the right. Equations must be **balanced** (made to show the same number of atoms of each kind on both sides) by using the smallest possible whole numbers for **coefficients.** Equations may show the states of the reactants and products and the conditions necessary for reaction.

ADDITIONAL PRACTICE PROBLEMS

The first Additional Practice Problems follow the order of the material presented in the chapter; the numbers in parentheses at the ends of these problems show the number of the section that the problems are related to. Later Additional Practice Problems do not follow the chapter order and are not classified by section because learning to identify different types of problems is an important part of learning to solve problems. Answers to problems with blue numbers appear in Appendix I. Solutions to these problems are in the Student Solutions Manual. Solutions to Additional Practice Problems with black numbers may be available at the option of your instructor.

1.42 What is chemistry? (Chapter opener)

1.43 Classify each of the following as an observation or a conclusion: (a) The solid is cold. (b) The solid is ice. (c) The

liquid is milk. (d) The liquid is orange. (e) There was a loud, sudden noise. (1.1)

1.44 Identify the phases present in a cardboard box containing chocolate-covered raisins. (1.2)

1.45 Which of the following statements about phosphorus describe chemical properties? Which describe physical properties? (a) Phosphorus is never found free in nature. (b) The common form of phosphorus is white. (c) Phosphorus melts at 44.1 °C. (d) Phosphorus catches fire spontaneously in air. (e) Phosphorus can be stored under water. (f) Phosphorus dissolves in carbon disulfide. (1.2)

1.46 Define *decomposition* and give an example of a reaction that is decomposition. (1.2)

1.47 Define *combination* and give an example of a reaction that is combination. (1.3)

1.48 (a) What is the difference between a *heterogeneous mixture* and a *homogeneous mixture?* (b) In what ways are they alike? (1.3)

1.49 Classify each of the following as compound, element, or mixture. Tell whether mixtures are solutions or heterogeneous mixtures: (a) aluminum foil, (b) blueberry muffins, (c) hot tea without any tea leaves in it, (d) salt, (e) carbon dioxide. (1.3)

1.50 Give one contribution that the alchemists made to chemistry. (RT)

1.51 Summarize Dalton's atomic theory in your own words. (1.5)

1.52 For each of the following elements, give the number of protons in the nucleus, the charge on the nucleus, and the number of electrons outside the nucleus: (a) bromine, (b) gold, (c) iron, (d) phosphorus, (e) sodium. (1.5)

1.53 What is the name and symbol of each element? (a) An atom of the element has 88 protons in its nucleus. (b) The nuclear charge of an atom is $9+$. (c) An atom of the element has 28 electrons around its nucleus. (1.5)

1.54 Give the names of the elements represented by the following symbols: (a) I, (b) Si, (c) Mg, (d) Pb, (e) Fe. (1.7)

1.55 Write symbols for (a) antimony, (b) bromine, (c) manganese, (d) potassium, (e) calcium. (1.7)

1.56 Of the following elements, Ca, He, I, Li, Sn, which is (a) an alkali metal? (b) an alkaline earth metal? (c) in Group IVA? (d) a halogen? (e) a noble gas? (1.8)

1.57 How many elements are in (a) the fourth period? (b) The sixth period? (1.8)

1.58 List the symbols and names of the elements that have chemical properties similar to chlorine. (1.8)

1.59 Give three physical properties of metals. (1.8)

1.60 Of the elements in Group IVA, which are (a) metals? (b) Semimetals? (c) Nonmetals? (1.8)

1.61 Answer the following questions for these elements (numbers in parentheses are atomic numbers): europium (63), potassium (19), sulfur (16), tungsten (74), and xenon (54). (a) Which are main group elements? (b) Which is a transition element? (c) Which are nonmetals? (d) Which is in the lanthanide series? (e) Which are in the sixth period? (1.8)

1.62 How many atoms of each kind are represented by the following formulas? (a) NH_3 (b) NH_4NO_3 (c) $(NH_4)_3PO_4$ (d) $Ca(C_2H_3O_2)_2$ (1.9)

1.63 A molecule of phosphoric acid is composed of three hydrogen atoms, four oxygen atoms, and one phosphorus atom. What is the formula for phosphoric acid? (1.9)

1.64 Which of the following formulas represent elements and which represent compounds? (a) $MgCl_2$ (b) P_4 (c) CH_4 (d) NO (e) C_6H_6 (1.9)

1.65 Classify each of the following as either a molecule or an ion: (a) H_3O^+, (b) H_2O, (c) H_2O_2, (d) OH^-. (1.9)

1.66 Classify each of the following ions as either an anion or a cation: (a) Ca^{2+}, (b) H^-, (c) I_3^-, (d) K^+, (e) SO_3^{2-}. (1.9)

1.67 Write formulas for (a) a magnesium ion, (b) a chloride ion. (1.9)

1.68 Write formulas for (a) the compound barium oxide. Units are barium ions, Ba^{2+}, and oxide ions, O^{2-}. (b) The compound ammonium carbonate. Units are ammonium ions, NH_4^+, and carbonate ions, CO_3^{2-}. (1.9)

1.69 Classify each of the following compounds as ionic or molecular: (a) K_2CO_3, (b) CO_2, (c) $CaSO_4$, (d) NH_4Cl, (e) SO_3. (1.9)

1.70 Write formulas for (a) perbromate ion, (b) hydrogen sulfite ion, (c) hydrogen phosphate ion. (1.9)

1.71 Tell whether each of the following compounds is inorganic or organic: (a) $CoBr_2$, (b) $C_4H_{10}O$, (c) Cr_2O_3, (d) $HC_2H_3O_2$, (e) NH_3. (1.10)

1.72 What is each of the following called? (a) The compound of calcium and oxygen (b) The compound of sodium and sulfur (1.10)

1.73 Write formulas for the following compounds: (a) manganese(II) bromide, (b) manganese(II) sulfide, (c) manganese(IV) sulfide. (1.10)

1.74 Write formulas for (a) phosphorus pentachloride, (b) carbon disulfide, (c) oxygen. (1.10)

1.75 Name the following compounds: (a) FeO, (b) Fe_2O_3, (c) N_2O_3, (d) SF_6, (e) TiN. (1.10)

1.76 Given the equation

$$(NH_4)_2S(aq) + 2HCl(aq) \longrightarrow H_2S(g) + 2NH_4Cl(aq)$$

(a) List the formulas of the reactants. (b) List the formulas of the products. (c) How many atoms of each element appear on each side of the equation? (d) Which substances are in aqueous solution? (e) Which substance is a gas? (1.11)

1.77 Supply the coefficients needed to make each of the following expressions into an equation:

(a) $Ag_2O(s) \xrightarrow{\text{heat}} Ag(s) + O_2(g)$

(b) $H_2(g) + F_2(g) \longrightarrow HF(g)$

(c) $CaCO_3(s) \xrightarrow{\text{heat}} CaO(s) + CO_2(g)$

(d) $Ca(s) + H_2O(l) \longrightarrow Ca(OH)_2(s) + H_2(g)$

(e) $Cu(s) + AgNO_3(aq) \longrightarrow$
$$Ag(s) + Cu(NO_3)_2(aq) \quad (1.11)$$

1.78 State the information given in the following equation in words:

$$S(s) + O_2(g) \longrightarrow SO_2(g) \quad (1.11)$$

1.79 In a reaction, the reactants are magnesium and oxygen. The product is magnesium oxide, which is a white solid. Write the equation for the reaction. (1.11)

1.80 The reaction

$$HCl(aq) + NaOH(aq) \longrightarrow NaCl(aq) + H_2O(l)$$

is well known. Write equations for the reactions that will probably take place when the following aqueous solutions are mixed: (a) HBr(aq) + NaOH(aq), (b) HCl(aq) + KOH(aq), (c) HBr(aq) + KOH(aq). (1.12)

1.81 What is combustion? (RT)

1.82 The formula of a compound formed between silicon and oxygen is SiO_2. Predict the formula of the compound formed between (a) carbon and oxygen, (b) silicon and sulfur, (c) carbon and sulfur.

1.83 Write formulas for the following compounds: (a) manganese(III) oxide, (b) dibromine monoxide, (c) strontium nitrate, (d) sodium perchlorate, (e) ammonium carbonate.

1.84 Name the following compounds: (a) LiCl, (b) SO_2, (c) V_2O_5, (d) $Ca(C_2H_3O_2)_2$, (e) Al_2S_3.

1.85 (a) Can the original mixture of iron filings and powdered sulfur be separated with a magnet? (b) What can you conclude happened to the mixture when it was heated? Explain your answer.

A mixture of iron filings and powdered sulfur before (left) and after (right) heating.

1.86 (a) $Mg(OH)_2$ does not dissolve in water at room temperature. $Ca(OH)_2$ and $Sr(OH)_2$ dissolve to a small extent; more $Sr(OH)_2$ than $Ca(OH)_2$ dissolves in a cup of water. Will more or less $Ba(OH)_2$ than $Sr(OH)_2$ dissolve in a cup of water? Explain your answer. (b) $MgSO_4$ dissolves in water at room temperature. $CaSO_4$ only dissolves to a small extent, and $SrSO_4$ does not dissolve. Will $BaSO_4$ dissolve? Explain your answer.

1.87 Magnesium metal does not react with water at room temperature. Calcium metal reacts slowly with water at room temperature. Predict how strontium metal behaves toward room temperature water.

1.88 Explain why (a) a potassium ion has a 1+ charge, (b) an oxide ion has a 2− charge.

1.89 Explain why sodium sulfide must have two sodium ions for each sulfide ion.

1.90 For the formation of hydrogen chloride by the reaction of hydrogen with chlorine, show the predicted result of each of the following combinations of molecules by completing the diagram:

(a)

(b)

(c)

STOP & TEST YOURSELF

Before going on to problems that require combining ideas and skills, test your mastery of individual skills and ideas by choosing the best answer for each of the following questions. Then check your responses in Appendix I. If you have answered a question incorrectly, review the section identified by number. The answers are discussed in the Student Solutions Manual.

1. Which of the following changes is (are) chemical changes?
   (i) Heat from a stove flows through an aluminum pan to the water inside.
   (ii) The water boils.
   (iii) An egg is hard-boiled.
   (iv) The shell is peeled off the egg.
   (v) The silver spoon used to eat the egg turns black.
   (a) i, ii, and iv   (b) iii and v   (c) only iii   (d) only v
   (e) None of the preceding answers is correct.

2. Which of the following statements is (are) true of solutions?
   (i) The properties of a single sample are the same everywhere in the sample.
   (ii) The properties of a single sample are different at different points in the sample.
   (iii) The composition of different samples may be different.
   (iv) A sample can only be separated into its components by a chemical reaction.
   (v) A sample can be separated into its components by physical means.
   (a) i, iii, and v   (b) ii and iv   (c) iii and v   (d) ii, iii, and v
   (e) only i

3. Which of the following is (are) elements?
   (i) CO   (ii) Co   (iii) C   (iv) Cl_2 (v) CaC_2
   (a) only iii   (b) iii and iv   (c) i and v   (d) ii, iii, and iv
   (e) ii and iii

4. Which of the following is (are) metals? (Numbers in parentheses are atomic numbers.)
   (i) K (19)   (ii) Ne (10)   (iii) P (15)   (iv) Th (90)   (v) Zr (40)
   (a) i and iv   (b) i and v   (c) only i   (d) only v   (e) i, iv, and v

5. Which of the following has properties similar to phosphorus (15)? (Numbers in parentheses are atomic numbers.)
   (a) As (33)   (b) Si (14)   (c) S (16)   (d) O (8)   (e) Pb (82)

6. The name of NO_2 is
   (a) nitrogen(II) oxide   (b) nitrite ion   (c) dioxygen nitride
   (d) nitrogen dioxide   (e) nitric oxide.

7. In one unit of $Co(NO_3)_2$ there are
   (a) 1 C, 1 N, 7 O   (b) 1 C, 2 N, 7 O   (c) 2 Co, 2 N, 6 O
   (d) 1 Co, 1 N, 6 O   (e) 1 Co, 2 N, 6 O

8. Which of the following statements is (are) true of an atom of the element with atomic number 20?
   - (i) The symbol for the element is Ne.
   - (ii) The element is an alkali metal.
   - (iii) The element forms a $2+$ ion.
   - (iv) The element is in Group IIA.
   - (v) The element is in the second period.
   - (a) i and iv (b) i and v (c) iii and iv (d) ii, iii, and iv (e) only i

9. The formula for potassium sulfate is
   (a) K_2SO_3 (b) K_2S (c) $KHSO_4$ (d) $K(SO_4)_2$ (e) K_2SO_4

10. When steam is passed over hot iron, hydrogen gas and an oxide of iron that has the formula Fe_3O_4 are produced. Steam is water in the form of a gas. The *best* equation for this reaction is
    (a) $3Fe(s) + 4H_2O(g) \longrightarrow Fe_3O_4(s) + 8H(g)$
    (b) $3Fe(s) + 4H_2O(l) \longrightarrow Fe_3O_4(s) + 4H_2(g)$
    (c) $6Fe(s) + 8H_2O(g) \longrightarrow 2Fe_3O_4(s) + 8H_2(g)$
    (d) $3Fe(s) + 2H_2O_2(g) \longrightarrow Fe_3O_4(s) + 2H_2(g)$
    (e) $3Fe(s) + 4H_2O(g) \longrightarrow Fe_3O_4(s) + 4H_2(g)$

11. In the equations in Question 10, (s) indicates
    (a) stirring (b) a solid (c) steam (d) a solution (e) a substance

12. If an equation is written for the reaction of carbon disulfide with chlorine gas to produce carbon tetrachloride and disulfur dichloride, the coefficient of chlorine gas in the equation is
    (a) 2 (b) 3 (c) 4 (d) 5 (e) 6

13. Which one of the following is a molecular compound?
    (a) KI (b) $Fe(NO_3)_3$ (c) H_2S (d) Na_2S (e) O_2

14. In a bromide ion, Br^-, there are
    (a) 34 protons and 35 electrons
    (b) 35 protons and 34 electrons
    (c) 35 protons and 35 electrons
    (d) 35 protons and 36 electrons
    (e) 36 protons and 35 electrons

15. Which of the following is a conclusion? (a) The liquid is hot. (b) The gas is chlorine. (c) The liquid has no odor. (d) The gas is yellow-green. (e) Touching a burning splint to the mouth of a test tube containing the gas produces a loud noise.

PUTTING THINGS TOGETHER

Answers to problems with blue numbers appear in Appendix I. Solutions to these problems are in the Student Solutions Manual.

1.91 Define and give an example of (a) a law, (b) a theory.

1.92 Write formulas for molecules of (a) helium gas, (b) bromine liquid, (c) nitrogen dioxide gas, (d) hydrogen chloride gas, (e) dinitrogen pentoxide solid.

1.93 Write formulas for the following ionic compounds:
(a) sodium carbonate, (b) sodium hydrogen carbonate, (c) potassium nitrite, (d) magnesium oxide, (e) calcium chloride, (f) vanadium(II) sulfide, (g) iron(III) sulfate.

1.94 When heated, magnesium reacts with nitrogen to give magnesium nitride:

$$3Mg(s) + N_2(g) \xrightarrow{heat} Mg_3N_2(s)$$

Write the equations for the reactions of calcium with nitrogen and of barium with nitrogen.

1.95 Classify each of the following reactions as combination, decomposition, or neither:
(a) $Cu(s) + 2AgNO_3(aq) \longrightarrow 2Ag(s) + Cu(NO_3)_2(aq)$
(b) $HCl(aq) + NaOH(aq) \longrightarrow H_2O(l) + NaCl(aq)$
(c) $2H_2O_2(aq) \longrightarrow 2H_2O(l) + O_2(g)$
(d) $Na_2S(s) + 2HCl(aq) \longrightarrow H_2S(g) + 2NaCl(aq)$
(e) $P_4(s) + 5O_2(g) \longrightarrow P_4O_{10}(s)$

1.96 (a) What does the formula O_2 mean? (b) How does it differ from 2 O?

1.97 Which of the following formulas represents an element? Which represent ionic compounds? Which represent molecular compounds? Which represents an organic compound?
(a) MgO (b) P_4 (c) $Ni(NO_3)_2$ (d) HBr (e) C_2H_4

1.98 A molecule of ethyl alcohol is composed of two carbon atoms, six hydrogen atoms, and one oxygen atom. (a) What is the formula for ethyl alcohol? (b) How many carbon atoms are needed to make two molecules of ethyl alcohol? (c) Is ethyl alcohol an organic or an inorganic compound?

1.99 How many electrons are there in each of the following species? (a) an atom of chlorine (b) a molecule of chlorine (c) a chloride ion (d) a ClO_4^- ion

1.100 (a) How is the charge of a monatomic cation related to the group number of the element? (b) How is the charge of a monatomic anion related to the group number of the element?

1.101 (a) What is the atomic number of tin? (b) In what group is tin? (c) In what period is tin? (d) How many protons are in the nucleus of a tin atom? (e) How many electrons are in a tin atom? (f) How many electrons are there in a Sn^{2+} ion?

1.102 Complete the following table:

Species	Atomic Number	Number of Protons	Number of Electrons
K			
K^+			
Ba			
Ba^{2+}			
Br			
Br^-			
N			
N^{3-}			

1.103 Natural gas is composed mostly of methane, CH_4. The products of combustion of methane are carbon dioxide gas and water. Write the equation for the combustion of methane.

1.104 Make diagrams similar to Figure 1.5 for (a) compounds, (b) elements.

1.105 Suppose ○ represents an atom of sulfur. (a) Sketch an atom of phosphorus. (b) Sketch an atom of selenium.

1.106 This chapter contains a number of references to science and how scientists do science. Summarize them.

1.107 In chemistry, everything is interconnected. You need to learn to use this text as a reference. First, examine it noting what information is inside the front and back covers and in the appendices. Then use the index to find page numbers where more information about the following topics can be found: (a) atomic number, (b) nomenclature of inorganic compounds, (c) solutions.

APPLICATIONS

Answers to problems with blue numbers appear in Appendix I. Solutions for these problems are in the Student Solutions Manual.

1.108 Most rocks are masses of one or more minerals joined together. The widely used building stone granite is described in a field book of rocks and minerals as a combination of at least three minerals having a texture coarse enough so that the individual minerals can be recognized with the unaided eye. What term would a chemist use to describe granite?

Granite

1.109 Gold was probably the first metal to be used by people. It does not dissolve in water so is often washed down into streambeds where its bright color attracts the eye. It is soft and is easily worked into various shapes. A piece of gold smaller than the head of a pin can be drawn into a wire 500 feet long or beaten into a thin leaf covering a square 7.5 inches on a side. Gold forms few compounds. It does not tarnish and is mostly found as the metal. Which of the preceding properties of gold are physical properties, and which are chemical properties?

1.110 A wire service report of an underground explosion in a gold mine suggested that the cause might be "a large shipment of explosives, including 50 bags of powdered nitrogen." What is wrong with this suggestion?

1.111 A joint Russian–American project to gather data on Earth's atmosphere will measure nitrogen dioxide, chlorine dioxide, and ozone. Ozone is a form of the element oxygen that has triatomic molecules. Write formulas for these substances.

1.112 Write equations for the reaction of hydrochloric acid, HCl(aq), in your stomach with the following common antacids in tablet form: (a) calcium carbonate (the products are water, carbon dioxide gas, and an aqueous solution of calcium chloride), (b) magnesium hydroxide (water and an aqueous solution of magnesium chloride are formed), (c) aluminum hydroxide (the products are water and an aqueous solution of aluminum chloride).

1.113 In geology a mineral is defined as a natural inorganic substance of definite chemical composition that has its units arranged in an orderly way. For example, the mineral, halite, is sodium chloride. (Figure 1.18 shows the arrangement of Na^+ and Cl^- ions in NaCl.) (a) What is meant by the term *inorganic?* The names and formulas of some common minerals are (b) chalcocite, Cu_2S, (c) hematite, Fe_2O_3, (d) siderite, $FeCO_3$, (e) corundum, Al_2O_3 (clear and perfect crystals of this mineral are the highly prized gems, sapphire and ruby), (f) molybdenite, MoS_2, (g) quartz, SiO_2, (h) calcite, $CaCO_3$. Give the names a chemist would use for these compounds. (i) What is wrong with the statement that the atoms composing corundum make a molecule?

1.114 A number of substances are added to tobacco for various reasons, such as to improve the flavor of cigarettes. The Bureau of Alcohol, Tobacco, and Firearms, which regulates the tobacco industry, assumes that these additives are safe if they are approved by the Food and Drug Administration for use in foods and beverages. What is wrong with this assumption?

Science and Public Policy

RONALD V. DELLUMS

Congressman

B. A. Psychology
San Francisco State College

M. A. Social Welfare
University of California, Berkeley

As a youngster growing up in west Oakland's African-American community, my dreams had little to do with science. For that matter, most boys and girls in my neighborhood did not think of science as a career option. Structurally, society discouraged young African Americans from going to college, much less pursuing a science career. Our neighborhood high school did not even offer many of the math and science classes needed to meet college admission requirements.

Despite my social ties to my community high school, my mother wanted to ensure that the avenues to college were open for me. She therefore transferred me from my neighborhood high school into one of the city's college preparatory high schools. Not only did this school offer the math, science, history, and language courses required for college admission, but the coach did not believe that Black youth could play baseball. My mother was pleased, but I saw the transfer as the end of my baseball career.

I approached my high school science classes with some consternation. Biology, chemistry, and physics were the standards at the time, and for the life of me I could not figure out how they would benefit me in the future.

After high school, a hitch in the Marine Corps, and two years at Oakland Community College, I found myself at San Francisco State College. Throughout college, I gravitated toward the social sciences, receiving a bachelor's degree in psychology and a master's degree in social welfare at the University of California, Berkeley.

At first, my indifference to science persisted in college. But, as I began to find my way, I became interested in what I regarded as the related sciences of the development of our planetary system and the evolution of life on Earth. These were personal interests, however; I did not think these subjects would have any direct application to my chosen career as a psychiatric social worker.

As the 1960s were ending, I decided to seek election to the United States Congress. By this time, I had worked as a social worker and manpower development and training specialist and had served on the Berkeley City Council. All of these endeavors convinced me that reordering our priorities at the federal level was essential to the betterment and well-being of our nation and its various communities.

My concerns were political, not scientific: I went to Washington to achieve civil rights for our citizens, end the war in Southeast Asia, and redirect funds from the military budget to our urgent social needs. Two years later, I wound up on the House Armed Services Committee—and ultimately became chair of the very technologically oriented Subcommittee on Research and Development.

I learned very quickly that the debate over the military budget was conducted in terms of technology, not just priorities. At first, I was overwhelmed by the testimony of the "rocket scientists," who appeared before the committee to discuss the "yields" of various nuclear weapons, the "circular error probability" of a missile landing near a target, "exchange" rates, computer "architecture," and all of the other esoterica of the Strangelovian world of nuclear war planning. I wanted to discuss whether or not we should build weapons per se; I found myself having to discuss whether or not these systems would work, alone or in combination with each other. To do that, I had to know the science behind them.

Physics, chemistry, and math all came into the dialogue. I found that my credibility in the debate over nuclear weapons systems was enhanced when I could discuss them in terms of scientific principles. I therefore forced myself to return to those lessons I had so painfully learned during my high school days. At first, summoning up that half-forgotten scientific knowledge was difficult. But once I had remembered it, I found that it played a significant role in my daily work as a public policy maker—work that, unfortunately, requires me almost constantly to contemplate systems that might one day destroy our planet.

No longer could my arguments be attacked as wistful liberalism or naive pacifism. Now those who favored the development of more nuclear weapons systems had to confront my challenge to the internal logic of the systems themselves. They were forced to defend the inconsistency of developing weapons that would, if used, destroy the very planet and society they were ostensibly designed to protect. They also had to acknowledge the instability created when massive destructive power is linked to computer technology and put on a short fuse. Time, distance, yield, instability, symbiosis, synergy, and other elements of weapons technology became my allies.

In my more reflective and philosophical moments, I also recall my exploration of planetary development and species evolution in college. I believe profoundly that the human spirit, given its current state of evolution, is capable of reaching beyond violence and war. The longer I am in public life, the more I hope this vision will be realized. Therefore, I not only apply the hard sciences to challenge these insane weapons systems, but I try to do my part to hasten our evolution toward a more peaceful and cooperative species that lives in harmony with the planet rather than as its principal threat and antagonist.

Our future as a species depends on our ability to understand the chemical and physical impact of our actions and the threats to the environment posed by our conduct. Accordingly, our public policies must be devoted to enhancing the quality of life on this planet. Without an understanding of science, I, as a public policy maker, would be unable to serve future generations who rely upon us to leave them a healthy and vibrant legacy—the Earth.

2

MEASUREMENT

Reprinted from Scientific American, August 1885.

Instruments used to make measurements can be beautiful as well as useful.

Which is the tallest of the three figures in the picture at the left? Number 3 looks tallest to most people. But if you measure each of the three figures with a ruler, you will find that Number 1 is really tallest.

— Why is a study of measurement a part of general chemistry?

The simple example above shows one advantage of **quantitative observations,** that is, observations *involving numbers,* over qualitative observations. **Qualita-tive observations** *do not involve num-bers.* Quantitative observations or mea-surements are made with instruments that extend the human senses of sight, smell, taste, touch, and hearing. These instru-ments may be as simple as a ruler or as complicated as a mass spectrometer (Section 2.11). They are less likely to be fooled, however, than the human senses alone.

Another advantage of measurements is that they are usually reproducible. One of the characteristics of a science is that observations can be repeated both by the original scientist and by other scientists.

The results of measurements can be stored in a laboratory notebook or a computer and can be compared easily with similar observations. For example, if one person records the fact that the pages of this book measure ten inches from top to bottom, the observation can easily be proved right or wrong. Qualitative observations are difficult to check. A sample may look blue to one person and green to another. Even a single person may have difficulty remembering the exact shade of blue or green.

Measurement is central to the sciences and engineering; modern technology and development are based on measurements. Measurements made possible by newly developed instruments are the basis for many recent advances in medicine, environmental science, materials science, and other areas. With these instruments, an ever-increasing variety of measurements can be carried out faster on less sample, and smaller and smaller quantities can be detected.

— What do you already know about measurements?

Everyone knows how to count the pennies in their pockets and the number of runs in a baseball game. You can measure your weight with a scale, your height with a ruler, time with a watch or clock, volume with a measuring cup or the meter on a gas pump, pressure with a tire gauge, and the speed of a car with a speedometer. You know that every measurement has two parts, a number and a unit. You would not just say "It is three from my house to school" because your listener would have to ask "Three what?"

Some everyday measuring devices

2.1   SI UNITS

Five thousand years ago in Egypt, the cubit, the length of a person's arm from the elbow to the tip of the middle finger, was the unit for the measurement of length. Different-sized people had different rulers. For people to be able to reproduce each other's measurements, they must agree on standard units. In their everyday life, people in the United States use the English system of units—inches, feet, ounces, pounds, and so forth. Scientists all over the world use **SI units,** the units of the *International System of Units.* (SI comes from the French, Le **S**ystème **I**nternational d'Unités.) The units of the International System of Units were adopted as recommended units for use in science and technology by the General Conference of Weights and Measures* in 1960. The purpose of adopting SI units was to make communication easier among scientists and engineers working in different scientific subjects and in different countries.

In the English system, there are a number of units for each physical quantity. For example, inches, feet, yards, and miles are all units of length. In SI, each physical quantity has a single base unit. For example, the unit of length is the meter. The seven **base units** are listed in Table 2.1. Of the units listed in Table 2.1, those of length, mass, time, electric current, temperature, and amount of substance will be used in general chemistry. You should memorize the names and symbols of the

The Omnibus Trade Act of 1988 made the SI system the preferred measurement system for U.S. trade and commerce and gave federal agencies until 1992 to begin using the SI system in their business dealings. Labels now give both SI and English units.

---

*The General Conference of Weights and Measures is the governing body of the International Bureau of Weights and Measures. The International Bureau of Weights and Measures was formed, at the suggestion of the French government, by a treaty signed in 1875 by representatives of 17 countries including the United States. It is now supported by 40 countries.

A meter is a little longer than a yard, and a kilogram is slightly more than two pounds. A comfortable room temperature of 73 °F is 296 K.

TABLE 2.1	SI Base Units	
Physical Quantity	Name of Unit	Symbol
length	**meter**[a]	**m**
mass	**kilogram**[b]	**kg**
time	**second**	**s**
electric current	**ampere**	**A**
temperature	**kelvin**	**K**
luminous intensity	candela	cd
amount of substance	**mole**	**mol**

[a]The spelling given is for the United States. The official spelling is metre.
[b]The official spelling is kilogramme.

SI units for these quantities, which are in dark type. To use SI units, follow these rules:

1. Use only the singular form of units; do not put a period after the symbol. Separate the symbol from the number by a space. For example, for eight meters, write 8 m (not 8 ms, 8 m., or 8m).
2. Use a dot on the baseline for the decimal point; that is, write 8.6, not 8·6. If a number is less than 1, write a 0 to the left of the decimal point. For example, write 0.62 m, not .62 m.
3. Group digits in threes around the decimal point to make reading long numbers easier. Do *not* use commas to space digits in numbers. For example, write 1 650 763.000 03, not 1,650,763.00003. Do *not* use either a space or a comma in four-digit numbers such as 1651 and 0.1651.
4. Omit the degree sign when the Kelvin scale is used for temperature; write 298 K, not 298 °K.

PRACTICE PROBLEM

2.1 (Remember that answers for Practice Problems are in Appendix I. Detailed solutions are in the Student Solutions Manual.) Rewrite each of the following according to the SI rules: (a) 6·39 kgs., (b) .000015 °K, and (c) 299,792,458 m/s.

The size of each SI base unit is defined exactly. For example, the standard of mass is the mass of a platinum–iridium cylinder kept in a vault near Paris. Each country has one or more copies. ➡Figure 2.1 is a photograph of a U.S. copy at the National Institute of Standards and Technology (known as the National Bureau of Standards until 1988) in Gaithersburg, Maryland.

Objects like the platinum–iridium cylinder can be damaged or stolen. They can only be in one place at a time. Therefore, all base units except the kilogram are now defined in terms of reproducible physical measurements that can be made anywhere by a skilled person with the right instruments. For example, in 1983 the meter was defined as the distance that light travels through empty space during 1/(299 792 458) of a second.

The symbols for SI units can be multiplied and divided like other mathematical symbols to obtain **derived units.** For example, the SI unit for volume is cubic meters, m^3. *All other SI units are derived from the seven base units (and two supplementary units of angle).*

➡ FIGURE 2.1   U.S. copy of the standard kilogram. The cylinder is about 90% platinum and 10% iridium, and the metal alone is worth about $36,000.

When added to meats such as hot dogs, ham, and bacon, sodium nitrite, $NaNO_2$, prevents the growth of the bacterium *Clostridium botulinum,* which causes botulism—a sometimes fatal form of food poisoning. Sodium nitrite, however, can react in the human body to form nitrosamines, which have been found to produce cancer in laboratory animals. Should the addition of sodium nitrite to meats be prohibited? This question is typical of the choices that must be made in modern life. The risk of dying of botulism must be balanced against the risk of dying of cancer.

Quantitative estimates of risk show that some of the ways humans think qualitatively about risk don't make much sense. Delayed effects, such as developing cancer, are feared more than immediate effects such as poisoning. People tend to consider risks they do not personally choose, such as pesticide residues on food, as greater than risks they do choose. For example, would you believe that drinking a single cup of coffee yields the same risk of cancer as eating a year's worth of fruits and vegetables that have man-made pesticide residues? We think of man-made risks as greater than natural risks even if the risks are actually smaller. All fruits

Cured meats, such as bacon, usually contain sodium nitrite.

and vegetables contain natural pesticides that the plants have developed to protect themselves from being eaten by insects or infected by fungi. The risk from these naturally occurring pesticides is 9999 times the risk from man-made pesticide residues. Catastrophies are seen as riskier than ordinary events. For example, some people are afraid of flying, but few people refuse to ride in cars; yet the risk of traveling by car is over three times as high as the risk of traveling by jet. Most people think that chemicals in the environment are more of a risk than are personal habits, despite research showing that the reverse is true. A study published in the *Journal of the American Medical Association* in

1993 showed that the main contributors to premature and preventable deaths are use of tobacco, poor diet, and lack of exercise. Less than 3% of premature and preventable deaths result from toxic materials in the environment and workplace.

Some things that can be done to reduce risks are simple and inexpensive, such as reading labels and following the directions on them or painting clearer lines down the center of roads to reduce the number of automobile accidents. However, in the past a great deal of money has been spent on projects of little benefit. For example, removal of asbestos from schools has cost billions of dollars although the death rate that results from the asbestos is only 0.005–0.03 per 1 million people. In comparison, the following everyday activities produce *one* additional death per million people exposed to the risk: eating 40 tablespoons of peanut butter, traveling ten minutes by bicycle, having one chest X ray. We do not have enough money and resources to deal with all our problems. Therefore, we need to obtain the best possible information and use it to decide what to do. Spending huge amounts of money and effort on small risks takes resources away from real hazards and hurts the economy.

PRACTICE PROBLEM

2.2   (a) What is the SI base unit for mass? (b) What is the SI derived unit for area?

2.2   CONVERTING UNITS

SI Prefixes

The sizes of the SI base units are not always convenient. For example, a meterstick is a little longer than a yardstick. Using a meterstick to measure the diameter of a penny or the distance between two cities would be awkward. The SI solves the problem of awkward units by using prefixes to make larger and smaller units from the base units. SI rulers, for example, are usually labeled in centimeters. One centimeter is 10^{-2} meter, or one-hundredth (0.01) of a meter. There are 100 centimeters in a meter. The symbol for a centimeter is cm. Table 2.2 shows the SI prefixes with their symbols. In Table 2.2, the SI prefixes are arranged in order from very

● See Appendix B.2 for a review of exponential (scientific) notation.

The material on exponential notation in Appendix B.2 provides students with Sample Problems and Practice Problems.

The equals sign means "describes the same amount or quantity as."

● Always check your answer to be sure that it is reasonable.

large to very small. The information given in dark type is most important in chemistry, and you should memorize it. One exagram is 10^{18}, or 1 000 000 000 000 000 000 g; one attogram is 10^{-18}, or 0.000 000 000 000 000 001 g. Sample Problem 2.1 shows how to read Table 2.2. You need to practice using the information in Table 2.2 until you can do it in your head.

SAMPLE PROBLEM

2.1 (a) Fill in the blank: One micrometer = _____ meter. (b) How many micrometers are there in one meter?

Solution (a) From Table 2.2, 1 micrometer equals 10^{-6} meter.

(b) Thinking a problem through qualitatively before you do any arithmetic helps prevent careless mistakes. A micrometer is very small (0.000 001 m). Therefore, there must be a great many micrometers in a meter. The answer should be a large number.

At this point, you are probably more used to decimal notation than to exponential notation. Let's work the problem in both. The relationship between micrometers and meters is

Decimal notation	*Exponential notation*
1 μm = 0.000 001 m	1 μm = 10^{-6} m
Division of both sides of the equation by 0.000 001 gives	Division of both sides of the equation by 10^{-6} gives
$\dfrac{1\ \mu m}{0.000\ 001} = 1\ m = 1\ 000\ 000\ \mu m$	$\dfrac{1\ \mu m}{10^{-6}} = 1\ m = 10^{6}\ \mu m$

The answer makes sense because it is a large number.

PRACTICE PROBLEMS

2.3 Write the name of each of the following units: (a) Ms, (b) pmol, and (c) cA.

2.4 Write the symbol for each of the following units: (a) milligram, (b) nanosecond, and (c) kilokelvin.

2.5 (a) Fill in the blank: One megagram = _____ grams. (b) How many grams are there in one megagram?

2.6 (a) How many millimeters are there in a meter? (b) Which is smaller, a millimeter or a meter? (c) Which is smaller, a micrometer or a nanometer? (d) How many nanometers are there in a meter?

Unit Conversion Factors

Now suppose that we need to convert a measurement in centimeters to the SI base unit of length, the meter, so that we can publish it. According to Table 2.2, 10^{-2} meter (0.01 m) equals one centimeter:

$$0.01\ m = 1\ cm \tag{2.1}$$

From equation 2.1, we can obtain a factor for converting centimeters to meters. Dividing both sides of equation 2.1 by 1 cm gives

$$\frac{0.01\ m}{1\ cm} = \frac{1\ cm}{1\ cm} = 1$$

Multiplying by 1 does not change the value of an expression. Measurements in centimeters can be converted to measurements in meters by multiplying by

TABLE 2.2

SI Prefixes

Factor	Prefix	Symbol
10^{18}	Exa	E
10^{15}	Peta	P
10^{12}	Tera	T
10^{9}	**Giga**	**G**
10^{6}	**Mega**	**M**
10^{3}	**Kilo**	**k**
10^{2}	Hecto	h
10^{1}	Deka	da
10^{-1}	**Deci**	**d**
10^{-2}	**Centi**	**c**
10^{-3}	**Milli**	**m**
10^{-6}	**Micro**	**μ**
10^{-9}	**Nano**	**n**
10^{-12}	**Pico**	**p**
10^{-15}	Femto	f
10^{-18}	Atto	a

0.01 m/1 cm. Centimeters will cancel:

$$\cancel{cm} \times \frac{0.01 \text{ m}}{1 \cancel{cm}} = m$$

Expressions such as 0.01 m/1 cm *that are equal to 1 and can be used to change one unit into another* are called **unit conversion factors.** *Every equality gives two unit conversion factors.* Dividing both sides of equation 2.1 by 0.01 m gives a unit conversion factor for converting meters to centimeters:

$$\frac{0.01 \text{ m}}{0.01 \text{ m}} = \frac{1 \text{ cm}}{0.01 \text{ m}} = 1$$

Sample Problem 2.2 shows the use of a unit conversion factor.

SAMPLE PROBLEM

2.2 The distance from New York to Philadelphia is 130.5 km. What is this distance in the SI base unit, meters?

Solution The question gives the distance in kilometers and asks for the distance in meters. To convert kilometers to meters, what unit conversion factor is needed?

$$km \times ? = m$$

According to Table 2.2, the relation between kilometers and meters is 10^3 m = 1 km or 1000 m = 1 km. The two unit conversion factors that can be obtained from this relation are

$$\frac{1000 \text{ m}}{1 \text{ km}} \quad \text{and} \quad \frac{1 \text{ km}}{1000 \text{ m}}$$

To get km to cancel, multiply by 1000 m/1 km:

$$130.5 \cancel{km} \left(\frac{1000 \text{ m}}{1 \cancel{km}} \right) = 130\ 500 \text{ m}$$

The distance from New York to Philadelphia is 130 500 meters. This answer is reasonable because meters are smaller than kilometers. The number of meters should be larger than the number of kilometers.

If the wrong unit conversion factor is used, the units of the answer do not make sense; in this case, the numerical value of the answer is not reasonable either:

$$130.5 \text{ km} \left(\frac{1 \text{ km}}{1000 \text{ m}} \right) = 0.1305 \frac{\text{km}^2}{\text{m}}$$

The question asks for distance in meters, not km²/m, which is not an appropriate unit for distance. In addition, the numerical value of the answer, 0.1305, is smaller than the original 130.5, not larger.

Because unit conversion factors between fractions and multiples of SI base units are powers of 10, conversion between fractions and multiples of SI units simply results in moving the decimal point. You need to practice conversions like this until you can do them by moving the decimal point. For now, write the unit conversion factor and use your calculator if you need to.

▶ Remember to check your answer to be sure that it is reasonable.

▶ Changing one SI unit to another is like converting dollars to cents: 1 cent = 10^{-2} dollars ($0.01) or 1 dollar = 10^2 cents (100¢).

PRACTICE PROBLEMS

2.7 What conversion factor should you multiply milligrams by to convert to grams?

2.8 Fill in the blanks: (a) 6.4 ns = _____ s and
(b) 872 g = _____ kg.

If neither the original unit nor the unit wanted is a base unit, two (or more) unit conversion factors are needed to accomplish a conversion. Sample Problem 2.3 shows this type of problem.

SAMPLE PROBLEMS

2.3 Convert 6.8 μm to cm.

Solution The question gives length in micrometers and asks for length in centimeters. Because neither micrometers nor centimeters are base units, two relationships from Table 2.2 are needed to solve this problem:

$$10^{-6} \text{ m} = 1 \ \mu\text{m} \qquad \text{and} \qquad 10^{-2} \text{ m} = 1 \text{ cm}$$

Length in micrometers must be changed to length in meters using one unit conversion factor:

$$6.8 \ \mu\text{m} \left(\frac{10^{-6} \text{ m}}{1 \ \mu\text{m}} \right) = 6.8 \times 10^{-6} \text{ m}$$

Then length in meters must be converted to length in centimeters using a second unit conversion factor:

$$6.8 \times 10^{-6} \text{ m} \left(\frac{1 \text{ cm}}{10^{-2} \text{ m}} \right) = 6.8 \times 10^{-4} \text{ cm}$$

The length 6.8 μm is equal to 6.8×10^{-4} cm.

Always reason through problems in steps. However, *after* you have reasoned through a problem, you may combine steps before doing the arithmetic:

$$6.8 \ \mu\text{m} \left(\frac{10^{-6} \text{ m}}{1 \ \mu\text{m}} \right) \left(\frac{1 \text{ cm}}{10^{-2} \text{ m}} \right) = 6.8 \times 10^{-4} \text{ cm}$$

Combining steps reduces the possibility of making careless mistakes. You do not have to read numbers from your calculator and then reenter them. However, when a calculation is done stepwise, you can see whether the answer for each step is reasonable, and it is sometimes easier to find a mistake when one has been made.

Check: The units of the answer to this problem are correct. The number is reasonable: micrometers are smaller than centimeters, so 6.8 μm should only equal a fraction of a centimeter.

2.4 Convert 0.645 m^2 to cm^2.

Solution The question gives area in square meters, m^2, and asks for area in square centimeters, cm^2. From Table 2.2,

$$10^{-2} \text{ m} = 1 \text{ cm}$$

To convert m^2 to cm^2, the conversion factor

$$\frac{1 \text{ cm}}{10^{-2} \text{ m}}$$

must be squared. Unit conversion factors may be squared, cubed, or raised to any power because

$$1^2 = 1^3 = 1^n = 1$$

$$0.645 \text{ m}^2 \left(\frac{1 \text{ cm}}{10^{-2} \text{ m}} \right)^2 = \frac{0.645 \text{ cm}^2}{10^{-4}} = 6450 \text{ cm}^2 = 6.45 \times 10^3 \text{ cm}^2$$

🔵 Checking your work is very important in the real world. If a bridge falls down and kills a number of people, the engineer responsible for building the bridge can't very well say to the grieving relatives, "I'm sorry—it was just a little mistake in the decimal point." A misplaced decimal point once credited spinach with ten times its actual iron content.

2.9 What conversion factors should you multiply micrograms, μg, by to convert to kilograms?

2.10 Fill in the blanks: (a) 4.9 km = _____ cm, (b) 2648 mg = _____ Mg, and (c) 341 mm² = _____ m².

Conversion factors can also be used to change measurements made in English units to SI units (or measurements made in SI units to English units). The necessary definitions are given inside the back cover.

SAMPLE PROBLEM

2.5 Convert 34.5 miles to kilometers, km.

Solution The question gives length in miles and asks for length in kilometers. From the table inside the back cover,

$$1 \text{ mile (mi)} = 1.6093 \text{ km}$$

The proper unit conversion factor is

$$\frac{1.6093 \text{ km}}{1 \text{ mi}} \quad \text{and} \quad 34.5 \text{ mi} \left(\frac{1.6093 \text{ km}}{1 \text{ mi}} \right) = 55.5 \text{ km}$$

Check: The answer is reasonable because one mile is equal to more than one kilometer. The numerical value of the distance in kilometers should be greater than the numerical value of the distance in miles.

PRACTICE PROBLEM

2.11 Fill in the blanks: (a) 63.8 lb = _____ g and (b) 44.7 km = _____ mi.

2.3 UNCERTAINTY IN MEASUREMENT

Making measurements often involves reading a scale. For example, to measure the length of the black rectangle in ►Figure 2.2 you would use a ruler. The object being measured is likely to fall between subdivisions on the scale as the rectangle in Figure 2.2 does. In everyday life, people usually read the nearest subdivision on a scale. Using the bottom ruler, they would say the length of the rectangle is 10.1 cm. In scientific work, people estimate what decimal fraction of the way from one mark to the next a reading is. The right end of the black rectangle appears to be

○ Always remember to read between the lines.

► FIGURE 2.2 A rectangle and rulers.

- FIGURE 2.3   The length of the rectangle should be recorded as 9.0 cm to show that the ruler is divided into centimeters and tenths of a centimeter can be estimated.

about four-tenths of the way from 10.1 cm to 10.2 cm. Thus, the length of the rectangle is 10.14 cm. Because the last digit, 4, is estimated, a person might read the length as 10.13 cm one time and 10.15 cm another time. The last digit is uncertain. Except when small numbers of objects are counted, *one uncertain or estimated digit should always be recorded.*

Digits recorded according to this rule are called **significant figures.** The number of digits in measurements depends on the size of the subdivisions on the instrument scale. A person using the top ruler in Figure 2.2 would say that the length of the rectangle is 10.1 cm. This time the one-tenth is estimated because the distance between marks is 1 cm instead of 0.1 cm. From the number of digits recorded for a measurement, the reader can tell how the scale was divided. A length of 10.1 cm was measured with a ruler subdivided into centimeters; a length of 10.14 cm was measured with a ruler subdivided into tenths of centimeters. The same thing is true of measurements from more complicated instruments. The last digit recorded is always understood to be estimated or uncertain. *In science, a measurement makes known three pieces of information: quantity, units, and uncertainty.*

Because the value for a measurement should tell the uncertainty of the measurement, a zero must be written if the measured value is exactly at a line of the scale. If the rectangle in -Figure 2.3 were measured with a ruler marked in centimeters, the length should be recorded as 9.0 cm.

PRACTICE PROBLEMS

2.12   How long are the rectangles?

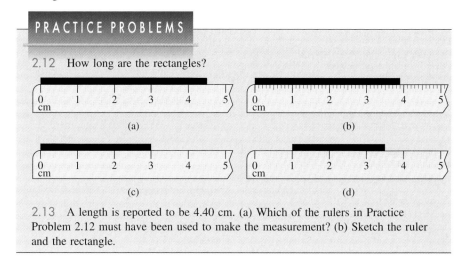

(a)

(b)

(c)

(d)

2.13   A length is reported to be 4.40 cm. (a) Which of the rulers in Practice Problem 2.12 must have been used to make the measurement? (b) Sketch the ruler and the rectangle.

Number of Significant Figures

In counting the number of significant figures in a correctly written measurement—that is, a measurement with one uncertain digit—all nonzero numbers are significant. There are two significant figures in the length 6.8 cm, and three in the length 6.83 cm.

Zeros are sometimes significant and sometimes not significant. How can you tell if a zero is significant?

1. Zeros between nonzero numbers are significant; for example, the zeros in 2.003 are significant. There are four significant figures in 2.003.
2. Zeros at the end of a number to the right of the decimal point are significant. The zero in 9.0 is significant. If it were not, it would not be written. The number would be written 9. There are two significant figures in 9.0.
3. Zeros in numbers like 0.0032 are *not* significant. The easiest way to see this is to write the number in exponential notation:

$$0.0032 = 3.2 \times 10^{-3}$$

There are just two significant figures in 3.2. The zeros in 0.0032 only show the position of the decimal point.

4. The last possibility is to have zeros at the end of a number but to the left of the decimal point, as in 4200. In this case, there is no way of telling whether the zeros are significant or not. There may be two, three, or four significant figures in 4200. The number of significant figures can be shown by writing the number in exponential notation. If four significant figures are meant, 4200 should be written 4.200×10^3. If three are meant, it should be written 4.20×10^3, and if only two, 4.2×10^3. *It is the responsibility of the person writing a number to show the number of significant figures.*

The *numbers obtained by counting,* such as the number of people in a car, *and the numbers in definitions* are exact. For example, the definition of the centimeter as 10^{-2} meter means that exactly one centimeter equals exactly 10^{-2} meter (1 cm = 0.01 m). **Exact numbers** are considered to have an infinite number of significant figures. *Exact numbers never limit the number of significant figures in the results of calculations.*

Some general chemistry texts have a "house rule" that writing a dot at the end of a number (for example, 4200.) means that the trailing zeros are significant. We think this rule does students a disservice because they forget that it is a "house rule" and expect it to be followed elsewhere, which it frequently is not. In addition, a period at the end of a sentence is apt to be mistaken for a dot.

PRACTICE PROBLEMS

2.14   Write 321 000 to show (a) three significant figures, (b) four significant figures, and (c) six significant figures.

2.15   How many significant figures are there in each of the following measurements? (a) 5269 m (b) 4.07 m (c) 42.300 m (d) 0.0065 m (e) 320 m (f) 4.8×10^{-9} m

2.16   Which of the following can be determined exactly, and which must be measured with some uncertainty? (a) The number of students in a classroom (b) The number of people in the United States (c) Your height

---

People often check measurements by repeating them. To check the length of the black rectangle in Figure 2.2, a scientist would remove and replace the ruler each time he or she measured the length of the rectangle. If the ruler was left in place, an error in lining up the zero of the scale with the left end of the rectangle would affect every reading. The scientist would probably also use different parts of the ruler, for example, the space between 1 and 12 cm instead of the space between 0 and 11 cm, to guard against errors in the scale. He or she would measure the length of the rectangle at the top and at points between the top and the bottom as well as at the bottom. A series of measurements might give the readings shown in Table 2.3.

Median and Average

What value should be reported for the length of the black rectangle? For a small set of observations, the median is usually the best value. If a set of measurements is arranged in order of increasing or decreasing value, the **median** is the *middle one of an odd number of measurements or the average of the two middle measurements for an even number of measurements.* Arranging the lengths from Table 2.3 in order of increasing value gives the series

| 1 | 2 | 3 | 4 | 5 | 6 | 7 | 8 | 9 | 10 | 11 | 12 |
| 10.13, | 10.13, | 10.14, | 10.14, | 10.14, | 10.14, | 10.14, | 10.14, | 10.14, | 10.15, | 10.15, | 11.14 |

Because 12 measurements (an even number) were made, the sixth and seventh

TABLE 2.3

Measurements of Length of Rectangle

Observation Number	Length, cm
1	10.14
2	10.15
3	10.13
4	10.14
5	10.14
6	11.14
7	10.14
8	10.13
9	10.15
10	10.14
11	10.14
12	10.14

TABLE 2.4

Frequency of Observations

Length, cm	Times Observed
10.12	0
10.13	2
10.14	7
10.15	2
10.16	0
11.14	1

values in this series, 10.14 and 10.14, are averaged to obtain the median. The median is 10.14 cm. The best value to report for the length of the black rectangle is 10.14 cm.

You are probably more familiar with calculating the average of a series of measurements than with finding the median. The **average** of the lengths in Table 2.3 *(obtained by adding all the individual values and dividing by the number of values)* is 10.22 cm. The one observation (observation number 6 in Table 2.3) that is much higher than all the others makes the average high. This observation is probably a mistake. The advantage of using the median instead of the average is that a mistake does not change the median as much as it does the average. If observation number 6 is omitted from the calculation, the average is 10.14 cm. Why not just leave out observation number 6 and calculate the average? Observations that you know are mistakes should be omitted. But it is dishonest to leave out observations simply because they are not what you expect or like.

Normal Distribution

Table 2.4 shows how many times each value for length was observed. The data in Table 2.4 (except for observation number 6, which does not fit on the page) are shown by small circles in ▪Figure 2.4.

When a large number of measurements are made, the distribution of the individual values is often shown by a normal distribution curve. Imagine that the numbers of observations in Figure 2.4 are multiplied by 10. Then the solid line in Figure 2.4 shows the **normal distribution** curve for measurements of the length of the black rectangle. From the normal distribution curve in Figure 2.4, 57 measurements would probably be 10.14 cm, 20 would be 10.13, and 20 would be 10.15 cm. The probabilities of measurements of 10.12 cm or less and measurements of 10.16 cm or more are zero except for mistakes.

For measurements that have a normal distribution, the median and the average are the same. The normal distribution curve shows that most observations are close to the median. Observations that are much larger or much smaller than the median are unlikely unless a mistake is made. It is usually a waste of time to make a large number of measurements. For example, the median of any set of 3 successive measurements in Table 2.3 is the same as the median of all 12 measurements.

▪ FIGURE 2.4   A normal distribution curve. Frequencies from Table 2.4 are shown by red circles and numbers.

 See Appendix B.6 for directions for making and reading graphs.

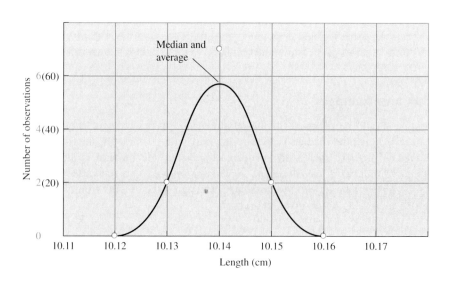

2.17 (a) What is the median of the following measurements? 62.4 cm, 61.4 cm, and 62.6 cm (b) What is the average of the measurements in part (a)? (c) Which is probably the better value to report, the median or the average? Explain your answer. (d) What is the median of the following measurements? 63.6 m, 62.4 m, 62.6 m, and 62.4 m

2.18 What does the normal distribution curve show about the chance of measurements greater than the median compared to the chance of measurements less than the median? Explain your answer.

Precision and Accuracy

The *repeatability of measurements* is called **precision.** Precision refers to how closely two or more measurements of the same property agree with one another. To get good precision, a scientist tries to make a measurement in exactly the same way each time.

The *correctness of measurements* is called **accuracy.** Accuracy refers to how close a measurement is to the true value of a property. Measurements can be precise without being accurate.* For example, if the black rectangle in Figure 2.2 had been measured with the ruler shown in ▬Figure 2.5, measurements would have been precise but not accurate. (What's wrong with the ruler in Figure 2.5?) However, if measurements are accurate without being precise, the person making the measurements is very lucky. To make sure their measurements are as accurate as possible, scientists calibrate their instruments. **Calibrate** means *to check the readings of an instrument.* The instrument can be compared with an instrument that is known to be correct or it can be used to measure a known sample. Accuracy and precision depend both on the instrument and on the skill of the person using the instrument. In the real world, the true result is almost never known.

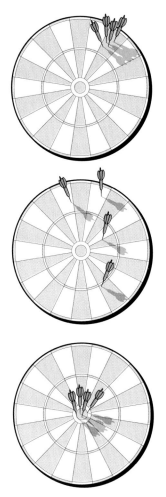

Dartboards showing the difference between precision and accuracy.
Top: Precise but not accurate.
Middle: Neither precise nor accurate.
Bottom: Both precise and accurate.

 2.4 SIGNIFICANT FIGURES IN CALCULATIONS

Using measurements in calculations does not make them any more or less precise or accurate. The uncertainty in the results of adding, subtracting, multiplying, and dividing measurements depends on the uncertainty in the measurements. The results of calculations should *always* be rounded to the correct number of significant figures. Digits that are not significant should be removed.

▬ FIGURE 2.5 An inaccurate ruler.

---

*Failure to distinguish between precision and accuracy on the part of the people who made the mirror for the Hubble Space Telescope cost their company $25 million. Measurements of the surface of the mirror that were precise to ±0.006 nm but accurate to only ±0.06 nm were responsible for the fuzzy images of the stars obtained with the telescope after it was launched in 1990. Only one method was used to test the mirror although there are a number of simple and inexpensive experiments that could have been used to check the results. After successful repairs in December 1993, the telescope now "sees" as well as it was supposed to.

Rounding

In rounding numbers to the correct number of significant figures, use the following rules:

Because there are five digits (0, 1, 2, 3, and 4) that are less than 5 and five digits (5, 6, 7, 8, and 9) that are 5 or more, these rules result in half of a large number of numbers being rounded up and half being rounded down.

*See Guare, C. J. J. Chem. Educ. **1991**, 68, 818 for a discussion of the inadequacy of the other common method for rounding 5.*

1. If the first digit to be removed is less than 5, simply remove the unwanted digits. For example, 6.7495 rounded to two digits is 6.7 because 4 is less than 5.
2. If the first digit to be removed is 5 or more, increase the preceding digit by 1. For example, 3.350 rounded to two digits is 3.4 because the first digit to be removed is 5, and 6.7938 rounded to two digits is 6.8 because 9 is greater than 5.

PRACTICE PROBLEM

2.19 Round each of the following numbers to three significant figures: (a) 3.432, (b) 3.438, and (c) 3.435.

Addition and Subtraction

To find the number of significant figures in the answer to addition and subtraction, you must set up the calculation. For example, consider the sum of 78.012 16 and 2.96:

$$\begin{array}{r} 78.012\ 16 \\ +\ \ 2.96 \\ \hline \end{array}$$

The answer should have the same number of digits to the right of the decimal point as there are in the number having the fewest digits to the right of the decimal point:

$$\begin{array}{r} 78.012\ 16 \\ +\ \ 2.96\ \boxed{?} \\ \hline 80.97\_\_ \end{array}$$ ⟵ You do not know these digits.
⟵ Therefore, you cannot know these digits.

There are four significant figures in 80.97, the sum of 78.012 16, which has seven significant figures, and 2.96, which has three. *In addition and subtraction, the number that has the smallest number of digits to the right of the decimal point determines the number of significant figures in the answer.* Note that significant figures are often lost by subtraction:

$$\left.\begin{array}{r} 78.0128 \\ -78.0121 \end{array}\right\}$$ Both these numbers have six significant figures.
$$0.0007$$ The difference between them has only one significant figure.

Multiplication and Division

How do you know how many significant figures there should be in the answer to multiplication or division? To find out, let's consider calculation of the area of the black rectangle in Figure 2.2. The area of a rectangle is found by multiplying the length by the width:

$$\text{area}_{\text{rectangle}} = \text{length} \times \text{width}$$

Measurements of width ranged from 0.66 cm to 0.68 cm; the median width was 0.67 cm. The best value for the area would be found by multiplying the median length by the median width

$$\text{area} = (10.14\ \text{cm})(0.67\ \text{cm}) = 6.7938\ \text{cm}^2$$

according to a calculator. How many of the digits displayed by the calculator are significant? The lowest measured value of length was 10.13 cm and the highest (except for the wrong value), 10.15 cm. The lowest value that could be obtained for the area would be found by multiplying the lowest length by the lowest width:

$$\text{lowest area} = (10.13 \text{ cm})(0.66 \text{ cm}) = 6.6858 \text{ cm}^2$$

The highest value that could be obtained for the area would be found by multiplying the highest length by the highest width:

$$\text{highest area} = (10.15 \text{ cm})(0.68 \text{ cm}) = 6.902 \text{ cm}^2$$

The difference between the lowest and highest values for area is 0.216 cm². The calculated areas differ in the first decimal place; we do not know whether the number in the first decimal place should be 6 or 9 (or any number in between). Therefore, the first digit to the right of the decimal point is uncertain. The best value for the area should be rounded to one decimal place and written 6.8 cm². There are two significant figures in the area. Notice that the number of significant figures in the product is the same as in the measurement with the smaller number of significant figures. This is generally true for multiplication and division. *For multiplication and division, the answer should be rounded to the same number of significant figures as there are in the quantity having the smallest number of significant figures.*

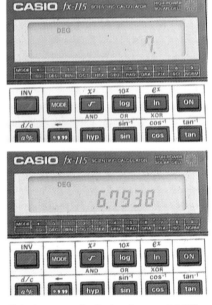 Finding powers and roots is similar to multiplication and division.

PRACTICE PROBLEM

2.20   How many significant figures are there in the answers to each of the following calculations (assume that the numbers are the results of measurements)?
(a) The quotient of 9.546 and 2.31—that is, $(9.546 \div 2.31)$
(b) The difference between 53.674 and 7.26—that is, $(53.674 - 7.26)$

Calculators and Computers

Your calculator does not know anything about significant figures. Sometimes it will display too many digits and sometimes too few. *It is your job to write answers to the correct number of significant figures.* If your calculator displays too many digits, you must round off. If it displays too few digits, you must add zeros. For example, according to many calculators,

$$4.362 + 2.638 = 7.$$

The correct answer is 7.000. *Zeros must be written when they are significant.* The programmer should make a computer display or print answers to the correct number of digits. If a program you are using does not, *you* must write the answers with the correct number of digits.

In Conclusion

The rules given here are useful rules of thumb. They must be applied thoughtfully. For example, the sum of ten or more numbers, each having uncertainty in the second decimal place, is uncertain in the first decimal place. If you take more advanced chemistry courses like quantitative analysis and physical chemistry, you will learn better (but more time-consuming) ways of dealing with the uncertainty in measurements. The question of whether differences are significant is a very important one in both physical and social sciences. Does one experimental drug really cure more patients than another experimental drug? Does a change in the design of a car really make a difference in the gas mileage? Do students taught by one method really learn more than students taught by another method?

Calculators usually display the sum of 4.362 and 2.638 as 7. (too few digits) and the product of 0.67 and 10.14 as 6.7938 (too many digits).

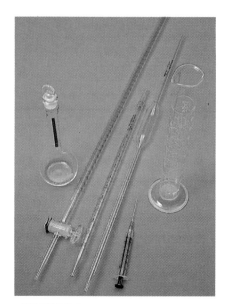

FIGURE 2.6 Glassware for measuring volumes of liquids. *Left to right:* volumetric flask, buret, graduated pipet, volumetric pipet, syringe, graduated cylinder.

⬤ The official spelling is litre. Sometimes l is used for liter. L is less likely to be mistaken for the number 1 or the capital letter I.

2.5 MEASURING VOLUME

Volume is the amount of space that an object occupies. Volumes of liquids must often be measured in chemical laboratories. Glassware used for measuring the volumes of liquids is shown in ▪Figure 2.6. The volumes of the glassware are shown in milliliters (mL) and microliters (μL). A **liter*** **(L)** is defined as $10^{-3}\ m^3$ or $1\ dm^3$. The prefixes m and μ have their usual meanings (see Table 2.2). One milliliter is exactly equal to one cubic centimeter; that is,

$$1 \text{ mL} = 1 \text{ cm}^3 \quad \text{or} \quad 1 \text{ cc}$$

and one liter is equal to 1000 cubic centimeters,

$$1 \text{ L} = 1000 \text{ cm}^3 \quad \text{or} \quad 1000 \text{ cc}$$

One thousand cubic centimeters is the volume of a cube that has edges ten centimeters, or one decimeter, long—hence the definition of the liter as one cubic decimeter. Liters and milliliters are *not* SI units. The SI unit for volume, m^3, is derived from the base unit for length, the meter. However, a cubic meter is too large to be a convenient unit for laboratory use, and chemists usually measure volumes of liquids in liters and milliliters.

2.6 MEASURING MASS

Mass is the *quantity of matter that an object contains.* Many people confuse mass and weight. **Weight** *depends both on the quantity of matter in an object and on the force of gravity on the object.* An astronaut on the moon weighs only one-sixth as much as the same astronaut weighs on Earth. The mass of the astronaut is the same on the moon as on Earth.

Mass is measured by comparing the weights of unknown objects with the weights of known masses by means of a balance. A simple balance is shown in ▪Figure 2.7. Because both sides of the balance are the same distance from the center of

FIGURE 2.7 A simple mechanical balance.

⬤ Note that the SI base unit for mass, the kilogram, has a prefix in its name.

---

*The original choice of the liter as a unit of volume was based on the fact that one kilogram of pure water (at 4 °C) has a volume of one liter. A liter is slightly larger than a quart, and one teaspoon has a volume of about five milliliters.

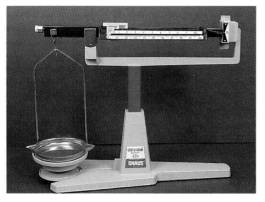

Cent-O-Gram triple-beam balance

Electronic toploading balance

Electronic microbalance

- FIGURE 2.8 Some modern laboratory balances. Electronic balances can be connected to computers for direct entry of measured masses into computer memory.

Earth, the force of gravity on them is the same. When the two sides balance, the masses on the pans must be equal. The known masses are calibrated by comparison with standard masses.

-Figure 2.8 shows some modern laboratory balances. With these balances, masses of 2.5 mg to 6 kg can be measured to four to six significant figures. Mass is usually measured in grams because a gram is a more convenient-sized unit than a kilogram for laboratory work.

Doctors use balances to determine their patients' masses. Bathroom scales measure weight. To lose weight quickly, go to the moon!

PRACTICE PROBLEM

2.21 Which of the following are matter? (a) sound (b) wood (c) light (d) rainwater (e) a dime

2.7 EXTENSIVE AND INTENSIVE PROPERTIES

Length, volume, and mass are all **extensive properties.** Extensive properties *depend on the quantity of sample measured.* **Intensive properties** are *independent of sample size.* For example, the temperature at which a liquid boils is an intensive property. A cup of water boils at the same temperature as a gallon of water. -Figure 2.9 shows the difference between an extensive property and an intensive property.

Intensive properties are often characteristic of substances. For this reason, they are frequently used for identifying substances, checking their purity, and deciding whether a substance is suitable for a given use. The fact that ice melts at 0 °C, for example, means that ice is useful for building houses (igloos) in cold places but not in hot places.

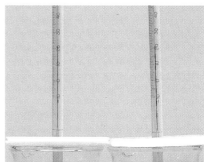

- FIGURE 2.9 The volume of a solution is an extensive property that depends on the amount of solution. The temperature of a solution is an intensive property. Both large and small samples are at the same temperature.

PRACTICE PROBLEM

2.22 Which of the following are extensive properties and which are intensive properties? (a) The number of calories in a chocolate bar, (b) the color of gold coins, (c) the area of a table top, (d) the fact that wood burns, (e) the fact that iron rusts.

Ice is less dense than water and floats on water. Gold is more dense than water and sinks.

DENSITY

The *ratio of the extensive property, mass, to the extensive property, volume,* is an intensive property called **density:**

$$\text{density} = \frac{\text{mass}}{\text{volume}} \tag{2.2}$$

The masses of samples do not change if the temperature is raised or lowered. The volumes of solids and of most liquids, however, increase slightly if the temperature is raised. The volumes of all gases held at constant pressure increase greatly when temperature is raised. If the denominator of a fraction is increased while the numerator remains constant, the value of the fraction decreases. Density usually decreases with increasing temperature. When the density of a material is given, the temperature at which the density was measured should also be given.

The SI unit for density, kg/m³ (read "kilogram per cubic meter"), is derived from the base units for mass and length. In the laboratory, mass is usually measured in grams and volume in cm³ (solids), mL (liquids), or L (gases). Density is usually expressed in g/cm³, g/mL, or g/L. You may also see g/cm³ written as

$$\frac{\text{g}}{\text{cm}^3} \qquad \text{g cm}^{-3} \qquad \text{or} \qquad \text{g} \cdot \text{cm}^{-3}$$

Sample Problem 2.6 shows how to calculate density from mass and volume.

SAMPLE PROBLEM

2.6 A block of wood has a mass of 492 g and a volume of 746 cm³. What is the density of the wood (a) in g/cm³ and (b) in SI base units of kg/m³?

Solution (a) The question gives the mass in g and volume in cm³ and asks for the density in g/cm³. By definition,

$$\text{density} = \frac{\text{mass}}{\text{volume}} \tag{2.2}$$

Substituting the mass and volume given in the problem in the definition of density gives

$$\text{density} = \frac{492 \text{ g}}{746 \text{ cm}^3} = 0.659\,5174\,\frac{\text{g}}{\text{cm}^3} = 0.660\,\frac{\text{g}}{\text{cm}^3}$$

Check: The unit, g/cm³, is suitable for density and is the unit asked for in the problem. Is the number 0.660 reasonable? Most samples of wood float on water because most kinds of wood are less dense than water. At room temperature, the density of water is 1.00 g/cm³. The number 0.660 is less than 1. Therefore, a density of 0.660 g/cm³ is reasonable.

(b) The question asks for the density in kg/m³. The density in g/cm³ is known from part (a). To convert g/cm³ to kg/m³, two unit conversion factors are needed, one to convert g to kg and the other to convert cm³ to m³. The relations required are (from Table 2.2)

$$10^3 \text{ g} = 1 \text{ kg} \qquad \text{and} \qquad 10^{-2} \text{ m} = 1 \text{ cm}$$

To decide which of the two unit conversion factors from each relationship to use, start to set up the calculation. Units must cancel to give the correct units for the answer, kg/m³:

$$\left(\frac{0.660 \text{ g}}{\text{cm}^3}\right)\left(\underline{\qquad}\right)\left(\underline{\qquad}\right) = \frac{\text{kg}}{\text{m}^3}$$

Wooden logs float on water because they are less dense than water.

(a)

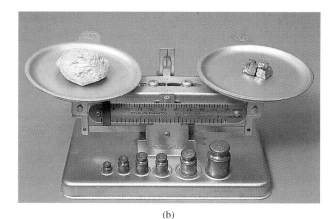

(b)

For the units to cancel properly, the unit conversion factors shown below must be used:

$$\left(\frac{0.660\ \text{g}}{\text{cm}^3}\right)\left(\frac{1\ \text{kg}}{10^3\ \text{g}}\right)\left(\frac{1\ \text{cm}}{10^{-2}\ \text{m}}\right)^3 = 0.660\ \frac{\text{kg}}{10^{-3}\ \text{m}^3} = 660\ \frac{\text{kg}}{\text{m}^3} = 6.60 \times 10^2\ \frac{\text{kg}}{\text{m}^3}$$

Notice how the answer is written in exponential notation to show that it has three signifcant figures.

PRACTICE PROBLEM

2.23   A piece of lead has a mass of 864 g and a volume of 76.2 cm³. What is the density of lead (a) in g/cm³ and (b) in kg/m³?

Just as you must be careful not to confuse mass and weight, you must be careful to distinguish between mass and density. If you could compare the masses of the two objects in ▬Figure 2.10(a) by holding first one and then the other in your hand, you would probably think that the smaller one has the greater mass. But as you can see from Figure 2.10(b), the masses of the two objects are equal. This is another example of the advantage of quantitative observations. The density of the smaller object is greater. Because its mass is in a smaller volume, the smaller object feels heavier.

The densities of some familiar materials are given in Table 2.5.

▬ FIGURE 2.10   Pumice and galena. (a) Pumice *(left)* has a density of 0.5 g/cm³; galena *(right)* has a density of 7.5 g/cm³. (b) Although the volume of the pumice is much larger than the volume of the galena, the masses are equal.

⬤ Remember that the ones in the conversion factors are exact numbers and do not limit the number of significant figures in the answer. Don't forget to cube both the number and the unit.

TABLE 2.5	Densities of Some Familiar Materials at Ordinary Temperature (20–30° C)[a]		
Material	Density, g/cm³	Material	Density, g/cm³
Air(dry)	0.001 18	Gold	19.3
Alcohol	0.79	Helium	0.000 164
Aluminum	2.7	Ice	0.92
Butter	0.86	Iron	7.9
Carbon dioxide	0.001 80	Mercury	13.5
Cork	0.24	Sugar	1.58
Glass	2.6	Water	1.00

[a]Most data are from *Handbook of Chemistry and Physics,* 75th ed., Lide, D. R., Ed.; CRC Press: Boca Raton, FL, 1994.

Why are "most" data referenced in tables? It was sometimes necessary to correct mistakes or add a little data from another source, and we think that detailed references are inappropriate to a general chemistry text. However, we strongly believe that the chief sources of information should be given to set a good example for students, to teach them where to look for information, and to enable either instructors or students to check the correctness of information if they wish.

Calculation of Mass and Volume from Density

If the density of a material is known, it can be used to calculate the mass of a sample from its volume or the volume of a sample from its mass as shown in Sample Problem 2.7.

See Appendix B.4 for a review of how to solve equations.

You should memorize the definition of density and use it to solve all problems about relationships between density, mass, and volume. Do not try to memorize the relationship

$$\text{volume} = \frac{\text{mass}}{\text{density}}$$

Why more than one method for solving the same problem? Students need to learn how to solve problems, not just how to get the answer to a particular problem. Also, different students have different learning styles.

The symbol ⇌ means "is equivalent to."

Why no Sample Problem for finding mass from volume and density? Students do not learn how to solve problems if examples of all the easy ones are worked out for them. A student who understands Sample Problem 2.7 should be able to do Practice Problem 2.25.

SAMPLE PROBLEM

2.7 What is the volume of 454 g of butter?

Solution The problem gives the mass of a sample of butter. The density of butter is given in Table 2.5 (0.86 g/cm³). The problem asks for the volume of the sample. One way to solve problems like this is to begin with the definition of density:

$$\text{density} = \frac{\text{mass}}{\text{volume}} \tag{2.2}$$

The definition of density is an equation and can be rearranged to give volume from mass and density:

$$\text{volume} = \frac{\text{mass}}{\text{density}} \tag{2.2a}$$

Substitution of the information given in the problem in equation 2.2a gives

$$\text{volume butter} = \frac{454 \text{ g butter}}{0.86 \dfrac{\text{g butter}}{1 \text{ cm}^3 \text{ butter}}} = 527.906\,97 \text{ cm}^3 \text{ butter}$$

$$= 5.3 \times 10^2 \text{ cm}^3 \text{ butter}$$

Check: The units of the answer, cm³ butter, are appropriate for volume of butter. The number 5.3×10^2 (530) is a little larger than the number for mass, 454. This is reasonable. The density of butter is 0.86 g/cm³. Volume should be a little larger than mass.

Another way to solve this type of problem is to use density to write a conversion factor for changing mass to volume. According to Table 2.5, the density of butter is 0.86 g/cm³; that is, 0.86 g of butter occupies a volume of 1 cm³, or 0.86 g of butter is **equivalent to** (*contains the same quantity* of butter *as*) 1 cm³ of butter:

$$0.86 \text{ g butter} \rightleftharpoons 1 \text{ cm}^3 \text{ butter}$$

Division of both sides of this expression by 0.86 g butter gives a unit conversion factor for converting mass of butter to volume of butter:

$$454 \text{ g butter} \left(\frac{1 \text{ cm}^3 \text{ butter}}{0.86 \text{ g butter}} \right) = 527.906\,97 \text{ cm}^3 \text{ butter} = 5.3 \times 10^2 \text{ cm}^3 \text{ butter}$$

Both methods lead to the same arithmetic and give the same answer.

PRACTICE PROBLEMS

2.24 What is the volume of 56.3 g of ice?

2.25 What is the mass of 4.36 cm³ of iron?

The Discovery of the Noble Gases

The story of the discovery of the noble gases shows the importance of careful experimental work and of knowing when a difference between two measurements is significant. In the 1880s, English physical scientist Lord Rayleigh measured the densities of some common gases and found to his surprise that the density of nitrogen obtained from air by removal of oxygen, carbon dioxide, and water was 1.2561 g/L, whereas the density of nitrogen obtained by burning ammonia

$$4NH_3(g) + 3O_2(g) \longrightarrow 2N_2(g) + 6H_2O(l)$$

was 1.2498 g/L under the same conditions.

Lord Rayleigh asked chemists to explain the difference in density between atmospheric nitrogen and nitrogen obtained from a nitrogen compound by chemical reaction. British chemist William Ramsay suggested that nitrogen from the atmosphere must contain an unknown dense gas.

Rayleigh and Ramsay then treated the nitrogen obtained from air with hot magnesium metal. When hot, magnesium metal reacts with nitrogen forming magnesium nitride:

$$3Mg(s) + N_2(g) \longrightarrow Mg_3N_2(s)$$

After all the nitrogen had reacted, a small amount of a denser gas was left. The denser gas indeed proved to be a new element that Rayleigh and Ramsay named argon. The name "argon" is derived from the Greek word *argós,* which means "inactive."

There was no place in the periodic table of that time for an unreactive gaseous element, and Ramsay suggested addition of another group between the halogens (Group VIIA) and the alkali metals (Group IA). In 1898 Ramsay separated three more noble gases from the atmosphere: neon, krypton, and xenon. This left room in Group 0 for two more elements, and Ramsay discovered both of them. In 1903 he showed that helium is produced by the radioactive decay of radium, and in 1910 he detected radon in the radioactive material given off by radium.

Rayleigh received the Nobel Prize for physics in 1904 for his discovery of argon, and Ramsay received the 1904 Nobel Prize for chemistry for his discovery of the other noble gases—neon, krypton, xenon, and helium—and their places in the periodic table.

Incandescent light bulbs are filled with argon to prevent reaction of the hot filament with oxygen. The chief use of argon is as an inert atmosphere for arc-welding metals such as aluminum and stainless steel.

Summary of Steps for Solving Problems

1. Read the question carefully, noting what's given and what's asked for.
2. Figure out the relationships between what's given and what's asked for.
3. Reason through the problem stepwise from what's given to what's asked for.
4. Find the answer.
5. Check the answer: Are the units correct? Is the value reasonable? Is the answer expressed to the correct number of significant figures?

2.9 MEASURING TEMPERATURE

Temperature is *a measure of how hot or cold something is.* The difference in temperature between two objects controls the direction in which heat flows. For example, your hand is warmer than ice. If you pick up a piece of ice, your warm hand heats the cold ice. Your hand feels cold, and the ice melts. If you continue

The relationship between kinetic energy and temperature is discussed in Chapter 5; heat is correctly defined in Chapter 6.

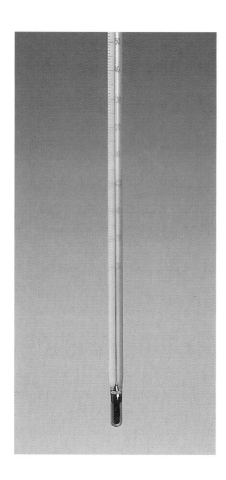

The Kelvin scale is named for the British scientist Lord Kelvin, who was a pioneer in understanding temperature.

■ FIGURE 2.11   A mercury thermometer. Mercury freezes at −38.842 °C. Therefore, mercury thermometers cannot be used below −38.842 °C. The temperature at which the glass begins to melt determines the highest temperature that can be measured with a mercury-in-glass thermometer. When a thermometer is used to measure the temperature of a sample, the sample must not gain a significant amount of heat from the thermometer (or lose a significant amount of heat to the thermometer). Therefore, the part of the thermometer that is in the sample must be small compared to the size of the sample. When the thermometer is at a low temperature, most of the mercury is in the bulb. The space above the mercury contains dry nitrogen gas. As the thermometer is heated, the volume of the mercury increases. Mercury rises up the narrow tube. The tube is so narrow that a small increase in the volume of the mercury results in a large increase in the height of the mercury column. Except for the bulb, the tube has a uniform diameter so that the same temperature change will produce the same increase in the height of the mercury over the whole temperature range of the thermometer. (The change in the volume of mercury per degree is the same over the whole temperature range.)

holding the ice, your hand will continue to feel cold until all the ice has melted and the water has become as warm as your hand.

Temperature cannot be measured directly by comparison with a standard as can length and mass. In everyday life and in the laboratory, temperature is usually measured with a thermometer. ■Figure 2.11 shows a mercury thermometer. You will probably use a mercury thermometer in your general chemistry laboratory.

In everyday life, Americans generally use the Fahrenheit scale to measure temperature. In scientific work, the Celsius scale is usually used. Kelvin temperature, the SI scale, must often be used in equations that involve temperature. ■Figure 2.12 shows the relationship between the Fahrenheit, Celsius, and Kelvin scales.

Interconverting Celsius and Kelvin Temperatures

Zero on the Celsius scale is defined as the temperature at which ice melts, and 100 is defined as the temperature at which water boils. On the Kelvin scale, the lowest possible temperature is defined as zero; there are no negative temperatures on the Kelvin scale (see Figure 2.12). The kelvin is the same size as the Celsius degree. There are 100 degrees Celsius and 100 kelvin between the melting point of ice and the boiling point of water. But 0 on the Kelvin scale equals −273 °C

■ FIGURE 2.12   A comparison of the Fahrenheit, Celsius, and Kelvin temperature scales. Note that although the symbol for kelvin is a capital K, the name is written with a lowercase k when used as a noun.

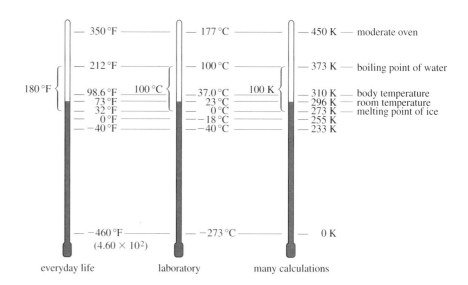

(-273.15 °C to be exact). The relationship between temperature on the Kelvin scale and temperature on the Celsius scale is

$$T\,K = (t\,°C + 273.15\,°C)\left(\frac{1\,K}{1\,°C}\right) \qquad (2.3)$$

In equation 2.3, capital T means Kelvin temperature, and lowercase t means Celsius temperature.

◉ Because kelvins are the same size as Celsius degrees, Kelvin temperature is equal to Celsius temperature plus 273.15.

SAMPLE PROBLEM

2.8   Convert 233 K to °C.

Solution   The problem gives temperature in K and asks for temperature in °C. Rearrange equation 2.3 to give Celsius temperature as a function of Kelvin temperature:

$$\left(\frac{1\,°C}{1\,K}\right)T\,K - 273.15\,°C = t\,°C \qquad (2.3a)$$

Substitute the Kelvin temperature given in the problem in the rearranged equation and solve for the Celsius temperature:

$$\left(\frac{1\,°C}{1\,K}\right)233\,K - 273.15\,°C = t\,°C = -40\,°C$$

◉ Numerically, the Celsius temperature is equal to the Kelvin temperature minus 273.15.

PRACTICE PROBLEM

2.26   Convert (a) 299 K to °C and (b) 19.5 °C to K.

Interconverting Celsius and Fahrenheit Temperatures

From Figure 2.12, you can see that the Fahrenheit and Celsius scales differ both in zero point and in the number of degrees between the melting point of ice and the boiling point of water. On the Fahrenheit scale, the temperature at which ice melts is defined as 32 °F. The temperature at which water boils is defined as 212 °F. Thus, on the Fahrenheit scale, there is a difference of 180 degrees between the melting point of ice and the boiling point of water. On the Celsius scale, there is only 100 degrees difference between the melting point of ice and the boiling point of water. If Δ is used to mean difference, then

$$\Delta 180\,°F = \Delta 100\,°C$$

or, simplifying by dividing by 20,

$$\Delta 9\,°F = \Delta 5\,°C$$

From this relationship, conversion factors for converting temperature in Fahrenheit to Celsius or Celsius to Fahrenheit can be obtained. The difference in zero point must also be taken into account. At the freezing point of ice, the temperature is 0 on the Celsius scale and 32 on the Fahrenheit scale. For this to be true, the relationship between the two scales must be

$$t_F\,°F = \left(\frac{9\,°F}{5\,°C}\right)t\,°C + 32\,°F \qquad (2.4)$$

where t_F means Fahrenheit temperature and t means Celsius temperature.

PRACTICE PROBLEMS

2.27 Which unit is larger, a Celsius degree or a Fahrenheit degree? Explain your answer.

2.28 Convert (a) 103 °F to °C and (b) 52 °C to °F.

2.10 MEASURING TIME

In the laboratory as in everyday life, time is measured with clocks, watches, and timers. The SI unit of time is the second, s. For periods much longer than a second, either the usual minutes (min), hours (h), days (day), and years (year or a), or SI prefixes like kilo and mega, can be used.

PRACTICE PROBLEMS

2.29 Fill in the blanks: (a) 1 day = _____ s, (b) 2.7 ns = _____ ps, and (c) 1 min = _____ ns.

2.30 Something happened at 9:22:34, and something else happened at 12:13:27. How long was it between these two events in (a) hours, minutes, and seconds and (b) seconds?

2.11 ATOMIC MASSES

To understand the SI unit of amount of substance, the mole, you first need to learn about atomic masses and formula masses. Individual atoms are extremely small; the heaviest atoms have masses of about 10^{-22} g. Even an ultramicrobalance, which can detect a difference in mass of 0.1 μg (10^{-7} g), cannot measure the mass of a single atom. However, finding the relative masses of atoms of different elements is possible. The following example should help you to understand the idea of relative masses better.

The masses of individual nails are too low to be measured with a student balance [see ▪Figures 2.13(a) and (b)]. But the masses of groups of nails can be determined [see Figure 2.13(c)]. If the mass of 100 large nails is 4.74 g and the mass of 100 small nails is 3.32 g, 100 large nails are 1.43 times as heavy as 100 small nails:

$$\frac{4.74 \text{ g}}{3.32 \text{ g}} = 1.43$$

Because the groups of nails contain equal numbers of nails, the mass of one large nail must be 1.43 times as large as the mass of one small nail. If the mass of a small nail is defined as one unit, the mass of a large nail will be 1.43 units. The relative masses of atoms of different elements can be found similarly by measuring the relative masses of large groups of atoms. Because the masses of atoms are so very small, the numbers of atoms in the group must be extremely large.

The measurement of the relative atomic masses of the elements occupied many chemists during the nineteenth and early twentieth centuries. At first, the mass of the lightest atom, hydrogen, was chosen as the unit of mass. On this scale, the atomic mass of oxygen was 15.87. Later, the atomic mass of oxygen was set equal to 16 (exactly), and the unit of mass was changed to 1/16 the mass of an oxygen

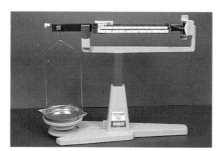

(a)

(b)

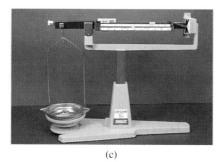

(c)

▪ FIGURE 2.13 (a) Student balance with nothing on the pan and masses and pointer at zero. (b) Addition of one nail to the pan does not move the pointer from zero. (c) The mass of 100 nails can be measured.

(a)

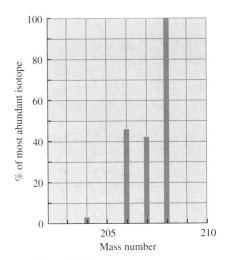

Electric field makes ions move faster and faster.

Field of magnet bends path of ions. Path of light ions is bent more than path of heavy ions. Ions are spread out into "rainbow" of masses.

Sample of gaseous atoms enters mass spectrometer.

Heavy ions

Beam of electrons knocks electrons off outside of atoms forming mostly 1+ ions.

Light ions

Recorder records mass spectrum.

(b)

■ FIGURE 2.14 (a) A modern mass spectrometer. (b) Schematic diagram of mass spectrometer. A mass spectrometer sorts ions according to their masses. The sample may consist of gaseous molecules instead of gaseous atoms.

atom. Oxygen was chosen as a standard for comparison because almost all elements form compounds with oxygen.

Then, early in the twentieth century, the American chemist T. W. Richards observed, to his amazement, that lead from samples collected in different places differed in atomic mass. This suggested that, contrary to Dalton's atomic theory, all atoms of an element are not exactly alike. Although atoms of an element have the same chemical properties, they can differ in mass.

Richards' surprising observation led to the discovery of isotopes. **Isotopes** are *atoms of an element that differ in mass.* In 1919 British physicist F. W. Aston invented the **mass spectrograph,** *an instrument that separates ions of different masses.* The results are recorded on a photographic plate. The modern version records results electrically and is called a **mass spectrometer.** ■Figure 2.14(b) is a schematic diagram of a mass spectrometer. With his mass spectrograph, Aston showed that most elements are mixtures of isotopes (see ■Figure 2.15).

The existence of isotopes was explained when the English physicist James Chadwick discovered neutrons in 1932. **Neutrons** are *particles with no charge* and are present in the nuclei of all atoms except one isotope of hydrogen. All atoms of an element have the same number of protons and electrons, but nuclei of different isotopes of the same element have different numbers of neutrons. Compared to neutrons and protons, electrons have negligible mass. This does not mean that electrons have no mass. An electron has mass, but its mass is so small compared to the masses of protons and neutrons that it can usually be neglected. The properties of the particles that make up atoms are summarized in Table 2.6.

The *sum of the number of protons and neutrons in the nucleus of an atom* is called its **mass number, A:**

$$\text{mass number (A)} = \text{number of protons (Z)} + \text{number of neutrons (N)} \qquad (2.5)$$

The *number of protons in the nucleus* is equal to the **atomic number, Z.** The *number of neutrons in the nucleus* is sometimes called the **neutron number, N.** All isotopes of an element have the same atomic number. However, different isotopes have different neutron numbers and different mass numbers.

The term **nuclide** is used to indicate a *specific nuclear species* such as an atom of lithium that contains four neutrons (all of lithium contain three protons). The symbol for a nuclide shows the mass number as a left superscript and the atomic number as a left subscript. For example, naturally occurring lithium consists mostly of a nuclide with mass number equal to 7. The atomic number of lithium is 3. The symbol for the nuclide of lithium with mass number equal to 7 is

$$^{7}_{3}\text{Li}$$

■ FIGURE 2.15 Mass spectrum of a sample of lead. Samples of lead from different places have different isotopic compositions.

TABLE 2.6

Particles Making Up Atoms

Particle	Mass[a]	Charge[b]
Proton	1	1+
Neutron	1	0
Electron	0	1−

[a] Rounded to nearest whole number. The unit of mass is defined later in this section.
[b] In chemistry, the charge on an electron is usually used as the unit of charge.

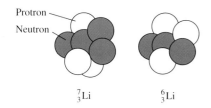

Protron
Neutron

$^{7}_{3}$Li $^{6}_{3}$Li

Diagram of nuclei of lithium-7 and its isotope, lithium-6.

The atomic number is sometimes omitted from the symbol for a nuclide because the atomic number is the same for all isotopes of an element:

$$^{7}Li$$

Another way of representing this nuclide is by writing lithium-7 (read "lithium seven").

Remember that a right superscript shows the charge on an ion and a right subscript shows the number of atoms of a given kind in an ion or molecule. For example, the symbol for the mercury(I) ion formed from mercury-202 is

$$^{202}_{80}Hg_2{}^{2+}$$

In this symbol, the right superscript shows that the total charge on the ion is $2+$ and the right subscript shows that the ion is formed from two mercury atoms. Therefore, the average charge on each mercury is $1+$, and the name of the ion is mercury(I) ion.

The Atomic Weights module of the Reactions section shows a working mass spectrometer. It teaches students to determine the masses of atoms relative to C-12 from the mass spectra and to calculate weighted average masses from isotropic masses and abundances. It also shows the relationships between nuclide symbols, mass numbers, and numbers of protons, neutrons, and electrons.

SAMPLE PROBLEM

2.9 (a) In an atom of mercury-202, how many protons are there? (b) How many neutrons are there in an atom of mercury-202? (c) How many electrons are there in one mercury(I) ion?

Solution   (a) The question gives the mass number and asks for the number of protons and neutrons. The atomic number of mercury is 80; therefore, there are 80 protons in the nucleus of a mercury atom.

(b) According to the definition of mass number (equation 2.5),

$$\text{mass number} = \text{number of protons} + \text{number of neutrons}$$

From this relationship,

$$\text{mass number} - \text{number of protons} = \text{number of neutrons}$$

and

$$202 \quad - \quad 80 \quad = \text{number of neutrons} = 122$$

(c) According to the formula for the mercury(I) ion ($Hg_2{}^{2+}$), two mercury atoms combine to form one mercury(I) ion. In two mercury atoms, there are

$$2(80 \text{ protons}) = 160 \text{ protons}$$

In an atom, the number of electrons around the nucleus is equal to the number of protons in the nucleus. In two mercury atoms, there are 160 electrons. The charge on the mercury(I) ion is $2+$. The combination of two mercury atoms has lost two electrons in forming the ion:

$$(160 - 2) \text{ electrons} = 158 \text{ electrons}$$

There are 158 electrons in one mercury(I) ion.

If a mouse creeps onto the scale while you are weighing yourself, the reading on the dial will not change because the mouse's mass is negligible compared to your mass just as the mass of an electron is negligible compared with the masses of protons and neutrons.

PRACTICE PROBLEMS

2.31   In an atom of tin-118, (a) how many protons are there? (b) How many neutrons? (c) Write the symbol for the nuclide tin-118. (d) How many electrons are there in one Sn^{2+} ion?

2.32   In an atom of chlorine-35, (a) how many protons are there? (b) How many neutrons? (c) Write the symbol for the nuclide chlorine-35. (d) How many electrons are there in one Cl^- ion? (e) How many electrons are there in one molecule of Cl_2?

Today, *the unit chosen for atomic masses by international agreement between both chemists and physicists is 1/12 the mass of an atom of carbon-12, which is defined to have a mass of exactly 12.* This unit is called a **universal atomic mass unit** and its symbol is **u** (or amu). For example, the mass of an atom of lithium-7 is 7.016 003 u; that is, an atom of lithium-7 is 7.016 003 times as massive as 1/12 the mass of an atom of carbon-12. Because atomic masses are relative masses, they are pure numbers and are sometimes given without units.

Biochemists call a universal atomic mass unit a dalton, D.

The Atomic Masses of the Elements

Some elements consist of a single nuclide. However, about three-quarters of the naturally occurring elements are mixtures of two to ten isotopes. If the naturally occurring element is a mixture of just two isotopes, the isotopic composition can be estimated from the atomic mass given in the table inside the front cover.

SAMPLE PROBLEM

2.10   Naturally occurring chlorine is a mixture of chlorine-35 and chlorine-37. Estimate the isotopic composition of naturally occurring chlorine.

Solution   From the table inside the front cover, the atomic mass of naturally occurring chlorine (rounded to one decimal place) is 35.5 u. Because 35.5 is closer to 35 than it is to 37, naturally occurring chlorine must contain more chlorine-35 than chlorine-37.

Since 35.5 is about one-quarter of the way from 35 to 37, naturally occurring chlorine must be about one-quarter (25%) chlorine-37 and three-quarters (75%) chlorine-35. (The composition calculated from the results of mass spectroscopy is 24.23% chlorine-37 and 75.77% chlorine-35.)

PRACTICE PROBLEM

2.33   Naturally occurring bromine is a mixture of bromine-79 and bromine-81. Estimate the isotopic composition of naturally occurring bromine.

The atomic mass of a mixture of isotopes depends both on the atomic masses of the individual isotopes and on the percentage of each isotope in the mixture. The atomic masses shown in the table of the elements inside the front cover are the most accurate values known at the present time. They are the results of experiments by many scientists. All were determined by measuring the number percents and the atomic masses of the isotopes of an element in naturally occurring samples with mass spectrometers (Figure 2.14). The **number percent** of a given isotope is the *percentage of the total number of atoms in the sample that are atoms of the given isotope:*

$$\text{number \%} = \frac{\text{number of atoms of isotope}}{\text{total number of atoms in sample}} \times 100$$

See Appendix B.5 for a review of percent.

The atomic masses of nuclides can be determined to six or more significant figures with a mass spectrometer. The number percent of each nuclide in the sample can also be obtained from the mass spectrum. From this information, the atomic mass of the sample can be calculated.

Calculation of Atomic Mass

Sample Problem 2.11 shows how the atomic mass of a naturally occurring mixture of isotopes is calculated from the isotopic composition of the naturally occurring mixture and the atomic masses of the nuclides that make up the naturally occurring mixture. The atomic mass must be "weighted" for the fact that the naturally occurring mixture is *not* composed of equal numbers of atoms of the different isotopes. To help you understand what "weighted" means, let's consider a much simpler and more familiar calculation, calculating a student's quiz average. Suppose a student receives the following scores on a set of ten quizzes: 5, 8, 9, 10, 9, 10, 9, 10, 10, and 10. The average of 5, 8, 9, and 10 calculated in the usual way is 8.0:

$$\frac{5 + 8 + 9 + 10}{4} = 8.0$$

Would the student be satisfied with this average? No! The student's average is

$$\frac{5 + 8 + 9 + 10 + 9 + 10 + 9 + 10 + 10 + 10}{10}$$

$$= \frac{1}{10}(5) + \frac{1}{10}(8) + \frac{3}{10}(9) + \frac{5}{10}(10) = 9.0$$

The correct average is "weighted" for the fact that the one of the ten quizzes was a 5, one was an 8, three were 9s, and five were 10s.

SAMPLE PROBLEM

2.11 Calculate the atomic mass of naturally occurring neon, which is a mixture of 90.48% neon-20 (atomic mass = 19.992 435 u), 0.27% neon-21 (atomic mass = 20.993 843 u), and 9.25% neon-22 (atomic mass = 21.991 383 u).

Solution The statement that naturally occurring neon contains 90.48% neon-20 means that 90.48 of 100 atoms have an atomic mass of 19.992 435 u. The weighted average for the atomic mass of neon is

$$\frac{90.48}{100}(19.992\ 435\ u) + \frac{0.27}{100}(20.993\ 843\ u) + \frac{9.25}{100}(21.991\ 383\ u) =$$

$$18.09\ u \qquad + \qquad 0.057\ u \qquad + \qquad 2.03\ u \qquad = 20.18\ u$$

Check: An atomic mass of 20.18 u (slightly greater than 20) is reasonable because naturally occurring neon is mostly (90.48%) neon-20 with smaller amounts of the heavier isotopes.

○ The statement that "90.48% neon-20 means that 90.48 of 100 atoms have an atomic mass of 19.992 435 u" does *not* mean that a fraction of an atom can exist any more than the statement that "9.21% of a class of 76 students received a grade of A" means that a fraction of a student can exist.

PRACTICE PROBLEMS

2.34 Naturally occurring silicon is a mixture of 92.2% silicon-28 (atomic mass = 27.976 927 u), 4.67% silicon-29 (atomic mass = 28.976 495 u), and 3.10% silicon-30 (atomic mass = 29.973 770 u). Calculate the atomic mass of naturally occurring silicon.

2.35 (a) From the atomic masses of the naturally occurring isotopes of neon and silicon given in Sample Problem 2.11 and Practice Problem 2.34, what relation is there between atomic mass and mass number? (b) What is the mass number of the isotope of nitrogen having atomic mass = 14.003 074 u?

2.36 Suggest an explanation for the fact that the atomic mass of carbon is 12.011, not 12.000 u.

In the table inside the front cover, some atomic masses are given to more significant figures than others because the isotopic composition of some elements varies more from sample to sample than the isotopic composition of other elements. Usually the number of significant figures in measurements increases as time passes because better instruments are invented. The number of significant figures in atomic masses may well decrease in the future for two reasons:

1. Modern processing methods are changing the isotopic composition of some elements to a small extent. For example, large quantities of heavy water (^2H_2O) have been separated from ordinary water for use in nuclear power plants.

2. Materials from space have different isotopic compositions from materials found on Earth. Xenon found in meteorites may have twice as much of the heavy isotopes (xenon-131, xenon-132, xenon-134, and xenon-136) as xenon found on Earth. When meteorites rich in the heavy isotopes of xenon strike Earth, the percentage of heavy isotopes in the xenon on Earth increases. If space mining ever becomes a reality, the isotopic composition of some elements may change significantly.

The photo shows a meteorite. Some elements in meteorites have isotopic compositions different from those on Earth.

Symbols without left superscripts are used to represent naturally occurring elements whether they consist of a single isotope or a mixture of isotopes. For example, the symbol F means naturally occurring fluorine and the symbol Cl, naturally occurring chlorine.

Some nuclides are radioactive. Nuclei of **radioactive nuclides** *change spontaneously (decay) into nuclei of other elements and give off energy in the form of radiation and particles* such as electrons and helium nuclei. Some nuclei, such as beryllium-8 and boron-9, decay in less than a femtosecond (10^{-15} s). Others, such as potassium-40, take more than a billion (10^9) years. *Nuclides that are not radioactive* are said to be **stable.** The numbers in parentheses in the table inside the front cover are the mass numbers of the *most* **stable isotopes** (longest-lived isotopes) of elements that have no stable isotopes. All nuclides having an atomic number greater than 83 and all nuclides of technetium (atomic number 43) and promethium (atomic number 61) are radioactive.

2.12 FORMULA MASSES

When atoms combine to form compounds, some of their electrons arrange in a new way around the nuclei. However, the number of nuclei of each kind and the number of electrons are the same in the compounds as they are in the elements. Therefore, the **formula masses** *of compounds are equal to the sum of the atomic masses of the atoms forming the compound.* The formula masses of molecular compounds are often called **molecular masses.**

SAMPLE PROBLEM

2.12   Calculate the formula mass of $Pb(NO_3)_2$ to as many significant figures as possible.

Solution   The formula mass of $Pb(NO_3)_2$ is the sum of the masses of the atoms combined to form lead nitrate. In one unit of $Pb(NO_3)_2$, one lead atom, two nitrogen atoms, and six oxygen atoms are combined. Atomic masses are listed in the table

$$
\begin{array}{lrl}
1Pb @ 207.2\ u & = & 207.2\ u \\
2N @ \ 14.006\ 74\ u & = & 28.013\ 48\ u \\
6O @ \ 15.9994\ u & = & 95.996\ 4\ u \\
\hline
1Pb(NO_3)_2 & = & 331.209\ 88\ u = 331.2\ u
\end{array}
$$

The formula mass of lead nitrate is rounded to one decimal place because it is obtained by addition from the atomic mass of lead, which has only one decimal place.

PRACTICE PROBLEM

2.37   Calculate the formula masses of each of the following compounds to as many significant figures as possible: (a) $CaBr_2$, (b) Al_2O_3, (c) $(NH_4)_2S$, and (d) CCl_4.

Left: *Copper(II) sulfate pentahydrate CuSO₄ · 5H₂O(s). Right: Addition of water to anhydrous (without water) copper(II) sulfate CuSO₄(s), which is white, yields the blue pentahydrate. Anhydrous copper(II) sulfate is used for detecting and removing trace amounts of water from alcohols and other organic compounds. The pentahydrate is used as an agricultural fungicide, algicide, bactericide, and herbicide.*

Hydrates

Some compounds form *crystals with a definite proportion of water* called **hydrates.** Zinc sulfite is an example. The formula for solid zinc sulfite is $ZnSO_3 \cdot 2H_2O$. Note the centered dot between $ZnSO_3$ and $2H_2O$. The 2 in front of the formula for water multiplies *both* H_2 and O by 2, although there are no parentheses. Greek prefixes are used to show the number of water molecules in hydrates. This compound is called zinc sulfite dihydrate. As usual, 1 is understood if no number is written. The compound $MgSO_4 \cdot H_2O$ has one molecule of water for each unit of $MgSO_4$ and is called magnesium sulfate monohydrate.

SAMPLE PROBLEM

2.13   Calculate the formula mass of $ZnSO_3 \cdot 2H_2O$ to as many significant figures as possible.

Solution   In one unit of $ZnSO_3 \cdot 2H_2O$, there are 1 Zn, 1 S, 3 O (from $ZnSO_3$) + 2 O from $2H_2O$, or a total of 5 O and 4 H:

$$
\begin{array}{lrl}
1Zn @ 65.39\ u & = & 65.39\ u \\
1S @ 32.066\ u & = & 32.066\ u \\
5O @ 15.9994\ u & = & 79.997\ 0\ u \\
4H @ \ 1.007\ 94\ u & = & 4.031\ 76\ u \\
\hline
1ZnSO_3 \cdot 2H_2O & = & 181.484\ 76\ u = 181.48\ u
\end{array}
$$

PRACTICE PROBLEM

2.38   (a) Calculate the formula mass of $CaCl_2 \cdot 6H_2O$ to as many significant figures as possible. (b) Fill in the blank: $CaCl_2 \cdot 6H_2O$ is called calcium chloride _____ . (c) Write formulas for calcium chloride monohydrate and anhydrous calcium chloride.

○ Arranging calculations in an orderly way helps prevent mistakes.

2.13 AMOUNT OF SUBSTANCE

So far we have usually taken a **microscopic view** of chemical reactions; we have talked about individual atoms, molecules, and ions. Samples in the laboratory and the world outside contain enormous numbers of these units. A **macroscopic view** of chemical reactions is necessary for practical work.

Just as a pair and a dozen are convenient-sized packages for objects like socks and eggs, the mole, the SI unit of amount of substance, is a convenient-sized package for atoms, molecules, and ions. Using the mole as a unit of amount of substance, we can count microscopic particles such as atoms, molecules, and ions by measuring mass.

Officially, one **mole** (abbreviated mol) is defined as *the amount of substance in a sample that contains as many units as there are atoms in exactly 0.012 kg (12 g) of carbon-12.* As a result, *the mass of one mole in grams is numerically equal to the formula mass.* The *mass of one mole in grams* is called the **molar mass.** For example, one hydrogen atom H has a mass of 1.007 94 u. Therefore, one mole of hydrogen atoms has a mass of 1.007 94 g. The formula mass of hydrogen gas, H_2, is

$$2H @ 1.007\ 94\ u = 2.015\ 88\ u$$

One molecule of hydrogen gas has a mass of 2.015 88 u, and one mole of hydrogen gas has a mass of 2.015 88 g. The **molar mass** of hydrogen gas is 2.015 88 g.

The *number of units in a mole* has been found by experiment to be $6.022\ 137 \times 10^{23}$ units/mol. This number is called **Avogadro's number.*** Avogadro's number is usually rounded to 6.022×10^{23} units/mol because measurements made in chemistry laboratories do not often have more than four significant figures. The units may be atoms, molecules, monatomic or polyatomic ions, electrons, even formula units for ionic compounds such as NaCl. *One mole of any units contains Avogadro's number of those units.* The units must be named; for example, one mole of hydrogen molecules, H_2, contains twice as many hydrogen atoms as one mole of hydrogen atoms, H. More examples are given in Table 2.7. ▪Figure 2.16 shows one mole of each of the substances in Table 2.7.

▪ FIGURE 2.16 *Clockwise starting at lower left:* One-mole samples of iron (Fe), sulfur (S_8), water (H_2O), and sodium chloride (NaCl). All samples contain the same number of formula units, 6.022×10^{23} formula units. Under the same conditions, a 1-mol sample of hydrogen gas (H_2) would occupy about 24 L (more than 1300 times the volume of the 1-mole sample of water).

For the moment, you will just have to take it on faith that counting such tiny particles to seven significant figures is possible.

TABLE 2.7	Relationship Between Formula Mass, Mass of One Mole, and Number of Particles

Name of Substance	Formula	Formula Mass, u	Mass of One Mole, g	Kind of Particle and Number in One Mole of Substance	
Iron	Fe	55.847	55.847	Fe atoms,	6.022×10^{23}
Sulfur	S_8	256.53	256.53	S_8 molecules,	6.022×10^{23}
				S atoms,	48.177×10^{23}
Water	H_2O	18.0153	18.0153	H_2O molecules,	6.022×10^{23}
				H atoms,	12.044×10^{23}
				O atoms,	6.022×10^{23}
Sodium chloride	NaCl	58.4425	58.4425	NaCl formula units,	6.022×10^{23}
				Na^+ ions,	6.022×10^{23}
				Cl^- ions,	6.022×10^{23}

---

*Avogadro was an Italian physicist who lived from 1776 to 1856. He was the first person to distinguish between atoms and molecules, and his work made possible the determination of a table of atomic masses.

Avogadro's number is an enormous number. If Avogadro's number of pennies were placed side-by-side, they would circle the world 270 trillion (2.7×10^{14}) times.

If you were to measure a mole of iron atoms by counting them at a rate of one atom per second, it would take you 1.9×10^{16} years—too long! However, one mole of iron atoms has a mass of 55.847 g. You could measure 55.847 g of iron with a balance in less than a minute. Therefore, in the laboratory, chemists usually measure amounts of substances in grams. Amount of substance expressed in grams, moles, and number of particles can be interconverted by calculation as shown in Sample Problems 2.14–2.16. The mass of a single atom can be obtained by calculation (Sample Problem 2.17).

SAMPLE PROBLEMS

2.14 How many moles of NaCl is 2.735 g NaCl?

Solution The problem gives the number of grams of NaCl and asks for the number of moles of NaCl. We must find the formula mass for NaCl and use it to write a conversion factor for changing grams of NaCl to moles of NaCl.

To save time and let you concentrate on new skills, a table of frequently used formula masses is printed inside the back cover. From the table inside the back cover, the formula mass for NaCl is 58.4425 u. Because the formula mass for NaCl is 58.4425 u,

$$1 \text{ mol NaCl} \leftrightarrow 58.4425 \text{ g NaCl}$$

The conversion factor for converting g NaCl to mol NaCl is 1 mol NaCl/58.44 g NaCl (the mass of 1 mol NaCl is rounded to four digits because there are four significant figures in 2.735 g NaCl, the quantity given in the problem):

$$2.735 \text{ g NaCl}\left(\frac{1 \text{ mol NaCl}}{58.44 \text{ g NaCl}}\right) = 0.046\,80 \text{ mol NaCl}$$

Check: The answer is reasonable because 2.735 g NaCl is about 3/60 or 0.05 mol NaCl.

2.15 What is the mass in grams of 3.47 mol NaCl?

Solution The problem gives the number of moles of NaCl and asks for the number of grams of NaCl. As in Sample Problem 2.14, the conversion factor is obtained from the relationship

$$1 \text{ mol NaCl} \leftrightarrow 58.4425 \text{ g NaCl}$$

This time there are three significant figures in the number given in the problem, 3.47.

$$3.47 \text{ mol NaCl}\left(\frac{58.4 \text{ g NaCl}}{1 \text{ mol NaCl}}\right) = 203 \text{ g NaCl}$$

Check: The answer is reasonable because 203 g is between 3 and 4 times 60, the approximate mass of 1 mol NaCl, and 3.47 mol is between three and four moles.

2.16 Calculate the number of atoms in 2.00 g N₂.

Solution The problem asks for the number of atoms in 2.00 g N_2. From the table inside the back cover, the formula mass of nitrogen gas, N_2, is 28.013 48 u.

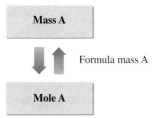

Mass A

Formula mass A

Mole A

Masses and moles can be interconverted by means of formula masses.

Therefore, one mole of nitrogen gas has a mass of 28.013 48 g. The relationships needed to solve this problem are

$$1 \text{ mol N}_2 \,\Leftrightarrow\, 28.0 \text{ g N}_2 \qquad 1 \text{ N}_2 \text{ molecule} \,\Leftrightarrow\, 2 \text{ N atoms}$$

$$1 \text{ mol N}_2 \text{ molecules} \,\Leftrightarrow\, 6.02 \times 10^{23} \text{ N}_2 \text{ molecules}$$

Using these relationships to write conversion factors gives

$$2.00 \text{ g N}_2 \left(\frac{1 \text{ mol N}_2}{28.0 \text{ g N}_2} \right) \left(\frac{6.02 \times 10^{23} \text{ N}_2 \text{ molecules}}{1 \text{ mol N}_2} \right) \left(\frac{2 \text{ N atoms}}{1 \text{ N}_2 \text{ molecule}} \right)$$

$$= 8.60 \times 10^{22} \text{ N atoms}$$

Remember that $\Leftrightarrow$ means "is equivalent to."

Check: The answer is reasonable because it is a very large number.

2.17 Calculate the mass in grams of a single nitrogen atom to three significant figures.

Solution The problem asks for the mass of a single N atom to three significant figures. From the table inside the front cover, the atomic mass of nitrogen is 14.006 74 u. Therefore, the mass of one mole of N atoms is 14.006 74 g. The relationships needed to calculate the mass of a single N atom to three significant figures are

Always check to be sure your work is reasonable.

$$1 \text{ mol N atoms} \,\Leftrightarrow\, 14.0 \text{ g N} \qquad 1 \text{ mol N atoms} \,\Leftrightarrow\, 6.02 \times 10^{23} \text{ N atoms}$$

Using these relationships to write conversion factors gives

$$1 \text{ N atom} \left(\frac{1 \text{ mol N atoms}}{6.02 \times 10^{23} \text{ N atoms}} \right) \left(\frac{14.0 \text{ g N}}{1 \text{ mol N atoms}} \right) = 2.33 \times 10^{-23} \text{ g N}$$

The mass of one N atom is 2.33×10^{-23} g.

Check: The answer is reasonable because it is a very small number.

PRACTICE PROBLEMS

2.39 (a) What is the mass in grams of a 1.000-mol sample of gold? (b) How many atoms of gold are there in 1.000 mol of gold? (c) How many atoms of gold are there in 2.00 mol of gold? (d) How many atoms of gold are there in 0.500 mol of gold?

2.40 (a) What is the mass in grams of a 1.000-mol sample of phosphorus, P_4? (b) How many molecules are there in a 1.000-mol sample of phosphorus? (c) How many phosphorus atoms are there in a 1.000-mol sample of phosphorus?

2.41 (a) How many moles is 17.031 g of ammonia? (b) How many nitrogen atoms are there in 17.031 g of ammonia? How many hydrogen atoms?

2.42 How many nitrate ions are there in 1.0000 mol $Ca(NO_3)_2$?

2.43 What is the mass in grams of 0.54 mol H_2SO_4?

2.44 How many moles of H_2SO_4 are there in 22.36 g H_2SO_4?

2.45 (a) Calculate the number of atoms in 5.000 g P_4. (b) Calculate the mass in grams of a single phosphorus atom to four significant figures.

Remember that Practice Problems are continued at the end of the chapter under the heading Additional Practice Problems. To emphasize the importance of the in-chapter Practice Problems, the Additional Practice Problems at the end of the chapter continue the sequence of numbering begun in the chapter.

Remember that if you need to know the meaning of any of the terms introduced in other chapters, you should use the index at the back of the book (blue-edged pages).

Observations may be **qualitative** or **quantitative.** Quantitative observations (measurements) involve numbers and are made with instruments. Qualitative observations do not involve numbers. The advantages of quantitative observations are as follows: (1) Instruments are less likely to be fooled than humans; (2) observations made with instruments are usually reproducible; (3) quantitative observations can easily be stored and compared. Measurements give three pieces of information: quantity, units, and uncertainty.

Scientists all over the world use **SI units,** both **base** and **derived,** to report their results. **Unit conversion factors** are used to convert measurements from one unit to another.

In general chemistry, uncertainty is shown by writing one uncertain or estimated digit in measurements and in the results of calculations involving measurements. Digits recorded according to this rule are called **significant figures.** The reproducibility of a measurement is called **precision.** The **normal distribution curve** shows that most observations are close to average. Once you have learned how to make a measurement, repeating the measurement more than two or three times is usually a waste of time. **Accuracy** refers to the correctness of measurements. To obtain results that are as accurate as possible, scientists **calibrate** their instruments and repeat measurements two or three times. The **median** is the best value to report for a small set of observations.

Mass is the quantity of matter that an object contains. **Weight** depends both on the quantity of matter in an object and on the force of gravity on the object. Mass is measured with a balance and is an **extensive property.** Extensive properties depend on the quantity of sample measured. **Intensive properties** are independent of sample size. Intensive properties are often characteristic of substances; therefore, they are useful for identifying substances, checking their purity, and deciding whether a substance is suitable for a given use.

The ratio of the extensive property, mass, to the extensive property, volume, is the intensive property called **density,**

$$\text{density} = \frac{\text{mass}}{\text{volume}}$$

which can be used to convert mass to volume or volume to mass. Density depends on temperature (and, for gases, on pressure).

Temperature is a measure of hotness or coldness and determines the direction in which heat flows; heat flows from a hotter to a colder body. In general chemistry laboratories, temperature is usually measured with a thermometer in degrees Celsius. To convert temperature in degrees Celsius to temperature in the SI base unit kelvin, add 273.15; to convert temperature in kelvin to temperature in degrees Celsius, subtract 273.15. **Kelvin temperature** must often be used in equations involving temperature.

The nuclei of atoms are composed of **protons** and **neutrons.** All atoms of an element have the same number of protons in the nucleus and electrons outside the nucleus. This number is the **atomic number, Z,** of the element. All atoms of an element do not have the same number of neutrons. Atoms of an element that have different numbers of neutrons are called **isotopes.** Isotopes differ in mass and can be separated with a **mass spectrometer.** The sum of the number of protons and neutrons in the nucleus of an atom is called the **mass number, A,** of the atom.

The term **nuclide** is used to indicate a specific nuclear species. Nuclides are represented by writing the symbol for the element with the mass number as a left superscript and the atomic number as a left subscript. Some naturally occurring elements consist of a single nuclide; others are mixtures of two to ten isotopes.

The masses of individual atoms are too small to measure. However, the masses of packages of large numbers of atoms can be compared. A package of 6.022×10^{23} particles is one **mole.** The number 6.022×10^{23} is called **Avogadro's number. Universal atomic mass units, u,** are used for atomic masses. By definition, one universal atomic mass unit is 1/12 the mass of an atom of carbon-12; the mass of one mole of carbon-12 is defined as exactly 0.012 kg (12 g). Atomic masses in the table inside the front cover are given to different numbers of significant figures because the isotopic compositions of some elements vary more from sample to sample than the isotopic compositions of other elements. The numbers in parentheses in the table inside the front cover are the mass numbers of the most stable isotopes of radioactive elements. Nuclei of **radioactive elements** change spontaneously into nuclei of other elements and give off energy in the form of radiation and particles.

The **formula mass** of a compound is equal to the sum of the atomic masses of the atoms forming the compound. If the compound is a molecular compound, the formula mass is frequently referred to as the **molecular mass.** The mass of one mole of any substance in grams has the same numerical value as the formula mass. The mass of one mole in grams is often called the **molar mass.**

The importance of the mole as a unit of amount of substance is that it supplies a conversion factor between the **microscopic** world of atoms, molecules, and ions and the real **(macroscopic)** world where every sample contains enormous numbers of particles. The mole concept provides a way to count atoms, molecules, and ions by measuring mass.

The following steps should be followed in solving problems: (1) Read the question carefully, noting what's given and what's asked for. (2) Figure out the relationships between what's given and what's asked for. (3) Reason through the problem stepwise. (4) Find the answer. (5) Check the answer. If the answer is numerical: Are the units correct? Is the value reasonable? Is the answer expressed to the correct number of significant figures?

Remember that the first Additional Practice Problems follow the order of the material presented in the chapter; the numbers in parentheses at the ends of these problems show the number of the section that the problems are related to. Later Additional Practice Problems do not follow the chapter order and are not classified by section because learning to identify different types of problems is an important part of learning to solve problems. Answers to problems with blue numbers appear in Appendix I. Solutions for these problems are in the Student Solutions Manual. Solutions to Additional Practice Problems with black numbers may be available at the option of your instructor.

2.46 Rewrite each of the following according to the SI rules: (a) 3,116mol., (b) 224·3 ks, (c) .014865 m. (2.1)

2.47 What is the SI base unit for each of the following? (a) temperature (b) amount of substance (2.1)

2.48 Which is the only SI base unit that is still defined in terms of a physical object? (2.1)

2.49 Write the name of each of the following units: (a) kK, (b) cg, (c) μmol. (2.2)

2.50 Write the symbol for each of the following units: (a) decimeter, (b) megagram, (c) millimole. (2.2)

2.51 (a) How many grams are there in a kilogram? (b) Which is smaller, a gram or a kilogram? (c) Which is smaller, a kilogram or a megagram? (d) How many kilograms are there in a megagram? (2.2)

2.52 (a) What conversion factor should you multiply Mm by to convert to m? (b) What conversion factors should you multiply μm by to convert to nm? (2.2)

2.53 Carry out the following conversions: (a) 0.342 mol to mmol, (b) 7.2×10^{-8} cm to pm, (c) 6.4 kg to g, (d) 72.4 Mm to km. (2.2)

2.54 Fill in the blank with the correct unit:
(a) 6.4 _____ = 6.4×10^{-9} m
(b) 3.87 _____ = 3.87×10^{6} mol
(c) 0.0457 _____ = 4.57 mg (2.2)

2.55 Use exponential notation to express the following in SI base units: (a) 8.9 km, (b) 26 mmol, (c) 37.81 g, (d) 3.4 pm, (e) 7.8 Ms. (2.2)

2.56 Convert each of the following to the SI unit that will give the smallest number greater than 1: (a) 3.35×10^{-6} m, (b) 5.05×10^{7} m. (2.2)

2.57 The liter is defined as 1 dm^{3}. (a) How many liters are in 1 (exact number) m^{3}? (b) How many m^{3} are in 1 (exact number) L? (2.2)

2.58 Fill in the blanks: (a) 673 mL = _____ qt, (b) 247 ft = _____ m. (2.2)

2.59 Write each of the following in decimal notation: (a) 1/8, (b) 6.72×10^{5}, (c) 8.3×10^{-4}. (2.2)

2.60 Write each of the following in exponential notation: (a) 24 362, (b) 0.035, (c) 78, (d) 6.4. (2.2)

2.61 Which member of each of the following pairs of numbers is larger?
(a) 6.02×10^{23} or 6.02×10^{24}
(b) 3.4×10^{-8} or 3.4×10^{-9}

(c) 3.4×10^{-8} or 4.4×10^{-8}
(d) 3.4×10^{7} or 4.4×10^{6}
(e) 44×10^{-7} or 4.2×10^{-6} (2.2)

2.62 Fill in the blanks using the correct number of digits in your answers:
(a) 6.83×10^{-9} = _____ $\times 10^{-8}$
(b) $(8.4 \times 10^{-12}) - (6.2 \times 10^{-13})$ = _____
(c) $(5.26 \times 10^{8})(4.1 \times 10^{12})$ = _____
(d) $3.26 \times 10^{8}/4.1 \times 10^{-12}$ = _____
(e) $26.83 - (7.2 \times 0.473)$ = _____ (2.2)

2.63 Enlargements of parts of the scales of three thermometers are shown below. All three are Celsius thermometers. What temperature is shown by each thermometer?

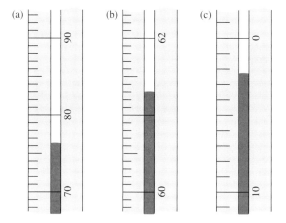

(d) Which thermometer was used to measure another temperature that was recorded as 48.23 °C? (2.3)

2.64 Write 670 000 to show (a) two significant figures, (b) four significant figures, (c) six significant figures. (2.3)

2.65 How many significant figures are there in each of the following measured quantities? (a) 6 thermometers (b) 42.3100 g (c) 42 300 s (d) 0.003 24 m (e) 3.4×10^{8} mg (2.3)

2.66 Old textbooks state that 1 in. = 2.540 005 cm. New textbooks define one inch as equal to exactly 2.54 cm. Are there more, fewer, or the same number of significant figures in the new definition? Explain your answer. (2.3)

2.67 What is the difference between a length of 2 cm and a length of 2.0 cm? (2.3)

2.68 Which of the following can be determined exactly, and which must be measured with some uncertainty?
(a) your weight
(b) the number of fingers on your left hand
(c) the number of hairs on your left arm (2.3)

2.69 (a) What is the median of the following measurements? 22.18 g, 22.29 g, 22.17 g, 22.16 g (b) What is the average of the measurements in part (a)? (c) What is the median of the following measurements? 4.6835 g, 4.6837 g, 4.6832 g (2.3)

2.70 (a) What is the difference between *accuracy* and *precision?* (b) Define *calibrate*. (2.3)

2.71 Round each of the following to two significant figures: (a) 43.481, (b) 536.5, (c) 6.7426, (d) 3.48×10^{-8}, (e) 88.98. (2.4)

2.72 How many significant figures are there in the answers to each of the following calculations?
(a) (23.9684 g) − (23.9680 g)
(b) (2.7 cm) + (6.84 cm) + (69.4 cm)
(c) (54.3286 g)/(25.4 mL)
(d) (2.7 cm)(26.8 cm)(79.4 cm) (2.4)

2.73 Carry out the following conversions: (a) 645 mL to cm^3, (b) 645 mL to L, (c) 0.072 L to mL. (2.5)

2.74 What is the difference between *mass* and *weight*? (2.6)

2.75 Which of the following are extensive properties, and which are intensive properties? (a) the volume of a can of orange juice (b) the density of aluminum (c) the freezing point of water (d) the weight of a steak (e) the fact that silver tarnishes (2.7)

2.76 A piece of chalk having a mass of 27.6 g is found to have a volume of 12.55 cm^3. What is the density of the chalk? (2.8)

2.77 The density of a sample of wood is 0.32 g/cm^3. (a) What is the mass in grams of 59.8 cm^3 of this wood? (b) What is the volume in cubic centimeters of 224 g of this wood? (2.8)

2.78 The density of a sample of marble is 2.73 g/cm^3.
(a) Convert this density to the SI units of kg/m^3.
(b) What is the density of marble in $kg \cdot L^{-1}$? (2.8)

2.79 If a sample is heated from 296 K to 373 K, what is the change in temperature in °C? (2.9)

2.80 Carry out the following conversions: (a) 167 K to °C, (b) 186 °C to K, (c) 125 °C to °F, (d) −15 °F to °C. (2.9)

2.81 Fill in the blanks: (a) 1 h = _____ s
(b) 4.3 Ms = _____ ks
(c) 2 min 35 s = _____ s (2.10)

2.82 If a student started an experiment at 2:43:39 and finished it at 3:23:09, how many seconds did it take to do the experiment? (2.10)

2.83 The path of which of the following cations is bent most in a mass spectrometer: $^2_1H^+$ or $^1_1H^+$? Explain your answer. (2.11)

2.84 How many protons, neutrons, and electrons are there in one unit of each of the following? (2.11)
(a) lead-208 (b) $^{238}_{92}U$ (c) $^{56}_{26}Fe^{3+}$ (d) $^{127}_{53}I^-$

2.85 Write symbols for (a) an atom of beryllium-9, (b) the 1+ ion formed from an atom of lithium-7, (c) the 2− ion formed from an atom of oxygen-16, (d) an atom with 58 protons and 83 neutrons in its nucleus, (e) the species with 50 protons and 68 neutrons in its nucleus and 48 electrons around its nucleus. (2.11)

2.86 Naturally occurring boron is a mixture of boron-10 and boron-11. Estimate the isotopic composition of naturally occurring boron. (2.11)

2.87 Naturally occurring chromium is a mixture of 4.35% chromium-50 (atomic mass = 49.9461 u), 83.79% chromium-52 (atomic mass = 51.9405 u), 9.50% chromium-53 (atomic mass = 52.9407 u), and 2.36% chromium-54 (atomic mass = 53.9389 u). What is the atomic mass of naturally occurring chromium? (2.11)

2.88 (a) What is the mass number of the isotope of tin having atomic mass = 117.9018 u? (b) Explain why the atomic

mass of the element chlorine is so far from being a whole number. (2.11)

2.89 Calculate the formula mass of each of the following compounds to two decimal places: (a) PH_3, (b) $ZnCl_2$, (c) $(NH_4)_2HPO_4$, (d) $Al_2(SO_4)_3$. (2.12)

2.90 (a) Calculate the formula mass of $Co_3(PO_4)_2 \cdot 8H_2O$ to one decimal place. (b) What is $Co_3(PO_4)_2 \cdot 8H_2O$ called? (2.12)

2.91 (a) How many of each kind of atom are there in one mole of $Al_2(SO_4)_3$? (b) How many aluminum ions are there in exactly half a mole of $Al_2(SO_4)_3$? (2.13)

2.92 Fill in the blanks:
(a) 3.47 g KI = _____ mol KI
(b) 4.8 mol CO_2 = _____ g CO_2
(c) 0.467 mol KNO_3 = _____ g KNO_3
(d) 67.2 g NH_3 = _____ mol NH_3 (2.13)

2.93 For each of the following substances, calculate the number of moles in a 25.0-g sample and the number of grams in 3.5 mol: (a) SF_6, (b) H_3PO_4, (c) $Fe(IO_3)_3$, (d) $(NH_4)_2Cr_2O_7$, (e) $CoCl_2 \cdot 2H_2O$. (2.13)

2.94 For a sample of 1 billion atoms of Lr-256, (a) calculate the mass in u. (b) Calculate the mass in g. (c) How many atoms are there in a 0.0010-g sample of Lr-256? (2.13)

2.95 (a) Calculate the number of hydrogen atoms in 3.52 g of H_2S molecules. (b) Calculate the mass in grams of a single molecule of H_2S. (2.13)

2.96 List five common mistakes that people make in thinking about risks.

2.97 In what way is the nucleus of hydrogen-1 unique?

2.98 A brick measures 20.5 cm × 9.8 cm × 6.4 cm. The brick has a mass of 2.0483 kg. What is the density of the brick in g/cm^3?

2.99 A rubber stopper has a mass of 7.50 g. When the rubber stopper is carefully added to 19.7 mL of water in a graduated cylinder, it sinks. The surface of the water rises to 26.0 mL. What is the density of the rubber stopper in g/mL?

2.100 Naturally occurring lithium is a mixture of lithium-6 (atomic mass = 6.015 12 u) and lithium-7 (atomic mass = 7.016 00 u). What is the percent composition of naturally occurring lithium?

2.101 The mass of one electron is 9.11×10^{-28} g. What is the mass of 1.00 mol of electrons?

2.102 Atomic nuclei have radii of about 0.5×10^{-12} cm and masses of the order of 1×10^{-22} g. Assuming that nuclei are spherical, what is the density of atomic nuclei? (Formula for calculating the volume of a sphere is in Appendix B.7; value of π is inside the back cover.)

2.103 From the bar graph of the mass spectrum of naturally occurring magnesium (at the right), calculate the percent of each isotope.

2.104 The isotopic composition of naturally occurring boron is 19.9% boron-10 and 80.1% boron-11. Make a bar graph showing the mass spectrum. The relative abundance of the more abundant isotope should be set equal to 100.

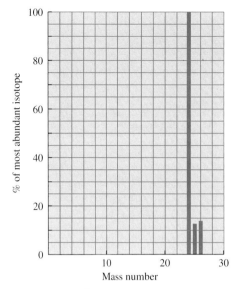

Mass spectrum of naturally occurring magnesium

STOP & TEST YOURSELF

Remember, before going on to problems that require combining ideas and skills, test your mastery of the individual ideas and skills in this chapter by choosing the best answer for each of the following questions. Then check your responses in Appendix I. If you have answered a question incorrectly, review the section identified by number. The answers are discussed in the Student Solutions Manual.

1. Which of the following is the symbol for the SI base unit of mass?
   (a) m   (b) g   (c) kg   (d) gr   (e) mol

2. One millimeter is defined as
   (a) 10^{-3} meter   (b) 10^{-6} meter   (c) 10^3 meters
   (d) 10^6 meters   (e) one mill

3. The period of time one hour, eighteen minutes, and forty-seven seconds equals how many seconds?
   (a) 5×10^3 (b) 4727   (c) 66   (d) 1487   (e) 3665

4. Fill in the blank: 4.3 Mm = _____ m.
   (a) 4.3×10^{-6} (b) 4.3×10^{-3} (c) 4.3×10^3
   (d) 4.3×10^6 (e) 4.3×10^9

5. Fill in the blank: 324 ng = _____ μg.
   (a) 3.24×10^{-13} (b) 3.24×10^{-5} (c) 3.24×10^{-1}
   (d) 3.24×10^5 (e) 3.24×10^{17}

6. What is the length of the block of wood?

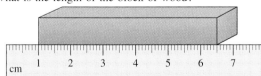

   (a) 6.5 cm   (b) 6.50 cm   (c) 5.5 cm   (d) 7.50 cm
   (e) 5.50 cm

7. How many significant figures are there in 0.060 30 m?
   (a) 6   (b) 5   (c) 4   (d) 3   (e) 2

8. The sum of 3.71×10^8 and 4.62×10^7 is
   (a) 4.991×10^7 (b) 8.33×10^{15} (c) 4.172×10^8
   (d) 4.99×10^7 (e) 4.17×10^8

9. The density of a sample of glass is 2.48 g/cm^3. What is the volume of 7.44 g of this glass?
   (a) 3 cm^3 (b) 18.4 cm^3 (c) 18.5 cm^3 (d) 3.00 cm^3
   (e) 18.45 cm^3

10. What is the temperature in K that corresponds to a temperature of -12.0 °C?
    (a) -285.2 (b) $+285.2$ (c) 261.15   (d) 261.2
    (e) 285.15

11. An atom of molybdenum-98 contains
    (a) 42 protons, 56 neutrons, and 42 electrons
    (b) 42 protons, 98 neutrons, and 42 electrons
    (c) 42 protons, 56 neutrons, and 56 electrons
    (d) 56 protons, 42 neutrons, and 42 electrons
    (e) 56 protons, 42 neutrons, and 56 electrons

12. The symbol for the ion containing 24 protons, 28 neutrons, and 21 electrons is
    (a) $^{45}_{21}\text{Sc}^{3+}$ (b) $^{45}_{21}\text{Sc}^{3-}$ (c) $^{45}_{21}\text{Sc}$ (d) $^{52}_{24}\text{Cr}^{3+}$
    (e) $^{52}_{24}\text{Cr}^{3-}$

13. The formula mass of $(\text{NH}_4)_2\text{SO}_4$ is
    (a) 118.13 u   (b) 132.17 u   (c) 164.19 u   (d) 214.31 u
    (e) 228.19 u

14. In 1.000 mol $(\text{NH}_4)_2\text{SO}_4$ there are
    (a) 6.022×10^{23} ammonium sulfate molecules
    (b) 1.204×10^{22} ammonium ions
    (c) 1.204×10^{24} ammonium ions
    (d) 6.022×10^{23} nitrogen atoms
    (e) 24.08×10^{23} hydrogen atoms

15. The mass in grams of 0.333 33 mol of oxygen gas is
    (a) 0.333 33 g   (b) 5.3331 g   (c) 10.666 g   (d) 32.000 g
    (e) 96.000 g

Remember that answers to problems with blue numbers appear in Appendix I. Solutions for these problems are in the Student Solutions Manual.

2.105 If the volume of a liquid (45.3 mL) can only be measured to the nearest 0.1 mL, can the precision and accuracy of the density measurement be increased by measuring mass (about 41 g) with an analytical balance instead of with a student balance? An analytical balance measures mass to the nearest 0.0001 g; mass can only be determined to ± 0.01 g with a student balance. Explain your answer.

2.106 (a) How many significant figures are there in the atomic mass of manganese? (b) To how many decimal places is the atomic mass of manganese known?

2.107 For which of the following compounds is the term *molecular mass* appropriate? Explain your answer.
(a) NaCl (b) CO_2 (c) H_2O (d) $Ca(NO_3)_2$ (e) CH_4

2.108 Identify each of the following: (a) an element of Group IVA that has 50 electrons in its atom, (b) a very reactive nonmetal with 9 protons in its nucleus, (c) an element that forms a $2-$ ion having 18 electrons, (d) a very unreactive gas with atomic mass between 30 and 40.

2.109 Explain why the chemical properties of carbon-12 and carbon-13 are almost the same.

2.110 Is atomic mass a periodic function of atomic number? Explain your answer.

2.111 In 1993, 80.31 billion lb of sulfuric acid and 34.50 billion lb of ammonia were produced, more pounds of sulfuric acid than pounds of ammonia. How does the number of moles of ammonia produced compare with the number of moles of sulfuric acid produced?

2.112 Silver is much more common than gold. However, it tarnishes when exposed to air or any other material containing sulfur compounds. Therefore, it did not catch people's eyes as early as gold. The largest known silver nugget, which has a mass of 363 kg, was found in Peru. A piece of silver smaller than the head of a pin can be drawn into a wire 400 ft long or beaten into leaves 0.000 01 in. thick. At room temperature, silver conducts electricity better than any other metal. Silver forms compounds with most nonmetals. In nature it is usually found as argentite, Ag_2S. From the preceding information about silver, give one example of each of the following: (a) extensive property, (b) intensive property, (c) physical property, (d) chemical property, (e) qualitative statement, (f) quantitative statement. (g) What is the mass of the largest nugget of silver in pounds?

2.113 Classify each of the following as length, area, volume, mass, density, or amount of substance: (a) 223 mL, (b) 6.78 g/cm^3, (c) 27.4 cm^2, (d) 5.6 Mmol, (e) 0.896 kg.

2.114 The densities of the elements in Group IIA are (all in g/cm^3): Be, 1.85; Mg, 1.74; Ca, 1.55; Sr, 2.63; and Ba, 3.62. (a) List the members of Group IIA in order of increasing density. (Write the element with the lowest density on the left.) (b) Make a graph showing the densities of these elements as a function of atomic number. (See Appendix B.6 for directions for making a graph.) (c) Use the graph to estimate the density of radium. (d) Calcium reacts with water at room temperature, but magnesium does not. Could you determine the density of radium by the method used for rubber stoppers in Problem 2.99? Explain your answer.

2.115 Samples of two compounds of carbon and hydrogen, **A** and **B**, and samples of three compounds of nitrogen and oxygen, **C, D,** and **E,** have the following compositions: **A,** 3.00 g C/1.00 g H; **B,** 6.00 g C/1.00 g H; **C,** 0.57 g O/1.00 g N; **D,** 1.14 g O/1.00 g N; **E,** 2.28 g O/1.00 g N. (a) These two series of compounds are typical of series of compounds formed from the same elements. What generalization can be made about the ratios of the masses of an element that combine with one gram of another element? (b) Choose two compounds of another pair of elements and see if this pair follows your generalization. (c) Does your generalization support Dalton's atomic theory? Explain your answer.

2.116 Three students measured the mass of 25-mL samples of water and obtained the following results. The laboratory temperature was 20.0 °C.

| | Mass, g | | |
Student	First Determination	Second Determination	Third Determination
A	24.59	24.80	24.77
B	23.95	24.44	24.96
C	24.99	24.95	24.97

(a) Which student's results were most precise? (b) Which student's results were most accurate? (*Hint:* Use the index to find the information you need to answer this question.) (c) To measure the water, one student used a graduated cylinder, one used a volumetric pipet, and one used a beaker (all of the appropriate size). Which student most likely used the beaker? Explain your answers.

Remember that answers to problems with blue numbers appear in Appendix I. Solutions for these problems are in the Student Solutions Manual.

2.117 The Sahara Desert is between 7 000 000 and 9 000 000 km^2 in area. Write the area of the Sahara showing the correct number of significant figures.

2.118 A bottle of vinegar is labeled "32 FL. OZ., 1 QT., 946 mL." From this information, (a) how many mL are there in a fluid ounce? (b) How many quarts are equal to one liter?

2.119 What is the molar mass of a protein that has a molecular mass of 73 kD (kilodalton)?

2.120 Many drugs have very complex formulas. For example, the formula for one of the penicillins is $C_{29}H_{38}N_4O_6S \cdot H_2O$. What is the formula mass of this penicillin to one decimal place?

2.121 The neutron-to-proton ratio in an atom is one factor that determines whether the atom is radioactive. (a) How many neutrons are there per proton in each of the following atoms? (Two significant figures are enough.)

carbon-12 $^{32}_{16}S$ $^{56}_{26}Fe$ $^{107}_{47}Ag$ $^{197}_{79}Au$ $^{238}_{92}U$

(b) How does the ratio of neutrons to protons change as atomic number increases?

2.122 Archaeologists use differences in density (d) to separate the mixtures that they dig up. (a) If a sample containing bone ($d = 1.7$–2.0 g/cm^3), charcoal ($d = 0.3$–0.6 g/cm^3), rock ($d = 2.6$–5 g/cm^3 and up), sand ($d = 2.2$–2.4 g/cm^3), and soil ($d = 2.6$–2.8 g/cm^3) is added to a solution that has a density of 2.1 g/cm^3, which materials will float, and which will sink? (b) What density of solution should be used to separate a mixture of bone and charcoal? (c) What would have to be true of the chemical properties of the solutions? Explain your answers.

2.123 With our present calendar, one year has an average length of 365.2425 days. Actually, an average year lasts 365.242 1934 days. This error amounts to an extra day in how many years?

2.124 (a) The temperature of the surface of Venus is 475 °C. What is this temperature in kelvin? (b) At the South Pole, the average temperature is -56 °C. What is this temperature in kelvin? (c) Earth's temperature at its core is about 6150 K. What is this temperature in degrees Celsius? (d) A magazine article states that temperatures in the centers of stars are of the order of 50 million degrees without telling what kind of degrees. Were the author and editor careless, or does it not matter whether the Celsius or the Kelvin scale is meant? How about the Fahrenheit scale? Explain your answers.

2.125 The label gives the volume of cola in a 12-oz can as 355 mL. A "full" can probably has some "head space" (empty space above the liquid). Experimentally, how could you find the volume of the head space?

2.126 The silicon chips used to make smart cards are 0.011 in. thick. Express this thickness in the SI unit that will give the smallest number greater than 1.

2.127 The units of the chain system of measure used by surveyors are 7.92 in. = 1 link, 100 links or 66 ft = 1 chain, 10 chains = 1 furlong, 80 chains = 1 mi. A length of 3.8 furlongs is equal to _____ chains, _____ links, or _____ miles.

2.128 To predict the risks of earthquakes, geologists need to measure small changes in distances between points far apart on Earth. Using Earth-orbiting satellites, the distance between points 1000 km apart can be measured with an uncertainty of 1 cm (or less). How many significant figures are there in these measurements?

2.129 (a) Flows of fragments of volcanic rock, such as one produced during an eruption of Mount St. Helens in 1980, have temperatures approaching 1000 °C and move at speeds of up to 300 m/s. What is this temperature in K and this speed in mi/h? (b) The eruption of Mt. Pinatubo in the Philippines in 1991 added 15–30 million tons of sulfur dioxide to the stratosphere. How many moles of SO_2 is this? (c) The number of people killed by volcanoes increased from 315 per year for the years 1600–1900 to 845 per year currently. What information would you need to tell whether the risk of being killed by a volcano has increased, decreased, or stayed the same?

2.130 Unfermented grape juice used to make wine is called a "must." The sugar content of the must determines whether a dry or a sweet wine will be produced. The sugar content is found by measuring the density of the must. If the density is lower than 1.070 g/mL, sugar syrup is added until the density reaches 1.075 g/mL. (a) If a 44.6-mL sample of must has a mass of 47.28 g, will sugar syrup have to be added? (b) During fermentation the temperature must be kept below 30 °C (zero is significant). What is this temperature in °F?

2.131 A microsensor that detects tiny amounts of the bioregulator and neurotransmitter molecule NO in a single cell has been developed. (a) In a typical cell volume, the sensor can detect 10^{-20} mol NO, much less than the amount produced by a single cell. How many molecules of NO can the microsensor detect? (b) A carbon fiber with a tip diameter of about 0.5 μm is used to make the sensor. If the radius of a carbon atom is 77 pm, how many atoms wide is the tip?

2.132 Radioactive nuclides are widely used in medicine for diagnosis, treatment, and research. (a) The radioactive nuclide ^{51}Cr is used in diagnosis to study blood volume and red cell survival. How many protons, neutrons, and electrons are there in an atom of ^{51}Cr? (b) The radioactive nuclide ^{18}F is given by mouth or injected into a vein in the form of sodium fluoride (for diagnostic bone scans). In a fluoride ion formed from this isotope, how many protons, neutrons, and electrons are there? (c) The radioactive nuclide ^{59}Fe is given by mouth or injected into a vein in the form of iron(II) sulfate to study iron metabolism. What is the formula for iron(II) sulfate? In the iron(II) ion formed from iron-59, how many protons, neutrons, and electrons are there?

Chemistry: The Unknown Adventure

M. CECILIA OLAVARRIETA-KUHN

Product Manager
American Trans-Chem Corporation

B. S. Chemical Engineering
Texas A&M University

When I came to the United States from the tip of South America, I had little idea of the adventures that awaited me. My experiences with chemistry started in December 1979 when my family arrived in the United States and I attended my first day of high school. At that time I did not know how to speak English, and I was terrified of walking into the classroom.

I remember my first chemistry class vividly. I walked into the room and felt the whole world was staring at me. The teacher, Mr. Stockton, approached me, looked me straight in the eyes, and in slow motion blurted out the words: Do you understand me? Actually, I didn't, but soon he found someone who could communicate a little with me and he became my lab partner. Everyone was very kind and helpful and some of my fears started to vanish.

At the time we were learning how to name organic chemicals, and I was very impressed by the long chains of hydrocarbons that stretched from east to west and north to south on the blackboard. Needless to say, I did not understand much then, but I copied every single carbon, hydrogen, nitrogen, and so on that Mr. Stockton put on the board. Naming them was as challenging as trying to fit them into my notebook.

I always enjoyed the lectures, but had a harder time with the lab experiments. It was fascinating to see chemicals turning different colors under the flame of the Bunsen burner or crystals growing, but I was very slow and never had enough time to finish my experiments.

In spite of my problems with lab experiments, math and chemistry were my favorite subjects in high school, and I went on to study chemical engineering at Texas A&M University. Feeling more mature and having become familiar with the English language, I felt confident and ready to tackle any new challenges my chemistry classes might offer.

My newfound confidence was short-lived. After a few weeks, I was having many problems in my laboratory classes. In fact, for the first year and a half—during my freshman chemistry and my organic chemistry labs—my performance was a disaster. I would read and study the instructions for the experiment before class, but I was never really prepared. I was too shy to seek help from my professors and was afraid I would cause an explosion and break all the expensive equipment. (Notice how teachers act as inhibitors when they mention the cost!)

I discovered that the problem was my attitude, and I started the second half of my sophomore year determined to be more enthusiastic about labs. I approached my teachers before class and asked questions until they were tired of listening to me. Now I was determined to succeed. To my amazement, when I opened up and began to talk about my difficulties, I discovered that my case was not unique and that other students were having similar problems, problems that could be resolved only by communicating with others.

Graduation came and I found a job as a product manager for an international chemical exporting company. I look for chemicals, mainly from the United States, for immediate export to South America, Europe, Africa, the Far East, and Egypt. Our main products are polymers such as polyethylene, polypropylene, polyvinyl chloride, acrylonitrile-butadiene-styrene, and the like. My background in chemistry helps me better understand the products required by our clients, the proper grades, and their use as raw materials in manufacturing processes. I know how the chemicals are produced, and knowing the basic raw materials also helps me predict market prices. I have worked for this company for three years and have definitely found a new and unexpected way to use the knowledge acquired in my chemistry classes.

Chemistry, therefore, is not limited to the four walls of your chemistry lab. A background in chemistry opens a variety of opportunities in different fields. Goggles, test tubes, gas chromatographs, Bunsen burners, and the sweet smell of aromatic chemicals may be requirements for an exciting job, but they are not necessarily permanent features of your future working environment. So I welcome you to the exciting world of chemistry and urge you to join in the adventure.

STOICHIOMETRY

3

Adjusting the air:fuel ratio in a racing car is a stoichiometry problem.

Stoichiometry is the *study of quantitative relationships between substances undergoing chemical changes.*

▬ Why is stoichiometry a part of general chemistry?

Stoichiometry is a very important subject from the point of view of both theory and practice. Dalton proposed his atomic theory to explain quantitative observations of amounts of substances undergoing

chemical changes. Analytical chemistry, the branch of chemistry concerned with finding the composition of matter, is based on stoichiometry. Stoichiometry is also very important in the chemical industry. For example, suppose you were in the business of making ammonia for use as a fertilizer by reacting nitrogen and hydrogen gases. You would need to know how much nitrogen and how much hydrogen to use and how much ammonia you could expect.

▬ What do you already know about stoichiometry?

You are certainly familiar with the products of cooking. You have probably also cooked. The reasoning used in solving stoichiometry problems is exactly like the reasoning used in cooking. Let's take baking cupcakes as an example. Suppose the recipe says "Take $2\frac{1}{4}$ cups of flour and two eggs (plus other things like sugar, baking powder, butter, milk, and flavor-

The shorter, lighter green rows of corn in the upper half of the photo are growing in nitrogen-deficient soil. The taller, darker rows in the foreground have been fertilized with urea, an organic compound that contains a high percentage of nitrogen.

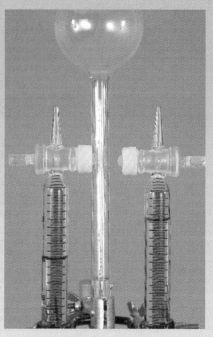

The volume of hydrogen produced by the electrolysis of water is always twice the volume of the oxygen.

ing) to make 30 cupcakes." If, when you go to the refrigerator, you find that you have only one egg, you can still make cupcakes. You do not have to go to the store. But you can only make half a recipe of cupcakes. You will need $1\frac{1}{8}$ cups of flour and will only get fifteen cupcakes. If there are 3 cups of flour in the bag, you will have $1\frac{7}{8}$ cups of flour left.

LAW OF CONSERVATION OF MATTER

The Russian chemist Lomonosov was the first person to study chemical reactions quantitatively. In the 1750s he found that if reactions are carried out in closed containers so that gases cannot escape, the total mass of the products equals the total mass of the reactants (■Figure 3.1). In other words, the quantity of matter is not changed by chemical reactions; matter is neither created nor destroyed by

■ FIGURE 3.1   *Left:* An unused flashbulb containing magnesium and oxygen. *Right:* When the flashbulb is used, magnesium reacts with oxygen to form magnesium oxide, but the mass of the flashbulb does not change.

Waste Disposal and Resources

Our waste-disposal problems are a direct result of the conservation of matter. You cannot really throw anything away. All you can do is to put it somewhere else. For example, most people dispose of unwanted materials by putting them in the trash. The trash is collected and deposited in a dump; it does not disappear. Even if the trash is burned in an incinerator, it is not destroyed, only converted by chemical reaction into other forms. Usually these other forms are less useful than the original form. For example, trash has some value as a fuel. (However, the carbon dioxide formed by burning trash escapes into the atmosphere where it may be changing Earth's climate for the worse.) Metals, glass, paper, and plastics can be recovered from trash and recycled; the use of recycled materials usually saves energy as well as natural resources.

Another result of the conservation of matter is that the number of atoms of each kind available for people's use on Earth is, for all practical purposes, fixed. If the world's population continues to grow, the

Photo shows the world's largest dump, which occupies 5000 acres of tidal wetlands near New York City. We are rapidly running out of land for dumps.

number of atoms of a given kind available to each person will decrease. Even if we figure out how to obtain materials from parts of Earth that we cannot reach now and from space, more energy will probably be required to obtain them. We will also have to develop new sources of energy.

chemical changes. The number of atoms of each kind must be the same after reaction as it was before reaction (Section 1.11). This general rule is called the **law of conservation of matter.**

Law of Conservation of Matter

In chemical reactions, the quantity of matter does not change. The total mass of the products equals the total mass of the reactants.

Credit for the discovery of the law of conservation of matter is usually given to the French chemist Lavoisier* because Lomonosov's work is not well known outside Russia. Lavoisier's work provided the foundation for modern chemistry. It led other chemists to examine chemical reactions quantitatively and resulted in the discovery of the law of constant composition (Section 1.3). Dalton suggested his atomic theory to account for the law of conservation of matter and the law of constant composition.

---

*Lavoisier was secretary and treasurer of a commission appointed in 1790 to make weights and measures used throughout France uniform. The metric system on which SI is based resulted. Lavoisier's many contributions to science and society ended when he was only 51. He was guillotined during the French Revolution.

Remember that Section 1.11 tells how to write chemical equations. When balancing equations, it pays to be systematic.

3.2 MORE ABOUT BALANCING EQUATIONS

Consideration of the amounts of substances reacted and formed in chemical reactions must begin with the equation for the reaction. Many equations are easy to balance. However, when you cannot see what to do to balance an equation, a system is helpful:

1. Begin with the compound that has the most atoms or the most kinds of atoms and use *one* of these atoms as a starting point.
2. Balance elements that appear only once on each side of the arrow first.
3. Then balance elements that appear more than once on a side.
4. Balance free elements last.

Sample Problem 3.1 shows how to balance an equation that is not as easy to balance as the ones we met in Chapter 1.

SAMPLE PROBLEM

3.1 When methyl alcohol, CH_4O, burns in air, it reacts with oxygen from air to form carbon dioxide gas and water. Write an equation for this reaction.

Solution The word equation for burning methyl alcohol in air is

$$\text{methyl alcohol} + \text{oxygen} \longrightarrow \text{carbon dioxide} + \text{water}$$

Writing formulas for the reactants and products gives

$$CH_4O + O_2 \xrightarrow{\text{spark}} CO_2 + H_2O$$

Start with CH_4O because it has the most atoms and the most kinds of atoms. Balance C and H first because C and H appear in only one substance on each side.

Balance O last because O appears twice on each side and because O_2 is an element.

Because the number of carbons in CH_4O is the same as the number of carbon atoms in CO_2, whatever the coefficient of CH_4O, the coefficient of CO_2 must be the same. One is all right for the time being. To balance the hydrogens, multiply H_2O by 2:

$$CH_4O + O_2 \xrightarrow{\text{spark}} CO_2 + 2H_2O$$

There are now four hydrogens on each side. But there are three oxygens on the left and four on the right. It would be convenient to be able to use one oxygen atom at a time. Imagine that you can use half an oxygen molecule at a time. You would need to use three of these half molecules to get a total of four oxygens on the left:

$$CH_4O + 3(\tfrac{1}{2})O_2 \xrightarrow{\text{spark}} CO_2 + 2H_2O$$

or

$$CH_4O + \tfrac{3}{2}O_2 \xrightarrow{\text{spark}} CO_2 + 2H_2O$$

But you cannot really use half an oxygen molecule. A molecule is the smallest unit into which a substance that is composed of molecules can be divided. Coefficients should usually be whole numbers. Multiply both sides of the equation by 2 to clear fractions:

$$2(CH_4O + \tfrac{3}{2}O_2 \xrightarrow{\text{spark}} CO_2 + 2H_2O)$$

A spark is necessary to start the reaction.

This camp stove is fueled by methyl alcohol.

or

$$2CH_4O + 3O_2 \xrightarrow{\text{spark}} 2CO_2 + 4H_2O$$

The equation is balanced. There are two carbons, eight hydrogens, and eight oxygens on each side. Now add symbols showing whether substances are solids, liquids, gases, or aqueous solutions. Methyl alcohol and water are liquids under ordinary conditions; oxygen and carbon dioxide are gases:

$$2CH_4O(l) + 3O_2(g) \xrightarrow{\text{spark}} 2CO_2(g) + 4H_2O(l)$$

Equations with fractional coefficients such as

$$CH_4O + \tfrac{3}{2}O_2 \longrightarrow CO_2 + 2H_2O$$

are acceptable on a mole scale.

Polyatomic ions often react as units and can be balanced as units. Sample Problem 3.2 shows how.

SAMPLE PROBLEM

3.2   Balance the following:

$$Ba(NO_3)_2(aq) + (NH_4)_2SO_4(aq) \longrightarrow BaSO_4(s) + NH_4NO_3(aq)$$

Solution   There are two ammonium ions on the left. Multiply NH_4NO_3 by 2 to get two ammonium ions on the right:

$$Ba(NO_3)_2(aq) + (NH_4)_2SO_4(aq) \longrightarrow BaSO_4(s) + 2NH_4NO_3(aq)$$

There are now two ammonium ions and two nitrate ions on each side. The numbers of barium atoms and sulfate ions are also the same (one each on both sides of the arrow). The equation is balanced.

PRACTICE PROBLEM

3.1   Balance each of the following expressions:

(a) $HNO_3(aq) \longrightarrow NO_2(aq) + H_2O(l) + O_2(g)$
(b) $B_2O_3(s) + H_2O(l) \longrightarrow H_3BO_3(aq)$
(c) $Cu(s) + AgNO_3(aq) \longrightarrow Ag(s) + Cu(NO_3)_2(aq)$
(d) $NH_3(g) + O_2(g) \longrightarrow NO(g) + H_2O(l)$
(e) $C_3H_6O(l) + O_2(g) \longrightarrow CO_2(g) + H_2O(l)$

If more than one reaction takes place between the reactants, an equation must be written for each reaction. For example, if carbon (charcoal) is burned inside, where the supply of oxygen from air is limited, some of the carbon reacts to form carbon monoxide, and some reacts to form carbon dioxide. *Because two different reactions take place, two equations are necessary, one for each reaction:*

$$2C(s) + O_2(g) \xrightarrow{\text{heat}} 2CO(g) \tag{3.1}$$

and

$$C(s) + O_2(g) \xrightarrow{\text{heat}} CO_2(g) \tag{3.2}$$

The proportions of carbon reacting by reactions 3.1 and 3.2 depend on the supply of oxygen. If there is less oxygen, more of the carbon reacts to form carbon monoxide. If there is more oxygen, more of the carbon reacts to form carbon dioxide.

Safety Note: Carbon monoxide is a poison. Never charcoal broil indoors.

Although both these reactions give off heat, considerable heat must be supplied to start them, as you know if you have ever started a charcoal fire. Therefore, heat is shown over the arrow.

3.3 EQUATIONS ON A MACROSCOPIC SCALE

Reactions in the laboratory and manufacturing plant involve enormous numbers of atoms. Let's consider the decomposition of water by sunlight, which has been suggested as a way of using the sun's energy, as an example of scaling up a reaction from a microscopic to a macroscopic level. A catalyst* must be present to make the decomposition of water by sunlight take place at a practical rate. The equation for this reaction is

$$2H_2O(l) \xrightarrow{\text{sunlight, catalyst}} 2H_2(g) + O_2(g) \tag{3.3}$$

On a microscopic scale, equation 3.3 is interpreted as meaning

$$\begin{array}{ccc}
\text{2 molecules} & \xrightarrow{\text{sunlight, catalyst}} & \text{2 molecules of} \\
\text{of water} & & \text{hydrogen gas}
\end{array} + \begin{array}{c}\text{1 molecule of} \\ \text{oxygen gas}\end{array}$$

or, using models to represent the reaction,

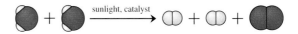

Like other equations, chemical equations can be multiplied by any number as long as both sides are multiplied by the same number. For example, equation 3.3 could be multiplied by 2, giving

$$4H_2O(l) \xrightarrow{\text{sunlight, catalyst}} 4H_2(g) + 2O_2(g)$$

$$\begin{array}{ccccc}
\text{4 molecules} & \xrightarrow{\text{sunlight, catalyst}} & \text{4 molecules} & + & \text{2 molecules of} \\
\text{of water} & & \text{hydrogen gas} & & \text{oxygen gas}
\end{array}$$

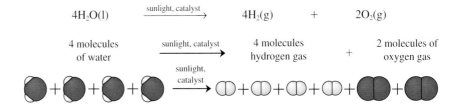

or multiplied by 3, giving

$$6H_2O(l) \xrightarrow{\text{sunlight, catalyst}} 6H_2(g) + 3O_2(g)$$

or multiplied by 6.022×10^{23}, giving

$$(12.04 \times 10^{23})H_2O(l) \xrightarrow{\text{sunlight, catalyst}} (12.04 \times 10^{23})H_2(g) + (6.022 \times 10^{23})O_2(g) \tag{3.4}$$

Because 6.022×10^{23} is the number of molecules in 1 mol (Section 2.13), equation 3.4 says

$$2 \text{ mol } H_2O(l) \xrightarrow{\text{sunlight, catalyst}} 2 \text{ mol } H_2(g) + 1 \text{ mol } O_2(g)$$

In other words, the original equation, equation 3.3, can be interpreted as meaning

---

*A catalyst is a substance that makes a reaction take place faster without being used up in the reaction. You are most likely to be familiar with the word *catalyst* in connection with catalytic converters in cars. Catalytic converters reduce air pollution by speeding the reaction of unburned gasoline and carbon monoxide with air. They also make nitrogen oxides formed in the engine change back to nitrogen and oxygen at a useful rate.

moles instead of molecules. *The mole is the key to quantitative relations between substances undergoing chemical changes on a practical scale.*

SAMPLE PROBLEM

3.3   Answer the following questions, given the equation

$$N_2(g) + O_2(g) \xrightarrow{\text{energy}} 2NO(g)$$

(a) How many moles of NO can be formed from 2 mol O_2, assuming enough N_2 is available? (b) How many moles of N_2 are needed to react with 3 mol O_2? (c) How many moles of N_2 are needed to make 1 mol NO, assuming enough O_2 is available?

Solution   The cupcake recipe gives the relationships between the quantities of eggs, flour, etc., and the number of cupcakes made. Similarly, the balanced chemical equation for a reaction gives relationships between quantities of reactants and products. The balanced chemical equation for a reaction can be used to write conversion factors between quantities. According to the coefficients in the equation,

$$N_2(g) + O_2(g) \xrightarrow{\text{energy}} 2NO(g)$$

1 mol N_2 ⇌ 1 mol O_2 1 mol N_2 ⇌ 2 mol NO   1 mol O_2 ⇌ 2 mol NO

These three relations can be used to write six conversion factors. As usual, choose the conversion factor that leads to the right units for the answer.

   (a) This part of the question asks about the number of moles of NO that can be formed from 2 mol O_2. We want to convert 2 mol O_2 to the equivalent number of moles of NO:

$$2 \text{ mol } O_2 \left(\frac{}{} \right) = \text{? mol NO}$$

The relationship between moles of O_2 and moles of NO is

$$1 \text{ mol } O_2 \ ⇌ \ 2 \text{ mol NO}$$

The correct conversion factor for converting moles of O_2 to moles of NO is 2 mol NO/1 mol O_2:

$$2 \text{ mol } O_2 \left(\frac{2 \text{ mol NO}}{1 \text{ mol } O_2} \right) = 4 \text{ mol NO}$$

   (b) $3 \text{ mol } O_2 \left(\dfrac{}{} \right) = \text{? mol } N_2$. The correct conversion factor is 1 mol N_2/1 mol O_2:

$$3 \text{ mol } O_2 \left(\frac{1 \text{ mol } N_2}{1 \text{ mol } O_2} \right) = 3 \text{ mol } N_2$$

   (c) $1 \text{ mol NO} \left(\dfrac{}{} \right) = \text{? mol } N_2$. The correct conversion factor is 1 mol N_2/2 mol NO:

$$1 \text{ mol NO} \left(\frac{1 \text{ mol } N_2}{2 \text{ mol NO}} \right) = 0.5 \text{ mol } N_2$$

You must include formulas for substances in the labels—that is, you must write "mol NO" not just "mol." Complete labeling is necessary whenever you are dealing with more than one similar thing. Without it, you may get mixed up. For example, you might confuse moles of NO with moles of O_2.

Remember that the symbol ⇌ means "is equivalent to."

Complete labeling is necessary whenever you are dealing with more than one similar thing.

3.2   Answer the following questions, given the equation

$$3Mg(s) + N_2(g) \xrightarrow{heat} Mg_3N_2(s)$$

(a) How many moles of Mg_3N_2 can be formed by the reaction of 6 mol Mg, assuming that enough N_2 is available? (b) How many moles of N_2 are needed to react with 1 mol Mg? (c) How many moles of Mg are needed to make 3 mol Mg_3N_2, assuming that enough N_2 is available?

3.4   MASS RELATIONSHIPS IN CHEMICAL REACTIONS

The information in Sections 2.13 and 3.3 can be combined to solve problems about mass relationships in chemical reactions. Sample Problems 3.4 and 3.5 show how. Stoichiometric calculations like the ones in Sample Problems 3.4 and 3.5 are *very* important. Much of your work in chemistry will involve stoichiometry.

3.4   If 37.6 g of water are decomposed to hydrogen and oxygen, (a) how many grams of hydrogen should be formed and (b) how many grams of oxygen should be formed?

Solution   (a) *Always begin solving problems about quantitative relationships in reactions by writing an equation for the reaction. The coefficients in the equation describe the relationships between moles of products and moles of reactants* (see Sample Problem 3.3). The equation for the decomposition of water is

Equation	$2H_2O(l) \xrightarrow{\text{sunlight, catalyst}} 2H_2(g) + O_2(g)$

Next write the information given in the problem and what's asked for above the formulas in the equation:

Problem	37.6 g		? g
Equation	$2H_2O(l)$	$\xrightarrow{\text{sunlight, catalyst}}$	$2H_2(g) + O_2(g)$

Formula masses of reactants and products are needed to convert grams to moles and moles to grams. Below the formula of each substance that is involved in the question, write the formula mass of the substance:

Problem	37.6 g		? g
Equation	$2H_2O(l)$	$\xrightarrow{\text{sunlight, catalyst}}$	$2H_2(g) + O_2(g)$
Formula mass, u	18.0		2.02

At the beginning, addition of another line to the table giving the "recipe" for the reaction in moles, which is obtained from the coefficients in the equation, is a good idea. The equation for the reaction says that 2 mol H_2O decompose to 2 mol H_2 (and 1 mol O_2). (Later, after you've become experienced, you will probably simply refer to the coefficients in the equation for the "recipe.")

Problem	37.6 g		? g
Equation	$2H_2O(l)$	$\xrightarrow{\text{sunlight, catalyst}}$	$2H_2(g) + O_2(g)$
Formula mass, u	18.0		2.02
Recipe, mol	2		2

▶ Remember that an equation must always be balanced.

▶ The term *recipe* is used in the chemical process industries as well as in the kitchen.

▶ One of the advantages of a table is that labels for a whole column need only be written once. All of the items in the $H_2O(l)$ column, for example, refer to $H_2O(l)$: 37.6 g is the mass of H_2O, 18.0 is the formula mass of H_2O, and 2 is the number of moles of H_2O.

Now we have all the information needed to solve part (a) of the problem.

First, grams of water must be converted to moles of water (see Sample Problem 2.14):

$$37.6 \text{ g } H_2O \left(\frac{1 \text{ mol } H_2O}{18.0 \text{ g } H_2O} \right) = 2.09 \text{ mol } H_2O$$

Second, moles of water must be converted to moles of hydrogen (see Sample Problem 3.3):

$$2.09 \text{ mol } H_2O \left(\frac{2 \text{ mol } H_2}{2 \text{ mol } H_2O} \right) = 2.09 \text{ mol } H_2$$

Third, moles of hydrogen must be converted to grams of hydrogen (see Sample Problem 2.15):

$$2.09 \text{ mol } H_2 \left(\frac{2.02 \text{ g } H_2}{1 \text{ mol } H_2} \right) = 4.22 \text{ g } H_2$$

The 1s and 2s are exact numbers and do not affect the number of significant figures in the answer. There are three significant figures in the answer because there are three significant figures in the quantity, 37.6 g H_2O.

Once you've figured out what to do, you can combine the steps to avoid the possibility of entering a wrong number in your calculator:

$$37.6 \text{ g } H_2O \left(\frac{1 \text{ mol } H_2O}{18.0 \text{ g } H_2O} \right) \left(\frac{2 \text{ mol } H_2}{2 \text{ mol } H_2O} \right) \left(\frac{2.02 \text{ g } H_2}{1 \text{ mol } H_2} \right) = 4.22 \text{ g } H_2$$

With the steps combined, it is easy to see that the units cancel to give the correct units for the answer, g H_2. Combining steps also avoids accumulation of rounding errors.

(b) The table of information for this part of the problem is

Problem	37.6 g		? g
Equation	$2H_2O(l)$	$\xrightarrow{\text{sunlight, catalyst}}$	$2H_2(g)$ + $O_2(g)$
Formula mass, u	18.0		32.0
Recipe, mol	2		1

With a little practice, you should be able to set up the combined calculation right away:

$$37.6 \text{ g } H_2O \left(\frac{1 \text{ mol } H_2O}{18.0 \text{ g } H_2O} \right) \left(\frac{1 \text{ mol } O_2}{2 \text{ mol } H_2O} \right) \left(\frac{32.0 \text{ g } O_2}{1 \text{ mol } O_2} \right) = 33.4 \text{ g } O_2$$

↑   ↑   ↑

First, use formula mass   *Second,* use coefficients   *Third,* use formula mass
H_2O to convert   from balanced O_2 to convert
g H_2O to mol   equation to   mol O_2 to g O_2.
H_2O convert mol
H_2O to mol O_2.

The answers are reasonable because the sum of the masses of hydrogen and oxygen formed equals the mass of the water decomposed:

$$4.22 \text{ g } + 33.4 \text{ g } = 37.6 \text{ g}$$

Mass is conserved. The mass of the hydrogen is much less than the mass of the oxygen, as would be expected because the formula mass of hydrogen is much smaller than the formula mass of oxygen.

Remember to include formulas for substances in labels.

Exact numbers do not affect the number of significant figures in the results of a calculation.

The reaction

$$4NH_3(g) \ + \ 5O_2(g) \xrightarrow{\text{Pt, 750–900 °C}} 4NO(g) \ + \ 6H_2O(g)$$

is a step in the production of nitric acid, an important industrial chemical. (a) How many grams of O_2 are needed to react with 374 g NH_3? (b) How many moles of NO can be formed from 374 g NH_3 if plenty of O_2 is present? (c) How many grams of NH_3 are needed to make 74.4 g NO if plenty of O_2 is present? (d) How many pounds of NH_3 are needed to make 74.4 lb of NO if plenty of O_2 is present?*
(e) How many tons of NH_3 are needed to make 74.4 tons of NO if plenty of O_2 is present?

Solution   (a) The table of information for this part is

Problem	*374 g*	*? g*	
Equation	**$4NH_3(g)$ + $5O_2(g)$** $\xrightarrow{\text{Pt, 750–900 °C}}$		**$4NO(g)$ + $6H_2O(g)$**
Formula masses, u	17.0	32.0	
Recipe, mol	4	5	

The setup for converting 374 g NH_3 to g O_2 is

$$374 \text{ g NH}_3 \left(\frac{1 \text{ mol NH}_3}{17.0 \text{ g NH}_3} \right) \left(\frac{5 \text{ mol O}_2}{4 \text{ mol NH}_3} \right) \left(\frac{32.0 \text{ g O}_2}{1 \text{ mol O}_2} \right) = 8.80 \times 10^2 \text{ g O}_2$$

The answer must be written in exponential notation to show that it has three significant figures.

(b) The table of information for this part is

Problem	*374 g*		*? mol*
Equation	**$4NH_3(g)$ + $5O_2(g)$** $\xrightarrow{\text{Pt, 750–900 °C}}$	**$4NO(g)$ + $6H_2O(g)$**	
Formula masses, u	17.0		
Recipe, mol	4		4

The setup is

$$374 \text{ g NH}_3 \left(\frac{1 \text{ mol NH}_3}{17.0 \text{ g NH}_3} \right) \left(\frac{4 \text{ mol NO}}{4 \text{ mol NH}_3} \right) = 22.0 \text{ mol NO}$$

Note that, because this question asks for moles of NO, not grams, the formula mass of NO is not needed, and one less conversion factor is required.

(c) The table of information for this part is

Problem	*? g*		*74.4 g*
Equation	**$4NH_3(g)$ + $5O_2(g)$** $\xrightarrow{\text{Pt, 750–900 °C}}$	**$4NO(g)$ + $6H_2O(g)$**	
Formula masses, u	17.0		30.0
Recipe, mol	4		4

The setup is

$$74.4 \text{ g NO} \left(\frac{1 \text{ mol NO}}{30.0 \text{ g NO}} \right) \left(\frac{4 \text{ mol NH}_3}{4 \text{ mol NO}} \right) \left(\frac{17.0 \text{ g NH}_3}{1 \text{ mol NH}_3} \right) = 42.2 \text{ g NH}_3$$

(d) and (e) We found in part (c) that 42.2 g NH_3 are needed to make 74.4 g NO. *Because atomic masses are relative masses, different units of mass can be used in stoichiometric calculations without changing the arithmetic as long as the same units are used throughout a calculation.* Therefore, 42.2 lb NH_3 are needed to make 74.4 lb NO, and 42.2 tons NH_3 are needed to make 74.4 tons NO.

---

*A gram is too small a unit for industrial scale reactions. In the United States, chemical engineers often work with pounds and tons.

⬤ As in any table, the column headings refer to all the entries in the column.

⬤ If you do not follow the reasoning in the solution for parts (d) and (e), do part (d) by converting pounds of NO to grams of NO and then grams of NH_3 to pounds of NH_3. Look carefully at your setup. Have you wasted time multiplying by a number early in the calculation and then dividing by the same number later?

3.3 Answer the following questions, given the reaction

$$3Na(s) + P(s) \longrightarrow Na_3P(s)$$

(a) How many grams of P are needed to react with 63.5 g Na? (b) How many grams of Na_3P can be formed from 63.5 g Na, provided enough P is used?

3.4 Potassium nitrate is decomposed by heat to potassium nitrite and oxygen gas. Potassium nitrate and potassium nitrite are solids. (a) How many moles of oxygen gas are formed by the decomposition of 3.91 g of potassium nitrate? (b) How many grams of oxygen gas are formed by the decomposition of 25.63 g of potassium nitrate?

3.5 Carbon disulfide is manufactured by the reaction of carbon with sulfur at high temperatures. At these high temperatures, carbon is a solid, but sulfur and carbon disulfide are both gases. (a) How many grams of sulfur are needed to react with 487 g of carbon? (b) How many grams of sulfur are needed to make 236 g of carbon disulfide, provided that enough carbon is used? (c) How many tons of sulfur are needed to react with 487 tons of carbon? (d) How many pounds of sulfur are needed to make 236 lb of carbon disulfide, provided that enough carbon is used?

> ● Always begin solving problems about quantitative relationships in reactions by writing an equation for the reaction.

3.5 LIMITING REACTANTS

When chemical reactions take place in a laboratory, manufacturing plant, or in the environment, the reactants are *not* usually present in the proportions shown by the equation for the reaction. The quantity of *one reactant controls the amount of products that can be formed.* This reactant is called the **limiting reactant** or **limiting reagent.** In the cupcake example at the beginning of this chapter, the egg is the limiting reactant. The recipe calls for two eggs and $2\frac{1}{4}$ cups of flour to make 30 cupcakes. If you have only one egg, you can only make half a recipe; you will get 15 cupcakes. The egg controls the amount of product (cupcakes). An **excess** of the other reactant or reactants is present. To make half a recipe of cupcakes, you will need $1\frac{1}{8}$ cups of flour. If there are 3 cups of flour in the bag, you will have $1\frac{7}{8}$ cups of flour left. An excess of flour is present.

When you are making cupcakes, you leave the excess of flour in the bag (unless you like very dry cupcakes). However, when a reaction is carried out in a laboratory, manufacturing plant, or in the atmosphere, an excess of one reactant is often present in the reaction mixture. For example, hydrogen gas made by using the sun's energy, instead of electricity, to decompose water has been suggested as a nonpolluting fuel because the only product of burning hydrogen gas in air is water. Remember that when a fuel burns in air, it reacts with the oxygen gas in the air. The equation for the burning of hydrogen gas in air is

$$2H_2(g) + O_2(g) \xrightarrow{\text{spark}} 2H_2O(l)$$

or, using models to show the reaction on a molecular scale,

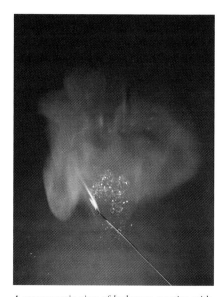

*Chemicals used in the laboratory are often referred to as **reagents.***

2 hydrogen molecules + 1 oxygen molecule ⟶ 2 water molecules

When hydrogen gas burns in air, many more oxygen molecules than hydrogen molecules are usually present. However, no more than two water molecules can

A macroscopic view of hydrogen reacting with oxygen. The soap bubbles are filled with H_2 (g); air contains O_2 (g). The burning splint provides the spark needed to start reaction. Which reactant is in excess?

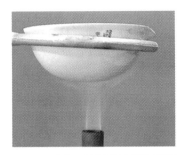

Top: *The flame of a properly adjusted laboratory burner is almost invisible.* Bottom: *If too little air is present when methane burns, carbon, not carbon dioxide, is formed. Hot carbon glows, making the flame luminous, and a deposit of soot forms on whatever is being heated. What is the limiting reactant in the bottom photo?*

❶ If quantities are given for more than one reactant, a problem may be a limiting-reactant problem.

❷ Remember, always begin solving problems about quantitative relationships in reactions by writing an equation for the reaction.

be formed from two molecules of hydrogen gas, even if extra oxygen molecules are present:

2 hydrogen molecules + 3 oxygen molecule $\xrightarrow{\text{spar}}$ 2 water molecule + 2 leftover oxygen

Hydrogen gas is the limiting reactant and is all used up in the reaction. An excess of oxygen molecules is present; oxygen molecules are left over.

The idea of a limiting reactant is very important. For example, plants need nitrogen and phosphorus to grow. If the plants do not have enough of either one, they will not be able to grow well. If a farmer applies nitrogen when his fields need phosphorus, the plants won't grow any faster or bigger. The farmer will have wasted his money and time only to pollute the water that runs off his fields. Babies and children need compounds called amino acids in their diets to make the proteins that their bodies need. The amino acid present in smallest amount compared to need limits their growth (especially mental growth). Carburetors in cars mix gasoline with the air needed to burn the gasoline. If the mixture contains too much gasoline, air (O_2) is limiting; some gasoline does not burn. The unburned gasoline goes into the exhaust and pollutes the atmosphere. If the mixture contains too little gasoline, then the gasoline is limiting, and the car stalls. These are but three of many practical applications of the idea of a limiting reactant.

Sample Problem 3.6 shows how to work problems that involve a limiting reactant. You can recognize a limiting-reactant problem by the fact that quantities are given for more than one reactant.

SAMPLE PROBLEM

3.6 A mixture of 3.50 g of hydrogen and 26.0 g of oxygen is made to react to form water. (a) Which reactant is limiting? (b) How many grams of water will be formed? (c) Which reactant will be left over? (d) How many grams of this reactant will be left over?

Solution As usual, begin by writing the equation for the reaction and making a table of information. The table of information for this problem is

Problem	3.50 g	26.0 g	? g
Equation	$2H_2(g)$ +	$O_2(g)$ $\xrightarrow{\text{spark}}$	$2H_2O(l)$
Formula masses, u	2.02	32.0	18.0
Recipe, mol	2	1	2

(a) and (c) To find out which reactant is limiting and which will be left over, reason exactly as you would if you were baking cupcakes. You'd say to yourself: "One egg, that's half a recipe. I have 3 cups of flour but that's more than a recipe of flour. I'll only be able to make half a recipe of cupcakes; I'll have some flour left over." Because the "recipe" for a chemical reaction is in moles, you must convert the quantities of reactants to moles before finding the fraction or multiple of a recipe available:

$$3.50 \text{ g } H_2 \left(\frac{1 \text{ mol } H_2}{2.02 \text{ g } H_2} \right) = 1.73 \text{ mol } H_2$$

and

$$26.0 \text{ g } O_2 \left(\frac{1 \text{ mol } O_2}{32.0 \text{ g } O_2} \right) = 0.813 \text{ mol } O_2$$

Now 1.73 mol H_2 is

$$\frac{1.73 \text{ mol } H_2}{2 \text{ mol } H_2} = 0.865$$

▶ The fraction 0.865 is a pure number—that is, it has no units.

of a recipe of H_2, and 0.813 mol O_2 is

$$\frac{0.813 \text{ mol } O_2}{1 \text{ mol } O_2} = 0.813$$

of a recipe of O_2. Because the fraction of a recipe of oxygen used (0.813) is less than the fraction of a recipe of hydrogen used (0.865), O_2 is limiting, and hydrogen will be left over. The amount of the limiting reactant, O_2, that is available determines how much hydrogen can react and how much water can be formed.

▶ None of the limiting reactant is left over; all the limiting reactant is used up in the reaction.

(b) Calculation of the amount of water that can be formed is similar to Sample Problem 3.5(b):

$$26.0 \text{ g } O_2 \left(\frac{1 \text{ mol } O_2}{32.0 \text{ g } O_2}\right)\left(\frac{2 \text{ mol } H_2O}{1 \text{ mol } O_2}\right)\left(\frac{18.0 \text{ g } H_2O}{1 \text{ mol } H_2O}\right) = 29.3 \text{ g } H_2O$$

(d) The problem states that 3.50 g of hydrogen were present at the beginning of the reaction. To find out how much hydrogen will be left over, we need to determine how much hydrogen will be used up by reaction with 26.0 g of oxygen. This calculation is similar to Sample Problem 3.5(a):

$$26.0 \text{ g } O_2 \left(\frac{1 \text{ mol } O_2}{32.0 \text{ g } O_2}\right)\left(\frac{2 \text{ mol } H_2}{1 \text{ mol } O_2}\right)\left(\frac{2.02 \text{ g } H_2}{1 \text{ mol } H_2}\right) = 3.28 \text{ g } H_2$$

Subtract the mass of H_2 needed to react with 26.0 g O_2 from the mass of H_2 that was present at the beginning of the reaction to find out how much hydrogen is left over:

$$(3.50 \text{ g } H_2 \text{ present}) - (3.28 \text{ g } H_2 \text{ reacted}) = 0.22 \text{ g } H_2 \text{ left over}$$

Summary of Steps for Solving Stoichiometry Problems

1. Write an equation for the reaction.
2. Note what's given and asked for above the equation.
3. Write the formula masses needed and recipe in moles below the equation.
4. If quantities of more than one reactant are given, determine which reactant is limiting by finding the fraction of a recipe of each reactant present.
   The quantity of the limiting reactant determines the amount of product formed and the amounts of other reactants that react.
5. Set up and carry out the calculation:
   a. Use formula masses to convert grams to moles.
   b. Use the equation (recipe) for the reaction to write conversion factors for converting moles of one substance to moles of other substances.
   c. Use formula masses to convert moles to grams. Be sure to include formulas of substances in labels.
6. Round the answer to the correct number of digits.
7. Check work: Is answer reasonable? Do units cancel to give correct units for answer? Is arithmetic correct? Is answer expressed to correct number of digits?

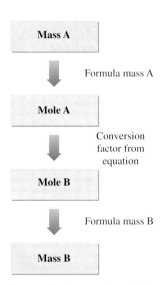

Flow diagram showing steps in stoichiometric calculations.

3.6 The mixture of hydrogen and chlorine shown in the molecular-level picture is made to react to form hydrogen chloride. Sketch a molecular-level picture of the mixture that will result.

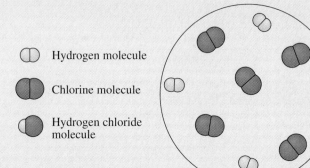

Hydrogen molecule

Chlorine molecule

Hydrogen chloride molecule

3.7 The reaction

$$2NH_3(g) + H_2SO_4(aq) \longrightarrow (NH_4)_2SO_4(aq)$$

is used in the commercial production of fertilizer. If 22.7 g NH_3 and 54.8 g H_2SO_4 are used, (a) which reactant is limiting? (b) How many grams of $(NH_4)_2SO_4$ can be formed? (c) Which reactant will be left over and how many grams of it will be left?

3.6 THEORETICAL, ACTUAL, AND PERCENT YIELDS

The *amount of product calculated to be formed* by a reaction is called the **theoretical yield.** In some cases, the *calculated amount of product actually is formed.* Reactions like this are called **quantitative reactions.** In other cases, less product is obtained, usually for one or more of the following reasons:

1. A reactant is impure.
2. Some product is lost **mechanically** (for example, because it sticks to the container like cupcake batter sticks to the bowl and spoon).
3. The reactants undergo other reactions in addition to the one you want to take place.

The *reaction you want to take place* is called the **main reaction.** In many cases, other reactions take place in addition to the main reaction. These *unwanted reactions* are called **side reactions;** the *products of side reactions* are called **by-products.***
For example, carbon dioxide is the product of the combustion of carbon in excess

An everyday example of mechanical loss.

---

*Products of the main reaction other than the product wanted are also called by-products. For example, water is a by-product of the reaction of NH_3 and O_2 to make NO:

$$4NH_3(g) + 5O_2(g) \xrightarrow{\text{Pt, heat}} 4NO(g) + 6H_2O(g)$$

This reaction is a step in the manufacture of nitric acid, an important industrial chemical that is used to make fertilizers and explosives.

oxygen. The equation is

$$C(s) + O_2(g) \longrightarrow CO_2(g) \qquad (3.5)$$

If the amount of oxygen is limited, some carbon monoxide is also formed:

$$2C(s) + O_2(g) \longrightarrow 2CO(g)$$

This second reaction is a side reaction if the first reaction is the main (wanted) reaction. Carbon monoxide is a by-product. Because part of the carbon is converted to carbon monoxide, the amount of carbon dioxide actually formed is less than the amount calculated according to equation 3.5. The **actual yield** of carbon dioxide is less than the theoretical yield. The larger the amount of reactants wasted in side reactions, the less product is formed (the lower the actual yield).

The **percent yield** or **% yield** of a reaction is the *amount of product actually formed, divided by the amount of product calculated to be formed, times 100:*

$$\% \text{ yield} = \frac{\text{actual yield}}{\text{theoretical yield}} \times 100$$

The equilibria involved in most synthetic reactions are shifted so that reaction is, for practical purposes, complete. Therefore, yields less than theoretical that result from incomplete reaction are not considered until later (Chapter 14).

SAMPLE PROBLEM

3.7   If 68.5 g of carbon are burned in air, (a) what is the theoretical yield of carbon dioxide? (b) If 237 g of carbon dioxide are actually formed, what is the percent yield?

Solution   When a substance burns in air it reacts with the oxygen in the air. The table of information is

Problem	68.5 g		? g
Equation	C(s)	$+ O_2(g) \longrightarrow$	$CO_2(g)$
Formula masses, u	12.0		44.0
Recipe, mol	1		1

(a) The theoretical yield is

$$68.5 \text{ g C} \left(\frac{1 \text{ mol C}}{12.0 \text{ g C}} \right) \left(\frac{1 \text{ mol CO}_2}{1 \text{ mol C}} \right) \left(\frac{44.0 \text{ g CO}_2}{1 \text{ mol CO}_2} \right) = 251 \text{ g CO}_2$$

(b) The percent yield is

$$\% \text{ yield} = \frac{\text{actual yield}}{\text{theoretical yield}} \times 100$$

$$= \frac{237 \text{ g CO}_2}{251 \text{ g CO}_2} \times 100 = 94.4\%$$

PRACTICE PROBLEM

3.8   Answer the following questions, given the reaction

$$C_2H_2(g) + H_2O(l) \xrightarrow{\text{HgSO}_4, \text{ H}_2\text{SO}_4} C_2H_4O(g)$$

(a) What is the theoretical yield of C_2H_4O from 5.67 g C_2H_2? (Assume that an excess of water is present.) (b) If 8.38 g C_2H_4O are actually obtained, what is the percent yield?

Green Chemistry

Green chemistry is preventing pollution rather than cleaning it up. Until the 1990s, industrial chemists and chemical engineers were trained to aim for the highest possible yields and profits when they developed methods for manufacturing compounds. In the future, they will probably be trained to consider environmental effects and environmental costs first.

What can be done to prevent pollution? To begin with, changes that can be brought about quickly and are relatively inexpensive should be made. The most obvious are housekeeping improvements such as stopping leaks. Solvents, leftover reactants, by-products, and catalysts can be recovered and reused. (Catalysts are substances that make reactions take place faster, without being used up in the reaction.) Changes like these often increase profits as well as prevent pollution.

Changing processes in the following ways takes longer and is more expensive because these steps usually involve building new plants: (1) considering the product's entire life beginning with the raw materials and energy used to make it, its use, and how to get rid of it when it is no longer needed; (2) substituting less toxic reactants, solvents, and catalysts wherever

possible; (3) making toxic reactants, solvents, and catalysts, for which substitutes are not available, as they are needed and where they are needed to avoid having to store and transport them; (4) using renewable resources, such as cellulose from plants, as raw materials instead of nonrenewable resources like petroleum; (5) substituting solar energy and wind energy for energy produced from fossil fuels such as coal.

However, the right decisions are often not simple or obvious. For example, clothes made from the natural fiber cotton might seem to be "greener" than clothes made from synthetic fibers. However, when the environmental costs of raising and harvesting the cotton, such as tractor fuel and fertilizers, of ironing the clothes, and of having to replace them sooner are included, clothes made from synthetic fibers turn out to be better for the environment. What's best in the United States and other developed countries may not be best in less developed countries. Much research, education of the public and lawmakers, and wise laws will be necessary if people all over the world are to enjoy a better life both now *and* in the future.

A high proportion of chemical spills occur during transportation.

3.7 QUANTITATIVE ANALYSIS

One important use of stoichiometry is in quantitative analysis. **Quantitative analysis** is *finding out **how much** of a given substance is present in a sample.* (*Finding out **what** substances are present in a sample* is called **qualitative analysis.**) Some typical applications of quantitative analysis include determination of pollutants in air and water, the quantity of metals in ores, and the purity of drugs. A wide range of methods is used, for example, both combination and decomposition reactions can be used. The results of quantitative analysis are often expressed as percent by mass:

$$\text{percent by mass of } A \text{ in sample} = \frac{\text{mass } A \text{ in sample}}{\text{total mass of sample}} \times 100$$

where A is any substance in the sample.

3.8   Magnesium metal burns in air; the magnesium metal reacts with oxygen gas from the air to form solid magnesium oxide. If 0.5684 g of magnesium metal forms 0.9426 g of magnesium oxide, (a) what is the percent (by mass) of magnesium in magnesium oxide? (b) The percent of oxygen?

Solution   (a) Substituting the information given in the problem in the definition of percent by mass,

$$\text{Mass Mg in magnesium oxide} = \frac{0.5684 \text{ g magnesium}}{0.9426 \text{ g magnesium oxide}} \times 100$$

$$= 60.30\%$$

Because there are just two elements, magnesium and oxygen, in magnesium oxide, the percent by mass of oxygen can be found by subtracting the percent magnesium from 100%:

$$100.00\% - 60.30\% = 39.70\% \text{ oxygen}$$

Check: If 0.5684 g Mg forms 0.9426 g of magnesium oxide, the increase in mass that results from the combination of oxygen with magnesium is

$$0.9426 \text{ g} - 0.5684 \text{ g} = 0.3742 \text{ g}$$

and the mass percent O in magnesium oxide is

$$\frac{0.3742 \text{ g}}{0.9426 \text{ g}} \times 100 = 39.70\%$$

The fact that the same answer is obtained by two methods of calculation is evidence that the answer is correct.

3.9   An oxide of mercury decomposes to mercury and oxygen gas when heated. If a 0.8349-g sample of the oxide gives 0.7732 g of mercury, (a) what is the percent by mass of mercury in the oxide? (b) The percent of oxygen?

Analysis for Carbon and Hydrogen in Organic Compounds

Because a majority of the over 13 million compounds that are known are organic (contain carbon and hydrogen), analysis for carbon and hydrogen is a common procedure. The sample is rapidly and completely burned; the carbon is quantitatively converted to carbon dioxide and the hydrogen to water. Modern analyzers are automated and computerized. The amounts of carbon dioxide and water are measured instrumentally. Other elements that occur frequently in organic compounds, such as nitrogen and sulfur, can be determined at the same time as carbon and hydrogen. ▬Figure 3.2 shows a typical modern analyzer.

▬ FIGURE 3.2   A modern elemental analyzer. With this analyzer, carbon, hydrogen, nitrogen, and sulfur can be determined in 3 minutes using only 2 mg of sample. Determination of oxygen takes just 12 minutes more.

3.9   A 0.024 54-g sample of an organic compound gave 0.074 84 g CO_2 and 0.036 75 g H_2O on combustion. (a) What is the percent carbon in the organic compound? (b) The percent hydrogen? (c) Are there any elements besides carbon and hydrogen in this organic compound?

Solution (a) Because we do not know the formula of the organic compound, we cannot begin by writing a balanced equation for the combustion. From the formula, CO_2, you can see that in one molecule of carbon dioxide, there is one carbon atom. In 1 mol of carbon dioxide, there is 1 mol of carbon atoms. The formula mass of carbon dioxide is 44.01. In 44.01 g of carbon dioxide, there are 12.01 g of carbon:

$$44.01 \text{ g } CO_2 \Leftrightarrow 12.01 \text{ g C}$$

This relationship can be used to write a conversion factor from grams of carbon dioxide to grams of carbon:

$$0.074\ 84 \text{ g } CO_2 \left(\frac{12.01 \text{ g C}}{44.01 \text{ g } CO_2} \right) = 0.020\ 42 \text{ g C}$$

By definition,

$$\% \text{ by mass of C in sample} = \frac{\text{g C}}{\text{g sample}} \times 100 = \frac{0.020\ 42 \text{ g C}}{0.024\ 54\text{-g sample}} \times 100$$

$$= 83.21\% \text{ C}$$

(b) The grams of hydrogen and the percent hydrogen in the sample can be calculated similarly:

$$0.036\ 75 \text{ g } H_2O \left(\frac{2.016 \text{ g H}}{18.02 \text{ g } H_2O} \right) = 0.004\ 111 \text{ g H}$$

$$\% \text{ H} = \frac{\text{g H}}{\text{g sample}} \times 100 = \frac{0.004\ 111 \text{ g H}}{0.024\ 54\text{-g sample}} \times 100$$

$$= 16.75\% \text{ H}$$

(c) The percent carbon and percent hydrogen total 100:

$$83.21\% + 16.75\% = 99.96\%$$

Thus, the compound can contain only carbon and hydrogen. (Remember that the last digit in measured numbers is uncertain, and 99.96% = 100.00%.)

PRACTICE PROBLEM

3.10 A 0.018 69-g sample of an organic compound that contained only C, H, and O gave 0.044 38 g CO_2 and 0.022 72 g H_2O on combustion. (a) What is the percent carbon in the compound? (b) The percent hydrogen? (c) The percent oxygen?

3.8 EMPIRICAL FORMULAS FROM PERCENT COMPOSITION

Empirical means derived from experience or observation alone.

The **empirical formula** for a compound is the *simplest formula that shows the ratios of the numbers of atoms of each kind in the compound.* For example, the formula for a molecule of hydrogen peroxide is H_2O_2, showing that a molecule of hydrogen peroxide consists of two hydrogen atoms and two oxygen atoms. One mole of hydrogen peroxide contains 2 mol of hydrogen atoms and 2 mol of oxygen atoms. The ratio of hydrogen atoms to oxygen atoms or moles hydrogen atoms to moles oxygen atoms is 2 : 2 or 1 : 1. Both the subscripts in the formula for hydrogen peroxide can be divided by 2. The simplest or empirical formula for hydrogen peroxide is HO. Sometimes the subscripts in the formula for a molecule do not have a common denominator. Water, H_2O, and ammonia, NH_3, are examples. For these compounds, the usual formulas are the simplest or empirical formulas.

PRACTICE PROBLEMS

3.11   The formula for a molecule of butane is C_4H_{10}. What is the empirical formula for butane?

3.12   The formula for a molecule of propene is C_3H_6. What is the empirical formula for propene?

The empirical formula for a compound can be calculated from the percent composition by mass of the elements in the compound. Sample Problems 3.10 and 3.11 show how.

See Appendix B.5 for a review of percent.

SAMPLE PROBLEMS

3.10   The composition of a compound is 36.4% Mn, 21.2% S, and 42.4% O. All percents are mass percents. What is the empirical formula of the compound?

Solution   Because the composition of the compound is given in mass percent—that is, in grams per 100 g—it is convenient to think of an exactly 100-g sample. A 100-g sample consists of 36.4 g Mn, 21.2 g S, and 42.4 g O. The problem is to find the subscripts x, y, and z in the formula, $Mn_xS_yO_z$. The subscripts tell the ratio of the number of atoms and moles of Mn, S, and O in the compound. The first step in finding the empirical formula is to find the number of moles of each of the elements in the 100-g sample (this calculation is like the one in Sample Problem 2.14):

Because percent composition is an intensive property and does not depend on sample size, we can think about any size sample we like.

$$36.4 \text{ g Mn}\left(\frac{1 \text{ mol Mn}}{54.9 \text{ g Mn}}\right) = 0.663 \text{ mol Mn}$$

$$21.2 \text{ g S}\left(\frac{1 \text{ mol S}}{32.1 \text{ g S}}\right) = 0.660 \text{ mol S}$$

$$42.4 \text{ g O}\left(\frac{1 \text{ mol O}}{16.0 \text{ g O}}\right) = 2.65 \text{ mol O}$$

The formula for the compound is $Mn_{0.663}S_{0.660}O_{2.65}$. But a formula can be interpreted in terms of atoms as well as in terms of moles. Therefore, the subscripts in formulas should not be fractions; they should be whole numbers. To convert the subscripts to whole numbers, divide each of them by the smallest:

Remember that, in formulas, an element that is to the left of another element in the periodic table is usually written on the left. If two elements are in the same group, the one that is lower in the periodic table is usually written on the left in the formula.

$$\frac{0.663}{0.660} = 1.00, \qquad \frac{0.660}{0.660} = 1.00, \qquad \text{and} \qquad \frac{2.65}{0.660} = 4.02$$

The formula for the compound is $Mn_{1.00}S_{1.00}O_{4.02}$ or, rounding to nearest whole numbers, $MnSO_4$.

Check: The best way to check the answer to a problem like this one is to calculate the percent composition from the formula (see Section 3.10). If the percent composition calculated from the formula is the same as the percent composition given in the problem, the answer must be correct.

3.11   The composition of a compound is 27.6% Mn, 24.2% S, and 48.2% O. What is the empirical formula of the compound?

Solution A 100-g sample consists of 27.6 g Mn, 24.2 g S, and 48.2 g O. The first step is to find the number of moles of each of the elements in the 100-g sample:

$$27.6 \text{ g Mn}\left(\frac{1 \text{ mol Mn}}{54.9 \text{ g Mn}}\right) = 0.503 \text{ mol Mn}$$

$$24.2 \text{ g S}\left(\frac{1 \text{ mol S}}{32.1 \text{ g S}}\right) = 0.754 \text{ mol S}$$

$$48.2 \text{ g O}\left(\frac{1 \text{ mol O}}{16.0 \text{ g O}}\right) = 3.01 \text{ mol O}$$

Dividing each of the subscripts by the smallest to change them to whole numbers gives

$$\frac{0.503}{0.503} = 1.00, \qquad \frac{0.754}{0.503} = 1.50, \qquad \text{and} \qquad \frac{3.01}{0.503} = 5.98$$

The formula for the compound is $Mn_{1.00}S_{1.50}O_{5.98}$. The subscript for sulfur is still not close to being a whole number. Multiplying it by 2 will give a number that is close to a whole number:

$$2(1.50) = 3.00$$

The other subscripts must also be multiplied by 2 to keep the proportions of Mn and O in the compound the same. The formula of this compound is $Mn_2S_3O_{12}$.

PRACTICE PROBLEMS

3.13 The composition of a compound is 28.73% K, 1.48% H, 22.76% P, and 47.03% O. What is the empirical formula of the compound?

3.14 The composition of a compound is 17.55% Na, 39.70% Cr, and 42.75% O. What is the empirical formula of the compound?

Compounds whose units are ions are always represented by empirical formulas. Compounds whose units are molecules are usually represented either by molecular or by structural formulas.

3.9   MOLECULAR AND STRUCTURAL FORMULAS

Molecular Formulas

A **molecular formula** *shows the number of atoms of each kind in a molecule.* Molecular formulas are multiples of empirical formulas. If the multiplier is 1, the molecular formula is the same as the empirical formula. For example, both the empirical and the molecular formulas for water are H_2O. Frequently, the multiplier is greater than 1. The molecular formula for hydrogen peroxide is H_2O_2, and the empirical formula is HO. The molecular formula is twice the empirical formula.

The molecular formula can be found if both the empirical formula and the molecular mass are known. The molecular mass need not be known very accurately. The fact that the molecular mass of oxygen gas is 32 instead of 16 is evidence that oxygen gas is composed of diatomic molecules, O_2, not single oxygen atoms.

This method is not suitable for deciding whether a formula is $C_{33}H_{68}$ or $C_{35}H_{72}$. $C_{33}H_{68}$ is 85.26% C; $C_{35}H_{72}$, 85.28% C.

The Mole Concept module of the Reactions section provides practice in determining empirical formulas by means of an animated chemical analyzer.

3.12   The empirical formula for benzene is CH. The molecular mass of benzene is 78 u. What is the molecular formula for benzene?

Solution   The mass of the unit, CH, is 13 u. We must find out how many times 13 u 78 u is:

$$n(13 \text{ u}) = 78 \text{ u}$$

$$n = \frac{78 \text{ u}}{13 \text{ u}} = 6$$

The molecular formula for benzene is six times the empirical formula

$$6(CH) = C_6H_6$$

The molecular formula for benzene is C_6H_6.

3.15   What is the molecular formula of a compound that has the empirical formula CCl_2O and a molecular mass of 297?

Another way of finding molecular formulas is to use a mass spectrometer (Figure 2.14). High-resolution mass spectrometers can separate ions differing in mass by just 1 part in 10 000. An ion with mass 500.00 u can be separated from an ion with mass 499.95 u. The composition of molecules can be found from these very accurate determinations of mass. For example, the molecular mass of $^{12}C^{16}O$ is 27.994 915 u, and the molecular mass of $^{14}N_2$ is 28.006 148 u:

$1\,^{12}C$ @ 12.000 000 u = 12.000 000 u $2\,^{14}N$ @ 14.003 074 u = 28.006 148 u

$1\,^{16}O$ @ 15.994 915 u = $\underline{15.994\ 915\ \text{u}}$

$1\,^{12}C^{16}O$ = 27.994 915 u

If the molecular mass found for a sample is 27.994 915 u, the formula of the sample is CO; if it is 28.006 148 u, the formula is N_2.

Structural Formulas

Finding the molecular formula is the first step in determining the structure of molecules. The **structure** of a molecule is *the way the atoms are joined together.* For example, in a water molecule, the oxygen is in the middle, and the two hydrogens are joined to it as shown in ▬Figure 3.3(a). The two hydrogens are not joined together with the oxygen fastened to one of them as shown in Figure 3.3(b).

▶ If you did not know the molecular formula, you would not know how many atoms of each kind to put together.

(a) Atoms connected correctly

(b) Atoms connected incorrectly

▬ FIGURE 3.3   In a water molecule, the oxygen is in the middle, and the two hydrogens are joined to it.

The *models shown in Figure 3.3* are called **ball-and-stick models.** They are more useful for showing the bonds that join atoms together to form molecules than the models we have been using. In ball-and-stick models, the balls represent atoms, and the sticks represent bonds. The *models we have been using* are called **space-filling models.** Space-filling models show the size and shape of molecules better than ball-and-stick models.

The properties of molecules depend on their structures. For example, ethyl alcohol [◾Figure 3.4(a)] is a liquid used in alcoholic beverages, as a fuel, and as a solvent in industry. Dimethyl ether is made of the same atoms arranged differently [Figure 3.4(b)]. Dimethyl ether is a gas used in refrigeration. *Different compounds that have the same molecular formula,* such as ethyl alcohol and dimethyl ether, are called **isomers.** Isomers have different structures; the same atoms are joined together in different ways.

◾ FIGURE 3.4   Ball-and-stick *(top)* and space-filling *(bottom)* models of the isomeric compounds ethyl alcohol and dimethyl ether. The molecular formula of both compounds is C_2H_6O but the atoms are joined together differently.

Remember that hydrogen attached to oxygen is smaller than hydrogen attached to carbon because the computer generated models in this text are boundary surface diagrams.

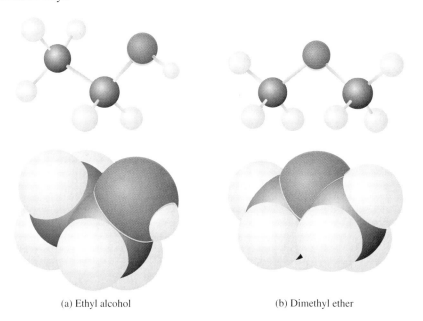

(a) Ethyl alcohol   (b) Dimethyl ether

Structural formulas are used to *show structures without models.* In structural formulas, lines are used to represent bonds that hold atoms together in molecules. For example, the structural formulas of ethyl alcohol and dimethyl ether are

$$\begin{array}{cccc} & H & H & \\ & | & | & \\ H- & C- & C- & O-H \\ & | & | & \\ & H & H & \end{array} \qquad and \qquad \begin{array}{cccc} & H & & H \\ & | & & | \\ H- & C- & O- & C-H \\ & | & & | \\ & H & & H \end{array}$$

ethyl alcohol   dimethyl ether

In the structural formulas for ethyl alcohol and dimethyl ether, notice that carbon forms four bonds and oxygen forms two bonds. Hydrogen only forms one bond. These are typical numbers of bonds for C, O, and H to form.

To see how these structural formulas are related to the ball-and-stick models, imagine that the models are at eye level and that you are looking at each of them from straight ahead. ◾Figure 3.5 shows the models viewed in this way.

Structural formulas for organic compounds like ethyl alcohol and dimethyl ether are often written in more compact form. The more *compact structural formulas* are called **condensed structural formulas.** The condensed structural formulas for ethyl alcohol and dimethyl ether are CH_3CH_2OH and CH_3OCH_3. Condensed structural formulas can be thought of as directions for writing complete structural formulas or putting models together. Sample Problem 3.13 shows how to write a complete structural formula from a condensed structural formula.

(a)

(b)

◾ FIGURE 3.5   Ball-and-stick models of ethyl alcohol CH_3CH_2OH (a) and dimethyl ether CH_3OCH_3 (b) viewed at eye level so that the models resemble the structural formulas.

3.13 Write complete structural formulas for compounds having the condensed structural formulas (a) $CH_3CH_2CH_3$, (b) $CH_3(CH_2)_2CH_2Cl$, and (c) $(CH_3)_2CHNH_2$.

Solution (a) The condensed structural formula $CH_3CH_2CH_3$ means "Write the symbol for a carbon atom:

$$C$$

To it, attach symbols for three hydrogens and a second carbon atom:

$$\begin{array}{c} H \\ | \\ H-C-C \\ | \\ H \end{array}$$

To the second carbon atom, attach two hydrogens and the third carbon:

$$\begin{array}{c} H \quad H \\ | \quad | \\ H-C-C-C \\ | \quad | \\ H \quad H \end{array}$$

Finally, attach the last three hydrogens to the third carbon." The complete structural formula is

$$\begin{array}{c} H \quad H \quad H \\ | \quad | \quad | \\ H-C-C-C-H \\ | \quad | \quad | \\ H \quad H \quad H \end{array}$$

(b) In the condensed structural formula $CH_3(CH_2)_2CH_2Cl$, $(CH_2)_2$ means "Write the carbon with two hydrogens attached twice." The complete structural formula is

$$\begin{array}{c} H \quad H \quad H \quad H \\ | \quad | \quad | \quad | \\ H-C-C-C-C-Cl \\ | \quad | \quad | \quad | \\ H \quad H \quad H \quad H \end{array}$$

(c) In the condensed structural formula $(CH_3)_2CHNH_2$, $(CH_3)_2C$ means "Write two carbons, each with three hydrogens, and attach these two groups of symbols to the next carbon. This next carbon also has a hydrogen and a nitrogen attached to it. The last two hydrogens are attached to the nitrogen." The complete structural formula is

$$\begin{array}{c} H \quad H \\ | \quad | \\ H-C-C-N-H \\ | \quad | \quad | \\ H \quad | \quad H \\ \quad H-C-H \\ \quad \quad | \\ \quad \quad H \end{array}$$

Structural formulas give more information about a compound than empirical and molecular formulas. They tell not only how many of each kind of atom there are in a molecule of the compound but also the order in which the atoms are attached to each other. Both the molecular formula and the empirical formula for a compound can be written from a structural formula. A molecular formula gives the number of each kind of atom in the molecule. The molecular mass can be calculated from it. Empirical formulas show only the relative numbers of each kind of atom in a compound.

PRACTICE PROBLEMS

3.16   Write complete structural formulas for compounds having the condensed structural formulas (a) H_2NNH_2, (b) $CH_3(CH_2)_3CH_2OH$, and (c) $(CH_3)_3CCl$.

3.17   The condensed structural formula for a compound is $CH_3(CH_2)_4CH_3$. (a) What is the molecular formula of the compound? (b) What is the empirical formula of the compound?

3.10   PERCENT COMPOSITION FROM FORMULAS

Why is percent composition at the end instead of at the beginning of this chapter? The title and main subject of the chapter are stoichiometry. Therefore, stoichiometry is treated first.

The Mole Concept module of the Reactions section provides practice in calculating percent composition.

Once the empirical or molecular formula is known, the percent composition can be calculated from it.

SAMPLE PROBLEM

3.14   The empirical formula for iron(III) chloride is $FeCl_3$. (a) What is the percent by mass of iron in iron(III) chloride to one decimal place? (b) What is the percent by mass of chlorine in iron(III) chloride to one decimal place?

Solution   (a) The formula mass of $FeCl_3$ is

$$
\begin{array}{r}
1Fe @ 55.8\ u = 55.8\ u \\
+\ 3Cl @ 35.5\ u = 106.5\ u \\
\hline
1FeCl_3 = 162.3\ u
\end{array}
$$

Of 162.3 u $FeCl_3$, 55.8 u is Fe, and 106.5 u is Cl. Therefore,

$$
\text{Percent by mass of Fe in } FeCl_3 = \frac{55.8\ u}{162.3\ u} \times 100 = 34.4\%\ Fe
$$

$$
\text{Percent by mass of Cl in } FeCl_3 = \frac{106.5\ u}{162.3\ u} \times 100 = 65.62\%\ Cl
$$

Always check to be sure your work is reasonable.

Calculating the percent Cl as shown above is better than finding it by subtracting the percent Fe from 100.0% because you can check your work by seeing if the percents total 100:

$$
34.4\% + 65.62\% = 100.0\%
$$

PRACTICE PROBLEM

3.18   The molecular formula for hydrazine is N_2H_4. (a) What is the percent by mass of hydrogen in hydrazine to one decimal place? (b) What is the percent by mass of nitrogen in hydrazine to one decimal place? (c) Would it make any difference in the percent composition found if the empirical formula for hydrazine was used instead of the molecular formula? Explain your answer.

Hydrazine is used as a rocket fuel in guided missiles and in the space program.

Percent compositions calculated from formulas are used for identifying compounds. Agreement between percent composition found by experiment and percent composition calculated from a proposed formula provides one piece of evidence that the formula proposed for a new compound is correct.

PRACTICE PROBLEM

3.19   A green compound is formed on burning chromium metal in oxygen. The percent composition is 68.42% Cr and 31.58% O. Is the compound CrO_2, CrO, Cr_2O_3, or CrO_3?

SUMMARY

Stoichiometry is the study of quantitative relationships between substances undergoing chemical changes. In chemical reactions, the total mass of the products equals the total mass of the reactants. This summary of many observations is called the **law of conservation of matter.** On a practical scale, equations are interpreted as meaning moles. However, scientists and engineers usually measure quantity of matter by measuring mass. Conversion factors between moles and mass are obtained from formula masses. The reactants are not usually present in the proportions shown by the equation for the reaction. The amount of one reactant, which is called the **limiting reactant,** controls the amount of products that can be formed.

The amount of product calculated to be formed by the **main reaction** (desired reaction) is called the **theoretical yield.** Reactions in which the theoretical yield is actually obtained are called **quantitative reactions.** The **actual yield** of many reactions is less than the theoretical yield as a result of (1) impurities in reactants, (2) **mechanical losses,** and (3) formation of **by-products** by side reactions. **Side reactions** are unwanted reactions that take place in addition to the main reaction. If more than one reaction takes place between reactants, an equation must be written for *each* reaction. The **percent yield** of a reaction is the actual yield divided by the theoretical yield times 100:

$$\% \text{ yield} = \frac{\text{actual yield}}{\text{theoretical yield}} \times 100$$

Qualitative analysis is the testing of something to find out what substance or substances are present in it. **Quantitative analysis** is finding out how much of a substance is present in a sample. The results of quantitative analysis are often expressed as percent by mass:

$$\% \text{ by mass } A \text{ in sample} = \frac{\text{mass } A \text{ in sample}}{\text{total mass of sample}} \times 100$$

The **empirical formula** for a compound is the simplest formula that shows the ratios of the numbers of atoms of each kind in the compound; the subscripts in empirical formulas do not have a common divisor other than 1. Empirical formulas are calculated from the percent composition of compounds and are always used for ionic compounds. For molecular compounds, the **molecular formula** shows the number of each kind of atom in a molecule as well as the ratios of the numbers of atoms of each kind in the compound. Molecular formulas are often multiples greater than 1 of empirical formulas; the multiplier is found by determining molecular mass. Once an empirical or molecular formula is known for a compound, it can be used to calculate the percent composition of the compound. **Structural formulas** show how the atoms are joined together in molecules. The **structures** of molecules can also be shown by **ball-and-stick** and **space-filling models.** The properties of molecules depend on their structures. Compounds that have the same molecular formula but different structures are called **isomers.**

For information about the organization of Additional Practice Problems, Stop & Test Yourself, Putting Things Together, and Applications, see the beginnings of these sections in Chapter 1.

3.20 Which of the following is or are contrary to the law of conservation of matter? (a) $4FeS_2(s) + 11O_2(g) \longrightarrow 2Fe_2O_3(s) + 8SO_2(g)$ (b) A 2.40-g sample of carbon reacts with 12.28 g of sulfur to form 15.22 g of carbon disulfide. (c) $2C_4H_{10} + 9O_2 \longrightarrow 8CO + 10H_2O$ (d) Decomposition of a 5.68-g sample of solid potassium chlorate, $KClO_3$, by heat yields 3.46 g of solid potassium chloride and 2.22 g of oxygen gas. (3.1)

3.21 Balance each of the following expressions:

(a) $CO_2(g) + H_2O(l) \longrightarrow C_6H_{12}O_6(aq) + O_2(g)$

(b) $S(s) + H_2SO_4(aq) \longrightarrow SO_2(g) + H_2O(l)$

(c) $Ag(s) + H_2S(g) + O_2(g) \longrightarrow Ag_2S(s) + H_2O(g)$

(d) $Na_3PO_4(aq) + CoCl_2(aq) \longrightarrow$
$Co_3(PO_4)_2(s) + NaCl(aq)$

(e) $C_8H_{18}(l) + O_2(g) \longrightarrow CO_2(g) + H_2O(l)$ (3.2)

3.22 Balance each of the following expressions:

(a) $FeTiO_3(s) + Cl_2(g) + C(s) \longrightarrow$
$FeCl_2(s) + TiO_2(s) + CO_2(g)$

(b) $Na_2O_2(s) + H_2O(l) \longrightarrow NaOH(aq) + H_2O_2(aq)$

(c) $Ca_3(PO_4)_2(s) + H_2SO_4(aq) \longrightarrow CaSO_4(s) + H_3PO_4(aq)$

(d) $Cu(NO_3)_2(aq) + NH_3(aq) + H_2O(l) \longrightarrow$
$Cu(OH)_2(s) + NH_4NO_3(aq)$

(e) $(NH_4)_2Cr_2O_7(s) \longrightarrow N_2(g) + Cr_2O_3(s) + H_2O(g)$ (3.2)

3.23 How many moles of C atoms are in one mole of C_4H_{10}? (3.3)

3.24 Given the equation $2KClO_3(s) \longrightarrow 2KCl(s) + 3O_2(g)$, (a) how many moles of $O_2(g)$ are formed by the decomposition of 0.35 mol $KClO_3(s)$? (b) How many moles of $KClO_3(s)$ must decompose to form 4.8 mol $KCl(s)$? (3.3)

3.25 In sunlight, hydrogen and chlorine react to form hydrogen chloride, which is a gas under ordinary conditions. (a) How many moles of chlorine are needed to react with 3.8 mol of hydrogen? (b) How many moles of hydrogen chloride will be formed by the reaction in part (a)? (3.3)

3.26 Given the equation $2Na(s) + Cl_2(l) \longrightarrow 2NaCl(s)$, (a) how many grams of NaCl can be formed from 4.68 g Na, assuming enough Cl_2 is present to react with all of the Na? (b) How many grams of Cl_2 are needed to react with 4.68 g Na? (c) To make 39.6 g NaCl, how many grams of Na should you start with, assuming the reaction is quantitative and enough chlorine is available? (d) To make 39.6 kg NaCl, how many kilograms of Na are needed? (3.4)

3.27 Given the equation $N_2O_4(l) + 2N_2H_4(l) \longrightarrow 3N_2(g) + 4H_2O(g)$, (a) how many grams of N_2H_4 are required to react with 25.49 g N_2O_4? (b) How many grams of N_2 will be formed from the reaction of 89.7 g N_2H_4, assuming that the reaction is quantitative and that enough N_2O_4 is available? (3.4)

3.28 Given the equation $4HCl(aq) + MnO_2(s) \longrightarrow 2H_2O(l) + MnCl_2(aq) + Cl_2(g)$, (a) how many moles of HCl are required to react with 31.8 g MnO_2? (b) How many grams of MnO_2 must react for 0.56 mol Cl_2 to be formed? (3.4)

3.29 When heated, mercury(II) oxide decomposes to oxygen gas and mercury vapor. (a) How many grams of mercury(II) oxide must be decomposed to form 0.567 mol oxygen gas? (b) How many grams of mercury will be formed by the reaction in part (a)? (3.4)

3.30 The mixture of sulfur dioxide and oxygen shown in the molecular-level picture is made to react to form sulfur trioxide. Sketch a molecular-level picture of the result. (3.4)

Oxygen

Sulfur dioxide

Sulfur trioxide

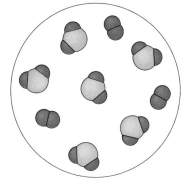

3.31 Given the equation $2AgNO_3(aq) + BaCl_2(aq) \longrightarrow 2AgCl(s) + Ba(NO_3)_2(aq)$, how many grams of AgCl will be formed if a solution that contains 41.6 g $AgNO_3$ is mixed with a solution that contains 35.4 g $BaCl_2$? (3.5)

3.32 The reaction $Cr_2O_3(s) + 2Al(l) \longrightarrow 2Cr(l) + Al_2O_3(l)$ takes place at high temperatures. If 42.7 g Cr_2O_3 and 9.8 g Al are mixed and reacted until one reactant is used up, which reactant will be left over? How much of it will be left over? (3.5)

3.33 (a) What is the difference between a *main reaction* and *side reactions*? (b) What is a *by-product*? (c) What is a *quantitative reaction*? (d) Give three reasons that many reactions are not quantitative. (3.6)

3.34 Given the equation

$$4Fe(s) + 3O_2(g) \xrightarrow{\text{heat}} 2Fe_2O_3(s)$$

(a) What is the theoretical yield of Fe_2O_3 from the reaction of 4.86 g Fe with excess O_2? (b) If 6.76 g Fe_2O_3 are actually obtained from 4.86 g Fe, what is the percent yield? (c) Under other conditions, the percent yield is 75.6%. How many grams of Fe_2O_3 are formed from 4.86 g Fe under these conditions? (3.6)

3.35 Lead(II) chloride can be prepared by the reaction of aqueous solutions of lead(II) nitrate and sodium chloride. The equation is $Pb(NO_3)_2(aq) + 2NaCl(aq) \longrightarrow PbCl_2(s) + 2NaNO_3(aq)$. If a solution that contains 49.1 g $Pb(NO_3)_2$ is treated with an excess of NaCl, (a) what is the theoretical yield of $PbCl_2$? (b) If 38.2 g $PbCl_2$ are obtained, what is the percent yield? (3.6)

3.36 (a) Calculate the mass of Ag in 34.8 g Ag_2S. (b) How many grams of Ag could be formed by the decomposition of 1 mol Ag_2S? (3.7)

3.37 If 2.85 g of thallium chloride is formed by the reaction of 2.43 g of thallium metal with an excess of chlorine gas, (a) what is the percent by mass of thallium in thallium chloride? (b) The percent chlorine? (3.7)

3.38 Qualitative analysis of an unknown organic compound showed that it contained only carbon, hydrogen, and chlorine. If combustion analysis of a 0.018 69-g sample of the unknown compound gave 0.009 68 g CO_2 and 0.003 96 g H_2O, (a) what was the percent by mass of carbon in the compound? (b) The percent hydrogen? (c) The percent chlorine? (3.7)

3.39 A molecule of an organic compound is composed of 9 carbon atoms, 12 hydrogen atoms, and 3 oxygen atoms. (a) Write the molecular formula. (b) Write the empirical formula. (3.8)

3.40 A sample of an unknown compound contains 43.95% Mo, 7.33% O, and 48.72% Cl. What is the empirical formula? (3.8)

3.41 What is the empirical formula for the compound with each of the following percent compositions? (a) 72.4% Fe, 27.6% O (b) 24.2% Cu, 27.0% Cl, 48.8% O (c) 28.7% K, 1.5% H, 22.8% P, 47.0% O (3.8)

3.42 The empirical formula of a compound is CH. The molecular mass of the compound is 26 u. What is the molecular formula of the compound? (3.9)

3.43 The empirical formula of a compound is CClN. The molecular mass of the compound is 184 u. What is the molecular formula of the compound? (3.9)

3.44 Classify each of the following formulas as empirical, molecular, or structural. (a) N_2H_4 (b) O H (c) CH_4O (d) $MgBr_2$ (e) P_4 (3.9) H O

3.45 Classify each of the following formulas as empirical, molecular, or structural. (3.9)

$$
\begin{array}{cc}
\text{H} & \text{H} \\
| & | \\
\text{(a) } O_2 \quad \text{(b) S} \quad \text{(c) H} \; \text{C} & \text{C} \; \text{H} \quad \text{(d) } C_2H_6 \quad \text{(e) } CH_3 \\
| & | \\
\text{H} & \text{H}
\end{array}
$$

3.46 Write complete structural formulas for compounds having the following condensed structural formulas: (a) CH_3CCl_3, (b) $CH_3CH_2OCH_2CH_3$, (c) $(CH_3)_2NH$. (3.9)

3.47 Write complete structural formulas for compounds having the following condensed structural formulas: (a) $(CH_3)_2CHOH$, (b) CH_3SH, (c) $(CH_3)_3N$, (d) $CH_3CH_2CH_2Cl$, (e) $(CH_3)_4C$. (3.9)

3.48 The condensed structural formula for ethane is CH_3CH_3. (a) Write the complete structural formula for ethane. (b) What is the molecular formula for ethane? (c) What is the empirical formula for ethane? (3.9)

3.49 Write structural, condensed structural, and molecular formulas for the substances represented by the ball-and-stick models. (See Figure 1.8 for color code.) (3.9)

3.50 Calculate the percent by mass of each element in the following compounds to one decimal place: (a) $CoCl_3$, (b) N_2O_5, (c) $Al(C_2H_3O_2)_3$, (d) $(NH_4)_2S$, (e) $NiSO_4 \cdot 7H_2O$. (3.10)

3.51 Calculate the percent by mass of each element in the following compounds to two decimal places: (a) KI, (b) $Ca(NO_3)_2$, (c) C_2H_4, (d) $CuSO_4 \cdot 5H_2O$, (e) $(NH_4)_2B_4O_7 \cdot 4H_2O$. (3.10)

3.52 An oxide of nitrogen is 63.65% nitrogen. Is the compound NO, NO_2, N_2O, or N_2O_5? (3.10)

3.53 The reaction of element A with element B to form compound C is represented by the diagram. (a) Write the equation for this reaction. (b) Which reactant is limiting?

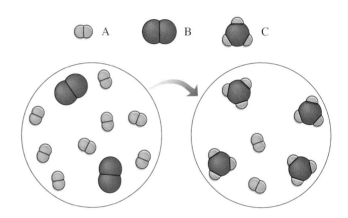

3.54 Given the equation $H_2SO_4(aq)$ + $2NaOH(aq)$ $\longrightarrow$ $Na_2SO_4(aq) + 2H_2O(l)$, (a) how many grams of Na_2SO_4 will be formed if a solution that contains 44.14 g H_2SO_4 is mixed with a solution that contains 36.00 g NaOH? (b) How many grams of which reactant will be left over?

3.55 A 41.5-g sample of an unknown compound contains 0.27 mol C and 1.08 mol Cl. What is the empirical formula of the compound?

3.56 Which contains the greater mass of copper, 23.0 g Cu_2O or 24.0 g Cu_2S?

3.57 Calculate the percent of water in $Mg_3(PO_4)_2 \cdot 22H_2O$ to four significant figures.

3.58 Careful heating of 2.69 g $NaHCO_3$ in a test tube yields bubbles of colorless gas. Drops of a colorless liquid are observed near the top of the test tube, and 1.71 g of a white solid remains in the bottom of the test tube. Which of the following reactions takes place? (a) $NaHCO_3(s) \longrightarrow NaOH(s)$ + $CO_2(g)$ (b) $2NaHCO_3(s) \longrightarrow Na_2O(s) + 2CO_2(g) + H_2O(l)$ (c) $2NaHCO_3(s) \longrightarrow Na_2CO_3(s) + CO_2(g) + H_2O(l)$

3.59 Given the equation $Zn(s) + 2HCl(aq) \longrightarrow ZnCl_2(aq) + H_2(g)$, (a) how many moles of HCl are needed to react with 24.8 g Zn? (b) How many moles of H_2 will be formed from 24.8 g Zn, assuming enough HCl is used to react with all of the Zn? (c) How many grams of Zn are needed to make 3.4 mol H_2, assuming enough HCl is used to react with all of the Zn? (d) How many grams of $ZnCl_2$ will be formed along with the 3.4 mol H_2 in part (c)?

3.60 If a mixture of 4.6 g Na(s) and 7.8 g $Cl_2(g)$ reacts, what is the total mass of the product and leftover reagent?

3.61 "If, in a reaction between A and B, more moles of A than moles of B are present, then B must be the limiting reactant." What is wrong with this statement?

3.62 For the reaction $CH_4(g) + 4S(g) \longrightarrow CS_2(g) + 2H_2S(g)$, a 90% yield (zero is significant) is usually obtained. To make 75 g CS_2, how many grams of sulfur are needed?

3.63 A metal oxide decomposes to the metal and oxygen gas when heated. From the decomposition of 3.48 g of metal oxide, 3.24 g of metal are obtained. How many grams of metal are combined with 1.00 g of oxygen in this metal oxide?

3.64 The SI mole is a gram-mole. One mole of water molecules is 6.02×10^{23} water molecules and has a mass of 18.0 g. (a) What is the mass in kilograms of one kilogram-mole of water molecules, and how many water molecules are in one kilogram-mole? (b) What is the mass in pounds of one pound-mole of water molecules, and how many water molecules are in one pound-mole?

3.65 Calcium carbide is produced by the reaction $CaO(s) + 3C(s) \longrightarrow CaC_2(l) + CO(g)$. To make 153 tons of CaC_2, an average day's production, how many tons of C are needed?

3.66 Ammonia is produced by the following sequence of reactions:

Step 1: $CH_4(g) + H_2O(g) \longrightarrow 3H_2(g) + CO(g)$

Step 2: $N_2(g) + 3H_2(g) \longrightarrow 2NH_3(g)$

How many grams of CH_4 must be used to produce 25 g NH_3, assuming 100% yield in each step?

3.67 If each step in a synthesis gives a 90% yield, what is the overall yield for   a) a two-step synthesis?   (b) A three-step synthesis?   (c) As the number of steps in a synthesis increases, what happens to the overall yield?   (d) What is the significance of this conclusion for the planning of syntheses?

3.68 An impure sample of NaCl contains 32.4% Na. Assuming the impurities do not contain any sodium, what percentage of the impure sample is NaCl?

3.69 The complete structural formula for a compound is

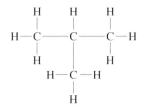

(a) Write a condensed structural formula for this compound. (b) What is the molecular formula? (c) The empirical formula?

3.70 Consider the reaction

$$7H_2SO_4(aq) + Na_2Cr_2O_7(aq) + 6NaCl(aq) \longrightarrow$$
$$Cr_2(SO_4)_3(aq) + 3Cl_2(g) + 4Na_2SO_4(aq) + 7H_2O(l)$$

Assume the reaction is quantitative. If a mixture of 16.57 g H_2SO_4, 3.46 g $Na_2Cr_2O_7$, and 4.23 g NaCl is caused to react, (a) which will be the limiting reactant? (b) How many grams of Cl_2 can be formed? (c) How many grams of which reactants will be left over?

STOP & TEST YOURSELF

1. In the equation for the reaction, $C_4H_{10}O(l) + O_2(g) \longrightarrow CO_2(g) + H_2O(l)$, the coefficient of $O_2(g)$ is
   (a) 2   (b) 3   (c) 4   (d) 5   (e) 6

2. In the equation for the reaction $Na_3PO_4(aq) + BaCl_2(aq) \longrightarrow Ba_3(PO_4)_2(s) + NaCl(aq)$, the coefficient of NaCl(aq) is
   (a) 2   (b) 3   (c) 4   (d) 5   (e) 6

3. The diagram shows the reaction of molecule A with molecule B to form molecule C. Which equation is best?
   (a) $3A + B \longrightarrow 4C$ (b) $6A + 4B \longrightarrow 4C$
   (c) $3A + 2B \longrightarrow 2C$ (d) $6A + 2B \longrightarrow 4C$ (e) $3A + B \longrightarrow 2C$

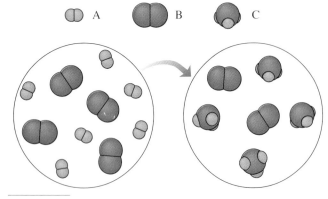

Use the following information for questions 4 through 7. Chlorine can be prepared by the reaction

$$MnO_2(s) + 4HCl(aq) \longrightarrow Cl_2(g) + MnCl_2(aq) + 2H_2O(l)$$

This reaction goes to completion. Formula masses in u are MnO_2, 86.9; HCl, 36.5; $MnCl_2$, 125.8; Cl_2, 70.9; H_2O, 18.02.

4. How many grams of Cl_2 can be produced from 25.0 g MnO_2? Assume enough HCl is used to react with all of the MnO_2.
   (a) 0.0326   (b) 10.2   (c) 20.4   (d) 30.6   (e) 246

5. How many moles of HCl are required to react with 25.0 g MnO_2?
   (a) 1.15   (b) 3.48   (c) 10.5   (d) 13.9   (e) 42.0

6. How many moles of Cl_2 can be prepared from 0.75 mol MnO_2 and 2.0 mol HCl?
   (a) 2.0   (b) 1.3   (c) 1.0   (d) 0.75   (e) 0.50

7. If 0.75 mol MnO_2 and 2.0 mol HCl are used, how many moles of which reactant will be left over?
   (a) 1.3 mol HCl   (b) 0.75 mol HCl   (c) None of either
   (d) 0.25 mol MnO_2 (e) 0.50 mol MnO_2

8. Consider the reaction of sodium metal with chlorine gas to form sodium chloride. If 1.5 mol NaCl is to be prepared, how many moles of chlorine must be used?
   (a) 0.75   (b) 1.0   (c) 1.3   (d) 1.5   (e) 3.0

9. What is the percent yield of a reaction in which 762 g CH_4 is converted to 2048 g CH_3Cl? The equation for the reaction is

$$CH_4(g) + Cl_2(g) \xrightarrow{light} CH_3Cl(g) + HCl(g)$$

(a) 2.69 (b) 26.9 (c) 37.2 (d) 85.4 (e) 117

10. The formula for glucose (corn sugar) is $C_6H_{12}O_6$. What is the percent by mass of carbon in glucose to two decimal places? (a) 4.00 (b) 25.00 (c) 40.00 (d) 41.39 (e) None of the preceding answers is correct.

11. What is the percent by mass of nitrogen in $Ca(NO_3)_2\cdot 3H_2O$ to two decimal places?
(a) 6.42 (b) 9.27 (c) 12.84 (d) 15.05 (e) 24.43

12. A compound has the following composition (percent by mass): 38.37% C, 1.49% H, 52.28% Cl. What is its empirical formula?

(a) C_2HCl (b) $C_{6.5}H_3Cl_3O$ (c) $C_{12}H_5Cl_5$ (d) $C_{13}H_6Cl_6O_2$ (e) $C_{26}H_{12}Cl_{12}O_4$

13. The empirical formula for a compound is C_2HCl_2. Which of the following is a possible molecular formula for the compound?
(a) $CH_{0.5}Cl$ (b) $CHCl$ (c) C_2HCl (d) $C_4H_2Cl_4$ (e) $C_6H_3Cl_3$

14. The empirical formula for a compound having a molecular mass of 70 u is CH_2. The molecular formula for this compound is
(a) CH_2 (b) C_2H_4 (c) C_3H_6 (d) C_4H_8 (e) C_5H_{10}

15. A sample having a mass of 0.015 32 g was burned in pure oxygen and 0.005 65 g CO_2 and 0.001 16 g H_2O were obtained. What is the percent C and the percent H in the compound?
(a) 10.1% C, 0.847% H (b) 10.1% C, 0.424% H (c) 10.1% C, 89.9% H (d) 15.8% C, 0.847% H (e) 15.8% C, 0.424% H

PUTTING THINGS TOGETHER

3.71 Magnesium metal and oxygen react to form magnesium oxide, which is a white solid. If 36.5 g Mg and 27.0 g O_2 are caused to react, (a) which reactant is limiting? (b) How many grams of which reactant will be left over? (c) What is the theoretical yield of magnesium oxide? (d) If 59.1 g of magnesium oxide are obtained, what is the percent yield?

3.72 A 0.2536-g sample of a compound of iron and sulfur was reacted with oxygen. The products were 0.1688 g Fe_2O_3 and 0.2708 g SO_2. What is the empirical formula of the compound of iron and sulfur?

3.73 A 6.38-g sample of a hydrate of calcium chloride was heated until all of the water of hydration was removed. The residue of anhydrous calcium chloride had a mass of 4.82 g. In the formula for the hydrate, $CaCl_2\cdot nH_2O$, what is the value of n?

3.74 A molecule of an organic compound has twice as many hydrogen atoms as carbon atoms and the same number of oxygen atoms as carbon atoms. The molecular mass of this compound is 120 u. What is its molecular formula?

3.75 An impure sample of iron is treated with an excess of hydrochloric acid. The equation for the reaction that takes place is $Fe(s) + 2HCl(aq) \longrightarrow FeCl_2(aq) + H_2(g)$; the reaction is quantitative. Assuming that the impurities do not react with hydrochloric acid to give hydrogen gas, if 9.54×10^{-3} mol of hydrogen gas is formed from 0.5483 g of the impure iron, what is the percent of iron in the impure sample?

3.76 When sodium metal reacts with water, hydrogen gas and aqueous sodium hydroxide are formed. The heat given off by the reaction melts the remaining sodium metal. The hydrogen gas and some of the remaining sodium metal burn. (a) How many different chemical reactions take place when sodium metal reacts with water? (b) How many equations should be written?

3.77 Sodium forms a series of compounds with chlorine and oxygen. The general formula for this series is $NaClO_n$. If one of these compounds was found to contain 28.96% Cl, what is the value of n for this compound?

3.78 Oxygen gas is often prepared in the laboratory by decomposing a compound of potassium, chlorine, and oxygen. The other product is KCl. Complete decomposition of a 1.6422-g sample of this compound gives 0.020 10 mol O_2. What is the empirical formula of the compound?

3.79 Combustion analysis of a compound containing only carbon and hydrogen gave 0.049 77 g CO_2 and 0.024 44 g H_2O. (a) What is the empirical formula of the compound? (b) The molecular mass is 72 u. What is the molecular formula? (c) What was the mass of the sample that was burned?

3.80 The mineral dolomite is a mixed carbonate of calcium and magnesium. Calcium and magnesium carbonates both decompose on heating. The products are the metal oxides and carbon dioxide. If a 4.87-g residue is left when an 8.76-g sample of dolomite is heated until decomposition is complete, what percent by mass of the sample is calcium carbonate?

3.81 (a) At red heat, 0.9264 g of copper reacted with oxygen to give 1.1596 g of copper oxide. How much oxygen reacted? How many grams of copper reacted per gram of oxygen? (b) At a higher temperature, 1.7498 g of copper reacted with oxygen to give 1.9701 g of a different oxide of copper. How much oxygen reacted? How many grams of copper reacted per gram of oxygen? (c) Are the different masses of copper that reacted with 1 g of oxygen in the ratio of small whole numbers? (d) What are the formulas and names of the two oxides of copper?

3.82 In a modern elemental analyzer, a 2.485-mg sample of an antibacterial drug composed of carbon, hydrogen, nitrogen, oxygen, and sulfur gave 4.500 mg CO_2, 1.089 mg H_2O, 0.391 mg N_2, and 0.596 mg SO_2. (a) What was the percent composition of the drug? (b) What was the empirical formula of the drug? (c) The molecular mass of the drug was 267 u. What was the molecular formula?

3.83 Tetraphosphorus hexoxide, a powdery white solid, can be made by reacting phosphorus with oxygen gas. The molecular mass of phosphorus is found by experiment to be 124 u. Phosphorus is a soft, white, waxy solid that must be stored underwater because it catches fire in air. (a) If 87.8 g of phosphorus and 45.3 g of oxygen are used, what is the theoretical yield? (b) If 50.8 g of tetraphosphorus hexoxide are obtained, what is the percent yield? (c) How many grams of which reactant will be left over?

3.84 Suppose you need to make 350.0 g of phosphorus tribromide by the reaction of phosphorus with bromine. (a) How many grams of phosphorus and how many grams of bromine should you start with if a 100% yield of phosphorus tribromide is expected? (b) It is easier to measure a liquid like bromine by volume. The density of bromine is 3.12 g/mL. How many milliliters of bromine should be used in part (a)? (c) Phosphorus catches fire spontaneously in air. Suppose that a 22% excess of bromine should be used in the reaction so that all the phosphorus is used up. How many grams of bromine should be used? (d) How many grams of phosphorus should you start with if only a 95% yield of phosphorus tribromide (based on phosphorus) is expected?

3.85 Dinitrogen monoxide gas is made by heating molten ammonium nitrate. The other product is water vapor. Some nitrogen monoxide and a little nitrogen gas are also formed. (a) Write the equation for the main reaction. (b) What are the reactions that form nitrogen monoxide and nitrogen gas called? (c) If 1.7300 g of ammonium nitrate are decomposed, what is the theoretical yield of dinitrogen monoxide in moles? (d) If 0.018 22 mol of dinitrogen monoxide is obtained, what is the percent yield? (e) "The ammonium nitrate must be free of chloride ion because chloride ion catalyzes decomposition to nitrogen gas." What does this statement mean? How is a catalyst shown in the equation for a reaction?

3.86 When heated in air, an element, E, burns to form an oxide with empirical formula E_2O_3. A 0.5386-g sample of the element gave 0.7111 g of oxide. (a) What is the atomic mass of the element E? (b) What is the symbol for the element? (c) What is the oxide called?

3.87 A sample of silver ore was analyzed by converting the silver to silver chloride; from 35.40 g of ore, 0.2940 g AgCl were obtained. Calculate the percent by mass of Ag in the ore.

APPLICATIONS

3.88 (a) The coal burned to generate the electricity needed to operate a single refrigerator for a year produces about 1.00 ton CO_2. If coal is 85% C by mass, how many tons of coal must be burned to operate a refrigerator for a year? (b) How many tons per year of sulfur dioxide does a power plant burning 3 million tons of 3% sulfur coal per year give off? Assume all sulfur is burned.

3.89 The condensed structural formula for methyl *tertiary*-butyl ether, a gasoline additive, is $CH_3OC(CH_3)_3$. Write the complete structural formula for methyl *tertiary*-butyl ether.

3.90 Write condensed structural, molecular, and empirical formulas for the compounds represented by the ball-and-stick models:

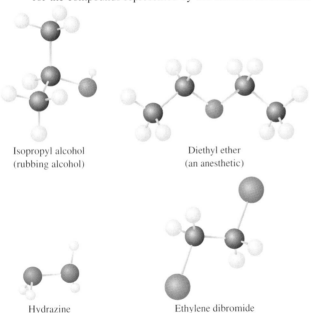

Isopropyl alcohol
(rubbing alcohol)

Diethyl ether
(an anesthetic)

Hydrazine
(a rocket fuel)

Ethylene dibromide
(a fumigant)

3.91 Ammonium nitrate and urea, CH_4N_2O, are both used as sources of nitrogen in fertilizers. If ammonium nitrate sells for $7.82/kg and urea for $8.11, which gives you more nitrogen for your dollar?

3.92 Future historians may well refer to our times as the "Polymer Age." Polymers are very large molecules. A number of them are composed of a single simple unit repeated many times. The structural formula for a part of a molecule of a polymer that is used to make phonograph records is shown below:

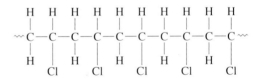

(a) What is the repeating unit (structural formula) of the polymer? (b) If the average molecular mass of a sample of the polymer is 1.5×10^6 u, how many times is the unit repeated in an average molecule?

3.93 Hemoglobins (the major components of red blood cells) from mammals have molecular masses of 6.45×10^4 u. They contain 0.35% Fe by mass. How many iron atoms are there in each hemoglobin molecule?

3.94 A newspaper article says that "serious damage is being done to the Earth by removing and destroying over 12 million pounds of oil every day. . . ." The reason given is that as a result of the combustion of petroleum in power plants and vehicles, Earth is losing so much weight that gravity will move it out of its orbit. Discuss the writer's reasoning.

3.95 A sample of white powder seized by the police was suspected of being cocaine. Combustion analysis of a 0.018 32-g sample gave 0.048 04 g CO_2 and 0.010 99 g H_2O. The molecular formula for cocaine is $C_{17}H_{21}NO_4$. Can the white powder be cocaine? Explain your reasoning.

3.96 The most important process for making atmospheric nitrogen into N-containing compounds is the Haber process. In the Haber process, nitrogen gas is reacted with hydrogen gas at high temperature and pressure to produce ammonia. There are no side reactions. The best yields are obtained when the reactants are mixed in the proportions shown by the balanced equation. A catalyst is necessary to make the reaction take place in a practical length of time. After enough time has passed so that no further increase in the amount of ammonia formed is observed, 0.45 mol of ammonia is obtained from 1.00 mol N_2. Does the reaction go to completion? Explain your answer.

3.97 Sulfuric acid, H_2SO_4, is one of the most widely used of all manufactured chemicals. Increasing amounts of it are being made from sulfur dioxide, which is a by-product of production of metals from ores and refining of crude oil. Conversion of sulfur dioxide to sulfuric acid is a two-step process. In the first step, sulfur dioxide is converted to sulfur trioxide by heating with excess oxygen in the presence of a vanadium catalyst. In the second step, sulfur trioxide reacts with water, forming sulfuric acid. Typical percent yields for the two steps are 96.5 and 98.0, respectively. (a) How many grams of sulfuric acid will be obtained from 454 g of sulfur dioxide? (b) The density of sulfuric acid is 1.84 g/mL. How many milliliters of sulfuric acid will be obtained from 454 g of sulfur dioxide?

3.98 The following reactions take place in some termites. Write the equation for each. (a) Glucose, $C_6H_{12}O_6$ (a solid), and water form acetic acid, $C_2H_4O_2$ (a liquid), carbon dioxide, and hydrogen. (b) Hydrogen and carbon dioxide form acetic acid and water. (c) Acetic acid and oxygen form carbon dioxide and water.

3.99 (a) Write equations for the reactions of diesel oil with air to form (i) carbon dioxide and water, (ii) carbon monoxide and water, (iii) carbon and water. Use the formula $C_{15}H_{32}$ to represent diesel oil, which is a mixture. (b) Diesel smoke is finely divided carbon. Explain why using "new-and-improved air" that contains 35% oxygen instead of the usual 20% in diesel engines prevents the formation of smoke.

3.100 The density of plutonium metal is 19.84 g/cm³, and the density of plutonium(IV) oxide is 11.46 g/cm³. Over a period of 7–10 years, a 2.5-kg sample of plutonium metal stored at Los Alamos National Laboratory in New Mexico was converted to plutonium(IV) oxide by reaction with oxygen from air. (a) What happened to the volume of the sample? (b) What happened to the container?

3.101 The recommended daily allowance (RDA) for calcium is 1000 mg·day^{-1}. (a) If a person's diet provides 650 mg Ca (two glasses of milk plus a serving of brocolli), how many Tums tablets are required to supply the missing Ca? A Tums tablet contains 500 mg $CaCO_3$. (b) An article on diet warns: "Always read the label to make sure you're not getting more carbonate than elemental calcium." Are you? Is there anything you can do about it? Explain your answers.

3.102 Metallic lead is used to make storage batteries. The first step in obtaining lead from its chief ore, galena (PbS), is roasting in limited air. The products are sulfur dioxide gas and a solid oxide containing 92.83% Pb. (a) What is the empirical formula of the oxide of lead? (b) Write the equation for the reaction involved in roasting galena. (c) For 1.000 kg of galena, what would be the largest number of moles of oxygen gas that could be present before galena, rather than air, became the limiting reactant? (d) What is the maximum number of grams of lead that could be obtained from 1.000 kg of galena, and how much SO_2, a potential air pollutant, will be produced?

3.103 Sodium hydroxide and chlorine are manufactured by passing an electric current through an aqueous solution of sodium chloride. Hydrogen gas is also formed. If the salt solution is 36.5% by mass NaCl, how many kilograms of each product will be obtained from each 1.000 kg of salt solution?

3.104 (a) If a car gets 25 mi/gal and travels 26 000 mi/year, how many tons of carbon dioxide does it produce in a year? Assume that gasoline, which is a mixture, is C_8H_{18} ($d = 0.70$ g/mL). (b) A young apple tree produces about 44 lb of fructose (fruit sugar), $C_6H_{12}O_6$, each growing season by the reaction of carbon dioxide from air with water. Oxygen is also formed. How many apple trees must be planted to use up the carbon dioxide from one car?

3.105 Automobiles are a major source of air pollutants. At the high temperatures inside a car's engine, nitrogen and oxygen from air react to form nitrogen monoxide gas. Nitrogen monoxide reacts with oxygen in air, forming nitrogen dioxide, a very poisonous red-brown gas with an irritating odor. (a) and (b) Write equations for these two reactions. (Solutions of a little nitrogen dioxide in a lot of air are yellow. You may have noticed that smog often has a faint yellow color.) Catalytic converters cause carbon monoxide gas and unburned fuel in automotive exhaust to react rapidly with oxygen in the air, forming carbon dioxide gas and water. (c) and (d) Write equations for these two reactions. Use the formula C_8H_{18} to represent unburned gasoline. Unfortunately, catalytic converters also cause sulfur dioxide gas in exhaust to react with oxygen in air, forming sulfur trioxide gas and thus contributing to acid rain. (e) Write the equation for this reaction.

3.106 Ethyl alcohol is a member of the family of organic compounds known as alcohols. Alcohols have an OH group attached to carbon. The water molecule also includes an OH group and, like water, alcohols react with Group IA metals forming hydrogen gas. Write equations for the reactions of each of the following compounds with sodium: (a) water (HOH, although the molecular formula H_2O is generally used); (b) wood alcohol or methyl alcohol, CH_3OH; (c) grain alcohol or ethyl alcohol, CH_3CH_2OH; (d) rubbing alcohol or isopropyl alcohol, $(CH_3)_2CHOH$.

Green Chemist, Rock Star, and Entrepreneur

JAMES A. CUSUMANO

Chairman of the Board, Catalytica, Inc.

B.A. Math and Chemistry
Rutgers–The State University of New Jersey

Ph.D. Physical Chemistry
Rutgers–The State University of New Jersey

As a young boy, I had a somewhat impetuous curiosity for discovering how things work. I liked to take toys apart, and frequently I asked adults questions that I am told were difficult to answer. As the father of two inquisitive daughters, I can now understand this situation.

My father was an "old school," professional, Italian emigrant who wanted his first of what were to be ten children to become a physician. I was made well aware of my "chosen" profession before the age of nine! Although intrigued with medicine, somehow it just didn't fit for me.

And then it happened! Without knowing it or planning it, I guess I decided to become a chemist. The single event that started me down this path was the most exciting gift I remember receiving as a youngster—a top-of-the-line Gilbert chemistry set—for my ninth Christmas! I was captivated by the experiments described in the accompanying instructions book—disappearing inks, spot removers, changing magical colors, the synthesis of pleasant aromas, and, of course, the power and intrigue of pyrotechnics.

My fascination with chemistry began in the early 1950s, when science was much more respected by the press and society than it is today. But I believe this respect can and will return when we scientists do a better job of explaining in understandable terms that science is part of the solution of society's problems and not the problem. Only the misuse of science can be a problem. It was therefore easy for me to become enthralled with a profession that was viewed by most as important and promising. The excitement of space exploration in the 1960s only reinforced this aura.

Chemistry intrigued me to the point that I spent most of my free time in a dark, damp, dingy basement laboratory that I had set up in our home in Elizabeth, New Jersey. At age eleven, I used money I had earned as a newsboy to buy chemicals, chemistry books, and apparatus for my increasingly more complex experiments. At first I was interested in trying to uncover the secrets of nature, but then I was bitten by the entrepreneurial "bug."

I learned to make practical products in my lab, including detergents, inks, spot removers, perfumes, cosmetics, and more. I packaged these products and sold them in our neighborhood. In 1954, I introduced "all natural" perfumes made by extraction of flower petals gathered in the evenings from Mom's and the neighbors' gardens—without their or the FDA's approval, I'm afraid! Whether folks on our block were impressed with my entrepreneural spirit or just felt sorry for me, I was an overnight success! I used the profits to increase the size and scope of my lab. My parents were amazingly accepting and encouraging of this endeavor, particularly in light of several narrow escapes I had in my lab over the years.

I ordered letterhead and named my "company" O & O Research Laboratories, Inc. I ordered free samples of chemicals from manufacturers from all over the U.S. They obviously thought that I was a serious potential customer. Upon finding out that I was a young teenager doing research and development, and making small-scale amounts of products, most of them were amused, and some were impressed.

There was, however, the occasional exception. Once, a salesman, hoping to sell me several tons of ammonium dichromate that his company had sampled me, arrived at my front door. He had traveled all the way from Philadelphia without calling first. He rang the bell and, believing he had the wrong address, proceeded to ask Mom if she had ever heard of O & O Research Laboratories. Embarrassed, she confided that the proprietor was thirteen years old and would not be home from school until 3:30 P.M. He was one of the few who was not pleased with my approach.

In retrospect, I guess the common thread in my initial scientific undertakings was a preoccupation with using chemistry to solve problems and to make things of value. One of my most memorable experiences was convincing Mrs. Semereski, who ran the neighborhood grocery store, to carry my new brand of ink. In those days, we used fountain pens in school, the old type with a rubber bladder inside that you filled by pulling and then releasing a small lever on the side of the pen. I labeled the ink with my logo and priced it at one half that of Waterman's, the leading brand at the time. The kids bought it like "hotcakes." But then disaster struck.

All of my customers' pens clogged. Mrs. Semereski was in an uproar. The customers demanded their money back, and a new pen as well. I immediately looked in my lab notebook only to discover that I had mistakenly added five times the required amount of gum Arabic, an ingredient that enhances the way ink sticks to paper. After collecting the defective pens, cleaning them by boiling in water, and providing a free bottle of properly formulated ink to each customer, I was back in business. I did not make a profit from that episode, but I learned a couple of important things about business: quality control is critical, and responsive customer technical service is a must.

From age ten I had two passions in life—chemistry and music. During my early teenage years, I was absorbed (as were most teenagers at the time) in rock 'n' roll. I devoted all of my waking hours to chemistry and to writing and singing. At age sixteen, I had the good fortune to become a successful recording artist and traveled the country, performing on television, in concert, and in nightclubs.

My life was often put in perspective by my parents, who would say, "The money is great, the experience is wonderful, but you need to decide if you want to be on stage at age fifty doing the same thing." I didn't think much of their counsel at the time, but eventually it sunk in and I had to come to terms with what I wanted to do with my life—which fork in the road should I take?

I chose chemistry, and for the most part I have never looked back, except for the memories. I went on to study chemistry, physics and mathematics at Rutgers–The State University of New Jersey. I developed a strong interest in physical chemistry. Although fortunately awarded a full scholarship, I still needed financial support,

so I continued to work as a musician in New York City throughout undergraduate school.

In 1964, during my senior year, I received several offers of fellowships, including an attractive one at Johns Hopkins University. I turned them down and decided to stay at Rutgers. I was doing too well financially to give up music. This also enabled me to complete my Ph.D. in two and one-half years. However, I do not recommend this approach. It is better to receive a new perspective by attending graduate school at a different institution. After leaving graduate school I went to work for Exxon, where, under the guidance and inspiration of internationally acclaimed catalytic scientist Dr. John H. Sinfelt, I became Director of Research for the Catalysis Research and Development Department. I learned that catalysis could have a profound positive economic and environmental impact on many of the chemical products that are made by industry.

In 1974 I left Exxon with a friend and colleague, Dr. Ricardo B. Levy. Together with our mutual friend, Professor Michel Boudart, a world authority in catalytic science from Stanford University, we formed Catalytica Associates, a worldwide consulting and contract research and development firm that focused on energy and the environmental issues. On these subjects we were somewhat ahead of our time. We built a substantial and profitable business, having worked on more than two hundred projects for one hundred companies in twenty-two countries during our first decade. The experience was incredibly rewarding!

In 1985 we refocused our efforts on building a technology and manufacturing operation, Catalytica, Inc. Our driving force is currently still the same. Through fundamental, practical chemical research on catalytic technologies, we develop new processes to manufacture products that exhibit significant cost benefits and are environmentally benign. We do this by de-

signing catalysts at the molecular level that are highly selective to the desired product and that eliminate the formation of unwanted toxic by-products. This, in our view, is the future—pollution prevention.

We operate in two diverse markets, power generation and pharmaceuticals. In the former area, we have invented a means to catalytically combust fuels with formation of virtually no toxic pollutants—only carbon dioxide and water. By catalytically controlling the chemical kinetics or rates of the key reactions, toxins such as nitrogen oxides, unburned hydrocarbons, and carbon monoxide just do not form! We are working with a major automobile manufacturer to adapt our technology to produce a Zero Emission Vehicle (ZEV) that will be mandated by law in several states beginning in 1998. At this point the only ZEVs are electric cars. Our technology would enable the consumer to use gasoline and the existing conventional fuels distribution and infrastructure.

In the area of pharmaceuticals, we use advanced catalytic technologies to manufacture the "building blocks" for modern drugs that are sold to major pharmaceutical firms that formulate the final therapeutic to fight cancer, Alzheimer's disease, AIDS, cardiovascular disease, depression, schizophrenia, and other disease states.

All of this is a long way from bottled ink in Mrs. Semereski's grocery store, but the basic operating principles of science and business are the same.

As I look over the horizon, it is clear that Planet Earth is faced with significant economic and environmental challenges to provide the four basic needs of our developing society: food, energy, health care, and materials. At Catalytica, we intend to make a difference through catalytic chemistry. As Count Antoine Saint Exuperey, author of my favorite book as a child, *The Little Prince,* once said, "We have not inherited the land of our ancestors, we are borrowing the land of our children."

REACTIONS IN SOLUTION

Life in the largest solution on Earth.

This chapter is an introduction to solutions and the reactions that take place in aqueous solution. Water is usually available, relatively cheap, nontoxic, does not burn, and dissolves many materials. Most of the reactions that you will carry out in general chemistry laboratory take place in aqueous solution. (Because only a few organic compounds are soluble in water, organic solvents such as alcohol are usually used to carry out reactions involving organic compounds.)

You may not think of a brass doorknob as a solution but it is one.

— Why is a study of reactions in solution a part of general chemistry?

For a chemical reaction to take place between two substances, the ions or molecules of the reactants must be close to each other. Molecules must be close enough to each other for the electrons to rearrange around the nuclei of the atoms and form new bonds. Ions must be close enough to each other for the attraction between opposite charges to be important. In solids, particles are fixed in

- FIGURE 4.1   *Left:* A drop of ink immediately after addition to water. *Right:* The ink drop rapidly mixes with the water.

- FIGURE 4.2   *Top:* In the solid state, no evidence of reaction between sodium hydrogen carbonate, $NaHCO_3$, and citric acid, $H_3C_6H_5O_7$, can be observed. *Bottom:* When aqueous solutions of the two compounds are mixed, rapid formation of bubbles of carbon dioxide gas, $CO_2(g)$, shows that a reaction is taking place. The same reaction occurs when an Alka-Seltzer tablet is dissolved in water.

position. When two solids are mixed, only the ions or molecules at the surfaces of the particles are close to each other and can react. In liquids and gases, particles are free to move. When two liquids or gases are mixed, the molecules or ions are mixed and free to move. ▬Figure 4.1 shows two liquids mixing spontaneously.

Chemical reactions can take place throughout the whole volume of a solution. In ▬Figure 4.2, you can see the difference in rate of reaction of two substances in the solid state and in solution.

Gases are hard to handle because you can't see them unless they are colored, and they fill the container and escape easily. In the laboratory, chemists use liquid solutions to carry out reactions whenever possible. Chemical engineers often use liquid solutions in manufacturing plants.

Another reason for studying reactions in solution is that most **biochemical reactions**—that is, most *reactions that take place in plants and animals*—take place in aqueous solution.* Many reactions of interest to geologists also take place in aqueous solution.

— What do you already know about reactions in solution?

Many familiar materials are solutions. Air is a solution of nitrogen, oxygen, argon, water vapor, carbon dioxide, and small amounts of other gases. Seawater is an aqueous solution of salt (and other substances). Brass is a solid solution formed by freezing a melted mixture of copper and zinc (containing 37% or less zinc).

You know something about making solutions from your everyday experience. You know that solids dissolve faster if the solvent is hot. Sweeteners dissolve faster in hot coffee and tea than in iced coffee and tea. Stirring makes sweeteners dissolve more rapidly.

Formation of rings in bathtubs is an everyday example of reaction in solution. The ring is the product of the reaction between soap and calcium and magnesium ions in hard water. The scale that builds up in hot water heaters and pipes is the product of the reaction of calcium and magnesium ions with carbonate ions in hard water.

---

*Only species in aqueous solution can affect your tastebud receptors. If you dry your tongue with a clean cloth and then sprinkle salt or sugar on it, you will not be able to taste the salt or sugar. However, as soon as you add a few drops of water to the crystals on your tongue, you will be able to taste them.

■ FIGURE 4.3 When phenol and menthol, which are both solids, are mixed *(left)*, a liquid solution soon forms *(right)*. The liquid solution is dispensed by pharmacists for the relief of itching.

■ FIGURE 4.4 The upper beaker contains a saturated solution of common table salt, NaCl. Some undissolved salt remains. The solution in the lower beaker is still saturated after the undissolved salt has been removed by filtration.

⬤ Water would probably be considered the solvent in a solution containing equal quantities of water and alcohol because water is so common.

4.1 SOME IMPORTANT DEFINITIONS

Like the composition of a compound, the composition of a given solution is the same no matter where you take a sample. Solutions are homogeneous on a macroscopic scale. However, one important difference between compounds and solutions is that two (or more) substances can be mixed in an infinite number of different proportions to make solutions. For example, alcohol–water solutions containing from the smallest fraction to 99.9% alcohol can be prepared. Only certain proportions of elements are possible for compounds. For example, water is 11.19% hydrogen and 88.81% oxygen. The only other compound of hydrogen and oxygen is hydrogen peroxide, H_2O_2, which is 5.93% hydrogen and 94.07% oxygen.

The other important difference between compounds and solutions is that chemical reactions are needed to decompose compounds into the elements that make them up. For example, decomposition of water to hydrogen gas and oxygen gas is a chemical change. Solutions can be separated into the substances they are composed of by physical means. Salt, NaCl, can be separated from seawater by evaporation of water (see Figure 1.4). The salty taste of seawater shows that the compound NaCl is present.

Liquid solutions can be made from two or more liquids, from solids and liquids, and from gases and liquids. White wine is a solution of ethyl alcohol, which is a liquid, in water. Salt water is a solution of salt in water, and soda water is a solution of carbon dioxide gas in water. Liquid solutions can even occasionally be made from two solids as shown in ■Figure 4.3.

Solutions are composed of solvents and solutes. If a *solution is in the same phase (solid, liquid, or gas) as one of the substances that compose the solution, that substance is usually called* the **solvent.** For example, seawater is a liquid mixture of salt (and other solids) and water. Water is a liquid; therefore, water is the solvent in seawater.

The *other substance in the solution* is called the **solute.** Salt is the solute in seawater. A solution may have more than one solute. Seawater contains magnesium sulfate (and other substances) as well as sodium chloride.

If all the substances in a solution are in the same state when pure, the substance present in greatest amount in the solution is usually called the solvent. For example, in a solution containing water and alcohol, which are both liquids, water is the solvent if the solution contains more water than alcohol. Alcohol is the solvent if the solution contains more alcohol than water.

PRACTICE PROBLEMS

4.1 In a liquid solution containing sugar and water, (a) which substance is the solvent, and (b) which is the solute?

4.2 In a solution containing 70% alcohol and 30% water, (a) which substance is the solvent, and (b) which is the solute?

A **saturated solution** *contains as much solute as will dissolve at a given temperature in the presence of undissolved solute.* The *quantity of solute that will dissolve in a given quantity of solvent or solution* is called the **solubility** of the solute. For example, 35.7 g of sodium chloride will dissolve in 100 mL of water at 0 °C; the solubility of sodium chloride in water at 0 °C is 35.7 g per 100 mL. A solution containing 35.7 g of sodium chloride in 100 mL of water at 0 °C is a saturated solution. If more solute is added to a saturated solution, the additional solute does not dissolve. ■Figure 4.4 shows a saturated solution. The solubilities of some representative substances are given in Table 4.1.

TABLE 4.1	Solubilities of Some Substances[a]	
Substance	Temperature, °C[b]	Solubility, g/100 mL Water
NaCl(s)	0	35.7
NaCl(s)	100	39.12
PbCl$_2$(s)	20	0.99
PbCl$_2$(s)	100	3.34
AgCl(s)	10	0.000 089
AgCl(s)	100	0.0021
KC$_2$H$_3$O$_2$(s)	20	253
KC$_2$H$_3$O$_2$(s)	62	492
CH$_3$CH$_2$OH(l) (ethyl alcohol)	0–100	∞^c
CH$_3$CH$_2$CH$_2$CH$_2$CH$_3$(l) (pentane)	16	0.036
CH$_3$CH$_2$OCH$_2$CH$_3$(l) (diethyl ether)	15	8.43
O$_2$(g)	0	0.0070^d
O$_2$(g)	60	0.0023
CO$_2$(g)	0	0.35
CO$_2$(g)	40	0.097
SO$_2$(g)	0	22.8
SO$_2$(g)	40	5.41

^{a}Most data are from *CRC Handbook of Chemistry and Physics,* 75th ed.; Lide, D. R., Ed. CRC Press: Boca Raton, FL, 1994.
^{b}In this table, all zeros at the end of a number are significant.
^{c}The symbol ∞ means that there is no limit to the solubility of the solute (alcohol) in the solvent (water).
^{d}The solubility of a gas depends on the pressure of the gas. All measurements of gas solubility were made at the same pressure.

(a)

(b)

PRACTICE PROBLEM

4.3   Assume that the behavior of the compounds in Table 4.1 is typical. (a) Are solids more or less soluble at high temperatures than at low temperatures? (b) What happens to the solubility of gases in water as temperature increases?

A *solution that contains less dissolved solute than a saturated solution* is called **unsaturated.** For example, a solution of 8.93 g NaCl in 100 mL of water at 0 °C is unsaturated. If more solute is added to an unsaturated solution, the additional solute dissolves.

With some substances, it is possible to make *a solution that contains more solute than a saturated solution*. A solution like this is called a **supersaturated** solution. A solution containing 35.8 g NaCl in 100 mL of water at 0 °C would be a supersaturated solution. *Supersaturated solutions are not stable.* If more solute is added to a supersaturated solution, solute separates from solution until the solution is only saturated. ▪Figure 4.5 shows the different behaviors of unsaturated, saturated, and supersaturated solutions when more solute is added.

A *solid that separates from a solution* is called a **precipitate.** The solute is precipitating from the supersaturated solution in Figure 4.5(c).

(c)

▪ FIGURE 4.5   (a) When a crystal of solute is added to a solution, the crystal dissolves if the solution is unsaturated. (b) If the solution is saturated, the crystal falls to the bottom unchanged. (c) If the solution is supersaturated, the crystal grows until the solution is only saturated.

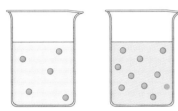

- FIGURE 4.6 Macroscopic *(top)* and microscopic *(bottom)* views of dilute *(left)* and concentrated *(right)* solutions of the same solute. The concentrated solution contains more solute in the same volume of solution. (In the photograph the quantities of solute in each solution are shown in front of the solutions.) What property of the solutions indicates that the concentrations are different?

In a metal, the current consists of moving electrons.

▶ The definitions of the terms *soluble, slightly soluble,* and *insoluble* that are used in this book are given in the first footnote of Table 4.4.

PRACTICE PROBLEMS

4.4   At 25 °C, 0.61 g $HgBr_2$ will dissolve in 100 mL of water. Classify each of the following solutions as unsaturated, saturated, or supersaturated. Assume that all solutions are at 25 °C: (a) a solution containing 0.61 g $HgBr_2$ in 100 mL of water, (b) a solution containing 0.46 g $HgBr_2$ in 100 mL of water, (c) a solution containing 0.62 g $HgBr_2$ in 100 mL of water, and (d) a solution containing 0.61 g $HgBr_2$ in 200 mL of water.

4.5   At 100 °C the solubility of $HgBr_2$ is 4.0 g/100 mL water. Predict what will be observed if a hot solution containing 3.5 g $HgBr_2$/100 mL water is cooled to 25 °C.

The **concentration** of a solution is the *amount of solute dissolved in a given quantity of solvent or solution.* A **dilute solution** *contains only a low concentration of solute.* For example, a solution of a few crystals of sugar in a glass of water is a dilute solution of sugar. A **concentrated solution** *contains a high concentration of solute.* For example, syrups such as Karo are concentrated solutions of corn sugar. *Dilute* and *concentrated* are relative, qualitative terms. ▪Figure 4.6 shows dilute and concentrated solutions of a colored substance.

PRACTICE PROBLEM

4.6   If 50 mL of water at 10 °C contain 0.000 0445 g of dissolved silver chloride, (a) is the solution unsaturated, saturated, or supersaturated, and (b) is the solution dilute or concentrated?

4.2   ELECTROLYTES

Electrolytes are *compounds that conduct electricity when dissolved or melted.* An electric current is a flow of charge. In a solution of an electrolyte or in a molten electrolyte, the current consists of moving ions (see ▪Figure 4.7).

The *ability to conduct electricity* is called **conductivity;** conductivity can be measured quantitatively as well as observed qualitatively. ▪Figure 4.8 shows a simple apparatus for observing conductivity qualitatively.

Strong and Weak Electrolytes

Electrolytes can be classified as strong electrolytes or weak electrolytes. In a solution of a **strong electrolyte,** *most of the solute is in the form of ions.* There are few, if any, molecules of solute in a solution of a strong electrolyte. Sodium chloride dissolved in water is an example of a strong electrolyte. If a strong electrolyte is soluble, its solutions are good conductors of electricity unless they are very dilute. However, saturated solutions of insoluble strong electrolytes do not conduct electricity very well because too little solute is dissolved. Barium sulfate, $BaSO_4$, which has a solubility of only 0.0002 g/100 mL of water at 25 °C, is an example of an insoluble strong electrolyte.

In a solution of a **weak electrolyte,** *most of the solute is in the form of molecules;* only a little of the solute is in the form of ions. Solutions of weak electrolytes do not conduct electricity as well as solutions of strong electrolytes of the same

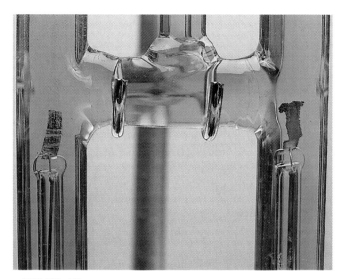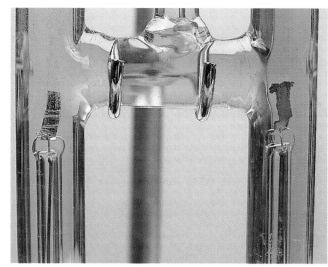

- FIGURE 4.7 *Left:* An aqueous solution of copper(II) nitrate, $Cu(NO_3)_2$, and potassium permanganate, $KMnO_4$. *Right:* When electricity is passed through the solution, blue Cu^{2+} ions move to the left of the apparatus and purple MnO_4^- ions to the right. Movement of ions carries charge through the solution. (Charge is also carried by K^+ and NO_3^- ions, but these ions are colorless and therefore invisible.)

concentration because there are too few ions in solution. Acetic acid and ammonia are examples of weak electrolytes. Water is also a weak electrolyte but it is so weak that its conductivity cannot be detected with the apparatus shown in Figure 4.8.

If *solutions of a compound do not conduct an observable amount of electricity,* the compound is called a **nonelectrolyte.** In a solution of a nonelectrolyte, all of the solute is in the form of molecules; none of the solute is in the form of ions. Alcohol and sugar are examples of nonelectrolytes. Figure 4.8 shows aqueous

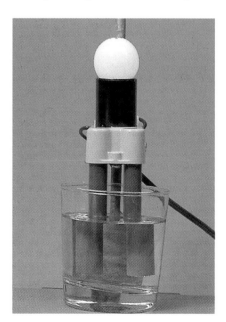

- FIGURE 4.8 Simple conductivity apparatus containing aqueous solutions. The concentrations of the three solutes are the same. *Left:* Solution of strong electrolyte (sodium chloride, NaCl) conducts well and bulb lights brightly. *Middle:* Solution of weak electrolyte (acetic acid, $HC_2H_3O_2$) conducts poorly and bulb lights dimly. *Right:* Solution of nonelectrolyte (alcohol, C_2H_6O) does not conduct and bulb does not light. Does pure water conduct electricity?

solutions of a strong electrolyte, a weak electrolyte, and a nonelectrolyte in the simple conductivity apparatus.

Acids, Bases, and Salts

Three types of compounds are electrolytes: acids, bases, and salts. The simplest definition of an acid is that an **acid** is *a compound that increases the concentration of hydrogen ions when dissolved in water.* Vinegar and lemon juice are familiar examples of solutions of acids. The sour taste of vinegar and of lemon juice is a characteristic property of acids. The names and molecular formulas of some common acids are given in Table 4.2.

The unit particles of acids are molecules. Hydrogen chloride gas, for example, is composed of HCl molecules. Hydrogen ions are formed when an acid dissolves in water.

$$HCl(g) \xrightarrow{H_2O} H^+(aq) + Cl^-(aq)$$

Note that an aqueous solution of hydrogen chloride is called hydrochloric acid. Hydrochloric acid is a strong electrolyte. There are almost no molecules of hydrogen chloride in water solution; almost all of the hydrogen chloride is in the form of hydrogen ions and chloride ions.

Some acids, such as sulfuric acid, H_2SO_4, contain more than one hydrogen that forms hydrogen ions in solution. Some acids contain hydrogens that do not form hydrogen ions in solution. Acetic acid has three hydrogen atoms that do not form hydrogen ions. Hydrogens like this are written separately in the formula, for example, the formula for acetic acid is often written $HC_2H_3O_2$. Hydrogens that form ions in solution are written at the left of a formula.

Arrhenius is given credit for his definition in Chapter 15. This chapter is full without any information that is not absolutely necessary.

○ Never taste anything in a chemistry laboratory. However, you can safely experience the sour taste of acids by tasting a lemon.

| TABLE 4.2 | Some Common Acids and Their Sodium Salts |

Acid		Sodium Salt	
Formula	Name	Formula	Name
$HC_2H_3O_2$	Acetic acid	$NaC_2H_3O_2$	Sodium acetate
$CO_2 + H_2O$	Carbonic acid[a]	$NaHCO_3$	Sodium hydrogen carbonate
		Na_2CO_3	Sodium carbonate
HCl	Hydrogen chloride[b]	NaCl	Sodium chloride
H_2S	Hydrogen sulfide[c]	NaHS	Sodium hydrogen sulfide
		Na_2S	Sodium sulfide
HNO_3	Nitric acid	$NaNO_3$	Sodium nitrate
$HClO_4$	Perchloric acid	$NaClO_4$	Sodium perchlorate
H_3PO_4	Phosphoric acid	NaH_2PO_4	Sodium dihydrogen phosphate
		Na_2HPO_4	Sodium hydrogen phosphate
		Na_3PO_4	Sodium phosphate
H_2SO_4	Sulfuric acid	$NaHSO_4$	Sodium hydrogen sulfate
		Na_2SO_4	Sodium sulfate
$SO_2 + H_2O$	Sulfurous acid[d]	$NaHSO_3$	Sodium hydrogen sulfite
		Na_2SO_3	Sodium sulfite

[a] Carbonic acid decomposes spontaneously to carbon dioxide and water, $H_2CO_3(aq) \rightarrow CO_2(g) + H_2O(l)$.
[b] An aqueous solution of the gas hydrogen chloride is called hydrochloric acid.
[c] An aqueous solution of the smelly gas hydrogen sulfide is called hydrosulfuric acid.
[d] Sulfurous acid H_2SO_3 probably does not exist. However, $SO_2(aq)$ is commonly referred to as sulfurous acid.

PRACTICE PROBLEMS

4.7 How many hydrogens that form hydrogen ions in aqueous solution are shown by each of the following formulas? (a) H_3PO_4 (b) $HC_7H_5O_2$ (c) $H_2C_8H_4O_4$ (d) CH_4O

4.8 An aqueous solution of hydrogen chloride gas is called hydrochloric acid. What is an aqueous solution of hydrogen bromide gas called?

4.9 Name each compound: (a) HNO_3, (b) $Mg(ClO_4)_2$, (c) $KHCO_3$, and (d) H_3PO_4.

4.10 Write the formula for each compound: (a) calcium sulfide, (b) sulfuric acid, (c) carbonic acid, and (d) aluminum sulfate.

The simplest definition of a base is that a **base** is *a compound that increases the concentration of hydroxide ions when dissolved in water.* Sodium hydroxide, NaOH, barium hydroxide, $Ba(OH)_2$, and ammonia, NH_3, are common bases. Concentrated solutions of strong bases feel slippery. The hydroxides of the Group IA and IIA metals such as NaOH and $Ba(OH)_2$ are ionic compounds; the solids are composed of metal ions and hydroxide ions. In the solid state, the ions cannot move, but when the solid is melted or dissolved in water, the ions become free to move and conduct electricity. The ions in an aqueous solution of ammonia are formed by a chemical reaction between water and ammonia:

$$NH_3(g) + H_2O(l) \longrightarrow NH_4^+(aq) + OH^-(aq)$$

The oxides of the Group IA and IIA metals, such as Na_2O and BaO, also react with water to form hydroxide ions and thus are bases:

$$Na_2O(s) + H_2O(l) \longrightarrow 2Na^+(aq) + 2OH^-(aq)$$
$$BaO(s) + H_2O(l) \longrightarrow Ba^{2+}(aq) + 2OH^-(aq)$$

Salts are *compounds of metals (or polyatomic cations such as the ammonium ion, NH_4^+) and nonmetals (including polyatomic anions such as the nitrate ion, NO_3^-).* Sodium chloride, NaCl, ammonium bromide, NH_4Br, and magnesium nitrate, $Mg(NO_3)_2$, are typical examples of salts. *Hydroxides, such as sodium hydroxide, NaOH, and oxides, such as calcium oxide, CaO, are usually classified as bases rather than as salts.* The units of many salts are ions (remember Figure 1.18). In the solid state, the ions are held in their positions by the strong attractions that exist between positive and negative charges. When a salt is melted or dissolved in water, the ions become free to move and carry an electric current. Even salts such as $AlBr_3$, which are molecular compounds, form ions in solution. Almost all salts are strong electrolytes.

⬤ Bottles containing aqueous ammonia are often labeled "Ammonium Hydroxide, NH_4OH." This is probably incorrect. At present, there is no evidence that ammonium hydroxide exists; $NH_3(aq) + H_2O(l)$ is the best "formula" for aqueous ammonia.

PRACTICE PROBLEMS

4.11 Which of the following compounds are bases? (a) KOH (b) NaCl (c) SrO (d) $Ca(NO_3)_2$ (e) HI

4.12 Which of the following compounds are salts? (a) KI (b) CCl_4 (c) $Ca(OH)_2$ (d) NH_4NO_3 (e) Ag_2SO_4 (f) MgO

Salts are formed when acids and bases react. For example, hydrochloric acid and sodium hydroxide react to form the salt, sodium chloride, and water:

$$HCl(aq) + NaOH(aq) \longrightarrow NaCl(aq) + H_2O(l)$$

The water is formed by combination of H^+ and OH^- ions.

$$H^+(aq) + OH^-(aq) \longrightarrow HOH \text{ [usually written } H_2O(l)]$$

The *reaction of an acid with a base to form a salt and water* is called **neutralization.** If the acid and base are mixed in the proportions shown by the equation for the reaction, both the sharp taste of the acid and the slippery feeling of the base disappear. The solution formed has a salty taste because it is an aqueous solution of a salt such as NaCl. Notice that in a salt formed by the neutralization of an acid with a base, the cation comes from the base and the anion from the acid. The names and formulas of the salts formed by the neutralization of some common acids with sodium hydroxide are given in Table 4.2.

Remember, never taste anything in a chemistry laboratory.

The Solubility Rules module of the Solutions section helps students learn to decide whether solutions of two reagents will react to form a precipitate. Then students can play a video selection to get feedback about their predictions.

PRACTICE PROBLEMS

4.13 Write the formulas and names of each of the following: (a) potassium salt of perchloric acid, (b) calcium salt of nitric acid, (c) potassium salts of carbonic acid, (d) calcium salts of carbonic acid, and (e) aluminum salt of hydrochloric acid.

4.14 Write equations for the neutralization of the following: (a) an aqueous solution of sulfuric acid with an aqueous solution of sodium hydroxide to form an aqueous solution of sodium sulfate and (b) an aqueous solution of nitric acid with an aqueous solution of barium hydroxide (both hydroxide ions react).

4.3 REACTIONS BETWEEN IONS IN SOLUTION

Neutralization is an example of a reaction between ions in solution. If you measured the temperatures of the hydrochloric acid and sodium hydroxide solutions before mixing and the temperature of the mixture, you would observe an increase in temperature. From this observation, you could conclude that a reaction had taken place.

The reaction between silver nitrate solution and sodium chloride solution is another typical example of a reaction between ions in solution. If you mixed silver nitrate and sodium chloride solutions, you would observe the formation of a white precipitate (see ▪Figure 4.9) and conclude that a reaction had taken place. The equation for this reaction is

$$AgNO_3(aq) + NaCl(aq) \longrightarrow AgCl(s) + NaNO_3(aq)$$

If you mixed solutions of sodium carbonate and nitric acid, you would observe gas bubbles escaping from the solution (see ▪Figure 4.10) and conclude that a reaction had taken place. The equation for this reaction is

$$Na_2CO_3(aq) + 2HNO_3(aq) \longrightarrow 2NaNO_3(aq) + H_2O(l) + CO_2(g)$$

If you mix solutions of potassium nitrate and sodium chloride, nothing happens. There is no change in temperature, no precipitate, no bubbles of gas, no sign of reaction. In this case, there is no evidence that a reaction has taken place. *On the basis of this experiment,* you would have to conclude that:

$$KNO_3(aq) + NaCl(aq) \longrightarrow \text{no reaction}$$

This conclusion is correct; however, not all reactions show easily observable signs of reaction.

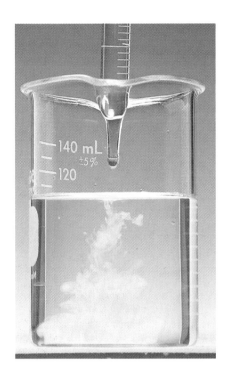

▪ **FIGURE 4.9** When $AgNO_3(aq)$ is added to NaCl(aq), a white precipitate of AgCl(s) forms because silver chloride is insoluble.

The following rule summarizes observations of many different combinations of electrolyte solutions:

◐ Keep in mind that writing an equation suggests that reaction takes place. Do not write an equation when no reaction occurs.

Rule for Predicting Reactions Between Electrolyte Solutions

When solutions of electrolytes are mixed, a reaction will take place if one or both of the possible products is either insoluble or a weak or nonelectrolyte.

Predicting the Products of Reactions Between Electrolyte Solutions

How do you know what the possible products are? Let's consider the reaction between hydrochloric acid and sodium hydroxide solution. Hydrochloric acid contains hydrogen ions, H^+, and chloride ions, Cl^-. Sodium hydroxide solution contains sodium ions, Na^+, and hydroxide ions, OH^-. The mixture of the two solutions contains H^+, Cl^-, Na^+, and OH^- ions. Possible new combinations of ions are H^+ with OH^- to form water, HOH or H_2O, and Na^+ with Cl^- to form NaCl. Ions that have charges of the same sign, such as H^+ and Na^+ or Cl^- and OH^-, repel each other; therefore, there is no need to consider combinations of them. Only combinations of oppositely charged ions, which attract each other, need be considered.

The rule says that if either of the possible products is insoluble or a weak or nonelectrolyte, a reaction will take place. Water is a very weak electrolyte. Therefore, reaction will take place to form water.

To use the rule to predict whether a reaction will take place when two electrolyte solutions are mixed, you need to know which compounds are weak electrolytes and which are insoluble. Tables 4.3 and 4.4 summarize this information. Sample Prob-

TABLE 4.3

Weak and Strong Electrolytes

1. Water is a weak electrolyte.
2. The acids HCl, HBr, HI, $HClO_4$, HNO_3, and H_2SO_4 are strong electrolytes. All other common acids are weak electrolytes.
3. The hydroxides of all IA metals and of Ca, Sr, Ba (and Ra) in Group IIA are strong electrolytes. Ammonia is a weak electrolyte.
4. Salts are strong electrolytes.[a]

[a]$CdSO_4$, $CdCl_2$, $CdBr_2$, CdI_2, $HgCl_2$, and $Pb(C_2H_3O_2)_2$ are exceptions to this rule.

TABLE 4.4 Water Solubility[a] of Common[b] Inorganic Compounds

1. All acids are soluble. Ammonia is soluble.
2. All Na^+, K^+, and NH_4^+ compounds are soluble.
3. All nitrates and acetates are soluble.
4. All chlorides are soluble except AgCl and Hg_2Cl_2; $PbCl_2$ is only slightly soluble. All bromides are soluble except AgBr and Hg_2Br_2; $PbBr_2$ and $HgBr_2$ are only slightly soluble. All iodides are soluble except AgI, Hg_2I_2, PbI_2, and HgI_2.
5. All sulfates are soluble except $PbSO_4$, Hg_2SO_4, $SrSO_4$, and $BaSO_4$; Ag_2SO_4 and $CaSO_4$ are only slightly soluble.
6. All sulfides are insoluble except those of the Group IA and IIA metals and ammonium sulfide.
7. All hydroxides are insoluble except those of the Group IA metals and $Ba(OH)_2$; $Sr(OH)_2$ and $Ca(OH)_2$ are slightly soluble.
8. All carbonates, phosphates, and sulfites are insoluble except those of the Group IA metals and the ammonium ion.
9. Carbon dioxide (CO_2), sulfur dioxide (SO_2), and hydrogen sulfide (H_2S) gases are slightly soluble.

[a]Soluble means that 1 g or more of the compound dissolves in 100 mL of water. Insoluble means that less than 0.1 g of the compound dissolves in 100 mL of water. Slightly soluble means that the solubility of the compound is between 0.1 g and 1 g per 100 mL of water.
[b]Compounds of some elements are common because the elements are abundant, like aluminum and iron. Compounds of other elements are common because the elements are easily obtained, like copper and silver. Compounds of the following cations were considered in writing the solubility rules: Ag^+, Al^{3+}, Ba^{2+}, Ca^{2+}, Cd^{2+}, Co^{2+}, Cr^{3+}, Cu^{2+}, Fe^{2+}, Fe^{3+}, Hg_2^{2+}, Hg^{2+}, K^+, Mg^{2+}, Mn^{2+}, Na^+, NH_4^+, Ni^{2+}, Pb^{2+}, Sn^{2+}, Sr^{2+}, and Zn^{2+}.

■ FIGURE 4.10   When $HNO_3(aq)$ is added to $Na_2CO_3(aq)$, bubbles of $CO_2(g)$ form rapidly.

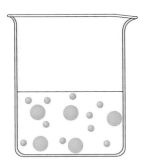

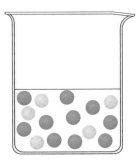

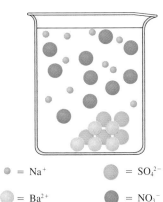

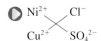 = Na⁺   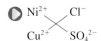 = SO₄²⁻

 = Ba²⁺   = NO₃⁻

When solutions of Na₂SO₄ *(top)* and Ba(NO₃)₂ *(middle)* are mixed *(bottom)*, BaSO₄(s) precipitates; Na⁺ ions and NO₃⁻ ions remain in solution.

A solution containing Ni²⁺, Cu²⁺, Cl⁻, and SO₄²⁻ ions would also be obtained by mixing solutions of nickel(II) sulfate and copper(II) chloride.

lems 4.1 through 4.3 show how to use the information in Tables 4.3 and 4.4 to predict reactions.

SAMPLE PROBLEMS

4.1 Will a reaction take place if solutions of sodium sulfate and barium nitrate are mixed? If a reaction will take place, write its equation.

Solution A solution of sodium sulfate contains Na^+ and SO_4^{2-} ions. A solution of barium nitrate contains Ba^{2+} and NO_3^- ions. First, list the ions that are present in the mixture before any reaction takes place:

$$Na^+ \qquad SO_4^{2-}$$

$$Ba^{2+} \qquad NO_3^-$$

Then consider possible new combinations of ions:

$$Na^+ \quad\diagdown\diagup\quad SO_4^{2-}$$
$$Ba^{2+} \quad\diagup\diagdown\quad NO_3^-$$

Possible products are $NaNO_3$ and $BaSO_4$.

The rule says that if either of the possible products is insoluble or a weak or nonelectrolyte, a reaction will take place. $BaSO_4$ is insoluble; therefore, a reaction will take place. A precipitate of $BaSO_4$ will be formed. Sodium nitrate is soluble and the equation for the reaction that will take place is

$$Na_2SO_4(aq) + Ba(NO_3)_2(aq) \longrightarrow BaSO_4(s) + 2NaNO_3(aq)$$

4.2 Will a reaction take place if solutions of potassium hydroxide and ammonium chloride are mixed? If a reaction will take place, write its equation.

Solution A solution of potassium hydroxide contains K^+ and OH^- ions. A solution of ammonium chloride contains NH_4^+ and Cl^- ions. The ions that are present in the mixture before any reaction takes place are

$$K^+ \qquad OH^-$$

$$NH_4^+ \quad Cl^-$$

Possible new combinations of ions are KCl and NH_4OH or $NH_3 + H_2O$. (Remember that NH_4OH probably does not exist.)

Ammonia and water are weak electrolytes. A reaction will take place, and NH_3 and H_2O will be formed. The equation for the reaction that will take place is

$$KOH(aq) + NH_4Cl(aq) \longrightarrow NH_3(aq) + H_2O(l) + KCl(aq)$$

4.3 Will a reaction take place if solutions of nickel(II) chloride and copper(II) sulfate are mixed? If a reaction will take place, write its equation.

Solution A solution of nickel(II) chloride contains Ni^{2+} and Cl^- ions. A solution of copper(II) sulfate contains Cu^{2+} and SO_4^{2-} ions. The ions that are present in the mixture before any reaction takes place are

$$Ni^{2+} \quad Cl^-$$

$$Cu^{2+} \quad SO_4^{2-}$$

Possible new combinations of ions are $NiSO_4$ and $CuCl_2$. Both these compounds are soluble salts and are strong electrolytes. Therefore, no reaction will take place:

$$NiCl_2(aq) + CuSO_4(aq) \longrightarrow \text{no reaction}$$

4.15   Classify each of the following compounds as strong electrolyte, weak electrolyte, or nonelectrolyte: (a) HF, (b) KOH, (c) $MgCl_2$, (d) C_3H_6O (acetone), and (e) $(NH_4)_2SO_4$.

4.16   Classify each of the following compounds as soluble, slightly soluble, or insoluble: (a) Ag_2SO_4, (b) $CaCO_3$, (c) $CoCl_2$, (d) HNO_3, and (e) PbS.

4.17   Predict whether a reaction will take place when each of the following pairs of aqueous solutions is mixed. If a reaction will take place, write its equation: (a) sodium hydroxide and magnesium chloride, (b) potassium sulfite and hydrochloric acid, (c) ammonium iodide and zinc chloride, (d) sodium acetate and hydrobromic acid, and (e) sodium carbonate and calcium chloride.

4.4   IONIC EQUATIONS

The equations we have written so far *show all compounds as if they exist in solution in the form of molecules* and are called **molecular equations.** For example, the molecular equation for the reaction between sodium sulfate and barium nitrate solutions (Sample Problem 4.1) is

$$Na_2SO_4(aq) + Ba(NO_3)_2(aq) \longrightarrow BaSO_4(s) + 2NaNO_3(aq)$$

However, strong electrolytes are present as ions, not molecules, in aqueous solution. In **ionic equations,** *compounds that exist mostly as ions in solution are shown as ions. Molecular formulas are used for weak electrolytes and insoluble compounds.* For example, the ionic equation for the reaction between sodium sulfate solution and barium nitrate solution is

$$2Na^+(aq) + SO_4^{2-}(aq) + Ba^{2+}(aq) + 2NO_3^-(aq) \longrightarrow$$
$$BaSO_4(s) + 2Na^+(aq) + 2NO_3^-(aq) \quad (4.1)$$

To save time in writing ionic equations, (aq) after formulas for ions is often omitted. Ions are assumed to be in aqueous solution and equation 4.1 is written

$$2Na^+ + SO_4^{2-} + Ba^{2+} + 2NO_3^- \longrightarrow BaSO_4(s) + 2Na^+ + 2NO_3^- \quad (4.2)$$

In equations 4.1 and 4.2, Na^+ ions and NO_3^- ions are present in solution both before and after reaction takes place. Ions like this are called **spectator ions** because they *are present but do not take part in the reaction.*

A **net ionic equation** *shows only the species that take part in the reaction.* The net ionic equation for the reaction in equations 4.1 and 4.2 is

$$Ba^{2+} + SO_4^{2-} \longrightarrow BaSO_4(s) \quad (4.3)$$

Net ionic equations must be balanced both materially and electrically—that is, they must show both the same number of each kind of atom *and* the same net charge on each side. The net charge is zero on both sides of equation 4.3.

Equations, such as 4.1 and 4.2, that *show all the ions that are present including spectator ions* are called **complete ionic equations.** The net ionic equation is obtained from the complete ionic equation by subtracting the spectator ions from both sides. The net ionic equation directs the reader's attention to the change that takes place. More than one reaction can have the same net ionic equation. For example, equation 4.3 is also the net ionic equation for the reaction of a solution of potassium sulfate with a solution of barium chloride. The molecular equation for this reaction is

$$K_2SO_4(aq) + BaCl_2(aq) \longrightarrow BaSO_4(s) + 2KCl(aq)$$

The Balancing Equations module of the Reactions section teaches students a novel method of writing equations for a wide variety of inorganic reactions. It does this by classifying the reactions into 10 fundamental types based on the nature of the reactants.

From here on, (aq) will often be omitted after the formulas for ions when ionic equations for reactions in aqueous solution are written.

Our experience is that insisting that students write (aq) after formulas for ions antagonizes the better students. In addition, it is not usually done in texts for related fields such as biology and geology, so students need to get used to the omission.

and the complete ionic equation is

$$2K^+ + SO_4^{2-} + Ba^{2+} + 2Cl^- \longrightarrow BaSO_4(s) + 2K^+ + 2Cl^-$$

The net charge on each side of a complete ionic equation is always zero. Sample Problem 4.4 shows how to write an ionic equation that involves a weak electrolyte.

You do not get an electric shock when you touch a solution (unless the solution is connected to an electric outlet or other source of electric current).

Sodium phosphate, Na_3PO_4, is used in scouring powders and paint removers.

SAMPLE PROBLEM

4.4 (a) Write the complete ionic equation for the reaction that takes place when aqueous phosphoric acid reacts with an excess of sodium hydroxide solution. The molecular equation is

$$H_3PO_4(aq) + 3NaOH(aq) \longrightarrow Na_3PO_4(aq) + 3H_2O(l)$$

(b) Which ions are spectator ions? (c) Write the net ionic equation. What is the net charge on each side? (d) Write the molecular equation for another reaction that has the same net ionic equation.

Solution (a) In a solution of a weak electrolyte, most of the solute is in the form of molecules. Therefore, molecular formulas should be used for weak electrolytes in ionic equations. Strong electrolytes should be written as ions because, in a solution of a strong electrolyte, most of the solute is in the form of ions. The complete ionic equation for the reaction of phosphoric acid with an excess of sodium hydroxide in aqueous solution is

$$H_3PO_4(aq) + 3Na^+ + 3OH^- \longrightarrow 3Na^+ + PO_4^{3-} + 3H_2O(l)$$

(b) Only Na^+ is a spectator ion. It is the only ion that is present both before and after reaction.

(c) The net ionic equation is

$$H_3PO_4(aq) + 3OH^- \longrightarrow PO_4^{3-} + 3H_2O(l)$$

The net charge on each side is $3-$.

(d) Another reaction with the same net ionic equation is

$$H_3PO_4(aq) + 3KOH(aq) \longrightarrow K_3PO_4(aq) + 3H_2O(l)$$

PRACTICE PROBLEMS

4.18 How should each of the following be represented in a complete ionic equation? (a) solid calcium carbonate (b) aqueous nitric acid (c) aqueous acetic acid (d) aqueous magnesium chloride (e) solid iron(II) hydroxide

4.19 The molecular equation for the reaction between hydrogen sulfide gas and a solution of zinc(II) nitrate in water is $H_2S(g) + Zn(NO_3)_2(aq) \longrightarrow ZnS(s) + 2HNO_3(aq)$. (a) Write the complete ionic equation. (b) Which ions are spectator ions? (c) Write the net ionic equation. (d) Is there another soluble zinc salt that could be used instead of zinc(II) nitrate but that will still give the same net reaction?

4.20 Write complete and net ionic equations for the reactions in Practice Problem 4.17.

Ionic equations are better descriptions of reactions between electrolytes than molecular equations. However, molecular equations are necessary in practical work in the laboratory and industry. You cannot measure out barium ions separately from some kind of negative ions; you cannot measure out sulfate ions separately from some kind of positive ions. Molecular equations are useful when you are deciding what chemicals to obtain from the stockroom to carry out a reaction and are usually used for stoichiometric calculations.

Reactions between ions in solution to form insoluble substances and weak or nonelectrolytes are sometimes called **metathesis** or **double-replacement reactions.** *Metáthesis* is a Greek word that means "changing of position." In molecular equations such as

$$HCl(aq) + NaOH(aq) \longrightarrow HOH(l) + NaCl(aq)$$

and

$$AgNO_3(aq) + NaCl(aq) \longrightarrow AgCl(s) + NaNO_3(aq)$$

 ## 4.5 SINGLE-REPLACEMENT REACTIONS

Metals

Single-replacement reactions are one member of a large and important class of reactions called oxidation–reduction reactions. In **single-replacement reactions,** *one element takes the place of another element in a compound.* Very reactive metals react with water at room temperature. The reactive metal takes the place of hydrogen in water, which can be thought of as hydrogen hydroxide, HOH. Hydrogen gas and the metal hydroxide are formed; the reaction of sodium metal with water is an example. The molecular equation for this reaction is

$$2Na(s) + 2H_2O(l) \longrightarrow 2NaOH(aq) + H_2(g)$$

The complete ionic equation is

$$2Na(s) + 2H_2O(l) \longrightarrow 2Na^+ + 2OH^- + H_2(g)$$

Because there are no spectator ions, the net ionic equation is the same as the complete ionic equation.

■Figure 4.11 shows what happens when samples of calcium, magnesium, and sodium metals are added to room-temperature water. Observation of Figure 4.11 leads to the conclusion that magnesium does not react with water *at room temperature.* Calcium and sodium do react, and sodium reacts more vigorously than calcium. Equations for the reactions that take place are:

$$2Na(s) + 2H_2O(l) \longrightarrow 2NaOH(aq) + H_2(g)$$
$$Ca(s) + 2H_2O(l) \longrightarrow Ca(OH)_2(s) + H_2(g)$$
$$Mg(s) + H_2O(l) \longrightarrow \text{no reaction}$$

These observations lead to the conclusion that the order of reactivity of these metals toward water is

$$Na > Ca > Mg$$

Sodium and magnesium are in the same row of the periodic table. Thus, *reactivity of metals in Groups IA and IIA (and aluminum)* appears to *decrease from left to right across a row in the periodic table.* Observations of other metals support this general rule. For example, potassium reacts faster with water than calcium.

Calcium and magnesium are in the same group in the periodic table. Comparison of their reactivity suggests that *for metals in Groups IA and IIA, reactivity increases from top to bottom of a group.* Again, observations of other metals confirm the general rule. Potassium reacts with water more vigorously than sodium and sodium reacts more vigorously than lithium (remember Table 1.1).

 Combustion reactions are another type of oxidation–reduction reaction. Oxidation–reduction reactions are the subject of Chapter 11.

■ FIGURE 4.11 *Upper left:* When a piece of calcium metal is added to room-temperature water, bubbles of gas are formed showing that a reaction occurs. *Upper right:* When a piece of magnesium metal is placed in room-temperature water, no change is observed indicating that no reaction is taking place. *Bottom:* Sodium metal reacts vigorously with room-temperature water.

- FIGURE 4.12 When pieces of different metals stand in dilute hydrochloric acid at room temperature, tiny bubbles of gas form slowly if the metal is cobalt *(left)*. No change is observed if the metal is copper *(middle)*. Many larger bubbles of gas form rapidly if the metal is iron *(right)*. What other change can you observe in the right-hand test tube?

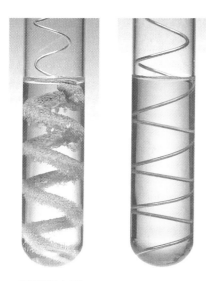

- FIGURE 4.13 *Left:* When copper is placed in a silver nitrate solution, copper replaces silver. The silver crystallizes on the copper and the originally colorless solution becomes blue as copper(II) ions form. *Right:* When silver is placed in a copper(II) nitrate solution, which is blue, nothing happens.

Only a few metals—the Group IA metals and calcium, strontium, and barium from Group IIA—react with water at room temperature. Many metals react with dilute acids. The metals replace the hydrogen of the acid, and hydrogen gas and a salt are formed. Reaction of iron metal with hydrochloric acid is an example. The molecular equation for this reaction is

$$Fe(s) + 2HCl(aq) \longrightarrow FeCl_2(aq) + H_2(g)$$

The complete ionic equation is

$$Fe(s) + 2H^+ + 2Cl^- \longrightarrow Fe^{2+} + 2Cl^- + H_2(g)$$

and the net ionic equation is

$$Fe(s) + 2H^+ \longrightarrow Fe^{2+} + H_2(g)$$

Dilute acid can be used to compare the reactivity of metals that do not react with water. -Figure 4.12 shows what happens when samples of cobalt, copper, and iron are placed in dilute hydrochloric acid at room temperature. From Figure 4.12, you can conclude that copper does not react with dilute hydrochloric acid. Iron reacts moderately fast. Cobalt reacts slowly. Of the elements of the **first transition series**—scandium, titanium, vanadium, chromium, manganese, iron, cobalt, nickel, copper, and zinc—titanium and vanadium are like copper and do not react with dilute acids. Chromium, manganese, and nickel are like iron and react moderately fast with dilute acids. The reactivity of transition metals does not change in a regular way across a row of the periodic table.

Reactivity of metals that do not react with water can also be compared by seeing whether one metal will replace another metal from its salt. -Figure 4.13 shows what happens when copper wire is placed in a solution of silver nitrate and silver wire is placed in a solution of copper nitrate. Observation of Figure 4.13 leads to the conclusion that copper reacts with silver nitrate solution but silver does not react with copper(II) nitrate solution:

$$Cu(s) + 2AgNO_3(aq) \longrightarrow Cu(NO_3)_2(aq) + 2Ag(s)$$
$$Ag(s) + Cu(NO_3)_2(aq) \longrightarrow \text{no reaction}$$

Copper is able to replace silver, but silver does not replace copper. Copper is more reactive than silver. *For transition metals (and the metals that follow the transition metals in the periodic table such as tin and lead), reactivity decreases from top to bottom of a column in the periodic table.* This general rule is also supported by many other observations. For example, most people know that silver is more reactive than gold; silver tarnishes, but gold does not (see -Figure 4.14).

Comparing the reactivity of metals toward oxygen gives the same general rules as are obtained by single-replacement reactions in solution. The **activity series** in

- FIGURE 4.14 *Left:* Silver tarnishes in days or weeks. *Right:* Gold remains bright and shining for hundreds of years. This Greek death mask from about 1525 B.C. was made by hammering a thin plate of metal into a mold from the back.

Table 4.5 is a convenient summary of the results of many experiments. Each metal in Table 4.5 that does not react with water will take the place of any metal below it in salt solutions. For example, zinc is above nickel. If a strip of zinc is placed in a solution of nickel(II) nitrate, a solution of zinc nitrate is formed. Nickel metal is deposited. If a strip of nickel is placed in a solution of zinc nitrate, nothing happens.

PRACTICE PROBLEM

4.21   Predict whether a reaction will take place between the following combinations. If a reaction will take place, write molecular, complete ionic, and net ionic equations. If no reaction will take place, write "no reaction": (a) aluminum metal and water, (b) strontium metal and water, (c) aluminum metal and dilute hydrochloric acid, (d) gold metal and dilute hydrochloric acid, (e) chromium metal and copper(II) nitrate solution, and (f) iron metal and zinc chloride solution.

Nonmetals

Changes in reactivity of nonmetals across rows and down groups in the periodic table can also be investigated by single replacement reactions. ■Figure 4.15 shows what happens when a dilute solution of chlorine in water is added to a dilute solution of sodium sulfide. Sodium chloride is soluble. The precipitate in Figure 4.15 must be sulfur. You can conclude that chlorine takes the place of sulfur:

$$Cl_2(aq) + Na_2S(aq) \longrightarrow S(s) + 2NaCl(aq)$$

Chlorine is more reactive than sulfur. *Except for the noble gases, reactivity of nonmetals increases from left to right across a row of the periodic table.*

■Figure 4.16 shows what happens when a dilute solution of bromine in water is added to sodium chloride solution. It also shows the addition of a dilute solution of chlorine in water to sodium bromide solution. Dilute solutions of bromine in

TABLE 4.5

Activity Series for Metals

K ⎫
Na ⎬ These metals react with
Ca ⎭ water.

Mg ⎫
Al ⎪
Mn ⎪
Zn ⎪
Cr ⎬ These metals react with
Fe ⎪ dilute acids.[a]
Ni ⎪
Sn ⎪
Pb ⎭

Cu ⎫
Ag ⎬ These metals do not react
Au ⎭ with dilute acids.

[a]Mn, Zn, Fe, Ni, Sn, and Pb all give 2+ ions. Cr gives Cr^{3+}.

■ FIGURE 4.15   When chlorine water (greenish yellow solution) is added to aqueous sodium sulfide (colorless solution), sulfur precipitates. A similar reaction is used to purify spring water. Spring water often contains hydrogen sulfide, H_2S, which is very poisonous. The hydrogen sulfide is removed by treating the water with chlorine gas. The sulfur that precipitates is separated by filtration (see Figure 4.17).

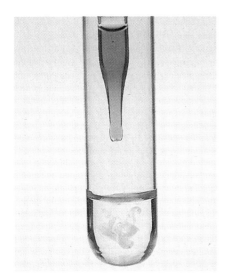
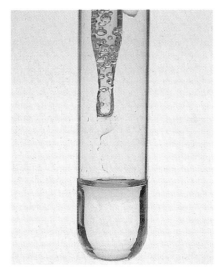

■ FIGURE 4.16   *Left:* When bromine water, which is brownish yellow, is added to aqueous sodium chloride, the color of bromine remains, showing that no reaction takes place. *Right:* When chlorine water is added to aqueous sodium bromide, which is colorless, chlorine replaces bromine as shown by the appearance of the yellow color characteristic of dilute aqueous solutions of bromine. This reaction is used to produce bromine from seawater. Salts of bromine are used in medicine for their sedative properties.

4.5 SINGLE-REPLACEMENT REACTIONS

121

water are yellow. When bromine water is added to sodium chloride solution, the yellow color of the bromine remains. Bromine is *not* used up by reaction with sodium chloride:

$$Br_2(aq) + NaCl(aq) \longrightarrow \text{no reaction}$$

When chlorine water is added to sodium bromide solution, the yellow color of bromine appears. A reaction takes place and chlorine takes the place of bromine:

$$Cl_2(aq) + 2NaBr(aq) \longrightarrow 2NaCl(aq) + Br_2(aq)$$

Chlorine is more reactive than bromine. *The reactivity of nonmetals decreases from top to bottom of a group in the periodic table.* The physical and chemical properties of nonmetals are more varied than those of metals. A single activity series like the one in Table 4.5 is not possible for nonmetals. However, comparing the reactivity of nonmetals toward magnesium gives the same general rules as obtained by single-replacement reactions in solution.

The arrows show the direction of increasing reactivity. More reactive elements replace less reactive elements from compounds.

PRACTICE PROBLEMS

4.22 Write complete ionic and net ionic equations for the reaction of chlorine water with sodium bromide solution.

4.23 Predict whether a reaction will take place if bromine water is added to a solution of potassium iodide. If so, write molecular, complete ionic, and net ionic equations. Iodine is a water-insoluble solid.

Not all single-replacement reactions take place in aqueous solution. The reaction

$$2Mg(l) + TiCl_4(g) \xrightarrow{800\ °C,\ \text{argon atmosphere}} 2MgCl_2(l) + Ti(s)$$

is used to make pure titanium metal.

Sometimes a nonmetal takes the place of a metal in an oxide:

$$3C(s) + 2Fe_2O_3(s) \xrightarrow{\text{heat}} 4Fe(s) + 3CO_2(g) \tag{4.4}$$

Many metal ores are oxides and other ores can be changed to oxides. Single-replacement reactions like the one shown by equation 4.4 are very important in obtaining metals from ores because carbon in the form of coke is cheap and the gaseous by-product, CO_2, is easily separated from the metal. However, recent rapid increases in the quantity of carbon dioxide in Earth's atmosphere are suspected of causing global warming (the greenhouse effect), and the production of carbon dioxide may have to be reduced in the future.

Titanium metal is used in turbine engines and in industrial, chemical, aircraft, and marine equipment. Titanium is less dense than other metals of similar strength and is also unusually resistant to corrosion.

Some texts cover balancing equations for oxidation–reduction reactions here. We find that students do better if they have a little time to get used to writing ionic equations for relatively simple reactions before tackling equations for which the half-reaction method (or change-in-oxidation-number method) is necessary.

4.6   SOME USES OF REACTIONS IN SOLUTION

Dissolving Insoluble Compounds

Reactions between ions are often used to dissolve insoluble compounds. An insoluble compound will dissolve if a reaction between ions to form a weak electrolyte will take place. For example, one of the most important uses of hydrochloric acid is to remove rust from steel. Rust is iron(III) oxide, Fe_2O_3. Hydrogen ions from HCl(aq) react with oxide ions from Fe_2O_3 to form the weak electrolyte, water, and the rust dissolves. Molecular, complete ionic, and net ionic equations for this reaction are as follows:

Rusty iron and iron that has been cleaned with hydrochloric acid. The largest single use of hydrochloric acid is for removing rust from steel.

Molecular:	$Fe_2O_3(s) + 6HCl(aq) \longrightarrow 2FeCl_3(aq) + 3H_2O(l)$
Complete ionic:	$Fe_2O_3(s) + 6H^+ + 6Cl^- \longrightarrow 2Fe^{3+} + 6Cl^- + 3H_2O(l)$
Net ionic:	$Fe_2O_3(s) + 6H^+ \longrightarrow 2Fe^{3+} + 3H_2O(l)$

Both growing plants and manufacturing plants synthesize large quantities of an enormous variety of compounds.

SAMPLE PROBLEM

4.5   (a) Is ZnS(s) soluble in HCl(aq)? (b) Is AgCl(s) soluble in HNO$_3$(aq)? Explain your answers.

Solution   (a) H$^+$ from HCl(aq) will react with S^{2-} from ZnS to form the weak electrolyte H$_2$S(g). Zinc sulfide is soluble in hydrochloric acid.

(b) Both AgNO$_3$ and HCl are soluble, strong electrolytes. No reaction will take place. Silver chloride is not soluble in nitric acid.

PRACTICE PROBLEM

4.24   (a) Is BaSO$_4$ soluble in HCl(aq)? (b) Is BaCO$_3$ soluble in HCl(aq)? Explain your answers.

Synthesis of Inorganic Compounds

Synthesis, the *making of a compound* in a laboratory or manufacturing plant, is an important activity of chemists and chemical engineers. Reactions between ions to form insoluble substances and weak electrolytes and single-replacement reactions can be used to prepare many compounds. Sample Problems 4.6–4.8 show some representative syntheses.

SAMPLE PROBLEMS

4.6   Silver bromide is widely used in photography. Outline a method for synthesizing pure silver bromide.

Solution   Silver bromide is insoluble in water. It can be made by precipitation. The reactants and the other product must all be soluble. One reactant must supply silver ion and the other, bromide ion.

Silver nitrate is a soluble silver salt. Sodium bromide is a soluble bromide. The other product of the reaction between silver nitrate and sodium bromide

$$AgNO_3(aq) + NaBr(aq) \longrightarrow AgBr(s) + NaNO_3(aq)$$

is sodium nitrate, a soluble salt. Pure silver bromide can be made by mixing a solution of silver nitrate with a solution of sodium bromide. Silver nitrate is much more expensive than sodium bromide. An excess of NaBr(aq) should be used so that AgNO$_3$ is the limiting reactant and is all used in the reaction. The precipitate of AgBr(s) can be separated by filtration, washed with water to remove NaNO$_3$(aq) and leftover NaBr(aq), and allowed to dry. (■Figure 4.17 shows how to separate a precipitate by filtration, wash, and air-dry it.) Silver bromide darkens on standing in light. It should be kept in the dark while it is being dried.

4.7   Zinc sulfate is used in the manufacture of rayon. Outline a method for synthesizing pure zinc sulfate.

Solution   Zinc sulfate is soluble. It is harder to make soluble inorganic compounds than insoluble compounds. One way of making zinc sulfate is to dissolve zinc metal in dilute sulfuric acid:

$$Zn(s) + H_2SO_4(aq) \longrightarrow ZnSO_4(aq) + H_2(g)$$

The H$_2$(g) will bubble off into the air as it is formed. Leftover zinc metal can be separated easily from ZnSO$_4$(aq) by filtration. Therefore, an excess of zinc metal should be used. Solid zinc sulfate can be obtained by evaporating the water from

Hydrogen gas forms an explosive mixture with air. Use good ventilation and avoid sparks and flames.

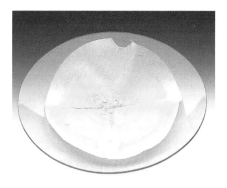

- FIGURE 4.17 *Left:* Separating a precipitate from the supernatant liquid or mother liquor by filtration. *Middle:* Washing a precipitate. *Right:* Air drying a precipitate.

◯ Hydrofluoric acid dissolves glass. This reaction would have to be carried out in a plastic or wax container. Hydrofluoric acid causes painful sores on the skin that are not usually noticed until the next day.

$ZnSO_4(aq)$. Separation of a soluble salt from aqueous solution by evaporation is shown in ▪Figure 4.18. The solid zinc sulfate will be hydrated. The water of hydration can be removed by heating in an oven.

4.8 Sodium fluoride is used as a pesticide and for fluoridation of water. Outline a method for synthesizing pure sodium fluoride. Sodium fluoride is soluble in water.

Solution Dissolving sodium metal in dilute hydrofluoric acid would *not* be a good way to make sodium fluoride. Sodium metal reacts rapidly with water. Its reaction with dilute hydrofluoric acid would probably be dangerously fast.

Neutralization of dilute hydrofluoric acid with dilute sodium hydroxide

$$NaOH(aq) + HF(aq) \longrightarrow NaF(aq) + H_2O(l)$$

would be satisfactory. The other product of this reaction is water that could be removed by evaporation along with the water used to dissolve the reactants. The reactants should be used in the proportion shown by the balanced equation so that both are used up by the reaction. Leftovers would make the product impure.

- FIGURE 4.18 Soluble salts can be separated from solution by evaporation of the solvent. This picture shows copper(II) sulfate pentahydrate, $CuSO_4 \cdot 5H_2O$, starting to crystallize. Why is the evaporating dish shaped the way it is?

PRACTICE PROBLEM

4.25 Tell how to make each of the following compounds. Include a molecular equation for each reaction: (a) $Mg(OH)_2$, (b) $MgCl_2$, and (c) KI.

The reactants for syntheses can, of course, be measured by mass, dissolved in water, and the solutions mixed. However, measurement of the volume of a liquid is much easier than measurement of the mass of a solid. A quantitative way of expressing concentration, especially one that would make possible measurement of moles by measurement of volume, would be very convenient.

4.7 CONCENTRATION EXPRESSED AS PERCENT

In everyday life, concentration is often expressed as percent. For example, the label on a bottle of rubbing alcohol reads "70% isopropyl alcohol." Although the concentration is expressed quantitatively, the meaning of the label is not clear. Sometimes percent composition is mass percent and sometimes volume percent. If

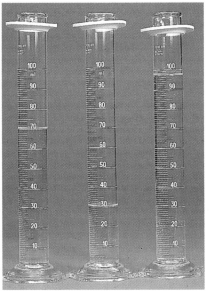
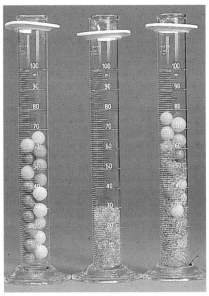

- FIGURE 4.19 *Left:* When isopropyl alcohol (rubbing alcohol) and water are mixed, 70 mL + 30 mL ≠ 100 mL. *Right:* A possible model to explain why 70 mL + 30 mL ≠ 100 mL. What is the total volume of the solution?

the rubbing alcohol is 70% by mass isopropyl alcohol, it is a mixture of isopropyl alcohol and water in the ratio 70 g of isopropyl alcohol to 30 g of water:

$$\text{mass \% } A = \frac{\text{mass component } A}{\text{total mass solution}} \times 100$$

If the rubbing alcohol is 70% by volume isopropyl alcohol, it is made by diluting 70 mL of isopropyl alcohol to a volume of 100 mL with water.

$$\text{volume \% } A = \frac{\text{volume component } A}{\text{total volume solution}} \times 100$$

Diluting 70 mL of isopropyl alcohol to a volume of 100 mL with water does not produce a solution of the same concentration as mixing 70 mL of isopropyl alcohol with 30 mL of water. The total volume of a mixture of 70 mL of isopropyl alcohol and 30 mL of water is *not* 100 mL. It is only 97.4 mL, as shown in -Figure 4.19. How is it possible that the sum of 70 mL and 30 mL is not 100 mL? Adding glass beads to marbles shows one way—see Figure 4.19.

Figuring out how to make solutions containing given volume and mass percents and preparing the solutions are simple.

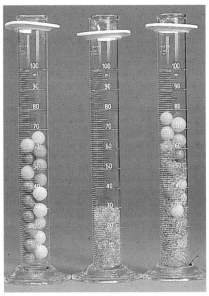 The Units of Concentration module of the Solutions section interactively helps students learn concentration units—including percent, molarity, molality, and mole fraction—either singly or in combination, and provides individualized problems to test their understanding.

Clear labels are very important in chemistry laboratories. Poor labels can be dangerous.

Notice that concentration, like density, is an intensive property that is the ratio of two extensive properties.

SAMPLE PROBLEMS

4.9 Write directions for preparing 100 mL of a 25% by volume solution of ethyl alcohol in water.

Solution Measure 25 mL of ethyl alcohol in a 100-mL graduated cylinder. Add water until the volume of liquid in the graduated cylinder is 100 mL and mix.

4.10 Write directions for preparing 525 g of aqueous solution containing 6.80% by mass NaCl.

Solution Before writing directions, we need to calculate how many grams of NaCl and water to use. The statement that the solution is to contain 6.80% by mass NaCl

A 25% by volume aqueous solution of ethyl alcohol is 50 proof.

means that it must contain 6.80 g NaCl for each 100 g of solution or, in other words,

$$6.80 \text{ g NaCl} \Leftrightarrow 100 \text{ g solution}$$

This relationship can be used to write a conversion factor for converting grams of solution to grams of NaCl:

$$525 \text{ g solution}\left(\frac{6.80 \text{ g NaCl}}{100 \text{ g solution}}\right) = 35.7 \text{ g NaCl}$$

The rest of the solution is water:

$$525 \text{ g solution} - 35.7 \text{ g NaCl} = 489 \text{ g H}_2\text{O}$$

The easiest way to measure a liquid is to measure volume with a graduated cylinder. The density of water at room temperature is 0.998 g/mL:

$$489 \text{ g H}_2\text{O}\left(\frac{1 \text{ mL H}_2\text{O}}{0.998 \text{ g H}_2\text{O}}\right) = 490 \text{ mL H}_2\text{O} \ (4.90 \times 10^2 \text{ mL})$$

To make 525 g of aqueous solution containing 6.80% by mass NaCl, measure out 35.7 g NaCl(s) and dissolve it in 490 mL (4.90×10^2 mL) of water.

PRACTICE PROBLEMS

4.26 Write directions for preparing 500 mL (5.00×10^2 mL) of 12.0% by volume solution of ethyl alcohol in water.

4.27 Write directions for preparing 75 g of aqueous solution containing 8.2% by mass KI.

Concentration Expressed as Parts per Million or Parts per Billion

One inch in a 16-mile trip is one part per million. One part per billion is one inch in a trip of 16 000 miles (about two-thirds of the way around Earth at the equator).

Very low concentrations such as the concentrations of pollutants in air and water are usually expressed in parts per million (ppm) or parts per billion (ppb). One percent is one part per hundred (pph). Just as

$$\% \text{ by volume} = \text{pph (volume)} = \frac{\text{volume of solute}}{\text{volume of solution}} \times 10^2$$

$$\text{ppm (volume)} = \frac{\text{volume of solute}}{\text{volume of solution}} \times 10^6$$

and

$$\text{ppb (volume)} = \frac{\text{volume of solute}}{\text{volume of solution}} \times 10^9$$

● As long as the same units are used for the volume of both solute and solution units cancel and any units can be used.

PRACTICE PROBLEMS

4.28 If a 100.0-mL sample of air contains 3.30×10^{-5} mL N_2O, what is the concentration of N_2O in ppb (volume)?

4.29 Write the definition for concentration in parts per trillion by volume. (A trillion is 10^{12}.)

MOLARITY

Using concentration expressed in percent by volume or percent by mass to solve stoichiometry problems is *not* simple. Therefore, in chemistry, concentration is usually expressed as molarity. **Molarity, M,** is defined as the *number of moles of solute per liter of solution:*

$$\text{molarity, M} = \frac{\text{number of moles of solute}}{\text{volume of solution in liters}}$$

The ACS Style Guide recommends the use of M *rather than M for molarity.*

For example, a solution of salt in water containing 1.000 mol or 58.44 g NaCl in 1.000 L of solution is a 1.000 M solution. Molarity is a much more convenient unit of concentration for solving stoichiometry problems than are mass and volume percent, because the "recipes" for reactions are in moles.

SAMPLE PROBLEM

4.11 What is the molarity of a solution prepared by dissolving 44.86 g NaCl in water and diluting to 250.0 mL?

Solution The problem gives the amount of NaCl in grams and the volume of solution to be prepared in milliliters. To use the definition of molarity to calculate molarity, the amount of NaCl must first be converted to moles and the volume of solution to liters. From the table inside the back cover, the formula mass of NaCl is 58.44 u; 1 mol NaCl has a mass of 58.44 g:

$$44.86 \text{ g NaCl}\left(\frac{1 \text{ mol NaCl}}{58.44 \text{ g NaCl}}\right) = 0.7676 \text{ mol NaCl}$$

$$250.0 \text{ mL soln}\left(\frac{1 \text{ L soln}}{1000 \text{ mL soln}}\right) = 0.2500 \text{ L soln}$$

⬤ Conversion of grams to moles was shown in Sample Problem 2.14.

Substituting mol NaCl and volume solution in liters in the definition of molarity,

$$\text{molarity, M} = \frac{\text{number of moles of solute}}{\text{volume of solution in liters}} = \frac{0.7676 \text{ mol NaCl}}{0.2500 \text{ L soln}}$$

$$= 3.070 \frac{\text{mol NaCl}}{1 \text{ L soln}} = 3.070 \text{ M}$$

As usual, you may combine steps before doing the arithmetic:

$$\left(\frac{44.86 \text{ g NaCl}}{250.0 \text{ mL soln}}\right)\left(\frac{1000 \text{ mL soln}}{1 \text{ L soln}}\right)\left(\frac{1 \text{ mol NaCl}}{58.44 \text{ g NaCl}}\right) = 3.070 \frac{\text{mol NaCl}}{1 \text{ L soln}}$$

$$= 3.070 \text{ M}$$

Check: Almost one mole of NaCl is dissolved in only one-quarter of a liter. The molarity should be somewhat less than 4 M: 3.070 M is somewhat less than 4 M.

PRACTICE PROBLEM

4.30 What is the molarity of a solution prepared by dissolving 57.81 g KI in water and diluting to 500.0 mL?

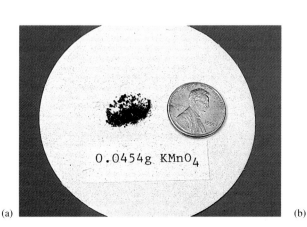
(a)

(b)

0.0454g KMnO₄

(c)

(d)

■ FIGURE 4.20 Steps in the preparation of 500 mL 5.75×10^{-4} M $KMnO_4$ from $KMnO_4(s)$. (a) A 0.0454-g sample of $KMnO_4$ is measured using an analytical balance. (b) The sample is dissolved in less than 500 mL of deionized water. (c) The solution is transferred to a 500-mL volumetric flask. (d) Transfer of the solution from the beaker to the volumetric flask is completed by rinsing beaker, stirring rod, and funnel with deionized water. (e) The solution is mixed by swirling the flask. (f) The solution is allowed to stand until the temperature reaches the temperature marked on the flask. (g) The volume of the solution is adjusted to exactly 500 mL by adding deionized water with a dropper until the bottom of the meniscus is even with the line etched on the neck of the flask. (h) The solution is mixed thoroughly by inverting the flask about 60 times.

Water is a good example of the fact that whether chemicals are safe or not depends on how they are used. Drowning is one of the leading causes of accidental death.

Preparation of Solutions of Known Molarity from Pure Solute

One way of preparing solutions of known molarity is to start with pure solute. In this book, the solvent will be assumed to be water unless another solvent is mentioned. Water is a good solvent for many inorganic compounds and for some organic compounds. In addition, water is readily available, cheap, and nonpoisonous.

The simplest way to solve molarity problems is to realize that the molarity of a solution provides a relationship between number of moles of solute and volume of solution that can be used to write conversion factors between moles of solute and volume of solution. For example, the fact that the solution in Sample Problem 4.11 is 3.070 M in NaCl means that 1 L of the solution contains 3.070 mol NaCl:

1 L soln ⟺ 3.070 mol NaCl

Sample Problem 4.12 shows how to calculate the quantity of solute required to make a solution of known molarity, and ■Figure 4.20 shows how to prepare the solution.

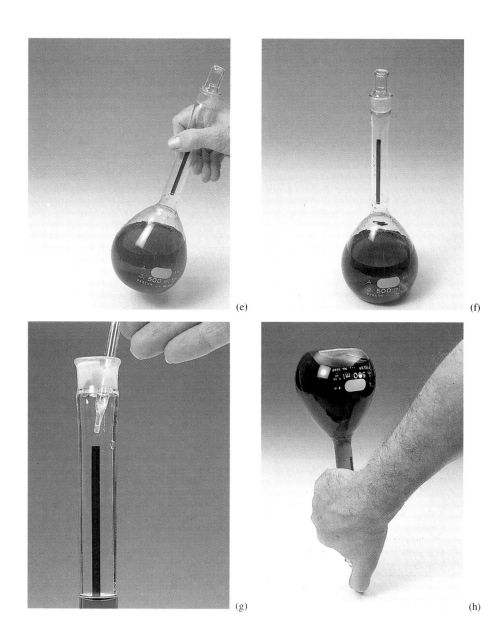

(e)

(f)

(g)

(h)

4.12   How many grams of KMnO₄ (potassium permanganate) are needed to make 500.0 mL of 5.75×10^{-4} M KMnO₄?

Solution   The relationship between moles of solute and volume of solution is

$$1 \text{ L soln} \Leftrightarrow 5.75 \times 10^{-4} \text{ mol KMnO}_4$$

The problem gives the volume and molarity of solution to be prepared and asks for the grams of solute needed. The relationship between moles of solute and volume of solution can be used to write a conversion factor for calculating the number of moles of KMnO₄ required to make 500.0 mL of 5.75×10^{-4} M KMnO₄. (The volume, which is given in milliliters, must be converted to liters.) The formula mass can then be used to write a conversion factor for converting moles of KMnO₄ to grams of KMnO₄.

Potassium permanganate solutions are used as bleaches, deodorizers, and dyes as well as in analysis.

Sample Problem 2.12 shows how to calculate a formula mass.

The answer obtained by stepwise calculation was slightly higher as a result of the accumulation of rounding errors.

Step 1. Calculation of moles of $KMnO_4$ required to make 500.0 mL of 5.75×10^{-4} M $KMnO_4$:

$$500.0 \text{ mL soln}\left(\frac{1 \text{ L soln}}{1000 \text{ mL soln}}\right)\left(\frac{5.75 \times 10^{-4} \text{ mol } KMnO_4}{1 \text{ L soln}}\right)$$

                    Convert   Use molarity to write a
                    volume to L.   conversion factor between
                                          volume in L and mol solute.

$$= 2.88 \times 10^{-4} \text{ mol } KMnO_4$$

Step 2. Conversion of mol solute to mass solute: The formula mass for $KMnO_4$ is 158 u. One mole of $KMnO_4$ has a mass of 158 g.

$$(2.88 \times 10^{-4}) \text{ mol } KMnO_4\left(\frac{158 \text{ g } KMnO_4}{1 \text{ mol } KMnO_4}\right) = 0.0455 \text{ g } KMnO_4$$

Once you have figured out how to solve the problem, the steps in the calculation can, of course, be combined:

$$500.0 \text{ mL soln}\left(\frac{1 \text{ L soln}}{1000 \text{ mL soln}}\right)\left(\frac{5.75 \times 10^{-4} \text{ mol } KMnO_4}{1 \text{ L soln}}\right)\left(\frac{158 \text{ g } KMnO_4}{1 \text{ mol } KMnO_4}\right)$$

$$= 0.0454 \text{ g } KMnO_4$$

Check: The concentration of the solution is much lower than 1 M, and the volume is less than a liter. The mass of $KMnO_4$ should be much lower than the molar mass: $0.0454 \text{ g} \ll 158 \text{ g}$.

PRACTICE PROBLEM

4.31 (a) How many grams of $NaNO_3$ are needed to make 250.0 mL 0.325 M $NaNO_3$? (b) Write directions for preparing 250.0 mL 0.325 M $NaNO_3$ from $NaNO_3(s)$.

Using Solutions of Known Molarity

Once you have a solution of known molarity, you can use it to measure known amounts of solute by measuring volume. Measuring the volume of a liquid is much easier and quicker than measuring mass.

SAMPLE PROBLEM

4.13 What volume of solution is required to obtain 2.35 mol H_2SO_4 from a 6.55 M solution of H_2SO_4?

Solution The relationship between mol H_2SO_4 and volume is

$$1 \text{ L soln} \Leftrightarrow 6.55 \text{ mol } H_2SO_4$$

$$2.35 \text{ mol } H_2SO_4\left(\frac{1 \text{ L soln}}{6.55 \text{ mol } H_2SO_4}\right) = 0.359 \text{ L soln}$$

Check: The quantity 2.35 mol is less than the 6.55 mol of solute that are in a liter of a 6.55 M solution. The volume required should be less than 1 L; 0.359 L is less than 1 L and is a reasonable answer.

These checks may seem unnecessary, but we can't count the times we've had students tell us they were going to use 500 g or more of solute to make 50 mL 10^{-5} M solution.

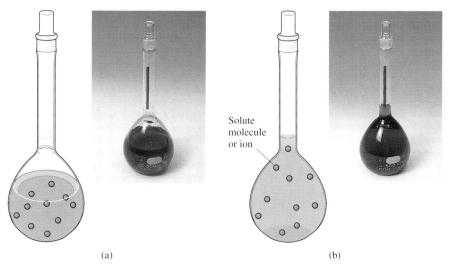

FIGURE 4.21 Microscopic-level view of dilution. When a concentrated solution *(left)* is diluted *(right)*, the solute is spread throughout a larger volume. The quantity of solute does not change.

PRACTICE PROBLEM

4.32 What volume of solution (in milliliters) is required to obtain 0.0750 mol NaOH from a 0.325 M solution of NaOH?

Preparation of Solutions of Known Molarity from More Concentrated Solutions

A second method of preparing solutions of known molarity is to dilute a more concentrated solution. Dilution results in spreading a given quantity of solute throughout a larger solution volume. The quantity of solute does not change. (See ▬Figure 4.21.)

SAMPLE PROBLEM

4.14 Write directions for preparing 200.0 mL 2.25×10^{-5} M $KMnO_4$ from 5.75×10^{-4} M $KMnO_4$.

Solution In this problem, the 2.25×10^{-5} M solution is the more dilute solution; it will be symbolized by "dil soln." The more concentrated solution, the 5.75×10^{-4} M solution, will be symbolized by "conc soln." First, calculate the number of moles of $KMnO_4$ needed to make 200.0 mL of the dilute (2.25×10^{-5} M) solution. Then calculate the volume of concentrated (5.75×10^{-4} M) solution that is required to provide this many moles of $KMnO_4$. Finally, write directions.

Step 1. Calculate the number of moles of solute needed to prepare the dilute solution:

$$200.0 \text{ mL dil soln} \left(\frac{1 \text{ L dil soln}}{1000 \text{ mL dil soln}} \right) \left(\frac{2.25 \times 10^{-5} \text{ mol } KMnO_4}{1 \text{ L dil soln}} \right)$$

$$= 4.50 \times 10^{-6} \text{ mol } KMnO_4$$

The dilution formula is not used because we believe that a logical rather than a plug-the-numbers-in-a-formula approach to problem solving is worth the extra effort required in terms of greater comprehension and better retention.

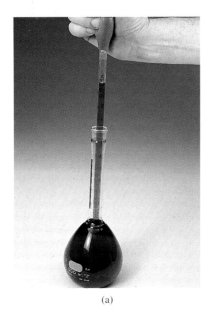

(a)

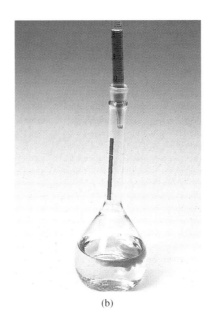

(b)

(c)

▪ FIGURE 4.22   Steps in the preparation of 200 mL 2.25 × 10⁻⁵ M KMnO₄ from 5.75 × 10⁻⁴ M KMnO₄. (a) Using a rubber bulb, draw the concentrated solution into the pipet past the zero mark and allow the solution to run out until the bottom of the meniscus is level with the zero line on the pipet. (b) Allow 7.83 mL of the concentrated solution to run into about 150 mL of deionized water in a 200.0-mL volumetric flask. Stir by swirling and adjust the temperature of the solution if necessary. Then dilute to the mark (c) and mix thoroughly [see Figure 4.20(e)–(h)].

A volumetric pipet should be used if available. Volumetric pipets are commonly made in the following sizes: 0.500, 1.000, 2.000, 3.00, 4.00, 5.00, 10.00, 15.00, 20.00, 25.00, 50.00, and 100.0 mL.

Step 2. Calculate the volume of concentrated solution required to obtain the necessary number of moles:

$$4.50 \times 10^{-6} \text{ mol KMnO}_4 \left(\frac{1 \text{ L conc soln}}{5.75 \times 10^{-4} \text{ mol KMnO}_4} \right) \left(\frac{1000 \text{ mL conc soln}}{1 \text{ L conc soln}} \right)$$

$$= 7.83 \text{ mL conc soln}$$

Combining steps, the setup is

$$200.0 \text{ mL dil soln} \left(\frac{1 \text{ L dil soln}}{1000 \text{ mL dil soln}} \right) \left(\frac{2.25 \times 10^{-5} \text{ mol KMnO}_4}{1 \text{ L dil soln}} \right)$$

$$\left(\frac{1 \text{ L conc soln}}{5.75 \times 10^{-4} \text{ mol KMnO}_4} \right) \left(\frac{1000 \text{ mL conc soln}}{1 \text{ L conc soln}} \right) = 7.83 \text{ mL conc soln}$$

Now we have the information that we need to write directions. To measure 7.83 mL, either a 25-mL buret or a 10-mL graduated pipet should be used. Most of the steps in making a solution of known concentration by diluting a concentrated solution are the same as the steps in making a solution from pure solute. The ones that are different are shown in ▪Figure 4.22.

PRACTICE PROBLEM

4.33 Write directions for preparing 1.000 L 0.0500 M NaOH from 0.325 M NaOH.

Sometimes the concentration of the concentrated solution is given in mass percent instead of in molarity (see ▪Figure 4.23). The concentration must be converted to molarity before the solution can be used to prepare a more dilute solution of known molarity. To convert concentration in mass percent to concentration in molarity, the density of the solution must be known.

4.15 The label on a bottle of concentrated ammonia says that the contents are 28.0% (mass) NH_3 and have a density of 0.898 g/mL. What is the molarity of the concentrated ammonia?

Solution By definition, the molarity of a solution is the number of moles of solute in a liter of solution. We need to figure out how many moles of ammonia there are in 1.00 L of the concentrated ammonia. We can use density to find the mass of 1.00 L of concentrated ammonia. Then we can use the definition of percent by mass to find out how many grams of ammonia there are in 1.00 L. Finally, we can convert grams of ammonia to moles of ammonia.

Use density to find the mass of 1.00 L of concentrated ammonia:

$$0.898 \frac{\text{g soln}}{\text{mL soln}} \left(\frac{1000 \text{ mL soln}}{1 \text{ L soln}} \right) = 898 \frac{\text{g soln}}{1.00 \text{ L soln}}$$

But the solution only contains 28.0% (by mass) NH_3:

$$898 \text{ g soln} \left(\frac{28.0 \text{ g } NH_3}{100 \text{ g soln}} \right) = 251 \text{ g } NH_3$$

Finally, convert grams NH_3 to moles NH_3. The formula mass of NH_3 is 17.0 u:

$$251 \text{ g } NH_3 \left(\frac{1 \text{ mol } NH_3}{17.0 \text{ g } NH_3} \right) = 14.8 \text{ mol } NH_3$$

Because 14.8 mol NH_3 is the amount of NH_3 in 1 L, the molarity of the concentrated ammonia is 14.8. Combining steps into a single series of conversion factors gives

$$0.898 \frac{\text{g soln}}{\text{mL soln}} \left(\frac{1000 \text{ mL soln}}{1 \text{ L soln}} \right) \left(\frac{28.0 \text{ g } NH_3}{100 \text{ g soln}} \right) \left(\frac{1 \text{ mol } NH_3}{17.0 \text{ g } NH_3} \right)$$

$$= 14.8 \frac{\text{mol } NH_3}{1 \text{ L soln}} = 14.8 \text{ M}$$

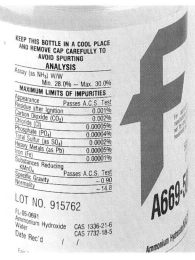

■ FIGURE 4.23 Concentrated ammonia is usually labeled "ammonium hydroxide." It is 28.0–30.0% NH_3 by mass and has a density of 0.898–0.892 g/mL.

4.34 The label on a bottle of concentrated hydrochloric acid, HCl(aq), says that the contents are 36.0% HCl by mass and have a density of 1.179 g/mL. What is the molarity of HCl in the concentrated hydrochloric acid?

When pure liquids like isopropyl alcohol and water are mixed, the total volume of the solution is not equal to the sum of the volumes mixed (remember Figure 4.19). However, dilute solutions are mostly solvent. When 1 M or less concentrated solutions in the same solvent are mixed, the total volume is usually, within our ability to measure, equal to the sum of the volumes of the individual solutions. Sample Problem 4.16 shows a calculation of the molarity of a solution made by mixing two dilute solutions.

4.16 What is the molarity of the solution formed by mixing 25.0 mL 0.375 M NaCl with 42.0 mL 0.632 M NaCl?

Nature's Solutions

Earth's oceans, lakes, and rivers are one of our planet's most striking features. Earth's neighbors in space—the moon, Mars, and Venus—do not have water on their surfaces.

The oceans, which cover about 70% of Earth's surface and contain more than 1.4×10^{18} kg of water, are all connected to each other and contain the same proportions of the major solutes. The concentrations of Na^+ and Cl^- are about ten times as high as the concentrations of Mg^{2+} and SO_4^{2-}. Seawater is a moderately concentrated aqueous electrolyte solution. It is about 0.5 M in Na^+ and Cl^- and 0.05 M in Mg^{2+} and SO_4^{2-} with a trace of just about every other solute imaginable. Although the proportions of the major solutes are constant, the concentrations at different locations may vary as a result of dilution by rain, runoff, and melting ice.

Variations, both in the proportions and in the concentrations of minor solutes, such as HCO_3^-, NO_3^-, $H_2PO_4^-$, HPO_4^{2-}, and silicon-containing anions, are extremely large. (Silicon-containing anions come from the action, both mechanical and chemical, of water on rock.) These ions serve as food for plants and algae growing near the surface. The phosphorus-containing ions are usually the limiting reactant.

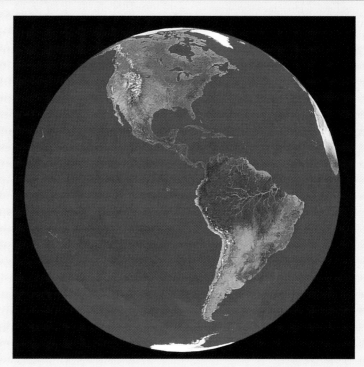

The oceans cover about 70% of Earth's surface.

Most of the reactions that take place in seawater occur at or near boundaries between phases. Many occur near the interface between water and air. Others take place on the bottom and some on the surfaces of small particles suspended in the water.

Although the oceans have a considerable natural capacity for disposing of pollutants, in recent years the rate at which materials have been dumped into the oceans has come to exceed this capacity. Ocean dumping both pollutes the seas and dilutes resources so that they are lost to

Solution To calculate the molarity of the mixture, we need to know how many moles of sodium chloride it contains and what its volume is. The number of moles of sodium chloride will be equal to the sum of the moles of sodium chloride in each of the original solutions. Because the solutions being mixed are less than 1 M, the volume of the mixture will be equal to the sum of the volumes of the original solutions:

$$25.0 \text{ mL soln} \left(\frac{1 \text{ L soln}}{1000 \text{ mL soln}} \right) \left(\frac{0.375 \text{ mol}}{1 \text{ L soln}} \right)$$

$$= 9.38 \times 10^{-3} \text{ mol NaCl in 25.0 mL 0.375 M NaCl}$$

$$42.0 \text{ mL soln} \left(\frac{1 \text{ L soln}}{1000 \text{ mL soln}} \right) \left(\frac{0.632 \text{ mol}}{1 \text{ L soln}} \right)$$

$$= 2.65 \times 10^{-2} \text{ mol NaCl in 42.0 mL 0.632 M NaCl}$$

future use. In addition, wastes have usually been dumped close to shore where most fisheries are located, thus reducing our food supply.

Freshwater lakes, rivers, ponds, and streams are also solutions of electrolytes, although much more dilute. The total concentration of salts in fresh water is only about 1/2000 the concentration of salts in seawater. The major solute ions in fresh water are Ca^{2+}, HCO_3^-, SO_4^{2-}, and silicon-containing anions.

Fresh water that contains Ca^{2+} and HCO_3^- is called temporary hard water because it can be softened by boiling. The reaction that takes place is

$$Ca^{2+} + 2HCO_3^- \xrightarrow{\text{heat}}$$
$$CaCO_3(s) + CO_2(g) + H_2O(l) \quad (4.5)$$

Fresh water that contains Ca^{2+} and SO_4^{2-} is called permanent hard water because it cannot be softened by boiling.

Hardness is a common and expensive nuisance. The reaction shown in equation 4.5 is responsible for the scale that deposits in boilers, hot-water heaters, and tea kettles. The $CaCO_3$ deposit is a poor conductor of heat and hinders the transfer of heat to water. In addition, it clogs pipes and contributes to corrosion problems.

The reaction of equation 4.5 takes place slowly when groundwater that contains calcium ions and hydrogen carbonate ions seeps through the ceilings of caves. Beautiful stalactites and stalagmites are formed. (The colors are caused by the presence of ions such as Fe^{3+}.)

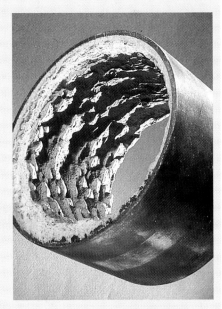

When the reaction of equation 4.5 takes place inside a pipe, scale deposits and eventually clogs the pipe completely.

Hard water is also a problem in the bathroom because calcium ion forms a precipitate with soap—that is, soap forms a bathtub ring instead of a lather. The net ionic equation for this reaction is

$$Ca^{2+} + 2C_{17}H_{35}COO^- \longrightarrow$$
$$Ca(C_{17}H_{35}COO)_2(s)$$

There is a total of
$$(9.38 \times 10^{-3} \text{ mol NaCl}) + (2.65 \times 10^{-2} \text{ mol NaCl}) = 3.59 \times 10^{-2} \text{ mol NaCl}$$
in the new solution. The volume of the new solution is
$$25.0 \text{ mL} + 42.0 \text{ mL} = 67.0 \text{ mL}$$
Therefore, the molarity of the new solution is
$$\frac{3.59 \times 10^{-2} \text{ mol NaCl}}{67.0 \text{ mL}\left(\frac{1 \text{ L}}{1000 \text{ mL}}\right)} = 0.536 \frac{\text{mol NaCl}}{1 \text{ L}} = 0.536 \text{ M}$$

Check: This answer is reasonable because it is greater than 0.375 M and less than 0.632 M. It is closer to 0.632 M, as would be expected because more of the 0.632 M solution was used to make the mixture.

4.35   What is the molarity of the solution formed by mixing 247 mL 0.325 M NaOH with 538 mL 0.249 M NaOH?

4.9   STOICHIOMETRY OF REACTIONS IN SOLUTION

Using concentration in molarity simplifies solving stoichiometry problems involving solutions. Measuring volumes of solutions of known molarity is an easy way of counting moles.

SAMPLE PROBLEMS

4.17   How many milliliters of 0.495 M NaOH are needed to neutralize 63.5 mL 0.368 M HCl? The equation for the reaction is

$$HCl(aq) + NaOH(aq) \longrightarrow NaCl(aq) + H_2O(l)$$

Solution   The equation for the reaction gives the relation between moles of HCl and moles of NaOH reacting; in this reaction, 1 mol HCl $\Leftrightarrow$ 1 mol NaOH. The moles of HCl present before reaction can be calculated from the volume and molarity of HCl. Because 1 mol HCl $\Leftrightarrow$ 1 mol NaOH, the same number of moles of NaOH are needed. The volume of NaOH needed can be calculated from the number of moles needed and the molarity:

$$63.5 \text{ mL HCl soln} \left(\frac{1 \text{ L HCl soln}}{1000 \text{ mL HCl soln}} \right) \left(\frac{0.368 \text{ mol HCl}}{1 \text{ L HCl soln}} \right)$$
$$= 2.34 \times 10^{-2} \text{ mol HCl used}$$

$$2.34 \times 10^{-2} \text{ mol HCl} \left(\frac{1 \text{ mol NaOH}}{1 \text{ mol HCl}} \right) = 2.34 \times 10^{-2} \text{ mol NaOH needed}$$

$$2.34 \times 10^{-2} \text{ mol NaOH} \left(\frac{1 \text{ L NaOH soln}}{0.495 \text{ mol NaOH}} \right) \left(\frac{1000 \text{ mL NaOH soln}}{1 \text{ L NaOH soln}} \right)$$
$$= 47.3 \text{ mL NaOH soln}$$

❍ The answer obtained by combining steps to avoid accumulation of rounding errors is 47.2 mL NaOH soln.

Check: This answer is reasonable. The concentration of the NaOH solution is about 4/3 the concentration of the HCl solution. One mole of NaOH reacts with 1 mol HCl. Therefore, 3/4 as much NaOH solution will be needed.

4.18   If 21.4 g of solid zinc are treated with 3.13 L 0.200 M HCl, how many grams of hydrogen gas will theoretically be formed? How much of which reactant will be left after reaction is over? The equation for the reaction is

$$Zn(s) + 2HCl(aq) \longrightarrow H_2(g) + ZnCl_2(aq)$$

Solution   The quantities of two reactants are given. This is a limiting-reactant problem. The number of moles of HCl used can be calculated from the volume and the molarity:

$$3.13 \text{ L soln} \left(\frac{0.200 \text{ mol HCl}}{1 \text{ L soln}} \right) = 0.626 \text{ mol HCl}$$

The table of information for this problem is

Problem	21.4 g Zn	0.626 mol HCl	? g
Equation	Zn(s) +	2HCl(aq)	$\longrightarrow$ H$_2$(g) + ZnCl$_2$(aq)
Formula mass, u	65.4		2.02
Recipe, mol	1	2	1

Because the "recipe" for the reaction is in moles, the quantity of zinc must be converted to moles:

$$21.4 \text{ g Zn}\left(\frac{1 \text{ mol Zn}}{65.4 \text{ g Zn}}\right) = 0.327 \text{ mol Zn}$$

The fractions of a recipe of the reactants are

$$\frac{0.327 \text{ mol Zn}}{1 \text{ mol Zn}} = 0.327 \qquad \text{and} \qquad \frac{0.626 \text{ mol HCl}}{2 \text{ mol HCl}} = 0.313$$

Because $0.313 < 0.327$, HCl is limiting. The amount of HCl must be used to calculate the quantity of $H_2(g)$ that will be formed and the amount of $Zn(s)$ that will react. The quantity of $H_2(g)$ that will be formed is

$$0.626 \text{ mol HCl}\left(\frac{1 \text{ mol } H_2}{2 \text{ mol HCl}}\right)\left(\frac{2.02 \text{ g } H_2}{1 \text{ mol } H_2}\right) = 0.632 \text{ g } H_2$$

The amount of $Zn(s)$ that will react is

$$0.626 \text{ mol HCl}\left(\frac{1 \text{ mol Zn}}{2 \text{ mol HCl}}\right)\left(\frac{65.4 \text{ g Zn}}{1 \text{ mol Zn}}\right) = 20.5 \text{ g Zn}$$

The amount of zinc that will be left over is

$$21.4 \text{ g Zn present} - 20.5 \text{ g Zn reacted} = 0.9 \text{ g Zn left over}$$

PRACTICE PROBLEMS

4.36   Make a flow diagram similar to the one on page 85 showing the steps used in solving Sample Problem 4.17.

4.37   How many milliliters of 0.236 M KOH are needed to react with 24.96 mL of 0.1254 M H_2SO_4? The equation for the reaction is

$$H_2SO_4(aq) + 2KOH(aq) \longrightarrow K_2SO_4(aq) + 2H_2O(l)$$

4.38   If 35.4 g of aluminum are treated with 721 mL of 5.86 M HCl, how many grams of hydrogen gas will theoretically be formed? How much of which reactant will be left after reaction is over? The equation for the reaction is

$$2Al(s) + 6HCl(aq) \longrightarrow 3H_2(g) + 2AlCl_3(aq)$$

4.10   TITRATION

Reactions in solution are used for analysis as well as synthesis. **Titration** is an important *method for determining the amount of a substance present in solution. A solution of known concentration,* called a **standard solution,** is added from a buret to the solution being analyzed. Addition is stopped when the quantity of reactant called for by the equation for the reaction has been added. This point is signaled by a color change or change in an electrical property such as conductivity. It is called the **end point** of the titration. *A substance whose change in color shows that the end point has been reached* is called an **indicator.** For example, the concentration of acetic acid in vinegar can be determined by titration with standard sodium hydroxide solution. The equation for the reaction is

$$HC_2H_3O_2(aq) + NaOH(aq) \longrightarrow NaC_2H_3O_2(aq) + H_2O(l) \qquad (4.6)$$

A sample of vinegar is measured with a balance or a pipet and dissolved in water.

Equivalence point will be defined in Chapter 15, when students can understand the difference between end point and equivalence point.

A few drops of phenolphthalein indicator are added. Phenolphthalein is an organic compound that is colorless in acidic solutions and pink in basic solutions. Because vinegar is acidic, the solution is colorless at the beginning of the titration. It remains colorless until all the acetic acid has been neutralized. Then the next drop of sodium hydroxide solution makes the solution basic. The phenolphthalein turns pink, and addition of sodium hydroxide is stopped. ■Figure 4.24 shows the titration of a sample of vinegar with sodium hydroxide solution.

From the volume of sodium hydroxide solution added and its concentration, the quantity of acetic acid in the sample can be calculated. Sample Problem 4.19 shows how to calculate the concentration of acetic acid in vinegar from the results of titration. Sample Problem 4.19 also illustrates the combination of skills from several chapters.

SAMPLE PROBLEM

4.19 A 15.00-mL sample of vinegar was titrated with 0.500 M NaOH and 22.59 mL 0.500 M NaOH were required to reach the end point. The density of the vinegar was 1.005 g/mL. (a) What was the molarity of acetic acid in the vinegar? (b) How many grams of acetic acid did the 15.00-mL sample of vinegar contain? (c) What was the mass % acetic acid in the vinegar?

Solution (a) First calculate the number of moles of NaOH required to neutralize the acetic acid in the sample of vinegar from the volume used and the molarity. From the balanced equation, figure out how many moles of acetic acid were in the vinegar sample. Then use the definition of molarity to calculate the molarity of the vinegar:

$$22.59 \text{ mL NaOH soln} \left(\frac{1 \text{ L NaOH soln}}{1000 \text{ mL NaOH soln}} \right) \left(\frac{0.500 \text{ mol NaOH}}{1 \text{ L NaOH soln}} \right)$$

$$= 0.0113 \text{ mol NaOH}$$

(a)

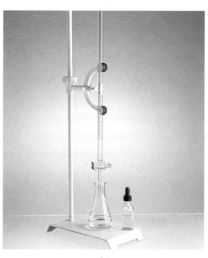
(b)

■ FIGURE 4.24 Titration of a sample of vinegar with standard sodium hydroxide solution. (a) A measured volume of vinegar is added to deionized water in an Erlenmeyer flask. (b) The buret is filled with the standard solution. The surface under the sample should be white so that the color change that signals the end point of the titration will be easily visible. (c) Just before the end point, the solution is still colorless. (d) Addition of one more drop of standard solution produces a definite color change. The end point has been reached.

(c)

(d)

According to equation 4.6, 1 mol NaOH neutralizes 1 mol of acetic acid:

$$0.0113 \text{ mol NaOH} \left(\frac{1 \text{ mol HC}_2\text{H}_3\text{O}_2}{1 \text{ mol NaOH}} \right) = 0.0113 \text{ mol HC}_2\text{H}_3\text{O}_2$$

There was 0.0113 mol $HC_2H_3O_2$ in the vinegar sample. The molarity of acetic acid in the vinegar was

$$M = \frac{0.0113 \text{ mol HC}_2\text{H}_3\text{O}_2}{0.015\ 00 \text{ L HC}_2\text{H}_3\text{O}_2 \text{ soln}} = 0.753 \frac{\text{mol HC}_2\text{H}_3\text{O}_2}{\text{L HC}_2\text{H}_3\text{O}_2 \text{ soln}} = 0.753 \text{ M}$$

(b) This part of the question asks how many grams of acetic acid were in the sample and is similar to Sample Problem 2.15. From the table in the back of the book, the formula mass of acetic acid is 60.1 u. The mass of 1 mol $HC_2H_3O_2$ is 60.1 g:

$$0.0113 \text{ mol HC}_2\text{H}_3\text{O}_2 \left(\frac{60.1 \text{ g HC}_2\text{H}_3\text{O}_2}{1 \text{ mol HC}_2\text{H}_3\text{O}_2} \right) = 0.679 \text{ g HC}_2\text{H}_3\text{O}_2$$

(c) This part of the question asks for the mass % acetic acid in the vinegar. From part (b), we know the mass of acetic acid in the sample. From the volume of the sample and the density, we can calculate the mass of the sample:

$$15.00 \text{ mL} \left(\frac{1.005 \text{ g}}{1 \text{ mL}} \right) = 15.08 \text{ g}$$

This calculation is similar to Practice Problem 2.25.

Now we can calculate the mass % from the definition for mass %:

$$\text{mass \%} = \frac{\text{mass solute}}{\text{total mass solution}} \times 100$$

$$\text{mass \% HC}_2\text{H}_3\text{O}_2 = \frac{0.679 \text{ g HC}_2\text{H}_3\text{O}_2}{15.08 \text{ g sample}} \times 100 = 4.50 \text{ mass \% HC}_2\text{H}_3\text{O}_2$$

PRACTICE PROBLEM

4.39 A 3.00-mL sample of sulfuric acid was dissolved in water and titrated with 0.250 M NaOH. The equation for the neutralization of sulfuric acid by sodium hydroxide is

$$H_2SO_4(aq) + 2NaOH(aq) \longrightarrow Na_2SO_4(aq) + 2H_2O(l)$$

To reach the end point, 24.62 mL 0.250 M NaOH were required. The density of the sulfuric acid was 1.0626 g/mL. (a) What was the molarity of the sulfuric acid? (b) How many grams of sulfuric acid did the 3.00-mL sample contain? (c) What was the mass % sulfuric acid in the sample?

Reactions other than neutralization can be used in titrations. For example, the concentration of chloride ion in water can be determined by titration with standard silver nitrate solution, using potassium chromate as indicator. Silver chloride is white and is much less soluble than silver chromate, which is red. When standard silver nitrate is added to a chloride solution that contains a little chromate ion, all the chloride ion is precipitated before silver chromate begins to precipitate. Formation of a red precipitate of silver chromate signals the end point of the titration:

$$Cl^- + Ag^+ \longrightarrow AgCl(s)$$
white

and then

$$2Ag^+ + CrO_4^{2-} \longrightarrow Ag_2CrO_4(s)$$
red

The reaction that takes place in a titration must be both quantitative and rapid as neutralization and the precipitation of silver chloride both are.

SUMMARY

Solutions are homogeneous mixtures. If a solution is in the same state as one of the materials that make up the mixture, that material is usually called the **solvent.** If all the substances in a solution are in the same state when pure, the substance present in greatest amount in the solution is called the solvent. The other substances are called **solutes.** Chemists use solutions to carry out reactions because they are easy to handle and the solute particles can move around and hit each other so that reaction can take place at a practical rate.

A **saturated solution** contains as much solute as will dissolve in the presence of undissolved solute at a given temperature; the quantity of solute that will dissolve is called the **solubility** of the solute. **Unsaturated solutions** contain less dissolved solute than a saturated solution. **Supersaturated solutions** contain more solute than a saturated solution and are unstable. A solid that separates from a solution is called a **precipitate.**

Electrolytes are compounds that conduct an electric current when dissolved or melted. The ability to conduct electricity is called **conductivity.** In a solution of an electrolyte or in a molten electrolyte, the current consists of moving ions. In solutions of **strong electrolytes,** most of the solute is in the form of ions; there are few molecules of solute. In solutions of **weak electrolytes,** most of the solute is in the form of molecules; only a little is in the form of ions. Solutions of **nonelectrolytes** do not contain any ions; all of the solute is in the form of molecules. Acids, bases, and salts are electrolytes. **Acids** increase the concentration of hydrogen ions when dissolved in water. **Bases** increase the concentration of hydroxide ions when dissolved in water. Acids and bases **neutralize** each other and form water and **salts,** which are compounds composed of metals (or polyatomic cations such as the ammonium ion, NH_4^+) and nonmetals (including polyatomic anions such as the nitrate ion, NO_3^-). Metal hydroxides and oxides are classified as bases, not salts.

Reactions take place between ions in solution whenever at least one of the possible products is either insoluble or a weak or nonelectrolyte. Reactions in solution can be represented by molecular, complete ionic, and net ionic equations. **Molecular equations** show all substances as if they exist in solution as molecules. **Ionic equations** show compounds that exist as ions in solution as ions. Molecular formulas are used for weak electrolytes and insoluble compounds. **Net ionic equations** show only the species that take part in the reaction. Ions that are present but do not take part in the reaction are called **spectator ions.** Ionic equations must show the same number of atoms of each kind and the same charge on both sides.

In a **single-replacement reaction,** one element takes the place of another element in a compound. There is a change in properties from metallic to nonmetallic across rows in the periodic table. The reactivity of metals in Groups IA and IIA (and aluminum) decreases across a row and increases going down a group. The reactivity of transition metals and nonmetals decreases down a group.

Reactions in solutions can be used for dissolving insoluble compounds, for making compounds **(synthesis),** and for qualitative and quantitative analysis. **Titration** is a method of determining the amount of substance present in a solution by adding a solution of known concentration, called a **standard solution,** from a buret until the **end point** is reached. **Indicators** are used to signal end points.

The **concentration** of a solution is the amount of solute dissolved in a given quantity of solvent or solution. **Dilute solutions** have relatively low concentrations of solute; **concentrated solutions** have relatively high concentrations of solute. In everyday life, concentration is often expressed as mass percent or volume percent. Molarity is the unit most used in laboratory work because it provides a conversion factor between volume of solution and moles of solute. **Molarity, M,** is the number of moles of solute per liter of solution:

$$\text{molarity, M} = \frac{\text{number of moles of solute}}{\text{volume of solution in liters}}$$

Solutions of known molarity can be prepared either from pure solute or by diluting a concentrated solution of known molarity. Once a solution of known molarity is available, it can be used to measure known amounts of solute simply by measuring volume. When pure liquids or concentrated solutions are mixed, the total volume is *not* necessarily equal to the sum of the individual volumes. However, the total volume of mixtures of dilute solutions is equal to the sum of the volumes mixed. Using molarity as a unit of concentration simplifies solving stoichiometry problems involving solutions.

For information about the organization of Additional Practice Problems, Stop & Test Yourself, Putting Things Together, and Applications, see the beginnings of these sections in Chapter 1.

4.40 Why are liquid solutions used for carrying out chemical reactions whenever possible? (chapter opener)

4.41 A mixture of 125.0 g of acetone, C_3H_6O, and 99.8 g H_2O is a clear, colorless liquid. The melting point of acetone is $-95.3\ °C$, and its boiling point is $56.2\ °C$. Which substance is the solvent? (4.1)

4.42 A mixture of 125.0 g $NH_4C_2H_3O_2$ and 99.8 g H_2O is a clear, colorless liquid. The melting point of $NH_4C_2H_3O_2$ is $114\ °C$. Which substance is the solvent? (4.1)

4.43 What term is used to describe a solid that separates from a solution? (4.1)

4.44 The solubility of lead chloride, $PbCl_2$, is 0.99 g/100 mL of water at 20 °C and 3.34 g/100 mL of water at 100 °C. (a) A 3.09-g sample of $PbCl_2$ was stirred overnight at 20 °C with 250 mL of water. What was the final concentration of $PbCl_2$ in g/100 mL of water? How much $PbCl_2$ dissolved? (b) Another 3.09-g sample of $PbCl_2$ was stirred overnight at 100 °C with 250 mL of water. How much $PbCl_2$ dissolved? What was the final concentration of $PbCl_2$ in g/100 mL of water? (c) The solution from part (b) was cooled to 20 °C, and a tiny crystal of $PbCl_2$ was added to it. What was observed? (d) Classify each of the solutions in parts (a), (b), and (c) as unsaturated, saturated, or supersaturated. (4.1)

4.45 (a) Are all saturated solutions concentrated solutions? Back up your answer with data from Table 4.1. (b) Are all unsaturated solutions dilute solutions? Again back up your answer with data from Table 4.1. (4.1)

4.46 (a) Experimentally, how can you tell the difference between strong electrolytes, weak electrolytes, and nonelectrolytes? (b) What are present in solutions of electrolytes that are not present in solutions of nonelectrolytes? (4.2)

4.47 Which of the following would conduct electricity? Explain your answers. (a) NaCl(s) (b) NaCl(l) (c) NaCl(aq) (d) C_2H_6O(l) (e) C_2H_6O(aq) (4.2)

4.48 Classify each of the following compounds as acid, base, or salt and name each compound. (a) KOH (b) NH_4Cl (c) HBr (d) $Ca(NO_3)_2$ (e) HNO_2 (f) Li_2O (4.2)

4.49 Write molecular equations for the neutralization of (a) an aqueous solution of perchloric acid with an aqueous solution of potassium hydroxide, (b) an aqueous solution of barium hydroxide with an aqueous solution of hydrochloric acid. (4.2)

4.50 Classify each of the following acids as weak or strong: (a) HF, (b) $HClO_4$, (c) HI, (d) $HC_2H_3O_2$, (e) H_2S. (4.3)

4.51 For aqueous solutions of each of the following compounds, tell whether the solute exists as molecules, ions, or a mixture of molecules and ions: (a) KOH, (b) $HC_7H_5O_2$, (c) $C_4H_{10}O$, (d) $FeSO_4$, (e) HNO_3. (4.3)

4.52 Classify each of the following compounds as soluble, slightly soluble, or insoluble in water: (a) $Ca(C_2H_3O_2)_2$, (b) NH_4Cl, (c) H_2SO_4, (d) $Fe(OH)_3$, (e) ZnS. (4.3)

4.53 Predict whether a reaction will take place. If a reaction will take place, tell why and write a molecular equation. If a reaction will not take place, write "no reaction" and tell why not. (a) Aqueous solutions of sodium carbonate and nickel nitrate are mixed. The formula for nickel nitrate is $Ni(NO_3)_2$. (b) Aqueous solutions of sodium acetate and magnesium chloride are mixed. (c) Aqueous solutions of nitric acid and potassium hydroxide are mixed. (d) Aqueous solutions of acetic acid and potassium hydroxide are mixed. (4.3)

4.54 Write complete ionic and net ionic equations for the reactions in the previous problem. (4.4)

4.55 Arrange the elements of each of the following sets in order of increasing reactivity: (a) Cd, Hg, Zn; (b) Al, Mg, Na; (c) Ba, Ca, Sr; (d) Cr, Mn, Ni; (e) F, Ne, O; (f) Br, Cl, F. (4.5)

4.56 Predict whether a reaction will take place between the following combinations. If a reaction will take place, write molecular, complete ionic, and net ionic equations. If no reaction will take place, write "no reaction." (a) zinc metal and dilute hydrochloric acid (b) zinc metal and aqueous nickel(II) chloride (c) mercury metal and aqueous zinc nitrate (d) potassium metal and water (e) zinc metal and water (f) copper metal and dilute hydrochloric acid (4.5)

4.57 Predict whether a reaction will take place if bromine water is added to aqueous potassium chloride. If so, write molecular, complete ionic, and net ionic equations. (4.5)

4.58 The compound $BaSO_3$ is insoluble in water but soluble in dilute hydrochloric acid. Write molecular, complete ionic, and net ionic equations for the reaction that takes place when $BaSO_3$ is dissolved by hydrochloric acid. (4.6)

4.59 (a) Is MnO_2(s) soluble in HCl(aq)? (b) Is $Mg(OH)_2$ soluble in HNO_3(aq)? Explain your answers. (4.6)

4.60 Tell how to synthesize each of the following compounds: (a) $Ni(OH)_2$, (b) $MgSO_4$, (c) KNO_3. (4.6)

4.61 Describe how to make (a) 1.00 L of a 25% by volume solution of acetone in water, (b) 1000 g (1.00×10^3 g) of a 25% by mass solution of acetone in water. (4.7)

4.62 The surface concentration of F in seawater is 1.3 mg/kg. Express this concentration of F in ppm (mass). (4.7)

4.63 What is the molarity of a solution prepared by dissolving 22.0 g NaOH in water and diluting to 250.0 mL? (4.8)

4.64 How many grams of $NH_4C_2H_3O_2$ are needed to make 750 (7.50×10^2) mL of 0.225 M $NH_4C_2H_3O_2$? (4.8)

4.65 Describe how to make 500.0 mL of 3.6 M acetone (C_3H_6O). (4.8)

4.66 What volume of solution is required to obtain 3.47 mol $NH_4C_2H_3O_2$ from a 0.225 M solution? (4.8)

4.67 How does the number of moles of solute in a solution change when the solution is diluted? (4.8)

4.68 (a) How many milliliters of 0.225 M $NH_4C_2H_3O_2$ are needed to make 750.0 mL of 0.1667 M $NH_4C_2H_3O_2$? (b) Describe how to make 750.0 mL of 0.1667 M $NH_4C_2H_3O_2$ from 0.225 M $NH_4C_2H_3O_2$. (4.8)

4.69 The label on a bottle of concentrated nitric acid, $HNO_3(aq)$, says that the contents are 71.2% HNO_3 by mass and have a density of 1.420 g/mL. What is the molarity of HNO_3 in the concentrated nitric acid? (4.8)

4.70 What is the molarity of a solution formed by mixing 37.2 mL of 0.225 M HNO_3 with 67.8 mL of 0.578 M HNO_3? (4.8)

4.71 How many milliliters of 0.0487 M $Ba(OH)_2$ are needed to react with 35.67 mL of 0.0748 M HCl? The equation for the reaction is $Ba(OH)_2(aq) + 2HCl(aq) \longrightarrow 2H_2O(l) + BaCl_2(aq)$. (4.9)

4.72 How many milliliters of 6.0 M HCl are needed to dissolve 3.78 g of magnesium metal? The equation for the reaction is $Mg(s) + 2HCl(aq) \longrightarrow MgCl_2(aq) + H_2(g)$. (4.9)

4.73 A 25.00-mL sample of 0.1253 M HCl was titrated with 0.0768 M KOH. How many milliliters of KOH were required to reach the end point? (4.10)

4.74 A 21.67-mL sample of H_2SO_4 was titrated with 0.254 M NaOH. If 45.87 mL of 0.254 M NaOH were required to reach the end point, what was the molarity of the sulfuric acid? (Na_2SO_4 is formed.) (4.10)

4.75 In answer to the question "If the density of 6.00% by mass $BaCl_2$ is 1.063 g/mL, what is the molarity?", a student gave the following solution:

$$1.063\ \frac{g}{mL}\left(\frac{1\ mol}{208.24\ g}\right)\left(\frac{1000\ mL}{1\ L}\right) = 5.105\ \frac{mol}{L}$$

The molarity of the solution is actually 0.306 M. Explain what's wrong with the student's solution and show how to solve the problem correctly.

4.76 (a) What is the difference between a molecular equation and an ionic equation? (b) What is one advantage of each?

4.77 Suppose that in preparing a 0.200 M solution of lithium bromide from pure solid you make one of the following mistakes. Tell whether the concentration of the solution will be 0.200 M, less than 0.200 M, or greater than 0.200 M LiBr and explain your reasoning. (a) When you go to the stockroom to get the lithium bromide, you can only find $LiBr \cdot H_2O$, but you use the amount calculated for LiBr. (b) You spill some of the solid LiBr while pouring it into the volumetric flask. (c) You forget to let the solution of lithium bromide warm to room temperature before diluting it to the mark. (d) After the solution is made, you spill some while transferring it from the volumetric flask to a bottle for storage. (e) You do not notice that the inside of the bottle is wet.

4.78 In a 0.10 M solution of $CaCl_2$, (a) what is the molarity of Ca^{2+}? (b) the molarity of Cl^-? (c) the total concentration of ions?

4.79 If a solution suspected of containing Ca^{2+}, K^+, or Zn^{2+} gives a precipitate when treated with hydrogen sulfide gas, which of the three ions is present? Explain your answer.

STOP & TEST YOURSELF

1. The solubility of potassium carbonate in water is 112 g/100 mL at 20 °C. A solution of 25 g of potassium carbonate in 25 mL of water at 20 °C would be described as (a) saturated but dilute, (b) saturated and concentrated, (c) unsaturated and dilute, (d) unsaturated but concentrated, (e) supersaturated.

2. How many moles of NH_3 are in 4.6 L 2.3 M NH_3?
   (a) 0.50 (b) 2.0 (c) 2.3 (d) 6.9 (e) 11

3. Calculate the molarity of a solution of 4.21 g NH_3 in 3.50 L of solution.
   (a) 7.05×10^{-5} (b) 0.0706 (c) 0.864 (d) 1.16 (e) 1.20

4. Calculate the mass in grams of HCl present in 0.748 L 0.52 M HCl.
   (a) 94 (b) 52 (c) 25 (d) 14 (e) 0.011

5. To prepare 250.0 mL 0.200 M KNO_3, what volume of 1.000 M KNO_3 must be diluted with water?
   (a) 5.00×10^{-2} L (b) 8.00×10^{-1} L (c) 1.25 L
   (d) 20.0 L (e) 50.0 L

Use the equation $CaCO_3(s) + 2HCl(aq) \longrightarrow CaCl_2(aq) + H_2O(l) + CO_2(g)$ to answer questions 6 through 8.

6. What volume of 6.0 M HCl is required to react with 8.65 g $CaCO_3$?
   (a) 1.4 mL (b) 14 mL (c) 29 mL (d) 52 mL
   (e) 1.9×10^3 mL

7. If 23.32 g $CaCO_3$ and 65.3 mL 6.0 M HCl are mixed, how many grams of CO_2 will be formed?
   (a) 0.48 (b) 8.6 (c) 10 (d) 13 (e) 17

8. If 23.32 g $CaCO_3$ and 65.3 mL 6.00 M HCl are mixed, how much of which reactant will be left over?
   (a) 0.074 mol HCl (b) 0.16 mol HCl (c) 0.32 mol HCl
   (d) 3.7 g $CaCO_3$ (e) 7.4 g $CaCO_3$

9. Electricity is carried through a solution of an electrolyte by (a) electrons only, (b) anions only, (c) cations only, (d) both cations and anions, (e) both electrons and ions.

10. Which of the following is a weak acid?
    (a) NH_3 (b) H_2SO_4 (c) HNO_3 (d) HCl (e) $HC_2H_3O_2$

11. Which of the following is an insoluble salt?
    (a) NiS (b) $NiSO_4$ (c) $Fe(OH)_3$ (d) Na_3PO_4 (e) $(NH_4)_2CO_3$

12. An aqueous solution of which of the following compounds will not react with aqueous KOH?
    (a) HNO_3 (b) $NaNO_3$ (c) NH_4Cl (d) $NiCl_2$
    (e) All of them will react.

13. Which of the following is a net ionic equation?
    (a) $Cl_2 + I^- \longrightarrow Cl^- + I_2$
    (b) $Cl_2 + 2I^- \longrightarrow 2Cl + I_2$
    (c) $Cl_2 + 2I^- \longrightarrow 2Cl^- + I_2$
    (d) $Cl_2 + 2KI \longrightarrow 2KCl + I_2$
    (e) $Cl_2 + 2K^+ + 2I^- \longrightarrow 2K^+ + 2Cl^- + I_2$

14. Which of the following pairs of reactants would you use to prepare manganese(II) sulfide by a reaction between ions in solution?
    (a) $MgCl_2$ and Na_2S (b) $MnSO_4$ and ZnS
    (c) $MnCO_3$ and Na_2S (d) $MnCl_2$ and Na_2S
    (e) $MnCl_2$ and Na_2SO_4

15. Which of the following terms can correctly be applied to the reaction $Zn(s) + 2HCl(aq) \longrightarrow ZnCl_2(aq) + H_2(g)$?
(a) double replacement   (b) neutralization
(c) decomposition   (d) combination
(e) single replacement

PUTTING THINGS TOGETHER

4.80 Write the net ionic equation for the neutralization of a strong acid with a strong base.

4.81 Suggest simple tests for determining whether   (a) a clear, colorless liquid is water or HCl(aq),   (b) a white solid is NaOH or NaCl,   (c) a white solid is $CaCl_2$ or $Ca(NO_3)_2$,   (d) a white solid is NH_4Cl or NaCl,   (e) a white solid is Na_2CO_3 or $CaCO_3$.

4.82 The labels have fallen off the bottles of potassium sulfate and sodium chloride in the stockroom. (a) Describe a simple chemical test that could be used to find out which bottle is which. Write net ionic, complete ionic, and molecular equations for any reaction that takes place.   (b) The density of potassium sulfate is 2.662 g/cm^3; the density of NaCl is 2.165 g/cm^3. How could the method of displacement of water (Problem 2.99) be adapted to the determination of the densities of potassium sulfate and sodium chloride?

4.83 Write the formula for each compound: (a) copper(II) sulfide,   (b) iron(III) phosphate dihydrate,   (c) magnesium bromide hexahydrate,   (d) tin(II) acetate,   (e) barium dihydrogen phosphate.

4.84 Name each compound: (a) $Cr_2(SO_3)_3$,   (b) $Cr(NO_3)_3 \cdot 9H_2O$,   (c) $LiClO_4 \cdot 3H_2O$,   (d) $PbCl_4$,   (e) $Al(C_2H_3O_2)_3$.

4.85 A new method of determining mass values may make possible a definition of the kilogram not based on a physical object. By this method, the mass of silicon-28 is 27.976 926 5324 $\pm$ 20. What is the uncertainty in this measurement in ppt (parts per trillion)?

4.86 By titration, 24.68 mL of 0.1017 M NaOH is required to neutralize 0.1482 g of an unknown organic acid. What is the formula mass of the acid (a) assuming that each molecule contains one acidic hydrogen—that is, that the formula is $HC_nH_mO_p$?   (b) Assuming that each molecule contains two acidic hydrogens—that is, that the formula is $H_2C_nH_mO_p$?

4.87 A 2.00-g sample of zinc was dissolved in 50.0 mL of 6.0 M HCl. The final volume of solution was 50.2 mL. (a) What was the concentration of Cl^- after the zinc was dissolved?   (b) What was the concentration of hydrogen ion after the zinc was dissolved?

4.88 The solubility of $CaSO_4$ is 0.209 g/100 mL of water at 30 °C. If 0.500 g $CaSO_4$ is stirred overnight at 30 °C with 100.0 mL of water, (a) how many grams of $CaSO_4$ remain undissolved?   (b) Calculate the molarity of $CaSO_4$ in solution.

4.89 In an experiment to measure the solubility of $Ba(OH)_2 \cdot 8H_2O$ in water, an excess of solid $Ba(OH)_2 \cdot 8H_2O$ was stirred with water overnight. The saturated solution was allowed to stand until all the undissolved solid had settled. A 10.00-mL sample of the clear supernatant liquid (the liquid above an undissolved solid is described as supernatant) was titrated with 0.1250 M HCl. If 31.24 mL of 0.1250 M HCl was required to neutralize

the hydroxide ion in the saturated solution, what was the molarity of $Ba(OH)_2 \cdot 8H_2O$ in the saturated solution?

4.90 A 25.00-mL sample of a dilute solution of HCl required 27.96 mL of 0.0477 M NaOH for neutralization. The dilute solution was prepared by diluting 50.0 mL of a more concentrated solution to 1.000 L. What was the concentration of the more concentrated solution?

4.91 If 22.91 mL of NaOH are required to exactly neutralize 0.1285 g of pure benzoic acid, $HC_7H_5O_2$, what is the molarity of the NaOH solution?

4.92 A sample was analyzed for Ba^{2+} by adding a small excess of $H_2SO_4(aq)$ to an aqueous solution of the sample. The $BaSO_4(s)$ that precipitated was collected by filtration, washed, and thoroughly dried. If the mass of the sample was 7.8417 g and 0.4318 g of pure, dry $BaSO_4(s)$ was obtained, what was the mass percent Ba in the sample?

4.93 When a solution of barium hydroxide is placed in a conductivity apparatus and titrated with dilute sulfuric acid, the light grows dimmer, goes out completely, and then grows brighter again. Can the light be used to signal the end point of the titration? Explain your answer.

4.94 The density of pure water is 0.998 g/mL at 20 °C. (a) What is the molarity of water in pure water at 20 °C?   (b) The density of 0.60% by mass aqueous NaCl is 1.0025 g/mL. What is the molarity of NaCl and of H_2O in this solution? Is the molarity of water in this concentration sodium chloride solution significantly different from the molarity of water in pure water?

4.95 What is the difference between six molecules of HCl, six moles of HCl, and 6 M HCl?

4.96 Write directions for preparing 250.0 g of a 5.0% by mass solution of $MnCl_2$ from $MnCl_2 \cdot 4H_2O$.

4.97 In an aqueous solution of ammonia, (a) what molecules are present (give names and formulas)?   (b) What anion?   (c) What cation?

4.98 If the concentration of a solution is 3.4 M, (a) what is the numerical value of the concentration in mmol/mL?   (b) In the SI base units of mol/m^3?

4.99 How many molecules of CH_4O are dissolved in 1.00 mL of a 1.00×10^{-4} M solution of CH_4O?

4.100 A 0.2726-g sample of a metal was dissolved in 50.00 mL of 0.5000 M HCl. After all of the metal had dissolved, the leftover acid was titrated with 0.1054 M NaOH. If 24.36 mL of 0.1054 M NaOH were required to neutralize the leftover acid, what was the atomic mass of the metal? The metal dissolved to form an M^{2+} ion.

4.101 (a) Write molecular, complete ionic, and net ionic equations for the reaction that takes place when sodium carbonate solution is mixed with a solution of acetic acid.   (b) Name the products of this reaction.

4.102 Classify each of the following reactions as combination, decomposition, single replacement, or metathesis:

(a) $Ca(s) + 2H_2O(l) \longrightarrow Ca(OH)_2(s) + H_2(g)$

(b) $CaCO_3(s) \xrightarrow{\text{heat}} CaO(s) + CO_2(g)$

(c) $Ca(s) + Cl_2(g) \longrightarrow CaCl_2(s)$

(d) $CaCl_2(aq) + Na_2CO_3(aq) \longrightarrow$
$$CaCO_3(s) + 2NaCl(aq)$$

(e) $CaS(aq) + Cl_2(aq) \longrightarrow CaCl_2(aq) + S(s)$

4.103 (a) Write the molecular equation for a typical single-replacement reaction. (b) Write the molecular equation for a typical combustion reaction. (c) Both single-replacement and combustion reactions are oxidation-reduction. Some combination reactions—for example, $H_2(g) + Cl_2(g) \longrightarrow 2HCl(g)$—are oxidation-reduction, but others—for example, $CaO(s) + H_2O(l) \longrightarrow Ca(OH)_2(s)$—are not. Some decomposition reactions, for example, $2HgO(s) \longrightarrow 2Hg(l) + O_2(g)$ are oxidation-reduction, but others, for example, $CaCO_3(s) \longrightarrow CaO(s) + CO_2(g)$ are not. What do all the reactions that are oxidation-reduction have in common?

APPLICATIONS

4.104 Indium is used in the manufacture of transistors. The purest grade is 99.999 99% indium. Express the concentration of impurities in this grade of indium in units that will make the number between 1 and 999.

4.105 Quartz-crystal watches lose or gain about a second a week. To three significant figures, how many ppm is this?

4.106 The salinity of coastal ocean water is normally just over 33 parts per thousand by mass, of which 85% is sodium chloride. What is the concentration of sodium chloride in molarity? The density of seawater is 1.025 g/cm³.

4.107 (a) The concentration of trimethyl lead in French wines was 460 pg/g in 1978. (It is now only 40 pg/g as a result of the reduction of the levels of lead additives in gasoline.) What are these concentrations expressed in ppt (parts per trillion)? (b) In brandy stored for over five years in a lead-crystal decanter, the lead concentration reached 21 530 μg/L. The density of the brandy was 0.930 g/mL. Express the lead concentration in mass/mass units such that it is between 1 and 999.

4.108 The Clean Air Act amendments of 1990 require that gasoline contain 2.7% oxygen in 100 cities with unacceptably high levels of carbon monoxide and ozone at ground level. One method of achieving this is to add methyl *tert*-butyl ether (MTBE), $CH_3OC(CH_3)_3$. (a) What percentage of MTBE must gasoline contain? (b) Write a complete structural formula for MTBE.

4.109 Pure water is such a weak electrolyte that, for all practical purposes, it is a nonelectrolyte. Why is it dangerous to talk on the telephone while sitting in the bathtub or to swim during a thunderstorm?

4.110 An average cup (250 mL) of coffee or tea contains 125 mg of caffeine, an average cup of cocoa, 50 mg, and a 12-oz Coke, 50 mg. The molecular formula for caffeine is $C_8H_{10}N_4O_2$. (a) What is the molarity of caffeine in tea or coffee? (b) Doses of caffeine greater than one gram may cause alarming symptoms in humans. How many cups of coffee or tea provide one gram of caffeine?

4.111 The minerals anhydrite ($CaSO_4$), barite ($BaSO_4$), and celestite ($SrSO_4$) are formed by precipitation reactions. Write a net ionic equation for each of these reactions.

4.112 The compound MoS_2 is an important lubricant. Making it from the elements takes five days at 900 °C. However, it can be made in less than five minutes at room temperature by the reaction

$$MoCl_5(s) + Na_2S(s) \longrightarrow MoS_2(s) + NaCl(s) + S(s)$$

(a) Write the equation for the synthesis of MoS_2 from the elements. What type of reaction is this reaction? (b) How many grams of MoS_2 will be obtained from 18.03 g $MoCl_5$ and 12.88 g Na_2S if the percent yield is 80%? (c) The sulfur boils away during the reaction. How can the sodium chloride be removed from the product? (d) The purity of the product is established by analysis for sulfur. The sulfur content is within 0.1% of the theoretical value. What is the theoretical percent sulfur for MoS_2?

4.113 A mail-order catalog advertises a tarnish remover for silver. The ad says "Fill the plastic pan with tap water, add liquid detergent, place the magnesium bars in the bottom, put in the silver and *watch the miracle*. The tarnish is transferred from the silver piece to the bar by electrolytic action. No chemical odor, easy on the hands, and safe for silver. Reusable too!" Tarnish is Ag_2S. The tarnish remover works because magnesium takes the place of silver in silver sulfide. (a) Write the molecular equation for this single-replacement reaction. (b) Criticize the statement that the tarnish is transferred from the silver piece to the bar. (c) Will the bar be reusable forever?

4.114 Tell how to make each of the following compounds: (a) $CaBr_2$. Calcium bromide is used to make high-density fluids employed in drilling for oil. (b) $CaCO_3$. Pure calcium carbonate is synthesized for antacids and dietary calcium supplements. (c) ZnS. Zinc sulfide is used as a white pigment in paints. (d) NH_4NO_3. Ammonium nitrate is used to make fertilizers and explosives. (e) AgI. Silver iodide is used in photography and in attempts to manage rainfall by seeding clouds.

4.115 Explain why carbonated beverages go flat if they are allowed to warm up.

4.116 Too little iron in the diet leads to anemia and too little calcium to osteoporosis. Spinach is rich in iron and calcium, but spinach is also rich in oxalate ion, $C_2O_4^{2-}$. Typically, a 100-g sample of spinach contains 99.4 mg Ca^{2+}, 2.6 mg Fe^{2+}, and 571 mg $C_2O_4^{2-}$. Both calcium oxalate and iron(II) oxalate are insoluble. (a) Write net ionic equations for the precipitation of calcium oxalate and iron(II) oxalate. Iron(II) oxalate separates as the dihydrate. (b) How many milligrams of oxalate ion are required to precipitate 99.4 mg Ca^{2+}? (c) How many milligrams of oxalate ion are required to precipitate 2.6 mg Fe^{2+}? (d) How many milligrams of Ca^{2+} and Fe^{2+} will be left for the people who eat their spinach? (A serving is half a pound.)

4.117 Solutions for intravenous infusion (IV) must have concentrations such that water does not move into or out of the red blood cells. Glucose, $C_6H_{12}O_6$, must be 0.31 M for IV use. (a) Explain how to prepare 1.00 L of a 0.31 M solution of glucose. (b) What concentration of NaCl would have the same concentration of particles as a 0.31 M solution of glucose?

4.118 A biochemical researcher investigating an allergy-related occurrence reported that a 10^{-120} M solution of an antibody produced a biological effect. (a) How many liters of a 10^{-120} M solution contain one molecule of solute? (b) All the world's oceans put together have a volume of 1.37×10^9 km^3. Do you believe the researcher? Explain your answer.

4.119 In lists of the principal components of seawater, no compounds (except water) are listed. The lists consist of ions such as chloride ion, sulfate ion, sodium ion, and magnesium ion. Explain why.

4.120 One commercial source of bromine is ocean water. The average concentration of Br^- in seawater is 0.065 g/kg. Bromine is obtained from seawater by replacement with the more active chlorine. (a) Write the net ionic equation for the preparation of bromine from seawater. (b) How many kilograms of seawater must be processed to obtain 454 g of bromine?

4.121 A 5.00-mL sample of seawater that had a density of 1.0116 g/mL was titrated with 0.0998 M $AgNO_3$ solution using K_2CrO_4 to indicate the end point. If 26.35 mL of 0.0998 M $AgNO_3$ were required to reach the end point, (a) what was the concentration of Cl^- in the seawater in molarity? (b) What was the mass percent Cl^- in the seawater?

4.122 Milk of magnesia is a mixture of magnesium hydroxide, $Mg(OH)_2(s)$, and a saturated aqueous solution of magnesium hydroxide. Stomach acid is HCl(aq). (a) Write molecular, complete ionic, and net ionic equations for the reaction that takes place between $Mg(OH)_2(aq)$ and stomach acid. (b) If taken internally, sodium hydroxide burns all tissues. Explain why magnesium hydroxide does not burn tissue but is able to neutralize stomach acid.

4.123 The explosion of the Soviet nuclear reactor at Chernobyl is believed to have involved single-replacement reactions between carbon and steam and zirconium and steam. Partial equations for these reactions are

$$C(s) + H_2O(g) \longrightarrow \text{_____} + H_2(g)$$
$$Zr(s) + 2H_2O(g) \longrightarrow \text{_____} + 2H_2(g)$$

(a) Fill in the blanks with the formulas of the products.
(b) What are the products called?

4.124 According to the label, the contents of a bottle of wine have a volume of 750 mL and are 12.0% alcohol, C_2H_6O, by volume. At 20 °C, the density of alcohol is 0.789 g/mL, and the density of water is 0.998 g/mL. If the density of the wine is 0.982 g/mL, (a) how many grams of alcohol are there in the bottle? (b) What is the percent by mass of alcohol in the wine? (c) The alcohol is made by fermentation of grape sugar,

$C_6H_{12}O_6$. The other product is carbon dioxide gas. Write the equation for the reaction. How many grams of grape sugar must be fermented to produce the alcohol in one bottle of wine? How many grams of carbon dioxide are also formed? The density of carbon dioxide at 20 °C and normal barometric pressure is 1.831 g/L. What is the volume of the carbon dioxide formed?

4.125 In the 1850s, an English physician noticed that when he mashed some preserved gooseberries with a steel fork, the tongs became coated with copper. Write the net ionic equation for a reaction that explains the observation. Steel is composed mostly of iron.

4.126 Increased use of filler in papermaking could result in the use of 20% fewer trees. The space inside the fibers' cell walls can be filled by dipping the wet wood pulp first in a solution of calcium chloride and then in a solution of sodium carbonate. Write net ionic, complete ionic, and molecular equations for the reaction that takes place.

4.127 (a) Barium chloride is poisonous. Which ion is toxic? Explain your answer. (b) Explain why barium sulfate, which is opaque to X-rays, can safely be taken before intestinal X-rays.

4.128 Chromates are poisonous. The water solubility of strontium chromate, $SrCrO_4$, is 0.12 g/100 mL; the water solubility of lead chromate, $PbCrO_4$, is 0.000 0058 g/100 mL. (a) Explain why strontium chromate is no longer used as a yellow pigment, but lead chromate is still used in traffic stripe paints for highways and airports. (b) What is the charge on the chromate ion?

4.129 Imagine that you visited one of your company's customers and they told you that they were using an acidic solution to remove zinc from galvanized parts and getting vigorous "boiling." You noticed electrical motors and people smoking nearby. (a) Write the net ionic equation for the reaction between zinc and an acidic solution. (b) What would you (tactfully) tell the customer?

4.130 Dutch scientists are exploring the possibility of raising the level of the ground in low-lying areas of Holland by treating the subsurface limestone ($CaCO_3$) with waste sulfuric acid forming gypsum ($CaSO_4 \cdot 2H_2O$). They say that "gypsum occupies twice the volume of a corresponding amount of calcium carbonate." The density of $CaCO_3$ is 2.71 g/cm^3 and the density of $CaSO_4 \cdot 2H_2O$ is 2.32 g/cm^3. Sulfuric acid is a strong acid, but the hydrogen sulfate ion is a weak acid. (a) What names would a chemist use for limestone and gypsum? (b) Write the molecular, complete ionic, and net ionic equations for the reaction of limestone with sulfuric acid to form gypsum. (c) Explain the sources of the twofold increase in volume.

4.131 The best method for diagnosing diabetes is measurement of the elevation of fasting glucose level. Normal plasma glucose levels are 70–110 mg/dL. A plasma glucose level of 140 mg/dL or greater on two or more occasions indicates that a patient has diabetes. Standard glucose solutions are needed to check the analytical method. Write directions for preparing 100.0 mL of a solution that contains 150.0 mg of glucose per deciliter from a stock solution that contains 2.5000 g/L. The formula for glucose is $C_6H_{12}O_6$.

A Scientist Serves the Global Environment

ELVIA NIEBLA

National Coordinator, National Global Change Research Program, U. S. Forest Service

B.S. Zoology
University of Arizona

Ph.D. Soil Chemistry
University of Arizona

I first became interested in science as a child when, encouraged by my mother, I would explore the world around me. That world was Nogales, Arizona, where there were innumerable bugs and other natural phenomena to explore. I recall one particular incident that sparked my interest. Near our house ran a creek where I spent many hours playing and exploring. One day at the creek I used a tin can to scoop out a tiny "fish," which I brought home in an old Coke bottle. Excitedly I told my mother all about the fish I'd caught and how I wanted to keep it until it grew into a big fish. She humored me, and we set up a world for my fish in a fishbowl half-filled with water and with a rock large enough so that part of it was above water, as my mother instructed. We set the bowl in our living room where every day I monitored my fish's growth. I spent endless afternoons observing the fish. To my amazement, one day as I watched it peck at the rock, I noticed tiny legs sprouting where its gills had been and its tail all but disappearing. I couldn't believe my eyes—my "fish" was turning into a frog! That did it. I was hooked on the wonders of nature and from then on pursued the natural transformations around me with even greater awe.

In high school I gravitated toward math and science courses. My high school chemistry laboratory, in particular, was where I felt I belonged. I loved the transformations that I could effect in lab and the way that I could identify unknowns through simple procedures, like exposing salts to fire and watching the colors emerge. In college I considered and chose multiple majors (Math, Physics, Zoology), but was finally drawn to chemistry because that's where I could see tangible, immediate results; I guess I felt it was the most practical science, too. I enjoyed organic chemistry; it was like solving a puzzle. I loved my analytical chemistry classes, too; they required exactness, so much so that for some experiments I had to measure unknown substances to accuracies of 1/1000th of a gram.

Coming from the desert country in Arizona instilled in me an appreciation for the preciousness of water. While still an undergraduate at the University of Arizona, I went to a picnic at nearby Patagonia Creek and saw that the creek water was polluted. I began a research project on the creek, assuming that the pollutants were there because of the sewage plant from the nearby town. In the course of my research, I found out that the real pollutants came from the mines up in the mountains. Thus, I was ultimately drawn to water and soil chemistry because of its applicability to environmental issues: I could investigate the pollutants in water and soils; I could also measure the exact nutrients in fertilizers needed to affect plant growth.

The deterioration of our environment also challenged me. I had the confidence that something could be done. We had the technology, and I felt that I could help meet the challenge somehow. I worked as a University professor in soil chemistry, performed countless hours of laboratory research, and published a number of substantive articles in the field. Although I enjoyed teaching and doing research, I eventually left that very rewarding job to work with the National Park Service as director of the Materials Testing Program for the Western Archaeological Center. From there I moved to a position with the Environmental Protection Agency that required my soils research experience to develop national regulations on sludge used for soil amendment.

Currently, I work in the U.S. Forest Service as National Coordinator of the National Global Change Research Program, a position that allows me to work closely with both policy makers and scientists. I am a science administrator for global change. As such, I am meeting the challenge posed by the deterioration of our environment. I coordinate a multimillion-dollar research program and have had many opportunities to make substantial contributions to overall U.S. policy, including the U.S. Global Change Research Program plan, "Our Changing Planet."

Although I did see myself working in the environmental field after I graduated from college, I never imagined I would be working at the national level with policy makers and scientists. I guess I saw myself as a scientist, but never as an administrator.

What I learned in my first general chemistry course was discipline, as well as the basic principles that remain with me to this day: the value of memorization, problem solving, assessing a situation, and seeking solutions through the scientific method. Even today, I use those same basic skills when I evaluate a proposal or solve a problem.

My advice if you are contemplating a career in chemistry is to realize that although it takes time, it is an infinitely rewarding pursuit. I remember that when I was in college, I sometimes envied other students who weren't majoring in chemistry. They didn't seem to spend as much time on their studies, while I spent endless hours in the lab. But then I remember, too, how rewarding the work in the lab was, how intellectually stimulating it was to solve a problem. I would advise you to go with your passion, and if chemistry is what motivates, intrigues, and awes you, then you should stick to it.

GASES

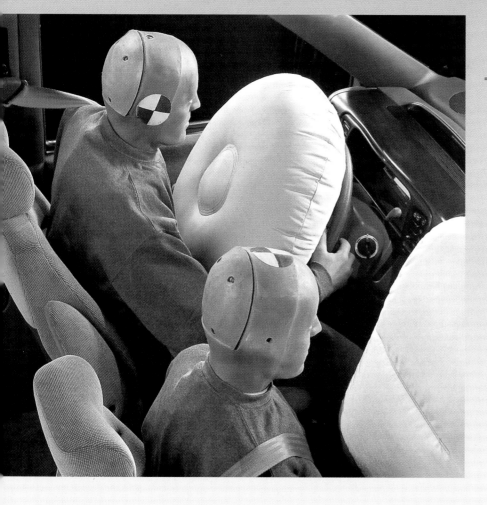

Automobile "air bags" are filled with nitrogen gas. During development, they were thoroughly tested with dummies.

I n this chapter, we will see how the laws that describe the way gases behave were developed and how the study of gases during the nineteenth century contributed to the growth of the atomic theory. We will also learn about the kinetic-molecular theory, a model that is used to explain why gases behave as they do. The history of the study of gases provides an excellent example of how science progresses from observations to laws to theories and of how theories must be changed to explain new observations.

— *Why is a study of gases a part of general chemistry?*

Many elements and compounds are gases under everyday conditions. The elements that are gases under ordinary conditions are hydrogen (H_2), nitrogen (N_2), oxygen (O_2) and ozone (O_3), fluorine (F_2), chlorine (Cl_2), and the noble gases—helium (He), neon (Ne), argon (Ar), krypton (Kr), xenon (Xe), and radon (Rn). Some important compounds that are gases are carbon dioxide (CO_2), carbon monoxide

(CO), ammonia (NH_3), nitric oxide (NO), nitrogen dioxide (NO_2), sulfur dioxide (SO_2), hydrogen chloride (HCl), hydrogen sulfide (H_2S), and methane (CH_4). Substances that are usually liquids or solids, such as water and mothballs, can also exist in the gas phase. *Gases formed from*

◗ Notice that all the elements that are gases under ordinary conditions are nonmetals; the compounds that are gases under ordinary conditions are compounds of nonmetals.

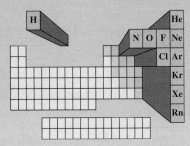

These elements are gases under ordinary conditions.

substances that are usually liquids or solids are called **vapors.**

Many chemical reactions, both in the laboratory and in manufacturing plants, involve gases as reactants or products or both. The air you breathe and the carbon dioxide you exhale are both gases. Plants use the gas carbon dioxide to synthesize complex organic materials such as sugar and starch, using sunlight as the source of energy. Combustion cannot take place without oxygen. All the reactions that take place in Earth's atmosphere, such as destruction of ozone in the ozone layer, involve gases.

Earth's weather is largely the result of changes in the properties of the mixture of gases called air. Changes in temperature lead to pressure changes and to changes in the amount of water vapor that air can hold. Clouds, rain, snow, winds, and storms can be explained in terms of the properties of gases. The properties of the gaseous mixture, air, have also affected the evolution of lungs, organs that can handle only gases, not liquids or solids. An important function of blood is to carry the gas oxygen from lungs to cells and

to carry the gas carbon dioxide from cells to lungs. Decreases and increases in volumes of gases are used in all sorts of ways in everyday life and in industry—for example, in car and airplane engines and in guns.

What do you already know about gases?

The odor of coffee brewing or dinner cooking soon spreads through the entire house. A sample of gas completely fills its container and takes the shape of the container (▬Figure 5.1). An enclosed sample of a gas can be **compressed** *(made smaller) by pressure*—for example, by pushing down the piston of a bicycle pump. When a gas is heated, its volume increases. For example, when a cake is baked, the volume of the air and carbon dioxide in the batter increases, and the cake rises. Gases are less dense than liquids or solids. Compare how it feels to walk through air and the water in a swim-

A hot-air balloon.

ming pool! Mixtures of gases are homogeneous; that is, they are solutions. "Pure" air is a mixture of nitrogen, oxygen, argon, carbon dioxide, water vapor, and a number of other gases. If a vapor is cooled, it condenses or deposits, like water vapor from damp air on a glass containing a cold drink.

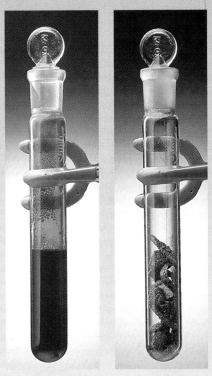

▬ FIGURE 5.1 *Left:* Gases, such as the bromine vapor, fill their containers. Liquids, such as the liquid bromine, take the shape of the bottom of the container but do not expand to fill the container. *Right:* Solids, such as these iodine crystals, have the same shape and volume regardless of the shape and volume of the container.

<table>
<tr><td>5.1</td><td></td></tr>
</table>

5.1   PRESSURE

Sucking soda through a straw makes use of the pressure of Earth's atmosphere. You suck the air out of the straw above the surface of the liquid, and atmospheric pressure then pushes the liquid up the straw against gravity to your mouth.

Pressure is defined as *force per unit area.* (A **force** is *that which causes a change in the motion of a body that is free to move.* Gravity is a force.) The air in Earth's atmosphere is attracted to Earth by gravity and pushes against every surface it touches. You are not aware of the atmosphere pushing down on you because you are used to its pressure, but the pressure of the atmosphere can crush a metal can (see ▬Figure 5.2).

- FIGURE 5.2 If the air is pumped out of a can, the pressure of the atmosphere crushes the can.

The pressure of the atmosphere changes from day to day, usually increasing when the weather is clearing and decreasing when a storm is coming. The pressure of the atmosphere also changes from place to place. It is highest below sea level and lowest on mountaintops for two reasons. The major reason is that the height of the column of air pushing down on an area below sea level is greater than the height of the column of air pushing down on an equal area on a mountaintop. The decrease in the force of gravity as distance from the center of Earth increases is much less important.

In chemistry laboratories, the pressure of Earth's atmosphere is usually measured with a mercury **barometer.** A simple mercury barometer is shown in -Figure 5.3. The mercury barometer was invented in 1643 by the Italian scientist Evangelista Torricelli, a student of Galileo. With a mercury barometer, pressure is measured in millimeters of mercury, **mmHg.** The *unit of pressure, mmHg,* is sometimes called a **torr** in honor of Torricelli.

In a mercury barometer, the weight of the column of mercury balances the weight of the atmosphere. To understand how a mercury barometer works, consider the pressure of the mercury in the column on the surface of the mercury in the container at the bottom. When the level of the mercury in the barometer is not moving, the pressure of the mercury in the column on the surface of the mercury in the container and the pressure of the atmosphere on the surface of the mercury in the container must be equal. If the pressure of the atmosphere were greater, mercury would be pushed up into the column [see Figure 5.3(b)]; if the pressure of the atmosphere were less, mercury would flow out of the column. Thus, the height of the mercury in the column is proportional to the atmospheric pressure:

$$\text{height of mercury} \propto \text{atmospheric pressure}$$

or

$$\text{height of mercury} = \text{a constant} \times \text{atmospheric pressure}$$

The height of the mercury in the column depends only on the atmospheric pressure, not on the diameter of the column*; at the same pressure, the height of the mercury in a fat barometer is the same as the height of the mercury in a thin barometer.

_____

*In a barometer,

$$\text{pressure at bottom of column} = \frac{\text{weight of mercury column}}{\text{area of bottom of column}} \propto \frac{\text{mass of mercury}}{\text{area}}$$

$$\propto \frac{\text{volume of mercury} \times \text{density of mercury}}{\text{area}}$$

$$\propto \frac{\text{height} \times \text{area} \times \text{density of mercury}}{\text{area}}$$

$$\propto \text{height} \times \text{density of mercury}$$

Mercury is very toxic and can be absorbed through the skin as well as by breathing mercury vapor.

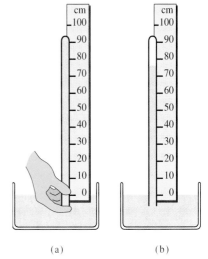

(a)   (b)

- FIGURE 5.3 A simple mercury barometer (not to scale). (a) Assembling a mercury barometer. A long tube sealed at one end is completely filled with mercury and carefully turned upside down with the open end under the surface of a pool of mercury. (b) A mercury barometer in use. The pressure of the atmosphere pushing down on the surface of the mercury holds up a column of mercury about 76 cm (760 mm) high. Pressing gently on the surface of the mercury makes the mercury rise in the column.

The *SI unit of pressure* is the **pascal, Pa.*** The pascal is a derived unit:

$$1 \text{ Pa} = \frac{1 \text{ kg}}{\text{m} \cdot \text{s}^2}$$

To give you an idea of the size of the pascal, one pascal is the pressure exerted by a layer of water that is 0.1 mm thick at sea level. One pascal is a very small pressure! For high pressures, the standard atmosphere (atm), which is defined as exactly $1.013\ 25 \times 10^5$ Pa, is frequently used. By definition, 1 atm = 760 torr (exactly), which is the average barometric pressure at sea level:

$$1 \text{ atm} = 1.013\ 25 \times 10^5 \text{ Pa} = 760 \text{ torr} = 760 \text{ mmHg}^\dagger$$

Mercury was chosen as the liquid for barometers because mercury is very dense (density = 13.6 g/mL). A barometer filled with water (density = 1.00 g/mL) would have to be almost 34 feet high!

PRACTICE PROBLEMS

5.1   (a) Which is smaller, 1 Pa or 1 mmHg? (b) Will a pressure in pascals be a smaller or a larger number than the same pressure in millimeters of mercury?

5.2   Convert (a) 473.1 mmHg to torr, (b) 467.3 mmHg to atm, (c) 4.83 atm to mmHg, (d) 473.1 mmHg to Pa, and (e) 2.3 atm to Pa.

5.2   RELATION BETWEEN PRESSURE AND VOLUME OF A GAS

The apparatus shown in Figure 5.4 can be used to study the relation between the pressure and the volume of a sample of gas at the constant temperature of the laboratory (usually around 20 °C). A *device used to measure the pressure of a sample of gas in a container* (such as the sample of air in Figure 5.4) is called a **manometer.**

To begin a typical experiment, the leveling bulb was placed so that the volume of air in the gas buret was exactly 5.0 mL, as shown in Figure 5.4(a). Because the stopcock was open, the atmosphere was able to push down on the surface of the mercury in the gas buret as well as on the surface of the mercury in the leveling bulb. The heights of the two mercury surfaces were the same. Both the pressure on the mercury in the gas buret and the pressure on the mercury in the leveling bulb were equal to the atmospheric pressure, which was 760.1 mmHg (76.01 cmHg) when the experiment was done.

---

*The pascal is named after Blaise Pascal, a French mathematician, physicist, and religious philosopher and writer, who was the father of the modern theory of probability. Pascal invented the first digital calculator in about 1643. He also invented the syringe and discovered the principle that a pressure change at one point in a fluid is transmitted to all points of the fluid and to the walls of the container. This principle is applied in hydraulic brakes.
†Other units of pressure that you may run into are "psi" and bar. American engineers often express pressure in psi (pounds per square inch). Meteorologists (scientists who study the atmosphere and its phenomena) generally use millibars (1 millibar = 0.750 mmHg) in discussing atmospheric pressure. The relationships between these units of pressure and the atmosphere are

$$1 \text{ atm} = 1.013\ 25 \text{ bar} = 14.70 \text{ psi}$$

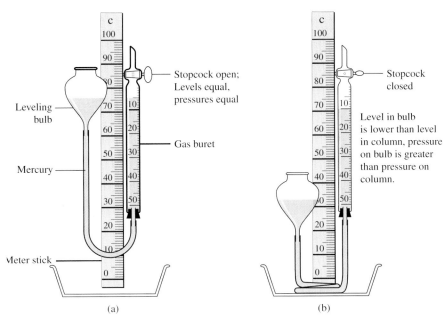

FIGURE 5.4 Apparatus for determination of pressure–volume relation of a gas.

In the figure (a):
- Stopcock open; Levels equal, pressures equal
- Leveling bulb
- Gas buret
- Mercury
- Meter stick

In figure (b):
- Stopcock closed
- Level in bulb is lower than level in column, pressure on bulb is greater than pressure on column.

We prefer this apparatus to the classical J-tube because we think it is safer. Ventilation in our classrooms is not always good, and we don't like to pour mercury. Students usually love to manipulate mercury and can safely experiment with this apparatus on the way out of class. If a picture is worth a thousand words, a personal experience must be worth a million.

The stopcock was closed to trap the sample of air in the gas buret. The leveling bulb was then lowered until the volume of the sample of air in the gas buret was 10.00 mL, as shown in Figure 5.4(b). In Figure 5.4(b), the pressure on the mercury in the leveling bulb is still equal to the atmospheric pressure. The surface of the mercury in the leveling bulb, however, is lower than the surface of the mercury in the gas buret. The atmosphere is able to push the surface of the mercury in the gas buret higher than the surface of the mercury in the leveling bulb. Therefore, the pressure of the atmosphere must be greater than the pressure of the gas in the gas buret. The difference in levels is equal to the difference in pressures measured in centimeters of mercury. The levels of the mercury in the leveling bulb and the gas buret were recorded, and the pressure of the sample of air in the gas buret was calculated from them in the following way.

In Figure 5.4(b), the level of the mercury in the gas buret is 70.9 cm. The level of the mercury in the leveling bulb is 33.0 cm. The difference in pressure between the two levels is

$$70.9 \text{ cmHg} - 33.0 \text{ cmHg} = 37.9 \text{ cmHg}$$

The pressure of the sample of air in the gas buret is 37.9 cmHg less than the pressure of the atmosphere. Because the atmospheric pressure was 76.01 cmHg, the pressure of the gas sample was

$$76.01 \text{ cmHg} - 37.9 \text{ cmHg} = 38.1 \text{ cmHg}$$

The leveling bulb was then lowered until the volume of the sample of air in the gas buret was 5.00 mL larger (15.00 mL). The heights of the mercury in the leveling bulb and the gas buret were recorded, and the pressure of the sample of air in the gas buret was calculated from them as before. This process was repeated until the volume of the sample of air in the gas buret reached 45.00 mL. The volumes and pressures observed are shown in Table 5.1(a).

○ You've undoubtedly heard of a tug-of-war. In a mercury manometer, there is a push-of-war.

TABLE 5.1(a)

Pressure of a Sample of Air[a] at Different Volumes

Volume, mL	Pressure, cmHg
5.00	76.0
10.00	38.1
15.00	25.3
20.00	19.1
25.00	15.1
30.00	12.7
35.00	10.9
40.00	9.4
45.00	8.4

[a]Fixed number of moles of air trapped in a closed container (see Figure 5.4).

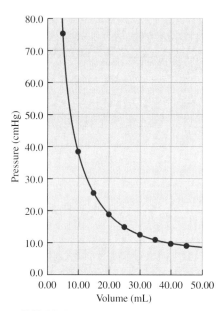

Relationship between
pressure and volume of a sample of air at
constant temperature. [Data are given in Table
5.1(a)].

TABLE 5.1(b)	Pressure of a Sample of Air at Different Volumes	
Volume, mL	Pressure, cmHg	Pressure × Volume, mL·cmHg
5.00	76.0	3.80×10^2
10.00	38.1	3.81×10^2
15.00	25.3	3.80×10^2
20.00	19.1	3.82×10^2
25.00	15.1	3.78×10^2
30.00	12.7	3.81×10^2
35.00	10.9	3.82×10^2
40.00	9.4	$3.7 \ \times 10^2$
45.00	8.4	$3.7 \ \times 10^2$

PRACTICE PROBLEMS

5.3   As the pressure of a sample of air decreases at constant temperature, what
happens to the volume?

5.4   At a pressure of 80.0 cmHg, will the volume of the sample in the experiment
be less than 5.0 mL, equal to 5.0 mL, or greater than 5.0 mL if the temperature
does not change? Explain your reasoning.

Graphs often help you to see relations between variables better than tables do.
■Figure 5.5 is a graph of the information in Table 5.1(a). Remembering your math
courses, you may already have recognized that the curve in Figure 5.5 looks like
a graph of $xy = $ a constant or $y = $ a constant$/x$. Let's test the information in Table
5.1(a) to see if the product of pressure and volume is constant. Beginning with the
first line,

$$5.00 \text{ mL} \times 76.0 \text{ cmHg} = 3.80 \times 10^2 \text{ mL·cmHg}$$

Continuing with the second line,

$$10.00 \text{ mL} \times 38.1 \text{ cmHg} = 3.81 \times 10^2 \text{ mL·cmHg}$$

All the pressure–volume products from Table 5.1(a) are shown in Table 5.1(b).
    As you can see from Table 5.1(b), the product of pressure and volume is indeed
constant. (Remember that the last digit in numbers calculated from measurements
is uncertain.) We have found a mathematical equation for stating the relation between
the volume of a sample of a gas and its pressure at constant temperature:

$$\text{pressure} \times \text{volume} = \text{a constant}$$

In other words, we have discovered a scientific law. Experiments at different temperatures show that the pressure–volume product is still constant provided that the
temperature is constant. However, the pressure–volume product has different values
at different temperatures. When similar experiments were carried out in which other
gases were substituted for air, the same results were obtained.
    The relation between the volume of a sample of a gas and its pressure at constant
temperature was first studied by Towneley, Power, and Hooke in England in 1660–
1661. The law we have "discovered" was first published by the Irish scientist, Robert
Boyle, in 1662 and is called **Boyle's law.** Mariotte, who published his results in
1679 in France, confirmed the earlier work, as have many other chemists and
physicists since that time. However, later work has shown that Boyle's law is true

▶ The simplest way to examine the effects
of different variables is to change one variable
at a time and keep all others the same.

▶ Remember that one characteristic of a
science is that observations can be repeated.
Experiments that lead to important conclusions
are always checked in different laboratories.

only at relatively high temperatures (around room temperature and above) and low pressures (a few atmospheres or less).

Boyle's Law

At constant temperature, the volume of a sample of gas is inversely proportional to the pressure of the gas.

$$V = \frac{a\ constant}{P} \qquad or \qquad P \times V = a\ constant$$

In everyday language, increasing the pressure on a sample of a gas at constant temperature decreases the volume of the gas.

Boyle's law can be used to calculate what the new volume of a sample of gas will be if the pressure is changed. It can also be used to calculate what change in pressure is needed to bring about a given change in volume (provided that the temperature does not change). It is a lot easier to push the buttons on a calculator than it is to do an experiment!

If a sample of a gas has a volume, V_1, at a pressure, P_1, and the same sample of gas has a volume, V_2, at a pressure, P_2, then

$$P_1V_1 = a\ constant \qquad and \qquad P_2V_2 = the\ same\ constant$$

providing that the temperature does not change. Since P_1V_1 and P_2V_2 are both equal to the same constant,

$$P_1V_1 = P_2V_2 \qquad\qquad (5.1)$$

The relationship in equation 5.1 makes possible calculation of any one of the four variables, P_1, P_2, V_1, or V_2, if the values of the other three are known. For example, dividing both sides of equation 5.1 by P_2, we find that

$$\frac{P_1V_1}{P_2} = V_2 \qquad or \qquad \left(\frac{P_1}{P_2}\right)V_1 = V_2$$

If the pressure on a sample of gas is changed while the temperature is kept constant, the old volume, V_1, can be converted to the new volume, V_2, by multiplying the old volume by (P_1/P_2). Dividing both sides of equation 5.1 by V_2, we get

$$\frac{P_1V_1}{V_2} = P_2 \qquad or \qquad P_1\left(\frac{V_1}{V_2}\right) = P_2$$

If the volume of a sample of gas is changed while the temperature is kept constant, the old pressure, P_1, can be converted to the new pressure, P_2, by multiplying the old pressure by (V_1/V_2). Sample Problem 5.1 illustrates this type of calculation.

SAMPLE PROBLEM

5.1 A sample of gas has a volume of 54 mL at a pressure of 452 mmHg. What will the volume be if the pressure is changed to 649 mmHg while the temperature is kept constant?

Solution Begin by making a table of the information given and asked for in the problem:

	Volume, mL	Pressure, mmHg
Old(1)	54	452
New(2)	?	649

To convert the old volume to the new volume, multiply by a factor obtained from the old and new pressures. Do *not* try to memorize whether to multiply by (P_1/P_2) or by (P_2/P_1). Think—is the new pressure greater or less than the old pressure? In this problem, it is greater.

If you increase the pressure on a sample of gas, will the volume become larger or smaller? You should know by now that it will become smaller.

To make the volume smaller, the old volume must be multiplied by a number less than 1. Should the old volume be multiplied by

$$\frac{452 \text{ mmHg}}{649 \text{ mmHg}} \qquad \text{or by} \qquad \frac{649 \text{ mmHg}}{452 \text{ mmHg}}?$$

The ratio with the lower pressure as numerator and the higher pressure as denominator is less than 1. To make the volume smaller, multiply by

$$\frac{452 \text{ mmHg}}{649 \text{ mmHg}}$$

The new volume will be

$$54 \text{ mL}\left(\frac{452 \text{ mmHg}}{649 \text{ mmHg}}\right) = 38 \text{ mL}$$

Check: The new volume is smaller than the old volume as it should be if the new pressure is greater than the old pressure.

Notice two things about this problem:

1. It does not matter what the gas is. The same results were obtained in the pressure–volume experiment when other gases were substituted for air.
2. It does not matter what units of pressure are used as long as both old and new pressures have the same units. One pressure unit is converted into another pressure unit by multiplication and the conversion factors cancel. For example,

$$54 \text{ mL}\left[\frac{452 \text{ mmHg}\left(\dfrac{1 \text{ atm}}{760 \text{ mmHg}}\right)}{649 \text{ mmHg}\left(\dfrac{1 \text{ atm}}{760 \text{ mmHg}}\right)}\right] = 38 \text{ mL}$$

(Similarly, it does not matter what the units of volume are as long as both old and new volumes have the same units.)

⬤ Think about a bicycle pump. Pushing down on the piston makes the volume of the air inside the pump smaller. Associating a new situation with a familiar one helps you to understand new ideas.

⬤ Always check to be sure your work is reasonable. This answer is reasonable because 38 mL is less than 54 mL, the original volume.

PRACTICE PROBLEMS

5.5 From the graph of pressures and volumes of a gas (Figure 5.5), what was the volume of the gas sample at 58.0 cmHg?

5.6 If the pressure on a sample of gas is doubled at constant temperature, what happens to the volume?

5.7 If the volume of a sample of gas at constant temperature is increased to three times the original volume, what happens to the pressure?

5.8 A sample of a gas has a volume of 325 mL at a pressure of 538 mmHg. What will the volume be if the pressure is changed to 274 mmHg at constant temperature?

5.9 (a) A sample of neon gas has a volume of 9.80 mL at a pressure of 437 Pa. What pressure is needed to change the volume to 6.00 mL at constant temperature? (b) A sample of argon gas has a volume of 9.80 mL at a pressure of 437 Pa. What pressure is needed to change the volume to 6.00 mL at constant temperature?

5.3 RELATION BETWEEN VOLUME AND TEMPERATURE OF A GAS

Heating or cooling the apparatus shown in Figure 5.4 would not be very convenient. Instead, let's put our gas sample in a syringe as shown in ▬Figure 5.6. Table 5.2(a) and ▬Figure 5.7 summarize the results of typical experiments at 760.0 and 570.0 mmHg. In Figure 5.7, the "curves" are straight lines. However, although volume increases as temperature increases, volume is *not* directly proportional to the Celsius temperature:

$$V \neq \text{a constant} \times t$$

For example, increasing the temperature from 20.0 to 40.0 °C, a doubling of the Celsius temperature, does not double the volume. At 760.0 mmHg, volume only increases from 26.9 to 28.6 mL.

If both lines in Figure 5.7 are extrapolated to very low temperatures (extrapolations are shown by dashed lines), both intersect the temperature axis at −273 °C; that is, volume has decreased to zero at −273 °C. The volume–temperature lines for other pressures also intersect the *x*-axis at −273 °C. Because volume can't very well become less than zero, the temperature, −273 °C, must be the lowest temperature possible; therefore, in 1848, Lord Kelvin proposed a new temperature scale. On this new temperature scale, now called the Kelvin scale, the *lowest possible temperature,* which is called **absolute zero,** is defined as zero. More accurate measurements give −273.15 °C for this lowest possible temperature.* The degrees

 To find an atmospheric pressure as low as 570.0 mmHg, we would have to move the apparatus to the top of an 8000-foot mountain.

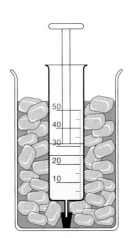

(a) Syringe in ice-water bath.
Temperature = 0.0 °C

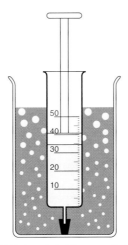

(b) Syringe in boiling water bath.
Temperature = 100.0 °C

▬ FIGURE 5.6 Apparatus for the determination of the volume–temperature relationship of a gas at the constant pressure of the atmosphere. Part (b) is at sea level.

---

*Determination of absolute zero by measurement of gas volumes at very low temperatures is impossible because all gases condense to liquids before the temperature reaches −273.15 °C. However, the value −273.15 °C has been checked by another method. You will learn about this method if you study thermodynamics in physics. At present, the record low temperature is 700 nK, attained with cesium in 1994 at the National Institute of Standards and Technology.

Although changing one variable at a time and keeping all others the same is the simplest way to examine the effects of different variables, with modern statistical methods and computers it is not the most efficient. Today, if experiments are carefully designed, the effects of a number of variables can be investigated at one time.

TABLE 5.2(a)	Volume of a Sample of Gas at Different Temperatures Under Constant Pressure	
Temperature, °C	760.0 mmHg[a] Volume, mL	570.0 mmHg Volume, mL
0.0	25.0	33.4
20.0	26.9	35.7
40.0	28.6	38.2
60.0	30.4	40.7
80.0	32.4	43.1
100.0	34.2	—[b]

[a]Barometric pressure plus pressure from force of gravity on plunger.
[b]At 570.0 mmHg, water boils at about 92 °C, making it impractical to get a value at this temperature and pressure *with this kind of experiment.*

on the Kelvin scale are the same size as the degrees on the Celsius scale. Temperature on the Kelvin scale equals temperature on the Celsius scale plus 273.15°:

$$T\ \text{K} = (t\ °\text{C} + 273.15\ °\text{C})\left(\frac{1\ \text{K}}{1\ °\text{C}}\right) \tag{2.3}$$

as we learned in Section 2.9.

If the Kelvin temperature scale is used instead of the Celsius scale, the volume of a sample of gas is directly proportional to the temperature:

$$V = \text{a constant} \times T \qquad \text{or} \qquad \frac{V}{T} = \text{a constant}$$

◗ An everyday example of the meaning of proportional is the cost of oranges. The cost of three oranges is proportional to the price of one orange.

For example, if $t = 0.0\ °\text{C}$, then

$$T = 0.0 + 273.2 = 273.2\ \text{K}$$

At $t = 0.0\ °\text{C}$ and 760.0 mmHg, volume = 25.0 mL [Table 5.2(a)]:

$$\frac{V}{T} = \frac{25.0\ \text{mL}}{273.2\ \text{K}} = 0.0915\ \text{mL} \cdot \text{K}^{-1}$$

If $t = 20.0\ °\text{C}$, then

$$T = 20.0 + 273.2 = 293.2\ \text{K}$$

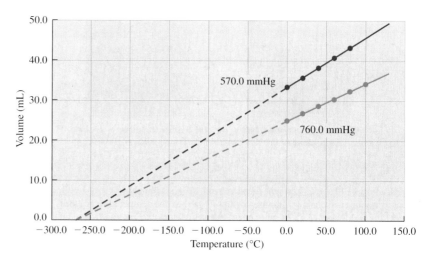

▬ FIGURE 5.7 Volume of a sample of gas at different temperatures (pressure constant).

At $t = 20.0$ °C and 760.0 mmHg, volume $= 26.9$ mL:

$$\frac{V}{T} = \frac{26.9 \text{ mL}}{293.2 \text{ K}} = 0.0917 \text{ mL} \cdot \text{K}^{-1}$$

All volume–temperature ratios for 760.0 mmHg are shown in Table 5.2(b). As you can see, the ratio of volume to Kelvin temperature is constant within experimental error. When similar experiments were carried out, substituting other gases for air, the same results were obtained.

The law that, at constant pressure, the volume of a sample of a gas is directly proportional to the Kelvin temperature is called **Charles's law.** Charles was the French chemist who first observed, in 1787, that gases expand at constant rates as the temperature rises.

Charles's Law

At constant pressure, the volume of a sample of a gas is directly proportional to the Kelvin or absolute temperature:

$$V = a \text{ constant} \times T \qquad \text{or} \qquad \frac{V}{T} = a \text{ constant}$$

In everyday language, increasing the temperature of a sample of a gas at constant pressure increases the volume of the gas.

Charles's law makes possible calculation of any one of the four variables, V_1, V_2, T_1, or T_2, if the values of the other three are known. For example, if $(V_1/T_1) = $ a constant and $(V_2/T_2) = $ the same constant, then

$$\frac{V_1}{T_1} = \frac{V_2}{T_2}$$

Rearrangement gives

$$T_1\left(\frac{V_2}{V_1}\right) = T_2 \qquad \text{and} \qquad V_1\left(\frac{T_2}{T_1}\right) = V_2$$

If the temperature of a sample of gas is changed while the pressure is kept constant, the old volume, V_1, can be converted to the new volume, V_2, by multiplying the old volume by (T_2/T_1). Again, reason from the fact that when a sample of a gas is heated at constant pressure, volume increases. Do not try to memorize equations. Sample Problem 5.2 illustrates a Charles's law calculation.

SAMPLE PROBLEM

5.2   A sample of gas has a volume of 364 mL at a temperature of 25.0 °C. What will the volume be if the temperature is changed to 100.0 °C while the pressure is kept constant?

Solution   First, convert the temperatures to kelvin so that you do not forget to do so. Then make a table of the information given and asked for in the problem:

	Volume, mL	Temperature, °C	Temperature, K
Old	364	25.0	298.2
New	?	100.0	373.2

TABLE 5.2(b)

Volume–Temperature Ratios Under Constant Pressure of 760.0 mmHg

Temperature, °C	K	Volume, mL	V/T, mL/K
0.0	273.2	25.0	0.0915
20.0	293.2	26.9	0.0917
40.0	313.2	28.6	0.0913
60.0	333.2	30.4	0.0912
80.0	353.2	32.4	0.0917
100.0	373.2	34.2	0.0916

To convert the old volume to the new volume, multiply by a factor obtained from the old and new temperatures. Do *not* try to memorize whether to multiply by (T_1/T_2) or by (T_2/T_1). Think—is the new temperature higher or lower than the old temperature? In this problem, it is higher. If you increase the temperature at constant pressure, will the volume become larger or smaller? You should know by now that it will become larger. To make the volume larger, should the old volume be multiplied by

$$\frac{298.2 \text{ K}}{373.2 \text{ K}} \quad \text{or by} \quad \frac{373.2 \text{ K}}{298.2 \text{ K}} ?$$

Multiplication by

$$\frac{373.2 \text{ K}}{298.2 \text{ K}}$$

will make the volume larger:

$$\text{new volume} = 364 \text{ mL} \left(\frac{373.2 \text{ K}}{298.2 \text{ K}} \right) = 456 \text{ mL}$$

Again, it does not matter what the gas is. The same results were obtained in the temperature–volume experiment when other gases were substituted for air. But temperature *must* be in kelvin; addition is involved in the conversion of one temperature unit into another temperature unit.

PRACTICE PROBLEMS

5.10   Convert (a) 0.0 °C to K, (b) 40.0 °C to K, (c) −37.4 °C to K, (d) 365 K to °C, and (e) 205 K to °C.

5.11   If the Kelvin temperature of a sample of helium is decreased by one-half at constant pressure, what will happen to the volume of the sample?

5.12   To increase the volume of a sample of gas fourfold at constant pressure, what must be done to the Kelvin temperature?

5.13   A sample of gas has a volume of 3.7 L at 40.0 °C. What will the volume be if the temperature is changed to 0.0 °C while the pressure is kept constant?

5.14   A sample of gas has a volume of 6.82 mL at 21.0 °C. For the volume to become 10.00 mL at constant pressure, what will the Celsius temperature have to be?

5.4   STANDARD TEMPERATURE AND PRESSURE

Because the volumes of samples of gases depend on their temperature and pressure, it is convenient to define a standard temperature and pressure for reporting volumes so that observations from different laboratories around the world can be compared. **Standard temperature** is *0 °C or 273.15 K.* **Standard pressure** is *760 mmHg or 1 atm (1.013 25 × 10⁵ Pa).** Observed volumes can be converted to volumes at standard temperature and pressure (STP) by applying both Boyle's and Charles's laws.

---

*In 1982 the IUPAC Commission on Thermodynamics recommended the use of 1 bar (10⁵ Pa) as standard pressure. The old definition is used in this book because a majority of American chemists, including the editors of handbooks containing tables of data, still use the old definition.

5.3 A sample of gas has a volume of 2.51 L at 26.3 °C and 756.2 mmHg. What would its volume be at STP?

Solution First convert the temperature to kelvin. Then make a table of the information given and asked for in the problem:

	Volume, L	Pressure, mmHg	Temperature, °C	Temperature, K
Old	2.51	756.2	26.3	299.5
New	?	760.0	0.0	273.2

To convert the old volume to the new volume, multiply by factors obtained from the old and new pressures and temperatures. To figure out these factors, think—is the new pressure higher or lower than the old pressure? The new pressure is higher than the old pressure. As a result of the change in pressure, will the new volume be larger or smaller than the old volume? The new volume will be smaller than the old volume as a result of the change in pressure. The old volume should be multiplied by

$$\frac{756.2 \text{ mmHg}}{760.0 \text{ mmHg}}$$

Then ask yourself, is the new temperature higher or lower than the old temperature? The new temperature is lower than the old temperature. As a result of the change in temperature, will the new volume be larger or smaller than the old volume? The new volume will be smaller than the old volume as a result of the change in temperature. The old volume should be multiplied by

$$\frac{273.2 \text{ K}}{299.5 \text{ K}}$$

Combination of the two steps gives

$$V_{new} = 2.51 \text{ L} \left(\frac{756.2 \text{ mmHg}}{760.0 \text{ mmHg}}\right)\left(\frac{273.2 \text{ K}}{299.5 \text{ K}}\right) = 2.28 \text{ L}$$

5.15 A sample of gas has a volume of 419 mL at 19.8 °C and 765.4 mmHg. What will its volume be at STP?

5.16 A sample of oxygen has a volume of 274 mL at STP. What will its volume be at 758.4 mmHg and 27.3 °C?

5.5 **GAY-LUSSAC'S LAW OF COMBINING VOLUMES AND AVOGADRO'S LAW**

Gay-Lussac and Charles were both balloonists.

The French chemist Joseph Gay-Lussac studied the volumes of gases reacted and formed in chemical reactions. He published his results in 1808: For reactions taking place at constant temperature and pressure, the volumes of the gases involved are in the ratios of small whole numbers. For example,

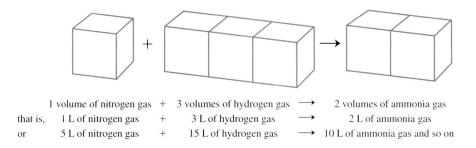

	1 volume of nitrogen gas	+	3 volumes of hydrogen gas	$\longrightarrow$	2 volumes of ammonia gas
that is,	1 L of nitrogen gas	+	3 L of hydrogen gas	$\longrightarrow$	2 L of ammonia gas
or	5 L of nitrogen gas	+	15 L of hydrogen gas	$\longrightarrow$	10 L of ammonia gas and so on

Gay-Lussac's Law of Combining Volumes

At constant temperature and pressure, the volumes of gases involved in chemical reactions are in the ratios of small whole numbers.

To explain why the volumes of gases involved in chemical reactions are in the ratios of small whole numbers, Italian physicist Amedeo Avogadro suggested, in 1811, that *equal volumes of all gases at the same temperature and pressure contain the same number of molecules.* In other words, 100 molecules of any gas have the same volume. This idea at first seems contrary to common sense: 100 oranges do not have the same volume as 100 peas or 100 grapefruit. However, remember that gases have very low densities compared to liquids and solids (Table 2.5). The densities of liquids and solids are of the order of grams per cubic centimeter, whereas the densities of gases are of the order of 0.001 grams per cubic centimeter. Picturing gases as mostly empty space, with the molecules occupying only a small part of the total volume of the gas, would account for the low densities of gases and explain how equal volumes of all gases at the same temperature and pressure can contain the same number of molecules (see ▬Figure 5.8). You may wonder how the gas molecules in Figure 5.8 can occupy so much space. Think of a small boy swinging a stone on the end of a string around himself. Although the stone is not very big, no one would want to come very close to the little boy. The small stone is able to occupy a large volume because it is moving fast. The small gas molecules are able to occupy a large volume because they are moving rapidly. This *microscopic picture of a gas as composed of small molecules moving rapidly about in a large volume* is called the **kinetic-molecular theory.** We will come back to it in Section 5.10 after we have finished discussing the properties of macroscopic samples of gases.

▬ FIGURE 5.8 Picturing a gas as mostly empty space accounts for the low densities of gases and explains how equal volumes of different gases at the same temperature and pressure can contain the same number of molecules.

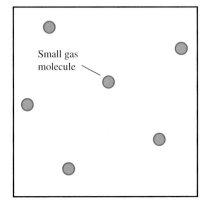

Small gas molecule

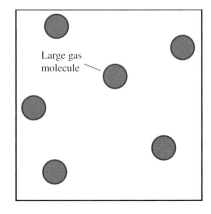

Large gas molecule

Avogadro was the first person to distinguish between atoms and molecules. He used Gay-Lussac's observations to show that the unit particles of gaseous elements must be molecules made up of more than one atom. For example, from the observation that

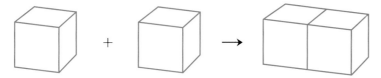

1 volume of hydrogen gas   +   1 volume chlorine gas ⟶ 2 volumes of hydrogen chloride gas

it is obvious that, if the simplest particles of hydrogen gas and of chlorine gas are atoms, then for one atom of hydrogen to combine with one atom of chlorine to give two molecules of hydrogen chloride, the hydrogen and chlorine atoms would have to be divided between two molecules of hydrogen chloride:

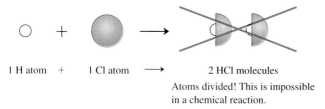

1 H atom   +   1 Cl atom   ⟶   2 HCl molecules
Atoms divided! This is impossible
in a chemical reaction.

However, division of atoms is impossible if atoms of elements are the smallest particles of elements and cannot be divided in chemical reactions.

If the smallest particle of hydrogen gas is a molecule made up of two atoms of hydrogen and the smallest particle of chlorine gas is a molecule made up of two atoms of chlorine, however, then the molecules can be divided, and two molecules of hydrogen chloride can be formed.

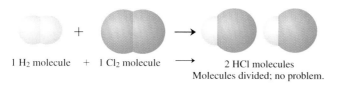

1 H₂ molecule   +   1 Cl₂ molecule   ⟶   2 HCl molecules
Molecules divided; no problem.

Now let's see how Avogadro's idea explains Gay-Lussac's law of combining volumes. We'll use the formula H_2 for hydrogen gas, the formula Cl_2 for chlorine gas, and the formula HCl for hydrogen chloride, although actually these formulas weren't known for sure until 1858. The equation for the reaction is then

$$H_2(g) + Cl_2(g) \longrightarrow 2HCl(g)$$

Be sure to notice that the coefficient of each substance in the balanced chemical equation is the same as the volume of each gas that reacts:

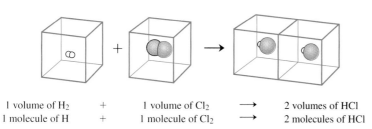

| 1 volume of H_2 | + | 1 volume of Cl_2 | ⟶ | 2 volumes of HCl |
| 1 molecule of H | + | 1 molecule of Cl_2 | ⟶ | 2 molecules of HCl |

In this case, the volume of the product, hydrogen chloride, is equal to the sum of the volumes of the reactants. This is not always true; for example, we saw earlier that one volume of nitrogen gas reacts with three volumes of hydrogen gas to form two volumes of ammonia, not four:

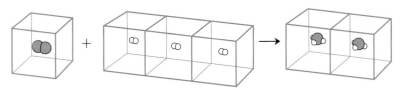

1 volume of nitrogen gas + 3 volumes of hydrogen gas $\longrightarrow$ 2 volumes of ammonia gas

If equal volumes of gases at the same temperature and pressure contain equal numbers of molecules, then the volume of a sample of a gas must depend on the number of particles. For example, two molecules of hydrogen chloride gas occupy twice as much volume as one molecule of hydrogen chloride gas, and three molecules of hydrogen gas occupy three times as much volume as one molecule of hydrogen gas. This has been proved experimentally and is now known as **Avogadro's law.**

Avogadro's Law

The volume of a gas at constant temperature and pressure is directly proportional to the number of molecules of the gas, n:

$$V = \text{a constant} \times n$$

In everyday language, at constant temperature and pressure, more gas molecules take up more space.

Because the volume of a gas is proportional to the number of molecules of the gas, volumes of gases can be used to count numbers of molecules of gases. Avogadro's law connects the properties of gases to the stoichiometry of chemical reactions that involve gases.

SAMPLE PROBLEM

5.4 (a) How many liters of chlorine gas are required to react with 6.3 L of hydrogen gas? (b) How many liters of hydrogen chloride gas will be formed by the reaction in part (a), assuming that reaction goes to completion? (All volumes are measured under the same conditions of temperature and pressure.)

Solution As for all stoichiometry problems, begin by writing the equation for the reaction:

$$H_2(g) + Cl_2(g) \longrightarrow 2HCl(g)$$

According to the equation, one molecule of hydrogen gas reacts with one molecule of chlorine gas to form two molecules of hydrogen chloride gas. In other words, one volume of hydrogen gas reacts with one volume of chlorine gas to form two volumes of hydrogen chloride gas. The relationships between volumes of hydrogen and chlorine reacting and volume of hydrogen chloride formed can be used to write conversion factors.
  (a) According to the equation

1 volume $H_2(g)$ ⇌ 1 volume $Cl_2(g)$

To react with 6.3 L $H_2(g)$,

$$6.3 \text{ L } H_2(g) \left[\frac{1 \text{ L } Cl_2(g)}{1 \text{ L } H_2(g)} \right] = 6.3 \text{ L } Cl_2(g)$$

are required.

(b) According to the equation

$$1 \text{ volume } H_2(g) \Leftrightarrow 2 \text{ volumes } HCl(g)$$

From 6.3 L $H_2(g)$,

$$6.3 \text{ L } H_2(g) \left[\frac{2 \text{ L } HCl(g)}{1 \text{ L } H_2(g)} \right] = 12.6 \text{ L } HCl(g)$$

will be formed.

PRACTICE PROBLEMS

5.17 (a) If 1.204×10^{24} molecules of $H_2(g)$ occupy a volume of 50.6 L, what volume will 1.204×10^{24} molecules of $N_2(g)$ occupy if both gases are at the same temperature and pressure? (b) What volume will 2.408×10^{24} molecules of $H_2(g)$ occupy (still at the same temperature and pressure)?

5.18 The equation for the complete combustion of methane is

$$CH_4(g) + 2O_2(g) \longrightarrow 2H_2O(g) + CO_2(g)$$

(a) How many liters of oxygen gas are required to react with 0.874 L of $CH_4(g)$? (b) How many liters of water vapor and how many liters of carbon dioxide gas will be formed by the reaction in part (a), assuming that reaction is quantitative? (All volumes are measured under the same conditions of temperature and pressure.)

5.19 Carbon monoxide gas reacts with oxygen gas to form carbon dioxide gas. (a) What volume of oxygen is needed to convert 21.6 L of carbon monoxide gas to carbon dioxide gas? (b) How many liters of carbon dioxide gas will be formed by the reaction in part (a), assuming that reaction goes to completion? (All volumes are measured under the same conditions of temperature and pressure.)

Remember that the mole, a package that contains 6.022×10^{23} molecules, is the connection between the microscopic world of molecules and the real world of macroscopic samples. The volume of gas measured at STP that contains Avogadro's number of molecules—that is, one mole of gas—is called the **STP molar volume** and has been found by experiment to be *22.4141 L*. To give you an idea of this volume, a cube 11.1 in. (28.2 cm) on an edge has a volume of 22.4 L. *One mole of any gaseous sample contains 6.022×10^{23} molecules and has a volume of 22.4 L at STP* whether the sample is a pure substance such as oxygen gas or a gaseous mixture such as air.

Three basketballs have a total volume of about 22.4 L.

5.6 THE IDEAL GAS EQUATION

In Sections 5.2, 5.3, and 5.5, we saw that the behavior of gases can be described by three laws:

Boyle's law	Charles's law	Avogadro's law
$V = \dfrac{\text{constant}_B}{P}$ (T, n constant)	$V = \text{constant}_C \times T$ (P, n constant)	$V = \text{constant}_A \times n$ (T, P constant)

Part of the label from an aerosol can. An "empty" aerosol can is filled with vapors from the contents and has a constant volume. If the temperature of the empty aerosol can is increased by incinerating the can, what will happen to the pressure and, as a result, to the can?

Because V is proportional to $(1/P)$, T, and n, it would seem that V should be proportional to all three,

$$V = \text{a constant}\left(\frac{T \cdot n}{P}\right) \tag{5.2}$$

and experiments show that equation 5.2 is correct.* Equation 5.2 is called the **ideal gas equation** or **ideal gas law** and is usually written

$$PV = nRT$$

where P, V, and T have their usual meanings; n is the number of moles of gas; and R is a proportionality constant.

Ideal Gas Equation

$$PV = nRT$$

where P is the pressure of the gas, V is the volume of the gas, n is the number of moles of gas, R is a proportionality constant, and T is the Kelvin temperature.

The value of the proportionality constant, R, which is called the **gas constant,** can be calculated from the STP molar volume, 22.4141 L/mol, which is the volume of 1 mol of gas at 0 °C (273.15 K) and 1 atm of pressure. Solution of the ideal gas equation for R gives

$$R = \frac{PV}{nT} = \frac{(1\text{ atm})(22.4141\text{ L})}{(1\text{ mol})(273.15\text{ K})} = 0.082\ 058\ \frac{\text{atm} \cdot \text{L}}{\text{mol} \cdot \text{K}}$$

Note that the value of R depends on the units used for pressure and volume. Table 5.3 lists the most useful values for R.

A *mathematical relation between the temperature, pressure, and volume of a given quantity of material* such as the ideal gas equation is called an **equation of state.** A *gas that behaves exactly as described by the ideal gas equation* is called an **ideal gas.** No real gases are ideal. But this one simple equation of state describes the behavior of most real gases to within a few percent at temperatures around room temperature and above, and under pressures of about one atmosphere or less. (The differences between real and ideal gases will be discussed in more detail in Section 5.11.) Liquids and solids are different from gases in this respect. The equations of state for liquids and solids are more complicated and differ from one substance to another.

TABLE 5.3

Values of the Gas Constant, R, in Different Units

$0.082\ 06$ atm $\cdot$ L $\cdot$ mol^{-1} $\cdot$ K^{-1}
62.36 mmHg $\cdot$ L $\cdot$ mol^{-1} $\cdot$ K^{-1}
8.315 kPa $\cdot$ dm^3 $\cdot$ mol^{-1} $\cdot$ K^{-1}
8.315 J $\cdot$ mol^{-1} $\cdot$ K$^{-1\ a}$

^{a}The joule, J, is the SI unit of energy.

5.7 USING THE IDEAL GAS EQUATION TO SOLVE PROBLEMS

The ideal gas equation can be used to solve a variety of problems about gases. If the values of any three of the four variables, P, V, n, and T, are known, the value of the fourth variable can be calculated by means of the ideal gas equation. Stoichiometry problems involving gases as reactants or products can be solved, the

---

*Mathematically, Boyle's, Charles's, and Avogadro's laws cannot simply be combined into one equation because the three laws are true for different conditions: constant temperature and number of moles for Boyle's law; constant pressure and number of moles for Charles's law; and constant temperature and pressure for Avogadro's law. However, equation 5.2 simplifies to the three gas laws; for example, if n and P are constant, then $V = \text{a constant} \times T$, which is Charles's law. Boyle's, Charles's, and Avogadro's laws are special cases of the more general ideal gas law.

densities of gases can be calculated if their molecular formulas are known, and the molecular masses of gases can be determined by measuring their densities. The following sample problems illustrate these types of calculation.

SAMPLE PROBLEMS

5.5 What is the volume of 0.683 mol of nitrogen gas at 25.2 °C and 767.3 mmHg?

Solution As usual, first convert the temperature to kelvin:

$$25.2 + 273.2 = 298.4 \text{ K}$$

In using the ideal gas equation to solve problems, you must be very careful to use the value for the gas constant that has the right units for pressure and volume. The amount of substance must always be in moles and the temperature in kelvin. In this problem, the amount of substance is given in moles. Because pressure is given in mmHg, we must use the value, $62.36 \text{ mmHg} \cdot \text{L} \cdot \text{mol}^{-1} \cdot \text{K}^{-1}$, for R. The volume calculated will then be in liters.

To avoid copying numbers, begin by solving the ideal gas equation for the unknown quantity, which is V in this problem, and then substitute the numbers in the equation that results and do the necessary arithmetic:

$$V = \frac{nRT}{P} = \frac{(0.683 \text{ mol})(62.36 \text{ mmHg} \cdot \text{L} \cdot \text{mol}^{-1} \cdot \text{K}^{-1})(298.4 \text{ K})}{(767.3 \text{ mmHg})} = 16.6 \text{ L}$$

5.6 How many moles of gas are there in a sample having a volume of 6.8 L at 20.6 °C and 763.4 mmHg?

Solution Solution of this problem is similar to solution of Sample Problem 5.5 except that the unknown quantity is the number of moles, n. On the Kelvin scale, the temperature is 293.8 K:

$$\frac{PV}{RT} = n = \frac{(763.4 \text{ mmHg})(6.8 \text{ L})}{(62.36 \text{ mmHg} \cdot \text{L} \cdot \text{mol}^{-1} \cdot \text{K}^{-1})(293.8 \text{ K})} = 0.28 \text{ mol}$$

5.7 Oxygen gas is sometimes made in the laboratory by heating solid potassium chlorate, $KClO_3(s)$. The other product is potassium chloride. What volume of oxygen gas at 21.6 °C and 763.8 mmHg will be obtained from 29.6 g of potassium chlorate?

Solution This is a stoichiometry problem. Remember that the first step in solving stoichiometry problems is to write the equation for the reaction. The equation for this reaction is

$$2KClO_3(s) \xrightarrow{\text{heat}} 2KCl(s) + 3O_2(g)$$

The second step in solving stoichiometry problems is to write, above the equation, the information given and asked for in the problem and, below the equation, the formula masses of the substances whose quantities are given in grams. The third step is to write the "recipe" for the reaction (from the coefficients in the equation) below the formula masses. The completed table for this problem is

Problem	29.6 g		? L at 21.6 °C and 763.8 mmHg
Equation	$2KClO_3(s) \xrightarrow{\text{heat}}$	$2KCl(s) +$	$3O_2(g)$
Formula mass, u	123	—	—
Recipe, mol	2	2	3

Now we are ready to solve the problem by converting the mass of $KClO_3$ used to moles, using the "recipe" to convert moles of $KClO_3$ used to moles of $O_2(g)$

The Importance of Gas Densities

From popping popcorn to predicting the weather, gas densities have many real-world results. Popcorn pops because the density of water vapor, like the densities of all gases, is of the order of 0.001 g/mL, whereas the density of liquid water is 1.00 g/mL. When heat from the hot air in a popcorn popper converts water in the corn kernel to vapor, the volume of the water increases about a thousandfold almost instantaneously, and the kernel blows up or pops.

Explosives make use of similar rapid large changes in volume. Nitroglycerin, the active ingredient of dynamite,* is an example. When nitroglycerin, which is a liquid under ordinary conditions, explodes, each mole of nitroglycerin is converted to eight moles of gaseous products:

$$2C_3H_5(ONO_2)_3(l) \xrightarrow[\text{or mechanical shock}]{\text{rapid heating}}$$

$$3CO(g) + 2CO_2(g) + 4H_2O(g)$$
$$+ 6NO(g) + H_2CO(g)$$

*Dynamite was invented by the Swedish chemist Alfred Nobel, who thought that explosives would make war so terrible that there would be no more wars. When he discovered he was wrong, he left his fortune for the prizes that bear his name.

Enough heat is given off to raise the temperature to about 5000 °C (9000 °F). The result is an almost instantaneous increase of 2.4×10^4-fold in volume. It is no wonder that things are blown to bits. Explosives are used in industry to save billions of man-hours of work each year. For example, explosives are used to mine coal and metal ores, to quarry rock, and to dig tunnels. Explosives, of course, are also used in weapons. Like all chemicals, whether explosives are used for constructive or destructive purposes depends on people.

A knowledge of gas densities might save your life if you are ever involved in an accident, such as the overturn of a truck carrying liquid chlorine or ammonia, that involves the release of a gas. If the density of the gas is greater than the density of air, like the density of chlorine, the gas will tend to settle into any low spots, and you should head for the highest place available (run uphill or climb a tree). If the density of the gas is lower than the density of air, like the density of ammonia, the gas will rise, and you should dive into the nearest ditch or run downhill. Most industrial gases, except hydrogen, ammonia, and methane, are more dense than air. In ▪Figure 5.9, bromine vapor is used to

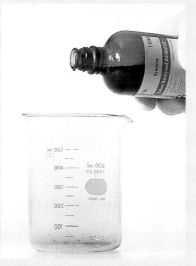

▪ FIGURE 5.9 Bromine vapor is denser than air and flows downward from the bottle into the beaker. Notice that the color is darkest at the bottom of the beaker because the concentration of bromine vapor is greatest there.

show how a gas denser than air flows downward. Methyl isocyanate, H_3CNCO, the gas that was accidentally released at Bhopal in India in 1984, is denser than air and spread out over the flat ground, killing about 3100 people and injuring many thousands.

The equation for the decomposition of nitroglycerin is from the Kirk-Othmer Encyclopedia of Chemical Technology.

formed, and, finally, using the ideal gas equation to find the volume of the oxygen:

$$29.6 \text{ g KClO}_3 \left(\frac{1 \text{ mol KClO}_3}{123 \text{ g KClO}_3} \right) \left(\frac{3 \text{ mol O}_2}{2 \text{ mol KClO}_3} \right) = 0.361 \text{ mol O}_2 \text{ formed}$$

$$V = \frac{nRT}{P} = \frac{(0.361 \text{ mol})(62.36 \text{ mmHg} \cdot \text{L} \cdot \text{mol}^{-1} \cdot \text{K}^{-1})(294.8 \text{ K})}{763.8 \text{ mmHg}} = 8.69 \text{ L}$$

5.8 What is the density of oxygen gas at 62.4 °C and 3.45 atm?

Solution The volume of one mole of oxygen gas at 62.4 °C and 3.45 atm can be calculated from the ideal gas equation:

$$V = \frac{nRT}{P} = \frac{(1 \text{ mol})(0.082 \, 06 \text{ atm} \cdot \text{L} \cdot \text{mol}^{-1} \cdot \text{K}^{-1})(335.6 \text{ K})}{3.45 \text{ atm}} = 7.98 \text{ L}$$

The mass of 1 mol O_2 is 32.0 g. Substitution of the mass and volume of 1 mol of oxygen gas in the definition of density gives

$$\text{density} = \frac{\text{mass}}{\text{volume}} = \frac{32.0 \text{ g}}{7.98 \text{ L}} = 4.01 \text{ g/L}$$

FIGURE 5.10 Weather balloons *(left)* and the Goodyear blimp *(right),* like party balloons, are filled with helium.

Carbon dioxide, which is a by-product of the formation of alcohol by fermentation, is denser than air and collects at low spots in wineries and breweries. Although carbon dioxide is not poisonous, workers have been asphyxiated. The fact that carbon dioxide gas is denser than air is put to use in carbon dioxide fire extinguishers. Carbon dioxide gas is nonflammable and forms a protective layer over a fire keeping air out.

If you take an organic chemistry laboratory, you will be warned that the vapors of organic solvents are denser than air and flow along the benchtop. Most organic solvents are flammable, and their vapors form explosive mixtures with air. Many fires and explosions have been caused by seemingly remote flames.

Water vapor is less dense than air. Therefore, moist air is slightly less dense than dry air. The barometric pressure is

low when air is moist, and decreasing barometric pressure is an indicator of wet weather.

Hot-air balloons (see page 148) float about in the sky because the density of hot air is lower than the density of cool air. Blimps and weather balloons (Figure 5.10) are filled with helium, which is much less dense than air.

5.9 A 2.78-g sample of a gas occupies a volume of 4.24 L at 23.6 °C and 755.1 mmHg. What is the molecular mass of the gas?

Solution We can use the ideal gas equation to calculate how many moles of gas there are in the sample:

$$n = \frac{PV}{RT} = \frac{(755.1 \text{ mmHg})(4.24 \text{ L})}{(62.36 \text{ mmHg}\cdot\text{L}\cdot\text{mol}^{-1}\cdot\text{K}^{-1})(296.8 \text{ K})} = 0.173 \text{ mol}$$

This number of moles has a mass of 2.78 g. From this relationship, we can calculate the number of grams in one mole:

$$1 \text{ mol}\left(\frac{2.78 \text{ g}}{0.173 \text{ mol}}\right) = 16.1 \text{ g in 1 mol}$$

The molecular mass of the gas is 16.1 u.

Notice that the density of the gas could easily be calculated from the information given in this problem, "A 2.78-g sample of a gas occupies a volume of 4.24 L." However, calculation of the density is not necessary to solve the problem.

In the Dalton's Law module of the Gases section, students can open the stopcock between two connected containers of gas and watch the gases mix; they can then calculate the partial pressure of each gas in the mixture. The module asks a wide variety of questions, most of them written using randomized variables so that a different answer is found each time the student uses the exercise.

PRACTICE PROBLEMS

5.20 What is the volume in liters of 2.8 mol of nitrogen gas at 22.1 °C and 753.4 mmHg?

5.21 How many moles of gas are there in a sample having a volume of 5.7 L at 25.9 °C and 759.4 mmHg?

5.22 When mercury(II) oxide is heated, it decomposes to mercury metal and oxygen gas. What volume of oxygen gas at 18.7 °C and 764.3 mmHg will be obtained from 27.65 g of mercury(II) oxide?

5.23 What is the density of helium gas at 22.4 °C and 57.6 atm? (Assume ideal gas behavior.)

5.24 At the same temperature and pressure, which has the higher density, fluorine gas or chlorine gas? Give your reasoning.

5.25 A 3.29-g sample of a gas occupies a volume of 2.87 L at 24.7 °C and 758.3 mmHg. What is its molecular mass?

The Haber process was developed in Germany during World War I to make explosives and fertilizers. Most fertilizers responsible for recent increases in the world's food supply are made from ammonia manufactured by the Haber process.

5.8   DALTON'S LAW OF PARTIAL PRESSURES

Gaseous mixtures are as important as pure gases. For example, dry air is a mixture of 75.5% (by mass) nitrogen, 23.1% oxygen, and 1.3% argon. The remaining 0.1% is mainly carbon dioxide. Mixtures of gases are also important in industry. For example, the Haber process for making ammonia, the most important process for making nitrogen from air into useful products such as fertilizers, synthetic fibers, and explosives, involves mixtures of nitrogen, hydrogen, and ammonia gases.

Under ordinary conditions* all gaseous mixtures are solutions—that is, gaseous mixtures consist of just one phase. If enough time has passed for the gases in a mixture to become thoroughly mixed, the composition is the same everywhere in a sample. Smoke, which is a mixture of small solid particles with the gaseous mixture air, is not classified as a gaseous mixture. Neither is fog, which is a mixture of tiny drops of liquid water with air. Smoke and fog are colloids; colloids are discussed in Chapter 13.

The *pressure of each gas in a mixture* is called the **partial pressure** of that gas. In 1802, just before he proposed his atomic theory, John Dalton observed that, if gases do not react with each other, the individual gases in a mixture do not have any effect on each other's pressures. The total pressure of a mixture of gases is equal to the sum of the partial pressures of the gases in the mixture.

Dalton's Law of Partial Pressures

The total pressure of a mixture of gases is equal to the sum of the partial pressures of the individual gases in the mixture:

$$P_{total} = p_{gas\ 1} + p_{gas\ 2} + \cdots$$

Sample Problem 5.10 shows how Dalton's law can be used to calculate the pressure of a mixture of gases.

———————

*At very high pressures (greater than 200 atm), some gaseous mixtures, for example, a mixture of helium and xenon, consist of more than one phase.

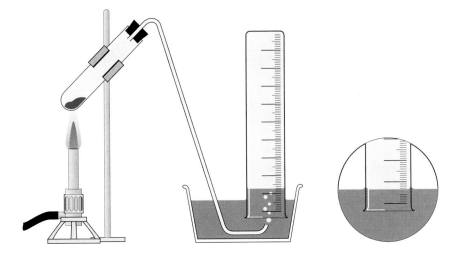

■ FIGURE 5.11 Collection of a gas over water. The detail at the right shows the graduated cylinder positioned so that the water levels inside and outside the graduated cylinder are the same. This adjustment makes the pressure of the gas in the cylinder on the water inside the cylinder equal to the pressure of the atmosphere on the water outside the cylinder. Atmospheric pressure is easily measured with a barometer. Thus, the total pressure of gas and water vapor can be obtained.

SAMPLE PROBLEM

5.10 (a) What is the partial pressure of each gas in a mixture of 2.43 mol nitrogen gas and 3.07 mol oxygen gas in a 5.00-L container at 298 K? (b) What is the total pressure of the mixture of gases in part (a)? (Assume ideal behavior.)

Our students seem to have enough to learn in this chapter without mole fractions. If you wish to introduce mole fractions and their use to calculate partial pressures of gases here, assign this part of Section 13.5.

Solution The pressure of each gas can be calculated using the ideal gas equation:

$$p_{N_2} = \frac{n_{N_2}RT}{V} = \frac{(2.43 \text{ mol})(0.082\ 06 \text{ atm} \cdot L \cdot mol^{-1} \cdot K^{-1})(298 \text{ K})}{5.00 \text{ L}} = 11.9 \text{ atm}$$

$$p_{O_2} = \frac{n_{O_2}RT}{V} = \frac{(3.07 \text{ mol})(0.082\ 06 \text{ atm} \cdot L \cdot mol^{-1} \cdot K^{-1})(298 \text{ K})}{5.00 \text{ L}} = 15.0 \text{ atm}$$

(b) According to Dalton's law, $P_{total} = p_{N_2} + p_{O_2}$. Therefore,

$$P_{total} = 11.9 \text{ atm} + 15.0 \text{ atm} = 26.9 \text{ atm}$$

Note that if the question had not asked for the partial pressure of each gas, the total pressure could have been found by one ideal gas equation calculation using the total number of moles of gas in the mixture. The total number of moles of gas in the mixture in part (a) is

$$2.43 \text{ mol} + 3.07 \text{ mol} = 5.50 \text{ mol}$$

Using the total number of moles of gas in the mixture for n in the ideal gas equation,

$$P_{total} = \frac{(5.50 \text{ mol})(0.082\ 06 \text{ atm} \cdot L \cdot mol^{-1} \cdot K^{-1})(298 \text{ K})}{5.00 \text{ L}} = 26.9 \text{ atm}$$

In the laboratory, gases are more difficult to handle than liquids and solids because they tend to escape from their containers just as gas escapes from a balloon. Also many gases are colorless so you can't see them. A container filled with a colorless gas looks the same as an empty one. (What to you looks like an "empty" container is actually filled with air!) Because gases tend to escape from their containers and are invisible, gases are often collected over water. A typical apparatus for the collection of a gas over water is shown in ■Figure 5.11. To be collected over water, a gas must not react with water and must be relatively insoluble in water.

If the level of the water inside the cylinder is the same as the level of the water outside the cylinder (Figure 5.11 detail), the total pressure of the gas above the

TABLE 5.4	Pressure of Water Vapor over Water in mmHg[a]				
Temp., °C	0.0	0.2	0.4	0.6	0.8
15	12.788	12.953	13.121	13.290	13.461
16	13.634	13.809	13.987	14.166	14.347
17	14.530	14.715	14.903	15.092	15.284
18	15.477	15.673	15.871	16.071	16.272
19	16.477	16.685	16.894	17.105	17.319
20	17.535	17.753	17.974	18.197	18.422
21	18.650	18.880	19.113	19.349	19.587
22	19.827	20.070	20.316	20.565	20.815
23	21.068	21.324	21.583	21.845	22.110
24	22.377	22.648	22.922	23.198	23.476
25	23.756	24.039	24.326	24.617	24.912
26	25.209	25.509	25.812	26.117	26.426
27	26.739	27.055	27.374	27.696	28.021
28	28.349	28.680	29.015	29.354	29.697
29	30.043	30.392	30.745	31.102	31.461
30	31.824	32.191	32.561	32.934	33.312

[a] Reprinted with permission from Weast, R. C. *CRC Handbook of Chemistry and Physics*, 65th ed.; 1984; p D-193. Copyright CRC Press, Inc., Boca Raton, FL.

water in the cylinder must be equal to the atmospheric pressure. The gas in the graduated cylinder in Figure 5.11 is a mixture of gas and water vapor from evaporation of the water. The total pressure is the sum of the pressure of the gas and the pressure of the water vapor:

$$P_{total} = p_{gas} + p_{water\ vapor}$$

The pressure of the water vapor must be subtracted from the total pressure to find the partial pressure of the gas:

$$P_{total} - p_{water\ vapor} = p_{gas}$$

The *pressure of* water *vapor formed by evaporation* of water *in a closed container* depends only on the temperature of the water (see ▪Figure 5.12) and is called the **vapor pressure** of the water. Tables of the vapor pressure of water at different temperatures, such as Table 5.4, are found in handbooks. Sample Problem 5.11 shows how to use Table 5.4 to obtain the vapor pressure of water at a given temperature. An example of how to correct the pressure of a gas for water vapor is shown in Sample Problem 5.12.

SAMPLE PROBLEMS

5.11   What is the vapor pressure of water at 21.5 °C?

Solution   The temperature, 21.5 °C, is between 21.4 and 21.6 °C.

In Table 5.4, find the line that begins with 21. In this line, find the values under 0.4 and 0.6. The value under 0.4 is 19.113 and the value under 0.6 is 19.349.

The temperature, 21.5 °C, is halfway between 21.4 and 21.6 °C. From the vapor pressure curve for water (Figure 5.12), you can see that the "curve" for such a small temperature range is very close to a straight line. The vapor pressure at 21.5 °C will be halfway between the vapor pressure at 21.4 °C and the vapor pressure at 21.6 °C:

$$(19.113 + 19.349)/2 = 19.231$$

The vapor pressure at 21.5 °C is 19.231 mmHg.

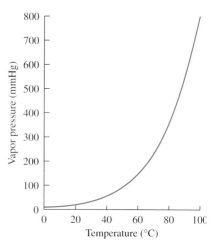

▪ FIGURE 5.12   Vapor pressure of water at different temperatures.

◐ The process of finding a value that is between two given values is called interpolation (see Appendix B.6).

5.12   In lab, a student made oxygen gas by heating $KClO_3$ and collected the oxygen over water using the apparatus shown in Figure 5.11. The student was careful to measure the volume of the gas when the levels of water inside and outside the graduated cylinder were the same. Suppose that the student collected 94.8 mL of gas, the barometric pressure was 764.7 mmHg, and the laboratory temperature was 21.5 °C. Calculate the volume the oxygen in the gas would have occupied at the same temperature and pressure if it had been dry.

Solution   Because the student measured the volume of the gas when the levels of water inside and outside the graduated cylinder were the same, the total pressure of the gas collected was equal to the barometric pressure of 764.7 mmHg. In Sample Problem 5.11, we found that the vapor pressure of water at 21.5 °C is 19.231 mmHg. Therefore, according to Dalton's law,

$$p_{oxygen} = P_{total} - p_{water\ vapor} = 764.7\ mmHg - 19.231\ mmHg = 745.5\ mmHg$$

The problem asks us to calculate the volume the oxygen would have occupied if it had been dry. If the oxygen had been dry, its pressure would have been equal to the barometric pressure, 764.7 mmHg. Now we are ready to tabulate the information given and asked for:

	Volume, mL	Pressure, mmHg	Temperature, °C
Oxygen, wet	94.8	745.5	21.5
Oxygen, dry	?	764.7	21.5

If the pressure of the oxygen increases from 745.5 to 764.7 mmHg, the volume should decrease. Multiply the volume of the wet oxygen by (745.5 mmHg/764.7 mmHg) to find the volume of the dry oxygen:

$$94.8\ mL\left(\frac{745.5\ \text{mmHg}}{764.7\ \text{mmHg}}\right) = 92.4\ mL$$

PRACTICE PROBLEMS

5.26   What is the total pressure of a mixture of nitrogen and oxygen if the partial pressure of nitrogen is 0.362 atm and the partial pressure of oxygen is 0.279 atm?

5.27   (a) What is the partial pressure in atmospheres of each gas in a mixture of 3.7 mol of nitrogen and 4.6 mol of hydrogen in a 2.5-L container at 500 (5.00 $\times$ 10^2) K? (b) What is the total pressure of the mixture of gases in (a)?

5.28   What is the total pressure in atmospheres of a mixture of 5.83 mol of nitrogen and 2.76 mol of hydrogen in a 3.75-L container at 752 K?

5.29   What is the vapor pressure of water (a) at 16.0 °C and (b) at 19.7 °C?

5.30   A student made hydrogen gas by reacting zinc with hydrochloric acid. The other product was a solution of zinc chloride. (The formula for zinc chloride is $ZnCl_2$.) The hydrogen gas was collected over water at 24.4 °C and a barometric pressure of 753.8 mmHg. The volume of the wet hydrogen gas was 0.567 L. Calculate the volume the hydrogen in the gas would have occupied at the same temperature and pressure if it had been dry.

All liquids have vapor pressures. The vapor pressures of some liquids, such as ethyl alcohol, are higher than the vapor pressure of water. The vapor pressure of ethyl alcohol is 40 mmHg at 19.0 °C; the vapor pressure of water is only 16.5 mmHg at 19.0 °C. The vapor pressures of other liquids, such as mercury, are lower than the vapor pressure of water. The vapor pressure of mercury does not reach 1

Top: *A balloon freshly filled with helium.*
Bottom: *The same balloon the morning after.*
Helium atoms diffuse rapidly through rubber.

Separation of H-1 and H-2 is an exception to the generalization that mixtures of isotopes cannot be separated by chemical means. Because the mass of H-2 is twice the mass of H-1, these isotopes can be separated chemically.

The Task Force on the General Chemistry Curriculum has pointed out that Graham's laws of diffusion and effusion do not apply to the cases usually discussed in general chemistry texts. (See Hawkes, S. J. J. Chem. Educ. 1993, 70, 836–837.) Therefore, Graham's laws do not appear in this text.

mmHg until the temperature is 126.2 °C. Mercury's low vapor pressure makes mercury a good liquid for barometers and manometers because the vapor pressure of mercury can be neglected at ordinary temperatures.

5.9 DIFFUSION AND EFFUSION RATES OF GASES

Diffusion means *spreading out*. For example, the odor of coffee brewing or dinner cooking soon spreads through the entire house. All gases diffuse rapidly to fill their containers. In the early 19th century, the English chemist Thomas Graham studied the rates of diffusion of different gases under the same conditions and found that gases having low molecular masses diffuse faster than gases having high molecular masses. Therefore, light gases diffuse from one point to another in a shorter time than heavy gases. In a given time, a gas that has a low molecular mass moves farther than a gas that has a high molecular mass (see ➡Figure 5.13).

Graham also studied the rates of effusion of gases. **Effusion** means the *flow of a gas through a small hole under conditions such that the molecules pass through the hole without colliding with each other.* The difference in the rates of effusion of two gases was used to separate the mixture of isotopes in naturally occurring uranium to make the atomic bomb dropped on Hiroshima and is still being used to make fuel for nuclear power plants. Naturally occurring uranium is a mixture of greater than 99.27 mass % U-238 and 0.72 mass % U-235; only U-235 is fissionable. Because isotopes have almost identical chemical properties, mixtures of isotopes cannot be separated by chemical means.

For separation of U-235 from U-238, uranium is converted to uranium(VI) fluoride. Naturally occurring fluorine is 100% F-19; therefore, only two kinds of uranium(VI) fluoride, $^{238}UF_6$ and $^{235}UF_6$, are formed. Uranium(VI) fluoride, UF_6, which is a solid under ordinary conditions, has a high vapor pressure and sublimes (changes from a solid to a gas) readily. The lighter $^{235}UF_6$ molecules pass through a porous barrier slightly faster than the heavier $^{238}UF_6$ molecules. Thus, the gas that passes through the barrier is slightly richer in $^{235}UF_6$ than the gas that remains behind. If the process is repeated enough times, $^{235}UF_6$ and $^{238}UF_6$ can be separated. To

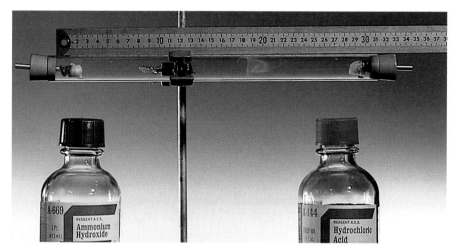

➡ FIGURE 5.13   The cotton ball at the left end of the glass tube is saturated with concentrated $NH_3(aq)$ and the cotton ball at the right end with concentrated HCl(aq). The puff of $NH_4Cl(s)$ formed by the reaction of $NH_3(g)$ with HCl(g) begins at about 18.5 cm (16.0 cm from the left cotton ball and 10.0 cm from the right cotton ball). Thus, $NH_3(g)$, which has a molecular mass of 17.0 u, has traveled about 1.6 times as far as HCl(g), which has a molecular mass of 36.5 u.

prepare uranium-235 concentrated enough for the bomb, the mixture of $^{235}UF_6$ and $^{238}UF_6$ was pumped through miles of pipe containing a total of about 2 million porous barriers.

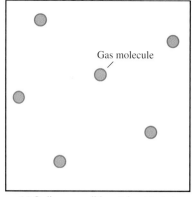

Gas molecule

(a) Ordinary conditions (about 1 atm)

PRACTICE PROBLEM

5.31   Which would diffuse faster, CO(g) or CO_2(g)? Give your reasoning.

5.10   THE KINETIC-MOLECULAR THEORY

The behavior of gases is summarized by the gas laws. It is explained by the kinetic-molecular theory, first stated by the German physicist Rudolf Clausius in 1857. According to this theory:

1. Gases consist of very small particles called molecules. The distances between molecules are large compared to the diameters of the molecules. ▪Figure 5.14 shows how the kinetic-molecular theory pictures a real gas. The molecules of an ideal gas are pictured as points; that is, they have no volume.

2. The molecules of a gas are constantly moving at very high speeds. For example, the average speed of an oxygen molecule at 25 °C is 444 m·s^{-1}, almost 1000 miles an hour. The average speed of a hydrogen molecule at 25 °C is 1768 m·s^{-1}, almost 4000 miles an hour.

3. Pressure is the result of molecules hitting the container. Each molecule of a real gas moves in a straight line until it hits another molecule or the container. At 25 °C and atmospheric pressure, one molecule of a typical real gas hits another molecule about 10^{10} times per second, so its path is completely random as shown in ▪Figure 5.15. Molecules of real gases travel only about 10^{-5} cm between collisions.

If you look at a ray of sunlight entering a window, you can see little pieces of dust in it moving suddenly as gas molecules hit them. This *random movement*, which was first observed in 1827 by the Scotch botanist Robert Brown, is called **Brownian motion.** Early in the twentieth century, theoretical work on Brownian motion by Albert Einstein* and quantitative studies of it by French physical

(b) 8 atmospheres   (c) 64 atmospheres

▪ **FIGURE 5.14**   The kinetic-molecular model of a real gas. Each picture shows the view through a cube of gas at 25 °C (298 K). The pictures are approximately to scale. The edges of the largest cube are actually 6.25 × 10^{-7} cm (6.25 nm) long, and the molecules are actually about 4 × 10^{-8} cm (400 pm) in diameter. These dimensions are magnified 8 million times in the picture. (A light microscope magnifies 1000 times.) Remember that the molecules are not really as close together as they appear to be because you are looking *through* the cube. The molecules occupy only about 0.04% of the volume occupied by the gas in part (a).

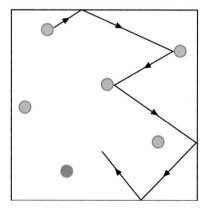

▪ **FIGURE 5.15**   Random path of a molecule of a real gas. The word *gas* is derived from the Greek word *cháos*, which originally meant "atmosphere" and now means "complete disorder."

*Einstein was a German physicist who developed the theory of relativity. He gave up his German citizenship and left Germany in 1933, soon after Hitler became chancellor. Einstein became an American citizen in 1940. Fearing that Nazi scientists would develop an atomic bomb, he wrote to President Roosevelt suggesting that the United States start work on one. Although his letter resulted in the Manhattan Project, he never worked on the bomb himself. After Hiroshima, he joined with other scientists to try to prevent further use of the bomb.

When you drop a ball, it bounces lower and lower each time until it finally stops bouncing completely. If balls were like ideal gas molecules, balls would keep on bouncing to the same height forever.

 Energy is discussed in Chapter 6.

chemist Jean-Baptiste Perrin put an end to the last doubts about the existence of atoms and molecules.

4. No attractive forces exist between the molecules of an ideal gas or between the molecules and their container; repulsive forces exist only during collisions. When one molecule hits another, energy can be transferred from one molecule to the other, but the *total energy of motion of the molecules remains unchanged.* Collisions like collisions between ideal gas molecules are called **elastic collisions.**

5. *Energy of motion* is called **kinetic energy.** The average kinetic energy of the molecules is proportional to the Kelvin temperature:

$$\text{average kinetic energy} = \text{a constant} \times T$$

At any given temperature, the molecules of all gases have the same average kinetic energy. The average kinetic energy does not change as time passes as long as the temperature is constant.

Kinetic energy depends on mass and speed. For example, a heavy freight train has much more kinetic energy than a small car moving at the same speed. A car moving fast has more kinetic energy than the same car moving slowly. Quantitatively,

$$\text{kinetic energy} = \tfrac{1}{2}mv^2 \tag{5.3}$$

where m is mass and v is speed.

Because a gas molecule hits other gas molecules so often, its speed changes frequently. The speed of a gas molecule varies from almost zero to very high, and the direction in which it is moving changes often (see Figure 5.15). Since any sample of gas contains a very large number of molecules (about 10^{22} per liter at room temperature and atmospheric pressure), it is not possible to predict the speed of any one molecule. But the distribution of molecular speeds can be predicted. In 1860 the English physicist James Clerk Maxwell calculated that the fraction of molecules having a given speed varies with the speed as shown in ▪Figure 5.16. This distribution of molecular speeds has been proven to be correct by experiments with beams of molecules.

Looking at the curve for oxygen at 25 °C in Figure 5.16, you can see that most molecules have speeds close to the average speed. Some have speeds much less than the average speed, and some have speeds much greater than the average speed. Comparing the curve for oxygen at 700 °C with the one for oxygen at 25 °C, you

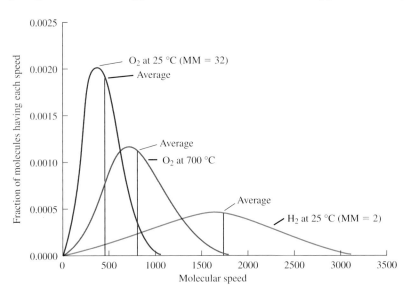

▪ FIGURE 5.16   Distribution of molecular speeds.

can see that, as temperature increases, the average speed increases, speeds are distributed over a wider range, and the number of molecules moving at very high speeds increases. The number of molecules moving at very high speeds increases much more than the average speed increases. This large increase in the number of molecules moving at very high speeds with increasing temperature is extremely important because the very fast-moving, high-energy molecules are the ones that are most likely to undergo chemical reactions. The average speed of the lighter hydrogen molecules at 25 °C is much greater than that of the oxygen molecules, and there is a wide range of speeds. The curve for hydrogen at 25 °C looks the way we might expect the curve for oxygen to look at very high temperatures.

All the gas laws can be derived from the kinetic-molecular theory. However, derivation of the gas laws from the kinetic-molecular theory is beyond the scope of this text because collisions between molecules are involved. Both the speeds and the directions of movement of the molecules are constantly changing as the molecules hit each other and the container. Nevertheless, the kinetic-molecular theory provides a qualitative explanation for the gas laws. For example, according to the kinetic-molecular theory, pressure on the walls of the container is the result of gas molecules hitting the walls. The pressure depends on the number of times gas molecules strike a unit area of the wall and their speed when they hit. The larger the number of molecules in the sample, the more molecules strike the walls and the higher the pressure. If the volume of the container is increased, molecules must move farther between collisions with the walls of the container, so they do not strike the walls as often and the pressure decreases. Increasing the temperature increases the speed of the molecules. Molecules strike the walls more often and at a higher speed; therefore, pressure increases. The total pressure of a mixture of gases is equal to the sum of the pressures of the individual gases because there are no attractive forces between molecules. The molecules of each gas in the mixture keep on hitting the walls of the container just as if the molecules of the other gases were not there. Mixtures of gases are homogeneous because the molecules of one gas can move about in the empty spaces between the molecules of the other gas or gases.

The kinetic-molecular theory also explains your everyday experience with gases. When you walk through the air in a room, you do not realize you are walking through anything because the air is mostly empty space. The odor of coffee brewing or dinner cooking quickly spreads through the whole house or apartment because the molecules are moving very fast. From the average speed of molecules at room temperature, which is of the order of m/s, you might expect that the odor would reach the far corners of your home almost immediately. But it does not because each molecule hits so many other molecules on the way.

PRACTICE PROBLEM

5.32 Use the kinetic-molecular theory to explain Charles's law.

5.11 REAL GASES

Behavior of Real Gases

One useful way of presenting the experimental facts about real gases is to graph the **compressibility factor,** PV/nRT. For an ideal gas, $PV = nRT$ and the compressibility factor equals 1. Figure 5.17(a) is a graph of compressibility factor against pressure for several common gases at 0 °C; Figure 5.17(b) is an enlargement of the low pressure part of Figure 5.17(a). In Figure 5.17(c), the curves for one gas, nitrogen,

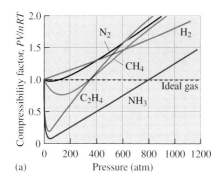

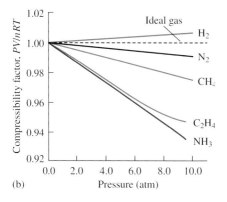

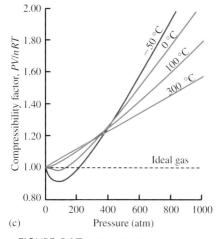

FIGURE 5.17 The behavior of real gases differs from the behavior of an ideal gas; the difference depends on the identity of the gas and on the temperature. (a) Compressibility factor for several common gases at 0 °C. Walter J. Moore, *Physical Chemistry*, 4e, © 1972; p. 20. Adapted by permission of Prentice Hall, Inc.: Englewood Cliffs, NJ. (b) Enlargement of low pressure region of part (a). (c) Effect of temperature on the compressibility factor for nitrogen (Modified from Alberty, A.; Daniels, F. *Physical Chemistry*, 5th ed.; Wiley: New York, 1980; p 23.)

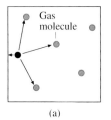 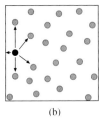

(a)   (b)

- FIGURE 5.18 The pressure of a real gas is lower than the pressure of an ideal gas as a result of attractions between molecules. The molecule that is colored black is about to hit the wall of the container. The black molecule in part (a) will not be able to hit the wall of the container as hard as a molecule of an ideal gas because it is attracted to the molecules near it. At the lower pressure shown in part (a), there are only three molecules near the black molecule; at the higher pressure shown in part (b), there are four molecules near it and they are closer. The black molecule in part (b) will not be able to hit the wall of the container as hard because it is more strongly attracted to a greater number of other molecules.

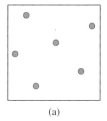 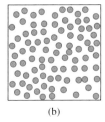

(a)   (b)

- FIGURE 5.19 (a) At low pressure, a real gas is mostly empty space. (b) At high pressure, the molecules of a real gas occupy a significant part of the total volume of the gas. Therefore, to obtain the same amount of empty space, a larger volume of high-pressure gas is required. The total volume of a real gas, the sum of the volumes of the molecules and empty space, is greater than the total volume of an ideal gas, which is entirely empty space.

Ⅾ Attractions between molecules are discussed in Chapter 12. For now, the important point is that attractions between molecules exist.

Ⅾ Remember that at 25 °C and 1 atm, the molecules occupy only about 0.04% of the volume occupied by a gas.

are shown for different temperatures. The behavior of these real gases is typical of the behavior of real gases.

In Figure 5.17, notice that *the behavior of all gases approaches ideal behavior as pressure approaches zero.* At very high pressures, the deviation from ideal behavior is large for all gases. The lower the temperature, the more a gas deviates from ideal behavior [Figure 5.17(c)]. As temperature increases, real gas behavior approaches ideal gas behavior.

At low temperatures, the compressibility factors for most gases decrease as pressure increases, as the compressibility factor for nitrogen does at -50 and 0 °C [Figure 5.17(c)]. At high temperatures, the compressibility factors for most gases increase as pressure increases, as the compressibility factor for nitrogen does at 100 and 300 °C. Compressibility factors of real gases are usually lower than the compressibility factor of an ideal gas at low pressures and higher than the compressibility factor of an ideal gas at high pressures [Figure 5.17(a)]. However, the pressure at which the minimum in the graph of compressibility factor occurs depends on the identity of the gas and on the temperature.

How can we change the kinetic-molecular model to explain the differences between the behavior of real and ideal gases shown in Figure 5.17? *All the observations can be explained by the existence of attractive forces between molecules of real gases and the fact that molecules of real gases are not points but have volume.*

A compressibility factor of less than 1 means that the pressure of the real gas is smaller than the pressure calculated by means of the ideal gas law. The existence of attractive forces between the molecules of real gases accounts for compressibility factors less than 1. Because the molecules of a real gas attract each other, they exert less pressure than the molecules of an ideal gas, which do not attract each other [see -Figure 5.18(a)]. The compressibility factor decreases with increasing pressure because, at higher pressures, a molecule is closer to its nearest neighbors and has more nearest neighbors [see Figure 5.18(b)].

A compressibility factor greater than 1 shows that the volume of the real gas is larger than the volume calculated by means of the ideal gas equation. The fact that real molecules are not points but have volume accounts for a compressibility factor greater than 1. In an ideal gas, where the molecules have no volume, the entire volume of the gas consists of the space between molecules. At low pressure, the volume of the molecules of a real gas is negligible compared to the space between the molecules. The molecules themselves are incompressible. As pressure increases, the molecules are pushed closer together. The volume of the space between molecules decreases, and a higher percentage of the total volume of the gas is occupied by gas molecules. At high pressures, the volume of the molecules of a real gas is significant compared to the space between the molecules (see -Figure 5.19).

Attractions between molecules make the compressibility factors of real gases decrease with increasing pressure. The volume occupied by the molecules of real gases makes the compressibility factor increase with increasing pressure. At low pressures, attractions between molecules are relatively more important than the volume occupied by the molecules; the compressibility factor is less than 1 and decreases as pressure increases. As pressure increases, the space between molecules decreases, and, at high pressures, the volume occupied by the molecules becomes more important than attractions between molecules. The compressibility factor is greater than 1 and increases with increasing pressure. Minimums (and maximums) in graphs, such as the minimums in the graphs of compressibility factor against pressure, usually result from opposing changes.

According to the kinetic-molecular theory, the higher the temperature of a gas, the faster the molecules of the gas move. Increasing the temperature of the real gas gives the molecules more kinetic energy to overcome the attractive forces between molecules. Therefore, the behavior of real gases approaches the behavior of an ideal gas as temperature increases. In addition, gases expand on heating. However, the volumes of the molecules themselves do not change; the increase in volume caused by heating is a result of increasing distance between molecules. The decrease in the fraction of the total volume of a gas that is occupied by gas molecules as temperature increases is another factor that causes the behavior of real gases to approach the behavior of an ideal gas as temperature increases.

PRACTICE PROBLEMS

5.33 At 0 °C, is molecular volume or intermolecular attraction (attraction between molecules) more important in making the behavior of the real gas CH_4 different from the behavior of an ideal gas (a) at 200 atm? (b) At 800 atm? Give your reasoning.

5.34 At 100 atm, is the difference between the compressibility factor for the real gas nitrogen and an ideal gas greater at high temperatures or at low temperatures?

The Plasma State

As we have just seen, the behavior of real gases becomes more and more like that of ideal gases as pressure decreases and temperature increases. However, at very high temperatures, molecules are ionized (become ions) and decomposed to atoms, and the atoms are stripped of some or all their electrons. The *mixture of positively charged ions and electrons* that results is called a **plasma.** Because each particle in a plasma has an electric charge, the properties of a plasma are altogether different from those of a gas. A plasma is classified as a fourth state of matter. Bright stars such as the sun have their matter in the plasma state; indeed, it is estimated that 99% of the matter in the universe exists in the plasma state. Neon in neon lights is also in the plasma state. In this case, the conversion of neon gas to the plasma state is brought about by electrical energy. Plasmas now have a wide range of applications from light sources to analytical instruments to municipal solid-waste disposal to rocket thrusters.

If you wish to cover the van der Waals equation, a handout is provided in the Instructor's Resource Manual.

SUMMARY

Gases completely fill whatever container they are in and take the shape of the container; gases can be **compressed,** that is, their volume can be made smaller by pressure. The densities of gases under ordinary conditions of temperature and pressure are of the order of grams per liter, whereas the densities of liquids and solids are of the order of grams per milliliter. Gases **diffuse** (spread out through space) and **effuse** (flow through small holes) rapidly. Gases that have low molecular masses diffuse and effuse faster than gases having high molecular masses; effusion can be used to separate mixtures of isotopes. Mixtures of gases are homogeneous under ordinary conditions.

Pressure is defined as force per unit area. In laboratories, the pressure of Earth's atmosphere is usually measured with a mer-

cury **barometer.** A **manometer** is used to measure the pressure of a sample of a gas. Common units of pressure are

$$1 \text{ atm} = 1.013\,25 \times 10^5 \text{ Pa} = 760 \text{ torr} = 760 \text{ mmHg}$$

According to **Boyle's law,** at constant temperature, the volume of a sample of gas is inversely proportional to pressure:

$$V = \frac{\text{a constant}}{P} \qquad (T, n \text{ constant})$$

Charles's law says that at constant pressure the volume of a sample of gas is directly proportional to the Kelvin temperature:

$$V = \text{a constant} \times T \qquad (P, n \text{ constant})$$

Zero on the Kelvin scale is the lowest possible temperature and is called **absolute zero.**

Because the volume of a sample of gas depends on its temperature and pressure, a **standard temperature** of 0 °C or 273.15 K and a **standard pressure** of 1 atm, 760 mmHg, or 1.013 25 × 10⁵ Pa have been defined for gases so that results obtained under different conditions can easily be compared.

Avogadro's law says that the volume of a gas at constant temperature and pressure is directly proportional to the amount of gas:

$$V = \text{a constant} \times n \qquad (T, P \text{ constant})$$

Equal volumes of all gases at the same temperature and pressure contain the same number of molecules. The volume of one mole of ideal gas at standard temperature and pressure, which is called the **STP molar volume,** has been found by experiment to be 22.4141 L·mol⁻¹. Avogadro's law connects the properties of gases to stoichiometry.

The **ideal gas equation**

$$PV = nRT$$

relates the pressure, volume, temperature, and number of moles of a gas and describes the behavior of most real gases to within a few percent under ordinary conditions. The value of the **gas constant,** R, depends on the units used for pressure and volume; for example, one common value is 62.36 mmHg·L·mol⁻¹·K⁻¹. The ideal gas equation is an **equation of state.** Gases that behave exactly as described by the ideal gas equation are called **ideal gases.**

The ideal gas equation can be used to solve problems about volume, temperature, pressure, and quantity of a gas. Gas densities can be calculated if the molecular formula is known or the molecular mass of an unknown gas can be determined by measuring the density of the gas. In problems about gases, Kelvin temperatures must be used, and the amount of substance must be in moles although temperature is usually measured in °C and amount of substance in grams or kilograms.

The pressure of each gas in a mixture is called the **partial pressure** of the gas. **Dalton's law of partial pressures** says that the total pressure of a mixture of gases is equal to the sum of the partial pressures of the individual gases in the mixture:

$$P_{\text{total}} = p_{\text{gas 1}} + p_{\text{gas 2}} + \cdots$$

Gases formed from substances that are usually liquids or solids are called **vapors.** The pressure of the vapor formed by evaporation of a given liquid in a closed container is called **vapor pressure** and depends only on temperature.

The **kinetic-molecular theory** explains why gases behave as they do. According to the kinetic-molecular theory, gases consist of very small particles called molecules. The distances between molecules are very large compared to the diameters of the molecules. Each molecule of a gas moves in a straight line at very high speed until it hits another molecule or the container. Pressure is the result of molecules hitting the container. The average **kinetic energy** (energy of motion) of the molecules is proportional to the Kelvin temperature and does not change as time passes as long as the temperature is constant. At any given temperature, the molecules of all gases have the same average kinetic energy; light molecules move faster than heavy molecules. In a sample of gas, some molecules have very low speeds, and some have very high speeds; most have speeds around average. There are no attractive or repulsive forces between molecules of an ideal gas.

Real gases behave as described by the gas laws to within a few percent under ordinary conditions of temperature and pressure. The molecules of real gases, however, have volume and attract each other. The attractions between molecules affect how the molecules behave at very low temperatures, and the volumes of the molecules become important at very high pressures.

ADDITIONAL PRACTICE PROBLEMS

For information about the organization of Additional Practice Problems, Stop & Test Yourself, Putting Things Together, and Applications, see the beginnings of these sections in Chapter 1.

5.35 At the bottom of a mine that is 475 ft deep, is the atmospheric pressure greater than, equal to, or less than the barometric pressure at the surface? (5.1)

5.36 (a) 7.83 torr = _____ mmHg. (b) 0.76 atm = _____ mmHg. (c) 25.6 mmHg = _____ atm. (d) 247.5 Pa = _____ mmHg. (e) 31.59 MPa = _____ atm. (f) 247.5 Pa = _____ kPa. (g) 31.59 MPa = _____ Pa. (h) 2.36 GPa = _____ kPa. (5.1)

5.37 From Figure 5.5, what was the volume of the gas sample at 31.2 cmHg? (5.2)

5.38 (a) If the pressure on a sample of gas is decreased to one-quarter of its original value at constant temperature, what happens to the volume? (b) By what factor must the pressure on a sample of gas be changed to make the volume only one-eighth of the original volume? Assume that the temperature is constant. (5.2)

5.39 The volume of a sample of nitrogen gas is 4.6 L at a pressure of 0.74 atm. What will the volume be if the pressure is changed to 1.86 atm, assuming that there is no change in temperature? (5.2)

5.40 A sample of sulfur dioxide gas has a volume of 562 mL at a pressure of 2.43 kPa. What pressure is required to change the volume to 893 mL at constant temperature? (5.2)

5.41 A sample of helium gas has a volume of 21.6 L at 0.750 atm. What pressure is required to change the volume to 4.65 L at the same temperature? (5.2)

5.42 A sample of oxygen gas has a volume of 387 mL at a temperature of 15.2 °C. What temperature (in °C) is required to change the volume to 439 mL under constant pressure? (5.3)

5.43 The volume of a sample of carbon monoxide gas is 295 mL at 19.6 °C. What will the volume be if the temperature is changed to 45.3 °C, assuming no change in pressure? (5.3)

5.44 A sample of helium gas occupies a volume of 654 mL at −25.3 °C and 747.2 mmHg. What will its volume be at STP? (5.4)

5.45 The volume of a sample of chlorine gas is 98.6 mL at 31.4 °C and 769.6 mmHg. What volume will the sample occupy at 23.2 °C and 748.8 mmHg? (5.4)

5.46 Hydrogen cyanide gas, HCN(g), is manufactured from methane and ammonia by the reaction

$$2CH_4(g) \ + \ 2NH_3(g) \ + \ 3O_2(g) \xrightarrow{\text{Pt, 1100 °C}}$$
$$2HCN(g) \ + \ 6H_2O(g)$$

If the volumes of both reactants and products are measured at the same temperature, (a) how many liters of ammonia are needed to react with 3.0 L of methane? (b) How many liters of oxygen are needed to react with 3.0 L of methane? (c) How many liters of hydrogen cyanide and how many liters of water vapor will be formed? (Assume reaction is complete.) (5.5)

5.47 (a) If 0.25 mol $N_2(g)$ occupies a volume of 5.8 L, what volume will 0.25 mol $O_2(g)$ occupy if both gases are at the same temperature and pressure? (b) What volume will 1.00 mol $N_2(g)$ occupy (still at the same temperature and pressure)? (c) If 3.33×10^{-2} mol $N_2(g)$ occupy a volume of 747 mL, what volume will 4.02×10^{21} molecules of $N_2(g)$ occupy at the same temperature and pressure? (5.5)

5.48 Water can be decomposed into hydrogen gas and oxygen gas by passing an electric current through it. The equation for this reaction is

$$2H_2O(l) \xrightarrow{\text{electricity}} 2H_2(g) \ + \ O_2(g)$$

If 18.46 mL of $H_2(g)$ are collected, (a) what volume of $O_2(g)$, measured at the same temperature and pressure, will also be formed? (b) Was the volume of $H_2O(l)$ decomposed 18.46 mL? Explain your answer. (5.5)

5.49 What is the temperature (in °C) of a 0.536-g sample of neon gas, Ne(g), that occupies a 500.0-mL container at a pressure of 762.3 mmHg? (5.7)

5.50 How many grams of carbon monoxide gas, CO(g), are present in a 750.0-mL container if the temperature is 22.3 °C and the pressure is 2.6 atm? (5.7)

5.51 (a) What is the density of neon gas, Ne(g), to three decimal places at STP? At 24.3 °C and 751.2 mmHg? (b) What volume will 3.82 g of neon occupy at 24.3 °C and 751.2 mmHg? (5.7)

5.52 What is the molecular mass of a gas if 4.26 g occupy 2.36 L at 23.2 °C and 756.1 mmHg? (5.7)

5.53 Small amounts of pure iron are made by treating iron(III) oxide with hydrogen gas:

$$Fe_2O_3(s) \ + \ 3H_2(g) \longrightarrow 2Fe(s) \ + \ 3H_2O(g)$$

How many liters of hydrogen gas at 754.3 mmHg and 723 K are required to react with 34.21 g of iron(III) oxide? (5.7)

5.54 What is the total pressure of a mixture of hydrogen and chlorine if the partial pressure of hydrogen is 245 mmHg and the partial pressure of chlorine is 463 mmHg? (5.8)

5.55 (a) What is the total pressure in mmHg of a mixture of 0.563 mol $H_2(g)$ and 0.841 mol $Cl_2(g)$ in a 789-mL container at 84.2 °C? (b) What is the partial pressure of each gas in the mixture in part (a)? (5.8)

5.56 (a) What is the vapor pressure of water at (i) 18.4 °C? (ii) 25.5 °C? (b) At what temperature is the vapor pressure of water equal to 18.997 mmHg? (5.8)

5.57 A sample of $N_2(g)$ was collected over water at a total pressure of 742.5 mmHg and a temperature of 22.9 °C. What was the partial pressure of the nitrogen? (5.8)

5.58 What is the difference between *diffusion* and *effusion*? (5.9)

5.59 Which would effuse faster, He(g) or Ne(g)? Explain your answer. (5.9)

5.60 Use the kinetic-molecular theory to explain Boyle's law. (5.10)

5.61 Use the kinetic-molecular theory to explain why heating a gas sample at constant volume results in an increase in pressure. (5.10)

5.62 Suppose you have two 5.0-L containers at the same temperature and pressure. One contains He and the other contains Ar. Compare (a) numbers of molecules, (b) masses of samples, (c) average kinetic energies of molecules, (d) average speeds of molecules. (5.10)

5.63 On the following sketch of the distribution of molecular speeds for O_2 at 25 °C, sketch the distributions expected for He and Kr at the same temperature. (5.10)

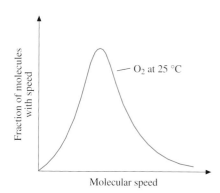

5.64 (a) What is meant by saying that real gases do not behave ideally? (b) Why do real gases not behave ideally? (c) Under what conditions do real gases behave most like ideal gases? (5.11)

5.65 A sample of bromine vapor is at 20.0 °C and 1 atm. Would you expect it to behave more or less ideally if (a) the temperature is increased to 95.0 °C? (b) The pressure is increased to 95 atm? (5.11)

5.66 Consider $F_2(g)$ and $Cl_2(g)$. (a) Which gas would move from point A to point B in the shorter time? (b) Which gas would move farther in 30 s? Explain your answers.

5.67 Bulbs that contain Ne at 237 mmHg and Kr at 372 mmHg are connected by a stopcock:

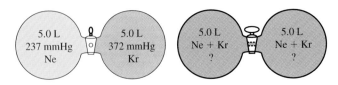

A sufficient time after the stopcock is opened so that the composition of the gaseous mixture has become uniform, what is the partial pressure of each gas and the total pressure?

5.68 Mercury barometers must be at least 80 cm high. However, some commercial mercury manometers are only 30–40 cm high. Two types are available, open-end and closed-end. One type is used to measure pressures that are much less than atmospheric pressure and the other to measure pressures close to atmospheric pressure. (a) Which is which? Explain your answer. (b) If the barometric pressure is 753.2 mmHg and the difference between the levels of the mercury is 4.87 cmHg in both manometers, what is the pressure of the sample attached to the closed-end manometer? The open-end manometer?

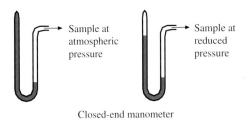

Closed-end manometer

Open-end manometer

5.69 The pressure of natural gas in the line in your laboratory is only about six inches of water greater than atmospheric pressure. What will happen if the gas is turned on in the apparatus shown (*not* to scale)? Explain your answer.

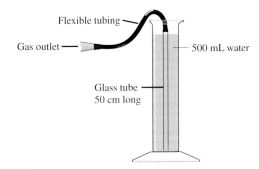

5.70 What is the experimental evidence that gases are mostly empty space?

5.71 (a) At what pressure does $NH_3(g)$ behave ideally at 0 °C? (b) According to Figure 5.17, around what other pressure does $NH_3(g)$ *appear* to behave ideally at 0 °C? (c) Explain why $NH_3(g)$ appears to behave ideally at this other pressure.

5.72 Draw pictures showing molecular-level views of the reddish brown gas in the two photographs in Figure 12.1(c). Use small circles to represent the molecules.

5.73 Suppose the tube in a simple barometer has a larger diameter than the one shown in Figure 5.3. Will the height of the mercury be greater, the same as, or less than the height of the mercury in the tube in Figure 5.3(b)? Give your reasoning.

5.74 What is the pressure (in atmospheres) of a 0.867-g sample of methane gas, $CH_4(g)$, that occupies a 250.0-mL container at 20.6 °C?

5.75 The volume of a sample of neon gas is 3.68 L at 16.8 °C and 742.3 mmHg. If the temperature of the neon is changed to 29.7 °C, to what value must the pressure be changed to keep the volume constant at 3.68 L?

5.76 Calculate the molecular mass of a gas that has a density of 1.143 g/L at 21.8 °C and 749.4 mmHg.

5.77 What volume would a sample of nitrogen gas that occupied a volume of 254 mL when dry at a laboratory temperature of 20.0 °C and an atmospheric pressure of 762.3 mmHg occupy at the same temperature and barometric pressure if collected over water?

5.78 Explain why a balloon filled with helium gets smaller as time passes, but a balloon filled with sulfur hexafluoride gets bigger and bursts.

5.79 (a) If the Celsius temperature of a sample of oxygen gas is decreased by one-half (from 200 to 100 °C) at constant pressure, what will happen to the volume? (b) If the Celsius temperature of the sample is again decreased by one-half (from 100 to 50 °C) at constant pressure, what will happen to the volume?

5.80 Given the following molecular-level view of a sample of gas contained in a rigid steel sphere at 25 °C, draw another molecular-level picture of the sample of gas in the steel sphere after the sphere is cooled in an ice bath.

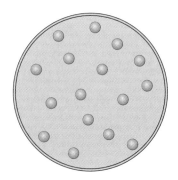

1. 745 mmHg = _____ Pa
   (a) 5.59 (b) 9.93×10^4 (c) 9.93×10^6 (d) 5.74×10^{10}

2. A sample of gas occupies a volume of 39.0 cm^3 at 125 °C and 0.568 atm pressure. What will the pressure of the gas be if the volume is changed to 74.0 cm^3 at constant temperature?
   (a) 0.928 atm (b) 1.08 atm (c) 0.953 atm (d) 0.299 atm

3. A sample of gas occupies a volume of 39.0 cm^3 at 125 °C and 0.568 atm pressure. What will the volume of the gas be if the temperature is changed to −23 °C at constant pressure?
   (a) 62.1 cm^3 (b) 7.18 cm^3 (c) 212 cm^3 (d) 24.5 cm^3

4. A sample of gas occupies a volume of 8.60 L at 67 °C and 748 mmHg. What volume will the sample occupy at STP?
   (a) 6.80 L (b) 7.02 L (c) 10.5 L (d) 10.9 L

5. Consider a 0.50-L container of H$_2$(g) at 1.00 atm and 0.0 °C and a 0.50-L container of NH$_3$(g) at 1.00 atm and 0.0 °C. Which of the following statements is true?
   (a) Both contain the same mass of gas. (b) Both contain the same number of hydrogen atoms. (c) The volume of the molecules of the gas is large compared with the total volume of the container. (d) Both contain the same number of molecules of gas.

6. Sulfur dioxide gas reacts with oxygen gas to form sulfur trioxide gas. What volume of oxygen gas is needed to convert 9.8 L of sulfur dioxide gas to sulfur trioxide gas if all gases are measured at the same temperature and pressure?
   (a) 4.9 L (b) 9.8 L (c) 20 L (d) 22 L

7. What is the density of ammonia gas at 745 mmHg and 65 °C?
   (a) 457 g/L (b) 0.184 g/L (c) 0.602 g/L
   (d) Impossible to calculate from information given.

8. A 0.602-g sample of a gas occupies 448 mL at STP. What is the molecular mass of the gas?
   (a) 0.0301 (b) 6.04 (c) 13.5 (d) 30.1

9. A sample of gas was collected over water at 22.1 °C when the barometric pressure was 748.6 mmHg. What was the partial pressure of the gas in mmHg? (You may use Table 5.4 to answer this one.)
   (a) 728.6 mmHg (b) 728.7 mmHg (c) 728.8 mmHg
   (d) 768.5 mmHg

10. At the same temperature and pressure, molecules of which of the following have the highest average speed?
    (a) Br$_2$(g) (b) Cl$_2$(g) (c) F$_2$(g) (d) I$_2$(g)

11. Which of the following statements is not part of the kinetic-molecular theory?

(a) There are no attractive or repulsive forces between molecules. (b) The molecules of a gas are small compared with the space between the molecules. (c) The molecules of a gas are attracted to the walls of the container. (d) The movement of the molecules of a gas is random.

12. The sketch below shows the fraction of molecules with a certain speed of a gas sample versus speed at various temperatures. Which curve represents the lowest temperature?

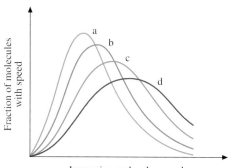

Increasing molecular speed

13. Which sample of a gas would follow the ideal gas equation most closely?
    (a) One at 1 atm and 0 °C (b) One at 100 atm and 0 °C
    (c) One at 1 atm and 50 °C (d) One at 100 atm and 50 °C

14. Real gases behave differently from ideal gases because
    (i) The molecules of real gases are in constant motion.
    (ii) Molecules of real gases collide with the walls of the container.
    (iii) Molecules of real gases have volume
    (iv) Molecules of real gases attract each other.
    (a) i and ii (b) iii only (c) iii and iv (d) All of the above.

15. What does the expression $P_1V_1/T_1 = P_2V_2/T_2$ reduce to if the gas volume is a constant? Another way of expressing this pressure–temperature relationship is _____. (C stands for a constant.)
    (a) $P_1/T_1 = T_2/P_2$, $P = CT$
    (b) $P_1/T_1 = T_2/P_2$, $PT = C$
    (c) $P_1/T_1 = P_2/T_2$, $P = CT$
    (d) $P_1/T_1 = P_2/T_2$, $PT = C$

5.81 A sample of gas has a volume of 3.62 L at 4.98 kPa. What pressure (in atmospheres) is needed to change the volume to 768 mL at constant temperature?

5.82 Dry air is 78.084% N$_2$, 20.946% O$_2$, and 0.934% Ar by volume. (a) What is the molecular mass of dry air? (b) The percent water vapor in air varies from 0 to 7%. To how many significant figures should the molecular mass of air be written?

5.83 (a) Which would diffuse faster, Cl$_2$(g) at 0 °C or Cl$_2$(g) at 35 °C? (b) Would Cl$_2$(g) diffuse faster through a vacuum or through air? Explain your answers.

5.84 The behavior of samples of five gases is displayed graphically in the sketches below. Which of the five gases behave ideally?

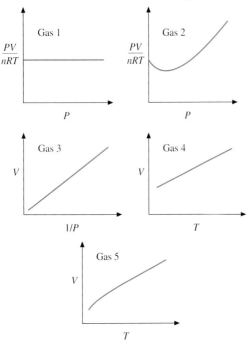

5.85 In the laboratory, hydrogen gas is usually made by reacting zinc metal with hydrochloric acid. The other product is zinc chloride, $ZnCl_2$. How many milliliters of $H_2(g)$, collected over water at a barometric pressure of 751.8 mmHg and a laboratory temperature of 21.2 °C, can theoretically be formed from 0.783 g Zn and an excess of hydrochloric acid?

5.86 How many significant figures are there in the pressure shown by the simple barometer in Figure 5.3(b)?

5.87 Industrially, hydrogen chloride gas is made by the combination of hydrogen and chlorine:

$$H_2(g) + Cl_2(g) \longrightarrow 2HCl(g)$$

Reaction is complete. If a mixture of 362 mL of hydrogen and 345 mL of chlorine is caused to react, (a) how many milliliters of H_2 will be present after reaction has taken place? (b) How many milliliters of Cl_2? (c) How many milliliters of HCl? (All volumes are measured under the same conditions of temperature and pressure.)

5.88 Measuring the volume of a gas may seem like a poor way to measure the quantity of a gas because the volume of a gas depends on its temperature and pressure. However, gases are usually measured by volume. What is one practical problem with measuring gases by mass?

5.89 A sample of carbon dioxide gas has a volume of 438 mL at 739.5 mmHg and 22.4 °C. (a) How many CO_2 molecules are there in the sample? (b) How many O atoms are there in the sample?

5.90 In industry, sulfuric acid is made by the contact process, which consists of the following steps:

$$S(s) + O_2(g) \longrightarrow SO_2(g)$$
$$2SO_2(g) \xrightarrow{V_2O_5} 2SO_3(g)$$
$$SO_3(g) + H_2O(l) \longrightarrow H_2SO_4(l)$$

How many liters of oxygen gas, measured at STP, are needed to convert 2.50×10^3 kg of sulfur to sulfuric acid?

5.91 An unknown compound contains 80.0% C and 20.0% H. A 0.2367-g sample has a volume of 191.7 mL at 22 °C and 756.8 mmHg. What is the molecular formula of the compound?

5.92 Chlorine gas is made in industry by passing an electric current through an aqueous solution of sodium chloride. The other products are hydrogen gas and an aqueous solution of sodium hydroxide. (a) How many liters of dry chlorine gas at STP can be made from 50.0 L of 5.085 M sodium chloride? (b) How many liters of dry hydrogen gas, also at STP, will be formed?

5.93 In the laboratory, hydrogen chloride gas is made by heating solid sodium chloride with concentrated sulfuric acid. The other product is solid sodium hydrogen sulfate. (a) If 5.68 g of sodium chloride and 6.46 mL of concentrated sulfuric acid are used, how many liters of hydrogen chloride gas at 24.3 °C and 762.4 mmHg will be formed? Concentrated sulfuric acid is 17.8 M. (b) Could the hydrogen chloride gas be collected over water? Explain.

5.94 (a) Calculate the mass of $CO_2(g)$ formed by the reaction of 5.37 g $CaCO_3(s)$ with 275 mL of 1.000 M HCl. (b) When the reaction was carried out in an unstoppered flask, the loss of mass was 1.96 g. Explain why the loss of mass is *not* equal to the mass of $CO_2(g)$ formed. (c) When the reaction was carried out in a flask closed with a balloon so that the $CO_2(g)$ was collected in the balloon, the balloon was inflated to a volume of 1070 cm³. A mass loss of 1.34 g was observed! Explain this observation. (*Hint:* The density of air is about 0.0012 g/mL at 25 °C and 760 mmHg pressure.) (d) How could this experiment be carried out to prove the law of conservation of matter? (e) How could the volume of the inflated balloon be measured?

5.95 Methanol, CH_3OH, is manufactured by reaction of carbon monoxide gas with hydrogen gas. If a 95% yield is usually obtained, how many liters of hydrogen gas at 350 (3.5×10^2) °C and 300 (3.0×10^2) atm are usually required to make 1.00 L of methanol measured at 20.0 °C? The density of methanol at 20.0 °C is 0.7914 g/mL.

5.96 Reaction of a 0.5328-g sample of iron with excess hydrochloric acid yields 232 mL $H_2(g)$ at 755.8 mmHg and 21.9 °C. Write the equation for the reaction.

5.97 Which of the following elements and compounds are gases under ordinary conditions? Explain your answer. (a) KI (b) NO (c) O_3 (d) S_8 (e) SO_2 (f) Ti

5.98 Given the microscopic view of a gas in a cylinder with a movable piston,

redraw the picture in two ways, each showing a new pressure half as large as the original pressure.

5.99 Like salad oil and vinegar, pentane ($CH_3CH_2CH_2CH_2CH_3$), which is a liquid under ordinary conditions, and water do not form a solution on mixing. Do their vapors form a solution? Explain your answer.

APPLICATIONS

5.101 (a) Normal blood pressure is 120 mmHg at heartbeat and 80 mmHg between heartbeats. What are these pressures in kPa? (b) The Martian atmosphere, which is mostly carbon dioxide gas, exerts a pressure of 6.1 mbars on the surface of the planet. What is this pressure in atmospheres? (c) The pressure exerted in a human hip when walking is 6 MPa. What is this pressure in mmHg?

5.102 The surface temperature near the Martian equator at noon is about 20 °C. However, the pressure of the Martian atmosphere [which consists largely of CO_2(g)] is only about 5 mmHg. Will the CO_2(g) at the bottom of the Martian atmosphere behave more or less ideally than CO_2(g) behaves at Earth's surface? Explain your answer.

5.103 (a) You are all ready to paint your room when you find that the lid on the can of paint is stuck and you cannot open the can. Explain why warming the can in a pail of hot water will make the lid easier to pry out. (b) Will the same trick make a screw-cap container easier to open?

5.104 "Empty" aerosol cans are still filled with gas at atmospheric pressure. (a) If an empty aerosol can is thrown into a fire (temperature = 475 °C), what will the pressure (in atmospheres) inside the can become? (b) Why do aerosol cans carry a label warning the user not to incinerate them?

5.105 Tennis balls are usually filled with air at a pressure of 2 atm and sealed in a can that is also filled with air at 2 atm. (a) Explain why, once a new can of tennis balls is opened, the quality of the tennis balls begins to deteriorate. (b) Suggest two ways of slowing the loss of quality. (c) Better quality tennis balls that are filled with SF_6(g) last longer. Why?

5.106 To study the effects of volcanoes on atmospheric chemistry, a team of researchers plans to measure the concentrations of carbon monoxide, carbon dioxide, hydrogen, hydrogen chloride, hydrogen fluoride, hydrogen sulfide, nitrogen, oxygen, and sulfur dioxide gases and sulfuric acid vapor in volcanic vents. (a) Write formulas for these substances. (b) Which of the substances are elements, and which are compounds? (c) What is the difference between a gas and a vapor?

5.107 Among the gases offered for sale are ammonia, argon, boron trichloride, carbon tetrafluoride, chlorine, hydrogen bromide, hydrogen iodide, hydrogen selenide, nitric oxide (nitrogen monoxide), nitrogen dioxide, nitrous oxide (dinitrogen monoxide), phosphorus pentafluoride, silicon tetrachloride, and sulfur hexfluoride. (a) Write formulas for these substances. (b) Which of these gases have triatomic molecules? (c) What do the elements that compose all these gases have in common?

5.108 The space between the stars contains 1500 (1.5×10^3) atoms per liter, of which 90% are hydrogen atoms and 10% are

helium atoms. If the temperature is 50 K, calculate the pressure. (Assume the mixture of hydrogen atoms and helium atoms behaves like an ideal gas.)

5.109 How many molecules per cubic centimeter does air at Earth's surface contain? Assume a temperature of 25 °C.

5.110 Standard atmospheric pressure is 14.69 psi. The area of Earth's surface is 1.970×10^8 mi². What is the mass of the entire atmosphere (in pounds)?

5.111 An adult human breathes in about 500 mL of air at 1 atm with each breath. At a temperature of 20 °C, how many molecules of oxygen is this? Air is 21% oxygen by volume and has a molecular mass of 29 u.

5.112 The ability of cross-country skiers to absorb oxygen into their blood is higher than that of any other Olympic athletes, 80 mL/kg of body weight·minute at 1.00 atm and 20 °C. How long will it take a 150-lb cross-country skier to absorb 1.00 mol O_2(g)?

5.113 Alex Jones got a barometer for his birthday. It read 30.05 when the atmospheric pressure was 763.3 mmHg. But there are no units on the dial. Please help Alex figure out what the units are because he needs to know them for his school science project.

5.114 The old Soviet Union produced more natural gas than the United States. About 60% of that country's gas was in northern Siberia (part of the Russian Republic), where the average temperature is −10 °C. (a) What volume (in cubic feet) would a sample of gas that has a volume of 1.00 cubic feet at −10 °C (zero is significant) occupy at a comfortable room temperature of 21 °C? (b) The Soviets used compressed natural gas instead of gasoline in many of their cars and trucks. What would be the disadvantage of using compressed natural gas as a fuel for motor vehicles?

5.115 Deep earthquakes, such as the one that occurred in Bolivia in 1994, take place at depths greater than 300 km where the pressure is greater than 10 gigapascals. How many atmospheres is 10 gigapascals?

5.116 The helium in the tanks sold for filling party balloons is at a pressure of 265 psi at 70 °F. The tanks have a volume of 14.0 ft³. How many balloons can be filled from a tank if the balloons have a volume of 1.5 L at 70 °F and the pressure inside the balloons is 1.02 atm?

5.117 An ad for Honda Scooters says that the underseat storage compartment will hold 3.34×10^{23} molecules of krypton gas at STP. What is the volume of the compartment?

5.118 A cylinder of an automobile engine has a volume of 0.500 L. If the cylinder is filled with a mixture of gasoline vapor

5.100 Consider methane gas, CH_4(g), and neopentane gas, $C(CH_3)_4$(g). (a) Write complete structural formulas for each gas. (b) At the same temperature and pressure, which is more likely to behave ideally? Explain your answer.

and air at 1.00 atm, what pressure is necessary to compress the gasoline vapor–air mixture to 57 mL before it is ignited by the spark plug? Assume the temperature is constant and the gas behaves as an ideal gas.

5.119 A motorist checked his tires at the start of a trip and the tire gauge read 25 psi. (A tire gauge reads 0 at atmospheric pressure, 14.7 psi.) The temperature was 21 °C. After driving for several hours, the motorist checked his tires again. The gauge pressure was up to 31 psi. To what temperature had the air in the tires been heated? Assume the volume was constant.

5.120 An instrument ad states that the instrument provides precision low-pressure measurements in ranges from 0.25 to 200 in. of water. (a) Convert these ranges to mmHg. At 20 °C, the density of water is 0.99821 g/cm^3, and the density of mercury is 13.5462 g/cm^3. (b) Why is water used instead of mercury for measuring very small differences in pressure?

5.121 The Rankine scale is a temperature scale used by engineers. The Rankine scale is an absolute scale—that is, 0 °R = 0 K. Rankine degrees are the same size as Fahrenheit degrees (1 °R = 1 °F). (a) What is 0 °R on the Fahrenheit scale? (b) What is the value of the gas constant R in $psi \cdot ft^3 \cdot mol^{-1} \cdot °R^{-1}$?

5.122 A magazine article reports that the hearts of mammals produce pressures from 10 000 to 40 000 Pa. (a) What are these pressures in atmospheres? (b) Can the pressures inside mammals really be less than atmospheric pressure? Explain your answer. (c) Does a mouse or a giraffe's heart produce the higher pressure? Explain your answer.

5.123 Dinitrogen monoxide, which is usually called nitrous oxide, is used as an anesthetic and as a propellant gas in whipped-cream dispensers. It is made by heating ammonium nitrate, which melts at 170 °C, to 250–260 °C. The other product is water vapor. From 36.4 g of ammonium nitrate, (a) how many liters of dinitrogen monoxide gas, measured at 255 °C and 1.00 atm, can be made? (b) How many liters of water vapor, also at 255 °C and 1.00 atm, will be formed?

5.124 The most important uses of ammonium nitrate are in fertilizer and explosives. When ammonium nitrate is heated at 800 (8.00×10^2) °C, the products are nitrogen gas, oxygen gas, and water vapor. (a) What total volume of gas, measured at 800 °C and 1.00 atm, will be formed from 36.4 g of ammonium nitrate? (b) If this gas is confined in a 25.0-mL space, what will be the pressure? (c) What do you think will happen?

5.125 The gas in a laser discharge cell is prepared by mixing 2.0 mol CO_2, 1.0 mol N_2, and 16 mol He. The total pressure is 3.0 atm. What is the partial pressure of each gas in the mixture?

5.126 Gaseous sodium atoms can be slowed with the pressure of the light from a laser beam until their average temperature is only 43 μK. (a) What is this temperature in K? (b) The average speed in meters per second of the molecules in a sample of a gas can be calculated from the formula

$$average\ speed = \left(\frac{8RT}{\pi \, MM}\right)^{1/2}$$

where R is $8.315\ J \cdot mol^{-1} \cdot K^{-1}$, T is the temperature in kelvin, $\pi = 3.1416$, and MM is the molecular mass in kilograms. Joules, J, are the SI unit of energy and have the dimensions $kg \cdot m^2 \cdot s^{-2}$. What is the average speed of sodium atoms at 43 μK? (c) At what temperature is the average speed of sodium atoms 20 m/s (zero is significant)? (d) What do you think the speed of sodium atoms will be at 0 K? (e) Describe how the distribution of molecular speeds curves (Figure 5.16) for Na(g) at 43 μK will look.

5.127 The gas formed by anaerobic decomposition of bat droppings is about 40% CO_2 and 60% CH_4 by volume. (*Anaerobic* means "in the absence of oxygen.") What is the density of this mixture at 25 °C and 1 atm?

5.128 Gases are now available in refillable cylinders that hold 104 L of gas at 1800 psig and 70 °F. The symbol psig means the gauge reading; gauges read zero at 1 atm. How many kilograms of nitrogen gas does one of these cylinders hold?

Chemistry: An Introduction to Medicine

YVONNE A. MALDONADO

Assistant Professor of Medicine

B. A. Bacteriology
University of California, Los Angeles

M.D.
Stanford University

L ike many kids, I suppose I first became interested in chemistry in grammar school when my parents bought me a junior chemistry set. I had fun tinkering with all the home experiments in the set and decided that I enjoyed science.

High school reinforced that idea. However, it wasn't until I started taking biochemistry in college that I began to consider pursuing my interest in science through medicine. I was in college in the 1970s—biochemistry and its cousin, molecular biology, were exploding with new developments. I became caught up in the excitement of the biochemical discoveries and their applications in biology. Unfortunately, my university did not offer a biochemistry undergraduate major, so I was advised to major in bacteriology and then pursue a graduate degree in biochemistry.

I was thrilled with my major in bacteriology, which included quite a bit of virology as well. My classes were small, I enjoyed the opportunity to do lab research, and the sense of discovery was overwhelming. I learned that biochemistry was really part of every living thing, from humans to the smallest living particles, the viruses. My real dilemma was deciding what I would do after graduation. I loved laboratory research and wanted to pursue graduate work in biochemistry, but I was also intrigued by the interconnections among biochemistry, biology, and physiology. I took the plunge and decided to apply to medical school in the hope that I could pursue the medical applications of biochemistry.

I have never regretted that decision. I am now on the faculty of the Stanford University School of Medicine. My area of specialty is pediatric infectious diseases. My research has taken two directions. First, I am involved in research on the diagnosis and treatment of pediatric HIV infection and AIDS. Much of my work in this field involves the biochemical interactions of HIV, the virus that causes AIDS, with the developing fetus and growing child. In this work, I am studying a new disease but am using known biochemical concepts to diagnose and treat this as yet incurable infection.

My other area of research involves the biochemical aspects of poliovirus infection and their implications for global polio vaccination strategies. Contrary to what we see in developed countries, poliovirus infection still occurs in many parts of the developing world. Therefore, many scientists have suggested that new vaccination strategies may be a solution to the persistence of poliovirus worldwide. Since polioviruses live and replicate only in the human intestinal tract, my work depends heavily on understanding the structure and function of poliovirus proteins and their interaction with human cells and other viruses. It is fascinating to piece together the interactions on a microscopic level and apply them on a global scale to the control of poliovirus infection.

Another important part of my job is to see pediatric patients with infectious diseases and to teach medical students and doctors in training about the spectrum and therapy of human infection. Most of this teaching also requires me to integrate my basic understanding of biochemistry with the material I learned in medical school.

In retrospect, it is interesting to see how I have come full circle in taking my interest in science and turning it into a career. Throughout my graduate medical training, it was clear to me that biochemistry is highly important for understanding the human body, but I never thought I would be using the biochemistry of both viruses and humans to pursue my medical interests. Thus, I started out with two choices and ended up combining them in one career.

6

CHEMICAL THERMODYNAMICS: THERMOCHEMISTRY

Getting a Concorde off the ground requires the conversion of a large amount of chemical potential energy to kinetic energy.

The word *thermodynamics* is derived from Greek words meaning "heat" and "power." Thermodynamics was developed during the nineteenth century by physicists who were interested in the work that could be done by steam engines. Now thermodynamics is an essential part of all the sciences. **Thermochemistry** *is the part of thermodynamics that deals with the relationship between chemical reactions and heat.*

— *Why is a study of chemical thermodynamics and thermochemistry a part of general chemistry?*

A knowledge of thermodynamics gives scientists and engineers the ability to predict which physical and chemical changes are possible. (However, thermodynamics does not provide any information about how fast possible changes take place.) We begin our study of thermodynamics with thermochemistry for two reasons:

Heat is useful, and heat is easy to measure.

— *What do you already know about chemical thermodynamics and thermochemistry?*

From your everyday life, you know that some changes take place naturally: spilled peas roll all over the floor; at room temperature, ice cubes melt, and the water spreads out; outside on a cold winter's day, water freezes to ice; if set on fire with a match, natural gas burns, providing far more heat than was supplied by the match. Other changes must be made to take place: spilled peas must be picked up; indoors, water must be put in the freezer to make ice cubes; a cake must be heated in order to bake it.

You also know that many chemical reactions are carried out to obtain heat. For example, the burning of fuels supplies warmth in winter and hot water to keep us and our possessions clean. Heat from the burning of fuels can be converted into mechanical energy; it supplies power to move us around and to manufacture all the things we can't get along without. Power stations convert heat from the burning of fuels into electrical energy to light our lights, run our computers and

Heat flows spontaneously from a hot burner to a cooler pan and its contents.

A natural gas burner.

TVs, and operate our appliances. However, supplies of fuels are limited, and, also, we are beginning to find that the burning of fuels leads to problems such

as acid rain, thermal pollution, and the greenhouse effect. Thus, an understanding of thermochemistry is important to everyone.

 ## 6.1 SYSTEM, SURROUNDINGS, AND UNIVERSE

In thermodynamics, terms are defined very precisely. The *part of the universe a scientist is interested in* is called the **system.** For example, if you were studying the reaction of $AgNO_3(aq)$ with $NaCl(aq)$ (■Figure 6.1), the system would consist of the mixture of $AgNO_3(aq)$ and $NaCl(aq)$ and the precipitated $AgCl(s)$. *The rest of the universe* is referred to as the **surroundings.** For practical purposes, the surroundings are the apparatus used and the space around it; in Figure 6.1, the test tube, stopper, clamp, and the air around them are the surroundings. Together, *system and surroundings* compose the **universe.**

■ FIGURE 6.1   The system consists of $AgNO_3(aq)$, $NaCl(aq)$, and $AgCl(s)$. The test tube, stopper, clamp, and the air around them make up the surroundings.

6.2 WHY CHANGES TAKE PLACE

Physical and chemical changes that take place naturally are said to be **spontaneous.** If you drop this book, it will fall to the floor. That is a spontaneous change. The *opposite of a spontaneous change* is **nonspontaneous.** If the book is on the floor, it will not rise up to your hand. That is a nonspontaneous change. The fact that a change is nonspontaneous does not mean that it cannot take place. You can lift your book back up from the floor. A nonspontaneous change does not take place unless someone or something makes it happen.

A change may be spontaneous under one set of conditions and nonspontaneous under another set of conditions. For example, at a comfortable room temperature, solid ice spontaneously melts to liquid water. But outside in the winter, liquid water may spontaneously freeze to solid ice. *As far as the system is concerned, a combination of two factors determines whether a change is spontaneous or nonspontaneous. These two factors are energy and disorder.*

A dam and power station that convert the energy of falling water into electrical energy. The seven operating units each provide $3.17 \times 10^7 \, J \cdot s^{-1}$ or 42 500 horsepower.

Energy

Scientists define **energy** as the *ability to do work.* **Work** is done when *a force applied to some object moves the object.* For example, lifting a heavy box is work. There are a number of types of energy: **thermal energy,** which is *commonly called heat,* electrical energy, radiant energy (including light), chemical energy, mechanical energy (including sound), and nuclear energy.

Energy can also be classified as kinetic energy or potential energy. As we saw in Section 5.10, kinetic energy is energy of motion. **Potential energy** is *stored energy.* For example, a book on your desk has greater potential energy than the same book on the floor. When you drop the book, it loses potential energy and gains kinetic energy as it falls. You may become painfully aware of the kinetic energy if the book happens to land on your foot.

Energy can be transferred from one object to another. For example, thermal energy can be transferred from a burner on your stove to a kettle of water. One form of energy can be changed into another form. For example, when magnesium and oxygen from air react to form magnesium oxide, chemical energy stored in magnesium and oxygen is changed into thermal energy and light energy. ■Figure 6.2 shows magnesium metal reacting with oxygen to form magnesium oxide.

The purpose of carrying out many chemical reactions is to change chemical energy into other forms of energy. For example, the motor in your car changes energy from burning gasoline into mechanical energy; the battery in your flashlight changes chemical energy into electrical energy.

Exothermic and Endothermic Changes

At room temperature, most spontaneous changes give off energy. The energy given off is usually in the form of thermal energy. *Changes that give off thermal energy— that is, changes that heat their surroundings—*are called **exothermic.** The burning of natural gas in a laboratory burner is an example of an exothermic reaction. The flame feels hot and heats anything that is near. Spontaneous reactions may need some help getting started. The gas in your laboratory burner doesn't burn unless you light it; once started, however, spontaneous reactions continue all by themselves.

The decomposition of water is different than the burning of natural gas. You must continue to pass electricity through the water or reaction stops. The energy needed to decompose water can be supplied by heating water to a very high temperature instead of by passing electricity through it. *Changes that remove thermal energy from their surroundings—that is, changes that cool their surroundings—*are called **endothermic.** The decomposition of water is nonspontaneous and endothermic. A few changes that take place spontaneously at room temperature are endothermic. For example, when some solids dissolve in water, the solution feels cold. This effect

■ FIGURE 6.2 A little energy is needed to start the reaction between magnesium metal and oxygen from air. But once started, reaction takes place spontaneously, releasing thermal energy and light energy. The magnesium oxide formed is a white solid. This reaction was used in camera flashbulbs for many years.

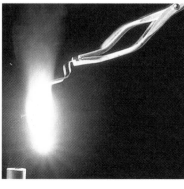

is used in instant cold packs. ▬Figure 6.3 shows an instant cold pack in use. An ice cube melting is another example of a spontaneous endothermic change. As an ice cube melts, it feels cold because it is removing thermal energy (heat) from your hand.

In ice, water molecules are arranged in order; they are fixed in position. ▬Figure 6.4(a) is a picture of a model of ice, and Figure 6.4(b) shows a model of liquid water. As ice is heated, the added energy makes the molecules move about more in their places. If enough energy is added, they are able to move out of their places and to move around. Solid ice has become liquid water, which can flow. The attraction between water molecules is not nearly as strong as the attraction between positively charged sodium ions and negatively charged chloride ions in sodium chloride. Therefore, much less thermal energy is needed to melt ice than to melt sodium chloride. The melting point of ice is low (32 °F or 0 °C). Even at room temperature, enough thermal energy is available from air and table top to melt ice.

Entropy

As you can see from Figure 6.4(b), in liquid water the water molecules are no longer arranged in order. (If you could shake the beaker, you could also observe that they can move around.) *If a change leads to a large enough increase in disorder of the system, it takes place spontaneously despite being endothermic* as the melting of ice at room temperature does. The scientific word that describes disorder is **entropy.** Entropy is *a measure of disorder;* as disorder increases, entropy increases. Entropy is treated in more detail in Chapter 17.

Rates of Changes

Not all spontaneous changes take place in a useful period of time. Some spontaneous changes are slow. For example, the reaction of hydrogen and oxygen to form water

$$2H_2(g) + O_2(g) \longrightarrow 2H_2O(l)$$

is very slow under ordinary conditions. Mixtures of hydrogen and oxygen can sit around for years without reacting. But if a little energy (such as a spark) is added to start the reaction, reaction takes place in an instant; an explosion results.

Sometimes a material that is not used up in a reaction makes the reaction take place faster. *A material that makes a reaction take place faster without being used up in the reaction* is called a **catalyst.** Catalysts are shown above or below the arrow in chemical equations. For example, iron is a catalyst for the reaction of nitrogen with hydrogen to form ammonia:

$$N_2(g) + 3H_2(g) \xrightarrow{\text{iron}} 2NH_3(g)$$

This reaction has played a major part in increasing food production throughout the world. It does not take place at a useful rate without a catalyst. *To predict whether*

▬ FIGURE 6.3   An instant cold pack in use. Instant cold packs usually contain a bag of water and ammonium nitrate [$NH_4NO_3(s)$]. When the bag of water is broken open by squeezing the outer bag, the ammonium nitrate dissolves, providing instant cold. Is the dissolving of ammonium nitrate in water exothermic or endothermic?

You are most likely to be familiar with the word catalyst *in connection with catalytic converters in cars. Catalytic converters reduce air pollution by speeding the reaction of unburned gasoline and carbon monoxide with air. They also make nitrogen oxides formed in the engine change back to nitrogen and oxygen at a useful rate.*

▬ FIGURE 6.4   (a) A space-filling model of ice. Note the orderly arrangement and fixed positions of the water molecules. (b) A space-filling model of water. The arrangement of the molecules is no longer orderly, and the molecules are free to move around.

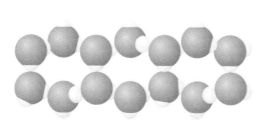

(a)

(b)

An everyday example of the spontaneity of an increase in disorder is your room. Have you ever known it to straighten itself up?

a change will take place, you need to know how the energy and entropy of the system will change as a result of the change, and whether the change will take place at a practical speed. Rates of reaction are discussed in Chapter 18.

PRACTICE PROBLEMS

6.1   Which of the following changes are spontaneous? (a) A damp towel hung on a rack dries; (b) spilled peas gather themselves into a neat pile; (c) the odor of cookies baking spreads through a house; (d) on a damp day, drops of water form on a glass that contains a cold drink.

6.2   Tell which of the following changes are exothermic and which are endothermic: (a) Water boils, (b) paper burns, and (c) rubbing alcohol evaporates from your skin.

6.3   Knowing that sodium metal reacts spontaneously with water to form a solution of sodium hydroxide and hydrogen gas, would you expect the reaction

$$2NaOH(aq) + H_2(g) \longrightarrow 2Na(s) + 2H_2O(l)$$

to take place spontaneously? Explain your answer.

⊙ Remember Figure 2.9.

6.3   TEMPERATURE, THERMAL ENERGY, AND HEAT

An understanding of the difference between temperature, thermal energy, and heat is essential for understanding thermochemistry. As we saw in Section 2.9, temperature is a measure of the hotness or coldness of an object or substance. Temperature is an intensive property—that is, temperature does not depend on sample size. Both a few drops of boiling water and a panful of boiling water are at the same temperature.

Thermal energy is *energy of motion of molecules, atoms, or ions,* whichever are the unit particles of a substance. All objects that are not at absolute zero (zero kelvin) have thermal energy. In a solid, the unit particles can only move back-and-forth in their places and rotate. Although, in a liquid, the unit particles still touch each other, they can move around: Liquids flow. Gases are mostly empty space, and the unit particles move freely through the space. The higher the temperature, the faster the unit particles move. Thermal energy is an extensive property—that is, thermal energy depends on sample size. A few drops of boiling water spattered on your skin do not hurt very much, but a panful of boiling water spilled on yourself can cause a painful burn.

See Peckham, G. D.; McNaught, I. J. J. Chem. Educ. 1993, 70, 103–104 for a discussion of the definition of heat.

Heat is the *thermal energy transfer that results from a difference in temperature.* Net transfer of thermal energy always takes place from a hot to a cold object. If no net transfer of thermal energy takes place between two objects, the two objects are at the same temperature. Energy can be transferred to an object by doing work on the object as well as by heating it. For example, water can be heated either by vigorous stirring or by adding a hot stone. Unfortunately, the term *heat* is often used when thermal energy is meant. Misuse of the term *heat* goes back to the eighteenth century when scientists thought that heat was a substance, a fluid that could flow. Heat is not a substance; heat is a process.

6.4   LAW OF CONSERVATION OF ENERGY

The work of many scientists and engineers has shown that energy, like matter, is neither created nor destroyed in chemical reactions. However, energy can be transferred from one body to another and changed from one form to another. For example, thermal energy can be transferred from the "burner" on your stove to the contents of a pot. Electrical energy can be changed into thermal energy in the burner. But the total quantity of energy remains the same—no energy is produced or lost. *The general rule that energy can neither be created nor destroyed* is called the **law of conservation of energy.**

Law of Conservation of Energy

Energy cannot be created or destroyed.

The law of conservation of energy is often called the first law of thermodynamics.

6.5 ENERGY UNITS

In Section 5.10, we saw that kinetic energy depends on mass and speed. The mathematical relationship between kinetic energy, mass, and speed is

$$\text{kinetic energy} = \tfrac{1}{2}mv^2 \qquad\qquad (5.3,\ 6.1)$$

where m is mass and v is speed. From equation 5.3, the kinetic energy of a 2-kg mass moving at a speed of 1 m/s can be calculated:

$$\text{kinetic energy} = \frac{1}{2}\,(2\ \text{kg})\left(\frac{1\ \text{m}}{1\ \text{s}}\right)^2 = 1\ \frac{\text{kg}\cdot\text{m}^2}{\text{s}^2}$$

> Remember that $\text{kg}\cdot\text{m}^2$ means $\text{kg} \times \text{m}^2$.

Thus, the SI unit of energy is $1\ \text{kg}\cdot\text{m}^2\cdot\text{s}^{-2}$; this *SI unit of energy and work* is called a **joule, J.*** The joule is a derived unit; it is *not* a base unit.

To give you an idea of how much energy a joule is, if you drop a textbook that has a mass of 5 lb from a height of 3.5 ft (a little over a meter), it will hit your toe with a kinetic energy of about 24 joules. Another way of thinking about the joule is that it is the amount of work that must be done to lift a 2-lb bag of sugar to a height of about 10 cm or 4 in. The bag of sugar has then gained 1 J of gravitational potential energy.

> *Each beat of your heart uses about 1 J of energy to drive blood through your body.*

You are probably more used to thinking of energy in calories. Before 1925, one calorie was defined as the amount of energy required to heat one gram of water from 15 to 16 °C. Today, one calorie is defined as 4.184 joules:

$$1\ \text{calorie (cal)} = 4.184\ \text{J (exactly)}$$

> You should memorize the relationship between calories and joules.

This definition was chosen to keep the quantity of energy represented by the calorie the same. The "calories" in a candy bar or a brownie are really kilocalories. To convert the energy gained by eating a 120-"calorie" brownie to gravitational potential energy, you would have to lift the 2-lb bag of sugar about 31 miles!

PRACTICE PROBLEM

6.4 Fill in the blanks: (a) 23.5 cal = _____ J,
(b) 642 J = _____ kJ, (c) 778 kcal = _____ kJ, and
(d) 3.86 kJ = _____ J.

---

*The joule is named after English physicist James Prescott Joule. Joule showed that mechanical, thermal, and electrical energy can be changed into one another. His discovery that, when gases expand at ordinary temperatures and pressures, their temperatures decrease, led to the development of the refrigeration industry. Joule was, at one time, a student of John Dalton's at the University of Manchester.

TABLE 6.1

Specific Heats of Some Common Substances at 25 °C and 1 atm[a]

Substance	Specific Heat, $J \cdot g^{-1} \cdot °C^{-1}$
Al(s)	0.90
Br_2(l)	0.47
C(diamond)	0.51
C(graphite)	0.71
CH_3CH_2OH(l)	2.42
$CH_3(CH_2)_6CH_3$(l)	2.23
Fe(s)	0.45
H_2O(s)	2.09
H_2O(l)	4.18
H_2O(g)	1.86
N_2(g)	1.04
Na(s)	1.23
O_2(g)	0.92

[a]Most data are from Atkins, P. W. *Physical Chemistry,* 5th ed.; Freeman: New York, 1994; pp C8–C15.

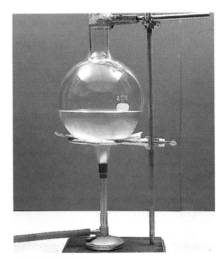

- FIGURE 6.5 Thermal energy released by the combustion of natural gas enters the water, and the temperature of the water increases. The final temperature minus the initial temperature is >0—that is, the temperature change, Δt, is +. The energy of the system (the water) increases; the energy change of the system is also +.

Thermodynamics is concerned with the energy and entropy changes that accompany physical and chemical changes. Although there are a number of forms of energy, such as radiant energy, electrical energy, mechanical energy, and chemical energy, all can be converted to thermal energy. Thermal energy is useful, and thermal energy changes are easy to measure; therefore, we begin our study of thermodynamics with thermochemistry.

6.6 HEAT CAPACITY AND SPECIFIC HEAT

The amount of energy involved in heating and cooling is very important from a practical point of view. A good part of the money spent for energy in an average household goes for heating and cooling air and water. Costs of energy for heating and cooling make up a significant portion of manufacturing costs in most processes. It seems reasonable that the quantity of thermal energy required to change the temperature of an object should depend on the size of the object, the material that the object is made of, and the change in temperature. Quantitatively, the relationship is given by equation 6.2:

$$\text{thermal energy change} = \underbrace{\text{mass} \times \text{specific heat}}_{\text{heat capacity}} \times \text{temperature change} \qquad (6.2)$$

The **specific heat** of a substance is the *amount of thermal energy required to heat one gram of the substance one degree.* The product of mass and specific heat is the **heat capacity** of the object, the *quantity of thermal energy needed to raise the temperature of the object one degree.* Heat capacity is an extensive property; specific heat is an intensive property. Some typical values for specific heats are shown in Table 6.1.

Specific heats can be used to calculate how much thermal energy must be added or removed to change the temperature of a sample from one temperature to another. Sample Problems 6.1 and 6.2 show how. Scientists and engineers have agreed to give the *energy changes associated with exothermic changes a negative sign because energy is lost by the system. The energy changes associated with endothermic changes have positive signs because energy is added to the system* (see ▪Figure 6.5).

SAMPLE PROBLEMS

6.1 How many joules must be added to a 63.43-g sample of Fe(s) to raise the sample's temperature from 19.7 to 54.2 °C? Assume that the specific heat of iron is constant over this temperature range.

Solution From Table 6.1, the specific heat of Fe(s) is 0.45 $J \cdot g^{-1} \cdot °C^{-1}$. This means that 0.45 J are required to raise the temperature of 1 g Fe 1 °C. The temperature of the 63.43-g sample of Fe in the problem must be raised from 19.7 to 54.2 °C:

$$(54.2 - 19.7) °C = 34.5 °C$$

The increase in temperature is 34.5 °C. Notice that *the initial temperature is subtracted from the final temperature to find the temperature change Δt:*

$$\Delta t = t_{\text{final}} - t_{\text{initial}} = (54.2 - 19.7) °C = 34.5 °C$$

The sign of the temperature change should be positive because temperature increases. According to equation 6.2,

$$\text{thermal energy change} = \text{mass} \times \text{specific heat} \times \text{temperature change} \qquad (6.2)$$

Substitution in equation 6.2 gives

$$\text{thermal energy change} = 63.43 \text{ g Fe}\left(\frac{0.45 \text{ J}}{\text{g Fe}\cdot{}^\circ\text{C}}\right)34.5 \,^\circ\text{C} = 9.8 \times 10^2 \text{ J}$$

The question asks for the number of joules required. In the setup for the calculation, the units cancel to give joules. The sign of the thermal energy change is positive because thermal energy must be added to the system (the 63.43-g sample of Fe) in order to increase the temperature of the system.

6.2   How many joules must be removed from a 4.51-g sample of Al(s) to lower the sample's temperature from 27.6 to 5.2 °C? Assume that the specific heat of aluminum is constant over this temperature range.

Solution   From Table 6.1, the specific heat of Al(s) is 0.90 $\text{J}\cdot\text{g}^{-1}\cdot{}^\circ\text{C}^{-1}$. The temperature of the 4.51-g sample of Al in the problem must be lowered from 27.6 to 5.2 °C:

$$\Delta t = t_{\text{final}} - t_{\text{initial}} = (5.2 - 27.6)\,^\circ\text{C} = -22.4\,^\circ\text{C}$$

This time, subtraction of the initial temperature from the final temperature gives a negative sign for the temperature change because temperature decreases. Substitution in equation 6.2 gives

$$\text{thermal energy change} = 4.51 \text{ g Al}\left(\frac{0.90 \text{ J}}{\text{g Al}\cdot{}^\circ\text{C}}\right)(-22.4\,^\circ\text{C}) = -91 \text{ J}$$

The sign of the thermal energy change is negative because thermal energy must be removed from the system in order to decrease the temperature of the system; 91 J must be removed.

PRACTICE PROBLEMS

6.5   How many joules must be added to a 7.92-g sample of $CH_3CH_2OH(l)$ to raise the sample's temperature from 6.8 to 75.2 °C?

6.6   How many joules must be removed from a 287-g sample of $CH_3CH_2OH(l)$ to lower the sample's temperature from 56.2 to 19.8 °C?

In Table 6.1, notice how high the specific heat of liquid water is compared to that of all the other substances in the table. The unusually high specific heat of water has very important results. The huge quantity of water in the world's oceans can absorb or release enormous amounts of thermal energy by means of a small temperature change. Thus, Earth is protected from sudden large changes in temperature that would be harmful to both plant and animal life. Cities located near large bodies of water are warmer in winter and cooler in summer than inland cities. Human beings, whose bodies average 60% water by mass, are able to maintain a constant body temperature of about 37 °C (99 °F).

6.7   MEASUREMENT OF THERMAL ENERGY GAINED OR LOST DURING CHANGES

Thermal energy itself cannot be measured. Only differences in energy can be measured. An *apparatus used to measure the quantity of thermal energy gained or lost during changes* is called a **calorimeter** (a calorie meter). In general chemistry labs,

The high specific heat of water can be used to store energy from the sun. This solar pond in El Paso, Texas supplies a food-processing plant with process hot water and is used experimentally to generate electric power and to desalt water. The high specific heat of water is also responsible for the fact that hot water causes serious burns. Many small children are badly injured by burns from hot water every year.

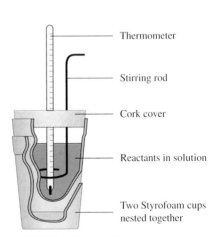

You mix hot and cold water to get water of comfortable temperature in your shower or bathtub.

calorimeters usually consist of covered Styrofoam cups with a thermometer and stirrer as shown in ▪Figure 6.6.

Now let's imagine that we carry out a very simple experiment. Suppose that we mix 25.00 g of water at 20.0 °C with 75.00 g of water at 40.0 °C. What will be the temperature of the mixture? Qualitatively, you should be able to predict that the temperature of the mixture will be between 20.0 and 40.0 °C and closer to 40.0 °C than to 20.0 °C because more warm water than cool water was used. From Table 6.1, the specific heat of water is 4.18 J·g^{-1}·°C^{-1}. If we let t_{mixt} equal the temperature of the mixture, then $(t_{mixt} - 40.0)$ °C is the change in temperature of the warmer water and

thermal energy change of warmer water = mass$_{water}$ × specific heat

× change in temperature

$$= 75.00 \text{ g H}_2\text{O}\left(\frac{4.18 \text{ J}}{\text{g H}_2\text{O} \cdot °\text{C}}\right)(t_{mixt} - 40.0) °\text{C}$$

This change must have a negative sign because the warmer water will lose thermal energy and become cooler. The change in temperature of the cooler water is $(t_{mixt} - 20.0)$ °C and

thermal energy change of cooler water = mass$_{water}$ × specific heat

× change in temperature

$$= 25.00 \text{ g H}_2\text{O}\left(\frac{4.18 \text{ J}}{\text{g H}_2\text{O} \cdot °\text{C}}\right)(t_{mixt} - 20.0) °\text{C}$$

This change must have a positive sign because the cooler water will gain thermal energy and become warmer.

According to the law of conservation of energy, energy cannot be created or destroyed. If no thermal energy is lost to the surroundings and all the thermal energy stays in the water, the thermal energy change of the warmer water must be equal but opposite in sign to the thermal energy change of the cooler water. Quantitatively,

thermal energy change of warmer water ($-$) = thermal energy change of cooler water ($+$)

and

$$-75.00 \text{ g H}_2\text{O}\left(\frac{4.18 \text{ J}}{\text{g H}_2\text{O} \cdot °\text{C}}\right)(t_{mixt} - 40.0) °\text{C}$$

$$= +\left[25.00 \text{ g H}_2\text{O}\left(\frac{4.18 \text{ J}}{\text{g H}_2\text{O} \cdot °\text{C}}\right)(t_{mixt} - 20.0) °\text{C}\right] \quad (6.3)$$

Solution of equation 6.3 for t_{mixt} gives

$$t_{mixt} = 35.0 °\text{C}$$

The temperature of the mixture is 35.0 °C, a reasonable answer because 35.0 °C is between 20.0 and 40.0 °C but closer to 40.0 °C.

A coffee-cup calorimeter can be used to measure the gain or loss of thermal energy that accompanies many kinds of physical and chemical changes. Sample Problem 6.3 illustrates the use of a coffee-cup calorimeter to determine the energy change that takes place during a chemical reaction. In accurate work, calorimeters are very thoroughly insulated, and a correction is made for the thermal energy absorbed by the calorimeter. A coffee cup is designed to keep coffee hot; it is a good insulator and does not absorb much thermal energy. Therefore, to simplify the calculations, we will assume that the amount of thermal energy lost to the surrounding air and the amount of thermal energy absorbed by the coffee-cup calorimeter are negligible.

— Thermometer

— Stirring rod

— Cork cover

— Reactants in solution

— Two Styrofoam cups nested together

▪ FIGURE 6.6 A coffee-cup calorimeter.

6.3 A 1.00 M aqueous solution of NaOH, a 0.50 M aqueous solution of H_2SO_4, and a coffee-cup calorimeter were allowed to stand at a room temperature of 25.4 °C until the temperature of all three reached 25.4 °C. A 50.0-mL sample of the 1.00 M NaOH was then placed in the calorimeter, 50.0 mL of the 0.50 M H_2SO_4 was added as rapidly as possible, and the two solutions were mixed thoroughly. The temperature rose to 31.9 °C. From this experiment, what is the heat of neutralization of one mole of sulfuric acid? For simplicity, assume that the densities of the NaOH and H_2SO_4 solutions were 1.00 g/mL and that the specific heat of the solution after reaction was 4.18 $J \cdot g^{-1} \cdot {}^\circ C^{-1}$ (the same as the density and specific heat of water).

○ Remember that the reaction of an acid with a base to form water and a salt is called neutralization (Section 4.2).

Solution In this experiment, the temperature rose; thermal energy was given off by the reaction and absorbed by the solution. To find the heat of neutralization of one mole of sulfuric acid, we have to calculate how many joules were given off by the reaction and how many moles of sulfuric acid were neutralized. We can then calculate the number of joules that would be given off by the neutralization of one mole of sulfuric acid.

The thermal energy given off had to warm 100.0 g of solution:

thermal energy absorbed by solution (+)

$$= 100.0 \text{ g soln} \left(\frac{4.18 \text{ J}}{\text{g soln} \cdot {}^\circ C} \right) (31.9 - 25.4) \, {}^\circ C$$

$$= 2.7 \times 10^3 \text{ J}$$

To find how many moles of sulfuric acid were neutralized in the experiment, use the definition of molarity:

$$M = \frac{\text{moles solute}}{\text{vol soln in liters}}$$

Solve the definition of molarity for moles solute and substitute the information given in the problem:

$$M \times (\text{vol soln in liters}) = \text{moles solute}$$

$$\left(0.50 \, \frac{\text{mol } H_2SO_4}{1 L \text{ soln}} \right) 50.0 \text{ mL soln} \left(\frac{1 L \text{ soln}}{1000 \text{ mL soln}} \right) = 0.025 \text{ mol } H_2SO_4$$

Because thermal energy was given off by the reaction, the heat of neutralization has a negative sign. Thus, the heat of neutralization per mole of sulfuric acid is

$$\frac{-2.7 \times 10^3 \text{ J}}{0.025 \text{ mol } H_2SO_4} = -1.1 \times 10^5 \, \frac{\text{J}}{\text{mol } H_2SO_4}$$

This large a quantity of energy is usually expressed in kilojoules. The heat of neutralization of sulfuric acid is

$$-1.1 \times 10^5 \, \frac{\text{J}}{\text{mol } H_2SO_4} \left(\frac{1 \text{ kJ}}{1000 \text{ J}} \right) = -1.1 \times 10^2 \, \frac{\text{kJ}}{\text{mol } H_2SO_4}$$

(The value for the heat of neutralization of sulfuric acid in the chemical literature is −112 kJ/mol.)

6.7 A 1.00 M solution of NaOH, a 1.00 M solution of HCl, and a calorimeter were allowed to stand at a room temperature of 19.5 °C until the temperature of all three reached 19.5 °C. A 200.0-mL sample of the 1.00 M NaOH was then placed in the calorimeter, 200.0 mL of the 1.00 M HCl was added as rapidly as possible, and the two solutions were mixed thoroughly. The temperature rose to 25.8 °C. (a) From this experiment, what is the heat of neutralization of one mole of hydrochloric acid? Assume that the densities of the NaOH and HCl solutions were 1.00 g/mL and that the specific heat of the solution after reaction was 4.18 $J \cdot g^{-1} \cdot °C^{-1}$. (b) Compare the heat of neutralization of one mole of hydrochloric acid with the heat of neutralization of one mole of sulfuric acid in Sample Problem 6.3 and explain the difference.

6.8 ENTHALPY

Like changes studied in a coffee-cup calorimeter, most changes in laboratories and manufacturing plants take place under the constant (approximately) pressure of Earth's atmosphere. Most biological reactions also take place under atmospheric pressure. The *thermal energy gained or lost when a change takes place under constant pressure* is called the **enthalpy change, ΔH,** for the change. The Greek letter delta, Δ, means change; H is the symbol for enthalpy. The symbol ΔH represents the difference between the final and the initial enthalpies:

$$\Delta H = H_{final} - H_{initial}$$

Fusion means melting.

Subscripts are used to show the type of change. For example, the enthalpy of fusion (heat of fusion) is symbolized by ΔH_{fus} and the enthalpy of vaporization (heat of vaporization) by ΔH_{vap}. The enthalpy of neutralization is represented by ΔH_{neut}, and so forth. In general, $\Delta H_{reaction}$, ΔH_{rxn}, or simply ΔH, stands for the enthalpy change accompanying a chemical reaction.

Stoichiometry

For many reactions, the thermal energy released is the chief reason for carrying out the reaction. For example, when natural gas is burned in a furnace or water heater, thermal energy is the desired product. When gasoline is burned in a car's engine, the thermal energy given off is converted to mechanical energy that moves the car. In an electric power plant, the thermal energy released by burning coal is converted to electrical energy.

Thermal energy given off can be shown as a product of the reaction in a **thermochemical equation,** for example,

$$CH_4(g) + 2O_2(g) \longrightarrow CO_2(g) + 2H_2O(l) + 890.32 \text{ kJ} \tag{6.4}$$

Commercial natural gas is mostly methane, CH_4, which is odorless. A small amount of smelly sulfur compound is added so that leaks will be noticed.

Thermochemical equations are always interpreted in terms of moles. Equation 6.4 says "One mole of methane gas reacts with two moles of oxygen gas. The products are one mole of carbon dioxide gas, two moles of liquid water, and 890.32 kJ of thermal energy." (Thermal energy absorbed can be shown as a reactant.) More commonly ΔH_{rxn} is given to the right of the equation for a reaction:

$$CH_4(g) + 2O_2(g) \longrightarrow CO_2(g) + 2H_2O(l) \qquad \Delta H_{rxn} = -890.32 \text{ kJ}$$

The $-$ sign for ΔH_{rxn} shows that the reaction is exothermic. Thermal energy is lost by the system to the surroundings. *When ΔH_{rxn} is given for a reaction, the equation for the reaction must be interpreted in terms of moles (the number of moles shown in the equation).*

Sample Problem 6.4 shows how to calculate the amount of thermal energy given off or absorbed during a change.

6.4 For the burning of methane gas,

$$CH_4(g) + 2O_2(g) \longrightarrow CO_2(g) + 2H_2O(l)$$

$\Delta H_{rxn} = -890.32$ kJ. How many kilojoules will be given off by the burning of 451 g of methane?

Solution The equation must be interpreted in terms of moles. But the quantity of methane is given in grams. We must convert the quantity of methane to moles. The formula mass of CH_4 is 16.043 u (see the table on the inside back cover); 1 mol CH_4 has a mass of 16.043 g. Therefore, 451 g CH_4 equals 28.2 mol CH_4 as shown below:

$$451 \text{ g } CH_4\left(\frac{1 \text{ mol } CH_4}{16.0 \text{ g } CH_4}\right) = 28.2 \text{ mol } CH_4$$

When 1 mol CH_4 is burned, 890.32 kJ are given off. When 28.2 mol CH_4 are burned,

$$28.2 \text{ mol } CH_4\left(\frac{890 \text{ kJ}}{1 \text{ mol } CH_4}\right) = 2.51 \times 10^4 \text{ kJ}$$

will be given off.

   Remember that once you have figured out how to solve a problem, combining steps avoids mistakes in entering numbers in your calculator and rounding errors. The two steps in this problem can be combined to give

$$451 \text{ g } CH_4\left(\frac{1 \text{ mol } CH_4}{16.0 \text{ g } CH_4}\right)\left(\frac{890 \text{ kJ}}{1 \text{ mol } CH_4}\right) = 2.51 \times 10^4 \text{ kJ}$$

Check: Don't forget to check to be sure that units cancel to give the correct units for the answer, that the answer is reasonable, and that you have rounded the answer off to the right number of significant figures. Our setup leads to an answer in kilojoules, the units asked for in the problem. The answer we have obtained is reasonable because the quantity of CH_4, 451 g, is almost 30 times the mass of 1 mol CH_4; the number of kilojoules given off, 2.51×10^4 kJ, is almost 30 times the number of kilojoules given off by the combustion of 1 mol CH_4. The three digits given for the mass of CH_4 limit the number of significant figures in the answer to 3.

▶ Always check your work.

6.8 For the burning of isooctane,

$$2(CH_3)_2CHCH_2C(CH_3)_3(l) + 25O_2(g) \longrightarrow 16CO_2(g) + 18H_2O(l)$$

$\Delta H_{rxn} = -10\,930.9$ kJ. How many kilojoules will be given off by the burning of 369 g of isooctane?

Isooctane is used to define the octane rating of gasoline.

6.9 Classify each of the following changes as exothermic or endothermic:
(a) $CaCO_3(s) \longrightarrow CaO(s) + CO_2(g)$ $\Delta H_{rxn} = +178.3$ kJ
(b) $H_2(g) + Cl_2(g) \longrightarrow 2HCl(g) + 184.62$ kJ
(c) $NH_3(g) + HCl(g) \longrightarrow NH_4Cl(s)$ $\Delta H_{rxn} = -176.01$ kJ

6.10 For the reactions in 6.9(a) and (c), rewrite the equations showing thermal energy as a reactant or product.

6.11 (a) After the reaction in 6.9(b) has taken place, will the walls of the container feel hot or cold? Explain your answer. (b) What is ΔH_{rxn} for the reaction in 6.9(b)?

Fortunately, it is not necessary to measure ΔH for all changes that may be of interest by calorimetry. Once a few enthalpy changes have been determined experimentally, many enthalpy changes can be calculated by means of Hess's law.

6.9 HESS'S LAW

In 1840 the Swiss-born chemist Germain Henri Hess* stated an empirical law:

Hess's Law

If the net change in energy by different series of steps were different, energy could be created. Raising a system from a low-energy to a high-energy state by a path requiring little energy and then letting the system return to the low-energy state by a path releasing more energy would result in the creation of energy.

The thermal energy given off or absorbed in a given change is the same whether the change takes place in a single step or in several steps.

For example, addition of hydrogen to acetylene to form ethane can be carried out in one step

$$HC{\equiv}CH(g) + 2H_2(g) \longrightarrow H_3CCH_3(g) \qquad \Delta H_{rxn} = -311.42 \text{ kJ} \qquad (6.5)$$

acetylene   ethane

or in two steps

$$HC{\equiv}CH(g) + H_2(g) \longrightarrow H_2C{=}CH_2(g) \qquad \Delta H_{rxn} = -174.47 \text{ kJ} \qquad (6.6)$$

and

$$H_2C{=}CH_2(g) + H_2(g) \longrightarrow H_3CCH_3(g) \qquad \Delta H_{rxn} = -136.95 \text{ kJ} \qquad (6.7)$$

Chemical equations can be combined like any other equations. The equation for the one-step reaction, equation 6.5, is the sum of the equations for the two steps, equations 6.6 and 6.7:

$$(6.5) = (6.6) + (6.7)$$

$$HC{\equiv}CH(g) + H_2(g) \longrightarrow H_2C{=}CH_2(g) + 174.47 \text{ kJ} \qquad (6.6)$$

$$+ H_2C{=}CH_2(g) + H_2(g) \longrightarrow H_3CCH_3(g) + 136.95 \text{ kJ} \qquad (6.7)$$

$$HC{\equiv}CH(g) + \cancel{H_2}(g)^2 + \cancel{H_2C{=}CH_2}(g) + \cancel{1H_2}(g) \longrightarrow \cancel{H_2C{=}CH_2}(g) + H_3CCH_3(g) + 311.42 \text{ kJ}$$

○ Terms that appear on both sides of an equation, such as H_2C $CH_2(g)$, can be canceled, and terms that appear more than once on a side, such as $H_2(g)$, can be combined.

or

$$HC{\equiv}CH(g) + 2H_2(g) \longrightarrow H_3CCH_3(g) + 311.42 \text{ kJ} \qquad (6.5)$$

The enthalpy change for the one-step reaction, -311.42 kJ, is the sum of the ΔHs for the two steps:

$$[(-174.47) + (-136.95)] \text{ kJ} = -311.42 \text{ kJ}$$

These energy relationships are shown graphically in ▪Figure 6.7.

In an energy diagram such as Figure 6.7, a horizontal line represents the sum of the enthalpies of the substances shown above the line in the physical states shown. The higher a line on the diagram, the greater the enthalpy of the substances shown above the line. The vertical arrows show differences in enthalpy. Arrows pointing

○ Hess's law is another way of stating the law of conservation of energy.

*Hess spent most of his life in Russia. He practiced medicine in Siberia before he became a professor of chemistry at the Technological Institute of the University of Saint Petersburg. His studies of thermal energy changes accompanying chemical reactions formed a foundation for thermodynamics.

down represent exothermic changes. (Arrows pointing up represent endothermic changes.) If the direction of an arrow is reversed, the arrow then shows a change that is the reverse of the original change and the sign of ΔH must be changed. For example,

$$HC{\equiv}CH(g) + 2H_2(g) \longrightarrow H_3CCH_3(g) \qquad \Delta H_{rxn} = -311.42 \text{ kJ}$$
$$H_3CCH_3(g) \longrightarrow HC{\equiv}CH(g) + 2H_2(g) \qquad \Delta H_{rxn} = +311.42 \text{ kJ}$$

State Functions

Properties such as enthalpy *that depend only on the initial and final states of the system and not on how the system gets from one state to another* are called **state functions.** Pressure, volume, and temperature are also state functions. A good analogy for a state function is the distance between New York and San Francisco measured on a map or globe (■Figure 6.8). The straight-line distance between New York and San Francisco is always the same. However, neither the distance traveled in going from New York to San Francisco nor the time required to get from New York to San Francisco are state functions. Both the distance traveled and the time required depend on how you get from one city to the other. For example, the distance traveled in getting from New York to San Francisco depends on whether you go by way of Chicago or by way of Houston. The time required depends on whether you travel by car or by plane as well as on the route you take.

In thermodynamics, the term **state** has a very precise meaning. *In thermodynamics, a system is said to be in a certain state when its properties have certain values.* For example, a sample of acetylene gas is in one state at 298 K and 1 atm and in another state at 273 K and 1 atm. Usually only a few properties of a system need be given to define the state of the system. The values of most properties depend on the values of a few properties; for example, the volume of the sample of acetylene gas is determined by its temperature and pressure.

The enthalpy changes accompanying many reactions are difficult or impossible to measure by calorimetry. Some reactions, for example, geological changes, take place too slowly; other reactions are accompanied by side reactions (see the example that follows). Enthalpy changes that cannot be measured by calorimetry can often be found by using Hess's law.

Calculation of Enthalpies of Reaction by Combination of Thermochemical Equations

We have just seen that, when the equation for a reaction is the sum of the equations for two other reactions, ΔH_{rxn} for the reaction is the sum of the ΔH_{rxn}s for the two other reactions. In general, if the equation for a chemical reaction can be obtained by combining the equations for other chemical reactions, ΔH_{rxn} for the reaction can be calculated by combining the ΔH_{rxn}s for the reactions in the same way.

For example, the enthalpy change for the reaction

$$2C(graphite) + O_2(g) \longrightarrow 2CO(g) \tag{6.8}$$

cannot be determined directly because part of the graphite is always converted to carbon dioxide. However, the enthalpies of reaction of both graphite and carbon monoxide with oxygen to form carbon dioxide can be measured:

$$C(graphite) + O_2(g) \longrightarrow CO_2(g) \qquad \Delta H_{rxn} = -393.51 \text{ kJ} \tag{6.9}$$
$$2CO(g) + O_2(g) \longrightarrow 2CO_2(g) \qquad \Delta H_{rxn} = -565.98 \text{ kJ} \tag{6.10}$$

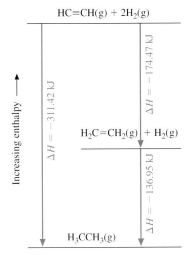

■ FIGURE 6.7 Change in enthalpy does not depend on pathway. The thermal energy given off (or absorbed) during a change is the same whether the change takes place in a single step or in more than one step.

■ FIGURE 6.8 The straight-line distance between New York and San Francisco is always the same. However, the distance traveled to get from New York to San Francisco by way of Houston is longer than the distance traveled to get from New York to San Francisco by way of Chicago.

Equation 6.8 can be obtained by combining equations 6.9 and 6.10. The product of the wanted equation, equation 6.8, is 2 mol CO(g). To get 2 mol CO(g) on the product side, we can reverse equation 6.10 and change the sign of ΔH:

$$2CO_2(g) \longrightarrow 2CO(g) + O_2(g) \qquad \Delta H_{rxn} = +565.98 \text{ kJ} \qquad (6.11)$$

Two moles of C(graphite) are a reactant in the wanted equation, equation 6.8. To get 2 mol C(graphite) on the reactant side, we must multiply equation 6.9 by 2. Because enthalpy is an extensive property, we must also multiply ΔH_{rxn} for equation 6.9 by 2:

$$2[C(graphite) + O_2(g) \longrightarrow CO_2(g)] \qquad \Delta H_{rxn} = 2(-393.51 \text{ kJ})$$

that is,

$$2C(graphite) + 2O_2(g) \longrightarrow 2CO_2(g) \qquad \Delta H_{rxn} = -787.02 \text{ kJ} \qquad (6.12)$$

Now, if we add thermochemical equations 6.11 and 6.12 and simplify, we get the wanted equation, equation 6.8, and its ΔH value:

$$2CO_2(g) \longrightarrow 2CO(g) + O_2(g) \qquad \Delta H_{rxn} = +565.98 \text{ kJ} \quad (6.11)$$
$$+\ 2C(graphite) + 2O_2(g) \longrightarrow 2CO_2(g) \qquad \Delta H_{rxn} = -787.02 \text{ kJ} \quad (6.12)$$

$$2C(graphite) + 2O_2(g) + 2CO_2(g) \longrightarrow 2CO_2(g) + 2CO(g) + O_2(g) \qquad \Delta H_{rxn} = -221.04 \text{ kJ} \quad (6.8)$$

These energy relationships are shown graphically in ▪Figure 6.9.

To figure out how to combine equations, start with the equation that you want to obtain (equation 6.8 in the example) and work backward. Compare the given equations (6.9 and 6.10 in this example) with the wanted equation and reverse and multiply them as needed. When you think you have figured out a way to combine equations, try it and see if it works (does it give you the equation you want?). Especially at first, some trial and error will probably be required; there is almost always more than one way to figure out how to combine the given equations. Sample Problem 6.5 is another example of the calculation of a ΔH_{rxn} by combination of thermochemical equations.

 Subtracting has the same result as reversing the sign and adding.

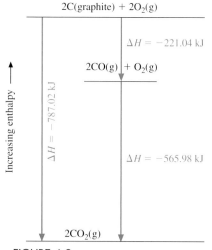

▪ FIGURE 6.9 The enthalpy change for the reaction $2C(graphite) + 2O_2(g) \rightarrow 2CO_2(g)$ is equal to the sum of the enthalpy change for the reaction $2C(graphite) + O_2(g) \rightarrow 2CO(g)$ and the enthalpy change for the reaction $2CO(g) + O_2(g) \rightarrow 2CO_2(g)$.

SAMPLE PROBLEM

6.5 Calculate ΔH_{rxn} for the reaction

$$2Ca(s) + 2C(graphite) + 3O_2(g) \longrightarrow 2CaCO_3(s) \qquad \Delta H_1 = ? \text{ kJ} \qquad (1)$$

from the following information:

$$2Ca(s) + O_2(g) \longrightarrow 2CaO(s) \qquad \Delta H_2 = -1270.18 \text{ kJ} \qquad (2)$$
$$C(graphite) + O_2(g) \longrightarrow CO_2(g) \qquad \Delta H_3 = -393.51 \text{ kJ} \qquad (3)$$
$$CaO(s) + CO_2(g) \longrightarrow CaCO_3(s) \qquad \Delta H_4 = -178.32 \text{ kJ} \qquad (4)$$

Solution Start with the equation for the wanted reaction (equation 1) and work backward. The product of reaction 1 is 2 mol $CaCO_3(s)$. To obtain 2 mol $CaCO_3(s)$, we must multiply equation 4 by 2. As reactants in equation 1 we need 2Ca(s) and 2C(graphite). To get 2Ca(s) and 2C(graphite), we must add equation 2 and twice equation 3:

	ΔH, kJ
$2Ca(s) + O_2(g) \longrightarrow 2CaO(s)$	−1270.18
$+\ 2[C(graphite) + O_2(g) \longrightarrow CO_2(g)]$	2(−393.51)
$+\ 2[CaO(s) + CO_2(g) \longrightarrow CaCO_3(s)]$	2(−178.32)
$2Ca(s) + 2C(graphite) + 3O_2(g) \longrightarrow 2CaCO_3(s)$	−2413.84 kJ

6.12   Given the thermochemical equation

$$2Hg(l) + O_2(g) \longrightarrow 2HgO(s) \qquad \Delta H_{rxn} = -181.66 \text{ kJ}$$

(a) Write the thermochemical equation for the decomposition of 2 mol HgO(s) to Hg(l) and $O_2(g)$. (b) Write the thermochemical equation that shows the formation of 1 mol HgO from Hg(l) and $O_2(g)$.

6.13   Calculate ΔH_{rxn} for the reaction

$$Sn(s) + Cl_2(g) \longrightarrow SnCl_2(s)$$

from the following information

$$Sn(s) + 2Cl_2(g) \longrightarrow SnCl_4(l) \qquad \Delta H_{rxn} = -545.2 \text{ kJ}$$
$$SnCl_2(s) + Cl_2(g) \longrightarrow SnCl_4(l) \qquad \Delta H_{rxn} = -195.4 \text{ kJ}$$

⬤ The equations and their ΔHs should be numbered (as was done in Sample Problem 6.5) so that you can identify them easily.

Calculation of Enthalpies of Reaction from Enthalpies of Formation

The real problem in using Hess's law is figuring out which equations to combine. The reactions that are most often combined are formation reactions. **Formation reactions** are *reactions in which compounds are formed from elements.* For example, the reactions shown by equations 6.13 and 6.14 are formation reactions. Equation 6.13

$$Ca(s) + O_2(g) + H_2(g) \longrightarrow Ca(OH)_2(s) \qquad \Delta H_{rxn} = -986.09 \text{ kJ} \qquad (6.13)$$

is the equation for the formation of the compound calcium hydroxide from the elements calcium, oxygen, and hydrogen. Equation 6.14

$$C(graphite) + O_2(g) \longrightarrow CO_2(g) \qquad \Delta H_{rxn} = -393.51 \text{ kJ} \qquad (6.14)$$

is the equation for the formation of the compound carbon dioxide from the elements carbon and oxygen. A reaction in which a compound is formed from another compound is *not* a formation reaction. For example, although $Ca(OH)_2$ is formed by the reaction

$$CaO(s) + H_2O(l) \longrightarrow Ca(OH)_2(s) \qquad (6.15)$$

the reaction shown by equation 6.15 is *not* a formation reaction because CaO and H_2O are compounds, not elements.

The enthalpies of individual substances cannot be measured; only the changes in enthalpy that accompany physical and chemical changes can be measured. The values of the changes in enthalpy that accompany physical and chemical changes depend on the conditions under which the changes take place. Therefore, a set of standard conditions has been defined so that results can be compared easily. The International Union of Pure and Applied Chemistry now recommends using a pressure of 10^5 Pa (1 bar) as standard pressure although most tables still use the old standard pressure of 1 atm. Because 10^5 Pa is 0.987 atm, differences in ΔH between 10^5 Pa and 1 atm are small and will be ignored in this text. For solids and liquids, the **standard state** is *the pure solid or pure liquid under standard pressure.* The standard state for a gas is *the ideal gas at a partial pressure of 10^5 Pa.** If *all*

---

*Because no real gas is ideal, the standard states for gases are imaginary states. As Moore says, "The definition of the standard state is subtle and curious, since it is not any real state of the gas, but an imaginary (or hypothetical) state." (Moore, W. J. *Basic Physical Chemistry;* Prentice Hall: Englewood Cliffs, NJ, 1983; p 199.)

Appendix D gives values of ΔH_f° for additional compounds.

TABLE 6.2	Some Standard Enthalpies of Formation, ΔH_f°, at 25 °C[a]		
Substance	ΔH_f°, kJ/mol	Substance	ΔH_f°, kJ/mol
C(s, diamond)	1.895	$(H_2N)_2CO(s)$	−333.51
$CaCO_3(s)$	−1206.92	$HNO_3(l)$	−174.1
CaO(s)	−635.09	$H_2O(l)$	−285.83
$CH_4(g)$	−74.85	$H_2O(g)$	−238.92
$C_2H_6(g)$	−84.67	$I_2(g)$	62.24
$CH_3OH(l)$	−238.64	KCl(s)	−436.7
$CH_3OH(g)$	−201.2	N(g)	472.70
$CH_3CH_2OH(l)$	−277.63	NaCl(s)	−411.12
$CH_3CH_2OH(g)$	−235.3	$NaClO_3(s)$	−358.69
CO(g)	−110.52	$NH_3(g)$	−46.11
$CO_2(g)$	−393.51	NO(g)	90.25
H(g)	217.97	$NO_2(g)$	33.2
HBr(g)	−36.40	O(g)	249.17
HCl(g)	−92.31	$SO_2(g)$	−296.83
HCN(g)	131	$SO_3(s)$	−454.5
HF(g)	−271.12	$SO_3(l)$	−441.0
HI(g)	25.94	$SO_3(g)$	−395.72

[a]Most of the thermodynamic data in this book are from Bard, A. J.; Parsons, R.; Jordan, J. *Standard Potentials in Aqueous Solution;* Dekker: New York, 1985.

the reactants and products are in their standard states, the enthalpy change that accompanies a reaction is called the **standard enthalpy change, ΔH°,** for the reaction. Thus, the **standard enthalpy of formation of a substance, ΔH_f°,** is the enthalpy change that results when one mole of the substance is formed from its elements with all substances in their standard states.

The standard enthalpies of formation of the elements in their standard states are zero. Formation of an element in its standard state from the element in its standard state, for example,

$$O_2(g) \longrightarrow O_2(g) \qquad \Delta H^\circ = 0$$

is *not* a change. If an element exists in more than one form, only the most common or most stable (lowest energy) form has zero enthalpy of formation. For example, graphite is the most common and the most stable form of the element carbon under ordinary conditions. Therefore, under ordinary conditions, the enthalpy of formation of carbon in the form of graphite is zero. Diamond is also a form of carbon. The enthalpy of formation of carbon in the form of diamond is 1.895 kJ/mol at 25 °C.

$$C(graphite) \longrightarrow C(diamond) \qquad \Delta H^\circ = 1.895 \text{ kJ}$$

Tables of large numbers of standard enthalpies of formation are available; Table 6.2 shows a few standard enthalpies of formation at 25 °C. (More values are given in Appendix D.) There is *no* standard temperature for thermodynamics. However, values of thermodynamics quantities such as enthalpy of formation are usually tabulated for 25 °C. There are several points you should notice in Table 6.2. The standard enthalpies of formation of most compounds are negative. Only a few compounds have positive values for the standard enthalpy of formation. The combination of elements to form compounds is usually exothermic; the decomposition of compounds to elements is usually endothermic.

The value of the standard enthalpy of formation of a compound depends on the physical state of the compound. For example, ΔH_f° for $H_2O(l)$ is −285.83 kJ/mol, while ΔH_f° for $H_2O(g)$ is −238.92 kJ/mol. The enthalpy change accompanying a

The annual production of graphite is 450 million times the annual production of diamonds, which is only about 9000 kilograms.

Fortunately for diamond lovers, the rate at which diamonds are converted to graphite under ordinary conditions is extremely low.

reaction depends on whether the substances involved are gases, liquids, or solids. For example, the amount of thermal energy given off when hydrogen gas and oxygen gas react to form liquid water is greater than the amount of thermal energy given off when the product is water vapor:

$$2H_2(g) + O_2(g) \longrightarrow 2H_2O(l) \qquad \Delta H^\circ_{rxn} = -571.66 \text{ kJ}$$

$$2H_2(g) + O_2(g) \longrightarrow 2H_2O(g) \qquad \Delta H^\circ_{rxn} = -477.84 \text{ kJ*}$$

Physical states must always be given in thermochemical equations.

For reactions that involve solids that can exist in more than one crystalline form, such as carbon, the enthalpy change depends on the crystalline form. For graphite, the most stable form of carbon at standard conditions and 25 °C, $\Delta H^\circ_f = 0$ by definition; for diamond, $\Delta H^\circ_f = 1.895 \text{ kJ/mol}$. The crystalline form must be given in thermochemical equations; for example, graphite must be represented by C(graphite) not just C(s).

Also note that the standard enthalpies of formation given in Table 6.2 are for the formation of *one* mole of compound. Therefore, equations for formation reactions are usually written to show the formation of 1 mol of the compound. For example, the equation for the formation of ammonia is generally written

$$\tfrac{1}{2}N_2(g) + \tfrac{3}{2}H_2(g) \longrightarrow NH_3(g) \qquad \Delta H^\circ_f = -46.11 \text{ kJ}$$

Instead of

$$N_2(g) + 3H_2(g) \longrightarrow 2NH_3(g) \qquad \Delta H^\circ_{rxn} = -92.22 \text{ kJ}$$

Fractional coefficients can be used because thermochemical equations are understood to refer to moles not molecules.

Hess's law can be used to calculate enthalpies of reaction for many reactions from enthalpies of formation given in tables. The equation for any reaction can be obtained by combining equations for formation reactions. For example, the equation for the reaction

$$2NH_3(g) + CO_2(g) \longrightarrow (H_2N)_2CO(s) + H_2O(l) \qquad (6.16)$$

which is used to manufacture urea, can be obtained by combining the equations for the formation reactions for the reactants and products.

Equations 6.17–6.20 show the formation reactions for all the substances involved in the synthesis of urea:

$$N_2(g) + 2H_2(g) + C(graphite) + \tfrac{1}{2}O_2(g) \longrightarrow (H_2N)_2CO(s) \qquad \Delta H^\circ_f(H_2N)_2CO = -333.51 \text{ kJ} \qquad (6.17)$$

$$H_2(g) + \tfrac{1}{2}O_2(g) \longrightarrow H_2O(l) \qquad \Delta H^\circ_f(H_2O) = -285.83 \text{ kJ} \qquad (6.18)$$

$$\tfrac{1}{2}N_2(g) + \tfrac{3}{2}H_2(g) \longrightarrow NH_3(g) \qquad \Delta H^\circ_f(NH_3) = -46.11 \text{ kJ} \qquad (6.19)$$

$$C(graphite) + O_2(g) \longrightarrow CO_2(g) \qquad \Delta H^\circ_f(CO_2) = -393.51 \text{ kJ} \qquad (6.20)$$

▶ The value found for ΔH° is for whatever quantities (in moles) of reactants and products are given by the coefficients in the equation for the reaction.

Urea is used as a fertilizer and to make plastics.

---

*These two equations can be combined to find the thermal energy required to vaporize water:

$$2H_2O(l) \longrightarrow 2H_2(g) + O_2(g) \qquad \Delta H^\circ_{rxn} = 571.66 \text{ kJ}$$

$$+\ 2H_2(g) + O_2(g) \longrightarrow 2H_2O(g) \qquad \Delta H^\circ_{rxn} = -477.84 \text{ kJ}$$

$$\overline{2H_2O(l) \longrightarrow 2H_2O(g) \qquad \Delta H^\circ = 93.82 \text{ kJ}}$$

This value is for 25 °C and 2 mol of water. Half this amount of thermal energy, 46.91 kJ, is required to vaporize 1 mol of water at 25 °C. This is *not* the standard heat of vaporization of water because, by definition, the standard heat of vaporization is the thermal energy required to vaporize a liquid at its boiling point; the boiling point of water is 100 not 25 °C. The standard heat of vaporization of water is 40.7 kJ/mol (Table 12.2).

In the equation we're trying to find, equation 6.16, the products are 1 mol $(H_2N)_2CO(s)$ and 1 mol $H_2O(l)$. Equation 6.17 yields 1 mol $(H_2N)_2CO(s)$, and equation 6.18 yields 1 mol $H_2O(l)$; therefore, equations 6.17 and 6.18 must be added. Two moles of $NH_3(g)$ react in the wanted equation; equation 6.19 must be reversed, multiplied by 2, and added. One mole of $CO_2(g)$ is a reactant in the wanted reaction. Equation 6.20 must be reversed and added:

$$\Delta H°, \text{kJ}$$

$$N_2(g) + 2H_2(g) + C(\text{graphite}) + \tfrac{1}{2}O_2(g) \longrightarrow (H_2N)_2CO(s) \qquad -333.51$$

$$H_2(g) + \tfrac{1}{2}O_2(g) \longrightarrow H_2O(l) \qquad -285.83$$

$$2[NH_3(g) \longrightarrow \tfrac{1}{2}N_2(g) + \tfrac{3}{2}H_2(g)] \qquad 2(46.11)$$

$$CO_2(g) \longrightarrow C(\text{graphite}) + O_2(g) \qquad 393.51$$

$$2NH_3(g) + CO_2(g) \longrightarrow (H_2N)_2CO(s) + H_2O(l) \qquad -133.61$$

Now, let's look closely at how we have combined the standard enthalpies of formation to obtain $\Delta H°_{rxn}$:

$$\Delta H°_{rxn} = \quad (-333.51\text{ kJ}) \quad + (-285.83\text{ kJ}) - \quad 2(-46.11\text{ kJ}) \quad - (-393.51\text{ kJ})$$

$$= \Delta H°_f[(H_2N)_2CO(s)] + \Delta H°_f[H_2O(l)] - 2\,\Delta H°_f[NH_3(g)] - \Delta H°_f[CO_2(g)]$$

$$(6.21)$$

We have added $\Delta H°_f$s for the substances that are products of the wanted equation and subtracted $\Delta H°_f$s for the substances that are reactants in the wanted equation. Because $\Delta H°_f$ is an extensive property, we have multiplied $\Delta H°_f(NH_3)$ by 2 since the coefficient of NH_3 in the equation for the reaction is 2 and the equation for the formation of NH_3 was multiplied by 2. Equation 6.21 can be rearranged to

$$\Delta H°_{rxn} = \{\Delta H°_f[(H_2N)_2CO] + \Delta H°_f(H_2O)\} - [2\,\Delta H°_f(NH_3) + \Delta H°_f(CO_2)] \quad (6.22)$$

that is,

$$\Delta H°_{rxn} = \quad \begin{matrix}\text{sum of standard enthalpies} \\ \text{of formation of products}\end{matrix} \quad - \quad \begin{matrix}\text{sum of standard enthalpies} \\ \text{of formation of reactants}\end{matrix}$$

where each standard enthalpy of formation is multiplied by the coefficient of the substance in the equation for the reaction.

This statement is true in general; for the reaction

$$aA + bB + \cdots \longrightarrow eE + fF + \cdots$$

where a is the coefficient of the reactant A, b is the coefficient of the reactant B, e is the coefficient of the product E, and so forth,

$$\Delta H°_{rxn} = [e \cdot \Delta H°_f(E) + f \cdot \Delta H°_f(F) + \cdots] - [a \cdot \Delta H°_f(A) + b \cdot \Delta H°_f(B) + \cdots]$$

or

▶ Equation 6.23 provides a shortcut to combining thermochemical equations for formation reactions.

$$\Delta H°_{rxn} = \Sigma n_p\,\Delta H°_f(\text{product}) - \Sigma n_r\,\Delta H°_f(\text{reactant}) \qquad (6.23)$$

where Σ means sum of, n_p is the number of moles of each product, $\Delta H°_f(\text{product})$ is the standard enthalpy of formation of the product, n_r is the number of moles of each reactant, and $\Delta H°_f(\text{reactant})$ is the standard enthalpy of formation of the reactant.

Equation 6.23 is the most useful form of Hess's law. The standard enthalpies of reaction of thousands of reactions can be calculated from tabulated values of ΔH_f° by means of equation 6.23. Sample Problem 6.6 shows how to use Hess's law in the form of equation 6.23 to calculate a standard enthalpy of reaction from the standard enthalpies of formation in Table 6.2. (Remember that more standard enthalpies of formation are given in Appendix D.)

SAMPLE PROBLEM

6.6   Use standard enthalpies of formation to calculate ΔH_{rxn}° for the reaction

$$2SO_2(g) + O_2(g) \longrightarrow 2SO_3(g)$$

Solution   Assuming an ordinary laboratory pressure of 1 atm, oxygen gas (O_2) is an element in its standard state, and its standard enthalpy of formation is zero by definition. The standard enthalpies of formation of the compounds SO_2 and SO_3 are given in Table 6.2. As usual, organizing the information needed to solve the problem in a table helps avoid careless mistakes:

Equation	$2SO_2(g)$	$+ O_2(g) \longrightarrow$	$2SO_3(g)$
ΔH_f°, **kJ/mol**	-296.83	0	-395.72

According to equation 6.23,

$$\Delta H_{rxn}^\circ = \Sigma n_p \, \Delta H_f^\circ(\text{product}) - \Sigma n_r \, \Delta H_f^\circ(\text{reactant}) \qquad (6.23)$$

For the reaction in this problem,

$$\Delta H_{rxn}^\circ = [2 \text{ mol } SO_3 \cdot \Delta H_f^\circ(SO_3)] - [2 \text{ mol } SO_2 \cdot \Delta H_f^\circ(SO_2) + 1 \text{ mol } O_2 \cdot \Delta H_f^\circ(O_2)]$$

Therefore,

$$\Delta H_{rxn}^\circ = \left[2 \text{ mol } SO_3 \left(-395.72 \, \frac{\text{kJ}}{\text{mol } SO_3} \right) \right]$$
$$- \left[2 \text{ mol } SO_2 \left(-296.83 \, \frac{\text{kJ}}{\text{mol } SO_2} \right) + 1 \text{ mol } O_2 \left(0 \, \frac{\text{kJ}}{\text{mol } O_2} \right) \right]$$
$$= (-791.44 \text{ kJ}) - [(-593.66 \text{ kJ}) + 0 \text{ kJ}] = -197.78 \text{ kJ}$$

Notice how mol cancels so that ΔH_{rxn}° is in kilojoules as it should be.

PRACTICE PROBLEMS

6.14   Write the formation equations for (a) $NaClO_3(s)$, (b) $HNO_3(l)$, and (c) $CH_3CH_2OH(l)$.

6.15   Use standard enthalpies of formation to calculate ΔH_{rxn}° for the following reactions. Don't forget to multiply each ΔH_f° by the coefficient of the substance in the balanced equation for the reaction. Be very careful not to make a mistake in sign.

(a) $2NaClO_3(s) \longrightarrow 2NaCl(s) + 3O_2(g)$

(b) $3NO_2(g) + H_2O(l) \longrightarrow 2HNO_3(l) + NO(g)$

(c) $CH_3CH_2OH(g) + 3O_2(g) \longrightarrow 2CO_2(g) + 3H_2O(l)$

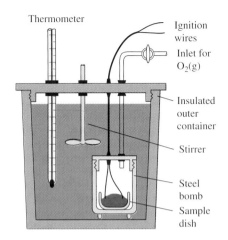

- **FIGURE 6.10** *Top:* Schematic diagram of a bomb calorimeter. *Bottom:* Bomb from a modern bomb calorimeter.

Labels for the top schematic diagram:

- Thermometer
- Ignition wires
- Inlet for $O_2(g)$
- Insulated outer container
- Stirrer
- Steel bomb
- Sample dish

Calculation of Enthalpies of Formation from Enthalpy of Reaction

Equation 6.23 can be used to obtain the standard enthalpy of formation of a compound if ΔH°_{rxn} for a reaction involving the compound can be measured and the standard enthalpies of formation of all the other reactants and products are known. The standard enthalpies of formation of many compounds, especially organic compounds, have been obtained from measurements of standard enthalpies of combustion. **Standard enthalpy of combustion** or **standard heat of combustion, $\Delta H^\circ_{combustion}$** or **$\Delta H_{comb}$**, is the *thermal energy given off when one mole of a substance burns in an excess of oxygen. (All reactants and products must be in their standard states.)* For example, the thermochemical equation for the combustion of methane under standard conditions at 25 °C is

$$CH_4(g) + 2O_2(g) \longrightarrow CO_2(g) + 2H_2O(l) \qquad \Delta H^\circ_{rxn} = -890.32 \text{ kJ}$$

The standard enthalpy of combustion of methane, $\Delta H^\circ_{comb}(CH_4)$, is -890.32 kJ/mol.

High temperatures are needed for combustion. In addition, the products of combustion of all organic compounds include carbon dioxide gas. An open calorimeter like the coffee-cup calorimeter would let carbon dioxide gas escape. Enthalpies of combustion cannot be determined in a calorimeter that is open to the atmosphere. Enthalpies of combustion must be determined in a more elaborate type of calorimeter called a bomb calorimeter. -Figure 6.10 shows a modern bomb calorimeter.

A bomb calorimeter has a constant volume. *If* the number of moles of gaseous products is the same as the number of moles of gaseous reactants, the thermal energy given off or absorbed in a constant-volume calorimeter is the same as the thermal energy given off or absorbed in a constant-pressure calorimeter and is equal to the enthalpy change. The combustion of graphite

$$C(graphite) + O_2(g) \longrightarrow CO_2(g)$$

is an example of this type of reaction.

However, if the number of moles of gaseous products is *not* the same as the number of moles of gaseous reactants, the energy change measured in a constant-volume calorimeter must be corrected to obtain the enthalpy. To understand why, imagine that a reaction is carried out at constant temperature in a cylinder with a piston that is fastened in position as shown in -Figure 6.11(a). If the number of moles of gaseous products is greater than the number of moles of gaseous reactants, pressure increases. Now compare what happens if the piston is not fastened but pressure is constant [Figure 6.11(b)]. As reaction takes place and the number of molecules of gas increases, the piston is pushed up.* Pushing up a piston is work; energy is required to push a piston up. Therefore, the energy given off by an exothermic reaction under constant pressure is less than the energy given off at constant volume. The energy absorbed by an endothermic reaction under constant pressure is greater than the energy absorbed at constant volume. At ordinary temperatures, however, the difference between energy changes measured in constant-volume and constant-pressure calorimeters is usually small compared to the magnitude of the energy change. For example, for the combustion of methane, it amounts to only 0.3% of the value of ΔH.

Sample Problem 6.7 shows the calculation of the standard enthalpy of formation of acetylene, $HC\equiv CH$, from the standard enthalpy of combustion of acetylene.

---

*The work done by expanding gases in the cylinders of your car propels your car. The gases are formed by the reaction of gasoline with oxygen from air. In a gun, extremely rapidly expanding gases from the combustion of the propellant push the projectile down the barrel.

Equation 6.24 represents the combustion of 1 mol of acetylene:

$$HC{\equiv}CH(g) + \tfrac{5}{2}O_2(g) \longrightarrow 2CO_2(g) + H_2O(l) \qquad (6.24)$$

At 25 °C, the temperature for which heats of formation are tabulated, water is a liquid. Of course, the temperature of the flame is much higher than 25 °C. But because enthalpy is a state function, as long as the reactants are at 25 °C at the start of the reaction and the products are cooled to 25 °C after combustion has taken place, the enthalpy change will be the standard enthalpy change.

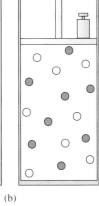

(a)

(b)

▬ FIGURE 6.11 (a) At constant temperature and constant volume, if the number of molecules of gas increases, pressure increases. (b) At constant temperature and constant pressure, if the number of molecules of gas increases, volume increases.

SAMPLE PROBLEM

6.7 The standard enthalpy of combustion of acetylene, $HC{\equiv}CH$, is -1299.60 kJ/mol. What is ΔH_f° for acetylene?

Solution The equation for the combustion of 1 mol of acetylene is:

Equation	$HC{\equiv}CH(g)$	$+ \tfrac{5}{2}O_2(g) \longrightarrow$	$2CO_2(g)$	$+ H_2O(l)$
ΔH_f°, **kJ/mol**	?	0	-393.51	-285.83

According to equation 6.23,

$$\Delta H_{rxn} = \Sigma n_p\, \Delta H_f^\circ(\text{product}) - \Sigma n_r\, \Delta H_f^\circ(\text{reactant})$$

For the reaction in this problem,

$$\Delta H_{rxn}^\circ = [2\text{ mol }CO_2 \cdot \Delta H_f^\circ(CO_2) + 1\text{ mol }H_2O \cdot \Delta H_f^\circ(H_2O)]$$
$$- [1\text{ mol }C_2H_2 \cdot \Delta H_f^\circ(C_2H_2) + \tfrac{5}{2}\text{ mol }O_2 \cdot \Delta H_f^\circ(O_2)]$$

Solving this equation for $\Delta H_f^\circ(C_2H_2)$ gives

$$\Delta H_f^\circ(C_2H_2)$$
$$= \frac{2\text{ mol }CO_2 \cdot \Delta H_f^\circ(CO_2) + 1\text{ mol }H_2O \cdot \Delta H_f^\circ(H_2O) - \tfrac{5}{2}\text{mol }O_2 \cdot \Delta H_f^\circ(O_2) - \Delta H_{rxn}^\circ}{1\text{ mol }C_2H_2}$$

and substitution of the values for ΔH_{rxn}°, $\Delta H_f^\circ(CO_2)$, $\Delta H_f^\circ(H_2O)$, and $\Delta H_f^\circ(O_2)$ gives

$$\Delta H_f^\circ(C_2H_2) = \frac{-787.02\text{ kJ} - 285.83\text{ kJ} + 1299.60\text{ kJ}}{1\text{ mol }C_2H_2} = +226.75\ \frac{kJ}{\text{mol }C_2H_2}$$

PRACTICE PROBLEMS

6.16 The heat of combustion of acetic acid, $CH_3COOH(l)$, is -871.7 kJ/mol. What is ΔH_f° for acetic acid?

6.17 Classify each of the following reactions as formation, combustion, both formation and combustion, or neither formation nor combustion:

(a) $MgO(s) + H_2O(l) \longrightarrow Mg(OH)_2(s)$

(b) $H_2(g) + Br_2(l) \longrightarrow 2HBr(g)$

(c) $2C_2H_2(g) + 5O_2(g) \longrightarrow 4CO_2(g) + 2H_2O(l)$

(d) $2HgO(s) \longrightarrow 2Hg(l) + O_2(g)$

(e) $H_2(g) + N_2(g) + 3O_2(g) \longrightarrow 2HNO_3(l)$

(f) $C(graphite) + O_2(g) \longrightarrow CO_2(g)$

6.18 If the number of moles of gaseous products is less than the number of moles of gaseous reactants, is the amount of thermal energy given off by an exothermic reaction under constant pressure greater or less than the thermal energy given off under constant volume? Explain your answer.

Acetylene is the fuel in oxyacetylene torches used for welding and cutting metals. Burning acetylene in oxygen produces an unusually high flame temperature (3300 °C).

The Energy Problem

Each year, Earth receives 5×10^{21} kJ of radiant energy from the sun, 10 000 times as much energy as is needed to meet all the world's energy needs. The energy problem is not a shortage of energy, it is a shortage of useful forms of energy. The sun's energy is distributed over a wide area and is very dilute. In addition, the sum is not a steady source of energy—nights and cloudy days interrupt the supply of energy from the sun, and less energy is received from the sun in winter than in summer.

At present, the United States obtains about 90% of its energy from fossil fuels: 42% from petroleum, 24% from natural gas, and 24% from coal. With only 6% of the world's population, the United States accounts for 30% of the world's total energy consumption. Our consumption of petroleum increased an average of 3.3% per year between 1985 and 1988 and is expected to continue to increase. And while consumption of petroleum increased an average of 3.3% per year from 1985 to 1988, net imports of petroleum increased an average of 17.8% per year. Until 1950, all the petroleum used in the United States was produced in the United States; now about half is imported. If we do not do something soon, we are heading for an enormous crisis.

Besides the fact that we have used up a large part of our own supplies of oil and natural gas and are rapidly using up other people's, burning of fossil fuels is a major contributor to environmental pollution. Petroleum and coal often contain sulfur, which is converted to sulfur dioxide gas

by combustion:

$$S(s) + O_2(g) \longrightarrow SO_2(g)$$

When sulfur dioxide dissolves in water, sulfurous acid, a component of acid rain, is formed:

$$SO_2(g) + H_2O(l) \longrightarrow H_2SO_3(aq)$$

Also, sulfur dioxide is oxidized to sulfur trioxide by oxygen in the atmosphere

$$2SO_2(g) + O_2(g) \longrightarrow 2SO_3(g)$$

and sulfur trioxide reacts with water to form sulfuric acid, another component of acid rain:

$$SO_3(g) + H_2O(l) \longrightarrow H_2SO_4(aq)$$

Moreover, sulfur dioxide is very irritating to the eyes and respiratory tract and sulfur dioxide in the air is believed to be responsible for the deaths of several thousand chronic bronchitis, asthma, and emphysema victims every year.

At the high temperatures reached in combustion reactions, reaction of nitrogen and oxygen from air results in the formation of nitric oxide:

$$N_2(g) + O_2(g) \longrightarrow 2NO(g)$$

This reaction is the first step in the formation of smog.

All three fossil fuels contain carbon. Natural gas is mostly methane, CH_4, and petroleum is a complex mixture consisting largely of hydrocarbons. Hydrocarbons are organic compounds that contain only carbon and hydrogen. Methane is the simplest example of a hydrocarbon; propane (C_3H_8), butane (C_4H_{10}), octane (C_8H_{18}), benzene (C_6H_6), ethylene (C_2H_4), and acetylene (C_2H_2) are all common and important hydrocarbons. Coal contains carbon and high molecular mass solid hydrocarbons. All three fossil fuels yield carbon dioxide on combustion. More and more observations seem to indicate that the concentration of carbon dioxide in the atmosphere is increasing as a result of the burning of fossil fuels and that this increase is causing Earth's climate to become warmer (the greenhouse effect). The possible effects of a warmer climate on life on Earth are so complicated that it is impossible to predict them with any certainty.

What can we do to decrease our use of fossil fuels, especially imported petroleum? Each of us can try to be less wasteful of energy, for example, by turning off the lights when we leave a room, turning the thermostat down in the winter and up in the summer, and planning the trips we make in our car. In addition, each of us can urge our lawmakers to encourage the substitution of other sources of energy such as wind energy and solar energy for fossil fuels by tax deductions and to discourage the use of fossil fuels by tax increases. (The money raised by the tax increases could be used to fund research on alternative energy sources.) Some experts estimate that the United States *could* replace half its fossil fuel use with renewable and nonpolluting sources of energy such as solar energy, wind energy, and energy from waste plant products by the year 2020. *You* can make a difference if you will.

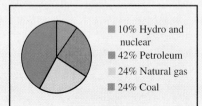

- ■ 10% Hydro and nuclear
- ■ 42% Petroleum
- ☐ 24% Natural gas
- ■ 24% Coal

A forest destroyed by acid rain.

The **system** is the part of the universe in which a scientist is interested. The rest of the universe is referred to as the **surroundings.** For practical purposes, the surroundings are the apparatus used and the space around it. Together, system and surroundings compose the **universe.**

Spontaneous changes take place all by themselves; the opposites of spontaneous changes are **nonspontaneous.** As far as the system is concerned, a combination of disorder and energy determines whether a change is spontaneous or nonspontaneous. The scientific word that describes disorder is **entropy. Energy** is the ability to do work; **work** is done when a force applied to some object moves the object. **Exothermic** changes release thermal energy to the surroundings. **Endothermic** changes remove thermal energy from their surroundings. Kinetic energy is energy of motion; **potential energy** is stored energy. Not all spontaneous changes take place in a useful period of time; a knowledge of reaction rate is necessary to predict whether a spontaneous change will actually take place. **Catalysts** are substances that make reactions take place faster without being used up themselves.

Thermochemistry deals with the relationship between chemical reactions and thermal energy. **Thermal energy** is the energy of motion of molecules, atoms, or ions, whichever are the units of a substance. Thermal energy is an extensive property. **Heat** is the thermal energy transfer that results from a difference in temperature. According to the **law of conservation of energy,** energy can neither be created nor destroyed. The SI unit of energy is the **joule.** One joule is $1 \ kg \cdot m^2 \cdot s^{-2}$. By definition

$$1 \ \text{calorie (cal)} = 4.184 \ \text{J (exactly)}$$

The **specific heat** of a substance is the amount of thermal energy required to heat one gram of the substance one degree. Water has an unusually high specific heat. The relationship between the mass of an object, the specific heat of the object, the temperature change that the object undergoes, and the thermal energy required to bring about the temperature change is

$$\text{thermal energy} = \text{mass} \times \text{specific heat} \times \text{temperature change}$$

The product of mass and specific heat is the **heat capacity** of the object. Specific heats can be used to calculate how much thermal energy must be added or removed to change the temperature of a sample from one temperature to another. Scientists and engineers have agreed to give the energy changes associated with exothermic changes a negative sign and the energy changes associated with endothermic changes a positive sign.

Energy cannot be measured; only differences in energy can be measured. An apparatus used to measure the quantity of thermal energy gained or lost during changes is called a **calorimeter.** The thermal energy gained or lost when a change takes place under constant pressure is called the **enthalpy change, ΔH,** for the change. The Greek letter Δ means change; H is the symbol for enthalpy. Subscripts are used to show the type of change.

For many reactions, the thermal energy given off is the chief reason for carrying out the reaction. The thermal energy change that accompanies a reaction can be shown by a **thermochemical equation.** Thermochemical equations are always interpreted in terms of moles and often involve fractional coefficients. When ΔH_{rxn} is given for a reaction, the equation for the reaction must be given.

According to **Hess's law,** the thermal energy given off or absorbed in a given change is the same whether the change takes place in a single step or in several steps. Hess's law is a consequence of the law of conservation of energy. The energy changes that accompany reactions can be summarized graphically.

Properties such as enthalpy that depend only on the initial and final states of the system and not on how the system gets from one state to another are called **state functions.** The enthalpy change for a reaction can be calculated by combining thermochemical equations in such a way that the thermochemical equation for the desired reaction is obtained. The equations that are combined may have to be multiplied by a factor or reversed. If a thermochemical equation is reversed, the magnitude of ΔH is the same but the sign is changed.

The equation

$$\Delta H_{rxn}^{\circ} = \Sigma n_p \, \Delta H_f^{\circ}(\text{product}) - \Sigma n_r \, \Delta H_f^{\circ}(\text{reactant})$$

where Σ means sum of, n_p is the number of moles of each product, $\Delta H_f^{\circ}(\text{product})$ is the standard enthalpy of formation of the product, n_r is the number of moles of each reactant, and $\Delta H_f^{\circ}(\text{reactant})$ is the standard enthalpy of formation of the reactant, is the most useful form of Hess's law. The **standard enthalpy of formation of a substance, ΔH_f°,** is the enthalpy change that results when one mole of the substance is formed from its elements with all substances in their standard states. The standard enthalpies of formation of the elements in their standard states are zero. For solids and liquids, the **standard state** is the pure solid or pure liquid under standard pressure of 1 atm. The standard state for a gas is the ideal gas at a partial pressure of 1 atm. Tables of standard enthalpies of formation of compounds are available; the values in the tables are usually obtained either by Hess's law calculations or from standard enthalpies of combustion determined in a bomb (constant-volume) calorimeter.

For information about the organization of Additional Practice Problems, Stop & Test Yourself, Putting Things Together, and Applications, see the beginnings of these sections in Chapter 1.

6.19 In Sample Problem 6.3, of what does the system consist? (6.1)

6.20 Define the following terms: (a) spontaneous change, (b) energy, (c) potential energy, (d) entropy, (e) exothermic, (f) catalyst. (6.2)

6.21 What three factors determine whether a reaction takes place? (6.2)

6.22 Classify each of the following changes as spontaneous or nonspontaneous: (a) spilled water running all over the floor, (b) a teaspoon of sugar dissolving in a cup of water, (c) a car rolling uphill, (d) a rotten egg turning into a fresh egg. (6.2)

6.23 From your everyday life, give one example of each of the following kinds of energy: (a) thermal, (b) electrical, (c) chemical, (d) light, (e) mechanical, (f) kinetic, (g) potential. (6.2)

6.24 From your everyday life, give an example of a change that takes place spontaneously because the entropy of the system increases as a result of the change. (6.2)

6.25 Classify each of the following changes as exothermic or endothermic: (a) perspiration evaporates, (b) natural gas burns, (c) water freezes. (6.2)

6.26 (a) Knowing that electricity (or some other energy source) is necessary to decompose water to hydrogen and oxygen, would you expect the reaction $2H_2(g) + O_2(g) \longrightarrow 2H_2O(l)$ to take place spontaneously? Explain your answer. (b) How would you explain the fact that mixtures of $H_2(g)$ and $O_2(g)$ can stand around for years without reacting? (6.2)

6.27 (a) In your own words, explain the difference between the meanings of the terms *heat, thermal energy,* and *temperature.* (6.3)

6.28 Is thermal energy conserved? Explain your answer. (6.4)

6.29 Carry out each of the following conversions of units: (a) 495 J to kJ, (b) 8.9 cal to J, (c) 208 cal to kJ, (d) 7.31 kJ to J, (e) 895 J to cal. (6.5)

6.30 How many kilojoules are required to raise the temperature of 33.4 g $CH_3CH_2OH(l)$ from 19.6 to 49.8 °C? (6.6)

6.31 How many kilojoules must be removed from a 325-g sample of $CH_3(CH_2)_6CH_3(l)$ to lower the sample's temperature from 82.4 to 31.6 °C? (6.6)

6.32 If 822 J are required to heat a 32.7-g sample of NaCl(s) from 21.2 to 50.3 °C, what is the specific heat of sodium chloride? (6.6)

6.33 If 41.2 g of water at 18.7 °C are mixed with 27.6 g of water at 38.4 °C, what will be the temperature of the mixture? (Assume that no thermal energy is lost to the surroundings.) (6.7)

6.34 If you drop 25.0 g of warm iron nails (initial temperature = 65.4 °C) into a beaker that contains 100.0 g of water at 20.0 °C, what will the temperature be when the nails and water have reached the same temperature? Assume that the beaker and thermometer do not absorb any thermal energy and that

no thermal energy escapes into the surrounding air or into the desktop. (6.7)

6.35 When 5.63 g LiBr(s) were added to 75.0 g H_2O in a calorimeter, the temperature rose from 21.6 to 31.7 °C. What is the enthalpy of solution of LiBr(s) in kJ/mol? (Assume that the initial temperature of the LiBr(s) was 21.6 °C. Neglect the contribution of the LiBr to both the mass *and* the specific heat of the water—the resulting errors cancel.)(6.7)

6.36 A 3.46-g sample of zinc heated to 95.4 °C was dropped into 50.0 g of water at 20.7 °C in the calorimeter. The temperature rose to 21.2 °C. What is the specific heat of zinc? (6.7)

6.37 The enthalpy of solution of LiBr(s) is -48.8 kJ/mol. If 4.82 g of LiBr(s) is dissolved in 100.0 g of water in a calorimeter at 19.8 °C, what temperature will be reached? Neglect the heat capacity of the calorimeter and the contribution of the LiBr to the mass and specific heat of the water. (6.7)

6.38 For the reaction

$$2CH_3CH_2CH_2CH_3(g) + 13O_2(g) \longrightarrow 8CO_2(g) + 10H_2O(g)$$
$$\text{butane}$$

$\Delta H = -5285.0$ kJ. (a) How many kilojoules are released when 27.86 g of butane are burned? Assume that an excess of oxygen is present. (b) If 327 J are released, how much butane was burned? (6.8)

6.39 (a) Classify each of the following reactions as exothermic or endothermic:

(i) $H_2(g) + Cl_2(g) \longrightarrow 2HCl(g) + 184.62$ kJ

(ii) $CH_4(g) + Cl_2(g) \longrightarrow CH_3Cl(g) + HCl(g),$
$$\Delta H = -99.5 \text{ kJ}$$

(iii) 181.66 kJ $+ 2HgO(s) \longrightarrow 2Hg(l) + O_2(g)$

(b) What is ΔH for reaction (i)? (c) Rewrite reaction (ii) showing thermal energy as a reactant or a product as appropriate. (d) Continued heating will be necessary to bring about which reaction? (6.8)

6.40 Given the following information

$$P_4(s) + 6Cl_2(g) \longrightarrow 4PCl_3(l) \qquad \Delta H = -1277.2 \text{ kJ}$$
$$PCl_3(l) + Cl_2(g) \longrightarrow PCl_5(s) \qquad \Delta H = -123.8 \text{ kJ}$$

Use a diagram like Figures 6.7 and 6.9 to find ΔH for the reaction

$$P_4(s) + 10Cl_2(g) \longrightarrow 4PCl_5(s) \quad (6.9)$$

6.41 Under what conditions can the value of ΔH for a reaction be symbolized by $\Delta H°$? (6.9)

6.42 Why does the table of standard enthalpies of formation (Table 6.2) not include the common forms of the elements, for example, C(graphite), $Cl_2(g)$, and $I_2(s)$? (6.9)

6.43 (a) What is a *state function?* (b) What state function is discussed in this chapter? (c) Give one other example of a state function. (6.9)

6.44 Which of the following depend on path? (a) The distance between the ground floor and the tenth floor of a building (b)

The time required to go from the ground floor to the tenth floor (6.9)

6.45 For the reaction

$$2KClO_3(s) \longrightarrow 2KCl(s) + 3O_2(g)$$

$\Delta H° = -91.0$ kJ. (a) What is $\Delta H°$ for the reaction

$$2KCl(s) + 3O_2(g) \longrightarrow 2KClO_3(s)$$

(b) What is $\Delta H°$ for the reaction

$$KClO_3(s) \longrightarrow KCl(s) + \tfrac{3}{2}O_2(g) \quad (6.9)$$

6.46 Calculate ΔH_{rxn} for the reaction

$$MnO_2(s) + 2C(graphite) \longrightarrow Mn(s) + 2CO(g)$$

from the following information (6.9):

$$Mn(s) + O_2(g) \longrightarrow MnO_2(s) \qquad \Delta H = -520.03 \text{ kJ}$$
$$2C(graphite) + O_2(g) \longrightarrow 2CO(g) \qquad \Delta H = -221.04 \text{ kJ}$$

6.47 Given the following data,

$$S(s) + O_2(g) \longrightarrow SO_2(g) \qquad \Delta H = -296.83 \text{ kJ}$$
$$2S(s) + 3O_2(g) \longrightarrow 2SO_3(g) \qquad \Delta H = -791.44 \text{ kJ}$$

use Hess's law to calculate ΔH for the reaction

$$2SO_2(g) + O_2(g) \longrightarrow 2SO_3(g) \quad (6.9)$$

6.48 Calculate the enthalpy change for the reaction

$$WO_3(s) + 3H_2(g) \longrightarrow W(s) + 3H_2O(g)$$

from the following information (6.9):

$$2W(s) + 3O_2(g) \longrightarrow 2WO_3(s) \qquad \Delta H = -1685.4 \text{ kJ}$$
$$2H_2(g) + O_2(g) \longrightarrow 2H_2O(g) \qquad \Delta H = -477.84 \text{ kJ}$$

6.49 From the following heats of reaction,

$$2H_2(g) + O_2(g) \longrightarrow 2H_2O(l) \qquad \Delta H = -571.66 \text{ kJ}$$
$$N_2O_5(g) + H_2O(l) \longrightarrow 2HNO_3(l) \qquad \Delta H = -92 \text{ kJ}$$
$$N_2(g) + 3O_2(g) + H_2(g) \longrightarrow 2HNO_3(l) \qquad \Delta H = -348.2 \text{ kJ}$$

calculate the heat of reaction for

$$2N_2(g) + 5O_2(g) \longrightarrow 2N_2O_5(g) \quad (6.9)$$

6.50 Calculate the enthalpy change for the reaction (6.9)

$$3C(graphite) + 4H_2(g) \longrightarrow CH_3CH_2CH_3(g)$$

given that

$$C(graphite) + O_2(g) \longrightarrow CO_2(g) \qquad \Delta H = -393.51 \text{ kJ}$$
$$2H_2(g) + O_2(g) \longrightarrow 2H_2O(l) \qquad \Delta H = -571.66 \text{ kJ}$$
$$CH_3CH_2CH_3(g) + 5O_2(g) \longrightarrow 3CO_2(g) + 4H_2O(l)$$
$$\Delta H = -2220 \text{ kJ}$$

6.51 What is meant by the term *standard state?* (6.9)
6.52 For the formation of methyl alcohol, $CH_3OH(l)$, $\Delta H_f° = -238.64$ kJ/mol. (a) Write the equation that corresponds to $\Delta H_f°$ for methyl alcohol. (b) Why are fractional coefficients

used for writing thermochemical equations but not for writing ordinary molecular equations? (c) Why is it especially important to show phases in thermochemical equations? (6.9)

6.53 Use standard enthalpies of formation to calculate standard enthalpies of reaction for the following reactions:

(a) $C(graphite) + 2H_2O(l) \longrightarrow CO_2(g) + 2H_2(g)$

(b) $2CH_4(g) + O_2(g) \longrightarrow 2CO(g) + 4H_2(g)$

(c) $2CO(g) + 2Cl_2(g) \longrightarrow 2COCl_2(g)$

(d) $Na_2SO_4(s) + 4CO(g) \longrightarrow Na_2S(s) + 4CO_2(g)$

(e) $4HCl(g) + O_2(g) \longrightarrow 2Cl_2(g) + 2H_2O(l)$ (6.9)

6.54 Use the information given in Problem 6.45 and standard enthalpies of formation to calculate the standard enthalpy of formation for $KClO_3(s)$. (6.9)

6.55 The standard enthalpy of combustion for acetone, $(CH_3)_2C\,O(l)$, is -1790.4 kJ/mol. What is $\Delta H_f°$ for acetone? (6.9)

6.56 (a) When 50.0 g of water at 41.6 °C was added to 50.0 g of water at 24.3 °C in a calorimeter, the temperature rose to 32.7 °C. What is the heat capacity of this calorimeter? (b) When 4.82 g $KClO_3(s)$ was added to 100.0 g of water in the calorimeter of part (a), the temperature dropped to 20.6 °C. What is the enthalpy of solution of $KClO_3(s)$ in kJ/mol? (Assume that the initial temperature was 24.3 °C. Neglect the contribution of the $KClO_3$ to both the mass *and* the specific heat of the water— the resulting errors cancel.)

6.57 Classify each of the following reactions as formation, combustion, both, or neither:

(a) $H_2C\,CH_2(g) + Br_2(l) \longrightarrow BrCH_2CH_2Br(l)$

(b) $2Na(s) + N_2(g) + 3O_2(g) \longrightarrow 2NaNO_3(s)$

(c) $P_4(s) + 5O_2(g) \longrightarrow P_4O_{10}(s)$

(d] $2CO(g) + O_2(g) \longrightarrow 2CO_2(g)$

(e) $NH_3(g) + HCl(g) \longrightarrow NH_4Cl(s)$

6.58 Imagine that a piece of iron and a piece of aluminum each absorb the same quantity of radiant energy from sunlight. The masses of the two pieces of metal are the same. Which piece of metal will be warmer? Explain your answer.

6.59 Of the following changes,

(i) $2CH_3OH(g) + 3O_2(g) \longrightarrow 2CO_2(g) + 4H_2O(l)$

(ii) $2CH_3OH(l) + 3O_2(g) \longrightarrow 2CO_2(g) + 4H_2O(l)$

(a) Which is more exothermic? Explain your answer. (b) Is the value of ΔH for the more exothermic reaction more or less negative than the value of ΔH for the less exothermic reaction?

6.60 Use the following information

$$CH_4(g) + 2O_2(g) \longrightarrow$$
$$CO_2(g) + 2H_2O(g) \qquad \Delta H = -796.50 \text{ kJ}$$
$$C(graphite) + O_2(g) \longrightarrow CO_2(g) \qquad \Delta H = -393.51 \text{ kJ}$$
$$2H_2(g) + O_2(g) \longrightarrow 2H_2O(g) \qquad \Delta H = -477.84 \text{ kJ}$$

to calculate ΔH for the reaction

$$CH_4(g) \longrightarrow C(graphite) + 2H_2(g)$$

6.61 (a) For which of the following reactions is the energy change at constant pressure different from the energy change at constant volume?

$$N_2(g) + 3H_2(g) \longrightarrow 2NH_3(g)$$
$$S(s) + O_2(g) \longrightarrow SO_2(g)$$
$$2SO_2(g) + O_2(g) \longrightarrow 2SO_3(g)$$
$$2HgO(s) \longrightarrow 2Hg(l) + O_2(g)$$
$$H_2(g) + Cl_2(g) \longrightarrow 2HCl(g)$$

(b) Why can the difference usually be neglected?

6.62 Given the following information,

$$N_2(g) + O_2(g) \longrightarrow 2NO(g) \qquad \Delta H = 180.50 \text{ kJ}$$
$$N_2(g) + 2O_2(g) \longrightarrow 2NO_2(g) \qquad \Delta H = 66.4 \text{ kJ}$$

use a diagram like Figures 6.7 and 6.9 to find ΔH for the reaction $2NO(g) + O_2(g) \longrightarrow 2NO_2(g)$.

6.63 Calculate standard enthalpies of reaction from standard enthalpies of formation:

(a) $2CH_4(g) + 3O_2(g) \longrightarrow 2CO(g) + 4H_2O(l)$

(b) $CH_4(g) + NH_3(g) \longrightarrow HCN(g) + 3H_2(g)$

(c) $4NH_3(g) + 5O_2(g) \longrightarrow 4NO(g) + 6H_2O(l)$

(d) $N_2(g) \longrightarrow 2N(g)$

(e) $NaHCO_3(s) \longrightarrow CO_2(g) + NaOH(s)$

6.64 Write equations that correspond to ΔH_f° for each of the following compounds: (a) $K_2SO_4(s)$, (b) $CH_3NH_2(g)$, (c) $HCONH_2(l)$, (d) $CH_3Cl(g)$, (e) $H_3PO_4(s)$.

6.65 For the reaction $S(s) + O_2(g) \longrightarrow SO_2(g)$, $\Delta H = -296.83 \text{ kJ}$. If 49.1 g of sulfur are burned, how many kilojoules are released?

6.66 The heat of combustion of carbon disulfide, a liquid, is -1085.1 kJ/mol. The products are $CO_2(g)$ and $SO_2(g)$. What is ΔH_f° for carbon disulfide in kJ/mol?

STOP & TEST YOURSELF

1. 346 J = _____ kJ
   (a) 3.46×10^{-5} (b) 0.346   (c) 3.46   (d) 3.46×10^4
   (e) 3.46×10^5

2. 0.266 kJ = _____ cal
   (a) 6.36×10^{-5} (b) 1.11×10^{-3} (c) 63.6   (d) 266
   (e) 1.11×10^3

3. Calculate the standard enthalpy change ΔH_{rxn}° for the reaction

   $$2HNO_3(l) + NO(g) \longrightarrow 3NO_2(g) + H_2O(l)$$

   (a) -168.8 kJ   (b) -71.7 kJ   (c) -59.1 kJ   (d) 118.6 kJ
   (e) 71.7 kJ

4. Calculate the standard enthalpy change ΔH_{rxn}° for the reaction to the nearest kilojoule.

   $$CH_3CH_3(g) + \tfrac{7}{2}O_2(g) \longrightarrow 2CO_2(g) + 3H_2O(l)$$

   (a) -1560 kJ   (b) -1428 kJ   (c) -764 kJ   (d) $+1560$ kJ
   (e) Not enough information is available to answer this question.

Use the following information to answer questions 5 through 9.

(a) $SnCl_4(l) \longrightarrow Sn(s) + 2Cl_2(g)$

(b) $SnCl_4(l) \longrightarrow SnCl_2(s) + Cl_2(g) \qquad \Delta H_{rxn}^\circ = 195.4 \text{ kJ}$

(c) $SnCl_2(s) \longrightarrow Sn(s) + Cl_2(g) \qquad \Delta H_{rxn}^\circ = 349.8 \text{ kJ}$

(d) $Cl_2(g) + \tfrac{1}{2}O_2(g) \longrightarrow Cl_2O(g) \qquad \Delta H_{rxn}^\circ = 80.3 \text{ kJ}$

(e) $Cl_2(g) + H_2O(l) \longrightarrow$
$$HCl(aq) + HOCl(aq) \qquad \Delta H_{rxn}^\circ = 1.92 \text{ kJ}$$

5. The ΔH_{rxn}° for which reaction above is a standard enthalpy of formation?
   (a)   (b)   (c)   (d)   (e)

6. All of the reactions are _____.
   (a) endothermic   (b) exothermic

7. If you carried out these reactions in the laboratory, the container would feel _____.
   (a) cool   (b) warm

8. What is the value of ΔH_{rxn}° for reaction (a)?
   (a) -545.2 kJ   (b) -154.4 kJ   (c) 154.4 kJ   (d) 545.2 kJ
   (e) Not enough information is given to answer this question.

9. What is the value of ΔH_{rxn}° for the reaction

   $$2Cl_2O(g) \longrightarrow 2Cl_2(g) + O_2(g)$$

   (a) -160.6 kJ   (b) -160.6 kJ/mol   (c) -80.3 kJ
   (d) -40.2 kJ   (e) 160.6 kJ

10. For the reaction

    $$C_3H_8(g) + 5O_2(g) \longrightarrow 3CO_2(g) + 4H_2O(l)$$

    $\Delta H_{rxn}^\circ = -2220.00$ kJ. How many grams of $C_3H_8(g)$ must be burned to supply 3169 kJ of thermal energy?
    (a) 6.295×10^{-6} (b) 3.237×10^{-2} (c) 62.95
    (d) 1.595×10^5 (e) 3.102×10^8

11. Given that ΔH_{rxn}° for the reaction

    $$CH_4(g) + 4Cl_2(g) \longrightarrow CCl_4(g) + 4HCl(g)$$

    is -402 kJ, what is ΔH_f° for $CCl_4(g)$?
    (a) -696 kJ/mol   (b) -294 kJ/mol   (c) -116 kJ/mol
    (d) -108 kJ/mol   (e) 384 kJ/mol

12. What quantity of thermal energy must be removed from an aluminum pan having a mass of 289 g to cool the pan from 100.0 to 25.0 °C? Assume that the specific heat of aluminum is constant over this temperature range.
    (a) 0.23 kJ   (b) 3.4 kJ   (c) 6.5 kJ   (d) 20 kJ   (e) 26 kJ

13. For the reaction

$$2Al(s) + Fe_2O_3(s) \longrightarrow$$
$$2Fe(s) + Al_2O_3(s) \qquad \Delta H° = -851.4 \text{ kJ}$$

how much thermal energy is released when 36.0 g Al reacts with excess Fe_2O_3?
(a) 568 kJ (b) 639 kJ (c) 1.14×10^3 kJ
(d) 1.27×10^3 kJ
(e) None of the preceding answers is correct.

14. Given the energy diagram,

$$H \uparrow \quad \frac{H_2(g) + Cl_2(g)}{2HCl(g)}$$

which of the following statements is true of the reaction $H_2(g) + Cl_2(g) \longrightarrow 2HCl(g)$?
(a) It is endothermic.
(b) It is exothermic.

PUTTING THINGS TOGETHER

6.67 Draw molecular-level pictures to show what happens when (a) the system is heated, (b) work is done on the system at constant temperature.

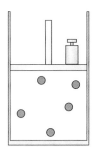

6.68 (a) What is the specific heat of water in $\text{cal} \cdot \text{g}^{-1} \cdot °\text{C}^{-1}$? (b) What is the molar heat capacity of water in $\text{J} \cdot \text{mol}^{-1} \cdot °\text{C}^{-1}$?

6.69 If a real gas is allowed to expand into a vacuum (empty space), the temperature of the gas usually decreases. Explain why.

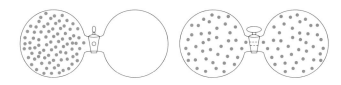

6.70 The enthalpy of combustion of ethyl alcohol, CH_3CH_2OH, is 326.68 kcal/mol. How many kilojoules are released by the combustion of one pint of ethyl alcohol if the density of ethyl alcohol is 0.789 g/cm^3?

6.71 Given the following information:

$$4Fe(s) + 3O_2(g) \longrightarrow 2Fe_2O_3(s) \qquad \Delta H = -1648.4 \text{ kJ}$$
$$2CO(g) + O_2(g) \longrightarrow 2CO_2(g) \qquad \Delta H = -565.98 \text{ kJ}$$

What is ΔH_{rxn} for the reaction between iron(III) oxide and carbon monoxide to give iron metal and carbon dioxide gas?

6.72 If the potential energy of a particle increases as it is moved away from another particle, do the two particles attract or repel each other? Explain your answer.

6.73 Nitrogen dioxide gas reacts with water to give nitric acid, which is a liquid, and nitric oxide (nitrogen monoxide) gas. At constant pressure, 69.9 kJ of thermal energy are released for each mole of nitric oxide that is formed. Write a thermochemical equation for the reaction that shows the formation of one mole of nitric acid.

6.74 Complete combustion of 8.62 g of benzoic acid, $C_6H_5COOH(s)$, at constant pressure released 227.8 kJ of thermal energy. (The reactants were at 25 °C at the beginning of the reaction, and the products were returned to 25 °C after the reaction had taken place.) What is $\Delta H_f°$ for benzoic acid at 25 °C?

6.75 What change takes place in the thermal energy of the system when a mixture of 1.25 g $H_2(g)$ and 11.00 g $O_2(g)$ reacts to form water?

6.76 If 6.92 kJ of thermal energy are used to decompose ammonium chloride to ammonia and hydrogen chloride, (a) how many grams of ammonium chloride will be decomposed? (b) How many milliliters of ammonia (measured at STP) will be formed? The standard enthalpy of formation of ammonium chloride is -314.43 kJ/mol.

6.77 How many kilojoules are released when 3.6 L $H_2(g)$ and 2.4 L $F_2(g)$ react to form HF(g) at a constant pressure of 1 atm and initial and final temperatures of 25 °C?

6.78 The enthalpies of combustion for the series of compounds with molecular formula C_nH_{2n+2} are (all in kJ/mol): $CH_4(g)$, 889.7; $C_2H_6(g)$ 1559.1; $C_3H_8(g)$, 2217.0; $C_4H_{10}(g)$, 2874.9; $C_5H_{12}(l)$ 3505.8; $C_6H_{14}(l)$, 4159.5; $C_7H_{16}(l)$, 4812.8; $C_8H_{18}(l)$, 5465.7. (a) On a sheet of $8\frac{1}{2} \times 11$-in. graph paper, plot enthalpy of combustion as a function of number of carbon atoms. (b) Use your graph to predict the enthalpy of combustion for the nine-carbon member of the series. What is its molecular formula? (c) Is more thermal energy released by burning $C_4H_{10}(g)$ or $C_4H_{10}(l)$? Explain your reasoning. (d) Is the difference between enthalpies of combustion of these compounds in the gaseous and liquid states significant to the number of figures allowed by the size of the graph paper? Explain your answer. (e) Why is enthalpy of combustion not zero for zero carbon atoms? (f) A typical condensed structural formula for this series of compounds is $CH_3(CH_2)_3CH_3$. Write a complete structural formula for this compound.

6.79 An aluminum pan that has a mass of 276 g is accidentally heated to 218 °C from room temperature, which is 22 °C. How many kilojoules are wasted?

6.80 The meat in a "quarter-pounder" has a calorie count of 236 cal. (The "calories" in food are actually kilocalories.) Playing tennis or bicycling uses up about 7.5 cal/min. For how many minutes would you have to play tennis or ride a bike to work off a quarter-pounder?

6.81 Explain how spraying plants with water keeps the plants from freezing when the temperature goes slightly below 0 °C.

6.82 The reaction

$$C_2H_6(g) + O_2(g) \longrightarrow 3H_2(g) + 2CO(g)$$

is one of the major reactions in a new process for making transport fuels such as gasoline from natural gas. From the following information, calculate ΔH_{rxn}° for this reaction.

$$2C_2H_6(g) + 7O_2(g) \longrightarrow$$
$$4CO_2(g) + 6H_2O(g) \qquad \Delta H^\circ = -3119.7 \text{ kJ}$$
$$2H_2(g) + O_2(g) \longrightarrow 2H_2O(g) \qquad \Delta H^\circ = -478.84 \text{ kJ}$$
$$2CO(g) + O_2(g) \longrightarrow 2CO_2(g) \qquad \Delta H^\circ = -565.98 \text{ kJ}$$

6.83 A cold pack (Figure 6.3) consists of a small inner bag that contains water inside another larger bag that contains solid ammonium nitrate. When the cold pack is needed, the inner bag is broken, and the water and ammonium nitrate are shaken together. If the inner bag contains 117 mL of water, how much ammonium nitrate must the outer bag contain to lower the temperature of the water from 23.0 to 5.0 °C? The heat of solution of ammonium nitrate is 2.569×10^4 J/mol. (Neglect the contribution of the ammonium nitrate to both the mass and the specific heat of the water—the resulting errors cancel.)

6.84 (a) What is ΔH° for the reaction NaOH(aq) + HCl(aq) $\longrightarrow$ NaCl(aq) + H$_2$O(l)? (b) How many kilojoules of thermal energy are released when 0.1 mol each of NaOH(aq) and HCl(aq) react? When 10.0 mol of each react? (c) What is the area and the volume of a cube that is 1.0 cm on a side? What is the area and the volume of a cube that is 100 cm on a side? What is the surface area per cubic centimeter for each cube? (d) What safety problem occurs when exothermic reactions are scaled up from a laboratory scale to an industrial scale?

6.85 Hydrogen cyanide, a poisonous gas which is used to make Lucite and Plexiglas, is manufactured by the reaction

$$2CH_4(g) + 2NH_3(g) + 3O_2(g) \longrightarrow 2HCN(g) + 6H_2O(g)$$

(a) Calculate ΔH° for this reaction. (b) What is ΔH° for the formation of 1 mol HCN? (c) Is the reaction exothermic or endothermic?

6.86 Two important methods for producing hydrogen chloride gas are

$$H_2(g) + Cl_2(g) \longrightarrow 2HCl(g)$$

and

$$2NaCl(s) + H_2SO_4(l) \longrightarrow 2HCl(g) + Na_2SO_4(s)$$

Is heating or cooling needed when each of these reactions is carried out?

6.87 The nutrition information on a package of cookies says that each cookie contains 1 g of protein, 7 g of carbohydrate, and 4 g of fat. For carbohydrates, the average calorie count is 4 kcal/g. For fats, the average calorie count is 9 kcal/g, and for protein, the average calorie count is 4 kcal/g. (a) What is the energy value of a cookie in kilocalories? (b) What does the label give for the calorie count per cookie? Remember that a nutritional calorie is a kilocalorie. (c) What is the energy value of a cookie in kilojoules?

6.88 Much of air pollution, of the greenhouse effect, and of the unfavorable U.S. balance of trade result from the combustion of gasoline in cars. Suppose the automobile industry had developed cars that used alcohol produced from biomass instead of gasoline produced from petroleum. (a) Calculate the number of kilograms of energy released per liter of fuel burned for alcohol, C$_2$H$_5$OH, and octane, C$_8$H$_{18}$ ($\Delta H_f^\circ = -250.1$ kJ/mol), a typical component of gasoline. The density of alcohol is 0.79 g/mL and the density of octane is 0.70 g/L. (b) How many grams of carbon dioxide would be formed by the combustion of one liter of each of the fuels? Does octane produce significantly more carbon dioxide per kilojoule than alcohol? (c) What advantages would the petroleum industry claim for gasoline? (d) What advantages would the alcohol and automobile industries claim for alcohol?

6.89 (a) How many pounds of water are in a waterbed measuring 60 in. × 84 in. × 9 in.? (b) How many kilocalories are required to heat the water in the waterbed from 65 °F, the temperature of the tapwater used to fill the waterbed, to 85 °F, the operating temperature of the waterbed?

6.90 Sodium chloride is obtained by the evaporation of seawater. Because the evaporation of water is an endothermic process, it must be coupled with an exothermic process such as the burning of a fuel (or the collection of solar energy). A typical sample of seawater is 2.90% by mass NaCl. (a) How many grams of NaCl(s) can be obtained by the evaporation of 454 g (1 lb) of seawater? (b) How many grams of methane must be burned to evaporate the water from 454 g of seawater? The enthalpy of combustion of methane is −890 kJ/mol.

6.91 A 75-watt incandescent light bulb and an 18-watt compact fluorescent bulb are equally bright. How much warmer will an insulated, closed 12 by 11 by 8.0-ft room get in 1.0 h if it is lighted by the incandescent bulb than if it is lighted by the compact fluorescent bulb? At room temperature, the specific heat of air is $1.007 \text{ J} \cdot \text{g}^{-1} \cdot {}^\circ\text{C}^{-1}$, and its density is $1.2 \text{ g} \cdot \text{L}^{-1}$. The watt (W) is the SI unit of power; 1 W = 1 J/1 s.

6.92 A mixture of powdered aluminum and powdered iron(III) oxide is called Thermit and is used in fire bombs and as a source of thermal energy for welding. The reaction of aluminum and iron(III) oxide is very exothermic, and the temperature reaches 2400 °C (4400 °F). How many kilojoules of thermal energy are released by the reaction of a mixture of 5.82 g of aluminum and 18.33 g of iron(III) oxide under standard conditions? The products are iron and aluminum oxide, which is a solid under ordinary conditions.

6.93 Water, which is usually not considered a pollutant, is the product of the combustion of hydrogen gas in air. Combustion of hydrogen gas provides more energy per unit of mass than any other fuel, and the combustion product of hydrogen is water. (a) How many kilojoules of thermal energy are released by combustion of one gram of hydrogen? (b) What is the volume of 1 g of $H_2(g)$ at 25 °C and 1 atm? (c) For isooctane, $(CH_3)_2CHCH_2C(CH_3)_3$, ΔH_f° is −255.1 kJ/mol. What is the enthalpy of combustion for isooctane, and how many kilojoules of thermal energy are released by the combustion of one gram of isooctane? (d) The density of isooctane is 0.692 g/mL at 25 °C. What is the volume of one gram of isooctane at 25 °C. (e) What is one problem with using hydrogen gas as a fuel for cars?

6.94 Natural gas produced in Pennsylvania is a mixture of methane, CH_4, ethane, CH_3CH_3, and 1.1% by volume N_2 and has a heating value of 1232 Btu·ft^{-3}. The Btu (British thermal unit) is a unit of energy used by American engineers and is the quantity of thermal energy required to raise the temperature of 1 lb water 1 °F (1 Btu = 1055 J). The heating value of methane is 1013 Btu·ft^{-3}, and the heating value of ethane is 1786 Btu·ft^{-3}. Nitrogen gas has no heating value. What is the composition in percent by volume of natural gas from Pennsylvania?

6.95 (a) Does hydrazine, H_2NNH_2, or dimethylhydrazine, $(CH_3)_2NNH_2$, release more thermal energy per gram on com-

bustion? The N-containing product of combustion of these compounds is $N_2(g)$. Both compounds are liquids; for $(CH_3)_2NNH_2(l)$, $\Delta H_f^\circ = 48.9$ kJ/mol. (b) Which compound is the better fuel for spacecraft? Explain your answer.

6.96 The thermal conductivity of stainless steel is 0.3 J·s^{-1}·cm^{-1}·K^{-1}, and that of copper is 7.1 J·s^{-1}·cm^{-1}·K^{-1}. Explain why several brands of stainless steel pans have copper bottoms.

6.97 In 1947 the French freighter *Grandcamp,* which was loaded with ammonium nitrate fertilizer, exploded in the harbor of Texas City, Texas. Nearly 600 people were killed. The products of the explosion of ammonium nitrate are nitrogen gas, oxygen gas, and water vapor. (a) For each pound (454 g) of ammonium nitrate that exploded, how many kilojoules of thermal energy were released, and how many liters of gas (measured at 1 atm pressure and 25 °C) were formed? According to the *Handbook of Chemistry and Physics,* ΔH_f° for ammonium nitrate is −87.37 kcal/mol. (b) The density of ammonium nitrate is 1.725 g/cm^3. How many liters did the volume of the system increase when the pound of ammonium nitrate exploded?

6.98 Suppose you have two pans the right size for baking brownies. One is made of aluminum and the other of Pyrex glass. (a) How many kilojoules are required to heat each pan from 22 to 177 °C (room temperature to baking temperature)? (b) If you were in a hurry to eat the brownies, which pan would you use? Explain your answer. (c) Why is the Pyrex pan so much heavier than the aluminum pan?

	Aluminum	Pyrex
Density, g·cm^{-3}	2.8	2.6
Mass of pan, g	213	1.89×10^3
Specific heat, cal·g^{-1}·°C^{-1}	0.216	0.176

Applying Chemistry as a Technical Illustrator

BRIAN C. BETSILL

Graphic Designer and Technical Illustrator

B. A. Computer Science
DePauw University

The second semester of my junior year in college, I changed my major from chemistry to computer science. The previous semester I had tried to muddle through a course in inorganic chemistry. During the course we had conducted analytical experiments to find out how much iron there was in spinach and how much calcium there actually was in powdered milk. I could not get excited about these or any other such experiments and could not fathom another year and a half as a chemistry major. So I made what seemed the logical choice and changed majors.

After changing majors, I thought I would never have to apply chemistry in a practical sense again. However, as a technical illustrator of math and science textbooks, I have frequently found myself trying to recall the molecular structures of organic compounds and the various laboratory apparatuses. Although I did not pursue chemistry as a major, let alone a career, I have found it useful in my chosen profession.

As a technical illustrator, I am responsible for supplying art for math and science textbooks. I have recently illustrated general chemistry books (including this one) and chemistry laboratory manuals. In preparing the illustrations, I must work very closely with the authors who provide the general guidelines for the art. My background in chemistry has helped me communicate with the authors and understand the subjects of my illustrations.

In my line of work, it is important to establish a good rapport with the authors. We need to be able to communicate comfortably with one another in order to complete the complex process of publishing a book. I find that when I am working on these projects, I have more credibility with the authors once they discover that I have a background in chemistry. We are speaking the same language, as it were, and they are relieved to find they can discuss problems with me directly without constantly having to search for a lay expression to make a point.

Just as a medical illustrator would find it difficult to do his or her job without understanding how the human body works, I would have problems doing chemistry drawings without a basic understanding of chemistry. When I am asked to draw a given molecule or chemical compound, I often begin with only a rough sketch or a brief description. I must then transform this sketch into a detailed illustration. Sometimes I find it helpful to construct visual aids such as Styrofoam models to assist in the process. Since I do the vast majority of my drawings on the computer, I am also able to take advantage of software programs that are designed specifically for molecular modeling. They require the user to input formulas, atoms, or the various substructures needed to create the desired compounds. Without these visual aids or actual computer-generated models, it would be difficult to arrive at a final product, and without a general understanding of chemistry, it would be virtually impossible for me to make use of these tools.

Although I thought I would not be using chemistry once I finished school, I have found it useful time and again in my career as a technical illustrator. I suppose the lesson to be learned is that academic subjects such as chemistry can be helpful in the "real world." Biochemists and pharmacists certainly find chemistry directly relevant to their work, but chemistry can also be applied indirectly in a number of other careers.

ATOMIC STRUCTURE

Neon (and incandescent) lights.

This chapter describes some of the many experiments that have contributed to our ideas about atomic structure. It traces the development of theories of atomic structure from their beginnings in the nineteenth century until the present. The basic structure of the atom is discussed first in Sections 7.1–7.4. The remainder of the chapter deals with the electronic structure of atoms. The term **electronic structure** refers to *the way the electrons are arranged around the nucleus*. Most of the properties of the elements and their compounds depend on electronic structure.

— *Why is a study of atomic structure a part of general chemistry?*

Almost all properties of matter depend on atomic structure. Atomic structure provides the answers to such questions as: Why do the properties of metals differ from the properties of nonmetals? Why does sodium react vigorously with water, whereas gold coins in sunken ships can be recovered in good condition after hundreds of years underwater? Why is radon radioactive although it is unreactive chemically? A knowledge of atomic structure will help you understand why

substances behave as they do and remember their properties.

In addition, the development of the modern atomic theory is an excellent example of how science progresses. Many people contribute and check each other's work. Theories are invented to explain observations; the theories suggest new experiments, and the new experiments lead to changes in the old theories or even to new theories. Theories are useful in providing the basis for further work even if later proved wrong. Advances in one field, such as physics, contribute to advances in other fields, such as chemistry and biology, and practical applications are based on new understanding.

— What do you already know about atomic structure?

An atom is made up of a small, dense, positively charged nucleus surrounded by rapidly moving electrons (Section 1.5). The positive charge on the nucleus results from the presence of protons in the nucleus; protons are small particles with 1+ charge. The number of protons in the nucleus is called the atomic number of the element. Atomic numbers distinguish atoms of one element from atoms of other elements and show which element an atom is. The nuclei of all atoms except hydrogen-1 also contain neutrons (Section 2.11). Neutrons have about the same mass as protons (1 u) but are electrically neutral.

The nucleus is surrounded by electrons. An electron is an extremely small particle with a 1 − charge. Compared to protons and neutrons, electrons have negligible mass. Atoms do not have a net electric charge. The positive charge of the protons in the nucleus is exactly balanced by the negative charge of the electrons outside the nucleus. Thus, the number of electrons outside the nucleus of an atom is equal to the number of protons in the nucleus (the atomic number of the element). All atoms of a given element have the same number of electrons around the nucleus; atoms of different elements have different numbers of electrons. The number of electrons around the nucleus determines the properties of an atom.

This picture of atomic structure was presented in the first two chapters so that we could use it. No description of the experiments on which it is based was given. Most of these experiments involve physical, not chemical, changes and were carried out by physicists. They are described in the following four sections.

| 7.1 | DISCOVERY OF THE ELECTRON |

As early as 1807, the English chemist Sir Humphry Davy suggested that the forces holding elements together in compounds are electrical.* He discovered five elements—sodium, potassium, calcium, strontium, and barium—by using electricity to decompose compounds. In 1832–1833 Michael Faraday, another English chemist, carried out a series of electrolysis experiments. He found a relationship between atomic mass and the quantity of electricity needed to free a given amount of an element from a compound. On the basis of Faraday's experiments, the English physicist George Stoney proposed that electricity exists in units associated with atoms. He suggested that atoms have no net charge because they contain both positive and negative units of electricity. In 1891 he proposed the name "electron" for the unit of negative electricity.

During the second half of the nineteenth century, a number of physicists studied electric discharges through gases. ▪Figure 7.1 is a diagram of a simple gas discharge tube. The purpose of the slits in the anode is to let a beam of "cathode rays" pass

Davy also invented the safety lamp used by miners until about 1900. Although electric lamps have been used since that time, flame safety lamps are still used to detect the flammable, explosive gas methane, CH_4. The flame gets longer when methane is present. Faraday began his career as an assistant in Davy's laboratory.

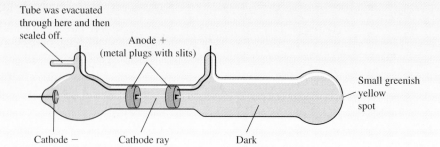

▪ FIGURE 7.1   A simple gas discharge tube.

---

*The force between the electron and the nucleus (a single proton) in an atom of hydrogen-1 is 8.2×10^{-8} N (1 Newton, N, equals 1 kg·m·s^{-2}), but the gravitational force between the electron and the nucleus in an atom of hydrogen-1 is only 3.6×10^{-47} N. Therefore, the force that holds the electron and the nucleus together cannot be gravity. Gravitational attraction between bodies is only important when the masses of the bodies are very large compared with the masses of the particles that make up a nucleus.

through. All particles except those traveling in the desired direction are screened out. When a high-voltage direct current is applied to a gas discharge tube, no current flows as long as the tube is filled with gas. Gases at ordinary temperatures and atmospheric pressure are very poor conductors of electricity. If the pressure is lowered below 2 mmHg, the tube begins to glow with a steady light. The light follows the tube from cathode (negative electrode) to anode through any curves and angles (see ▬Figure 7.2). Its color depends on what gas is in the tube.

When the pressure of the gas in the discharge tube is lowered below 10^{-5} mmHg, light is replaced by darkness except for a small greenish-yellow spot at the end of the tube opposite the cathode. The position of the anode makes no difference in the position of the spot. The "rays" causing the spot are called **cathode rays.**

The English physicist J. J. Thomson is usually given credit for discovering the electron because he was, in 1897, the first to measure its properties quantitatively. ▬Figure 7.3 is a diagram of Thomson's apparatus. Thomson's value for the mass-to-charge ratio of the cathode ray particles was -6×10^{-9} g/C or -6×10^{-12} kg/C.

The **coulomb, C,** is the *SI unit of charge.* The coulomb is defined as *the amount of charge that flows past a fixed point in a wire in one second when the current is one ampere.* You may wonder why charge is defined in such a complicated way. The International System tries to make definitions of units as precise as possible. Current can be measured more precisely than any natural unit of charge such as the charge on an electron.

Thomson obtained the same mass-to-charge ratio for the cathode ray particles regardless of the gas that originally filled the tube and the metal used to make the cathode. He concluded that cathode rays must be units common to all atoms. They must be the particles of electricity that Stoney called electrons.

Thomson's experiments could not measure the mass or charge of the electron separately. They could only measure the mass-to-charge ratio. A mass-to-charge ratio of -6×10^{-12} kg/C is very low. The smallest value for a mass-to-charge ratio observed before Thomson's experiments with cathode rays was that observed for the hydrogen ion. This value, which was obtained by electrolysis experiments, was 1837 times as large as the mass-to-charge ratio of the cathode ray particles. Either

If the electric service to your house is 115 volts, it takes 1.15 seconds for one coulomb of electrons to pass through a 100-watt light bulb. It takes one day, six hours, and 50 minutes for one mole of electrons to pass through.

Thomson's apparatus is a cathode ray tube. Television picture tubes and computer displays are cathode ray tubes.

▬ FIGURE 7.3   (a) Thomson's apparatus for measuring the mass-to-charge ratio of the electron. The position where the small greenish-yellow spot at the end of the tube opposite the cathode appears when there is no charge on the plates is labeled 1. If the plates are charged as shown, the greenish-yellow glow appears at spot 2. Bending of the cathode rays toward the positive plate indicates that the particles composing them must have a negative charge. If the plates are discharged so that there is no electric field between them, and the poles of a magnet are placed on either side of the tube as shown in (b), the greenish-yellow glow appears at spot 3. The glow can be moved back to spot 1 by applying both electric and magnetic fields at the same time and adjusting the field strengths. From the field strengths, when the glow is at spot 1 and the position of spot 3, Thomson calculated the mass-to-charge ratio of the particles composing the cathode rays. He checked his work by using a different method—measuring the increase in temperature of a body struck by cathode rays.

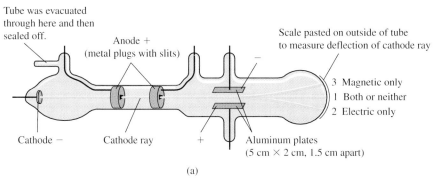

Tube was evacuated through here and then sealed off.

Anode + (metal plugs with slits)

Scale pasted on outside of tube to measure deflection of cathode ray

3 Magnetic only
1 Both or neither
2 Electric only

Cathode −   Cathode ray   +   Aluminum plates (5 cm × 2 cm, 1.5 cm apart)

(a)

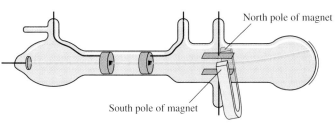

North pole of magnet

South pole of magnet

(b)

the mass of the particles making up cathode rays must be very low or the charge of the particles must be very high.

DETERMINATION OF THE CHARGE AND MASS OF THE ELECTRON

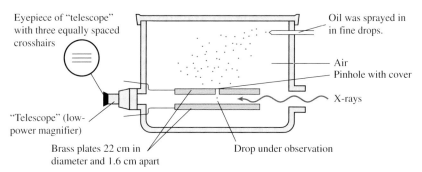

Although making a large number of measurements is usually a waste of time, very important measurements are carried out many times.

The American physicist R. A. Millikan determined the unit of charge by observing electrically charged oil drops in the apparatus shown in ▪Figure 7.4. From several thousand measurements, the charges on the oil drops, whether positive or negative, were always found to be whole number multiples of 1.5924×10^{-19} C. After obtaining the same value using other liquids for the drops, Millikan concluded that 1.5924×10^{-19} C must be the unit of electric charge. The modern value for the charge on the electron, the unit of electric charge, is $-1.602\ 1773 \times 10^{-19}$ C.

Once the charge on the electron was known, the mass of the electron, m_e, could be calculated from Thomson's mass-to-charge ratio for the electron, $m_e/e = -(6 \times 10^{-12})$ kg/C:

$$-(6 \times 10^{-12})\frac{\text{kg}}{\text{C}} = \frac{m_e}{-(1.5924 \times 10^{-19}\ \text{C})}$$

Solving for m_e,

$$m_e = \left(6 \times 10^{-12}\frac{\text{kg}}{\text{C}}\right)(1.5924 \times 10^{-19}\ \text{C}) = 1 \times 10^{-30}\ \text{kg}$$

The modern value for the mass of the electron is $9.109\ 390 \times 10^{-31}$ kg or $5.485\ 799 \times 10^{-4}$ u.

▪ FIGURE 7.4   Millikan's apparatus for determining the unit of electric charge. An oil drop was allowed to fall through the pinhole in the top plate and the hole was closed. The air around this drop was ionized by the X-rays. Soon one of the ions from the air hit the oil drop and stuck to its surface. The drop, which looked like a bright star, was observed through the telescope. When the plates were connected so that they could not be charged with respect to each other and there was no electric field between them, the drop was pulled down by gravity and fell slowly through the air. The time required for the drop to fall from the top crosshair of the telescope to the bottom crosshair was measured. After the drop had fallen past the bottom crosshair, the plates were charged so that the drop rose, and the time needed for the drop to rise past the space between the crosshairs was measured. Then the plates were discharged and the drop was observed again while falling. The process was repeated over and over again; a single drop could be observed for over an hour. From the time and the distance traveled by the drop, the speed of the drop could be calculated. The mass of the drop could be calculated from the speed of the falling drop. The charge on the the the drop could be calculated from the speed of the rising drop. According to his report of his work (Millikan, R. A. *Phys. Rev.* **1913**, *2(2)*, 109–143), Millikan included all observations except those he knew to be wrong because some mistake was made in carrying out the experiment. He also said that he did not do any calculations until all of the observations had been made so that his observations could not possibly be influenced by his idea of what the result should be.

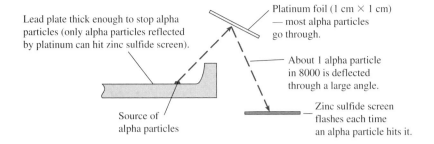

Lead plate thick enough to stop alpha particles (only alpha particles reflected by platinum can hit zinc sulfide screen).

Platinum foil (1 cm × 1 cm) — most alpha particles go through.

About 1 alpha particle in 8000 is deflected through a large angle.

Source of alpha particles

Zinc sulfide screen flashes each time an alpha particle hits it.

◾ FIGURE 7.5 Marsden's apparatus for observing what happens to alpha particles aimed at thin pieces of metal foil. (Adapted from Geiger, H.; Marsden, E. *Proceedings of the Royal Society of London* **1909,** *82,* 495–500.)

Why have we used platinum in Marsden's apparatus when discussions in other texts use gold? We have used platinum because that is what Geiger and Marsden say they used.

7.3 THE NUCLEAR ATOM

Thomson thought that the positive charge occupied most of the volume of the atom. He pictured an atom as a uniform ball of positive electricity and imagined the electrons as enclosed in this ball like blueberries in a spherical blueberry muffin.

Discovery of the Nucleus

In 1909 Rutherford, another English physicist, had one of his students, an undergraduate named Ernest Marsden, observe what happens to alpha particles aimed at a thin piece of metal foil. (Alpha particles are helium ions, He^{2+}. The mass of an alpha particle is about 8000 times the mass of an electron; the velocity of the alpha particles was about 2×10^7 m/s.) Rutherford thought the alpha particles would pass through the foil in a straight line (like a bullet would go through a blueberry muffin). However, he told his student to check to be sure no particles were deflected through large angles. ◾Figure 7.5 is a diagram of Marsden's apparatus.

By comparing the number of alpha particles that hit the zinc sulfide screen with the total number of alpha particles emitted, Marsden estimated that the path of about 1 in 8000 alpha particles was deflected through an average angle of 90°. Most of the alpha particles did go through the foil in a straight line. It took Rutherford until 1911 to figure out an explanation. ◾Figure 7.6 shows this explanation. Platinum, like most solids, is nearly incompressible. Therefore, it seems reasonable to picture the atoms as touching as shown in Figure 7.6. If the atom has a nearly uniform distribution of charge as Thomson thought, alpha particles should pass straight through, as most of them did. For an alpha particle to be turned aside through a large angle, it must pass close to something with a very concentrated positive charge that repels it. For an alpha particle to be bounced back toward its source, it must be deflected by a very strong electric field. Rutherford pictured the atom as consisting

If you saw someone shoot a bullet at a blueberry muffin, you would be amazed if the bullet did not go straight through.

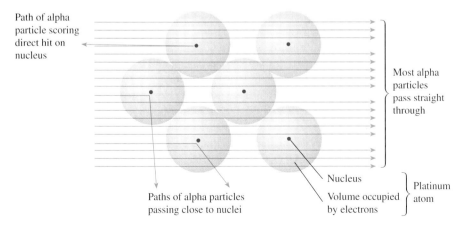

Path of alpha particle scoring direct hit on nucleus

Most alpha particles pass straight through

Paths of alpha particles passing close to nuclei

Nucleus

Volume occupied by electrons

Platinum atom

◾ FIGURE 7.6 Rutherford's picture of platinum atoms bombarded by alpha particles. If the nucleus was drawn to the same scale as the platinum atoms, it would be invisible because it would be only about 0.0001 cm in diameter. If the nucleus was the size shown (about 1 mm), the atom would be about 10 m in diameter.

◐ Remember the small boy swinging a stone on the end of a string around him (Section 5.5). The stone is able to occupy a large volume because it is moving rapidly. The electrons in an atom are able to occupy a large volume because they are moving very fast.

In 1915 Moseley was killed in World War I at the age of 28.

◐ The terms *wavelength* and *frequency* are explained in Section 7.5. For now, the important thing is that the graph is a straight line.

When you have your teeth or chest x-rayed, the X-rays are made by bombarding a tungsten anode in a cathode ray tube.

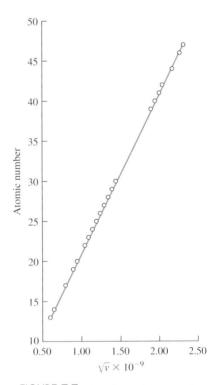

■ FIGURE 7.7 Atomic number versus the square root of the frequency of the X-rays given off by metals. The square roots of the frequencies of the X-rays in this graph range from 0.50×10^9 to 2.50×10^9. To avoid having to write $\times 10^9$ five times when labeling the *x*-axis, all values have been multiplied by 10^{-9}.

of a small, dense, positively charged nucleus containing most of the mass of the atom with the electrons in the space outside the nucleus. According to Rutherford's model, the moving electrons occupied most of the volume of the atom. Because the electrons, which are very small, occupy most of the volume of the atom, they must be moving very rapidly in the space around the nucleus.

Determination of Nuclear Charges

One of the questions raised by Rutherford's picture of the atom is "What are the charges on the nuclei?" From his observations, Rutherford estimated that the charge on the nucleus of an atom is numerically equal to about half the atomic mass, but he knew that this was only a very rough approximation. A young English physicist, H. G. J. Moseley, working in Rutherford's laboratory in 1913–1914, discovered a way to determine nuclear charge accurately.

When a metal is made the anode in a cathode ray tube, the metal gives off X-rays. Moseley invented a method of measuring the wavelengths of X-rays given off by metals when they are bombarded with a stream of high-energy electrons. Moseley found that a graph of atomic numbers against the square root of the frequency of the X-rays is a straight line as shown in ■Figure 7.7. Moseley noted that if the elements were not characterized by the atomic numbers, the graph of atomic number against square root of frequency would not be a straight line. He concluded:

> Now Rutherford has proved that the most important constituent of an atom is its central positively charged nucleus, and van den Broek has put forward the view that the charge carried by this nucleus is in all cases an integral multiple of the charge on the hydrogen nucleus. There is every reason to suppose that the integer which controls the X-ray spectrum is the same as the number of electrical units in the nucleus, . . .

7.4 DISCOVERY OF THE PROTON AND THE NEUTRON

If electrons are knocked off gas atoms or molecules in a cathode ray tube, positively charged particles must be left. The positively charged particles will move toward the cathode because it is negatively charged. They can be detected and studied by making a hole in the cathode. The charge-to-mass ratios of the positive particles were measured by the same method used for electrons. The masses of the positively charged particles were found to depend on the gas in the tube, with the particles from hydrogen gas having the lowest mass. The particles from hydrogen gas were assumed to be another unit particle common to all atoms and were given the name proton.

Because atoms have no net charge, they must have the same number of protons as electrons. However, the total mass of atoms was found to be about twice the mass of the protons. There must be uncharged particles in nuclei to account for the missing mass. Uncharged particles are hard to detect, and it was not until 1932 that they were discovered by English physicist James Chadwick. Chadwick observed that when beryllium-9 was exposed to alpha particles (He^{2+} ions) from the radioactive decay of radium, particles with about the same mass as protons, but with no charge, were given off.

Neutrons are present in the nuclei of all atoms except hydrogen-1. Protons repel each other because of their like charges; more than one proton can't be packed into a small volume to form a stable nucleus unless neutrons are present. Neutrons reduce the repulsive force between positively charged protons and contribute to the force that holds the particles in the nucleus together. In light atoms (up to calcium in the periodic table), one neutron per proton is enough. Heavier atoms with more protons in the nucleus need more neutrons in the nucleus for the nucleus to be stable.

Since the 1960s, physicists have found that even protons and neutrons have structure, and the investigation of the detailed structure of these particles is an active field of research in physics today. However, because the chemical and physical properties of the elements and their compounds can be explained by protons, neutrons, and electrons, chemists continue to regard atoms as composed of these particles.

Chadwick was another of Rutherford's students.

7.5   TRAVELING WAVES

Much of our information about the electronic structure of atoms comes from observations of the action of visible light and atoms on each other. To understand the experiments that have led to our present understanding of the electronic structure of atoms, you first need to know something about waves and electromagnetic radiation.

When you think of waves, you probably think of water. If a pebble is dropped into a pool of water, a wave spreads in circles from the point where the pebble hit the surface of the water, as shown in ▪Figure 7.8. If you were actually looking at the water, you would think that the water was moving outward from the center. But the water is only moving up and down; the *wave motion* is moving outward. A ball on the surface just bobs up and down as the wave passes; the ball does not move outward with the wave. Wave motion transfers energy, not matter, from one place to another.

If you could look at the wave from the side at water level, it would look something like the ideal wave shown in ▪Figure 7.9. A wave consists of *repeating units* called **cycles.** A wave is said to be in the same phase at similar points such as A and B. The distance between similar points such as A and B is called **wavelength, λ;** the wavelength is the *length of one cycle*. The **amplitude** of a wave is *half the vertical distance from the top of the wave to the bottom of the wave.* In Figure 7.9, notice that the mathematical sign of the wave function changes from + to − (or from − to +) when the amplitude is zero. *Points* such as C and D *where the amplitude of a wave is zero* are called **nodes.**

The **frequency, ν,** of wave motion is the *number of cycles passing a fixed point in one second.* Frequency depends both on the speed at which a wave is moving and its wavelength. If two waves traveling at the same speed are compared, the wave with the shorter wavelength has the higher frequency, as shown in ▪Figure 7.10.

▪ FIGURE 7.8   If a pebble is dropped into a pool of water, a wave spreads in circles from the point where the pebble hit the surface of the water.

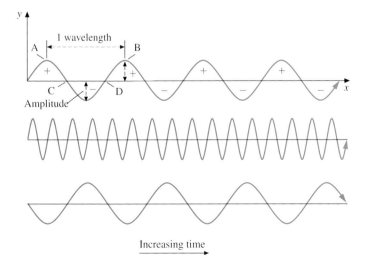

Increasing time

▪ FIGURE 7.9   An ideal or perfect wave. A wave is said to be in the same phase at similar points such as A and B. The + and − signs show the mathematical sign of the wave; the mathematical sign changes at points like C and D where the wave crosses the *x*-axis.

▪ FIGURE 7.10   The two waves are traveling at the same speed. The top wave is of shorter wavelength and higher frequency than the bottom wave.

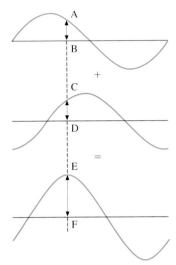

▬ FIGURE 7.11   Addition of two waves; EF = AB + CD.

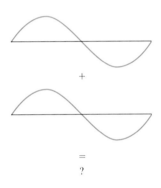

PRACTICE PROBLEM

7.1   Assuming that the two waves are traveling at the same speed, which has the higher frequency? (a) A wave with a wavelength of 1 m or one with a wavelength of 2 m? (b) A wave with a wavelength of 10^{-7} m or one with a wavelength of 10^{-10} m?

Wave amplitudes can be added algebraically, which leads to two important properties of waves: interference and diffraction. **Interference,** the *addition of two waves,* is shown in ▬Figure 7.11. *If the signs of the waves that are added are the same,* as they are in the example in Figure 7.11 (both +), the height (or depth) of the new wave is greater than the height or depth of either old wave. *The interference is said to be* **constructive.**

PRACTICE PROBLEM

7.2   Sketch the wave that will be formed by the addition of two waves of equal amplitude in phase as shown below at the left. How will the amplitude of the new wave compare with the amplitude of the original waves?

If the signs of the waves that are added are different, the height or depth of the new wave is less than the height or depth of either old wave. *The interference is said to be* **destructive.** When two waves having the same wavelength and amplitude that are in opposite phases are added, the two waves completely cancel each other. The sum is zero as shown in ▬Figure 7.12. Interference between two sets of water waves is shown in ▬Figure 7.13.

Because waves spread, *waves bend around the edge of any object in their path that is about the same size or larger than their wavelength.* This phenomenon is called **diffraction.** Diffraction is shown in ▬Figure 7.14.

Although you may not think of them as waves, you are also familiar with a number of kinds of waves besides water waves, such as sound waves, microwaves, radiowaves, and visible light. All these waves except sound waves are examples of electromagnetic radiation.

7.6   ELECTROMAGNETIC RADIATION

Wave Properties

Electromagnetic radiation is *a form of energy that consists of perpendicular electric and magnetic fields that change, at the same time and in phase, with time.* The

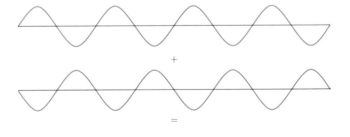

▬ FIGURE 7.12   Addition of two waves with the same wavelength and amplitude that are in opposite phases. The two waves completely cancel each other.

intensity of electromagnetic radiation is *proportional to the square of the amplitude of the wave.* Electromagnetic radiation exhibits many of the characteristic properties of waves such as interference and diffraction. The changing electric and magnetic fields spread from their source in much the same way as the waves made by the pebble in the pool of water. However, unlike water waves, electromagnetic radiation can move through empty space. *Empty space* is called a **vacuum.** All kinds of electromagnetic radiation travel through a vacuum at the same speed, *2.998 × 10⁸ m/s,* which is often referred to as the **speed of light, c.**

The *SI unit of frequency* is called a **hertz, Hz,** in honor of the nineteenth-century German physicist Heinrich Hertz, who was the first person to broadcast and receive radio waves. One hertz equals one cycle per second:

$$1 \text{ Hz} = \frac{1}{s} \qquad \text{or} \qquad 1 \text{ Hz} = 1 \text{ s}^{-1}$$

For example, in Figure 7.10 if a point in the wave moves from left to right in one second, then the upper wave has a frequency of 16 Hz, and the lower wave has a frequency of 4 Hz.

Wavelength and frequency are related. The product of the wavelength (λ) and the frequency (v) of electromagnetic radiation is equal to the speed of light (c), as shown by equation 7.1:

$$\lambda v = c \qquad\qquad (7.1)$$

In equation 7.1, if the speed of light is in meters per second, then wavelength must be in meters and frequency in hertz.

Because the product of the wavelength and frequency of electromagnetic radiation is constant (see equation 7.1), long-wavelength radiation must have low frequency, and short-wavelength radiation must have high frequency. The frequency of electromagnetic radiation can be calculated from wavelength or wavelength can be calculated from frequency by means of equation 7.1.

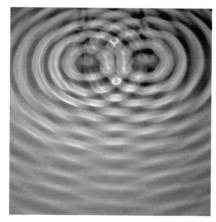

■ FIGURE 7.13   Pattern formed by interference between two waves.

(a)

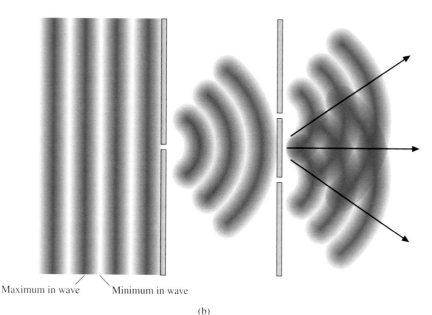

Maximum in wave   Minimum in wave

(b)

■ FIGURE 7.14   (a) Waves are diffracted (bent) by an object in their path. (b) A diffraction pattern is produced by the interference of the two waves formed when a single wave passes through two slits. Intensity maxima occur along the lines indicated by the arrows.

SAMPLE PROBLEM

7.1 What is the wavelength in meters of a radiowave having a frequency of 92.1 MHz?

Solution First we must convert 92.1 MHz to Hz. Then we can use equation 7.1 and the speed of light, 2.998×10^8 m/s, to calculate wavelength in meters.
Mega means 10^6 (Table 2.2):

$$92.1 \text{ MHz}\left(\frac{10^6 \text{ Hz}}{1 \text{ MHz}}\right) = 92.1 \times 10^6 \text{ Hz} \qquad \text{or} \qquad 9.21 \times 10^7 \text{ s}^{-1}$$

Solution of equation 7.1 for wavelength and substitution gives

$$\lambda = \frac{c}{v} = \frac{2.998 \times 10^8 \text{ m} \cdot \text{s}^{-1}}{9.21 \times 10^7 \text{ s}^{-1}} = 3.26 \text{ m}$$

PRACTICE PROBLEMS

7.3 Which travels faster through a vacuum, electromagnetic radiation with a frequency of 10^9 s^{-1} or electromagnetic radiation with a frequency of 10^3 s^{-1}?

7.4 Which has the longer wavelength, electromagnetic radiation with a frequency of 10^9 s^{-1} or electromagnetic radiation with a frequency of 10^3 s^{-1}?

7.5 What is the wavelength of electromagnetic radiation that has a frequency of 6.10 kHz?

7.6 What is the frequency in MHz of a wave with a wavelength of 5.6 m?

Ultraviolet radiation is used to sterilize the air in operating rooms. Gamma rays are used to sterilize packages of medical supplies.

Electromagnetic radiation is described qualitatively by name. For example, people speak of a red light, a microwave oven, and so forth. In a vacuum, light and other electromagnetic radiation can be identified quantitatively by either wavelength or frequency. The whole **spectrum** (range) **of electromagnetic radiation** is shown in ▬Figure 7.15(a) and the visible part in 7.15(b). As you can see, visible light is just a small part of the electromagnetic spectrum.

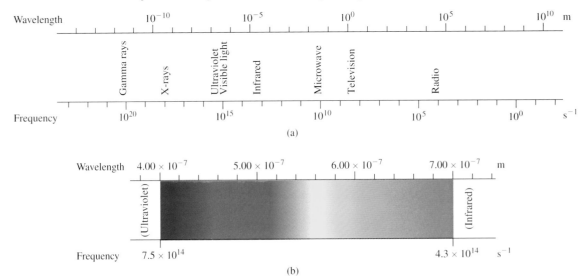

▬ FIGURE 7.15 (a) The whole electromagnetic spectrum. (b) The visible part of the electromagnetic spectrum, which lies between the ultraviolet and infrared regions, is shown in color. *Ultraviolet and infrared radiation are invisible.*

Light from the sun and from light bulbs is composed of light of many colors that can be separated with a prism. ▬Figure 7.16 shows a prism separating light from a lamp. Electromagnetic radiation is slowed by passing through matter. Some materials slow it more than others; the speed of light through water or glass is less than the speed of light through air. Some wavelengths are slowed more than others. A prism separates light into a spectrum because the speed of blue light through glass is less than the speed of red light through glass.

Empty space is the only completely "clear" medium. If no matter is present, all wavelengths of electromagnetic radiation are **transmitted** *(passed through)*; different kinds of matter **absorb** *(take up)* different wavelengths. Some aqueous solutions appear colored because the solutes absorb some wavelengths of light and transmit others. The color you see is the color of the transmitted wavelengths (see ▬Figure 7.17). Although glass transmits visible light, it absorbs most ultraviolet light. Sunburn is caused by ultraviolet light. Therefore, you are less likely to get sunburned riding in a car when the windows are closed. Earth's atmosphere is **transparent** to *(transmits)* visible light and part of the radio region of the electromagnetic spectrum.

The leaves of green plants obtain energy to carry out **photosynthesis,** the *process by which plant cells make sugars and starch from carbon dioxide and water,* by absorbing red and orange light from sunlight. Violet light is also absorbed. Yellow, green, and blue light are reflected and the leaves appear green, as shown in ▬Figure 7.18.

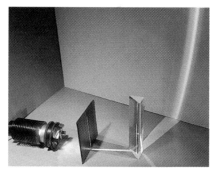

▬ FIGURE 7.16 The white light from the lamp is composed of many colors. The colors are separated by the prism.

Because the atmosphere absorbs X-rays, detectors for X-ray astronomy must be placed above the atmosphere. The "X" in X-ray stands for unknown. When X-rays were discovered, their nature was unknown.

Particle Properties

Although electromagnetic radiation shows wave properties such as diffraction and interference, describing electromagnetic radiation as waves does not account for all

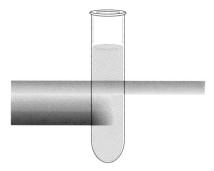

▬ FIGURE 7.17 An orange solution absorbs green, blue, and violet light. Only yellow, orange, and red light is transmitted to the observer's eyes.

▬ FIGURE 7.18 When sunlight falls on a green leaf, light of red, orange, and violet wavelengths is absorbed. Light of green, yellow, and blue wavelengths is reflected to the observer's eyes, and the leaf looks green.

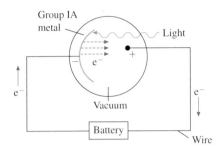

FIGURE 7.19 The photoelectric effect. The battery pumps electrons through the wire in the direction shown by the arrows. The metal becomes negatively charged, and the other end of the wire becomes positively charged, as shown in the diagram. However, no current flows in the dark. When light shines on the metal, electrons are knocked off the surface. The electrons are attracted by the positive charge and travel through the empty space completing the circuit. A door can be made to open when the light is shut off and to close when the light is turned on. (Why must the metal in the photocell be surrounded by a vacuum?)

Einstein was awarded the Nobel Prize in physics in 1921 for his work in theoretical physics, including his explanation of the photoelectric effect.

the properties of light. The simplest and most familiar phenomenon that requires a different model of light is the photoelectric effect used in some automatic door openers.* When visible light falls on the active metals of Group IA (the alkali metals—Li, Na, K, Rb, and Cs), electrons, called **photoelectrons,** are **emitted** *(given off)* from the metal, as shown in ▪Figure 7.19.

Electrons do not spontaneously escape from metals; thermal energy or light energy is needed to knock electrons out of metals. The brighter the light (that is, the greater the intensity of the light), the more electrons are knocked out of a metal. However, the kinetic energy (speed) of the electrons depends on the color (frequency) of the light, not on brightness. In addition, a certain minimum frequency is required or no electrons at all are knocked out (see ▪Figure 7.20). For example, the dimmest of violet lights produces a photoelectric effect with potassium but the brightest of red lights does not; the frequency of violet light is higher than the frequency of red light [see Figure 7.15(b)]. A bright violet light produces more photoelectrons than a dim violet light; however, the maximum kinetic energy of the photoelectrons is the same.

The photoelectric effect was first observed by Heinrich Hertz in 1887 (the same Hertz for whom the SI unit of frequency is named), but it was not explained until 1905 (by Albert Einstein). To help you understand Einstein's reasoning, let's imagine a vending machine that accepts nickels, dimes, and quarters, but not pennies. No matter how many pennies you have, you cannot buy anything from the machine; you must have a nickel, dime, or quarter—a piece composed of 5, 10, or 25 cents—to make a purchase. To knock electrons out of a metal, light must have a certain minimum frequency. This fact suggests that light energy must come in pieces that have different energies comparable to the pieces of money called nickels, dimes, and quarters.

If the object you want from the vending machine costs only a nickel but you put in a dime, you will get back a nickel in change. If a packet of light that has more energy than needed to knock an electron out hits a metal, the extra energy is converted to kinetic energy. The electron that is knocked out comes out at high speed. The more nickels you put in the machine, the more objects you will get from the machine just as a bright light, which contains more packets of light, produces more photoelectrons than a dim light of the same frequency.

The *small pieces or particles of electromagnetic radiation* are called **photons;** each photon carries a definite amount of energy, just as each coin is equal in value to a certain number of cents. The energy of a photon is proportional to the frequency of the electromagnetic radiation and inversely proportional to the wavelength of the electromagnetic radiation:

$$\text{energy of a photon} = h\nu \qquad (7.2a)$$

or

$$\text{energy of a photon} = \frac{hc}{\lambda} \qquad (7.2b)$$

where h is a proportionality constant called **Planck's constant,** c is the speed of light in a vacuum (2.998×10^8 m·s^{-1}), ν is the frequency of the radiation in s^{-1}, and λ is the wavelength of the radiation in meters. Planck's constant has the value

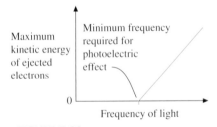

FIGURE 7.20 The photoelectric effect. No matter how bright the light, no electrons are knocked out if the frequency of the light is too low. The minimum frequency is characteristic of the kind of metal. At frequencies above this minimum frequency, the maximum kinetic energy of the electrons is directly proportional to the frequency of the light.

---

*Movie soundtracks and burglar alarms are among the many other practical applications of the photoelectric effect. A movie soundtrack is a strip along the edge of the film that causes the light passing through to vary in brightness depending on the frequency of the recorded sound. A photoelectric cell converts the light to a varying electric current that is amplified and played through speakers.

6.626 × 10^{-34} J·s. Qualitatively, equation 7.2a says that the higher the frequency, the greater the energy of a photon of electromagnetic radiation. Equation 7.2b tells us that the longer the wavelength, the lower the energy of a photon of electromagnetic radiation.* Equations 7.2a and b can be used to calculate the energy of photons of light of different frequencies and wavelengths.

Planck, a German physicist, was the first person to propose particles of energy, which he called "quanta." However, he was one of the last physicists to believe in them.

SAMPLE PROBLEM

7.2   Calculate the energy in kJ/mol of photons of red light of wavelength 656 nm.

Solution   Equation 7.2b gives the relationship between the energy of a photon and the wavelength of electromagnetic radiation. Substitution of the wavelength given (656 nm) in equation 7.2b gives

$$\text{energy of a photon} = \frac{hc}{\lambda} = \frac{(6.626 \times 10^{-34}\ \text{J} \cdot \text{s})(2.998 \times 10^{8}\ \text{m} \cdot \text{s}^{-1})}{\left(656\ \text{nm} \times \dfrac{10^{-9}\ \text{m}}{1\ \text{nm}}\right)}$$

$$= 3.03 \times 10^{-19}\ \text{J}$$

The quantity of energy, 3.03 × 10^{-19} J, is the energy for a single photon of red light. For a mole of photons, the energy is

$$\left(3.03 \times 10^{-19}\ \frac{\text{J}}{\text{photon}}\right)\left(6.022 \times 10^{23}\ \frac{\text{photon}}{\text{mol}}\right)\left(\frac{1\ \text{kJ}}{10^{3}\ \text{J}}\right) = 182\ \frac{\text{kJ}}{\text{mol}}$$

PRACTICE PROBLEMS

7.9   Calculate the energy in kJ/mol of photons of blue-green light of wavelength 486 nm.

7.10   (a) Calculate the energy in kJ/mol of photons of electromagnetic radiation that has a frequency of 3.72 × 10^{17} s^{-1}. (b) Which type of electromagnetic radiation has a frequency of 3.72 × 10^{17} s^{-1}?

7.11   What is the wavelength of photons of electromagnetic radiation that have an energy of 1722 kJ/mol?

In conclusion, a certain minimum frequency of electromagnetic radiation, which depends on the material used, is needed to knock electrons out of a material. When the frequency of the electromagnetic radiation is less than this minimum frequency, nothing happens. When the frequency is equal to or greater than this minimum frequency, electrons are knocked out of the material. If the frequency of the electromagnetic radiation is greater than the minimum frequency, the extra energy makes the electrons that are given off move faster, and the maximum speed of the electrons depends on the frequency of the electromagnetic radiation. The number of electrons emitted depends on the intensity (brightness for visible light) of the electromagnetic radiation.

Photons of visible light only have enough energy to knock electrons out of active metals such as sodium, potassium, and cesium (the metals of Group IA of the periodic table). Ultraviolet light, which has shorter wavelength, higher frequency,

---

*If you have ever played the children's game where one person moves the end of a rope back and forth and another child jumps over the waves in the rope as they go by, you know that it takes more energy to make short quick waves than to make long, slow waves.

■ FIGURE 7.21   Young lady and old woman.

and more energy per photon than visible light, is needed to knock electrons out of materials other than the active metals.

PRACTICE PROBLEM

7.12   Which have more energy, photons of X-rays or photons of gamma rays?

Twofold Character of Electromagnetic Radiation

To understand all the properties of electromagnetic radiation, *both* a wave model and a particle model must be used. Some things are better explained by a wave model, and others are better explained by a particle model. However, photons are not ordinary particles; photons have no mass, and all photons move at the same speed. The energies of photons do not depend on their speeds. There is no analogy for the twofold character of electromagnetic radiation in everyday life. However, the picture in ■Figure 7.21 may help you to understand that the way something looks can depend on the way you look at it. Sometimes, when you look at Figure 7.21, you will see the young lady; other times you will see the old woman.

7.7   THE BOHR MODEL OF THE HYDROGEN ATOM

Spectra is the plural of spectrum.

🔘 The Models of the Atom module of the Atoms section provides an interactive hydrogen atom that students can use to reinforce their understanding of the emission and absorption of light.

Atomic Spectra

A second question raised by Rutherford's nuclear model for an atom is "Why are the negatively charged electrons not pulled into the positively charged nucleus?" At the same time that Moseley was investigating the X-ray spectra of atoms, a young Danish physicist, Niels Bohr, proposed an answer to this second question. Bohr was also working in Rutherford's laboratory at the time.

Besides explaining why the electrons do not fall into the nucleus, Bohr wanted to account for the spectra emitted by atoms in gas discharge tubes and flames. Atomic spectra extend from the infrared through the visible to the ultraviolet regions of the electromagnetic spectrum. A simple spectroscope (device for producing and observing spectra) and the visible parts of some typical atomic emission spectra are shown in ■Figure 7.22. As you can see, atomic spectra consist of lines of different

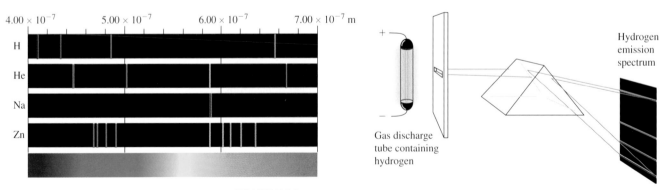

■ FIGURE 7.22   The spectrum of white light is continuous (*bottom*); the spectra of atoms consist of lines of different colors. For simplicity only the brightest lines are shown. The spectra of atoms of different elements are different. Helium in the sun was discovered before helium was found on Earth as a result of observation of a new yellow line in the spectrum of the sun's chromosphere during a solar eclipse in 1868.

colors. The spectra of different elements are different. For example, the spectrum of sodium is different from the spectrum of hydrogen. The spectrum of sodium has two closely spaced yellow lines at 590 and 589 that are so bright no other lines are visible; the spectrum of hydrogen has a red line at 656 nm, a green line at 486 nm, and purple lines at 434 and 410 nm. Atomic spectra are characteristic properties that can be used to identify elements. They are like a fingerprint for each element.

Atomic spectra were used to identify 27 or more elements in rock samples from the moon.

Toward the end of the nineteenth century, a Swiss high school teacher, Johann Balmer, found an empirical equation that accurately describes the spectrum of the hydrogen atom in the visible region:

◗ Remember that empirical means based on experiment and observation.

$$\frac{1}{\lambda} = 1.097 \times 10^7 \text{ m}^{-1} \left(\frac{1}{2^2} - \frac{1}{n^2} \right) \qquad n = 3, 4, 5, \dots \qquad (7.3)$$

The spectra of many other atoms can be represented by similar equations. However, no one could figure out any explanation for the spectral lines or for the equations.

SAMPLE PROBLEM

7.3   Calculate the wavelength in nm of the line in the spectrum of the hydrogen atom for which $n = 3$.

Solution   Solve equation 7.3 for wavelength and substitute the value given for n:

$$\lambda = \frac{1}{1.097 \times 10^7 \text{ m}^{-1} \left(\frac{1}{2^2} - \frac{1}{n^2} \right)} = \frac{1}{1.097 \times 10^7 \text{ m}^{-1} \left(\frac{1}{2^2} - \frac{1}{3^2} \right)}$$

$$= 6.563 \times 10^{-7} \text{ m}$$

The problem asks for the wavelength in nm:

$$6.563 \times 10^{-7} \text{ m} \left(\frac{1 \text{ nm}}{10^{-9} \text{ m}} \right) = 656.3 \text{ nm}$$

The wavelength of the red line in the spectrum of the hydrogen atom is 656.3 nm (see Figure 7.23).

PRACTICE PROBLEM

7.13   Calculate the wavelength in nm of the line in the spectrum of the hydrogen atom for which $n = 4$.

The Energy of Electrons in Atoms Is Quantized

To explain why the electron is not pulled into the nucleus and why hydrogen atoms only emit certain wavelengths of light, Bohr proposed that the electron moves around the nucleus in one of many possible orbits similar to the orbits of the planets about the sun. Each of the orbits has a certain energy associated with it, that is, *only certain values* of energy *are possible*. The energy of the electron is said to be **quantized.**

◗ Quantized energy is like the energy of someone or something going down stairs. The potential energy changes in steps. Continuous energy is like the energy of someone or something going down a ramp.

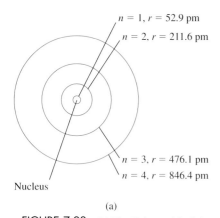

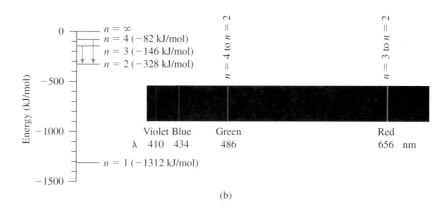

FIGURE 7.23 (a) The Bohr model of the hydrogen atom showing relative radii of the orbits. The nucleus is shown at the center of the picture of the Bohr model, although it would be invisible if drawn to the same scale as the orbits. (b) The allowed energy levels of the orbits in the Bohr atom and the spectrum of the hydrogen atom.

In Bohr's model, each orbit was assigned a number called the **principal quantum number, n,** that could have any whole number value from 1 to infinity. ◄Figure 7.23 shows Bohr's model of the hydrogen atom and the energy levels of the orbits. In Figure 7.23(a) notice how rapidly the radii of the orbits increase as n increases. Also notice, in Figure 7.23(b), how successive energy levels become more and more closely spaced. As energy (either electromagnetic, thermal, or electrical) is added to the atom, the electron is raised to higher and higher energy levels farther and farther from the nucleus. Finally, it is so far from the nucleus that it is no longer attracted by the positive charge on the nucleus. The electron is, for all practical purposes, infinitely far from the nucleus and is called a **free electron.** The energy of a free electron is not quantized but is continuous. The hydrogen atom has become a hydrogen ion. The *energy necessary to remove the electron* from the first energy level of the hydrogen atom and form a hydrogen ion, H^+, is called the **ionization energy** of the hydrogen atom.

Remember that energy itself cannot be measured; only differences in energy can be measured. The energy of a free electron—that is, an electron at an infinite distance from the nucleus—is set equal to zero, and other energies are measured relative to the energy of the free electron. From Figure 7.23(b), the ionization energy of hydrogen is $+1312$ kJ/mol. (For comparison, the heat of combustion of methane gas is -890.3 kJ/mol.)

The energy of the electron depends on the orbit it occupies. The smaller the radius of the orbit, the lower the energy of the electron in the orbit and the more stable the system of the nucleus and one electron. According to the Bohr model, the electron does not fall into the nucleus because it cannot have a radius smaller than the radius of the first orbit or an energy lower than the energy of the first orbit. *When the electron is in the lowest energy orbit,* the hydrogen atom is said to be in its **ground state.** *When the electron is in any higher energy level,* the hydrogen atom is said to be in an **excited state.**

Emission and Absorption Spectra

When an electron falls from one orbit to another lower energy orbit, energy, in the form of electromagnetic radiation, is given off. The energy can only be given off in definite amounts because the difference in energy between the two orbits is fixed. Each spectral line corresponds to a certain energy change. The red line at 656 nm in the hydrogen spectrum results from the electron falling from the $n = 3$ to the $n = 2$ level. The green line at 486 nm results from the electron falling from the $n = 4$ to the $n = 2$ level. The greater the difference in energy between the two levels, the higher the energy of the electromagnetic radiation given off and the

Lightning, neon signs, and fireworks are examples of emission of visible light resulting from electronic transitions.

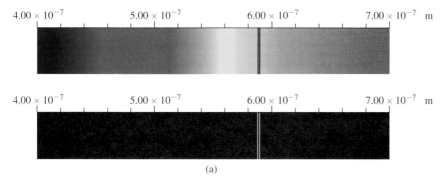

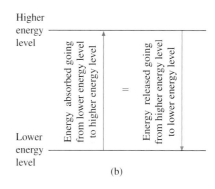

(a)

(b)

■ FIGURE 7.24 (a) When white light passes through a sample of sodium atoms, light of some wavelengths is absorbed, and dark lines are observed. The dark lines are at the same wavelengths as the bright lines in the emission spectrum of sodium. The quantity of energy that is absorbed when an electron is raised from a lower energy level to a higher energy level is the same as the quantity of energy that is emitted when an electron falls from the same higher energy level to the same lower energy level. (b) The energy absorbed going from a lower energy level to a higher energy level is equal to the energy released going from the higher energy level to the lower energy level. When the continuous light from inside a star such as the sun passes through the cooler vapors on the outside of the star, light is absorbed. Which wavelengths of light are absorbed depends on which elements are present in the vapors. The compositions of the stars can be determined from the wavelengths or frequencies of the dark lines in their spectra.

shorter its wavelength. The spectrum of a sample of gaseous hydrogen atoms contains all possible lines because of the large number of atoms in a macroscopic sample. The brightness of a line depends on how many photons of that wavelength are given off. [Remember that a photon is a particle of electromagnetic radiation (Section 7.6)]. The quantity of energy given off when an electron falls from one energy level to another is proportional to the frequency of the electromagnetic radiation and inversely proportional to the wavelength (equations 7.2a and b).

When an electron is raised from one energy level to another, electromagnetic radiation is absorbed. The same amount of energy is absorbed as is given off when the electron falls from the higher level to the lower level. The same wavelength of electromagnetic radiation is absorbed as is emitted when the electron falls from the higher level to the lower level. ■Figure 7.24 shows the absorption spectrum of sodium and how it compares with the emission spectrum of sodium and with a continuous spectrum.

Bombarding an atom that has many electrons, such as a sodium atom, with high-speed electrons may, if the energy of the electrons is high enough, knock an electron out of the lowest energy shell closest to the nucleus. An electron from a higher energy level then falls down into the empty space in the lowest level. The difference in energy between the lowest level and the other levels is large, and the electromagnetic radiation given off is of X-ray wavelength. Fall of an electron from a higher level into the lowest energy level is the origin of the X-ray spectra used by Moseley to determine atomic numbers.

Mercury vapor (top) *and sodium vapor* (bottom) *lights. The characteristic colors of these lights result from the wavelengths of the light emitted by excited mercury and sodium atoms. Low pressure sodium lights emit at a single wavelength leaving the rest of the visible spectrum dark for astronomers.*

SAMPLE PROBLEM

7.4 The blue line in the spectrum of the hydrogen atom, wavelength = 434 nm, results from the electron dropping from the $n = 5$ level to the $n = 2$ level.
(a) Calculate the energy in kJ/mol of photons of light of wavelength 434 nm.
(b) Draw a diagram like the one in Figure 7.23(b) showing the energy levels from $n = 2$ to $n = \infty$ that includes the $n = 5$ level.

Solution Part (a) of this problem is similar to Sample Problem 7.2:

$$\text{energy of a photon} = \frac{hc}{\lambda} = \frac{(6.626 \times 10^{-34}\ \text{J} \cdot \text{s})(2.998 \times 10^{8}\ \text{m} \cdot \text{s}^{-1})}{\left(434\ \text{nm} \times \dfrac{10^{-9}\ \text{m}}{1\ \text{nm}}\right)}$$

$$= 4.58 \times 10^{-19}\ \text{J}$$

$$\left(4.58 \times 10^{-19}\ \frac{\text{J}}{\text{photon}}\right)\left(6.022 \times 10^{23}\ \frac{\text{photon}}{\text{mol}}\right)\left(\frac{1\ \text{kJ}}{10^{3}\ \text{J}}\right) = 276\ \frac{\text{kJ}}{\text{mol}}$$

When a mole of electrons drops from the $n = 5$ level to the $n = 2$ level, 276 kJ are emitted.

(b) The $n = 5$ level must be 276 kJ/mol above the $n = 2$ level, which is at -328 kJ/mol:

$$-328\ \text{kJ/mol} + 276\ \text{kJ/mol} = -52\ \text{kJ/mol}$$

The $n = 5$ level must be at -52 kJ/mol.

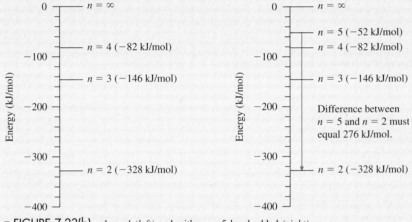

• FIGURE 7.23(b) enlarged *(left)* and with $n = 5$ level added *(right)*.

PRACTICE PROBLEM

7.14 The violet line in the spectrum of the hydrogen atom, wavelength = 410 nm, results from the electron dropping from the $n = 6$ level to the $n = 2$ level. (a) Calculate the energy in kJ/mol of photons of light of wavelength 410 nm. (b) Draw a diagram like the one in Figure 7.23(b) showing the energy levels from $n = 2$ to $n = \infty$ that includes the $n = 6$ level.

From his model, Bohr was able to derive an equation for the wavelengths of light emitted by hydrogen that fit the observed spectrum of the hydrogen atom. Bohr's theoretical equation was identical to the empirical equation developed earlier (equation 7.3). Bohr also was able to calculate the radius of the hydrogen atom in its ground state (52.9 pm) and its ionization energy (-1312 kJ/mol) from his theory. The *radius of the hydrogen atom in its ground state, 52.9 pm,* is called the **Bohr radius;** the Bohr radius is sometimes used as a unit.

Atoms with more than one electron are more complicated than the hydrogen atom. All efforts to extend Bohr's model to atoms with more than one electron, such as a helium atom or a lithium atom, failed. In addition, the Bohr model did not explain *why* energy should be quantized. The picture of electrons moving in fixed orbits was abandoned in 1927 after German physicist Werner Heisenberg

Atomic Absorption Spectroscopy

Atomic absorption spectroscopy (AAS) is one of the most widely used analytical techniques. AAS is a very sensitive and specific method of analysis and is the method of choice for most commonly determined metals. It is applicable to any problem involving low concentrations of metal ions in aqueous solution. Solid samples such as steel and hair can be dissolved by appropriate chemical reactions. Thus, AAS is applicable to a large variety of samples.

Determination of air and water quality is often carried out by atomic absorption spectroscopy. Cadmium and lead have been measured in cigarette smoke, and mercury levels in offices and labs. AAS is also used to measure trace elements in food. For example, the concentration of zinc (Zn^{2+}) in cow's milk and the concentration of lead (Pb^{2+}) in wine have been determined. Biological fluids such as blood and urine can be analyzed, and atomic absorption spectroscopy is becoming a routine diagnostic tool. AAS is used to establish the presence or absence of essential trace metals in human and animal tissue. For example, the metal content of hair varies with the individual and his or her diet and environment. As a result, hair can be used for identification for forensic purposes. Crime labs also analyze soil taken from suspects' shoes and compare it with soil collected from the scene of the crime. AAS is used in agriculture, archaeology, art, criminal investigations, environmental studies, medicine, nutrition, and many other fields.

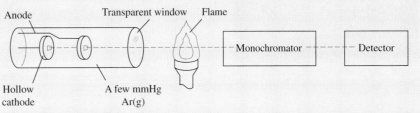

■ FIGURE 7.25   Schematic diagram of an atomic absorption spectrometer.

Atomic absorption spectroscopy can be used to analyze qualitatively for the presence of different metals in a sample and quantitatively to establish how much of a given metal is present. It is also often used for the metalloids arsenic, antimony, and boron. More than 60 elements can be determined with accuracies of 1–2%; detection limits for most elements are of the order of 10^{-7} M. AAS is fast and can be automated.

■Figure 7.25 is a schematic diagram of an atomic absorption spectrometer. The hollow cathode in the gas discharge tube is coated or filled with the element to be determined; application of 600–1000 V to the gas discharge tube causes a glow discharge. The argon gas is ionized:

$$Ar + e^- \longrightarrow Ar^+ + 2e^-$$

The argon ions are attracted to the cathode, where they knock metal atoms off into the gas phase. The gaseous metal atoms are excited by collisions with argon ions and emit electromagnetic radiation of characteristic wavelengths. Thus, a beam of intense radiation consisting of narrow bands of the wavelengths characteristic of the metal to be determined is produced.

The sample is introduced into the flame, where water is evaporated and metal ions are reduced to gaseous metal atoms in the ground state. Only those ground-state metal atoms in the flame that are atoms of the same metal that is in the hollow cathode can absorb radiation from the gas discharge tube. AAS is very specific because atomic spectra are like fingerprints; each element absorbs and emits only certain characteristic wavelengths.

The monochromator allows only radiation of a single wavelength to pass through to the detector where the fraction of the radiation from the discharge tube that was absorbed by the sample is measured. The fraction of the radiation from the discharge tube that is absorbed depends on the concentration of the metal in the sample. The instrument must be calibrated with solutions of known concentrations. The chief disadvantage of atomic absorption spectroscopy is that a different discharge tube, at several hundred dollars each, is usually needed for each element that is to be determined.

---

pointed out that both the position and the speed of a submicroscopic particle like an electron cannot be known at the same time (see Section 7.9). However, Bohr's model was important because it introduced the ideas of quantum numbers and of quantized energy states for electrons in atoms.

7.8   THE WAVE THEORY OF THE ELECTRON

To answer the question of why the energy of the electron in the hydrogen atom should be quantized, a French graduate student, Louis de Broglie, suggested in 1924 that electrons have wave properties as well as particle properties. De Broglie reasoned that the electron in the hydrogen atom is "fastened" in the space around the nucleus

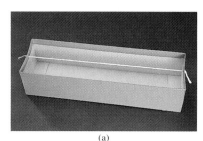

(a)

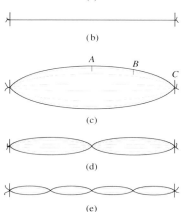

(b)

(c)

(d)

(e)

■ FIGURE 7.26 Standing waves in a string. If a tightly stretched string fastened to the sides of a box at both ends as shown in the photo (a) and the diagram (b) is plucked, it will vibrate as shown in (c). If the vibrating string is touched lightly at its center A, the string will vibrate as shown in (d). If the vibrating string is touched at point B, instead of at point A, the string will vibrate as shown in (e).

⊙ Remember that points such as C, where the amplitude of the wave is zero, are called nodes.

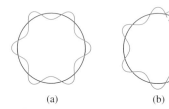

(a)   (b)

■ FIGURE 7.27 (a) The number of waves around the orbit is a whole number (6) times the wavelength. (b) The number of waves around the orbit is *not* a whole number multiple of the wavelength. When the wave goes around the orbit many times, it cancels itself out.

by the attraction of the positive charge on the nucleus. He argued that the electron's wave behavior should be that of a standing wave. The water waves and electromagnetic waves we have talked about are all traveling waves. **Traveling waves** *move from one place to another.* **Standing waves** *do not move from one place to another;* they are confined to a region of space. For example, the vibrations of a guitar string are standing waves. Guitar strings are fastened to the guitar at both ends as the tightly stretched string in ■Figure 7.26 is fastened to the sides of a box at both ends.

If the string is plucked it will vibrate as shown in Figure 7.26(c). (The amplitudes of the vibrations in Figure 7.26 are enlarged so that you can see them.) The wavelength of the vibration shown in Figure 7.26(c) is twice the length of the box. There are two nodes, one at each end of the box where the string is fastened to the box. If the vibrating string is touched lightly at its center, A, the string will vibrate as shown in Figure 7.26(d) with an additional node in the center. Two half-wavelengths of the vibration in Figure 7.26(d) are equal to the length of the box. Point B is halfway between the center and one end of the string. If the string vibrating as shown in Figure 7.26(c) is touched at point B instead of at point A, the string will vibrate as shown in Figure 7.26(e). In Figure 7.26(e), there are three nodes besides the nodes at the ends. Four half-wavelengths are equal to the length of the box. Because the string is fastened to the box at both ends, vibrations are confined to inside the box. There must be nodes at both ends of the string where it is fastened to the box and cannot move. Only certain wavelengths are possible for vibrations. The wavelengths of the allowed vibrations must be such that a whole number of half-wavelengths are equal to the length of the box.

PRACTICE PROBLEM

7.15 Suppose the string in Figure 7.26(c) is touched at a point that is one-third of the way from one end of the box to the other. (a) Make a sketch showing how the string will vibrate. (b) How many nodes are there besides the nodes at the ends? (c) How many half-wavelengths are equal to the length of the box?

If the electron's wave behavior is that of a standing wave, each orbit must be equal to a whole number times the wavelength, as shown in ■Figure 7.27(a). If the orbit is not equal to a whole number times the wavelength, the wave will interfere with itself and eventually be destroyed as shown in Figure 7.27(b). According to de Broglie's theory, only orbits with lengths equal to a whole number times the wavelength of the electron are possible. Orbits whose lengths are not equal to a whole number times the wavelength of the electron are impossible.

The energy of the hydrogen atom depends on the electron's orbit. Because only certain orbits are possible, only certain energies are possible. The energy of the hydrogen atom can have some values but not others; that is, the energy is quantized.

De Broglie proposed that the wavelength of a particle, λ, and its mass, m, and speed, v, are related by equation 7.4:

$$\lambda = \frac{h}{mv} \tag{7.4}$$

In equation 7.4, h is Planck's constant. From equation 7.4, you can see that either high mass or high speed or both result in short wavelengths. Equation 7.4 can be used to calculate the wavelength of the electron in the hydrogen atom. Substitution of Planck's constant (6.6×10^{-34} kg·m²/s), the mass of the electron (9.1×10^{-31} kg), and Bohr's value for the speed of the electron (2.2×10^{6} m/s) in equation 7.4 gives

$$\lambda = \frac{(6.6 \times 10^{-34}\,\text{kg·m}^2/\text{s})}{(9.1 \times 10^{-31}\,\text{kg})(2.2 \times 10^{6}\,\text{m/s})} = 3.3 \times 10^{-10}\,\text{m}$$

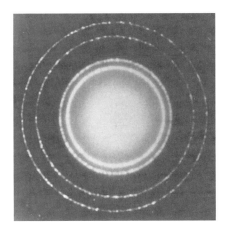

The wavelength associated with the electron in the hydrogen atom is similar to the wavelength of X-rays. The wavelength associated with macroscopic objects like cars is too small to be observed.

At the time de Broglie suggested that electrons have wave properties, wave behavior of electrons had not been observed experimentally. Soon afterward (1928), C. J. Davisson and L. H. Germer at Bell Labs in the United States and G. P. Thomson in England showed that electrons do indeed have wave properties. They showed that electrons can be diffracted like light; diffraction is a property of waves (remember Figure 7.14). ■Figure 7.28 compares an electron-diffraction pattern with the diffraction pattern from X-rays. Davisson and Germer also found that the wavelengths of electrons calculated from de Broglie's equation, equation 7.4, were correct.

G. P. Thomson was the son of J. J. Thomson, who is given credit for discovering the electron.

The wave properties of electrons are applied in the electron microscope. The electron microscope is an instrument that uses the wave characteristics of electrons to obtain pictures of tiny objects like molecules. Ordinary light microscopes cannot be used to examine objects smaller than the wavelength of visible light (about $4 - 7 \times 10^{-7}$ m). X-rays have shorter wavelengths than visible light but are hard to focus. As we have seen, electrons have about the same wavelength as X-rays. Because of their charge, electrons can be focused by means of electric and magnetic fields. In addition, the wavelengths of the electrons can be varied by changing the voltage used to produce them. A modern electron microscope can resolve details as small as 5×10^{-10} m in diameter; an image obtained with an electron microscope is shown in ■Figure 7.29.

■ FIGURE 7.29 Electron micrograph of AIDS virus (yellow-orange) on a lymphocyte (light blue). Electron microscopy, which provides images at magnifications of 50 000 times and more, has made major contributions to the biological sciences by revealing details that could not have been observed with ordinary microscopes (maximum magnification about $1400\times$).

THE HEISENBERG UNCERTAINTY PRINCIPLE

We see objects by noting their interaction with visible light. The object must be at least as large as the wavelength of the light. Electromagnetic radiation of an extremely short wavelength would be needed to locate an object as small as an electron. Photons of radiation having a short wavelength and a high frequency have a great deal of energy. When a photon of high-energy radiation strikes an electron, the collision causes the direction of motion and the speed of the electron to change. The attempt to locate the electron changes the speed of the electron.

Photons with longer wavelengths are less energetic and would have a smaller effect on the electron's speed. However, because of their longer wavelength, these photons would not show the position of the electron very precisely. According to Heisenberg, the uncertainty in the position of an object times the uncertainty in the speed of the object is equal to or greater than $h/4\pi m$:

Equation 7.5 is correct as long as the speed of the particle does not approach the speed of light and m is constant. We do not want to deal with special relativity here.

$$\Delta x \cdot \Delta v \geq \frac{h}{4\pi m} \tag{7.5}$$

In equation 7.5, Δx is the uncertainty in position of the particle and Δv, the uncertainty in its speed; h is Planck's constant, 6.626×10^{-34} kg · m²/s; and m is the mass of the particle in kg.

From equation 7.5, you can see that the smaller the mass of an object, the greater the product of the uncertainty of its position and speed. The uncertainty in measuring position and speed is very important for objects as small as an electron. The uncertainty in measuring position and speed is not significant for objects of ordinary size. Police officers have no difficulty measuring both the position and the speed of cars. The world of submicroscopic particles like the electron is very different from the ordinary macroscopic world of our everyday life.

Uncertainty in measuring the position and speed of a submicroscopic particle is different from ordinary uncertainty in measurements. Ordinary uncertainty in measurements can be reduced by using more precise and more accurate instruments; uncertainty in measuring the position and speed of a submicroscopic particle is unavoidable. An exact description of the path of an electron in an atom like the orbits in Bohr's model is impossible and attempts to patch up Bohr's theory are useless. Looking at an electron as a wave, it is not surprising that we are uncertain about its position.

THE QUANTUM MECHANICAL OR WAVE MECHANICAL MODEL OF THE ATOM: THE SCHRÖDINGER EQUATION

Schrödinger was 39 years old in 1926 when he published his wave equation, his first important work. Most scientists are much younger than this when they begin making major contributions.

On the basis of the idea that electrons have wave properties, Austrian physicist Erwin Schrödinger wrote an equation called the **Schrödinger equation** to *describe the behavior and energies of electrons in atoms.* The Schrödinger equation is similar to equations used to describe electromagnetic waves. It is too complicated to write here. Schrödinger used his equation to calculate a number of the properties of the electron in the hydrogen atom; the calculated values agreed well with the properties observed by experiment. We can use the results of solutions of the Schrödinger equation even if we can't carry out the calculations. Solutions of the Schrödinger equation contain all the information about an atom that is allowed by the uncertainty principle; this information is usually in the form of probabilities.

Probability means chance or odds of something happening.

For atoms with more than one electron, the calculations are very complicated. The total energy of the atom depends on the positions of all the electrons. Each of the electrons repels all the others and is repelled by all the others. Even with the world's most advanced supercomputer system, the Schrödinger equation has not been solved exactly for atoms containing more than one electron. However, results in good agreement with experiment have been obtained for a number of atoms by using approximations to simplify the calculations.*

Solutions of the Schrödinger equation are functions (not numbers) called **wave functions, ψ.** The wave function of an electron can have either a positive or a negative sign. The wave function of an electron has no physical meaning, but the square of the wave function ψ^2,† is a mathematical expression of how the probability of finding an electron in a small volume varies from place to place. The probability of finding an electron must be positive because the electron must be found somewhere. A negative probability has no meaning. The *volume in space where an electron with a particular energy is likely to be found* is called an **orbital.**

The quantized energy and quantum numbers proposed by Bohr are natural results of Schrödinger's theory. The electrons in atoms are attracted by the positive charge of the nucleus and are confined to a small volume of space near the nucleus. The electron waves in atoms are standing waves; their energy is quantized. If an electron escapes from the attractive force of the nucleus (the atom is ionized), the energy of the electron is no longer quantized but is continuous. Free electrons are traveling waves.

The Schrödinger equation has an infinite family of solutions. Each solution is identified by three quantum numbers. A set of three quantum numbers is needed to describe an electron because the electrons in atoms are moving in three-dimensional space.

The **principal quantum number, n,** *tells the size of an orbital and largely determines its energy.* It can have any positive whole number value. The larger the principal quantum number, the greater the average distance of an electron in the orbital from the nucleus and the higher the energy of the electron. *All the orbitals with the same principal quantum number* are referred to as a **shell.**

All shells except the first shell are divided into **subshells.** All orbitals in a subshell have the same **angular momentum quantum number, l,** as well as the same principal quantum number. The angular momentum quantum number *tells the shape of the orbitals.* It can have any whole number value from zero to $n - 1$ where n

A simple example of a solution that is a number is $y = 2$. A simple example of a solution that is a function is $y^2 = x^2$.

Be careful to distinguish an orbital from an orbit. An orbit is a curved path such as Earth's path around the sun. According to the Heisenberg uncertainty principle, the path of an electron is unknowable.

The Models of the Atom module of the Atoms section introduces students to orbitals and their characteristics and lets them explore the four quantum numbers and their restrictions.

---

*Calculation of a square root with a cheap calculator that can only add, subtract, multiply, and divide is a simple example of the use of approximations. Suppose you need to find the square root of 7.698. Your scientific calculator is broken, and you must get along with a cheap calculator that can only add, subtract, multiply, and divide. The number 7.698 is between 4 and 9, the squares of 2 and 3. Therefore, the square root of 7.698 must be between 2 and 3. To get started, guess that the square root of 7.698 is 2.500. Find the square of 2.500 and compare it with 7.698. The square of 2.500 is 6.250. Because 6.250 is less than 7.698, 2.500 is too low a value for the square root of 7.698. Try a higher number, say, 2.750. The square of 2.750 is 7.5625, still a little low but a lot closer. Suppose we try 2.780; $(2.780)^2 = 7.7284$, very close but slightly high. Next try 2.775. The square of 2.775 is 7.701, only 0.003 too high. The square of 2.774 is 7.695, 0.003 too low. If 7.698 is a measured number so that the last digit is uncertain, either 2.775 or 2.774 is a satisfactory answer, as good an answer as could be obtained with a scientific calculator. (A scientific calculator gives 2.774 526 987 for the square root of 7.698.) A computer can be programmed to make approximations and test them until a satisfactory answer is obtained.

†The square of the wave function is like the square of the amplitude of an electromagnetic wave. The intensity of electromagnetic radiation is proportional to the square of the amplitude. The square of the wave function for an electron gives the probability that the electron will be found at a certain point in space.

1s
2s 2p   Not allowed
3s 3p 3d
4s 4p 4d 4f
5s 5p 5d 5f 5g
6s 6p 6d 6f 6g 6h
7s 7p 7d 7f 7g 7h 7i

Principal Quantum Number, n (Shell)	Angular Momentum Quantum Number, l (Subshell)	Subshell Label	Magnetic Quantum Number, m_l	Number of Orbitals in Subshell
1	0	1s	0	1
2	0	2s	0	1
	1	2p	−1, 0, +1	3
3	0	3s	0	1
	1	3p	−1, 0, +1	3
	2	3d	−2, −1, 0, +1, +2	5
4	0	4s	0	1
	1	4p	−1, 0, +1	3
	2	4d	−2, −1, 0, +1, +2	5
	3	4f	−3, −2, −1, 0, +1, +2, +3	7

TABLE 7.1 Allowed Combinations of Quantum Numbers for n = 1–4

is the principal quantum number. Thus, if the principal quantum number, n, is 3, the angular momentum quantum number, l, can be 0, 1, or 2. To avoid getting numbers mixed up, angular momentum quantum numbers are usually shown by letter as follows:

Angular momentum quantum number	0	1	2	3	4	5
Letter used	s	p	d	f	g	h*

Principal quantum numbers, angular momentum quantum numbers, and magnetic quantum numbers come from the mathematical solution of the Schrödinger equation.

The third quantum number is called the **magnetic quantum number, m_l.** The magnetic quantum number *describes the direction that the orbital projects in space.* The magnetic quantum number can have whole number values from $-l$ to $+l$ where l is the angular momentum quantum number. For example, if $l = 2$, m_l can have values of -2, -1, 0, $+1$, and $+2$. The possible values of the three quantum numbers for the first four shells are summarized in Table 7.1.

In Table 7.1, you should notice the following points: There is one subshell in the first shell, two subshells in the second shell, three in the third, and four in the fourth. Thus, the number of subshells in a shell is the same as the principal quantum number of the shell. The number of possible m_l values for a given value of l is equal to the number of orbitals present in that subshell. There is one orbital in each s subshell, three orbitals in each p subshell, five in each d, and seven in each f. The number of orbitals in subshells is given by the series of odd numbers 1, 3, 5, ... There are $(2l + 1)$ orbitals with a given l value.

PRACTICE PROBLEM

7.16   Add a section for $n = 5$ to Table 7.1.

---

*The letters *s, p, d,* and *f* describe lines in the atomic emission spectra of the alkali metals: **s**harp, **p**rincipal, **d**iffuse, and **f**undamental. Following *f*, other letters are assigned in alphabetical order.

7.11 PICTURES OF ORBITALS

Because orbitals have shapes and positions in space, pictures and models of orbitals are very useful. *Imagine* that we could take photographs of an electron as it moves about the nucleus. Each exposure would show the position of the electron at some point in time. (Remember, the Heisenberg uncertainty principle tells us that we cannot *really* know the position of an electron.) If we could make a very large number of exposures on the same film, we would end up with a picture like ▪Figure 7.30(a) for an electron in a 1s orbital. The 1s orbital is spherical; Figure 7.30(a) shows a cross section cut through the center of the atom. The nucleus is too small to be shown in Figure 7.30(a). It is at the center of the darkest part of the **charge cloud** in the picture.

Charge density is *the amount of electronic charge per unit volume of space.* From Figure 7.30(a), you can see that the charge density in a 1s orbital is greatest in the center of the atom close to the nucleus. The 1s electron spends most of its time close to the nucleus. The charge density decreases as distance from the nucleus increases. The chance of finding the electron farther than about 106 pm from the nucleus is very small.

Figure 7.30(b) is a graph of the probability per unit volume of finding the electron (ψ^2) against distance from the nucleus plotted to the same scale as Figure 7.30(a). Figure 7.30(b) shows the same information as Figure 7.30(a) in a different way.

Figure 7.30(c) shows the radial distribution function for a 1s electron. The **radial distribution function** is the *probability of finding the electron in thin spherical shells of uniform thickness at distances r from the nucleus.* Maximums in radial distribution functions occur at radii where an electron is most likely to be found. For the 1s electron of the hydrogen atom, the maximum is at 5.29×10^{-11} m or 52.9 pm, the Bohr radius.

You may wonder why, if a 1s electron spends most of its time close to the nucleus as shown in Figures 7.30(a) and (b), the radial distribution function is low close to the nucleus and reaches a maximum at one Bohr radius, as shown in Figure 7.30(c). The volume of thin spherical shells of uniform thickness increases as distance from the nucleus increases [Figure 7.30(d)]. Charge density decreases as distance from the nucleus increases [Figures 7.30(a) and (b)]. Very close to the nucleus, although charge density is high, the volume of thin spherical shells is so small that the probability that the electron is in a shell is small. Far from the nucleus, although the volume of thin spherical shells is large, charge density is very low, and radial probability is very small. At one Bohr radius, volume is becoming significant, and charge density is still high; therefore, radial probability is at a maximum. Radial probability is an example of a property that has a maximum because it is a combination of two properties that change in opposite directions.

Figure 7.30(e) shows a surface such that the electron is found inside this surface 90% of the time. A **boundary surface** like this is the easiest type of orbital picture to draw; we will usually use boundary surfaces to show orbitals. The most important features of an orbital are its shape, its relative size, and its location in space. All these are shown by boundary surface diagrams.

▪ FIGURE 7.30   (a) Charge cloud of the 1s orbital of a hydrogen atom. The greater the concentration of dots, the higher the charge density. (b) ψ^2 plotted against distance from the nucleus. (c) Radial probability plotted against distance from the nucleus. (d) Volume of thin spherical shells of uniform thickness as a function of distance from the nucleus. (e) Boundary surface diagram for 1s orbital. The electron is found inside the volume shown 90% of the time.

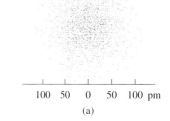

100 50 0 50 100 pm
(a)

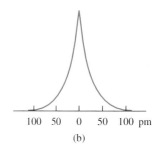

100 50 0 50 100 pm
(b)

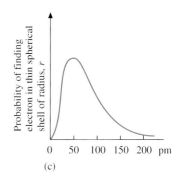

Probability of finding electron in thin spherical shell of radius, r

0 50 100 150 200 pm
(c)

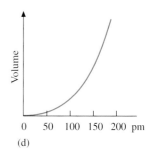

Volume

0 50 100 150 200 pm
(d)

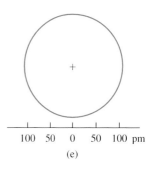

100 50 0 50 100 pm
(e)

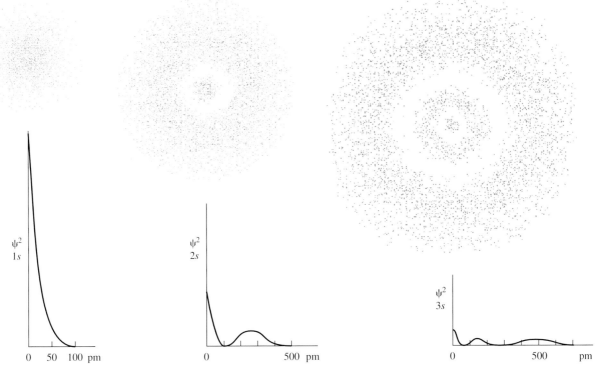

▪ FIGURE 7.31 Charge clouds and probabilities of finding an electron as a function of distance from the nucleus, ψ^2, for s orbitals of a hydrogen atom.

⬤ Remember that the boundary "surface" is not real.

In the boundary surface diagram, Figure 7.30(e), the plus sign is the mathematical sign of the wave function. The sign of the wave function does not change in a 1s orbital. There are no nodes in 1s orbitals; that is, there is no place in a 1s orbital where the probability of finding an electron is zero. Although the probability of finding an electron far from the nucleus is very small, it is *not* zero.

There is just one s orbital in each shell. All s orbitals are spherical; however, different s orbitals differ in size and energy. ▪Figure 7.31 shows charge clouds and probabilities of finding an electron as a function of distance from the nucleus for 1s, 2s, and 3s electrons.

Electrons in s orbitals with high principal quantum numbers tend to stay farther away from the nucleus than electrons in s orbitals with low principal quantum numbers. Electrons in s orbitals with high principal quantum numbers have higher energies than electrons in s orbitals with low principal quantum numbers. The 2s orbital has one node, and the 3s orbital has two nodes. The higher the energy of an orbital, the more nodes; the more nodes, the higher the energy. The relationship between energy and number of nodes is analogous to the relationship between energy and number of nodes in the standing wave in a stretched string (remember Figure 7.26).

PRACTICE PROBLEM

7.17 (a) How many nodes would you predict for a 4s orbital? (b) What shape would the nodes be?

The part of the electron cloud close to the nucleus does not take part in bond formation. Only electrons near the outside of the atom are involved in forming bonds. The nodes in *s* orbitals are not near the outside of the atom; therefore, the fact that they do not show in boundary surface diagrams is not important.

The first shell (principal quantum number 1) has only one *s* orbital (see Table 7.1). The first *p* orbitals occur when the principal quantum number is 2. ▪Figure 7.32 shows *one p* orbital. The nucleus, which is too small to show, is at the origin (where the axes intersect). A *p* orbital is not spherically symmetric about the nucleus like an *s* orbital; the electron density is concentrated along one axis. The *p* orbital shown in Figure 7.32 is called a p_x orbital because its electron density is concentrated along the *x*-axis.

A *p* orbital has a node close to the nucleus. Students often ask how an electron gets from one lobe of a *p* orbital to the other if there is a node—a volume where the electron doesn't spend any time—around the nucleus between the lobes. They are thinking of the electron as a particle. But an electron also has wave properties. A wave has no trouble passing through a node. The + and − signs show the mathematical sign of the wave function. The sign of the wave function for an electron changes when it passes through a node just as the sign for the simple wave shown in Figure 7.9 changes sign at each node.

The electron density of an *s* orbital is greatest close to the nucleus, whereas *p* orbitals have nodes at the nucleus. This is sometimes described by saying that *s* electrons have greater ability to **penetrate** to the nucleus than do *p* electrons.

The 2*p* orbitals have no nodes other than the one between the lobes. The 3*p* orbitals have one other node, the 4*p* two other, and so on. The extra nodes are not near the outside of the atom and are usually ignored. The outer parts of the 3*p* and 4*p* orbitals are similar to 2*p* orbitals. ▪Figure 7.33 compares 2*p* and 3*p* orbitals.

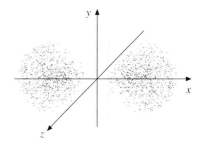

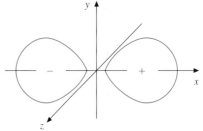

▪ FIGURE 7.32 Charge cloud and boundary surface diagram for one *p* orbital. The nucleus, which is too small to be seen, is at the origin. Remember that the + and − signs show the mathematical sign of the wave and have nothing to do with electrical charge.

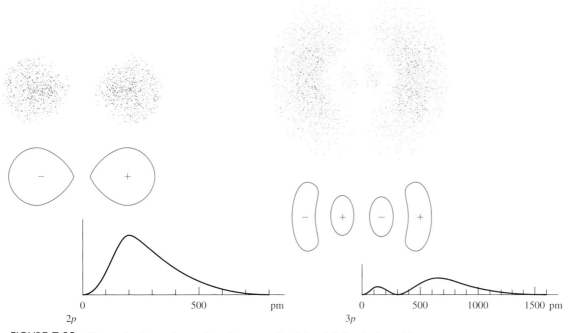

▪ FIGURE 7.33 Charge cloud, boundary surface diagram, and radial probability for 2*p* and 3*p* orbitals of a hydrogen atom.

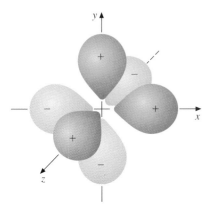

FIGURE 7.34 Boundary surface model of a set of three *p* orbitals. Lobes that have a + sign are red, and lobes that have a − sign are lighter red.

If you wish to cover them now, the shapes of d *orbitals are shown in Chapter 24 where coordination compounds are discussed.*

Some reviewers objected to the use of the terms spin up and spin down. Spin up and spin down are used by physicists (see, for example, Science *1994, 265, 185). We prefer these terms to clockwise and counterclockwise because the picture of an electron as a spinning top is questionable. In addition, students easily connect spin up and spin down with the arrows used in energy-level diagrams.*

Why include the Dirac equation? (See Related Topic.) This material has two messages that we think are important to students: Failure (to get a job in Dirac's case) is not the end of the world, and many problems have more than one solution. In addition, for students who take physics, it shows another connection between subjects.

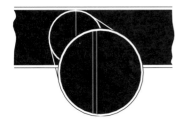

FIGURE 7.35 The red line at 656 nm in the spectrum of the hydrogen atom [see Figure 7.23(b)] is actually two very closely spaced lines at 656.272 and 656.285 nm.

There are three *p* orbitals in each shell. The three *p* orbitals in a shell are identical in size, shape, and energy. They differ only in their location in space. ▪Figure 7.34 shows a model of a set of three *p* orbitals. You can see that the three *p* orbitals are perpendicular to each other. Therefore, if the axis that passes through one of the *p* orbitals is selected as the *x*-axis, the other two will lie along the *y*- and *z*-axes. The three *p* orbitals are often distinguished from each other by labeling them p_x, p_y, and p_z depending on which axis the electron density is concentrated along.

The shapes of the five *d* and seven *f* orbitals are harder to show. Fortunately, a knowledge of the shapes and positions in space of *s* and *p* orbitals will be enough for most of the topics that we will discuss. The shapes and locations in space of the *d* orbitals are shown later (in Figures 24.9 and 24.10) when we need to use them.

7.12 ELECTRON SPIN AND THE PAULI EXCLUSION PRINCIPLE

The quantum numbers derived from the Schrödinger equation explain a great deal of experimental data, but they do not account for the fact that some atomic spectral lines actually consist of two closely spaced lines as shown in ▪Figure 7.35. Austrian physicist Wolfgang Pauli suggested that the two lines could be explained by the electron having two states available to it, either one of which it can occupy. These states were later identified with electron spin. An electron is pictured as spinning like a top about its axis. Like a top, it can only spin in one of two directions. You would describe a top's spin as clockwise or counterclockwise; an electron's spin is said to be up or down. A fourth quantum number, the **spin quantum number, m_s,** had to be added to the three quantum numbers obtained by solving the Schrödinger equation. Only two values, which are equal but opposite in sign, are possible for the spin quantum number. The two values, + 1/2 and − 1/2, result from the mathematics used to describe the electron's magnetism.

Pauli also proposed that *no two electrons in an atom can have all four quantum numbers alike.* This proposal is called the **Pauli exclusion principle.** As a result of the Pauli exclusion principle, an orbital can hold only two electrons; the two electrons must have opposite spins. An orbital occupied by two electrons with opposite spins is filled. The exclusion principle is a statement of an experimental fact with no explanation according to the Schrödinger model of the atom, but with very important effects in systems that have more than one electron. All electrons tend to avoid each other because all electrons are negatively charged and like charges repel each other. However, electrons with the same spin have an especially low probability of being close to each other that has nothing to do with their charges.

SAMPLE PROBLEM

7.5 (a) How many electrons can occupy the 2*p* subshell? (b) How many electrons can be in the second shell?

Solution (a) In a 2*p* subshell, there are three orbitals (see Table 7.1). According to the Pauli exclusion principle, an orbital can hold two electrons. Therefore, six electrons can occupy the 2*p* subshell.

(b) In the second shell, there is one 2*s* orbital and three 2*p* orbitals making a total of 4 orbitals in the second shell. Each orbital can hold two electrons. Eight electrons can be in the second shell.

The Dirac Equation: A Relativistic Model of the Atom

Four quantum numbers and the Pauli exclusion principle arise naturally from the solution of *another wave equation,* the **Dirac equation,** published in 1928 by the English physicist P. A. M. Dirac.* Dirac applied the ideas of Einstein's special theory of relativity to quantum mechanics. According to the special theory of relativity, events must be described by four numbers: *x-, y-,* and *z-*coordinates telling where the event took place and a fourth coordinate, *t,* telling when. Four quantum numbers are needed to describe the electrons in atoms because the electrons have four coordinates.

Also according to the special theory of relativity, the masses of objects moving at speeds approaching the speed of light increase as their speed increases. The mass of an object can be defined as a measure of the object's resistance to the change in its state of motion caused by a given force. The larger the mass, the smaller the acceleration caused by the same force. (A push that is hard enough to start a small car rolling will not have much effect on a freight train.) Objects with mass can't move faster than light (and other electromagnetic radiation). If an object is already traveling at a speed approaching the speed of light, it must

offer increasing resistance to further increases in speed. If it does not, it will soon go faster than light, and this is impossible according to the special theory of relativity. In large atoms, the speed of the 1*s* electrons approaches the speed of light, and relativistic effects are important. Relativistic effects are unimportant for objects moving much slower than the speed of light, just as quantum effects are unimportant for objects of ordinary size.

The principal quantum number, *n,* determines the size and energy of an orbital in the Dirac model as it does in the Schrödinger model. The other three quantum numbers have different meanings; the third and fourth, instead of the second, determine the shape of the orbitals. The shapes of the Dirac orbitals are different from the shapes of the Schrödinger orbitals, and there are no nodes like the node between the two lobes of a *p* orbital. There is no such thing as electron spin in the Dirac treatment.

Dirac's equation predicted the existence of positrons. **Positrons** are *particles having the same mass as electrons but with one unit of positive charge.* The discovery of positrons in 1932 supported Dirac's model.

The Schrödinger equation is satisfactory for dealing with elements having low atomic numbers. It can be corrected for relativistic effects for elements of intermediate atomic number. Calculations based on the Schrödinger equation are *not* adequate for a discussion of the chemical

The discovery photograph of the positron, the first antimatter particle to be discovered.

properties of the atoms that follow the lanthanide series in the periodic table. For these heavy elements, the Dirac equation is required.

Relativistic effects increase very rapidly as atomic number increases and are partly responsible for the chemical differences that exist between the elements of the fifth and sixth periods. Predicting properties of elements in the seventh period from the properties of elements in the sixth period is risky as a result of relativistic effects.

The math involved in solving the Dirac equation is even more complicated than the math involved in solving the Schrödinger equation. The Dirac equation is used only when the Schrödinger equation is inadequate.

---

*Dirac studied electrical engineering as an undergraduate. He became a theoretical physicist because, after graduation, he was unable to get a job as an electrical engineer.

PRACTICE PROBLEM

7.18 (a) How many electrons can occupy the 3*d* subshell? (b) How many electrons can be in the third shell?

SUMMARY

A wave is a disturbance that moves transmitting energy from one place to another. A wave consists of repeating units called **cycles.** The shortest distance between similar points on a wave

is called **wavelength.** A wavelength is the length of one cycle. The number of cycles passing a fixed point in one second is called the **frequency** of the wave. The SI unit of frequency is the **hertz, Hz;** one Hertz equals one cycle per second. The height or depth of a wave is called its **amplitude.** Points where the mathematical sign of a wave's amplitude changes from + to − are called **nodes. Interference** and **diffraction** are properties characteristic of waves. Matter **transmits** (lets through) some wavelengths and is said to be **transparent** to these wavelengths. It **absorbs** (takes up) other wavelengths. **Traveling waves** move from one place to another; **standing waves** are confined to a region of space. All wavelengths are possible for traveling waves; only certain wavelengths are possible for standing waves.

Electromagnetic radiation consists of perpendicular electric and magnetic fields that change, at the same time and in phase, with time. The wavelength and frequency of electromagnetic radiation are related by the equation

$$\lambda v = c$$

where λ stands for wavelength, v for frequency, and c for the speed of electromagnetic radiation in a vacuum, 2.998×10^8 m/s. A **vacuum** is empty space without even air in it. The intensity (brightness) of electromagnetic radiation is proportional to the square of the amplitude of the wave. Electromagnetic radiation has particle properties as well as wave properties. A particle of electromagnetic radiation is called a **photon.** The energy of a photon is proportional to the frequency and inversely proportional to the wavelength of the electromagnetic radiation.

Rutherford found that most of the mass of an atom is concentrated in a small, dense nucleus. The electrons are outside the nucleus and occupy most of the volume of the atom. The nucleus of a hydrogen atom is a proton. The nuclei of all other atoms are composed of protons and neutrons. The number of protons in the nucleus is equal to the number of electrons outside the nucleus in an atom and is called the atomic number of the atom.

To explain why the negatively charged electrons do not fall into the positively charged nucleus, **Bohr** proposed that the energies of electrons in atoms are **quantized**—that is, they can have only certain values. Bohr pictured the electrons as circling the nucleus in orbits. This latter part of Bohr's model of the atom was abandoned after **Heisenberg** proposed his **uncertainty principle.** According to the uncertainty principle, it is impossible to measure both the position and the speed of submicroscopic particles like electrons at the same time.

To explain why the energies of the electrons in atoms are quantized, **de Broglie** suggested that electrons have properties of standing waves. **Schrödinger** wrote a wave equation for an electron in an atom; solutions of the Schrödinger equation are called **wave functions, ψ.** The square of the wave function, ψ^2, expresses the probability of finding an electron at a given point in space; the volume in space where an electron with a particular energy is likely to be found is called an **orbital.** The **principal quantum number, n,** tells the size of an orbital and largely determines its energy. The **angular momentum quantum number, l,** tells the shape of an orbital and the **magnetic quantum number, m_l,** describes the position in space of the orbital. A fourth quantum number, the **spin quantum number, m_s,** and the **Pauli exclusion principle** must be added to the three quantum numbers from solutions to the Schrödinger equation to explain the details of atomic spectra and complete the description of electrons in an atom. According to the Pauli exclusion principle, no two electrons in an atom can have all four quantum numbers alike. As a result of the Pauli exclusion principle, only two electrons can occupy an orbital, and the two electrons must have opposite spins. An orbital occupied by two electrons with opposite spins is filled.

Charge density pictures, **radial distribution functions,** and **boundary surfaces** show where the electron density is greatest in an orbital. Electrons in s orbitals **penetrate** closer to the nucleus than electrons in p orbitals.

ADDITIONAL PRACTICE PROBLEMS

For information about the organization of Additional Practice Problems, Stop & Test Yourself, Putting Things Together, and Applications, see the beginnings of these sections in Chapter 1.

7.19 Describe how Thomson measured the mass-to-charge ratio of the electron. (7.1)

7.20 Summarize Millikan's method for determining the charge on an electron. (7.2)

7.21 In your own words explain how (a) Marsden's experiment disproved Thomson's model of the atom, (b) Moseley determined nuclear charges. (7.3)

7.22 Sketch a wave and use the sketch to define the following terms: (a) cycle, (b) wavelength, (c) amplitude, (d) node. (7.5)

7.23 (a) Convert a wavelength of 6.3 pm to m.   (b) Convert a frequency of 8.91 MHz to Hz. (7.6)

7.24 What is the frequency of electromagnetic radiation that has a wavelength of 3.56×10^3 m? (7.6)

7.25 What is the wavelength of electromagnetic radiation that has a frequency of 5.72×10^{21} s^{-1}? (7.6)

7.26 If a wave has a wavelength of 0.34 m and a frequency of 0.75 s^{-1}, (a) what is the speed of the wave?   (b) Is the wave electromagnetic radiation? Explain your answer. (7.6)

7.27 (a) Which has the longer wavelength, X-rays or ultraviolet light?   (b) Which has the higher frequency, red light or green light? (7.6)

7.28 What is the color of a leaf that absorbs blue, green, and violet light and reflects yellow, orange, and red light? Explain your answer. (7.6)

7.29 (a) What is a photon? (b) Explain how the photoelectric effect suggests that photons exist. (7.6)

7.30 What is the energy (in joules) of a photon of electromagnetic radiation that has a wavelength of 5.75 μm? (7.6)

7.31 What is the energy (in kilojoules) of 57 photons of electromagnetic radiation that has a frequency of 6.83×10^{19} Hz? (7.6)

7.32 (a) Distinguish between a continuous spectrum and a line spectrum. (b) What conclusion is drawn from the observation that the emission and absorption spectra of atoms are line spectra? (7.7)

7.33 What is meant by saying that the energy of the electron in a hydrogen atom is quantized? (7.7)

7.34 (a) Explain what happens in an atom when the atom absorbs electromagnetic radiation. (b) Explain the process of emission of light by an atom. (7.7)

7.35 According to the Bohr model, is electromagnetic radiation emitted or absorbed when the electron in a hydrogen atom undergoes each of the following transitions? (a) from $n = 1$ to $n = 2$ (b) from the orbit with radius = 476.1 pm to the orbit with radius = 211.6 pm (c) from $n = 4$ to $n = 3$ (7.7)

7.36 Which of the transitions in Problem 7.35 involves the largest quantity of energy? (7.7)

7.37 Compare the emission spectrum of an unidentified sample given in this problem with the emission spectra for Zn, H, He, and Na in Figure 7.22. Is the unidentified sample Zn, H, He, or Na? (7.7)

7.38 Compare the absorption spectrum of the unidentified sample below with the emission spectra for Zn, H, He, and Na in Figure 7.22. Is this unidentified sample Zn, H, He, or Na? (7.7)

7.39 The wavelength of the yellow line in the emission spectrum of helium is 588 nm. What is the energy difference between the two energy levels involved in kJ/mol? (7.7)

7.40 Explain why (a) Bohr's model of the hydrogen atom was important, (b) Bohr's model is no longer used. (7.7)

7.41 Draw a line that is 6.00 cm long. (a) Using the ends of this line as the ends of the standing wave, draw a standing wave that has two nodes in addition to the nodes at the ends. (b) What is the wavelength of the standing wave in part (a)? (c) How many waves of wavelength 1.00 cm will fit between the ends of the line? (7.8)

7.42 Summarize de Broglie's contribution to our understanding of atoms. (7.8)

7.43 State the Heisenberg uncertainty principle in your own words. (7.9)

7.44 How does the product of the uncertainty in the position and velocity of an object change as the mass of the object increases? (7.9)

7.45 (a) Define *wave function*. (b) What is the physical significance of the square of the wave function? (c) What is an *orbital*, and how does it differ from an orbit? (7.10)

7.46 According to Schrödinger, what does each of the following quantum numbers describe? (a) principal quantum number (b) angular momentum quantum number (c) magnetic quantum number (7.10)

7.47 How many orbitals are there in each of the following subshells? (a) $3d$ (b) $2s$ (c) $4f$ (d) $5p$ (7.10)

7.48 (a) How many electrons can occupy an orbital? (b) How many electrons can occupy each of the subshells in Problem 7.47? (c) What is the total number of electrons that can occupy the fourth shell? (d) What is the total number of electrons in an atom that has the first two shells filled? (7.10)

7.49 (a) What is the first shell in which a g subshell can occur? (b) How many electrons can occupy a g subshell? (7.10)

7.50 According to the quantum mechanical model, do atoms really have definite surfaces like the surfaces of space-filling models of atoms? (7.11)

7.51 In what ways are all s orbitals alike, and in what ways are s orbitals that have different principal quantum numbers (for example, $1s$, $2s$, and $3s$ orbitals) different? (7.11)

7.52 (a) Sketch boundary surface diagrams of the three $2p$ orbitals showing clearly their shapes and directions in space. (b) What is meant by saying that a $2s$ orbital penetrates more to the nucleus than a $2p$ orbital? (7.11)

7.53 (a) State the Pauli exclusion principle in your own words. (b) What is an important result of this principle? (7.12)

7.54 A hyperactive small child is put in a circular playpen. Suppose you observed the child's location at 1-min intervals for 30 min. (a) Make a diagram showing the child's position at the time of each observation with a black dot. (b) Your diagram probably looks like the charge cloud for an atomic orbital. Which one? (c) No analogy is perfect. In what ways is the child's location in the playpen *not* like the position of an electron in an atomic orbital?

7.55 How many photons of electromagnetic radiation of frequency 4.56 MHz have a total energy of 5.4 J?

7.56 How many photons of electromagnetic radiation of wavelength 6.2 km have a total energy of 6.32×10^{-18} J?

7.57 A minimum frequency of 5.56×10^{14} Hz is needed for observation of a photoelectric effect with potassium metal. (a) What color is light of this frequency? (b) With which colors of light will a photoelectric effect be observed for potassium? (c) With which colors of light will a photoelectric effect *not* be observed for potassium? (d) What is the minimum energy needed to knock an electron out of an atom of potassium metal?

7.58 Figure 7.30(a) shows that the charge density of a $1s$ orbital is greatest at the nucleus. Explain why the radial probability [Figure 7.30(c)] is zero at the nucleus and is at a maximum at about 50 pm from the nucleus.

7.59 How does the uncertainty in measuring the position and speed of submicroscopic particles such as electrons differ from ordinary uncertainty in measurements?

7.60 What is the wavelength of a neutron moving at a speed of 2200 m/s (zeros are significant)?

7.61 In his first published work on the unit of charge, Millikan used water drops instead of oil drops in a much simpler apparatus. (Millikan, R. A. *Phil. Mag. Series 6* **19,** *110,* Feb. 1910, p 209) Results of this earlier experiment are summarized in the table:

Determination of Charge on Electron	
Expt No.	Charge, C
1	4.60×10^{-19}
2	6.09×10^{-19}
3	3.10×10^{-19}
4	8.06×10^{-19}
5	3.25×10^{-19}
6	9.40×10^{-19}

(a) Can the smallest charge obtained be the unit of charge? Explain your answer. (b) From these data, what is the unit of charge? (c) How many units of charge does the drop in experiment 1 have?

STOP & TEST YOURSELF

1. Of photons of electromagnetic radiation with the wavelengths below, which has the highest frequency?
(a) 321 nm (b) 21.6 nm (c) 436 nm (d) 6.32 nm (e) 896 nm

2. Of photons of electromagnetic radiation with the frequencies below, which has the highest energy?
(a) 1.091×10^{10} Hz (b) 1.091×10^{11} Hz
(c) 1.091×10^{-11} Hz (d) 2.182×10^{11} Hz
(e) 1.091×10^{12} Hz

3. If the frequency of electromagnetic radiation is 1.29×10^{15} Hz, what is the wavelength in meters?
(a) 2.32×10^{23} (b) 2.32×10^{-7} (c) 3.87×10^{23}
(d) 4.31×10^{6} (e) 4.31×10^{8}

4. If the wavelength of electromagnetic radiation is 3.4 μm, what is the frequency in Hz?
(a) 1.1×10^{-14} (b) 1.0×10^{3} (c) 88 (d) 8.8×10^{13}
(e) 8.8×10^{15}

5. What is the energy in joules of a photon of electromagnetic radiation of wavelength 4.50×10^{-7} m?
(a) 9.94×10^{-49} (b) 4.41×10^{-19} (c) 2.98×10^{-40}
(d) 2.04×10^{35} (e) 135

6. How many photons of electromagnetic radiation of frequency 2.17×10^{18} Hz have a total energy equal to 4.73×10^{-14} J?
(a) 32.9 (b) 1.44×10^{3} (c) 33 (d) 5.17×10^{29} (e) 30

7. Which of the following types of electromagnetic radiation has the longest wavelength?
(a) infrared (b) microwaves (c) radio waves
(d) ultraviolet light (e) X-rays

8. Electromagnetic radiation with wavelength 550 nm (a) has a higher frequency than radiation with wavelength 450 nm. (b) is in the ultraviolet region of the electromagnetic spectrum. (c) has a higher speed in vacuum than does radiation of wavelength 450 nm. (d) has a greater energy per photon than does

radiation with wavelength 450 nm. (e) has a lower frequency than radiation with wavelength 450 nm.

9. Schrödinger
(a) measured the charge-to-mass ratio of cathode ray particles. (b) measured the charge on the electron. (c) proposed the nuclear model of the atom to explain the scattering of alpha particles by thin metal foils. (d) suggested that the energy of the electron is quantized. (e) proposed the theory that we call wave mechanics or quantum mechanics.

10. Light is given off by a sodium vapor streetlight when (a) electrons move from a given energy level to a higher energy level. (b) electrons are removed from atoms and cations are formed. (c) electrons move from a given energy level to a lower energy level. (d) electrons are added to atoms and anions are formed. (e) atoms combine to form molecules.

11. The charge cloud shown represents an orbital indicated by
(a) $2p_x$ (b) $1s$ (c) $3p_x$ (d) $3s$ (e) $2s$

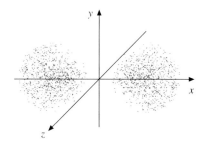

12. In the $3d$ subshell the maximum possible number of electrons is _____.
(a) 2 (b) 6 (c) 10 (d) 14 (e) 18

13. In the second shell the maximum possible number of electrons is _____.

(a) 2 (b) 4 (c) 6 (d) 8 (e) 10

14. Which quantum number is shared by all the orbitals in a shell? (a) principal (b) angular momentum (c) magnetic (d) spin (e) none of the preceding

PUTTING THINGS TOGETHER

7.62 Match each of the scientists in the lefthand column with their contribution to our understanding of atomic structure in the righthand column.

Bohr

de Broglie

Heisenberg

Millikan

Moseley
Pauli

Rutherford

Schrödinger

Thomson

1. concluded that atoms are composed of a nucleus and electrons.
2. suggested that both the velocity and the position of small particles can't be known exactly.
3. is credited with discovery of the electron.
4. determined charge and mass of electron.
5. determined the charges on nuclei.
6. proposed that energies of electrons in atoms are quantized.
7. proposed that there is only room for two electrons in an orbital.
8. suggested that electrons behave like waves.
9. wrote an equation that describes electrons in atoms.

7.63 Distinguish between (a) *wavelength* and *frequency,* (b) *interference* and *diffraction,* (c) *traveling waves* and *standing waves,* (d) *ground state* and *excited state,* (e) an *orbit* and an *orbital.*

7.64 Classify each behavior as wavelike or particlelike: (a) An electron moves from the positive lobe of a *p* orbital to the negative lobe. (b) The path of an electron is bent by a magnet. (c) Electrons form diffraction patterns.

7.65 What two questions were raised by Rutherford's nuclear model for the atom?

7.66 Discuss the similarities and differences between traveling waves and standing waves.

7.67 Consider the first four energy levels of the hydrogen atom shown in Figure 7.23(b). (a) How many spectral lines result from transitions between these four energy levels? (b) Which transition will give the highest frequency radiation? (c) Which transition will give the longest wavelength radiation?

7.68 One mole of photons is called an einstein. For yellow light of wavelength 560 nm, how many kilojoules does one einstein represent?

7.69 According to the Bohr model, when an electron in a hydrogen atom falls from the orbit for which $n = 2$ to the orbit with $n = 1$, (a) what is the energy emitted in kJ/mol? (b) What is the wavelength of the radiation that is emitted? (c) What is the frequency of the radiation that is emitted? (d) In

what region of the electromagnetic spectrum is the emitted radiation? Explain your answer.

7.70 Consider the two orbitals shown by the boundary surface diagrams below:

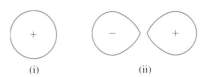

(i)   (ii)

(a) What is the maximum number of electrons that can be in orbital (i)? In orbital (ii)? (b) What is the minimum principal quantum number, *n*, for an orbital such as (ii)? (c) How many orbitals such as (ii) can have the same principal quantum number?

7.71 In equation 7.3, *n* is the higher energy orbit of the electron, and 2 is its lower energy orbit. Suitably modify equation 7.3 and use the modified equation to calculate the ionization energy of the electron in the hydrogen atom.

7.72 A line in the spectrum of the hydrogen atom, wavelength = 397 nm, results from the electron falling from the $n = 7$ to the $n = 2$ level. (a) In what region of the electromagnetic spectrum does this line occur? (b) Calculate the energy in kilojoules per mole of photons of electromagnetic radiation of wavelength 397 nm. (c) Make a graph like the one in Sample Problem 7.4 showing the energy levels from $n = 2$ to $n = \infty$ that includes the $n = 7$ level.

7.73 Compare the behavior of very small particles such as electrons with the behavior of everyday-size objects such as baseballs.

7.74 The minimum energy needed to force an electron out of an atom of rubidium is 3.46×10^{-19} J. (a) What is the kinetic energy of the electron emitted if light of the following wavelengths shines on rubidium? (i) 675 nm (ii) 425 nm? (b) What is the maximum speed of the electron emitted? [kinetic energy = (1/2)mass · speed²] (c) Compare what will happen if the light is dim with what will happen if the light is bright.

7.75 If Marsden had used electrons instead of alpha particles to investigate the structure of the nucleus, what would he have observed?

7.76 (a) Some wavelengths used to be given in angstrom units. By definition 1 angstrom unit, Å $= 10^{-8}$ cm. What is the wavelength, 7600 Å (zeros significant), expressed in nanometers? (b) Other wavelengths used to be given in micrometers. What is the wavelength, 3.40 μm, expressed in nanometers?

7.77 The minimum energy needed to knock an electron out of an atom in a metal is called the work function. The work functions for a number of elements are given in the table:

Element	Work Function, J × 10^{19}
Al	6.86
Cs	3.43
Li	4.6
Mg	5.86
Na	4.40
Rb	3.46

Assuming that the data in the table are typical, (a) how does the value of the work function change down a group in the periodic table? (b) How does the value of the work function change across a row in the periodic table? (c) What is the minimum energy needed to knock an electron off an atom of magnesium metal? (d) In what region of the electromagnetic spectrum is the energy in part (c)?

7.78 For the determination of the minimum energies required to remove electrons from metal atoms, a photocell evacuated to 10^{-9} to 10^{-10} torr is used. If the laboratory temperature is 20.0 °C, how many molecules of gas are there in 1 (exact number) cm^3 inside the photocell?

7.79 The proton has a mass of 1.0073 u, a diameter of the order of 1×10^{-15} m, and a charge of $+1.602 \times 10^{-19}$ C. (a) Assuming that the proton is spherical, what is the density of the proton in g/cm^3? (See Appendix B.7 for the formula for calculating the volume of a sphere.) (b) What is the charge-to-mass ratio of the proton in C/kg? (c) The neutron has a mass of 1.0087 u. What is the charge-to-mass ratio of the neutron in C/kg?

7.80 (a) Of what three types of particles are atoms composed? (b) Compare the masses and charges of these particles.

APPLICATIONS

7.81 Give one practical application of the wave properties of electrons.

7.82 Why does the gas in a fluorescent light bulb emit light?

7.83 A popular AM radio station broadcasts at a frequency of 650 kHz. What is the wavelength (in meters) of the electromagnetic radiation emitted by this station?

7.84 The compound known as Sunbrella, which is the active ingredient in some sunscreens, absorbs strongly around 266 nm. (a) What is the energy of a photon of electromagnetic radiation of this wavelength? (b) What is the frequency of the absorption? (c) In which region of the electromagnetic spectrum does Sunbrella absorb?

7.85 Arrange the following types of electromagnetic radiation in order of increasing energy per photon: (a) gamma rays from a nuclear reactor, (b) microwaves from an oven, (c) red-orange light from a neon sign, (d) ultraviolet light from a sunlamp, (e) X-rays used by a dentist to examine teeth.

7.86 Explain how the photoelectric effect could be used to operate a burglar alarm.

7.87 Lasers are devices that produce an intense beam of light in which all photons have identical wavelengths, are in phase, and travel in parallel paths. Lasers are used in microscopic surgery and welding, in surveying, and to make holograms (three-dimensional pictures), to name but a few of their many applications. Movements of continents of only a few centimeters per year have been measured with lasers on satellites. (a) Sketch a beam of laser light showing the paths of three photons. (b) On your sketch, label one wavelength. (c) Explain why laser beams are intense (bright).

7.88 Imagine that you are the editor of a magazine. The following statements appeared in an article scheduled for publication. Correct them. (a) Below 400 nm per second is infrared. This is heat. (b) Between 400 and 700 nm per second is ultraviolet light. Color, or the rainbow, falls into this category. (c) Light traveling more than 700 nm per second is gamma radiation—the stuff in science fiction movies.

7.89 (a) Compare X-rays and visible light with respect to: (i) speed in a vacuum, (ii) wavelength, (iii) frequency, (iv) energy per photon, (v) ability to penetrate flesh. (b) X-ray technicians wear lead aprons to protect them from the harmful effects of overexposure to X-rays. What conclusion can you draw about the ability of lead to transmit electromagnetic radiation of X-ray frequencies?

7.90 Light from distant stars is very weak, and astronomers often need to detect just a few photons. What is the total energy of 23 photons of light having a wavelength of 575 nm?

7.91 A ruby laser, one of the earliest types of laser, "lases" at 694.3 nm. If a ruby laser emits 4.50×10^{22} photons, what is the total energy emitted by the laser (in kilojoules)?

7.92 Orange-colored lenses are used in sunglasses that allow airplane pilots and air traffic controllers to see better in bright light, haze, and fog. What colors of light do these lenses absorb?

7.93 Almost the only means of learning about the stars is through analysis of the electromagnetic radiation they emit. Astronomers are concerned with wavelengths from about 20 to 10^{-7} cm. With what regions of electromagnetic radiation are astronomers concerned?

7.94 Cesium metal is used in the photocells of automatic door openers. The minimum energy needed to knock an electron out of an atom of cesium is 3.43×10^{-19} J. (a) What is the maximum wavelength of light (in nanometers) that can be used to operate an automatic door opener with a cesium photocell? (b) What colors of light could be used? (c) Why does the cesium, which is very reactive, not react with the oxygen and water vapor in the air? (*Hint:* See Figure 7.19.)

7.95 The intensity of the light that enters the pupil of the eye must be at least 1.5×10^{-11} J·m^{-2}·s^{-1} for objects to be visible. If the diameter of the pupil is 0.6 cm, what is the minimum rate (in photons per second) at which photons must enter the eye in order for a person to be able to see? Assume that the pupil is circular and that the wavelength of the light is 550 nm, the average wavelength of visible light. (See Appendix B.7 for formula for calculating the area of a circle.)

7.96 Explain why photographic film can be handled in a "dark-room" lighted with red light but is ruined if it is exposed to the same intensity of yellow light.

7.97 The light-sensitive substance on most black-and-white photographic films is silver bromide, AgBr. The energy required to darken silver bromide is 100.0 kJ·mol^{-1}. (a) Calculate the energy (in joules) that a photon must have to darken a unit of silver bromide. (b) Explain the fact that light from a match can expose photographic film whereas the signal (radiation) from a TV station cannot.

7.98 Newborn babies usually cry at about 450 Hz. (a) Explain why the wavelength of a baby's cry cannot be calculated from the equation $\lambda v = c$. (b) What information is needed to calculate the wavelength of a baby's cry?

7.99 A new detector that uses atomic emission can be used to determine quantitatively picogram quantities of almost any element in a wide variety of samples including soil samples from chemical dumps, feedstocks from oil refineries, and body fluids. A wavelength of 180.7 nm is used for analysis for sulfur. (a) In which region of the electromagnetic spectrum is this wavelength? (b) What is the energy of a photon of this radiation in joules?

7.100 In 1967 the second was redefined as 9 192 631 770 cycles of radiation resulting from an energy-level change in an atom of cesium-133. This particular band was chosen to make the new second as close as possible to the astronomically determined second (1/31 556 925.974 of the length of the year 1900), which had been used previously. What is the difference in energy between the two energy levels in cesium-133?

7.101 In 1993 an unusually bright supernova was observed. As astronomers watched, its spectrum changed from the upper spectrum to the lower spectrum. (a) What element was present at first? (b) What element was present later?

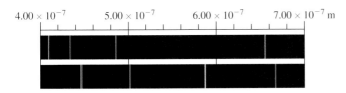

7.102 Chemical changes brought about by light are called photochemical reactions. The photochemical reaction

$$O_3(g) \xrightarrow[\text{radiation}]{\text{electromagnetic}} O_2(g) + O(g)$$

which takes place in the stratosphere, protects Earth from harmful radiation. (a) What is $\Delta H^\circ_{\text{rxn}}$ for this reaction? (b) What is the minimum frequency of photons that have this energy?

7.103 In a microwave oven, radiation is absorbed by water in the food and the food is heated. How many photons of wavelength 4.00 mm are required to heat 250.0 g of water (about a cup) from faucet temperature of 10.6 to 100.0 °C (the boiling point of water)? Assume that all the energy from the radiation is used to heat the water.

7.104 The closest that Pluto, the outermost planet, comes to Earth is 5.8022×10^9 km. How long does it take light from Pluto to reach telescopes on Earth? (Assume that the space between Pluto and Earth is a vacuum and express time in the usual units for the length of time required.)

7.105 The distance from an observation point on Earth to a mirror placed on the moon by the Apollo astronauts can be determined by bouncing laser signals off the mirror and measuring the time required for the signal to go to the moon and return. (a) If 2.564 s are required, how far is the moon from Earth (in kilometers)? (b) By this method, distance can be measured to ±2.5 cm (1 in.). To how many significant figures can the distance to the moon be measured?

7.106 (a) The sun and Earth radiate most of their energy at 0.5 μm and 10 μm, respectively. (a) What is the wavelength of the sun's radiation in units such that the number is between 1 and 999? (b) In which regions of the electromagnetic spectrum do the sun and Earth radiate?

7.107 The lenses of sunglasses that darken in light contain a little AgCl(s). Light causes the reaction AgCl(s) ⟶ Ag(s) + Cl(g) to take place, and the silver atoms darken the glass. The chlorine atoms cannot escape from the glass; therefore, the reverse reaction takes place in the absence of light. If 310 kJ/mol of energy are needed to make reaction take place, what is the maximum wavelength of electromagnetic radiation that has enough energy to cause reaction to occur?

7.108 In 1960 the meter was defined as 1 650 763.73 wavelengths of a line in the spectrum of krypton-86. (a) What is the wavelength of this line in nanometers? (b) What color is the line? (c) Write the symbol for krypton-86. (d) In an atom of krypton-86, how many protons, neutrons, and electrons are there?

7.109 The lasers in light shows are usually argon-ion and krypton-ion lasers. The ions are formed by passing an electric discharge through the gas at about 1 torr. The Ar$^+$ ion has two strong emissions at 488 and 514 nm. (a) What is the color of these emissions? (b) What are their frequencies? (c) In an Ar$^+$ ion, how many protons are there? How many electrons? (d) To four significant figures, what is a pressure of 1 torr in mmHg? In atm? In Pa?

Chemistry and the Stars

SALLIE BALIUNAS

Astrophysicist
Harvard-Smithsonian Center for
Astrophysics

B. A. Astronomy
Villanova University

A. M., Ph.D. Astronomy
Harvard University

A tabloid newspaper might have run the headline "Chemistry Stops Girl's Moon Journey." Here's how it happened. As a five year old, I caught the "mission to the Moon" fever. Humanity was going to the Moon. The next step would be Mars, then the stars! And I was angling to get in on the adventure.

I soon learned that only *boys* could become astronauts because they were pilots with many hours of flying time in advanced jet aircraft. I chose a backup career in aeronautical engineering, a field where women did work. Then along came tenth-grade chemistry. I remember Chapter 7, "Spectroscopy," in our chemistry textbook. Imagine

studying the spectrum of a gas and being able to deduce physical properties about the gas, for example, the speed of its atoms and the relative abundances of different elements. The gas need not be present in the lab to infer its properties, only the spectrum. Thus, spectroscopy could be used as a remote-sensing device—distant stars and galaxies could yield their physical properties in our terrestrial laboratories where the spectra could be collected and analyzed. Here at last was a way to study space *without leaving Earth*.

Chemistry also taught me about a fabulous scientist, Marie Curie. She was the first person to be awarded two Nobel prizes (one in chemistry, one in physics). Her example taught me that a woman could become a research scientist.

The preparation for my job as an astrophysicist at the Smithsonian Institution wasn't easy (many years of study were required), but the end result is *fun*. The Smithsonian has several research labs (in addition to its museums), and I work at its Astrophysical Observatory in Cambridge, Massachusetts. I study the changes in the sun's magnetic activity (the 11-year sunspot cycle) by observing the magnetic phenomena of nearby stars and contrasting their behavior to the sun's. The sun's changing magnetism seems to be connected to its total energy output, which, in turn, could affect Earth's climate. Although my work focuses on understanding the complicated engine that produces solar magnetism, it may also help us understand global change.

My research tools are telescopes both on the ground and in space. I use satellite observatories to do spectroscopy of the ultraviolet radiation from nearby stars. The ultraviolet spectrum is rich in emission lines that detail the behavior of magnetic activity and hot plasma on these stars. This information must be acquired above Earth's atmosphere, which blocks the ultraviolet light.

From Earth, I study the spectrum of singly ionized calcium in stars. The spectrum of starlight contains a pair of emission lines, the "H" and "K" doublet near 393 nm, which changes in intensity as the number of magnetic "starspots" waxes and wanes.

In this way, I can see changes in the surface magnetic activity of stars without having to resolve their surfaces (and without traveling there!).

I have been working recently in a program concerned with the search for extra-terrestrial intelligence (ETI). We are seeking evidence of extraterrestrial civilizations by listening for their radio communications. However, such a technically advanced civilization took over four billion years to develop in our solar system, so we should be targeting star systems at least as old as ours. I am determining the ages of over a thousand stars in the solar neighborhood in order to identify the stars old enough to have ETI. Because the H and K lines of calcium weaken with age, their intensities tell us how old the stars are.

My research has taken me from arid mountaintops (where most observatories are located) to satellite control stations at NASA labs. In the course of my research, I have run into hungry bears, poisonous rattlesnakes, and deranged computers!

I like the thrill of working on the edge of the unknown. Because the arena of scientific exploration is serendipitous, an experiment is always *slightly* out of control. Not only the unexpected, but the *unexpectable* may happen—and sometimes does—whether or not I am prepared for it. Scientific experimentation is often depicted as a character in a lab coat shouting "Eureka" after the lab blows up. Although a bit exaggerated, this element of science is real and adds adventure to the quest.

Beyond research, my job has another enjoyable function that is very important—communication. If the results of an experiment are never explained to others, then the experiment might as well not have been conducted in the first place!

Now—what does your everyday astrophysicist do in her spare moments? I give older cars a new life as modern hot rods. I've rebuilt a 1933 Ford, a 1957 Chevy, and a 1970 Corvette. At any time, I'm overhauling or upgrading at least one car. This is as close to speeding through the universe as I've come!

ELECTRONIC STRUCTURE AND THE PERIODIC TABLE

8

Shells are composed largely of calcium carbonate.

T his chapter tells you how to show the distribution of electrons in the orbitals of atoms. It relates electronic structures, the periodic table, and some key properties of atoms.

— *Why is a study of electronic structure and the periodic table a part of general chemistry?*

The original periodic table was based on observations of chemical reactions and physical properties. The electronic structure of atoms explains the form of the periodic table.

— *What do you already know about electronic structure and the periodic table?*

You know that the electrons move about rapidly in the space around the nucleus. They are arranged in shells of varying energy and average distance from the nucleus. Because only certain values are allowed for quantum numbers, only the following subshells are possible:

$$1s$$
$$2s, 2p$$
$$3s, 3p, 3d$$
$$4s, 4p, 4d, 4f$$

and so on. There is one orbital in each *s* subshell, three orbitals in each *p* subshell, five in each *d* subshell, and seven in each *f* subshell. The number of orbitals in subshells is given by the series of odd numbers 1, 3, 5, 7, ... Each orbital can hold two electrons that have opposite spins. The higher the principal quantum number, the higher the energy of a shell, although the difference in energy between successive shells becomes small at high principal quantum numbers [remember Figure 7.23(b)]. The distance of a shell from the nucleus increases rapidly as the principal quantum number increases [remember Figure 7.23(a)].

You know that in the periodic table the elements are arranged in order of increasing atomic number and that the atomic number tells the number of electrons around the nucleus of an atom. Thus, the elements are arranged in the periodic table in order of increasing number of electrons. The periodic table is constructed so that elements in a vertical column have similar properties. The vertical columns in the periodic table are called groups, and the horizontal rows are called periods. There is a gradual change in properties going from the top to the bottom of a group or across a period from left to right. The elements in gray spaces in the periodic table inside the front cover are metals, the elements in purple spaces are nonmetals, and the elements in blue spaces are semimetals. The elements in Groups IIIB–IIB are called transition elements, and those in the lanthanide series and actinide series at the bottom of the table are called inner-transition elements. All other elements are called main group elements. Before reading any further in this chapter, you should review the introduction to the periodic table in Section 1.8 and the material on how the reactivity of the elements changes across rows and down groups in Section 4.5.

8.1   ELECTRON CONFIGURATIONS

The Hydrogen Atom

In the hydrogen atom, the value of the principal quantum number alone determines the energy of the orbitals. All subshells of a shell have the same energy, as shown in ▪Figure 8.1. In this figure, each line represents one orbital. An infinite number of energy levels exists. The energy levels get closer and closer together as n increases. Finally, they merge into a continuum. There is not room in Figure 8.1 to show any more energy levels.

If more than 1312 kJ of energy is added to a mole of hydrogen atoms in the ground state, the lone electron is removed from each hydrogen atom. The hydrogen atoms become hydrogen ions.

The $1s$ orbital is smallest and closest to the nucleus. In the ground state of the hydrogen atom, the lowest energy state, the electron is in the $1s$ orbital, the lowest energy orbital. In orbital energy level diagrams, electrons are represented by arrows. The first arrow in an orbital is usually drawn pointing up representing an electron with spin up. Figure 8.1 shows the energy level diagram for the hydrogen atom in its ground state.

Drawing an energy level diagram is time-consuming. The *way the electrons are distributed among the atomic orbitals of an atom* can be *shown* more easily *by* writing the **electron configuration** of the atom. To write an electron configuration, write the principal quantum number of the main shell and the letter for the subshell. Show the number of electrons in the subshell by a right superscript. For example, the electron configuration for the hydrogen atom in its ground state is $1s^1$:

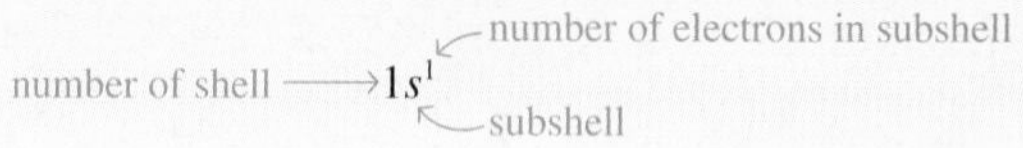

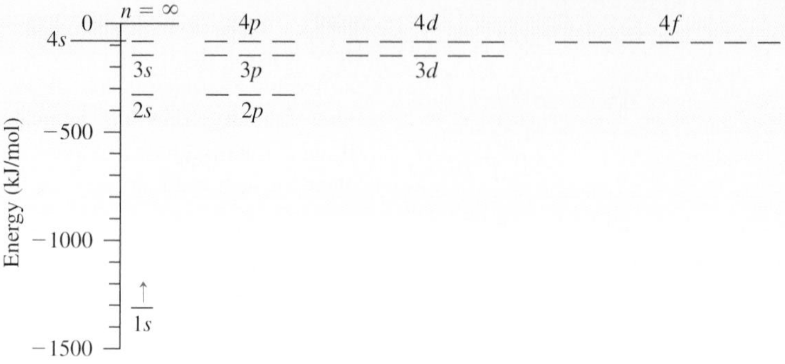

number of shell ⟶ $1s^1$ ⟵ number of electrons in subshell

⟵ subshell

▪ FIGURE 8.1   Energy level diagram for the hydrogen atom in its ground state.

This electron configuration means that the hydrogen atom's one electron is in the $1s$ orbital.

Any other electron configuration of the hydrogen atom is an excited state. For example, the electron configuration $2s^1$ (meaning that the hydrogen atom's one electron is in the $2s$ orbital) represents an excited state of the hydrogen atom. An atom has an infinite number of excited states.

PRACTICE PROBLEMS

8.1 What does the symbol $3s^2$ mean?

8.2 Write the electron configuration for another excited state of the hydrogen atom (*not* the one in the paragraph above).

Atoms with More Than One Electron

In atoms with more than one electron (that is, all atoms other than hydrogen), the orbital energy levels are assumed to be similar to the orbital energy levels in the hydrogen atom except that *the energy of orbitals in different subshells of the same shell is different.* Why is the energy of orbitals in different subshells of the same shell different in atoms with more than one electron? Each electron repels every other electron and is repelled by every other electron. The attractive force that each electron feels from the nucleus is affected by all the other electrons in the atom. Also, the *inner electrons screen* or **shield** *the outer electrons from the positive charge on the nucleus.* As a result the **effective nuclear charge**—that is, the *charge actually felt by the outer electrons*—is less than the actual nuclear charge. Because s electrons penetrate to the nucleus more than p electrons (see Section 7.11), they feel the attractive force of the nucleus more than p electrons. Therefore, the energy of s orbitals is lower than the energy of p orbitals of the same shell. ▪Figure 8.2 shows the relative orbital energy levels of hydrogen and some second-period elements.

In Figure 8.2, notice how the energies of similar orbitals—for example, $1s$ orbitals—decrease as atomic number increases. The energy of a $1s$ orbital in a fluorine atom, atomic number 9, is lower than the energy of a $1s$ orbital in a carbon atom, atomic number 6. The lower energy is not surprising. The nucleus of a fluorine atom contains three more protons than the nucleus of a carbon atom. As a result, the positive charge on the nucleus of a fluorine atom is greater than the positive charge on the nucleus of a carbon atom. Although a fluorine atom also contains three more electrons than a carbon atom, the three additional electrons are in the same subshell. Electrons in the same subshell do not shield each other very effectively from the nucleus. The effective nuclear charge of a fluorine atom, as well as its actual nuclear charge, is larger than the effective nuclear charge of a carbon atom.

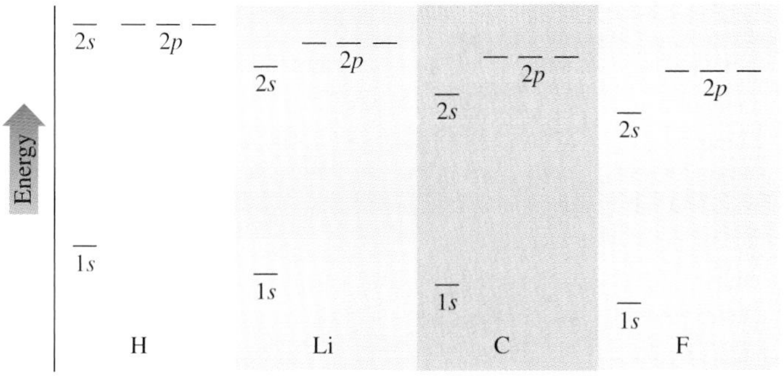

Remember that the Schrödinger equation has not been solved exactly for atoms containing more than one electron.

▪ FIGURE 8.2 Relative orbital energy levels of hydrogen and some second period elements (energies are not to scale).

The greater the effective nuclear charge, the more strongly the nucleus attracts the negatively charged electrons around it.

Just as s electrons penetrate closer to the nucleus than p electrons, electrons in p orbitals penetrate closer to the nucleus than electrons in d orbitals. The p electrons feel the attractive force of the nucleus more than d electrons do. The energy of p orbitals is lower than the energy of d orbitals of the same shell. Similarly, electrons in d orbitals penetrate closer to the nucleus than electrons in f orbitals. The d electrons feel the attractive force of the nucleus more. The energy of d orbitals is lower than the energy of f orbitals of the same shell. In other words, for orbitals with the same principal quantum number, the order of increasing energy is

$$s < p < d < f$$

The electron arrangements of the elements can be considered to be built up by putting electrons in energy-level diagrams like the ones in Figure 8.2. This *building up process* is called the **aufbau** approach. *Aufbau* is a German word that means "building up." The atomic number of the element tells how many electrons to place. For example, the atomic number of carbon is 6; there are six electrons around the nucleus of a carbon atom. To build up the electron configuration of the carbon atom, begin at the bottom of the diagram for the carbon atom in Figure 8.2. Put electrons in the lowest energy orbital available, the $1s$ orbital. Because an orbital can hold only two electrons and the $1s$ orbital is of much lower energy than the $2s$ orbital, two electrons go into the $1s$ orbital before any electrons go into the $2s$ orbital. Remember that two electrons in an orbital must have opposite spins, one up and one down. Only two electrons are allowed in the $1s$ orbital.

Moving up the energy-level diagram, the next orbital is the $2s$ orbital. Put two electrons with opposite spins in the $2s$ orbital. Four electrons have now been placed. A carbon atom has six electrons, so two electrons are left. These go into the $2p$ orbitals, the next orbitals in energy. The three $2p$ orbitals all have the same energy. *Orbitals that have the same energy* are called **degenerate orbitals.** Should we put both electrons in one $2p$ orbital or one in each of two $2p$ orbitals? Remember that an orbital is a volume in space where an electron with a particular energy is likely to be found. Electrons repel each other because they all have negative charges. Therefore, electrons get as far apart as possible. The two electrons should go into two different p orbitals.

The electron configuration of the carbon atom is written $1s^2 2s^2 2p^2$, which means "two $1s$, two $2s$, and two $2p$ electrons." If you want to show that the two $2p$ electrons are in different orbitals, you can write the electron configuration $1s^2 2s^2 2p_x^1 2p_y^1$. The subscript x indicates the p orbital that is symmetric around the x-axis. The subscript y indicates the p orbital that is symmetric around the y-axis. (Since the three $2p$ orbitals are degenerate, any two of the three could have been used. Thus, $2p_x^1 2p_z^1$ or $2p_y^1 2p_z^1$ would be equally acceptable electron configurations, although $2p_x^1 2p_y^1$ is customary.)

⬤ Remember that, according to the Pauli exclusion principle (Section 7.12), an orbital occupied by two electrons is filled. The two electrons must have opposite spins.

Energy level diagram for the carbon atom in its ground state. The numbers above the arrows representing the electrons show the order in which the electrons are placed in the orbitals by the aufbau process.

⬤ For a picture of a set of three p orbitals, see Figure 7.34.

PRACTICE PROBLEMS

8.3 The energy-level diagrams of all the atoms in the second period are similar to those shown in Figure 8.2. (a) Sketch the energy-level diagram including electrons for the nitrogen atom in its ground state. (b) Write the electron configuration for the ground state of the nitrogen atom.

8.4 (a) Sketch the energy-level diagram for an excited state of the beryllium atom. (b) Write the electron configuration for the excited state of the beryllium atom in part (a).

To save time and space and to focus attention on the outer electrons, **abbreviated electron configurations** are often used. The symbol of the nearest noble gas with lower atomic number is written in square brackets to represent the core electrons. The **core electrons** are *all the electrons in an atom of the nearest noble gas with lower atomic number.* For example, the abbreviated electron configuration for carbon is $[He]2s^22p^2$ or $[He]2s^22p_x^12p_y^1$. The symbol $[He]$ represents the two $1s$ electrons.

Another way of showing the arrangement of electrons that calls attention to the unpaired electrons is the **orbital diagram.** The orbital diagram for the carbon atom in its ground state is usually written

C ⇅ ⇅ ↑ ↑   or   C [He] ⇅ ↑ ↑
    $1s$ $2s$ $2p$ $2s$ $2p$

However, orbital diagrams such as

C [He] ⇅ ↑ ↓   or   C [He] ⇅ ↑ ↑
        $2s$ $2p$ $2s$ $2p$

would be equally acceptable. Orbital diagrams show clearly that there are two unpaired electrons in a ground-state carbon atom.

*See Campbell, M. L. J. Chem. Educ. **1991**, 68, 134–135 for the reason for omitting the common statement that the electrons should have the same spin from Hund's rule.*

PRACTICE PROBLEMS

8.5 Write abbreviated electron configurations for (a) lithium, atomic number 3, (b) oxygen, atomic number 8, and (c) neon, atomic number 10.

8.6 Write orbital diagrams for each of the atoms in Practice Problem 8.5 and tell how many unpaired electrons each atom has.

8.7 Write an orbital diagram for the excited state of the carbon atom that has the electron configuration $[He]2s^12p_x^12p_y^12p_z^1$.

Hund's Rule

The rule which says that *when putting electrons into orbitals with the same energy, you should put one electron in each orbital before putting two in any one* is called **Hund's rule.** Hund's rule is based on the results of measurements of magnetic properties. *Atoms with unpaired (odd) electrons are paramagnetic.* **Paramagnetic materials** are *weakly magnetized when brought near a magnet and are attracted into the magnetic field.* The magnetism of most paramagnetic materials disappears when the magnet is removed. *Atoms with all electrons paired are diamagnetic.* **Diamagnetic materials** are *pushed out of a magnetic field.* ▪Figure 8.3 shows the difference between a paramagnetic sample and a diamagnetic sample.

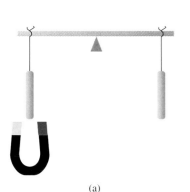

(a)

(b)

(c)

▪ FIGURE 8.3   (a) Electromagnet off. Samples balance. (b) Electromagnet on. Paramagnetic material is drawn into magnetic field. (c) Electromagnet on. Diamagnetic material is pushed out of magnetic field.

Core Period
1

																	0
								1 **H** $1s^1$									2 **He** $1s^2$ [He]

	IA	**IIA**											**IIIA**	**IVA**	**VA**	**VIA**	**VIIA**	
[He] 2	3 **Li** $2s^1$	4 **Be** $2s^2$											5 **B** $2s^22p^1$	6 **C** $2s^22p^2$	7 **N** $2s^22p^3$	8 **O** $2s^22p^4$	9 **F** $2s^22p^5$	10 **Ne** $2s^22p^6$ or [Ne]
[Ne] 3	11 **Na** $3s^1$	12 **Mg** $3s^2$	**IIIB**	**IVB**	**VB**	**VIB**	**VIIB**	(VIII)		**IB**	**IIB**		13 **Al** $3s^23p^1$	14 **Si** $3s^23p^2$	15 **P** $3s^23p^3$	16 **S** $3s^23p^4$	17 **Cl** $3s^23p^5$	18 **Ar** $3s^23p^6$ or [Ar]

	IA	**IIA**	**IIIB**	**IVB**	**VB**	**VIB**	**VIIB**	VIII			**IB**	**IIB**	**IIIA**	**IVA**	**VA**	**VIA**	**VIIA**	**0**
[Ar] 4	19 **K** $4s^1$	20 **Ca** $4s^2$	21 **Sc** $3d^14s^2$	22 **Ti** $3d^24s^2$	23 **V** $3d^34s^2$	24 **Cr** $3d^54s^1$	25 **Mn** $3d^54s^2$	26 **Fe** $3d^64s^2$	27 **Co** $3d^74s^2$	28 **Ni** $3d^84s^2$	29 **Cu** $3d^{10}4s^1$	30 **Zn** $3d^{10}4s^2$	31 **Ga** $3d^{10}4s^2 4p^1$	32 **Ge** $3d^{10}4s^2 4p^2$	33 **As** $3d^{10}4s^2 4p^3$	34 **Se** $3d^{10}4s^2 4p^4$	35 **Br** $3d^{10}4s^2 4p^5$	36 **Kr** $3d^{10}4s^2 4p^6$ or [Kr]
[Kr] 5	37 **Rb** $5s^1$	38 **Sr** $5s^2$	39 **Y** $4d^15s^2$	40 **Zr** $4d^25s^2$	41 **Nb** $4d^35s^2$	42 **Mo** $4d^55s^1$	43 **Tc** $4d^55s^2$	44 **Ru** $4d^75s^1$	45 **Rh** $4d^85s^1$	46 **Pd** $4d^{10}$	47 **Ag** $4d^{10}5s^1$	48 **Cd** $4d^{10}5s^2$	49 **In** $4d^{10}5s^2 5p^1$	50 **Sn** $4d^{10}5s^2 5p^2$	51 **Sb** $4d^{10}5s^2 5p^3$	52 **Te** $4d^{10}5s^2 5p^4$	53 **I** $4d^{10}5s^2 5p^5$	54 **Xe** $4d^{10}5s^2 5p^6$ or [Xe]
[Xe] 6	55 **Cs** $6s^1$	56 **Ba** $6s^2$ ∗	71 **Lu** $4f^{14}5d^1 6s^2$	72 **Hf** $4f^{14}5d^2 6s^2$	73 **Ta** $4f^{14}5d^3 6s^2$	74 **W** $4f^{14}5d^4 6s^2$	75 **Re** $4f^{14}5d^5 6s^2$	76 **Os** $4f^{14}5d^6 6s^2$	77 **Ir** $4f^{14}5d^7 6s^2$	78 **Pt** $4f^{14}5d^9 6s^1$	79 **Au** $4f^{14}5d^{10} 6s^1$	80 **Hg** $4f^{14}5d^{10} 6s^2$	81 **Tl** $4f^{14}5d^{10} 6s^26p^1$	82 **Pb** $4f^{14}5d^{10} 6s^26p^2$	83 **Bi** $4f^{14}5d^{10} 6s^26p^3$	84 **Po** $4f^{14}5d^{10} 6s^26p^4$	85 **At** $4f^{14}5d^{10} 6s^26p^5$	86 **Rn** $4f^{14}5d^{10} 6s^26p^6$ or [Rn]
[Rn] 7	87 **Fr** $7s^1$	88 **Ra** $7s^2$ ◆	103 **Lr** $5f^{14}6d^1 7s^2$	104 **Rf** $5f^{14}6d^2 7s^2$	105 **Ha** $5f^{14}6d^3 7s^2$	106 **Sg** $5f^{14}6d^4 7s^2$	107 **Ns** $5f^{14}6d^5 7s^2$	108 **Hs** $5f^{14}6d^6 7s^2$	109 **Mt** $5f^{14}6d^7 7s^2$	110	111							

Metal   Semimetal   Nonmetal

∗ **Lanthanide series**

[Xe]

57 **La** $5d^16s^2$	58 **Ce** $4f^26s^2$	59 **Pr** $4f^36s^2$	60 **Nd** $4f^46s^2$	61 **Pm** $4f^56s^2$	62 **Sm** $4f^66s^2$	63 **Eu** $4f^76s^2$	64 **Gd** $4f^75d^1 6s^2$	65 **Tb** $4f^96s^2$	66 **Dy** $4f^{10}6s^2$	67 **Ho** $4f^{11}6s^2$	68 **Er** $4f^{12}6s^2$	69 **Tm** $4f^{13}6s^2$	70 **Yb** $4f^{14}6s^2$

◆ **Actinide series**

[Rn]

89 **Ac** $6d^17s^2$	90 **Th** $6d^27s^2$	91 **Pa** $5f^26d^1 7s^2$	92 **U** $5f^36d^1 7s^2$	93 **Np** $5f^46d^1 7s^2$	94 **Pu** $5f^67s^2$	95 **Am** $5f^77s^2$	96 **Cm** $5f^76d^1 7s^2$	97 **Bk** $5f^97s^2$	98 **Cf** $5f^{10}7s^2$	99 **Es** $5f^{11}7s^2$	100 **Fm** $5f^{12}7s^2$	101 **Md** $5f^{13}7s^2$	102 **No** $5f^{14}7s^2$

- FIGURE 8.4 Periodic table showing electron configurations.

For the evidence in favor of placing lutetium and lawrencium instead of lanthanum and actinium in Group IIIB see Jensen, W. B. J. Chem. Educ. 1982, 59, 634–636. The outer-shell configurations for the lanthanides are from Scerri, E. R. J. Chem. Educ. 1991, 68, 122–126.

The magnetism of a paramagnetic species depends on the number of unpaired electrons. Using a sensitive analytical balance and a powerful electromagnet, the number of unpaired electrons in a species can be found by measuring the force produced on the sample by the magnetic field. A carbon atom in the ground state does indeed have two unpaired electrons.

8.2 THE PERIODIC TABLE AND ELECTRON CONFIGURATIONS

If we had to draw the energy-level diagram for an atom to write its electron configuration, writing electron configurations would be a lot of work. Fortunately, it is possible to write the electron configurations for most atoms from the periodic table. In -Figure 8.4, the abbreviated electron configurations for all the elements are shown in their spaces in the periodic table. Because the core is the same for all the elements in a period, the symbol for the core is shown at the beginning of each row. Most of the electron configurations were found by both experiment and calculation.

You may wonder why potassium and calcium have electrons in the $4s$ orbital instead of in the $3d$ orbital. Chemists differ in their explanations for why two electrons enter the $4s$ level before any electrons enter the $3d$ level. However, they all agree that potassium has electron configuration $[Ar]4s^1$ and calcium has electron configuration $[Ar]4s^2$.

Much of the experimental evidence for electron configurations comes from atomic spectra.

Electron configurations make clear the reason for the structure of the periodic table. *Hydrogen, helium, and the elements in Groups IA and IIA* are called **s-block elements;** the outer electrons of these elements are all in *s* subshells. The *elements in Groups IIIA–0 are called* **p-block elements.** Going across a row of the periodic table, the difference between *p*-block elements lies in the number of *p* electrons in the outer shell. As atomic number increases by one, the number of *p* electrons in the outer shell also increases by one. All *p*-block elements have filled *s* subshells in their outer shells. Some also have filled *d* and *f* subshells outside the core. The *s*-block and *p*-block elements are the elements that we have previously called main group elements.

*See Pilar, F. L. J. Chem. Educ. **1978,** 55, 2–6, Scerri, E. R. J. Chem. Educ. **1989,** 66, 481–483, and Vanquickenborne, L. G.; Pierloot, K.; Devoghel, D. J. Chem. Educ. **1994,** 71, 469–471 for expositions of the opposing points of view regarding the relative energies of 4s and 3d orbitals.*

The *elements in Groups IIIB–IIB* are called **d-block elements.** The *d*-block elements are the elements that we have been referring to as transition elements. In most cases, going across a row in the periodic table, the difference between *d*-block elements lies in the number of *d* electrons in the next-to-the-outermost or $(n - 1)$ shell. As atomic number increases by one, the number of *d* electrons in the $(n - 1)$ shell also increases by one. The *d-block elements in the fourth period (Sc–Zn) are* called the **first transition series.** The *d-block elements in the fifth period (Y–Cd)* are referred to as the **second transition series** and the *d-block elements in the sixth period (Lu–Hg) as the* **third transition series.**

Some chemists do not consider the elements of Group IIB to be transition elements.

If we consider the elements in order of increasing atomic number, the first element that does not have the electron configuration expected from its place in the periodic table is chromium (atomic number 24). From its position between vanadium $[Ar]3d^34s^2$ and manganese $[Ar]3d^54s^2$, chromium would be expected to have the electron configuration $[Ar]3d^44s^2$. Instead, chromium has the electron configuration $[Ar]3d^54s^1$. At the present time, chemists do not agree on a reason for the unexpected electron configuration of chromium. However, note that both the 3*d* and the 4*s* subshells in the chromium atom are half-filled. Molybdenum, copper, silver, and gold, which also have unexpected electron configurations, also have half-filled subshells.

The Instructor's Resource Manual includes a periodic table with elements having unexpected electron configurations highlighted to help you avoid accidentally using these elements for test questions.

Only in the case of palladium does the electron configuration predicted by position in the periodic table differ by more than one electron from the experimental electron configuration. The predicted configuration for palladium is $4d^85s^2$. The observed configuration is $4d^{10}5s^0$. Because orbitals with high principal quantum numbers have similar energies, the difference between the predicted and observed electron configurations for these elements is not important as far as their chemical properties and the chemical properties of their compounds are concerned. Chemical and physical properties are readily explained in terms of predicted electron configurations.

Students can reinforce their understanding of electron configurations in the Atoms section by building electron configurations with the aid of the periodic table.

PRACTICE PROBLEM

8.8 (a) From its position between nickel and zinc in the periodic table, what electron configuration would you predict for copper? (b) From Figure 8.4, what is the electron configuration of copper?

The *elements* in the two rows at the bottom of the periodic table, those *in the lanthanide and actinide series,* are called **f-block elements.** In most cases, the difference between *f*-block elements going across a row in the periodic table lies in the number of *f* electrons in the next-to-the-next-to-the-outermost or $(n - 2)$ shell. All *f*-block elements in a period have the same outer electron configuration, that is, $6s^2$ or $7s^2$. The *f*-block elements are sometimes called inner-transition elements because the difference between elements in the same period is one more shell in from the outside of the atom than the difference between transition elements.

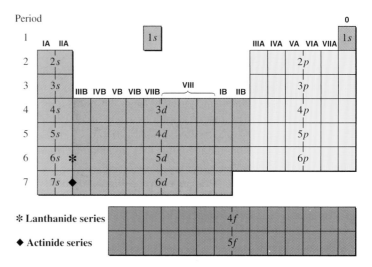

■ FIGURE 8.5   Periodic table showing orbitals being filled.

■Figure 8.5 shows the orbitals that are being filled for elements in different parts of the periodic table. ■Figure 8.6 shows the periodic table with the lanthanides and actinides or *f*-block elements in their proper place between the *s*-block elements and the *d*-block elements. We do not ordinarily use the expanded form of the periodic table because it is too wide.

Notice the general shape of the expanded form. Every two periods, a new block of elements follows the *s*-block elements. Each new block is wider than the ones before. The *s*-block is only two elements wide because there is only a single *s* orbital in each shell. A single *s* orbital can only contain two electrons. The first period is only two elements wide because the first shell only has an *s* orbital. The *p*-block is six elements wide because there are three *p* orbitals, each of which can contain two electrons in each shell. The *d*-block is 10 elements wide because there are five *d* orbitals that can contain a total of ten electrons in each shell that has a *d* subshell. The *f*-block is 14 elements wide because there are seven *f* orbitals that can contain a total of 14 electrons in each shell that has an *f* subshell.

PRACTICE PROBLEM

8.9   If more synthetic elements are made, for example, atomic numbers 112–121,
(a) predict the atomic number of the next element that will be in Group 0.
(b) Predict the atomic number of the first element in the new block of elements that will follow the *s*-block elements in the eighth period.

■ FIGURE 8.6   Expanded form of the periodic table.

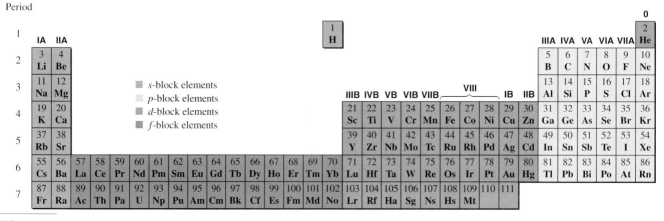

From Figure 8.4, you can see that all the elements in each column of the periodic table have similar outer electron configurations except for a few elements in Groups VIB and VIII.* For example, all members of Group IA have one *s* electron in their outermost shell, and all elements in Group IVA have four electrons—two *s* and two *p* electrons—in their outermost shell. *The physical and chemical properties of the elements and their compounds depend on the outer electron configuration. The members of a group have similar properties because they have the same outer electron configuration.*

For elements in A groups and Group 0, the number of the group is the same as the total number of *s* and *p* electrons outside the core. The noble gases (Group 0) are sometimes labeled VIIIA because, except in the case of helium, there are eight *s* and *p* electrons outside the core. For elements in Groups IIIB–VIIB and in the lefthand column of Group VIII, the number of the group is the same as the total number of *d* and *s* electrons outside the core. For Groups IB and IIB, the number of the group is the same as the total number of *s* electrons outside the core.

The energies of filled subshells are so low that the electrons in filled subshells aren't usually used in forming chemical bonds. Therefore, the chemistry of Ga is very similar to the chemistry of Al, the chemistry of Ge is very similar to the chemistry of Si, and so forth. Only the electrons of filled *s* subshells in the outermost shell such as the 3*s* electrons of Mg and the 2*s* electrons of C are commonly used to form chemical bonds.

When *f* electrons are present, they are buried so deep inside the atom that they do not play much part in forming chemical bonds. The properties of all the members of the lanthanide series, which all have an outer electron configuration of $6s^2$ but differing numbers of 4*f* electrons, are so similar that these elements usually occur together and are difficult to separate from each other.

 8.3 ## USING THE PERIODIC TABLE TO WRITE ELECTRON CONFIGURATIONS

Atoms

The periodic table is the best memory aid for writing electron configurations. From Figure 8.4, you can see that the principal quantum number, *n*, of the outer electrons (the *s* and *p* electrons) is always the same as the number of the period. Thus, the outer electrons of fluorine, which is in the second period, are in the second shell; the outer electrons of radium, which is in the seventh period, are in the seventh shell. The principal quantum number of *d* electrons outside the core is $(n - 1)$. For example, nickel is in the fourth period. Its *d* electrons are in the third shell. The principal quantum number of *f* electrons outside the core is $(n - 2)$. Cerium is in the sixth period. Its *f* electrons are in the fourth shell.

To write the ground-state electron configuration of an element using the periodic table as a memory aid:

1. Begin with hydrogen.
2. Go through the elements in order of increasing atomic number.
3. As you move across a period:
   a. Add electrons to the *ns* orbital as you pass through Groups IA and IIA.
   b. Add electrons to the *np* orbitals as you pass through Groups IIIA to 0.

*The outer electron configurations of W and Sg in Group VIB [$(n - 1)d^4ns^2$ where *n* is the principal quantum number] are different from the outer electron configurations of Cr and Mo [$(n - 1)d^5ns^1$].

c. Add electrons to $(n - 1)d$ orbitals as you pass through the Groups IIIB to IIB and add electrons to $(n - 2)f$ orbitals as you pass through the f-block elements.

4. Continue until you reach the element whose electron configuration you are writing.

Sample Problems 8.1–8.3 show how to write electron configurations for atoms from the periodic table.

SAMPLE PROBLEMS

8.1 The atomic number of sulfur is 16. For the sulfur atom in its ground state, (a) write the complete electron configuration, (b) write the abbreviated electron configuration, and (c) give the orbital diagram.

Solution (a) The fact that the atomic number of sulfur is 16 tells us that a sulfur atom has 16 electrons.

Refer to a periodic table and begin with hydrogen (atomic number 1). Going across the first row of the periodic table, put two electrons in the $1s$ orbital. Then go to the left end of the second row. Put the third and fourth electrons in the $2s$ orbital. Continuing across the second row, put the next six electrons in the $2p$ orbitals. Go to the third row and put two electrons in the $3s$ orbital. So far we have put 12 electrons in orbitals. There are four more electrons to place. Continuing across the third row, follow Hund's rule and place one electron in each of the three $3p$ orbitals. The 16th and last electron must go into one of the $3p$ orbitals that already contains one electron. The electron configuration of the ground state of the sulfur atom is

$$1s^2 2s^2 2p^6 3s^2 3p_x^2 3p_y^1 3p_z^1$$

The sum of the right superscripts must equal the number of electrons in the atom. Always add the superscripts to check your work:

$$2 + 2 + 6 + 2 + 2 + 1 + 1 = 16$$

(b) Going backward from atomic number 16 in the periodic table, we find that the nearest noble gas with lower atomic number is neon. In the abbreviated electron configuration, the core electrons (the first ten electrons) are represented by [Ne]. The abbreviated electron configuration for sulfur is

$$[Ne]3s^2 3p_x^2 3p_y^1 3p_z^1$$

The symbol [Ne] stands for $1s^2 2s^2 2p^6$. The atomic number of an element is equal to the number of electrons in an atom of the element. To check an abbreviated electron configuration, add the right superscripts to the atomic number of the next lower noble gas. The atomic number of neon is 10:

$$10 + 6 = 16$$

(c) In writing an orbital diagram, remember that the first shell has only an s orbital. The second shell has one s and three p orbitals. The third shell has one s, three p, and five d orbitals, and the fourth shell has one s, three p, five d, and seven f orbitals. The number of subshells in a shell is the same as the principal quantum number. The number of orbitals in subshells is given by the odd numbers 1, 3, 5, 7, ... Use the electron configuration to write the orbital diagram. When there is more than one orbital in a subshell, follow Hund's rule. Put one electron in each orbital before putting two electrons in any orbital. Each orbital can hold only two electrons, one with spin up and the other with spin down. The orbital diagram for sulfur in its ground state is

S   ⇅   ⇅   ⇅ ⇅ ⇅   ⇅   ⇅ ↑ ↑
    $1s$ $2s$ $2p$ $3s$ $3p$

or

S   [Ne] $\boxed{\uparrow\downarrow}$ $\boxed{\uparrow\downarrow\,\vert\,\uparrow\,\vert\,\uparrow}$
            $3s$ $3p$

8.2   The atomic number of iron is 26. For the iron atom in its ground state, (a) write the complete electron configuration, (b) write the abbreviated electron configuration, (c) give the orbital diagram, and (d) tell how many unpaired electrons are present.

Solution   (a) Again refer to a periodic table. Going across the first row of the periodic table, put two electrons in the $1s$ orbital. Then go to the left end of the second row. Put the third and fourth electrons in the $2s$ orbital. Continuing across the second row, put the next six electrons in the $2p$ orbitals. Go to the third row and put two electrons in the $3s$ orbital and six in the $3p$ orbitals. Then put two electrons in the $4s$ orbital.

So far we have put 20 electrons in orbitals. There are six more electrons to place. Continuing across the fourth row of the periodic table, place one electron in each of the five $3d$ orbitals. The principal quantum number of d electrons is always one less than the number of the row in the periodic table; d electrons go into the next-to-outermost shell. The 26th and last electron must go into one of the $3d$ orbitals that already contains one electron. Its spin must be opposite to the spin of the electron that is already in the orbital. The electron configuration for the ground state of the iron atom is

$$1s^2 2s^2 2p^6 3s^2 3p^6 4s^2 3d^6$$

or, grouping orbitals with the same principal quantum number together,

$$1s^2 2s^2 2p^6 3s^2 3p^6 3d^6 4s^2$$

The superscripts total 26, the number of electrons in an iron atom.

(b) The abbreviated electron configuration for an iron atom in its ground state is

$$[Ar]4s^2 3d^6 \qquad \text{or} \qquad [Ar]3d^6 4s^2.$$

(c) The orbital diagram for an iron atom in its ground state is

Fe $\boxed{\uparrow\downarrow}$ $\boxed{\uparrow\downarrow}$ $\boxed{\uparrow\downarrow\,\vert\,\uparrow\downarrow\,\vert\,\uparrow\downarrow}$ $\boxed{\uparrow\downarrow}$ $\boxed{\uparrow\downarrow\,\vert\,\uparrow\downarrow\,\vert\,\uparrow\downarrow}$ $\boxed{\uparrow\downarrow\,\vert\,\uparrow\,\vert\,\uparrow\,\vert\,\uparrow\,\vert\,\uparrow}$ $\boxed{\uparrow\downarrow}$
     $1s$ $2s$ $2p$ $3s$ $3p$ $3d$ $4s$

or

Fe   [Ar] $\boxed{\uparrow\downarrow\,\vert\,\uparrow\,\vert\,\uparrow\,\vert\,\uparrow\,\vert\,\uparrow}$ $\boxed{\uparrow\downarrow}$
             $3d$ $4s$

(d) There are four unpaired electrons in an iron atom in its ground state. [Count them in the orbital diagrams in part (c).]

8.3   The atomic number of plutonium is 94. For the plutonium atom in its ground state, (a) write the complete electron configuration, (b) write the abbreviated electron configuration, and (c) give the orbital diagram (abbreviated).

Solution   (a) As usual, refer to a periodic table. Going across the first row of the periodic table, put two electrons in the $1s$ orbital. Then go to the left end of the second row. Put the third and fourth electrons in the $2s$ orbital. Continuing across the second row, put the next six electrons in the $2p$ orbitals. Go to the third row and put two electrons in the $3s$ orbital and six in the $3p$ orbitals. Going across the fourth row, put two electrons in the $4s$ orbital, ten in the $3d$ orbitals, and six in the $4p$ orbitals. Similarly, put two electrons in the $5s$ orbital, ten in the $4d$ orbitals, and six in the $5p$ orbitals. Next put two electrons in the $6s$ orbital. Then, following the atomic numbers in the periodic table, place fourteen electrons in $4f$ orbitals, ten in the $5d$ orbitals, and six in the $6p$ orbitals. Then place two electrons in the $7s$ orbital. So far we have put 88 electrons in orbitals. There are six more electrons to place. Continuing across the seventh row of the periodic table, place one electron in each of six $5f$ orbitals. The electron configuration for the ground state of plutonium is

$$1s^22s^22p^63s^23p^64s^23d^{10}4p^65s^24d^{10}5p^66s^24f^{14}5d^{10}6p^67s^25f^6$$

or, grouping orbitals of the same principal quantum number together,

$$1s^22s^22p^63s^23p^63d^{10}4s^24p^64d^{10}4f^{14}5s^25p^65d^{10}5f^66s^26p^67s^2$$

The superscripts total 94, the atomic number of plutonium.

(b) The abbreviated electron configuration is

$$[\text{Rn}]5f^67s^2$$

(c) The abbreviated orbital diagram for plutonium is

$$[\text{Rn}] \quad \boxed{\uparrow\ \uparrow\ \uparrow\ \uparrow\ \uparrow\ \uparrow\ \ \ } \quad \boxed{\uparrow\downarrow}$$
$$5f \qquad\qquad 7s$$

PRACTICE PROBLEM

8.10   For each of the following atoms in its ground state, (a) write the complete electron configuration, (b) write the abbreviated electron configuration, (c) give the orbital diagram (abbreviated), and (d) tell how many unpaired electrons are present. (i) Ne (ii) Cl (iii) Zr (iv) Pm

Ions

The ground-state electron configurations of ions can be written from the ground-state electron configurations of atoms.

1. To write the ground-state electron configuration of an anion, add electrons to the ground-state electron configuration of the atom following the usual rules for placing electrons in orbitals (see Sample Problem 8.4).

2. To write the ground-state electron configuration of a cation, remove electrons from the ground-state electron configuration of the atom. Start with the *highest* occupied energy level. The form of the ground-state electron configuration of the atom with the orbitals of the same principal quantum number grouped together is most convenient. Starting from the right, remove enough electrons to give the correct charge for the cation (see Sample Problem 8.5).

To form a cation, remove the electrons with the highest principal quantum number first. The results of measurements of paramagnetism support this rule.

Note that the last electrons added in the building up (aufbau) process are not always the first electrons removed when a positive ion is formed. Electrons are placed in $3d$ orbitals after electrons are placed in $4s$ orbitals, but electrons are removed from $4s$ orbitals before electrons are removed from $3d$ orbitals. *In an atom, the electrons occupy the orbitals that give the atom the lowest total energy.* Similarly, *in an ion, the electrons occupy the orbitals that give the ion the lowest total energy.*

SAMPLE PROBLEMS

8.4   For the sulfide ion S^{2-} in its ground state, (a) write the complete electron configuration and (b) give the orbital diagram (abbreviated).

Solution   (a) From Sample Problem 8.1, the ground-state electron configuration of the sulfur atom is

$$1s^22s^22p^63s^23p_x^{\ 2}3p_y^{\ 1}3p_z^{\ 1}$$

A sulfide ion has a 2− charge; therefore, a sulfide ion has two more electrons than a sulfur atom. The electron configuration of a sulfide ion is

$$1s^22s^22p^63s^23p_x^23p_y^23p_z^2$$

Note that a sulfide ion has the same number of electrons (18) and the same electron configuration as the next noble gas, argon.

(b) From Sample Problem 8.1, the abbreviated orbital diagram for the sulfur atom is

[Ne] | ↑↓ | | ↑↓ | ↑ | ↑ |
     3s   3p

The abbreviated orbital diagram for the sulfide ion is

[Ne] | ↑↓ | | ↑↓ | ↑↓ | ↑↓ |
     3s   3p

Note that although a sulfur atom is paramagnetic with two unpaired electrons, a sulfide ion is diamagnetic with no unpaired electrons.

8.5 For the iron(II) ion, Fe^{2+}, and the iron(III) ion, Fe^{3+}, in their ground states, (a) write the complete electron configuration and (b) give the orbital diagram (abbreviated).

Solution (a) *Use the ground-state electron configuration for an atom that has orbitals with the same principal quantum number grouped together* to write the electron configuration of a cation. From Sample Problem 8.2, the ground-state electron configuration of the iron atom written showing orbitals with the same principal quantum number grouped together is

$$1s^22s^22p^63s^23p^63d^64s^2$$

Starting at the right side, remove electrons until enough electrons have been removed to give the required positive charge.

Two electrons must be removed from an iron atom to form an iron(II) ion. The electron configuration for the ground state of the Fe^{2+} ion is

$$1s^22s^22p^63s^23p^63d^6$$

Three electrons must be removed from an iron atom to form an iron(III) ion. The electron configuration for the ground state of the Fe^{3+} ion is

$$1s^22s^22p^63s^23p^63d^5$$

(b) From Sample Problem 8.2, the abbreviated orbital diagram for the iron atom is

Fe   [Ar] | ↑↓ | ↑ | ↑ | ↑ | ↑ | | ↑↓ |
              3d   4s

The abbreviated orbital diagram for the Fe^{2+} ion is

Fe^{2+} [Ar] | ↑↓ | ↑ | ↑ | ↑ | ↑ | | |
              3d   4s

and the abbreviated orbital diagram for the Fe^{3+} ion is

Fe^{3+} [Ar] | ↑ | ↑ | ↑ | ↑ | ↑ | | |
              3d   4s

PRACTICE PROBLEMS

8.11 For the chloride ion Cl^- in its ground state, (a) write the complete electron configuration and (b) give the orbital diagram (abbreviated).

Many trends in physical and chemical properties of the elements and their compounds across rows and down columns in the periodic table are easily explained by their electron configurations. Important examples are atomic and ionic radii, ionization energies, and electron affinities.

8.4 ATOMIC AND IONIC RADII

Atomic Radii

Real atoms and ions are not hard balls. There is no sharp boundary that defines the outside of an atom or ion. However, charge density diagrams and radial distribution curves similar to those in Figures 7.30(a) and (c), 7.31, and 7.33 suggest that the charge clouds of atoms occupy fairly well-defined volumes. The sizes of the circles in ▬Figure 8.7 are proportional to the best available set of estimates of radii of atoms in metals, in molecules of elements, and in molecular compounds. For metals, the atomic radius shown in Figure 8.7 is half the distance between the centers of

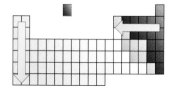

◗ The noble gases are absent because noble gas atoms don't combine to form diatomic molecules.

◗ In general, the atomic radii of main group elements increase from right to left across periods and from top to bottom down groups.

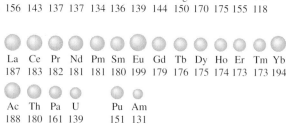

FIGURE 8.7 Periodic table showing relative sizes of atoms (magnified about 10 million times). Numbers are radii in pm.* (Data are from Moeller, T. *Inorganic Chemistry: A Modern Introduction;* Wiley: New York, 1982; pp. 70–72.)

*You may wonder why circles are used to show the radii of all atoms when you have just learned (Section 7.11) that only *s* orbitals are spherical. *Sets* of orbitals are roughly spherical (remember Figure 7.34). In addition, atoms are not motionless but rotate continuously. Rotation of even a single *p* orbital in all directions yields a sphere.

the atoms in the metal [■Figure 8.8(a)]. For nonmetals, the radius shown in Figure 8.7 is half the distance between the centers of the atoms in diatomic molecules of the elements [Figure 8.8(b)].

The attraction of the positively charged nucleus for the negatively charged electrons pulls the electrons toward the nucleus. The repulsion between negatively charged electrons pushes the electrons apart. The combination of attraction by the nucleus and repulsion by other electrons determines the size of an atom.

As you can see from Figure 8.7 in general, the *atomic radii of main group elements decrease from left to right across rows in the periodic table.* The principal quantum number of the outer electrons is the same all across a period; the outer electrons are in the same shell. Electrons added to the outer shell do not shield other outer shell electrons from the positive charge on the nucleus very well. On the other hand, the charge on the nucleus increases from left to right across a row in the periodic table. As a result, the outer electrons are attracted more strongly and pulled in toward the nucleus, making the atom smaller.

Going down main groups, atoms get bigger. The number of shells of electrons increases from top to bottom of a group. The higher the principal quantum number of a shell, the larger the radius of the shell. Although the charge on the nucleus increases going down a group, the effective nuclear charge—that is, the charge that the outer electrons actually feel—does not increase as much. For example, the atomic number of lithium is 3, and the atomic number of potassium is 19; the positive charge in the nucleus of a potassium atom is over six times as large as the positive charge in the nucleus of a lithium atom. However, the effective nuclear charge that the outer electron in a potassium atom feels is less than three times as large as the effective nuclear charge that the outer electron of a lithium atom feels. In a potassium atom, there are more electrons between the nucleus and the outer electrons to shield the outer electrons from the positive charge in the nucleus. The increase in size that accompanies an increase in principal quantum number predominates over increasing effective nuclear charge and atomic radius increases going down a group.

The radii of transition metals generally first decrease and then increase across each series. Changes in radius are small compared to the changes in size of atoms of main group elements across a period. The smallest main group atom in a period is only about half as big as the biggest main group atom in the period. The smallest transition metal atom in a period is three-quarters or more the size of the largest transition metal atom in the period. The small change in size across a series of transition metals is explained by the fact that electrons are being added to an inner shell. These additional inner electrons screen the outer electrons. Effective nuclear charge does not increase very much, so size does not decrease very much. Toward the end of each series, repulsion between the increasing number of electrons makes the atomic radius increase.

Although atoms usually get bigger going down main groups, notice that gallium (atomic number 31) is smaller than aluminum (atomic number 13). The small size of the gallium atom is a result of the large increase in nuclear charge between aluminum and gallium. As a result of the occurrence of the first transition series between calcium and gallium, gallium has 18 more protons in its nucleus than does aluminum. Potassium and calcium each have only eight more protons in the nucleus than the elements in the preceding period, sodium and magnesium.

Going down groups, the radii of transition elements increase from the fourth period to the fifth period. However, *the sixth-period transition elements, the elements of the third transition series, are about the same size as the fifth-period transition elements.* The third transition series follows the lanthanide series in the periodic

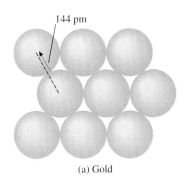

144 pm

(a) Gold

37 pm

74 pm

(b) H₂ molecule

■ FIGURE 8.8 How atomic radii are measured.

The more layers of clothing you put on in winter to protect yourself from the cold, the bigger you get.

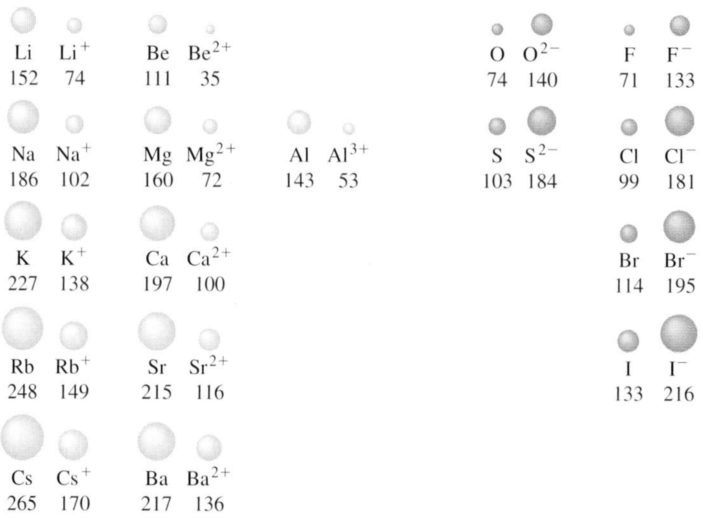

| Li | Li$^+$ | | Be | Be^{2+} | | | | | O | O^{2-} | | F | F$^-$ |
| 152 | 74 | | 111 | 35 | | | | | 74 | 140 | | 71 | 133 |

| Na | Na$^+$ | | Mg | Mg^{2+} | | Al | Al^{3+} | | S | S^{2-} | | Cl | Cl$^-$ |
| 186 | 102 | | 160 | 72 | | 143 | 53 | | 103 | 184 | | 99 | 181 |

| K | K$^+$ | | Ca | Ca^{2+} | | | | | | | | Br | Br$^-$ |
| 227 | 138 | | 197 | 100 | | | | | | | | 114 | 195 |

| Rb | Rb$^+$ | | Sr | Sr^{2+} | | | | | | | | I | I$^-$ |
| 248 | 149 | | 215 | 116 | | | | | | | | 133 | 216 |

| Cs | Cs$^+$ | | Ba | Ba^{2+} |
| 265 | 170 | | 217 | 136 |

■ FIGURE 8.9   Scale drawings of some main group atoms and ions (enlarged about 10 million times). Numbers are radii in pm. (Data are from Moeller, T. *Inorganic Chemistry: A Modern Introduction;* Wiley: New York, 1982; pp. 141–144.)

table. Each element in the second transition series has 18 more protons in its nucleus than the element above it. Each element in the third transition series has 32 more protons in its nucleus than the element above it, and the increase in radius expected from an increase in principal quantum number does not take place. This phenomenon is called the **lanthanide contraction.***

PRACTICE PROBLEMS

8.13   Explain why (a) a beryllium atom is smaller than a lithium atom, (b) a chlorine atom is larger than a fluorine atom, and (c) a gold atom is about the same size as a silver atom.

Use the periodic table inside the front cover, not Figure 8.8, to answer the following Practice Problems.

8.14   Which atom has the larger radius? (a) Na or K (b) K or Ca

8.15   Arrange the atoms of the following set in order of increasing radius: B, Be, Mg.

Ionic Radii

Many of the properties of ions are explained by their charges and sizes. In ■Figure 8.9, the sizes of some monatomic ions formed from main group elements are compared with the sizes of the atoms from which the ions are formed. As you can see, *cations (positive ions) are smaller than the atoms from which they are formed.*

Differences in ionic size often affect biological activity. For example, the toxicity of beryllium is believed to result from the substitution of the smaller Be^{2+} for Mg^{2+} in enzymes (biochemical catalysts). Substitution of ions of similar size for each other in minerals is important in geology.

For an excellent review of the importance of relativistic effects in atoms with atomic number 29 and beyond, see Pykkö, P. Chem. Rev. *1988,* 88, *563–594.*

*Another reason why atomic radius does not increase from the second transition series to the third is a relativistic effect. The speed of the 1s electrons in atoms of elements of the third transition series approaches the speed of light as does the speed of other s electrons when they penetrate close to the nucleus. As a result, the masses of the s electrons increase significantly, and the radii of the s orbitals become smaller. The outer electrons in the third transition series are 6s electrons. The elements in the third transition series are smaller than expected because the radii of their 6s orbitals have been reduced.

When positive ions are formed from main group elements, the outer shell of electrons is removed. It is not surprising that the cations are smaller than the atoms from which they are formed. *Anions (negative ions) are larger than the atoms from which they are formed.* Adding electrons to the outermost shell results in increased repulsion between electrons. Therefore, the electron cloud spreads out.

Species that have the same electron configurations are called **isoelectronic.** For example, O^{2-}, F^-, Ne, Na^+, Mg^{2+}, and Al^{3+} are isoelectronic because all have the electron configuration $1s^2 2s^2 2p^6$. For isoelectronic species, the higher the nuclear charge, the smaller the species. For example, Mg^{2+} is smaller than Na^+ (see Figure 8.9).

Students can test their understanding of the electronic structure of ions and isoelectronic series in the Electron Configurations module of the Atoms section.

PRACTICE PROBLEMS

8.16   Explain why (a) a sodium atom, Na, is larger than a sodium ion, Na^+, and (b) a chloride ion, Cl^-, is larger than a chlorine atom, Cl.

8.17   Which is smaller? (a) Al^{3+} or Mg^{2+} (b) O^{2-} or F^- (c) F^- or Na^+

8.18   Write symbols for a positive ion and a negative ion that are isoelectronic with argon.

All of the properties of atoms that measure the ease of loss or gain of electrons are related to their size. Two such properties are ionization energy and electron affinity.

8.5   IONIZATION ENERGY

To promote an electron in an atom from one energy level to the next higher energy level, energy must be supplied. If enough energy is available, the electron is removed from the atom, and a cation with a $1+$ charge is formed. The *amount of energy needed to remove the least tightly bound electron from a mole of gaseous atoms* is called the **first ionization energy,** I_1. Unlike atoms in liquids and solids, atoms in the gaseous state are not influenced by neighboring atoms. For atoms in the gaseous state, the first ionization energy is equal to the energy with which an atom binds its least tightly bound electron:

$$A(g) + \text{first ionization energy} \longrightarrow A^+(g) + e^- \qquad (8.1)$$

where A stands for an atom of any element. Because the electron is attracted by the positive charge of the nucleus, removing an electron from an atom requires energy. Ionization in the gas phase is always an endothermic change. The value of the first ionization energy depends on a combination of effective nuclear charge, atomic radius, and electron configuration. The first ionization energies of most of the elements are shown in the periodic table in ▪Figure 8.10. Notice how the elements with the lowest first ionization energies are in the lower left corner of the periodic table. Elements with low ionization energies lose electrons relatively easily and form cations. They are reactive metals.

Elements with high first ionization energies are in the upper right corner. They are nonmetals. Elements with high ionization energies do not lose electrons easily. In fact, they tend to gain electrons when they form compounds.

From Figure 8.10, the first ionization energy for sodium, which is relatively low, is 496 kJ/mol. A mole of sodium has a mass of 23 grams or less than an ounce. The energy needed to remove one electron from all the atoms in a mole of sodium atoms is equivalent to the energy in a brownie or to the energy needed to lift a 2-

Period   **0**

1   **IA** **IIA**   **IIIA** **IVA** **VA** **VIA** **VIIA** | 2 He 2370

| | 1 H 1311 | | | | | | |

Period	IA	IIA	IIIB	IVB	VB	VIB	VIIB	VIII			IB	IIB	IIIA	IVA	VA	VIA	VIIA	0
1																		2 He 2370
2	3 Li 521	4 Be 899											5 B 799	6 C 1087	7 N 1404	8 O 1314	9 F 1682	10 Ne 2080
3	11 Na 496	12 Mg 737											13 Al 576	14 Si 786	15 P 1052	16 S 1000	17 Cl 1245	18 Ar 1521
4	19 K 419	20 Ca 590	21 Sc 646	22 Ti 660	23 V 647	24 Cr 650	25 Mn 717	26 Fe 755	27 Co 756	28 Ni 736	29 Cu 745	30 Zn 906	31 Ga 576	32 Ge 784	33 As 1013	34 Se 939	35 Br 1135	36 Kr 1351
5	37 Rb 402	38 Sr 549	39 Y 627	40 Zr 671	41 Nb	42 Mo 681	43 Tc	44 Ru 724	45 Rh 743	46 Pd 782	47 Ag 731	48 Cd 867	49 In 559	50 Sn 704	51 Sb 834	52 Te 865	53 I 1007	54 Xe 1170
6	55 Cs 375	56 Ba 503	✳71 Lu	72 Hf	73 Ta	74 W 766	75 Re 772	76 Os 839	77 Ir 888	78 Pt 859	79 Au 891	80 Hg 1007	81 Tl 590	82 Pb 716	83 Bi 849	84 Po 791	85 At 926	86 Rn 1037
7	87 Fr	88 Ra 508	◆103 Lr	104 Rf	105 Ha	106 Sg	107 Ns	108 Hs	109 Mt	110	111							

◻ Metal   ▨ Semimetal   ▦ Nonmetal

✳ **Lanthanide series**

57 La 540	58 Ce 631	59 Pr 560	60 Nd 608	61 Pm	62 Sm 540	63 Eu 544	64 Gd 646	65 Tb 646	66 Dy 656	67 Ho	68 Er	69 Tm	70 Yb 600

◆ **Actinide series**

89 Ac	90 Th	91 Pa	92 U 386	93 Np	94 Pu	95 Am	96 Cm	97 Bk	98 Cf	99 Es	100 Fm	101 Md	102 No

▬ FIGURE 8.10   First ionization energies, kJ/mol. (Data are from Bard, A. J.; Parsons, R.; Jordan, J. *Standard Potentials in Aqueous Solution;* Dekker: New York, 1985; pp. 24–27.)

lb bag of sugar 31 miles. This amount of energy is enough to heat six cups of coffee from room temperature to boiling. These figures should give you an idea of how tightly atoms hold on to electrons.

▬Figure 8.11 is a graph of first ionization energies as a function of atomic number. Note that first ionization energies are periodic functions of atomic number. The elements in Group 0, the noble gases, have the highest first ionization energies of all the groups. Helium, the noble gas with the smallest atomic radius, has the highest first ionization energy of all the elements. The high ionization energies of the noble gases are convincing evidence that the electron configurations of the noble gases are unusually stable.

The elements in Group IA, the alkali metals, have the lowest first ionization energies of all the groups. Cesium, the alkali metal with the largest atomic radius, has the lowest first ionization energy of all the elements. Removal of one electron from an atom of a Group IA element gives the atom the electron configuration of the next lower noble gas. For example, the electron configuration of sodium is [Ne]$3s^1$. Removal of the $3s$ electron gives [Ne] for the electron configuration of the Na^+ ion.

Trends Across Periods

In general, first ionization energies increase going across rows in the periodic table from left to right. Going across a row, each successive atom has one more proton in its nucleus. The positive charge on the nucleus increases by one unit. Each successive atom also has one more electron. However, electrons added to the outer

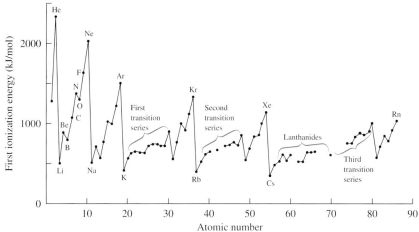

FIGURE 8.11 First ionization energies as a function of atomic number. Although there is a general increase in first ionization energy across periods, some irregularities occur. For example, the first ionization energy of boron is lower than the first ionization energy of beryllium, and the first ionization energy of oxygen is lower than the first ionization energy of nitrogen. The electron that is removed from boron is in a 2p orbital, whereas the electron that is removed from beryllium is in a 2s orbital. Because a 2p orbital is of higher energy than a 2s orbital, an electron is more easily removed from boron than from beryllium. An oxygen atom has two electrons in one of its 2p orbitals. Because these paired electrons are in the same orbital, they have more repulsion between them than electrons in the same subshell but different orbitals. This causes the first ionization energy of oxygen to be lower than that of nitrogen.

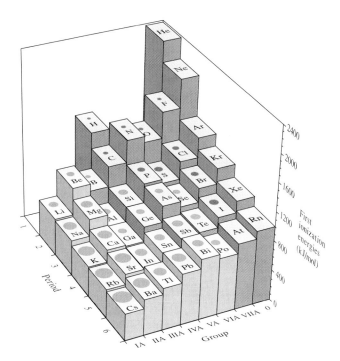

The three-dimensional graph shows how both first ionization energies and atomic radii (circles) of main group elements change across periods and down groups.

shell don't shield other electrons in the outer shell very well from the positive charge on the nucleus. Effective nuclear charge increases proceeding from left to right, and the radii of the atoms decrease from left to right. This means that the outer electrons of elements on the right side of a row in the periodic table are closer to a larger positive charge. Thus, it is not surprising that more energy is needed to remove them.

Transition Elements

From left to right across a row in the periodic table, the first ionization energies of the transition elements do not increase as fast as those of the main group elements. Going from one element to the next, the additional electron is added to a d orbital of the next-to-outermost shell. The additional d electrons shield the outer electrons from the positive charge of the nucleus. Effective nuclear charge does not increase as rapidly as actual nuclear charge.

The first ionization energies of the lanthanides increase very slowly across the row. The added electron is going into the $4f$ subshell deep inside the atom. Increased shielding of the outer electrons almost cancels the effect of increased nuclear charge.

Group Trends

For most main group elements, first ionization energy decreases going down a group. The elements in a group have the same outer electron configuration but different principal quantum numbers. As principal quantum number increases, so does the average distance of the electrons in the shell from the nucleus. Because the outer electrons are farther from the positive charge in the nucleus, they are less strongly attracted by it.

For almost all third transition series elements, the first ionization energy is higher than the first ionization energy of the element in the same column in the second transition series. The high first ionization energies of the elements in the third transition series are a result of their high effective nuclear charge. Atoms of elements in the third transition series are about the same size as atoms of elements in the second transition series, but the nuclear charges of atoms in the third transition series are 32 units higher. As a result, effective nuclear charges are greater in atoms of the third transition series and first ionization energies of these elements are higher.

PRACTICE PROBLEMS

8.19   Explain why (a) the first ionization energy of K is lower than the first ionization energy of Na, (b) the first ionization energy of F is higher than the first ionization energy of O, (c) the first ionization energy of N is higher than the first ionization energy of O, and (d) the first ionization energy of Au is greater than the first ionization energy of Ag.

8.20   Use the periodic table inside the front cover to answer the following questions. Which element has the higher first ionization energy? (a) H or He (b) K or Na (c) Cd or Hg

Successive Ionization Energies

All atoms other than hydrogen have more than one electron. *The energy needed to remove the second most loosely bound electron from a mole of gaseous cations* is called the **second ionization energy,** I_2:

$$A^+(g) \; + \; \text{second ionization energy} \longrightarrow A^{2+}(g) \; + \; e^-$$

Because of its positive charge, the cation A^+ attracts electrons more strongly than the atom A does. Also, in the cation, there is one less electron than in an atom of the element and electron–electron repulsions are reduced. As a result, more energy is required to remove the second electron than the first; second ionization energies are always greater than first ionization energies.

The energy needed to remove the third most loosely bound electron is called the **third ionization energy,** I_3. Higher ionization energies are defined similarly. The larger the positive charge on the cation, the stronger the attraction between the cation and the electron being removed and the fewer electron–electron repulsions there are. The larger the positive charge on the cation, the more energy is required to remove an electron. For any element, the value of the third ionization energy is always higher than the value of the second ionization energy, the fourth ionization energy is higher than the third, and so on.

Successive ionization energies I_1 through I_8 of the first 18 elements are shown in Table 8.1. Ionization energies to the right of the stepwise lines involve removal of electrons from the core of the atom. Notice how, for each atom, there is a large increase in ionization energy after all the outer electrons have been removed. For sodium, for example, I_1 is 496 kJ/mol and I_2 is 4564 kJ/mol, almost ten times as much. For magnesium, I_1 is 737 kJ/mol and I_2, 1447 kJ/mol, about twice as much. However, I_3 is 7738 kJ/mol, over five times as much as I_2. After all electrons in the outer shell have been removed, the next electron must be removed from the next-to-outermost shell—that is, it must be removed from the core. The core has a noble gas electron configuration; in the core, all subshells that contain electrons are filled. Values for successive ionization energies support the idea that only outer electrons outside the noble gas core are involved in chemical change. Inner electrons in the core are too tightly bound to the nucleus.

Ionization energies provide strong evidence for the existence of electron shells in atoms. ▬Figure 8.12 shows the values of all the successive ionization energies of calcium. The energy required to remove the first electron from a calcium atom is 590 kJ/mol. The energy required to remove the last (20th) electron from a Ca^{19+} ion is 5.28×10^5 kJ/mol. The difference between these two energies is so large that the logarithms of the successive ionization energies are plotted instead of the successive ionization energies themselves. As you can see from Figure 8.12, the

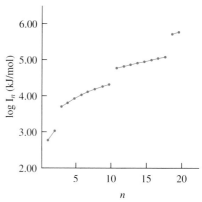

▬ FIGURE 8.12   Log ionization energy, I_n, vs. n for calcium.

TABLE 8.1	Successive Ionization Energies, kJ/mol[a]							
	1st	2nd	3rd	4th	5th	6th	7th	8th
H	1 311							
He	2 370	5 220						
Li	521	7 304	11 752					
Be	899	1 756	14 849	20 899				
B	799	2 422	3 657	25 019	32 660			
C	1 087	2 393	4 622	6 223	37 822	46 988		
N	1 404	2 856	4 573	7 468	9 446	53 250	63 970	
O	1 314	3 396	5 297	7 468	10 990	13 325	71 312	83 652
F	1 682	3 367	6 050	8 423	11 028	15 167	17 869	91 950
Ne	2 080	3 946	6 165	9 301	12 138	15 148	19 972	22 963
Na	496	4 564	6 918	9 542	13 373	16 644	20 175	25 501
Mg	737	1 447	7 738	10 546	13 624	18 033	21 767	25 742
Al	576	1 814	2 750	11 578	14 820	18 361	23 465	27 575
Si	786	1 582	3 232	4 361	16 007	19 693	23 658	29 110
P	1 052	1 901	2 914	4 959	6 272	21 516	25 858	30 489
S	1 000	2 258	3 387	4 544	6 947	8 500	27 112	31 734
Cl	1 245	2 287	3 850	5 162	6 542	9 359	11 028	33 442
Ar	1 521	2 653	3 927	5 886	7 526	8 587	11 964	13 778

[a]Data are from Bard, A. J.; Parsons, R.; Jordan, J. *Standard Potentials in Aqueous Solution;* Dekker: New York, 1985; pp 24–27.

successive ionization energies fall into groups of 2, 8, 8, and 2 points corresponding to the number of electrons in occupied shells in the calcium atom. Graphs of logarithms of successive ionization energies of other elements are similar to the graph for calcium.

PRACTICE PROBLEMS

8.21   What would be the problem with plotting the successive ionization energies themselves?

8.22   (a) Why is the second ionization energy for Be greater than the first ionization energy for Be? (b) Why is the third ionization energy for Be much greater than the second ionization energy for Be?

The large difference in ionization energy between outer and core electrons is why only one kind of cation is formed from elements in Groups IA and IIA and from aluminum. Atoms of transition elements have d electrons of the $(n - 1)$ shell as well as s electrons of the n shell outside the core. The energy of electrons in $(n - 1)d$ orbitals is not very different from the energy of electrons in ns orbitals. As a result, atoms of transition elements often form more than one kind of cation, for example, Fe^{2+} and Fe^{3+}.

8.6   ELECTRON AFFINITY

Electron affinities can now be measured as accurately as ionization energies, and values are available for most elements except the inner transition elements. Therefore, we feel that electron affinities should be given roughly equal time with ionization energies. A number of general chemistry texts give the opposite signs for electron affinities. This makes sense from a thermodynamic viewpoint, but is confusing from a linguistic viewpoint because it gives the elements with the strongest tendency to gain electrons the lowest instead of the highest electron affinities. The signs used in this text are the ones used in advanced texts and in the chemical literature.

Atoms not only lose electrons to form positive ions but also gain electrons to form negative ions. **Electron affinity,** as the name indicates, is *a measure of an atom's tendency to gain an electron.* The higher an atom's electron affinity, the more likely it is to gain an electron. Quantitatively, electron affinity is the *negative* of the enthalpy change associated with the addition of an electron to a mole of gaseous atoms to form gaseous ions with a $1-$ charge. Thermal energy is released when an electron is added to most atoms to form a $1-$ ion:

$$A(g) + e^- \longrightarrow A^-(g) + \text{thermal energy} \qquad (8.2)$$

where A represents an atom of an element. The change shown in equation 8.2 is exothermic. The *energy change* associated with an exothermic change always has a $-$ sign. The higher an element's electron affinity, the more thermal energy is given off when an electron is added to an atom of the element. The most recent values for electron affinities are given in ▬Figure 8.13 and shown graphically in ▬Figure 8.14.

Figure 8.14 shows that electron affinity is a rather irregular periodic function of atomic number. In general, electron affinity increases going from left to right across a row of the periodic table, reaching a maximum with the halogens in Group VIIA. The halogens have a much greater tendency to gain an electron than the alkali metals in Group IA. Both actual and effective nuclear charges increase across a period. Atomic radius decreases, and an electron added to the outer shell is closer to a larger positive charge in atoms on the right side of the periodic table. Thus, more energy is released when an electron is added.

As you can see from Figure 8.13, only nitrogen and the elements of Groups IIA, IIB, and 0 have negative electron affinities. Energy is required to force an electron onto atoms of nitrogen, the alkaline earth metals (Group IIA), the elements of Group IIB, and the noble gases (Group 0). Addition of an electron to nitrogen and to atoms

Period

	IA	IIA											IIIA	IVA	VA	VIA	VIIA	0
1	1 H 73																	2 He −21
2	3 Li 60	4 Be −19											5 B 27	6 C 122	7 N −7	8 O 141	9 F 328	10 Ne −29
3	11 Na 53	12 Mg −19	IIIB	IVB	VB	VIB	VIIB	VIII			IB	IIB	13 Al 43	14 Si 134	15 P 72	16 S 200	17 Cl 349	18 Ar −35
4	19 K 48	20 Ca −10	21 Sc 18	22 Ti 8	23 V 51	24 Cr 64	25 Mn	26 Fe 16	27 Co 64	28 Ni 112	29 Cu 118	30 Zn −47	31 Ga 29	32 Ge 116	33 As 78	34 Se 195	35 Br 325	36 Kr −39
5	37 Rb 47	38 Sr	39 Y −30	40 Zr 41	41 Nb 86	42 Mo 72	43 Tc 53	44 Ru 101	45 Rh 110	46 Pd 54	47 Ag 126	48 Cd −32	49 In 29	50 Sn 116	51 Sb 103	52 Te 190	53 I 295	54 Xe −41
6	55 Cs 45	56 Ba	71 Lu	72 Hf	73 Ta 31	74 W 79	75 Re 14	76 Os 106	77 Ir 151	78 Pt 205	79 Au 223	80 Hg −61	81 Tl 20	82 Pb 35	83 Bi 91	84 Po 183	85 At 270	86 Rn −41
7	87 Fr 44	88 Ra	103 Lr	104 Rf	105 Ha	106 Sg	107 Ns	108 Hs	109 Mt	110	111							

Metal   Semimetal   Nonmetal

■ FIGURE 8.13   Electron affinities of the elements, kJ/mol. (Values are from Hotop, H.; Lineberger, W. C. *J. Phys. Chem. Ref. Data* **1985**, *14*, 731–750.)

Note that the numerical values of electron affinities are small compared to the numerical values of ionization energies. Electron affinities range from −61 kJ/mol for Hg to 349 kJ/mol for Cl; first ionization energies range from 375 kJ/mol for Cs to 2370 kJ/mol for He.

of the elements of the groups listed is an endothermic process. Because the noble gases have both high ionization energies and low electron affinities, the noble gases are unreactive.

In atoms of elements of Groups IIA, IIB, and 0, all subshells that contain electrons are filled; the next electron must go into a higher energy subshell. The irregularities of the Group VA elements, such as N and P, are also explained by electron configuration. An electron added to an atom of a Group VA element must go into a p orbital that already contains an electron. The electron already present in the orbital repels the second electron. As a result, less energy than expected is released when an electron is added to a Group VA element.

In general, electron affinity decreases going down groups in the periodic table, for example, Cl > Br > I > At (see Figure 8.14). Going down a group, the radii of the shells increases because the principal quantum number is higher. The added electron is farther from the positive charge in the nucleus. Therefore, the quantity of energy released when an electron is added is smaller.

The Periodic Trends module of the Atoms section lets students explore trends in atomic and ionic radii (including those of isoelectronic series), ionization energy, and electron affinity.

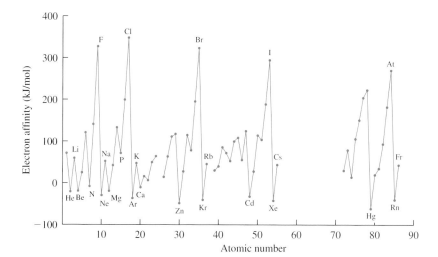

■ FIGURE 8.14   Variation of electron affinities of the elements with atomic number. Unfortunately, the electron affinities of the lanthanides and of a few other elements have not yet been measured.

Mendeleev and the Periodic Table

The periodic table was originally based on experimental observations of the chemical and physical properties of the elements. The Russian chemist Dmitri Ivanovich Mendeleev, who was professor of chemistry at the University of St. Petersburg, was the most important contributor to the early development of the periodic table. Mendeleev could not find a textbook that suited him and decided to write his own. In the course of writing his book, Mendeleev examined relationships between the properties of the elements and their compounds trying to find a system of organization that would help his students to learn chemistry. He discovered the periodic law, which he published in 1869, and constructed a periodic table soon afterward.

According to Mendeleev's periodic law, "the properties of the elements are a periodic function of their atomic weights." Mendeleev realized that, when placed according to their atomic weights (masses),

Reihen	Gruppe I. — R^2O	Gruppe II. — RO	Gruppe III. — R^2O^3	Gruppe IV. RH^4 RO^2	Gruppe V. RH^3 R^2O^5	Gruppe VI. RH^2 RO^3	Gruppe VII. RH R^2O^7	Gruppe VIII. — RO^4
1	H=1							
2	Li=7	Be=9,4	B=11	C=12	N=14	O=16	F=19	
3	Na=23	Mg=24	Al=27,3	Si=28	P=31	S=32	Cl=35,5	
4	K=39	Ca=40	—=44	Ti=48	V=51	Cr=52	Mn=55	Fe=56, Co=59, Ni=59, Cu=63.
5	(Cu=63)	Zn=65	—=68	—=72	As=75	Se=78	Br=80	
6	Rb=85	Sr=87	?Yt=88	Zr=90	Nb=94	Mo=96	—=100	Ru=104, Rh=104, Pd=106, Ag=108.
7	(Ag=108)	Cd=112	In=113	Sn=118	Sb=122	Te=125	J=127	
8	Cs=133	Ba=137	?Di=138	?Ce=140	—	—	—	— — — —
9	(—)	—	—	—	—	—	—	
10	—	—	?Er=178	?La=180	Ta=182	W=184	—	Os=195, Ir=197, Pt=198, Au=199.
11	(Au=199)	Hg=200	Tl=204	Pb=207	Bi=208	—	—	
12	—	—	—	Th=231	—	U=240	—	— — — —

MENDELEEF'S PERIODIC TABLE FROM HIS PAPER IN LIEBIG'S ANNALEN SUPP. **8, 133.**

- **FIGURE 8.15** Mendeleev's periodic table was published in a German journal in 1872. In German, "Reihen" means row, "Gruppe" means group, and "J" is the symbol for iodine. In the formulas for oxides and hydrides at the tops of the columns, the letter "R" is used to represent any of the elements in a group. Notice that the numbers that are now subscripts in formulas were superscripts in Mendeleev's day. The numbers after the equal signs in the body of the table are the atomic masses of the elements; the elements are arranged in order of increasing atomic mass. Mendeleev placed elements with similar properties in the same group leaving spaces for elements that were not known, such as the elements with atomic masses 44, 68, and 72.

Fluorine appears out of line compared to the other members of Group VIIA. The electron affinities of oxygen and nitrogen also seem low. These three atoms are small and have high effective nuclear charges. It is surprising that so little energy is given off when an electron is added to nitrogen, oxygen, and fluorine atoms. However, the second shell is small, and repulsions are large between the electrons already present in the second shell and the electron that is entering. In third- (and higher) period elements, the electrons already present in the outer shell are spread out through a larger volume. Repulsions between them and the electron that is entering are smaller.

several elements were out of place. He concluded that the atomic weights must be wrong and put the elements where they belonged on the basis of their properties. ▪Figure 8.15 shows Mendeleev's periodic table of 1871.

Because a number of elements had not yet been discovered, there were several blank spaces (for example at 44, 68, and 72) in Mendeleev's table. On the basis of the properties of neighboring elements, Mendeleev predicted the properties of these unknown elements. Three of these missing elements were discovered within the next 15 years. The fact that the properties of the newly discovered elements were very similar to the properties predicted by Mendeleev (see Table 8.2) provided convincing evidence for both the correctness and the usefulness of the periodic table. The last missing element to be found was francium, which was not discovered until 1939.

A number of changes have been made in the periodic table since 1871. When the noble gases helium and argon were isolated from air in 1894, another column was added. The existence of neon, krypton, xenon, and radon was then predicted from the periodic law, and these gases were soon discovered. The position of the lanthanides, an early problem, was worked out by Bohr in 1913. American chemist Glenn Seaborg realized in 1944 that the elements beginning with actinium (atomic number 89) form a second f-block series like the lanthanides. Chemists had previously thought that these elements were transition elements. The ability to predict the properties of the transuranium elements (the elements with atomic num-

TABLE 8.2	Some Properties of Germanium	
Property	Predicted by Mendeleev in 1871	Found by Winkler in 1886
Atomic mass	72	72.32
Density, g/cm^3	5.5	5.47
Molar volume, cm^3/mol	13.1	13.22
Color	Dirty gray	Grayish white
Heating in air gives	EO$_2$	GeO$_2$
Properties of EO$_2$	High melting point, density = 4.7 g/cm^3	Melting point, 1086 °C^a density = 4.703 g/cm^3
Properties of chloride	Formula, ECl$_4$, boiling point a little under 100 °C, and density = 1.9 g/cm^3	Formula, GeCl$_4$; boiling point, 86 °C; density = 1.887 g/cm^3

^{a}Modern value.

ber greater than 92) was of great help in working with these elements when they were synthesized. All transuranium elements are radioactive, and some have only been made in minute amounts (a few atoms).

The problem of the elements that were out of order according to their masses (Ar and K, Co and Ni, Te and I, and Th and Pa) was solved in 1913 when Moseley discovered atomic numbers. According to the modern periodic law, the properties of the elements are a periodic function of their *atomic numbers,* not of their atomic masses. The atomic number of an element describes the number of electrons outside the nucleus. The arrangement of the electrons in shells and subshells, the electron configuration of the elements, determines the properties of the elements. Why did atomic mass work so well as a basis for organizing the properties of the elements? The answer to this question lies in the

aufbau process—as protons are added to the nucleus, the mass of the atom increases. As protons are added to the nucleus, neutrons must also be added, and mass increases faster than atomic number. However, the order according to atomic number is about the same as the order according to atomic mass. A few elements are out of order when arranged according to atomic mass because naturally occurring samples of these elements contain unusually high proportions of heavy isotopes.

Mendeleev, a very practical man, used his scientific knowledge to improve the yields and quality of Russian crops and contributed to the development of the chemical and petroleum industries. However, his forward-looking political views were not popular with the czar. In 1890 he carried a request for the relief of unjust conditions from the students at the university to the administration and was retired early.

SAMPLE PROBLEM

8.6 For the addition of one mole of electrons to one mole of gaseous chlorine atoms,

$$Cl(g) + e^- \longrightarrow Cl^-(g) \qquad (8.3)$$

what is ΔH?

Solution From Figure 8.13, the electron affinity of Cl is 349 kJ/mol. Electron affinity is defined as the *negative* of the enthalpy change associated with the addition of an electron to a mole of gaseous atoms to form gaseous ions:

$$\text{electron affinity} = -\Delta H$$
$$349 \text{ kJ/mol} = -\Delta H$$
$$-349 \text{ kJ/mol} = \Delta H$$

PRACTICE PROBLEMS

8.23 For the addition of one mole of electrons to one mole of gaseous helium atoms,

$$He(g) + e^- \longrightarrow He^-(g)$$

what is ΔH?

8.24 Explain why the order of electron affinities for the elements in Group VIA is

$$O < S > Se > Te > Po$$

8.25 Explain why the electron affinity of Mg is negative.

8.26 Predict whether each of the following changes will be exothermic or endothermic and explain your reasoning: (a) $Na(g) + e^- \longrightarrow Na^-(g)$ and (b) $Mg(g) + e^- \longrightarrow Mg^-(g)$.

The idea of electron affinity as a measure of tendency to gain an electron can be extended to ions. Electron affinities of anions are negative. An electron is repelled by the negative charge of an anion, and energy must be added to force an electron onto a negative ion. Addition of electrons to negatively charged ions is always endothermic. To add a second electron to an O^- ion to form an O^{2-} ion requires 708 kJ/mol of energy:

$$O^-(g) + e^- + 708 \text{ kJ} \longrightarrow O^{2-}(g) \tag{8.4}$$

The energy needed to force an electron onto an O^- ion is greater than the energy given off when an electron is added to an oxygen atom to form the O^- ion:

$$O(g) + e^- \longrightarrow O^-(g) + 141 \text{ kJ} \tag{8.5}$$

Addition of equations 8.5 and 8.4 gives equation 8.6:

$$[O(g) + e^- \longrightarrow O^-(g) + 141 \text{ kJ}] \tag{8.5}$$
$$+ \; [O^-(g) + e^- + 708 \text{ kJ} \longrightarrow O^{2-}(g)] \tag{8.4}$$
$$\overline{O(g) + 2e^- + 567 \text{ kJ} \longrightarrow O^{2-}(g)} \tag{8.6}$$

Thus, the formation of an oxide ion, O^{2-}, in the gas phase is endothermic. In addition, entropy (disorder) decreases when one oxygen atom and two electrons combine to form an oxide ion. Because of the combined unfavorable effects of both energy and entropy changes, oxide ions, O^{2-}, are not formed spontaneously in the gas phase.

However, oxide ions are common in solids. In solids, the negatively charged oxide ions are close to positively charged ions. The attraction between oppositely charged ions is large and the combination of anions and cations is very exothermic.

8.27   Predict whether each of the following changes will be exothermic or endothermic and explain your reasoning: (a) $Mg^{2+}(g) + e^- \longrightarrow Mg^+(g)$ and (b) $S^-(g) + e^- \longrightarrow S^{2-}(g)$.

8.7   CHEMICAL PROPERTIES AND THE PERIODIC TABLE

Electron configurations help us understand changes in atomic radii, ionization energies, and electron affinities across rows and down groups in the periodic table. In turn, ionization energies and electron affinities help us understand chemical reactions. It is easier to remember material that makes sense. Let's see how our new knowledge of electron configurations, ionization energies, and electron affinities helps make sense out of some of the chemical properties we talked about earlier.

Across rows of the periodic table, atomic radii decrease. Both actual and effective nuclear charge, ionization energy, and electron affinity increase. Elements lose electrons with increasing difficulty and, except for the noble gases, gain electrons more readily. In chemical reactions, metals tend to lose electrons and nonmetals to gain them. The elements change from metals to semimetals to nonmetals from left to right across every row of the periodic table.

The reactivity of main group metals increases down groups as ionization energies decrease. The reactivity of nonmetals decreases down groups of main group elements as electron affinities decrease. Elements become more metallic and less nonmetallic proceeding down groups. In Groups IVA, VA, and VIA, there is a transition from nonmetal to semimetal to metal going down the group.

The reactivity of transition metals decreases down groups. Atoms of elements in the third transition series have about the same radii as atoms of elements in the second transition series. However, nuclear charge is much higher in the third transition series. Therefore, ionization energies of elements in the third transition series are higher than ionization energies of elements in the second transition series.

Hydrogen

Like the elements of Group IA, hydrogen has a single electron in its outermost occupied shell and forms a $1+$ ion, H^+ (hydrogen ion). However, hydrogen is a nonmetal under ordinary conditions; the elements of Group IA are all metals. Like the elements of Group VIIA, hydrogen needs only one electron to reach the electron configuration of the next noble gas. Also, like the elements of Group VIIA, hydrogen is a nonmetal and forms a $1-$ ion, H^- (called a hydride ion), when it reacts with the metals in Groups IA and IIA (except beryllium). For example,

$$2Na(l) + H_2(g) \xrightarrow{>200\,°C} 2NaH(s)$$

Hydrogen occurs in most organic compounds and has a half-filled outer shell. In these respects, hydrogen is similar to carbon, which is in Group IVA. In many periodic tables, hydrogen is placed either in Group IA or in both IA and VIIA. In this book, hydrogen is not placed in any group; it is placed in the middle of the first row of the periodic table.

Industrially, most hydrogen is produced by the reaction of hydrocarbons, compounds of carbon and hydrogen such as methane (CH_4), and carbon monoxide with steam:

$$CH_4(g) + H_2O(g) \xrightarrow{\text{catalyst, 900 °C}} 3H_2(g) + CO(g)$$

$$CO(g) + H_2O(g) \xrightarrow{\text{catalyst, 200–400 °C}} H_2(g) + CO_2(g)$$

More hydrogen is used to make ammonia than for any other purpose. Ammonia is used mainly for fertilization of crops. It makes possible production of enough food to feed the world's ever-increasing population on the available land.

The Noble Gases (Group 0)

Helium has a filled s subshell in its outermost occupied energy level. The other noble gases all have filled s and p subshells in their outermost occupied energy levels. These electron configurations are very stable as shown by the high ionization energies of the noble gases, compared with other elements in the same period, and their negative electron affinities. The noble gases are unreactive and occur uncombined as monatomic gases in the atmosphere and trapped in small pockets in rocks. The lack of reactivity of helium and argon, the most common noble gases, is the basis for their main use—as an inert atmosphere for high-temperature processes such as arc welding. Although helium is denser than hydrogen, it is used to fill weather balloons and the Goodyear blimp because it is nonflammable.

No compounds of helium, neon, or argon are known. However, xenon does react with fluorine:

$$Xe(g) + F_2(g) \xrightarrow{>250 °C} XeF_2(g)$$

Other compounds of xenon with fluorine have also been prepared, as well as a few compounds of xenon with oxygen and nitrogen and some krypton compounds.

Alkali Metals (Group IA)

The alkali metals have the outer electron configuration ns^1 where n is the principal quantum number. Loss of the single s electron from the outer shell leaves the electron configuration of the preceding noble gas, a very stable electron configuration. For example,

$$[\text{Ne}]\,3s^1 \longrightarrow [\text{Ne}]^+ + e^-$$
$$\text{Na} \longrightarrow \text{Na}^+ + e^-$$

Loss of an electron to form a $1+$ ion is the basis of almost all the reactions of the alkali metals.

Compared with other elements in the same period, the alkali metals have large radii, low nuclear charges, and low ionization energies. As a result, the alkali metals are very reactive. They react with most nonmetals and with many compounds. For example, all react with cold water forming a solution of the metal hydroxide and hydrogen gas (Section 1.8). Using M to represent any of the alkali metals, the ionic equation for this reaction is

$$2M(s) + 2H_2O(l) \longrightarrow 2M^+(aq) + 2OH^-(aq) + H_2(g)$$

The reactivity of the elements in Group IA increases from top to bottom of the group in the periodic table because the ionization energies of the elements decrease from top to bottom in the group. Less energy is required to remove an electron from Na than from Li; less energy is required to remove an electron from K than from Na, and so forth.

About 10 million tons of salt are applied to U.S. roads each winter to prevent icing.

As a result of the great reactivity of the alkali metals, their compounds are much more important than the metals themselves. Most other sodium compounds are made from sodium chloride, which occurs naturally as rock salt and in brine (very salty water), and is third on the list of the most widely used natural inorganic materials in the United States (air is first and water, second). Other sodium compounds that are produced in billion-pound quantities every year* are (1) sodium hydroxide, which is manufactured by the electrolysis of brine and used as a base and for the manufacture of other compounds; (2) sodium carbonate, which occurs naturally and is used to make glass and to remove sulfur dioxide from the gases given off by power plants; (3) sodium sulfate, which is a by-product of hydrogen chloride manufacture and is used to make brown paper and corrugated boxes; and (4) sodium silicate, which is used in detergents. In most reactions that involve sodium compounds, sodium ions are spectator ions (Section 4.4). Potassium compounds would do as well but are more expensive. The only other alkali metal compound produced in billion-pound quantity is potassium chloride, which occurs naturally and is used as fertilizer.

Information about potassium chloride is from Chenier, P. J. Survey of Industrial Chemistry, 2nd ed.; VCH: New York, 1992; p 111.

In Earth's crust, there are about twice as many sodium ions as potassium ions; in seawater, there are 47 times as many sodium ions as potassium ions, although the solubilities of sodium and potassium salts in water are usually of the same order of magnitude. Potassium ions are essential to plant growth, and plants use up a large part of the potassium ions in groundwater.

Although the human body contains only a few tenths of a percent of sodium and potassium, the presence of both sodium ions and potassium ions is essential. Inside most human (and animal) cells, the concentration of potassium ions is high (about 0.15 M), and the concentration of sodium ions is low (about 0.005 M). The situation is reversed outside cells—for example, in blood plasma—where the concentration of potassium ions is low, and the concentration of sodium ions is high. The transport of Na^+ and K^+ into and out of cells is very important in physiology. The ratio of the concentrations of Na^+ and K^+ in animal cells controls the volume of the cells, makes possible the excitation of nerve and muscle cells by electrons, and drives the transport of sugars and amino acids. (Amino acids are organic acids similar to acetic acid that also contain amino groups NH_2. Amino acids are the building blocks for proteins.)

The Alkaline Earth Metals (Group IIA)

The alkaline earth metals are not as reactive as the alkali metals. They have smaller radii and higher nuclear charges than alkali metals in the same row of the periodic table. Also, they must lose two electrons to attain a noble gas electron configuration.

Like the reactivity of the alkali metals, the reactivity of the alkaline earth metals increases going from top to bottom of the group in the periodic table and for the same reasons. Calcium and the elements below it react with water at room temperature, forming the metal hydroxide and hydrogen gas:

$$M(s) + 2H_2O(l) \longrightarrow M(OH)_2(s) + H_2(g)$$

However, the reactions are not as vigorous as the reactions of the alkali metals. Magnesium requires hot water or steam to react. Magnesium is the only alkaline earth metal that is produced on a large scale. It is made by electrolysis of magnesium chloride from seawater. Magnesium is the lightest metal that can be used for construc-

---

*See Appendix E for a table showing names, formulas, U.S. production data, and uses for the "Top 50 Chemicals for 1993."

tion, and its major use is as a structural metal in aircraft, cars, and intercontinental ballistic missiles.

Calcium carbonate occurs in nature as limestone, chalk, marble, eggshells, pearls, coral, stalagtites and stalagmites, and shells, such as oyster shells. You are undoubtedly familiar with uses of some of these forms of calcium carbonate in your everyday life. Limestone is the fourth most widely used natural inorganic material in the United States. Two calcium compounds, calcium oxide, which is commonly called lime, and calcium chloride are made in billion-pound quantities. Lime is made by heating limestone:

$$CaCO_3(s) \xrightarrow{1200-1300\ °C} CaO(s)\ +\ CO_2(g)$$

Lime is used in the steel industry and in water treatment. Calcium chloride is obtained from brine, which contains a number of other cations besides Na^+, and is used to melt ice and snow on highways.

Properties of the first members of groups of main group elements are often out of line compared with the properties of the larger members of the group. In several cases, the properties of elements in the second period are similar to the properties of the third-period element in the next-higher-numbered group. For example, lithium is more like magnesium than sodium in several respects. Both lithium and magnesium react with oxygen to form oxides:

$$4Li(s)\ +\ O_2(g) \longrightarrow 2Li_2O(s)$$
$$2Mg(s)\ +\ O_2(g) \longrightarrow 2MgO(s)$$

However, sodium forms sodium peroxide, Na_2O_2, with oxygen; and potassium, rubidium, and cesium form superoxides, MO_2, where M represents the metal. (Remember Section 1.12.) Although the alkali metals are generally more reactive than the alkaline earth metals, the alkaline earth metals, but not the alkali metals, react with nitrogen to form nitrides:

$$3Mg(s)\ +\ N_2(g) \xrightarrow{heat} Mg_3N_2(s)$$

However, lithium, the first member of Group IA, reacts with nitrogen:

$$6Li(s)\ +\ N_2(g) \xrightarrow{heat} 2Li_3N(s)$$

Both calcium and magnesium compounds are abundant at Earth's surface and are involved in many biochemical processes in animals and plants. Calcium is the most abundant metallic element in the human body. Calcium phosphate, $Ca_3(PO_4)_2$, in the form of hydroxyapatite, $Ca_5(PO_4)_3OH$, is the chief inorganic material in teeth and bones, and calcium is needed to maintain the heart's rhythm, to clot blood, and in vision. The major source of calcium in the diet is dairy products. Magnesium is needed for many important enzymes (biochemical catalysts) to be active, for nerve impulse transmission, for muscle contraction, and for carbohydrate metabolism. The best sources of magnesium in the diet include cereals, peas and beans, nuts, meats, and dairy products. Chlorophyll, which is necessary for photosynthesis in plants, contains magnesium.

Beryllium and barium are both toxic. The toxicity of beryllium and barium is probably the result of the ability of Be^{2+} and Ba^{2+} to take the place of Ca^{2+} and Mg^{2+}, yet not function properly in their place.

Group IIIA

Aluminum is the most abundant metal in Earth's crust, making up 8.3% by mass of the crust. Aluminum-containing minerals, such as feldspars, weather to give clay, which forms a major part of soil. Geologists regard this weathering as the most important chemical reaction on Earth because without soil, no plants could grow.

Aluminum metal is made by electrolysis of bauxite, a mixture that typically contains 40–60% aluminum oxide. It is almost an ideal metal. Although its density is low, many alloys of aluminum are very strong. In addition, aluminum resists corrosion because the surface is protected by a thin coating of aluminum oxide. Aluminum is a good conductor of thermal energy and electricity and is very malleable and ductile. It is not attracted by magnets and does not spark. It can be shaped by all known metal-working methods, joined to itself or other metals by all procedures, and finished easily and cheaply by many processes. You are undoubtedly familiar with aluminum metal; like plastics, aluminum is everywhere in modern life.

Commercially, the most important compound of aluminum is aluminum sulfate, which is made by the reaction of aluminum oxide in bauxite with sulfuric acid:

$$Al_2O_3(s) + 3H_2SO_4(aq) \longrightarrow Al_2(SO_4)_3(aq) + 3H_2O(l)$$

Aluminum sulfate is used in water treatment and in the coating of paper.

Aluminum is generally considered to be nontoxic. Studies suggesting that abnormal concentrations of aluminum in brain tissue are the cause of Alzheimer's disease have been shown to be in error.

In addition to the familiar uses of aluminum shown in the photo, aluminum is used for construction and in transportation. More than 90% of all the overhead electric transmission lines in the United States are made of aluminum.

The Transition Metals (Groups IIIB–IIB)

All transition elements except palladium have either one or two s electrons in the outermost shell. Across a period, the transition elements differ from each other in the number of d electrons in the next-to-the-outermost $(n - 1)$ shell. Because the transition elements have similar outer electron configurations, they have some properties in common. All transition elements are metals. Most of the transition metals have high melting and boiling points and high densities and are hard and strong. Transition metals are less reactive than aluminum, and their main uses are as metals. Many are of major importance. Iron forms the skeleton of all large buildings, and most machinery is made of iron. Gold has been by far the most important means of payment between nations.

Many enzymes contain zinc ions, Zn^{2+}. Hemoglobin, which picks up oxygen in the lungs and delivers it to tissues throughout the body, contains iron(II) ions. Vitamin B_{12} contains cobalt(III) ions.

Most of the transition metals form more than one kind of ion—for example, iron forms iron(II) ion, Fe^{2+}, and iron(III) ion, Fe^{3+}. Most compounds that contain transition-metal ions are colored (see ►Figure 8.16). In addition, colored anions usually contain a transition element; the permanganate ion, MnO_4^-, which is purple, and the dichromate ion, $Cr_2O_7^{2-}$, which is orange-red, are examples. Most compounds of elements in Groups IA, IIA, and IIIA are colorless or white except for salts that have colored anions. Many of the transition metals and their compounds have interesting and useful magnetic properties. Iron, and to a lesser extent, cobalt and nickel, are **ferromagnetic,** that is, they *can be permanently magnetized.*

The only transition-metal compound that is produced in billion-pound quantities in the United States is titanium(IV) oxide, which occurs naturally but must be purified. Pure titanium(IV) oxide is white and is the most important white pigment. It is used in the manufacture of paint, as a surface coating on paper, and as a filler

◄ FIGURE 8.16 Solutions of compounds of the elements of the first transition series. The transition metal ions in the solutions are (left to right): Sc^{3+}, Ti^{3+}, V^{3+}, Cr^{2+}, Mn^{2+} (very pale pink), Fe^{3+}, Co^{2+}, Ni^{2+}, Cu^{2+}, and Zn^{2+}.

in plastics and rubber. Titanium(IV) oxide has the greatest covering power of all inorganic white pigments and is unreactive and therefore nontoxic.

The Lanthanides and Actinides

In general, lanthanides differ from each other in the number of electrons in the $4f$ subshell, and actinides differ from each other in the number of electrons in the $5f$ subshell. Because the energies of the $(n - 2)f$ orbitals and the $(n - 1)d$ orbitals are very similar, the ground-state electron configurations of some lanthanide and actinide atoms are slightly different than those predicted from the periodic table (see Figure 8.4).

All lanthanides have the same outer electron configuration, $6s^2$. Therefore, the chemical properties of all lanthanide elements are very much alike. The lanthanides occur together in nature and are difficult to separate from each other. All lanthanides form $3+$ ions with outer electron configuration $4f^n5d^06s^0$, where n can be 0 or any number from 1 to 13. In reactivity, the lanthanides are similar to magnesium; they react with hot water forming hydrogen gas, for example,

$$2La(s) + 6H_2O(l) \xrightarrow{heat} 2La(OH)_3(s) + 3H_2(g)$$

The characteristic property of the actinides is radioactivity. Only thorium and uranium occur naturally in useful amounts; all other actinides are man-made. In some cases, only a few atoms have been synthesized. During early work on the chemistry of the synthetic actinides, prediction of the properties of the actinides by analogy to the properties of the lanthanides was very helpful. Without an idea of what to expect, isolation of only a few atoms of radioactive material would have been impossible. Uranium and plutonium are used in nuclear power plants and in nuclear weapons.

The Halogens (Group VIIA)

Compared with other elements in the same period, the halogens have small radii and high nuclear charges. The outer electron configuration of the halogens,

$$ns^2np^5 \qquad \text{(neglecting filled } d \text{ and } f \text{ subshells)}$$

is just one electron short of the electron configuration of the noble gases, ns^2np^6 (again neglecting filled d and f subshells). The halogens have the highest electron affinities of any group in the periodic table and react with elements that lose electrons easily—that is, with active metals having low ionization energies—to form $1-$ ions. The reaction

$$2Na(s) + Cl_2(g) \longrightarrow 2NaCl(s)$$

is a typical example.

The reactivity of the halogens decreases from top to bottom of the group. Fluorine is the most reactive nonmetal and reacts with all other elements except helium, neon, and argon. Reactions involving fluorine often take place so fast and are so exothermic that explosions occur. Even asbestos and steam catch fire in fluorine gas! Fluorine is much more reactive than chlorine and bromine despite the fact that

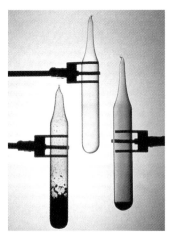

The halogens. Left to right: *Iodine is a black solid that sublimes to a purple vapor. Chlorine is a greenish yellow gas. (Fluorine, which is not shown because of its great reactivity, is also a greenish yellow gas.) Bromine is a red-brown liquid with a high vapor pressure.*

the electron affinity of fluorine is not as high as the electron affinity of chlorine but is about the same as the electron affinity of bromine. The great reactivity of fluorine is a result of the weakness of the F—F bond in molecules of F_2. The fluorine molecule is so small (•Figure 8.17) that repulsions between electrons weaken the bond. Beginning with chlorine, electron affinity decreases going down the group.

Chlorine is the most used halogen. About 24 billion pounds of chlorine are produced in the United States each year by electrolysis of brine or molten sodium chloride:

$$2NaCl(l) \xrightarrow{\text{electrolysis}} 2Na(l) + Cl_2(g)$$

Manufacture of organic compounds such as $ClCH_2CH_2Cl$ and H_2C $CHCl$, which are used in the manufacture of the plastic "vinyl," is the main use of chlorine. Large quantities are also used as a bleach and to purify water. Calcium chloride and hydrogen chloride are billion pound per year halogen compounds. Calcium chloride has already been discussed in connection with Group IIA; 90% of the hydrogen chloride produced is a by-product of the organic chemicals industry. Hydrogen chloride is used to manufacture both organic and inorganic chemicals.

Chloride ion, Cl^-, is the major anion outside cells. The average human body contains about 14 mg of iodine, mostly in the thyroid gland. The thyroid gland produces iodine-containing hormones that are necessary for normal metabolism in all the body's cells. A slight deficiency of iodine in the diet results in an enlarged thyroid gland. Severe and prolonged deficiency is more serious and leads to weak muscles, weight increase, loss of energy, and slow thinking. In a baby, lack of iodine may prevent normal physical and mental development. Iodine deficiency is prevented by use of iodized salt.

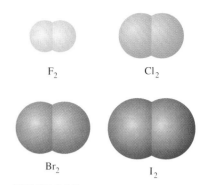

• FIGURE 8.17   Scale models of halogen molecules.

PRACTICE PROBLEMS

8.28   Write formulas for (a) sodium hydroxide, (b) sodium carbonate, (c) sodium sulfate, (d) sulfur dioxide, and (e) hydrogen chloride.

8.29   Name (a) NaH, (b) XeF_2, (c) KOH, (d) CO_2, and (e) Mg_3N_2.

8.30   Rank in order of increasing reactivity: K, Mg, Na.

8.31   Write molecular equations for the reaction of each of the following. If no reaction will take place, write "no reaction." (a) calcium with hydrogen (b) potassium with water (c) magnesium with steam (d) calcium with nitrogen (e) sodium with nitrogen (f) potassium with chlorine (g) calcium with chlorine (h) chlorine with aqueous sodium bromide (i) bromine with aqueous sodium chloride (j) ethane, CH_3CH_3, which is a gas, with steam

 8.8   A NEW WAY OF NUMBERING GROUPS IN THE PERIODIC TABLE

The method of numbering the groups of the periodic table that is used in this book is the one that is preferred by most American chemists. However, you need to be aware of two other systems because they are used in many books and articles. The

Arabic numbers are often used instead of Roman numerals in the preferred U.S. and previous IUPAC systems.

Period

	IA	IIA	Previous IUPAC									IIIB	IVB	VB	VIB	VIIB	VIIIB
	1	2	Current ACS and IUPAC									13	14	15	16	17	18
1	IA	IIA	Preferred U.S.		1 H							IIIA	IVA	VA	VIA	VIIA	0
																	2 He

| 2 | 3 Li | 4 Be | | | | | | | | | | 5 B | 6 C | 7 N | 8 O | 9 F | 10 Ne |

IIIA IVA VA VIA VIIA — VIIIB — IB IIB
3 4 5 6 7 — 8 9 10 — 11 12
IIIB IVB VB VIB VIIB — VIII — IB IIB

3	11 Na	12 Mg	13 Al	14 Si	15 P	16 S	17 Cl	18 Ar
4	19 K	20 Ca	21 Sc	22 Ti	23 V	24 Cr	25 Mn	26 Fe	27 Co	28 Ni	29 Cu	30 Zn	31 Ga	32 Ge	33 As	34 Se	35 Br	36 Kr
5	37 Rb	38 Sr	39 Y	40 Zr	41 Nb	42 Mo	43 Tc	44 Ru	45 Rh	46 Pd	47 Ag	48 Cd	49 In	50 Sn	51 Sb	52 Te	53 I	54 Xe
6	55 Cs	56 Ba	*71 Lu	72 Hf	73 Ta	74 W	75 Re	76 Os	77 Ir	78 Pt	79 Au	80 Hg	81 Tl	82 Pb	83 Bi	84 Po	85 At	86 Rn
7	87 Fr	88 Ra	◆103 Lr	104 Rf	105 Ha	106 Sg	107 Ns	108 Hs	109 Mt	110	111

□ Metal   ▨ Semimetal   ▨ Nonmetal

* Lanthanide series

57 La	58 Ce	59 Pr	60 Nd	61 Pm	62 Sm	63 Eu	64 Gd	65 Tb	66 Dy	67 Ho	68 Er	69 Tm	70 Yb

◆ Actinide series

89 Ac	90 Th	91 Pa	92 U	93 Np	94 Pu	95 Am	96 Cm	97 Bk	98 Cf	99 Es	100 Fm	101 Md	102 No

▪ FIGURE 8.18   Different methods used to number groups in the periodic table.

one labeled "Previous IUPAC" in ▪Figure 8.18 was used in Europe for some time. The one labeled "Current ACS and IUPAC" was recommended by the American Chemical Society in 1983 and is also recommended by the International Union for Pure and Applied Chemistry. The new numbering system was proposed to end the use of different systems by chemists in the United States and Europe. It is easier to learn than either of the old systems because the columns are numbered continuously from left to right. Notice how the units digit of the group number for most groups in the new numbering system is the same as the group number in both the old systems. For groups 1–12 and for groups 13–18 in the fourth period and beyond, the number of the group is the same as the total number of s, p, and d electrons outside the core. For main group elements, the units digit of the group number tells the number of electrons in the outermost shell except for helium. Because helium only has two electrons, it cannot have eight electrons in its outermost shell.

The names and symbols for elements 104–108 used in this book are those adopted by the ACS. The IUPAC has recommended different names and symbols for these elements: 104 dubnium, Db; 105 joliotium, Jl; 106 rutherfordium, Rf; 107, bohrium, Bh; 108, hahnium, Hn. The differences are the result of a long argument about who discovered elements 102, 104, 105, and 106. (There is no disagreement over the name and symbol for element 109.)

In the hydrogen atom, all subshells of a shell have the same energy. In atoms with more than one electron (that is, all atoms other than hydrogen), the energy of the orbitals in different subshells of the same shell is different. The inner electrons screen or shield the outer electrons from the positive charge on the nucleus. The **effective nuclear charge**—that is, the charge actually felt by the electrons—is less than the actual nuclear charge. The more the electrons in a subshell penetrate to the nucleus, the more they feel its attractive force, and the lower the energy of the subshell. For orbitals with the same principal quantum number, the order of increasing energy is $s < p < d < f$.

The electron arrangements of the elements can be considered to be built up by putting electrons into orbitals beginning with the lowest energy $1s$ orbital closest to the nucleus. This building up process is called **aufbau.** Orbitals that have the same energy are said to be **degenerate.** When putting electrons into degenerate orbitals, follow **Hund's rule:** Put one electron into each orbital before putting two in any one. The electrons in doubly occupied orbitals must have opposite spins.

Either a **complete electron configuration,** an **abbreviated electron configuration,** or an **orbital diagram** can be used to show the arrangement of the electrons. In abbreviated electron configurations, the inner or **core** electrons are represented by the symbol of the nearest noble gas with lower atomic number in square brackets. Orbital diagrams call attention to unpaired electrons. Atoms with one or more unpaired electrons are **paramagnetic** and are attracted into a magnetic field. Atoms with all electrons paired are **diamagnetic** and are slightly pushed out of a magnetic field. **Ferromagnetic** elements can be permanently magnetized.

Electron configurations make clear the reason for the structure of the periodic table. The elements in Groups IA and IIA are called **s-block elements,** and the elements in Groups IIIA–0 are called **p-block elements.** The **d-block elements** are the elements in Groups IIIB–IIB. The d-block elements in the fourth period are called the **first transition series.** The d-block elements in the fifth period are referred to as the **second transition series** and the d-block elements in the sixth period as the **third transition series.** The elements in the lanthanide and actinide series are called **f-block elements.**

The periodic table is the best memory aid for writing electron configurations. The ground-state electron configurations of ions can be written from the ground-state electron configurations of atoms. Ions and atoms that have the same electron configuration are called **isoelectronic.**

All elements in each group have similar outer electron configurations except for a few elements in Groups VIB and VIII. The physical and chemical properties of the elements and their compounds depend on the outer electron configuration; therefore, the members of a group have similar properties.

The **first ionization energy** is the amount of energy needed to remove the most loosely bound electron from a gaseous atom.

The energy needed to remove the second most loosely bound electron is called the **second ionization energy.** Successive ionization energies are named similarly. For all atoms, values of successive ionization energies are higher and higher. For each atom, there is a very large increase in ionization energy after all outer electrons have been removed. Values for successive ionization energies provide evidence for the existence of electron shells in atoms and support the idea that only outer electrons are involved in chemical change.

Electron affinity is a measure of an atom's tendency to gain an electron. The greater an atom's electron affinity, the more likely it is to gain an electron. Quantitatively, electron affinity is the negative of the enthalpy change associated with the addition of an electron to a mole of gaseous atoms to form gaseous ions with a $1-$ charge.

Although there is no sharp boundary that defines the outside of an atom or ion, we can think of the size of the atom as the volume containing about 90% of the total electron density around the nucleus and can represent this volume by a boundary surface. Positive ions are smaller than the atoms from which they are formed, and negative ions are larger than the atoms from which they are formed.

Across rows of the periodic table, atomic radii decrease. Actual and effective nuclear charges and ionization energy increase, and electron affinities become higher. Therefore, elements lose electrons with increasing difficulty and, except for the noble gases, gain electrons more readily. In chemical reactions, metals tend to lose electrons and nonmetals to gain them. The elements change from metals to semimetals to nonmetals from left to right across every complete row of the periodic table except the first.

The reactivity of main group metals increases down groups as ionization energies decrease. The reactivity of nonmetals decreases down groups as electron affinities become lower. Elements become more metallic and less nonmetallic proceeding down groups. In Groups IVA, VA, and VIA, there is a transition from nonmetal to semimetal to metal going down the group.

The reactivity of transition metals decreases down groups. Atoms of elements in the third transition series have about the same radii as atoms of elements in the second transition series. However, nuclear charge is much higher in the third transition series. Therefore, ionization energies of elements in the third transition series are higher than ionization energies of elements in the second transition series.

The high ionization energies of the noble gases compared with other elements in the same period and their negative electron affinities are evidence for the great stability of the electron configurations of the noble gases. The noble gases are unreactive and occur uncombined as monatomic gases.

Alkali metals have the outer electron configuration ns^1, where n is the principal quantum number. Loss of the single s electron from the outer shell leaves the stable electron configuration of the preceding noble gas. Therefore, loss of an electron to form a 1+ ion is the basis of almost all reactions of the alkali metals. Alkali metals are very reactive, and their compounds are much more important than the metals themselves. Alkaline earth metals are not as reactive as the alkali metals.

In reactions with active metals, the halogens achieve a noble gas electron configuration by gaining one electron to form 1− ions. Chlorine is the most used halogen; fluorine is the most reactive nonmetal.

Properties of the first members of groups of main group elements are often out of line compared with the properties of the larger members of the group. In several cases, the properties of elements in the second period are similar to the properties of the third-period element in the next-higher-numbered group. For example, lithium is more like magnesium than sodium in several respects.

Because the transition elements have similar outer electron configurations, they have some properties in common. All transition elements are metals. Most transition metals have high melting and boiling points and high densities and are hard and strong. Transition metals are less reactive than aluminum, and their major uses are as metals. Most transition metals form more than one kind of ion, and many compounds that contain transition elements are colored. The properties of all the lanthanides are very similar; the characteristic property of the actinides is radioactivity.

ADDITIONAL PRACTICE PROBLEMS

For information about the organization of Additional Practice Problems, Stop & Test Yourself, Putting Things Together, and Applications, see the beginnings of these sections in Chapter 1.

8.32 What does the symbol $4f^{11}$ mean? (8.1)

8.33 (a) Write the electron configuration for an excited hydrogen atom with the electron in the $4s$ orbital. (b) Would the energy of the hydrogen atom be any different with the electron in a $4p$ orbital? Explain your answer. (8.1)

8.34 In your own words, explain why the energy of the $2s$ subshell in a boron atom is lower than the energy of a $2p$ orbital in a boron atom. (8.1)

8.35 (a) Sketch the energy-level diagram including electrons for the oxygen atom in its ground state. (b) Write the complete electron configuration for the ground state of the oxygen atom. (8.1)

8.36 (a) What is meant by the term *core electrons?* (b) How are the core electrons represented in abbreviated electron configurations? (8.1)

8.37 (a) Define the term *degenerate* as applied to orbitals. (b) State Hund's rule in your own words. (8.1)

8.38 (a) What is paramagnetism? (b) What is the cause of paramagnetism? (8.1)

8.39 Using fluorine as an example, illustrate the four methods of showing electron configurations. Give one advantage and one disadvantage of each. (8.1)

8.40 (a) From its position in the periodic table, what electron configuration would you predict for platinum? (b) From Figure 8.4, what is the actual electron configuration of platinum? (8.2)

8.41 Which group has each of the following outer electron configurations in the ground state? (a) ns^2np^6 (b) ns^2 (c) ns^2np^1 (d) $(n-1)d^5ns^2$ (e) ns^2np^4 (8.2)

8.42 (a) How many completely filled subshells are there in the ground state of selenium? (b) How many completely filled shells are there in the ground state of nobelium? (8.2)

8.43 Use elements from the first transition series to illustrate the characteristics of the ground-state electron configurations of transition elements. (8.2)

8.44 In which block is each of the following elements? (a) Ce (b) Fe (c) K (d) P (e) U (8.2)

8.45 Explain why (a) there are 2 elements in the first period of the periodic table; (b) there are 8 elements in the second and third periods; (c) there are 18 elements in the fourth and fifth periods. (8.3)

8.46 How many elements are there in the sixth period? (8.3)

8.47 Write the abbreviated orbital diagram for the ground state of each of the following elements: (a) Al, (b) K, (c) Sn (d) Th, (e) Ti. (8.3)

8.48 Write the abbreviated electron configuration for the ground state of each of the following elements: (a) Cl, (b) Eu, (c) Ir, (d) Li, (e) Sr. (8.3)

8.49 Write the abbreviated orbital diagram for the ground state of each of the following elements: (a) I, (b) P, (c) Rn, (d) Si, (e) Zn. (8.3)

8.50 Write the abbreviated electron configuration and orbital diagram for the ground state of each of the following ions: (a) Ca^{2+}, (b) H^+, (c) H^-, (d) O^{2-}, (e) Ti^{4+}. (8.3)

8.51 How many unpaired electrons are there in the ground state of each of the following species? (a) a cobalt atom (b) a Co^{2+} ion (8.3)

8.52 Write the electron configuration for the first excited state of the sodium atom. (The first excited state is the lowest energy excited state.) (8.3)

8.53 The elements of which main group form (a) 2+ ions? (b) 2− ions? (c) 1− ions? (d) 3+ ions? (8.3)

8.54 Explain what is meant by the size of an atom. (8.4)

8.55 What is the distance between the centers of the atoms in a Br_2 molecule? (8.4)

8.56 (a) What is the lanthanide contraction? (b) Why does the lanthanide contraction occur? (8.4)

8.57 In each pair, which species has the larger radius? Explain your answers. (a) Mg or Na (b) K or Na (c) Na or Na^+ (d) Cl or Cl^- (e) Cu^+ or Cu^{2+} (8.4)

8.58 Which of the following is (are) isoelectronic with Kr? (a) Se^{2-} (b) Sr^{2+} (c) Ar (d) Zn^{2+} (8.4)

8.59 Which of the following ions are isoelectronic with a noble gas? (a) Al^{3+} (b) Ca^{2+} (c) F^- (d) Fe^{2+} (e) Pb^{2+} (8.4)

8.60 (a) Define *first ionization energy.* (b) Is the process exothermic or endothermic? Explain your answer. (c) Is ionization energy + or −? (d) How does the magnitude of the second ionization energy compare with the magnitude of the first ionization energy? Explain. (8.5)

8.61 In general, how does the magnitude of the first ionization energy change (a) across a row in the periodic table? (b) Down a group in the periodic table? (8.5)

8.62 Explain how ionization energies provide evidence for the existence of shells of electrons in atoms. (8.5)

8.63 Explain why Cu forms both Cu^+ and Cu^{2+} ions but K forms only K^+ ion. (8.5)

8.64 Explain why the first ionization energy of Cs is less than the first ionization energy of Rb, but the first ionization energy of Au is greater than the first ionization energy of Ag. (8.5)

8.65 The first, second, and fourth ionization energies for Mg and Al are of the same order of magnitude. However, the third ionization energy for Mg is much larger than the third ionization energy for Al. Explain why. (8.5)

8.66 (a) Define *electron affinity*. (b) In general, how does electron affinity change across a row in the periodic table? (c) In general, how does electron affinity change down a group in the periodic table? (8.6)

8.67 Explain why (a) the electron affinities of the elements of Groups IIA and 0 are negative; (b) the electron affinity of N is less than the electron affinities of C and O. (8.6)

8.68 Why are the electron affinities of N, O, and F low compared with the electron affinities of P, S, and Cl? (8.6)

8.69 Predict whether each of the following changes will be exothermic or endothermic and explain your reasoning:
(a) $Al(g) + e^- \longrightarrow Al^-(g)$
(b) $Ne(g) + e^- \longrightarrow Ne^-(g)$
(c) $Cl(g) + e^- \longrightarrow Cl^-$ (8.6)

8.70 From the properties of aluminum and indium given in the table below, estimate the properties of gallium. (8.7)

Element	Density, g/cm³	Boiling Point, °C[a]	Formula Oxide	Formula Nitride
Aluminum	2.7	2327	Al_2O_3	AlN
Indium	7.3	2070	In_2O_3	InN

[a]Although the melting points of aluminum and indium are 660 and 157 °C, respectively, the melting point of gallium is only 30 °C. There is no simple explanation for the unexpectedly low melting point of gallium.

8.71 Write formulas for (a) barium oxide, (b) lithium carbonate, (c) calcium nitride, (d) lanthanum chloride, (e) potassium permanganate. (8.7)

8.72 Name (a) Zn_2SiO_4, (b) XeF_6, (c) $NdBr_3$, (d) $Na_2Cr_2O_7$, (e) AlN. (8.7)

8.73 Rank in order of increasing reactivity: F, O, and S. (8.7)

8.74 Write molecular equations for the reaction of each of the following. If no reaction will take place, write "no reaction." (a) Propane, $CH_3CH_2CH_3$, which is a gas, with steam (b) Strontium and nitrogen (c) Aluminum and water (d) Chlorine and aqueous sodium sulfide (e) Sodium and hydrogen (8.7)

8.75 (a) In terms of the ground-state electron configuration of the hydrogen atom, explain why hydrogen could be placed either in Group IA or in Group VIIA of the periodic table. (b) What does the ground-state electron configuration of hydrogen have in common with the ground-state electron configuration of carbon?

8.76 Which of the following electron configurations describe atoms in excited states? For those that describe atoms in excited states, write the electron configuration of the ground state. (a) $[Kr]4d^{10}5s^2$ (b) $[Xe]4f^76s^2$ (c) $[He]2s^22p^53s^1$ (d) $[Ne]3s^23p^1$ (e) $[Xe]4f^{14}5d^{10}6s^16p^4$

8.77 Identify each of the following atoms from its ground-state electron configuration. (a) $1s^22s^22p^3$ (b) $[Xe]5d^16s^2$ (c) $[Rn]6d^17s^2$

8.78 Second electron affinities

$$A^-(g) + e^-(g) \longrightarrow A^{2-}(g)$$

and higher electron affinities are always negative. Explain why.

8.79 Explain why (a) atomic radii change less across the ten elements of the first transition series than they do across the eight elements of the third period; (b) across each of the three transition series, atomic radius first decreases slightly and then increases slightly; (c) the radii of the elements in the third transition series are about the same as the radii of the elements of the same group in the second transition series.

8.80 Explain briefly why (a) the first ionization energy of Al is less than the first ionization energy of Mg; (b) the first ionization energy of S is lower than the first ionization energy of P; (c) although the first ionization energy of Na is lower than the first ionization energy of Mg, the second ionization energy of Na is higher than the second ionization energy of Mg; (d) the first ionization energy of Ga is about the same as the first ionization energy of Al.

8.81 A gaseous ion, A^+, has two unpaired electrons in its ground state. If A is a main group element, to which group does A belong?

8.82 (a) From its position in the periodic table, what electron configuration would you predict for molybdenum? (b) From Figure 8.4, what is the actual electron configuration of molybdenum? (c) How could the predicted and actual electron configurations be distinguished experimentally?

8.83 Identify each of the following ions from its ground-state electron configuration. (a) $[Ar]3d^4$, charge 3+ (b) $[Kr]4d^{10}$, charge 1+ (c) $[Kr]4d^{10}5s^25p^6$, charge 1−

8.84 Removal of one electron from an atom of gaseous oxygen yields a gaseous oxygen ion, O^+:

$$O(g) \longrightarrow O^+(g) + e^-(g)$$

The electron can be removed from any occupied orbital:

$$1s^22s^22p^4 \longrightarrow 1s^12s^22p^4 + e^-$$
$$1s^22s^22p^4 \longrightarrow 1s^22s^12p^4 + e^-$$
$$1s^22s^22p^4 \longrightarrow 1s^22s^22p^3 + e^-$$

(a) Which process will require the least energy? (b) Which process will require the most energy? (c) In which process is the energy required the first ionization energy for oxygen?

8.85 Write a paragraph summarizing what Section 1.8 says about the properties of metals.

8.86 Make an outline of what Section 8.7 says about transition metals.

8.87 (a) Do the lanthanides occur free in nature? Explain your answer. (b) Lanthanide ores are mixtures that contain all the lanthanides. Why?

8.88 What is the characteristic property of the actinides?

8.89 How many kilojoules are required to remove three electrons from one mole of gaseous aluminum atoms to form one mole of gaseous Al^{3+} ions?

8.90 Describe the results of dietary iodine deficiency.

8.91 Which substances are colored? (a) N_2 (b) $Cr(NO_3)_3$ (c) $KMnO_4$ (d) $Ca(NO_3)_2$ (e) $Na_2Cr_2O_7$

8.92 Given the successive ionization energies for magnesium in Table 8.1, would you expect any large jumps beyond those shown? If so, how many? Explain your answer.

STOP & TEST YOURSELF

1. The ground-state electron configuration of neodymium, atomic number 60, is
   (a) $1s^22s^22p^63s^23p^64s^23d^{10}4p^65s^24d^{10}5p^66s^24f^5$.
   (b) $1s^22s^22p^63s^23p^64s^23d^{10}4p^65s^24d^{10}5p^66s^24f^4$.
   (c) $1s^22s^22p^63s^23p^64s^23d^{10}4p^65s^24d^{10}5p^66s^14f^5$.
   (d) $1s^22s^22p^63s^23p^64s^23d^{10}4p^65s^24d^{10}5p^64f^6$.
   (e) None of the preceding answers is correct.

2. Which of the following statements is true of the ground state of the Co^{2+} ion? (a) The number of unpaired electrons is 0 and the Co^{2+} ion is paramagnetic. (b) The number of unpaired electrons is 0 and the Co^{2+} ion is *not* paramagnetic. (c) The number of unpaired electrons is 3 and the Co^{2+} ion is paramagnetic. (d) The number of unpaired electrons is 3 and the Co^{2+} ion is *not* paramagnetic. (e) The number of unpaired electrons is 1 and the Co^{2+} ion is paramagnetic.

3. The core for I is (a) [At], (b) [Br], (c) [Kr], (d) [Te], (e) [Xe].

4. Choose the electron configuration that corresponds to an excited state. (a) $[Ar]3d^34s^2$ (b) $[Kr]5s^1$ (c) $[Ar]3d^54s^1$ (d) $[Ne]3s^13p^6$ (e) $[Xe]4f^{14}5d^{10}6s^2$

5. Choose the orbital diagram or electron configuration that is possible.
   (a) $1s^22s^22p^63s^23p^73d^64s^2$

   (b)

1s	2s	2p

3s	3p	3d	4s

   (c)

1s	2s	2p

3s	3p	3d	4s

   (d) $1s^22s^22p^63s^23p^63d^74s^2$
   (e) $1s^22s^22p^62d^73s^23p^64s^2$

6. The electrons in which subshell experience the largest effective nuclear charge in a multielectron atom?

(a) $4s$ (b) $4p$ (c) $4d$ (d) $4f$ (e) All of the preceding experience the same effective nuclear charge.

7. How many electrons are required to fill a d subshell?
   (a) 2 (b) 6 (c) 10 (d) 14 (e) 18

8. Arrange the following in order of decreasing radii:

$$K \quad Mg \quad Mg^{2+} \quad Na$$

(a) $Mg^{2+} > Mg > Na > K$ (b) $K > Na > Mg^{2+} > Mg$
(c) $K > Mg > Na > Mg^{2+}$ (d) $K > Na > Mg > Mg^{2+}$
(e) None of the preceding answers is correct.

9. Which of the following sets is arranged in order of increasing radii? (a) $Br^- < Br < Cl$ (b) $Cl < Br^- < Br$ (c) $Cl < Br < Br^-$ (d) $Br^- < Cl < Br$ (e) $Br < Br^- < Cl$

10. In which of the following series are the atoms arranged in order of decreasing ionization energy?
    (a) $Sr > Ca > Mg$ (b) $Li > Be > B$ (c) $O > F > Ne$
    (d) $Ne > Na > Mg$ (e) $Cl > Br > I$

11. The following successive values of ionization energy (in kJ/mol) are for the atom _____?

$$I_1 \ 799 \quad I_2 \ 2422 \quad I_3 \ 3657 \quad I_4 \ 25\,019 \quad I_5 \ 32\,660$$

(a) B (b) Be (c) C (d) Li (e) N

12. Among the following elements the one with the highest electron affinity is _____?
    (a) Ar (b) Br (c) Cl (d) S (e) Se

13. Which of the following elements has no known compounds?
    (a) Xe (b) N (c) Na (d) Ne (e) Ni

14. Which of the following elements does not react with water at room temperature?
    (a) Ca (b) K (c) Li (d) Mg (e) Na

15. Vanadium is a(an)
    (a) *s*-block element.   (b) *p*-block element.
    (c) *d*-block element.   (d) *f*-block element.
    (e) *g*-block element.

8.93 What is the electron affinity of the following ions in kJ/mol? (a) Na^+ (b) Mg^{2+}

8.94 (a) The Group IB elements have the same outer electron configuration (ns^1) as the Group IA elements. Why are the Group IB elements much less reactive than the Group IA elements? (b) Why is K more reactive than Na? (c) Why is Ag more reactive than Au?

8.95 State the modern periodic law.

8.96 Which of the following ions are likely to be formed, and which are unlikely to be formed? (a) Br^- (b) Fe^{2+} (c) Fe^{8+} (d) Pb^{2+} (e) Rb^+ (f) S^{6+} (g) Xe^-
Explain your answers. Do all the ions that are likely to be formed have noble gas electron configurations?

8.97 Which of the following species is paramagnetic in the ground state? (a) Zn (b) Zn^{2+} (c) Cl (d) Cl^-

8.98 Use orbital diagrams to explain why the third ionization energy for Fe is lower than the third ionization energy for Mn.

8.99 (a) Use complete electron configurations to show the first three steps in the ionization of calcium. (b) Explain why the second ionization energy for calcium is greater than the first ionization energy. (c) Explain why the third ionization energy for calcium is *much* greater than the second ionization energy. (d) In the ionization of aluminum, between which steps would the biggest increase in ionization energy take place? Explain your answer.

8.100 Some chemists think that element number 114 will be synthesized. (a) Write the abbreviated ground-state electron configuration for element number 114. (b) What type of element will 114 be? (c) In which period will 114 be? (d) In which group will 114 be? (e) Which known element will 114 be most like? (f) Using Ele as the symbol for element 114, write the probable formula or formulas for the compound or compounds that the element would be expected to form with oxygen.

8.101 Explain why Mg has a higher first ionization energy than both Na and Al and a negative electron affinity.

8.102 The energy levels in a hydrogen-like ion (a hydrogen-like ion is an ion that contains only one electron) can be calculated from the equation

$$energy_n = -\frac{(2.18 \times 10^{-18}\ J)Z^2}{n^2}$$

where $energy_n$ is the energy of the nth shell, Z is the atomic number of the element, and n is the principal quantum number. (a) Calculate the ionization energy (in kJ/mol) of the He^+ ion. (b) Qualitatively, how would the size of the $1s$ orbital in the He^+ ion compare with the size of the $1s$ orbital in the hydrogen atom?

8.103 In general, elements that have high first ionization energies also have high electron affinities. (a) Explain why. (b) Which group is an exception?

8.104 In which block are each of the following found? (a) nonmetals (b) semimetals (c) the halogens (d) the transition metals (e) the actinides

8.105 Of the following elements

$$Ca\quad Cl_2\quad H_2\quad Na\quad Ne$$

(a) Which will be gases under ordinary conditions?
(b) Which will be liquids under ordinary conditions?
(c) Which will be solids under ordinary conditions?

8.106 The formula for the simplest compound of carbon and hydrogen, methane, is CH_4. What are the formulas for the simplest compounds of each of the following elements with hydrogen? (a) silicon (b) germanium (c) tin

8.107 Figure 7.31 shows the probabilities of finding the electron as a function of distance from the nucleus for the $1s$, $2s$, and $3s$ orbitals of the hydrogen atom. Which other orbitals are degenerate with each of these three orbitals in the hydrogen atom?

8.108 For each of the following nuclides, (a) tell how many protons and neutrons are in the nucleus and how many electrons are outside the nucleus. (b) Write the electron configuration. (i) potassium-41 (ii) bromine-81 (iii) silver-107 (iv) lead-208

8.109 Write formulas for and name each of the following.
(a) a common cation that is isoelectronic with argon
(b) a common anion that is isoelectronic with argon
(c) a common cation that is not isoelectronic with a noble gas

8.110 No known elements have electrons in a g subshell in the ground state. If an element with electrons in the g subshell in the ground state is synthesized, (a) what is the lowest numbered period that the element can be in? (b) How many orbitals will there be in the g subshell?

8.111 (a) Predict the state (solid, liquid, or gas) under ordinary conditions of At_2. (b) Predict the atomic number of the next halogen after astatine and give the expected ground-state electron configuration.

bromine   iodine

8.112 The first ionization energy for potassium is given in Figure 8.10. What type of electromagnetic radiation is needed to ionize potassium?

8.113 Write molecular equations for each of the following reactions. If no reaction will take place, write "no reaction." (a) magnesium metal with room-temperature water (b) barium metal with room-temperature water (c) potassium metal with bromine vapor (d) neon with fluorine (e) sodium metal with nitrogen

8.114 In general, how do the following properties compare for metals and nonmetals? (a) atomic radius (b) radius of commonly found ions (c) first ionization energies (d) electron affinities (e) For electron affinities, which group is a marked exception to the general rule?

8.115 The German chemist Lothar Meyer, who was educated as a physician, developed the periodic law a year before Mendeleev's periodic law was published but did not publish his work until after Mendeleev and did not use the periodic law to predict the existence and properties of unknown elements. Meyer based the periodic law largely on measurements of physical properties such as molar volume. The molar volume of an element is the volume occupied by one mole of the element. (a) Calculate the molar volume of sodium. The density of sodium is 0.97 g/cm^3. (b) Meyer found that the molar volumes of elements that are solids under ordinary conditions are periodic functions of atomic mass. Are molar volumes of elements that are gases under ordinary conditions periodic functions of atomic mass? Explain your answer.

8.116 (a) Write the formulas and names for the hydrides (binary compounds with hydrogen such as HCl) of the elements of the second period. (Use common names if you know them.) (b) Classify the hydrides of the elements of Groups IA, IVA, VA, VIA, and VIIA as predominantly ionic or predominantly molecular compounds.

8.117 From the data in the accompanying table and information given elsewhere in this book, estimate the following properties for francium. (a) density (b) melting point (What will be the state of francium under ordinary conditions?) (c) radius of atom (d) formula of ion (e) radius of ion (f) first ionization energy (g) reactivity toward water (h) formula of chloride (i) solubility of francium compounds.

Element	Density, g/cm^3	Melting Point, °C
Lithium	0.53	181
Sodium	0.97	98
Potassium	0.86	63
Rubidium	1.53	39
Cesium	1.88	28

8.118 (a) From your everyday experience, which metal is more reactive, aluminum or iron? Explain your answer. (b) According to the activity series for metals in Table 4.5, which is more reactive, aluminum or iron? Explain your answer. (c) The explanation is that both metals react with oxygen in air to form oxides. However, aluminum oxide is white and forms a smooth layer that sticks to the surface of the metal. Hydrated iron(III) oxide (rust) is red and forms a rough layer that flakes off the surface of the metal. Write equations for the reactions of aluminum and iron with oxygen.

8.119 The metathesis reaction between aqueous sodium nitrate and aqueous potassium chloride is used to manufacture potassium nitrate. (a) What can you conclude about the relative solubilities of sodium nitrate, potassium chloride, potassium nitrate, and sodium chloride from this information? (b) Should the aqueous sodium nitrate and aqueous potassium chloride be concentrated or dilute? (c) Write molecular, complete ionic, and net ionic equations.

8.120 Write molecular equations: (a) Potassium hydroxide is made by electrolysis of aqueous potassium chloride; (b) potassium dihydrogen phosphate, potassium hydrogen phosphate, and potassium phosphate are made from aqueous potassium hydroxide.

8.121 The sulfuric acid used to make $Al_2(SO_4)_3$ is 77.67% by mass H_2SO_4 and has a density of 1.706 g/mL. What is its molarity?

APPLICATIONS

8.122 Metals are usually found in nature together with other metals that have similar properties and similar radii. With which metal would you expect to find hafnium? Explain your answer.

8.123 The nuclide strontium-90 is considered the most dangerous part of radioactive fallout because it can replace calcium in foods and become concentrated in bones and teeth. Explain why strontium can replace calcium.

8.124 Mercury(I) chloride, Hg_2Cl_2, is commonly known as calomel and is used in antiseptic salves. Mercury(II) chloride, $HgCl_2$, is used as an agricultural fungicide and a topical antiseptic and also as a catalyst in the manufacture of "vinyl." Write electron configurations for the (a) mercury atom, (b) mercury(I) ion, (c) mercury(II) ion.

8.125 The orange-yellow color of sodium-vapor streetlights results from electrons in sodium atoms falling from $3p$ to $3s$ orbitals. The wavelength of one orange-yellow line in the spectrum of sodium is 589 nm (see Figure 7.24). (a) Write the electron configuration of the excited state of the sodium atom that is involved in this change in energy level. (b) Calculate the difference in energy between the $3p$ and $3s$ orbitals of the sodium atom.

8.126 (a) Production of sodium hydroxide by the electrolysis of brine also yields chlorine. For each pound (1.00 lb) of sodium hydroxide that is produced by the electrolysis of brine, how many pounds of chlorine are produced? (b) Demand for sodium hydroxide is increasing because of its environmental uses, such as neutralization of acidic waste water. Demand for chlorine is decreasing because chlorofluorocarbons have been banned as ozone depleters and because use of chlorine to bleach paper is being phased out due to the dioxin discharges that result. Plants that produce sodium hydroxide by reaction of lime with aqueous sodium carbonate, which were closed because they were unprofitable in the early 1990s, are being reopened in 1995. Write the molecular equation for this reaction. What compound is a by-product? Can it be recycled if there is no market for it? Explain your answer.

8.127 Demand for anhydrous metal halides, high-tech materials used in solid-state synthesis, is increasing. A recent ad lists aluminum iodide, barium bromide, cesium iodide, cobalt(II) chloride, scandium(III) chloride, and many others. (a) Write formulas for these compounds. (b) In laboratory quantities, aluminum iodide costs $167.40 for 25 g according to the ad. To how many significant figures must the mass really be accurate before the customers start complaining? How much does one mole cost? (c) These anhydrous halides generally contain less than 100 ppm of water. What percent water is 100 ppm? (d) They are packaged in sealed ampules under argon. Why?

8.128 Magnesium hydroxide is being used to replace sodium hydroxide for the neutralization of acidic waste water. (a) Write molecular, complete ionic, and net ionic equations for the neutralization of HCl(aq) by magnesium hydroxide. (b) What is the advantage of magnesium hydroxide over sodium hydroxide? (c) Another new use for magnesium hydroxide is as a flame-retardant additive for plastics. This use is based on the decomposition of the hydroxide to oxide and water. Explain how magnesium hydroxide slows the burning of plastics. (*Hint:* Calculate ΔH_{rxn}.) (d) The halogen-containing materials previously used as flame retardants release toxic and corrosive gases when heated. Does magnesium hydroxide?

8.129 (a) What is the electron affinity of chlorine? (b) What is the first ionization energy of the chloride ion? (c) Electrons can be removed from chloride ions with laser light. What is the maximum wavelength of electromagnetic radiation that has enough energy per photon to remove one electron from a gaseous chloride ion? In what region of the electromagnetic spectrum is this wavelength?

8.130 The iron in some cereals is in the form of iron(III) phosphate; in others it is tiny particles of iron metal. Iron capsules taken as dietary supplements contain iron(II) sulfate. (a) Write formulas for the two iron compounds. (b) Which of these forms of iron is soluble in water? (c) Iron(III) phosphate dissolves rapidly in the stomach, which contains HCl(aq). Write molecular, complete ionic, and net ionic equations for this reaction. (d) Iron metal also dissolves in hydrochloric acid. [Iron(II)

chloride is formed.] Write molecular, complete ionic, and net ionic equations for this reaction. (e) Very little metallic iron, however, dissolves in stomach acid in the time that cereal is in the stomach. Outline a simple experiment that you could do to separate the iron from a sample of your cereal if it is in the form of metallic iron. (f) The label of a common iron supplement says that each capsule contains 159 mg of dried ferrous sulfate. [Ferrous sulfate means iron(II) sulfate.] How many milligrams of iron are provided by the capsule?

8.131 After iron, aluminum is the most widely used metal. The density of iron is 7.9 g/cm^3; the density of aluminum is 2.7 g/cm^3. (a) If the mass of an engine part is 1.73 kg when made of iron, what is the mass of an otherwise identical part made of aluminum? (b) What percent of the mass of the heavier part is the mass of the lighter part? (c) If the strengths of the parts are the same, what is one advantage of the aluminum part?

8.132 Hydrogen fluoride is produced by the reaction of calcium fluoride with sulfuric acid. Calcium sulfate is a by-product. (a) What is ΔH°_{rxn} for this reaction? Is the reaction exothermic or endothermic? (b) How many liters of HF(g) at 20 °C and 760 mmHg can theoretically be obtained by the reaction of 20.52 g CaF_2 and 28.36 g H_2SO_4? (c) Until the early 1990s, production of chlorofluorocarbons, compounds such as CCl_3F and CCl_2F_2, which were used as refrigerants, blowing agents for plastic foams, and solvents was a major use of HF(g). Write complete structural formulas for CCl_3F and CCl_2F_2. (Today the production of chlorofluorocarbons is prohibited by international agreement because they are believed to be important contributors to the destruction of the ozone layer.)

8.133 Special fire extinguishers must be used for burning metals. (a) Write the equation for the combustion of magnesium metal. (b) If water is poured on burning magnesium, an explosion results. Write equations for the reactions leading to explosion. (c) Calculate ΔH°_{rxn} for the reaction of magnesium oxide with carbon dioxide. What will happen if a carbon dioxide extinguisher is used on burning magnesium?

Chemistry and Microscopic Marine Life

JUDY WILLIAMS-HOWZE

Research Scientist,
Marine Science Program/Baruch
Institute for Marine Biology
and Coastal Research,
University of South Carolina

B.S. Zoology
University of Southern Mississippi

M.S. Zoology
Louisiana State University

Ph.D. Biology
University of South Carolina

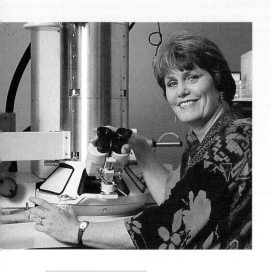

In all honesty, I never used to think of using chemistry in my chosen field of biology. I considered myself a classical biologist, interested in the whole organism, as I studied taxonomy (how organisms are classified into groups). Before going to college, I worked as a field hand on a marine science research boat, where I helped collect and identify many marine animals in the Gulf of Mexico, including my favorites—tiny invertebrate organisms. Later, I got a job as a trainee in histology at a veterinary school. There I learned how to prepare animal tissue for research and study of diseases. This was extremely interesting work, and I became a certified technician in histology. I learned that you must add specific chemicals to "fix" the tissue so it will not decay and the cells will look normal. Dyes are used to stain the tissues so that pathologists can analyze them and detect any abnormalities. In order for the dyes to work on particular types of tissues (for example, on muscles rather than fats), specific chemical compounds are used that allow the dyes to bind only to the tissue under study (it all has to do with the molecular structure).

Through this work, my initiation into chemistry began. In a roundabout way, I learned a lot of chemistry without ever studying it specifically. Because I came into chemistry through the "back door," it never seemed to scare me as it does so many students. I just accepted that you did things a certain way because, chemically, that is the only way you could get them to work!

When I went to college I once again pursued my interest in marine organisms. For my master's thesis and doctoral dissertation, I combined my fascination with how tiny marine organisms work and my interest in tissue histology to study the functional morphology of a group of tiny marine invertebrates called copepods (about the size of a flea). Functional morphology is the study of how organisms are structurally put together to allow them to do what they can do. For instance, a frog has webbed feet to aid it in swimming, and a bird has hollow bones so it will be lighter in the air. Despite their small size, copepods have very complex life histories and can do some incredible things in seawater, such as building a mucus tube to live in, or they can secrete a cocoon-like cyst around themselves, in which they undergo a dormant or resting state for three to four months. This is amazing when you consider how small these creatures are and how few cells they have.

Because of their size, I could not study copepods using the standard microscopes and tissue techniques that I had used before for animal tissue. For copepods, I needed the more powerful electron microscopes, which allow you to view cells and the organelles within them at magnifications of 1000 to 100,000 times their normal size. This is wonderful when your subjects are less than a millimeter to begin with! There are two standard types of these microscopes, scanning and transmission electron microscopes, both of which use electron beams rather than light beams. The scanning microscope does just as its name implies: it scans the outer surface of the organism only. The transmission microscope, on the other hand, allows us to look at individual cells within organisms by examining thin slices of their tissues.

As you might have guessed, there is a LOT of chemistry involved in preparing samples for these types of analysis. First, one must again "fix" the tissue so it won't degrade or decay. This involves directly linking up the proteins with a special chemical. Then, the tissues must be completely dehydrated (removing all water from cells) because water molecules interfere with the electrons coming down the microscope. If the specimen is to be scanned, it must be coated with a special thin layer of gold. The gold causes the electrons to be bounced off the sample surface. The electrons are then collected, so to speak, and viewed on something similar to a TV screen. For transmission microscopy, the tissue must be totally embedded in a very hard plastic that will then be cut very thin (about five millionths of a centimeter), stained (again with special dyes), and viewed on the microscope.

My current position is with the Marine Science Program and the Baruch Institute for Marine Biology and Coastal Research, where I continue my work on the tiny copepods. However, I study the copepods on even a smaller scale by investigating the molecules within their nervous system that allows these animals to know when to begin and when to stop their dormant state (called diapause). It is suspected that these molecules are neuropeptides or neurohormones. The suite of neurohormones that control functions such as growth, molting (shedding of old shells), and mate finding are well described for insect species. My goal now is to see how biochemically related the land-based insects are to the sea-based copepods, which exhibit many of the same complex behaviors as the insects.

As I pursue this area of research, I am reminded daily that all living things have a chemical basis, from your cells, to the processes of growth and development, to the hormones that control behavior. P.S. I did finally take formal chemistry and cell biology/biochemistry courses in college!

CHEMICAL BONDS

This trellis is made of PVC and will never rot, be eaten by termites, or need painting.

O nly the noble gases occur naturally as separate atoms. In most materials, atoms are joined by chemical bonds. **Chemical bonds** are the *forces that hold atoms together in molecules of elements* such as O_2 and N_2, *in compounds, and in metals.* In this chapter, we will discuss the bonds that hold atoms together in ionic compounds and in molecules. The bonding between metal atoms in metals is discussed in Chapters 10 and 12.

— *Why is a study of chemical bonds a part of general chemistry?*

When chemical reactions take place, bonds break in the reactants and form in the products. Electrons rearrange around the nuclei. An understanding of chemical bonds is basic to an understanding of chemistry and answers questions such as the following: Why do sodium metal and chlorine gas react to form sodium chloride (salt)? Why is salt a high melting solid that is quite soluble in water? Why do aqueous solutions of salt conduct electricity whereas neither pure water nor aqueous solutions of sugar conduct electricity?

— *What do you already know about chemical bonds?*

Some compounds, such as sugar and water, are molecular compounds that have molecules as units (Section 1.9). A molecule is the smallest unit of an element or molecular compound that can exist. Molecules of elements consist of one or more atoms of the same element; molecules of compounds are composed of two or more atoms of at least two elements. Molecules do not have net electric charges. The bonds in molecules are represented by sticks in ball-and-stick models and by lines in complete structural formulas (Section 3.9). Compounds composed of two or more nonmetals, such as sugar, $C_{12}H_{22}O_{11}$, and water, H_2O, and compounds composed of unreactive metals (metals with high ionization energies) and nonmetals, such as $HgCl_2$, are molecular.

When atoms of metals that have low ionization energies, such as the elements of Group IA, and the lower members of Group IIA, combine with nonmetals, especially nonmetals with high electron affinities such as oxygen and the halogens, ionic compounds form. The units of ionic compounds are ions, for example, sodium ions, Na^+, and chloride ions, Cl^-, in the case of NaCl. In an ionic solid, each ion is surrounded by ions of opposite charge (remember Figure 1.18); there are no molecules in ionic compounds. Because compounds are electrically neutral, the proportions of the positive and negative ions must be such that the total number of positive charges equals the total number of negative charges. The formulas of ionic compounds give the simplest ratios of + and − ions; that is, they are empirical formulas. In an ionic compound, the oppositely charged ions are held together

by the attraction between positive and negative charges. Because the attraction between positive and negative charges is strong, ionic compounds are usually hard and have high melting points.

Monatomic ions are formed by transfer of electrons from metal to nonmetal. Often, enough electrons are transferred to give each atom the same number of electrons as the noble gas with the closest atomic number. Loss of an electron or electrons leaves an atom with a positive charge; positively charged ions are called cations. Metals usually form cations. Gain of an electron or electrons gives an atom a negative charge; negatively charged ions are called anions. Nonmetals usually form anions. Some ions are

This wok is heated by the combustion of natural gas, which is mainly methane, a typical molecular compound.

polyatomic, that is, consist of more than one atom; the ammonium ion, NH_4^+, the hydroxide ion, OH^-, and the permanganate ion, MnO_4^-, are examples of polyatomic ions.

A salt mine under Lake Erie three miles out from Cleveland, Ohio. Common table salt (sodium chloride) is a typical example of an ionic compound.

9.1 VALENCE ELECTRONS

Only the outer electrons of an atom are influenced to a significant extent by the approach of another atom. The core electrons are usually not affected much. Neither are electrons in *filled d* and in *f* subshells because they are in the interior of the atom, not on the surface. Evidence for the noninvolvement of inner electrons comes from X-ray electron spectroscopy. The energy required to remove inner electrons from an atom is almost independent of whether the atom is in a compound or is in an uncombined element. The energy required to remove outer electrons depends markedly on the state of combination of the atom.

The *outer electrons* are called **valence electrons,** and the *outer shell* is called the **valence shell.** All elements in a group usually have the same number of electrons in the valence shell. Possession of the same number of valence electrons by all members of a group is the reason that the members of a group have similar properties. For main group elements, the number of valence electrons is equal to the group number. For example, the elements in Group IA have one valence electron, and the elements in Group VIIA have seven valence electrons.

Some ionic compounds. Left: *Cobalt(II) chloride hexahydrate.* Center: *Sodium chloride.* Right: *Copper(II) sulfate pentahydrate.*

9.2 IONIC BONDS

The *attraction between positively charged ions and negatively charged ions in ionic compounds* is called an **ionic bond.** When the ionic compound sodium chloride is formed from a sodium atom and a chlorine atom, one electron is transferred from the sodium atom to the chlorine atom:

$$[Ne]3s^1 + [Ne]3s^23p^5 \longrightarrow [Ne] + [Ne]3s^23p^6 \text{ or } [Ar] \qquad (9.1)$$
$$Na \quad + \quad Cl \quad \longrightarrow Na^+ + \quad Cl^-$$

Removal of one electron from a sodium atom to form a sodium ion gives the sodium atom the outer electron configuration of the noble gas neon, which has eight electrons in its outermost shell. Addition of one electron to a chlorine atom to form a chloride ion gives the chlorine atom the outer electron configuration of the noble gas argon, which has eight electrons in its outermost shell.

The best evidence for the presence of ions in solids such as sodium chloride comes from X-ray diffraction. Measurements of electron density show that sodium chloride is made up of spherical groups of 10 and 18 electrons. Sodium ions have 10 electrons, and chloride ions have 18 electrons.

SAMPLE PROBLEM

9.1 (a) Use abbreviated electron configurations to show the formation of the ionic compound, magnesium fluoride, from magnesium atoms and fluorine atoms.
(b) What is the formula for magnesium fluoride?

Solution (a) The abbreviated electron configurations for magnesium and fluorine are

$$Mg \quad [Ne]3s^2 \qquad \text{and} \qquad F \quad [He]2s^22p^5$$

Two electrons must be removed from a magnesium atom to leave it with a noble gas electron configuration. A fluorine atom has seven electrons in its outermost shell (the second shell). Addition of one electron to a fluorine atom gives the electron configuration of the noble gas neon. A fluorine atom can only accept one electron. Therefore, two fluorine atoms are needed to accept the two electrons from one magnesium atom:

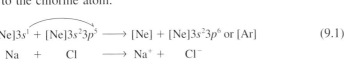

$$[Ne]3s^2 + \begin{matrix} [He]2s^22p^5 \\ [He]2s^22p^5 \end{matrix} \longrightarrow [Ne] + \begin{matrix} [He]2s^22p^6 \text{ or } [Ne] \\ [He]2s^22p^6 \text{ or } [Ne] \end{matrix}$$
$$Mg \quad + \quad 2F \quad \longrightarrow Mg^{2+} + \quad 2F^-$$

(b) The formula for magnesium fluoride is MgF_2.

PRACTICE PROBLEM

9.1   (a) Use abbreviated electron configurations to show the formation of (i) potassium iodide from potassium atoms and iodine atoms, (ii) calcium bromide from calcium atoms and bromine atoms, and (iii) sodium sulfide from sodium atoms and sulfur atoms. (b) What are the formulas of potassium iodide, calcium bromide, and sodium sulfide?

Transition Metal Ions

All monatomic *anions* have noble gas electron configurations. However, not every monatomic *cation* has a noble gas electron configuration. For example, all transition elements except palladium have either one or two electrons in their outermost shell. The difference in electron configuration between one transition atom and the next going across a row in the periodic table is in the next-to-outermost shell. Most of the elements in the first transition series form ions with a $2+$ charge. Consider the formation of iron(II) oxide as an example:

$$[Ar]3d^64s^2 + [He]2s^22p^4 \longrightarrow [Ar]3d^6 + [He]2s^22p^6 \text{ or } [Ne]$$
$$\text{Fe} \quad + \quad \text{O} \quad \longrightarrow \quad \text{Fe}^{2+} \quad + \quad \text{O}^{2-}$$

Remember (from Sample Problem 8.5) that the two $4s$ electrons are removed before the $3d$ electrons because removal of the $4s$ electrons gives an ion of lower energy. For Fe^{2+}, the electron configuration $[Ar]3d^6$ is of lower energy than the electron configuration $[Ar]4s^23d^4$. The iron(II) ion, Fe^{2+}, does not have the electron configuration of a noble gas. Because each successive ionization energy is larger (Table 8.1), much more energy than is available under ordinary conditions would be required to remove all eight outer electrons from the iron atom.

However, there is not much difference in energy between the $3d$ orbitals and the $4s$ orbital. In the presence of oxygen, the iron(II) ion rapidly reacts to form iron(III) ion. The third electron must be removed from the $3d$ subshell because there are no more electrons left in the fourth shell:

$$[Ar]3d^6 \longrightarrow [Ar]3d^5 + e^-$$
$$\text{Fe}^{2+} \longrightarrow \text{Fe}^{3+} + e^-$$

Formation of more than one kind of cation is a characteristic property of transition metals because the s orbital of the outer (n) shell and the d orbitals of the next-to-outermost ($n - 1$) shell have similar energies.

There is no simple rule for predicting which cation will be more common. For iron, the $3+$ cation is more common on Earth because Earth's atmosphere is 21% by volume oxygen. Removal of three electrons from an iron atom gives a half-filled subshell and none of the d electrons are paired. For cobalt and nickel, the $2+$ ion is more common than the $3+$ ion. In the case of the cobalt(II) and nickel(II) ions, magnetic measurements can prove that the $4s$ electrons, not $3d$ electrons, are removed. Taking cobalt as an example,

$$[Ar]3d^74s^2 \longrightarrow [Ar]3d^7 \text{ or } [Ar]3d^54s^2? + 2e^-$$
$$\text{Co} \longrightarrow \text{Co}^{2+} + 2e^-$$

The electron configuration $[Ar]3d^7$ has three unpaired electrons, and the electron configuration $[Ar]3d^54s^2$ has five unpaired electrons:

Co²⁺ [Ar] ⤷ 3d

or [Ar] 3d   4s

[Ar]3d⁷   [Ar]3d⁵4s²

Magnetic measurements show that the cobalt(II) ion has three unpaired electrons. Therefore, the electron configuration of the cobalt(II) ion is $[Ar]3d^7$.

PRACTICE PROBLEMS

9.2 (a) How many unpaired electrons are there in a Ni^{2+} ion with electron configuration $[Ar]3d^8$? (b) With electron configuration $[Ar]3d^64s^2$? (c) Magnetic measurements show that the nickel(II) ion has two unpaired electrons. What is the electron configuration of the nickel(II) ion?

9.3 Why can't magnetic measurements tell the difference between the electron configurations $[Ar]3d^6$ and $[Ar]3d^44s^2$ for the iron(II) ion?

9.4 In one formula unit of iron(III) oxide, how many iron(III) ions and how many oxide ions are there? (Remember that a compound cannot have a charge.)

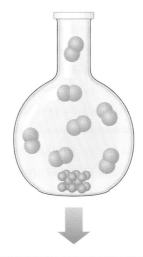

A number of elements near the end of the transition series have filled d subshells. For example, the electron configuration of zinc is

$$[Ar]3d^{10}4s^2$$

and the electron configuration of the zinc ion, Zn^{2+}, is

$$[Ar]3d^{10}$$

There is a filled d subshell in Zn^{2+}, and zinc does not form a $3+$ ion.

Metals that follow the transition elements in the periodic table such as thallium and lead have filled f, d, and s subshells and form ions by losing electrons from the unfilled p subshell:

$$[Xe]4f^{14}5d^{10}6s^26p^1 \longrightarrow [Xe]4f^{14}5d^{10}6s^2 + e^-$$
$$Tl \longrightarrow Tl^+ + e^-$$
$$[Xe]4f^{14}5d^{10}6s^26p^2 \longrightarrow [Xe]4f^{14}5d^{10}6s^2 + 2e^-$$
$$Pb \longrightarrow Pb^{2+} + 2e^-$$

PRACTICE PROBLEM

9.5 Of the following ions

$$Al^{3+} \quad Bi^{3+} \quad Cu^+ \quad Cu^{2+} \quad Hg^{2+} \quad S^{2-}$$

(a) Which have the outer electron configuration of a noble gas? (b) Which have a filled d subshell?

Energy and Ionic Bond Formation: The Born–Haber Cycle

The reaction between sodium and chlorine is spontaneous and energy is released as you can see in ▪Figure 9.1. Let's analyze the situation to see where the energy comes from. To focus attention on the transfer of electrons that takes place between metal and nonmetal, we used sodium atoms and chlorine atoms in equation 9.1. When we actually carry out the reaction, we must use sodium metal and chlorine gas (see Figure 9.1). Suppose we use 1 mol Na(s):

$$Na(s) + \tfrac{1}{2}Cl_2(g) \longrightarrow NaCl(s) \tag{9.2}$$

▪ FIGURE 9.1 Although you cannot see from the photo that the reaction between sodium and chlorine, $2Na(s) + Cl_2(g) \longrightarrow 2NaCl(s)$, is exothermic, you can see that energy in the form of yellow-colored light is released.

You should recognize the reaction shown by equation 9.2 as the formation reaction for sodium chloride (Section 6.9). The enthalpy change that accompanies this reaction under standard conditions is ΔH_f° for NaCl(s). The reaction between sodium metal and chlorine gas to form solid sodium chloride can be broken down into the following five steps:

Step 1. Change of sodium metal to vapor:

$$Na(s) \longrightarrow Na(g) \qquad \Delta H^\circ = 92 \text{ kJ}$$

Energy must be supplied to change a solid to a gas—that is, this change is endothermic. Therefore, the change in enthalpy that that accompanies it has a positive sign.

Step 2. Decomposition of chlorine molecules into chlorine atoms:

$$\tfrac{1}{2}Cl_2(g) \longrightarrow Cl(g) \qquad \Delta H^\circ = 121 \text{ kJ}$$

Step 3. Ionization of sodium vapor:

$$Na(g) \longrightarrow Na^+(g) + e^-(g) \qquad \Delta H^\circ = 496 \text{ kJ}$$

The quantity, 496 kJ, is the first ionization energy of sodium vapor (Figure 8.10).

Step 4. Addition of an electron to Cl(g):

$$Cl(g) + e^-(g) \longrightarrow Cl^-(g) \qquad \Delta H^\circ = -349 \text{ kJ}$$

See the instructor's marginal note on page 274 regarding the definition of electron affinity.

Energy is released when an electron is added to a gaseous atom of chlorine—that is, addition of an electron to a gaseous atom of chlorine is an exothermic process. Therefore, the change in enthalpy that accompanies this change has a negative sign. The quantity, 349 kJ, is the electron affinity of chlorine (Figure 8.13).

Step 5. Combination of gaseous sodium ions with gaseous chloride ions to form solid sodium chloride:

$$Na^+(g) + Cl^-(g) \longrightarrow NaCl(s) \qquad \Delta H^\circ = -771 \text{ kJ}$$

Addition of the thermochemical equations for the five steps gives the thermochemical equation for the formation of sodium chloride:

	ΔH°, kJ
$Na(s) \longrightarrow Na(g)$	92
$\tfrac{1}{2}Cl_2(g) \longrightarrow Cl(g)$	121
$Na(g) \longrightarrow Na^+(g) + e^-(g)$	496
$Cl(g) + e^-(g) \longrightarrow Cl^-(g)$	-349
$Na^+(g) + Cl^-(g) \longrightarrow NaCl(s)$	-771
$Na(s) + \tfrac{1}{2}Cl_2(g) \longrightarrow NaCl(s)$	-411

The energy changes in the formation of solid sodium chloride from solid sodium metal and chlorine gas are summarized graphically in ▪Figure 9.2. *A cycle of changes like the one shown in Figure 9.2* is called a **Born–Haber cycle.** Born–Harber cycles are applications of Hess's law (Section 6.9). In Figure 9.2, notice that, although sodium has a relatively low ionization energy compared to most elements (see Figure 8.10) and chlorine has the highest electron affinity of any element (see Figure 8.13), more energy is requred to remove an electron from a gaseous sodium atom than is released when an electron is added to a gaseous chlorine atom [compare steps (3) and (4)]. Most of the energy given off when NaCl(s) is formed (771 kJ out of a total of 1120 kJ or over two-thirds) is due to the formation of solid sodium chloride from the gaseous ions [step (5)]. In NaCl(s), each positive sodium ion is

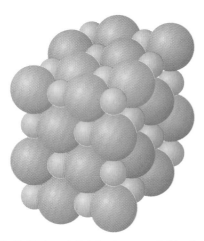

In NaCl(s), each Na^+ ion is surrounded by six Cl^- ions, and each Cl^- ion is surrounded by six Na^+ ions. There are no molecules of NaCl in NaCl(s).

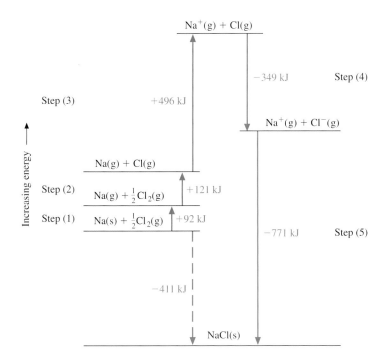

Step (4) −349 kJ

Na⁺(g) + Cl(g)

Step (3) +496 kJ

Na⁺(g) + Cl⁻(g)

Na(g) + Cl(g)

Step (2) Na(g) + $\frac{1}{2}$Cl$_2$(g) +121 kJ

Step (1) Na(s) + $\frac{1}{2}$Cl$_2$(g) +92 kJ

−771 kJ Step (5)

−411 kJ

NaCl(s)

Increasing energy →

- FIGURE 9.2 Born–Haber cycle for NaCl(s). Energy changes are for the number of moles shown. Haber is even better known as the developer of the commercial process for making ammonia from nitrogen in air.

surrounded by six negative chloride ions, and each chloride ion is surrounded by six sodium ions (remember Figure 1.18 which is reproduced on the opposite page); attractions between opposite charges are very strong.

The energy required to separate the ions of an ionic solid to an infinite distance [the reverse of step (5)] is called the **lattice energy** of the solid. The lattice energy for sodium chloride is +771 kJ/mol. The lattice energy of an ionic solid is a measure of the force of attraction between the ions in the solid. The higher the lattice energy, the stronger the ionic bonds, and, in general, the harder and higher melting the solid.

If the values of $\Delta H°$ for all but one of the changes in a Born–Haber cycle are known, the value of the remaining one can be calculated. Standard enthalpies of formation are readily available in tables. *Born–Haber cycles are usually used to calculate lattice energies, which are difficult to measure experimentally.* The lattice energies for some typical ionic compounds are given in Table 9.1. From this table, you can see that lattice energies depend on the charges and sizes of the ions involved. Increasing charge while size is kept constant increases the lattice energy. For example, the lattice energy of MgCl$_2$ is much greater than the lattice energy of LiCl; the radii of Mg^{2+} and Li$^+$ are about the same (72 and 74 pm, respectively). The lattice energy of Na$_2$S is much greater than the lattice energy of NaCl; the radii of S^{2-} and Cl$^-$ are about the same (184 and 181 pm, respectively). Increasing the size of the cations while charges are kept constant lowers the lattice energy (compare LiCl, NaCl, and KCl). Increasing the size of the anions also results in a decrease in lattice energy (compare NaCl and NaBr or Na$_2$O and Na$_2$S). Compounds of 2+ and 2− ions have very high lattice energies and are very stable as shown by the value for MgO in Table 9.1.

Since compounds of 2+ and 2− ions have very high lattice energies, you may wonder why sodium chloride is not made up of Na^{2+} and Cl^{2-} ions instead of Na$^+$ and Cl$^-$ ions. To form a Na^{2+} ion, the second electron must be removed from the next-to-the-outermost shell against the attraction of a 1+ charge. This second ionization energy for sodium is 4564 kJ/mol, almost ten times the first ionization energy. The energy needed to force a second electron onto a chlorine atom is also very large. To form a Cl^{2-} ion from a Cl$^-$ ion, the additional electron must go into

TABLE 9.1

Lattice Energies of Some Typical Ionic Compounds[a]

Compound	Lattice Energy, kJ/mol
LiCl	834
NaCl	769
KCl	701
NaBr	732
Na$_2$O	2481
Na$_2$S	2192
MgCl$_2$	2326
MgO	3795

[a]Most values are from Lide, D. R., Ed. *CRC Handbook of Chemistry and Physics,* 75th ed.; CRC Press: Boca Raton, FL, 1994, p **12**-13 to **12**-23.

Sodium chloride (common table salt) is a typical high-melting ionic compound (melting point about 800 °C). Both aqueous solutions of sodium chloride and molten sodium chloride conduct electricity; solid sodium chloride is a nonconductor.

Sodium chloride occurs naturally in the solid state and in aqueous solution. The mineral halite, which is commonly called rock salt, has been known for several thousand years. Deposits of halite are products of the evaporation of ancient seas. Sodium chloride makes up almost 80% of the salts in seawater, which averages about 3% salts. On average, there is a little more than a quarter of a pound of salt in each gallon of seawater. The ancient Egyptians produced salt by evaporation of salt water, and salt wells have been in existence in Szechuan, China, for more than 2000 years. Small quantities of salt are present in blood and in milk. Crude oil also contains sodium chloride, which must be removed in refining if more than 0.7% by mass is present.

An adequate intake of salt is essential to the health of both people and animals. Salt makes up about 0.9% of blood and body cells. As a result and also because of its preservative and seasoning properties, salt has been prized for thousands of years. In the ancient world, salt had religious significance; for example, sacrifices of grain were usually accompanied by salt. Cakes of salt have been used for money in Ethiopia and other parts of Africa and in Tibet. The word *salary* is derived from the Latin word *salarium,* which originally referred to the payment of salt as wages to Roman soldiers. The economic importance of salt is indicated by the prevalence of taxes on salt and the fact that the production of salt has often been a government monopoly. The transport of salt was important in the development of early trade routes.

About 23% of the salt produced in the world in 1980—some 3.7×10^{10} tons worth more than $600 million—was produced in the United States. Louisiana and Texas are the leading salt-producing states. In the United States, 55% of the salt produced is obtained from brine and 32% from rock salt.

Sodium chloride was the starting point for the chemical industry; the LeBlanc process for making Na_2CO_3 from NaCl, which is now obsolete, was patented in 1791. Today more NaCl than any other material is used for the manufacture of inorganic chemicals. In the United States, about half the annual production of sodium chloride is used to make chlorine gas and sodium hydroxide by electrolysis. (Electrolysis is discussed in Chapter 19.) Another 20% is used to make other chemicals. Highway snow clearance uses 21%; only about 3% is used for seasoning and food processing.

In the Middle Ages and during the Renaissance, large and elaborate salt cellars such as this one, made by the Renaissance goldsmith Benvenuto Cellini in the sixteenth century, were used as centerpieces.

the next higher energy level:

$$[Ne]3s^23p^6 + e^- \longrightarrow [Ne]3s^23p^64s^1$$
$$Cl^- + e^- \longrightarrow Cl^{2-}$$

Moreover, the additional electron is repelled by the negative charge on the Cl^- ion. The second electron affinity for chlorine is very negative. As a result of the high second ionization energy for sodium and the negative second electron affinity for chlorine, the formation of $Na^{2+}Cl^{2-}$ is very endothermic—that is, the energy change accompanying the formation of $Na^{2+}Cl^{2-}$ is very unfavorable. Therefore, $Na^{2+}Cl^{2-}$ does not form despite its greater lattice energy.

PRACTICE PROBLEM

9.6 In each of the following pairs of ionic compounds, which compound has the higher lattice energy? Explain your answers. (a) $CaCl_2$ or $MgCl_2$ (b) CaO or NaF (c) FeO or Fe_2O_3 (d) NaBr or NaI

SHOWING MOLECULAR STRUCTURE WITH LEWIS FORMULAS

Although some ionic compounds, such as salt (NaCl), lime (CaO), sodium hydroxide (NaOH), sodium carbonate (Na_2CO_3), and ammonium nitrate (NH_4NO_3), are common and important commercially, most compounds are molecular. Water (H_2O), sulfuric acid (H_2SO_4), methane (CH_4), ethylene (C_2H_4), ammonia (NH_3), ethyl alcohol (C_2H_6O), and sugar ($C_{12}H_{22}O_{11}$) are examples of common molecular compounds that are used in huge quantities.

After the discovery of the electron and the nuclear atom, many attempts were made to develop an explanation of the chemical bonding in molecules based on electrons. In 1916 the American chemist Gilbert Newton Lewis proposed that chemical bonds in molecules are formed by atoms sharing pairs of outer electrons. Lewis assumed that unshared electrons are also paired. He also put forth the idea that *groups of eight electrons* (**octets**) around atoms have a special stability. A little later (1919) another American chemist, Irving Langmuir, pointed out how well Lewis's idea organized the facts of chemistry and suggested the name **covalent bond** for a *shared pair of electrons.* Although Lewis's description of the covalent bond as a pair of shared electrons was very useful, it did not explain why or how the electrons were shared. A genuine theory of the covalent bond was not possible until after the development of the quantum theory.

Theories of covalent bonding are discussed in Chapter 10.

Lewis invented a simple and convenient way of showing the valence electrons. In **Lewis** (or **electron-dot**) **symbols,** the *core of the atom—that is, the nucleus and the inner electrons—is represented by the symbol for the element. A dot is used to represent each valence electron.* The Lewis symbol for the chlorine atom is

$$\cdot \ddot{\underset{..}{Cl}} :$$

Lewis symbols are used mainly for s- and p-block elements. For these elements, the number of valence electrons is the same as the group number. The Lewis symbols for the atoms in the second period are

$$\text{Li} \quad \cdot\text{Be}\cdot \quad \cdot\dot{\text{B}}\cdot \quad \cdot\dot{\underset{.}{\text{C}}}\cdot \quad \cdot\dot{\underset{..}{\text{N}}}\cdot \quad \cdot\ddot{\underset{..}{\text{O}}}: \quad \cdot\ddot{\underset{..}{\text{F}}}: \quad :\ddot{\underset{..}{\text{Ne}}}:$$

In writing Lewis symbols, one dot (representing one electron) is placed on each of the four sides of the symbol for the element before two dots are placed on any side. Which side you start on makes no difference. The symbol for lithium could equally well have been written

$$\text{Li}\cdot \qquad \text{or} \qquad \text{L}\underset{.}{\text{i}} \qquad \text{or} \qquad \cdot\text{Li}$$

The idea that two of the electrons of beryllium, boron, and carbon are paired was not developed until later (1924). Lewis placed one dot on each of the four sides of the symbol for the element before placing two on any side so that the number of unpaired electrons in the Lewis symbol would be equal to the number of bonds that an element *usually* forms in compounds. Beryllium usually forms two bonds; boron forms three; carbon, four; nitrogen, three; oxygen, two; and fluorine one. Neon, with all electrons paired, does not form any bonds (see Table 9.2).

TABLE 9.2

Number of Bonds Usually Formed

Lewis Symbol	Number of Bonds
H·	1
·Be·	2
·Ḃ·	3
·Ç·	4
·N̈·	3
·Ö:	2
·F̈:	1
:N̈e:	0

PRACTICE PROBLEM

9.7 Write the Lewis (electron-dot) symbols for the elements in the third period.

Lewis symbols are combined in **Lewis** (or **electron-dot**) **formulas.** The Lewis formula for hydrogen chloride is

$$H\!:\!\ddot{\underset{..}{C}}l\!:$$

The Lewis formula for hydrogen chloride is formed by combining the Lewis symbol for a hydrogen atom with the Lewis symbol for a chlorine atom:

$$H\cdot \; + \; \cdot\ddot{\underset{..}{C}}l\!: \; \longrightarrow \; H\!:\!\ddot{\underset{..}{C}}l\!:$$

In Lewis formulas, two dots between the symbols for two atoms represent *a pair of electrons that is shared between two atoms*—that is, a **bonding pair** or covalent bond. *Electrons that are not shared* are called **unshared pairs, lone pairs,** or **nonbonding electrons.** The chlorine atom in hydrogen chloride has three unshared pairs. In the following Lewis formula, circles have been drawn to show how to count the number of electrons belonging to each atom. A pair of electrons in the area where two circles overlap is shared between two atoms. This Lewis formula shows two valence electrons around H (like He) and eight valence electrons around Cl (like Ar):

unshared pairs—belong only to Cl

shared pair—belongs to both H and Cl

In HCl, both H and Cl have obtained the same number of valence electrons as the nearest noble gas by sharing electrons.

The Lewis formulas for ammonia and water are

$$\begin{array}{cc} H\!:\!\ddot{N}\!:\!H & H\!:\!\ddot{\underset{..}{O}}\!: \\ \;\;H & \;\;H \end{array}$$

ammonia   water

The atom in the center of a molecule, such as the N in ammonia and the O in water, is referred to as a **central atom.** There may be more than one central atom in a molecule. Both carbons in ethane and both the carbon and the oxygen in methyl alcohol are regarded as central atoms:

$$\begin{array}{cc} \begin{array}{cc} H & H \\ H\!:\!C\!:\!C\!:\!H \\ H & H \end{array} & \begin{array}{cc} H \\ H\!:\!C\!:\!\ddot{\underset{..}{O}}\!:\!H \\ H \end{array} \end{array}$$

ethane   methyl alcohol

Notice that there are octets (eight electrons) around each of the atoms except hydrogen in the Lewis formulas for hydrogen chloride, ammonia, water, ethane, and methyl alcohol. Eight electrons in the valence shell is the electron configuration of the noble gases (except helium, which has only two). Each of the elements, Cl, N, O, and C, has reached the electron configuration of a noble gas by sharing electrons with hydrogen atoms in return for a share of the hydrogen atoms' electrons. Most compounds follow the **octet rule** that *there should be eight electrons around each atom except hydrogen* although there are a number of common and important exceptions (see Section 9.10).

Each hydrogen atom in all the Lewis formulas shown has a share in two electrons. Two electrons is the maximum number of electrons that can be placed in the first shell. By sharing its electron with another atom in return for a share in one of the other atom's electrons, each hydrogen atom has gained the electron configuration of the noble gas with the closest atomic number, helium.

Remember the discussion of electron spin and the Pauli exclusion principle in Section 7.12.

CHAPTER 9 CHEMICAL BONDS

306

9.8   (a) How many covalent bonds are there in a molecule of ammonia? (b) How many unshared pairs are there in a water molecule?

9.9   Copy the Lewis formula for methyl alcohol and use circles to show that there are eight valence electrons around C and O and two around each H.

9.10   Write the Lewis (electron-dot) formula for each of the following:
(a) hydrogen fluoride, HF, (b) phosphine, PH_3, (c) hydrogen sulfide, H_2S,
(d) methane, CH_4, and (e) a fluorine molecule, F_2.

To save time and make the formulas simpler, a line is often used to represent each shared pair of electrons; that is, the Lewis formulas for hydrogen chloride, ammonia, water, ethane, and methyl alcohol are written:

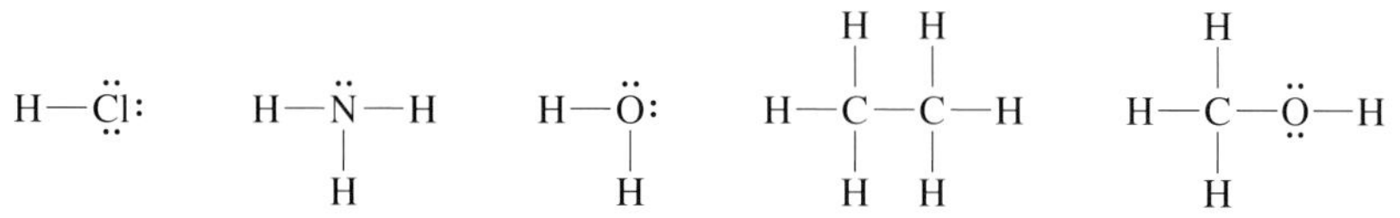

$$
\text{H}-\ddot{\underset{..}{\text{C}}}\text{l:} \qquad
\text{H}-\underset{|}{\overset{..}{\text{N}}}-\text{H} \qquad
\text{H}-\ddot{\underset{..}{\text{O}}}: \qquad
\underset{\underset{\text{H}}{|}}{\overset{\overset{\text{H}}{|}}{\text{H}-\text{C}}}-\underset{\underset{\text{H}}{|}}{\overset{\overset{\text{H}}{|}}{\text{C}}}-\text{H} \qquad
\underset{\underset{\text{H}}{|}}{\overset{\overset{\text{H}}{|}}{\text{H}-\text{C}}}-\ddot{\underset{..}{\text{O}}}-\text{H}
$$

You should recognize these Lewis formulas as being the same as the structural formulas we wrote in Section 3.9 except that unshared pairs of valence electrons are shown.

9.11   Rewrite the Lewis formulas in Practice Problem 9.10 using lines to represent shared electron pairs.

9.4   NONPOLAR AND POLAR COVALENT BONDS

If the two atoms sharing an electron pair are the same, they will share the electron pair equally. *Bonds in which electron pairs are shared equally* are called **nonpolar bonds.** The bonds in hydrogen molecules and in chlorine molecules are nonpolar:

$$\text{H:H} \quad \text{or} \quad \text{H}-\text{H} \qquad :\ddot{\underset{..}{\text{C}}}\text{l}:\ddot{\underset{..}{\text{C}}}\text{l}: \quad \text{or} \quad :\ddot{\underset{..}{\text{C}}}\text{l}-\ddot{\underset{..}{\text{C}}}\text{l}:$$

If the two atoms sharing the electron pair are different, they will usually not share the electron pair equally. One of them will have more and the other less than an equal share. For example, the electron affinity of chlorine is almost five times as large as the electron affinity of hydrogen (Figure 8.13) while the first ionization energies of hydrogen and chlorine are about the same (Figure 8.10). Thus, chlorine atoms have a greater tendency to gain an electron than hydrogen atoms do; hydrogen atoms and chlorine atoms give up electrons with equal difficulty. As a result, in a hydrogen chloride molecule, the chlorine has more than half of the shared pair, and the hydrogen has less than half. *A covalent bond in which the electron pair is not shared equally* is called a **polar bond.** The bond in hydrogen chloride is polar. Because it has a polar bond, the hydrogen chloride molecule is polar.

There are several ways of showing a polar bond. The electron pair can be drawn closer to the atom that has the larger share of it instead of in the middle:

$$\text{H} :\ddot{\underset{..}{\text{C}}}\text{l}:$$

Because the hydrogen originally had one electron and now has less than half of two, it has a part of a positive charge. Because the chlorine originally had seven electrons and now has a share in another one, it has a part of a negative charge. Partial positive and negative charges are usually shown by the lowercase Greek letter delta:

$$\overset{\delta+}{H}\overset{\delta-}{:\overset{..}{\underset{..}{Cl}}}:$$

Sometimes a crossed arrow is used to show which atom has a part of a positive charge and which a part of a negative charge. The head of the arrow points toward the atom with the partial negative charge. The crossed part of the arrow, which looks like a + sign, is located near the partially positive atom.

$$\overset{+\longrightarrow}{H:\overset{..}{\underset{..}{Cl}}:}$$

A pair of equal and opposite charges separated by a distance (like the pair in HCl) is called a **dipole.**

9.5 ELECTRONEGATIVITY

Electronegativity is *the ability of an atom that is bonded to another atom or atoms to attract electrons to itself.* For example, the electronegativity of chlorine is greater than the electronegativity of hydrogen. Electronegativity is related to ionization energy and electron affinity. Atoms that have high ionization energies and high electron affinities—that is, atoms that lose electrons with difficulty and gain electrons readily—are very electronegative. However, ionization energies and electron affinities are properties of separate gaseous atoms whereas electronegativities refer to atoms that are bonded to other atoms.

Ionization energies and electron affinities are determined by experiment. Electronegativities cannot be measured directly. However, several methods that have been used to estimate relative electronegativities all give similar results. Because electronegativities are relative, they have no units. The difference in electronegativity of the atoms joined by a bond is what's important. The concept of electronegativity was originally suggested by the American chemist Linus Pauling.* ▬Figure 9.3 shows modern values for electronegativities.

The electronegativity of an element varies slightly from compound to compound. For example, the electronegativity of carbon in CH_3Cl is 2.4, and the electronegativity of carbon in CCl_4 is 2.6. The values in Figure 9.3 are average values useful for making *qualitative* predictions.

▬Figure 9.4 is a graph of electronegativity as a function of atomic number. From Figure 9.4, it is obvious that electronegativity is a periodic property. In general, electronegativity increases from left to right across rows of the periodic table and decreases from top to bottom of groups. It is constant across rows for *f*-block elements. The most electronegative elements are in the upper right corner of the periodic table; the least electronegative elements are in the lower left corner.

In the Bond Polarity module of the Molecules section, students learn to estimate relative bond polarities from electronegativity differences. A 3-dimensional periodic table showing electronegativity values is provided; some questions pertaining to the trends shown by the table are posed.

We have used Allred–Rochow electronegativities because, according to several recent inorganic texts (e.g., Huheey), these values are the most frequently used.

---

*Linus Pauling was among the first to apply the principles of quantum mechanics to molecules. He won the Nobel Prize for chemistry in 1954 for his studies of the nature of the chemical bond and the Nobel Prize for peace in 1962 for his work toward stopping the testing of nuclear weapons and establishing international control over them. His work on sickle cell anemia and his campaign for taking large doses of vitamin C to prevent the common cold are also well known. Pauling was still active professionally almost until he died in 1994 at the age of 93.

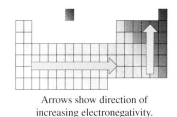

Arrows show direction of
increasing electronegativity.

- FIGURE 9.3   Electronegativities of the
elements. (Allred–Rochow scale from Huheey,
J. E.; Keiter, E. A.; Keiter, R. L. *Inorganic
Chemistry: Principles of Structure and
Reactivity*, 4th ed.; HarperCollins: New York,
1993; pp 187–190. Why are no values listed
for He, Ne, and Ar?

Periodic Table of Electronegativities

	0
	2 He

| 1 H 2.2 | | | | | | | | | | | | | | | | | 2 He |

IA	IIA											IIIA	IVA	VA	VIA	VIIA	
3 Li 1.0	4 Be 1.5											5 B 2.0	6 C 2.5	7 N 3.1	8 O 3.5	9 F 4.1	10 Ne
11 Na 1.0	12 Mg 1.2	IIIB IVB VB VIB VIIB				VIII			IB	IIB		13 Al 1.5	14 Si 1.7	15 P 2.1	16 S 2.4	17 Cl 2.8	18 Ar

IA	IIA	IIIB	IVB	VB	VIB	VIIB	VIII			IB	IIB	IIIA	IVA	VA	VIA	VIIA
19 K 0.9	20 Ca 1.0	21 Sc 1.2	22 Ti 1.3	23 V 1.5	24 Cr 1.6	25 Mn 1.6	26 Fe 1.6	27 Co 1.7	28 Ni 1.8	29 Cu 1.8	30 Zn 1.7	31 Ga 1.8	32 Ge 2.0	33 As 2.2	34 Se 2.5	35 Br 2.7	36 Kr 2.9
37 Rb 0.9	38 Sr 1.0	39 Y 1.1	40 Zr 1.2	41 Nb 1.2	42 Mo 1.3	43 Tc 1.4	44 Ru 1.4	45 Rh 1.5	46 Pd 1.4	47 Ag 1.4	48 Cd 1.5	49 In 1.5	50 Sn 1.7	51 Sb 1.8	52 Te 2.0	53 I 2.2	54 Xe 2.4
55 Cs 0.9	56 Ba 1.0	71 Lu 1.1	72 Hf 1.2	73 Ta 1.3	74 W 1.4	75 Re 1.5	76 Os 1.5	77 Ir 1.6	78 Pt 1.4	79 Au 1.4	80 Hg 1.4	81 Tl 1.4	82 Pb 1.5	83 Bi 1.7	84 Po 1.8	85 At 1.9	86 Rn 2.1
87 Fr 0.9	88 Ra 1.0	103 Lr	104 Rf	105 Ha	106 Sg	107 Ns	108 Hs	109 Mt	110	111							

☐ Metal   ■ Semimetal   ■ Nonmetal

✳ **Lanthanide**
✿ **series**

◆ **Actinide**
◇ **series**

57 La 1.1	58 Ce 1.1	59 Pr 1.1	60 Nd 1.1	61 Pm 1.1	62 Sm 1.1	63 Eu 1.0	64 Gd 1.1	65 Tb 1.1	66 Dy 1.1	67 Ho 1.1	68 Er 1.1	69 Tm 1.1	70 Yb 1.1
89 Ac 1.0	90 Th 1.1	91 Pa 1.1	92 U 1.2	93 Np 1.2	94 Pu 1.2	95 Am 1.2	96 Cm 1.2	97 Bk 1.2	98 Cf 1.2	99 Es 1.2	100 Fm 1.2	101 Md 1.2	102 No 1.2

Differences in electronegativity are helpful in predicting the direction of polarization of covalent bonds. The negative end of a bond is the end having the more
electronegative atom. For example, chlorine (electronegativity 2.8) is more electronegative than hydrogen (electronegativity 2.2). The bond between hydrogen and
chlorine in hydrogen chloride is polar with a partial positive charge on hydrogen
and a partial negative charge on chlorine as we saw in Section 9.4.

PRACTICE PROBLEM

9.12   Show the direction of polarization of each of the following bonds, using
either partial charges or an arrow (see Section 9.4): (a) H—O, (b) H—N,
(c) C—Cl, and (d) C—Mg.

The greater the difference in electronegativity between bonded atoms, the more
polar the bond. If the difference in electronegativity is large enough, electrons are
transferred from the less electronegative to the more electronegative atom, and the
bond is ionic. Bonds vary from nonpolar to slightly polar to very polar to mainly
ionic depending on the difference in electronegativity between the bonded atoms.

SAMPLE PROBLEM

9.2   Of the following bonds

Al—Cl   Cl—Cl   H—Cl   K—Cl

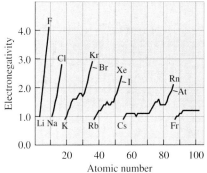

- FIGURE 9.4   Electronegativity as a
function of atomic number.

(a) Which bond is nonpolar? (b) Only one bond is ionic. Which one is it? (c) Arrange the bonds in order of increasing polarity.

Solution (a) Electrons are shared equally in bonds between two identical atoms. The Cl—Cl bond is nonpolar.

(b) From Figure 9.3, the electronegativity of Cl is 2.8. The electronegativities of the other atoms are: Al, 1.5; H, 2.2; K, 0.9. Differences in electronegativity between bonded atoms are

$$\text{for Al—Cl, } 2.8 - 1.5 = 1.3$$
$$\text{for H—Cl, } 2.8 - 2.2 = 0.6$$
$$\text{for K—Cl, } 2.8 - 0.9 = 1.9$$

Electronegativity difference is largest for the K—Cl bond. The K—Cl bond must be the one ionic bond.

(c) The order of increasing polarity is the same as the order of difference in electronegativity:

$$\text{Cl—Cl} < \text{H—Cl} < \text{Al—Cl} < \text{K—Cl}$$

PRACTICE PROBLEM

9.13 Of the following bonds

Ba—O   Be—O   C—O   O—O

(a) Which bond is nonpolar? (b) Only one bond is ionic. Which one is it? (c) Arrange the bonds in order of increasing polarity.

Aluminum chloride is an example of a compound that is ionic in the solid state but molecular in the liquid state. (See Greenwood, N. N., Earnshaw, A. Chemistry of the Elements; Pergamon: Oxford, 1984; p 263.)

As a general rule, ionic compounds have high melting points and boiling points and are hard in the solid state. Solid ionic compounds do not conduct electricity, but liquids formed by melting ionic compounds are usually conductors. Some ionic compounds are soluble in water and other polar solvents. Molecular compounds are soft. Decrease in polarity of compounds is accompanied by decreases in boiling point, in melting point, in conductivity in the liquid state, and in solubility in polar solvents. Molecular compounds are nonconductors of electricity in both solid and liquid states.

The following empirical rules, which are based on observations of melting points and boiling points and on conductivity measurements, are useful for predicting whether a compound is mainly ionic or mainly covalent:

We have not given the common rule that compounds with electronegativity difference greater than 1.7 are ionic because we find that it leads to too many wrong predictions. Also it gives the students the idea that whether a compound is predominantly ionic or predominantly molecular is a simple matter. As mentioned in Section 9.2, electron density plots from X-ray diffraction seem to provide the best evidence.

1. Most metal fluorides and oxides (except beryllium) are mainly ionic. Compounds such as WF_6, OsF_6, and OsO_4 are exceptions to this rule. Too much energy is required to remove enough electrons to form W^{6+}, Os^{6+}, and Os^{8+} ions for compounds such as WF_6, OsF_6, and OsO_4 to be ionic.
2. Compounds of metals in Group IA, in Group IIA (except beryllium and magnesium), and in the lanthanide series with nonmetals are mainly ionic.
3. Compounds consisting solely of nonmetals are mainly covalent (except for ammonium compounds, which are ionic).

PRACTICE PROBLEM

9.14 Use the rules to classify each of the following compounds as ionic or molecular. (a) CaO, (b) NaI, (c) CS_2, (d) AlF_3, (e) NH_4Cl, and (f) NO.

Although compounds that contain polyatomic ions such as the ammonium ion, NH_4^+, and the hydroxide ion, OH^-, are ionic compounds, these compounds contain covalent bonds as well as ionic bonds. In polyatomic ions, two or more atoms are bound together by covalent bonds. For example, the Lewis formula for the ammonium ion is

$$\left[\begin{array}{c}H \\ H:\overset{\cdot\cdot}{N}:H \\ \overset{\cdot\cdot}{H}\end{array}\right]^{+} \quad \text{or} \quad \left[\begin{array}{c}H \\ | \\ H-N-H \\ | \\ H\end{array}\right]^{+}$$

Thus, the ionic compound ammonium chloride, NH_4Cl, has four covalent bonds and one ionic bond:

$$\left[\begin{array}{c}H \\ H:\overset{\cdot\cdot}{N}:H \\ \overset{\cdot\cdot}{H}\end{array}\right]^{+} :\overset{\cdot\cdot}{\underset{\cdot\cdot}{Cl}}:^{-} \quad \text{or} \quad \left[\begin{array}{c}H \\ | \\ H-N-H \\ | \\ H\end{array}\right]^{+} :\overset{\cdot\cdot}{\underset{\cdot\cdot}{Cl}}:^{-}$$

Notice that a line is not used to show the ionic bond between the ammonium ion and the chloride ion. In Lewis formulas, a line represents a pair of shared electrons. The ammonium ion and the chloride ion are held together by the attraction between opposite charges, not by a shared pair of electrons.

PRACTICE PROBLEM

9.15 Write both forms (electron-dot and line) of the Lewis formula for the hydroxide ion.

9.6 MORE ABOUT WRITING LEWIS FORMULAS

The Lewis Structures and Resonance module of the Molecules section leads students through each step in the writing of Lewis formulas.

Writing Lewis formulas is an important skill because Lewis formulas are very useful. For example, the shapes of molecules and ions can often be predicted from Lewis formulas. Shape is an important factor that affects both physical and chemical properties.

Although the Lewis formulas for very simple species can be written by combining Lewis symbols (remember Section 9.3), a system is helpful for more complex species. *The first step is to write the symbols for the elements in the correct structural order.*

Hydrogen usually forms only one covalent bond. A hydrogen atom with two electrons has the electron configuration of the noble gas helium. Additional electrons would have to be placed in the $2s$ orbital, which is of much higher energy than the $1s$ orbital. Because hydrogen usually forms only one covalent bond, hydrogen isn't commonly bonded to two other atoms. Arrangements with hydrogen as a central atom such as

$$H = H = O$$

Chemists use the term species *when they mean either an atom, a molecule, or an ion.*

are rare. Oxygen–oxygen bonds, O—O, occur only in peroxides such as hydrogen peroxide (H—O—O—H).

In species with formulas of the type A_1N_n, A is usually the central atom with the B atoms arranged around it. Ammonia (NH_3), ammonium ion (NH_4^+), nitrate ion (NO_3^-), and water (H_2O) are examples. In all three of the N-containing species, N is the central atom; O is the central atom in H_2O (▪Figure 9.5).

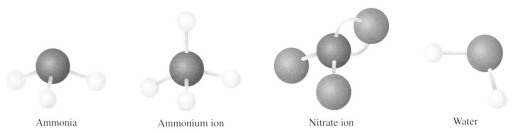

| Ammonia | Ammonium ion | Nitrate ion | Water |

▪ FIGURE 9.5 Nitrogen (blue) is the central atom in ammonia, the ammonium ion, and the nitrate ion. Oxygen (red) is the central atom in water.

In the real world, the positions of the atoms in a molecule can be found by experiment. In problems, the positions of the atoms are often shown by structural formulas. For example, suppose you need to write the Lewis formula for formaldehyde and are told that the condensed structural formula for formaldehyde is H_2CO. The correct arrangement of atoms is

$$
\begin{array}{l}
\text{H} \\
\quad \text{C O} \\
\text{H}
\end{array}
$$

The second step in writing a Lewis formula is to calculate how many valence electrons to show. Hydrogen has only one electron. For Group A elements, the group number in the periodic table tells the number of valence electrons. If the species is an anion, enough extra electrons must be added to account for the charge; if it is a cation, enough electrons must be removed to account for the charge. Formaldehyde is a molecule. Formaldehyde, H_2CO, has

$$
\begin{array}{rcl}
2 \text{ hydrogens @ 1 valence electron} & = & 2 \text{ valence electrons} \\
1 \text{ carbon @ 4 valence electrons} & = & 4 \text{ valence electrons} \\
1 \text{ oxygen @ 6 valence electrons} & = & 6 \text{ valence electrons} \\
\hline
\text{total} & = & 12 \text{ valence electrons}
\end{array}
$$

The Lewis formula for formaldehyde must show 12 electrons (six pairs).

The third step in writing a Lewis formula is to place the electrons around the symbols. First, place one pair of electrons between each of the symbols to hold the atoms together.

$$
\begin{array}{l}
\text{H} \\
\quad \text{:C:O} \\
\text{H}
\end{array}
$$

Next, beginning at the outside of the formula and working toward the center, place pairs of electrons on the atoms (except hydrogen) until each has eight electrons around it or you run out of electrons, whichever happens first. Six electrons (three shared pairs) are needed to hold the atoms in formaldehyde together. Subtracting 6 from 12 (the total number of valence electrons) leaves 6 more (three pairs) to place. These all fit around the oxygen:

$$
\begin{array}{l}
\text{H} \\
\quad \text{:C:Ö:} \\
\text{H}
\end{array}
$$

Now each hydrogen has a share in two electrons, and the oxygen has eight electrons around it; but the carbon only has six electrons around it. *If an atom other than hydrogen has less than eight electrons around it; move unshared pairs to form multiple bonds.** Moving one of the unshared pairs on the oxygen between carbon and oxygen

$$
\begin{array}{l}
\text{H} \\
\quad \text{:C:O:} \\
\text{H}
\end{array}
$$

to form a second bond between carbon and oxygen

$$
\begin{array}{l}
\text{H} \\
\quad \text{:C::O:} \quad\quad \text{or} \quad\quad \text{H}{\diagdown}\!\!\!\!\!\!\!\! \\
\text{H} \quad\quad\quad\quad\quad\quad\quad\quad \text{C}{=}\text{O:} \\
\quad\quad\quad\quad\quad\quad\quad\quad\quad \text{H}{\diagup}
\end{array}
$$

---

*You will learn in Section 9.10 that there are some exceptions to this rule. Few simple rules work *all* of the time. If a rule works 90 to 95% of the time, it is a useful rule. (You do not have to get 100% on *every* test to receive a grade of A.)

gives a Lewis formula with eight electrons around both oxygen and carbon and two around each hydrogen. Formation of a covalent bond, like formation of an ionic bond, is exothermic and leads to a state of lower energy and greater stability.

Finally, check your work. The Lewis formula we have written for formaldehyde shows 2 H atoms, 1 C atom, 1 O atom, and 12 electrons as it should. Each hydrogen atom has two electrons around it, and the carbon and oxygen atoms each have eight. Each H atom has one bond, the C atom has four bonds, and the O atom has two bonds, as predicted by the number of unpaired electrons in the Lewis symbols.

Sample Problems 9.3 and 9.4 further illustrate how to write Lewis formulas.

▶ Always check your work.

SAMPLE PROBLEMS

9.3   Write the Lewis formula for the compound having molecular formula PCl_3.

Solution   In species with formulas of the type A_1B_n, A is usually the central atom with the B atoms arranged around it. Therefore, in a molecule of PCl_3, P is probably in the middle:

Phosphorus trichloride is used to make oil and fuel additives, plasticizers, flame retardants, and insecticides.

$$Cl \quad P \quad Cl$$
$$Cl$$

The number of valence electrons to be shown is

$$
\begin{array}{rl}
1\ P\ @\ 5\ \text{valence electrons} = & 5\ \text{valence electrons} \\
3\ Cl\ @\ 7\ \text{valence electrons} = & 21\ \text{valence electrons} \\
\hline
\text{total} = & 26\ \text{valence electrons}
\end{array}
$$

Six (three pairs) of the 26 valence electrons are needed to hold the atoms together:

$$Cl:P:Cl$$
$$\ddot{C}l$$

This leaves 20 more valence electrons (ten pairs). Putting three unshared pairs around each Cl gives each Cl 8 electrons and uses up 18 more electrons:

$$:\ddot{C}l:P:\ddot{C}l:$$
$$:\ddot{C}l:$$

The remaining pair of electrons is placed on the central atom, phosphorus, and completes its octet of 8 electrons. The Lewis formula for PCl_3 is

▶ Place any extra electrons on the central atom.

$$:\ddot{C}l:\ddot{P}:\ddot{C}l: \quad \text{or} \quad :\ddot{C}l—\ddot{P}—\ddot{C}l:$$
$$:\ddot{C}l: \qquad\qquad\qquad :\ddot{C}l:$$

The Lewis formula shows the correct number of each kind of atom and the right number of valence electrons (26). Each atom has 8 electrons around it. Phosphorus has three bonds, and each Cl has one bond.

9.4   Write the Lewis formula for the hypochlorite ion ClO^-

Solution   In the hypochlorite ion, chlorine must be bonded to oxygen:

The hypochlorite ion is the active ingredient in many household bleaches such as Clorox.

$$Cl:O$$

The total number of electrons that should be shown in the Lewis formula is

$$
\begin{array}{rl}
1\ Cl\ @\ 7\ \text{valence electrons} & =\ 7\ \text{valence electrons} \\
1\ O\ @\ 6\ \text{valence electrons} & =\ 6\ \text{valence electrons} \\
1\ \text{extra electron for}\ 1-\ \text{charge} & =\ 1\ \text{valence electron} \\
\hline
\text{total} & =\ 14\ \text{valence electrons}
\end{array}
$$

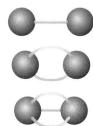

- FIGURE 9.6 Double bonds are shorter than single bonds between the same kinds of atoms, and triple bonds are even shorter.

Fourteen valence electrons less the 2 used to hold the chlorine and oxygen together leaves 12 electrons or six unshared pairs. The Lewis formula for the hypochlorite ion is

$$\left[:\ddot{\text{Cl}}:\ddot{\text{O}}:\right]^{-} \quad \text{or} \quad \left[:\ddot{\text{Cl}}-\ddot{\text{O}}:\right]^{-}$$

(You should check this Lewis formula to be sure it shows the correct number of each kind of atom, the right number of electrons, and has eight electrons around each atom.)

Ethylene (IUPAC name, ethene) is one of the most important raw materials of the organic chemical industry. It is also a plant hormone that ripens fruit. Acetylene (IUPAC name, ethyne) is widely used as a fuel in welding and cutting of metals. The flame in an oxygen–acetylene torch reaches the highest temperature of any known mixture of gases—about 3300 °C.

Multiple Bonds

Do double bonds, such as the one we drew in the Lewis structure for formaldehyde, really exist? The answer to this question is yes. Triple bonds are also quite common. For example, not only formaldehyde but also ethylene and acetylene

$$\text{H}_2\text{C}=\text{CH}_2 \qquad \text{HC}\equiv\text{CH}$$
ethylene   acetylene

have multiple bonds. Double bonds are shorter than single bonds between the same kinds of atoms, and triple bonds are even shorter as shown in -Figure 9.6.

9.7 BOND LENGTH, BOND ENERGY, AND BOND ORDER

Measurements of bond lengths and strengths can be used to distinguish between multiple bonds and single bonds. **Bond length** is defined as the *distance between the centers of two bonded atoms* (see -Figure 9.7). Bond lengths are equal to the sums of the radii of the atoms joined by the bond. The radius of a hydrogen atom is 37 pm (Figure 8.8); the distance between the centers of the two hydrogen atoms in a hydrogen molecule is

$$37 \text{ pm} + 37 \text{ pm} = 74 \text{ pm}$$

The lengths of single, double, and triple bonds are compared in Table 9.3 using carbon as an example. A triple bond is shorter than a double bond, which is shorter than a single bond.

74 pm

- FIGURE 9.7 The length of the bond between the nuclei of the two hydrogen atoms in the H_2 molecule (74 pm) is shown by the broken line.

TABLE 9.3		Bond Lengths and Strengths		
			Example	
Bond	Length, pm	Bond Energy, kJ/mol	Name	Structural Formula
C—C	154	347	Ethane	H_3C—CH_3
C=C	134	615	Ethylene (ethene)	H_2C=CH_2
C≡C	120	812	Acetylene (ethyne)	HC≡CH

Table 9.3 also shows the bond energies of carbon–carbon single, double, and triple bonds. The stronger a bond is, the more energy is required to break it. In polyatomic species, bond lengths and bond strengths vary slightly from one molecule to another and from one position in a molecule to another. **Bond energy** is the *average energy needed to break a given type of bond.* From Table 9.3, you can see that a double bond is not only shorter than a single bond but also stronger. A triple bond is even shorter and stronger than a double bond. Double bonds are most common between carbon, nitrogen, and oxygen atoms; triple bonds are most common between carbon atoms and between carbon and nitrogen atoms. Quadruple bonds are not common.

The *number of bonds between two atoms in a molecule or ion* is called **bond order.** For a single bond, the bond order is one, for a double bond, the bond order is two, and for a triple bond the bond order is three.

◖ Bond energies and their uses are discussed in more detail at the end of this chapter (Section 9.12).

A quadruple bond was first recognized in 1964 between rhenium atoms. Molybdenum forms hundreds of compounds that contain Mo–Mo quadruple bonds. Chromium, tungsten, and technetium also form quadruple bonds.

PRACTICE PROBLEM

9.18 Of the following nitrogen–nitrogen bonds,

(a) Which is longest? (b) Which is strongest? (c) Which has a bond order of two?

9.8 FORMAL CHARGES

In some Lewis formulas, one or more atoms have fewer or more bonds than usual. (Table 9.2 shows the usual number of bonds for some common elements.) The Lewis formula for carbon monoxide is an example. Measurements of the length and strength of the carbon–oxygen bond in carbon monoxide show that this bond is a triple bond; the Lewis formula for carbon monoxide must be

$$:C⋮⋮O:$$

This Lewis formula shows the correct number of each kind of atom, the correct number of electrons, and eight electrons around both C and O. However, the C atom only has three bonds instead of the usual four, and the oxygen atom has three bonds instead of the usual two.

Whenever one or more atoms have fewer or more bonds than usual, formal charges should be shown with the Lewis formula. **Formal charges** *are a bookkeeping device for electrons. They show the approximate distribution of electron density in molecules and polyatomic ions. In this book, formal charges will be circled so that you do not mistake them for ionic charges.* A carbon monoxide molecule with its formal charges is shown below:

$$\overset{\scriptsize{⊖} \; {⊕}}{:C⋮⋮O:}$$

To calculate the formal charge on an atom, assume that all shared pairs of electrons are shared *equally* between the two atoms joined by the bond. Then

1. Assign each atom half of the electrons in pairs it shares and all the electrons in any unshared pairs it has.
2. Subtract the number of electrons assigned to each atom from the number of valence electrons in an uncombined atom of the element.

Why cover formal charge in a general chemistry text? We instructors write Lewis formulas easily as a result of practice and experience. Students find rules for judging their Lewis formulas very helpful.

◓ In one of the activities in the Lewis Structures and Resonance module of the Molecules section, Lewis formulas are provided, and students must determine the formal charge on each step.

The net formal charge for a molecule must be zero; the net formal charge for an ion must be equal to the charge on the ion. Sample Problem 9.5 shows the calculation of formal charges for each atom in the carbon monoxide molecule. Sample Problem 9.6 shows how to calculate formal charges for each atom in the hypochlorite ion.

SAMPLE PROBLEMS

9.5 Calculate the formal charge on each atom in the Lewis formula for carbon monoxide:

$$:C::O:$$

Solution Draw a line around all the electrons assigned to each atom:

Five electrons are assigned to C. Carbon is in Group IVA and has four valence electrons:

$$\text{formal charge on C} = 4 - 5 = -1$$

Five electrons are also assigned to O. Oxygen is in Group VIA and has six valence electrons:

$$\text{formal charge on O} = 6 - 5 = +1$$

Including formal charges, the Lewis formula for carbon monoxide is

as shown previously. The sum of the formal charges equals zero

$$(-1) + (+1) = 0$$

as it should for a molecule.

9.6 Calculate the formal charge on each atom in the hypochlorite ion. The Lewis formula for the hypochlorite ion is

$$\left[:\ddot{C}l:\ddot{O}:\right]^-$$

Solution Draw a line around all the electrons assigned to each atom:

Seven electrons are assigned to Cl. Chlorine is in Group VIIA and has seven valence electrons:

$$\text{formal charge on Cl} = 7 - 7 = 0$$

Seven electrons are also assigned to O. Oxygen is in Group VIA and has six valence electrons:

$$\text{formal charge on O} = 6 - 7 = -1$$

Including formal charges, the Lewis formula for the hypochlorite ion is

$$\left[:\ddot{C}l:\ddot{O}:\right]^-$$

The sum of the formal charges equals the charge on the ion:

$$(0) + (-1) = -1$$

● Always check to be sure your work is reasonable.

9.19 Show the calculation of the formal charges on nitrogen and hydrogen in the ammonium ion. The Lewis formula for the ammonium ion is

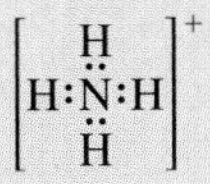

$$\left[\begin{array}{c} H \\ H \!:\! \ddot{N} \!:\! H \\ H \end{array} \right]^{+}$$

9.20 In the hydroxide ion, OH^-, which atom carries the formal charge?

Formal charges are useful for helping decide which of several possible Lewis formulas is best.

1. A Lewis formula with formal charges of zero is preferable to one with nonzero formal charges. Small formal charges are preferable to large formal charges.
2. Lewis formulas with negative formal charges on the more electronegative atom are more likely than Lewis formulas with negative formal charges on the less electronegative atom.
3. Lewis formulas with unlike charges close together are more likely than Lewis formulas with opposite charges widely separated.
4. Lewis formulas with like charges on adjacent atoms are very unlikely.

Sample Problems 9.7 and 9.8 show the use of formal charge to choose between Lewis formulas.

9.7 Which of the following structures for carbon dioxide is the better structure?

$$:\ddot{O}:C\!::\!O: \qquad \text{or} \qquad :O\!::\!C\!::\!O:$$

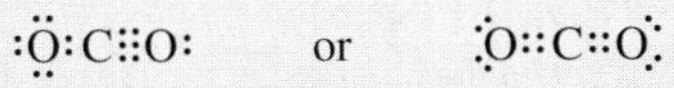

Solution The usual procedure for calculating formal charges shows that the formal charges on the two structures are

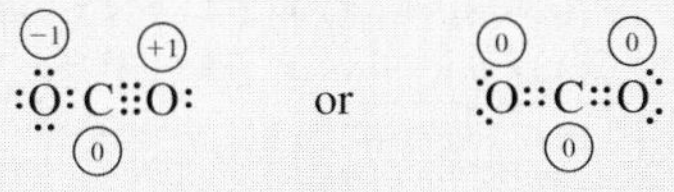

$$\overset{\scriptsize{\textcircled{-1}}}{:\ddot{O}:}\overset{}{C}\overset{\scriptsize{\textcircled{+1}}}{::O:} \qquad \text{or} \qquad \overset{\scriptsize{\textcircled{0}}}{:O::}\overset{}{C}\overset{\scriptsize{\textcircled{0}}}{::O:}$$
$$\underset{\scriptsize{\textcircled{0}}}{} \qquad\qquad \underset{\scriptsize{\textcircled{0}}}{}$$

The structure with all formal charges equal to zero is the better structure.

Formal charges of zero are often not shown; if no formal charge is shown on an atom in a species with formal charges, you should assume that the formal charge on the unlabeled atom is zero.

9.8 Suppose the positions of the atoms in formaldehyde (molecular formula CH_2O) were not known by experiment. Use formal charge to choose between the three possible arrangements of atoms in formaldehyde:

(1) H C O	(2) C O H	(3) H C O H
H	H	

Solution As we saw earlier, formaldehyde has 12 valence electrons. Following the rules for writing Lewis formulas gives these Lewis formulas:

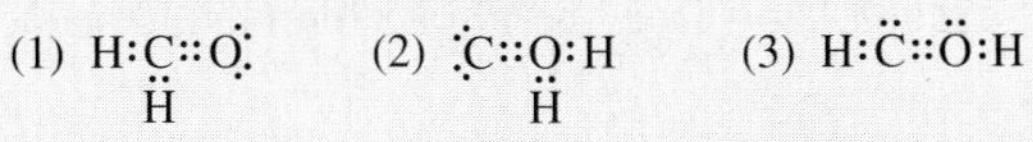

(1) $H\!:\!C\!::\!\ddot{O}:$ (2) $:\ddot{C}\!::\!\ddot{O}\!:\!H$ (3) $H\!:\!\ddot{C}\!::\!\ddot{O}\!:\!H$
      $\ddot{H}$ $\ddot{H}$

Following the procedure for calculating formal charges shows that the formal charges on the three Lewis formulas are

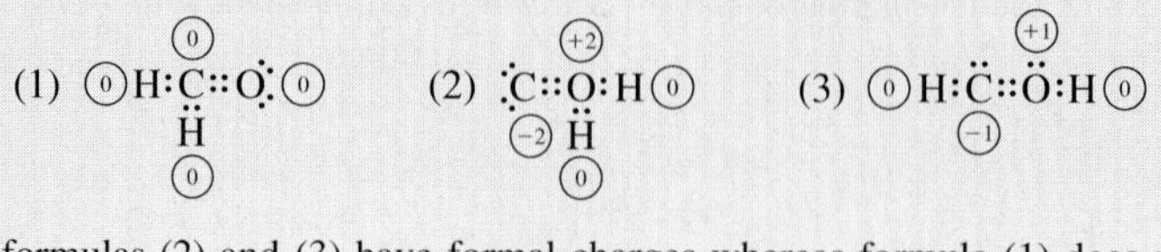

(1) ⓪H:C::O:⓪
　　　　Ḧ
　　　　⓪

(2) :C::O:H⓪
　⊖₂ Ḧ
　　　⓪

(3) ⓪H:C::O:H⓪
　　　⊖₁

Because formulas (2) and (3) have formal charges whereas formula (1) does not, formula (1) is best. As we saw earlier, this is the arrangement of atoms found by experiment.

PRACTICE PROBLEMS

9.21　Use formal charges to show which of the following two Lewis structures for the cyanate ion is better:

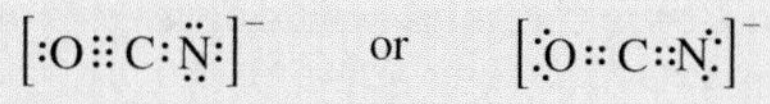

$$\left[:O::C:\ddot{N}:\right]^{-} \quad \text{or} \quad \left[:\ddot{O}::C::N:\right]^{-}$$

9.22　Use formal charges to show why the other possible arrangement of atoms in carbon dioxide, C O O, is unlikely.

Appropriate formulas in the Lewis Structures and Resonance module of the Molecules section take students to another screen where they are asked to identify correct resonance structures and to identify the most important resonance structure from formal charge.

9.9　RESONANCE STRUCTURES

Each molecule and polyatomic ion has *one* structure. But sometimes more than one Lewis formula must be written to represent the bonding in a molecule or polyatomic ion. For example, the Lewis formula for the nitrate ion, NO_3^{-}

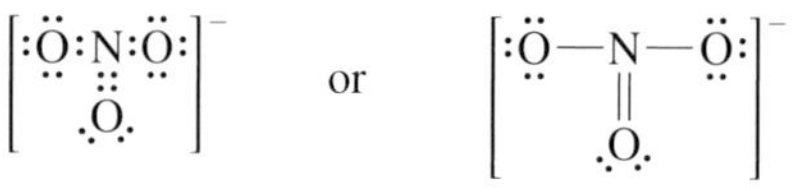

$$\left[:\ddot{O}:N:\ddot{O}: \atop \quad :\ddot{O}: \right]^{-} \quad \text{or} \quad \left[:\ddot{O}-N-\ddot{O}: \atop \quad \overset{\|}{:\ddot{O}:}\right]^{-}$$

follows all the rules. However, measurements show that all the N—O bonds in the nitrate ion have the same length and strength. In the structure shown above, one nitrogen–oxygen bond would be expected to be shorter and stronger than the other two since one is a double bond and the other two are single bonds. Two other structures that follow all the rules can also be written for the nitrate ion:

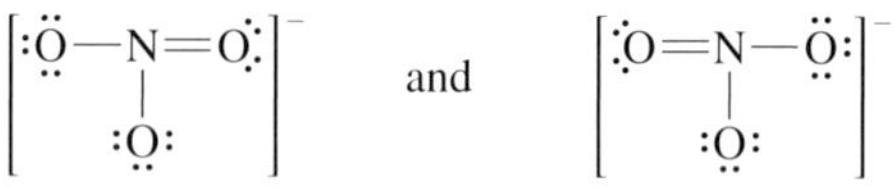

$$\left[:\ddot{O}-N=\ddot{O}: \atop \quad :\ddot{O}: \right]^{-} \quad \text{and} \quad \left[:\ddot{O}=N-\ddot{O}: \atop \quad :\ddot{O}: \right]^{-}$$

The different Lewis formulas for the nitrate ion are called **resonance structures.** In resonance structures, all *atoms* must be shown *in the same positions;* only the *positions of the electrons are different.* A double-headed arrow is used between resonance structures:

$$\left[:\ddot{O}-N-\ddot{O}: \atop \overset{\|}{:\ddot{O}:}\right]^{-} \longleftrightarrow \left[:\ddot{O}=N-\ddot{O}: \atop \quad :\ddot{O}: \right]^{-} \longleftrightarrow \left[:\ddot{O}-N=\ddot{O}: \atop \quad :\ddot{O}: \right]^{-}$$

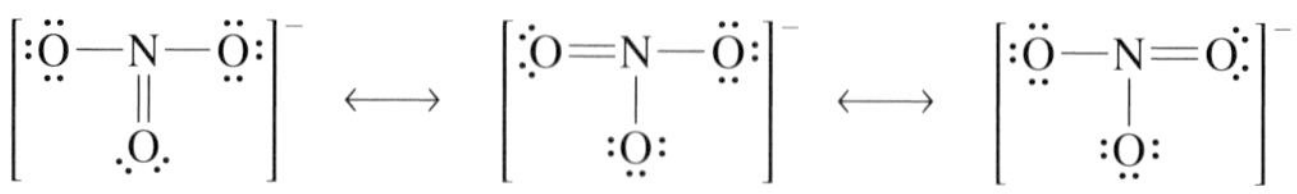

No one of the three Lewis formulas correctly describes the bonding in the nitrate ion. All three are needed. However, the nitrate ion has the same structure all the time. It does *not* change back-and-forth from one structure to another. The nitrate ion is said to be a **hybrid** of the structures shown by the three Lewis formulas.*

In resonance structures, the atoms must be shown in the same positions because resonance structures are pictures of the same thing (a nitrate ion in the example given). The positions of atoms can be determined by experiment. The positions of electrons cannot be determined; only the probability of finding an electron in a certain region of space can be known. More than one resonance structure must be drawn to describe bonding because each Lewis formula shows the electrons in a certain position. In this respect, a Lewis formula is not a completely accurate way of describing bonding.

The second nitrogen–oxygen bond in the nitrate ion is shared equally between the three nitrogen–oxygen bonds. Each nitrogen–oxygen bond can be regarded as consisting of a single bond and a one-third share of the second bond. Thus, the bond order of the nitrogen–oxygen bonds in the nitrate ion is $1\frac{1}{3}$ (1.33). The nitrogen–oxygen bonds in the nitrate ion are shorter and stronger than single bonds but longer and weaker than double bonds.

◗ Species for which resonance structures must be written have fractional bond orders.

Steps in Writing Resonance Structures

The first step in writing the resonance structures for a species is to write one correct Lewis formula for the species. The second step is to recognize that more than one Lewis formula is needed. More than one Lewis formula is needed when there is more than one way to place unshared pairs to form multiple bonds. For example, 24 electrons should be shown in the Lewis formula for nitric acid HNO_3. In working out the Lewis formula for nitric acid, HNO_3, you would arrive at the formula

The most important use of nitric acid is for the manufacture of NH_4NO_3 for fertilizers.

$$:\ddot{O}:N:\ddot{O}:$$
$$:\ddot{O}:$$
$$H$$

In this formula, the nitrogen atom has only six electrons around it. An unshared pair from *either* of the two upper oxygens could be moved to form a double bond:

$$:\ddot{O}:N:\ddot{O}:\quad or \quad :\ddot{O}:N:\ddot{O}:$$
$$:\ddot{O}:\qquad\qquad :\ddot{O}:$$
$$H\qquad\qquad\quad H$$

Thus, there are two resonance structures for nitric acid:

$$\left[\begin{array}{c}:\ddot{O}=N-\ddot{O}:\\ |\\ :O:\\ |\\ H\end{array}\right] \quad and \quad \left[\begin{array}{c}:\ddot{O}-N=\ddot{O}:\\ |\\ :O:\\ |\\ H\end{array}\right]$$

The structure

$$\left[\begin{array}{c}:\ddot{O}-N-\ddot{O}:\\ \|\\ O:\\ |\\ H\end{array}\right]$$

is not as good a structure for nitric acid because it has more formal charges.

Resonance structures are needed to account for the properties of certain organic molecules and ions too. For example, two resonance structures must be written for

---

*You are probably familiar with the term *hybrid* in connection with plants and animals. For example, a mule is the hybrid offspring of a male donkey and a female horse. A mule resembles a horse in height and a donkey in the shape of its head and the length of its ears. A mule is a mule all of the time. It does not change back-and-forth from a donkey to a horse. An even better analogy to a resonance hybrid is to think of a rhinoceros, which is a real animal, as a hybrid of two imaginary animals, a unicorn and a dragon.

Benzene is a component of gasoline and is used to make drugs, dyes, and many other organic compounds.

benzene, C_6H_6.

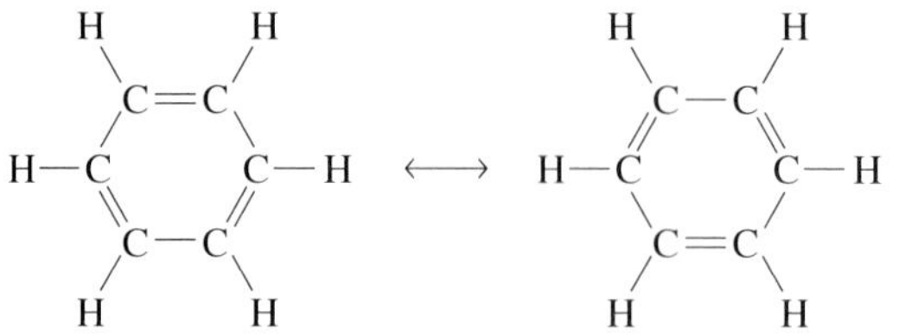

to explain the experimentally observed fact that all the C—C bonds in benzene are the same length. All the C—C bonds in benzene are 139 pm long, a length between the C—C single bond length of 154 pm and the C=C double bond length of 134 pm. The resonance structures for benzene are usually abbreviated

What is the bond order of the carbon–carbon bonds in benzene?

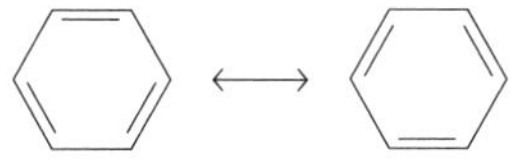

In the abbreviated resonance structures for benzene, each corner of the hexagons represents a carbon atom. The hydrogens attached to the ring carbons are not shown.*

In writing resonance structures, all atoms must be left in the same place. Only the positions shown for electrons are different. For example, the two structures

Benzene is toxic and carcinogenic.

$$[\,:\!\ddot{S}::C::\ddot{N}\!:\,]^- \longleftrightarrow [\,:\!\ddot{S}\!:C::N\!:\,]^-$$

are resonance structures for the thiocyanate ion SCN⁻ but the structure

$$[\,:\!\ddot{S}\!:N::C\!:\,]^-$$

is *not* a resonance structure for the thiocyanate ion because the order in which the atoms are attached to each other is different. Nitrogen, not carbon, is the central atom in this Lewis formula.

PRACTICE PROBLEMS

9.23 One resonance structure for the cyanate ion, OCN⁻, is

$$[\,:\!N\equiv C-\ddot{O}\!:\,]^-$$

Of the following structures,

$$[\,:\!C\equiv N-\ddot{O}\!:\,]^- \quad [\,:\!N=C=\ddot{O}\,]^- \quad [\,:\!\ddot{N}-C\equiv O\!:\,]^-$$

(a) Which one is *not* a resonance structure for the cyanate ion? Explain your answer. (b) Use formal charge to decide which of the other two structures is the better structure.

9.24 One resonance structure for the acetate ion, $C_2H_3O_2^-$, is

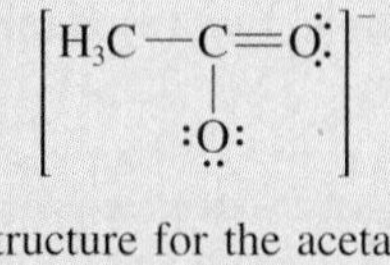

$$\left[\begin{array}{c} H_3C-C=\ddot{O}\!: \\ | \\ :\ddot{O}\!: \end{array}\right]^-$$

(a) Write the other resonance structure for the acetate ion. (b) What is the bond order of the carbon–oxygen bond in the acetate ion?

---

*In reading, you may also find the symbols 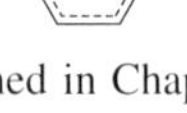 and 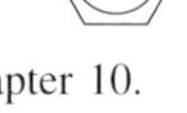 used for benzene. The meaning of the dotted line and the circle is explained in Chapter 10.

9.25 Write resonance structures for (a) the carbonate ion, CO_3^{2-} and (b) the ozone molecule, O_3. Use the correct symbol between resonance structures.

Ozone is much more reactive than O_2. As far as damage to plants and people is concerned, ozone is one of the worst components of smog. Ozone is used to purify drinking water.

9.10 EXCEPTIONS TO THE OCTET RULE

The Lewis formulas for some species do not obey the octet rule. There are three types of species involving main group elements that do not follow the octet rule:

1. Species with more than eight electrons around an atom
2. Species with fewer than eight electrons around an atom
3. Species with an odd total number of electrons

Species with More Than Eight Electrons Around an Atom

Except for species that contain hydrogen, the most common type of species that does not obey the octet rule has more than eight electrons around the central atom. Phosphorus pentafluoride, PF_5, sulfur tetrafluoride, SF_4, sulfur hexafluoride, SF_6, and the hexafluorosilicate ion, SiF_6^{2-}, are typical examples:

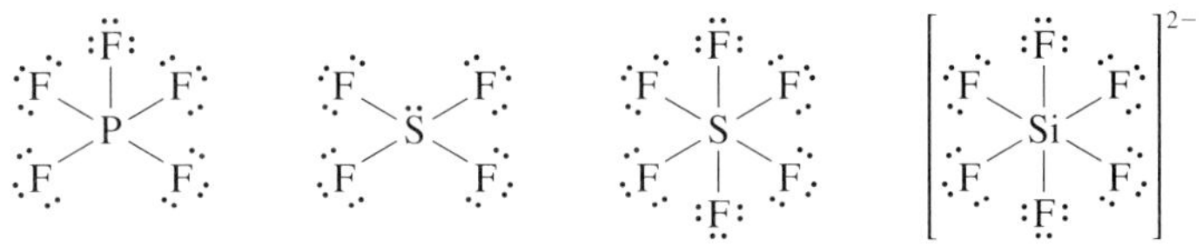

The Lewis formulas for phosphorus pentafluoride and sulfur tetrafluoride each have 10 electrons around the central atom, and the Lewis formulas for sulfur hexafluoride and the hexafluorosilicate ion each have 12. This type of species is not observed for elements in the second row of the periodic table like nitrogen; NF_3 is known but not NF_5. The valence electrons of nitrogen are in the second shell, which has only s and p orbitals. When the $2s$ and $2p$ orbitals of a second-period element are filled and the second-period element has the electron configuration of neon, additional electrons have to go into the $3s$ orbital. The $3s$ orbital of second-period elements is of significantly higher energy than the $2p$ orbitals. Elements of the third period and beyond, however, have $(n - 1)d$ orbitals available that are of about the same energy as the np orbitals. Thus, elements of the third period and beyond can have more than eight electrons around an atom. (In addition, atoms of elements in the third period and beyond are larger and have more room around them than atoms of elements in the second period.)

For elements of the third period and beyond, showing more than eight electrons around an atom can sometimes avoid the need for Lewis formulas with high formal charges. Sulfur dioxide* is a good example. Following the octet rule, the resonance structures

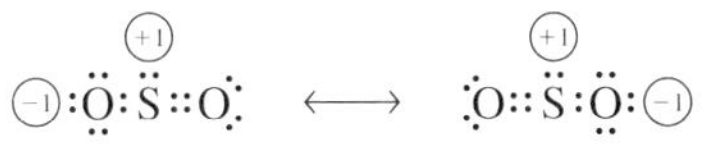

would be written to describe the bonding in sulfur dioxide. However, sulfur is in the third period and can have more than eight electrons around it, making possible

---

*Sulfur dioxide is formed in large quantities by the burning of sulfur-containing coal and oil. It also occurs naturally in volcanic gases. Sulfur dioxide is one of the most harmful air pollutants because it is toxic to both plants and animals and leads to acid rain. Sulfur dioxide is also a very useful compound. It is used to make sulfuric acid and as a preservative in dried fruits and wine.

a third resonance structure

$$\text{:O::S::O:}$$

with ten electrons around sulfur but no formal charges. The experimentally determined length of the sulfur–oxygen bond supports the conclusion that the structure with ten electrons around S but no formal charges best represents the structure of sulfur dioxide.

Species with Fewer Than Eight Electrons Around an Atom

Beryllium chloride in the gas phase at high temperatures (above 900 °C) is an example of a nonhydrogen-containing compound with fewer than eight electrons around the central atom. The Lewis formula

$$\text{:Cl:Be:Cl:}$$

fits the properties observed for beryllium chloride, such as bond length, much better than the structure

$$\text{:Cl::Be::Cl:}$$

which has octets around all atoms. In the former structure, none of the atoms has a formal charge. In the latter structure, all of the atoms have formal charges:

$$\overset{+1}{\text{:Cl::}}\underset{-2}{\text{Be::}}\overset{+1}{\text{Cl:}}$$

This structure is a poor structure because of its high formal charge and the location of the high negative charge on the less electronegative beryllium atom and the positive charges on the more electronegative chlorine atoms. However, for the halides of boron, such as boron trifluoride (BF_3), measurements of bond lengths indicate that both structures

$$\text{:F:B:F:}\qquad \text{and} \qquad \overset{+1}{\text{:F::}}\overset{-1}{\text{B:F:}}$$
$$\text{:F:}\qquad\qquad\qquad\qquad\text{:F:}$$

contribute significantly to the resonance hybrid.

In the singly bonded structures, there are only four electrons around the beryllium atom and six around the boron atom. *Atoms other than hydrogen and helium that have fewer than eight valence electrons around them* are said to be **electron deficient.** Species with electron-deficient atoms are usually very reactive. The central atom tends to achieve an octet by combining with a species with an unshared electron pair such as ammonia. For example, boron trifluoride gas reacts rapidly with ammonia gas to form a white crystalline solid:

$$\text{:F:}\quad\text{H}\qquad\qquad\qquad \text{:F:}\quad\text{H}$$
$$\text{:F—B + :N—H} \longrightarrow \text{:F—B—N—H}$$
$$\text{:F:}\quad\text{H}\qquad\qquad\qquad \text{:F:}\quad\text{H}$$

Species with an Odd Total Number of Electrons

Species with an odd total number of electrons cannot have eight electrons around each atom. They must have one unpaired electron and are usually very reactive (see ▬Figure 9.8). Not many compounds of this type are known. One example is mononitrogen monoxide, NO, which is commonly called nitric oxide. Nitric oxide

For a discussion of the bonding in sulfur dioxide see Purser, G. H. J. Chem. Educ. 1989, 66, 710–713. However, Suidan, L.; Badenhoup, J. K.; Glendening, E. D.; Weinhold, F. J. Chem. Educ. 1995, 72, 583–586 have recently shown, using the powerful ab initio quantum-chemistry programs now available for solving the Schrödinger equation, that the best structures for sulfur dioxide are those that conform to the octet rule. The shortened bond lengths result from the highly ionic character of the S O bonds.

◑ Remember that you must always make a theoretical explanation fit the experimental facts. You cannot make the experimental facts fit your explanation.

For a discussion of Lewis formulas of boron compounds involving multiple bonding see Straub D.K., J. Chem. Educ. 1995, 72, 494–497.

▬ FIGURE 9.8 Most species with an odd number of electrons are very reactive. When it meets the air, colorless NO(g) reacts instantly with O_2(g) to form NO_2(g), which is red-brown. Nitrogen dioxide also has an odd number of electrons. It reacts further to give N_2O_4, which has an even number of electrons. The equation for the reaction taking place in the flask is $3Cu(s) + 8HNO_3(aq) \longrightarrow 3Cu(NO_3)_2(aq) + 2NO(g) + 4H_2O(l)$. Why is the solution blue?

has a total of 11 valence electrons, 5 from the nitrogen and 6 from the oxygen. The best Lewis structure for nitric oxide is

$$\ddot{:}N::\ddot{O}:$$

Species that have one or more unpaired electrons are called **radicals** or **free radicals.** Radicals are believed to play important parts in the aging process and in cancer.

Although important exceptions to the octet rule exist, the octet rule is useful because it works for second-period elements. Second-period elements are more common than elements in lower periods and form more compounds than elements in lower periods.

Why are hydrogens that have only two electrons around them such as the hydrogens in water

$$H:\ddot{O}:$$
$$H$$

not electron deficient?

▶ Species that have odd electrons, such as NO, are also electron deficient.

In Conclusion

At this point, we can list a series of steps to be followed in writing Lewis formulas.

Summary of Procedure for Writing Lewis Formulas

1. Using whatever information is available, write the symbols for the elements in the correct arrangement.
2. Calculate the total number of valence electrons.
3. Place one pair of electrons between each pair of bonded atoms.
4. Beginning at the outside of the formula, place the remaining electrons in pairs until there are eight electrons around each atom (two for hydrogen) or all electrons have been used. If there are extra electrons, place them on the central atom. Remember that atoms in the third period and beyond can have more than eight electrons around them but elements in the second period cannot.
5. If not enough electrons are available to give all atoms except hydrogen an octet, move unshared pairs to form double or triple bonds. However, remember that Be, B, and other Group IIIA elements may have fewer than eight electrons.
6. Examine your Lewis formula to see if resonance structures are needed. (Is more than one position possible for multiple bonds?)
7. Check your answer: Does the Lewis formula show the correct number of atoms? The correct number of electrons? The right number of electrons around each atom? The minimum number of formal charges?

PRACTICE PROBLEMS

9.26 Write Lewis formulas for (a) AsF_5 and (b) XeF_4.

9.27 (a) Write the best Lewis formula for sulfuric acid that follows the octet rule. (b) Write the best Lewis formula for sulfuric acid in which sulfur has more than eight electrons around it. (c) Use formal charge to decide which Lewis formula is better.

9.28 Which of the following molecules has an odd number of electrons?

$$ClO_2 \quad Cl_2O \quad Cl_2O_7 \quad ClN_3 \quad N_2O_5$$

To "demonstrate" the synthesis of a polymer, before class put a long chain of paper clips in an opaque container, such as an oatmeal box. During class, show students a box of "monomer" and then dump contents into the opaque container. Stir with a glass rod and draw out the polymer chain. Models of cross-linked chains and networks like those in gels can also be made from paper clips.

9.11 MACROMOLECULES

The molecules that we have discussed so far are composed of from 2 to 50 atoms, a relatively small number of atoms. **Macromolecules** are *molecules that are com-*

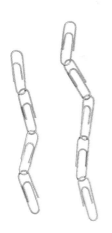

posed of thousands of atoms. *Substances whose molecules are macromolecules* are called **polymers.** Plastics, fibers, rubber, and the biologically important proteins, polysaccharides (such as cellulose and starch), and nucleic acids are polymers.

Polymers are made up of repeating units called **monomers.** A chain of paper clips (■Figure 9.9) is a simple model for a polymer molecule; each paper clip represents a monomer molecule. The monomer molecules can be the same [Figure 9.9(a)] or different [Figure 9.9(b)].

The simplest example of a polymer is polyethylene. In the presence of a catalyst, the monomer ethylene, $H_2C{=}CH_2$, is converted to polymer by heat and pressure:

$$x H_2C{=}CH_2 \xrightarrow[\text{7–20 atm}]{\text{catalyst, 100 °C}} \sim CH_2{-}CH_2{-}CH_2{-}CH_2{-}CH_2{-}CH_2{-}CH_2{-}CH_2 \sim$$

monomer   part of polyethylene molecule

■ **FIGURE 9.9** *Left:* Model of a simple polymer molecule, a chain of identical units. *Right:* The polymer molecule represented by this model has two different units and is a copolymer.

Molecules of polyethylene have between 50 and 3500 $CH_2{-}CH_2$ units; the molecular masses are between 1500 and 100 000 u. Different molecules of polyethylene have different numbers of $CH_2{-}CH_2$ units and different molecular masses. Most synthetic polymers are made up of a mixture of molecules with different molecular masses. Polyethylene is used to make bottles for milk and juices, bread wrappers, trash bags, storage drums, fuel tanks for trucks and cars, squeeze bottles, and many other familiar objects. Without the Eastern bloc, for which no data are available, world production of polyethylene was about 2.5×10^{10} lb in 1988. ■Figure 9.10 shows some things made of polyethylene.

Other compounds with carbon–carbon double bonds, such as $F_2C{=}CF_2$ and $H_2C{=}CHCl$, also polymerize. The polymer formed from $F_2C{=}CF_2$ is called Teflon and has the structure

$$\sim CF_2{-}CF_2{-}CF_2{-}CF_2{-}CF_2{-}CF_2{-}CF_2{-}CF_2{-}CF_2{-}CF_2{-}CF_2{-}CF_2{-}CF_2{-}CF_2 \sim$$

Teflon is very unreactive and very slippery and is used to line kitchenware, greaseless bearings, and artificial joints. The polymer formed from $H_2C{=}CHCl$ is called poly(vinyl chloride) (PVC) and is used to make pipes, siding, floor tile, computer keyboards, and the sheathing on your telephone cord. Its structure is

$$\sim CH_2{-}\underset{\underset{Cl}{|}}{CH}{-}CH_2{-}\underset{\underset{Cl}{|}}{CH}{-}CH_2{-}\underset{\underset{Cl}{|}}{CH}{-}CH_2{-}\underset{\underset{Cl}{|}}{CH}{-}CH_2{-}\underset{\underset{Cl}{|}}{CH}{-}CH_2{-}\underset{\underset{Cl}{|}}{CH} \sim$$

Like PVC, the polymers formed from unsymmetrically substituted ethylenes usually have head-to-tail structures, not tail-to-tail structures such as

$$\sim CH_2{-}\underset{\underset{Cl}{|}}{CH}{-}\underset{\underset{Cl}{|}}{CH}{-}CH_2{-}CH_2{-}\underset{\underset{Cl}{|}}{CH}{-}\underset{\underset{Cl}{|}}{CH}{-}CH_2{-}CH_2{-}\underset{\underset{Cl}{|}}{CH}{-}\underset{\underset{Cl}{|}}{CH}{-}CH_2 \sim$$

or random structure such as

$$\sim CH_2{-}\underset{\underset{Cl}{|}}{CH}{-}\underset{\underset{Cl}{|}}{CH}{-}CH_2{-}\underset{\underset{Cl}{|}}{CH}{-}CH_2{-}\underset{\underset{Cl}{|}}{CH}{-}CH_2{-}CH_2{-}\underset{\underset{Cl}{|}}{CH}{-}CH_2{-}\underset{\underset{Cl}{|}}{CH} \sim$$

Two or more different monomers can be polymerized together, for example,

■ **FIGURE 9.10** Polyethylene containers.

$$x H_2C{=}CHCl + y H_2C{=}CCl_2 \xrightarrow{\text{catalyst, heat}}$$

$$\sim CH_2{-}\underset{\underset{Cl}{|}}{CH}{-}CH_2{-}\underset{\overset{Cl}{|}}{\underset{\underset{Cl}{|}}{C}}{-}CH_2{-}\underset{\overset{Cl}{|}}{\underset{\underset{Cl}{|}}{C}}{-}CH_2{-}\underset{\underset{Cl}{|}}{CH}{-}CH_2{-}\underset{\overset{Cl}{|}}{\underset{\underset{Cl}{|}}{C}}{-}CH_2{-}\underset{\underset{Cl}{|}}{CH} \sim$$

The subunits are arranged head-to-tail but occur in a random order along the chain. *The products of polymerization of more than one monomer are called* **copolymers.**

By the proper choice of monomers, polymerization conditions, and additives, an almost infinite variety of materials having properties exactly suited for different purposes can be produced. (**Additives** are *materials that are added to polymers to improve their properties.*) Like most things, synthetic polymers have both advantages and disadvantages. Synthetic polymers can be made-to-order for different purposes and don't decay as rapidly as natural polymers. Clothes made of synthetic fibers wear better, as well as needing less care. Unfortunately, plastic cups and food containers and other items made of plastic also resist decay. Plastic litter lasts much longer than paper litter and is a danger to birds, animals, and fish, as well as being an eyesore.

Many wild creatures are killed by improperly disposed of plastic.

Biodegradable plastics *(plastics that can be broken down by the action of living organisms)* are a solution to the litter problem. However, although biodegradable plastics disappear when exposed to sunlight and rain, they do not usually decay rapidly when buried in landfills and are *not* an answer to the solid waste problem. Fortunately, efforts to invent ways of recycling plastics are beginning to succeed. Articles made of recyclable plastics are labeled with a recycling symbol that identifies the kind of plastic so that things made of different plastics can be separated. The label is usually on the bottom of the object but may be inside the lid or on the side of a plastic bag. Each area has its own list of acceptable and unacceptable materials as well as requirements for preparing them for recycling.

The first synthetic polymer, Bakelite, was introduced in 1909. The use of synthetic polymers increased rapidly after World War II, and by 1976 synthetic polymers had become the most widely used materials in the United States. More industrial chemists work in the field of polymer chemistry than in any other area of chemistry. Many engineers and all biologists and biochemists work with polymeric materials.

PRACTICE PROBLEMS

9.29   For the polymer of $CH_3CH{=}CH_2$, show a part of the polymer chain that contains three monomer units.

9.30   Write the formula for the monomer of the polymer

$$\sim \underset{\underset{F}{|}}{\overset{\overset{Cl}{|}}{C}} - \underset{\underset{F}{|}}{\overset{\overset{F}{|}}{C}} - \underset{\underset{F}{|}}{\overset{\overset{Cl}{|}}{C}} - \underset{\underset{F}{|}}{\overset{\overset{F}{|}}{C}} - \underset{\underset{F}{|}}{\overset{\overset{Cl}{|}}{C}} - \underset{\underset{F}{|}}{\overset{\overset{F}{|}}{C}} \sim$$

9.12   MORE ABOUT BOND ENERGIES

Why do compounds with carbon–carbon double bonds polymerize? Not because entropy increases. Just as a chain of paper clips is a more orderly arrangement of paper clips than the same paper clips loose in a box, polymers are more orderly than their monomers. Polymerization results in a decrease in disorder or entropy.

When ethylene and compounds derived from it polymerize, carbon–carbon double bonds are converted to carbon–carbon single bonds. Carbon–carbon double bonds are stronger than carbon–carbon single bonds; the bond energy for a $C{=}C$ is 615 kJ/mol, whereas the bond energy for a $C{-}C$ is only 347 kJ/mol (Table 9.3). However, the bond energy for a $C{=}C$ is *less than twice as large* as the bond energy for a $C{-}C$; 2(347 kJ/mol) = 694 kJ/mol. Therefore, thermal energy is released

Recycling symbols. The number 2 shows that the object is made of polyethylene and the number 3 that the object is made of PVC.

Methane, CH_4, is a typical example of a covalently bonded compound. Methane occurs abundantly in nature as the chief constituent of natural gas and in coal. Mixtures of methane and air are explosive, and methane has caused many explosions in coal mines. Although it is not poisonous, it can cause death by asphyxiation. Methane is formed by the action of bacteria on freshly deposited sediments of plant materials at the bottoms of lakes and seas. Gas rich in methane is also produced by activated sludge processes of sewage disposal, and processes have been developed to recover methane from animal feedlot wastes.

The use of natural gas was recorded in China about 900 B.C.; wells were drilled, and the gas was piped through bamboo tubes. But natural gas was not known in Europe until much later; it was discovered in England in 1659. There methane from coal began to be used as fuel for internal combustion engines and to light streets and houses in 1790. The town of Fredonia, New York, was first illuminated by natural gas in 1821. However, it was not until the 1920s that natural gas was transported by pipeline. The period of tremendous expansion in the use of natural gas that continues at the present time did not begin until after World War II.

Before it can be used, natural gas must be processed to remove compounds with higher molecular masses that are liquids under ordinary conditions, sulfur compounds such as hydrogen sulfide, and carbon dioxide. Natural gas is usually transported over land by buried pipelines. Although a pressure of about 70 atm is used to reduce the volume of the gas, a gas pipeline can only carry from one-third to one-half as much energy as an oil pipeline in a given time. Beginning in 1959, liquefied natural gas (LNG) has been transported overseas by ship. Liquefying natural gas reduces the volume of the gas about 600-fold.

About a quarter of the total energy required in North America comes from the burning of natural gas. The combustion of natural gas causes far less air pollution than the burning of coal and gasoline. Although the major use of natural gas is as a fuel, increasing amounts are being used as a raw material for the chemical industry. Methyl alcohol, CH_3OH, acetylene, $HC{\equiv}CH$, and hydrogen are all made from natural gas. In addition, all the helium in commerce comes from natural gas.

A gas light. The chief constituent of natural gas is methane, a typical example of a covalently bonded compound.

when an ethylene molecule adds to the end of the polymer chain because the carbon–carbon double bond is converted to two carbon–carbon single bonds. One of the new single bonds connects the two carbon atoms to each other, and the second connects them to the chain. Thus, polymerization is exothermic and, as a result, spontaneous despite the decrease in entropy.

Bond energies can be used to estimate values of ΔH_{rxn} for reactions involving species so reactive that they cannot be isolated or enthalpies of formation for compounds that have not yet been prepared. For example, the reaction

$$CO(g) + OH(g) \longrightarrow CO_2(g) + H(g)$$

is the most important step in the conversion of carbon monoxide to carbon dioxide in flames from burning hydrocarbons. The species, OH, is a radical. It is very reactive and cannot be isolated; therefore, the heat of formation of OH cannot be measured. However, the enthalpy of reaction can be estimated from bond energies. Table 9.4 shows some typical values for bond energies. Sample Problem 9.9 shows how to estimate the enthalpy of reaction from bond energies.

TABLE 9.4 Average Bond Energies at 25 °C (298 K)[a]

Bond	Bond Energy, kJ/mol	Bond	Bond Energy, kJ/mol
C—C	347	H—Cl	432
C=C	615	H—F	568
C≡C	812	H—I	298
C—Cl	326	H—N	389
C—N	305	H—O	464
C—O	360	H—S	368
C=O	728	N≡N	946
C≡O	1075	O=O	498
Cl—Cl	244	P—H	322
Cl—O	205	Si—Si	192
F—F	158	Si—C	305
H—H	436	Si—H	318
H—Br	366	Si—O	464
H—C	414		

[a]Most data are from Jolly, W. L. *Modern Inorganic Chemistry,* 2nd ed.; McGraw-Hill: New York, 1991; pp. 58, 62.

In the Bond Energies module of the Thermodynamics section students deconstruct and reassemble molecules in order to investigate reaction enthalpies.

SAMPLE PROBLEM

9.9 Use bond energies to estimate ΔH_{rxn} at 25 °C for the reaction

$$CO(g) + OH(g) \longrightarrow CO_2(g) + H(g)$$

Solution To use bond energies to estimate ΔH_{rxn}, we must know which bonds are broken and which bonds are formed in the reaction. *The first step in solving a problem of this type is to write Lewis formulas for the reactants and products:*

$$:C{\equiv}O: + \ddot{O}{-}H \longrightarrow \ddot{O}{=}C{=}\ddot{O}: + \cdot H$$

As you can see from the Lewis formulas, the reaction involves breaking one carbon–oxygen triple bond and one hydrogen–oxygen bond and forming two carbon–oxygen double bonds.

Next, look up the bond energies in Table 9.4 and record them under the Lewis formulas for easy reference:

$$:C{\equiv}O: + \ddot{O}{-}H \longrightarrow \ddot{O}{=}C{=}\ddot{O}: + \cdot H$$

Bond energies, kJ/mol   1075   464   728

Breaking one carbon–oxygen triple bond requires 1075 kJ and breaking one hydrogen–oxygen bond, 464 kJ. A total of

$$(1075 + 464) \text{ kJ} = 1539 \text{ kJ}$$

must be supplied to break bonds. When two carbon–oxygen double bonds are formed,

$$2(728 \text{ kJ}) = 1456 \text{ kJ}$$

are released. The difference between the energy that must be supplied to break bonds and the energy that is released by bond formation is

$$(1539 - 1456) \text{ kJ} = 83 \text{ kJ}$$

More energy must be supplied than is released. The enthalpy of reaction is positive; the estimated $\Delta H_{rxn} = 83$ kJ.

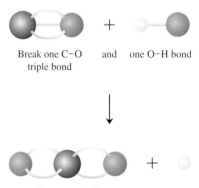

Break one C–O triple bond   and   one O–H bond

Form two C–O double bonds

When a molecule of carbon monoxide reacts with a hydroxyl radical to form carbon dioxide and a hydrogen atom, a carbon–oxygen triple bond and a hydrogen–oxygen bond must be broken. Two carbon–oxygen double bonds form.

These rocks at Acadia National Park are composed of compounds with Si—O bonds.

When molecular solids melt and liquids vaporize, no chemical bonds are broken. However, energy must be added to overcome the intermolecular attractions. Therefore, simple calculations of heats of reaction and heats of formation from bond energies such as the calculation in Sample Problem 9.9 should only be carried out if all reactants and products are gases.

Bond strengths help explain the chemical forms of the elements in nature. For example, nitrogen gas is unreactive because of the great strength of the nitrogen–nitrogen triple bond. The strength of carbon–carbon bonds compared with bonds between carbon and other elements is an important factor in why there are so many more compounds of carbon than compounds of all the other elements together. Compounds with Si—O bonds, which are very strong, make up 95% of Earth's crust and upper mantle.

SUMMARY

The forces that hold atoms together in compounds are called **chemical bonds.** Two kinds of bonds connect atoms in compounds: ionic bonds and covalent bonds. Formation of both ionic and covalent bonds is exothermic and leads to a state of lower energy and greater stability.

Ionic bonds are the attraction between positively and negatively charged ions. Most of the energy released by the formation of ionic bonds is lattice energy. **Lattice energy** is the energy required to separate the ions of an ionic solid to an infinite distance.

Covalent bonds consist of pairs of electrons shared between two atoms. One, two, or three pairs of electrons can be shared between two atoms leading to single, double, and triple bonds. The average energy needed to break a given type of bond is called the **bond energy** of that type of bond. If bonds between the same two atoms are compared, triple bonds are shorter and stronger than double bonds, which are shorter and stronger than single bonds. Bond energies can be used to estimate ΔH_{rxn}° for reactions involving species so reactive that they cannot be isolated or ΔH_f° for compounds that have not yet been prepared.

The number of bonds between two atoms in a molecule or ion is called **bond order.** A pair of electrons that is shared between two atoms is called a **bonding pair.** Only the outer electrons are involved in bond formation; the outer electrons used in bond formation are called **valence electrons,** and the outer shell is called the **valence shell.** Valence electrons that are not shared are referred to as **unshared pairs, lone pairs,** or **nonbonding electrons.**

Lewis (electron-dot) formulas are used to represent molecules and polyatomic ions composed of main group elements. Many compounds follow the **octet rule** that there should be eight electrons around each atom except hydrogen. However, although the octet is a very stable grouping of electrons, not all atoms have octets in compounds. Three types of species involving main group elements are exceptions to the octet rule: (1) species with more than eight electrons around the central atom, (2) species with fewer than eight electrons around the central atom, and (3) species with an odd total number of electrons. Species that have one or more unpaired electrons are called **radicals** or **free radicals.**

Electronegativity is the ability of an atom that is bonded to another atom or atoms to attract electrons to itself. Electronegativity is a periodic property. It usually increases from left to right across rows of the periodic table and from bottom to top of groups. Differences in electronegativity are useful in predicting the direction of polarization of covalent bonds and whether bonds are slightly polar, very polar, or ionic. The greater the difference in electronegativity between bonded atoms, the more polar the bond. Bonds vary from nonpolar to slightly polar to very polar to mainly ionic, depending on the difference in electronegativity between the bonded atoms.

Sometimes more than one Lewis formula must be written to describe a molecule or polyatomic ion. The different Lewis formulas are called **resonance structures.** The real structure of the molecule or ion is a **hybrid** of the resonance structures.

Formal charges show the approximate distribution of electron density in molecules and polyatomic ions. They are helpful in deciding which of several possible Lewis formulas is best.

Macromolecules are molecules that are composed of thousands of atoms. Substances whose molecules are macromolecules are called **polymers.** Plastics, fibers, rubber, and the biologically important proteins, polysaccharides (such as cellulose and starch),

and nucleic acids are polymers. Polymers are made up of repeating units called **monomers.** Different molecules of most synthetic polymers contain different numbers of monomer units and thus have different molecular masses. Polymers made from more than one kind of monomer are called **copolymers.** Polymers can be made-to-order for different purposes by the proper choice of monomers, polymerization conditions, and **additives** (materials added to polymers to improve their properties). Some polymers are **biodegradable**—that is, they can be broken down by living organisms. Synthetic polymers are the most widely used materials in the United States at the present time, and more industrial chemists work in the field of polymer chemistry than in any other area of chemistry.

ADDITIONAL PRACTICE PROBLEMS

For information about the organization of Additional Practice Problems, Stop & Test Yourself, Putting Things Together, and Applications, see the beginnings of these sections in Chapter 1.

9.33 How many valence electrons are in each of the following atoms? (a) Al (b) C (c) Cl (d) K (e) N (f) S (9.1)

9.34 How many valence electrons are in each of the following atoms? (a) Ba (b) Na (c) H (d) He (e) O (f) Sn (g) P (9.1)

9.35 Use electron configurations to describe the formation of each of the following ions from the atom. (a) Ag^+ (b) Br^- (c) Cr^{3+} (d) Mn^{2+} (e) O^{2-} (f) Sn^{2+} (g) Sr^{2+} (9.2)

9.36 Use electron configurations to describe the formation of each of the following ionic compounds: (a) aluminum oxide from aluminum atoms and oxygen atoms, (b) lithium hydride from lithium atoms and hydrogen atoms, (c) magnesium nitride from magnesium atoms and nitrogen atoms. Write formulas for each of the compounds formed. (9.2)

9.37 (a) How many unpaired electrons are in a Cu^{2+} ion with electron configuration $[Ar]3d^84s^1$? (b) With an electron configuration $[Ar]3d^9$? (c) Magnetic measurements show that the copper(II) ion has one unpaired electron. What is the electron configuration of the copper(II) ion? (9.2)

9.38 In one formula unit of chromium(IV) oxide, how many chromium(IV) ions and how many oxide ions are there? (9.2)

9.39 Which of the following compounds contain ions that do *not* have noble gas electron configurations? (a) AlF_3 (b) Fe_2O_3 (c) $PbCl_2$ (d) TiO_2 (e) $ZnCl_2$. For each ion that does not have a noble gas electron configuration, write the electron configuration. (9.2)

9.40 Arrange the compounds of the following set in order of increasing lattice energy:

$$BaCl_2 \quad BaS \quad KBr \quad KCl \ (9.2)$$

9.41 (a) Given the following information,

$$K(s) \longrightarrow K(g) \qquad \Delta H° = 77 \text{ kJ}$$
$$Br_2(l) \longrightarrow Br_2(g) \qquad \Delta H° = 30 \text{ kJ}$$

$$Br_2(g) \longrightarrow 2Br(g) \qquad \Delta H° = 194 \text{ kJ}$$
$$K(g) \longrightarrow K^+(g) \qquad \Delta H° = 419 \text{ kJ}$$
$$Br(g) + e^- \longrightarrow Br^-(g) \qquad \Delta H° = -324 \text{ kJ}$$
$$2K(s) + Br_2(l) \longrightarrow 2KBr(s) \qquad \Delta H° = -788 \text{ kJ}$$

use a Born–Haber cycle to calculate the energy change associated with the reaction

$$K^+(g) + Br^-(g) \longrightarrow KBr(s)$$

All equations are written on a mole basis. (b) What is the energy change associated with the change

$$KBr(s) \longrightarrow K^+(g) + Br^-(g)$$

called? (c) Is the change in part (b) exothermic or endothermic? (9.2)

9.42 Tell which group of the periodic table X belongs to if the Lewis symbol for X is:

(a) $\cdot \ddot{X} \cdot$ (b) $:\ddot{X}:$ (c) $\cdot \ddot{X}:$ (9.3)

9.43 Referring only to the periodic table, write Lewis formulas for each of the following by combining Lewis symbols: (a) an ammonium ion, NH_4^+, (b) a bromine molecule, Br_2, (c) ethane, C_2H_6, (d) hydrogen iodide, HI, (e) a hydrogen molecule, H_2, (f) hydrogen selenide, H_2Se, (g) silane, SiH_4, (h) stibine, SbH_3. (9.3)

9.44 Use the periodic table to arrange the atoms of the following set in order of increasing electronegativity:

$$Br \quad Cl \quad Fe \quad K \quad Rb \ (9.5)$$

9.45 Use the periodic table to label the atoms joined by the following bonds with partial positive and negative charges and to predict which member of each pair of bonds is more polar.
(a) Al—F or Al—Cl (b) C—Si or C—C
(c) K—Br or Cu—Br (d) Mg—Cl or Ca—Cl
(e) Si—O or Si—S (9.5)

9.46 Use the empirical rules from Section 9.5 to answer this question. Of the following compounds, (a) which are mainly ionic? (b) Which are mainly covalent? (i) AgF (ii) Fe_2O_3 (iii) BeO (iv) MoF_6 (v) KI (vi) N_2H_4 (9.5)

9.47 (a) Calculate the difference in electronegativity between each of the pairs of atoms in Problem 9.46. (b) Does the often-stated rule of thumb that "if the difference in electronegativity between the elements in a compound composed of two elements is greater than 1.7, the compound is predominantly ionic" lead to correct predictions? Explain your answer. (9.5)

9.48 How many valence electrons should be shown in the Lewis formulas for each of the following? (a) H_3O^+ (b) IF_7 (c) SO_4^{2-} (9.6)

9.49 The Lewis formula for the dichromate ion, $Cr_2O_7^{2-}$, is

$$\left[\begin{array}{c} \ddot{O} \qquad \ddot{O} \\ | \qquad | \\ \ddot{O}-Cr-\ddot{O}-Cr-\ddot{O} \\ | \qquad | \\ \ddot{O} \qquad \ddot{O} \end{array} \right]^{2-}$$

How many valence electrons does each chromium atom contribute? (9.6)

9.50 Write Lewis formulas for each of the following: (a) acetaldehyde, CH_3CHO, (b) hydronium ion, H_3O^+, (c) propene, C_3H_6, (d) sulfate ion, SO_4^{2-}, (e) sulfuric acid, H_2SO_4. (9.6)

9.51 Write Lewis formulas for each of the following: (a) HOI, (b) SiF_4, (c) PCl_4^+, (d) Cl_2O, (e) NH_2^-. (9.6)

9.52 Of the following C N bonds:

$$H_3C-C\equiv N: \qquad H_3CHC=\ddot{N}H \qquad H_3CH_2C-\ddot{N}H_2$$

(a) Which is shortest? (b) Which is weakest? (c) Which has a bond order of three? (9.7)

9.53 Add formal charges to the following Lewis formulas:

(a)
$$\begin{array}{c} \ddot{O} \\ \| \\ H-\ddot{O}-P-\ddot{O}-H \\ | \\ \ddot{O} \\ | \\ H \end{array}$$

(b)
$$\begin{array}{c} \ddot{O} \\ \| \\ H-\ddot{O}-P-\ddot{O}-H \\ | \\ H \end{array}$$

(c)
$$\begin{array}{c} \ddot{O} \\ \| \\ H-P-H \\ | \\ \ddot{O} \\ | \\ H \end{array}$$

(d)
$$H-\ddot{O}-N=\ddot{O} \\ | \\ \ddot{O}$$

(e) $H-\ddot{O}-\ddot{N}=\ddot{O}$ (9.8)

9.54 (a) Write the best Lewis formula, including formal charges if any, for the cyanide ion, CN^-. (b) If a hydrogen ion, H^+, attaches to a cyanide ion, does it attach to the C or to the N? Explain your answer. (9.8)

9.55 Use formal charge to choose between different arrangements of atoms. (a) Carbon disulfide, CSS or SCS? (b) Hypochlorous acid, HOCl, HClO, or OHCl? (c) Nitrosyl chloride, NOCl, ClNO, or NClO? (9.8)

9.56 (a) What are resonance structures? (b) Why is the concept of resonance used? (c) How do you know when you need to write resonance structures? (d) Is it possible to isolate a resonance structure? Explain your answer. (9.9)

9.57 One resonance structure for pyridine is ⬡N. (a) Write another resonance structure for pyridine. (b) What is the bond order of the carbon–nitrogen bonds? (9.9)

9.58 Phosphorus, sulfur, and chlorine have expanded octets in many compounds. (a) Do nitrogen, oxygen, and fluorine have expanded octets in their compounds? (b) Do arsenic, selenium, and bromine? Explain your answer. (9.10)

9.59 Which of the following species have an odd number of electrons? (a) Br (b) OH (c) OH^- (d) NO_2 (e) PCl_2 (f) PCl_3 (9.10)

9.60 Explain why the octet rule works for most compounds although exceptions exist. (9.10)

9.61 (a) Why doesn't hydrogen have an octet in its compounds? (b) How many covalent bonds does hydrogen form? (9.10)

9.62 Write Lewis formulas for each of the following: (a) BBr_3, (b) BrF_4^-, (c) OH. (9.10)

9.63 Write the complete structural formula for a piece of the polymer chain from the polymerization of H_2C CHF that contains four monomer units. (9.11)

9.64 For the polymer

$$\begin{array}{ccccccc} CH_3 & & CH_3 & & CH_3 & \\ | & H & | & H & | & H \\ | & | & | & | & | & | \\ \sim C & - C & - C & - C & - C & - C \sim \\ | & | & | & | & | & | \\ | & H & | & H & | & H \\ CH_3 & & CH_3 & & CH_3 & \end{array}$$

write a structural formula for the monomer. (9.11)

9.65 (a) What is a copolymer? (b) What is the advantage of copolymerization? (9.11)

9.66 Use bond energies to estimate ΔH_{rxn} for the reaction

$$CH_4(g) + 2H_2O(g) \longrightarrow CO_2(g) + 4H_2(g) \quad (9.12)$$

9.67 (a) Use bond energies to estimate ΔH_{rxn} for the reaction

$$H_2(g) + Cl_2(g) \longrightarrow 2HCl(g)$$

(b) The value for ΔH_f° [HCl(g)] in Table 6.2 is -92.31 kJ/mol. What percent of the accurate value is the error in the estimated value? (9.12)

9.68 The most stable oxide of chlorine is Cl_2O_7 (condensed structural formula $O_3ClOClO_3$). Write the best Lewis formula.

9.69 Briefly describe some properties of ionic compounds and explain how ionic bonding accounts for these properties.

9.70 Why are there no molecules of NaCl in salt crystals?

9.71 Which of the following formulas represent stable ionic compounds? Explain your answers. (a) BaF_2 (b) CaO (c) HCl (d) K_2Br (e) $RbCl_2$

9.72 Use the following information

$$\begin{array}{ll} Mg(s) \longrightarrow Mg(g) & \Delta H^\circ = 146.4 \text{ kJ} \\ F_2(g) \longrightarrow 2F(g) & \Delta H^\circ = 158.2 \text{ kJ} \\ Mg(s) + F_2(g) \longrightarrow MgF_2(s) & \Delta H^\circ = -1123.4 \text{ kJ} \end{array}$$

and the appropriate information given elsewhere in this book to calculate the lattice energy of $MgF_2(s)$.

9.73 (a) Write two resonance structures including formal charges for hydrazoic acid HN_3. (b) What are the bond orders of the N N bonds?

9.74 Given the following information,

$$NH_3(g) \longrightarrow H(g) + NH_2(g) \qquad \Delta H° = 453 \text{ kJ}$$
$$NH_2(g) \longrightarrow H(g) + NH(g) \qquad \Delta H° = 360 \text{ kJ}$$
$$NH(g) \longrightarrow H(g) + N(g) \qquad \Delta H° = 356 \text{ kJ}$$

calculate the bond energy of the N H bond.

9.75 Which group of the periodic table does X belong to if the Lewis formula for a compound is

$$:\ddot{F}-\overset{..}{X}-\ddot{F}:$$
$$:\ddot{F}:$$

9.76 (a) What is a polymer such as

called? (b) What are the monomers?

STOP & TEST YOURSELF

1. The formula for aluminum nitride is
   (a) AlN (b) AlN_3 (c) Al_2N_3 (d) Al_3N (e) None of the preceding answers is correct.

2. The electron configuration of the Cr^{3+} ion is:
   (a) [Ar] ↑↑ □□□ | ↑ 3d 4s
   (b) [Ar] ↑↓ □□□ | ↑ 3d 4s
   (c) [Ar] ↑↑↑ □□ | □ 3d 4s
   (d) [Ar] ↑↓↑ □□ | □ 3d 4s
   (e) [Kr] ↑↑↑ □□ | □ 3d 4s

3. Which of the following elements cannot have more than eight electrons around it?
   (a) Cl (b) C (c) Si (d) Sn (e) S

4. Which chloride should exhibit the most covalent (least ionic) type of bond?
   (a) $AlCl_3$ (b) BCl_3 (c) KCl (d) $MgCl_2$ (e) NaCl

5. For a ClF_4^+ ion, how many valence electrons should be shown in the Lewis formula?
   (a) 32 (b) 33 (c) 34 (d) 35 (e) 36

6. Which of the following compounds has the highest lattice energy?
   (a) $CaBr_2$ (b) $CaCl_2$ (c) CaO (d) CaS (e) RbF

7. Which species has more than eight electrons around the central atom?
   (a) BF_3 (b) BF_4^- (c) BrF_3 (d) NF_3 (e) PF_3

8. Use bond energies to estimate ΔH_{rxn} for the reaction $O(g) + H_2O(g) \longrightarrow 2OH(g)$. (a) -928 kJ (b) -464 kJ (c) 0 kJ (d) $+464$ kJ (e) $+928$ kJ

9. How many unshared pairs of valence electrons are there in the Lewis formula for $AsCl_3$?
   (a) 1 (b) 3 (c) 10 (d) 13 (e) 26

10. One resonance structure for the nitrite ion is

Which of the following is another resonance structure for the nitrite ion?

(e) None of them is.

11. In this Lewis formula for sulfurous acid,

$$H-\overset{..}{O}-\overset{..}{S}-\overset{..}{O}-H$$
$$:\overset{..}{O}:$$

what is the formal charge on sulfur?
   (a) -2 (b) -1 (c) 0 (d) $+1$ (e) $+2$

12. Given the electronegativities, C = 2.5, N = 3.1, and S = 2.4, which of the following Lewis formulas is best for the thiocyanate ion?
   (a) $\left[:\ddot{C}=N=\ddot{S}:\right]^-$
   (b) $\left[:\ddot{C}=S=\ddot{N}:\right]^-$
   (c) $\left[:\ddot{S}=C=\ddot{N}:\right]^-$
   (d) $\left[:\ddot{S}-C\equiv N:\right]^-$
   (e) $\left[:S\equiv C-\ddot{N}:\right]^-$

13. The compound shown on the next page forms part of a safflower plant's natural defenses against parasites. Of the bonds given,

the carbon–carbon bond that begins at which carbon is shortest and strongest?

$$\overset{13}{CH_3}-\overset{12}{CH}=\overset{11}{CH}-\overset{10}{C}\equiv\overset{9}{C}-\overset{8}{C}\equiv\overset{7}{C}-\overset{6}{C}\equiv\overset{5}{C}-\overset{4}{CH}=\overset{3}{CH}-\overset{2}{CH}=\overset{1}{CH_2}$$

(a) 1   (b) 2   (c) 3   (d) 4   (e) 5

14. The monomer from which the polymer

$$\sim CH_2-\underset{\underset{COOCH_3}{|}}{\overset{\overset{CH_2CH_3}{|}}{C}}-CH_2-\underset{\underset{COOCH_3}{|}}{\overset{\overset{CH_2CH_3}{|}}{C}}-CH_2-\underset{\underset{COOCH_3}{|}}{\overset{\overset{CH_2CH_3}{|}}{C}}\sim$$

is made is
(a) $CH_3CH{=}CHCOOCH_3$.
(b) $H_2C{=}C(CH_3)COOCH_3$.
(c) $CH_3CH{=}CHCOOCH_2CH_3$.
(d) $H_2C{=}C(CH_2CH_3)COOCH_3$.
(e) $H_2C{=}C(CH_3)COOCH_2CH_3$.

PUTTING THINGS TOGETHER

9.77   (a) Explain what an ionic bond is.   (b) Explain the difference between an ionic and a covalent bond.

9.78   (a) What combination of a metal and a nonmetal will form the most ionic compound?   (b) Why doesn't ionic bonding take place when two nonmetals combine to form a compound? (c) Give an example of an ionic compound that contains only nonmetals.

9.79   (a) Add formal charges to the Lewis formulas for acrylonitrile, C_3H_3N:

$$H-\underset{\underset{H}{|}}{C}=\underset{\underset{H}{|}}{C}-C\equiv N\colon \qquad H-\underset{\underset{H}{|}}{C}=\underset{\underset{H}{|}}{C}-C{=}C{=}\ddot{N}{\cdot}$$

$$H-\underset{\underset{H}{|}}{\ddot{C}}-\underset{\underset{H}{|}}{C}{=}C{=}N\colon$$

(b) What are different Lewis formulas for the same substance, such as those in part (a), called?   (c) What symbol should be used between the different Lewis formulas?   (d) Which of the three Lewis formulas contributes most to the physical properties of acrylonitrile?   (e) Which contributes least? Explain your answers to parts (d) and (e).

9.80   (a) Draw two resonance structures for toluene, $C_6H_5CH_3$, a derivative of benzene in which a methyl group, CH_3, is substituted for one of the hydrogens.   (b) What is the bond order of the carbon–carbon bonds in the benzene ring? In the bond between the methyl group and the ring?

9.81   Does the following Lewis formula represent an anion, a cation, or a molecule? If it represents an ion, what is the charge on the ion?

$$\colon\ddot{O}{=}\ddot{S}{=}\ddot{O}\colon$$
$$\underset{\underset{\cdot\ddot{O}\cdot}{\|}}{}$$

9.82   The H Cl bond in HCl can be broken in three ways.
(1) Each atom gets one electron of the shared pair:

$$H\colon\!\ddot{\underset{\cdot\cdot}{Cl}}\colon(g) \longrightarrow H{\cdot}(g) + {\cdot}\ddot{\underset{\cdot\cdot}{Cl}}\colon(g)$$

(2) The chlorine atom gets both electrons of the shared pair:

$$H\colon\!\ddot{\underset{\cdot\cdot}{Cl}}\colon(g) \longrightarrow H^+(g) + \colon\ddot{\underset{\cdot\cdot}{Cl}}\colon^-(g)$$

(3) The hydrogen atom gets both electrons of the shared pair:

$$H\colon\!\ddot{\underset{\cdot\cdot}{Cl}}\colon(g) \longrightarrow H\colon^-(g) + \ddot{\underset{\cdot\cdot}{Cl}}\colon^+(g)$$

(a) Predict whether each reaction will be exothermic or endothermic. Explain your answers.   (b) Predict whether reaction 2 or 3 will involve the larger enthalpy change.   (c) Check your predictions by calculating ΔH_{rxn} for each of the reactions.

9.83   In the Lewis formula for nitric oxide (NO), why is the odd electron shown on nitrogen instead of on oxygen?

9.84   Use formal charge to argue against structures with octets for BF_3.

9.85   What properties distinguish an ionic compound such as NaCl from a molecular compound such as sugar?

9.86   Which properties of atoms determine whether compounds have ionic or covalent bonds?

9.87   Explain why the formula C_2H_2 is written for acetylene, but it is incorrect to write the formula Na_2Cl_2 for salt.

9.88   Which of the following are paramagnetic? (a) NO   (b) NO^+ (c) NO_2 (d) NO_2^- (e) N_2O_4 (f) ClO_2 (g) ClO_2^-

9.89   Write Lewis formulas for the four species: CN^-, CO, N_2, and NO^+.   (a) What do these four species have in common?   (b) What is the proper term to describe species that are related as these four species are related?   (c) Could C_2H_2 be considered another member of the series? Explain your answer.

9.90   For the polymer

$$\sim CH_2-\underset{\underset{\text{(ring)}}{|}}{\overset{\overset{H}{|}}{C}}-CH_2-\underset{\underset{\text{(ring)}}{|}}{\overset{\overset{H}{|}}{C}}-CH_2-\underset{\underset{\text{(ring)}}{|}}{\overset{\overset{H}{|}}{C}}\sim$$

(a) What is the monomer?   (b) If the mass of one typical molecule is 2.6×10^5 u, how many monomer units combined to form the molecule?

9.91   Which member of each of the following pairs would you expect to be more stable? Explain your choices.
(a) BH_3 or BH_4^- (b) CaF or CaO   (c) KCl or KCl_2 (d) NF_5 or PF_5 (e) NO_2 or NO_2^-

9.92 (a) Write two Lewis formulas for thionyl chloride, $SOCl_2$. One should involve an expanded octet, and one should not. (b) On the basis of formal charge, which structure is better?

9.93 Given the following condensed structural formulas, write Lewis formulas. Include formal charges, if any, and important resonance structures, if any.
(a) $(CH_3)_2CHOH$ (b) $(CH_3)_3COCH_2CH_3$ (c) C_6H_5OH
(d) $H_2C{=}C(CH_3)COOCH_3$ (e) CH_3NO_2

9.94 Use bond energies to estimate ΔH for

$$CH_3COO(g) \longrightarrow CH_3(g) + CO_2(g)$$

Is this reaction exothermic or endothermic?

9.95 An unknown compound was 38.67% C, 16.23% H, and 45.10% N and had a molecular mass of 32. Write the Lewis formula of this compound.

9.96 A 0.036 82-g sample of an unknown compound gave 0.115 53 g CO_2 and 0.047 31 g H_2O on combustion. A 0.3864-g sample of the unknown compound occupied a volume of 223 mL at a laboratory temperature of 24.8 °C and a barometric pressure of 765 mmHg. Write Lewis formulas for the two compounds that fit the data.

9.97 Use electron configurations to describe the formation of the following compounds. Write chemical equations for formation of the compounds by combination of the elements and name the compounds. (a) compound of cesium and fluorine   (b) compound of barium and oxygen   (c) compound of barium and chlorine   (d) compound of calcium and nitrogen   (e) compound of calcium and hydrogen

9.98 Use Lewis symbols to figure out the formulas of the following compounds and write chemical equations for formation of the compounds by combination of the elements. (a) simplest compound of fluorine and chlorine (name this compound) (b) simplest compound of germanium and hydrogen   (c) simplest compound of arsenic and hydrogen   (d) simplest compound of arsenic and chlorine (name this compound)

9.99 When CCl_4 is added to water, no reaction takes place. The CCl_4, which is a liquid, is insoluble in water and sinks to the bottom. When PCl_3, which is a liquid, or PCl_5, which is a solid, is added to water, a violent reaction occurs. Hydrochloric acid and a solution of phosphorus acid, H_3PO_3 (from PCl_3), or phosphoric acid, H_3PO_4 (from PCl_5), are formed. (a) Write molecular and ionic equations for the two reactions.   (b) In terms of Lewis formulas, explain the difference between CCl_4 and the phosphorus chlorides.   (c) Predict what will be observed if $SiCl_4$ is added to water.   (d) What conclusion can your draw concerning the relative densities of water and CCl_4?

9.100 A compound is 82.66% C and 17.34% H. A 0.472-g sample occupies a volume of 196 mL at 756 mmHg and 19.1 °C. Write Lewis formulas for the two possible isomers.

9.101 For the reaction,

$$CH_4(g) + Cl_2(g) \longrightarrow CH_3Cl(g) + HCl(g)$$

(a) Estimate ΔH from bond energies (Table 9.4).   (b) Calculate $\Delta H°$ from standard enthalpies of formation. For CH_3Cl, $\Delta H_f° = -82$ kJ/mol.   (c) Explain why values of ΔH from bond energies are only estimates.

9.102 For acetonitrile, $CH_3C{\equiv}N(g)$, $\Delta H_f° = 87.9$ kJ/mol, and for $C(g)$, $\Delta H_f° = 718.384$ kJ/mol. Use this information and bond energies from Table 9.4 to calculate the bond energy for $C{\equiv}N$.

9.103 Which species are isoelectronic with carbon dioxide? (a) C_3H_4 (b) ClO_2 (c) $NO_2{}^+$ (d) SCN^- (e) SO_2

APPLICATIONS

9.104 Phosgene, which has molecular formula CCl_2O, is a very poisonous gas that was used in gas warfare in World War I and is now used in large quantities to make resins. Write the best Lewis formula for phosgene.

9.105 Cyanates, salts containing the cyanate ion, OCN^-, are stable. Fulminates, salts containing the fulminate ion, CNO^-, are explosive. Suggest an explanation for the difference. (*Hint:* Compare formal charges of Lewis formulas.)

9.106 Saturn's largest moon, Titan, is about the same size as Earth and, like Earth, has an atmosphere that consists largely of nitrogen. Methane plays the same role on Titan as water does on Earth—existing as solid, liquid, and gas at the surface temperature of 93 K. Titan's atmosphere acts like a chemical manufacturing plant that uses nitrogen and methane as raw materials. Among the organic molecules produced are C_4H_2 and C_3HN. (a) Write the Lewis formulas for all five molecules. (b) Label the longest and weakest C—C bond in the Lewis formula for C_4H_2.

9.107 In the Middle Ages, alchemists obtained formic acid by distilling red ants; formic acid is the irritant in ant bites. The Lewis formula for formic acid is

$$H{-}C{-}\overset{\cdot\cdot}{\underset{\underset{\overset{\cdot\cdot}{\cdot O \cdot}}{\parallel}}{O}}{-}H$$

In formic acid, one carbon–oxygen bond length is 134 pm, and the other is 120 pm. Label each carbon–oxygen bond with the appropriate length.

9.108 Paper mills can greatly reduce their release of toxic chemicals by using chlorine dioxide instead of chlorine to bleach paper. For chlorine dioxide, (a) write the molecular formula.   (b) Write four Lewis formulas, two of which involve an expanded octet and two of which do not.   (c) Can you choose between the four Lewis formulas on the basis of formal charge?   (d) What experimental evidence could be obtained to compare the importance of the Lewis formulas to the resonance hybrid (e) What is a species such as chlorine dioxide called?

9.109 Heavy metal azides such as barium azide, $Ba(N_3)_2$, are explosive and are used in detonation caps. One Lewis formula for the azide ion is

$$\left[\ddot{\text{N}}=\text{N}=\ddot{\text{N}}\ddot{}\right]^{-}$$

(a) Write two other resonance structures for the azide ion. (b) Add formal charges to all three resonance structures for the azide ion. (c) Which structure contributes most to the hybrid? Explain your answer.

9.110 Aluminum chloride, which is used as a catalyst and for the synthesis of other aluminum compounds, is a solid that changes to a gas without melting at 177.8 °C. At temperatures just above 177.8 °C, aluminum chloride vapor consists of Al_2Cl_6 molecules. At higher temperatures, the vapor consists of $AlCl_3$ molecules. (a) Write a Lewis formula for $AlCl_3$. (b) Does $AlCl_3$ follow the octet rule? If not, in what way does it not follow the octet rule? (c) Suggest a Lewis formula for Al_2Cl_6.

9.111 The decomposition of arsine, AsH_3, to arsenic, which is deposited on hot surfaces as a mirror, is used to test for arsenic in crime labs. (a) Write the Lewis formula for arsine. (b) Write the equation for the decomposition of arsine. A temperature of 250–300 °C is required, and the other product is hydrogen gas. At 250–300 °C, arsenic is a solid, and arsine is a gas.

9.112 Sodium diphosphate is used in instant puddings, sodium dihydrogen diphosphate in baking powder, and calcium diphosphate in toothpaste. The condensed structural formula for the diphosphate ion is $[O_3POPO_3]^{4-}$. (a) Write the formulas of the three salts mentioned and of diphosphoric acid. (b) Write the best Lewis formula for the diphosphate ion.

9.113 Most people can synthesize tyrosine in their bodies, but a few who have an inherited condition known as phenylketonuria (PKU) cannot. The diets of people with phenylketonuria must contain some tyrosine. A condensed structural formula for tyrosine is

HO—⟨benzene ring⟩—$CH_2CH(NH_2)COOH$

(a) Write a Lewis formula for tyrosine including resonance structures and formal charges, if any. (b) Briefly discuss the relative lengths of the various bonds.

9.114 The reaction

$$ClO(g) + O(g) \longrightarrow Cl(g) + O_2(g)$$

plays a part in the destruction of stratospheric ozone. Use bond energies to estimate ΔH for this reaction. The bond dissociation energy of the Cl—O bond in ClO is 64.3 kcal/mol.

9.115 The phosphorus compounds that are used as oil and fuel additives, plasticizers, flame retardants, and for the manufacture of insecticides are all made from phosphorus trichloride. The ΔH_f° for $PCl_3(g)$ is −287.0 kJ/mol, and ΔH_f° for P(g) is 314.25 kJ/mol. What is the average bond energy of the P—Cl bond?

9.116 Hydroxyurea $H_2NCONHOH$ is used to treat (unfortunately not cure) sickle-cell anemia. (a) What is the molecular formula? (b) Write the best Lewis formula.

9.117 Write the Lewis formula for ammonium perchlorate, which is used in rocket propellants.

9.118 Polyacrylamide is used as a matrix for electrophoresis, a method of sorting proteins. The condensed structural formula for acrylamide is $H_2C{=}CHCONH_2$. Write a structural formula for polyacrylamide showing three monomer units.

9.119 Poly(vinyl acetate)

$$\sim CH_2\!-\!\underset{\underset{OOCCH_3}{|}}{\overset{\overset{H}{|}}{C}}\!-\!CH_2\!-\!\underset{\underset{OOCCH_3}{|}}{\overset{\overset{H}{|}}{C}}\!-\!CH_2\!-\!\underset{\underset{OOCCH_3}{|}}{\overset{\overset{H}{|}}{C}}\sim$$

is used to make chewing gum. Write a condensed structural formula for vinyl acetate.

A Chemist Turns Historian

WILLIAM B. JENSEN

Oesper Professor of History of Chemistry
and Chemical Education

B.S. Chemistry; M.S. Education;
Ph.D. Chemistry
University of Wisconsin

I became interested in chemistry very early, probably around the fifth grade. I would like to claim that I was attracted by the theories and ideas of chemistry or that I had some burning desire to understand nature, but, in fact, I was attracted by the laboratory equipment. Those wonderful arrays of tubes and flasks filled with mysterious liquids that populated the laboratories of mad scientists on the Friday night reruns of old Frankenstein movies were infinitely more exciting to me than the coils of wire and falling weights of the physicist or the plants and animals of the biologist. Chemical reactions—where materials were completely transformed into something entirely different—seemed more basic and more wondrous than the subjects studied by other branches of science. To my 11-year-old mind, a person who could understand and manipulate reactions at will possessed a magical power. Needless to say, I soon acquired a Chemcraft chemistry set. I quickly abandoned its toy test tubes and plastic, undersized equipment in favor of the real laboratory apparatus I ordered from a catalog supplied by the company after I sent in a dollar to become an official "Chemcraft Chemist." This, I suppose, was my first chemical "degree." I was an avid home chemist throughout my school years. By the

time my basement laboratory was dismantled during my second year of college (much to my mother's relief), it had become a respectable private chemical laboratory.

In junior high school, I went to work for a local drugstore under the mistaken impression that this occupation had something to do with chemistry. Though I soon discovered otherwise, the job did prove a useful source of chemicals for my ever-expanding home laboratory. In addition, the science coordinator for the local schools introduced me to the chemists at a local industrial laboratory, and they gave me chemicals and equipment when I needed them. This pattern continued through high school. At each step, interested adults and teachers were willing to go out of their way to encourage my fascination, or perhaps I should say obsession, with chemistry.

My interest in the history of chemistry also developed quite early, largely as a result of reading Mary Elvira Weeks's *Discovery of the Elements* when I was 14. This book was packed full of pictures of old laboratories and famous chemists. As my understanding of chemical theory matured, I became more interested in learning about the origins of the ideas I was studying and the personalities of the chemists who had discovered them. During undergraduate training, I used every available free elective to take history of science courses.

After finishing my B.S. in chemistry, I did not immediately continue on to the Ph.D. I felt that I needed a break and wanted to teach. I entered a two-year masters program in chemistry and education and became certified as a high school chemistry teacher. When I finally decided to return to graduate school, I was offered fellowships in both chemistry and history of science and had to decide whether I was a chemist interested in history, or a historian interested in chemistry. I knew I would have many more career options with a chemistry degree than with a history degree, and I also realized that my interest in history was an extension of my more fundamental interest in chemistry. I have never regretted the choice.

After finishing my Ph.D. in inorganic chemistry, I took a faculty position in chemistry at a college in New York. I continued to pursue history of chemistry as a hobby but

assumed it would never be anything more. In 1986, however, a position in history of chemistry and chemical education became available at the University of Cincinnati. This position was unique in that it was located in the chemistry department. It required someone qualified to teach both chemistry and history of chemistry. Since few professional historians of science have science degrees beyond the B.S. level, most were not qualified and so, almost by default, I was given an opportunity to pursue my hobby professionally.

As a historian of chemistry, I study the older chemical literature to trace the origins of our current knowledge. I also edit a journal for the history of chemistry and manage a library of more than 4000 books relating to the history of chemistry as well as a large collection of illustrations of old laboratories and famous chemists, and a museum of chemical apparatus.

Besides the courses I teach at Cincinnati, I travel to a number of schools to lecture on the history of chemistry. In the course of these visits, I go through the storerooms and basements of the chemistry departments looking for old equipment to add to the apparatus museum. I also visit the local used bookstores to purchase old textbooks for the book collection. My woodworking hobby has enabled me to restore some of the old equipment and build reproductions of hard-to-find items. Likewise, my interest in caricature and art has proven valuable in doing layouts and design work for the journal and in constructing museum displays. The topics of these displays have ranged from the history of the chemistry set to the development of the modern spectrophotometer.

As a historian, I also plan for the future by interviewing retiring members of the chemistry department and editing the departmental newsletter (a form of current history). I also see that important documents are saved for use by future historians. In short, my interests in history and art, combined with the excellent college education I received as a chemistry major, have all fused to produce an unusual hybrid career that I never imagined on that fateful day 30 years ago when I found my first Chemcraft chemistry set under the Christmas tree.

10

MOLECULAR SHAPE AND THEORY OF CHEMICAL BONDING

An Electronically Programmable Read-Only Memory (EPROM) chip magnified 180-times.

L ewis formulas for molecules and polyatomic ions are very useful because they help organize a large number of chemical facts. But Lewis formulas do not give any information about the shapes of molecules, nor do they explain how and why atoms share electrons. In the first part of this chapter, we will learn a very simple method for predicting shape from Lewis formulas that gives the correct shape for most molecules. In the second part of the chapter, two theories that explain how and why covalent bonds form will be discussed.

— *Why is a study of molecular shape and theories of chemical bonding a part of general chemistry?*

The shape of a molecule or polyatomic ion is one of its most important characteristics. You cannot understand molecules or ions or the reactions that they undergo without a knowledge of their shapes. Reactions that take place in living systems are especially sensitive to the shapes of the species involved.

As we have seen before, theories are explanations of observations; theories help make sense out of information. What

makes sense is easier to remember. With the aid of theories, you can extend what you know by making predictions about new situations.

— *What do you already know about molecular shape and theories of chemical bonding?*

From the pictures of models in Figure 9.5, you know that an ammonia molecule is shaped like a three-legged stool. The hydrogens in an ammonium ion are located at the four corners of a tetrahedron surrounding the nitrogen atom. A water molecule is bent like a boomerang. If you could turn the model of the nitrate ion so that you could look at it from directly above, you would be able to see that it is planar and that the oxygen atoms are located at the three corners of an equilateral triangle around the nitrogen atom. (However, you would not be able to see the double bond very well.)

You also know that covalent bonds consist of pairs of electrons shared between two atoms. The number of bonds between two atoms in a molecule or ion is called bond order. Bonding involves only the valence electrons and often results in the formation of octets of electrons, the outer electron configuration of the noble gases, around atoms. Lewis formulas are used to represent molecules and polyatomic ions of main group elements. Sometimes more than one Lewis formula must be written to describe a molecule or polyatomic ion. The different Lewis formulas are called resonance structures, and the real structure of the molecule or ion is a hybrid of the structures represented by the resonance structures.

10.1 SHAPES OF MOLECULES AND POLYATOMIC IONS: THE VALENCE SHELL ELECTRON PAIR REPULSION (VSEPR) MODEL

When there are four or more pairs of electrons around a central atom, as there are in an ammonium ion, the pairs of electrons often do not lie in a plane. **Stereochemical formulas** must be used to show the positions of the electron pairs on flat paper (or on a flat chalkboard or screen). In stereochemical formulas, both atoms and unshared pairs of electrons are shown attached to the central atom by a bond. In one common type of stereochemical formula,

1. Ordinary lines are used to show bonds that lie in the plane of the paper.
2. Solid wedges are used to show bonds that stick out toward the viewer in front of the plane of the paper.
3. Dashed lines are used to show bonds that go away from the viewer behind the plane of the paper.

Compare the shapes of the bunches of balloons in ◾Figure 10.1 with the stereochemical formulas to be sure that you can picture the shapes from the stereochemical formulas.

The shapes of molecules and polyatomic ions can be observed experimentally and bond lengths and angles measured precisely. However, it is convenient to be able to predict shapes qualitatively. The **valence shell electron pair repulsion** model, which is often abbreviated by the first letters, **VSEPR** (pronounced ves'per), is a very simple way of predicting the shapes of species *that have main group elements as central atoms.* According to the VSEPR model, *electron pairs will be as far apart from each other in three-dimensional space as possible.* The two electrons of an electron pair can occupy the same region of space because they have opposite spins. But according to the Pauli exclusion principle, more than one pair of electrons cannot occupy the same region of space. Pairs of electrons repel each other because of their negative charges. Therefore, pairs of electrons get as far apart from each other as possible. Bunches of balloons that are tied together naturally take up positions as far apart as possible and provide a good model for the results of electron pair repulsion as shown in Figure 10.1.

Shapes of Polyatomic Species Without Unshared Pairs on the Central Atom

If there are no unshared pairs of electrons attached to the central atom, the molecules or ions have the shapes shown by the balloons in Figure 10.1. Figure 10.1 also

 If you have trouble picturing three-dimensional shapes in your mind, try making models using gumdrops or marshmallows for atoms and toothpicks for bonds.

◗ Remember that main group elements are the elements in A groups and Group 0.

① From the Lewis Structures and Resonance module of the Molecules section students can move directly to the Geometries module to determine the shape of a molecule or ion whose Lewis formula they have written.

Stereochemical formula	:—A—:				
Number of sets of electron pairs around central atom, A	2	3	4	5	6
Shape	Linear	Trigonal planar	Tetrahedral	Trigonal bipyramidal	Octahedral
Bond angle,°	180	120	109.5	120 and 90	90
Example					
Model				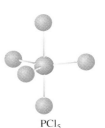	
	$BeCl_2$	BF_3	CH_4	PCl_5	SF_6

▪ FIGURE 10.1 If balloons of the same size and shape are tied together, they take the positions shown. Sets of electron pairs take up similar positions around a central atom. The compounds shown as examples are all species with no unshared pairs of electrons on the central atom.

shows stereochemical formulas and models of some molecules with no unshared electron pairs on the central atom.

Shapes of Species with Unshared Pairs on the Central Atom

Now let's consider what happens when some of the electron pairs are unshared pairs instead of bonding pairs. An ammonia molecule has one unshared pair of electrons, and a water molecule has two:

$$H:\overset{\cdot\cdot}{\underset{H}{N}}:H \qquad H:\overset{\cdot\cdot}{\underset{\cdot\cdot}{O}}:H$$

According to experiment, an ammonia molecule is shaped like a three-legged stool:

(Scientists describe the shape of the ammonia molecule as trigonal pyramidal.) The angle between the H—N bonds in ammonia is 107.3°. A water molecule is shaped like a boomerang with an angle between the H—O bonds of 104.5°:

(Scientists describe the shape of the water molecule as bent or angular.) How does valence shell electron pair repulsion account for the shapes of the ammonia and water molecules?

If we define the **total coordination number** of an atom as the *total number of atoms and sets of unshared electron pairs around a central atom,* both ammonia and water have total coordination numbers of 4. According to the VSEPR model, four sets of electron pairs should be located at the corners of a tetrahedron about the central atom. The VSEPR model pictures both ammonia and water as having tetrahedral arrangements of electron pairs about the central atom. However, the positions of the atoms determine the molecular shape observed.

According to the VSEPR model, the fact that the angles between H—N bonds in ammonia (107.3°) and the angles between the H—O bonds in water (104.5°) are less than the angles between H—C bonds in methane (109.5°) is explained by assuming that unshared pairs of electrons repel shared pairs of electrons more than shared pairs repel each other. This seems reasonable because the unshared pair is only attracted by the positive charge on one nucleus, but the shared pair is attracted by the positive charges on two nuclei. Because it is concentrated between two nuclei, the shared pair takes up less space than the unshared pair. An unshared pair pushes the shared pairs closer together (see ◾Figure 10.2).

The angle between H—O bonds in water (104.5°) is even smaller than the angles between H—N bonds in ammonia because repulsion between unshared pairs is greater than repulsion between an unshared pair and a shared pair. Two unshared pairs push the remaining shared pairs even closer to each other than one unshared pair does.

VSEPR can be used to predict the shapes of species with unshared pairs of electrons and total coordination numbers greater than 4, of species with multiple bonds, and of resonance hybrids. *The shape of the molecule is determined by the positions of the atoms that are bonded to the central atom. The positions of the atoms that are bonded to the central atom depend both on total coordination number and on the number of unshared pairs on the central atom.*

The term, total coordination number, *was introduced by Jolly (Jolly, W. L.* Modern Inorganic Chemistry; *McGraw-Hill: New York, 1984; p 78). It is a very convenient term.*

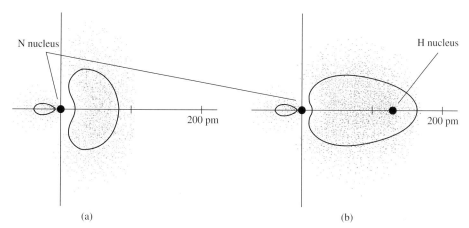

◾ FIGURE 10.2   Comparison of the charge clouds of an unshared pair and a bonding pair: (a) the unshared pair of electrons on the nitrogen atom in an ammonia molecule, NH_3, and (b) one of the three bonding pairs of electrons in the ammonia molecule. Boundary lines that enclose equal percentages of each charge cloud have been drawn. Note that the unshared pair (a) takes up more space (is "fatter") near the nitrogen nucleus than the bonding pair (b). (Computer generated.) (Copyright © 1975 by W. G. Davies and J. W. Moore. Reprinted with permission.)

Shapes of Species with Multiple Bonds

According to experiment, a carbon dioxide molecule is linear:

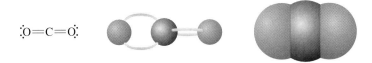

There are two oxygen atoms and no unshared pairs attached to the carbon atom in carbon dioxide; the total coordination number of carbon is 2. The beryllium atom in a beryllium chloride molecule, which is linear

$$:\ddot{C}l — Be — \ddot{C}l:$$

also has a total coordination number of 2. Because a carbon dioxide molecule is the same shape as a beryllium chloride molecule, we can conclude that the effect of a double bond on shape is similar to the effect of a single bond. This seems reasonable because the four electrons of a double bond must be located between the atoms joined by the bond (or the bond wouldn't be a double bond). Thus, *as far as the shape predicted by VSEPR is concerned, double bonds are no different than single bonds. Triple bonds are also similar to single bonds in their effect on shape.*

Using VSEPR to Predict Shape

Sample Problems 10.1–10.3 show how to use VSEPR to predict shape.

SAMPLE PROBLEMS

10.1 Use VSEPR to predict the shape of the SF_4 molecule.

Solution   The first step in predicting the shape of molecules by VSEPR is to write the correct Lewis formula for the molecule. The Lewis formula for SF_4 is

$$\ddot{\underset{\ddot{F}}{F}} \overset{\ddot{F}}{\underset{\ddot{F}}{S}}$$

There are four fluorine atoms and one unshared pair of electrons around S. The total coordination number for S in SF_4 is 5, and a trigonal bipyramidal arrangement of atoms and unshared pairs would be expected (see Figure 10.1):

$$: — A \overset{\cdot\cdot}{\underset{\cdot\cdot}{}}$$

Because unshared electron pairs take up more space than shared pairs, the unshared pair should be in one of the positions where it is 120° away from the other pairs, not in one of the positions where it is just 90° away from the other pairs. Placing the four fluorines in the other four positions gives the structure

$$: — \overset{F}{\underset{F}{S}} \overset{F}{\underset{F}{}}$$

However, the unshared pair will push the shared pairs closer together, as happened in ammonia, and the SF_4 molecule will have the shape

$$:-\underset{\underset{F}{|}}{\overset{F}{\underset{|}{S}}}\diagup F \qquad or \qquad \underset{\underset{F}{|}}{\overset{F}{S}}\diagup F$$

This shape is called seesaw.

10.2 Use **VSEPR** to predict the shape of the ethylene molecule, $H_2C{=}CH_2$.

Solution The Lewis formula for ethylene is

$$H-\underset{\underset{H}{|}}{C}=\underset{\underset{H}{|}}{C}-H$$

Each carbon in the ethylene molecule has two hydrogens and one carbon attached to it, making a total coordination number of 3. The atoms will lie in a plane, and the angles between the bonds will be 120°:

$$\underset{H}{\overset{H}{\diagdown}}C{=}C \qquad or \qquad C{=}C\underset{\diagdown H}{\overset{\diagup H}{}}$$

> Remember that the fact that the carbon is attached by a double bond does not make any difference in the total coordination number.

Putting the two halves of the molecule together gives

$$\underset{H}{\overset{H}{\diagdown}}C{=}C\underset{\diagdown H}{\overset{\diagup H}{}}$$

for the shape of the ethylene molecule. ■Figure 10.3 shows models of ethylene. All the atoms lie in a plane, and all bond angles are 120°.

For species with more than one "central atom" such as ethylene, the stereochemistry around each central atom must be predicted individually and the results combined to find the shape of the whole species.

10.3 Use **VSEPR** to predict the shape of the nitrate ion.

Solution As we saw earlier, three contributing structures are needed to describe the nitrate ion:

$$\left[\overset{..}{:}\overset{..}{O}-N\overset{O}{\underset{\overset{||}{\underset{:O:}{..}}}{=}}\overset{..}{O}:\right]^{-} \longleftrightarrow \left[\overset{..}{:}O{=}N\underset{\underset{:O:}{|}}{-}\overset{..}{O}:\right]^{-} \longleftrightarrow \left[\overset{..}{:}\overset{..}{O}-N\underset{\underset{:O:}{|}}{=}\overset{..}{O}:\right]^{-}$$

In any one of the three contributing structures for the nitrate ion, *three* oxygen atoms and no unshared pairs are attached to the nitrogen atom. Thus, the total coordination number of N is 3, and the nitrate ion is planar with bond angles of 120°:

$$\left[\underset{\underset{:O:}{|}}{\overset{:\overset{..}{O}:\quad\overset{..}{O}:}{N}}\right]^{-} \longleftrightarrow \left[\underset{\underset{:O:}{|}}{\overset{\overset{..}{O}:\quad\overset{..}{O}:}{N}}\right]^{-} \longleftrightarrow \left[\underset{\underset{:O:}{|}}{\overset{:\overset{..}{O}:\quad\overset{..}{O}}{N}}\right]^{-}$$

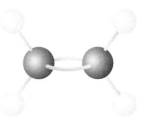

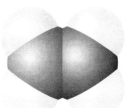

■ **FIGURE 10.3** Ball-and-stick and space-filling models of ethylene (ethene). The planarity of the ethylene molecule is not predicted by VSEPR. We will learn why ethylene is planar in Section 10.4.

Summary of Procedure for Predicting Shapes by VSEPR

1. Write the Lewis formula.
2. From the Lewis formula, find the total coordination number by counting the atoms and unshared pairs that are attached to the central atom.
3. Place the total coordination number of electron pairs around the central atom in

| TABLE 10.1 | Shapes of Molecules Based on VSEPR |

Total Coordination Number	Number of Bonding Pairs	Number of Unshared Pairs	General Formula	Stereochemical Formula	Shape	Example	Model[a]
2	2	0	AB_2	B—A—B	Linear	$BeCl_2$	
3	3	0	AB_3		Trigonal planar	BF_3	
	2	1	AB_2		Bent or angular	GeF_2	
4	4	0	AB_4		Tetrahedral	CH_4	
	3	1	AB_3		Trigonal pyramidal	NH_3	
	2	2	AB_2		Bent or angular	H_2O	
	1	3	AB		Linear	HF	

[a] Sizes of different species are not to scale.
[b] The distortion that results from the larger volume occupied by unshared pairs is not shown in this table.

such a way that they are as far from each other as possible in three-dimensional space.

4. Remembering that unshared pairs occupy more space than shared pairs, place the atoms around the central atom. The positions of the atoms determine the shape of the molecule or polyatomic ion.

Table 10.1 summarizes the molecular shapes predicted by VSEPR.

TABLE 10.1 Shapes of Molecules Based on VSEPR (continued)

Total Coordination Number	Number of Bonding Pairs	Number of Unshared Pairs	General Formula	Stereochemical Formula	Shape	Example	Model
5	5	0	AB_5		Trigonal bipyramidal	PCl_5	
	4	1	AB_4		Seesaw	SF_4	
	3	2	AB_3		T-shaped	ClF_3	
	2	3	AB_2		Linear	ICl_2^-	
6	6	0	AB_6		Octahedral	SF_6	
	5	1	AB_5		Square pyramidal	IF_5	
	4	2	AB_4		Square planar	IF_4^-	

^{c}Note that the two lone pairs are opposite each other.

PRACTICE PROBLEMS

10.1 For the ion $(SnCl_6)^{2-}$, (a) write the Lewis formula, (b) predict the shape, and (c) draw a stereochemical formula.

10.2 Predict the shape of beryllium fluoride, BeF_2.

10.3 Draw stereochemical formulas for (a) SiH_4 and (b) AsF_5.

Although the VSEPR model is very simple, its predictions are usually quite accurate. Naturally, the exceptions that prove the rule exist, and VSEPR occasionally predicts the wrong shape for a molecule. For example, according to VSEPR, hydrogen sulfide, H_2S, should be bent like water, H_2O, and have a bond angle slightly less than the tetrahedral angle of 109.5°. However, although hydrogen sulfide is indeed bent, the bond angle is actually only 90°.

10.2   POLAR AND NONPOLAR MOLECULES

Polarity is an important property of molecules because both physical properties, such as melting point, boiling point, and solubility, and chemical properties depend on molecular polarity. **Dipole moment, μ,** is *a quantitative measure of the polarity of a molecule.* Dipole moments can be determined by studying the behavior of molecules between charged plates. A polar molecule turns in the electric field between charged plates so that the negative end of the molecule is toward the positive plate and the positive end is toward the negative plate as shown in ▬Figure 10.4(a). Nonpolar molecules are not affected by the electric field [see Figure 10.4(b)].

Because oxygen (electronegativity, 3.5) is more electronegative than carbon (electronegativity, 2.5), C—O bonds are polar with a partial positive charge on carbon and a partial negative charge on oxygen:

$$\overset{\delta+ \quad \delta-}{C—O}$$

Why are carbon dioxide molecules, which contain two C—O bonds, not polar? Molecules with just one polar bond like the hydrogen chloride molecule must be polar and have a nonzero dipole moment. If a molecule contains more than one polar bond, however, whether the molecule as a whole is polar depends on the shape of the molecule. The two triatomic molecules CO_2 and H_2O provide a good example of the effect of molecular shape on the polarity of molecules. Carbon dioxide is linear, whereas water is bent. Both carbon–oxygen and hydrogen–oxygen

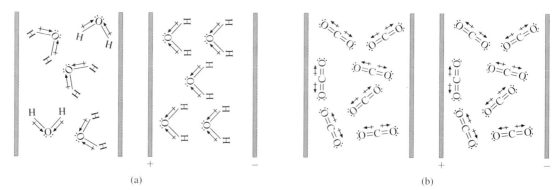

(a)   (b)

▬ FIGURE 10.4   Behavior of (a) polar and (b) nonpolar molecules between uncharged (*left*) and charged (*right*) plates.

bonds are polar because oxygen is more electronegative than carbon and hydrogen (see Figure 9.3):

$$\ddot{\text{O}} \overset{\rightharpoonup}{=} \text{C} \overset{\leftharpoonup}{=} \ddot{\text{O}}$$ (water structure with H—O—H)

Under ordinary conditions, carbon dioxide is a gas because molecules of carbon dioxide are nonpolar; water is a liquid as a result of the great polarity of water molecules.

Dipole moments are vectors, that is, they have direction as well as magnitude. Because carbon dioxide is linear, the dipole moments of the two C—O bonds cancel each other. The carbon dioxide molecule is nonpolar despite the fact that it has two polar bonds. The dipole moments of the two H—O bonds in water do not cancel because water is bent. A water molecule has a net dipole moment in the direction shown by the boldface arrow:*

(water molecule with bold upward arrow between H and H)

The nonzero dipole moment of water and the zero dipole moment of carbon dioxide confirm the prediction of VSEPR that water molecules are bent and carbon dioxide molecules are linear.

The four compounds that can be derived from methane by substitution of chlorine for hydrogen—chloromethane, dichloromethane, trichloromethane (chloroform), and tetrachloromethane (carbon tetrachloride)—

(Lewis structures of the four chloromethanes)

provide another illustration of the importance of molecular shape in determining the polarity and dipole moments of molecules. The stereochemical formulas for these four compounds are

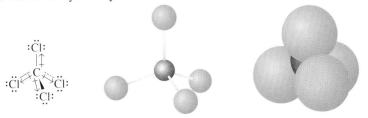

Although each of the four C—Cl bonds in tetrachloromethane is polar because chlorine is more electronegative than carbon, the carbon tetrachloride molecule is nonpolar due to its symmetry:

(structural formula of CCl₄ with dipole arrows, and two ball-and-stick/space-filling models)

The four dipoles cancel each other. In the other three molecules, the dipoles do not cancel, and the other three molecules are polar. Notice that you must think about three-dimensional shape to recognize that dichloromethane is polar. From the Lewis formula for dichloromethane,

$$:\ddot{\text{Cl}}:\ddot{\text{C}}:\ddot{\text{Cl}}:$$ (with H above and H below)

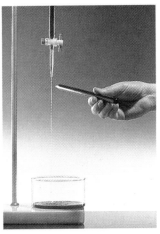

Top: *A stream of polar molecules is attracted by a charged rod.* Bottom: *A stream of nonpolar molecules is not affected. (Both liquids are dyed so that you can see them.) The polar molecules are* cis-*1,2-dichloroethylene*

(structure of cis-1,2-dichloroethylene)

and the nonpolar molecules are trans-*1,2-dichloroethylene*

(structure of trans-1,2-dichloroethylene)

These compounds have similar densities.

*The situation in the carbon dioxide molecule is like a tug-of-war over a third person between twins. Because the twins are pulling in opposite directions with the same force, the person in the middle will not move. The water molecule is like twins pushing on a third person from the same side. The third person will be pushed sideways away from the twins (but not toward either twin).

the wrong conclusion that the polarities of the C—Cl bonds cancel and the molecule is nonpolar is easily drawn.

Comparison of the dipole moments of NH_3 and NF_3 shows the effect of unshared pairs of electrons in the valence shell on dipole moment. Both molecules are trigonal pyramidal:

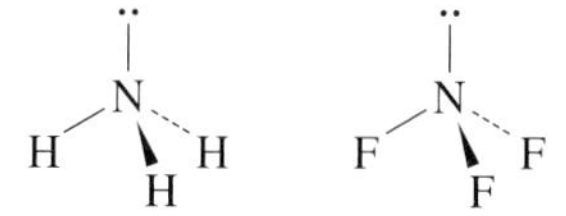

The difference in electronegativity between N and F (1.0) is about the same as the difference in electronegativity between N and H (0.9). Therefore, we might expect that the dipole moment of NF_3 would be about the same as the dipole moment of NH_3. However, measurements of dipole moment show that the dipole moment of NH_3 is over seven times as large as the dipole moment of NF_3. Moreover, the dipole moment of NF_3 is very small. How is the small dipole moment of NF_3 explained?

Nitrogen is more electronegative than hydrogen. As a result, the dipole of the N—H bond points toward N. However, fluorine is more electronegative than nitrogen, and the dipole of the N—F bond points away from nitrogen toward F. Unshared pairs of valence electrons, such as the unshared pairs in NH_3 and NF_3, contribute to the dipole moments of molecules:

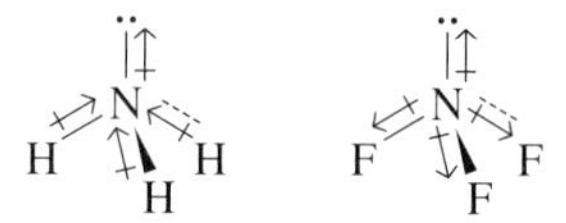

In NH_3, the polarity that results from the unshared pair reinforces the polarity of the N—H bonds. In NF_3, the polarity that results from the unshared pair cancels most of the polarity of the N—F bonds. Therefore, NF_3 only has a small dipole moment.

The polarity of molecules that have formal charges is affected by the distribution of electron density shown by the formal charge. For example, because oxygen is more electronegative than carbon, carbon monoxide would be expected to be polar with the negative charge on oxygen. However, because oxygen has a +1 formal charge and carbon has a −1 formal charge in carbon monoxide (Section 9.8), the dipole moment of carbon monoxide is very small, and the negative end of the dipole is on carbon.

In Conclusion

Molecules in which the central atom is symmetrically substituted by identical atoms, such as CO_2 and CCl_4 are nonpolar. Molecules with only one polar bond and molecules in which the central atom is not symmetrically substituted are usually polar. Formal charges must be considered in predicting polarity.

PRACTICE PROBLEM

10.6 Of the compounds given as examples in Table 10.1, (a) which are nonpolar? (b) In which will the dipole moment of the unshared pair or pairs act in the opposite direction to the dipole moment of the A—B bonds?

INTRODUCTION TO BONDING THEORY

The Schrödinger equation can be used to describe the behavior and energies of electrons in molecules as well as the behavior and energies of electrons in atoms. In principle, the Schrödinger equation can be solved to give the wave function for each electron in a molecule, and all the properties of the molecule can be calculated. However, in practice, solving the wave equations for molecules is even more complicated than solving the wave equations for atoms because there are more variables. Even simple molecules such as CH_4, NH_3, H_2O, and HF contain ten electrons as well as two or more nuclei, and approximations must be used.

Two different methods of approximation are commonly used: the valence bond method and the molecular orbital method. In the **valence bond method,** which is discussed in Section 10.4, *bonds are assumed to be formed by overlap of atomic orbitals.* According to the **molecular orbital method,** which is discussed in Section 10.6, *when atoms form compounds, their orbitals combine to form new orbitals called molecular orbitals.* After enough approximations have been made, both methods give identical results. However, each has advantages and disadvantages.

The fact that liquid oxygen is suspended between the poles of a magnet shows that oxygen is paramagnetic. The molecular orbital method easily explains the paramagnetism of oxygen, whereas the valence bond method does not.

The Hydrogen Molecule

Just as we began our consideration of atomic structure with the simplest atom, the hydrogen atom, let's begin our consideration of the theory of chemical bonding with the simplest common molecule, the hydrogen molecule, H_2. Picture two hydrogen atoms, originally an infinite distance apart and imagine that the two hydrogen atoms move toward each other. Let's think about what happens to the energy of the system when the two hydrogen atoms get close to each other. First let's suppose that the electrons in the two hydrogen atoms have opposite spins. Charge densities for the two hydrogen atoms are shown in ▪Figure 10.5.

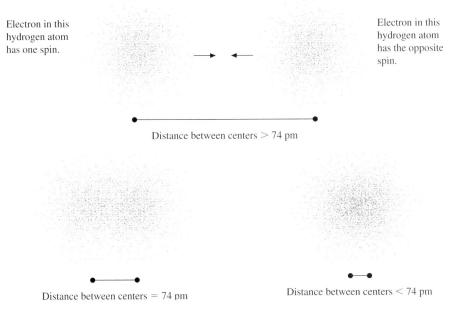

Electron in this hydrogen atom has one spin.

Electron in this hydrogen atom has the opposite spin.

Distance between centers $>$ 74 pm

Distance between centers $=$ 74 pm

Distance between centers $<$ 74 pm

▪ FIGURE 10.5   Charge densities as two hydrogen atoms with electrons having opposite spins approach each other. The experimental bond length for the H—H bond in the H_2 molecule is 74 pm. Notice how electron density is concentrated between the two nuclei in the hydrogen molecule.

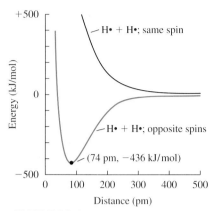

■ FIGURE 10.6 Energy of two hydrogen atoms as a function of the distance between them. The lower curve is the binding energy curve for the hydrogen molecule and is typical of binding energy curves.

The **binding energy curve** for the hydrogen molecule (see lower curve in ■Figure 10.6) *shows how the energy of the system changes as the two* hydrogen *atoms move closer together.* In Figure 10.6, the energy of two hydrogen atoms at infinite distance from each other is set equal to zero. As the distance between the two hydrogen atoms having electrons with opposite spins decreases, energy decreases, reaching a minimum at 74 pm, the length of the H—H bond in a H_2 molecule. As the two hydrogen atoms approach each other, the electron of one hydrogen atom begins to be attracted by the positive charge on the nucleus of the other hydrogen atom, as well as by the positive charge on its own nucleus. At the same time, the electron of one hydrogen atom begins to repel the electron on the other hydrogen atom; the two nuclei also begin to repel each other. As long as the distance between the centers of the two hydrogen atoms is greater than 74 pm, attraction increases faster than repulsion, and energy decreases. If the two hydrogen atoms are pushed closer than 74 pm to each other, the nuclei start to repel each other strongly. Repulsions become more important than attractions and energy increases very rapidly. Energy is at a minimum when the hydrogen atoms are the measured bond length of 74 pm apart. The observed bond length is the result of a balance between electron–nucleus attractions and electron–electron and nucleus–nucleus repulsions.

The minimum energy reached is −436 kJ/mol. The difference in energy between two moles of gaseous hydrogen atoms that are infinitely far from each other and one mole of gaseous hydrogen molecules is 436 kJ. When two moles of gaseous hydrogen atoms combine to form one mole of gaseous hydrogen molecules, 436 kJ are given off:

$$H\cdot \quad + \quad \cdot H \quad \longrightarrow \quad H\!:\!H + 436\ kJ$$

Electron in this   Electron in this
hydrogen atom   hydrogen atom has
has one spin.   the opposite spin.

To decompose a mole of gaseous hydrogen molecules into atoms, 436 kJ are required:

$$H\!:\!H + 436\ kJ \quad \longrightarrow \quad H\cdot \quad + \quad \cdot H$$

Electron in this   Electron in this
hydrogen atom   hydrogen atom has
has one spin.   the opposite spin.

The *energy required to break a specific bond in a certain molecule* is called **bond dissociation energy;** the bond dissociation energy for the H—H bond is 436 kJ/mol.

If the electrons of two hydrogen atoms have the same spin, the electrons cannot pair to form a bond. According to the Pauli exclusion principle, two electrons with the same spin cannot be in the same region of space. In addition, electrons repel each other because of their negative charges. As the two hydrogen atoms move closer together, repulsion between electrons is greater than the attraction of the electrons for the nucleus of the other hydrogen atom. Electron density moves away from the space between the nuclei; ■Figure 10.7 shows how charge density changes

▶ Bond dissociation energy is not the same as bond energy except for diatomic molecules. Bond energy is the average energy needed to break a given type of bond (Section 9.7).

■ FIGURE 10.7 Charge densities as two hydrogen atoms with electrons having the same spin approach each other. Notice that there is a node between the two nuclei when the atoms are a bond length apart.

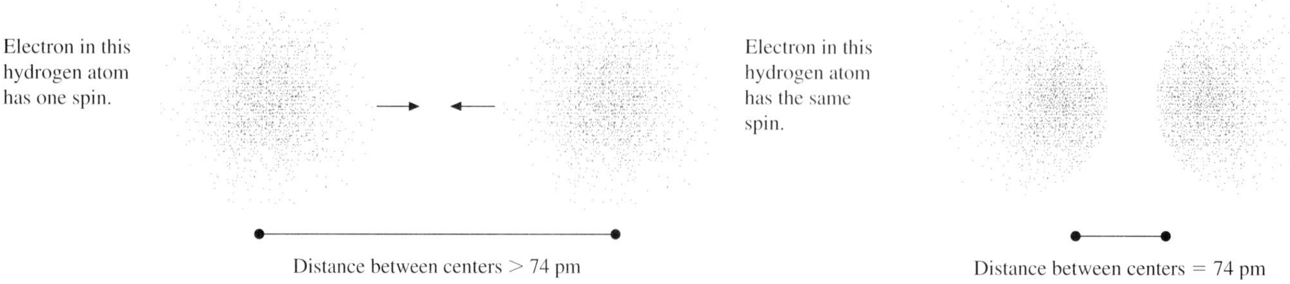

Electron in this
hydrogen atom
has one spin.

Electron in this
hydrogen atom
has the same
spin.

Distance between centers > 74 pm

Distance between centers = 74 pm

as two hydrogen atoms with electrons having the same spin approach each other. The nuclei are not shielded from each other and repel each other because they both have positive charges. Energy increases at first slowly and then rapidly as shown by the upper curve in Figure 10.6. No bond is formed.

Just as the model developed for the structure of the hydrogen atom in Chapter 7 was extended to other atoms in Chapter 8, the picture of the formation of the hydrogen–hydrogen bond can be extended to other bonds. The binding energy curves for the formation of all bonds are similar in shape to the binding energy curve for the hydrogen molecule (Figure 10.6).* The bond length and the energy at the minimum depend on the atoms joined by the bond.

10.4 THE VALENCE BOND METHOD

According to the valence bond method, the H H bond in the H_2 molecule results from overlap of the $1s$ orbitals of the hydrogen atoms so that the two electrons move in the space between the two nuclei (-Figure 10.8). In isolated hydrogen atoms, each electron moves in the field of a single nucleus.

- FIGURE 10.8   The valence bond method pictures the formation of the covalent bond between hydrogens in a hydrogen molecule as resulting from overlap of the $1s$ orbitals of the hydrogen atoms. The distance between the centers of the nuclei is 74 pm.

The explanation of covalent bonds as formed by overlap of atomic orbitals can be applied to polyatomic molecules and ions as well as to diatomic molecules, such as the H_2 molecule. However, the idea of hybrid orbitals is needed to explain the number of bonds that many atoms form and the directions in space of the bonds. *Hybrid orbitals are models used to account for molecular shapes that are observed by experiment.* Hybrid orbitals allow better overlap and the formation of more bonds than unhybridized orbitals. Thus, the molecules that result from the overlap of hybrid orbitals have lower energies and are more stable than the molecules that would result from overlap of unhybridized orbitals.

The compounds of carbon provide more examples of the use of the concept of hybrid orbitals than the compounds of any other element. The carbon atom has the ground-state electron configuration $1s^2 2s^2 2p_x^1 2p_y^1$:

C atom in ground state   [He] $\underset{2s}{\uparrow\downarrow}$ $\underset{2p}{\uparrow\ \ \uparrow}$ ___

Because the carbon atom has two unpaired electrons, we might expect it to form two bonds. But carbon almost always forms four bonds, and in compounds such as methane (CH_4) and tetrachloromethane (CCl_4), the four bonds are identical. Because a theory is an explanation of observed facts, we must change our picture of the carbon atom to fit the facts that carbon almost always forms four bonds and that the four bonds in CH_4 and CCl_4 are identical.

Tetrachloromethane, CCl_4, is commonly known as carbon tetrachloride. Today, carbon tetrachloride is used mainly as a solvent. Its use for dry cleaning and for extinguishing fires has been discontinued because of its toxicity.

---

*There is one difference between the formation of bonds between hydrogen atoms and the formation of bonds between all other atoms. All other atoms (except helium, which is not known to form any bonds) have cores consisting of filled subshells of electrons between the valence electrons and the nucleus. For all atoms except hydrogen, exclusion principle repulsion between core electrons is the main cause of the strong repulsion between atoms at distances less than the bond length.

sp^3 Hybrid Orbitals

To explain why carbon forms four identical single bonds, we imagine that the s orbital combines with the three p orbitals to form four new orbitals that are degenerate:

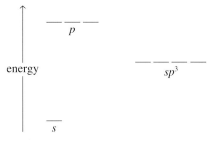

Atomic orbitals pictured as resulting from combinations of atomic orbitals on the same atom are called **hybrid orbitals.** The hybrid orbitals are called sp^3 (read as "ess-pee-three") orbitals because they are formed from one s and three p orbitals. (As usual, 1 is assumed when no superscript is written.) There are four sp^3 orbitals because four orbitals were combined. *The number of hybrid orbitals formed must be the same as the number of atomic orbitals combined.* One of the four orbitals used to form an sp^3 orbital is an s orbital, and the other three are p orbitals. Therefore, each sp^3 orbital has one-quarter s character and three-quarters p character. Because each sp^3 orbital has one-quarter s character and three-quarters p character, the energies of the sp^3 orbitals are lower than the energies of the p orbitals but higher than the energy of the s orbital.

In a singly bonded carbon, the four electrons of the carbon atom are placed in the four degenerate sp^3 orbitals according to Hund's rule:

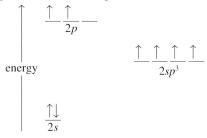

Thus, the sp^3 hybridized carbon atom has four identical unpaired electrons to use for bond formation.

Each sp^3 orbital has the shape shown in ▬Figure 10.9, with a large lobe and a small lobe. The large lobes are the ones that are important for forming bonds. In

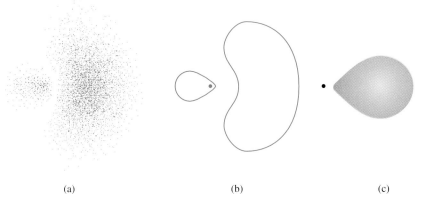

▬ FIGURE 10.9   One sp^3 orbital: (a) charge cloud, (b) boundary surface diagram, (c) simplified diagram.

(a)   (b)   (c)

the simplified diagram that is usually used for showing the formation of molecules, the small lobe is omitted, and the large lobe is made thinner to make drawing the molecule easier.

The angles between the C—H bonds in CH₄ and the C—Cl bonds in CCl₄ are 109.5°. Therefore, the large lobes of the four sp^3 orbitals are pictured as being at 109.5° angles to each other (that is, they are directed to the four corners of a tetrahedron) as shown in ►Figure 10.10.

Single bonds to carbon are pictured as formed by overlap of singly occupied orbitals from other atoms with the large lobes of the four singly occupied sp^3 hybridized orbitals of carbon. For example, ►Figure 10.11(a) shows the overlap of the large lobes of the four sp^3 orbitals from a carbon atom with the 1s orbitals of four hydrogen atoms to form methane, CH₄.

As we saw in Sections 9.7 and 9.12, formation of bonds is exothermic. More energy is given off when four bonds are formed than when only two bonds are formed. Enough energy is produced by forming four bonds instead of just two bonds to more than make up for the energy required to promote the two 2s electrons to $2sp^3$ orbitals.

The electron clouds of sp^3 orbitals are concentrated in smaller volumes than the electron clouds of p orbitals and s orbitals. The large lobes are more directed in space than unhybridized p or s orbitals and overlap better with other orbitals forming stronger bonds. Hybrid orbitals are the way chemists picture the distortion of the electron cloud of one atom by another atom to form a chemical bond.

► FIGURE 10.10   Four sp^3 orbitals attached to a carbon atom. Note that the four sp^3 orbitals are directed to the four corners of a tetrahedron.

sp^2 Hybrid Orbitals

To account for the observed shape of molecules with double bonds, a second type of hybrid orbital must be pictured. Ethylene, $H_2C=CH_2$ (see Figure 10.3) is the simplest compound with a carbon–carbon double bond. In the second type of hybridization, the s orbital is combined with *two* of the p orbitals. The remaining p orbital is not hybridized. Combining one s and two p orbitals gives three hybrid orbitals. Each of the three hybrid orbitals and the remaining unhybridized p orbital contains a single electron:

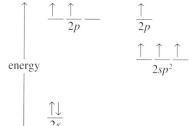

unhybridized C atom   hybridized C atom

⊙ The energy of the p orbital must be close enough to the energies of the sp^2 orbitals so that more energy is gained by forming two additional bonds than would be gained by putting carbon's fourth valence electron in an sp^2 orbital.

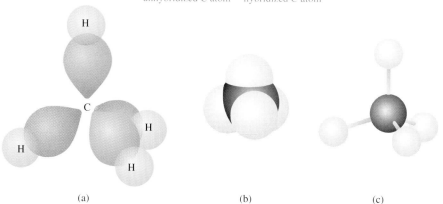

(a)   (b)   (c)

► FIGURE 10.11   (a) The four C—H bonds in methane are pictured as formed by overlap of the 1s orbital of a hydrogen atom with each of the four identical sp^3 orbitals of a carbon atom. (b) Space-filling model of methane. (c) Ball-and-stick model of methane.

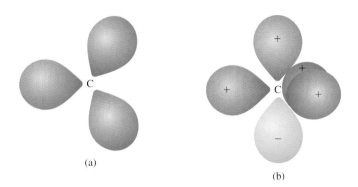

FIGURE 10.12 (a) Three sp^2 orbitals (attached to a carbon atom). The three sp^2 orbitals lie in a plane. The angles between them are 120°. (b) Three sp^2 orbitals and one p orbital (vertical) attached to carbon. The p orbital is perpendicular to the plane of the three sp^2 orbitals. Remember that + and − are the mathematical signs of the wave function and have nothing to do with electrical charge.

These hybridized orbitals are called sp^2 orbitals because they are a combination of one s and two p orbitals. Each sp^2 orbital looks about like an sp^3 orbital (see Figure 10.9). The large lobes of the three sp^2 orbitals lie in a plane at 120° angles to each other as shown in ▪Figure 10.12(a). The unhybridized p orbital is perpendicular to the plane of the three sp^2 orbitals. Figure 10.12(b) shows three sp^2 and one p orbital attached to a carbon atom.

The formation of a double bond between two carbon atoms is pictured as follows. The large lobe of one sp^2 orbital from one carbon atom overlaps the large lobe of one sp^2 orbital from the other carbon atom as shown in ▪Figure 10.13(a). At the same time, the p orbital from one carbon atom overlaps the p orbital from the other carbon atom sidewise, as shown in Figure 10.13(b). Sidewise overlap of p orbitals allows more than one pair of electrons to be shared between atoms. So that sidewise overlap of p orbitals can take place, the carbon atoms joined by a double bond move closer to each other than the normal C—C single bond distance of 154 pm; the length of C=C double bonds is only 134 pm (Table 9.3). The atoms joined by the double bond and the four atoms attached to them all lie in a plane.

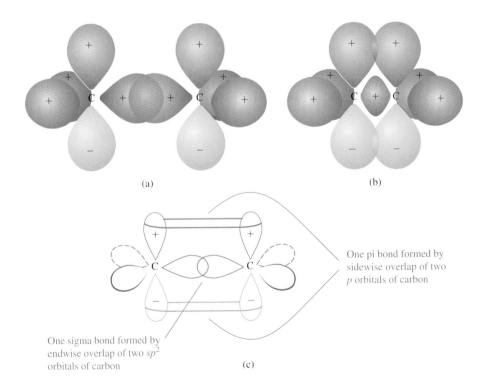

FIGURE 10.13 (a) Two sp^2 orbitals overlapping endwise (head-on). (b) If close enough together and aligned as shown, p orbitals overlap sidewise. (c) Sketch showing endwise overlap of sp^2 orbitals and sidewise overlap of p orbitals. The simplified boundary surface diagrams in the sketch have been made thinner so that you can see the sigma bond better. What would happen if one carbon atom was rotated so that the − lobe of one p orbital overlapped the + lobe of the other p orbital?

One pi bond formed by sidewise overlap of two p orbitals of carbon

One sigma bond formed by endwise overlap of two sp^2 orbitals of carbon

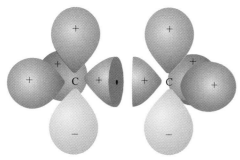

■ FIGURE 10.14   A cross section of a sigma bond is circular. The internuclear axis (an imaginary line connecting the centers of two atoms joined by a sigma bond) is represented by the black dot in the center of the sigma bond.

Sigma and Pi Bonds

Bonds formed by endwise (head-on) overlap are called **sigma (σ) bonds** and are symmetrical about the internuclear axis. This means that if you cut the bond in half, the cross section is a circle with the line between the centers of the two nuclei in its center, as shown in ■Figure 10.14. Single bonds are always sigma bonds.

Bonds formed by sidewise overlap (see Figure 10.13) are called **pi (π) bonds.** A pi bond is *not* symmetrical about the internuclear axis. In a pi bond, the electron density is in two separate regions, one above and the other below the plane of the molecule. Both parts make *one* pi bond. A double bond consists of one sigma bond and one pi bond.

Rotation around single bonds is free as shown in ■Figure 10.15. *The different arrangements of parts of a molecule in space that result from rotation about single bonds* are called **conformations.** The number of possible conformations is infinite, although some conformations are more stable than others and a molecule spends most of its time in the lowest energy, most stable conformation.

Rotation around double bonds is not free. If one doubly bonded carbon were to be rotated relative to the other, the *p* orbitals could no longer overlap sidewise (see ■Figure 10.16). Free rotation about single bonds and lack of free rotation around double bonds are important factors in the properties of many molecules, including physiologically important molecules like proteins.

All points on a circle are the same distance from the center of the circle.

The biological functions of molecules such as proteins in living organisms depend on the conformations of the molecules.

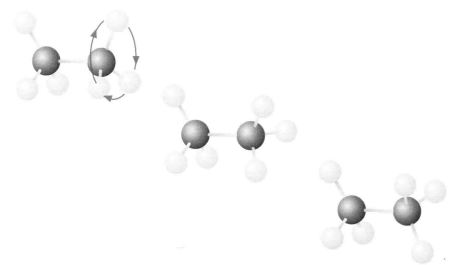

■ FIGURE 10.15   Some conformations of CH_3CH_3 (ethane) resulting from rotation of the right-hand carbon.

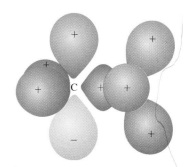

■ FIGURE 10.16   Rotation of a doubly bonded carbon destroys the sidewise overlap of the *p* orbitals.

The Chemistry of Vision

The rod cells in the human eye contain a molecule called rhodopsin that consists of a protein and 11-*cis*-retinal.

11-*cis*-retinal

light →

11-*trans*-retinal

light →

Light that strikes the rod cells is absorbed by rhodopsin. (The human eye can detect as few as 100 photons of light.) The light energy apparently is able to break the bond marked with the dashed line so that rotation can take place. Then the bond reforms; 11-*cis*-retinal is converted to

11-*trans*-retinal in only 200 fs.

As you can see from the structural formulas and the models, the shape of 11-*trans*-retinal is very different from the shape of 11-*cis*-retinal. The energy of the photon of light is converted into kinetic energy of the atoms in retinal. The change

in shape is the beginning of the process of seeing; it starts a nerve impulse that is detected by the brain as a visual image. Retinal is so perfectly suited for its job that the visual systems of all known forms of life are based on retinal.

sp Hybrid Orbitals

A third type of hybridization is needed to explain the shapes observed for compounds with triple bonds. The carbon–carbon triple bond in acetylene, HC≡CH, is an example. Figure 10.17 shows models of acetylene. In the third type of hybridization, one *s* orbital is combined with one *p* orbital to give two *sp* orbitals:

energy

unhybridized C atom   hybridized C atom

- FIGURE 10.17 Ball-and-stick and space-filling models of acetylene (ethyne).

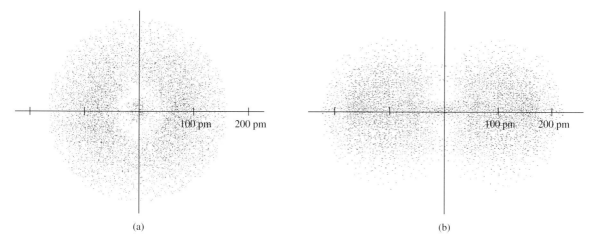

(a)   (b)

- FIGURE 10.18   Electron-density distribution for the valence electron configuration $2s^2 2p_x^2$. (a) Color coded to show s (red) and p_x (blue) electron densities. (b) Color coded to show left-pointing and right-pointing sp hybrid orbital electron densities. (Computer generated.) (Copyright © 1975 by W. G. Davies and J. W. Moore. Reprinted with permission.)

In **-**Figure 10.18, combination of one s and one p orbital to form two sp orbitals is shown using electron density diagrams produced by a computer. Unfortunately, it's difficult to show the formation of sp^2 and sp^3 hybrid orbitals in this way.

An sp orbital looks very much like an sp^3 orbital and a similar simplified diagram is used. The two sp orbitals lie on a straight line [**-**Figure 10.19(a)]. Two uncombined p orbitals are left. The two sp orbitals are perpendicular to the plane of the two leftover p orbitals. Figure 10.19(b) shows the orbitals of an sp hybridized carbon atom.

The formation of a triple bond between two carbon atoms is pictured as follows. The large lobe of one sp orbital from one carbon atom overlaps the large lobe of one sp orbital from the other carbon atom head-on. At the same time, the two p orbitals from one carbon atom overlap the two p orbitals from the other carbon atom sidewise as shown in **-**Figure 10.20. One bond of the triple bond is a sigma bond formed by endwise overlap of an sp orbital from each carbon. The other two bonds are pi bonds formed by the sidewise overlap of the unhybridized p orbitals.

⊙ For central atoms that obey the octet rule, any p orbitals that are not hybridized are involved in the formation of pi bonds.

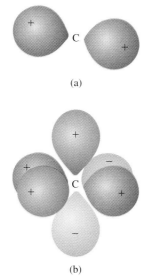

(a)

(b)

- FIGURE 10.19   (a) Two sp orbitals. b) Orbitals of an sp hybridized carbon atom. The two sp orbitals are perpendicular to the plane of the two p orbitals.

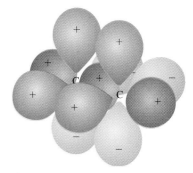

- FIGURE 10.20   Orbital picture showing the two pi bonds formed by sidewise overlap of two p orbitals. The sigma bond is barely visible in the center of the picture.

TABLE 10.2 Hybrid Orbitals

Hybrid	Formed from	Number	Geometry			Example
			Shape	Angle, °		
sp	One s and one p	2	Linear	180		HC≡CH
sp^2	One s and two p	3	Trigonal planar	120		H_2C=CH_2
sp^3	One s and three p	4	Tetrahedral	109.5		CH_4
sp^3d	One s, three p, and one d	5	Trigonal bipyramidal	120 and 90		PCl_5
sp^3d^2	One s, three p, and two d	6	Octahedral	90		SF_6

Elements Other Than Carbon

The idea of hybrid orbitals is also applied to elements other than carbon. For example, hybrid orbitals are often used to account for the observed shapes of water and ammonia. Both water and ammonia have a total coordination number of 4 (Section 10.1), and the VSEPR model predicts that the four electron pairs of water and ammonia will be tetrahedrally arranged around the central atom:

Both the nitrogen in ammonia and the oxygen in water can be pictured as sp^3 hybridized. According to this picture, the bond angles are less than the ideal tetrahedral angle of 109.5° because unshared pairs occupy more space than shared pairs (remember Figure 10.2). The bond angle in water is smaller than the bond angle in ammonia because water has two unshared pairs.

As we saw in Section 9.10, elements in the third period and beyond can form compounds in which the central atom has more than four covalent bonds such as PF_5 and SF_6. To form five or six hybrid orbitals, d orbitals must be used because the total number of s and p orbitals is four. To form five hybrid orbitals, one d orbital must be used; a hybrid orbital formed by combination of one s, three p, and one d orbital is called an sp^3d orbital. To form six hybrid orbitals, two d orbitals must be used; a hybrid orbital formed by combination of one s, three p, and two d orbitals is called an sp^3d^2 orbital. Individual sp^3d and sp^3d^2 orbitals look about like the sp^3 orbital shown in Figure 10.9. The trigonal bipyramidal shape observed for PF_5 requires that the five sp^3d orbitals be arranged around the central atom in a trigonal bipyramid. The octahedral shape observed for SF_6 requires that the six sp^3d^2 orbitals be arranged around the central atom in an octahedron. ▪Figure 10.21 shows models of five sp^3d and six sp^3d^2 orbitals. Information about five common types of hybrid orbitals is summarized in Table 10.2.

For a different explanation of the shapes of water and ammonia see Laing, M., J. Chem. Educ. 1987, 64, 124–129.

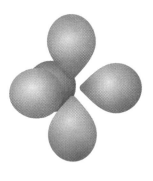

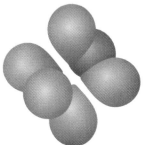

▪ FIGURE 10.21   *Top:* Five sp^3d orbitals. *Bottom:* Six sp^3d^2 orbitals.

SAMPLE PROBLEMS

10.4 Sketch a picture like Figure 10.11(a) showing the formation of AlH_4^-.

Solution   First, write the Lewis formula. The Lewis formula for the AlH_4^- ion is

Then use VSEPR to predict the shape of the species from the Lewis structure. There are four shared pairs of electrons around Al, and a tetrahedral arrangement would be predicted.

From Table 10.2, species with tetrahedral geometry are pictured as having sp^3 hybridization. Draw an Al at the center of four sp^3 orbitals:

Three valence electrons from the aluminum atom plus one extra electron to account for the negative charge on the ion makes a total of four valence electrons; there is one electron in each of the four sp^3 orbitals. Overlap the s orbitals of the four hydrogen atoms, each of which contains one electron, with the four sp^3 lobes:

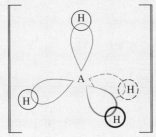

10.5 (a) Predict the shape of the formaldehyde molecule, $H_2C{=}O$. (b) Which type of hybridization is used to describe C in $H_2C{=}O$? (c) What is the total number of covalent bonds in formaldehyde? (d) How many pi bonds are there? (e) How many sigma bonds are there? (f) What kinds of orbitals overlap to form each of the bonds? (g) What is the angle between the H—C bonds?

Solution   (a) The Lewis formula for the formaldehyde molecule is

$$H{-}C{=}\ddot{O}\!:$$
$$\underset{\displaystyle H}{|}$$

The total coordination number of C is 3; C has no unshared pairs. According to VSEPR, the formaldehyde molecule is trigonal planar.

(b) The C should be pictured as being sp^2 hybridized.

(c) There are four covalent bonds.

(d) There is one pi bond.

(e) There are three sigma bonds.

(f) To form the H—C bonds, the s orbitals of the hydrogens each overlap an sp^2 orbital of C. The C$=$O bond is formed by endwise overlap of an sp^2 orbital from C with a p orbital from O and sidewise overlap of p orbitals, one from each atom.

(g) The four electrons of the carbon–oxygen double bond probably take up more room than the two electrons of the single bonds. The angle between the H—C bonds is probably slightly less than 120°. (The experimentally observed bond angle is 116.5°.)

PRACTICE PROBLEMS

10.7   Are the carbon–hydrogen bonds in CH_4 pi bonds or sigma bonds? Explain your reasoning.

10.8 The Lewis structure for SF_4 is

(a) How many d orbitals are there in the hybrid orbitals used to describe the S—F bonds in SF_4? (b) What are these hybrid orbitals called? (c) How many of them are there?

10.9 Sketch pictures like Figure 10.11(a) showing the formation of (a) $BeCl_2$ and (b) SiH_4.

10.10 In a molecule of $CH_3CH{=}CH_2$, (a) what is the total number of covalent bonds? (b) How many pi bonds are there? (c) How many sigma bonds?

10.11 What kinds of orbitals overlap to form each of the bonds in (a) $HC{\equiv}N$ and (b) $CH_3CH{=}CH_2$?

10.12 In $H_2C{=}CHCl$, what is the approximate angle between (a) the C—Cl bond and the C—H bond that is attached to the same carbon as chlorine? (b) The C—Cl bond and the carbon–carbon double bond?

10.13 Do all the atoms in $CH_3CH{=}CH_2$ lie in the same plane? Explain your answer.

The common name for $CH_3CH{=}CH_2$ is propylene. Propylene is polymerized (see Sections 9.11 and 23.1) to make plastics and synthetic fibers.

The compound $H_2C{=}CHCl$ is called vinyl chloride. In 1993, 13.75 billion pounds of vinyl chloride were made. It was used to make "vinyl" plastics, water pipes, and films such as Saran Wrap.

2p

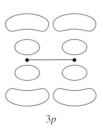

3p

Good sidewise overlap of 2p orbitals is possible, but 3p and larger p orbitals are too far apart to overlap well sidewise. As a result, carbon, nitrogen, and oxygen frequently form multiple bonds but silicon, phosphorus, and sulfur do not.

Pi bonding is very important in the chemistry of the elements in the second period. It is not as important for third row and larger main group elements. The larger p orbitals of the third-period elements do not overlap sidewise very well with the smaller p orbitals of second-period elements. Two larger atoms of third- and lower-period elements can't get close enough together for good sidewise overlap.

10.5 SOLUTION OF THE SCHRÖDINGER EQUATION FOR THE HYDROGEN MOLECULE

The Schrödinger equation for the electrons in the simplest molecule, the hydrogen molecule, was first solved by the valence bond method by the German physicists Heitler and London,* in 1927, just one year after Schrödinger published his equation. Heitler and London's first approximation was that the wave function for the electrons in the hydrogen–hydrogen bond can be pictured as a combination of the atomic wave functions of the electrons in the hydrogen atoms joined by the bond. This first approximation gave a wave function that had the right shape; the binding energy curve had a minimum. However, the minimum came at a bond length 1.7 times the observed H—H bond length of 74 pm, and the calculated bond dissociation energy was only 5% of the experimental value. Heitler and London's second approximation included a correction for the fact that two electrons cannot be distinguished from each other. The second approximation gave a bond length only 1.2 times the observed bond length and a bond dissociation energy 66% of the experimental value. Heitler and London and other physicists improved the valence bond method further by adding more corrections until, in the mid-1950s, a value for the bond dissociation energy for the hydrogen molecule was obtained that later proved to be more accurate than the best experimental value available at the time!

*London also developed an early theory of superconductivity, the loss of all resistance to the passage of an electric current by some solids when cooled below a characteristic temperature. He emigrated to the United States at the beginning of World War II.

The valence bond method supposes that electrons in molecules occupy the atomic orbitals of the individual atoms that compose the molecule. Bonds between atoms are formed by overlapping atomic orbitals; the idea of hybrid orbitals was a product of valence bond calculations. In the valence bond method, the structures of molecules and polyatomic ions are represented by Lewis formulas. However, as we saw in Section 9.9, not all molecules and polyatomic ions can be described by a single Lewis formula; two or more Lewis formulas are required to describe some molecules and polyatomic ions. In addition, simple valence bond theory fails to account for the fact that oxygen, O_2, is paramagnetic with two unpaired electrons.

Since World War II, the molecular orbital method has come to be used more and more for calculations until today almost all calculations are done by the molecular orbital method. The molecular orbital method supposes that electrons in molecules occupy molecular orbitals that may extend over the whole molecule. The energies of the molecular orbitals are calculated using the Schrödinger equation. Electrons are then assigned to the molecular orbitals using the aufbau approach and following the same rules (Hund's rule and the Pauli exclusion principle) followed in placing electrons in atomic orbitals.

After enough approximations have been made, the molecular orbital and valence bond methods give identical results. However, the molecular orbital method gives a better first approximation, and the math is simpler. The molecular orbital method also explains easily some observations that the valence bond method explains only with difficulty such as the paramagnetism of oxygen. The idea of resonance is unnecessary in the molecular orbital method; one picture is enough to describe any molecule or polyatomic ion. On the other hand, valence bond structures are simpler to picture than molecular orbitals (compare the Lewis formula for water with the molecular orbital contour map for the ground state of water in ▬Figure 10.22) and will be used often in the remainder of this book (and in organic and inorganic texts if you study these subjects later).

As a first approximation, molecular orbitals are formed by mathematically combining atomic orbitals from different atoms. Although the math required for the molecular orbital method is simpler than the math required for the valence bond method, it is still far from easy. We will simply use molecular orbital energy level diagrams (such as the molecular orbital energy level diagram for the hydrogen molecule shown in ▬Figure 10.23) obtained as the results of other people's calculations.

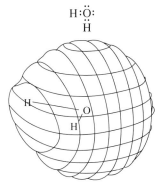

▬ FIGURE 10.22 *Top:* Lewis formula for water. *Bottom:* Molecular orbital contour map for water. A molecular orbital contour map consists of a series of boundary surface diagrams. The boundary surface diagrams show the volumes inside which the electrons are found different percentages of the time. (The molecular orbital contour map is reproduced from Jorgensen, W. L.; Salem, L. *The Organic Chemist's Book of Orbitals;* Academic: New York, 1973; p 70 with permission of the publisher.)

Identical results are expected because approximations are made until the answers agree with experiment.

10.6   THE MOLECULAR ORBITAL METHOD

Again, let's begin by considering the simplest common molecule, the hydrogen molecule, H_2. *When atomic orbitals are combined to form molecular orbitals, the number of molecular orbitals formed must be the same as the number of atomic orbitals mathematically combined.* Thus, when the $1s$ atomic orbitals of two hydrogen atoms are combined, two molecular orbitals are formed.

When two atomic orbitals are combined to form two molecular orbitals, one molecular orbital has lower energy than the atomic orbitals combined. The other molecular orbital is of higher energy than the atomic orbitals. *The molecular orbital with energy lower than the atomic orbitals that were combined* is called a **bonding molecular orbital.** The charge density in the bonding molecular orbital of the hydrogen molecule, which is labeled σ_{1s} in Figure 10.23, is shown at the bottom of Figure 10.23. Notice how the charge density in the bonding molecular orbital is concentrated between the nuclei of the two hydrogen atoms. The attraction of the negative charge of the electrons between the two nuclei for the positive charges of

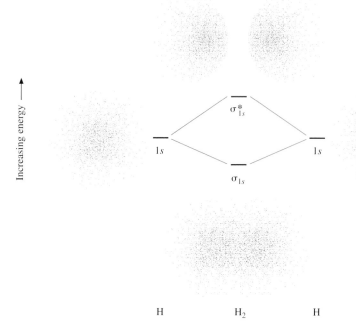

FIGURE 10.23 Bonding and antibonding molecular orbitals of the hydrogen molecule.

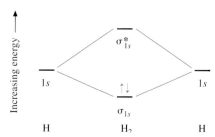

FIGURE 10.24 Molecular orbital energy level diagram for the ground state of the hydrogen molecule.

the nuclei holds the nuclei together. Also observe that the charge density is similar to the charge density resulting from two hydrogen atoms that have electrons with opposite spins approaching each other (Figure 10.5).

The *molecular orbital with energy higher than the energy of the atomic orbitals that were combined* is called an **antibonding molecular orbital.** The charge density in the antibonding molecular orbital of the hydrogen molecule, which is labeled σ_{1s}^{*}, is shown at the top of Figure 10.23. In the antibonding molecular orbital, there is a node in charge density between the nuclei. Charge density in the antibonding orbital is concentrated outside of the space between the nuclei. Without much charge density between them, the hydrogen nuclei repel each other. The atoms tend to fly apart; a bond is not formed. Antibonding orbitals are usually indicated by superscripting an asterisk as shown in Figure 10.23. Both the bonding and the antibonding molecular orbitals of the hydrogen molecule are *symmetric about the internuclear axis.* They are both **sigma orbitals.** When a bonding sigma orbital contains a pair of electrons, a sigma bond results.

In molecular orbital energy level diagrams, the green lines leading from each molecular orbital to two atomic orbitals show the atomic orbitals that were combined to form the molecular orbital. They also show which molecular orbitals were formed from a given atomic orbital.

Most of what we have learned about writing electron configurations for atoms applies to molecules as well. Electrons are placed in orbitals in a molecular orbital energy level diagram beginning with the lowest energy orbital at the bottom (aufbau approach). No more than two electrons can be put in one orbital, and the two electrons in an orbital must have opposite spins (Pauli exclusion principle). One electron is put in each of a set of degenerate orbitals before two are put in any one orbital (Hund's rule).

There are two electrons in a hydrogen molecule. In the ground state, both electrons are in the lowest energy σ_{1s} orbital. ▪Figure 10.24 shows the molecular orbital energy level diagram for the hydrogen molecule in its ground state. The electron configuration of the hydrogen molecule in its ground state is written $(\sigma_{1s})^2$. In the electron configurations of molecules, the superscript numbers show the number of electrons in each orbital. Thus, the symbol $(\sigma_{1s})^2$ says that there are two electrons in the σ_{1s} molecular orbital. A molecule is in its ground state unless electrons are promoted to an excited state by collision, thermal energy, or electromagnetic radiation.

All electron configurations except the ground state are excited states. Promoting one electron to the σ_{1s}^* orbital gives the first excited state for the hydrogen molecule as shown in ▬Figure 10.25. Molecular orbital calculations for the hydrogen molecule give the result that, relative to the energies of the $1s$ atomic orbitals of the hydrogen atoms, the energy of the σ_{1s}^* orbital is raised more than the energy of the σ_{1s} orbital is lowered. As a result, one electron in the antibonding molecular orbital more than cancels the bonding effect of one electron in the bonding orbital. A hydrogen molecule decomposes into two hydrogen atoms when enough energy is added to promote an electron to the σ_{1s}^* orbital.

Molecular orbital energy level diagrams like Figures 10.23–10.25 include only the lowest energy molecular orbitals. (Remember that the principal quantum number, n, can have any whole number value from 1 to infinity.) More high-energy molecular orbitals can be obtained by combining two $2s$ atomic orbitals or two $2p_x$ atomic orbitals. The possibility of higher energy excited states is important in explaining the visible and ultraviolet spectra of molecules as well as some of their reactions. *The atomic orbitals that are combined to form a molecular orbital must have similar energies.* For example, a $1s$ orbital from one hydrogen atom and a $2s$ orbital from another hydrogen atom cannot be combined to form a molecular orbital.

Also, to be combined to form a molecular orbital, the charge clouds of atomic orbitals must occupy the same region in space. For example, a p_y orbital can overlap with another p_y orbital either endwise or sidewise as shown in ▬Figures 10.26(a) and (b). But a p_y orbital cannot overlap well either endwise or sidewise with a p_x orbital or with a p_z orbital [see Figure 10.26(c)]. Orbitals are electron waves. Constructive interference yields bonding molecular orbitals; destructive interference yields antibonding molecular orbitals.

Homonuclear Diatomic Molecules of Elements Other Than Hydrogen

Diatomic molecules in which both atoms are of the same kind, such as H_2, are called **homonuclear diatomic molecules.** The molecular orbital energy level diagram for He_2 is similar to the molecular orbital energy level diagram for H_2. The molecular

▬ FIGURE 10.25 Molecular orbital energy level diagram for the first excited state of the hydrogen molecule.

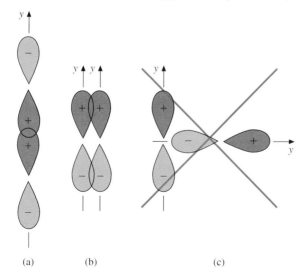

(a)   (b)   (c)

▬ FIGURE 10.26 (a) A p_y orbital can overlap with another p_y orbital either endwise to form a σ molecular orbital or (b) sidewise to form a π molecular orbital. (c) A p_y orbital cannot overlap either endwise or sidewise with a p_x orbital (or with a p_z orbital). If two p_y orbitals overlap endwise, the p_x and p_z orbitals overlap sidewise.

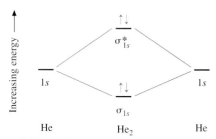

FIGURE 10.27 Molecular orbital energy level diagram for the ground state of the He$_2$ molecule.

Existence of a He$_2$ molecule at <0.001K was reported in Science *1993, 259, 1537. The He—He is so weak that He$_2$ is as large as a small protein.*

orbital energy level diagram for He$_2$ can be used to show why helium does not form a diatomic molecule. Two helium atoms have a total of four electrons in 1s orbitals. Putting four electrons in the molecular orbital energy level diagram gives the electron configuration shown in ▪Figure 10.27 for the He$_2$ molecule.

The bonding effect of the two electrons in the bonding molecular orbital is more than canceled by the antibonding effect of the two electrons in the antibonding molecular orbital. The energy of a helium molecule is higher than the energy of two separate helium atoms. In addition, two separate helium atoms are more disorderly than two helium atoms neatly combined into a helium molecule. Entropy as well as energy favors two separate helium atoms over a diatomic helium molecule. Therefore, the He$_2$ molecule is less stable than two helium atoms, and covalently bonded He$_2$ molecules do not exist.

The molecular orbital energy level diagrams for homonuclear diatomic molecules of the elements in the second row of the periodic table are shown in ▪Figure 10.28. In principle, similar diagrams apply to homonuclear diatomic molecules of third- and higher period elements. However, because pi bonds between the larger atoms of third- and higher period elements are not very strong, only the halogens normally form diatomic molecules. **Pi molecular orbitals, π and π*,** are *not symmetric about the internuclear axis.*

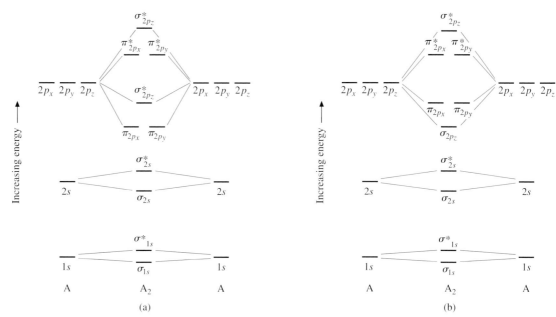

FIGURE 10.28 Molecular orbital energy level diagram for molecular orbitals formed from 1s, 2s, and 2p atomic orbitals: (a) Li$_2$ through N$_2$ and (b) O$_2$ through Ne$_2$. (These diagrams show the order of filling only. Energies are not to scale.) In (a), the energy of the π_{2p_x} and π_{2p_y} orbitals is lower than the energy of the σ_{2p_z} orbital, whereas in (b), the energy of the σ_{2p_z} orbital is lower than the energies of the π_{2p_x} and π_{2p_y} orbitals. The order of increasing energy of molecular orbitals in (b) is easier to understand than the order in (a). Endwise overlap of p orbitals to form a sigma orbital lowers energy more than sidewise overlap of p orbitals to form a pi orbital. The order shown in (b) is correct for O$_2$ and F$_2$ but leads to wrong predictions about the magnetic properties of B$_2$ and C$_2$. Why should the order of energies for O$_2$ and F$_2$ be different than the order for Li$_2$ through N$_2$? Oxygen and fluorine have very high effective nuclear charges; the difference in energy between the 2s and 2p orbitals in oxygen and fluorine atoms is large, 2500 kJ/mol for F. Compared with oxygen and fluorine, the elements Li through N have low effective nuclear charges and the difference in energy between the 2s and 2p orbitals is small, only 200 kJ/mol for lithium (remember Figure 8.2). For the elements Li through N, interaction between the 2s and 2p orbitals is significant and makes the energy of the π_{2p_x} and π_{2p_y} orbitals lower than the energy of the σ_{2p_z} orbital.

Sample Problem 10.6 illustrates the use of a molecular orbital energy level diagram for writing the electron configuration of a molecule and describing the bonding in the molecule.

SAMPLE PROBLEM

10.6 (a) Draw the molecular orbital energy level diagram for the ground state of the nitrogen molecule, N_2, and write the electron configuration. (b) How many sigma bonds are there in a N_2 molecule? (c) How many pi bonds? (d) Is the nitrogen–nitrogen bond a single bond, a double bond, or a triple bond? What is the bond order of the nitrogen–nitrogen bond?

Solution (a) From Figure 10.28, the appropriate molecular orbital energy level diagram is (a). In a N_2 molecule, there are 14 electrons. The electron configuration of the N_2 molecule in its ground state is

$$
\begin{array}{c}
\underline{} \quad \sigma^*_{2p_z} \\[4pt]
\pi^*_{2p_x} \;\underline{} \quad \underline{}\; \pi^*_{2p_y} \\[10pt]
\underline{\uparrow\downarrow} \quad \sigma_{2p_z} \\[4pt]
\pi_{2p_x}\; \underline{\uparrow\downarrow} \quad \underline{\uparrow\downarrow}\; \pi_{2p_y} \\[10pt]
\underline{\uparrow\downarrow} \quad \sigma^*_{2s} \\[4pt]
\underline{} \quad \sigma_{2s} \\[14pt]
\underline{\uparrow\downarrow} \quad \sigma^*_{1s} \\[4pt]
\underline{\uparrow\downarrow} \quad \sigma_{1s}
\end{array}
$$

(Increasing energy →, vertical axis)

or $(\sigma_{1s})^2(\sigma^*_{1s})^2(\sigma_{2s})^2(\sigma^*_{2s})^2(\pi_{2p_x})^2(\pi_{2p_y})^2(\sigma_{2p_z})^2$.

(b) The bonding effect of the pair of electrons in the σ_{1s} orbital is canceled by the antibonding effect of the pair of electrons in the σ^*_{1s} orbital, and the bonding effect of the pair of electrons in the σ_{2s} orbital is canceled by the antibonding effect of the pair of electrons in the σ^*_{2s} orbital. According to the molecular orbital method, only valence electrons are involved in bonding because *both* bonding and antibonding orbitals are filled in lower energy levels. There is one pair of electrons in the σ_{2p_z} bonding orbital and none in the $\sigma^*_{2p_z}$ antibonding orbital; therefore, there is one sigma bond in a N_2 molecule.

(c) There are two pairs of electrons in pi bonding orbitals (π_{2p_x} and π_{2p_y}) and none in pi antibonding orbitals; therefore, there are two pi bonds in a N_2 molecule.

(d) Two pi bonds plus one sigma bond make three bonds between nitrogens; the bond between the nitrogens is a triple bond. The bond order of the nitrogen–nitrogen bond is three (Section 9.7).

Why no summary table of electron configurations of diatomic molecules of second-period elements? If one is included, there is little left for students to do (except exotic species such as O_2^+). Without a summary table, writing the electron configuration for F_2 (Practice Problem 10.14) can be used as a problem.

PRACTICE PROBLEM

10.14 (a) Draw the molecular orbital energy level diagram for the ground state of the fluorine molecule, F_2, and write the electron configuration. (b) How many pi bonds are there in a F_2 molecule? (c) How many sigma bonds? (d) Is the fluorine–fluorine bond a single bond, a double bond, or a triple bond? What is the bond order of the fluorine–fluorine bond?

The Oxygen Molecule

One of the major successes of the molecular orbital method is that it explains the electronic structure of the oxygen molecule. Oxygen is paramagnetic (see ▪Figure 10.29), and measurements of paramagnetism show that the oxygen molecule has two unpaired electrons. The valence bond method does not predict a structure with two unpaired electrons for oxygen. The Lewis formula for O_2 would be written as follows:

$$:\!\ddot{O}\!=\!\ddot{O}\!:$$

Thus, according to the valence bond method, the oxygen molecule should be diamagnetic and be pushed out of a magnetic field.

The molecular orbital method correctly predicts that oxygen molecules have two unpaired electrons and a double bond between the oxygen atoms in agreement with experimental measurements. Two oxygen atoms have a total of 16 electrons. The molecular orbital diagram for the oxygen molecule in its ground state is

▪ FIGURE 10.29   Oxygen is paramagnetic and sticks to the poles of a magnet. If you poured a diamagnetic liquid, such as water, between the poles of a magnet, you would get your feet wet.

Measurements of paramagnetism are used to determine the concentration of oxygen in the blood of surgical patients while they are anesthetized.

The ground-state electron configuration of the oxygen molecule is

$$(\sigma_{1s})^2(\sigma_{1s}^*)^2(\sigma_{2s})^2(\sigma_{2s}^*)^2(\sigma_{2p_z})^2(\pi_{2p_x})^2(\pi_{2p_y})^2(\pi_{2p_x}^*)^1(\pi_{2p_y}^*)^1$$

The oxygen molecule has two more electrons in sigma bonding orbitals than it has in sigma antibonding orbitals and two more electrons in pi bonding orbitals than it has in pi antibonding orbitals. According to the molecular orbital method, the oxygen molecule has one sigma bond and one pi bond; there is a double bond between the oxygen atoms. There are also two unpaired electrons.

Bond Order from Molecular Orbital Energy Level Diagrams

As we saw in Section 9.7, the number of bonds between two atoms in a molecule is often referred to as the bond order. The bond order of the carbon–carbon bond in the ethylene molecule is 2, and the bond order of the carbon–carbon bond in the acetylene molecule is 3. In the molecular orbital method, the bond order of a bond is equal to the net number of pairs of bonding electrons in the electron configuration of the molecule:

bond order = $\frac{1}{2}$(number of bonding electrons − number of antibonding electrons)

Why does bond order follow the discussion of diatomic molecules? Students need to be familiar with molecular orbital energy level diagrams for diatomic molecules to understand the discussion of bond order.

For example, in the molecular orbital energy level diagram for the ground state of the hydrogen molecule (Figure 10.24) there are two bonding electrons and no antibonding electrons.

$$\text{bond order} = \tfrac{1}{2}(2 - 0) = 1$$

The bond order of the hydrogen–hydrogen bond in H_2 is 1; there is one single bond between the hydrogen atoms in the hydrogen molecule.

In the molecular orbital energy level diagram for the ground state of the helium molecule He_2 (Figure 10.27), there are two bonding electrons and two antibonding electrons:

$$\text{bond order} = \tfrac{1}{2}(2 - 2) = 0$$

The bond order of the helium–helium bond in He_2 is 0. There is no bond between the helium atoms in the helium molecule; He_2 is not a stable molecule.

As we saw in Section 9.9, fractional bond orders are possible. The species H_2^+, which has been observed although it is not common under ordinary conditions, is another example. In the molecular orbital energy level diagram for the ground state of H_2^+, the single electron is in the σ_{1s} orbital:

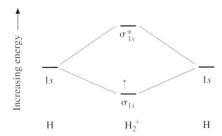

The bond order of the hydrogen–hydrogen bond in H_2^+ is $\tfrac{1}{2}$ or 0.5:

$$\text{bond order} = \tfrac{1}{2}(1 - 0) = \tfrac{1}{2} = 0.5$$

PRACTICE PROBLEM

10.15   (a) Draw the molecular-orbital-energy-level diagram for the ground state of the boron molecule, B_2, and write the electron configuration. (b) What is the bond order of the boron–boron bond? (c) Is B_2 diamagnetic or paramagnetic? If it is paramagnetic, how many unpaired electrons are there?

Heteronuclear Diatomic Molecules

The bonds we have considered so far in this part of the chapter have all been between identical atoms. *Bonds between nonidentical atoms,* which are called **heteronuclear,** are much more common.

If the two different atoms are only one or two places apart in the periodic table, the calculated molecular orbital energy level diagram is not much different than the molecular orbital energy level diagram for two identical atoms. For example, ▪Figure 10.30 shows the molecular orbital energy level diagram for the NO molecule. As you can see, the bonding molecular orbitals for NO are closer in energy to the atomic orbitals of the more electronegative atom, O. The antibonding molecular orbitals for NO are closer in energy to the atomic orbitals of the less electronegative atom, N.

If the energies of the atomic orbitals from which the molecular orbitals are formed are quite different, the molecular orbital energy level diagram will *not* resemble the

Remember that NO is commonly called nitric oxide. Nitric oxide is formed from nitrogen and oxygen in air at high temperatures, such as those in automobile engines. It is a major air pollutant. Since 1987, NO has also been found to play a role, often as a biological messenger, in an astonishing variety of physiological processes in humans and animals.

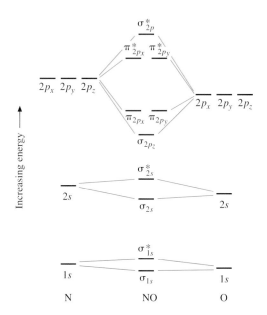

◾ FIGURE 10.30 Atomic orbital energy level diagrams for N and O *(sides)* and molecular orbital energy level diagram for NO *(center).*

molecular orbital energy level diagrams for homonuclear diatomic atoms. A different diagram must be constructed for each molecule by calculating energies from the Schrödinger equation.

PRACTICE PROBLEM

10.16 (a) Copy the molecular-orbital-energy-level diagram for NO and place the electrons in the copy. (b) Write the electron configuration of the NO molecule in its ground state. (c) What is the bond order of the N O bond in NO? (d) Is this the order predicted by the valence bond method? (e) What experimental evidence would be used to decide which method better describes NO?

10.7 DELOCALIZED ELECTRONS

Calculations of molecular orbital energy level diagrams for polyatomic species are often simplified by assuming that the electrons in all sigma orbitals and in some pi orbitals are **localized** between and around the bonded atoms just as they are according to the valence bond model. For example, the electrons in the pi bond in ethene, $H_2C{=}CH_2$, are localized between the two carbon atoms (Figure 10.3). However, *the electrons in pi bonds of species for which resonance structures must be written* are pictured as **delocalized,** that is, *free to move around three or more atoms.* An important advantage of the molecular orbital method of describing the bonding in molecules is that a single electron configuration is enough to describe any species and the idea of resonance is unnecessary.

The formate ion, $HCOO^-$, is an example of a species that has delocalized electrons. In the formate ion, the two carbon–oxygen bonds are the same length and strength, and two structures must be written to represent the bonding:

Experimental observations of the formate ion also show that all four atoms lie in a plane and that the angles between bonds are approximately 120°. The carbon atom in the formate ion can be pictured as sp^2 hybridized with the unhybridized p orbital perpendicular to the plane formed by the atoms. If a p orbital of each oxygen atom is also perpendicular to the plane, the three p orbitals overlap sidewise as shown in ▪Figure 10.31, and electrons are delocalized over three atoms. *For electrons to be delocalized, three or more* p *orbitals must be located on adjacent atoms that lie in the same plane. The* p *orbitals must be perpendicular to the plane formed by the atoms.*

The organic compound benzene, C_6H_6, is the traditional example of a species that has delocalized electrons. Experimental observations of the benzene molecule have established that the six carbons form a ring and that all carbon–carbon bonds are the same length and strength. All the atoms, both carbons and hydrogens, lie in a plane and all the bond angles are 120°. ▪Figures 10.32(a) and 10.32(b) show ball-and-stick and space-filling models of benzene. The valence bond method pictures benzene as a hybrid of two equivalent resonance structures:

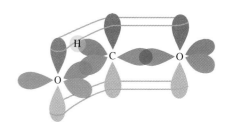

▪ FIGURE 10.31 Sketch showing sidewise overlap of p orbitals in the HCOO$^-$ ion.

Because of the 120° bond angles and the fact that all the atoms lie in a plane, the carbons of benzene are pictured as sp^2 hybridized with the unhybridized p orbitals perpendicular to the plane of the ring. The p orbitals overlap sidewise all around the ring as shown in Figure 10.32(c). Combination of the six p orbitals gives six pi molecular orbitals. Figure 10.32(c) shows the lowest energy pi molecular orbital of the benzene molecule. From Figure 10.32(c), it is easy to see why the pi electrons in benzene are referred to as delocalized. There are two "doughnuts" of charge density above and below the ring of carbon atoms.

Structures with delocalized electrons, such as the benzene molecule and the formate ion, always have lower energy (greater stability) than similar structures without delocalized electrons. For example, although pi bonds are weaker than sigma bonds between carbon atoms and are often the site of chemical reactivity in molecules, benzene is unreactive toward substances that react with carbon–carbon double bonds. In solution in CCl_4,

◗ In the symbols ⬡ and ⬡ which are sometimes used to represent benzene, the dashed line and the circle represent the delocalized pi electrons.

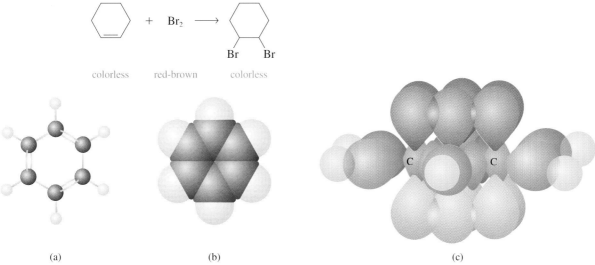

(a)   (b)   (c)

▪ FIGURE 10.32 (a) Ball-and-stick model of benzene. (b) Space-filling model of benzene. (c) Orbital picture of benzene.

FIGURE 10.33 *Left:* Substances that have carbon–carbon double bonds decolorize a solution of bromine in carbon tetrachloride. *Center:* Substances that do not have carbon–carbon double bonds do not decolorize bromine. *Right:* Benzene does not decolorize bromine showing that the carbon–carbon double bonds in the resonance structures for benzene are not ordinary carbon–carbon double bonds.

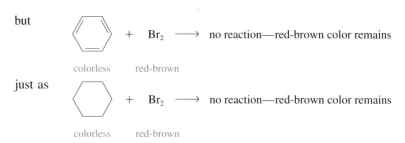

but   + Br₂ ⟶ no reaction—red-brown color remains

just as   + Br₂ ⟶ no reaction—red-brown color remains

Although the Lewis formula for each resonance structure for benzene has three double bonds, benzene reacts like a compound with no double bond rather than like a compound with a double bond (see ▪Figure 10.33).

PRACTICE PROBLEM

10.17   (a) Which of the following isomers of C_5H_8

$$H_2C=CHCH=CHCH_3 \qquad H_2C=CHCH_2CH=CH_2$$

has delocalized electrons? Explain your answer. (b) Over how many atoms are electrons delocalized?

10.8   BAND THEORY OF BONDING IN SOLIDS

Band theory is an extension of delocalized orbital ideas that accounts for the properties of solids. At the beginning of our discussion of chemical bonding, we considered what happens when two atoms are brought together to form a diatomic molecule. Bonding in solids can be understood by thinking about what happens when a large number of atoms are brought together to form a crystal. Let's take sodium metal as an example. The smallest crystal of sodium metal that can be seen with the naked eye is about 0.1 mm on a side. A crystal this size contains 10^{16} atoms of sodium. Each sodium atom has one valence electron in a $3s$ orbital. If 10^{16} atoms of sodium are brought together, 10^{16} $3s$ orbitals must be combined to form "molecular orbitals" for the crystal.

Just as the energy levels for the bonding and antibonding molecular orbitals of diatomic molecules can be found by solving the Schrödinger equation, the energy levels of the bonding and antibonding molecular orbitals of clusters of atoms or particles of 10^{16} atoms can be calculated. ▪Figure 10.34 shows how the molecular orbital energy level diagram for a cluster of ten sodium atoms

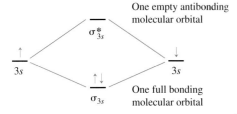

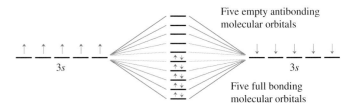

FIGURE 10.34 (a) Molecular orbital energy level diagram for Na₂. Two half-filled $3s$ atomic orbitals combine to give one bonding molecular orbital, which is full, and one antibonding molecular orbital, which is empty. The energy gap between the bonding and antibonding molecular orbitals is large. (b) Molecular orbital energy level diagram for Na₁₀. Ten half-filled $3s$ atomic orbitals combine to give five bonding molecular orbitals, which are full, and five antibonding molecular orbitals, which are empty. The energy gap between the bonding and antibonding molecular orbitals is small.

compares with the molecular orbital energy level diagram for a molecule composed of two sodium atoms.

The molecular orbital energy levels are closer together when 3s orbitals from ten sodium atoms are combined than when 3s orbitals from only two sodium atoms are combined. If 10^{16} atomic orbitals are combined, the molecular orbital energy levels are so closely spaced that they appear to be continuous. *A group of very closely spaced energy levels* is called a **band.**

The *difference in energy between the bonding and the antibonding molecular orbitals* is called an **energy gap.** Notice how the energy gap is smaller when atomic orbitals from ten sodium atoms are combined than when atomic orbitals from only two sodium atoms are combined. When atomic orbitals from 10^{16} sodium atoms are combined, the energy gap disappears (see ▪Figure 10.35).

Bands are separated by spaces, called **forbidden bands,** where there are no allowed energies. Figure 10.35 compares energy-level diagrams for the valence shells of solid sodium and the sodium atom.

Each energy band is formed by combining a certain number of orbitals (10^{16} for our 0.1-mm cube of sodium, for example). According to the Pauli exclusion principle, each orbital can hold only two electrons. Therefore, there is only room in each energy band for a certain number of electrons. If the highest energy band that has some electrons is unfilled, as it is in solid sodium (see Figure 10.35), electrons may be excited from a lower to a higher energy level *within* the band. The separation of levels within the band is so small that the addition of a small amount of energy can excite the electrons. An electron in an unfilled level will move under the influence of an electric field or if heated. *A material with a partly filled energy band* is a **conductor;** a conductor conducts both electricity and thermal energy by movement of electrons.

All metals are conductors—that is, all metals have partly filled energy bands. Transition-metal atoms have unfilled *d* subshells, and lanthanide and actinide atoms have unfilled *f* subshells; therefore, crystals of lanthanide and actinide elements have partly filled bands. Even the metals of Group IIA, which have filled 2s subshells, are conductors. Metals of Group IIA can conduct because *s* and *p* bands overlap as shown in ▪Figure 10.36. The overlapping energy bands are only partly filled so that there are many closely spaced levels available within the band.

Metals are extreme examples of delocalized electrons; the valence electrons are free to move throughout the entire crystal. However, the metal ions are fixed in position and can only vibrate. As a metal is heated, the metal ions vibrate more vigorously about their fixed positions. Each metal ion has a positive charge, and the valence electrons are attracted by the positive charges of the cores. Motion of the valence electrons through the crystal is hindered by vibrations of the cores, and the electrical conductivity of metals decreases with increasing temperature.

Most nonmetals and most compounds are **insulators**—that is, they are *poor conductors of electricity and thermal energy.* The energy-level diagram for a typical insulator is shown in ▪Figure 10.37. For insulators, the highest occupied band is always completely (or almost completely) filled. The forbidden band just above the highest filled band is wide. An electric field billions of times as strong as the electric field needed to make a current flow in sodium is needed to excite electrons in an insulator.

Semiconductors like silicon make up a third class of materials. (Elements along the borderline between metals and nonmetals in the periodic table and their compounds are semiconductors.) The difference between insulators and semiconductors is that *the gap between the highest filled band and the next higher permitted band is relatively narrow* as shown in ▪Figure 10.38. The thermal energy available at room temperature is enough to raise some electrons from the highest occupied band

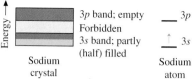

▪ **FIGURE 10.35** Energy levels in solid sodium and the sodium atom (not to scale). When a very large number of atomic orbitals are combined to form the molecular orbitals of a crystal, the energy gap between bonding and antibonding molecular orbitals disappears. The 3s band in the sodium crystal is only half filled because this band is formed by combining 3s orbitals of sodium atoms, which are only half filled. The empty 3p band is formed by combining empty 3p orbitals of sodium atoms.

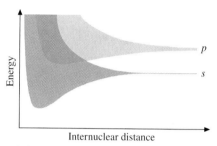

▪ **FIGURE 10.36** The *s* and *p* bands of Group IIA metals overlap.

▪ **FIGURE 10.37** Energy bands of an insulator.

▪ **FIGURE 10.38** Energy bands of a semiconductor.

■ FIGURE 10.39 *Left:* Ultraviolet-visible absorption spectrum of the permanganate ion, MnO_4^-. There are two bands in the spectrum, one at about 310 nm in the ultraviolet and the other at about 535 nm in the visible region. Remember that visible light is composed of light of all colors (Section 7.6). The permanganate ion absorbs light of wavelengths around 535 nm strongly. Light of wavelengths around 425 nm (violet-blue) and 650 nm (orange-red) is almost completely transmitted. As a result, solutions containing the permanganate ion appear purple in color. *Right:* A solution that contains permanganate ion.

Data are from Manahan, S. E. Quantitative Chemical Analysis; Brooks/Cole: Monterey, 1986; p 569.

(called the **valence band**) to the next permitted band (called the **conduction band**). The conduction band is then only partly filled, and conduction can take place. The number of electrons in the conduction band increases with increasing temperature; therefore, the conductivity of semiconductors usually increases as temperature increases.

When electrons are transferred to the conduction band from the valence band, the valence band is no longer filled. *The space left by the removal of an electron is a small electron-deficient region* called a **hole.** Some charge is carried by the movement of holes. An electron moves into a hole leaving a new hole; electrons move forward, and holes move backward.

10.9 MOLECULAR SPECTRA

Besides predicting correctly the structure of the oxygen molecule and explaining the properties of solids, the molecular orbital method gives the energies of unoccupied orbitals. Excited electronic states can be described and related to the ultraviolet-visible spectra of molecules and polyatomic ions. Many experimental bond lengths, bond angles, and binding energy curves have been obtained from molecular spectra.

Absorption spectra are usually used to study molecules. ■Figure 10.39, which shows the ultraviolet-visible absorption spectrum of the permanganate ion, MnO_4^-, is a typical ultraviolet-visible absorption spectrum. Molecular spectra and spectra of polyatomic ions such as the permanganate ion, MnO_4^-, consist of bands of closely spaced lines although, in most cases, the individual lines making up the bands are not visible (see Figure 10.39). The *peaks* in an ultraviolet-visible absorption spectrum are called **bands.**

Absorption of electromagnetic radiation of ultraviolet and visible wavelengths by molecules and polyatomic ions in their ground states results in promotion of valence electrons to higher energy levels. If the difference in energy between the highest molecular orbital that is occupied in the ground state and the lowest unoccupied molecular orbital is small, a molecule or polyatomic ion absorbs some wavelengths of visible light. The other wavelengths are transmitted, and the species is colored. If the difference in energy between the highest molecular orbital that is occupied in the ground state and the lowest unoccupied molecular orbital is high, the species absorbs in the ultraviolet. All wavelengths of visible light are transmitted, and the species is colorless.

Instruments for measuring how much ultraviolet and visible radiation is absorbed at different wavelengths are relatively inexpensive. The quantity of energy absorbed by a given amount of sample is usually large so that low concentrations can be measured accurately. Ultraviolet-visible spectroscopy is ideally suited to the routine

The photoelectric effect (Section 7.6) is not limited to visible light and ultraviolet radiation, which knock valence electrons out of materials. X-ray photons have more energy than photons of ultraviolet and visible light and can knock core electrons out of a sample. If the energy of an X-ray photon is greater than the minimum energy required to knock an electron out of a sample, the ejected electron will move with a kinetic energy equal to the excess energy of the photon:

energy of photon

= energy required to knock electron out of atom

+ kinetic energy of the electron

The ejected electrons can be sorted according to their kinetic energies, and the number of electrons having each kinetic energy can be counted. A graph of the number of electrons having a given kinetic energy against kinetic energy is called a photoelectron spectrum. ➡Figure 10.40

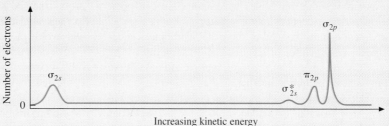

➡ FIGURE 10.40   Photoelectron spectrum of N_2. There is not room on the page to include the σ_{1s} and σ_{1s}^* peaks. Photoelectron spectra were used to analyze rocks brought back from the moon by the Apollo mission.

shows the photoelectron spectrum of a sample of nitrogen molecules, N_2.

All the photons used to produce the photoelectron spectrum of nitrogen had the same energy. From Figure 10.40, you can see that all photoelectrons had one of a small number of kinetic energies. Therefore, in the nitrogen molecule, all electrons must have had one of a small number of energies. Thus, the photoelectron spectrum provides direct experimental evidence for the existence of electronic energy levels in the nitrogen molecule.

Each peak in Figure 10.40 is labeled with the symbol of the molecular orbital that the electrons came from. The kinetic energies observed for photoelectrons can be used to calculate the energies of molecular orbitals, and the results agree well with energies calculated by means of the Schrödinger equation. Photoelectron spectra can also be used to measure orbital energies of electrons in atoms.

analyses used to inspect raw materials and finished products and to control industrial processes. For example, ultraviolet-visible spectroscopy is used in the pharmaceutical industry to measure the amount of penicillin in antibiotics, in the food industry to measure the amount of nitrite in meat, and in steel production to determine the amount of manganese in steel. Environmental chemists use ultraviolet-visible spectroscopy to measure the concentration of lead and sulfur dioxide in air and the concentration of nitrate in groundwater. Ultraviolet-visible spectroscopy is also one of the basic tools of the clinical chemist, who uses it to determine glucose and cholesterol, and of the forensic scientist. Development of automated recording instruments linked with computers has made analysis by ultraviolet-visible spectroscopy fast and accurate as well as relatively inexpensive.

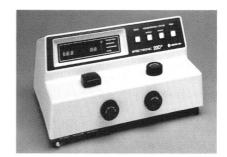

Spectronic 20s, which have a wavelength range of 340–625 nm, are widely used for routine testing and teaching.

SUMMARY

The shapes of molecules and ions that have main group elements as central atoms can be predicted from their Lewis structures by means of **valence shell electron pair repulsion (VSEPR).** According to the VSEPR model, electron pairs around a central atom tend to be located as far apart from each other in space as possible. The **total coordination number** of an atom is the total number of atoms *and* unshared electron pairs around a central atom. The shapes of molecules depend on the positions occupied by the atoms attached to the central atom. **Stereochemical formulas** must be used to show the shapes of three-dimensional species.

Bonds in which electron pairs are shared equally are called **nonpolar bonds.** Bonds in which electron pairs are not shared equally are called **polar bonds.** A pair of equal and opposite charges separated by a distance is called a **dipole.** The **dipole moment, μ,** is a quantitative measure of the polarity of a molecule. Dipole moments are **vectors**—that is, they have both magnitude and direction. All molecules that have one polar covalent bond are polar. Whether molecules with more than one polar bond are polar or not depends on their shape.

Approximations must be used to solve the Schrödinger equation for the wave functions of electrons in all molecules. The **valence bond method** is one method of approximating, and the **molecular orbital method** is another.

In the valence bond method, formation of bonds is pictured as resulting from overlap of atomic orbitals so that electrons move in the region of space between two nuclei. The valence bond method uses Lewis formulas including resonance structures to describe molecules and polyatomic ions. **Hybrid orbitals,** orbitals formed by combination of atomic orbitals on the same atom, are needed to explain the number of bonds that some atoms form and the directions in space of the bonds. The number of hybrid orbitals formed must be the same as the number of atomic orbitals combined.

Sigma (σ) bonds are pictured as formed by endwise (head-on) overlap of atomic orbitals on different atoms. All single bonds are sigma bonds. Rotation about single bonds is free; the different arrangements of parts of a molecule in space that result from rotation about sigma bonds are called **conformations. Pi (π) bonds** are pictured as formed by sidewise overlap of atomic orbitals on different atoms. Multiple bonds consist of one sigma bond and one or more pi bonds. Rotation about pi bonds is not possible unless the pi bond is broken. Pi bonding is very important in the chemistry of the elements in the second period of the periodic table.

Binding energy curves show how the energy of the system changes as atoms move toward each other. For a stable bond to form, there must be a minimum in the binding energy curve; the distance between the atoms at the minimum is the bond length. The difference in energy between the energy at the minimum and the energy of the separate atoms is the **bond dissociation energy,** the energy required to break a specific bond in a certain molecule.

The molecular orbital method supposes that electrons in molecules occupy molecular orbitals that may extend over the whole molecule. The energies of the molecular orbitals are calculated using the Schrödinger equation. Electrons are then assigned to the molecular orbitals using the aufbau approach and following the same rules (Hund's rule and the Pauli exclusion principle) followed in placing electrons in atomic orbitals. The molecular orbital method successfully explains the paramagnetism of the oxygen molecule, the properties of solids, and molecular spectra and avoids the need for resonance structures.

The atomic orbitals that are combined to form a molecular orbital come from different atoms. The atomic orbitals that are combined to form a molecular orbital must occupy the same region in space and have similar energies; the number of molecular orbitals formed must be equal to the number of atomic orbitals combined. The energy of a **bonding molecular orbital** is lower than the energies of the atomic orbitals that were combined to form the molecular orbital. Electron density is concentrated between the bonded atoms. The energy of an **antibonding molecular orbital** is higher than the energies of the atomic orbitals that were combined to form the molecular orbital. There are nodes in electron density between atoms and the atoms are not bonded.

If three or more p orbitals are perpendicular to adjacent atoms that lie in the same plane, the electrons of the pi bonds formed by sidewise overlap are **delocalized** and move over three or more atoms. Structures with delocalized electrons always have lower energy (greater stability) than similar structures without delocalized electrons.

Bonding in solids can be understood by thinking about what happens when many atoms are brought together to form a crystal. The molecular orbital energy levels are so closely spaced that they appear continuous. A group of closely spaced molecular orbital energy levels is called a **band.** Bands are separated by spaces, called **forbidden bands,** where there are no allowed energies. A material with a partly filled energy band is a **conductor** of electricity and thermal energy. All metals are conductors. Most nonmetals and most compounds are **insulators**—that is, they are poor conductors of electricity and thermal energy. For insulators, the highest occupied band is filled. The forbidden band just above the highest filled band is wide. In a **semiconductor,** the forbidden band just above the highest filled band is narrow. The thermal energy available at room temperature is enough to raise some electrons from the highest occupied band (called the **valence band**) to the next permitted band (called the **conduction band**).

ADDITIONAL PRACTICE PROBLEMS

For information about the organization of Additional Practice Problems, Stop & Test Yourself, Putting Things Together, and Applications, see the beginnings of these sections in Chapter 1.

10.18 Why can the shape of a species for which two or more resonance structures must be drawn be predicted from any one of the resonance structures? (10.1)

10.19 Without referring to Table 10.1, predict the arrangement of the electron pairs around the central atom and the molecular shape of each of the following: (a) $AsCl_3$, (b) AsF_5, (c) BrF_3, (d) BrF_5, (e) ClF_4^+, (f) ClO_4^-, (g) $CuCl_2^-$ (this ion is diamagnetic), (h) HCl, (i) ICl_2^+, (j) PCl_6^-, (k) $SnCl_2$, (l) SO_3, (m) XeF_2, (n) XeF_4. (10.1)

10.20 BF_3, NH_3, and ClF_3 all have the general formula AB_3, but the shapes of these molecules are quite different. (a) Without referring to Table 10.1, predict the shapes of these molecules. (b) Name each molecular shape. (10.1)

10.21 The F—Cl—F bond angles in ClF_3 are only 80° instead of the 90° shown in Table 10.1. Why? (10.1)

10.22 For a molecule to be polar, what must be true of the structure of the molecule? (10.2)

10.23 Of what use is a knowledge of the dipole moments of molecules? (10.2)

10.24 Indicate, with a crossed arrow $\mapsto$, the polarity of each polar bond and the net polarity of the molecule, if any, for each of the following: (a) Cl_2, (b) ClF, (c) ClF_3, (d) XeF_4, (e) $TeCl_4$. (10.2)

10.25 How do the dipole moments of CS_2 and OCS compare? (10.2)

10.26 Hydrogen peroxide, H_2O_2, has a dipole moment. (a) Write a Lewis formula for H_2O_2. (b) Which of the bonds in H_2O_2 are polar? (c) Sketch a possible shape for the H_2O_2 molecule. (10.2)

10.27 (a) Which of the following compounds is nonpolar? (b) For the polar molecules, show the direction of the dipole moment with an arrow. (10.2)

(a)   (b)   (c)

10.28 In binding energy curves (see, for example, the lower curve in Figure 10.6), why does energy first decrease, then reach a minimum, and finally increase rapidly as two atoms move toward each other? (10.3)

10.29 (a) What is a hybrid orbital? In your answer, use a hybrid orbital of carbon as an example. (b) Why is the idea of hybrid orbitals used? (10.4)

10.30 Without referring to this book or to your notes, tell which type of hybrid orbital is used to describe each of the following shapes: (a) linear, (b) octahedral, (c) tetrahedral, (d) trigonal bipyramidal, (e) trigonal planar. (10.4)

10.31 Without referring to this book or to your notes, answer the following questions: (a) If an atom is *sp* hybridized, how many pure *p* orbitals remain in the shell? (b) How many pi bonds can an *sp* hybridized atom form? (c) How many sigma bonds? (d) What is the maximum number of types of hybrid orbitals that a carbon atom can form? Explain your answer. (10.4)

10.32 Use orbital diagrams to show the silicon atom in its ground state and hybridized as it is pictured when bonded to form SiH_4. (10.4)

10.33 (a) Predict the angles between H—N bonds in the ammonium ion. (b) What type of hybrid orbital is used to picture this geometry? (c) Sketch a picture like Figure 10.11(a) showing the formation of an ammonium ion from an ammonia molecule and a hydrogen ion. (10.4)

10.34 The Lewis structure for $[SnCl_6]^{2-}$ is

(a) How many *d* orbitals are there in the hybrid orbitals used to describe the Sn—Cl bonds? (b) What are these hybrid orbitals called? (c) How many of them are there? (10.4)

10.35 What hybridization would be pictured for the underlined atom in each of the following species? (a) $\underline{C}O_2$ (b) $\underline{I}F_6^+$ (c) $\underline{Sn}Cl_5^-$ (d) $\underline{Ge}F_4$ (10.4)

10.36 Sketch pictures like the one in Sample Problem 10.4 showing the formation of each of the species in the previous problem. (10.4)

10.37 For each of the species in Problem 10.35, tell (a) the number of pi bonds, (b) the number of sigma bonds. (10.4)

10.38 What kinds of orbitals overlap to form each of the bonds in the species in Problem 10.35? (10.4)

10.39 Predict the angles between the bonds in each of the species in Problem 10.35. (10.4)

10.40 Of the species in Problem 10.35, which has all atoms in a single plane? (10.4)

10.41 Why is rotation about pi bonds not free? (10.4)

10.42 (a) Why are the energies of electrons in bonding molecular orbitals formed by combination of two atomic orbitals, such as the energy of the two electrons in the bonding molecular orbital of the hydrogen molecule, lower than the energies of the electrons in their original atomic orbitals? (b) Why is the higher energy molecular orbital formed by combination of two atomic orbitals referred to as an antibonding orbital? (10.6)

10.43 (a) How many molecular orbitals must result from combination of six atomic orbitals? (b) What is the maximum number of electrons that one molecular orbital can hold? (10.6)

10.44 Without referring to this book or to your notes, (a) draw the molecular orbital energy level diagram for the He_2^+ ion in its ground state. (b) Write the electron configuration of the He_2^+ ion in its ground state. (c) What is the bond order of the helium–helium bond? (d) Is He_2^+ paramagnetic or diamagnetic? (If paramagnetic, how many unpaired electrons are there?) (10.6)

10.45 (a) Using the appropriate molecular orbital energy level sequence from Figure 10.28, draw the molecular orbital energy level diagram for the C_2 molecule in its ground state. (b) Write the electron configuration of the C_2 molecule in its ground state. (c) What is the bond order of the carbon–carbon bond? (d) Is C_2 paramagnetic or diamagnetic? (If paramagnetic, how many unpaired electrons are there?) (10.6)

10.46 According to the molecular orbital theory, why is neon gas monatomic? (10.6)

10.47 According to the molecular orbital theory, which species has the shorter fluorine–fluorine bond, F_2 or F_2^{2+}? (10.6)

10.48 According to the molecular orbital theory, which species has the strongest bond, B_2, B_2^-, or B_2^{2-}? (10.6)

10.49 Which homonuclear diatomic species would have each of the following electron configurations? (a) $(\sigma_{1s})^2(\sigma_{1s}^*)^2$ (b) $(\sigma_{1s})^2(\sigma_{1s}^*)^2(\sigma_{2s})^2(\sigma_{2s}^*)^2(\pi_{2p_x})^1(\pi_{2p_y})^1$ (10.6)

10.50 Measurements of the magnetic properties of B_2 show that it is paramagnetic with two unpaired electrons. How does this indicate that the π_{2p_x} and π_{2p_y} orbitals are lower in energy than the σ_{2p_z} orbital in the B_2 molecule? (10.6)

10.51 (a) Sketch the molecular orbital energy-level diagram expected for the CN^- ion in the ground state. (b) Write the electron configuration of the CN^- ion in the ground state. (c) What is the bond order of the carbon–nitrogen bond? (d) Is the CN^- ion in its ground state paramagnetic or diamagnetic? (If paramagnetic, how many unpaired electrons are there?) (10.6)

10.52 (a) What are delocalized electrons? (b) What structural feature must be present for delocalization? (10.7)

10.53 What effect does the presence of delocalized electrons have on the stability of a molecule or ion? (10.7)

10.54 Over how many atoms are the pi electrons in *cis*-retinal delocalized? (10.7)

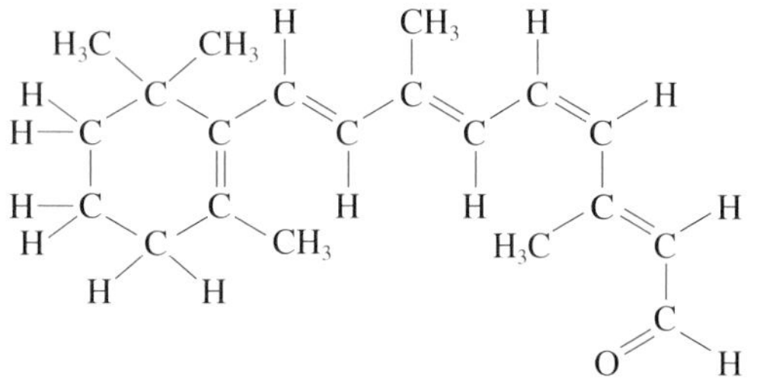

cis-retinal

10.55 According to the band theory, what is the difference between conductors, semiconductors, and insulators? (10.8)

10.56 As temperature increases, how do the conductivities of metals and semiconductors change? Explain why they change as they do. (10.8)

10.57 Are the electrons in allene, $H_2C{=}C{=}CH_2$, delocalized? Explain your answer.

10.58 Sketch a picture like the one in Sample Problem 10.4 showing the formation of an H_3O^+ ion from a water molecule and a hydrogen ion. (Note that *both* electrons in the new H—O bond come from the oxygen atom.)

10.59 Draw the molecular orbital energy–level diagrams for the Li_2 molecule, the Be_2^+ ion, and the Be_2 molecule in their ground states. Compare the stabilities of these three species.

10.60 (a) What is the angle between bonds in propane, $CH_3CH_2CH_3$? (b) In cyclopropane (sketched below), the C—C—C bond angles are only 60°. In terms of orbital overlap, explain why the carbon–carbon bonds in cyclopropane are weaker than the carbon–carbon bonds in propane.

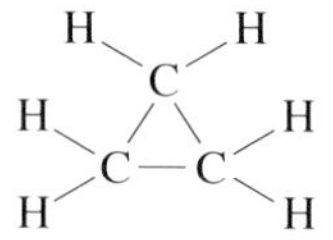

cyclopropane

10.61 What change, if any, in hybridization of the underlined atom takes place in each of the following reactions?

(a) $HC{\equiv}\underline{C}H(g) + 2H_2(g) \xrightarrow{\text{catalyst}} H_3C{-}\underline{C}H_3(g)$

(b) $CH_3\underline{C}H_2OH(l) \xrightarrow{\text{acid, heat}} H_2\underline{C}{=}CH_2(g) + H_2O(l)$

(c) $CH_3\underline{C}H_2OH(l) + HBr(\text{conc}) \longrightarrow$
$\qquad\qquad\qquad\qquad CH_3CH_2Br(l) + H_2O(l)$

(d) $\underline{P}Cl_3(l) + Cl_2(g) \longrightarrow PCl_5(s)$

(e) $\underline{Xe}F_2(s) + F_2(g) \longrightarrow XeF_4(s)$

10.62 A node is one of the characteristics of an antibonding molecular orbital. What kind of bonding orbital also has a node?

10.63 Comparing the stereochemical formulas in each set, which are the same?

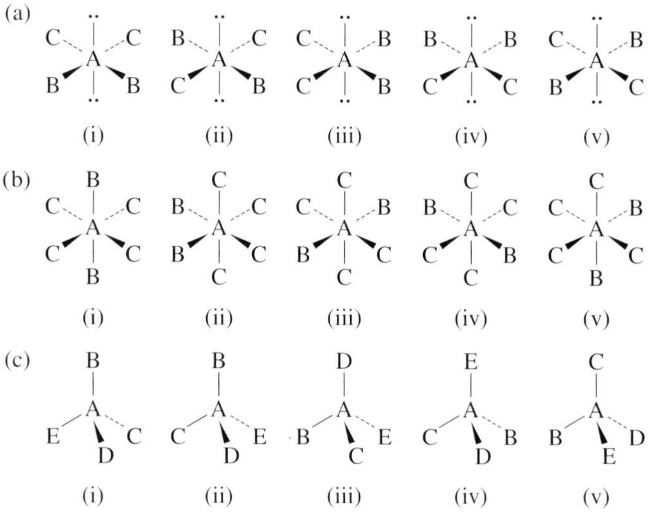

(a)

(i)   (ii)   (iii)   (iv)   (v)

(b)

(i)   (ii)   (iii)   (iv)   (v)

(c)

(i)   (ii)   (iii)   (iv)   (v)

10.64 Why does silicon not commonly form double bonds with oxygen as carbon does?

Use the Lewis formula for xenon tetrafluoride below to answer questions 1 and 2.

$$F-\overset{..}{\underset{/\ \ \backslash}{Xe}}-F$$
F   F

1. In a molecule of xenon tetrafluoride, the arrangement of electron pairs would be described as (a) octahedral   (b) square planar   (c) square pyramidal   (d) tetrahedral   (e) trigonal bipyramidal

2. The shape of a molecule of xenon tetrafluoride would be described as (a) octahedral   (b) square planar   (c) square pyramidal   (d) tetrahedral   (e) trigonal bipyramidal

3. Which of the following molecules has zero dipole moment? (a) CH_2Cl_2 (b) H_2S (c) NH_3 (d) SO_2 (e) SO_3

4. A set of sp^2 orbitals is (a) linear   (b) octahedral   (c) tetrahedral   (d) trigonal bipyramidal   (e) trigonal planar

5. The number of orbitals in a set of sp^3d hybrid orbitals is (a) 2   (b) 3   (c) 4   (d) 5   (e) 6

6. The central atom in $CHCl_3$ is pictured as _____ hybridized.
(a) sp (b) sp^2 (c) sp^3 (d) sp^3d (e) sp^3d^2

Use the structural formula below to answer questions 7–10.

$$H-\overset{1}{C}\equiv\overset{2}{C}-\overset{3}{\underset{\underset{H}{|}}{C}}=\overset{4}{\underset{\underset{H}{|}}{C}}-\overset{5}{\underset{\underset{H}{|}}{\overset{\overset{H}{|}}{C}}}-H$$

7. The total number of pi bonds in the molecule is
(a) 1   (b) 2   (c) 3   (d) 4   (e) 5

8. The total number of sigma bonds in the molecule is
(a) 8   (b) 9   (c) 10   (d) 11   (e) 12

9. The H C H bond angle on C-5 is closest to
(a) $60°$ (b) $90°$ (c) $120°$ (d) $109.5°$ (e) $180°$

10. The shortest carbon–carbon bond is the one between
(a) C-1 and C-2   (b) C-2 and C-3
(c) C-3 and C-4   (d) C-4 and C-5
(e) All the carbon–carbon bonds are the same length.

11. According to the molecular orbital theory, which of the following is *not* a stable species?
(a) Be_2 (b) C_2^{2-} (c) Li_2 (d) NO   (e) O_2^+

12. According to the molecular orbital theory, the species with bond order three is
(a) B_2 (b) C_2 (c) N_2 (d) O_2 (e) F_2

13. Which of the following compounds does *not* have delocalized electrons?

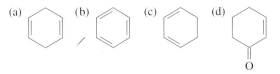

(e) All of them do.

14. All metals are _____; their conductivity _____ with decreasing temperature.
(a) conductors, increases   (b) conductors, decreases
(c) conductors, does not change   (d) semiconductors, increases
(e) insulators, increases

PUTTING THINGS TOGETHER

10.65 (a) Write the Lewis formula, including formal charges, for the triiodide ion, I_3^-.   (b) Explain why F_3^- does not exist under normal conditions although Cl_3^- and Br_3^- are known. (c) What shape is the triiodide ion?

10.66 (a) Use stereochemical formulas to show the changes in molecular shape that accompany the reaction

$$2ClF(g) + AsF_5(g) \longrightarrow [FCl_2]^+[AsF_6]^-$$

(b) What is the physical state of the product? Explain your answer.

10.67 The oxygen molecule, O_2, can gain one electron to form the superoxide ion, two electrons to form the peroxide ion, or lose one electron to form the dioxygenyl ion. For each of these ions, (a) write the formula.   (b) Draw the molecular

orbital energy-level diagram for the ground state.   (c) Write the electron configuration of the ground state.   (d) What is the bond order of the oxygen–oxygen bond?   (e) Tell whether the species is paramagnetic or diamagnetic and, if it is paramagnetic, the number of unpaired electrons.

10.68 How does the molecular orbital energy level diagram for a heteronuclear diatomic molecule differ from the diagram for a homonuclear diatomic molecule?

10.69 Describe the bonding in the formate ion, $HCOO^-$, using a combination of localized and delocalized orbitals.

10.70 Make a diagram similar to Figure 10.38 showing the energy bands of solid (a) Bi, (b) P_4, (c) Sb, (d) NaCl.

10.71 Use bond energies from Table 9.4 to estimate the energy needed for rotation about a carbon–carbon double bond.

10.72 Discuss the structure of the carbonate ion in terms of (a) resonance, (b) molecular orbital theory. (Use a combination of localized and delocalized electrons.)

10.73 What is the difference between a resonance hybrid and a hybrid orbital?

10.74 What is the difference between bond dissociation energy and bond energy?

10.75 With what two common molecules is the carbide ion, C_2^{2-}, isoelectronic?

10.76 In ammonium nitrate, (a) how many covalent bonds are there? (b) How many ionic bonds? (c) How many pi bonds? (d) How many sigma bonds?

10.77 Compare the structures of O_2 predicted by the valence bond and molecular orbital theories. Which structure agrees with the properties observed by experiment? Explain your answer.

10.78 (a) How is bond order defined in terms of the molecular orbital theory? (b) How are bond order, bond length, and bond strength related?

10.79 (a) If three atomic orbitals are combined to form a hybrid orbital, how many hybrid orbitals are formed? (b) If three atomic orbitals are combined to form molecular orbitals, how many molecular orbitals are formed? (c) What is the maximum number of electrons that one hybrid orbital can hold? One molecular orbital? (d) What is the difference between a molecular orbital and a hybrid orbital?

10.80 Compare the valence bond theory and the molecular orbital theory.

10.81 Describe the bonding in O_3 using both the valence bond and the molecular orbital theories.

10.82 Arrange the three species of Problem 10.67 and the oxygen molecule in order of increasing bond length.

10.83 Arrange the three species of Problem 10.67 and the oxygen molecule in order of increasing bond strength.

10.84 How do the members of each of the following pairs differ, and in what ways are they similar? (a) molecular orbitals and atomic orbitals (b) bonding orbitals and antibonding orbitals (c) sigma bonds and pi bonds (d) a ground state and an excited state (e) sigma bonds and sigma orbitals

10.85 In Section 10.7, we learned that structures with delocalized electrons, such as benzene,

always have lower energies (greater stabilities) than similar structures without delocalized electrons. For the change,

$$C(s) \longrightarrow C(g) \qquad \Delta H° = 718.384 \text{ kJ/mol}$$

(a) Use this information and bond energies from Table 9.4 to estimate $\Delta H°$ for the reaction

$$6C(s) + 3H_2(g) \longrightarrow C_6H_6(g)$$

(b) The standard enthalpy of formation for gaseous benzene is 82.927 kJ/mol; that is, $6C(s) + 3H_2(g) \longrightarrow C_6H_6(g)$ $\Delta H° = 82.927$ kJ/mol. What is the difference between the estimated and the experimental values for $\Delta H°$? This difference is referred to as the resonance energy of the benzene molecule.

(c) Draw the other resonance structure for benzene.

(d) Sketch the molecular orbital model of benzene.

10.86 Under ordinary conditions, barium metal reacts with oxygen from the air to form barium oxide. At 500 °C, barium peroxide is formed, and there is some evidence for the existence of a superoxide of barium. (a) Use electron configurations to show the formation of barium oxide from barium and oxygen atoms. (b) Write Lewis formulas for the peroxide and superoxide ions (see Problem 10.67). (c) Which of the species in this problem has an odd electron? (d) For which species should resonance structures be written? (e) Draw molecular orbital energy-level diagrams for the peroxide and superoxide ions and write their electron configurations. (f) What type of measurement should show which theory is better? (g) Write molecular equations for the formation of the three oxides of barium from barium metal and oxygen gas.

APPLICATIONS

10.87 The industrial production of nitric acid, HNO_3, involves reaction of nitrogen from air with hydrogen to form ammonia and reaction of ammonia with oxygen from air. Use hybrid orbitals to describe the bonding in N_2, NH_3, and HNO_3.

10.88 Until about 1950, the reaction of calcium carbide with water

$$CaC_2(s) + 2H_2O(l) \longrightarrow HC \equiv CH(g) + Ca(OH)_2(s)$$

was the only industrial method for making acetylene. (a) Draw the molecular orbital energy-level diagram for the carbide ion in its ground state. (b) Write the electron configuration of the carbide ion in its ground state. (c) What is the bond order of the carbon–carbon bond in the carbide ion? (d) How many sigma bonds and how many pi bonds are there?

(e) How is the bond order of C_2 changed by the addition of two electrons? (f) Is the carbide ion diamagnetic or paramagnetic? (If paramagnetic, how many unpaired electrons are there?)

10.89 The ions O_2^+ and N_2^+ are found in Earth's upper atmosphere. For each of these ions, (a) sketch the molecular orbital energy level diagram. (b) Write the electron configuration. (c) Predict the bond order. (d) Tell whether the ion is paramagnetic or diamagnetic. If it is paramagnetic, how many unpaired electrons are there?

10.90 The compound carbon monoxide is a common air pollutant, especially in cities and tunnels. (a) Assuming that the molecular orbital energy-level diagram for the carbon monoxide molecule is similar to that of NO, sketch a molecular orbital energy level diagram for carbon monoxide. (b) According to the molecular orbital theory, what is the bond order in carbon monoxide? (c) How does the bond order predicted by molecular orbital theory compare with the bond order predicted by valence bond theory? (d) Qualitatively, in what respect would you expect the molecular orbital energy level diagram for carbon monoxide to differ from that of NO? Explain your answer.

10.91 The compound $FClO_3$ has been considered for use as a rocket propellant. (a) Use a sketch to show the bonding in this compound in terms of hybrid orbitals. (b) Use formal charges to show why this compound is very reactive.

10.92 There is only one compound with structural formula $HOOCCH_2CH_2COOH$ (A), but two with structural formula $HOOCCH{=}CHCOOH$ (B). The ability of living systems to detect differences in molecular shape is shown by the fact that one of the two compounds (B) is essential to vegetable and animal tissue respiration and is used in soft drinks and baking powders while the other is a strong irritant. (a) What hybridization is pictured for each carbon in (B)?

$$HOCCH{=}CHCOH$$
$$\overset{\|}{O} \qquad \overset{\|}{O}$$

(b) How many pi bonds are there in $HOOCCH{=}CHCOOH$? How many sigma bonds? (c) Explain the existence of two compounds with structural formula $HOOCCH{=}CHCOOH$ by writing stereochemical formulas for them. Show approximate bond angles and relative bond lengths. (d) What type of measurement would distinguish between the two compounds in part (c)? (e) Either one of the two compounds, $HOOCCH{=}CHCOOH$, can be converted to a mixture of both of them by heating. Explain. (f) Over how many atoms are electrons delocalized in the two compounds $HOOCCH{=}CHCOOH$? (g) Describe the bonding in one of the two compounds $HOOCCH{=}CHCOOH$ using a combination of localized and delocalized electrons.

10.93 The colored material in the atmosphere of Jupiter is a polymer of carbon suboxide, C_3O_2. (a) Write the Lewis formula for carbon suboxide. (b) What is the hybridization of the carbon atoms? (c) What are the bond angles? (d) What is a polymer? (e) What is the building block of a polymer, such as carbon suboxide, called?

10.94 The product of the reaction

$$2HF + SbF_5 \longrightarrow [H_2F]^+[SbF_6]^-$$

which takes place when antimony pentafluoride is dissolved in liquid HF, is known as a "superacid." Write Lewis formulas for and predict the shapes of the reactants and products.

10.95 How fast greenhouse warming develops will depend on how fast the surface layers of the ocean mix with the deeper regions and carry away atmospheric thermal energy and carbon dioxide. Oceanographers used SF_6, which can be easily and quickly detected at concentrations of less than 1 g in 1 km^3 of seawater, as a tracer to measure mixing in the eastern Atlantic. (a) Draw a stereochemical formula for SF_6. (b) What concentration in ppb by mass can be detected? How many molecules per milliliter of seawater is this? Use 1.02 g/mL as the density of seawater. (c) A total of 139 kg SF_6 were used. This compound, which is a gas, is made by reacting sulfur with fluorine. Assuming an excess of sulfur is used and a 100% yield is obtained, how many kilograms of fluorine are required to make 139 kg SF_6? (d) What is SF_6 called? (e) What chemical properties must SF_6 have to be suitable for this application?

10.96 Peroxyacetylnitrate, $C_2H_3NO_5$ (PAN), is one of the most irritating compounds in smog. The arrangement of the atoms in PAN is

(a) Finish writing the Lewis formula for PAN including resonance structures, if any, and formal charges.

(b) Sketch a stereochemical formula and show the bond angles on it. Show the relative lengths of the various bonds.

10.97 Polar molecules are responsible for the absorption of microwave energy by food items. (a) Which of the following substances absorb microwaves? CS_2, CCl_4, H_2O, $CH_3CH_2CH_2OH$, $CH_3(CH_2)_{14}COOCH_3$ (a fatlike molecule)? (b) Under similar conditions in a microwave oven, the temperature of $CH_3CH_2CH_2OH$ rises much faster than the temperature of water. Explain why. (c) Write the Lewis formula for $CH_3CH_2CH_2OH$.

10.98 Comets are the least changed samples of early solar system ices available for study. Estimates of the abundance of molecular nitrogen in comets are based largely on detection of N_2^+ in the comae. (a) Draw the molecular orbital energy-level diagram for the ground state of N_2^+. (b) Write the electron configuration for N_2^+. (c) Is N_2^+ diamagnetic or paramagnetic? If it is paramagnetic, how many unpaired electrons are there? (d) What is the bond order?

10.99 The compound, β-carotene

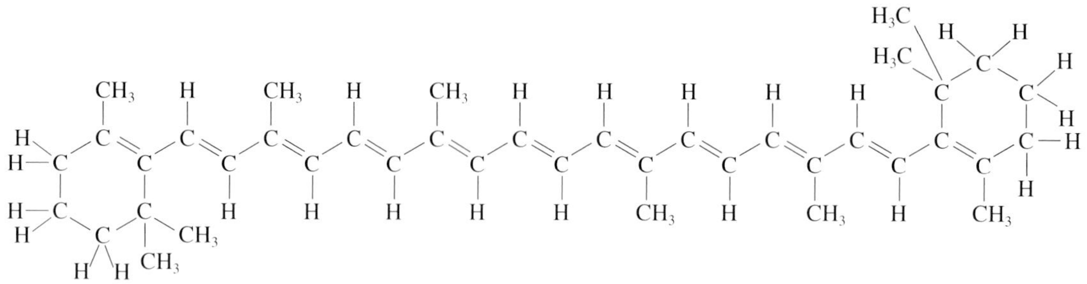

β-carotene

which can be obtained from carrots, is used as a yellow coloring in foods and as a sunscreen agent and is a precursor of vitamin A in all species except cats.

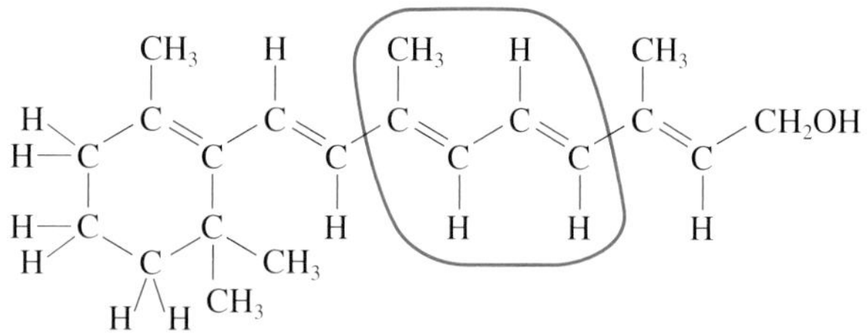

vitamin A

(a) To which retinal, *cis-* or *trans-*, are vitamin A and β-carotene related? (b) Over how many atoms are pi elec- trons delocalized in vitamin A? In β-carotene? (c) How many atoms in vitamin A must lie in a plane? (d) Describe the bonding in the circled part of the vitamin A structure using a combination of localized and delocalized electrons. Tell which hybrid orbitals overlap to form each of the bonds. (e) The weakest pulse of light that a human eye can detect sends about 100 photons through the pupil and converts the *cis*-retinal in about 15 molecules of rhodopsin to *trans*-retinal. If the wavelength of the light is 550 nm, how much energy is available from 100 photons?

A Biologist Looks at Chemistry

DENOLA M. BURTON

Associate Toxicologist
Eli Lilly Research Laboratories

B.S. Biology
Jarvis Christian College

M.S. Biology
Texas Southern University

My mother was a nurse, and from the time I was a little girl, I can remember wanting to work in the medical profession. I worked hard toward that goal—health careers curriculum in high school; jobs in hospitals and nursing homes; pre-med courses in college, including biology, chemistry, and physics; and special research programs (I was a student collaborator at Brookhaven National Laboratory and a student researcher at Woods Hole Marine Biological Laboratory). All were chosen to guide me in a scientific direction.

Over the years, my goals in life changed, and I started looking at other career options. While I was in college, I began exploring other possible scientific careers, which led me to expand my pre-med studies to include more chemistry and physics. At the time of my graduation, I had enough chemistry hours to have been either a chemistry or a biology major. Nevertheless, I was accepted into the graduate program in biology at Texas Southern University and that is the area I pursued.

I completed my master's degree in biology with an emphasis on immunogenetics. Then came the real world. I began working in a histocompatibility laboratory in Houston where I did blood studies for patients with leukemia. Then I worked in a rheumatology laboratory where I conducted research on various diseases similar to arthritis. Subsequently, I moved back to Indianapolis to become the coordinator of the Reproductive Endocrinology Clinical Programs where we studied male and female infertility.

I know my previous work experiences didn't involve chemistry, but look where I am now—associate toxicologist for one of the largest pharmaceutical companies in the world, Eli Lilly and Company in Greenfield, Indiana! Chemistry is a part of everyday life here at Eli Lilly. The development of new chemicals or medicines is based on chemistry and chemical reactions, assays and properties of substances, synthesis and degradations.

In the Toxicology Division, we test the compounds produced by Eli Lilly chemists to determine the toxic effects of high doses. My area of expertise lies in the reproduction phase where we test compounds to determine if they have any effect on fertility. I design studies involving laboratory animals to provide information on gonadal function, estrous cycles, mating behavior, conception, parturition, lactation, and growth and development of progeny. I then analyze the data and write reports that are reviewed and submitted to various regulatory agencies before the compound becomes available to physicians and their patients.

When I design reproduction studies, I need to determine the amount of test chemical needed for a particular study. This is where chemistry can be involved. Formulas must be used to determine the amount of test chemical needed as well as the concentration of that chemical. I need to be familiar with the compounds, their structures, and their properties in order to draw some conclusions about the specifics of a study. Sometimes a dose-ranging study must be conducted before an actual reproduction study can begin. All this sounds like chemistry, doesn't it? A basic background in chemistry is necessary just to identify the compounds we work with daily.

Another way that we continue to learn more about toxicology, pharmaceuticals, and chemistry in general is through our seminars. Project Reviews, the Toxicology Monthly Seminar, and Tox Topics are training lectures where group leaders, project leaders, and scientist supervisors review the status of ongoing projects, learn more about new technologies, and become aware of new developments in the world of pharmaceuticals. These lectures and discussions are very interesting and feature some of the top scientists from Lilly and around the country. They challenge us and give us ideas about our own involvement in the Toxicology Division and the scientific community.

Because my educational efforts were centered on biology, my major job responsibilities still lie in that area. But thanks to my chemistry background, I am able to function at a higher level where chemistry is involved as well. As I look back, I am very glad that I chose my courses well and planned the direction of my career.

11 OXIDATION–REDUCTION REACTIONS

Wrought iron (iron shaped by hammering) has been used for decoration since the Middle Ages.

The term *oxidation* originally referred to reaction with oxygen. Oxygen gas is very reactive, and most substances react with it. For example, the metal magnesium, the nonmetal sulfur, and the compound methyl alcohol all react with oxygen in air:

$$2Mg(s) + O_2(g) \longrightarrow 2MgO(s)$$

$$S(s) + O_2(g) \longrightarrow SO_2(g)$$

$$2CH_3OH(l) + 3O_2(g) \longrightarrow$$

$$2CO_2(g) + 4H_2O(l)$$

All of these are oxidation reactions. Oxygen gas, the *substance that brings about oxidation,* is referred to as an **oxidizing agent.**

Reduction at first meant removal of oxygen. For example, the reactions

$$NiO(s) + C(s) \longrightarrow Ni(s) + CO(g)$$

$$WO_3(s) + 3H_2(g) \longrightarrow W(s) + 3H_2O(g)$$

$$5CaO(s) + 2Al(l) \longrightarrow$$

$$3Ca(g) + Ca_2Al_2O_5(s)$$

which are used to obtain the metals nickel, tungsten, and calcium from their ores, are reduction of the metal oxides. *Substances that bring about reduction,* such as C, H_2, and Al, are called **reducing agents.** Notice that in each reaction the reducing agent is oxidized (combined with oxygen).

— *Why is a study of oxidation–reduction reactions a part of general chemistry?*

All the combustion reactions by which we convert energy stored in fossil fuels into thermal energy to heat our houses and our dinners and into mechanical energy to move our cars are oxidation–reduction reactions. Both the production of metals from their ores and the corrosion of metals involve oxidation–reduction reactions. In addition, the original definitions of oxidation and reduction have been broadened to include a large number of reactions of great importance in chemistry, geology, and biology. For example, photosynthesis, the process by which green plants use the sun's energy to convert carbon dioxide to carbohydrates*

$$x\text{CO}_2(g) + y\text{H}_2\text{O}(l) \xrightarrow{\text{sunlight}}$$
$$\text{C}_x(\text{H}_2\text{O})_y(aq) + x\text{O}_2(g)$$

is an oxidation–reduction reaction. Metabolism also involves oxidation–reduction; many biochemical reactions are oxidation–reduction reactions.

Spontaneous oxidation–reduction reactions, such as those that take place in batteries, are used to supply electrical energy. Electrical energy can be used to make nonspontaneous oxidation–reduction reactions, such as the electrolysis of water (Section 1.2), occur. Electrolysis is used to produce some metals, such as aluminum, from their ores and to purify other metals, such as copper. The *production of electricity by chemical changes*

and the chemical changes produced by electricity make up an important branch of chemistry called **electrochemistry** (Chapter 19).

— What do you already know about oxidation–reduction reactions?

Not only combustion reactions but also single-replacement reactions such as the reactions between metals and water,

$$2\text{Na}(s) + 2\text{H}_2\text{O}(l) \longrightarrow$$
$$\text{H}_2(g) + 2\text{NaOH}(aq)$$

between metals and acids,

$$\text{Zn}(s) + 2\text{HCl}(aq) \longrightarrow$$
$$\text{H}_2(g) + \text{ZnCl}_2(aq)$$

between metals and the ions of less active metals,

$$\text{Cu}(s) + 2\text{Ag}^+ \longrightarrow 2\text{Ag}(s) + \text{Cu}^{2+}$$

and between nonmetals and the ions of less reactive nonmetals,

$$\text{Cl}_2(aq) + 2\text{Br}^- \longrightarrow 2\text{Cl}^- + \text{Br}_2(aq)$$

that we studied in Section 4.5 are oxidation–reduction reactions. Combination

Rusted iron. About 20% of the iron and steel produced in the United States each year is used to replace objects ruined by oxidation of iron to rust.

reactions that have elements as reactants such as the reaction between nitrogen gas and hydrogen gas to form ammonia gas

$$\text{N}_2(g) + 3\text{H}_2(g) \longrightarrow 2\text{NH}_3(g)$$

and decomposition reactions that have elements as products such as the electrolysis of water

$$2\text{H}_2\text{O}(l) \xrightarrow{\text{electricity}} 2\text{H}_2(g) + \text{O}_2(g)$$

are also oxidation–reduction reactions.

Oxidation–reduction reactions occur every time you metabolize your food, as well as every time you drive your car. The reactions in the batteries that power your calculator and start your car are oxidation–reduction reactions.

---

*Compounds that have the general formula $\text{C}_x(\text{H}_2\text{O})_y$ and related compounds are called *carbohydrates* (carbo- for carbon and hydrate for water). Both starch and ordinary table sugar are carbohydrates. Carbohydrates are not really hydrates of carbon. For example, the simplest sugar has the structure

$$\text{HOCH}_2\text{CH(OH)CHO}$$

and is called glyceraldehyde. Photosynthesis is one of the most important chemical reactions on Earth. More than 10^{17} kJ of energy is stored by photosynthesis every year, and more than 10^{10} tons of carbon are converted into carbohydrates and other organic substances. Not only the food we eat, but all the fossil fuels, wood, paper, and natural fibers such as cotton also result from photosynthesis. Photosynthesis produces the oxygen that we breathe and that is needed to burn fuels.

OXIDATION NUMBERS

Many oxidation–reduction reactions can be recognized as such by the fact that an element is formed from a compound or a compound is formed from an element. To be able to classify other reactions as oxidation–reduction or not oxidation–reduction, we need to learn about oxidation numbers. Like formal charges (Section 9.8), oxidation numbers are tools for keeping track of electrons.

The **oxidation number** or **oxidation state** of a covalently bound element is the *charge the element would have if all the shared pairs of electrons in the Lewis formula for the species were transferred to the more electronegative atom.* For example, the Lewis formula for carbon monoxide is

$$:C \equiv O:$$

The electronegativity of oxygen (3.5) is greater than the electronegativity of carbon (2.5). If all the shared pairs of electrons in carbon monoxide were transferred to the more electronegative oxygen atom,

$$:C \overset{\equiv}{\longrightarrow} O:$$

the oxygen atom would have eight valence electrons, and the carbon would be left with just two valence electrons:

$$:C \quad ::O:$$

The oxygen atom, which originally had six valence electrons (oxygen is in Group VIA) and now has eight, has gained two electrons; the oxygen atom has an oxidation number of -2. The carbon atom, which originally had four valence electrons (carbon is in Group IVA) and now has two, has lost two electrons; the carbon atom has an oxidation number of $+2$.

Oxidation numbers of covalently bound elements are not real charges like the charges on ions; therefore, they cannot be measured experimentally. Only for simple monatomic ions such as the magnesium ion, Mg^{2+}, and the oxide ion, O^{2-}, and for elements such as O_2 and O_3 are oxidation numbers real charges that can be determined by experiment.

Although oxidation numbers, like formal charges, are bookkeeping devices for electrons, oxidation numbers are not the same as formal charges. For example, the oxidation number of carbon in carbon monoxide is $+2$, and the oxidation number of oxygen in carbon monoxide is -2. The formal charge on carbon in carbon monoxide is -1; the oxygen in carbon monoxide has a formal charge of $+1$.

Although oxidation numbers are not real charges, they are very useful because they help chemists and students of chemistry to organize a great many facts. Fortunately, it is not usually necessary to write a Lewis formula and look up the electronegativities of the bonded atoms in order to assign oxidation numbers. The oxidation numbers of the elements in many species can be calculated from the empirical formula of the species by means of four simple rules, which are based on relative electronegativities. Only if a species contains more than one element not covered by the rules is it sometimes necessary to use Lewis structures and electronegativities to assign oxidation numbers. The rules, which *must* be applied in order, are:

◗ The electronegativities of the elements are given in Figure 9.3.

◗ You should check the formal charges given for carbon and oxygen in carbon monoxide to be sure that they are correct.

Rules for Assigning Oxidation Numbers

Rules must be applied in order.
1. The sum of the oxidation numbers of all the atoms in a species must equal the net charge on the species.

2. In compounds, the oxidation numbers of the elements in Group IA are $+1$. The oxidation numbers of the elements in Group IIA are $+2$, and the oxidation numbers of boron and aluminum are $+3$. The oxidation number of fluorine is -1.

3. In compounds, the oxidation number of hydrogen is $+1$.

4. In compounds, the oxidation number of oxygen is -2.

Sample Problem 11.1 shows how to use the rules to assign oxidation numbers.

SAMPLE PROBLEM

11.1 What is the oxidation number of each atom in each of the following species? (a) O_3 (b) $HClO_4$ (c) H_2O_2 (d) AlH_4^- (e) KO_2 (f) $FePO_4$

Solution (a) The O_3 molecule has no net charge. Therefore, according to rule 1, the sum of the oxidation numbers of the oxygens in the O_3 molecule must be zero. Therefore, for O_3,

$$3(\text{oxidation number of oxygen}) = 0 \qquad \text{and}$$
$$\text{oxidation number of oxygen} = 0/3 = 0$$

(b) According to rule 1, the sum of the oxidation numbers of hydrogen, oxygen, and chlorine in the $HClO_4$ molecule must be zero. By rule 3, the oxidation number of hydrogen is $+1$ and by rule 4, the oxidation number of oxygen is -2. Therefore, for $HClO_4$,

$$(+1) + \text{oxidation number of chlorine} + 4(-2) = 0 \qquad \text{and}$$
$$\text{oxidation number of chlorine} = -4(-2) - (+1) = +7$$

Once you understand the process, you can shorten the procedure as follows. Write the formula for the species and, over the symbol for each element, write the oxidation numbers given by the rules:

$$\overset{+1}{H} \quad \overset{}{Cl} \quad \overset{-2}{O_4}$$

Then write the total oxidation number for each element under the symbol for the element:

$$\overset{+1}{\underset{+1}{H}} \quad \overset{}{Cl} \quad \overset{-2}{\underset{-8}{O_4}}$$

The sum of the oxidation numbers must equal the charge on the species, zero in this case. You should be able to see that the total oxidation number for chlorine equals $+7$:

$$\overset{+1}{\underset{+1}{H}} \quad \overset{}{\underset{+7}{Cl}} \quad \overset{-2}{\underset{-8=0}{O_4}}$$

Because the formula shows just one chlorine atom, the oxidation number of chlorine is equal to $+7$.

(c) The oxidation number of hydrogen is $+1$ by rule 3 (which must be applied before rule 4). Using the shortcut method,

$$\overset{+1}{\underset{+2}{H_2}} \quad \overset{}{O_2}$$

According to rule 1, the sum of the oxidation numbers of hydrogen and oxygen in H_2O_2 must equal zero. Therefore, the total oxidation number for oxygen must equal -2:

$$\overset{+1}{\underset{+2}{H_2}} \quad \overset{}{\underset{-2=0}{O_2}}$$

When not combined with atoms of another element, an element has an oxidation number of zero.

Note that oxidation numbers are on a per atom basis. The oxidation number of each atom in a species must be multiplied by the number of the particular kind of atom in the species to find the sum of the oxidation numbers.

Because the formula shows two oxygen atoms, the oxidation number of oxygen in hydrogen peroxide is equal to

$$-2/2 = -1$$

(d) The oxidation number of aluminum is $+3$ by rule 2 (which must be applied before rule 3):

$$\underset{+3}{(\overset{+3}{Al}\quad H_4)^-}$$

By rule 1, the sum of the oxidation numbers of aluminum and hydrogen in the AlH_4^- ion must equal -1; therefore, the total oxidation number for hydrogen must equal -4:

$$\underset{+3\qquad -4=-1}{(\overset{+3}{Al}\quad H_4)^-}$$

The formula shows four hydrogen atoms, and the oxidation number of hydrogen in AlH_4^- is equal to $-4/4 = -1$.

(e) The oxidation number of potassium is $+1$ by rule 2 (which must be applied before rule 4):

$$\underset{+1}{\overset{+1}{K}\quad O_2}$$

By rule 1, the sum of the oxidation numbers of potassium and oxygen in KO_2 must equal zero, and the total oxidation number for oxygen must equal -1:

$$\underset{+1\qquad -1=0}{\overset{+1}{K}\quad O_2}$$

The formula shows two oxygen atoms; the oxidation number of oxygen in KO_2 is equal to $-1/2$. Oxidation numbers may be fractions because they are just bookkeeping devices. (Ionic charges must be whole numbers because electrons can't be split.)

(f) The compound, $FePO_4$, contains two elements, Fe and P, that are not covered by the rules. However, you should recognize the group PO_4 as the phosphate ion, PO_4^{3-}. Because the charge on the phosphate ion is $3-$, the charge on the iron ion must be $3+$ in order for the compound to have zero charge. Thus, the oxidation number of iron is $+3$ by rule 1.

The oxidation number of oxygen is -2 by rule 4, and for the phosphate ion the total oxidation number for oxygen is -8:

$$\underset{-8}{(P\quad \overset{-2}{O_4})^{3-}}$$

For the sum of the oxidation numbers to equal -3, the charge on the phosphate ion, the oxidation number of phosphorus must be $+5$:

$$\underset{+5\qquad -8=-3}{(P\quad \overset{-2}{O_4})^{3-}}$$

◗ Because no metal is as electronegative as hydrogen, the oxidation number of H is always -1 in binary compounds with metals.

◗ Note that, although the oxidation number of oxygen is usually -2, the oxidation number of oxygen in peroxides is -1 and the oxidation number of oxygen in superoxides is $-(1/2)$.

◗ Notice that the sign for the charge on an ion is written after the number as in Fe^{3+} and PO_4^{3-}. The sign for an oxidation number is written in front of the number as in $+3$, $+5$, and -2 so that you can tell the difference between ionic charges and oxidation numbers.

PRACTICE PROBLEMS

11.1   Use the rules for assigning oxidation numbers to find the oxidation number of each of the elements in each of the following species: (a) N_2, (b) NO_3^-, (c) NH_4^+, (d) Mg_3N_2, (e) HNO_2, (f) $FeSO_4$, and (g) $Cr_2(SO_4)_3$.

11.2   Use the definition of oxidation number to assign the oxidation number of each atom in (a) PCl_3 and (b) CN^-.

OXIDATION NUMBERS AND THE PERIODIC TABLE

The oxidation numbers observed for the elements in their common compounds are related to the positions of the elements in the periodic table as shown in ▪Figure 11.1. In general, metals have positive oxidation numbers in compounds. Most transition metals appear in more than one positive oxidation state. Most nonmetals and semimetals show more than one oxidation number in their compounds and have both positive and negative oxidation numbers, although the number of negative oxidation numbers is small compared with the number of positive oxidation numbers. Except for second-period elements, the oxidation numbers of nonmetals are either all even or all odd.

No compounds containing elements in an oxidation state higher than $+8$ are known. Except for Cu, Au, and the noble gases, the highest oxidation number

Group																		
IA	IIA											IIIA	IVA	VA	VIA	VIIA	0

1 H +1, −1

2 He (0)

3 Li +1 — **4 Be** +2

5 B +3 — **6 C** +4, +2, −1, −4 — **7 N** All from +5 to −3 — **8 O** −1, −2 — **9 F** −1 — **10 Ne**

11 Na +1 — **12 Mg** +2

13 Al +3 — **14 Si** +4, −4 — **15 P** +5, +3, −3 — **16 S** +6, +4, +2, −2 — **17 Cl** +7, +5, +3, +1, −1 — **18 Ar**

Transition groups	IIIB	IVB	VB	VIB	VIIB	VIII			IB	IIB

19 K +1 — **20 Ca** +2 — **21 Sc** +3 — **22 Ti** +4, +3, +2 — **23 V** +5, +4, +3, +2 — **24 Cr** +6, +3, +2 — **25 Mn** +7, +6, +4, +3, +2 — **26 Fe** +3, +2 — **27 Co** +3, +2 — **28 Ni** +2 — **29 Cu** +2, +1 — **30 Zn** +2 — **31 Ga** +3, +2 — **32 Ge** +4, +2, −4 — **33 As** +5, +3, −3 — **34 Se** +6, +4, −2 — **35 Br** +7, +5, +3, +1, −1 — **36 Kr** +2

37 Rb +1 — **38 Sr** +2 — **39 Y** +3 — **40 Zr** +4, +3 — **41 Nb** +5, +4, +2 — **42 Mo** +6, +5, +4, +3 — **43 Tc** +7, +5, +4 — **44 Ru** +8, +5, +4, +3 — **45 Rh** +4, +3 — **46 Pd** +4, +2 — **47 Ag** +1 — **48 Cd** +2 — **49 In** +3, +2, +1 — **50 Sn** +4, +2, −4 — **51 Sb** +5, +3, −3 — **52 Te** +6, +4, −2 — **53 I** +7, +5, +3, +1, −1 — **54 Xe** +6, +4, +2

55 Cs +1 — **56 Ba** +2 — **71 Lu** +3 — **72 Hf** +4, +3 — **73 Ta** +5, +4, +3 — **74 W** +6, +5, +4 — **75 Re** +8, +6, +5, +4 — **76 Os** +8, +6, +4, +3, +2 — **77 Ir** +4, +3, +1 — **78 Pt** +4, +2 — **79 Au** +3, +1 — **80 Hg** +2, +1 — **81 Tl** +3, +1 — **82 Pb** +4, +2 — **83 Bi** +3 — **84 Po** +6, +5, +4, +2, −2 — **85 At** +7, +5, +3, +1, −1 — **86 Rn** +2

87 Fr +1 — **88 Ra** +2 — **103 Lr** +3

Metal   Semimetal   Nonmetal

▪ **FIGURE 11.1** Important oxidation numbers of the elements in their compounds. (Information is from Bard, A. J.; Parsons, R.; Jordan, J. *Standard Potentials in Aqueous Solution;* Dekker: Basel, 1985.)
* Lanthanide series: La–Yb, $+3$; Ce also $+4$.
♦ Actinide series: Ac, $+3$; Th, $+4$; Pa, $+5$; U, $+6$; Np, $+5$; Pu, $+4$; Am, $+4$ and $+3$; Cm–No, $+3$.

The products of aerobic (in presence of air) metabolism are species with the central atom in a high oxidation state such as CO_2 and $CO_3{}^{2-}$. The products of anaerobic (in absence of air) metabolism are species with the central element in a low oxidation state such as NH_3 and CH_4.

commonly observed is never greater than the group number. For example, the maximum oxidation number for osmium (atomic number 76), which is in Group VIII, is $+8$ and the maximum oxidation number for chlorine (atomic number 17), which is in Group VIIA, is $+7$.

The most negative oxidation number observed for an element is equal to the group number minus eight, the number of electrons that must be gained to complete an octet by transfer or sharing. For example, an oxygen atom has six electrons in its outer shell; two electrons are needed to complete an octet. The most negative oxidation number observed for oxygen is -2.

Oxidation numbers make it easier to remember the chemical properties of the elements and their compounds. For example, the chemistry of the elements that can exist in more than one oxidation state in addition to zero, such as nitrogen, chromium, and manganese, involves a large number of different oxidation–reduction reactions. The chemistry of compounds of elements that can only exist in a single oxidation state (in addition to zero), such as sodium and magnesium, is much simpler and less varied.

An increase in acidity with increasing oxidation number is common. For example, sulfuric acid, H_2SO_4, in which sulfur has an oxidation number of $+6$, is a stronger acid than sulfurous acid, H_2SO_3, in which sulfur has an oxidation number of $+4$.

Compounds in which a metal has a high oxidation number are more covalent than compounds in which the metal has a low oxidation number. For example, the compound lead(IV) chloride, $PbCl_4$, in which lead has an oxidation number of $+4$, is predominantly covalent. The evidence for this statement is that $PbCl_4$ is a yellow, oily liquid that does not conduct electricity. The compound, lead(II) chloride, $PbCl_2$, in which lead has an oxidation number of $+2$, is predominantly ionic. Lead(II) chloride is a white crystalline solid with a high melting point (501 °C); the melt conducts electricity. Ionic charges greater than $3+$ are unlikely because successive ionization energies increase rapidly (Section 8.5).

PRACTICE PROBLEMS

11.3 Use the periodic table inside the front cover to predict the most positive and the most negative oxidation numbers expected for phosphorus. (Do not refer to Figure 11.1.)

11.4 Which is the stronger acid, HNO_3 or HNO_2? Explain your prediction.

11.3 OXIDATION NUMBERS AND NOMENCLATURE

The Roman numeral system of nomenclature (Section 1.10), which is called the Stock system, is based on oxidation numbers. Remember that Roman numerals are used to show which of several possible compounds of a metal with a given nonmetal or polyatomic ion is meant. For example, the two kinds of iron nitrate, $Fe(NO_3)_2$ and $Fe(NO_3)_3$, are called iron(II) nitrate and iron(III) nitrate; the oxidation number of iron in iron(II) nitrate is $+2$, and the oxidation number of iron in iron(III) nitrate is $+3$. The names of the two oxides of copper, Cu_2O and CuO, are copper(I) oxide and copper(II) oxide because the oxidation number of copper is $+1$ in Cu_2O and $+2$ in CuO.

A knowledge of oxidation numbers can be used to write formulas from names. Consider, for example, lead(IV) oxide. The Roman numeral IV tells us that the lead

Copper(I) oxide (left) is red. Copper(II) oxide (right) is black.

TABLE 11.1	Names of Oxo Acids and Their Salts				
Formula of Acid	Oxidation Number of Cl	Name of Acid	Formula of Salt	Name of Salt
$HClO_4$	+7	Perchloric acid	$NaClO_4$	Sodium perchlorate
$HClO_3$	+5	Chloric acid	$NaClO_3$	Sodium chlorate
$HClO_2$	+3	Chlorous acid	$NaClO_2$	Sodium chlorite
$HClO$	+1	Hypochlorous acid	$NaClO$	Sodium hypochlorite

in this compound has an oxidation number of +4. The sum of the oxidation numbers of the atoms in a compound must be zero; therefore, the sum of the oxidation numbers of the oxygens in lead(IV) oxide must equal −4. According to rule 4 for assigning oxidation numbers, the oxidation number of oxygen is −2, and there must be two oxygen atoms in a unit of lead(IV) oxide. The formula for lead(IV) oxide must be PbO_2.

The *inorganic oxygen-containing acids,* such as nitric acid (HNO_3), phosphoric acid (H_3PO_4), and sulfuric acid (H_2SO_4), are often called **oxo acids.** The *anions of the oxo acids* are called **oxo anions.** A knowledge of oxidation numbers is useful for naming oxo acids and their salts. The name of the most common oxo acid of an element usually ends in -ic; nitric acid, phosphoric acid, and sulfuric acid are examples. If the element has a lower oxidation number than it does in the most common oxo acid, the ending -ic is changed to -ous. Thus, the acid HNO_2 is called nitrous acid, H_3PO_3 is called phosphorous acid, and H_2SO_3 is called sulfurous acid. An oxo acid in which the element has a still lower oxidation number is named hypo-_____-ous acid as in hypochlorous acid. An oxo acid in which the element has a higher than usual oxidation number is named per-_____-ic acid as in perchloric acid. Table 11.1 shows the names of the complete series of chlorine-containing oxo acids and their salts. Notice how each acid has one fewer oxygen than the acid above it. The oxidation numbers of Cl in these compounds differ by 2 and are all odd numbers. Also notice that the names of the salts of -ic acids end in -ate and the names of salts of -ous acids end in -ite.

The endings -ous and -ic are also sometimes used to name metal salts, oxides, and hydroxides. For example, iron(II) nitrate, $Fe(NO_3)_2$, is called ferrous nitrate, and iron(III) nitrate, $Fe(NO_3)_3$, is called ferric nitrate; copper(I) oxide, Cu_2O, is called cuprous oxide, and copper(II) oxide, CuO, is called cupric oxide. Notice how the ending -ous is used to show the lower oxidation number of the metal and the ending -ic is used to show the higher oxidation number. However, in these nonsystematic names, you have to remember what the ionic charges are. The charge on a cuprous ion is 1+, the charge on a ferrous ion is 2+, and the charge on a cerous ion is 3+! In addition, the names are often derived from the Latin rather than the English names of the elements. Another disadvantage is that the method does not provide for metals such as titanium, vanadium, chromium, and manganese that have more than two oxidation numbers in compounds. The most important oxide of manganese is neither manganous oxide, MnO, nor manganic oxide, Mn_2O_3, but MnO_2. In chemistry, the nonsystematic names are not used much any more, but unfortunately you still need to know about them because you may come across them in reading and in the labels on bottles. For reference, the old nonsystematic names of the common cations are given in Table A.2 in Appendix A.

Prefixes are preferred to Roman numerals for naming compounds composed entirely of nonmetals (Section 1.10) because Roman numerals do not show the difference between compounds such as NO_2 and N_2O_4 that have the same empirical formula but different molecular formulas.

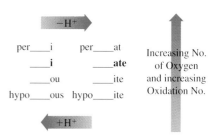

Naming oxo acids and oxo anions.

Manganese(IV) oxide, MnO_2, is the most plentiful ore of manganese. Geologists call it pyrolusite. Manganese(IV) oxide is used to produce steel and to make dry cells, bricks, and glass. Most other manganese compounds are made from MnO_2.

11.5 Name each of the following: (a) CrO_3, (b) $Ca(ClO_2)_2$, (c) ClO^-, (d) $KClO_3$, and (e) HIO_4.

11.6 Write the formula of each of the following: (a) lithium chlorite, (b) calcium hypochlorite, (c) sodium sulfite, (d) manganese(II) chloride, and (e) iron(III) oxide.

Magnesium metal burning in air.

🕐 The real world examples in the Oxidation and Reduction module of the Electrochemistry section help students learn to identify species that undergo oxidation and reduction.

In case anyone asks, the oxidation number of the metal atom is zero in metal carbonyls, and the oxidation number of carbon is zero in formaldehyde, in acetic acid, and in dichloromethane.

▶ In the reaction shown by equation 11.1, which element is oxidized and which element is reduced?

11.4   IDENTIFYING OXIDATION–REDUCTION REACTIONS

When an element is oxidized, its oxidation number increases. When an element is reduced, its oxidation number decreases. For example, when magnesium burns in air, magnesium metal reacts with oxygen in air forming magnesium oxide:

$$2Mg(s) + O_2(g) \longrightarrow 2MgO(s)$$

Magnesium is oxidized, and the oxidation number of magnesium increases from 0 to $+2$. Oxygen is reduced, and the oxidation number of oxygen decreases from 0 to -2. *One definition of oxidation is that oxidation is an increase in oxidation number and reduction is a decrease in oxidation number. If the oxidation number of any element changes in the course of a reaction, the reaction is oxidation–reduction.*

Consider the following reactions of potassium dichromate, $K_2Cr_2O_7$:

$$K_2Cr_2O_7(aq) + 6KCl(aq) + 7H_2SO_4(aq) \longrightarrow$$
$$Cr_2(SO_4)_3(aq) + 3Cl_2(g) + 4K_2SO_4(aq) + 7H_2O(l) \quad (11.1)$$

$$K_2Cr_2O_7(aq) + KI(aq) + 4H_2SO_4(aq) \longrightarrow$$
$$Cr_2(SO_4)_3(aq) + KIO_3(aq) + K_2SO_4(aq) + 4H_2O(l) \quad (11.2)$$

$$K_2Cr_2O_7(aq) + 2KOH(aq) \longrightarrow 2K_2CrO_4(aq) + H_2O(l) \quad (11.3)$$

If an element appears in a compound on one side of an equation and as a free element on the other side, a reaction can usually immediately be identified as an oxidation–reduction reaction. The oxidation numbers of atoms in all free elements are zero (rule 1). The oxidation number of an atom in a compound is very rarely zero. Therefore, when a free element is formed or used up in a reaction, the reaction is almost always oxidation–reduction. In the first reaction (equation 11.1), the element chlorine, which is combined with potassium in KCl on the reactant side, is a product; therefore, this reaction is oxidation–reduction. The oxidation number of chlorine changes from -1 in KCl to 0 in Cl_2.

If no reactant or product is a free element, the oxidation number of each element in the reactants must be compared with the oxidation number of the same element in the products to determine whether the reaction is oxidation–reduction. The oxidation number of each atom is shown over the symbol for the atom in the equation for the second reaction (equation 11.2):

$$\overset{+1\ +6\ \ -2}{K_2Cr_2O_7}(aq) + \overset{+1\ -1}{KI}(aq) + \overset{+1\ +6\ -2}{4H_2SO_4}(aq) \longrightarrow$$

$$\overset{+3\ \ +6\ -2}{Cr_2(SO_4)_3}(aq) + \overset{+1\ +5\ -2}{KIO_3}(aq) + \overset{+1\ +6\ -2}{K_2SO_4}(aq) + \overset{+1\ -2}{4H_2O}(l)$$

The oxidation number of chromium decreases from $+6$ to $+3$. Therefore, chromium is reduced in the reaction. The oxidation number of iodine increases from -1 to $+5$; iodine is oxidized. This reaction is oxidation–reduction. The complete ionic equation is

$$2K^+ + Cr_2O_7^{2-} + K^+ + I^- + 8H^+ + 4SO_4^{2-} \longrightarrow$$
$$2Cr^{3+} + 3SO_4^{2-} + K^+ + IO_3^- + 2K^+ + SO_4^{2-} + 4H_2O(l)$$

and the net ionic equation is

$$Cr_2O_7^{2-} + I^- + 8H^+ \longrightarrow 2Cr^{3+} + IO_3^- + 4H_2O(l)$$

Potassium ions and sulfate ions are spectator ions. The dichromate ion, $Cr_2O_7^{2-}$, is the oxidizing agent (species that brings about oxidation), and the iodide ion, I^-, is the reducing agent (species that brings about reduction). Notice that *the oxidizing agent is reduced and the reducing agent is oxidized.*

For the third reaction (equation 11.3), the oxidation number of each atom is shown over the symbol for the atom in the equation below:

$$\overset{+1 +6 \ -2}{K_2Cr_2O_7}(aq) + 2\overset{+1 -2 +1}{KOH}(aq) \longrightarrow 2\overset{+1 +6 -2}{K_2CrO_4}(aq) + \overset{+1 -2}{H_2O}(l)$$

Because the oxidation numbers of all the elements are the same in the products as in the reactants, this reaction is *not* oxidation–reduction.

Notice that *both* oxidation and reduction take place in each oxidation–reduction reaction. Oxidation and reduction always take place at the same time; reduction cannot take place without oxidation, and oxidation cannot take place without reduction. Because oxidation and reduction always take place at the same time, *oxidation–reduction reactions* are often called **redox** reactions for short.

▶ All reactions are either redox or not redox.

Although it seems obvious, the idea that all reactions are either redox or not redox does not occur to many students unless it is pointed out to them.

PRACTICE PROBLEMS

11.7   In each of the following reactions, which species is oxidized and which species is reduced?

(a) $Zn(s) + 2H^+ + 2Cl^- \longrightarrow Zn^{2+} + 2Cl^- + H_2(g)$

(b) $2Na(s) + Cl_2(g) \longrightarrow 2NaCl(s)$

(c) $2HgO(s) \longrightarrow 2Hg(l) + O_2(g)$

11.8   Classify each of the following reactions as either redox or not redox. For each reaction that is redox, tell which element is oxidized and which element is reduced and write complete ionic and net ionic equations. Identify the oxidizing agent and the reducing agent.

(a) $CaCO_3(s) + 2HCl(aq) \longrightarrow CaCl_2(aq) + CO_2(g) + H_2O(l)$

(b) $6FeCl_2(aq) + 6HCl(aq) + NaClO_3(aq) \longrightarrow 6FeCl_3(aq) + NaCl(aq) + 3H_2O(l)$

(c) $Na_2S(aq) + 4I_2(s) + 4H_2O(l) \longrightarrow 8HI(aq) + Na_2SO_4(aq)$

11.5 WRITING EQUATIONS FOR OXIDATION–REDUCTION REACTIONS

◯ *Students balance half-reactions interactively in the Half-Reactions module of the Electrochemistry section. Real world examples of redox reactions are used in this module.*

The equations for most reactions that are not oxidation–reduction reactions can be balanced by inspection (Section 3.2). The equations for many redox reactions can also be balanced by inspection. For example, the equation for the reaction between hydrogen sulfide gas and aqueous hydrogen peroxide to form solid sulfur and liquid water

$$H_2S(g) + H_2O_2(aq) \longrightarrow S(s) + 2H_2O(l)$$

is easily balanced.

11.9 Balance the equations for the following redox reactions by inspection:
(a) $Sb(s) + Cl_2(g) \longrightarrow SbCl_3(s)$
(b) $HBr(aq) + H_2SO_4(aq) \longrightarrow Br_2(aq) + SO_2(g) + H_2O(l)$

However, the equations for many redox reactions that take place in aqueous solution are difficult if not impossible to balance by inspection. In addition, for some redox reactions, the expressions obtained by inspection are not correct. The reaction between potassium permanganate and hydrogen peroxide is an example. For this reaction the following expression is obtained by inspection:

$$2KMnO_4(aq) + H_2O_2(aq) + 3H_2SO_4(aq) \longrightarrow$$
$$2MnSO_4(aq) + K_2SO_4(aq) + 3O_2(g) + 4H_2O(l)$$

Although this expression shows 2 K, 2 Mn, 22 O, 8 H, and 3 S on each side, it does *not* show the right relationship between moles of $KMnO_4$ and moles of H_2O_2 reacting, as we will see shortly.

A method is needed for balancing equations for redox reactions. Unfortunately, there is no one best method. Two methods are commonly used: The change-in-oxidation-number method and the half-reaction method. Both methods yield the same equation.

Change-in-Oxidation-Number Method

A simple example, the reaction of magnesium metal with oxygen to form magnesium oxide, should help you understand the basis of the change-in-oxidation-number method. The equation for the reaction of magnesium with oxygen is

$$2Mg(s) + O_2(g) \longrightarrow 2MgO(s)$$

The product (magnesium oxide) is an ionic compound that is composed of magnesium ions, Mg^{2+}, and oxide ions, O^{2-}. In magnesium oxide, the oxidation number of Mg is $+2$. The oxidation number of magnesium in magnesium metal is 0; thus, the oxidation number of each magnesium increases by 2 in the reaction. Since two magnesium atoms are oxidized, the total increase in oxidation number is 4. In magnesium oxide, the oxidation number of O is -2. The oxidation number of oxygen in oxygen gas is 0 and the oxidation number of each oxygen decreases by 2 in the reaction. Since two oxygen atoms are reduced, the total decrease in oxidation number is also 4. *In any oxidation–reduction reaction, the total increase in oxidation number equals the total decrease in oxidation number.* This fact is the basis for the change-in-oxidation-number method of balancing the equations for redox reactions. For detailed directions for balancing equations that cannot be balanced by inspection by the change-in-oxidation-number method, see Appendix F.

Examination of the change in oxidation number for the expression

$$\overset{+1 \ +7 \ -2}{2KMnO_4}(aq) + \overset{+1 \ -1}{H_2O_2}(aq) + \overset{+1 \ +6 \ -2}{3H_2SO_4}(aq) \longrightarrow$$
$$\overset{+2 \ +6 \ -2}{2MnSO_4}(aq) + \overset{+1 \ +6 \ -2}{K_2SO_4}(aq) + \overset{0}{3O_2}(g) + \overset{+1 \ -2}{4H_2O}(l)$$

shows why this expression is *not* the molecular equation for the reaction between potassium permanganate and hydrogen peroxide. According to this expression, oxygen is oxidized from -1 in H_2O_2 to 0 in O_2. However, the number of oxygens with oxidation number -1 on the left is not the same as the number of oxygens with oxidation number 0 on the right. If oxygen with oxidation number -2 in either

The element oxidized is shown in blue and the element reduced in red. The element reduced is in the species that is the oxidizing agent. Thus, the element reduced is red like models of oxygen atoms. Oxygen is the commonest oxidizing agent.

Some users of the first edition taught balancing equations for complicated redox equations here. Others postponed it until just before electrochemistry. Still others did not teach it at all. Therefore, and because it is a skill with little conceptual content like using exponential numbers, we have moved it to the Appendix in this edition.

KMnO₄ or H₂SO₄ is oxidized, one equation has been written for two reactions, which is wrong. An equation must be written for *each* reaction. The equation for the reaction between is potassium permanganate and hydrogen peroxide in acidic solution is

$$2\overset{+1\,+7\,-2}{KMnO_4}(aq) + 5\overset{+1\,-1}{H_2O_2}(aq) + 3\overset{+1+6-2}{H_2SO_4}(aq) \longrightarrow$$

$$2\overset{+2\,+6-2}{MnSO_4}(aq) + \overset{+1+6\,-2}{K_2SO_4}(aq) + 5\overset{0}{O_2}(g) + 8\overset{+1\,-2}{H_2O}(l)$$

which not only shows the same number of each kind of atom and the same charge on both sides but also has the same numerical values for the total changes in oxidation number.

PRACTICE PROBLEM

11.10   Check your answers to Practice Problem 11.9 by comparing total increases in oxidation numbers with total decreases in oxidation numbers.

Half-Reaction Method

The half-reaction method does not require assignment of all oxidation numbers. Therefore, it can be used when assignment of all oxidation numbers is difficult or impossible. In addition, the half-reactions that are used in this method will be useful later for discussing electrochemistry in Chapter 19.

The displacement of silver from a solution of a silver salt by copper is a simple and familiar example:

$$Cu(s) + 2Ag^+ \longrightarrow 2Ag(s) + Cu^{2+}$$

In this reaction, each copper atom loses two electrons to form a copper(II) ion in solution:

$$Cu(s) \longrightarrow Cu^{2+} + 2e^- \qquad (11.4)$$

The oxidation number of copper increases from 0 to +2; copper is oxidized. The reaction shown by equation 11.4 is called a **half-reaction of oxidation.** Notice that *electrons are a product of a half-reaction of oxidation;* indeed, **oxidation** is often defined as *loss of electrons.* Each silver ion gains one electron and becomes a silver atom:

$$Ag^+ + e^- \longrightarrow Ag(s) \qquad (11.5)$$

The oxidation number of silver decreases from +1 to 0; silver is reduced. The reaction shown by equation 11.5 is called a **half-reaction of reduction.** Notice that *electrons are a reactant in a half-reaction of reduction;* **reduction** is often defined as *gain of electrons.*

The equations for a half-reaction of oxidation and a half-reaction of reduction can be added to give the equation for the overall reaction. However, electrons must cancel when half-reactions are combined because electrons are matter and cannot be created or destroyed in a chemical reaction. *The number of electrons released in the oxidation half-reaction must be the same as the number of electrons used up in the reduction half-reaction.* Thus, two silver atoms must be reduced for each copper atom that is oxidized. For electrons to cancel when the half-reactions are

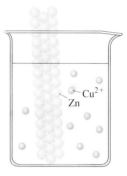

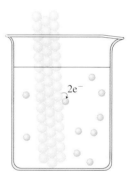

Macroscopic (top) *and microscopic views of another displacement reaction, Zn(s) + Cu²⁺ → Zn²⁺ + Cu(s). For this reaction, what is the half-reaction of oxidation? the half-reaction of reduction?*

Remember the law of conservation of matter, Section 3.1.

added, equation 11.5 must be multiplied by 2:

$$\begin{array}{rr} Cu \longrightarrow Cu^{2+} + 2e^- & (11.4) \\ 2(Ag^+ + e^- \longrightarrow Ag) & +2(11.5) \\ \hline Cu + 2Ag^+ + \cancel{2e^-} \longrightarrow Cu^{2+} + \cancel{2e^-} + 2Ag & \end{array}$$

PRACTICE PROBLEMS

11.11 For the oxidation-reduction reaction

$$Zn(s) + 2H^+ + 2Cl^- \longrightarrow Zn^{2+} + 2Cl^- + H_2(g)$$

(a) Write the net ionic equation for the oxidation half-reaction. (b) Write the net ionic equation for the reduction half-reaction.

11.12 If the net ionic equation for one half-reaction is

$$2I^- \longrightarrow I_2(s) + 2e^-$$

and the net ionic equation for the other half-reaction is

$$Br_2(aq) + 2e^- \longrightarrow 2Br^-$$

(a) What is the net ionic equation for the overall reaction of oxidation–reduction? (b) Which half-reaction is the oxidation half-reaction?

Some users of the first edition taught balancing equations for complicated redox equations here. Others postponed it until just before electrochemistry. Still others did not teach it at all. Therefore, and because it is a skill with little conceptual content like using exponential numbers, we have moved it to the Appendix in this edition.

Thus, the half-reaction method for balancing oxidation–reduction equations that cannot be balanced by inspection is based on division of the equation for the overall reaction into an oxidation half-reaction and a reduction half-reaction. Each half-reaction is then balanced separately, and the equation for the overall reaction is obtained by combining the half-reactions in such a way that electrons cancel. For detailed directions for balancing equations that cannot be balanced by inspection by the half-reaction method, see Appendix F.

11.6 DISPROPORTIONATION REACTIONS

Sometimes the same species is both oxidized and reduced. The reaction,

$$3Br_2(aq) + 6OH^- \longrightarrow BrO_3^- + 5Br^- + 3H_2O(l)$$

is an example. In this reaction, bromine, which has an oxidation number of 0 in Br_2, is oxidized to bromine in the +5 oxidation state and reduced to bromine in the −1 oxidation state. *Reactions in which the same species is both oxidized and reduced* are called **disproportionation reactions.** Only species that contain an element in an intermediate oxidation state, such as Br_2, can disproportionate. Bromine in Br_2 has an oxidation number of 0 and can be oxidized to +1, +3, +5, or +7 or reduced to −1 (see Figure 11.1). Species that contain an element in its most positive oxidation state, such as Fe^{3+}, can only be reduced and act only as oxidizing agents. Species that contain an element in its most negative oxidation state, such as I^-, can only be oxidized and act only as reducing agents. Species with elements in intermediate oxidation states can be either oxidized or reduced and can act either as reducing agents or as oxidizing agents.

11.13 Of the following species, (a) which can only be oxidized? (b) Which can only be reduced? (c) Which can undergo disproportionation? Explain your answers. (i) Fe^{2+} (ii) Ca (iii) F_2 (iv) SO_2 (v) Cu^+

11.14 (a) Write a balanced net ionic equation for the disproportionation reaction

$$HS_2O_4^- + H_2O(l) \longrightarrow S_2O_3^{2-} + H_2SO_3(aq)$$

(b) Which element disproportionates?

11.15 If the source of OH^- is NaOH(aq), write complete ionic and molecular equations for the disproportionation reaction

$$3Br_2(aq) + 6OH^- \longrightarrow BrO_3^- + 5Br^- + 3H_2O(l)$$

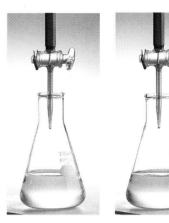

- FIGURE 11.2 Titration with potassium permanganate. (a) Potassium permanganate solution (in the buret) is purple. (b) At the end point, addition of one drop of permanganate solution gives a pink color.

11.7 OXIDATION–REDUCTION TITRATIONS

Remember that titration is an important method for determining the amount of a substance present in solution. In a titration, a solution of known concentration, called a standard solution, is added from a buret to the solution being analyzed. Addition is stopped when the quantity of reactant called for by the equation for the reaction has been added. This point is called the end point of the titration and is signaled by a color change or change in an electrical property. A substance whose change in color shows that the end point has been reached is called an indicator.

More titration methods involve oxidation–reduction reactions than any other type of reaction. Standard solutions of oxidizing agents are usually used for redox titrations because solutions of reducing agents often react with oxygen in the air. All the species being analyzed for must be in the same oxidation state at the beginning of the titration. Therefore, all species that result from dissolving the sample are reduced to a single oxidation state before titration. For example, after a sample of iron ore is dissolved, both iron(II) ion (Fe^{2+}) and iron(III) ion (Fe^{3+}) are present. Iron(III) ion is reduced to iron(II) ion with zinc. The net ionic equation is

$$2Fe^{3+} + Zn(s) \longrightarrow 2Fe^{2+} + Zn^{2+}$$

Potassium permanganate, $KMnO_4$, is the commonest oxidizing agent for oxidation–reduction titrations because the purple-colored permanganate ion acts as its own indicator. During a titration with permanganate, the solution remains colorless until the end point is reached. Then, addition of one drop of permanganate solution gives a pink color (see -Figure 11.2).

The half-reaction for the reduction of the permanganate ion in acidic solution is

$$MnO_4^- + 8H^+ + 5e^- \longrightarrow Mn^{2+} + 4H_2O(l)$$

In the determination of iron in ores, the overall reaction is

$$5Fe^{2+} + MnO_4^- + 8H^+ \longrightarrow 5Fe^{3+} + Mn^{2+} + 4H_2O(l)$$

Potassium permanganate is readily available, inexpensive, and reacts with a large number of organic and inorganic substances. Unfortunately, pure potassium permanganate is not available for the preparation of standard solutions of permanganate. The concentration of standard solutions must be determined by using the solution to titrate oxalic acid, $H_2C_2O_4 \cdot 2H_2O$, which is available pure. *Substances* such as oxalic acid, *which are available pure and are used to standardize (determine the concentration of) solutions of other substances,* are called **primary standards.** The

Equivalence point will be defined in Chapter 15 when students can understand the difference between end point and equivalence point.

Titration was discussed in Section 4.10. In general chemistry labs, students are more likely to do an acid–base titration than an oxidation–reduction titration.

The new Random House Unabridged Dictionary *prefers commonest to most common.*

See, for example, Kennedy, J. H. Analytical Chemistry: Principles; *Harcourt Brace Jovanovich: San Diego, 1984; p 350.*

RELATED TOPIC

The Breathalyzer*

Each year some 24 000 people die in alcohol-related automobile accidents, and more than 2 million are injured seriously enough to require treatment in a hospital. Although establishing and administering laws that regulate the operation of motor vehicles is a state responsibility, uniform laws have been enacted nationwide. In most states, any person with a blood-alcohol level of one gram per liter (1.0 g/L) or more is legally intoxicated. However, research seems to be indicating that the level should probably be lower, and many states have reduced the limit to 0.8 g/L.

The effect of alcohol on a person depends on the concentration of alcohol in his or her brain, which is difficult to measure. However, the concentration of alcohol in the blood is proportional to the concentration in the brain, and the concen-

tration in the breath is proportional to the concentration in the blood. Thus, the concentration in the breath, which is easily measured, provides a reasonably accurate measure of intoxication.

One of the most popular breath-testing instruments is the Breathalyzer, which was developed in 1954 by R. F. Burkenstein, then a captain in the Indiana State Police. A Breathalyzer collects a sample of breath and measures its alcohol content by means of an oxidation–reduction reaction. The sample of breath passes into 3.0 mL of 0.025% potassium dichromate ($K_2Cr_2O_7$) and 0.025% silver nitrate ($AgNO_3$) in sulfuric acid and water. Any alcohol present in the breath dissolves and is oxidized to acetic acid:

$$2K_2Cr_2O_7(aq) + 3CH_3CH_2OH(aq)$$
red

$$+ 8H_2SO_4(aq) \xrightarrow{AgNO_3 \text{ catalyst}}$$
$$2Cr_2(SO_4)_3(aq) + 2K_2SO_4(aq)$$
$$+ 3CH_3COOH(aq) + 11H_2O(l)$$

The amount of dichromate that is used up depends on the amount of alcohol in the

A suspect providing a sample of breath to a Breathalyzer.

sample of breath and is determined by measuring the disappearance of the red color of the dichromate ion quantitatively.

*Information is from Saferstein, R. *Criminalistics: An Introduction to Forensic Science,* 4th ed.; Prentice Hall: Englewood Cliffs, NJ, 1990; pp 247–267.

---

reaction involved in the standardization of permanganate solutions with oxalic acid is

$$2MnO_4^- + 5H_2C_2O_4(aq) + 6H^+ \longrightarrow 2Mn^{2+} + 10CO_2(g) + 8H_2O(l)$$

The calculations for redox titrations are similar to other stoichiometric calculations. Sample Problem 11.2 shows an example.

SAMPLE PROBLEM

11.2 In an experiment to standardize a $KMnO_4$ solution, a 0.3396-g sample of $H_2C_2O_4 \cdot 2H_2O$ required 21.63 mL of $KMnO_4$ solution to titrate it to the first detectable pink color. What was the molarity of the $KMnO_4$ solution? Titration with potassium permanganate is usually carried out in a solution that has been acidified with sulfuric acid. The equation for the reaction that takes place is

$$2KMnO_4(aq) + 5H_2C_2O_4(aq) + 3H_2SO_4(aq) \longrightarrow$$
$$2MnSO_4(aq) + K_2SO_4(aq) + 10CO_2(g) + 8H_2O(l)$$

Solution To calculate the molarity of the $KMnO_4$ solution, we need to know the number of moles of $KMnO_4$; the volume of $KMnO_4$ solution is given (21.63 mL). Finding the number of moles of $KMnO_4$ is a stoichiometry problem. Remember that the first step in solving stoichiometry problems is to write the equation for the reaction. The second step is to write, above the equation, the information given and asked for in the problem, and, below the equation, the formula masses of the substances whose quantities are given in grams. The third step is to write the "recipe" for the reaction (from the coefficients in the equation) below the formula masses. The completed table for this problem is as follows:

Problem $21.63 \ mL \ ?M \quad 0.3396 \ g \ H_2C_2O_4 \cdot 2H_2O$

Equation $2KMnO_4 + 5H_2C_2O_4 + 3H_2SO_4 \longrightarrow 2MnSO_4 + K_2SO_4$
$$+ \ 10CO_2 + 8H_2O$$

FM, u   126.1

Recipe, mol   2   5

First, we must calculate the number of moles of $KMnO_4$ reacted. Then we must use the definition of molarity to calculate the molarity of the $KMnO_4$ solution.

Step 1. Calculation of moles of $KMnO_4$ reacted:

$$0.3396 \ g \ H_2C_2O_4 \cdot 2H_2O \left(\frac{1 \ mol \ H_2C_2O_4 \cdot 2H_2O}{126.1 \ g \ H_2C_2O_4 \cdot 2H_2O} \right)$$

$$\times \left(\frac{2 \ mol \ KMnO_4}{5 \ mol \ H_2C_2O_4 \cdot 2H_2O} \right) = 0.001 \ 077 \ mol \ KMnO_4$$

Step 2. Calculation of molarity:

$$M = \frac{mol \ solute}{vol \ soln \ in \ L} = \frac{0.001 \ 077 \ mol \ KMnO_4}{21.63 \ mL \left(\dfrac{1 \ L}{1000 \ mL} \right)} = 0.049 \ 79 \ M$$

11.16   A 25.00-mL sample of $FeSO_4$ solution was titrated with $KMnO_4$ in a solution that had been acidified with H_2SO_4. If 22.43 mL of 0.0995 M $KMnO_4$ were required to reach the end point, what was the molarity of the $FeSO_4$ solution? The equation for the reaction is

$$10FeSO_4(aq) + 2KMnO_4(aq) + 8H_2SO_4(aq) \longrightarrow$$
$$2MnSO_4(aq) + 5Fe_2(SO_4)_3(aq) + K_2SO_4(aq) + 8H_2O(l)$$

11.8   OXIDATION BY OXYGEN

Oxygen from the atmosphere is the most important oxidizing agent. Oxygen reacts with all other elements except the noble gases, the halogens, and a few "noble" metals such as gold. Oxygen also reacts with many inorganic and most organic compounds. Organic compounds in which halogen atoms have been substituted for a number of the hydrogens are exceptions. For example, CCl_4, $CHCl_3$, and CH_2Cl_2 do not burn. However, CH_3Cl is flammable. Some examples of reactions of oxygen are the following:

$$2Mg(s) + O_2(g) \xrightarrow{\text{heat}} 2MgO(s)$$
$$C(s) + O_2(g) \xrightarrow{\text{heat}} CO_2(g)$$
$$2SO_2(g) + O_2(g) \xrightarrow{\text{catalyst}} 2SO_3(g)$$
$$2CH_3OH(l) + 3O_2 \xrightarrow{\text{heat}} 2CO_2(g) + 4H_2O(l)$$

For an inorganic compound to react with oxygen, the compound must contain an element that forms compounds with oxygen. This element must *not* be in its highest oxidation state. For example, the oxidation number of S is -2 in H_2S, $+4$ in SO_2, and $+6$ (the maximum possible) in SO_3; H_2S and SO_2 react with oxygen, but SO_3 does not react with oxygen.

Everyone knows to put food in the refrigerator to slow the rate at which it spoils.

Although oxidation by oxygen is often slow at room temperature, once started, reaction is fast because thermal energy is released by the reaction and raises the temperature. (The higher the temperature, the faster most reactions take place.) For example, mixtures of hydrogen and oxygen can stand for long periods, but a spark or flame results in an explosion.

The oxidation–reduction reaction that is carried out on the largest scale is **combustion,** *rapid oxidation accompanied by heat and usually by light also.* In your everyday life, you are used to heating your home, when it's cold outside, by the combustion of fossil fuels such as natural gas, oil, and coal. You are also used to using electricity for running air conditioners when it's hot outside. In the United States, about 90% of electricity is generated by burning fossil fuels. Most cars and trucks are powered by internal combustion engines that burn gasoline or diesel fuel. Rockets, such as those used to launch the space shuttle, are often powered by combustion involving liquid oxygen as the oxidizer.

Side reactions that accompany these combustion reactions are a major cause of air pollution. At the high operating temperatures of car, truck, and airplane engines, nitrogen from air is oxidized by oxygen from air forming nitric oxide:

$$N_2(g) \ + \ O_2(g) \xrightarrow{\text{heat}} 2NO(g)$$

Nitric oxide reacts further with oxygen forming nitrogen dioxide:

$$2NO(g) \ + \ O_2(g) \longrightarrow 2NO_2(g)$$

Nitrogen dioxide is red-brown and is responsible for the color of smog; it reacts with water in the air forming nitric and nitrous acids, components of acid rain. When the coal that is burned in power plants contains sulfur as an impurity, combustion yields sulfur dioxide, another major air pollutant and contributor to acid rain.

Carbon dioxide is a product of the combustion of fossil fuels, all of which contain carbon. The burning of natural gas, which is largely methane, is an example:

$$CH_4(g) \ + \ 2O_2(g) \xrightarrow{\text{heat}} CO_2(g) \ + \ 2H_2O(l)$$

Large quantities of carbon dioxide are also being formed by the burning of tropical rain forests. Many scientists believe that higher levels of carbon dioxide in the atmosphere are leading to "global warming," an increase in the temperature of the atmosphere that may be hazardous to life on Earth. This warming effect is called the "greenhouse effect."

Besides being used for heat and power, combustion reactions are steps in some important industrial syntheses. For example, sulfur dioxide, which is used to make sulfuric acid, is produced by burning either sulfur or hydrogen sulfide:

$$S(s) \ + \ O_2(g) \xrightarrow{\text{heat}} SO_2(g)$$
$$2H_2S(g) \ + \ 3O_2(g) \xrightarrow{\text{heat}} 2SO_2(g) \ + \ 2H_2O(l)$$

More sulfuric acid than any other compound is manufactured in the United States.* Sulfuric acid is widely used because it is a strong and cheap inorganic acid. It is

The color of smog is due to NO_2. Dilute solutions of NO_2 are yellow.

---

*See Appendix E for a table that shows names, formulas, U.S. production data, and uses for the "Top 50 Chemicals for 1993."

Although fire is very useful, it can also be extremely destructive. Each year, 5000 Americans die and 30 000 are hospitalized for long periods as a result of about 5 million unwanted fires.* North Americans lose life and property to fire at about twice the rate of people in other industrial countries such as Japan.

Two-thirds of all fatal fires take place in houses. Only 20% of deaths result from burns and heart attacks combined; most deaths (80%) result from breathing combustion products. Combustion in unwanted fires is never complete, and considerable carbon monoxide is formed. In addition wool, silk, and nitrogen-containing synthetic fibers such as nylon yield hydrogen cyanide on combustion. Hydrogen chloride, which is similar in toxicity to hydrogen cyanide, forms when poly(vinyl chloride) burns. Aldehydes, such as formaldehyde, are produced by combustion of organic materials and cause the eyes to fill with tears, making escape difficult.[†]

Most fires are started by faulty wiring, electrical devices, or cigarettes. Some smolder for hours, filling the house with toxic vapors; others grow.

Full-scale experiments have shown what happens when a fire grows. Fires do not grow at a constant rate but instead increase in size at an exponential rate because of the energy liberated by the fire. Hot gases are less dense than cool gases and rise. As a result, a layer of hot gas and soot from incomplete combustion builds up just below the ceiling. Everything inside the room is heated by radiation from the flames and from the layer of hot gas and soot. In a few minutes, steam or smoke rising from objects and surfaces far from the flame signal that the rest of the room is ready to ignite. Suddenly every combustible surface and object in the room bursts into flames; this event is called flashover.

Air from outside the burning room enters through the doorway near the floor. Before flashover, escape is possible by keeping the head low to breathe. As the layer of hot gas and soot just below the ceiling gets thicker, it starts to flow out through the top of the doorway into the next room. When flashover takes place, smoke and flame shoot through the doorway into the next room and displace cool air below the ceiling. Flames on the floor of the fire room start to come through the doorway into the next room, heating the incoming air and making it rise to the ceiling. As incoming air is no longer flowing into the fire room, burning stops. However, unburned combustibles in the fire room remain very hot and continue to decompose to combustible gases hot enough to burn. The hot combustible gases keep flowing out into the next room. When they meet the rising air at the door,

■ FIGURE 11.3   The result of flashover in a 25-story department store/office building in downtown São Paulo, Brazil (February 24, 1972).

the next room flashes over. This type of flashover can take place over large distances, and a common report about big fires is that the fire was suddenly everywhere. Although buildings may be made of noncombustible materials, their contents are not; ■Figure 11.3 shows the results of flashover in a tall modern building.

*Half of the fire victims between 30 and 58 years of age have a blood-alcohol level above the legal level for intoxication.
[†]Many research chemists are active today trying to find ways to keep plastics and other polymers from burning and from producing harmful gases when heated by fires.

---

used to make fertilizers, leather, and tin plate; in petroleum refining; and in the dyeing of fabrics.

Oxidation with oxygen can also be carried out catalytically. An important example is the production of ethylene oxide, another billions-of-pounds-a-year compound.

$$2H_2C\!=\!CH_2(g) + O_2(g) \xrightarrow{\text{catalyst}} 2H_2C\!-\!CH_2(g)$$
$$\text{O}$$

ethylene   ethylene oxide

Why must a catalyst be used instead of heat to make ethylene oxide?

Ethylene oxide is a member of the class of organic compounds known as ethers. An ether has a C—O—C group. Ethylene oxide is a cyclic ether because the C—O—C group is part of a ring. It is used to make antifreeze and detergents.

Reaction of oxygen with the protein hemoglobin is the basis of oxygen transport

11.8 OXIDATION BY OXYGEN

in blood. Arterial blood carries oxygen from the lungs to the muscles where it is transferred to myoglobin, another protein similar to hemoglobin. Myoglobin stores oxygen until it is needed to release energy by oxidizing the sugar glucose, $C_6H_{12}O_6$ or $HOCH_2CH(OH)CH(OH)CH(OH)CH(OH)CHO$.

PRACTICE PROBLEMS

11.17 Which of the following compounds will probably react with oxygen?
(a) CH_3COCH_3 (b) $ClCF_2C(Cl)F_2$ (c) NH_3 (d) FeS (e) $Fe_2(SO_4)_3$

11.18 Write equations for the reactions that will take place. If no reaction is expected, write "no reaction."

(a) $Ca(s) + O_2(g) \xrightarrow{\text{heat}}$

(b) $Si(s) + O_2(g) \xrightarrow{\text{heat}}$

(c) $C_6H_6(l) + O_2(g) \xrightarrow{\text{heat}}$

(d) $Au(s) + O_2(g) \xrightarrow{\text{heat}}$

(e) $Li(s) + O_2(g) \xrightarrow{\text{heat}}$

(f) $CH_3CH\ CH_2(g) + O_2(g) \xrightarrow{\text{heat}}$

(g) $CH_3CH\ CH_2(g) + O_2(g) \xrightarrow{\text{catalyst}}$

11.9 METALLURGY

The form in which a metal occurs in nature depends on its reactivity and the solubility of its compounds. The periodic chart in ▬Figure 11.4 shows the most important sources of the elements. The elements for which no source is given in Figure 11.4 are usually obtained as by-products of the production of other elements with similar properties. In Figure 11.4, notice how the nature of the principal source of an element is related to position in the periodic table. For example, elements that are usually obtained from sulfates are at the bottom of Group IIA, and metals that occur free in nature are all fifth- and sixth-period transition elements in Groups VIII and IB.

Only nine metals occur free in nature, the seven shown in Figure 11.4 and copper and silver. Extremely pure copper is found near Lake Superior. However, most copper is obtained from ores that contain only about 1% copper. An **ore** is *a rock or mineral that serves as a source of an element.* Silver metal is sometimes found with the sulfide ore, Ag_2S, which is called argentite or silver glance, but the free metal is not an important source of silver.

Only two metals—Li and Be—are obtained from silicate minerals. Although more than 90% of Earth's crust is composed of silicate minerals, silicate minerals are not usually economical sources of metals. The proportion of silicon and oxygen in silicate minerals is high, and the concentration of metals is low. Metals in silicate minerals are also hard to convert to the free element. Fortunately for people, geological processes have produced relatively concentrated deposits of most metals in more easily reduced forms.

The United States is dependent on foreign sources for most of the commercially important metals. Only Cu, Fe, Mg, Mo, Pb, and W are available in sufficient quantities in the United States. The metals Al, Ag, Co, Cr, Hg, Mn, Ni, Sn, U, and V must be imported.

A mineral is a naturally occurring, inorganic, crystalline solid. Minerals are substances; they may be either elements or compounds. Rocks are usually heterogeneous mixtures of minerals.

		IA	IIA		IIIB	IVB	VB	VIB	VIIB	VIII	VIII	VIII	IB	IIB	IIIA	IVA	VA	VIA	VIIA	0
										H_2O CH_4										He
		$LiAl-Si_2O_6$	$Be_3Al_2Si_6O_{18}$												$Na_2[B_4O_5(OH)_4]\cdot 2H_2O$	Coal	N_2	O_2	CaF_2	Ne
		NaCl	$MgCO_3$												$AlO_x(OH)_{3-2x}$ $(0<x<1)$	SiO_2	$Ca_3(PO_4)_2$	S_8	NaCl	Ar
		KCl	$CaCO_3$			$FeTiO_3$ TiO_2		$Fe-Cr_2O_4$	MnO_2	Fe_2O_3 Fe_3O_4 $FeCO_3$	$(Ni,Fe)_9S_8{}^a$		$CuFeS_2$	ZnS			$FeAs_2$ $FeAsS$		NaBr	Kr
			$SrSO_4$					MoS_2	—	Ru	Rh	Pd	Ag_2S			SnO_2	Sb_2S_3		NaI $NaIO_3$	Xe
			$BaSO_4$					$CaWO_4$ (Fe, Mn) WO_4		Os	Ir	Pt	Au	HgS		PbS			—	Rn

^{a} Also $(Ni, Mg)_6Si_4O_{10}(OH)_8$ and $(Fe, Ni)O(OH)\cdot mH_2O$

- FIGURE 11.4 Principal sources of the elements. The elements in boldface are the elements that are obtained from the ores. A dash, —, in a space means that the element does not occur naturally. Blank spaces are left for elements that are by-products of the production of other elements.

Because metals have positive oxidation numbers in their compounds, metals must be obtained from ores by reduction. The *separation of metals from their ores* by reduction is part of **metallurgy.** Metallurgy also includes the *making of alloys* and the *working or heating of metals so as to give them certain properties.* **Alloys** are *materials,* such as steel or brass, *that have metallic properties but are not pure metals.* They are composed of two or more metals or of a metal or metals and a nonmetal. Alloys are discussed in Section 24.11; in this section, we will concentrate on the separation of metals from their ores.

The art of metallurgy dates from about 4000 B.C. Copper is believed to have been the first metal obtained from an ore. A close relationship exists between the discovery that metals can be obtained from ores and made into useful objects, such as weapons and tools, and the rise of civilization. Around 3000 B.C. in Asia Minor (where Turkey is today), the Hittites learned how to obtain iron from iron ore by reduction with charcoal. (Charcoal is carbon obtained from wood.) They also learned how to make weapons from the iron, which gave them such an advantage over their enemies that they kept the process a secret for about 1800 years. When the Hittite Empire fell around 1200 B.C., everybody learned how to make and use iron, and the Iron Age began.

About 1773, wood became so scarce in England that the use of charcoal to reduce iron ore was forbidden. Abraham Darby substituted coke, which is carbon obtained from coal, for charcoal. The quantity of iron that could be produced increased enormously, and the price dropped. For the first time, iron cylinders could be made for steam engines and iron rails for railroads, and iron could be used for building ships and bridges. The availability of iron played an important part in the Industrial Revolution. In the past 100 years, metallurgy has gradually developed from an art to a science.

Production of metals from ores involves three steps:

1. *Preliminary treatment.* In preliminary treatment, impurities are removed and the desired component of the ore is concentrated. If necessary, the desired component is changed, by chemical reaction, to a form better suited for reduction.
2. *Reduction.* The desired component is reduced to the free metal.
3. *Refining.* Undesirable impurities are removed from the metal; desirable impurities may be added.

Panning for gold is a very simple method of concentration.

Preliminary treatment usually involves crushing and grinding the ore and isolating the desired component from the ore. The desired component can sometimes be separated by gravity; for concentration by gravity, the desired component and the impurities must have different densities. The iron ore magnetite, Fe_3O_4, can be separated with a magnet. Chemical reactions can also be used for concentration.

The processes by which metals from ores are reduced to the free metal are redox reactions on a large scale. Only a few methods are used for reducing metals to the free element; the method used for reduction depends on the reactivity of the metal. The most vigorous method is electrolytic reduction, which is used to produce the active metals Al, Mg, and Na. These active metals are used to produce other active metals.

Metal oxides are usually reduced with carbon or sometimes with hydrogen:

$$MnO_2(s) + 2C(s) \xrightarrow{heat} Mn(s) + 2CO(g)$$

$$WO_3(s) + 3H_2(g) \xrightarrow{heat} W(s) + 3H_2O(g)$$

Reductions of metal oxides with either carbon or hydrogen are usually endothermic. The enthalpy change, ΔH, is positive and reaction is not spontaneous at room temperature. However, reaction becomes spontaneous at high temperatures and also takes place more rapidly.

Sulfide ores are usually roasted in air. Roasting converts some metal sulfides to metal oxides, for example,

$$2ZnS(s) + 3O_2(g) \xrightarrow{heat} 2ZnO(s) + 2SO_2(g)$$

The oxide is then reduced to the free metal. The free metal is produced directly by roasting other sulfide ores such as CuS, HgS, and PbS:

$$CuS(s) + O_2(g) \xrightarrow{heat} Cu(s) + SO_2(g)$$

Electrolysis is often used to refine metals. For example, the pure copper required for use as an electrical conductor is produced from impure copper by electrolysis.

Iron and Steel

Production of 1 ton of steel requires 1700 lb of iron ore, 1050 lb of scrap steel, 1600 lb of coal (to make coke), 800 lb of $CaCO_3$, 150 kWh of electricity, and 1600 gal of water. One ton of steel is enough to make about nine washing machines.

Hematite is red. It was used for war paint by the Indians and today is used to make red paint. The red color of many rocks is due to hematite.

Iron has played a more important part in people's material progress than any other element. Today most iron is used in the form of a variety of **steels,** *alloys that contain more than 50% iron and 0.03–1.5% carbon.* Other elements are added to give the properties desired for different purposes. For example, steel containing more than 5% chromium resists staining and is used for tableware. The world's annual production of steel is of the order of 1.4×10^{12} lb.

The most important ores of iron are hematite (Fe_2O_3), magnetite (Fe_3O_4), and siderite ($FeCO_3$). Iron pyrite (FeS_2), which is also a common iron-containing mineral, is not used to produce iron because complete removal of sulfur from iron is difficult. Sulfur makes iron brittle.

Iron ore is reduced to metallic iron in a **blast furnace** (■Figure 11.5). Sand and gravel (SiO_2) usually occur with the iron ore. Limestone is added to form the basic metallic oxide, CaO, and neutralize the acidic nonmetallic oxide, SiO_2. The product, $CaSiO_3$, is called **slag.** Molten slag is less dense than molten iron and floats on the iron, protecting it from oxidation by the hot air. The slag and molten iron are run off at regular intervals.

A blast of hot air passes through the furnace in about ten seconds. Coke is converted to CO(g) at the bottom of the furnace, and carbon monoxide reduces the iron ore, which is usually either Fe_2O_3 or Fe_3O_4, in steps to impure iron. Coke is

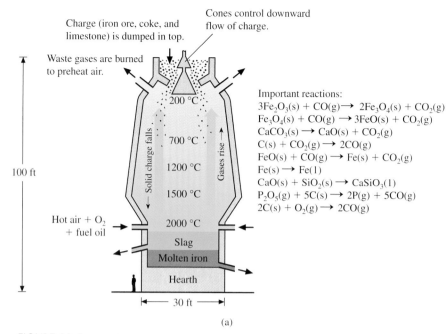

Charge (iron ore, coke, and limestone) is dumped in top.

Cones control downward flow of charge.

Waste gases are burned to preheat air.

100 ft

200 °C

Solid charge falls

Gases rise

700 °C

1200 °C

1500 °C

2000 °C

Hot air + O_2 + fuel oil

Slag

Molten iron

Hearth

30 ft

Important reactions:
$$3Fe_2O_3(s) + CO(g) \longrightarrow 2Fe_3O_4(s) + CO_2(g)$$
$$Fe_3O_4(s) + CO(g) \longrightarrow 3FeO(s) + CO_2(g)$$
$$CaCO_3(s) \longrightarrow CaO(s) + CO_2(g)$$
$$C(s) + CO_2(g) \longrightarrow 2CO(g)$$
$$FeO(s) + CO(g) \longrightarrow Fe(s) + CO_2(g)$$
$$Fe(s) \longrightarrow Fe(l)$$
$$CaO(s) + SiO_2(s) \longrightarrow CaSiO_3(l)$$
$$P_2O_5(g) + 5C(s) \longrightarrow 2P(s) + 5CO(g)$$
$$2C(s) + O_2(g) \longrightarrow 2CO(g)$$

(a)

(b)

(c)

▪ FIGURE 11.5   (a) Diagram of a blast furnace. Iron ore is reduced to iron in blast furnaces. (b) A blast furnace skyline. (c) Workers at a blast furnace.

both the fuel and the source of the reducing agent (CO). Oxygen and fuel oil are added to the air so that only a minimum quantity of coke is needed to reach the proper temperatures.

A modern blast furnace is operated continuously and produces about 10 000 tons of iron in 24 hours. The molten iron from a blast furnace contains 4–5% C and variable amounts of Mn, Si, P, and S. If the impure iron is poured into molds and allowed to cool, it is called pig iron or cast iron. Pig iron is brittle, hard to weld, and not strong. Most of the impure iron is converted into steel. The slag is used as a building material and for the manufacture of cement.

Today most pig iron is converted to steel by the **basic oxygen process.** ▪Figure 11.6 shows a basic oxygen furnace. A jet of pure oxygen is directed at the surface of a charge of molten pig iron, scrap iron, and limestone. The exothermic reaction between carbon and oxygen to form carbon dioxide takes place so fast that the temperature rises nearly to the boiling point of iron without any external heating. The impurities are oxidized to acidic oxides, for example, silicon is converted to silicon dioxide:

$$Si(l) + O_2(g) \longrightarrow SiO_2(l)$$

The acidic oxides react with basic CaO formed by decomposition of limestone to form salts

$$SiO_2(l) + CaO(l) \longrightarrow CaSiO_3(l)$$

and the salts form a slag. When the percent carbon has been reduced to the desired level, addition of oxygen is stopped. (Too much oxygen would oxidize the iron.) Other metals such as nickel are added and the furnace is tipped to pour off slag and steel. The entire process takes less than an hour; about 7200 tons of steel can be produced in a day.

The highest quality steels are made in **electric arc furnaces.** In an electric arc furnace, an electric arc between graphite electrodes and the metal charge heats the furnace to about 3500 °C. The charge consists of scrap steel to which alloying

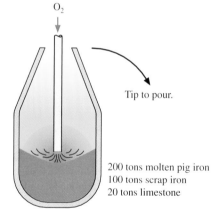

O_2

Tip to pour.

200 tons molten pig iron
100 tons scrap iron
20 tons limestone

▪ FIGURE 11.6   *Top:* Diagram of a basic oxygen furnace. *Bottom:* Photo shows workers at a basic oxygen furnace.

The electric discharge is called an arc because, if the discharge is horizontal, it is bent upward in an arc by heated air rising.

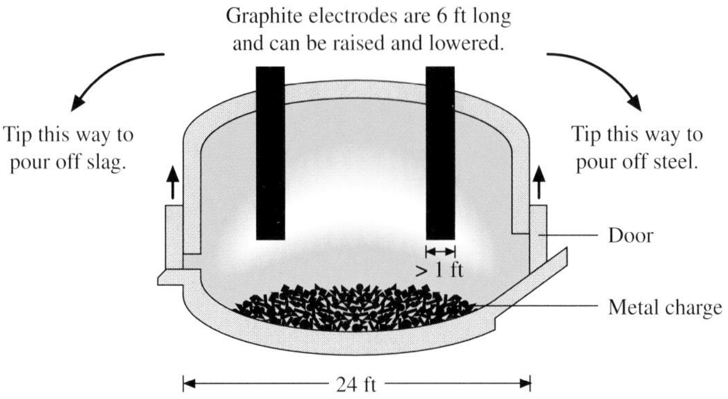

Graphite electrodes are 6 ft long and can be raised and lowered.

Tip this way to pour off slag.

Tip this way to pour off steel.

> 1 ft

Door

Metal charge

24 ft

■ FIGURE 11.7 Diagram of an electric arc furnace.

materials have been added. ■Figure 11.7 is a schematic drawing of an electric arc furnace. The electrodes can be moved down as the charge melts so that all of the thermal energy from the arc goes into the metal charge and no thermal energy is wasted. Because no fuel is needed in an electric arc furnace, no impurities from a fuel are added to the steel. In addition, at the very high temperatures that are reached, sulfur and phosphorus impurities can be completely removed. Steels for demanding uses such as jet engine parts are made in an electric arc furnace.

PRACTICE PROBLEMS

11.19 From the observation that iridium occurs free in nature, what can you conclude about the reactivity of iridium?

11.20 Write molecular equations for each of the following reactions:

(a) $MoO_3(s) + H_2(g) \xrightarrow{heat}$

(b) $PbO(s) + C(s) \xrightarrow{heat}$

SUMMARY

Oxidation numbers are a bookkeeping device that is useful for remembering the chemical properties of the elements and their compounds, naming compounds, and identifying oxidation–reduction reactions. The **oxidation number** or **oxidation state** of a covalently bound element is the charge the element would have if all the shared pairs of electrons in the Lewis formula for the species were transferred to the more electronegative atom. The oxidation numbers of elements are zero; the oxidation number of an atom in a monatomic ion is equal to the charge on the ion. The oxidation numbers of the elements in many common species can be calculated from the formulas of the species by means of a few simple rules. The oxidation numbers observed for the elements in their compounds are related to the positions of the elements in the periodic table.

Oxidation is defined as an increase in oxidation number or a loss of electrons. **Reduction** is defined as a decrease in oxidation number or a gain of electrons. In oxidation–reduction reactions, at least one element must undergo a change in oxidation number. Oxidation and reduction always take place together, and oxida-

tion–reduction reactions are often called **redox** reactions for short. A species that brings about oxidation is called an **oxidizing agent** and is reduced. A species that brings about reduction is called a **reducing agent** and is oxidized.

In any oxidation–reduction reaction, the total increase in oxidation number is the same as the total decrease in oxidation number. The equation for the overall reaction can be divided into a **half-reaction of oxidation** and a **half-reaction of reduction.** Electrons appear on the product side of the equation for a half-reaction of oxidation and on the reactant side of the equation for a half-reaction of reduction. The half-reactions are combined so that electrons cancel to give the equation for the overall reaction. Many titration methods involve oxidation–reduction reactions.

Species with elements in their most positive oxidation states can only be reduced and act only as oxidizing agents; species with elements in their most negative oxidation states can only be oxidized and act only as reducing agents. Species with elements in intermediate oxidation states can be either oxidized or reduced and can act either as reducing agents or as oxidizing agents.

Reactions in which the same species is both oxidized and reduced are called **disproportionation reactions.**

Oxygen from the atmosphere is the most important oxidizing agent. The oxidation–reduction reaction that is carried out on the largest scale is **combustion,** rapid oxidation accompanied by heat and usually by light also. Side reactions that accompany the combustion of fossil fuels are the cause of a number of environmental problems. Besides being used for heat and power, combustion reactions are steps in some important industrial syntheses. Controlled catalytic oxidations with oxygen are also very useful in the manufacture of synthetic chemicals.

The principal source of an element is related to the element's position in the periodic table. Because metals have positive oxidation numbers in their compounds, metals must be obtained from ores by reduction. An **ore** is a rock or mineral that serves as a source of an element. Silicate minerals, although common, are rarely used as ores. Three steps are involved in the production of a metal from its ore: (1) preliminary treatment, (2) reduction, and (3) refining. The separation of metals from their ores, the making of alloys, and the working or heating of metals so as to give them certain properties are all parts of **metallurgy. Alloys** are materials that have metallic properties but are not pure metals. They are composed of two or more metals or of a metal or metals and a nonmetal. **Steels** are alloys that contain more than 50% iron and 0.03–1.5% carbon. Most iron is used in the form of one of the many varieties of steel. Iron has played a more important part in people's material progress than any other element.

ADDITIONAL PRACTICE PROBLEMS

For information about the organization of Additional Practice Problems, Stop & Test Yourself, Putting Things Together, and Applications, see the beginnings of these sections in Chapter 1.

11.21 Use the rules for assigning oxidation numbers to find the oxidation number of each atom in each of the following species: (a) $CaCrO_4$, (b) $Cr(NO_3)_3$, (c) H_2SeO_3, (d) P_4, (e) VO_2, (f) OF_2. (11.1)

11.22 Use the definition of oxidation number to assign oxidation numbers to each atom in each of the following molecules: (a) NCl_3, (b) SbH_3, (c) $SOCl_2$. (11.1)

11.23 Use the periodic table inside the front cover to predict the most positive and the most negative oxidation numbers expected for (a) Cl, (b) F, (c) K, (d) Mn, (e) S. (11.2)

11.24 Use the periodic table inside the front cover to predict the most positive and the most negative oxidation numbers expected for (a) Al, (b) C, (c) Cr, (d) N, (e) O. (11.2)

11.25 The two most important oxides of chromium are Cr_2O_3 and CrO_3. The former melts at 2266 °C and the latter at 197 °C. Suggest an explanation for this difference. (11.2)

11.26 Which is the stronger acid, $HClO_3$ or $HClO_4$? Explain your prediction. (11.2)

11.27 Name each of the following: (a) $FeSO_4$, (b) Cr_2O_3, (c) $HBrO_2$, (d) $Mg(ClO_4)_2$, (e) TlOH. (11.3)

11.28 Name each of the following: (a) KIO_3, (b) HIO_4, (c) V_2O_5, (d) $Ce(SO_4)_2$, (e) Cu_2S. (11.3)

11.29 Write the formula for each of the following: (a) aluminum sulfate, (b) manganese(IV) oxide, (c) potassium nitrite, (d) cobalt(III) acetate, (e) vanadium(II) sulfide. (11.3)

11.30 Write the formula for each of the following: (a) osmium(VIII) oxide, (b) lithium sulfite, (c) sodium nitrite, (d) mercury(II) sulfide, (e) hypobromous acid. (11.3)

11.31 Which of the following reactions are oxidation-reduction? For those that are redox, tell which element is oxidized, which element is reduced, which substance is the oxidizing agent, and which substance is the reducing agent. (11.4)
(a) $2Ag_2CrO_4(s) + 4HNO_3(aq) \longrightarrow$
$$4AgNO_3(aq) + H_2Cr_2O_7(aq) + H_2O(l)$$

(b) $CaCO_3(s) + SiO_2(s) \longrightarrow CaSiO_3(l) + CO_2(g)$
(c) $CH_3OH(aq) + 2CrO_3(aq) \longrightarrow$
$$Cr_2O_3(s) + CO_2(g) + 2H_2O(l)$$
(d) $3H_2SO_3(aq) + 2HNO_3(aq) \longrightarrow$
$$3H_2SO_4(aq) + 2NO(g) + H_2O(l)$$
(e) $Mn(OH)_2(s) + H_2O_2(aq) \longrightarrow MnO_2(s) + 2H_2O(l)$

11.32 Which of the following reactions are oxidation-reduction? For those that are redox, tell which element is oxidized, which element is reduced, which substance is the oxidizing agent, and which substance is the reducing agent. (11.4)
(a) $Ba(OH)_2(aq) + H_2SO_4(aq) \longrightarrow BaSO_4(s) + 2H_2O(l)$
(b) $CaCO_3(s) \longrightarrow CaO(s) + CO_2(g)$
(c) $2Fe_2O_3(l) + 3C(s) \longrightarrow 4Fe(l) + 3CO_2(g)$
(d) $2NH_3(aq) + NaOCl(aq) \longrightarrow$
$$H_2NNH_2(aq) + NaCl(aq) + H_2O(l)$$
(e) $3NO_2(g) + H_2O(l) \longrightarrow 2HNO_3(aq) + NO(g)$

11.33 Write the equation for each of the following redox reactions. (11.5)
(a) $Fe_2O_3(s) + C(s) \longrightarrow Fe(l) + CO_2(g)$
(b) $Fe(s) + O_2(g) + H_2O(l) \longrightarrow Fe(OH)_3(s)$
(c) $C_3H_8(g) + O_2(g) \longrightarrow CO_2(g) + H_2O(g)$
(d) $Cr_2O_3(s) + Si(s) \longrightarrow Cr(s) + SiO_2(s)$

11.34 Check your answer to problem 11.33 by comparing total increase in oxidation number to total decrease in oxidation number. (11.5)

11.35 A student wrote $Cu(s) + Ag^+(aq) \longrightarrow Cu^{2+}(aq) + Ag(s)$ as the equation for the replacement of Ag^+ by Cu. Is this expression an equation? Explain your answer. (11.5)

11.36 For the oxidation–reduction reaction
$$Cl_2(aq) + Na_2S(aq) \longrightarrow S(s) + 2NaCl(aq)$$
(a) write the net ionic equation for the half-reaction of oxidation. (b) Write the net ionic equation for the half-reaction of reduction. (11.5)

11.37 If the net ionic equation for one half-reaction is $Cu^{2+} + 2e^- \longrightarrow Cu(s)$ and the net ionic equation for the other half-reaction is $Al(s) \longrightarrow Al^{3+} + 3e^-$, (a) what is the net ionic equation for the overall reaction? (b) Which half-reaction is the half-reaction of reduction? (11.5)

11.38 Of the following species, (a) which can only be oxidized? (b) Which can only be reduced? (c) Which can undergo disproportionation? Explain your answers.

$$Al \quad Cr^{3+} \quad HNO_2 \quad I_2 \quad S^{2-} \quad (11.6)$$

11.39 Of the following species, (a) which can act only as an oxidizing agent? (b) Which can act only as a reducing agent? (c) Which can act either as an oxidizing agent or as a reducing agent? Explain your answers.
(i) H_2 (ii) H_2SO_4 (iii) I^- (iv) Li (v) $SnCl_2$ (11.6)

11.40 Which of the following can disproportionate?
(a) H_2O_2 (b) Na (c) Cl^- (d) OCl^- (e) P_4 (11.6)

11.41 Write the net ionic equation for the disproportionation reaction:

$$ClO^- \longrightarrow Cl^- + ClO_3^- \quad (11.6)$$

11.42 For the reaction

$$Cu(s) + Cu^{2+} + Cl^- \longrightarrow CuCl(s)$$

(a) write the net ionic equation. (b) Write the molecular equation. (c) Is this reaction disproportionation? (11.6)

11.43 A 10.00-mL sample of Na_2SO_3 solution was titrated with $K_2Cr_2O_7$ in a solution that had been acidified with H_2SO_4. If 19.21 mL of 0.0498 M $K_2Cr_2O_7$ were required to reach the end point, what was the molarity of the Na_2SO_3 solution? The equation for the reaction is

$$3Na_2SO_3(aq) + K_2Cr_2O_7(aq) + 4H_2SO_4(aq) \longrightarrow$$
$$3Na_2SO_4(aq) + K_2SO_4(aq) + Cr_2(SO_4)_3(aq) + 4H_2O(l)$$
$$(11.7)$$

11.44 How many milliliters of 0.0523 M $Na_2S_2O_3$ are required to react with 0.2268 g NaI_3? The equation for the reaction is $NaI_3(aq) + 2Na_2S_2O_3(aq) \longrightarrow 3NaI(aq) + Na_2S_4O_6(aq)$. (11.7)

11.45 Which of the following substances will probably react with oxygen? (a) $C_{12}H_{22}O_{11}$ (b) CuCl (c) He (d) Li (e) $Zn(NO_3)_2 \cdot 6H_2O$ (11.8)

11.46 Write equations for the reactions that will take place. If no reaction is expected, write "no reaction." (11.8)

(a)
$(l) + O_2(g) \xrightarrow{\text{catalyst}}$

(b)
$(l) + O_2(g) \xrightarrow{\text{heat}}$

(c) $Cl_2(g) + O_2(g) \xrightarrow{\text{heat}}$

(d) $CS_2(l) + O_2(g) \xrightarrow{\text{heat}}$

(e) $Fe(s) + O_2(g) \xrightarrow{\text{heat}}$

11.47 (a) What does *metallurgy* mean? (b) What are *ores?* (c) List the three steps involved in the production of metals from ores. (11.9)

11.48 Which will give the more acidic solution if the concentrations are equal, SO_2 or SO_3? Explain your prediction.

11.49 What is the oxidation number of each atom in each of the following substances?
(a) AlN (b) NaOCN (c) $(NH_4)_2Mo_2O_7$ (d) S_8 (e) TiO_2

11.50 What is the oxidation number of each atom in each of the following compounds?
(a) $CaWO_4$ (b) CS_2 (c) FeS (d) $KMnO_4$ (e) $Mg_2P_2O_7$

11.51 Arrange the following manganese-containing species in order of increasing oxidation number of manganese:
(a) MnO_2, (b) Mn_2O_3, (c) MnO_4^-, (d) $MnSO_4$.

11.52 Explain why HCl(aq) is referred to as a nonoxidizing acid whereas HNO_3(aq) is referred to as an oxidizing acid.

11.53 Are the most positive and the most negative oxidation numbers shown in Figure 11.1 periodic functions of atomic number? Explain your answer. [*Hint:* Plot the most positive and the most negative oxidation numbers for the second- and third-row elements against atomic number.]

11.54 (a) On what two factors does the form in which a metal occurs in nature depend? (b) Sketch the periodic table and label the areas where each of the following are located: (i) metals that occur free in nature (ii) metals that are obtained from silicates (iii) metals that are obtained from chlorides (iv) metals that are obtained from sulfides (v) metals that are obtained from sulfates (vi) metals that are obtained from carbonates (vii) metals that are obtained from oxides. (c) Why are silicates *not* usually economical sources of metals?

11.55 In the production of pig iron in a blast furnace, what is the function of each of the following? (a) limestone (b) blast of hot air (c) coke

11.56 Should an oxidizing agent or a reducing agent be used to bring about the following change? $H_3PO_3 \longrightarrow H_4P_2O_6$

STOP & TEST YOURSELF

1. The oxidation number of S in $NaHSO_3$ is
   (a) -2 (b) 0 (c) $+2$ (d) $+4$ (e) $+6$

2. The oxidation number of I in ICl is
   (a) -1 (b) 0 (c) $+1$ (d) $+3$ (e) $+7$

3. The highest oxidation number expected for sulfur in its compounds is
   (a) -2 (b) $+2$ (c) $+4$ (d) $+6$ (e) $+8$

4. Which of the following reactions is not oxidation-reduction?
   (a) $4H_3PO_3(l) \longrightarrow PH_3(g) + 3H_3PO_4(l)$
   (b) $2Na_2S_2O_3(aq) + 4H_2O_2(aq) \longrightarrow$
   $$Na_2S_3O_6(aq) + Na_2SO_4(aq) + 4H_2O(l)$$
   (c) $3Na_2S_2O_4(aq) + 6NaOH(aq) \longrightarrow$
   $$5Na_2SO_3(aq) + Na_2S(aq) + 3H_2O(l)$$
   (d) $3H_2S(aq) + 2HNO_3(aq) \longrightarrow 3S(s) + 2NO(g) + 4H_2O(l)$

(e) $PF_3(g) + 3H_2O(l) \longrightarrow H_3PO_3(aq) + 3HF(aq)$

5. In Problem 4, which reactions are disproportionation?
6. Which of the following can disproportionate?
   (a) Al (b) Cl_2 (c) $Fe_2(SO_4)_3$ (d) HNO_3 (e) ZnS
7. The chemistry of which of the following elements involves the greatest variety of oxidation–reduction reactions?
   (a) Ca (b) Cl (c) F (d) Na (e) Ne
8. The preferred name for MnO_2 is
   (a) dioxygen manganide (b) manganese(II) oxide
   (c) manganese(IV) oxide (d) manganous oxide
   (e) manganese dioxide
9. The formula for iron(II) sulfate is
   (a) $FeSO_4$ (b) Fe_2SO_4 (c) $Fe(SO_4)_2$ (d) $Fe_2(SO_4)_3$
   (e) $FeSO_3$
10. The products of the reaction of carbon with concentrated nitric acid are carbon dioxide, nitrogen dioxide, and water. In the equation for the reaction, the coefficient of nitric acid is
    (a) 1 (b) 2 (c) 3 (d) 4 (e) 5

Use the equation

$2KMnO_4(aq) + 5H_2C_2O_4(aq) + 3H_2SO_4(aq) \longrightarrow$
$$2MnSO_4(aq) + K_2SO_4(aq) + 10CO_2(g) + 8H_2O(l)$$

to answer problems 11 and 12.

11. The element that is reduced is
    (a) carbon (b) hydrogen (c) manganese (d) oxygen
    (e) sulfur
12. The substance that is the oxidizing agent is
    (a) $H_2C_2O_4$ (b) H_2SO_4 (c) $KMnO_4$ (d) K_2SO_4 (e) $MnSO_4$
13. Which substance will be oxidized by oxygen?
    (a) CH_3Cl (b) CCl_4 (c) $Mg(ClO_4)_2$ (d) Ne (e) Pt
14. Titration of a 25.00-mL sample of a solution containing Sn^{2+} ion required 24.12 mL of 0.0987 M $Ce(SO_4)_2$ solution. What is the molarity of Sn^{2+} in the solution? The equation for the reaction is $Sn^{2+} + 2Ce^{4+} \longrightarrow Sn^{4+} + 2Ce^{3+}$
    (a) 0.001 19 (b) 0.002 38 (c) 0.0476 (d) 0.0952
    (e) 0.191
15. Which is *not* a part of preliminary treatment of an ore?
    (a) Addition of desirable impurities
    (b) Removal of undesirable impurities
    (c) Concentration of the desired component
    (d) Conversion of the desired component to a form better suited for reduction
    (e) Any of the above may be a part of preliminary treatment.

PUTTING THINGS TOGETHER

11.57 The molecular equation for the reaction between sodium sulfite and potassium permanganate is

$3Na_2SO_3(aq) + 2KMnO_4(aq) + H_2O(l) \longrightarrow$
$$3Na_2SO_4(aq) + 2MnO_2(s) + 2KOH(aq)$$

(a) Write the net ionic equation for this reaction. (b) Which element is oxidized, and which element is reduced? (c) Which substance is the oxidizing agent, and which substance is the reducing agent?

11.58 The net ionic equation for the reaction of iodine with chlorate ion is

$$3I_2(s) + 5ClO_3^- + 3H_2O(l) \longrightarrow 6HIO_3(aq) + 5Cl^-$$

(a) Write the complete ionic equation and the molecular equation for this reaction if $NaClO_3$ is the source of the ClO_3^- ion. (b) Which element is oxidized, and which element is reduced? (c) Which substance is the oxidizing agent, and which substance is the reducing agent?

11.59 Copper(II) sulfide dissolves in concentrated nitric acid. The molecular equation is

$3CuS(s) + HNO_3(conc) \longrightarrow$
$$3CuSO_4(aq) + 8NO(g) + 4H_2O(l)$$

If 35.4 g copper(II) sulfide and 41.3 mL of concentrated nitric acid (16.0 M) are used, how many grams of which reactant will be left over?

11.60 Suppose that the source of the Fe^{2+} ion in the reaction

$$Fe^{2+} + Cr_2O_7^{2-} \longrightarrow Fe^{3+} + Cr^{3+}$$

is $FeSO_4$, the source of the $Cr_2O_7^{2-}$ is $K_2Cr_2O_7$, and the acid used is H_2SO_4. The net ionic equation for the reaction that takes place in acidic solution is

$$6Fe^{2+} + 14H^+ + Cr_2O_7^{2-} \longrightarrow 6Fe^{3+} + 2Cr^{3+} + 7H_2O(l)$$

Write the molecular equation for the reaction.

11.61 The net ionic equation for the reaction of glycerol with permanganate in basic solution is

$14MnO_4^- + 20OH^- + C_3H_8O_3(aq) \longrightarrow$
$$14MnO_4^{2-} + 3CO_3^{2-} + 14H_2O(l)$$

If the source of the MnO_4^- ion is potassium permanganate, $KMnO_4$, and the source of the OH^- is potassium hydroxide, KOH, (a) write the complete ionic equation for the reaction. (b) Write the molecular equation for the reaction.

11.62 Camphor ($C_{10}H_{16}O$) can be made by the oxidation of borneol ($C_{10}H_{18}O$) with hypochlorous acid. The equation is

$C_{10}H_{18}O(s) + HOCl(aq) \longrightarrow$
$$C_{10}H_{16}O(s) + HCl(aq) + H_2O(l)$$

If 6.16 g of borneol are treated with an excess of the oxidizing agent, (a) what is the theoretical yield of camphor? (b) If 3.28 g of camphor are obtained, what is the percent yield?

11.63 Chlorine forms several oxides. Write the formula and name for the oxide of chlorine in which chlorine has an oxidation number of (a) $+1$ (b) $+4$ (c) $+6$. (The molecular mass of this oxide is 167.)

11.64 For each of the following compounds,

$$SnCl_4 \qquad PCl_3 \qquad PCl_5$$

(a) write the Lewis formula. (b) Describe the shape of the molecule. (c) Calculate the formal charge on each atom. (d) Tell the oxidation number of each atom. (e) Give the systematic name.

11.65 Arrange the oxo acids of chlorine in order of decreasing acidity.

11.66 Titration of the iron(II) ion from a 0.5811-g sample of iron ore required 22.49 mL of 0.053 47 M $K_2Cr_2O_7$ to reach the end point (which was signaled by an indicator). What was the percentage Fe in the ore? In acidic solution, dichromate ion, $Cr_2O_7^{2-}$, is reduced to Cr^{3+}. The net ionic equation is

$$6Fe^{2+} + Cr_2O_7^{2-} + 14H^+ \longrightarrow 6Fe^{3+} + 2Cr^{3+} + 7H_2O(l)$$

What was the mass percentage iron in the ore?

11.67 Predict the lowest and highest oxidation numbers likely to be observed for element 114.

11.68 Distinguish clearly between ionic charge, partial charge, formal charge, and oxidation number.

11.69 (a) What is the oxidation number of O in hydrogen peroxide, H_2O_2? (b) Can H_2O_2 act as an oxidizing agent? If so, predict the product. (c) Can H_2O_2 act as a reducing agent? If so, predict the product.

11.70 One way to organize the mass of information available about reactions is to classify them. Of the following reactions, which are (a) redox? (b) Combination? (c) Combustion? (d) Decomposition? (e) Double replacement? (f) Formation? (g) Neutralization? (h) Single replacement?

(i) $S(s) + O_2(g) \longrightarrow SO_2(g)$
(ii) $H_2(g) + Cl_2(g) \longrightarrow 2HCl(g)$
(iii) $2HgO(s) \longrightarrow 2Hg(l) + O_2(g)$
(iv) $CaO(s) + CO_2(g) \longrightarrow CaCO_3(s)$
(v) $CaCO_3(s) \longrightarrow CaO(s) + CO_2(g)$
(vi) $CH_4(g) + 2O_2(g) \longrightarrow CO_2(g) + 2H_2O(l)$
(vii) $Zn(s) + 2HCl(aq) \longrightarrow ZnCl_2(aq) + H_2(g)$
(viii) $Ca(s) + 2H_2O(l) \longrightarrow Ca(OH)_2(s) + H_2(g)$
(ix) $HNO_3(aq) + NaOH(aq) \longrightarrow NaNO_3(aq) + H_2O(l)$
(x) $AgNO_3(aq) + NaCl(aq) \longrightarrow AgCl(s) + NaNO_3(aq)$
(xi) $Na_2CO_3(aq) + 2HCl(aq) \longrightarrow$
$$2NaCl(aq) + H_2O(l) + CO_2(g)$$
(xii) $3Cu(s) + 8HNO_3(aq) \longrightarrow$
$$3Cu(NO_3)_2(aq) + 2NO(g) + 4H_2O(l)$$

11.71 Which will probably react with oxygen?

(a) $\sim CF_2-CF_2-CF_2-CF_2-CF_2-CF_2 \sim$

(b) $\sim CH_2-CH_2-CH_2-CH_2-CH_2-CH_2 \sim$

(c) $\sim CH_2-CH-CH_2-CH-CH_2-CH \sim$
$\qquad \qquad | \qquad \qquad | \qquad \qquad |$
$\qquad \quad OH \qquad \quad OH \qquad \quad OH$

11.72 Write equations for the following:

(a) $Al(powder) + O_2(g) \xrightarrow{heat}$

(b) $Na(s) + O_2(g) \xrightarrow{heat}$ (*Hint:* Remember Section 1.12.)

(c) $K(s) + O_2(g) \xrightarrow{heat}$

(d) $N_2(g) + O_2(g) \xrightarrow{heat}$

(e) $\sim CH-CH_2-CH-CH_2 \sim (s) + O_2(g) \xrightarrow{heat}$
$\qquad | \qquad \qquad \quad |$
$\qquad CH_3 \qquad \qquad CH_3$

11.73 Heating the sulfide ore, CuS, in air results in *reduction* of copper to the metal

$$CuS(s) + O_2(g) \longrightarrow Cu(s) + SO_2(g)$$

Explain this surprising result.

11.74 The reactions of organic compounds are often organized on the basis of functional groups. For example, all organic compounds that contain the group —COOH, such as acetic acid (CH_3COOH), belong to a family called carboxylic acids and react similarly. Two characteristic properties of carboxylic acids are odor and reaction with bases. (a) Write molecular equations for the neutralization of aqueous acetic acid with aqueous sodium hydroxide; aqueous propionic acid, CH_3CH_2COOH, with aqueous sodium hydroxide; aqueous butyric acid, $CH_3CH_2CH_2COOH$, with aqueous potassium hydroxide. (b) The class of organic compounds introduced in this chapter is ethers. Cyclic ethers with three-membered rings, such as ethylene oxide, are very reactive; for example, ethylene oxide reacts with water to form ethylene glycol:

$$H_2C-CH_2(g) + H_2O(l) \longrightarrow HOCH_2CH_2OH(l)$$
$$\qquad \backslash \; O \; /$$

Write the equation for the reaction of propylene oxide,

$$CH_3CH-CH_2(g)$$
$$\qquad \backslash O /$$

with water. What do you think the product is called? Write complete structural formulas for propylene oxide and the product. (c) Acyclic ethers such as CH_3OCH_3 and $CH_3CH_2OCH_2CH_3$ are very unreactive, and the latter is often used as a solvent. What is the meaning of the prefix *a*-? Why is ethylene oxide more reactive than the acyclic ethers?

11.75 When copper is dissolved in nitric acid, copper(II) nitrate, nitrous acid, nitric oxide gas, and water are formed. Explain why more than one "equation," for example,

$$5Cu(s) + 14HNO_3(aq) \longrightarrow$$
$$5Cu(NO_3)_2(aq) + 2HNO_2(aq) + 2NO(g) + 6H_2O(l)$$

and

$$9Cu(s) + 26HNO_3(aq) \longrightarrow$$
$$9Cu(NO_3)_2(aq) + 6HNO_2(aq) + 2NO(g) + 10H_2O(l)$$

can be written for this reaction.

11.76 For the disproportionation reaction

$$Cl_2(g) + Ca(OH)_2(s) \longrightarrow Ca(OCl)_2(s) + CaCl_2(s) + H_2O(l)$$

which is used to make bleaching powder, (a) write the molecular equation. (b) Which element disproportionates? (c) What is $Ca(OCl)_2$ called?

11.77 Give the oxidation number of the atom that is in italics in each of the following formulas: (a) Fe_3O_4, an ore of iron that gives many rocks and soils a reddish color (b) $Na_2S_2O_3$, "hypo" used as a fixer in photography (c) MoS_2, used as a dry lubricant and as a hydrogenation catalyst (d) Cl_2O, used as a bleach for wood pulp and textiles (e) $AlSb$, used in semiconductor research

11.78 In seawater, NH_4^+ from decay and metabolic processes is oxidized first to NO_2^- and then to NO_3^-. What is the oxidation number of nitrogen in each of these three species?

11.79 The following compounds are on the Food and Drug Administration (FDA) list of substances that may be added to foods. Write the name of each of these compounds. (a) CuI (b) $Fe_3(PO_4)_2$ (c) $FeSO_4$ (d) H_2SO_4 (e) $KHSO_3$ (f) K_2SO_4 (g) Na_2SO_3 (h) SO_2

11.80 The iron from a 29.99-g sample of soil from the surface of the moon was brought into solution, and all the iron was converted to Fe^{2+}. The solution was then analyzed for iron by titration with standard 0.020 00 M $KMnO_4$ in acidic solution. If 26.32 mL of 0.020 00 M $KMnO_4$ were required to reach the end point, what was the percent by mass iron in the sample of lunar soil?

11.81 The half-reactions

$$Fe(s) \longrightarrow Fe^{2+} + 2e^-$$
$$Fe^{2+} \longrightarrow Fe^{3+} + e^-$$
$$O_2(g) + 2H_2O(l) + 4e^- \longrightarrow 4OH^-$$

result in the corrosion (rusting) of iron. (a) Which half-reaction is reduction? (b) Write the net ionic equation for the overall reaction obtained by combining the second and third half-reactions.

11.82 Some typical reactions that are involved in the weathering of Earth's crust are
(a) $CaCO_3(s) + H_2O(l) + CO_2(aq) \longrightarrow Ca(HCO_3)_2(aq)$
(b) $MnS(s) + 2O_2(aq) \longrightarrow MnSO_4(aq)$
(c) $SiO_2(s) + 2H_2O(l) \longrightarrow H_4SiO_4(aq)$
Which is redox?

11.83 Emission of sulfur dioxide warned of the eruption of Mount Pinatubo, thereby allowing areas potentially in the path of lava flows to be evacuated and saving many lives. In an analysis of air for sulfur dioxide, 5.00×10^2 L of air at 25 °C and 760 mmHg was bubbled through 250.0 mL of acidic 0.004 98 M $KMnO_4$. The net ionic equation for the reaction that occurred is

$$5SO_2(g) + 2MnO_4^- + 2H_2O(l) \longrightarrow$$
$$2Mn^{2+} + 5SO_4^{2-} + 4H^+$$

The MnO_4^- remaining in solution was titrated with 0.0998 M $FeSO_4(aq)$. The reaction that took place is

$$8H^+ + MnO_4^- + 5Fe^{2+} \longrightarrow Mn^{2+} + 5Fe^{3+} + 4H_2O(l)$$

and 37.26 mL were required to reach the end point. What was the concentration of SO_2 in the air in ppm (volume)?

11.84 (a) What is the molarity of $K_2Cr_2O_7$ in the Breathalyzer? Assume that the density of the solution is 1.00 g/mL as it is very dilute. (b) The volume of the sample of breath analyzed is 52.5 mL. If the concentration of alcohol in a sample of breath is 1.3 g/L, how many moles of $K_2Cr_2O_7$ will be used up?

11.85 Metabolism of foods such as carbohydrates provides energy for maintaining body temperature, moving muscles, and synthesizing more complex organic substances. Oxygen from air is the preferred oxidizing agent. (a) Write the molecular equation for the reaction of glucose, $C_6H_{12}O_6$, with oxygen in aqueous solution. (b) Oxidation of one mole of glucose releases 2.87×10^3 kJ. Oxidation takes place in many small steps so that energy is released gradually. Explain why the total quantity of energy released is independent of the steps involved. (c) If the supply of O_2 is limited, as it is in some water and in sediments, mud, and soil, anerobic organisms can use nitrate as an oxidizing agent,

$$NO_3^- \longrightarrow N_2(g)$$

and when all nitrates are used up, sulfate,

$$SO_4^{2-} \longrightarrow S^{2-}$$

Write net ionic equations for these half-reactions of reduction. (Hint: H^+ is a reactant, and H_2O is a product. First, balance materially; then use electrons to balance charge.)

11.86 Silver is expensive and is usually recovered after use in industrial operations. During recovery of silver in a plant, a violent explosion took place as a result of the reaction between finely divided titanium metal and aqueous silver nitrate, which yields titanium(IV) oxide, silver, and nitrogen. (a) Write the molecular equation for the reaction. (b) Which element is oxidized? (c) Which element is reduced? (d) Which substance is the oxidizing agent? (e) Which substance is the reducing agent? (f) What is ΔH_{rxn}° for the reaction? (g) Which reactant was probably used in excess, and why? (h) The material that exploded contained 40% by mass titanium and 60% by mass silver nitrate. Which substance was the limiting reactant?

11.87 Lead azide, $Pb(N_3)_2$, which is used in detonating caps, is made by the reaction $Pb(NO_3)_2 + 2NaN_3 \longrightarrow Pb(N_3)_2 + 2NaNO_3$ in aqueous solution. (a) Write the Lewis formula (including formal charges) for the azide ion. (b) What is the oxidation number of nitrogen in the azide ion? (c) Is the reaction used to prepare lead azide a redox reaction? Explain your answer. (d) From the method of preparation, what can you conclude about the solubility of lead azide in water? (e) Write the complete ionic and net ionic equations for the reaction used to prepare lead azide. (f) When lead azide explodes, the products are nitrogen gas and lead metal. Write a molecular equation for this reaction. What element is oxidized? What element is reduced?

11.88 Imagine that you are a photographer and have a bottle of hypo ($Na_2S_2O_3$) that you are going to use to make a solution to fix a black-and-white picture. You want to check the purity of the hypo by titration with iodine. If a 2.858-g sample of the hypo requires 22.48 mL of 0.251 M I_2 to reach the end point, what is the mass percent of $Na_2S_2O_3$ in the hypo? Reaction of thiosulfate ion with iodine yields tetrathionate ion ($S_4O_6^{2-}$) and iodide ion.

11.89 Nitric acid is made by the oxidation of ammonia with oxygen from air. The steps in the process are

$$4NH_3(g) + 5O_2(g) \longrightarrow 4NO(g) + 6H_2O(l)$$

$$2NO(g) + O_2(g) \longrightarrow 2NO_2(g)$$

$$3NO_2(g) + H_2O(l) \longrightarrow 2HNO_3(aq) + NO(g)$$

(a) Which reactions are redox? (b) Which reaction is disproportionation? (c) What is the theoretical yield of HNO_3 from 100.0 lb of a mixture of 10.0 lb NH_3 and 90.0 lb air? Air is 23.1% O_2 by mass. (d) If the percent yield is 95%, what is the actual yield in pounds? (e) The process was originally operated at atmospheric pressure, but modern plants operate at 10 MPa. Why?

11.90 In the production of zinc metal, which is used as an anticorrosion coating and to make alloys and batteries, the reaction

$$Zn(powder) + CdSO_4(aq) \longrightarrow Cd(s) + ZnSO_4(aq)$$

is used to separate cadmium from zinc. (a) What compound is the principal source of zinc? (b) Why does zinc ore usually contain cadmium? (c) Where must cadmium be located in the activity series (Table 4.5)? (d) Write the net ionic equation for the half-reaction of oxidation. (e) Write the net ionic equation for the half-reaction of reduction.

A Geochemist Puts the Pressure on Minerals

CAROLYN REBBERT

Geochemist, American Museum of Natural History, New York

B.S. and M.S. Geological Sciences
Virginia Polytechnic Institute and State University

Ph.D. Geological Sciences
University of Oregon

I come from a family of chemists. My Mexican grandmother and grandfather both worked as chemical engineers, and my parents met during their graduate studies in physical chemistry. As the youngest of four children, I felt the need to choose a career that would distinguish me from my parents and siblings. I enjoyed science among other things, but my brother was studying physics, one of my sisters was in a forestry program, and my parents were too well known in the local chemistry community for my liking. The natural sciences were attractive to me with their mixture of field and laboratory work. I had volunteered one summer for a geologist at the Smithsonian museum, and while I enjoyed the work it also made me realize how little I really knew about geology. But one thing I did know was that geologists could teach me more about minerals, which had fascinated me since I was a child. So, between that and the allure of an otherwise mysterious science, I chose to major in geology in college.

I took chemistry my freshman year as part of the requirements for a geology major. I was a little intimidated, since it was an advanced general chemistry course and my professor had been a student of my father's. In our first laboratory session, the professor had given his speech on safety, warning that despite his efforts, one of us would injure ourselves; I kept him honest by fulfillng his prophecy that very day. In actuality, the course went well, but I recall thinking that I probably wasn't practical enough to be a very good lab scientist and hoped that rocks would be more forgiving.

I currently classify myself as a high-temperature geochemist and am involved in investigating the interactions that occur between rocks and the fluids that migrate through them along mineral-grain boundaries. The fluids, which are often aqueous solutions, may originate from a number of sources. For example, they may be the result of volatiles exsolved from a deeply buried crystallizing magma (molten rock), or they may be produced by the decomposition of volatile-bearing minerals as temperatures and pressures increase during burial of a sediment. Whatever the origin and subsequent history of these fluids, they present a dilemma to the geologist studying them: by the time geologic processes have uplifted and exposed a rock that was originally formed deep below the surface of the earth, whatever fluids it may have interacted with are no longer present.

One way of addressing this problem (and one which I use in my research) is by studying volatile-bearing minerals that supply clues to the composition of the fluids with which they equilibrated. These are minerals that contain variable amounts of anions such as OH^-, Cl^-, F^-, CO_3^{2-}, or SO_4^{2-}. In particular, I have investigated the hydrous mineral annite, $KFe_3AlSi_3O_{10}(OH)_2$, to determine what partial pressures of H_2 are required in an aqueous fluid to produce different degrees of oxidation of the iron in the mineral. Most of the iron in annite occurs as Fe^{2+}, but other geologists had observed variable amounts of Fe^{3+} in natural annites. It has been widely assumed that this oxidation occurs by the reaction

$$Fe^{2+} + OH^- \longrightarrow Fe^{3+} + O^{2-} + 0.5H_2(g)$$

where the mineral would lose some hydrogen from its structure as the iron oxidized. By applying the principle of Le Châtelier, it is expected that a decrease in the partial pressure of H_2 in a coexisting fluid will result in an increase in the amount of oxidation in annite.

Rather than using naturally occurring annites that are chemically much more complex than the above formula, I synthesized my own minerals. Once I had synthesized annite, I encapsulated it with distilled water in 2.5-cm-long capsules of $Ag_{70}Pd_{30}$ and welded the capsules shut to prevent water loss. I put the capsule in a pressure vessel called a bomb and placed the bomb in a furnace at temperatures of 500–750 °C. A total pressure of 1 kbar was maintained by an Ar-H_2 gas mixture. An interesting property of AgPd alloys is that H_2 can diffuse through them at high temperatures. Thus, I can control the partial pressure of H_2 by varying its amount in the initial Ar-H_2 mix. I did a series of experiments with different initial partial pressures of H_2 and then analyzed the Fe^{3+} contents of the annite by a titration technique.

As expected, annites that had been heated in a more reduced atmosphere had less Fe^{3+} than those heated with a lower H_2 partial pressure. By correlating the amount of Fe^{3+} to the H_2 pressure, I was able to develop a hydrogen barometer; for example, geologists with the mineral annite occurring in a rock sample can analyze the Fe^{3+} contents of the annite and then refer to my calibration to determine the partial pressure of H_2 in the fluid that coexisted with their rock.

Chemistry is an integral part of many geological studies. Application of thermodynamics to mineral compositions provides clues not just to fluid compositions, but also to the temperatures and pressures at which the rock formed. Correlation of crystal structures with periodic variations in ion size, bond strength, and bond type helps us understand how compositional changes in a given mineral enhance its stability or affect other properties of the mineral. Other geologists are concerned with the distribution of elements throughout the earth and the means by which some elements are concentrated to produce economic deposits that provide the raw materials for many of our manufactured goods. The periodic table is virtually spread out beneath our feet.

12

LIQUIDS, SOLIDS, AND CHANGES IN STATE

OUTLINE

Like the water in this stream, liquids flow and take the shape of the bottom of their container. Solids, such as the rocks, do not flow and have shapes of their own.

In this chapter, we will extend the kinetic-molecular theory to liquids and solids. We will learn about the intermolecular attractions that cause gases to condense to liquids and liquids to freeze to solids, and we will study the properties of liquids and solids and the changes that take place between the different states.

— *Why is a study of liquids, solids, and changes in state a part of general chemistry?*

The physical properties of substances and their reaction rates depend on state. Seventy-one percent of Earth's surface is covered with liquid water; if water were not liquid, life as we know it would be impossible. However, most materials are solids under the conditions that exist on the surface of Earth, and a large part of modern technology is based on the characteristics of different types of solids. Changes in state are also important; for example, evaporation (change from liquid to gas) of sweat keeps our bodies from overheating.

— *What do you already know about liquids, solids, and changes in state?*

From your everyday experience, you know that liquids have definite volumes but take the shape of the lower part of their container. Solids not only have fixed volumes but also have shapes of their own. ◄Figure 12.1 compares the behavior of liquids and solids with the behavior of gases. Liquids and solids are not nearly as compressible as gases. (Have you ever kicked a rock or a football filled with water?) You know that samples of liquids of any size can be poured (provided that you can lift them). Samples of solids cannot be poured unless the solid is first divided into small pieces. You have probably noticed that some liquids can be poured faster than others. Water, for example, flows more easily than molasses or motor oil. You may have discovered that, if you are careful, you can heap water up a little above the top of the sides of a container.

Many liquids evaporate if left in an open container. So do some solids—mothballs, for instance. Evaporation of rubbing alcohol from your skin cools your skin. Vapors condense to liquids when cooled; drops of water from warm, humid air collect on the outside of a glass containing a cold drink. Sometimes vapors deposit as solids; crystals of ice form on the inside of the lid of a box of ice cream if you keep the box of ice cream in the freezer too long.

The liquid, water, freezes if cooled and boils if heated. The solid, ice, melts when heated; if an ice cube melts in your hand, your hand feels cold. Ice cubes float at the top of a glass of water but do not stick up very much above the surface of the water.

From your work in general chemistry laboratory, you know that the meniscus at the top of water in a glass graduated cylinder is concave and the meniscus at the top of mercury in a barometer is convex. You may also have observed that the surface at the top of water in a plastic graduated cylinder is flat.

Top: *Frost on grass (and maple leaf).*
Bottom: *Rock split by freezing water.*

(a)

(b)

▬ FIGURE 12.1   (a) Solids have shapes of their own. (b) Liquids take the shape of the lower part of their container. All the samples of water shown have a volume of 100 mL. (c) Gases fill their containers taking the shape of the container.

(c)

A KINETIC-MOLECULAR VIEW OF LIQUIDS AND SOLIDS

All real gases can be condensed to liquids by lowering the temperature and increasing pressure. According to the kinetic-molecular theory, lowering the temperature of a sample of gas decreases the average speed at which the molecules move. A smaller fraction of the molecules have very high speeds at the lower temperature. Increasing the pressure on a sample of gas pushes the molecules closer together. Molecules of real gases attract each other; therefore, if molecules of a real gas are close together and not moving very fast, they stick to each other and form a drop of liquid. In the

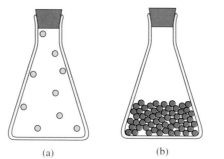

FIGURE 12.2 Kinetic-molecular view of gases, liquids, and solids: (a) gas, (b) liquid or glass, and (c) crystalline solid. A useful analogy for the kinetic-molecular view is going to the movies. On the way to the theater, people are far apart like the molecules of a gas. Going into the theater, if there is a crowd, people are touching but still moving like the molecules of a liquid. Once in their seats, like the molecules of a solid, they are fixed in position in an orderly arrangement, but can still wiggle.

liquid state, the molecules are still free to move but they touch each other; the molecules are still arranged in a disorderly way [compare Figures 12.2(a) and (b)].

Under ordinary conditions, the densities of liquids and solids are of the order of g/cm³ (10³ kg/m³). The densities of gases are of the order of g/L (kg/m³), only about one-thousandth as large as the densities of liquids and solids. A sample of a gas is mostly empty space.

If the temperature of a liquid is lowered, the average speed at which the molecules move decreases. When the temperature becomes low enough, the molecules no longer have sufficient energy to move around and can only vibrate. If the temperature is lowered very quickly, the *molecules get stuck in the disorderly arrangement of the liquid* as shown in Figure 12.2(b). A **glass** or **amorphous solid** is formed. If the temperature is lowered slowly, the *molecules have time to arrange themselves in an orderly way* as shown in Figure 12.2(c), and a **crystalline solid** is formed. Figure 12.3 shows sugar solidified in both glassy and crystalline forms.

Amorphous means having no definite shape.

Note that the terms *gas, liquid,* and *solid* refer to collections of a number of molecules. You can't melt or boil a single molecule; melting and boiling and other *properties of collections of molecules* are called **bulk properties.**

Figures 12.2(b) and (c) explain why liquids and solids are not nearly as compressible as gases. The molecules are touching; they cannot be pushed much closer together. The density of a substance in the solid phase is *usually* a little greater than the density in the liquid phase. Most solids sink to the bottom of the liquid

FIGURE 12.3 *Left:* The sugar in a lollipop has solidified rapidly in a glassy form. Glasses can be molded into different shapes. *Right:* The sugar in rock candy has solidified slowly in a crystalline form. Crystals of sugar have a characteristic shape. Crystals are formed in many natural processes; for example, snowflakes are crystalline water. Glasses are relatively rare in nature.

instead of floating as ice does (▪Figure 12.4). The same volume generally holds more molecules when the molecules are arranged in an orderly way. Water is exceptional; it is unusual in many ways.

The kinetic-molecular picture also explains why liquids take the shape of the lower part of the container and why liquids can be poured. In liquids the molecules are free to move. As one molecule moves, it leaves a space. Another molecule moves into the space leaving another space as shown in ▪Figure 12.5.

▪ FIGURE 12.4   Ice cubes float on water *(left)*. However, most substances are denser in the solid state than in the liquid state. For example, cubes of *t*-butyl alcohol, $(CH_3)_3COH(s)$, sink to the bottom of liquid *t*-butyl alcohol, $(CH_3)_3COH(l)$ *(right)*.

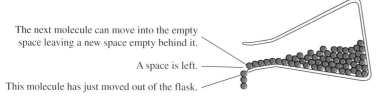

The next molecule can move into the empty space leaving a new space empty behind it.

A space is left.

This molecule has just moved out of the flask.

▪ FIGURE 12.5   Pouring a liquid. As one molecule moves, it leaves a space. The next molecule moves into the empty space and leaves another empty space. The molecules in a liquid move like cars in a traffic jam. When cars are bumper-to-bumper, a car can only move forward when the car ahead moves. Cars move forward into empty spaces, and the empty space moves back.

Solids keep their shape; they cannot be poured because the molecules are fixed in position. (A finely divided solid like dry sand can be poured because the particles can move like the molecules of a liquid.) ▪Figure 12.6 shows macroscopic views of some crystalline solids and a glass. The smooth faces and characteristic angles between the faces of crystalline solids result from the orderly arrangement of the units of the solid [compare Figures 12.2(c) and 12.6(a) and (b)]. The properties of gases and liquids are the same in all directions; the properties of most crystalline solids are different in different directions. The directional nature of the properties of mica and asbestos can be seen with the naked eye. Going from top to bottom of the crystal of mica in Figure 12.6(c), for example, layers are encountered. Going from front to back or left to right, no layers are encountered.

(a)

(b)

(c)

(d)

(e)

▪ FIGURE 12.6   (a) Quartz, (b) calcite, (c) mica, and (d) asbestos are minerals—solid substances that occur in nature, are usually inorganic, and have definite composition and crystal form. Crystals break (cleave) along planes. (e) Obsidian is a naturally occurring glass. Glasses break along curves.

Bond holding two
bromine atoms together
in a molecule is an
intramolecular attraction.

Force holding two
bromine molecules
together is an
intermolecular
attraction.

- FIGURE 12.7 Intermolecular and intramolecular attractions. The difference between intermolecular and intramolecular is like the difference between intercollegiate and intramural.

van der Waals was a Dutch physicist.

In the Interparticle Attractions module of the Liquids section, students are given a pair of compounds. They must describe the type(s) of interparticle attraction for each, decide which compound has the stronger attraction between particles, and predict the effect of the stronger attractions on properties such as vapor pressure and boiling point.

Glassy solids can be molded in almost any shape (see lollipop in Figure 12.3); the molecules get stuck in the disorderly arrangement of the liquid from which the glass was formed. When glasses break, they must split between atoms because atoms cannot be divided in physical and chemical changes. However, since the atoms are not arranged in planes in a glass, the surfaces of broken pieces of glass are curved as shown in Figure 12.6(e).

12.2 INTERMOLECULAR ATTRACTIONS

Before we consider how the kinetic-molecular theory explains the properties of liquids and solids in more detail, we first need to think about the nature of the attractive forces between molecules. *Attractions between molecules* are called **intermolecular attractions.** You must be very careful to distinguish intermolecular attractions from intramolecular attractions. **Intramolecular attractions** are the forces *inside* molecules that hold the atoms making up the molecule together; intramolecular attractions are *chemical bonds.* Intermolecular attractions hold two or more molecules together. -Figure 12.7 shows the difference.

Intermolecular attractions must exist between *all* molecules because all gases condense to liquids if the temperature is low enough and the pressure high enough. *Intermolecular attractions* are often called **van der Waals forces.** Intermolecular attractions are not nearly as strong as intramolecular forces. The energies required to break covalent bonds ordinarily range from 150 to 800 kJ/mol, whereas the energies required to overcome intermolecular attractions are normally of the order of 1–40 kJ/mol. The strength of intermolecular attractions decreases rapidly with increasing distance between molecules. Therefore, intermolecular attractions are not very important in gases, where the molecules are far apart, but are very important in liquids and solids, where the molecules are close together. There are three kinds of intermolecular attractions: dipole–dipole interactions, hydrogen bonds, and London forces.

Dipole–Dipole Attractions

In Section 9.5, you learned how to use differences in electronegativity to predict the polarities of bonds; and in Section 10.2, you learned how to predict from molecular shape whether molecules that have polar bonds are polar. Consider hydrogen chloride as an example. The electronegativity of chlorine (2.8) is greater than the electronegativity of hydrogen (2.2); therefore, the bond between hydrogen and chlorine in a hydrogen chloride molecule is polar. Because the hydrogen chloride molecule has only one polar bond, the hydrogen chloride molecule is polar. In the liquid and solid states, molecules are touching. The positive ends of dipoles are attracted to the negative ends of dipoles in other molecules as shown in -Figure 12.8.

In Section 9.2, you learned that the attraction between positive and negative ions is very strong. In the hydrogen chloride molecule, the positive and negative charges

- FIGURE 12.8 The positive end of the dipole in one HCl molecule is attracted by the negative end of the dipole in another HCl molecule. (a) In the solid, the molecules are fixed in position and arranged in a regular order. (b) In the liquid, the molecules are free to move and are not arranged in an orderly way.

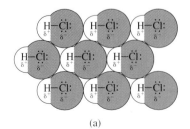

(a)

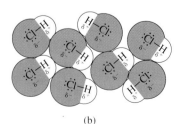

(b)

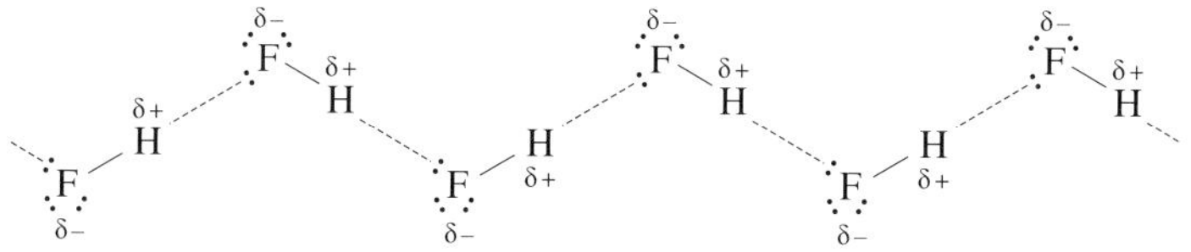

■ FIGURE 12.9 Hydrogen-bonded structure of HF(s). The chain is zigzag rather than linear because the four electron pairs around fluorine are at the four corners of a tetrahedron and the hydrogen bond is directed toward one of the unshared pairs.

on hydrogen and chlorine are only partial charges; therefore, the dipole–dipole attraction between hydrogen chloride molecules is not nearly as strong as the attraction between sodium ions and chloride ions, which have whole charges. As a result, hydrogen chloride is a gas under ordinary conditions, whereas sodium chloride is a high melting solid.

Hydrogen Bonding

Molecules that have hydrogen bonded to nitrogen, oxygen, or fluorine have *unusually strong dipole–dipole attractions* called **hydrogen bonds.** Attraction between opposite charges is greater the higher and closer together the charges. Fluorine, oxygen, and nitrogen are the most electronegative elements. Covalent bonds between these elements and hydrogen are more polar than bonds between hydrogen and any other elements. The partial positive charge on hydrogen bonded to fluorine, oxygen, or nitrogen is greater than the partial positive charge on hydrogen bonded to any other elements. The partial negative charges on fluorine, oxygen, or nitrogen bonded to hydrogen are greater than the partial negative charge on any other elements bonded to hydrogen. In addition, hydrogen is the smallest atom, and atoms of fluorine, oxygen, and nitrogen, which are at the right side of the second row of the periodic table, are smaller than atoms of any other elements except hydrogen. As a result of the large partial charges and the small sizes of these atoms, dipole–dipole attractions between molecules with hydrogen–fluorine, hydrogen–oxygen, and hydrogen–nitrogen bonds are unusually strong.

Attraction between opposite charges is like attraction between opposite magnetic poles. The stronger the magnets and the closer the opposite poles, the stronger the attraction.

The hydrogen bonding of hydrogen fluoride is the easiest hydrogen bonding to show on flat paper because the hydrogen-bonded structure is planar. Hydrogen bonds are commonly indicated by a dashed line as shown in ■Figure 12.9. In hydrogen-bonded substances such as hydrogen fluoride, hydrogen is attached to one fluorine, oxygen, or nitrogen atom by a covalent bond and to another fluorine, oxygen, or nitrogen atom by a hydrogen bond. The three atoms joined by a hydrogen bond usually lie on a straight line. The hydrogen is about normal bonding distance from the atom to which it is covalently bonded and farther than normal from the other atom. Each HF molecule forms two hydrogen bonds. For one, the HF molecule provides the hydrogen; for the other, it accepts a hydrogen from another molecule. The hydrogen bond that is accepted is directed toward one of the unshared pairs of electrons on fluorine.

Hydrofluoric acid, an aqueous solution of hydrogen fluoride, dissolves glass and is used to polish, frost, and etch glass.

Only the very electronegative atoms, F, O, and N, can accept hydrogen bonds. Fluorine, oxygen, and nitrogen atoms in molecules that do not have a hydrogen attached to fluorine, oxygen, or nitrogen can accept hydrogen bonds from molecules that do. For example, although the hydrogens in formaldehyde are attached to carbon (not F, O, or N) so that formaldehyde molecules can't hydrogen bond to other formaldehyde molecules, the oxygen atom in formaldehyde can accept a hydrogen bond from a water molecule:

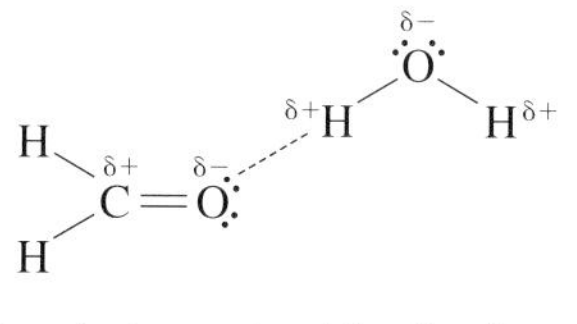

This type of hydrogen bonding is important in the formation of aqueous solutions.

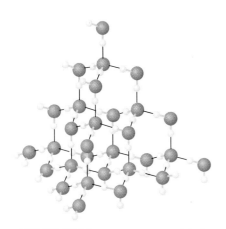

FIGURE 12.10 Ball-and-stick model of ice.

Many of the properties of water that make life as we know it possible are a result of hydrogen bonding. For example, hydrogen sulfide, hydrogen selenide, and hydrogen telluride are all gases under ordinary conditions. Water is, of course, a liquid. In most series of similar compounds, intermolecular attractions are stronger the larger the atoms. Intermolecular attractions increase in the order

$$H_2S < H_2Se < H_2Te$$

and water should be a gas. The fact that it is a liquid is explained by hydrogen bonding. Each hydrogen in a water molecule can form a hydrogen bond to oxygen in another water molecule, and the oxygen can accept a hydrogen bond from two other water molecules:

The hydrogen-bonded structure of water is three-dimensional and is best shown by a ball-and-stick model. ▪Figure 12.10 shows a model of an ice crystal, which has an orderly arrangement of water molecules. Because the four electron pairs of oxygen are located at the four corners of a tetrahedron, the hydrogen bonds in ice are arranged tetrahedrally around oxygen. Liquid water has regions that have ice-like structures and regions in which the arrangement of water molecules is much less orderly.

Fluids like blood, urine, and the sap in trees are mostly water. Life, if it existed at all, would be very different if water were not liquid. The fact that water is a liquid is also very important in geological processes such as erosion.

Hydrogen bonding is significant in almost all biologically important molecules; for example, hydrogen bonds occur in both carbohydrates and proteins. The shapes of large biological molecules such as proteins are determined largely by hydrogen bonding. Shape is a major factor in the properties of biological molecules.

PRACTICE PROBLEMS

12.1 Which of the following molecules can hydrogen bond to other molecules of the same kind? Explain your answer. (a) CH_3CH_2OH (b) CH_3OCH_3 (c) CH_4 (d) NH_3 (e) F_2

12.2 Which of the molecules in Practice Problem 12.1 can accept hydrogen bonds from water?

12.3 Using the method of Figure 12.9, show two molecules of methanol, CH_3OH, joined by a hydrogen bond.

12.4 Which member of each of the following pairs of molecules has the stronger intermolecular attractions? Explain your answers. (a) O_2 or NO (b) CH_4 or NH_3

London Forces (Dispersion Forces)

Even substances with nonpolar molecules such as H_2, Cl_2, CH_4, and He condense to liquids if the temperature is low enough and the pressure is high enough. What is the nature of the attractive forces between nonpolar molecules? The attractive forces that exist between nonpolar molecules are called London forces* or dispersion forces. London forces result from the movement of electrons in molecules. The helium atom is the simplest example. On *average,* the charge density distribution of a helium atom is spherically symmetrical as shown in ▪Figure 12.11(a). The electrons are moving very fast in a spherical volume around the nucleus. At any *instant,* the charge density distribution can be unsymmetrical as shown in Figure 12.11(b). At the next instant, it can be unsymmetrical as shown in Figure 12.11(c).

At any time when the charge density distribution is unsymmetrical, a molecule has a temporary dipole moment. The positive end of this temporary dipole attracts the electrons of nearby atoms. The negative end of the temporary dipole repels the electrons of nearby atoms. Nearby atoms temporarily become polar. ▪Figure 12.12 shows the creation of temporary dipoles in nearby atoms. The *temporary dipoles* are called **induced dipoles.** The oppositely charged ends of the temporary dipoles attract each other. Thus **London forces** are the *attractions between molecules that result from temporary dipoles caused by the movement of electrons.*†

London forces between helium atoms are very weak. Helium gas must be cooled to 4.3 K (−268.9 °C) at atmospheric pressure before it condenses to liquid helium. However, helium atoms have only two electrons in a 1s orbital. The electrons in helium atoms are close to the nucleus with no shells of electrons between them and the nucleus. They are very strongly attracted to the nucleus. In addition, the 1s orbital is very small, and the electrons in the 1s orbital don't have much space to move in. The charge distribution in a helium atom can't ever be very unsymmetrical because the two electrons repel each other.

The situation is different in larger atoms with more electrons. For example, a xenon atom has 54 electrons, and the outer electrons are in the fifth shell, far from the nucleus compared with helium's electrons. The attraction between the positive charge on the nucleus and the negatively charged electrons decreases very rapidly as the distance between the unlike charges increases. In addition, the fifth shell is much larger than the first shell, and outer electrons have room to get off-center without getting too close to each other. The outer part of the *charge cloud* of a xenon atom is much more *easily pulled out of shape* than the charge cloud of a helium atom; xenon is said to be more **polarizable** than helium. The temporary dipoles in xenon are much stronger than the temporary dipoles in helium, and as a result, xenon gas must be cooled to only 166.1 K (−107.1 °C) at atmospheric pressure to make it condense to liquid xenon. The greater the number of electrons and the farther the outer electrons are from the nucleus, the more polarizable the outer electrons and the higher the temperature at which a gas can be condensed to a liquid. The outer electrons in many molecules are sufficiently polarizable that the substances are liquids or solids at room temperature. Bromine and iodine are examples.

PRACTICE PROBLEM

12.5 (a) Which substance has stronger London forces, neon gas or argon gas? Explain your answer. (b) One of the gases in part (a) must be cooled to 87.5 K (−185.7 °C) and the other to 27.3 K (−245.9 °C) to condense to a liquid at atmospheric pressure. Which is which? Explain your answer.

*London forces are named after the London who, with Heitler, thought up the first quantum mechanical treatment of the hydrogen molecule.

†Some authors use the term *van der Waals force* to refer to the forces that we call London forces. In reading, you must be careful to see how a particular author defines van der Waals force.

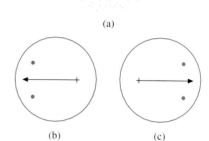

▪ FIGURE 12.11 (a) Charge cloud of a helium atom. (b) and (c) Temporary dipoles result from movement of electrons. [Remember that the nucleus, which has a 2+ charge, is at the center of the atom although it is much too small to show in the boundary surface diagrams (b) and (c).]

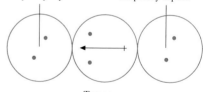

These electrons are repelled by the negative end of the temporary dipole.   These electrons are attracted by the positive end of the temporary dipole.

Temporary dipole

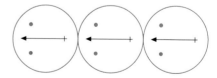

Temporary dipoles are formed in the adjacent atoms. The temporary dipoles attract each other.

▪ FIGURE 12.12 Creation of temporary dipoles in adjacent atoms.

London forces depend on the sizes of molecules as well as on the polarizability of the outer electrons of the atoms in the molecule. Consider the series of molecules, CH_4, CH_3CH_3, $CH_3CH_2CH_3$, and $CH_3(CH_2)_2CH_3$. ▬Figure 12.13 shows space-filling models of these molecules. Note that the four molecules become longer as the number of carbon atoms increases. The longer molecules can touch each other over a larger area, and as a result, the strength of the London forces between them increases as the number of carbon atoms increases. The temperatures to which the compounds must be cooled in order to condense them increase as the number of carbon atoms increases.

London forces also depend on the shapes of molecules. For example, two isomers of molecular formula C_4H_{10} exist; one has the structural formula $CH_3(CH_2)_2CH_3$, and the other, $(CH_3)_3CH$. A molecule of $CH_3(CH_2)_2CH_3$ is shaped like a hot dog, whereas a molecule of $(CH_3)_3CH$ is very nearly spherical. Contact between $CH_3(CH_2)_2CH_3$ molecules is much better than contact between $(CH_3)_3CH$ molecules (see ▬Figure 12.14). As a result, London forces between molecules of $CH_3(CH_2)_2CH_3$ are stronger than London forces between molecules of $(CH_3)_3CH$. Both these substances are gases at room temperature and atmospheric pressure. However, the temperature must be lowered to 261.6 K (-11.6 °C) to condense $(CH_3)_3CH(g)$ to liquid but only to 272.7 K (-0.5 °C) to condense $CH_3(CH_2)_2CH_3(g)$ to a liquid.

○ Remember that isomers are different compounds that have the same molecular formula (Section 3.9).

PRACTICE PROBLEMS

12.6 (a) Which substance has stronger London forces, $CH_3CH_2CH_2CH_2Cl$ or $CH_3CH_2CH_2Cl$? Explain your answer. (b) One of the substances in part (a) condenses from a gas to a liquid at 47 °C and the other at 78 °C at atmospheric pressure. Which is which? Explain your answer.

12.7 (a) Which substance has stronger London forces, $CH_3CH_2CH_2CH_2OH$ or $(CH_3)_3COH$? Explain your answer. (b) One of the substances in part (a) condenses from a gas to a liquid at 82.3 °C and the other at 117.2 °C at atmospheric pressure. Which is which? Explain your answer.

London forces are *not* a result of gravitational attraction between molecules. Masses of individual molecules are so small that gravitational attraction between molecules is negligible. Gravitational attraction is negligible even for macroscopic objects like marbles. Two marbles on a smooth flat surface do *not* roll together. Gravity is only an important force for objects with very large masses like planets. *The relationship between molecular mass and intermolecular attraction is a good example of a relationship that is not cause and effect.* The greater strength of intermolecular attractions between molecules having higher mass is *not* caused

Compound	Methane, CH_4	Ethane, CH_3CH_3	Propane, $CH_3CH_2CH_3$	Butane, $CH_3(CH_2)_2CH_3$
Temperature to condense, °C	-164	-89	-42	-0.5

▬ FIGURE 12.13 Space-filling models of different length molecules. London forces are greater the longer a molecule.

Compound	Butane, $CH_3(CH_2)_2CH_3$	Isobutane or Methylpropane, $(CH_3)_3CH$
Temperature to		
condense, °C	−0.5	−11.6

- FIGURE 12.14 Space-filling models of differently shaped isomers. London forces are stronger for linear than for spherical molecules as shown by the lower temperature required to condense isobutane. Note that *two* molecules of each shape are shown to illustrate the difference in contact area.

by the higher mass. Intermolecular attractions usually increase as molecular mass increases because molecules that have greater mass usually have either larger, more polarizable atoms or longer molecules or both.

All molecules are attracted to other molecules by London forces—that is, all molecules have temporary dipoles caused by the movement of electrons. Polar molecules such as hydrogen chloride molecules are attracted to each other by *both* London forces and dipole–dipole attractions. *If London forces are about the same, polar molecules are attracted to each other more strongly than molecules without dipoles.* A temperature of 188 K (−85 °C) is low enough to cause hydrogen chloride gas to condense to a liquid, but a temperature of 87 K (−186 °C) is required to cause argon gas to condense. Chlorine atoms and argon atoms are about the same size and have similar polarizabilities and London forces. The additional attraction between hydrogen chloride molecules is dipole–dipole attraction.

However, for most molecules, London forces are more important than dipole–dipole attractions. For example, although the dipole moment of hydrogen chloride is over two and a half times as large as the dipole moment of hydrogen iodide, a temperature of 188 K (−85 °C) is needed to cause hydrogen chloride gas to condense to a liquid, whereas a temperature of 238 K (−35 °C) is low enough to cause hydrogen iodide gas to condense. The larger iodine atom is more polarizable than the smaller chlorine atom, and London forces are much greater for hydrogen iodide than for hydrogen chloride.

PRACTICE PROBLEM

12.8 (a) Which substance has dipole–dipole attractions in addition to London forces, HCl or F_2? Explain your answer. (b) One of the gases in part (a) must be cooled to 85 K (−188 °C) and the other to 188 K (−85 °C) to condense to a liquid at atmospheric pressure. Which is which? Explain your answer.

12.3 PROPERTIES OF LIQUIDS

Diffusion

Diffusion takes place in liquids as well as in gases (see - Figure 12.15). As you might expect from the kinetic-molecular picture of liquids and gases [Figures 12.2(a)

- FIGURE 12.15 Diffusion takes place in liquids, although more slowly than in gases. In time, all the water will be uniformly colored.

(a)

(b)

(c)

(d)

- FIGURE 12.16   (a) Liquids composed of spherical molecules flow easily like peas. (b) Liquids composed of long flexible molecules that get tangled up in each other like spaghetti are viscous. (c) The viscosity of a liquid (fudge sauce) is lower when the liquid is hot than when (d) the liquid is cold.

and (b)], diffusion in a liquid is slower than diffusion in a gas. In a gas, a molecule can travel a significant distance before striking another molecule; in a liquid, the molecules are touching. Diffusion is faster in hot liquids than in cold liquids because the molecules in a hot liquid are, on average, moving faster than the molecules in a cold liquid. The curves that show the distribution of molecular speeds in liquids and gases are similar (remember Figure 5.16).

Viscosity

Viscosity is *resistance to flow;* molasses and motor oil are more viscous than water. The viscosity of a liquid depends on intermolecular attractions. The stronger the intermolecular attractions between molecules in a liquid, the harder it is for the molecules to move past each other. For example, ethylene glycol ($HOCH_2CH_2OH$), which has two —OH groups that can form hydrogen bonds, is more viscous than alcohol (CH_3CH_2OH), which has only one. Liquids composed of long, flexible molecules that can get twisted with each other are also more viscous (see - Figure 12.16). The viscosity of a liquid usually decreases as the temperature is raised. At higher temperatures, the molecules have more kinetic energy to overcome the attractive forces between them. As a result, most liquids flow more easily when heated [Figure 12.16(c)].

Motor oils are an example of the importance of viscosity. If an oil is too thin, it runs out from between the engine parts and does not lubricate them. If an oil is too thick, it lowers engine efficiency by creating greater resistance to motion and does not flow into the spaces between the engine parts as it should. Modern motor oils contain additives to keep viscosity the same at high and low temperatures so that the engine is properly lubricated when it is first started and after it has warmed up in both summer and winter.

Surface Tension

Surface tension is *the force in the surface of a liquid that makes the area of the surface as small as possible.* For a given volume, a sphere has the smallest surface, and surface tension makes small drops of liquid spherical (see - Figure 12.17).

Ethylene glycol is the major component of antifreezes.

- FIGURE 12.17   Surface tension makes beads of water on a freshly waxed car form rounded drops. Surface tension is the most important force in liquids in zero gravity. It holds the mashed potatoes together on space flights.

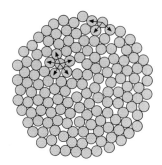

- FIGURE 12.18 The molecules in the interior of a drop of liquid are attracted by their neighbors on all sides. The molecules at the surface are not attracted from outside the drop. As a result, the surface molecules are pulled toward the interior, and the surface becomes rounded. The sample behaves as if it were enclosed in a thin sheet of rubber (like air in a balloon).

The molecules in the interior of a sample of liquid are attracted by their neighbors from all sides. The molecules at the surface of a liquid do not have neighbors on all sides. Therefore, they are attracted toward the liquid as shown in ▪Figure 12.18.

Surface tension allows some insects to walk on water and water to be heaped up a little above the sides of a container. Needles, paper clips, and razor blades do not sink if placed *carefully* on a smooth water surface, although the metal of which they are made is denser than water (see ▪Figure 12.19). Although the surface of water stretches a little when an insect walks on it or a needle is placed on it, the surface would have to break to let the insects' feet or the needle go through.

- FIGURE 12.19 The surface tension of water makes possible the floating of a needle although needles are denser than water and sink once they have passed through the surface.

Capillary Action

Another result of surface tension is capillary action. **Capillary action** *raises the surface of water in a glass tube having a slender bore and lowers the surface of mercury* as shown in ▪Figure 12.20. The molecules of a liquid attract each other and are also attracted by the walls of the container. The *forces between the molecules of a substance* are called **cohesive forces,** and the *forces between the molecules of different substances* are called **adhesive forces.** The relative strengths of adhesive and cohesive forces determine whether the meniscus curves up like the meniscus of water in glass or down like the meniscus of mercury in glass and whether a

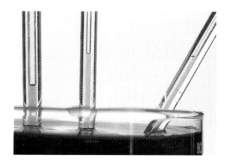

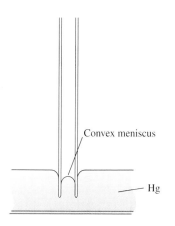

Convex meniscus

Hg

- FIGURE 12.20 *Left:* Water has a concave meniscus and rises in glass capillaries as a result of surface tension and adhesive forces between glass and water. The smaller the capillary, the higher the water rises. The angle between the capillary and the surface of the water does not matter. *Middle:* Water (dyed blue) rises in the capillaries of a stalk of celery. *Right:* Mercury, which has high surface tension but does not adhere to (wet) glass, has a convex meniscus and falls in a capillary as a result of gravity.

liquid rises or falls in a capillary. If the adhesive forces are stronger than the cohesive forces, as is the case for glass and water, the meniscus is concave, and the liquid rises in a capillary. The attraction of glass for water pulls water that touches glass up, and surface tension makes the rest of the surface of the water follow. Gravity pulls down on the water, limiting the height to which it rises. The smaller the diameter of the capillary, the higher the water rises. The capillary can be slanted (as shown in Figure 12.20) or even horizontal.

Capillary action is responsible for the capacity of paper towels and cotton to absorb water. Water is pulled into capillaries in the paper and the cotton. Capillary action is also involved in blood circulation and the rise of water from roots to leaves in plants. Surfactants or wetting agents in dishwasher and laundry detergents lower the surface tension of water, and the adhesive forces are better able to make water spread out over (wet) the surfaces being cleaned.

If the cohesive forces are stronger than the adhesive forces, as is the case for mercury and glass, the meniscus is convex, and the level of the liquid falls in a capillary [see Figure 12.20, right]. The cohesive forces in water are stronger than the adhesive forces between water and wool. As a result, water does not spread through wool fabrics, and wool clothing does not get as wet in rainy weather as cotton clothing.

Vaporization

Vaporization is the *formation of a gas from a liquid.* (*Formation of a gas from a solid* is called **sublimation.**) At temperatures below the boiling point of a liquid, vaporization takes place at the surface of the liquid. Molecules in a liquid, like molecules in a gas, do not all move at the same speed. Some molecules move faster than average and others slower. If a fast-moving molecule at the surface of a liquid has enough kinetic energy, it escapes from the attractive force of the other molecules in the surface of the liquid and becomes a gaseous molecule as shown in ▬Figure 12.21. If vaporization takes place in an open container, the gas molecules diffuse away from the surface of the liquid, and the liquid evaporates.

As the molecules having the most energy escape, the liquid that remains has less kinetic energy than the original liquid. For the moment, the remaining liquid becomes slightly cooler than its surroundings, and thermal energy flows into the liquid from the surroundings, making the surroundings feel cold. Animals that can sweat like humans are cooled by the evaporation of the water. Dogs can only evaporate water from the surface of their tongues, not as efficient a cooling process as sweating because of the relatively small surface area of the tongue.

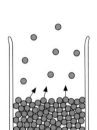

■ FIGURE 12.21   The most rapidly moving molecules at the surface of a liquid have enough energy to escape from the attractive forces of the molecules around them. A liquid in an open container evaporates.

Dynamic Equilibrium

Now let's consider what happens if a container is stoppered after a liquid is placed in it. After a number of molecules have escaped from the surface of the liquid and part of the liquid has become gaseous, the molecules in the gas phase will start bumping into each other and into the walls of the container. Some will bounce back and hit the surface of the liquid. Some molecules will be moving faster than average and others slower (remember Figure 5.16). If a slow-moving molecule hits the surface of the liquid it will get stuck—held to the surface by the attractive forces of the molecules around it. *Conversion of a gas to a liquid* is called **condensation.**

As the number of molecules in the gas phase increases, the number hitting the surface of the liquid and getting stuck in a given period of time (the rate of condensation) will increase. In the meantime, fast molecules will continue to escape from the surface. The number escaping in a given period of time (the rate of vaporization) will depend on the temperature of the liquid and its surface area. The rate of vaporization will be constant. After enough time has gone by, the rate at which molecules are returning to the liquid will become equal to the rate at which molecules are escaping from the liquid. ▪Figure 12.22 is a graph of the rates of vaporization and condensation as a function of time. In formal terms, the rate of condensation will become equal to the rate of vaporization:

$$\text{rate}_{condensation} = \text{rate}_{vaporization}$$

The concentration of molecules in the gas phase will remain constant because as many molecules are condensing as are being vaporized in a given period of time, as shown in ▪Figure 12.23. However, the same molecules will not be in the gas phase all the time. Molecules will keep going back-and-forth between the gas phase and the liquid phase. *A situation where nothing appears to be happening because two opposing changes are taking place at the same rate* is called a **dynamic equilibrium.*** The idea of dynamic equilibrium is one of the most important ideas in chemistry. We will refer to it again and again.

Shifting Equilibria

Now suppose temperature increases. More molecules will have sufficient energy to escape from the surface of the liquid, and the rate of vaporization will increase. As a result, the concentration of molecules in the gas phase will increase. At the higher temperature, molecules in the gas phase will move faster. Because there are more molecules in the gas phase and these molecules are moving faster, more molecules will hit the surface in a given time period. The rate of condensation will also increase and will again become equal to the rate of vaporization, although both rates will be greater than before. There will be more molecules in the gas phase and fewer molecules in the liquid phase; the equilibrium is said to have **shifted.**†

If temperature decreases, fewer molecules will have enough energy to escape from the surface of the liquid, and the rate of vaporizaton will decrease. Molecules in the gas phase will move slower. Gas molecules will not hit the surface of the liquid as often, and the rate of condensation will also decrease. The rate of condensation will again become equal to the rate of vaporization although, at the lower temperature, both rates will be slower than before. There will be fewer molecules in the gas phase and more molecules in the liquid phase.

Raising the temperature shifts the equilibrium between a liquid and its vapor in favor of vapor. More molecules are in the vapor phase, and fewer molecules are in the liquid phase. The concentration of molecules in the vapor phase is greater, and

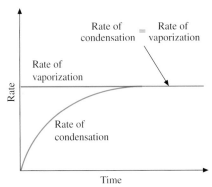

▪ FIGURE 12.22 Rates of vaporization and condensation as a function of time.

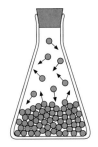

▪ FIGURE 12.23 In a closed container, liquid and vapor reach equilibrium. The rate at which molecules return to the liquid phase is equal to the rate at which molecules escape from the liquid phase.

---

*An everyday example of a dynamic equilibrium takes place at a beach on a hot summer afternoon. At any time, a certain number of people will be on the beach, and a certain number of people will be in the water. Some people will be entering the water, and about the same number of people will be leaving it each minute. After a few minutes have passed, about the same number of people will be on the beach and in the water, but they will not be the same people.

†If a cold breeze comes up, the number of people entering the water in a given time will decrease. Only a few hardy souls will enter the water. At first, the rate of people coming out of the water will remain the same because you do not feel a breeze when you are immersed in water. The number of people left in the water will decrease. Then because the number of people left in the water is smaller, the rate of people leaving the water will decrease until, once again, the rates of people entering the water and leaving the water are equal. Both rates will be lower. The concentration of people on the beach will be larger, and the concentration of people in the water will be smaller than before. The cold breeze has shifted the equilibrium in favor of people on the beach.

the volume of the liquid phase is smaller at the higher temperature. (The concentration of the liquid phase does not change very much with changing temperature. As long as the liquid remains a liquid, molecules must continue to touch each other.) Cooling shifts the equilibrium between a liquid and its vapor in favor of the liquid. There is more liquid, and the concentration of the vapor phase is smaller at the lower temperature.

Conversion of liquid to vapor requires heating—the transfer of thermal energy to the liquid; thermal energy is given off when a vapor condenses to liquid. Condensation is the opposite of vaporization. *Changes that can take place in either of two directions* are said to be **reversible.** When the two opposing changes are taking place at the same rate, the system is in equilibrium. Equilibria are shown in equations by writing two arrows, one pointing to the right and the other to the left:

$$\text{liquid + thermal energy} \leftrightarrows \text{vapor}$$

When thermal energy is added to a liquid and its vapor at equilibrium, the additional energy converts liquid to vapor:

$$\text{liquid + thermal energy} \rightarrow \text{vapor}$$

When a mixture of a liquid and its vapor at equilibrium is cooled, vapor condenses to liquid. Thermal energy is given off to replace the thermal energy removed:

$$\text{vapor} \rightarrow \text{liquid + thermal energy}$$

The thermal energy released by the formation of raindrops drives hurricanes. If converted to electrical energy, the energy released by a typical hurricane in one day would supply U.S. needs for six months.

Le Châtelier's Principle

What happens to the equilibrium between a liquid and its vapor when the temperature is raised is typical of the behavior of equilibria. If a change is made in a system at equilibrium, the equilibrium shifts in such a way as to reduce the effect of the change. This summary of experimental observations is known as *Le Châtelier's principle* (pronounced luh shatél yays) after the French chemist, metallurgist, and mining engineer who first proposed it in 1884.

Le Châtelier's principle has been very useful in the development of efficient processes in the chemical industry (see Section 14.6). When you dry your laundry in a dryer, you are applying Le Châtelier's principle. Liquid water changes to water vapor to use up the added thermal energy, and the water vapor is removed from the equilibrium by the fan. Le Châtelier's principle is discussed further in Section 14.5.

Le Châtelier's Principle

If a change is made in a system at equilibrium, the equilibrium will shift in such a way as to reduce the effect of the change.

Le Châtelier's principle can be used to make qualitative predictions about the effect of changes in temperature, pressure, and concentration on a system in equilibrium, as shown in Sample Problem 12.1.

SAMPLE PROBLEM

12.1 (a) Use Le Châtelier's principle to predict the effect of decreasing the volume of the space above a liquid in equilibrium with its vapor at constant temperature. (b) Use the kinetic-molecular theory to explain your answer to part (a).

Solution (a) Sketches are often helpful in reasoning. This problem can be pictured by imagining liquid in equilibrium with vapor in a cylinder closed by a piston as shown in figure (a).

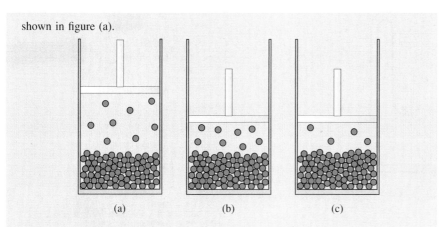

(a)   (b)   (c)

The volume of the vapor can be decreased by pushing the piston down as shown in figure (b). The volume of the vapor in figure (b) is half the volume of the vapor in figure (a). As a result of the decrease in volume, the concentration of molecules in the vapor phase in figure (b) is twice as large as the concentration of molecules in the vapor phase in figure (a). Vapor will condense to liquid to reduce the increased concentration.

(b) Because the concentration of molecules in the vapor phase in figure (b) is twice the concentration of molecules in the vapor phase in figure (a), more molecules will hit the surface of the liquid in a given period of time. The rate of condensation will increase. The rate of vaporization will not change because the surface area of the liquid has not changed. Therefore, the rate of condensation will be greater than the rate of vaporization. The concentration of molecules in the vapor phase will decrease until the rate of condensation has slowed down to again equal the rate of vaporization. At the new equilibrium, more molecules will be in the liquid phase. The volume of the liquid phase will be greater. The concentration of molecules in the vapor phase will be the same as that before the volume was lowered [see figure (c)]. Decreasing the volume of the vapor phase will shift the equilibrium in favor of the liquid phase as predicted by Le Châtelier's principle.

PRACTICE PROBLEM

12.9   (a) Use the kinetic-molecular theory to predict the effect of decreasing the volume of a liquid in equilibrium with its vapor at constant temperature. [*Hint:* What will happen if the layer of liquid in Sample Problem 12.1(a) is only half as deep?] (b) State your conclusion in terms of Le Châtelier's principle.

12.4   VAPOR PRESSURE AND BOILING POINT

Equilibrium Vapor Pressure

The *pressure of the vapor in equilibrium with a liquid* is called the **equilibrium vapor pressure** of the liquid. The equilibrium vapor pressure of a specific liquid depends on the intermolecular forces in the liquid and on temperature. It does *not* depend on the volume of the vapor or on the volume or surface area of the liquid. However, some liquid must be present for an equilibrium to exist, and the container must be closed. ▪Figure 12.24 shows the vapor pressures of several common liquids at different temperatures. As you can see, the higher the temperature, the greater the vapor pressure. Greater vapor pressure at higher temperatures seems reasonable because raising the temperature increases the number of molecules having enough

Br₂ ——   CH₃CH₂OH ——
H₂O ——   CH₃(CH₂)₆CH₃ ——

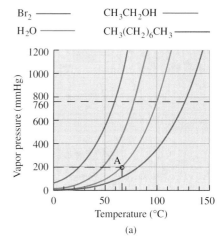

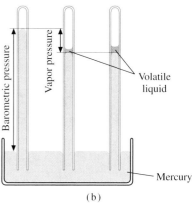

(b)

▪ FIGURE 12.24   (a) Vapor pressure of some common liquids as a function of temperature. (b) The vapor pressure of a liquid is the same whether the volume of the sample of liquid is small or large.

The word equilibrium *is often omitted. Vapor pressure is frequently used to mean equilibrium vapor pressure.*

Differences in vapor pressures between various liquids can easily be demonstrated by comparing the evaporation times for spots of liquid dabbed on the chalkboard with cotton balls.

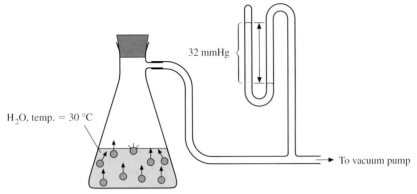

32 mmHg

H_2O, temp. = 30 °C

To vacuum pump

Pressure inside bubbles
at the surface = 32 mmHg

■ FIGURE 12.25 A liquid boils when its vapor pressure equals the pressure on its surface. The pressure on bubbles *below* the surface is equal to 32 mmHg plus the pressure from the water above the bubble. (This figure is *not* drawn to scale!)

Equilibrium vapor pressure is an intensive property.

The relatively low vapor pressure of water is a result of the strong hydrogen bonding in water.

energy to escape from the liquid. At a higher temperature, fewer molecules will be in the liquid phase, and more molecules will be in the gas phase. The concentration of molecules in the gas phase will be greater. The molecules in the gas phase will move faster at the higher temperature. Because there are more molecules in the gas phase and the molecules are moving faster, the gas will exert more pressure.

The equilibrium vapor pressure is the maximum vapor pressure a specific liquid can exert at a given temperature. Before liquid and vapor have reached equilibrium at constant temperature, vapor pressure is less than the equilibrium vapor pressure. Equilibrium vapor pressure is a characteristic property of a substance and is independent of the quantity of liquid present. *As long as some liquid is present, the liquid will exert its characteristic vapor pressure providing that the container is closed and that enough time has passed for the system to reach equilibrium. The vapor pressure of a liquid indicates the strength of the intermolecular attractions between molecules of the liquid.* The fact that the vapor pressure of bromine is higher than the vapor pressure of water at all temperatures shows that intermolecular attractions between water molecules are stronger than intermolecular attractions between bromine molecules (even though bromine has many more electrons and thus stronger London forces than water).

Boiling Point

The **boiling point** of a liquid is the *temperature at which the liquid's vapor pressure equals the pressure on the sample.* When the liquid's vapor pressure equals the pressure on the sample, bubbles of vapor form throughout the liquid, rise to the surface, and burst. (If the vapor pressure inside bubbles were less than the pressure on the bubbles, the bubbles could not form.) The liquid boils (see ■Figure 12.25).

The boiling point of a liquid depends on the pressure on the liquid. The boiling points of a liquid at different pressures can be read from the vapor pressure curve for the liquid by finding the temperature at which the vapor pressure is equal to the pressure on the liquid. For example, from Figure 12.24(a), the boiling point of water at 200 mmHg is 67 °C (point A). As you can see from Figure 12.24(a), the lower the pressure on a liquid, the lower its boiling point.

Because boiling point depends on pressure, pressure should be given along with temperature for a boiling point. One way to show pressure is by a right superscript. For example, to show that the boiling point of water at 200 mmHg is 67 °C, the boiling point is written, boiling point = 67^{200} °C.

Pressure cookers raise pressure above atmospheric pressure. Water boils at a higher temperature in a pressure cooker, and food cooks faster. The record high barometric pressure adjusted to sea level is 813 mmHg recorded in the middle of Siberia. At this pressure, water boils at 102 °C. The record low barometric pressure at sea level is 658 mmHg recorded during a typhoon in the South Pacific. At this pressure, water boils at 96 °C.

Cake mixes must have high-altitude directions because the boiling point of water is lower at high altitude where barometric pressure is lower than it is at sea level.

Normal Boiling Point

Normal boiling point refers to the *boiling point at standard atmospheric pressure (760 mmHg).* For example, from Figure 12.24, the normal boiling point of water is 100 °C, and the normal boiling point of octane [$CH_3(CH_2)_6CH_3$] is 126 °C. If no information about pressure is given with a boiling point, the boiling point is assumed to be the normal boiling point. *The normal boiling point of a liquid indicates the strength of the intermolecular attractions between molecules of the liquid.* The fact that water boils at 100 °C and octane at 126 °C shows that intermolecular attractions between octane molecules are stronger than intermolecular attractions between water molecules. Although water is strongly hydrogen-bonded, octane molecules have many more electrons and a greater volume than water molecules and are much longer than water molecules.

Liquids that have high vapor pressures and low normal boiling points are said to be **volatile.** Volatile liquids evaporate readily. *Liquids that have low vapor pressures and high normal boiling points* are said to be **nonvolatile** and do not evaporate readily.

PRACTICE PROBLEMS

12.10   From Figure 12.24, what is the normal boiling point of ethyl alcohol?

12.11   What is the boiling point of ethyl alcohol at 900 mmHg?

12.12   What must the pressure be if ethyl alcohol boils at 50 °C?

12.13   Which has stronger intermolecular attractions, CH_3CH_2OH or $CH_3(CH_2)_6CH_3$? Explain your answer.

12.5   MELTING POINTS AND FREEZING POINTS

The **normal melting point** of a solid is the *temperature at which the solid changes to liquid under atmospheric pressure.* (*Melting* is sometimes called **fusion.**) The **freezing point** of a liquid is the *temperature at which liquid changes to solid.* The freezing point of a substance in the liquid state is the same as the melting point of the same substance in the solid state.

The melting point does not change very much as the pressure on a solid is changed. (The melting point of water is lowered from 0 to −1 °C by increasing pressure from 1 atm to 125 atm.) At the melting point, solid is in equilibrium with liquid. The rate at which solid is forming from liquid is the same as the rate at which solid is melting to liquid. Molecules that are in the solid phase at one time are in the liquid phase at another time.

The equilibrium between a solid and its melt is the result of a balance between the energy (enthalpy) and the entropy of the system. Conversion of solid to liquid is endothermic. The reverse change, conversion of liquid to solid, is exothermic. Thus, energy favors the solid phase. However, conversion of solid to liquid leads to an increase in disorder. In a solid, molecules are arranged in order. A liquid does not have an orderly arrangement; therefore, entropy favors the liquid phase. At low temperatures (for a given substance, a low temperature is any temperature below the melting point of the solid), the solid phase of a substance is favored. At high

■ FIGURE 12.26 After almost 3000 years, the glass in the bowl is starting to crystallize. The white spots are partly crystalline. Does entropy increase or decrease when a glass crystallizes?

temperatures (for a given substance, a high temperature is any temperature above the melting point), the liquid phase is favored. At low temperatures, energy is more important than entropy. At high temperatures, the reverse is true—entropy is more important than energy. The terms *high* and *low* are relative. Ordinary room temperature is a "high" temperature for water, which melts at 0 °C but a low temperature for iodine (melting point = 113.5 °C). *All* equilibria result from a balance between the energy and entropy of the system.

For the conversion of a liquid to a solid, the rate of the change is also important. At low temperatures, the rate at which molecules are able to move into the ordered arrangement of the crystalline solid may be so low that the substance solidifies as a glass if cooled rapidly. A glass is *not* in equilibrium with liquid. (Glasses gradually crystallize. However, the process may take a long time—hundreds or even thousands of years. ■Figure 12.26 shows a piece of very old glass that has partly crystallized.)

SAMPLE PROBLEM

12.2 Does the entropy of the sample increase or decrease when a liquid changes to a vapor? Explain your answer.

Solution When a sample of a liquid changes to a vapor, the volume of the sample increases about a thousandfold. Because the sample occupies a much larger volume as a vapor, many more disorderly arrangements of the molecules are possible, and the entropy of the sample increases.

PRACTICE PROBLEM

12.14 Does the entropy of the sample increase or decrease when a liquid changes to a solid? Explain your answer.

12.6 HEATING AND COOLING CURVES

Let's look at the results of a very simple experiment that you can easily check in lab. Suppose a 75.0-g sample of water, originally at room temperature, is heated at a uniform rate (with an electric heater) until it boils. The water is stirred gently so that the temperature is the same throughout the whole sample. The temperature is measured at intervals until after the water has boiled for six minutes. The results of this experiment are tabulated in Table 12.1. As you can see, the temperature rose

Although our students do a boiling point determination in lab early in the year, a number of them still expect the temperature of the water to keep on rising as long as the water is heated.

TABLE 12.1	Temperatures Observed when Heating 75.0 g of H_2O	
Time, s	Temperature, °C	Temperature Change, °C
0	22.2	—
180	45.5	23.3
360	68.9	23.4
540	92.2	23.3
600	100.0[a]	7.8
780	100.0	0.0
960	100.0	0.0

[a]Water began to boil.

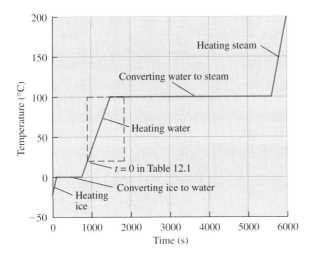

■ FIGURE 12.27 Temperature vs. time curve for heating 75.0 g of ice at −23 °C for 6000 seconds (1 hour and 40 minutes). The area inside the dashed rectangle shows the temperature range in Table 12.1. Determination of the complete heating curve for the conversion of ice to steam requires a more complicated apparatus than a hot plate, beaker, and thermometer. The fact that the specific heats (quantity of thermal energy required to heat one gram of a substance one degree) of ice and steam are about the same is a coincidence. How do the times required to melt the ice, heat the water to boiling, and boil the water compare?

23.3 °C each 180 seconds until the water began to boil. After the water began to boil, the temperature was constant. Temperature is constant during the phase change from liquid to vapor because the energy added is used up separating molecules from the liquid.

A graph of temperature vs. time for an experiment like the one just described is called a **heating curve.** ■Figure 12.27 shows the heating curve for a sample of ice from the time it was taken from the freezer until after all of the sample had been converted to steam. As you can see, the temperature is constant while ice melts to water, as well as while water is converted to steam. A longer time (more thermal energy) is required to convert a quantity of water to steam than to convert the same quantity of ice to water. If you look carefully at Figure 12.27, you can also see that the slopes of the lines showing ice being warmed to 0 °C and steam being heated above 100 °C are about the same. About the same amount of thermal energy is required to heat one gram of ice one degree as is needed to heat one gram of steam one degree. The line showing the temperature of liquid water as it is warmed from 0 to 100 °C is less steep. More thermal energy is required to heat one gram of water one degree than is needed to heat one gram of ice or steam one degree.

The *amount of thermal energy necessary to melt one mole of a substance at the melting point* is called the **heat of fusion** or **enthalpy of fusion, ΔH_{fus},** of the substance. The *amount of thermal energy required to convert one mole of a substance to vapor at the boiling point* is called the **heat of vaporization** or **enthalpy of vaporization, ΔH_{vap},** of the substance. Some typical values for heats of fusion and heats of vaporization are shown in Table 12.2. Notice that each substance's heat of vaporization is much larger than its heat of fusion. Much more energy is required

Turning up the heat under a pot of boiling water will make the water evaporate faster but will not make the water any hotter.

The high specific heat of water is another result of hydrogen bonding. The high specific heat of water and the enormous quantity of water in seas and lakes combine to prevent wild rises and falls in temperature on Earth.

For molecular substances, specific heat for the liquid is usually greater than specific heat for the solid and gaseous states. For metals, the specific heats of solid and liquid are about the same and greater than the specific heat of the gaseous state.

TABLE 12.2	Heats of Fusion and Heats of Vaporization of Some Common Substances*			
Substance	Melting Point, °C	ΔH_{fus}, kJ/mol	Boiling Point, °C	ΔH_{vap}, kJ/mol
Br_2	−7.3	10.57	59.2	29.5
CH_3CH_2OH	−117	4.60	79	43.5
$CH_3(CH_2)_6CH_3$	−56.8	20.65	125.7	38.6
H_2O	0.0	6.01	100.0	40.7
Na	97.8	2.60	883	98.0

*Most of the data are from Atkins, P. W. *Physical Chemistry*, 5th ed.; Freeman: New York, 1994, p. C5.

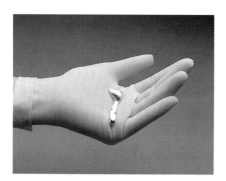

A human hand is warm enough to melt gallium (mp 29.8 °C).

to separate molecules from each other than to free them from fixed positions in a solid. Like vapor pressure and normal boiling point, heat of vaporization is a good indicator of the strength of intermolecular attractions. Melting points and heats of fusion are poorer indicators of the strength of intermolecular attractions because they depend on how efficiently molecules pack as well as on intermolecular attractions.

Heats of fusion and vaporization can be used to calculate how much energy is needed to melt or vaporize a sample. Sample Problem 12.3 shows how.

SAMPLE PROBLEM

12.3 How many kilojoules are needed to melt 674 g of sodium at its melting point?

Solution From Table 12.2, ΔH_{fus} for Na at its melting point is 2.60 kJ/mol. In the problem, the amount of substance is given in grams. Because the heat of fusion is per mole, the amount of substance must first be converted to moles. From the table inside the front cover, the atomic mass of sodium is 22.989 768 u; therefore, 1 mol Na has a mass of 22.989 768 g, and we can write a conversion factor for converting mass to moles.

$$\text{kJ needed} = \text{moles sample} \times \text{heat of fusion}$$

$$= 674 \text{ g Na} \left(\frac{1 \text{ mol Na}}{23.0 \text{ g Na}}\right)\left(\frac{2.60 \text{ kJ}}{1 \text{ mol Na}}\right) = 76.2 \text{ kJ}$$

PRACTICE PROBLEM

12.15 How many kilojoules are needed to vaporize 25.64 g of ethyl alcohol, CH_3CH_2OH, at its boiling point?

Cooling is the reverse of heating. Specific heats tell how much thermal energy must be removed to cool one gram of a substance one degree. Heats of fusion tell how much thermal energy must be removed to freeze one mole of a substance, and heats of vaporization tell how much thermal energy must be removed to condense one mole of a substance. ▪Figure 12.28 shows a cooling curve for water. As you can see by comparing Figure 12.28 with Figure 12.27, a cooling curve is the reverse of a heating curve except at the freezing point. Many liquids can be cooled below

▪ FIGURE 12.28 Temperature vs. time curve for cooling 75.0 g of steam at 200 °C.

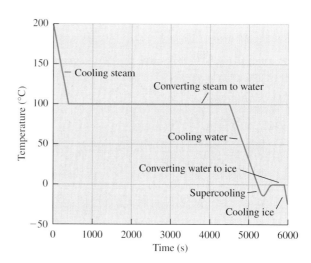

proper place in the ordered crystalline pattern. A *liquid cooled below its freezing point* is called a **supercooled liquid.** Supercooled liquids are not stable. Once a supercooled liquid starts to crystallize, the whole sample solidifies very rapidly. Thermal energy is given off, and the temperature rises to the melting point as shown in Figure 12.28. Once the whole sample has solidified, the temperature falls normally as thermal energy is removed.

12.7 PHASE DIAGRAMS

Information about phase changes is often summarized by **phase diagrams.** ▪Figure 12.29 shows the phase diagram for water. Phase diagrams are drawn by plotting the pressures and temperatures at which solid and liquid, liquid and vapor, and solid and vapor are in equilibrium. A phase diagram like that for water is said to be for a one-component system. All of the pressure is due to vapor; no part of the pressure is due to air. However, phase diagrams can usually be used to predict the results of experiments carried out under normal atmospheric pressure (such as the determination of the heating and cooling curves for water, Figures 12.27 and 12.28) without significant error.

In the phase diagram for a single substance like water, one area represents the solid, another area the liquid, and a third area, the vapor or gas phase. The areas corresponding to ice, liquid water, and water vapor are labeled in Figure 12.29.

Phase diagrams are graphs (to scale), not sketches. The sketches in most texts give students an exaggerated idea of how fast melting point changes with pressure.

Because it is less dense than liquid water, ice floats on lakes and rivers, insulating the water below. If ice were denser than water and sank, the water would freeze solid from bottom to top and fish could not survive the winter.

The Phase Diagrams module of the Liquids section asks students to identify the points and areas in phase diagrams and then asks questions that can be answered from phase diagrams.

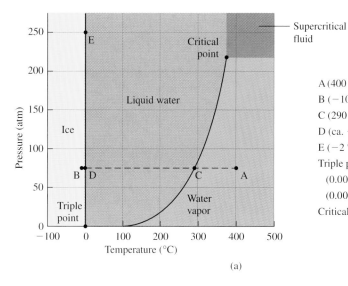

A (400 °C, 75 atm)

B (−10 °C, 75 atm)

C (290 °C, 75 atm)

D (ca. −0.5 °C, 75 atm)

E (−2 °C, 250 atm)

Triple point
  (0.0098 °C, 0.006 03 atm)
  (0.0098 °C, 4.58 mmHg)

Critical point (374 °C, 218 atm)

(a)

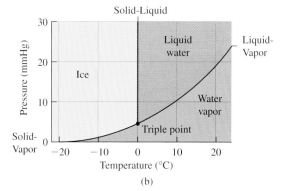

(b)

▪ FIGURE 12.29   (a) Phase diagram for water. (b) Enlargement of area around the triple point.

FIGURE 12.30 *Left to right:* As a sample of a liquid is heated in an evacuated sealed tube, the density of the liquid decreases, and the density of the vapor increases. At the critical temperature, the densities become equal, and the meniscus disappears.

Supercritical fluids are sometimes classified as a fifth state of matter (plasma is the fourth).

Icicles gradually disappear even on days when the temperature never goes above freezing because ice sublimes.

Freeze-drying is often used to preserve foods and reduce their mass and volume. Foods to be freeze-dried are first frozen. Then the pressure is reduced below 1 mmHg, and water sublimes. Without water, foods don't spoil; their mass and bulk are much lower.

Lines in phase diagrams show equilibria between two phases. Ice and liquid water are in equilibrium along the straight line between the ice and liquid water areas in Figure 12.29. This line shows the melting point of ice (or the freezing point of water) at different pressures. The melting point "curve" is usually an almost vertical straight line because melting points are not very much affected by pressure. The melting point curve for ice is unusual because it slopes to the left—that is, its slope is negative; ice is less dense than water. Why is this so? Most solids are denser than their melts. A given sample of water occupies a larger volume as ice than it occupies as liquid water as a result of the open structure of ice crystals. Ice crystals contain an unusually large amount of empty space because of the hydrogen bonding in ice (see Figure 12.10). Increasing the pressure on ice melts the ice as predicted by Le Châtelier's principle. The melting point of ice decreases about 1 °C for each 125 atm of increase in pressure. Note that liquid water and water vapor are in equilibrium along the curved line between the liquid water and water vapor areas in Figure 12.29. This curve shows the vapor pressure of liquid water at different temperatures or the boiling point of water at different pressures.

At the **triple point,** all *three phases are in equilibrium; temperature and pressure are fixed.* At the triple point for water shown in Figure 12.29, ice, liquid water, and water vapor are in equilibrium. The triple point of water is more reproducible than the melting point of ice in air that is used to define zero on the Celsius scale. The lowering of the triple point from 0.0098 °C to 0 is caused by dissolved air and the increase in pressure from 0.006 03 atm (the partial pressure of water vapor) to 1 atm. The triple point of water, 273.1600 K, is used to define the Kelvin scale. For most purposes, the difference between 0 and 0.0098 °C is not important.

The **critical point** is the *end of the vapor pressure curve.* The *temperature at the critical point* is called the **critical temperature, t_c (T_c** if temperature is in K). The *pressure at the critical point* is called the **critical pressure, P_c.** At temperatures above the critical temperature, a substance cannot be liquefied no matter how great the pressure. At temperatures above the critical temperature, the molecules have so much kinetic energy that no amount of pressure is enough to hold them in contact with each other. The critical pressure is the minimum pressure required to condense gas to liquid at the critical temperature. As a liquid is heated, it expands; the density of the liquid decreases. As a gas is compressed, its density increases. At the critical point, the densities of liquid and gas are the same. There is no longer any difference between liquid and gas (see **Figure 12.30).

At temperatures higher than the critical temperature and pressures greater than the critical pressure, a substance is still fluid. Because it is neither a gas nor a liquid, it is referred to as a **supercritical fluid.** Supercritical fluids have low viscosities like gases but are as dense as liquids and are good solvents. Supercritical carbon dioxide is used in large quantities to recover crude oil left in reservoirs by standard methods.

The line between the ice area and the water vapor area in Figure 12.29(b) is the vapor pressure curve for ice. (Note that the vapor pressure of ice is less than the vapor pressure of liquid water.) Along the solid line between the ice area and the water vapor area, ice and water vapor are in equilibrium. Below the triple point, ice *changes directly to* water *vapor without melting*—that is, ice **sublimes.** The opposite of sublimation, *change from vapor to solid,* is called **deposition.** **Figure 12.31 summarizes the names of the various types of phase changes and shows the relative energies of the three ordinary phases of matter.

The phase diagram for carbon dioxide (**Figure 12.32) illustrates several other points about phase diagrams. The solid–liquid line in the phase diagram for carbon dioxide has a positive slope (slopes to the right) as it does for most substances [see Figure 12.32(a)]. Solid carbon dioxide is denser than liquid carbon dioxide. Solid carbon dioxide does not float on liquid carbon dioxide; it sinks. Carbon dioxide is

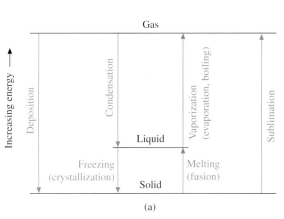

(a)   (b)

▪ FIGURE 12.31   (a) Energy diagram for phase changes. (b) On standing, iodine crystals, which
were originally all at the bottom of the container, sublime. New crystals of iodine deposit at the top
of the container.

unusual in another respect, however. Liquid carbon dioxide does not exist at normal
atmospheric pressure [see Figure 12.32(b)]. At pressures up to 5.11 atm, solid carbon
dioxide (dry ice) sublimes instead of melting.

A phase diagram for a substance tells the state in which the substance exists
under different combinations of temperature and pressure. Phase diagrams can be
used to predict what will happen if the temperature of a substance is changed at
constant pressure or if the pressure on a substance is changed at constant temperature.
For example, if a sample of water vapor that is originally at 400 °C and 75 atm
[point A in Figure 12.29(a)] is cooled to −10 °C at a constant pressure of 75 atm
[point B in Figure 12.29(a)], the temperature will decrease as shown by the broken
horizontal line. When the temperature reaches 290 °C, point C on the liquid–vapor

*Dry ice is used as a refrigerant during
shipment of products that must be kept
cold, such as ice cream and meat. The
atmosphere of carbon dioxide gas produced
by sublimation reduces spoilage. Liquid
carbon dioxide is used to make carbonated
drinks and in fire extinguishers.*

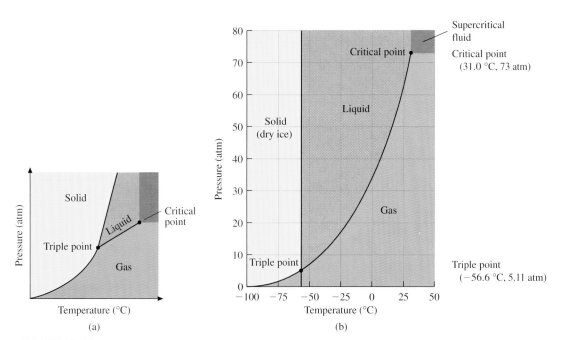

(a)   (b)

▪ FIGURE 12.32   Phase diagram for carbon dioxide. (a) Sketch with slope of solid–liquid line
exaggerated so that you can see it. (b) Scale drawing.

Top: *An empty paper cup burns when heated with a laboratory burner.* Bottom: *A paper cup that contains water does not burn because the thermal energy supplied by the burner is used to convert liquid water to water vapor. The temperature remains at 100 °C, the normal boiling point of water.*

The Crystal Types module of the Solids section provides material for exploration and application of the ideas that link bonding in solids to their properties.

curve, water vapor will start to condense to liquid water. The temperature will remain constant at 290 °C until all of the water vapor has condensed to liquid water. As soon as all the water vapor has condensed to liquid water, the temperature will again start to fall gradually as the liquid water is cooled. When crystals of ice start to form, the temperature will remain constant (at about −0.5 °C, point D) until all of the water has frozen. Then the temperature will again fall gradually. The cooling curve will be similar to the cooling curve in Figure 12.28. The design of many industrial operations, such as the manufacture of ceramics, alloys, and glass, depends on a knowledge of phase diagrams, and geological studies of the history of minerals in rocks are based on phase equilibria.

SAMPLE PROBLEMS

12.4   From the phase diagram for water, Figure 12.29, will water be a solid, a liquid, or a gas (a) at 300 °C and 240 atm? (b) At 450 °C and 240 atm?

Solution   (a) Find the point (300 °C, 240 atm) on the phase diagram for water. This point is in the area labeled "Liquid water." Water will be a liquid at 300 °C and 240 atm.

(b) Find the point (450 °C, 240 atm) on the phase diagram for water. This point is in the area labeled "Supercritical fluid." Water will be a supercritical fluid at 450 °C and 240 atm.

12.5   Describe what will happen if: (a) the pressure on a sample of water at 300 °C and 240 atm is gradually lowered to 25 atm at a constant temperature of 300 °C; (b) the pressure on a sample of water at 450 °C and 240 atm is gradually lowered to 25 atm at a constant temperature of 450 °C.

Solution   (a) On the phase diagram for water [Figure 12.29(a)], lightly draw a broken vertical line from (300 °C, 240 atm) to (300 °C, 25 atm). Beginning at 240 atm, the pressure will gradually decrease, as shown by the broken vertical line, until it reaches the liquid–vapor curve at 83 atm. Then liquid water will begin to evaporate and pressure will remain constant at 83 atm, the vapor pressure of water at 300 °C, until all of the water has evaporated. The pressure of the water vapor will then decrease, as shown by the broken vertical line, until it reaches 25 atm.

(b) On the phase diagram for water [Figure 12.29(a)], lightly draw a broken vertical line from (450 °C, 240 atm) to (450 °C, 25 atm). Beginning at 240 atm, the pressure will gradually decrease as shown by the broken vertical line; the supercritical fluid will gradually expand.

PRACTICE PROBLEMS

12.16   From the phase diagram for carbon dioxide, Figure 12.32, will carbon dioxide be a solid, a liquid, or a gas at 0 °C and 10 atm?

12.17   (a) Describe what will happen if a sample of carbon dioxide at −75 °C and 13 atm is warmed to 0 °C at a constant pressure of 13 atm. (b) Sketch the heating curve for part (a).

12.18   Sketch the phase diagram for O_2 from the following information: triple point (54.4 K, 1.14 mmHg); critical point (154.6 K, 37 823 mmHg); normal melting point 54.8 K; and normal boiling point 90.2 K.

12.8   TYPES OF CRYSTALS

The physical properties of crystalline solids vary widely. Some solids are hard like diamonds; others are soft like sodium. Some have high melting points like iron, and others can be melted by the heat of your hand like ice. Some conduct heat and electricity like copper, whereas others are insulators like asbestos. The physical properties of solids depend on the nature of the particles that make up the solid and on the attractive forces between them. Crystals can be divided into four types on the basis of their units and the attractive forces that hold them together: molecular, ionic, covalent network, and metallic.

Snowflakes are molecular crystals.

Molecular Crystals

In a **molecular crystal,** the *units are molecules.* For example, in a crystal of methane, the units are CH_4 molecules. In an argon crystal, the units are argon atoms. The units of nonpolar molecular crystals such as these are held together only by London forces. If the molecules are polar, dipole–dipole attractions as well as London forces hold the crystals together. The molecules in solid hydrogen chloride (melting point -114.8 °C) are held together by both London forces and dipole–dipole attractions. The molecules in solid methyl alcohol (CH_3OH, melting point -93.9 °C) are held together by both London forces and hydrogen bonds. Because intermolecular attractions are relatively weak, it is easy to move one layer of molecules past another; it is also easy to free molecules from the solid so that they can move. As a result, molecular crystals are soft and have low melting points. Neither the solid nor the liquid is a good conductor of electricity or heat.

Although no one who has fallen on ice would consider ice to be soft, you can scratch an ice cube with your fingernail.

For compounds that have molecules as units, shape is an important factor in packing. For example, $(CH_3)_4C$, which is approximately spherical, is a gas under ordinary conditions, but $CH_3(CH_2)_3CH_3$, which is shaped like a hot dog, is a liquid that boils at 36.1 °C (remember Figure 12.14). Intermolecular attractions are stronger in $CH_3(CH_2)_3CH_3$ than in $(CH_3)_4C$ because molecules of $CH_3(CH_2)_3CH_3$ can interact with one another over a much larger surface area than molecules of $(CH_3)_4C$. However, $(CH_3)_4C$ has a much higher melting point than $CH_3(CH_2)_3CH_3$; $(CH_3)_4C$ melts at -16.5 °C, whereas $CH_3(CH_2)_3CH_3$ melts at -130 °C. Spherical molecules of $(CH_3)_4C$ pack better than the longer, more flexible molecules of $CH_3(CH_2)_3CH_3$.

Ionic Crystals

In an **ionic crystal,** the *units are ions.* Sodium chloride is a typical example. Attraction between oppositely charged ions is strong (Section 9.2), and the melting points of ionic crystals are high—usually greater than 300 °C. If one layer of an ionic crystal is pushed past another, ions with like charges are moved closer together as shown in ▪Figure 12.33. Like charges repel each other, and a large amount of

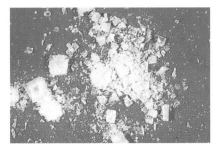

Top: *Sodium chloride crystals.* Bottom: *Sodium chloride crystals after being hit with a hammer.*

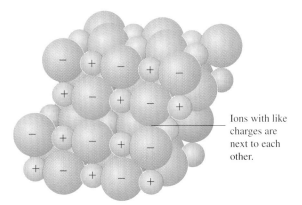

Ions with like charges are next to each other.

▪ FIGURE 12.33   Pushing layers of an ionic crystal past one another places ions with like charges next to each other, and the crystal shatters.

(a)

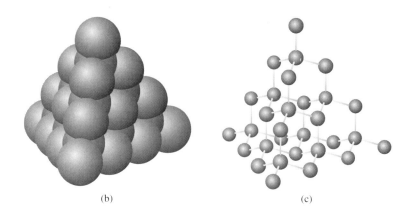

(b)   (c)

▪ FIGURE 12.34 *Left:* A macroscopic view of a diamond. Pure diamonds are colorless and transparent. *Middle:* Space-filling model of diamond structure. *Right:* Ball-and-stick model of diamond structure.

energy is required to push one layer past another. If enough force is applied to push one layer past another, so that ions with like charges are near, the crystal flies apart. Ionic crystals are hard and brittle. Because the ions are fixed in position in the crystal and can only vibrate, ionic crystals do not usually conduct electricity. However, if the crystals are melted, the melt is a good conductor. (*Liquids near their freezing points* are called **melts.**)

Covalent Network Crystals

In a **covalent network crystal,** the *units are held together by covalent bonds.* Carbon (diamond) is the classic example of a covalent network crystal; in a crystal of diamond, each carbon is covalently bonded to four other carbons arranged at the four corners of a tetrahedron as shown in ▪Figure 12.34.

Diamond is one of the hardest known substances because each crystal is a single giant molecule. To split a diamond, covalent bonds must be broken. The energy required to break covalent bonds is much greater than the energy required to overcome London forces and dipole–dipole attractions. Not only are covalent network crystals very hard, but they also have extremely high melting points. If covalent network crystals are melted, the melts do not conduct electricity.

Carbon more commonly occurs in another crystalline form called graphite. In graphite, the carbons are arranged in flat sheets of rings of six carbons similar to benzene rings (see ▪Figure 12.35). Each layer of graphite is a covalent network of carbon atoms; the distance between bonded carbon atoms in a layer is 141.5 pm, intermediate between the average carbon–carbon single-bond length of 154 pm and

(a)

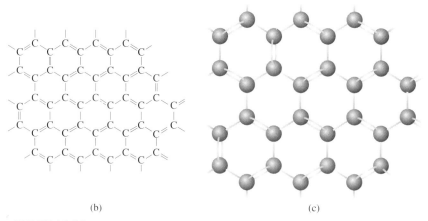

(b)   (c)

▪ FIGURE 12.35 (a) A macroscopic view of a piece of graphite. (b) One resonance structure for a small part of a layer of graphite. (c) Ball-and-stick model of a layer of graphite.

the double-bond length of 134 pm. Two resonance structures are needed to describe the properties of benzene (Section 9.9). Many resonance structures are required to describe the properties of graphite. One resonance structure for a small part of one layer of graphite is shown in Figure 12.35(b). In graphite, the pi electrons are delocalized over entire sheets, and graphite conducts electricity along the sheets.

A crystal of graphite consists of stacked layers. The distance between layers is 335 pm, much greater than the distance between carbon atoms in a layer. The layers are held together by London forces, which are weak compared with covalent bonds. Not much energy is needed to slide one layer of graphite past another or rub off some layers on a piece of paper. The "lead" in pencils is graphite (mixed with clay), and graphite is also used as a lubricant, for example, for locks.

Different forms of the same element that differ in bonding are called **allotropes;** diamond and graphite are allotropes. Allotropes differ in both physical and chemical properties. For example, diamond is denser than graphite. The density of diamond is 3.51 g/cm^3, and the density of graphite is 2.22 g/cm^3. Graphite is more stable than diamond by 2.9 kJ/mol under ordinary conditions. However, a great many carbon–carbon bonds must be broken to change diamond into graphite. Breaking bonds requires energy, and an extremely long time would be required for diamond to change into graphite under ordinary conditions.

In 1985 a third allotrope of carbon was discovered at Rice University in Houston, Texas. Crystals of this allotrope, which is yellow, are composed of spherical C$_{60}$ molecules (see ▬Figure 12.36). Because C$_{60}$ molecules are shaped like the geodesic domes invented by Buckminster Fuller, this carbon allotrope was named buckminsterfullerene. It is often called buckyball for short. Groups of U.S. and German scientists found a way to make C$_{60}$ in relatively large quantities in 1990. Since that time, it has become one of the most popular subjects for research.

Buckyballs have turned out to be a member of a large family of carbon molecules known as fullerenes, which have many interesting and potentially useful properties. For example, a water-soluble compound made from buckyballs has general viricidal activity and has been shown to have in-vitro activity against the AIDS virus. However, buckyballs are still very expensive, costing up to $1000 per gram in 1993, and have no commercial uses yet.

In 1991 another form of carbon, nanotubes, was discovered. Carbon nanotubes are concentric tubes formed from graphitelike sheets of carbon capped by fullerenelike hemispheres of carbon. These tubes are only nanometers in diameter but up to a few micrometers in length. A fourth allotrope of carbon consisting of chains of *sp* hybridized carbon atoms, ⁓ C≡C—C≡C—C≡C ⁓ , was reported early in 1995.

Metallic Crystals

The bonding in metals, which is called metallic bonding, is neither covalent nor ionic. The simplest model that explains the properties of metals pictures *metals as consisting of metal ions at fixed positions in a sea of mobile valence electrons* (see ▬Figure 12.37). The metal ions are held together in the crystal by the attraction between the positive charges of the metal ions and the negative charges on the mobile valence electrons, which are delocalized over the entire crystal. Because the valence electrons are free to move, metals are good conductors of thermal energy and electricity in both solid and liquid states. Both thermal energy and electricity are conducted through metals by the movement of electrons.

When a metal is heated, the positively charged metal ions vibrate more vigorously about their fixed positions in the crystal. The metal ions effectively occupy more

The lead in pencils is graphite. Graphite conducts electricity well and is often used for electrodes.

Oxygen, O$_2$, and ozone, O$_3$, are also allotropes. A number of nonmetals exist in allotropic forms. However, only three metals—tin, bismuth, and polonium—have allotropes. These three metals are all near the metal–nonmetal dividing line in the periodic table.

*In case someone asks how buckyballs are formed, one set of possible steps is outlined in Dravid, V. P., et al. Science **1993,** 259, 1601–1604, another in Chem. Eng. News **1994,** 72 (37), 27–28, and another in Hunter, J. M.; Fye, J. L.; Roskamp, E. J.; Jarrold, M. F. J. Phys. Chem. **1994,** 98, 1810–1818.*

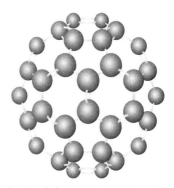

▬ **FIGURE 12.36** Ball-and-stick model of a buckminsterfullerene molecule. There are 20 benzene rings and 12 five-membered rings per molecule. Substitution of the five-membered rings for the six-membered rings of graphite makes the flat sheet curve into a ball. (However, this is *not* how buckminsterfullerene molecules are formed.) What is the hybridization of the carbon atoms in buckminsterfullerene?

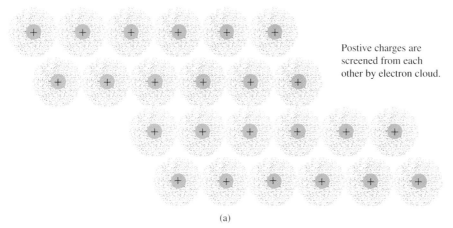

Positive charges are
screened from each
other by electron cloud.

(a)   (b)

■ FIGURE 12.37   (a) Microscopic view of
one layer of a metallic crystal moving past
another. (b) Macroscopic view of steel being
rolled into a sheet.

space in the crystal and get in the way of the mobile valence electrons. As a result,
the electrical conductivity of metals decreases with increasing temperature.

The electron-sea model also explains why metals are ductile and malleable.
Ductile means *capable of being drawn out into wire*, and **malleable** means *capable
of being shaped by hammering or spread out into sheets by pressure from rollers.*
When one layer of a metal crystal is moved past another, the positive charges on
the metal ions are shielded from each other by the valence electrons as shown in
Figure 12.37. Absorption of photons of electromagnetic radiation by the mobile
valence electrons and reemission of the energy as light accounts for the characteristic
shininess of metals.

The Crystal Structure module of the Solids
section allows students to manipulate 3-D
models to explore cubic unit cells. Interactive
problems help students connect unit cell
structure to compound stoichiometry.

In Conclusion

Table 12.3 summarizes information about the four types of crystals.

TABLE 12.3	Types of Crystals			
Type	Units	Attractive Forces	Properties	Examples
Molecular	Molecules	London forces Dipole–dipole including hydrogen bonding	Soft, low melting, neither solid nor melt conduct	O_2, Ar H_2S CH_3OH, $C_{12}H_{22}O_{11}$ (sugar)
Ionic	Ions	Attraction between oppositely charged ions	Hard, brittle, high melting, melt conducts (solid does not conduct)	NaCl, $CaCO_3$
Covalent network	Atoms	Covalent bonds	Very hard, very high melting, neither solid nor melt conducts	C (diamond), SiO_2
Metallic	Metal ions	Attraction between + charge on metal ions and − charge on mobile electrons	Shiny, ductile and malleable, may be soft or hard, low melting or high melting, both solid and liquid conduct	Na, Fe

SAMPLE PROBLEMS

12.6 Classify crystals of each of the following substances by type. Explain your reasoning. (a) Cu (b) Si (c) $CaCl_2$ (d) NH_3

Solution (a) Cu is a metal, and crystals of copper are metallic.

(b) Si is in Group IVA like C. Therefore, crystals of silicon are probably covalent network like diamond.

(c) $CaCl_2$ is a compound of a metal with low ionization energy and a nonmetal with high electron affinity and is probably mainly ionic.

(d) NH_3 is a compound of two nonmetals and is a molecular compound; crystals of solid ammonia are molecular.

12.7 Boron nitride, BN, is a white crystalline solid under ordinary conditions. Boron nitride does not melt but sublimes if heated to about 3000 °C. It is extremely hard. What type of crystal does BN form?

Solution The facts that boron nitride is extremely hard and that it does not melt or sublime until heated to a very high temperature suggest that BN probably forms covalent network crystals.

PRACTICE PROBLEMS

12.19 Classify crystals of each of the following substances by type: (a) P_4, (b) HCl, (c) Mn, and (d) KI.

12.20 Describe the attractive forces holding the units in place in crystals of each of the substances in Practice Problem 12.19.

12.21 Predict the properties expected of crystals of each of the substances in Practice Problem 12.19. (*Hint:* For part (c), you will need to use information from everyday life and the periodic table.)

12.22 $SnCl_4$ is a liquid under ordinary conditions. What type of crystal will $SnCl_4$ form if cooled below its freezing point?

12.9 ARRANGEMENT OF UNITS IN CRYSTALS

As far as the arrangement of the units in a crystal is concerned, crystals of metals are the simplest crystals for two reasons: All the atoms in a crystal of a metal are the same, and metal atoms are spherical. Let's begin by considering crystals of metals.

Simple Cubic Crystals

The easiest arrangement of identical spheres to understand is the simple cubic arrangement shown in ▬Figure 12.38. Figure 12.38(a) shows one layer of a simple cubic structure. In a simple cubic structure, layers are stacked with atoms of one layer directly on top of atoms of the other layers as shown in Figure 12.38(b). In a crystal that has a simple cubic structure, each atom has six nearest neighbor atoms—four in the same layer, one in the layer above, and one in the layer below. The *number of nearest neighbors* that an atom has is called the **coordination number** of the atom. In the simple cubic structure, the coordination number of each atom is 6.

Scientists usually use unit cells to describe crystal structures. The **unit cell** of a crystalline solid is the *smallest unit of the structure which, if repeated in three dimensions, would produce the crystal.* In a simple cubic unit cell, one-eighth of

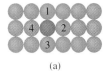

(a)

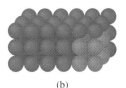

(b)

(c)

(d)

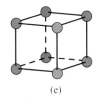

(e)

▬ **FIGURE 12.38** Simple cubic packing: (a) One layer of simple cubic packing. In a layer, each atom has four nearest neighbors. (b) In a crystal, layers are stacked with atoms of one layer directly on top of atoms of the other layers. In a crystal, each atom has six nearest neighbors—four in the same layer, one in the layer above, and one in the layer below. (c) Simple cube. (d) Unit cell. (e) Expanded model. The advantage of an expanded model is that you can see inside the model.

■ FIGURE 12.39   *Left:* One close-packed layer of oranges. *Right:* A close-packed arrangement of cannon balls.

each of the eight corner atoms of the simple cube lies inside the unit cell. Thus, the simple cubic unit cell, which has eight corners, contains a total of one atom. Simple cubic packing is rare. Polonium is the only metal, indeed the only element, known to crystallize in a simple cubic form.

Liquids freeze to solids because solids have lower energies than liquids. The decrease in energy when liquid changes to solid is a result of attractive forces between the particles making up the solid. The attractive forces are greatest when the particles are close together and touch as many other particles as possible. Therefore, the unit particles of solids usually tend to pack into as small a volume as possible. In a crystal with the simple cubic structure, only 52% of the space is occupied, and the coordination number of each particle is 6. A simple cubic arrangement is not very stable. If you were to stack oranges (or cannon balls) in a simple cubic arrangement, the pile would be very likely to collapse. Instead, oranges and cannon balls are usually stacked as shown in ■Figure 12.39. The arrangements shown here are called *close-packed* because there is a minimum of empty space. In a close-packed arrangement, 74% of the space is occupied, and each particle has a coordination number of 12, the highest possible coordination number.

○ Note that the coordination number of an atom does not include unshared pairs of valence electrons as the total coordination number does.

Close Packing

The crystals of most metals are close-packed. The arrangements of the atoms in the crystals can be described by picturing the building up of layers of close-packed spheres. ■Figure 12.40 shows one layer of close-packed spheres. In a close-packed layer, each atom has six nearest neighbors.

In close-packed structures, layers are stacked with atoms of one layer nested in the holes in the two touching layers. All the holes in a layer of close-packed spheres are identical. But if a sphere in the second layer is nested in one of the holes labeled A in Figure 12.40, a sphere cannot be nested in either of the two nearest holes labeled B [see ■Figure 12.41(a)]. You can see that in a close-packed crystal, an atom has three nearest neighbors in the next layer. Since an atom in a close-packed structure has six nearest neighbors in the same layer, atoms in close-packed crystals have a coordination number of 12.

All the spheres in the second layer must be nested either in holes of set A or in holes of set B. Suppose we nest spheres of the second layer in holes of set A as shown in Figure 12.41(b). The holes in the second layer are *not* identical. Half are

A picture of a close-packed layer of spherical buckyball molecules taken with a scanning tunneling microscope. Photo is courtesy of Y. Z. Li, M. Chandler, J. C. Patrin, J. H. Weaver, Department of Materials Science and Chemical Engineering, University of Minnesota. From Science, 26 July 1991.

Hole belonging to set A

Hole belonging to set B

■ FIGURE 12.40   One layer of close-packed spheres.

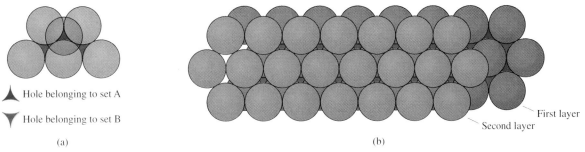

Hole belonging to set A

Hole belonging to set B

(a)

First layer
Second layer

(b)

FIGURE 12.41 (a) If a second-layer sphere is nested in a set A hole, spheres cannot be nested in the two nearest set B holes. (b) Second layer of close-packed spheres placed over holes in set A.

over spheres in the first layer, and half are over holes in the first layer. Spheres in the third layer can be placed either above spheres in the first layer or above holes in the first layer. Suppose that spheres in the third layer are placed above holes in the first layer as shown in ▪Figures 12.42(a) and (b), and the next (fourth layer of spheres) is placed over the spheres in the first layer. The fourth layer is a repeat of the first layer. The sequence of three layers is repeated over and over again, ABCABCABC . . . .

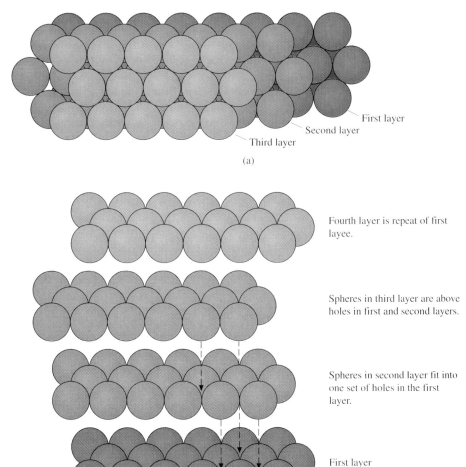

First layer
Second layer
Third layer

(a)

Fourth layer is repeat of first layee.

Spheres in third layer are above holes in first and second layers.

Spheres in second layer fit into one set of holes in the first layer.

First layer

(b)

FIGURE 12.42 (a) Top view of third layer of close-packed spheres over holes in first layer. (b) Side view of third layer of close-packed spheres over holes in first layer. Three layers are repeated over and over again.

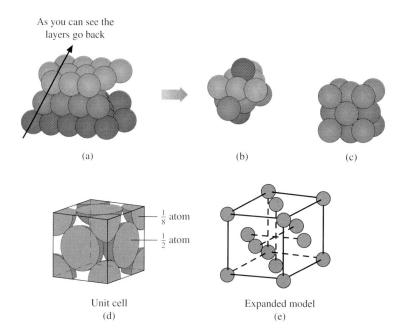

As you can see the
layers go back

(a)   (b)   (c)

$\frac{1}{8}$ atom

$\frac{1}{2}$ atom

Unit cell
(d)

Expanded model
(e)

- FIGURE 12.43 Relationship between cubic close packing and face-centered cubic unit cell. (a) Cubic close-packed arrangement. (b) Cube of 14 spheres separated from cubic close-packed arrangement. (c) Cube of 14 spheres rotated to usual position. (d) Face-centered cubic unit cell. (e) Expanded model.

Cubic Close Packing: Face-Centered Cubic Unit Cell

The relationship between the unit cell usually used for the close-packed arrangement of spheres with three repeating layers [Figure 12.42(a) and (b)] and the close-packed arrangement is shown in -Figure 12.43. From Figure 12.43(c), you can easily see why this arrangement of spheres is called **face-centered cubic.** Another name for this arrangement is **cubic close-packed.** From the usual unit cell, it is not easy to see that the layers of spheres in a face-centered cubic crystal are close-packed. The model of the unit cell must be rotated so that the cube is standing on one corner like the cube in Figure 12.43(b).

Placing spheres in the third layer above spheres in the first layer instead of above holes in the first layer gives another type of close packing called hexagonal close packing. In both close-packed arrangements, the maximum amount of space (74%) is occupied by spheres. The minimum amount of space (26%) is empty. In both close-packed arrangements, each sphere has the maximum possible number of spheres touching it. The coordination number of each unit in a close-packed arrangement is 12. We will not discuss hexagonal close packing further because the unit cell is not a cube. In this book, we will only discuss crystal structures for which the unit cell is a cube.

SAMPLE PROBLEM

12.8 How many spheres does the cubic close-packed unit cell contain?

Solution From Figure 12.43(d), you can see that the cubic close-packed unit cell has half a sphere at the center of each face and an eighth of a sphere at each corner. A cube has six faces and eight corners. Therefore, there are

$$6(\tfrac{1}{2} \text{ sphere}) + 8(\tfrac{1}{8} \text{ sphere}) = 4 \text{ spheres}$$

in the cubic close-packed unit cell.

Body-Centered Cubic Unit Cell

Almost all metals (except polonium) that do not crystallize in one of the two close-packed arrangements crystallize in a **body-centered cubic** arrangement. The unit cell for the body-centered cubic arrangement is shown in ▬Figure 12.44. The spheres in a body-centered cubic structure occupy almost as large a percentage of the volume as the spheres in the close-packed arrangements (see Table 12.4). However, you can see from Figure 12.44 that an atom in a body-centered cubic crystal only has a coordination number of 8. Why do a number of metals crystallize in a body-centered cubic arrangement? The six atoms in the centers of the six unit cells that surround a body-centered cubic unit cell are quite close to the atom in the center of the body-centered cubic unit cell. Thus, an atom in a body-centered cubic unit cell is close to a total of 14 other atoms, 8 in the unit cell and 6 in surrounding unit cells.

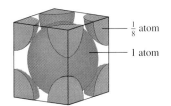

$\frac{1}{8}$ atom

1 atom

PRACTICE PROBLEM

12.23   How many spheres does the body-centered cubic unit cell contain?

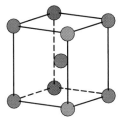

▬ FIGURE 12.44   *Top:* Body-centered cubic unit cell. *Bottom:* Expanded model of body-centered cubic cell.

In Conclusion

Table 12.4 summarizes information about the various types of arrangements. Substances that have spherical units tend to crystallize either in one of the close-packed arrangements or in the body-centered cubic arrangement. The noble gases and hydrogen have the cubic close-packed (face-centered cubic) structure. Almost all metals except polonium have either one of the close-packed arrangements or a body-centered cubic structure.

Ionic Compounds

The arrangements of the units in crystals of ionic compounds are complicated by the facts that two (or more) kinds of particles are involved, the particles usually differ in size, and the particles are charged. In addition, not all ions are even approximately spherical; for example, nitrate ions are trigonal planar.

In ionic compounds, the major attractive force between particles is the attraction between opposite charges. The attraction of an ion for ions of opposite charge is the same in all directions. The lowest energy structure lets the largest possible number of oppositely charged ions touch. Ions with like charges repel each other; therefore, ions with like charges should not touch. The relative sizes of the ions are also very important.

Many ionic compounds can be pictured as derived from close-packed arrangements of anions by placing small cations in the holes. Let's consider the cubic close-packed arrangement as an example. There are two sizes of holes in the cubic close-packed arrangement. Referring to Figures 12.41 and 12.42, the smaller holes are

Insulin crystals. Top: *Formed on Earth.* Bottom: *Grown in zero gravity on the Space Shuttle during a mission.*

TABLE 12.4	Crystal Structures		
Name	Coordination Number	Percentage of Space Occupied	Example
Face-centered cubic (cubic close-packed)	12	74	Al
Body-centered cubic	8	68	Na
Simple cubic	6	52	Po

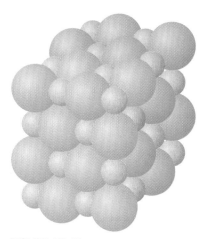

■ FIGURE 12.45   Model of a sodium chloride crystal. The large green balls represent Cl⁻ ions; the small silver balls represent Na⁺ ions.

the ones formed from the holes in set A when a sphere from the second layer is placed on top of them:

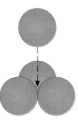

Because they are surrounded by four spheres, the smaller holes are called **tetrahedral holes.** There are two tetrahedral holes per sphere. The larger holes can be seen if face-centered cubes are viewed from the front:

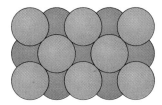

They are surrounded by six spheres and are called **octahedral holes.** There is one octahedral hole per sphere. In sodium chloride crystals (■Figure 12.45), the large chloride ions (radius = 181 pm) are in a face-centered (close-packed) cubic arrangement. The small sodium ions (radius = 102 pm) occupy the octahedral holes. The chloride ions must spread out a little bit to make room for the sodium ions. Each sodium ion is surrounded by six chloride ions, and each chloride ion is surrounded by six sodium ions. This structure results in the lowest energy because every positive ion is touching as many negative ions as possible. Every negative ion is touching as many positive ions as possible. Positive ions are not touching positive ions, and negative ions are not touching negative ions. Many other compounds have a structure similar to the sodium chloride structure; $LiCl$, $NaBr$, MgO, NiO, and NH_4I are some examples.

The structures of many other compounds can also be derived from close-packed arrangements. However, substances with highly directed bonds, such as diamond and water, and substances with nonspherical molecules, such as butane (Figure 12.13), are *not* close-packed in crystals; the crystal structures of substances with nonspherical molecules are difficult to predict.

12.10   CRYSTAL STRUCTURE FROM X-RAY DIFFRACTION PATTERNS

The arrangement of the particles in a crystal determines the shape of a crystal, its most obvious property. For example, the relationships between the cubic arrangement of ions in sodium chloride and the cubic shape of sodium chloride crystals and between the hexagons of atoms in ice crystals and the characteristic hexagonal shapes of snowflakes are easy to see (■Figure 12.46). Most of what we know about the arrangement of the particles in crystals comes from X-ray diffraction patterns.

The diameters of atoms are about the same as the wavelengths of X-rays, around 100 pm, and diffraction patterns are produced by the interaction of X-rays with the atoms in a crystal. [Figure 7.14(b) explains how diffraction patterns are formed.]

(a)

(b)

(c)

(d)

► FIGURE 12.46 (a) Space-filling model of sodium chloride. (b) The cubic shape of sodium chloride crystals results from the structure of the crystals. (c) Ball-and-stick model of ice. The oxygen atoms are located at the corners of hexagons. (d) The hexagonal shapes of snowflakes result from the structure of ice.

► Figure 12.47(a) shows a typical X-ray diffraction pattern. The X-rays interact with the electrons in the atoms, not with the nuclei of the atoms. From measurements of the positions and intensities of the dark spots, electron density can be calculated and shown as a contour map such as the one in Figure 12.47(b). Because electron density is highest close to the nucleus of an atom, an electron-density contour map shows the positions of the centers of the atoms, which determine the bond lengths and bond angles.

With the automated instruments developed in recent years, the structures of most molecules can be determined in a few days, and X-ray diffraction has become the accepted way of proving that a proposed structure for a new compound is correct. X-ray diffraction patterns can also be used as "fingerprints" for small samples of

Unusual molecules take longer and still require special skill on the part of the scientist.

► FIGURE 12.47 (a) An X-ray diffraction pattern. (b) An electron-density contour map of salicylic acid,

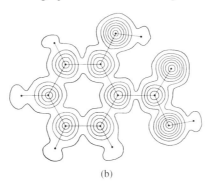

which is used to make aspirin.

(a)

(b)

RELATED TOPIC   Something to Bragg About

A Nobel Prize is something to be proud of at any age, but British physicist W. L. Bragg was only 25 years old when he received one. The work for which he received it was done while he was still an undergraduate at Cambridge University. During summer vacation, Bragg's father, W. H. Bragg, who was professor of physics at the University of Leeds, discussed with him a book on the German physicist Max von Laue's work on the diffraction of X-rays by crystals.

When the younger Bragg returned to college, using only algebra and trigonometry, he derived the equation relating the wavelength of the X-rays, the angle at which the X-rays strike the crystal, and the distance between planes of atoms in a crystal that is the basis for the use of X-ray diffraction to analyze crystal structures.

Meanwhile, Bragg senior designed an instrument, which used X-rays of a single wavelength and rotated the crystal, that was to become the prototype of all modern X-ray diffractometers. The two Braggs spent vacations using the diffractometer to determine crystal structures.

The younger Bragg showed that, in a crystal of sodium chloride, sodium and

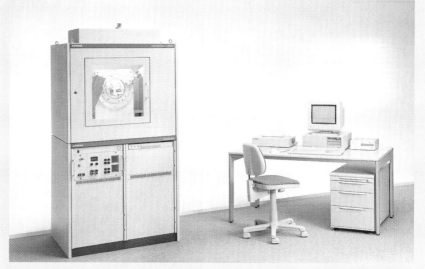

A modern X-ray diffractometer.

chlorine alternate. All six chlorines surrounding a sodium are the same distance from the sodium, and all six sodiums surrounding a chlorine are the same distance from the chlorine. As a result, he concluded that there are no molecules of NaCl in a crystal of sodium chloride. The conclusion that solid sodium chloride is not composed of sodium chloride molecules upset some of Bragg's elders consider-

ably. Father and son were awarded a Nobel Prize in 1915.

Deducing a crystal structure from an X-ray diffraction pattern was a time-consuming process in precomputer days. Beginning in 1937, Max Perutz, an Austrian-born British biochemist, spent *22 years* determining the structure of hemoglobin, the oxygen-carrying coloring material of red blood cells.

---

crystalline solids. Identification by X-ray diffraction has been used to detect forged works of art and to identify cars involved in hit-and-run accidents from bits of paint left at the scene of the accident.

X-ray diffraction is a powerful tool in biochemistry. For example, Watson, Crick, and Wilkins deduced the famous double-helix structure of DNA, which stores genetic information, from Rosalind Franklin's X-ray data. One problem with X-ray diffraction as a method for studying physiologically important molecules is that the sample must be crystalline. Glasses and liquids do not give sharp X-ray diffraction patterns because they do not have an orderly structure. Another problem is that the structures of complex molecules, such as proteins, in their natural environments may well be different from their structures as crystalline solids.

A number of manuscript reviewers identified as optional the calculations in this section, and the discussion of crystal defects in the next section. Other reviewers thought they were useful. These topics are placed at the end of the chapter so you may easily omit them if you choose.

 ### 12.11   CALCULATION OF ATOMIC AND IONIC RADII AND AVOGADRO'S NUMBER

The information obtained by X-ray diffraction can be used to calculate atomic and ionic radii and Avogadro's number. Sample Problems 12.9 and 12.10 illustrate these calculations.

12.9 Iron crystallizes in a body-centered cubic arrangement. The edge of a unit cell of iron is 286 pm long. What is the radius of the iron atom?

Solution In a body-centered cubic arrangement (Figure 12.44), spheres touch across the diagonal of the cube, AB.

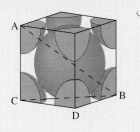

This iron meteorite has been sliced, polished, and etched with acid to show its crystallinity. Most of the metallic iron that occurs naturally on Earth is in meteorites. Iron rarely occurs uncombined because it rusts so easily.

The diagonal of the cube AB is $4r$, where r is the radius of the sphere that represents an iron atom. If we can find the length of AB,

$$r = \frac{AB}{4} \tag{12.1}$$

AB is the hypotenuse of the right triangle ACB. According to the Pythagorean theorem, the square of the hypotenuse of a right triangle is equal to the sum of the squares of the sides:

$$(AC)^2 + (CB)^2 = (AB)^2 \tag{12.2}$$

CB is the hypotenuse of the right triangle CDB. Therefore,

$$(CD)^2 + (DB)^2 = (CB)^2 \tag{12.3}$$

The sides of the cube, AC, CD, and DB, all equal 286 pm. Substitution in equation 12.3 gives

$$(286 \text{ pm})^2 + (286 \text{ pm})^2 = (CB)^2 \tag{12.4}$$

Solving equation 12.4 for CB,

$$CB = 404 \text{ pm}$$

Substituting this value for CB and 286 pm for AC in equation 12.2 gives

$$(286 \text{ pm})^2 + (404 \text{ pm})^2 = (AB)^2 \tag{12.5}$$

Solving equation 12.5 for AB,

$$AB = 495 \text{ pm}$$

According to equation 12.1, $r = AB/4$; therefore,

$$r = \frac{495 \text{ pm}}{4} = 124 \text{ pm}$$

12.10 Copper crystallizes in a face-centered (close-packed) cubic arrangement (Figure 12.43). The edges of the unit cell are 361.6 pm long; the density of copper is 8.92 g/cm³. Calculate Avogadro's number.

Native copper in calcite.

Solution   Avogadro's number, N, is the number of atoms in a mole

$$N = \frac{?\ \text{atoms}}{\text{mol}}$$

On a macroscopic scale, we can calculate the volume occupied by one mole of copper atoms from the atomic mass and density of copper. On a microscopic scale, we can calculate the volume occupied by one copper atom from the volume of the unit cell and the number of copper atoms in one unit cell. By calculating how many times larger the volume occupied by one mole is than the volume occupied by a single atom, we can find the number of atoms in one mole.

From the table inside the front cover, the atomic mass of copper is 63.546; therefore, 1 mol of copper has a mass of 63.546 g. The density of copper, 8.92 g/cm³, is given in the problem. We can use the density to find the volume occupied by one mole of copper atoms:

$$\text{volume occupied by 1 mol Cu} = \left(\frac{63.5\ \text{g Cu}}{\text{mol Cu}}\right)\left(\frac{1\ \text{cm}^3}{8.92\ \text{g Cu}}\right) = \frac{7.12\ \text{cm}^3}{\text{mol Cu}}$$

Because copper crystallizes in a face-centered cubic arrangement, the unit cell is a cube (Figure 12.43). The volume of a cube is equal to the cube of the length of one edge:

$$V_{\text{unit cell}} = (361.6\ \text{pm})^3 = 4.728 \times 10^7\ \text{pm}^3$$

The unit cell has 6 faces and 8 corners. Each face contains 1/2 of an atom; each corner contains 1/8 of an atom. The unit cell contains a total of

$$[6(\tfrac{1}{2}) + 8(\tfrac{1}{8})]\ \text{atoms Cu} = 4\ \text{atoms Cu}$$

The volume occupied by 1 atom is

$$\frac{4.728 \times 10^7\ \text{pm}^3}{4\ \text{atoms Cu}} = 1.182 \times 10^7\ \frac{\text{pm}^3}{\text{atom Cu}}$$

Next we must convert the volume occupied by one mole and the volume occupied by one atom to the same units. It doesn't matter which units we use. Suppose we choose pm³:

$$7.12\ \frac{\text{cm}^3}{\text{mol}}\left[\left(\frac{1\ \text{m}}{10^2\ \text{cm}}\right)\left(\frac{10^{12}\ \text{pm}}{\text{m}}\right)\right]^3 = 7.12 \times 10^{30}\ \frac{\text{pm}^3}{\text{mol}}$$

Dividing the volume occupied by one mole by the volume occupied by one atom gives the number of atoms in a mole:

$$\frac{7.12 \times 10^{30} \frac{pm^3}{mol\ Cu}}{1.182 \times 10^7 \frac{pm^3}{atom\ Cu}} = 6.02 \times 10^{23} \frac{atoms\ Cu}{mol\ Cu}$$

Avogadro's number is 6.02×10^{23}.*

This is one of the most accurate methods for determining Avogadro's number.

PRACTICE PROBLEM

12.24   Nickel crystallizes in a face-centered cubic arrangement. The edge of a unit cell of nickel is 352 pm long. What is the radius of the nickel atom?

12.12   DEFECTS IN CRYSTALS

Thus far, we have assumed that all crystals are perfect or ideal crystals. Real crystals are almost always imperfect, and defects in crystal structure have important effects on the properties of real crystals.

X-ray diffraction gives the average arrangement of units in crystals. To study defects in crystals, individual atoms must be observed. Individual atoms were first viewed by means of an electron microscope, but only very large atoms could be "seen." With the scanning tunneling microscope, which was invented in 1981, even atoms as small as carbon can be observed. ▪Figure 12.48 shows a picture of graphite produced by a scanning tunneling microscope; the six-membered rings of carbon atoms can be seen clearly.

There is only one right way for a crystal to form but many wrong ways. Most crystals have many defects. Foreign atoms or ions can take the place of the units of a crystal or be trapped in the holes. Atoms or ions may be missing or misplaced. Defects interfere with the normal attractive forces between the units of a crystal and usually weaken a crystal. For example, perfect crystals of metals are much stronger and much more resistant to corrosion than ordinary samples of metals.

Considerable care is necessary to grow crystals without defects and techniques for making more perfect crystals are the subject of a great deal of research. Strangely enough, much research is also directed toward introducing defects because defects sometimes result in desirable properties; for example, impurities increase the conductivity of some semimetals. Semimetals have electrical conductivity less than that of metals but greater than that of nonmetals. Therefore, they are often referred to as semiconductors. *Addition of impurities to semiconductors* is called **doping.**

Murphy's law, "If something can go wrong, it will," applies to crystallization as well as to human activities.

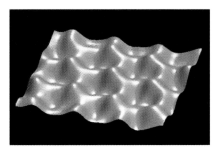

▪ **FIGURE 12.48**   Scanning tunneling micrograph showing the surface of a crystal of graphite. The hexagonal rings of six carbon atoms, which are 250 pm across, are clearly visible. Three of the carbon atoms in each ring are slightly higher than the other three.

*The reasoning used in answering Sample Problem 12.10 is like the reasoning used in solving the problem, "One dozen eggs cost $1.20. One egg costs 10¢. How many eggs are in a dozen?" You would immediately realize that both prices must be stated in the same units and convert units: One dozen eggs costs 120¢. Then you would probably think to yourself, "The price for one dozen eggs divided by the price of one egg gives the number of eggs in a dozen."

$$\frac{120¢/dozen\ eggs}{10¢/egg} = \frac{12\ eggs}{dozen\ eggs}$$

Scanning Probe Microscopes

$\mathcal{S}$ canning probe microscopes are a family of instruments that are used to measure the properties of surfaces. Surfaces are very important because changes, both physical and chemical, usually take place at surfaces. For example, when iron rusts, it does so at the surface where the metal is in contact with air.

All scanning probe microscopes use a sharp tip to scan the surface, usually from a distance of less than one nanometer (see ■Figure 12.49). They measure the surface in three dimensions and can "see" individual atoms and molecules, sometimes in new ways. For example, one member of the family has a tip that is sensitive to magnetic fields and thus can be used to examine the magnetic patterns on computer hard disks.

The first scanning probe microscope was the scanning tunneling microscope (STM), which was invented by Gerd Binnig and Heinrich Rohrer of IBM's Zurich (Switzerland) Research Laboratory in 1981. Only five years later, Binnig and Rohrer won the Nobel Prize for physics for their invention. For scanning tunneling microscopy, both the tip and the sample must conduct electricity, although nonconductors can sometimes be examined on a conducting surface. For example, the important biological molecule DNA was imaged under water on a gold surface. When a small voltage is applied to the sample, electrons "tunnel" through the space between the sample and the tip.

Tunneling is a quantum phenomenon. According to quantum theory, although electrons spend most of their time around the nucleus in a volume shown by the boundary surface diagram, the probability of an electron being farther away from the nucleus, although small, is not zero. Thus, an electron that belongs to an atom in the sample surface may be near the atom at the tip of the probe or an electron that belongs to an atom in the tip may be near a surface atom. The probability of tunneling occurring is increased by applying a small

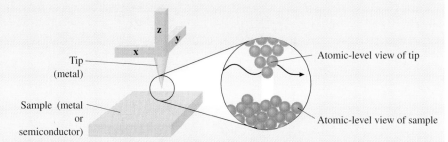

■ FIGURE 12.49   Diagram of a scanning probe microscope.

voltage (about one volt) to either the tip or the surface.

When the tip of a STM is moved across the sample in parallel lines, it moves up-and-down to keep the current constant. Because the current depends exponentially on the distance between the tip and the surface, movement of the tip is detectable. Measurements of the movement of the tip are converted to a relief map of the surface by a computer and displayed on the computer screen (see Figure 12.48).

The STM can be used to move atoms or molecules around on a surface. The person in ■Figure 12.50 was drawn with carbon monoxide molecules on a platinum surface. The person is only five nanometers tall! The ability to move individual atoms makes possible the construction of tiny tools such as a battery so small that 100 of them can fit into a red blood cell.

In 1989 the atomic force microscope (AFM) was introduced to better image nonconducting samples. For example, an AFM was used to examine pits in the window of the space shuttle. In an AFM, the tip touches the sample but with a force low enough (about 10^{-9} N) that the surface is usually not damaged. The AFM maps the surface by "feeling" interatomic forces similarly to the way blind people tap their canes to investigate the ground in front of them. Biological samples, such as chromosome clusters, can be observed under

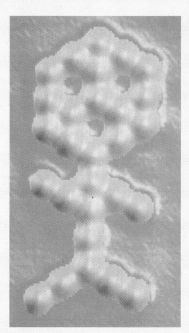

■ FIGURE 12.50   Person drawn with CO molecules (magnified 15 million times). A period magnified this much would be 3 mi (4.5 km) in diameter. The carbon atoms are attached to the platinum surface (blue) and the oxygen atoms point up.

water, so that the samples do not dry out, and even in vivo (in living systems). The motion associated with catalysis by an enzyme (a biological catalyst) was observed with an AFM in 1994.

How do impurities increase the conductivity of semiconductors? To answer this question, let's consider silicon, the most widely used semiconductor. Silicon has the same structure as diamond (Figure 12.34). An atom of arsenic (radius = 125 pm) is about the same size as an atom of silicon (radius = 118 pm), and atoms of arsenic can take the place of atoms of silicon in a crystal of silicon. But an atom of arsenic has five valence electrons, one more than silicon; the extra electrons can move through the crystal if an electric field is applied. A *semiconductor in which the current is carried by the movement of negatively charged electrons* is called an **n-type semiconductor.**

If an atom of boron is substituted for an atom of silicon, there is one too few electrons. There is an electron-deficient hole in the crystal, and in an electric field, an electron moves from a neighboring bond to fill the hole. A new hole is left; holes move through the semiconductor carrying current. This type of semiconductor is called a **p-type semiconductor** because the current is, in effect, carried by the movement of "positive" (electron-deficient) holes.* In modern electronic equipment, compact, durable semiconductor devices have largely replaced bulky, fragile vacuum tubes. They have made possible pocket calculators, personal computers, microwave ovens, video games, heart pacemakers, industrial robots, and space exploration.

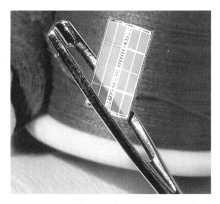

A circuit printed on a silicon wafer is small enough to go through the eye of a needle.

There is no microscopic view of a semiconductor because the usual ones show far too high a concentration of dopant.

Replacement of as few as ten atoms of silicon per million atoms of silicon by boron makes silicon a semiconductor.

Cathode ray tubes (Figure 7.3) and photoelectric cells (Figure 7.19) are vacuum tubes.

PRACTICE PROBLEM

12.25 Classify each of the following as an *n*-type or a *p*-type semiconductor: (a) Ge doped with As and (b) Ge doped with Ga.

SUMMARY

All real gases can be condensed to liquids by lowering temperature and increasing pressure. Formation of liquids from gases is due to **intermolecular attractions,** which are called **van der Waals forces.** There are three kinds of intermolecular attractions: **London forces, dipole–dipole attractions,** and **hydrogen bonds.** London forces are due to temporary dipoles, called induced dipoles, caused by the movement of electrons, and they exist between all molecules. The strength of London forces depends on polarizability, molecular size, and molecular shape. **Polarizability** refers to the ease with which the outer part of an atom's charge cloud is distorted. Polar molecules are also attracted to each other by dipole–dipole attractions, attractions between permanent dipoles. Hydrogen bonds are unusually strong dipole–dipole attractions that occur when hydrogen is bonded to nitrogen, oxygen, or fluorine.

In liquids, molecules are touching but free to move. There is no long-range order, and liquids have the same properties in all directions. Some important properties of liquids are **viscosity, surface tension, capillary action,** and **vapor pressure.**

If a liquid vaporizes in a closed container, a **dynamic equilibrium** is established when the rate of condensation equals the rate of vaporization. The pressure of the vapor over the liquid is called the **equilibrium vapor pressure.** The **boiling point** is the temperature at which the vapor pressure of the liquid is equal to the pressure on the liquid. The **normal boiling point** is the boiling point at standard pressure (1 atm or 760 mmHg). Liquids that have high vapor pressures and low normal boiling points are said to be **volatile.** Volatile liquids evaporate readily. Liquids that have low vapor pressures and high normal boiling points are said to be **nonvolatile** and do not evaporate readily.

Le Châtelier's principle can be used to predict the effect of changes in temperature, pressure, and concentration on a system in equilibrium. According to Le Châtelier's principle, if a change is made in a system at equilibrium, the equilibrium will shift in such a way as to reduce the effect of the change.

The amount of thermal energy necessary to melt one mole of a substance at its melting point is called the **heat of fusion,** ΔH_{fus}, and the amount of thermal energy needed to vaporize one

*According to the band theory of bonding in solids (Section 10.8), addition of arsenic to silicon provides energy levels in the forbidden band below the empty band. Addition of boron provides energy levels in the forbidden band above the filled band.

mole of a substance at its normal boiling point is called the **heat of vaporization, ΔH_{vap}. Heating curves** are graphs showing how temperature changes with time as a system is heated at a constant rate; **cooling curves** are the reverse of heating curves except for supercooling.

Information about phase changes is often summarized by **phase diagrams.** Phase diagrams are drawn by plotting the pressures and temperatures at which solid and liquid, liquid and vapor, and solid and vapor are at equilibrium. In the phase diagram for a single substance, areas represent phases, and lines show equilibria between two phases. At **triple points,** three phases are in equilibrium. The **critical point** is the end of the vapor pressure curve; at temperatures above the critical point, a substance cannot be liquefied, no matter how great the pressure. Substances at temperatures and pressures above their critical point are called **supercritical fluids.**

When liquids are cooled slowly, the molecules become fixed in an orderly arrangement known as a **crystalline solid.** Minimization of energy is the controlling factor in the existence of crystalline solids. The properties of most crystalline solids are different in different directions. The **normal melting point** of a solid is the temperature at which the solid changes to liquid under standard pressure and is the same as the **freezing point,** the temperature at which liquid changes to solid. Many liquids can be **supercooled** to temperatures below their normal freezing point. Supercooled liquids change spontaneously to crystalline solids when disturbed.

The physical properties of solids depend on the units that make up the solid and on the attractive forces between them. Crystals can be divided into four types on the basis of their units: **molecular crystals, ionic crystals, covalent network crystals,** and **metallic crystals. Allotropes** are different forms of the same element that differ in bonding.

Many substances crystallize in one of the two **close-packed** arrangements, **face-centered cubic (cubic close-packed)** or hexagonal close-packed. The arrangement of units in a crystal can be shown by a **unit cell.** A unit cell is a small volume of a crystal that can be repeated in three dimensions to show the whole structure of the crystal. Other ordered arrangements are **body-centered cubic** and **simple cubic.** Many ionic compounds can be pictured as derived from close-packed arrangements of anions by placing small cations in the holes. There are two kinds of holes, **tetrahedral holes** and **octahedral holes.** The crystal structures of ionic compounds depend on the relative sizes of the ions and the number of each kind of ion. For compounds that have molecules as units, shape is an important factor in crystal structure.

Much of our knowledge of the structure of crystals comes from X-ray diffraction. Contour maps showing electron density in a molecule can be plotted and the positions of atoms and the angles between bonds determined by X-ray diffraction.

Real crystals almost always have defects. Defects interfere with the normal attractive forces between the units of a crystal and usually weaken the crystal. However, defects sometimes result in desirable properties. For example, defects increase the conductivity of semiconductors, substances that are better conductors than nonmetals but not as good conductors as metals. If the current in a semiconductor is carried by movement of negatively charged electrons, the semiconductor is called an *n*-type **semiconductor.** If the current is carried by the movement of "positive" (electron-deficient) holes, the semiconductor is called a *p*-type **semiconductor.**

When liquids are cooled rapidly, the molecules can become fixed in the disorderly arrangement of the liquid before they have time to arrange themselves in order. The disorderly solid is called a **glass** or **amorphous** solid.

ADDITIONAL PRACTICE PROBLEMS

For information about the organization of Additional Practice Problems, Stop & Test Yourself, Putting Things Together, and Applications, see the beginnings of these sections in Chapter 1.

12.26 Tell how each of the following physical properties differs for the three states of matter: (a) density   (b) compressibility   (c) fluidity (ability to flow)   (d) rate of diffusion. (12.1)

12.27 Use the kinetic-molecular theory to explain why: (a) Liquids have definite volumes but take the shape of the lower part of their container.   (b) Solids have both fixed volumes and shapes.   (c) Gases are compressible, but liquids and solids are almost incompressible.   (d) Diffusion takes place rapidly in gases, slowly in liquids, and extremely slowly in solids.   (e) Diffusion takes place faster in hot liquids than in cold liquids. (12.1)

12.28 (a) In what ways is a glass like a crystalline solid?   (b) In what ways is a glass different from a crystalline solid? (12.1)

12.29 For a perfect crystal to form, should the crystal be formed from the melt slowly or rapidly? Explain your answer. (12.1)

12.30 What is the difference between intramolecular and intermolecular attractions? (12.2)

12.31 Why are intermolecular attractive forces important in liquids and solids but unimportant in gases under ordinary conditions? (12.2)

12.32 Using the method of Figure 12.9, (a) show how a hydrogen bond can form between acetone, $(CH_3)_2C{=}O$, and water. (b) Show all the types of hydrogen bonds that can form in a mixture of water, H_2O, and methyl alcohol, CH_3OH. (12.2)

12.33 Explain why: (a) the hydrogen bonds in HF are stronger than the hydrogen bonds in water.   (b) Hydrogen bonding is more important in water (boiling point, 100 °C) than in liquid HF (boiling point, 19.5 °C). (12.2)

12.34 Which of the following molecules can hydrogen bond to other molecules of the same kind? Explain your answer. (a) CH_3F (b) $CH_3CH_2CH_2NH_2$ (c) CH_3COOH (d) HF (e) H_2S (f) $(CH_3)_3N$ (12.2)

12.35 Which of the molecules in Problem 12.34 can accept a hydrogen bond from water? (12.2)

12.36 (a) What is polarizability? (b) How does the polarizability of argon compare with the polarizability of neon? Explain your answer. (c) How is polarizability related to the strength of intermolecular attractions? (12.2)

12.37 What is the difference between (a) induced dipoles and permanent dipoles? (b) London forces and dipole–dipole attractions? (12.2)

12.38 The melting points of the halogens are F_2, -219.6 °C; Cl_2, -101.0 °C; Br_2, -7.2 °C; I_2, 113.5 °C. Explain why the melting points of the halogens change as they do going down the group. (12.2)

12.39 Which member of each of the following pairs has the stronger London forces? Explain your answers. (a) CH_3CH_2Cl or CH_3CH_2Br (b) CH_3CH_2Cl or $CH_3CH_2CH_2Cl$ (c) $CH_3CH_2CH_2Cl$ or $(CH_3)_2CHCl$ (12.2)

12.40 (a) What is *viscosity?* (b) Explain why the following order of increasing viscosity is observed:

$$CH_3CH_2OH < HOCH_2CH_2OH < HOCH_2CH(OH)CH_2OH$$

ethyl alcohol   ethylene glycol   glycerine

(c) Explain why viscosity decreases as temperature increases. (12.3)

12.41 (a) What is surface tension? (b) Why do molecules at the surface of a liquid behave differently from molecules in the interior of a liquid? (c) Why does a 500-mL sample of water not form a spherical ball although a 0.05-mL sample of water forms a spherical drop? (d) Arrange the three liquids in Problem 12.40 in order of increasing surface tension. Explain your answer. (e) How would you expect surface tension to change as temperature is increased? Explain your answer. (12.3)

12.42 (a) What is capillary action? (b) Give an example or application of capillary action from your everyday life. (12.3)

12.43 Explain what is meant by the term *dynamic equilibrium* using a phase change as an example. (12.3)

12.44 (a) State Le Châtelier's principle. (b) Use Le Châtelier's principle to predict the effect of increasing temperature on the equilibrium between a solid and a vapor below the triple point. (12.3)

12.45 (a) How does the boiling point of a liquid change as atmospheric pressure decreases? (b) Why does the boiling point of a liquid vary with atmospheric pressure? (12.4)

12.46 Which member of each of the following pairs has the higher vapor pressure at 25 °C? Explain your answers. (a) CoO or CO_2 (b) A liquid with normal boiling point 35 °C or a liquid with normal boiling point 50 °C (c) $CH_3(CH_2)_3CH_3$ or $CH_3(CH_2)_4CH_3$ (d) $(CH_3)_4C$ or $CH_3(CH_2)_3CH_3$ (12.4)

12.47 Use Figure 12.24 to answer the following questions:
(a) What is the vapor pressure of bromine at 50.0 °C?
(b) What is the normal boiling point of bromine?
(c) What is the boiling point of bromine at 400 mmHg?
(d) What must the pressure be if bromine boils at 30 °C?
(e) Which has the stronger intermolecular attractions, Br_2 or CH_3CH_2OH? Explain your answer. (12.4)

12.48 (a) Suggest a reason why the boiling point of isooctane, $(CH_3)_3CCH_2CH(CH_3)_2$, is lower than the boiling point of octane, $CH_3(CH_2)_6CH_3$. (b) Make a rough copy of the curve for octane from Figure 12.24. On the copy, sketch in the curve for isooctane. (12.4)

12.49 (a) How is the vapor pressure of a liquid affected by each of the following? (i) the volume of the liquid (ii) the volume of the container (iii) the surface area of the liquid (iv) the strength of the intermolecular attractions in the liquid (v) the temperature of the liquid. (b) How is the time required to reach the equilibrium vapor pressure related to the surface area of the liquid? (12.4)

12.50 Does the entropy of the sample increase or decrease when a solid changes to a vapor? Explain your answer. (12.5)

12.51 The water in a beaker on a hot plate is boiling. (a) Would adjusting the temperature control of the hot plate to a higher setting increase the temperature of the water? Explain your answer. (b) What would be the result of adjusting the temperature control to a higher setting? (12.6)

12.52 On a molecular level, explain why temperature remains constant as a heated liquid is converted to vapor at its boiling point. (12.6)

12.53 Explain why heat of vaporization is always greater than heat of fusion. (12.6)

12.54 (a) How many kilojoules are needed to melt 52.4 g of ice at 0 °C? (b) How many kilojoules must be removed from a 52.4-g sample of water at 0 °C to make it freeze? (12.6)

12.55 (a) How many kilojoules are needed to boil 52.4 g of water at 100 °C? (b) How many kilojoules must be removed from a 52.4-g sample of steam at 100 °C to make it condense? (12.6)

12.56 How many kilojoules must be removed from a 34.9-g sample of sodium at its melting point to make it solidify? (12.6)

12.57 Use the phase diagram for water to answer the following questions. (a) What is the phase of water at 2 mmHg and -3 °C? (b) What term describes water at 400 °C and 225 atm? (c) Describe what will happen if a sample of water at -20 °C and 3 mmHg is heated to $+20$ °C at constant pressure. (d) Sketch the heating curve for part (c). (e) Describe what will happen if the pressure on a sample of water at 200 °C and 1 atm is increased to 50 atm at constant temperature. (12.7)

12.58 (a) Explain why the melting point "curve" in the phase diagram for water slopes one way and the melting point curve in the phase diagram for carbon dioxide slopes the other way. (b) Which slope is more common? (12.7)

12.59 (a) Why does ice float on water? (b) Benzene freezes at 5.5 °C. Would you expect "ice cubes" made of benzene to sink or float on liquid benzene? Explain your answer. (12.7)

12.60 Use the following data to sketch a phase diagram for xenon: triple point -121 °C, 0.37 atm; critical point 17 °C, 58 atm; normal melting point -112 °C; normal boiling point -107 °C. (12.7)

12.61 (a) What does the term *critical temperature* mean? (b) What happens to the phase boundary between liquid and vapor above the critical temperature? (c) Use the kinetic-molecular theory to explain what is observed at the critical temperature. (12.7)

12.62 (a) The critical temperature for ammonia is 132.5 °C, and the critical pressure is 112.5 atm. Is it possible to liquefy ammonia at 20 °C?   (b) For fluorine the critical point is −128.9 °C and 51.5 atm. Can fluorine be liquefied at 20 °C? (12.7)

12.63 Give an example of each of the six types of phase changes (see Figure 12.31) from your everyday life. (12.7)

12.64 Classify crystals of each of the following substances by type: (a) S₈ (b) Ni (c) MgO (d) Ge (e) Xe. For each substance, describe the attractive forces holding the units in place in crystals and predict the properties expected. (12.8)

12.65 What type of crystal (molecular, ionic, covalent network, or metallic) are crystals of each of the following? (a) An element melts at 44.1 °C; neither solid nor melt conduct electricity. (b) An element conducts heat and electricity well.   (c) An element melts at 1410 °C; neither the solid nor the melt conduct electricity. (12.8)

12.66 Define and give an example of allotropes. (12.8)

12.67 How many spheres does the simple cubic unit cell contain? (12.9)

12.68 (a) Explain what is meant by close packing.   (b) Sketch the unit cell for cubic close packing and label an octahedral hole. (12.9)

12.69 Lithium ions and magnesium ions have similar radii. The crystal structure of lithium chloride is similar to the crystal structure of sodium chloride (Figure 12.45). Can magnesium chloride have a sodium chloride structure? Explain your answer. (12.9)

12.70 The unit cell of *perovskite,* a compound of calcium, titanium, and oxygen, is shown below. (a) How many calcium atoms are in the unit cell? How many titanium atoms? How many oxygen atoms?   (b) What is the formula of perovskite? (12.9)

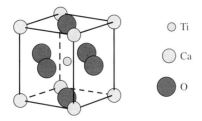

- ○ Ti
- ○ Ca
- ● O

12.71 The unit cell of a compound of uranium and oxygen can be described as cubic close-packed uranium ions with oxide ions in all the tetrahedral holes. What is the formula of this compound? (12.9)

12.72 Which of the following materials would give sharp X-ray diffraction patterns? (a) KI (b) sugar, C₁₂H₂₂O₁₁ (c) ice (d) water (e) window glass (12.10)

12.73 Chromium metal has a body-centered cubic structure; the density of chromium metal is 7.20 g/cm³. What is the length of the edge of the unit cell of chromium metal? (12.11)

12.74 Silver crystallizes in a cubic structure. The density of silver is 10.50 g/cm³, and the edge of a unit cell is 408 pm long. (a) How many atoms of silver are there in a unit cell? (b) What type of a cubic unit cell does silver form? (12.11)

12.75 The crystal structure of lithium bromide, LiBr, is similar to the crystal structure of sodium chloride (Figure 12.45) with the bromide ions in contact. The radius of a Br⁻ ion is 195 pm (Figure 8.9). (a) Calculate the length of the edge of the unit cell.   (b) The radius of a Li⁺ ion is 74 pm. Do the lithium ions in lithium bromide touch the bromide ions, or do they rattle around in the octahedral holes a little? Explain your answer. (12.11)

12.76 Sodium chloride melts at 801 °C, sodium bromide melts at 747 °C, and sodium iodide melts at 661 °C. (a) Suggest an explanation for the trend observed in the melting points of the sodium halides going from top to bottom of the periodic table.   (b) Arrange KCl, NaCl, and RbCl in order of increasing melting points.

12.77 (a) Does the entropy of a sample increase or decrease when a glass crystallizes? Explain your answer.   (b) What happens to the energy of the sample when it crystallizes?   (c) Sometimes a very viscous liquid crystallizes rapidly when scratched with a glass stirring rod. What temperature change, if any, will be observed? Explain your answer.

12.78 The boiling points of the monohalo methanes are as follows: CH₃Br, 3.6 °C; CH₃Cl, −24.2 °C; CH₃I, 42.4 °C. Explain the observed order of boiling points.

12.79 How many kilojoules will be released when a 6.37-g sample of bromine condenses from vapor to liquid?

12.80 The heat of vaporization of CH₃CH₂OH(l) is 43.5 kJ/mol. Would you expect the heat of vaporization of CH₃OH(l) to be 35.3, 43.5, or 51.7 kJ/mol? Explain your answer.

12.81 Explain the following observations of a flask containing water: (a) The water must be heated to make it boil at atmospheric pressure.   (b) If the open flask is allowed to stand at room temperature and atmospheric pressure, the temperature of the water does not change, but the water gradually disappears.   (c) If the flask is connected to a vacuum pump (a thick-walled flask must be used), the water boils, and the flask feels very cold. The water left in the flask freezes.

12.82 An element crystallizes in a cubic close-packed structure. The edge of the unit cell is 408 pm, and the density of the element is 19.27 g/cm³. (a) What is the atomic mass of the element?   (b) Which element is it?

12.83 The density of silicon is 2.33 g/cm³, and a silicon atom has a radius of 118 pm. (a) What is the volume occupied by 1 mol of silicon?   (b) What is the volume of 1 mol of silicon atoms?   (c) What percent of the space in silicon is empty?

12.84 The heat of fusion of ice is 6.01 kJ/mol. If all of the heat of fusion goes toward breaking hydrogen bonds and 3.5×10^{-20} J are required to break each hydrogen bond, what percent of the hydrogen bonds in ice is broken when ice melts to water?

12.85 The table of data for Cl₂ gives the temperature at which the vapor has the pressure indicated:

Pressure, mmHg	Temperature, °C
100	−72
300	−53
500	−43
700	−36
900	−31

(a) Graph pressure as a function of temperature. (b) From your graph, what is the normal boiling point of Cl_2? (c) What is the vapor pressure of $Cl_2(l)$ at $-50\ °C$?

12.86 (a) Suppose that, instead of heating a sample of room-temperature (20.0 °C) water at a uniform rate, a student heated the sample with a constant-temperature bath set at 80.0 °C. Sketch the temperature vs. time curve. (b) If acetone (bp 56.2 °C) was used instead of water, how would the curve differ?

12.87 The critical temperatures for ethane, CH_3CH_3, and ethene, $H_2C{=}CH_2$, are 32.2 °C and 9.9 °C, respectively. The quan-tity of one of these gases that remains in a cylinder must be determined by weighing; the quantity of the other can be determined by reading the pressure gauge. Which is which? Explain your answer.

12.88 People learn when they are shown that their ideas are wrong. Many people think that temperature always decreases when a sample is cooled. Describe a demonstration to show that this idea is not correct.

12.89 Make a sketch of the phase diagram for water similar to Figure 12.32(a).

STOP & TEST YOURSELF

1. The molecular level picture to the right represents
   (a) a gas.
   (b) a liquid.
   (c) a crystalline solid.
   (d) an amorphous solid or glass.
   (e) either an amorphous solid or a liquid.

2. Viscosity
   (a) is the resistance of fluids to flow. (b) is the force in the surface of a liquid that makes the area of the surface as small as possible. (c) is the resistance of solids to flow. (d) raises the surface of the water in a glass tube having a slender bore. (e) is the formation of a gas from a liquid.

3. The boiling point of $HCl(l)$ is higher than the boiling point of $Ar(l)$ as a result of
   (a) hydrogen bonding. (b) higher formula mass. (c) stronger London forces. (d) permanent dipole forces. (e) ion–ion forces.

4. From the graph of the equilibrium vapor pressures of some common liquids below, which of the following statements is true?

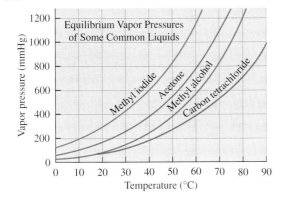

   (a) The normal boiling point of acetone is 59 °C. (b) The normal boiling point of acetone is 75 °C. (c) The normal boiling point of acetone is 100 °C. (d) The boiling point of acetone is lower than the boiling point of methyl iodide. (e) Acetone has a higher vapor pressure than methyl iodide.

5. A liquid is in equilibrium with its vapor. If some of the vapor escapes, what is the immediate result?
   (a) Vaporization rate decreases. (b) Condensation rate decreases. (c) Vaporization rate increases. (d) Condensation rate increases. (e) None of these answers is correct.

6. Which of the following substances has the strongest intermolecular attractions?
   (a) He, boiling point $-268.6\ °C$ (b) Ne, boiling point $-245.9\ °C$ (c) Ar, boiling point $-185.7\ °C$ (d) Kr, boiling point $-152.3\ °C$ (e) All have equally strong intermolecular attractions.

7. To convert a sample of air into a liquid, you would probably have to
   (a) increase temperature and increase pressure (b) decrease temperature and increase pressure (c) decrease temperature and decrease pressure (d) increase temperature and decrease pressure (e) Air cannot be converted into a liquid.

8. Which of the following gases can be liquefied by compressing it at room temperature?
   (a) methane, CH_4, critical temperature $-82.1\ °C$ (b) silicon tetrafluoride, SiF_4, critical temperature $-14.06\ °C$ (c) ethylene, C_2H_4, critical temperature 9.9 °C (d) xenon, Xe, critical temperature 16.6 °C (e) acetylene, C_2H_2, critical temperature 35.5 °C

9. What type of unit cell is represented by the figure at the right?
   (a) simple cubic
   (b) body-centered cubic
   (c) end-centered cubic
   (d) face-centered cubic
   (e) cubic close-packed

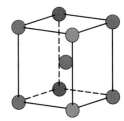

10. Boron nitride has the formula BN. Its cubic form is so hard that it will scratch diamond. What type of crystal does BN form?
    (a) ionic (b) molecular (c) covalent network (d) metallic (e) None of these is correct.

11. A liquid will "wet" a surface if
    (a) the liquid has a lower density than the surface. (b) the liquid has a high vapor pressure. (c) the attractive forces between the liquid and the surface are greater than the attractive forces between the molecules of the liquid. (d) the attractive forces

between the liquid and the surface are lower than the attractive forces between the molecules of the liquid. (e) the forces between the liquid molecules are weak.

12. The molar heat of vaporization of ethyl acetate, $CH_3COOCH_2CH_3$, is 32.5 kJ/mol. How much energy is required to vaporize 2.57 g of ethyl acetate?
(a) 0.897 J   (b) 6.97 kJ   (c) 83.5 kJ   (d) 948 J   (e) 7360 kJ

13. The Lewis formulas for five molecules are given below. Which molecule can hydrogen bond to other molecules of the same kind?

(a)
$$H-\overset{\overset{\displaystyle H}{|}}{\underset{\underset{\displaystyle H}{|}}{C}}-H$$

(b)
$$H-\overset{\overset{\displaystyle H}{|}}{\underset{\underset{\displaystyle H}{|}}{C}}-\ddot{O}:$$

(c)
$$\overset{\displaystyle H}{\underset{\displaystyle H}{\diagdown}}C=\ddot{O}:$$

(d) $H-H$

(e) $H-\ddot{\underset{..}{C}l}:$

Given the phase diagram below:

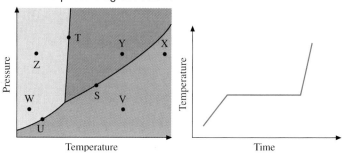

14. Which phase(s) is(are) present at point T?
(a) gas   (b) liquid   (c) solid   (d) both liquid and gas   (e) both solid and liquid

15. Referring to the phase diagram, the graph of temperature against time shows which change?
(a) X to Y   (b) W to V   (c) V to W   (d) X to Z   (e) Z to X

PUTTING THINGS TOGETHER

12.90 Only two elements are liquids under ordinary conditions: bromine, Br_2, and mercury, Hg. Which would you expect to have the higher surface tension? Explain your answer.

12.91 A molecule of a sugar has a number of —OH groups. For example, the condensed structural formula for glucose is $HOCH_2CH(OH)CH(OH)CH(OH)CH(OH)CHO$. Explain why sugars are notoriously difficult to crystallize.

12.92 Benzene, C_6H_6, melts at 5.5 °C and boils at 80.1 °C. The specific heat of liquid benzene is 1.742 J g^{-1} °C^{-1}, the heat of fusion is 10.59 kJ/mol, and the heat of vaporization is 30.8 kJ/mol. For a 5.68-g sample of benzene, calculate the number of kilojoules required to melt the sample, heat the sample to its boiling point, and vaporize the sample.

12.93 Rubbing alcohol on your skin makes you feel cool because the alcohol evaporates. How many kilojoules would the evaporation of 1.00 mL of alcohol [ethanol, $CH_3CH_2OH(l)$] remove from your skin? The density of alcohol is 0.79 g/mL.

12.94 (a) In a mercury barometer, what is in the space above the mercury in the column? (b) Although it would seem that a more sensitive barometer (one that would give a greater change in the height of the liquid column for a given change in atmospheric pressure) could be made by using water instead of mercury, water barometers are not only inconvenient because of their height (greater than 34 ft) but also inaccurate. Explain why water barometers are inaccurate. (c) Suppose that when a mercury barometer was put together, some bubbles of air got into the space at the top of the column of mercury. Would barometric pressures measured with this barometer be too high, correct, or too low? Explain your reasoning.

12.95 (a) What volume will be occupied by a 1.00-g sample of water vapor at 20.0 °C and 760.0 mmHg? (b) If a 1.00-g sample of liquid water is placed in an *empty* 2.0-L container

under reduced pressure. (a) Why does iodine sublime more rapidly when heated? (b) What is the purpose of using reduced pressure? (c) Make a sketch similar to Figure 12.23 showing the equilibrium between a solid and its vapor. (d) Explain how some molecules in a solid can have enough energy to sublime although the solid is not hot enough to melt.

12.97 From Table 5.4, what is the boiling point of water at 20.00 mmHg?

12.98 What is the hybridization of carbon in (a) diamond? (b) Graphite? (c) Which electrons in graphite are delocalized?

12.99 A compound of iron and oxygen has O^{2-} ions in a cubic close-packed crystal structure with iron ions in all the octahedral holes. What is the percent iron in this compound?

12.100 Sketch curves comparing the distributions of molecular speeds in water at 20 and 80 °C.

12.101 The surface tension of water at temperatures from 20 to 100 °C is given in the table:

Temperature, °C	Surface Tension, J/m²
20	7.28×10^{-2}
40	6.96×10^{-2}
60	6.62×10^{-2}
80	6.26×10^{-2}
100	5.89×10^{-2}

(a) Plot surface tension as a function of temperature. (b) What kind of curve is the graph (approximately)? (c) Write an equation that describes the relationship between surface tension and temperature for water in this temperature range quite well (although not exactly).

12.102 Suppose that a sample of ethyl alcohol, CH_3CH_2OH, is placed in a stoppered flask at 25 °C. (a) From Figure 12.24, what is the equilibrium vapor pressure of alcohol at 25 °C? (b) If you cool the flask in an ice-water bath (temperature approximately 0 °C), what will happen to the vapor pressure of the alcohol? To the pressure of the air inside the flask? To the total pressure inside the flask? If you have used a stopper that is too small, what will happen to the stopper? (c) Now suppose that, instead of cooling the flask containing the alcohol, you heat the flask on a hot plate. What will happen to the vapor pressure of the alcohol and the total pressure inside the flask? What will happen to the stopper?

12.103 A 0.500-g sample of water was placed in an empty (evacuated) 250-mL flask (zero is significant) at 23.6 °C. After enough time had passed for equilibrium to be reached, how many grams of water had evaporated?

12.104 (a) A mixture of 0.050 mol H_2 and 0.025 mol O_2 in a 10.00-L container was caused to react. What was the pressure after the products had cooled to room temperature (20.0 °C)? (b) Suppose that 0.100 mol H_2 was used instead of 0.050 mol. What was the pressure?

12.105 The specific heat of $CH_3CH_2OH(s)$ is 0.97 J g^{-1} °C^{-1}, the specific heat of $CH_3CH_2OH(g)$ is 1.42 J g^{-1} °C^{-1}, the melting point of $CH_3CH_2OH(s)$ is -117 °C, and the boiling point of $CH_3CH_2OH(l)$ is 79 °C. Plot the heating curve for an experiment in which a 4.72-g sample of CH_3CH_2OH was heated from -125 to 85 °C at a rate of 0.50 J/s.

12.106 There are three isomers of dichlorobenzene:

ortho-
dichlorobenzene

meta-
dichlorobenzene

para-
dichlorobenzene

(a) Which of the three isomers of dichlorobenzene is nonpolar? (b) Two of the three are liquids under ordinary conditions and one is a solid. Which one is the solid? Explain your answers.

12.107 Which of the following compounds have permanent dipole–dipole attractions? (a) CO_2 (b) SO_2 (c) PCl_3 (d) PCl_5 (e) ICl_3.

12.108 One equation of state, the ideal gas law, describes the behavior of all gases to within a few percent under ordinary conditions. However, a different equation of state must be written for each liquid or solid. Explain why the behavior of liquids and solids depends on what the liquid or solid is but the behavior of gases does not.

12.109 Comparing the same capillary and the same liquid, will the height of the liquid be the same, higher, or lower on the moon than on Earth? Explain.

12.110 (a) In the electrolysis of water (Figure 1.3), what observations prove that the gas bubbles are *not* water vapor or air? (b) Design an experiment to show that the gas bubbles in Figure 12.25 are water vapor.

12.111 Of each pair, which member (a) has the stronger intramolecular attraction? (b) The stronger intermolecular attraction? Explain your answers. (i) H_2NNH_2 or N_2 (ii) Br_2 or Cl_2 (iii) NH_3 or PH_3

APPLICATIONS

12.112 Water at about 300 atm and 500 °C can be used as a solvent for the destruction of toxic wastes. (a) Give two advantages of using water as a solvent. (b) What is the state of water under the conditions given? (c) An advantage of this state is that physical properties can be changed easily. If pressure is increased at constant temperature, how will density change? How will viscosity change? How will the rates of diffusion of solutes through the solvent change? Explain your answers.

12.113 In colonial days, bullets were made by dropping molten lead into water. Explain why the bullets were round.

12.114 Explain why you can get the last drops out of a bottle of molasses by warming the bottle.

12.115 Dishwashing detergents always contain wetting agents. What do wetting agents do to the surface tension of dishwater?

12.116 Explain why ice cubes get smaller and smaller and finally disappear if left in an open container in the freezer for a long time.

12.117 Explain why the sharp edges of glass tubing are rounded by fire-polishing.

12.118 Why do people who live in areas where the temperature drops below zero Celsius for long periods of time in the winter have to drain their swimming pools in the fall?

12.119 The butane used to fill cigarette lighters has a normal boiling point of -0.5 °C. (a) Why doesn't the liquid butane in the lighters boil at room temperature? (b) What can you deduce about the critical temperature of butane?

12.120 Suppose that you are camping in the mountains and the atmospheric pressure is only 495 mmHg. At what temperature will water boil?

12.121 Normal body temperature is about 37 °C for humans. If the air in the lungs is saturated with water vapor when it leaves the lungs, what is the vapor pressure of water in the air?

12.122 The same mass of water causes a worse burn as steam at 100 °C than as liquid water at 100 °C. Explain why.

12.123 The unit cell of the most widely used white pigment is shown below. What is the formula of this pigment?

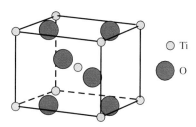

Ti
O

12.124 A compound of gallium and arsenic is an important semiconductor. The structure of this compound can be described as cubic close-packed arsenide ions with gallium ions in half the tetrahedral holes. What is the formula of this compound?

12.125 The compound OsO_4 is used as a biological stain. It is a soft solid that melts at 39.5 °C. Neither the solid nor the melt conducts electricity. What type of crystal are crystals of OsO_4?

12.126 Ethyl chloride, CH_3CH_2Cl, which boils at 12.3 °C, is used as a local anesthetic. When ethyl chloride is sprayed on a small area of skin, the area is cooled enough to numb it. Explain.

12.127 A new solid that, when pure, may be harder than diamond was predicted theoretically in 1989 and first synthesized in 1993. This solid may be useful for making ultrahard coatings for machine tools and glass windshields; it is 39.13% C and 60.87% N. (a) Write the empirical formula. (b) Name the compound. (c) What type of crystal does it form?

12.128 Before the days of refrigerators and ice makers, drinking water in hot countries was often cooled by storage in clay pots. Evaporation of water through the pot cooled the remaining water. How much water must evaporate to cool 3.79 L (1 gal) of water from 32.2 to 5.0 °C? Use 1.00 g/mL for the density of water.

12.129 The moisture content of air is usually expressed in terms of relative humidity:

relative humidity $=$

$$\frac{\text{partial pressure of water vapor in the air}}{\text{equilibrium vapor pressure at same temperature}} \times 100$$

(a) Calculate the relative humidity if the partial pressure of water vapor in the air is 14.4 mmHg and the air temperature is 22.4 °C. (b) Qualitatively, what happens to the relative humidity of a sample of cold air that is warmed to room temperature?

12.130 Explain each of the following observations: (a) A can filled with steam at 100 °C and atmospheric pressure collapses when sealed and then cooled. (b) A "three-minute egg" must be cooked more than three minutes in Denver (altitude 1 mi). (c) Spilled mercury forms tiny balls that roll all over. (d) Water stored in a porous clay pot or canvas bag stays cool. (e) Clothes dry faster in a dryer than when hung up to dry.

12.131 Current prices for small quantities of the dichlorobenzenes (see Problem 12.106) are: *o*-dichlorobenzene, $9.43/100 g; *m*-dichlorobenzene, $10.75/100 g; *p*-dichlorobenzene, $7.80/100 g. Naturally prices for industrial quantities are lower, but relative prices are similar. (a) One of the dichlorobenzenes is used as a solvent. Which one? (b) One of them is sold for use as a moth repellent in bureau drawers and closets. Which one? Explain your choices. (c) What must be true of the triple point of the moth repellent?

12.132 Ammonia, NH_3, Freon 12, CCl_2F_2, and sulfur dioxide, SO_2, have all been used as the fluids in refrigerators. In a refrigerator, a gas is compressed and condenses to a liquid. Evaporation of the liquid provides the cooling. What is the nature of the intermolecular attractions between the molecules in each of these compounds?

12.133 Ammonia is used for refrigeration (see Problem 12.132). The heat of vaporization of ammonia is 23.35 kJ/mol. How many grams of ammonia would have to vaporize to freeze one tray of ice cubes [500 g (5.00 $\times$ 10^2 g) water] if the water was originally at 40.0 °C?

12.134 Aluminum is used in airplanes as a result of its low density. Aluminum crystallizes in a cubic close-packed structure (Figure 12.43). The radius of an aluminum atom is 143 pm (Figure 8.8). Calculate the density of aluminum.

12.135 Diamond is a very wide gap semiconductor. Because it resists damage from radiation, heat, chemicals, and stress, diamond would be ideal for use in car or jet engine microcircuits. Which element should be used to dope diamond to make (a) An *n*-type semiconductor? (b) A *p*-type semiconductor?

12.136 Like Earth's atmosphere, the atmosphere of Triton, Neptune's largest satellite, is largely nitrogen gas. However, at the temperature of Triton's surface, 38 K, the partial pressure of $N_2(g)$ in the atmosphere is only 16 μbar, the vapor pressure of $N_2(s)$ at 38 K. (a) What is the vapor pressure in atmospheres? (b) What is the temperature in °C? (c) Explain why Triton's surface is isothermal, that is, has the same constant temperature everywhere.

12.137 Arctic ground squirrels hibernate for eight months each year, underground but above the permafrost. Their body temperatures remain below 0 °C for weeks and sometimes fall as low as -2.9 °C. Explain how this can be. (There is no evidence that their blood contains "antifreeze" molecules as the blood of cold-water fishes does.)

12.138 Highly fluorinated polymers such as the polymer from $H_2C=CHCOOCH_2(CF_2)_6CF_3$ (FOA) are used as lubricants in computer disk drives, protective coatings, and aircraft-fuel sealants. These fluoropolymers are insoluble in most conventional solvents except chlorofluorocarbons (CFCs) such as CCl_2F_2. Use of CFCs is being phased out because their release into the atmosphere has been identified as one of the main causes of the destruction of the ozone layer. However, supercritical CO_2 at 59.4 °C and 207 bar appears to be an excellent substitute. (a) Write a structural formula showing three monomer units for the polymer of FOA. (b) How many atmospheres is 207 bar? (c) Substitution of supercritical CO_2 for chlorofluorocarbons is an example of green chemistry. What is *green chemistry?*

Chemistry and Planetary Interiors

RAYMOND JEANLOZ

Geologist, Professor of Geophysics,
University of California, Berkeley

B.A. Geology
Amherst College

Ph.D. Geology, Geophysics
California Institute of Technology

I study materials at very high pressures and temperatures—up to millions of atmospheres of pressure and thousands of degrees Celsius in temperature. Why do I do this? I am interested in knowing what happens deep inside planets, and these are the conditions of planetary interiors. The pressure and temperature at Earth's center, for example, are about 3.6×10^6 atm and 6000 to 7000 °C, respectively.

To understand how the planets have evolved over their billions of years of geological history, it is necessary to examine the materials and processes of the deep planetary interiors. And to do this, one must study what happens at the appropriate conditions of pressure and temperature.

Such experiments are carried out with a combination of lasers and diamonds. Specifically, materials are taken to high pressures by squeezing them between the points of two gem-quality diamonds. Because the points of the diamonds are very small, usually less than 0.5 mm across, and because pressure is equal to force divided by area, we can achieve pressures of millions of atmospheres just by pushing the diamonds together with only a relatively modest force.

Why use diamond? It is the strongest material known and can therefore support the high pressures at the points without crushing. In addition, because diamond is transparent, we can actually observe our sample—not just by eye, but also with X-rays, lasers, and spectrometers—and analyze it in detail while it is at high pressure.

To simulate the high temperatures existing inside planets, we use a powerful laser beam that is focused right through the diamonds. The sample absorbs the continuous (not pulsed) infrared light from the laser, and in this way it becomes heated to the thousands of degrees that we are trying to reach. The temperature is then determined by measuring the thermal radiation from the sample: that is, by measuring how "red" hot or "white" hot the sample is.

Before I describe the insides of planets, I should mention a more basic reason that pressures of millions of atmospheres are interesting. If you calculate the change in the energy of a sample that is caused by such pressures, you obtain a value that is comparable in magnitude to bonding energies. Put another way, the work expended in squeezing the atoms together under a million atmospheres of pressure has a very large effect on the bonding energies. As a result, the nature of bonding—that is, the chemical properties of matter—can be greatly altered by pressure.

Many examples of the effect of pressure on chemistry are now known. For example, xenon is normally an unreactive "inert" element, yet it becomes a metal at pressures of about 1.3×10^6 atm. Similarly, hydrogen has recently been converted into a metal by compression to pressures above 1.7 to 2.0 $\times 10^6$ atm. Because hydrogen is the most abundant element in the universe, its high-pressure metallic form is thought to be the primary material in the interiors of the giant planets (such as Jupiter and Saturn) and stars. Thus, metallic hydrogen is not only a novel material that is unavailable at normal conditions—low pressures—but it is also one of the most important materials for astrophysicists and planetary scientists.

Changes in chemical bonding due to pressure are important inside the earth as well. One way to see this is to think of the deep interior, where the rocky outer shell (mantle and crust) of our planet is in contact with the metallic core. Most of the core is made of liquid iron alloy, and this is where the earth's magnetic field is produced.

Rock is composed of oxides, such as the compounds typically making up ceramics. In fact, rock consists of more than 50% oxygen on an atomic basis. At normal conditions, it is very unlike a metal. At the boundary between the mantle and the core, nearly 3000 km beneath our feet, however, the pressure is over 1.35×10^6 atm, and the situation is quite different. From our experiments, we have discovered that oxygen combines with iron quite easily at high pressures, forming a truly metallic iron oxide alloy.

Qualitatively, pressure has the effect of making oxygen behave chemically more like sulfur does at zero pressure (remember that sulfur is just below oxygen in the periodic table); that is, sulfur easily forms metal alloys with iron, such as FeS_2 pyrite ("fool's gold"). Similarly, oxygen alloys with iron at the pressures of the deep interior.

As a result, when we simulate the conditions existing at the boundary between the mantle and the core, we observe dramatic chemical reactions occurring between liquid iron or iron alloy (corelike material) and the high-pressure minerals of the deep mantle. Some of the oxide minerals dissolve into (alloy with) the liquid metal, leaving a heterogeneous mixture of metal alloys and nonmetallic minerals as reaction products.

We expect that the same reactions may take place inside Earth, with the rocky mantle dissolving into the liquid metal of the core over geological history. In fact, seismologists, who study the deep interior using the sound waves created by earthquakes, find that just such a zone of heterogeneous materials exists at the mantle-core boundary inside Earth. A simple explanation for why the core is an iron alloy—it does not have exactly the same properties as pure liquid iron at high pressures and temperatures—is that this region has reacted with, and become contaminated by, the mantle above it. Thus, based on the high-pressure experiments, we may have identified what is the most chemically active region inside our planet, the boundary between the mantle and core, and may have determined one of the main ways in which Earth evolves chemically over geological time.

13

SOLUTIONS REVISITED

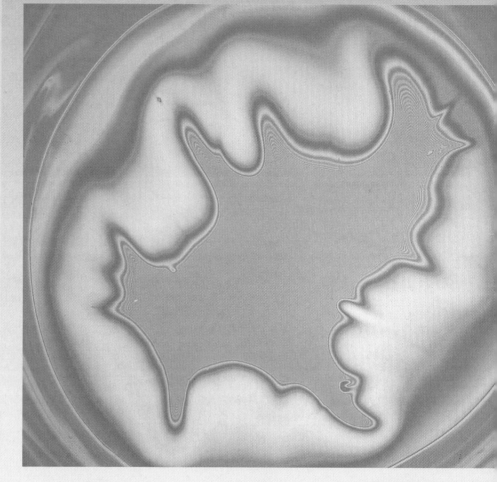

Oil and water do not mix.

Solutions can be gases, liquids, or solids. An example of each possible type is given in Table 13.1. Mixtures that consist of a liquid in a gas (such as fog) and a solid in a gas (such as smoke) are colloidal dispersions. Colloidal dispersions are discussed in the last section of this chapter.

Solutions in which the solvent is a liquid, such as seawater, are the most common type of solution, and the ability to act as a solvent is an important property of a liquid. Most of this chapter will be devoted to solutions in which the solvent is a liquid.

— *Why is a further study of solutions a part of general chemistry?*

In Chapter 4, the emphasis was on reactions in solution; in this chapter, the emphasis is on the physical properties of solutions, which are also of great practical importance. For example, the coolant/water solution that keeps your car's engine from overheating costs more than plain water. But it is worth the extra money because it does not freeze and crack the block in cold weather or boil over on hot days as pure water does.

— *What do you already know about solutions?*

Everyone uses solutions all the time without thinking about them as solutions. Coca Cola, apple juice, beer, and clear soup are aqueous solutions and so are mouthwash, cough syrup, and some shampoos and laundry detergents. Your blood plasma, the water in the ocean and in swimming pools, and the sap in plants are also aqueous solutions. Gasoline, lubricating oil, and antifreeze are all nonaqueous solutions.

TABLE 13.1 Types of Solutions

State of Solute	State of Solvent	Example
Gas	Gas	Air
Gas	Liquid	Carbonated beverages [$CO_2(g)$ in $H_2O(l)$]
Gas	Solid	$H_2(g)$ in $Pd(s)$
Liquid	Liquid	Vinegar [$CH_3COOH(l)$ in $H_2O(l)$]
Liquid	Solid	Dental amalgam [$Hg(l)$ in $Ag(s)$]
Solid	Liquid	Seawater [$NaCl(s)$ in $H_2O(l)$]
Solid	Solid	Brass [$Zn(s)$ in $Cu(s)$]

 Mixtures of gases were discussed in Section 5.8.

The dissolving of hydrogen in palladium is used to separate hydrogen gas from mixtures with other gases, none of which is soluble in palladium.

Solutions are homogeneous mixtures (Section 1.3). If the solution has the same state (solid, liquid, or gas) as one of the pure components, that component is usually called the solvent (Section 4.1). If more than one (or none) of the components is in the same state as the solution, the component that is present in greatest quantity is usually regarded as the solvent. The other substances in the solution are called solutes. The concentration of a solution is the amount of solute dissolved in a given quantity of solvent or solution. Concentration can be expressed as mass or volume percent (Section 4.7) or in molarity (Section 4.8):

$$\text{molarity, M} = \frac{\text{mol solute}}{\text{vol soln L}}$$

A concentrated solution is one that contains a high concentration of solute,* and a dilute solution is one that contains only a low concentration of solute. A saturated solution is a solution containing as much solute as will dissolve at a given temperature in the presence of excess solute. The quantity of solute that will dissolve in a given quantity of solvent or solution is called the solubility of the solute. A solution that contains less dissolved solute than a saturated solution is called unsaturated. A solution that contains more dissolved solute than a saturated solution is called supersaturated; supersaturated solutions are

not stable. A solid that separates from a solution is called a precipitate.

Electrolytes are materials that conduct an electric current when dissolved or melted (Section 4.2) and can be classified as either strong or weak electrolytes. In solutions of strong electrolytes, almost all the solute is in the form of ions. Solutions of strong electrolytes are good conductors of electricity if they are sufficiently concentrated. In solutions of weak electrolytes, most of the solute is in the form of molecules, and only a little of the solute is in the form of ions. Solutions of weak electrolytes are poor conductors of electricity.

*In solutions that contain water, water is often considered to be the solvent even if more of another component is present. For example, the name "concentrated sulfuric acid" suggests that sulfuric acid is the solute. However, concentrated sulfuric acid is 95% by mass sulfuric acid and only 5% by mass water.

13.1 A KINETIC-MOLECULAR VIEW OF THE SOLUTION PROCESS

From your everyday life, you know that solubilities vary enormously. Salt, sugar, and ethyl alcohol are very soluble in water. Marble, people, and plants are very insoluble in water. Salad oil is not soluble in vinegar; however, salad oil, if spilled on your clothes or the rug, can be dissolved by cleaning fluid. How are these great differences in solubility explained?

Oil and vinegar dressing must be shaken each time it is used.

Ideal Solutions

A solution looks uniform even under a microscope, but at the level of atoms, molecules, and ions, a solution appears heterogeneous. At this submicroscopic level,

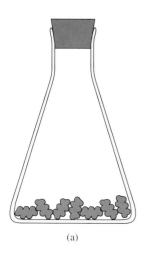

(a)

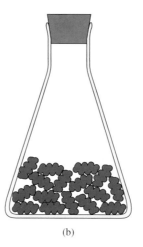

(b)

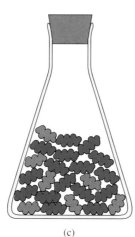

(c)

■ FIGURE 13.1 Pentane dissolves in hexane—microscopic and macroscopic views. (a) Pentane. (b) Hexane. (c) Solution of pentane in hexane.

In the Solution Theory module of the Solutions section, three solvents and three solutes are displayed. Students are asked to drag any solvent-solute pair into a beaker; a molecular-level animation then shows how the solute is solveted and tells the type of interparticle attraction operating between solute and solvent. A second activity asks students to predict solubilities and identify interparticle attractions.

London forces were discussed in Section 12.2.

there are individual atoms, molecules, ions, and empty space. ■Figure 13.1 shows models of pentane, $CH_3CH_2CH_2CH_2CH_3$, and hexane, $CH_3CH_2CH_2CH_2CH_2CH_3$, and a solution of pentane in hexane. The models in Figure 13.1 are about 10^7 times as large as real molecules.

When pentane is dissolved in hexane, three things happen at the microscopic level: (1) molecules of pentane separate from each other; (2) molecules of hexane move apart to make room for the molecules of pentane; and (3) molecules of pentane and molecules of hexane come together to form a solution. Molecules of pentane are uniformly distributed throughout the hexane in the solution. London forces hold molecules of pentane together in liquid pentane; London forces hold molecules of hexane together in liquid hexane. Energy is required to separate molecules of pentane and to separate molecules of hexane. But there is not much difference between the sizes and shapes of molecules of pentane and molecules of hexane. Hexane molecules are slightly longer than pentane molecules, but the difference is so small it is hard to see in Figure 13.1. London forces between molecules of pentane and hexane are about the same as those between molecules of pentane and those between molecules of hexane. When the molecules of pentane and hexane come together to form the solution, the amount of energy given off by attractions between pentane and hexane molecules is about the same as the amount of energy used up separating molecules of each component. There is no significant difference in energy between the solution and the solute and solvent separately. *A solution that forms without a significant change in energy* is called an **ideal solution.** The volume occupied by an ideal

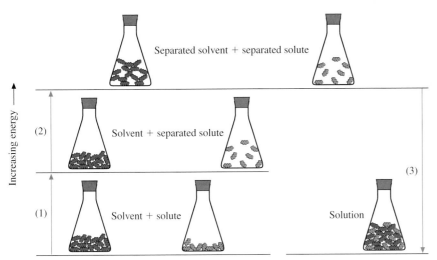

Energy diagram for formation of an ideal solution.

Remember Hess's law, Section 6.9.

solution is equal to the sum of the volumes of the components. The entropy of the mixture is greater than the entropy of the separate liquids. *Increased entropy of the system is the driving force that makes an ideal solution form:*

Nonideal Solutions

Now let's think about what happens if ethyl alcohol, CH_3CH_2OH, and water, H_2O (or HOH), are mixed. The —OH part of the alcohol molecule is like the —OH part of the water molecule. The —OH group in the alcohol molecule and the —OH group in the water molecule can hydrogen bond

$$CH_3CH_2 - \overset{\delta-}{\ddot{O}}: \cdots \cdots \overset{\delta+}{H} - \overset{H^{\delta+}}{\underset{\delta-}{\overset{|}{\ddot{O}}}}:, \qquad CH_3CH_2 - \overset{\delta-}{\ddot{O}}:$$

$$CH_3CH_2 - \overset{\delta-}{\ddot{O}}: \cdots \cdots \overset{\delta+}{H} \qquad , \qquad \text{and so forth.}$$

When samples of ethyl alcohol and water that are at the same temperature are mixed, the temperature of the solution is higher than the original temperature. In other words, thermal energy is given off when ethyl alcohol and water are mixed; formation of an alcohol–water solution is exothermic. Solutions of ethyl alcohol and water are not ideal solutions. The energy released by formation of hydrogen bonds between alcohol and water is greater than the energy used up in separating alcohol molecules and water molecules. Energy as well as entropy favors solutions of water and alcohol, and water and ethyl alcohol are soluble in each other in all proportions. *Liquids that are soluble in each other in all proportions* are said to be **miscible.**

The CH_3CH_2— part of the ethyl alcohol molecule is similar to the ends of the hexane molecule, $CH_3CH_2CH_2CH_2CH_2CH_3$. Like pentane and hexane, ethyl alcohol and hexane are soluble in each other in all proportions. However, when ethyl alcohol and hexane are mixed, temperature decreases; an ethyl alcohol–hexane solution is

The ethyl alcohol must be dry (free of water).

not ideal. Hexane has no hydrogens bonded to nitrogen, oxygen, or fluorine and cannot hydrogen bond to ethyl alcohol. The only intermolecular forces possible between ethyl alcohol molecules and hexane molecules are London forces. More energy is needed to separate the hydrogen-bonded alcohol molecules from each other than is gained from London forces between alcohol molecules and hexane molecules. However, the increase in entropy on mixing is large enough that a solution forms spontaneously:

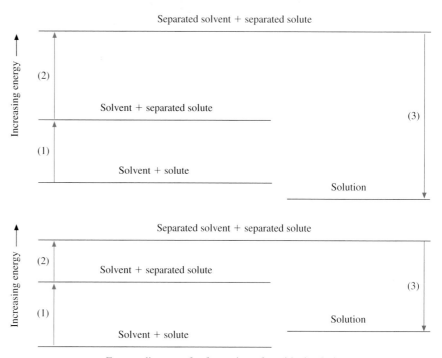

Energy diagrams for formation of nonideal solutions.
Top: Formation of solution is exothermic like solution of ethyl alcohol in water.
Bottom: Formation of solution is endothermic like solution of ethyl alcohol in hexane.

Immiscible Mixtures

Next let's consider what happens if water and hexane are mixed. Water is strongly hydrogen-bonded, and considerable energy is required to separate water molecules from each other. Although a solution of hexane and water would have greater entropy than pure hexane and pure water, so much energy is needed to separate water molecules from each other and so little energy is gained from the interaction between water molecules and hexane molecules that solution does not take place. Water and hexane are not soluble in each other.

Suppose hexane and water are mixed by vigorous stirring or shaking. As water molecules move through the mixture and bump into other water molecules, water molecules stick together. Water separates from hexane. Because the density of water (1.00 g/mL) is greater than the density of hexane (0.660 g/mL), water sinks to the botton as shown in ◾Figure 13.2(c). *Liquids* like water and hexane, *which are not soluble in each other,* are said to be **immiscible.** The energy of a solution of hexane and water would be so much higher than the energy of hexane and water separately that the increase in entropy on mixing is not large enough for a solution to form.

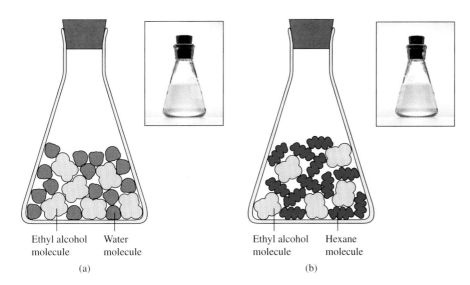

Ethyl alcohol molecule	Water molecule		Ethyl alcohol molecule	Hexane molecule		Water molecule	Hexane molecule
(a)			(b)			(c)	

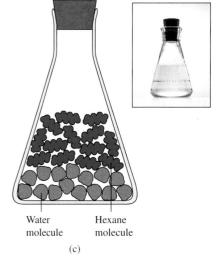

FIGURE 13.2 Like dissolves like: (a) Ethyl alcohol is soluble in water. (b) Ethyl alcohol is soluble in hexane. (c) Water is not soluble in hexane. Because the density of hexane (0.660 g/mL) is lower than the density of water (1.00 g/mL), hexane floats on water.

The common saying, "Oil and water do not mix," is another way of stating this rule of thumb.

Predicting Solubilities

Like dissolves like is a good rule of thumb for predicting solubility where "like" refers to substances with similar forces between particles.* Nonpolar solvents like hexane and carbon tetrachloride, CCl_4, dissolve nonpolar substances like iodine, I_2. Polar solvents like water are not good solvents for nonpolar substances. Iodine is very soluble in carbon tetrachloride but only slightly soluble in water. Alcohols with long carbon chains like 1-octanol, $CH_3(CH_2)_6CH_2OH$, are soluble in hexane. Most of a 1-octanol molecule is like hexane; only a very small part of the 1-octanol molecule, the —OH group, however, is like water. Therefore, 1-octanol and similar long-chain alcohols are not significantly soluble in water.

Molecules that contain fluorine, oxygen, or nitrogen but do not have hydrogen bonded to fluorine, oxygen, or nitrogen can accept hydrogen bonds although they cannot form hydrogen bonds to themselves. For example, diethyl ether, $CH_3CH_2OCH_2CH_3$, molecules cannot hydrogen bond to other diethyl ether molecules. The boiling point of diethyl ether is only 35°C, about the same as the boiling point of pentane, $CH_3CH_2CH_2CH_2CH_3$. The boiling point of 1-butanol, $CH_3CH_2CH_2CH_2OH$, an isomer of diethyl ether that can form hydrogen bonds to itself, is 117°C. However, the water solubilities of 1-butanol and diethyl ether are similar (8 g/100 g water at room temperature). Both 1-butanol and diethyl ether can accept hydrogen bonds from water:

$$CH_3CH_2CH_2CH_2 \overset{\delta-}{\underset{\delta+H}{-\ddot{\underset{\cdot\cdot}{O}}:}} \cdots\cdots \overset{H^{\delta+}}{\underset{\delta-}{\overset{|}{H}}} - \overset{\delta+}{\underset{\cdot\cdot}{\ddot{O}}:} \qquad CH_3CH_2 \overset{\delta+}{\underset{\delta+CH_2}{-\overset{\delta-}{\ddot{\underset{\cdot\cdot}{O}}}:}} \cdots\cdots \overset{H^{\delta+}}{\underset{\delta-}{\overset{|}{H}}} - \overset{\delta+}{\underset{\cdot\cdot}{\ddot{O}}:}$$

---

*"Like dissolves like" applies to solutions that are solid as well as to solutions that are liquid. For example, potassium and rubidium, which are similar in size and have similar ionization energies, form a complete series of solid solutions ranging in composition from 0% K–100% Rb to 100% K–0% Rb. Sodium does not form a complete series of solid solutions with potassium because sodium is smaller than potassium and has a higher ionization energy. (Remember that the units of metallic crystals are metal ions.) The minerals siderite, $FeCO_3$, and rhodochrosite, $MnCO_3$, form a complete series of solid solutions; Fe^{2+} ions and Mn^{2+} ions are similar in size and charge.

13.1   Which of the following substances would you expect to be soluble in

benzene, ? Explain your reasoning. (a) acetone, CH_3CCH_3 (b) NaCl
$\overset{\|}{O}$

Solution   (a) Benzene is nonpolar. The C=O group in acetone is polar but cannot hydrogen bond to itself. The two CH_3 groups in acetone are nonpolar like benzene. Acetone is probably soluble in benzene.

(b) NaCl is an ionic compound. It is probably insoluble in the nonpolar solvent benzene. The energy required to separate oppositely charged ions from each other is very large, but little energy would be gained from the interaction between the ions and nonpolar benzene molecules. The solution process would be too endothermic to take place despite the entropy increase that would result.

13.1   Which of the following solvents would you expect to be miscible with pentane, $CH_3CH_2CH_2CH_2CH_3$? Explain your reasoning. (a) H_2O
(b) Heptane, $CH_3(CH_2)_5CH_3$ (c) 1-Propanol, $CH_3CH_2CH_2OH$ (d) CCl_4

Chemists often call the process of dissolving "dissolution."

Reviewers were unanimous in preferring solution to dissolution in a general chemistry text. Solution is correct according to the new Random House unabridged dictionary.

*For evidence that the water molecules around halide ions are oriented as shown, see Sharpe, A. G. J. Chem. Educ. **1990**, 67, 309.*

Macroscopic view of ionic solid $CuSO_4 \cdot 5H_2O$ dissolving in water. The blue solution is below the crystal because the solution is denser than water.

Solution of Solids

Now let's think about what happens when an ionic solid like salt dissolves in water. Some of the ions in a sodium chloride crystal are vibrating more vigorously than others. Some of the ions, for example, the ones on corners, are more exposed to being hit by water molecules than others, and some of the water molecules are moving faster than others. If a fast-moving water molecule hits a vigorously vibrating, exposed ion with the oppositely charged end of its dipole toward the ion as shown in ▪Figure 13.3, the ion is attracted to the water molecule. The ion leaves the crystal and goes into solution.

In solution the ions become surrounded by shells of water molecules. Positively charged sodium ions are surrounded by water molecules with their negative sides

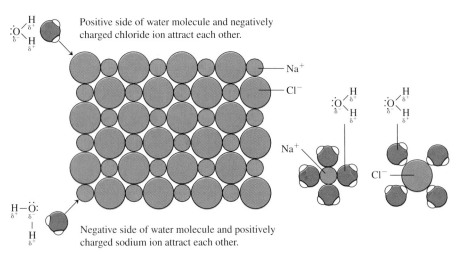

▪ FIGURE 13.3   Microscopic view of ionic solid NaCl dissolving in water. A NaCl crystal dissolves in water when a fast-moving water molecule hits a vigorously vibrating ion in an exposed position. In solution, sodium ions are surrounded by water molecules with the − end in and chloride ions are surrounded by water molecules with the + end in.

in, and negatively charged chloride ions are surrounded by water molecules with their positive sides in. In both cases, the ions are surrounded by spherical shells of water molecules, which separate the positively charged ions from the negatively charged ions. The sodium ions and chloride ions are said to be **solvated.** The term **hydrated** is often used to refer to *ions (or molecules) that are solvated by water.* Solvation of ions is an exothermic process; energy released by solvation helps make up for the energy that is required to separate the oppositely charged ions in a crystal of an ionic solid from each other.

Dynamic Equilibria in Saturated Solutions

If the amount of sodium chloride is smaller than the amount needed to form a saturated solution, all the sodium chloride will dissolve. If the amount of sodium chloride is larger than the amount needed to form a saturated solution and enough time is allowed, an equilibrium will be reached, and some undissolved sodium chloride will remain. When the sodium chloride is first added to the water, no sodium ions or chloride ions will be in solution. As time passes, the concentration of sodium ions and chloride ions in solution will increase until, after awhile, there will be a significant number of sodium ions and chloride ions in solution. The ions and their shell of water will move around the solution. Some of them will bump into water molecules or the sides of the container and be turned back toward the crystal. When an ion and its shell of water hits the crystal near an oppositely charged ion, the shell of water molecules may be knocked off, and the ion may stick to the crystal. The ion may precipitate. The greater the concentration of ions in solution, the greater the rate of precipitation. If enough solid is present, the rate of precipitation will become equal to the rate of solution:

$$\text{rate}_{\text{precipitation}} = \text{rate}_{\text{solution}}$$

A dynamic equilibrium will be established; the undissolved solid will be in equilibrium with ions in solution:

$$NaCl(s) \rightleftharpoons Na^+(aq) + Cl^-(aq)$$

The concentration of sodium chloride in solution will remain constant although the same sodium ions and the same chloride ions will not be in solution all the time. The quantity of solid sodium chloride will not change as long as the temperature does not change. The constant equilibrium concentration is equal to the solubility; the solution is saturated. The same constant concentration will be reached at equilibrium whether you start with only a little more sodium chloride than can dissolve or with a lot more sodium chloride than can dissolve. If you start with a lot more sodium chloride than can dissolve, a large quantity of undissolved solid will be left at equilibrium.

Solution of molecular solids like sugar is similar to solution of ionic solids except that molecules, not ions, go into solution. The molecules are solvated in solution. If the amount of sugar is larger than the amount needed to form a saturated solution and enough time is allowed, an equilibrium will be reached. Undissolved solid will be in equilibrium with molecules in solution:

$$C_{12}H_{22}O_{11}(s) \rightleftharpoons C_{12}H_{22}O_{11}(aq)$$

Experimental Test for Equilibrium

An experimental test for equilibrium is that the same situation must be reached no matter which side you start from. For example, the same concentration of sodium chloride is reached by forming solid sodium chloride by precipitation from aqueous solution

$$Na^+(aq) + Cl^-(aq) \longrightarrow NaCl(s)$$

as by dissolving enough solid sodium chloride in water to form a saturated solution:

$$NaCl(s) \longrightarrow Na^+(aq) + Cl^-(aq)$$

The temperatures must, of course, be the same and enough time must be allowed for the system to reach equilibrium.

Rate of Reaching Equilibrium

At a given temperature, the rate at which a solid dissolves and precipitates depends on the surface area of the solid. The larger the surface area, the more particles escape in a given period of time and the more particles hit the surface and precipitate in a given time. Therefore, the larger the surface area, the faster equilibrium is reached. However, the same constant concentration is reached whether the surface area of the solid is large or small (providing that enough solid is available to form a saturated solution).

The same quantity of material has a larger surface area if it is finely divided than if it is in large pieces. Consider what happens to the surface area if a cube 1 cm on a side is cut into 1000 little cubes, each 0.1 cm on a side (see ▪Figure 13.4). The smaller the particles of a solid, the larger the surface area and the faster the solid will dissolve.

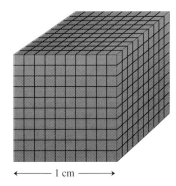

$$Vol_{\text{large cube}} = Total\ vol_{\text{small cubes}}$$
$$= (1\ cm)^3 = 1\ cm^3$$

$$Area_{\text{large cube}} = 6\ faces\left(\frac{1\ cm^2}{face}\right) = 6\ cm^2$$

$$Total\ area_{\text{small cubes}} = 10^3\ cubes\left(\frac{6\ faces}{cube}\right)\left(\frac{0.01\ cm^2}{face}\right)$$

$$= 60\ cm^2$$

←——— 1 cm ———→

▪ FIGURE 13.4 If a cube of material is divided into 1000 small cubes, the total surface area is increased by a factor of 10.

The higher the temperature, the faster a solid will dissolve. At a higher temperature, the particles of a solid are vibrating more vigorously, and the average speed of the solvent particles is greater. More solvent particles will hit the surface of the solute in a given time, and the collisions will be more violent.

Stirring also speeds up solution. The rate of diffusion of dissolved particles through the liquid away from the solid is not very high. Stirring moves the dissolved particles away from the solid so they cannot hit it and stick once again.

PRACTICE PROBLEMS

13.2 Describe the solution of sugar, $C_{12}H_{22}O_{11}$, in water as pictured by the kinetic-molecular theory.

13.3 List three things you could do to dissolve a lump of $CuSO_4 \cdot 5H_2O$ faster.

Ionic substances like sodium chloride are not soluble in nonpolar solvents such as hexane. Considerable energy is needed to separate the oppositely charged ions in an ionic crystal from each other. A nonpolar solvent can't solvate ions, and no energy is released to make up for the energy required to separate the oppositely charged ions.

It is difficult to predict whether ionic solids will be soluble in water except by the empirical rules in Table 4.4 because a number of competing factors are involved when an ionic solid dissolves in water. Attractions between opposite charges are stronger when the charges are higher and the distance between the charges is smaller. As a result, the smaller an ion and the higher its charge, the stronger the attractive force between the ion and the molecules of a polar solvent and the more energy is released by solvation. For example, a lithium ion, Li^+, is more strongly hydrated than a sodium ion, Na^+, because a lithium ion (radius $= 74$ pm) is smaller than a sodium ion (radius $= 102$ pm). A doubly positively charged magnesium ion, Mg^{2+}, is more strongly hydrated than a singly charged lithium ion. (These two ions have similar radii.) Anions are usually less strongly hydrated than cations because anions are usually larger than cations.

However, the same factors that result in strongly hydrated ions—small size and high charge—also make attractions between ions in the crystalline solid strong and the energy required to separate ions from the solid large. In addition, *prediction of the entropy change that accompanies solution is difficult when water is the solvent.* Usually entropy increases when a solution is formed. But solvation decreases the entropy of a solution because the solvent molecules are clustered around the solute ions or molecules. Solvation by water molecules is unusually strong, and entropy may either increase or decrease when an aqueous solution is formed.

Substances that crystallize in covalent networks are very insoluble. Too much energy is needed to break the covalent bonds for these substances to dissolve. Metals are also insoluble in ordinary liquids. Again, the forces holding the atoms together in crystals are too strong. Metals can only be "dissolved" in ordinary liquids by chemical reactions that change the metals into soluble compounds. For example, zinc can only be "dissolved" in water by converting it to a soluble salt such as zinc chloride:

$$Zn(s) \ + \ 2HCl(aq) \longrightarrow ZnCl_2(aq) \ + \ H_2(g)$$

Sodium can only be "dissolved" in water by converting it to a soluble compound such as sodium hydroxide:

$$2Na(s) \ + \ 2H_2O(l) \longrightarrow 2NaOH(aq) \ + \ H_2(g)$$

In the solution produced by the reaction of zinc with hydrochloric acid, the solute is zinc chloride, not zinc. In the solution produced by the reaction of sodium with water, the solute is sodium hydroxide, not sodium. Zinc and sodium metals are not soluble in water.

Crystalline Hydrates

Many compounds crystallize from aqueous solution with a definite proportion of water. For example, evaporation of a water solution of $CuSO_4$ yields beautiful blue crystals of $CuSO_4 \cdot 5H_2O$. Compounds of small, highly positively charged ions such as Cu^{2+} and Mg^{2+} commonly form hydrates, whereas compounds of larger, singly positively charged ions such as Na^+ and K^+ do not.

▶ Remember that *empirical* means "derived from experience or observation alone."

Metals are often soluble in melted salts.

Solution is a physical process. The solute and solvent can be separated by evaporation of the solvent, a physical change.

Large crystal of $CuSO_4 \cdot 5H_2O$ formed by slow evaporation of a saturated solution.

13.4 Which member of each of the following pairs of ions would you expect to be more strongly hydrated? Explain your answers. (a) Mg^{2+} or Ca^{2+} (b) K^+ or Ba^{2+} (Radii are similar.) (c) Na^+ or Cl^-

13.5 Which of the following substances would you expect to dissolve in water (to the extent of 1 g of solute/100 g H_2O or more)? Explain your reasoning. (a) $KI(s)$ (b) $Ti(s)$ (c) $CCl_4(l)$ (d) $CH_3OH(l)$ (e) C (diamond)

13.6 Which would be more likely to crystallize as a hydrate, $BaSO_4$ or $CaSO_4$?

13.3 EFFECT OF TEMPERATURE ON SOLUBILITY

Liquids and Solids

The solubilities of most liquids and solids are greater at high temperatures than at low temperatures. Applying Le Châtelier's principle, you might conclude that solution is an endothermic process and the temperature often does fall noticeably when solids are dissolved in water. The cold packs used in hospitals and often found in first-aid kits put the large absorption of thermal energy that accompanies the dissolving of certain solids to practical use. Some cold packs contain ammonium nitrate and water separated by thin plastic. Squeezing a cold pack breaks the thin plastic and allows the ammonium nitrate and water to mix; the cold pack instantly becomes cold.

However, in a number of cases, the temperature rises when solids are dissolved in water, although the solids are more soluble in hot water than in cold water. Sodium hydroxide is a good example. The solubility of sodium hydroxide increases from 109 g/100 g water at 20 °C to 174 g/100 g water at 60 °C. But dissolving sodium hydroxide in water is a very exothermic process. How is this apparent contradiction explained? Separating positively charged sodium ions from negatively charged hydroxide ions requires energy. Making space between water molecules also requires energy. However, hydrating sodium ions and hydroxide ions releases energy. *When sodium hydroxide dissolves in water,* the amount of energy released to the surroundings by the solvation process is greater than the amount of energy absorbed from the surroundings by separating oppositely charged ions and making space between water molecules. *This process is exothermic.* However, *when sodium hydroxide dissolves in a nearly saturated sodium hydroxide solution, the process is endothermic.* In a nearly saturated sodium hydroxide solution, more energy is absorbed separating ions than is released by the hydration of ions. For this endothermic process, Le Châtelier's principle correctly predicts that the solubility of sodium hydroxide will increase as the temperature is raised. It is the dissolving of sodium hydroxide (and other solutes) *in a nearly saturated solution* that determines how solubility changes with changing temperature.

Le Châtelier's principle should not be used to predict the effect of temperature on the solubility of liquids and solids. *The effect of temperature on the solubility of liquids and solids must be determined by experiment.* ▪Figure 13.5 shows how the solubilities of some solids change as temperature changes. As you can see, the solubilities of some solids (sugar, $C_{12}H_{22}O_{11}$, and potassium nitrate, KNO_3, for example) increase more rapidly as temperature increases than the solubilities of other solids (for example, sodium chloride, NaCl, and potassium chloride, KCl). The solubility of a few salts decreases with increasing temperature like the solubility of cerium(III) sulfate, $Ce_2(SO_4)_3$.

The greater solubility of most solids at higher temperatures accounts for the formation of supersaturated solutions. Suppose 150.0 g KNO_3 are dissolved in

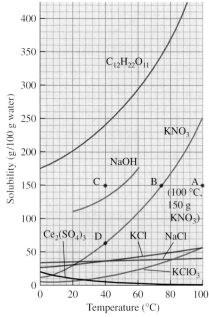

▪ FIGURE 13.5 Solubilities of some solids at different temperatures. Along the curves, solid is in equilibrium with saturated solution. Most of the data are from Dean, J. A., *Lange's Handbook of Chemistry,* 14th ed.; McGraw-Hill: New York, 1992, pp. 5.9–5.23.

100.0 g H_2O by heating the mixture to 100 °C. This solution is represented by point A in Figure 13.5. If the solution is cooled, KNO_3 should start to precipitate at 76 °C (point B). However, if no dust particles* are available for crystals to start growing on, the solution can be cooled below 76 °C without precipitation. Below 76 °C, the solution is supersaturated. Once crystallization starts—say, at 40 °C (point C)—KNO_3 will precipitate until the solution is only saturated (point D). Then, as cooling continues, the concentration will change along the KNO_3 solubility curve. The solubility curve ends at about 0 °C, the freezing point of the solvent, water. If cooling is continued much past 0 °C, water will begin to crystallize.

Differences in solubility can be used to separate mixtures. As you might expect, the greater the difference in solubility between the components of a mixture, the easier the mixture is to separate. A mixture of salt and sand can be separated simply by stirring the mixture with water to dissolve the salt. Filtering the solution leaves sand on the filter paper. Salt can be recovered from the filtrate by evaporation of the water.

When compounds are synthesized in laboratories or manufacturing plants, they usually contain impurities (by-products of side reactions, leftover starting materials, and so forth). Impure samples can be purified by crystallization even if the impurities and the product have similar solubilities. If the mixture contains much more of one component than the other, a solution that is saturated with respect to the more abundant component is not saturated with respect to the less abundant component. For example, suppose you need to prepare pure sodium chloride from a mixture of 39 g NaCl and 1 g KCl obtained by the evaporation of seawater. According to Figure 13.5, at 100 °C the solubility of NaCl is 40 g/100 g of water, and the solubility of KCl is 55 g/100 g of water. For simplicity, we will assume that NaCl and KCl do not change each other's solubilities. According to the solubility data, the mixture can be completely dissolved in 100 g of water at 100 °C. At 0 °C, the solubility of NaCl is 35 g/100 g of water or 3.5 g/10 g, and the solubility of KCl is 28 g/100 g of water or 2.8 g/10 g. If the solution is evaporated until only 10 g of water are left and cooled to 0 °C, all of the KCl will remain in solution, but 35 g of pure NaCl will precipitate. Crystallization is the most frequently used method for purifying solids.

Gases

The solubilities of some typical gases in water at different temperatures are given in Table 13.2, where you can see that the solubilities of these gases decrease as temperature increases. For aqueous solutions, decreasing solubility with increasing temperature is typical of most gases at ordinary pressures. The only common gas that is an exception to this rule is helium. The solubility of helium is lowest at about 30 °C; above this temperature it increases as temperature increases.

TABLE 13.2

Solubilities of Some Typical Gases in Water[a]

Gas	Temperature, °C		
	0	25	50
CO_2	0.33	0.145	0.076
N_2	0.0029	0.00175	0.00122
O_2	0.0069	0.0039	0.0027
SO_2	23	9.4	4.3

[a]In g gas/100 g H_2O at a total pressure of 1 atm ($P_{total} = p_{gas} + p_{water\ vapor}$). Most data are from Dean, J. A. *Lange's Handbook of Chemistry*, 14th ed.; McGraw-Hill: New York, 1992, pp 5.3–5.8.

Making coffee or tea, if you do not care for instant, is an example of separation of fairly soluble materials from very insoluble grounds or leaves.

Granulated sugar is purified by crystallization.

When you have boiled water, you may have noticed that bubbles form on the bottom and sides of the container before the water boils. These bubbles are air coming out of solution because oxygen and nitrogen and the other gases in air are less soluble at higher temperatures.

*See Mysels, K. J. J. Chem. Educ. **1955**, 32, 399 for references concerning the greater solubility of gases in organic liquids at higher temperatures. We are grateful to Donald B. Alger, California State University, Chico, for calling this paper to our attention.*

PRACTICE PROBLEMS

Use Figure 13.5 to answer Practice Problems 13.7–13.9.

13.7   What is the solubility of $KClO_3$ in water at 60 °C?

---

*If the solution does not contain any particles at all, there is nothing in the solution for the hydrated ions to bump into to knock off their shells of water molecules. The shells of water molecules prevent combination of a K^+ ion and a NO_3^- ion to form a KNO_3 particle. Dust particles on which precipitation begins are referred to as nucleation centers.

Thermal Pollution

In the United States, about 10^{14} gallons of water are used by industry for cooling each year, mostly in power plants. This water is usually returned to lakes and rivers at a higher temperature. The increase in the temperature of the water results in a decrease in the solubility of oxygen gas in the water (see Table 13.2), which may be fatal to fish. The problem is worse in lakes because warmer water is less dense than cooler water (above 4 °C) and the warmer water forms a layer on top of the cooler water preventing normal absorption of oxygen from the air.

One solution to the problem is to use less power so that less cooling water will be required and less thermal energy will be released to the environment. Another is to find ways to put the waste thermal energy to work such as using hot water

Fish killed by thermal pollution of water.

from power plants for heating. The water can be cooled and treated with air to return oxygen to it before returning it to lakes and rivers, but this is expensive.

On hot days, expert fishermen choose deep spots to fish in because the fish go where the water is deeper and cooler and contains more oxygen.

13.8   Will a solution of 198 g of sucrose in 100 g of water at 25 °C be saturated, unsaturated, or supersaturated?

13.9   What is the minimum quantity of water needed to dissolve 62 g of KNO_3 at 100 °C?

13.10   Suggest a method for increasing the concentration of dissolved oxygen gas in water.

13.4   EFFECT OF PRESSURE ON SOLUBILITY

Pressure does not have much effect on the solubility of liquids and solids. Qualitatively, the effect of pressure on the solubility of gases can be predicted from the kinetic-molecular theory. Imagine the equilibrium between a gas and a solution of the gas in a cylinder with a piston shown in ▪Figure 13.6(a). If the pressure is doubled while the temperature is kept constant, the volume of the gas phase will

▪ FIGURE 13.6   At constant temperature, increasing the pressure of a gas in equilibrium with solution increases the concentration of the gas in solution. (For simplicity, assume that the liquid solvent has a negligible vapor pressure.) (a) Original equilibrium. (b) Pressure doubled, volume halved. (c) New equilibrium.

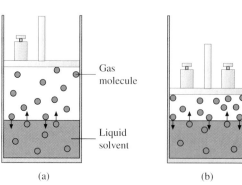

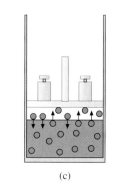

Gas molecule

Liquid solvent

(a)   (b)   (c)

be halved as shown in Figure 13.6(b). The concentration of molecules in the gas phase will be doubled. The number of gas molecules hitting the surface of the liquid and dissolving in a given time will increase. The concentration of gas molecules in the liquid phase will increase. The number of gas molecules leaving the liquid will increase until it is again equal to the rate of molecules entering the liquid. When equilibrium is reached at the higher pressure, the concentration of gas molecules in the liquid phase will be greater as shown in Figure 13.6(c).

▪Figure 13.7 shows quantitatively how the equilibrium concentrations of gases in solution depend on the pressure of the gas at constant temperature. Note that the higher the pressure, the higher the solubility of gases in water. Because the graphs of concentration against pressure are straight lines, the relation between solubility and pressure is given by the equation

$$c_g = kp_{gas}$$

where c_g is the concentration of the dissolved gas, k is a constant characteristic of the gas, and p_{gas} is the pressure of the gas above the solution. This equation is named Henry's law after William Henry, who discovered the relationship between pressure and solubility of gases in 1803.

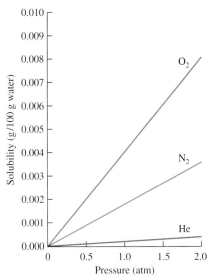

▪ FIGURE 13.7 Dependence of solubility of gases in water on pressure at 25°C.

Henry's Law

At constant temperature, the solubility of a gas is directly proportional to the pressure of the gas above the solution:

$$c_g = kp_{gas}$$

Henry's law is accurate to within 1–3% for slightly soluble gases at pressures up to 1 atm; the gas must not react with the solvent. Henry's law makes possible the calculation of the solubility of a gas at a given temperature and any pressure from a single experimental measurement of the solubility of the gas at one pressure at the given temperature.

William Henry was an English doctor. When ill health forced him to give up his medical practice, he turned to chemistry.

The Solubility of Gases module of the Solutions section allows students to see what happens when the temperature and pressure of a gas/liquid system are varied. Animations show how changing temperature or pressure influences the equilibrium; Henry's law is demonstrated.

SAMPLE PROBLEM

13.2 The solubility of O_2 gas in water is 0.0041 g O_2/100 g H_2O when the pressure of the oxygen gas is 1.00 atm and the temperature is 25 °C. What is the solubility of O_2 gas in water at 0.20 atm (the usual pressure of oxygen gas in Earth's atmosphere at sea level) and 25 °C?

Solution   The pressure in the problem, 0.20 atm, is 1/5 of 1.00 atm. According to Henry's law, concentration of a gas is directly proportional to pressure at constant temperature. Therefore, the solubility of oxygen at 0.20 atm is 1/5 the solubility at 1.00 atm:

$\frac{1}{5}$(0.0041 g O_2/100 g H_2O)

$= 0.000\ 82$ g O_2/100 g H_2O or 8.2×10^{-4} g O_2/100 g H_2O

If you do not follow this reasoning, you can use Henry's law to calculate the value of the constant k for oxygen at 25 °C. Then you can use the value of k and Henry's law to calculate the concentration of oxygen at 0.20 atm. From Henry's law,

$$k = \frac{c_g}{p_{gas}} = \frac{0.0041 \ \frac{g\ O_2}{100\ g\ H_2O}}{1.00\ atm} = 0.0041 \ \frac{g\ O_2}{100\ g\ H_2O\ atm}$$

Using the value found for k in Henry's law,

$$c_{\text{oxygen}} = 0.0041 \left(\frac{\text{g } O_2}{100 \text{ g } H_2O \text{ atm}} \right) 0.20 \text{ atm} = 0.000\,82 \, \frac{\text{g } O_2}{100 \text{ g } H_2O}$$

13.11 The concentration of N_2 gas in water is 0.0018 g N_2/100 g H_2O when the pressure of the nitrogen gas is 1.00 atm and the temperature is 25 °C. What is the concentration of N_2 gas in water at 0.80 atm (the usual pressure of nitrogen gas in Earth's atmosphere at sea level) and 25 °C?

Henry's law is applied to filling bottles and cans of carbonated beverages; the pressure of carbon dioxide is increased to about 4 atm. When the container is opened and the pressure is reduced, gas fizzes out of the liquid. Henry's law is also important to deep-sea divers, who must work under high pressure; 100 feet under the sea the total pressure is 4.0 atm (3.0 atm due to seawater and 1.0 atm due to Earth's atmosphere). Divers' blood is saturated with air at high pressure. Oxygen is used up by the cells, but nitrogen accumulates; if a diver comes to the surface too rapidly, nitrogen comes out of solution fast and forms bubbles that block blood capillaries and cause the bends. The bends got their name because they are so painful that they cause people to double up; a severe case can cause death.

To help prevent the bends, a mixture of helium and oxygen is used in scuba diving tanks. Helium is inert and is much less soluble than nitrogen at high pressures (see Figure 13.7).

13.5 TWO MORE CONCENTRATION UNITS: MOLALITY AND MOLE FRACTION

Before we discuss the physical properties of solutions any further, we need to learn about two more concentration units, molality and mole fraction.

If you wish to cover normality, the Instructor's Resource Manual contains a complete text section available for reproduction and distribution as a handout.

Molality

Molality, *m*, is defined as *the number of moles of solute dissolved in one kilogram of solvent:*

$$\text{molality, } m = \frac{\text{number of moles of solute}}{\text{mass of solvent in kilograms}}$$

One advantage of molality as a unit of concentration is that *molality does not change with temperature* because masses do not vary with temperature. Molarity decreases with increasing temperature because the volumes of liquids and liquid solutions increase as temperature increases and volume of solution is in the denominator of the definition of molarity:

$$\text{molarity, M} = \frac{\text{number of moles of solute}}{\text{volume of solution in liters}}$$

However, molal solutions must be measured by mass whereas molar solutions can be measured by volume, which is quicker and easier. For dilute aqueous solutions (0.1 M and lower), molality and molarity are equal for all practical purposes; the difference between molarity and molality is of the order of 1–3% for 1 M aqueous solutions.

PRACTICE PROBLEM

13.12   What is the molality of each of the following solutions? (a) 0.53 mol NaCl in 2.1 kg H_2O (b) 6.00 g NaCl in 996.5 g of water

Mole Fraction

The **mole fraction** of a component in solution X_A is the *fraction of the total number of moles in the solution that is A*. For a solution containing components A, B, C, . . . ,

$$X_A = \frac{n_A}{n_A + n_B + n_C + \cdots}$$

where n_A is the number of moles of component A in the solution, n_B is the number of moles of component B in the solution, n_C is the number of moles of component C in the solution, and so forth. Because the units of both numerator and denominator in the definition of mole fraction are moles, units cancel and mole fractions are pure numbers without units. The sum of the mole fractions of all components of a mixture must equal 1. A mole is a package of 6.02×10^{23} particles; therefore, the mole fraction of a component, A, tells what fraction of the particles present are particles of A.

PRACTICE PROBLEMS

13.13   For a mixture of 2.9 mol of methyl alcohol, 3.6 mol of ethyl alcohol, and 7.2 mol of water, what are the mole fractions of methyl alcohol, ethyl alcohol, and water? Check your answer by adding the three mole fractions to be sure the total is 1.

13.14   For a solution of 6.00 g NaCl in 996.5 g of water [the solution of Practice Problem 13.12(b)], calculate the mole fractions of NaCl and water.

Use of Mole Fractions to Calculate Partial Pressures of Gases in Gaseous Mixtures

Mole fractions can be used to calculate the partial pressures of the individual gases in a mixture of gases from the total pressure of the mixture. The partial pressure of a component of the mixture, p_A, is equal to the mole fraction of the component, X_A, times the total pressure, P_{total}:

$$p_A = X_A P_{total}* \tag{13.1}$$

---

*The relationship of equation 13.1 is easily derived from the ideal gas equation (Section 5.6). The ideal gas equation, $PV = nRT$, applies to mixtures of gases as well as to each gas in the mixture:

$$P_{total} = n_{total}\frac{RT}{V} \qquad p_{gas\ 1} = n_{gas\ 1}\frac{RT}{V} \qquad p_{gas\ 2} = n_{gas\ 2}\frac{RT}{V}$$

and so forth. Dividing the expression for one gas by the expression for P_{total} gives, using gas 1 as an example,

$$\frac{p_{gas\ 1}}{P_{total}} = \frac{n_{gas\ 1}\dfrac{RT}{V}}{n_{total}\dfrac{RT}{V}}$$

Simplifying, $\dfrac{p_{gas\ 1}}{P_{total}} = \dfrac{n_{gas\ 1}}{n_{total}}$

But $\dfrac{n_{gas\ 1}}{n_{total}} = X_{gas\ 1}$ so $\dfrac{p_{gas\ 1}}{P_{total}} = X_{gas\ 1}$ and $p_{gas\ 1} = X_{gas\ 1}\,P_{total}$

13.3 What is the partial pressure of each gas in a mixture of 2.43 mol of nitrogen gas and 3.07 mol of oxygen gas if the total pressure is 26.9 atm?

Solution If we first calculate the mole fraction of each gas in the mixture, we can then use equation 13.1 to find the partial pressure of each gas:

$$X_{N_2} = \frac{n_{N_2}}{n_{N_2} + n_{O_2}} = \frac{2.43 \text{ mol}}{2.43 \text{ mol} + 3.07 \text{ mol}} = 0.442$$

and

$$X_{O_2} = \frac{n_{O_2}}{n_{N_2} + n_{O_2}} = \frac{3.07 \text{ mol}}{2.43 \text{ mol} + 3.07 \text{ mol}} = 0.558$$

Substitution of these mole fractions in equation 13.1 gives

$$
\begin{array}{ll}
p_{N_2} = X_{N_2} P_{total} & \quad \text{and} \quad & p_{O_2} = X_{O_2} P_{total} \\
\quad = 0.442(26.9 \text{ atm}) & & \quad = 0.558(26.9 \text{ atm}) \\
\quad = 11.9 \text{ atm} & & \quad = 15.0 \text{ atm}
\end{array}
$$

Check: According to Dalton's law of partial pressures (Section 5.8),

$$P_{total} = p_{gas\ 1} + p_{gas\ 2} = 11.9 \text{ atm} + 15.0 \text{ atm} = 26.9 \text{ atm}$$

Always check your work.

13.15 What is the partial pressure of each gas in a mixture of 4.36 mol of nitrogen gas and 0.215 mol of water vapor if the total pressure is 752.8 mmHg?

Interconversion of Concentration Units

Sample Problem 13.4 shows how to calculate molality from mole fraction. To interconvert concentrations expressed in molarity with concentrations expressed in either molality or mole fraction, the density of the solution is needed (Sample Problem 13.5).

13.4 If the mole fraction of methyl alcohol, CH_3OH, in an aqueous solution of CH_3OH is 0.1508, what is the molality of the solution?

Solution The definitions of the concentration units are the key to solving all problems about concentrations of solutions. Begin by writing down the definitions of the units involved in the problem:

$$X_{CH_3OH} = \frac{\text{mol } CH_3OH}{\text{mol } CH_3OH + \text{mol } H_2O} \qquad m = \frac{\text{mol } CH_3OH}{\text{kg } H_2O}$$

The problem gives X_{CH_3OH} and asks for m.

The sum of the mole fractions of the components of a mixture must equal 1:

$$1 = X_{CH_3OH} + X_{H_2O} \qquad \text{and} \qquad 1 - X_{CH_3OH} = X_{H_2O}$$

Therefore, for the solution in this problem,

$$X_{H_2O} = 1 - 0.1508 = 0.8492$$

If we consider a 1-mol sample of the mixture, a convenient-sized sample for this type of problem, the sample is composed of 0.1508 mol CH_3OH and 0.8492 mol H_2O. If we find the mass in kg of 0.8492 mol H_2O, we will have the information we need to calculate molality, m, from the definition of molality.

From the table inside the back cover, the formula mass of H_2O is 18.0153 u; therefore, 1 mol H_2O has a mass of 18.0153 g, and we can write a conversion factor for converting mol H_2O to g H_2O. By definition, 1 kg $= 10^3$ g, and this information can be used to convert grams to kilograms. The setup is

$$0.8492 \text{ mol } H_2O \left(\frac{18.02 \text{ g } H_2O}{\text{mol } H_2O}\right)\left(\frac{1 \text{ kg } H_2O}{10^3 \text{ g } H_2O}\right) = 0.015\ 30 \text{ kg } H_2O$$

Now we are ready to use the definition of molality to calculate the molality of the mixture:

$$m_{CH_3OH} = \frac{\text{mol}_{CH_3OH}}{\text{kg } H_2O} = \frac{0.1508 \text{ mol } CH_3OH}{0.015\ 30 \text{ kg } H_2O} = 9.856\ m$$

13.5 An aqueous solution of CH_3OH in which the mole fraction of CH_3OH is 0.1508 has a density of 0.9606 g/cm^3. What is the molarity of the solution?

Solution Begin by writing down the definitions of the units involved in the problem:

$$X_{CH_3OH} = \frac{\text{mol } CH_3OH}{\text{mol } CH_3OH + \text{mol } H_2O} \qquad M = \frac{\text{mol } CH_3OH}{\text{vol soln L}}$$

The problem gives X_{CH_3OH} and asks for M. From Sample Problem 13.4, we know that a 1-mol sample of the mixture is composed of 0.1508 mol CH_3OH and 0.8492 mol H_2O. To calculate M, we need to know the volume in liters of this sample.

Density provides a conversion factor between mass and volume; we need to find the mass of the 1-mol sample of mixture, which we can do by calculating the sum of the masses of the 0.1508 mol CH_3OH and 0.8492 mol H_2O in the 1-mol sample of mixture. From the table inside the back cover, the formula masses of CH_3OH and H_2O are 32.042 and 18.0153 u, respectively:

$$0.1508 \text{ mol } CH_3OH\left(\frac{32.04 \text{ g } CH_3OH}{\text{mol } CH_3OH}\right) = 4.832 \text{ g } CH_3OH$$

$$0.8492 \text{ mol } H_2O\left(\frac{18.02 \text{ g } H_2O}{\text{mol } H_2O}\right) = 15.30 \text{ g } H_2O$$

The total mass of the 1-mol sample of mixture is

$$4.832 \text{ g } CH_3OH + 15.30 \text{ g } H_2O = 20.13 \text{ g}$$

The volume of the 1-mol sample is

$$20.13 \text{ g}\left(\frac{1 \text{ cm}^3}{0.9606 \text{ g}}\right)\left(\frac{1 \text{ L}}{10^3 \text{ cm}^3}\right) = 0.020\ 96 \text{ L}$$

and the molarity is

$$M = \frac{0.1508 \text{ mol } CH_3OH}{0.020\ 96 \text{ L}} = 7.195 \text{ M}$$

PRACTICE PROBLEMS

13.16 If the mole fraction of ethyl alcohol, CH_3CH_2OH, in an aqueous solution of CH_3CH_2OH is 0.640, what is the molality of the solution?

13.17 An aqueous solution of CH_3CH_2OH in which the mole fraction of CH_3CH_2OH is 0.640 has a density of 0.8387 g/cm^3. What is the molarity of the solution?

13.18 If the molality of sucrose, $C_{12}H_{22}O_{11}$, in an aqueous solution of sucrose is 6.80, what is the mole fraction of sucrose in the solution?

13.6   RAOULT'S LAW

● Remember that substances with similar intermolecular attractions form ideal solutions (Section 13.1). Most dilute solutions (<1 M) are close to ideal.

The dependence of the vapor pressure of a volatile component of an ideal solution on the vapor pressure of the pure component was first described by French chemist François-Marie Raoult in about 1886.

Raoult's Law

The vapor pressure of a component A of an ideal solution, $VP_{A\ in\ solution}$, is equal to the product of the mole fraction of the component in the solution, $X_{A\ in\ solution}$, and the vapor pressure of the pure component, $VP_{pure\ A}$, at the temperature of the solution:

$$VP_{A\ in\ solution} = X_{A\ in\ solution} \times VP_{pure\ A}$$

The total vapor pressure of an ideal solution is equal to the sum of the vapor pressures of the volatile components of the solution:

$$VP_{solution} = VP_{A\ in\ solution} + VP_{B\ in\ solution} + VP_{C\ in\ solution} + \cdots$$

where A, B, C, . . . are volatile components of the ideal solution. Sample Problem 13.6 shows how to use Raoult's law to calculate the vapor pressure of an ideal solution and ▬Figure 13.8 is a graph of the vapor pressures of pentane and hexane, which form an ideal solution (Section 13.1), and of the total vapor pressure of the mixture as a function of composition in mole fractions.

SAMPLE PROBLEM

13.6 The mole fraction of pentane in a mixture of pentane and hexane is 0.5443. The vapor pressure of pentane is 420.8 mmHg, and the vapor pressure of hexane is 101.9 mmHg at 20.0 °C. What is the vapor pressure of the mixture at 20.0 °C?

Solution A mixture of pentane and hexane is an ideal solution. The total vapor pressure of the solution is the sum of the vapor pressures of the components:

$$VP_{solution} = VP_{pentane\ in\ solution} + VP_{hexane\ in\ solution}$$

We need to use Raoult's law to find the vapor pressures of pentane and hexane and add them. According to Raoult's law,

$$VP_{pentane\ in\ solution} = X_{pentane}VP_{pure\ pentane}$$
$$VP_{hexane\ in\ solution} = X_{hexane}VP_{pure\ hexane}$$

The mole fraction of pentane is given as 0.5443. The mixture contains only pentane and hexane and the mole fraction of hexane is $(1 - 0.5443) = 0.4557$. Substitution in Raoult's law gives

$$VP_{pentane\ in\ solution} = 0.5443(420.8\ mmHg) = 229.0\ mmHg$$
$$VP_{hexane\ in\ solution} = 0.4557(101.9\ mmHg) = 46.44\ mmHg$$

Therefore,

$$VP_{solution} = 229.0\ mmHg + 46.44\ mmHg) = 275.4\ mmHg$$

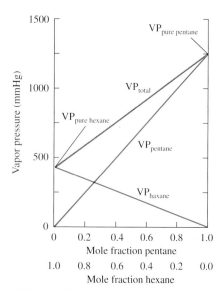

▬ FIGURE 13.8   Vapor pressure of an ideal solution. This graph is for a temperature of 50 °C.

In Figure 13.8, the $VP_{pentane}$ line shows how the vapor pressure of pentane changes as the mole fraction of pentane in the solution changes, and the VP_{hexane} line shows how the vapor pressure of hexane changes as the mole fraction of hexane in the solution changes. The total vapor pressure at any mole fraction is equal to the sum of the vapor pressures of the pentane and hexane.

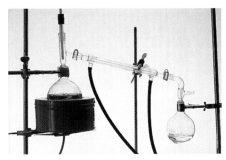

- FIGURE 13.9   Simple distillation. The colored solute (on left in heater) is nonvolatile and remains in the pot when the solution is distilled. Cold water is run through the condensor jacket to cause vapor to condense to liquid.

PRACTICE PROBLEM

13.19   The mole fraction of hexane in a mixture of pentane and hexane is 0.2674. The vapor pressure of pentane is 420.8 mmHg, and the vapor pressure of hexane is 101.9 mmHg at 20.0 °C. (a) What is the vapor pressure of the mixture at 20.0 °C? (b) Qualitatively, if the mole fraction of the component having the higher vapor pressure increases, does the vapor pressure of the mixture increase or decrease? Give evidence in support of your answer.

Distillation

Like a pure liquid, a solution boils when its vapor pressure equals the pressure on it. The boiling points of ideal solutions are between the boiling points of the lowest and highest boiling components, that is, higher than the boiling point of the component with the lowest boiling point but lower than the boiling point of the component with the highest boiling point (see Figure 13.11).

Mixtures of substances with different vapor pressures can be separated by **distillation.** If the boiling points of two substances are very different, they can be separated by **simple distillation;** a laboratory apparatus for simple distillation is shown in -Figure 13.9. The sample to be separated is placed in the pot and heated until it boils. The vapor, which consists almost entirely of the more volatile component, goes up into the head and then into the condenser where it is condensed to liquid by the cold water in the condenser jacket. The condensed liquid drips down through the adapter into the receiver. The less volatile material is left behind in the pot.

The composition of the vapor in equilibrium with a liquid solution is not the same as the composition of the liquid solution. For example, in Sample Problem 13.6 we found that, for a mixture of pentane and hexane in which $X_{pentane} = 0.5443$, $p_{pentane} = 229.0$ mmHg and $P_{total} = 275.4$ mmHg. The mole fraction of pentane in the vapor can be calculated from this information by means of equation 13.1:

$$p_A = X_A P_{total}$$

To distinguish between the mole fractions of pentane in the liquid phase and in the vapor phase, we will use the symbols $X_{pentane(g)}$ for the mole fraction of pentane in the vapor phase and $X_{pentane(l)}$ for the mole fraction of pentane in the liquid phase. Solution of equation 13.1 for $X_{pentane(g)}$ gives

$$\frac{p_{pentane}}{P_{total}} = X_{pentane(g)}$$

Substitution of the information concerning the vapor above a pentane–hexane solution in which $X_{pentane(l)} = 0.5443$ gives

$$\frac{229.0 \text{ mmHg}}{275.4 \text{ mmHg}} = X_{pentane(g)} = 0.8315$$

The mole fraction of pentane in the liquid is only 0.5443; thus, the mole fraction of pentane in the vapor, 0.8315, is greater than the mole fraction of pentane in the liquid. *The mole fraction of the component with the higher vapor pressure is always*

Freshwater can be obtained from seawater by distillation. However, lots of energy is required due to the high specific heat and high heat of vaporization of water.

Distillation should always be stopped before the pot is completely empty.

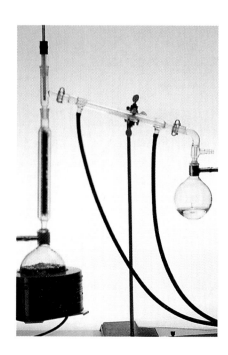

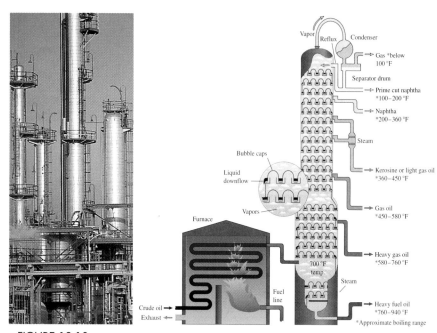

FIGURE 13.10 Fractional distillation. *Left:* Laboratory apparatus. The column is packed with a stainless steel "scrubber." *Middle:* Fractionating column in an oil refinery. *Right:* Diagram of fractionating column in oil refinery.

higher in the vapor above a liquid than in the liquid. Condensation of the vapor above a liquid gives a liquid that contains more of the component with the higher vapor pressure than the original liquid.

When a mixture of two volatile substances with similar boiling points is distilled, the first liquid to distill is richer in the more volatile component than the original mixture. The material remaining in the pot is enriched in the less volatile component. If the first liquid to distill is distilled again, the first liquid to distill from it is even richer in the more volatile component. If the process is repeated enough times, a pure (for all practical purposes) sample of the more volatile component can be obtained. Fractionating columns, such as the ones shown in ▬Figure 13.10, carry out repeated distillations automatically. The vapor condenses on the first layer of packing and is vaporized again. It then condenses on the second layer of packing and is vaporized again; the process is repeated over and over. With an efficient fractionating column, mixtures of substances boiling only 1° apart can be separated; heavy water, D_2O (bp 101.42 °C), can be separated from ordinary water, H_2O (bp 100.00 °C), in this way.

Fractional distillation must be used to separate mixtures of substances that have similar boiling points. ▬Figure 13.11 compares the good separation of a mixture of trichloromethane, $CHCl_3$, and toluene obtained by fractional distillation through the laboratory fractionating column shown in Figure 13.10, with the very poor separation that results from simple distillation. Boiling points of trichloromethane and toluene differ by 49 °C; trichloromethane boils at 62 °C and toluene at 111 °C at 1 atm.

In Figure 13.11, notice how the boiling point in the simple distillation increases gradually from beginning to end of the distillation. At the beginning, the distillate was richer in trichloromethane than the material in the pot, although it contained some toluene. As the distillation progressed, the distillate became richer and richer in toluene. In the fractional distillation, the boiling point stays almost constant at the boiling point of trichloromethane until most of the trichloromethane has distilled. The boiling point then rises rapidly to the boiling point of pure toluene and remains constant until distillation is stopped.

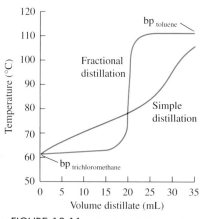

FIGURE 13.11 Simple and fractional distillation of a 1:1 by volume mixture of trichloromethane (bp = 62 °C) and toluene (bp = 111 °C). The data from which this graph was plotted were obtained by Abdul Ghafoor, a student at the University of Houston–Downtown, using the laboratory apparatus shown in Figures 13.9 and 13.10.

Toluene is ⬡ — CH_3.

Distillation is the most frequently used procedure for purifying liquids. Larger quantities of liquids are separated by fractional distillation than by any other method. For example, refineries use fractional distillation to separate crude oil into gasoline, kerosene, lubricating oils, greases, wax, and asphalt. If liquids have high normal boiling points or decompose when heated, the temperature required for distillation can be lowered by reducing the pressure and performing a vacuum distillation.

Oxygen, nitrogen, and the noble gases are obtained by fractional distillation of liquid air.

13.7 COLLIGATIVE PROPERTIES

Most properties of solutions depend on what the solute is. The identity of the solute determines the chemical properties of a solution; for example, a solution of sodium hydroxide, NaOH, turns red litmus blue and neutralizes acids. A solution of potassium iodide, KI, does neither. The identity of the solute is also a factor in many physical properties. Solutions of copper(II) sulfate are blue; solutions of cobalt(II) sulfate are pink. A 25% by mass solution of ethyl alcohol in water has a density of 0.962 g/mL at 20 °C; a 25% by mass solution of sodium chloride in water has a density of 1.189 g/mL at 20 °C.

Four physical properties of solutions are the same for all *nonvolatile* solutes: *vapor pressure lowering, boiling point elevation, freezing point depression, and osmotic pressure.* These four properties are called **colligative properties.** *The colligative properties of solutions depend only on the number of solute particles. The identity of the solute particles does not matter.* A 0.1 *m* solution of sugar, $C_{12}H_{22}O_{11}$, has the same vapor pressure, the same boiling point, the same freezing point, and the same osmotic pressure as a 0.1 *m* solution of urea, H_2NCONH_2,* in the same solvent.

In the Colligative Properties module of the Solutions section, students can see how the equilibrium between a liquid and its vapor shifts as a nonvolatile solute is added. Randomly generated questions about colligative properties are posed; feedback designed to lead the student to the correct answer is provided if necessary.

Vapor Pressure Lowering

Because the vapor pressure of a nonvolatile component is, for all practical purposes, zero, a nonvolatile solute does not contribute to the vapor pressure of a solution. The vapor pressure of a solution of a nonvolatile substance in a volatile solvent is equal to the mole fraction of solvent in the solution times the vapor pressure of the pure solvent. Because the mole fraction of solvent in a solution must be less than one, the vapor pressure of a solution of a nonvolatile solute is lower than the vapor pressure of the pure solvent. Sample Problem 13.7 shows the calculation of the vapor pressure of a solution of a nonvolatile substance and the vapor pressure lowering due to the nonvolatile solute.

Remember Raoult's law.

SAMPLE PROBLEM

13.7 (a) What is the vapor pressure of an aqueous solution of sucrose, $C_{12}H_{22}O_{11}$, in which $X_{sucrose} = 0.017\ 378$ if the temperature is 22.6 °C? (b) What is the vapor pressure lowering caused by the dissolved sucrose?

---

*Urea is the final product of the body's use of protein. The preparation of urea, a typical organic compound, from the inorganic salt, ammonium cyanate, NH_4OCN,

$$NH_4OCN(s) \xrightarrow{heat} H_2NCONH_2(s)$$

by a German chemist, Friedrich Wöhler, in 1828 was the first synthesis of a naturally occurring organic compound in the laboratory. The percent nitrogen in urea is very high. About 10^{10} pounds of urea are produced in the United States every year, mostly for use as a fertilizer.

▪ FIGURE 13.12 When a solution of copper(II) sulfate is cooled below its freezing point, the solid that separates is pure ice. The copper(II) sulfate, which is blue, remains in solution.

Solution (a) According to Raoult's law,

$$VP_{A \text{ in solution}} = X_{A \text{ in solution}} \times VP_{\text{pure A}}$$

The mole fraction of water in solution is equal to 1 minus the mole fraction of sucrose:

$$X_{H_2O} = 1 - 0.017\ 378 = 0.982\ 622$$

The vapor pressure of water at different temperatures is given in Table 5.4. From Table 5.4, the vapor pressure of water at 22.6 °C is 20.565 mmHg.

Because sucrose is a nonvolatile solute, the vapor pressure of the solution is equal to the vapor pressure of water over the solution. Substitution in Raoult's law gives

$$VP_{\text{solution}} = 0.982\ 622(20.565 \text{ mmHg}) = 20.208 \text{ mmHg}$$

(b) The vapor pressure lowering is the difference between the vapor pressure of pure water and the vapor pressure of the solution:

$$\text{vapor pressure lowering} = (20.565 \text{ mmHg} - 20.208 \text{ mmHg}) = 0.357 \text{ mmHg}$$

PRACTICE PROBLEM

13.20 (a) What is the vapor pressure of an aqueous solution of urea, H_2NCONH_2, in which $X_{\text{urea}} = 0.015\ 425$ if the temperature is 24.8 °C? (b) What is the vapor pressure lowering caused by the dissolved urea?

Boiling Point Elevation and Freezing Point Depression

Because a nonvolatile solute lowers the vapor pressure of the solvent, a higher temperature must be reached for the vapor pressure of the solvent to be equal to atmospheric pressure. The boiling point of a solution of a nonvolatile solute is higher than the boiling point of the pure solvent.

The solid that freezes out of a dilute solution is usually pure solvent. The vapor pressure of the solid solvent is not changed by the presence of solute because the solute stays in the solution that remains (see ▪Figure 13.12). A nonvolatile solute, such as sugar or urea, lowers the freezing point of the solvent, as well as raising the boiling point, as can be seen from ▪Figure 13.13. The solution becomes more concentrated as the solvent freezes out and the melting point decreases further; therefore, mixtures do not have constant melting points.

Freezing is the best method of producing concentrated fruit juices.

Quantitatively, the sizes of the boiling point elevation and freezing point depression are proportional to the concentration of solute particles expressed in molality. The boiling point elevation, Δbp, is given by equation 13.2:

$$\Delta bp = K_{bp}m \tag{13.2}$$

In equation 13.2, m is the molality of solute particles in the solution, and K_{bp} is a constant. The value of the boiling point elevation constant, K_{bp}, is characteristic of the solvent; Table 13.3 shows some typical boiling point elevation constants.

Potassium carbonate and potassium acetate are used to depress the freezing point of water in fire extinguishers for unheated warehouses.

The freezing point depression, Δfp, is given by equation 13.3:

$$\Delta fp = K_{fp}m \tag{13.3}$$

In equation 13.3, m is the molality of the solution, and K_{fp} is a constant. The value of the freezing point depression constant, K_{fp}, is characteristic of the solvent; Table 13.3 also shows some typical freezing point depression constants.

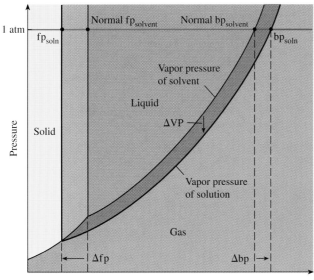

- FIGURE 13.13 A nonvolatile solute lowers the vapor pressure of a solvent. The boiling point is raised, and the freezing point is lowered.

Equations 13.2 and 13.3 can be used to calculate the boiling points and freezing points of solutions as shown in Sample Problem 13.8.

SAMPLE PROBLEM

13.8   For a 0.222 m aqueous solution of sucrose, $C_{12}H_{22}O_{11}$, (a) what is the boiling point? (b) The freezing point?

Sucrose is the common chemical name for ordinary table sugar.

Solution   (a) According to equation 13.2,

$$\Delta bp = K_{bp}m$$

Substitution of the boiling point constant for water from Table 13.3 and the molality given in the problem in equation 13.2 gives

$$\Delta bp = 0.512 \frac{°C}{m}(0.222\ m) = 0.114\ °C$$

The boiling point of the solution is equal to the normal boiling point of pure water, 100.00 °C, plus the boiling point elevation:

$$bp_{soln} = (100.00\ °C + 0.114\ °C) = 100.11\ °C$$

TABLE 13.3	Some Boiling Point Elevation and Freezing Point Depression Constants[a]			
Substance	Normal bp, °C	K_{bp}, °C/m	Normal fp, °C	K_{fp} °C/m
Water	100.0	+0.512	0.0	−1.86
Benzene	80.1	+2.53	5.5	−5.12
Camphor	207.4	+5.61	178.8	−39.7
Ethyl alcohol	78.3	+1.22	−117.3	−1.99

[a]Most data are from Dean, J. A. *Lange's Handbook of Chemistry,* 14th ed.; McGraw-Hill: New York, 1992, pp 11.4 and 11.13. The molality m is the molality of solute particles in the solution.

(b) According to equation 13.3, $\Delta fp = K_{fp}m$. Substitution of the freezing point constant for water from Table 13.3 and the molality given in the problem in equation 13.3 gives

$$\Delta fp = -1.86\,\frac{°C}{m}\,(0.222\,m) = -0.413\,°C$$

The freezing point of the solution is equal to the normal freezing point of pure water, 0.00 °C plus the freezing point depression:

$$fp_{soln} = [0.00\,°C + (-0.413\,°C)] = -0.41\,°C$$

PRACTICE PROBLEMS

13.21   For a 1.344 m solution of urea, H_2NCONH_2, in ethyl alcohol, CH_3CH_2OH, (a) what is the boiling point? (b) The freezing point?

13.22   If the freezing point of a solution of urea in ethyl alcohol is $-121.0\,°C$, what is the molality of the solution?

Because a nonvolatile solute lowers the freezing point of a solvent, the melting points of mixtures of two compounds having similar melting points are usually lower than the melting points of the pure compounds. Impurities, unless they have very high melting points, usually lower melting points and widen the melting range. The melting range is the difference between the temperature at which the first liquid appears and the temperature at which the last solid disappears when a solid is melted slowly. A narrow melting range indicates that a solid sample is pure.

Electrolyte Solutions

If a solute is an electrolyte, each formula unit of solute yields more than one particle in aqueous solution. For example, sodium chloride is a strong electrolyte and is believed to be completely ionized in water solution. Therefore, 1 formula unit of NaCl gives 1 Na^+ ion and 1 Cl^- ion or a total of two particles. Each formula unit of NaCl should lower the vapor pressure, raise the boiling point, and lower the freezing point twice as much as a formula unit of a nonelectrolyte such as sugar, $C_{12}H_{22}O_{11}$, or urea, H_2NCONH_2. However, although strong electrolytes like NaCl are completely dissociated (separated into ions) in water solution, the ions are not completely independent of each other. Hydrated ions of opposite charge attract each other [◀Figure 13.14(a)], and ions combine to form ion pairs [Figure 13.14(b)].

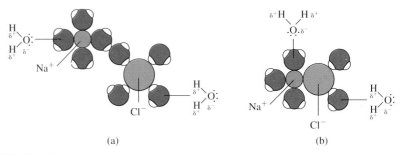

(a)   (b)

◀ FIGURE 13.14   (a) Hydrated ions of opposite charge attract each other because the outside of the shell of water molecules surrounding the cation is positively charged and the outside of the shell of water molecules surrounding the anion is negatively charged. (b) Ions of opposite charge form ion pairs in aqueous solution.

A glassblower at work. Glassblowing is simplified by the fact that glass is a mixture with a broad melting point range. Some laboratory apparatus is made of quartz. Blowing quartz is difficult because quartz is a compound with a narrow melting point range. One minute you have solid quartz, and the next minute you have a puddle of molten quartz.

The effects of interionic attractions and ion-pair formation on colligative properties are small. For example, the freezing point depression observed for 0.100 m NaCl is -0.348 °C. Because each formula unit of NaCl gives two particles, a Na^+ ion and a Cl^- ion, the freezing point depression expected for 0.100 m NaCl if interionic attractions and ion-pair formation are neglected is

If you wish to cover the van't Hoff i factor, the Instructor's Resource Manual contains a complete text section available for reproduction and distribution as a handout.

$$2(0.100 \ m) \frac{-1.86 \ °C}{m} = -0.372 \ °C$$

A freezing point depression of 0.372 °C is only 0.024 °C greater than the experimentally observed value of 0.348 °C. *As far as colligative properties are concerned, we will assume that interionic attraction and ion-pair formation are not significant.*

PRACTICE PROBLEMS

13.23   What will be the total concentration of particles in each of the following solutions? (a) 0.1 m $C_{12}H_{22}O_{11}$ (b) 0.1 m NaCl (c) 0.1 m $CaCl_2$ (d) 0.05 m HCl

13.24   Arrange the solutions in each of the following sets in order of increasing freezing point: (a) 0.1 m $CaCl_2$, 0.1 m $C_{12}H_{22}O_{11}$, and 0.1 m NaCl; (b) 0.05 m HCl, 0.1 m HCl, and 0.1 m $HC_2H_3O_2$.

13.25   For a 0.0100 m aqueous solution of K_2SO_4, what is the freezing point?

Applications of Vapor Pressure Lowering, Boiling Point Elevation, and Freezing Point Depression

Vapor pressure lowering, boiling point elevation, and freezing point depression have many practical applications. Saturated solutions of some strong electrolytes—$CaCl_2$ for example—have vapor pressures lower than the vapor pressure of water in damp air. The solids **deliquesce**—that is, they *absorb water from moist air and form solutions.* Solid calcium chloride is spread on dirt roads in the summer to keep the roads from becoming dusty. In the winter, solid calcium chloride is spread on roads

Why is there no worked example of the calculation of molecular mass from freezing point depression? To make room for current material, something had to go, and we chose this because low molecular masses are usually determined by mass spectroscopy today.

▪ FIGURE 13.15 If a hollowed-out carrot filled with molasses is immersed in water, water passes through the carrot into the molasses by osmosis. The aqueous solution of molasses that results has a larger volume than the molasses and rises in the tube. In one author's (JBU) most successful performance of this demonstration (from the students' point of view), the carrot blew up.

(a)   (b)

to keep ice from forming because dissolved calcium chloride lowers the freezing point of water.

The time necessary to process filled cans of food in boiling water was an early obstacle to commercial canning. Then, in 1866, it was discovered that addition of $CaCl_2$ to boiling water raised the boiling temperature from 100 to 116 °C and reduced the time necessary to protect the contents of the can against spoiling from 4 to 5 hours to 25 to 40 minutes. The average production of canneries increased almost tenfold.

Permanent antifreeze/coolant for car radiators is composed largely of ethylene glycol, $HOCH_2CH_2OH$. Ethylene glycol has a high boiling point and low vapor pressure. When mixed with water (usually about 1 : 1 by volume), ethylene glycol lowers the freezing point and raises the boiling point of water. Use of ethylene glycol–water mixtures in car radiators prevents freezing in the winter and boiling in the summer.

Osmotic Pressure

If a pure solvent and a solution of a nonvolatile solute in the solvent are allowed to stand separated by a semipermeable membrane as shown in ▪Figure 13.15(a), there is a net movement of solvent molecules through the semipermeable membrane into the solution as shown by Figure 13.15(b). A **semipermeable membrane** is *a membrane that will let only solvent molecules through;* solute particles can't pass through a semipermeable membrane. The carrot is the semipermeable membrane in Figure 13.15. Many plant and animal tissues are semipermeable membranes. In Figure 13.15, the solvent is water; molasses is a solution of sugar (and small amounts of other materials) in water.

When the hollowed-out carrot filled with molasses is placed in the beaker of water, water molecules pass through the carrot into the molasses. Water molecules also pass out through the carrot into the beaker. However, in a given period of time, more water molecules move into the carrot than move out of the carrot. The molasses is diluted, and its volume increases; the diluted molasses rises into the glass tube. The *spontaneous net movement of solvent molecules through a semipermeable membrane from a dilute solution (or pure solvent) into a more concentrated solution* is called **osmosis.** In osmosis, the solvent always moves from the less concentrated to the more concentrated solution.

As the column of molasses rises in the glass tube, the pressure exerted by the column of liquid on the molasses in the carrot increases. As the pressure increases, the rate at which water moves out of the carrot increases while the rate at which

Cellophane is also a semipermeable membrane.

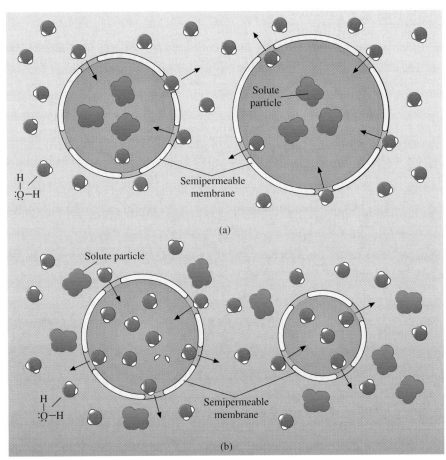

(a)

(b)

FIGURE 13.16 (a) If the concentration of solution outside a cell is less than the concentration of solution inside the cell, a net movement of water through the cell membrane into the cell takes place. The cell expands. (b) If the concentration of solution outside a cell is greater than the concentration inside the cell, there is a net movement of water through the cell membrane out of the cell. The cell shrinks.

water is moving into the carrot remains the same. Eventually, the rate at which water moves out of the carrot becomes equal to the rate at which water is moving into the carrot, and the system is at equilibrium. Although water molecules are still moving into and out of the carrot, there is no further net movement of water into the carrot. The level of the column of dilute molasses in the glass tube is constant. The *pressure that is needed to stop the net movement of solvent molecules through a membrane* is called the **osmotic pressure** of the solution.

If *a pressure greater than the osmotic pressure is applied to a solution in contact with a semipermeable membrane, water can be forced out of the solution.* This process is called **reverse osmosis;** reverse osmosis is used to make freshwater from seawater. Reverse osmosis requires much less energy than distillation because no phase change is involved and the cost of freshwater produced by reverse osmosis is much lower than the cost of freshwater produced by distillation.

Osmosis is very important in medicine and biology. For example, if the concentration of solution outside a cell is less than the concentration of solution inside the cell, a net movement of water through the cell membrane into the cell takes place as shown in ■Figure 13.16(a). The amount of water inside the cell increases, and the cell swells and bursts. If the concentration of solution outside a cell is greater than the concentration inside the cell, there is a net movement of water through the cell membrane out of the cell [see Figure 13.16(b)]. The cell shrinks, wrinkles like

Reverse osmosis is also used to concentrate egg whites, fruit juices, and maple syrup.

Freshwater fish and aquatic plants die when placed in salt water, and saltwater fish and aquatic plants die when placed in freshwater largely as a result of the effects of osmotic pressure.

13.7 COLLIGATIVE PROPERTIES

a prune, and dies. Solutions for intravenous use must have the same osmotic pressure as blood so that there is no net movement of water into or out of cells.

People who eat a lot of salt have a high concentration of salt in their cells. Water passes into the cells by osmosis, causing water retention and swelling. Sugar and salt used in preserving food cause water to move out of bacterial cells, and the bacterial cells shrink and die. Carrots and celery go limp because they have lost water to the air; they can be crisped by covering them with water. Water passes back in by osmosis. Osmotic pressure is the major factor that forces water up from the soil into the leaves of plants. Plants lose water to air through their leaves; water moves back into the leaves by osmosis. (Capillary action can only raise water a few centimeters, but mature trees are many meters tall.)

Experimental measurements on dilute solutions of known concentration have shown that the relation between osmotic pressure π and concentration is

$$\pi = MRT \qquad (13.4)$$

In equation 13.4,* M is the molarity of the solution, R is the gas constant, and T is the Kelvin temperature.† Osmotic pressure does not depend on the identity of the solvent as do vapor pressure lowering, boiling point elevation, and freezing point depression. The relationship of equation 13.4 can be used to find the molecular masses of biologically important molecules that have high molecular masses and low solubilities. Sample Problem 13.9 shows how.

SAMPLE PROBLEM

13.9   A solution of 2.04 g of hemoglobin in 100.0 mL of solution has an osmotic pressure of 5.83 mmHg at 22.5°C. What is the molecular mass of hemoglobin?

Solution   In the problem, concentration is expressed as g hemoglobin in 100.0 mL of solution. A mole is a package of particles such that the mass in grams is numerically equal to the molecular mass. Molecular mass is the same number as grams per mole. From molarity, M, and concentration in g/100.0 mL solution, g/mol can be calculated.

M can be found from the osmotic pressure law:

$$\pi = MRT$$

Solving the osmotic pressure law for M gives

$$\frac{\pi}{RT} = M \qquad (13.5)$$

Because the pressure is given in mmHg, the value, 62.36 mmHg·L·mol^{-1}·K^{-1}, should be used for the gas constant, R. Temperature must be converted to K:

$$K = 273.2 + 22.5 = 295.7$$

---

*The osmotic pressure law was discovered by the Dutch chemist van't Hoff, who was the first winner of the Nobel Prize for chemistry. van't Hoff also proposed the three-dimensional nature of molecules. His work is important in biology and geology as well as in chemistry. Petrology, the field of geology that deals with rocks, began with van't Hoff's work on the precipitation of salts by the evaporation of seawater.

†Note the similarity between the osmotic pressure equation, $\pi = MRT$, and the ideal gas equation, $PV = nRT$. Dividing both sides of the ideal gas equation by V gives

$$P = \frac{n}{V}RT$$

But $(n/V) = M$ when V is in liters; the ideal gas equation can be written $P = MRT$.

Substitution in equation 13.5 then gives

$$\frac{5.83 \text{ mmHg}}{(62.36 \text{ mmHg} \cdot \text{L} \cdot \text{mol}^{-1} \cdot \text{K}^{-1})295.7 \text{ K}} = M = 3.16 \times 10^{-4}$$

Using M and the concentration given, 2.04 g/100.0 mL or 20.4 g/L, to find g/mol:

$$20.4 \frac{\text{g}}{\text{L}} \left(\frac{1 \text{ L}}{3.16 \times 10^{-4} \text{ mol}} \right) = 6.45 \times 10^4 \frac{\text{g}}{\text{mol}}$$

The molecular mass of hemoglobin is 6.45×10^4 u.

PRACTICE PROBLEMS

13.26 The molecular mass of thrombin is 33 580. What is the osmotic pressure of a solution of 0.887 g of thrombin in 50.0 mL of solution at 20.8 °C?

13.27 A solution of 0.3521 g of pepsin in 25.0 mL of solution has an osmotic pressure of 7.46 mmHg at 19.8 °C. What is the molecular mass of pepsin?

Thrombin is an enzyme that plays a part in blood clotting. Pepsin is the principal digestive enzyme of gastric juice.

Molecular masses can also be determined by measuring freezing point depression and boiling point elevation. However, osmotic pressure measurements are much more sensitive. The solution of pepsin in Practice Problem 13.27 that produces an osmotic pressure of 7.46 mmHg would only raise the boiling point of an aqueous solution 0.000 21 °C and lower the freezing point 0.000 76 °C. A pressure of 7.46 mmHg is much easier to measure than temperature differences of less than 0.001 °C. Freezing point depression and boiling point elevation measurements are useful only for measuring small molecular masses. Historically, measurements of freezing point depressions were very important in establishing the theory of electrolytes. However, today small molecular masses are usually determined by mass spectroscopy.

Cause of Colligative Properties

All solutions, both ideal and nonideal, that have the same value of one colligative property have the same value for all four. This fact leads to two conclusions: (1) all four colligative properties have a common cause; (2) the cause is an entropy effect because no energy change takes place when an ideal solution is formed. The entropy of the pure liquid solvent is lower than the entropy of its vapor because gases are more disordered than liquids. The solvent vaporizes because it can become more disordered by vaporizing. Because a solution is more disordered than the pure solvent, dissolving a solute in the solvent decreases the difference in entropy between the liquid and gaseous phases. Therefore, the tendency of the solvent to vaporize decreases lowering the vapor pressure and raising the normal boiling point. Similarly, the greater disorder of the solution compared to the pure solvent opposes the tendency of the solvent to freeze and lowers the normal freezing point.

For the arguments against the common explanation of vapor pressure lowering as a dilution effect see Mysels, K. J. J. Chem. Educ. 1978, 64, 21–22.

13.8 COLLOIDS

Matter exists in a range of particle sizes. Some particles, such as grains of sand, can be seen with the naked eye. Others, like red blood cells and bacteria, cannot be seen with the naked eye but are visible under an ordinary microscope. Particles like the hemoglobin molecule are not visible under an ordinary microscope but can

The maximum magnification obtainable with a light microscope is about 1000. With an electron microscope, magnification as high as 50 000 is possible.

be observed with an electron microscope. With a scanning tunneling microscope, even small atoms can be viewed.

Particles like the hemoglobin molecule *that are larger than ordinary molecules and ions but too small to be seen with a light microscope* are said to be **colloidal.** Colloidal particles are composed of 10^3-10^9 atoms and have masses between 10^4 and 10^{10} u. Viruses, proteins, clays, and synthetic polymers are all colloids. Colloids are important in biology and medicine; in the food, chemical, and pharmaceutical industries; in geology; and in many other fields.

Colloidal particles are usually found dispersed in a homogeneous medium like water or air. Colloidal dispersions have some properties in common with solutions. Like particles in solution, colloidal particles do not settle out on standing as do macroscopic-sized particles like sand and raindrops. Colloidal particles also pass through ordinary filters like particles in solution (see ▪Figure 13.17).* On the other hand, colloidal particles diffuse much more slowly than particles in solution. Vapor pressure lowering, boiling point elevation, and freezing point depression due to colloidal particles are negligible; even osmotic pressure due to very high molecular mass particles is negligible.

Light scattering is a characteristic property of colloids; concentrated colloidal dispersions look milky. Light scattered by colloidal dispersions that appear transparent can be seen from the side as shown in Figure 13.17. *Light scattering by colloidal dispersions* is called the **Tyndall effect.** You can often observe a Tyndall effect in the sky (see Figure 13.17).

Because of their small size, colloidal particles have very large surface-to-volume ratios. In macroscopic particles, most of the units are in the interior. In colloidal particles, a high proportion of the units is at the surface. At a surface, there are always unsatisfied attractive forces and colloidal particles **adsorb** *(gather on their surface)* molecules and ions from solution.

Classification of Colloids

One very useful classification of colloidal dispersions is based on the physical states of the colloidal particles and of the homogeneous medium. Table 13.4 shows the names of the various types of colloidal dispersions and gives examples of each type. The dispersed particles can be either inorganic or organic materials; almost all materials can exist as colloids. Water is the most important dispersion medium, especially in biology and medicine. Air is another common dispersion medium. Other solvents and gases are used as dispersion mediums in industry.

Another important way of classifying colloids is by the shape of the colloidal particles. Extremes of shape are three-dimensional like a ball, two-dimensional like film, or one-dimensional like thread. Differences in shape make classification of particles by length or diameter difficult. For example, a rod-shaped particle may have a diameter similar to the diameter of an ordinary molecule but be extremely long. A film may have a large area but be no thicker than an ordinary molecule.

The cytoplasm of a cell (the substance between the membrane and the nucleus) is colloidal.

Colloidal dispersions of gold prepared by Michael Faraday in 1857 have not settled.

▪ FIGURE 13.17   *Top:* Like particles in solution, colloidal particles pass through ordinary filters. However, although a colloidal suspension may look like a solution, the colloidal particles scatter light; the scattered light can be seen from the side. *Bottom:* A Tyndall effect produced by light from the setting sun. You have probably also observed the Tyndall effects produced by scattering light from headlights in fog and from a slide projector in a darkened room.

*Colloidal particles can be separated from solvent molecules and small dissolved molecules and ions by means of a semipermeable membrane. In an artificial kidney, blood is brought into contact with one side of a membrane that is permeable to small solute molecules such as urea, H_2NCONH_2, and ions of inorganic salts but not to colloidal particles such as red and white blood cells, platelets, and proteins. A sterile solution of sugars, amino acids, and salts needed by the body is circulated on the other side of the membrane. Urea and ions of inorganic salts not needed by the body pass through the membrane out of the blood. Sugars, amino acids, ions of salts needed by the body, and water pass through the membrane into the blood. Dilution of the blood by water is prevented by pressure. This process is called dialysis. Dialysis is similar to osmosis except that the membrane used for dialysis is permeable to small molecules and ions as well as to solvent molecules.

TABLE 13.4 Types of Colloidal Dispersions

State of Colloidal Particle	State of Medium	Name	Examples
Gas	Liquid	Foam	Suds on beer or soapy water,[a] whipped cream, shaving cream, fire-fighting foam
Gas	Solid	Foam	Foam rubber, Styrofoam, marshmallows, sponge, Ivory soap, pumice
Liquid	Gas	Aerosol	Fog, clouds, aerosol sprays
Liquid	Liquid	Emulsion	Homogenized milk, mayonnaise
Liquid	Solid	Emulsion	Butter
Solid	Gas	Aerosol	Smoke, airborne viruses
Solid	Liquid	Sol[b]	Cream, milk of magnesia, mud, detergents, many paints, Jell-O, toothpaste
Solid	Solid	—	Many alloys, ruby glass

[a]Materials that behave like colloids are often classified as colloids even if particles can be seen with a microscope or with the eye, as can many soap bubbles.
[b]A semisolid sol such as gelatin is called a gel.

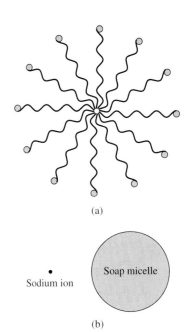

(a)

Sodium ion   Soap micelle

(b)

■ FIGURE 13.18 (a) Two-dimensional model of soap micelle. A real micelle is, of course, three- rather than two-dimensional. A real soap micelle contains about 50 soap ions (some more, some less), is approximately 4600 pm in diameter, and has a molecular mass of about 13 000 u. (b) A sodium ion and a soap micelle drawn to the same scale.

The mechanical properties of a colloidal dispersion depend primarily on the shape of the particles. For example, dispersions of spherical particles have low viscosities, and dispersions of linear particles, which can get twisted with each other, have high viscosities. All structural tissues in plants and animals are built of linear colloids. Proteins circulating in body fluids (hemoglobin, for example) have spherical molecules. When blood clots, some spherical particles are converted into linear particles, which can then form a semisolid jelly or gel.

Colloids are also classified as lyophobic and lyophilic. **Lyophobic** means *solvent hating,* and **lyophilic** means *solvent loving.* If the dispersion medium is water, the colloids are referred to as **hydrophobic** and **hydrophilic.** Gelatin is an example of a hydrophilic colloid; very insoluble substances such as silver chloride and sulfur form hydrophobic colloids.

Soaps and Detergents

Many colloidal particles are composed of clusters of small particles that are held together by intermolecular attractions. (Some colloidal particles are composed of single polymeric molecules. Polymeric molecules are discussed in Section 9.11.) Soaps and detergents provide examples of colloidal particles made up of clusters of small particles. Molecules of soaps and detergents have a polar hydrophilic end that is soluble in water and a nonpolar hydrophobic end that is not soluble in water. Sodium palmitate, $CH_3(CH_2)_{14}COO^-Na^+$, is a typical soap:

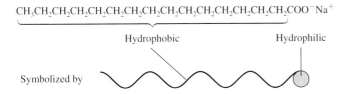

$$CH_3CH_2CH_2CH_2CH_2CH_2CH_2CH_2CH_2CH_2CH_2CH_2CH_2CH_2CH_2COO^-Na^+$$

Hydrophobic   Hydrophilic

Symbolized by

When dispersed in water, molecules of soaps and detergents form clusters called **micelles** as shown in ■Figure 13.18. Attractions between water molecules squeeze

Although soap was probably discovered by the ancient Egyptians, the art of soapmaking was lost during the Dark Ages. Soapmaking was reinvented during the Renaissance, but the use of soap did not become common until the eighteenth century.

Micelles are very important in biology. For example, bile salts emulsify the products of the digestion of fats by combining with them to form micelles.

The dyes used to color fabrics also form micelles.

soap ions out of the water into the micelle. In the micelle, the hydrophobic ends of the soap ions in the center are held together by London forces. The negatively charged hydrophilic ends of the soap ions are solvated by water molecules with their positive ends in. The sodium ions are in solution and can move through the solution relatively rapidly. The micelles, which are much larger, can only move slowly.

Soaps and detergents can remove dirt from people and clothes because most dirt is hydrophobic and is soluble in the hydrophobic ends of the soap or detergent particles. The hydrophilic end of the soap or detergent particles is soluble in water. ▬Figure 13.19 shows how soaps and detergents remove dirt.

Detergents and soaps are **surfactants** or *surface-active agents.* Surfactants concentrate at the surface of liquid water:

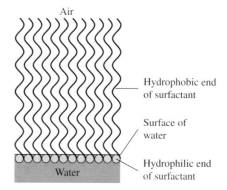

Surface tension tends to reduce the surface area; concentration of surfactant ions at the surface tends to reduce surface tension and expand the surface area. Thus, surfactants make water wet surfaces better. Surfactants are used in dips for animals, in insecticides and other plant sprays, and in the textile industry for even results in bleaching and dyeing. A wetting agent must be present to make stable foams and emulsions.

Loss of water from lakes and reservoirs in hot dry countries can be reduced by covering the surface with cetyl alcohol, $CH_3(CH_2)_{14}CH_2OH$. Cetyl alcohol molecules have a hydrophobic end and a hydrophilic end. The cetyl alcohol spreads to form a layer one molecule thick with the —OH groups in the water and the $CH_3(CH_2)_{14}CH_2$— groups in the air. Water cannot easily evaporate through the cetyl alcohol layer. However, oxygen from air can diffuse through the monomolecular layer into the water so that fish and plants survive.

Formation and Precipitation of Colloidal Dispersions

Many colloids are produced naturally, for example, by biochemical reactions or geological processes such as erosion. Chemical reactions yield colloidal dispersions when precipitation takes place rapidly from very dilute solution. In the laboratory or manufacturing plant, colloids are made by dividing large particles into smaller particles, for example, by grinding a sample in a mortar or by forming large particles from small ones as in the formation of micelles of soap.

Like solutions, dispersions of lyophilic colloids form because the attractive forces between the colloidal particles and the medium are greater than colloid particle/colloid particle and solvent molecule/solvent molecule attractions. The entropy increase due to mixing also favors formation of colloidal dispersions. Colloidal dispersions of micelle formers and macromolecules are lyophilic.

Colloidal dispersions of very insoluble materials such as silver chloride and metals are lyophobic. Dispersions of lyophobic colloids form because the colloidal

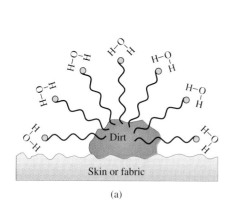

(a)

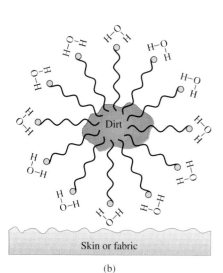

(b)

▬ FIGURE 13.19 (a) Hydrophobic ends of soap ions dissolve in dirt. Water molecules attract hydrophilic ends of soap ions. (b) Dirt leaves skin or fabric and becomes center of micelle. Micelle can be washed away with water.

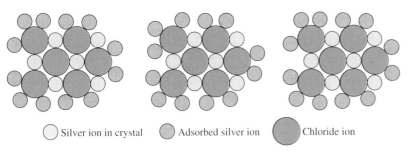

○ Silver ion in crystal　　● Adsorbed silver ion　　● Chloride ion

- FIGURE 13.20 Colloidal silver chloride formed in the presence of an excess of Ag^+ adsorbs a layer of Ag^+ ions. The outsides of all the colloidal particles are positively charged. Positively charged particles repel each other, and AgCl(s) does not precipitate but instead forms a colloidal dispersion.

Top: *Rounded drops of water on a freshly waxed car.* Bottom: *Water to which a surfactant has been added has lower surface tension and wets the surface better.*

particles adsorb ions. All the colloidal particles adsorb ions having the same charge as shown in **-Figure 13.20.** The adsorbed ions repel each other, preventing collision and precipitation of the colloidal particles. The colloidal particles of silver chloride are very small pieces of crystals of AgCl.

Colloidal dispersions of lyophobic colloids can be precipitated by the addition of electrolytes. Ions of the added electrolyte neutralize the charge of the adsorbed ions as shown in **-Figure 13.21.** When the colloidal particles no longer have the same charge, they precipitate. The higher the charge of the added ions, the more effective the added ions are at precipitating the colloid. For example, sulfate ions, SO_4^{2-}, are more effective than nitrate ions, NO_3^-, for precipitating a positively charged colloid. Solvation, as well as adsorbed ions, helps keep colloidal particles suspended.

Both the making and breaking of colloidal dispersions are of great practical importance. For example, 10^{12} tons of metal ore are concentrated every year by froth (foam) flotation (4×10^8 tons in the United States alone). Separation by froth flotation depends on the relative wettability of surfaces. The ore is finely ground and treated with a flotation collector. A flotation collector is a compound that sticks to the metal-containing component of the ore and makes it hydrophobic. The unwanted components of the ore are hydrophilic. The treated ore is mixed with water that contains a foaming agent, and air is bubbled through the mixture to form a foam. The hydrophilic components are wetted and sink; the wanted component, which is hydrophobic, sticks to the bubbles in the foam and floats.

Styptic pencils used to stop bleeding from small cuts are a practical application of the precipitation of a colloidal dispersion by addition of an electrolyte (Al^{3+} ion).

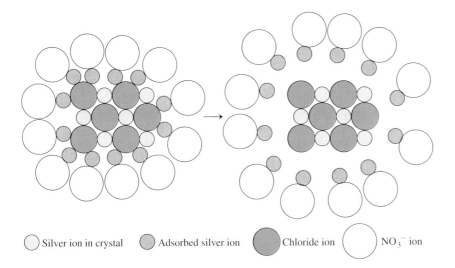

○ Silver ion in crystal　　● Adsorbed silver ion　　● Chloride ion　　○ NO_3^- ion

- FIGURE 13.21 Added electrolyte (NO_3^-) neutralizes charge of adsorbed ions. The next time particles of AgCl collide, they will stick together and form a precipitate.

Separation of Mixtures

The more than 13 million chemical compounds that are known exist mainly in the form of mixtures. People have used methods of separating and purifying materials since ancient times. Today the oil industry separates crude oil into products used as fuels, lubricants, and raw materials for the chemical industry. The mining industry is based on the separation of metals (and some nonmetals) from their ores. Pharmaceutical companies separate and purify natural and synthetic drugs. Modern medicine depends on purifications, such as those performed by artifical kidneys, and separations, such as those involved in the analysis of biological fluids like blood and urine. Many advances in the life sciences are based on the development of new separation methods.

We have already discussed some methods for separating mixtures in this chapter. Separations based on differences in solubility were treated in Section 13.3, and distillation was discussed in Section 13.6.

The most powerful approach for separating mixtures is chromatography. The ability of chromatography to separate complex mixtures of similar substances is unequaled among separation methods; for example, more than 100 components can be separated from complex mixtures such as urine and gasoline. Chromatography can be used to separate samples as small as a picogram and as large as several tons.

All methods of chromatography involve the distribution of solutes between a moving phase and a nonmoving phase. Experimentally, the simplest chromatographic method is paper chromatography, which is sometimes used in general chemistry laboratories and often used in organic and biochemistry laboratories.

To understand how paper chromatography works, we first need to discuss liquid–liquid extraction. Consider a colored compound that is soluble in both water and a solvent, such as dichloromethane, CH_2Cl_2, that is not miscible with water.

Suppose a crystal of a colored compound is dissolved in 10.0 mL of water and the aqueous solution is mixed thoroughly with 10.0 mL of dichloromethane as shown in ▬Figure 13.22. The colored compound is distributed between the two immiscible solvents [see Figure 13.22(d)].

If, in a second experiment, the same-size crystal of the colored compound is dissolved in 10.0 mL of dichloromethane and the dichloromethane solution is shaken with 10.0 mL of water, the intensities of the color in the two layers will be the same as in the first experiment. The fact that the same intensities are obtained in both experiments indicates that the distribution of a solute between two immiscible solvents is another example of equilibrium. The rate at which solute molecules are moving from water to dichloromethane must be equal to the rate at which solute molecules are moving from dichloromethane to water. The intensities of the color in the two layers do not change if

(a)

(b)

(c)

(d)

▬ FIGURE 13.22   (a) Dichloromethane (*left*) is colorless. A solution of crystal violet in water is violet (*right*). (b) If dichloromethane and aqueous crystal violet are placed in a separatory funnel, the aqueous solution floats on the dichloromethane. (c) After the separatory funnel has been shaken so that the two layers are thoroughly mixed, the layers separate slowly. (d) After the layers have separated, both are colored. The crystal violet has been distributed between the two solvents.

■ FIGURE 13.23 Separation of black ink by paper chromatography. A line of black ink was chromatographed instead of a spot so that you can see the separation better.

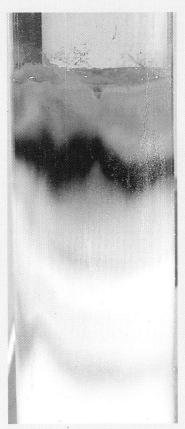

the two solutions are mixed again.

The process shown in Figure 13.22 is called liquid–liquid extraction. As you can see from Figure 13.22, a single extraction is not enough to remove the colored solute from water; repeated extractions with fresh portions of dichloromethane are needed. Carrying out extraction after extraction is a time-consuming process. Repeated extractions can be carried out much more efficiently by chromatography.

In paper chromatography, the stationary solvent is water that has been absorbed from air by the paper. If a spot of the mixture to be separated is placed near the bottom edge of a piece of filter paper and the bottom of the filter paper is dipped into a solvent that wets it, the solvent rises up the paper by capillary action. The solutes are distributed between the moving solvent and the stationary water adsorbed on the filter paper just as the purple dye was distributed between the two layers in the extraction shown in Figure 13.22—except that the process is repeated hundreds of times. A solute that is very soluble in the moving solvent and only weakly adsorbed by the stationary phase moves up the filter paper very quickly. A solute that is not very soluble in the moving solvent but strongly adsorbed by the stationary phase moves slowly. ■Figure 13.23 shows the separation of black ink on filter paper. Paper chromatographic separation of inks is used in crime labs to compare inks used on checks suspected of having been altered.

In other types of chromatography, the stationary phase is packed in a column as shown in ■Figure 13.24. There are a number of different types of chromatography, and species ranging in size from hydrogen gas molecules to viruses can be separated. Chromatographic separations are used on samples as different as exhaust gas from cars, the atmospheres of Mars and Venus, peanut butter, and aspirin and have been a major factor in the recent rapid growth of biochemistry. They are also used routinely for analysis and process control in oil refineries and chemical plants.

■ FIGURE 13.24 Separation of black ink by column chromatography. The colors move down the column from the top. As a result, the rose-colored band, which moves slowest and is at the bottom of the paper chromatogram, is at the top.

Formation of deltas at the mouths of rivers is an example of precipitation of lyophobic colloidal particles by the addition of an electrolyte. Rivers contain colloidal particles of mud; the salt in the ocean causes the mud particles to settle.

Particulate matter from smoke produced by cars and power stations is a major form of air pollution. Smoke from power plants is passed through electrostatic precipitators to remove particulate matter. The particles are charged in an electric field, and the charged particles move to the surface of opposite charge where they deposit. Electrostatic precipitation is the most energy-efficient way of removing colloidal particles from smoke; however, the equipment is very expensive.

SUMMARY

On a microscopic scale, three things happen when a solution forms: (1) solute particles separate from each other; (2) solvent molecules move apart to make room for solute particles; (3) solute particles and solvent molecules mix. A solution that forms without any change in energy is called an **ideal solution.** The increased entropy of the mixture makes an ideal solution form.

Liquids that are soluble in each other in all proportions are said to be **miscible,** and liquids that are not soluble in each other are described as **immiscible. Like dissolves like** is a good rule of thumb for predicting solubility; "like" refers to intermolecular attractions.

A dynamic equilibrium exists between undissolved solid and ions or molecules in solution. Ions and polar molecules in solution in polar solvents are **solvated**—that is, they are surrounded by spherical shells of solvent molecules. The term **hydrated** refers to ions or molecules that are solvated by water. The higher an ion's charge and the smaller it is, the more strongly an ion is hydrated.

The solubilities of most liquids and solids are greater at high temperatures than at low temperatures; however, the quantitative effect of temperature on the solubility of liquids and solids cannot be predicted but must be found experimentally. The solubilities of most gases in water are smaller at high temperatures.

Pressure does not have much effect on the solubility of liquids and solids. Increasing the pressure of a gas in equilibrium with a solution increases the concentration of gas in solution. According to **Henry's law,** at constant temperature, the concentration of a gas in solution is directly proportional to the pressure of the gas above the solution:

$$c_g = kp_{gas}$$

Raoult's law describes how the vapor pressure of a component of an ideal solution depends on the vapor pressure of the pure component at the temperature of the solution. The vapor pressure of a component of a solution is equal to the product of the mole fraction of the component in the solution and the vapor pressure of the pure component at the temperature of the solution:

$$VP_{A\ in\ solution} = X_{A\ in\ solution} \cdot VP_{pure\ A}$$

The **mole fraction** of a component A in a solution is the fraction of the total number of moles in the solution that is A:

$$X_A = n_A/(n_A + n_B + n_C + \cdots)$$

The total vapor pressure of an ideal solution is equal to the sum of the vapor pressure of the volatile components of the solution:

$$VP_{solution} = VP_{A\ in\ solution} + VP_{B\ in\ solution} + VP_{C\ in\ solution} + \cdots$$

The partial pressure of a component of a gaseous mixture is equal to the mole fraction of the component times the total pressure of the mixture:

$$p_A = X_A P_{total}$$

Nonvolatile solutes lower the vapor pressure of a volatile solvent; therefore, they raise the boiling point and lower the freezing point of the solvent. Vapor pressure lowering, boiling point elevation, freezing point depression, and osmotic pressure are known as **colligative properties.** Colligative properties are the same for all solutes; they depend on the concentration of solute particles. A formula unit of a strong electrolyte AB yields two particles. A formula unit of a strong electrolyte AB_2 or A_2B yields three particles, and so forth.

Boiling point elevation and freezing point depression are proportional to the molality of the solute; the value of the proportionality constant is characteristic of the solvent. **Molality** is defined as the number of moles of solute in one kilogram of solvent,

$$molality, m = (moles\ solute)/(mass\ solvent\ in\ kg)$$

and does not change with temperature. For dilute aqueous solutions, molality is identical with molarity for all practical purposes.

The boiling points of mixtures of two volatile substances are usually between the boiling points of the two pure components. Mixtures of substances with very different vapor pressures can be separated by **simple distillation. Fractional distillation** must be used to separate mixtures of substances that have similar boiling points. When a mixture of two volatile substances is distilled, the first liquid to distill is richer in the more volatile component that the original mixture. Distillation is the most frequently used method for purifying liquids; larger quantities of liquids are separated by fractional distillation than by any other method.

Osmosis refers to the movement of solvent molecules through a semipermeable membrane from a dilute to a concentrated solution. A **semipermeable membrane** is a membrane that lets only solvent molecules pass through. The pressure needed to stop the net motion of solvent particles through a semipermeable membrane is called the **osmotic pressure** of the solution:

$$\text{osmotic pressure, } \pi = MRT$$

Molecular masses of biologically important molecules that have moderately high molecular masses and low solubilities can be found by osmotic pressure measurements.

Colloids are particles that are larger than ordinary molecules and ions but too small to be seen with a light microscope; colloidal particles are composed of 10^3–10^9 atoms and have masses between 10^4 and 10^{10} u. Colloidal particles are usually found dispersed in a homogeneous medium like water or air. Like solutions, colloidal dispersions are stable and pass through filters. The **Tyndall effect** is characteristic of colloidal dispersions.

Colloids are classified according to the physical states of the colloidal particles and the medium, by shape, or as **lyophobic** (solvent hating) or **lyophilic** (solvent loving). If the dispersion medium is water, lyophobic colloids are referred to as **hydrophobic** and lyophilic colloids as **hydrophilic.** Colloidal particles have very large surface-to-volume ratios, and as a result, colloidal particles **adsorb** molecules and ions from solution. Hydrophobic colloids settle if the charge of the layer of adsorbed ions is neutralized.

Some colloidal dispersions are made up of clusters of particles called **micelles.** Other colloidal particles are composed of a single polymeric molecule.

ADDITIONAL PRACTICE PROBLEMS

For information about the organization of Additional Practice Problems, Stop & Test Yourself, Putting Things Together, and Applications, see the beginnings of these sections in Chapter 1.

13.28 Classify each of the following pairs of solvents as probably miscible or probably immiscible. Explain your answers.
(a) $HOCH_2CH_2OH$ and water
(b) $HOCH_2CH(OH)CH_2OH$ and $CH_3(CH_2)_4CH_3$
(c) $CH_3CH_2CH_2OH$ and CH_3CH_2OH
(d) $CH_3(CH_2)_6CH_2OH$ and water (13.1)

13.29 Which member of each of the following pairs is more soluble in water? Explain your answers. (a) CH_4 or NH_3 (b) $CH_3(CH_2)_3CH_3$ or $CH_3(CH_2)_2CH_2OH$ (c) $MgCl_2$ or CH_2Cl_2 (d) CH_3CH_2OH or CH_3CH_2SH (13.1)

13.30 (a) Compare the meanings of the terms *solvation* and *hydration.* (b) What factors cause an ion to be strongly hydrated? (c) Why is hydration of ions an important factor in the solubility of ionic compounds? (13.1)

13.31 Which member of each of the following pairs of ions is more strongly hydrated? (a) Li^+ or Be^{2+} (b) Li^+ or Na^+ (c) Fe^{2+} or Fe^{3+} (d) H^+ or NH_4^+ (e) F^- or K^+ (radii are similar) (13.2)

13.32 (a) The solubility of NaCl in g/100 g solvent at 25 °C is 36 for H_2O, 1.3 for CH_3OH, and 0.0 for CCl_4. Explain why. (b) Suggest an explanation for the fact that Br_2 is more soluble in CCl_4 than I_2 is. (13.2)

13.33 Consider glucose, $HOCH_2(CHOH)_4CHO$, and biphenyl,

⬡—⬡ . (a) Which is more soluble in water? (b)

Which is more soluble in benzene, ⬡ ?
Explain your answers. (13.2)

13.34 Explain why: (a) some salts, such as NaCl, are quite soluble in water but other salts, such as AgCl, are quite insoluble; (b) ionic compounds are quite insoluble in pentane, $CH_3(CH_2)_3CH_3$; (c) glucose, $HOCH_2(CHOH)_4CHO$, is soluble in water but insoluble in pentane. (13.2)

13.35 Which would be more likely to crystallize as a hydrate, NaCl or $CaCl_2$? (Sodium ions and calcium ions have similar radii.) (13.2)

13.36 Use Figure 13.5 to answer the following questions: (a) If a solution of 95 g of KNO_3 in 100 g of water is cooled from 100 °C to 0 °C, at what temperature will KNO_3 start to precipitate provided that no supersaturation takes place? (b) At 80 °C, what is the solubility of KNO_3 in water and what is the total mass of a sample that contains 100 g water? (c) Suppose that the total mass of sample from part (b) is cooled to 20 °C, how many grams of KNO_3 will separate? (d) If only 50.0 g of a solution that is saturated with KNO_3 at 80 °C is cooled to 20 °C, how many grams of KNO_3 will separate? (13.3)

13.37 Qualitatively, how does increasing pressure affect the solubilities of solids, liquids, and gases? (13.4)

13.38 The Henry's law constant for helium is 3.7×10^{-4} M atm^{-1}. (a) What is the molarity of He(g) in water if the pressure of the helium is 0.80 atm? (b) If the concentration of an aqueous solution of helium is 0.10 M, what must the pressure of He(g) be? (13.4)

13.39 Would you expect HCl(g) to follow Henry's law? Explain your answer. (13.4)

13.40 In a mixture of pentane, hexane, and heptane, the mole fraction of pentane is 0.18, and the mole fraction of hexane is 0.36. What is the mole fraction of heptane? (13.5)

13.41 What is the mole fraction of each component in a mixture that contains 0.38 mol pentane, 0.27 mol hexane, 0.45 mol heptane, and 0.51 mol octane? (13.5)

13.42 What is the mole fraction of each component in a mixture of 22.6 g pentane, C_5H_{12}, 42.8 g hexane, C_6H_{14}, and 37.5 g heptane, C_7H_{16}? (13.5)

13.43 How many moles of CH_3OH are present in 362 g of 2.50 *m* solution? (13.5)

13.44 What is the molality of each of the following solutions? (a) 0.367 mol NH_4Cl in 4.85 kg water (b) 22.9 g NH_4Cl in 812 g water (c) 92.4 g $CaCl_2 \cdot 2H_2O$ in 2.79 kg water (13.5)

13.45 (a) How many grams of water must be used to dissolve 85.4 g NaCl to make a 3.46 m solution? (b) How many grams of NaCl must be dissolved in 724 g water to make a 2.95 m solution? (c) To make 675 g of 3.29 m NaCl, how many grams of NaCl and how many grams of water should be used? Write directions for preparing this solution. (13.5)

13.46 (a) What is the molality of NH_4Cl in an aqueous solution that has mole fraction $NH_4Cl = 0.0823$? (b) What is the mole fraction of NH_4Cl in a 3.76 m aqueous solution of NH_4Cl? (13.5)

13.47 A solution of CH_3OH made by dissolving 16.24 g of CH_3OH in 41.76 g H_2O has a density of 0.955 g/cm³. What is (a) the molality? (b) The mole fraction of each component? (c) The molarity? (13.5)

13.48 The mole fraction of CH_3CH_2OH in an aqueous solution of CH_3CH_2OH is 0.4538; the density of the solution is 0.872 g/cm³. What is the molarity of the solution? (13.5)

13.49 The density of a 3.54 M solution of NH_4Cl is 1.0512 g/cm³. What is the molality of the solution? (13.5)

13.50 The density of a 21.15 M solution of CH_3OH in H_2O is 0.847 g/cm³. What is the mole fraction of each component in the solution? (13.5)

13.51 What is the partial pressure of each gas in a mixture of 0.163 mol N_2, 0.482 mol H_2, and 0.375 mol NH_3 if the total pressure is 599.3 mmHg? (13.5)

13.52 In a mixture of $CO(g)$ and $CO_2(g)$, $p_{CO} = 222.6$ mmHg and $p_{CO_2} = 437.1$ mmHg. What is the mole fraction of each gas in the mixture? (13.5)

13.53 The mole fraction of trichloromethane, $CHCl_3$, in a mixture of trichloromethane and tetrachloromethane, CCl_4, is 0.453. The vapor pressure of trichloromethane is 193 mmHg, and the vapor pressure of tetrachloromethane is 108 mmHg at 24 °C. What is the vapor pressure of the mixture at 24 °C? (13.6)

13.54 For the mixture in Problem 13.53, what is the mole fraction of $CHCl_3$ in the vapor? (13.6)

13.55 Why must the addition of a nonvolatile solute to a volatile solvent lower the vapor pressure of the solvent? (13.7)

13.56 At ordinary temperatures, the vapor pressure of glycerol, $HOCH_2CH(OH)CH_2OH$, is negligible. (a) What is the vapor pressure at 22.2 °C of an aqueous solution of glycerol in which the mole fraction of glycerol is 0.294? (b) What is the vapor pressure lowering caused by the dissolved glycerol? (13.7)

13.57 For an 11.77 m aqueous glycerol solution, (a) what is the freezing point? (b) What is the boiling point? (13.7)

13.58 Assuming that strong electrolytes are completely ionized in aqueous solution, what will the concentration of particles be in each of the following solutions? (a) 0.05 m $HOCH_2CH_2OH$ (b) 0.05 m $Ce(NO_3)_3$ (c) 0.05 m Na_2SO_4 (d) 0.05 m KI (e) 0.05 m $MgCl_2$ (13.7)

13.59 Consider 0.1 m aqueous solutions of NaCl and glucose, $C_6H_{12}O_6$. (a) Which has the lower freezing point? (b) Which has the lower boiling point? Explain your answers. (13.7)

13.60 For an aqueous solution of $(NH_4)_2SO_4$ at 20.8 °C in which the mole fraction of $(NH_4)_2SO_4$ is 0.0712, (a) what is the vapor pressure? (b) What is the vapor pressure lowering caused by the dissolved $(NH_4)_2SO_4$? (13.7)

13.61 For a 0.0354 m aqueous solution of $CaCl_2$, (a) what is the freezing point? (b) What is the boiling point? (13.7)

13.62 What is the molality of an aqueous solution of a nonelectrolyte that boils at 100.087 °C at 1 atm? (13.7)

13.63 What is the osmotic pressure at 19.6 °C of a 2.752 M solution of glycerol? (13.7)

13.64 For a 0.0263 M solution of $FeCl_3$ at 24.9 °C, what is the osmotic pressure? (13.7)

13.65 The osmotic pressure at 23.6 °C of 500.0 mL of a solution containing 0.302 g of an antibiotic has an osmotic pressure of 8.34 mmHg. What is the molecular mass of the antibiotic? (13.7)

13.66 (a) Tell how colloids differ from solutions and in what respects they are the same as solutions. (b) What is the difference between a colloid and a heterogeneous mixture? (13.8)

13.67 (a) What is the Tyndall effect? (b) What causes the Tyndall effect? (13.8)

13.68 Given a clear liquid, how could you determine whether it is a solution or a colloidal dispersion? (13.8)

13.69 In each of the following colloidal dispersions, what is the state of the colloidal particle, and what is the state of the medium? (a) a foam (b) an aerosol (c) an emulsion (d) a sol (13.8)

13.70 For each of the following examples of colloidal dispersions, give the name of the type of colloidal dispersion: (a) butter (b) clouds (c) homogenized milk (d) marshmallows (e) mud (f) ruby glass (g) smoke (h) whipped cream (13.8)

13.71 What is a micelle? (13.8)

13.72 Explain how soap removes oily dirt. (13.8)

13.73 Explain why colloidal particles adsorb molecules and ions from solution. (13.8)

13.74 Give two methods by which colloidal particles can be formed and two methods by which colloidal particles can be precipitated. (13.8)

13.75 (a) The particles of a colloidal dispersion of As_2S_3 are negatively charged. Which of the following compounds would be most effective for precipitating the colloid? NaCl, $CaCl_2$, $Al(NO_3)_3$, Na_2SO_4, or Na_3PO_4 (b) The particles of a colloidal dispersion of $Al(OH)_3$ are positively charged. Which of the compounds in part (a) would be most effective for precipitating the colloid? (13.8)

13.76 Sketch a micelle formed from a typical synthetic detergent:

$$CH_3(CH_2)_{10}CH_2SO_4^- Na^+$$

13.77 To make 2.50 mol of a mixture of pentane, hexane, and heptane that has mole fraction pentane $= 0.18$ and mole fraction hexane $= 0.36$, how many grams of each component should be used?

13.78 Use Figure 13.5 to answer the following questions: If a sample that contains 25.0 g KCl and 5.0 g NaCl is heated long enough with 25.0 g water to dissolve all the NaCl and then cooled to 5 °C, how many grams of KCl will crystallize? Will the KCl be pure or will it still be contaminated with NaCl? (Assume that NaCl and KCl don't affect each other's solubility.)

13.79 (a) Under what conditions is M = *m* in aqueous solutions? (b) Is M = *m* under these conditions in all solvents? Explain your answer.

13.80 A 0.0953 *m* solution of naphthalene, $C_{10}H_8$, in acetic acid, CH_3COOH, froze at 16.328 °C. The freezing point of the acetic acid was 16.700 °C. What is the value of K_{fp} for acetic acid?

13.81 What is the difference between (a) a lyophobic colloid and a lyophilic colloid? (b) A hydrophobic colloid and a hydrophilic colloid? (c) A sol and a gel? (d) A foam and an emulsion? (e) An aerosol and a foam?

13.82 A 0.0100 *m* aqueous solution of Na_2HPO_4 has a freezing point of −0.050 °C. Which of the following equations best represents what happens when $Na_2HPO_4(s)$ dissolves?
(a) $Na_2HPO_4(s) \longrightarrow Na_2HPO_4(aq)$
(b) $Na_2HPO_4(s) \longrightarrow Na^+(aq) + NaHPO_4^-(aq)$
(c) $Na_2HPO_4(s) \longrightarrow 2Na^+(aq) + HPO_4^{2-}(aq)$
(d) $Na_2HPO_4(s) \longrightarrow 2Na^+(aq) + H^+(aq) + PO_4^{3-}(aq)$
(e) $Na_2HPO_4(s) \longrightarrow$
  $2Na^+(aq) + H^+(aq) + P^{5+}(aq) + 4O^{2-}(aq)$

13.83 (a) A mixture of 0.0263 g of an unknown organic compound and 0.9476 g of benzene melted at 4.4 °C. What is the molecular mass of the unknown organic compound? (b) Give two reasons why it is better to use camphor than benzene.

13.84 Arrange the solutions in each of the sets in order of increasing boiling point: (a) 0.1 *m* $Al(NO_3)_3$, 0.1 *m* $Ca(NO_3)_2$, and 0.1 *m* $NaNO_3$ (b) 0.1 *m* HNO_2, 0.1 *m* HNO_3, 0.2 *m* HNO_3.

13.85 Using circles to represent molecules, draw microscopic-level pictures showing (a) 12 molecules of a nonvolatile solute dissolved in 10 mL of solvent A in a stoppered 50-mL Erlenmeyer flask (b) the flask after addition of 10 mL of solvent B, which is immiscible with and less dense than solvent A (c) the flask after the two layers have been thoroughly mixed and then allowed to separate; 8 molecules of solute should now be in the upper layer (d) 10 mL of solvent A in a stoppered 50-mL Erlenmeyer flask (e) the flask after addition of a solution of 12 molecules of solute in 10 mL of solvent B (f) the flask after the two layers have been thoroughly mixed and then allowed to separate.

STOP & TEST YOURSELF

1. Which of the following substances is most soluble in water?
(a) C (b) $CH_3(CH_2)_6CH_2OH$ (c) CH_4 (d) CH_3OH (e) Cu

2. The solubility of a gas is 0.0010 g/100 g H_2O at 25 °C and 1 atm. Would you expect the solubility of the gas to be less at (a) lower temperature and higher pressure? (b) Higher temperature and lower pressure? (c) Lower temperature and same pressure? (d) Same temperature and higher pressure? (e) The identity of the gas must be known to answer this question.

3. Which of the following mixtures is most likely to be an ideal solution?
(a) CH_3CH_2OH and $CH_3(CH_2)_3CH_3$
(b) $CH_3(CH_2)_3CH_3$ and $CH_3(CH_2)_4CH_3$
(c) CH_3CH_2OH and H_2O
(d) NaOH and H_2O
(e) $CH_3(CH_2)_3CH_3$ and H_2O

4. Which of the following has the lowest freezing point?
(a) Pure H_2O (b) 0.10 *m* NaCl (c) 0.10 *m* $CaCl_2$ (d) 0.10 *m* CH_3COOH (e) 0.10 *m* CH_3CH_2OH

5. The best method for determining the molecular mass of a high molecular mass substance is measurement of (a) vapor density (b) vapor pressure lowering (c) freezing point depression (d) boiling point elevation (e) osmotic pressure

6. In which of the following solutions is molality most nearly equal to molarity?
(a) 0.30 M aqueous KI (b) 0.030 M aqueous KI (c) 0.30 M KI in CH_3OH (The density of CH_3OH is 0.791 g/cm³.) (d) 0.030 M KI in CH_3OH (e) Molality is never nearly equal to molarity.

7. Use Figure 13.5 to answer this question. Consider an aqueous solution that is saturated with $KClO_3$ at 86 °C. If a sample of saturated solution containing 100 g H_2O is cooled to 22 °C, how many grams of $KClO_3$ will precipitate?
(a) 8 (b) 9 (c) 36 (d) 40 (e) 44

8. What is the osmotic pressure of a 0.0050 M aqueous solution of NaCl at 19.7 °C?
(a) 0.12 mmHg (b) 6.1 mmHg (c) 12.3 mmHg (d) 91 mmHg (e) 183 mmHg

9. A solution of 0.072 65 g of a hormone in 100.00 mL of solution had an osmotic pressure of 12.60 mmHg at 21.6 °C. What is the molecular mass of the hormone?
(a) 77.7 (b) 521 (c) 943 (d) 1060 (e) 2.32×10^5

10. If the mole fraction of glucose, $C_6H_{12}O_6$, in an aqueous solution is 0.247 and the temperature is 18.6 °C, what is the vapor pressure of the solution?
(a) 3.97 mmHg (b) 12.1 mmHg (c) 16.1 mmHg
(d) 188 mmHg (e) 572 mmHg

11. What is the normal boiling point of a 0.122 *m* aqueous solution of KI?
(a) 99.88 °C (b) 99.94 °C (c) 100.00 °C (d) 100.12 °C (e) 100.23 °C

12. For a solution of 35.2 g KI in 92.4 g H_2O, what is molality?
(a) 0.002 29 (b) 0.381 (c) 0.436 (d) 2.29 (e) 381

13. For a solution of 35.2 g KI in 92.4 g H_2O, what is mole fraction KI?
(a) 0.0397 (b) 0.0413 (c) 0.212 (d) 0.276 (e) 0.381

14. A colloidal dispersion of a liquid in a liquid such as homogenized milk is called a(an)
(a) aerosol (b) emulsion (c) foam (d) gel (e) sol

15. A negatively charged colloidal dispersion can be precipitated most effectively by addition of
(a) $CaCl_2(s)$ (b) $C_{12}H_{22}O_{11}(s)$ (c) NaCl(s)
(d) $Na_2SO_4(s)$ (e) $Na_3PO_4(s)$

13.86 For the solvent in which a reaction is to be carried out, (a) what properties are essential? (b) What properties are desirable?

13.87 (a) What four factors enter into solubility? (b) Why can't the solubility of ionic compounds in water be predicted from these four factors, that is, why must the empirical rules of Table 4.4 be used to predict the solubility of ionic compounds in water?

13.88 At ordinary temperatures, the vapor pressure of ethylene glycol, $HOCH_2CH_2OH$, is negligible. (a) What is the vapor pressure of a solution of 462 g ethylene glycol in 523 g H_2O at 24.6 °C? (b) What is the vapor pressure lowering caused by the dissolved ethylene glycol? (c) What is the boiling point? (d) The freezing point?

13.89 For a solution of 22.5 g NaCl in 83.9 g H_2O at 18.6 °C, (a) what is the vapor pressure of the water above the solution? (b) What is the vapor pressure lowering caused by the dissolved NaCl? (c) What is the freezing point of the solution? (d) The boiling point of the solution?

13.90 Arrange the substances in each of the following sets in order of decreasing solubility in water: (a) AgCl, CH_3CH_2OH, and NaCl (b) CH_3OH, CuS, and $CuSO_4$ (c) $CH_3CH_2CH_2OH$, $CH_3CH_2CH_2CH_2OH$, and $CH_3CH_2CH_2CH_2CH_3$

13.91 Which would be more likely to crystallize as a hydrate, $Ba(OH)_2$ or KOH?

13.92 What is the osmotic pressure at 24.8 °C of a solution containing 0.5216 g of oxytocin, $C_{43}H_{66}N_{12}O_{12}S_2$, per liter?

13.93 Given the equilibrium

$$N_2(g) \rightleftarrows N_2(aq)$$

(a) Will increasing the pressure of $N_2(g)$ shift the equilibrium to the right (increase solubility of N_2) or to the left? (b) Will cooling shift the equilibrium to the right or to the left? Explain your answers.

13.94 (a) Write an ionic equation for the equilibrium that exists in a saturated aqueous solution of KCl(s). (b) Use Le Châtelier's principle to predict what will happen when chloride ions are added to a saturated solution of KCl in water.

13.95 (a) Write a molecular equation for the equilibrium that exists in a saturated aqueous solution of glucose, $C_6H_{12}O_6$. (b) What will be observed if a little more water is added to the saturated solution?

13.96 The density of a 6.39 m solution of CH_3OH in H_2O is 0.971 g/cm³. What is the percent by mass of each component in the mixture?

13.97 A solution of 0.2168 g benzoic acid, C_6H_5COOH, in 23.65 g of benzene, C_6H_6, freezes at 5.317 °C. A sample of the benzene freezes at 5.513 °C. (a) What is the molecular mass of benzoic acid found by this experiment? (b) How does the experimentally determined molecular mass compare with the molecular mass calculated for C_6H_5COOH? (c) Suggest an explanation. (Write a structural formula.)

13.98 The density of gold is 19.3 g/cm³. (a) What is the surface area of a cube of gold that has a mass of 1.00 g? (b) An average colloidal particle has a mass of 1.00×10^7 u. Assume that a colloidal particle is spherical and that the gold atoms in a colloidal particle are close-packed. What is the total surface area of the 1.00-g sample of gold if the sample is divided into spheres of mass 1.00×10^7 u? (c) How many times as large as the surface area of the gold in a cube is the surface area of the same quantity of gold in average colloidal particles? (d) Explain how it is possible for Faraday's colloidal gold dispersions to have remained dispersed for over 100 years.

13.99 According to the *Merck Index,* water at 25 °C holds 31% by mass $NH_3(g)$. (a) At 25 °C, what is the solubility of $NH_3(g)$ in units of g gas/100 g H_2O? (b) What can you conclude takes place when $NH_3(g)$ dissolves in water? (c) According to the *CRC Handbook,* the solubility of $NH_3(g)$ is 89.9 g/100 cm³ water, but no temperature is given. Is the temperature <25 °C, = 25 °C, or >25 °C? Explain your answer.

13.100 For each of the following solutions, tell which type or types of attraction is (are) important between solute and solvent. (a) NaCl in water (b) CH_3CH_2OH in water (c) $(CH_3)_2C{=}O$ in water (d) $CHCl_3$ in CCl_4 (e) CH_2Cl_2 in $CHCl_3$ (f) CCl_4 in $CH_3(CH_2)_3CH_3$.

13.101 A solution of NH_4Cl made by dissolving 3.16 g NH_4Cl in 30.14 g H_2O has a density of 1.0272 g/cm³. What is (a) the molality? (b) The molarity? (c) The mole fraction of each component? (d) The percent by mass of each component?

13.102 A white crystalline solid contained 71.62% C, 6.01% H, and 10.44% N according to the results of analysis. (a) Assuming that the missing material was oxygen, what was the % O in the compound? (b) A mixture of 0.1234 g of the unknown with 0.9501 g of camphor, $C_{10}H_{16}O$, melted at 160.6 °C. What was the molecular formula of the unknown?

13.103 (a) Which would dissolve buckyballs, C_{60}, water or benzene, ⬡? (b) To make hexane, $CH_3(CH_2)_4CH_3$, more soluble in water, which group should be substituted for hydrogen, NH_2 or CH_3? (c) Which is more soluble in water, benzoic acid, ⬡—COOH, or sodium benzoate, ⬡—COONa? Explain your answers.

13.104 Solutions that are not ideal deviate significantly from Raoult's law. For example, solutions of chloroform, $CHCl_3$, and acetone, CH_3COCH_3, deviate as shown by the graph. In the graph, solid curves are observed vapor pressures; dashed lines show the vapor pressures predicted by Raoult's law. (a) How does the strength of acetone–chloroform attractions compare with acetone–acetone and chloroform–chloroform attractions? (b) Is formation of an acetone–chloroform solution exothermic or endothermic? (c) Is the enthalpy change that accompanies formation of the solution positive or negative?   (d) Sketch a graph showing vapor pressure as a function of composition for solutions of ethyl alcohol and hexane.

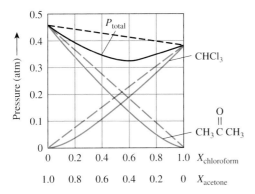

13.105 Explain why carbonated beverages must be stored in sealed containers.

13.106 When smoking was allowed in movie theaters, you could see the beam of light passing through the air from the projector to the screen. Explain.

13.107 Write directions for obtaining drinkable water by freezing seawater.

13.108 Give two practical applications of freezing point depression from everyday life.

13.109 Explain why a green salad wilts if the greens are allowed to stand in an oil and vinegar dressing that contains salt or a salt substitute.

13.110 A student bought an expensive tropical fish for a pet. To make sure the fish didn't get sick, the student sterilized the water used to fill the aquarium by boiling it. Although the student was careful to cool the water to the same temperature as the water in the aquarium in the fish store before putting the fish in it, the fish soon died. Explain.

13.111 Why does the temperature have to drop below 0 °C to freeze plants?

13.112 Pickles are made by soaking cucumbers in salt water. Why are pickles wrinkled like prunes?

13.113 Why are deltas formed at the mouths of rivers?

13.114 Diarrhea is the leading killer of small children in developing countries. Oral rehydration therapy (ORT) is a simple, low-cost treatment that replaces lost fluid and electrolytes. ORT now saves the lives of about 1 million children every year. To make the solution for ORT, a package containing 3.5 g NaCl, 2.9 g HOC(CH$_2$COONa)$_2$COONa (sodium citrate), 1.5 g KCl, and 20.0 g C$_6$H$_{12}$O$_6$ (glucose) is dissolved in enough water to make 1.0 L of solution. (a) What is the osmotic pressure of this solution in mmHg at body tempera-

ture (37.0 °C)?   (b) The osmotic pressure of the solution is the same as the osmotic pressure of blood. Putting more glucose in the mixture would increase the calorie content. Why is increasing the concentration of glucose *not* a good thing to do?   (c) An alternative to adding more glucose is to add a polymer of glucose. Why is a polymer of glucose preferable to glucose?

13.115 Soft drinks are carbonated by treatment with CO$_2$(g) under pressure. The solubility of CO$_2$(g) in water is 0.145 g/100 g of water at 25 °C and 1 atm. What is the solubility of CO$_2$(g) in water at the same temperature if the pressure is 4 atm?

13.116 A sample of gas from a volcano analyzed 97.3% H$_2$O, 1.2% CO$_2$, and 1.5% SO$_2$ by mass. What is the partial pressure of each gas in the mixture if the total pressure is 1 atm?

13.117 Streptomycin is an antibiotic produced by soil. If 50.0 mL of a solution containing 0.013 70 g of streptomycin has an osmotic pressure of 8.72 mmHg at 23.6 °C, what is the molecular mass of streptomycin?

13.118 The boiling point of water can be raised by adding salt as well as by using a pressure cooker. At a mountain camp where water boils at 95.2 °C, how many grams of NaCl would have to be added to 1.0 L of water to raise the boiling point to 100.0 °C? Assume that the salt is completely ionized.

13.119 Reverse osmosis can be used to obtain freshwater from seawater. In reverse osmosis, a pressure greater than the osmotic pressure of the solution is applied, and salt-free water is pushed out through the semipermeable membrane. The salt can't pass through the semipermeable membrane and remains behind. Assume that seawater is a 0.604 M solution of salt, NaCl. (a) What is the osmotic pressure of seawater at 25 °C?   (b) What is the minimum pressure that must be applied for reverse osmosis of seawater?   (c) Why must the applied pressure be greater than this minimum to obtain a significant quantity of freshwater from seawater? (d) If the applied pressure is 100.0 atm, how many liters of seawater must be processed to obtain 4.0 L of freshwater, the quantity one person in a lifeboat in the hot sun needs to drink in a day?

13.120 Values of Henry's law constants for water at 25 °C are O_2, 1.28×10^{-3} M atm^{-1}; N_2, 6.5×10^{-4} M atm^{-1}; He, 3.7×10^{-4} M atm^{-1}. Under water, pressure increases by 1 atm for each 34 ft of depth. High concentrations of nitrogen gas in blood and body tissues leads to drowsiness and stupor; high concentrations of oxygen are also dangerous, and a solution of oxygen in helium is better for breathing than air at high pressures. This solution has $X_{O_2} = 0.036$ and $X_{He} = 0.908$; the remainder is N_2. (a) At Earth's surface, the solubility of $O_2(g)$ in water is 2.6×10^{-4} M. At what depth will the concentration of oxygen from the oxygen–helium–nitrogen solution be the same as the solubility at Earth's surface? (b) At the depth found in part (a), what will the concentrations of He and N_2 total? (c) Compressed air has $X_{N_2} = 0.80$. At the depth found in part (a), what will be the concentration of N_2 from compressed air? (d) Will use of the oxygen–helium–nitrogen solution also help to avoid the bends? Explain your answer.

13.121 If NaCl or $CaCl_2$ is used to keep water on sidewalks from freezing, the runoff kills grass and other plants. Substitution of urea, H_2NCONH_2, which is a fertilizer, has been suggested. (a) How many grams of each of these three substances are needed to lower the freezing point of 1 kg of water to -5.0 °C (23 °F)? (b) If the prices are urea, $4.32/kg; NaCl, $9.25/kg; and $CaCl_2$, $7.59/kg; what is the cost of the quantity of each needed to lower the freezing point of 1 kg of water to -5.0 °C? (c) What is the chief disadvantage of using the cheapest one?

13.122 Ethylene glycol, $HOCH_2CH_2OH$, is used as antifreeze. Ethylene glycol is nonvolatile and has a density of 1.109 g/cm^3 at 20 °C. (a) To protect the water in your car radiator against freezing down to -34 °F (-36.7 °C), how many liters of ethylene glycol should be used per liter of water? (b) What is the percent by mass ethylene glycol in the mixture in part (a)? (c) The label on the antifreeze container says to use 1 L of antifreeze per liter of water for protection to -34 °F. What's the problem with the calculation in part (a)? (d) Calculate the boiling point of the mixture in part (a).

13.123 Methyl alcohol, CH_3OH, has been used as antifreeze. Methyl alcohol is volatile and has a normal boiling point of 65.0 °C and a density of 0.791 g/cm^3 at 20 °C. To protect a car radiator against freezing down to -34 °F (-36.7 °C), a 38.8% by mass solution of methyl alcohol in water is needed. (a) Why can't the freezing points of mixtures of methyl alcohol and water be calculated? (b) How many liters of methyl alcohol must be used per liter of water? (c) The price of ethylene glycol is $4.60/kg and the price of methyl alcohol is $3.58/kg. The density of the ethylene glycol–water mixture in Problem 13.124 is 1.071 g/cm^3, and the density of the methyl alcohol–water mixture in this problem is 0.935 g/cm^3. What is the cost of 1 L of each mixture? (Assume that water is free.) (d) The boiling point of the ethylene glycol–water mixture is >100 °C. How will the boiling point of the methyl alcohol–water mixture compare with the boiling point of pure water? Why do you think that methyl alcohol is no longer used as antifreeze?

13.124 Solutions for intravenous injection (IV) must have the same osmotic pressure as blood. According to the label on the bottle, each 100 mL (zeros significant) of normal saline solution used for IVs contains 900 mg NaCl (zeros significant). (a) What is the molarity of NaCl in normal saline solution? (b) What is the osmotic pressure of normal saline solution at the average body temperature of 37 °C? (c) What concentration of glucose, $C_6H_{12}O_6$, has the same osmotic pressure as normal saline solution?

13.125 Air and water are the major components of the part of Earth's atmosphere in which breathing is possible. Therefore, the Henry's law constant for the air–water equilibrium is one of the most important factors governing the distribution of chemicals in the environment. The Henry's law constants H, where $H = p_{gas}/c_{gas}$, for carbon tetrachloride, CCl_4, and heptane, $CH_3(CH_2)_5CH_3$, are 3.31 and 91.3 kPa·m^3·mol^{-1}, respectively, at a little over 25 °C. (a) Which compound will tend to accumulate in water and which in air? (b) At 5 °C, will the constants be larger or smaller? Explain your answers.

13.126 Large crystals free of defects are necessary for X-ray crystallography. In a new process, a dilute salt solution containing the substance to be crystallized is separated from a concentrated salt solution by a semipermeable membrane through which only water can pass. Draw a microscopic-level picture showing how this process works.

13.127 The determination of the average molecular masses of crude oils and hydrocarbon fractions by freezing point depression is widely used in the petroleum industry. A new instrument permits the use of a wide variety of solvents. (a) Why is the ability to use a wide variety of solvents desirable? (b) In an experiment to determine the molal freezing point depression constant, K_{fp}, for the solvent $Cl_2C(F)C(Cl)_2F$, the following data were obtained:

Mass empty container:	52.3679 g
Mass container + $CH_3(CH_2)_{10}CH_3$:	53.0060 g
Mass container + $CH_3(CH_2)_{10}CH_3$ +	
$Cl_2C(F)C(Cl)_2F$:	83.4172 g

The freezing point of the mixture was 33.06 °C; the freezing point of $Cl_2C(F)C(Cl)_2F$ is 37.70 °C. What is the value of K_{fp}?

13.128 Each cylinder contains 10 mL of 2,2,4-trimethylpentane, $(CH_3)_3CCH_2CH(CH_3)_2$, a typical component of gasoline, sitting on top of 1 mL of water representing water at the bottom

of an automobile gas tank. A trace of crystal violet has been dissolved in the water so that you can easily distinguish the two layers. One brand of "dry gas" was added to the left-hand cylinder and another to the right-hand cylinder. (Both brands are colorless.) The contents of each of the two cylinders were thoroughly mixed and then allowed to stand.

(a) Describe the contents of each cylinder. (b) Which brand of dry gas would you use? Explain your answer. (c) One dry gas contains methyl alcohol, CH_3OH, and the other isopropyl alcohol, $(CH_3)_2CHOH$. Write a complete structural formula for each alcohol and for 2,2,4-trimethylpentane. (d) Which cylinder contains the methyl alcohol? Explain your answer.

13.129 In a helium–neon laser, the active medium is a mixture of helium and neon in a 5:1 mole ratio. (a) What is the mole fraction of helium in this mixture? (b) This laser generates 633-nm radiation. What color is this radiation? (c) What is the frequency of the radiation in hertz? (d) What is the energy of one photon in joules?

13.130 The water in cells does not begin to freeze until the temperature is significantly below 0 °C. If cells are frozen slowly, damage occurs. If cells are frozen rapidly, however, they can be preserved indefinitely in a form capable of living. Explain.

Chemical Needs of a Solar Engineer

JIM WILLIAMSON

Solar Engineer, National Renewable
Energy Laboratory, A Division of Midwest
Research Institute

B.S. Mathematics
Montana State University

M.S. Mathematics
University of California, Berkeley

Solar energy is an emerging field that is growing rapidly in the 1990s. Since few universities offer a field of study specifically pointed toward solar technologies, it is one of those fields that you tend to grow into. I started in mathematics for the atomic energy technologies and moved into a specialty energy field of power for satellites and then gradually into solar systems. This pattern is not unusual; a large number of satellites used solar energy before many solar systems were built on the ground. The common threads of my career have been math and science, with chemistry being particularly significant because of the importance of materials.

The leader of the solar technologies is photovoltaics, a process of converting sunlight to electricity through solar cells. The reliance of solar technology on chemistry is characterized by this special group of semiconductor materials that Nature has blessed with the property of producing electricity when exposed to sunlight. The trick in developing the technology is to select the combination of materials that will produce the most electricity at the cheapest cost. This combination is Si, amorphous Si, GaAr, CdTe, or special synthesized binary and ternary semiconductors of type V-VI and I-II-V. Research into different materials is being conducted at the Solar Energy Research Institute (SERI) in Golden, Colorado. SERI, a national laboratory of the U.S. Department of Energy, is also studying a variety of other solar systems, each of which relies on a basic understanding of chemistry. The chemistry foundation of all research into solar energy technologies cannot be overstated. My high school chemistry teacher told us that "Chemistry is the queen of the sciences because it is the basis for so much of our understanding." To me, that understanding flows from basic chemistry to the interactions of more complex materials to final applications in solar devices. Chemistry reactions have always been a major part of my work and have enabled me to take an analytical approach to problem solving. Problem solving in research, though structured and exacting, can still be fun and may include travel to exotic places.

One of my favorite research projects involved the design and operation of a pilot plant for desalting seawater using solar energy. The project was jointly sponsored by the national solar research programs in the United States and Saudi Arabia. Most of the design and engineering was done in the United States, and the plant construction and testing were done on the Red Sea coast near Yanbu, Saudi Arabia. The desalting process used an innovative freezing technique wherein the seawater was frozen into ice crystals and the salt was merely rinsed off the ice. This process was developed to replace more energy-intensive osmotic membrane separations or conventional distillation techniques. The basic chemistry principle at work is that less energy is required in the latent heat of fusion of water than in the higher heat of evaporation. The solar technology in the project was an array of large solar thermal collectors that used mirrors to concentrate heat into a receiver fluid that operated a LiBr absorption chiller, fairly standard air-conditioning equipment.

The project went very smoothly through design, construction, and initial operation. We all enjoyed the Red Sea area with its colorful reefs and fishes. The relatively rural community of Yanbu provided a glimpse of Bedouin lifestyles and cuisine. Then a problem with the "queen of sciences" occurred.

The individual mirrors of the solar collectors began to turn black almost before our eyes. "Black-lace" degradation of mirrors is very common in old or antique mirrors that do not have adequate backing or protection for the silvering. The black, nonreflective appearance results when the silver reacts with salts, forming a black AgCl substance. However, the mirrors for our experiment had been carefully manufactured with a special low-iron glass to improve reflectivity and a thick anticorrosion paint to protect the silvering. The project for making fresh drinking water from the sea quickly turned into a giant experiment in mirror coatings as we looked for a way to stop the chemical reaction of the mirrors.

We finally identified the source of the corrosive salts as coming from the exhaust plume of a nearby petrochemical plant in combination with the natural saltation of the air near the seacoast. The solution was a thick, resistant epoxy coating that slowed the penetration of the airborne salt into the mirror coatings. Thus, the project was not only successful in demonstrating an innovative method of desalting but also had the unexpected benefit of developing a new means of mirror protection for the mirror industry. Saudi Arabia has continued to develop desalination plants and now has the largest systems in the world; the plants are the principal source of drinking water for even the inland cities.

When I first took chemistry in high school and college, I was intrigued by the investigative nature of experimentation. The scientific approach to problem solving that I learned in those classes has been invaluable in my career in solar technology research. In addition, I am always on the lookout for materials problems and reactions that underlie engineering challenges.

CHEMICAL EQUILIBRIUM

People on the beach and in the water on a hot day provide a good analogy for dynamic equilibria.

When pure substances are mixed, a chemical reaction often takes place, and the original substances are, for all practical purposes, completely converted to different substances. In other cases, no detectable reaction takes place or some, but not all, of the reactants is converted to products. For example, when a weak electrolyte like acetic acid is dissolved in water, a few of its molecules ionize; however, most remain in the form of molecules.

Why do some molecules of a weak electrolyte ionize while others do not? A dynamic equilibrium exists between the molecules of the weak electrolyte and the ions in solution. In a solution of a weak electrolyte, molecules are continually forming ions, and ions are continually combining to form molecules. The rates of the two opposing changes are equal. When the two opposing changes that are taking place at equal rates are chemical reactions, the dynamic equilibrium is called a **chemical equilibrium.**

■ *Why is a study of chemical equilibrium a part of general chemistry?*

Chemical equilibria are important in all fields of chemistry, in engineering, in the health sciences, in geology—in any field where chemistry is used. Chemical equilibria control the solution of limestone in groundwater and the formation of stalactites and stalagmites in caves and of scale in teakettles. The oxygen-carrying capacity of the hemoglobin in your blood depends on chemical equilibria. Understanding the factors that determine the composition of systems at equilibrium helps people predict the results of changes in systems at equilibrium and allows them to control reactions. For example, chemical engineers can choose the operating conditions for a manufacturing plant that will give the highest yield of product at the lowest cost. A quantitative knowledge of chemical equilibria is needed to calculate how much product to expect from reactions that go to equilibrium.

— *What do you already know about equilibrium?*

You first met the idea of dynamic equilibrium in connection with the vaporization of a liquid in a closed container (Section 12.3). When a system is in a state of dynamic equilibrium, nothing appears to be happening because two opposing changes are taking place at the same rate.

The concentration of molecules in the gas phase over a liquid in a closed container is constant *once enough time has passed for equilibrium to be reached.* The pressure of the gas phase is constant as long as some liquid is present. However, the same molecules are not in the gas phase all the time. Molecules move back-and-forth between the gas and liquid phases.

You also learned that solid is in equilibrium with liquid at the melting point of the solid (the freezing point of the liquid) (Section 12.5). The equilibrium between a solid and its melt is the result of a balance between the energy (enthalpy) and the entropy of the system.

Indeed, *all* equilibria result from this balance. At low temperatures, energy is usually more important than entropy. At high temperatures, entropy is more important than energy. Equilibria between undissolved solute and ions or molecules of solute in solution are yet another type of dynamic equilibrium (Section 13.1).

Qualitative predictions about the effect of changes in temperature, pressure, and concentration on a system at equilibrium can be made by means of Le Châtelier's principle: If a change is made in a system at equilibrium, the equilibrium will shift in such a way as to reduce the effect of the change (Section 12.3).

14.1 INTRODUCTION TO CHEMICAL EQUILIBRIA

In Section 4.2, we saw that when a weak acid such as acetic acid is dissolved in water, some ions are formed

$$HC_2H_3O_2(aq) \longrightarrow H^+(aq) + C_2H_3O_2^-(aq) \tag{14.1}$$

although most of the acetic acid remains in the form of molecules. In Section 4.3, we learned that when solutions of electrolytes are mixed, a reaction will take place if one of the possible products is a weak electrolyte. If an aqueous solution of sodium acetate is mixed with an aqueous solution of hydrochloric acid, acetate ions react with hydrogen ions to form molecules of the weak acid, acetic acid. The complete ionic equation is

$$Na^+(aq) + C_2H_3O_2^-(aq) + H^+(aq) + Cl^-(aq) \longrightarrow$$

$$HC_2H_3O_2(aq) + Na^+(aq) + Cl^-(aq)$$

Sodium ion and chloride ion are spectator ions. The net ionic equation is

$$H^+(aq) + C_2H_3O_2^-(aq) \longrightarrow HC_2H_3O_2(aq) \tag{14.2}$$

The reaction in equation 14.2 is the reverse of the reaction in equation 14.1. Both reactions take place almost instantaneously. Therefore, any solution that contains hydrogen ions and acetate ions also contains acetic acid molecules; any solution that contains acetic acid molecules also contains hydrogen ions and acetate ions. In any solution containing hydrogen ions, acetate ions, and acetic acid molecules, an equilibrium exists between the three species:

$$HC_2H_3O_2(aq) \rightleftharpoons H^+(aq) + C_2H_3O_2^-(aq) \tag{14.3}$$

⬤ The double arrows are often abbreviated by the symbol ⇌.

The *two arrows* in equation 14.3, *one pointing to the right and the other to the left, indicate an equilibrium.* At equilibrium, reaction is taking place in both directions.

In a solution prepared by dissolving acetic acid in water, there are mostly acetic acid molecules at equilibrium. For example, in a 0.1000 M solution of acetic acid at 25 °C, 98.7% of the solute is in the form of molecules, and only 1.3% is ionized. The fact that there are more molecules than ions at equilibrium can be shown qualitatively by making one arrow longer than the other:

$$HC_2H_3O_2(aq) \xrightleftharpoons{} H^+(aq) + C_2H_3O_2^-(aq)$$

The longer arrow points to the most numerous species. Strong electrolytes, such as hydrochloric acid, are usually assumed to be ionized completely in aqueous solution, and the equation for the ionization of hydrochloric acid is commonly written with only a single arrow:

$$HCl(aq) \longrightarrow H^+(aq) + Cl^-(aq)$$

The strengths of weak acids (and the positions of other equilibria) are shown quantitatively by numbers called equilibrium constants.

14.2   EQUILIBRIUM CONSTANTS AND EQUILIBRIUM CONSTANT EXPRESSIONS

Homogeneous Equilibria

Let's begin our quantitative discussion of chemical equilibria by considering a gaseous equilibrium, the equilibrium between hydrogen gas, iodine vapor, and hydrogen iodide gas:

$$H_2(g) + I_2(g) \xrightleftharpoons{} 2HI(g) \qquad (14.4)$$

Equilibria that involve only a single phase like the gaseous hydrogen–iodine–hydrogen iodide equilibrium and the equilibrium between acetic acid molecules and hydrogen ions and acetate ions in aqueous solution are called **homogeneous equilibria.**

If a mixture of hydrogen gas and iodine vapor is heated, hydrogen iodide is formed. The concentrations of H_2 and I_2 decrease, and the concentration of HI, which is zero at the beginning of the experiment, increases. ▪Figure 14.1(a) shows

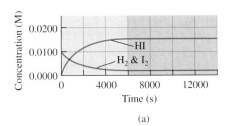

(a)

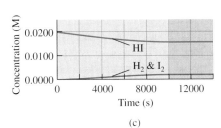

(c)

For ease of comparison, all time scales are the same; the concentration scales of (a) and (c) and the rate scales of (b) and (d) are the same.

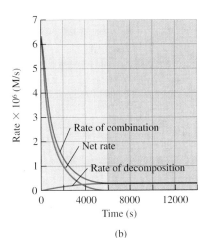

(b)

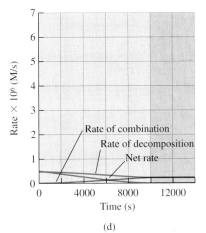

(d)

▪ FIGURE 14.1   Approach to the $H_2(g) + I_2(g) \xrightleftharpoons{} 2HI(g)$ equilibrium at 425.4 °C. Starting with $[H_2] = [I_2] = 0.0100$ and $[HI] = 0.0000$, where the formulas in square brackets indicate the concentration of a species in molarity, (a) shows concentration as a function of time, and (b) shows rate of reaction as a function of time. The net rate is the difference between the rate of combination and the rate of decomposition and is zero at equilibrium. Starting with $[H_2] = [I_2] = 0.0000$ and $[HI] = 0.0200$, (c) shows concentration as a function of time, and (d) shows rate of reaction as a function of time. The net rate is the difference between the rate of decomposition and the rate of combination.

Although rates of reaction have not yet been treated quantitatively, students were introduced qualitatively in Section 1.8 to the idea that some reactions take place faster than others. Rate as a factor in whether or not a reaction will be observed was discussed in Section 6.2.

how concentrations change as equilibrium is approached at 425.4 °C starting with concentrations of both hydrogen and iodine equal to 0.0100 M and a concentration of hydrogen iodide equal to 0.0000 M. As soon as some HI has been formed, it begins to decompose. In Figure 14.1(a), notice how the concentrations change rapidly at first and then more slowly until they become constant after about 6000 seconds. Figure 14.1(b) shows how the rates of combination and decomposition change with time. Rates of reaction are the subject of Chapter 18. For now, notice that reaction rates are high when concentrations are changing rapidly (reaction is fast) and low when concentrations are changing slowly (reaction is slow). After about 6000 seconds, the rate of decomposition has become equal to the rate of combination. The system has reached equilibrium. After about 6000 seconds, the concentrations of H_2, I_2, and HI are constant because hydrogen iodide is decomposing as fast as it is forming. The net rate of change in concentration is zero, and from a macroscopic point of view, nothing appears to be happening.

We can also approach this equilibrium from the other side. When heated, hydrogen iodide gas decomposes to hydrogen and iodine. The concentration of hydrogen iodide decreases, and the concentrations of hydrogen and iodine, which are zero to begin with, increase. As soon as some hydrogen and iodine have been formed, they begin to react to form hydrogen iodide. Figure 14.1(c) shows how the concentrations of hydrogen iodide, hydrogen, and iodine change with time and become constant after about 10 000 seconds. Figure 14.1(d) shows how the rates of decomposition and combination change with time. After about 10 000 seconds, the rates of decomposition and combination become equal. The system has reached equilibrium; the net rate of change in concentration is zero, and from a macroscopic point of view, nothing appears to be happening.

Now compare Figures 14.1(a) and (c). Notice that the same constant concentrations are reached whether the experiment begins with 0.0100 M H_2 and I_2 or with 0.0200 M HI. *The same equilibrium is reached starting from either side providing that temperature and number of each kind of atom are the same.* The same constant rates are reached. [Compare Figures 14.1(b) and (d).]

From Figures 14.1(a) and (c), the concentration of hydrogen iodide in the equilibrium mixture is 0.0158 M. The concentrations of hydrogen and iodine in the equilibrium mixture are only 0.0021 M. The concentration of hydrogen iodide is about 7.5 times as large as the concentrations of hydrogen and iodine. In the equation for the equilibrium, the arrow pointing to the right should be longer than the arrow pointing to the left:

$$H_2(g) \ + \ I_2(g) \ \underset{\longleftarrow}{\overset{\longrightarrow}{\rightleftharpoons}} \ 2HI(g)$$

▪Figure 14.2 shows a molecular-level view of the hydrogen–iodine–hydrogen

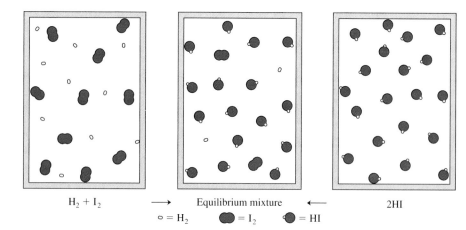

▪ FIGURE 14.2   Providing that temperature and number of hydrogen atoms and iodine atoms are the same, the same equilibrium mixture is obtained whether an experiment begins with $H_2(g)$ and $I_2(g)$ or with HI(g).

$$H_2 + I_2 \qquad \longrightarrow \qquad \text{Equilibrium mixture} \qquad \longleftarrow \qquad 2HI$$

$\circ = H_2$ ⬤ $= I_2$ ⬤ $= HI$

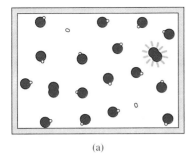
(a)

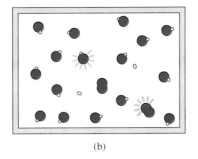
(b)

- FIGURE 14.3 A molecular-level view of the hydrogen–iodine–hydrogen iodide equilibrium. (a) One of the iodine molecules is composed of iodine-131, which is radioactive (shown by yellow rays). (b) Soon a molecule of HI contains an atom of iodine-131.

iodide equilibrium. Equal volumes of 0.0100 M I_2 and H_2 and 0.0200 M HI at the same temperature contain the same numbers of iodine atoms and hydrogen atoms and yield the same equilibrium mixture. In the equilibrium mixture, hydrogen iodide molecules are decomposing to hydrogen and iodine, and hydrogen and iodine molecules are reacting to form hydrogen iodide. The rates of the two opposing reactions are equal. On a microscopic scale, chemical equilibria are dynamic equilibria.

The dynamic nature of chemical equilibria can be shown by experiments with isotopic tracers. Naturally occurring iodine is all iodine-127, which is not radioactive. If a little iodine-131, which is radioactive, is substituted for some of the iodine-127 in the equilibrium mixture of hydrogen, iodine, and hydrogen iodide, radioactive iodine-131 can soon be detected in the hydrogen iodide as shown in ▬Figure 14.3. If a little hydrogen iodide that contains iodine-131 is substituted for some of the ordinary hydrogen iodide, iodine-131 can soon be detected in the iodine. Although concentrations are constant at equilibrium, both forward and reverse reactions are still taking place. Iodine atoms are moving back-and-forth between I_2 and HI all the time. Hydrogen atoms are also continuously moving back-and-forth between H_2 and HI.

Stoichiometry

In Figures 14.1(a) and (c), the concentrations of H_2, I_2, and HI are related by the equation for the reaction (equation 14.4). To keep the arithmetic simple, let's think about one liter of gas. In one liter of gas, the number of moles of each component is the same as the concentration of the component. From Figure 14.1(a), the concentrations of both hydrogen and iodine were 0.0100 M, and the concentration of hydrogen iodide was zero at the beginning of the experiment (time = 0 s). In one liter of gas, there were 0.0100 mol H_2, 0.0100 mol I_2, and 0.0000 mol HI at the beginning of the experiment.

After 2000 s, the concentrations of both hydrogen and iodine were 0.0043 M [Figure 14.1(a)]. The number of moles of hydrogen and iodine that reacted in 2000 s is equal to the difference between the number of moles present at the start and the number of moles left after 2000 s:

$$\begin{array}{ccc} \text{mol } H_2 \text{ (or } I_2) & \text{mol } H_2 \text{ (or } I_2) \text{ left} & \text{mol } H_2 \text{ (or } I_2) \\ \text{present at start} & - \quad \text{after 2000 s} & = \quad \text{reacted in 2000 s} \\ 0.0100 \text{ mol} & - \quad 0.0043 \text{ mol} & = \quad 0.0057 \text{ mol} \end{array}$$

According to the equation for the reaction (equation 14.4), 1 mol H_2 reacts with 1 mol I_2 to form 2 mol HI. For each 0.0057 mol H_2 and I_2 reacted, 0.0114 mol HI should be formed:

$$0.0057 \text{ mol } H_2 \left(\frac{2 \text{ mol HI formed}}{1 \text{ mol } H_2 \text{ reacted}} \right) = 0.0114 \text{ mol HI formed}$$

The concentration of HI after 2000 s should be 0.0114 M. From Figure 14.1(a), [HI] at 2000 s is indeed 0.0114 M.

We can think about any volume that is convenient because concentration is an intensive property and therefore does not depend on volume.

Practice Problem 14.1 is an exercise in graph reading and stoichiometry. We find that our students need constant practice in both these skills.

14.1 (a) From Figure 14.1(c), what was the concentration of HI at the beginning of the experiment? The concentrations of H_2 and I_2? (b) From Figure 14.1(c), what is the concentration of HI after 6000 s? (c) How many moles of HI have reacted (in 1 L of solution) after 6000 s? (d) From the equation for the decomposition of HI,

$$2HI(g) \longrightarrow H_2(g) + I_2(g)$$

how many moles of H_2 and how many moles of I_2 should have formed in 6000 s (in 1 L of solution)? (e) From Figure 14.1(c), what was the concentration of H_2 and I_2 after 6000 s?

Equilibrium Constants and Equilibrium Constant Expressions

This research was carried out by Taylor to fulfill part of the requirements for a Ph.D.

The hydrogen–iodine–hydrogen iodide equilibrium was studied by Taylor and Crist who published their work in 1941.* A typical sample of the data they obtained is shown in Table 14.1. In experiments 1–4, Taylor and Crist started with a mixture of $H_2(g)$ and $I_2(g)$. In experiments 5 and 6, they began with HI(g). Their experiments were carried out at 425.4 °C because equilibrium is reached in a convenient time at this temperature. At room temperature, the rate of reaction of $H_2(g)$ and $I_2(g)$ and the rate of decomposition of HI(g) are extremely slow.

The data in Table 14.1 show that a number of different combinations of equilibrium concentrations were observed at 425.4 °C. The equilibrium concentrations depend on the initial concentrations; different initial concentrations give different equilibrium concentrations. *A set of equilibrium concentrations* is referred to as an **equilibrium position;** an infinite number of equilibrium positions is possible.

The highest value obtained was 54.75 and the lowest, 53.96. The difference between these values is 0.79. Thus, the first decimal place is uncertain.

However, the quantity, $[HI]_{eq}^2/([H_2]_{eq}[I_2]_{eq})$, in the far right column is constant to three significant figures. The median value for this constant from Table 14.1 is 54.5. This constant is called the **equilibrium constant** for the equilibrium

$$H_2(g) + I_2(g) \rightleftharpoons 2HI(g)$$

and is represented by the symbol $\boldsymbol{K_c}$. The subscript c indicates that concentrations in molarity were used to calculate the value of the constant.

The relationship

● Remember that square brackets indicate concentration in M.

$$K_c = \frac{[HI]_{eq}^2}{[H_2]_{eq}[I_2]_{eq}} \qquad (14.5)$$

TABLE 14.1	Equilibrium Between H_2(g), I_2(g), and HI(g) at 425.4 °C			
Experiment	$[H_2]_{eq}^a$	$[I_2]_{eq}$	$[HI]_{eq}$	$[HI]_{eq}^2/([H_2]_{eq}[I_2]_{eq})$
1	2.9070×10^{-3}	1.7069×10^{-3}	1.6482×10^{-2}	54.75
2	3.5600×10^{-3}	1.2500×10^{-3}	1.5588×10^{-2}	54.60
3	2.2523×10^{-3}	2.3360×10^{-3}	1.6850×10^{-2}	53.96
4	1.8313×10^{-3}	3.1292×10^{-3}	1.7671×10^{-2}	54.49
5	1.1409×10^{-3}	1.1409×10^{-3}	8.410×10^{-3}	54.34
6	4.953×10^{-4}	4.953×10^{-4}	3.655×10^{-3}	54.46

^{a}The subscript "eq" means equilibrium concentration.

*Taylor, A. H.; Crist, R. H. *J. Am. Chem. Soc.* **1941**, *63*, 1377–1386.

is called the **equilibrium constant expression** for the equilibrium. The square brackets stand for concentration in molarity, and the subscript "eq" indicates an equilibrium concentration. In writing equilibrium constant expressions, the "eq" subscripts are usually omitted because the concentration in equilibrium constant expressions *must* be equilibrium concentrations. The value of the equilibrium constant, 54.5, is the same for all of the infinite number of equilibrium positions possible for the H_2–I_2–HI system as long as the temperature is not changed. There is just one value for an equilibrium constant for a given reaction at a given temperature.

In principle, all chemical reactions are reversible, and all reactions have equilibrium constants. For the general reaction,

$$aA + bB + \cdots \rightleftarrows eE + fF + \cdots$$

where a is the coefficient of the reactant A, b is the coefficient of the reactant B, e is the coefficient of the product E, and so forth, the equilibrium constant expression is

$$K_c = \frac{[E]^e[F]^f \ldots}{[A]^a[B]^b \ldots}$$

The value of the equilibrium constant, K_c, is equal to the product of the equilibrium molarities of the products of reaction (right side of equation), each raised to a power given by the coefficient in the equation for the reaction, divided by the product of the equilibrium molarities of the reactants (left side of equation), each raised to a power given by the coefficient in the equation for the reaction. Sample Problem 14.1 shows how to write equilibrium constant expressions from the equations for equilibria.

The first person to write equilibrium constant expressions was the Dutch chemist van't Hoff who discovered the osmotic pressure law and proposed the three-dimensional nature of molecules.

The fraction $[E]^e[F]^f \ldots /[A]^a[B]^b \ldots$ is sometimes known as the mass action expression.

SAMPLE PROBLEM

14.1   Write the equilibrium constant expression for each of the following equilibria:

(a) $4NH_3(g) + 3O_2(g) \rightleftarrows 2N_2(g) + 6H_2O(g)$

(b) $HC_2H_3O_2(aq) \rightleftarrows H^+(aq) + C_2H_3O_2^-(aq)$

Solution   (a) The equilibrium is

$$4NH_3(g) + 3O_2(g) \rightleftarrows 2N_2(g) + 6H_2O(g)$$

Write the product of the equilibrium concentrations of the products, N_2 and H_2O, in the numerator and the product of the equilibrium concentrations of the reactants, NH_3 and O_2, in the denominator. The exponent of each concentration should be the same as the species' coefficients in the equation for the equilibrium:

$$K_c = \frac{[N_2]^2[H_2O]^6}{[NH_3]^4[O_2]^3}$$

(b) The equilibrium is

$$HC_2H_3O_2(aq) \rightleftarrows H^+(aq) + C_2H_3O_2^-(aq)$$

Write the product of the equilibrium concentrations of the products, H^+ and $C_2H_3O_2^-$, in the numerator and the equilibrium concentration of the single reactant, $HC_2H_3O_2$, in the denominator. Because the coefficients of all these species are ones, no exponents are written in the equilibrium constant expression. When no exponent is shown for a concentration in an equilibrium constant expression, an exponent of 1 is understood:

$$K_c = \frac{[H^+][C_2H_3O_2^-]}{[HC_2H_3O_2]}$$

The symbol K_a is introduced in Chapter 15, Acids and Bases.

14.2   Write the equilibrium constant expression for each of the following equilibria:

(a) $2SO_3(g) \rightleftharpoons 2SO_2(g) + O_2(g)$

(b) $Sn^{4+}(aq) + 2Fe^{2+}(aq) \rightleftharpoons Sn^{2+}(aq) + 2Fe^{3+}(aq)$

(c) $HNO_2(aq) \rightleftharpoons H^+(aq) + NO_2^-(aq)$

(d) $N_2(g) + 3H_2(g) \rightleftharpoons 2NH_3(g)$

(e) $3O_2(g) \rightleftharpoons 2O_3(g)$

Heterogeneous Equilibria

Many equilibria involve more than one phase. The equilibrium between calcium carbonate, calcium oxide, and carbon dioxide is an example:

$$CaCO_3(s) \rightleftharpoons CaO(s) + CO_2(g) \qquad (14.6)$$

Equilibria that involve more than one phase, like the calcium carbonate–calcium oxide–carbon dioxide equilibrium are called **heterogeneous equilibria.**

For the equilibrium shown in equation 14.6, the equilibrium constant expression found by experiment is simply

$$K_c = [CO_2]$$

The $CaCO_3$–CaO–CO_2 equilibrium is typical of heterogeneous equilibria. *Equilibrium constant expressions for heterogeneous equilibria do not involve the concentration of the pure solids (or pure liquids) present.*

Why don't equilibrium constant expressions for equilibria involving pure liquids or solids show the concentration of the pure liquid or solid? The concentrations of pure liquids and solids—that is, the quantities in a given volume or densities—are constant. Density is a characteristic property of a liquid or solid. The densities of pure liquids and solids do not vary much with changing temperature and hardly at all with pressure. When a chemical equilibrium involves a pure liquid or solid, the constant concentration of the pure liquid or solid is included in K_c. As long as some of each solid is present, the concentration of carbon dioxide gas in equilibrium with solid calcium carbonate and solid calcium oxide does not depend on how much solid calcium carbonate or how much solid calcium oxide is present (see ▪Figure 14.4).

▪ FIGURE 14.4   As long as some of each solid is present and temperature is constant, the concentration of $CO_2(g)$ in equilibrium with CaO(s) and $CaCO_3(s)$ is the same whether a lot of solid is present or only a little.

The concentration of carbon dioxide gas in equilibrium with solid calcium carbonate and solid calcium oxide depends only on the temperature.*

In dilute solutions, the concentration of the solvent is practically constant. For example, the molarity of water in 0.100 M NaCl is 55.3, and the molarity of water in 0.200 M NaCl is 55.2.[†] Like the constant concentrations of pure liquids and pure solids, the essentially constant concentration of the solvent in dilute solutions is included in K_c.

To Write an Equilibrium Constant Expression

1. Write an equation for the equilibrium.
2. Put the product of the equilibrium concentrations of the products in the numerator. Omit pure liquids, pure solids, and the solvents in dilute solutions (concentration around 0.1 M or less).
3. Write the product of the equilibrium concentrations of the reactants in the denominator. Omit pure liquids, pure solids, and the solvents in dilute solutions (concentration around 0.1 M or less).
4. The exponent of each concentration should be the same as the coefficient of the species in the equation. Values of 1 are understood, not written.

▶ Remember that "equations" must be balanced.

SAMPLE PROBLEMS

14.2 Write the equilibrium constant expression for the following equilibrium:

$$CaCO_3(s) + H_2O(l) + CO_2(aq) \rightleftharpoons Ca^{2+}(aq) + 2HCO_3^-(aq)$$

Solution The concentrations of $CaCO_3(s)$ and $H_2O(l)$ are constant and are included in K_c:

$$K_c = \frac{[Ca^{2+}][HCO_3^-]^2}{[CO_2]}$$

14.3 Which of the following can reach the equilibrium shown by the equation

$$CaCO_3(s) \rightleftharpoons CaO(s) + CO_2(g)$$

if placed in a closed container and allowed to stand for a sufficient length of time?
(a) Pure $CaCO_3(s)$ (b) Pure $CaO(s)$ (c) A mixture of $CaCO_3(s)$ and $CaO(s)$

---

*The situation is similar to the equilibrium between a liquid and its vapor. The equilibrium vapor pressure of a liquid does *not* depend on the volume of the vapor or on the volume or surface area of the liquid. Some liquid must be present for an equilibrium to exist, and the container must be closed. The equilibrium vapor pressure of a liquid is different at different temperatures.
[†]The molarity of water in 0.100 M NaCl is calculated as follows: The density of 0.100 M NaCl is 1.002 g/cm³ at 20 °C; one liter (1000 cm³) of solution has a mass of 1002 g and contains 0.100 mol NaCl. The mass of 0.100 mol NaCl is

$$0.100 \text{ mol NaCl} \left(\frac{58.4 \text{ g NaCl}}{\text{mol NaCl}} \right) = 5.84 \text{ g NaCl}$$

Therefore, the quantity of water in one liter of 0.100 M NaCl is

$$1002 \text{ g solution} - 5.84 \text{ g NaCl} = 996 \text{ g H}_2\text{O} \qquad \text{or}$$

$$996 \text{ g H}_2\text{O} \left(\frac{1 \text{ mol H}_2\text{O}}{18.0 \text{ g H}_2\text{O}} \right) = 55.3 \text{ mol H}_2\text{O}$$

The concentration of water in 0.100 M NaCl solution is 55.3 M.

(d) $CaCO_3(s)$ and a concentration of $CO_2(g)$ greater than the equilibrium concentration (e) $CaO(s)$ and a concentration of $CO_2(g)$ greater than the equilibrium concentration

Solution Think about each possibility individually.

(a) Pure $CaCO_3(s)$ can decompose

$$CaCO_3(s) \longrightarrow CaO(s) + CO_2(g)$$

If enough $CaCO_3$ is present at the start so that $[CO_2]$ can reach the equilibrium value before all the $CaCO_3$ has decomposed, equilibrium can be reached.

(b) Equilibrium cannot be reached starting with pure $CaO(s)$. Calcium carbonate contains carbon; there is no carbon in pure CaO.

(c) As in part (a), if enough $CaCO_3$ is used, equilibrium can be reached. The extra CaO will not make any difference (except to make the pile of solid at the bottom of the container larger).

(d) Because the $[CO_2]$ is already greater than the equilibrium concentration, no more $CaCO_3$ can decompose. No CaO is available to react with the excess CO_2. Nothing will happen. Equilibrium cannot be reached.

(e) If enough CaO is present so that $[CO_2]$ can reach the equilibrium value before all the CaO has reacted, equilibrium can be reached.

In summary, equilibrium can be reached in situations (a), (c), and (e).

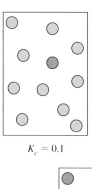
$K_c = 0.1$

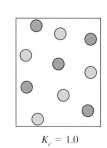
$K_c = 1.0$

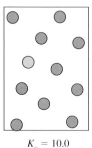
$K_c = 10.0$

For the equilibrium A $\rightleftarrows$ B, $K_c = \dfrac{[B]}{[A]}$

PRACTICE PROBLEMS

14.3 Classify each of the following equilibria as homogeneous or heterogeneous:

(a) $2SO_2(g) + O_2(g) \rightleftarrows 2SO_3(g)$

(b) $AgCl(s) \rightleftarrows Ag^+(aq) + Cl^-(aq)$

(c) $2Mg(s) + O_2(g) \rightleftarrows 2MgO(s)$

(d) $NH_3(aq) + H_2O(aq) \rightleftarrows NH_4^+(aq) + OH^-(aq)$

14.4 Write the equilibrium constant expression for each of the following equilibria:

(a) $(NH_4)_2CO_3(s) \rightleftarrows 2NH_3(g) + CO_2(g) + H_2O(g)$

(b) $AgCl(s) \rightleftarrows Ag^+(aq) + Cl^-(aq)$

(c) $2Mg(s) + O_2(g) \rightleftarrows 2MgO(s)$

(d) $O_2(g) + 2SH^-(aq) \rightleftarrows 2OH^-(aq) + 2S(s)$

14.5 Can the equilibrium in equation 14.6 be reached if $CaCO_3(s)$ is heated in an open container? Explain your answer.

Position of Equilibrium and Rate of Reaching Equilibrium

In equilibrium constant expressions, the products are written in the numerator so that the size of the equilibrium constant for an equilibrium shows how complete the reaction is. If the equilibrium constant is very large, reactants are, for all practical purposes, completely converted to products at equilibrium. The position of the equilibrium is said to lie to the right. The reaction between hydrogen gas and oxygen gas to form water vapor

$$2H_2(g) + O_2(g) \rightleftarrows 2H_2O(g)$$

is an example; the value of K_c for this reaction is 2.9×10^{81} at 25.0 °C. If the hydrogen and oxygen are mixed in the proportions shown by the equation—two moles of hydrogen for each mole of oxygen—the quantity of hydrogen and oxygen left at equilibrium is insignificant.

However, the combination of hydrogen and oxygen to form water illustrates a very important point about equilibrium constants. *There is no relationship between the magnitude of the equilibrium constant and the rate of reaction.* The reaction between hydrogen and oxygen is very slow. At room temperature, mixtures of hydrogen and oxygen can stand around for years without reacting.

If the equilibrium constant is very small, little reaction takes place even if reaction is fast and equilibrium is reached rapidly. Most of the starting materials remain unreacted at equilibrium. The position of the equilibrium is said to lie to the left. The ionization of hydrocyanic acid, HCN(aq), is an example of this type of reaction:

$$\text{HCN(aq)} \rightleftharpoons \text{H}^+\text{(aq)} + \text{CN}^-\text{(aq)} \qquad K_c = 5 \times 10^{-10} \text{ at } 25\ ^\circ\text{C}$$

If the equilibrium constant is neither very large nor very small—that is, its value is between 10^{-2} and 10^2—then the concentrations of products and reactants are of the same order of magnitude at equilibrium, and equilibrium is easily observed experimentally. The reaction between hydrogen and iodine to form hydrogen iodide, which has $K_c = 54.5$ at $425.4\ ^\circ\text{C}$, is an example of this type of reaction.

The value of the equilibrium constant depends on the particular equilibrium and the temperature. Different equilibria have different equilibrium constants as you can see from the examples discussed thus far. The equilibrium constant for a reaction usually changes when temperature changes. For example, the equilibrium constant for the hydrogen–iodine–hydrogen iodide equilibrium is 54.5 at $425.4\ ^\circ\text{C}$ and 45.6 at $490.6\ ^\circ\text{C}$. *Chemical reactions tend to go to equilibrium providing that reaction takes place at a significant rate.*

 Mixtures of H_2(g) and O_2(g) are very dangerous because, if ignited by a spark, they react so rapidly that an explosion takes place. Many reactions between gases are slow at low temperatures and very fast at high temperatures and are difficult to control.

PRACTICE PROBLEM

14.6 Of the following reactions, (a) which goes practically to completion at equilibrium? (b) Which essentially does not take place? (c) Which go to an easily detectable equilibrium providing that reaction takes place at a practical rate?

   (i) $\text{N}_2\text{(g)} + \text{O}_2\text{(g)} \rightleftharpoons 2\text{NO(g)}$ $K_c = 1 \times 10^{-30}$

   (ii) $2\text{NH}_3\text{(g)} \rightleftharpoons \text{N}_2\text{(g)} + 3\text{H}_2\text{(g)}$ $K_c = 9.5$

   (iii) $\text{CO}_2\text{(g)} + \text{H}_2\text{(g)} \rightleftharpoons \text{CO(g)} + \text{H}_2\text{O(g)}$ $K_c = 0.11$

   (iv) $\text{H}_2\text{(g)} + \text{Cl}_2\text{(g)} \rightleftharpoons 2\text{HCl(g)}$ $K_c = 1.33 \times 10^{34}$

14.3 DETERMINATION OF VALUES OF EQUILIBRIUM CONSTANTS

Concentration Equilibrium Constants

The numerical values of equilibrium constants can be found by experiment. If the concentrations of all reactants and products at equilibrium are known, the numerical value of an equilibrium constant can be calculated by substituting the equilibrium concentrations in the equilibrium constant expression. Sample Problem 14.4 shows the calculation of the numerical value of the equilibrium constant for the equilibrium

$$\text{H}_2\text{(g)} + \text{I}_2\text{(g)} \rightleftharpoons 2\text{HI(g)}$$

from the data from experiment 1 in Table 14.1.

The hydrogen–iodine–hydrogen iodide equilibrium is used repeatedly in this chapter so that students can focus their attention on equilibrium concepts without being distracted by a variety of examples. Real data are used throughout.

14.4 Use the data from experiment 1 in Table 14.1 to calculate the numerical value of the equilibrium constant K_c for the equilibrium

$$H_2(g) + I_2(g) \rightleftharpoons 2HI(g) \qquad \text{at 425.4 °C}$$

Solution For this equilibrium, the equilibrium constant expression is

$$K_c = \frac{[HI]^2}{[H_2][I_2]}$$

From Table 14.1, the equilibrium concentrations are: $[H_2] = 2.9070 \times 10^{-3}$, $[I_2] = 1.7069 \times 10^{-3}$, and $[HI] = 1.6482 \times 10^{-2}$. Substitution of the equilibrium concentrations in the equilibrium constant expression gives

$$K_c = \frac{(1.6482 \times 10^{-2})^2}{(2.9070 \times 10^{-3})(1.7069 \times 10^{-3})} = 54.748*$$

From experiment 1, the numerical value of the equilibrium constant at 425.4 °C is 54.748.

14.7 What is the numerical value of K_c for the equilibrium

$$N_2O_4(g) \rightleftharpoons 2NO_2(g)$$

at 100 °C if $[N_2O_4] = 1.40 \times 10^{-3}$ and $[NO_2] = 1.72 \times 10^{-2}$ at equilibrium?

Determination of the concentrations of all reactants and products is not necessary. If the concentrations of the reactants at the beginning of the experiment are known, the concentrations of all the components of the equilibrium mixture can be calculated from the equilibrium concentration of any one component and the equation for the reaction, as shown in Sample Problem 14.5. Although the calculations are more complicated, the experimental work is much simpler. The *concentrations at the beginning of the experiment* are called the **initial concentrations, [ ]₀.**

14.5 A mixture that was initially 0.005 00 M in $H_2(g)$ and 0.012 50 M in $I_2(g)$ and contained no HI(g) was heated at 425.4 °C until equilibrium was reached. The equilibrium concentration of $I_2(g)$ was found by experiment to be 0.007 72 M. What is the numerical value of K_c for the following equilibrium?

$$H_2(g) + I_2(g) \rightleftharpoons 2HI(g)$$

Both HI(g) and H₂(g) are colorless; I₂(g) is purple. The concentration of I₂(g) can be determined easily and continuously by measuring the purple color quantitatively.

*Units are not usually given for equilibrium constants because the more accurate ways of treating equilibrium constants used in thermodynamics and physical chemistry define equilibrium constants in terms of activities, not molarities. Activities, which are numerically equal to effective concentrations corrected for nonideal behavior, are dimensionless numbers; that is, they have no units. For dilute solutions (0.1 M and lower), use of molarities usually gives accurate enough results for general chemistry (within 5% of the results obtained by more accurate but more complicated methods). Pure liquids and solids can be disregarded when writing equilibrium constant expressions because their activities are one.

Solution *The concentrations in K_c must be equilibrium concentrations.* Therefore, the first step is to calculate the equilibrium concentrations of $H_2(g)$ and $HI(g)$ from the initial concentrations and the equilibrium concentration of $I_2(g)$, which are given in the problem; this is a *stoichiometry problem*. Once we have the equilibrium concentrations of all species, the *equilibrium problem* is just like the one in Sample Problem 14.4.

Divide and conquer applies to chemistry too!

In a relatively complicated problem like this one, a summary table helps to keep information organized and to avoid careless mistakes. You should make the summary table when you begin solving the problem so that you can easily see what information is available and what is needed and so that you have a place to record information as you obtain it. Use the equation for the equilibrium as column headings for the summary table. The summary table for this problem is given below with the information provided shown in boldface:

	$H_2(g)$	+	$I_2(g)$	$\rightleftarrows$	$2HI(g)$
Initial concentration, M	**0.005 00**		**0.012 50**		**0.000 00**
Change in concentration due to reaction, ΔM					
Equilibrium concentration, M			**0.007 72**		

▶ Notice that the first letters of the key words in the three lines of the summary table spell "ICE."

Step 1. *Calculate the equilibrium concentrations of $H_2(g)$ and $HI(g)$.* For convenience, consider a sample that has a volume of exactly one liter. In a 1-L sample, the number of moles of each solute is numerically equal to the molarity of the solute, which simplifies the calculations. The initial concentration of I_2 was 0.012 50 M; a 1-L sample contained 0.012 50 mol $I_2(g)$. The equilibrium concentration of $I_2(g)$ is given in the problem and is 0.007 72 M. At equilibrium, 0.007 72 mol I_2 was left in the one-liter sample. The number of moles of I_2 that reacted equals the difference between the number of moles present at the beginning and the number present at equilibrium:

The person doing the experiment chooses the initial concentrations. The equation for the equilibrium and the value of the equilibrium constant control the changes in concentration and the equilibrium concentrations.

$$\text{moles } I_2 \text{ present at start} \; - \; \begin{array}{c}\text{moles } I_2 \text{ present at}\\ \text{equilibrium}\end{array} = \text{moles } I_2 \text{ reacted}$$

$$0.012\,50 \text{ mol } I_2 \; - \; 0.007\,72 \text{ mol } I_2 \; = \; 0.004\,78 \text{ mol } I_2 \text{ reacted}$$

Because we are considering a 1-L sample, the number of moles is the same numerically as the molarity. The change in molarity of I_2 is $-0.004\,78$ M ①. The sign is negative because iodine is used up; the concentration of iodine at equilibrium is less than the initial concentration of iodine.

▶ The numbers in circles, such as the ① after $-0.004\,78$ M, are keys to the position of the quantity in the completed summary table.

According to the equation for the equilibrium, 1 mol I_2 reacts with 1 mol H_2 to form 2 mol HI. The reaction of 0.004 78 mol I_2 will use up 0.004 78 mol H_2; the change in molarity of hydrogen is $-0.004\,78$ M ②. The equilibrium concentration of H_2 is equal to the initial concentration of H_2 plus the change in concentration of H_2:

$$\begin{array}{ccccc}\begin{array}{c}\text{initial concentration}\\ \text{of } H_2\end{array} & + & \begin{array}{c}\text{change in concentration}\\ \text{of } H_2\end{array} & = & \begin{array}{c}\text{equilibrium concentration}\\ \text{of } H_2\end{array}\\ 0.005\,00 & + & (-0.004\,78) & = & 0.000\,22 \; ③\end{array}$$

From the equation for the equilibrium, twice as many moles of HI will be formed as moles of H_2 and I_2 are used up. Therefore,

$$0.004\,78 \text{ mol } I_2 \left(\frac{2 \text{ mol HI}}{1 \text{ mol } I_2} \right) = 0.009\,56 \text{ mol HI}$$

will be formed. The change in concentration of HI is $+0.009\,56$ M ④. The sign is positive because hydrogen iodide is formed. The concentration of hydrogen iodide at equilibrium is greater than the initial concentration of hydrogen iodide. The equilibrium concentration of HI is equal to the initial concentration of HI plus the change in concentration of HI:

initial concentration of HI		change in concentration of HI		equilibrium concentration of HI
	+		=	
0.000 00	+	(+0.009 56)	=	0.009 56 ⑤

The completed summary table is shown below:

	$H_2(g)$	+	$I_2(g)$	$\rightleftharpoons$	$2HI(g)$
Initial concentration, M	**0.005 00**		**0.012 50**		**0.000 00**
Change in concentration due to reaction, ΔM	−0.004 78 ②		−0.004 78 ①		+0.009 56 ④
Equilibrium concentration, M	0.000 22 ③		**0.007 72**		0.009 56 ⑤

⬤ Remember that the information given in the problem is in boldface and the calculated quantities are numbered to correspond with their calculation.

⬤ The quantities in the change of concentration line are always in the same relationship to each other as the coefficients in the equation; in this example, $1:1:2$.

⬤ The answer is only expressed to two significant figures because the equilibrium concentration of H_2 is only known to two significant figures.

Step 2. *Calculation of value of K_c from equilibrium concentrations.* Substitution of the equilibrium concentrations in the equilibrium constant expression gives

$$K_c = \frac{[HI]^2}{[H_2][I_2]} = \frac{(0.009\ 56)^2}{(0.000\ 22)(0.007\ 72)} = 54$$

PRACTICE PROBLEM

14.8 In an experiment starting with $[N_2O_4]_0 = 0.020\ 00$ and $[NO_2]_0 = 0.000\ 00$, $[N_2O_4]_{eq}$ was found to be 0.004 52. (a) What is $[NO_2]_{eq}$? (b) From this experiment, what is the numerical value of K_c for the equilibrium

$$N_2O_4(g) \rightleftharpoons 2NO_2(g)$$

Partial Pressure Equilibrium Constants

At constant temperature, the pressure of a gas is proportional to its molarity:

$$P = (RT)M*$$

For equilibria that involve gases, partial pressures can be used instead of concentrations in equilibrium constant expressions. For the equilibrium

$$aA(g) + bB(g) + \cdots \rightleftharpoons eE(g) + fF(g) + \cdots$$

the quotient

$$\frac{p_E{}^e p_F{}^f \cdots}{p_A{}^a p_B{}^b \cdots}$$

is a constant when the pressures are equilibrium partial pressures. If partial pressures are in atmospheres, the constant is symbolized by K_p:

$$K_p = \frac{p_E{}^e p_F{}^f \cdots}{p_A{}^a p_B{}^b \cdots}$$

In general, $K_p \neq K_c$. Instead

$$K_p = K_c(RT)^{\Delta n_g} \qquad (14.7)$$

---

*This relationship was derived from the ideal gas equation in the discussion of osmotic pressure (Section 13.7).

where Δn_g is the number of moles of gaseous products minus the number of moles of gaseous reactants:

$$\Delta n_g = (e + f + \cdots) - (a + b + \cdots)$$

If $\Delta n_g = 0$, that is, the number of moles of products is the same as the number of moles of reactants, then $K_p = K_c$. The equilibrium

$$H_2(g) + I_2(g) \rightleftharpoons 2HI(g)$$

is an example of a reaction where there is no change in the number of moles of gas.

▶ Remember that any number n raised to the zeroth power equals one, $n^0 = 1$.

SAMPLE PROBLEM

14.6   For the equilibrium

$$NH_4HS(s) \rightleftharpoons NH_3(g) + H_2S(g)$$

$K_c = 1.81 \times 10^{-4}$ at 25 °C, (a) what is the value of K_p? (b) Write the expression for K_p.

Solution   (a) Equation 14.7 gives the relationship between K_p and K_c:

$$K_p = K_c(RT)^{\Delta n_g}$$

For the reaction in this problem, the change in the number of moles of gas, Δn_g, is 2:

$$\Delta n_g = 2 - 0 = 2$$

If pressure is to be expressed in atmospheres, the value 0.082 06 atm·L·mol^{-1}·K^{-1} must be used for R because concentration is expressed in molarity (mol·L^{-1}). For this equilibrium, the concentration equilibrium constant expression is

$$K_c = [NH_3][H_2S]$$

The units of K_c, mol^2·L^{-2}, must be used for units to cancel. The temperature in K is

$$(25 + 273) = 298 \text{ K}$$

Substitution of these values in equation 14.7 gives

$$K_p = \left(1.81 \times 10^{-4} \frac{mol^2}{L^2}\right)\left[\left(0.082\ 06 \frac{atm \cdot L}{mol \cdot K}\right)(298\ K)\right]^2$$

$$= 1.08 \times 10^{-1} \text{ atm}^2 \text{ or } 0.108 \text{ atm}^2$$

The value of K_p is 0.108.
   (b) $K_p = p_{NH_3} \cdot p_{H_2S}$

PRACTICE PROBLEMS

14.9   For the equilibrium

$$2H_2(g) + S_2(g) \rightleftharpoons 2H_2S(g)$$

$K_c = 1.10 \times 10^7$ at 700 °C (zeros are significant). What is the value of K_p?

14.10   Write the expressions for K_p for the following equilibria:
(a) $2H_2(g) + S_2(g) \rightleftharpoons 2H_2S(g)$
(b) $CaCO_3(s) \rightleftharpoons CaO(s) + CO_2(g)$

Relationship Between the Equation for an Equilibrium and the Value of the Equilibrium Constant

The equation for the equilibrium must be written whenever an equilibrium constant expression is written. You cannot simply say, "The equilibrium constant for the equilibrium between hydrogen gas, iodine vapor, and hydrogen iodide gas is 54.5 at 425.4 °C." The equilibrium between hydrogen gas, iodine vapor, and hydrogen iodide gas can be written either

$$H_2(g) + I_2(g) \rightleftharpoons 2HI(g) \qquad \text{or} \qquad 2HI(g) \rightleftharpoons H_2(g) + I_2(g)$$

For the equation on the left, the equilibrium constant expression is

$$K_c = \frac{[HI]^2}{[H_2][I_2]} = 54.5$$

For the equation on the right, the equilibrium constant expression is

$$K_c' = \frac{[H_2][I_2]}{[HI]^2} = \frac{1}{K_c} = \frac{1}{54.5} = 0.0183 = 1.83 \times 10^{-2}$$

When an equation is written in the opposite direction, the value of the new equilibrium constant is the reciprocal of the original value.

In the equation for an equilibrium, the species on the left side of the double arrows are, by definition, the reactants. The species on the right side of the arrows are, by definition, the products. For the equilibrium

$$H_2(g) + I_2(g) \rightleftharpoons 2HI(g)$$

hydrogen iodide is considered the product whether an experiment begins with a mixture of hydrogen and iodine or with pure hydrogen iodide.

The equation for the hydrogen–iodine–hydrogen iodide equilibrium may also be written

$$\tfrac{1}{2}H_2(g) + \tfrac{1}{2}I_2(g) \rightleftharpoons HI(g)$$

For this equilibrium,

$$K_c'' = \frac{[HI]}{[H_2]^{\frac{1}{2}}[I_2]^{\frac{1}{2}}} = \sqrt{K_c} \qquad \text{or} \qquad (K_c)^{\frac{1}{2}} = 7.38$$

In general, when the equation for an equilibrium is multiplied by a factor, the equilibrium constant must be raised to a power equal to the factor.

PRACTICE PROBLEM

14.12 For the equilibrium

$$2H_2(g) + S_2(g) \rightleftharpoons 2H_2S(g)$$

$K_c = 1.105 \times 10^7$ at 700 °C. (a) What is the value of K_c for the equilibrium $2H_2S(g) \rightleftharpoons 2H_2(g) + S_2(g)$? (b) What is the value of K_c for the equilibrium $H_2(g) + \tfrac{1}{2}S_2(g) \rightleftharpoons H_2S(g)$?

CALCULATIONS INVOLVING EQUILIBRIUM CONSTANTS

Once the value of an equilibrium constant has been determined, the results of any experiments carried out at the same temperature can be predicted. Doing calculations is a lot less work than doing experiments.

Predicting the Direction of Reaction

For the general reaction,

$$aA + bB + \cdots \rightleftharpoons eE + fF + \cdots$$

the *product of the concentrations of the products divided by the product of the concentrations of the reactants, each raised to a power equal to the coefficient of the species in the equation for the equilibrium,* is called the **reaction quotient, Q.**

$$Q = \frac{[E]^e[F]^f \cdots}{[A]^a[B]^b \cdots}$$

In the Equilibrium Constants module of the Equilibrium section animation is used to reinforce the concepts of reaction quotients and equilibrium constants. Students can work ICE problems of various types and complexity.

The quotient in the reaction quotient has the same form as the quotient in the equilibrium constant expression. However, the concentrations in the reaction quotient can be *any* concentrations; the concentrations in the equilibrium constant expression must be equilibrium concentrations.

Comparison of the number obtained by substitution of any experimental set of concentrations in the reaction quotient with the equilibrium constant tells whether a system is at equilibrium. If $Q = K_c$, the system is at equilibrium. For example, consider a mixture in which $[H_2] = 3.62 \times 10^{-2}$, $[I_2] = 5.49 \times 10^{-3}$, and $[HI] = 1.041 \times 10^{-1}$ at 425.4 °C. Substitution of these concentrations in the reaction quotient

$$Q = \frac{[HI]^2}{[H_2][I_2]}$$

for the equilibrium

$$H_2(g) + I_2(g) \rightleftharpoons 2HI(g) \quad \text{gives}$$

$$Q = \frac{(1.041 \times 10^{-1})^2}{(3.62 \times 10^{-2})(5.49 \times 10^{-3})} = 54.5$$

The value of K_c for this equilibrium is 54.5 at 425.4 °C. The value of Q for the experiment is equal to the value of K_c; therefore, the system is at equilibrium, and no net change will take place.

A value of Q for an experiment that is smaller than the value of the equilibrium constant shows that the concentrations of the products are too low and the concentrations of the reactants are too high for the system to be at equilibrium. Net reaction will take place in the forward direction (from left to right). The concentrations of the products will increase, and the concentrations of the reactants will decrease until Q becomes equal to K_c. The system will then have reached equilibrium, and no further change will take place.

A value of Q for an experiment that is larger than the value of the equilibrium constant shows that the concentrations of the products are too high and the concentrations of the reactants are too low for the system to be at equilibrium. Net reaction will take place in the reverse direction (from right to left). The concentrations of the products will decrease, and the concentrations of the reactants will increase until Q becomes equal to K_c. The system will then have reached equilibrium, and no further change will take place.

$Q < K_c$ Net forward reaction will take place.

$Q = K_c$ No net change will take place (system is at equilibrium).

$Q > K_c$ Net reverse reaction will take place.

Sample Problem 14.7 shows the use of Q to predict the direction of a chemical reaction.

14.7 Is a mixture in which $[H_2] = 7.69 \times 10^{-2}$, $[I_2] = 3.45 \times 10^{-4}$, and $[HI] = 5.21 \times 10^{-3}$ at 425.4 °C at equilibrium, or will reaction take place to reach equilibrium? If reaction will take place, in which direction will reaction take place? The value of K_c for the equilibrium

$$H_2(g) + I_2(g) \rightleftharpoons 2HI(g)$$

is 54.5 at 425.4 °C.

Solution   First, write the reaction quotient. Then, substitute the concentrations given in the reaction quotient, calculate the value of Q, and compare the value of Q with K_c. The reaction quotient for this equilibrium is

$$Q = \frac{[HI]^2}{[H_2][I_2]}$$

Substitution of the concentrations given in the reaction quotient gives

$$Q = \frac{(5.21 \times 10^{-3})^2}{(7.69 \times 10^{-2})(3.45 \times 10^{-4})} = 1.02$$

Because 1.02 is less than 54.5, net reaction will take place in the forward direction. Net reaction in the forward direction increases [HI] and decreases $[H_2]$ and $[I_2]$, making Q larger. Net reaction in the forward direction will take place until Q becomes equal to K_c. The system will then be at equilibrium.

14.13   Use the reaction quotient to determine whether each of the following mixtures is at equilibrium at 425.4 °C. (Refer to Sample Problem 14.7 for the equation for the equilibrium and for the value of K_c.) If reaction is predicted, tell in which direction reaction will take place.

(a) $[H_2] = 8.48 \times 10^{-3}$, $[I_2] = 9.91 \times 10^{-4}$, and $[HI] = 2.57 \times 10^{-2}$

(b) $[H_2] = 6.42 \times 10^{-5}$, $[I_2] = 5.36 \times 10^{-2}$, and $[HI] = 1.369 \times 10^{-2}$

Calculating Equilibrium Concentrations from K_c

In Chapters 3 and 4, you learned how to solve stoichiometry problems such as "If a mixture of 3.50 g of hydrogen and 26.0 g of oxygen is made to react to form water, how many grams of water will be formed?" The methods you learned in Chapters 3 and 4 can only be used if the reaction goes to completion ($K_c \approx \infty$) as the reaction between hydrogen and oxygen to form water does. If a reaction reaches equilibrium before all the reactants have been used up, as the reaction between hydrogen and iodine to form hydrogen iodide does, the equilibrium constant must be used to calculate how much product will be formed.

Usually, only the initial concentrations and the value of K_c are known. The general method of calculating equilibrium concentrations from initial concentrations and the value of the equilibrium constant is to let x equal one of the unknown equilibrium concentrations. All other equilibrium concentrations are then expressed in terms of x. The equilibrium concentrations in terms of x are substituted in the equilibrium constant expression to give an equation in one unknown. The equation is solved for x, and the equilibrium concentrations calculated from the value found for x.

Calculations are checked by substituting the answers in the equilibrium constant expression and comparing the result with the equilibrium constant. Sample Problems 14.8 and 14.9 illustrate the application of this method to some common types of problems.

SAMPLE PROBLEM

14.8   A sample of $COCl_2$ is allowed to decompose. The value of K_c for the equilibrium

$$COCl_2(g) \rightleftarrows CO(g) + Cl_2(g)$$

is 2.2×10^{-10} at 100 °C. If the initial concentration of $COCl_2$ is 0.095 M, what will be the equilibrium concentrations of $COCl_2$, CO, and Cl_2?

Solution   The problem says "A sample of $COCl_2$ is allowed to decompose." It does not say anything about CO or Cl_2 being present at the start. *When a problem does not say that a species is present, you should assume that it is not.* In the solution of this problem, the initial concentrations of CO and Cl_2 are assumed to be zero.

Suppose we let $x = [CO]_{eq}$ and think about 1 L of solution so that the number of moles and concentration are numerically equal. According to the equation for the equilibrium, when 1 mol CO is formed, 1 mol Cl_2 is also formed. Therefore, $[Cl_2]_{eq} = x$ too. To form x mol CO and x mol Cl_2, x mol $COCl_2$ must decompose, and $[COCl_2]_{eq} = (0.095 - x)$. The summary table looks like this:

	$COCl_2(g)$	$\rightleftarrows$ $CO(g)$ +	$Cl_2(g)$
Initial concentration, M	**0.095**	**0.000**	**0.000**
Change in concentration due to reaction, ΔM	$-x$	$+x$	$+x$
Equilibrium concentration, M	$(0.095 - x)$	x	x

The equilibrium constant expression is

$$K_c = \frac{[CO]_{eq}[Cl_2]_{eq}}{[COCl_2]_{eq}} = 2.2 \times 10^{-10}$$

Substitution of the equilibrium concentrations in terms of x in the equilibrium constant expression gives

$$\frac{x^2}{(0.095 - x)} = 2.2 \times 10^{-10} \qquad (14.8)$$

An equation in one unknown can be solved for the unknown. Equation 14.8 is a second-order equation and can be solved by means of the quadratic formula. However, we can avoid having to use the quadratic formula by noting that, because the value of K_c is very small, not much of the $COCl_2$ is decomposed at equilibrium. If we assume that x is negligible compared to 0.095, equation 14.8 simplifies to

$$\frac{x^2}{0.095} = 2.2 \times 10^{-10}; \qquad \text{rearrangement gives}$$

$$x^2 = (0.095)(2.2 \times 10^{-10}) \qquad \text{or} \qquad x^2 = 0.21 \times 10^{-10} \qquad (14.9)$$

⊙ If no initial concentration is given for a species, assume that the species is absent and its initial concentration is zero.

⊙ Remember that because molarity is an intensive property, we can think about any size sample we choose.

For a review of how to take the square root of an exponential number, see Appendix B.2.

Always check to be sure your work is reasonable.

We can find x by taking the square root of both sides of equation 14.9:

$$x = \pm 0.46 \times 10^{-5} \quad \text{or} \quad \pm 4.6 \times 10^{-6}$$

The root, -4.6×10^{-6}, is physically impossible. A concentration can't be negative. The answer is $x = +4.6 \times 10^{-6}$

Next, we must check our assumption that x is negligible compared with 0.095:

$$[COCl_2]_{eq} = (0.095 - x) = [0.095 - (4.6 \times 10^{-6})] = 0.095$$

Subtracting 4.6×10^{-6} from 0.095 doesn't make any difference in the value of $[COCl_2]_{eq}$. The assumption that x is negligible compared with 0.095 is correct. Now we can use the value of x to find the other two equilibrium concentrations:

$$[CO]_{eq} = [Cl_2]_{eq} = x = 4.6 \times 10^{-6}$$

Finally, we should check our answers by substitution of the values found in the equilibrium constant expression:

$$\frac{(4.6 \times 10^{-6})^2}{0.095} = 2.2 \times 10^{-10}$$

The value given for K_c in the problem is 2.2×10^{-10}. Our answers are correct.

PRACTICE PROBLEM

14.14   The value of K_c for the equilibrium

$$COCl_2(g) \rightleftharpoons CO(g) + Cl_2(g)$$

is 2.2×10^{-10} at 100 °C. If the initial concentration of $COCl_2$ is 0.074 M, what will be the equilibrium concentrations of $COCl_2$, CO, and Cl_2?

SAMPLE PROBLEM

14.9   The value of K_c for the equilibrium

$$N_2O_4(g) \rightleftharpoons 2NO_2(g)$$

is 4.64×10^{-3} at 25 °C. If the initial concentration of N_2O_4 is 0.0367 M and the initial concentration of NO_2 is 0.0000, what will be the concentrations of both gases at equilibrium?

Solution   Follow a procedure similar to the one used in Sample Problem 14.8. The equilibrium constant expression is

$$K_c = \frac{[NO_2]_{eq}^2}{[N_2O_4]_{eq}} = 4.64 \times 10^{-3}$$

Sample Problem 14.9 shows what to do if the value of K is not very low.

If we think about a 1-L sample and let $x =$ mol N_2O_4 that have decomposed when the system reaches equilibrium, then $2x =$ mol NO_2 that have formed when the system reaches equilibrium. Thus $[NO_2]_{eq} = 2x$, and $[N_2O_4]_{eq} = 0.0367 - x$. The summary table is

Using x for a species that has a coefficient of 1 in the equation for an equilibrium usually simplifies the algebra.

	$N_2O_4(g)$	$\rightleftharpoons$	$2NO_2(g)$
Initial concentration, M	**0.0367**		**0.0000**
Change in concentration due to reaction, ΔM	$-x$		$+2x$
Equilibrium concentration, M	$(0.0367 - x)$		$2x$

Substitution of the equilibrium concentrations in terms of x in the equilibrium constant expression gives

$$\frac{(2x)^2}{(0.0367 - x)} = 4.64 \times 10^{-3} \qquad (14.10)$$

If we assume that x is negligible compared with 0.0367, equation 14.10 simplifies to

$$\frac{(2x)^2}{(0.0367)} = 4.64 \times 10^{-3}; \qquad \text{rearranging}$$

$$4x^2 = (0.0367)(4.64 \times 10^{-3}),$$

or

$$x^2 = \frac{(0.0367)(4.64 \times 10^{-3})}{4} = 4.26 \times 10^{-5}$$

and

$$x = 6.53 \times 10^{-3} \quad \text{or} \quad 0.006\ 53$$

In this case, the value found for x is not negligible compared with 0.0367:

$$0.0367 - 0.006\ 53 = 0.0302 \neq 0.0367$$

The value found for x, 0.006 53, is almost 20% of 0.0367:

$$\frac{0.006\ 53}{0.0367} \times 100 = 17.8\%$$

What do we do now? Make a second approximation. Instead of assuming that x is 0 compared with 0.0367 and $[N_2O_4]_{eq} = 0.0367$, use the value of x obtained by the first approximation to calculate a better value for $[N_2O_4]_{eq}$:

$$[N_2O_4]_{eq} = 0.0367 - 0.006\ 53 = 0.0302$$

Repeat the calculation using this better value for $[N_2O_4]_{eq}$:

$$\frac{4x^2}{0.0302} = 4.64 \times 10^{-3}$$

$$x^2 = \frac{0.0302(4.64 \times 10^{-3})}{4} = 3.50 \times 10^{-5} \qquad \text{and} \qquad x = 5.92 \times 10^{-3}$$

Continue repeating the calculation, each time using the improved value for x obtained from the previous approximation, until the answer does not differ significantly from the one obtained by the previous calculation. In this problem, the third approximation gives $x = 5.98 \times 10^{-3}$ and the fourth approximation gives $x = 5.97 \times 10^{-3}$. Because the last digit is uncertain, the value of x obtained by the fourth approximation is the same as the value obtained by the third approximation. No further approximations are necessary; we have obtained as good a value for x as possible and are ready to use it to calculate the equilibrium concentrations:

$$[N_2O_4]_{eq} = (0.0367 - x) = (0.0367 - 0.005\ 97) = 0.0307$$

$$[NO_2]_{eq} = 2x = 2(0.005\ 97) = 0.011\ 94$$

Check:

$$\frac{(0.011\ 94)^2}{0.0307} = 4.64 \times 10^{-3}$$

This is the value given in the problem for K_c. The answer is correct.

The answer could, of course, be obtained by using the quadratic formula. Equation 14.10

$$\frac{(2x)^2}{(0.0367 - x)} = 4.64 \times 10^{-3}$$

The assumption that x is negligible compared with a number is often correct if x is added to or subtracted from the number and is usually worth trying. The assumption that x is negligible compared with a number must not be made if x is to be multiplied or divided by the number.

A graphing calculator provides another way to solve equations. (See Ruch, D. K.; Chasteen, T. G. J. Chem. Educ. 1993, 70, A184–A185.)

Quadratic equations and their solution are reviewed in Appendix B.4.

must first be put in the standard form for a quadratic equation, $ax^2 + bx + c = 0$.

Multiplication of both sides of equation 14.10 by $(0.0367 - x)$ gives

$$(2x)^2 = (0.0367 - x)(4.64 \times 10^{-3}) = (1.70 \times 10^{-4}) - (4.64 \times 10^{-3})x$$

Rearrangement gives

$$4x^2 + (4.64 \times 10^{-3})x - (1.70 \times 10^{-4}) = 0 \qquad (14.11)$$

Substitution of the coefficients from equation 14.11 in the quadratic formula

$$x = \frac{-b \pm \sqrt{b^2 - 4ac}}{2a} \qquad \text{gives}$$

$$x = \frac{-(4.64 \times 10^{-3}) \pm \sqrt{(4.64 \times 10^{-3})^2 - 4(4)(-1.70 \times 10^{-4})}}{2(4)}$$

The two roots of equation 14.11 are

$$x = 5.96 \times 10^{-3} \qquad \text{and} \qquad x = -7.12 \times 10^{-3}$$

The negative answer is impossible because $[NO_2]_{eq}$, which is equal to $2x$, can't be negative. The answer is

$$x = 5.96 \times 10^{-3}$$

If you do not use a programmable calculator to solve quadratic equations, be especially careful not to round off until you reach the final answer or significant figures may be lost.

The first method used to solve Sample Problem 14.9 is called the **method of successive approximations.** The number of approximations that must be made depends on how good the first guess is. If the value of the equilibrium constant is close to 1, a first guess that x is negligible will not be very good. For second-order equations, use of the quadratic formula will be simpler than successive approximations *if you use a programmable calculator to solve the quadratic equation.* However, if you must solve the quadratic equation by hand, once you have practiced using successive approximations, you will probably find that successive approximations is faster and more accurate. An approximation method is usually used to solve third-order and higher equations. Fortunately a computer can be programmed to try various approximations until a satisfactory answer is obtained.

Approximation methods are very useful for solving complex problems. For example, approximations must be used to solve the Schrödinger equation for atoms and molecules with more than one electron. The ability to use appropriate approximations to solve problems is an important skill that you should practice.

PRACTICE PROBLEM

14.15 The value of K_c for the equilibrium

$$PCl_5(g) \rightleftharpoons PCl_3(g) + Cl_2(g)$$

is 1.21×10^{-2} at 500 K. If the initial concentration of PCl_5 is 0.0405 M and the initial concentrations of PCl_3 and Cl_2 are 0, what will be the concentrations of all gases at equilibrium?

Calculations involving K_p are very similar to calculations involving K_c. Sample Problem 14.10 illustrates a K_p calculation in which the total pressure is given. Total pressure is often much easier to measure than partial pressures.

14.10 For the equilibrium

$$N_2O_4(g) \rightleftharpoons 2NO_2(g)$$

$K_p = 0.113$ at 25 °C. If the total pressure at equilibrium $= 2.00$ atm, what are the partial pressures of $N_2O_4(g)$ and $NO_2(g)$ at equilibrium?

Solution The equilibrium constant expression is

$$K_p = \frac{p_{NO_2}^2}{p_{N_2O_4}} = 0.113$$

Let $x = p_{N_2O_4}$ at equilibrium. According to Dalton's law of partial pressures, the total pressure of a mixture of gases is equal to the sum of the partial pressures of the individual gases in the mixture (Section 5.8). Therefore,

$$x + p_{NO_2} = 2.00 \text{ atm}; \quad \text{and rearranging,} \quad p_{NO_2} = (2.00 - x)$$

Substitution of the equilibrium partial pressures in terms of x in the equilibrium constant expression gives

$$\frac{(2.00 - x)^2}{x} = 0.113$$

Two positive roots are obtained from the quadratic formula, $x = 2.54$ and $x = 1.58$. However, the former is impossible because $p_{NO_2} = (2.00 - x)$ would be negative if x were equal to 2.54. The value of x must be greater than 0 but less than 2.00. Successive approximations give $x = 1.58$. At equilibrium, $p_{N_2O_4} = 1.58$ atm, and $p_{NO_2} = (2.00 - 1.58) = 0.42$ atm.

Check: $[(0.42)^2/1.58] = 0.112$

14.16 The value of K_p is 0.113 at 25 °C for the equilibrium

$$N_2O_4(g) \rightleftharpoons 2NO_2(g)$$

If the total pressure $= 3.0$ atm, what are the partial pressures of $N_2O_4(g)$ and $NO_2(g)$ at equilibrium?

Summary of Method of Calculating Equilibrium Concentrations and Pressures

1. Write an equation for the equilibrium.
2. Write the equilibrium constant expression.
3. Express all unknown concentrations or partial pressures in terms of a single variable, x. (Remember, to simplify the algebra, if possible x should be small compared to any quantities to which x is to be added or from which x is to be subtracted.)
4. Substitute the equilibrium concentrations or partial pressures in terms of the single variable, x, in the equilibrium constant expression.
5. Solve for x.
6. Use the value found for x to calculate equilibrium concentrations or equilibrium partial pressures.
7. Check work by substituting answers in equilibrium constant expression and comparing the value obtained with the value of the equilibrium constant.

The smaller the value of K_c and the higher the concentration with which x is compared, the better the assumption that x is negligible.

K_c and K_p are truly constant only for ideal solutions and ideal gases. However, for real solutions and real gases, calculations involving K_c and K_p usually give results that are accurate to $\pm 5\%$ or better providing that solutions are dilute (0.1 M or less) and gas pressures are low (1 atm or less). These results are accurate enough to be very useful.

Remember that, according to Le Châtelier's principle (Section 12.3), if a change is made in a system at equilibrium, the equilibrium will shift in such a way as to reduce the effect of the change.

The Le Châtelier's Principle module of the Equilibrium section uses an animated equilibrium system to depict shifts resulting from changes made by the user.

14.5 USING LE CHÂTELIER'S PRINCIPLE TO PREDICT SHIFTS IN CHEMICAL EQUILIBRIA

Once a reaction has reached equilibrium, no further change occurs on a macroscopic scale. On a microscopic scale, both forward and reverse reactions continue to take place; the rates of forward and reverse reactions are equal. Many changes in experimental conditions, such as changes in concentration and temperature, change the rate of one of the reactions more than the rate of the other reaction, and the position of the equilibrium shifts.

Chemists and chemical engineers would usually like to obtain as much product as possible from a given quantity of reactants. Sometimes formation of the minimum quantity of a substance, such as an air pollutant, is desirable. In either case, the factors that affect the composition of the equilibrium mixture must be known to choose the best conditions for carrying out a reaction. (Information about the factors affecting the *rate* at which equilibrium is reached must also be available; reaction rates are the subject of Chapter 18.) The ability to make qualitative predictions about equilibria is helpful in planning and in checking the results of calculations to see if they are reasonable. Le Châtelier's principle provides a simple way to make qualitative predictions about the direction in which chemical equilibria shift as a result of changes in concentration, pressure, and temperature.

Effect of Changes in Concentration

Changes in concentration do not change the value of the equilibrium constant. As long as temperature is constant, the value of the equilibrium constant remains the same. However, changes in concentration do make a difference in the composition of the equilibrium mixture, that is, in the position of equilibrium. Table 14.2 and

TABLE 14.2	Effect of Changing Concentration on the Equilibrium $H_2(g) + I_2(g) \rightleftharpoons 2HI(g)$ at 425.4 °C			
	$[H_2]$	$[I_2]$	$[HI]$	$\dfrac{[HI]^2}{[H_2][I_2]}$
Original equilibrium	4.562×10^{-3}	0.7384×10^{-3}	1.355×10^{-2}	54.5
New equilibrium after 1.000×10^{-3} mol I_2 added to 1 L original equilibrium	3.877×10^{-3}	1.053×10^{-3}	1.492×10^{-2}	54.5
New equilibrium after 1.000×10^{-3} mol H_2 removed from 1 L original equilibrium	3.698×10^{-3}	0.8748×10^{-3}	1.328×10^{-2}	54.5

Figure 14.5. compare the effects of adding iodine gas to and removing hydrogen gas from an equilibrium mixture of $H_2(g)$, $I_2(g)$, and $HI(g)$.

As you can see from Figure 14.5(a), addition of $I_2(g)$ to the hydrogen–iodine–hydrogen iodide equilibrium results at first in a considerable increase in the concentration of $I_2(g)$. The concentrations of $I_2(g)$ and of $H_2(g)$ then gradually decrease, and the concentration of $HI(g)$ gradually increases until a new equilibrium is reached. At the new equilibrium, the concentration of HI is larger than the concentration of HI at the original equilibrium. The concentration of H_2 is smaller than the concentration of H_2 at the original equilibrium. Although the concentration of I_2 at the new equilibrium is higher than the concentration of I_2 at the old equilibrium, it is not as high as it was immediately after the addition of the $I_2(g)$ to the original equilibrium mixture. Thus, the equilibrium has shifted in such a way as to use up some of the added $I_2(g)$ as predicted by Le Châtelier's principle.

Removal of $H_2(g)$ from the original equilibrium mixture results in decreases in the concentration of $H_2(g)$ [see Figure 14.5(b)] and $HI(g)$ (see Table 14.2) and an increase in the concentration of $I_2(g)$. The concentration of $H_2(g)$ at the new equilibrium is not as low as the concentration of $H_2(g)$ immediately after the removal of $H_2(g)$. Thus, the original equilibrium has shifted in such a way as to replace some of the $H_2(g)$ that was removed as predicted by Le Châtelier's principle.

The positions of the new equilibria could also have been predicted by means of the reaction quotient. For example, the reaction quotient for the equilibrium

$$H_2(g) + I_2(g) \rightleftharpoons 2HI(g) \qquad \text{is}$$

$$Q = \frac{[HI]^2}{[H_2][I_2]}$$

Originally, $Q = K_c$ because the system was at equilibrium. Addition of $I_2(g)$ to the original equilibrium mixture would increase $[I_2]$, which is in the denominator. An increase in $[I_2]$ would result in a decrease in the value of Q. For the value of Q again to become equal to K_c, net reaction would have to take place in the forward direction so that $[HI]$ would increase and $[H_2]$ and $[I_2]$ would decrease.

Sample Problem 14.11 shows how the new equilibrium concentrations in Table 14.2 were calculated.

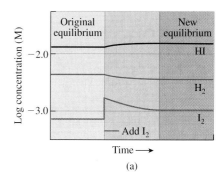

(a)

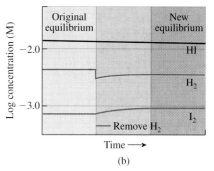

(b)

■ FIGURE 14.5 Effect of changing concentration on the hydrogen–iodine–hydrogen iodide equilibrium. A logarithmic scale must be used for concentration so that the graph fits on a page; the highest concentration (1.4920×10^{-2}) is more than 20 times as large as the lowest concentration (7.384×10^{-4}). The disadvantage of a logarithmic scale is that small differences such as the difference between $[HI]$ = 1.355×10^{-2} and $[HI]$ = 1.328×10^{-2} can't be seen. (a) I_2 added. (b) H_2 removed.

The logarithmic scale that you are most likely to be familiar with is the Richter scale used for measuring the magnitude of earthquakes. A magnitude 8 earthquake is ten times as strong as a magnitude 7 earthquake.

○ Sample Problem 14.11 shows how to calculate new equilibrium concentrations after a shift in equilibrium brought about by a change in concentration.

14.5 USING LE CHÂTELIER'S PRINCIPLE TO PREDICT SHIFTS IN CHEMICAL EQUILIBRIA

SAMPLE PROBLEM

14.11 A mixture containing 4.562×10^{-3} mol $H_2(g)$, 7.384×10^{-4} mol $I_2(g)$ and 1.355×10^{-2} mol $HI(g)$ in a 1.000-L container at 425.4 °C is at equilibrium. If 1.000×10^{-3} mol $I_2(g)$ are added, what will be the concentrations of $H_2(g)$, $I_2(g)$, and $HI(g)$ after the system has again reached equilibrium?

Solution The equation for the equilibrium is

$$H_2(g) + I_2(g) \rightleftharpoons 2HI(g)$$

and the equilibrium constant expression is

$$K_c = \frac{[HI]^2}{[H_2][I_2]}$$

First, we must use Le Châtelier's principle to predict the direction of the shift in equilibrium that will occur. Then, we must calculate the initial concentrations, the concentrations immediately after the addition of $I_2(g)$ but before any reaction has had time to occur. From this point on, the problem is similar to Sample Problem 14.9.

According to Le Châtelier's principle, addition of I_2 will shift the equilibrium to the right. Net reaction will take place to form HI and use up H_2 and I_2.

Immediately after the addition of the $I_2(g)$, before any reaction has had time to take place,

$$[H_2] = 4.562 \times 10^{-3} = [H_2]_0, \quad [HI] = 1.355 \times 10^{-2} = [HI]_0, \text{ and}$$

$$[I_2] = (7.384 \times 10^{-4}) + (1.000 \times 10^{-3}) = 1.738 \times 10^{-3} = [I_2]_0.$$

If x = mole of $H_2(g)$ = mole of $I_2(g)$ that react to form $HI(g)$ in order to reach the new equilibrium (eq'), then $2x$ = mol $HI(g)$ that form.
$[H_2]_{eq'} = [(4.562 \times 10^{-3}) - x]$, $[I_2]_{eq'} = [(1.738 \times 10^{-3}) - x]$, and $[HI]_{eq'} = [(1.355 \times 10^{-2}) + 2x]$. The completed summary table is

	$H_2(g)$	$+$	$I_2(g)$	$\rightleftharpoons$	$2HI(g)$
Eq. conc., M	4.562×10^{-3}		7.384×10^{-4}		1.355×10^{-2}
Init. conc., M	4.562×10^{-3}		1.738×10^{-3}		1.355×10^{-2}
Change, ΔM	$-x$		$-x$		$+2x$
Eq.' conc., M	$[(4.562 \times 10^{-3}) - x]$		$[(1.738 \times 10^{-3}) - x]$		$[(1.355 \times 10^{-2}) + 2x]$

Substitution of the new equilibrium concentrations in terms of x in the equilibrium constant expression gives

$$\frac{[(1.355 \times 10^{-2}) + 2x]^2}{[(4.562 \times 10^{-3}) - x][(1.738 \times 10^{-3}) - x]} = 54.5 \qquad (14.12)$$

The value of the equilibrium constant, 54.5, is not very small. Therefore, $2x$ is not negligible compared with (1.355×10^{-2}), nor is x negligible compared with (4.562×10^{-3}) and (1.738×10^{-3}). Either successive approximations or the quadratic formula must be used to find the value of x. By either method

$$x = 6.85 \times 10^{-4}$$

The new equilibrium concentrations can now be calculated:

$$[H_2]_{eq'} = [(4.562 \times 10^{-3}) - x] = [(4.562 \times 10^{-3}) - (6.85 \times 10^{-4})]$$
$$= 3.877 \times 10^{-3}$$

$$[I_2]_{eq'} = [(1.738 \times 10^{-3}) - x] = [(1.738 \times 10^{-3}) - (6.85 \times 10^{-4})]$$
$$= 1.053 \times 10^{-3}$$

$$[HI]_{eq'} = [(1.355 \times 10^{-2}) + 2x] = [(1.355 \times 10^{-2}) + 2(6.85 \times 10^{-4})]$$
$$= 1.492 \times 10^{-2}$$

○ Converted to standard quadratic form, equation 14.12 is $50.5x^2 - 0.398x + 2.49 \times 10^{-4} = 0$

○ Always check your work.

These answers can be checked by substitution into the equilibrium constant expression:

$$\frac{(1.492 \times 10^{-2})^2}{(3.877 \times 10^{-3})(1.053 \times 10^{-3})} = 54.53 = 54.5$$

The answers are correct.

PRACTICE PROBLEM

14.17 A mixture containing 4.562×10^{-3} mol $H_2(g)$, 7.384×10^{-4} mol $I_2(g)$, and 1.355×10^{-2} mol $HI(g)$ in a 1.0000-L container at 425.4 °C is at equilibrium. If 0.005 000 mol $HI(g)$ is removed from the container, what will be the concentrations of $H_2(g)$, $I_2(g)$, and $HI(g)$ after the system has again reached equilibrium?

The addition of reactants or products that are pure solids or pure liquids does not shift equilibria because the concentrations of pure solids and pure liquids are constant. The removal of reactants or products that are pure solids or pure liquids does not affect equilibria *as long as some of the solid or liquid remains*. If *all* of a solid or liquid is removed, the system is no longer at equilibrium. (Remember Figure 14.4 and Sample Problem 14.3.)

Sample Problem 14.12 provides another example of the qualitative prediction of the effect of change in concentration on an equilibrium.

SAMPLE PROBLEM

14.12   Predict the effect of the following changes on the equilibrium

$$CaCO_3(s) \rightleftharpoons CaO(s) + CO_2(g)$$

(a) adding $CO_2(g)$ to the container (b) removing some of the $CaCO_3$ from the container

Solution   (a) When $CO_2(g)$ is added to the container, the equilibrium shifts to the left. [Note that, in this case, the equilibrium concentration returns to its original value if enough CaO(s) is present.] Calcium carbonate is formed. The quantity of $CaCO_3$ increases, and the quantity of CaO present decreases. (The concentrations of CaO and $CaCO_3$ do not change because the concentrations of pure solids are constant.)

(b) $CaCO_3$ is a solid, and its concentration is constant. Removing some of the $CaCO_3$ from the container will not shift the equilibrium. The concentration of CO_2 at equilibrium will remain the same as long as there is some $CaCO_3$ (and some CaO) in the container and the temperature is not changed.

PRACTICE PROBLEM

14.18   Predict the effect of the following changes on the equilibrium

$$CO(g) + H_2O(g) \rightleftharpoons CO_2(g) + H_2(g)$$

(a) increasing [CO] (b) decreasing $[H_2]$ (c) decreasing $[H_2O]$ (d) increasing $[CO_2]$

The effect of adding a reactant or removing a product on equilibrium is often used to make reactions go to completion. For example, a large excess of an inexpensive reactant can be used to shift an equilibrium to the right so that, for all practical purposes, an expensive reactant is completely converted to product. Complete conversion of reactants to products can also be brought about by removing a product as it is formed. The equilibrium shifts to the right; more product is formed to replace the product removed.

On a microscopic level, an increase in the concentration of a reactant increases the number of collisions between that reactant and the other reactants, and the rate of the forward reaction increases. As a result of the increase in the rate of the forward reaction, the concentrations of the products increase. The rate of the reverse reaction then also increases. As time passes, the rate of the forward reaction decreases because reactants are used up by the formation of products. After enough time has passed, the rates of the forward and reverse reactions again become equal; a new equilibrium has been reached.

Effect of Changes in Pressure

Changing the pressure does not change the value of the equilibrium constant. As long as the temperature is constant, the value of the equilibrium constant remains

TABLE 14.3	Effect of Changing Pressure by Changing Volume on Equilibrium $N_2O_4(g) \rightleftharpoons 2NO_2(g)$ at 25 °C			
P_{tot}, atm		$p_{N_2O_4}$, atm	p_{NO_2}, atm	% Decomposition
1.0		0.72	0.28	16
2.0		1.58	0.42	12

the same. Pressure can be changed by addition of a gaseous reactant or product to an equilibrium mixture or by removal of a gaseous reactant or product from the equilibrium mixture. Addition or removal of a gaseous reactant or product changes the concentration of the reactant or product. Shifts in equilibrium that result from change in concentration have already been discussed.

The pressure of a gaseous reaction mixture can also be changed by changing the volume of the container. A decrease in the volume of the container increases the concentrations and partial pressures of both reactants and products and the rates of both forward and reverse reactions. Table 14.3 shows the effect of increasing the pressure by decreasing the volume of the container on the equilibrium

$$N_2O_4(g) \rightleftharpoons 2NO_2(g)$$

(The value of K_p for this equilibrium is 0.113 at 25 °C.) The data in Table 14.3 show that doubling the pressure by decreasing the volume increases the partial pressures of both $N_2O_4(g)$ and $NO_2(g)$ at equilibrium. However, the partial pressure of $N_2O_4(g)$ is more than doubled while the partial pressure of $NO_2(g)$ is less than doubled. The percent decomposition is less at the higher pressure. Increasing pressure by decreasing volume shifts the equilibrium in favor of N_2O_4 (to the left).

According to the balanced equation for the equilibrium, 1 mol $N_2O_4(g)$ decomposes to form 2 mol $NO_2(g)$. Therefore, according to Avogadro's law (Section 5.5), one volume of $N_2O_4(g)$ yields two volumes of $NO_2(g)$. Increasing pressure by decreasing volume shifts the equilibrium to the side that occupies the smaller volume as predicted by Le Châtelier's principle.

In general, increasing pressure by decreasing volume shifts equilibria toward the side that has the smaller number of moles of gas. If the number of moles of gas is the same on both sides of an equilibrium, changing the pressure by changing the volume does not have any effect on an equilibrium. For example, changing pressure does not shift the equilibrium

$$H_2(g) + I_2(g) \rightleftharpoons 2HI(g)$$

because two moles of gaseous reactants are converted to two moles of gaseous product. ▪Figure 14.6 shows the effect of increasing pressure by decreasing volume on the nitrogen dioxide–dinitrogen tetroxide equilibrium:

$$\underset{\text{colorless}}{N_2O_4(g)} \rightleftharpoons \underset{\text{red-brown}}{2NO_2(g)}$$

Pressure can also be changed by addition of a gas that does not take part in the reaction. *A gas that does not react either with the reactants or with the products of a reaction* is called an **inert gas.** Increasing the total pressure by adding an inert gas at constant volume does not shift gaseous equilibria (assuming ideal behavior). There is plenty of room for the molecules of the inert gas in the space between the molecules of the gases that are in equilibrium. (Remember that a gas under ordinary conditions of temperature and pressure is mostly empty space.) There are no intermolecular attractions between molecules of ideal gases. Therefore, the presence of the inert gas molecules has no effect on the equilibrium.

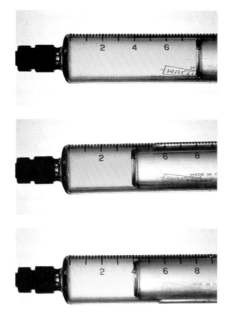

▪ FIGURE 14.6 *Top:* Dinitrogen tetroxide (colorless) and nitrogen dioxide (red-brown) are at equilibrium in a syringe. *Middle:* The plunger is pushed. *Bottom:* After some time has passed, N_2O_4 and NO_2 are again at equilibrium. Why is the red-brown color darkest immediately after the plunger has been pushed down? Is the equilibrium concentration of red-brown NO_2 greater at high pressure or at low pressure? Does the N_2O_4–NO_2 equilibrium shift as predicted by Le Châtelier's principle?

Effect of Changes in Temperature

Changing the temperature usually changes the value of the equilibrium constant. The values of some equilibrium constants increase as temperature is increased. The values of other equilibrium constants decrease as temperature is increased. Table 14.4 shows how the equilibrium constants for two industrially important reactions change with temperature.

The reaction of methane with steam,

$$CH_4(g) + H_2O(g) \longrightarrow 3H_2(g) + CO(g)$$

which is used to make hydrogen gas for the manufacture of ammonia, is endothermic:

$$203.25 \text{ kJ} + CH_4(g) + H_2O(g) \longrightarrow 3H_2(g) + CO(g)$$

As you can see from Table 14.4(a), the value of the equilibrium constant for the equilibrium

$$203.25 \text{ kJ} + CH_4(g) + H_2O(g) \rightleftharpoons 3H_2(g) + CO(g)$$

increases as temperature increases. Increasing temperature shifts the equilibrium to the right as predicted by Le Châtelier's principle. Remember that, by definition, the species on the left side of the arrows in the equation for an equilibrium are the reactants and the species on the right are the products (Section 14.3). The thermochemical equation for the methane–steam–hydrogen–carbon monoxide equilibrium is usually written

$$CH_4(g) + H_2O(g) \rightleftharpoons 3H_2(g) + CO(g) \qquad \Delta H^\circ = +203.25 \text{ kJ}$$

Reaction of nitrogen and hydrogen to form ammonia is exothermic:

$$N_2(g) + 3H_2(g) \longrightarrow 2NH_3(g) + 92.22 \text{ kJ}$$

As you can see from the values for K_p in Table 14.4(b), the higher the temperature, the smaller the equilibrium constant for the equilibrium

$$N_2(g) + 3H_2(g) \rightleftharpoons 2NH_3(g) + 92.22 \text{ kJ}$$

or

$$N_2(g) + 3H_2(g) \rightleftharpoons 2NH_3(g) \qquad \Delta H^\circ = -92.22 \text{ kJ}$$

Increasing temperature shifts the equilibrium to the left as predicted by Le Châtelier's principle.

■Figure 14.7 shows the effect of temperature on the nitrogen dioxide–dinitrogen tetroxide equilibrium:

$$N_2O_4(g) \rightleftharpoons 2NO_2(g)$$

colorless   red-brown

TABLE 14.4

(a) Values of K_p for Equilibrium $CH_4(g) + H_2O(g) \rightleftharpoons 3H_2(g) + CO(g)$ at Different Temperatures[a]

Temperature, °C	K_p
649	2.679×10^0
760	6.343×10^1
871	8.166×10^2
982	6.755×10^3

(b) Values of K_p for Equilibrium $N_2(g) + 3H_2(g) \rightleftharpoons 2NH_3(g)$ at Different Temperatures[a]

Temperature, °C	K_p
227	9.02×10^{-2}
427	8.12×10^{-5}
627	1.28×10^{-6}
827	9.71×10^{-8}

[a]Data are from Kent, J. A. *Riegel's Handbook of Industrial Chemistry,* 8th ed.; van Nostrand Reinhold: New York, 1983; pp 150 and 165.

Heating a closed container results in increased pressure. The container should be shielded in case it explodes.

■ FIGURE 14.7   Is the equilibrium concentration of red-brown NO_2 greater at high temperature (*right*) or at low temperature (*left*)? Is the decomposition of N_2O_4 to NO_2 exothermic or endothermic?

If no thermal energy is released or absorbed when a reaction takes place (which is very rare), the equilibrium constant for the reaction is the same at all temperatures. Equilibrium is not shifted by changing temperature.

Effect of Catalysts

Catalysts are substances that make reactions take place faster but are not used up in the reaction (Section 6.2). Catalysts are shown over the arrow in equations for chemical reactions and do not appear in equilibrium constant expressions or in reaction quotients. Catalysts increase the rates of *both* forward and reverse reactions by the same factor; therefore, addition of a catalyst does not shift an equilibrium. A catalyst makes a reaction reach equilibrium faster but does not change either the value of the equilibrium constant or the equilibrium concentrations.

SAMPLE PROBLEM

14.13 Predict the effect of each of the following changes on the equilibrium

$$2SO_2(g) + O_2(g) \rightleftharpoons 2SO_3(g) \qquad \Delta H° = -197.78 \text{ kJ}$$

(a) increasing pressure by decreasing volume at constant temperature (b) increasing pressure by adding an inert gas at constant temperature and volume (c) increasing temperature at constant pressure (d) adding a catalyst

Solution (a) Picture the equilibrium in a cylinder with a movable piston. The cylinder is kept at a constant temperature as the piston is pushed down:

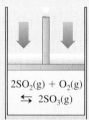

According to the equation for the equilibrium 2 mol $SO_2(g)$ and 1 mol $O_2(g)$ form 2 mol $SO_3(g)$. The number of moles of gas and, therefore, the volume, are decreased by reaction to form SO_3. Increasing pressure by decreasing volume will shift the equilibrium to the right.

(b) Adding an inert gas will not shift the equilibrium.

(c) Picture the equilibrium in a cylinder with a movable piston. The contents of the cylinder are kept at a constant pressure as the cylinder is heated:

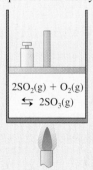

According to the equation for the equilibrium, thermal energy is released when SO_2 and O_2 form SO_3. Increasing the temperature will cause the equilibrium to shift to the left.

(d) Adding a catalyst will not shift the equilibrium, only the *speed* with which equilibrium is reached.

14.19   Predict the effect of each of the following changes on the equilibrium

$$CaCO_3(s) \rightleftharpoons CaO(s) + CO_2(g) \qquad \Delta H° = +178.32 \text{ kJ}$$

(a) increasing pressure by decreasing volume at constant temperature (b) increasing pressure by adding an inert gas at constant temperature and volume (c) increasing temperature at constant pressure

14.20   Predict the effect of each of the following changes on the equilibrium

$$CH_4(g) + H_2O(g) \rightleftharpoons 3H_2(g) + CO(g) \qquad \Delta H° = +203.25 \text{ kJ}$$

(a) increasing pressure by decreasing volume at constant temperature (b) increasing pressure by adding an inert gas at constant temperature and volume (c) increasing temperature at constant pressure (d) adding a catalyst

14.21   (a) Which of the changes in Practice Problem 14.20 will change the value of the equilibrium constant? (b) For each change that will affect the value of the equilibrium constant, tell whether the value of the equilibrium constant will increase or decrease.

14.6   SOME INDUSTRIALLY IMPORTANT CHEMICAL EQUILIBRIA

The chemical industry is a major part of all industrial economies. It provides fertilizers, fibers, dyes, building materials, paints, detergents, plastics, adhesives, drugs, cosmetics, explosives, and so forth, and employs over a million people in the United States. Many important processes in the chemical industry involve shifting equilibria to make the most product at the lowest cost.

Although the trade balance of the United States is negative, the balance of trade in chemicals is favorable.

Contact Process for Sulfuric Acid

Almost twice as much sulfuric acid as any other compound is manufactured in the United States each year. In 1993, 80.31 billion pounds of sulfuric acid were produced—over 300 pounds for each person. About two-thirds of the sulfuric acid produced is used to make fertilizers. The rest is used for a variety of purposes—for example, to make other chemicals; in petroleum refining; to produce iron, steel, and other metals; to make paints, pigments, dyes, rayon, cellulose film, and explosives; and for battery acid. The key step in the production of sulfuric acid is the **contact process.** In the contact process, *sulfur dioxide is reacted with oxygen to form sulfur trioxide:*

You are most likely to be familiar with sulfuric acid as the battery acid in your car, although battery acid is not a very important use of sulfuric acid as far as quantity is concerned. Don't forget that there is a table showing names, formulas, U.S. production data, and major uses for the "Top 50 Chemicals for 1993" in Appendix E.

$$2SO_2(g) + O_2(g) \longrightarrow 2SO_3(g)$$

The sulfur dioxide used in the contact process is made by burning sulfur in dry air,

$$S(s) + O_2(g) \longrightarrow SO_2(g)$$

or from hydrogen sulfide or metal sulfides such as FeS_2:

$$2H_2S(g) + 3O_2(g) \longrightarrow 2SO_2(g) + 2H_2O(g)$$
$$4FeS_2(s) + 11O_2(g) \longrightarrow 8SO_2(g) + 2Fe_2O_3(s)$$

The sulfur trioxide produced by the contact process is absorbed in sulfuric acid to form disulfuric acid, $H_2S_2O_7$:

Sulfuric acid absorbs sulfur trioxide better than water does. The reaction between sulfuric acid and water produces a fog that is hard to condense.

$$SO_3(g) + H_2SO_4(l) \longrightarrow H_2S_2O_7(l)$$

TABLE 14.5	Values of Equilibrium Constant for Equilibrium $2SO_2(g) + O_2(g) \rightleftharpoons 2SO_3(g)$ at Different Temperatures[a]			
Temperature, °C	K_p		Temperature, °C	K_p
400	1.576×10^5		800	8.37×10^{-1}
500	2.31×10^3		900	1.475×10^{-1}
600	9.08×10^1		1000	3.40×10^{-2}
700	6.92×10^0		1100	9.60×10^{-3}

[a]Data are from Austin, G. T. *Shreve's Chemical Process Industries,* 5th ed.; McGraw-Hill: New York, 1984, p 333.

and disulfuric acid is diluted with water to form sulfuric acid:

$$H_2S_2O_7(l) + H_2O(l) \longrightarrow 2H_2SO_4(l)$$

The reaction between sulfur dioxide and oxygen that takes place in the contact process is exothermic and reversible:

$$2SO_2(g) + O_2(g) \rightleftharpoons 2SO_3(g) \qquad \Delta H° = -197.78 \text{ kJ}$$

Table 14.5 shows the values of the equilibrium constant for this reaction at different temperatures. Judging by the large values of K_p at the lower temperatures in Table 14.5, sulfur dioxide should not exist in air at room temperature. Instead, sulfur dioxide should be completely converted to sulfur trioxide. However, sulfur dioxide does exist in air at room temperature. The rate of the reaction between sulfur dioxide and oxygen must be very low at room temperature.

The rates of almost all reactions are higher at high temperatures than at low, but even at 400 °C in the presence of a catalyst, reaction of sulfur dioxide with oxygen is slow. Increasing the temperature to 600 °C speeds the reaction but lowers the percent conversion of sulfur dioxide to sulfur trioxide. Therefore, the temperature used to make sulfur trioxide must be a compromise. The temperature must be high enough for reaction to take place in a reasonable time but low enough that most of the sulfur dioxide is converted to sulfur trioxide. In a modern sulfuric acid plant, as much sulfur dioxide as possible must be converted to sulfur trioxide to minimize atmospheric pollution by sulfur dioxide, which is an important factor in acid rain. The equilibrium between $SO_2(g)$, $O_2(g)$, and $SO_3(g)$ is shifted to the right both by using an excess of oxygen and by removing sulfur trioxide as it is formed. An increase in pressure would also shift the equilibrium to the right (3 mol of gas are converted to 2 mol of gas), but the increased conversion is not worth the cost of high-pressure equipment.

Vanadium(V) oxide, V_2O_5, is the catalyst used for the contact process.

Although the contact process was patented in England in 1831, the technology necessary to make it work well was not developed until about 1900. The success of the contact process led to the development of other catalytic processes such as the Haber–Bosch process for the fixation of nitrogen gas from air.

Haber–Bosch Process for Ammonia

Fixation of nitrogen means *conversion of unreactive nitrogen gas from air into reactive nitrogen compounds such as ammonia.* Nitrogen in compounds is known as "fixed" nitrogen. Nitrogen gas is unreactive because the strong triple bond between nitrogen atoms in molecular nitrogen must be broken for molecular nitrogen to react.

Fixed nitrogen is almost always the limiting reactant for the production of food crops. Nitrogen is fixed naturally by microorganisms in soil. However, around 1900, demand for fixed nitrogen to grow food for the world's rapidly increasing population began to exceed supply from natural sources. In addition, Germany was beginning to prepare for World War I and needed a source of nitrates to make explosives. In 1904–1908 Fritz Haber,* a German physical chemist, determined equilibrium data for the $N_2(g)$, $H_2(g)$, $NH_3(g)$ system and developed a laboratory method for synthesizing ammonia from nitrogen. A German industrial chemist, Karl Bosch, then adapted Haber's laboratory method to an industrial scale. The Haber–Bosch process is the most economical industrial method for nitrogen fixation and is one of the basic processes of the chemical industry worldwide. Large quantities of ammonia are used to make explosives and plastics as well as fertilizers, and in the paper, rubber, textile, food, and drug industries. In 1993 in the United States, 34.50 billion pounds of ammonia were synthesized, over 130 pounds per person.

Growth of the shorter corn in the rectangle in the upper half of the photo was limited by a nitrogen deficiency.

The reaction between nitrogen and hydrogen to form ammonia is exothermic and reversible:

$$N_2(g) + 3H_2(g) \rightleftharpoons 2NH_3(g) \qquad \Delta H^\circ = -92.22 \text{ kJ}$$

[Values of K_p at different temperatures are shown in Table 14.4(b).] At room temperature in the absence of a catalyst, the rates of formation and decomposition of ammonia are negligible. However, increasing temperature to increase the rate of reaction also shifts the equilibrium to the left and decreases the proportion of nitrogen that is converted to ammonia. Therefore, the temperature used to make ammonia must be a compromise. The temperature must be high enough for reaction to take place in a reasonable time but low enough that a significant fraction of the nitrogen is converted to ammonia.

One volume of nitrogen gas and three volumes of hydrogen gas combine to give only two volumes of ammonia gas. Increasing pressure shifts the equilibrium to the right and increases the proportion of nitrogen that is converted to ammonia. Increasing pressure also increases the rate of reaction. In modern plants, a temperature of 500 °C is used, and the pressure is 150–300 atm. Even at 500 °C, a catalyst (iron) must be used to obtain a practical rate of reaction.

The Haber–Bosch process was the first industrial process carried out at high temperature and high pressure. The original reactors were made from cannons!

The best yields are obtained by using the stoichiometric proportions of nitrogen and hydrogen, 1 mol N_2 : 3 mol H_2. At 150 atm, the percent yield of ammonia is only 14%. However, the equilibrium is shifted to the right by removing the ammonia as it is formed. (Ammonia condenses to a liquid under conditions at which nitrogen and hydrogen are still gases.) More nitrogen and hydrogen are added to the leftover nitrogen and hydrogen to replace that used up in the reaction. The mixture of nitrogen and hydrogen is then recycled. In the end, almost all the hydrogen, which is more expensive than nitrogen, is converted to ammonia. Hydrogen is made from natural gas by the reactions

$$CH_4(g) + H_2O(g) \longrightarrow CO(g) + 3H_2(g) \qquad \text{and}$$
$$2CH_4(g) + O_2(g) \longrightarrow 2CO(g) + 4H_2(g)$$

Production of Lime

Another important industrial reaction that involves an equilibrium is the production of lime, CaO, by heating calcium carbonate, the major component of limestone.

---

*Haber was a very patriotic German citizen and played a leading part in the development of poison gas as a weapon. After World War I, he tried to find a way of recovering gold from seawater to pay the reparations imposed on Germany by the victorious Allies. He died a refugee from Hitler's anti-Jewish policy. Bosch's contributions to the nitrogen fixation process were to find a suitable catalyst and to invent methods for carrying out reactions at high pressures (150–300 atm). Haber received the Nobel Prize for chemistry in 1918 and Bosch in 1931.

The Oxygen–Hemoglobin Equilibrium

At high altitudes, the concentration of oxygen is low.

Transport of oxygen by blood depends on the reversible combination of oxygen with hemoglobin. In blood, hemoglobin, oxygen, and oxyhemoglobin are in equilibrium:

$$hemoglobin\ (Hb)\ +\ O_2(g) \rightleftharpoons$$
$$oxyhemoglobin\ [Hb(O_2)]$$

At high altitudes, the concentration of oxygen is low, and the equilibrium shifts to the left. Dizziness and a tired feeling result; in some cases, altitude sickness, which can be a severe disease, occurs. The body gradually adapts to high altitudes by producing more hemoglobin, which shifts the equilibrium to the right.

Carbon monoxide forms a complex with hemoglobin [Hb(CO)] that is more stable than the oxygen–hemoglobin complex [Hb(O₂)]:

$$Hb(O_2)\ +\ CO \rightleftharpoons Hb(CO)\ +\ O_2$$

$$K_c = \frac{[Hb(CO)][O_2]}{[Hb(O_2)][CO]} = 2.1 \times 10^2$$

Formation of the carbon monoxide–hemoglobin complex prevents hemoglobin from carrying oxygen. As a result, the toxicity of carbon monoxide is very high. The first noticeable effect of carbon monoxide poisoning is reduced alertness, which is probably the cause of many automobile accidents. First aid is fresh air. Recovery from mild carbon monoxide poisoning is rapid and complete; the effects of carbon monoxide poisoning are not cumulative.

Increasing the concentration of CO in air shifts the equilibrium to the right. High concentrations of CO result in unconsciousness and death; exposure for one hour to a carbon monoxide concentration of 0.1% (by volume) in air is often fatal. Carbon monoxide is a colorless, odorless gas, which makes it especially dangerous because you can't smell it or see it.

Carbon monoxide formed in the engines of cars is a major air pollutant; police officers who direct traffic in big cities have unusually high concentrations of carbon monoxide in their blood. Smokers also have unusually high concentrations of carbon monoxide in their blood because carbon monoxide is formed when tobacco burns.

The Romans used the same reaction to make lime for mortar with which they built roads and buildings 2000 years ago.

The production of lime was one of the first manufacturing processes carried out by the settlers in colonial America.

The difference between continuous process and batch process is like the difference between manufacture of cars on an assembly line and production of individual, custom-made cars.

The equation for the equilibrium is

$$CaCO_3(s) \rightleftharpoons CaO(s)\ +\ CO_2(g) \qquad \Delta H° = +178.32\ kJ$$

The carbon dioxide, which is a by-product of this reaction, is also sold. Below 600 °C, the pressure of CO_2 at equilibrium is very small. Above 600 °C, the equilibrium pressure of CO_2 increases rapidly and reaches 1 atm at about 900 °C. The equilibrium is shifted to the right by removing CO_2 as it is formed.

Lime is used for making mortars and plasters, as a cheap base in industry, for removing excess acid from soil in farming, in steel making, in the manufacture of glass, and in water purification. The annual production of lime in the United States in 1993 was 36.80 billion pounds, slightly greater than the production of ammonia.

Continuous Processes

Modern large-scale industrial processes such as the contact process, the Haber process, and the production of lime are usually carried out continuously. The reactants flow steadily through the processing units; reaction conditions are optimized with the aid of computers. In the laboratory, reactions are carried out in batches. Batch processing is used in industry for the preparation of small-volume chemicals (materials that are only produced in small quantities) and for the manufacture of dangerous materials like explosives.

In principle, all chemical reactions are reversible. The positions of chemical equilibria are shown by the numerical values of the equilibrium constants. The value of the **equilibrium constant K_c** for the reaction

$$aA + bB + \cdots \rightleftharpoons eE + fF + \cdots$$

where a is the coefficient of the reactant A in the equation for the reaction, b is the coefficient of the reactant B, e is the coefficient of the product E, and so forth, is given by the expression

$$K_c = \frac{[E]^e[F]^f \cdots}{[A]^a[B]^b \cdots}$$

which is called the **equilibrium constant expression** for the reaction. The square brackets stand for concentration in molarity; concentrations are *equilibrium* concentrations. The concentrations of pure liquids and solids and of the solvents in dilute solutions, which are constant, are included in the value of K_c.

The chemical equation for an equilibrium must be written in order to write an equilibrium constant expression. When the equation for an equilibrium is written in the opposite direction, the quotient in the equilibrium constant expression is the reciprocal of the quotient in the original equilibrium constant expression, and the numerical value of the equilibrium constant is the reciprocal of the original value. When the equation for an equilibrium is multiplied by a factor, the equilibrium constant must be raised to a power equal to the factor.

The numerical value of an equilibrium constant can be found by experiment. The concentrations at the beginning of an experiment are called the **initial concentrations.** If the initial concentrations of all reactants and products are known, the concentrations of all the components of the equilibrium mixture can be calculated from the equilibrium concentration of any one component and the chemical equation for the reaction.

For equilibria that involve gases, partial pressures can be used instead of concentrations in equilibrium constant expressions, and $K_p = K_c(RT)^{\Delta n_g}$ where Δn_g is the difference between the number of moles of gaseous products and reactants.

If the value of K_c (or K_p) is very large, reaction is complete at equilibrium. If the value of K_c (or K_p) is very small, no significant reaction takes place. For intermediate values of K_c (or K_p), measurable quantities of both reactants and products are present at equilibrium. Once the value of the equilibrium constant has been determined, the results of any experiments carried out at the same temperature can be calculated.

For the reaction

$$aA + bB + \cdots \rightleftharpoons eE + fF + \cdots$$

the **reaction quotient, Q,** is defined as

$$Q = \frac{[E]^e[F]^f \cdots}{[A]^a[B]^b \cdots}$$

where concentrations may be any concentrations. Comparison of the value of the reaction quotient, Q, for an experiment with the value of the equilibrium constant tells whether the system is at equilibrium and, if not, in which direction the net reaction will take place. If $Q < K_c$, net reaction takes place in the forward direction. If $Q = K_c$, the system is at equilibrium and no net change takes place. If $Q > K_c$, net reaction takes place in the reverse direction.

Concentration and partial pressure equilibrium constants are truly constant only for ideal solutions and ideal gases. For real solutions and real gases, calculations involving K_c and K_p are usually accurate enough to be useful at concentrations of less than 0.1 M and partial pressures of less than 1 atm.

The effects of changes in concentration on systems at equilibrium can be predicted from Le Châtelier's principle or from the reaction quotient, Q. If the concentration of a component of an equilibrium mixture is increased, the equilibrium will shift in the direction that uses up the added component. If the concentration of a component of an equilibrium mixture is decreased, the equilibrium will shift in the direction that forms the component. The effect of pressure and temperature changes on systems at equilibrium can also be predicted from Le Châtelier's principle. Increasing temperature favors the endothermic reaction; decreasing temperature favors the exothermic reaction. Increasing pressure causes an equilibrium to shift in the direction that reduces volume; decreasing pressure causes an equilibrium to shift in the direction that increases volume. A catalyst makes a system reach equilibrium faster but does not change the position of an equilibrium.

Equilibria that involve only a single phase are called **homogeneous equilibria.** Equilibria that involve more than one phase are called **heterogeneous equilibria.**

The **contact process** is the step in the manufacture of sulfuric acid that involves reaction of sulfur dioxide with oxygen to form sulfur trioxide. The **Haber-Bosch process** is a method of reacting nitrogen from air with hydrogen to form ammonia. Conversion of nitrogen to reactive compounds such as ammonia is called **nitrogen fixation.**

For information about the organization of Additional Practice Problems, Stop & Test Yourself, Putting Things Together, and Applications, see the beginnings of these sections in Chapter 1.

14.22 What is meant by the term *dynamic equilibrium?* Use a chemical equilibrium as an example in your explanation. (14.1)

14.23 Classify each of the following equilibria as homogeneous or heterogeneous:

(a) $CS_2(g) + 4H_2(g) \rightleftharpoons CH_4(g) + 2H_2S(g)$

(b) $Ni(s) + 4CO(g) \rightleftharpoons Ni(CO)_4(g)$

(c) $2HgO(s) \rightleftharpoons 2Hg(l) + O_2(g)$

(d) $4HCl(g) + O_2(g) \rightleftharpoons 2H_2O(l) + 2Cl_2(g)$

(e) $4HCl(g) + O_2(g) \rightleftharpoons 2H_2O(g) + 2Cl_2(g)$ (14.2)

14.24 Classify each of the following equilibria as homogeneous or heterogeneous:

(a) $2H_2S(g) + 3O_2(g) \rightleftharpoons 2H_2O(g) + 2SO_2(g)$

(b) $HSO_4^-(aq) \rightleftharpoons H^+(aq) + SO_4^{2-}(aq)$

(c) $Mg(OH)_2(s) \rightleftharpoons Mg^{2+}(aq) + 2OH^-(aq)$

(d) $2Cu^{2+}(aq) + Sn^{2+}(aq) \rightleftharpoons Sn^{4+}(aq) + 2Cu^+(aq)$

(e) $CO_2(aq) + H_2O(l) \rightleftharpoons H^+(aq) + HCO_3^-(aq)$ (14.2)

14.25 For each of the equilibria in Problem 14.23, write the concentration equilibrium constant expression. (14.2)

14.26 For each of the equilibria in Problem 14.24, write the concentration equilibrium constant expression. (14.2)

14.27 Explain why the concentrations of pure liquids and solids are included in the value for the equilibrium constant. (14.2)

14.28 (a) What qualitative information is given by the magnitude of the equilibrium constant for an equilibrium? (b) Explain why no reaction takes place when $H_2(g)$ and $O_2(g)$ are mixed under ordinary conditions despite the fact that K_c for the equilibrium

$$2H_2(g) + O_2(g) \rightleftharpoons 2H_2O(g)$$

is extremely large. (14.2)

14.29 Assuming that the system is at equilibrium, (a) which of the following reactions goes essentially to completion? (b) Which does not take place to a significant extent? (c) Which equilibria are easily detectable? (14.2)

(i) $Br_2(g) + Cl_2(g) \rightleftharpoons 2BrCl(g)$ $K_c = 7.0$

(ii) $C(graphite) + O_2(g) \rightleftharpoons CO_2(g)$ $K_c = 1.3 \times 10^{69}$

(iii) $2CH_4(g) \rightleftharpoons HC{\equiv}CH(g) + 3H_2(g)$
$K_c = 0.154$

(iv) $Br_2(g) \rightleftharpoons 2Br(g)$ $K_c = 4 \times 10^{-18}$

(v) $2HF(g) \rightleftharpoons H_2(g) + F_2(g)$ $K_c = 1.0 \times 10^{-95}$

14.30 If the equilibrium concentrations are $[O_2] = 0.21$ and $[O_3] = 6.0 \times 10^{-8}$, what is the value of K_c for the following equilibrium? (14.3)

$$2O_3(g) \rightleftharpoons 3O_2(g)$$

14.31 For the equilibrium

$$CO(g) + 2H_2(g) \rightleftharpoons CH_3OH(g)$$

if $[CO]_{eq} = 1.00$, $[CH_3OH]_{eq} = 1.50$, and $K_c = 14.5$, what is $[H_2]_{eq}$? (14.3)

14.32 For the equilibrium

$$2CH_4(g) \rightleftharpoons HC{\equiv}CH(g) + 3H_2(g)$$

if the initial concentration of CH_4 is 0.0300 M and the concentration of C_2H_2 at equilibrium is 0.013 75 M, (a) what are the equilibrium concentrations of CH_4 and H_2? (b) What is the numerical value of K_c? (14.3)

14.33 Explain why either partial pressures or concentrations can be used to write equilibrium constant expressions for gaseous reactions. (14.3)

14.34 If the equilibrium partial pressures are $p_{CO_2} = 0.0387$ and $p_{NH_3} = 0.0774$, what is the value of K_p for the following equilibrium? (14.3)

$$H_2NCO_2NH_4(s) \rightleftharpoons 2NH_3(g) + CO_2(g)$$

14.35 What is K_c for the equilibrium

$$CS_2(g) + 4H_2(g) \rightleftharpoons CH_4(g) + 2H_2S(g)$$

if the molarity of CH_4 at equilibrium is 2.50×10^{-2} when the initial molarities of both CS_2 and H_2 are 0.1635? (14.3)

14.36 For the following equilibrium

$$N_2(g) + 3H_2(g) \rightleftharpoons 2NH_3(g)$$

(a) write the expression for K_p. (b) The value of K_p is 9.02×10^{-2} at 227 °C. What is the value of K_c at 227 °C? (14.3)

14.37 The value of the equilibrium constant, K_c, for the equilibrium

$$CO(g) + 3H_2(g) \rightleftharpoons CH_4(g) + H_2O(g)$$

is 0.1764 at 1500 °C. What is the value of K_c for the equilibrium

$$CH_4(g) + H_2O(g) \rightleftharpoons CO(g) + 3H_2(g)$$

at 1500 °C? (14.3)

14.38 The value of the equilibrium constant for the equilibrium

$$2H_2(g) + S_2(g) \rightleftharpoons 2H_2S(g)$$

is 2.5×10^2 at a certain temperature. What is the value of the equilibrium constant for the equilibrium

$$H_2(g) + \tfrac{1}{2}S_2(g) \rightleftharpoons H_2S(g)$$

at the same temperature? (14.3)

14.39 (a) In what way are reaction quotients and equilibrium constant expressions alike? (b) In what way are they different? (c) What are reaction quotients used for? (14.4)

14.40 If $K_c = 6.4 \times 10^{-7}$ for the equilibrium

$$2CO_2(g) \rightleftharpoons 2CO(g) + O_2(g)$$

tell whether each of the following mixtures is at equilibrium. If the mixture is *not* at equilibrium, tell whether reaction will take place to the left or to the right to reach equilibrium. (a) $[CO_2]$ = 5.3 × 10⁻²; $[CO]$ = 3.6 × 10⁻⁴; $[O_2]$ = 2.4 × 10⁻³   (b) $[CO_2]$ = 1.78 × 10⁻¹; $[CO]$ = 2.1 × 10⁻²; $[O_2]$ = 5.7 × 10⁻⁵   (c) $[CO_2]$ = 1.03 × 10⁻¹; $[CO]$ = 2.4 × 10⁻²; $[O_2]$ = 1.18 × 10⁻⁵ (14.4)

14.41 If K_c = 2.5 × 10² for the equilibrium

$$2H_2(g) + S_2(g) \rightleftharpoons 2H_2S(g)$$

when a mixture with $[H_2]$ = 7.6 × 10⁻⁴, $[S_2]$ = 6.8 × 10⁻⁵, and $[H_2S]$ = 3.2 × 10⁻⁵ reaches equilibrium, which concentrations will have increased and which concentrations will have decreased? Explain your answer. (14.4)

14.42 For the equilibrium

$$H_2(g) + I_2(g) \rightleftharpoons 2HI(g)$$

K_c = 54.5 at 425.4 °C. If $[I_2]_{eq}$ = 0.0346 and $[HI]_{eq}$ = 0.0287, what is $[H_2]_{eq}$? (14.4)

14.43 The value of K_c for the equilibrium

$$NO_2(g) + O_2(g) \rightleftharpoons NO(g) + O_3(g)$$

is 7.58 × 10⁻¹¹. If the initial concentration of NO_2 is 0.072 M and the initial concentration of O_2 is 0.083 M, what are the equilibrium concentrations of all four substances? (14.4)

14.44 The value of K_c for the equilibrium

$$2H_2S(g) \rightleftharpoons 2H_2(g) + S_2(g)$$

is 1.00 × 10⁻⁶. If the initial concentration of H_2S is 0.075 M, what are the equilibrium concentrations of all three substances? (14.4)

14.45 The value of K_c for the equilibrium

$$SO_2Cl_2(g) \rightleftharpoons SO_2(g) + Cl_2(g)$$

is 6.4 × 10⁻³ at a certain temperature. If the initial concentration of SO_2Cl_2 is 0.044 M and the initial concentrations of $SO_2(g)$ and $Cl_2(g)$ are both zero, what will be the equilibrium concentrations of all three substances? (14.4)

14.46 If the value of K_c for the equilibrium

$$N_2O_4(g) \rightleftharpoons 2NO_2(g)$$

is 4.64 × 10⁻³ and the equilibrium concentration of NO_2 formed by the decomposition of pure $N_2O_4(g)$ is 1.54 × 10⁻², (a) what is the equilibrium concentration of N_2O_4? (b) What was the initial concentration of N_2O_4? (14.4)

14.47 The equilibrium constant, K_c, for the equilibrium

$$CO(g) + 3H_2(g) \rightleftharpoons CH_4(g) + H_2O(g)$$

is 0.1764 at 1500 °C. If the initial concentration of CO is 0.1000 M and the initial concentration of H_2 is 0.300 M, what are the concentrations of all species at equilibrium? (14.4)

14.48 For the equilibrium

$$H_2(g) + I_2(g) \rightleftharpoons 2HI(g)$$

K_c = 54.5 at 425.4 °C. If 0.020 000 M HI(g) at 425.4 °C is allowed to come to equilibrium, what will be the concentrations of $H_2(g)$, $I_2(g)$, and HI(g)? (14.4)

14.49 For the equilibrium

$$H_2(g) + I_2(g) \rightleftharpoons 2HI(g)$$

K_p = 54.5 at 425.4 °C. If p_{HI} = 0.0453 atm and p_{H_2} = 0.0216 atm at equilibrium, what is p_{I_2} at equilibrium? (14.4)

14.50 The value of K_p for the equilibrium

$$COCl_2(g) \rightleftharpoons CO(g) + Cl_2(g)$$

is 6.7 × 10⁻⁹ at 100 °C. If the initial partial pressure of $COCl_2$ is 0.87 atm, (a) what will be the equilibrium partial pressures of $COCl_2$, CO, and Cl_2?   (b) What will be the total pressure at equilibrium? (14.4)

14.51 For the equilibrium

$$NH_4Cl(s) \rightleftharpoons NH_3(g) + HCl(g)$$

K_p = 6.0 × 10⁻⁹ at a certain temperature. What is the vapor pressure of $NH_4Cl(s)$ at this temperature? (14.4)

14.52 For the equilibrium

$$COCl_2(g) \rightleftharpoons CO(g) + Cl_2(g)$$

K_p = 6.7 × 10⁻⁹ at a certain temperature. If, starting with pure $COCl_2(g)$, the total pressure is 1.25 atm, what are the partial pressures of each of the three gases at equilibrium? (14.4)

14.53 Give three means by which the equilibrium composition of a reaction mixture can be changed. (14.5)

14.54 Consider the equilibrium

$$Br_2(g) + Cl_2(g) \rightleftharpoons 2BrCl(g)$$

At a certain temperature, a mixture in which $[Br_2]$ = 0.0337, $[Cl_2]$ = 0.0200, and $[BrCl]$ = 0.0687 is at equilibrium. (a) What is the value of K_c?   (b) If 0.0137 mol Br_2 is removed from one liter of the equilibrium mixture, what will be the concentrations of all substances after the system again reaches equilibrium? (14.5)

14.55 Tell how the equilibrium

$$2O_3(g) \rightleftharpoons 3O_2(g) \qquad \Delta H° = -285.4 \text{ kJ}$$

would be shifted by each of the following changes. (a) increasing temperature   (b) increasing pressure by decreasing volume   (c) adding a catalyst   (d) removing O_3 (e) increasing pressure by adding helium (14.5)

14.56 Tell how the equilibrium

$$\text{thermal energy} + CO_2(g) + H_2(g) \rightleftharpoons CO(g) + H_2O(g)$$

would be shifted by each of the following changes. (a) increasing temperature   (b) increasing pressure by decreasing volume   (c) adding a catalyst   (d) removing CO_2 (e) adding H_2 (14.5)

14.57 Tell how the equilibrium

$$C(s) + 2H_2(g) \rightleftharpoons CH_4(g) \qquad \Delta H° = -74.85 \text{ kJ}$$

would be shifted by: (a) adding C;   (b) cooling;   (c) increasing the partial pressure of H_2;   (d) decreasing the pressure by increasing the volume. Which of these changes would change the value of the equilibrium constant? (14.5)

14.58 How does each of the following affect the numerical value of the equilibrium constant for the equilibrium

thermal energy + $CH_4(g)$ + $2H_2S(g)$ $\rightleftharpoons$ $CS_2(g)$ + $4H_2(g)$

(a) adding a catalyst   (b) removing H_2 (c) increasing the concentration of CH_4 (d) increasing the temperature   (e) increasing the pressure by decreasing the volume (14.5)

14.59 What is the value of K_p for the equilibrium

$$H_2S(g) + I_2(s) \rightleftharpoons 2HI(g) + S(s)$$

if the equilibrium partial pressures of H_2S and HI are 9.90×10^{-1} and 3.50×10^{-3} atm respectively?

14.60 The value of K_c for the equilibrium

$$CO(g) + Cl_2(g) \rightleftharpoons COCl_2(g)$$

is 4.6×10^9 at 100 °C. In an experiment with $[CO]_0 = 0.0350$, $[Cl_2]_0 = 0.0270$, and $[COCl_2]_0 = 0.000$, what will be the equilibrium concentrations?

14.61 Write the equation for the equilibrium that corresponds to each of the following equilibrium constant expressions:

(a) $K_c = \dfrac{[F]^2}{[F_2]}$ (b) $K_c = \dfrac{[NO]^4[H_2O]^6}{[NH_3]^4[O_2]^5}$

(c) $K_c = \dfrac{[C_6H_5COO^-][H^+]}{[C_6H_5COOH]}$ (d) $K_c = [H_2]^2[O_2]$

14.62 Consider an equilibrium mixture containing 4.562×10^{-3} mol $H_2(g)$, 7.384×10^{-4} mol $I_2(g)$, and 1.355×10^{-2} mol HI(g) in a 1.000-L container at 425.4 °C (the mixture of Sample Problem 14.11). How many moles of $I_2(g)$ must be added to adjust the equilibrium concentration of HI to exactly 1.500×10^{-2} M?

14.63 The value of K_c for the equilibrium

$$COCl_2(g) \rightleftharpoons CO(g) + Cl_2(g)$$

is 2.2×10^{-10}. If the concentration of Cl_2 formed by the decomposition of pure $COCl_2(g)$ at equilibrium is 3.8×10^{-6} M, (a) what are the equilibrium concentrations of the other two substances?   (b) What was the initial concentration of $COCl_2$?

14.64 Which of the following equilibria would be shifted more by increasing the concentration of O_2? Explain your answer.

$$4NH_3(g) + 3O_2(g) \rightleftharpoons 2N_2(g) + 6H_2O(g)$$
$$4NH_3(g) + 5O_2(g) \rightleftharpoons 4NO(g) + 6H_2O(g)$$

14.65 Given the equilibrium

$$2C(s) + O_2(g) \rightleftharpoons 2CO(g) \qquad \Delta H° = -221.04 \text{ kJ}$$

(a) is the forward reaction favored at high or low temperature?   (b) Is the forward reaction favored by high or low pressure?   (c) If equilibrium is reached too slowly, what can be done?

14.66 For the equilibrium

$$C(s) + H_2O(g) \rightleftharpoons CO(g) + H_2(g)$$

$K_c = 6.5 \times 10^{-23}$ at 298 K and 0.111 at 1100 K. Is the forward reaction exothermic or endothermic? Explain your answer.

14.67 (a) For the equilibrium in Problem 14.66, what is the value of K_p at 298 K?   (b) If, at 900 K, $K_p = 2.2$, what is the value of K_c at 900 K?

14.68 A sample of $COCl_2$ is allowed to decompose. The value of K_p for the equilibrium

$$COCl_2(g) \rightleftharpoons CO(g) + Cl_2(g)$$

is 6.7×10^{-9} at 100 °C. If the initial partial pressure of $COCl_2$ is 0.087 atm, what will be the equilibrium concentrations of $COCl_2$, CO, and Cl_2?

14.69 For the equilibrium

$$N_2O_4(g) \rightleftharpoons 2NO_2(g)$$

$K_c = 0.21$ at 100 °C. (a) If 0.100 M $N_2O_4(g)$ is allowed to decompose at 100 °C, what will $[N_2O_4]$ and $[NO_2]$ be at equilibrium?   (b) What will $[N_2O_4]$ be when $[NO_2] = 0.050$?   (c) What will $[NO_2]$ be when $[N_2O_4] = 0.060$?   (d) Use the concentrations you have calculated to sketch a graph like Figure 14.1(a) showing how concentrations change with time.

14.70 The decomposition of HI to H_2 and I_2 was followed by measuring the $[I_2]$ colorimetrically and the following data were obtained:

Time, h	$[I_2]$	Time, h	$[I_2]$
0.00	0.00	8.00	5.53×10^{-4}
0.50	2.20×10^{-4}	12.00	5.77×10^{-4}
2.00	3.93×10^{-4}	20.00	5.75×10^{-4}
4.00	4.91×10^{-4}	40.00	5.76×10^{-4}

(a) Graph $[I_2]$ against time.   (b) From your graph, about how long was required for the system to reach equilibrium?   (c) On your graph, show the effect of adding a catalyst with a broken line.   (d) The decomposition of HI is endothermic. On your graph, show the effect of increasing temperature with a dotted line.

14.71 The value of K_c for the equilibrium

$$N_2O_4(g) \rightleftharpoons 2NO_2(g)$$

is 4.64×10^{-3} at 25 °C. If the initial concentration of NO_2 is 0.0256 M and the initial concentration of N_2O_4 is 0.0000, what will be the concentrations of both gases at equilibrium?

14.72 If $[NH_3]_0$ is 0.250 when the value of K_c for the equilibrium

$$N_2(g) + 3H_2(g) \rightleftharpoons 2NH_3(g)$$

is 1.05×10^{-1}, what are the equilibrium concentrations of all three components?

14.73 If the initial pressure of PCl_5 is 1.65 atm when the value of K_p for the equilibrium

$$PCl_5(g) \rightleftharpoons PCl_3(g) + Cl_2(g)$$

is 0.497, what are the equilibrium pressures of all three components?

14.74 For the equilibrium A(g) $\rightleftharpoons$ B(g), use white circles to represent molecules of A and black circles to represent molecules of B. (a) If $K_c = 5$, (i) draw a molecular-level view of 12 molecules of A in a closed container; (ii) draw the reaction mixture after enough time has passed to reach equilibrium; (iii) show the reaction mixture after enough time has passed to observe that reaction is taking place but before equilibrium is reached; (iv) draw the reaction mixture after twice the time required to reach equilibrium has passed; (v) repeat parts (i) and (ii) starting with molecules of B instead

of A. (b) Starting with any number of A *or* B you like, sketch the equilibrium mixture if $K_c = 0.5$.

14.75 At a temperature where $K_c = 3.6 \times 10^8$ for the equilibrium

$$N_2(g) + 3H_2(g) \rightleftharpoons 2NH_3(g)$$

what is the value of K_c for the equilibrium

$$NH_3(g) \rightleftharpoons \tfrac{1}{2}N_2(g) + \tfrac{3}{2}H_2(g)?$$

STOP & TEST YOURSELF

1. Assuming that the system is at equilibrium, which of the following reactions goes most nearly to completion?

(a) $CH_3CH_2CH_2CH_3(g) \longrightarrow$
$\qquad CH_3CH(CH_3)_2(g) \qquad K_c = 2.5$
(b) $C(s) + 2H_2(g) \longrightarrow CH_4(g) \qquad K_c = 27.5$
(c) $2C(s) + O_2(g) \longrightarrow 2CO(g) \qquad K_c = 1 \times 10^{16}$
(d) $2Cl_2(g) + 2H_2O(g) \longrightarrow$
$\qquad 4HCl(g) + O_2(g) \qquad K_c = 1.88 \times 10^{-15}$
(e) $FeO(s) + CO(g) \longrightarrow$
$\qquad Fe(s) + CO_2(g) \qquad K_c = 0.403$

2. The equation for the equilibrium between nitrogen, hydrogen, and ammonia is

$$N_2(g) + 3H_2(g) \rightleftharpoons 2NH_3(g)$$

If, in the preparation of ammonia from nitrogen and hydrogen, the initial concentration of N_2 is 0.250 M, the initial concentration of H_2 is 0.750 M, and the equilibrium concentration of NH_3 is 0.350 M, what are the equilibrium concentrations of N_2 and H_2? (a) N_2, 0.075 M and H_2, 0.225 M (b) N_2, 0.075 M and H_2, 0.400 M (c) N_2, 0.075 M and H_2, 0.575 M (d) N_2, 0.100 M and H_2, 0.400 M (e) It is impossible to calculate the equilibrium concentrations of N_2 and H_2 because the value of K_c is not given.

3. The equilibrium constant expression for the equilibrium

$$2NH_3(g) + 2O_2(g) \rightleftharpoons N_2O(g) + 3H_2O(g)$$

is: (a) $K_c = \dfrac{[N_2O][H_2O]^3}{[NH_3][O_2]}$ (b) $K_c = \dfrac{[N_2O][H_2O]^3}{[NH_3]^2[O_2]^2}$

(c) $K_c = \dfrac{[NH_3]^2[O_2]^2}{[N_2O][H_2O]^3}$ (d) $K_c = \dfrac{[N_2O][H_2O]}{[NH_3][O_2]}$

(e) $K_c = \dfrac{[NH_3][O_2]}{[N_2O][H_2O]}$

4. The equilibrium constant expression for the equilibrium

$$C(s) + CO_2(g) \rightleftharpoons 2CO(g)$$

is: (a) $K_c = \dfrac{[CO]^2}{[C][CO_2]}$ (b) $K_c = \dfrac{[C][CO_2]}{[CO]^2}$

(c) $K_c = \dfrac{[CO]}{[C][CO_2]}$ (d) $K_c = \dfrac{[CO]^2}{[CO_2]}$ (e) $K_c = \dfrac{[CO_2]}{[CO]^2}$

5. Starting with pure $COCl_2$, if the initial concentration of $COCl_2$ is 0.0350 M and the equilibrium concentration of CO is

0.0232 M, what is the value of K_c for the following equilibrium?

$$COCl_2(g) \rightleftharpoons CO(g) + Cl_2(g)$$

(a) 1.54×10^{-2} (b) 4.56×10^{-2} (c) 1.97 (d) 2.19×10^1 (e) 6.50×10^1

6. For the equilibrium

$$N_2(g) + O_2(g) \rightleftharpoons 2NO(g)$$

if the initial concentration of N_2 is 0.036 M and the initial concentration of O_2 is 0.075 M, what is the equilibrium concentration of NO? The value of K_c is 4.59×10^{-7}. (a) 6.2×10^{-10} (b) 3.8×10^{-6} (c) 8.8×10^{-6} (d) 1.8×10^{-5} (e) 3.5×10^{-5}

7. For the equilibrium

$$Cl_2(g) \rightleftharpoons 2Cl(g)$$

$K_c = 3.6 \times 10^{-2}$. If the initial concentration of Cl_2 is 0.049 M, what is the concentration of Cl at equilibrium? (a) 0.017 (b) 0.021 (c) 0.034 (d) 0.044 (e) 0.052

Use the following information to answer Problems 8 and 9. For the equilibrium

$$CO(g) + 3H_2(g) \rightleftharpoons CH_4(g) + H_2O(g)$$

$K_c = 3.89$.
(a) $[CH_4] = 0.047$, $[H_2O] = 0.058$, $[CO] = 0.033$, $[H_2] = 0.156$
(b) $[CH_4] = 0.068$, $[H_2O] = 0.072$, $[CO] = 0.029$, $[H_2] = 0.251$
(c) $[CH_4] = 0.052$, $[H_2O] = 0.034$, $[CO] = 0.296$, $[H_2] = 0.145$
(d) $[CH_4] = 0.236$, $[H_2O] = 0.029$, $[CO] = 0.368$, $[H_2] = 0.124$
(e) $[CH_4] = 0.025$, $[H_2O] = 0.036$, $[CO] = 0.041$, $[H_2] = 0.178$

8. Which of the mixtures is at equilibrium?
9. In which of the mixtures will the net reaction take place in the forward direction to reach equilibrium?
10. If 0.0250 mol Cl_2 is added to an equilibrium mixture of 0.0275 mol PCl_3, 0.0275 mol Cl_2, and 0.0195 mol PCl_5 in a 1-L container, what are the concentrations of all three substances after the new equilibrium has been reached? The equation for this equilibrium is

$$PCl_5(g) \rightleftharpoons PCl_3(g) + Cl_2(g)$$

and the value of K_c is 3.9×10^{-2}. A correct setup for solving this problem is

(a) $\dfrac{(0.0195 + x)}{(0.0275 - x)(0.0525 - x)} = 3.9 \times 10^{-2}$

(b) $\dfrac{(0.0195 - x)}{(0.0275 + x)(0.0525 + x)} = 3.9 \times 10^{-2}$

(c) $\dfrac{(0.0195 + x)}{(0.0275 - x)^2} = 3.9 \times 10^{-2}$

(d) $\dfrac{(0.0275 - x)^2}{(0.0195 + x)} = 3.9 \times 10^{-2}$

(e) $\dfrac{(0.0275 - x)(0.0525 - x)}{(0.0195 + x)} = 3.9 \times 10^{-2}$

11. Which of the following changes would cause the equilibrium

$$SiF_4(g) + 2H_2O(g) \rightleftharpoons$$
$$SiO_2(s) + 4HF(g) \qquad \Delta H^\circ = +148.9 \text{ kJ}$$

to shift to the left? (a) adding more SiO_2 (b) increasing the temperature (c) adding a catalyst (d) decreasing the pressure by increasing the volume (e) removing water

12. Which of the changes in Problem 11 will result in a change in the value of the equilibrium constant?
13. For which of the following equilibria is $K_c = K_p$?
    (a) $C(s) + CO_2(g) \rightleftharpoons 2CO(g)$
    (b) $Cl_2(g) \rightleftharpoons 2Cl(g)$
    (c) $2H_2(g) + O_2(g) \rightleftharpoons 2H_2O(g)$
    (d) $CH_4(g) + H_2O(g) \rightleftharpoons CO(g) + 3H_2(g)$
    (e) $N_2(g) + O_2(g) \rightleftharpoons 2NO(g)$
14. For the equilibrium

$$CO(g) + 3H_2(g) \rightleftharpoons CH_4(g) + H_2O(g)$$

the value of K_c is 0.176 at 1500 °C. What is the value of K_p at 1500 °C? (a) 5.00×10^{-9} (b) 8.31×10^{-6} (c) 1.16×10^{-5} (d) 2.67×10^{-3} (e) 3.73×10^3

15. For the equilibrium

$$Cl_2(g) \rightleftharpoons 2Cl(g)$$

$K_p = 6.71$. If the total pressure is 0.987 atm, what is the partial pressure of Cl? (a) 0.114 atm (b) 0.316 atm (c) 0.389 atm (d) 0.873 atm (e) 2.57 atm

PUTTING THINGS TOGETHER

14.76 In the molecular-level view of the left photo in Figure 14.7, white circles represent N_2O_4 molecules and brown circles represent NO_2 molecules.

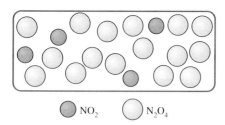

NO₂   N₂O₄

(a) Draw a molecular-level view of the right photo. (b) Explain why the pressure in the tube in the right photo may be dangerously high.

14.77 (a) At 3000 K, a 1.00-mol sample of CO_2 in a 1.00-L container is 54.8% decomposed to carbon monoxide and oxygen at equilibrium:

$$2CO_2(g) \rightleftharpoons 2CO(g) + O_2(g)$$

What is the value of K_c? (b) No decomposition of carbon dioxide into carbon monoxide and oxygen can be detected at room temperature. Explain the difference.

14.78 A mixture of 0.074 mol H_2 and 0.151 mol CH_4 in a 1.00-L container is at equilibrium with solid carbon. The equation for the equilibrium is

$$C(s) + 2H_2(g) \rightleftharpoons CH_4(g)$$

(a) What is the value of K_c? (b) If 0.015 mol H_2 is removed, what will be the concentrations of H_2 and CH_4 at the new equilibrium?

14.79 For the equilibrium

$$2SO_2(g) + O_2(g) \rightleftharpoons 2SO_3(g)$$

$K_p = 9.08 \times 10^1$ at 600 °C. In an experiment in which the initial concentration of SO_2 is 9.31×10^{-3} M, the initial concentration of O_2 is 4.65×10^{-3} M, and the initial concentration of SO_3 is 0, what will be the concentrations of all three species at equilibrium?

14.80 From the graph below, (a) write the equation for the equilibrium using whole-number coefficients. (b) What is the value of K_c?

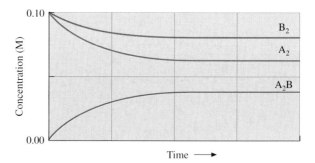

(c) How would the graph be changed if a catalyst was used?

14.81 A mixture with $[N_2O_4] = 0.090$ and $[NO_2] = 0.134$ is at equilibrium. (a) What is the value of K_c for the following equilibrium?

$$N_2O_4(g) \rightleftharpoons 2NO_2(g)$$

(b) If the volume is decreased to half the original volume by increasing the pressure on the equilibrium mixture, what happens to the concentrations of N_2O_4 and NO_2? (c) Is the mixture in half the original volume still at equilibrium? If not, in which direction will reaction have to take place to restore equilibrium? (d) What will be the concentrations of N_2O_4 and NO_2 at the new equilibrium?

14.82 The forward reaction in the equilibrium

$$CH_3COOH(l) + CH_3CH_2OH(l) \rightleftharpoons$$

$$CH_3COOCH_2CH_3(l) + H_2O(l)$$

which is a homogeneous equilibrium, is an example of an esterification reaction, a type of reaction that is important in organic chemistry. The equilibrium constant expression for this equilibrium is

$$K_c = \frac{[CH_3COOCH_2CH_3][H_2O]}{[CH_3COOH][CH_3CH_2OH]}$$

The value of K_c at 25 °C is 4.00. (a) Why is the concentration of water not included in the value of the equilibrium constant K_c for this equilibrium—that is, why does the concentration of water appear in the equilibrium constant expression? (b) Suppose 0.500 mol CH_3COOH and 1.000 mol CH_3CH_2OH are used and you need to calculate how much $CH_3COOCH_2CH_3$ will be formed at equilibrium. The CH_3COOH and CH_3CH_2OH are dissolved in a suitable solvent (one that dissolves both reactants and both products) and the volume of the solution is one liter. What value do you get for x if you make the usual approximation that x is negligible? Is this value possible? Explain your answer. (c) What is the maximum value that is possible for x? (d) Assume a possible value for x and calculate how much $CH_3COOCH_2CH_3$ will be formed at equilibrium by successive approximations. (e) Explain why esterification can be made to go to completion by removing water as it is formed.

14.83 (a) Write an equation for the equilibrium between liquid water and water vapor. (b) Write the equilibrium constant expression for K_p for this equilibrium. (c) What is the numerical value of K_p for this equilibrium at 28.6 °C? (d) What is the numerical value of K_c for this equilibrium at 28.6 °C? (e) At equilibrium, what is the molarity of water vapor in air at 28.6 °C?

14.84 Imagine that the quantities of $CO_2(s)$ (dry ice), $CaCO_3(s)$, and $CaO(s)$ given below were placed in an empty 2.5-L container at a low temperature and the temperature was then raised to 800 °C. (At 800 °C, $K_p = 0.236$, and the carbon dioxide is gaseous.) For each mixture, tell whether equilibrium can be reached if the mixture is held at 800 °C. If equilibrium can be reached, will the quantity of solid calcium carbonate increase, decrease, or remain the same? For simplicity, assume that the volume occupied by the solid $CaCO_3$ and CaO is negligible.
(a) Experiment 1: 0.75 g $CaCO_3$; 0.50 g CaO; 0.50 g CO_2
(b) Experiment 2: 0.75 g $CaCO_3$; 0.50 g CaO; 0.295 g CO_2
(c) Experiment 3: 0.34 g $CaCO_3$; 0.40 g CaO; 0.10 g CO_2

14.85 In an experiment to determine the equilibrium constant for the equilibrium

$$2Cl_2(g) + 2H_2O(g) \rightleftharpoons 4HCl(g) + O_2(g)$$

8.4 g of Cl_2 and 2.7 g of H_2O were placed in a 4.2-L container. After equilibrium was reached, the concentration of O_2 was found to be 2.3×10^{-5} M. From this experiment, what is the value of K_c?

14.86 For the equilibrium

$$\underset{\text{butane}}{CH_3CH_2CH_2CH_3(g)} \rightleftharpoons \underset{\text{isobutane}}{CH_3CH(CH_3)_2(g)}$$

if the value of K_c is 2.5, what is the percent by mass of isobutane in the equilibrium mixture?

14.87 An equilibrium mixture has the following composition: $[CO] = 0.022$, $[H_2O] = 0.022$, $[CO_2] = 0.019$, and $[H_2] = 0.034$. (a) What is the value of K_c for the following equilibrium?

$$CO_2(g) + H_2(g) \rightleftharpoons CO(g) + H_2O(g)$$

(b) What are the concentrations of the four substances in mole fractions? (c) How does the value of the equilibrium constant calculated using mole fractions compare with the value calculated using molarities in part (a)? (d) Would the relationship between K_c calculated using molarities and K_c calculated using mole fractions be the same for the following equilibrium?

$$CO(g) + 3H_2(g) \rightleftharpoons CH_4(g) + H_2O(g)$$

Explain your answer.

14.88 A 50.0-mL sample of the equilibrium mixture from an experiment to determine the value of K_c for the equilibrium

$$H_2(g) + I_2(g) \rightleftharpoons 2HI(g)$$

was cooled very rapidly so that no reaction took place. The HI was absorbed in 25.00 mL of 0.0502 M NaOH, and the excess NaOH was titrated with 0.0498 M HCl. If 9.55 mL of 0.0498 M HCl were required to neutralize the excess NaOH, what was the concentration of HI in the original 50.0-mL sample of equilibrium mixture?

14.89 (a) What is the theoretical yield of HI from the reaction of 0.003 55 g H_2 with 0.556 g I_2? (b) How many grams of HI are present in the equilibrium mixture from the reaction of 0.003 55 g H_2 with 0.556 g I_2 in a 1.000-L container if the value of K_c for the equilibrium

$$H_2(g) + I_2(g) \rightleftharpoons 2HI(g)$$

is 55.2? (c) How could the fraction of the I_2 converted to HI be increased?

14.90 Write molecular equations and equilibrium constant expressions for each of the following equilibria: (a) Carbon disulfide gas and chlorine gas in equilibrium with disulfur dichloride gas and carbon tetrachloride gas (b) Liquid silicon tetrachloride and gaseous water in equilibrium with solid silicon dioxide and hydrogen chloride gas (c) Solid tin(IV) oxide and hydrogen gas in equilibrium with solid tin and water vapor (d) Solid copper and a water solution of silver nitrate in equilibrium with solid silver and a water solution of copper(II) nitrate

14.91 The density of $CaCO_3(s)$ is 2.71 g/cm^3. What is the molarity of $CaCO_3(s)$?

14.92 The mixture of hydrogen sulfide and hydrogen gases in equilibrium with diantimony trisulfide and antimony at 440 °C and 765.0 mmHg is 92.71% by mass H_2S. The equation for the equilibrium is

$$Sb_2S_3(s) + 3H_2(g) \rightleftharpoons 2Sb(s) + 3H_2S(g)$$

What is the value of K_c for this equilibrium at 440 °C?

14.93 A 6.45×10^{-2} M sample of N_2O_4 is allowed to stand until $[N_2O_4]$ is constant. If the value of K_c for the equilibrium

$$N_2O_4(g) \rightleftharpoons 2NO_2(g)$$

is 4.61×10^{-3}, calculate (a) $[N_2O_4]_{eq}$ and (b) the percent decomposition of N_2O_4 at equilibrium.

14.94 (a) Can you tell whether the equilibrium mixture in the top photo in Figure 14.6 was prepared from NO_2 or from N_2O_4? Explain your answer. (b) Explain why the reactions shown in Figure 4.12 do not reach equilibrium.

14.95 If a 40.0-L reaction vessel contains 0.800 mol N_2, 2.40 mol H_2, and 0.400 mol NH_3 and K_c for the equilibrium

$$N_2(g) + 3H_2(g) \rightleftharpoons 2NH_3(g)$$

is 0.500, will more ammonia decompose or be formed as the system moves toward equilibrium?

14.96 For the equilibrium ●● + ●● $\rightleftharpoons$ 2●●, K_c = 4. Given the molecular-level drawing for the initial mixture,

sketch a molecular-level picture of the equilibrium mixture.

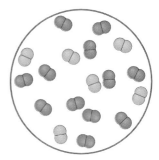

Initial mixture

14.97 If 0.0220 mol SO_3 decomposes in a 1.50-L container at 800 K, 0.0150 mol SO_3 remains at equilibrium. What is the value of K_c for the following equilibrium?

$$2SO_3(g) \rightleftharpoons 2SO_2(g) + O_2(g)$$

14.98 If 7.44 g $N_2O_4(g)$ and 1.64 g $NO_2(g)$ are present at equilibrium in a 3.00-L container, what is the value of K_c for the following equilibrium?

$$N_2O_4(g) \rightleftharpoons 2NO_2(g)$$

APPLICATIONS

14.99 Rainwater contains carbon dioxide gas from the atmosphere. Explain the following observations in terms of shifting the equilibrium

$$CaCO_3(s) + H_2O(l) + CO_2(g) \rightleftharpoons$$
$$Ca^{2+}(aq) + 2HCO_3^-(aq) \qquad \Delta H° = -321.7 \text{ kJ}$$

(a) When rainwater passes through limestone, $CaCO_3$, caves are formed. The rainwater becomes hard water. (Water containing calcium ions and magnesium ions in solution is called hard water.) (b) When hard water is heated in kettles, hot water heaters, or boilers, scale forms. (c) When hard water seeps slowly through the roofs of caves, stalactites and stalagmites form.

14.100 In an industrial plant, synthesis of ammonia from nitrogen and hydrogen is carried out at 500 °C (zeros are significant) and a total pressure of 300.0 atm. The value of K_p for the equilibrium

$$N_2(g) + 3H_2(g) \rightleftharpoons 2NH_3(g)$$

is 1.35×10^{-5} at 500 °C. The mole ratio of reactants given by the equation is used. What are the equilibrium partial pressures of all three substances?

14.101 Imagine that you are a newspaper reporter assigned to interview an inventor. The inventor claims to have discovered a new catalyst that increases the percent of the nitrogen and hydrogen that is converted to ammonia by the Haber process. Would you advise your editor to publish the story or not? Explain your answer.

14.102 The equilibrium

$$4NH_3(g) + 5O_2(g) \rightleftharpoons$$
$$4NO(g) + 6H_2O(g) + \text{thermal energy}$$

is part of the industrial process for making nitric acid from ammonia. What conditions should be used to obtain the maximum possible conversion of NH_3 to NO?

14.103 The forward reaction of the equilibrium

$$CH_4(g) + H_2O(g) \rightleftharpoons CO(g) + 3H_2(g)$$

is called reforming. The mixture of CO and H_2 produced is known as synthesis gas and is used to make ammonia and methyl alcohol. The reverse reaction, which is called methanation, is used to remove CO from H_2 before the H_2 is converted to NH_3. (a) Calculate $\Delta H°_{rxn}$ for the forward reaction. (b) Compare the conditions that would be used to favor the forward reaction with the conditions required to favor the reverse reaction.

14.104 The value of K_c for the equilibrium

$$N_2(g) + O_2(g) \rightleftharpoons 2NO(g)$$

is 4.8×10^{-31} at 25 °C. However, at the temperatures that exist in automobile engines, the value of K_c is much higher (of the order 10^{-4} to 10^{-3}). When the exhaust leaves an engine, it cools very rapidly, and the NO formed at the higher temperature does not decompose and is a serious air pollutant. (a) Is the forward reaction in this equilibrium endothermic or exothermic? Give your reasoning.   (b) Suggest a way to reduce the amount of NO formed in automobile engines.   (c) What can you conclude about the rate of the reverse reaction at ordinary temperatures?

14.105 For the equilibrium

$$Na_2SO_4 \cdot 10H_2O(s) \rightleftharpoons Na_2SO_4(s) + 10H_2O(g)$$

the value of K_p is 9.99×10^{-17} at 25 °C, and for the equilibrium

$$CaCl_2 \cdot 6H_2O(s) \rightleftharpoons CaCl_2(s) + 6H_2O(g)$$

the value of K_p is 5.090×10^{-44} at 25 °C. (a) Calculate the vapor pressure of water above each of these salt hydrates. (b) Which would be the more effective drying agent?

14.106 Besides the equilibrium in Problem 14.102, the following equilibria also exist when ammonia is oxidized in air:

$$4NH_3(g) + 3O_2(g) \rightleftharpoons 2N_2(g) + 6H_2O(g)$$
$$2NH_3(g) + 2O_2(g) \rightleftharpoons N_2O(g) + 3H_2O(g)$$
$$4NH_3(g) + 7O_2(g) \rightleftharpoons 4NO_2(g) + 6H_2O(g)$$
$$2NH_3(g) \rightleftharpoons N_2(g) + 3H_2(g)$$

(a) Write the equilibrium constant expressions for these four equilibria.   (b) From the value of K_c in Problem 14.111, what is the value of K_c for the fourth equilibrium at 427 °C?

14.107 The product of the reaction

$$C(coal) + H_2O(g) \longrightarrow CO(g) + H_2(g)$$

which is called water gas, was used for illumination and heating until it was displaced by natural gas in the 1950s. (a) If $[H_2O]_0 = 0.8$ and the value of K_c for the equilibrium

$$C(s) + H_2O(g) \rightleftharpoons CO(g) + H_2(g)$$

is 0.2 around 1000 °C, the temperature at which the reaction is carried out, what are the equilibrium concentrations?   (b) At 25 °C, the value of K_c is 7×10^{-23}. Is the production of water gas an exothermic or an endothermic process?   (c) To obtain the maximum conversion of H_2O to water gas, should pressure be high or low? Explain your answer.   (d) Give one practical problem associated with using low pressures for gaseous reactions and one practical problem associated with using high pressures.

14.108 Sodium hydrogen carbonate is commonly known as baking soda and is also used in dry-powder fire extinguishers. The value of K_p for the equilibrium

$$2NaHCO_3(s) \rightleftharpoons Na_2CO_3(s) + CO_2(g) + H_2O(g)$$

is 0.25 at 125 °C. (a) Over $NaHCO_3(s)$ in a closed container at 125 °C, what is the partial pressure of CO_2? The partial pressure of water vapor? The total gas pressure?   (b) Does the size of the container make any difference? Explain your answer.   (c) From the fact that $NaHCO_3$ can be kept indefinitely on a closet shelf at room temperature but releases CO_2 and water vapor when baked or thrown on a fire, what can you conclude about on which side of the equation thermal energy belongs?

14.109 Cobalt can be produced by heating cobalt(II) oxide with carbon monoxide. If the value of K_c for the equilibrium

$$CoO(s) + CO(g) \rightleftharpoons Co(s) + CO_2(g)$$

is 490 at 550 °C, (a) what is the value of K_p at 550 °C?   (b) If the total pressure is 12.4 atm, what is the partial pressure of CO_2? Of CO?   (c) What type of reaction are the forward and reverse reactions?

14.110 The concentrations of trace substances in Earth's atmosphere are sometimes expressed in molecules/cm³. If K_c for the equilibrium

$$HO_2NO_2(g) \rightleftharpoons HO_2(g) + NO_2(g)$$

is 1.32×10^{-10} at a certain temperature, (a) what are the units of K_c?   (b) What would be the numerical value of the equilibrium constant if the concentrations were in molecules/cm³?

14.111 For the equilibrium in the Haber–Bosch process for the synthesis of ammonia,

$$N_2(g) + 3H_2(g) \rightleftharpoons 2NH_3(g)$$

$K_c = 2.68 \times 10^{-1}$ at 427 °C. If the initial concentration of $N_2 = 0.653$, the initial concentration of $H_2 = 1.958$, and the initial concentration of NH_3 is 0, (a) what will be the equilibrium concentration of NH_3?   (b) Considering a 1.000-L sample, how many moles of NH_3 would be formed if the reaction went to completion—that is, what is the theoretical yield of NH_3?   (c) What is the percent yield of NH_3 at equilibrium?

14.112 (a) In addition to O_2 and CO, hemoglobin can also combine with CO_2 and H^+. Although the equilibrium is complicated, it can be summarized as

$$[Hb(O_2)] + H^+ + CO_2 \rightleftharpoons [Hb(H^+)(CO_2)] + O_2$$

(a) Predict the effect of the production of lactic acid, $CH_3CH(OH)COOH$, and carbon dioxide by muscles during exercise on the bonding of oxygen by hemoglobin.   (b) Exercising muscle also releases thermal energy, which causes the release of oxygen from oxyhemoglobin. Which reaction of the hemoglobin–oxygen–oxyhemoglobin equilibrium is exothermic?

14.113 The fastest-growing use of methyl alcohol, CH_3OH, is to make the octane enhancer methyl *tert*-butyl ether, $CH_3OC(CH_3)_3$. Methyl alcohol is also being considered as a fuel for automobiles and as a starting material for synthetic gasoline. Today all methyl alcohol is produced by the reaction of carbon monoxide and hydrogen. For the equilibrium

$$CO(g) + 2H_2(g) \rightleftharpoons CH_3OH(l)$$

$K_p = 1.3 \times 10^{-4}$ at 300 °C (zeros significant). (a) What is the value of K_c?   (b) In which direction will this equilibrium shift if the temperature is raised?   (c) In the industrial process, the stoichiometric ratio of CO to H_2 is used. If reaction is carried out at a total pressure of 30 atm, what partial pressures of CO and H_2 are used? (Assume that the vapor pressure of $CH_3OH(l)$ is negligible.)

Forensic Science and Chemistry

DONNA R. REES

Forensic Scientist, Illinois State Police

B.S. Chemistry
Illinois State University

M.S. Forensic Chemistry
University of Pittsburgh

As a forensic scientist, I apply science to the law, providing the courts with information to bring them to the truth. I have not always wanted to be a forensic scientist—in fact, when I was a kid, I did not even know that they existed! I liked science in high school, and when I thought about the future, I decided to find some field I could work in that involved science. My father, a dentist, encouraged me to consider a medical field—like a dental hygienist. In that way I could achieve another one of my career goals: helping people. But the healing arts just did not appeal to me. I thought I would become a teacher instead, and entered college intending to double major in biology and French.

Once in college, I discovered chemistry: here was something that I liked, at which I was pretty good, and had a job market that looked promising! I had also decided that the life of a teacher was not for me, so I changed my major to chemistry. When I was a junior, I received a pamphlet from the American Chemical Society on nontraditional jobs in chemistry. One of the fields described was forensic chemistry—I was fascinated! I had found a career that sounded interesting, used science, and allowed me to serve my fellow man. One of the professors at Illinois State worked in forensic science on the side, and he had developed a course in arson investigation. I immediately enrolled to get a feel of what forensic science was about, and I was hooked.

After graduation, I enrolled in a graduate program in forensic science. There I got an overview of all of the different specialities found in a crime lab: from photography to firearms examination, from document examination to drug analysis, and from typing body fluids to developing fingerprints. We learned about relatively mundane subjects such as how a case proceeds through the court system, as well as more exciting subjects such as the construction of a working letter bomb. We also were able to do internships and work on research projects—an important facet of any science.

Much of forensic science is based on Locard's principle, which basically states that when two objects come together, some trace of each is left with the other as the result of that event. For example, if Mr. X stabs Mr. Z, there may be some transfer of hair, blood, fibers, soil, etc., that can connect the two people, as well as connect them with the place where the stabbing occurred. Items of evidence are examined for class characteristics, which would include the item in a group of similar things, and then individual characteristics are looked for to establish uniqueness. For example, a shoe print left at a robbery scene may be determined to have been caused by a Nike size 9 shoe (class characteristics), and it may later be narrowed down to the Nike size 9 shoes of John Doe by its unique (individual) characteristics.

Every section in our laboratory uses chemistry to some extent in looking for these transfers and characteristics, as well as for establishing simple facts. My use of chemistry is obvious when I examine a powder for the presence of cocaine. I may take a small amount of the sample that the police have submitted and dissolve it in methanol (CH_3OH). I can then inject it with a syringe into a gas chromatograph/mass spectrometer (GCMS), where the sample enters a heated tube called a column. This column is lined with a coating that attracts the components of a sample to differing degrees. Thus, if a sample contains both cocaine and heroin, they will arrive at the end of the column at different times. As the separated components of the sample emerge, they are bombarded with a beam of electrons that breaks

up the molecules and ionizes the fragments, which can then be analyzed by their charge and mass. The computer on the GCMS produces a graph of these sizes versus the amounts of each fragment. Different graphs are produced for each compound. Comparing the graph of an "unknown" drug to graphs made of known drugs (like cocaine or heroin) may allow me to identify the unknown drug. Similar equipment is used to identify the components in a fire scene sample being tested for gasoline, or to prepare profiles to compare the lipstick found on a dead body to that found on a suspect's shirt.

I also use chemistry in the forensic biology section. When a piece of clothing is submitted for examination as evidence in a homicide case, I study the garment carefully, looking for any signs of a bloodstain. I can then cut out a sample of a stain and test it with Kastle-Meyer (KM) reagent. The heme found in blood will react with the components of the reagent to produce a pink color. A piece of the garment that has not been stained is also tested to make sure the fabric itself does not cause a color. When I find the development of a pink color for the stain and none for the unstained portion, I know that there may be blood present. I will then proceed to positively identify the stain as blood, as well as test to determine which species produced the blood. I use the Ouchterlony double diffusion method for this test: I punch holes in a gel, then fill them with knowns, my samples, and antiserum (antihuman serum when I am trying to show that the sample has human blood). The compounds diffuse through the gel, and where human blood and anti-human serum meet, an opaque band develops. When I see this reaction, I know I am dealing with human blood. Once the blood is identified as being human in origin, I can proceed to type it and compare it to the blood types of the people involved in the case.

Being a forensic scientist/chemist requires a clean background, good communication skills, good laboratory skills, and scientific inquisitiveness—the ability to look for those hidden links in a case. While the job is not all fun (there is a lot of paperwork and record keeping to do, and some of the evidence I work with may be stinky, gory, and hazardous to my health), I have found forensic science to be a rewarding career.

ACIDS AND BASES

15

The tart taste of fruits results from organic acids.

This chapter will introduce one of the most frequently used definitions of acids and bases, the Brønsted–Lowry definitions. It will illustrate the various common types of calculations that involve acids and bases and discuss some of the major uses of these calculations.

◼ Why is a study of acids and bases a part of general chemistry?

Of the 50 compounds manufactured in largest quantity in the United States, 13, or 26%, are acids or bases. Acids and bases are used as either reactants or catalysts in many industrial processes. For example, ammonium nitrate, which is used to make fertilizers and explosives, is manufactured by neutralizing ammonia with nitric acid,

$$NH_3(g) + HNO_3(aq) \longrightarrow NH_4NO_3(aq)$$

and phosphoric acid is used as a catalyst for the production of ethyl alcohol,

$$H_2C\ CH_2(g) + H_2O(g) \xrightarrow{\quad H_3PO_4,\ 300\ °C,\ 70\ atm \quad} CH_3CH_2OH(g)$$

Many biological processes also involve acids and bases—for example, buildup of lactic acid in muscle tissue during heavy exercise leads to the feeling of tiredness, and formation of lactic acid in milk during souring causes curdling. Hydrochloric acid is a component of the digestive juices in your stomach; too much causes ulcers, and too little sometimes results in anemia. The acidity and basicity of soil and water are of great importance to plants and animals living in them. Geological processes such as the weathering of rocks are also very much affected by the acidity of water.

A number of familiar materials, such as aspirin and vitamin C, are acids. Vinegar is a dilute solution of acetic acid, lemonade contains citric and ascorbic acids (ascorbic acid is vitamin C), and the battery in your car contains sulfuric acid. Milk of magnesia is basic as are baking soda and washing soda. Household ammonia, whitewash, oven cleaners,

549

and most drain cleaners are also basic. Acid–base reactions form one of the most important classes of chemical reactions, and an understanding of acids and bases is necessary in biology and geology as well as in chemistry.

— What do you already know about acids and bases?

Acids are compounds that increase the concentration of hydrogen ions when dissolved in water. Acids taste sour (Section 4.2). The names and formulas of some common acids are given in Table 4.2.

When an acid dissolves in water, hydrogen ions are formed, for example,

$$HCl(aq) \longrightarrow H^+(aq) + Cl^-(aq)$$

Hydrochloric acid is a strong electrolyte. In dilute aqueous solutions (concentration 0.01 M or less), almost all the hydrochloric acid is in the form of hydrogen ions and chloride ions. Other acids, such as acetic acid, are weak electrolytes; most of a weak acid is in the form of molecules in aqueous solution.

Some acids contain more than one hydrogen that can form hydrogen ions in solution. Sulfuric acid, H_2SO_4, is an example. Some acids contain hydrogens that do *not* form hydrogen ions in solution. For example, acetic acid, $HC_2H_3O_2$ or CH_3COOH, has three hydrogen atoms—the three which are bonded to carbon—that do not form hydrogen ions:

Only this H forms H^+ in aqueous solution.

Hydrogens that form hydrogen ions are written separately on the left side of formulas such as $HC_2H_3O_2$. In the Lewis formula for acetic acid, notice that it is the hydrogen bonded to the strongly electronegative oxygen atom that is acidic. The hydrogens that are bonded to the less electronegative carbon atom are not acidic.

Bases are compounds that increase the concentration of hydroxide ion when dissolved in water. Sodium hydroxide, NaOH, barium hydroxide, $Ba(OH)_2$, and ammonia, NH_3, are common bases. Solutions of bases feel slippery. The unit particles of the hydroxides of the Group IA and IIA metals, such as NaOH and $Ba(OH)_2$, are metal ions and hydroxide ions. The ions in an aqueous solution of ammonia are formed by a chemical reaction between ammonia and water:

$$NH_3(g) + H_2O(l) \longrightarrow NH_4^+(aq) + OH^-(aq)$$

Acids and bases react to form water and a salt; for example, hydrochloric acid and sodium hydroxide react to form water and sodium chloride:

$$HCl(aq) + NaOH(aq) \longrightarrow H_2O(l) + NaCl(aq)$$

The red coloring material extracted from a rose with methyl alcohol is an indicator that turns blue in basic solution.

The reaction of an acid with a base to form water and a salt is called neutralization. If the acid and base are mixed in the proportions shown by the equation for the reaction, both the sour taste of the acid and the slippery feeling of the base disappear, and the solution formed has a salty taste. The driving force for neutralization is the formation of the weak electrolyte, water. The net ionic equation for the neutralization of a strong acid with a strong base is

$$H^+(aq) + OH^-(aq) \longrightarrow H_2O(l)$$

Another reaction characteristic of acids is reaction with metals to form hydrogen gas and a salt of the metal (Section 4.5), for example,

$$2HCl(aq) + Zn(s) \longrightarrow H_2(g) + ZnCl_2(aq)$$

This type of reaction is oxidation-reduction.

15.1 THE BRØNSTED–LOWRY DEFINITIONS

The simple definitions of acids and bases given in Section 4.2 are usually referred to as the **classical** or **Arrhenius* definitions**. Why do we need other definitions? The Arrhenius definitions are limited to aqueous solutions; acids increase the concentration of hydrogen ions, and bases increase the concentration of hydroxide ions *when dissolved in water*. However, reactions that resemble acid–base reactions take place in the gas phase and in solvents other than water. For example, ammonia and hydrogen chloride react in the gas phase to form the salt ammonium chloride:

$$NH_3(g) + HCl(g) \longrightarrow NH_4Cl(s)$$

as shown in ▪Figure 15.1. When benzene solutions of acetic acid and ammonia are

---

*Arrhenius was a Swedish physical chemist whose definitions were first presented in his Ph.D. thesis in 1884. The faculty of the University of Uppsala thought Arrhenius's idea wrong and barely passed him. However, in this case, the faculty turned out to be wrong; in 1903 Arrhenius received the Nobel Prize for chemistry for his work on the theory of electrolytes.

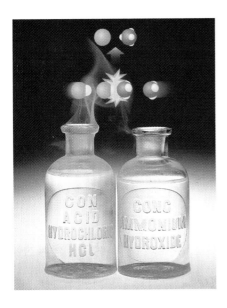

■ FIGURE 15.1   Macroscopic and microscopic views of the reaction of ammonia gas and hydrogen chloride gas to form ammonium chloride, which is a white solid. This reaction is the source of the white film on the reagent bottles (and possibly on the windows) in your general chemistry lab. Why is the white cloud closer to the hydrochloric acid bottle?

The Acid-Base Reactions module of the Acids and Bases section asks students to predict the products of acid-base reactions, identify conjugate acid/base pairs, and predict the position of equilibrium.

mixed, acetic acid reacts with ammonia to form the salt ammonium acetate:

$$CH_3COOH(benzene) + NH_3(benzene) \longrightarrow CH_3COONH_4(s)$$

The ammonium acetate precipitates because it is insoluable in benzene.

A more general definition of acids and bases that includes reactions in the gas phase and in nonaqueous solvents is desirable.

According to the **Brønsted–Lowry**[*] definitions, an **acid** is *any species that can donate a proton* (H^+, a hydrogen atom without its electron). A **base** is *any species that can accept a proton.* An **acid–base reaction** *consists of the transfer of a proton from an acid to a base.* According to the Brønsted–Lowry definitions, the reaction between ammonia and hydrogen chloride in the gas phase is an acid–base reaction:

Brønsted–Lowry base + Brønsted–Lowry acid ⟶

The curved arrow shows the shift of electrons that results in the transfer of a proton from the acid to the base. Brønsted–Lowry bases might better be described as species that *take* a proton from an acid and Brønsted–Lowry acids as species that *give up* a proton to a base.

The curved arrows show how electrons must rearrange to result in the transfer of the hydrogen atom from the acid to the base. The unshared pair of electrons on the nitrogen of ammonia is attracted by the partial positive charge on the hydrogen in the polar hydrogen chloride molecule. Comparison of the structures of the reactants and products shows that the hydrogen atom, which was bonded to chlorine in the reactants, is bonded to nitrogen in the products. Thus, a proton has been transferred from chlorine to nitrogen. To act as a Brønsted–Lowry base, a species must have at least one unshared pair of valence electrons so that it can form a covalent bond to the proton. To act as a Brønsted–Lowry acid, a species must contain at least one hydrogen that is attached to an atom that is more electronegative than hydrogen.

Many of our general chemistry students will take organic and have to learn to use electron-shift arrows to write resonance structures and mechanisms. For the arguments in favor of starting students off using curved arrows to represent movement of electrons rather than atoms, see Loudon, G. M. Organic Chemistry, 2nd ed.: Benjamin/Cummings, Menlo Park, CA, 1988; p. 15; and Dawood, A.; Delaware, D. L.; Fountain, K. R. J. Chem. Educ. 1990, 67, 1011–1018.

PRACTICE PROBLEM

15.1   Use Lewis formulas to show that the reaction between ammonia molecules and acetic acid molecules to form the ionic solid ammonium acetate, which takes

---

[*]Brønsted was a Danish physical chemist, and Lowry was an English chemist. In 1923 they both introduced the same definitions for acids and bases. Occurrence of the same idea to two or more people at about the same time is fairly common in science.

Organic chemists usually use the condensed structural formula CH_3COOH to represent acetic acid and write formulas such as CH_3COONH_4 for salts of acetic acid. Inorganic chemists often write $HC_2H_3O_2$ for acetic acid and formulas such as $NH_4C_2H_3O_2$ for salts of acetic acid.

place in benzene solution, is a Brønsted–Lowry acid–base reaction. Remember that the Lewis formula for an acetic acid molecule is

$$
\begin{array}{c}
\quad\quad H \\
\quad\quad | \quad\quad\quad \ddot{\text{O}}\text{:} \\
H - C - C \\
\quad\quad | \quad\quad\quad \ddot{\text{O}} - H \\
\quad\quad H
\end{array}
$$

and that the hydrogen which is bonded to oxygen is the one transferred.

The Brønsted–Lowry acid–base reactions that take place in aqueous solution are the most important from a practical point of view. Most biological and many geological reactions take place in aqueous solution; many reactions that are carried out in the laboratory and in manufacturing plants are also carried out in aqueous solution.

When a Brønsted–Lowry acid dissolves in water, the water acts as a base. For example, when a molecule of hydrogen chloride gas dissolves in water, a proton is rapidly transferred from the hydrogen chloride molecule to a water molecule. The water molecule acts as a Brønsted–Lowry base; the hydrogen chloride molecule acts as a Brønsted–Lowry acid:

The vapor pressure of HCl(g) over HCl(conc) is considerable. The gaseous HCl molecules react with H_2O molecules in humid air to form a white cloud of hydrochloric acid vapor.

$$
\delta + H \overset{\cdot\cdot}{\text{O}} \text{:} \delta - \quad + \quad \overset{\delta+}{\text{H}} \text{:} \overset{\delta-}{\ddot{\text{Cl}}}\text{:} \quad \longrightarrow \quad H \text{:} \overset{\cdot\cdot}{\text{O}} \text{:} H \;+\; \text{:}\ddot{\text{Cl}}\text{:}^{-} \tag{15.1}
$$
$$
\overset{\text{H}}{\underset{\delta+}{}} \qquad\qquad\qquad\qquad\qquad\qquad\qquad \overset{}{\text{H}}
$$

Brønsted–Lowry   Brønsted–Lowry
base   acid

Where does the energy needed to break the H—Cl bond come from? We saw in Section 13.1 that ions in aqueous solutions are solvated (hydrated) by spherical shells of water molecules. The *hydrated H_3O^+ ion* is called a **hydronium ion.** The exact number of water molecules that solvates a hydronium ion in aqueous solution is not known; therefore, the hydronium ion is written $H_3O^+(aq)$ in ionic equations. More than enough energy is released by the hydration of hydronium ions and chloride ions to supply the energy needed to break the H—Cl bond; the reaction shown by equation 15.1 is exothermic.

Because the water molecule that gains a proton forming a hydronium ion can't be distinguished from water molecules that solvate the H_3O^+ ion, the symbol $H^+(aq)$ (or H^+) is frequently used as an abbreviation for the hydrated hydronium ion, $H_3O^+(aq)$,* and the ion is often referred to simply as a hydrogen ion. *In the rest of this book, the symbol H^+ and the name* hydrogen ion *will be used except when it is important to think about the part played by the water molecules in acid–base reactions.*

In aqueous solutions of weak acids, such as hydrofluoric acid, an equilibrium exists between the molecules and ions of the weak acid:

As Cotton and Wilkinson say (Advanced Inorganic Chemistry, 5th ed.; Wiley: New York, 1988; p. 98), "We shall usually write H^+ for the hydrogen ion and assume it to be understood that the ion is aquated since other ions, for example, Na^+ and Fe^{2+}, are customarily so written although it is understood that the ions in solution are hydrated." The symbol H^+ and the name hydrogen ion are commonly used in books on subjects related to chemistry, such as biology and geology. Therefore, students need to become accustomed to these abbreviations.

$$
HF(aq) \;+\; H_2O(l) \;\rightleftharpoons\; H_3O^+(aq) \;+\; F^-(aq) \tag{15.2}
$$

Proton transfers take place very rapidly, and equilibrium is reached almost immediately. Remember that when the arrow pointing to the left is longer than the arrow pointing to the right, equilibrium lies to the left; there are more hydrogen fluoride molecules than hydrogen ions and fluoride ions in an aqueous solution of hydrofluoric

*The species H_3O^+ [$H^+(H_2O)$] and $H_5O_2^+$ [$H^+(H_2O)_2$] are known to exist in crystals. The species H_3O^+ is a rather flat pyramid.

acid. For example, in a 0.010 M solution of hydrofluoric acid at 25 °C, 83% of the hydrofluoric acid is in the form of hydrogen fluoride molecules, and only 17% is in the form of hydrogen ions and fluoride ions. Hydrogen fluoride molecules are called the **predominant species** (of solute) in an aqueous solution of hydrofluoric acid.

In equation 15.2, notice that the reverse reaction is also a Brønsted–Lowry acid–base reaction. In the reverse reaction in equation 15.2, the $H_3O^+(aq)$ ion is the Brønsted–Lowry acid (loser of H^+), and the fluoride ion is the Brønsted–Lowry base (taker of H^+):

Naturally, more water molecules than hydrogen fluoride molecules are present in an aqueous solution of hydrofluoric acid because water is the solvent.

$$:\!\ddot{F}\!: \;+\; H\!:\!\overset{+}{\underset{H}{\ddot{O}}}\!:\!H \;\longrightarrow\; :\!\ddot{F}\!:\!H \;+\; :\!\underset{H}{\ddot{O}}\!:\!H$$

Brønsted–Lowry   Brønsted–Lowry
base   acid

Acids and bases that are related by loss or gain of H^+, as H_3O^+–H_2O and HF–F^- are related, are called **conjugate acid–base pairs.** Water is the conjugate base of H_3O^+; H_3O^+ is the conjugate acid of H_2O. Hydrogen fluoride is the conjugate acid of F^-; F^- is the conjugate base of the HF molecule. The acid member of a conjugate acid–base pair always has one more hydrogen and a charge one unit more positive than the base member of the pair. The members of conjugate acid–base pairs are often marked with the same subscript as shown in equation 15.3 where the HF–F^- pair is labeled 1 and the H_3O^+–H_2O pair is labeled 2:

The adjective conjugate *is derived from the Latin word that means "to yoke together."*

$$\text{HF(aq)} \;+\; \text{H}_2\text{O(l)} \;\rightleftharpoons\; \text{H}_3\text{O}^+\text{(aq)} \;+\; \text{F}^-\text{(aq)} \qquad (15.3)$$
$$\text{acid}_1 \qquad \text{base}_2 \qquad\quad \text{acid}_2 \qquad \text{base}_1$$

A Brønsted–Lowry acid–base reaction is a competition between bases for a proton. The stronger base ends up with more protons at equilibrium. A fluoride ion is a stronger base than a water molecule.

When combined with acids, such as hydrochloric acid and hydrofluoric acid, water reacts as a base (see equation 15.1 and the forward reaction of equation 15.3). However, when water is mixed with a base, such as ammonia, water reacts as an acid as shown in equation 15.4a and 15.4b:

$$^{\delta+}H\!:\!\underset{\underset{\delta+}{H}}{\overset{\overset{\delta+}{H}}{N}}\!:\!^{\delta-} \;+\; ^{\delta+}H\!:\!\underset{\underset{\delta+}{H}}{\ddot{O}}\!:^{\delta-} \;\longrightarrow\; H\!:\!\underset{H}{\overset{\overset{H}{+}}{N}}\!:\!H \;+\; :\!\underset{H}{\ddot{O}}\!:^{-} \quad \text{or} \qquad (15.4a)$$

$$\text{NH}_3\text{(aq)} \;+\; \text{H}_2\text{O(l)} \;\rightleftharpoons\; \text{NH}_4^+\text{(aq)} \;+\; \text{OH}^-\text{(aq)} \qquad (15.4b)$$
$$\text{base}_1 \qquad\quad \text{acid}_2 \qquad\quad \text{acid}_1 \qquad\quad \text{base}_2$$

A *species* like water *that can react either as an acid or as a base* is said to be **amphoteric.**

As you might expect from the fact that water is amphoteric, water can undergo a Brønsted–Lowry acid–base reaction with itself, as shown in equation 15.5:

$$^{\delta+}H\!:\!\underset{\underset{\delta+}{H}}{\ddot{O}}\!:^{\delta-} \;+\; ^{\delta+}H\!:\!\underset{\underset{\delta+}{H}}{\ddot{O}}\!: \;\rightleftharpoons\; H\!:\!\underset{H}{\overset{+}{\ddot{O}}}\!:\!H \;+\; :\!\underset{H}{\ddot{O}}\!:^{-} \qquad (15.5)$$

Brønsted–Lowry   Brønsted–Lowry   Brønsted–Lowry   Brønsted–Lowry
base   acid   acid   base

A reaction in which a substance reacts with itself to form ions, such as the one shown by equation 15.5, is called **autoionization.** In equation 15.5, the arrow that points to the left is very much longer than the arrow that points to the right because water is a very weak acid and a very weak base. The fraction of water molecules

The prefix auto- *means "self." An automobile does not need a horse to pull it.*

The hydrogen sulfate ion, HSO_4^-, is the active ingredient of Sani-Flush.

The acidity of the hydrated aluminum ion is discussed in more detail in Section 15.6 and treated quantitatively in Section 16.2.

that is ionized is much smaller than the fraction of hydrofluoric acid or ammonia molecules that is in the form of ions in aqueous solution.

Brønsted–Lowry acids can be molecules, such as HCl, CH_3COOH, and HF; anions such as the hydrogen sulfate ion, HSO_4^-; or cations such as NH_4^+. Hydrated cations also act as Brønsted–Lowry acids if the cation is small and has a high charge. For example, aqueous solutions of the aluminum ion, Al^{3+}, are acidic as a result of a Brønsted–Lowry acid–base reaction between the hydrated ion and water:

$$Al(H_2O)_6^{3+}(aq) + H_2O(l) \rightleftharpoons Al(H_2O)_5(OH)^{2+}(aq) + H_3O^+(aq)$$

Brønsted–Lowry bases can be molecules like NH_3 and H_2O, anions such as the acetate ion, CH_3COO^-, and hydroxide ion, OH^-, or cations like the $Al(H_2O)_5(OH)^{2+}$ ion.

PRACTICE PROBLEMS

15.2   For the Brønsted–Lowry acid–base reaction,

$$H_2PO_4^-(aq) + H_2O(l) \rightleftharpoons HPO_4^{2-}(aq) + H_3O^+(aq)$$

(a) classify each of the species as an acid or a base. (b) What is the formula of the predominant solute species?

15.3   For each of the following Brønsted–Lowry acids, write the formula of the conjugate base: (a) HPO_4^{2-} (b) NH_3 (c) $Cu(H_2O)_4^{2+}$

15.4   For each of the following Brønsted–Lowry bases, write the formula of the conjugate acid. (a) NO_2^- (b) CH_3OH (c) $Al(H_2O)_5(OH)^{2+}$

15.5   Like water, liquid ammonia can undergo a Brønsted–Lowry acid–base reaction with itself:

$$NH_3(l) + NH_3(l) \longrightarrow NH_2^- + NH_4^+$$

(a) What are reactions of this type called? (b) What are substances such as water and ammonia that can act both as acids and as bases called?

15.2   THE ION PRODUCT FOR WATER, K_w

In water, water molecules, hydrogen ions, and hydroxide ions are always in equilibrium; the water equilibrium is shown, using ordinary formulas, in equation 15.6:

$$H_2O(l) + H_2O(l) \rightleftharpoons H_3O^+(aq) + OH^-(aq) \qquad (15.6)$$

Following the usual custom of including the constant concentrations of pure substances in the equilibrium constant, the equilibrium constant expression for the autoionization of water is

$$K_c = [H_3O^+][OH^-]$$

or, using the symbol H^+ for the hydrogen ion,

$$K_c = [H^+][OH^-]$$

This constant is called the **ion product for water** and, because of its importance, is given the special symbol, K_w. Some values of K_w at different temperatures are shown in Table 15.1. These values were determined by careful measurements of the conductivity of pure water. As you can see, the value of K_w increases as temperature increases. In the problems in this book, a temperature of slightly less than 25 °C

⊙ Section 14.2 tells how to write equilibrium constant expressions. Remember that the square brackets, [ ], stand for concentration in molarity, M. Brønsted–Lowry acid–base reactions (proton transfers) take place very rapidly. Equilibrium is reached almost immediately. In this chapter, [ ] will be understood to mean [ ]$_{eq}$.

will be assumed and the easily remembered value, 1.0×10^{-14}, will be used for K_w:

$$K_w = [H^+][OH^-] = 1.0 \times 10^{-14}$$

You should memorize this value because you will be using it often.

The equation for the ionization of water, equation 15.6, shows that for each hydrogen ion that is formed, one hydroxide ion is also formed. In pure water,

$$[H^+] = [OH^-]$$

Therefore, in pure water,

$$1.0 \times 10^{-14} = [H^+]^2 \qquad (\text{or } 1.0 \times 10^{-14} = [OH^-]^2)$$

Solving for $[H^+]$ (or $[OH^-]$),

$$[H^+] = [OH^-] = 1.0 \times 10^{-7}$$

A *solution that has the same concentration of hydrogen ions and hydroxide ions as pure water* is called a **neutral solution.** A *solution that has a higher concentration of hydrogen ions (lower concentration of hydroxide ions) than pure water* is called an **acidic solution.** A *solution that has a lower concentration of hydrogen ions (higher concentration of hydroxide ions) than pure water* is called a **basic solution.**

In dilute aqueous solutions, the concentration of water is constant for all practical purposes. For example, the concentration of water in pure water at 20 °C is 55.4 M; the concentration of water in 1 M HCl is 54.4 M, a difference of less than 2%. Thus, the value 1.0×10^{-14} can be used for the ion product of water in dilute aqueous solutions as well as for pure water. The constant value of the ion product means that, *in any dilute aqueous solution (at 25 °C), the product of the concentrations of hydrogen and hydroxide ions must always equal 1.0×10^{-14}.* If the $[H^+]$ is greater than 1.0×10^{-7}, the $[OH^-]$ must be lower than 1.0×10^{-7}. If the $[H^+]$ is less than 1.0×10^{-7}, the $[OH^-]$ must be greater than 1.0×10^{-7}. Note that the $[OH^-]$ is *not* zero in an acidic solution; the $[H^+]$ is *not* zero in a basic solution. Neither $[H^+]$ nor $[OH^-]$ is zero in a neutral solution. In summary,

acidic solution	$[H^+] > 1.0 \times 10^{-7} > [OH^-]$
neutral solution	$[H^+] = 1.0 \times 10^{-7} = [OH^-]$
basic solution	$[H^+] < 1.0 \times 10^{-7} < [OH^-]$

The ion product for water can be used to calculate the $[OH^-]$ if the $[H^+]$ is known or to calculate the $[H^+]$ if the $[OH^-]$ is known, as shown in Sample Problem 15.1.

TABLE 15.1

Ion Product for Water[a]

Temperature, °C	K_w
0	1.153×10^{-15}
20	6.87×10^{-15}
25	1.012×10^{-14}
30	1.459×10^{-14}
50	5.31×10^{-14}

[a]Data are from Light, T.S. *Anal. Chem.* **1987**, *59*, 2327–2330.

▶ Remember that for negative powers of 10, the greater the numerical value of the exponent, the smaller the value of the power of 10.

▶ Remember that the information in this table applies only at temperatures around 25 °C.

SAMPLE PROBLEM

15.1 In a solution that has $[H^+] = 5.0 \times 10^{-4}$, (a) what is the $[OH^-]$? (b) Is the solution acidic, neutral, or basic?

Solution In any dilute aqueous solution at about 25 °C,

$$[H^+][OH^-] = 1.0 \times 10^{-14}$$

Thus,

$$[OH^-] = \frac{(1.0 \times 10^{-14})}{[H^+]} = \frac{(1.0 \times 10^{-14})}{(5.0 \times 10^{-4})} = 2.0 \times 10^{-11}$$

(b) The $[H^+] = 5.0 \times 10^{-4}$, which is greater than 1.0×10^{-7}. Therefore, the solution is acidic.

15.6 In a solution that has $[H^+] = 6.0 \times 10^{-9}$, (a) what is the $[OH^-]$? (b) Is the solution acidic, neutral, or basic?

15.7 In a solution that has $[OH^-] = 8.0 \times 10^{-4}$, (a) what is the $[H^+]$? (b) Is the solution acidic, neutral, or basic?

In the pH module of the Acids and Bases section, students practice calculations involving pH, pOH, $[H_3O^+]$, and $[OH^-]$.

15.3 THE pH AND OTHER "p" SCALES

The concentrations of hydrogen ions commonly found in aqueous solutions range over more than 14 orders of magnitude. As a result, a logarithmic scale called the pH scale is often used to express the hydrogen ion concentrations of *aqueous solutions*. An additional advantage of the pH scale is that it combines the number and the power of 10 of an exponential number into a single number. The pH scale is widely used in medicine and biology, environmental science, and geology as well as in chemistry.

The **pH** of a solution is usually defined as the *negative of the base 10 logarithm of the molar concentration of hydrogen ion:*

$$pH = -\log [H^+]$$

Logarithms are explained more fully in Appendix B.3.

A logarithm to the base 10 is the power to which 10 must be raised to give the number. For example, $\log (100) = 2$ because $10^2 = 100$; $\log (1000) = 3$ because $10^3 = 1000$. The logarithm of 10^{-7} is -7,

$$\log (10^{-7}) = -7$$

and the logarithm of 1 is 0 ($1 = 10^0$):

$$\log (1) = 0$$

Seven is an exact number. Remember that exact numbers are considered to have an infinite number of significant figures and do not limit the number of significant figures in the results of calculations.

Logarithms are exponents. The logarithm of the product of two numbers is equal to the sum of the logarithms of the numbers. The logarithm of 1.0×10^{-7} is

$$\log (1.0) + \log (10^{-7}) = 0.00 + (-7) = -7.00$$

Therefore, the pH of a solution that has $[H^+] = 1.0 \times 10^{-7}$ is 7.00:

$$pH = -\log [H^+] = -(-7.00) = 7.00$$

In addition to being able to find pH from hydrogen ion concentration, you must also be able to find the hydrogen ion concentration from the pH. For example, if $pH = 3.00$, $[H^+] = 1.0 \times 10^{-3}$.

Consult the instruction book that came with your calculator to learn how to find logarithms and antilogarithms with your calculator.

If the number part of the exponential number is not 1 or the pH is not an integer, you can use a scientific calculator to find the logarithm or antilogarithm. Sample Problems 15.2 and 15.3 will let you check your skill in using your scientific calculator to find logarithms and antilogarithms.

How do you know how many digits there should be in the pH? The *number to the left of the decimal place in a logarithm* is called the **characteristic** of the logarithm. The characteristic of a logarithm is related to the power of 10 in the exponential number. The number to the left of the decimal point in a pH has nothing to do with significant figures. *The number of digits to the right of the decimal place in a pH should be the same as the number of digits in the number part of the hydrogen ion concentration.* The *number to the right of the decimal place in a logarithm* is called the **mantissa** of the logarithm. The mantissa of a logarithm is related to the number part of the exponential number and should be written to the correct number of significant figures (see Sample Problems 15.2 and 15.3).

15.2   If $[H^+] = 3.0 \times 10^{-3}$, what is the pH?

Solution   If $[H^+] = 3.0 \times 10^{-3}$ M, then log $[H^+] = -2.52$ and

$$pH = -(-2.52) = 2.52$$

There are two digits to the right of the decimal place in the pH because $[H^+]$ is given to two significant figures.

15.3   In a solution with pH = 9.20, what is $[H^+]$?

Solution   Because pH = $-$log $[H^+]$,

$$\log [H^+] = -pH \qquad \text{and}$$
$$[H^+] = \text{antilog } (-pH) = \text{antilog } (-9.20) = 6.3 \times 10^{-10}$$

The answer has two significant figures because there are two digits to the right of the decimal place in the pH.

15.8   (a) What is the pH of a solution that has $[H^+] = 1.0 \times 10^{-5}$? (b) What is the pH of a solution that has $[H^+] = 8.3 \times 10^{-4}$?

15.9   (a) If a solution has pH = 8.7, what is the $[H^+]$? (b) If the pH of a solution is 5.64, what is $[H^+]$ in the solution?

Because pH is a logarithmic scale, a difference of 1 in pH corresponds to a factor of 10 in $[H^+]$. For example, a solution with $[H^+] = 1.0 \times 10^{-2}$ (0.010 M) has pH = 2.00; a solution with pH = 1.00 has $[H^+] = 1.0 \times 10^{-1}$ (0.10 M), ten times as high. Also note that the *lower pH* corresponds to the *more acidic solution*. Pure water has a pH of 7.00 at 25 °C. At 25 °C, a solution with pH < 7.00 is acidic; a solution with pH > 7.00 is basic.

> pH < 7.00 acidic
> pH = 7.00 neutral
> pH > 7.00 basic

The pH of a solution can be estimated with test paper or measured with a pH meter. ▪Figure 15.2 shows test paper and a pH meter. Table 15.2 shows the pH of some liquids that you are probably familiar with in your everyday life.

The pH scale is commonly used only for hydrogen ion concentrations between 1 M (pH = 0) and 1×10^{-14} M (pH = 14). A solution that has $[H^+] > 1$ has a negative pH; for example, if $[H^+] = 3$, pH = (-0.5). Hydrogen ion concentrations greater than 1 M and less than 1.0×10^{-14} M are usually expressed by giving the molarity.

Logarithmic scales similar to the pH scale are convenient for expressing other quantities that vary over a large range. For example, the concentration of hydroxide ions is often expressed as pOH:

$$pOH = -\log [OH^-]$$

Because $K_w = [H^+][OH^-] = 1.0 \times 10^{-14}$ (at 25 °C),

$$\log K_w = \log [H^+] + \log [OH^-] = -14.00 \tag{15.7}$$

◯ Be sure you use the "LOG" not the "LN" button on your calculator. (The "LN" button is for logarithms to the base e.)

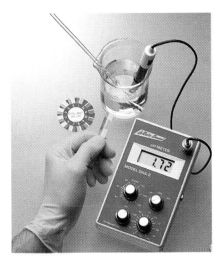

▪ FIGURE 15.2   Both the test paper and the pH meter show that the solution is strongly acidic. Before use, a pH meter must be calibrated with solutions of accurately known pH. pH meters are made with probes small enough to be inserted into a single living cell!

TABLE 15.2

pH of Some Common Liquids

Liquid	pH
Gastric juices	1.0–3.0
Lemon juice	2.2–2.4
Classic Coke	2.5
Saliva	6.5–7.5
Pure water	7.0
Blood	7.3–7.5
Milk of magnesia	10.5
Household ammonia	12.0

15.3 THE pH AND OTHER "p" SCALES

Multiplication of both sides of equation 15.7 by -1 gives

$$(-\log K_w) = (-\log [H^+]) + (-\log [OH^-]) = 14.00 \qquad \text{or}$$
$$pK_w = pH + pOH = 14.00$$

In any aqueous solution at 25 °C, the sum of the pH and the pOH is 14.00.

PRACTICE PROBLEM

15.10 (a) What is the pOH of a solution with $[OH^-] = 1.0 \times 10^{-4}$ M?
(b) What is the pOH of a solution with $[OH^-] = 6.2 \times 10^{-9}$? (c) What is the
$[OH^-]$ of a solution that has pOH $= 11.00$? (d) What is the $[OH^-]$ of a solution
that has pOH $= 8.55$? (e) What is the pOH of a solution that has pH $= 7.99$?

 The Acid-Base Equilibria module of the
Acids and Bases section provides randomly
generated problems involving calculation of K_a
or K_b from equilibrium concentrations and
equilibrium concentrations from one initial
concentration and the value of K_a or K_b.

15.4 CONCENTRATIONS OF HYDROGEN IONS IN AQUEOUS SOLUTIONS OF ACIDS

Strong Acids

The pH of a 0.010 M solution of HCl, measured with a pH meter, is 2.00; that is,
$[H^+]$ in a 0.010 M solution of HCl is 0.010. The equation for the ionization of HCl is

$$HCl(aq) \longrightarrow H^+(aq) + Cl^-(aq)$$

According to this equation, 1 mol H^+ should be formed from 1 mol HCl or 0.010
mol H^+ from 0.010 mol HCl. The fact that 0.010 mol H^+ is actually observed shows
that HCl is completely ionized in 0.010 M solution. *For moderately dilute solutions
(1.0×10^{-6} to 0.01 M) of strong acids that have only one ionizable hydrogen,
$[H^+]$ = stoichiometric concentration of the strong acid.* The concentration of
hydrogen ion formed by the autoionization of water is negligible in 0.010 M HCl.
The fact that the contribution of the autoionization of water to the concentration of
hydrogen ion is negligible is not surprising because the value of K_w is very small.
Moreover, the H^+ from the strong acid causes the equilibrium

$$H_2O(l) \xrightleftharpoons{} H^+(aq) + OH^-(aq)$$

to shift to the left.

How dilute can the strong acid be before the concentration of hydrogen ion
formed by the autoionization of water becomes significant? The answer to this
question can be found by a calculation similar to the calculations in Chapter 14.
Let's calculate the $[H^+]$ formed by the autoionization of water in a 1.0×10^{-6} M
solution of HCl as an example. The equations for the ionization of hydrochloric
acid and the autoionization of water are

$$HCl(aq) \longrightarrow H^+(aq) + Cl^-(aq) \qquad \text{and}$$
$$H_2O(l) \xrightleftharpoons{} H^+(aq) + OH^-(aq)$$

The equilibrium constant expression for the autoionization of water is

$$K_w = [H^+][OH^-] = 1.0 \times 10^{-14}$$

If, as usual, we consider one liter of solution, the number of moles is numerically
equal to concentration. If we let $x = [H^+]$ from water, according to the equation
for the autoionization of water, $[OH^-]$ is also equal to x. The total concentration
of H^+ is equal to the sum of the concentration from HCl and the concentration

from water:

$$[H^+] = [(1.0 \times 10^{-6}) + x]$$

Substitution of the concentrations in terms of x in the equilibrium constant expression for the ionization of water gives

$$[(1.0 \times 10^{-6}) + x]x = 1.0 \times 10^{-14} \qquad (15.8)$$

Rearrangement of equation 15.8 gives the quadratic equation

$$x^2 + (1.0 \times 10^{-6})x - (1.0 \times 10^{-14}) = 0$$

and, either by the quadratic formula or by successive approximations,

$$x = 1.0 \times 10^{-8}$$

Calculation of $[H^+]$ gives

$$[H^+] = [(1.0 \times 10^{-6}) + x] = [(1.0 \times 10^{-6}) + (1.0 \times 10^{-8})] = 1.0 \times 10^{-6}$$

Thus, in 1.0×10^{-6} M HCl, the concentration of hydrogen ion formed by the autoionization of water is not significant. *The concentration of hydrogen ion formed by the autoionization of water can be neglected unless the acid is very dilute or very weak.* However, no matter how dilute the HCl, $[H^+]$ in a solution of HCl can never be less than 1.0×10^{-7}. A solution of HCl (or any other acid) can *never* be basic.

If the stoichiometric concentration of HCl is much less than 1.0×10^{-6}, the contribution of the H^+ from water will be significant. For example, in 1.0×10^{-7} M HCl, $[H^+]$ from water is 0.6×10^{-7} or almost 40% of the total $[H^+]$.

PRACTICE PROBLEM

15.11 What is $[H^+]$ in a 3.1×10^{-4} M solution of HCl?

Weak Acids

The equation for the ionization of acetic acid, CH_3COOH, is

$$CH_3COOH(aq) \longrightarrow H^+(aq) + CH_3COO^-(aq) \qquad (15.9)$$

If acetic acid were completely ionized, $[H^+]$ in a 0.010 M solution would be 0.010 or 1.0×10^{-2}, and pH would be 2.00. However, the pH of a 0.010 M solution of acetic acid is 3.38; that is, $[H^+] = 4.2 \times 10^{-4}$ or 0.000 42. Only a small fraction of the dissolved acetic acid is ionized. Acetic acid is a weak acid. The symbol K_a is used for the equilibrium constant for the ionization of a weak acid. For acetic acid,

$$K_a = \frac{[H^+][CH_3COO^-]}{[CH_3COOH]}$$

K_a is called the **acid ionization constant** or **acid dissociation constant** for the weak acid.

PRACTICE PROBLEM

15.12 (a) Write the equation for the ionization of the weak acid HCN. (b) Write the expression for K_a for HCN.

The numerical value of K_a for a weak acid can be calculated from the percent ionization determined by conductivity measurements as illustrated by Sample Problem 15.4.

■ FIGURE 15.3 Addition of ethyl alcohol to water does not change the pH of the water. The beaker on the left contains pure water. The liquid in the beaker on the right is a solution of ethyl alcohol in water.

15.4 A 0.10 M aqueous solution of acetic acid, CH_3COOH, is 1.3% ionized at 25 °C. What is the value of K_a for acetic acid at this temperature?

Solution From the percent ionization, we can calculate the equilibrium concentrations. Then we can find the value of K_a by substituting the equilibrium concentrations in the equilibrium constant expression.

Percent is always equal to the part divided by the whole times 100; therefore, percent ionization is equal to the fraction of the solute that is ionized times 100 and

$$\% \text{ ionization}_{CH_3COOH} = \frac{[CH_3COO^-]_{eq}}{[CH_3COOH]_0} \times 100 \qquad (15.10)$$

Solution of equation 15.10 for $[CH_3COO^-]_{eq}$ gives

$$\frac{\% \text{ ionization}_{CH_3COOH} \times [CH_3COOH]_0}{100} = [CH_3COO^-]_{eq}$$

and, substituting the information given in the problem.

$$\frac{1.3 \times 0.10}{100} = [CH_3COO^-]_{eq} = 0.0013 = 1.3 \times 10^{-3}$$

The initial concentrations at the moment the acetic acid was dissolved in water but before any ionization had time to take place were:

$$[CH_3COOH]_0 = 0.10 \qquad [CH_3COO^-]_0 = 0.00$$
$$[H^+]_0 = 1.0 \times 10^{-7} = 0.00$$

Because $[CH_3COO^-]_{eq} = 1.3 \times 10^{-3}$, $[CH_3COO^-]$ has increased from 0.00 to 1.3×10^{-3}, and the change in $[CH_3COO^-]$ is $+ (1.3 \times 10^{-3})$. All coefficients in the equation for the equilibrium are 1; therefore, the change in $[H^+]$ is also $+(1.3 \times 10^{-3})$, and the change in $[CH_3COOH]$ is $-(1.3 \times 10^{-3})$. The completed summary table is as follows:

	$CH_3COOH(aq) \rightleftharpoons$	$H^+(aq)$	$+ CH_3COO^-(aq)$
Init. conc., M	**0.10**	**0.00**	**0.00**
Change, ΔM	$-(1.3 \times 10^{-3})$	$+(1.3 \times 10^{-3})$	$+(1.3 \times 10^{-3})$
Eq. conc., M	0.10	1.3×10^{-3}	1.3×10^{-3}

Substitution of the equilibrium concentrations in the expression for K_a

$$K_a = \frac{[H^+]_{eq}[CH_3COO^-]_{eq}}{[CH_3COOH]_{eq}} \qquad \text{gives} \qquad K_a = \frac{(1.3 \times 10^{-3})^2}{0.10} = 1.7 \times 10^{-5}$$

15.13 A 0.050 M aqueous solution of formic acid, HCOOH, is 5.8% ionized at 25 °C. What is the value of K_a for formic acid at this temperature?

The stronger the acid, the greater the value of K_a. The values of K_a for a large number of acids have been determined. Table 15.3 shows the values for some common weak acids listed in order of decreasing strength. Acids that are weaker acids than water are neutral compared to water. For example, the value of K_a for ethyl alcohol, CH_3CH_2OH, is 1.3×10^{-16} and aqueous solutions of ethyl alcohol are neutral (see ■Figure 15.3). From the value of K_a, the concentration of hydrogen ions in any dilute solution of a weak acid (at 25 °C) can be calculated as illustrated by Sample Problem 15.5.

Because ionization of a weak acid involves transfer of a proton to water, K_a values apply only to aqueous solutions.

TABLE 15.3 Ionization Constants of Some Common Weak Acids at 25 °C[a]

Acid: Name	Acid: Formula	K_a	Conjugate Base: Name	Conjugate Base: Formula
Hydrogen sulfate ion	HSO_4^-	1.2×10^{-2}	Sulfate ion	SO_4^{2-}
Iron(III) ion (hexaaquairon(III) ion)	$Fe(H_2O)_6^{3+}$	6×10^{-3}	Pentaaquahydroxo-iron(III) ion	$Fe(H_2O)_5(OH)^{2+}$
Chloroacetic acid	$ClCH_2COOH$	1.4×10^{-3}	Chloroacetate ion	$ClCH_2COO^-$
Nitrous acid	HNO_2	4.6×10^{-4}	Nitrite ion	NO_2^-
Hydrofluoric acid	HF	3.5×10^{-4}	Fluoride ion	F^-
Formic acid	$HCOOH$	1.8×10^{-4}	Formate ion	$HCOO^-$
Benzoic acid	C_6H_5COOH[b]	6.5×10^{-5}	Benzoate ion	$C_6H_5COO^-$
Acetic acid	CH_3COOH	1.8×10^{-5}	Acetate ion	CH_3COO^-
Propionic acid	CH_3CH_2COOH	1.3×10^{-5}	Propionate ion	$CH_3CH_2COO^-$
Hypochlorous acid	$HOCl$	3.2×10^{-8}	Hypochlorite ion	ClO^-
Zinc ion (hexaaquazinc ion)	$[Zn(H_2O)_6]^{2+}$	1×10^{-9}	Pentaaquahydroxo-zinc ion	$Zn(H_2O)_5(OH)^+$
Ammonium ion	NH_4^+	5.6×10^{-10}	Ammonia	NH_3
Hydrocyanic acid	HCN	4.9×10^{-10}	Cyanide ion	CN^-
Phenol	C_6H_5OH[b]	1.3×10^{-10}	Phenoxide ion	$C_6H_5O^-$
Hydrogen carbonate ion	HCO_3^-	5.6×10^{-11}	Carbonate ion	CO_3^{2-}
Magnesium ion	$Mg(H_2O)_x^{2+}$	4×10^{-12}	—	$Mg(H_2O)_{x-1}(OH)^+$
Water	H_2O[c]	2×10^{-16}	Hydroxide ion	OH^-

↑ Increasing strength of acids

↓ Increasing strength of bases

[a]Most values are from Lide, D. R., Ed. *CRC Handbook of Chemistry and Physics,* 75th ed. CRC Press: Boca Raton, FL, 1994; pp. 8-43–8-55. As usual, no units are given for values of equilibrium constants.

[b]C_6H_5 means 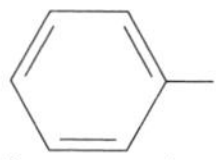 .

[c]Note that K_a for water is not the same as K_w. The values given for K_a's are for the weak acids as solutes. For comparison, K_a for the hydronium ion, H_3O^+, the strongest acid that can exist in aqueous solution, is 55.3.

SAMPLE PROBLEM

15.5 (a) What is the concentration of hydrogen ions in a 0.050 M solution of acetic acid? (b) How does the percentage of the 0.050 M acetic acid that is ionized compare with the 1.3% of the 0.10 M acetic acid that is ionized (Sample Problem 15.4)? (c) How does the $[H^+]$ in a solution of acetic acid compare with the $[H^+]$ in a solution of formic acid that has the same initial concentration of weak acid?

Solution (a) The method used to solve this problem is similar to the method used to solve problems about gaseous equilibria (Section 14.4). This problem is like Sample Problem 14.8.

Step 1. *Write the equation for the equilibrium.* For the ionization of acetic acid, the equation is

$$CH_3COOH(aq) \rightleftharpoons H^+(aq) + CH_3COO^-(aq)$$

Step 2. *Write the expression for K_a and find the value of K_a in a table.* The expression for K_a is

$$K_a = \frac{[H^+][CH_3COO^-]}{[CH_3COOH]} = 1.8 \times 10^{-5} \qquad \text{(from Table 15.3)}$$

Step 3. *Express the quantities of all the species in terms of a single variable, x, and make a summary table.* If we let x equal mol/L H^+ formed, then x also equals mol/L CH_3COO^- formed, and $-x$ equals mol/L CH_3COOH reacted. The completed

See Campbell, M. L.: Waite, B. A. J. Chem. Educ. ***1990,*** *67, 386–388 for the arguments in favor of the value 2×10^{-16} for K_a for H_2O and 55.3 for H_3O^+. As an example of the reasoning, K_a for acetic acid is for the equilibrium $CH_3COOH(aq) + H_2O(l)$ $H_3O^+ + CH_3COO^-$ not for the equilibrium $2CH_3COOH(l)$ $CH_3COOH_2^+ + CH_3COO^-$, which is analogous to the equilibrium $2H_2O(l)$ $H_3O^+ + OH^-$, $K_w = 1.0 \times 10^{-14}$.*

summary table is as follows:

$$CH_3COOH(aq) \rightleftharpoons H^+(aq) + CH_3COO^-(aq)$$

	$CH_3COOH(aq)$	$H^+(aq)$	$CH_3COO^-(aq)$
Init. conc., M	**0.050**	**0.000**	**0.000**
Change, ΔM	$-x$	$+x$	$+x$
Eq. conc., M	$(0.050 - x)$	x	x

Step 4. *Substitute the equilibrium concentrations in terms of x in the expression for K_a and solve for x.* Substitution of the concentrations in terms of x in the expression for K_a gives

$$\frac{x^2}{(0.050 - x)} = 1.8 \times 10^{-5} \qquad (15.11)$$

The number 1.8×10^{-5} is small; not much acetic acid is ionized at equilibrium. If we assume that x is negligible compared with 0.050, equation 15.11 simplifies to

$$\frac{x^2}{0.050} = 1.8 \times 10^{-5}; \text{ rearranging}$$

$$x^2 = 0.050(1.8 \times 10^{-5}) = 9.0 \times 10^{-7} \qquad \text{and} \qquad x = 9.5 \times 10^{-4}$$

Is the assumption that x is negligible compared with 0.050 correct?

$$[CH_3COOH] = (0.050 - x) = [0.050 - (9.5 \times 10^{-4})]$$
$$= (0.050 - 0.000\,95) = 0.049)$$

Because the last digit is uncertain, 0.049 does not differ significantly from 0.050; therefore, the assumption that x is negligible compared with 0.050 is correct. (The value of x found by successive approximations or the quadratic formula is 9.4×10^{-4}.)

(b) In Sample Problem 15.4, we saw that the percent ionization of acetic acid is 1.3% in a 0.10 M solution. We must calculate the percent ionization in the 0.050 M solution so that we can compare it with 1.3%:

$$\% \text{ ionization} = \frac{[CH_3COO^-]_{eq}}{[CH_3COOH]_0} \times 100$$

Substitution of values for $[CH_3COO^-]_{eq}$ and $[CH_3COOH]_0$ from this problem gives

$$\% \text{ ionization} = \frac{(9.4 \times 10^{-4})}{0.050} \times 100 = 1.9\%$$

The percent ionization in the more dilute (0.050 M) solution, 1.9%, is greater than the percent ionization in the more concentrated solution (0.10 M), which is 1.3%. Greater ionization in more dilute solution seems reasonable because the more solvent molecules there are between the solute ions, the less the chance of oppositely charged solute ions bumping into each other and forming molecules. The rate of the reverse reaction should be lower in the more dilute solution, and equilibrium should be shifted to the right by dilution.

(c) In the first part of this problem we found that $[H^+] = 9.5 \times 10^{-4}$ in 0.050 M acetic acid. According to Practice Problem 15.13, $[H^+] = 2.9 \times 10^{-3}$ in 0.050 M formic acid. Thus, the concentration of hydrogen ion in acetic acid is less than the concentration of hydrogen ion in formic acid of the same concentration ($9.5 \times 10^{-4} < 2.9 \times 10^{-3}$). This is reasonable because K_a for acetic acid is smaller than K_a for formic acid ($1.8 \times 10^{-5} < 1.8 \times 10^{-4}$).

▶ Always check your work to be sure that it is reasonable.

PRACTICE PROBLEM

15.14 What is the concentration of hydrogen ions in a 0.10 M solution of formic acid?

The percent ionization of a weak acid depends on the value of K_a for the acid and the stoichiometric concentration of the acid. The larger the value of K_a, the greater the percent ionization at the same concentration. As you can see from Table 15.3, the values of K_a for common weak acids range from 10^{-2} to 10^{-16}; therefore, pK_a's are often used instead of K_a's:

$$pK_a = -\log K_a$$

Values for pK_a and K_a are converted to each other in the same way as pH and [H^+]. For example, K_a for acetic acid is 1.8×10^{-5}, and pK_a for acetic acid is 4.74. The value of K_a for formic acid is 1.8×10^{-4}, and its pK_a is 3.74. Notice that *the stronger the acid, the lower its pK_a.* ▪Figure 15.4 shows the relationship between percent ionization and the value of pK_a for 0.10 M solutions of weak acids.

The percent ionization at different molarities for an acid that has $K_a = 1.0 \times 10^{-5}$ is shown in ▪Figure 15.5. Note how the percent ionization of a strong acid is 100% in all solutions that are 0.0100 M or less. The percent ionization of a weak acid increases as concentration decreases. However, the ionization of a weak acid is aqueous solution is never complete even at infinite dilution. In aqueous solution, the percent ionization of weak acids is limited by the hydrogen ion formed by the autoionization of water. The H^+ from the water shifts the equilibrium between the weak acid and its ions in solution to the left. A moderately weak acid such as acetic acid ($K_a = 1.8 \times 10^{-5}$) is more than 99% ionized at infinite dilution. However, only 0.13% of the much weaker acid, phenol ($K_a = 1.3 \times 10^{-10}$), is ionized at infinite dilution.

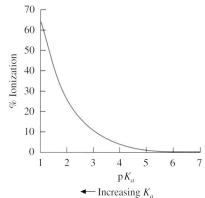

▪ **FIGURE 15.4** Percent ionization in 0.10 M solution as a function of pK_a.

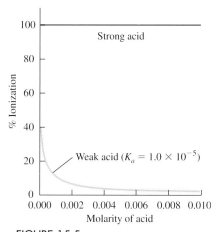

▪ **FIGURE 15.5** Percent ionization as a function of molarity.

PRACTICE PROBLEM

15.15 Why is percent ionization plotted against pK_a rather than against K_a in Figure 15.4?

▪Figure 15.6 will give you an idea of the relative numbers of solvent molecules, solute molecules, and ions in solutions of weak electrolytes.

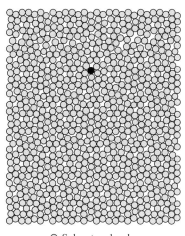

○ Solvent molecule
● Solute molecule

▪ **FIGURE 15.6** A microscopic view of a 0.1 M solution. The area of the solution in the picture is about 1/20 of the area of the page. If the solvent was water and the solute a weak acid that was 10% ionized ($K_a = 1 \times 10^{-3}$), there would be 20 solute molecules and two hydrogen ions on a page. To show one hydroxide ion, a picture the size of 5×10^9 pages would be required!

15.5 CONCENTRATIONS OF HYDROXIDE IONS IN AQUEOUS SOLUTIONS OF BASES

Strong Bases

The most important water-soluble strong bases are sodium hydroxide and potassium hydroxide. The hydroxides of the other Group IA elements are also soluble strong bases. However, lithium hydroxide, rubidium hydroxide, and cesium hydroxide are too expensive to be used simply as soluble strong bases. Of the Group IIA hydroxides, only calcium, strontium, and barium hydroxides are soluble enough to be used as strong bases, and strontium hydroxide is not available commercially.* All other basic hydroxides except thallium(I) hydroxide are insoluble.

In dilute (0.1 M or less) aqueous sodium hydroxide, all of the solute is present in the form of sodium ions and hydroxide ions. One mole of sodium hydroxide gives one mole of sodium ions and one mole of hydroxide ions in solution:

$$NaOH(s) \longrightarrow Na^+(aq) + OH^-(aq)$$

In pure water, $[OH^-] = 1.0 \times 10^{-7}$. Addition of sodium hydroxide to water shifts the equilibrium

$$H_2O(l) + H_2O(l) \rightleftharpoons H_3O^+(aq) + OH^-(aq)$$

Only if the concentration of hydroxide ion from a dissolved base is less than 1.0×10^{-6} M does the autoionization of water make a significant contribution to the $[OH^-]$.

to the left. Thus, water does not make a significant contribution to the hydroxide ion concentration in a 0.10 M solution of NaOH; all hydroxide ions come from the sodium hydroxide. Therefore, in a 0.10 M aqueous solution of NaOH, $[Na^+] = [OH^-] = 0.10$.

SAMPLE PROBLEM

15.6   What is the concentration of hydroxide ions in a 0.035 M solution of $Ba(OH)_2$?

Solution   Barium hydroxide is a strong base. Each formula unit yields two hydroxide ions in solution:

$$0.035 \left(\frac{\text{mol } Ba(OH)_2}{\text{L soln}} \right) \left(\frac{2 \text{ mol } OH^-}{\text{mol } Ba(OH)_2} \right) = 0.070 \, \frac{\text{mol } OH^-}{\text{L soln}}$$

The concentration of OH^- in 0.035 M $Ba(OH)_2$ is 0.070 M.

PRACTICE PROBLEM

15.16   What is the concentration of hydroxide ions in a 0.050 M solution of KOH?

Weak Bases

Ammonia is the only important soluble *inorganic* molecular weak base. Organic compounds derived from ammonia by substitution of carbon-containing groups for

---

*The low solubility of magnesium hydroxide is an advantage in the antacid milk of magnesia. A high concentration of hydroxide ion in solution would burn the tissues of the mouth, esophagus, and stomach. As hydroxide ions are used up by reaction with stomach acid, the equilibrium between magnesium hydroxide solid and magnesium hydroxide in solution shifts, and more magnesium hydroxide dissolves to replace the hydroxide ions used.

the hydrogens of ammonia make up the only important type of soluble *organic molecular weak base. Organic compounds derived from ammonia* are called **amines.** Some simple examples of amines are methylamine, dimethylamine, trimethylamine, and aniline:

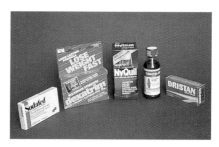

NH_3	CH_3NH_2	$(CH_3)_2NH$	$(CH_3)_3N$	aniline
ammonia	methylamine	dimethylamine	trimethylamine	

Many amines have powerful effects on the mind and on the body and are important medically. For example, epinephrine (adrenaline), the hormone released into the bloodstream when an animal senses danger that prepares the animal for fight or flight, is an amine. Caffeine and nicotine are amines; most abused drugs and some vitamins are amines.

The equations for Brønsted–Lowry acid–base equilibria between molecules of weak bases, water molecules, the cations from the weak base, and hydroxide ions are usually written with the water molecules on the left (reactant) side as shown in equation 15.12:

$$:NH_3(aq) + H_2O(l) \rightleftharpoons NH_4^+(aq) + :\ddot{O}H^-(aq) \qquad (15.12)$$

The *equilibrium constants for the reactions of weak bases with water* are called **base ionization constants** or **base dissociation constants, K_b.** For the equilibrium shown by equation 15.12,

$$K_b = \frac{[NH_4^+][OH^-]}{[NH_3]}$$

Over-the-counter drugs that are salts of weak bases.

The nitrogen atoms in ammonia and amines each have an unshared pair of electrons.

⊙ Remember that proton transfers take place very rapidly and equilibrium is reached almost immediately.

PRACTICE PROBLEM

15.17 (a) Write the equation for the equilibrium between molecules of the weak base, CH_3NH_2, water, $CH_3NH_3^+$, and hydroxide ions. (b) Write the expression for K_b.

The numerical value of K_b for a weak base can be calculated from the percent ionization determined by conductivity measurements just as the numerical value of K_a for a weak acid can be calculated from percent ionization (remember Sample Problem 15.4). Sample Problem 15.7 shows the calculation of K_b for ammonia from percent ionization.

SAMPLE PROBLEM

15.7 A 0.075 M solution of ammonia is 1.5% ionized at 25 °C. What is the value of K_b for ammonia at this temperature?

Solution The equation for the ionization of ammonia is

$$NH_3(aq) + H_2O(l) \rightleftharpoons NH_4^+(aq) + OH^-(aq)$$

and

$$\% \text{ ionization} = \frac{[NH_4^+]_{eq}}{[NH_3]_0} \times 100 \qquad (15.13)$$

Solution of equation 15.13 for $[NH_4^+]$ and substitution gives

$$\frac{\% \text{ ionization} \times [NH_3]_0}{100} = [NH_4^+]_{eq} = \frac{1.5 \times 0.075}{100} = 1.1 \times 10^{-3}$$

The initial concentrations at the moment the ammonia was dissolved in water but before any reaction had time to take place were

$$[NH_3]_0 = 0.075 \qquad [NH_4^+]_0 = 0.000 \qquad [OH^-]_0 = 1.0 \times 10^{-7} = 0.000$$

Because $[NH_4^+]_{eq} = 1.1 \times 10^{-3}$, $[NH_4^+]$ has increased from 0.000 to 1.1×10^{-3}, and the change in $[NH_4^+]$ is $+(1.1 \times 10^{-3})$. All coefficients in the equation for the equilibrium are 1; therefore, the change in $[OH^-]$ is also $+(1.1 \times 10^{-3})$, and the change in $[NH_3]$ is $-(1.1 \times 10^{-3})$. The completed summary table is as follows:

	$NH_3(aq)$	$+ H_2O(l) \rightleftharpoons$	$NH_4^+(aq)$	$+ OH^-(aq)$
Init. conc., M	**0.075**		**0.000**	**0.000**
Change conc., ΔM	$-(1.1 \times 10^{-3})$		$+(1.1 \times 10^{-3})$	$+(1.1 \times 10^{-3})$
Eq. conc., M	0.074		(1.1×10^{-3})	(1.1×10^{-3})

Substitution of the equilibrium concentrations in the expression for K_b

$$K_b = \frac{[NH_4^+][OH^-]}{[NH_3]} \qquad \text{gives} \qquad K_b = \frac{(1.1 \times 10^{-3})^2}{0.074} = 1.6 \times 10^{-5}$$

PRACTICE PROBLEM

15.18 A 0.090 M solution of aniline, $C_6H_5NH_2$, is 0.0069% ionized. (a) What is $[OH^-]$ formed by the ionization of aniline? (b) Can the $[OH^-]$ formed by the autoionization of water be neglected? Explain your answer. (c) What is the value of K_b for aniline?

Table 15.4 shows values of K_b and pK_b for a few molecular weak bases. The concentration of hydroxide ions in any dilute solution of a weak base (at 25 °C) can be calculated from the value of K_b for the weak base. The method used is similar to that used to find the concentration of hydrogen ions in a dilute solution of a weak acid from K_a (see Sample Problem 15.5).

PRACTICE PROBLEM

15.19 (a) What is the concentration of hydroxide ions in a 0.075 M solution of aniline? (b) What is the concentration of anilinium ions in a 0.075 M solution of aniline? (c) The concentration of aniline molecules?

Most inorganic weak bases are anions, the conjugate bases of weak acids. The acetate ion, CH_3COO^-, is a typical example of an anionic weak base. When a salt that contains one of these basic anions is dissolved in water, a Brønsted–Lowry acid–base reaction takes place between the anion and water. Equilibrium between the anionic weak base, water, the molecular weak acid, and hydroxide ion is reached almost instantly. For example, when an acetate such as sodium acetate is dissolved in water, the equilibrium

$$CH_3COO^-(aq) + H_2O(l) \rightleftharpoons CH_3COOH(aq) + OH^-(aq) \qquad (15.14)$$

is established almost immediately. The expression for K_b for the equilibrium shown by equation 15.14 is

$$K_b = \frac{[CH_3COOH][OH^-]}{[CH_3COO^-]}$$

TABLE 15.4 Ionization Constants of Some Molecular Weak Bases at 25 °C[a]

Name	Base Formula	K_b	pK_b	Name	Conjugate Acid Formula	K_a	pK_a
Ammonia	NH_3	1.8×10^{-5}	4.75	Ammonium ion	NH_4^+	5.6×10^{-10}	9.25
Methylamine	CH_3NH_2	4.3×10^{-4}	3.37	Methylammonium ion	$CH_3NH_3^+$	2.3×10^{-11}	10.63
Dimethylamine	$(CH_3)_2NH$	4.8×10^{-4}	3.32	Dimethylammonium ion	$(CH_3)_2NH_2^+$	2.1×10^{-11}	10.68
Trimethylamine	$(CH_3)_3N$	6.3×10^{-5}	4.20	Trimethylammonium ion	$(CH_3)_3NH^+$	1.6×10^{-10}	9.80
Aniline	$C_6H_5NH_2$[b]	4.3×10^{-10}	9.37	Anilinium ion	$C_6H_5NH_3^+$	2.3×10^{-5}	4.63

[a]Data are from Lide, D. R., Ed. *CRC Handbook of Chemistry and Physics*, 75th ed. CRC Press: Boca Raton, FL, 1994; pp **8**-43–**8**-55.

[b]C_6H_5 means ⬡ .

In aqueous solutions of acetic acid or of an acetate such as sodium acetate, all three equilibria

$$H_2O(l) \rightleftharpoons H^+(aq) + OH^-(aq)$$

$$CH_3COOH(aq) \rightleftharpoons H^+(aq) + CH_3COO^-(aq)$$

$$CH_3COO^-(aq) + H_2O(l) \rightleftharpoons CH_3COOH(aq) + OH^-(aq)$$

exist together. Examination of the equilibrium constant expressions for the three equilibria

$$K_w = [H^+][OH^-] \qquad K_a = \frac{[H^+][CH_3COO^-]}{[CH_3COOH]} \qquad K_b = \frac{[CH_3COOH][OH^-]}{[CH_3COO^-]}$$

shows that

$$K_a \times K_b = K_w \qquad (15.15)$$

$$\frac{[H^+][CH_3COO^-]}{[CH_3COOH]} \times \frac{[CH_3COOH][OH^-]}{[CH_3COO^-]} = [H^+][OH^-]$$

Another useful way of stating this relationship is

$$pK_a + pK_b = pK_w = 14.00$$

Many references list only the values of K_a and pK_a for the conjugate acids of weak bases. The values of K_b and pK_b are not given.

Thus, the value of K_b can be calculated from the values of K_a and K_w. It is not necessary to have a separate table of K_b values for weak bases.

From equation 15.15 note that, if the value of K_a is large, the value of K_b must be small because the value of K_w is always 1.0×10^{-14}. The stronger an acid, the weaker its conjugate base. Since Table 15.3 lists the weak acids in order of decreasing strength, the conjugate bases are listed in order of increasing strength. The conjugate bases of strong acids are weak bases. For example, Cl^-, the conjugate base of the strong acid HCl, is an extremely weak base.

The smaller the value of K_a, the larger the value of K_b; that is, the weaker an acid, the stronger its conjugate base. For example, hydrocyanic acid, HCN, is a weaker acid than acetic acid, and the cyanide ion, CN^-, is a stronger base than the acetate ion. However, even if the value of K_a is small, the value of K_b is still small because K_w is small. The cyanide ion is *not* a strong base; the value of K_b for the cyanide ion is only 2.0×10^{-5}. Only the conjugate bases of extremely weak acids, substances so weakly acidic that they are not usually considered to be acids at all (such as methane), are strong bases.

Sample Problem 15.8 shows how to use equation 15.15 to calculate K_b from K_a and K_w.

For methane, CH_4, K_a = ca. 10^{-49}.

15.8 (a) Use the value of K_a for formic acid from Table 15.3 to calculate the value of K_b for the formate ion, the equilibrium constant for the equilibrium

$$HCOO^-(aq) + H_2O(l) \rightleftharpoons H_3O^+(aq) + HCOOH(aq)$$

(b) Is the formate ion a weaker or a stronger base than aniline?

Solution (a) From Table 15.3, the value of K_a for formic acid is 1.8×10^{-4}. You should have memorized the fact that $K_w = 1.0 \times 10^{-14}$. According to equation 15.15,

$$K_a \cdot K_b = K_w \qquad \text{or} \qquad K_b = \frac{K_w}{K_a} = \frac{1.0 \times 10^{-14}}{1.8 \times 10^{-4}} = 5.6 \times 10^{-11}$$

(b) Because the value of K_b for the formate ion is smaller than the value of K_b for aniline (4.3×10^{-10} according to Table 15.4), the formate ion is a weaker base than aniline.

15.20 (a) From Table 15.3, which is the stronger acid, chloroacetic acid or acetic acid? (b) Which is the stronger base, the chloroacetate ion or the acetate ion? Explain your answers.

15.21 (a) Use the value of K_a for phenol from Table 15.3 to calculate the value of K_b for the phenoxide ion. (b) Is the phenoxide ion a weaker or a stronger base than ammonia? Explain your answer.

Sample Problem 15.9 illustrates the calculation of the concentration of hydroxide ions in a solution of a salt with a basic anion.

15.9 (a) What is the concentration of hydroxide ions in a 0.075 M solution of sodium acetate? (b) Is a 0.075 M solution of sodium acetate acidic, basic, or neutral?

Solution (a) First, write the net ionic equation for the equilibrium that exists in an aqueous solution of an acetate:

$$CH_3COO^-(aq) + H_2O(l) \rightleftharpoons CH_3COOH(aq) + OH^-(aq)$$

Then write the expression for K_b,

$$K_b = \frac{[CH_3COOH][OH^-]}{[CH_3COO^-]}$$

and calculate the value of K_b for the acetate ion from the value for K_a for the conjugate acid, acetic acid, given in Table 15.3 as we calculated the value of K_b for the formate ion in Sample Problem 15.8:

$$K_b = \frac{K_w}{K_a} = \frac{1.0 \times 10^{-14}}{1.8 \times 10^{-5}} = 5.6 \times 10^{-10}$$

Then express the changes in concentration of all the species in the K_b expression in terms of a single variable. If x equals mol/L OH^- formed, mol/L CH_3COOH formed also equals x, and mol/L CH_3COO^- reacted equals $-x$. The completed summary table is as follows:

$$CH_3COO^-(aq) + H_2O(l) \rightleftharpoons CH_3COOH(aq) + OH^-(aq)$$

	CH_3COO^-		CH_3COOH	OH^-
Init. conc., M	**0.075**		**0.000**	**0.000**
Change, ΔM	$-x$		$+x$	$+x$
Eq. conc., M	$0.075 - x$		x	x

Substitute the equilibrium concentrations in terms of x in the K_b expression,

$$K_b = \frac{x^2}{(0.075 - x)} = 5.6 \times 10^{-10}$$

and solve for x. Always begin by assuming that x is negligible compared to the initial concentration. If x is negligible compared to 0.075, then

$$\frac{x^2}{0.075} = 5.6 \times 10^{-10}$$

and

$$x^2 = (0.075)(5.6 \times 10^{-10}) = 4.2 \times 10^{-11}$$

If $x^2 = 4.2 \times 10^{-11}$, then

$$x = \sqrt{4.2 \times 10^{-11}} = 6.5 \times 10^{-6}$$

Because $6.5 \times 10^{-6} = 0.000\,0065$, it is indeed negligible compared to 0.075. The concentration of hydroxide ion in 0.075 M sodium acetate solution is 6.5×10^{-6} M.

(b) Because $[OH^-] = 6.5 \times 10^{-6} > 1.0 \times 10^{-7}$, the $[OH^-]$ in pure water, the solution is basic.

Because the $[OH^-]$ found is greater than 1.0×10^{-6}, the formation of OH^- by the autoionization of water can be neglected as usual.

PRACTICE PROBLEM

15.22 (a) What is the concentration of hydroxide ions in a 0.075 M solution of sodium phenoxide? (b) Is a 0.075 M solution of sodium phenoxide more or less basic than a 0.075 M solution of sodium acetate? Explain your answer.

15.6 HYDROLYSIS

Although the reactions of basic anions with water are ordinary Brønsted–Lowry acid–base reactions, they are so important that they are given a special name, hydrolysis. The word *hydrolysis* is derived from Greek words meaning "water" and "loosening." Hydrolysis reactions split one part of a water molecule from the other. The reaction of acetate ion with water is a typical example.

$$CH_3COO^-(aq) + H_2O(l) \rightleftharpoons CH_3COOH(aq) + OH^-(aq)$$

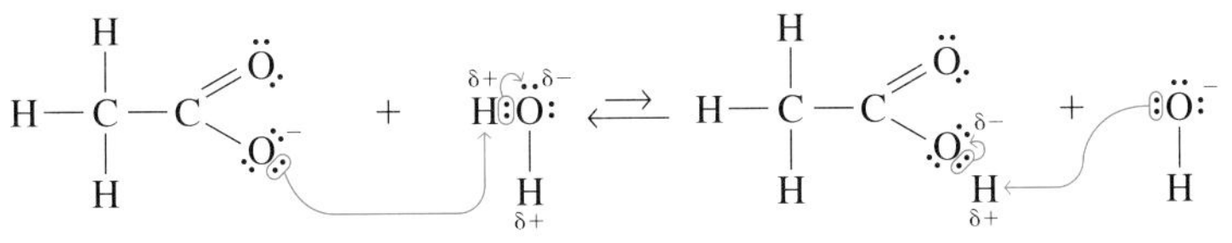

Hydrated cations also hydrolyze as was noted in Section 15.1. In hydrated cations, the water molecules are arranged around the cation with the negative ends of the water molecules toward the metal ion:

$$H-\overset{\overset{\displaystyle H}{|}}{\underset{\cdot\cdot}{O}}: \quad :\overset{\overset{\displaystyle H^{\delta+}}{|}}{\underset{\cdot\cdot}{O}}-H^{\delta+}$$

$$H-\overset{\overset{\displaystyle H}{|}}{\underset{\cdot\cdot}{O}}: \quad :\overset{\overset{}{}}{\underset{|}{O}}-H$$

• Metal ion M^{n+}

If the metal ion is small and has a high positive charge, electrons on the oxygen and in the O—H bonds are pulled toward the metal ion away from the hydrogens. The partial positive charges on the hydrogens increase. If the attraction of the metal ion for electrons is strong enough, one or more of the water molecules is split. One of the surrounding water molecules takes a proton from the hydrated metal ion:

• Metal ion M^{n+} $\qquad\qquad\qquad\qquad M^{(n-1)+}$

In this reaction, the hydrated metal ion acts as a Brønsted–Lowry acid. Hydrolysis of the hydrated aluminum ion is an example:

$$Al(H_2O)_6{}^{3+}(aq) + H_2O(l) \rightleftharpoons Al(H_2O)_5(OH)^{2+}(aq) + H_3O^+(aq)$$

The acidity of the hydrated aluminum ion is treated quantitatively in Section 16.2.

Thus, **hydrolysis** is the *reaction of an anion with water to produce the conjugate acid of the anion and hydroxide ion or the reaction of a cation with water to produce the conjugate base of the cation and hydronium ion.*

Relatively large cations with low positive charges such as Na^+ and K^+ do not attract electrons as strongly as small cations with high positive charges. Hydrated sodium and potassium ions do not hydrolyze and are not acidic.

Although salts are a product of the neutralization of acids by bases, only a few salts form neutral solutions when dissolved in water. Solutions of most salts are either acidic or basic as a result of hydrolysis (see ▬Figure 15.7).

Solutions of salts formed by the reaction of weak acids with strong bases such as sodium acetate and sodium phenoxide are basic. The cations of these salts do not hydrolyze; they are spectator ions. The anions are bases and react with water to form the conjugate acid of the anion and hydroxide ion:

$$Na^+(aq) + CH_3COO^-(aq) + H_2O(l) \rightleftharpoons Na^+(aq) + CH_3COOH(aq) + OH^-(aq)$$
$$Na^+(aq) + C_6H_5O^-(aq) + H_2O(l) \rightleftharpoons Na^+(aq) + C_6H_5OH(aq) + OH^-(aq)$$

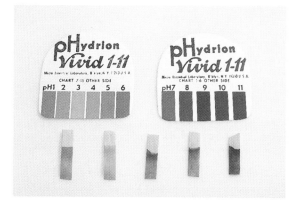

▬ FIGURE 15.7   The center strip of test paper shows the effect of pure water on test paper. Compared with pure water, 0.1 M $Fe(NO_3)_3(aq)$ is strongly acidic *(far left)*, and 0.1 M $NH_4Cl(aq)$ is weakly acidic *(second from left)*. A 0.1 M aqueous solution of sodium formate, HCOONa, is weakly basic *(second from right)*, and 0.1 M aqueous sodium phenoxide, C_6H_5ONa, is strongly basic *(far right)*.

Solutions of salts of weak molecular bases, such as ammonia, and strong acids are acidic. Ammonium chloride is an example of this type of salt. The cations of these salts are acids and react with water to form hydronium ions. The anions are extremely weak bases because they are the conjugates of strong acids and do not react with water:

The pHs of salt solutions can be shown with an overhead projector using samples in Petri dishes with Universal Indicator.

$$NH_4^+(aq) + Cl^-(aq) + H_2O(l) \rightleftharpoons NH_3(aq) + H_3O^+(aq) + Cl^-(aq)$$

The net ionic equation for this equilibrium is

$$NH_4^+(aq) + H_2O(l) \rightleftharpoons NH_3(aq) + H_3O^+(aq)$$

Solutions of many metal salts of strong acids are acidic because the hydrated cations of the metals are acids. For example, a 0.1 M solution of iron(III) perchlorate is strongly acidic as a result of the equilibrium

Aqueous solutions of the salts of most metals (except Group IA and IIA metals) with strong acids are acidic.

$$Fe(H_2O)_6^{3+}(aq) + 3ClO_4^-(aq) + H_2O(l) \rightleftharpoons$$
$$Fe(H_2O)_5(OH)^{2+}(aq) + H_3O^+(aq) + 3ClO_4^-(aq)$$

The perchlorate ions are the conjugates of the strong acid, perchloric acid, and are spectator ions.

Salts of weak acids and weak bases can give either acidic, neutral, or basic solutions. If the weak acid is stronger than the weak base, a solution of the salt is acidic. Anilinium acetate, $CH_3COONH_3C_6H_5$, is an example of this type of salt. If the weak base is stronger than the weak acid, the solution of the salt is basic. An example is ammonium phenoxide, $C_6H_5ONH_4$. If the acid dissociation constant of the weak acid is of the same order of magnitude as the base dissociation constant of the weak base, a solution of the salt will be neutral. Ammonium acetate, CH_3COONH_4, is an example of a salt that is neutral because the anion and the cation hydrolyze to the same extent.

Solutions of salts formed by neutralization of strong acids with strong bases are neutral. For example, solutions of sodium iodide, potassium nitrate, and calcium chloride are neutral. Neither the cation nor the anion of these salts reacts with water. No hydrolysis occurs:

$$Na^+(aq) + I^-(aq) + H_2O(l) \longrightarrow \text{no reaction}$$

The handling and uses of salt solutions are strongly influenced by their acidic and basic properties. For example, acidic salt solutions corrode metals and can't be stored in uncoated metal containers. Strongly basic salts cannot be used as medicines because they would burn the patient's mouth and esophagus.

SAMPLE PROBLEM

15.10 Predict whether an aqueous solution of each of the following salts will be acidic, basic, or neutral: (a) NaCl, (b) NH$_4$I, (c) Cr(NO$_3$)$_3 \cdot$ 9H$_2$O, and (d) KCN. Refer to Tables 15.3 and 15.4 for values of ionization constants of weak acids and bases.

Solution   (a) NaCl is the salt of a strong base, NaOH, and a strong acid, HCl. Neither the Na$^+$ ion nor the Cl$^-$ ion will hydrolyze. An aqueous solution of NaCl will be neutral.

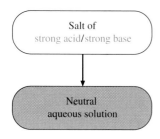

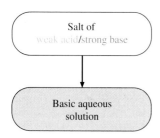

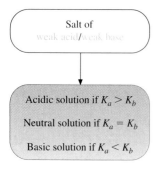

(b) NH₄I is the salt of a weak base, NH₃, and a strong acid, HI. The I⁻ ion will not hydrolyze but the NH₄⁺ ion will:

$$NH_4^+(aq) + H_2O(l) \rightleftharpoons NH_3(aq) + H_3O^+(aq)$$

An aqueous solution of NH₄I will be acidic.

(c) Cr(NO₃)₃·9H₂O is the salt of a small, highly charged metal ion, Cr³⁺, and a strong acid, HNO₃. The NO₃⁻ ion will not hydrolyze but the hydrated Cr³⁺ ion will:

$$Cr(H_2O)_x^{3+}(aq) + H_2O(l) \rightleftharpoons Cr(H_2O)_{(x-1)}(OH)^{2+}(aq) + H_3O^+(aq)$$

An aqueous solution of Cr(NO₃)₃·9H₂O will be acidic.

(d) KCN is the salt of a strong base, KOH, and a weak acid, HCN. The K⁺ ion will not hydrolyze but the CN⁻ ion will:

$$CN^-(aq) + H_2O(l) \rightleftharpoons HCN(aq) + OH^-(aq)$$

An aqueous solution of KCN will be basic.

PRACTICE PROBLEM

15.23 Predict whether an aqueous solution of each of the following salts will be acidic, basic, or neutral: (a) NH₄Br, (b) NaClO, (c) AlCl₃·6H₂O, and (d) KI.

15.7 THE COMMON ION EFFECT

In Section 15.4, we learned how to calculate the hydrogen ion concentration in a solution of a weak acid and in Section 15.5 how to calculate the hydrogen ion concentration in a solution of a salt of a weak acid. In this section, we learn how to calculate the concentration of hydrogen ions in a solution that contains both a weak acid and a salt of the weak acid.

First, let's consider a solution that contains both a weak acid and a salt of the weak acid qualitatively so that we will know whether an answer is reasonable. A solution that contains a mixture of acetic acid and sodium acetate is an example. The acetate ion is present both in solutions of acetic acid and in solutions of sodium acetate. The acetate ion is the common ion in a solution containing a mixture of acetic acid and sodium acetate. A **common ion** is an *ion that is produced by more than one solute.*

According to Le Châtelier's principle, the equilibrium between acetic acid molecules and acetate ions in an aqueous solution of acetic acid,

$$CH_3COOH(aq) \rightleftharpoons CH_3COO^-(aq) + H^+(aq)$$

will be shifted to the left by the presence of CH₃COO⁻ ions from sodium acetate. The concentration of H⁺(aq) ions will be decreased, and the concentration of CH₃COOH(aq) molecules will be increased. The *shift in equilibrium caused by the addition of an ion formed from the solute* is called the **common ion effect.** Sample Problem 15.11 illustrates the calculation of concentrations of species present in a solution that contains two solutes with a common ion.

15.11   What is the concentration of hydrogen ions in a solution formed by the addition of 0.097 mol (8.0 g) $CH_3COONa(s)$ to 1.00 L of 0.099 M acetic acid? Assume that the volume of the acetic acid solution is not changed by the addition of the solid sodium acetate.

We have used 0.097 mol and 0.099 M instead of 0.100 to make it easier for students to follow the numbers through the calculation.

Solution   The equation for the equilibrium between molecules of acetic acid and hydrogen ions and acetate ions is

$$CH_3COOH(aq) \rightleftharpoons CH_3COO^-(aq) + H^+(aq)$$

and the expression for K_a for the equilibrium is

$$K_a = \frac{[CH_3COO^-][H^+]}{[CH_3COOH]} = 1.8 \times 10^{-5} \quad \text{(from Table 15.3)}$$

The completed summary table is as follows:

	$CH_3COOH(aq) \rightleftharpoons$	$CH_3COO^-(aq) +$	$H^+(aq)$
Init. conc., M	**0.099**	**0.097**	**0.000**
Change, ΔM	$-x$	$+x$	$+x$
Eq. conc., M	$0.099 - x$	$0.097 + x$	x

Solution of the equilibrium constant expression for $[H^+]$ and substitution of the equilibrium concentrations gives

$$x = \frac{K_a[CH_3COOH]}{[CH_3COO^-]} = \frac{(1.8 \times 10^{-5})(0.099 - x)}{(0.097 + x)}$$

which simplifies to

$$x = \frac{(1.8 \times 10^{-5})0.099}{0.097} = 1.8 \times 10^{-5}$$

● Remember that if $[H^+]$ from a dissolved acid is greater than 1×10^{-6}, the quantity of H^+ formed by the autoionization of water is negligible.

if we make the usual assumption that x is negligible compared with 0.099 and 0.097.

   The assumption that x is negligible compared with 0.099 and 0.097 is correct, and $[H^+]$ is equal to 1.8×10^{-5}. According to Sample Problem 15.4, $[H^+]$ in 0.100 M acetic acid is 1.3×10^{-3}. Addition of sodium acetate has lowered $[H^+]$ (from 1.3×10^{-3} to 1.8×10^{-5}) as we predicted qualitatively that it should.

● Notice that in this solution, which has equal concentrations of the weak acid and its salt, $[H^+] = K_a$.

15.24   What is the concentration of hydrogen ions in a solution formed by the addition of 0.085 mol of solid potassium formate, $HCOOK(s)$, to 1.0 L of 0.083 M formic acid, $HCOOH(aq)$? Assume that the volume of the formic acid solution is not changed by the addition of the solid potassium formate.

   Calculation of the concentration of hydroxide ion in a solution containing a weak base and the salt of the weak base is similar to calculation of the concentration of hydrogen ion in a solution containing a weak acid and the salt of the weak acid, as illustrated by Sample Problem 15.12.

15.12   What is the concentration of hydroxide ions in a solution formed by the addition of 0.073 mol $NH_4Cl(s)$ to 1.0 L of 0.075 M $NH_3(aq)$? Assume that the volume of the ammonia solution is not changed by the addition of the solid ammonium chloride.

Solution   The equation for the equilibrium is

$$NH_3(aq) + H_2O(l) \rightleftharpoons NH_4^+(aq) + OH^-(aq)$$

and the expression for K_b is

$$K_b = \frac{[NH_4^+][OH^-]}{[NH_3]} = 1.8 \times 10^{-5} \qquad \text{(from Table 15.4)}$$

The completed summary table is as follows:

	$NH_3(aq)$ + $H_2O(l)$ $\rightleftharpoons$	$NH_4^+(aq)$ +	$OH^-(aq)$
Init. conc., M	**0.075**	**0.073**	**0.000**
Change, ΔM	$-x$	$+x$	$+x$
Eq. conc., M	$0.075 - x$	$0.073 + x$	x

Solution of the equilibrium constant expression for $[OH^-]$ and substitution of the equilibrium concentrations gives

$$x = \frac{K_b[NH_3]}{[NH_4^+]} = \frac{(1.8 \times 10^{-5})(0.075 - x)}{(0.073 + x)}$$

which simplifies to

$$x = \frac{(1.8 \times 10^{-5})0.075}{0.073} = 1.8 \times 10^{-5}$$

if we make the usual assumption that x is negligible compared with 0.075 and 0.073.

   The assumption that x is negligible compared with 0.075 and 0.073 is correct, and $[OH^-]$ is equal to 1.8×10^{-5}. According to Sample Problem 15.7, $[OH^-]$ in 0.075 M NH_3 is 1.1×10^{-3}. The presence of the common ion, NH_4^+, has shifted the equilibrium to the left and reduced $[OH^-]$, as would be predicted by Le Châtelier's principle.

◗ Notice that in this solution, which has equal concentrations of the weak base and its salt, $[OH^-] = K_b$.

◗ Always check your work to be sure that it is reasonable.

15.25   If a solution is prepared by addition of 0.082 mol $CH_3NH_3Cl(s)$ to 1.0 L of 0.074 M $CH_3NH_2(aq)$, what is the concentration of hydroxide ions in the solution? Assume that the volume of the solution is not changed by the addition of the solid.

◔ In the Acids and Bases section, a pH meter records the pH of water and a buffered solution as acid or base is added. An animation with narration explains the behavior observed. Students are asked to identify acid-base pairs that buffer solutions and solve numerical problems about buffers.

15.8   BUFFER SOLUTIONS

The most interesting thing about solutions that have approximately equal concentrations of weak acids and a salt of the weak acid (or solutions that have approximately equal concentrations of weak bases and a salt of the weak base) is that they resist significant changes in hydrogen ion concentration. Neither addition of small quantities of acids or bases nor dilution results in much change in pH. *Solutions that resist changes in hydrogen ion concentration* are called **buffered solutions** or **buffer**

TABLE 15.5	Comparison of Buffered and Unbuffered Solutions	
	Unbuffered Solution: 1.78×10^{-5} M HCl	Buffered Solution: 0.099 M CH_3COOH and 0.097 M CH_3COONa
Initial pH of 1.00-L sample	4.8	4.8
pH after addition of 0.010 mol NaOH	12.0	4.8
pH after addition of 0.010 mol HCl	2.0	4.7

solutions. The data in Table 15.5 and ■Figure 15.8 show the difference between unbuffered and buffered solutions.

Buffered solutions occur frequently in nature; for example, seawater and blood are buffered. The oxygen-carrying capacity of hemoglobin in blood depends on pH. The normal pH of human blood is 7.4 (measured at 25 °C). At pH < 7.3, blood cannot remove carbon dioxide from cells; at pH > 7.7, blood cannot release carbon dioxide to the lungs. Variation in the pH of blood of greater than 0.6 pH unit for more than a few seconds results in death. Culture media used in bacteriology are usually buffered at the pH needed for the growth of the bacteria being studied, and many reactions are carried out in buffered solutions in chemistry laboratories and chemical manufacturing plants. Buffer solutions are also used to calibrate pH meters.

How does a buffer solution keep the hydrogen ion concentration from changing? Consider the equilibrium

$$CH_3COOH(aq) \rightleftharpoons H^+(aq) + CH_3COO^-(aq)$$

In a buffer solution, there is a large supply of both acid molecules and anions:

$$\underset{\text{large supply}}{CH_3COOH(aq)} \rightleftharpoons H^+(aq) + \underset{\text{large supply}}{CH_3COO^-(aq)}$$

If an acid is added to the buffer solution, the anions react with the added hydrogen ions and use them up so that the concentration of hydrogen ions does not increase very much. If a base is added to the buffer solution, the hydrogen ions react with the added hydroxide ions and form water; the equilibrium shifts to the right to form more hydrogen ions. The concentration of hydrogen ions does not decrease very much. In a buffer solution, the concentrations of both acid molecules and anions must be large compared to the quantities of acid and base that are likely to be added to the buffer in order for the buffer to have enough capacity. The **capacity** of a buffer solution is the *quantity of acid or base that can be added before the pH changes significantly.*

We have already seen that, in a solution that contains both a weak acid, such as acetic acid, and a salt of the acid, such as sodium acetate, the concentration of weak acid molecules is, for all practical purposes, the same as the initial concentration of weak acid:

$$[\text{acid molecules}]_{eq} = [\text{acid molecules}]_0$$

The concentration of anions formed by ionization of the weak acid is very small, and the concentration of anions is, for all practical purposes, the same as the concentration of the salt:

$$[\text{anions}]_{eq} = [\text{anions}]_0$$

■ FIGURE 15.8 *Top:* Samples of unbuffered and buffered hydrochloric acid solution containing bromocresol green indicator. Addition of sodium hydroxide *(middle)* and hydrochloric acid *(bottom)* changes the pH and color of the indicator in the unbuffered solution *(left)* but not in the buffered solution *(right).*

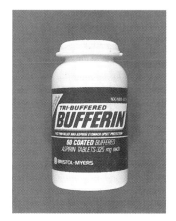

Some aspirin is buffered.

See Sample Problem 15.11.

(assuming that the salt is a 1:1 salt like sodium acetate, CH_3COONa). Substitution of these values for equilibrium concentrations in the K_a expression gives

$$K_a = \frac{[\text{anions}]_0[H^+]_{eq}}{[\text{acid molecules}]_0}$$

or, solving for $[H^+]_{eq}$,

$$[H^+]_{eq} = K_a \frac{[\text{acid molecules}]_0}{[\text{anions}]_0} \tag{15.16}$$

Thus, the hydrogen ion concentration of a buffer solution depends on the ionization constant of the weak acid and on the ratio of the concentrations of the weak acid and its salt. If equal concentrations of weak acid and salt are used to prepare a buffer, $[\text{acid molecules}]_0 = [\text{anions}]_0$ and $[H^+] = K_a$. The concentration of hydrogen ion in a buffer can be adjusted somewhat by varying the ratio of the concentrations of molecules to anions. However, neither concentration can be very low, or the buffer will not have sufficient capacity. A weak acid with an ionization constant as close as possible to the $[H^+]$ desired for the buffer should be chosen to make a weak acid–salt buffer.

Solutions containing weak bases and salts of the weak bases can also be used to make buffers. For example, a solution that is 0.10 M in both NH_3 and NH_4Cl is a buffer and has a pH of 9.3. The hydroxide ion concentration of a weak base–salt buffer solution depends on K_b for the weak base and on the ratio of the concentrations of the weak base and its salt. If equal concentrations of weak base and salt are used to prepare a buffer, $[\text{base molecules}]_0 = [\text{cations}]_0$ and $[OH^-] = K_b$. Neither of the components of a buffer should react with any components of the solution to be buffered, or the buffer will be destroyed. Calculations that involve weak base–salt buffers are similar to calculations for weak acid–salt buffers.

Sample Problem 15.11 showed the calculation of the concentration of hydrogen ions in a buffer; Sample Problem 15.13 illustrates calculation of the change in pH that occurs when base is added to a buffer. Calculation of change in pH when acid is added to a buffer is similar.

Again, refer to Sample Problem 15.11.

SAMPLE PROBLEM

15.13 A buffer solution that is 0.099 M in CH_3COOH and 0.097 M in CH_3COO^- has $[H^+] = 1.8 \times 10^{-5}$ (pH = 4.74). What will the $[H^+]$ and pH be after 0.010 mol NaOH(s) has been added to 1.0 L of the buffer solution? Assume no volume change takes place.

Solution Consider the equilibrium

$$CH_3COOH(aq) \rightleftharpoons H^+(aq) + CH_3COO^-(aq)$$

for which

$$K_a = \frac{[H^+][CH_3COO^-]}{[CH_3COOH]} = 1.8 \times 10^{-5}$$

The ionic equation for the reaction beween the weak acid, CH_3COOH, and the strong base, NaOH, is

$$CH_3COOH(aq) + Na^+(aq) + OH^-(aq) \longrightarrow CH_3COO^-(aq) + Na^+(aq) + H_2O(l)$$

The net ionic equation is

$$CH_3COOH(aq) + OH^-(aq) \longrightarrow CH_3COO^-(aq) + H_2O(l)$$

This reaction is the reverse of the hydrolysis of an acetate. Therefore, the equilibrium constant is the reciprocal of the equilibrium constant for the hydrolysis of an acetate. In Sample Problem 15.9, we found that the equilibrium constant for the hydrolysis of an acetate is 5.6×10^{-10}; the equilibrium constant for the reaction between CH_3COOH and OH^- is

⊙ Remember that, when an equation is written in the opposite direction, the value of the new equilibrium constant is the reciprocal of the original value (Section 14.3).

$$\frac{1}{5.6 \times 10^{-10}} = 1.8 \times 10^9$$

Thus, K_c for the reaction between CH_3COOH and OH^- is very large, and reaction is complete.

According to the net ionic equation, 0.010 mol OH^- will react with 0.010 mol CH_3COOH. After reaction has taken place,

$$[CH_3COOH] = [(0.099) - (0.010)] = 0.089$$

$$[CH_3COO^-] = [(0.097) + (0.010)] = 0.107$$

Substitution in equation 15.16 gives

$$[H^+] = K_a \frac{[CH_3COOH]}{[CH_3COO^-]} = 1.8 \times 10^{-5} \frac{(0.089)}{(0.107)} = 1.5 \times 10^{-5}$$

and

$$pH = 4.82$$

Sample Problem 15.14 shows how to choose the best weak acid for a buffer and the right ratio of acid to base to obtain the desired pH. The ratio of the concentration of acid molecules to the concentration of anions must be between 0.1 and 10 in order for the buffer to have sufficient capacity. As a result, a given combination of weak acid and salt can only be used to buffer solutions to within ± 1 pH unit of pK_a.

SAMPLE PROBLEM

15.14 (a) To prepare a buffer having a hydrogen ion concentration of 1.5×10^{-3}, which of the acid–base pairs in Table 15.3 should be used? (b) What should the ratio of [acid molecules]/[anions] be to obtain a hydrogen ion concentration of 1.5×10^{-3}?

Solution (a) The acid used for a buffer should have K_a as close as possible to the desired concentration of hydrogen ion. The only acid in Table 15.3 with K_a close to the desired hydrogen ion concentration (1.5×10^{-3}) is chloroacetic acid ($K_a = 1.4 \times 10^{-3}$). A mixture of chloroacetic acid and sodium chloroacetate should be used to prepare the buffer.

(b) According to equation 15.16,

$$[H^+] = K_a \frac{[\text{acid molecules}]_0}{[\text{anions}]_0}$$

Solution for the ratio $[\text{acid molecules}]_0/[\text{anions}]_0$ and substitution of the values for $[H^+]$ and K_a gives

$$\frac{[\text{acid molecules}]_0}{[\text{anions}]_0} = \frac{[H^+]}{K_a} = \frac{(1.5 \times 10^{-3})}{(1.4 \times 10^{-3})} = 1.1$$

Any pair of concentrations of acid molecules and anions that are in the ratio 1.1:1 would give a buffer solution of the correct pH. However, for the buffer to have a reasonable capacity, the concentrations should not be too low. A solution that is 0.11 M in chloroacetic acid and 0.10 M in sodium chloroacetate would have a reasonable capacity for most purposes. However, concentrations of 0.000 11 M and 0.000 10 M would have too little capacity.

PRACTICE PROBLEMS

15.26 Which of the following combinations would be useful as buffers? Explain your answer. (a) (1.0×10^{-1}) M NH_3 and (1.0×10^{-1}) M NH_4NO_3
(b) (1.0×10^{-1}) M HCOOH and (1.0×10^{-1}) M HCOOK
(c) (1.0×10^{-5}) M HCOOH and (1.0×10^{-5}) M HCOOK

15.27 (a) What is the concentration of hydrogen ions and pH in a solution that is 0.075 M HCOOK and 0.080 M HCOOH? (b) What will be the $[H^+]$ and pH after 0.0050 mol HCl(g) has been added to 1.00 L of the buffer solution? Assume no change in volume.

15.28 To prepare a buffer having a hydroxide ion concentration of 3.0×10^{-4}, (a) which of the base–conjugate acid pairs in Table 15.4 should be used? (b) What should the ratio of [base molecules]/[cations] be?

Use of the Henderson–Hasselbalch equation is convenient for technicians who often make up buffer solutions. However, plugging numbers into the formula does not contribute to students' understanding of buffer solutions. Therefore, we encourage our students to reason through problems such as Practice Problem 15.27 rather than to use the Henderson–Hasselbalch equation.

Biologists and biochemists use buffer solutions a great deal. For routine calculations involving the pH of buffers, the Henderson–Hasselbalch equation

$$pH = pK_a + \log \frac{[base]_0}{[acid]_0} \qquad (15.17)$$

is very convenient.* In using the Henderson–Hasselbalch equation, it is important to remember that the acid must be weak enough and the concentrations high enough so that the assumption that the equilibrium concentrations are the same as the initial concentrations is correct. In addition, the concentrations must not be so high that K_a is not constant and the temperature must be the temperature to which K_a applies.

PRACTICE PROBLEM

15.29 Use the Henderson–Hasselbalch equation (equation 15.17) to calculate the pH of a solution that has [HF] = 0.110 and $[F^-]$ = 0.090.

15.9 HOW INDICATORS WORK

An indicator is a substance used in a titration to signal the point at which reaction is complete (Section 4.10). Most indicators used for acid–base titrations are weak

---

*L. J. Henderson, an American biochemist, was professor of biological chemistry at Harvard Medical School. Henderson developed the equation that bears his name; Danish biochemist Karl Hasselbalch modified it. The Henderson–Hasselbalch equation is easily derived from the relationship of equation 15.16 stated in general form:

$$[H^+]_{eq} = K_a \frac{[acid]_0}{[base]_0}$$

Taking logarithms of both sides of the above relationship gives

$$\log[H^+]_{eq} = \log K_a + \log \frac{[acid]_0}{[base]_0} \qquad (15.18)$$

Multiplication of both sides of equation 15.18 by -1 gives

$$-\log[H^+]_{eq} = -\log K_a - \log \frac{[acid]_0}{[base]_0}$$

$$pH = pK_a + \log \frac{[base]_0}{[acid]_0} \qquad (15.17)$$

organic acids with complex structures represented by HIn. The molecules of indicators, HIn, are one color and the conjugate bases, In⁻, are another color. For example, molecules of the indicator bromocresol green are yellow; the anions that are the conjugate base of bromocresol green are blue. In aqueous solution, indicator molecules are in equilibrium with hydrogen ions and indicator ions:

$$\underset{\substack{\text{yellow in the case of} \\ \text{bromocresol green}}}{HIn(aq)} \; \rightleftharpoons \; H^+(aq) \; + \; \underset{\substack{\text{blue in the case of} \\ \text{bromocresol green}}}{In^-(aq)}$$

■ FIGURE 15.9   Color of the indicator bromocresol green at different pHs. *Left:* pH = 3.0. *Middle:* pH = 4.5. *Right:* pH = 6.0.

Addition of an acid shifts the equilibrium to the left; the color changes to the color of the indicator molecules (yellow in the case of bromocresol green). Addition of a base removes hydrogen ions by reaction to form water, and the equilibrium shifts to the right. The color changes to the color of the indicator ions (blue in the case of bromocresol green). ■Figure 15.9 shows a solution containing bromocresol green at different hydrogen ion concentrations. The human eye sees a solution containing bromocresol green as yellow when about 90% (or more) of the indicator is in the form of molecules. It sees the solution as blue when about 90% (or more) of the bromocresol green is in the form of ions. When approximately half the bromocresol green is in the form of molecules and half is in the form of ions, the solution appears green, a mixture of yellow and blue. The quantity of indicator used must be very small, or the indicator itself will affect the concentration of hydrogen ion significantly. Therefore, the color of an indicator must be very strong so that an extremely small quantity of the indicator gives a visible color.

The ionization constant expression for an indicator can be written

$$K_a = \frac{[H^+][In^-]}{[HIn]}$$

Rearrangement of the ionization constant expression gives

$$\frac{K_a}{[H^+]} = \frac{[In^-]}{[HIn]}$$

Thus, the ratio [In⁻]/[HIn] depends on the value of K_a for the indicator and on the [H⁺]. If the ratio [In⁻]/[HIn] is ≥ 10, the human eye sees the color of the anion In⁻. If the ratio is ≤ 0.1, the human eye sees the color of the molecule HIn. In other words, the human eye can detect a change of 100-fold in the ratio of [In⁻] to [HIn]. For a given indicator, K_a is a constant; therefore, a change of 100-fold in the ratio of [In⁻] to [HIn] corresponds to a change of 100-fold in [H⁺]. Because the pH scale is a logarithmic scale, a change of 100-fold in [H⁺] corresponds to a change of 2 in pH. Indicators are observed to change from their acid color to their base color over a pH range of about 2 units. When [In⁻] = [HIn], the color of a solution containing the indicator is a mixture of the colors of the acid and base forms. For example, a solution of bromocresol green is green, a mixture of yellow and blue. When [In⁻] = [HIn], the ratio [In⁻]/[HIn] is equal to 1, and [H⁺] = K_a.

Different indicators have different values of K_a. As a result, different indicators change color at different pH ranges. Table 15.6 shows the colors and pH ranges of some common indicators.

If the intensities of the colors of the acid and base forms are very different, the color will not be a mixture of the colors when [In⁻] = [HIn]. However, we think this detail should be left for analytical chemistry.

PRACTICE PROBLEM

15.30   (a) What color is thymol blue at a pH of 7.0? (b) At about what pH would thymol blue be orange? (Orange is a mixture of red and yellow.)

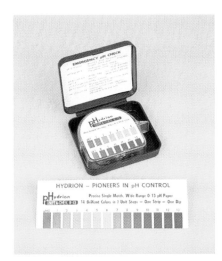

FIGURE 15.10 One brand of pH paper.

Many indicators are obtained from plants; for example, litmus is extracted from lichen. Other indicators, such as phenolphthalein, which is also used in laxatives, are synthetic.

TABLE 15.6	Some Common Indicators[a]		
Name	Acid Color	pH Range of Color Change	Base Color
Methyl violet	Yellow	0.0–1.6	Blue
Thymol blue[b]	Red	1.2–2.8	Yellow
Methyl orange	Red	3.2–4.4	Yellow
Bromocresol green	Yellow	3.8–5.4	Blue
Methyl red	Red	4.8–6.0	Yellow
Litmus	Red	5.0–8.0	Blue
Bromothymol blue	Yellow	6.0–7.6	Blue
Thymol blue[b]	Yellow	8.0–9.6	Blue
Phenolphthalein	Colorless	8.2–10.0	Pink
Thymolphthalein	Colorless	9.4–10.6	Blue
Alizarin yellow R	Yellow	10.1–12.0	Red

[a]Most information is from Lide, D. R., Ed. *CRC Handbook of Chemistry and Physics,* 75th ed.; CRC Press: Boca Raton, FL, 1994, pp **8**-17–8-18.
[b]Thymol blue has two end points.

pH test papers are often used to estimate pH rapidly, for example, to test soil. ▬Figure 15.10 shows one brand of pH test paper. pH test papers contain a mixture of indicators. As a simple example, paper soaked in a mixture of bromocresol green and methyl red and then dried would be orange (a mixture of red and yellow) at pH 3, purple (a mixture of blue and red) at pH 5, and green (a mixture of blue and yellow) at pH 7.

15.10   TITRATION REVISITED

In Section 4.10, we saw that titration is a method of determining the quantity of a substance present in a solution. A solution of known concentration called a standard solution is added from a buret to the solution being analyzed (see ▬Figure 15.11). Addition is stopped when the end point of the titration is reached. The end point is signaled by a color change of an indicator or by a change in an electrical property such as conductivity. *If the means of measuring the end point has been chosen correctly, the end point occurs at the equivalence point.* The **equivalence point** is the *point at which the quantity of reactant called for by the equation for the reaction has been added.*

A titration curve gives the clearest picture of what happens during a titration. The **titration curve** for an acid–base titration is a *graph of pH against the number of milliliters of acid or base added in the titration.* ▬Figure 15.12 shows titration curves for the titration of the strong acid, HCl, with the strong base, NaOH. The data used to plot titration curves can be obtained either by computation or by measuring the pH during titration with a pH meter. The black line in Figure 15.12 shows the calculated titration curve for the titration of 25.00 mL of 0.1000 M HCl with 0.100 M NaOH. The purple line in Figure 15.12 shows the calculated titration curve for the titration of 25.00 mL of 0.01000 M HCl with 0.01000 M NaOH. Note how use of the logarithmic pH scale makes possible inclusion of the whole titration in a single graph. The hydrogen ion concentration changes by a factor of 10^{12} during the titration of 0.1000 M HCl.

The titration curves in Figure 15.12 are typical of titration of a strong acid with a strong base. During the first part of the titration, the pH increases slowly and gradually. The concentration of hydrogen ion is large at the beginning of the titration,

FIGURE 15.11 In titration, a standard solution is added from a buret to the solution being analyzed, which is stirred with a magnetic stirrer.

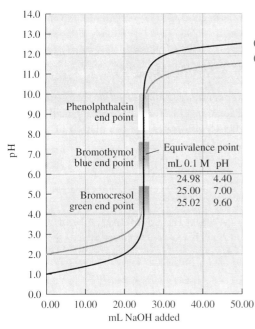

The curve for the titration of a strong base with a strong acid is similar except that it begins at high pH and ends at low pH.

The Acid-Base Titration curves module of the Acids and Bases section allows students to choose the type of titration curve they wish to study: strong acid/strong base, weak acid/strong base, or strong acid/weak base. They can move a ball along the titration curve and see how pH is calculated for any volume of titrant. Then the student must calculate the pH at various points along the curve. Sample calculations and on-line help are provided.

▪ FIGURE 15.12 Titration curves for titration of 25.00 mL of strong acid with strong base.

and addition of hydroxide ions produces only a small change in $[H^+]$. Near the equivalence point, the concentration of hydrogen ion is very small; addition of hydroxide ions produces a large proportional change in $[H^+]$. There is a very rapid change in pH around the equivalence point. At the equivalence point, addition of one drop (about 0.05 mL) of 0.1000 M NaOH results in a change in pH of more than 5.2 units, a change in $[H^+]$ of about 100 000-fold.

In Section 15.9 you learned that indicators usually change color over a pH range of about 2 units. If an indicator is to signal the equivalence point of a titration, *an indicator must be chosen that will change color when the right quantity of standard solution has been added.* For example, in the titration of 25.00 mL of 0.1000 M HCl, 25.00 mL of 0.1000 M NaOH must be added to reach the equivalence point. The pH changes rapidly around the equivalence point. Any indicator that changes color in the pH range of 3.5–10.5 is suitable for measuring the equivalence point of the titration of a 0.1000 M solution of a strong acid with a 0.1000 M solution of a strong base. Bromocresol green, methyl red, litmus, bromothymol blue, thymol blue, phenolphthalein, and thymophthalein are all suitable (see Table 15.6). Neither methyl violet nor alizarin yellow R are suitable. Methyl violet changes from yellow to blue between pH 0.0 and 1.6. Methyl violet would have changed color before the equivalence point. Alizarin yellow R changes from yellow to red between pH 10.1 and 12.0. The equivalence point would have been passed before alizarin yellow R changed color.

The purple line in Figure 15.12 shows the effect of dilution on a titration curve. Although a very rapid change is evident in pH around the equivalence point, the vertical portion of the curve is shorter. Only indicators that change color between 5.0 and 9.5, such as litmus, bromothymol blue, and thymol blue, can be used for the titration of 0.01000 M HCl with 0.01000 M NaOH. Because the vertical portion of the titration curve becomes shorter as the solutions become more dilute, titration is not suitable for analyzing very dilute solutions.

The methods we have learned in this chapter for calculating the concentration of hydrogen ions in solutions of acids, bases, salts, and combinations of acids or bases and salts can be used to calculate the concentration of hydrogen ion and the pH at any point in a titration. Calculation of the concentration of hydrogen ion and the pH at different points in titrations is a good review of calculations involving acid–base equilibria in aqueous solution.

Titration of a Strong Acid with a Strong Base

The hydrogen ion concentration *at the beginning of the titration,* before any base has been added, is simply the hydrogen ion concentration of the strong acid solution (see Section 15.4).* Sample Problem 15.15 illustrates the calculation of pH *after addition of some base but before the equivalence point.*

SAMPLE PROBLEM

Burets can be read to the nearest 0.01 mL. The concentrations of commercial standard solutions are given to the nearest part per thousand; for example, 0.1 M HCl is 0.0950–0.1050 M. The exact molarity is given on the label.

15.15 In the titration of 25.00 mL of 0.1000 M HCl with 0.1000 M NaOH, what is the pH after 15.00 mL of the standard NaOH solution have been added?

Solution The equation for the neutralization reaction is

$$HCl(aq) + NaOH(aq) \longrightarrow H_2O(l) + NaCl(aq) \qquad (15.19)$$

The reaction between a strong acid and a strong base takes place almost instantly and is complete. However, the 15.00 mL of 0.1000 M NaOH added is not enough to neutralize all of the HCl in 25.00 mL of 0.1000 M HCl. Some HCl is left over. The H^+ from the leftover HCl will shift the equilibrium between water molecules and hydrogen and hydroxide ions,

$$H_2O(l) \rightleftharpoons H^+(aq) + {}^-OH(aq)$$

to the left. We probably do not have to worry about hydrogen ions from water.

Neither sodium ions nor chloride ions hydrolyze. No hydrogen ions or hydroxide ions will be formed by hydrolysis. All hydrogen ions in solution must come from HCl. First, we must figure out how many moles of HCl are left over. Then we must find the total volume of the solution and calculate the $[H^+]$ and the pH.

Figuring out how many moles of HCl are left over is a limiting-reactant stoichiometry problem. We must calculate how many moles of HCl were present at the start of the titration and how many moles of HCl were neutralized by the 15.00 mL of 0.1000 M NaOH:

○ Limiting reactant stoichiometry problems were discussed in Section 3.6.

$$\left[25.00 \text{ mL HCl} \left(\frac{1 \text{ L HCl}}{1000 \text{ mL HCl}} \right) \left(\frac{0.1000 \text{ mol HCl}}{1 \text{ L HCl}} \right) \right.$$
$$\left. = 0.002\ 500 \text{ mol HCl present at start} \right]$$

$$\left[15.00 \text{ mL NaOH} \left(\frac{1 \text{ L NaOH}}{1000 \text{ mL NaOH}} \right) \left(\frac{0.1000 \text{ mol NaOH}}{1 \text{ L NaOH}} \right) \right.$$
$$\left. = 0.001\ 500 \text{ mol NaOH added} \right]$$

According to the equation for the neutralization (equation 15.19), 1 mol NaOH neutralizes 1 mol HCl. Therefore, 0.001 500 mol NaOH neutralizes 0.001 500 mol HCl. The number of moles of HCl left is equal to the difference between the

---

*Measurement of the pH of a 0.1000 M solution of HCl with a pH meter shows that the pH is 1.1 not 1.0. The difference is a result of the nonideality of a 0.1 M solution. In a 0.1 M solution of HCl, there is some attraction between oppositely charged ions and some ion-pair formation. Note in Figure 15.12 that the difference between a pH of 1.0 and a pH of 1.1 at the beginning of the titration is not important.

number of moles of HCl present at the start and the number neutralized:

$$0.002\ 500 \text{ mol HCl present at start}$$
$$\underline{-\ 0.001\ 500 \text{ mol HCl neutralized}}$$
$$0.001\ 000 \text{ mol HCl left}$$

We began with 25.00 mL of HCl solution and added 15.00 mL of NaOH solution. The total volume of the solution is now

$$(25.00 \text{ mL} + 15.00 \text{ mL}) = 40.00 \text{ mL, or } 0.040\ 00 \text{ L}$$

and the molarity of HCl is now

$$M = \frac{0.001\ 000 \text{ mol HCl}}{0.040\ 00 \text{ L}} = 2.50 \times 10^{-2}$$

The assumption that all hydrogen ions in solution come from HCl (and none from the autoionization of water) is correct. The $[H^+] = 2.500 \times 10^{-2}$, and pH = 1.6021.

⬤ This part is a dilution problem (Section 4.8).

PRACTICE PROBLEM

15.31   In the titration of 25.00 mL of 0.1000 M HCl with 0.1000 M NaOH, what is the pH after 20.00 mL of the standard NaOH solution has been added?

At the equivalence point, all the acid has been neutralized, but no excess base has been added. The solution contains only water and a salt. In the neutralization of a strong acid with a strong base, the pH at the equivalence point is 7.00 (the pH of pure water) because salts of strong acids and strong bases do not hydrolyze. Solutions of salts of strong acids and strong bases are neutral.

After the equivalence point, excess base is added to the neutral salt solution. The pH of the solution at any point after the equivalence point can be calculated as shown in Sample Problem 15.16.

SAMPLE PROBLEM

15.16   In the titration of 25.00 mL of 0.1000 M HCl with 0.1000 M NaOH, what is the pH after 40.00 mL of the standard NaOH solution has been added?

Solution   In Sample Problem 15.15, we found that there is 0.002 500 mol HCl in 25.00 mL of 0.1000 M HCl. By a similar calculation, we can find that there is 0.004 000 mol NaOH in 40.00 mL of 0.1000 M NaOH.

According to the equation for the neutralization reaction (equation 15.19), 1 mol NaOH neutralizes 1 mol HCl. Therefore, 0.002 500 mol NaOH is required to neutralize 0.002 500 mol HCl. An excess of NaOH over the amount required to neutralize all HCl has been added. The number of moles of NaOH in excess is equal to the difference between the total number of moles of NaOH added and the number used up neutralizing HCl:

$$0.004\ 000 \text{ mol NaOH added}$$
$$\underline{-\ 0.002\ 500 \text{ mol NaOH reacted}}$$
$$0.001\ 500 \text{ mol NaOH in excess}$$

We began with 25.00 mL of HCl solution and added 40.00 mL of NaOH solution. The total volume of the solution is now

$$(25.00 \text{ mL} + 40.00 \text{ mL}) = 65.00 \text{ mL, or } 0.065\ 00 \text{ L}$$

and the molarity of NaOH is

$$M = \frac{0.001\ 500\ mol\ NaOH}{0.065\ 00\ L} = 2.308 \times 10^{-2}$$

The $[OH^-] = 2.308 \times 10^{-2}$, and pOH = 1.6368.
  In any aqueous solution, (pH + pOH) must equal 14.00. Therefore,

  pH + 1.6368 = 14.00   and   pH = (14.00 − 1.6368) = 12.36

PRACTICE PROBLEM

15.32 In the titration of 25.00 mL of 0.1000 M HCl with 0.1000 M NaOH, what is the pH after 30.00 mL of the standard NaOH solution has been added?

Titration of a Weak Acid with a Strong Base

The hydrogen ion concentration *at the beginning of the titration,* before any base has been added, is simply the hydrogen ion concentration of the weak acid solution. (See Sample Problem 15.5 for the method used to calculate the hydrogen ion concentration of a weak acid.) As base is added, the weak acid is converted to its salt. The solution contains both a weak acid and the salt of the weak acid, that is, it is *buffered.* (See Sample Problem 15.11 for the method used to calculate the hydrogen ion concentration of a buffered solution.)

At the equivalence point, all the weak acid has reacted to form salt. Because the salt is a salt of a weak acid and a strong base, it hydrolyzes. The pH at the equivalence point is *not* 7.00. Salts of weak acids and strong bases give basic solutions; the pH at the equivalence point is greater than 7.00. (See Sample Problem 15.9 for the method used to calculate the hydroxide ion concentration of a solution of a salt of a weak acid and a strong base.)

After the equivalence point has been passed, base is being added to a solution of the salt of the weak acid. The added hydroxide ion shifts the hydrolysis equilibrium to the left. All the hydroxide ions present in the solution come from sodium hydroxide. After the equivalence point, calculation of the pH of the solution is exactly like calculation of the pH of the solution from the titration of a strong acid with a strong base (Sample Problem 15.16).

In the laboratory, buffer solutions are often prepared by adding a solution of a strong base to an excess of weak acid solution.

PRACTICE PROBLEM

15.33 If 25.00 mL of 0.1000 M CH_3COOH is titrated with 0.1000 M NaOH, what is the pH (a) after 2.00 mL of 0.1000 M NaOH has been added? (b) After 25.00 mL of 0.1000 M NaOH has been added? (c) After 40.00 mL of 0.1000 M NaOH has been added?

The calculated titration curve for the titration of 25.00 mL of 0.1000 M CH_3COOH with 0.1000 M NaOH is shown by the purple line in ►Figure 15.13. Comparison of the titration curve for a weak acid with a strong base with the titration curve for a strong acid with a strong base (black line) shows some similarities and some differences. At the beginning of the titration, although both acids are 0.1000 M, the pH of the weak acid is higher. Almost all the molecules of a strong acid are ionized, but only a small percentage of the molecules of a weak acid are ionized. The

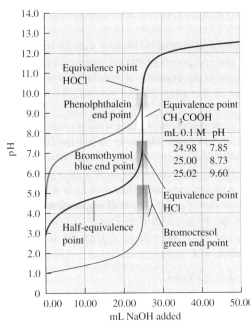

0.1 M CH₃COOH with 0.1 M NaOH ———
0.1 M HOCl with 0.1 M NaOH ———
0.1 M HCl with 0.1 M NaOH ———

Equivalence point HOCl

Phenolphthalein end point

Bromothymol blue end point

Half-equivalence point

Equivalence point CH₃COOH

mL 0.1 M	pH
24.98	7.85
25.00	8.73
25.02	9.60

Equivalence point HCl

Bromocresol green end point

■ FIGURE 15.13 Titration curves for titration of 25.00-mL samples of weak acids with strong bases. (The strong acid–strong base curve is shown for comparison.) The value of K_a for CH₃COOH is 1.8×10^{-5}; the value of K_a for HOCl is 3.2×10^{-8}.

anions formed by neutralization of the weak acid are basic; the anions formed by neutralization of the strong acid are neutral. As a result, the pH increases rapidly at the beginning of the weak acid titration, although slowly at the beginning of the strong acid titration. After the initial rapid rise in pH in the weak acid titration, the rate of increase of pH becomes smaller as the concentration of anions of the weak acid builds up and the solution becomes buffered.

Near the equivalence point of the weak acid, the pH rises very rapidly, although not quite as rapidly as in the titration of the strong acid. The nearly vertical portion of the titration curve is shorter in the weak acid curve; one drop of base only produces a change of a little more than 1.75 pH units. Choice of an indicator for a weak acid–strong base titration is more limited than choice of an indicator for a strong acid–strong base titration.

The pH at the equivalence point is greater than 7.00 in the titration of the weak acid; the solution is basic at the equivalence point. However, the number of milliliters required to reach the equivalence point is the same for both the strong and weak acids. After the steep portions of the curves have been passed, the two curves are identical.

PRACTICE PROBLEM

15.34 Of the indicators listed in Table 15.6, which could be used for the titration of 0.1000 M CH₃COOH with 0.1000 M NaOH?

Hypochlorous acid, HOCl, is weaker than acetic acid. The value of K_a for HOCl is 3.2×10^{-8}; K_a for CH₃COOH is 1.8×10^{-5}. The green line in Figure 15.13 shows the calculated titration curve for the titration of 25.00 mL of 0.1000 M HOCl with 0.1000 M NaOH. Notice how short and how far from vertical the steep portion of the titration curve for hypochlorous acid is. Of the indicators in Table 15.6, only thymolphthalein changes color in the right pH range. About 1 mL of 0.1000 M NaOH is required to raise the pH from 9.4 to 10.6; therefore, about 1 mL of 0.1000

M NaOH is needed to change the color of thymolphthalein from colorless to blue. The end point is not sharp.

Very weak acids such as hypochlorous acid cannot be titrated with an indicator. A pH meter must be used to determine the end points of the titrations of very weak acids. The pH is read after the addition of small portions of standard solution, and a titration curve is plotted. The middle of the steep part of the titration curve is taken as the equivalence point.

Unless you are lucky enough to have an automatic titrator, titration using a pH meter takes longer than titration using an indicator. However, a pH meter can be used for colored solutions such as tomato juice, which would hide the color change of an indicator. In addition, a titration curve obtained by titration with a pH meter gives more information than titration with an indicator. The value of K_a can be determined from the $[H^+]$ at the half-equivalence point. The **half-equivalence point** is the *point in a titration where half the base needed to completely neutralize the acid has been added.* In the titration of 25.00 mL of 0.1000 M acid with 0.1000 M base, the half-equivalence point is at

$$25.00 \text{ mL/2} = 12.50 \text{ mL}$$

At the half-equivalence point, half the acid remains, and half has been neutralized. The concentration of the anion of the weak acid is equal to the concentration of molecules of the weak acid. From the expression for the ionization constant of a weak acid, for example, acetic acid,

$$K_a = \frac{[H^+][CH_3COO^-]}{[CH_3COOH]} = 1.8 \times 10^{-5}$$

you can see that if $[CH_3COO^-] = [CH_3COOH]$, then

$$K_a = [H^+] = 1.8 \times 10^{-5}$$

The value of the ionization constant is an important property of weak acids. The ionization constant must be known in order to calculate a titration curve and choose a suitable indicator.

Around the half-equivalence point, the solution contains approximately equal concentrations of weak acid and weak acid salt and is buffered. Addition of more (or less) base does not result in much change in pH, and the titration curve has only a slight slope as you can see in Figure 15.13. The slope of the titration curve for strong acid with strong base is small at the beginning (and end) of the titration. Solutions of strong acids are buffer solutions with low pH, and solutions of strong bases are buffer solutions with high pH. A solution of a strong acid is buffered at a low pH because the concentration of H^+ is high and addition of a little acid or base does not change the $[H^+]$ very much. Similarly, a solution of a strong base is buffered at high pH because the concentration of OH^- is high and addition of a little acid or base does not change the $[OH^-]$ very much. Mixtures of weak acids and their salts and weak bases and their salts are used to make buffer solutions having intermediate pHs.

In laboratories where many titrations are carried out, automatic computerized titrators such as the one shown in ▬Figure 15.14 are used. All you have to do is add the sample. The automatic titrator displays the titration curve and does all the calculations.

Titration of a Weak Base

Calculation of the titration curve for the titration of a weak base with a strong acid from K_b is similar to calculation of the titration curve of a weak acid with a strong

Although solutions of strong acids and bases are buffered toward the addition of acid or base, they are not *buffered toward dilution.*

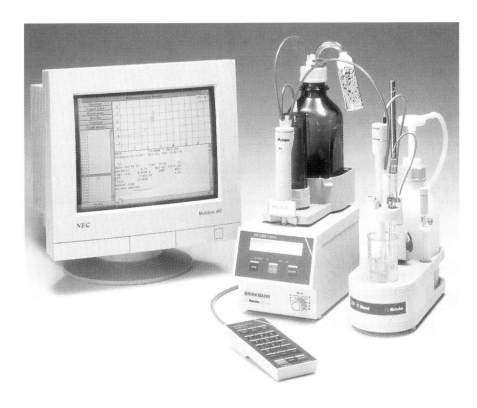

■ FIGURE 15.14 An automatic computerized titrator.

base from K_a. ■Figure 15.15 shows the titration curve for the titration of 25.00 mL of 0.1000 M NH_3 with 0.1000 M HCl.

Titrations in which both acid and base are weak are rarely carried out. The titration curves for weak acid–weak base titrations have no portion that is even close to vertical.

15.11 POLYPROTIC ACIDS

The acids that we have been dealing with thus far are all **monoprotic acids,** acids that *have only one ionizable hydrogen.* For example, hydrochloric acid and acetic acid are both monoprotic acids:

$$HCl(aq) \longrightarrow H^+(aq) + Cl^-(aq)$$

$$CH_3COOH(aq) \rightleftharpoons CH_3COO^-(aq) + H^+(aq)$$

Many monoprotic acids, such as acetic acid, contain more than one hydrogen atom. However, only one of the hydrogens ionizes in aqueous solution.

Polyprotic acids such as phosphoric acid, H_3PO_4, *have more than one ionizable hydrogen* and ionize stepwise. Phosphoric acid has three ionizable hydrogens and ionizes in three steps:

$$H_3PO_4(aq) \rightleftharpoons H^+(aq) + H_2PO_4^-(aq) \qquad K_{a1} = 7.1 \times 10^{-3}$$

$$H_2PO_4^-(aq) \rightleftharpoons H^+(aq) + HPO_4^{2-}(aq) \qquad K_{a2} = 8.0 \times 10^{-8}$$

$$HPO_4^{2-}(aq) \rightleftharpoons H^+(aq) + PO_4^{3-}(aq) \qquad K_{a3} = 4.8 \times 10^{-13}$$

The number 1 in the subscript of K_{a1} shows that the ionization constant is for the ionization of the first hydrogen. The ionization constant for the first proton is often

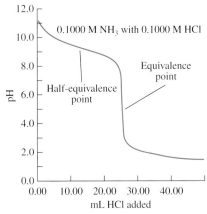

■ FIGURE 15.15 Titration curve for titration of a weak base with a strong acid.

TABLE 15.7 Ionization Constants of Some Polyprotic Acids at 25 °C[a]

	Acid			Conjugate Base	
Name	Formula	K_a	Name		Formula
Adipic acid	$H_2C_6H_8O_4$	3.7×10^{-5}	Hydrogen adipate ion		$HC_6H_8O_4^-$
Hydrogen adipate ion	$HC_6H_8O_4^-$	3.9×10^{-6}	Adipate ion		$C_6H_8O_4^{2-}$
Carbonic acid	$H_2CO_3^{b}$	4.3×10^{-7}	Hydrogen carbonate ion		HCO_3^-
Hydrogen carbonate ion	HCO_3^-	5.6×10^{-11}	Carbonate ion		CO_3^{2-}
Hydrogen sulfide	H_2S	8.9×10^{-8} [c]	Hydrogen sulfide ion		HS^-
Hydrogen sulfide ion	HS^-	3.1×10^{-19c}	Sulfide ion		S^{2-}
Oxalic acid	$H_2C_2O_4$	5.9×10^{-2}	Hydrogen oxalate ion		$HC_2O_4^-$
Hydrogen oxalate ion	$HC_2O_4^-$	6.4×10^{-5}	Oxalate ion		$C_2O_4^{2-}$
Phosphoric acid	H_3PO_4	7.1×10^{-3}	Dihydrogen phosphate ion		$H_2PO_4^-$
Dihydrogen phosphate ion	$H_2PO_4^-$	8.0×10^{-8}	Hydrogen phosphate ion		HPO_4^{2-}
Hydrogen phosphate ion	HPO_4^{2-}	4.8×10^{-13}	Phosphate ion		PO_4^{3-}
Sulfuric acid	H_2SO_4	—[d]	Hydrogen sulfate ion		HSO_4^-
Hydrogen sulfate ion	HSO_4^-	1.2×10^{-2}	Sulfate ion		SO_4^{2-}

[a]Most values are from Lide, D. R., Ed. *CRC Handbook of Chemistry and Physics,* 75th ed.; CRC Press: Boca Raton, FL, 1994; pp 8-43–8-55.
[b]Relatively few carbonic acid molecules actually exist in aqueous solution. The first step in the ionization of dissolved carbon dioxide should be written $CO_2(aq) + H_2O(l) \rightleftharpoons H^+(aq) + HCO_3^-(aq)$ instead of $H_2CO_3(aq) \rightleftharpoons H^+(aq) + HCO_3^-(aq)$. However, the latter equation is often used to be consistent with the equations for the ionization of other weak acids. Carbon dioxide is not a Brønsted–Lowry acid.
[c]At 20 °C, Brouwer, H. *J. Chem. Educ.* **1995,** *72,* 182–183.
[d]Ionization of the first hydrogen of sulfuric acid is complete in dilute solution.

Terephthalic acid is used to make polyesters and adipic acid to make nylon. A structural formula for terephthalic acid is

and a structural formula, for adipic acid is HOOCCH₂CH₂CH₂CH₂COOH.

represented simply by K_1. In addition to phosphoric acid, other industrially important polyprotic acids are sulfuric acid, H_2SO_4, terephthalic acid, $H_2C_8H_4O_4$, and adipic acid, $H_2C_6H_8O_4$. Many of the buffer systems that occur naturally and those used in laboratories and manufacturing plants involve polyprotic acids and their salts. The ionization constants for some common polyprotic acids are listed in Table 15.7.

As you can see from the formulas in Table 15.7, the conjugate base from one step is the conjugate acid for the next step. The second ionization constant is always smaller than the first. If there is a third ionization constant, it is even smaller than the second. The main reason for the decrease in successive ionization constants is electrostatic; removal of a proton requires more energy the higher the negative charge on the conjugate acid. Addition of a proton is easier the higher the negative charge on the conjugate base.

The product of neutralization of polyprotic acids depends on the quantity of base used relative to the quantity of acid used. For example, the molecular equation for the reaction of one mole of carbonic acid with one mole of sodium hydroxide is

$$H_2CO_3(aq) + NaOH(aq) \longrightarrow H_2O(l) + NaHCO_3(aq)$$

The molecular equation for the reaction of one mole of carbonic acid with two moles of sodium hydroxide is

$$H_2CO_3(aq) + 2NaOH(aq) \longrightarrow 2H_2O(l) + Na_2CO_3(aq)$$

The hydrogen-containing anions are amphoteric. For example, the HCO_3^- ion reacts with a H^+ as a base. The net ionic equation is

$$HCO_3^-(aq) + H^+(aq) \longrightarrow [H_2CO_3] \longrightarrow H_2O(l) + CO_2(g)$$

The HCO_3^- ion reacts with the base OH^- as an acid. The net ionic equation is

$$HCO_3^-(aq) + OH^-(aq) \longrightarrow H_2O(l) + CO_3^{2-}(aq)$$

The hydrogen carbonate ion reacts with the weak acid and base, H_2O, as a base:

$$HCO_3^-(aq) + H_2O(l) \rightleftharpoons H_2CO_3(aq) + OH^-(aq) \qquad K_b = 2.3 \times 10^{-8}$$

Because K_b for the hydrogen carbonate ion (2.3×10^{-8}) is larger than K_a for the hydrogen carbonate ion (5.6×10^{-11}), aqueous solutions that contain hydrogen carbonate ions are slightly basic (pH = 8.4 for a 0.1 M solution). That a species can contain an ionizable hydrogen and neutralize bases like a typical acid yet give a basic solution may seem surprising, but it is a fact.

From Table 15.7, you can see that the difference between successive ionization constants can be very large (hydrogen sulfide), large (phosphoric acid), or small (adipic acid). While there is a difference of 10^{11} between the ionization constants of hydrogen sulfide, the first ionization constant of adipic acid, 3.7×10^{-5}, is only nine times as large as the second ionization constant, 3.9×10^{-6}. When successive ionization constants differ by a factor of 10^3 or more, the concentration of hydrogen ion in a solution of a polyprotic acid can be calculated as easily as the concentration of hydrogen ion in a solution of a monoprotic acid. Sample Problem 15.17 shows how to calculate the concentrations of species in a solution of a diprotic acid.

SAMPLE PROBLEM

15.17 In a 0.10 M solution of carbonic acid, (a) what are $[H^+]$ and $[HCO_3^-]$? (b) What is $[CO_3^{2-}]$?

Solution Two equilibria exist at the same time in a solution of carbonic acid in addition to the autoionization of water that takes place in all aqueous solutions:

$$H_2CO_3(aq) \rightleftharpoons H^+(aq) + HCO_3^-(aq) \qquad K_{a1} = 4.3 \times 10^{-7}$$

$$HCO_3^-(aq) \rightleftharpoons H^+(aq) + CO_3^{2-}(aq) \qquad K_{a2} = 5.6 \times 10^{-11}$$

The equilibrium constant expressions are

$$K_{a1} = \frac{[H^+][HCO_3^-]}{[H_2CO_3]} \qquad \text{and} \qquad K_{a2} = \frac{[H^+][CO_3^{2-}]}{[HCO_3^-]}$$

(a) Because the value of K_{a1} is approximately 10^4 times as large as the value of K_{a2}, we can assume that all the hydrogen ion in the solution is formed by the first ionization. As usual, we will assume that the $[H^+]$ from the autoionization of water is negligible so that, at the moment the carbonic acid dissolves, before it has time to ionize, $[H^+] = [HCO_3^-] = 0.00$ and $[H_2CO_3] = 0.10$. If we let $x = $ mol/L of H^+ formed, then x mol/L of HCO_3^- will also be formed, and x mol/L H_2CO_3 will ionize because the coefficients of all the species in the equilibrium equation are 1. The completed summary table is as follows:

	$H_2CO_3(aq) \rightleftharpoons$	$H^+(aq) +$	$HCO_3^-(aq)$
Init. conc., M	0.10	0.00	0.00
Change, ΔM	$-x$	$+x$	$+x$
Eq. conc., M	$(0.10 - x)$	x	x

Substitution of the equilibrium concentrations in the equilibrium constant expression for the first ionization gives

$$\frac{x^2}{(0.10 - x)} = 4.3 \times 10^{-7}$$

which, if we assume that x is negligible compared with 0.10, simplifies to

$$\frac{x^2}{0.10} = 4.3 \times 10^{-7} \qquad \text{or} \qquad x^2 = 4.3 \times 10^{-8} \qquad \text{and} \qquad x = 2.1 \times 10^{-4}$$

Carbonic acid solutions play an important part in environmental chemistry. Carbon dioxide released into the air by burning fossil fuels and tropical rain forests dissolves in rain and in lakes, rivers, and oceans, making them acidic.

Sample Problem 15.17 is a good example of the use of approximations to simplify a calculation.

Acids and Bases in the Human Body

Small changes in hydrogen ion concentration have large effects on normal cell function, and only a narrow pH range is suitable for life. For example, if the pH of arterial blood, which is normally 7.45, becomes less than 6.8 or greater than 8.0 for more than a few seconds, death results. The major clinical effect of a hydrogen ion concentration that is too high (pH too low) is depression of the central nervous system. Too low a hydrogen ion concentration (pH too high) results in overexcitability.

Carbonic, sulfuric, phosphoric, lactic, $CH_3CH(OH)COOH$, and other organic acids are continuously formed by metabolism and added to the body fluids. The hydrogen ion from these acids is almost instantaneously removed from solution by buffers. So that the capacity of the buffers is not exceeded, the acids are then eliminated. Carbon dioxide is eliminated by the lungs in minutes and sulfuric, phosphoric, and organic acids by the kidneys in hours. The kidneys also supply hydrogen carbonate ions for the H_2CO_3–HCO_3^- buffer, which is an important buffer system in the fluids outside the cells such as blood plasma.

When H^+ is added to the H_2CO_3–HCO_3^- buffer,

$$CO_2(aq) + H_2O(l) \rightleftharpoons H^+ + HCO_3^-$$

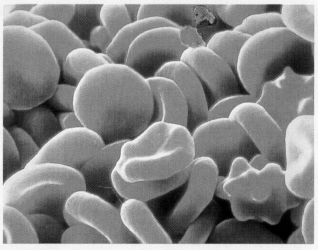

Scanning electron micrograph of human red blood cells. The biconcave disks are normal cells. The star-shaped cells are abnormal cells resulting from too concentrated plasma, which causes water to move out of the cells by osmosis so that they shrivel up.

the added hydrogen ion is used up by reaction with HCO_3^- ions. When H^+ is removed, the equilibrium shifts to the right, forming more hydrogen ion.

The H_2CO_3–HCO_3^- buffer also buffers saliva, removing hydrogen ion produced by bacteria in the mouth and from acids in foods. This buffering of saliva helps prevent cavities.

The H_2CO_3–HCO_3^- buffer cannot buffer against acidity resulting from carbon dioxide. This function is carried out mainly by hemoglobin. A $H_2PO_4^-$–HPO_4^{2-} buffer system

$$H_2PO_4^- \rightleftharpoons H^+ + HPO_4^{2-}$$

is important inside cells and in the urine. Urine may also be buffered by an ammonia–ammonium ion buffer

$$NH_3(aq) + H^+ \rightleftharpoons NH_4^+$$

The assumptions that H^+ from the autoionization of water is negligible and that x is negligible compared with 0.10 are both correct. Therefore, $[H^+] = [HCO_3^-] = 2.1 \times 10^{-4}$, and $[H_2CO_3] = 0.10$.

(b) Carbonate ion is formed by the second ionization. The concentrations of H^+ and HCO_3^- from part (a) are the initial concentrations of these species. Suppose we use y to represent the mol/L of CO_3^{2-} formed. Then y also equals the mol/L of H^+ formed by the second ionization and the mol/L of HCO_3^- ionized:

	$HCO_3^-(aq)$	$\rightleftharpoons$	$H^+(aq)$	$+ CO_3^{2-}(aq)$
Init. conc., M	2.1×10^{-4}		2.1×10^{-4}	0.0
Change, ΔM	$-y$		$+y$	$+y$
Eq. conc., M	$[(2.1 \times 10^{-4}) - y]$		$[(2.1 \times 10^{-4}) + y]$	y

Substitution of the equilibrium concentrations in the equilibrium constant expression for the second ionization gives

$$5.6 \times 10^{-11} = \frac{[(2.1 \times 10^{-4}) + y]y}{[(2.1 \times 10^{-4}) - y]}$$

which, if we assume that y is negligible compared with 2.1×10^{-4}, simplifies to

$$5.6 \times 10^{-11} = \frac{(2.1 \times 10^{-4})y}{(2.1 \times 10^{-4})} = y$$

Thus, $[CO_3^{2-}] = 5.6 \times 10^{-11}$.

The assumption that y can be neglected compared with 2.1×10^{-4} is correct. The quantities of H^+ formed and HCO_3^- used up by the second ionization are indeed negligible.

▶ Note that $[CO_3^{2-}] = K_{a2}$. In any solution that contains a weak polyprotic acid as the only solute, the concentration of the anion formed by the second ionization is equal to K_{a2}.

PRACTICE PROBLEM

15.35 For ascorbic acid, $K_{a1} = 7 \times 10^{-5}$ and $K_{a2} = 3 \times 10^{-12}$. In a 0.05 M solution of ascorbic acid, $H_2C_6H_6O_6$, (a) what are $[H^+]$ and $[HC_6H_6O_6^-]$? (b) What is $[C_6H_6O_6^{2-}]$?

Ascorbic acid is commonly known as vitamin C.

You may wonder why the basic counterparts of polyprotic acids, polyhydroxy bases, are not discussed. Most polyhydroxy bases are quite insoluble. A saturated solution of barium hydroxide, $Ba(OH)_2$, the most soluble of the Group IIA hydroxides, is less than 0.2 M. The hydroxides of the transition metals, such as nickel(II) hydroxide, $Ni(OH)_2$, and iron(III) hydroxide, $Fe(OH)_3$, are even less soluble in water.

SUMMARY

According to the **Brønsted–Lowry definitions,** an **acid** is any species that can donate a proton. Acids must have at least one hydrogen attached to an atom that is more electronegative than hydrogen. A **base** is any species that can accept a proton. Bases must have at least one unshared pair of electrons. Proton transfers take place very rapidly, and equilibrium is reached almost immediately. The Brønsted–Lowry definitions can be used for reactions that take place in the gas phase and nonaqueous solvents as well as in aqueous solution.

The equations for Brønsted–Lowry acid–base equilibria involving weak acids and water molecules are usually written with the water molecules on the reactant side:

$$HB(aq) + H_2O(l) \rightleftharpoons H_3O^+(aq) + B^-(aq)$$

The product of the addition of a proton to a water molecule, H_3O^+, is hydrated in aqueous solution; the hydrated H_3O^+ ion is called a **hydronium ion** and is written $H_3O^+(aq)$. The formula H^+ is often used as an abbreviation for the hydrated hydronium ion, which is commonly referred to simply as a hydrogen ion. The equilibrium constants for ionization equilibria of weak acids in water are called **acid ionization constants** or **acid dissociation constants, K_a:**

$$K_a = [H^+][B^-]/[HB]$$

The concentration of water in dilute aqueous solutions is constant for all practical purposes and is included in K_a.

The equations for Brønsted–Lowry acid–base equilibria involving weak bases and water are also usually written with the water molecules on the reactant side:

$$B^-(aq) + H_2O(l) \rightleftharpoons HB(aq) + OH^-(aq)$$

The equilibrium constants for equilibria involving weak bases and water are called **base ionization constants** or **base dissociation constants, K_b:**

$$K_b = [HB][OH^-]/[B^-]$$

Acids and bases that are related by loss or gain of H^+, as H_3O^+ and H_2O are related, are called **conjugate acid–base pairs.** Water is the conjugate base of H_3O^+; H_3O^+ is the conjugate acid of H_2O. The acid member of a conjugate acid–base pair always has one more hydrogen and a charge one unit more positive than the base member of the pair.

Substances that can react either as an acid or as a base, such as water, are called **amphoteric.** A reaction in which a cation and an anion are formed from two molecules of the same substance is called **autoionization.** The autoionization of water is an example:

$$H_2O(l) + H_2O(l) \rightleftharpoons H_3O^+(aq) + OH^-(aq)$$

In any aqueous solution, water molecules, hydronium ions (hydrogen ions), and hydroxide ions are in equilibrium. The product of the concentrations of hydrogen ions and hydroxide ions is called the **ion product,** K_w, for water:

$$K_w = [H^+][OH^-] = 1.0 \times 10^{-14} \quad \text{at } 25\ °C$$

The **pH** of an aqueous solution is defined as $-\log [H^+]$ and the **pOH** as $-\log [OH^-]$. In any aqueous solution at 25 °C, pH + pOH = 14.00. In pure water, $[H^+] = [OH^-] = 1.0 \times 10^{-7}$, and pH = 7.00. Pure water is defined as **neutral.** Solutions that have $[H^+] > 1.0 \times 10^{-7}$ (pH < 7.00) are **acidic;** solutions that have $[H^+] < 1.0 \times 10^{-7}$ (pH > 7.00) are **basic.**

In dilute solutions of strong monoprotic acids, $[H^+]$ is the same as the stoichiometric concentration of the acid; in dilute solutions of strong bases, $[OH^-]$ can be obtained from the stoichiometric concentration of the base. The $[H^+]$ in dilute solutions of weak acids can be calculated from the stoichiometric concentration of the weak acid and the value of K_a for the weak acid. The $[OH^-]$ in dilute solutions of weak bases can be calculated from the stoichiometric concentration of the weak base and the value of K_b for the weak base. The $[H^+]$ and $[OH^-]$ from water can be neglected except in very dilute solutions or if the value of K_a or K_b is very small.

The acid and base ionization constants of conjugate acids and bases and K_w are related:

$$K_a \cdot K_b = K_w$$

The stronger an acid is, the weaker its conjugate base; the weaker an acid, the stronger its conjugate base.

The Brønsted–Lowry acid–base reactions of ions with water are called **hydrolysis.** Salts of weak acids and strong bases are basic. Salts of strong acids and weak bases are acidic. Salts of most metals, except the metals of Groups IA and IIA, with strong acids are also acidic. Only salts formed from acids and bases of equal strength are neutral.

A **common ion** is an ion that is produced by more than one solute in a solution. The presence of a common ion shifts equilibria; the shift in equilibrium that results from the addition of a common ion is called the **common ion effect.**

Buffer solutions resist changes in hydrogen ion concentration. Mixtures of a weak acid and a salt of the weak acid or of a weak base and a salt of the weak base buffer solutions at intermediate pHs; the hydrogen ion concentration of the solution depends on the ionization constant of the weak acid or base and on the ratio of the concentrations of the weak acid or base and the salt. The concentrations of the weak acid or base and salt must be fairly high for the buffer to have a useful capacity. The **capacity** of a buffer solution is the quantity of acid or base that can be added before the hydrogen ion concentration changes significantly.

Titration can be used to analyze moderately dilute solutions of strong and moderately weak acids and bases. The **equivalence point** is the point in a titration at which the proportions of the reactants in moles are the same as the proportions in the balanced equation for the reaction. **Titration curves** for acid–base reactions are graphs of pH against the number of milliliters of standard acid or base added in a titration. Titration curves can be used to choose a suitable indicator for a titration or to find the value of K_a or K_b.

Most indicators for acid–base reactions are weak organic acids. The molecules of the weak acid are one color, and the anions are a different color.

Monoprotic acids have only one ionizable hydrogen; **polyprotic acids** have more than one ionizable hydrogen ion. Successive ionization constants for polyprotic acids always get successively smaller.

ADDITIONAL PRACTICE PROBLEMS

For information about the organization of Additional Practice Problems, Stop & Test Yourself, Putting Things Together, and Applications, see the beginnings of these sections in Chapter 1.

15.36 Use Lewis formulas to show the Brønsted–Lowry acid–base reaction between water and formic acid, HCOOH. (15.1)

15.37 Write equations similar to equation 15.2 for the Brønsted–Lowry acid–base reactions of each of the following acids with water: (a) $HClO_4$ (b) CH_3COOH (c) HCN (d) HOCl (e) HCl (15.1)

15.38 For the Brønsted–Lowry acid–base reaction

$$PO_4^{3-}(aq) + H_3O^+(aq) \rightleftarrows HPO_4^{2-}(aq) + H_2O(l)$$

(a) classify each of the species as an acid or base.
(b) What is the formula of the predominant solute species? (15.1)

15.39 For each of the following Brønsted–Lowry acids, write the formula of the conjugate base. (a) HSO_4^- (b) HBr (c) $Zn(H_2O)_6^{2+}$ (15.1)

15.40 For each of the following Brønsted–Lowry bases, write the formula of the conjugate acid. (a) HSO_4^- (b) CN^- (c) $Ni(H_2O)_5(OH)^+$ (15.1)

15.41 Write the equation for the autoionization of HSO_4^-. (15.1)

15.42 (a) Which of the following species are amphoteric? (i) H_3PO_4 (ii) $H_2PO_4^-$ (iii) HPO_4^{2-} (iv) PO_4^{3-} (b) For those that are amphoteric, write equations to show the amphoteric behavior. (15.1)

15.43 In a solution that has $[H^+] = 2.9 \times 10^{-3}$, (a) what is $[OH^-]$? (b) Is the solution acidic, basic, or neutral? (15.2)

15.44 In a solution that has $[OH^-] = 4.1 \times 10^{-8}$, (a) what is $[H^+]$? (b) Is the solution acidic, basic, or neutral? (15.2)

15.45 What is pH of solutions with each of the following $[H^+]$? (a) 1.0×10^{-12} (b) 5.3×10^{-10} (c) 2.4 (d) 6.6×10^{-2} (15.3)

15.46 What is the $[H^+]$ in solutions with each of the following pHs? (a) 4.0 (b) 3.7 (c) 10.5 (d) 14.0 (15.3)

15.47 Given a solution with a pH of 4.2, (a) is the solution acidic, basic, or neutral? (b) What is the $[H^+]$ in the solution? (c) What is the pOH? (15.3)

15.48 Given a solution with a pH of 10.2, (a) is the solution acidic, basic, or neutral? (b) What is the $[H^+]$ in the solution? (c) What is the pOH? (15.3)

15.49 What is the pOH of solutions with each of the following $[OH^-]$? (a) 0.16 (b) 1.0×10^{-7} (c) 3.4×10^{-5} (d) 6.2×10^{-9} (e) 6.0 (15.3)

15.50 What is $[OH^-]$ of solutions with each of the following pOH? (a) 3.9 (b) 11.5 (c) 0.6 (d) −4.6 (15.3)

15.51 Rewrite each of the equations in Problem 15.37 using the symbol H^+ for the hydrated hydronium ion (see equation 15.9) and write the expressions for K_a for the weak acids. (15.4)

15.52 Arrange the following acids in order of increasing strength: (a) benzoic acid (b) chloroacetic acid (c) hydrocyanic acid. (15.4)

15.53 A 0.050 M aqueous solution of hypobromous acid (HOBr) is 0.020% ionized at 25 °C. What is the value of K_a for hypobromous acid at this temperature? (15.4)

15.54 A 0.075 M aqueous solution of hypoiodous acid (HOI) is 0.001 75% ionized at 25 °C. What is the value of K_a for hypoiodous acid at this temperature? (15.4)

15.55 What is the $[H^+]$ in (a) 2.9×10^{-3} M HNO_3? (b) 3.5×10^{-4} M HOCl? (c) 4.1×10^{-2} M $ClCH_2COOH$? (15.4)

15.56 What is the $[H^+]$ in each of the following solutions? (a) 5.8×10^{-3} M $HClO_4$ (b) 6.7×10^{-2} M propionic acid (c) 7.9×10^{-4} M HF (15.4)

15.57 Arrange the acids of the following set in order of increasing percent ionization of 0.100 M solutions at 25 °C: (a) HNO_2 (b) HNO_3 (c) HCN (15.4)

15.58 Arrange the solutions of the following set in order of increasing percent ionization at 25 °C: (a) 0.100 M HOCl (b) 0.0100 M HOCl (c) 5.0×10^{-2} M HOCl (15.4)

15.59 In a 0.050 M aqueous solution of phenol at 25 °C, (a) what is $[H^+]$? $[C_6H_5O^-]$? $[C_6H_5OH]$? (b) What is the percent ionization? (15.4)

15.60 In 0.032 M HCN at 25 °C, (a) what is $[H^+]$? $[CN^-]$? [HCN]? (b) What is the percent ionization? (15.4)

15.61 Arrange the following bases in order of increasing strength: (a) benzoate ion (b) chloroacetate ion (c) cyanide ion. (15.5)

15.62 What is the $[OH^-]$ in (a) 6.8×10^{-4} M NaOH? (b) 8.7×10^{-2} M $(CH_3)_3N$? (c) 7.6×10^{-3} M C_6H_5COONa? (15.5)

15.63 If, in a 0.050 M aqueous solution of hydroxylamine, $HONH_2$, at 20 °C, $[OH^-] = 2.3 \times 10^{-5}$, (a) what is K_b for $HONH_2$ at 20 °C? (b) What percent of the $HONH_2$ molecules are ionized? (15.5)

15.64 Calculate what percent of the trimethylamine molecules in a 0.075 M aqueous solution at 25 °C are ionized. (15.5)

15.65 In 0.050 M aqueous solution, NH_3 is 1.72% ionized at 5 °C. What is the value of K_b for NH_3 at this temperature? (15.5)

15.66 For a conjugate acid–base pair, what is the relationship between the values of pK_a and pK_b? (15.5)

15.67 From the value of K_b given for methylamine in Table 15.4, calculate the value of K_a for the methylammonium ion, $CH_3NH_3^+$. (15.5)

15.68 What is $[OH^-]$ in 0.047 M $(CH_3)_2NH$ at 25 °C? $[(CH_3)_2NH_2^+]$? $[(CH_3)_2NH]$? (15.5)

15.69 Predict whether aqueous solutions of each of the following salts will be acidic, basic, or neutral. For those solutions that are not neutral, write equations for the reactions that take place. (a) $BaCl_2$ (b) CH_3NH_3Cl (c) C_6H_5COONa (d) $KHSO_4$ (e) $ZnCl_2$ (15.6)

15.70 Calculate (a) $[OH^-]$ in 0.050 M KCN (b) $[H^+]$ in 0.025 M NH_4Cl (c) $[OH^-]$ in 3.8×10^{-4} M KI. (15.6)

15.71 Will $[H^+]$ increase, decrease, or remain the same if (a) HCl is added to a solution of $NaNO_2$? (b) NH_4Cl is added to a solution of NH_3? (c) $NaNO_2$ is added to a solution of HNO_2? (d) KOH is added to a solution of CH_3COOK? (e) NaCl is added to a solution of NH_4Cl? (15.6)

15.72 Explain what is meant by the term *common ion effect*. Use an aqueous solution of NH_4Cl as an example. (15.7)

15.73 What is the $[H^+]$ in a solution formed by the addition of 5.1×10^{-4} mol NH_4NO_3 to 0.25 L of 0.0076 M NH_3? Assume that the volume of the ammonia solution is not changed by the addition of the solid ammonium nitrate. (15.7)

15.74 (a) Explain what is meant by the term *buffered solution*. (b) Using a solution that is 0.1 M in NH_4Cl and 0.1 M in NH_3 as an example, explain how a buffer solution works. (c) What two factors determine the pH of a buffer solution? (d) What is the capacity of a buffer solution, and what factor controls the capacity of a buffer solution? (15.8)

15.75 To prepare a buffer solution that has pH = 7.0, what should you do? (15.8)

15.76 What restrictions are there on the use of the Henderson–Hasselbalch equation to calculate the pH of a buffered solution? (15.8)

15.77 In a solution of CH_3COOH and CH_3COOK, what is the ratio of $[CH_3COO^-]/[CH_3COOH]$ if $[H^+] = 3.6 \times 10^{-4}$? (15.8)

15.78 (a) To prepare a buffer solution having a hydrogen ion concentration of 2.0×10^{-4}, which of the acid–base pairs in Table 15.3 should be used? (b) What should the ratio of [acid molecules]/[anions] be? (15.8)

15.79 Why should only a small quantity of indicator be used? (15.9)

15.80 What is the approximate pH of a colorless solution that remains colorless when thymolphthalein is added but turns pink when phenolphthalein is added? (15.9)

15.81 What is the approximate value of K_a for the indicator thymolphthalein? (15.9)

15.82 In a titration, (a) what is the equivalence point? (b) What is the end point? (c) What should be the relationship between the equivalence point and the end point? (15.10)

15.83 Of the indicators listed in Table 15.6, which could be used for the titration of 0.1 M NH_3 with 0.1 M HCl? (15.10)

15.84 Titration of a 0.23-mL sample of household ammonia required 25.62 mL of 0.049 M HCl to reach the end point. What was the molarity of the ammonia solution? (15.10)

15.85 Calculate the pH at the following points in the titration of 20.00 mL of 0.0500 M KOH with 0.0500 M HNO_3 and sketch the titration curve (a) before any HNO_3 has been added (b) after addition of 5.00 mL 0.0500 M HNO_3 (c) after addition of 20.00 mL (d) after addition of 20.05 mL (e) after addition of 25.00 mL. (15.10)

15.86 Calculate the pH at the following points in the titration of 30.00 mL of 0.075 M propionic acid with 0.075 M NaOH. (a) before any NaOH has been added   (b) after addition of 10.00 mL 0.075 M NaOH   (c) after addition of 30.00 mL (d) after addition of 40.00 mL. (15.10)

15.87 If pH = 5.2 after addition of 10.00 mL of 0.050 M HCl to 20.00 mL of 0.050 M pyridine,

$$\begin{array}{ccc} & H & \quad H \\ & \backslash & / \\ & C=C & \\ H-C & & N \\ \backslash & & / \\ & C-C & \\ / & & \backslash \\ H & & H \end{array}$$

what is the value of K_b for pyridine? (15.10)

15.88 In Figure 15.13, the curves for the titrations of 0.1 M HCl and 0.1 M CH_3COOH with 0.1 M NaOH are different before the equivalence point but identical after the equivalence point. Explain why. (15.10)

15.89 What are the concentrations of all solute species present in 3.8×10^{-2} M H_2SO_4? (15.11)

15.90 (a) Calculate the concentrations of all solute species present in 0.050 M maleic acid,

$$\begin{array}{ccc} HOOC & & COOH \\ \backslash & & / \\ & C=C & \\ / & & \backslash \\ H & & H \end{array}$$

For maleic acid, $K_{a1} = 1.42 \times 10^{-2}$ and $K_{a2} = 8.57 \times 10^{-7}$.   (b) List the species in order of decreasing concentration. (15.11)

15.91 (a) What is the concentration of OH^- in 1.00×10^{-7} M NaOH?   (b) What percent of the OH^- in 1.00×10^{-7} M NaOH comes from water?

15.92 What is the percent ionization at infinite dilution of an acid having $K_a = 1.00 \times 10^{-8}$?

15.93 For a solution to have a negative pH, what must be true of the $[H^+]$?

15.94 The following titration curve is for the titration of 25.00 mL of 0.1000 M weak acid with 0.1000 M NaOH(aq). (a) What is the value of K_a for this weak acid?   (b) Which of the indicators listed in Table 15.6 could be used?

15.95 (a) From the value of K_w at 50 °C given in Table 15.1, what is $[H^+]$ in pure water at 50 °C? $[OH^-]$?   (b) Is a solution with $[H^+] = 1.7 \times 10^{-7}$ (measured at 50 °C) acidic, basic, or neutral?

15.96 In a solution that is 2.8×10^{-3} M in both CH_3COOH and C_6H_5OH, (a) what is the $[H^+]$?   (b) $[CH_3COOH]$?   (c) $[CH_3COO^-]$?   (d) $[C_6H_5OH]$?   (e) $[C_6H_5O^-]$?

15.97 A 0.072 M solution of formic acid that also contains potassium formate has a $[H^+]$ of 2.0×10^{-4}. What is the concentration of potassium formate in the solution?

15.98 If a weak acid, HB, is 2.3% ionized in 0.025 M solution, (a) what is the percent ionization in 0.050 M HB?   (b) At what concentration is the weak acid 4.6% ionized?

15.99 The concentrations of standard solutions of bases are often determined by using the solution of base to titrate a sample of pure potassium hydrogen phthalate

$$\begin{array}{ccc} & H & \quad H \\ & \backslash & / \\ & C=C & \\ H-C & & C-H \\ \backslash & & / \\ & C-C & \\ / & & \backslash \\ KOOC & & COOH \end{array}$$

of accurately known mass. If 26.32 mL of NaOH solution are required to neutralize 0.5690 g of potassium hydrogen phthalate, what is the molarity of the NaOH solution?

15.100 (a) Use your answer to Problem 15.70 (a) to calculate the percent hydrolysis for 0.050 M KCN.   (b) Calculate $[OH^-]$ and percent hydrolysis for 0.0050 M KCN.   (c) How does percent hydrolysis change as concentration decreases?

15.101 From the curve for the titration of 0.1 M HOCl with 0.1 M NaOH in Figure 15.13, make a table of pH and milliliters of NaOH added that covers the pH range from 8.0 to 12.0 in 1.0-mL increments. Graph the absolute value of the change in pH per milliliter of NaOH added. A graph like this is called a differential plot. What is the advantage of using a differential plot for the titration of a weak acid or a weak base?

15.102 Use the data from Table 15.1 to graph pK_w against $1/T$ and read pK_w for water at normal body temperature of 37 °C from the graph.

15.103 Use the following symbols

HCl or HOCl   Cl⁻ or OCl⁻ H_3O^+

to draw microscopic-level pictures of (a) HCl(aq) (b) HOCl(aq). To keep your drawings as simple as possible, do *not* show molecules of the solvent, water.

15.104 Draw a series of three microscopic-level pictures (before, during, and after collision) showing formation of (a) H_3O^+

and Br$^-$ from HBr(g) and H$_2$O(g)   (b) NH$_4^+$ and OH$^-$ from NH$_3$(g) and H$_2$O(g).

15.105 Draw a microscopic-level picture of an H$_2$PO$_4^-$–HPO$_4^{2-}$ buffer. Use the symbols shown at the right:

H$_2$PO$_4^-$ HPO$_4^{2-}$

STOP & TEST YOURSELF

1. Which of the following reactions is a Brønsted–Lowry acid–base reaction?
   (a) Ca^{2+} + 2OH$^-$ ⟶ Ca(OH)$_2$
   (b) (CH$_3$)$_3$C$^+$ + H$_2$O ⟶ (CH$_3$)$_3$COH$_2^+$
   (c) (CH$_3$)$_3$COH + H$_3$O$^+$ ⟶ (CH$_3$)$_3$COH$_2^+$ + H$_2$O
   (d) CO$_2$ + H$_2$O ⟶ H$^+$ + HCO$_3^-$
   (e) H$_2$C=CH$_2$ + H$_2$ ⟶ H$_3$CCH$_3$
2. Which of the following reactions is autoionization?
   (a) 2(CH$_3$)$_2$C=CH$_2$ ⟶ (CH$_3$)$_3$CCH$_2$C(CH$_3$)=CH$_2$
   (b) HCl + H$_2$O ⟶ H$_3$O$^+$ + Cl$^-$
   (c) H$_3$PO$_4$ + PO$_4^{3-}$ ⟶ H$_2$PO$_4^-$ + HPO$_4^{2-}$
   (d) 2HPO$_4^{2-}$ ⟶ H$_2$PO$_4^-$ + PO$_4^{3-}$
   (e) 2NO$_2$ ⟶ N$_2$O$_4$
3. The conjugate acid of H$_2$PO$_4^-$ is
   (a) H$_3$O$^+$ (b) PO$_4^{3-}$ (c) HPO$_4^{2-}$ (d) H$_2$PO$_3^-$ (e) H$_3$PO$_4$
4. If, in an aqueous solution, [H$^+$] = 3.8 × 10^{-5}, then [OH$^-$] = _____ and the solution is _____.
   (a) 2.6 × 10^{-10}, acidic   (b) 2.6 × 10^{-10}, basic   (c) 2.6 × 10^{-8}, acidic   (d) 2.6 × 10^{-8}, basic   (e) 2.6 × 10^{-20}, acidic
5. If pH = 9.7, then [H$^+$] = _____.
   (a) 5 × 10^9 (b) 2 × 10^4 (c) 6 × 10^{-5} (d) 7 × 10^{-9}
   (e) 2 × 10^{-10}
6. In 0.025 M ClCH$_2$COOH, [H$^+$] = _____.
   (a) −0.0067   (b) 3.5 × 10^{-5} (c) 0.0053   (d) 0.0059
   (e) 0.025
7. If a 0.040 M solution of weak acid, HB, is 1.3% ionized, what is the value of K_a for the weak acid?
   (a) 6.8 × 10^{-6} (b) 1.4 × 10^{-5} (c) 6.8 × 10^{-4} (d) 1.3 × 10^{-2} (e) 1.4 × 10^5
8. For CH$_3$CH$_2$NH$_3^+$, K_a = 1.56 × 10^{-11}. The equilibrium constant expression is

   (a) $K_a = \dfrac{[CH_3CH_2NH_3^+][OH^-]}{[CH_3CH_2NH_2]}$

   (b) $K_a = \dfrac{[CH_3CH_2NH_2]}{[CH_3CH_2NH_3^+][OH^-]}$

   (c) $K_a = \dfrac{[CH_3CH_2NH_2][H_3O^+]}{[CH_3CH_2NH_3^+][H_2O]}$

   (d) $K_a = \dfrac{[CH_3CH_2NH_3^+]}{[CH_3CH_2NH_2][H^+]}$

   (e) $K_a = \dfrac{[CH_3CH_2NH_2][H^+]}{[CH_3CH_2NH_3^+]}$

9. In which solution is percent ionization the lowest?
   (a) 0.1 M solution of weak acid with K_a = 4 × 10^{-5}
   (b) 0.1 M solution of weak acid with K_a = 8 × 10^{-5}
   (c) 0.1 M solution of weak acid with K_a = 4 × 10^{-4}
   (d) 0.05 M solution of weak acid with K_a = 4 × 10^{-5}
   (e) The percent dissociation is the same in (a) and (d).
10. Which of the following salts gives a basic solution when dissolved in water?
    (a) Ba(NO$_3$)$_2$ (b) NaNO$_2$ (c) KBr   (d) NH$_4$I   (e) ZnCl$_2$
11. In a 0.050 M solution of C$_6$H$_5$ONa, [OH$^-$] = _____.
    (a) 6.5 × 10^{-12} (b) 2.6 × 10^{-6} (c) 3.9 × 10^{-6} (d) 7.7 × 10^{-5} (e) 1.9 × 10^{-3}
12. In a solution that is 0.020 M in HCOOK and 0.040 M in HCOOH,
    (a) [H$^+$] = 1.1 × 10^{-14}
    (b) [H$^+$] = 3.6 × 10^{-4}
    (c) [H$^+$] = 2.6 × 10^{-3}
    (d) [H$^+$] = 3.0 × 10^3
    (e) [OH$^-$] = 1.1 × 10^{-6}.
13. To buffer a solution at pH = 10.06, which combination would be best?
    (a) 0.025 M CH$_3$NH$_2$ and 0.025 M CH$_3$NH$_3$Cl
    (b) 0.025 M CH$_3$NH$_2$ and 0.100 M CH$_3$NH$_3$Cl
    (c) 0.100 M CH$_3$NH$_2$ and 0.025 M CH$_3$NH$_3$Cl
    (d) 0.0025 M CH$_3$NH$_2$ and 0.0100 M CH$_3$NH$_3$Cl
    (e) 0.0100 M CH$_3$NH$_2$ and 0.0025 M CH$_3$NH$_3$Cl
14. Which titration would give the sharpest end point?
    (a) 0.1 M CH$_3$COOH with 0.1 M (CH$_3$)$_3$N
    (b) 0.1 M HCl with 0.1 M (CH$_3$)$_3$N
    (c) 0.1 M HNO$_2$ with 0.1 M KOH
    (d) 0.01 M HNO$_3$ with 0.01 M KOH
    (e) 0.1 M HNO$_3$ with 0.1 M KOH
15. Which of the following substances is a polyprotic acid?
    (a) CH$_3$NH$_2$ (b) C$_3$H$_6$O$_2$ (c) HC$_2$H$_3$O$_2$ (d) H$_2$C$_2$O$_4$
    (e) H$_2$NCH$_2$CH$_2$NH$_2$

PUTTING THINGS TOGETHER

15.106 What are the predominant solute species in aqueous solutions of each of the following?   (a) NaCl   (b) NaOH   (c) NH$_3$
   (d) HF   (e) NaC$_2$H$_3$O$_2$

15.107 (a)   In pure water, what percentage of the water molecules are ionized?   (b) How many hydrogen ions are there in 1.0 L of pure water at 25 °C?

15.108 All of the following solutions have pH = 4.7. Which are buffered at this pH? Explain your answers. (a) 25.00 mL 0.100 M HCl + 24.99 mL 0.100 M NaOH (b) 25.00 mL 0.100 M CH_3COOH + 12.50 mL 0.100 M NaOH (c) 25.00 mL 0.100 M NH_3 + 25.02 mL 0.100 M HCl (d) 25.00 mL 0.100 M HOCl + 0.04 mL 0.1 M NaOH (e) 10.00 mL 0.100 M H_3PO_4 + 10.20 mL 0.100 M NaOH.

15.109 What is the pH of a 1.0×10^{-12} M solution of HCl?

15.110 If the pH of a 0.020 M solution of aniline is reduced to 5.50 by the addition of HCl, what percentage of the aniline is present in the form of anilinium ions?

15.111 (a) What is the pH of a solution that is 0.025 M in HCl and 0.032 M in CH_3COOH? (b) What is the pH of a solution prepared by mixing 50.0 mL of 0.025 M HCl with 50.0 mL of 0.032 M CH_3COOH?

15.112 At the point in the titration of 25.00 mL of 0.100 M CH_3COOH with 0.100 M NaOH at which pH = 6.0, what percentage of the CH_3COOH is present as CH_3COO^- ions?

15.113 What is the pH of each of the following solutions? (a) 0.035 M $HClO_4$ (b) 0.027 M $(CH_3)_3N$ (c) 0.041 M $C_6H_5NH_3Cl$ (d) 0.052 M $ZnCl_2$ (e) 0.024 M HF

15.114 What is the pH of a solution formed by the addition of 228 mL of hydrogen chloride gas measured at 25.0 °C and 755.2 mmHg to 500.0 mL of 0.0340 M aqueous NH_3? Assume that the volume of the ammonia solution is not changed by the addition of the hydrogen chloride.

15.115 What are the formulas and names of the conjugate bases of the following acids? (a) HSO_3^- (b) HI (c) NH_4^+ (d) OH^- (e) CH_3COOH

15.116 Write complete ionic, net ionic, and molecular equations for each of the following reactions: (a) Sodium formate reacts with water. (b) Perchloric acid solution reacts with potassium hydroxide solution. (c) Acetic acid solution reacts with barium hydroxide solution. (d) Aqueous ammonia reacts with hydrobromic acid solution. (e) Solid ammonium chloride reacts with water.

15.117 What is the pH of a solution prepared by dissolving 5.32 g of concentrated sulfuric acid in water and diluting to 1.00 L at 25 °C? Concentrated sulfuric acid is 95% by mass H_2SO_4 and has density = 1.840 g/cm³.

15.118 (a) If the freezing point of a 0.070 M solution of dichloroacetic acid, $Cl_2CHCOOH$, is −0.19 °C, what is the value of K_a for dichloroacetic acid at this temperature? (b) Briefly discuss the disadvantages of using freezing point depression to determine the value of K_a. (c) How does the value of K_a for dichloracetic acid compare with the values of K_a for acetic acid and chloroacetic acid in Table 15.3? Predict a value of K_a for trichloroacetic acid, Cl_3CCOOH.

15.119 (a) Explain why hydrochloric acid conducts electricity but hydrogen chloride gas does not. (b) In hydrochloric acid, how is charge carried? (c) Why doesn't a solution of HCl in benzene conduct?

15.120 Titration of a solution of a 0.1243-g sample of an unknown acid with 0.0500 M NaOH required 26.31 mL to reach the end point. Was the unknown acid $CH_3(CH_2)_{15}COOH$, $ClCH_2COOH$, $BrCH_2CH_2COOH$, or $CH_3(CH_2)_{14}COOH$?

15.121 When 0.1 M aqueous solutions of NH_3 and CH_3COOH are placed in a conductivity apparatus such as the one in Figure 4.8, the light bulb only lights dimly. However, if a mixture of equal volumes of the two solutions is placed in the apparatus, the light bulb lights brightly. Explain.

15.122 In the species $H_5O_2^+$, which is known to exist in crystals, water is believed to be hydrogen bonded to an H_3O^+ ion. Draw a Lewis formula for $H_5O_2^+$.

15.123 A solution of sodium dihydrogen phosphate neutralizes both acids and bases. (a) Write molecular, net ionic, and complete ionic equations for the reaction of sodium dihydrogen phosphate with hydrochloric acid. (b) Write molecular, net ionic, and complete ionic equations for the reaction of sodium dihydrogen phosphate with sodium hydroxide solution.

15.124 Use Le Châtelier's principle to predict the effect of the following changes on the equilibrium

$$C_6H_5COOH(aq) + H_2O(l) \rightleftharpoons H_3O^+ + C_6H_5COO^-$$

(a) evaporation of water (b) addition of hydrochloric acid (c) addition of sodium benzoate (d) addition of sodium hydroxide (e) increased pressure

15.125 What is the pH of a solution prepared by diluting 22.3 mL of concentrated CH_3COOH that contains 38.0% by mass CH_3COOH and has a density of 1.0454 g/cm³ to 1.000 L?

15.126 Values of K_w for water at different temperatures are given in Table 15.1. (a) Write the equation for the autoionization of water showing thermal energy on the correct side. (b) If pH is measured at 50 °C, what is the pH of pure water?

15.127 A researcher obtained the following titration curve. Suggest two possible explanations.

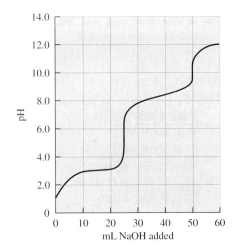

15.128 The following microscopic-level picture shows $H_3PO_4(aq)$. (a) Draw new pictures showing the solution after the addition of (i) 10 OH^- (ii) 20 OH^- (total). (b) How many OH^- must be added for PO_4^{3-} to be the only phosphorus-containing species?

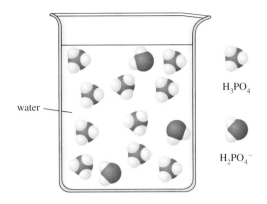

water —

H_3PO_4

$H_2PO_4^-$

15.129 A buffer solution was prepared by adding 10.7 mL of 0.10 M NaOH to 50.0 mL of 0.050 M $NaHCO_3$. What is the pH of this buffer?

15.130 All organic compounds that contain the group —COOH are members of a family having similar properties called carboxylic acids. (a) List the condensed structural formulas of the carboxylic acids in Table 15.3. (b) Using condensed structural formulas, write molecular equations for the neutralization in aqueous solution of each acid in part (a) with KOH. (c) Two of the polyprotic acids in Table 15.7, adipic acid and oxalic acid, are carboxylic acids. Write Lewis formulas for these acids. (*Hint:* Oxalic acid is easier because it is composed of fewer atoms. Do it first.) (d) Using a condensed structural formula, write the molecular equation for the reaction of 1 mol of oxalic acid with: 1 mol NaOH; 2 mol NaOH.

15.131 The general formula R_3N where R = H or a C-containing group, such as the methyl group CH_3, can be written for amines. (a) In aniline, what is R? (b) Write the formulas and names of the series of amines similar to methyl-, dimethyl-, and trimethylamine in which R = CH_3CH_2, an ethyl group. (c) Reaction of amines with acids is similar to reaction of ammonia with acids. Use Lewis formulas to show the reaction between a methylamine molecule and a hydrogen chloride molecule. (d) Write the net ionic equation for the reaction of methylamine with hydrochloric acid using a condensed structural formula for methylamine.

15.132 The group C_6H_5 is called an aromatic group because many compounds that contain it have odors. (a) Which of the weak acids in Table 15.3 are aromatic? (b) How does the strength of these acids compare with the strength of acetic and propionic acids? (c) Which of the weak bases in Table 15.4 is aromatic? (d) How does the strength of this base compare with the strength of the other weak organic bases?

APPLICATIONS

15.133 Industrial wastes often contain acidic metal ions such as $Al(H_2O)_6^{3+}$ and $Fe(H_2O)_6^{3+}$. Write an equation showing why each of these metal ions is acidic.

15.134 The amine trimethylamine, $(CH_3)_3N$, is partly responsible for the odor of fish. Explain, by writing an equation, why lemon juice, which contains citric acid,

$$HOOCCH_2C(OH)(COOH)CH_2COOH,$$

reduces the fishy odor.

15.135 The body's acid–base condition is determined by comparison to the normal blood pH of 7.4 (measured at 25 °C), not to the pH of pure water. Acidosis occurs when the blood pH falls below 7.35 and alkalosis when the blood pH rises above 7.45. Choose the correct terms to complete the following sentence: Blood with a pH of 7.25 is (*acidotic* or *alkalotic*) but (*acidic* or *basic*).

15.136 Soap consists mainly of salts such as sodium stearate, $CH_3(CH_2)_{16}COONa$. For acids such as stearic acid, K_a = 1.3×10^{-5} at 25 °C. (a) Are soap solutions acidic, basic, or neutral? Explain your answer. (b) What is pK_a for acids such as stearic? (c) What is pK_b for the anions from these acids?

15.137 Glycine, H_2NCH_2COOH, is one of the building blocks of proteins; gelatin and silk fibroin are the best sources. Although $H_2NCH_2CH_2CH_2CH_3$ and CH_3CH_2COOH, which have molecular masses similar to the molecular mass of glycine, are typical organic compounds (both are liquids under ordinary conditions and are soluble in ether), glycine behaves more like a salt (it is a high melting solid that is insoluble in ether). Suggest an explanation for this behavior.

15.138 When microorganisms reproduce, they release waste products that may change the pH and prevent further reproduction. Therefore, culture mediums for growing microorganisms are usually buffered. The medium for the culture of lactobacilli includes 250.0 mL of a buffer prepared by dissolving 25.00 g K_2HPO_4 and 25.00 g KH_2PO_4 in water and diluting to 250.0 mL. (a) What is the pH of this buffer solution? (b) Before the culture medium is used, it is diluted to 1.000 L with distilled water. What is the pH after dilution?

15.139 Urine is buffered by a NaH_2PO_4–Na_2HPO_4 buffer. If the pH of a urine specimen is 6.6, what is the ratio of $[HPO_4^{2-}]/[H_2PO_4^-]$?

15.140 For aspirin,

OOCCH$_3$

—COOH

$K_a = 3 \times 10^{-5}$ at body temperature (37 °C). There are 325 mg of aspirin in an aspirin tablet. If two aspirin tablets are dissolved in a full stomach (volume, 1 L) in which the pH is 2, what percentage of the aspirin is in the form of molecules?

15.141 For morphine

the Merck index gives $K_b = 7.5 \times 10^{-7}$ and $K_a = 1.4 \times 10^{-10}$ at 20 °C. (a) Write the structural formula for the conjugate acid of morphine. (b) Is the value given for K_a the value for the conjugate acid of morphine? Explain your answer. (c) If the answer to part (b) is no, which H is the acidic H? Explain your answer.

15.142 The compound, 2,4-dinitrophenol

which is used as a wood preservative and insecticide, is also an indicator that changes from colorless to yellow at pH 2.6–4.4. (a) Estimate the value of K_a for 2,4-dinitrophenol. (b) Although 2,4-dinitrophenol itself is very sparingly soluble in cold water, the sodium salt is soluble in water. What is the pH of a 0.050 M solution of the sodium salt?

15.143 For pseudoephedrine hydrochloride

$$\text{—CH(OH)CH(CH}_3\text{)N(CH}_3\text{)H}_2\text{Cl}$$

which is used in cough syrup, $pK_a = 9.22$. (a) What is K_b for the free base? (b) What is the pH of a 0.030 M aqueous solution of pseudoephedrine hydrochloride?

15.144 The capacity of natural waters to neutralize acids is called alkalinity. Alkalinity is used by biologists as a measure of water fertility. The species responsible for alkalinity are HCO_3^-, CO_3^{2-}, and OH^-. (a) Write an equation for the reaction of each of these species with H^+. (b) In pH 7.0 water having an alkalinity equivalent to 1.00×10^{-3} mol OH^-/L, (i) what is $[H^+]$? (ii) What is $[OH^-]$? (iii) Which is the predominant solute species? (iv) Calculate $[CO_2]$, $[HCO_3^-]$, and $[CO_3^{2-}]$.

15.145 Cocaine

is a weak base. The osmotic pressure of a 2.5000×10^{-3} M solution of cocaine is 46.32 mmHg at 15.0 °C. What is the value of K_b for cocaine at 15.0 °C?

15.146 The salt calcium propionate, which is often added to foods to slow spoilage, is most effective at acid pHs. (a) What is the formula for calcium propionate? (b) The average pH of white bread is 5.5. Is white bread acidic, basic, or neutral? (c) At pH = 5.5, what is the ratio of $[CH_3CH_2COOH]/[CH_3CH_2COO^-]$?

15.147 In a human stomach, the pH may be as low as 2 as a result of the production of HCl. While an empty stomach has a volume of only 50 mL, a full stomach has a volume of 1 L. (a) What is $[H^+]$ in a stomach that has pH = 2.0? (b) How many grams of HCl does a full stomach contain? (c) Vomiting results in a decrease in plasma $[H^+]$ (as well as in dehydration). Does the pH of the plasma increase or decrease as a result of vomiting? Explain your answer. (d) Why do doctors often recommend that patients drink cola drinks when they have upset stomachs?

15.148 Hydrogen carbonate ion buffers blood toward most acids and bases but cannot buffer it toward acidity from carbon dioxide. Blood is buffered toward acidity from carbon dioxide produced by metabolic processes by the hemoglobin buffer system. Venous blood (pH = 7.35) is only slightly more acidic than arterial blood (pH = 7.45), despite the CO_2 the venous blood carries back to the lungs. Three equilibria are involved in the hemoglobin buffer system:

$$Hb + H^+ \rightleftharpoons HHb^+$$
$$Hb + O_2 \rightleftharpoons HbO_2$$
$$CO_2 + H_2O \rightleftharpoons H^+ + HCO_3^-$$

(a) Tell how these three equilibria will be shifted by (i) diffusion of CO_2 produced in tissue cells into the blood (ii) addition of HCO_3^- from the kidneys (iii) removal of CO_2 by lungs. (b) Calculate the ratio of $[H^+]_{\text{venous blood}}/[H^+]_{\text{arterial blood}}$.

15.149 Pure water in equilibrium with air has pH = 5.7 as a result of dissolved CO_2. (a) What is the $[CO_2]$ in this water? (b) Water used to prepare standard base solutions should be free of CO_2. Explain why and tell how to obtain water free of CO_2.

15.150 The results of recent research suggest that nicotine consumed in foods should be considered when testing people for passive inhalation of tobacco smoke. For example, ripe tomatoes contain 4.1 ng of nicotine per gram. (a) What is the concentration of nicotine in ripe tomatoes in "ppb" by mass? (b) How many grams of tomatoes must be eaten to consume 1 μg of nicotine, the quantity absorbed by a nonsmoker who spends three hours in an environment with a minimum concentration of smoke? (c) For nicotine

$pK_{b1} = 6.16$ and $pK_{b2} = 10.96$. What is the pH of a 0.050 M solution?

15.151 What effect does the use of ammonium nitrate as a fertilizer have on soil pH? Explain your answer.

15.152 The titration curve at the right is that of ephedrine,

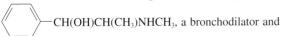CH(OH)CH(CH$_3$)NHCH$_3$, a bronchodilator and

decongestant: (a) What is the value of K_b? (b) Write a complete structural formula for ephedrine. (c) What is the molecular formula? (d) What is the hybridization of the carbons in the ring? Of the carbons attached to nitrogen?

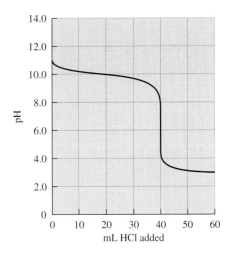

A Chemist Escapes from the Lab

ROBERTA FRIEDMAN

Freelance Science Journalist

B.A. Neuroscience
University of Rochester

Ph.D. Pharmacology
Vanderbilt University

Chemistry was far from my favorite subject. I took it the first time because it was a prerequisite. The first semester of my freshman year, I struggled to follow the lectures delivered by a nationally respected physical chemist. Three hundred fellow classmates were not as patient. After the Schrödinger equation and a second exam that was even harder than the first, they greeted the professor with boos and paper airplanes. More disgusted by the grade-conscious premedical students than by the ponderous style of my teacher (I actually had enjoyed the challenge posed by ol' Schrödinger), I swore off chemistry, biology, and anything that would put me in a pre-med's path.

Of course, I wouldn't be writing this if I hadn't come back to the sciences. After taking a few psychology courses the following year, I wanted to study how the brain works, and the scientific method seemed mightier than Freud. Back I went to the required chemistry. The second half of the course, taken in the calm of summer, had a more sedate tone. Though I still fumbled through the lab, trashing my smock and botching the procedures, I could take all the time that acid-base equations required (for me—a lot!) as I studied with the dorm window open to the warm evenings.

Organic was the next step toward the chemistry of the brain. Labs were still a nightmare, with memorable events such as my neighbor's experiment smoking out of control and forcing a temporary evacuation into the hallway. Not just my coat, but even my lab notebook were graced by awkward spills. There was the aniline dye I'd tried to synthesize—it was supposed to be bright red, I remember, but instead it came out a muddy brick color—which ate through half the pages of my notebook.

But organic was where I learned the secret to labs: no experiment works the first time. Halfway into the "unknown" exercise, I ruined my sample. Given a fresh sample, I found that repeating the first steps was surprisingly simple and clear—I could rise above the performance and understand where I was going. It was a thrill finally to be able to hand the properly processed sample to the teaching assistant, get the printout back from the mysterious NMR machine, decipher its code, and identify the unknown correctly.

Success at last?—not so fast. Then came graduate school and biochemistry, a prerequisite for the courses I would take in pharmacology, the science of how drugs work. Biochem lab rarely kept normal hours. We were lucky to get home by six or seven at night. I spent forever on a step that should have taken five minutes, waiting for a decidedly low-tech machine called a rotary evaporator to evaporate my extract of rat liver down to a drop. It remained a stubborn puddle for over an hour.

With my courses over and my qualifying exams passed with praise from all examiners, I was in the lab for real, working to publish my own data. I started each day mixing buffers, those magical acid-base balanced solutions that keep biological samples happy. Pulverized brain tissues need those buffers, or they will stop living in the test tubes. I was supposed to have learned the skills of titration and pipetting in all those chemistry labs I'd taken, but I only became adept at them when I did them day in, day out for that first month as a neuro-pharmocologist-in-training.

Neuropharmacology was my thesis subject. Specifically, I was studying the cell molecules that recognize and receive a nerve cell messenger, serotonin. The receptor molecules sit in the nerve cell membranes, where they act as switches for modulating brain activity. Give them a dose of psychedelic LSD, or the latest drug that treats depression, called Prozac, or a migraine reliever called methysergide, and these serotonin receptors will recognize them all because their molecular shapes all resemble that of the natural messenger, serotonin. To understand exactly how this works—guess what—you need to know chemistry, pi orbitals, and the Schrödinger equation.

It was satisfying to master molecular pharmacology and use everything I'd learned. It was also rewarding to have my experiments finally start to work and to set up and run successfully a new piece of equipment for our lab called a high-performance liquid chromatograph with an electrochemical detector. I'd come a long way from a rotovap.

But once I left the lab to write my Ph.D. thesis, I discovered another secret. I liked writing about what I was doing, far more than I enjoyed fumbling in the lab. Finally, I decided to give up labwork. Let's face it, I'm a klutz.

My clothes no longer cower under a lab coat. I report science rather than perform it. I write about medical high technology, people diagnosed by NMR, and mystery street drugs deciphered by mass spectrometry. I write simply, for the layperson, so that chemistry is not a prerequisite.

Still, as a freelancer who contracts with many different outlets, I never know what I could be assigned next. I've written on astronomy for Time-Life books and mathematics for a children's encyclopedia, as well as more typical news articles on medicine and biotechnology that draw on my biology background. I followed the debate over cold fusion with special interest when I found the person heading a national review panel was my college chemistry professor. If I had covered that story, I'd have interviewed him!

MORE ABOUT EQUILIBRIA

16

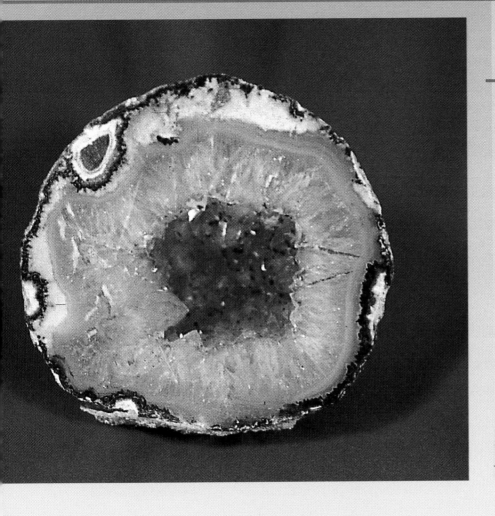

*The beautiful crystals inside this geode were
formed by slow growth from a saturated
solution.*

In this chapter, you will learn how to combine equilibrium constants to obtain the equilibrium constants for additional equilibria. The relationship between the acidity and basicity of an element's compounds and the position of the element in the periodic table will be shown, and another set of definitions of acids and bases—the Lewis definitions—will be introduced. The Lewis definition for an acid includes compounds that do not contain ionizable hydrogens. Equilibria between complexes and their parts will be considered, and solubility equilibria will be discussed.

■ Why is a further study of equilibria a part of general chemistry?

The ability to obtain many equilibrium constants by calculation from a few experimentally determined or tabulated equilibrium constants saves labor. The ability to deduce the acidity or basicity of a compound from its structure and the periodic table is a memory-saver. The Lewis definitions of acids and bases extend the concepts of acid–base chemistry to many reactions of great practical importance. For example, the reaction between calcium oxide and silicon dioxide,

$$CaO(s) + SiO_2(s) \longrightarrow CaSiO_3(l)$$

which is used to remove sand and gravel from iron ore in the production of pig iron, is a Lewis acid–base reaction.

The formation of complexes is important in many chemical and biological systems. Complexes are used to separate

The small dark object is a kidney stone. Kidney stones form when the concentrations in the kidneys of ions that can combine to form insoluble solids become too high. Kidney stones are composed mainly of phosphates and oxalates of calcium and magnesium.

metals from their ores and as catalysts in a number of industrial processes. Many biologically important molecules such as hemoglobin, chlorophyll, and vitamin B_{12} are complexes. In this chapter, we will be concerned mainly with the effect of complex formation on solubility equilibria. Complexes are discussed further in Chapter 24.

Equilibria that involve solution or precipitation of insoluble substances are also very important from a practical point of view. The separation of metals from their ores, the manufacture of many chemicals, and a number of the most interesting questions in geochemistry, such as the order in which minerals form during the evaporation of seawater and weathering, involve solubility equilibria. The precipitation of sodium urate in joints results in gouty arthritis, and excess salts precipitated as solid particles in the kidneys cause kidney stones.

— What do you already know about equilibria?

In Chapter 14, you learned that the product of the equilibrium molarities of the products of a reaction, each raised to a power given by the coefficient in the reaction's equation, divided by a similar product for the reactants is a constant. That is, for the general reaction

$$a\text{A} + b\text{B} + \cdots \rightleftharpoons e\text{E} + f\text{F} + \cdots$$

the equilibrium constant expression is

$$K_c = \frac{[\text{E}]^e[\text{F}]^f \cdots}{[\text{A}]^a[\text{B}]^b \cdots}$$

where a is the coefficient of the reactant A in the equation for the reaction, b is the coefficient of the reactant B, e is the coefficient of the product E, and so forth. If the equilibrium constant is very large, reactants are, for all practical purposes, completely converted to products, providing that the rate of reaction is high enough so that equilibrium is reached. If the equilibrium constant is very small, little reaction takes place even if reaction is fast and equilibrium is reached rapidly. If the equilibrium constant is neither very large nor very small, the equilibrium concentrations of reactants and products are of the same order of magnitude, and equilibrium is easily observed experimentally.

In Chapter 14, you also learned that the expression for the reaction quotient Q has the same form as the equilibrium constant expression. However, the concentrations in the reaction quotient can be any concentrations, not just equilibrium concentrations. If the value of Q is smaller than K_c, net reaction will take place in the forward direction until Q becomes equal to K_c. If the reaction quotient is larger than K_c, net reaction will take place in the reverse direction until Q becomes equal to K_c. If $Q = K_c$, the reaction is at equilibrium, and no net change will occur.

Equilibrium concentrations can be calculated from the initial concentrations and the value of K_c by letting x equal one of the unknown equilibrium concentrations. All other equilibrium concentrations are then expressed in terms of x and substituted in the equilibrium constant expression to give an equation in one unknown. This equation is solved for x, and the equilibrium concentrations are calculated from the value found for x. In Chapter 14, you learned how to solve problems about gas phase equilibria by this general method; in Chapter 15, you learned how to solve problems about acid–base equilibria by the same general

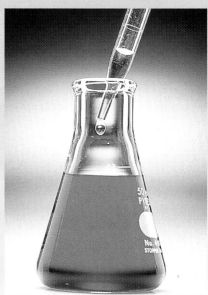

Top: An aqueous solution of $Fe_2(SO_4)_3$. Bottom: Addition of colorless $NH_4SCN(aq)$ to the yellow solution of $Fe_2(SO_4)_3$ yields an equilibrium mixture that contains the red complex ion $[Fe(SCN)(H_2O)_5]^{2+}$. This highly colored ion is used in both qualitative and quantitative analysis for iron.

method. In this chapter, the method will be applied to more problems about acid–base equilibria, equilibria between complexes and their parts, and equilibria between undissolved solid and solute in saturated solutions.

16.2 ACIDITY, BASICITY, AND THE PERIODIC TABLE

The strengths of acids and bases in aqueous solutions are a result of a combination of factors of which bond polarity, bond strength, and solvation are the most important. Fortunately, a few empirical rules that relate acidity to molecular structure work very well at room temperature.

Strengths of Binary Acids of Elements in the Same Group

Binary acids are *acids that consist of just two elements,* hydrogen and some other nonmetal. Hydrogen chloride, HCl, and hydrogen sulfide, H_2S, are examples. How do the acidities of binary acids compare going down a group and across a period in the periodic table? Table 16.1 shows the first ionization constants for the binary acids of the Group VIA elements. As you can see, the acidity of the binary acids of Group VIA increases going down the group, that is, as the radius of the nonmetal and the length of the bond increase. This trend is easy to remember because long bonds are weaker and more easily broken than short bonds (Section 9.7).

The acidity of the binary acids of Group VIIA also increases down the group in the periodic table, that is, HF < HCl < HBr < HI. However, the difference in acidity between HCl, HBr, and HI cannot be observed in aqueous solution. All three of these acids are stronger acids than the hydronium ion, H_3O^+. Water is a stronger base than the halide ions Cl^-, Br^-, and I^-. As we saw in Section 16.1, in Brønsted–Lowry acid–base reactions, the stronger acid and the stronger base react to form the weaker acid and the weaker base. The reactions

$$HCl(aq) + H_2O(l) \longrightarrow H_3O^+(aq) + Cl^-(aq)$$
$$HBr(aq) + H_2O(l) \longrightarrow H_3O^+(aq) + Br^-(aq)$$
$$HI(aq) + H_2O(l) \longrightarrow H_3O^+(aq) + I^-(aq)$$

all go to completion in dilute aqueous solution. Water is said to **level** the acidity of HCl, HBr, and HI because *no difference can be observed in the acidity* of HCl, HBr, and HI in aqueous solution. In aqueous solution, any acid stronger than the hydronium ion instantly and completely reacts with water to form hydronium ions. The hydronium ion is the strongest acid that can exist in aqueous solution; therefore, the strengths of acids that are stronger than the hydronium ion can't be compared in aqueous solution.

The difference in acid strength between HCl, HBr, and HI can be detected by using a weaker base than water, such as acetone, $(CH_3)_2C{=}O$, as the solvent:

In acetone, HCl, HBr, and HI are not completely ionized, and the order of acidity is observed to be HCl < HBr < HI. *A solvent that reveals the difference in acidity between two acids* is called a **differentiating solvent.**[*]

TABLE 16.1

First Ionization Constants of Binary Acids of Group VIA Elements at 25 °C[a]

Formula	K_1	pK_1
H_2O	2×10^{-16}	15.7
H_2S	1×10^{-7}	7.0
H_2Se	3×10^{-4}	3.6
H_2Te	3×10^{-3}	2.6

[a] Myers, R. J., *J. Chem. Educ.* **1986,** *63,* 687–690.

The value of K_a for HCl is estimated to be about 1×10^7 ($pK_a = -7$).

In some texts, the relationships between molecular structure and acid strength are treated at the beginning of the discussion of acid–base equilibria. We have placed this material after acid–base equilibria calculations so that students will have a quantitative idea of relative acid strengths when they read it.

HF is probably a weak acid because of entropy effects. See Meyers, T. J. J. Chem. Educ. ***1976,*** *53, 17–19 and Lessley, S. D.; Ragsdale, R. O. J. Chem. Educ.* ***1976,*** *53, 19–20.*

---

[*] An everyday analogy to the leveling effect of a strong base on acids is two little children, one smaller than the other, each with a candy bar. Any normal ten-year-old can take the candy bars away from both little children. The ten-year-old is so much stronger than both little children that the additional strength of the stronger little child does not enable him or her to hold on to the candy bar. The ten-year-old levels the strengths of the little children. On the other hand, another little child who is midway in strength between the two children with the candy bars is like a differentiating solvent. The third child can only take a candy bar away from the weaker child.

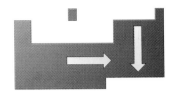

The strengths of binary acids increase from left to right across periods and from top to bottom in groups.

◐ The difference in acidity between H bonded to C and H bonded to O explains why the Hs of the CH_3 group in acetic acid, CH_3COOH, are not acidic.

TABLE 16.2

First Ionization Constants of Binary Acids of Some Second Period Elements at 25 °C

Formula	K_a	pK_a
CH_4^a	ca. 10^{-49}	ca. 49
NH_3	ca. 10^{-35}	ca. 35
H_2O	2×10^{-16}	15.7
HF	3.5×10^{-4}	3.46

^{a}The solubility of methane in water is 3.5 mL/ 100 mL H_2O at 17 °C and 1 atm. A saturated solution is 1.5×10^{-3} M.

Calcium hydride reacting with water to yield hydrogen gas and calcium hydroxide. The reaction of calcium hydride with water is rapid but not violent. Calcium hydride is used to dry organic solvents and gases and is a convenient source of hydrogen gas. Sodium hydride reacts violently with water.

Strengths of Binary Acids of Elements of the Same Period

The first ionization constants for the binary acids of some elements in the second row of the periodic table are given in Table 16.2. Going across a row in the periodic table from left to right, the acidity of binary acids increases. This trend is easy to remember because the electronegativities of the elements also increase from left to right (Figure 9.3). As a result, the polarity of the nonmetal-hydrogen bond increases. The partial positive charge on hydrogen becomes larger, and the hydrogen becomes more like a hydrogen ion. The atomic radius and bond length do not change much across a row. Across a row, the bonds are all about the same length and strength.

Methane, CH_4, and ammonia, NH_3, are such weak acids that no acidity can be observed in aqueous solution. The conjugate bases of methane and ammonia, CH_3^- and NH_2^-, are such strong bases that they react completely with water forming hydroxide ions:

$$CH_3^- + H_2O \longrightarrow CH_4 + OH^- \qquad K_b = \text{ca. } 10^{35}$$
$$NH_2^- + H_2O \longrightarrow NH_3 + OH^- \qquad K_b = \text{ca. } 10^{21}$$

Hydroxide ions are the strongest base that can exist in aqueous solution. Water levels the basicity of the very strong bases, CH_3^- and NH_2^-. A solvent that is less acidic than water, such as liquid ammonia, must be used to differentiate between the basicity of CH_3^- and the basicity of NH_2^-.

Metal Hydrides

The binary compounds of the reactive metals of Groups IA and IIA at the left side of the periodic table with hydrogen, such as sodium hydride, NaH and calcium hydride, CaH_2, are very strong bases. The reactive metals have low ionization energies and give up electrons relatively easily. As a result, sodium hydride and calcium hydride are ionic compounds. Sodium hydride consists of sodium ions, Na^+, and hydride ions, H^-, and calcium hydride is composed of calcium ions, Ca^{2+}, and hydride ions. The hydride ion is the conjugate base of the hydrogen molecule, H_2, an extremely weak acid. The hydride ion is an extremely strong base and immediately and quantitatively reacts with water:

$$H^-(aq) + H_2O(l) \longrightarrow H_2(g) + OH^-(aq)$$

Oxo Acids and Organic Acids

The hydroxides of the reactive metals are also ionic compounds and are strong bases. For example, potassium hydroxide is composed of potassium ions, K^+, and hydroxide ions, OH^-. (Remember, there are no molecules of ionic compounds.) A potassium atom, which has a low ionization energy, gives an electron to a combination of the nonmetals oxygen and hydrogen, which have high electron affinities.

However, the hydroxides of nonmetals are molecular compounds and are acidic. As a result of the combined electronegativities of the nonmetal and oxygen, the H—O bond is polar with the partial positive charge on hydrogen. Hypochlorous acid is the simplest example:

$$\overset{\text{+}\longrightarrow}{H - O - Cl}$$

When hypochlorous acid ionizes, the H O bond breaks, and a hydrogen ion is formed:

$$HOCl(aq) \underset{\longleftarrow}{\overset{\longrightarrow}{\rightleftharpoons}} H^+(aq) + OCl^-(aq)$$

The industrially important inorganic acids, sulfuric acid, phosphoric acid, and nitric acid, as well as most organic acids, such as acetic acid, are nonmetal hydroxides:

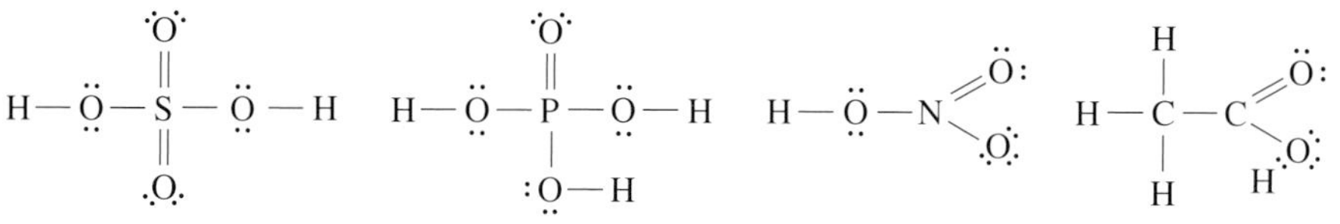

⬤ Remember that the inorganic oxygen-containing acids, such as sulfuric acid, phosphoric acid, and nitric acid, are often referred to as oxo acids and the anions that are their conjugate bases are called oxo anions.

$$H-\overset{\cdot\cdot}{\underset{\cdot\cdot}{O}}-\overset{\overset{\cdot\cdot}{O}}{\underset{\cdot\cdot}{\overset{\|}{S}}}-\overset{\cdot\cdot}{\underset{\cdot\cdot}{O}}-H \qquad H-\overset{\cdot\cdot}{\underset{\cdot\cdot}{O}}-\overset{\overset{\cdot\cdot}{O}}{\underset{\underset{H}{\overset{|}{O}}}{\overset{\|}{P}}}-\overset{\cdot\cdot}{\underset{\cdot\cdot}{O}}-H$$

sulfuric acid   phosphoric acid   nitric acid   acetic acid

The stronger the attraction of the central atom in an oxo acid for electrons, the more polar the O—H bonds, and the more acidic the oxo acid. Therefore, the acidity of oxo acids is related to the electronegativity of the central atom (see Figure 9.3). To see the effect of differences in the electronegativity of the central atom, acids with similar structures must be compared. For example, the hypohalous acids, HOCl, HOBr, and HOI, have similar structures; none of the hypohalous acids have an extra oxygen that is not part of an —OH group like sulfuric, phosphoric, nitric, and acetic acids do. The acidities of the hypohalous acids decrease as the electronegativities of the halogens decrease going down the group in the periodic table:

⬤ To see the effect of a change easily, keep all other possible variables constant.

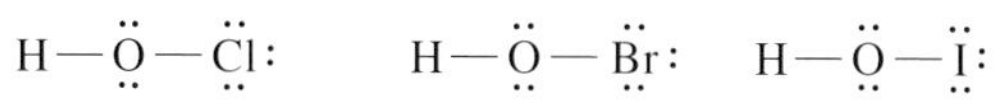

$$H-\overset{\cdot\cdot}{\underset{\cdot\cdot}{O}}-\overset{\cdot\cdot}{\underset{\cdot\cdot}{Cl}}: \qquad H-\overset{\cdot\cdot}{\underset{\cdot\cdot}{O}}-\overset{\cdot\cdot}{\underset{\cdot\cdot}{Br}}: \qquad H-\overset{\cdot\cdot}{\underset{\cdot\cdot}{O}}-\overset{\cdot\cdot}{\underset{\cdot\cdot}{I}}:$$

$$K_a = 3.2 \times 10^{-8} \qquad 2.1 \times 10^{-9} \qquad 2.3 \times 10^{-11}$$

Hypofluorous acid, HOF, is unstable.

The acidities of oxo acids are also related to the number of extra oxygens (oxygens that are not part of —OH groups) attached to the central atom. Oxygen is a very electronegative element. The more extra oxygens there are in a molecule of an acid, the stronger an acid the molecule is. The chlorine-containing oxo acids provide a good example:

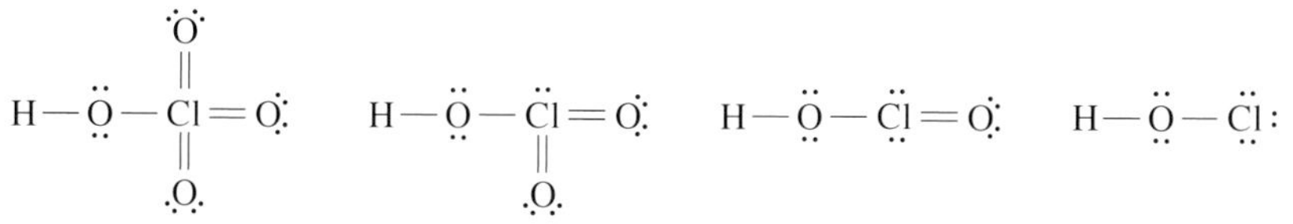

According to electron diffraction (Greenwood, N. N.; Earnshaw, A. Chemistry of the Elements; Pergamon: Oxford, 1986, p. 1016), the Cl O bonds in $HClO_4$ are 140.8 pm long, whereas the Cl OH bond is 163.5 pm. Thus, the former bonds appear to have double-bond character as shown. (For an alternate explanation of the bond lengths in this and similar compounds see Suidan, L.; Badenhoop, J.K.; Glendening, E.D.; Weinhold, F. J. Chem. Educ. 1995, 72, 583–586.

perchloric acid,   chloric acid,   chlorous acid,   hypochlorous acid,
three extra oxygens   two extra oxygens   one extra oxygen   no extra oxygens
$K_a =$ very large   large 1.1×10^{-2} 3.2×10^{-8}

Perchloric acid has three extra oxygens and is a very strong acid; chloric acid with two extra oxygens is a strong acid. Chlorous acid, with one extra oxygen, is a moderately strong weak acid, and hypochlorous acid, with no extra oxygens, is a very weak acid. *In general, oxo acids with no extra oxygens are weak ($K_a = 10^{-8}$ to 10^{-9}); oxo acids with one extra oxygen are moderately strong ($K_a = 10^{-2}$ to 10^{-4}), and oxo acids with two or more extra oxygens are strong.*

Like the strengths of inorganic oxo acids, the strengths of organic acids are related to the polarity of the O—H bond. For example, chloroacetic acid, $ClCH_2COOH$ ($K_a = 1.4 \times 10^{-3}$), is a stronger acid than acetic acid, CH_3COOH ($K_a = 1.8 \times 10^{-5}$). Substitution of the more electronegative Cl atom for a H close to the O—H group makes the O—H bond more polar.

⬤ Note the o in the ending of the name of phosphorous acid, an oxo acid of the element phosphorus, which does not have an o in the last syllable of its name.

SAMPLE PROBLEM

16.1   The first ionization constant for phosphorous acid, H_3PO_3, is of the same order of magnitude as the first ionization constant for phosphoric acid, H_3PO_4.

16.2 ACIDITY, BASICITY, AND THE PERIODIC TABLE

Which Lewis structure shows phosphorous acid, (a) or (b)? Explain your answer.

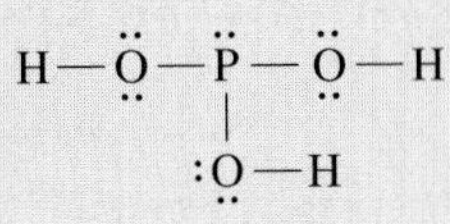

(a) 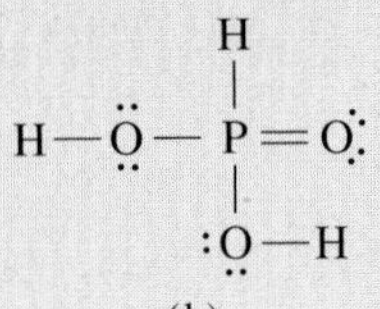   (b)

Solution The Lewis structure of phosphoric acid is

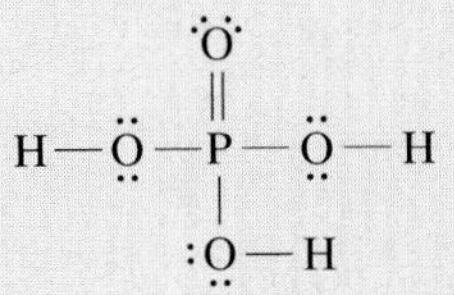

with one extra oxygen. The fact that the first ionization constant for phosphorous acid is of the same order of magnitude as that for phosphoric acid suggests that both acids have the same number of extra oxygens. Structure (b) has one extra oxygen like the structure of phosphoric acid. Therefore, (b) is probably the correct structure for phosphorous acid. Structure (a) has no extra oxygens and should be a weaker acid.

The conclusion that (b) is the correct structure for phosphorous acid is borne out by the fact that phosphorous acid is a diprotic acid; 1 mol of phosphorous acid requires 2 mol of sodium hydroxide for neutralization. Structure (a) should be triprotic and require 3 mol of sodium hydroxide per mole of acid for neutralization.

PRACTICE PROBLEMS

16.3 The first ionization constant for hypophosphorous acid, H_3PO_2, is of the same order of magnitude as that for phosphoric acid, H_3PO_4. Which Lewis structure shows hypophosphorous acid, (a) or (b)? Explain your answer.

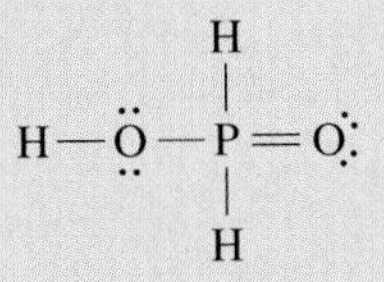

(a) 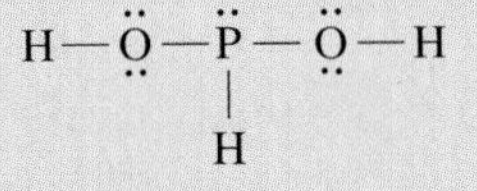   (b)

Is hypophosphorous acid a monoprotic, a diprotic, or a triprotic acid?

16.4 Which member of each of the following pairs is the stronger acid (has the higher value for K_1)? Give your reasoning. (a) CH_3OH or CH_3SH (b) H_2S or PH_3 (c) H_3PO_4 or H_2SO_3 (d) H_2SO_3 or H_2SO_4 (e) CH_3OH or CH_3COOH

16.5 What is HOBr called?

The hydroxides of elements near the borderline between metals and nonmetals in the periodic table are amphoteric—that is, they can react either as acids or as bases. With acids, they react as bases; with bases, they react as acids. Aluminum hydroxide is an example. To understand why aluminum hydroxide is amphoteric, we must consider the hydrated cation, $Al(H_2O)_6^{3+}$.

Hydrated Cations

In Sections 15.1 and 15.6, we saw that if a cation is small and has a high positive charge like the Al^{3+} ion, the hydrated cation acts as a Brønsted–Lowry acid. As a

The acidity of hydrated ions is always a difficult topic for our students. Therefore, we have discussed it in several places: qualitatively at the end of Section 15.1 and again in Section 15.6, and quantitatively here in Section 16.2.

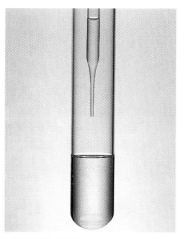
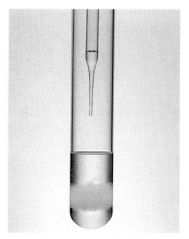
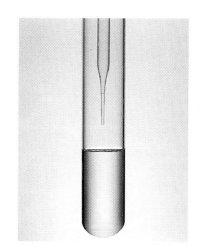

■ FIGURE 16.1 As a solution of sodium hydroxide is added to a solution of aluminum nitrate in a test tube *(top)*, a precipitate of $Al(OH)_3$ forms *(center)* and then dissolves again *(bottom)* as a result of the formation of $Al(OH)_4^-$ ion.

strong base such as hydroxide ion is added to a solution of the $Al(H_2O)_6^{3+}$ ion, protons are transferred from the hydrated ion to the base. The following reactions take place:

$$Al(H_2O)_6^{3+}(aq) + OH^-(aq) \rightleftharpoons Al(H_2O)_5(OH)^{2+}(aq) + H_2O(l)$$
$$Al(H_2O)_5(OH)^{2+}(aq) + OH^-(aq) \rightleftharpoons Al(H_2O)_4(OH)_2^+(aq) + H_2O(l)$$
$$Al(H_2O)_4(OH)_2^+(aq) + OH^-(aq) \rightleftharpoons Al(H_2O)_3(OH)_3(s) + H_2O(l)$$
$$Al(H_2O)_3(OH)_3(s) + OH^-(aq) \rightleftharpoons Al(OH)_4^-(aq) + 3H_2O(l)$$

Addition of OH^- ions shifts all four equilibria to the right. As hydroxide ions are added to a solution containing $Al(H_2O)_6^{3+}$ ion, insoluble $Al(H_2O)_3(OH)_3(s)$ first precipitates and then dissolves again (see ■Figure 16.1). If a solution of a strong acid is added to a solution containing $Al(OH)_4^-$ ions, H^+ reacts with OH^- to form water. As OH^- is removed, all four equilibria shift to the left. Insoluble $Al(H_2O)_3(OH)_3(s)$ precipitates and then redissolves. $Al(H_2O)_3(OH)_3(s)$ reacts with either acid or base and is amphoteric. The equations for these reactions are usually simplified by omitting the water molecules that hydrate the aluminum ion:

$$Al^{3+}(aq) + OH^-(aq) \rightleftharpoons Al(OH)^{2+}(aq)$$
$$Al(OH)^{2+}(aq) + OH^-(aq) \rightleftharpoons Al(OH)_2^+(aq)$$
$$Al(OH)_2^+(aq) + OH^-(aq) \rightleftharpoons Al(OH)_3(s)$$
$$Al(OH)_3(s) + OH^-(aq) \rightleftharpoons Al(OH)_4^-(aq)$$

The hydroxides of chromium [$Cr(OH)_3$], lead [$Pb(OH)_2$], tin [$Sn(OH)_2$ and $Sn(OH)_4$], and zinc [$Zn(OH)_2$] are also amphoteric.

Equilibrium constants for the hydrolysis of hydrated cations

The amphoterism of aluminum hydroxide is used in the production of aluminum to separate aluminum from its ore.

● Metal ion, M^{n+} ● Metal ion, $M^{(n-1)+}$

can be measured. (Note that when protons are transferred from hydrated cations,

TABLE 16.3 Values of K_a for Some Hydrated Cations[a]

Cation	Radius, pm	Electronegativity of Metal	K_a	pK_a
K^+	138	0.9	3×10^{-15}	14.5
Na^+	102	1.0	6×10^{-15}	14.2
Li^+	74	1.0	3×10^{-14}	13.6
Ca^{2+}	100	1.0	2×10^{-13}	12.8
Mg^{2+}	72	1.2	4×10^{-12}	11.4
Zn^{2+}	75	1.7	1.0×10^{-9}	9.0
Al^{3+}	53	1.5	1.0×10^{-5}	5.0

[a]Values for pK_a are from Wulfsberg, G. *Principles of Descriptive Inorganic Chemistry;* Brooks/Cole: Monterey, CA, 1987; p. 25.

an O—H bond breaks just as it does when protons are transferred from oxo acids and organic acids.) However, the uncertainty in the value of pK_a may be as large as one unit so small differences are not significant. Table 16.3 shows values of K_a and pK_a for some typical hydrated cations.

From a comparison of the data for K^+, Na^+, and Li^+ in Table 16.3, you can see that if charge and electronegativity are similar, then the smaller the radius of the ion, the more acidic the ion is:

$$\text{acidity} \quad K^+ < Na^+ < Li^+$$

The acidity of hydrated cations of the alkali and alkaline earth metals decreases down groups of the periodic table.

If radii and electronegativities are similar, as they are for Na^+ and Ca^{2+}, then the higher the charge on the cation, the more acidic the hydrated ion. Going across a row of the periodic table from left to right, ionic radius decreases and charge increases. As a result, acidity of hydrated cations increases across a row, for example,

$$Na^+ < Mg^{2+} < Al^{3+}$$

If charge and radius are similar, the greater the electronegativity of the metal, the more acidic the hydrated cation (compare Mg^{2+} and Zn^{2+}). The hydrated Al^{3+} ion with high charge, small radius, and relatively high electronegativity (for a metal) is by far the most acidic cation listed in Table 16.3.

PRACTICE PROBLEM

16.6 For Sr^{2+}, radius = 132 pm, electronegativity = 1.0, and pK_a = 13.3. Which of the following is the value of pK_a for Ba^{2+} (radius = 149 pm, electronegativity = 1.0)? Explain your answer. (a) 12.8 (b) 13.3 (c) 13.5

Weak Bases

Metal hydroxides are either soluble strong bases such as sodium and potassium hydroxides or insoluble like magnesium hydroxide. Ammonia is the only important soluble molecular inorganic weak base. A discussion of the basicity of amines, organic compounds derived from ammonia by substitution of groups, such as methyl-CH_3— or ethyl CH_3CH_2—, for one or more of the hydrogens of ammonia is a part of organic, not general, chemistry.

Most inorganic weak bases are anions such as the hypochlorite ion, ClO^-, or cations such as $Al(H_2O)_5(OH)^{2+}$. The basicity of anionic and cationic weak bases can be deduced from the acidity of their conjugate acids. *The stronger an acid, the weaker its conjugate base; the stronger a base, the weaker its conjugate acid.* For example, the CH_3COO^- ion is the conjugate base of the weak acid, CH_3COOH ($K_a = 1.8 \times 10^{-5}$). The weak acid, $HOCl$ ($K_a = 3.2 \times 10^{-8}$), is weaker than acetic acid. Therefore, the conjugate base of $HOCl$, the hypochlorite ion, OCl^-, is a stronger base than the acetate ion. However, the hypochlorite ion is *not* a strong base; only the conjugate bases of extremely weak acids such as CH_4, NH_3, and H_2 are strong bases.

PRACTICE PROBLEM

16.7   Which is the stronger base, CH_3COO^- or $ClCH_2COO^-$? Explain your answer.

Acidic and Basic Oxides

Like metal hydroxides, most metal oxides are basic. The oxide ion, O^{2-}, is the conjugate base of the very weak acid OH^-. Therefore, the oxide ion is a very strong base. Metal oxides that are soluble in water, such as sodium oxide, immediately react with water forming hydroxide ions:

$$Na_2O(s) + H_2O(l) \longrightarrow 2NaOH(aq) \qquad or$$
$$O^{2-}(aq) + H_2O(l) \longrightarrow OH^-(aq) + OH^-(aq)$$

Only the oxides of the metals of Group IA and calcium, strontium, and barium in Group IIA are significantly soluble in water. However, water-insoluble metal oxides react with acids to form salts and water:

$$MgO(s) + 2HCl(aq) \longrightarrow MgCl_2(aq) + H_2O(l)$$

Reaction with acids to form a salt and water is the characteristic reaction of Arrhenius bases; both soluble and insoluble metal oxides behave like bases.

On the other hand, most nonmetal oxides are acidic and react with water to form oxo acids, for example,

$$CO_2(g) + H_2O(l) \longrightarrow H_2CO_3(aq)$$
$$SO_3(g) + H_2O(l) \longrightarrow H_2SO_4(aq)$$

The formulas of the oxo acids formed by reaction of nonmetal oxides with water aren't always easy to predict. For example, $2NO_2(g) + H_2O(l) \longrightarrow HNO_3(aq) + HNO_2(aq)$.

Some nonmetals form more than one kind of oxide; for example, sulfur forms both SO_2 and SO_3. When an element forms more than one oxide, the oxide with the higher proportion of oxygen reacts with water to form a stronger acid:

$$SO_2(g) + H_2O(l) \longrightarrow H_2SO_3(aq) \qquad K_1 = 1.5 \times 10^{-2}$$

Sulfuric acid is a strong acid as far as the first ionization is concerned. A few nonmetal oxides, such as NO, N_2O, and CO, fail to react with water to form acids; these oxides do not react with acids or bases either.

Oxides of some metals near the metal–nonmetal border are amphoteric and react with both acids and bases. Aluminum oxide, Al_2O_3, is an example:

$$Al_2O_3(s) + 6HCl(aq) \longrightarrow 2AlCl_3(aq) + 3H_2O(l)$$
$$Al_2O_3(s) + 2NaOH(aq) + 3H_2O(l) \longrightarrow 2NaAl(OH)_4(aq)$$

Other common amphoteric oxides are chromium(III) oxide (Cr_2O_3), lead(II) oxide (PbO), tin(II) and tin(IV) oxides (SnO and SnO_2), and zinc oxide (ZnO).

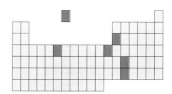

The oxides and hydroxides of the elements with red squares are amphoteric.

16.2 ACIDITY, BASICITY, AND THE PERIODIC TABLE

16.8   Classify each of the following oxides as acidic, basic, or amphoteric. Explain your answers. (a) BaO (b) P_4O_{10} (c) ZnO

16.9   Write net ionic, complete ionic, and molecular equations for each of the following reactions: (a) $K_2O(s)$ with water; (b) $Fe_2O_3(s)$ with dilute HCl; (c) $N_2O_5(s)$ with water. The product is nitric acid.

Basic and acidic oxides often react directly to form salts, for example,

$$Na_2O(s) + SO_3(g) \longrightarrow Na_2SO_4(s)$$

Many such reactions are of great practical importance. For example, silica, SiO_2, is one of the chief impurities in iron ore. In the production of pig iron, SiO_2 is removed from iron by reaction with CaO:

$$CaO(s) + SiO_2(s) \xrightarrow{\text{about 1500 °C}} CaSiO_3(l)$$

The liquid $CaSiO_3$ floats on top of the liquid iron at the bottom of the furnace and protects it from air. The $CaSiO_3$ is easily separated from the iron when too much collects. To include reactions between basic and acidic oxides as acid–base reactions, another set of definitions for acids and bases is needed.

16.3   LEWIS ACIDS AND BASES

The Brønsted–Lowry acid–base definitions classify more species as bases than the Arrhenius definitions do and can be applied to reactions in the gas phase and in nonaqueous solvents. However, acids are limited to species that contain ionizable hydrogens. There are still some reactions that resemble acid–base reactions, such as the reaction between solid sodium oxide and gaseous sulfur trioxide to form the solid sodium sulfate

$$Na_2O(s) + SO_3(g) \longrightarrow Na_2SO_4(s)$$

that are not covered by the Brønsted–Lowry definitions. Sodium oxide is a basic oxide, and sulfur trioxide is an acidic oxide. Reaction of one with the other neutralizes the acidic and basic properties; a salt is formed. The salt, sodium sulfate, also results from the reaction of sodium hydroxide solution with sulfuric acid solution

$$2NaOH(aq) + H_2SO_4(aq) \longrightarrow 2H_2O(l) + Na_2SO_4(aq)$$

which is a typical neutralization reaction.

In 1923, the same year that Brønsted and Lowry proposed their definitions, Lewis (of Lewis formulas) proposed an even more general acid–base definition that does not limit acids to species that contain hydrogen. According to the **Lewis** definitions, an **acid** is *a species that can attach to an electron pair to form a covalent bond.* A **base** is *a species that has an electron pair that can form a covalent bond.*

A Lewis acid must have an empty valence orbital for the electron pair to go into. For example, boron trifluoride is a typical Lewis acid. Its Lewis formula is

Boron trifluoride is used as a catalyst for many industrially important organic reactions.

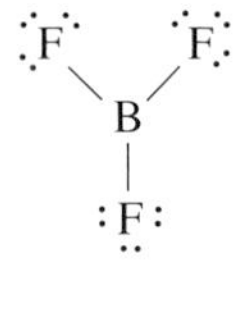

The boron atom in boron trifluoride only has six electrons around it. The boron atom is pictured as sp^2 hybridized with an empty p orbital perpendicular to the plane of the molecule.

A Lewis base must have an unshared pair of valence electrons. Thus, *there is no difference between Lewis bases and Brønsted–Lowry bases;* ammonia, NH_3, is a typical Lewis base as well as a typical Brønsted–Lowry base. The reaction between ammonia gas and boron trifluoride gas

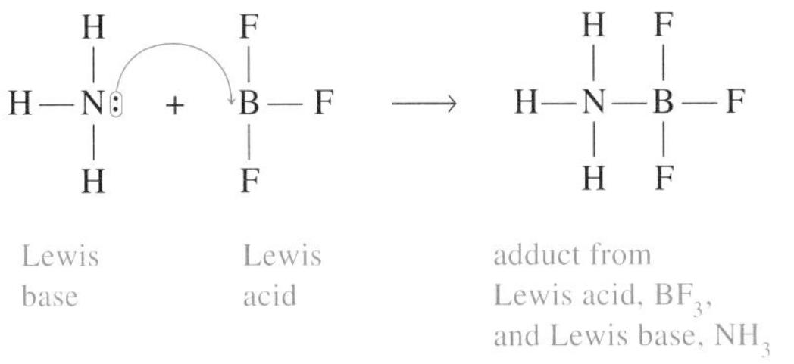

Boron trifluoride is a Lewis acid; it is not a Lewis base.

is a typical Lewis acid–base reaction. The Lewis acid BF_3 attaches to an electron pair from the Lewis base NH_3, and a covalent bond is formed between N and B. The product is a white crystalline solid similar in physical properties to salts. *A shared electron pair of which both electrons originally belonged to the same atom,* such as the pair that forms the N—B bond in the adduct formed from NH_3 and BF_3, is called a **coordinate covalent bond.**

In addition to species with incomplete octets of electrons like boron trifluoride, small cations with high charges such as the copper(II) ion, Cu^{2+}, act as Lewis acids:

$$Cu^{2+}(aq) + 4\,{:}NH_3(aq) \rightleftharpoons Cu({:}NH_3)_4^{\,2+}$$

Lewis acid — Lewis base — adduct from Lewis acid, Cu^{2+}, and Lewis base, NH_3

The reaction used at the beginning of this section to show the need for another set of acid–base definitions is pictured as a Lewis acid–base reaction as follows. Sulfur trioxide is a resonance hybrid with three equally important contributing structures:

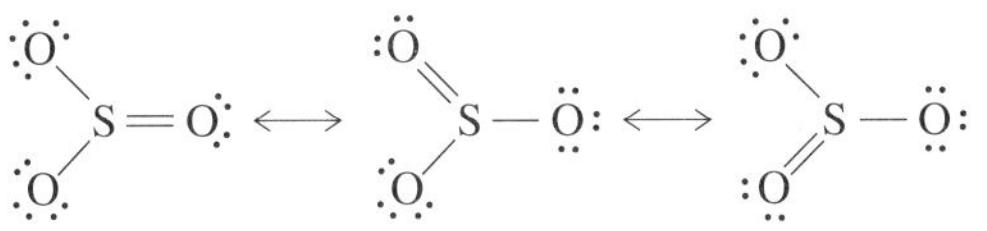

The electronegativity of oxygen, 3.5, is greater than the electronegativity of sulfur, 2.4. Therefore, the S—O bonds are polar with a partial positive charge on sulfur. Sulfur trioxide functions as a Lewis acid and attaches to a pair of electrons from the oxide ion. The oxide ion, O^{2-}, acts as a Lewis base. The sodium ion is a spectator ion:

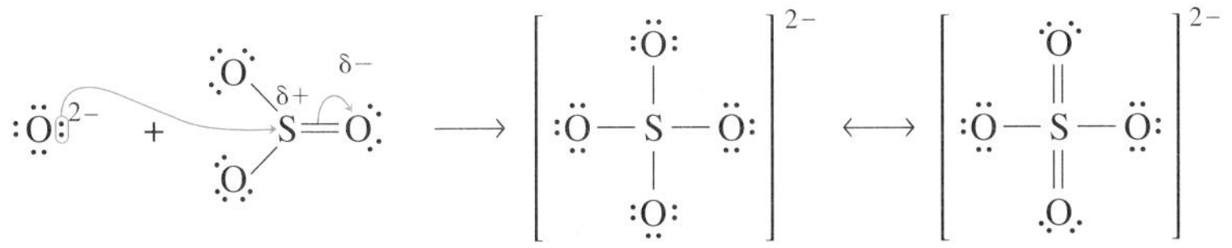

Sulfur and nitrogen oxides in the atmosphere, which are Lewis acids, react with the Lewis base, water, to form acid rain (see Related Topic elsewhere in this chapter).

Notice how we picture an empty orbital being created on the sulfur atom by moving a pair of electrons from the double bond onto the oxygen atom.

16.2 Picture the formation of H_2CO_3 as a Lewis acid–base reaction between H_2O and CO_2.

Solution The electronegativity of oxygen, 3.5, is greater than the electronegativity of carbon, 2.5. Therefore, the C—O bonds are polar with a partial positive charge on carbon. Carbon dioxide functions as a Lewis acid and attaches to the oxygen of a water molecule. The water molecule acts as a Lewis base:

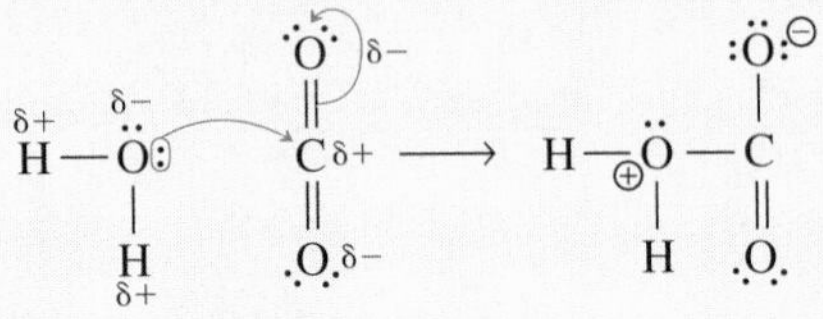

A proton shifts from the positively charged oxygen to the negatively charged oxygen to give a carbonic acid molecule:

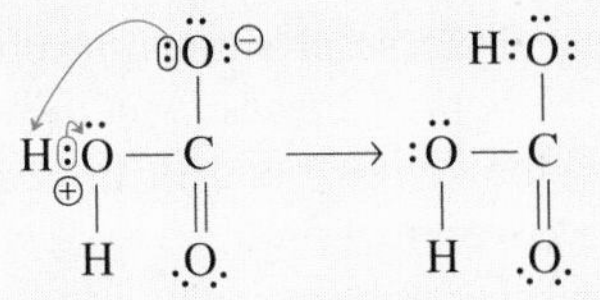

16.10 Of the following species,

$$Cr^{3+} \quad CH_3NH_2 \quad Na^+ \quad BeCl_2 \quad CH_4$$

(a) which will probably act as Lewis acids? (*Hint:* Write Lewis formulas for molecules.) (b) Which will probably act as Lewis bases? (c) Which will not react either as a Lewis acid or as a Lewis base?

16.11 Each of the following reactions is a Lewis acid–base reaction. For each, identify the Lewis acid and the Lewis base.
(a) $Al(OH)_3(s) + OH^-(aq) \longrightarrow Al(OH)_4^-(aq)$
(b) $Ag^+(aq) + 2NH_3(aq) \longrightarrow Ag(NH_3)_2^+(aq)$
(c) $(CH_3CH_2)_2O(l) + BF_3(g) \longrightarrow (CH_3CH_2)_2OBF_3(l)$

16.12 Picture the formation of H_2SO_3 as a Lewis acid–base reaction between H_2O and SO_2.

In the Student Solutions Manual, students are reminded that $BeCl_2$ does not act as a Lewis base.

Reaction of $CO_2(g)$ and $SO_2(g)$ with water is the reason that carbon dioxide and sulfur dioxide are much more soluble in water than $N_2(g)$ and $O_2(g)$ (Table 13.2).

One of the disadvantages of the Lewis definitions is that typical hydrogen-containing acids such as hydrogen chloride (HCl) and acetic acid (CH_3COOH) are not, strictly speaking, Lewis acids. Typical acids are included as Lewis acids in one of two ways. One way is to regard the regular acids as sources of protons. The proton, H^+, is a Lewis acid because it has an empty $1s$ orbital. Hydrogen-containing acids are classified as Lewis acids because they are sources of protons.

The other approach to classifying hydrogen-containing acids as Lewis acids is to regard Lewis acid–base reactions as substitution reactions: One Lewis base takes the place of another Lewis base. For example, water displaces chloride ion from hydrogen chloride:

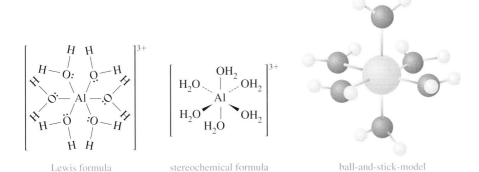

$$\underset{\substack{\text{Lewis}\\\text{base}}}{\overset{\delta+}{H}-\overset{\delta-}{\underset{\underset{H^{\delta+}}{|}}{\overset{..}{O}:}} \ + \ \underset{\substack{\text{adduct from}\\\text{Lewis acid, }H^+,\\\text{and Lewis base, }Cl^-}}{\overset{\delta+}{H}-\overset{\delta-}{\overset{..}{\underset{..}{Cl}}:} \ \longrightarrow \ \underset{\substack{\text{adduct from}\\\text{Lewis acid, }H^+,\\\text{and Lewis base, }H_2O}}{H-\overset{+}{\underset{\underset{H}{|}}{\overset{..}{O}}}-H} \ + \ \underset{\substack{\text{Lewis}\\\text{base}}}{:\overset{..}{\underset{..}{Cl}}:^-}$$

In this way, all Brønsted–Lowry acid–base reactions can be classified as Lewis acid–base reactions.

A second disadvantage of the Lewis definitions is that no uniform scale of acid–base strengths exists for Lewis acids and bases. Brønsted–Lowry acids or bases can be arranged easily in order of increasing or decreasing strength by the values of their ionization constants.

The different definitions of acids and bases help us organize experimental observations. The definition that is most convenient for a particular situation should be used. For example, for quantitative calculations involving aqueous solutions, the Brønsted–Lowry definitions are best. For consideration of reactions that do not include proton transfer, the Lewis definitions must be used. The Brønsted–Lowry and the Lewis definitions are used in both inorganic and organic chemistry. No one set of definitions is best in all circumstances.

16.4 EQUILIBRIA BETWEEN COMPLEXES AND THEIR PARTS

As we saw in Section 15.6, small, highly charged cations such as Al^{3+} attract electrons strongly. When small, highly charged cations are hydrated, the attraction of the cation for the unshared pair of electrons of the water molecules is so strong that the cation actually shares the "unshared" pair. Coordinate covalent bonds are formed between the metal ion and the oxygens. A definite number of water molecules is involved in the hydration of each cation. The species that is formed is called a **complex ion.** The hydrated aluminum ion $Al(H_2O)_6^{3+}$ is an example:

Lewis formula   stereochemical formula   ball-and-stick-model

Although most general chemistry texts say that the central atom must be a metal, the authors of recent inorganic texts include species such as PF_6^- and SiF_6^{2-} as complexes.

Nomenclature of complexes is discussed in Chapter 24.

Complexes are *species in which a central atom is surrounded by a set of Lewis bases that are covalently bonded to the central atom.* Complexes can be ions like $Al(H_2O)_6^{3+}$ and $Al(OH)_4^-$ or molecules such as $Al(H_2O)_3(OH)_3$. *The Lewis bases in complexes are usually referred to as* **ligands.** The ligands in the hexaaquaaluminum ion, $Al(H_2O)_6^{3+}$, are water molecules. Other common ligands include ammonia molecules, and halide, hydroxide, and cyanide ions. Because small, highly charged

Acid Rain

Rain normally has a pH close to 5.6 because it contains small amounts of both weak and strong acids of natural origin including carbon dioxide. The term *acid rain* refers to precipitation with pH substantially less than 5.6 (sometimes as low as 2.2–2.7). The additional acidity is due mainly to sulfuric and nitric acids formed by the reaction of sulfur trioxide and nitrogen dioxide with water:

$$SO_3(g) + H_2O(l) \longrightarrow H_2SO_4(aq)$$
$$2NO_2(g) + H_2O(l) \longrightarrow$$
$$HNO_3(aq) + HNO_2(aq)$$

Sulfur trioxide and nitrogen dioxide are formed by the reaction of sulfur dioxide and nitric oxide with oxygen from the air:

$$2SO_2(g) + O_2(g) \xrightarrow{\text{dust}} 2SO_3(g)$$
$$2NO(g) + O_2(g) \longrightarrow 2NO_2(g)$$

The nitric oxide results from reaction of nitrogen and oxygen in air at high temperatures:

$$N_2(g) + O_2(g) \xrightarrow{\text{high temperature}} 2NO(g)$$

Both sulfur dioxide and nitric oxide are produced naturally; for example, volcanoes give off SO_2, and NO is formed in lightning bolts. But since the middle of

Acid rain has severely damaged these marble gargoyles at Notre Dame in Paris.

the nineteenth century, and especially in the twentieth century, people have been responsible for a great increase in the production of both gases in cities and industrial areas. The sulfur dioxide is formed mainly by the burning of fossil fuels that contain sulfur. Nitric oxide is formed from nitrogen and oxygen in air at the high temperatures involved in the combustion of fossil fuels.

Acid rain attacks buildings made of limestone and marble, which both consist

---

cations tend to form complex ions, *formation of complex ions is a characteristic property of transition metal ions.*

Formation of complex ions other than aqua ions in aqueous solution always involves substitution of other ligands for water. For example, addition of concentrated ammonia to a solution that contains copper(II) ions results in substitution of ammonia molecules for water molecules. The net ionic equation is

$$4NH_3(aq) + Cu(H_2O)_6{}^{2+}(aq) \longrightarrow Cu(H_2O)_2(NH_3)_4{}^{2+}(aq) + 4H_2O(l)$$
$$\text{pale blue} \qquad\qquad\qquad \text{deep blue}$$

Complexes are often intensely colored, and the colors of complex ions in solution are frequently used to detect low concentrations of metal ions such as the Cu^{2+} ion.

The color of the solution changes from the pale blue of the $Cu(H_2O)_6{}^{2+}$ ion to the deep blue of the $Cu(H_2O)_2(NH_3)_4{}^{2+}$ ion (see ■Figure 16.2). Replacement of water in hydrated metal ions by other ligands usually takes place very rapidly, and equilibrium is reached almost immediately. The only common ion that usually undergoes substitution slowly is $Cr^{3+}(aq)$. (Many of the rarer second and third transition series ions react slowly.)

A forest damaged by acid rain.

largely of calcium carbonate:

$$CaCO_3(s) + H_2SO_4(aq) \longrightarrow$$
$$CaSO_4(s) + H_2O(l) + CO_2(g)$$

$$CaCO_3(s) + 2HNO_3(aq) \longrightarrow$$
$$Ca(NO_3)_2(aq) + H_2O(l) + CO_2(g)$$

Calcium nitrate, of course, is soluble in water; calcium sulfate is much more soluble than calcium carbonate and also washes away with time. Nitric and sulfuric

acids also corrode metals from which cars, machinery, and railroad rails are made, for example,

$$Fe(s) + H_2SO_4(aq) \longrightarrow$$
$$FeSO_4(aq) + H_2(g)$$

Breathing these acids is definitely hazardous to your health.

Acid rain damages plants by interfering with nitrogen fixation and extracting nutrients from leaves. Acid rain also leaches aluminum from the soil and lake sediments, and Al^{3+} in the water damages the roots of trees and kills fish. Young fish and species low in the food chain are killed when lakes and streams become too acidic, and adult fish starve. The harmful effect of nitric acid is increased by the fact that nitrate ion fertilizes the growth of algae using up the oxygen dissolved in the water so that fish can't breathe. The algal bloom shades plants living in the water so that they cannot grow.

The fact that many lakes have limestone basins delayed discovery of the harmful effects of acid rain on lakes and streams because the water was buffered with carbonate and hydrogen carbonate ions. Although rain was gradually becoming more acidic, the pH of many lakes did not start to decrease until the buffer capacity was used up.

Algal bloom shades plants living in the water so that they cannot grow.

Acid rain is a worldwide problem because sulfur dioxide, sulfur trioxide, and the nitrogen oxides are carried across national boundaries by the movement of air. Governments are beginning to work together to reduce emissions, but solution of the problem will be difficult and expensive. For example, one method of removing sulfur dioxide from the gases given off by power plants is reaction with lime:

$$SO_2(g) + CaO(s) \longrightarrow CaSO_3(s)$$

However, not only does this process cost money and use energy but, in addition, about one ton of calcium sulfite is produced each year for each person who uses electricity from the power plant. The $CaSO_3$ must be disposed of in landfills because no uses have been found for it.

- FIGURE 16.2 *Left:* A solution containing $Cu(H_2O)_6^{2+}$. *Middle:* Addition of concentrated ammonia results in precipitation of $Cu(OH)_2(s)$ and formation of the deep blue $Cu(H_2O)_2(NH_3)_4^{2+}$ ion. *Right:* Continued addition of ammonia dissolves the precipitate; more $Cu(H_2O)_2(NH_3)_4^{2+}$ is formed.

Substitution of other ligands for water takes place in steps and is reversible. For example, the first step in the formation of the $Cd(CN)_4^{2-}$ ion by displacement of water from the $Cd(H_2O)_6^{2+}$ ion by CN^- is

$$CN^-(aq) + Cd(H_2O)_6^{2+}(aq) \longrightarrow Cd(CN)(H_2O)_5^+(aq) + H_2O(l)$$

This equation is commonly written without showing water,

$$CN^-(aq) + Cd^{2+}(aq) \longrightarrow Cd(CN)^+(aq)$$

and the equilibrium constant expression for the first step is usually written

$$K_1 = \frac{[Cd(CN)^+]}{[CN^-][Cd^{2+}]}$$

The equations for all four steps in the formation of the $Cd(CN)_4^{2-}$ ion and their equilibrium constant expressions are

$$Cd^{2+}(aq) + CN^-(aq) \rightleftharpoons Cd(CN)^+(aq) \qquad K_1 = \frac{[Cd(CN)^+]}{[Cd^{2+}][CN^-]} = 1.0 \times 10^6$$

$$Cd(CN)^+(aq) + CN^-(aq) \rightleftharpoons Cd(CN)_2(aq) \qquad K_2 = \frac{[Cd(CN)_2]}{[Cd(CN)^+][CN^-]} = 1.3 \times 10^5$$

$$Cd(CN)_2(aq) + CN^-(aq) \rightleftharpoons Cd(CN)_3^-(aq) \qquad K_3 = \frac{[Cd(CN)_3^-]}{[Cd(CN)_2][CN^-]} = 3.4 \times 10^4$$

$$Cd(CN)_3^-(aq) + CN^-(aq) \rightleftharpoons Cd(CN)_4^{2-}(aq) \quad K_4 = \frac{[Cd(CN)_4^{2-}]}{[Cd(CN)_3^-][CN^-]} = 1.9 \times 10^2$$

$$Cd^{2+}(aq) + 4CN^-(aq) \rightleftharpoons Cd(CN)_4^{2-}(aq) \qquad K_f = \frac{[Cd(CN)_4^{2-}]}{[Cd^{2+}][CN^-]^4} = K_1 \cdot K_2 \cdot K_3 \cdot K_4$$
$$= 8.4 \times 10^{17}$$

The last equation, the equation for the overall reaction for the formation of $Cd(CN)_4^{2-}$, is the sum of the other four. The subscript f stands for formation. When equations are added, the equilibrium constant expression is the product of the equilibrium constant expressions for the individual equations (Section 16.1). The equilibrium constant for the overall reaction is the product of the equilibrium constants for the individual reactions. The large magnitude of the formation constant for the $Cd(CN)_4^{2-}$ ion shows that cyanide ions displace water molecules from $Cd^{2+}(aq)$ even though the concentration of cyanide ions is much lower than the concentration of water molecules.

The value of the equilibrium constant for each of the four steps is smaller than the value of the equilibrium constant for the preceding step. As each water molecule is displaced, fewer water molecules remain to be displaced. Also, the charge on the reactant complex ion becomes less positive with each step; the reactant complex ion has less attraction for negative CN^- ions. However, the differences between equilibrium constants for successive steps are small. As cyanide ions are added to a solution of cadmium(II) ions, the solution usually contains a mixture of the various cadmium-containing ions. Only when the concentration of cyanide ions is very low or relatively high is a single cadmium-containing species present. Because differences between equilibrium constants for the steps in the formation of most complex ions are small, calculations of the concentrations of the species present in solution are too complicated for a course in general chemistry *except* when the equilibrium constant for the overall reaction is high and an excess of the complexing ion is present. Sample Problem 16.3 shows how to calculate the concentration of hydrated ion remaining after the addition of an excess of complexing ion. As you will see, the calculation is similar to all the other equilibrium calculations we have done.

16.3 The value of K_f for $Cd(CN)_4^{2-}$ is 8.3×10^{17}. If 0.40 mol NaCN(s) is added to 1.0 L of 0.010 M $Cd(NO_3)_2(aq)$, what will $[Cd^{2+}]$ be at equilibrium? [Assume that no change in the volume of the solution takes place when NaCN(s) is added.]

Solution Although CN^- is the conjugate base of the weak acid HCN, K_b for the hydrolysis of the CN^- ion is only 2.0×10^{-5}. The change in $[CN^-]$ resulting from hydrolysis is negligible in 0.40 M NaCN(aq).

The equilibrium constant for the equilibrium

$$Cd^{2+}(aq) + 4CN^-(aq) \rightleftharpoons Cd(CN)_4^{2-}(aq)$$

is 8.3×10^{17}, a very large number, and reaction is essentially complete at equilibrium. So that the change in concentration will be small compared to the initial concentration, assume that formation of the complex is complete and consider the reverse reaction, the dissociation of the complex

$$Cd(CN)_4^{2-}(aq) \rightleftharpoons Cd^{2+}(aq) + 4CN^-(aq)$$

For the dissociation of the complex,

$$K_d = \frac{[Cd^{2+}][CN^-]^4}{[Cd(CN)_4^{2-}]} = \frac{1}{K_f} = \frac{1}{(8.3 \times 10^{17})} = 1.2 \times 10^{-18}$$

According to the overall equation for the formation of $Cd(CN)_4^{2-}$,

$$Cd^{2+}(aq) + 4CN^-(aq) \longrightarrow Cd(CN)_4^{2-}(aq)$$

4 mol CN^- react with 1 mol Cd^{2+} to form 1 mol of the complex ion. Therefore, 0.040 mol CN^- will react with 0.010 mol Cd^{2+}, and the number of moles of cyanide ion left in a liter of solution will be

0.40 mol CN^- originally present − 0.04 mol CN^- reacted = 0.36 mol CN^- left

Because we are considering 1.0 L of solution, $[CN^-] = 0.36$.

According to the overall equation for the dissociation of $Cd(CN)_4^{2-}$, 1 mol $Cd(CN)_4^{2-}$ yields 1 mol Cd^{2+} and 4 mol CN^-. If we let x equal mol/L Cd^{2+} formed, then $4x$ equals mol/L CN^- formed, and $-x$ equals mol/L $Cd(CN)_4^{2-}$ dissociated. The completed summary table is:

	$Cd(CN)_4^{2-}(aq) \rightleftharpoons$	$Cd^{2+}(aq) +$	$4CN^-(aq)$
Init. conc., M	**0.010**	**0.000**	**0.36**
Change, ΔM	$-x$	$+x$	$+4x$
Eq. conc., M	$(0.010 - x)$	x	$(0.36 + 4x)$

Substitution of the equilibrium concentrations in the equilibrium constant expression gives

$$\frac{x(0.36 + 4x)^4}{(0.010 - x)} = 1.2 \times 10^{-18}$$

The usual assumption that x is negligible compared to 0.010 and $4x$ is negligible compared to 0.36 simplifies the preceding equation to

$$\frac{x(0.36)^4}{(0.010)} = 1.2 \times 10^{-18} \qquad \text{and} \qquad x = 7.1 \times 10^{-19}$$

The assumption that x is negligible compared to 0.010 and $4x$ is negligible compared to 0.36 is correct and $[Cd^{2+}]$ at equilibrium is 7.1×10^{-19}, an extremely low concentration. (The best answer, obtained with no intermediate rounding, is 7.2×10^{-19}.)

○ You should check this statement. Sample Problem 15.9 shows how to calculate the change in concentration of an anion that results from hydrolysis.

○ If K ≫ 1, always assume that reaction goes to completion and then consider the reverse reaction.

According to the new *Random House Unabridged Dictionary, both* solution *and* dissolution *mean the process by which a solid, liquid, or gas is dispersed homogeneously in a solid, liquid, or gas. We have used* solution *because it is among the 5000 commonest words, whereas* dissolution *is in the 8th thousand. Also "dis-" has a reverse meaning in many words (for example, distrust) so that* dissolution *implies the opposite of solution or precipitation to some students and, as a result, is confusing. Reviewers unanimously preferred* solution. *Students who take more advanced chemistry courses can learn to use* dissolution *later.*

TABLE 16.4	Formation Constants for Some Common Complex Ions at 25 °C[a]		
Complex Ion	K_f	Complex Ion	K_f
$AgBr_2^-$	5.0×10^7	$Cu(NH_3)_4^{2+}$	5.6×10^{11}
$AgCl_2^-$	1.8×10^5	$Fe(CN)_6^{4-}$	2.5×10^{35}
$Ag(CN)_2^-$	3.0×10^{20}	$Fe(CN)_6^{3-}$	4.0×10^{43}
AgI_2^-	1.1×10^9	$FeSCN^{2+}$	1.0×10^3
$Ag(NH_3)_2^+$	1.7×10^7	$Hg(CN)_4^{2-}$	9.3×10^{38}
$Ag(S_2O_3)_2^{3-}$	5.2×10^{12}	$Ni(CN)_4^{2-}$	1.7×10^{30}
AlF_6^{3-}	6.3×10^{19}	$Ni(NH_3)_4^{2+}$	4.7×10^7
$Al(OH)_4^-$	1.0×10^{33}	$Ni(NH_3)_6^{2+}$	2.0×10^8
$Cd(CN)_4^{2-}$	8.3×10^{17}	$Zn(CN)_4^{2-}$	4.2×10^{19}
$Cd(NH_3)_4^{2+}$	5.5×10^6	$Zn(NH_3)_4^{2+}$	7.8×10^8
$Co(NH_3)_6^{2+}$	2.5×10^4	$Zn(OH)_4^{2-}$	6.3×10^{14}

[a]Data are from Smith, R. M.; Martell, A. E. *Critical Stability Constants;* Plenum: New York, 1976; Vol. 4. "Critical" means that the values have been selected by experts as the most reliable values available.

◗ Formation constants for complex ions are of interest only if they are very large. Dissociation constants for weak acids and bases are of interest only if they are small.

Table 16.4 lists the formation constants for some common complex ions. Because we will not be using the equilibrium constants for the individual steps, only the equilibrium constants for the overall reactions are given.

PRACTICE PROBLEM

16.13   If 0.50 mol $NH_3(g)$ is added to 1.00 L of 0.020 M $Cu(NO_3)_2(aq)$, what will $[Cu^{2+}]$ be at equilibrium? (Assume that no change in the volume of the solution takes place when $NH_3(g)$ is added.)

16.5 THE SOLUBILITY PRODUCT CONSTANT AND SOLUBILITY

◗ What can be done to speed up the solution process—that is, what do you do to make the sweetener in your iced tea dissolve faster?

Dynamic equilibria exist in saturated solutions as long as some undissolved solute is present. The rates of solution and precipitation are equal. Once enough time has passed for equilibrium to be reached, concentrations are constant as long as the temperature does not change (Section 13.1). The equations for equilibria between slightly soluble solids and ions in solution are usually written with the slightly soluble solid on the left (reactant) side and the ions on the right (product) side:

$$CaF_2(s) \rightleftharpoons Ca^{2+}(aq) + 2F^-(aq) \tag{16.5}$$

The equilibrium constant expression for the equilibrium in equation 16.5 is

$$K_c = [Ca^{2+}][F^-]^2$$

◗ Remember that the constant concentrations of pure solids are included in the equilibrium constant. The concentrations of pure solids do not appear in the equilibrium constant expression.

Because the constant concentration of the pure solid is included in the value of K, K is equal to the product of the concentrations of the ions in solution (each raised to the power of the coefficient in the balanced equation) and is usually referred to as the **solubility product constant** or **solubility product** and represented by the symbol K_{sp}.

TABLE 16.5

TABLE 16.5 Calculated and Observed Solubilities

Substance	Temperature, °C	$K_{sp}{}^a$	Solubility, g/100 cm³ water From K_{sp}	Observed[b]
CaCO₃	25	5×10^{-9}	0.0007	0.0014
PbCO₃	20	1.4×10^{-13}	0.000010	0.00011
Li₂CO₃	0	2×10^{-3}	0.6	1.5
NiCO₃	25	1.2×10^{-7}	0.004	0.009
CuI	18	7×10^{-13}	0.000016	0.008
PbBr₂	20	9×10^{-6}	0.5	0.8
PbCl₂	20	1.2×10^{-5}	0.4	1.0
HgI₂	25	3×10^{-29}	0.000000009	0.01
Hg₂Br₂	25	6×10^{-23}	0.0000014	0.000004
Hg₂Cl₂	25	1.4×10^{-18}	0.00003	0.0002
AgBr	100	5×10^{-10}	0.0004	0.0004
AgCl	10	4×10^{-11}	0.00009	0.00009
AgI	25	9×10^{-17}	0.0000002	0.0000003
Ca(OH)₂	0	9×10^{-6}	0.10	0.19
Fe(OH)₂	18	7×10^{-16}	0.00005	0.00015
Pb(OH)₂	20	2×10^{-15}	0.00019	0.016
Mn(OH)₂	18	1.7×10^{-13}	0.0003	0.0002
BaSO₄	25	1.1×10^{-10}	0.0002	0.0002
CaSO₄	30	6×10^{-5}	0.11	0.2
Hg₂SO₄	25	7×10^{-7}	0.04	0.06
Ag₂SO₄	0	5×10^{-6}	0.3	0.6
SrSO₄	0	4×10^{-7}	0.012	0.011
CdS	18	4×10^{-30}	3×10^{-14}	0.00013
MnS	18	3×10^{-14}	0.0000015	0.0005
HgS	18	1.5×10^{-53}	9×10^{-26}	0.000001
CuS	18	2×10^{-37}	4×10^{-18}	0.00003

[a]Values for K_{sp} tabulated in references such as the *CRC Handbook of Chemistry and Physics* are usually calculated from thermodynamic data. The values for K_{sp} given in this table were calculated using thermodynamic data from Bard, A. J.; Parsons, R.; Jordan, J. *Standard Potentials in Aqueous Solution;* Dekker: New York, 1985. The method of calculation is shown in Appendix G.
[b]Most solubilities are from Lide, D. R. *CRC Handbook of Chemistry and Physics,* 75th ed.; CRC Press: Boca Raton, FL, 1994.

In Table 16.5 and in the rest of Chapter 16, numbers that begin with 1 are written with an extra digit. Numbers that begin with 1 are often written with an extra digit so that they have the same uncertainty as other numbers—for example, in some states any person with a blood-alcohol level of 1.0 g/L or more is legally intoxicated. Other states have reduced the limit to 0.8 g/L. Both limits are uncertain to about one part in ten.

Why are solubilities written in decimal notation? Solubilities are written in decimal notation to be consistent with the CRC Handbook. Solubilities in decimal notation are less likely to be confused with K_{sp} values, and it is easier to see at a glance which are low and which are very low.

If data on the solubility of a compound at 25 °C were not available, the value of K_{sp} for a temperature for which the solubility was available was obtained by estimating $\Delta G°$ for this temperature from thermodynamic data.

PRACTICE PROBLEM

16.14 For each of the following slightly soluble compounds, write the equation for the equilibrium that exists in a saturated solution and the solubility product expression. (a) AgCl (b) Fe(OH)₂ (c) PbBr₂ (d) Li₂CO₃ (e) Ag₂SO₄

For a recent discussion of the problems with the usual general chemistry treatment of solubility equilibria, see Slade, P. W., Rayner-Canham, G. W. J. Chem. Educ. 1990, 67, 316–317.

One might think that solubility and solubility product are related so that solubility can be calculated from the solubility product or the solubility product calculated from solubility data. Let's see if that is true. Sample Problem 16.4 shows how to calculate solubility from solubility product. The solubilities calculated from K_{sp} for a number of typical salts and hydroxides are compared with the observed solubilities in Table 16.5.

16.4   The value of K_{sp} for AgBr is 5×10^{-13} at 25 °C. What is the solubility of AgBr in g/100 cm³ of water at 25 °C?

Solution   The equation for the equilibrium that exists in a saturated solution of AgBr is

$$AgBr(s) \rightleftharpoons Ag^+(aq) + Br^-(aq) \qquad (16.6)$$

and the expression for the solubility product constant is

$$K_{sp} = [Ag^+][Br^-] = 5 \times 10^{-13}$$

Before any AgBr has dissolved, $[Ag^+] = [Br^-] = 0.0$. According to the equation for the solubility equilibrium (equation 16.6), if x mol Ag^+ go into solution in a liter of water, then x mol Br^- also dissolve. The summary table is:

$$AgBr(s) \rightleftharpoons Ag^+(aq) + Br^-(aq)$$

	Ag^+	Br^-
Init. conc., M	0.0	0.0
Change, ΔM	$+x$	$+x$
Eq. conc., M	x	x

Substitution in the solubility product expression gives

$$x^2 = 5 \times 10^{-13} \qquad (16.7)$$

Solution of equation 16.7 for x gives

$$x = 7 \times 10^{-7}$$

The solubility of AgBr is 7×10^{-7} mol/L at 25 °C.

But the problem asks for solubility in g/100 cm³ of water. We must convert units. The formula mass of AgBr is 188 u; 1 mol AgBr has a mass of 188 g. Because saturated silver bromide solution is very dilute, the volume of the solution is equal to the volume of water in the solution:

$$7 \times 10^{-7} \frac{\text{mol AgBr}}{\text{L}} \left(\frac{1 \text{ L}}{1000 \text{ cm}^3} \right) 100 \text{ cm}^3 \left(188 \frac{\text{g AgBr}}{\text{mol AgBr}} \right)$$

$$= 1.3 \times 10^{-5} \text{ g AgBr in 100 cm}^3$$

Thus, the calculated solubility of AgBr is 1.3×10^{-5} g/100 cm³ of water at 25 °C. According to the Merck Index, the measured solubility of AgBr at 25 °C is 1.35×10^{-5} g/100 cm³ of water.

16.15   The value of K_{sp} for AgCl is 1.6×10^{-10} at 25 °C. What is the solubility of AgCl in g/100 cm³ of water at 25 °C?

16.16   The solubility of thallium iodide, TlI, is 0.006 g/100 cm³ of water at 20 °C. What is the value of K_{sp} for TlI at 20 °C?

From the data in Table 16.5, you can see that although calculated and observed solubilities are equal for a few compounds such as silver bromide, silver chloride, and barium sulfate, actual solubilities are higher than solubilities calculated from K_{sp} for most salts and hydroxides. For a few salts, such as HgI_2 and all of the sulfides except MnS, the observed solubility is many powers of 10 greater than the calculated solubility. Why? To answer this question, we need to consider the factors that influence the solubility of salts and hydroxides.

(a)

(b)

(c)

(d)

■ FIGURE 16.3 Common ion and salt effects. (a) A cloud of AgCl(s) is formed when dilute AgNO₃(aq) and KCl(aq) are mixed in pure water. (b) No cloud forms when the same volumes of AgNO₃(aq) and KCl(aq) are mixed in potassium nitrate solution, an example of the salt effect. When the same volumes of AgNO₃(aq) and KCl(aq) are mixed in (c) silver nitrate solution and (d) potassium chloride solution, precipitates of AgCl(s) form. The latter are examples of the common ion effect.

16.6 FACTORS INFLUENCING THE SOLUBILITY OF SALTS AND HYDROXIDES

Four effects are important in solubility equilibria: the common ion effect, the salt effect, side reactions such as hydrolysis and complex ion formation, and incomplete dissociation or ion-pair formation. We discuss each of them in turn.

Common Ion and Salt Effects

■Figure 16.3 shows the effect of added electrolytes on solubility. In Figure 16.3(a), solutions of silver nitrate, $AgNO_3$, and potassium chloride, KCl, have just reacted to form solid silver chloride. The net and complete ionic equations are

Net: $Ag^+(aq) + Cl^-(aq) \longrightarrow AgCl(s)$

Complete: $Ag^+(aq) + NO_3^-(aq) + K^+(aq) + Cl^-(aq) \longrightarrow$
$$AgCl(s) + K^+(aq) + NO_3^-(aq)$$

Potassium ions and nitrate ions are spectator ions.

From a comparison of Figure 16.3(b) with (a) you can see that less precipitate is formed in the presence of KNO_3. Comparing Figure 16.3(c) and (d) with (a) you can see that more precipitate is formed in the presence of $AgNO_3$ and KCl. Because less precipitate is formed in the presence of KNO_3, formation of more precipitate in the presence of $AgNO_3$ and KCl must be due to silver ions and chloride ions, not to nitrate ions and potassium ions. If we consider the equilibrium between solid silver chloride and silver ions and chloride ions in solution,

$$AgCl(s) \rightleftharpoons Ag^+(aq) + Cl^-(aq)$$

a shift to the left on addition of either Ag^+ or Cl^- would be predicted either by Le Châtelier's principle or from the equilibrium constant expression,

$$K_{sp} = [Ag^+][Cl^-]$$

For $[Ag^+][Cl^-]$ to be constant, if $[Ag^+]$ increases, $[Cl^-]$ must decrease. Chloride ion must be used up by reaction with silver ion to form solid silver chloride. On a microscopic level, the higher the concentration of silver ion, the greater the chance of a chloride ion bumping into a silver ion and forming silver chloride. The effect of silver ion on the equilibrium between solid silver chloride and silver ions and

The common ion effect is discussed in Section 15.7.

The salt effect is sometimes called the uncommon ion effect.

chloride ions in solution is an example of the common ion effect. The effect of chloride ion on the equilibrium is also an example of the common ion effect. The presence of an ion in common with one of the ions in a solubility equilibrium shifts the equilibrium in favor of undissolved solid.

Now, how do we explain the effect of potassium nitrate (which has no ions in common with silver chloride) on the equilibrium between solid silver chloride and silver ions and chloride ions in solution? The added potassium ions are attracted by the negative charge on the chloride ions in solution and cluster around the chloride ions. The added nitrate ions are attracted by the positive charge on the silver ions and cluster around the silver ions. The clusters of oppositely charged ions around each silver ion and chloride ion keep the silver ions and chloride ions from combining to form solid silver chloride. *The increase in solubility of a slightly soluble salt caused by the presence of ions in solution* is called a **salt effect.**

Hydrolysis

If the anion is the conjugate base of a weak acid or the cation is the conjugate acid of a weak base, salts are more soluble than expected from the values of K_{sp} as a result of hydrolysis. Let's consider silver cyanide, AgCN, as an example. Because the cyanide ion is the conjugate base of the weak acid HCN, the cyanide ion is moderately basic. Reaction of cyanide ion with water,

$$CN^-(aq) + H_2O(l) \longrightarrow HCN(aq) + OH^-(aq)$$

uses up CN^- ions and shifts the solubility equilibrium,

$$AgCN(s) \rightleftharpoons Ag^+(aq) + CN^-(aq)$$

to the right.

From Table 15.3, K_a for acetic acid is 1.8×10^{-5}, and K_a for HCN is 4.9×10^{-10}.

Dilution has the same effect on percent hydrolysis that it has on the percent ionization of weak acids (Figure 15.5).

Percent hydrolysis of anions in dilute solution is limited by the hydroxide ion formed by the autoionization of water just as percent ionization of weak acids is limited by the hydrogen ion formed by the autoionization of water (Section 15.4).

The extent to which a salt of a weak acid is hydrolyzed depends both on the value of K_a for the weak acid and on the concentration of the salt. The weaker an acid, the stronger its conjugate base and the greater the percent hydrolysis of salts of the weak acid. For example, a 0.1 M solution of sodium acetate is only 0.007% hydrolyzed, but a 0.1 M solution of sodium cyanide is 1.4% hydrolyzed. The lower the concentration of a weak acid salt, the greater the percent hydrolysis. For example, whereas a 0.1 M solution of sodium cyanide is 1.4% hydrolyzed, a 0.01 M solution of sodium cyanide is 4% hydrolyzed. The effect of dilution on the extent of hydrolysis is predicted by Le Châtelier's principle; water is on the reactant side of the equation for the hydrolysis equilibrium. Because the concentrations of solute ions in saturated solutions of slightly soluble salts are low, hydrolysis is significant for salts of weak acids and salts of weak bases.

Complex Ion Formation

The solubility of slightly soluble salts and hydroxides is also often increased by complex ion formation. For example, a small excess of Cl^- shifts the equilibrium between undissolved AgCl and ions in solution

$$AgCl(s) \rightleftharpoons Ag^+(aq) + Cl^-(aq)$$

to the left, and more silver chloride precipitates, an example of the common ion effect. However, AgCl(s) dissolves in the presence of a large excess of Cl^- as a result of the formation of the dichloroargentate(I) ion:

$$Ag^+(aq) + 2Cl^-(aq) \rightleftharpoons AgCl_2^-(aq)$$

dichloroargentate(I) ion

Formation of the $AgCl_2^-$ ion removes Ag^+ from solution and shifts the solubility equilibrium to the right. More solid AgCl dissolves. The higher the value of the formation constant, K_f, for the complex ion, the lower the concentration of uncomplexed cation that remains in solution.

Solution of Insoluble Compounds by Brønsted–Lowry Acid–Base Reactions

We have just seen that some slightly soluble salts and metal hydroxides can be dissolved by complex ion formation, a Lewis acid–base reaction. Brønsted–Lowry acid–base reactions can also be used to dissolve some insoluble compounds. For example, $CaCO_3(s)$ is insoluble in water but soluble in acids. The net ionic equation for the solution of calcium carbonate in an acid is

$$CaCO_3(s) + 2H^+(aq) \longrightarrow Ca^{2+}(aq) + H_2O(l) + CO_2(g)$$

Formation of the weak electrolyte H_2O and the gas CO_2 removes carbonate ions from solution and shifts the equilibrium,

$$CaCO_3(s) \rightleftharpoons Ca^{2+}(aq) + CO_3^{2-}(aq)$$

to the right.

However, not all insoluble compounds can be dissolved by Brønsted–Lowry acid–base reactions. Silver chloride, for example, is not dissolved by nitric acid because chloride ion is the conjugate base of the strong acid hydrochloric acid and silver nitrate is a soluble, strong electrolyte. No reaction takes place either between a hydrogen ion and a chloride ion or between a silver ion and a nitrate ion, the two possible combinations of oppositely charged ions (see Section 4.3):

$$H^+(aq) + Cl^-(aq) \longrightarrow \text{no reaction}$$
$$Ag^+(aq) + NO_3^-(aq) \longrightarrow \text{no reaction}$$

The equilibrium

$$AgCl(s) \rightleftharpoons Ag^+(aq) + Cl^-(aq)$$

is not shifted by the addition of nitric acid (except by the salt effect, which is relatively small unless the concentration of ions is high).

Calcium carbonate, which is not soluble in water, dissolves in hydrochloric acid because carbon dioxide gas escapes.

PRACTICE PROBLEMS

16.17   Which of the following slightly soluble compounds are more soluble in dilute nitric acid than they are in pure water? (a) $MgCO_3$ (b) $Ca(OH)_2$ (c) AgBr (d) ZnS

16.18   Write net ionic, complete ionic, and molecular equations for (a) solution of $Mg(OH)_2(s)$ in HCl(aq), (b) solution of AgCl(s) in NH_3(aq) (*Hint:* Refer to Table 16.4.) (c) solution of AgCl(s) in excess HCl(aq).

Incomplete Dissociation and Ion-Pair Formation

If a salt or hydroxide is not completely dissociated into ions in solution, the salt or hydroxide will be more soluble than predicted from the solubility product. The situation is similar to the situation that exists in a saturated solution of an insoluble weak acid such as benzoic acid. Benzoic acid, C_6H_5COOH, is a molecular compound; the units of solid benzoic acid are benzoic acid molecules. In a saturated solution of benzoic acid, undissolved benzoic acid molecules are in equilibrium with both dissolved benzoic acid molecules and with hydrogen ions and benzoate ions, $C_6H_5COO^-$, in solution. In addition, dissolved benzoic acid molecules are in equilibrium with hydrogen ions and benzoate ions:

$$C_6H_5COOH(s) \rightleftarrows C_6H_5COOH(aq)$$

$$C_6H_5COO^-(aq) + H^+(aq)$$

A solution that is saturated with benzoic acid at 25 °C is 0.028 M. However, only 5% of the dissolved benzoic acid is ionized; 95% is in the form of molecules. The concentrations of hydrogen ions and benzoate ions are 0.0014 M, and the concentration of benzoic acid molecules is 0.027 M. Thus, the solubility measured by the product of the concentration of the ions in solution would be much lower than the actual solubility.

Mercury(II) iodide, HgI_2, provides a good example of excess solubility resulting from incomplete dissociation. It is a predominantly molecular compound and is a weak electrolyte. The value of K_{sp} for HgI_2 is 3×10^{-29}. From the value of K_{sp}, the solubility of HgI_2 is calculated to be 2×10^{-10} M. Actually, the solubility of HgI_2 is 2×10^{-4} mol/L, about a million times the calculated value.

The formation of ion pairs in solution can cause the solubility of ionic compounds to be greater than predicted by K_{sp}. An **ion pair** is *two touching oppositely charged ions that are held together by the attraction between the unlike charges.* Ion-pair formation takes place to a greater extent the higher the charge on the oppositely charged ions and the smaller the ions are. Therefore, ion pairing is not important for salts consisting of singly charged ions such as silver bromide, and the solubilities of these salts are quantitatively related to the value of K_{sp} provided that neither anion nor cation hydrolyzes significantly.

See Martin, R. B. J. Chem. Educ. 1986, 63, 471–472.

Data are from Russo, S. O.; Hanania, G. I. H. J. Chem. Educ. 1989, 66, 148–153.

The most studied example of a salt that is more soluble than predicted from its solubility product as a result of ion pairing is calcium sulfate. For $CaSO_4$, $K_{sp} = 7 \times 10^{-5}$ at 25 °C, and the solubility calculated from K_{sp} is 0.008 M. But a solution that is saturated with $CaSO_4$ at 25 °C is actually 0.016 M, twice as concentrated as predicted. In a saturated solution of calcium sulfate at 25 °C, only 70% of the dissolved calcium sulfate is ionized; the remaining 30% of the dissolved calcium sulfate is in the form of ion pairs.

In solutions of slightly soluble salts (and metal hydroxides) such as calcium sulfate, at least three equilibria exist at the same time—one equilibrium between undissolved solid and ions in solution, another between undissolved solid and ion pairs in solution, and a third equilibrium between ion pairs in solution and ions in solution. For example, in an aqueous solution of calcium sulfate, solid calcium sulfate is in equilibrium with ion pairs in solution. Molecular formulas with charges will be used to represent ion pairs:

$$CaSO_4(s) \rightleftarrows \underset{\text{ion pair}}{Ca^{2+}SO_4{}^{2-}(aq)}$$

Ion pairs in solution are in equilibrium with ions in solution:

$$Ca^{2+}SO_4{}^{2-}(aq) \rightleftarrows Ca^{2+}(aq) + SO_4{}^{2-}(aq)$$

Undissolved solid is also in equilibrium with ions in solution:

$$CaSO_4(s) \rightleftarrows Ca^{2+}(aq) + SO_4{}^{2-}(aq)$$

In summary, the following three equilibria exist together in a saturated solution of $CaSO_4(s)$:

$$CaSO_4(s) \rightleftharpoons Ca^{2+}SO_4^{2-}(aq)$$
$$Ca^{2+}(aq) + SO_4^{2-}(aq)$$

*For calculations showing that hydrolysis equilibria are negligible for $CaSO_4$, see Meites, L., Pode, J., Thomas H. J. J. Chem. Educ. **1966,** 43, 667–672.*

For calcium sulfate, only these three equilibria are important. However, for many solutes, hydrolysis equilibria and complex ion equilibria must also be considered.

PRACTICE PROBLEM

16.19 What is the most likely reason why the observed solubility is many powers of 10 greater than the solubility calculated from K_{sp} for most sulfides?

The solubility product constant K_{sp} really describes the concentration of ions in solution, not the total solubility. However, *although the relationship between solubility and solubility product constant is quantitative for only a few salts, calculations involving K_{sp} are still useful in many cases when only the order of magnitude need be known and are widely used.*

16.7   CALCULATIONS INVOLVING K_{sp}

In the Solubility Equilibria module of the Equilibrium section students can explore behavior of slightly soluble ionic compounds. They engage in manipulations involving K_{sp} and Q_{sp} and predict precipitate formation.

Ionic substances are not soluble in nonpolar solvents such as hexane (Section 13.2). But it is difficult to predict whether or not salts and metal hydroxides will be soluble in water except by the empirical rules given in Table 4.4 because of the competing factors—interionic attractions in the solid, hydration of ions, entropy of solution—that are involved when an ionic solid dissolves in water.

In Section 4.3, we learned that a precipitation reaction will take place if, when solutions of electrolytes are mixed, at least one of the possible products is insoluble. The solubility rules in Table 4.4 were used to make predictions. The solubility product constant can be used to calculate answers to questions such as "Will a precipitate form if 20.0 mL of 0.032 M $AgNO_3$ are mixed with 15.0 mL of 0.041 M NaBr?" more precisely. The reaction quotient Q (Section 14.4) is used. Remember:

If $Q < K_c$, net forward reaction will take place.

If $Q = K_c$, no net change will take place (system is at equilibrium).

If $Q > K_c$, net reverse reaction will take place.

Because the "reaction quotient" for a solubility equilibrium is a product, Q *values for solubility equilibria* are given the special name **ion product** and represented by the symbol Q_{sp}:

If $Q_{sp} < K_{sp}$, solid will dissolve until $Q_{sp} = K_{sp}$.

If $Q_{sp} = K_{sp}$, no change will take place (solution is saturated).

If $Q_{sp} > K_{sp}$, solid will precipitate until $Q_{sp} = K_{sp}$.

For example, in a solution in which $[Ag^+] = 7 \times 10^{-5}$ and $[Br^-] = 6 \times 10^{-5}$,

$$Q_{sp} = [Ag^+][Br^-] = (7 \times 10^{-5})(6 \times 10^{-5}) = 4 \times 10^{-9}$$

A precipitate will form because $4 \times 10^{-9} > 5 \times 10^{-13}$, the solubility product for AgBr at 25 °C. In practice, the value of Q_{sp} must be about 1000 times K_{sp} for the

A Tyndall effect (Section 13.8) can be observed when the ion product is only ten times K_{sp}.

TABLE 16.6	Solubility Product Constants at 25 °C		
Compound	K_{sp}	Compound	K_{sp}
AgCl	1.6×10^{-10}	Fe(OH)$_2$	8×10^{-16}
AgBr	5×10^{-13}	Hg$_2$Br$_2$	6×10^{-23}
AgC$_2$H$_3$O$_2$	1.8×10^{-3}	Hg$_2$Cl$_2$	1.4×10^{-18}
Ag$_2$C$_2$O$_4$	5×10^{-12}	Hg$_2$SO$_4$	7×10^{-7}
AgI	9×10^{-17}	Li$_2$CO$_3$	1.1×10^{-3}
AgSCN	1.0×10^{-12}	Mn(OH)$_2$	1.7×10^{-13}
Ag$_2$SO$_4$	1.0×10^{-5}	NiCO$_3$	1.2×10^{-7}
BaCrO$_4$	8×10^{-11}	PbBr$_2$	1.2×10^{-5}
BaSO$_4$	1.1×10^{-10}	PbCl$_2$	1.5×10^{-5}
CaCO$_3$	5×10^{-9}	PbCO$_3$	1.2×10^{-13}
CaF$_2$	1.4×10^{-10}	SrSO$_4$	3×10^{-7}
Ca(OH)$_2$	5×10^{-6}	SrF$_2$	4×10^{-9}
CaSO$_4$	7×10^{-5}		

precipitate to be visible. Both precipitation and solution may take place slowly, and enough time must be allowed for equilibrium to be reached.

The values of K_{sp} for some salts and hydroxides for which solubilities in water calculated from K_{sp} are within an order of magnitude of observed solubilities are given in Table 16.6. Sample Problems 16.5–16.7 illustrate three of the types of calculations that can be done using K_{sp}.

SAMPLE PROBLEMS

16.5   Will a precipitate form if 20.0 mL of 0.032 M AgNO$_3$ are mixed with 15.0 mL of 0.041 M NaBr at 25 °C?

Solution   For a precipitate to form, at least one of the possible new combinations of the ions, Ag$^+$, NO$_3^-$, Na$^+$, and Br$^-$, must be insoluble. Silver bromide, AgBr, is insoluble (Table 4.4); sodium nitrate, NaNO$_3$, is soluble. If a precipitate is to form, the precipitate will have to be AgBr. The equilibrium involved in this question is

$$AgBr(s) \rightleftharpoons Ag^+(aq) + Br^-(aq) \qquad K_{sp} = 5 \times 10^{-13}$$

We must calculate the concentrations of Ag$^+$ and Br$^-$ in the mixture and see if the ion product is greater than, equal to, or less than K_{sp}.

Step 1.   *Calculation of moles of AgNO$_3$ and NaBr used:*

$$20.0 \text{ mL AgNO}_3(aq)\left(\frac{1 \text{ L AgNO}_3(aq)}{1000 \text{ mL AgNO}_3(aq)}\right)\left(\frac{0.032 \text{ mol AgNO}_3}{\text{L AgNO}_3(aq)}\right)$$
$$= 6.4 \times 10^{-4} \text{ mol AgNO}_3$$

$$15.0 \text{ mL NaBr}(aq)\left(\frac{1 \text{ L NaBr}(aq)}{1000 \text{ mL NaBr}(aq)}\right)\left(\frac{0.041 \text{ mol NaBr}}{\text{L NaBr}(aq)}\right)$$
$$= 6.2 \times 10^{-4} \text{ mol NaBr}$$

Step 2.   *Calculation of [Ag$^+$] and [Br$^-$] in the mixture:* The total volume of the mixture is

$$20.0 \text{ mL} + 15.0 \text{ mL} = 35.0 \text{ mL} = 0.0350 \text{ L}$$

$$[Ag^+] = \frac{6.4 \times 10^{-4} \text{ mol Ag}^+}{0.0350 \text{ L}} = 0.018$$

$$[Br^-] = \frac{6.2 \times 10^{-4} \text{ mol Br}^-}{0.0350 \text{ L}} = 0.018$$

Silver bromide precipitating as solutions of silver nitrate and sodium bromide are mixed.

Step 3. *Calculation of the ion product and comparison with K_{sp}:*

$$\text{ion product} = [Ag^+][Br^-] = (0.018)(0.018) = 3.2 \times 10^{-4}$$

Since $3.2 \times 10^{-4} > 5 \times 10^{-13}$, a precipitate will form.

16.6 (a) If $CaF_2(s)$ is precipitated from a solution that contains Ca^{2+} by adding F^-, what $[Ca^{2+}]$ will remain in solution when $[F^-]$ reaches 1.0×10^{-4}? Assume a temperature of 25 °C.

Solution The equilibrium involved in this question is

$$CaF_2(s) \rightleftharpoons Ca^{2+}(aq) + 2F^-(aq) \qquad K_{sp} = 1.4 \times 10^{-10}$$

In any saturated solution of CaF_2,

$$[Ca^{2+}][F^-]^2 = 1.4 \times 10^{-10}$$

Solving for $[Ca^{2+}]$ and substituting gives

$$[Ca^{2+}] = \frac{(1.4 \times 10^{-10})}{(1.0 \times 10^{-4})^2} = 1.4 \times 10^{-2}$$

▶ No matter how high the $[F^-]$, $[Ca^{2+}]$ cannot be reduced to zero because $[Ca^{2+}][F^-]^2$ must equal K_{sp}.

16.7 What is the molar solubility of AgBr in 0.10 M NH_3 at 25 °C? The solution is buffered with NH_4^+ to keep $[OH^-]$ formed by the reaction of NH_3 with water

$$NH_3(aq) + H_2O(l) \rightleftharpoons NH_4^+(aq) + OH^-(aq)$$

low so that AgOH does not precipitate.

Solution Two equilibria are important to the solution of this problem:

$$AgBr(s) \rightleftharpoons Ag^+(aq) + Br^-(aq) \qquad K_{sp} = 5 \times 10^{-13}$$
$$Ag^+(aq) + 2NH_3(aq) \rightleftharpoons Ag(NH_3)_2^+(aq) \qquad K_f = 1.7 \times 10^7$$

▶ When a large excess of ligand is present such that the steps in the formation of the complex ion can be neglected, calculations involving complex ions are greatly simplified.

Addition of these two equilibria gives the overall equation for the solution of $AgBr(s)$ in NH_3 (aq)

$$AgBr(s) + 2NH_3(aq) \rightleftharpoons Ag(NH_3)_2^+(aq) + Br^-(aq) \qquad (16.8)$$

Therefore, the equilibrium constant for the overall reaction is equal to the product of K_{sp} and K_f (Section 16.1):

$$K = K_{sp} \cdot K_f = (5 \times 10^{-13})(1.7 \times 10^7) = 9 \times 10^{-6}$$

Now proceed as usual and let x equal mol/L $Ag(NH_3)_2^+$ and Br^- formed. Then $2x$ equals the mol/L of NH_3 reacted. The summary table is:

	$AgBr(s) +$	$2NH_3(aq) \rightleftharpoons$	$Ag(NH_3)_2^+(aq) +$	$Br^-(aq)$
Init. conc., M		**0.10**	**0.00**	**0.00**
Change, ΔM		$-2x$	$+x$	$+x$
Eq. conc., M		$(0.10 - 2x)$	x	x

Substitution in the equilibrium constant expression for the equilibrium shown by equation 16.8 gives

$$9 \times 10^{-6} = \frac{x^2}{(0.10 - 2x)^2} \qquad (16.9)$$

Because both numerator and denominator on the right are perfect squares, the easiest way to solve equation 16.9 is to take the square root of both sides:

$$\sqrt{9 \times 10^{-6}} = \frac{x}{(0.10 - 2x)}$$

Solution for x gives

$$x = 3 \times 10^{-4}$$

A precipitate of silver bromide can be dissolved by adding a concentrated solution of ammonia.

and $[Ag(NH_3)_2^+] = [Br^-] = 3 \times 10^{-4}$. The molar solubility of AgBr in 0.10 M NH_3 at 25 °C is 3×10^{-4}. This answer is reasonable because 3×10^{-4} M is larger than the solubility of AgBr in pure water, 7×10^{-7} M (Sample Problem 16.4).

PRACTICE PROBLEMS

16.20 What $[Br^-]$ must be used to reduce $[Ag^+]$ to 6×10^{-10}?

16.21 Will a precipitate form if 25.0 mL of 2×10^{-4} M $(NH_4)_2C_2O_4$ are mixed with 75.0 mL of 3×10^{-5} M $AgNO_3$ at 25 °C?

16.22 What is the molar solubility of AgI in each of the following at 25 °C?
(a) Pure water (b) 0.010 M HNO_3 (c) 0.50 M NH_3 (d) 0.025 M $AgNO_3$

16.8 SOME PRACTICAL APPLICATIONS OF SOLUBILITY EQUILIBRIA

Analytical Chemistry

The concentration of Cl^- in water can be determined by precipitating the Cl^- by addition of an excess of $AgNO_3(aq)$:

$$Cl^-(aq) + Ag^+(aq) + NO_3^-(aq) \longrightarrow AgCl(s) + NO_3^-(aq)$$

An excess of $AgNO_3(aq)$ is used to shift the equilibrium

$$AgCl(s) \; \underset{\longleftarrow}{\overset{\longrightarrow}{}} \; Ag^+(aq) + Cl^-(aq)$$

in favor of AgCl(s) by the common ion effect so that all but a negligible amount of the Cl^- is precipitated. The AgCl(s) is collected by filtration, washed to remove excess Ag^+ and NO_3^-, and dried. From the mass of AgCl(s) formed, the amount of Cl^- in the sample can be calculated.

Geology

Most groundwater is acidic because it contains acids formed by the decay of vegetable matter. When acidic groundwater flows through limestone, $CaCO_3(s)$, the limestone dissolves:

$$CaCO_3(s) + 2H^+(aq) \longrightarrow Ca^{2+}(aq) + H_2O(l) + CO_2(g)$$

If the limestone deposit is near the surface, a sinkhole results. If the limestone deposit is well below the surface, a cave is formed.

Water that contains dissolved CO_2 is acidic (see Section 16.3) and reacts with $CaCO_3(s)$. This reaction is reversible:

$$CaCO_3(s) + CO_2(aq) + H_2O(l) \rightleftharpoons Ca^{2+}(aq) + 2HCO_3^-(aq) \qquad (16.10)$$

As groundwater containing Ca^{2+} and HCO_3^- ions drips through the ceiling of a cave, the water and carbon dioxide evaporate. Equilibrium is shifted to the left, and $CaCO_3(s)$ precipitates; stalactites and stalagmites are formed. ◾Figure 16.4 shows a cave with stalactites and stalagmites.

Hard water contains dissolved $Ca(HCO_3)_2$. When hard water is heated, carbon dioxide is lost because gases are less soluble in hot water than in cold water. The

◾ **FIGURE 16.4** Stalactites and stalagmites are formed when water and carbon dioxide evaporate from groundwater that contains Ca^{2+} and HCO_3^- ions.

equilibrium shown in equation 16.10 is shifted to the left, and a deposit of $CaCO_3(s)$ is formed inside steam generators, hot water pipes, hot water heaters, and teakettles.

Industrial Chemistry

Sodium hydrogen carbonate that is 99.9% pure (baking soda) is manufactured by bubbling carbon dioxide gas through a saturated solution of sodium carbonate:

$$Na_2CO_3(satd) + H_2O(l) + CO_2(g) \longrightarrow 2NaHCO_3(s)$$

The $NaHCO_3$ is collected by filtration, washed, and dried. Drying must be carried out at a low temperature so that the reaction

$$2NaHCO_3(s) \longrightarrow Na_2CO_3(s) + H_2O(g) + CO_2(g)$$

does not take place.

Pure barium sulfate is also made by precipitation:

$$BaCl_2(aq) + Na_2SO_4(aq) \longrightarrow BaSO_4(s) + 2NaCl(aq)$$

Barium sulfate is a white pigment used in the paint industry.

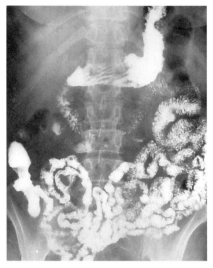

An abdominal X-ray taken after the patient has drunk a suspension of barium sulfate. The stomach and intestines are white and can be seen clearly because barium sulfate is opaque to X-rays.

Medicine

Barium sulfate is also used when the digestive tract is x-rayed. The patient is given a suspension of $BaSO_4(s)$ in $Na_2SO_4(aq)$ by mouth long enough before the X-rays are taken for the $BaSO_4(s)$ to reach the digestive tract. Because $BaSO_4(s)$ is opaque to X-rays, the digestive tract can be seen clearly in the photographs. Although barium ion, Ba^{2+}, is very poisonous, the solubility of $BaSO_4(s)$ in $Na_2SO_4(aq)$ is so low that the $[Ba^{2+}]$ in solution is not dangerous to the patient. The excess of sulfate ions from the sodium sulfate shifts the barium sulfate solubility equilibrium

$$BaSO_4(s) \rightleftharpoons Ba^{2+}(aq) + SO_4^{2-}(aq)$$

to the left by the common ion effect.

Metallurgy

You are undoubtedly familiar with aluminum metal; like plastics, aluminum is everywhere in modern life. It is the most widely used metal after iron. The principal ore of aluminum is bauxite, a mixture of hydrated aluminum oxides, $Al_2O_3 \cdot nH_2O$. Bauxite is usually open-pit mined and reduced to aluminum electrolytically. However, bauxite commonly contains Fe_2O_3 and SiO_2 as well as Al_2O_3 and must be purified before reduction. Treatment of bauxite with hot concentrated sodium hydroxide solution dissolves the acidic oxide, SiO_2, and the amphoteric oxide, Al_2O_3,

$$SiO_2(s) + 2OH^- + 2H_2O(l) \longrightarrow Si(OH)_6^{2-}$$
$$Al_2O_3(s) + 2OH^- + 3H_2O \longrightarrow 2Al(OH)_4^-$$

but not the basic oxide, Fe_2O_3, which is separated by filtration. Treatment of the filtrate with carbon dioxide gas precipitates Al_2O_3 but leaves $Si(OH)_6^{2-}$ in solution:

$$2Al(OH)_4^-(aq) + CO_2(g) \longrightarrow Al_2O_3(s) + CO_3^{2-}(aq) + 4H_2O(l)$$

After separation by filtration, the purified aluminum oxide is ready for reduction to aluminum metal.

Whether the equilibrium between an acid and a base lies to the right or to the left can be predicted from the positions of the acid and base in a table of K_a values. The stronger acid and stronger base react to form the weaker acid and the weaker base. The value of the equilibrium constant can be calculated from the values of the equilibrium constants for the two weak acids by making use of the fact that the equilibrium constant for the sum of two or more equations is the product of the equilibrium constants for the original equations.

The strengths of acids and bases in aqueous solutions are a result of a combination of factors of which bond polarity, bond strength, and solvation are the most important. Fortunately, a few empirical rules that relate acidity to structure happen to work reasonably well at room temperature. The acidity of the binary compounds (compounds consisting of two elements) of hydrogen increases from left to right across rows and from top to bottom down groups in the periodic table. The hydrides of the reactive metals of Groups IA and IIA are strong bases; the hydrides of many nonmetals are acids.

The hydroxides of the reactive metals are also strong bases. The hydroxides of nonmetals are acidic, and the hydroxides of elements near the metal–nonmetal borderline are amphoteric. The acidity of oxo acids is related to the electronegativity of the central atom and to the number of extra oxygens attached to the central atom. The acidity of organic acids is increased by substitution of more electronegative atoms for hydrogens close to the —COOH group.

Hydrated cations also act as Brønsted–Lowry acids. If charge and electronegativity are similar, the smaller the radius of the ion, the more acidic the hydrated cation is. The acidity of hydrated cations decreases down groups of the periodic table. If radii and electronegativities are similar, the higher the charge, the more acidic the hydrated cation. If charge and radius are similar, the greater the electronegativity of the metal, the more acidic the hydrated cation.

The basicity of anionic and cationic weak bases can be deduced from the acidity of their conjugate acids. Anions that contain one or more ionizable hydrogens, such as HCO_3^-, are amphoteric as are species like $Al(H_2O)_5(OH)^{2+}$.

Water **levels** the acidity of strong acids and bases. A solvent that shows differences in acidity between acids or differences in basicity between bases is called a **differentiating solvent.**

Most metal oxides are basic. Oxides of some metals near the metal–nonmetal borderline are amphoteric. Most nonmetal oxides are acidic, and many nonmetal oxides react with water to form oxo acids. When an element forms more than one oxide, the oxide with the higher proportion of oxygen reacts with water to form a stronger acid. A few nonmetal oxides are neither acidic nor basic. Acidic and basic oxides often react directly to form salts.

Lewis acids are not limited to compounds that contain hydrogen. A **Lewis acid** is a species that can attach to an electron pair to form a covalent bond. A Lewis acid must have an empty valence orbital for the electron pair to go into. Species with incomplete octets, small cations with high charges, and species with multiple bonds are Lewis acids. A **Lewis base** is a species that has an unshared electron pair that can form a covalent bond. A Lewis base must have an unshared pair of valence electrons. Lewis acid–base reactions can be regarded as substitution reactions in which one Lewis base takes the place of another Lewis base.

Complexes are species in which a central atom is surrounded by Lewis bases that are covalently bonded to the central atom. The Lewis bases are usually referred to as **ligands.** Complexes can be either ions or molecules. Formation of complexes other than aqua complexes in water solution involves substitution of other ligands for water; substitution of other ligands for water takes place in steps and is reversible. The values of the equilibrium constants for successive steps decrease although differences between equilibrium constants are small and solutions usually contain mixtures of complexes. Complex ion equilibria are too complicated to deal with quantitatively in general chemistry unless an excess of the ligand is present so that a single species is predominant.

The **solubility product constant** or **solubility product, K_{sp},** is equal to the product of the concentrations of the ions in a saturated solution of a slightly soluble salt, each raised to the power of the coefficient in the balanced equation. For a few slightly soluble salts and hydroxides, solubility is quantitatively related to the value of the equilibrium constant. For most, actual solubilities are higher than solubilities calculated from K_{sp}. For some, the observed solubility is many powers of 10 greater than the calculated solubility. The common ion effect, the salt effect, side reactions such as hydrolysis and complex ion formation, and incomplete dissociation and ion-pair formation are important in solubility equilibria. A **salt effect** is an increase in solubility caused by the presence in the solution of ions that are not common ions.

Some slightly soluble salts and metal hydroxides can be dissolved by complex ion formation. Others can be dissolved by Brønsted–Lowry acid–base reactions.

For information about the organization of Additional Practice Problems, Stop & Test Yourself, Putting Things Together, and Applications, see the beginnings of these sections in Chapter 1.

16.23 (a) Arrange the following weak acids in order of decreasing strength:

(i) $HC_3H_5O_3$, $K_a = 1.4 \times 10^{-4}$
(ii) $HC_4H_7O_2$, $K_a = 1.5 \times 10^{-5}$
(iii) $HClO_2$, $K_a = 1.2 \times 10^{-2}$
(iv) $HOBr$, $K_a = 2 \times 10^{-9}$
(v) $HOCN$, $K_a = 1.2 \times 10^{-4}$

(b) Of the five acids, which has the strongest conjugate base? (16.1)

16.24 For each of the following equilibria, predict whether equilibrium lies to the right or to the left. Explain your answers.

(a) $NO_2^-(aq) + C_6H_5COOH(aq) \rightleftharpoons$
$$HNO_2(aq) + C_6H_5COO^-(aq)$$
(b) $HSO_4^-(aq) + HCOO^-(aq) \rightleftharpoons$
$$HCOOH(aq) + SO_4^{2-}(aq)$$
(c) $CH_3CH_2COO^-(aq) + C_6H_5OH(aq) \rightleftharpoons$
$$CH_3CH_2COOH(aq) + C_6H_5O^-(aq)$$
(d) $HCl(aq) + OH^-(aq) \rightleftharpoons H_2O(l) + Cl^-(aq)$
(e) $CH_3CH_2O^-(aq) + CH_3COOH(aq) \rightleftharpoons$
$$CH_3CH_2OH(aq) + CH_3COO^-(aq)$$
(f) $NH_3(aq) + H_2O(l) \rightleftharpoons NH_4^+(aq) + OH^-(aq)$ (16.1)

16.25 For equilibria (a) and (b) in Problem 16.24, calculate the values of the equilibrium constants. (16.1)

16.26 For each of the following equilibria, predict whether equilibrium lies to the right or to the left:

(a) $H_2CO_3(aq) + SO_4^{2-}(aq) \rightleftharpoons$
$$HCO_3^-(aq) + HSO_4^-(aq)$$
(b) $H_2CO_3(aq) + CO_3^{2-}(aq) \rightleftharpoons 2HCO_3^-(aq)$
(c) $CH_3COOH(aq) + HPO_4^{2-} \rightleftharpoons$
$$CH_3COO^-(aq) + H_2PO_4^-(aq)$$
(d) $H_2SO_4(aq) + OH^-(aq) \rightleftharpoons HSO_4^-(aq) + H_2O(l)$
(e) $H_2S(aq) + HPO_4^{2-}(aq) \rightleftharpoons HS^-(aq) + H_2PO_4^-(aq)$
(f) $C_6H_5NH_3^+(aq) + CN^-(aq) \rightleftharpoons$
$$C_6H_5NH_2(aq) + HCN(aq)$$ (16.1)

16.27 For equilibria (b) and (f) in Problem 16.26, calculate the values of the equilibrium constants. (16.1)

16.28 From Table 16.3, for Ca^{2+}, radius = 100 pm, electronegativity = 1.0 and pK_a = 12.8. Which of the following is the value of pK_a for Hg^{2+} (radius = 102 pm, electronegativity = 1.4)? Explain your answer. (a) 3.4 (b) 12.8 (c) 13.5 (16.2)

16.29 Which member of each of the following pairs would you expect to be more acidic? Explain your choices.

(a) $HBrO_3$ or $HOBr$ (b) HOI or $HOBr$ (c) H_2SO_3 or HSO_3^- (d) H_2O or H_2S (e) HF or H_2O (f) $Fe(H_2O)_6^{3+}$ or $Zn(H_2O)_6^{2+}$ (g) $CH_3NH_3^+$ or $C_6H_5NH_3^+$ (h) $Al(OH)_3$ or $ClOH$ (16.2)

16.30 Classify each of the following oxides as acidic, basic, or neutral. (a) CO (b) CO_2 (c) MgO (d) SO_3 (e) SnO (16.2)

16.31 Would you expect bismuth hydroxide, $Bi(OH)_3$, and bismuth oxide, Bi_2O_3, to be acidic, basic, or neutral? Explain your answer. (16.2)

16.32 (a) Explain what is meant by the statement "Water levels the acid strength of strong acids such as HCl and HBr and of strong bases such as H^- and NH_2^-." (b) To detect a difference in basicity between H^- and NH_2^-, should you use a substance that is a stronger or a weaker acid than water as solvent? Explain your answer. (c) What is a solvent like the one in part (b) called? (16.2)

16.33 Which member of each of the following pairs would you expect to be more basic? Explain your choices.

(a) NaH or ClH (b) CH_3COO^- or CN^- (c) CH_3COO^- or ClO_4^- (d) $Al(H_2O)_5OH^{2+}$ or $Zn(H_2O)_5OH^+$ (e) $ClCH_2COO^-$ or $BrCH_2COO^-$ (16.2)

16.34 In aqueous solution, copper(II) ion exists as $Cu(H_2O)_6^{2+}$. Write an equation showing why solutions of copper(II) ion are acidic. (16.2)

16.35 What are each of the following called? (a) $HBrO_3$ (b) HIO_4 (c) HOF (16.2)

16.36 Write molecular equations for each of the following reactions: (a) $P_2O_5(s)$ with water (the product is phosphoric acid) (b) $ZnCO_3(s)$ with dilute HNO_3 (c) $CaO(s)$ with water (d) $Na_2S(s)$ with water. If appropriate, also write complete and net ionic equations. (16.2)

16.37 (a) Define the terms *Lewis acid* and *Lewis base* and give an example of each. (b) What must be present for a species to act as a Lewis acid? A Lewis base? (16.3)

16.38 Explain why F^- can act as a Lewis base but not as a Lewis acid. (16.3)

16.39 Of the following species—H_3CCH_3, Cl^-, Fe^{3+}, $AlCl_3$, $(CH_3)_2O$—(a) which will probably act as Lewis acids? (b) Which will probably act as Lewis bases? (c) Which will probably not act as either a Lewis acid or a Lewis base? (16.3)

16.40 Each of the following reactions is a Lewis acid–base reaction. For each identify the Lewis acid and the Lewis base.

(a) $BF_3(g) + H_2O(l) \longrightarrow BF_3 \cdot H_2O(s)$
(b) $SO_3(g) + K_2O(s) \longrightarrow K_2SO_4(s)$
(c) $Ag^+ + 2CN^- \longrightarrow Ag(CN)_2^-$
(d) $Zn(OH)_2(s) + 2OH^- \longrightarrow Zn(OH)_4^{2-}$
(e) $(CH_3)_3C^+ + OH^- \longrightarrow (CH_3)_3COH(aq)$ (16.3)

16.41 Picture the formation of H_2SO_4 as a Lewis acid–base reaction between H_2O and SO_3. (16.3)

16.42 (a) Define the term *complex* and give an example of a complex ion. (b) In the complex ion in part (a), label the central atom and the ligand. (16.4)

16.43 Which of the following ions form complex ions? Explain your answer. (a) Ba^{2+} (b) Cu^{2+} (c) Fe^{3+} (d) K^+ (e) Ni^{2+} (16.4)

16.44 For the formation of the complex ion, $Ag(NH_3)_2^+$, (a) write the equation for each step and its equilibrium constant expression (b) Write the equation for the overall reaction and its equilibrium constant expression. (16.4)

16.45 (a) What is the value of the equilibrium constant for the decomposition of $Ag(NH_3)_2{}^+$?

$$Ag(NH_3)_2{}^+(aq) \rightleftharpoons Ag^+(aq) + 2NH_3(aq)$$

(b) What is the value of the equilibrium constant for the decomposition of $Ag(CN)_2{}^-$? (c) Is the $Ag(NH_3)_2{}^+$ ion more or less stable than the $Ag(CN)_2{}^-$ ion? Explain your answer. (16.4)

16.46 The equilibrium constants for the steps in the formation of $Cd(NH_3)_4{}^{2+}$ are step 1: 4×10^2; step 2: 1×10^2; step 3: 2×10^1; and step 4: 7×10^0. What is the overall formation constant? (16.4)

16.47 Given a 0.1 M solution of Ag^+, should you add Cl^-, CN^-, or NH_3 to lower $[Ag^+]$ the most? Explain your answer. (16.4)

16.48 Explain the following observations by writing equations for all reactions that take place: Addition of a few drops of concentrated $NH_3(aq)$ to a solution of $CuSO_4$ yields a light blue precipitate. As more $NH_3(aq)$ is added, the precipitate dissolves and the solution becomes dark blue. (16.4)

16.49 When sodium hydroxide solution is added gradually to a solution that contains Zn^{2+}, a precipitate forms and then redissolves. Explain. (16.4)

16.50 What is $[Ag^+]$ in a solution that is 0.050 M in $Ag(NH_3)_2{}^+$? (16.4)

16.51 If 0.50 mol $NH_3(g)$ is added to 1.0 L 0.021 M $AgNO_3(aq)$, what will $[Ag^+]$ be at equilibrium? Assume that the volume does not change. (16.4)

16.52 (a) What $[NH_3]$ is required to reduce $[Ag^+]$ in 0.030 M $AgNO_3$ to 1.0×10^{-5}? (b) What $[CN^-]$ is required to reduce $[Ag^+]$ in 0.030 M $AgNO_3$ to 1.0×10^{-5}? (16.4)

16.53 What is $[Cu^{2+}]$ in a solution made by mixing 20.0 mL of 0.034 M $CuSO_4$ with 30.0 mL of 0.150 M NH_3? (16.4)

16.54 In a solution that is 0.045 M in Ag^+, what $[CN^-]$ is needed to reduce $[Ag^+]$ to 4.5×10^{-5}? (16.4)

16.55 Explain why the concentration of the slightly soluble solid does not appear in the equilibrium constant expressions for solubility equilibria. (16.5)

16.56 If the solubility product expression is $K_{sp} = [Ca^{2+}]^3[PO_4{}^{3-}]^2$, what is the equation for the solubility equilibrium? (16.5)

16.57 How many liters of water are needed to dissolve 1.00 g of AgCl at 25 °C? (16.5)

16.58 For each of the following slightly soluble compounds, write the equation for the equilibrium that exists in a saturated solution and the solubility product expression: (a) $Cd(IO_3)_2$ (b) Cu_2S (like the oxide ion, the sulfide ion is the conjugate of a very weak acid and is completely hydrolyzed in aqueous solution) (c) $Fe(OH)_3$ (d) $BaCO_3$ (e) $Ca_3(PO_4)_2$ (16.5)

16.59 In a saturated solution of AgCl at 25 °C, $[Cl^-] = 3.0 \times 10^{-7}$. What is $[Ag^+]$? (16.5)

16.60 At 25 °C, (a) what is the molar solubility of $Ca(OH)_2$? (b) The solubility in g/100 cm³ of water? (16.5)

16.61 According to the *CRC Handbook,* the solubility of TlCl is 0.29 g/100 cm³ water at 15.6 °C. What is the value of K_{sp} for TlCl at 15.6 °C? (16.5)

16.62 From the data in Table 16.5, (a) how would you expect the salt effect in a saturated solution of $BaSO_4$ to compare with the salt effect in a saturated solution of $CaSO_4$? (b) Is ion pairing as important in aqueous solutions of $BaSO_4$ as it is in aqueous solutions of $CaSO_4$? Explain your reasoning. (c) Suggest two reasons for the difference in the extent of ion pairing between $CaSO_4$ and $BaSO_4$. (16.6)

16.63 Which of the following compounds will be more soluble in acid than in pure water? Explain your answers. (a) $BaCO_3$ (b) $Cd(OH)_2$ (c) $PbCl_2$ (d) MnS (e) AgI (16.6)

16.64 For which of the following compounds will hydrolysis of the anion contribute to solubility? Explain your answer. (a) MnS (b) $Mn(OH)_2$ (c) $PbCl_2$ (d) Ag_3PO_4 (16.6)

16.65 Write net ionic, complete ionic, and molecular equations for the solution of (a) $Al(OH)_3$ in excess NaOH(aq) (b) ZnO(s) in HCl(aq) (c) $Cu(OH)_2$(s) in NH_3(aq) (16.6)

16.66 How does the solubility of $CaCO_3$ in pure water compare with the solubility of $CaCO_3$ in (a) a solution containing a fairly high concentration of Ca^{2+}? (b) A solution buffered at pH = 3? (c) A solution buffered at pH = 11? (d) A solution containing a fairly high concentration of NaCl? Explain your answers. (16.6)

16.67 Explain the following facts about the solubility of AgCN: (a) Why is the solubility in acidic solutions greater than the solubility in pure water? (b) Why is the solubility in solutions containing a fairly high concentration of cyanide ion greater than the solubility in pure water? (16.6)

16.68 Which is more soluble, AgCl or CaF_2? (16.7)

16.69 If a solution is 4×10^{-1} M in Pb^{2+} and 2×10^{-3} M in Cl^- at 25 °C, will a precipitate of $PbCl_2$ form? (16.7)

16.70 How many moles of NH_3 must be added to 1.0 L of water to dissolve 0.010 mol AgCl(s)? (16.7)

16.71 Calculate the value of K for the equilibrium

$$AgI(s) + 2CN^-(aq) \rightleftharpoons Ag(CN)_2{}^-(aq) + I^-(aq)$$

from the value of K_{sp} for AgI and K_f for $Ag(CN)_2{}^-$. (16.7)

16.72 (a) Will AgCl precipitate from a 0.023 M solution of $AgNO_3$ that is also 0.020 M in NaCl? (b) What is $[Ag^+]$ in a 0.023 M solution of $AgNO_3$ that is also 0.50 M in NH_3? (c) Will AgCl precipitate from a 0.023 M solution of $AgNO_3$ that is also 0.020 M in NaCl *and* 0.50 M in NH_3? (16.7)

16.73 Picture each of the following reactions as a substitution reaction in which one Lewis base takes the place of another Lewis base. Label the Lewis bases. (a) $O^{2-} + H_2O \rightarrow 2OH^-$ (b) $NH_2{}^- + H_2O \rightarrow NH_3 + OH^-$ (c) $CH_3I + OH^- \rightarrow CH_3OH + I^-$ (d) $CH_3COOH + H_2O \rightarrow CH_3COO^- + H_3O^+$.

16.74 Would the equilibrium similar to

$$C_6H_5OH(aq) + H_2O(l) \rightleftharpoons$$
$$C_6H_5O^-(aq) + H_3O^+(aq) \qquad K_a = 1.3 \times 10^{-10}$$

that exists when phenol is dissolved in NH_3(l) have K greater than, equal to, or less than K_a? Explain your answer.

16.75 If $AgNO_3$(s) is added slowly to a solution that is 0.010 M in KI and 0.015 M in NaCl, (a) which will precipitate first, AgCl or AgI? (b) What will $[I^-]$ be when AgCl begins to precipitate? What percentage of the original $[I^-]$ is still in solution? Assume that addition of $AgNO_3$(s) does not change the volume of the solution.

1. Which of the following equilibria lies to the left?
   (a) $CH_3COOH(aq) + CN^-(aq) \rightleftharpoons$
   $$CH_3COO^-(aq) + HCN(aq)$$
   (b) $CH_3CH_2COOH(aq) + SO_4^{2-}(aq) \rightleftharpoons$
   $$CH_3CH_2COO^-(aq) + HSO_4^-(aq)$$
   (c) $ClCH_2COOH(aq) + OCl^-(aq) \rightleftharpoons$
   $$ClCH_2COO^-(aq) + HOCl(aq)$$
   (d) $HF(aq) + NH_3(aq) \rightleftharpoons F^-(aq) + NH_4^+(aq)$
   (e) All of them lie to the right.

2. What is the value of K for the following equilibrium?
   $$HNO_2(aq) + CH_3CH_2COO^-(aq) \rightleftharpoons$$
   $$CH_3CH_2COOH(aq) + NO_2^-(aq)$$
   (a) 6.0×10^{-9} (b) 4.6×10^{-4} (c) 2.3×10^1
   (d) 3.5×10^1 (e) 7.7×10^4

3. Arrange the acids HCl, HF, and H_2O in order of decreasing
   strength: (a) $HCl > H_2O > HF$
   (b) $HF > H_2O > HCl$ (c) $HF > HCl > H_2O$
   (d) $H_2O > HF > HCl$ (e) $HCl > HF > H_2O$

4. Arrange the acids H_2SO_3, HSO_3^-, and H_2SO_4 in order of decreasing strength:
   (a) $H_2SO_4 > HSO_3^- > H_2SO_3$
   (b) $HSO_3^- > H_2SO_4 > H_2SO_3$
   (c) $H_2SO_3 > HSO_3^- > H_2SO_4$
   (d) $HSO_3^- > H_2SO_3 > H_2SO_4$
   (e) $H_2SO_4 > H_2SO_3 > HSO_3^-$

5. Arrange the bases $ClCH_2COO^-$, CH_3COO^-, and FCH_2COO^- in order of increasing strength:
   (a) $ClCH_2COO^- < FCH_2COO^- < CH_3COO^-$
   (b) $CH_3COO^- < ClCH_2COO^- < FCH_2COO^-$
   (c) $FCH_2COO^- < ClCH_2COO^- < CH_3COO^-$
   (d) $CH_3COO^- < FCH_2COO^- < ClCH_2COO^-$
   (e) $FCH_2COO^- < CH_3COO^- < ClCH_2COO^-$

6. Which of the following solutions is most acidic? All are 0.1 M.
   (a) $NaNO_3$ (b) $Cr(NO_3)_3$ (c) $Ni(NO_3)_2$
   (d) $Ba(NO_3)_2$ (e) All are neutral.

7. Which of the following oxides is basic? (a) BaO (b) CO
   (c) CO_2 (d) SO_2 (e) Cl_2O

8. Which of the following *cannot* act as a Lewis acid?
   (a) O^{2-} (b) $Al(OH)_3$ (c) CO_2 (d) Fe^{3+} (e) BBr_3

9. If 0.45 mol NaCN is added to 1.0 L of 0.025 M $FeSO_4$, what
   will be the $[Fe^{2+}]$? Assume that no change in volume occurs.
   (a) 2.8×10^{-37} (b) 1.4×10^{-35} (c) 1.4×10^{-34}
   (d) 1.4×10^{-2} (e) 2.8×10^{43}

10. If the value of K_{sp} for $Zn(OH)_2$ is 8×10^{-17} at 25 °C, $Zn(OH)_2$ is
    completely ionized in solution, and the solution behaves ideally,
    what is the molar solubility of $Zn(OH)_2$ in pure water at 25 °C?
    (a) 6×10^{-9} (b) 3×10^{-6} (c) 4×10^{-6}
    (d) 9×10^{-6} (e) 1.3×10^{-5}

11. What is the molar solubility of AgCl in 0.050 M $AgNO_3$ solution?
    (a) 8.0×10^{-12} (b) 3.2×10^{-11} (c) 1.6×10^{-10}
    (d) 3.2×10^{-9} (e) 1.3×10^{-5}

12. What is the molar solubility of AgCl in 0.15 M HCl?
    (a) 0.15 (b) 0.075 (c) 1.1×10^{-9} (d) 1.3×10^{-5}
    (e) 4.3×10^{-6}

13. Which of the following slightly soluble compounds is *not* soluble
    in dilute nitric acid? (a) $PbCO_3$ (b) $PbCl_2$
    (c) $Pb(OH)_2$ (d) PbS (e) All of them are soluble in dilute
    nitric acid.

14. If equal volumes of the following pairs of solutions are mixed,
    which mixture will give a precipitate?
    (a) (1.6×10^{-5}) M $AgNO_3$ and (4.0×10^{-5}) M NaCl
    (b) (8.0×10^{-6}) M $AgNO_3$ and (4.0×10^{-5}) M NaCl
    (c) (1.0×10^{-4}) M $AgNO_3$ and (1.4×10^{-4}) M NaCl
    (d) none of them (e) all of them

16.76 Referring to the data for the sulfides in Table 16.5, (a) arrange the four sulfides listed in order of decreasing value of K_{sp} at 18 °C. (b) For each of the four sulfides, calculate the ratio of observed solubility to solubility calculated from K_{sp}. (c) Explain the trend observed for the ratios.

16.77 Which of the following hydroxides are soluble in sodium hydroxide solution as well as in hydrochloric acid solution?
   (a) $Zn(OH)_2$ (b) $Mg(OH)_2$ (c) $Al(OH)_3$ (d) $Fe(OH)_3$
   Explain your answers.

16.78 Use the Lewis definitions of acids and bases to explain the formation of the $Ag(NH_3)_2^+$ ion.

16.79 What is the solubility of AgCl in a solution made by dissolving 5.0 g $CaCl_2$ in water and diluting to 1.00 L?

16.80 Describe a simple test for distinguishing between
   (a) AgCl(s) and $AgNO_3$(s) (b) $AgNO_3$(aq) and
   $NaNO_3$(aq) (c) HCl(aq) and HNO_3(aq).

16.81 Write an equation for the reaction that will take place between each of the following pairs of substances. If no reaction will take place, write "no reaction."
   (a) NaH(s) + H_2O(l) (b) CaH_2(s) + H_2O(l)
   (c) $NaNH_2$(s) + H_2O(l) (d) Na_2O(s) + H_2O(l)
   (e) SO_2(g) + H_2O(l) (f) NO(g) + H_2O(l)

16.82 Write an equation for the reaction that will take place between each of the following pairs of substances. If no reaction will take place, write "no reaction."
   (a) CH_4(g) + H_2O(l) (b) NH_3(g) + H_2O(l)
   (c) HCl(g) + H_2O(l) (d) CaO(s) + H_2O(l)
   (e) ZnO(s) + H_2O(l) (f) ZnO(s) + HCl(aq)

16.83 Write an equation for the reaction that will take place between each of the following pairs of substances. If no reaction will take place, write "no reaction."
   (a) $Al(OH)_3$(s) + HCl(aq) (b) $Al(OH)_3$(s) + NaOH(aq)

(c) $NaHCO_3(s) + H_2O(l)$ (d) $NaCl(s) + H_2O(l)$

(e) $NH_4Cl(s) + H_2O(l)$ (f) $CaO(s) + SO_3(g)$

16.84 Write a molecular equation for the reaction that might occur if aqueous solutions of sodium cyanide and acetic acid are mixed. Does the reaction take place to any significant extent? Explain your answer.

16.85 Use Lewis formulas to show that each of the following reactions is a Lewis acid–base reaction and identify the Lewis acid and the Lewis base in each case.

(a) $Ni^{2+} + 6NH_3(aq) \longrightarrow Ni(NH_3)_6^{2+}$

(b) $CaO(s) + CO_2(g) \longrightarrow CaCO_3(s)$

(c) $(CH_3)_2S(l) + BBr_3(l) \longrightarrow (CH_3)_2S \cdot BBr_3(s)$

(d) $2PCl_5(g) \longrightarrow [PCl_4^+][PCl_6^-](s)$

(e) $2AlCl_3(g) \longrightarrow Al_2Cl_6(l)$

16.86 (a) In the formation of the triiodide ion, $I^-(aq) + I_2(aq) \longrightarrow I_3^-(aq)$, which species acts as the Lewis acid? As the Lewis base? (b) In the triiodide ion, how are the electron pairs distributed around the central I? Explain your answer. (c) What is the shape of the triiodide ion?

16.87 Compare the advantages and disadvantages of the Lewis and Brønsted–Lowry acid–base definitions.

16.88 Use Le Châtelier's principle to explain each of the following observations: (a) $AgCl(s)$ is soluble in $NH_3(aq)$. (b) $AgCl(s)$ precipitates when a solution that contains $Ag(NH_3)_2^+$ and Cl^- is acidified with dilute nitric acid. (c) $AgCl(s)$ precipitates when a solution that contains $AgCl_2^-$ is diluted.

16.89 When $Al(NO_3)_3$ dissolves in water, both a Lewis acid–base reaction and a Brønsted–Lowry acid–base reaction take place. Write the equation for each reaction.

16.90 Explain the meaning of saturated, unsaturated, and supersaturated solutions in terms of solubility equilibria and K_{sp}.

16.91 If equal volumes of 4×10^{-6} M $AgNO_3$ and 8×10^{-8} M NaBr are mixed, will a precipitate form? If so, write net ionic, complete ionic, and molecular equations.

16.92 Explain why $Mg(OH)_2$, which is insoluble in water, is soluble in $NH_4Cl(aq)$.

16.93 If a solution is 4.2×10^{-6} M in Ag^+ and 5.5×10^{-5} M in Cl^-, what will be the concentration of each ion after precipitation is complete?

16.94 (a) What is the molar solubility of $Ca(OH)_2$ in pure water at 25 °C? (b) What is the pH of a saturated solution of $Ca(OH)_2$? (c) What is the molar solubility of $Ca(OH)_2$ in a pH 2.0 buffer at 25 °C?

16.95 (a) How many milliliters of $HCl(g)$ at 764 mmHg and 21.8 °C must be added to 254 mL of a 3.6×10^{-8} M solution of $AgNO_3$ in order for $AgCl(s)$ to begin to precipitate? Assume that the $HCl(g)$ behaves ideally and the volume of the solution does not change significantly. (b) How many milliliters of 0.050 M $HCl(aq)$? (c) How many milligrams of $NaCl(s)$?

16.96 For each of the following slightly soluble compounds, write the equation for the equilibrium that exists in a saturated solution and the solubility product expression: (a) iron(II) sulfide (b) lead(II) bromide (c) silver sulfide (d) chromium(III) phosphate (e) cobalt(II) phosphate (the sulfide ion is completely hydrolyzed in aqueous solution).

16.97 Explain why, when CO_3^{2-} is added to a solution containing both Ca^{2+} and Mg^{2+} that is buffered with NH_3–NH_4Cl, $CaCO_3$ precipitates but $MgCO_3$ does not precipitate. According to the *CRC Handbook*, K_{sp} for $MgCO_3$ is 1.15×10^{-5}. The value of K_{sp} for $CaCO_3$ is 5×10^{-9}.

16.98 The compound Hg_2Br_2 is composed of Hg_2^{2+} ions and Br^- ions. (a) What is the molar solubility of this compound in 0.10 M NaBr? (b) What is the name of this compound? (c) Suppose that a very conscientious student who is asked to find the molar solubility of AgBr in 0.10 M NaBr wants to get an exact instead of an approximate answer and uses a programmable calculator set to disply ten digits (the maximum number possible) to solve the quadratic formula. What will happen?

16.99 Write the formula for each of the following: (a) potassium chlorite (b) calcium hypochlorite (c) ammonium perchlorate (d) sodium chlorate (e) aluminum bromate nonahydrate.

16.100 (a) For selenic acid, H_2SeO_4, and telluric acid, H_6TeO_6, write Lewis formulas. (b) One of these acids is a strong acid like sulfuric acid, and the other is weak. Which is which? Explain your answer. (c) Predict the shapes of the selenic acid and telluric acid molecules.

16.101 (a) Write Lewis formulas for periodic acid, H_5IO_6, and iodic acid, HIO_3. (b) Which is the stronger acid? Explain your answer. (c) Predict the shapes of the periodic acid and iodic acid molecules.

16.102 From Table 16.5, the observed solubility of $CaCO_3(s)$ is 0.0014 g/100 cm³ water at 25 °C. Calculate $[HCO_3^-]$ and $[CO_3^{2-}]$ in a saturated solution of $CaCO_3(s)$.

16.103 In Section 15.4, you learned that the ionization of a weak acid is never complete even at infinite dilution. Silver chloride is a weak electrolyte yet the solubility of silver chloride calculated from K_{sp} is approximately the same as the observed solubility. Explain.

16.104 (a) Write the electron configuration for the Al^{3+} ion. (b) Predict the geometry of the $Al(OH)_4^-$ ion. (Draw a stereochemical formula.)

16.105 Write net ionic, complete ionic, and molecular equations for the following reactions: (a) Nickel(II) hydroxide dissolves in hydrochloric acid. (b) Iron(II) carbonate dissolves in hydrochloric acid. (c) Zinc sulfide dissolves in hydrochloric acid. (d) Aqueous cobalt(III) chloride is treated with an excess of aqueous ammonia. (e) Solid aluminum hydroxide dissolves in excess aqueous sodium hydroxide. (f) Mixing of solutions of silver nitrate and potassium chromate, K_2CrO_4, gives a red precipitate.

16.106 When 3.09 g $KOH(s)$ is added to 500 (5.00×10^2) mL of 0.050 M $FeCl_2$, (a) what mass of $Fe(OH)_2$ will precipitate? (b) After the $Fe(OH)_2$ is collected by filtration, what $[Fe^{2+}]$ remains in the filtrate? (c) What is the pH of the filtrate?

16.107 Bismuth sulfide, Bi_2S_3, has an extremely small value for K_{sp}, 1.6×10^{-72} at 25 °C. (a) How many ions does 1.0 mL of a saturated solution of Bi_2S_3 contain? (Remember that the sulfide ion is completely hydrolyzed in aqueous solution.) (b) Mercury(II) sulfide has a much larger (although still very small) value for K_{sp}. The value of K_{sp} for HgS is 4×10^{-54} at 25 °C. How many ions does 1.0 mL of a saturated solution of HgS solution contain? (c) The strengths of weak acids can easily be compared by comparing the values of K_a—for example, acetic acid with $K_a = 1.8 \times 10^{-5}$ is stronger than phenol with $K_a = 1.3 \times 10^{-10}$. Can the total concentrations of ions in saturated solutions of slightly soluble solids be easily judged by comparing the values of K_{sp}? Explain your answer.

16.108 A calorimeter had a heat capacity of 46 J/°C. When 50.0 mL of 0.500 M $AgNO_3$(aq) and 50.0 mL of 0.500 M NaBr(aq) were mixed in the calorimeter, an off-white precipitate formed, and the temperature rose to 27.1 °C. The original temperature of the calorimeter and both solutions was 22.6 °C. (a) Write ionic, net ionic, and molecular equations for the reaction that took place. (b) Assuming that the specific heat of the solution after reaction took place was 4.18 J/g·°C and that the densities of the two solutions were 1.00 g/mL, what was ΔH_{rxn} in kJ/mol?

16.109 What is the difference between the members of each of the following pairs of terms? (a) solubility product and ion product (b) concentration and activity (c) saturated solution and supersaturated solution (d) common ion effect and salt effect (e) Brønsted–Lowry acid and Lewis acid

16.110 (a) Draw a microscopic view showing the equilibrium between NaCl(s) and NaCl(aq) in a saturated solution of sodium chloride. To keep your drawing simple, do *not* show molecules of the solvent, water. (b) Why are K_{sp} values not tabulated for soluble salts such as sodium chloride? (c) If undissolved solid is separated from the saturated solution by filtration, is the filtrate still a saturated solution? Does an equilibrium exist in the filtrate? If HCl(aq) is added to the filtrate, what will be observed? If a crystal of NaCl(s) is added to the filtrate, what will be observed?

APPLICATIONS

16.111 Lime (calcium oxide) is fifth on the Top 50 list for 1993. (a) The chief use of lime is for steelmaking. In what kind of reaction is lime involved? (b) Because the thermal energy released when lime is slaked (reacts with water) can set fire to paper, wood, and plastic containers, slaked lime is sold for many uses. Calculate ΔH_{rxn}° for the slaking of lime. (c) A growing use of slaked lime is removal of sulfur dioxide from the gases produced by burning high-sulfur coal. Calcium sulfite is formed. Write the equation for this reaction.

16.112 In the production of sulfuric acid, sulfur trioxide is absorbed in sulfuric acid. The product is disulfuric acid, $H_2S_2O_7$. Show this reaction as a Lewis acid–base reaction.

16.113 The reaction between Ca(OH)$_2$(s) in mortar and CO$_2$(g) from air is involved in the setting of mortar. Picture this reaction as a Lewis acid–base reaction.

16.114 Formation of the soluble cyanide complex, $Au(CN)_2^-$, is used to extract gold from its ores. Gold(I) cyanide is only very slightly soluble. Write the equations for the steps in the formation of $Au(CN)_2^-$ and the formation constant expressions.

16.115 The recovery of magnesium from seawater depends on the position of the equilibrium

$$Ca(OH)_2(s) + Mg^{2+} \rightleftharpoons Mg(OH)_2(s) + Ca^{2+}$$

What is the value of K for this equilibrium? According to the *CRC Handbook*, the value of K_{sp} for Mg(OH)$_2$ is 5.66×10^{-12}.

16.116 Chickens do not have sweat glands and cool themselves by panting, which lowers the level of carbon dioxide in the chicken's blood. Eggshells are 95% calcium carbonate and too much panting results in thin egg shells that break easily. Explain in terms of the equilibrium

$$H_2O(l) + CO_2(g) \rightleftharpoons 2H^+(aq) + CO_3^{2-}(aq)$$

and suggest a simple solution to the problem.

16.117 In addition to Cr_2O_3, which is amphoteric, an oxide of chromium, CrO_3, exists. (a) Is this other oxide of chromium probably acidic or basic? Explain your answer. (b) Name both these oxides of chromium. (c) A third oxide of chromium, called chromium(IV) oxide, also is known. Write the formula for chromium(IV) oxide. Chromium(IV) oxide is used to make magnetic recording tapes, and Cr_2O_3 is widely used as a green pigment.

16.118 Chloride ion can be determined quantitatively by titration with standard silver nitrate solution using chromate ion, CrO_4^{2-}, as indicator. Silver chloride is white; silver chromate is brick red. Silver chromate is more soluble than silver chloride and does not start to precipitate until after "all" the chloride ion has been precipitated. The red color of silver chromate signals the end point. (a) Can all the chloride ion really be precipitated—that is, can [Cl$^-$] be reduced to zero? Explain your answer. (b) At the equivalence point in the titration of Cl$^-$ with Ag$^+$, what is the relationship between [Cl$^-$] and [Ag$^+$]? (c) What is [Cl$^-$] at the equivalence point? (d) What [CrO_4^{2-}] is required for precipitation of Ag_2CrO_4 to start at the equivalence point? According to the *CRC Handbook*, K_{sp} for silver chromate is 1.11×10^{-12} at 25 °C.

16.119 Silver ion can be determined quantitatively by titration with a standard solution of thiocyanate, SCN$^-$. A solution of iron(III) ion is used as indicator; the red complex ion, $FeSCN^{2+}$, forms at the end point. (a) At the equivalence point in the titration of Ag$^+$ with SCN$^-$, what is the relationship between [SCN$^-$] and [Ag$^+$]? (b) What is [SCN$^-$] at the equivalence point? (c) To be visible, [Fe(SCN)$^{2+}$] must equal 6.5×10^{-6}. What [Fe^{3+}] should be used?

16.120 In the hardest natural waters, [Ca^{2+}] = 4×10^{-4}. When water is fluoridated to prevent tooth decay, not more than 1 ppm of F$^-$ is added. Remember that a "ppm" is one part per million by mass; 1 mass % = 1 pph (part per hundred).

Will calcium fluoride precipitate when hard water is fluoridated?

16.121 The silver nitrate solution in a qualitative analysis laboratory was prepared by dissolving 17 g AgNO₃ in water and diluting to one liter. For a precipitate to be visible, the ion product must be at least 1000 times as large as K_{sp}. If the tapwater in the laboratory turns cloudy when the silver nitrate solution is added, the molarity of Cl⁻ in the tapwater must be at least what?

16.122 The safe disposal of many chemicals requires chemical reactions. The directions for the disposal of barium acetate say to add an excess of dilute sulfuric acid, let stand overnight, separate insoluble material by filtration, and bury it in a landfill site approved for hazardous-waste disposal. (a) Write molecular, complete ionic, and net ionic equations for the reaction involved in this procedure. (b) What do you think is the purpose of letting the mixture stand overnight? (c) How could you test to be sure you had used an excess of dilute sulfuric acid?

16.123 Boric acid, H₃BO₃, which is used as an astringent and antiseptic, is not a Brønsted–Lowry acid. Instead, boric acid is a Lewis acid. (a) Use Lewis structures to show the reaction of boric acid with water. (b) pK_a for boric acid is 9.2. What is K_a? (c) Calculate the pH of a 0.020 M aqueous solution of boric acid. (d) Boric acid–borate buffers occur in nature and are used as pH standards and in detergents. What is the pH of a buffer that is 0.020 M in boric acid and 0.020 M in borate ion?

16.124 Limewater, a saturated solution of Ca(OH)₂, is often added to cream that has been separated from whole milk to reduce acidity before pasteurization and conversion to butter. Limewater is usually prepared by stirring Ca(OH)₂(s) with water until no more will dissolve. The undissolved Ca(OH)₂ is then allowed to settle and the clear supernatant liquid is siphoned off. (a) Does an equilibrium exist in the saturated solution that has been siphoned off? Explain your answer. (b) Explain why, when CO₂(g) is bubbled through limewater, a precipitate forms that then redissolves.

16.125 Calcite and aragonite are two different crystalline forms of calcium carbonate. Coral is calcite, and pearls are aragonite. For calcite $K_{sp} = 5 \times 10^{-9}$, and for aragonite, $K_{sp} = 7 \times 10^{-9}$ at 25 °C. (a) Which is more soluble at 25 °C, calcite or aragonite? Explain your answer. (b) What is the value of K for the equilibrium

$$CaCO_3(\text{calcite}) \rightleftharpoons CaCO_3(\text{aragonite})$$

at 25 °C? (c) Cheap "coral" beads are made of red gypsum, CaSO₄·2H₂O. Suggest a simple test for distinguishing between real and fake coral.

16.126 One of the simplest methods for determining Cl⁻ quantitatively is precipitation of AgCl by gradual addition of AgNO₃ to an aqueous solution of the sample, collection of the AgCl by filtration, washing, drying, and determining the mass of the precipitate. (a) If the [Cl⁻] remaining after precipitation of AgCl must be 2.99×10^{-7} M or less, what is the minimum [Ag⁺] needed? (b) If not more than 9.65×10^{-5} g AgCl may be lost by solution in the wash water,

what is the maximum volume of water (in mL) that should be used for washing? (Assume that equilibrium will be reached.) (c) During the addition of much of the AgNO₃, the precipitate is colloidal. Around the equivalence point, the precipitate coagulates so that it can be separated by filtration, but if too large an excess of AgNO₃ is added, the precipitate again becomes colloidal. The colloidal particles at the end have the opposite charge from the colloidal particles at the beginning. What ions are absorbed on the colloidal AgCl particles at the beginning? At the end?

16.127 When photographic film is fixed, unexposed silver bromide is removed by solution in sodium thiosulfate solution. (The formula for sodium thiosulfate is Na₂S₂O₃.) A soluble complex ion, Ag(S₂O₃)₂³⁻, is formed. (a) What mass of Na₂S₂O₃ is needed to prepare 1.0 L of a solution that will dissolve 0.14 g AgBr, the quantity left on a roll of film after exposure? (b) "Hypo" is actually Na₂S₂O₃·5H₂O. How many grams of Na₂S₂O₃·5H₂O are needed?

16.128 Make a diagram similar to the one in Figure H.1 in the Appendix showing how purified aluminum oxide is separated from bauxite.

16.129 Acid rain has been recognized as an environmental problem for a number of years. Recently, acid aerosols have been identified as a health hazard. (a) What is an aerosol? (b) List the five acids present in acid rain. Acid aerosols contain the same five acids. (c) Why is the acidity of acid rain and acid aerosols due mainly to just two of the five acids? (d) The reactions of CO₂(g), SO₂(g), and SO₃(g) with water are Lewis acid–base reactions. What type of reaction is the reaction of NO₂(g) with water?

16.130 Phenol was the first antiseptic used by Lister in 1865–1869. During this period, the surgical mortality in his ward dropped from 45 to 15%. (a) The value of K_a for ethyl alcohol, CH₃CH₂OH, is 1.3×10^{-16}. What is the value of K for the following equilibrium?

$$C_6H_5OH(aq) + CH_3CH_2O^- \rightleftharpoons$$
$$C_6H_5O^- + CH_3CH_2OH(aq)$$

(b) Write Lewis formulas for the two most important resonance structures for phenol. (c) What is the hybridization of the carbon atom to which the —OH is attached in phenol? In ethyl alcohol?

16.131 Hydrothermal vents are places under the oceans where fluids rise through Earth's crust. (a) At hydrothermal vents relatively near the ocean surface, water coming out of vents contains bubbles of gas. At deep vents, there are no gas bubbles. Explain this observation. (b) Snail shells are made of calcium carbonate. Vent fluids can have pH as low as 2.8; near such vents snails are "naked." Explain why.

16.132 Tooth enamel consists mainly of Ca₅(PO₄)₃OH. Cavities are caused by acids:

$$Ca_5(PO_4)_3OH(s) + 4H^+ \longrightarrow 5Ca^{2+} + 3HPO_4^{2-} + H_2O(l)$$

If fluoride is present, Ca₅(PO₄)₃OH is converted to Ca₅(PO₄)₃F, and decay is prevented. Why is the latter compound less soluble in acids?

A Lawyer Uses Chemistry

L. GENE SPEARS, JR.

Associate, Arnold, White, & Durkee,
Houston, Texas

B.A. Chemistry and Chemical Physics
Rice University

J.D., *University of Texas College of Law*

My father is a chemistry professor and my mother is a high school chemistry teacher. Although neither of them ever directly suggested that I become a chemist, I was naturally drawn in that direction. In high school, my math and science courses were generally more challenging, and thus more interesting, than the other courses I took. I felt that I was actually learning something in my math and science classes.

My first chemistry course in college was organic chemistry. I found it somewhat boring and very difficult. I believe that many college students become disenchanted with chemistry because the freshman- and sophomore-level courses are so dry. But it is important to recognize that these courses are not "real chemistry." Real chemistry is in the laboratory and is best appreciated through undergraduate research. Fortunately, Rice had a number of professors who included undergraduates in their research groups.

I started research as soon as possible—the summer after my freshman year. While I was an undergraduate, I worked with two different professors in two very different areas of chemistry. Some of my research concerned proton and methyl transfers to and from organic phosphonates. My other work was a theoretical study of the reaction dynamics of trans-diimide. These research projects proved to be very rewarding. The faculty became "real people," rather than just figures in front of a classroom. Both of my research areas generated publications in major journals. The thrill of discovering something new and then later seeing your name in print attached to this discovery really cannot be equaled by any conventional undergraduate experience.

I decided that I wanted to become a professor of chemistry at a Texas university. Accordingly, I went to California for graduate school. Shortly after I arrived, I began to have second throughts about my chosen career. I returned to Texas and enrolled in law school.

My law school class was filled with business, economics, political science, and even engineering majors—but very few natural scientists. Therefore I was something of a curiosity. I have often been asked how my background in chemistry has helped or hurt me in my nascent legal career. I believe that it has helped.

Some have suggested that a scientist would make an effective lawyer because law, like science, involves some sort of objective, analytical search for truth. This is nonsense. One need only spend a few hours reading through some of the decisions that the U.S. Supreme Court has produced over the past two hundred years to conclude that law is more than the product of objective, systematic reasoning. Scientists follow the tradition of Aristotle. Law professors, and often judges as well, follow the tradition of the sophists. The law is something that must be accepted. Those who attempt to explain it in a rational, scientific manner only end up looking foolish.

Instead, it is in the nuts and bolts application of the law that I believe a chemistry background is helpful. We live in a world of microcomputers, nuclear reactors, airplanes, and countless human-made carcinogens. We also live in a nation where the regulatory bureaucracy is growing and the scope of regulation is expanding through private tort law. As a result, lawyers are increasingly being asked to give advice or to litigate in fairly technical areas.

I worked one summer for a law firm that specialized in defending so-called toxic tort lawsuits. In these suits, the plaintiffs seek recovery for injuries caused by exposure to toxic chemicals. I felt that I had an advantage in this work simply because I could look at terms like "butadiene" and "methyl ethyl ketone" without blinking. I could also identify possible sources of exposure and, with the assistance of some introductory texts, could understand the epidemiological and toxicological studies that are at the center of this sort of litigation.

The law firm where I work specializes in the law of intellectual property, which includes patents, copyrights, trademarks, and trade secrets. In prosecuting a patent application I must explain the technology in language that an educated nonspecialist can understand, distinguish it from earlier technologies, provide sufficient detail to create a strong patent, and avoid disclosing matters that the client would prefer to protect as a trade secret. To do this, I must understand the technology—almost as thoroughly as the inventor. When the validity of a patent is challenged in a lawsuit, the level of scrutiny can be higher than that encountered in the most rigorous scientific peer review. Millions, even billions of dollars may be at stake. Catalysts, polymers, pharmaceuticals, and industrial synthetic processes may all be the subject matter of patents. And the lawyer who works with such patents, if not a chemist, must often work with someone who is.

Patents and toxic torts are not the only legal areas where my chemistry background will be helpful. Because I have experience in one technical field, I am not intimidated by others. Although I could never work as a biologist, geologist, or electrical engineer, I can at least navigate the technical jargon of those fields with a reasonable effort. The areas of law that intersect with some technical field are numerous—oil and gas, environmental law, medical malpractice, engineering malpractice, and product liability are but a few examples. I look forward to always using in my legal career what I learned as a student of chemistry.

17

CHEMICAL THERMODYNAMICS REVISITED: A Closer Look at Enthalpy, Entropy, and Equilibrium

Some fires, like this forest fire, are very destructive.

Chemical thermodynamics is concerned with whether physical and chemical changes will take place spontaneously and how far a change will proceed before reaching equilibrium. *Thermodynamics does not provide any information about how fast spontaneous changes take place or about how reactions take place.*

■ *Why is a study of chemical thermodynamics a part of general chemistry?*

A key question in chemistry is "Will a physical or chemical change take place?"

Chemical thermodynamics provides part of the answer. Chemical thermodynamics tells whether a change can take place spontaneously and how far a change can go before reaching equilibrium. (Chemical kinetics, the subject of Chapter 18, tells whether a change will take place at a practical rate and deals with how reactions take place.)

■ *What do you already know about thermodynamics?*

You know that a spontaneous change is one that takes place by itself (Section 6.2). Energy may be needed to start a

spontaneous change—for example, natural gas must be lighted in order to burn. However, once started, a spontaneous change continues until the materials are used up or an equilibrium is reached. Once lighted, natural gas continues to burn until the gas is turned off (or the supply of oxygen runs out).

The opposite of a spontaneous change is nonspontaneous. A nonspontaneous change can be made to take place by supplying energy, but a nonspontaneous change stops if energy is no longer supplied. For example, liquid water can be made to boil under atmospheric pressure by heating, but as soon as the heating stops, the boiling stops. Water vapor spontaneously condenses to liquid water under ordinary laboratory conditions until the partial pressure of the water vapor has decreased to the equilibrium vapor pressure of water.

Most changes that take place spontaneously under ordinary conditions are exothermic—that is, they heat their surroundings. An increase in the entropy of the system must occur for an endothermic change to take place spontaneously. The system is the part of the universe in which you are interested (Section 6.1). The rest of the universe is referred to as the surroundings. For practical purposes, the surroundings are the apparatus used and the laboratory around it.

A large part of thermodynamics deals with the energy changes that accompany almost all physical and chemical processes. Chapter 6 dealt mainly with thermal energy changes. In Chapter 6, you learned that energy itself cannot be measured. Only differences in energy can be measured. The thermal energy gained or lost when a change takes place under constant pressure is called the enthalpy change, ΔH. Thermal energy is given off by the system in an exothermic change and gained by the system in an endothermic change. The thermal energy change that takes place in an exothermic change has a negative sign; the thermal energy change that takes place in an endothermic change has a positive sign. Enthalpy is a state function—that is, enthalpy changes depend only on the initial and final states of the system and not on how the system gets from one state to another.

The changes in enthalpy that accompany many physical and chemical changes can be determined experimentally in a calorimeter. Because, according to the law of conservation of energy, energy cannot be created or destroyed, enthalpy changes can be calculated by combining thermochemical equations (Hess's law). The most frequently used form of Hess's law is

$$\Delta H^{\circ}_{\text{rxn}} = \Sigma n_p \, \Delta H^{\circ}_f \text{ (product)} - \Sigma n_r \, \Delta H^{\circ}_f \text{ (reactant)}$$

where Σ means sum of, n_p is the number of moles of each product, ΔH°_f (product) is the standard enthalpy of formation of the product, n_r is the number of moles of each reactant, and ΔH°_f (reactant) is the standard enthalpy of formation of the reactant.

The standard enthalpy of formation of a substance, ΔH°_f, is the enthalpy change that results when one mole of the substance is formed from its elements with all substances in their standard states. For liquids and solids, the standard state is the pure liquid or pure solid under standard pressure of 1 atm. The standard state for a gas is the ideal gas at a partial pressure of 1 atm. The standard enthalpies of formation of the elements in their standard states are zero. Values of standard enthalpies of formation of compounds are obtained from tables (see Appendix D).

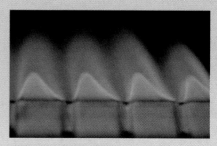

Other fires, such as this natural gas flame in a furnace, are very useful. Once started, all fires are spontaneous and exothermic.

17.1   THE LAWS OF THERMODYNAMICS

The law of conservation of energy (Section 6.4) is the first law of thermodynamics. A simple statement of the first law is that energy cannot be created or destroyed but only transferred from one body to another and changed from one form to another; that is, the energy of the universe is constant.

According to the second law of thermodynamics, every spontaneous change increases the entropy of the universe. The second law does not mean that local decreases in entropy cannot take place. However, if the entropy of a system decreases, the entropy of the surroundings must increase more. The second law makes possible prediction of the direction of change.

The first two laws are based on observations of properties of *macroscopic* samples of matter such as temperature, pressure, and volume. Thermodynamics was developed in the second half of the nineteenth century and the beginning of the twentieth century and, at that time, some scientists still did not believe in atoms. Therefore, the first two laws of thermodynamics are not based on any model of the microscopic nature of matter.

The third law was not stated in its modern form until 1923 (after the atomic theory was accepted and the quantum theory had been invented). The third law of thermodynamics says that the entropy of perfect crystalline substances (those with

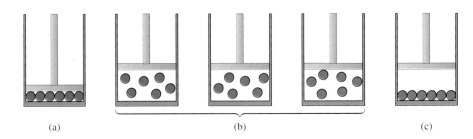

■ FIGURE 17.1 (a) Ideal gas at 0 K and 1 atm. (b) Ideal gas at temperature > 0 K and 1 atm. (c) One of many possible arrangements of molecules in larger volume. Is this arrangement likely?

no disorder) is zero at the absolute zero of temperature (0 K); however, it is impossible to reach absolute zero.

17.2 ENTROPY

German mathematician and physicist Rudolf Clausius introduced the idea of entropy as a measure of driving force for change in 1850. Every spontaneous change increases the entropy of the universe according to the second law of thermodynamics. Like the first law, the second law is accepted because conclusions based on it agree with experience. No exceptions have been observed.

A Kinetic-Molecular View of Entropy

Although the second law of thermodynamics was based on observations of properties of macroscopic samples of matter and was stated before the atomic and kinetic-molecular theories were widely accepted, the kinetic-molecular theory is very helpful in understanding entropy. Up to this point we have considered entropy to be a measure of disorder. The idea that an increase in entropy is a measure of an increase in disorder is very useful. However, it must be applied thoughtfully.

Let's begin by considering a sample of an ideal gas in a container with a movable piston. The piston is under the constant pressure of the atmosphere. If the temperature is lowered, the volume of the gas will decrease as described by Charles's law (Section 5.3). At absolute zero, the molecules will have no energy of motion. The only arrangement possible for the molecules at absolute zero is shown in ■Figure 17.1(a).* Now, if the temperature is raised, the molecules will begin to move. Because there is no attraction between the molecules of an ideal gas, the molecules will move apart and occupy a larger volume. As soon as the molecules occupy a larger volume, many different arrangements of the molecules are possible. Three of these arrangements are shown in Figure 17.1(b). Because many different arrangements are possible, the chance of the molecules being found in any one of them is very small. The original arrangement is only one of many possible arrangements. Therefore, the probability of the molecules returning to the original arrangement [see Figure 17.1(c)] is very small.

As temperature becomes higher, the volume will become still larger. The number of possible arrangements will increase even more. The probability of finding the molecules in any one of the arrangements will decrease. The probability of finding the molecules in a disordered arrangement will increase because so many disorderly arrangements are possible. The entropy of an ideal gas at constant pressure increases with increasing temperature because the volume of the gas increases.

Spreading anything throughout a larger volume increases the entropy of the system. For example, litter spread over the countryside has higher entropy than litter deposited in a container.

We have not used the popular analogy of a deck of new cards after being dropped to explain increase in entropy because it depends on the fact that each card is different from every other card, whereas all molecules of a substance are identical.

___

*The molecules of an ideal gas would not have any volume. But it is impossible to draw a picture of a particle with zero volume.

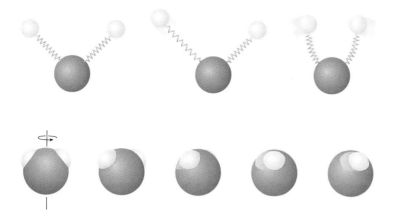

- FIGURE 17.2   Covalent bonds are really more like springs than sticks and stretch and bend.

- FIGURE 17.3   At any temperature above absolute zero, molecules rotate.

Another reason entropy increases as temperature increases is because the distribution of molecular speeds at high temperatures is broader than the distribution of molecular speeds at low temperatures (remember Figure 5.16). In addition, more energy levels in atoms and molecules are occupied at higher temperatures. Both atoms and molecules have many electronic energy levels (Sections 8.1 and 10.6). Molecules also have vibrational and rotational energy levels. The bonds in molecules are actually more like springs than sticks and can vibrate as shown in ▪Figure 17.2. Molecules can also rotate as shown in ▪Figure 17.3.

The entropy of a substance in the liquid state is lower than the entropy of the same substance in the gaseous state. The number of possible arrangements of molecules in the liquid state is smaller because the volume of the sample is smaller. In crystalline solids, the unit particles are fixed in an ordered arrangement and can only vibrate about their average positions. As a result, the entropy of a substance in the solid state is less than the entropy of the same substance in the liquid state.*

When two liquids are mixed to form an ideal solution, entropy increases. The volume of the solution is larger than the volumes of each of the pure substances. Therefore, the number of possible arrangements of molecules is greater in the solution. If two ideal gases are mixed in such a way that both occupy the same volume as each did before mixing [see ▪Figure 17.4(a)], there is no change in entropy. Only if the volume of the mixture is different from the volumes of the individual gases as shown in Figure 17.4(b) is there an entropy change when two ideal gases are mixed.

For a discussion of the common misconception that mixing causes an increase in entropy, see Meyer, E. J. Chem. Educ. 1987, 64, 676.

- FIGURE 17.4   (a) At constant temperature, if a mixture of ideal gases occupies the same volume as each of the individual gases occupied, entropy does *not* increase on mixing. (b) At constant temperature, if the volume of a mixture of ideal gases is greater than the volumes of each of the individual gases, entropy increases.

---

*For substances that can crystallize in more than one arrangement, such as carbon, the entropies of the different crystalline forms are often slightly different. For example, the entropy of carbon in the diamond structure is lower than the entropy of carbon in the graphite structure. The crystalline form that is harder usually has the lower entropy. The fact that a solid is hard shows that moving the unit particles is difficult.

Entropy of the System

According to the second law of thermodynamics, all spontaneous changes increase the entropy of the universe. However, the entropy of a system can either increase or decrease in a spontaneous change. If the entropy of the system decreases, the entropy of the surroundings must increase more so that the total entropy of the universe increases as required by the second law. For example, consider condensation of a gas to a liquid. The entropy of the system decreases when a gas condenses to a liquid. However, thermal energy is given off to the surroundings (remember Section 12.6), and the temperature of the surroundings increases, which results in an increase in the entropy of the surroundings. If the increase in the entropy of the surroundings is greater than the decrease in the entropy of the system, the entropy of the universe increases and condensation is spontaneous.

If a change results in an increase in the number of moles of gas as the reaction

$$CaCO_3(s) \longrightarrow CaO(s) + CO_2(g)$$

does, the change will bring about an increase in the entropy of the system. If a change results in a decrease in the number of moles of gas as the reverse reaction

$$CaO(s) + CO_2(g) \longrightarrow CaCO_3(s)$$

does, the change will bring about a decrease in the entropy of the system. If no change in the number of moles of gas accompanies a physical or chemical change, entropy may either increase or decrease. However, the change in entropy will be small. The reaction

$$H_2(g) + I_2(g) \longrightarrow 2HI(g)$$

is an example of a reaction that takes place without any change in the number of moles of gas.

Experimentally, entropies are usually determined by measuring the increase in temperature that results from the addition of a known amount of energy to a known quantity of sample in a calorimeter beginning at a temperature close to absolute zero. The determination of entropies is more complicated than the determination of enthalpy changes, and we will simply use the results of other people's work.

Standard Entropy

The *entropy of a substance in a standard state* is called **standard entropy, $S°$**. Standard entropy is the difference between the entropy of perfect crystals of the substance at absolute zero and the entropy of the substance in a standard state. (Remember that, for liquids and solids, the standard state is the pure liquid or pure solid under standard pressure of 1 atm. The standard state for a gas is the ideal gas at a partial pressure of 1 atm.) Tables of standard entropies for a large number of substances are available; some standard entropies at 25 °C are shown in Table 17.1 and more are given in Appendix D.

There are a number of points you should notice in Table 17.1. The units of standard entropy are J/K·mol. The magnitudes of entropy changes are usually small compared to the magnitudes of enthalpy changes; therefore, joules, not kilojoules, are used for tabulating entropies. Standard entropies depend on temperature as well as sample size.

The standard entropies of all substances are positive; elements as well as compounds have nonzero entropies. As long as substances are in the same physical state, entropy increases down groups in the periodic chart. For example, the entropies of the noble gases are in the order He < Ne < Ar < Kr, and the entropies of the

The reason why entropy increases with increasing atomic mass is beyond the scope of a general chemistry text.

TABLE 17.1	Some Standard Entropies at 25 °C (298 K)[a]						
Substance	$S°$, J/K·mol	Substance	$S°$, J/K·mol	Substance	$S°$, J/K·mol	Substance	$S°$, J/K·mol
Ar(g)	154.84	$CH_3OH(g)$	238	HI(g)	206.33	Na(s)	51.45
C(diamond)	2.37	$CH_3CH_2OH(l)$	161	$H_2O(l)$	69.91	Na(l)	57.85
C(graphite)	5.69	$CH_3CH_2OH(g)$	282	$H_2O(g)$	188.72	Na(g)	153.61
CaO(s)	39.75	$Cl_2(g)$	222.96	He(g)	126.15	NaCl(s)	72.12
$CaCO_3(s)$	92.2	CO(g)	197.91	$I_2(g)$	260.58	Ne(g)	146.33
$CH_4(g)$	186.2	$CO_2(g)$	213.64	K(s)	64.68	$NH_3(g)$	192.3
$C_2H_2(g)$	200.82	$H_2(g)$	130.68	Kr(g)	164.08	$O_2(g)$	205.03
$C_2H_4(g)$	219.4	HBr(g)	198.59	Li(s)	29.10	$SO_2(g)$	248.12
$C_2H_6(g)$	229.5	HCl(g)	186.80	$N_2(g)$	191.50	$SO_3(g)$	256.72
$CH_3OH(l)$	127	HF(g)	173.67				

[a]Most values are from Bard, A. J.; Parsons, R.; Jordan, J. *Standard Potentials in Aqueous Solution;* Dekker: New York, 1985.

hydrogen halides are in the order HF < HCl < HBr < HI. If species having similar structures are compared, the more atoms that are joined together, the greater the entropy. The entropy of $CO_2(g)$, for example, is greater than the entropy of CO(g), and the entropy of ethyl alcohol, $CH_3CH_2OH(l)$, is greater than the entropy of methyl alcohol, $CH_3OH(l)$.* The more atoms there are in a molecule, the more bonds there are and the more rotational and vibrational motions are possible.

Thomas, T. J. Chem. Educ. 1995, 72, 16 discusses periodic trends in the entropy of the elements.

SAMPLE PROBLEM

17.1   Predict the sign of the entropy change for each of the following changes. Explain your answers.
(a) $O_2(g)$ (1 atm, 25 °C) $\longrightarrow$ $O_2(g)$ (0.1 atm, 25 °C)
(b) $NH_4Cl(s) \longrightarrow NH_3(g) + HCl(g)$
(c) $CO(g) + H_2O(g) \longrightarrow CO_2(g) + H_2(g)$

Solution   (a) For the pressure of $O_2(g)$ to decrease from 1 to 0.1 atm at constant temperature, the volume of the sample must increase. Therefore, the entropy increases and the sign of ΔS is +.
    (b) When $NH_4Cl(s)$ decomposes to $NH_3(g)$ and HCl(g), volume increases and ΔS is +.
    (c) When carbon monoxide gas and water vapor react to form carbon dioxide gas and hydrogen gas, no change takes place in the number of moles of gas. The entropy change will be relatively small, and its sign cannot be predicted.

PRACTICE PROBLEMS

17.1   Predict the sign of the entropy change for each of the following changes. Explain your answers.
(a) $CH_4(g) + 2O_2(g) \longrightarrow CO_2(g) + 2H_2O(g)$
(b) $CH_4(g) + 2O_2(g) \longrightarrow CO_2(g) + 2H_2O(l)$
(c) $N_2(g)$ (1 atm, 25 °C) $\longrightarrow$ $N_2(g)$ (1 atm, 0 °C)
(d) $N_2(g)$ (1 atm, 25 °C) $\longrightarrow$ $N_2(g)$ (2 atm, 25 °C)
(e) $2H_2(g) + O_2(g) \longrightarrow 2H_2O(g)$

---

*The relationship between entropy and structure is quantitative. Entropies of gaseous species can be calculated from properties of individual molecules by statistical thermodynamics.

17.2   Predict whether the entropy of the system will increase or decrease when each of the following physical changes takes place at constant temperature and pressure. Explain your answers. (a) Liquid benzene freezes to solid benzene. (b) Liquid benzene evaporates. (c) Pentane, a liquid, is dissolved in hexane, also a liquid.

17.3   Indicate which member of each of the following pairs has the higher standard entropy at 25 °C. Explain your answers. (a) Ca(s) or Mg(s) (b) $H_2O(g)$ or $H_2S(g)$ (c) $PCl_3(g)$ or $PCl_5(g)$ (d) $Cl_2(g)$ or $F_2(g)$ (e) $Br_2(l)$ or $I_2(s)$

Calculation of Standard Entropy Changes

Like enthalpy, entropy is a state function. The entropy of a substance in a given state does not depend on how the substance arrived at the given state. A **standard entropy change** is the *change in entropy that accompanies the conversion of reactants in their standard states to products in their standard states.* Using a method similar to the method used to calculate standard enthalpy changes from standard heats of formation, values of standard entropies from tables can be used to calculate the standard entropy changes that accompany physical and chemical changes.

$$\Delta S^{\circ}_{rxn} = \Sigma n_p S^{\circ} \text{ (product)} - \Sigma n_r S^{\circ} \text{ (reactant)} \tag{17.1}$$

where Σ means sum of, n_p is the number of moles of each product, S° (product) is the standard entropy of the product, n_r is the number of moles of each reactant, and S° (reactant) is the standard entropy of the reactant.

Sample Problems 17.2 and 17.3 show how to use equation 17.1 to calculate ΔS°_{rxn} from standard entropies.

SAMPLE PROBLEMS

17.2   Use standard entropies to calculate ΔS°_{rxn} at 25 °C for the reaction

$$2SO_2(g) + O_2(g) \longrightarrow 2SO_3(g)$$

Solution   Remember that S° is *not* zero for elements. For both elements and compounds, S° must be obtained from Table 17.1 or Appendix D. Use the equation for the reaction as column headings for a table of the standard entropies:

Equation	$2SO_2(g)$	+	$O_2(g)$	$\longrightarrow$	$2SO_3(g)$
S°, J/K·mol	248.12		205.03		256.72

According to equation 17.1,

$$\Delta S^{\circ}_{rxn} = \Sigma n_p S^{\circ} \text{ (product)} - \Sigma n_r S^{\circ} \text{ (reactant)}$$

Therefore, for the reaction in this problem,

$$\Delta S^{\circ}_{rxn} = \left[2 \text{ mol } SO_3 \left(256.72 \frac{J}{K \cdot mol \ SO_3} \right) \right]$$

$$- \left[2 \text{ mol } SO_2 \left(248.12 \frac{J}{K \cdot mol \ SO_2} \right) + 1 \text{ mol } O_2 \left(205.03 \frac{J}{K \cdot mol \ O_2} \right) \right]$$

$$= 513.44 \frac{J}{K} - \left[496.24 \frac{J}{K} + 205.03 \frac{J}{K} \right] = -187.83 \text{ J/K}$$

Is this answer reasonable? Yes, because, according to the equation for the reaction, 3 mol of gas combine to form only 2 mol of gas, and entropy should decrease considerably.

17.3 Calculate ΔS_f° for NaCl(s) at 25 °C.

Solution The symbol ΔS_f° means entropy of formation. We must begin by writing the equation for the formation of 1 mol NaCl(s) from the elements. Then the problem is similar to Sample Problem 17.2. The formation equation for NaCl(s) is

Equation	Na(s) + $\frac{1}{2}$Cl$_2$(g) $\longrightarrow$ NaCl(s)		
S°, J/K·mol	51.45	222.96	72.12

$$\Delta S_{rxn}^\circ = \left[1 \text{ mol NaCl} \left(72.12 \frac{J}{K \cdot \text{mol NaCl}} \right) \right]$$

$$- \left[1 \text{ mol Na} \left(51.45 \frac{J}{K \cdot \text{mol Na}} \right) + \left(\frac{1}{2} \text{ mol Cl}_2 \right) \left(222.96 \frac{J}{K \cdot \text{mol Cl}_2} \right) \right]$$

$$= 72.12 \frac{J}{K} - \left[51.45 \frac{J}{K} + 111.48 \frac{J}{K} \right] = -90.81 \text{ J/K}$$

Thus, ΔS_f° for NaCl(s) = -90.81 J/K·mol. A negative value is reasonable because a gas has reacted to form a solid.

> ⊙ Always check to be sure your work is reasonable.

PRACTICE PROBLEMS

17.4 Use standard entropies to calculate ΔS_{rxn}° at 25 °C for each of the following reactions. Compare your answers with the qualitative predictions you made about these same reactions in Practice Problem 17.1.
(a) CH$_4$(g) + 2O$_2$(g) $\longrightarrow$ CO$_2$(g) + 2H$_2$O(g)
(b) CH$_4$(g) + 2O$_2$(g) $\longrightarrow$ CO$_2$(g) + 2H$_2$O(l)
(c) 2H$_2$(g) + O$_2$(g) $\longrightarrow$ 2H$_2$O(g)

17.5 Calculate ΔS_f° for Li$_2$CO$_3$(s) at 25 °C. [S° Li$_2$CO$_3$(s) = 90.37 J/K·mol]

Remember that, according to the second law of thermodynamics, the entropy of the universe is increasing. However, the universe consists of the system *and* its surroundings. The entropy change of the universe is the sum of the entropy change of the system and the entropy change of the surroundings:

$$\Delta S_{universe} = \Delta S_{system} + \Delta S_{surroundings}$$

It is not always easy to observe entropy changes in the surroundings. Therefore, it is simpler to think of change as the result of a combination of energy and entropy changes in the system as we have been doing from the beginning. A single thermodynamic function combining the enthalpy and entropy of the system would be convenient.

Because the difference between ΔH and ΔE is usually small compared to the magnitude of ΔH, this difference is disregarded in this book, and enthalpy is referred to as energy. The error introduced is usually less than the error introduced by using molarity instead of activity, as is commonly done in general chemistry texts, and we think that the simplification that results is justified in both cases.

17.3   FREE ENERGY

The problem of combining the enthalpy and entropy of the system into a single thermodynamic function was solved by the American theoretical physicist and

Top: *Aqueous hydrogen peroxide does not decompose at room temperature although decomposition is spontaneous at all temperatures because it is exothermic and the entropy of the system increases.* Bottom: *When a lump of manganese(IV) oxide, one of many substances that catalyzes the decomposition of hydrogen peroxide, is lowered into the solution, decomposition takes place rapidly.*

By the 1860s, the French chemist Marcellin Berthelot had observed that all very exothermic reactions are spontaneous. Berthelot was the first to use the terms exothermic *and* endothermic.

chemist J. Willard Gibbs in the late nineteenth century.* In his honor, the official name of the function is Gibbs energy, and its symbol is *G*. Gibbs energy is usually referred to as **Gibbs free energy** or simply **free energy.**

As usual, it is the *changes* in free energy that accompany physical and chemical changes that are of interest. The change in free energy accompanying a physical or chemical change at constant temperature and pressure is equal to the change in enthalpy minus the product of the temperature in kelvin and the entropy change:

$$\Delta G = \Delta H - T\,\Delta S \qquad (T \text{ and } P \text{ constant})$$

(17.2)

17.4 TEMPERATURE AND DIRECTION OF SPONTANEOUS CHANGE

What is the sign of the free-energy change for spontaneous changes? We already know that decreases in enthalpy and increases in entropy of the system favor spontaneous change. In other words, spontaneous changes tend to be associated with a − sign for ΔH and a + sign for ΔS. Temperature in kelvin, T, is always positive; as a result, if ΔS is +, the term, $-T\,\Delta S$ in equation 17.2 is −. If both ΔH and $-T\,\Delta S$ are −, ΔG must be −. Therefore, ΔG *is negative for spontaneous changes.* The decomposition of aqueous hydrogen peroxide

$$2H_2O_2(aq) \longrightarrow 2H_2O(l) + O_2(g)$$

is an example of a reaction with a negative enthalpy change and a positive entropy change.

Remember that spontaneous changes can take place either slowly or rapidly. A 3% solution of hydrogen peroxide in water is sold as an oral antiseptic in drugstores and supermarkets. This solution can be kept for months in a bathroom closet. However, as soon as you swish it around your mouth, an enzyme in your saliva catalyzes the decomposition, and you can feel the little bubbles of oxygen forming.

A change accompanied by an enthalpy increase and an entropy decrease is nonspontaneous. If ΔS is negative, the term $-T\,\Delta S$ must be positive. If ΔH is positive and $-T\,\Delta S$ is also positive, ΔG must be positive. *A positive value of ΔG means that a change will be nonspontaneous.* The reaction of oxygen gas with water to form hydrogen peroxide

$$2H_2O(l) + O_2(g) \longrightarrow 2H_2O_2(aq)$$

is an example of this type of reaction. Oxygen gas does not react spontaneously with water to form hydrogen peroxide.

Now suppose ΔH and ΔS have the same sign—for example, ΔH is negative and ΔS is also negative. If ΔS is negative, the term $-T\,\Delta S$ is positive. Whether ΔG is + or − depends on which term is larger, ΔH or $T\,\Delta S$. For most changes, ΔH is of the order of kJ/mol, but ΔS is only of the order of J/mol, 1000 times smaller. At low temperatures, the ΔH term is more important than the $T\,\Delta S$ term; if ΔH is negative, ΔG is negative, and the change is spontaneous. As a result, most exothermic changes are spontaneous at room temperature. But the higher the temperature, the larger the $T\,\Delta S$ term becomes. At high temperatures, the $T\,\Delta S$ term is more important

---

*Gibbs received the first doctorate of engineering to be conferred in the United States from Yale University in 1863. He was a professor of mathematical physics at Yale from 1871 to 1903 and, during this time, laid the foundations for many areas of thermodynamics.

TABLE 17.2		Factors Determining Direction of Change (at constant T and P)	
ΔH	ΔS	ΔG	Comments
$-$	$+$	$-$	Spontaneous at all temperatures
$+$	$-$	$+$	Nonspontaneous at all temperatures
$-$	$-$	$-$ low $+$ high	Spontaneous at low temperatures but nonspontaneous at high temperatures
$+$	$+$	$+$ low $-$ high	Nonspontaneous at low temperatures but spontaneous at high temperatures

than the ΔH term, ΔG is positive, and reaction is nonspontaneous. If ΔH and ΔS are both negative, a change is spontaneous at low temperatures but nonspontaneous at high temperatures.

Which temperatures are high and which temperatures are low depend on the relative values of ΔH and ΔS. The freezing of water

$$H_2O(l) \longrightarrow H_2O(s)$$

is a familiar example of a change that has both ΔH and ΔS negative. At temperatures below 0 °C, the freezing point of water, water spontaneously freezes to ice. At temperatures above 0 °C, water does not freeze; the reverse change is spontaneous, and ice melts to water. Any temperature lower than 0 °C is a low temperature for water, and any temperature above 0 °C is high. The freezing point of ethyl alcohol, -117.3 °C, is much lower than the freezing point of water. For ethyl alcohol, any temperature above -117.3 °C is a high temperature!

If ΔH and ΔS are both positive, as they are for the melting of ice, ΔG is positive at low temperatures but negative at high temperatures. Change is nonspontaneous at low temperatures but becomes spontaneous at high temperatures. *If ΔH and ΔS both have the same sign, temperature determines the direction of spontaneous change.* Table 17.2 summarizes the effects of the various possible combinations of signs for ΔH and ΔS.

The value of ΔG must be positive, negative, or zero. The value of ΔG is zero when $\Delta H = -T\Delta S$. *If $\Delta G = 0$, spontaneous change does not take place in either direction; the system is at equilibrium.* For example, a mixture of ice and water at 0 °C under atmospheric pressure in a closed container is at equilibrium:

$$H_2O(s) \rightleftharpoons H_2O(l)$$

Under these conditions, the free energies of the ice and water are equal, and $\Delta G = 0$. As long as the mixture of ice and water is kept at 0 °C—that is, as long as no thermal energy flows into or out of the system—the quantities of ice and water do not change. The rates of melting and freezing are equal. At any time, some ice is melting and some water is freezing, but the total quantities of ice and water are constant.

If the temperature of the mixture of ice and water is raised above 0 °C, $-T\Delta S$ becomes greater than ΔH. Because ΔS for the conversion of ice to water is $+$ and $-T\Delta S$ is $-$, ΔG becomes $-$ at the higher temperature, and melting becomes spontaneous. The equilibrium is shifted, and all the ice melts. If the temperature is lowered below 0 °C, equilibrium is shifted in the opposite direction. All the water freezes. If the sign of ΔG for a change is known, the direction of spontaneous change (but not the rate) can be predicted.

The $N_2O_4(g) \rightleftharpoons 2NO_2(g)$ equilibrium is an example of a chemical equilibrium with both ΔH and ΔS positive. At low temperatures, ΔG is positive, and equilibrium lies to the left. The concentration of red-brown $NO_2(g)$ is low (top). Increasing the temperature shifts the equilibrium to the right. The concentration of red-brown $NO_2(g)$ increases (bottom). At high temperatures, ΔG is negative, and equilibrium lies to the right.

The idea that the value of ΔG must either be positive, negative, or zero seems obvious, but it helps a lot of students to remind them of it at this point.

The Free Energy module of the Thermodynamics section allows investigation of the quantitative aspects of free energy. Animated graphs help reinforce qualitative concepts of enthalpy, entropy, and free energy.

PRACTICE PROBLEM

17.6   For the reaction

$$NH_3(g) + HCl(g) \longrightarrow NH_4Cl(s)$$

both ΔH and ΔS are negative. *Without referring to Table 17.2,* use the definition of free-energy change, $\Delta G = \Delta H - T\Delta S$, to figure out whether the reaction will be (a) spontaneous at all temperatures, (b) not spontaneous at any temperature, (c) spontaneous at low temperatures but nonspontaneous at high temperatures, or (d) nonspontaneous at low temperatures but spontaneous at high temperatures. Then check your answer with Table 17.2.

Standard free-energy changes, $\Delta G°$, are *free-energy changes that accompany changes from reactants in their standard states to products in their standard states.* Standard states for free-energy changes are defined in the same way as standard states for enthalpy and entropy changes. There are several ways to find the value of $\Delta G°$ for a change.

17.5   CALCULATION OF $\Delta G°$ FROM $\Delta H°$ AND $\Delta S°$

The definition of free-energy change, equation 17.2, can be used to calculate $\Delta G°$ if $\Delta H°$, $\Delta S°$, and T are known:

$$\Delta G° = \Delta H° - T\Delta S° \qquad (T \text{ and } P \text{ constant}) \qquad (17.3)$$

Sample Problem 17.4 illustrates calculation of the standard free-energy change for a reaction from tabulated values of standard heats of formation and standard entropies.

SAMPLE PROBLEM

17.4   Use standard heats of formation and standard entropies to calculate $\Delta G°$ for the reaction

$$N_2(g) + 3H_2(g) \longrightarrow 2NH_3(g)$$

at 25.0 °C and 1 atm partial pressure of each gas.

Solution   First look up $\Delta H_f°$ (NH_3) and the standard entropies of $N_2(g)$, $H_2(g)$, and $NH_3(g)$ in Tables 6.2 and 17.1 or in Appendix D and record their values:

Equation	$N_2(g)$	$+ 3H_2(g)$	$\longrightarrow 2NH_3(g)$
$\Delta H_f°$, kJ/mol	0.00	0.00	−46.11
$S°$, J/K·mol	191.50	130.68	192.3

Then use the values found to calculate $\Delta H°$ and $\Delta S°$ for the reaction. Finally calculate $\Delta G°$ for the reaction using equation 17.3, $\Delta G° = \Delta H° - T\Delta S°$.

Step 1.   *Calculation of $\Delta H°$:* To calculate $\Delta H°$, use Hess's law in the form

$$\Delta H° = \Sigma n_p \, \Delta H_f° \text{ (product)} - \Sigma n_r \, \Delta H_f° \text{ (reactant)}$$

$$= \left[2 \text{ mol } NH_3 \left(-46.11 \, \frac{kJ}{mol \, NH_3} \right) \right]$$

$$- \left[1 \text{ mol N}_2 \left(0.00 \frac{\text{kJ}}{\text{mol N}_2} \right) + 3 \text{ mol H}_2 \left(0.00 \frac{\text{kJ}}{\text{mol H}_2} \right) \right]$$

$$= -92.22 \text{ kJ}$$

Step 2. *Calculation of $\Delta S°$:* To calculate $\Delta S°$, use equation 17.1:

$$\Delta S° = \Sigma n_p \, S° \text{ (product)} - \Sigma n_r \, S° \text{ (reactant)}$$

$$= \left[2 \text{ mol NH}_3 \left(192.3 \frac{\text{J}}{\text{K} \cdot \text{mol NH}_3} \right) \right]$$

$$- \left[1 \text{ mol N}_2 \left(191.50 \frac{\text{J}}{\text{K} \cdot \text{mol N}_2} \right) + 3 \text{ mol H}_2 \left(130.68 \frac{\text{J}}{\text{K} \cdot \text{mol H}_2} \right) \right]$$

$$= -198.9 \frac{\text{J}}{\text{K}}$$

Step 3. *Calculation of $\Delta G°$:* In equation 17.3,

$$\Delta G° = \Delta H° - T \, \Delta S°$$

T stands for temperature in kelvin. The values for standard entropies in Table 17.1 are for 25 °C. We must convert temperature in °C to K:

$$K = (t + 273.2) \, °C \left(\frac{1 \text{ K}}{1 \, °C} \right) = (25.0 + 273.2) \, °C \left(\frac{1 \text{ K}}{1 \, °C} \right) = 298.2 \text{ K}$$

We must also convert the units of $\Delta S°$ to kJ because $\Delta H°$ is expressed in kJ and $\Delta G°$ is usually given in kJ:

$$\left(-198.8 \frac{\text{J}}{\text{K}} \right) \left(\frac{1 \text{ kJ}}{1000 \text{ J}} \right) = -0.1989 \frac{\text{kJ}}{\text{K}}$$

Now substitution in the equation $\Delta G° = \Delta H° - T \, \Delta S°$ gives

$$\Delta G° = (-92.22 \text{ kJ}) - 298.2 \text{ K} \left(-0.1989 \frac{\text{kJ}}{\text{K}} \right) = -32.91 \text{ kJ}$$

> ▶ You must be careful not to add J to kJ; you would not expect to get two dollars by adding one cent to one dollar.

PRACTICE PROBLEM

17.7 Use standard heats of formation and standard entropies to calculate $\Delta G°$ for the reaction

$$CH_4(g) + 2O_2(g) \longrightarrow CO_2(g) + 2H_2O(g)$$

at 25.0 °C and 1 atm partial pressure of each gas.

17.6 CALCULATION OF $\Delta G°$ FROM $\Delta G_f°$

Another way to calculate standard free-energy changes accompanying physical and chemical changes is to use tabulated values of standard free energies of formation. The **standard free energy of formation, $\Delta G_f°$,** of a substance is the *free-energy change that results when one mole of the substance is formed from its elements*

Enthalpy, Entropy, and Acidity

For most chemical changes at room temperature, the enthalpy term $\Delta H°$ is more important than the entropy term $T \Delta S°$. But for the ionization of many weak acids in water, $\Delta H°$ is small, $\Delta S°$ is large, and $T \Delta S°$ is larger than $\Delta H°$ under ordinary conditions. Some examples from the weak acids in Tables 15.3 and 15.7 are shown in Table 17.3.

Why is the $T \Delta S°$ term more important than the $\Delta H°$ term for the ionization of many weak acids at room temperature? Ionization of most organic acids and many inorganic acids involves breaking an O—H bond in the acid and forming an O—H bond with a water molecule. The ionization of acetic acid is an example:

TABLE 17.3	Thermodynamic Functions for the Ionization of Some Weak Acids at 25 °C[a]		
Acid			
Name	Formula	$\Delta H°$, kJ/mol	$T \Delta S°$, kJ/mol
Formic acid	HCOOH	−0.04	−21.4
Acetic acid	CH_3COOH	−0.5	−27.6
Chloroacetic acid	$ClCH_2COOH$	−4.9	−21.2
Propionic acid	CH_3CH_2COOH	−0.7	−28.6
Hypochlorous acid	HOCl	+13.9	−28.4
Carbonic acid	$CO_2 + H_2O$	+7.7	−28.6
Phosphoric acid	H_3PO_4	−7.9	−19.9
Nitrous acid	HNO_2	+14	−3.9
Ammonium ion	NH_4^+	+52.2	−0.6

[a]Most data are from Jolly, W. L. *Modern Inorganic Chemistry*, 2nd ed.; McGraw-Hill: New York, 1991; p 225.

Although the strength of an O—H bond depends somewhat on the species the bond is part of, the strengths of most O—H bonds are similar. The amount of energy used up in breaking the O—H bond in the acid is about the same as the amount of energy given off when the O—H bond in the hydronium ion is formed. On the other hand, the ionic products are solvated more strongly than the molecular reactants. Clustering of water molecules around the ions orders the water molecules, and entropy decreases. The product of the entropy decrease and the absolute temperature is greater than the enthalpy change. Entropy changes, not energy changes, are the major factor that determines the strength of these weak acids in aqueous solutions.

Unfortunately, entropy changes in aqueous solution are difficult to predict. The larger volume available to the solute in solution results in an increase in entropy. In addition, introduction of a solute destroys some of the order due to hydrogen bonding in water and also results in an increase in entropy. But solvation of ions arranges water molecules around solute ions in spherical shells and results in a decrease in entropy. Thus, the net entropy change that results is hard to foretell.

Although the entropy term, $T \Delta S°$, is more important than the enthalpy term, $\Delta H°$, for some weak acids, the enthalpy term is more important than the entropy term for other weak acids. For some pairs of acids, the acid that is stronger at one temperature is weaker at another, not much different, temperature. For example, CH_3COOH is weaker than $(CH_3CH_2)_2CHCOOH$ at 20 °C but stronger than $(CH_3CH_2)_2CHCOOH$ at 40 °C. It is a wonder that we are able to predict the acidities of weak acids as well as we can!

with all substances in their standard states. Just as the standard enthalpies of formation for elements in their most stable or most common form are zero because changing a pure element into itself does not involve any change, the standard free energies of formation for elements in their most stable or most common form are zero. Tables of large numbers of standard free energies of formation for compounds are available; some standard free energies of formation at 25 °C are given in Table 17.4, and more are given in Appendix D.

Free energy is a state function like enthalpy and entropy. The method used to calculate standard free-energy changes from standard free energies of formation is similar to the methods used to calculate standard enthalpy changes from standard

TABLE 17.4   Some Standard Free Energies of Formation, ΔG_f°, at 25 °C[a]

Substance	ΔG_f°, kJ/mol	Substance	ΔG_f°, kJ/mol	Substance	ΔG_f°, kJ/mol	Substance	ΔG_f°, kJ/mol
C(s, diamond)	2.832	$CH_3OH(g)$	−161.9	HF(g)	−273.22	$Na_2SiO_3(s)$	−1467.4
$CaCO_3(s)$	−1128.84	$CH_3CH_2OH(l)$	−174.8	HI(g)	1.30	$NH_3(g)$	−16.5
CaO(s)	−604.04	$CH_3CH_2OH(g)$	−168.6	$H_2O(l)$	−237.18	O(g)	231.75
$CH_4(g)$	−50.79	CO(g)	−137.27	$H_2O(g)$	−228.59	$SO_2(g)$	−300.19
$C_2H_2(g)$	209	$CO_2(g)$	−394.38	$I_2(g)$	19.37	$SO_3(s)$	−369.0
$C_2H_4(g)$	86.12	H(g)	203.25	N(g)	340.9	$SO_3(l)$	−368.4
$C_2H_6(g)$	−32.89	HBr(g)	−53.43	NaCl(s)	−384.04	$SO_3(g)$	−371.08
$CH_3OH(l)$	−166.3	HCl(g)	−95.30				

[a]Most values are from Bard, A. J.; Parsons, R.; Jordan, J. *Standard Potentials in Aqueous Solution;* Dekker: New York, 1985.

heats of formation and standard entropy changes from standard entropies.

$$\Delta G_{rxn}^\circ = \Sigma n_p \Delta G_f^\circ \text{ (product)} - \Sigma n_r \Delta G_f^\circ \text{ (reactant)} \qquad (17.4)$$

where Σ means sum of, n_p is the number of moles of each product, ΔG_f° (product) is the standard free energy of formation of the product, n_r is the number of moles of each reactant, and ΔG_f° (reactant) is the standard free energy of formation of the reactant.

Sample Problem 17.5 illustrates the calculation of a standard free-energy change from standard free energies of formation.

SAMPLE PROBLEM

17.5   Use standard free energies of formation to calculate ΔG° for the reaction

$$2SO_2(g) + O_2(g) \longrightarrow 2SO_3(g)$$

at 25 °C and 1 atm partial pressure of each gas.

Solution   Record the standard free energies from Table 17.4 or Appendix D under the equation for the reaction:

	Equation		$2SO_2(g)$	+ $O_2(g)$ $\longrightarrow$	$2SO_3(g)$
	ΔG_f°, **kJ/mol**		−300.19	0	−371.08

[Because O_2 is the most stable (and the most common) form of the element oxygen at 25 °C and 1 atm partial pressure, $\Delta G_f^\circ(O_2) = 0$.] Substitution of these values in equation 17.4 gives

$$\Delta G^\circ = \left[2 \text{ mol SO}_3\left(-371.08 \, \frac{kJ}{\text{mol SO}_3} \right) \right]$$

$$- \left[2 \text{ mol SO}_2\left(-300.19 \, \frac{kJ}{\text{mol SO}_2} \right) + 1 \text{ mol O}_2\left(0.00 \, \frac{kJ}{\text{mol O}_2} \right) \right]$$

$$= -141.78 \text{ kJ}$$

PRACTICE PROBLEM

17.8   Use standard free energies of formation to calculate ΔG_{rxn}° for the following reaction at 25 °C: $CH_4(g) + 2O_2(g) \longrightarrow CO_2(g) + 2H_2O(g)$.

17.7 ESTIMATION OF $\Delta G°$ AT DIFFERENT TEMPERATURES

The rates of many reactions at 25 °C are too low to be useful. We need to be able to find $\Delta G°$ at other temperatures. Accurate calculations of $\Delta G°$ at temperatures other than 25 °C are beyond the scope of a general chemistry course. However, if we assume that $\Delta H_f°$ and $\Delta S°$ do not change with temperature,* we can use values of $\Delta H_f°$ and $S°$ from tables and equation 17.3 to *estimate* $\Delta G°$ at temperatures other than 25 °C. Sample Problem 17.6 shows how to do this.

SAMPLE PROBLEM

17.6 Estimate $\Delta G°$ for the reaction

$$N_2(g) + 3H_2(g) \longrightarrow 2NH_3(g)$$

at 427 °C.

Solution In Sample Problem 17.4, we calculated $\Delta H°$ and $\Delta S°$ for this reaction at 25 °C from standard heats of formation and standard entropies and found that $\Delta H° = -92.22$ kJ and $\Delta S° = -0.1989$ kJ/K. If we assume that $\Delta H°$ and $\Delta S°$ have the same values at 427 °C as they do at 25 °C, substitution in equation 17.3,

$$\Delta G° = \Delta H° - T\,\Delta S°$$

gives

$$\Delta G° = (-92.22 \text{ kJ}) - (427 + 273) \text{ K} (-0.1989 \text{ kJ/K})$$

$$= (-92.22 \text{ kJ}) - (700 \text{ K})(-0.1989 \text{ kJ/K})$$

$$= (-92.22 \text{ kJ}) + (139 \text{ kJ}) = +47 \text{ kJ}$$

PRACTICE PROBLEM

17.9 Estimate $\Delta G°$ for the reaction

$$CH_4(g) + 2O_2(g) \longrightarrow CO_2(g) + 2H_2O(g)$$

at 250 °C. (You calculated values for $\Delta H°$ and $\Delta S°$ for this reaction in Practice Problem 17.7.)

17.8 ESTIMATION OF TEMPERATURE AT WHICH DIRECTION OF SPONTANEOUS CHANGE REVERSES

Earlier we saw that if both ΔH and ΔS have the same sign, temperature determines the direction of spontaneous change. Equation 17.3 can be used to estimate the

---

*Standard entropies, $S°$, of course, depend markedly on temperature especially at low temperatures (close to 0 K). However, because increasing temperature increases the entropy of all substances and the increase per degree is less at high temperatures, standard entropy changes, $\Delta S°$, often do not change greatly with temperature at ordinary temperatures. The situation is similar to that of two children, one of whom is 2'7" tall and the other of whom is 2'8" tall when both are two years old. If, when the two are both twenty, the shorter one is 5'7" and the taller, 5'8", the latter is still one inch taller than the former although both have grown three feet. Standard enthalpy changes, $\Delta H°$, are also often quite constant as temperature changes. For example, $\Delta H°$ for the reaction, C(graphite) + O_2(g) $\longrightarrow$ CO_2(g), changes from -393.51 kJ/mol at 25 °C to -393.67 kJ/mol at 125 °C, a change of only 0.04% for a 100° change in temperature. At 1025 °C, the value of $\Delta H°$ is -391.41 kJ/mol.

temperature at which a change that is spontaneous in one direction becomes spontaneous in the other direction.

SAMPLE PROBLEMS

17.7 Estimate the normal boiling point of methyl alcohol, CH_3OH.

Solution At the normal boiling point of a liquid, liquid and vapor are at equilibrium:

$$CH_3OH(l) \rightleftharpoons CH_3OH(g)$$

and the partial pressure of the vapor equals 1 atm. The pressure on the liquid is 1 atm. Both liquid and vapor are in their standard states. At equilibrium $\Delta G = 0$; if the values of $\Delta H°$ and $\Delta S°$ are known, substitution in equation 17.3 gives an equation in one unknown, T, which can be solved for T. From Tables 6.2 and 17.1, values of the standard enthalpy of formation and standard entropy of methyl alcohol liquid and gas are

	$\Delta H_f°$, kJ/mol	$S°$, J/K·mol
$CH_3OH(l)$	-238.64	127
$CH_3OH(g)$	-201.2	238

Step 1. *Calculation of $\Delta H°$:* Hess's law, equation 6.23,

$$\Delta H° = \Sigma n_p \Delta H_f° \text{ (product)} - \Sigma n_r \Delta H_f° \text{ (reactant)}$$

can be used to calculate $\Delta H°$ for the phase change from liquid to vapor:

$$\Delta H° = \left[1 \text{ mol } CH_3OH(g) \left(-201.2 \frac{kJ}{\text{mol } CH_3OH(g)} \right) \right]$$
$$- \left[1 \text{ mol } CH_3OH(l) \left(-238.64 \frac{kJ}{\text{mol } CH_3OH(l)} \right) \right]$$
$$= (-201.2 \text{ kJ}) - (-238.64 \text{ kJ}) = 37.4 \text{ kJ}$$

Step 2. *Calculation of $\Delta S°$:* Equation 17.1,

$$\Delta S° = \Sigma n_p S° \text{ (product)} - \Sigma n_r S° \text{ (reactant)}$$

can be used to calculate $\Delta S°$ for the phase change:

$$\Delta S° = \left[1 \text{ mol } CH_3OH(g) \left(238 \frac{J}{K \cdot \text{mol } CH_3OH(g)} \right) \right]$$
$$- \left[1 \text{ mol } CH_3OH(l) \left(127 \frac{J}{K \cdot \text{mol } CH_3OH(l)} \right) \right]$$
$$= (238 \text{ J/K}) - (127 \text{ J/K}) = 111 \text{ J/K or } 0.111 \text{ kJ/K}$$

Step 3. *Calculation of T:* At equilibrium, $\Delta G° = 0$. If the assumption is made that the values of $\Delta H°$ and $\Delta S°$ at the normal boiling point of methyl alcohol are the same as the values at 25 °C, equation 17.3 can be used to estimate the normal boiling point. Solution of equation 17.3

$$\Delta G° = \Delta H° - T \Delta S° \qquad \text{for } T \text{ gives}$$
$$T = (\Delta H° - \Delta G°)/\Delta S°$$

and substitution of the values for $\Delta G°$, $\Delta H°$, and $\Delta S°$ gives

$$(T \text{ K}) = (+37.4 \text{ kJ} - 0.0 \text{ kJ})/(0.111 \text{ kJ/K}) = 337 \text{ K}$$

The estimated boiling point in °C is

$$337 - 273 = 64 \text{ °C}$$

The actual boiling point of methyl alcohol is 65 °C. When you consider that boiling points tabulated in the *CRC Handbook* vary from -268 °C for He(l) to $+6000$ °C for tungsten carbide, WC(l), the estimated boiling point of 64 °C agrees very well with the actual boiling point of 65 °C.

17.8 The normal boiling point of benzene, C_6H_6, is 80.0 °C (353.2 K) and ΔH_{vap} is 30.8 kJ/mol. Calculate ΔS_{vap} for benzene.

Solution Solution of equation 17.3

$$\Delta G° = \Delta H° - T \Delta S° \qquad \text{for } \Delta S° \text{ gives}$$
$$\Delta S° = (\Delta H° - \Delta G°)/T$$

At the boiling point of a liquid, liquid and vapor are in equilibrium and at equilibrium, $\Delta G° = 0$. Substitution gives

$$\Delta S° = (30.8 \text{ kJ/mol} - 0.0 \text{ kJ/mol})/353.2 \text{ K} = 0.0872 \text{ kJ/K·mol or } 87.2 \text{ J/K·mol}$$

Conversion of liquid to vapor results in an increase in the entropy of the system. The answer is reasonable.

○ Always check to be sure your work is reasonable.

PRACTICE PROBLEMS

17.10 Estimate the normal boiling point of ethyl alcohol, CH_3CH_2OH.

17.11 The normal boiling point of bromine is 58.8 °C and ΔH_{vap} is 29.5 kJ/mol. Calculate ΔS_{vap} for bromine.

17.9 CALCULATION OF ΔG FOR NONSTANDARD CONDITIONS

Remember that $\Delta G_f°$, *the standard free energy of formation, is for the formation of products in their standard states from reactants in their standard states*. The standard state for a gas is the ideal gas at a partial pressure of 1 atm. However, *reactions are often carried out with partial pressures of gases that are different from standard partial pressure*. For example, in Sample Problem 17.6, we estimated that $\Delta G°$ for the reaction

$$N_2(g) + 3H_2(g) \longrightarrow 2NH_3(g)$$

at 427 °C is $+47$ kJ. From the positive value of $\Delta G°$ at 427 °C, we might predict that ammonia would not be formed from nitrogen and hydrogen at 427 °C. However, temperatures of 427 °C and higher are used for the commercial synthesis of ammonia. The synthesis of ammonia is carried out at pressures of 133 to 667 atm. The partial pressures of the three gases are *not* 1 atm. (In addition, the gases, especially ammonia, are far from ideal at these high pressures.) To predict the direction of spontaneous change, we need to be able to find ΔG, the free-energy change under nonstandard conditions.

The free-energy changes, ΔG, accompanying changes carried out under nonstandard conditions can be calculated from equation 17.5 (the derivation of equation 17.5 is beyond the scope of this text):

$$\Delta G = \Delta G° + RT \ln Q \qquad (17.5)$$

Equation 17.5 is derived in most physical chemistry and thermodynamics texts.

In equation 17.5, $\Delta G°$ is the standard free-energy change, R is the gas constant (8.315 J/K·mol or 0.008 315 kJ/K·mol), T is the Kelvin temperature, and Q is the reaction quotient (Section 14.4). The symbol ln means natural or base e logarithm (Appendix B.3). The value of ΔG can then be used to predict the direction of spontaneous change.

The more accurate ways of treating equilibrium constants used in thermodynamics and physical chemistry define equilibrium constants in such a way that they are pure numbers and have no units (Section 14.3). Therefore, units are not usually given for equilibrium constants and for Q. If ln Q has no units, then the units of RT ln Q are J/mol or kJ/mol:

$$RT \ln Q = \left(8.315 \frac{J}{K \cdot mol}\right) K = 8.315 \frac{J}{mol}$$

We have been writing J or kJ for the units of $\Delta G°$ and ΔG. But quantities with different units cannot be added; J/mol cannot be added to J or kJ/mol to kJ. However, $\Delta G°$ and ΔG are extensive properties and really do include units of mol^{-1} because, in thermodynamics, equations are always interpreted in terms of moles. *In calculations that involve both $\Delta G°$ or ΔG and R, the unit J/mol (or kJ/mol) must be used for $\Delta G°$ and ΔG.* Sample Problem 17.9 illustrates the calculation of ΔG for a change carried out under nonstandard conditions.

▶ The unit mol^{-1} for $\Delta G°$ and ΔG will still be omitted when it is not needed.

SAMPLE PROBLEM

17.9   Calculate ΔG at 427 °C (700 K) for the reaction

$$N_2(g) + 3H_2(g) \longrightarrow 2NH_3(g)$$

if p_{N_2} = 33.0 atm, p_{H_2} = 99.0 atm, and p_{NH_3} = 2.0 atm. Predict the direction of spontaneous change.

Solution   For this reaction and the partial presures given

$$Q = p_{NH_3}^2/p_{N_2} \cdot p_{H_2}^3 = (2.0)^2/(33.0)(99.0)^3 = 1.3 \times 10^{-7}$$

$\Delta G°$ for this reaction at 427 °C is estimated to be +47 kJ (Sample Problem 17.6). Substitution of these values for Q and $\Delta G°$ in equation 17.5

$$\Delta G = \Delta G° + RT \ln Q \qquad \text{gives}$$

$$\Delta G = +47 \frac{kJ}{mol} + \left(0.008\ 315 \frac{kJ}{K \cdot mol}\right)(700\ K)\ln(1.3 \times 10^{-7}) = -45 \frac{kJ}{mol}$$

Reaction between hydrogen and nitrogen to form ammonia will be spontaneous *under these conditions* because ΔG is negative.

▶ Remember that ΔG is $-$ for spontaneous changes and $+$ for nonspontaneous changes. If $\Delta G = 0$, the system is at equilibrium.

PRACTICE PROBLEM

17.12   Calculate ΔG at 250 °C (523 K) for the reaction

$$CH_4(g) + 2O_2(g) \longrightarrow CO_2(g) + 2H_2O(g)$$

if p_{CH_4} = 0.56 atm, p_{O_2} = 0.45 atm, p_{CO_2} = 2.10 atm, and p_{H_2O} = 3.40 atm. (You estimated $\Delta G°$ in Practice Problem 17.9.)

17.10   STANDARD FREE ENERGIES AND EQUILIBRIUM CONSTANTS

At equilibrium, $\Delta G = 0$ and $Q = K$, the equilibrium constant. Substitution of these values in equation 17.5 gives the equation

$$0 = \Delta G° + RT \ln K \qquad\qquad (17.6)$$

Rearrangement of equation 17.6 gives a quantitative relationship between $\Delta G°$ and the equilibrium constant:

$$\Delta G° = -RT \ln K \qquad (17.7)$$

Because $\Delta G°$ is the difference in free energy between products in their standard states and reactants in their standard states, K means K_p for gaseous reactions and K_c for reactions that involve solutions.

Equation 17.7 can be used to calculate the value of the equilibrium constant from $\Delta G°$.* Equilibrium constants that are difficult to measure experimentally because equilibrium is reached very slowly can be calculated to see if it is worthwhile to look for a catalyst. Also, equilibrium constants that are difficult to measure accurately because the value of K is very large or very small can be found. (If K is very large, equilibrium partial pressures or concentrations of reactants are so low that they are hard to measure. If K is very small, equilibrium partial pressures or concentrations of products are so low that they are difficult to measure.) Sample Problem 17.10 illustrates calculation of an equilibrium constant from $\Delta G°$.

SAMPLE PROBLEM

17.10   For the equilibrium

$$H_2(g) + I_2(g) \rightleftharpoons 2HI(g)$$

at 425.4 °C, $\Delta G°$ is estimated, from tabulated values of $\Delta H_f°$ and $S°$, to be -25.31 kJ. Calculate the value of K_p at 425.4 °C.

Solution   According to equation 17.7,

$$\Delta G° = -RT \ln K$$

where K means K_p because the reactants and products are all gases. Solution of equation 17.7 for $\ln K$ gives

$$\frac{\Delta G°}{-RT} = \ln K$$

Substitution in this expression gives

$$\ln K_p = \frac{(-25.31 \text{ kJ} \cdot \text{mol}^{-1})}{-(0.008\ 315 \text{ kJ} \cdot \text{K}^{-1} \cdot \text{mol}^{-1})(425.4 + 273.2) \text{ K}}$$

$$= 4.35 \text{ and } K_p = 78.0$$

Because the number of moles of gaseous product in the H_2–I_2–HI equilibrium is the same as the number of moles of gaseous reactants, $K_p = K_c$. The experimental value for K_c for this equilibrium at 425.4 °C is 54.5 (Section 14.2). You may think that the agreement between a calculated value of 78.0 and an experimental value of 54.5 is not very good. Actually, it is satisfactory for many uses. The values for $\Delta G_f°$ in the *CRC Handbook* range from -2055 kJ/mol for $Mg_2Al_4Si_5O_{18}(s)$ to $+198.7$ kJ/mol for $TbC_2(g)$. This range in $\Delta G_f°$ values corresponds to a change in

---

*Don't forget that the value of the equilibrium constant depends on how the equation for an equilibrium is written. For example, the equilibrium constant for the equilibrium

$$\tfrac{1}{2}N_2(g) + \tfrac{3}{2}H_2(g) \rightleftharpoons NH_3(g)$$

is equal to the square root of the equilibrium constant for the equilibrium

$$N_2(g) + 3H_2(g) \rightleftharpoons 2NH_3(g)$$

The value of $\Delta G_f°$ in Table 17.4 is for the former equilibrium.

the value of the equilibrium constant from 10^{1506} to 10^{-146}! Equilibrium constants that are of the same order of magnitude (power of 10) are the same for many practical purposes. As a result of the logarithmic relationship between $\Delta G°$ and K, a difference of a power of 10 in K corresponds to a difference of 5.7 kJ in $\Delta G°$. If $\Delta G°$ is 114 kJ, a difference of 5.7 kJ is only a 5% difference in $\Delta G°$; if $\Delta G°$ is greater than 114 kJ, a difference of 5.7 kJ is less than 5% of $\Delta G°$.

We made a number of assumptions in our calculations. We assumed that the gases were ideal and used partial pressures to calculate the experimental value of the equilibrium constant. In our calculation of $\Delta G°$, we assumed that $\Delta H°$ and $\Delta S°$ have the same values at 424.4 °C as they do at 25 °C. In addition, calculations of both $\Delta H°$ and $\Delta S°$ involved finding a small difference between two large numbers, a process that always results in a loss of precision. It is not surprising that slightly different values for the equilibrium constant resulted from different methods of obtaining the equilibrium constant.

PRACTICE PROBLEM

17.13   Calculate K if $\Delta G° = -80.0$ kJ/mol and $T = 298$ K.

Table 17.5 shows values of the equilibrium constant calculated from equation 17.7 for different values of $\Delta G°$. If the equilibrium constant can be measured, equation 17.7 can be used to calculate the value of $\Delta G°$. Practice Problem 17.14 illustrates this type of calculation.

PRACTICE PROBLEM

17.14   If the equilibrium constant for a reaction is 2.7×10^4 at 25 °C, what is $\Delta G°$ for the reaction at 25 °C?

The difference in free energy between reactants and products is the major factor in the position of equilibrium. ▪Figure 17.5 shows how the percent completion of reaction at equilibruium changes with the difference in free energy between reactants and products for reactions of the simplest possible type, A → B. As the difference

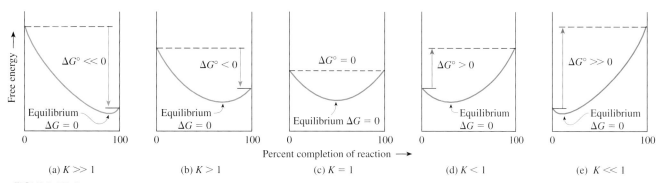

▪ FIGURE 17.5   If $\Delta G°$ is very negative (a), equilibrium is close to the product side of the diagram, and reactants are completely or almost completely converted to products at equilibrium. The reaction is said to be complete. On the other hand, if $\Delta G°$ is very positive (e), equilibrium is close to the reactant side of the diagram. Little or no reaction takes place. Only if $\Delta G°$ is small, either negative or positive (or even zero), are significant quantities of both reactants and products present at equilibrium under standard conditions (b)–(d).

An oil field fire. No useful work is being obtained from the combustion of this petroleum.

in free energy between reactants and products changes from very negative to zero to very positive, the percent completion of reaction changes from almost complete to almost no reaction at equilibrium. If $\Delta G°$ is more negative than about -20 kJ/mol, reaction is more than 99.9% complete at equilibrium. If $\Delta G°$ is more positive than about 20 kJ/mol, little reaction takes place; more than 99.9% of the starting materials remains unreacted at equilibrium. *Only if $\Delta G°$ is between -20 and $+20$ kJ/mol can significant quantities of both reactants and products be present at equilibrium under standard conditions.*

Note carefully the difference between ΔG and $\Delta G°$: *$\Delta G°$ is the free-energy change that accompanies a change from reactants in their standard states to products in their standard states, while ΔG is the free-energy change that accompanies a change from reactants in nonstandard states to products in nonstandard states.* For any system of equilibrium, $\Delta G = 0$ but $\Delta G°$ is very rarely equal to zero.

17.11 FREE ENERGY AND USEFUL WORK

Chemical reactions are often carried out in order to obtain useful work. For example, the purpose of burning gasoline in a car's engine is to convert chemical potential energy to thermal energy. The thermal energy released causes the gases formed by combustion (CO_2 and H_2O) to expand and push the pistons so that the car moves and people and things are carried from one place to another. Free energy is called free energy because the free energy lost by the system during a spontaneous chemical (or physical) change that takes place at constant temperature and pressure is a measure of the maximum amount of energy that is free to do work. The work actually done is always less than this maximum amount. The surroundings (the coolant in your car's engine, for example) are heated. Some energy is always wasted raising the temperature of the surroundings; this wasted energy is the cause of thermal pollution. Thus, energy, which cannot be created or destroyed, is changed from a concentrated, useful form into energy that is less concentrated and less useful. The entropy of the surroundings increases.

The quantity of work done by a spontaneous change depends on how the change is carried out. For example, if a gallon of gasoline is burned in an open container, no useful work is done. A gallon of gasoline burned in the engine of a car moves the pistons which in turn move the car and its contents 15–30 miles. Work is *not* a state function. Usually, the faster a change takes place, the less work is obtained and the more energy is wasted heating the surroundings. However, if a change is carried out too slowly, it is of no practical value.

To bring about a nonspontaneous change, it must be coupled with a spontaneous change. For example, you know from your everyday experience that iron rusts spontaneously. Rust is hydrated iron(III) oxide; the reaction involved in the rusting of iron is

$$4Fe(s) + 3O_2(g) \longrightarrow 2Fe_2O_3(s) \qquad \Delta G° = -1484.4 \text{ kJ}$$

The reverse reaction

$$2Fe_2O_3(s) \longrightarrow 4Fe(s) + 3O_2(g) \qquad \Delta G° = +1484.4 \text{ kJ}$$

is *not* spontaneous. Yet many tons of iron are produced each year from iron(III) oxide. The conversion of $Fe_2O_3(s)$ to $Fe(s)$ is coupled with a spontaneous reaction that releases more free energy than is needed to reduce iron(III) oxide to iron:

$$
\begin{array}{ll}
2Fe_2O_3(s) \longrightarrow 4Fe(s) + 3O_2(g) & \Delta G° = +1484.4 \text{ kJ} \\
\underline{6CO(g) + 3O_2(g) \longrightarrow 6CO_2(g)} & \Delta G° = -1542.7 \text{ kJ} \\
2Fe_2O_3(s) + 6CO(g) \longrightarrow 4Fe(s) + 6CO_2(g) & \Delta G° = -58.3 \text{ kJ}
\end{array}
$$

Many familiar examples exist of the tendency of systems to become disordered. However, systems that are far from equilibrium can proceed spontaneously to order and organization. ▪Figure 17.6 shows organization created by a chemical reaction. Organization of a system, of course, results in an increase in entropy of the surroundings. The entropy of the universe must increase in any spontaneous change.

A bottle of water provides a simple example of a system near and far from equilibrium that may help you to understand the difference. A bottle of water that is standing up [▪Figure 17.7(a)] is in a state of equilibrium. As long as the bottle of water is not disturbed, it will remain standing forever. (This equilibrium is called a static equilibrium in contrast to the physical and chemical equilibria we have been talking about, which are dynamic equilibria.) If the bottle is tipped to a position not too far from the equilibrium position, water flows from the bottle in a steady stream as shown in Figure 17.7(b). But if the bottle is placed on its side, a position far from the vertical equilibrium position, water flows from the bottle in spurts as shown in Figure 17.7(c). The flow of water under conditions far from equilibrium has a periodic property.

Organization and periodic repeating changes are characteristic of biological systems. Think of even the simplest cell or a beating heart. An understanding of how organization takes place by means of chemical reactions may help to explain the origin of life.

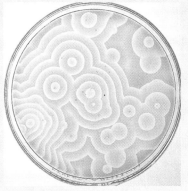

▪ FIGURE 17.6   In the Belousov–Zhabotinskii reaction, striking patterns of concentric circles develop from an initially homogeneous mixture.

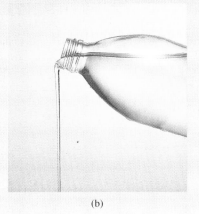

| (a) | (b) | (c) |

▪ FIGURE 17.7   If a bottle of water is far enough from the vertical equilibrium position, water flows from it in spurts. Note the periodicity of wide and narrow portions in the lower part of the stream of water in (c).

or simplifying by dividing by 2:

$$Fe_2O_3(s) + 3CO(g) \longrightarrow 2Fe(s) + 3CO_2(g) \qquad \Delta G° = -29.2 \text{ kJ}$$

Coupled reactions are very important in biochemistry. For example, in cells the energy required to bring about nonspontaneous reactions, such as the synthesis of proteins, is provided by spontaneous reactions that are involved in the metabolism of foods.

The free energy gained by a system during a nonspontaneous change is a measure of the minimum amount of work needed to bring about the nonspontaneous change. More work than this minimum amount is always required because some energy is always wasted raising the temperature of the surroundings.

Directions for demonstrating the Belousov–Zhabotinskii reaction are in Chem. Eng. News *30 March* **1987,** *28–29.*

17.11 FREE ENERGY AND USEFUL WORK

The law of conservation of energy is the first law of thermodynamics. According to the first law, energy cannot be created or destroyed. The second law makes possible prediction of the direction of change. According to the second law, although the entropy of a system may decrease, the entropy of the universe is increasing. The third law says that the entropy of perfect crystalline substances is zero at absolute zero; however, it is impossible to reach absolute zero.

Entropy is a measure of the disorder of a system and of the probability that a system will be in a given state. The more arrangements that are possible for the molecules of a sample or the more energy levels that can be occupied, the greater the entropy. Entropy increases with increasing temperature. Although all spontaneous changes increase the entropy of the universe, the entropy of a system can either increase or decrease in a spontaneous change. If the entropy of the system decreases, the entropy of the surroundings must increase more. **Standard entropy, $S°$,** is the entropy of a substance in a standard state. Entropy is a state function; calculation of standard entropy changes from standard entropies is similar to calculations of standard enthalpy changes from standard enthalpies of formation.

Free energy, G, combines enthalpy and entropy in a single state function. The equation relating standard free-energy change to changes in standard enthalpy and standard entropy is

$$\Delta G° = \Delta H° - T\,\Delta S°$$

Standard free-energy changes can be calculated from the definition of standard free-energy changes or from standard free-energy changes of formation. The free-energy changes accompanying changes carried out under nonstandard conditions can be calculated from the equation

$$\Delta G = \Delta G° + RT \ln Q$$

For spontaneous changes, ΔG is negative. If ΔH and ΔS both have the same sign, temperature determines the direction of spontaneous change. If ΔH and ΔS are assumed not to change with temperature, the definition of free-energy change can be used to estimate free-energy changes at different temperatures.

The standard free-energy change is related to the equilibrium constant by the equation

$$\Delta G° = -RT \ln K$$

This relationship can be used to find the value of K or, if K is known, to calculate $\Delta G°$. The free energy lost by the system during a spontaneous change at constant temperature and pressure is the maximum amount of energy that is free to do useful work. The free energy gained by the system during a nonspontaneous change at constant temperature is the minimum amount of work needed to bring about a nonspontaneous change.

Thermodynamics tells whether a change can take place spontaneously and how far a change goes before reaching equilibrium. It does not provide any information about how fast a change will take place or about how changes take place.

For information about the organization of Additional Practice Problems, Stop & Test Yourself, Putting Things Together, and Applications, see the beginnings of these sections in Chapter 1.

17.15 State the three laws of thermodynamics in your own words. (17.1)

17.16 From your everyday life, give an example of an increase in entropy. (17.2)

17.17 (a) Explain why $S°$ for an element in its standard state at 25 °C is not zero. (b) Why does $S°$ increase with increasing temperature? (17.2)

17.18 When an exothermic change occurs, what happens to the entropy of the surroundings? (17.2)

17.19 What property of entropy makes possible the calculation of $\Delta S°_{rxn}$ by equation 17.1? (17.2)

17.20 Which member of each of the following pairs has the higher standard entropy at 25 °C? Explain your answers.   (a) $H_2O(l)$ or $H_2O_2(l)$ (b) $N_2O(g)$ or $NO(g)$ (c) $NaCl(s)$ or $NaBr(s)$ (d) $NH_3(g)$ or $PH_3(g)$ (17.2)

17.21 Predict whether the entropy of the system will increase, decrease, or remain about the same when each of the following changes takes place at constant temperature and pressure. Explain your answers.
(a) $I_2(s) \longrightarrow I_2(g)$
(b) $2C(graphite) + O_2(g) \longrightarrow 2CO(g)$
(c) $C(graphite) + O_2(g) \longrightarrow CO_2(g)$
(d) $3Fe_2O_3(s) + CO(g) \longrightarrow 2Fe_3O_4(s) + CO_2(g)$
(e) $4NH_3(g) + 3O_2(g) \longrightarrow 2N_2(g) + 6H_2O(g)$ (17.2)

17.22 Predict whether the entropy of the system will increase, decrease, or remain about the same when each of the following changes takes place at constant temperature and pressure:
(a) 0.5 mol $H_2(g)$ is added to a container that already holds 0.5 mol $Cl_2(g)$ (b) $CaO(s) + CO_2(g) \longrightarrow CaCO_3(s)$
(c) $H_2O(g) \longrightarrow H_2O(l)$ (d) $2HgO(s) \longrightarrow 2Hg(l) + O_2(g)$
(e) $N_2(g) + O_2(g) \longrightarrow 2NO(g)$ (17.2)

17.23 Calculate the standard entropy changes that accompany each of the following reactions:

(a) $2NH_3(g) + 2O_2(g) \longrightarrow N_2O(g) + 3H_2O(g)$

(b) $2NaHCO_3(s) \longrightarrow Na_2CO_3(s) + H_2O(g) + CO_2(g)$

(c) $CaO(s) + CO_2(g) \longrightarrow CaCO_3(s)$

(d) $2Na(l) + Cl_2(g) \longrightarrow 2NaCl(s)$

(e) $2SO_2(g) + O_2(g) \longrightarrow 2SO_3(g)$ (17.2)

17.24 Calculate the standard entropy changes that accompany each of the following reactions: (a) $H_2(g) + Cl_2(g) \longrightarrow 2HCl(g)$

(b) $N_2(g) + O_2(g) \longrightarrow 2NO(g)$

(c) $C_2H_2(g) + 2H_2(g) \longrightarrow C_2H_6(g)$ (d) $Na(l) \longrightarrow Na(s)$

(e) $CH_3CH_2OH(l) + 3O_2(g) \longrightarrow 2CO_2(g) + 3H_2O(l)$ (17.2)

17.25 Calculate ΔS_f° for $CH_3OH(l)$ at 25 °C. (17.2)

17.26 Why is the value of ΔG much more dependent on temperature than the values of ΔH and ΔS? (17.3)

17.27 Explain why, under ordinary conditions, most changes that are spontaneous are exothermic. (17.4)

17.28 From the following equations and standard enthalpies, predict whether the reaction will be spontaneous at all temperatures, spontaneous at low temperatures but nonspontaneous at high temperatures, nonspontaneous at low temperatures but spontaneous at high temperatures, or not spontaneous at any temperature.

(a) $2HgO(s) \longrightarrow 2Hg(l) + O_2(g)$ $\Delta H^\circ = 90.8$ kJ

(b) $4Fe(s) + 3O_2(g) \longrightarrow 2Fe_2O_3(s)$ $\Delta H^\circ = -1648.4$ kJ

(c) $N_2(g) + 3Cl_2(g) \longrightarrow 2NCl_3(g)$ $\Delta H^\circ = 460$ kJ

(d) $2NH_4NO_3(s) \longrightarrow 2N_2(g) + 4H_2O(g) + O_2(g)$
$\Delta H^\circ = -225.5$ kJ

(e) $H_2C\!=\!CH_2(g) + H_2(g) \longrightarrow H_3CCH_3(g)$
$\Delta H^\circ = -137.0$ kJ (17.4)

17.29 From the following equations and standard enthalpies, predict whether the changes will be spontaneous at all temperatures, spontaneous at low temperatures but nonspontaneous at high temperatures, nonspontaneous at low temperatures but spontaneous at high temperatures, or not spontaneous at any temperature:

(a) $I_2(s) \longrightarrow I_2(g)$ $\Delta H^\circ = 62.24$ kJ

(b) $2SO_2(g) + O_2(g) \longrightarrow 2SO_3(g)$ $\Delta H^\circ = -197.78$ kJ

(c) $CaCO_3(s) \longrightarrow CaO(s) + CO_2(g)$ $\Delta H^\circ = 178.32$ kJ

(d) $CH_3OH(l) + CO(g) \longrightarrow CH_3COOH(l)$
$\Delta H^\circ = -37.8$ kJ

(e) $2HCN(g) + 6H_2O(g) \longrightarrow 2CH_4(g) + 2NH_3(g) + 3O_2(g)$
$\Delta H^\circ = 929.60$ kJ (17.4)

17.30 A reaction is spontaneous at 975 °C but nonspontaneous at 25 °C. Without looking at Table 17.2, what are the signs of ΔH° and ΔS°? Explain your answer. (17.4)

17.31 For each of the following reactions, calculate the value of ΔG° at 25 °C:

(a) $CH_4(g) + NH_3(g) \longrightarrow HCN(g) + 3H_2(g)$

(b) $MnO_2(s) + 2C(graphite) \longrightarrow Mn(s) + 2CO(g)$

(c) $3Fe_2O_3(s) + CO(g) \longrightarrow 2Fe_3O_4(s) + CO_2(g)$

(d) $WO_3(s) + 3H_2(g) \longrightarrow W(s) + 3H_2O(g)$

(e) $2H_2S(g) + 3O_2(g) \longrightarrow 2H_2O(l) + 2SO_2(g)$ (17.6)

17.32 For each of the following reactions, calculate the value of ΔG° at 25 °C:

(a) $2Fe_2O_3(s) + 3C(graphite) \longrightarrow 4Fe(s) + 3CO_2(g)$

(b) $4NH_3(g) + 5O_2(g) \longrightarrow 4NO(g) + 6H_2O(l)$

(c) $2NO(g) + O_2(g) \longrightarrow 2NO_2(g)$

(d) $CaO(s) + H_2O(l) \longrightarrow Ca(OH)_2(s)$

(e) $2H_2S(g) + 3O_2(g) \longrightarrow 2H_2O(g) + 2SO_2(g)$ (17.6)

17.33 Estimate the value of ΔG° at 995 °C for the reaction in Problem 17.32(a). (17.7)

17.34 The reaction in 17.32(a) is nonspontaneous at low temperature and spontaneous at high temperature. Estimate the temperature (in °C) at which the change from nonspontaneous to spontaneous takes place. (17.8)

17.35 For $SO_3(l)$,   (a) estimate the normal boiling point.   (b) Estimate the normal freezing point. (17.8)

17.36 According to Table 12.2, ΔH_{fusion} for sodium metal, which melts at 97.8 °C, is 2.60 kJ/mol. Calculate ΔS_{fusion} for sodium metal. (17.8)

17.37 Under what conditions can a reaction be spontaneous if ΔG° is positive? (17.9)

17.38 Calculate ΔG for the reaction

$$2SO_2(g) + O_2(g) \longrightarrow 2SO_3(g)$$

at 25 °C if $p_{SO_2} = 0.153$ atm, $p_{O_2} = 0.164$ atm, and $p_{SO_3} = 0.683$ atm. The value of ΔG° at 25 °C is -141.78 kJ. Predict the direction of spontaneous change. (17.9)

17.39 The value of K_p for the equilibrium

$$C(graphite) + CO_2(g) \rightleftharpoons 2CO(g)$$

is 1.81 at 1000 K (1.000×10^3 K). From this information, calculate the value of ΔG° for this equilibrium at 1000 K. (17.10)

17.40 For the equilibrium

$$C(graphite) + 2H_2(g) \rightleftharpoons CH_4(g)$$

the value of ΔG° at 1000 °C (1.000×10^3 °C) estimated from ΔH° and ΔS° at 25 °C is $+28.24$ kJ. Calculate the value of K_p for this equilibrium at 1000 °C. (17.10)

17.41 Will both Br_2 and I_2 react spontaneously with methane

$$CH_4(g) + Br_2(g) \longrightarrow CH_3Br(g) + HBr(g)$$

$$CH_4(g) + I_2(g) \longrightarrow CH_3I(g) + HI(g)$$

under standard conditions at 25 °C?

17.42 Entropy has been called the "arrow of time." Discuss the meaning of this description of entropy.

17.43 For the reaction

$$2CH_4(g) \longrightarrow CH_3CH_3(g) + H_2(g)$$

(a) what is the value of ΔG°?   (b) Is this reaction spontaneous under standard conditions?   (c) Is it spontaneous under any conditions?   (d) Would it be worth looking for a catalyst?

17.44 When stretched rubber contracts rapidly, it cools. (You can observe this for yourself by stretching a wide rubber band or a balloon, placing it across your lips, and letting it contract rapidly.)   (a) What must be the sign of ΔS when rubber contracts? (b) Are the molecules in stretched rubber more or less ordered than the molecules in unstretched rubber?

17.45 If temperatures are the same, which member of each pair has higher entropy? Explain your answers.

(a)

(b)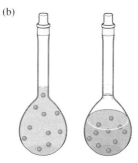

(c)

STOP & TEST YOURSELF

1. Consider the following diagram for a gas:

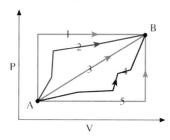

The change in entropy ΔS in going from state A to state B is (a) the same for paths 1 through 5. (b) different for each of the paths. (c) smallest for path 3 because it is the shortest path. (d) the same for paths 1 and 5 but different for paths 2, 3, and 4. (e) None of the above statements is correct.

2. As observed in the world around us, each of the following processes is spontaneous except (a) a rock rolling downhill. (b) formation of hydrogen and oxygen by boiling water. (c) rusting of an iron nail. (d) burning of natural gas. (e) dissolving of salt (NaCl) in water.

3. Which of the following changes (at constant temperature and pressure) would not result in a substantial increase in the entropy of the system?
(a) $2NH_3(g) \longrightarrow N_2(g) + 3H_2(g)$
(b) $2HF(g) \longrightarrow H_2(g) + F_2(g)$
(c) $2NaNO_3(s) \longrightarrow 2NaNO_2(s) + O_2(g)$
(d) $CH_3OH(l) \longrightarrow CH_3OH(g)$
(e) $CH_3OH(s) \longrightarrow CH_3OH(l)$

4. Calculate $\Delta S°$ for the reaction $2NH_3(g) + 3Cl_2(g) \longrightarrow N_2(g) + 6HCl(g)$ at 25 °C. (a) -37.0 J/K (b) 37.0 J/K (c) 258.8 J/K (d) 736.2 J/K (e) 2365.8 J/K

5. The ΔH for a reaction was -20 kJ. On the basis of this fact, we can conclude that the reaction (a) is spontaneous. (b) is fast. (c) is slow. (d) absorbs thermal energy. (e) gives off thermal energy.

6. Which of the following changes is spontaneous at low temperatures but nonspontaneous at high temperatures?
(a) $C_6H_5OH(s) \longrightarrow C_6H_5OH(l)$
(b) $CS_2(g) + 3Cl_2(g) \longrightarrow CCl_4(g) + S_2Cl_2(g)$
$$\Delta H°_{rxn} = -238 \text{ kJ}$$
(c) $C_6H_{12}O_6(s) + 6O_2(g) \longrightarrow 6CO_2(g) + 6H_2O(l)$
$$\Delta H°_{rxn} = -2816 \text{ kJ}$$
(d) $CaO(s) + 3C(s) \longrightarrow CaC_2(s) + CO(g)$ $\Delta H°_{rxn} = 462$ kJ
(e) $N_2(g) + 3Cl_2(g) \longrightarrow 2NCl_3(l)$ $\Delta H°_{rxn} = 230$ kJ

7. Calculate $\Delta G°$ at 25 °C for the reaction

$$3NO_2(g) + H_2O(l) \longrightarrow 2HNO_3(l) + NO(g)$$

(a) 8.07 kJ (b) 639.43 kJ (c) 331.63 kJ (d) -466.29 kJ (e) None of the preceeding answers is correct.

8. Calculate $\Delta G°$ at 552 °C for the reaction

$$N_2(g) + O_2(g) \longrightarrow 2NO(g)$$

(a) -2.02×10^4 kJ (b) -167.1 kJ (c) 69.9 kJ (d) 160.1 kJ (e) 166.9 kJ

9. A reaction mixture has concentrations such that ΔG is zero. However, $\Delta G°$ for the reaction is $+10$ kJ. This information means that (a) further reaction will take place to form more products. (b) the reverse reaction will take place to form more reactants. (c) the reaction mixture is at equilibrium but the concentration of products is small. (d) the reaction mixture is at equilibrium and the concentration of products is large. (e) None of the above statements is correct.

10. Which of the following statements is (are) true? (i) The entropy of 1 mol Ar(g) at 25 °C is greater than the entropy of 1 mol HCl(g) at 25 °C. (ii) The entropy of 1 mol HCl(g) at 25 °C is greater than the entropy of 1 mol Ar(g) at 25 °C. (iii) The entropy of 1 mol Ar(g) at 25 °C is greater than the entropy of 1 mol Ar(g) at 25 K. (iv) The entropy of 1 mol Ar(g) at 25 K is greater than the entropy of 1 mol Ar(g) at 25

°C. (v) The entropy of 1 mol HCl dissolved in enough water to make 1 M HCl(aq) is greater than the entropy of 1 mol HCl dissolved in enough water to make 0.1 M HCl(aq). (a) Only ii (b) Only iii (c) ii and iii (d) i, iv, and v (e) None of the above statements is correct.

11. Which of the following statements is true of a spontaneous process? (a) The entropy of the universe increases. (b) The energy of the universe increases. (c) The process is fast. (d) The reverse process cannot be made to take place. (e) ΔG is +.

12. For the reaction $CO(g) + Cl_2(g) \longrightarrow COCl_2(g)$, $\Delta H° = -108.28$ kJ and $\Delta S° = -131.63$ J/K at 25 °C. At what temperature does the reverse reaction become spontaneous? (a) 0.823 °C (b) 823 °C (c) 549 °C (d) -272 °C (e) None of the preceeding answers is correct.

13. If, for the reaction $A + B \longrightarrow C + D$, $\Delta G° = 28.5$ kJ at 25 °C, what is the numerical value of the equilibrium constant? (a) 1×10^5 (b) 1×10^{-5} (c) 1×10^{60} (d) 1×10^{-60} (e) None of the preceeding answers is correct.

14. Which of the following statements is true? (a) Thermal energy cannot be completely converted into work. There is always some thermal energy that is transferred to the surroundings. (b) When a process takes place spontaneously, the system tends to end up in a more ordered state. (c) When a process takes place spontaneously, the system tends to end up in a higher energy state. (d) For processes that are not spontaneous, the free-energy change is a measure of the maximum amount of work that must be done to cause the process to take place. (e) How much work is obtained from a process does not depend on how the process is carried out.

PUTTING THINGS TOGETHER

17.46 Calculate $\Delta G_f°$ for CoO(s) at 25 °C.

17.47 Estimate ΔG at 525 K for the reaction

$$CoO(s) + H_2(g) \longrightarrow Co(s) + H_2O(g)$$

if $p_{H_2} = 2.1$ atm and $p_{H_2O} = 4.9$ atm.

17.48 Estimate the value of K_p for the equilibrium

$$2CH_4(g) \rightleftharpoons HC\equiv CH(g) + 3H_2(g)$$

at 2000 K (2.000×10^3 K) from thermodynamic data.

17.49 Would you predict the products of the reaction of carbon monoxide gas and hydrogen gas at 25 °C to be chiefly formaldehyde, $H_2C{=}O(g)$, or methyl alcohol, $CH_3OH(l)$?

17.50 The reaction

$$2Cl_2(g) + 2H_2O(g) \longrightarrow 4HCl(g) + O_2(g)$$

is nonspontaneous at low temperatures but spontaneous at high temperatures. (a) Estimate the temperature (in °C) at which the change from nonspontaneous to spontaneous takes place. (b) Explain what the change from nonspontaneous to spontaneous means (that is, at a temperature 5 °C lower than the transition temperature, does no reaction take place, and is reaction complete at a temperature 5 °C higher than the transition temperature?).

17.51 What is the vapor pressure of iodine at 25 °C?

17.52 At 0 K, a crystal of carbon monoxide is found to have an entropy greater than zero. Suggest two possible explanations.

17.53 The normal melting point of benzene is 5.5 °C. For the melting of benzene at 1 atm, what is the sign of (a) ΔH? (b) ΔS? (c) ΔG at 5.5 °C? (d) ΔG at 0.0 °C? (e) ΔG at 25.0 °C?

17.54 Calculate the value of $\Delta S°$ at 25 °C for the reaction of silver with hydrogen sulfide gas and oxygen from air to form silver sulfide and water.

17.55 Calculate the value of $\Delta G°$ at 25 °C for the decomposition of hydrogen peroxide to water and oxygen gas.

17.56 Distinguish between members of each of the following pairs of terms: (a) bond dissociation energy and bond energy (b) Q and K (c) ΔG and $\Delta G°$

17.57 A student is upset because the instructor has marked the equation

$$CH_4(g) + H_2O(g) \longrightarrow CH_3OH(g) + H_2(g)$$

wrong on a test. If you were the instructor, how would you show the student that this reaction is impossible?

17.58 (a) What are allotropes? (b) Which allotrope of phosphorus is more stable, white or red? (c) Which allotrope must be the more common allotrope? Explain your answers.

17.59 For the equilibrium

$$SO_2(g) + NO_2(g) \rightleftharpoons NO(g) + SO_3(g)$$

what is the value of K_c at 25.0 °C?

17.60 Use a Born–Haber cycle (Section 9.2) to calculate the lattice energy for $CaCl_2$. You will need the following information in addition to that given in tables in this book: For calcium, the enthalpy of sublimation is 178.2 kJ/mol, and the second ionization energy is 1145 kJ/mol.

17.61 For the equilibrium

$$PCl_5(g) \rightleftharpoons PCl_3(g) + Cl_2(g)$$

(a) estimate the value of $\Delta G°$ at 250 °C. (b) Estimate the value of K_p at 250 °C. (c) If the initial pressure of PCl_5 is 0.99 atm, what will be the partial pressures of all three gases at equilibrium? (d) If some PCl_3 is removed from the equilibrium mixture, in which direction will the equilibrium shift? (e) How will the value of ΔG immediately after the PCl_3 is removed compare with the value $\Delta G°$?

17.62 (a) Calculate $\Delta S°$ for the dissolving of sodium chloride in water. (b) Calculate $\Delta S°$ for the dissolving of magnesium sulfate in water. (c) Explain the difference between the answers to parts (a) and (b). (A good explanation will include a microscopic picture.)

17.63 If two liquids are miscible and form an ideal solution, would a positive, a negative, or a zero value be expected for ΔH, ΔS, and ΔG? Explain your answers.

17.64 (a) Which of the four types of reactions in Table 17.2 is shown by the graph? (b) On the graph, label the point that shows ΔH°. (c) Sketch graphs showing the other three types of reactions.

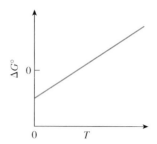

APPLICATIONS

17.65 In living cells, small molecules are joined together to form large molecules. Explain how this can happen.

17.66 From your everyday life, give an example of a chemical reaction that is used to do work.

17.67 Hydrazine, H_2NNH_2, is used as a rocket fuel. For liquid hydrazine at 25 °C, $\Delta H_f^\circ = 50.63$ kJ/mol and $S^\circ = 121.2$ J/K·mol. Can hydrazine be prepared by the following reaction? Explain your answer.

$$N_2(g) + 2H_2(g) \longrightarrow H_2NNH_2(l)$$

17.68 Hydrogen cyanide, which is used to make Lucite and Plexiglas, is manufactured by the reaction

$$2CH_4(g) + 2NH_3(g) + 3O_2(g) \longrightarrow 2HCN(g) + 6H_2O(g)$$

(a) Is this reaction spontaneous at all temperatures, spontaneous only at low temperatures, or spontaneous only at high temperatures? (b) A temperature of 1100 °C is used to carry out the reaction industrially. Suggest a reason why such a high temperature is used. (Heating is expensive.)

17.69 The reaction between 1,1-dimethylhydrazine, $(CH_3)_2NNH_2$, and dinitrogen tetroxide to form carbon dioxide, nitrogen, and water vapor was used to power the Apollo Lunar Landers. The reactants were chosen because both are liquids under ordinary conditions (so are easy to store) and reaction between them starts spontaneously. They were used in the proportions shown by the equation because the cost of putting materials in space is enormous. Launching an excess of either reactant would be very wasteful. (a) How many grams of dinitrogen tetroxide are required to react with each gram of 1,1-dimethylhydrazine? (b) For this reaction, $\Delta H_{rxn}^\circ = -29$ kJ/g of 1,1-dimethylhydrazine. Give two reasons why reaction is spontaneous.

17.70 (a) Is the formation of ozone from molecular oxygen, $3O_2(g) \longrightarrow 2O_3(g)$, spontaneous under standard conditions? Explain your answer. (b) Explain why the characteristic odor of ozone is observed after lightning has hit.

17.71 Uranium isotopes are usually separated by effusion of $UF_6(g)$ (Section 5.9). Effusion takes place spontaneously although the enthalpy change involved is close to zero. What must be the sign of the entropy change?

17.72 Gasohol is a mixture of about 90% gasoline and 10% ethyl alcohol, CH_3CH_2OH. Addition of ethyl alcohol to gasoline increases the octane number, gives farmers a market for sur-

plus corn (ethyl alcohol can be made by fermentation of the sugars in corn), and reduces the need to import oil. Recently, substitution of butyl alcohol, $CH_3CH_2CH_2CH_2OH$, which can also be prepared by fermentation, for ethyl alcohol has been suggested. (a) For $CH_3CH_2CH_2CH_2OH(l)$, ΔH_f° is -327 kJ/mol. What are the enthalpies of combustion of ethyl alcohol and butyl alcohol? (b) How many kilojoules of thermal energy are released by burning 1 g of each of these two alcohols? (c) How would you expect the vapor pressure of butyl alcohol to compare with the vapor pressure of ethyl alcohol? Explain your answer. (d) In an automobile engine, the gasoline must vaporize. What is one advantage of ethyl alcohol over butyl alcohol? (e) How would you expect the solubility of water in ethyl alcohol and in butyl alcohol to compare? (f) One problem with gasohol is that it absorbs water from air. What is one advantage of butyl alcohol as an additive for gasoline?

17.73 The formation of smog begins with the reaction of nitrogen and oxygen in automobile engines:

$$N_2(g) + O_2(g) \longrightarrow 2NO(g)$$

The nitric oxide then reacts with oxygen to form red-brown nitrogen dioxide:

$$2NO(g) + O_2(g) \longrightarrow 2NO_2(g)$$

The nitrogen dioxide is decomposed by light to atomic oxygen and nitric oxide,

$$NO_2(g) \xrightarrow{\text{light}} NO(g) + O(g)$$

and the atomic oxygen reacts with oxygen to form ozone:

$$O(g) + O_2(g) \longrightarrow O_3(g)$$

Ozone damages plants and animals including people and reacts with unburned hydrocarbons from auto exhaust in a complex series of reactions yielding a number of irritants. (a) Calculate ΔH° for each of the four reactions given. (b) For which of the four reactions can the sign of ΔS° be predicted from the equation? (c) For the other(s), calculate ΔS°. (d) For each reaction, tell the temperature conditions, if any, under which the reaction is spontaneous or nonspontaneous. (e) Write Lewis formulas for O_3, NO, and NO_2. (f) Explain why NO and NO_2 are very reactive species.

17.74 Reaction of NO_2, formed as described in Problem 17.73, with water to form nitric acid

$$3NO_2(g) + H_2O(l) \longrightarrow 2HNO_3(aq) + NO(g)$$

would contribute to the formation of acid rain. Calculate $\Delta G°$ for the reaction of NO_2 with water at 25 °C and tell whether reaction is spontaneous or nonspontaneous under standard conditions.

17.75 Considerable effort is being made to apply anaerobic digestion (fermentation) to the conversion of glucose from plant biomass to fuel:

$$C_6H_{12}O_6(aq) \longrightarrow 3CH_4(g) + 3CO_2(g) \qquad \Delta G° = -418 \text{ kJ}$$

(a) What is $\Delta G_f°$ for $C_6H_{12}O_6(aq)$? (b) How many kilojoules of free energy are released by the combustion of the methane formed from one mole of glucose?

17.76 Hydrogen sulfide must be removed from natural gas before the natural gas is piped to consumers. One process for the removal of hydrogen sulfide from natural gas involves the equilibrium

$$Na_2CO_3(s) + H_2S(g) \rightleftharpoons NaHCO_3(s) + NaHS(s)$$

(a) Write the equilibrium constant expression for this equilibrium. (b) What is $\Delta G°$ for this equilibrium at 25 °C? (c) Calculate the value of K_p at 25 °C. (d) What is the value of K_c at 25 °C? (e) Should the temperature used for this process be higher or lower than 25 °C? Explain your answer. (f) Hydrogen sulfide is used as the source of sulfur or of sulfur dioxide for sulfuric acid plants. How could H_2S be recovered? (g) The value of $S°$ for NaHS is *not* given in the table in the appendix. Calculate the value of $S°$ for NaHS at 25 °C.

17.77 Deep in the sea where there is no light, bacteria use the oxidation of hydrogen sulfide,

$$2H_2S(aq) + O_2(aq) \longrightarrow 2H_2O(l) + 2S(s)$$

to provide energy for the synthesis of glucose:

$$6CO_2(g) + 6H_2O(l) \longrightarrow C_6H_{12}O_6(aq) + 6O_2(g)$$
$$\Delta G° = 2870 \text{ kJ at } 25 \text{ °C}$$

The overall reaction is

$$24H_2S(aq) + 6CO_2(aq) + 6O_2(aq) \longrightarrow$$
$$C_6H_{12}O_6(aq) + 18H_2O(l) + 24S(s)$$

Show that this overall reaction is spontaneous at 25 °C.

17.78 (a) If a tank truck carrying liquid hydrogen cyanide is involved in a highway accident and develops a leak, what happens to the liquid hydrogen cyanide? (b) Which requires less thermodynamic work, cleaning up a spill or preventing it (the green chemical approach)? (c) Suggest some steps for preventing spills.

The Pharmacist Who Hated Arithmetic

JEANETTE K. WIEGAND

Pharmacist

B.S. Psychology
Idaho State University

Pharm. D.
Idaho State University

As was typical with young girls in the 1950s and 1960s, I thought I hated math and science. In the 7th grade I finally had a real introduction to science—a genuine science teacher and a good-looking one at that. He had just graduated from college and was motivated (inspired) and made science fun and interesting. I had this same science teacher the next year, and he convinced me to take as much science as possible in high school. My small, rural high school in Nevada didn't offer much variety in the sciences, but chemistry and physics were options that were available. I am sorry to say that math remained uninteresting to me, and no similar mentors surfaced.

My first job was in a drugstore when I was sixteen. The pharmacist suggested I try pharmacy as a career goal. Upon graduation from high school I entered pharmacy school in Idaho, but other things got in my way of a pharmacy degree (marriage and a child to raise). After living in the East for several years, I returned to Idaho and earned a Bachelor of Science degree in psychology. An undergraduate liberal arts degree wasn't very helpful in the job market though, and I wasn't interested in pursuing a graduate level education at that time.

After many years as a bookkeeper (interpretation: doing arithmetic every day), I went back to college to finish my pharmacy education. It had been 13 years since I had graduated from college and almost 20 years since I had taken any chemistry, but I completed organic chemistry and all of the other classes I was required to take. So at age forty-five, and 28 years after starting pharmacy school, I received a Doctor of Pharmacy degree.

The curriculum in pharmacy is based on chemistry, biology, and math. Most of the classes use chemistry either directly or indirectly. My current job as a pharmacist in a pediatric setting also uses chemistry (and biology) either directly or indirectly. Much of my job also requires communications skills—dealing with parents and answering their questions. Many of these questions involve "chemicals." All drugs, even the natural ones, are chemicals. Drugs work in the body or on body invaders (bacteria, fungi, viruses, and cancer cells). The job of the health professional is to know the targets for the drug (such as receptor sites, organs, and bacteria) and to know and inform the patients of the effects of their medications, both desired and undesired (side effects). I must be aware of (or look up) the pH of drugs and of the body systems they target (e.g., acidic or basic). I have to know about protein binding (chemical bonding). I have to recognize "functional groups" to be able to predict what a particular drug's mechanism of action is.

Occasionally I have to "compound" a drug, that is, combine two or more ingredients to make a special medication or change the form of an existing medication. For example, I often make suspensions from tablets, prepare special medicated creams, or make small doses out of larger ones. Laboratory skills such as weighing and measuring are frequently used by pharmacists and pharmacy technicians. The most frightening aspect of my job is preparing the chemotherapy for cancer patients. There is a very fine line between killing the cancer and killing the patient. Obviously we want to do as much damage as possible to the cancer cells while causing the least amount of damage to the patient.

So, although to the average customer or outside observer it may appear as though pharmacists only transfer medications from larger bottles to smaller ones and put labels on them, there is much more involved. For the most part, I really enjoy my profession and the challenges that are presented to me every day.

CHEMICAL KINETICS: A Closer Look at Reaction Rates

18

After being heated in the flame of a laboratory burner, steel wool reacts rapidly with oxygen in air releasing thermal energy and light energy.

To predict whether a reaction will occur, you need to know whether it will take place at a practical speed as well as about how completely reactants are converted to products at equilibrium. **Chemical kinetics** is the *study of how fast chemical reactions take place.*

Studies of reaction rates are an important source of information about reaction mechanisms. A **reaction mechanism** is a *detailed molecular-level picture of how a reaction takes place;* the equation for a reaction simply identifies the reactants and products. Reaction mechanisms are theories that explain experimental observations. Although new observations can prove them wrong, they are very useful in organizing the huge amount of information that exists about chemical reactions.

■ *Why is a study of kinetics a part of general chemistry?*

A knowledge of rates of reactions and pathways from reactants to products is important from both practical and theoretical viewpoints. For example, the success of chemical processes in industry depends on rates of chemical reactions. A reaction that needs years to take place is ordinarily not very useful. One that takes place so fast that it cannot be controlled and blows up the plant is a disaster. There is no relation between the position of equilibrium and how fast a reaction reaches equilibrium. Many reactions that should, according to thermodynamics, go to completion take place too slowly to be useful. The chemical industry spends millions of dollars every year trying to develop catalysts to speed these reactions.

To describe how rapidly a reaction takes place, kinetics is needed. In chemical manufacturing, an understanding of reaction mechanisms and relative rates of reaction may make it possible to choose reaction conditions leading to higher yields of desired products and lower yields of by-products, thereby increasing profits and decreasing pollution. The changes that take place in sediments, which are of great importance in petroleum exploration, are controlled by kinetics. The rates of biochemical reactions—for example, how long the concentration of a drug stays high enough to produce the desired effect—are important in the health sciences. An understanding of reaction mechanisms may make it possible to design a more powerful drug, or one with fewer side effects.

The explosion of dust in a grain elevator started this fire. Dust is the most dangerous form of solid matter.

— What do you already know about kinetics?

From your everyday experience, you know that the rates of chemical reactions depend on what the reactants are. For example, tuna fish salad sandwiches spoil faster than peanut butter and jelly sandwiches. You also know that chemical reactions take place faster at high temperatures. Every cook puts bread dough in a warm place to make it rise faster and the remains of the Thanksgiving turkey in the refrigerator to keep them from spoiling before they get eaten. If you have ever had any experience with an oxygen tent or flown on an airliner, you also know that the rates of chemical reactions depend on concentration. Smoking is forbidden when oxygen is in use because a cigarette bursts into flames in pure oxygen (–Figure 18.1). Knowing that stirring makes sugar dissolve faster in coffee or tea, you probably could figure out that, in a heterogeneous mixture, reaction takes place faster if the mixture is stirred. You also very likely know that lumps of sugar are slow to dissolve. Reactions in heterogeneous mixtures also take place more rapidly if reactants are finely divided.

The rates of many reactions are increased by catalysts. Although catalysts called enzymes are necessary to make biological processes such as digestion take place at body temperature, you are more likely to be aware of the catalytic converter in the exhaust system of your car. Catalytic converters reduce discharge of pollutants, such as carbon monoxide, nitrogen oxides, and unburned hydrocarbons, into the atmosphere and require use of unleaded gas. Unfortunately, most familiar reactions are very complex, and we will have to use simpler but less familiar reactions as examples.

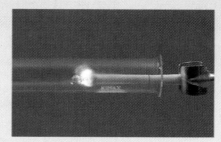

▪ FIGURE 18.1 Combustion of a cigarette is more rapid in pure oxygen than in air!

Hydrocarbons are compounds composed of carbon and hydrogen and no other elements. Methane (CH_4), ethylene ($H_2C{=}CH_2$), and benzene (⬡) are hydrocarbons.

RATES OF REACTIONS

The rate you are probably most familiar with is the speed of your car. If your speedometer reads "55 MPH" (or "88 KPH"), it means that in an hour at that speed you will travel a distance of 55 miles (or 88 kilometers). In a minute, you will travel almost a mile (1.5 km). In other words, the rate at which your car travels is expressed in terms of the change in the position of the car that takes place in a period of time:

$$\text{speed of car} = \frac{\text{position}_{\text{later}} - \text{position}_{\text{earlier}}}{\text{time}_{\text{later}} - \text{time}_{\text{earlier}}} = \frac{\Delta \text{ position}}{\Delta \text{ time}}$$

The rate of a chemical reaction is expressed in terms of the change in the quantity of a reactant or product that takes place in a period of time. If the reaction takes place in a constant volume in the gas phase or in solution, as most reactions studied in the laboratory do, the rate is usually expressed in terms of change in concentration with time:

$$\text{rate of reaction} = \frac{\text{concentration}_{\text{later}} - \text{concentration}_{\text{earlier}}}{\text{time}_{\text{later}} - \text{time}_{\text{earlier}}} = \frac{\Delta[\ \ ]}{\Delta t}$$

Studying the Rate of a Chemical Reaction

Studying the rate of a chemical reaction is harder than studying a chemical equilibrium because the system is changing all the time. Significant additional reaction must not take place during analysis. Temperature must be accurately and precisely controlled because the rates of most reactions change very rapidly as temperature changes. Also, analysis must not disturb the reaction. For example, a cold pipet must not be inserted into a warm reaction mixture to withdraw a sample for analysis.

To study the rate of a reaction, first identify the reactants and products. Sometimes the same reactants can form more than one set of products. Either the relative rates of the reactions or the relative stability of the products can determine the product composition. Products should be identified as soon as possible after reaction is complete because occasionally the products that are formed fastest are converted to thermodynamically more stable products as time passes (see ▪Figure 18.2).

Then carry out the reaction and measure the amounts or concentrations of one of the reactants or products at intervals as reaction takes place. If the reaction takes place slowly, a sample can be removed, the reaction stopped by cooling or dilution, and analyzed, for example, by titration. For fast reactions, continuous methods, such as measuring color, pH, or pressure of a mixture of gases at constant volume, are essential. Even for slow reactions, continuous methods of following the reaction are better. Sensitive automated instruments for following reactions and computers for handling data have made studying reaction rates much easier and have made possible the investigation of very fast and very slow reactions.

A Case Study: The Decomposition of N_2O_5

Let's consider the decomposition of dinitrogen pentoxide as an example of a reaction rate study. In the gas phase or when dissolved in inert solvents like carbon tetrachloride, dinitrogen pentoxide decomposes completely at a rate that is conveniently measured at room temperature or a little above. The products are oxygen gas and dinitrogen tetroxide:

$$2N_2O_5 \longrightarrow 2N_2O_4 + O_2(g)$$

▪ FIGURE 18.2   In dilute solution (right beaker), the less stable yellow form of HgI_2 is formed faster than the more stable red form and then slowly changes to the red form. Similarly, when first formed, the hard parts of star fishes, sea urchins, sand dollars, and sea cucumbers are aragonite, a form of calcium carbonate that is less stable than the common form, calcite, and is slowly converted to calcite.

The time required for changes to take place range from femtoseconds for some biological reactions such as the binding of oxygen to hemoglobin to billions of years for the decay of some radioactive nuclides.

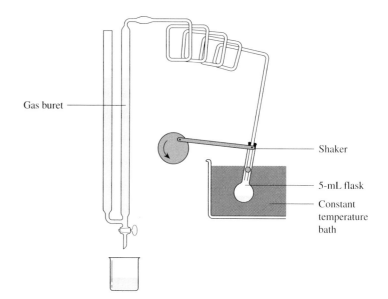

Gas buret

Shaker

5-mL flask

Constant
temperature
bath

◾ FIGURE 18.3   Apparatus for the study of
the rate of decomposition of dinitrogen
pentoxide in carbon tetrachloride solution.
(Reproduced from Eyring, H.; Daniels, F.
J. Amer. Chem. Soc. **1930**, *52*, 1472. © 1930
by The American Chemical Society,
Washington, D.C.)

When a solution of dinitrogen pentoxide in carbon tetrachloride decomposes, the
dinitrogen tetroxide* remains in solution in the carbon tetrachloride while the oxygen, which is not soluble in carbon tetrachloride, escapes.

The easiest substance to measure is the oxygen gas. It can be collected in a gas
buret and its volume measured at intervals. ◾Figure 18.3 shows the apparatus used
to study the rate of decomposition of dinitrogen pentoxide in inert solvents like carbon
tetrachloride. The N_2O_5 was dissolved in carbon tetrachloride. The temperature of
the reaction mixture was kept constant to within ±0.01 °C, and the reaction mixture
was shaken to keep the oxygen from forming a supersaturated solution. The results
of a typical experiment are summarized in ◾Figure 18.4. As you can see, the volume
of oxygen increased quite rapidly at first. As reaction took place, the rate of formation
of oxygen gradually slowed down. By the time all of the dinitrogen pentoxide had
decomposed (infinite time), the rate of formation of oxygen had, of course, decreased
to zero. This behavior is typical: *Most reactions take place rapidly at first and then
slow down as the reactants are used up.*

Average Rates

The **average rate** of formation of oxygen during any interval of time can be
calculated by dividing the change in the volume of oxygen by the change in time:

$$\text{average rate of formation of oxygen} = \frac{\Delta V_{O_2}}{\Delta t}$$

This calculation is similar to calculating your average speed for a trip. If it takes
you 3 hours and 20 minutes to drive from Houston to San Antonio, a distance of

---

*Dinitrogen tetroxide is the dimer of nitrogen dioxide. (A dimer is a molecule composed of two
identical, simpler molecules.) Because nitrogen dioxide is soluble in carbon tetrachloride, the
position of the equilibrium

$$N_2O_4 \rightleftharpoons 2NO_2$$

makes no difference in the volume of oxygen given off; we will not take any further notice of
this equilibrium.

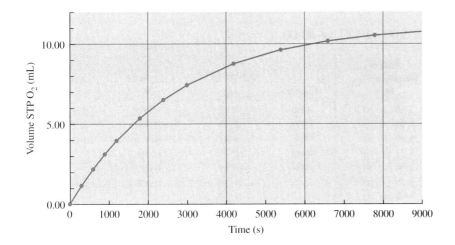

■ FIGURE 18.4 Volume of STP oxygen formed by decomposition of dinitrogen pentoxide in carbon tetrachloride at 40.00 °C. Zeros in times are significant. At infinite time—that is, after enough time had passed for reaction to be complete—the volume of STP oxygen formed was 11.20 mL. There is not room on the page to show infinite time.

183 miles, your average speed is 55 mi/h:

$$\text{average speed} = \frac{183 \text{ miles}}{3.33 \text{ hours}} = 55.0 \text{ mi/h}$$

Sample Problem 18.1 shows a calculation of the average rate of formation of oxygen gas.

SAMPLE PROBLEM

18.1   What was the average rate of formation of STP oxygen gas during the first 300 s? According to the data used to plot Figure 18.4, 1.15 mL of STP oxygen gas were formed in the first 300 s.

Solution

$$\text{Average rate of formation of oxygen} = \frac{1.15 \text{ mL}}{300 \text{ s}} = 0.003\ 83 \text{ mL/s}$$

PRACTICE PROBLEM

18.1   (a) What was the average rate of formation of STP oxygen gas during the period from 6600 to 7800 s? According to the data used to plot Figure 18.4, 10.17 mL of STP oxygen gas were formed in 6600 s and 10.53 in 7800 s. (b) Did the average rate of formation of STP oxygen gas decrease as time passed?

Instantaneous Rates

A speedometer shows **instantaneous speed.** If you stopped for a hamburger during your trip from Houston to San Antonio, your instantaneous speed was 0 mi/h. While you were passing a slow-moving car, your instantaneous speed may have been 65 mi/h. The instantaneous rate of reaction—that is, the *rate at any point in time*— can be found by drawing a tangent to the graph of quantity or concentration vs. time. The instantaneous rate of reaction is equal to the slope of the tangent. Sample Problem 18.2 shows the determination of the initial rate of formation of oxygen by decomposition of dinitrogen pentoxide in carbon tetrachloride at 40.00 °C from Figure 18.4. **Initial rate of formation** means the *rate of formation at time zero when the reactants are mixed and reaction starts.*

A tangent is a straight line that touches a curve at only one point.

TABLE 18.1	Decomposition of N_2O_5 in 5.00 mL CCl_4 at 40.00 °C							
Time, s	Vol STP O_2, mL	Conc N_2O_5, M	Rate of Decomposition, M/s	Time, s	Vol STP O_2, mL	Conc N_2O_5, M	Rate of Decomposition, M/s
0	0.00	0.200	7.29×10^{-5}	3000	7.42	0.068	2.44×10^{-5}
300	1.15	0.180	6.46×10^{-5}	4200	8.75	0.044	1.59×10^{-5}
600	2.18	0.161	5.80×10^{-5}	5400	9.62	0.028	1.03×10^{-5}
900	3.11	0.144	5.21×10^{-5}	6600	10.17	0.018	—
1200	3.95	0.130	4.69×10^{-5}	7800	10.53	0.012	—
1800	5.36	0.104	3.79×10^{-5}	∞	11.20	0.000	—
2400	6.50	0.084	3.04×10^{-5}				

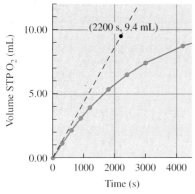

Detail of Figure 18.4 for Sample Problem 18.2.

The decomposition of N_2O_5 is used repeatedly in this chapter so that students can focus their attention on kinetic concepts without being distracted by a variety of examples.

SAMPLE PROBLEM

18.2 What is the initial rate of formation of oxygen by the decomposition of dinitrogen pentoxide in carbon tetrachloride at 40.00 °C?

Solution Draw a tangent to the curve at time 0 as shown by the broken line in the detail of the first part of Figure 18.4 at the left. Remember that the slope of a line, λ, is given by the formula

$$\lambda = \frac{(y_2 - y_1)}{(x_2 - x_1)}$$

(See Appendix B.6 for more information about the slope of curves.) Be sure to make the tangent long enough for there to be as many digits as possible in the differences between y_2 and y_1 and x_2 and x_1. Choose a point near the end of the tangent like the point (2200 s, 9.4 mL) for (x_2, y_2). Use the point (0, 0) for (x_1, y_1). Substitution of these values in the formula gives

$$\text{slope of the tangent} = \frac{(9.4 - 0.0)\ \text{mL}}{(2200 - 0)\ \text{s}} = 0.0043 \text{ or } 4.3 \times 10^{-3} \text{ mL/s}$$

Although volume can be read from the graph to two significant figures, different people might draw tangents with slightly different slopes. The initial rate of formation of oxygen gas should probably be reported as 4×10^{-3} mL/s.

PRACTICE PROBLEM

18.2 From Figure 18.4, what was the instantaneous rate of formation of oxygen gas at 3000 s?

The Decomposition of N_2O_5 Again

The equation for a reaction relates the quantities of all substances involved in the reaction. In experiments to determine reaction rates, you can measure whichever reactant or product is easiest to measure. The quantities of all other reactants and products can then be calculated from the one experimentally determined quantity by means of the equation for the reaction. Table 18.1 shows the concentrations of N_2O_5 calculated from the volumes of oxygen used to plot Figure 18.4. ▪Figure 18.5 shows how the concentration of N_2O_5 changed with time.

● The calculation of the quantities of other reactants and products from one experimentally determined quantity by means of the equation for the reaction is a problem in stoichiometry.

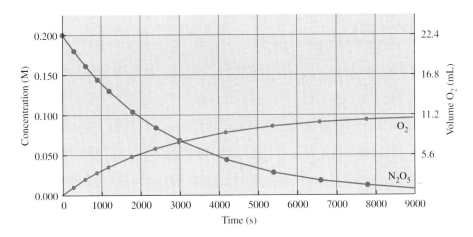

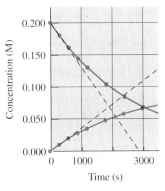

■ FIGURE 18.5   Decomposition of dinitrogen pentoxide in carbon tetrachloride at 40.00 °C. So that you can easily compare the rate of disappearance of N_2O_5 with the rate of appearance of O_2, the left scale shows what the concentration of oxygen would have been *if oxygen had remained in solution in carbon tetrachloride.* The right scale, like Figure 18.4, shows the volume of oxygen formed. Notice how hard the awkward labels of the right scale make reading the volume of oxygen. This shows the importance of the way a graph is labeled.

PRACTICE PROBLEM

18.3   From Figure 18.5, (a) What is the initial rate of disappearance of N_2O_5 in M/s? (b) What is the initial rate of appearance of O_2 in M/s? (c) How does the initial rate of disappearance of N_2O_5 compare with the initial rate of formation of O_2? The tangents to the curves at time zero are drawn for you in the detail of Figure 18.5 at the right because it is hard to draw tangents accurately.

Detail of Figure 18.5 for Practice Problem 18.3.

Like quantities of reactants and products, the rates at which different reactants are used up and products formed are related by the stoichiometry of the reaction. Once you know one rate, you can calculate all others from it. For example, you can see from the equation for the decomposition of dinitrogen pentoxide,

$$2N_2O_5 \longrightarrow 2N_2O_4 + O_2$$

that for each mole of oxygen formed, 2 mol of dinitrogen pentoxide are used. The rate of formation of oxygen will be only one-half the rate of disappearance of dinitrogen pentoxide just as you found in Practice Problem 18.3. This information is usually expressed like this:

$$\text{rate of reaction} = \frac{\Delta[O_2]}{\Delta t} = -\frac{1}{2}\frac{\Delta[N_2O_5]}{\Delta t} \qquad (18.1)$$

The sign is minus for N_2O_5 because the concentration of N_2O_5 decreases as reaction takes place.* The sign is plus (understood) for O_2 because the amount of O_2 increases as reaction takes place. Equation 18.1 is read as "rate of reaction is defined as change in concentration of oxygen with change in time or minus half the change in concentration of dinitrogen pentoxide with change in time." A helpful memory aid is

for a reaction $aA + bB + \cdots \longrightarrow eE + fF + \cdots$

$$\text{rate} = -\frac{1}{a}\frac{\Delta[A]}{\Delta t} = -\frac{1}{b}\frac{\Delta[B]}{\Delta t} = \frac{1}{e}\frac{\Delta[E]}{\Delta t} = \frac{1}{f}\frac{\Delta[F]}{\Delta t} \qquad (18.2)$$

◗ As usual, square brackets mean concentration in moles per liter and Δ means change.

---

*Remember that

$$\text{rate of reaction} = \frac{\text{concentration}_{later} - \text{concentration}_{earlier}}{\text{time}_{later} - \text{time}_{earlier}} = \frac{\Delta[\ ]}{\Delta t}$$

For reactants, concentration$_{later}$ is lower than concentration$_{earlier}$. Thus, (concentration$_{later}$ − concentration$_{earlier}$) is negative. A negative rate makes no physical sense, and $\Delta[\ ]/\Delta t$ must be multiplied by −1 so that the rate is positive.

The Rates module of the Kinetics section presents the concept of reaction rate using graphs and supporting video. Students can explore stoichiometric relations of instantaneous rates.

SAMPLE PROBLEM

18.3 For the reaction

$$2HgCl_2(aq) + C_2O_4^{2-}(aq) \longrightarrow 2Cl^-(aq) + 2CO_2(g) + Hg_2Cl_2(s)$$

(a) show how rate of reaction can be expressed in terms of the rates of disappearance of the reactants, $HgCl_2$ and $C_2O_4^{2-}$, and the rate of formation of the product, Cl^-. (b) How is the rate of formation of Cl^- related to the rate of disappearance of $C_2O_4^{2-}$? (c) If the rate of disappearance of $C_2O_4^{2-}$ is 5.6×10^{-5} M/min, what is the rate of formation of Cl^-?

Solution (a) From the equation for the reaction and equation 18.2,

$$\text{rate of reaction} = -\frac{1}{2}\frac{\Delta[HgCl_2]}{\Delta t} = -\frac{\Delta[C_2O_4^{2-}]}{\Delta t} = \frac{1}{2}\frac{\Delta[Cl^-]}{\Delta t} \qquad (18.3)$$

(b) From equation 18.3, you can see that $C_2O_4^{2-}$ disappears half as fast as Cl^- is formed or Cl^- is formed twice as fast as $C_2O_4^{2-}$ disappears.

(c) If the rate of disappearance of $C_2O_4^{2-}$ is 5.6×10^{-5} M/min, the rate of formation of Cl^- is

$$2(5.6 \times 10^{-5}\ \text{M/min}) = 11.2 \times 10^{-5} \text{ or } 1.12 \times 10^{-4}\ \text{M/min}$$

Another way of answering part (c) is to use the coefficients in the equation for the reaction to write a conversion factor:

$$\left(\frac{5.6 \times 10^{-5}\ \text{mol}\ \cancel{C_2O_4^{2-}}}{\text{L min}}\right)\left(\frac{2\ \text{mol}\ Cl^-}{1\ \text{mol}\ \cancel{C_2O_4^{2-}}}\right)$$
$$= 11.2 \times 10^{-5} \text{ or } 1.12 \times 10^{-4}\ \text{mol}\ Cl^-/\text{L}\cdot\text{min}$$

PRACTICE PROBLEMS

18.4 Show how the rates of each of the following reactions can be expressed in terms of the rates of disappearance of reactants and rates of appearance of products:
(a) $5Br^-(aq) + BrO_3^-(aq) + 6H^+(aq) \longrightarrow 3Br_2(aq) + 3H_2O(l)$
(b) $H_2O_2(aq) + 3I^-(aq) + 2H^+(aq) \longrightarrow I_3^-(aq) + 2H_2O(l)$

18.5 For the reaction in Practice Problem 18.4(a), (a) how is the rate of formation of Br_2 related to the rate of disappearance of BrO_3^-? (b) How is the rate of formation of Br_2 related to the rate of disappearance of Br^-? (c) Do you think that measuring the concentration of water at intervals would be a good way to determine the rate of this reaction? Explain your answer. (*Hint:* Remember that the equation shows that the reaction takes place in aqueous solution.)

18.6 For the reaction,

$$H_2O_2(aq) + 2S_2O_3^{2-}(aq) + 2H^+(aq) \longrightarrow 2H_2O(l) + S_4O_6^{2-}(aq)$$

if the rate of disappearance of $S_2O_3^{2-}$ is 2.56×10^{-4} M/min, what is the rate of formation of $S_4O_6^{2-}$?

18.2 RATE AND CONCENTRATION

Let's see if we can find a quantitative relationship between instantaneous rate of reaction and concentration of the reactant, N_2O_5. By drawing tangents to the curve for the disappearance of N_2O_5 in Figure 18.5 and calculating their slope, we can find the rate at each of the experimental points. Results are summarized in Table 18.1. After 5400 seconds, tangents are almost horizontal, and the slope is too small to be measured precisely. Therefore, in Table 18.1, no rates are given for times greater than 5400 seconds. From Table 18.1, you can see that as the concentration of dinitrogen pentoxide decreases, its rate of disappearance decreases. This suggests that, in this case, perhaps rate may be proportional to concentration, that is,

$$\text{rate} = k[N_2O_5]$$

where k is a constant. If this is so, then rate/concentration will be constant:

$$\text{rate}/[N_2O_5] = k$$

Let's see if this is true. The values obtained by dividing the rates of decomposition by the concentrations of dinitrogen pentoxide are shown in Table 18.2. The quotient (rate/$[N_2O_5]$) is indeed constant.

The relationship rate $= k[N_2O_5]$ is called the **rate law** for the reaction, and the constant, k, is called the **rate constant** for the reaction. *Note that the value of the rate constant, k, is only constant as long as the temperature is constant and, for reactions in solution, the solvent is the same.* In Section 18.9, we will see how the value of k varies with temperature; in Chapter 22, we will study an example of the effect of changing solvent on rate of reaction.

The rate laws for many reactions have been determined. For a reaction

$$aA + bB + \cdots \longrightarrow eE + fF + \cdots$$

the rate law often has the form

$$\text{rate} = k[A]^x[B]^y \ldots$$

where the square brackets mean molar concentrations as usual.

Let's consider a specific example. The rate law for the reaction with net ionic equation

$$5Br^-(aq) + BrO_3^-(aq) + 6H^+(aq) \longrightarrow 3Br_2(aq) + 3H_2O(l) \qquad (18.4)$$

is

$$\text{rate} = k[Br^-][BrO_3^-][H^+]^2$$

The *power to which each concentration is raised in the rate law* is called the **order of the reaction with respect to that substance.** For example, in the above rate law, the order with respect to Br^- and BrO_3^- is one, and the order with respect to H^+ is two. The reaction is first order with respect to Br^- and BrO_3^- and second order with respect to H^+.

If *changing the concentration of a reactant does not change the rate of a reaction*, the reaction is **zero order** with respect to that reactant. For example, the rate of decomposition of hydrogen iodide on a gold catalyst,

$$2HI(g) \xrightarrow{\text{Au}} H_2(g) + I_2(g)$$

is independent of the concentration of hydrogen iodide. The rate law is

$$\text{rate} = k[HI]^0$$

Since any number to the zero power equals 1, another way of writing this rate law is

$$\text{rate} = k$$

The concentration of the reactant does not appear in the rate law.

TABLE 18.2

Rates of Decomposition Divided by Concentration for Decomposition of N_2O_5 in CCl_4 at 40.00 °C[a]

Time, s	Rate/Concentration, s^{-1}
0	3.65×10^{-4}
300	3.59×10^{-4}
600	3.60×10^{-4}
900	3.62×10^{-4}
1200	3.61×10^{-4}
1800	3.64×10^{-4}
2400	3.6×10^{-4}
3000	3.6×10^{-4}
4200	3.6×10^{-4}
5400	3.7×10^{-4}

[a]Rates and concentrations are given in Table 18.1.

A and B are usually reactants but may be products or catalysts.

For the reasons for omitting overall order from this edition, see Reeve, J. C. J. Chem. Educ. 1991, 68, 728–730.

The equations and experimentally determined rate laws for a number of reactions are given below. Use this information to answer Practice Problems 18.7–18.9.

Equation	**Rate Law**
(1) $2N_2O_5(g) \longrightarrow 4NO_2(g) + O_2(g)$	rate $= k[N_2O_5]$
(2) $2NO_2(g) + F_2(g) \longrightarrow 2NO_2F(g)$	rate $= k[NO_2][F_2]$
(3) $2NO(g) + O_2(g) \longrightarrow 2NO_2(g)$	rate $= k[NO]^2[O_2]$
(4) $NO(g) + N_2O_5(g) \longrightarrow 3NO_2(g)$	rate $= k[N_2O_5]$
(5) $3NO(g) \longrightarrow N_2O(g) + NO_2(g)$	rate $= k[NO]^2$

(6) $2Ce^{4+}(aq) + Tl^+(aq) \xrightarrow{Mn^{2+}} 2Ce^{3+}(aq) + Tl^{3+}(aq)$
$$\text{rate} = k[Ce^{4+}][Mn^{2+}]$$

(7) $NO_2(g) + 2HCl(g) \longrightarrow NO(g) + H_2O(g) + Cl_2(g)$
$$\text{rate} = k[NO_2][HCl]$$

(8) $2NO_2Cl(g) \longrightarrow 2NO_2(g) + Cl_2(g) \qquad$ rate $= k[NO_2Cl]$

(9) $NO_2(g) + CO(g) \longrightarrow NO(g) + CO_2(g)$
$$\text{rate} = k[NO_2]^2$$

(10) $2NO_2(g) + O_3(g) \longrightarrow N_2O_5(g) + O_2(g)$
$$\text{rate} = k[NO_2][O_3]$$

(11) $C_3H_6Br_2 + 3KI \longrightarrow C_3H_6 + 2KBr + KI_3$
$$\text{rate} = k[C_3H_6Br_2][KI]$$

(12) $14H_3O^+ + 2HCrO_4^- + 6I^- \longrightarrow 2Cr^{3+} + 3I_2 + 22H_2O$
$$\text{rate} = k[HCrO_4^-][I^-][H_3O^+]^2$$

18.7 (a) As a general rule, are the exponents in the rate law for a reaction the same as the coefficients of the same substance in the equation for the reaction? Give an example to prove your point. (b) Can you see any general relationship between the equation for a reaction and its rate law that could be used to write the rate law from the equation instead of determining the rate law by experiment?

18.8 For reactions 1–5, give the order with respect to each substance.

18.9 Assuming that these twelve reactions are a representative sample of reactions, what can you conclude about the relative frequency of (a) Reactions that are third order with respect to a species? (b) Reactions that are zero order with respect to a species?

18.10 The reaction, $3I^-(aq) + H_3AsO_4(aq) + 2H^+(aq) \longrightarrow I_3^-(aq) + H_3AsO_3(aq) + H_2O(l)$, is first order with respect to I^-, first order with respect to H_3AsO_4, and first order with respect to H^+. Write the rate law.

The rate law for a reaction must be determined by experiment. The exponents, x, y, . . . , cannot be assumed to be the same as the coefficients of the species in the equation for the reaction. Both the exponents, x, y, . . . , and the numerical value of the rate constant, k, must be found experimentally.

In the Rate Laws module of the Kinetics section students can use initial rate data from real world systems to determine rate laws.

18.3 FINDING RATE LAWS

Method of Initial Rates

Finding the rate law for a reaction by drawing tangents to the graph of concentration vs. time is very time-consuming and not very accurate. Better ways have been developed. As far as analysis of data is concerned, the simplest is the **method of**

initial rates. In this method, *the order with respect to each reactant is found by changing the concentration of that reactant while keeping the concentrations of all other reactants constant.*

Finding the rate law for the decomposition of N_2O_5 by the method of initial rates provides a simple example. The necessary data are

Experiment Number	[N₂O₅]	Initial Rate, M/s
1	0.100	3.62×10^{-5}
2	0.200	7.29×10^{-5}

Substituting the information given in the table in the general form of the rate law

$$\text{rate} = k[A]^x[B]^y \ldots$$

gives for experiment 2: $7.29 \times 10^{-5} \text{ M/s} = k(0.200 \text{ M})^x$

and for experiment 1: $3.62 \times 10^{-5} \text{ M/s} = k(0.100 \text{ M})^x$

Dividing the equation for experiment 2 by the equation for experiment 1 gives

$$\frac{7.29 \times 10^{-5} \text{ M/s}}{3.62 \times 10^{-5} \text{ M/s}} = \frac{k(0.200 \text{ M})^x}{k(0.100 \text{ M})^x} \quad \text{or}$$

$$2.01 = (2.00)^x \text{ and, by inspection, } x = 1$$

Comparing experiment 2 with experiment 1, the initial rate was doubled when [N₂O₅] was doubled. What is true of this pair of experiments is true in general:

$$\left(\begin{array}{c}\text{factor by which} \\ \text{initial rate changes}\end{array}\right) = \left(\begin{array}{c}\text{factor by which} \\ \text{concentration changes}\end{array}\right)^{\text{reaction order}} \quad (18.5)$$

⊙ If the reaction rate doubles when the concentration of a reactant is doubled, the reaction is first order with respect to that reactant.

SAMPLE PROBLEM

18.4   The initial rate of a hypothetical reaction, A + B → C, was measured using different initial concentrations of A and B. Results are summarized in the table below. What is the rate law for the reaction?

Experiment Number	[A]	[B]	Initial Rate, M/s
1	0.030	0.010	1.7×10^{-8}
2	0.060	0.010	6.8×10^{-8}
3	0.030	0.020	4.9×10^{-8}

While we prefer to use real reactions, we have used hypothetical reactions occasionally in this chapter so that we could illustrate a combination of points with a single example.

Solution   Comparing experiment 2 with experiment 1, [A] was doubled in experiment 2 while [B] remained the same. The initial rate increased fourfold. From equation 18.5,

$$4.0 = (2.0)^x$$

where x is the exponent of [A] in the rate law. By inspection, $x = 2$.

Now compare experiment 3 with experiment 1 in the same way. In this pair of experiments, [A] remains the same while [B] is changed. If y is the exponent of [B]

⊙ If the reaction rate increases fourfold when the concentration of one reactant is doubled, the reaction is second order with respect to that reactant.

in the rate law then

$$\left(\frac{4.9 \times 10^{-8}}{1.7 \times 10^{-8}}\right) = (2.0)^y \quad \text{or} \quad 2.9 = (2.0)^y$$

If the value of the exponent is not obvious by inspection, take logarithms of both sides of the equation and solve for y:

$$\ln 2.9 = y \ln 2.0$$

$$\frac{\ln 2.9}{\ln 2.0} = y = 1.54 = 1\frac{1}{2} = \frac{3}{2}$$

The rate law is

$$\text{rate} = k[A]^2[B]^{3/2}$$

○ Remember that the symbol ln means natural logarithm (logarithm to the base e).

Once the rate law is known, the value of the rate constant, k, can be found using data from any experiment as shown in Sample Problem 18.5.

SAMPLE PROBLEM

18.5   What is the value of the rate constant, k, for the reaction in Sample Problem 18.4?

Solution   In Sample Problem 18.4, the rate law was found to be rate $= k[A]^2[B]^{3/2}$. Solving for k and substituting the concentrations and initial rate from experiment 1 gives

$$\frac{1.7 \times 10^{-8}\,\text{M/s}}{(0.030\,\text{M})^2(0.010\,\text{M})^{3/2}} = k = 1.9 \times 10^{-2}\,\text{s}^{-1}\,\text{M}^{-5/2}$$

Units of rate constants often don't make much sense physically, but including them helps prevent mistakes in problem setups. The data from either of the other two experiments can be used to check the value found for k.

○ Use the instruction book that came with your calculator to learn how to find the value of $(0.010)^{3/2}$.

Once you know the rate law and the value of k, you can sit back in your chair and calculate the results of any other experiment involving the same reactants *at the same temperature (and in the same solvent if reactants are in solution)* instead of doing the experiment.

SAMPLE PROBLEM

18.6   What will be the initial rate of the reaction in Sample Problem 18.4 if the initial concentration of A $= 0.050$ M and the initial concentration of B $= 0.020$ M?

Solution   Substitute the concentrations given and the numerical value of the rate constant from Sample Problem 18.5 ($k = 1.9 \times 10^{-2}\,\text{s}^{-1}\,\text{M}^{-5/2}$) in the rate law from Sample Problem 18.4, (rate $= k[A]^2[B]^{3/2}$) and solve for the rate:

$$\text{rate} = (1.9 \times 10^{-2}\,\text{s}^{-1}\,\text{M}^{-5/2})(0.050\,\text{M})^2(0.020\,\text{M})^{3/2}$$

$$= 1.3 \times 10^{-7}\,\text{M·s}^{-1}$$

18.11   The initial rate of the reaction,

$$2HgCl_2(aq) + C_2O_4^{2-} \longrightarrow 2Cl^- + 2CO_2(g) + Hg_2Cl_2(s)$$

was measured using different initial concentrations of $HgCl_2$ and $C_2O_4^{2-}$. Results are summarized in the table below:

Experiment Number	$[HgCl_2]$	$[C_2O_4^{2-}]$	Initial Rate, M/s
1	0.100	0.20	3.1×10^{-5}
2	0.100	0.40	1.2×10^{-4}
3	0.050	0.40	6.2×10^{-5}

(a) Find the order of the reaction with respect to $HgCl_2$ by comparing experiments with the same $[C_2O_4^{2-}]$. (b) Find the order of the reaction with respect to $C_2O_4^{2-}$ by comparing experiments with the same $[HgCl_2]$. (c) What is the rate law for the reaction? (d) What is the numerical value of k? (e) Qualitatively, how will the initial rate in an experiment using an initial concentration of $HgCl_2 = 0.20$ M and an initial concentration of $C_2O_4^{2-} = 0.30$ M compare with the initial rate in experiment 1? (f) Quantitatively, what will be the initial rate in the experiment using an initial concentration of $HgCl_2 = 0.20$ M and an initial concentration of $C_2O_4^{2-} = 0.30$ M?

18.12   (a) If a reaction is second order with respect to a reactant, what will happen to the rate if the concentration of that reactant is doubled? (b) If a reaction is zero order with respect to a reactant, what will happen to the rate if the concentration of that reactant is doubled? (c) If a reaction is first order with respect to a reactant, what will happen to the rate if the concentration of that reactant is tripled?

18.13   For a reaction with the rate law, rate $= k[A][B]^2$, what will happen to the rate if the concentrations of both A and B are doubled?

In the method of initial rates, the reaction is followed until it is about 10% complete. The problem with this method is that it takes time to mix the reactants and time for them to reach the temperature of the constant temperature bath. Also the analytical method used must be very sensitive to detect the relatively small changes that take place during the first 10% of the reaction. On the other hand, not much product is present, so the reverse reaction and other reactions involving the product are not important.

Graphical Method

The graphical method of finding the order of a reaction with respect to a reactant A is based on integrated rate laws. To derive integrated rate laws, which relate time and concentration, from the rate laws relating rate and concentration that we have discussed so far, calculus is necessary. However, you do not have to know calculus to use integrated rate laws. For a first-order reaction, the integrated rate law is

$$\ln \frac{[A]_t}{[A]_0} = -kt \tag{18.6a}$$

In the Reaction Order module of the Kinetics section students use graphs to determine reaction order and half-life.

TABLE 18.3　Rate Laws for Common Reaction Orders

Order with Respect to A	Rate Law	Integrated Rate Law	Graph of ___ vs. t Is a Straight Line	Slope of Linear Graph =
0	rate = k	$[A]_t = -kt + [A]_0$	$[A]_t$	$-k$
1	rate = $k[A]$	$\ln [A]_t = -kt + \ln [A]_0$	$\ln [A]_t$	$-k$
2	rate = $k[A]^2$	$\dfrac{1}{[A]_t} = kt + \dfrac{1}{[A]_0}$	$1/[A]_t$	k

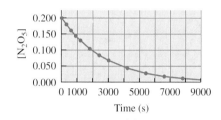

(a)

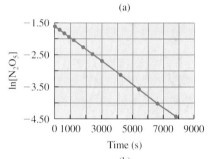

(b)

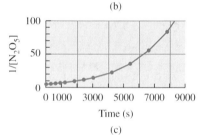

(c)

- FIGURE 18.6　Decomposition of N_2O_5 in CCl_4 at 40.00 °C. (a) Concentration vs. time. (b) ln concentration vs. time. (c) 1/concentration vs. time.

⊙ A computer with graphics software makes graphing easy.

where $[A]_t$ is the concentration of reactant A at any time t, $[A]_0$ is the initial concentration of A, and k is the rate constant. The integrated rate law for a first-order reaction can also be written in the form

$$\ln [A]_t = -kt + \ln [A]_0 \qquad (18.6b)$$

Equation 18.6b is of the form $y = mx + b$; that is, it is the equation of a straight line of slope, $-k$, and y-intercept, $\ln [A]_0$. Table 18.3 summarizes information about rate laws for the common reaction orders.

- Figure 18.6 shows graphs of $[A]_t$, $\ln [A]_t$, and $1/[A]_t$ against t for the decomposition of dinitrogen pentoxide in carbon tetrachloride at 40.00 °C. The graph of $\ln [N_2O_5]$ against time is a straight line; therefore, the decomposition of dinitrogen pentoxide in carbon tetrachloride at 40.00 °C is first order with respect to dinitrogen pentoxide. The other two graphs are not straight lines. However, note that the first parts of all three curves in Figure 18.6 resemble straight lines. A short enough part of *any* curve looks like a straight line. Concentration must be measured until the reaction is more than 75% complete for the graphical method to distinguish between zero-, first-, and second-order reactions.

PRACTICE PROBLEM

18.14　In an experiment to determine the order of a reaction with respect to A, the following data were obtained:

Time, s	[A]
0	2.00×10^{-3}
1.00×10^3	1.39×10^{-3}
2.00×10^3	0.78×10^{-3}
3.00×10^3	0.17×10^{-3}

(a) Find the order with respect to A by plotting [A], ln [A], and 1/[A] against time.
(b) From the slope of the graph that is a straight line, what is the value of k?

In using the graphical method, you must investigate the order with respect to each reactant separately by keeping the concentrations of all other reactants constant. High concentrations are usually used to keep the concentrations of other species constant. The concentrations of reactants present in very large excess are not changed significantly by the reaction. Their concentration is, for all practical purposes, constant so that the rate of reaction depends only on the concentration of the reactant being studied. Because a solvent is, by definition, present in large excess, the order of the reaction with respect to the solvent cannot be determined.

TABLE 18.4	Reactions with Complex Rate Laws	

Equation	Rate Law
$CO(g) + Cl_2(g) \longrightarrow COCl_2(g)$	$rate = k[CO][Cl_2]^{3/2}$
$Cr(H_2O)_6^{3+} + SCN^- \longrightarrow Cr(H_2O)_5NCS^{2+} + H_2O(l)$	$rate = [Cr(H_2O)_6^{3+}][SCN^-]\left(k + \dfrac{k'}{[H^+]} + \dfrac{k''}{[H^+]^2}\right)$

Buffers may be used to maintain a constant concentration of hydrogen and hydroxide ions.

The value of k is the rate constant for the reaction only if the rate of the reaction is independent of the concentrations of all other components of the reaction mixture. If the rate of the reaction depends on the concentration of any other component of the reaction mixture, such as the solvent, the value of k found is not the true rate constant. It is the product of the true rate constant and the concentrations of the substances that affect the rate of reaction and is called a **pseudo rate constant.** After the orders with respect to all species that affect the rate have been found and the rate law for the reaction can be written, the true value of k can be calculated by substituting concentrations and rate in the rate law and solving for k as shown in Sample Problem 18.5.

Pseudo rate constants include constant concentrations, just as some equilibrium constants include constant concentrations. For example, K_{sp} includes the concentration of the slightly soluble solid, and K_a includes the concentration of water.

Reactions with Complex Rate Laws

Reactions that are third order with respect to a reactant are rare. But reactions with fractional and negative powers in the rate law and rate laws with more than one term are common. Some examples of reactions with complex rate laws are shown in Table 18.4. In general, the only practical way to find complex rate laws is to use a computer to fit the rate law to the experimental data. Like simple rate laws, complex rate laws can be changed into equations that give concentrations at any time. *Rate laws and rate constants are summaries of enormous amounts of information about reaction rates.*

18.4 FIRST-ORDER REACTIONS

Reactions that are first order with respect to a reactant are of great practical importance. For example, the rates at which many drugs pass into the bloodstream and the rates at which they are used by the body are first order. First-order kinetics are often useful in geochemistry; the rate of removal of magnesium from seawater at midocean ridges is one example. The decay reactions of all radionuclides are first order. These reactions are used in nuclear medicine and for determining the age of ancient objects (Chapter 20).

The **half-life** of a reaction, $t_{1/2}$, is the *time required for one-half of the quantity of reactant originally present to react.* The half-life is an important quantity for first-order reactions because it is constant. For first-order reactions, the half-life does not depend on the concentration of the reactant. You can see this if you look at Figure 18.5. The original concentration of N_2O_5 was 0.200 M. It took 1900 seconds for the concentration to decrease to 0.100 M, half the original concentration. It took another 1900 seconds for the concentration to decrease to 0.050 M, half of 0.100 M, and another 1900 seconds for it to decrease to 0.025 M, half of 0.050 M.

The half-life and the value of the rate constant for first-order reactions are related:*

$$t_{1/2} = \frac{0.693}{k}$$

(18.7)

Equation 18.7 can be used to calculate k from the half-life of a first-order reaction.

Because the half-life for a first-order reaction does not depend on initial concentration, the concentration (or amount) remaining after a whole number of half-lives can be calculated very simply, as shown in Sample Problem 18.7.

SAMPLE PROBLEM

18.7 The half-life for a reaction is 726 s. Starting with a concentration of 0.600 M, what will the concentration be after 1452 s *if the reaction is first order?*

Solution The time, 1452 s, is two half-lives. In one half-life, half the material originally present will react, and half will be left. The concentration remaining will be half the original concentration or 0.300 M.

After a second half-life, half the material remaining after one half-life will react, and half will be left. The concentration remaining will be 0.150 M. You are less likely to make a careless mistake if you work in a table:

Time, s	M
0	0.600
726	0.300
1452	0.150

It is equally simple to calculate the number of half-lives needed to reach a given concentration, such as the minimum concentration needed for a drug to be effective.

If the time you are interested in is not a whole number of half-lives, you can use the integrated rate law. From the integrated rate law, given the value of the rate constant, you can calculate: (1) the time required for an initial concentration to decrease to a specified concentration, (2) the concentration that will be reached in a given time, and (3) the initial concentration if the concentration after some period of time is known. Sample Problem 18.8 illustrates this type of calculation.

*The relationship of equation 18.7 can be derived from the integrated rate law, equation 18.6(a). By definition, at time $t_{1/2}$,

$$[A]_t = \frac{[A]_0}{2}$$

Substitution in equation 18.6(a) gives

$$\ln \frac{\frac{[A]_0}{2}}{[A]_0} = -kt_{1/2}$$

Clearing of fractions and simplifying gives

$$\ln \frac{1}{2} = -kt_{1/2} \qquad \text{or} \qquad -0.693 = -kt_{1/2}$$

and, solving for $t_{1/2}$,

$$\frac{0.693}{k} = t_{1/2}$$

(18.7)

The formula

$$\frac{A_t}{A_0} = \left(\frac{1}{2}\right)^n$$

can, of course, be used to solve Sample Problem 18.7. However, we find that if we give our students a formula, they tend to mindlessly plug numbers into it. They learn more by having to reason a problem out. If a student figures out the formula and tells us about it, as occasionally happens, it provides an opportunity to share the excitement of discovery with the whole class.

18.8   In the experiment in Sample Problem 18.7, how long will it take for the concentration to reach 0.100 M?

Solution   First we will have to find the value of k. Then we can use the integrated rate law to solve the problem. According to equation 18.7, for a first-order reaction,

$$t_{1/2} = \frac{0.693}{k}$$

Solving for k and substituting the half-life given in Sample Problem 18.7,

$$k = \frac{0.693}{726\ \text{s}} = 9.55 \times 10^{-4}\ \text{s}^{-1}$$

The integrated rate law for a first-order reaction is

$$\ln \frac{[A]_t}{[A]_0} = -kt \qquad (18.6a)$$

Solving for t and substituting the data from the problem gives

$$\frac{\ln \left(\dfrac{0.100}{0.600} \right)}{-(9.55 \times 10^{-4}\ \text{s}^{-1})} = t = 1.88 \times 10^3\ \text{s}$$

This answer looks reasonable because, in a third half-life or 2178 s, the concentration would be 0.075 M. One-tenth molar is between 0.150 and 0.075 M, and 1.88×10^3 s is between 1452 and 2178 s.

⬤ Remember that the number of digits to the right of the decimal point in the natural logarithm of a number and the number of digits in the number should be the same.

⬤ Always check to be sure your work is reasonable.

18.15   The half-life for a first-order reaction is 234 days. Starting with a concentration of 0.816 M, what will the concentration be after 702 days?

18.16   A first-order reaction has a half-life of 76.2 min. How long will it take for the concentration to decrease to 6.25% (1/16) of its original value?

18.17   The half-life for one first-order reaction is 273 s and the half-life for another first-order reaction is 541 s. (a) Which reaction is faster? (b) For which reaction does k have the larger numerical value?

18.18   The half-life for a first-order reaction is 2.6 years. Starting with a concentration of 0.25 M, what will the concentration be after 9.9 years?

18.5   RATE AND IDENTITY OF REACTANTS

The numerical value of the rate constant depends on the identity of the reactants. If the numerical value of the rate constant, k, is large, reaction is rapid, and not much time is required for reactants to be changed to products unless concentrations are very low. If the numerical value of the rate constant is small, the reaction is slow even if concentrations are high; a long time is required for complete reaction. From your experience in lab and study of how reactivity changes across rows and down groups in the periodic table, you should be getting some feel for which reactions are fast and which reactions are slow. For example, you should know that, of metals that react spontaneously with water, some (for example, sodium) react rapidly with water at room temperature, but others (like calcium) react slowly. (But

- FIGURE 18.7 Finely divided iron burns in air but a massive piece of iron such as a frying pan does not.

remember that most metals do not react spontaneously with water because the free-energy change for the reaction is positive.) Nonspontaneous reactions that are made to take place by supplying energy to the system are usually slow.

Spontaneous reactions between oppositely charged ions in solution are usually very fast. For example, the neutralization reaction,

$$H^+ + OH^- \longrightarrow H_2O(l)$$

is one of the fastest reactions whose rate has been measured. The rate constant for this reaction is $1.4 \times 10^{11} \text{ M·s}^{-1}$ at 25 °C! Automated instrumental methods must be used to measure the rates of very fast reactions like this.

Reactions between molecules involve breaking covalent bonds and are usually slower. Reactions of this kind commonly have half-lives greater than the human response time of a few seconds. Their rates can be measured by people using ordinary methods like titration. However, many important biological reactions between molecules are catalyzed by enzymes and are very fast.

One of the slowest reactions known is the radioactive decay of zirconium-96. This reaction has a half-life of more than 3.6×10^{17} years!

18.6 RATE AND SOLVENT

For reactions that take place in solution, one of the most important factors affecting rate of reaction is the nature of the solvent. Both physical properties of the solvent, such as viscosity, and chemical properties, such as acidity and basicity, can be important. Changing the solvent may increase the rate of some reactions as much as 10 million-fold. An example is discussed in Section 22.8. When more than one set of products is thermodynamically possible from a given set of reactants, different solvents may give different ratios of products or even entirely different products. By changing the solvent, a process may be made to yield a purer product faster with fewer by-products.

18.7 HETEROGENEOUS REACTIONS

In solutions, molecules and ions are free to move around. Reactions usually take place faster in solution than in heterogeneous mixtures because the reactants can get at each other easier. For example, nothing happens when solid sodium hydrogen carbonate is mixed with solid citric acid. But when aqueous solutions of these substances are mixed, carbon dioxide gas is given off rapidly. (Remember Figure 4.2.)

In heterogeneous reaction mixtures, reaction can only take place at boundaries between phases. The greater the area of these boundaries, the faster the reaction. Because the surface area of a given quantity of a substance increases rapidly as particle size decreases, small particles react faster than large particles (see -Figure 18.7). The effect of increasing surface area with decreasing particle size can be used to speed reactions you want to take place. It can also lead to undesirably fast reactions like grain elevator explosions. Dust from stored grain is very finely divided. When mixed with air, this dust burns extremely rapidly if ignited by a spark, and an explosion results (see chapter opening). Stirring is important for good contact between phases of a heterogeneous reaction mixture.

Remember Figure 13.4, which shows how surface area increases as particle size decreases. If you forget to get the potatoes started for dinner on time, you can make them cook faster by cutting them up.

18.8 CATALYSTS

The rates of many reactions can be greatly increased by catalysts, substances that make a reaction take place faster but are not used up in the reaction. Catalysts are useful for making slow reactions that have large equilibrium constants take place at useful rates. Although catalysts shorten the time needed to reach equilibrium, they do not change the composition of the equilibrium mixture.

Catalysts are usually very specific. For example, carbon monoxide and hydrogen react to form methane if a nickel catalyst is used, but they form methyl alcohol if the reaction is catalyzed by a mixture of zinc oxide and chromium(III) oxide:

$$CO(g) + 3H_2(g) \xrightarrow{\text{Ni catalyst}} CH_4(g) + H_2O(g)$$

$$CO(g) + 2H_2(g) \xrightarrow{\text{ZnO—Cr}_2\text{O}_3 \text{ catalyst}} CH_3OH(g)$$

Catalysts are classified as homogeneous or heterogeneous. **Homogeneous catalysts** are *present in the same phase as the reactants.* Catalysis of the decomposition of ozone to oxygen by nitric oxide in the upper atmosphere,

$$2O_3(g) \xrightarrow{\text{NO(g)}} 3O_2(g)$$

is an example of homogeneous catalysis. Ozone in the upper atmosphere protects the plants and animals on Earth from ultraviolet radiation from the sun. As a result, a decrease in the amount of ozone in the upper atmosphere is potentially harmful. Nitric oxide is an air pollutant formed by the reaction of nitrogen and oxygen from air at high temperatures such as the combustion temperatures in automobile and airplane engines. One of the purposes of the catalytic converter in the exhaust of your car (■Figure 18.8) is to change nitric oxide back to nitrogen.

The catalyst in a catalytic converter is a heterogeneous catalyst. A **heterogeneous catalyst** is *in a different phase from the reactants.* The catalyst in a catalytic converter is solid; the exhaust from the automobile engine is gaseous. Conversion of nitric oxide to nitrogen and oxygen takes place on the surface of the catalyst.

Lead in gasoline poisons the catalyst. *Substances* like lead *that interfere with reactions* are called **inhibitors.** Inhibitors can be useful, for example, in slowing tooth decay, corrosion, and aging. If you look at the lists of ingredients on most food products you will find one labeled "to retard spoilage."

Heterogeneous catalysts are used in many important industrial processes. Some examples are the Haber process to make ammonia from nitrogen from air, the contact process for making sulfuric acid, the hydrogenation of vegetable oils to solid fats as in the manufacture of margarine from corn oil, and the cracking of high-boiling fractions from petroleum to make gasoline. About $2 billion worth of catalysts, which are used mainly in petroleum processing and pollution control, are sold in the United States each year, and catalysts are the subject of a large part of industrial research.

Almost all biological reactions are catalyzed by enzymes. **Enzymes** are *complex substances (molecular mass 12 000 to 120 000 u and higher) produced in living cells.* Enzymes are extremely specific and are very efficient under the mild conditions in living animals and plants. Enzyme-catalyzed reactions in biological systems often convert reactants to products in milliseconds or even microseconds. Research on the structure of enzymes and how enzymes work is one of the most important fields of research in biochemistry. An understanding of catalysis by enzymes promises the ability to do many wonderful things, such as make ammonia for fertilizer from atmospheric nitrogen at low temperatures and pressures. Catalysis by enzymes is discussed in Section 18.12.

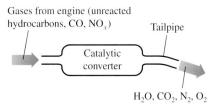

Gases from engine (unreacted hydrocarbons, CO, NO$_x$)

Catalytic converter

Tailpipe

H$_2$O, CO$_2$, N$_2$, O$_2$

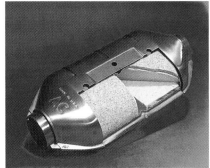

■ FIGURE 18.8 *Top:* In an automotive catalytic converter, CO is converted to CO$_2$, unburned hydrocarbons react to form CO$_2$ and H$_2$O, and nitrogen oxides (NO$_x$) are changed back to N$_2$ and O$_2$. *Bottom:* Cutaway of automotive catalytic converter showing outer cover, insulator, and ceramic catalyst, which consists of particles of platinum, palladium, and rhodium deposited in a ceramic honeycomb.

People have used enzymes since prehistoric times to make cheese, wine, and vinegar.

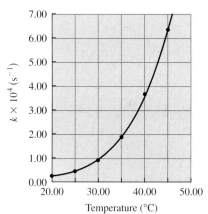

FIGURE 18.9 Graph of rate constant k vs. temperature for the decomposition of N_2O_5 in CCl_4. Notice how the vertical axis is labeled. To avoid repeating ... $\times 10^{-4}$, all values have been multiplied by 10^4. Values graphed range from 0.235×10^{-4} s^{-1} at 20.00 °C to 6.29×10^{-4} s^{-1} at 45.00 °C.

The rates of many reactions are increased two- to fourfold by a 10° increase in temperature.

This is the same Arrhenius who defined acids as substances that increase the concentration of hydrogen ions and bases as substances that increase the concentration of hydroxide ions when dissolved in water.

18.9 RATE AND TEMPERATURE

Figure 18.9 shows the relationship between rate constant and temperature for the decomposition of dinitrogen pentoxide in carbon tetrachloride. This relationship is typical. You can see that the rate constant increases as the temperature increases, but the relationship between rate constant and temperature is not linear. The rate constant increases faster and faster as the temperature increases. For example, from 20.00 to 25.00 °C, it increases from 0.235×10^{-4} s^{-1} to 0.469×10^{-4} s^{-1}, a difference of 0.234×11^{-4} s^{-1}. From 40.00 to 45.00 °C, it increases from 3.62×10^{-4} s^{-1} to 6.29×10^{-4} s^{-1}, a difference of 2.67×10^{-4} s^{-1}, or over ten times as much. This rapid increase suggests an exponential relationship between rate constant and temperature.

The mathematical equation that describes the relationship between temperature and rate constant is

$$k = Ae^{-E_a/RT} \tag{18.8a}$$

where A is a constant, called the frequency factor, which is characteristic of the reaction, e is the base of the natural logarithm system (2.718), R is the gas constant, T is the Kelvin temperature, and E_a is a constant called the **activation energy** of the reaction. The activation energy is the *energy the molecules must have in order to react*. We will learn more about activation energies in Section 18.10. Equation 18.8a is called the **Arrhenius equation** after Swedish chemist Svante Arrhenius, who showed that it applies to almost all kinds of reactions. The Arrhenius equation was based on experimental curves like the one in Figure 18.9 showing the dependence of rate constants of a number of reactions on reaction temperature.

Another form of the Arrhenius equation, which is obtained by taking the logarithms of both sides of equation 18.8a, is

$$\ln k = \left(-\frac{E_a}{R}\right)\left(\frac{1}{T}\right) + \ln A \tag{18.8b}$$

Equation 18.8b is the equation of a straight line with slope $= -E_a/R$ where the independent variable is $1/T$. If the value of k is determined at several temperatures, and $\ln k$ plotted against $(1/T)$, a straight line of slope $-E_a/R$ is obtained.

PRACTICE PROBLEM

18.19 The data graphed in Figure 18.9 are given in the table below.
(a) Determine the activation energy, E_a, for the decomposition of dinitrogen pentoxide in carbon tetrachoride by graphing $\ln k$ against $(1/T)$. (*Hint:* Determination of the slope of a line was explained in Sample Problem 18.2.)
(b) Use the graph from part (a) to estimate the value of k at 50.00 °C.

Temperature, °C	$k \times 10^4$, s^{-1}
20.00	0.235
25.00	0.469
30.00	0.933
35.00	1.820
40.00	3.62
45.00	6.29

The activation energy can also be calculated from the rate constants at two different temperatures using equation 18.9,*

$$\ln\left(\frac{k_2}{k_1}\right) = \frac{E_a}{R}\left(\frac{1}{T_1} - \frac{1}{T_2}\right) \tag{18.9}$$

although usually not as accurately as by graphing $\ln k$ against $(1/T)$. To obtain an accurate value for E_a by substitution into equation 18.9, the rate constants must be very accurate. Also, the two temperatures should be as widely separated as practical to avoid needless loss of significant figures by subtraction.

PRACTICE PROBLEM

18.20   Use equation 18.9 to calculate E_a for the decomposition of dinitrogen pentoxide in carbon tetrachloride from the data in the table in Practice Problem 18.19.

The reaction that makes a Cyalume light stick glow takes place more rapidly in hot water (left) than in ice water (right).

◗ Activation energy must be positive. If you obtain a negative value for E_a, you have probably interchanged T_1 and T_2 or k_1 and k_2.

Students are usually pretty good at plugging numbers into formulas. The absence of a sample problem here is intentional.

18.10   THEORIES OF REACTION RATES

The dependence of reaction rates on concentration and temperature is *described* by rate laws and the Arrhenius equation. It is *explained* by the collision and activated complex or transition state theories.

Collision Theory

The **collision theory** is based on the kinetic-molecular theory. Let's begin by considering a reaction between two molecules in the gas phase because this type of reaction is simplest. The reaction between ozone and nitric oxide (mononitrogen monoxide) to form nitrogen dioxide and oxygen

$$O_3(g) + NO(g) \longrightarrow NO_2(g) + O_2(g)$$

is a good example. According to the kinetic-molecular theory, a container "full" of ozone and nitric oxide molecules is mostly empty space. The ozone and nitric oxide molecules move about very rapidly in this space, colliding frequently with each other and with the container. *For the valence electrons to rearrange and form new bonds, the atoms to be joined by the new bonds must be within bonding distance of each other.* Therefore, it seems reasonable to suppose that reaction takes place when a molecule of ozone collides with a molecule of nitric oxide. This simple picture explains why the rate of reaction depends on the concentrations of the reactants. The more ozone and nitric oxide molecules there are in a given volume, the more collisions there will be between them just as a larger number of motorboats on a small lake means more collisions.

Under ordinary laboratory conditions, one molecule of gas hits another molecule of gas about 10^{10} times per second (Section 5.10). If all of these collisions resulted

The reaction between ozone and nitric oxide is important in smog formation.

⊕ The Activation Energy module of the Kinetics section presents the concepts of the collision theory using animation and graphics. Students can explore elementary reactions to learn about energy relations, activated complexes, and catalysts.

---

*Equation 18.9 is obtained from the Arrhenius equation as follows: If the rate constant at one temperature, T_1, is k_1 and the rate constant at a different temperature, T_2, is k_2, then

$$\ln k_1 = \ln A - \frac{E_a}{RT_1} \qquad \text{and} \qquad \ln k_2 = \ln A - \frac{E_a}{RT_2}$$

Then, because E_a and A are constant, subtraction of the former equation from the latter eliminates $\ln A$ and gives

$$\ln k_2 - \ln k_1 = \frac{E_a}{R}\left(\frac{1}{T_1} - \frac{1}{T_2}\right) \qquad \text{or} \qquad \ln\left(\frac{k_2}{k_1}\right) = \frac{E_a}{R}\left(\frac{1}{T_1} - \frac{1}{T_2}\right)$$

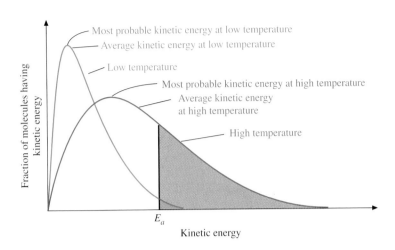

FIGURE 18.10 Distribution of kinetic energies. At the higher temperature, the increase in the fraction of the molecules that have energy greater than the activation energy is large compared with the increase in average kinetic energy.

Not only thermal energy, but also light, microwaves, and sound can supply the energy needed for chemical reactions to take place.

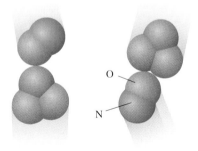

FIGURE 18.11 Some collisions between NO and O_3 molecules that cannot result in reaction because the geometry is wrong.

in reaction, reaction would be complete in an instant. Also, the rates of most reactions increase much faster with increasing temperature than would be expected on the basis of higher molecular speeds alone. The observations that not all collisions result in reaction and that the reaction rate increases very fast with increasing temperature are explained by proposing that molecules must collide with a certain minimum energy for reaction to take place. This is like removing an object from a shelf that has a rim to keep things from falling off accidentally. You have to put in some energy to lift the object over the rim. Then it will fall to the floor if you drop it. But if you do not put in some energy, the object can't fall to the floor.

If molecules collide with less than the minimum energy needed for reaction, they simply bounce apart again without reacting. ▪Figure 18.10 shows how kinetic energies are distributed among the huge number of particles in a sample. The distribution of kinetic energies is similar to the distribution of molecular speeds shown in Figure 5.16 because kinetic energy is proportional to the square of molecular speed. In Figure 18.10, the shaded areas represent the fraction of molecules having energy greater than the activation energy or, in other words, the fraction of molecules having enough energy to react. You can see that this fraction is much larger at the higher temperature. You can also see that the increase in the fraction of molecules having energy greater than the activation energy is large compared with the increase in average kinetic energy. Therefore, reaction rates increase much faster with increasing temperature than would be expected on the basis of average kinetic energy.

Figure 18.10 also shows that *the higher the activation energy, the smaller the fraction of the molecules that have enough energy to react. As a result, the higher the activation energy, the slower a reaction will be.*

Reactive collisions continuously remove the most energetic reactant molecules from the reaction mixture. But at constant temperature, the average kinetic energy remains constant. Nonreactive collisions of molecules with other molecules and with the container walls restore the usual distribution of energies.

Collisions between molecules are like collisions between people and cars. A person can walk into a parked car, or even a car that is moving slowly, without getting hurt. If the car is moving fast, the person is hurt or killed. But a fast-moving car doesn't hurt the person unless it hits him or her. *For a reaction to take place, molecules must collide with a certain minimum energy.*

However, not every collision with sufficient energy results in reaction. For example, ▪Figure 18.11 shows collisions that cannot result in reaction because the nitrogen atoms of the nitric oxide are not near enough to any of the oxygens of the ozone

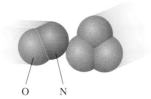

O   N

■ FIGURE 18.12   A collision between an NO and O_3 molecule that can result in reaction if the collision is energetic enough. The geometry is right.

molecule to become bonded to an oxygen and form nitrogen dioxide. A collision like the one shown in ■Figure 18.12 is necessary for a bond to be formed between nitrogen and oxygen.

Activated Complex Theory

The nucleus of each atom is surrounded by an electron cloud. When atoms combine to form molecules, the outside of the molecule is surrounded by an electron cloud. When two molecules approach each other, their electron clouds repel each other. For the nuclei to get close enough for the electron clouds to rearrange and form new bonds, the molecules must have unusually high kinetic energies. Three "still" shots in ■Figure 18.13 show an ozone molecule and a nitric oxide molecule moving toward each other, colliding, and then molecules of nitrogen dioxide and oxygen moving away from each other. Figure 18.13(b) shows the molecules while the electrons are rearranging after they have collided. A species like the one shown in Figure 18.13(b) is called an **activated complex.** The *state the activated complex is in* is called the **transition state.**

   ■Figure 18.14 shows how the potential energy of the system changes during the reaction shown in Figures 18.13 (a), (b), and (c). Graphs like Figure 18.14 are called **reaction profiles.** The **reaction coordinate** is the *lowest energy pathway from reactants to activated complex to products.* It is like a mountain pass as shown in

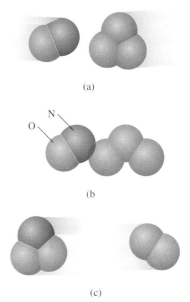

(a)

N

O

(b)

(c)

■ FIGURE 18.13   Reactive collision between NO and O_3 molecules. (a) Reactants. (b) Activated complex. (c) Products.

*See Paver, S. H.; Wilcox, C. F., Jr. J. Chem. Educ. **1995,** 72, 13–16 for a discussion of the common misuse of the term transition state.*

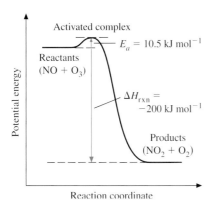

Activated complex

E_a = 10.5 kJ mol^{-1}

Reactants (NO + O_3)

$\Delta H_{rxn} = -200$ kJ mol^{-1}

Products (NO$_2$ + O$_2$)

Potential energy

Reaction coordinate

■ FIGURE 18.14   Reaction profile for reaction of NO and O_3*. This reaction has a low activation energy and is fast.

*Strictly speaking, the symbol $\Delta H^{\ddagger}$ should be used instead of E_a in reaction profiles. The value of E_a is an experimental quantity: It is the activation energy found from the temperature dependence of reaction rate. The value $\Delta H^{\ddagger}$ is a theoretical quantity: It is the difference between the enthalpy of the reactants and the enthalpy of the activated complex. Because E_a and $\Delta H^{\ddagger}$ are closely related and their values are usually similar, the difference is disregarded in this book.

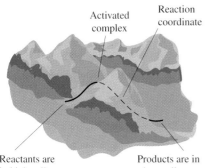

Activated complex   Reaction coordinate

Reactants are in high valley on near side of mountain.

Products are in low valley on other side of mountain.

▪ FIGURE 18.15   Mountain pass analogy for the reaction coordinate for the reaction of NO and O_3. The part of the reaction coordinate that is behind the front mountain is shown by a dashed line.

▪Figure 18.15. *The activated complex is the highest energy arrangement of atoms along the reaction coordinate.* The difference between the energy of the reactants and the energy of the activated complex is the activation energy for the reaction, E_a. The activation energy is the minimum energy the colliding molecules must have in order for reaction to take place. The difference in energy between the reactants and products is the enthalpy change for the reaction, ΔH_{rxn}.

PRACTICE PROBLEM

18.21   For the reaction

$$CO(g) + N_2O(g) \longrightarrow CO_2(g) + N_2(g),$$

$\Delta H_{rxn} = -365 \text{ kJ mol}^{-1}$ and $E_a = 96 \text{ kJ mol}^{-1}$. Sketch the reaction profile for this reaction.

In solution, reactant species collide not only with each other but also with the solvent molecules around them. The rates of very fast reactions, such as combination of oppositely charged ions, are controlled by the speed with which the ions can diffuse through the solvent to each other. A diffusion-controlled reaction is shown in ▪Figure 18.16. Just as in the gas phase, the reactant species must have enough energy to react and, unless they are very simple in structure, the correct orientation. However, reactions in solution are complicated by the solvent molecules that surround the activated complex. The solvent molecules that surround the activated complex act as a cage. Product species cannot separate immediately if reaction takes place; reactant species cannot separate immediately if a reaction does not occur. The reaction rate may either decrease or increase as a result.

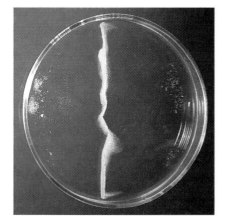

▪ FIGURE 18.16   If a solution of lead(II) nitrate is poured into a solution of potassium iodide in a test tube, a yellow precipitate of lead(II) iodide forms immediately. However, if a little $Pb(NO_3)_2(s)$ is added to one side of a thin layer of water in a Petri dish and then a little KI(s) is added to the opposite side, the solids dissolve and the Pb^{2+} and I^- ions diffuse toward each other. When they meet, a line of yellow $PbI_2(s)$ crystals forms.

18.11   REACTION MECHANISMS

A *detailed molecular-level picture,* like Figure 18.13, *of how a reaction might take place* is called a **reaction mechanism.** Reaction mechanisms are usually shown by means of formulas. For example, the reaction mechanism for the reaction between NO and O_3 can be shown as follows:

activated complex

The broken lines in the formula for the activated complex represent bonds breaking and forming in the reaction. This mechanism is simple, consisting of just one step. *Each step in a mechanism* is called an **elementary process.** The **molecularity** *of an elementary process is the number of particles that come together to form the activated complex.* The elementary process shown is **bimolecular,** that is, *two particles combine to form the activated complex.* If *just one particle is involved in the activated complex,* an elementary process is said to be **unimolecular.** If *three particles are involved,* the elementary process is **termolecular.** Termolecular processes are rare because the probability of three molecules colliding at the same time is very low. No elementary processes with molecularity greater than three are known.

For elementary processes, the exponents for each species in the rate law are the same as the coefficients in the equation for the step. For example, the rate law for the elementary process

$$NO + O_3 \longrightarrow NO_2 + O_2$$

is rate $= k[NO][O_3]$. Since one molecule of NO collides with one molecule of O_3 to form the activated complex, the rate is proportional to the concentrations of NO and O_3. *It is very important to distinguish between molecularity and order. Order must be determined by experiment and refers to the overall reaction. Molecularity refers to the elementary processes, that is, the steps that make up a mechanism.* The equation for an elementary process tells the molecularity. The equation for the overall reaction does not tell the order. Because the elementary process, $NO + O_3 \longrightarrow NO_2 + O_2$, is the only step in the reaction between nitric oxide and ozone to form nitrogen dioxide and oxygen, the rate of the elementary step is the rate of reaction and the rate law for the reaction is rate $= k[NO][O_3]$. The reaction is first order in nitric oxide and first order in ozone.

Notice the relationship between the rate law and the composition of the activated complex. In the rate law, both NO and O_3 appear to the first power. The activated complex consists of one molecule of nitric oxide and one molecule of ozone. In general, *the rate law gives the composition of the activated complex. The power of a species in the rate law is the same as the number of particles of the species in the activated complex.* For example, the experimentally observed rate law for the reaction

$$2NO_2(g) + F_2(g) \longrightarrow 2NO_2F(g) \qquad \text{is} \qquad \text{rate} = k[NO_2][F_2]$$

The activated complex for this reaction consists of a combination of one NO_2 molecule and one F_2 molecule and is composed of one N atom, two O atoms, and two F atoms.

The mechanisms of most reactions consist of more than one step. *If the exponents in the rate law are not the same as the coefficients in the equation for the reaction, the overall reaction must consist of more than one step.* For example, the decomposition of dinitrogen pentoxide is first order; that is, the experimentally determined rate law is rate $= k[N_2O_5]$. But the coefficient of dinitrogen pentoxide in the equation for the decomposition

$$2N_2O_5 \longrightarrow 2N_2O_4 + O_2$$

is 2, not 1. Therefore, a dinitrogen pentoxide molecule cannot simply fall apart in a single unimolecular step. Many apparently simple reactions such as this are actually quite complicated. The overall reaction takes place through a series of elementary processes.

Remember that, although thermochemical equations are always interpreted in terms of moles, ordinary equations can also be interpreted in terms of molecules. Therefore, the equation

$$N_2O_5(g) \longrightarrow N_2O_4(g) + \tfrac{1}{2}O_2(g)$$

is incorrect. Half a molecule of O_2 cannot be formed.

Figuring Out Reaction Mechanisms

The first steps in figuring out the mechanism of a reaction are always (1) to determine what the reactants and products are and (2) to find the rate law. To find out what the products are, it is important to keep reaction times short compared with the time needed to reach equilibrium so that you get information about reaction rates, not equilibrium (remember Figure 18.2). The elementary processes in a mechanism must add up to give the correct overall reaction. The mechanism must account for the observed rate law and any additional observations that may have been made of the reaction. Let's work through a simple example of how a possible mechanism is figured out. For the reaction

$$2NO_2(g) + F_2(g) \longrightarrow 2NO_2F(g)$$

the experimentally observed rate law is

$$\text{rate} = k[NO_2][F_2]$$

If this reaction took place in a single step, the rate law would be

$$\text{rate} = k[NO_2]^2[F_2]$$

Since this is not the rate law, the reaction must take place in more than one step.

Because the rate law gives the stoichiometry of the activated complex, collision of a molecule of NO_2 with a molecule of F_2 seems like a possible elementary process. But if a molecule of NO_2F were formed by collision of a molecule of NO_2 with a molecule of F_2, one fluorine atom would be left over:

$$NO_2 + F_2 \longrightarrow NO_2F + F$$

A fluorine atom is a very reactive species and, if formed, would probably react very fast with something. If the fluorine atom were to react with a molecule of NO_2, another molecule of NO_2F would be formed:

$$F + NO_2 \longrightarrow NO_2F$$

Let's see if these two steps are a possible way for the reaction to take place. When these two steps are added, the fluorine atom cancels, and the equation for the overall reaction is obtained:

$$
\begin{array}{rcl}
NO_2 + F_2 & \longrightarrow & NO_2F + F \\
+\,F + NO_2 & \longrightarrow & NO_2F \\
\hline
2NO_2 + F_2 & \longrightarrow & 2NO_2F
\end{array}
$$

The first step,

$$NO_2 + F_2 \longrightarrow NO_2F + F$$

would be expected to take place slowly because it involves breaking a fluorine–fluorine bond. Breaking bonds takes energy, so this reaction would have a significant activation energy.

In a series of steps, the slowest step determines the overall rate. Think about baking cupcakes again. There are three steps: (1) make batter, (2) bake, and (3) frost. Suppose that it takes 10 minutes to make the batter for a batch of 24 cupcakes, 20 minutes to bake them, and 15 minues to frost them. The rate at which you can make batter is

$$\frac{24 \text{ cupcakes}}{10 \text{ min}} \times \frac{60 \text{ min}}{h} = \frac{144 \text{ cupcakes}}{h}$$

The rate at which you can bake is 72 cupcakes per hour, and the rate at which you can frost is 96 cupcakes per hour. Suppose you are giving a big party and need to make several batches of cupcakes. You will only be able to make 72 cupcakes per hour. Baking is the slowest step in the sequence and limits the rate at which you can make cupcakes. If you keep on making batter as fast as you can, you will soon have a kitchen full of batter waiting to be baked. You can't frost cupcakes until after they are baked. *In the mechanism for a chemical reaction, the slow step* is called the **rate-determining step.** *Species* like the unfrosted cupcakes, *which are formed in one step and used up in another step,* are called **intermediates.**

Now that you understand what's meant by a rate-determining step, let's get back to the reaction between nitrogen dioxide and fluorine. This mechanism is usually written

Step 1 $NO_2 + F_2 \xrightarrow{k_1} NO_2F + F$ slow

Step 2 $F + NO_2 \xrightarrow{k_2} NO_2F$ fast

where k_1 is the rate constant for the first step and k_2 is the rate constant for the second step. (Both steps are elementary processes.) In this mechanism, the first step is the rate-determining step. Fluorine atoms are the intermediate. Once fluorine atoms are formed, they react immediately with NO_2 molecules. But they can't react until they are formed, so the rate of formation of the fluorine atoms determines the rate of formation of NO_2F molecules. The rate of formation of NO_2F molecules depends on the rate of the first step; the rate law for the first step is

$$rate = k_1[NO_2][F_2]$$

Therefore, this mechanism explains the experimentally observed rate law and also gives the correct overall reaction. The experimentally determined rate constant for the reaction, k, is equal to k_1 because the first step is slow and determines the rate of the overall reaction. Many reactions have simple rate laws because there is a single slow step in a sequence of elementary processes.

Steps after the rate-determining step do not affect the rate law, which explains how rate laws can be zero order with respect to a reactant. The reactant takes part in a fast step after the rate-determining step.

PRACTICE PROBLEMS

18.22   Which of the reactions given for Practice Problems 18.7–18.9 can take place in a single step?

18.23   Again referring to the reactions given for Practice Problems 18.7–18.9, what are the formulas for the activated complexes in reactions 1 and 3?

18.24   The mechanism,

Step 1 $NO_2 + O_3 \xrightarrow{k_1} NO_3 + O_2$ slow

Step 2 $NO_3 + NO_2 \xrightarrow{k_2} N_2O_5$ fast

has been proposed for a reaction. (a) Write the equation for the overall reaction. (b) Write the rate laws for steps 1 and 2. (c) Write the rate law predicted by the mechanism for the overall reaction. (d) Write the formulas for the activated complexes for each step and for the overall reaction. (e) Which species is an intermediate?

The reaction profile for the reaction of nitrogen dioxide with fluorine is shown in ▪Figure 18.17. Note carefully the difference between the reaction profile for a two-step reaction in Figure 18.17 and the reaction profile for a one-step reaction in Figure 18.14. The reaction profile for a two-step reaction looks like a combination of two profiles for one-step reactions. The energy level labeled "intermediate" includes both products of the first step—the product, NO_2F, as well as the intermediate, F. The activation energy E_a' for the second, fast step is very low compared to the activation energy E_a for the first, slow step.

Let's work through one more example. For the reaction

$$2NO(g) + O_2(g) \longrightarrow 2NO_2(g)$$

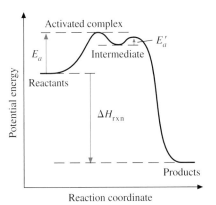

▪ FIGURE 18.17   Reaction profile for the reaction of NO_2 and F_2.

which is a step in the commercial production of nitric acid, the experimentally determined rate law is

$$\text{rate} = k[NO]^2[O_2]$$

For this reaction to take place in one step, two molecules of NO and one molecule of O_2 would have to collide. Collisions between three molecules are not very likely, just as simultaneous collisions of three cars are rare. What has usually happened in a three-car accident is that one car ran into another and a third ran into them. Similarly, collision between two of the molecules followed by a collision of a third molecule with the combination is much more probable than one collision between three molecules. The generally accepted mechanism for this reaction is

$$\text{Step 1} \qquad 2NO \quad \underset{k_{-1}}{\overset{k_1}{\rightleftarrows}} N_2O_2 \qquad \text{fast}$$

$$\text{Step 2} \qquad N_2O_2 + O_2 \overset{k_2}{\longrightarrow} 2NO_2 \qquad \text{slow}$$

Let's figure out the rate law that would be expected from this mechanism. The slow step is rate-determining. From the equation for step 2, the rate law is

$$\text{rate} = k_2[N_2O_2][O_2] \qquad (18.10)$$

But N_2O_2 is an intermediate. The experimentally determined rate law tells how the rate depends on the concentrations of the reactants. We must express the rate in terms of the concentrations of the reactants. The equilibrium in step 1 is reached rapidly. At equilibrium, the rate of the forward reaction is equal to the rate of the reverse reaction. The rate of the forward reaction $= k_1[NO]^2$ and the rate of the reverse reaction $= k_{-1}[N_2O_2]$. Setting these two rates equal gives

$$k_1[NO]^2 = k_{-1}[N_2O_2]$$

Solving for $[N_2O_2]$ gives

$$\frac{k_1}{k_{-1}}[NO]^2 = [N_2O_2]$$

Substituting this value for $[N_2O_2]$ in the rate law (equation 18.10) gives

$$\text{rate} = k_2\frac{k_1}{k_{-1}}[NO]^2[O_2]$$

This rate law agrees with the experimentally determined rate law. The experimental rate constant, k, is equal to $k_2(k_1/k_{-1})$.

The mechanism of a reaction is a theory developed to explain how the reaction takes place. It must account for all the observed facts. A mechanism can be proved wrong if a new experimental fact is discovered that is not explained by the mechanism. But it can never be proved right. There is always the possibility that someday another experiment will be done that will prove a mechanism wrong. For example, for years the reaction

$$H_2(g) + I_2(g) \longrightarrow 2HI(g)$$

which has the experimentally observed rate law, rate $= k[H_2][I_2]$, was given as an example of a one-step bimolecular gas phase reaction. However, it has been shown to have a much more complicated mechanism.* If the exponents in the rate law are

The effect of significant concentrations of an intermediate on calculations of equilibrium concentrations from initial concentrations and equilibrium constants was pointed out by Fainzilberg, V.E.; Karp, S. J. Chem. Educ. 1994, 71, 769–770.

─────────

*Calculations of equilibrium concentrations from initial concentrations and the value of the equilibrium constant, such as those in Section 14.4, assume that only reactants and products are present at equilibrium. This assumption is not always correct. For example, iodine atoms are an intermediate in the formation (and decomposition) of HI(g). At high temperatures, such as 1000 K, and low initial concentrations, such as $[H_2]_0 = [I_2]_0 = 5.00 \times 10^{-4}$, the concentration of iodine atoms is of the same order of magnitude as the concentrations of H_2, I_2, and HI.

the same as the coefficients in the equation for the reaction, the reaction *may* take place in one step. But it does not necessarily do so. A famous kineticist once said that the only simple reactions are those that have not been studied thoroughly. However, mechanisms are very useful in organizing the enormous amount of information that exists about chemical reactions.

PRACTICE PROBLEMS

18.25 Is a mechanism similar to the one developed for the reaction of NO_2 with F_2 possible for the reaction of NO and O_2? Explain your answer.

18.26 ΔH_{rxn} for the reaction $2NO(g) + O_2(g) \longrightarrow 2NO_2(g)$ is -114.1 kJ. Sketch the reaction profile for this reaction.

18.27 The experimentally observed rate law for the reaction, $2NO(g) + Br_2(g) \longrightarrow 2NOBr(g)$, is rate $= k[NO]^2[Br_2]$. (a) Which of the following mechanisms is (are) possible? (b) If more than one mechanism is possible, which is more probable?

Mechanism 1:	Step 1	$NO + NO + Br_2 \longrightarrow 2NOBr$	slow
Mechanism 2:	Step 1	$NO + Br_2 \xrightarrow{k_1} NOBr_2$	slow
	Step 2	$NOBr_2 + NO \xrightarrow{k_2} 2NOBr$	fast
Mechanism 3:	Step 1	$NO + Br_2 \underset{k_{-1}}{\overset{k_1}{\rightleftarrows}} NOBr_2$	fast
	Step 2	$NOBr_2 + NO \xrightarrow{k_2} 2NOBr$	slow

HOW CATALYSTS WORK

A catalyst works by providing another mechanism or pathway with lower activation energy for the reactants to follow in becoming products. Although catalysts are not used up in reactions, this does not mean that they do not take part in reactions. Catalysts are used up in one step and formed again in a later step.

Homogeneous Catalysis

The nitric oxide catalyzed reaction of sulfur dioxide with oxygen to form sulfur trioxide,

$$2SO_2(g) + O_2(g) \xrightarrow{NO(g)} 2SO_3(g)$$

is a simple example of homogeneous catalysis. Although reaction of sulfur dioxide and oxygen is spontaneous and exothermic, it is slow because the activation energy is high. Nitric oxide acts as a catalyst furnishing a mechanism with lower activation energy. The mechanism for the catalyzed reaction is believed to be

Step 1	$2NO + O_2 \longrightarrow 2NO_2$	fast
Step 2	$NO_2 + SO_2 \longrightarrow NO + SO_3$	fast

Step 2 must take place twice each time step 1 takes place once for the steps to add to give the overall equation for the reaction. In the mechanism, note how the catalyst, $NO(g)$, is used up in the first step and formed again in the second step. Also note the difference between a catalyst and an intermediate. A catalyst must be present at the beginning of the reaction; an intermediate is formed during the reaction.

The nitric oxide catalyzed reaction of sulfur dioxide with oxygen to form sulfur trioxide may play a part in the production of acid rain.

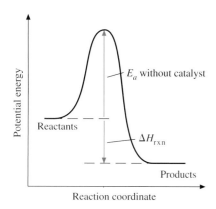

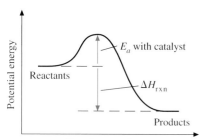

- FIGURE 18.18 Reaction profiles for catalyzed and uncatalyzed reactions of $SO_2(g)$ and $O_2(g)$. Notice that the catalyzed and uncatalyzed reactions have different reaction coordinates. The first step of the nitric oxide catalyzed reaction of sulfur dioxide with oxygen has such a low activation energy that this step does not show in the reaction profile.

*See Friend, C. M. Sci. Amer. **April 1993**, 74–79.*

*See Johnson, A. D.; Daley, S. P.; Utz, A. L.; Ceyer, S. T. Science **1992**, 257, 223–225.*

Reaction profiles for the catalyzed and uncatalyzed reactions are shown in ▬Figure 18.18. Notice that the catalyst does not change the energy of either the reactants or the products. Therefore, ΔH_{rxn} is not changed, and the same equilibrium is reached with and without the catalyst. However, equilibrium is reached faster with the catalyst because the energies of activation for both forward and reverse reactions are lower for the catalyzed reaction. A larger fraction of the molecules has enough energy to react. Also note that, in general, a catalyzed reaction must take place in at least two steps because the catalyst must be used up in one step and formed again in another.

Heterogeneous Catalysis

Nitric oxide-catalyzed conversion of sulfur dioxide to sulfur trioxide was discovered in 1746 and was used in the manufacture of sulfuric acid by the chamber process for more than a century. The contact process (Section 14.6) has now completely taken the place of the chamber process. The contact process uses a heterogeneous catalyst. Heterogeneous catalysis usually takes place at the surface of a solid. In addition to providing a mechanism with lower activation energy, heterogeneous catalysts may also increase the number of collisions with the right geometry.

Besides the contact process, another important application of heterogeneous catalysis is in catalytic converters for automobiles. The main purpose of a catalytic converter is to remove nitric oxide, NO, and carbon monoxide, CO, from the exhaust. As we have seen before, nitric oxide is involved in the formation of smog and contributes to acid rain. Carbon monoxide is very toxic. In the catalytic converter, nitric oxide and carbon monoxide react to form nitrogen and carbon dioxide:

$$2NO(g) + 2CO(g) \xrightarrow{\text{Pt, Pd, and Rh}} N_2(g) + 2CO_2(g)$$

This reaction is spontaneous, but reaction is slow, hence the need for a catalyst. The small size of the metal particles and the ceramic honeycomb in a catalytic converter (Figure 18.8) maximize contact of the exhaust with the catalyst and minimize the quantity of metals needed. Platinum, palladium, and rhodium metals are very expensive.

The mechanism of this reaction has been investigated in great detail by modern instrumental methods. Because atoms in the surface of the catalyst particles have fewer neighbors than atoms in the interior, the surface atoms have unsatisfied bonds, and reactant molecules are adsorbed. The interaction with the surface weakens the N—O bond so that it breaks. The oxygen atom combines with an adsorbed molecule of carbon monoxide, and the molecule of carbon dioxide that results is less strongly bonded to the surface than the other species and leaves the surface. ▬Figure 18.19 shows the complete series of events that takes place.

Yet another important application of heterogeneous catalysis is the conversion of liquid vegetable oils, such as corn oil, to margarine by catalytic hydrogenation. **Hydrogenation** is the *addition of hydrogen to a multiple bond.* The simplest example of catalytic hydrogenation is the addition of hydrogen to ethylene to form ethane:

$$H_2C{=}CH_2(g) + H_2(g) \xrightarrow{\text{Pt, Pd, or Ni catalyst}} H_3C{-}CH_3(g)$$

Until recently, this reaction was believed to take place at the surface of the catalyst. However, in 1992, a reaction involving hydrogen, atoms of which are very small, was shown to use hydrogen absorbed in the catalyst, not hydrogen adsorbed on the surface of the catalyst.

Hydrogenation reactions take place in living cells as well as in industrial plants and chemical laboratories. These biological reactions, like most biological reactions, are catalyzed by enzymes.

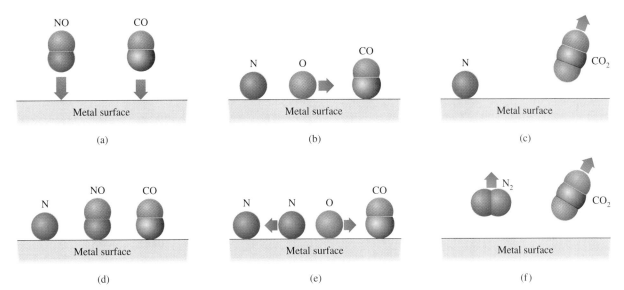

(a) (b) (c)

(d) (e) (f)

- FIGURE 18.19 (a) A molecule of nitric oxide and a molecule of carbon monoxide are adsorbed on the surface of the catalyst. (b) The nitric oxide molecule dissociates into a nitrogen atom and an oxygen atom, which reacts with the carbon monoxide molecule to form a molecule of carbon dioxide. (c) The carbon dioxide molecule leaves the surface of the catalyst. (d) If a second nitric oxide molecule and a second carbon monoxide molecule are adsorbed close to the remaining nitrogen atom, the second nitric oxide molecule dissociates. (e) The two nitrogen atoms combine to form a molecule of nitrogen. The second oxygen atom reacts with the second carbon monoxide molecule to form a second molecule of carbon dioxide. (f) The nitrogen molecule and the second molecule of carbon dioxide leave the surface of the catalyst.

Enzymes

The *reactant in an enzyme-catalyzed reaction* is known as the **substrate.** When an enzyme catalyzes a reaction, the substrate fits into a pocket or groove on the surface of the enzyme like a hand fits in a glove (see - Figure 18.20). The *pocket or groove* is called the **active site** of the enzyme.

Many inhibitors of enzyme-catalyzed reactions act by occupying the active site of the enzyme so that there is no room for the substrate. For example, the antibacterial action of sulfa drugs such as sulfanilamide is believed to involve inhibition of an enzyme in the bacteria, whose normal substrate is *p*-aminobenzoic acid. Molecules of sulfanilamide and *p*-aminobenzoic acid are about the same size and shape and have similar structures:

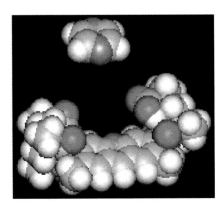

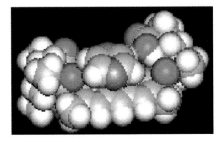

- FIGURE 18.20 Enzymes are very large, complex molecules (see Figure 23.7). The photos above show a relatively simple model of how a substrate fits into the active site of an enzyme. In these computer-generated models, carbon atoms are colored green so that they show up against the black background. As usual, hydrogen atoms are white, oxygen atoms are red, and nitrogen atoms are blue.

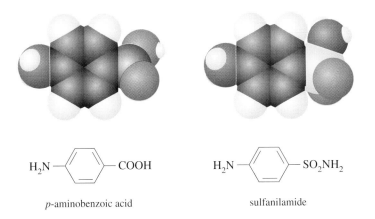

H_2N—⟨ ⟩—COOH

p-aminobenzoic acid

H_2N—⟨ ⟩—SO_2NH_2

sulfanilamide

The Kinetics of Drinking*

Alcohol does not have to be digested before it can be absorbed into the bloodstream. When an alcoholic drink is consumed, the alcohol is diluted in the mouth and stomach. A small fraction diffuses into the bloodstream from the stomach where the presence of food, especially fatty food, delays the process but carbonated drinks speed it up. Most of the alcohol diffuses into the bloodstream from the small intestine.

The rate of absorption of alcohol by the bloodstream from the gastrointestinal tract is proportional to the concentration of alcohol in the stomach and small intestine—that is, the movement of alcohol through the stomach and intestinal walls is a first-order process. The rate of absorption of alcohol from the gastrointestinal tract is so high that, within 30–60 min after ingestion, absorption is complete for all practical purposes; the rate constant k is about 10 h^{-1}.

Distribution of the alcohol throughout the various body fluids is also rapid. The alcohol is diluted by the body fluids. For

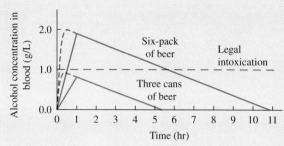

Many states have a legal limit of intoxication of 0.8 g/L instead of 1.0 g/L.

- FIGURE 18.21   Rate of alcohol metabolism in an average (150-lb) adult male. Strictly on a mass basis, a typical 110-lb female will become legally intoxicated by drinking about two beers. In addition, research indicates that females usually metabolize alcohol less rapidly than males.

example, an ounce (29.57 mL) of 100-proof liquor leads to a blood-alcohol concentration of 0.02% in an average (150-lb) adult male. In a larger person, the concentration is naturally lower and in a smaller person, higher.

The body begins to get rid of the alcohol immediately. A small part, 2–10%, is eliminated in breath, sweat, and urine. The remaining 90–98% is converted, first to acetaldehyde,[†] CH_3CHO, and then to acetic acid, CH_3COOH, by metabolic processes, mainly in the liver. The rate-determining step in the metabolism of alcohol is replacement of one of the enzymes that

catalyzes the formation of acetic acid from acetaldehyde. Like all catalysts, this enzyme is used up in one step and formed in another step. The steady-state concentration of the enzyme controls the rate of alcohol metabolism; therefore, the rate of metabolism does not depend on the concentration of alcohol and is zero order. In one individual, *the rate of elimination of alcohol* is constant although the rate *can*

[†]Acetaldehyde is toxic, especially to fetuses. Ingestion of alcohol by pregnant women is believed to be a major cause of mental retardation in Western countries.

*Most data are from Calder, G. V. *J. Chem. Educ.* **1974,** *51,* 19–22.

Figure 18.20 shows a model host-guest complex used in seminal studies probing the forces involved in nucleic acid recognition. The cleftlike structure is known as Rebek's cleft after Dr. Julius Rebek, Jr. of the Department of Chemistry at MIT. For a very interesting paper about molecular recognition with model systems see Rebek, J. Jr. Angew. Chem. Int. Ed. Engl. 1990, 29, 245–255.

If sulfanilamide molecules occupy the active sites of enzymes that catalyze reactions important to bacteria, there is no room for *p*-aminobenzoic acid molecules. The enzymes can't act as catalysts and the bacteria die. Many poisons are enzyme inhibitors, and a number of insecticides and herbicides work by inhibiting enzyme catalysis in insects and weeds.

Like all catalysts, enzymes act by lowering the activation energy. However, enzymes are much more efficient at increasing reaction rate than ordinary catalysts. Reactions that require long times at high temperatures in a laboratory or manufacturing plant take place rapidly at body temperature and physiological pH (about 7.4) in animals and in plants. For example, all plants and animals can synthesize proteins at ordinary temperature, but the manufacture of nylon, which is similar in structure to proteins, requires a temperature of 280–300 °C.

PRACTICE PROBLEMS

18.28   A possible mechanism for a reaction that takes place in solution is

Step 1	$Ce^{4+} + Mn^{2+} \longrightarrow Ce^{3+} + Mn^{3+}$		slow
Step 2	$Ce^{4+} + Mn^{3+} \longrightarrow Ce^{3+} + Mn^{4+}$		fast

vary by as much as ±30% from one individual to another. However, elimination of alcohol from the bloodstream is always much slower than absorption of alcohol into the bloodstream. A typical value of the rate constant for the elimination of alcohol from the bloodstream is 0.92 g/L·h or a total of about 8 g/h for an average adult male.

The dashed curve in ▪Figure 18.21 shows how the concentration of alcohol in the blood varies with time if the alcohol is all ingested at one time. At a party, alcohol is usually consumed over a period of time at a roughly constant rate. The solid lines in Figure 18.21 show how the concentration of alcohol varies with time under this more realistic condition. Naturally, the rate of consumption, like the rate of metabolism, varies from individual to individual.

From Figure 18.21, you can see that the concentration of alcohol in the blood increases rapidly, reaches a maximum, and then decreases gradually (assuming that no more is consumed). Whether the alcohol is ingested rapidly or slowly doesn't make much difference in the maximum concentration reached in the blood or in the time required to get below the

The "proof" is twice the alcohol content; 90 proof means 45% alcohol by volume.

legal level for intoxication or to eliminate alcohol completely from the body.

If more alcohol is consumed, the concentration again increases rapidly *starting from whatever concentration is already present.* Only about 8 g of alcohol an hour, a little over half a can of beer, is required to maintain a steady-state concentration. More alcohol adds to what's already there, and the concentration goes up. Every once in a while you hear about someone who,

as a result of a dare or as part of an initiation, drinks a large quantity of alcohol very rapidly and dies. A blood alcohol concentration of 3.5 g/L is sometimes fatal and 5.5 g/L almost always is.

For additional information on the kinetics of drinking, see Labianca, D. A. J. Chem. Educ. 1992, 69, 628–632.

| Step 3 | $Mn^{4+} + Tl^+ \longrightarrow Mn^{2+} + Tl^{3+}$ | fast |

(a) Write the equation for the overall reaction. (b) Write the predicted rate law. (c) Which species is the catalyst? (d) Which species is (are) intermediate(s)?

18.29 If E_a for a reaction is 23 kJ/mol and $\Delta H_{rxn} = -54$ kJ/mol, what is E_a for the reverse reaction?

18.13 THE STEADY-STATE APPROXIMATION

In the mechanisms we have discussed so far, one step has been slow and rate determining, and the rest have been fast. If two or more steps are slow, figuring out a mechanism and checking the rate law to be sure it has the same form as the experimental rate law can be complicated. An approximation called the **steady-state approximation** is very important in prediction of rate laws from mechanisms with more than one slow step. *In a steady state, the concentration of an intermediate is assumed to be constant because the intermediate is being formed in one step and used up in a later step.* Its concentration never becomes very high. You must be careful to distinguish between a steady state and a state of equilibrium. In a state

An understanding of the steady-state approximation is basic to an understanding of drug action.

of equilibrium, concentrations are constant because the rates of a reaction and its reverse are equal. In a steady state, the concentration of an intermediate is assumed to be constant because the intermediate is being formed by one elementary process and used up by a different elementary process. The later step is not the reverse of the earlier step. The concentration of the intermediate is not really constant. But the approximation that it is constant is accurate enough to be very useful.

In the natural world, constant concentrations are much more likely to be the result of steady states than of equilibria because the natural world does not exist in a closed container. For example, constant animal populations are steady states due to the equality of birth rates and death rates. The constant concentration of oxygen in the atmosphere results from formation of oxygen by plants and use of oxygen by animals. The metabolism of drugs in the body involves steady states, and steady states are important in the chemistry of air pollutants (see Problem 18.132 at the end of this chapter).

The constant concentration of H^+ in body fluids is another example of a steady state. Hydrogen ions are continuously added by metabolism and removed by the lungs and kidneys.

18.14 ANSWERS TO TWO COMMON QUESTIONS

Unimolecular Elementary Processes

One question that may have occurred to you is "If chemical reactions involve collisions, how can there be such a thing as a unimolecular elementary process?" According to the definition of a unimolecular elementary process given in Section 18.11, just one particle is involved in the activated complex.

How can energy be obtained from collisions even though a reaction takes place in a single unimolecular step and with a first-order rate law? The explanation is easier to understand if you think of reaction taking place in the presence of a high concentration of an inert gas such as nitrogen or argon. The reactant molecules get enough energy to react by colliding with inert gas molecules. (In practice, the molecules that give energy to the reactant molecules can be other reactant molecules instead of inert gas molecules.) Using A* to represent a reactant molecule with enough energy to react and M to represent an inert gas molecule,

*Now that "intermediates" have lifetimes on the order of a nanosecond whereas an "activated complex" should exist for no more than a picosecond, decomposition of cyclobutane gas to ethene gas can no longer be used as an example of a reaction that takes place by a single unimolecular step (Berson, J. A., Science **1994**, 266, 1338–1339).*

$$A + M \xrightarrow{k_1} A^* + M \tag{18.11}$$

The energetic reactant molecules can either lose their energy by colliding with another molecule of the inert gas (the reverse of the change in equation 18.11)

$$A^* + M \xrightarrow{k_{-1}} A + M$$

or they can form product

$$A^* \xrightarrow{k_2} product$$

Here's another place where the steady-state approximation is used.

The rate of formation of $A^* = k_1[A][M]$, and the rate of disappearance of $A^* = k_{-1}[A^*][M] + k_2[A^*]$. At the steady state,

rate of formation of A* = rate of disappearance of A*   or

$$k_1[A][M] = k_{-1}[A^*][M] + k_2[A^*] \tag{18.12}$$

Solution of equation 18.12 for the concentration of A* gives.

$$\frac{k_1[A][M]}{(k_{-1}[M] + k_2)} = [A^*]$$

But the rate of formation of products $= k_2[A^*]$. Substitution of the expression for the concentration of A* obtained by the steady-state approximation in the rate law for the formation of products gives

$$\text{rate of formation of products} = \frac{k_2 k_1 [\text{A}][\text{M}]}{(k_{-1}[\text{M}] + k_2)} \qquad (18.13)$$

From equation 18.13, you can see that if [M] is high, k_2 will be negligible compared to $k_{-1}[\text{M}]$, and equation 18.13 simplifies to

$$\text{rate of formation of products} = \frac{k_2 k_1 [\text{A}]}{k_{-1}}$$

a first-order rate law.

If [M] is low, $k_{-1}[\text{M}]$ will be negligible compared to k_2, and equation 18.13 simplies to

$$\text{rate of formation of products} = k_1 [\text{A}][\text{M}]$$

Thus, a second-order rate law is predicted for low concentrations. This change of rate law from first to second order at low concentrations is observed.

Reactions That Are Faster at Low Temperature

Something else you may have wondered about is whether there are any reactions that take place faster at lower temperatures. A few are known. For example, the rate constant for the reaction,

$$2\text{NO}(g) + \text{O}_2(g) \longrightarrow 2\text{NO}_2(g)$$

decreases with increasing temperature. Remember that the following mechanism is generally accepted for this reaction (Section 18.11):

Step 1 $2\text{NO} \rightleftharpoons \text{N}_2\text{O}_2$ fast
Step 2 $\text{N}_2\text{O}_2 + \text{O}_2 \longrightarrow 2\text{NO}_2$ slow

The forward reaction in step 1 is exothermic; therefore, equilibrium is shifted to the left by increasing temperature. The rapidly decreasing equilibrium concentration of N_2O_2 with increasing temperature results in a decrease in rate of formation of NO_2.

SUMMARY

Chemical kinetics deals with the rates of chemical reactions. Studies of reaction rates are an important source of information about **reaction mechanisms,** detailed pictures of how a reaction takes place. Mechanisms are theories and can never be proved to be correct although they can be shown to be wrong. They are useful in organizing information about chemical reactions.

Because the rates of formation and disappearance of the reactants and products are related by the equation for the reaction, the rate of a reaction can be followed by measuring the amount or concentration of any reactant or product at intervals. The rate of a chemical reaction depends on the identity of the reactants and their concentrations, temperature, presence of catalysts, the solvent, and, for heterogeneous reactions, particle size and mixing.

The dependence of reaction rate on concentration is described by the **rate law** for a reaction. Rate laws must be determined by experiment. Many rate laws have the form rate $= k[\text{A}]^x[\text{B}]^y \ldots$,

where k is a constant called the **rate constant.** The value of the rate constant depends on all the factors that affect the rate of a reaction except concentration. If the exponent of a concentration in the rate law is 0, the reaction is **zero order** with respect to that substance, and the rate of the reaction is independent of the concentration of that substance. If the exponent is 1, the reaction is **first order** with respect to that substance, and the rate doubles when the concentration of that reactant is doubled. If the exponent is 2, the reaction is **second order** with respect to that substance, and the rate increases fourfold when the concentration of that substance is doubled.

The rate law can be found by observing initial rates or by graphing. The **half-life** of a reaction, $t_{1/2}$, is the time required for one-half of the quantity of reactant originally present to react.

The rates of almost all reactions increase as the temperature is raised; quantitatively, the relationship is given by the **Arrhenius equation, $k = Ae^{-E_a/RT}$.** The dependence of reaction rates on concentration and temperature is explained by the collision and activated complex (transition state) theories.

The **collision theory** explains chemical reactions as involving collisions between molecules or ions. According to the collision theory, particles must collide with a minimum amount of energy, called the **activation energy,** and with the correct orientation in order for a reaction to take place. According to the **activated complex theory,** reactants follow the **reaction coordinate**—that is, the lowest energy path available—from reactants to activated complex to products. The **activated complex** is the highest energy arrangement of atoms along the reaction coordinate. **Reaction profiles** are graphs showing how potential energy changes as the reactants proceed along the reaction coordinate from reactants to activated complex to products.

Most reaction mechanisms consist of a series of simple steps called **elementary processes.** The **molecularity** of an elementary process is the number of particles that must collide to form the activated complex for the elementary process. For elementary processes, the exponents in the rate law are the same as the coefficients in the equation. If the exponents in the rate law for an overall reaction are different from the coefficients in the equation for the reaction, the mechanism must consist of more than one step. If the exponents in the rate law for the overall

reaction are the same as the coefficients in the equation, the reaction may take place in a single step, but it does not necessarily do so. The steps in a mechanism must add to give the net equation for the overall reaction, the mechanism must lead to the experimentally observed rate law, and the mechanism must account for any other observations that have been made of the reaction.

Intermediates are species that are formed in one step and used up in another step. In a **steady state,** the concentration of an intermediate is constant because the intermediate is being formed in one step and used up in a later step. A **rate-determining step** is a step in a sequence of elementary processes making up a reaction mechanism that is much slower than the other steps and determines the rate of the overall reaction.

Catalysts provide another mechanism with lower activation energy and/or increase the frequency of collisions with the right orientation. Catalysts may be homogeneous or heterogeneous. Catalysts are present both at the beginning and at the end of a reaction; they are used up in one step and formed again in another.

Enzymes are complex substances produced in living cells that catalyze biological reactions. The reactant in an enzyme-catalyzed reaction is known as the **substrate.** The substrate fits in the **active site** of the enzyme.

ADDITIONAL PRACTICE PROBLEMS

For information about the organization of Additional Practice Problems, Stop & Test Yourself, Putting Things Together, and Applications, see the beginnings of these sections in Chapter 1.

18.30 (a) What is the difference between average rate, instantaneous rate, and initial rate? (b) What is the difference between the rate of a reaction and the rate constant for a reaction? Is there any type of reaction for which they are the same? (18.1)

18.31 Use the data in Table 18.1 to calculate the average rate of formation of oxygen gas in the period from 1800 to 2400 s. (18.1)

18.32 From Figure 18.4, what was the instantaneous rate of formation of oxygen gas at 1500 s? (18.1)

18.33 Show how the rates of each of the following reactions can be expressed in terms of the rates of disappearance of the reactants and the rates of appearance of the products:
(a) $3I^- + H_3AsO_4(aq) + 2H^+ \longrightarrow$
$\qquad\qquad I_3^- + H_3AsO_3(aq) + H_2O(l)$
(b) $N_2O_5(g) + NO(g) \longrightarrow 3NO_2(g)$
(c) $2S_2O_8^{2-} + 2H_2O(l) \longrightarrow 4SO_4^{2-} + O_2(g) + 4H^+$ (18.1)

18.34 For the reaction in part (a) of Problem 18.33, (a) how is the rate of formation of I_3^- related to the rate of disappearance of I^-? (b) How is the rate of disappearance of H^+ related to the rate of disappearance of I^-? (18.1)

18.35 For the reaction $2NO_2Cl(g) \longrightarrow 2NO_2(g) + Cl_2(g)$, if the rate of disappearance of NO_2Cl is 8.5×10^{-9} M·s^{-1}, (a)

what is the rate of formation of NO_2? (b) What is the rate of formation of Cl_2? (18.1)

18.36 For the reactions below, give the order with respect to each species:
(a) $2Ce^{4+} + Tl^+ \xrightarrow{Mn^{2+}} 2Ce^{3+} + Tl^{3+}$
Rate $= k[Ce^{4+}][Mn^{2+}]$
(b) $14H_3O^+ + 2HCrO_4^- + 6I^- \longrightarrow$
$\qquad\qquad\qquad\qquad 2Cr^{3+} + 3I_2 + 22H_2O$
Rate $= k[HCrO_4^-][I^-][H_3O^+]^2$
(c) $NO_2(g) + 2HCl(g) \longrightarrow NO(g) + H_2O(g) + Cl_2(g)$
Rate $= k[NO_2][HCl]$ (18.2)

18.37 For the reactions below, give the order with respect to each species:
(a) $NO_2(g) + CO(g) \longrightarrow NO(g) + CO_2(g)$
Rate $= k[NO_2]^2$
(b) $2NO_2(g) + O_3(g) \longrightarrow N_2O_5(g) + O_2(g)$
Rate $= k[NO_2][O_3]$
(c) $CO(g) + Cl_2(g) \longrightarrow COCl_2(g)$
Rate $= k[CO][Cl_2]^{3/2}$ (18.2)

18.38 The reaction $2Br^- + 2H^+ + H_2O_2(aq) \longrightarrow Br_2(aq) + 2H_2O(l)$ is first order with respect to bromide ion, first order with respect to hydrogen ion, and first order with respect to hydrogen peroxide. Write the rate law. (18.2)

18.39 The initial rate of a reaction, $A + 3B \longrightarrow C + D$, was measured using different initial concentrations of A and B.

Results are summarized in the table below:

Experiment Number	[A]	[B]	Initial Rate, M·s⁻¹
1	0.100	0.100	5.40×10^{-4}
2	0.200	0.100	4.32×10^{-3}
3	0.200	0.200	4.32×10^{-3}

(a) What is the rate law for the reaction? (b) What is the numerical value of k? (c) Qualitatively, how will the initial rate in an experiment using an initial concentration of A = 0.250 M and an initial concentration of B = 0.300 M compare with the initial rate in experiment 3? What will be the initial rate in this experiment? (18.3)

18.40 The initial rate of a reaction, A + 2B $\longrightarrow$ C, was measured using different initial concentrations of A and B. Results are summarized in the table below:

Experiment Number	[A]	[B]	Initial Rate, M·s⁻¹
1	0.100	0.100	8.53×10^{-6}
2	0.100	0.200	3.41×10^{-5}
3	0.200	0.100	1.21×10^{-5}

(a) What is the rate law for the reaction? (b) What is the numerical value of k? (c) What will be the initial rate in an experiment using an initial concentration of A = 0.150 M and an initial concentration of B = 0.250 M? (18.3)

18.41 The initial rate of a reaction, A + B + C $\longrightarrow$ 2D + E, was measured using different initial concentrations of A, B, and C. Results are summarized in the table below:

Experiment Number	[A]	[B]	[C]	Initial Rate, M·s⁻¹
1	0.100	0.100	0.100	7.3×10^{-11}
2	0.200	0.100	0.100	1.2×10^{-9}
3	0.200	0.200	0.100	2.3×10^{-9}
4	0.200	0.100	0.300	1.1×10^{-8}

(a) What is the rate law for the reaction? (b) What is the numerical value of k? (c) What will be the initial rate in an experiment using initial concentrations A = 0.250 M, B = 0.300 M, and C = 0.125 M? (18.3)

18.42 The initial rate of a reaction, A + B $\longrightarrow$ C + D, was measured using different initial concentrations of A and B. Results are summarized in the table below:

Experiment Number	[A]	[B]	Initial Rate, M/min
1	0.050	0.060	1.8×10^{-2}
2	0.100	0.060	1.8×10^{-2}
3	0.050	0.030	7.1×10^{-2}

(a) What is the rate law for the reaction? (b) What is the numerical value of k? (c) What will be the initial rate in an experiment using an initial concentration of A = 0.075 M and an initial concentration of B = 0.050 M? (18.3)

18.43 If doubling the concentration of a reactant increases the rate of a reaction eightfold, what is the order of the reaction with respect to the reactant? (18.3)

18.44 For a reaction with the rate law of rate = $k[A]^2/[B]$, what will happen to the rate if (a) the concentration of A is doubled? (b) The concentration of B is doubled? (c) The concentrations of both A and B are doubled? (18.3)

18.45 In an experiment to determine the rate law for the decomposition of acetaldehyde, CH₃CHO(g) $\longrightarrow$ CH₄(g) + CO(g), the following data were obtained at 700 K:

Time, s	[CH₃CHO]
0	0.0500
1200	0.0300
2000	0.0240
6000	0.0120
10 000	0.0080
15 000	0.0056
20 000	0.0043

Use graphs to find the rate law for the decomposition of acetaldehyde and the value of k at 700 K. Assume that there are as many significant figures in the times as there are in the concentrations. (18.3)

18.46 In an experiment to determine the rate law and rate constant for the reaction A $\longrightarrow$ B, the following data were obtained:

Time, h	[A]
0.0	0.0400
40.0	0.0329
100.0	0.0245
200.0	0.0150
300.0	0.0092
400.0	0.0056
500.0	0.0034

What is the rate law and the value of k? (18.3)

18.47 The half-life for a first-order reaction is 2768 years. Starting with a concentration of 0.345 M, what will the concentration be after 11 072 years? (18.4)

18.48 A first-order reaction has a half-life of 4.48 months. How long will it take for the concentration to decrease of 25.0% of its original value? (18.4)

18.49 The half-life for one first-order reaction is 75 min, and the half-life for another first-order reaction is 322 min. Which reaction is faster? For which reaction does k have the larger numerical value? (18.4)

18.50 The half-life for a first-order reaction is 49 min. What is the rate constant per second? (18.4)

18.51 The rearrangement of methyl isonitrile to methyl nitrile,

$$H_3C\!-\!N\!\equiv\!C\!: \longrightarrow H_3C\!-\!C\!\equiv\!N\!:,$$

is first order. The rate constant is $3.02 \times 10^{-3}\,\text{s}^{-1}$ at 250.0 °C. (a) In an experiment at 250.0 °C, if the initial concentration of methyl isonitrile is 2.50×10^{-2} M, what will be the concentration of methyl isonitrile 172 s after the reaction begins? (b) In another experiment at 250.0 °C, if the initial concentration is 2.25×10^{-2} M, how many seconds will it take for the concentration to decrease to 1.125×10^{-3} M? (c) In a third experiment at 250.0 °C, if the concentration of methyl isonitrile is 3.65×10^{-3} M after 578 s, what was the initial concentration of methyl isonitrile? (18.4)

18.52 The decomposition of dimethyl ether, $(CH_3)_2O(g) \longrightarrow CH_4(g) + H_2(g) + CO(g)$, is first order with a half-life of 1733 s at 500 °C. In an experiment at this temperature, if $[(CH_3)_2O]$ is 0.033 after 865 s, what was the original concentration of dimethyl ether? (18.4)

18.53 (a) For a first-order reaction $A \longrightarrow$ products, the half-life is 726 s. Starting with a concentration of 0.600 M, what will the concentration be after 726 s? (b) What is the numerical value of k? (18.4)

18.54 Why are reactions usually carried out in solution? (18.7)

18.55 What can be done to increase the rate of a heterogeneous reaction? (18.7)

18.56 (a) What is meant by the term activation energy? (b) How is activation energy related to reaction rate? (18.9)

18.57 If the rate of a reaction doubles for every 10-degree increase in temperature, (a) how much faster will the reaction be at 40 °C than at 20 °C? (b) How much slower will the reaction be at 10 °C than at 20 °C? (c) How many degrees would the temperature have to be raised to increase the rate sixteen times? (18.9)

18.58 Calculate E_a for a reaction having $k = 2.61 \times 10^{-5}$ at 190.0 °C and $k = 3.02 \times 10^{-3}$ at 250.0 °C. (18.9)

18.59 (a) Calculate k at 30.0 °C for a reaction having $k = 4.59 \times 10^{-3}$ s^{-1} at 20.0 °C and $E_a = 51$ kJ mol^{-1}. (b) How many times faster does reaction take place at 30.0 °C than at 20.0 °C? (18.9)

18.60 For the decomposition of cyclobutane,

$$
\begin{array}{ccc}
H & & H \\
| & & | \\
H-C & - & C-H \\
| & & | \\
H-C & - & C-H \\
| & & | \\
H & & H
\end{array}
\quad (g) \longrightarrow 2H_2C = CH_2(g)
$$

$E_a = 262$ kJ/mol and $k = 6.1 \times 10^{-8}$ s^{-1} at 600 K. At what temperature is $k = 1.00 \times 10^{-4}$ s^{-1}? (18.9)

18.61 The value of k for the decomposition of acetaldehyde, $CH_3CHO(g) \longrightarrow CH_4(g) + CO(g)$, was determined at temperatures ranging from 750 to 1000 °C. Data are summarized in the following table:

Temperature, K	k
750	0.085
800	0.55
850	3.0
900	11.6
950	43
1000	140

Find the value of E_a graphically. (18.9)

18.62 For the reaction

$$
\begin{array}{ccc}
H & & H \\
\diagdown & & \diagup \\
 & C - C & \\
\diagup & & \diagdown \\
H & & H \\
 & \diagdown C \diagup & \\
 & \diagup \diagdown & \\
 & H \quad H &
\end{array}
\longrightarrow H_2C = CHCH_3
$$

$E_a = 271$ kJ/mol and $A = 1 \times 10^{15}$ s^{-1}. What is the value of k at 250 °C? (18.9)

18.63 Explain why increasing the temperature increases the rates of most reactions very rapidly. (18.10)

18.64 For the reaction

$$(CH_3)_3COH(l) \longrightarrow (CH_3)_2C=CH_2(g) + H_2O(l),$$

$\Delta H = +51.9$ kJ mol^{-1} and $E_a = 274.1$ kJ mol^{-1}. Sketch the reaction profile. (18.10)

18.65 For the reaction $2N_2O(g) \longrightarrow 2N_2(g) + O_2(g)$, $\Delta H_{rxn} = +78.0$ kJ mol^{-1} and $E_a = 240$ kJ mol^{-1}. This reaction is catalyzed by chlorine; E_a for the catalyzed reaction is 140 kJ mol^{-1}. What is E_a for the reverse of the catalyzed reaction? (18.10)

18.66 What two factors determine whether or not a collision leads to reaction? (18.10)

18.67 (a) What is a reaction mechanism? (b) What is an elementary process? (c) What does bimolecular mean? (d) Why are termolecular elementary processes rare? (e) What is a rate-determining step? (f) What is the difference between molecularity and order? (18.11)

18.68 Suppose an exothermic reaction takes place by a three-step mechanism in which the first step is slow and the second and third steps are fast. Sketch the reaction profile. (18.11)

18.69 Assuming that reactions (5) and (6) for Practice Problems 18.7–18.9 take place by mechanisms with a single slow step, what are the formulas for the activated complexes? (18.11)

18.70 The mechanism

Step 1 $\quad H_2O_2 + I^- \xrightarrow{k_1} H_2O + IO^- \quad$ slow

Step 2 $\quad H_2O_2 + IO^- \xrightarrow{k_2} H_2O + O_2 + I^- \quad$ fast

has been proposed for a reaction. (a) Write the equation for the overall reaction. (b) Write the rate laws for each step. (c) Write the rate law predicted by the mechanism for the overall reaction. (d) Which step is rate determining? (e) Write the formulas for the activated complexes for each step. (f) Which species is an intermediate? (g) Which species is a catalyst? (18.11)

18.71 The experimentally observed rate law for the reaction

$$H_2(g) + 2ICl(g) \longrightarrow 2HCl(g) + I_2(g)$$

is rate $= k[ICl][H_2]$. (a) Which of the following mechanisms is (are) possible?

Mechanism 1: $\quad H_2 + ICl + ICl \xrightarrow{k} 2HCl + I_2 \quad$ slow

Mechanism 2: Step 1 $\quad ICl + ICl \xrightarrow{k_1} Cl_2 + I_2 \quad$ slow

Step 2 $\quad Cl_2 + H_2 \xrightarrow{k_2} 2HCl \quad$ fast

Mechanism 3: Step 1 $H_2 + ICl \xrightarrow{k_1} HCl + HI$ slow

Step 2 $ICl + HI \xrightarrow{k_2} HCl + I_2$ fast

(b) What is the molecularity of each step in each mechanism? (18.11)

18.72 If a possible mechanism for a gas phase reaction is

Step 1	$2NO \rightleftharpoons N_2O_2$	fast
Step 2	$N_2O_2 + H_2 \longrightarrow H_2O + N_2O$	slow
Step 3	$N_2O + H_2 \longrightarrow N_2 + H_2O$	fast

(a) write the equation for the reaction. (b) Write the rate law. (18.11)

18.73 Catalysts used to be defined as substances that cause reaction to take place at a faster-than-normal rate but do not themselves react. Is this definition correct? If not, correct it. (18.12)

18.74 (a) What is the difference between a homogeneous catalyst and a heterogeneous catalyst? (b) Why are heterogeneous catalysts usually finely divided? (18.12)

18.75 (a) What are enzymes? (b) What is the reactant in an enzyme-catalyzed reaction called? (c) What is meant by the term "active site"? (18.12)

18.76 (a) What do steady states and equilibria have in common? (b) What is the difference between a steady state and a state of equilibrium? (c) From the everyday world, give one example of a steady state. (18.14)

18.77 For the reaction $2NOCl(g) \longrightarrow 2NO(g) + Cl_2(g)$, the rate law is rate $= k[NOCl]^2$ and $k = 5.9 \times 10^{-4}$ M^{-1} s^{-1} at 400 K. If the initial concentration of NOCl is 0.025 M, what [NOCl] will remain after 5.00 hours?

18.78 In an experiment to determine the order of a reaction, A + B $\longrightarrow$ C, the following data were obtained when [B] = 3.00×10^{-1}:

Time, s	[A]
0	2.00×10^{-4}
0.60×10^2	1.09×10^{-4}
1.20×10^2	0.72×10^{-4}
2.40×10^2	0.44×10^{-4}
3.60×10^2	0.31×10^{-4}
4.80×10^2	0.24×10^{-4}
6.00×10^2	0.20×10^{-4}

The following data were obtained when [A] = 3.00×10^{-1}:

Time, s	[B]
0	2.00×10^{-4}
0.60×10^2	1.82×10^{-4}
1.20×10^2	1.64×10^{-4}
2.40×10^2	1.28×10^{-4}
3.60×10^2	0.92×10^{-4}
4.80×10^2	0.57×10^{-4}
6.00×10^2	0.20×10^{-4}

(a) The order of the reaction did not depend on pH or anything else the investigator could think of to try. Write the rate law for the reaction. (b) Calculate the true value of k.

18.79 (a) For a zero-order reaction, how does the length of the first half-life (time for concentration to decrease to half its original value) compare with the length of the second half-life (time for concentration to decrease from one-half to one-quarter of its original value)? (b) For a second-order reaction, how does the length of the first half-life compare with the length of the second half-life?

18.80 For a reaction that is zero order with respect to a reactant A, what will [A] be after 26 min if $[A]_0 = 0.54$ and $k = 3.8 \times 10^{-3}$ M min^{-1}?

18.81 Explain why zero-order reactions are over much sooner than first-order reactions with similar numerical values for the rate constant.

18.82 Inhibitors used to be referred as "negative catalysts." If a negative catalyst is a substance that makes a reaction take place slower by providing another pathway with higher activation energy, what's wrong with the idea of a negative catalyst?

18.83 For a reaction that is third order with respect to a reactant, A, the equation for calculating the concentration of A at time t is

$$\frac{1}{[A]_t^2} = 2kt + \frac{1}{[A]_0^2}$$

(a) What are the units of the rate constant? (b) What should be plotted against t to give a straight line?

18.84 (a) For a reaction in which every collision between reactants leads to reaction, what determines the rate at which reaction takes place? (b) Why are reactions in which every collision leads to reaction relatively rare—that is, why do some collisions not result in reaction?

18.85 (a) Sketch the reaction profile for a reaction with $\Delta H = +91$ kJ/mol and $E_a = 101$ kJ/mol. (b) What is the minimum activation energy for an endothermic reaction? (c) Are reactions that are very endothermic likely to be fast? Explain your answer.

18.86 The initial rate of reaction $ClO^- + I^- \longrightarrow IO^- + Cl^-$, which takes place in basic solution, was measured using different initial concentrations of ClO^-, I^-, and OH^-. Results are summarized in the table below:

Experiment Number	$[I^-]$	$[OCl^-]$	$[OH^-]$	Initial Rate, $M \cdot s^{-1}$
1	0.0030	0.0010	1.00	1.8×10^{-4}
2	0.0030	0.0020	1.00	3.6×10^{-4}
3	0.0060	0.0020	1.00	7.2×10^{-4}
4	0.0030	0.0010	0.50	3.6×10^{-4}

(a) What is the rate law for the reaction? (b) What is the numerical value of k? (c) Qualitatively, how will the initial rate in an experiment using an initial concentration of $I^- = 0.0030$ M, an initial concentration of $OCl^- = 0.0010$ M, and an initial concentration of $OH^- = 0.25$ compare with the initial rate in experiment 4? What will be the initial rate in this experiment?

18.87 For the elementary process $Cl + NOCl \longrightarrow NO + Cl_2$, show microscopic views of (a) a collision that does *not* result in reaction even if it takes place with enough energy for reaction to occur (b) a collision that can result in reaction if energetic enough.

18.88 For a reaction $A(g) \longrightarrow B(g)$, the rate law is rate $= k[A]$.

If $t_{1/2} = 15$ min, draw microscopic pictures showing the system at (a) $t = 0$ (b) $t = 15$ min (c) $t = 30$ min (d) $t = 45$ min. Use ● to represent a molecule of A and ○ to represent a molecule of B; start with eight molecules of A. (e) What do you think will happen after 45 min? Explain your answer.

STOP & TEST YOURSELF

1. Given the graph of concentration as a function of time shown below, the equation for the reaction is
   (a) $A + B \longrightarrow 2C$ (b) $2A + 2B \longrightarrow C$
   (c) $A + B \longrightarrow C$ (d) $2A + B \longrightarrow 2C$

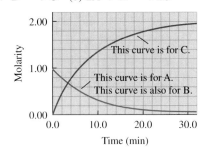

This curve is for C.
This curve is for A.
This curve is also for B.

Time (min)

2. What is the half-life of the reaction shown in the graph in Problem 1?
   (a) 3.0 min (b) 3.5 min (c) 6.0 min (d) 11.5 min

3. Again referring to the graph in Problem 1, what is the initial rate of formation of C?
   (a) -0.25 M/min (b) $+0.25$ M/min
   (c) $+0.093$ M/min (d) -0.093 M/min

4. Given the reaction

$$3KCN(aq) + 2KMnO_4(aq) + H_2O(l) \longrightarrow$$
$$3KCNO(aq) + 2KOH(aq) + 2MnO_2(s)$$

which of the following is true?

(a) $\dfrac{-2\Delta[KMnO_4]}{\Delta t} = \dfrac{3\Delta[KCNO]}{\Delta t}$

(b) $\dfrac{2\Delta[KMnO_4]}{\Delta t} = \dfrac{-3\Delta[KCNO]}{\Delta t}$

(c) $\dfrac{-3\Delta[KMnO_4]}{\Delta t} = \dfrac{2\Delta[KCNO]}{\Delta t}$

(d) $\dfrac{3\Delta[KMnO_4]}{\Delta t} = \dfrac{-2\Delta[KCNO]}{\Delta t}$

5. At low temperatures, the reaction $CO(g) + NO_2(g) \longrightarrow CO_2(g) + NO(g)$ is zero order in CO and second order in NO_2. The rate law is rate = _____.
   (a) $k[CO][NO_2]$ (b) $k[NO_2]$ (c) $k[CO]^2[NO_2]$ (d) $k[NO_2]^2$

6. For the reaction $NH_4^+ + NO_2^- \longrightarrow N_2(g) + 2H_2O(l)$, the rate law has been found to be rate $= k[NH_4^+][NO_2^-]$. In one experiment where initial $[NO_2^-]$ was 0.0400 and initial $[NH_4^+]$ was 0.200, the observed initial rate was 2.15×10^{-6} M/s. What is the value of k in $M^{-1} s^{-1}$?
   (a) 1.72×10^{-8} (b) 3.72×10^3 (c) 2.69×10^{-10}
   (d) 2.69×10^{-4}

7. For the reaction $2NO_2(g) \longrightarrow 2NO(g) + O_2(g)$, the rate law is rate $= k[NO_2]^2$. If $k = 0.543$ $M^{-1} s^{-1}$ and concentration of NO_2 is 0.0079 M, what is the rate in M/s?
   (a) 4.3×10^{-3} (b) 3.4×10^{-5} (c) 1.5×10^{-2} (d) 69

8. The following data were obtained for the reaction $A \longrightarrow$ products. What is the order, x, of the rate law, rate $= k[A]^x$?

Experiment Number	Initial [A]	Initial Rate, M/s
1	0.20	0.75×10^{-5}
2	0.40	3.0×10^{-5}

   (a) 0 (b) 1 (c) 2 (d) 3

9. The rate law for a reaction $B \longrightarrow$ products is rate $= k[B]$. Which of the following graphs will give a straight line?
   (a) ln[B] vs. time (b) [B] vs. time (c) 1/[B] vs. time
   (d) ln[B] vs. $1/T$

10. The reaction $C \longrightarrow$ products is a first-order reaction with $t_{1/2} = 1.3 \times 10^4$ s. If the initial concentration of C is 0.200 M, what will be the concentration of C after 2.6×10^4 s?
    (a) 0.025 M (b) 0.050 M (c) 0.100 M (d) 0.200 M

11. The major reason the rates of most chemical reactions increase very rapidly as temperature rises is (a) the fraction of the molecules with kinetic energy greater than the activation energy increases very rapidly as temperature increases. (b) the average kinetic energy increases as temperature rises. (c) the activation energy decreases as temperature increases. (d) more collisions take place with particles placed so that reaction can occur.

12. For the reaction mechanism

$$H_2 + 2NO \longrightarrow N_2 + H_2O_2 \qquad slow$$
$$H_2O_2 + H_2 \longrightarrow 2H_2O \qquad fast$$

the rate law is rate =
   (a) $k[H_2][NO]^2$ (b) $k[H_2O]^2$ (c) $k[N_2][H_2O_2]$ (d) $k[H_2O_2][H_2]$

For questions 13 and 14, consider the following reaction profile:

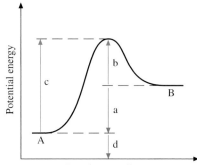

Reaction coordinate

13. For the reaction A $\longrightarrow$ B, the activation energy is
    (a) (b) (c) (d)
14. For the reaction A $\longrightarrow$ B in the presence of a catalyst, which of the following statements is true?

(a) Activation energy is lower, rate is faster, ΔH is different.
(b) Activation energy is lower, rate is slower, ΔH is same.
(c) Activation energy is lower, rate is faster, ΔH is same.
(d) Activation energy is higher, rate is faster, ΔH is same.

PUTTING THINGS TOGETHER

18.89 Explain each answer. (a) For which elementary process is the probability of a collision occurring with the correct orientation for reaction greater?

$$Cl\cdot + CH_4 \longrightarrow HCl + \cdot CH_3 \text{ or}$$
$$Cl\cdot + CHCl_3 \longrightarrow HCl + \cdot CCl_3$$

(b) Which elemental process(es) would have very low or zero activation energy?
  (i) $Cl\cdot + H\cdot \longrightarrow H{-}Cl$
  (ii) $Cl\cdot + CH_4 \longrightarrow H{-}Cl + \cdot CH_3$
  (iii) $CH_3{}^+ + Cl^- \longrightarrow CH_3Cl$

18.90 (a) The equilibrium constant expression for a reaction can be written from the equation for the reaction. Explain why rate laws can't be written from equations. (b) What two methods are used to find the rate laws of the form rate $= k[A]^a[B]^b \ldots$?

18.91 The rate constant for the reaction $Cl_2(aq) + H_2S(aq) \longrightarrow S(s) + 2HCl(aq)$ at 28 °C is $3.5 \times 10^{-2} M^{-1} s^{-1}$.
(a) What three rate laws are possible? (*Hint:* Look at the units of the rate constant.) (b) How could you find out which of these three rate laws is correct?

18.92 A first-order reaction has a half-life of 342 min at 23.0 °C; E_a for this reaction is 94.6 kJ mol^{-1}. Calculate the half-life at 37.0 °C.

18.93 (a) Reaction rates usually increase as temperature increases. Explain why. (b) Explain how a reaction rate can decrease as temperature increases.

18.94 What five factors affect the rate of a homogeneous reaction?

18.95 Suggest a method for measuring the rates of each of the following reactions:

(a) $NO_2(g) + CO(g) \xrightarrow{325\,°C} NO(g) + CO_2(g)$
    red-brown

(b) $2H_2O_2(aq) \xrightarrow{I^-(aq)} 2H_2O(l) + O_2(g)$

(c) and (d) $2HgCl_2(aq) + C_2O_4{}^{2-} \longrightarrow$
    $2Cl^- + 2CO_2(g) + Hg_2Cl_2(s)$ (two methods)

(e) $C_6H_{12}O_6(aq) \longrightarrow 2HC_3H_5O_3(aq)$
    glucose   lactic acid

18.96 For reactions taking place in the gas phase at constant volume and temperature, the change in pressure with time can be used to follow some reactions. For which of the following reactions is this method applicable? Explain your answer.

(a) $H_2C{=}CH_2(g) + H_2(g) \xrightarrow{Pd} H_3C{-}CH_3(g)$
(b) $H_3C{-}N{\equiv}C{:}(g) \longrightarrow H_3C{-}C{\equiv}N{:}(g)$

18.97 Neither of the following reactions takes place at a signficant rate under ordinary laboratory conditions. Which of them

fails to take place for kinetic and which for thermodynamic reasons? Give the evidence on which your conclusion is based.
(a) $H_2(g) + Cl_2(g) \longrightarrow 2HCl(g)$
(b) $N_2(g) + O_2(g) \longrightarrow 2NO(g)$

18.98 Formation of a bond in the activated complex causes a mole of the activated complex to become 10–15 cm^3 smaller than a mole of the reactants. What would you expect the effect of very high pressures (8000 to 20 000 atm) on rates of reaction involving bond formation to be?

18.99 In finding rate laws in Section 18.3, time was treated as the independent variable and concentration as the dependent variable. In some experiments, it is more convenient to treat concentration as the independent variable and time as the dependent variable. For example, the reaction

$$CH_3CCH_3 + Br_2 \xrightarrow{H^+} CH_3CCH_2Br + HBr$$
$$\quad\;\; \| \qquad\qquad\qquad\qquad\quad \|$$
$$\quad\;\; O \qquad\qquad\qquad\qquad\quad O$$
    acetone

can be studied by measuring the time required for the yellow color of bromine to disappear. Typical data for an experiment at 23.5 °C are shown in the table:

Experiment Number	Initial Concentration, M			Time, s
	Acetone	HCl	Br$_2$	
1	0.80	0.20	0.0010	2.9×10^2
2	0.80	0.20	0.0020	5.7×10^2
3	1.60	0.20	0.0010	1.5×10^2
4	0.80	0.40	0.0010	1.4×10^2

(a) Which species is limiting? (b) Does the concentration of acetone change significantly during each experiment? The concentration of HCl? Explain your answer. (c) What is the order of the reaction with respect to bromine? Explain your answer. (d) What is the order of the reaction with respect to acetone? To HCl? Explain your answers. (e) What is the rate law for the reaction? (f) If a fifth experiment were done with [acetone]$_0$ = 0.80, [HCl]$_0$ = 0.20, and [Br$_2$]$_0$ = 0.000 50, how many seconds would be required for the color of the bromine to disappear? (g) If a sixth experiment were done with [acetone]$_0$ = 0.80, [HCl]$_0$ = 0.80, and [Br$_2$]$_0$ = 0.0010, how many seconds would be required for the color of the bromine to disappear?

18.100 The rate of decomposition of di-*tert*-butyl peroxide (DTBP) to acetone and ethane,

$$(CH_3)_3COOC(CH_3)_3(g) \longrightarrow 2(CH_3)_2CO(g) + CH_3CH_3(g)$$
$$\quad\; DTBP \qquad\qquad\qquad\qquad acetone \qquad\quad ethane$$

can be followed by measuring the total pressure in a closed container at constant temperature. In one experiment, the following data were obtained:

Time, s	P_{total}, mmHg
0	700
0.50×10^4	1420
1.00×10^4	1770
1.50×10^4	1940
2.00×10^4	2022
2.50×10^4	2062

(a) What is p_{DTBP} at the beginning of the experiment before any decomposition has taken place? (b) What is p_{DTBP} at $t = \infty$? (Reaction goes to completion.) (c) What is P_{total} at $t = \infty$? (d) What is p_{DTBP} at $t = 0.50 \times 10^4$ s? (e) Calculate p_{DTBP} at the other times shown in the table and find the order of the reaction with respect to DTBP. (f) What is the value of k at the temperature of the experiment?

18.101 The reaction $(CH_3)_3CBr + OH^- \longrightarrow (CH_3)_3COH + Br^-$ is believed to take place by the following mechanism:

Step 1 $(CH_3)_3CBr \longrightarrow (CH_3)_3C^+ + Br^-$ slow

Step 2 $(CH_3)_3C^+ + OH^- \longrightarrow (CH_3)_3COH$ fast

(a) Write the rate law predicted by the mechanism. (b) What effect would doubling the concentration of $(CH_3)_3CBr$ have on the rate of reaction? (c) What effect would doubling the concentration of OH^- have on the rate of reaction? (d) This reaction takes place much faster in water than in alcohol. Explain why.

18.102 For the reaction of nitrogen gas with hydrogen gas to form ammonia, (a) show how the rate of disappearance of nitrogen gas, the rate of disappearance of hydrogen gas, and the rate of appearance of ammonia are related. (b) If the rate of disappearance of nitrogen gas is 6.3×10^{-4} M·s^{-1}, what is the rate of appearance of ammonia? (c) What is the bond energy of the $N{\equiv}N$ bond in $N_2(g)$? (d) Explain why the reaction of nitrogen gas with hydrogen gas to form ammonia is slow.

18.103 Nitric oxide gas reacts with chlorine gas to give ONCl. (a) Write the equation for this reaction. (b) Write the equation for the reaction you would predict between nitric oxide gas and fluorine gas. (c) The rate laws for the reactions in parts (a) and (b) are rate $= k[NO][Cl_2]^2$ and rate $= k[NO][F_2]$, respectively. What can you say about the mechanisms of these two reactions?

18.104 In an experiment to determine the order of the reaction

$$(CH_3)_3CCl + H_2O \longrightarrow (CH_3)_3COH + HCl$$
tertiary-butyl chloride

with respect to *tertiary*-butyl chloride, the following data were obtained:

Time, s	pH		Time, s	pH
0	5.00		1025	2.32
100	3.27		1625	2.20
305	2.74		2225	2.14
425	2.62		∞	2.02
725	2.43			

(a) What is $[H^+]_0$? (b) What is $[H^+]_\infty$? (c) What $[H^+]$ was formed by the reaction? (d) What was [*tertiary*-butyl chloride]$_0$? (e) Calculate the [*tertiary*-butyl chloride] for the times given in the table and add to the table. (f) What is the order of the reaction with respect to *tertiary*-butyl chloride? (g) What is k for the reaction?

18.105 Acetone is CH_3COCH_3. The solvent for the experiment in Problem 18.104 was prepared by mixing 75.0 mL of deionized water with 75.0 mL of acetone. (The volumes of both liquids were measured at the same temperature.) The solution felt warm compared to the pure solvents. After the temperature of the solution had returned to the original temperature, the volume of the solution was 141 mL. (a) Explain why the temperature increased when the acetone and water were mixed and why the total volume was not 150 mL. (b) Calculate the molarity of water in the solvent. (c) Did the concentration of water change significantly during the reaction?

18.106 For the reaction $2CH_3I(g) \longrightarrow CH_3CH_3(g) + I_2(g)$, the rate law is rate $= k[CH_3I]$. (a) Estimate the value of $\Delta G°$ for the reaction at 300 °C. (b) Is the reaction spontaneous under standard conditions at 300 °C? Explain your answer. (c) At what temperature does the reverse reaction become spontaneous? (d) For the equilibrium

$$2CH_3I(g) \rightleftharpoons CH_3CH_3(g) + I_2(g)$$

what is the relationship between K_p and K_c? (e) Estimate the value of K_p at 300 °C. (f) In studying the rate of the forward reaction, is it necessary to consider the reverse reaction at 300 °C?

18.107 The element phosphorus exists in a number of allotropic forms that melt to give the same liquid and burn in air forming $P_4O_{10}(s)$. The two most common allotropes are white and red. White phosphorus is formed when phosphorus vapor condenses or liquid phosphorus freezes. (a) What is meant by the term allotrope? (b) Which allotrope is more stable? Explain your answer. (c) Suggest an explanation for the fact that white phosphorus catches fire in moist air at about 30 °C and must be stored under water while red phosphorus must be heated to about 260 °C to burn and can be kept in an ordinary container even when powdered. (d) What is the compound P_4O_{10} called? Write an equation for its formation.

18.108 The reaction

$$ICN(g) \xrightarrow{light} I(g) + CN(g)$$

takes place by a single unimolecular step. Sketch the reaction profile labeling, where appropriate, with Lewis formulas.

18.109 Discuss the advantages and disadvantages of carrying out reactions in solution.

18.110 Draw microscopic views of a collision that leads to reaction when (a) hydrogen atoms react with iodine chloride. The main product is hydrogen iodide; little hydrogen chloride is formed. (b) Potassium atoms react with methyl iodide, CH_3I. The products are KI and methyl radicals. (Use stereochemical formula for CH_3I.) What are the shapes of CH_3I molecules and methyl radicals? What is the hybridization of carbon in each?

18.111 Explain why an explosion resulted when seven times the normal amount of gaseous reactants were charged to a reactor in a chemical plant.

18.112 Explain why a mixture of natural gas and air formed by leakage of gas from a pipe can stand for a long time without reacting but explodes if a spark is introduced.

18.113 Psychologists estimate that people forget half of what they've learned in six months if they don't use it or review it. If you take a course in your freshman year and don't think about the material again until graduation three years later, how much will you remember? (Assume that forgetting takes place by a first-order process.)

18.114 The rate at which fungi in soil remove carbon monoxide from the air is independent of the concentration of carbon monoxide in the air. What is the order of the reaction with respect to carbon monoxide?

18.115 People, other mammals, and birds are warm-blooded; their body temperature is approximately constant. On the other hand, fishes and reptiles are cold-blooded—that is, their body temperature depends on the temperature of the water, earth, or air around them. Explain why cold makes fishes and reptiles sluggish.

18.116 Why do the chemical and petroleum industries spend millions of dollars every year on research about catalysts?

18.117 Nitric oxide, which is formed from nitrogen and oxygen in air when fossil fuels are burned, reacts with oxygen to form nitrogen dioxide, $2NO(g) + O_2(g) \longrightarrow 2NO_2(g)$. The following data concerning this reaction were obtained at 25 °C:

Experiment Number	[NO]	[O$_2$]	Initial Rate, M/s
1	0.0020	0.0010	2.8×10^{-5}
2	0.0040	0.0010	1.1×10^{-4}
3	0.0020	0.0020	5.6×10^{-5}

(a) Write the rate law. (b) What is the value of k at 25 °C? (c) If the initial concentrations are [NO] = 0.0030 and [O$_2$] = 0.0015, what will be the initial rate?

18.118 Bacteria have been used in the manufacture of cheese and wine for thousands of years. Investigators have found that bacteria can also be used to break down a variety of toxic substances including polychlorinated biphenyls (PCBs). PCBs have the structure

where three to six of the Xs represent Cl and the others represent H. PCBs have been widely used in electrical equipment because they are good insulators and nonflammable. As a result, they have gotten into wastewater, which is cause for concern because PCBs are suspected of causing birth defects. A news item about the discovery of bacteria that can break down PCBs states that the bacteria should reduce the concentration of PCBs by 60–65% for every 20 days during which sewage sludge is exposed to them. What is the order of the reaction? Give your reasoning.

18.119 Researchers have created artificial red blood cells. These artificial red blood cells are cleared from circulation by a first-order reaction with a half-life of about 6 h. If it takes 1 h to get an accident victim, whose red blood cells have been replaced by the artificial red blood cells, to a hospital, what percentage of the artificial red blood cells will be left when the person reaches the hospital?

18.120 The table shows how concentration changes with time as a reaction A $\longrightarrow$ B takes place:

t, min	[A]	t, min	[A]
0	0.250	20	0.089
2	0.212	30	0.068
5	0.172	45	0.050
10	0.132	60	0.039

(a) Graph [A] as a function of time. (b) From the graph, how many minutes were required for one-half of the original material to react? (c) How many minutes were required for half of the remaining material to react? (d) Is the reaction zero, first, or second order with respect to A? Explain your answer. (e) If a toxic environmental pollutant disappears naturally by a reaction of this order, what problem arises?

18.121 A company producing photographic color print paper wanted to make a paper whose quality would last for 100 yr under ordinary conditions of temperature (75 °F), humidity (60%), and lighting. Since the chemist responsible for testing the paper did not want to wait around for 100 yr to observe the results, the chemist used accelerated aging and an Arrhenius plot (ln time vs. $1/T$) to extrapolate to the 100-yr time. (a) Assuming the rate of aging increased threefold for each 10 °C increase in temperature, estimate the temperature that would have to be used in the accelerated aging tests to reduce the reaction time from 100 yr to less than 1 yr. (b) Explain why the chemist used a ln time vs. $1/T$ plot instead of a time vs. T plot to extrapolate.

18.122 Children metabolize some drugs much more rapidly than adults; the elderly metabolize them slower than younger adults. For example, the half-life of the sedative and tranquilizer diazepam in hours is roughly equal to the patient's age in years. (a) How long will it take for 75% of a dose of diazepam to be metabolized in a 6-year-old patient? In a 60-year-old patient? (b) Doctors usually base their prescription of how much of a drug to give a patient in each dose on the patient's weight. Are young patients or old patients more likely to suffer from toxic side effects of drugs? Explain your answer.

18.123 Chlorofluorocarbons such as Freon-12 (CCl_2F_2)—which are used in refrigeration and air conditioning, for making plastic foams, as solvents, and for sterilizing medical equipment— are being phased out by international agreement although their economic value was estimated at $27 billion a year in the United States alone in 1986 (where they also provided 715 000 jobs). The reason for the phaseout is that chlorofluorocarbons are believed to be involved in the destruction of ozone in the stratosphere. Ozone in the stratosphere absorbs ultraviolet radiation that causes skin cancer and eye damage if it reaches Earth's surface. Dissociation of Freons produces chlorine atoms

$$CCl_2F_2(g) \xrightarrow{\text{sunlight}} CClF_2(g) + Cl(g)$$
$$\text{Freon-12}$$

The chlorine atoms are believed to destroy ozone by the following mechanism:

$$Cl + O_3 \longrightarrow ClO + O_2$$
$$ClO + O \longrightarrow Cl + O_2$$

(The O atoms are formed by the dissociation of oxygen molecules by light.)   (a) Is sunlight a catalyst for the dissociation of Freon-12? Explain your answer.   (b) In the mechanism for the destruction of ozone, how is Cl acting? ClO? Write the equation for the overall reaction.

18.124 Explain why   (a) when kept at the same temperature, beef ground under sterile conditions spoils more rapidly than steak.   (b) Steak left in a car in the sun spoils more rapidly than steak stored in the refrigerator.

18.125 When a drug is taken by mouth, it goes into the gastrointestinal (GI) tract. From there it is absorbed into the blood and used by the body. This process can be described by the following equations:

$$A \xrightarrow{k_1} B \qquad \text{and} \qquad B \xrightarrow{k_1} C$$

In the equations, A represents drug in the GI tract, B represents drug in blood, and C represents drug used by the body. A typical graph of the relative concentrations of A, B, and C as a function of time is shown below.

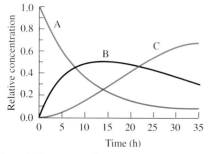

*(Reproduced from Calder, G. V. J. Chem. Educ. **1974**, 51, 19. © 1974 by Division of Chemical Education of the American Chemical Society, Washington, D.C. Used with permission of the publisher.)*

For most drugs, a minimum concentration characteristic of the drug must be present in the blood for the drug to be effective.   (a) If the minimum effective relative concentration in blood for the drug in the graph is 0.40, how long does it take for the drug to become effective?   (b) For how many hours does the relative concentration remain greater than 0.40?   (c) How long after the first dose should a second dose be taken?   (d) Does the second dose need to be as large as the first? Explain your answer.   (e) How can the time necessary to reach the minimum effective concentration be reduced?   (f) Explain why the directions on many prescriptions say "Take 2 capsules to start, then 1 every ___ hours."   (g) What is the order of the rate of absorption of the drug into the blood from the GI tract? This order is characteristic for drugs taken by mouth.

18.126 In a pressure cooker, the vapor pressure of water is 3.0 atm. Explain why food cooks much faster in a pressure cooker than in an ordinary pot.

18.127 Denver is a mile high, and water boils at 92 °C. A "3-minute egg" takes about 4½ minutes. What is the activation energy for boiling an egg?

18.128 In the early 1950s, people, cats, and seabirds around Minamata Bay in Japan suffered brain damage, which in some cases led to death, from eating fish contaminated with mercury from a local factory. Elimination of mercury from the body is first order with a half-life of 60 days (zero not significant). If a quarter-pound serving of fish containing 5.7×10^{-3} g of mercury (the concentration of mercury in the Minamata Bay fish) is eaten by a 150-lb (68-kg) person and the maximum normal level of mercury in people is 25 ppb by mass, how many days will be required for the mercury level to return to normal?

18.129 The reaction $NO(g) + O_3(g) \longrightarrow NO_2(g) + O_2(g)$ is important in the formation of photochemical smog. The rate law for this reaction is rate = $k[NO][O_3]$, and $k = 7.9 \times 10^1$ M^{-1} s^{-1} at 25 °C and 3.0×10^3 M^{-1} s^{-1} at 75 °C. (a) What is E_a?   (b) What is the value of the frequency factor, A?   (c) Can the reaction take place in a single step? Explain your answer.   (d) If [NO] = 1.0×10^{-8} and $[O_3] = 1.0 \times 10^{-9}$, what is the rate of the reaction at 25 °C? at 75 °C?   (e) How many times as fast is the reaction at 75 °C as at 25 °C?   (f) Calculate $\Delta H°$ for this reaction. (g) Sketch the reaction profile.   (h) What is E_a for the reverse reaction?   (i) The nitrogen dioxide reacts further with ozone to form dinitrogen pentoxide and oxygen gas. The rate law for this reaction is rate = $k[NO_2][O_3]$. Can this reaction take place in a single step?

18.130 The following mechanism has been proposed for another reaction that converts NO to NO_2 during smog formation:

Step 1	$CO + OH \longrightarrow CO_2 + H$
Step 2	$H + O_2 \longrightarrow HOO$
Step 3	$HOO + NO \longrightarrow OH + NO_2$

(a) What is the equation for this other reaction?
(b) Which species acts as a catalyst?   (c) Which species is (are) intermediates?   (d) Write the Lewis formula for OH. What is a species like this called?

18.131 Use of ethyl alcohol, CH_3CH_2OH, as a substitute for gasoline in cars has been suggested as a way of reducing air pollution by unburned hydrocarbons. (a) Write the equation for the combustion of ethyl alcohol. (b) Is it likely that combustion of alcohol takes place by a simple one-step mechanism? Explain your answer.

18.132 Concentration vs. time graphs from a computer simulation of photochemical smog formation are shown in the figure at the right. Propylene, $CH_3CH=CH_2$, and isobutylene, $(CH_3)_2C=CH_2$, are representative of hydrocarbons from unburned gasoline. PAN, peroxyacylnitrate ($RCOOONO_2$ where R is an organic group such as $H_3C—$), is a major cause of eye irritation from smog. Compare Figures 1(a) and 1(b) in the figure and answer the following questions: (a) What substances are reactants? (b) What substances are products? (c) What substance is an intermediate? (d) Which is more reactive, propylene or isobutylene? Compare Figures 1(a) and 3 and answer the following questions: (e) What is the difference between Figures 1(a) and 3? (f) What changes does this difference bring about? (g) What substances are intermediates? Which intermediate is shortest lived? Which is longest lived? (h) What can you conclude about the relative rates of formation and disappearance of NO_2 and O_3? Of O_3 and PAN? (i) Compare Figures 1(a) and 2 and answer the following question: Is severe smog likely on an overcast day? Explain your answer.

18.133 Researchers recently found that billions of HIV particles are continuously produced by newly infected cells and then rapidly destroyed leaving only a relatively low concentration. What is a state like this called?

18.134 Synthesis gas, a mixture of CO(g) and H_2(g), is used to make methyl alcohol, CH_3OH, an atom-economical product. Atom-economical products use all of the reactant atoms to form products. Other atom-economical products possible from synthesis gas, at least on paper, are HCHO, CH_3COOH, and OHCCHO. (a) Which of these reactions are thermodynamically possible? (b) How could a chemical engineer control the course of the reaction so as to make a particular product?

18.135 The reaction $O^+(g) + NO(g) \longrightarrow NO^+(g) + O(g)$, which takes place in the upper atmosphere, is believed to occur in a single elementary process. When the reaction is carried out in the laboratory using $O^+(g)$ enriched in the rare isotope O-18, the $NO^+(g)$ is not enriched in O-18. (a) Does the reaction involve breaking of the original N—O bond and formation of a new N—O bond or transfer of an electron from NO to O^+? Explain your answer. (b) How many protons, neutrons, and electrons are in an $^{18}O^+$ ion?

18.136 A new joint venture claims to have developed a fuel for gasoline engines that is 55% water and 45% low-grade gasoline. (The fuel contains a surfactant that permits gasoline to mix with water.) A small amount of nickel is added to each cylinder of the engine to catalyze the breakdown of water into hydrogen and oxygen. Would you invest in this company? Explain your answer.

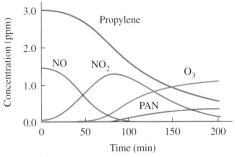

Figure 1(a) Sunny day

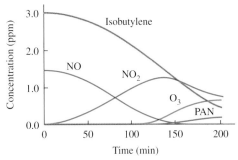

Figure 1(b) Sunny day

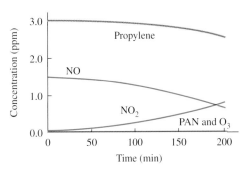

Figure 2 Cloudy day

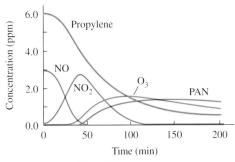

Figure 3 Sunny day

*(Reproduced from Huebert, B. J. J. Chem. Educ. **1974**, 51, 644. © 1974 by Division of Chemical Education of the American Chemical Society, Washington, D.C. Used with permission of the publisher.)*

Chemistry and the Search for Blood Substitutes

MIKE WONG

Graduate Student

B.S. Chemistry/Mathematics
University of Tulsa

Ph.D. Candidate in Chemistry
University of Illinois

As I was growing up, my parents, who were Chinese immigrants from Vietnam, strongly emphasized the importance of a good education. I always did well in school, especially in the sciences. When my elementary and junior high school teachers asked me what career I wanted to pursue, I told them I wanted to be a doctor. I guess that early decision came from the fact that my mom was convinced that a career in medicine would fit me well.

I took my first real chemistry course in high school. I was turned on to chemistry by my teacher, Mrs. Crowley, who herself was in the process of finishing her degree in chemical education. She made chemistry fun. Fifteen minutes before the end of every class, she would have a session entitled *Stump the Teacher*. The class could ask her any questions we wanted relating to chemistry (or just general science), and she would attempt to provide us with a simple answer. We submitted our questions on a sheet of paper so that no one would feel embarrassed in asking a question. She answered questions ranging from "Where did the periodic table come from?" to "Why do we have to study chemistry?" I asked questions that I had always been curious about but had been afraid to ask: Why is the sky blue? Where do the colors of a rainbow come from? How does an automobile engine work and why do we need gasoline? After two years of chemistry in Mrs. Crowley's class, I began to realize the important role that chemistry plays in our everyday life.

After high school, I attended the University of Tulsa as a pre-med chemistry major. I had always heard from teachers and friends that a pre-med chemistry major would better prepare me for medical school than the traditional biology major. My first chemistry course in college was general chemistry. I was in a class of about 100 students, most of them pre-med majors. Most students studied enough just to get by because they never wanted to see chemistry again after they finished the course. However, I felt the opposite. In fact, I wanted to learn even more.

I soon got out of the traditional pre-med curriculum and switched to a true chemistry major in my sophomore year. By this time, I was very intrigued by the chemistry-related work that various research groups were doing at the university, ranging from trying to find the gene for cancer to developing an improved catalytic converter for automobiles.

I joined the research group of Professor Dale Teeters, a physical chemist working in the area of polymer chemistry. We tried to design a battery that was made entirely of plastic. The advantages of a plastic battery over a traditional battery include a greater power capacity, less use of corrosive components, and recyclable parts. As a result, the plastic battery would reduce the pollution that came from the manufacturing and disposal of traditional batteries. My research work went very well, and eventually I was able to publish my work in a major journal and present the results at a chemical conference.

Having obtained my B.S., I entered graduate school in chemistry at the University of Illinois. My next decision was to choose which research group I wanted to join. The abundance and variation of research topics made this decision difficult. I wanted to do something with a biological theme (I guess I had a little bit of pre-med leftover in me), and eventually I joined the group headed by Professor Kenneth S. Suslick, a bioinorganic chemist.

After two years of classes, I was able to dedicate my full-time effort to research. I study the effects of high-intensity sound waves (or ultrasound) on proteins (major components of body tissues, muscles, organs, etc.). This work involves using both chemistry and biology—in fact, one might say that chemistry and biology have a symbiotic relationship in my research.

The specific project I am working on now involves the use of ultrasound waves and hemoglobin (the oxygen-carrying component in the body) to develop an alternative oxygen-carrying agent. With the current threat of AIDS and hepatitis, people are concerned about possibly contracting diseases while giving and receiving blood. There is a great need in the medical community for an alternative carrier of oxygen (or blood substitute). We have developed bubbles in the micrometer (10^{-6} m) size range (or microbubbles) from hemoglobin that seem to possess the characteristics of a blood substitute.

When I started the project, there were many problems that I had to overcome: designing the reaction apparatus, optimizing reaction conditions, and determining the possible chemical mechanisms. As I looked for solutions to these problems, I learned to ask many questions of my peers and, more importantly, I further developed my independent-thinking skills. Doing chemical research is like solving a puzzle. You are given a specific amount of time to find the solution to a problem. Along the way, you are given some help, but it is up to you to hunt down the final answer using various resources around you. Sometimes there might not be a good solution to the problem, but when you do find the solution or make a new discovery, the rewards are enormous.

Although the hemoglobin microbubbles are a long way from being developed commercially, I hope this process will stimulate additional scientific interest that could lead to the next generation of blood substitutes. I have recently been offered a position as a research scientist at a biotechnology company in California to further study the microbubbles. I will finish my thesis and obtain my doctorate in one year.

ELECTROCHEMISTRY

The chromeplated parts of this motorcycle are decorative. Plating with chromium also prevents corrosion.

Electrochemistry deals with the use of spontaneous redox reactions to supply electrical energy and the use of electrical energy to make nonspontaneous redox reactions take place. In this chapter, we will see how oxidation–reduction reactions are used to produce electric current and how they can be brought about by electricity. We will learn how the strengths of oxidizing and reducing agents are compared and how they depend on concentration. We will discuss how and why metals corrode and how corrosion can be prevented.

- *Why is a study of electrochemistry a part of general chemistry?*

The batteries that start your car and power your calculator and flashlight are devices that store energy and convert it to electrical energy by means of oxidation–reduction reactions. Electricity is used either to obtain or to purify most commonly used metals except iron and steel, and electroplating is used to decorate objects and increase their resistance to corrosion. Large quantities of electricity are used in the chemical industry to bring about

oxidation–reduction reactions. For example, electrolysis of brine, NaCl(aq), to produce sodium hydroxide, chlorine, and hydrogen is one of the largest volume processes in the chemical industry.

- *What do you already know about electrochemistry?*

The very first reaction that we discussed, the electrolysis of water (Section 1.2), was an electrochemical reaction. Electrical energy brought about the decomposi-

715

Getting ready to jump start a car that has a dead battery.

tion of water into hydrogen and oxygen. You have undoubtedly used the reactions that take place in batteries since you got your first battery-powered toy as a child. You are also used to drinking out of cans made of aluminum that is produced by electrolysis and perhaps to swimming in water purified by an end product of an electrolytic process.

In Chapter 11, you learned that oxidation–reduction or redox reactions involve the transfer of electrons from one species to another. Species that lose electrons (increase in oxidation number) are said to be oxidized, and species that gain electrons (decrease in oxidation number) are said to be reduced. Because electrons are matter and cannot be created or destroyed in chemical reactions, oxidation can't take place unless reduction also takes place;

reduction can't take place unless oxidation takes place. Redox reactions can be divided into a half-reaction of oxidation and a half-reaction of reduction.

The water in this public pool is purified with calcium hypochlorite powder, $Ca(OCl)_2 \cdot 2H_2O$, made by the reaction of chlorine with calcium hydroxide. The chlorine is produced by the electrolysis of aqueous sodium chloride.

19.1 VOLTAIC CELLS

Many oxidation–reduction reactions take place spontaneously. For example, if a strip of zinc metal is placed in a solution of copper(II) nitrate, copper metal immediately begins to plate out on the zinc strip. If the mixture is allowed to stand, the blue color of the copper(II) nitrate solution, which is due to hydrated copper(II) ions, $Cu(H_2O)_6^{2+}$, gradually fades. The molecular and net ionic equations for this reaction are, respectively,

$$Zn(s) + Cu(NO_3)_2(aq) \longrightarrow Cu(s) + Zn(NO_3)_2(aq) \qquad \text{and}$$

$$Zn(s) + Cu^{2+} \longrightarrow Cu(s) + Zn^{2+}$$

■Figure 19.1 shows pictures of this reaction. For the reaction between zinc metal

(a)

(b)

(c)

(d)

■ FIGURE 19.1   (a) Clockwise starting at lower left: zinc metal, copper(II) nitrate solution, zinc nitrate solution, copper metal. (b) When the strip of zinc is placed in the copper(II) nitrate solution, copper immediately begins to plate out. (c) As more copper plates out, the blue color of the solution, which is due to copper(II) ion, fades. (d) When the strip of copper is placed in the zinc nitrate solution, nothing happens. The reaction $Cu(s) + Zn(NO_3)_2(aq) \longrightarrow Zn(s) + Cu(NO_3)_2(aq)$, which is the reverse of the spontaneous reaction shown in (b) and (c), is not spontaneous.

and copper(II) nitrate solution, the half-reaction of oxidation is

$$Zn(s) \longrightarrow Zn^{2+} + 2e^-$$

and the half-reaction of reduction is

$$Cu^{2+} + 2e^- \longrightarrow Cu(s)$$

The overall reaction

$$Zn(s) + Cu^{2+} \longrightarrow Cu(s) + Zn^{2+}$$

amounts to the transfer of two electrons (per atom) from zinc to copper.

The reverse of a spontaneous change is nonspontaneous. No reaction takes place when a strip of copper metal is allowed to stand in a solution of zinc nitrate [Figure 19.1(d)].

When the spontaneous reaction between zinc metal and copper(II) nitrate solution is carried out in a single container as shown in Figures 19.1(b) and (c), the energy given off heats the water and its surroundings. If the two half-reactions of a spontaneous redox reaction are separated but connected by conductors, energy is given off in the form of an electric current as shown in ▪Figure 19.2. An **electric current** is a *flow of electric charge that takes place when there is a difference in electrical potential between two points connected by a conductor.* The *SI unit of electric current* is the ampere, A, one of the SI base units (Table 2.1). An **ampere** is *a coulomb per second*

$$1 \text{ A} = 1 \text{ C/s} \tag{19.1}$$

The coulomb, C, is the SI unit of charge (Section 7.1). Amperes are measured with an ammeter as shown in Figure 19.2.

A *device that uses a spontaneous oxidation–reduction reaction to produce an electric current,* such as the one shown in Figure 19.2, is called a **voltaic** or **galvanic cell.*** ▪Figure 19.3 is a schematic drawing of the voltaic cell shown in Figure 19.2. In an electrochemical cell such as the one shown in Figures 19.2 and 19.3, the metal strips are called **electrodes.** The *electrode at which oxidation takes place* is called the **anode,** and the *electrode at which reduction takes place* is called the **cathode.** (It is easy to remember that oxidation takes place at the anode and reduction at the cathode because both oxidation and anode begin with vowels; both reduction and cathode begin with consonants.) The *oxidation and reduction reactions that take place at the electrodes* are called **half-cell reactions.** You should recognize half-cell reactions as examples of the half-reactions used to balance equations for redox reactions by the half-reaction method (Section 11.5 and Appendix F.2). The overall cell reaction is the sum of the two half-reactions:

$$Zn(s) + Cu^{2+} \longrightarrow Cu(s) + Zn^{2+}$$

Notice that the half-reactions and the overall reaction are the same whether reaction takes place in a single container or in the two half-cells of a voltaic cell.

To understand how a voltaic cell works, picture the metal electrodes as made up of metal ions in a cloud of mobile electrons (Section 12.8). When a strip of zinc

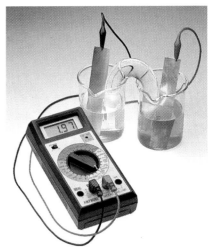

▪ FIGURE 19.2   A voltaic cell consisting of a zinc electrode in zinc nitrate solution and a copper electrode in copper(II) nitrate solution. Electrons flow from the zinc electrode to the copper electrode through the wire (and ammeter). Movement of ions through the salt bridge completes the circuit. Batteries are voltaic cells.

⏵ One ampere is about 6×10^{18} electrons per second.

ⓘ The Voltaic Cells module of the Electrochemistry section allows students to build cells and investigate various electrode materials. Animations help students connect macroscopic behavior to microscopic details.

*Luigi Galvani was an Italian doctor who discovered in 1780 that contact of two different metals with a frog's muscle resulted in an electric current. Galvani's observation led his friend, the Italian physicist Alessandro Volta, to invent the voltaic pile, a type of battery, in 1800. Volta's invention of the voltaic pile was the first step toward the modern age of electric power. Although not major users of electric power in terms of quantity, electronics and communications—telephones, radios, television, and computers—are among the most widespread products of the electrical age. Volta also discovered methane gas.

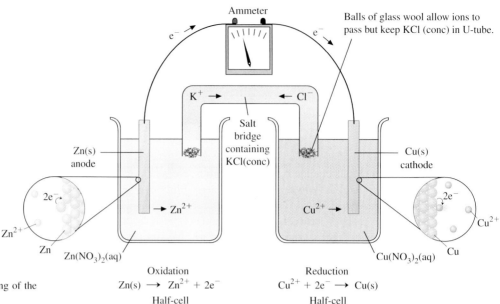

FIGURE 19.3 Schematic drawing of the voltaic cell shown in Figure 19.2.

Oxidation
$$Zn(s) \longrightarrow Zn^{2+} + 2e^-$$
Half-cell

Reduction
$$Cu^{2+} + 2e^- \longrightarrow Cu(s)$$
Half-cell

Balls of glass wool allow ions to pass but keep KCl (conc) in U-tube.

Salt bridge containing KCl(conc)

Zn(s) anode

Cu(s) cathode

Zn(NO$_3$)$_2$(aq)

Cu(NO$_3$)$_2$(aq)

Remember that someone's energy was required to charge the rod in Figure 1.6 by rubbing it with wool.

Zinc is more reactive than copper; zinc is above copper in the activity series for metals (Table 4.5).

For historical reasons, engineers regard current as flow of positive, not negative, charge.

is dipped in water, some zinc ions leave the surface of the strip and go into solution. A negative charge builds up on the strip as a result of the electrons that are left behind by the zinc ions. A positive charge that is due to the presence of Zn^{2+} ions builds up in the water around the strip. But the process soon stops because charge buildup requires that energy be supplied to a system. The same process takes place at the copper electrode but to a lesser extent.

Now, if the strip of zinc is connected to the strip of copper by a wire, the electrons can move away from the zinc electrode, which has the greater number of electrons, through the wire to the copper electrode. If a KCl salt bridge (shown in Figures 19.2 and 19.3) is used to connect the Zn/Zn^{2+} half-cell with the Cu^{2+}/Cu half-cell, negatively charged chloride ions can flow into the Zn/Zn^{2+} half-cell from the salt bridge to balance the positive charge of the zinc ions.

At the copper electrode, the negative charge on the electrons attracts the positively charged Cu^{2+} ions. Copper(II) ions move out of the solution onto the surface of the copper electrode to neutralize the negative charge on the electrode. Positively charged potassium ions flow from the salt bridge to the solution in the Cu^{2+}/Cu half-cell to replace the Cu^{2+} ions that have deposited on the electrode and maintain electrical neutrality.

Reaction takes place as long as the circuit is complete—that is, as long as there is a continuous path for electric current—and enough reactants remain. Current through the wire consists of the movement of electrons. Electrons travel through the wire from the anode to the cathode. Current through the solutions consists of the movement of ions. The positively charged ions, Cu^{2+}, K$^+$, and Zn^{2+}, travel toward the cathode. This is why positively charged ions are called cations. Negatively charged ions such as NO$_3^-$ and Cl$^-$ are called anions because they travel toward the anode.

The salt bridge permits movement of charge from one half-cell to the other but prevents mixing of the two solutions. If the copper nitrate and zinc nitrate solutions are allowed to mix, the situation will be just like that in the single container. Reaction will take place spontaneously, but there will be no current through the wire. On the other hand, if the circuit is not complete—that is, if there is not a continuous path for the current—reaction will soon stop because charges will build up.

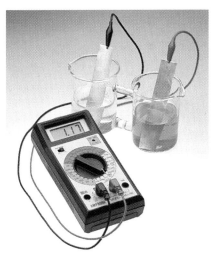

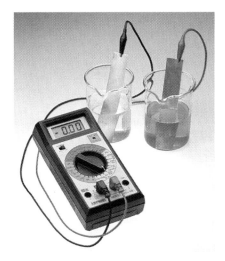

- FIGURE 19.4   A porous barrier, such as the fritted-glass disc in the tube that connects the two solutions, can be substituted for a salt bridge.

If no salt bridge or porous barrier connects the solutions, the circuit is not complete, and the current is zero.

A salt bridge contains a concentrated solution of a strong electrolyte. As a result of the high concentration of ions in the salt bridge, ions diffuse out of the salt bridge; diffusion of the less concentrated ions in the half-cells into the salt bridge is negligible. The cation and anion used in a salt bridge should be about the same size and have the same magnitude charge so that they diffuse at the same rate. The ions of the electrolyte in a salt bridge must be inert toward the solutions in the two half-cells and toward the electrodes. Loose balls of glass wool or cotton at the ends of the salt bridge keep the electrolyte solution from flowing down out of the salt bridge. A porous barrier such as the one shown in ▪Figure 19.4 can be used instead of a salt bridge.

Cell diagrams are often used instead of schematic drawings such as Figure 19.3 to describe voltaic cells. The cell diagram for the cell shown in Figure 19.3 is

$$Zn(s)|Zn^{2+}(aq)\|Cu^{2+}(aq)|Cu(s)$$

In a cell diagram, the anode is always written on the left and the cathode on the far right. A single vertical line shows a boundary between phases such as the boundary where the surface of the zinc meets the zinc nitrate solution.* A double vertical line shows a salt bridge or other barrier between solutions:

$$\underset{\text{anode}}{Zn(s)}|\underset{\substack{\text{boundary}\\\text{between}\\\text{phases}}}{Zn^{2+}(aq)}\|\underset{\substack{\text{salt}\\\text{bridge}}}{Cu^{2+}(aq)}|\underset{\text{cathode}}{Cu(s)}$$

species in contact
with electrodes

Cell diagrams are much easier to make than schematic drawings.

○ In both cell diagrams and schematic drawings of cells, the cathode, where reduction takes place, is always shown on the right. Both reduction and right begin with an r.

---

*Modern research in electrochemistry is concerned chiefly with how, at a molecular level, charge transfer takes place across the interface between electrodes and solutions. In other words, modern electrochemical research is about the mechanisms of the reactions that take place at the surfaces of electrodes, that is, with kinetics. The material in this chapter, which is a necessary foundation, deals with equilibria between electrodes and solutions, that is, with thermodynamics.

◄ FIGURE 19.5 A waterwheel. Waterwheels were a major source of power from about 100 B.C. until the invention of a practical steam engine by the Scottish instrument maker and inventor James Watt. Watt patented his steam engine in 1769.

Volta estimated the potential of his voltaic cells by connecting the electrodes with his finger.

A dead battery has reached chemical equilibrium.

19.2 STANDARD CELL POTENTIALS

Water at the top of a waterfall has greater potential energy than water at the bottom of a waterfall. As water runs down the waterfall, its potential energy is converted to kinetic energy. The kinetic energy can be made to do work, for example, to turn a waterwheel (see ◄Figure 19.5). Similarly, some half-cells have greater electric potential energy than other half-cells. When half-cells with different electric potential energies are connected in a complete circuit, electrons move through the wire from the half-cell of higher potential energy to the half-cell of lower potential energy. The moving electrons can be made to do work, for example, to turn a motor. The *difference in electric potential energy between half-cells is measured in* **volts, V.** By definition,

$$1 \text{ volt} = 1 \text{ joule/coulomb or } 1 \text{ V} = 1 \text{ J/C}$$

The *difference in electric potential energy between half-cells is called* the **electromotive force, E,** of the cell or the **cell potential** or **cell voltage.** The abbreviation **emf** is often used for electromotive force. The electromotive force is the driving force that pushes electrons through the wire from one electrode to the other. One volt is the emf required to give one joule of energy to a charge of one coulomb.

The cell potential for a voltaic cell can be measured by substituting a voltmeter for the ammeter; the voltmeter reading is the potential difference between the two half-cells. The cell potential depends on which reactions take place at the anode and at the cathode, on the concentrations of the solutions in the half-cells, and on the temperature. Because cell voltage depends on concentration and temperature, standard conditions for comparison of half-cells must be stated and temperature must be given. The same standard conditions that are used in thermodynamics are used for electrochemistry, that is, solutions are 1 M and partial pressures of gases are 1 atm.* The *cell potential under standard conditions* is called the **standard cell potential, $E°$.** The standard cell potential is a quantitative measure of the tendency of reactants in their standard states to form products in their standard states. It represents the driving force for reaction when all materials are in their standard states. The standard cell potential of a cell is a very useful and important quantity.

When the standard potential of a cell is measured, the current must be negligible. The current drawn by a modern electronic voltmeter is negligible. If current is drawn from the cell, reaction takes place, the concentrations in the half-cells change, and the cell potential decreases.† At equilibrium, the cell potential is zero, and no

---

*One molar solutions of salts do not usually behave ideally. For real solutions (and real gases), it is the activities or effective concentrations that should be equal to 1. However, we will have to assume that errors resulting from use of concentrations in molarity and partial pressures in atmospheres are not significant because, in general chemistry, we do not know how to do anything else.
†If current is drawn, the cell voltage will also be lowered as a result of the internal resistance of the cell.

more reaction takes place. (The situation is analogous to two ponds, one having a slightly higher water level than the other. If the two ponds are connected by a ditch, water will flow from the higher pond to the lower pond until the water level in the two ponds is the same.)

The Standard Hydrogen Electrode

Measurement of the potential of a single half-cell is impossible; only differences in potential between two half-cells can be measured. For this reason, some half-cell potential must be set equal to zero and all other half-cell potentials compared with this half-cell potential. By international agreement among scientists, the voltage for the half-reaction

$$2H^+(aq, 1 \text{ M}) + 2e^- \longrightarrow H_2(g, 1 \text{ atm})$$

is exactly zero. (The voltage for the reverse reaction

$$H_2(g, 1 \text{ atm}) \longrightarrow 2H^+(aq, 1 \text{ M}) + 2e^-$$

is also equal to zero.) A *half-cell that contains one molar hydrogen ion and hydrogen gas at a partial pressure of 1 atm* is called a **standard hydrogen electrode (SHE).** ▪Figure 19.6 shows a standard hydrogen electrode. Neither of the species in a standard hydrogen electrode, H^+ or $H_2(g)$, is a solid conductor. Therefore, a chemically inert (unreactive) conductor must be used for the electrode; platinum metal is very inert. The platinum electrode conducts electricity into or out of the solution, provides a surface where electron transfer can take place, and acts as a catalyst, but no platinum is used up in the reaction. The platinum electrode is coated with finely divided platinum to increase its surface area. Hydrogen gas must be bubbled through the standard hydrogen electrode so that the partial pressure of hydrogen gas is 1 atm if hydrogen gas is used up in the half-cell reaction.

In the cell shown in ▪Figure 19.7, a standard hydrogen electrode is combined with a copper(II) ion/copper metal half-cell. *By convention, the emf of voltaic cells is positive; voltages for spontaneous reactions have + signs. The experimentally measured potential of a cell is the difference between the reduction potentials of*

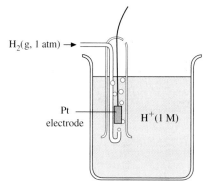

▪ FIGURE 19.6   Standard hydrogen electrode.

Setting one half-cell potential equal to zero is similar to choosing to measure temperature compared to the melting point of ice.

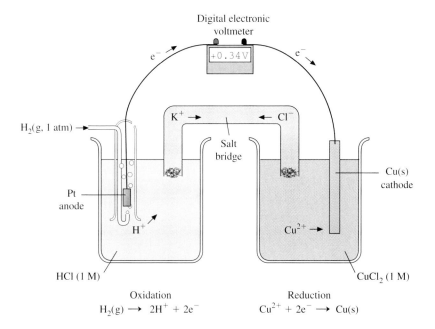

▪ FIGURE 19.7   Comparison of $Cu^{2+}/Cu(s)$ half-cell with SHE.

For a discussion of the conventions of electrochemistry and the reasons for them, see Birss, V. I.; Truax, D. R. J. Chem. Educ. *1990,* 67, 403–409.

the two half-cells that make up the cell:

$$E^\circ_{cell} = E^\circ_{\text{half-cell of reduction}} - E^\circ_{\text{half-cell of oxidation}} \qquad (19.2)$$

As you can see from the voltmeter reading in Figure 19.7, $E^\circ_{cell} = +0.34$ V for the cell

$$Pt|H_2(g, 1\ atm)|H^+(aq, 1\ M)||Cu^{2+}(aq, 1\ M)|Cu(s)$$

For this cell, the $Cu^{2+}/Cu(s)$ half-cell is the half-cell where reduction takes place, and the $H^+/H_2(g)$ half-cell is the half-cell where oxidation takes place. Thus, according to equation 19.2,

$$E^\circ_{cell} = E^\circ_{Cu^{2+}/Cu(s)} - E^\circ_{H^+/H_2(g)} \qquad (19.3)$$

If the cell potential of the standard hydrogen electrode is set equal to zero, then substitution in equation 19.3 gives

$$+0.34\ \text{V} = E^\circ_{Cu^{2+}/Cu(s)} - 0.00\ \text{V} \qquad \text{and} \qquad E^\circ_{Cu^{2+}/Cu(s)} = +0.34\ \text{V}$$

The salt bridge is assumed to have a negligible effect on E°_{cell}. The effect of the salt bridge is of the order of 0.05 V or less if the salt bridge is correctly made.

The standard potential for the reduction of Cu^{2+} ion is positive because the tendency of Cu^{2+} to be reduced is greater than the tendency of H^+ to be reduced. The greater tendency of Cu^{2+} to be reduced is the reason Cu^{2+}, not H^+, is reduced when the $Cu^{2+}/Cu(s)$ half-cell is paired with the standard hydrogen electrode. The SHE is the anode in the cell shown in Figure 19.7.

■Figure 19.8 shows a standard hydrogen electrode combined with a zinc ion/zinc metal half-cell. As you can see, $E^\circ_{cell} = +0.76$ V for the cell

$$Zn(s)|Zn^{2+}(aq, 1\ M)||H^+(aq, 1\ M)|H_2(g, 1\ atm)|Pt$$

For this cell, reduction takes place in the $H^+/H_2(g)$ half-cell. Oxidation takes place in the $Zn^{2+}/Zn(s)$ half-cell, and

$$E^\circ_{cell} = E^\circ_{H^+/H_2(g)} - E^\circ_{Zn^{2+}/Zn(s)} \qquad (19.4)$$

If the cell potential of the standard hydrogen electrode is set equal to zero, then substitution in equation 19.4 gives

$$+0.76\ \text{V} = 0.00\ \text{V} - E^\circ_{Zn^{2+}/Zn(s)} \qquad \text{and} \qquad E^\circ_{Zn^{2+}/Zn(s)} = -0.76\ \text{V}$$

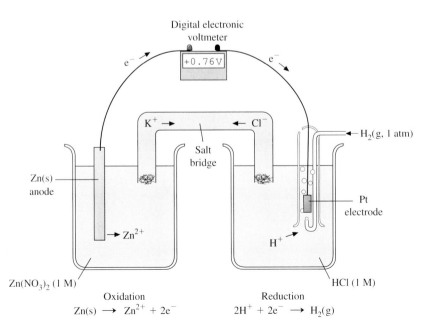

■ FIGURE 19.8 Comparison of $Zn^{2+}/Zn(s)$ half-cell with SHE.

The standard potential for the reduction of Zn^{2+} is negative because the tendency of Zn^{2+} to be reduced is less than the tendency of H^+ to be reduced. The lesser tendency of Zn^{2+} to be reduced is the reason H^+, not Zn^{2+}, is reduced when the $Zn^{2+}/Zn(s)$ half-cell is paired with the standard hydrogen electrode. Note how the direction in which a given half-reaction takes place depends on the other half-reaction with which it is combined. The standard hydrogen electrode is the anode when combined with the $Cu^{2+}/Cu(s)$ half-cell but the cathode when combined with the $Zn^{2+}/Zn(s)$ half-cell.

Standard Reduction Potentials

Once some cell potentials have been measured, the cell potentials for other cells can be calculated. For example, the standard cell potential for the cell

$$Zn(s)|Zn^{2+}(aq,\ 1\ M)||Cu^{2+}(aq,\ 1\ M)|Cu(s)$$

can be calculated from the standard potentials for the following half-cells:

$$Zn^{2+} + 2e^- \longrightarrow Zn(s) \qquad E° = -0.76\ V$$
$$Cu^{2+} + 2e^- \longrightarrow Cu(s) \qquad E° = +0.34\ V$$

Substitution in equation 19.2 gives

$$E°_{cell} = E°_{Cu^{2+}/Cu(s)} - E°_{Zn^{2+}/Zn(s)}$$
$$= +0.34\ V - (-0.76\ V) = +1.10\ V$$

The standard cell potential is an intensive property—that is, the standard cell potential does not depend on the size of the cell. Large C batteries and small AAA batteries both produce 1.5 V.

The standard cell potentials for a number of common half-reactions are given in Table 19.1. In tables of standard reduction potentials such as Table 19.1, all species in the lefthand column are oxidizing agents. Fluorine gas, MnO_4^- ion, and Li^+ are oxidizing agents. All species in the lefthand column accept electrons and are reduced when they undergo oxidation–reduction reactions. All species to the right of the arrows—for example, F^-, Mn^{2+}, and $Li(s)$—are reducing agents: They give up electrons and are oxidized when they undergo oxidation–reduction reactions. The potentials are potentials measured with respect to the half-reaction $2H^+(aq) + 2e^- \longrightarrow H_2(g)$ with all reactants and products in their standard states. Thus, a positive sign shows that a half-reaction has a greater tendency to take place than the half-reaction $2H^+(aq) + 2e^- \longrightarrow H_2(g)$ under standard conditions. A negative sign shows that a half-reaction has a lesser tendency to take place than the half-reaction $2H^+(aq) + 2e^- \longrightarrow H_2(g)$ under standard conditions.

In Table 19.1, the standard reduction potentials are listed in order of decreasing value. Tables in some books list standard reduction potentials in order of increasing value. If you use tables in other books, you will need to note which order is used.

The half-reaction at the top of Table 19.1 has the strongest tendency to take place of all the reactions in the table. Fluorine gas is the most easily reduced species to the left of the arrows in the equations for the half-reactions. Fluorine is the strongest oxidizing agent in the table. However, fluorine is rarely used as an oxidizing agent because its reactions with reducing agents often take place with explosive violence. Chlorine gas, oxygen gas, permanganate ion (MnO_4^-), dichromate ion ($Cr_2O_7^{2-}$), and hydrogen peroxide (H_2O_2) are commonly used oxidizing agents.

The half-reaction at the bottom of part A of Table 19.1, $Li^+ + e^- \longrightarrow Li(s)$, has the least tendency to take place of all the reactions. Lithium ion is the least easily reduced species to the left of the arrows in the equations for the half-reactions.

▶ Because $E°_{cell}$ for voltaic cells is always positive, the lower standard reduction potential is always subtracted from the higher standard reduction potential to obtain $E°_{cell}$ for a voltaic cell.

Oxygen gas is the strongest oxidizing agent found in the natural world because stronger oxidizing agents react with water to form $O_2(g)$. Only reducing agents that do not react with water to form $H_2(g)$ are found in nature.

TABLE 19.1   Standard Reduction Potentials at 25 °C[a]
A. Acidic Aqueous Solution

Half-reaction	$E°$, Volts
$F_2(g) + 2e^- \longrightarrow 2F^-$	2.87
$H_2O_2(aq) + 2H^+ + 2e^- \longrightarrow 2H_2O(l)$	1.76
$MnO_4^- + 8H^+ + 5e^- \longrightarrow Mn^{2+} + 4H_2O(l)$	1.51
$Cr_2O_7^{2-} + 14H^+ + 6e^- \longrightarrow 2Cr^{3+} + 7H_2O(l)$	1.36
$Cl_2(g) + 2e^- \longrightarrow 2Cl^-$	1.36
$O_2(g) + 4H^+ + 4e^- \longrightarrow 2H_2O(l)$	1.23
$Br_2(l) + 2e^- \longrightarrow 2Br^-$	1.07
$NO_3^- + 4H^+ + 3e^- \longrightarrow NO(g) + 2H_2O(l)$	0.96
$Ag^+ + e^- \longrightarrow Ag(s)$	0.80
$Fe^{3+} + e^- \longrightarrow Fe^{2+}$	0.77
$O_2(g) + 2H^+ + 2e^- \longrightarrow H_2O_2(aq)$	0.70
$I_2(s) + 2e^- \longrightarrow 2I^-$	0.54
$Cu^+ + e^- \longrightarrow Cu(s)$	0.52
$Cu^{2+} + 2e^- \longrightarrow Cu(s)$	0.34
$Cu^{2+} + e^- \longrightarrow Cu^+$	0.16
$2H^+ + 2e^- \longrightarrow H_2(g)$	0.00
$Fe^{3+} + 3e^- \longrightarrow Fe(s)$	-0.04
$Cr^{3+} + e^- \longrightarrow Cr^{2+}$	-0.42
$Fe^{2+} + 2e^- \longrightarrow Fe(s)$	-0.44
$Cr^{3+} + 3e^- \longrightarrow Cr(s)$	-0.74
$Zn^{2+} + 2e^- \longrightarrow Zn(s)$	-0.76
$Cr^{2+} + 2e^- \longrightarrow Cr(s)$	-0.90
$Al^{3+} + 3e^- \longrightarrow Al(s)$	-1.67
$Mg^{2+} + 2e^- \longrightarrow Mg(s)$	-2.36
$Na^+ + e^- \longrightarrow Na(s)$[b]	-2.71
$Ca^{2+} + 2e^- \longrightarrow Ca(s)$	-2.84
$Sr^{2+} + 2e^- \longrightarrow Sr(s)$	-2.89
$Ba^{2+} + 2e^- \longrightarrow Ba(s)$	-2.92
$Cs^+ + e^- \longrightarrow Cs(s)$	-2.92
$Rb^+ + e^- \longrightarrow Rb(s)$	-2.93
$K^+ + e^- \longrightarrow K(s)$	-2.93
$Li^+ + e^- \longrightarrow Li(s)$	-3.05

B. Basic Aqueous Solution

Half-reaction	$E°$, Volts
$OCl^- + H_2O(l) + 2e^- \longrightarrow Cl^- + 2OH^-$	0.89
$HO_2^- + H_2O(l) + 2e^- \longrightarrow 3OH^-$	0.87
$MnO_4^- + 2H_2O(l) + 3e^- \longrightarrow MnO_2(s) + 4OH^-$	0.60
$O_2(g) + H_2O(l) + 2e^- \longrightarrow OH^- + HO_2^-$	-0.06
$Ni(OH)_2(s) + 2e^- \longrightarrow Ni(s) + 2OH^-$	-0.72
$2H_2O(l) + 2e^- \longrightarrow H_2(g) + 2OH^-$	-0.83

[a]Data are from Bard, A. J.; Parsons, R.; Jordan, J. *Standard Potentials in Aqueous Solution;* Dekker: New York, 1985.
[b]For metals that react violently with water, such as sodium, an amalgam is used as the electrode. An amalgam is a solution of metal in mercury.

Lithium metal is the most easily oxidized species to the right of the arrows; lithium metal is the strongest reducing agent in the table. However, lithium metal and other very reactive metals cannot be used as reducing agents in aqueous solution because they reduce water:

$$2Li(s) + 2H_2O(l) \longrightarrow 2LiOH(aq) + H_2(g)$$

Hydrogen gas and moderately reactive metals such as zinc are commonly used reducing agents in aqueous solution.

You may wonder why the order of decreasing strength of the Group IA metals as reducing agents is

$$Li > K > Rb > Cs > Na \qquad \text{instead of} \qquad Cs > Rb > K > Na > Li$$

as you might expect from ionization energies. [The first ionization energies of the Group IA metals decrease down the group in the periodic table (Figure 8.10).] The explanation is that ionization energies are energies needed to remove electrons from *gaseous* atoms to form gaseous ions; for example,

$$Li(g) + \text{ionization energy} \longrightarrow Li^+(g) + e^-$$

Standard reduction potentials refer to solid metals and ions in aqueous solution. In aqueous solution, the situation is considerably more complicated than in the gas phase; the energy needed to remove a metal ion from the solid and *the energy released by solvation of the metal ion must be considered.*

If the standard reduction potential for a half-reaction has a + sign, the species to the left of the arrow is more easily reduced than H^+. For example, $E°$ for the half-reaction

$$Cu^{2+} + 2e^- \longrightarrow Cu(s)$$

is $+0.34$ V; Cu^{2+} is more easily reduced than H^+. If the standard reduction potential for a half-reaction has a − sign, the species to the left of the arrow is less easily reduced than H^+. For example, $E°$ for the half-reaction

$$Zn^{2+} + 2e^- \longrightarrow Zn(s)$$

is -0.76 V; Zn^{2+} is less easily reduced than H^+ (H^+ is more easily reduced than Zn^{2+}).

Qualitative Predictions from the Table of Standard Reduction Potentials

Under standard conditions, *any species to the left of the arrows will oxidize any species to the right of the arrows that is lower in the table.* This **diagonal relationship** can be used to predict whether a given redox reaction will be spontaneous. For example, consider the two half-reactions:

$$MnO_4^- + 8H^+ + 5e^- \longrightarrow Mn^{2+} + 4H_2O \qquad E° = +1.51 \text{ V}$$
$$Cl_2(g) + 2e^- \longrightarrow 2Cl^- \qquad E° = +1.36 \text{ V}$$

Permanganate ion is to the left of the arrow; chloride ion is to the right and below permanganate ion in the table. Therefore, reaction will take place. The standard potential for the overall cell reaction will be positive:

$$E°_{cell} = 1.51 \text{ V} - (1.36 \text{ V}) = +0.15 \text{ V}$$

Permanganate ion will oxidize Cl^- ion to $Cl_2(g)$ in acidic solution; the manganese will be reduced to Mn^{2+} ion. Note that the oxidation of Cl^- ion to $Cl_2(g)$

$$2Cl^- \longrightarrow Cl_2(g) + 2e^-$$

is the reverse of the half-reaction in the table. The half-reaction of reduction that

Solutions of reducing agents, such as the developer used in photography, can't be stored very long unless air is kept out because the oxygen in air oxidizes the reducing agents.

has the strongest tendency to take place (highest value for $E°$) will take place. The other half-reaction will be forced to go in the reverse direction.

The net ionic equation for the reaction of MnO_4^- and Cl^-

$$2MnO_4^- + 16H^+ + 10Cl^- \longrightarrow 2Mn^{2+} + 5Cl_2(g) + 8H_2O(l)$$

can be obtained by combining the equations for the two half-reactions so that electrons cancel. The half-reaction for the reduction of MnO_4^- must be multiplied by 2 and the half-reaction for the oxidation of Cl^- by 5.

Oxidation–Reduction and Conjugate Acid–Base Pairs

Note the similarity between the table of standard reduction potentials (Table 19.1) and the table of ionization constants of some common weak acids (Table 15.3). Both are tables of half-reactions. Reduction cannot take place without oxidation; a species cannot obtain an electron(s) unless some other species gives up the electron(s). A base cannot obtain a proton unless an acid gives up a proton. Table 15.3 lists acids in order of decreasing strength and bases in order of increasing strength. Table 19.1 lists oxidizing agents in order of decreasing strength and reducing agents in order of increasing strength. Any acid in Table 15.3 will react with any base that is lower in the table; any oxidizing agent in Table 19.1 will react with any reducing agent that is lower in the table. Strong acids and strong bases react to form weaker bases and weaker acids; strong oxidizing agents and strong reducing agents react to form weaker reducing agents and weaker oxidizing agents.

Species That Can Act Both as Oxidizing Agents and as Reducing Agents

Species that contain an element in its most positive oxidation state, such as Fe^{3+}, can only act as oxidizing agents and be reduced. Species that contain an element in its most negative oxidation state, such as I^-, can only act as reducing agents and be oxidized. However, species that contain an element in an intermediate oxidation state, such as Cu^+, can undergo either oxidation or reduction and can act either as reducing agents or as oxidizing agents. These species appear in *both* columns of tables of standard reduction potentials like Table 19.1. From the diagonal relationship in the table,

$$\mathbf{Cu^+} + e^- \longrightarrow Cu(s) \qquad E° = +0.52 \text{ V}$$
$$Cu^{2+} + e^- \longrightarrow \mathbf{Cu^+} \qquad E° = +0.16 \text{ V}$$

we would predict that Cu^+ should react with Cu^+ to form Cu^{2+} and $Cu(s)$ or, in other words, that copper(I) ion will autooxidize or disproportionate in aqueous solution

$$2Cu^+ \longrightarrow Cu(s) + Cu^{2+}$$

and this turns out to be the case. Disproportionation is very fast and takes place in less than a second. As a result, only very low concentrations of Cu^+ can exist in aqueous solution and no water-soluble Cu(I) compounds are known.

PRACTICE PROBLEMS

19.3 Use Table 19.1 to classify each of the following species as either an oxidizing agent or a reducing agent or both: (a) Ag^+ (b) NO_3^- (c) Al(s) (d) I^- (e) $Br_2(l)$ (f) Fe^{2+}

Calculation of Standard Cell Potentials from Standard Reduction Potentials

A table of standard reduction potentials is a very compact way of storing a great deal of chemical information. The 32 half-cell reactions in part A of Table 19.1 can be combined to form 496 cell reactions that are spontaneous under standard conditions! The potentials of the 496 cell reactions can be calculated from the 32 standard potentials. Much larger tables of standard reduction potentials are available. The standard potential of a cell is a quantitative measure of the tendency of reactants to form products under standard conditions and can be used to calculate the potential of a cell under nonstandard conditions as we will see in Section 19.3. Sample Problem 19.1 illustrates the calculation of a standard cell potential from standard half-cell reduction potentials in Table 19.1.

The most recent compilation is Standard Potentials in Aqueous Solution *edited by A. J. Bard, R. Parsons, and J. Jordan (Dekker, New York, 1985).*

SAMPLE PROBLEM

19.1 Calculate the standard cell potential, E°_{cell}, for the reaction

$$2Ag(s) + Cu^{2+} \longrightarrow 2Ag^+ + Cu(s)$$

and predict whether or not this reaction will take place spontaneously.

Solution The standard potential for a cell is the difference between the reduction potentials of the two half-cells that make up the cell (equation 19.2):

$$E^\circ_{cell} = E^\circ_{\text{half-cell of reduction}} - E^\circ_{\text{half-cell of oxidation}} \tag{19.2}$$

For the reaction in this problem, the half-cell of reduction is $Cu^{2+} + 2e^- \longrightarrow Cu(s)$, and the half-cell of oxidation is $Ag^+ + e^- \longrightarrow Ag(s)$ and

$$E^\circ_{cell} = E^\circ_{Cu^{2+}/Cu(s)} - E^\circ_{Ag^+/Ag(s)} \tag{19.5}$$

From Table 19.1, $E^\circ_{Cu^{2+}/Cu(s)} = +0.34$ V and $E^\circ_{Ag^+/Ag(s)} = +0.80$ V. Substitution of these values in equation 19.5 gives

$$E^\circ_{cell} = +0.34 \text{ V} - (+0.80 \text{ V}) = -0.46 \text{ V}$$

Because E°_{cell} is negative, reaction is nonspontaneous.

Check: The diagonal relationship can be used to check the sign of the value obtained for E°_{cell}:

$$Ag^+ + e^- \longrightarrow Ag(s) \qquad 0.80$$
$$Cu^{2+} + 2e^- \longrightarrow Cu(s) \qquad 0.34$$

In Table 19.1, Cu^{2+} in the left column is *below* Ag(s) in the right column; reaction between Cu^{2+} and Ag(s) is not spontaneous, and the sign of E°_{cell} should be negative, as it is.

▶ Notice that you can calculate a cell potential by changing the sign for the half-reaction that takes place in the reverse direction from the one shown in Table 19.1 and adding. Some general chemistry instructors prefer to calculate cell potentials in this way.

▶ Always check to be sure your work is reasonable.

When using a table of standard electrode potentials to predict whether a reaction will be spontaneous or nonspontaneous—either by the diagonal relationship or by calculation of E°_{cell}—*all possible half-reactions should be considered.* For example, to decide whether copper metal can be dissolved by nitric acid, reactions of both H^+ and NO_3^- must be considered. The standard reduction potential for the reaction

$$Cu(s) + 2H^+ \longrightarrow Cu^{2+} + H_2(g)$$

◖ You should check this value.

is -0.34 V. However, the standard reduction potential for the reaction

$$3Cu(s) + 8H^+ + 2NO_3^- \longrightarrow 3Cu^{2+} + 2NO(g) + 4H_2O(l)$$

is $+0.62$ V. Copper metal should dissolve in 1 M nitric acid at 25 °C providing that enough time is allowed. *Standard reduction potentials do not give any information about reaction rate.* In working the problems in this book, you will have to assume that reactions predicted to be spontaneous will take place at a useful rate and, for lack of other information, neglect the possibility of reactions that are not included in Table 19.1. In the real world, of course, both rate and the possibility of reactions not in tables must be considered.

Cell potentials for half-reactions of reduction not in the table cannot be obtained by adding cell potentials for half-reactions of reduction. For example, although the equation for the half-reaction

$$Cu^{2+} + 2e^- \longrightarrow Cu(s)$$

is the sum of the equations for the half-reactions

$$Cu^{2+} + e^- \longrightarrow Cu^+$$
$$\underline{+\ (Cu^+ + e^- \longrightarrow Cu(s))}$$
$$Cu^{2+} + 2e^- \longrightarrow Cu(s)$$

the cell potential for the half-reaction $Cu^{2+} + 2e^- \longrightarrow Cu(s)$ is *not* equal to the sum of the cell potentials for the other two half-reactions:

$$+0.34 \text{ V} \neq [(+0.16 \text{ V}) + (+0.52 \text{ V})]$$

Standard half-cell reduction potentials that are not in the table *can* be calculated from an experimentally measured standard cell potential and one standard half-cell reduction potential from the table. Sample Problem 19.2 shows how.

Top: *Copper metal immediately after being placed in 1 M HNO₃.* Bottom: *Eight hours later. The surface of the copper has been etched, and the solution shows the blue color of $Cu^{2+}(aq)$.*

SAMPLE PROBLEM

19.2 The standard cell potential for the reaction

$$2Ag^+ + Ni(s) \longrightarrow 2Ag(s) + Ni^{2+}$$

is $+1.06$ V at 25 °C. What is $E°$ for the half-reaction $Ni^{2+} + 2e^- \longrightarrow Ni(s)$ at 25 °C?

Solution Solve equation 19.2 for $E°_{\text{half-cell of oxidation}}$ and substitute the information given:

$$E°_{\text{half-cell of oxidation}} = E°_{\text{half-cell of reduction}} - E°_{\text{cell}}$$
$$= +0.80 \text{ V} - (+1.06 \text{ V}) = -0.26 \text{ V}$$

$E°$ for the half-reaction $Ni^{2+} + 2e^- \longrightarrow Ni(s)$ is -0.26 V.

19.6 Calculate the standard cell potential E°_{cell} for each of the following reactions and predict whether or not reaction will take place spontaneously under standard conditions:

(a) $2Fe(s) + 6H^+ \longrightarrow 2Fe^{3+} + 3H_2(g)$

(b) $O_2(g) + 4Fe^{2+} + 4H^+ \longrightarrow 4Fe^{3+} + 2H_2O(l)$

19.7 The standard cell potential for the reaction

$$Zn(s) + Sn^{2+} \longrightarrow Zn^{2+} + Sn(s)$$

is $+0.62$ V. What is E° for the half-reaction $Sn^{2+} + 2e^- \longrightarrow Sn(s)$?

19.8 The four half-reactions

$$
\begin{array}{ll}
F_2(g) + 2e^- \longrightarrow 2F^- & E^\circ = +2.87 \text{ V} \\
Cl_2(g) + 2e^- \longrightarrow 2Cl^- & E^\circ = +1.36 \text{ V} \\
Br_2(l) + 2e^- \longrightarrow 2Br^- & E^\circ = +1.07 \text{ V} \\
I_2(s) + 2e^- \longrightarrow 2I^- & E^\circ = +0.54 \text{ V}
\end{array}
$$

can be combined to make six different spontaneous reactions. Write net ionic equations for the six spontaneous reactions.

19.9 For the reaction $Ni^{2+} + 2e^- \longrightarrow Ni(s)$, $E^\circ = -0.26$ V. Will Ni^{2+} ions oxidize Zn to Zn^{2+} or will Zn^{2+} ions oxidize Ni to Ni^{2+}? Explain your answer.

19.3 EFFECT OF CONCENTRATION ON CELL POTENTIAL

Qualitative Consideration

In the real world, reactions are usually carried out using concentrations of solutions that are different from the standard concentration of one molar and often with partial pressures of gases that are different from one atmosphere. ▪Figure 19.9 shows how the cell potential of the cell

$$Cu(s)|Cu^{2+}(aq)||Ag^+(aq)|Ag(s)$$

in which the cell reaction is

$$Cu(s) + 2Ag^+ \longrightarrow Cu^{2+} + 2Ag(s) \qquad (19.6)$$

changes with concentration. From Figure 19.9, you can see that increasing the concentration of silver ion at constant copper(II) ion concentration increases the cell voltage. On the other hand, increasing the concentration of copper(II) ion at constant silver ion concentration decreases the cell voltage. Silver ion is a reactant in the cell reaction, and copper(II) ion is a product of the cell reaction. In general, an increase in the concentration of a reactant increases cell voltage; an increase in the concentration of a product decreases cell voltage. An increase in cell voltage means that the reaction has a greater tendency to take place. Thus, increasing the concentration of the reactant, Ag^+, increases the tendency of the reaction to take place; increasing the concentration of the product, Cu^{2+}, decreases the tendency of the reaction to take place. A shift in equilibrium to the right when the concentration of a reactant is increased or a shift to the left when the concentration of a product is increased is what you would expect from your study of chemical equilibrium and Le Châtelier's principle. However, a large relative change in concentration only results in a small change in cell potential. When the concentration of silver ion increases tenfold—for example, from 0.0100 to 0.100 M—the cell voltage increases a mere 0.06 V. The small change in cell potential that accompanies a large relative

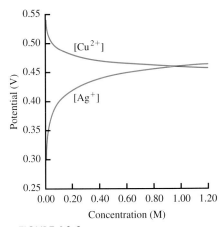

▪ FIGURE 19.9 Effect of concentration on cell potential. One curve shows how the potential changes as the concentration of silver ion changes while the concentration of copper(II) ion is constant (1 M). The other curve shows how the potential changes as the concentration of copper(II) ion changes while the concentration of silver ion is constant (1 M).

change in concentration suggests that a logarithmic relationship exists between cell potential and concentration. (Remember how pH only changes one unit when $[H^+]$ changes by a power of 10.) The fact that a large relative change in concentration only results in a small change in cell potential is the reason that use of concentrations instead of activities in electrochemistry is satisfactory.

A tenfold change in $[Cu^{2+}]$ only produces half as much change in cell potential, 0.03 V, as a tenfold change in $[Ag^+]$. Notice that the coefficient of Cu^{2+} in the equation for the cell reaction (equation 19.6) is 1 (understood) whereas the coefficient of Ag^+ is 2.

When the standard reduction potential for a cell reaction is small, the direction of spontaneous change can be reversed by changing concentrations. For example, $E°$ for the reaction

$$3Zn(s) + 2Cr^{3+} \longrightarrow 3Zn^{2+} + 2Cr(s) \qquad (19.7)$$

is only $+0.02$ V at 25 °C. If $[Cr^{3+}]$ is greater than 1 M or $[Zn^{2+}]$ is less than 1 M, reaction takes place spontaneously in the direction shown by equation 19.7. However, if $[Cr^{3+}]$ is less than 0.1 or $[Zn^{2+}]$ is greater than 5, the reverse reaction is spontaneous.

Quantitative Relationship Between Concentration and Cell Potential

Quantitatively, the relationship between cell potential and concentration is given by the equation

$$E = E° - \frac{RT}{nF} \ln Q \qquad (19.8)$$

Equation 19.8 is called the **Nernst equation** in honor of the German physical chemist Walther Nernst. In the Nernst equation, E is the potential of the cell under nonstandard conditions, $E°$ is the standard potential of the cell, R is the gas constant, 8.31 J/K·mol, T is the temperature in kelvin, Q is the reaction quotient (Section 14.4), n is the number of moles of electrons transferred, and F is the Faraday constant,* 9.6485×10^4 C/mol. The **Faraday constant** is the charge on one mole of electrons. The quantity, n, is the number of electrons that cancels when the two half-cell reactions are combined to give the cell reaction. For example, for the reaction, $Cu(s) + 2Ag^+ \longrightarrow Cu^{2+} + 2Ag(s)$, $n = 2$.

Because 1 V = 1 J/C, the term RT/nF in the Nernst equation is equal to 0.0257 V/n at 25 °C:

$$\frac{RT}{nF} = \frac{8.31 \text{ J} \cdot \text{K}^{-1} \cdot \text{mol}^{-1} (25 + 273) \text{ K}}{n(9.65 \times 10^4) \text{ C} \cdot \text{mol}^{-1}} = \frac{0.0257 \text{ J}}{n\text{C}} = \frac{0.0257 \text{ V}}{n}$$

Use of the expression†

$$E = E° - \frac{0.0257 \text{ V}}{n} \ln Q \qquad (19.8a)$$

instead of equation 19.8 simplifies calculations.

Why use natural logarithms? Almost all reviewers have said that their students have scientific calculators and that natural logarithms rather than base 10 logarithms should be used throughout the book. Although base 10 logs must, of course, be used for pH, students can learn 0.0257 just as easily as 0.0591, so we saw no reason not to be consistent here.

---

*The Faraday constant is named after Michael Faraday, an Englishman who was one of the greatest scientists of the nineteenth century. Faraday's work provided the experimental and much of the theoretical foundation for the classical theory of electromagnetic fields. In 1821, Faraday discovered the principle of the electric motor and built a model. He began his scientific career as an assistant in Sir Humphry Davy's chemistry laboratory and was the first person to liquefy chlorine gas. Faraday also discovered benzene.

†In older pre-scientific-calculator texts where logarithms to the base 10 are used, the equivalent expression is

$$E = E° - \frac{0.0591 \text{ V}}{n} \log Q$$

Like any other equation, the Nernst equation can be used to calculate any one of the variables if the values of all the other variables are known. Thus, the cell potentials for cell reactions and half-cell reactions under nonstandard conditions can be calculated from standard potentials. Standard potentials can be found from cell potentials measured using dilute solutions for which molarities are indeed equal to activities. Potential measurements can be used to find the concentration of an ion if the concentrations of all other ions are known and the standard potential is available. The Nernst equation *cannot* be used to calculate E at one temperature from E measured at another temperature because $E°$ depends on temperature. However, small differences in temperature, such as the difference in temperature between one day and the next, do not result in significant changes in potential. Changes in potential are of the order of mV/K around 25 °C. Sample Problem 19.3 illustrates a calculation using the Nernst equation.

19.3   What is the potential of the cell

$$Zn(s)|Zn^{2+}(aq, 0.35\ M)||Cu^{2+}(aq, 4.7 \times 10^{-5}\ M)|Cu(s)$$

at 25 °C?

Solution   The equation for the cell reaction is

$$Zn(s) + Cu^{2+} \longrightarrow Zn^{2+} + Cu(s)$$

For this reaction, the number of electrons transferred is 2, that is, $n = 2$.

Remember that the reaction quotient, Q, is the product of the concentrations of the products, each raised to a power equal to the coefficient in the balanced equation, divided by a similar product for the reactants. That is, for the general reaction

$$aA + bB + \cdots \rightleftharpoons eE + fF + \cdots, \qquad Q = \frac{[E]^e[F]^f \cdots}{[A]^a[B]^b \cdots}$$

The concentrations of pure liquids and solids are constant and are included in the value of Q. Therefore, for the reaction in this problem,

$$Q = \frac{[Zn^{2+}]}{[Cu^{2+}]}$$

In Section 19.2, we calculated that $E°$ for this cell is 1.10 V at 25 °C. Substitution of this value for $E°$ and the concentrations given in the problem in equation 19.8a gives

$$E = 1.10\ V - \frac{0.0257\ V}{2} \ln\left(\frac{0.35}{0.000\ 047}\right) = 0.99\ V$$

Check: Although both concentrations are lower than standard concentrations of 1 M, the concentration of the reactant Cu^{2+} is reduced much more than the concentration of the product Zn^{2+}. According to Le Châtelier's principle, equilibrium should shift to the left and the driving force for reaction should be decreased. Because 0.99 V is less than 1.10 V, the answer is reasonable.

19.10   What is the potential of the following cell at 25 °C?

$$Cu(s)|Cu^{2+}(aq, 0.100\ M)||Ag^+(aq, 0.0100\ M)|Ag(s)$$

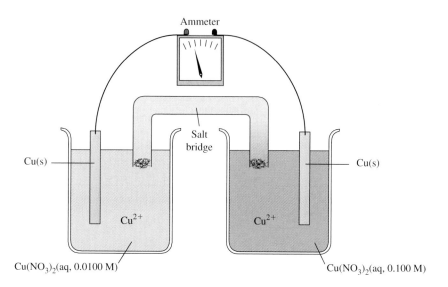

FIGURE 19.10 Cell with different concentrations in half-cells.

Another problem with hydrogen electrodes is that $H_2(g)$ forms explosive mixtures with oxygen in air; hydrogen electrodes can be dangerous under some circumstances.

Voltaic cells similar to the ones in Figures 19.7 and 19.8 can be used to measure the $[H^+]$ in solutions. In principle, pH meters are similar to these voltaic cells. A pH meter (Figure 15.2) actually measures voltage although the scale is marked in pH units for convenience. Glass electrodes that are sensitive to hydrogen ion concentration are substituted for the standard hydrogen electrode, which is not easy to use. Many pH meters appear to have only one electrode. However, this electrode is a combination of a glass electrode and another reference electrode.

Concentration Cells

The dependence of cell potential on concentration suggests that it should be possible to make a voltaic cell from similar electrodes that dip into solutions of different concentrations as shown in ▪Figure 19.10. Which half-cell will be the anode in the cell pictured in Figure 19.10? If 0.0100 and 0.100 M solutions of $Cu(NO_3)_2$ were mixed in a single container, the concentration of the new solution would be between 0.0100 and 0.100 M. A change is spontaneous whether it takes place in a single container or is separated in two half-cells; if reaction is allowed to take place in the cell in Figure 19.10, the concentrations of both half-cells will become equal. The concentration in the left half-cell will increase; the concentration in the right half-cell will decrease. For the concentration of Cu^{2+} in the left half-cell to increase, the half-reaction

$$Cu(s) \longrightarrow Cu^{2+} + 2e^-$$

must take place in the left half-cell. This half-reaction is oxidation; the left half-cell must be the anode. For the concentration of Cu^{2+} in the right half-cell to decrease, the half-reaction

$$Cu^{2+} + 2e^- \longrightarrow Cu(s)$$

must take place in the right half-cell. This half-reaction is reduction; the right half-cell must be the cathode. A *voltaic cell that produces an electric current as a result of a difference in concentration* is called a **concentration cell.** ▪Figure 19.11 shows a concentration cell with all its parts labeled. Concentration cells usually do not have very high voltages. However, even a very small difference in voltage causes an electric current, just as only a slight difference in level is needed to make water flow. Concentration cells are very important in biological systems; for example, the

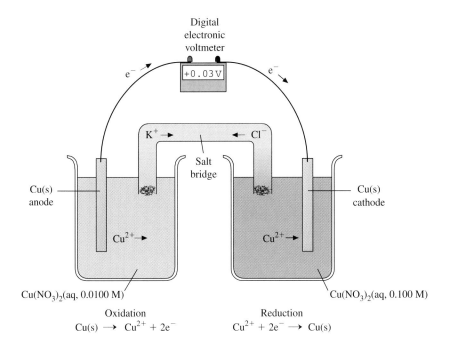

Cu(s) anode

Cu(s) cathode

$Cu(NO_3)_2(aq, 0.0100\ M)$

$Cu(NO_3)_2(aq, 0.100\ M)$

■ FIGURE 19.11   A concentration cell.

potential difference that results from differences in concentration of sodium ions and potassium ions inside and outside of cells plays a significant part in the transfer of nerve impulses. The Nernst equation can be used to calculate this potential.

19.4 FREE ENERGY AND CELL POTENTIAL

In Section 14.4, you learned how to use the value of Q, in Section 17.4 how to use the value of ΔG, and in Section 19.2 how to use the value of E to predict the direction of spontaneous change:

	Q	ΔG	E
Forward change spontaneous	$<K$	$-$	$+$
At equilibrium	$=K$	0	0
Reverse change spontaneous	$>K$	$+$	$-$

Quantitatively, you learned in Section 19.3 that E and Q are related by the Nernst equation. A relationship between $E°$ and equilibrium constant, K, is easily obtained from the Nernst equation

$$E = E° - \frac{RT}{nF} \ln Q \qquad (19.8)$$

as follows. When a system is at equilibrium, no further change in concentration takes place; there is no driving force for reaction, or, in other words, $E = 0$. Also, at equilibrium, $Q = K$. Thus,

$$0 = E° - \frac{RT}{nF} \ln K$$

and

$$\frac{RT}{nF} \ln K = E° \qquad (19.9)$$

Rearrangement gives

$$RT \ln K = nFE°$$

But from equation 17.7 (Section 17.10)

$$\Delta G° = -RT \ln K \qquad (17.7)$$

Therefore,

$$\Delta G° = -nFE° \qquad (19.10)$$

Equation 19.10 can be used to calculate $\Delta G°$ from $E°$. In addition, values for $E°$ that cannot be determined experimentally can be calculated from tabulated values of $\Delta G_f°$.

○ Equation 19.10 is a special case of the general relationship $\Delta G = -nFE$. From this relationship you can see that for ΔG to be negative (reaction spontaneous), E must be $+$.

◐ Students can predict voltages for common batteries in the Voltaic Cells module of the Electrochemistry section.

PRACTICE PROBLEMS

19.11   The standard cell potential for the cell

$$Zn(s)|Zn^{2+}(aq)||Cu^{2+}(aq)|Cu(s)$$

at 25 °C is $+1.10$ V. What is $\Delta G°$ for the following reaction?

$$Zn(s) + Cu^{2+} \longrightarrow Zn^{2+} + Cu(s)$$

19.12   The value of $\Delta G°$ for the reaction

$$Ni^{2+} + Cd(s) \longrightarrow Ni(s) + Cd^{2+}$$

is -31.2 kJ. What is $E°$ for the reaction?

19.13   For the reaction

$$3Au^+ \longrightarrow Au^{3+} + 2Au(s)$$

$E° = +0.47$ V. Write the equilibrium constant expression and calculate the value of the equilibrium constant.

■ FIGURE 19.12   This man is cranking the car to start it.

19.5   BATTERIES

Life without batteries is hard to imagine. No calculator, no portable radio or tape recorder, no battery-operated toys! In addition you would have to start your car each morning by cranking it (see ■Figure 19.12). The batteries that play such an important part in modern life are compact, portable voltaic cells; for example, the familiar "flashlight battery" or dry cell is a zinc–carbon cell. Batteries store energy so that it is ready to use whenever and wherever it is needed.

Dry Cells

Although zinc–carbon cells are called dry cells, they are not really dry. The electrolyte, aqueous NH_4Cl, is made into a paste that does not flow by adding an inert filler such as carbon black (finely divided, microcrystalline graphite). ■Figure 19.13 is a schematic drawing of a cross-section of a zinc–carbon cell. The seal keeps water from evaporating and prevents leaks. Most dry cells are guaranteed not to leak. The reactions that take place in zinc–carbon cells are complex but are usually represented by the following equations:

$$Zn(s) \longrightarrow Zn^{2+}(aq) + 2e^- \qquad \text{anode}$$

$$2MnO_2(s) + 2NH_4^+(aq) + 2e^- \longrightarrow Mn_2O_3(s) + 2NH_3(aq) + H_2O(l) \qquad \text{cathode}$$

$$Zn(s) + 2MnO_2(s) + 2NH_4^+(aq) \longrightarrow$$
$$Zn^{2+}(aq) + Mn_2O_3(s) + 2NH_3(aq) + H_2O(l) \qquad \text{cell reaction}$$

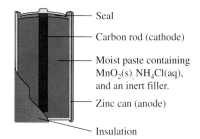

■ **FIGURE 19.13** Cross-section of a zinc–carbon "dry cell."

The graphite cathode is inert. Manganese(IV) oxide is the oxidizing agent; manganese(IV) is reduced to manganese(III) at the cathode. The zinc container is the anode; metallic zinc is oxidized to zinc ion at the anode. Zinc ions then move through the paste away from the anode toward the cathode. Near the cathode, zinc ions react with ammonia to form the complex ion $Zn(NH_3)_4^{2+}$:

$$Zn^{2+} + 4NH_3(aq) \longrightarrow Zn(NH_3)_4^{2+}$$

Formation of the complex ion uses up NH_3 molecules. Removal of ammonia molecules shifts the cell reaction, which is an equilibrium, to the right, as predicted by Le Châtelier's principle, and increases the tendency of the cell reaction to take place.

You may have noticed that if you keep a flashlight turned on for a few minutes, the light dims. However, the next night, the flashlight again gives a bright light when first turned on. This is because zinc ions can't diffuse away from the anode very rapidly through the viscous paste. The concentration of zinc ions builds up near the anode, and the cell potential decreases. *Electrodes with potentials that change greatly when current passes through them,* such as the electrodes in ordinary dry cells, are said to be **polarizable.** Given time, the zinc ions diffuse away from the anode and the flashlight is almost as good as new. Of course, with use, dry cells eventually go dead. When all of the manganese(IV) oxide has reacted or the concentration of ammonium ions has become very low, a dry cell must be replaced.

The potential of a battery depends on the materials used to make the battery, on their concentrations, and on the temperature. Each new zinc–carbon cell has a potential of about 1.5 V. In flashlights (and most other devices that use dry cells), two cells are connected in series, as shown in ■Figure 19.14. The potential of cells that are connected in series is the sum of the potentials of the individual cells. About 3 V are required to light a flashlight bulb brightly.

The amount of energy stored in a battery depends on its size. Big C batteries can deliver more amperes for a longer period of time than small AAA batteries before they go "dead." Big batteries contain larger quantities of the reactive species.

Ordinary dry cells are inexpensive. However, when used, they lose their ability to produce current quite rapidly because the reaction products can't diffuse away from the electrodes fast enough. In addition, ordinary dry cells cannot be recharged. So-called rechargers for dry cells just drive zinc ions away from the anode; attempted recharging voids the guarantee on ordinary dry cells.

The electrolyte solution in ordinary dry cells is acidic because the ammonium ion, NH_4^+, is a Brønsted–Lowry acid; ammonium chloride is the salt of a strong acid and a weak base and is acidic by hydrolysis. Alkaline dry cells also use zinc

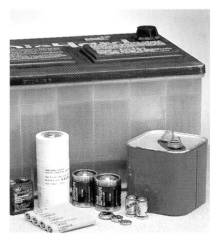

The photo shows an automotive lead storage battery (background), *rechargeable batteries* (left), *alkaline batteries* (center), *and a large zinc–carbon dry cell* (right).

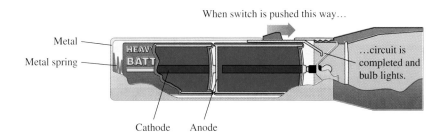

■ **FIGURE 19.14** When cells are connected in series, the cathode of one cell is connected to the anode of another cell. The voltage is the sum of the voltages of the individual cells. The current through all parts of the circuit is the same.

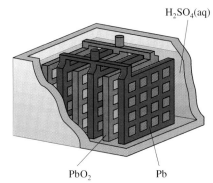

FIGURE 19.15 One cell of a lead storage battery. The 12-V batteries used in most cars have six cells connected in series. The lead storage battery was invented by Gaston Planté, a French physics professor, in 1859.

Before the invention of the electric self-starter eliminated the hand crank, only 22% of cars were gasoline-powered; 40% were powered by steam and 38% by electricity.

as the reducing agent and manganese(IV) oxide as the oxidizing agent, but under basic conditions. In alkaline batteries, the electrolyte is potassium hydroxide instead of ammonium chloride. Alkaline batteries have a longer shelf-life than ordinary zinc–carbon batteries and can deliver more current for longer periods of time, but they are more expensive. In alkaline batteries, the half-reactions are

$$Zn(s) + 2OH^- \longrightarrow ZnO(s) + H_2O(l) + 2e^- \quad \text{anode reaction}$$
$$\underline{2MnO_2(s) + H_2O(l) + 2e^- \longrightarrow Mn_2O_3(s) + 2OH^-} \quad \text{cathode reaction}$$
$$Zn(s) + 2MnO_2(s) \longrightarrow ZnO(s) + Mn_2O_3(s) \quad \text{cell reaction}$$

As you can see from the equation for the overall cell reaction, all substances involved in the cell reaction of an alkaline battery are solids. The concentrations of solids are constant; as a result, the cell potential is constant until all of the limiting reactant has been used up. Alkaline dry cells are not as polarizable as ordinary zinc–carbon batteries. However, like ordinary zinc–carbon batteries, alkaline batteries cannot be recharged.

Lead Storage Batteries

Lead storage batteries are used when a large capacity and moderately high currents are needed. Lead storage batteries have been used to start cars since about 1915. One cell of a lead storage battery is shown in ▪Figure 19.15. The electrolyte in lead storage batteries is aqueous sulfuric acid. One cell of a lead storage battery consists of a number of grids of a lead alloy. One set of alternate grids is packed with lead metal and the other set with lead(IV) oxide, PbO_2. Each set of grids is connected in parallel as shown in Figure 19.15. The larger the surface area of an electrode, the more current the electrode can deliver. When electrodes are connected in parallel, as they are in a cell of a lead storage battery, the voltage is the same as the voltage of a single electrode, but the current that can be delivered is increased. Each cell of a lead storage battery has a potential of about 2 V; the 12-V batteries used in most cars have six cells connected in series.

When a lead storage battery is producing current, the electrode reactions are

$$Pb(s) + HSO_4^- \longrightarrow PbSO_4(s) + H^+ + 2e^- \quad \text{anode (lead plates)}$$
$$\underline{PbO_2(s) + 3H^+ + HSO_4^- + 2e^- \longrightarrow PbSO_4(s) + 2H_2O(l)} \quad \text{cathode (PbO}_2 \text{ plates)}$$
$$Pb(s) + PbO_2(s) + 2H^+ + 2HSO_4^-(aq) \longrightarrow 2PbSO_4(s) + 2H_2O(l) \quad \text{cell reaction (discharging)}$$

The lead sulfate formed in both half-cell reactions sticks to the electrodes. Sulfuric acid is a reactant and water is a product; as a result, the concentration of sulfuric acid decreases during discharge. Dilute sulfuric acid is less dense than more concentrated sulfuric acid. Therefore the condition of a battery can be checked by measuring the density of the sulfuric acid solution in the battery.

One of the biggest advantages of lead storage batteries is that they can be recharged. When a car is running, the battery is recharged by the alternator, a generator powered by the engine. During recharging, the electrode reactions are the reverse of the reactions that take place during discharge:

$$PbSO_4(s) + H^+ + 2e^- \longrightarrow Pb(s) + HSO_4^- \quad \text{lead plates (cathode)}$$
$$\underline{PbSO_4(s) + 2H_2O(l) \longrightarrow PbO_2(s) + 3H^+ + HSO_4^- + 2e^-} \quad \text{PbO}_2 \text{ plates (anode)}$$
$$2PbSO_4(s) + 2H_2O(l) \longrightarrow$$
$$Pb(s) + PbO_2(s) + 2H^+ + 2HSO_4^-(aq) \quad \text{cell reaction (charging)}$$

Lead(II) sulfate is a reactant in both these half-cell reactions. Recharging would not be possible if $PbSO_4$ did not stick to the electrodes. However, solid $PbSO_4$ is

Fuel Cells

When electricity is generated by burning coal, natural gas, or oil in an electric power plant, only about 33% of the energy in the fuel is converted to electrical energy; the rest is wasted heating the surroundings. Fuel cells are twice as efficient; they convert about 70% of the energy in the fuel to electrical energy. Like batteries, fuel cells are voltaic cells. But in fuel cells the oxidizing and reducing agents are fed into the cell continuously. Reaction continues as long as reactants are fed in and the external circuit is complete.

Fuel cells usually use oxygen from air as the oxidizing agent. Hydrogen, carbon monoxide, methyl alcohol (CH_3OH), hydrazine (H_2NNH_2), and simple hydrocarbons such as methane (CH_4) are used as reducing agents. The electrodes are usually porous metals to which a catalyst has been added and the electrolytes are concentrated solutions of acids or bases. Figure 19.16 is a schematic diagram of the fuel cells used in Gemini spacecraft.* The potential of a single hydrogen–oxygen fuel cell is 1.23 V. The system of many hydrogen–oxygen fuel cells used in the Gemini spacecraft furnished an average power of 900 J s^{-1} and maximum power of 2000 J s^{-1}. It also provided freshwater for drinking.

The reactions that take place in a hydrogen–oxygen fuel cell are

$$2H_2(g) \longrightarrow 4H^+ + 4e^-$$
$$O_2(g) + 4H^+ + 4e^- \longrightarrow 2H_2O$$
$$\overline{2H_2(g) + O_2(g) \longrightarrow 2H_2O}$$

*The Gemini spacecraft program was a series of 12 two-man spacecraft launched into orbit around Earth by the United States in 1964–1967. The Gemini missions tested the astronauts' ability to control spacecraft by hand and helped develop the methods for orbital rendezvous and dockings with a target vehicle that were necessary for the Apollo moon landings.

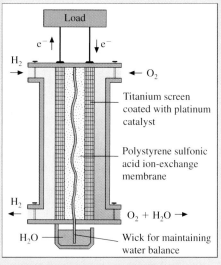

FIGURE 19.16 Schematic drawing of a single Gemini hydrogen–oxygen fuel cell. (Reproduced from Bockris, J. O'M.; Reddy, A. K. N. *Modern Electrochemistry;* Plenum: New York, 1970, p. 1388 by permission of the publisher.)

The polystyrene sulfonic acid ion-exchange membrane is the electrolyte. The hydrogen ions pass through this membrane from the anode to the cathode. The conductivity of the membrane depends on its water content; the wick removes or supplies water as needed by capillary action. Each side of the membrane is covered by a titanium screen that is coated with a platinum catalyst. The whole cell is only about 0.5 mm thick.

Besides being about twice as efficient as conventional power plants at converting chemical energy to electrical energy, fuel cells are quiet and require little maintenance. The reaction product from the hydrogen–oxygen fuel cell, water, does not pollute either air or water. Fuel cells have been used in New York and Tokyo to generate extra electricity at times of peak demand. Unfortunately, fuel cells are

During the Gemini IV mission (launched June 3, 1965), astronaut Edward H. White spent about 20 minutes outside the spacecraft. This picture was taken by astronaut James McDivitt.

too expensive for routine use at the present time. However, they represent a promising alternative energy source for the future. Perhaps you will be the one to discover less expensive materials for electrodes or a better catalyst.

A 200 kW fuel cell power plant that uses natural gas as the reducing agent and phosphoric acid as the electrolyte. This power plant satisfies California's environmental standards, which are the strictest in this country and possibly in the world; operating costs average 25-40% lower than those of conventional power plants. Power plants like this are now being used in North America, Europe, and Asia, and fuel cell power plants up to several megawatts in size are being demonstrated.

Aluminum forms an oxide coating (white) that protects it from further attack by oxygen. This coating is not visible to the naked eye but can be seen under magnification (8×).

not a good conductor of electricity, and a battery should not be allowed to run down because too thick a coating of lead sulfate on the electrodes prevents good contact between the sulfuric acid electrolyte and the electrodes. In addition, lead sulfate is less dense than lead(IV) oxide and lead; buildup of lead sulfate in the grids can crack the grids just as ice bursts frozen water pipes.

A dead battery can sometimes be recharged with a battery recharger. However, after repeated discharge–charge cycles and trips over bumpy roads, enough lead(II) sulfate falls to the bottom of the battery that the battery cannot be fully recharged and must be replaced.

Lead storage batteries are sturdy and relatively long-lived; some are guaranteed for as long as five years. They operate at temperatures ranging from -30 to more than 100 °F (-34 to >38 °C). Below -30 °F, the viscosity of the sulfuric acid electrolyte is so high that the ions can't move toward and away from the electrodes fast enough. Lead storage batteries can deliver a relatively large amount of energy for a short time. The disadvantages of the lead storage battery are the large mass of lead needed to deliver enough current and the fluidity and corrosive nature of the liquid electrolyte. Lead storage batteries leak if not kept upright.

Lithium Batteries

Lithium has the lowest density of any metal, and lithium is one of the strongest reducing agents known. The combination of these properties makes lithium batteries the most promising candidates for batteries that can power an electric car for 300 miles on one charge. Although lithium batteries are expensive, they last longer than other batteries and are already used in portable computers.

19.6 CORROSION

The spontaneous oxidation–reduction reactions that take place in batteries are very useful. However, some redox reactions are extremely destructive. About $200 billion are spent in the United States each year (about 4% of the gross domestic product) to prevent corrosion and to repair damage done by corrosion! **Corrosion** is the *deterioration of metals by chemical reaction.* Some metals, such as gold and platinum, are very unreactive and do not corrode. Many uses of gold and platinum, for example, in jewelry and for electrical contacts, depend on the lack of reactivity of these metals. Most metals are below and to the right of $O_2(g)$ in the table of standard reduction potentials and react spontaneously with oxygen. Fortunately, however, for many uses we make of metals, the rates of reaction of metals with oxygen tend to be low. In addition, many metals form oxide coatings that protect the metal from further attack by oxygen.

The rusting of iron and steel is certainly the most familiar and probably the most important example of corrosion. About 20% of the iron and steel produced in the United States every year is used to replace objects that have been ruined by corrosion. Results of the corrosion of iron and steel can be seen in junkyards and automobile graveyards everywhere.

Both oxygen and water must be present for corrosion to take place. Figure 19.17 shows an iron nail that has begun to corrode. The nail is embedded in agar, a gelatine-like material obtained from seaweed. The agar contains water, oxygen, phenolphthalein indicator, and potassium ferricyanide, $K_3[Fe(CN)_6]$. In Figure 19.17, you can see a blue color around the head and point of the nail and around the spot where it is bent. The blue color is ferric ferrocyanide, $Fe_4[Fe(CN)_6]_3$. Formation of

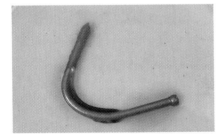

▪ FIGURE 19.17   A bent iron nail in agar containing water, oxygen, phenolphthalein indicator, and potassium ferricyanide. The blue color around the head, point, and bend is due to ferric ferrocyanide and indicates the presence of Fe^{2+} formed by the oxidation of iron. The pink color shows OH^- formed by reduction of oxygen from air in aqueous solution.

ferric ferrocyanide on addition of potassium ferricyanide is a test for Fe^{2+} ion:

$$4Fe^{2+} + 4Fe(CN)_6^{3-} + xH_2O(l) \longrightarrow Fe_4[Fe(CN)_6]_3 \cdot xH_2O(s) + Fe(CN)_6^{4-}$$
blue

At the points where the blue color is observed, iron(II) ions must be going into solution:

$$Fe(s) \longrightarrow Fe^{2+}(aq) + 2e^-$$

A pink color has developed along the nail between the blue spots. Phenolphthalein turns pink in the presence of hydroxide ions. Where the pink color is observed, hydroxide ions have been formed by reaction of oxygen from air with water:

$$O_2(g) + 2H_2O(l) + 4e^- \longrightarrow 4OH^-$$

▪Figure 19.18 shows the corrosion of iron on a microscopic scale. The iron(II) ions go into solution where the iron metal has the most defects. As a result, Fe^{2+} ions dissolve at the head and point of the nail and where it has been bent. These spots function as anodes; the spots in between function as cathodes. Rust forms where oxygen is available; the point where rust forms is often far from the point where iron dissolves. Very pure iron with a minimum of surface defects does not corrode nearly as fast as ordinary iron.

The picture of how iron corrodes shown in Figure 19.18 explains a number of observations that have been made of corrosion. It explains why both water and oxygen must be present for corrosion to occur; iron does not rust in the desert. Rusting takes place faster when water contains salt—iron and steel corrode more rapidly in salt water than in freshwater. Salt water acts like a salt bridge; the dissolved sodium ions and chloride ions help carry charge through the solution.

Rusting also takes place faster when the water is acidic. The standard potential for the reaction

$$Fe(s) + 2H^+ \longrightarrow Fe^{2+} + H_2(g) \qquad E° = +0.44 \text{ V}$$

is positive and this reaction takes place rapidly. In addition, the standard potential for the reaction

$$O_2(g) + 4H^+ + 4e^- \longrightarrow 2H_2O \qquad E° = +1.23 \text{ V}$$

is more positive than the standard potential for the reaction

$$O_2(g) + 2H_2O(l) + 4e^- \longrightarrow 4OH^- \qquad E° = +0.40 \text{ V}$$

Thus, oxygen has a much stronger tendency to be reduced in acidic solution. Reduction of oxygen gas takes place slowly. Reduction of oxygen in acidic solution probably has a lower activation energy than reduction in neutral solution so that oxygen is reduced more rapidly in acidic solution. Most water is weakly acidic because water usually contains dissolved carbon dioxide. Much water today also contains dissolved sulfur dioxide and sulfur trioxide and is quite acidic.

As you might expect from your knowledge of kinetics (Section 18.9), metals corrode faster in the tropics than they do in the arctic. The rates of almost all reactions increase as temperature increases.

Corrosion Prevention

Corrosion can be prevented or at least slowed down in several ways. Painting the metal protects the surface from air and water. (However, if the paint is chipped, the metal surface is exposed, and corrosion takes place.) In northern states, the

This structure for Prussian blue is the one given in the latest edition of Cotton and Wilkinson (Cotton, F. A.; Wilkinson, G., Advanced Inorganic Chemistry, 5th ed.; Wiley: New York, 1988; p 721).

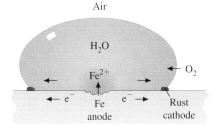

Anode:
$$Fe(s) \longrightarrow Fe^{2+} + 2e^-$$

Cathode:
$$Fe^{2+} \longrightarrow Fe^{3+} + e^-$$
$$O_2(g) + 2H_2O(l) + 4e^- \longrightarrow 4OH^-$$
$$Fe^{3+} + 3OH^- \longrightarrow FeO(OH)(s) + H_2O(l)$$
Red-brown
(rust)

▪ **FIGURE 19.18** Microscopic view of corrosion of iron. When iron corrodes, iron ions leave the surface of the iron and go into solution. The positive charges on the Fe^{2+} ions repel each other; the Fe^{2+} ions move away from each other through the solution. The negative charges of the electrons that are left behind in the iron metal when the Fe^{2+} ions go into solution repel each other. The electrons move away from each other through the iron. At the edge of the drop of water, oxygen from air dissolves in the water and oxidizes the iron(II) ions to iron(III) ions. The oxygens from the oxygen molecules are reduced from the zero oxidation state to the -2 oxidation state, and hydroxide ions are formed. The hydroxide ions and the iron(III) ions react to form a red-brown precipitate of FeO(OH) or rust.

After a collision, a car rusts both because paint has been removed and because of the defects in the metal where it has been dented.

A shining sheet of continuously galvanized steel leaves the molten zinc bath and enters the cooling chambers at Bethlehem Steel Corporation.

■ FIGURE 19.19 Protection of a pipeline with a sacrificial anode. The zinc wire is about 0.5 in. in diameter. It is connected to the pipeline at intervals of 500–1000 ft. Because zinc is more easily oxidized than iron, zinc is oxidized:

$$Zn(s) \longrightarrow Zn^{2+} + 2e^-$$

The electrons reduce oxygen,

$$O_2(g) + 2H_2O(l) + 4e^- \longrightarrow 4OH^-$$

preventing oxidation of the iron pipe.

undersides of cars are often coated to protect steel from corrosion by water containing sodium chloride and calcium chloride. These salts are sprinkled on roads in winter to prevent formation of ice.

Iron is frequently protected by galvanizing. **Galvanizing** means *coating a metal with zinc;* one method of galvanizing is to dip the metal in molten zinc. Although zinc is more easily oxidized than iron, in the presence of moisture, zinc metal reacts with oxygen and carbon dioxide in air to form a film of $Zn_2(OH)_2CO_3$:

$$2Zn(s) + O_2(g) + CO_2(g) + H_2O(l) \longrightarrow Zn_2(OH)_2CO_3(s)$$

This film sticks firmly to the surface of the zinc and protects the zinc from further reaction. Rust is more porous than $Zn_2(OH)_2CO_3$ and does not stick to iron metal very well. Unlike paint, a zinc coating still protects iron even if the zinc coating is chipped. Because zinc is more easily oxidized than iron, zinc around the hole in the coating oxidizes:

$$Zn(s) \longrightarrow Zn^{2+} + 2e^-$$

Zinc forms the anode, making iron the cathode and keeping iron reduced. Electrons formed by oxidation of zinc reduce any iron(II) ions formed before they have time to move away through the solution.

Electroplating is another method of preventing corrosion (it is discussed in Section 19.9). "Tin cans" are actually iron cans plated with tin. If the tin plate is chipped, the iron underneath rusts rapidly because iron is more easily oxidized than tin.

Although aluminum is more reactive than iron, aluminum cans do not corrode under neutral conditions. Aluminum metal that has been exposed to air is covered with a coating of aluminum oxide, Al_2O_3, which is nonporous and sticks firmly to the surface of the aluminum metal protecting the aluminum from air and water. However, because aluminum is amphoteric, aluminum is corroded by acids and bases.

Objects that cannot conveniently be painted, galvanized, or electroplated, such as buried pipelines and underground storage tanks, are often protected from corrosion by sacrificial anodes. **Sacrificial anodes** are *pieces of reactive metals such as zinc and magnesium that are connected to the object to be protected by a conductor* (see ■Figure 19.19). Sacrificial anodes act like the zinc around a hole in the coating of galvanized iron. Electrons from oxidation of the reactive metal keep iron reduced to iron metal. Sacrificial anodes must be replaced from time to time, but replacement of sacrificial anodes is cheaper than replacement of a pipeline or underground storage tank. In addition, the hazards of leaks are avoided.

Yet another method of preventing corrosion is to alloy a metal with a more reactive metal that forms a protective oxide coating. All types of stainless steel contain chromium; many also contain nickel.

PRACTICE PROBLEM

19.14   Predict the result of using iron nails to fasten an aluminum gutter to a house. Explain your answer.

19.7   ELECTROLYTIC CELLS

Oxidation–reduction reactions that are nonspontaneous can be made to take place by supplying electrical energy. Direct current (charge moves in only one direction), not ordinary alternating current, must be used. (The source of direct current can be thought of as a pump for electrons.) The *process of causing a nonspontaneous chemical reaction to take place by means of electrical energy* is called **electrolysis.*** An *electrochemical cell in which a nonspontaneous reaction is made to take place* is called an **electrolytic cell.** During charging, a lead storage battery acts as an electrolytic cell.

The direction of movement of charge reverses or alternates 60 times a second in standard U.S. current.

The minimum voltage necessary to bring about electrolysis is the calculated potential for the cell. However, the voltage actually needed is always greater than this minimum voltage. This extra voltage is required to overcome the resistance of the cell. In addition, some electrode reactions take place slowly; energy at least equal to the activation energy must be supplied in order for reaction to take place. The *extra voltage needed to make slow reactions take place at a practical rate* is called **overpotential.** Overpotential is often large when gases are formed in electrode reactions; overpotential is usually small when metals are deposited on electrodes.

Three types of reaction are possible in electrolytic cells: (1) A solute ion or molecule can be oxidized or reduced. (2) The solvent can be oxidized or reduced. (3) The electrode can react. As usual, which reaction actually takes place depends both on thermodynamics and on kinetics. In general, the most easily oxidized species will be oxidized and the most easily reduced species will be reduced. *Of the possible half-reactions, the ones with the most positive (least negative) potentials will usually take place.* However, if the difference between standard half-cell potentials is small, concentration may determine the result of electrolysis. Even if the difference between standard half-cell potentials is fairly large, predictions based on standard potentials may be wrong if the predicted reaction is slow.

Let's consider the electrolysis of a 1 M solution of sodium chloride as an example. Electrolysis of 1 M sodium chloride solution yields chlorine gas at the anode and hydrogen gas at the cathode as shown in ▬Figure 19.20. In making predictions based on standard potentials about which half-reaction will take place, all possible half-reactions should be considered. But you can't very well consider half-reactions you don't know about. To illustrate the method, we will assume as usual that only the half-reactions listed in Table 19.1 take place.

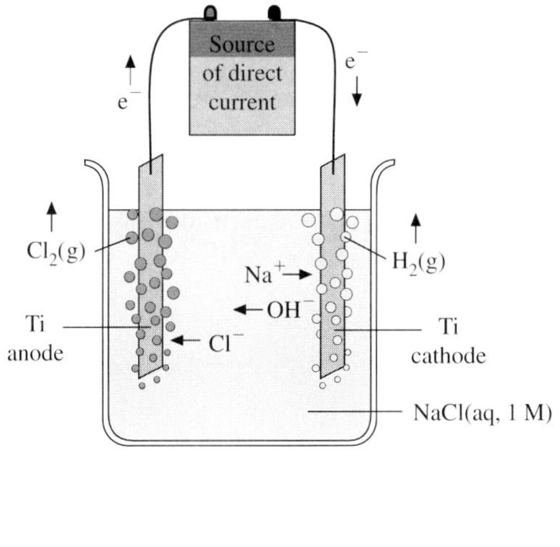

The solute anion, Cl^-, moves toward the anode.

The solute cation, Na^+, moves toward the cathode.

▬ FIGURE 19.20   Electrolysis of aqueous sodium chloride.
Anode reaction:
$$2Cl^- \longrightarrow Cl_2(g) + 2e^-$$
Cathode reaction:
$$2H_2O(l) + 2e^- \longrightarrow H_2(g) + 2OH^-$$
Overall reaction:
$$2Cl^- + 2H_2O(l) \longrightarrow$$
$$Cl_2(g) + H_2(g) + 2OH^-$$

---

*Remember that "-lysis" is derived from the Greek word meaning *loosening.* Electrolysis is the loosening of the parts of chemical compounds from each other by electricity. For example, electrolysis of water (Figure 1.3) splits water into hydrogen and oxygen:

$$2H_2O(l) \xrightarrow{\text{electricity}} 2H_2(g) + O_2(g)$$

A little electrolyte, such as sulfuric acid or potassium nitrate, must be dissolved in the water to make it a conductor.

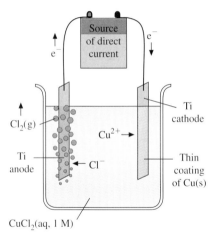

FIGURE 19.21 Electrolysis of aqueous copper(II) chloride.
Anode reaction:
$$2Cl^- \longrightarrow Cl_2(g) + 2e^-$$
Cathode reaction:
$$Cu^{2+} + 2e^- \longrightarrow Cu(s)$$
Overall reaction:
$$Cu^{2+} + 2Cl^- \longrightarrow Cu(s) + Cl_2(g)$$

In the electrochemical cell shown in Figure 19.20, the electrodes are made of titanium and are inert. The two remaining possible anode reactions are

oxidation of solute anion: $2Cl^- \longrightarrow Cl_2(g) + 2e^-$ $E° = -1.36$ V (Table 19.1)

oxidation of solvent: $2H_2O(l) \longrightarrow O_2(g) + 4H^+ + 4e^-$ $E = -0.82$ V*

The potential for the oxidation of water is less negative than that for the oxidation of chloride ion. From a comparison of the potentials of the two possible half-cell reactions, we would predict that oxidation of the water, not the chloride ions, should take place. However, formation of oxygen gas has an unusually high overpotential, and chlorine gas is formed instead, an example of kinetic rather than thermodynamic control of reaction product. As electrolysis is carried out, the concentration of chloride decreases because chloride ion reacts to form chlorine gas. When the concentration of chloride ion becomes much less than 1 M, the formation of oxygen gas at the anode becomes an important side reaction.

The two cathode reactions that are possible when aqueous sodium chloride is electrolyzed using inert electrodes are

reduction of solvent: $2H_2O(l) + 2e^- \longrightarrow H_2(g) + 2OH^-$ $E = -0.42$ V[†]

reduction of the solute cation: $Na^+ + e^- \longrightarrow Na(s)$ $E° = -2.71$ V

From the potentials of the two possible half-cell reactions, we would predict that reduction of water, not reduction of sodium ion, should take place. In this case, our prediction is correct. (We would hardly expect that metallic sodium would be formed in aqueous solution when we know that sodium metal reacts vigorously with water, forming hydrogen gas and sodium hydroxide solution.) As the reaction takes place and the concentration of hydroxide ions increases, a higher potential will be required to continue electrolysis.

If the solute cation is more easily reduced than water, the solute cation is reduced, and metal plates out on the cathode. The electrolysis of aqueous copper(II) chloride is an example (-Figure 19.21). From the potentials of the two possible half-reactions of reduction,

reduction of solute cation: $Cu^{2+} + 2e^- \longrightarrow Cu(s)$ $E° = +0.34$ V

reduction of solvent: $2H_2O(l) + 2e^- \longrightarrow H_2(g) + 2OH^-$ $E = -0.42$ V

we would correctly predict that copper(II) ion should be reduced.

If the anode is made of a moderately reactive metal such as copper, the anode may be oxidized. The cell shown in -Figure 19.22 is an example. In this cell, the possible anode reactions are

oxidation of electrode: $Cu(s) \longrightarrow Cu^{2+} + 2e^-$ $E° = -0.34$ V

oxidation of solvent: $2H_2O \longrightarrow O_2(g) + 4H^+(10^{-7} M) + 4e^-$ $E = -0.82$ V

---

*Table 19.1 gives $+1.23$ V for $E°$ for the half-reaction $O_2(g) + 4H^+ + 4e^- \longrightarrow 2H_2O(l)$. The standard potential for the reverse reaction $2H_2O(l) \longrightarrow O_2(g) + 4H^+ + 4e^-$ is -1.23 V. However, the symbol ° means that solutions are 1 M. In pure water, the concentration of H^+ is 10^{-7} M, not 1 M. The half-cell potential for neutral water, -0.82 V, was calculated by means of the Nernst equation (Section 19.3).
[†]According to Table 19.1, the standard reduction potential for water ($[OH^-] = 1$ M) is -0.83 V. The half-cell potential for neutral water, -0.42 V, was calculated by means of the Nernst equation (Section 19.3). You should note the potentials for the two half-reactions, $O_2(g) + 4H^+(10^{-7} M) + 4e^- \longrightarrow 2H_2O(l)$, $E = +0.82$ V, and $2H_2O(l) + 2e^- \longrightarrow H_2(g) + 2OH^-(10^{-7} M)$, $E = -0.42$ V, in their proper places according to the values of E in Table 19.1.

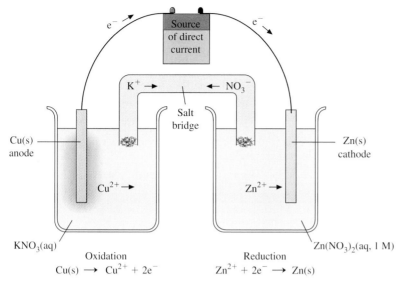

◉ The solute anion NO_3^- moves away from the cathode toward the anode.

Oxidation
$$Cu(s) \longrightarrow Cu^{2+} + 2e^-$$

Reduction
$$Zn^{2+} + 2e^- \longrightarrow Zn(s)$$

KNO₃(aq)

Zn(NO₃)₂(aq, 1 M)

- FIGURE 19.22 Electrolytic cell with reactive electrode.

Oxidation of the solute anion is not possible in this case because N in nitrate ion is already in its highest oxidation state, the +5 state. Even under standard conditions, copper is oxidized more easily than water. At the beginning of the electrolysis when the concentration of Cu^{2+} is less than 1 M, the tendency of copper to be oxidized will be greater still.

The possible cathode reactions are

reduction of solvent: $2H_2O(l) + 2e^- \longrightarrow H_2(g) + 2OH^-(10^{-7}\ M)$ $E = -0.42$ V

reduction of solute cation: $Zn^{2+} + 2e^- \longrightarrow Zn(s)$ $E° = -0.76$ V

Reduction of the zinc electrode is not possible because negative oxidation states are unknown for zinc (Section 11.2). From the half-cell potentials, we would predict that the solvent, water, should be reduced. However, the experimental fact is that zinc is deposited, which must be a result of the overpotential of hydrogen gas. The overall reaction is

$$Cu(s) + Zn^{2+} \xrightarrow{\text{electrolysis}} Cu^{2+} + Zn(s)$$

the reverse of the reaction in the voltaic cell in Figure 19.3.

To find the most easily reduced species in Table 19.1, start at the top of the left column and work down. For example, bromine is more easily reduced than iodine. To find the most easily oxidized species in Table 19.1, start at the bottom of the right column and work up. For example, Zn(s) is more easily oxidized than Fe(s). Sample Problem 19.4 gives another illustration of how to use the table of standard reduction potentials to predict the reactions that will take place during electrolysis.

SAMPLE PROBLEM

19.4 Predict the anode and cathode reactions for the electrolysis of aqueous H_2SO_4 of such concentration that $[H^+] = 1.00$ M using platinum electrodes.

Solution   The platinum electrodes are inert. Only reaction of solute ions and solvent need be considered. The hydrogen sulfate ion, HSO_4^-, is the predominant solute species. In HSO_4^-, the oxidation number of S is +6. Because +6 is the highest oxidation number observed for sulfur (Figure 11.1), sulfur in HSO_4^- ion

Top: *Hydrochloric acid reacting with a zinc strip.* Bottom: *Hydrochloric acid of the same concentration reacting with an iron strip. Hydrogen gas is formed much more rapidly with the zinc strip, showing that zinc is more easily oxidized than iron.*

A higher potential and carefully controlled conditions (high current density, temperature <30 °C, bright platinum electrodes, and a protected cathode) are necessary for HSO_4^- to be oxidized to peroxodisulfate ion.

◐ Why is reduction of HSO_4^- not a possibility?

cannot be oxidized any further. The probable anode reaction is oxidation of water:

$$2H_2O(l) \longrightarrow O_2(g) + 4H^+ + 4e^- \qquad E° = -1.23 \text{ V}$$

From Table 19.1, possible cathode reactions appear to be

$$2H^+ + 2e^- \longrightarrow H_2(g) \qquad \text{and} \qquad 2H_2O(l) + 2e^- \longrightarrow H_2(g) + 2OH^-$$

When $[H^+] = 1.00$ M, $[OH^-] = 1.00 \times 10^{-14}$, and $E°$ for both reactions is 0.00 V. Both lead to the same overall reaction, the electrolysis of water:

$$2H_2O(l) \xrightarrow{\text{electrolysis}} 2H_2(g) + O_2(g) \qquad E° = -1.23 \text{ V}$$

The negative sign of the cell potential for the predicted overall reaction shows that the reaction is not spontaneous. A potential of more than 1.23 V must be applied for electrolysis to take place. The only function of the sulfuric acid is to provide ions so that electricity can pass through the water.*

PRACTICE PROBLEM

19.15 Predict the anode, cathode, and overall reactions for each of the following cells: (a) 1.00 M LiBr, C electrodes (inert); (b) 1.00 M $AgNO_3$, Pt electrodes (inert); (c) 1.00 M KNO_3, Pt electrodes (inert); (d) 1.00 M $NiSO_4$, Ni electrodes. (For the reaction $Ni^{2+} + 2e^- \longrightarrow Ni(s)$, $E° = -0.26$ V.)

◉ In the Electrolysis module of the Electrochemistry section students can interactively examine the relationship between current, time, and mass of electrochemical products.

19.8 ## STOICHIOMETRY OF ELECTROCHEMICAL REACTIONS

The English chemist Michael Faraday's investigations of the relationship between atomic mass and the quantity of electricity needed to free a given amount of an element from a compound were an important step in the discovery of electrons (Section 7.1). Faraday found that the masses of the products formed by electrolysis and the masses of the reactants used up are directly proportional to the quantity of electricity transferred at the electrodes. The masses of the products and reactants also depend on the molar masses of the substances involved in the reaction.

The equations for the half-cell reactions show the relationship between quantity of electricity and quantities of reactants and products. For example, the equation for the half-reaction

$$Ag^+ + e^- \longrightarrow Ag(s)$$

says that one electron is needed to reduce one silver ion to a silver atom; 1 mol of electrons (6.02×10^{23} electrons) is needed to reduce 1 mol of silver ions to 1 mol of metallic silver. The equation for the half-reaction

$$Cu^{2+} + 2e^- \longrightarrow Cu(s)$$

says that two electrons are needed to reduce one copper(II) ion to a copper atom;

---

*Sulfuric acid is the electrolyte in the lead storage batteries used in cars. Electrolysis of water is a possible side reaction when a lead storage battery is charged. The mixture of hydrogen and oxygen produced by electrolysis of water can be exploded by sparks. If you have to jump start your car, you should make sure that the ground is connected to a part of the engine that is as far as possible from the battery and wear your safety goggles. In some modern batteries, the electrodes contain a little calcium metal, which prevents electrolysis of water. Because no hydrogen and no oxygen gas are formed, the battery can be sealed. No water needs to be added to replace water used up by electrolysis or lost by evaporation.

2 mol of electrons are needed to reduce 1 mol of copper(II) ions to 1 mol of metallic copper.

In Section 19.3, you learned that the charge on 1 mol of electrons is 9.65×10^4 C and that 9.65×10^4 C/mol is called a Faraday, F. Coulombs can either be measured directly or calculated from measurements of current and time. Equation 19.11 shows the relationship between coulombs, current, and time:

$$\text{coulombs} = \text{amperes} \times \text{seconds} \qquad \text{or} \qquad C = A \times s \qquad (19.11)$$

A current regulator is used to maintain a constant current.

Sample Problems 19.5 and 19.6 illustrate stoichiometric calculations for electrochemical reactions.

SAMPLE PROBLEMS

19.5 How many grams of chromium metal will be deposited from a solution that contains chromium(III) ions by a current of 1.24 A in 25.0 min? Assume that the theoretical yield of chromium metal will be obtained.

Solution The first step in solving any stoichiometry problem is always to write an equation for the reaction. The equation for the reduction of chromium(III) ion to chromium metal is

$$Cr^{3+} + 3e^- \longrightarrow Cr(s) \qquad (19.12)$$

Equation 19.12 says that 3 mol of electrons are needed to reduce 1 mol of Cr^{3+} ions to 1 mol of chromium metal. Next we must use equation 19.11, $C = A \times s$, and the relationship

$$1 \text{ mol } e^- \approx 9.65 \times 10^4 \text{ C}$$

to calculate how many moles of electrons are supplied by a current of 1.24 A in 25.0 min. From equation 19.11, A = C/s and C/s should be used instead of A so that units cancel. The time, 25.0 min, must be converted to seconds:

$$\text{mol } e^- = \left(1.24 \frac{C}{s}\right)\left(\frac{60 \text{ s}}{\text{min}}\right)25.0 \text{ min}\left(\frac{1 \text{ mol } e^-}{9.65 \times 10^4 \text{ C}}\right) = 0.0193$$

According to the equation for the reaction, 3 mol of electrons reduce 1 mol Cr^{3+} ions to 1 mol of chromium metal. From the table inside the front cover, the atomic mass of chromium is 51.9961 u. Therefore, 0.0193 mol e^- will reduce

$$0.0193 \text{ mol } e^- \left(\frac{1 \text{ mol Cr}}{3 \text{ mol } e^-}\right)\left(\frac{52.0 \text{ g Cr}}{\text{mol Cr}}\right) = 0.335 \text{ g Cr}$$

19.6 How long will it take to deposit 0.670 g of copper metal from a solution that contains copper(II) ions using a current of 1.24 A? Assume that a theoretical yield of copper metal will be obtained.

Solution We must use the equation for the half-reaction to calculate the number of moles of electrons needed and then use equation 19.11 to calculate the time required to supply this many moles of electrons.

The equation for the half-reaction is

$$Cu^{2+} + 2e^- \longrightarrow Cu(s)$$

From the table inside the front cover, the atomic mass of Cu is 63.546 u. Therefore,

$$0.670 \text{ g Cu}\left(\frac{1 \text{ mol Cu}}{63.5 \text{ g Cu}}\right)\left(\frac{2 \text{ mol } e^-}{1 \text{ mol Cu}}\right) = 0.0211 \text{ mol } e^- \qquad \text{and}$$

$$0.0211 \text{ mol } e^-\left(\frac{9.65 \times 10^4 \text{ C}}{\text{mol } e^-}\right) = 2.04 \times 10^3 \text{ C}$$

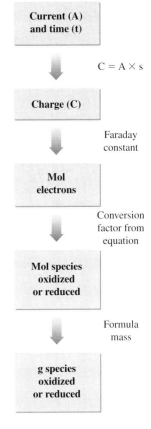

Current (A) and time (t)

$C = A \times s$

Charge (C)

Faraday constant

Mol electrons

Conversion factor from equation

Mol species oxidized or reduced

Formula mass

g species oxidized or reduced

19.8 STOICHIOMETRY OF ELECTROCHEMICAL REACTIONS

are needed. Solution of equation 19.11 for seconds and substitution gives

$$\frac{coulombs}{amperes} = seconds = \frac{2.04 \times 10^3 \, C}{1.24 \, C/s} = 1.65 \times 10^3 \, s$$

PRACTICE PROBLEMS

19.16   From a solution that contains nickel(II) ions, how many grams of nickel metal will be deposited by a current of 0.563 A in 1.25 h? Assume that the theoretical yield of nickel metal will be obtained.

19.17   How many amperes are necessary to deposit 0.324 g of silver metal from a solution that contains silver(I) ions in 5.00 min? Assume that a theoretical yield of silver metal will be obtained.

19.9   PRACTICAL APPLICATIONS OF ELECTROLYSIS

Reactive metals such as sodium, magnesium, and aluminum are produced by electrolysis; electrolysis is also used to purify less reactive metals such as copper. Electroplating of metals is an important method of preventing corrosion. In addition, many chemicals, both inorganic and organic, are manufactured by electrolytic processes.

Production of Active Metals

The most important method for the production of metallic sodium is electrolysis of molten sodium chloride. Lithium and the Group IIA metals are also produced by electrolysis of the molten chlorides. The reactions involved in the electrolysis of sodium chloride are

cathode reaction $Na^+(l) + e^- \longrightarrow Na(l)$

anode reaction $2Cl^-(l) \longrightarrow Cl_2(g) + 2e^-$

overall reaction $2Na^+(l) + 2Cl^-(l) \xrightarrow{electrolysis} 2Na(l) + Cl_2(g)$

Because liquid sodium metal and chlorine gas react vigorously to form sodium chloride (-Figure 19.23), the apparatus used for the electrolysis of sodium chloride must keep the liquid sodium and the chlorine gas apart. -Figure 19.24 is a schematic

- FIGURE 19.23   Liquid sodium metal and chlorine gas react vigorously.

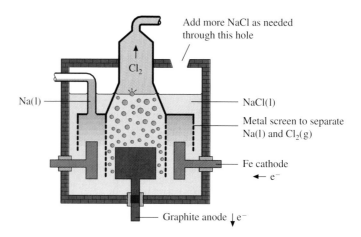

Add more NaCl as needed through this hole

Cl_2

Na(l)

NaCl(l)

Metal screen to separate Na(l) and $Cl_2(g)$

Fe cathode

$\leftarrow e^-$

Graphite anode $\downarrow e^-$

- FIGURE 19.24   Downs cell for production of sodium.

diagram of the apparatus, which is called a Downs cell after its inventor. Liquid sodium is insoluble in and less dense than molten sodium chloride. Sodium metal floats on molten sodium chloride and is drawn off continuously in the Downs cell. The liquid sodium metal must be protected from oxidation by oxygen in air.

The most familiar use of sodium metal is in sodium vapor street lights. Larger volume uses are as a reducing agent in the manufacture of dyes and drugs and in the production of less active metals from their ores. For example, titanium metal is obtained by the reaction

$$TiCl_4(g) + 4Na(l) \xrightarrow{heat} Ti(s) + 4NaCl(s)$$

Aluminum metal is produced by electrolysis of aluminum oxide, Al_2O_3. The melting point of pure aluminum oxide is greater than 2000 °C. The keys to successful electrolysis of aluminum oxide are a solvent that dissolves aluminum oxide to form a conducting solution at a temperature lower than 2000 °C and a powerful enough electric generator for direct current. The latter was available by the late nineteenth century. Two college students, Charles M. Hall at Oberlin College in Ohio and Paul-Louis-Toussaint Héroult in France, both became interested in producing aluminum inexpensively, and both solved the problem in the same way—use of molten cryolite, Na_3AlF_6, as a solvent for Al_2O_3—in 1886. Hall filed his patent application first and received the U.S. patent. The price of aluminum dropped from $100 a pound to $2 a pound, and an industry was born. Aluminum is light but strong and is a good conductor of heat and electricity. It does not corrode in air because a thin, tough, adherent surface layer of aluminum oxide quickly forms on fresh surfaces. Some familiar uses of aluminum are beverage cans, airplanes, pots and pans, and aluminum foil. Today more aluminum is used each year than any other metal except iron.

■Figure 19.25 shows a cell for the production of aluminum by the Hall–Héroult process. The electrode reactions are

cathode reaction $Al^{3+} + 3e^- \longrightarrow Al(l)$
anode reaction $2O^{2-} \longrightarrow O_2(g) + 4e^-$

The oxygen reacts with the carbon anode to yield mainly $CO_2(g)$ [and some $CO(g)$]

$$C(s) + O_2(g) \longrightarrow CO_2(g)$$

Thus, the overall reaction is chiefly

$$2Al_2O_3(l) + 3C(s) \longrightarrow 4Al(l) + 3CO_2(g)$$

Today about 5 million tons of 99.7% aluminum are produced in the United States each year. Production of aluminum uses about 5% of the total electricity generated.*

The Downs cell was patented in 1924.

Some reviewers questioned our use of sodium for the production of titanium. According to Greenwood, N. N.; Earnshaw, A. Chemistry of the Elements; Pergamon: Oxford, 1984, p 1113, the use of sodium instead of magnesium in the Kroll process requires little change in the basic process but gives a more readily leached product. This product yields titanium in a granular form that is preferred by some users. The most recent edition of the Kirk–Othmer Encyclopedia of Chemical Technology cites reduction of $TiCl_4$ as an important use of sodium metal.

Héroult also invented the electric-arc furnace, which is widely used in making steel, so he too was a very successful man.

Molten aluminum being poured.

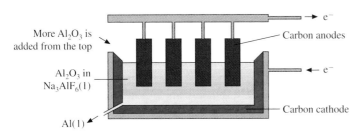

More Al_2O_3 is added from the top

Carbon anodes

Al_2O_3 in Na_3AlF_6(l)

e⁻

Carbon cathode

Al(l)

■ FIGURE 19.25   Aluminum reduction pot.

*Here are some data that may give you some idea of how much electrical energy is used to make aluminum. Electric bills are based on kilowatt-hours, kWh. A typical American household uses about 1000 kWh of electricity per month (at a cost of 10.5¢/kWh). One kWh is equivalent to 3.60×10^6 J or 8.60×10^5 cal. One kWh will light a 100-W light bulb for 10 h or run a hairdryer for about 50 min. About 11 kWh are required to make 1 lb of aluminum, the quantity in an average-sized roll of aluminum foil. Thus, some 110 billion kWh of electricity are used to produce aluminum in the United States each year.

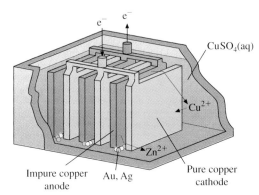

CuSO$_4$(aq)

Cu^{2+}

Zn^{2+}

Impure copper anode   Au, Ag   Pure copper cathode

- FIGURE 19.26   Purification of copper by electrolysis. At the present time, most of the silver produced is a by-product of the manufacture of other metals such as copper.

Much less electrical energy is needed to recycle aluminum cans and other scrap aluminum. Recycling also helps solve the solid-waste-disposal problem.

Purification of Less Active Metals

Although less reactive metals such as copper can be obtained from their ores by chemical reduction, the less reactive metals are often purified by electrolysis. Copper is an example. The largest use of copper is as an electrical conductor. Impurities reduce the conductivity of copper, and copper wiring made of impure copper gets dangerously hot. Impure copper (99%), which is to be purified, is made the anode in an electrolytic cell. Thin sheets of pure (99.98%) copper are used for the cathode; the electrolyte is an aqueous solution of copper(II) sulfate. -Figure 19.26 shows purification of copper by electrolysis. Copper(II) ions go into solution at the anode leaving electrons behind:

$$\text{anode reaction} \qquad Cu(s) \longrightarrow Cu^{2+} + 2e^-$$

Ions of impurities such as zinc that are easier to oxidize than copper also go into solution. Gold and silver are more difficult to oxidize than copper and do not dissolve. As copper in the anode dissolves, gold and silver are left behind and fall to the bottom of the cell. Electrons are pumped through the circuit from the anode to the cathode. The copper(II) ions move through the solution from the anode to the cathode where they are reduced to copper metal and deposited on the cathode:

$$\text{cathode reaction} \qquad Cu^{2+} + 2e^- \longrightarrow Cu(s)$$

Ions of more reactive metals such as Zn^{2+} also move to the cathode. However, if the voltage is correct, the ions of the more reactive metals are not reduced and remain in solution. Overall, copper does not undergo any reaction; it is simply transferred from the anode to the cathode. The impure copper anode becomes smaller and the pure copper cathode becomes larger as the electrolysis proceeds. Nickel can be purified in a similar way. Many other metals, for example, silver and gold, are also purified by electrolysis.

Electroplating

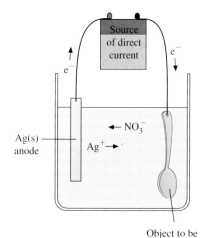

Source of direct current

e$^-$

e$^-$

Ag(s) anode

Ag$^+$ →

← NO$_3^-$

Object to be plated is made the cathode.

- FIGURE 19.27   Electroplating silver onto a spoon.

Deposition of a thin coating (of the order of 10^{-3} to 10^{-4} mm thick) of metal on the cathode is called **electroplating.** The thin coating of metal protects the metal underneath from corrosion and is decorative. For example, jewelry is often gold plated and table utensils, silver plated. -Figure 19.27 shows a spoon being electroplated. If a solution of silver nitrate alone is used as the electrolyte, silver plates out too rapidly and does not stick very well to the object being plated. Cyanide ion, CN^-, is usually added to the electrolyte solution to reduce the concentration of silver ion by complex formation:

$$Ag^+ + 2CN^- \rightleftharpoons Ag(CN)_2^-$$

A low concentration of silver ion results in a smooth plate that adheres firmly to the object being plated.

Synthesis

Electrolytic cells are used for the production of a number of compounds, the most important of which are sodium hydroxide and adiponitrile, $NC(CH_2)_4CN$. Adiponitrile is used in the manufacture of nylon.

Sodium hydroxide is made by electrolysis of aqueous sodium chloride with inert electrodes; chlorine and hydrogen are by-products. The electrolysis of sodium chloride solution was discussed earlier (Figure 19.20). The cell reactions are

anode reaction	$2Cl^- \longrightarrow 2e^- + Cl_2(g)$
cathode reaction	$2H_2O(l) + 2e^- \longrightarrow H_2(g) + 2OH^-$
overall reaction	$2Cl^- + 2H_2O(l) \xrightarrow{\text{electrolysis}} H_2(g) + Cl_2(g) + 2OH^-$

Because chlorine reacts with hydrogen to form hydrogen chloride and with hydroxide ion to form hypochlorite ion, OCl^-, the anode and cathode must be separated. In the most modern plants, a cation exchange resin that lets Na^+, but not Cl^- or OH^-, pass is used between anode and cathode compartments as shown schematically in ‑Figure 19.28. A concentrated solution of sodium chloride is used in commercial cells to favor oxidation of chloride ion over oxidation of water. The sodium hydroxide solution is drained from the cathode compartment, and water is evaporated to obtain solid sodium hydroxide.

The electrolysis of aqueous sodium chloride is second only to the production of aluminum as a consumer of electricity in the United States (about 0.5% of the electricity generated). Sodium hydroxide is used to neutralize acids in a number of industrial processes and to make soap, textiles, and paper. Hydrogen is used to make ammonia, to convert vegetable oils to margarine, and in the manufacture of organic chemicals. Manufacture of organic chemicals, such as vinyl chloride ($H_2C{=}CHCl$), which is used to make the plastic "vinyl," is the main use of chlorine. Large quantities of chlorine are also used as a bleach and to purify water. The value of the electrolysis of aqueous sodium chloride to the U.S. economy in 1980 was about $5 billion.

Silver-plated spoons immediately after being removed from the cyanide-containing plating solution.

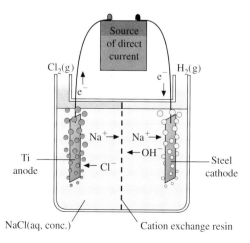

‑ FIGURE 19.28 Cell for production of NaOH, Cl_2, and H_2 by electrolysis of aqueous NaCl.

Electrochemistry deals with the use of spontaneous chemical reactions to supply electrical energy and the use of electrical energy to make nonspontaneous chemical reactions take place. In **voltaic** or **galvanic cells,** spontaneous chemical reactions are used to generate electricity. The oxidation and reduction reactions must be physically separated but connected by a wire and an electrolyte. An **electric current** is a flow of electric charge; charge is carried through an electrolyte by movement of ions and through a wire by movement of electrons. Reaction takes place at the surface of the **electrodes,** the conductors that carry charge into and out of an electrolyte. The electrode where oxidation takes place is named the **anode,** and the electrode where reduction takes place is named the **cathode.** The reactions that take place at electrodes are called **half-cell reactions** or **half-reactions.**

The **electromotive force (emf),** E, (or **cell potential** or **cell voltage**) of a cell is the difference in electric potential energy between the two half-cells that make up the cell. Potentials are measured in **volts;** $1\ V = 1\ J/C$. The **standard cell potential,** $E°$, is the cell potential when the cell is operating under standard conditions. By agreement among scientists, the voltage for the half-reaction

$$2H^+ + 2e^- \longrightarrow H_2(g)$$

or its reverse under standard conditions is exactly zero. A half-cell that contains one molar hydrogen ion and hydrogen gas at a partial pressure of 1 atm is called a **standard hydrogen electrode,** SHE. The potentials in tables of standard reduction potentials are measured with respect to the standard hydrogen electrode as the cathode:

$$2H^+ + 2e^- \longrightarrow H_2(g)$$

A positive sign for the voltage of a half-reaction shows that the half-reaction has a greater tendency to take place than the reduction of hydrogen ion under standard conditions; a negative sign shows that a half-reaction has a lesser tendency to take place than the reduction of hydrogen ion. Whether or not a given oxidation–reduction reaction will be spontaneous under standard conditions can be predicted on the basis of the **diagonal relationship** of the species to each other in the table of standard reduction potentials. When standard reduction potentials are listed in order of decreasing value, any species on the reactant side of a half-reaction equation will oxidize any species on the product side that is lower in the table *provided that reaction takes place at a significant rate.*

Standard reduction potentials can be used to calculate standard cell potentials:

$$E°_{cell} = E°_{half\text{-}cell\ of\ reduction} - E°_{half\text{-}cell\ of\ oxidation}$$

Cell potentials under nonstandard conditions can be calculated

from standard cell potentials by means of the **Nernst equation,**

$$E = E° - \frac{RT}{nF}\ln Q$$

where E is the potential of the cell under nonstandard conditions, $E°$ is the potential of the cell under standard conditions, R is the gas constant, T is the temperature in kelvin, Q is the reaction quotient, n is the number of moles of electrons transferred, and F is the Faraday constant. The **Faraday constant** is the charge on 1 mol of electrons, $9.65 \times 10^4\ C$. Reactions that have positive potentials are spontaneous. A **concentration cell** produces an electric current as a result of a difference in concentration between solutions in the two half-cells.

Free energy is related to cell potential by the equation

$$\Delta G = -nFE$$

which can be used to calculate $\Delta G°$ from experimental values for $E°$ and to calculate values for $E°$ that cannot be found experimentally from tabulated values of $\Delta G°_f$. By means of the relationship

$$E° = \frac{RT}{nF}\ln K$$

equilibrium constants can be found from measurements of $E°$.

Batteries are compact, portable voltaic cells that store energy ready to be used whenever and wherever needed. **Fuel cells** are voltaic cells that continuously convert chemical energy into electrical energy. The spontaneous redox reactions that take place in batteries and fuel cells are very useful. On the other hand, the spontaneous redox reactions involved in **corrosion,** the deterioration of metals by chemical reactions, are extremely destructive. Metals can be protected against corrosion by painting, plating, equipping with sacrificial anodes, or alloying.

Oxidation–reduction reactions that are nonspontaneous can be made to take place by supplying electrical energy. An electrochemical cell in which a nonspontaneous reaction is made to take place is called an **electrolytic cell.** The process that takes place in an electrolytic cell is called **electrolysis.** As usual, the reaction that takes place depends on both thermodynamics and kinetics. The cell potential for a cell is the minimum voltage necessary to bring about electrolysis; additional voltage is required to overcome the resistance of the cell and overpotential. **Overpotential** is the voltage needed to make slow reactions take place at a practical rate. Electrolytic cells are used to produce active metals, to refine metals, for electroplating, and for synthesis of both inorganic and organic compounds.

The masses of products formed and reactants used up in electrochemical reactions are directly proportional to the quantity of electricity transferred at the electrodes and to the molar masses of the substances involved in the reaction.

For information about the organization of Additional Practice Problems, Stop & Test Yourself, Putting Things Together, and Applications, see the beginnings of these sections in Chapter 1.

19.18 (a) Make a schematic drawing of a voltaic cell that uses the spontaneous reaction between metallic zinc and nickel(II) nitrate solution to form metallic nickel and zinc(II) nitrate solution. Label all parts of the cell and write net ionic equations for the half-reactions that take place at each electrode. (b) Write the cell diagram. (c) Write net ionic and molecular equations for the overall reaction. (19.1)

19.19 Classify each of the following as either an oxidizing agent or a reducing agent or both: (a) OCl^- (b) $Ca(s)$ (c) Cr^{2+} (d) H_2O_2 (e) HO_2^- (19.2)

19.20 Which member of each of the following pairs is the stronger oxidizing agent? (a) $Br_2(l)$ or $Cl_2(g)$ (b) MnO_4^- in acidic solution or MnO_4^- in basic solution (c) MnO_4^- in acidic solution or $Cr_2O_7^{2-}$ in acidic solution (19.2)

19.21 Arrange the following reducing agents in order of increasing strength:

$$H_2(g) \qquad I^-(aq) \qquad Na(s) \qquad Zn(s) \quad (19.2)$$

19.22 Will silver metal dissolve in 1 M HCl? Explain your answer. (19.2)

19.23 From the table of standard reduction potentials choose, (a) an oxidizing agent that can oxidize $Fe(s)$ to Fe^{2+} but not $Cu(s)$ to Cu^{2+} (b) a reducing agent that can reduce Cl_2 to Cl^- but not I_2 to I^-. (19.2)

19.24 From the fact that the following reactions are spontaneous at 25 °C with 1 M concentrations of all species in solution,

$$Pt(s) + Au^{3+} \longrightarrow Pt^{2+} + Au(s)$$
$$Ag(s) + Pd^{2+} \longrightarrow Ag^+ + Pd(s)$$
$$Pd(s) + Pt^{2+} \longrightarrow Pd^{2+} + Pt(s)$$

predict whether each of the following reactions will be spontaneous under the same conditions:
(a) $Ag(s) + Au^{3+} \longrightarrow Ag^+ + Au(s)$
(b) $Au(s) + Pd^{2+} \longrightarrow Au^{3+} + Pd(s)$
(c) $Ag(s) + Pt^{2+} \longrightarrow Ag^+ + Pt(s)$ (19.2)

19.25 Given the following standard reduction potentials

$$2HIO_3(aq) + 10H^+ + 10e^- \longrightarrow I_2(s) + 6H_2O(l)$$
$$E° = 1.20 \text{ V}$$
$$ClO_3^- + 6H^+ + 6e^- \longrightarrow Cl^- + 3H_2O$$
$$E° = 1.45 \text{ V}$$

(a) what is the standard potential for the following reaction?

$$3I_2(s) + 5ClO_3^- + 3H_2O(l) \longrightarrow 6HIO_3(aq) + 5Cl^-$$

(b) Is the reaction spontaneous? Explain your answer. (19.2)

19.26 Given the following equations for half-reactions with their standard potentials,

$$Cu^{2+} + 2e^- \longrightarrow Cu(s) \qquad E° = 0.34 \text{ V}$$
$$Ga^{3+} + 3e^- \longrightarrow Ga(s) \qquad E° = -0.53 \text{ V}$$
$$Pd^{2+} + 2e^- \longrightarrow Pd(s) \qquad E° = 0.92 \text{ V}$$

(a) write net ionic equations for all the spontaneous combinations and calculate $E°$ for each. (b) Which combination leads to the largest emf? (c) Which combination leads to the smallest emf? (19.2)

19.27 Calculate $E°$ for each of the following reactions and tell whether the forward or the reverse reaction will be spontaneous under standard conditions at 25 °C:
(a) $Cl_2(g) + Cu(s) \longrightarrow Cu^{2+} + 2Cl^-$
(b) $Ag(s) + Cu^{2+} \longrightarrow Ag^+ + Cu(s)$
(c) $O_2(g) + H^+ + Fe^{2+} \longrightarrow Fe^{3+} + H_2O(l)$
(d) $H_2(g) + Zn^{2+} \longrightarrow Zn(s) + 2H^+$
(e) $MnO_4^- + H^+ + Br^- \longrightarrow Mn^{2+} + Br_2(l)$ (19.2)

19.28 For the reaction $Ag(s) + Pt^{2+} \longrightarrow Ag^+ + Pt(s)$, $E° = 0.39$ V. What is $E°$ for the half-reaction $Pt^{2+} + 2e^- \longrightarrow Pt(s)$? (19.2)

19.29 How could the emf of the cell $Cu^{2+} + H_2(g) \longrightarrow Cu(s) + 2H^+$ be increased? (19.3)

19.30 The value of $E°$ for the decomposition of $H_2Se(aq)$ is $+0.03$ V under standard conditions at 25 °C. Predict the effect of each of the following changes on the equilibrium

$$H_2Se(aq) \rightleftharpoons Se(s) + H_2(g)$$

(a) increasing $[H_2Se]$ (b) decreasing the pressure of $H_2(g)$ (c) decreasing the quantity of $Se(s)$ on the bottom of the container (19.3)

19.31 As a voltaic cell operates, does E increase, decrease, or remain the same? Explain your answer. (19.3)

19.32 What is the potential of the following cell at 25 °C? (19.3)

$$Al(s) \mid Al^{3+}(aq, 1.25 \text{ M}) \parallel Ag^+(aq, 0.050 \text{ M}) \mid Ag(s)$$

19.33 For the reaction $MnO_4^- + 8H^+ + 5Ag(s) \longrightarrow Mn^{2+} + 5Ag^+ + 4H_2O(l)$, $E° = 0.71$ V. (a) What is E when $[H^+] = 0.100$ M and the concentrations of all other species remain at 1.00 M? (b) What is E when $[MnO_4^-] = 0.100$ M and the concentrations of all other species remain at 1.00 M? (19.3)

19.34 For the reaction

$$5SO_4^{2-} + 4H^+ + 2Mn^{2+} + 3H_2O(l) \longrightarrow 5H_2SO_3(aq) + 2MnO_4^-$$

$E = 0.11$ V when $[H_2SO_3] = [MnO_4^-] = 0.10$ and $[SO_4^{2-}] = [H^+] = [Mn^{2+}] = 1.0$ M. What is the value of $E°$ for this reaction? (19.3)

19.35 For the half-reaction $CuO(s) + H_2O(l) + 2e^- \longrightarrow Cu(s) + 2OH^-$, $E° = -0.29$ V. (a) What is the value of E in pure water when $[OH^-] = 1.00 \times 10^{-7}$? (b) At what $[OH^-]$ is $E = 0.00$ V? (19.3)

19.36 The $[H^+]$ in a solution was found by measuring the emf of a cell composed of a silver electrode in a 1.00 M solution of Ag^+ and a hydrogen electrode in the solution of unknown hydrogen ion concentration. If the emf was 1.28 V when the pressure of hydrogen gas was 1.00 atm, what was $[H^+]$? (19.3)

751

19.37 The cell described in Problem 19.36 can be used to determine $[Ag^+]$ if the hydrogen electrode is a standard hydrogen electrode. If the observed emf was 0.54 V, what was $[Ag^+]$? (19.3)

19.38 (a) Sketch a concentration cell with zinc electrodes in 0.100 and 0.001 00 M $Zn(NO_3)_2$(aq). (b) Calculate E for this cell. (19.3)

19.39 The standard cell potential for the cell

$$Cd(s) \mid Cd^{2+}(aq) \parallel Co^{2+}(aq) \mid Co(s)$$

at 25 °C is +0.126 V. What is $\Delta G°$ for the reaction Cd(s) + $Co^{2+} \longrightarrow Cd^{2+} + Co(s)$? (19.4)

19.40 For the reaction $2VO_2^+ + 4H^+ + Zn(s) \longrightarrow 2VO^{2+} + Zn^{2+} + 2H_2O(l)$, $\Delta G° = -340.3$ kJ. What is the value of $E°$? (19.4)

19.41 For the reaction $3Fe^{2+} \longrightarrow Fe(s) + 2Fe^{3+}$, $E° = -1.21$ V at 25 °C. What is the value of the equilibrium constant for this reaction? (19.4)

19.42 (a) What three factors determine the potential of a battery? (b) What does "connected in series" mean? (c) Why are two or more cells often connected in series in a battery? (d) What determines the total number of coulombs that a battery can deliver? (19.5)

19.43 (a) What is corrosion? (b) List four methods that are commonly used to protect metals from corrosion and briefly discuss each of them. (19.6)

19.44 (a) From the table of standard reduction potentials, which would you expect to corrode more easily, iron or aluminum? Explain your answer. (b) Why does aluminum foil not corrode as rapidly as sheets of iron would? (19.6)

19.45 When water that contains a little Na_2SO_4 is electrolyzed using inert electrodes, (a) what happens to the $[H^+]$ in the water around the anode? (b) What happens to the $[H^+]$ in the water around the cathode? (c) What is the minimum voltage required for the electrolysis of water at 25 °C? (d) Why would a higher voltage than this be required in practice? (e) What is meant by the term inert electrode? (19.7)

19.46 (a) Sketch a cell for the electrolysis of LiCl(l). Lithium chloride melts at 605 °C and lithium metal at 181 °C. (b) Would the same products be obtained by the electrolysis of LiCl(aq)? Explain your answer. (c) Why must LiCl be melted or dissolved in water to be electrolyzed? (19.7)

19.47 Predict the anode, cathode, and overall reactions for the electrolysis of the following aqueous solutions under standard conditions: (a) KI (inert electrodes) (b) $Ca(NO_3)_2$ (inert electrodes) (c) $Fe_2(SO_4)_3$ (inert electrodes) (d) $Cu(NO_3)_2$ with silver anode and copper cathode. (19.7)

19.48 How many grams of I_2(s) will be deposited from a solution that contains I^- by a current of 2.5 A in 45 min? Assume that a theoretical yield is obtained. (19.8)

19.49 How many seconds will be required to deposit 0.36 g of Au from a solution that contains Au^{3+} ions if a 0.25-A current is used? (19.8)

19.50 How many coulombs are required to form 2.8 g of Cl_2(g) from a solution that contains Cl^-? (19.8)

19.51 How many amperes are required to deposit 0.72 g of Cr from a solution that contains Cr^{3+} in 30.0 min? (19.8)

19.52 (a) What is electroplating? (b) In silver plating, why is CN^- added to the electrolyte solution that contains silver ion? (19.9)

19.53 Draw microscopic views showing what happens at the anode and cathode during electrolysis of NaCl(l) in a Downs cell. Show only a few representative particles. (19.9)

19.54 How could you bring about each of the following conversions? (a) Fe^{2+} to Fe^{3+} (b) Fe^{3+} to Fe^{2+} (c) Fe(s) to $Fe(OH)_3$(s) (d) Br^- to Br_2 (e) Cu^{2+} to Cu(s)

19.55 Three electrolysis cells are connected in series so that the same quantity of charge flows through each cell. If 3.68 g of silver are deposited from a solution that contains Ag^+ in the first cell, how many grams of copper will be deposited in a second cell containing Cu^{2+}, and how many grams of gold will be deposited in a third cell containing Au^{3+}?

19.56 Given the following half-reactions of reduction and their standard potentials,

$$2HNO_2(aq) + 4H^+ + 4e^- \longrightarrow$$
$$N_2O(g) + 3H_2O(l) \qquad E° = 1.30 \text{ V}$$
$$NO_3^- + 3H^+ + 2e^- \longrightarrow$$
$$HNO_2(aq) + H_2O(l) \qquad E° = 0.94 \text{ V}$$

(a) write the net ionic equation for the spontaneous reaction between HNO_2 and I^-. Is HNO_2 acting as an oxidizing agent or as a reducing agent in this reaction? (b) Write the net ionic equation for the spontaneous reaction between HNO_2 and MnO_4^-. Is HNO_2 acting as an oxidizing agent or as a reducing agent in this reaction? (c) Explain why HNO_2 can behave in this way.

19.57 For each of the following compounds, tell whether the compound will be stable in 1 M solution at 25 °C or will the two ions undergo a redox reaction with each other? (a) AgF (b) $FeBr_3$ (c) FeI_3

19.58 Electrolysis can be used to separate metals from each other— for example, gold and zinc are separated from copper by electrolysis (see Figure 19.26 and accompanying discussion). Which of the following pairs of ions can be separated by electrolysis of their aqueous solutions under standard conditions, and which cannot? Explain your answers. (a) Ca^{2+} and Fe^{3+} (b) Al^{3+} and Cu^{2+} (c) Cr^{3+} and Zn^{2+} (d) Li^+ and Na^+

19.59 Electrolysis of a solution containing an unknown metal ion with a 3+ charge with a current of 0.3001 A for 0.7500 h deposited 0.4036 g of the metal on the cathode. Identify the metal ion.

19.60 A solution containing iron was electrolyzed with a 0.75-A current for 30.0 min, and 0.3906 g of iron was deposited on the cathode. What was the charge on the iron ion?

19.61 If the following oxidation–reduction reactions involving A, B, and C are spontaneous

$$A + B^+ \longrightarrow A^+ + B$$
$$C + A^+ \longrightarrow C^+ + A$$
$$C + B^+ \longrightarrow B + C^+$$

what would be the order of the half-reactions of reduction in Table 19.1?

19.62 In an experiment to verify the formula of the mercury(I) ion, a student found that the voltage of the concentration cell

$$Hg(l) \mid Hg(I) \text{ (aq, } 10^{-5} \text{ M)} \parallel Hg(I) \text{ (aq, } 10^{-1} \text{ M)} \mid Hg(l)$$

is 115 mV. show that this result proves that the formula is Hg_2^{2+}, not Hg^+.

1. Which of the following is the strongest reducing agent?
   (a) $Cl_2(g)$ (b) $H_2(g)$ in acidic solution (c) $KI(aq)$ (d) $Zn(s)$
   (e) $ZnCl_2(aq)$

2. Sodium dichromate, $Na_2Cr_2O_7$, in acidic solution can oxidize all of the following except _____.
   (i) Br^- (ii) $Cu(s)$ (iii) Cu^+ (iv) Cu^{2+} (v) Mn^{2+}
   (a) i (b) ii (c) iv (d) v (e) iv and v

3. For the reaction $2MnO_4^- + 10Br^- + 16H^+ \longrightarrow$
   $2Mn^{2+} + 5Br_2(l) + 8H_2O(l)$, $E° =$ _____ V.
   (a) -2.33 (b) -1.62 (c) 0.44 (d) 2.58 (e) 8.37

4. The emf of the reaction in Problem 3 will be increased by
   (a) removing $Br_2(l)$ (b) adding OH^- (c) increasing $[Mn^{2+}]$
   (d) diluting the solution (e) increasing $[Br^-]$

5. For the cell
   $$Au(s) \mid Au^{3+}(aq, 0.025 \text{ M}) \parallel Au^{3+}(aq, 0.50 \text{ M}) \mid Au(s)$$
   $E =$ _____ V. (a) 0.0770 (b) 0.0257 (c) 0.0111
   (d) -0.0257 (e) -0.0770

6. For the reaction
   $$4Zn(s) + 10H^+ + NO_3^- \longrightarrow$$
   $$4Zn^{2+} + NH_4^+ + 3H_2O(l), \Delta G° = -1268.2 \text{ kJ}$$

   what is $E°$ (to three significant figures)?
   (a) 6.57 V (b) 1.64 V (c) $0.006\ 57$ V (d) $0.001\ 64$ V
   (e) -6.57 V

7. For the reaction $Cl_2(g) + H_2O(l) \longrightarrow H^+ + Cl^- + HOCl(aq)$,
   $E° = -0.272$ V at 25 °C. What is K?
   (a) 6×10^{-22} (b) 3×10^{-11} (c) 6×10^{-10}
   (d) 3×10^{-5} (e) 6×10^{-2}

8. If $CaCl_2(aq)$ is electrolyzed using inert electrodes,
   (a) $H_2(g)$ is produced at the cathode and $Cl_2(g)$ at the anode.
   (b) $Cl_2(g)$ is produced at the cathode and $H_2(g)$ at the anode.
   (c) $H_2(g)$ is produced at the cathode and $O_2(g)$ at the anode.
   (d) $Ca(s)$ is produced at the cathode and $Cl_2(g)$ at the anode.
   (e) $Cl_2(g)$ is produced at the cathode and $Ca(s)$ at the anode.

9. The minimum voltage required to electrolyze $CuCl_2(aq)$ under standard conditions at 25 °C is _____.
   (a) 1.70 V (b) 1.36 V (c) 1.02 V (d) 0.85 V (e) 0.34 V

10. How many grams of nickel will be plated out of a solution that contains nickel ion by a 0.25-A current in 37 min?
    (a) 0.0028 g (b) 0.11 g (c) 0.17 g (d) 0.33 g
    (e) Nickel cannot be plated out.

PUTTING THINGS TOGETHER

19.63 (a) Make a schematic drawing of the cell shown by the cell diagram. Label all parts of the cell and write net ionic equations for the half-reactions that take place at each electrode. (b) Write the net ionic equation for the overall reaction

$$Ni(s)|Ni^{2+}(aq, 1 \text{ M})\|Cu^{2+}(aq, 1 \text{ M})|Cu(s)$$

19.64 (a) What is the difference between a voltaic cell and an electrolytic cell? (b) Why must the cells in which the two half-reactions take place always be separated in a voltaic cell? (c) Why must the cells in which the two half-reactions take place sometimes be separated in an electrolytic cell? (The cell for the production of NaOH, Cl_2, and H_2 by the electrolysis of NaCl(aq) shown in Figure 19.28 is an example.) (d) What is the purpose of a salt bridge? (e) Why must direct current be used for electrolysis?

19.65 In Section 19.2 we saw that cell potentials for half-reactions of reduction (or half-reactions of oxidation) must not be added. However, the cell potential for a half-reaction that is the sum of two half-reactions can be obtained by adding the free energy changes for the two half-reactions. (a) Calculate $\Delta G°$ for each of the following half-reactions from the standard reduction potentials:

$$Cu^{2+} + e^- \longrightarrow Cu^+ \qquad E° = 0.16 \text{ V}$$
$$Cu^+ + e^- \longrightarrow Cu(s) \qquad E° = 0.52 \text{ V}$$

(b) Add the values of $\Delta G°$ from part (a) and use the total to calculate $E°$ for the half-reaction $Cu^{2+} + 2e^- \longrightarrow Cu(s)$, which is the sum of the other two half-reactions. (c) Look closely at the three voltages and see if you can find an easier way to obtain the third one from the first two.

19.66 If the measured value of E for the cell

$$Ga(s)|Ga^{3+}(aq, 0.0050 \text{ M})\|H^+(aq, 0.0100 \text{ M})|H_2(g, 1\text{atm})|Pt$$

is $+0.46$ V, what is $E°$ for the reaction $2Ga(s) + 6H^+ \longrightarrow 2Ga^{3+} + 3H_2(g)$?

19.67 What is ΔG in kJ for the reaction $Cl_2(g) + 2Br^- \longrightarrow Br_2(l) + 2Cl^-$ if $[Br^-] = 0.100$, $[Cl^-] = 0.50$, and $p_{Cl_2} = 1.00$ atm? The temperature is 25 °C.

19.68 Because use of a standard hydrogen electrode is inconvenient, other reference electrodes are commonly used. The saturated calomel electrode is a popular choice. In a saturated calomel electrode, mercury metal is in equilibrium with a solution that is saturated with both mercury(I) chloride (Hg_2Cl_2—commonly called calomel) and potassium chloride. A saturated calomel electrode has $E° = +0.24$ V at 25 °C. (a) Will the saturated calomel electrode be the anode or the cathode when compared with a standard hydrogen electrode? (b) What potential will be observed for a Cu^{2+}/Cu cell that is compared with a saturated calomel electrode?   (c) Will the Cu^{2+}/Cu half-cell be the anode or the cathode in the cell?   (d) If a potential of 0.28 V is observed for a M^{3+}/M half-cell that acts as the anode when compared with a saturated calomel electrode, what is $E°$ for the M^{3+}/M half-cell?   (e) Assuming that this half-cell is listed in Table 19.1, what element is M?

19.69 (a) For the reaction $Cr_2O_7^{2-} + Cl^- \longrightarrow Cr^{3+} + Cl_2(g)$, what is the value of $E°$ at 25 °C?   (b) Is this reaction a good way to oxidize Cl^- to Cl_2 under standard conditions? Explain your answer.   (c) How could you increase the percent conversion of Cl^- to Cl_2?

19.70 Suppose that a strip of iron is placed in a 1.00 M solution of Cd^{2+} at 25 °C and allowed to stand until equilibrium is reached. (a) What will $[Fe^{2+}]$ and $[Cd^{2+}]$ be at equilibrium? (b) What percentage of the Cd^{2+} originally present will have plated out on the iron strip? ($E°$ for the Cd^{2+}/Cd half-cell is -0.4025 V. Assume that $E°$ for the Fe^{2+}/Fe half-cell is -0.4400 V.)

19.71 A hydrogen electrode immersed in 1.00 M HCl and a hydrogen electrode immersed in 1.00 M NaOH are combined to form a voltaic cell. (a) What is E_{cell}?   (b) If 1.00 M NH_3 is substituted for the 1.00 M NaOH, will E_{cell} be lower or higher? Explain your answer.   (c) For the voltaic cell with the hydrogen electrode immersed in 1.00 M NH_3, what is the value of E_{cell}?

19.72 Suppose that two half-cells—an Ag electrode in 1.00 M Ag^+ and a Cu electrode in 1.00 M Cu^{2+}—are connected to make a voltaic cell. (a) What will the overall cell reaction be?   (b) What will $E°$ be?   (c) Sketch the cell and write the cell diagram.   (d) If there is 100.0 mL of solution in each cell and the cell is allowed to run for 8.00 h at 0.20 A, what will $[Ag^+]$, $[Cu^{2+}]$, and E be?   (e) If there is 100.0 mL of solution in each cell and the cell is allowed to run for 24.00 h at 0.20 A, what will $[Ag^+]$, $[Cu^{2+}]$, and E be?   (f) What will $[Ag^+]$, $[Cu^{2+}]$, and E be when the cell has reached equilibrium, and how long will it take for the cell to reach equilibrium?   (g) Sketch a graph of E against time and describe how E changes with time.

19.73 From the value of the Faraday constant and Avogadro's number, calculate the charge on an electron in coulombs.

19.74 In an experiment like the one shown in Figure 1.3, 20.4 mL of $H_2(g)$ and 10.2 mL of $O_2(g)$ were collected after water that contained a little sulfuric acid was electrolyzed for 45.0 min. The barometric pressure was 765.2 mmHg, and the laboratory temperature was 23.4 °C.   (a) What volumes would the $H_2(g)$ and $O_2(g)$ have occupied if they had been dry?   (b) How many coulombs of electricity passed through the apparatus? (c) How many amperes was the current?

19.75 Just as acids stronger than H_3O^+ and bases stronger than OH^- can't exist in aqueous solution, the half-reactions

$$2H_2O(l) \rightleftharpoons O_2(g) + 4H^+ + 4e^-$$
$$2H_2O(l) + 2e^- \rightleftharpoons H_2(g) + 2OH^-$$

limit the voltages of the reactions that can take place in aqueous solutions. Oxidizing agents stronger than oxygen gas oxidize water, and reducing agents stronger than water reduce

water. (a) What is $E°$ for the half-reaction that involves $O_2(g)$? (b) What is standard pressure for $O_2(g)$ and $H_2(g)$ in electrochemistry?   (c) What is E for the half-reaction that involves $O_2(g)$ when the partial pressure of $O_2(g)$ is only 0.21 atm (as it is in air) instead of standard pressure?   (d) Equations for the ionization of acids such as

$$CH_3COOH(aq) + H_2O(l) \rightleftharpoons H_3O^+ + CH_3COO^-$$

can be regarded as combinations of two half-reactions (from Table 15.3). Divide the equation for the ionization of acetic acid into half-reactions.   (e) What other analogies are there between Table 15.3 and Table 19.1?

19.76 Entropy changes are more difficult to measure experimentally than enthalpy changes. One way of finding entropy changes is to calculate them from the results of measurements of standard potentials and enthalpy changes. (a) For the reaction

$$Cr_2O_7^{2-} + 6Fe^{2+} + 14H^+ \longrightarrow 2Cr^{3+} + 6Fe^{3+} + 7H_2O(l)$$

what is the value of $E°$?   (b) Use the value of $E°$ from part (a) to calculate $\Delta G°$.   (c) For this reaction, $\Delta H° = -759$ kJ. What is $\Delta S°$ for the reaction?

19.77 (a) Derive an equation for $E°$ as a function of Kelvin temperature by combining the equations $\Delta G° = \Delta H° - T\Delta S°$ and $\Delta G° = -nFE°$.   (b) What property is needed to make a battery with a voltage that does not change much as temperature changes?

19.78 The sum of the two half-reactions

$$Ag(s) \longrightarrow Ag^+ + e^-$$
$$AgCl(s) + e^- \longrightarrow Ag(s) + Cl^- \qquad E° = 0.22 \text{ V}$$

is the solubility equilibrium for AgCl. (a) What is $E°$ for the AgCl solubility equilibrium?   (b) What is the value of K_{sp} for AgCl?

19.79 For the cell

$$Ag(s)|Ag^+(aq, \text{sat'd AgBr})\|Ag^+(aq, 0.100 \text{ M})|Ag(s)$$

the measured value of E is 0.305 V. (a) What is $[Ag^+]$ in saturated AgBr?   (b) What is the value of K_{sp} for AgBr? (c) Sketch the cell.

19.80 (a) How is charge transferred through NaCl(aq) [or through NaCl(l)]?   (b) Why must NaCl(s) be melted or dissolved in water to be electrolyzed?   (c) Why can't pure water be electrolyzed?   (d) Suggest some factors that might affect the rate of transfer of charge through NaCl(aq).   (e) The rate of transfer of charge through solutions of electrolytes usually increases as the temperature is raised. Explain why.   (f) How is charge transferred through Cu(s)?   (g) The rate of transfer of charge through metals decreases as the temperature is raised. Explain why.   (h) Why must reactions take place at the surfaces of the electrodes in order for a circuit to be complete?

19.81 Both aluminum and zinc have been considered as anode materials in air batteries—batteries that use oxygen from air as the oxidizing agent. The reactions that take place are

$$4Al(s) + 3O_2(g) \longrightarrow 2Al_2O_3(s)$$
$$2Zn(s) + O_2(g) \longrightarrow 2ZnO(s)$$

(a) Calculate ΔG° for these reactions.   (b) What is the maximum number of kilojoules that can be obtained from one pound (454 g) of each metal?   (c) What is the voltage, E°, of each battery to three significant figures?

19.82 Why do batteries not give voltages greater than about 3.6 V?

19.83 (a) Describe what you see in the photograph.   (b) What conclusions can you draw from your observations?   (c) How can your observations be explained?

19.84 (a) The voltage of a lithium battery is 3.4 V; the voltage of an otherwise similar zinc battery is 1.5 V. Why do lithium batteries have high voltages?   (b) How many times as many amperes can be obtained from 1.00 g Li as from 1.00 g Zn?

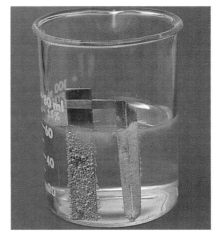

Photo shows copper connected to magnesium. The liquid is 0.1 M HCl(aq).

APPLICATIONS

19.85 Predict the result of using a copper strip to fasten an iron water pipe to the ceiling in a damp climate.

19.86 (a) To start a car, a current of 450 A is needed for 30 s. How many grams of lead are used up in the battery?   (b) Automobile batteries contain six lead storage cells connected in series. How is the total voltage of the battery related to the voltage of a single cell?

19.87 When electricity first became available for home use at the end of the nineteenth century, electric meters consisted of electrolytic cells in which zinc was deposited on the cathode. The meter reader measured the increase in mass of the cathode each month. If 2.36 g of zinc plated out in a month and 0.25% of the current entering the house passed through the coulometer, how many coulombs were used in the month?

19.88 A solution of $K[Au(CN)_2]$ is used for gold-plating baths. This solution is made by electrolysis using a gold electrode immersed in a solution of KCN. The other half-cell consists of an inert electrode immersed in KOH solution. The two half-cells are separated by a porous barrier (see Figure 19.4). Sketch the cell and write equations for the reactions that take place at the electrodes.

19.89 Nickel–cadmium batteries are more expensive than alkaline batteries, but they are light, rechargeable, and can be sealed to prevent leakage. The batteries in your calculator and in any cordless tools you may own are probably nickel–cadmium batteries. In a nickel–cadmium battery, the electrode reactions are

$$Cd(s) \longrightarrow Cd(OH)_2(s) \qquad E^\circ = 0.82 \text{ V}$$
$$NiO_2(s) \longrightarrow Ni(OH)_2(s) \qquad E^\circ = -0.49 \text{ V}$$

which take place in basic solution. (a) Which reaction is the anode reaction and which the cathode reaction?   (b) Write equations for each half-reaction and for the overall reaction.   (c) What is the standard emf of a nickel–cadmium battery?

19.90 The zinc–mercury battery, which was one of the earliest miniature batteries, was introduced in the 1940s. Although they are expensive, zinc–mercury batteries are used in hearing aids, pacemakers, photographic flashes, electric watches, and light meters because they have a high capacity for their mass and size and long lives. The electrode reactions take place in basic solution and are

$$HgO(s) \longrightarrow Hg(l) \qquad \text{and} \qquad Zn(s) \longrightarrow ZnO(s)$$

(a) Which reaction is the anode reaction and which the cathode reaction?   (b) Write equations for each half-reaction and for the overall reaction.   (c) Explain why the voltage (1.35 V) is nearly constant as the battery discharges.   (d) A zinc–mercury battery stops functioning when the zinc is used up. How many coulombs can be generated per gram of zinc, assuming 100% efficiency?

19.91 The aluminum in the cans in a six-pack has a mass of 106 g. (a) If a current of 1.00×10^5 A is used in a Hall–Héroult plant, how long does it take to make 106 g of aluminum?   (b) How many grams of C from the anode should theoretically be used up in making 106 g of aluminum if all the carbon is converted to CO_2? If all the carbon is converted to CO?   (c) If 45.0 g of C are actually used up, what percentage by mass of the gas is CO_2?   (d) To make 106 g of aluminum, 203 g Al_2O_3 must be electrolyzed. What is the percent yield of aluminum?

19.92 The skeleton reaction for the electrochemical production of adiponitrile, which is used in the manufacture of nylon (Section 19.9) is $H_2C=CHCN \longrightarrow NC(CH_2)_4CN$. (a) Write the net ionic equation for the half-reaction involved.   (b) Is the half-reaction a half-reaction of oxidation or a half-reaction of reduction? (c) If a current of 5.0×10^5 A is used, how many kilograms of adiponitrile will be produced per hour?

19.93 In the production of NaOH, H_2, and Cl_2 by the electrolysis of aqueous sodium chloride, (a) what current is required to produce 15.0 kg/h of NaOH, assuming 100% efficiency (that is, assuming that all the electrical energy goes to produce NaOH). (b) The actual efficiency is only about 65%. How many kg/h of NaOH will be produced? (c) What do you think happens to the rest of the electrical energy?

19.94 (a) From Table 19.1, what are two metals that will not dissolve in 1 M H^+ but will dissolve in 1 M HNO_3? (b) Gold, platinum, and iridium not only will not dissolve in dilute nitric acid but won't even dissolve in concentrated nitric acid. Aqua regia, a 3:1 by volume mixture of concentrated hydrochloric and nitric acids, must be used to dissolve gold, platinum, and iridium. Aqua regia is also used to dissolve very insoluble materials such as HgS and some ores for analysis. The nitrate ion in aqua regia is a powerful oxidizing agent, and the chloride ion complexes the metal ions. Write equations for the reactions that take place when gold, platinum, and mercury(II) sulfide dissolve in aqua regia. The skeleton reactions are

(i) $Au(s) + Cl^- + NO_3^- \longrightarrow HAuCl_4(aq) + NO_2(g)$
(ii) $Pt(s) + Cl^- + NO_3^- \longrightarrow H_2PtCl_6(aq) + NO(g)$
(iii) $HgS(s) + Cl^- + NO_3^- \longrightarrow$
$$HgCl_4^{2-} + NO_2(g) + S(s)$$

19.95 Consider the titration of 50.00 mL of 0.010 00 M Fe^{2+} with 0.020 00 M Ce^{4+}. The reaction is

$$Fe^{2+} + Ce^{4+} \longrightarrow Fe^{3+} + Ce^{3+}$$

This reaction is rapid and complete. The Nernst equation for the half-reaction

$$Fe^{3+} + e^- \longrightarrow Fe^{2+}$$

can be used to calculate *E before* the equivalence point when both $[Fe^{2+}]$ and $[Fe^{3+}]$ can be calculated from the stoichiometry of the reaction. (a) What is E after the addition of 10.00 mL of 0.020 00 M Ce^{4+}? (b) How many milliliters of 0.020 00 M Ce^{4+} are required to reach the half-equivalence point? (c) What is E at the half-equivalence point? (d) Suggest a reason why standard solutions of oxidizing agents are widely used but standard solutions of reducing agents are rarely used.

19.96 Sodium oxalate can be used to standardize potassium permanganate solution. The skeleton reaction is

$$MnO_4^- + C_2O_4^{2-} \longrightarrow Mn^{2+} + CO_2(g) \qquad \text{acidic solution}$$

(a) What is the half-reaction of oxidation? (b) The half-reaction of reduction? (c) The net ionic equation for the overall reaction? (d) The reaction is slow but is catalyzed by Mn^{2+}. If a sample of a solution of sodium oxalate is titrated with standard $KMnO_4(aq)$, what do you think will be observed? Explain your answer. (e) If 24.36 mL of

$KMnO_4(aq)$ are required to titrate 0.8120 g of sodium oxalate, what is the molarity of the solution? (f) Oxalic acid solution can be used to remove rust stains. A net ionic equation for the reaction is

$Fe_2O_3(s) + 6H_2C_2O_4(aq) \longrightarrow$
$$2Fe(C_2O_4)_3^{3-}(aq) + 3H_2O(l) + 6H^+$$

Are oxidation and reduction involved in this reaction? What are species like $Fe(C_2O_4)_3^{3-}$ called?

19.97 The concentrations of oxidizing agents, such as O_3, H_2O_2, and Cl_2, in air can be determined by bubbling the air through a solution of potassium iodide. In the presence of an excess of iodide ion, the product is the triiodide ion, I_3^-; the reaction for ozone is

$$O_3(g) + I^- \longrightarrow I_3^- + O_2$$

(a) The concentration of triiodide is determined spectrophotometrically by measuring the absorption at 352 nm. In what region of the electromagnetic spectrum is this wavelength? (b) Write the net ionic equation for the reaction. (c) Given the standard reduction potentials

$$O_3(g) + 2H^+ + 2e^- \longrightarrow O_2(g) + H_2O(l) \qquad E° = 2.08 \text{ V}$$
$$I_3^- + 2e^- \longrightarrow 3I^- \qquad E° = 0.54 \text{ V}$$

what is the value of $E°$? (d) The potassium iodide solution is buffered at pH 6.8. What is E at this pH? (e) Write the net ionic equations for the formation of triiodide ion by the reaction between Cl_2 and iodide ion and by the reaction between H_2O_2 and iodide ion. At pH 6.8, hydrogen peroxide is reduced to water and chlorine to chloride ion.

19.98 A number of chemicals are produced by electrolysis of salt. Sodium metal and chlorine gas are produced by the electrolysis of NaCl(l) in a Downs cell (Figure 19.24). Electrolysis of NaCl(aq) in the cell shown in Figure 19.28 yields NaOH(aq), $Cl_2(g)$, and $H_2(g)$, and electrolysis of NaCl(aq) in a cell without the partition between the anode and cathode compartments yields $NaClO_3(aq)$ and $H_2(g)$. (a) Is the production of sodium in a Downs cell a batch or a continuous process? (b) The Downs cell is operated at a temperature of 600 °C. The electrolyte is not pure NaCl; instead a mixture of NaCl (melting point, 801 °C) and $CaCl_2$ (melting point, 782 °C) is used. Explain the purpose of adding $CaCl_2$ to the NaCl. (c) The "Na(l)" is a mixture of Na(l) and Ca(l). The melting point of sodium metal is 97.8 °C, and the melting point of calcium metal is 839 °C. When the mixture is cooled, crystals form. The crystals are separated by filtration at 105–110 °C. Are the crystals sodium or calcium? (d) The filtrate is run into a nitrogen-filled tank car, allowed to solidify, and shipped. Why is the tank car filled with nitrogen? (e) Explain why different products are obtained by electrolysis of NaCl(l), NaCl(aq) with a partition between anode and cathode compartments, and NaCl(aq) without the partition.

An Art Conservator Looks at Chemistry

LYNDA A. ZYCHERMAN

Conservator of Sculpture

B. A. Art History
City College of the City University of New York

M. A. Art History, and Diploma,
Art Conservation
Institute of Fine Arts, New York University

I was nine years old when I discovered archaeology, ancient Egypt, and the Metropolitan Museum of Art in New York City. From then on, I was hooked on art. Almost every Saturday afternoon over the next ten years, I would spend a couple of hours in the museum until I knew the collection by heart. Because of my interest in art, I attended a special high school that offered fine arts as well as a rigorous academic program.

It was a foregone conclusion that I would be an art history major in college, but, of course, I also had to meet science requirements. My father, a chemical engineer, suggested that I take chemistry. After two weeks, the formulas and numbers proved too much for a freshman art historian to handle. I thought I could *never* master the science so I dropped chemistry as fast as I could.

In the middle of my sophomore year, I saw a description of a graduate program in the conservation and restoration of works of art. Fascinating, I thought, but chemistry was a requirement—and not just inorganic, but a year of organic too. I visited the graduate school to see if they were serious about the requirements (they were). I saw laboratories where paintings and sculptures were being treated to mitigate the ravages of time. The conservators were cleaning paintings or treating archaeological artifacts for "bronze disease." Medieval plaques were being examined microscopically for defects in the glassy enamel. Seeing these laboratories, I knew that this was the way I wanted to be involved with art and archaeology.

I was so enthusiastic about applying science to art that I registered for chemistry again, this time with a goal in mind and a will to succeed. Despite my earlier fiasco, I managed to do reasonably well, and later on I was accepted to the graduate program of my choice.

In the graduate program, we studied the materials and techniques of works of art from a chemical standpoint, i.e., the raw materials of a work of art, the way the materials were modified in making the object, and the changes it undergoes over time. Understanding these things can help us find a treatment that can arrest the deterioration of the object.

As a conservator for sculpture and objects, I am often required to examine archaeological bronzes. When I examine an ancient metal object, say, a weapon, ceremonial vessel, or figurine, in the laboratory, I study the corrosion carefully. With the aid of a simple chemical test, I can take further steps if necessary to arrest its physical deterioration and bring out the beauty that time and burial have concealed.

Bronze is an alloy of copper and tin that was first used around 3000 to 2500 B.C. in Mesopotamia. The alloy was cast into stone or clay molds, producing artifacts whose surfaces were subsequently worked with tools and abrasives to refine them and remove most of the defects produced by the casting.

When buried in damp soil, artifacts made of copper and its alloys lose their metallic appearance and begin to return to their oxides. The oxide layer increases in thickness, and cuprous oxide (Cu_2O) becomes compacted into a purplish red mineral known as cuprite. This in turn may become encrusted with "patina," beautiful basic green or blue carbonates corresponding to the minerals malachite ($CuCO_3 \cdot Cu(OH)_2$) and azurite ($2CuCO_3 \cdot Cu(OH)_2$). Not all green patinas are stable, however. Pale green, powdery spots of corrosion called "bronze disease" may flower up over the surface, disfiguring and weakening the object. The corrosion will continue until the hydrolysis is arrested or the artifact is a pile of green powder.

How does a conservator determine that green spots on a bronze are deleterious copper chloride and not desirable malachite? One simple, accurate way is to perform a chloride test. In a 10-ml graduated cylinder, a small sample of the green powder is dissolved in a weak nitric acid solution. A few drops of silver nitrate solution are added; if chloride is present, it will be precipitated as silver chloride, which appears opalescent when the cylinder is lit from the side in front of a dark background.

Physical and chemical methods have been developed that reach the deep cuprous chloride and render it innocuous without unduly altering the surface appearance of the bronze. In addition, since chloride activity is at a minimum when conditions are dry, objects with bronze disease should be kept in an atmosphere with as low a relative humidity as possible. Another method for stabilizing bronze disease uses benzotriazole (BTA). BTA forms an insoluble complex compound with cupric ions that precipitates over the cuprous chloride and forms a barrier that prevents the ingress of moisture.

Any detailed examination of archaeological artifacts can lead to surprises. While examining a Chinese painted pottery vessel from the Han Dynasty (208 B.C.–A.D. 220), a colleague and I observed blue and purple pigments that, when analyzed by microchemical and instrumental methods, did not correspond to any of the natural mineral pigments known to be in use in China around 200 B.C. Using scanning electron microscopy to identify the elements present (barium, copper, silicon) and optical mineralogy combined with X-ray diffraction to identify the crystalline structure, we identified the blue component as barium copper silicate, $BaCuSi_4O_{10}$, a hitherto unidentified synthetic pigment.

The purple pigment proved to be more elusive. Recently, scientists working in superconductor research synthesized a magenta compound (barium copper silicate, $Ba_3Cu_2Si_6O_{17}$) that proved to be identical to the purple we had found on that pottery.

The conservation and restoration of contemporary works of art presents a different sort of challenge. We can easily determine the technology of modern materials (plastic, Corten steel, acrylic paint) by consulting industry engineers, but no one knows much about the long-term stability of the materials because that is usually not a concern of the industry that produced them.

In the 1920s, when cellulose nitrate was a new plastic resin, some artists used it to create sculptures. We now know that cellulose nitrate is an unstable plastic. It deteriorates by loss of the plasticizer, which migrates out at the surface, followed by the emission of small amounts of nitric acid. The result is severe yellowing, opacification, and fragmentation into small pieces or powder. The nitric acid could also attack any metal fittings, causing the entire sculpture to self-destruct. Cellulose nitrate is also highly flammable and has spontaneously burst into flame in film collections (but, fortunately, never in an art museum). Although we cannot restore objects made of this material, we can try to arrest the deterioration by storing such works in a stable environment with a relatively low temperature because every 10-degree drop will halve the rate of a chemical reaction.

From the preservation and treatment of a deteriorated artifact to the analysis of ancient technology, I use chemistry in my work every day. If you are a museum visitor, art lover, or collector, you will see the results of chemistry in every work of art and artifact made by humans.

NUCLEAR CHEMISTRY

Organic material added to the pigments used in this cave painting, which is called the "White Shaman," were extracted with a low-temperature, low-pressure plasma and radio-carbon dated. The painting is about 4000 years old and is located near where the Pecos and Devils Rivers empty into the Rio Grande. The section shown is a little over one meter high.

The reactions we have studied so far involve electrons, which are outside the nucleus. In this chapter, we look at changes that affect the nucleus. First, we will examine the common types of nuclear reactions and learn how to write equations for them. Then some observations of nuclear reactions will be considered, and the models that have been developed to explain these observations will be discussed. Finally, some applications of nuclear chemistry will be described.

— *Why is a study of nuclear chemistry a part of general chemistry?*

For an atom to be electrically neutral, the number of electrons must equal the number of protons in the nucleus. Thus, the number of protons in the nucleus determines the number of electrons in an atom. Both the physical and chemical properties of the elements and their compounds depend on the number of electrons. Nuclear chemistry plays an essential part in determining what elements are available, and it is used in chemistry, in medicine and biology, and in geology. Nuclear reactions that take place in the sun release the energy that makes life possible. The use of energy from nuclear reactions to generate electricity and in nuclear weapons is the subject of continuing debate.

The second worst radiation accident in history shows why as many people as possible should know about nuclear chemistry. The doctors who owned a radiotherapy clinic in Goiania in central

Brazil moved to new quarters and left behind a machine containing cesium-137. (Cesium-137 is used to treat cancer.) Scavengers sold the cylinder containing the cesium-137 to a junk dealer, and an employee opened the capsule containing the cesium-137, a glowing blue substance. Several people took some of the fascinating material home with them, and a child rubbed it on her body like glitter. In all, 244 people were contaminated, 54 seriously enough to need hospitalization, and several died. The toll in terms of cancer and genetic defects has not yet been estimated. Cleanup of the 2000-m² area that was contaminated took six months. Brazil does not have a disposal site for nuclear waste and does not know what to do with the tons of radioactive material collected. None of this would have happened if the doctors had not abandoned the machine containing the cesium-137 or if the man who opened the capsule had recognized the danger.

▬ What do you already know about nuclear chemistry?

You have surely read or heard about nuclear power, radioactive waste, and nuclear weapons. You probably know someone who has had radiotherapy for cancer. If you are interested in medicine, you may have read about the growing use of positron emission tomography (PET) and many other methods involving the use of radioactive nuclides in diagnosis.

From previous chapters, you know that the nucleus of an atom is very small, extremely dense, positively charged, and composed of protons and neutrons. The number of protons in the nucleus is given by the atomic number, Z, of the element, and all atoms of an element have the same

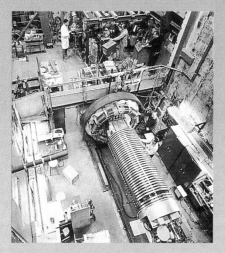

The Superhilac linear accelerator at the University of California at Berkeley.

Aerial view of the two-mile-long linear accelerator at Stanford Universtiy. Linear accelerators are discussed in Section 20.2.

number of protons in their nuclei. Nuclei of different isotopes of an element contain different numbers of neutrons. The sum of the number of protons and neutrons in the nucleus of an atom is called the mass number, A, of the atom. The symbol for a particular isotope of an element shows the mass number as a left superscript and the atomic number as a left subscript. The left subscript is sometimes omitted from the symbol because it is the same for all isotopes of an element. For example, naturally occurring fluorine consists of a single isotope with mass number 19. The atomic number of fluorine is 9. The symbol for the isotope of fluorine with mass number 19 is $^{19}_{9}F$ or ^{19}F. You can also write fluorine-19 or F-19. The term *nuclide* is used to indicate a specific nuclear species such as fluorine-19.

Most of the naturally occurring elements are mixtures of isotopes. The atomic masses shown in the table inside the front cover are weighted averages of

the masses of the isotopes that make up the naturally occurring element.

Nuclei of radioactive elements change spontaneously (decay) into nuclei of other elements and give off energy in the form of rays. Some nuclei decay in less than a femtosecond, whereas others take billions of years. Stable nuclides are nuclides that either do not decay or decay extremely slowly.

Protons repel each other because of their like charges. More than one proton can't be packed into a tiny volume like the nucleus unless neutrons are present. Neutrons reduce the repulsive forces between positively charged protons and contribute to the force that holds the particles in the nucleus together. In atoms that have low atomic numbers (up to calcium, atomic number 20, in the periodic table), one neutron per proton is enough; atoms with more than 20 protons in the nucleus need more neutrons in the nucleus for the nucleus to be stable.

20.1   RADIOACTIVE DECAY PROCESSES

Chemical reactions involve electrons, which are outside the nucleus. The nucleus does not undergo any change during chemical reactions. Radioactive decay *involves the protons and neutrons that make up the nucleus;* radioactive decays are **nuclear reactions.** Three types of rays are given off when nuclei decay: **alpha rays,** which are *He^{2+} ions;* **beta rays,** which are *electrons;* and **gamma rays,** which are *electromagnetic radiation of even shorter wavelength and greater energy and penetrating power than X-rays.*

Alpha Decay and Nuclear Equations

Nuclear reactions, like chemical reactions, are described by equations. Let's consider the alpha decay of uranium-238 as an example. A *radioactive nuclide* such as uranium-238 is called a **radionuclide.** ▬Figure 20.1 shows a microscopic model of the decay of the nucleus of an atom of uranium-238 to the nucleus of an atom of thorium-234 and an alpha particle (helium nucleus). The electrons outside the nucleus are not shown in Figure 20.1.

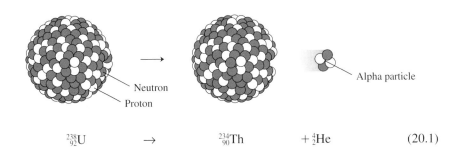

Neutron

Proton

Alpha particle

▬ FIGURE 20.1 Microscopic model of alpha decay of uranium-238.

$$^{238}_{92}U \longrightarrow {}^{234}_{90}Th + {}^{4}_{2}He \qquad (20.1)$$

Equation 20.1 is the nuclear equation for the alpha decay of uranium-238. **Nuclear equations** *show the changes that take place in nuclei.* In nuclear equations, as in chemical equations, the reactants are written to the left of the arrow and the products to the right. Nuclei are represented by nuclide symbols; the chemical form of the reactants and products is not shown because a nuclide undergoes the same nuclear reaction no matter what its state of chemical combination. Uranium-238 decays to thorium-234 and an alpha particle whether it is in the form of uranium metal, uranium hexafluoride, UF_6, or uranium(IV) sulfate octahydrate, $U(SO_4)_2 \cdot 8H_2O$.

The product nucleus has two fewer protons than the reactant nucleus and is actually a $^{234}_{90}Th^{2-}$ ion. The alpha particle is actually a $^{4}_{2}He^{2+}$ ion. However, sooner or later the $^{234}_{90}Th^{2-}$ ion loses two electrons and becomes a thorium atom; the $^{4}_{2}He^{2+}$ ion gains two electrons and becomes a helium atom. Ionic charges are never shown in equations for nuclear reactions.

The term **nucleon** is used to refer to *both protons and neutrons in the nucleus.* The equation for a nuclear reaction must show the same number of nucleons and the same nuclear charge on both sides. The sum of the mass numbers of the products must be equal to the sum of the mass numbers of the reactants; that is, the sum of the superscripts of the products must be equal to the sum of the superscripts of the reactants. The sum of the nuclear charges of the products must be equal to the sum of the nuclear charges of the reactants; that is, the sum of subscripts of the products must be equal to the sum of the subscripts of the reactants. The number of product atoms of each kind is *not* equal to the number of reactant atoms of each kind. In radioactive decay, an atom of one element changes into an atom of another element. An atom of uranium-238 becomes a thorium-234 atom when uranium-238 decays. The thorium atom has a mass number four units less than the uranium atom and an atomic number two units lower.

○ In chemical reactions, an atom of one element never changes into an atom of another element. The equation for a chemical reaction must show the same number of each kind of atom on both sides of the arrow.

Gamma rays are usually given off when a radionuclide decays. However, gamma rays are often not shown in equations for nuclear reactions because gamma rays are electromagnetic radiation and have no mass or charge.

SAMPLE PROBLEM

20.1   Write the equation for the alpha decay of mercury-180.

Solution   From the table inside the front cover, the atomic number of mercury is 80. The nuclide symbol for mercury-180 is $^{180}_{80}\text{Hg}$. The problem does not say what the product nucleus is; X can be used to represent the symbol of the product nucleus. The equation thus far is

$$^{180}_{80}\text{Hg} \longrightarrow {}^{A}_{Z}X + {}^{4}_{2}\text{He}$$

where A represents the mass number and Z the atomic number of the product nucleus. For the equation to be balanced,

$$180 = A + 4 \qquad \text{and} \qquad 176 = A$$

Also

$$80 = Z + 2 \qquad \text{and} \qquad 78 = Z$$

From the table inside the front cover, the element for which the atomic number $Z = 78$ is platinum, Pt. The equation for the alpha decay of mercury-180 is

$$^{180}_{80}\text{Hg} \longrightarrow {}^{176}_{78}\text{Pt} + {}^{4}_{2}\text{He}$$

After you have balanced the equation for a nuclear reaction, you should check to be sure that the sum of the superscripts on the right equals the sum of the superscripts on the left and the sum of the subscripts on the right equals the sum of the subscripts on the left.

Always check your work.

PRACTICE PROBLEM

20.1   Write the equation for the alpha decay of radon-222.

Beta Decay

The product of the radioactive decay of uranium-238, thorium-234, is also radioactive and emits beta rays. Beta rays are electrons; electrons have a mass number of zero because the mass of an electron is only 1/1850 of the mass of a proton or neutron, and a charge of $1-$. The equation for the decay of thorium-234 can be written either

$$^{234}_{90}\text{Th} \longrightarrow {}^{234}_{91}\text{Pa} + \beta^{-}$$

to emphasize that the electrons are *not* electrons from outside the nucleus or

$$^{234}_{90}\text{Th} \longrightarrow {}^{234}_{91}\text{Pa} + {}^{0}_{-1}\text{e}$$

to emphasize that beta particles are electrons. The simplest example of beta decay is the disintegration of hydrogen-3:

Particles called antineutrinos, $\bar{v}$, that have no charge and a mass less than 4×10^{-5} of the mass of an electron are also emitted along with beta particles, but chemists do not usually show antineutrinos in equations for nuclear reactions.

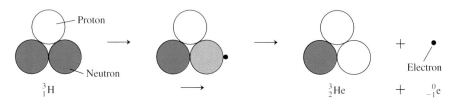

As you can see, electrons are formed from neutrons in the nucleus; beta decay converts a neutron in the nucleus into a proton. The mass number remains the same, and the atomic number is increased by one unit.

Three other types of radioactive decay—positron emission, electron capture, and fission—also take place spontaneously although they are not as common as alpha and beta emission.

Positron Emission

Positrons are represented either by the symbol β^+ or by $_{+1}^{0}e$. The simplest recorded example of positron emission is the disintegration of boron-8:

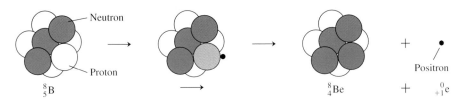

As you can see, positrons are formed from protons in the nucleus. Positron emission converts a proton in the nucleus into a neutron. The mass number remains the same; the atomic number decreases by one unit.

When a positron collides with an electron, both the positron and the electron disappear and gamma rays are given off:

$$_{-1}^{0}e + _{+1}^{0}e \longrightarrow _{0}^{0}\gamma$$

This process is called **annihilation.** Because electrons are everywhere in matter, a positron is bound to hit an electron soon after it is emitted; positrons don't last very long. Gamma rays are high-energy electromagnetic radiation; they are not matter. Matter is converted into energy when an electron annihilates a positron. The interconversion of matter and energy is discussed in Section 20.6.

A positron is referred to as **antimatter** because it destroys an electron, a particle of ordinary matter. The positron was the first example of antimatter to be discovered.* It is an antielectron.

Electron Capture

Electron capture, EC, like positron emission, results in a decrease of one unit in atomic number while the mass number remains the same. *One of the electrons, usually one from the first shell, becomes part of the nucleus and a proton becomes a neutron.* The simplest recorded example of electron capture is the formation of lithium-7 from beryllium-7:

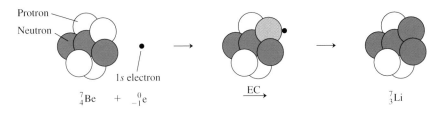

> ▶ Remember that positrons are particles that have the same mass as an electron but have one unit of positive (instead of one unit of negative) charge. Neutrinos, ν, are also emitted along with positrons, but chemists usually do not show neutrinos in equations for nuclear reactions. Neutrinos and antineutrinos are different particles but are practically identical.

> ▶ The neutrino and the antineutrino are another example of particle and antiparticle.

---

*The positron was discovered in 1932 by the American physicist Carl David Anderson in the course of a study of cosmic rays. (Cosmic rays are discussed in Section 20.9.) Positron emission was first observed in 1934.

An electron from a higher shell (the second shell in this case) falls down into the place left by the captured electron, and an X-ray is emitted. Electron capture is an exception to the general rule that nuclear reactions are not affected by the state of chemical combination of the atom of which the nucleus is a part. Because an electron, which is part of the electron cloud around the nucleus, is involved in capture, changes in electron density near the nucleus result in small changes in the rate of electron capture. Differences in the rate of electron capture of the order of 0.1% have been observed between beryllium-7 metal and beryllium compounds such as BeF_2.

Spontaneous Fission

Only atoms with atomic numbers greater than 92 undergo spontaneous fission; within this range of atomic numbers, the higher the atomic number, the more likely a nuclide is to undergo spontaneous fission. In **fission,** *an atom of high atomic number breaks into two approximately equal pieces and neutrons are emitted:*

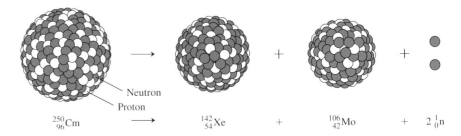

$$^{250}_{96}Cm \longrightarrow {}^{142}_{54}Xe + {}^{106}_{42}Mo + 2{}^{1}_{0}n$$

Different atoms break into different pieces so that a variety of products is formed. For example, a second atom of curium-250 might yield cerium-148, strontium-98, and four neutrons instead of xenon-142, molybdenum-106, and two neutrons:

$$^{250}_{96}Cm \longrightarrow {}^{148}_{58}Ce + {}^{98}_{38}Sr + 4{}^{1}_{0}n$$

Some nuclides decay by more than one type of reaction. Fermium-256, for example, undergoes both spontaneous fission and alpha decay.

PRACTICE PROBLEMS

20.2   Classify each of the following nuclear reactions as alpha decay, beta decay, positron emission, electron capture, or spontaneous fission:

(a) $^{96}_{41}Nb \longrightarrow {}^{96}_{42}Mo + {}^{0}_{-1}e$ (b) $^{90}_{41}Nb \longrightarrow {}^{90}_{40}Zr + {}^{0}_{+1}e$

(c) $^{228}_{93}Np \longrightarrow {}^{89}_{38}Sr + {}^{135}_{55}Cs + 4{}^{1}_{0}n$ (d) $^{196}_{84}Po \longrightarrow {}^{192}_{82}Pb + {}^{4}_{2}He$

(e) $^{90}_{41}Nb + {}^{0}_{-1}e \longrightarrow {}^{90}_{40}Zr$ (f) $^{223}_{86}Rn \longrightarrow {}^{223}_{87}Fr + {}^{0}_{-1}e$

20.3   Complete each of the following nuclear equations:

(a) $^{139}_{58}Ce + {}^{0}_{-1}e \longrightarrow$ _____   (b) $^{23}_{12}Mg \longrightarrow$ _____ $+ {}^{0}_{+1}e$

(c) _____ $\longrightarrow {}^{172}_{75}Re + {}^{4}_{2}He$

20.4   Write equations for each of the following nuclear reactions:
(a) Fermium-245 decays by alpha emission.
(b) Fermium-251 captures an electron from outside the nucleus.
(c) Nitrogren-18 decays by beta emission.   (d) Oxygen-14 emits a positron.
(e) Nobelium-258 spontaneously fissions to silver-120 and cesium-135. (*Hint:* How many neutrons must be released?)

INDUCED NUCLEAR REACTIONS

An **induced nuclear reaction** is a *nuclear reaction brought about by collision of a moving particle with a nucleus,* which is referred to as the **target.** The *moving particle* is called the **projectile.** The first induced nuclear reaction was carried out by Rutherford in 1919. Rutherford used alpha particles produced by radioactive decay as projectiles and nitrogen-14 as the target. A proton was expelled and oxygen-17 was formed:

$$^{14}_{7}\text{N} + {}^{4}_{2}\text{He} \longrightarrow {}^{17}_{8}\text{O} + {}^{1}_{1}\text{H}$$

This is the same Rutherford who showed that atoms consist of a nucleus surrounded by electrons.

Induced nuclear reactions are sometimes called **particle–particle reactions** because one particle is a reactant and another particle is a product. Another name for induced nuclear reactions is **transmutation;** *one element is changed or transmuted into another element.* (Elements are also transmuted by radioactive decay.)

Table 20.1 lists the symbols for particles commonly used to bring about induced nuclear reactions. Ordinary nuclide symbols are used for particles not listed in Table 20.1; for example, the symbol ${}^{12}_{6}\text{C}$ is used to represent the nucleus of a carbon-12 atom. *Projectiles with masses greater than the mass of an alpha particle* are called **heavy ions.**

The use of heavy ions for the treatment of cancer is being tested on patients in Japan.

Synthetic Radionuclides

In 1933 Irène and Frédéric Joliot-Curie prepared the first radionuclide not found in nature by bombarding aluminum-27, the only naturally occurring isotope of aluminum, with alpha particles:

$$^{27}_{13}\text{Al} + {}^{4}_{2}\text{He} \longrightarrow {}^{30}_{15}\text{P} + {}^{1}_{0}\text{n}$$

The Joliot-Curies were a husband and wife team of French physical chemists who won the Nobel Prize in chemistry in 1935 for their discovery of artificial radioactivity.

Phosphorus-30 decays by positron emission to silicon-30, which is not radioactive. Between 1933 and the present, over 2000 synthetic (artificial) radionuclides have been made.

More nuclides are being made all the time.

The higher the atomic number of the target nucleus, the higher its positive charge. The higher the positive charge on the target nucleus, the more kinetic energy a positively charged bombarding particle, such as an alpha particle, must have to overcome repulsion between like charges and hit the target nucleus. Alpha particles from radioactive decay do not have enough energy to bring about very many induced nuclear reactions. Accelerators such as linear accelerators, cyclotrons, and synchrotrons must be used to produce high-speed charged particles.

Accelerators *use an electric field to make charged particles go faster.* ◄Figure 20.2 is a diagram of a linear accelerator, the easiest type of accelerator to understand.

◗ Acceleration of a charged particle in an accelerator is similar to acceleration of a skateboard by repeated pushes.

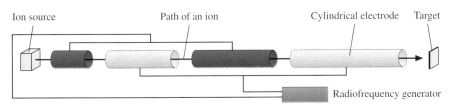

Ion source　　　Path of an ion　　　Cylindrical electrode　　Target

Radiofrequency generator

◀ **FIGURE 20.2**　A linear accelerator. One set of electrodes (red) has one charge; the other set of electrodes (gray) has the opposite charge. As a positively charged ion leaves the ion source, the first electrode has a negative charge, which attracts the positive ion and accelerates it. As the ion reaches the right-hand end of the first electrode, the signs of the electrodes are reversed so that the first electrode has a positive charge and the second electrode has a negative charge. The particle is repelled by the first electrode and attracted by the second and is accelerated some more. This process continues until the particle comes out of the accelerator and hits the target. The electrodes get longer from left to right because the ion is moving faster and faster, but the time between changes in sign of the electrodes is constant. The exit from the ion source, the electrodes, and the target are all enclosed in a vacuum chamber so that there are very few molecules in the accelerator for the particle to hit.

TABLE 20.1

Symbols for Particles

Name	Nuclide Symbol	Particle Symbol
Alpha	${}^{4}_{2}\text{He}$	α
Beta	${}^{0}_{-1}\text{e}$	β^{-}
Deuteron	${}^{2}_{1}\text{H}$	d
Gamma ray[a]	${}^{0}_{0}\gamma$	γ
Positron	${}^{0}_{+1}\text{e}$	β^{+}
Proton	${}^{1}_{1}\text{H}$	p
Neutron	${}^{1}_{0}\text{n}$	n

[a] A gamma ray is *not* a particle.

A cyclotron is similar to a linear accelerator except that the particle beam is curled into a spiral by a magnetic field. The particle beam also follows a curved path in synchrotrons.

A linear accelerator called the Superhilac was used by Seaborg (see related topic) and his co-workers at the University of California at Berkeley to synthesize seaborgium (atomic number 106) by the reaction

$$^{249}_{98}\text{Cf} + ^{18}_{8}\text{O} \longrightarrow ^{263}_{106}\text{Sg} + 4^{1}_{0}\text{n}$$

Hilac stands for *heavy-ion linear accelerator*. The Superhilac is 52 m long and can accelerate ions from atoms as heavy as uranium to an energy of 2×10^{11} kJ/mol. Although the Superhilac can accelerate 10^{12} particles per second, yields of seaborgium were only a few atoms per day. Yields of nuclides with high atomic numbers are very poor because these nuclei decay very rapidly.

Thermal (slow) neutrons are used as projectiles for the synthesis of large quantities of radionuclides. **Thermal neutrons (slow neutrons)** *have average kinetic energy about the same as neutrons would have if they existed as a monatomic gas at room temperature.* Because neutrons have no charge, neutrons are not repelled by the positive charge on target nuclei. Thermal neutrons have enough kinetic energy to react with nuclei. A typical example is

$$^{59}_{27}\text{Co} + ^{1}_{0}\text{n} \longrightarrow ^{60}_{27}\text{Co} + ^{0}_{0}\gamma$$

Such reactions are referred to as **neutron capture.** Isotopes of almost every element up to fermium-257 have been made by neutron capture.

You might think that if slow neutrons are good projectiles, fast neutrons would be better, but this is not the case. A good analogy is a hole in a miniature golf course. If the area around the hole is flat and you hit the ball gently, straight for the hole, the ball will fall into the hole. But if you hit the ball too hard, it will pass directly over the hole without dropping in. Hitting a ball into a hole located in the top of a hill is analogous to hitting a nucleus with a charged particle. To go into the hole, the ball must have enough kinetic energy to reach the top of the hill. If it does not, it will simply roll back down again. However, if the ball has too much energy, it will go over the hole.

To get close enough to a nucleus to combine with it, a positively charged particle must have enough energy to overcome the repulsion between like charges. ▪Figure 20.3 shows how potential energy changes with distance from the center of the nucleus for electrically neutral neutrons and for positively charged alpha particles.

Element 106 was first made in 1974. A linear accelerator was also used to synthesize the superheavy elements 107, 108, and 109 at the Institute for Heavy-Ion Research at Darmstadt, Germany, in the 1980s. Elements 110 and 111 were made at Darmstadt in 1994.

Cobalt-60 is radioactive and is used to treat malignant tumors.

◐ The reason for the large decrease in potential energy inside the nuclear surface is discussed in Section 20.6.

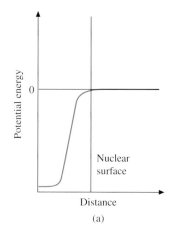

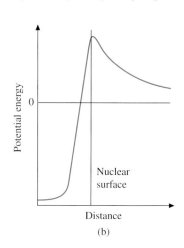

▪ FIGURE 20.3 Potential energy as a function of distance from the center of the nucleus: (a) neutron and (b) alpha particle.

Seaborg and the Transuranium Elements

Transuranium elements are elements that have atomic numbers greater than 92, the atomic number of uranium. The transuranium elements are all radioactive. They are made by induced nuclear reactions.

Glenn Seaborg is an American nuclear chemist who was a student of Lewis. His proposal in 1944 that the actinide series in the periodic table should begin with element 89 led to the discovery of most of the transuranium elements. Before 1944, element 89 was placed directly after element 88 in the main body of the periodic table.

Chemists thought that element 93 should resemble rhenium, element 94 should resemble osmium, and so forth. Because chemists were expecting the wrong

55	56	*71	72	73	74	75	76
Cs	Ba	Lu	Hf	Ta	W	Re	Os
87	88	89	90	91	92	93	94
Fr	Ra	Ac	Th	Pa	U	Np	Pu

properties for elements 95 and 96, they could not separate and identify them.

Of the nineteen transuranium elements that have been made, Seaborg and his group were unquestionably the first to make eight (elements 94–101). A Russian group at Dubna also claims credit for elements 102, 104, 105, and 106. Seaborg won the Nobel Prize for chemistry in 1951 for his discoveries about the chemistry of the transuranium elements. Element 106, seaborgium, is named for him.

Glenn Seaborg at the time of his Nobel Prize in 1951.

PRACTICE PROBLEMS

20.5 For the reaction

$$^{27}_{13}\text{Al} + ^{4}_{2}\text{He} \longrightarrow ^{30}_{15}\text{P} + ^{1}_{0}\text{n}$$

identify the target, the projectile, the particle ejected, and the product nuclide.

20.6 Of the following reactions, (a) which are induced nuclear reactions? (b) Which is a neutron capture reaction? (i) $^{239}_{94}\text{Pu} + ^{2}_{1}\text{H} \longrightarrow ^{240}_{95}\text{Am} + ^{1}_{0}\text{n}$
(ii) $^{169}_{76}\text{Os} \longrightarrow ^{165}_{74}\text{W} + ^{4}_{2}\text{He}$ (iii) $^{139}_{58}\text{Ce} + ^{0}_{-1}\text{e} \longrightarrow ^{139}_{57}\text{La}$
(iv) $^{140}_{58}\text{Ce} + ^{1}_{0}\text{n} \longrightarrow ^{141}_{58}\text{Ce} + \gamma$ (v) $^{239}_{94}\text{Pu} + ^{4}_{2}\text{He} \longrightarrow ^{242}_{96}\text{Cm} + ^{1}_{0}\text{n}$

20.3 RATES OF NUCLEAR REACTIONS

Why are some radionuclides such as uranium-238 found in nature whereas others such as technetium-98 are not? One reason is that some radionuclides decay much faster than others. This brings us to the question of how rates of radioactive decay are measured.

Activity of Radionuclides

The *rate at which radioactive decay takes place* is called the **activity** of the radioactive sample. The *SI unit of activity* is named the **becquerel, Bq,** in honor of the French physicist Henri Becquerel, who discovered radioactivity in 1896. One becquerel is equal to one event per second:

$$1 \text{ Bq} = 1 \text{ event} \cdot \text{s}^{-1}$$

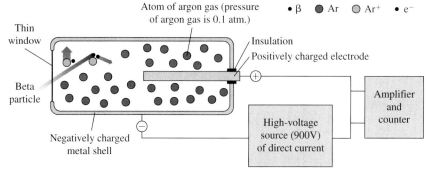

FIGURE 20.4 Diagram of a Geiger counter. Normally, a gas does not conduct electricity. When ionizing radiation, for example, a beta particle enters the Geiger counter through the window, which is made of a thin sheet of beryllium, aluminum foil, or a plastic film such as Mylar polyester, the beta particle causes the argon atoms in its path to ionize. The electrons formed by ionization move toward the positive electrode; they are accelerated by the positive charge on the electrode and ionize more argon atoms in their paths. The number of electrons increases rapidly, and a burst of current is recorded at the counter. The more massive positively charged argon ions move slowly toward the negative electrode (the metal shell of the Geiger counter). A second particle cannot be counted until the argon ions reach the metal shell.

The curie was originally defined as the activity of one gram of radium.

The *traditional unit of activity* is the **curie, Ci** (1 Ci = 3.70×10^{10} Bq); the curie is named after Pierre and Marie Curie, discoverers of the radioactive elements polonium and radium. Marie Curie introduced the term *radioactivity* in 1898.*

The **Geiger counter** (◾Figure 20.4) uses the ionization of a gas by radiation to measure activity. Geiger counters are slow and can only count the number of particles; no information is obtained about the energy of the particles. However, Geiger counters are relatively cheap.

A Geiger counter can be used to count neutrons if filled with boron trifluoride gas. The nuclide boron-10, which makes up about 20% of naturally occurring boron, absorbs low-energy neutrons and is converted to lithium-7 and alpha particles:

$$^{10}_{5}\text{B} + {}^{1}_{0}\text{n} \longrightarrow {}^{7}_{3}\text{Li} + {}^{4}_{2}\text{He}$$

One alpha particle is formed for each neutron absorbed; the alpha particles ionize argon gas and can be counted.

The apparatus used to detect alpha particles by Rutherford's student, Ernest Marsden, in the work that led Rutherford to the nuclear model of the atom was a primitive scintillation detector (Section 7.3). The student acted as the counter.

Scintillation detectors make use of the promotion of electrons to higher energy levels when rays from radioactive decay strike a fluorescent substance. When the electrons fall back down into lower energy levels, light is emitted. The light is converted to electricity by means of a photoelectric cell and the current measured. Scintillation detectors are used to count alpha particles and photons of gamma rays. Scintillation counters not only count but also measure the energies of particles.

The most modern devices for counting particles and measuring their energy are ionization detectors that use the solid semiconductor germanium instead of a gas. The atoms in a solid are much closer together than the atoms in a gas. Gamma ray photons, which pass right through gases without producing much ionization, are stopped by solid-state detectors. Solid-state detectors count particles accurately and

Marie and Pierre Curie in their laboratory.

*Marie and Pierre Curie were another husband and wife team who, with Becquerel, won the Nobel Prize for physics in 1903. Only three years later, Pierre was killed in a traffic accident. Marie continued their work and was awarded a second Nobel (for chemistry) in 1911; she is one of only four scientists who have received two Nobel prizes and the only one to win them for different scientific fields. She studied both the chemistry of radioactive substances and their medical applications. Irène Joliot-Curie was the daughter of Marie and Pierre; both Irène and Marie died of leukemia caused by radiation.

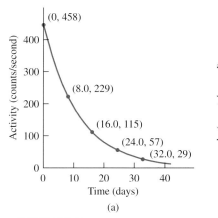
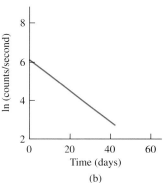
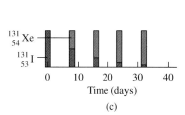

FIGURE 20.5 Radioactive decay of 0.100 pg of iodine-131. (a) Activity as a function of time. (b) Ln activity vs. time. (c) Mole fractions of iodine-131 and xenon-131.

can distinguish energies that differ only slightly. They can be made small enough to be inserted into a living animal and tough enough for measuring the radiation that hits satellites in orbit.

Kinetic Order of Radioactive Decay

*Figure 20.5 shows how the activity of iodine-131 changes with time. The equation for the decay of iodine-131 is

$$^{131}_{53}\text{I} \longrightarrow {}^{131}_{54}\text{Xe} + {}^{0}_{-1}\text{e}$$

For the radioactive decay of iodine-131, an event is the emission of a beta particle. From Figures 20.5(a) and (c), you can see that the time required for the activity of iodine-131 to decrease by one-half is constant and equal to 8.0 days. A constant half-life is characteristic of first-order reactions (Section 18.4). The straight-line graph of ln activity vs. time [Figure 20.5(b)] is also characteristic of a first-order reaction. The rate law for the radioactive decay of iodine-131 is

$$\text{rate} = kN \tag{20.2}$$

where k is the rate constant and N is number of atoms of iodine-131. The radioactive decay of iodine-131 is typical of radioactive decay; all radioactive decays are first-order processes.

Unlike the rates of most chemical reactions, the rates of spontaneous radioactive decay reactions do not change if temperature changes. The rates of induced nuclear reactions do depend on temperature. However, like the rates of radioactive decay, they are usually independent of physical state and of chemical combination.

Radioactive decay has a first-order rate law because it does not depend on any factor outside the nucleus. Radioactive decay depends on chance. Each nucleus in a sample of a radionuclide has a certain probability of decaying that is different for each radionuclide. However, there is no way of knowing which nucleus will decay during a certain period of time. The fact that iodine-131 has a half-life of 8.0 days means that every nucleus of iodine-131 has a 50% chance of decaying during an 8-day period. This does not mean that a given nucleus is sure to decay in 16.0 days. Instead 75% of the nuclei with half-lives of 8.0 days will probably decay in 16.0 days, 87.5% in 24.0 days (three half-lives), 93.75% in 32.0 days (four half-lives) and so on, *if the sample is large*. If a sample is too small, it may not be representative. For example, the chances of getting heads when you toss a coin is one in two. If

Iodine-131 is given intravenously or orally in the form of sodium iodide for treatment of cancer of the thyroid.

TABLE 20.2

Half-Lives of Some Radionuclides

Name	Half-Life[a]
Boron-9	8×10^{-19} s
Carbon-10	19.3 s
Nitrogen-13	9.97 m
Copper-64	12.701 h
Calcium-47	4.536 d
Cesium-137	30.3 y
Potassium-40	1.26×10^9 y[b]

[a]In this table and elsewhere in this chapter, m = minutes, d = days, and y = years to be consistent with the Table of Isotopes in the *CRC Handbook*.
[b]Nuclides with half-lives $> 3 \times 10^{10}$ y are usually classified as stable.

We chose these radionuclides to illustrate the range of half-lives observed.

you toss a coin 1000 times you will probably get heads about 500 times. But if you only toss the coin twice, you may very well get heads twice or not at all.

Half-Lives of Radionuclides

The rate constant k for a radioactive decay is a measure of the probability that a nucleus will decay in a unit of time. Information about the rates of radioactive decay reactions is usually tabulated in the form of the half-lives; half-life is a characteristic property of a nuclide. Because radioactive decay is a first-order process, the value of the rate constant k can be calculated from the half-life by means of the expression

$$t_{1/2} = \frac{0.693}{k} \qquad (18.7)$$

Thus, the value of the rate constant for the decay of iodine-131 is

$$k = \frac{0.693}{8.0 \text{ d}} = 0.087 \text{ d}^{-1}$$

for each atom in the sample.

The half-life of a radionuclide is *not* the average lifetime of the radionuclide; the average lifetime of a radionuclide is the reciprocal of k:

$$\text{average lifetime of a radionuclide } = \frac{1}{k}$$

For example, the average lifetime of the nucleus of an atom of iodine-131 is $(1/0.087 \text{ d}^{-1}) = 11.5$ d.

If the sample is large, that is, if many nuclei are present, the amount of radionuclide, N_t, left after time, t, has passed can be calculated by means of the integrated rate equation for first-order processes (Section 18.3):

$$\ln (N_t/N_0) = -kt \qquad (18.6a)$$

In equation 18.6a, N_0 is the amount of radionuclide present at the start of the experiment and the quotient, N_t/N_0, is the fraction of the original quantity of radionuclide left after time t. As long as the same units are used at time t as at time 0, units cancel from the ratio N_t/N_0. The quantity of radionuclide can be expressed in any one of a variety of units: number of atoms, counts, mass, moles, molarity.

This problem is similar to Sample Problems 18.7 and 18.8.

PRACTICE PROBLEM

20.7 The half-life for the radioactive decay of iridium-192 is 73.8 days. Calculate the amount in grams of Ir-192 that will be left from a 1.36-g sample after (a) 221.4 d and (b) 192.4 d.

Table 20.2 shows a sampling of half-lives ranging from the shortest that have been measured to the longest. You may wonder how half-lives as long as 1.26×10^9 years can be measured. Obviously, no one can sit around for 1.26×10^9 years making measurements. The answer is that any sample large enough to be handled contains a great many atoms. Even if the half-life is very long, enough disintegrations will take place in a short time period for a count to be taken. Sample Problem 20.2 shows the calculation of the number of disintegrations per second expected from the beta decay of rubidium-87, which has a half-life of 4.88×10^{10} years.

20.2 How many disintegrations (events) per second are expected from a sample of rubidium chloride, RbCl, that contains 0.100 g of rubidium-87, $t_{1/2} = 4.88 \times 10^{10}$ y?

Solution The first-order rate law, rate $= kN$ (equation 20.2), gives the relationship between the rate of radioactive decay (the number of disintegrations per second), the rate constant, k, and the number of atoms, N, of the radionuclide present. We must find the value of k from the half-life and the number of atoms from the mass of the sample. Then we can calculate the rate from the rate law.

Step 1. *Calculate the value of k.* Because the question asks about the number of disintegrations per second, we must begin by converting 4.88×10^{10} years to seconds:

$$4.88 \times 10^{10} \text{ y} \left(\frac{365 \text{ d}}{\text{y}} \right) \left(\frac{24 \text{ h}}{\text{d}} \right) \left(\frac{60 \text{ m}}{\text{h}} \right) \left(\frac{60 \text{ s}}{\text{m}} \right) = 1.54 \times 10^{18} \text{ s}$$

Solution of equation 18.7 for k and substitution of the half-life in seconds gives

$$k = \frac{0.693}{1.54 \times 10^{18} \text{ s}} = 4.59 \times 10^{-19} \text{ s}^{-1} \text{ (per atom)}$$

Step 2. *Calculate the number of atoms from the mass of the sample.* The atomic masses of many radionuclides are not readily available. Because the atomic mass of a nuclide is very close to its mass number, the mass number can be used in place of atomic mass when only two or three significant figures are needed. The number of atoms, N, of rubidium-87 in 0.100 g of rubidium-87 is

$$0.100 \text{ g Rb-87} \left(\frac{1 \text{ mol}}{87.0 \text{ g Rb-87}} \right) \left(\frac{6.02 \times 10^{23} \text{ atoms}}{\text{mol}} \right) = 6.92 \times 10^{20} \text{ atoms}$$

 This calculation is similar to the one in Sample Problem 2.16.

Step 3. *Calculate the rate.* Substitution of the values found for k and N in equation 20.2 gives

$$\text{rate} = \left(\frac{4.59 \times 10^{-19} \text{ events}}{\text{s} \cdot \text{atom}} \right) 6.92 \times 10^{20} \text{ atoms} = 3.18 \times 10^2 \text{ events} \cdot \text{s}^{-1}$$

The background radiation from naturally radioactive materials in the laboratory and cosmic rays is normally less than one disintegration per second. A count of 3.18×10^2 disintegrations per second can be measured easily with modern instruments.

20.8 The half-life for the beta decay of osmium-194 is 6.0 y. (a) How many disintegrations per second will be observed from a sample that contains 8.2 pg of osmium-194? (b) How many atoms of osmium-194 are required to give 5.0×10^7 disintegrations per second? (c) What mass of osmium-194 is required to give 5.0×10^7 disintegrations per second? Express the mass in the most appropriate units (without a power-of-10 term).

20.4 PREDICTING WHICH TYPE OF RADIOACTIVE DECAY WILL TAKE PLACE

Neutron : Proton Ratio

Whether a given nuclide can be synthesized, whether, once made, it will be radioactive, and, if it will be radioactive, how it will decay can often be predicted from

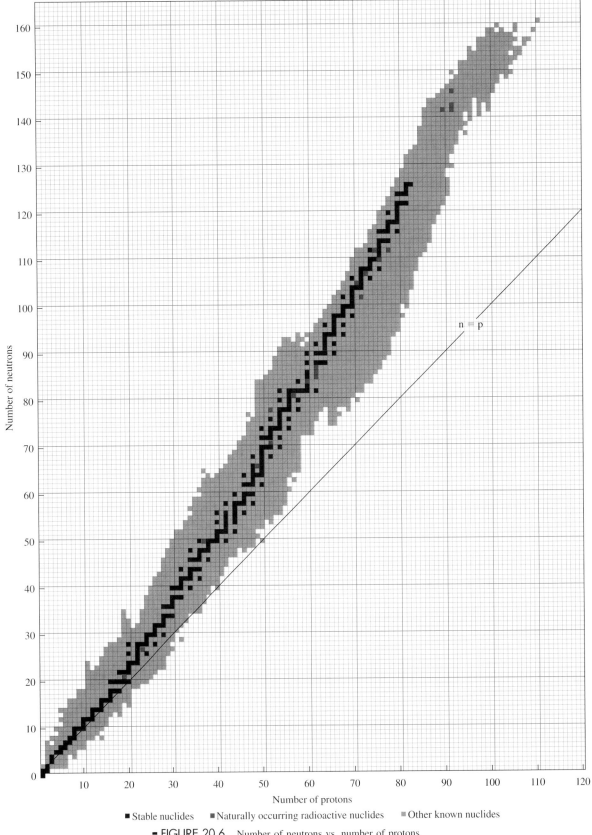

■ Stable nuclides   ■ Naturally occurring radioactive nuclides   ■ Other known nuclides

▪ FIGURE 20.6   Number of neutrons vs. number of protons.

the ratio of the number of neutrons to the number of protons in the nucleus. ▬Figure 20.6 is a graph of the number of neutrons vs. the number of protons in the nucleus for each of the nuclides known when the figure was drawn. In Figure 20.6, each black square represents a stable nuclide. Each red square represents a radioactive nuclide that makes up more than 0.1% of the naturally occurring element. Gray squares represent other known nuclides. The band of black squares, which shows the stable nuclides, is known as the **band of stability.**

In general, the further a nuclide is from the band of stability, the shorter the half-life of the nuclide. The half-lives for the eight longest-lived isotopes of carbon shown in Figure 20.6 are given in Table 20.3 and are typical.

From Figure 20.6, you can see that no stable isotopes are known for elements with atomic numbers higher than 83 (bismuth). Apparently, it is impossible to combine more than 83 protons, each with a positive charge, in a stable nucleus. There are no stable isotopes of either technetium (atomic number 43) or promethium (atomic number 61).

For elements with atomic number 20 and below, stable nuclides have approximately equal numbers of neutrons and protons; the neutron : proton ratio is close to 1. As the atomic number increases beyond 20, the ratio of neutrons to protons gradually increases to 1.52 for the only stable isotope of bismuth, bismuth-209. Additional neutrons in the nucleus both increase the attractive forces between nucleons (because there are more nucleons) and reduce the repulsive forces between protons by increasing the volume of the nucleus. (The volume of a nucleus is roughly proportional to the number of nucleons in the nucleus, that is, to the mass number.)

Nuclides that have a neutron : proton ratio either higher or lower than the stable ratio and nuclides beyond the band of stability are radioactive and decay in such a way as to gain a stable ratio.

1. Nuclides that lie above the band of stability have too many neutrons for stability, that is, the neutron : proton ratio is too high. These nuclides decay by emitting a beta particle. For example,

$$\,^{23}_{10}\text{Ne} \longrightarrow \,^{23}_{11}\text{Na} + \,^{\;0}_{-1}\text{e}$$

When a beta particle is emitted, a neutron in the nucleus is converted into a proton. The number of neutrons decreases by one and the number of protons increases by one. Thus, the neutron : proton ratio decreases.

2. Nuclides that lie below the band of stability have too few neutrons for stability and decay by positron emission or electron capture. Either increases the number of neutrons and decreases the number of protons. Nuclides with low atomic numbers such as neon usually increase their neutron : proton ratio by positron emission. The decay of neon-19 is an example:

$$\,^{19}_{10}\text{Ne} \longrightarrow \,^{19}_{9}\text{F} + \,^{\;0}_{+1}\text{e}$$

Nuclides with high atomic numbers such as tungsten (Z = 74) usually undergo electron capture. The higher the atomic number, the greater the positive charge on the nucleus and the more the inner electrons are pulled in toward the nucleus. The closer the inner electrons are to the nucleus, the more easily they can be captured. Nuclides with intermediate atomic numbers such as zirconium (Z = 40) undergo both positron emission and electron capture. ▬Figure 20.7 shows how beta decay, positron decay, and electron capture move a nuclide toward the band of stability.

3. Nuclides beyond the band of stability, such as astatine-213, commonly undergo alpha emission, which decreases both the number of protons and the number

TABLE 20.3

Half-Lives of Some Isotopes of Carbon

Nuclide	Half-Life
Carbon-9	0.127 s
Carbon-10	19.3 s
Carbon-11	20.3 m
Carbon-12	Stable
Carbon-13	Stable
Carbon-14	5715 y
Carbon-15	2.45 s
Carbon-16	0.75 s

Experts in nuclear chemistry and nuclear physics plot the number of protons against the number of neutrons. However, we think that graphing number of neutrons against number of protons makes the elementary discussion in this text easier to follow.

$\,^{23}_{10}$ Ne moves into band of stability by beta decay.

$\,^{19}_{10}$ Ne moves into band of stability by positron emission. Electron capture would also move $\,^{19}_{10}$ Ne into the band of stability.

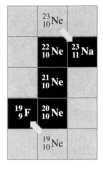

▬ **FIGURE 20.7** Enlargement of the part of Figure 20.6 around neon.

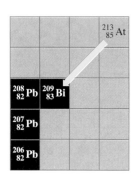

$^{213}_{85}$At moves into band of stability by emission of an alpha particle.

FIGURE 20.8 Enlargement of the part of Figure 20.6 around bismuth-209.

| TABLE 20.4 | | | |

Naturally Occurring Nuclides

Type	Number	Percent	Example
Even–even	168	59	$^{12}_{6}$C
Even–odd	57	20	$^{9}_{4}$Be
Odd–even	53	18	$^{7}_{3}$Li
Odd–odd	9	3	$^{14}_{7}$N
Total	287	100	

These magic numbers are from Greiner, W., Sandulescu, A. Sci. Amer. 1990, 262, 58–67.

⊙ Remember that species with the same number of electrons are referred to as isoelectronic species (Section 8.4).

of neutrons by two:

$$^{213}_{85}\text{At} \longrightarrow {}^{209}_{83}\text{Bi} + {}^{4}_{2}\text{He}$$

- Figure 20.8 shows how alpha decay moves a nuclide toward the band of stability.

4. Nuclides with atomic numbers greater than 92 may undergo spontaneous fission, for example,

$$^{252}_{98}\text{Cf} \longrightarrow {}^{140}_{54}\text{Xe} + {}^{108}_{44}\text{Ru} + 4{}^{1}_{0}\text{n}$$

Neutrons are almost always released when fission takes place because elements with lower atomic numbers have lower stable neutron : proton ratios than elements with higher atomic numbers.

Even–Even, Even–Odd, Odd–Even, and Odd–Odd Nuclei

Some other facts about nuclei can be seen from Figure 20.6. Elements with even atomic numbers usually have a greater number of stable isotopes than elements with odd atomic numbers. For example, tin (atomic number 50) has ten stable, naturally occurring isotopes whereas tin's neighbors in the periodic table (and in Figure 20.6), indium and antimony, each have only two stable isotopes. Elements with odd atomic numbers almost never have more than two stable isotopes; elements with even atomic numbers often have a large number.

Naturally occurring nuclei that have both an even number of protons and an even number of neutrons are most common. Stable nuclei with both an odd number of protons and an odd number of neutrons are rare. Table 20.4 summarizes statistics about the numbers of each type among naturally occurring nuclides.

If the even–odd character were a matter of chance, there would be about 25% of each type of nucleus. The excess of even numbers of protons and even numbers of neutrons is evidence for pairing of nucleons in the nucelus. Just as electrons outside the nucleus form pairs, protons inside the nucleus form pairs as do neutrons inside the nucleus.

Magic Numbers

Another similarity between nucleons and electrons is the existence of **magic numbers** *of protons and neutrons that occur more often that ordinary numbers.* Graphing data in a different way often makes different relationships apparent. The same data used for Figure 20.6 in the form of a plot of the number of naturally occurring nuclides having a certain number of neutrons against the number of neutrons (- Figure 20.9) reveals some of the magic numbers—20, 28, 50, and 82. The complete set of magic numbers observed for protons is 2, 8, 20, 28, 40, 50, and 82; for neutrons, 2, 8, 20, 28, 40, 50, 82, and 126. The "magic numbers" for electrons are 2, 10, 18, 36, 54, and 86, the numbers of electrons in atoms of the noble gases. These magic numbers of electrons occur not only in the noble gases but also in many common ions such as Na^+, Mg^{2+}, Al^{3+}, O^{2-}, and F^-, and in molecules like CH_4, NH_3, H_2O, and HF. All these different species have a total of ten electrons.

Existence of magic numbers for nucleons is evidence for energy levels in the nucleus similar to the energy levels of the electrons outside the nucleus. The gamma radiation that often accompanies other types of radioactive decay provides additional evidence for nuclear energy levels. Gamma rays have definite energies characteristic of the nuclide just as atomic spectra consist of lines characteristic of the element (Section 7.7). For example, when magnesium-27 undergoes beta decay,

$$^{27}_{12}\text{Mg} \longrightarrow {}^{27}_{13}\text{Al} + {}^{0}_{-1}\text{e}$$

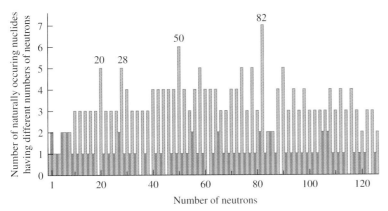

FIGURE 20.9 Number of naturally occurring nuclides having different numbers of neutrons.

Remember that the line spectrum of hydrogen furnished the data for calculation of the energy levels in Bohr's model of the hydrogen atom. Nuclear energy levels, like electron energy levels, are now calculated by means of the Schrödinger equation.

a total of 4.18×10^{-13} J is given off for *each* magnesium nucleus that decays. Emission of a beta particle from a magnesium-27 nucleus leaves an aluminum nucleus in one of two excited states. When an excited aluminum nucleus drops down into the ground state, gamma radiation of characteristic wavelength is emitted.

Figure 20.10 shows how the energy is distributed. Part goes into gamma rays having energy of either 1.62×10^{-13} or 1.35×10^{-13} J. Part goes into kinetic energy of the beta particle (2.55×10^{-13} or 2.80×10^{-13} J). The aluminum nucleus recoils just as a gun recoils when you shoot it. However, the aluminum atom does not move very fast or very far because its mass is so much greater than the mass of the electron. (Compare the speed and distance traveled by a bullet with the speed and distance traveled by the gun from which the bullet was fired.)

SAMPLE PROBLEM

20.3 Without referring to Figure 20.6, predict which of the following nuclides are probably stable. For the nuclides that are probably radioactive, predict which type of radioactive decay is most likely. Explain your answers. You may use the tables inside the front cover. (a) $^{17}_{9}F$ (b) $^{200}_{80}Hg$ (c) $^{20}_{8}O$ (d) $^{106}_{46}Pd$ (e) $^{210}_{84}Po$

Solution (a) The atomic mass of fluorine-17, which is close to 17, is lower than the atomic mass of naturally occurring fluorine given in the table inside the front cover (18.9984 u). Fluorine-17 has too few neutrons. Because fluorine has a low atomic number, fluorine-17 is more likely to decay by emitting positrons than by electron capture.

(b) The atomic mass of mercury-200, which is close to 200, is about the same as the atomic mass of naturally occurring mercury (200.59 u). Mercury-200 has even numbers of both protons and neutrons (80 p and 120 n) and is probably stable.

(c) The atomic mass of oxygen-20, which is close to 20, is higher than the atomic mass of naturally occurring oxygen (15.9994 u). Oxygen-20 has too many neutrons and will probably be radioactive and decay by beta emission.

(d) The atomic mass of palladium-106, which is close to 106, is about the same as the atomic mass of naturally occurring palladium (106.42 u). Palladium-106 has even numbers of both protons and neutrons (46 p and 60 n) and is probably stable.

(e) The atomic number of polonium (84) is greater than 83. No nuclides with atomic number > 83 are stable. Polonium is beyond the band of stability and will probably decay by alpha emission.

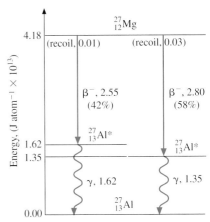

FIGURE 20.10 Magnesium-27 decays beta to two different excited states of aluminum-27. (Asterisks indicate excited states.) Gamma rays are emitted when excited aluminum-27 nuclei drop down to the ground state.

All of these predictions are correct.

20.4 PREDICTING WHICH TYPE OF RADIOACTIVE DECAY WILL TAKE PLACE

20.9 Without referring to Figure 20.6, predict which of the following nuclides are probably stable. For the nuclides that are probably radioactive, predict which type of radioactive decay is most likely. Explain your answers. You may use the tables inside the front cover. (a) $^{138}_{56}Ba$ (b) $^{24}_{10}Ne$ (c) $^{196}_{82}Pb$ (d) $^{214}_{82}Pb$ (e) $^{226}_{88}Ra$

20.10 Which has the shorter half-life, fluorine-17 or fluorine-18? Explain your answer.

Although consideration of the neutron : proton ratio, even–odd type, and magic numbers often results in correct predictions about radioactivity, it does not always do so. For example, consider the nuclides

$$^{8}_{4}Be \qquad and \qquad ^{14}_{7}N$$

Beryllium-8 has a neutron : proton ratio of 1 : 1 (usually the best neutron : proton ratio for an element with atomic number less than 20) and is even–even. But beryllium-8 is radioactive and beta decays with a half-life of only 7×10^{-17} s. On the other hand, although nitrogen-14 has a neutron : proton ratio of 1 : 1, it has an odd number of both protons and neutrons. Most odd–odd nuclides are radioactive. However, nitrogen-14 is not radioactive. Sometimes the energy change that will accompany a nuclear reaction must be calculated to predict correctly whether a nuclide will decay in a certain way. The energy changes that accompany nuclear reactions are discussed in Section 20.6.

20.5 RADIOACTIVE DECAY SERIES

Some nuclei can't gain stability by emission of a single particle. The radioactive decay of one nuclide leads to another radioactive nuclide; *a number of steps is required for the formation of a stable nuclide.* The thorium series shown in ▪Figure 20.11 is the simplest example of a naturally occurring **radioactive decay series.**

A *long-lived nuclide* is chosen as the first member or **parent** of a series. The decay of the parent is the slow, rate-determining step in the series (Section 18.11). *All the other nuclides in the series* are known as **daughters.** The radioactive daughters are constantly being produced and decaying; their concentration becomes constant, an example of a steady state (Section 18.13). In the natural radioactive decay series, alpha emission reduces the number of protons in the nucleus. The beta emissions are necessary because elements with lower atomic numbers have lower stable neutron : proton ratios than elements with higher atomic numbers.

The thorium series is one of three series that begin with a naturally occurring radionuclide and end with a stable isotope of lead, which has a magic number of protons (82). Most lead was formed by one of the three natural radioactive decay series. Because each series yields a different isotope of lead, the isotopic compositions of samples of lead from different places vary.

In addition to the elements in the three natural decay series, a few other radioactive nuclides occur naturally. The first to be discovered (1906) were rubidium-87, which has a half-life of 4.88×10^{10} y, and potassium-40 with a half-life of 1.26×10^{9} y. Rubidium-87 composes 27.83% of the element on Earth but only a very small percentage, 0.0117%, of potassium-40 remains in naturally occurring potassium. Either a very long half-life or very low concentration of a radionuclide makes radioactivity hard to detect. As better methods for detecting radioactivity are devel-

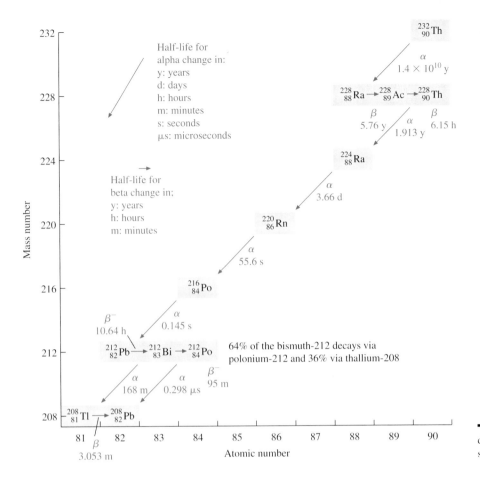

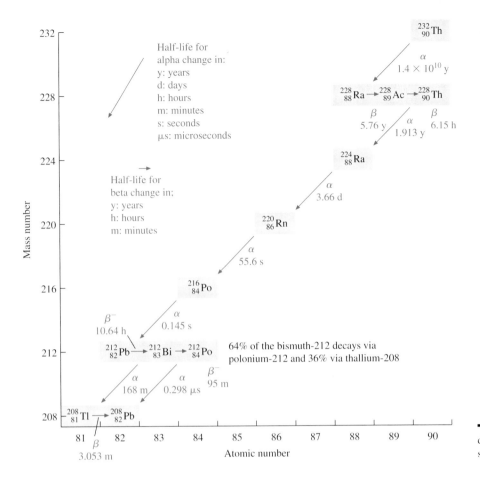

64% of the bismuth-212 decays via polonium-212 and 36% via thallium-208

■ FIGURE 20.11 The thorium radioactive decay series, the simplest naturally occurring series.

oped, some nuclides now classified as nonradioactive will probably be reclassified as radioactive.

20.6 ENERGY CHANGES ACCOMPANYING NUCLEAR REACTIONS

If ordinary chemical reactions are carried out in closed containers so that matter cannot escape, the total mass of the products equals the total mass of the reactants (Section 3.1). In nuclear reactions, the mass of the products is almost never equal to the mass of the reactants. The beta decay of carbon-14 is an example:

Carrying out reactions in closed containers is extremely dangerous. Suitable apparatus must be used, and proper safety precautions taken.

$$\underset{\substack{\text{mass nucleus, u} \quad 13.999\,950}}{^{14}_{6}\text{C}} \longrightarrow \underset{13.999\,234}{^{14}_{7}\text{N}} + \underset{(0.000\,549)}{^{0}_{-1}\text{e}}$$

The sum of the masses of the products

$$13.999\,234\ \text{u} + 0.000\,549\ \text{u} = 13.999\,783\ \text{u}$$

is not equal to the mass of the reactant, 13.999 950 u.

The explanation for the difference in mass between reactants and products of nuclear reactions is to be found in the equivalence of mass and energy. The equivalence of mass and energy was first proposed by Einstein in 1905 as a mathematical footnote to his special theory of relativity. Although Rutherford's nuclear model of the atom did not appear until 1911, the equivalence of mass and energy provides the key to understanding the energy changes that accompany nuclear reactions.

Mass-Energy and the Einstein Equation

According to Einstein's proposal, mass and energy are different forms of the same thing, which is called **mass-energy.** The conversion factor between mass and energy is the square of the speed of light:

$$E = mc^2 \qquad (20.3a)$$

If the speed of light, c, is in $m \cdot s^{-1}$ and mass, m, is in kg, then energy, E (where E stands for total potential and kinetic energy), will be in joules, J.* The Einstein equation is a combination of the law of conservation of mass and the law of conservation of energy:

total mass-energy before reaction = total mass-energy after reaction

Because we will usually be interested in energy changes and mass differences, a more useful form of the **Einstein equation** is

$$\boxed{\Delta E = \Delta mc^2} \qquad (20.3b)$$

From equation 20.3b you see that *if the mass of the products is less than the mass of the reactants (Δm is $-$), ΔE is negative. Nuclear reactions with negative values of ΔE are spontaneous and mass is converted into energy.* If the mass of the products is greater than the mass of the reactants (Δm is $+$), ΔE is positive. Nuclear reactions with positive values of ΔE are nonspontaneous. Energy must be supplied to make reaction take place; energy is converted into mass. Sample Problem 20.4 illustrates calculation of the energy change accompanying a typical nuclear reaction.

SAMPLE PROBLEM

20.4 For the reaction

$$^{50}_{22}\text{Ti} \longrightarrow {}^{50}_{23}\text{V} + {}^{0}_{-1}\text{e}$$

(a) Calculate the change in mass that will take place. The atomic mass of titanium-50 is 49.9448 u† and the atomic mass of vanadium-50 is 49.9472 u.
(b) Will this radioactive decay occur spontaneously? Explain your answer.
(c) Calculate the energy change in kJ/mol that will accompany this reaction.

Solution (a) Because the reaction involves nuclei, not atoms, it would seem as if nuclear masses, not atomic masses, should be used. To obtain the mass of a nucleus from the mass of an atom, the mass of the electrons outside the nucleus must be subtracted from the mass of the atom:

mass atom = mass nucleus + mass electrons

and

$$\text{mass atom} - \text{mass electrons} = \text{mass nucleus} \qquad (20.4)$$

The problem gives atomic masses because atomic masses are what you will find in tables of nuclide masses. From the table inside the back cover, the mass of an electron is 0.000 549 u.

---

*As you learned in Section 20.2, a few atoms are often all that are obtained by nuclear reactions, and nuclear equations are commonly interpreted in terms of individual particles, not moles. Nuclear chemists and physicists usually express the energy changes accompanying nuclear reactions in electron-volts, eV. One electron-volt is the energy gained by one electron when it is accelerated through a potential difference of one volt (1 eV = 1.602 1773 $\times$ 10^{-19} J). In this book, *all* energy changes are expressed in joules or kilojoules for easier comparison.

†Notice the small magnitude of the difference between accurate atomic masses and atomic masses estimated from mass numbers. The accurate atomic mass of titanium-50, for example, is 49.9448 u while the mass estimated from the mass number is 50.00 u, a difference of 0.06 u. Thus, the estimated atomic mass for titanium-50 is good to four significant figures.

The number of electrons outside the nucleus is the same as the number of protons inside the nucleus (given by the left subscript in the nuclide symbol). From equation 20.4, the nuclear masses of titanium-50 and vanadium-50 are

titanium-50: $[49.9448 \text{ u} - 22e^-(0.000\ 549 \text{ u/e}^-)]$

vanadium-50: $[49.9472 \text{ u} - 23e^-(0.000\ 549 \text{ u/e}^-)]$

The change in mass, Δm, from reactants to products is equal to the mass of the products minus the mass of the reactants:

$$\Delta m = m_{\text{products}} - m_{\text{reactants}}$$

The difference between the masses of the products and the mass of the reactant in this problem is

(mass vanadium-50 + mass electron) − mass titanium-50

$[49.9472 \text{ u} - 23e^-(0.000\ 549 \text{ u/e}^-)]$
$\qquad\qquad + 1e^-(0.000\ 549 \text{ u/e}^-) - [49.9448 \text{ u} - 22e^-(0.000\ 549 \text{ u/e}^-)]$

or

$49.9472 \text{ u} - 23e^-(0.000\ 549\ \text{u/e}^-) + 1e^-(0.000\ 549\ \text{u/e}^-)$
$\qquad\qquad\qquad\qquad\qquad - 49.9448 \text{ u} + 22e^-(0.000\ 549\ \text{u/e}^-)$

The terms for electrons cancel leaving just

$$49.9472 \text{ u} - 49.9448 \text{ u} = 0.0024 \text{ u}$$

the difference between the atomic masses of the two nuclides.

Not only do the masses of the electrons cancel but also, as we will see very soon, the energies involved in removing electrons from atoms are very small compared to the energies involved in nuclear reactions. The difference between the energy required to remove electrons from the products and the energy required to remove electrons from the reactants is negligible. *We could have saved ourselves some trouble and used atomic masses in the beginning without converting to nuclear masses.*

(b) Mass increases when the reaction

$$^{50}_{22}\text{Ti} \longrightarrow\ ^{50}_{23}\text{V} + \ ^{0}_{-1}e$$

takes place (Δm is +). The energy change is related to the mass change by the Einstein equation, $\Delta E = \Delta mc^2$. Because Δm is +, ΔE will be +. This radioactive decay will not occur spontaneously; titanium-50 does *not* undergo beta decay.

(c) From part (a), we know that $\Delta m = +0.0024$ u. This mass increase is for beta decay of a *single atom* of titanium-50. In other words, $\Delta m = +0.0024$ $\text{u}\cdot\text{atom}^{-1}$. The problem asks for the energy change in kJ/mol. We will need to use the following relationships from the inside back cover to write conversion factors: speed of light, $c = 3.00 \times 10^8 \text{ m}\cdot\text{s}^{-1}$; 1 u = 1.66×10^{-27} kg; and 1 mol of atoms = 6.02×10^{23} atoms. Substitution of c and Δm in the Einstein equation, $\Delta E = \Delta mc^2$, and conversion of units gives

$\Delta E = (+0.0024\ \text{u}\cdot\text{atom}^{-1})(1.66 \times 10^{-27}\ \text{kg}\cdot\text{u}^{-1})(6.02 \times 10^{23}\ \text{atom}\cdot\text{mol}^{-1})$
$\qquad\qquad\qquad\qquad\qquad\qquad\qquad\qquad \times (3.00 \times 10^8\ \text{m}\cdot\text{s}^{-1})^2$

$\quad = 2.2 \times 10^{11}\ \text{kg}\cdot\text{m}^2\cdot\text{s}^{-2}\cdot\text{mol}^{-1}$

In Section 6.5, we saw that, expressed in SI base units,

$$1 \text{ J} = 1\ \text{kg}\cdot\text{m}^2\cdot\text{s}^{-2}$$

Therefore,

$$\Delta E = 2.2 \times 10^{11}\ \text{J}\cdot\text{mol}^{-1} \text{ or } 2.2 \times 10^8\ \text{kJ}\cdot\text{mol}^{-1}$$

Atomic masses can usually be used to calculate mass changes for nuclear reactions; masses of electrons need not be considered. Positron emission is an exception to this rule. For positron emission, electron masses do not cancel and nuclear masses must be used.

20.11 For the reaction

$$^{14}_{6}\text{C} \longrightarrow \, ^{14}_{7}\text{N} + \, ^{0}_{-1}\text{e}$$

(a) Calculate the change in mass that will take place. The atomic mass of carbon-14 is 14.003 241 u and the atomic mass of nitrogen-14 is 14.003 074 u. (b) Will this radioactive decay occur spontaneously? Explain your answer. (c) Calculate the energy change in kJ/mol that will accompany this reaction.

The atomic masses used to calculate mass differences should have as many significant figures as possible because mass differences are small differences between large numbers. Atomic masses accurate to six or more decimal places can be obtained with a high-resolution mass spectrometer. Careful measurements of energy released can be used to determine atomic masses of nuclides that decay too rapidly for mass spectrometric determination of mass. For example, a neutron has a half-life of only 11 minutes outside a nucleus. Measurement of the energy given off by the formation of hydrogen-2 by addition of a neutron to hydrogen-1

$$^{1}_{1}\text{H} + \, ^{1}_{0}\text{n} \longrightarrow \, ^{2}_{1}\text{H}$$

was used to obtain the mass of the neutron. Agreement between calculated energies and energies measured by experiment provided some of the first evidence that Einstein's theory of the equivalence of mass and energy is correct.

When 1 mol of methane gas is burned, about 10^3 kJ of energy are given off. Beta decay of 1 mol of titanium-50 would adsorb 2.2×10^8 kJ. The energy changes accompanying nuclear reactions are about 10^5 times as large as the energy changes accompanying chemical reactions. Small changes in mass result in large changes in energy because the conversion factor between mass and energy, the square of the speed of light (9.00×10^{16} m$^2 \cdot$s^{-2}), is such a large number. The mass changes associated with ordinary chemical reactions are too small to be detected. For example, the heat of combustion of methane gas is -892.36 kJ mol^{-1}; loss of 892.36 kJ of thermal energy is accompanied by a decrease in mass of only 10^{-8} g. This mass decrease is too small to be detected with the balances that are currently available.

⦿ Although the mass lost when 1 mol of methane gas burns is undetectable, it would be hard to avoid noticing the thermal energy that is given off.

Mass Defect and Nuclear Binding Energy

The mass of the nucleus of an atom is always smaller than the sum of the masses of the nucleons that compose the nucleus. For example, consider the hydrogen-2 nucleus, which has a mass of 2.013 554 u. A hydrogen-2 nucleus is made up of one proton, mass = 1.007 276 u, and one neutron, mass = 1.008 665 u. The sum of the masses of the nucleons in a hydrogen-2 nucleus is

$$\begin{array}{cc} \text{mass one proton} + \text{mass one neutron} \\ (1.007\ 276\ \text{u}) \quad + \quad (1.008\ 665\ \text{u}) \quad = 2.015\ 941\ \text{u} \end{array}$$

The *difference between the observed mass of a nucleus and the sum of the masses of the nucleons* is called the **mass defect** because mass is always lost when nucleons combine to form a nucleus. The mass defect for hydrogen-2 is

$$\begin{array}{ccc} \text{observed nuclide mass} - (\text{sum of masses of nucleons}) = & \text{mass defect} \\ (2.013\ 554\ \text{u}) \quad - \quad (2.015\ 941\ \text{u}) & = -0.002\ 387\ \text{u} \end{array}$$

A loss in mass of 0.002 387 u for the formation of a hydrogen-2 nucleus corresponds to a release of energy of

$$\Delta E = \Delta mc^2$$
$$= (-0.002\ 387\ \text{u·atom}^{-1})(1.66 \times 10^{-27}\ \text{kg·u}^{-1})(6.02 \times 10^{23}\ \text{atom·mol}^{-1})$$
$$\times (3.00 \times 10^8\ \text{m·s}^{-1})^2$$
$$= 2.15 \times 10^{11}\ \text{kg·m}^2\text{·s}^{-2}\text{·mol}^{-1} = 2.15 \times 10^{11}\ \text{J·mol}^{-1}$$
$$= 2.15 \times 10^8\ \text{kJ·mol}^{-1}$$

When hydrogen-2 nuclei are formed from protons and neutrons, $2.15 \times 10^8\ \text{kJ·mol}^{-1}$ of energy are given off; therefore, a minimum of $2.15 \times 10^8\ \text{kJ·mol}^{-1}$ of energy must be supplied to break up an atom of hydrogen-2 into a proton and a neutron. The *energy required to break a nucleus apart into its nucleons* is called the **nuclear binding energy** of the nucleus.

The greater the number of nucleons in a nucleus, the greater the binding energy. The binding energies of the naturally occurring nuclides range from 3.74×10^{-13} J for hydrogen-2 to 2892×10^{-13} J for uranium-238. *Different nuclides are best compared by calculating the binding energy per nucleon,* as illustrated in Sample Problem 20.5.

SAMPLE PROBLEM

20.5 For the nuclide neptunium-239, which has an atomic mass of 239.052 933 u, (a) what is the mass defect? (b) Calculate the total binding energy in J. (c) What is the binding energy per nucleon in J?

Solution (a) The mass defect of an atom is the mass lost when the atom is formed from neutrons and protons. *Atomic masses usually given in tables can be used to calculate mass defects. For electrons to cancel, the atomic mass of hydrogen-1 must be used for the mass of the protons in the nucleus.* Hydrogen-1 (H) has an atomic mass of 1.007 825 u.

First, we must figure out how many protons and neutrons are in the nucleus of a neptunium-239 atom. From the table inside the front cover, the atomic number of neptunium is 93; there are 93 protons in the nucleus of a neptunium atom. The mass number is the total number of nucleons in the nucleus:

mass number = number of protons + number of neutrons

Solving for the number of neutrons gives

mass number − number of protons = number of neutrons

For Np-239   239   −   93   =   146

From the table inside the back cover, the mass of a neutron is 1.008 665 u. The difference between the atomic mass of neptunium-239 and the sum of the masses of the nucleons in the nucleus is

239.052 933 u − [93 H(1.007 825 u·H^{-1}) + 146 n(1.008 665 u·n^{-1})]
$$= 239.052\ 933\ \text{u} - 93.727\ 73\ \text{u} - 147.2651\ \text{u} = -1.9399\ \text{u}$$

The mass defect is −1.9399 u.

(b) The total binding energy can be calculated from the change in mass by means of the Einstein equation, $\Delta E = \Delta mc^2$. Substitution of the speed of light, $2.9979 \times 10^8\ \text{m·s}^{-1}$, and the change in mass, −1.9399 u, into the Einstein equation together with the conversion factor from the relationship

$$1\ \text{u} = 1.6605 \times 10^{-27}\ \text{kg gives}$$
$$\Delta E = (-1.9399\ \text{u})(1.6605 \times 10^{-27}\ \text{kg·u}^{-1})(2.9979 \times 10^8\ \text{m·s}^{-1})^2$$
$$= -2.8950 \times 10^{-10}\ \text{J}$$

▶ Notice that there are only five significant figures in the mass defect although there are eight in the atomic mass of neptunium-239 and seven in the atomic mass of hydrogen-1 and the neutron. Significant figures are always lost when one large number is subtracted from another large number to give a small difference.

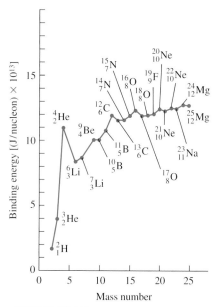

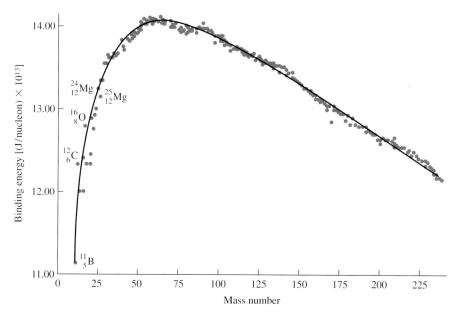

- FIGURE 20.12 Binding energies per nucleon as a function of mass number for naturally occurring nuclides. The enlargement at the left shows mass numbers 1–25.

This much energy is released when 93 protons and 146 neutrons combine to form a neptunium-239 nucleus. Therefore, $+2.8950 \times 10^{-10}$ J are required to break a neptunium-239 nucleus apart into its nucleons, and the binding energy of neptunium-239 is 2.8950×10^{-10} J.

(c) There are 239 nucleons in the nucleus of an atom of neptunium-239. The binding energy per nucleon is

$$\frac{2.8950 \times 10^{-10} \text{ J}}{239 \text{ nucleons}} = 1.2113 \times 10^{-12} \text{ J} \cdot \text{nucleon}^{-1}$$

PRACTICE PROBLEM

20.12   For the nuclide iron-56, which has an atomic mass of 55.934 939 u, (a) what is the mass defect? (b) Calculate the total binding energy in J. (c) What is the binding energy per nucleon in J? Hydrogen-1 (H) has an atomic mass of 1.007 825 u.

- Figure 20.12 is a graph of the average binding energy per nucleon as a function of mass number. Energies of the order of 10^{-13} J look like very small quantities of energy. However, remember that this amount of energy is *per nucleon*. You are used to thinking in terms of energy per mole. As we have seen, the energy changes associated with nuclear reactions are enormous compared to the energy changes associated with chemical reactions.

From Figure 20.12, you can see that, except for a few of the lightest nuclei, the average binding energy per nucleon ranges from 12×10^{-13} to 14×10^{-13} J. In other words, the average binding energy per nucleon is fairly constant. Maximum values of average binding energy occur for mass numbers 50–65, that is, around iron. From the maximum around iron, values of average binding energy per nucleon decrease slowly toward the high mass number side and rapidly toward the low mass number side.

The binding energy of helium-4, the alpha particle, is unusually high compared to the binding energies of neighboring nuclides. Other nuclides with mass number divisible by four such as carbon-12, oxygen-16, neon-20, and magnesium-24, are also at the top of peaks in Figure 20.12. However, beryllium-8 does not occur naturally; beryllium-8 has a half-life of 2×10^{-16} s. Decay of beryllium-8 gives *two* alpha particles:

$$^{8}_{4}\text{Be} \longrightarrow 2(^{4}_{2}\text{He})$$

The mass of the nucleus of an atom is always smaller than the sum of the masses of the nucleons that compose the nucleus (Δm is negative), and the energy of formation of nuclei from free nucleons is always negative. The fact that nuclei have lower potential energies than separate nucleons indicates that there is an attractive force between the nucleons in a nucleus. The large energy changes that accompany nuclear reactions show that *the attractive force between nucleons in a nucleus* is very strong; physicists call it the **nuclear strong force.** If the attractive force between nucleons were exerted over the whole volume of the nucleus, the attractive force per nucleon would increase as the number of nucleons increased. However, the potential energy per nucleon is quite constant for all but the lightest nuclei (see Figure 20.12). Therefore, the attractive force is exerted only on the nucleon's nearest neighbors like the intermolecular attractions that hold molecules together in a drop of liquid. The nuclear strong force is very short range.

Potential Energy Per Nucleon as a Function of Mass Number

Another way of showing attraction between nucleons is to graph potential energy per nucleon against mass number (■Figure 20.13). In Figure 20.13, a smooth curve has been drawn through the points; the minimum in the curve represents the lowest energy as is usually the case for energy diagrams. The curve of Figure 20.13 can be used to make some general predictions about nuclear reactions. For example, nuclides with high mass numbers can become more stable (reach lower energies) by decaying in such a way as to attain lower mass numbers. Fission and alpha decay result in lower mass numbers. Nuclides with high mass numbers will split into nuclides with mass numbers between 50 and 100 or alpha decay. Fission results in a large change in mass number and large energy change; emission of an alpha particle only results in a small change in mass number and a small energy change. More energy will be released by fission than by alpha decay. From the band of stability in Figure 20.6, one could predict that radium-226 would probably alpha decay and become radon-222, but aluminum-27 would be unlikely to alpha decay to form sodium-23.

Nuclides with low mass numbers (below about 60, the mass number of the potential energy minimum) can become more stable by combining to form nuclides with higher mass numbers. More energy will be released by combination of nuclides with low mass numbers than by fission of nuclides with high mass numbers.

All known nuclides, no matter how short their half-lives, have negative energies of formation (positive binding energies). For example, beryllium-8 with a half-life of only 7×10^{-17} s has a total binding energy of 9.0518×10^{-12} J or 11.31×10^{-13} J·nucleon^{-1}. If beryllium-8 occurred naturally, it would be at the top of a peak in Figure 20.12 like helium-4, carbon-12, oxygen-16, neon-20, and magnesium-24. The reason beryllium-8 is not stable is because the energy change for the alpha decay of beryllium-8 is negative. *A nucleus is unstable if a type of decay exists that results in release of energy.* A positive energy of formation (negative binding energy) would mean that a nuclide would spontaneously break up into the nucleons of which it is composed. This does not happen. Unstable nuclei of high mass number such as neptunium-239 have very high total binding energies because they are made up of lots of nucleons. *A nucleus always has lower energy and is more stable than the separate nucleons.*

A liquid-drop model for the nucleus was first proposed by the Danish physicist Niels Bohr, the same person who suggested that the energy of the electron in a hydrogen atom is quantized.

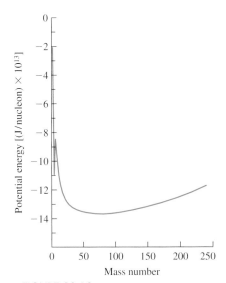

■ FIGURE 20.13 Potential energy per nucleon as a function of mass number.

20.7 FISSION

For practical purposes, the most important fission is fission of uranium-235 induced by thermal neutrons. Fission was discovered in 1938 by an Italian physicist, Enrico Fermi,† in the course of investigations of neutron capture. When Fermi used uranium as a target for neutrons, he obtained puzzling radioactive substances that he, a physicist, could not identify. In 1939 the German physical chemist Otto Hahn and Fritz Strassman, who was an expert analytical chemist, identified one of the products as barium, an element with an atomic number only 60% of the atomic number of uranium. Hahn sent news of this amazing result to his former co-worker, Lise Meitner, in Sweden who, with her nephew, Otto Frisch, figured out what had happened and named the process fission. ▬Figure 20.14 shows Meitner's explanation.

When a slow neutron hits a uranium-235 nucleus, the neutron becomes part of the nucleus forming a compound nucleus of uranium-236*. A **compound nucleus** is a *nucleus that is composed of a projectile and a target nucleus.* The asterisk shows that the compound nucleus is in an excited state. Because the compound nucleus of uranium-236 is in an excited state, it moves back-and-forth between various shapes, as you can imagine a drop of liquid doing, and splits into two parts:

The ratio of neutrons to protons is lower in nuclides of lower atomic number. Therefore, two or more neutrons are released when a nuclide of high atomic number fissions. Only one neutron is required to cause fission. *More neutrons are released than are used up and the number of neutrons increases rapidly* as shown in Figure

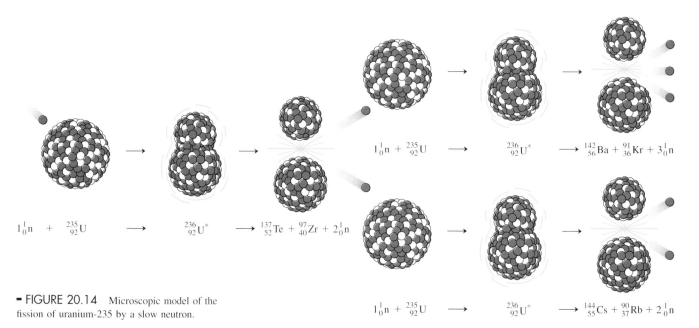

▬ FIGURE 20.14   Microscopic model of the fission of uranium-235 by a slow neutron.

---

†Fermi received the Nobel Prize in 1938 for his work with slow neutrons. Given permission by Mussolini's Fascist government to go to Sweden with his wife and family to receive the prize, he took advantage of the opportunity to escape Fascist Italy and emigrated to the United States. Fermi's use of slow neutrons as projectiles was an extension of the work of the Joliot-Curies, who you may remember, made the first synthetic radionuclides by using alpha particles as projectiles.

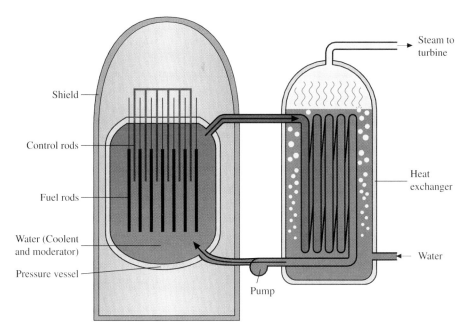

■ FIGURE 20.15 Pressurized water reactor. The fuel rods are about 1 cm in diameter and 4 m long. They contain UO_2 pellets enclosed in zirconium. The uranium in the fuel pellets is enriched to about 3% uranium-235 from the natural composition of 0.72% uranium-235 by gaseous effusion of UF_6 (Section 5.9). The "core" contains about 40 000 fuel rods and 1000 control rods. The tank containing the water that acts as the coolant and moderator is about 10 m high and 3 m in diameter with stainless steel walls 20 cm thick. The water enters this vessel at about 280 °C and leaves it at about 320 °C at a pressure of 150 atm. The temperature of the coolant–moderator water is high enough to boil the water in the heat exchanger producing steam. The steam is used to drive a turbine to generate electricity. The reactor tank is enclosed in a thick shield and the reactor building is surrounded by an outer containment (not shown) to prevent escape of radioactivity if an accident should occur.

20.14. If the neutrons released by fission are slowed down, they can cause further fission reactions to take place. A reaction like this is called a **branching chain** reaction.

What happens depends on the size of the piece of uranium that is undergoing fission. If the piece of uranium is small, the surface area is relatively large. Many neutrons escape from the surface, the chain is broken, and fission stops. *A sample that is too small to sustain a chain reaction* is called **subcritical.** If the piece of uranium is just the right size so that *the number of neutrons formed is exactly equal to the number of neutrons used up,* the chain continues in a controlled fashion. The sample is said to be **critical.** A *piece* of uranium *larger than critical* is referred to as **supercritical.** In a supercritical sample, *the number of neutrons increases very rapidly.* For example, as shown in Figure 20.14, one neutron brings about one fission and forms two neutrons. The two neutrons bring about two fissions and form five neutrons, and so on. An average of 2.5 neutrons is formed per fission. An enormous amount of energy is released in a very short period of time and a tremendous explosion—an atomic bomb—results.

On the other hand, energy from the controlled fission of a critical mass of uranium can be used to convert water to steam as shown in the diagram of the most common type of nuclear reactor in ■Figure 20.15. The steam can be used to generate electricity just as steam produced from water by burning coal or natural gas can be used to generate electricity.

The neutrons formed by fission are fast neutrons. Fast neutrons do not split uranium-235. Slow neutrons, which remain near a uranium-235 nucleus long enough to become a part of it, are required for fission. The **moderator** *slows the neutrons down.* The slow neutrons then pass back into the fuel rods and bring about more fission.

Together *the fuel rods and the control rods* are known as the **core.** The **control rods** are usually made of cadmium, which *absorbs slow neutrons very effectively.* If the number of neutrons begins to increase, the control rods are lowered between the fuel rods to absorb neutrons and slow the reaction. If the number of neutrons begins to decrease the control rods are raised, speeding up reaction. Thus, the nuclear reactor can be controlled. If the control rods stick or something interferes with the

A simple mechanical analogy for a branching chain can be constructed by standing several sets of dominoes on end.

The critical mass of uranium-235 is 22.8 kg or about 50 pounds, a sphere of uranium about 5.2 inches in diameter.

About 20% of the electricity generated in the United States is produced in nuclear power plants. Nuclear reactors are also used to power submarines and to provide power aboard satellites and space vehicles.

Top: *Three Mile Island nuclear power plant. In the accident at this plant, only about 2.5 MCi of radioactive krypton and xenon escaped, less than the 3 MCi of radiation added to the atmosphere by the eruption of the Mt. St. Helens volcano in 1980.* Bottom: *Measuring radiation inside the reactor at Chernobyl five years after the accident.*

Some reactors are run in order to make plutonium-239 for the manufacture of nuclear weapons.

Replacing fuel rods in a nuclear reactor. The blue glow is Cerenkov radiation and is caused by electrons from the reactor moving at speeds greater than the speed of light in water, which is 75% of the speed of light in a vacuum.

flow of the water that acts as both coolant and moderator, either reaction stops or the core can heat up and melt. However, a nuclear reactor can never explode like an atomic bomb; the concentration of uranium-235 in the fuel is too low.

The difference between the results of the accidents at Three Mile Island in Pennsylvania and at Chernobyl in the Ukraine shows the importance of containment. The nuclear reactor at Three Mile Island was enclosed in concrete, and cooling water failure resulted in only a small release of radioactive materials to the atmosphere,* mostly in the form of inert gases that quickly diffused so that no one was exposed to much radiation. The nuclear reactor at Chernobyl was not enclosed, and that accident, the worst radiation accident in history, killed a number of people and contaminated a large area.

After a few years, the concentration of fission products that capture neutrons builds up to the point where power can no longer be produced at a practical rate. The fuel rods must be replaced, leading to the problem of what to do with the used fuel. The fission products in the used fuel contain too many neutrons and are radioactive. They will continue to be dangerously radioactive for thousands of years. No one wants them stored near where they live. Concern about storing radioactive waste has just about stopped construction of nuclear power plants in the United States, although generation of electricity by nuclear power prevents atmospheric pollution by oxides of sulfur, nitrogen, and carbon formed by burning fossil fuels.

Another problem with nuclear power is that supplies of uranium-235 are limited. Almost all naturally occurring uranium (99.275%) is uranium-238, which is not split by thermal neutrons. Instead uranium-238 captures neutrons. The product of neutron capture by uranium-238 is plutonium-239:

$$^{238}_{92}U + ^{1}_{0}n \longrightarrow ^{239}_{92}U + \gamma$$

$$^{239}_{92}U \longrightarrow ^{239}_{93}Np + ^{0}_{-1}e$$

$$^{239}_{93}Np \longrightarrow ^{239}_{94}Pu + ^{0}_{-1}e$$

Plutonium-239 is formed as a side reaction in power reactors. Plutonium-239 is fissionable and could be recovered from used fuel rods and used as a fuel or reactors could be designed especially to make plutonium-239 from uranium-238. In fact, **breeder reactors** can be built that *not only produce thermal energy to generate electricity but also make more fuel for themselves.*

Fission of plutonium-239 is similar to fission of uranium-235 except that fast neutrons are used and no moderator is necessary. In breeder reactors, plutonium-239 is mixed with uranium-238. One of the neutrons produced by fission of plutonium-239 is used to continue the chain reaction. The other neutrons are used to convert uranium-238 to plutonium-239. In seven to ten years, the reactor not only produces enough fuel to refuel itself but also to fuel another reactor. All of the energy of the uranium can be used instead of only about 1% of it. In addition, because breeder reactors run at higher temperatures than ordinary nuclear reactors, breeders generate electricity more efficiently than ordinary nuclear reactors. However, control is more difficult and corrosion is a greater problem at higher temperatures.

The used fuel from breeder reactors must be reprocessed to recover plutonium. The plutonium can be used to make nuclear weapons as well as to fuel reactors, and many people are concerned about the possibility of theft of plutonium by

---

*Loss of cooling water allowed the core temperature to rise to more than 2200 °C. At about 1100 °C, the zirconium metal that enclosed the UO_2 pellets in the fuel rods began to react with steam, producing hydrogen gas, $Zr(s) + 2H_2O(g) \longrightarrow ZrO_2(s) + 2H_2(g)$. To prevent explosion of the hydrogen, gas was released into the atmosphere.

terrorists because only a few kilograms are required for a bomb. In addition plutonium is very poisonous and carcinogenic. Plutonium-239 has a half-life of 2.4×10^4 y; if any plutonium escapes into the biosphere, it will be around a long time. Research on breeder reactors has stopped in the United States, but other countries are continuing to use and develop them.

20.8 FUSION

The maximum in the binding energy curve (Figure 20.12) comes at a mass number of 50–65; combination of nuclides of low mass number as well as fission of nuclides of high mass number should be a source of energy. *Combination of nuclides of low mass number* is called **fusion.** A simple example of fusion is

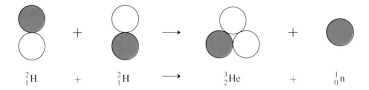

$$\ _{1}^{2}\text{H} \quad + \quad \ _{1}^{2}\text{H} \quad \longrightarrow \quad \ _{2}^{3}\text{He} \quad + \quad \ _{0}^{1}\text{n}$$

However, very high temperatures are required for fusion because nuclei have positive charges and repel each other. For fusion to take place, the nuclei must have enough kinetic energy to overcome the repulsion between like charges.

In fusion weapons, which are called hydrogen bombs or thermonuclear bombs, the kinetic energy needed to overcome repulsion between like charges is obtained by exploding a fission bomb. Although no official information is available, it is believed that modern hydrogen bombs are made of lithium hydride in which the hydrogen is hydrogen-2. Neutrons from the fission bomb react with lithium-6 to form hydrogen-3:

$$_{3}^{6}\text{Li} + \ _{0}^{1}\text{n(slow)} \longrightarrow \ _{1}^{3}\text{H} + \ _{2}^{4}\text{He}$$

$$_{3}^{7}\text{Li} + \ _{0}^{1}\text{n(fast)} \longrightarrow \ _{1}^{3}\text{H} + \ _{2}^{4}\text{He} + \ _{0}^{1}\text{n}$$

Hydrogen-2 and hydrogen-3 then fuse:

$$_{1}^{2}\text{H} + \ _{1}^{3}\text{H} \longrightarrow \ _{2}^{4}\text{He} + \ _{0}^{1}\text{n}$$

$$_{1}^{2}\text{H} + \ _{1}^{2}\text{H} \longrightarrow \ _{2}^{3}\text{He} + \ _{0}^{1}\text{n}$$

$$_{1}^{2}\text{H} + \ _{1}^{2}\text{H} \longrightarrow \ _{1}^{3}\text{H} + \ _{1}^{1}\text{p}$$

$$_{1}^{2}\text{H} + \ _{2}^{3}\text{He} \longrightarrow \ _{2}^{4}\text{He} + \ _{1}^{1}\text{p}$$

Fusion weapons are far more powerful than fission weapons. They have been tested by several countries (the U.S., the U.S.S.R., Great Britain, China, France, and India) but never used in war.

If fusion is to be used as a source of power, more energy must be released by fusion than is used in the acceleration of nuclei to high enough speeds so that they can collide and combine. Temperatures of 10^8 K are required. At temperatures of 10^8 K, all substances are in the plasma state; atoms are completely ionized to electrons and nuclei (Section 5.11). The problem with fusion power is to reach a temperature of 10^8 K and *at the same time* contain the sample in a small volume for a long enough time to permit nuclei to collide and combine. Strong magnetic fields are under investigation as containers; another possibility is use of intense beams of electromagnetic radiation or particles to heat and compress the sample.

The fireball that resulted from the detonation of a hydrogen bomb at Bikini Atoll, May 21, 1956. The picture was taken from an airplane flying at 3700 m at a distance of 80 km from the blast.

A technician positioning the target for an experiment using intense beams of electromagnetic radiation to initiate fusion.

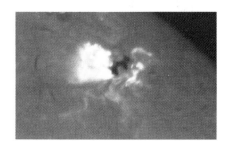

A solar flare. The fusion reactions that take place in the sun are the ultimate source of most of our energy.

The reaction

$$\text{}^{2}_{1}\text{H} + \text{}^{3}_{1}\text{H} \longrightarrow \text{}^{4}_{2}\text{He} + \text{}^{1}_{0}\text{n}$$

which releases 2.82×10^{-12} J per event (1.697×10^{9} kJ/mol of reaction or 3.37×10^{8} kJ/g fuel), appears most promising as a source of power because it has the lowest activation energy of any fusion reaction and takes place at the lowest temperature. The nuclide hydrogen-2 can be obtained from water. Although naturally occurring hydrogen contains only 0.015% hydrogen-2, 2.5×10^{13} tons of hydrogen-2 are available in the 3.6×10^{20} gallons of water in Earth's oceans. Hydrogen-3 does not occur naturally and would have to be made, probably by the reactions of lithium-6 and lithium-7 with neutrons shown previously. Supplies of lithium are estimated to be sufficient for a million years. However, whether technology for controlled fusion can be developed remains to be seen.*

In the meantime, the fusion reactions that take place in the interior of the sun,

$$2\text{}^{1}_{1}\text{H} \longrightarrow \text{}^{2}_{1}\text{H} + \text{}^{0}_{+1}\text{e} \qquad (20.5)$$

$$\text{}^{1}_{1}\text{H} + \text{}^{2}_{1}\text{H} \longrightarrow \text{}^{3}_{2}\text{He} + \gamma \qquad (20.6)$$

$$2\text{}^{3}_{2}\text{He} \longrightarrow \text{}^{4}_{2}\text{He} + 2\text{}^{1}_{1}\text{H} \qquad (20.7)$$

● Combination of four protons must take place in steps because the simultaneous collision of four particles is unlikely.

which combine [2(20.5) + 2(20.6) + (20.7)] to give the reaction (20.8),

$$4\text{}^{1}_{1}\text{H} \longrightarrow \text{}^{4}_{2}\text{He} + 2\text{}^{0}_{+1}\text{e} \qquad (20.8)$$

Global energy consumption is now about 6.5×10^{14} kJ per day.

will probably continue to shower down 1.5×10^{19} kJ of energy each day on the surface of Earth as they have done every day for millions of years. This quantity of energy is estimated to be approximately equal to the total amount of energy that humans have used since the beginning of time.

20.9 BIOLOGICAL EFFECTS OF RADIATION

Unfortunately, students now need to be familiar with both traditional and SI units. Although traditional units are still widely used, SI have started to appear in articles likely to be read by science students. See, for example, the article "Controlling Indoor Air Pollution" by A. V. Nero, Jr., in the May 1988 issue of Scientific American.

Why must nuclear reactors be contained, and why is radioactive waste such a problem? When radiation passes through matter, electrons are promoted to higher energy levels, and atoms and molecules are ionized. Excitation and ionization in nonliving matter are used in scintillation counters and ionization counters to measure radioactivity. In living matter, excitation and ionization break bonds in complex organic molecules and cause damage to cells. Interaction of radiation with water, both inside and outside of cells, produces radicals (Section 9.10). The radicals damage cells by causing oxidation–reduction reactions. A person, animal, or plant exposed to ionizing radiation may suffer either short-term or long-term injury. In addition, the offspring of the person, animal, or plant may be defective.

● Ionizing radiation is especially dangerous because you can't see it, hear it, smell it, or feel it.

The **dose** of ionizing radiation is the *quantity of ionizing radiation absorbed by a unit mass of matter.* Dose is measured in grays or rads. The **gray, Gy,** is the *SI unit of radiation dose* and is defined as joules of energy absorbed per kilogram of

---

*Because fusion offers the possibility of generating nearly unlimited power, reports that fusion had been achieved at room temperature in 1989 caused great excitement. The "cold fusion" was supposedly obtained by electrolyzing "heavy" water (^{2}H$_2$O) using a palladium cathode. Palladium absorbs hydrogen, and it was claimed that the hydrogen atoms in the palladium were close enough together to fuse. Unfortunately, hundreds of attempts to duplicate the work have failed. Cold fusion has probably been achieved using particles called muons as catalysts. Muons are similar to electrons but 207 times as massive. They are made by causing a beam of ions from an accelerator to collide with a sample of ordinary matter such as carbon. The problem with muon-catalyzed fusion is the short lifetime of muons; the average lifetime of a muon is only about 2 μs.

target material:

$$1 \text{ Gy} = 1 \text{ J/kg}$$

The *traditional unit for radiation dosage* is the **rad** (*r*adiation *a*bsorbed *d*ose); 1 Gy = 100 rad.

A dose in terms of absorbed energy is not enough. Different kinds and energies of radiation affect tissues differently. Before doses can be compared, they must be multiplied by a **quality factor, Q.** The quality factor is sometimes referred to as the *relative biological effectiveness* or RBE value. Table 20.5 lists the quality factors for common types of radiation.

The *SI unit that takes the type of radiation into account* is called the **sievert, Sv.** A sievert is a gray multiplied by the quality factor:

$$\text{Sv} = \text{Gy} \times Q$$

The *traditional unit that takes the type of radiation into account* is called the **rem** (*r*oentgen *e*quivalent *m*an):

$$\text{rem} = \text{rad} \times \text{RBE}$$

One sievert is equal to 100 rem.

To further complicate the matter, some tissues are more sensitive than others to the same radiation. Specialized cells and cells undergoing division are more affected by radiation. For example, radiation is 100 times as dangerous to a fetus during the third to seventh weeks of pregnancy than it is to the pregnant woman. Organs and tissues in which cells are replaced slowly, such as eyes, exhibit high radiation sensitivity. The age and health of the individual are also factors. Radiation damage to cells is shown in several ways. The rate of cell division is reduced, and the cell undergoes fewer divisions during its lifetime; cell mutations may take place. Precise evaluation of the possible biological effects of radiation is difficult.

External alpha radiation is not harmful because alpha radiation is not very penetrating. Alpha particles are stopped by the outer dead layer of skin. However if alpha particles are introduced into the body by breathing or eating or through a wound, they leave a stream of ions behind as they move through living tissue just as they do when they move through a Geiger counter. As a result, internal alpha particles are extremely harmful. Beta radiation is stopped by about 1 cm of tissue. Gamma rays, like X-rays, are very penetrating. Lead bricks or thick concrete are required to stop gamma rays.

The effect of different doses of radiation on people is shown in Table 20.6. The doses are instantaneous whole-body doses, that is, the whole dose is received within

TABLE 20.5	
Q-Values for Various Kinds of Radiation	
Radiation	Q (RBE)
X-rays, β, γ	1
Slow neutrons	3
Fast neutrons	10
α	20

The roentgen is a unit developed for X-rays.

TABLE 20.6	Effects of Instantaneous Whole-Body Radiation Doses on People
Dose, Sv (rem)	Effect
≥10 (1000)	Death within 24 h from destruction of the neurological system.
7.5 (750)	Death within 4–30 d from gastrointestinal bleeding.
1.5–7.5 (150–750)	Intensive hospital care required for survival. At the higher end of range, death through infection resulting from destruction of white blood cell–forming organs usually takes place 4–8 weeks after the accident. Those surviving this period usually recover.
<0.5 (50)	Only proven effect is decrease in white blood cell count.

A badge for measuring exposure to radiation.

A home radon detector.

The lifetime dose from the accident at the Chernobyl nuclear power plant was 100 mSv to evacuees from the area around the plant and 0.3 mSv to people around the world.

The international symbol warning of ionizing radiation and radiation sources.

TABLE 20.7	Sources of Human Radiation Doses for Nonsmokers[a]	
Source	Dose Rate, mSv/y (mrem/y)	Percent
Natural background		
Airborne radon	2.00 (200)	55
Cosmic rays[b] and terrestrial sources including K-40	1.00 (100)	28
Medical procedures		
X-ray diagnosis	0.39 (39)	11
Nuclear medicine	0.14 (14)	4
Consumer products other than tobacco products	0.09 (9)	2
All other sources (occupational, nuclear power, weapons testing fallout, etc.)	0.01 (1)	0
Total	3.63 (363)	100

[a]Data are from Moeller, D. W. *American Council on Science and Health News and Views,* **1988,** 9(2), 8.
[b]Cosmic rays are radiation of great penetrating power that falls continuously on Earth from somewhere outside Earth. The origin of cosmic rays is still under investigation. Near the top of Earth's atmosphere, cosmic rays produce violent nuclear reactions in which neutrons are formed. Most of the neutrons are slowed to thermal energies and react with nitrogen-14.

less than a day over the whole body. The dose required to produce a certain amount of damage depends on the period of time over which the dose is received. A dose that is deadly within a short period of time may lead to very few symptoms if spread out over the normal lifetime of the individual. Some effects of exposure to radiation may not be evident for many years.

Table 20.7 shows the sources of the radiation doses received by people. As you can see, the major part (83%) of the average annual dose of radiation *for nonsmokers* is from natural sources; only 17% is a result of human activities. Diagnostic X-rays are the largest source of exposure from artificial sources. The estimated dose for smokers from tobacco products is 13 mSv/y, over three times the average dose for nonsmokers.

Airborne radon is responsible for almost two-thirds of the radiation from natural sources. The isotope of radon that has the biggest effect on human health is radon-222. Radon-222 is formed in the decay series of uranium-238, which is present in most rocks and soils. It is a colorless, odorless, radioactive gas that enters houses and other buildings through cracks in the foundation and openings around pipes. Radon's daughters, polonium-214 and polonium-218, which are both alpha emitters, tend to lodge in the lungs. Radon is unquestionably a danger to uranium miners and other people who are exposed to high doses. Whether the levels found in homes increase the risk of cancer is still not known. A number of studies have been carried out, but the results do not agree.

Doses from cosmic radiation vary from 0.30 to 0.40 mSv at sea level to 1.00 to 1.50 mSv at an altitude of 3000 m. As a result of variations in natural background, determination of long-term effects of radiation is difficult. Scientists differ as to whether there is a threshold level of radiation below which radiation is not harmful or whether radiation effects at low dosages should be estimated by extrapolating the effects of high doses. There is even some evidence that low radiation doses similar to natural background radiation may reduce the cancer rate.

Another factor in the biological effects of radiation is the chemical properties of the radioactive source. The damage from internal radiation sources depends on how long the radiation source stays in the body. Radioactive strontium is particularly dangerous because strontium is similar in chemical properties to calcium and collects in bones; radioactive iodine is concentrated in the thyroid.

The fact that people who survive instantaneous whole-body doses of 1.5–7.5 Sv (150–700 rem) usually recover completely shows that the body can repair radiation damage. Because cells are constantly exposed to ionizing radiation, they have developed biochemical mechanisms to protect themselves from damage. Cancer is the main delayed effect of ionizing radiation.

The most important principle followed to keep occupational radiation exposure to a minimum is maintaining a safe distance between people and the radioactive material. *The intensity of radiation is proportional to $1/d^2$, where d is the distance from the source;* thus, intensity decreases rapidly with increasing distance. Remote controls or robots are often used to handle radioactive materials (■Figure 20.16). Work with radioactive materials is also usually separated from other work, workers are shielded from the radioactive materials (for example, by lead aprons), and exposure is carefully monitored.

■ FIGURE 20.16   Handling radioactive materials with remote controls.

20.10   USES OF RADIONUCLIDES

Radiotherapy

The biological damage caused by radiation has many useful applications. Radiotherapy is widely used to treat cancers and blood disorders such as leukemia. Because most cancer cells are rapidly growing and reproducing, cancer cells are usually more sensitive to radiation than normal cells. As a result, radiation kills more cancerous cells than healthy cells. Radionuclides can be enclosed in needles or small capsules and implanted to deliver a slow, continuous dose of radiation to small tumors that can be reached easily. The slow administration of radiation gives normal cells time to repair damage. Large tumors in organs deep in the body can be irradiated with gamma rays from cobalt-60. Damage to normal tissue is, of course, greater and side effects such as nausea and hair loss occur. Ten to twenty treatments are usually given over a period of several months.

Diagnosis

Radionuclides are also widely used to diagnose disease. Because the amount of energy released by decay of a single radioactive atom is relatively large, a few radioactive atoms can be used to label molecules composed mostly of ordinary atoms. The amount of energy released by decay of the radionuclide is too small to harm the patient but the molecules that contain radioactive atoms can be followed easily through the body by tracing their radioactivity. For example, human serum albumin labeled with ^{99}Tc can be injected intravenously and blood circulation followed with a scintillation counter to detect obstructions (such as blood clots), constrictions, and other circulatory disorders. The ideal nuclide for diagnosis decays by electron capture so that no particles are emitted, emits relatively low energy gamma rays, and has a half-life about as long as the study requires.

Preservation of Food

Bacteria in foods can be destroyed by radiation although changes in the flavor of the food are a problem. Preservation of food by radiation uses much less energy than preservation by heating or freezing. Radiation can also be used to kill bacteria in sewage. However, a combination of heat and gamma irradiation is more effective

A device used for generating ^{99}Tc. Over 80% of diagnostic imaging procedures use ^{99}Tc although technetium does not occur naturally.

Radionuclides used for both radiotherapy and diagnosis are products of nuclear reactors. Labeling with radionuclides is like belling a cat so that the cat's location is easily detected.

Left: *Mushrooms preserved by irradiation.* Right: *Untreated mushrooms after the same time.*

than radiation alone and is less expensive. Insects and fungi in grain can be wiped out by radiation, and medical supplies can be sterilized.

Radiation can also be used to control insect pests without harming useful insects or other living creatures. A large number of male insects are sterilized by radiation and released. As a result of the high proportion of sterile males, the probability of a female mating with a sterile male is high. Because no offspring result from mating with a sterile male, the population of the insect is rapidly reduced or eliminated.

Study of Reaction Pathways

Isotopic tracers have also been widely used to study the mechanisms of relatively simple reactions as well as the pathways for complex biochemical reactions that take place in many steps. For example, in photosynthesis,

$$x CO_2(g) + y H_2O(l) \xrightarrow{\text{sunlight}} C_x(H_2O)_y(s) + x O_2(g)$$

does oxygen in $O_2(g)$ come from H_2O? This question can be answered by labeling the oxygen in the water with O-18 and observing where the O-18 ends up in the products. It turns out that all the O-18 ends up in the oxygen gas:

$$x CO_2(g) + y H_2^{18}O(l) \xrightarrow{\text{sunlight}} C_x(H_2O)_y(s) + x {}^{18}O_2(g)$$

Oxygen in $O_2(g)$ must come from H_2O; all of the oxygen in the carbohydrate must come from CO_2. Oxygen-18 is not radioactive; the products must be analyzed with a mass spectrometer instead of a counter. Unfortunately, no radioactive isotopes of oxygen exist that have long enough half-lives to allow time for photosynthesis to take place and the products to be isolated.

The pathway from simple to complex molecules followed in photosynthesis was worked out using the radionuclide carbon-14 as a tracer. The understanding of biochemical pathways made possible by the use of isotopic tracers has been an important factor in recent rapid advances in the medical sciences.

Analysis

Several methods used in modern analytical chemistry such as neutron activation and isotope dilution involve nuclear reactions. In neutron activation analysis, the sample is bombarded with thermal neutrons. A typical capture reaction results; for example,

$$^{75}_{33}As + {}^{1}_{0}n \longrightarrow {}^{76}_{33}As^* \longrightarrow {}^{76}_{33}As + \gamma$$

The frequency of the gamma radiation is characteristic of the element and is used to identify the element. The intensity of the gamma radiation is proportional to the amount of the nuclide in the sample and is used to measure how much of the nuclide is present. Neutron activation can measure extremely small quantities; for example, 4 to $9 \times 10^{-4} \mu g$ of arsenic. As many as 30 different elements in a single sample can be determined in trace quantities without chemical separation. The sample is not destroyed. Neutron activation analysis is very useful in crime detection; a history of arsenic poisoning can be deduced from analysis of a single hair. Neutron activation analysis has also been used to detect art forgeries and to investigate trace components of air, water, biological, and geological samples. One of the few drawbacks is that a source of neutrons is needed.

Isotope dilution can be used to measure the volume of liquid in irregularly shaped containers such as the circulatory system and underground reservoirs of oil and water. For example, to determine the volume of a patient's blood, a small sample of blood is withdrawn and labeled with ^{99}Tc. The radioactivity of the labeled blood is measured and the labeled blood returned to the patient's circulatory system. After enough time has passed for the labeled blood to have become thoroughly mixed with the rest of the patient's blood, another sample is taken. The activity of the second sample is, of course, much lower than the activity of the original sample. From the ratio of the activities in the same volume of the first and second samples, the volume of the patient's circulatory system can be calculated.

Manufacturing

Radiation is also used in chemical manufacturing. Treatment of polyethylene with a dose of 10^7 rad increases the softening point from 90 to 150 °C and makes the polyethylene tougher with less tendency to crack. Very thin films can be formed that are used to make packaging that shrinks when heated.

Polyethylene is discussed in Section 9.11.

Dating of Objects of Geological, Archaeological, and Historical Interest

Because the rates of radioactive decay reactions are constant (not changing with temperature, pressure, physical state, or chemical form), they can be used as clocks. For example, the decay of carbon-14 is used to date objects containing once-living material, which has been of great value in archaeology, anthropology, and geology.

Carbon-14 is continuously being formed in the upper atmosphere by reaction of neutrons from cosmic rays with nitrogen-14:

$$^{14}_{7}N + ^{1}_{0}n \longrightarrow ^{14}_{6}C + ^{1}_{1}H$$

Carbon-14 is radioactive and decays beta with a half-life of 5715 years:

$$^{14}_{6}C \longrightarrow ^{14}_{7}N + ^{0}_{-1}e \qquad t_{1/2} = 5715 \text{ y}$$

Thus, carbon-14, which is continuously being formed in the upper atmosphere, is continuously decaying. The rate of formation of carbon-14 equals the rate of decay of carbon-14; a steady state exists. At the steady state, the radioactivity of carbon is 14.9 disintegrations per minute per gram of total carbon, an easily measurable activity level.

The carbon-14 is converted to $^{14}CO_2$ by oxygen in air. The $^{14}CO_2$ mixes with the ordinary carbon dioxide, a mixture of 99% $^{12}CO_2$ and 1% $^{13}CO_2$, in the atmosphere; the half-life of carbon-14 is long enough for complete mixing. The mixture that results contains about one atom of carbon-14 for every 10^{12} atoms of carbon-12. Plants use the carbon dioxide containing $^{14}CO_2$ for photosynthesis; animals and people eat the plants. Thus, the carbon-14 is distributed through all living matter. However, when a plant dies, it stops photosynthesizing; when an animal dies, it stops eating. A dead plant or animal no longer takes in carbon-14. However, the carbon-14 the plant or animal took in when alive continues to decay, and the radioactivity due to carbon-14 decreases. From a comparison of the radioactivity of the carbon in an ancient object with the radioactivity of carbon at the steady state and a knowledge of the half-life of carbon-14, the age of the object can be found. For example, suppose the activity of the carbon in a skeleton is 7.45 disintegrations per minute per gram of carbon, half the steady-state activity. For the activity to have decreased to half its original value, one half-life must have passed. The skeleton must be one half-life or 5715 years old. Sample Problem 20.6 illustrates calculation of the age of an object by carbon-14 dating.

The American chemist Willard Libby developed radiocarbon dating in 1947 and received the Nobel Prize for his work in 1960. Libby also helped to develop the method for separating uranium isotopes by gaseous effusion that was an essential step in the production of the atomic bomb (Section 5.9).

20.6 The carbon in a sample of charcoal from the remains of an ancient campfire found in a cave has an activity of four disintegrations per minute per gram of carbon. How old is the charcoal?

Solution Because four is less than half but more than a quarter of 14.9, the integrated rate equation for a first-order process, equation 18.6a,

$$\ln \frac{N_t}{N_0} = -kt$$

must be used to calculate t. We can use equation 18.7,

○ Sample Problem 18.8 showed how to calculate k from the half-life of a first-order reaction.

$$t_{1/2} = \frac{0.693}{k}$$

to find the value of k from the half-life. Solution of equation 18.7 for k and substitution of the half-life for carbon-14 gives

$$k = \frac{0.693}{5715 \text{ y}} = 1.21 \times 10^{-4} \text{ y}^{-1}$$

From the problem we know that

$$\frac{N_1}{N_0} = \frac{4 \text{ disintegrations}}{14.9 \text{ disintegrations}} = 0.3$$

(Remember that any unit can be used for quantity because units cancel out. The number of disintegrations measures the amount of a radioactive substance present.)

Now we are ready to substitute in equation 18.6a and find t:

$$t = \frac{\ln 0.3}{-(1.21 \times 10^{-4} \text{ y}^{-1})} = 1.0 \times 10^4 \text{ y}$$

20.13 The carbon in an ancient sample of bone has an activity of 11.0 disintegrations per minute per gram of carbon. How old is the bone?

Carbon-14 dating by measuring activity can't be used for times less than 1000 y because the change in number of disintegrations is too small. Times greater than 30 000 y are too large to be determined because almost all of the carbon-14 will have decayed. The count will be too low to measure accurately compared with the background count. However, if a mass spectrometer is used to measure the amount of carbon-14, times as long as 75 000 y can be measured and only a very small quantity (about 10^{-3} g) of the object is destroyed.

The accuracy of carbon-14 dating depends on the correctness of the assumption that the proportion of carbon-14 to carbon-12 in the atmosphere is constant. The proportion of carbon-14 to carbon-12 in the atmosphere varies from time to time as a result of changes in cosmic radiation accompanying sunspots. During the industrial age, burning of fossil fuels has released large quantities of carbon dioxide formed from old, nonradioactive carbon. This old carbon no longer contains any carbon-14 and thus decreases the proportion of carbon-14 in the atmosphere. On the other hand, aboveground nuclear explosions have added to atmospheric carbon-14. As a result, carbon-14 dates must be calibrated by comparison with dates obtained by other methods such as dendrochronology (dating by counting the annual rings of trees).

A mammoth foot. Carbon-14 dating has been used to determine the age of the remains of mammoths, an extinct genus of elephants often pictured in cave art.

20.11 SYNTHESIS OF THE ELEMENTS

In addition to the many applications of nuclear reactions and radionuclides discussed in Sections 20.7, 20.8, and 20.10, nuclear reactions are important because they are the source not only of all the energy available for use on Earth, but also of the chemical elements of which matter is composed. Except for relatively small quantities of synthetic nuclides, elements are made in stars and in the space between the stars. We have already seen how helium is made from hydrogen by fusion in the sun (Section 20.8).

From the spectra of the stars (and the clouds of dust and gas between the stars), the elements that compose the universe can be identified and their proportions determined. The part of the universe that has been observed by scientists must, of course, be assumed to be a representative sample. The *relative numbers of atoms of the various elements* are usually referred to as the **abundances** of the elements. The abundances of the elements in the universe vary by 12 orders of magnitude. Of all the atoms in the universe, 93.4% are estimated to be hydrogen atoms and 6.5% helium atoms; only 0.1% of the atoms in the universe are atoms of all the other elements together. The proportions of these other atoms vary by a factor of 10^9. ■Figure 20.17 is a graph of the logarithms of the abundances of each kind of atom in the universe.

In general, the proportion of an element in the universe decreases as atomic number increases. The decrease is rapid at first and then gradually becomes slower. The up–down character of most of the graph is striking. If elements with similar atomic numbers are compared, elements with even atomic numbers are more common than elements with odd atomic numbers. The abundances of elements with even and odd atomic numbers vary less for elements with atomic numbers higher than 30 than for elements with atomic numbers lower than 30. Lithium, beryllium, boron, fluorine, and scandium are much less common than would be expected from their atomic numbers; iron, nickel, and lead are more common.

The atomic abundances of the ten most common elements in the universe are listed in Table 20.8, which also shows the number of protons and neutrons in the nucleus and the total number of nucleons for the major isotope. In all but one case (Mg), the major isotope makes up more than 90% of the naturally occurring element. Because hydrogen, the simplest nucleus, is so abundant in the universe and because the atomic masses of most nuclides (and the mass of the neutron) are close to being

Most elements were made in stars, such as those in the Andromeda Galaxy, and in the space between stars.

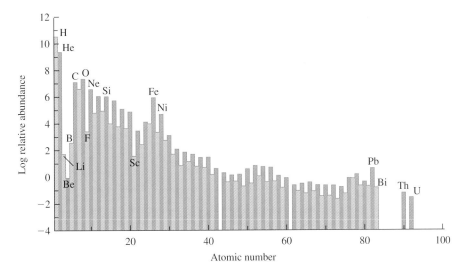

■ FIGURE 20.17 Relative abundances of the elements in the universe. The abundance of silicon is set equal to 1×10^6 atoms. (Data are from Moeller, T. *Inorganic Chemistry*; Wiley: New York, 1982; pp. 24–27.)

TABLE 20.8	Atomic Abundances of the Ten Most Common Elements in the Universe			
Element	Number of Protons	Number of Neutrons	Total Number of Nucleons	Number of Atoms/Si Atom
H	1	0	1	3.2×10^4
He	2	2	4	2.2×10^3
O	8	8	16	2.2×10^1
C	6	6	12	1.2×10^1
N	7	7	14	3.7
Ne	10	10	20	3.4
Mg	12	12	24	1.1
Si	14	14	28	1.0
Fe	26	30	56	8.3×10^{-1}
S	16	16	32	5.0×10^{-1}

multiples of the mass of a hydrogen atom, it has been suggested that hydrogen nuclei are the building blocks for other nuclei.

Examination of Table 20.8 reveals that six of the eight nuclides that have more protons than helium—oxygen, carbon, neon, magnesium, silicon, and sulfur—are multiples of helium-4; for example,

$$^{16}_{8}O = 4(^{4}_{2}He)$$

⦿ Remember that helium-4 is formed from hydrogen-1.

Thus, helium-4 appears to be a building block for many nuclides. The first step in the formation of carbon, oxygen, and so forth by combination of helium-4 nuclei must be the combination of two helium-4s to form beryllium-8:

$$2^{4}_{2}He \longrightarrow ^{8}_{4}Be$$

The sun is only an average-sized star.

However, beryllium-8 has a half-life of only 7×10^{-17} s. How can there be time for another helium to fuse with the beryllium-8 before the beryllium-8 decays? Only under the conditions of high density and high temperature that exist in large stars is enough beryllium-8 present at the steady state for the reaction

$$^{8}_{4}Be + ^{4}_{2}He \longrightarrow ^{12}_{6}C$$

to take place. Once carbon-12 has been formed, further helium capture reactions such as

$$^{12}_{6}C + ^{4}_{2}He \longrightarrow ^{16}_{8}O \qquad \text{and} \qquad ^{16}_{8}O + ^{4}_{2}He \longrightarrow ^{20}_{10}Ne$$

can take place. In the largest, hottest stars some of the fastest moving heavy ions have enough energy to fuse. Reactions such as

$$2^{12}_{6}C \longrightarrow ^{24}_{12}Mg \qquad \text{and} \qquad 2^{16}_{8}O \longrightarrow ^{32}_{16}S$$

occur. Helium capture and fusion explain the formation of even-numbered elements with mass numbers divisible by 4.

Because iron is at the minimum of the curve of potential energy per nucleon as a function of mass number (Figure 20.13), fusion of nuclei larger than iron is not spontaneous. Nuclei beyond iron cannot be formed by fusion. Nuclei beyond iron (and some lighter nuclei) are formed by neutron capture, for example,

$$^{56}_{26}Fe + ^{1}_{0}n \longrightarrow ^{57}_{26}Fe^{*} \longrightarrow ^{57}_{27}Co + ^{0}_{-1}e$$

The neutrons are formed by reactions such as

$$2^{2}_{1}H \longrightarrow ^{3}_{2}He + ^{1}_{0}n$$

Lithium, beryllium, and boron are rare in the universe because they are easily transmuted by collision with energic particles. They cannot survive in the hot, dense interiors of stars where elements are formed. The relatively small quantities of lithium, beryllium, and boron that exist are believed to be formed by collision of cosmic rays with low-temperature low-density clouds of gas and dust between the stars. A typical reaction is

$$^1_1H + {}^{12}_6C \longrightarrow {}^{11}_5B + 2\,^1_1H$$

The composition of the part of Earth that can be used by people is not the same as the composition of the universe. Table 20.9 gives the abundances of the ten most common elements in the crust, surface water, and atmosphere of Earth. Comparison of Tables 20.9 and 20.8 shows that the elements that are less common on Earth than in the universe—H, He, C, N, Ne, and S—are nonmetals. The nonmetals lost from Earth are either noble gases such as He and Ne or elements that form gaseous or low boiling compounds such as CH_4 and NH_3. Earth's gravitational attraction is not strong enough to hold these light gases. The elements that are more common on Earth than in the universe—Al, Na, Ca, K, and Ti—are metals that form nonvolatile compounds. No elements are common on Earth that are rare in the universe.

Use of the various elements depends on the availability of the element in Earth's crust, surface waters, and atmosphere. Availability is not the same as abundance. Some elements that are very abundant are not readily available because they are widely dispersed in small amounts and seldom, if ever, occur in large deposits. Titanium is a good example of a metal with many useful properties, such as corrosion resistance and strength, that is expensive because no concentrated deposits exist. Other elements are not used because of the difficulty of recovery from their sources. The importance of ease of extraction is illustrated by magnesium. The concentration of magnesium in dolomite ($CaCO_3 \cdot MgCO_3$), the chief ore of magnesium, is 22% by mass. The concentration of magnesium in seawater is only 0.13% by mass. However, a significant proportion of magnesium is produced from seawater because seawater is cheap and plentiful and magnesium is easily recovered from it by electrolysis. Gold was one of the first metals to be used by people because it occurs free in nature.

The chemical and physical properties of the elements and their compounds are also an important factor in their uses. For example, no one would use sodium metal instead of iron to build a bridge. Sodium metal is too soft and would melt in the sun—to say nothing of what would happen if it rained. But even the physical and chemical properties of an element and its compounds depend on the nucleus because the number of protons in the nucleus determines the number of electrons around the nucleus.

TABLE 20.9

Atomic Abundances of Common Elements in the Crust, Surface Water, and Atmosphere of Earth

Element	Number of Atoms/ Si Atom
O	3.4
Si	1.0
H	1.0
Al	0.30
Na	0.12
Fe	0.09
Ca	0.09
Mg	0.09
K	0.07
Ti	0.01

You can't use what you haven't got.

This picture shows why oxygen, silicon, and hydrogen are the most abundant elements. A large part of Earth's surface is covered with water. The silicate rocks that make up Earth's crust contain vast quantities of silicon and oxygen.

SUMMARY

Nuclear reactions are different from chemical reactions in six ways:

Nuclear Reactions	Chemical Reactions
(1) Involve protons and neutrons inside nucleus.	Involve electrons outside nucleus.
(2) Elements are transmuted into other elements.	Both reactants and products contain the same number of each kind of atom.
(3) Isotopes react differently.	Isotopes react similarly.
(4) Independent of state of chemical combination.	Depend on state of chemical combination.
(5) Energy changes of order of 10^8–10^9 kJ/mol.	Energies of order of 10–10^3 kJ/mol.
(6) Mass changes are detectable.	Mass products = mass reactants.

Radioactive decay reactions are spontaneous. Six types of radioactive decay are observed:

Type	Symbol	Mass Number	Atomic Number	n/p	Example
Alpha, α	^4_2He	-4	-2	Increases	$^{238}_{92}\text{U} \longrightarrow {}^{234}_{90}\text{Th} + {}^4_2\text{He}$
Beta, β^-	$^0_{-1}e$	0	$+1$	Decreases	$^{234}_{90}\text{Th} \longrightarrow {}^{234}_{91}\text{Pa} + {}^0_{-1}e$
Positron, β^+	$^0_{+1}e$	0	-1	Increases	$^8_5\text{B} \longrightarrow {}^8_4\text{Be} + {}^0_{+1}e$
Electron capture, EC	$^0_{-1}e$	0	-1	Increases	$^{181}_{74}\text{W} + {}^0_{-1}e \longrightarrow {}^{181}_{73}\text{Ta}$
Spontaneous fission, SF	—		About half	Average decreases	$^{250}_{96}\text{Cm} \longrightarrow {}^{142}_{54}\text{Xe} + {}^{106}_{42}\text{Mo} + 2{}^1_0n$
Gamma, γ	$^0_0\gamma$	0	0	No change	$^{27}_{13}\text{Al}^* \longrightarrow {}^{27}_{13}\text{Al} + {}^0_0\gamma$

Nuclides with atomic numbers greater than 83 usually decay alpha. Beta emission decreases the number of neutrons and increases the number of protons and is characteristic of nuclides with too high neutron : proton ratios. Both positron emission and electron capture increase the number of neutrons and decrease the number of protons of nuclides with neutron : proton ratios that are too low. Positron decay is characteristic of nuclides with low atomic numbers; electron capture is characteristic of nuclides with high atomic numbers. Spontaneous fission occurs only when the atomic number is greater than 92. Emission of gamma rays often accompanies other types of radioactive decay. Gamma rays are electromagnetic radiation of very short wavelength.

In equations for nuclear reactions, the sum of the mass numbers of the product nuclei must be equal to the sum of the mass numbers of the reactant nuclei. The sum of the atomic numbers of the product nuclei must be equal to the sum of the atomic numbers of the reactant nuclei.

Decay of some radionuclides to stable nuclides requires a number of steps. The steps make up a **radioactive decay series.** A long-lived nuclide is chosen as the first member or **parent** of a series. All the other nuclides in the series are known as **daughters.**

The **activity** of a radionuclide is the rate at which radioactive decay takes place. The SI unit of activity is named the **becquerel, Bq.** One becquerel equals one event per second. The traditional unit of activity is the **curie, Ci** ($1 \text{ Ci} = 3.70 \times 10^{10}$ Bq). Activity is usually measured with ionization counters such as Geiger counters and solid-state detectors, or with a scintillation detector. The rate of radioactive decay is independent of temperature as well as of chemical state.

Radioactive decays have first-order rate laws because radioactive decay depends on chance. Each nucleus in a sample of a radionuclide has a certain probability of decaying but there is no way of knowing which nucleus will decay during a given period of time.

The neutron : proton ratio, even–odd type, and magic numbers are considered in making qualitative predictions about radioactive decay. Stable nuclei have neutron : proton ratios in the **band of stability.** The further a nuclide is from the band of stability, the shorter the nuclide's half-life tends to be.

Qualitative predictions are not always correct; sometimes the energy change that will accompany a nuclear reaction must be calculated to predict whether a nuclide will decay in a certain way. For a nuclide to be radioactive and decay spontaneously, the mass of the products of decay must be less than the mass of the reactants. Mass and energy are related by the **Einstein equation:**

$$\Delta E = \Delta mc^2$$

The mass of the nucleus of an atom is always less than the sum of the masses of the nucleons that compose the nucleus. The difference between the observed mass of a nucleus and the sum of the masses of the nucleons is called the **mass defect.** The energy equivalent of the mass defect is referred to as the **nuclear binding energy** of the nucleus, the energy required to break a nucleus apart into its nucleons. Nuclear binding energies per nucleon are at a maximum (potential energy is at a minimum) around iron.

The **nuclear strong force** is the attractive force between nucleons. It is very strong but very short-range. A nucleus is stable if it can't be transmuted into a nucleus of another element without adding energy from outside the nucleus. Nucleons in a nucleus are paired and exist in different energy levels.

Induced nuclear reactions are brought about by collision of a moving nucleus called the **projectile** with a stationary nucleus called the **target nucleus.** Most nuclides **capture slow** or **thermal neutrons,** which have average kinetic energy about the same as neutrons would have if they existed as a monatomic gas at room temperature. Positively charged particles are repelled by the positive charge on a nucleus and must be accelerated before they have enough kinetic energy to react with the nucleus. Linear accelerators, cyclotrons, and synchrotrons are all accelerators. Projectiles with masses greater than the mass of an alpha particle are called **heavy ions.**

Some nuclides can be made to fission by supplying energy. The reaction is a **branching chain** reaction—more neutrons are released than are used up. A sample that is too small to sustain a chain reaction is called **subcritical.** In a **critical** sample, the number of neutrons formed is equal to the number of neutrons used up. A **supercritical** sample is larger than a critical sample and produces a nuclear explosion.

Controlled fission by slow neutrons may be used in **nuclear reactors** to heat water to generate electricity. **Breeder reactors** not only produce thermal energy to generate electricity but also produce fuel. **Fusion,** the combination of nuclei of low atomic number into nuclei of moderate atomic number, is a possible future source of power.

Radiation damages living tissue. Although this damage may be harmful, it can also be used to treat cancer. Nuclear chemistry is also applied in diagnosis, preservation of foods, for historical dating, and for analysis.

The **dose** of radiation is the quantity of radiation absorbed by a unit mass of matter and is measured in **grays** or **rads.** The dose required to produce a certain amount of damage depends on the period of time over which the dose is received. Different kinds and energies of radiation differ in their effect on a given tissue; before doses can be compared they must be multiplied by a **quality factor, Q.** The SI unit that takes type of radiation into account is called the **sievert, Sv.** The intensity of radiation is proportional to $1/d^2$, where d is the distance from the source. Cells undergoing division are affected more by radiation than other cells. The age and health of the individual are also factors.

Nuclear reactions are the source of all the energy used on earth. The chemical elements are formed in stars and in the space between the stars by nuclear reactions, mainly fusion and neutron capture.

ADDITIONAL PRACTICE PROBLEMS

For information about the organization of Additional Practice Problems, Stop & Test Yourself, Putting Things Together, and Applications, see the beginnings of these sections in Chapter 1.

20.14 What six types of radioactive decay take place spontaneously? Briefly discuss each of them. (20.1)

20.15 There are no electrons in atomic nuclei. Explain, then, how electrons (beta particles) can be emitted by nuclei. A good explanation will include a microscopic picture. (*Hint:* Use helium-6 as an example.) (20.1)

20.16 The nuclide bromine-82 beta decays. What is the daughter nuclide? (20.1)

20.17 The nuclide polonium-212 decays to lead-208. What kind of decay does polonium-212 undergo? (20.1)

20.18 Complete the following nuclear equations:
(a) $^{183}_{77}\text{Ir} + ^{0}_{-1}\text{e} \longrightarrow$ _____
(b) $^{172}_{68}\text{Er} \longrightarrow$ _____ $+ ^{0}_{-1}\text{e}$
(c) $^{256}_{100}\text{Fm} \longrightarrow ^{115}_{46}\text{Pd} + ^{138}_{54}\text{Xe} +$ _____
(d) $^{137}_{60}\text{Nd} \longrightarrow$ _____ $+ ^{0}_{+1}\text{e}$
(e) _____ $\longrightarrow ^{171}_{76}\text{Os} + ^{4}_{2}\text{He}$ (20.1)

20.19 Write equations for each of the following nuclear reactions:
(a) Antimony-116 emits a positron.
(b) Antimony-116 captures an electron.
(c) Cadmium-118 beta decays. (d) Californium-254 spontaneously fissions to molybdenum-106 and four neutrons.
(e) Thorium-224 decays alpha. (20.1)

20.20 Technetium is one of the two elements with atomic number 92 or less that is not found on Earth. Technetium-97 can be made by the reaction of molybdenum-96 with H-2. Write the equation for this reaction. (20.2)

20.21 Write the nuclear equation for the formation of iron-59 by neutron capture. (20.2)

20.22 Complete the following nuclear equations by filling in the blanks:
(a) _____ $+ ^{1}_{0}\text{n} \longrightarrow ^{239}_{93}\text{Np} + ^{0}_{-1}\text{e}$
(b) $^{238}_{92}\text{U} +$ _____ $\longrightarrow ^{238}_{93}\text{Np} + 2^{1}_{0}\text{n}$
(c) $^{239}_{94}\text{Pu} + ^{4}_{2}\text{He} \longrightarrow$ _____ $+ ^{1}_{0}\text{n}$
(d) $^{238}_{92}\text{U} +$ _____ $\longrightarrow ^{253}_{99}\text{Es} + 5^{1}_{0}\text{n}$
(e) $^{209}_{83}\text{Bi} + ^{58}_{26}\text{Fe} \longrightarrow ^{266}_{109}\text{Mt} +$ _____ (20.2)

20.23 Molybdenum-101 has a half-life of 14.6 m, and molybdenum-103 has a half-life of 62 s. (a) Which isotope decays more rapidly? (b) Which isotope is less stable? (20.3)

20.24 The half-life for the radioactive decay of iron-55 is 2.7 y. From a 2.00-g sample of iron-55, how many grams will be left after 13.5 y? (20.3)

20.25 If 1.25 mg of a 5.00-mg sample of arsenic-78 remain after 182 m, what is the half-life of arsenic-78? (20.3)

20.26 The half-life for the radioactive decay of zirconium-86 is 16.5 h. What is the value of k for this reaction? (20.3)

20.27 In exactly a week, 10.00 μg of radon-222 decays to 2.82 μg of polonium-218. (a) What is the value of the rate constant for this reaction in d^{-1}? (b) What is the half-life in days? (20.3)

20.28 The half-life of krypton-85 is 10.76 y. (a) What percentage of the original sample of krypton-85 will remain after 25.0 y? (b) How long will it take for 95.0% of a sample of krypton-85 to decay? (20.3)

20.29 What is the activity in Ci of a 5.00-μg sample of sodium-24, which has a half-life of 15.02 h? (20.3)

20.30 Sulfur-35 has a half-life of 87.4 d. What is the activity of sulfur-35 (a) in Bq/g? (b) In Ci/g? (20.3)

20.31 A 3.50-μg sample of calcium-47 has an activity of 7.92×10^{10} Bq. (a) What is the half-life of Ca-47 in days? (b) How many days will it take for the activity to decrease to 8.02×10^{8} Bq? (20.3)

20.32 What quantity (use appropriate units of mass) of cesium-137 is required to furnish an activity of 75 mCi? Cesium-137 has a half-life of 30.2 y. (20.3)

20.33 A sample that contains manganese-56 has an activity of 1.00 mCi at one time and an activity of 0.77 mCi 0.97 h later. What is the half-life of manganese-56? (20.3)

20.34 (a) What is meant by the term *band of stability*?
(b) Some nuclides, such as S-35, that lie within the band of stability are unstable. This being the case, what is the use of the band of stability? (20.4)

20.35 Both barium-123 and barium-140 are radioactive. Which is likely to decay more rapidly? Explain your answer. (20.4)

20.36 Arrange the following isotopes of magnesium in order of decreasing stability. Explain your answer. (20.4)

$$^{12}_{12}\text{Mg} \qquad ^{24}_{12}\text{Mg} \qquad ^{28}_{12}\text{Mg}$$

20.37 Which member of each pair of nuclides is probably more stable? Explain your answers. (a) $^{209}_{83}Bi$ or $^{213}_{83}Bi$ (b) $^{44}_{20}Ca$ or $^{45}_{20}Ca$ (c) $^{18}_{9}F$ or $^{19}_{9}F$ (d) $^{23}_{10}Ne$ or $^{23}_{11}Na$ (e) $^{201}_{81}Tl$ or $^{203}_{81}Tl$ (20.4)

20.38 Which member of each pair of elements has a greater number of stable isotopes? Explain your answers.
(a) Ba or Cs (b) Ba or Mg (c) Bi or Po (20.4)

20.39 (a) What is a magic number? (b) What are the magic numbers for protons? For neutrons? For electrons? (20.4)

20.40 The four stable isotopes of cerium that exist have mass numbers 136, 138, 140, and 142. What rule about nuclear stability is illustrated by the four stable isotopes of cerium? (20.4)

20.41 Give the evidence for thinking that (a) nucleons are paired in the nucleus, (b) the nucleons in the nucleus are arranged in shells with different energies. (20.4)

20.42 (a) Explain why electron capture is more common among elements with high atomic numbers than among elements with low atomic numbers. (b) Which electron is most likely to be captured, and why? (20.4)

20.43 Aluminum-27 is the only stable isotope of aluminum. Predict how each of the following radioactive isotopes of aluminum will decay and write equations for the decay reactions: (a) aluminum-26 (b) aluminum-28. (20.4)

20.44 Complete the diagram for the U-235 radioactive decay series by filling in the missing nuclides and particles. In the diagram, the numbers above or below the arrows are half-lives, and the dotted lines show minor branches. (20.5)

$^{235}_{92}U \xrightarrow[7.1 \times 10^8 \text{ y}]{\alpha} \underline{\quad} \xrightarrow[25.5 \text{ h}]{} {}^{231}_{91}Pa \xrightarrow[3.25 \times 10^4 \text{ y}]{}$

$^{227}_{89}Ac \quad ^{223}_{88}Ra \xrightarrow[11.43\text{ d}]{\alpha} \underline{\quad} \xrightarrow[4.0\text{ s}]{}$

$^{227}_{90}Th$

$^{215}_{84}Po \xrightarrow[1.78 \times 10^{-3}\text{ s}]{} \underline{\quad} \xrightarrow[36.1\text{ m}]{} {}^{211}_{82}Pb \xrightarrow{} {}^{211}_{83}Bi$

$^{207}_{81}Tl$

20.45 Why is mass conserved in ordinary chemical reactions but not in nuclear reactions? (20.6)

20.46 For the beta decay of carbon-13, (a) calculate the change in mass that will take place. The atomic mass of C-13 is 13.003 354 u, and the atomic mass of the product nuclide is 13.005 738 u. (b) Will this radioactive decay take place spontaneously? Explain your answer. (c) Calculate the energy change in kilojoules per mole that will accompany this reaction. (20.6)

20.47 When one atom of iodine-131 beta decays, 1.5541×10^{-13} J are released. The atomic mass of the product nuclide is 130.9051 u. What is the atomic mass of I-131? (20.6)

20.48 The atomic mass of No-254 is 254.090 953 u, and the atomic mass of H-1 is 1.007 825 u. (a) What is the mass defect for No-254? (b) Calculate the total binding energy in joules. (c) What is the binding energy per nucleon in joules? (20.6)

20.49 Explain why a critical mass is necessary for a nuclear chain reaction to take place and a supercritical mass for a nuclear explosion to occur. (20.7)

20.50 In a nuclear reactor, (a) what is in the core? (b) How is the rate of reaction controlled? (c) What is the function of the moderator? (20.7)

20.51 What does the term fusion mean in nuclear chemistry? (20.8)

20.52 If a worker moves twice as far from a source of radiation, how does the intensity of the radiation compare with the intensity at the worker's original position? (20.9)

20.53 Predict what would happen if a person absorbed a single dose of (a) 150 rad of fast neutrons (b) 150 rad of slow neutrons (c) 150 rad of gamma radiation. (20.9)

20.54 (a) If a person living near a nuclear power plant was exposed to a single dose of 5.0×10^{-3} rem, by what percentage would his or her average dose for the year be increased? (b) If a person had to have an extra X-ray and received a dose of 20 mrem, by what percentage would his or her average dose for the year be increased? (20.9)

20.55 What is the age of a sample of mummified skin from a prehistoric human that contains 12.5% of the original quantity of carbon-14? (20.10)

20.56 (a) What observations suggest that hydrogen nuclei are the building blocks for other nuclei? (b) Why can nuclei having atomic numbers greater than 26 not be formed by fusion? (20.11)

20.57 What factors affect people's use of the various elements? (20.11)

20.58 Beryllium-9 is the only stable isotope of beryllium. Beryllium-8, which the rules would predict to be stable, alpha decays with the extremely short half-life of 2×10^{-16} s. Atomic masses are Be-8, 8.005 308 u; Be-9, 9.012 186 u; and H-1, 1.007 825 u. (a) Calculate the mass defects for Be-9 and for Be-8. (b) Calculate the total binding energies in joules for both isotopes. (c) Calculate the binding energy per nucleon for each isotope. (d) Which isotope has the greater binding energy per nucleon? (e) Which isotope has the greater total binding energy and why? (f) Does either binding energy per nucleon or total binding energy determine whether a nuclide is radioactive or nonradioactive? Explain your answer.

20.59 The neutrons discovered by Chadwick (Section 7.4) were made by bombarding beryllium-9 with alpha particles. Write the equation for this reaction.

20.60 A rock sample contains 365 mg of Pb-206 (atomic mass 205.9745 u) and 526 mg of U-238 (atomic mass 238.0508 u). The half-life of U-238 is 4.47×10^9 y. How old is the rock?

20.61 Two structures have been proposed for the thiosulfate ion, $S_2O_3^{2-}$:

Studies with the radionuclide S-35 eliminated one of the structures. When thiosulfate prepared by treating $^{32}SO_3^{2-}$ with ^{35}S,

$$^{32}SO_3^{2-} + {}^{35}S(s) \longrightarrow {}^{35}S{}^{32}SO_3^{2-}$$

is decomposed by acid, all of the ^{35}S is in the sulfur, and none is in the SO_2:

$$^{35}S^{32}SO_3^{2-} + 2H^+ \longrightarrow {}^{35}S(s) + {}^{32}SO_2(g) + H_2O(l)$$

Which of the two proposed structures is eliminated by this observation? Explain your answer.

20.62 When the ester methyl acetate, CH_3COOCH_3, is made by the reaction of acetic acid and methyl alcohol using acetic

acid labeled with O-18, the O-18 ends up in the water:

$$CH_3C^{18}OH + HOCH_3 \longrightarrow CH_3COCH_3 + H_2^{18}O$$
$$\quad\; \overset{\|}{O} \qquad\qquad\qquad\qquad \overset{\|}{O}$$

Do the C—OH bond in the acid and the O—H bond of the alcohol break or do the C—OH bond in the alcohol and the O—H bond in the acid break? Explain your answer.

STOP & TEST YOURSELF

1. Which of the following statements is *not* true of nuclear reactions? (a) Nuclear reactions involve protons and neutrons inside the nucleus. (b) In nuclear reactions, elements are transmuted into other elements. (c) Nuclear reactions are independent of the state of chemical combination of the elements. (d) The energy changes that accompany nuclear reactions are of the order of 10 to 10^3 kJ/mol. (e) In nuclear reactions, the mass of the products does not equal the mass of the reactants.

2. The product of the alpha decay of $^{198}_{84}Po$ is _____.
   (a) $^{194}_{85}At$ (b) $^{198}_{85}At$ (c) $^{194}_{83}Bi$ (d) $^{198}_{83}Bi$ (e) $^{194}_{82}Pb$

3. Which of the following nuclides is most likely to decay alpha?
   (a) $^{20}_{12}Mg$ (b) $^{45}_{19}K$ (c) $^{52}_{26}Fe$ (d) $^{185}_{79}Au$ (e) $^{197}_{79}Au$

4. Which of the following nuclides is most likely to decay by electron capture?
   (a) $^{8}_{5}B$ (b) $^{13}_{5}B$ (c) $^{204}_{83}Bi$ (d) $^{209}_{83}Bi$ (e) $^{215}_{83}Bi$

5. The projectile in the induced nuclear reaction
   $$^{23}_{11}Na + {}^{2}_{1}H \longrightarrow {}^{24}_{11}Na + {}^{1}_{1}H \text{ is } \_\_\_\_\_.$$
   (a) p (b) $^{1}_{1}H$ (c) d (d) $^{23}_{11}Na$ (e) $^{24}_{11}Na$

6. The half-life for the beta decay of silicon-31 is 2.62 h. After 7.86 h, how many micrograms of phosphorus-31 will have formed from a 2.68-μg sample of Si-31?
   (a) 0.335 (b) 0.670 (c) 1.34 (d) 2.01 (e) 2.34

7. The half-life for the radioactive decay of fluorine-18 is 109.8 m. How many minutes will it take for 85.0% of a sample of F-18 to decay?
   (a) 11.2 (b) 25.8 (c) 131 (d) 301 (e) 352

8. How many disintegrations per second are expected from a sample that contains 3.91 mg of potassium-40 (the quantity of K-40 in 1.00 mol KCl)? For K-40, $t_{1/2} = 1.28 \times 10^9$ y.
   (a) 1.72×10^{-17} (b) 1.01×10^3 (c) 4.04×10^4
   (d) 3.19×10^{10} (e) 5.95×10^{19}

9. Which of the following nuclides is most likely to be stable?
   (a) $^{58}_{29}Cu$ (b) $^{63}_{29}Cu$ (c) $^{64}_{29}Cu$ (d) $^{68}_{29}Cu$ (e) $^{72}_{36}Kr$

10. What is the mass defect in atomic mass units for the nuclide sodium-22, which has an atomic mass of 21.994 437 u? The

atomic mass of H-1 is 1.007 825 u.
   (a) 0.000 840 (b) 0.180 925 (c) 0.186 953 (d) 0.192 988
   (e) 11.282 268

11. Calculate the energy change in joules that will accompany the reaction
    $$^{51}_{23}V + {}^{1}_{0}n \longrightarrow {}^{52}_{24}Cr + {}^{0}_{-1}e$$
    and predict whether the reaction will be spontaneous or nonspontaneous. The atomic mass of vanadium-51 is 50.9440 u, and the atomic mass of chromium-52 is 51.9405 u.
    (a) -6.07×10^{-13} J, spontaneous
    (b) -1.82×10^{-12} J, spontaneous
    (c) 1.82×10^{-12} J, nonspontaneous
    (d) -1.82×10^{-12} J, nonspontaneous
    (e) 1.82×10^{-12} J, spontaneous

12. Which of the following reactions is fusion?
    (a) $^{250}_{96}Cm \longrightarrow {}^{148}_{58}Ce + {}^{98}_{38}Sr + 4{}^{1}_{0}n$
    (b) $^{0}_{-1}e + {}^{161}_{69}Tm \longrightarrow {}^{161}_{68}Er$
    (c) $^{58}_{26}Fe + {}^{1}_{0}n \longrightarrow {}^{59}_{27}Co + {}^{0}_{-1}e$
    (d) $^{3}_{2}He + {}^{4}_{2}He \longrightarrow {}^{7}_{4}Be + \gamma$
    (e) $^{229}_{90}Th \longrightarrow {}^{225}_{88}Ra + {}^{4}_{2}He$

13. The biological effects of radiation depend on
    (a) the size of the dose.
    (b) the period of time over which the dose is received.
    (c) the type of radiation.
    (d) the kind of tissue exposed to the radiation.
    (e) all of the above.

14. The carbon in a sample of charcoal has an activity of 8.6 disintegrations per minute per gram of carbon. The half-life of carbon-14 is 5715 y and the steady state activity of C-14 is 14.9 disintegrations per minute per gram of carbon. How old is the charcoal?
    (a) 4.5×10^2 y (b) 2.0×10^3 y (c) 4.5×10^3 y
    (d) 7.1×10^3 y (e) 9.6×10^5 y

15. The most common element in the universe is _____.
    (a) C (b) Fe (c) H (d) O (e) Si

PUTTING THINGS TOGETHER

20.63 Briefly discuss the differences between chemical and nuclear reactions.

20.64 What two types of radionuclides occur naturally?

20.65 (a) Why is electron capture accompanied by the emission of X-rays? (b) Why are most nuclear reactions accompanied by the emission of gamma rays?

20.66 Both fission and fusion are nuclear reactions. (a) Why is a very high temperature required for fusion but not for fission? (b) Why are elements of low atomic number used as fuel for fusion whereas elements of high atomic number are used as fuel for fission? (c) What makes fission a branching chain reaction? (d) Why isn't it possible for a nuclear power reactor to blow up like an atomic bomb? (e) Why can't fission products be destroyed like other harmful chemicals?

20.67 (a) Like C-12, H-3 is formed by nuclear reactions brought about by cosmic rays in the upper atmosphere. About 1 atom of every 10^{18} H atoms in naturally occurring hydrogen is H-3. Complete the equation for one reaction by which H-3 is formed:

$$^{14}_{7}N + ^{1}_{0}n \longrightarrow ^{3}_{1}H + \underline{\hspace{1.5cm}}$$

(b) For the radioactive decay of H-3, $t_{1/2} = 12.26$ y. What is the average lifetime of an atom of H-3? (c) Write the equation for the radioactive decay of H-3. (d) Water in which 98% of the hydrogen atoms are H-3 is so radioactive that it glows at room temperature. How many disintegrations/s are expected from 1.00 g of $^{3}H_2O$ (all of the hydrogen atoms are H-3)? The atomic mass of H-3 is 3.016 05 u. (e) How many curies are contained in the sample of $^{3}H_2O$ in part (d)? (f) How many disintegrations /s are expected from 1.00 g of ordinary water?

20.68 (a) Write the equation for the alpha decay of beryllium-8. (b) Calculate the change in mass that will take place when Be-8 alpha decays. The atomic mass of Be-8 is 8.005 308 u, and the atomic mass of He-4 is 4.002 603 u. (c) Calculate the energy change in kilojoules per mole that will accompany this reaction. (d) Why is Be-8 radioactive?

20.69 A rock sample contains 627 atoms of uranium-235 for each 10 000 atoms of lead-207. The half-life of U-235 is 7.04×10^8 y. How old is the rock?

20.70 (a) The atomic mass of beryllium-7 is 7.0169 u, and the atomic mass of its decay product is 7.016 00 u. Which type of radioactive decay does beryllium-7 undergo? (b) For the radioactive decay of Be-7, the half-life is 53.37 days. How many kilojoules of energy are released per second by 1.00 g of beryllium-7?

20.71 Naturally occurring nitrogen is 99.63% ^{14}N and 0.37% ^{15}N; naturally occurring oxygen is 99.76% ^{16}O, 0.04% ^{17}O, and 0.20% ^{18}O. Calculate the percentages of NO molecules with masses 30, 31, 32, and 33.

20.72 Explain why fusion reactions have extremely high activation energies.

20.73 (a) If element 112 is synthesized, will it be a metal, a semimetal, or a nonmetal? (b) In which group will element 112 be? (c) Which known element will element 112 most resemble? (d) What will probably be the highest oxidation state of element 112 in its compounds? The lowest?

20.74 (a) Finish the seventh period of the expanded form of the periodic table (Figure 8.6) and add the eighth period. (b)

What group will element-168 be in? Describe the properties of element-168.

20.75 Describe an experiment with a radioactive nuclide that could be used to show that a dynamic equilibrium exists in a saturated solution of $BaSO_4$.

20.76 A 25.0-mL sample of 0.050 M $Ba(NO_3)_2$ solution was mixed with 25.0 mL of 0.050 M Na_2SO_4 solution labeled with the radionuclide S-35. The activity of the Na_2SO_4 solution was 1.22×10^6 Bq/mL. The temperature was adjusted to 25 °C, and the precipitate was separated by filtration. The filtrate had an activity of 250 Bq/mL. (a) Write net ionic, complete ionic, and molecular equations for the reaction that took place. (b) From this experiment, what is the value of K_{sp} for the insoluble salt that formed?

20.77 What is the difference between the following? (a) a nuclide and a nucleon (b) gamma rays and X-rays (c) atomic number and mass number (d) fission and fusion (e) beta particles and positrons (f) a rad and a rem (g) slow neutrons and fast neutrons

20.78 Elements 110 and 111, which have not yet been named, were synthesized by bombarding medium-weight atoms with medium-weight ions. For example, 110 was made using $^{208}_{82}Pb$ atoms as the target and $^{64}_{28}Ni^{9+}$ ions as projectiles. An experiment that lasted several weeks and in which 3×10^{12} ions/s hit the target yielded nine atoms of element 110. A compound nucleus in an excited state was formed that decayed by emission of a neutron. (a) Write equations for the nuclear reactions that took place using E as the symbol for element 110. (b) Write the abbreviated electron configuration of the Ni^{9+} ion. (c) What do you think happened to the Ni^{9+} ions that did not react? (d) Elements 110 and 111 were identified by their decay chains. Both decay by emitting a succession of alpha particles. Write equations showing the first three steps of the decay chain for $^{269}_{110}E$. (e) The researchers at Darmstadt are planning to make element 112 using $^{208}_{82}Pb$ as the target. What element will they use as projectile? (f) They also plan to make element 113 using the same projectile as in part (e). What element will they use as the target? (g) Tin-100 was made for the first time at Darmstadt in 1994. Why is tin-100 especially stable? (h) In addition to trying to make new elements, the researchers are also studying the chemical properties of elements 105 and 106 to see how these elements fit into the periodic table. Element 105 resembles tantalum. Which element should 106 behave like?

20.79 When IO_4^- is added to a solution that contains I^- labeled with I-128, which is radioactive, all of the radioactive iodine ends up in the I_2:

$$IO_4^- + 2\ ^{128}I^- + H_2O(l) \longrightarrow IO_3^- + ^{128}I_2(s) + 2OH^-$$

Is IO_3^- formed by the reduction of IO_4^- or by oxidation of I^-? Explain your answer.

20.80 The carbon in a sample of wood found in an Egyptian pyramid has an activity of 8.5 distintegrations per minute per gram of carbon. The half-life of C-14 is 5715 years, and the activity of wood from a recently cut tree is 14.9 disintegrations per minute per gram of carbon. How old is the wood? (Dates of Egyptian objects measured by C-14 dating are confirmed by records carved in stone.)

20.81 Radioisotopes such as aluminum-26, which is believed to have been present in the solar system when Earth was formed but can no longer be detected, are known as extinct radionuclides. The half-life for the decay of aluminum-26 is 7.20×10^5 years. (a) How many years did it take for 99.99% of the Al-26 to decay? (b) If Earth was formed 4.6×10^9 years ago, what percentage of the aluminum-26 present when the solar system was formed is still present today?

20.82 Strontium is similar to calcium and takes the place of calcium in bones. As a result, the radionuclide Sr-90 (half-life 28.1 y), which is formed by nuclear fission, is especially dangerous. What percentage of the Sr-90 that was released into the atmosphere when the first atomic bombs were exploded in 1945 still exists at the present time?

20.83 One figure for the estimated annual energy requirement of the world in 2025 is 1.98×10^{14} kJ (*Research & Development*, September 1988, p. 32). There are 1.36×10^{21} kg of water in the oceans. Naturally occurring hydrogen is 0.015% H-2. (a) What percentage of the hydrogen-2 in seawater would have to be fused to helium-4 to supply enough energy for one year if the estimated energy requirement is correct? The atomic mass of H-2 is 2.0140 u, and the atomic mass of He-4 is 4.002 60 u. (b) For how many years would the H-2 supply last at this rate of use?

20.84 The first nuclear reactors operated early in Earth's geological history. They were discovered in 1972 when an analyst in the French Atomic Energy laboratories found a uranium-235 concentration of only 0.7171% in a sample of uranium from a mine near Oklo in Gabon, Africa. The concentration of U-235 in naturally occurring uranium is usually 0.7202% at the present time. Naturally occurring nuclear reactors are impossible today because the proportion of U-235 in uranium is too low for a fission chain reaction to occur but were possible earlier in Earth's history when the proportion of U-235 in natural uranium was higher. The uranium in the fuel used in nuclear power plants is about 3.0% U-235. The half-life of U-235 is 7.1×10^8 y. How many years ago did natural uranium contain 3.0% U-235 so that a fission chain reaction was possible?

20.85 Naturally occurring phosphorus is 100% P-31. Six radioisotopes of phosphorus are known and are, with their half-lives, P-28, 0.28 s; P-29, 4.4 s; P-30, 2.50 m; P-32, 14.3 d; P-33, 25 d; P-34, 12.4 s. (a) Predict the type of radioactive decay expected for each of the radioisotopes. (b) Which isotope(s) is(are) suitable for use as radioactive tracers?

20.86 Analysis of a sample of rock from the moon shows that the ratio of the number of atoms of lead-208 to the number of

atoms of thorium-232 is 0.256. The half-life of Th-232 is 1.40×10^{10} y. How old is the moon rock?

20.87 To find uranium deposits, uranium prospectors usually use a Geiger counter (although a scintillation counter is better) to detect gamma radiation from potassium-40 and the daughter products of the uranium and thorium decay series. Why is uranium not found by looking for alpha and beta radiation?

20.88 Rubidium-87/strontium-87 dating is used for very old rocks. (a) What type of radioactive decay takes place when rubidium-87 decays to strontium-87? Write the nuclear equation. (b) Some Sr-87 was present in the material from which Earth was formed. Does this fact complicate calculations about Rb-87/Sr-87 dating? Explain your answer.

20.89 If only part of a substance can be separated from a sample, isotope dilution can be used to determine the amount of the substance in the sample. The analysis of vitamin tablets for vitamin B-12 is an example. Vitamin B-12 contains cobalt, and vitamin B-12 containing radioactive cobalt is available commercially. Suppose that 60 vitamin tablets, each having a mass of 777 mg, were dissolved in water. To the solution, 0.395 mg of B-12 labeled with radioactive cobalt were added. The tagged vitamin B-12 had an activity of 3.620 μCi/mg. The solution was mixed thoroughly. A sample of vitamin B-12 isolated from the solution had an activity of 1.282 μCi/mg. How many micrograms of vitamin B-12 were in each tablet? Decay of the radioactive cobalt during the time required for analysis is negligible.

20.90 The conversion of four hydrogen-1 nuclei to one helium-4 nucleus by the proton–proton chain (equations 20.5–20.7) powers the sun. (a) Write the nuclear equation for the overall reaction. (b) How much energy in kilojoules is released when four hydrogen-1 nuclei are converted to one helium-4 nucleus? Atomic masses you will need are H-1, 1.007 825 u; He-4, 4.002 603 u. The mass of an electron is 0.000 5486 u on the atomic mass scale. (*Hint:* Positrons are emitted, so nuclear masses, not atomic masses, should be used in the calculation.) (c) In the sun, 4.3×10^9 kg of matter are converted into energy each second. How many helium-4 nuclei are formed each second? How many H-1 nuclei are used up each second?

20.91 Briefly discuss the advantages and disadvantages of fusion, fission, and combustion of fossil fuels as energy sources.

20.92 Americium-241 is used in smoke detectors. It decays through a series of alpha and beta decays to bismuth-209, which is stable. The first step in the decay of americium-241, for which the half-life is 432 y, is emission of an alpha particle. The alpha particles ionize the air in the smoke detector and the ions in the air conduct an electric current. Smoke particles interfere with conductivity, and the alarm sounds. (a) Write the nuclear equation for the first step in the decay of americium-241. (b) How many alpha particles and how many beta particles are given off during the decay of americium-241 to bismuth-209? Give your reasoning. (c) One other isotope of americium, americium-243, can be made in useful quantities. The first step in the decay of americium-243, for which the half-life is 7.73×10^3 y, is also emission of an alpha particle. Why is americium-241, not americium-243, used in smoke detectors?

20.93 (a) One problem with another method of dating, potassium-40/argon-40 dating, is that argon is a gas and is apt to have escaped from the sample. If argon-40 has escaped from a sample, will the age found be higher or lower than the true age? Explain your reasoning. (b) Most of the K-40 (98.2%) is converted to Ar-40 by electron capture. A little (0.49%) undergoes positron emission. Write equations for both these nuclear reactions. (c) The remaining 1.35% of the K-40 beta decays. Write the equation for the beta decay of K-40. (d) The stable isotopes of potassium are K-39 and K-41. Why is K-40 unstable when these two isotopes are stable? (e) Explain why K-40 undergoes three different types of radioactive decay.

20.94 Plutonium-239, which is formed in nuclear reactors, is another dangerous nuclide. The half-life of plutonium-239 is 24 400 (2.44×10^4) y, and the product of its decay is U-235. (a) Write a nuclear equation for the decay of Pu-239. What type of radiation is emitted? (b) Why is inhalation and ingestion of Pu-239 hazardous? (c) How many years will be required for 99% of the plutonium-239 formed this year to decay? (To give you an idea of how very long a time this is, it is as far into the future as the Stone Age is into the past.)

20.95 Only one stable isotope of cobalt, Co-59, exists. Two radioactive isotopes of cobalt, Co-57 and Co-60, are used in medicine. Cobalt-57 has a half-life of 271 days; when an atom of cobalt-57 decays, a gamma-ray photon with an energy of 0.122 MeV is emitted. Cobalt-60 has a half-life of 5.27 y; when an atom of cobalt-60 decays, gamma-ray photons with energies of 1.173 and 1.332 MeV are emitted. (a) Predict the type of radioactive decay expected for each of the two radioisotopes. Explain your answers. (b) For diagnosis, which radioisotope would be used? Explain your answer. (c) For treatment of malignancies known to be responsive to gamma radiation, which radioisotope would be used? Explain your answer.

20.96 Solutions of nuclides that can be fissioned are produced in the reprocessing of spent fuel elements from nuclear reactors. Care must be taken that the critical size is not exceeded in any container in order to prevent an accidental chain reaction. For fission by thermal neutrons, the smallest critical size for a uranyl sulfate (UO_2SO_4) solution is 0.82 kg of U-235 in 6.3 L of solution. What is the molarity of this solution?

20.97 For the nuclide $^{99}_{43}Tc^*$, $t_{1/2} = 6.02$ h. The atomic mass of technetium-99 is 98.906 u. A 1.00-mL sample of a solution of $^{99}_{43}Tc^*$ with an activity of 1.481×10^8 Bq was injected intravenously. After 6.00 m (enough time for complete mixing to take place), a 1.00-mL sample of the person's blood had an activity of 2.79×10^4 Bq. (a) What was the dose given in millicuries? (b) After 6.00 m, what percentage of the original radioactivity would remain? (c) What was the volume of the person's blood? (d) What mass of $NaTcO_4$ was in the 1.00-mL sample of a solution of $^{99}_{43}Tc^*$ with an activity of 1.481×10^8 Bq? (e) The smallest mass that

can be measured with the necessary precision is of the order of 1.00×10^{-5} g. How could a solution of the concentration needed be obtained?

20.98 Thallium-201 is used to examine the myocardium (the muscular substance of the heart). The half-life of Tl-201 is 73 h. (a) If a 2.0-mg sample of Tl-201 is needed for some medical research and 24 h are required for delivery, how much $^{201}TlCl$ must be ordered? (b) The Merck Index notes that HCl decreases the solubility of TlCl in water. Explain why the solubility of TlCl is lowered by HCl.

20.99 In stars that are more massive and hotter than the sun, four H-1 nuclei are fused into a He-4 nucleus by a different series of steps, the carbon–nitrogen–oxygen (CNO) cycle, shown below:

$$^{12}_{6}C + ^{1}_{1}H \longrightarrow ^{13}_{7}N + ^{0}_{0}\gamma$$
$$^{13}_{7}N \longrightarrow ^{13}_{6}C + ^{0}_{+1}e$$
$$^{13}_{6}C + ^{1}_{1}H \longrightarrow ^{14}_{7}N + ^{0}_{0}\gamma$$
$$^{14}_{7}N + ^{1}_{1}H \longrightarrow ^{15}_{8}O + ^{0}_{0}\gamma$$
$$^{15}_{8}O \longrightarrow ^{15}_{7}N + ^{0}_{+1}e$$
$$^{15}_{7}N + ^{1}_{1}H \longrightarrow ^{12}_{6}C + ^{4}_{2}He$$

(a) In the CNO cycle, how is C-12 acting? (b) Which nuclides are intermediates? (c) How does the energy released compare with the energy released when four H-1 nuclei are converted into a He-4 nucleus by the proton–proton chain [see Problem 20.90(b)]? Explain your answer.

20.100 A typical induced fission reaction is

$$1^{1}_{0}n + ^{235}_{92}U \longrightarrow ^{137}_{52}Te + ^{97}_{40}Zr + 2^{1}_{0}n$$

(see Figure 20.14). (a) The nuclides produced by fission, such as Te-137 and Zr-97, are radioactive. What type of radioactive decay would be predicted for these nuclides? Explain your answer. (b) Assuming that Te-137 and Zr-97 decay until they become stable nuclides, what will the overall fission reaction be? (*Hint:* Use Figure 20.6 to find out which nuclides are stable.) Write the nuclear equation for the overall fission reaction. (c) Calculate the energy released by the overall fission reaction in kilojoules per gram of U-235 split. Use the Table of the Isotopes in a recent edition of the *CRC Handbook* to find nuclide masses. (d) Calculate the energy released by the combustion of methane, CH_4, in kilojoules per gram of methane burned. (e) Quantitatively, how do the energies released per gram of fuel by fission, fusion (Problem 20.90), and combustion compare?

20.101 The nuclide $^{99}_{43}Tc^*$ is an excellent nuclide for diagnosis. Only low-energy gamma rays are emitted when $^{99}_{43}Tc^*$ decays to $^{99}_{43}Tc$. The half-life is long enough that a dose can be delivered and a study carried out before the radioactivity disappears but short enough so that most of the radioactivity is given off during the study. Although the product nuclide $^{99}_{43}Tc$ beta decays, its half-life (2.12×10^5 y) is so long that its radioactivity is not a problem. (a) Write equations for the radioactive decay of $^{99}_{43}Tc^*$ and $^{99}_{43}Tc$. (b) How often, on average, is a beta particle emitted from the $^{99}_{43}Tc$ formed from the quantity of $^{99}_{43}Tc^*$ delivered in Problem 20.97? This radiation is of the same order of magnitude as normal background radiation.

(c) When an atom of $_{43}^{99}$Tc* decays to $_{43}^{99}$Tc, the photon of gamma radiation emitted has an energy of 0.141 MeV. (i) How much energy in kilojoules is emitted in the form of gamma rays when 1.00 mol $_{43}^{99}$Tc* decays? (ii) What is the wavelength of the gamma radiation given off? (iii) What is the difference between the mass of a mole of $_{43}^{99}$Tc* atoms and the mass of a mole of $_{43}^{99}$Tc atoms?

20.102 (a) In hospitals and clinics, $_{43}^{99}$Tc*solution for diagnostic studies is made from molybdenum-99 in the form of MoO_4^- ion, which is absorbed on finely divided alumina (Al_2O_3) in a column. Write an equation for the conversion of molybdenum-99 to $_{43}^{99}$Tc*.   (b) The TcO_4^- ion formed is removed from the column by passing a solution of NaCl of the same concentration as the solution inside the body's cells through the column. Why must the $_{43}^{99}$Tc* solution used for intravenous injection have the same concentration of NaCl as the solution inside the cells?   (c) The molybdenum-99 is made by bombarding molybdenum-98 (also in the form of MoO_4^-) with slow neutrons from a nuclear reactor. Gamma rays are emitted. For this nuclear reaction, write the nuclear equation. (d) Which of the following half-lives—2.0 × 10⁴ y, 66.69 h, or 40 s—must be the half-life of molybdenum-99? Explain your answer.

Chemistry and Paleontology

ELIZABETH GOMANI

Graduate Student

B.S. Geology/Chemistry
University of Malawi

M.S. Geology
Southern Methodist University

Ph.D. Candidate in Geology
Southern Methodist University

I became interested in chemistry when I was very young, going behind my mother's back to get some cooking oil for "little girl's cooking." Because I did not know how to cook, I just poured the cooking oil in a glass of water. I expected the two liquids to mix, but even after shaking the glass, the oil just formed globules that recombined to float on top of the water after standing for a bit. This puzzled me so much that I started mixing other liquids to see what happened. At that time, my grandmother would brew a local spirit from a fermented mixture of water, sugar, and corn flour. After brewing and prior to sales, the spirit would be diluted with water. One day I watched her out of curiosity to see if the alcohol was going to float on the water like the cooking oil did. When it failed to do so, I was very puzzled because I could not understand what was present in alcohol that allowed it to mix in water but was absent from cooking oil.

This curiosity continued through high school and my freshman year in college at the University of Malawi in East Africa. But after burning myself with sulfuric acid during one experiment, I began losing interest in chemistry. I wanted to eventually work with the Geological Survey of Malawi and was therefore advised to continue my studies in chemistry and geology, although my preferences at this point were biology and geology. I never thought that chemistry would be useful to me in geological work, but my subsequent experiences as a paleontologist have proven quite the opposite.

During 1989 I worked on a paleontology dig taking place in Malawi. An American researcher, Dr. Louis Jacobs, asked the Malawi Department of Antiquities to recommend a young science student to help work on the dig. Although I had received many discouragements during my geological studies, having been told by some that it was "scandalous for a woman to study geology," I was nonetheless chosen by the Department of Antiquities on the basis of my strong background in math and science. My experiences on this dig and additional digs in 1990 and 1992 helped me see a way in which I could combine my interests in geology and biology. During these digs we discovered bones and bone fragments of a relatively unknown species of dinosaur (now known as Malawisaurus), as well as an odd dwarf crocodile that had molarlike teeth and apparently lived in burrows. I have subsequently been involved in further research on both of these fossil specimens and am presently working toward my Ph.D. by investigating the relationship of Malawisaurus to its environment and to other species of dinosaurs.

In my studies, I work with the remains of past life, such as bones and shells. The bones are preserved because their chemical composition allows them to withstand the decay that destroys muscles and other soft tissues. The sedimentary rocks in which these fossils are usually found consist of rock particles (aggregates of minerals) that have been cemented together. Each mineral displays distinctive properties that reflect its characteristic microscopic structure. The chemical signatures of the cementing materials can also be used to determine the depositional environment of a rock. For example, in some cases, calcite, $CaCO_3$, concretions are formed. The stable carbon isotopic ratios of the concretions can then be used to determine the existing vegetation at the time of deposition and the oxyen isotopic ratios to estimate the temperature at the time of deposition. This information is important in understanding the environmental conditions during the time a particular animal lived.

Even before we try to interpret the paleoenvironment of an animal represented by fossil remains, the geologic time frame in which the animal lived must be determined. For fossils found in geologically recent deposits, radiocarbon dating can be used to date the fossil. However, most of geological time is far too distant for radiocarbon dating to be useful. To date fossils (such as those of dinosaurs) that occur in more geologically distant times on an absolute time scale requires other types of radiometric techniques. In such cases, heavier elements that possess long-lived radioactive isotopes are used to obtain an absolute age for the rock containing a fossil. This age can then be used to estimate the age of the fossil.

In paleontological research, chemistry is also involved in the preparation of samples. Chemicals are used in the laboratory to stabilize and clean bones. Stabilization is accomplished by applying a solution of polyvinylacetate in acetone to the bone sample. Acetone can also be used to separate bones that are stuck together if there is need to do that. If fossils are so strongly cemented together with calcite that regular preparation techniques will not work, acid is applied to the rock to dissolve the calcite. Care must be taken, however, because both the bone and the rock will dissolve in the acid.

Little did I know as I began working on that dig in 1989, my initiation into the world of paleontology, that the chemical principles and properties that so fascinated me as a young girl would prove so useful in my future research pursuits. Without the insights given by chemical knowledge, our understanding of the past worlds occupied by dinosaurs and other species would be limited. Even the practical problems of separating fossils from their rock surroundings and physically preparing them for laboratory study would be most difficult without a thorough understanding of chemical principles and reactions.

A CLOSER LOOK AT INORGANIC CHEMISTRY: NONMETALS AND SEMIMETALS AND THEIR COMPOUNDS

21

In lightning bolts, nitrogen and oxygen from air react to form nitric oxide, and oxygen is converted to ozone.

This chapter concerns hydrogen; water; the nonmetals of Groups IVA, VA, and VIA; and the most important semimetal, silicon, which is the second element in Group IVA. Both the free elements and their most important compounds are discussed.

— **Why is a study of nonmetals and semimetals a part of general chemistry?**

Although only 24 of the 111 elements are nonmetals and semimetals, the two commonest of these elements, oxygen and silicon, make up 74.5% by mass of Earth's crust. The water that covers much of Earth's surface is a compound of hydrogen and oxygen. Earth's atmosphere is composed mainly of nitrogen and oxygen. Five nonmetals—carbon, hydrogen, oxygen, nitrogen, and phosphorus—are the basic elements of the organic compounds that make up living systems. The nonmetals sulfur, fluorine, chlorine, and iodine and the semimetals boron, silicon, and selenium are also essential for life.

More compounds of hydrogen are known than compounds of any other element. Almost all organic compounds contain hydrogen, and it is present in all animal and vegetable materials. Most of the hydrogen on Earth is in the form of the compound water. After oxygen in air, water is the substance that is most essential to life. The majority of biological reactions, many industrial reactions, and most of the reactions carried out in general chemistry labs are carried out in aqueous solution. The reactions that take place in oceans, lakes, and rivers are important in environmental chemistry. Water has shaped the land on which we live. It is one of the reactants for photosynthesis and for many industrial and laboratory processes. Steam is used to drive the turbines that generate electricity in power plants, and water is used for heating and cooling, cleaning, and waste disposal. Swimming in water, boating on water, and skating on ice are all fun. Everything considered, water is probably the most important compound.

Carbon is the key element in the animal and vegetable kingdoms; silicon is the key element in the mineral kingdom. About 95% of Earth's crust consists of compounds of silicon, and most inorganic building materials are based on silicon. Air provides carbon dioxide for photosynthesis and oxygen for breathing. Oxygen is also necessary for the combustion of fossil fuels, the source of most of the energy used by people. The nitrogen that is essential for life also comes from air. Ammonia, nitric acid, ammonium nitrate, and ammonium sulfate are all produced in large quantities and are important in the fertilizer and explosive industries. Phosphoric acid is also produced in large quantities and is important in the fertilizer industry; bones and teeth are about 20% phosphorus. Sulfur and its compounds are so widely used in industry that a country's level of economic development is often measured by the quantity of sulfur it uses.

— *What do you already know about nonmetals and semimetals and their compounds?*

You are familiar with many of the properties of water and air from your everyday experience. You know chlorine as a disinfectant in drinking water and swimming pools and combined with sodium in table salt. In the United States, most table salt is "iodized" and contains about 1 part KI per 10 000 parts NaCl. Fluorine is familiar from toothpaste and fluoridated drinking water, which contains sodium fluoride. The nonstick Teflon lining on cooking utensils is polytetrafluoroethylene, $\{CF_2 - CF_2\}_n$. You are also undoubtedly familiar with diamonds and pencil leads (both carbon), chalk and marble (both $CaCO_3$), and possibly with washing soda (Na_2CO_3). Glass is probably the most widely known silicon-containing material.

The properties of the halogens and of the noble gases vary quite smoothly from top to bottom of the group in the periodic table. Therefore, these two groups and their compounds were considered at the end of Chapter 8. The reactions of oxygen as an oxidizing agent were discussed in Section 11.8. All Arrhenius and Brønsted–Lowry acids contain one or more ionizable hydrogens. From earlier chapters, you also know that, in general, nonmetals form molecular compounds with other nonmetals, and ionic compounds, in which the nonmetal is anionic, with active metals. Oxides and hydroxides of nonmetals, such as SO_3 and H_2SO_4, are acidic; nonmetal hydroxides are usually called oxo acids (Sections 11.3 and 16.2). You also learned about the allotropic forms of carbon (Section 12.8) and that addition of arsenic to silicon yields an n-type semiconductor and addition of boron gives a p-type semiconductor (Section 12.12). In Chapters 4 and 13, you learned about the chemical and physical properties of aqueous solutions. The contact process for manufacturing sulfuric acid and the uses of sulfuric acid were discussed in Section 14.6. The Haber–Bosch process for making ammonia was also considered in Section 14.6.

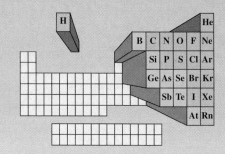

21.1 HYDROGEN

Isotopes

The differences between the isotopes of hydrogen are greater than the differences between the isotopes of any other element because the masses of hydrogen-2 and hydrogen-3 are two and three times the mass of hydrogen-1. The physical properties of hydrogen-2 and hydrogen-3 and their compounds differ significantly from those of hydrogen-1 and its compounds. Therefore, hydrogen-2 is often referred to as deuterium, D, and hydrogen-3, as tritium, T. Some physical properties of molecular hydrogen composed of the three isotopes are shown in Table 21.1. The chemical properties of the three isotopes and their compounds are similar; only rates and equilibrium constants differ.

Building deuterium into tissue results in the death of the test animal. However, mice containing 60 times the natural abundance of carbon-13 suffered no ill effects.

TABLE 21.1	Properties of Molecular Hydrogen		
Property	H_2	D_2	T_2
Melting point, K	14.1	18.6	20.6
Boiling point, K	20.4	23.5	25.0
Heat of fusion, kJ/mol	0.117	0.197	0.250
Heat of vaporization, kJ/mol	0.904	1.226	1.393
Heat of dissociation, kJ/mol (at 298.2 K)	435.88	443.35	446.9

Hydrogen gas was used to fill lighter-than-air dirigibles until 30 people were killed when the German airship Hindenberg burned while landing at Lakehurst, New Jersey, at the end of her first transatlantic voyage of the 1937 season.

Physical Properties of Molecular Hydrogen

Hydrogen, H_2, is a colorless gas with extremely low melting and boiling points (see Table 21.1). The low melting and boiling points of hydrogen indicate that the London forces, which are the only attractive forces that exist between H_2 molecules, are very weak. The temperature of most gases falls when the gas is allowed to expand because thermal energy is required to overcome the attractive forces between gas molecules. However, the temperature of hydrogen gas rises when the hydrogen gas is allowed to expand at room temperature. The molecules of hydrogen gas not only do not attract each other very strongly under ordinary conditions but actually repel each other. At 25 °C and 1 atm pressure, the density of hydrogen gas is 0.0824 g/L, lower than the density of any other substance at the same temperature and pressure.

Preparation of Molecular Hydrogen

Industrially, most hydrogen is produced by the reaction of hydrocarbons (compounds of carbon and hydrogen) such as methane with steam,

$$CH_4(g) + H_2O(g) \xrightarrow{\text{catalyst, 900 °C}} 3H_2(g) + CO(g)$$

followed by reaction of carbon monoxide with steam,

$$CO(g) + H_2O(g) \underset{\longleftarrow}{\xrightarrow{\text{catalyst, 200–400 °C}}} H_2(g) + CO_2(g) \qquad (21.1)$$

The reaction shown in equation 21.1 is called the **water–gas shift reaction.** The reaction between carbon monoxide and steam is exothermic, and the water–gas shift reaction is carried out at a relatively low temperature to shift the equilibrium in favor of hydrogen and carbon dioxide. Large plants make hydrogen at a rate of 1–8 tons·h^{-1}.

Hydrogen is also produced as a by-product of the manufacture of sodium hydroxide by the electrolysis of aqueous sodium chloride (Section 19.9):

$$2NaCl(aq) + 2H_2O(l) \xrightarrow{\text{electrolysis}} H_2(g) + Cl_2(g) + 2NaOH(aq)$$

Preparation of hydrogen by electrolysis of water,

$$2H_2O(l) \xrightarrow{\text{electrolysis}} 2H_2(g) + O_2(g)$$

is too expensive for large-scale production except where cheap electricity is available from water power.

About 10^{10} pounds of hydrogen are produced in the United States each year. Hydrogen is often made and used in the same plant; only about 5% of the hydrogen produced is sold. If necessary, hydrogen can be liquefied and transported in well-insulated tank cars and tank trucks. Steel cylinders are used for small quantities of hydrogen and other gases (see ▪Figure 21.1). When full, the cylinder shown in

▪ FIGURE 21.1 If a cylinder of compressed gas falls over and the valve at the top breaks off or springs a leak, the cylinder becomes jet propelled and extremely dangerous. Therefore, compressed gas cylinders should always be fastened as this one is (a chain is holding the cylinder to the carrier). Most of the energy needed to move hydrogen in cylinders around is required just to move the cylinder.

Figure 21.1 holds about 1.39 lb (631 g) of hydrogen at a pressure of 2640 pounds per square inch (179.7 atm). The empty cylinder has a mass of 125 lb.

PRACTICE PROBLEM

21.1   Write an equation for the synthesis of hydrogen gas by the reaction of propane, $CH_3CH_2CH_3$, with steam.

Chemical Properties of Molecular Hydrogen

Hydrogen gas is odorless and tasteless. Although hydrogen reacts with fluorine in the dark at room temperature,

$$H_2(g) + F_2(g) \longrightarrow 2HF(g)$$

hydrogen is relatively unreactive at low temperatures. (Fluorine gas is an unusually reactive substance.) Mixtures of hydrogen with oxygen or chlorine can stand indefinitely in the dark at room temperature without reacting. However, if energy is added in the form of a spark or light, reaction takes place with explosive violence:

$$2H_2(g) + O_2(g) \xrightarrow{\text{spark}} 2H_2O(l)$$

$$H_2(g) + Cl_2(g) \xrightarrow{\text{light}} 2HCl(g)$$

In other words, reaction of hydrogen with oxygen and chlorine is exothermic and spontaneous but has a high activation energy. The rate of reaction can be increased by a catalyst as is done in the synthesis of ammonia by reaction of hydrogen with nitrogen (Section 14.6).

The high bond dissociation energy of the H—H bond in the H_2 molecule,

$$\text{H:H} \longrightarrow \text{H·} + \text{·H} \qquad \text{or} \qquad H_2(g) \longrightarrow 2H(g) \qquad \Delta H° = 436 \text{ kJ/mol}$$

is an important factor in the lack of reactivity of hydrogen at low temperatures. *At high temperatures,* enough energy is available to break the H—H bond, and *hydrogen reacts vigorously with many metals and nonmetals.*

Hydrogen forms several types of binary compounds. The best method of classification of the hydrides seems to be on the basis of physical properties.

Hydrides

Reaction of hydrogen with active metals such as lithium (*s*-block elements) results in the formation of high-melting **saltlike hydrides.** Using abbreviated electron configurations to show the formation of lithium hydride from lithium atoms and hydrogen atoms, electron movement is as follows:

$$[\text{He}]2s^1 + 1s^1 \longrightarrow [\text{He}] + 1s^2$$
$$\text{Li} + \text{H} \longrightarrow \text{Li}^+ + \text{H}^-$$

The equation for the reaction of lithium with hydrogen gas is

$$2\text{Li}(s) + H_2(g) \xrightarrow{725\,°C} 2\text{LiH}(s)$$

The fact that molten lithium hydride is an electrolyte is experimental evidence for the ionic character of lithium hydride. Electrolysis of molten lithium hydride yields

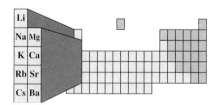

Elements that form saltlike hydrides.

hydrogen gas at the anode:

$$2H^-(l) \xrightarrow{\text{electrolysis}} H_2(g) + 2e^-$$

The hydride ion reacts with water to form hydrogen gas:

$$H^- + H_2O(l) \longrightarrow H_2(g) + OH^- \tag{21.2}$$

Reaction of water with either lithium hydride or calcium hydride is a convenient, easily moved source of small quantities of hydrogen gas. Calcium hydride is also used to dry organic solvents and gases.

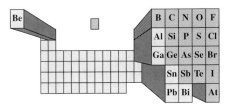

Elements that form molecular hydrides.

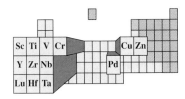

Elements that form metal-like hydrides. Hydrogen itself changes into a metallic form at 1.7–2.0 million atm. Metallic hydrogen probably exists in the interiors of the giant planets Jupiter and Saturn and of stars.

Information about hydrides is mostly from Greenwood, N. N.; Earnshaw, A. Chemistry of the Elements; Pergamon: Oxford, 1984, pp 69–74.

PRACTICE PROBLEMS

21.2 (a) Use abbreviated electron configurations to show the formation of calcium hydride from calcium atoms and hydrogen atoms. (b) Write an equation for the reaction of calcium metal with hydrogen gas. (c) Write net ionic, complete ionic, and molecular equations for the reaction of calcium hydride with water.

21.3 The reaction in equation 21.2 is a Brønsted–Lowry acid–base reaction. In this reaction, (a) which species acts as a Brønsted–Lowry acid? (b) Which species acts as a Brønsted–Lowry base?

21.4 The reaction in equation 21.2 is also a redox reaction. (a) Which element undergoes oxidation? (b) Which element undergoes reduction? (c) Which species is the oxidizing agent? (d) Which species is the reducing agent?

21.5 Use Lewis formulas to show that the reaction between AlH_3 and H^- is a Lewis acid–base reaction.

Most of the *p*-block elements form **molecular hydrides,** such as CH_4, NH_3, PH_3, H_2O, H_2S, and HCl, which are gases or low boiling liquids at room temperature. Hydrogen is a good reducing agent. As a result, hydrides in which the central atom has a high oxidation number such as PH_5 and SH_6 do not exist. Only the hydrides of low oxidation number, PH_3 and SH_2 (usually written H_2S), are known. Lewis formulas can be written for molecular hydrides and the shapes of most molecules predicted correctly by VSEPR (Section 10.1).

Many of the *d*-block metals form **metal-like hydrides,** which are solids and conduct electricity like metals. Metal-like hydrides are often nonstoichiometric. The *composition* of **nonstoichiometric compounds** is *not constant;* the subscripts in the formulas of nonstoichiometric compounds are not whole numbers and vary from sample to sample. The compounds $TiH_{1.7}$ and $ZrH_{1.9}$ are examples of nonstoichiometric hydrides. At the present time, chemists have not been able to agree on a simple model for the metal-like hydrides. No stable hydrides have yet been reported for a number of later *d*-block metals such as osmium and platinum.

Uses of Molecular Hydrogen

More hydrogen is used to make ammonia by the Haber process (Section 14.6) than is used for any other purpose. Large quantities of hydrogen are also used to make solid vegetable shortenings such as Crisco, Spry, and margarine from liquid vegetable oils like soybean and corn oil. Liquid vegetable oils contain carbon–carbon double

Molecules without double bonds (left model) *pack more efficiently than molecules with double bonds* (right model). *As a result, London forces are stronger in hydrogenated vegetable shortenings, and the shortenings are solid whereas the original vegetable oils, which contain double bonds, are liquid.*

bonds. In the presence of a catalyst, hydrogen adds to carbon–carbon double bonds:

$$CH_3(CH_2)_4CH{=}CHCH_2CH{=}CH(CH_2)_7COO\overset{\displaystyle CH_3(CH_2)_7CH{=}CH(CH_2)_7COOCH_2}{\underset{\displaystyle CH_3(CH_2)_{14}COOCH_2}{CH}}(l) + 3H_2(g) \xrightarrow{\text{catalyst}} CH_3(CH_2)_{16}COO\overset{\displaystyle CH_3(CH_2)_{16}COOCH_2}{\underset{\displaystyle CH_3(CH_2)_{14}COOCH_2}{CH}}(s)$$

The products are solids.

Manufacture of organic chemicals for use as solvents and as raw materials for the production of other organic chemicals also consumes huge amounts of hydrogen. The synthesis of methyl alcohol, CH_3OH,

$$2H_2(g) + CO(g) \underset{\xrightarrow{\hspace{1.5cm}}}{\overset{\text{catalyst, 300 °C, 300 atm}}{\rightleftharpoons}} CH_3OH(g) \qquad \Delta H° = -90.7 \text{ kJ}$$

is an example. Methyl alcohol plants are so similar to ammonia plants that one can be converted into the other as demand changes. The catalyst and high temperature are necessary to make the reaction take place at a practical rate. However, because the reaction between hydrogen and carbon monoxide is exothermic, the high temperature shifts the equilibrium in favor of reactants. A temperature of 300 °C is a compromise between rate and yield. High pressure shifts the equilibrium in favor of products because there are three moles of gaseous reactants and only one mole of gaseous product.

Large quantities of hydrogen are also used to obtain metals from their ores. Reduction of tungsten(VI) oxide is an example:

$$WO_3(s) + 3H_2(g) \xrightarrow{850\,°C} W(s) + 3H_2O(g)$$

Hydrogenation of liquid vegetable oils yields solid shortenings.

PRACTICE PROBLEM

21.6 Write equations for (a) the reaction of molecular hydrogen with $CH_3CH{=}CH_2(g)$ in the presence of a catalyst and (b) the reduction of molybdenum(VI) oxide with hydrogen.

Hydrogen as a Fuel

Substitution of hydrogen for coal and petroleum products as fuels is attractive for a number of reasons. The combustion product of hydrogen is water

$$2H_2(g) + O_2(g) \longrightarrow 2H_2O(l)$$

not carbon dioxide. Buildup of carbon dioxide from the burning of fossil fuels may be causing a disastrous increase in the temperature of Earth's atmosphere [the "greenhouse effect" (Section 21.2)]. With hydrogen as a fuel, there is no chance of accidental release of radioactivity and no radioactive waste to store as there would be if fossil fuels were replaced by nuclear fuels.

The energy content of a given mass of hydrogen is high. For example, burning of 1.00 g of hydrogen yields 142 kJ whereas burning of 1.00 g of octane [$CH_3(CH_2)_6CH_3$], a typical component of gasoline, gives only 48 kJ. An abundant supply of hydrogen is available in the 1.385×10^9 km³ of water in Earth's oceans, lakes, rivers and streams, glaciers, groundwater, and atmosphere.

Coal and petroleum occur naturally. Hydrogen, like electricity, must be generated. Thus, hydrogen is not really a source of energy but a means of transmitting energy. However, unlike electricity, hydrogen can be stored. The technology for stockpiling large quantities of hydrogen already exists. As much as 3.4×10^3 m³ (9×10^5 gal) of liquid hydrogen has been stored for the space program. The space shuttle's main engines are fueled by liquid hydrogen. Liquid hydrogen can be shipped in tank trucks and tank cars or transported by pipelines. The danger of leaks and explosions exists, of course, but natural gas and gasoline leaks have also

been known to result in explosions and fires. Metallic hydrides, which can be decomposed to hydrogen gas and the metals, may provide more compact and safer storage for hydrogen. The volumes needed to store 1.00 g of hydrogen are compared in Table 21.2. The most promising possibility is $FeTiH_{1.95}$ because it is relatively cheap.

Modified internal-combustion engines, for example, automobile engines, have already been operated successfully using hydrogen as a fuel. Fuel cells have been run on a commercial scale with an energy conversion efficiency of 70%.

The problem with using hydrogen as a fuel is that the production of hydrogen from water requires a large amount of energy. For the use of hydrogen as a fuel to avoid the problems associated with the burning of fossil fuels and with nuclear power, a way must be found to use energy from the sun to split water. Although a good deal of research has been done, a practical method has not yet been found. Perhaps you may be the one to invent a process for producing hydrogen from water by means of solar energy.

Launch of the space shuttle Discoverer. *Over two-thirds of the tank at the right is filled with $H_2(l)$. [The remainder contains $O_2(l)$.] Both are converted to water in a few minutes, providing the energy to lift the shuttle out of Earth's gravitational field.*

TABLE 21.2	Comparison of Volume Needed to Store Hydrogen[a]
Hydrogen Stored as	Volume Needed to Store 1.00 g Hydrogen, mL
$MgH_2(s)$	9.9
$LaNi_5H_7(s)$	11.2
$FeTiH_{1.95}(s)$	10.4
$H_2(l)$[b]	14.3

[a]Data are from Jolly, W. L. *Modern Inorganic Chemistry,* 2nd ed.; McGraw-Hill: New York, 1991, p 193.
[b]At normal boiling point of 20.4 K and 1 atm.

Some physical properties of water are given in Appendix C.

FIGURE 21.2 A small piece of an iceberg. Notice that most of the ice is under water.

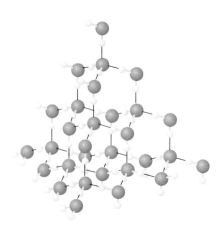

The crystal structure of ice, with its hexagonal rings of oxygen atoms, is responsible for the six-sided shapes always observed for snowflakes.

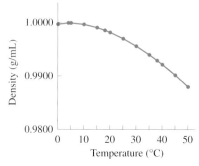

The density of water increases with increasing temperature from 0 to 3.98 °C. Above 3.98 °C, the usual decrease in density with increasing temperature is observed.

21.2 WATER

Most of the hydrogen on Earth is in the form of the compound water. Water is the only liquid that is common on Earth. In the liquid and solid states, in which the water molecules touch each other, water's ability to form three-dimensional hydrogen-bonded networks has no equal. (See Figure 12.10, which is reproduced here for convenience.) For example, although alcohols also have —OH groups and can hydrogen bond, they cannot form a three-dimensional hydrogen-bonded network. (Try it with models!)

Physical Properties

The three-dimensional hydrogen-bonded network structure explains the open crystal structure of ice. As a result of its open crystal structure, ice is less dense than liquid water. At 0 °C, the density of ice is 0.9168 g/mL, whereas the density of water is 0.9999 g/mL. Most substances are denser in the solid state than in the liquid state because orderly packing is more efficient than disorderly packing. (Remember Figure 12.4.)

The fact that ice is less dense than water has many important results. Ice floats on water (see ▪Figure 21.2) and insulates the water below; lakes, rivers, and oceans freeze from the top down. Fish and plants can survive winter's cold in the water under the ice. The increase in volume that takes place when water freezes breaks rocks into small particles of soil. As usual, with good news comes bad news. If icebergs did not float, the *Titanic* (and many other ships) would not have sunk. If a given mass of water did not occupy a larger volume in the form of solid ice than in the form of liquid water, the pipes in your house and the radiator in your car would not burst when they freeze (▪Figure 21.3). The freezing of aqueous solutions in plant cells causes the cells to break open and may destroy the plant.

When ice melts, some hydrogen bonds are broken. Disorderly groups of water molecules are formed between clusters of water molecules that still have an icelike structure. The regions of order and disorder shift constantly as some molecules attach to the edges of the orderly clusters and other molecules break away.

In the disorderly groups of water molecules, the water molecules are no longer held in the open ice structure. The open structure collapses, and density increases. Just above the melting point, the density of water increases as temperature increases. Above 3.98 °C, the normal increase in motion of water molecules with increasing temperature becomes more important than the collapse of the ice structure, and the density of water begins to decrease as temperature increases. Thus, the density of water reaches a maximum of 1.0000 g/mL at 3.98 °C. Near room temperature, liquid water has about half as many hydrogen bonds as ice. Even at the boiling point, much of the hydrogen bonding remains as shown by the high boiling point of water relative to the boiling points of H_2S, H_2Se, and H_2Te. All these compounds are gases under ordinary conditions; usually the boiling point of a series of similar compounds increases down a group in the periodic table.

Hydrogen bonding is also responsible for the high specific heat of water (Table 6.1). As a result of its high specific heat, water can absorb or give off a large amount of thermal energy with only a small change in temperature. The enormous quantities of water in lakes and oceans prevent large changes in the temperature of nearby land by absorbing thermal energy in summer and giving off thermal energy in winter. As a result, coastal cities and towns have milder climates than inland cities and towns. Water is widely used as a coolant (material used to remove thermal energy) as a result of its high specific heat and low cost. The coolant in the radiator of your car is about 50% water.

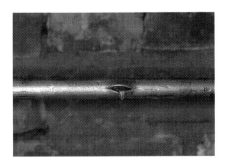 

Hydrogen bonding in water is the chief cause of the low solubility of nonpolar covalent compounds such as hexane, $CH_3(CH_2)_4CH_3$, in water. For one substance to be soluble in another, the effects of the higher entropy that results from increased volume and lower energy due to solute–solvent attraction must be large enough to overcome solute–solute and solvent–solvent attractions. Attraction between nonpolar compounds and water is low; solvent–solvent (water–water) attractions are unusually high as a result of strong hydrogen bonding. Therefore, nonpolar covalent compounds are not soluble in water.

Polar covalent compounds that can hydrogen bond with water are miscible with water (soluble in all proportions) as long as the molecule only has a small nonpolar part. For example, ethyl alcohol (CH_3CH_2OH), acetic acid (CH_3COOH), and acetone (CH_3COCH_3) are all miscible with water. However, octanol [$CH_3(CH_2)_6CH_2OH$], which has a larger nonpolar part, $CH_3(CH_2)_6CH_2$—, is not miscible with water (see ■Figure 21.4). Note that acetone has no hydrogen attached to oxygen and cannot form hydrogen bonds with itself. However, acetone can accept hydrogen bonds from water.

Some predominantly ionic compounds are soluble in water, and some are insoluble (Table 4.4). Solubility rules must be memorized because solubility is not easy to predict from thermodynamics. The free-energy change accompanying solution is the result of a combination of five factors: (1) and (2) the energies required to separate the ions from the crystal and to separate the solvent molecules from each other; (3) and (4) the energies released by solvation of cation and anion; and (5) the entropy change that accompanies solution. Usually, entropy increases when a solution is formed. However, because water is a polar molecule, when a strong electrolyte dissolves in water, both the positive ions and the negative ions become strongly hydrated, that is, surrounded by spherical shells of water molecules (Figure

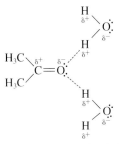

Acetone can accept hydrogen bonds from water.

The ability of water to dissolve a variety of materials makes water easily polluted.

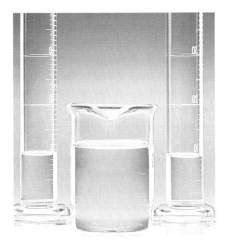 

FIGURE 21.5 Over 97% of the water on Earth is seawater. Two percent is ice in polar ice caps and glaciers.

13.3). Hydration of ions orders the water molecules in the shell of water molecules and decreases entropy. If the ions are small and highly charged like Ca^{2+}, the decrease in entropy can be a more important factor in the spontaneity of the solution process than the decrease in energy. Calcium sulfate is only slightly soluble because its enthalpy of solution is about -17 kJ/mol, whereas its entropy of solution is -0.1397 kJ/mol·K. For $CaSO_4$, the $T \Delta S$ term in the definition of free energy is equal to $+42$ kJ/mol at room temperature; the free energy of solution is positive and $CaSO_4$ is only slightly soluble.

Most reactions take place more rapidly in solution. Water is widely used as a solvent both in the laboratory and in industry because it is a good solvent for many materials. In addition, it is readily available, cheap, nonflammable, nontoxic,* and is a liquid over a convenient temperature range. The range of temperature over which water is a liquid, 0–100 °C, limits the temperatures at which reactions can take place in aqueous solution. The acidity and basicity of water limit the strengths of acids and bases that can exist in aqueous solution. The strength of water as an oxidizing and reducing agent limits the strength of the oxidizing and reducing agents that can be used in aqueous solution.

Water makes up 60–90% of all living matter. It is the solvent of life and life is believed to have begun in the sea, a complicated aqueous solution. Both animals and plants use aqueous solutions such as blood, digestive juices, and sap as mediums for carrying out biochemical reactions. Removal of water from fish, meat, milk, and fruit slows the growth of microorganisms, and drying in the sun is one of the oldest methods of preserving food. Salting, another method of preserving food that has been practiced for centuries, removes water by osmosis.

PRACTICE PROBLEM

21.7 Using the method of Figure 12.9, show a molecule of water joined to a molecule of ether, $CH_3CH_2OCH_2CH_3$, by a hydrogen bond.

There are about 7×10^{23} cm³ of water in the Pacific Ocean, a little over Avogadro's number of cm³.

FIGURE 21.6 The six cities in the United States that have populations of more than a million are all located near rivers or lakes.

Occurrence

About three-quarters of Earth's surface is covered with 1.4×10^{21} kg of water. Of this water, 97.3% is seawater, 2.0% is ice in polar ice caps and glaciers, and only 0.6% is freshwater (-Figure 21.5). Most of the freshwater is groundwater. Less than 0.1% of the water on Earth is in lakes and rivers. However, people obtain most of their freshwater from rivers and lakes; cities and towns tend to be located near rivers and lakes (see -Figure 21.6).

Seawater contains about 3.4% by mass of dissolved salts. Sodium ion, Na^+, and chloride ion, Cl^-, are by far the most common species (after water molecules!) and make up 86% by mass of the dissolved salts. Just ten species compose over 99.99% of the solutes in seawater; Table 21.3 lists the proportions of these species. At the present time, only water, sodium chloride, magnesium, and bromine can be obtained profitably from seawater.

---

*Nontoxic in the proper quantity. As is the case with many other substances that are essential to life, too much is bad. Drowning is the third leading cause of accidental death (after automobile accidents and falls) in the United States.

The Water Cycle

Earth's water passes through a great cycle. Water evaporates from oceans, lakes, rivers, soil, and plants and sublimes from snow and ice into the atmosphere. The sun supplies the energy needed for evaporation and sublimation. Winds move the water vapor around. Water then returns to Earth as snow or rain. Energy is released when water vapor condenses to liquid water or deposits as solid ice. The transfer of energy by evaporation and condensation of water are major factors in the weather. Release of energy by thunderstorms, tornadoes, and hurricanes is obvious (see ▬Figure 21.7). The quantity of water vapor in the atmosphere at any time is relatively small—only about 0.001% of the total water on Earth. However, the quantity of water that passes through the atmosphere in a year is large—greater than the total volume of freshwater on Earth.

The Greenhouse Effect

Although the amount of water vapor in the atmosphere is small, water vapor plays a major part in keeping Earth warm at night. Air and water vapor are transparent to the visible radiation from the sun that warms Earth's surface during the day. The warm Earth radiates infrared radiation. Water vapor absorbs infrared radiation and prevents loss of thermal energy at night. The lower atmosphere, which contains most of the water vapor in the atmosphere, acts like a blanket. Without its insulating blanket, Earth would be much cooler with an average surface temperature of about -20 °C instead of about 15 °C. This effect is known as the "greenhouse effect."

Water is the most important greenhouse gas because its concentration in the atmosphere is relatively high. However, carbon dioxide, methane, and many other gases are also greenhouse gases. The concentrations of these gases in the atmosphere are increasing as a result of people's activities. For example, the combustion of fossil fuels and the burning of tropical rain forests both release large amounts of carbon dioxide into the air. Many people believe these increasing quantities of greenhouse gases in the atmosphere will lead to an increase in the average temperature at Earth's surface with possible disastrous effects on climate.

Relative Humidity

The concentration of water vapor in air is usually expressed as relative humidity:

$$\text{relative humidity} = \frac{\text{partial pressure of water vapor in the air}}{\text{equilibrium vapor pressure of water at same temperature}} \times 100$$

The equilibrium vapor pressure is a measure of the maximum amount of water vapor that the air can hold at the given temperature. Thus, the relative humidity is the percent saturation of the air with water vapor.

Because the equilibrium vapor pressure of water increases as temperature increases, the relative humidity of a sample of air decreases as temperature increases. For example, a sample of air in which the partial pressure of water vapor is 2.8 mmHg has a relative humidity of 61% at 0 °C:

$$\text{relative humidity at 0 °C} = \frac{2.8 \text{ mmHg}}{4.6 \text{ mmHg}} \times 100 = 61\%$$

TABLE 21.3

Inorganic Solutes in Seawater

Ion	Percent of Solute (by Mass)
Cl^-	55.04
Na^+	30.61
SO_4^{2-}	7.68
Mg^{2+}	3.69
Ca^{2+}	1.16
K^+	1.10
HCO_3^-	0.41
Br^-	0.19
H_3BO_3	0.08
Sr^{2+}	0.04

Mars, which has very little atmosphere, has an average surface temperature of about -50 °C. The greenhouse effect explains why dry nights are cooler than humid nights.

The atmosphere of Venus consists mostly of CO_2. Venus has an average surface temperature of about 400 °C.

Although the relative humidity is much lower over tropical deserts than over polar ice caps, the air above tropical deserts contains more water than the air over polar ice caps!

▬ FIGURE 21.7 Trees in a hurricane.

■ FIGURE 21.8 When moist air cools at night, the relative humidity may become greater than 100%. If grass and other plants are warmer than 0 °C, liquid water (dew) condenses on them (*top*). If grass and other plants are cooler than 0 °C, ice (frost) deposits (*bottom*). The dates of the last frost in the spring and the first frost in the fall determine the length of the growing season.

At 20 °C, the relative humidity is only 16% if the partial pressure of water vapor is 2.8 mmHg (see ■Figures 21.8 and 21.9).

Relative humidities of 25–50% are comfortable. If the relative humidity is greater than 50%, you begin to feel uncomfortable. If it is hot, your perspiration doesn't evaporate fast enough to keep you cool. If it is cold (as in a cave), you feel clammy. On the other hand, if the relative humidity is less than 25%, the inside of your nose and mouth dry out.

PRACTICE PROBLEM

21.8 If the partial pressure of the water vapor in air is 6.5 mmHg at 20 °C, what is the relative humidity?

Purification

No naturally occurring water is pure. Some impurities, such as Ca^{2+} and particles of soil, get into water naturally. Other impurities are in water as a result of human activities, for example, nitrates from fertilizers used on crops and lawns. *Harmful substances present in water and air* are called **pollutants.** Almost any substance can be a pollutant if too much of it is present.

The method chosen to purify water depends on the original quality of the water, the amount of water needed, and the purpose for which the water is to be used. Fortunately, most humans can stand a considerable range of composition in their drinking water. The World Health Organization standards for some solutes in drinking water are given in Table 21.4. Good-quality drinking water contains few bacteria and no coliform bacteria.* The presence of coliform bacteria indicates that the water may contain disease-producing agents.

Water is usually disinfected both before and after other treatment. In the United States, liquid chlorine is usually used to disinfect water. Treatment with chlorine kills bacteria and makes viruses inactive. However, if organic compounds are present, chlorine may react with them. For example, one possible product of the reaction of methane, a common water pollutant, with chlorine is trichloromethane (chloroform):

$$CH_4(aq) + 3Cl_2(aq) \xrightarrow{\text{heat and/or light}} CHCl_3(aq) + 3HCl(aq)$$

Chloroform has been listed as a carcinogen (substance that tends to produce a cancer) by the Environmental Protection Agency. In Europe ozone, O_3, has been used to disinfect water for some time, and U.S. plants are beginning to switch to ozone. Use of ozone to disinfect water avoids pollution of water by chlorinated organic compounds such as trichloromethane. However, leftover chlorine remains in water long after treatment and prevents recontamination of the water. Leftover ozone decomposes to oxygen:

$$2O_3(aq) \longrightarrow 3O_2(aq)$$

■ FIGURE 21.9 Icing on aircraft, which caused this plane to crash, is usually produced by the freezing of supercooled water droplets from air.

---

*Coliform bacteria are found in the large intestine. Their presence in water indicates that the water is polluted by feces. People who drink such water often develop cholera, a bacterial infection that results in severe diarrhea with rapid loss of body fluids (3–4 gal or 15–20 L per day) and salts. Cholera killed 370 000 people in India between 1898 and 1907. In underdeveloped countries, people still die of cholera although treatment is simple if adequate medical facilities are available. The most recent outbreaks took place in Russia and Africa in 1993–1994.

TABLE 21.4	World Health Organization Standards for Drinking Water	
Solute	Maximum Desirable Concentration, mg/L	Maximum Allowable Concentration, mg/L
Ca^{2+}	75	200
Cl^-	20	60
Mg^{2+}	30	150
SO_4^{2-}	200	400
Total dissolved solids	500	1500

Always use water from the cold-water tap for drinking and food preparation. When you come home after a trip, be sure to run the water for a few minutes to flush out any contaminants that have formed in pipes and faucets.

Water treated with ozone is not protected against being contaminated again.

Water for drinking is usually subjected to other treatment besides disinfection. It is commonly filtered to remove solids and aerated by mixing air with the water. Aeration reduces unpleasant odors and tastes. In modern water-treatment plants, water is often softened—that is, calcium and magnesium ions are removed.

Laboratories and some industrial processes require water that is much purer than drinking water. For example, water used in high-pressure boilers must be 99.999 998% pure. Water of this purity is obtained by a combination of deionization, removal of dissolved organic materials with activated carbon (finely divided graphite with a very large surface area on which organic impurities are adsorbed), and reverse osmosis (Section 13.7).

Deionization of water is shown schematically in ■Figure 21.10. *Water from which ions have been removed with an ion-exchange resin is called* **deionized water.** * Water can also be freed of ions by distillation (evaporation and condensation) but distillation requires much more energy than ion exchange. (Distillation was discussed in Section 13.6.)

Home water softeners are often ion exchangers that exchange Na^+ for Ca^{2+} and Mg^{2+}. Sodium chloride is used to replace the calcium and magnesium ions with sodium ions so that the water softener can continue to be used.

Chemical Properties

Pure water is odorless and tasteless. The taste of drinking water is due to dissolved salts and gases. Water is thermodynamically stable toward decomposition to hydrogen and oxygen; even at 2700 °C, water is only 11% decomposed at equilibrium. However, water is a relatively reactive compound. Water can react either as an acid or a base or as an oxidizing agent or a reducing agent. How water reacts depends on the other substance (or substances) with which water is mixed. For example, with bases, water reacts as an acid:

$$NH_3(aq) + H_2O(l) \rightleftharpoons NH_4^+ + OH^-$$

Toward acids, water reacts as a base:

$$HCl(aq) + H_2O(l) \longrightarrow H_3O^+ + Cl^-$$

Water can act as a Lewis base as well as a Brønsted–Lowry base:

$$Al^{3+} + 6H_2O(l) \longrightarrow [Al(H_2O)_6]^{3+}$$

Reaction of water with aluminum ion could also be classified as complex formation with water as the ligand (Section 16.4).

---

*You probably use deionized water in your general chem lab. The ions present in ordinary tap water might interfere with your experiment. For example, you might obtain a positive test for chloride ion with an unknown that did not contain any chloride ion.

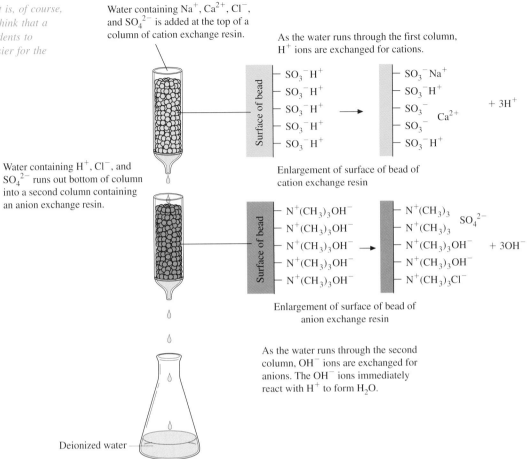

Water containing Na^+, Ca^{2+}, Cl^-, and SO_4^{2-} is added at the top of a column of cation exchange resin.

As the water runs through the first column, H^+ ions are exchanged for cations.

Surface of bead

$$\begin{array}{l} \text{—SO}_3^-\text{H}^+ \\ \text{—SO}_3^-\text{H}^+ \\ \text{—SO}_3^-\text{H}^+ \\ \text{—SO}_3^-\text{H}^+ \\ \text{—SO}_3^-\text{H}^+ \end{array} \longrightarrow \begin{array}{l} \text{—SO}_3^-\text{Na}^+ \\ \text{—SO}_3^-\text{H}^+ \\ \text{—SO}_3^- \\ \text{—SO}_3^- \quad \text{Ca}^{2+} \\ \text{—SO}_3^-\text{H}^+ \end{array} + 3\text{H}^+$$

Enlargement of surface of bead of cation exchange resin

Water containing H^+, Cl^-, and SO_4^{2-} runs out bottom of column into a second column containing an anion exchange resin.

Surface of bead

$$\begin{array}{l} \text{—N}^+(\text{CH}_3)_3\text{OH}^- \\ \text{—N}^+(\text{CH}_3)_3\text{OH}^- \\ \text{—N}^+(\text{CH}_3)_3\text{OH}^- \\ \text{—N}^+(\text{CH}_3)_3\text{OH}^- \\ \text{—N}^+(\text{CH}_3)_3\text{OH}^- \end{array} \longrightarrow \begin{array}{l} \text{—N}^+(\text{CH}_3)_3 \\ \text{—N}^+(\text{CH}_3)_3 \quad \text{SO}_4^{2-} \\ \text{—N}^+(\text{CH}_3)_3\text{OH}^- \\ \text{—N}^+(\text{CH}_3)_3\text{OH}^- \\ \text{—N}^+(\text{CH}_3)_3\text{Cl}^- \end{array} + 3\text{OH}^-$$

Enlargement of surface of bead of anion exchange resin

As the water runs through the second column, OH^- ions are exchanged for anions. The OH^- ions immediately react with H^+ to form H_2O.

Deionized water

■ FIGURE 21.10 Purification of water by deionization. The ion-exchange resins are plastic beads. The cation-exchange resin has —SO_3^- groups covalently bonded to the surface of the plastic beads. Hydrogen ions are ionicly bonded to the —SO_3^- groups. The anion-exchange resin has —$\overset{+}{N}(CH_3)_3$ groups covalently bonded to the surface of the plastic beads. Hydroxide ions are ionically bonded to the —$\overset{+}{N}(CH_3)_3$ groups. When water containing ions is run through the column of cation-exchange resin, hydrogen ions are exchanged for the cations in the water. Passage through the second column substitutes hydroxide ions for the anions in the water. The hydrogen ions and hydroxide ions immediately combine to form water: $H^+ + OH^- \longrightarrow H_2O(l)$. The cations on the cation-exchange resin can be replaced by hydrogen ions by passing dilute sulfuric acid through the column. The column can then be used again. The anions on the anion-exchange resin can be replaced by hydroxide ions by passing an aqueous solution of sodium hydroxide through the column and the column used again.

In the water–gas shift reaction,

$$\text{CO(g)} + \text{H}_2\text{O(g)} \underset{}{\overset{\text{catalyst, 200–400 °C}}{\rightleftharpoons}} \text{CO}_2\text{(g)} + \text{H}_2\text{(g)}$$

water acts as an oxidizing agent. Carbon is oxidized from oxidation number $+2$ to oxidation number $+4$, and H from water is reduced from oxidation number $+1$ to 0. Water also acts as an oxidizing agent with Group IA metals and with calcium, strontium, and barium from Group IIA, which are strong reducing agents (Section 19.2) and reduce water to hydrogen gas:

$$\text{Ca(s)} + 2\text{H}_2\text{O(l)} \longrightarrow \text{Ca(OH)}_2\text{(s)} + \text{H}_2\text{(g)}$$

Less active metals react with water only at high temperatures:

$$Mg(s) + H_2O(g) \longrightarrow H_2(g) + MgO(s)$$

At very high temperatures, even the nonmetal carbon and the semimetal silicon reduce water:

$$C(s) + H_2O(g) \xrightarrow{1500-1600\ °C} CO(g) + H_2(g)$$

$$Si(s) + 2H_2O(g) \xrightarrow{heat} SiO_2(s) + 2H_2(g)$$

The mixture of CO and H$_2$ produced by the reaction of carbon with steam is called water gas and is an important industrial fuel.

Fluorine is a very strong oxidizing agent and reacts vigorously with water, oxidizing oxygen from the -2 oxidation state to 0. In this reaction, water acts as a reducing agent:

$$2F_2(g) + 2H_2O(l) \longrightarrow 4HF(aq) + O_2(g)$$

Thermodynamically, chlorine should also oxidize water, but the reaction is slow. Instead, chlorine undergoes disproportionation:

$$Cl_2(g) + H_2O(l) \rightleftharpoons HOCl(aq) + H^+ + Cl^-$$

The mixture of $Cl_2(aq)$, $H_2O(l)$, $HOCl(aq)$, H^+, and Cl^- is called chlorine water. Notice that reaction between chlorine and water is not complete. The equilibrium can be shifted to the right by base:

$$Cl_2(g) + 2NaOH(aq) \longrightarrow NaOCl(aq) + NaCl(aq) + H_2O(l)$$

Reaction of Br_2 and I_2 with water and base is not simple. The elements of Groups VA and VIA do not react with water.

PRACTICE PROBLEM

21.9 Write equations for the reaction of water with (a) CaO(s), (b) HNO$_3$(l), (c) Na(s), and (d) CH$_3$COOH(l).

Hydrates

A number of compounds crystallize from aqueous solution as hydrates, crystals with a definite proportion of water (Section 2.12). Many hydrates lose some or all of their water when heated. For example, calcium sulfate dihydrate loses three-quarters of its water at 128 °C and the other quarter at 163 °C:

The nomenclature of hydrates was discussed in Section 2.12.

$$CaSO_4 \cdot 2H_2O(s) \xrightarrow{128\ °C} CaSO_4 \cdot \tfrac{1}{2}H_2O(s) + 1\tfrac{1}{2}H_2O(g)$$

$$CaSO_4 \cdot \tfrac{1}{2}H_2O(s) \xrightarrow{163\ °C} CaSO_4(s) + \tfrac{1}{2}H_2O(g)$$

$CaSO_4(s)$ is called anhydrous calcium sulfate. **Anhydrous** means *without water*. Anhydrous calcium sulfate is sold as a drying agent under the trade name Drierite; anhydrous calcium sulfate removes water from liquids and gases by forming hydrated calcium sulfate. Calcium sulfate hemihydrate, $CaSO_4 \cdot \tfrac{1}{2}H_2O$, is called plaster of Paris and is used to make plaster. When water is added to plaster of Paris, it sets—that is, calcium sulfate dihydrate, $CaSO_4 \cdot 2H_2O$ (gypsum), crystallizes.

You may have seen salt shakers with chunks of blue solid in their caps. The blue solid is anhydrous calcium sulfate to which an indicator that turns pink when wet has been added. Keeping salt dry prevents lumping.

Uses

People normally obtain almost as much water (about 1000 mL) by eating solid foods as they do by drinking. Another 350 mL of water per day are provided by the metabolic processes that take place in the body.

An average adult needs to drink about 1250 mL (over five 8-oz glasses) of water a day. A person can only survive around seven days without water. In the United States at the present time, about 1400 gallons of water are used per day for each person. Bathing, dishwashing, laundering, cooking, and heating and air conditioning of homes require an average of 90 gallons a day per person. Most of the 1400 gallons is consumed in the production of the goods we all use: food, electricity, gasoline, metals, synthetic fibers, and so forth. More water is used for irrigation to grow food than for any other single purpose, an average of more than 750 gallons per day per person. Eighty gallons are used to generate 1 kW of electricity, 10 gallons to produce 1 gallon of gasoline, 12.5 gallons to make 1 pound of paper, and so forth. In addition, ships travel on water, fish live in it, and people swim in it, sail boats on it, and skate on ice. Many animals use evaporation of water by sweating and panting to cool themselves.

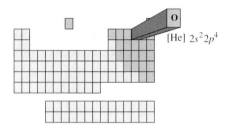

[He] $2s^2 2p^4$

21.3 OXYGEN

There are more atoms of oxygen in Earth's crust, surface waters, and atmosphere than there are atoms of any other element. Molecular oxygen makes up 23% by mass (21% by volume) of the atmosphere, and large quantities of molecular oxygen are dissolved in surface waters. Almost all of this molecular oxygen is a product of photosynthesis. Oxygen in compounds makes up 46% by mass of the crust and upper mantle and 86% of the surface waters of Earth. Water, the most important compound of oxygen as well as of hydrogen, is discussed in the previous section.

Production and Uses

The reactions of oxygen are discussed in section 11.8.

Industrially, oxygen is obtained by liquefying and fractionally distilling air. Although world production of oxygen is almost 2×10^{11} lb per year, this quantity is less than 0.000 01% of the oxygen in the atmosphere. Molecular oxygen is continuously being made by photosynthesis. Like water, oxygen cycles between the atmosphere, living organisms, surface waters, and Earth. Animals breathe in oxygen and breathe out carbon dioxide. Oxygen is also converted to carbon dioxide by the burning of fuels and tropical rain forests, in volcanoes, and by the decay of dead plants and animals. Plants and algae use carbon dioxide in photosynthesis and give off oxygen. Some oxygen is stored in minerals such as Fe_2O_3 and $CaCO_3$.

In many of the major industrial countries, 65–85% of the oxygen produced is used in steelmaking. Substitution of oxygen for air in steelmaking speeds up the reactions involved and increases productivity. Large quantities of oxygen are also used in the production of the white pigment, titanium(IV) oxide, TiO_2. Oxygen is also employed as an oxidizing agent to power the space shuttle, for sewage treatment, and in oxygen tents and hyperbaric oxygen chambers* in hospitals.

The bubbles are oxygen formed as a result of photosynthesis in the aquatic plant, waterweed.

*In hyperbaric oxygen chambers, patients are exposed to pure oxygen at 2–3 atm for about 2.5 hours. Hyperbaric oxygen treatment increases the concentration of oxygen in bone and body tissues. The growth of blood vessels and collagen (the fibrous protein in connective tissue) is promoted and the availability of white cells to fight infection is increased. People who might otherwise lose an arm or a leg do not. Rock singer Michael Jackson supposedly underwent hyperbaric oxygen treatment to prevent aging; however, there is *no* evidence that hyperbaric oxygen treatment really prevents aging.

Physical Properties of Molecular Oxygen

Molecular oxygen, O_2, is a colorless, odorless, tasteless gas that is slightly soluble in water. As is usually the case for gases, the solubility of oxygen decreases with increasing temperature. The solubility of oxygen in salt water is slightly less than in freshwater. However, the solubility of oxygen in some organic solvents is many times greater. Indeed, a mixture of perfluorodecalin, $C_{10}F_{18}$, and perfluorotripropylamine, $(CF_3CF_2CF_2)_3N$, is a good enough solvent for oxygen that an aqueous emulsion of the mixture called Fluosol DA can be used for blood transfusions. Fluosol DA can be used regardless of blood type, can be stored, and does not transmit disease.

Molecular oxygen is the only paramagnetic gaseous diatomic species that has an even number of electrons. Both the paramagnetism of the oxygen molecule and the strength of the O—O bond in molecular oxygen are explained by the molecular orbital theory (Section 10.6).

At 1 atm total pressure, the solubility of oxygen gas is 6.9×10^{-3} g/100 g $H_2O(l)$ at 0 °C and 3.9×10^{-3} g/100 g H_2O at 25 °C (only 57% as high). The lower solubility of oxygen in warm water is one reason fish die when water temperature increases.

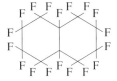

Perfluorodecalin

Ozone

Ozone, O_3, is an allotrope of molecular oxygen. It is prepared by the action of a silent electrical discharge or ultraviolet light on molecular oxygen:

$$3O_2(g) \xrightarrow{\text{silent electric discharge or UV}} 2O_3(g)$$

Ozone is a diamagnetic blue gas with a characteristic odor. Concentrations of ozone as low as 0.01 ppm can be detected by odor; you can often smell ozone after a severe thunderstorm.

Ozone is both unstable and very reactive. Either catalysts or ultraviolet light cause ozone to decompose:

$$2O_3(g) \xrightarrow{\text{UV or catalyst}} 3O_2(g)$$

Ozone is a much more powerful oxidizing agent than molecular oxygen. In acidic solution, it is one of the most powerful oxidizing agents known. For the half-reaction

$$O_3(g) + 2H^+ + 2e^- \longrightarrow O_2(g) + H_2O(l)$$

$E° = +2.07$ V.

Ozone absorbs ultraviolet radiation of wavelength 220–290 nm strongly. As a result, ozone in the stratosphere (the "ozone layer") protects Earth's surface from the sun's intense ultraviolet radiation. However, ozone in the troposphere (the layer of the atmosphere that touches Earth's surface) is involved in the formation of smog. In addition, oxidation by ozone in the troposphere causes rubber to crack, damaging automobile tires. Ozone at low concentrations is used to purify drinking water.

▶ Remember that allotropes are different forms of the same element that differ in bonding (Section 12.8).

A patient in a hyperbaric oxygen chamber.

It is strange that the presence of ozone in the stratosphere is essential to our survival when the presence of ozone in the lower atmosphere is so harmful.

A tire cracked by ozone.

PRACTICE PROBLEM

21.10 (a) Write resonance structures, including formal charges, for the ozone molecule. (b) Is ozone polar or nonpolar? Explain your answer.

Earth's Atmosphere

Many of the subjects that we have discussed thus far have involved Earth's atmosphere. The atmosphere is the source of the fixed nitrogen that is essential for life. Air also provides carbon dioxide for photosynthesis and oxygen for breathing; people can only live for about five minutes without oxygen. Oxygen is also necessary for the combustion of fossil fuels, the source of most of the energy used by people. Dilution of oxygen by nitrogen is necessary to prevent too rapid reaction. For example, even wet grass will burn in an atmosphere that contains 30% oxygen instead of 21%. Too high a concentration of oxygen is also harmful to people and leads to mental confusion, poor vision and hearing, and nausea. The atmosphere carries water necessary for life from the oceans to the land, protects life on Earth from cosmic rays and ultraviolet radiation from the sun, and keeps Earth's temperature within a range suitable for life. The formation of smog and acid rain take place in the atmosphere.

Only a millionth of the total mass of Earth is atmosphere. The total mass of Earth is 5.977×10^{27} g; the mass of the atmosphere is 5.1×10^{21} g. Earth's atmosphere is a thin layer of gases that has no sharp boundary. The concentration of matter decreases gradually with increasing altitude. Half of the mass of the atmosphere is within the first 5.5 km up from sea level, and three-fourths is below 11 km. Less than 1% is more than 40 km from Earth's surface. Thus, ■Figure 21.11 shows over 99% of Earth's atmosphere.

Earth's surface is warmed by energy from the sun and by decay of radioactive elements within Earth. Within the troposphere, the layer of the atmosphere that touches Earth's surface, temperature decreases as distance from the warm Earth increases. The troposphere is mixed constantly by rising masses of warm air so that its composition is quite uniform except for local air pollution and water content. The water content of the troposphere

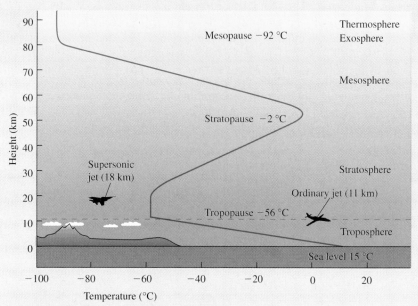

- FIGURE 21.11   The atmosphere.

changes as a result of cloud formation, rain and snow, and evaporation of water from lakes, rivers, and oceans.

The height of the troposphere varies by a kilometer or more even during one day at a single place. The troposphere is ended by a very cold layer called the tropopause. The low temperature of the tropopause causes water vapor to condense and prevents escape of water to higher altitudes.

Most of the ozone in the atmosphere is between 15 and 30 km from Earth's surface in the stratosphere. This region of the atmosphere is referred to as the ozone layer. Ozone in the stratosphere protects life on Earth from ultraviolet radiation (Section 21.3). Ultraviolet light is absorbed by ozone and causes dissociation of the ozone to molecular oxygen and atomic oxygen:

$$O_3(g) \xrightarrow{\text{ultraviolet light}} O_2(g) + O(g)$$

The molecular oxygen and atomic oxygen react exothermically to form ozone:

$$O_2(g) + O(g) \longrightarrow O_3(g)$$

No net chemical change takes place, but the energy of the ultraviolet light is converted to thermal energy.

The temperature of the stratosphere increases as altitude increases. Warmer, less dense air is above cooler, more dense air, and the stratosphere is not mixed by convection. Whatever gets into the stratosphere tends to stay there. However, the stratosphere is mixed horizontally by high-speed winds called jet streams that flow from west to east in the lower part of the stratosphere.

The troposphere and stratosphere have the same general composition except for water vapor (see Table 21.5). The percent water in the troposphere is variable. Very low concentrations of a number of other substances are also present; substances present in trace amounts are I_2, CO, SO_2, NH_3, NO, and NO_2.

Many of the reactions that take place in the atmosphere involve nitrogen oxides. Both nitric oxide, NO, and nitrogen dioxide, NO_2, are important components of polluted air. Together they are often represented by the symbol NO_x. In the troposphere, reaction of oxides of nitrogen with ozone and hydrocarbons results in the formation of smog. Hydrocarbons enter the atmosphere from three main sources: automobiles, plants such as trees, and decomposition of organic matter by bacteria in the absence of air. Hydrocarbons evaporate from gasoline, and exhaust from motor vehicles contains unreacted hydrocarbons. Green plants release huge quantities of hydrocarbons that contain double bonds, such as isoprene, $H_2C{=}C(CH_3)CH{=}CH_2$, to the atmosphere. In addition, methane is produced in large quantities by bacterial decomposition of organic matter in water, sediments, and soil.

Absorption of solar energy by the red-brown gas, NO_2, produces NO and atomic oxygen, O:

$$NO_2(g) \xrightarrow{\text{sunlight}} NO(g) + O(g)$$

Atomic oxygen reacts with molecular oxygen to form ozone; a third body, M, must be present to remove energy:

$$O_2(g) + O(g) + M \longrightarrow O_3(g) + M$$

Reaction of atomic oxygen and ozone with hydrocarbons forms radicals

$$O + RH \longrightarrow R\cdot + \text{other products}$$
$$O_3 + RH \longrightarrow$$
$$\qquad\qquad R\cdot \text{ and/or other products.}$$

and smog is then formed by radical reactions. The exact chemistry of smog formation is not known; however, hydroxyl radicals, $\cdot$OH, are believed to play a key role. Computers are widely used to simulate the effects of possible changes on the environment.

In the stratosphere, ultraviolet light makes the following reactions take

Trees are one of the main sources of hydrocarbons in the atmosphere.

place:

$$N_2O(g) \xrightarrow{\text{ultraviolet light}} N_2(g) + O(g)$$

$$N_2O(g) + O(g) \xrightarrow{\text{ultraviolet light}} 2NO(g)$$

Addition of the reactions of nitric oxide and ozone and of nitrogen dioxide with atomic oxygen shown below results in a net reaction that uses up ozone:

$$NO(g) + O_3(g) \longrightarrow$$
$$\qquad\qquad NO_2(g) + O_2(g)$$
$$+\ [NO_2(g) + O(g) \longrightarrow$$
$$\qquad\qquad NO(g) + O_2(g)]$$
$$\overline{\qquad O_3(g) + O(g) \longrightarrow 2O_2(g)\qquad}$$

Thus, nitric oxide catalyzes the destruction of ozone.

Notice that most reactions in the atmosphere involve radicals; three of the molecules that play an important part—O_2, NO, and NO_2—have unpaired electrons. Reaction by radical mechanisms is characteristic of reactions that take place in the gas phase because no solvent molecules are available to solvate ions.

TABLE 21.5	Present Composition[a] of the Dry Atmosphere at Earth's Surface[b]		
Component	Percent by Volume	Component	Percent by Volume
N_2	78.1	CH_4	0.000 2
O_2	21.0	Kr	0.000 114
Ar	0.934	H_2	0.000 05
CO_2	0.031 4	N_2O	0.000 025
Ne	0.001 82	Xe	0.000 0087
He	0.000 524	O_3[c]	0–trace

[a]When life on Earth began about 3.5 billion years ago, the atmosphere probably was made up largely of CO_2 and N_2. The atmosphere was not oxidizing as it is now; it was neither oxidizing nor reducing. The atmosphere has probably contained the same percent oxygen that it does now for about 50 million years, a long time compared with the existence of people (less than 1 million years).
[b]Data are from Manahan, S. E. *Environmental Chemistry;* Willard Grant: Boston, 1979; p 272.
[c]Increases with increasing altitude.

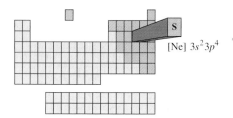

[Ne] $3s^2 3p^4$

SULFUR

Sulfur and its compounds are used so widely in industry that a country's economic development is often measured by the quantity of sulfur consumed. Sulfuric acid is first on the list of the Top 50 chemicals. Most proteins contain sulfur.

The Element

Sulfur has more allotropes than any other element. The S—S bond is the third strongest single bond between the same kind of atoms; only the H—H and C—C bonds are stronger. As a result of the strength of the S—S bond, elementary sulfur forms a variety of chains and rings. *At room temperature,* the commonest and most stable allotrope of sulfur is rhombic sulfur, which gets its name from the shape of the crystals. Rhombic sulfur consists of rings of eight sulfur atoms. ▪Figure 21.12(a) shows crystals of this allotrope of sulfur and Figure 21.12(b) shows a molecular model of the S_8 ring.* The packing of the molecules within a crystal of rhombic sulfur leads to a very complex structure.

When sulfur is heated, the packing of the S_8 molecules changes around 96 °C, and another allotrope of sulfur called monoclinic sulfur [Figure 21.12(c)] forms. Monoclinic sulfur is the most stable allotrope between 96 and 119 °C and forms when molten sulfur crystallizes. At room temperature, monoclinic sulfur gradually changes to rhombic sulfur over a period of several weeks.

The viscosity of liquids usually decreases as temperature increases (Section 12.3). When liquid sulfur is heated, however, an unusual change takes place, and from 160 to 195 °C the viscosity *increases* over 10 000 times (see ▪Figure 21.13). The change in viscosity can be explained by ring opening and polymerization of the chains of sulfur atoms that result from ring opening:

$$\begin{array}{c} \text{S—S} \\ \text{S} \qquad \text{S} \\ | \qquad | \\ \text{S} \qquad \text{S} \\ \text{S—S} \end{array} \longrightarrow \text{·S—S—S—S—S—S—S—S·} \longrightarrow$$

$$\text{·S—S—S—S—S—S—S—S—S—S—S—S—S—S—S·}$$

The longer the chains of sulfur atoms, the more the chains get tangled up in each other and the more viscous the sulfur becomes. Chains in liquid sulfur at 195 °C

▪ FIGURE 21.12 (a) Crystals of rhombic sulfur. (b) Ball-and-stick model of S_8 ring. (c) Crystals of monoclinic sulfur.

*Although the molecular formula of sulfur was determined by freezing-point depression in 1912 and sulfur was among the first substances to be examined by X-ray crystallography (by William Bragg in 1914), the crown structure of S_8 rings was not finally proved until 1935.

■ FIGURE 21.13 Left to right: When rhombic sulfur is heated, it melts to an orange liquid. As heating is continued, viscosity first increases as S₈ rings open and the chains that result polymerize, and then decreases as the chains of sulfur atoms are broken into shorter chains. The color changes to dark red. When liquid sulfur of maximum viscosity is poured into cold water, a rubberlike form of sulfur results. The sulfur atoms in the long tangled chains of sulfur atoms do not have time to rearrange themselves into crystals of rhombic sulfur.

are more than 200 000 sulfur atoms long. Above 195 °C, the chains begin to break up, and viscosity gradually decreases. The liquid sulfur becomes more colored because there are more unpaired electrons at the ends of shorter chains. At the boiling point of liquid sulfur (445 °C), liquid sulfur pours easily.

Species with unpaired electrons are often colored.

If liquid sulfur at about 195 °C is cooled suddenly by pouring it into cold water, a rubberlike form of sulfur called "plastic sulfur" results (see Figure 21.13). When plastic sulfur is not stretched, the long chains of sulfur atoms are coiled up. The coils straighten out when the sulfur is pulled and coil up again when it is released. Plastic sulfur is amorphous; it gradually changes to rhombic sulfur on standing at room temperature. Many other allotropes of sulfur that differ in molecular formula and in the way the molecules are packed have been prepared.

Sulfur is quite a reactive element and reacts with all elements except nitrogen, tellurium, iodine, iridium, platinum, gold, and the noble gases. Heating may be necessary to bring about reaction. Compounds of all elements except the noble gases with sulfur are known.

In spite of its reactivity, sulfur occurs free as well as combined. The most important commercial sources of sulfur are (in order of decreasing importance at the present time) sulfide minerals such as FeS_2, free sulfur in salt domes and around volcanoes, hydrogen sulfide in natural gas and crude oil, and organosulfur compounds in coal.

Production

Free sulfur is mined by the **Frasch process.** Free sulfur occurs in rock over salt domes near the coast in East Texas and Louisiana and under the Gulf of Mexico. The sulfur was probably produced by bacterial reduction of sulfur in $CaSO_4$. ■Figure 21.14 shows how sulfur is obtained by the Frasch process.

To convert hydrogen sulfide to sulfur, the H_2S is burned with a limited amount of oxygen; the products are sulfur, sulfur dioxide, and water vapor:

Herman Frasch was a German-born American chemist whose simple but clever process made possible use of large deposits of uncombined sulfur at a reasonable cost. The Frasch process was patented in 1891 and was first used successfully in Louisiana and East Texas.

To simplify equations, S will be used to represent S₈ molecules.

$$2H_2S(g) \ + \ 2O_2(g) \xrightarrow{400\,°C} S(l) \ + \ SO_2(g) \ + \ 2H_2O(g)$$

The SO_2 formed is then reacted with the excess H_2S in the presence of a catalyst:

$$SO_2(g) \ + \ 2H_2S(g) \xrightarrow{Fe_2O_3,\ 300\,°C} 3S(l) \ + \ 2H_2O(g)$$

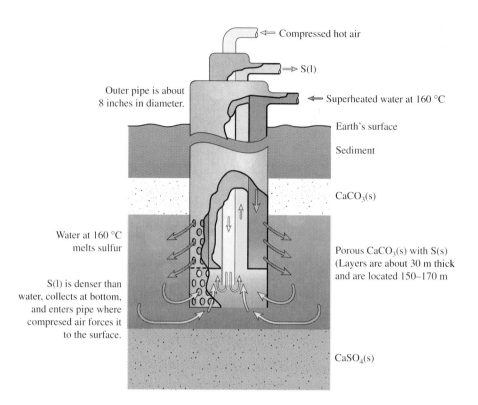

FIGURE 21.14 Schematic diagram of Frasch process for mining sulfur. Ordinary oil-well equipment is used to drill holes to the bottom of the sulfur-bearing layer. Three concentric pipes are then inserted. (Only the diameters of the three pipes are shown to scale in the diagram.) Pressurized hot water at about 160 °C is forced down the outside pipe and out through the upper set of holes into the sulfur-bearing layer. The sulfur melts and sinks to the bottom of the porous layer where it flows into the next pipe through the lower set of holes. The hot compressed air in the middle pipe then forces the mixture of liquid sulfur and air to the surface where the air is removed. The compressed air flow must not be too fast or sulfur will not be melted rapidly enough to fill the bottom of the pipe. Wells for removing water from the formation are located lower in the dome so that pressure does not build up. The liquid sulfur is either allowed to solidify in huge vats or kept liquid in steam-heated storage tanks. Most sulfur in the United States is shipped as a liquid.

⟸ Compressed hot air

⟹ S(l)

Outer pipe is about 8 inches in diameter.

⟸ Superheated water at 160 °C

Earth's surface

Sediment

$CaCO_3(s)$

Water at 160 °C melts sulfur

Porous $CaCO_3(s)$ with S(s) (Layers are about 30 m thick and are located 150–170 m

S(l) is denser than water, collects at bottom, and enters pipe where compresed air forces it to the surface.

$CaSO_4(s)$

Using the sulfur dioxide formed when sulfide minerals are roasted to make sulfuric acid reduces air pollution and conserves the sulfur in salt domes.

The sulfur dioxide gas formed when sulfide minerals are roasted

$$4FeS_2(s) + 11O_2(g) \xrightarrow{heat} 8SO_2(g) + 2Fe_2O_3(s)$$

is usually used directly for the manufacture of sulfuric acid.

Sulfuric Acid and Sulfates

The manufacture and uses of sulfuric acid were discussed in Section 14.6. Sulfuric acid is the cheapest strong acid. Hot concentrated sulfuric acid is an oxidizing agent. For example, hot concentrated sulfuric acid dissolves copper metal:

$$Cu(s) + 2H_2SO_4(conc) \xrightarrow{heat} CuSO_4(aq) + 2H_2O(l) + SO_2(g)$$

However, cold concentrated sulfuric acid does not react with copper and sulfuric acid is not often used as an oxidizing agent. Sulfuric acid is such a powerful dehydrating agent (water-removing agent) that it chars many organic substances (see ▬Figure 21.15).

Three salts of sulfuric acid—$Al_2(SO_4)_3$, Na_2SO_4, and $(NH_4)_2SO_4$—are among the Top 50 chemicals. Aluminum sulfate is produced by treating bauxite, $Al_2O_3 \cdot 2H_2O$, with sulfuric acid:

$$Al_2O_3 \cdot 2H_2O(s) + 3H_2SO_4(aq) \longrightarrow Al_2(SO_4)_3(aq) + 5H_2O(l)$$

Aluminum oxide acts as a base in this reaction. The aluminum sulfate is obtained by evaporation and is used to make water clear and to size (fill the pores of) paper. Sodium sulfate is made from common salt and sulfuric acid:

$$2NaCl(s) + H_2SO_4(l) \longrightarrow Na_2SO_4(s) + 2HCl(g)$$

The reaction is forced to go to the right by the escape of HCl(g). Sodium sulfate is used in the manufacture of brown wrapping paper and paper bags. Ammonium

◐ Concentrated sulfuric acid reacts with organic compounds in skin, producing painful burns. If you spill concentrated sulfuric acid on yourself, wash immediately with lots of water. When diluting concentrated acid, always add the concentrated acid to the water.

◑ Remember that sulfates are soluble except $PbSO_4$, Hg_2SO_4, $SrSO_4$, and $BaSO_4$. Ag_2SO_4 and $CaSO_4$ are only slightly soluble (Section 4.3).

■ FIGURE 21.15 *Left to right:* Crystals of sugar; sugar immediately after the addition of concentrated sulfuric acid; sugar reacting with concentrated sulfuric acid; sugar charred by concentrated sulfuric acid.

sulfate, which is made by the reaction of ammonia with sulfuric acid,

$$H_2SO_4(aq) + 2NH_3(g) \longrightarrow (NH_4)_2SO_4(aq)$$

is used in fertilizers.

Oxides

Although a number of oxides of sulfur are known to exist, SO_2 and SO_3 are by far the most stable and are the only important oxides of sulfur. Sulfur dioxide is a colorless gas with a sharp odor. It is harmful to both animal and vegetable life and is probably the most dangerous common air pollutant. Sulfur dioxide is formed when sulfur-containing fossil fuels are burned in power plants, homes, and cars, and when sulfide ores are roasted. Sulfur dioxide is also produced in large quantities by volcanoes and by the air oxidation of marsh gases. In polluted air, oxidation of SO_2 to SO_3 is catalyzed by dust particles and water droplets. Sulfur trioxide is a liquid under ordinary conditions and is hard to handle because it reacts vigorously with many materials.

Both SO_2 and SO_3 dissolve in water to form acidic solutions. Sulfur trioxide is known as the **anhydride** of sulfuric acid because it forms sulfuric acid when combined with water:

$$SO_3(g) + H_2O(l) \longrightarrow H_2SO_4(l)$$

Together with NO_2 and HCl, SO_2 and SO_3 account for the acidity of acid rain (see Chapter 16).

The highest oxidation number possible for sulfur is $+6$ and the lowest is -2. Sulfur has an intermediate oxidation number of $+4$ in SO_2 and can be either oxidized

INGREDIENTS: SUN-DRIED APRICOTS
PREPARED WITH SULPHUR DIOXIDE
FOR COLOR RETENTION

NUTRITIONAL INFORMATION PER SERVING
SERVING SIZE — 2 OZ.
SERVINGS PER CONTAINER — 4.

CALORIES 140
CARBOHYDRATE 35g
PROTEIN 2g
FAT 0g

Sulfur dioxide is often used as a bleach. The photographs on the left show a red rose before and after bleaching with sulfur dioxide. Sulfur dioxide is also often used to preserve dried fruit. Allergy to sulfur dioxide and sulfites is fairly common.

FIGURE 21.16 Tarnished and freshly polished silver spoons. Tarnish can be removed by either a chemical or a physical process. A chemical method consists of placing the tarnished silver in contact with aluminum (in dilute sodium carbonate solution). Aluminum is more reactive than silver and replaces silver ion: $Al(s) + 3Ag^+ \longrightarrow Al^{3+} + 3Ag(s)$. No silver is lost but *all* tarnish is removed including the tarnish that makes the design show. Removing tarnish by the physical process of polishing with a very fine abrasive leaves the tarnish that highlights the design in place. However, a little silver is lost each time an article is polished.

or reduced. Therefore, SO_2 can act either as a reducing agent or an oxidizing agent:

reducing agent: $H_2O_2(aq) + SO_2(g) \longrightarrow H_2SO_4(aq)$

oxidizing agent: $2H_2S(g) + SO_2(g) \xrightarrow{\text{catalyst}} 3S(s) + 2H_2O(l)$

Sulfur dioxide is used as a bleach and as an antibacterial agent to preserve fruit.

Sulfides

Sulfides are insoluble except for those of the Group IA and IIA metals and ammonium sulfide (Section 4.3). Many of the most important naturally occurring minerals and ores are sulfides. Zinc sulfide, which is called *sphalerite* by geologists, is an example; over 90% of the zinc produced comes from ZnS. The tarnish that forms on silver (see **Figure 21.16**) is silver sulfide formed by the reaction of traces of hydrogen sulfide gas in the air with silver metal.

$$4Ag(s) + 2H_2S(g) + O_2(g) \longrightarrow 2Ag_2S(s) + 2H_2O(l)$$
$$\text{black}$$

PRACTICE PROBLEMS

21.11 Write an equation for each of the following reactions: (a) A black precipitate forms when $H_2S(g)$ is bubbled into a solution that contains Pb^{2+}. (b) ZnS(s) is dissolved in HCl(aq).

21.12 Name each of the following: (a) $KHSO_4$, (b) FeS, (c) SO_3^{2-}.

21.13 Write the formula of each of the following: (a) magnesium sulfate heptahydrate, (b) calcium hydrogen sulfite, and (c) potassium sulfate.

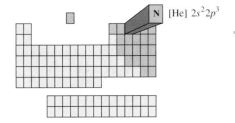

N [He] $2s^2 2p^3$

21.5 NITROGEN

Nitrogen gas, N_2, makes up 78% by volume of the atmosphere, and nitrogen is the fifth commonest element in the universe (Table 20.8). All living matter contains nitrogen in the form of nucleic acids and proteins. Eight of the Top 50 chemicals contain nitrogen. However, nitrogen is not very common in Earth's crust. The only important nitrogen-containing minerals are KNO_3 and $NaNO_3$, which may have come either from the decay of vegetable and animal material in the presence of air or from volcanoes. Deposits of KNO_3 and $NaNO_3$ occur only in dry areas because nitrates are soluble in water.

The Element

Nitrogen is a colorless, odorless, tasteless gas that is obtained industrially by liquefaction and fractional distillation of air. The free element exists as the diatomic gas, N_2, or

$$:N{\equiv}N:$$

because nitrogen, like its neighbors in the second row of the periodic table, C and O, readily forms multiple bonds. Nitrogen gas is not very reactive at room temperature. The nitrogen–nitrogen triple bond is very strong, *and* the N_2 molecule is nonpolar. Isoelectronic species that are polar, such as CO and CN^-, are much more reactive. At room temperature, nitrogen reacts only with lithium

$$6Li(s) + N_2(g) \longrightarrow 2Li_3N(s)$$

The bond dissociation energy of the nitrogen–nitrogen triple bond is 946 kJ/mol (Table 9.4).

$$[Ru(NH_3)_5(H_2O)]^{2+}(aq) + N_2(g) \rightleftharpoons H_2O + [Ru(NH_3)_5(N_2)]^{2+}$$

Carnations frozen by immersion in liquid nitrogen shatter like glass.

and with nitrogen-fixing bacteria. At high temperatures, nitrogen is more reactive and combines with many metals. For example, at high temperature, nitrogen reacts with magnesium metal to form magnesium nitride:

$$3Mg(s) + N_2(g) \xrightarrow{heat} Mg_3N_2(s)$$

Nitrogen also reacts with the nonmetal H_2 at high temperature in the presence of a catalyst to form ammonia (Section 14.6).

The chief use of nitrogen gas is as an inert atmosphere in processes where the presence of oxygen from air would involve danger of fire or explosion or result in the oxidation of the product. About 10% of the nitrogen produced is used as a refrigerant. Nitrogen boils at $-195.8\ °C$, and a liquid nitrogen bath has a temperature of about $-195.8\ °C$ just as a boiling water bath has a temperature of about $100\ °C$, the boiling point of water.*

The Nitrogen Cycle

Atmospheric nitrogen is fixed naturally by bacteria in nodules (small lumps) on the roots of peas, beans, and clover, and by fires and lightning, which produce nitrogen oxides. Atmospheric nitrogen is fixed industrially by the Haber process. At the present time, the quantity of nitrogen fixed industrially (about 2×10^{11} lb worldwide) is of the same order of magnitude as the quantity of nitrogen fixed naturally (about 4×10^{11} lb). Plants absorb the fixed nitrogen from soil and water and form plant proteins. Then, either the plants die and decay and nitrogen is returned to the atmosphere and seas or the plants are eaten by animals. The animals use the nitrogen to synthesize animal proteins, and the nitrogen is returned to the atmosphere or seas either by elimination or by death and decay of the animal.

Ammonia and Ammonium Salts

The nitrogen compound that is produced industrially in the largest quantities is ammonia, which is made by the Haber–Bosch process (Section 14.6). Ammonia is a colorless gas with a characteristic odor. It is very soluble in water; about 90 g of ammonia dissolve in 100 mL of water at room temperature and 1 atm. Ammonia is both a Brønsted–Lowry and a Lewis base, and aqueous solutions of ammonia are weakly basic as a result of the equilibrium

$$NH_3(aq) + H_2O(l) \rightleftharpoons NH_4^+(aq) + OH^-(aq)$$

Ammonium salts are made by neutralization of ammonia with the appropriate acid. For example, sulfuric acid is used to make ammonium sulfate:

$$2NH_3(g) + H_2SO_4(aq) \longrightarrow (NH_4)_2SO_4(aq)$$

Ammonium sulfate is a source of nitrogen in fertilizers.

Many ammonium salts decompose on heating; for example,

$$NH_4HCO_3(s) \xrightarrow{heat} NH_3(g) + CO_2(g) + H_2O(g)$$

An ammonia fountain. An ammonia fountain works because ammonia is very soluble in water. The flask is filled with dry $NH_3(g)$, and the beaker is filled with water containing phenolphthalein indicator. The bottom of the tube from which water is spouting is immersed in the water. When a little water is squirted into the flask from the syringe at the right, ammonia dissolves in the water. As a result, the pressure inside the flask decreases, and atmospheric pressure pushes water up the tube into the flask to make a fountain. Phosphine, PH_3, is only slightly soluble in water, and its aqueous solution is neutral.

---

*Liquid nitrogen is used to recover steel, copper, and aluminum from junked cars and other large steel or iron items such as washing machines and refrigerators. The liquid nitrogen is sprayed on the object. The steel becomes brittle and breaks in a shredder like glass breaks if dropped on a concrete floor. Copper and aluminum do not become brittle; they stick together and are easily separated from the steel.

If the anion of an ammonium salt is an oxidizing agent, the ammonium ion may be oxidized:

$$NH_4NO_3(s) \xrightarrow{200\ °C} N_2O(g) + 2H_2O(g) \tag{21.3}$$

The decomposition of ammonium nitrate can become explosively rapid. In 1947 an explosion of ammonium nitrate fertilizer on a ship in the harbor of Texas City killed nearly 600 people, injured 3500, and destroyed $33 million worth of property. When ammonium nitrate explodes, the reaction

$$2NH_4NO_3(s) \longrightarrow 2N_2(g) + 4H_2O(g) + O_2(g) \qquad \Delta H° = -236\ kJ$$

takes place extremely rapidly. The increase in volume that accompanies this reaction is enormous because the reaction is exothermic and 3.5 mol of gas are formed from each mole of solid ammonium nitrate. Formation of many moles of gas per mole of solid by an exothermic reaction is characteristic of explosive reactions.

Ammonium nitrate is an important fertilizer because of its high nitrogen content, the simplicity and low cost of its manufacture, and its combination of quick-acting nitrate nitrogen and slower-acting ammonium nitrogen. Ammonium nitrate is also used to make explosives.

Nitric Acid and Nitrates

Nitric acid, HNO_3, is the second most important compound of nitrogen. Industrially, nitric acid is made by the catalytic oxidation of ammonia with air. The steps in the manufacture of nitric acid are

$$4NH_3(g) + 5O_2(g) \xrightarrow{Pt\ catalyst,\ 920\ °C,\ 7\ atm} 4NO(g) + 6H_2O(g)$$

$$2NO(g) + O_2(g) \longrightarrow 2NO_2(g) \tag{21.4}$$

$$3NO_2(g) + H_2O(l) \longrightarrow 2HNO_3(aq) + NO(g)* \tag{21.5}$$

In the first step, 9 mol of gas react to form 10 mol of gas; the increase in volume is relatively small. Using a pressure of 7 atm does not shift the equilibrium to the left very much but does greatly increase the quantity of ammonia that can be oxidized in a reactor of a given volume.

A number of other oxidation products are possible from ammonia; for example, N_2 and N_2O could be formed instead of NO:

$$4NH_3(g) + 3O_2(g) \longrightarrow 2N_2(g) + 6H_2O(g)$$

$$2NH_3(g) + 2O_2(g) \longrightarrow N_2O(g) + 3H_2O(g)$$

However, the catalyst makes the desired reaction the fastest. The NO formed in the last step of the manufacture of nitric acid (equation 21.5) is used to make more NO_2 (equation 21.4).

The nitrate ion in nitric acid solutions acts as an oxidizing agent because nitrogen is small and electronegative. Nitric acid reacts with metals, such as copper (see equations 21.6 and 21.7 and ►Figure 21.17), that do not react with nonoxidizing acids like hydrochloric acid and phosphoric acid. Hot concentrated nitric acid dissolves nearly all metals except gold, platinum, iridium, and rhodium. Even nonmetals

---

Phosphine, PH_3, is only slightly soluble in water, and its aqueous solution is neutral.

To prevent the use of ammonium nitrate fertilizer for terrorist bombings such as the one in Oklahoma City, dilution with other salts such as $CaCO_3$, $NH_4H_2PO_4$, $(NH_4)_2HPO_4$, and $(NH_4)_2SO_4$ has been suggested. Unfortunately, tests show that none of these reduces the explosiveness of NH_4NO_3; in many cases the amount of thermal energy released increases.

See Rouhi, A. M. Chem. Eng. News 1995, 73(30), 10–19 for a very interesting article on combating terrorism.

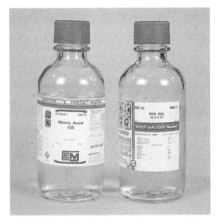

Nitric acid is colorless when pure (left) but turns yellow in light (right) as a result of the formation of NO_2. Nitrogen dioxide gas is red-brown but dilute solutions look yellow.

---

*Although NO_2 reacts with water to form nitric acid, NO_2 is not the anhydride of nitric acid. The reaction of NO_2 with water is a disproportionation reaction (equation 21.5); the oxidation number of nitrogen changes from +4 in NO_2 to +5 in HNO_3 and +2 in NO. The anhydride of nitric acid is dinitrogen pentoxide, N_2O_5: $N_2O_5(s) + H_2O(l) \longrightarrow 2HNO_3(aq)$. No change in oxidation number takes place when an acid anhydride reacts with water to form the acid.

such as carbon are oxidized (equation 21.8). When nitric acid reacts with metals such as zinc that also react with nonoxidizing acids, the product is not $H_2(g)$ (equations 21.9–21.12). The nitrate ion is a stronger oxidizing agent than the hydrogen ion, and nitrate ion, rather than hydrogen ion, is reduced. The nitrogen-containing products of nitric acid oxidation depend on the concentration of the nitric acid, the strength of the reducing agent, and the temperature. A mixture of nitrogen-containing products is usually obtained. Equations 21.6–21.12 show the predominant product of each these reactions:

$$Cu(s) + 4HNO_3(16\ M) \longrightarrow Cu(NO_3)_2(aq) + 2NO_2(g) + 2H_2O(l) \qquad (21.6)$$

$$3Cu(s) + 8HNO_3(6\ M) \longrightarrow 3Cu(NO_3)_2(aq) + 2NO(g) + 4H_2O(l) \qquad (21.7)$$

$$C(s) + 4HNO_3(16\ M) \longrightarrow CO_2(g) + 4NO_2(g) + 2H_2O(l) \qquad (21.8)$$

$$3Zn(s) + 8HNO_3(6\ M) \longrightarrow 3Zn(NO_3)_2(aq) + 2NO(g) + 4H_2O(l) \qquad (21.9)$$

$$4Zn(s) + 10HNO_3(3\ M) \longrightarrow 4Zn(NO_3)_2(aq) + N_2O(g) + 5H_2O(l) \qquad (21.10)$$

$$5Zn(s) + 12HNO_3(1\ M) \longrightarrow 5Zn(NO_3)_2(aq) + N_2(g) + 6H_2O(l) \qquad (21.11)$$

$$4Zn(s) + 10HNO_3(0.1\ M) \longrightarrow 4Zn(NO_3)_2(aq) + NH_4NO_3(aq) + 3H_2O(l) \ (21.12)$$

As you can see by comparing equations 21.6, 21.7, and 21.9–21.12, the more concentrated the nitric acid, the less the change in oxidation number of nitrogen. This seems reasonable because, in concentrated nitric acid, more nitrate ions are around to be reduced. Each nitrate ion can obtain only a small number of electrons. In dilute solution, fewer nitrate ions are competing for electrons, and each one gets more.

Nitric acid oxidizes compounds as well as elements. For example, copper(II) sulfide, which is very insoluble, can be dissolved by treatment with hot nitric acid, which oxidizes sulfide ion:

$$3CuS(s) + 8HNO_3(3\ M) \xrightarrow{heat} 3Cu(NO_3)_2(aq) + 3S(s) + 2NO(g) + 4H_2O(l)$$

Nitric acid also reacts with proteins, such as those in your skin, to give a yellow-colored stain. None of these reactions takes place with H^+ alone; NO_3^- is the oxidizing agent.

Surprisingly, neither aluminum nor iron dissolves in concentrated nitric acid although these metals dissolve readily in nonoxidizing acids such as hydrochloric acid. The explanation of this behavior is that aluminum and iron form coatings of Al_2O_3 and Fe_2O_3 that do not dissolve in concentrated nitric acid.

Like any other acid, nitric acid neutralizes bases forming salts. Neutralization of ammonia with nitric acid is an example:

$$NH_3(g) + HNO_3(aq) \longrightarrow NH_4NO_3(aq)$$

Oxidation Numbers

The number of oxidation numbers shown by nitrogen (10) is not matched by any other element. Table 21.6 gives an example of each. From Table 21.6, the importance of nitrogen's ability to form multiple bonds to the chemistry of nitrogen should be obvious.

Nitrogen gas from the decomposition of sodium azide

$$2NaN_3(s) \longrightarrow 3N_2(g) + 2Na(l)$$

is used to inflate air bags in cars (–Figure 21.18). The AIDS drug AZT contains a covalently bonded azide group. Hydroxylamine is made by the reduction of nitrites, nitric acid, or nitric oxide. It is used as a reducing agent in the manufacture of

– FIGURE 21.17   The reaction of copper metal with concentrated (16 M) nitric acid yields clouds of red-brown nitrogen dioxide gas and a green solution of copper(II) nitrate. Nitric acid is used for etching and photoengraving copper.

Pure nitric acid autoionizes. The equation is
$$3HNO_3(l) \longrightarrow H_3O^+ + NO_2^+ + 2NO_3^-.$$

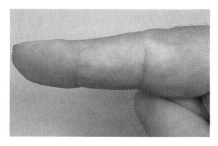

Finger stained yellow by nitric acid.

FIGURE 21.18 The pellets that inflate air bags contain not only NaN_3 but also MoS_2, S_8, and Fe_2O_3. The MoS_2 is a lubricant; its presence makes NaN_3 form pellets well. Sulfur makes the pellets burn smoothly. The Fe_2O_3 reacts with the liquid sodium:

$$6Na(l) + Fe_2O_3(s) \longrightarrow 3Na_2O(s) + 2Fe(s)$$

Why can Na(l) not be left in the used air bag?

TABLE 21.6	The Ten Oxidation Numbers of Nitrogen		
Oxidation Number	Formula	Lewis Formula	Name
+5	HNO_3		Nitric acid
+4	NO_2		Nitrogen dioxide
+3	HNO_2		Nitrous acid
+2	NO		Nitric oxide
+1	N_2O		Nitrous oxide
0	N_2		Nitrogen
−1/3	N_3^-		Azide ion
−1	H_2NOH		Hydroxylamine
−2	H_2NNH_2		Hydrazine
−3	NH_3		Ammonia

[a]Only one Lewis formula is given for each species. For some of the species, other resonance structures are needed to describe the species. For example, a second structure should be written for nitric acid to account for the fact that the bonds between nitrogen and the two oxygens on the right are identical.

polymers, as a component of photographic developers, and to make caprolactam, an intermediate for the polymer nylon-6. Hydrazine is a strong reducing agent and is used as a rocket fuel. Another important application is removal of O_2 from high-pressure boiler water in power plants:

$$H_2NNH_2(aq) + O_2(aq) \longrightarrow N_2(g) + 2H_2O(l)$$

Hydrazine, which is poisonous, is made by the reaction of ammonia with a basic solution of sodium hypochlorite:

$$2NH_3(aq) + NaOCl(aq) \longrightarrow H_2NNH_2(aq) + NaCl(aq) + H_2O(l)$$

This reaction is the reason you should be careful to keep ammonia and household bleach, which contains sodium hypochlorite, apart.

Oxides

Nitrogen forms an unusually large number of oxides—N_2O_5, NO_2 and its dimer, N_2O_4, NO, and N_2O. All of them are thermodynamically unstable (ΔG_f° positive), but the rates of decomposition are extremely slow under ordinary conditions. All have been discussed previously except dinitrogen monoxide, N_2O, which is usually called nitrous oxide. It is used as an aerosol propellant and aerating agent for whipped cream and is manufactured by the reaction shown in equation 21.3. Breathing nitrous

Nitrous oxide is the major form in which fixed nitrogen is returned to the atmosphere.

oxide causes insensitivity to pain; the anesthetic effect is preceded by mild hysteria sometimes including laughter, hence the nickname "laughing gas." Nitrous oxide has often been used as an anesthetic during short operations especially in dentistry. However, prolonged inhalation causes death.

Nitrogen-Containing Organic Compounds

Three of the nitrogen-containing compounds in the Top 50—urea, acrylonitrile, and caprolactam—are organic. Most of the urea produced is used as fertilizer. Urea is the leading nitrogen-containing fertilizer worldwide because it has the highest nitrogen content of any solid fertilizer, 47% N. The nitrogen is readily available to plants because urea is rapidly converted into ammonia and then into nitrate in soil. Urea is made from ammonia and carbon dioxide

Urea is of historical interest as the first synthetic organic compound. Wöhler's preparation of urea from the inorganic compound ammonium cyanate was a major blow to the idea that organic compounds result from a vital force peculiar to living organisms.

$$2NH_3(g) + CO_2(g) \underset{\text{pressure}}{\rightleftharpoons} H_2NCOONH_4(s) \underset{\text{heat}}{\rightleftharpoons} H_2NCONH_2(s) + H_2O(l)$$

and is also used for the manufacture of plastics.

Acrylonitrile, $H_2C{=}CHCN$, is made from propene, $CH_3CH{=}CH_2$, ammonia, and oxygen from the air:

$$2CH_3CH{=}CH_2(g) + 2NH_3(g) + 3O_2(g) \xrightarrow{\text{catalyst, heat, pressure}} 2H_2C{=}CHCN(l) + 6H_2O(l)$$

Acrylonitrile is polymerized to make acrylic fibers such as Acrilan, Creslon, and Orlon, which resemble wool and are resistant to attack by chemicals and weather.

Caprolactam, which is made from benzene by a series of steps, is used to make nylon-6:

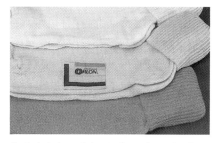

$$\underset{\text{caprolactam}}{\text{NH (s)}} \xrightarrow{\text{H}_2\text{O, heat}} HOOC(CH_2)_5NH_2(s) \xrightarrow{260\,°C} {-}{\Big(}C{-}(CH_2)_5{-}NH{\Big)}_{\!n}(s) + nH_2O(l)$$

nylon-6

Nylon-6 fibers are strong and flexible and resist wear. They are used to make ropes and tire cord.

The ropes used by mountain climbers are often made of nylon-6.

Socks knit from Orlon (polyacrylonitrile) fiber.

Nitric Oxide: Biochemical

Nitric oxide, NO, has been important as an intermediate in the production of nitric acid for almost 100 years and has been investigated as an air pollutant for about 50 years. Since 1987 it has also become one of the most studied molecules in biochemistry. Nitric oxide is synthesized in animals as different as barnacles, fruit flies, horseshoe crabs, chickens, trout, and humans. In bacteria, it plays a role in nitrogen fixation.

Nitric oxide takes part in an amazing variety of physiological processes in humans and animals. Its functions include carrying messages between nerve cells, dilating blood vessels, and controlling blood pressure, and it has a role in the immune system's ability to kill tumor cells and intracellular parasites. Understanding of the physiology and biochemistry of processes that involve NO has already be-

gun to produce improvements in the treatment of disease.

Unlike many other biochemicals, its functions depend on its chemical properties rather than its shape. Molecules of nitric oxide have an unpaired electron and thus are radicals. However, compared to most radicals, nitric oxide is not very reactive. It reacts mainly with other radicals, such as oxygen, and with transition-metal ions, with which it forms complexes. Because nitric oxide is toxic, it is not stored like other neurotransmitters but is produced as needed.

Nitric oxide is about twice as soluble in water as oxygen; a saturated solution is about 0.002 M under ordinary laboratory conditions. Because molecules of nitric oxide are small and are not charged, they diffuse easily through cell membranes. As a result, NO often serves as a messenger.

The diffusion of nitric oxide is believed to contribute not only to memory formation but also to the spatial organization of the brain during embryonic development.

Doctors began using nitroglycerin to treat angina about 1900 without knowing that the mechanism of its action involves nitric oxide. Most recently, NO has been found to be a powerful antiviral agent with activity against poxviruses and herpes simplex virus type 1, which causes cold sores. Perhaps, in the future, nitric oxide may provide a much-needed new approach to treating diseases caused by viruses.

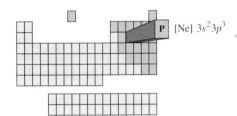

P [Ne] $3s^2 3p^3$

21.6 PHOSPHORUS

Occurrence

All living matter contains phosphorus. Phosphates are part of DNA (deoxyribonucleic acid), a polymeric molecule that transfers genetic characteristics in all forms of life, and of the molecules that store and transfer energy in living organisms. About two-thirds of bone and teeth are hydroxyapatite, $Ca_5(PO_4)_3OH$: the average 70-kg adult male contains about 3.5 kg of $Ca_5(PO_4)_3OH$. More than 1 g of phosphorus is eliminated daily and must be replaced. Phosphorus is usually the limiting reactant for plant growth.

Phosphorus is too reactive to be found free in nature. The chief source of phosphorus is phosphate rock, which has the approximate composition $Ca_5(PO_4)_3F$ or $CaF_2 \cdot 3Ca_3(PO_4)_2$. Because phosphorus is usually the limiting reactant for plant growth, most phosphate rock is used to make fertilizer.

The Element

Elemental phosphorus is obtained by heating phosphate rock with sand and coke in an electric furnace:

$$2Ca_3(PO_4)_2(s) + 6SiO_2(s) + 10C(s) \xrightarrow{1400-1500\,°C} 6CaSiO_3(l) + 10CO(g) + P_4(g)$$

Most of the phosphorus that is produced is used to make phosphoric acid; the remainder is used to make other phosphorus compounds.

Like oxygen and sulfur, phosphorus exists in allotropic forms. The commonest allotropic form is white phosphorus, which is made up of P_4 molecules. ▪Figure 21.19 shows models of the P_4 molecule. As you can see, the bonds that hold the

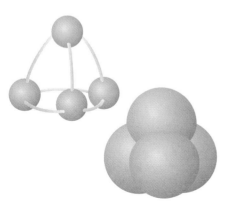

▪ FIGURE 21.19   Ball-and-stick and space-filling models of P_4.

■ FIGURE 21.20   White phosphorus must be stored under water because it ignites spontaneously in air.

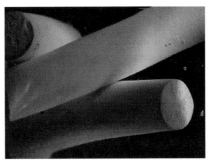

■ FIGURE 21.21   At room temperature, white phosphorus glows when exposed to air. It catches fire at about 35 °C. Substances that ignite spontaneously in air are said to be pyrophoric. Combustion of phosphorus is used to lay smoke screens.

For a recent review concerning nitric oxide as a biochemical, see Feldman, P. L.; Griffith, O. W.; Stuehr, D. J. Chem. Eng. News, *December 20,* **1993,** *26–38.*

P_4 molecule together are "bent." The orbitals that form the bonds overlap at an angle instead of endwise. Bonds formed by overlapping orbitals at an angle are not as strong as bonds formed by endwise overlap. The P_4 molecule is said to be **strained.** Weak bonds break easily, and white phosphorus is very reactive. White phosphorus catches fire spontaneously in air at about 35 °C and must be stored under water as shown in ■Figure 21.20.

The name "phosphorus" refers to the fact that white phosphorus glows in the dark when exposed to air (see ■Figure 21.21). Phosphorus exists as P_4 molecules both in the liquid phase and in the gas phase. The glow comes from reaction of the P_4 molecules in the vapor phase near the surface of the solid with oxygen from air:

$$P_4(g) + 5O_2(g) \longrightarrow P_4O_{10}(s) + light$$
excess

Although white phosphorus is the allotrope usually formed when phosphorus vapor condenses or liquid phosphorus freezes, it is the least stable and most reactive allotrope. Red phosphorus, which is made by heating white phosphorus in the absence of air, is much less reactive, is easier to handle, and is safer.* Red phosphorus has a very complex polymeric structure.

White phosphorus is very toxic. About 50 mg is a fatal dose. Even contact with the skin should be avoided.

Formation of white phosphorus by deposition of phosphorus vapor or freezing of liquid phosphorus is another example of kinetic rather than thermodynamic control.

Tetraphosphorus Decoxide

Tetraphosphorus decoxide, P_4O_{10} (■Figure 21.22), is usually called phosphorus pentoxide from its empirical formula P_2O_5. Phosphorus pentoxide is the commonest and most important oxide of phosphorus. The most outstanding property of P_4O_{10} is its eagerness to absorb water. Reaction of P_4O_{10} with water is one step in the synthesis of pure phosphoric acid (equation 21.13).

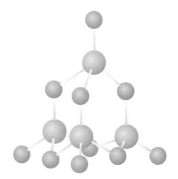

■ FIGURE 21.22   Ball-and-stick model of P_4O_{10}. The structure of P_4O_{10} can be derived from the structure of the P_4 molecule by inserting an oxygen in the middle of each P—P bond and adding an oxygen at each corner of the tetrahedron. As you can see, phosphorus pentoxide is made up of four {PO_4} tetrahedra linked through oxygen.

---

*White phosphorus was used in the heads of matches from 1831 to the end of the nineteenth century when serious toxic effects from the white phosphorus were observed in match-factory workers. Modern safety matches have a head that is largely $KClO_3$, an oxidizing agent. The surface provided for striking the matches consists mainly of red phosphorus, antimony(III) sulfide, Sb_2S_3, and an adhesive. When the match is struck, friction heats the $KClO_3$, red phosphorus, and Sb_2S_3, causing the red phosphorus and Sb_2S_3 to catch fire. The thermal energy released by the burning of red phosphorus and Sb_2S_3 in the presence of $KClO_3$ sets fire to the wood or cardboard, which serves both as fuel and as handle for the match.

When very dilute, phosphoric acid is nontoxic and odorless and is used to give a tart taste to many soft drinks such as Coca-Cola. About 0.05% H_3PO_4 is added; the pH of the soft drinks is about 2.3.

According to Cotton, F. A.; Wilkinson, G. Advanced Inorganic Chemistry, 5th ed.; Wiley-Interscience: New York, 1988; p 424, phosphoric acid has essentially no oxidizing properties below 350 to 400 °C.

Phosphoric Acid

The most important inorganic compound of phosphorus is phosphoric acid, H_3PO_4, which is sometimes called orthophosphoric acid. Some 23 billion lb of H_3PO_4 were produced in the United States in 1993, and phosphoric acid was number ten on the list of the Top 50 chemicals. Pure phosphoric acid is manufactured by burning phosphorus in a mixture of air and steam. The reactions that take place are

$$P_4(l) + 5O_2(g) \xrightarrow{\text{excess}} P_4O_{10}(s)$$

$$P_4O_{10}(s) + 6H_2O(g) \longrightarrow 4H_3PO_4(l) \tag{21.13}$$

Phosphorus pentoxide is the anhydride of phosphoric acid.

Phosphoric acid is a weak, triprotic acid. In contrast to nitric acid, it is *not* an oxidizing acid. Phosphoric acid solutions oxidize moderately active metals such as magnesium only because the hydrogen ion oxidizes the metal:

$$3Mg(s) + 2H_3PO_4(aq) \longrightarrow 3H_2(g) + Mg_3(PO_4)_2(s)$$

Salts of Phosphoric Acid

Salts of phosphoric acid are encountered everywhere in modern life. Because phosphoric acid is a triprotic acid, three types of salts can be made:

$$2H_3PO_4(aq) + Na_2CO_3(aq) \xrightarrow{\text{excess}} 2NaH_2PO_4(aq) + CO_2(g) + H_2O(l)$$

$$H_3PO_4(aq) + Na_2CO_3(aq) \xrightarrow{\text{excess}} Na_2HPO_4(aq) + CO_2(g) + H_2O(l)$$

$$H_3PO_4(aq) + 3NaOH(aq) \xrightarrow{\text{excess}} Na_3PO_4(aq) + 3H_2O(l)$$

Sodium hydroxide must be used to neutralize the third hydrogen of phosphoric acid because sodium carbonate is not a strong enough base. Sodium dihydrogen phosphate, NaH_2PO_4, is a solid, water-soluble weak acid and is used to adjust the pH of water for boilers and in laxative tablets that fizz, which also contain sodium hydrogen carbonate. The reaction that takes place when the laxative tablets are dropped into water is

$$NaH_2PO_4(aq) + NaHCO_3(aq) \longrightarrow CO_2(g) + Na_2HPO_4(aq) + H_2O(l)$$

Disodium hydrogen phosphate, Na_2HPO_4, is used as a basic buffer. Instant pudding mixes and quick-cooking cereals contain disodium hydrogen phosphate. Trisodium phosphate, Na_3PO_4, the salt of a very weak acid with a strong base, is very basic and is used in scouring powders and paint removers.

The Phosphorus Cycle

None of the phosphorus compounds that exist under ordinary conditions is volatile. Therefore, the phosphorus cycle does not involve the atmosphere. Phosphates are slowly removed from rocks by weathering and carried by rivers to lakes and seas. There they are precipitated as insoluble metal phosphates. As time passes, the precipitate becomes sediment. Eventually the sediment is uplifted to form a new land mass.

Some familiar products that contain phosphates.

Excess fertilizer is washed into lakes, rivers, and groundwater. Phosphorus-containing detergents and untreated sewage also add phosphorus to natural waters. Because phosphorus is usually the limiting element for the growth of plants, addition of large quantities of phosphorus from detergents and fertilizers to lakes and rivers makes algae grow. Decomposition of dead algae uses up the oxygen dissolved in the water; fish cannot breathe and die. This process is called **eutrophication**. To prevent eutrophication of our lakes, detergents that contain phosphorus have been prohibited in many places, and detergent manufacturers must use sodium carbonate, Na_2CO_3, (washing soda) as a builder again. At the present time, more phosphate is added to water by human activities than by natural processes, another example of significant interference by people in natural cycles.

PRACTICE PROBLEM

21.16 Write the molecular equation for each of the following reactions:

(a) $H_3PO_4(aq) + K_2CO_3(aq) \longrightarrow$
    excess

(b) $H_3PO_4(aq) + K_2CO_3(aq) \longrightarrow$
    excess

(c) $H_3PO_4(aq) + KOH(aq) \longrightarrow$
    excess

(d) $Zn(s) + H_3PO_4(aq) \longrightarrow$

21.7 CARBON

Occurrence and Production of Carbon and Its Compounds

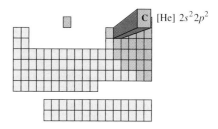

C [He] $2s^2 2p^2$

Although carbon is the fourth most abundant element in the universe, it is not especially plentiful on Earth. Diamonds, graphite, and buckyballs (Section 12.8) are found naturally,* but most carbon is combined in compounds. Most of the carbon on Earth is in carbonate minerals such as calcite ($CaCO_3$) and in coal, petroleum, and natural gas. Some carbon is dissolved in water, mostly in the form of hydrogen carbonate ion, HCO_3^-; a little is present in the atmosphere as carbon dioxide gas, CO_2.

Deposits of calcium carbonate occur in many places. Calcium ions and carbonate ions are present in solution in seawater and in most freshwater, and calcium carbonate is withdrawn from the water by many animals to make shells and bones (-Figure 21.23). These shells and bones fall to the bottom and become limestone. When subjected to heat and pressure, limestone is converted to marble. After quartz (SiO_2), $CaCO_3$ is the most abundant mineral in Earth's crust. Sodium carbonate also occurs naturally.

On an industrial scale, carbon dioxide is obtained as a by-product of the production of hydrogen for the synthesis of ammonia:

$$CH_4(g) + 2H_2O(g) \xrightarrow{\text{catalyst}} CO_2(g) + 4H_2(g)$$

Carbon dioxide is also obtained from the combustion products of coal, petroleum,

Remember that minerals are solid substances that occur in nature, are usually inorganic, and have definite composition and crystal structure.

- FIGURE 21.23   Both the pearl and the oyster shell are calcium carbonate.

---

*By far the largest diamond ever found was the Cullinan, which was discovered in South Africa in 1905. It measured approximately $10 \times 6.5 \times 5$ cm and had a mass of 605 g (about $1\frac{1}{3}$ lb). The Cullinan diamond was presented to King Edward VII of England, who had it cut into 11 stones. Four of these stones are each larger than any other diamond yet found.

and natural gas and as a by-product of fermentation*

$$C_6H_{12}O_6(aq) \xrightarrow{\text{yeast, O}_2 \text{ absent}} 2CH_3CH_2OH(aq) + 2CO_2(g)$$

glucose, a carbohydrate   ethyl alcohol

and of the production of lime from limestone (Section 14.6).

Three forms of carbon—coke, carbon black, and activated carbon, which are mostly microcrystalline forms of graphite—are manufactured on a large scale and used in industry. Coke is made by heating soft coal to about 1100 °C in the absence of air. A variety of liquid and gaseous substances—including water, coal tar, hydrogen, carbon monoxide, carbon dioxide, hydrogen sulfide, ammonia, nitrogen, methane and other hydrocarbons—distills, and coke is left behind. Coke is used in blast furnaces to produce iron from its ores; the carbon is both reducing agent and fuel (Section 11.9).

Carbon black is made by heating a mixture of high-boiling compounds from petroleum to 1400–1650 °C in a limited supply of air. The chief use of carbon black is in rubber products, especially tires; the pigments in black inks, paints, and plastics are usually carbon black.

Activation of carbon is a physical change. The surface of the carbon is greatly increased by the removal of adsorbed hydrocarbons; five pounds of activated carbon is estimated to have a surface area of one square mile. The atoms at the surface have unsatisfied attractive forces and can adsorb large quantities of gases and solutes. Activated carbon often adsorbs a mass of material equal to 25–100% of its own mass. It is used to purify sugar solutions in the production of white sugar, for the removal of unpleasant tastes and odors from drinking water, and for cleanup of many other solutions. Activated carbon is also used in gas masks, large air conditioning systems such as those in airport concourses, and in the control and recovery of vapors from manufacturing processes. Adsorbed materials can be desorbed by raising the temperature, the desorbed material recovered, and the activated carbon reused. Activated carbon is produced on a much smaller scale than coke and carbon black.

Very strong fibers of graphite can be made by heating organic polymer fibers to 1500 °C or more in the absence of air. These fibers are used to make plastics stronger; the reinforced plastics are both very light and very strong.

Diamond can be made from graphite at high pressure (10^5 atm). The temperature must be high enough to allow the atoms to move (about 2000 °C) and a catalyst such as Cr(l), Fe(l), Ni(l), or Pt(l) is needed to make the change from graphite to diamond take place in a practical length of time. As a result of their hardness, synthetic diamonds are used to make cutting tools such as drill points, glass cutters, and diamond saws (see ◄ Figure 21.24). The points of diamond cutting tools do not get hot because diamond conducts thermal energy better than any other known substance.

Chemical Properties of the Element Carbon

Both diamond and graphite are unreactive at room temperature; neither reacts with oxidizing or reducing agents or with acids or bases. At high temperatures, both

The frame of this tennis racket is reinforced with graphite fibers.

Carbon black is used to make carbon paper.

Solvent vapors that would otherwise escape during the manufacture of plastics are caught by activated carbon. The solvents can be recovered and reused, saving resources as well as decreasing air pollution.

In nature, diamonds are believed to have been formed over a long period of time under great heat and pressure deep in the lava in the necks of ancient volcanoes. Diamonds make up only 1 part in 15 million of the lava. No wonder diamonds are expensive!

In the early 1980s, researchers in Japan found that diamonds can be grown at normal atmospheric pressure from a mixture of 99% H_2 and 1% CH_4 at about 2200 °C. These diamonds are smoother than high-pressure diamonds and can be made in sheets.

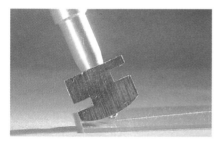

◄ FIGURE 21.24 Diamonds are harder than glass. The tip of this tool is a synthetic diamond.

*Fermentation is an enzyme-catalyzed, energy-producing reaction pathway in cells. Fuel molecules, such as glucose, are broken down anaerobically, that is, in the absence of air. The most familiar example of fermentation is the production of beer, wine, and other alcoholic beverages. Fermentation stops when the alcohol content reaches 15% (30 proof) because the alcohol denatures the enzyme. The alcohol content can be increased by distillation. If O_2 is present, the product is acetic acid, CH_3COOH, instead of ethyl alcohol. Fermentation was one of the first chemical reactions to be discovered and was known to the ancient Greeks. Penicillin G is prepared by a different fermentation.

diamond and graphite react with a number of elements and compounds; for example,

$$2C(graphite) + O_2(g) \xrightarrow{\text{thermal energy}} 2CO(g)$$
$$\quad\text{excess}$$

$$C(graphite) + O_2(g) \xrightarrow{\text{thermal energy}} CO_2(g)$$
$$\quad\text{excess}$$

$$2Fe_2O_3(l) + 3C(graphite) \xrightarrow{2000\ °C} 4Fe(l) + 3CO_2(g)$$

The same products are obtained from diamond as from graphite. However, although graphite is the thermodynamically more stable form of carbon at room temperature, graphite is more reactive than diamond. The bonds between the layers in graphite are much weaker than the bonds within the layers and the bonds between carbons in diamond. Reaction of diamond requires a higher temperature than reaction of graphite.

Carbon Dioxide and Carbonates

Carbon dioxide is a colorless gas; that is, it does not absorb visible light. Sunlight passes through the atmosphere, which contains 0.03% by volume CO_2, to the surface of Earth. Earth radiates energy to space in the form of infrared radiation; carbon dioxide absorbs infrared radiation and interferes with the escape of thermal energy. The concentration of carbon dioxide in the atmosphere is increasing as a result of the burning of fossil fuels; this may be causing Earth to warm up enough to change the climate and perhaps melt the polar ice caps (the greenhouse effect).

Carbon dioxide is odorless with a faintly acidic taste. High concentrations can cause death for lack of oxygen. When bubbled through limewater, a saturated solution of $Ca(OH)_2$, carbon dioxide turns the limewater milky:

$$CO_2(g) + Ca^{2+} + 2OH^- \longrightarrow CaCO_3(s) + H_2O(l)$$

This reaction is often used as a test for $CO_2(g)$.

Carbon dioxide is fairly soluble in water; a saturated solution at 1 atm and 25 °C is about 0.033 M. Carbon dioxide is a typical nonmetal oxide and is acidic. The predominant species in aqueous solutions of carbon dioxide is the CO_2 molecule. Only about 0.17% of the dissolved carbon dioxide is in the form of carbonic acid molecules, H_2CO_3, at equilibrium.

Over a third of the more than 9.84×10^9 lb of carbon dioxide produced in the United States each year is used as a refrigerant. *Solid carbon dioxide* is called **dry ice** because it sublimes rather than melts. Dry ice is used to refrigerate ice cream, meat, and frozen foods during shipment. A single batch of dry ice is enough for the trip from one coast to the other. Ordinary ice would have to be replaced several times during the trip and the water from the melted ice disposed of. Liquid carbon dioxide, which exists only under pressure, is used to make low-melting metals and hamburger meat easier to grind and to cool refrigerated trucks and railroad cars rapidly. It is also used for inflating life rafts, in blasting shells for coal mining, as the propellant in aerosol cans, and in fire extinguishers.

Carbonation of soft drinks and beer is another major use of carbon dioxide. In 1988, each American drank an average of 338 12-oz cans or bottles of soft drinks for a grand total of 8.32×10^{10} cans and bottles. Carbon dioxide gas is also used as an inert protective gas for welding and to neutralize basic wastewater. Manufacture of urea,

$$CO_2(g) + 2NH_3(g) \xrightarrow{185\ °C,\ 200\ atm} [H_2NCO_2NH_4] \longrightarrow H_2NCONH_2(s) + H_2O(l)$$
$$\text{urea}$$

Test for carbon dioxide gas: When carbon dioxide (generated by the reaction of an acid with a solid metal carbonate in the flask at the left) is bubbled through limewater, a saturated solution of calcium hydroxide, the limewater turns milky because calcium carbonate precipitates.

◗ The greenhouse effect is discussed in Section 21.2.

◗ Sample Problem 16.2 pictures CO_2 as a Lewis acid.

Information about soft drink consumption is from Predicasts' Basebook; Predicasts: Cleveland, 1990; p 230.

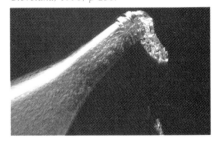

The bubbles in carbonated beverages are bubbles of carbon dioxide gas. Before the container is opened, the partial pressure of carbon dioxide may be as high as 4 atm. The acidity of carbon dioxide solutions is one reason carbonated drinks have a sharp, slightly acid taste. (Many carbonated beverages contain phosphoric acid as well.)

consumes significant amounts of carbon dioxide too. Urea is used as a fertilizer and to make plastics.

Over a third of the sodium carbonate produced each year is used in the glass industry. A new use is for removal of SO_2 from the gases formed by burning fossil fuels in power plants and other large furnaces. The reaction involved is

$$Na_2CO_3(s) + SO_2(g) \longrightarrow Na_2SO_3(s) + CO_2(g)$$

Sodium hydrogen carbonate (sodium bicarbonate) is used to make baking powder. The sodium hydrogen carbonate from baking powder reacts with hydrogen ion from vinegar, sour milk, or the hydrolysis of a salt to form carbon dioxide gas

$$NaHCO_3(s) + H^+ \longrightarrow CO_2(g) + Na^+ + H_2O(l)$$

The bubbles of gas make dough rise. Sodium hydrogen carbonate is also used in dry-powder fire extinguishers. When heated, sodium hydrogen carbonate decomposes, forming carbon dioxide gas:

$$2NaHCO_3(s) \xrightarrow{270\,°C} Na_2CO_3(s) + CO_2(g) + H_2O(g)$$

The decomposition is endothermic and cools the burning materials. In addition, carbon dioxide gas is denser than air and forms a blanket over the fire, keeping air away; with cooling and without oxygen from air, the fire goes out.

Do not use a CO_2 fire extinguisher on an active-metal fire. Active metals such as sodium and magnesium burn in carbon dioxide; for example,

$$CO_2(g) + Mg(s) \xrightarrow{heat} MgO(s) + CO(g)$$

Active-metal fires can be extinguished with powdered graphite.

The Carbon Cycle

Carbon, like water, oxygen, and nitrogen, cycles from Earth into the atmosphere and back again. Green plants and algae use the sun's energy to convert carbon dioxide (and water) into carbohydrates:

$$6CO_2(g) + 6H_2O(l) \xrightarrow{light,\ chlorophyll} \underset{glucose,\ a\ carbohydrate}{C_6H_{12}O_6(aq)} + 6O_2(g)$$

The plants are eaten by animals, including people and fish, who exhale carbon dioxide. Carbon dioxide is also formed by decomposition of dead animals and animal wastes by microorganisms. (If O_2 is scarce, as it is in sewage, marshes, and swamps, some carbon is released as methane gas, CH_4, another greenhouse gas.) The carbon dioxide passes into the atmosphere and is again used for photosynthesis. An equilibrium also exists between carbon dioxide in the atmosphere and dissolved carbon dioxide and HCO_3^- in oceans and lakes. In addition, much carbon is stored in Earth's crust in the form of fossil fuels (coal, petroleum, and natural gas) and in the form of limestone and coral. Since the middle of the nineteenth century, production of carbon dioxide by the combustion of fossil fuels and the decomposition of limestone has been increasing rapidly. In addition, destruction of tropical rain forests is reducing the quantity of carbon dioxide used up by photosynthesis. Human activities have now reached a scale where interference with the natural carbon cycle may well be significant. Unfortunately, being sure is difficult because the longest continuous records of the concentration of CO_2 in the atmosphere go back only to 1958.

PRACTICE PROBLEM

21.17 Write molecular equations for each of the following reactions:

(a) $C(graphite) + S(g) \xrightarrow{heat}$

(b) $MnO_2(s) + C(graphite) \xrightarrow{heat}$

(c) $CH_3COCH_3(l) + O_2(g) \xrightarrow{spark}$
    excess

(d) $Na_2CO_3(s) + HNO_3(aq) \longrightarrow$

SILICON

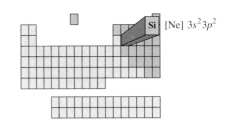

Si [Ne] $3s^2 3p^2$

Carbon, the first element in Group IVA, is the key element in living matter. Although silicon, the second element in Group IVA, is an essential element for humans, it is relatively unimportant in the plant and animal kingdoms. However, silicon is the key element in the mineral kingdom. About 95% of Earth's crust consists of compounds of silicon. Most of these silicon compounds are silica (SiO_2) and metal silicates. The soil, sand, and clay in Earth's crust are formed by the breakdown of silica and silicate minerals. Most inorganic building materials are based on silicon. The natural materials sandstone, granite, and slate are composed largely of silicates, and the synthetic materials cement, concrete, bricks, and glass are made from silica and silicates. Why is there such a great difference between silicon and carbon?

Differences Between Silicon and Carbon

Silicon–oxygen bonds are the key feature of silicon chemistry, whereas carbon–carbon bonds are the key feature of carbon chemistry. Silicon–oxygen bonds are much stronger than silicon–silicon bonds; carbon–carbon bonds are of about the same strength as carbon–oxygen bonds. The bond energies of the four bonds are compared in Table 21.7.

In carbon dioxide, each oxygen is linked to carbon by a double bond:

$$\ddot{O}{=}C{=}\ddot{O}$$

The units of carbon dioxide are small nonpolar molecules with weak intermolecular attractions. As a result, carbon dioxide is a gas under ordinary conditions. Carbon dioxide is fairly soluble in water because carbon dioxide acts as a Lewis acid (Section 16.3). Thus, carbon dioxide is available for photosynthesis and other reactions in aqueous solution. Because the larger silicon atom does not readily form double bonds, silicon dioxide forms a three-dimensional covalent network solid with a high melting point (about 1700 °C) that is insoluble in water. Therefore, silicon dioxide is unable to take part in reactions in aqueous solutions.

TABLE 21.7

Bond Energies

Bond	Bond Energy, kJ/mol
C—C	347
C—O	360
Si—Si	192
Si—O	464
Si—C	305

The solubility of $CO_2(g)$ in water at 20 °C and 1 atm pressure is 88 mL of CO_2 per 100 mL of water.

Silica and the Silicate Minerals

The structural unit of silica and silicates is the {SiO_4} tetrahedron [see ▬Figure 21.25(a)]. A few minerals, such as the semiprecious stone zircon ($ZrSiO_4$), are simple silicates, but most are more complex. Silicate tetrahedrons can be linked through one or more of the oxygens to form a variety of structures as shown in Figure 21.25(b)–(d). Fibrous minerals such as asbestos are derived from chains of {SiO_4} tetrahedrons similar to the one shown in Figure 21.25(c). Layer minerals such as mica are derived from sheets of {SiO_4} tetrahedrons similar to that shown in Figure 21.25(d).

Some crystalline silicate minerals. Left to right: *zircon embedded in zirconium silicate, beryl, asbestos, and mica.*

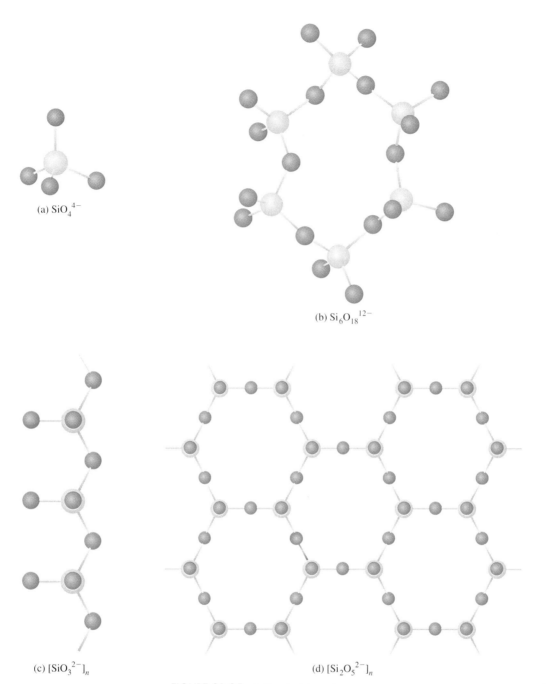

(a) SiO_4^{4-}

(b) $Si_6O_{18}^{12-}$

(c) $[SiO_3^{2-}]_n$

(d) $[Si_2O_5^{2-}]_n$

■ FIGURE 21.25   Ball-and-stick models: (a) The {SiO₄} tetrahedron. (b) Cyclic $Si_6O_{18}^{12-}$ ion. To find the charge on silicate anions, assume that the oxidation number of silicon is $+4$. The oxidation number of oxygen is, of course, -2. (c) Chain of {SiO₄} tetrahedrons. (d) Layer of {SiO₄} tetrahedrons.

The negative charges of the various silicon-containing anions such as $Si_6O_{18}^{12-}$ must be balanced by positive charges from metal ions. In crystals of silicates, the metal ions, which are relatively small, fit into the spaces between the larger anions, and any metal ion that is about the right size will do. For example, Be^{2+} (radius = 59 pm) and Al^{3+} (radius = 67 pm) fit the same holes. The total positive charge on the metal ions must equal the negative charge on the silicate anion:

$$3(2+) + 2(3+) = 12+$$

The strange formulas of silicate minerals—$Be_3Al_2Si_6O_{18}$ for beryl, for example—result.

Other silicate minerals such as quartz (see ▬Figure 21.26), which is the most common mineral, are derived from three-dimensional covalent networks of $\{SiO_4\}$ tetrahedrons. Quartz is the crystalline form of silica (SiO_2) that is stable under ordinary conditions. Crystal balls are carved from colorless (pure) quartz. Some varieties of quartz that are colored by impurities and some noncrystalline forms of SiO_2 are shown in ▬Figure 21.27.

Production of Pure Silicon

Silicon does not occur free in nature; the element is made from sand by reduction with carbon in an electric furnace:

$$SiO_2(l) + 2C(s) \xrightarrow{3000\,°C} Si(l) + 2CO(g)$$
$$\text{sand}$$

An excess of sand is used to prevent contamination of the product by silicon carbide, SiC. Silicon carbide is converted to silicon and carbon monoxide by the excess sand:

$$2SiC(l) + SiO_2(l) \longrightarrow 3Si(l) + 2CO(g)$$

World production of silicon is about 2×10^9 lb per year.

Manufacture of semiconductor devices such as transistors, solar cells, and microcomputer chips is an important use of silicon (Section 12.12). More than 99.999 999% of the atoms in silicon used for semiconductor devices must be silicon atoms—that is, there must be less than 1 atom of impurity for every 100 million silicon atoms. Silicon made by reduction of sand is only about 96–97% pure. Soluble impurities are removed by running water slowly through powdered silicon. Then the silicon is converted to silicon tetrachloride, which is a liquid:

$$Si(s) + 2Cl_2(g) \xrightarrow{heat} SiCl_4(l)$$

The silicon tetrachloride is purified by fractional distillation and reduced with very pure zinc or magnesium:

$$SiCl_4(l) + 2Zn(s) \longrightarrow Si(s) + 2ZnCl_2(s)$$

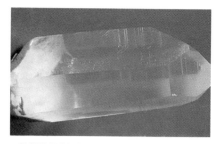

▬ FIGURE 21.26 Crystals of pure quartz are colorless. Quartz is used for frequency control in radios, TVs, clocks, and watches.

Crystals with three-dimensional covalent network structures are one giant molecule like diamond. Therefore, minerals with three-dimensional covalent network structures are hard and high melting.

▬ FIGURE 21.27 Some varieties of quartz that are colored by impurities and some noncrystalline forms of SiO_2. Clockwise starting at upper left: Rose quartz, amethysts, opals, flint, and agate.

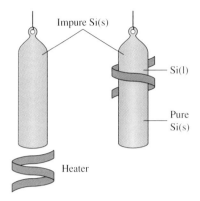

Impure Si(s)

Si(l)

Pure Si(s)

Heater

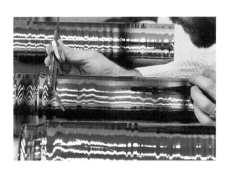

- FIGURE 21.28 *Left:* Schematic diagram showing purification by zone refining. *Right:* Cylinders of silicon purified by zone refining. These cylinders will be sliced into wafers and used for making computer chips.

When you were a child, did you ever suck the colored, flavored liquid out of a Popsicle, leaving almost colorless ice behind? This is an everyday example of zone refining. The impurities (flavoring and coloring) concentrate in the melt and purer ice remains.

Next the silicon is melted and grown into cylindrical crystals. Finally, the crystals of silicon are purified by zone refining.

In **zone refining,** the cylindrical crystal is heated near the bottom end, and a layer melts. The melted zone is then moved up the crystal by moving the heater up the crystal (or lowering the crystal through the heater). Below the heater, silicon crystallizes again. When silicon crystallizes, the impurities do not fit in the crystal structure as well as silicon atoms do; therefore, the impurities remain in solution in the melt. Thus, the impurities are carried along the crystal with the melted zone (see ▪Figure 21.28). When the melted zone and the impurities reach the top of the crystal, they are removed. Thin slices of the very pure silicon are used to produce chips for microcomputers. Transistors and solar cells are also made from very pure silicon.

Other Products from Sand: Sodium Silicate, Glass, Cement, and Ceramics

Sodium silicate, glass, cement, and ceramics are also manufactured from sand. Sodium silicate is made by melting a mixture of sodium carbonate and sand:

$$Na_2CO_3(l) \ + \ aSiO_2(l) \ \xrightarrow{1500\ °C} \ Na_2O \cdot aSiO_2(l) \ + \ CO_2(g)$$

The proportion of sand, a, can be varied to give sodium silicate with different properties; the value of a is usually between 0.5 and 4. On cooling, sodium silicate forms a glass that is soluble in water. The major use of sodium silicate is to make catalysts and gels for drying air.

To make ordinary glass for windows and bottles, a mixture that contains 70–74% pure sand (SiO_2), 13–16% soda ash (Na_2CO_3), and 10–13% lime (CaO) is melted. This composition gives glass that is viscous enough so that the glass does not crystallize yet not too viscous to be worked at reasonable temperatures. The chemical reactions involved in glass making are

$$Na_2CO_3(l) \ + \ aSiO_2(l) \ \xrightarrow{1200\ °C} \ Na_2O \cdot aSiO_2(l) \ + \ CO_2(g)$$
$$CaCO_3(l) \ + \ bSiO_2(l) \ \longrightarrow \ CaO \cdot bSiO_2(l) \ + \ CO_2(g)$$

where a and b are usually fractions. Glass is a homogeneous mixture, that is, a solution of silicates and silica. Because glass is a mixture, it softens gradually over a range of temperatures instead of having a sharp melting point as a pure solid does. Glass can be melted, solidified, and remelted. It was made in Egypt as early as 6000–5000 B.C. The chief difference between ancient and modern glass manufacture is that modern glass factories are automated.

Notice that silica acts as an acid in these reactions. Insolubility keeps silica from acting as an acid in aqueous solution.

A silicon wafer with one microchip removed.

Glass is attacked slowly by strong bases and rapidly by hydrofluoric acid. Representative equations for the reactions that take place are

$$CaSiO_3(s) + 6HF(aq) \longrightarrow CaF_2(s) + SiF_4(g) + 3H_2O(l)$$

$$SiO_2(s) + 2OH^- \longrightarrow SiO_3^{2-} + H_2O(l)$$

Hydrofluoric acid cannot be stored in glass bottles and concentrated sodium or potassium hydroxide should not be.

Borosilicate glasses such as Pyrex and Kimax usually contain about 13–28% B_2O_3 and 72–87% silica. Borosilicate glasses do not expand very much when heated, don't break easily, and are more resistant to chemicals than ordinary glass. They are used for baking dishes and laboratory glassware.

The disk of the 200-in. telescope at Mt. Palomar is made of borosilicate glass.

Safety glass is made by sandwiching a layer of flexible plastic between two layers of glass. If safety glass is broken, the polymer layer holds the pieces in place. Another type of safety glass called tempered glass is made by heating glass just below its softening point and then suddenly cooling it with jets of cold gas. Tempered glass breaks into many small pieces without sharp edges when damaged.

Tempered glass is used for automobile windshields in some countries. However, tempered glass is more rigid than the skulls of the car's occupants and sandwich-type safety glass is always used in American cars.

Cement, like glass, dates back to ancient times. Cement is now low in cost and is used everywhere for building roads, tunnels, dams, houses, and so forth. **Cement** is *made by heating sand, clay (hydrated aluminum silicate), limestone (CaCO₃), and gypsum (CaSO₄·2H₂O).* **Concrete** is *made by mixing cement with sand, pebbles, or gravel, and water and allowing the mixture to harden.* Cement and concrete are used in larger quantities than any other man-made materials.

The Egyptians used cement to make the pyramids.

In chemistry and engineering, the term **ceramic** refers to *an inorganic material that is not metallic.* Sand, clay, and feldspar (aluminum silicates) are the basic raw materials for ceramics. High temperatures are usually needed to process ("fire") ceramics. The reactions involved are complex. However, the overall change involves removal of water and conversion of silica to cristobalite, a crystalline form of silica that has a diamondlike structure:

The most important clay is kaolinite, $Al_2O_3 \cdot 2SiO_2 \cdot 2H_2O$. Clays are colloids.

$$3(Al_2O_3 \cdot 2SiO_2 \cdot 2H_2O)(s) \xrightarrow{1000\,°C} 3Al_2O_3 \cdot 2SiO_2(s) + 4SiO_2(s) + 6H_2O(g)$$

Phase diagrams (Section 12.7) have been of great importance in improving ceramics.

Ceramics are heterogeneous mixtures composed of a glassy phase with tiny crystals of silicates in it. Ceramics are nonflammable and resistant to heat and chemicals, although brittle. They are less dense than metals so articles made of ceramics are lighter. The raw materials for making ceramics are abundant and relatively inexpensive. Only 1/6 as much energy is required to make a ceramic as is needed to make the same volume of plastic and only 1/29 as much energy is required to make a ceramic as is needed to make the same volume of stainless steel. Many materials engineers are predicting that ceramics will become as important in the future as polymers have become in the recent past. The U.S. Office of Technology Assessment ranked ceramics with information technology and biotechnology as one of the three most important high-tech industries of the future.

Ceramic superconductors are discussed in Chapter 24.

Silicon and the Periodic Table

Just as lithium and magnesium resemble each other (Section 8.7), silicon resembles boron. Silicon and boron are both semimetals, whereas carbon is a nonmetal. Like SiO_2, boron oxide, B_2O_3, is a good glass former.

However, although silicon differs from carbon in many respects, the family resemblance is still there. Crystals of the free element silicon have the diamond structure. Like carbon, silicon is not very reactive under ordinary conditions. Silicon is usually tetravalent in compounds, and SiO_2 is a weak acid like CO_2.

Only carbon has a graphite structure. The larger members of Group IVA do not have carbon's ability to form double bonds.

Germanium is very similar to silicon. Germanium is also a semimetal, has the diamond crystal structure, and is used to make semiconducting devices. The lowest members of Group IVA, tin and lead, are both metals. The gradual change from nonmetallic to metallic properties down a group that is observed in Group IVA is typical of *p*-block groups.

SUMMARY

Hydrogen is unusual in several ways. The differences between the isotopes of hydrogen are greater than the differences between the isotopes of any other element. The density of hydrogen gas is lower than the density of any other substance at the same temperature and pressure. The temperature of hydrogen gas rises when the hydrogen gas is allowed to expand at room temperature. More compounds of hydrogen are known than compounds of any other element.

Hydrogen is produced industrially by the reaction of hydrocarbons with steam followed by the **water–gas shift reaction:**

$$CO(g) \ + \ H_2O(g) \xrightleftharpoons{\text{catalyst, 200–400 °C}} H_2(g) \ + \ CO_2(g)$$

Hydrogen is often made and used in the same plant; however, hydrogen can be liquefied and shipped in insulated tank cars and trucks. Small quantities of hydrogen are stored and shipped in steel cylinders.

Most of the hydrogen that is produced is used to make ammonia by the Haber process. Some day hydrogen may be used as a fuel instead of coal, natural gas, and oil.

Hydrogen is relatively unreactive at low temperatures. Reactions involving hydrogen have high activation energies as a result of the high bond dissociation energy of the H—H bond in the H_2 molecule. At high temperatures, hydrogen reacts vigorously with many metals and nonmetals. **Saltlike hydrides, molecular hydrides,** and **metal-like hydrides** are formed. Metal-like hydrides are often **nonstoichiometric**—that is, their composition is not constant.

Water is the most important chemical compound. About 74% of Earth's surface is covered with water; Earth's water cycles between Earth and its atmosphere. Water, which is strongly hydrogen-bonded, is the only liquid that is common on Earth.

Most reactions take place more rapidly in solution, and water is the most important solvent. All naturally occurring water contains impurities; harmful impurities are called **pollutants.** Impurities are removed by filtration, aeration, disinfection with chlorine or ozone, treatment with **ion-exchange resins** or distillation, and reverse osmosis. Organic impurities are removed with activated carbon. Water from which ions have been removed with an ion-exchange resin is called **deionized water.**

Water takes part in acid–base reaction, and oxidation-reduction reactions. It is a reactant for photosynthesis and for many industrial processes.

Hydrates are compounds that contain water. **Anhydrous** means "without water."

Oxygen makes up 23% by mass of the atmosphere, 86% of the surface waters of Earth, and 46% of the crust and upper mantle. Oxygen of the proper concentration is necessary for breathing and for the combustion of fossil fuels. Oxygen and nitrogen are obtained from the atmosphere. The ozone layer in the stratosphere protects life on Earth from ultraviolet radiation. Most reactions in the atmosphere involves species with odd electrons.

Sulfur and its compounds are so widely used in industry that a country's economic development is often measured by the quantity of sulfur consumed. Sulfur has more allotropes than any other element. It is a fairly reactive element. Sulfuric acid is the most important sulfur compound; large quantities of aluminum sulfate, sodium sulfate, and ammonium sulfate are also used.

Nitrogen gas, N_2, is relatively inert because the nitrogen–nitrogen triple bond is very strong. Nitrogen gas is used as an inert atmosphere and as a refrigerant. Nitrogen in compounds shows more oxidation numbers than any other element, and nitrogen forms an unusually large number of oxides. Ammonia and nitric acid are the most important inorganic nitrogen compounds. Nitric acid is an oxidizing acid.

All living matter contains phosphorus. The chief source of phosphorus is phosphate rock, which is used to make fertilizers and to produce elemental phosphorus. The commonest allotrope of phosphorus is white phosphorus, which is made up of strained P_4 molecules. White phosphorus is very reactive. Phosphoric acid is the most important inorganic phosphorus compound; P_2O_5 is the **anhydride** of phosphoric acid.

Even if organic compounds are not considered, people use more carbon (by mass) than any other element, mainly in the form of coke and carbon black, which are mostly microcrystalline forms of graphite. Coke is used as both reducing agent and fuel in the production of metals from their ores. The chief use of carbon black is in rubber products, especially tires. The most important inorganic carbon compounds are carbon dioxide and sodium carbonate.

Silicon is like carbon in crystal structure, lack of reactivity, and tetravalency. Germanium is very similar to silicon; tin and lead are metals. A gradual change from nonmetallic to metallic properties down a group is characteristic of *p*-block elements.

Silicon is the key element in the mineral kingdom but is relatively unimportant in the plant and animal kingdoms. Silicon–oxygen bonds are the most important feature of silicon chemistry. Most inorganic building materials are based on silicon; glass, **cement, concrete,** and **ceramics** are all manufactured from sand. Silicon is also made from sand. Silicon for semiconductors is purified by **zone refining.**

ADDITIONAL PRACTICE PROBLEMS

For information about the organization of Additional Practice Problems, Putting Things Together, and Applications, see the beginnings of these sections in Chapter 1.

21.18 (a) What are the three isotopes of hydrogen called, and what is the symbol for each? (b) Why does each have a different name and symbol?

21.19 Explain why there is very little $H_2(g)$ in Earth's atmosphere.

21.20 Compare the physical and chemical properties of the hydrides of (a) Li, C, N, O, and F (b) O and S.

21.21 What effect does hydrogenation have on vegetable oils?

21.22 Write equations for the reaction of hydrogen with (a) fluorine (b) oxygen (c) nitrogen (d) calcium (e) $H_2C=CH_2$.

21.23 Suggest a reaction that could be used to make each of the following: (a) NaOD (b) D_2 (c) DCl

21.24 Explain why (a) ice is less dense than water whereas most substances are denser when in the solid phase than when in the liquid phase, (b) the density of water first increases and then decreases with increasing temperature.

21.25 (a) Why is water often used as a solvent? (b) Why must solubility rules be memorized for ionic compounds? (c) Which of the following compounds are soluble in water? Explain your answers. (i) $AgNO_3$ (ii) CH_3OH (iii) $CH_3(CH_2)_3CH_3$ (iv) $CH_3(CH_2)_{16}COOH$ (v) ZnS

21.26 Write equations for the reaction of water with (a) HCl(g) (b) $NH_3(g)$ (c) BaO(s) (d) $SO_3(g)$ (e) Ca(s).

21.27 (a) What is the greenhouse effect? (b) Give the names and formulas of three greenhouse gases.

21.28 Compare ozone and chlorine as disinfectants for water.

21.29 (a) How is oxygen produced industrially? (b) What is the major industrial use of oxygen? (c) In the use in part (b), how is oxygen reacting?

21.30 Write equations for each of the following reactions. If no reaction will take place, write "No reaction."
(a) $H_2S(g)$ + excess $O_2(g)$ $\longrightarrow$
(b) Cu(s) + H_2SO_4(1 M) $\longrightarrow$
(c) Cu(s) + H_2SO_4(18 M) $\xrightarrow{heat}$
(d) NaOH(aq) + excess H_2SO_4 $\longrightarrow$
(e) H_2SO_4(aq) + excess NaOH(aq) $\longrightarrow$

21.31 (a) Nitric acid is colorless. Why are the contents of bottles of concentrated nitric acid in the lab usually yellow? (b) Why are copper and silver, which are not soluble in hydrochloric acid, soluble in nitric acid?

21.32 Write the molecular equation for each of the following reactions of nitric acid. Assume that the same predominant nitrogen-containing species will be formed as are shown in the equations illustrating oxidations by nitric acid in Section 21.5. If no reaction will take place, write "No reaction." (a) Ag(s) + HNO_3(16M) $\longrightarrow$ (b) Ag(s) + HNO_3(6 M) $\longrightarrow$ (c) $S_8(s)$ + HNO_3(16 M) $\longrightarrow$ (d) Cr(s) + HNO_3 (6 M) $\longrightarrow$ (e) Cr(s) + HNO_3(3 M) $\longrightarrow$ (f) Cr(s) + HNO_3(1 M) $\longrightarrow$ (g) Cr(s) + HNO_3(0.1 M) $\longrightarrow$ (h) CuS(s) + HNO_3(3 M) $\xrightarrow{heat}$ (i) Al(s) + HNO_3(conc) $\longrightarrow$ (j) NaOH(aq) + HNO_3(aq) $\longrightarrow$ (k) $CaCO_3$(s) + HNO_3(aq) $\longrightarrow$ (l) MgO(s) + HNO_3(aq) $\longrightarrow$

21.33 (a) Copy the Lewis formulas for nitric acid and nitrogen dioxide from Table 21.6 and write the other resonance structure for each using the appropriate symbol between structures. (b) Write the three resonance structures for the nitrate ion.

21.34 (a) Why is phosphorus very reactive? (b) Write the equation for the reaction of phosphorus with oxygen from air. (c) What is the most important property of the product of the reaction in part (b)? (Write the equation for the reaction that takes place.)

21.35 Explain how Na_2HPO_4 buffers a solution against an increase in $[H^+]$ on addition of H^+ and an increase in $[OH^-]$ on addition of OH^-.

21.36 (a) Write equations for four reactions carried out industrially that produce carbon dioxide. (b) What are two major uses of carbon dioxide?

21.37 Why is carbon the key element in the plant and animal kingdoms whereas silicon is the key element in the mineral kingdom?

21.38 How does the microscopic model of the structure of asbestos, mica, and quartz explain the physical properties of macroscopic samples of these minerals?

21.39 In chains of SiO_4 tetrahedrons, how many O does each tetrahedron share with neighboring tetrahedrons?

21.40 Discuss the similarities and differences between glass and quartz.

21.41 Why should glass bottles not be used to store the following? (a) hydrofluoric acid (b) concentrated sodium hydroxide solution

21.42 Which element most closely resembles silicon in properties? Explain your answer.

21.43 (a) What are allotropes? (b) For which of the elements in this chapter are allotropes known? (c) If allotropes are known, describe them briefly. (d) Why does silicon not have a graphitelike allotrope?

21.44 Industrially, hydrogen gas is usually produced by the reaction of coke (C), natural gas (CH_4), or other hydrocarbons, such as propane ($CH_3CH_2CH_3$), with steam. Which of these three raw materials for the production of hydrogen produces the highest ratio of product (H_2) to by-product (CO)? The lowest ratio?

21.45 Can ammonia gas be collected over water? Explain your answer.

21.46 The solubility of hydrogen gas in water is 1.91 mL/L at 25°C if the pressure of the hydrogen gas is 1.00 atm. (a) What is the molarity of the saturated solution? (b) How will the solubility change if the pressure of the hydrogen gas is increased? If the temperature is increased?

21.47 (a) Use Lewis formulas to show that the reaction between SO_3 and water is a Lewis acid–base reaction. (b) Write a net ionic equation showing that the reaction between $NaNH_2$ and water is a Brønsted–Lowry acid–base reaction.

21.48 (a) For the carbon dioxide molecule, write the Lewis formula. (b) How many pi bonds are shown by this formula? How many sigma bonds? (c) What is the hybridization of the carbon in carbon dioxide? (d) What is the angle between the carbon–oxygen bonds in carbon dioxide? (e) Is a carbon–oxygen bond polar or nonpolar? (f) Is the carbon dioxide molecule polar or nonpolar?

21.49 For each of the following species, draw the molecular orbital energy level diagram for the ground state, write the electron configuration of the ground state, give the bond order, and tell whether the species is paramagnetic or diamagnetic. If it is paramagnetic, give the number of unpaired electrons. (a) O_2 (b) N_2 (c) O_2^-

21.50 Silicon tetrafluoride reacts with fluoride ion to form the complex ion SiF_6^{2-}. (a) In the formation of this complex ion, how is silicon tetrafluoride acting? Fluoride ion? (b) Use VSEPR to predict the geometry of the SiF_6^{2-} ion. (c) Why does CF_4 not react with fluoride ion?

21.51 What type or types of intermolecular attraction must be overcome to (a) melt silicon? (b) Melt magnesium oxide? (c) Melt rhombic sulfur? (d) Melt ice? (e) Boil carbon disulfide?

21.52 (a) Write the Lewis formula for the hydrogen molecule. (b) Which type of intermolecular attraction exists between hydrogen molecules? (c) How do the intermolecular attractions between hydrogen molecules compare with the intermolecular attractions between D_2 molecules? Give evidence on which your conclusion is based.

21.53 Use the phase diagram for carbon dioxide, Figure 12.32, to answer the following questions. (a) If a sample of $CO_2(g)$ is compressed at -25 °C, at about what pressure will the gas condense to a liquid? (b) If a sample of carbon dioxide that is initially at 50 °C and 76 atm is cooled (keeping

the pressure constant), at what temperature will the sample solidify? (c) What is supercritical carbon dioxide?

21.54 (a) Why are P_2O_5 and concentrated H_2SO_4 good drying agents for air? (b) Magnesium perchlorate is also a good drying agent because it forms a hydrate, $Mg(ClO_4)_2 \cdot 6H_2O$. What is the name of this hydrate? (c) Anhydrous magnesium perchlorate deliquesces in humid air. What does this statement mean?

21.55 The acid

is usually called hypophosphorous acid. (a) How many ionizable hydrogens are there in this acid? (b) What is the salt $NaH_2PO_2 \cdot H_2O$, which is used in industry as a reducing agent, called? (c) What is the oxidation number of P in these two compounds?

21.56 (a) How does the bond dissociation energy of the D—D bond compare with the bond dissociation energy of the H—H bond? (b) The bond energy for the D—C bond is 419 kJ/mol. How does the bond energy of the D—C bond compare with the bond energy of the H—C bond? (c) These two examples are typical. Which are stronger, bonds to deuterium or bonds to hydrogen? Explain your answer. (d) How would you expect the rates of reactions involving breaking bonds to deuterium to compare with the rates of reactions involving breaking bonds to ordinary hydrogen? Explain your answer. This effect is called a kinetic isotope effect and is responsible for the deadly effect of building deuterium into tissue on animals.

21.57 Given the following enthalpies of hydration—K^+, -321 kJ/mol; Mg^{2+}, -1922 kJ/mol; Na^+, -405 kJ/mol—(a) explain why more energy is released when Na^+ is hydrated than when K^+ is hydrated. (b) Explain why more energy is released when Mg^{2+} is hydrated than when Na^+ is hydrated. (c) Of the following enthalpies of hydration (all in kJ/mol), -296, -515, -405, -1592, -4660, which is the enthalpy of hydration of Al^{3+}? of Rb^+? Explain your answers. (d) Explain why a solution of aluminum nitrate is acidic.

21.58 For the water–gas shift reaction

$$CO(g) + H_2O(g) \xrightleftharpoons{\text{catalyst, 200–400 °C}} H_2(g) + CO_2(g)$$
$$(21.1)$$

(a) What is $\Delta H°$? (b) Estimate the value of $\Delta G°$ at 300 °C. (c) Is the forward reaction spontaneous at 300°C? (d) What is the value of K_p at 300 °C? of K_c? (e) What would be the effect of raising the temperature on the position of equilibrium? (f) What would be the advantage of carrying out this reaction at high pressure? (g) What is the effect of the catalyst on the position of equilibrium? (h) What is the advantage of using a catalyst?   (i) What would be the effect of removing CO_2 on the equilibrium?

21.59 (a) Estimate the temperature at which each of the following reactions will become spontaneous.

$$SnO_2(s) + 2C(s) \longrightarrow Sn(s) + 2CO(g)$$
$$CaO(s) + C(s) \longrightarrow Ca(s) + CO(g)$$

For $SnO_2(s)$, $\Delta H_f^\circ = -580.7$ kJ/mol, $\Delta G_f^\circ = -519.9$ kJ/mol, $S^\circ = 52.3$ J/K·mol, for $Sn(s)$, $S^\circ = 51.5$ J/K·mol, and for $Ca(s)$, $S^\circ = 41.4$ J/K·mol.   (b) One reaction has been used to produce the metal since ancient times. Which one? Explain your answer.

21.60 (a) Calculate the standard free-energy change for the conversion of diamonds to graphite.   (b) Explain why diamonds don't spontaneously change to graphite.

21.61 For the forward reaction in the equilibrium

$$N_2O_3(g) \rightleftharpoons NO_2(g) + NO(g)$$

$\Delta G^\circ = -1.5$ kJ at 25 °C.   (a) What is the value of ΔG_f° for $N_2O_3(g)$ at 25 °C?   (b) What is the value of K_p for this equilibrium at 25 °C?   (c) What is the value of K_c for this equilibrium at 25 °C?   (d) For $N_2O_3(g)$ ΔH_f° at 25 °C = 83.72 kJ/mol. What is ΔH° for the forward reaction at 25 °C? ΔS°?

21.62 (a) Is the reaction $N_2(g) + O_2(g) \longrightarrow 2NO(g)$ spontaneous under standard conditions at 25 °C? Explain your answer. (b) What is ΔH° for this reaction at 25 °C? Is the reaction exothermic or endothermic?   (c) This reaction becomes spontaneous at very high temperatures. What must be the sign of ΔS° for this reaction? Give your reasoning.

21.63 The following mechanism is generally accepted for the reaction between hydrogen and chlorine to form hydrogen chloride:

Step (1)	$Cl_2 \xrightarrow{\text{light or thermal energy}} 2Cl$
Step (2)	$Cl + H_2 \longrightarrow HCl + H$
Step (3)	$H + Cl_2 \longrightarrow HCl + Cl$

(a) Show that the sum of steps (2) and (3) is the overall equation for the reaction between hydrogen and chlorine. (b) Which species shown in the mechanism are radicals? Rewrite the mechanism marking each of the radicals with a dot, ·, to show that it has an unpaired electron and is very reactive.   (c) For each radical used up in step (2), how many radicals are formed? For each radical used up in step (3), how many radicals are formed?   (d) Reactions in which one reactive species forms another reactive species are called chain reactions because, once started, reaction continues until something happens, such as the combination of two radicals,

$$H + H \longrightarrow H_2$$

to break the chain. Write equations for two other possible chain-terminating combinations of radicals.   (e) Why are reactions such as steps (2) and (3), which are called chain-propagating steps, much more likely than chain-terminating reactions? (*Hint:* How do the concentrations of H_2 and Cl_2 compare with the concentrations of H and Cl?)   (f) How many kilojoules are released by the formation of a mole of

HCl?   (g) Steps that start a chain, such as step (1), are called chain-initiating steps. Explain why, once chains are started by light, an explosively rapid reaction occurs.

21.64 A reaction typical of the upper atmosphere is

$$O + H_2O \longrightarrow 2OH$$

For this reaction, the value of the rate constant k is 2×10^{-3} L/mol·s at 25 °C, and the activation energy E_a is 77.4 kJ. What is the value of k at 40 km where the temperature is -13 °C?

21.65 Nitric oxide is believed to destroy stratospheric ozone by the following mechanism:

$$NO + O_3 \longrightarrow NO_2 + O_2$$
$$NO_2 + O \longrightarrow NO + O_2$$

(a) What is the overall reaction that results from this mechanism?   (b) In this mechanism, how is NO acting?   (c) What is a species such as NO_2 called?   (d) Write the rate law for each step.   (e) In a mechanism, what is each step called?

21.66 The value of E° for the half-reaction

$$O_3(g) + 2H^+ + 2e^- \longrightarrow O_2(g) + H_2O(l)$$

is 2.08 V. From Table 19.1 choose a species that should be oxidized by ozone but not by molecular oxygen and write the net ionic equation for the reaction.

21.67 A minimum of 1.24 V (neglecting overvoltage) is needed to electrolyze water that contains a little Na_2SO_4 as electrolyte. One kilowatt hour (kWh) of electricity, which will run a 100-W light bulb for 10 h, is equivalent to 3.60×10^6 J. How many grams of hydrogen gas can be produced by the electrolysis of water using 1.00 kWh of electricity?

21.68 If the low temperature of the tropopause did not prevent escape of water vapor to higher altitudes, water would be dissociated by ultraviolet light. Photons of what wavelength of ultraviolet light (in nanometers) have enough energy to break the O—H bonds in water molecules?

21.69 (a) Sodium metal is an example of a reducing agent that can't be used in aqueous solution because it is a stronger reducing agent than water and reacts by reducing water. Write the equation for the reaction of sodium metal with water.   (b) Fluorine gas is an example of an oxidizing agent that can't be used in aqueous solution because it is a stronger oxidizing agent than water and reacts by oxidizing water. Write the equation for the reaction of fluorine gas with water. (c) The hydride ion is an example of a base that can't be used in aqueous solution because it is a stronger base than water. Write the equation for the reaction of hydride ion with water.   (d) What is the strongest base that can exist in aqueous solution?   (e) In each of the reactions in parts (a)–(c), how is water acting?

21.70 Explain the following everyday observations of water: (a) If an ice cube tray is filled with water to the level of the top of the divider, the surface of the ice cubes is above the level of the top of the divider, and the ice cubes are all stuck together.   (b) A load of wet clothes forgotten in the washing machine is still wet the next time you go to use the machine, whereas the same clothes hung on a rack will soon be dry.

(c) Add three more observations of your own and explain them. Illustrate as many different properties of water as you can.

21.71 (a) Which air pollutant is formed by the reaction of nitrogen and oxygen from air in automobile engines? (b) Which air pollutant is formed by the combustion of high-sulfur fuels in power plants? (c) Which air pollutants are produced by lightning? (d) What type of air pollutant is produced by trees?

21.72 Explain why: (a) Carbon tetrachloride does not react with water, but silicon tetrachloride reacts vigorously with cold water. (b) Phosphorus pentachloride is a well-known compound, but nitrogen pentachloride does not exist. (c) Hydroxide ion is a stronger base than the hydrogen sulfide ion, HS^-. (d) Nitric acid is a strong acid, but nitrous acid is a weak acid. (e) Separation of boron from silicon is difficult.

21.73 How could you distinguish between the members of each of the following pairs of gases? (a) nitrogen and oxygen (b) carbon monoxide and carbon dioxide (c) sulfur dioxide and carbon dioxide (d) nitric oxide and nitrogen dioxide

21.74 (a) For the silicon atom, write the complete and abbreviated electron configurations and the orbital diagram. (b) What explanation is usually given for the fact that silicon forms four identical covalent bonds in most of its compounds?

21.75 How many kilojoules of thermal energy are released when 672 g SO_2 are formed by the combustion of sulfur?

21.76 Explain why: (a) Carbon dioxide is a gas, but silicon dioxide is a high melting solid. (b) The reaction between zinc and nitric acid does not yield hydrogen gas. (c) Many compounds exist with sulfur in positive oxidation states, but only a few compounds with oxygen in a positive oxidation state are known. (d) The normal boiling point of SiH_4 (-112 °C) is higher than the normal boiling point of methane (-164 °C), but the normal boiling point of phosphine (-88 °C) is lower than the normal boiling point of ammonia (-33 °C). (e) The element oxygen is more reactive than the element sulfur, but the element nitrogen is less reactive than the element phosphorus. (f) Carbon can't form more than four covalent bonds. (g) Solutions of sulfites develop a detectable concentration of sulfate ion on standing. (h) White phosphorus is much more reactive than red phosphorus. (i) $NaNO_3$ and KNO_3 occur naturally, but $NaNO_2$ and KNO_2 do not. (j) Nitrous acid can be both oxidized and reduced, but nitric acid can only be reduced.

21.77 (a) On a microscopic scale, what is the difference between crystalline forms of silica, SiO_2, such as quartz, and noncrystalline forms such as flint? (b) What type of solid (molecular, ionic, covalent network, or metallic) is quartz?

21.78 For the silicate ion, SiO_4^{4-}, (a) write the Lewis formula including formal charges. (b) What is the oxidation number of oxygen? Of silicon? (c) Use VSEPR to predict the shape and draw a stereochemical formula. (d) What is the hybridization of the silicon atom? (e) Draw stereochemical formulas for the silicon-containing anions in spudomene ($LiAlSi_2O_6$) and hardystonite ($Ca_2ZnSi_2O_7$).

21.79 Given the following half-reactions and standard reduction potentials for hydrogen peroxide in acidic solution,

$$H_2O_2(aq) + 2H^+ + 2e^- \longrightarrow 2H_2O(l) \qquad E° = 1.763 \text{ V}$$
$$O_2(g) + 2H^+ + 2e^- \longrightarrow H_2O_2(aq) \qquad E° = 0.685 \text{ V}$$

(a) should disproportionation of hydrogen peroxide be spontaneous in acidic solution? Explain your answer. (b) Calculate the value of $E°$ for the disproportionation. (c) Write the net ionic equation for the reaction of $H_2O_2(aq)$ and Cr^{3+}. In this reaction, is hydrogen peroxide acting as an oxidizing agent or as a reducing agent? (d) Write the net ionic equation for the reaction of $H_2O_2(aq)$ and $Cr_2O_7^{2-}$. In this reaction, is hydrogen peroxide acting as an oxidizing agent or as a reducing agent? (e) What is the predominant solute species in basic solutions of hydrogen peroxide?

APPLICATIONS

21.80 The size of gemstones is measured in carats (1 carat = 0.200 g). (a) The Cullinan diamond had a mass of 605 g. How many carats was the Cullinan? (b) The density of diamond is 3.514 g/cm³. What was the volume of the Cullinan? If the Cullinan was a cube, how long would each side have been? (c) What is the mass in grams of a 5.0-carat diamond?

21.81 Calcium monohydrogen phosphate is practically insoluble in water and is used as an abrasive in toothpaste. (a) Write net ionic, complete ionic, and molecular equations for the synthesis of calcium monohydrogen phosphate by a precipitation reaction. (b) The chief use of calcium monohydrogen phosphate is as a source of calcium in animal feeds. This use depends on the fact that calcium monohydrogen phosphate is soluble in acids. Explain why calcium monohydrogen phosphate is soluble in acids and write an equation for the dissolving of calcium monohydrogen phosphate by dilute hydrochloric acid.

21.82 Write molecular equations for each of the following reactions: (a) Iron(III) oxide and hydrogen gas, which is used to make small quantities of pure iron. (b) Calcium hydride and water, which is used to make hydrogen for inflating life rafts and weather balloons. (c) Iron metal and hydrochloric acid, which is used to make small quantities of hydrogen gas in the laboratory. The other product is iron(II) chloride. (d) Aluminum metal and potassium hydroxide, which was used to make hydrogen gas for observation balloons from scrapped airplane parts during World War II. The other product is a solution of $KAl(OH)_4$ (potassium aluminate).

21.83 The phosphorus content of commercial phosphate rock is usually expressed as BPL [bone phosphate of lime, $Ca_3(PO_4)_2$]. If the phosphorus content of a sample of phosphate rock is 71% expressed as BPL, (a) what is the phosphorus content expressed as percent P_2O_5? (b) Expressed as percent P?

21.84 Water from some springs contains hydrogen sulfide, which is very toxic. The hydrogen sulfide is removed by treating the water with chlorine gas; sulfur precipitates and can be separated by filtration. Calculate the number of milligrams of Cl_2 required to remove the hydrogen sulfide from the 1250 mL of water that an average adult needs to drink each day if the water contains 59 mg H_2S/L.

21.85 Reaction of a mixture of equal masses of methylhydrazine, CH_3NHNH_2, and dimethylhydrazine, $(CH_3)_2NNH_2$, which are both liquids, with N_2O_4(l) was used to decelerate the Apollo lunar modules on landing. The products were nitrogen gas, water vapor, and carbon dioxide gas. (a) Write equations for both reactions. (b) If 3.0 tons of the methylhydrazine–dimethylhydrazine mixture were used, how many tons of each of the products were formed, assuming that enough N_2O_4 was used to react with all of the hydrazines? (c) If N_2O_4 were the limiting reactant, could part (b) be solved? Explain your answer.

21.86 Hydrazine, H_2NNH_2, is used to remove O_2 from water for high-pressure boilers in power stations by the reaction

$$N_2H_4(aq) + O_2(aq) \longrightarrow N_2(aq) + 2H_2O(l)$$

The usual concentration of O_2 in boiler feedwater is about 0.01 ppm by mass. Twice the stoichiometric amount of hydrazine is used. (a) How many grams of N_2H_4 are needed to treat 12 500 tons (1.135×10^7 kg) of water, a day's supply? (b) Hydrazine hydrate, $N_2H_4 \cdot H_2O$, is usually used instead of anhydrous hydrazine because it is cheaper. What is the mass percent hydrazine in hydrazine hydrate? (c) How many grams of hydrazine hydrate are needed to treat a day's supply of water?

21.87 Many important minerals are aluminosilicates—compounds in which silicon has been replaced by aluminum. Aluminosilicates containing Na, K, or Ca are called feldspars and make up about 60% of Earth's crust. The feldspar orthoclase may be considered to be formed by replacement of one-fourth of the silicon atoms of SiO_2 with Al and maintenance of charge balance with K^+ ions. What is the empirical formula for orthoclase, which is used in the manufacture of porcelain?

21.88 The three compounds most often used as N-containing fertilizers are ammonia, urea (H_2NCONH_2), and ammonium nitrate. (a) Calculate the percent by mass nitrogen in each of these compounds. (b) If all of the nitrogen in a plant food that contains 15.0% N comes from urea, what percentage of the plant food is urea?

21.89 The buoyant force on an object is equal to the mass of the fluid (liquid or gas) displaced by the object. This summary of experimental observations is known as Archimedes's principle (after its discoverer) and was discovered in the third century B.C. (a) Assuming that air is 78.08% nitrogen gas, 20.95% oxygen gas, and 0.934% argon gas by volume, calculate the mass of 22.4 L of air at STP. (b) What is the mass of 22.4 L of hydrogen gas at STP? (c) The buoyant force on a 22.4-L balloon filled with hydrogen gas is approximately

equal to the difference between the mass of 22.4 L of air and the mass of 22.4 L of hydrogen gas (at same temperature and pressure). What is the buoyant force on a 22.4-L balloon filled with hydrogen gas? On a 22.4-L balloon filled with helium gas? (d) Give one reason why hydrogen is better for lifting balloons than helium. (e) What are two disadvantages of hydrogen compared to helium?

21.90 Like silicon, germanium is an important semiconductor. Germanium is obtained from the flue dusts of smelters for zinc ores and is separated from zinc by conversion to germanium tetrachloride, which is purified by fractional distillation. The purified germanium tetrachloride is hydrolyzed to germanium dioxide (and hydrochloric acid), and the germanium dioxide is reduced to germanium with hydrogen. Germanium for semiconductors is purified by zone refining. (a) Write the molecular equation for each reaction that is involved in the production of germanium. (b) Discuss the process of zone refining.

21.91 A diver working at a depth of 175 ft where the pressure is 6.06 atm and the temperature is 40 °F (4.4 °C) should breathe a mixture of helium and oxygen in which the partial pressure of O_2 is the same as at the surface, 0.21 atm. (a) What should the partial pressure be of the He in the mixture? (b) What percent by volume of the mixture should be O_2 and what He? (c) What is the density in grams per liter of a mixture that has this composition at STP? (d) What is the density at 40 °F (4.4 °C) and 6.06 atm?

21.92 Dilute gaseous nitrogen trichloride is used to bleach and sterilize flour. (a) Write the equation for the formation of nitrogen trichloride gas from nitrogen and chlorine. (b) The bond energy of the N—Cl bond is 188 kJ/mol. Use bond energies to estimate ΔH_f° for nitrogen trichloride. (c) Explain why the first person to prepare nitrogen trichloride lost three fingers and an eye studying its properties.

21.93 The Great Lakes make the area surrounding them cooler in the summer and warmer in the winter because of their ability to absorb thermal energy in the summer and release it in the winter. The total volume of water in the Great Lakes is 23 000 km³ (2.3×10^4 km³). How many kilojoules are required to raise the temperature of the water in the Great Lakes 1.00 °C?

21.94 The total volume of ice in polar ice and glaciers is 2.82×10^7 km³. The density of ice at 0 °C is 0.917 g/cm³, and the heat of fusion of ice is 6.01 kJ/mol. Calculate the number of kilojoules of thermal energy required to melt all the ice in polar ice and glaciers.

21.95 Citrus groves are sometimes sprayed with water to prevent damage by freezing weather. Explain how a thin coating of water keeps the trees and fruit from freezing.

21.96 For the combustion of H_2(g) and CH_4(g), calculate the following: (a) How many kilojoules are released per kilogram of fuel burned? (b) How many kilojoules are released per liter (STP) burned? (c) Which is the better fuel on a mass basis? (d) Which is the better fuel on a volume basis?

21.97 If a H_2-O_2 fuel cell such as the one shown in Figure 19.16 is operated at 105 °C so that the water formed is gaseous, how many grams of H_2(g) and how many grams of O_2(g) must react each second to provide power of 2.00 kJ/s (2.00 kW)? Remember that the free energy lost by a system during a spontaneous change is a measure of the maximum amount of energy that is free to do work (Section 17.11). Assume that the efficiency of the fuel cell is 75.0%—that is, that 75.0% of the free energy released is actually available to do useful work.

21.98 By means of radio astronomy, more than 70 molecules have been detected in the dark clouds in interstellar space where stars are formed. Three of these molecules are nitroxyl, HNO; C_3O; isocyanic acid, HNCO (formulas are condensed structural formulas). (a) What is the name of C_3O? (b) Write Lewis formulas for each of the three molecules. (c) Write the Lewis formula for isothiocyanic acid, which also occurs in the dark clouds in interstellar space. (d) The densest of the dark clouds contain about 10^6 particles per cubic centimeter. (At sea level on Earth, there are about 10^{19} molecules per cubic centimeter.) What is the volume of 1 mol of the particles in the dark cloud? What is the length in meters of each side of a cube having this volume? (e) If the temperature in the dense dark cloud in part (d) is 20 K, what is the pressure in atmospheres?

21.99 Ice, dry ice, and liquid nitrogen are all used for refrigeration. For each, (a) what phase change is involved? (b) What is the reverse phase change called?

21.100 Water spreads through soil by capillary action. (a) What is *capillary action*? (b) On what properties of water does capillary action depend?

21.101 A volcano's potential for exploding depends on the time since it last erupted and on the viscosity of its lava. Gases escape readily from less viscous lavas, but gas pressure builds up in viscous lavas. The viscosity of lava depends on temperature and composition; other factors being equal, the higher the SiO_2 content of a lava, the greater is its viscosity. (a) What does *viscosity* mean? (b) Explain why a high SiO_2 content results in a very viscous lava. (c) How and why does the viscosity of lava vary with temperature?

21.102 The hardness of water is often expressed as $[Ca^{2+}]$ or as mg $CaCO_3$ per liter of water. A 1.00-L sample of water was passed through a column of cation exchange resin (Figure 21.10) and the H^+ was titrated with 0.0500 M NaOH; 23.29 mL were required. (a) What was the hardness of the water expressed as $[Ca^{2+}]$? (b) What was the hardness expressed as mg $CaCO_3$ per liter of water? (c) Kits for testing water express hardness in grains; 1 grain = 0.065 g $CaCO_3$ per gallon of water. What was the hardness of the water in grains?

21.103 The concentration of gold in seawater is 4×10^{-6} mg/L. (a) How many kilograms of gold are there in the oceans? (b) How many liters of seawater must be processed to obtain 1.0 kg of gold?

21.104 (a) Why isn't phosphorus from $Ca_3(PO_4)_2$ in phosphate rock available to plants? (b) How can the phosphorus be made available?

21.105 A concentration of 1 ppm (by mass) Cl_2 in water kills bacteria without harming people. What is this concentration in molarity?

21.106 The osmotic pressure of a sample of seawater is 27 atm at 25 °C. (a) What is the minimum pressure needed for reverse osmosis to make freshwater from this seawater? (b) What is the total molarity of the solutes in the seawater? (c) Estimate the freezing point of the seawater. (d) How does the equilibrium vapor pressure of water over seawater compare with the vapor pressure of water over pure water at the same temperature?

21.107 Both $CaCl_2$ and NaCl are spread on roads in the winter to melt ice. (a) To obtain the same freezing point lowering, how does the mass of $CaCl_2$ required compare with the mass of NaCl required? How do the transportation and labor costs of using $CaCl_2$ compare with the transportation and labor costs of using NaCl? (b) If the cost of a given mass of $CaCl_2$ is 1.5 times as great as the cost of the same mass of NaCl, how does the cost of buying $CaCl_2$ compare with the cost of buying NaCl?

21.108 Minerals deposited more than 2×10^9 years ago contain iron(II) ion, whereas minerals deposited more recently contain iron(III) ion. What does this information suggest about the early atmosphere of Earth?

21.109 The reaction of ammonium perchlorate and powdered aluminum in the booster rockets provides the energy to launch the space shuttle. (a) The products of the reaction of ammonium perchlorate and powdered aluminum are aluminum oxide, aluminum chloride, nitric oxide, and water. Write the molecular equation for the reaction. (b) In the reaction, which elements are oxidized, and which element is reduced? Which species is the oxidizing agent? The reducing agent? (c) Why is the aluminum powdered? (d) Ammonium perchlorate is made by a precipitation reaction between solutions of sodium perchlorate and ammonium chloride. Write net ionic, complete ionic, and molecular equations for this reaction. According to the solubility rules (Table 4.4), all sodium and all ammonium salts are soluble. What can you conclude about the relative solubility of ammonium perchlorate? Should dilute or concentrated solutions of sodium perchlorate and ammonium chloride be used? Explain your answer. (e) In 1989 the ammonium perchlorate in a plant manufacturing ammonium perchlorate for the space shuttle exploded. The products were nitric oxide, chlorine, oxygen, and water vapor. How many moles of gaseous products were formed from each mole of ammonium perchlorate that exploded?

21.110 Write molecular equations for the synthesis of each of the following hydrogen-containing compounds from hydrogen gas: (a) Ammonia, which is used as a fertilizer. (b) High-purity hydrogen chloride, which is used to decompose the bones that are the raw material for the manufacture of gelatin. The hydrogen chloride must be of high purity because the gelatin is used in foodstuffs. (c) Hydrogen bromide, which is used to manufacture other compounds.

21.111 (a) Explain how water extinguishes burning paper. (b) Explain why water can't be used to extinguish a sodium fire. (c) Explain why water can't be used to extinguish burning oil. (d) If your clothing catches fire, why should you lie down and roll instead of running to the nearest faucet? (e) Treatment of fabrics with a solution of ammonium bromide under high pressure makes the fabrics burn less rapidly. Explain why the fabrics must be treated again after washing.

21.112 In northern states, people who have swimming pools empty them in the winter. Why?

21.113 Selenium is used to decolorize glass and in photocopying. (a) Which of the elements discussed in this chapter does selenium resemble most closely? (b) By analogy to this element, would you expect selenium to have allotropes? (c) What would you expect the units of selenium crystals to be? (d) Write the formula expected for the hydride of selenium. (e) Is the hydride acidic, basic, amphoteric, or neutral? (f) Write the formula expected for the compounds of sodium, magnesium, and zinc with selenium. (g) Nonstoichiometric compounds of selenium with transition elements are common. What are nonstoichiometric compounds?

21.114 Epsom salts, which are used as an anticonvulsive and as a laxative, are a hydrate of magnesium sulfate. When a 5.43-g sample of epsom salts was heated above 250 °C for long enough to drive off all water, the residue had a mass of 2.65 g. (a) What is the formula for epsom salts? (b) What is the systematic name for epsom salts?

21.115 Exposure for one hour to a carbon monoxide concentration of 0.1% by volume in air is often fatal. (a) What is this concentration of carbon monoxide expressed in ppm? (b) How many grams of carbon monoxide are required to produce this concentration in a room that measures $13.0 \times 14.5 \times 8.0$ ft if the temperature is 70 °F (21.1 °C) and the barometric pressure is 755.0 mmHg? (c) Air is 21% by volume $O_2(g)$. What is the molarity of oxygen gas in air at 25 °C and one atmosphere total pressure? (d) What is the molarity (at 25 °C and 1 atm total pressure) of CO in air that is 0.1% by volume CO? (e) At these molarities of O_2 and CO, what is the ratio $[O_2]/[CO]$? (f) At the ratio of $[O_2]$ to $[CO]$ found in part (e), what is $[Hb(CO)]/[Hb(O_2)]$ in blood? (g) What percentage of the hemoglobin in blood is tied up as $Hb(CO)$?

21.116 In the production of silicon from sand and coke, excess sand is used to prevent the formation of silicon carbide, SiC. Silicon carbide, which is commonly called carborundum, is almost as hard as diamond and is very strong and unreactive. About 10^9 lb of silicon carbide are made each year by the reaction of a slight excess of coke with sand for use as an abrasive to polish glass and granite and sharpen tools, as a refractory for furnace linings, and to make high-temperature semiconductors. (a) Why is the systematic name silicon carbide not carbon silicide? (b) What type of solid is silicon carbide? (c) Write the equations for the production of silicon carbide. An electric furnace at 2000–2500 °C is used to carry out the reaction, which takes place in two steps. Silicon formed in the first step reacts with carbon to form silicon carbide in the second step. (Approximate melting points in °C: SiO_2, 1600; Si, 1400; SiC 2700.) (d) If a "slight excess of coke" means a 10.0% excess of coke, how many kilograms of coke should be used per kilogram of sand? Assume that sand is pure SiO_2 and coke is pure C. (e) What element could be added to silicon carbide to make it an n-type semiconductor? A p-type semiconductor? How does charge move through an n-type semiconductor? A p-type semiconductor?

21.117 Calcium carbide, CaC_2, is an ionic solid. The reaction of calcium carbide with water is the major source of acetylene for the chemical industry and for oxyacetylene welding, although use of acetylene from petroleum is increasing especially in the chemical industry. (a) Explain why calcium and silicon form different types of carbides. (b) Write the electron configuration of the cation that is one unit of calcium carbide. (c) Write the Lewis formula including formal charges for the anion that is the other unit of calcium carbide. (d) What is the relationship between the anion and acetylene? (e) Write the molecular equation for the reaction of calcium carbide with water. (f) How many liters of dry acetylene at 20.0 °C and 1.00 atm will be formed by the reaction of 1.00 kg of calcium carbide with 1.00 kg of water? (g) What is $\Delta H°_{rxn}$? (h) How much of the thermal energy released will be used up heating unreacted water from 25.0 to 149.0 °C? (i) The cations in calcium carbide form a face-centered cubic structure with the anions in holes similar to the sodium chloride structure. What types of holes are the anions in? (j) Sketch the crystal structure of calcium carbide showing the anions as being vertical. (k) How many cations and how many anions are there in a unit cell of calcium carbide?

Biorganometallic Chemistry: Working at a Scientific Interface

FRANK A. GOMEZ

Assistant Professor of Chemistry
California State University, Los Angeles

B.S. Chemistry
California State University, Los Angeles

Ph.D. Chemistry
University of California, Los Angeles

My parents' love for teaching provided the impetus for a career in academics. Both secondary school teachers, their interest and dedication to teaching greatly influenced my decision. Their constant guidance and support bestowed upon me a desire to excel academically so I too could one day share this experience. They have always told me to strive to be the best I can possibly be. With this in mind, I have always tried to complement my research interests with a strong commitment to science education and teaching.

My interest in science was spawned by the chemistry and materials science-related courses I took while in high school. I attended a small private Catholic high school that was well suited to introduce young men to careers in science. I subsequently entered California State University, Los Angeles as an engineering major. As fate would have it, general chemistry was a requirement in the undergraduate engineering curriculum. I enjoyed the course so much that I changed my major. I was soon introduced to research by a wonderful teacher, Professor Thomas Onak. Our work focused on carbon- and boron-containing species called carboranes. The work was very interesting and provided me with my first real research experience. Professor Onak's enthusiasm for research and dedication to teaching were especially encouraging in helping me decide on a research and teaching career.

I enrolled in graduate school at the University of California, Los Angeles, where I did a great deal of organic and inorganic synthesis. My research advisor was a pretty hands-off kind of guy, which gave me the opportunity to really "play" in the lab. This playing should not be confused with what we commonly think of as playing; it is a chemist's way of saying that we worked hard, experimented a great deal, and had a lot of fun in the process. My research projects were varied and involved the synthesis of carborane-containing chelate systems called "venus flytrap ligands." Although the objective of the project was to synthesize compounds for radioimmunodiagnosis and radioimmunotherapy of cancer, I preferred the synthetic chemistry aspect of it more than anything else.

After graduation I went to Harvard University as a postdoctoral fellow to work with Professor George M. Whitesides. Up until then I had neither been involved nor had much of an interest in coupling chemistry to other fields of science like biology, medicine, and biochemistry. I soon saw the relationship. I owe a great deal of this "awakening" to my Harvard experience and my advisor. It was there that I was introduced to a new analytical technique called capillary electrophoresis (CE). CE is currently the fastest growing analytical technique in separation science and is an attractive technique for solving complex bioanalytical problems. While I was a postdoc I used CE to study protein-protein and protein-ligand interactions. We demonstrated CE to be a useful technique and applied it to several biological systems for which analysis by other techniques was difficult. CE is an alternative technique for measuring biomolecular noncovalent interactions and, in the future, will probably be the analytical technique of choice for evaluating biological problems.

My research group and I are currently working at the interface of several chemical areas, including synthetic inorganic chemistry, protein and peptide chemistry, and analytical chemistry. We call it bioorganometallic chemistry. The broad objective of our work is to examine the synthesis and molecular recognition properties of bioorganometallic complexes of peptides and proteins by the use of CE and other analytical techniques. This new field utilizes the diversity of organometallic materials that can now be syntheized for use in the solving of biological problems.

I have only been a college professor for a year but have found it a most exhilarating experience. Teaching has been challenging. They say "you never really know and understand the material until you have to teach it." Is that an understatement! But it is fun. I also enjoy the students that work in my research lab. It is a treat to see young eyes discover science through research. It reminds me of when I was in a similar boat; I know how they must be feeling.

Something else plays a big role in my life that I feel aids me in my work: my proud heritage. As a Mexican-American I feel an added responsibility to succeed and excel in chemistry. Latinos have historically been underrepresented in the science fields. Overall, the educational success of minorities is quite discouraging but it is improving. I am very committed to improving the state of education in the United States, especially where the Latino community is concerned. I currently sit on the board of directors of the Society for Advancement of Chicanos and Native Americans in Science (SACNAS) whose main goal is to increase minority representation in all sectors of science. As an academician I continue to help develop science policy for minorities in the hopes of encouraging more Latinos to consider a science career. Chemistry is fun, which is why people of all colors should be exposed to the feeling.

A CLOSER LOOK AT ORGANIC CHEMISTRY

Photo shows a star shell (Astraea Undosa) with a clockwise spiral.

Organic chemistry can be divided into three interrelated areas: determination of structure, reactions and synthesis, and the mechanisms by which reactions take place. Organic chemistry is too large a field to be summarized in one chapter. Instead, this chapter tries to give you an idea of what modern organic chemistry is like by considering typical examples of reactions, synthesis, and mechanisms. (Structure determination is discussed in connection with proteins in Chapter 23.)

■ *Why is a study of organic chemistry a part of general chemistry?*

About 91% of the more than 13 million known compounds are organic. As we saw in Chapter 21, water makes up 60–90% of all living matter; however, of the remaining 10–40% of living matter, less than 4% is inorganic. The remainder is organic. Biochemistry, the chemistry of living matter, is based on organic chemistry. The clothes we wear, the houses we live in, the furnishings inside them, the fuels we burn to keep us warm, and the medicines we take to keep us well are all largely organic.

Annual U.S. production of ten organic compounds—ethylene ($H_2C\!=\!CH_2$), methyl *tert*-butyl ether [$CH_3OC(CH_3)_3$], propylene ($CH_3CH\!=\!CH_2$), ethylene dichloride ($ClCH_2CH_2Cl$), urea (H_2NCONH_2), vinyl chloride ($H_2C\!=\!CHCl$), benzene (⬡),

ethylbenzene ($CH_3CH_2\!-\!$⬡),

methanol (CH₃OH), and styrene

$(H_2C{=}CH{-}\langle\!\!\!\bigcirc\!\!\!\rangle)$—is over 10

billion pounds each. The production of another 19 organic compounds is greater than 1 billion pounds per year.

— What do you already know about organic chemistry?

All organic compounds contain carbon. The name *organic* was given to organic compounds because, before 1828 when Wöhler made the organic compound urea, H_2NCONH_2, from the inorganic compound ammonium cyanate, NH_4OCN, people thought that all organic compounds came from plants and animals (Section 1.10).

Organic compounds are composed largely of nonmetals. As a result, organic compounds are usually molecular, and the atoms in organic compounds are generally held together by covalent bonds.

The properties of organic compounds depend on their structures. For example, ethyl alcohol, structural formula CH_3CH_2OH, is a liquid used in alcoholic

beverages and as a solvent in industry. Methyl ether, structural formula CH_3OCH_3, which is made of the same atoms arranged differently, is a gas used in refrigeration. Compounds that have the same molecular formula but different structures, such as ethyl alcohol and methyl ether, are called isomers (Section 3.9). The structures of substances can be shown by models as well as by structural formulas (remember Figure 3.4).

The shapes of species that contain carbon can be predicted by VSEPR (Section 10.1). Shape is a factor in London forces (Section 12.2) and therefore in boiling points and melting points (remember Figure 12.14).

The hybrid orbital model (Section 10.4) is one way of explaining the shapes of species that contain carbon. According to the hybrid orbital model, the carbon in methane (CH_4), which has only single bonds, is pictured as sp^3 hybridized [Figure 10.11(a)], the doubly bonded carbons in ethylene ($H_2C{=}CH_2$) are pictured as sp^2 hybridized (Figure 10.13), and the triply bonded carbons in acetylene (HC≡CH) are pictured as sp hybridized

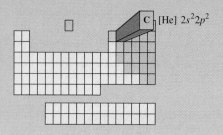

[Figure 10.19(b)]. Rotation around single bonds is free; rotation about double bonds is not possible because it destroys the sidewise overlap between *p* orbitals, as shown in Figure 10.16.

This cactus (Aconium tabulaeforme) *has a counterclockwise spiral. Handedness is characteristic of living creatures and plants.*

22.1   OCCURRENCE OF ORGANIC COMPOUNDS

Petroleum, natural gas, and coal are mixtures of organic compounds, and all living matter contains carbon compounds. Although carbon is not among the ten most abundant elements in Earth's crust, surface water, and atmosphere (Table 20.9), carbon is the third most abundant element in the human body in terms of number of atoms and the second most common in terms of mass. The composition of the human body is shown in Table 22.1. The ten elements listed account for 99.8% of the human body. The other 0.2% is made up of a number of elements that are called

Information about the essential elements is from Fraústo da Silva, J. J. R.; Williams, R. J. The Biological Chemistry of the Elements: The Inorganic Chemistry of Life; Clarendon: Oxford, 1991, pp 3–5 and Huheey, J. E.; Keiter, E. A.; Keiter, R. L. Inorganic Chemistry, 4th ed.; HarperCollins: New York, 1993; pp 941–954.

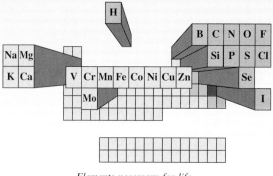

Elements necessary for life.

TABLE 22.1   Composition of the Human Body

Element	Percent by Mass	Role in the Body
O	64.6	Required for breathing. Present in water and many organic and inorganic compounds.
C	18.0	Present in all organic compounds and some inorganic compounds.
H	10.0	Present in water, many other inorganic compounds, and almost all organic compounds.
N	3.1	Present in proteins and nucleic acids.
Ca	1.9	Major part of bone and teeth.
P	1.1	Needed for energy transfer. Present in nucleic acids and a major part of bone and teeth.
Cl	0.4	Major anion outside cells.
K	0.4	Major cation inside cells.
S	0.2	Present in most proteins.
Na	0.1	Major cation in blood plasma.

Although the mass percent hydrogen in the human body is less than the mass percents of oxygen and carbon, there are about 6.6 times as many hydrogen atoms as carbon atoms (and 2.7 times as many oxygen atoms as carbon atoms).

essential elements because they *are necessary for life.* Boron, fluorine, magnesium, silicon, vanadium, chromium, manganese, iron, cobalt, nickel, copper, zinc, selenium, molybdenum, and iodine are essential elements. Only two elements that are abundant on or near Earth's surface are not essential—Al and Ti. Both of these elements form oxides that are very insoluble at the pH values existing in people. The amount of an essential element needed varies from a few hundredths of a percent for iron to less than one part in a million for molybdenum. The roles of some essential elements are not known. Some elements, such as selenium, which are essential at low concentrations, are toxic at higher concentrations.

22.2   HYDRIDES OF CARBON

Carbon forms an enormous number of compounds with hydrogen. The key feature of the chemistry of carbon is **catenation,** the *formation of chains of atoms of the same element.* For example, in ethane (C_2H_6), two carbon atoms are joined together and in propane (C_3H_8), three:

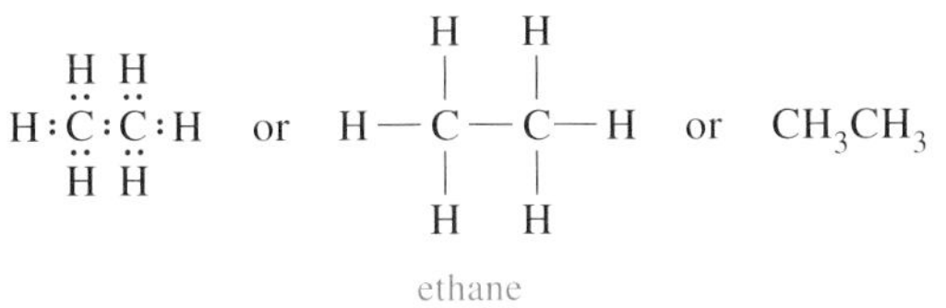

ethane

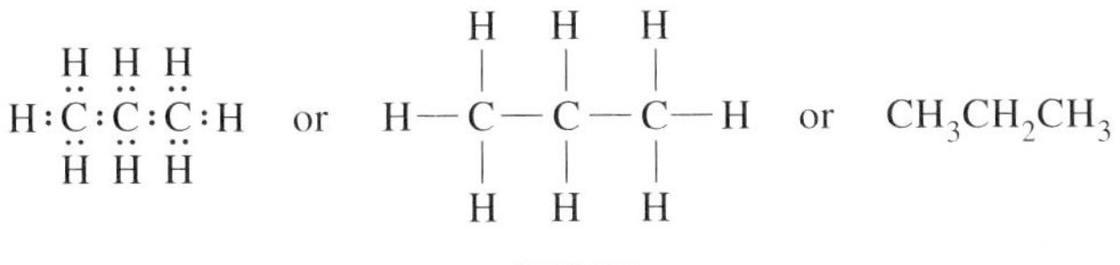

propane

There does not appear to be any limit to the number of carbon atoms that can be joined together in a chain. In polyethylene (Section 9.11), as many as 7000 carbons are linked.

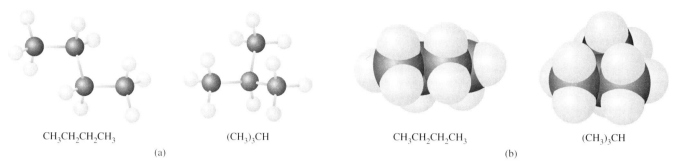

CH₃CH₂CH₂CH₃

(a)

(CH₃)₃CH

CH₃CH₂CH₂CH₃

(b)

(CH₃)₃CH

■ FIGURE 22.1   (a) Ball-and-stick models of butane and isobutane. The angles between bonds to carbon are 109.5°, not 90° as you might think from the complete structural formulas. (b) Space-filling models of butane and isobutane.

○ Remember that different compounds that have the same molecular formula are called isomers (Section 3.9).

Only one compound with molecular formula C_2H_6 and one compound with molecular formula C_3H_8 exist. However, there are two compounds with molecular formula C_4H_{10}. One boils at -0.5 °C and the other at -12 °C. Four carbons and 10 hydrogens can be put together in two ways:

$$H-\overset{\overset{\displaystyle H}{|}}{\underset{\underset{\displaystyle H}{|}}{C}}-\overset{\overset{\displaystyle H}{|}}{\underset{\underset{\displaystyle H}{|}}{C}}-\overset{\overset{\displaystyle H}{|}}{\underset{\underset{\displaystyle H}{|}}{C}}-\overset{\overset{\displaystyle H}{|}}{\underset{\underset{\displaystyle H}{|}}{C}}-H \quad \text{and} \quad H-\overset{\overset{\displaystyle H}{|}}{\underset{\underset{\displaystyle H}{|}}{C}}-\overset{\overset{\displaystyle H}{|}}{\underset{\underset{\displaystyle H-\overset{\overset{\displaystyle H}{|}}{\underset{\underset{\displaystyle H}{|}}{C}}-H}{|}}{C}}-\overset{\overset{\displaystyle H}{|}}{\underset{\underset{\displaystyle H}{|}}{C}}-H$$

CH₃CH₂CH₂CH₃ or CH₃(CH₂)₂CH₃
butane
boiling point, −0.5 °C

(CH₃)₃CH
isobutane (methylpropane)*
boiling point, −12 °C

The substance that boils at -0.5 °C (butane) is said to have a **continuous** or **straight chain.** The chain in the substance that boils at -12 °C (isobutane) is called a **branched chain.** ■Figure 22.1(a) shows ball-and-stick models of these two compounds so that you can see the difference in bonding easily and Figure 22.1(b) shows space-filling models. There are three isomers of C_5H_{12}.

Why is the boiling point of $CH_3(CH_2)_3CH_3$ higher than the boiling point of $CH_3(CH_2)_2CH_3$? Why do the boiling points of the isomers of C_5H_{12} increase in the order $(CH_3)_4C < (CH_3)_2CHCH_2CH_3 < CH_3(CH_2)_3CH_3$?

CH₃(CH₂)₃CH₃
pentane
boiling point, 36 °C

(CH₃)₂CHCH₂CH₃
isopentane (methylbutane)
boiling point, 28 °C

(CH₃)₄C
neopentane (dimethylpropane)
boiling point, 9.5 °C

The number of isomers that is possible increases rapidly as the number of carbon atoms in the molecule increases. For example, 75 different structures can be written

*The first name given, isobutane, is the common name. The name in parentheses is the IUPAC name.

for $C_{10}H_{22}$ and 62 481 801 147 341 for $C_{40}H_{82}$. Not all structures that can be written are possible because of the overcrowding that occurs in very branchy structures; for example, $C[C(CH_3)_3]_4$ has never been synthesized. However, even after allowing for the nonexistence of overcrowded molecules, the total number of possible organic compounds is still enormous.

Since only 13 million compounds are known, obviously not all of the isomers of $C_{40}H_{82}$ have been made!

For information about number of isomers, see Davies, R. E.; Freyd, P. J. J. Chem. Educ. **1989**, 66, 278–281.

In addition, carbon can share electrons with other elements such as oxygen, nitrogen, the halogens, and sulfur as well as with carbon and hydrogen. A few familiar examples are ethyl alcohol, CH_3CH_2OH, chloroform, $CHCl_3$, amphetamine, $C_6H_5CH_2CH(CH_3)NH_2$, and the compound $(CH_3)_2CHCH_2CH_2SH$, which is the chief defensive weapon of the skunk. There can be more than one atom other than carbon and hydrogen in a molecule. A simple example is ethylene glycol, $HOCH_2CH_2OH$, the major component of the antifreeze/coolant used in automobile radiators.

Further opportunities for more carbon compounds are introduced by the possibility of joining the ends of a chain together to form a ring, for example,

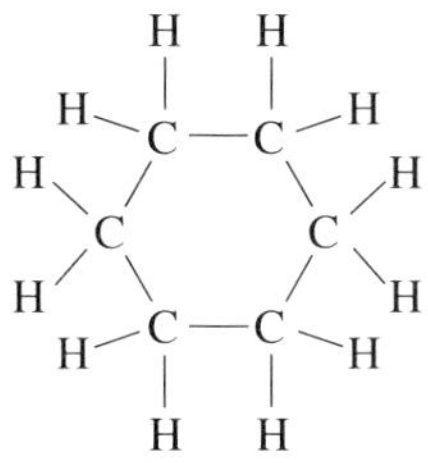

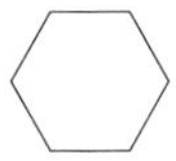

C_6H_{12}
cyclohexane

Cyclohexane is used to make nylon.

Line formulas like the one in the middle are often used in organic chemistry. In line formulas, lines represent bonds. Carbon atoms are assumed to be present at the end of each line segment, for example, at each corner of the hexagon. Each carbon is assumed to have enough hydrogen atoms attached to it to make a total of four bonds to each carbon. Nitrogen, oxygen, and halogen atoms are shown, but hydrogens are not shown unless bonded to an atom that is shown:

Remember the abbreviated resonance structures that we wrote for benzene in Section 9.9.

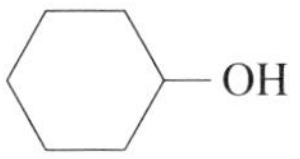

$C_6H_{11}OH$
cyclohexanol

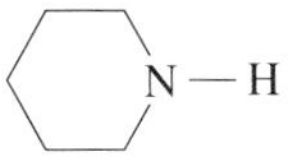

$C_5H_{10}NH$
piperidine

Cyclohexanol is an intermediate in the production of nylon. Piperidine is found in small quantities in black pepper.

PRACTICE PROBLEM

22.1 (a) Write the line formula for

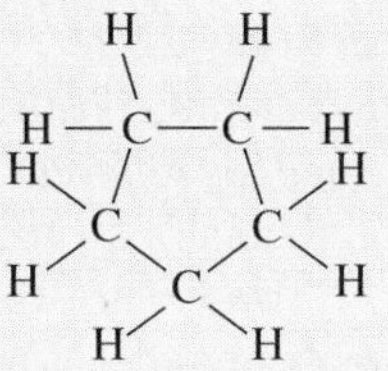

(b) Write complete structural formulas for 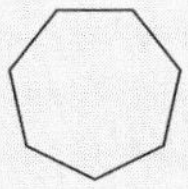 and —OH.

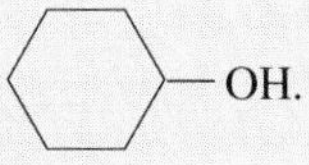

Another key feature of the chemistry of carbon that provides additional chances for more carbon compounds is the ability of carbon to form double and triple bonds.

A workman using an acetylene torch.

TABLE 22.2			Bonding Possibilities for Elements Commonly Found in Organic Compounds
Element	Number of Bonds	Number of Unshared Pairs	Bonding Possibilities
C	4	0	$-\overset{\|}{\underset{\|}{C}}-,\ \diagdown C=,\ -C\equiv,\ =C=^a$
H	1	0	$-H$
Cl^b	1	3	$-\overset{..}{\underset{..}{Cl}}:$
N	3^c	1	$-\overset{\|}{N}-,\ \ .\overset{..}{N}=,\ \equiv N:$
O	2	2	$-\overset{..}{\underset{..}{O}}-,\ =\overset{..}{O}:$

[a] This type is relatively uncommon and unimportant.
[b] The other halogens (Br, F, and I) are similar.
[c] In the ammonium ion, NH_4^+, and ions derived from it such as $CH_3NH_3^+$, N forms four covalent bonds.

Examples of carbon compounds with double and triple bonds are

$$\overset{H}{\underset{H}{\diagdown}}C=C\overset{H}{\underset{H}{\diagup}} \qquad HC\equiv CH \qquad \text{and} \qquad \overset{\displaystyle \overset{H}{\underset{H}{\diagup}}C\diagdown}{\underset{\displaystyle \overset{H}{\underset{H}{\diagdown}}C\diagup}{}}C=\overset{..}{O}:$$

$H_2C=CH_2$ $HC\equiv CH$ $(CH_3)_2C=O$
ethylene (ethene)   acetylene (ethyne)   acetone (propanone)

Ethylene is used to manufacture polyethylene

$$H_2C=CH_2(g) \xrightarrow{\text{catalyst, 100 °C, 7–20 atm}} -(CH_2-CH_2)_n\ (s) \qquad n = \text{about 50 to about 3500}$$

Many other organic chemicals, such as ethyl alcohol and ethylene glycol (antifreeze), are made from ethylene. In addition, ethylene is a plant hormone that ripens fruit. Acetylene is used as a fuel in oxyacetylene welding and cutting of metals. The combustion of acetylene produces the highest flame temperature (about 3300 °C) of any known gas. Acetylene is also a raw material for the manufacture of many organic chemicals and plastics. Acetone is an important solvent and is used as a starting material for the synthesis of a number of other organic compounds. The bonding possibilities for carbon and the other elements most commonly found in organic compounds are summarized in Table 22.2.

The isomers discussed in this section are all constitutional isomers; in **constitutional isomers** the *atoms are connected in different ways*. For example, in the isomer of C_4H_{10} that boils at -0.5 °C (butane), all four carbons are in a chain. In the isomer of C_4H_{10} that boils at -12 °C (isobutane), three carbons are connected in a chain; the fourth carbon is attached to the middle carbon of the three-carbon chain.

○ Constitutional isomers are sometimes referred to as structural isomers.

"Constitutional isomer" *is the preferred term in recent organic texts.*

Stereo- comes from Greek where it referred to three-dimensionality. You are probably most familiar with this prefix in stereophonic sound.

22.3   STEREOISOMERS

Constitutional isomers cannot be the only type of isomer because a number of pairs of compounds are known that have the same structural formula. For example, there are two kinds of 2-butene $CH_3CH=CHCH_3$ and two kinds of lactic acid,

CH₃CH(OH)COOH. Both kinds of lactic acid have identical physical properties except for the direction in which their solutions rotate the plane of polarized light. Both kinds of lactic acid are white, crystalline solids that melt at 53 °C and have $K_a = 1.6 \times 10^{-4}$. Under the same conditions, one kind of lactic acid rotates the plane of polarized light 2.6° clockwise (to the right) and the other rotates the plane of polarized light 2.6° in a counterclockwise direction (to the left).

Rotation of Plane of Polarized Light

What does "rotate the plane of polarized light 2.6° clockwise" mean? In a beam of ordinary light, the electromagnetic waves vibrate in all planes perpendicular to the direction in which the beam of light is moving. If the beam of light is passed through a polarizer, such as a lens from a pair of Polaroid sunglasses, only the waves that are vibrating in a single direction can get through [see ▪Figure 22.2(a)]. The *light beam with waves vibrating in just one direction* is said to be **plane-polarized.**

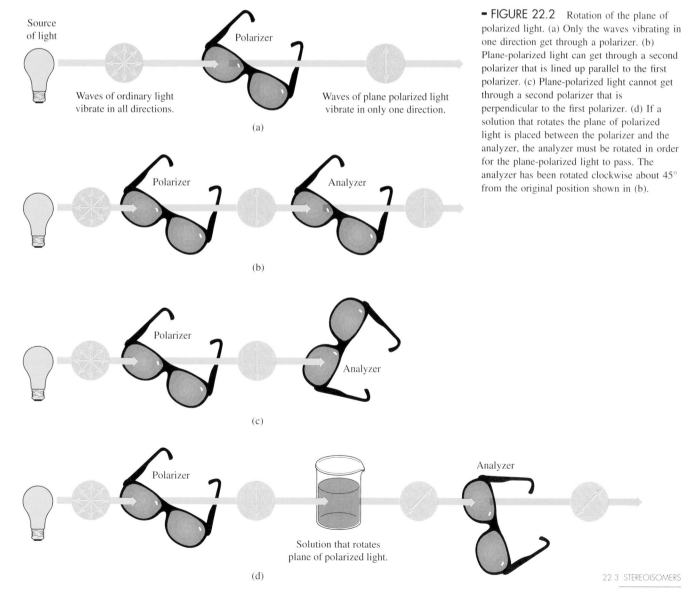

▪ FIGURE 22.2   Rotation of the plane of polarized light. (a) Only the waves vibrating in one direction get through a polarizer. (b) Plane-polarized light can get through a second polarizer that is lined up parallel to the first polarizer. (c) Plane-polarized light cannot get through a second polarizer that is perpendicular to the first polarizer. (d) If a solution that rotates the plane of polarized light is placed between the polarizer and the analyzer, the analyzer must be rotated in order for the plane-polarized light to pass. The analyzer has been rotated clockwise about 45° from the original position shown in (b).

Source of light

Waves of ordinary light vibrate in all directions.

Polarizer

Waves of plane polarized light vibrate in only one direction.

(a)

Polarizer

Analyzer

(b)

Polarizer

Analyzer

(c)

Polarizer

Solution that rotates plane of polarized light.

Analyzer

(d)

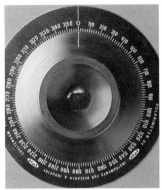

■ FIGURE 22.3 I²R student polarimeter. (a) When the polarimeter is in use, the lamp is positioned so that its light is reflected from the mirror at the bottom through an immovable polarizer, then lengthwise through the sample in the polarimeter tube, and finally through a rotatable polarizer with a dial at the top. (b) When the dial reads zero, the polarizers are perpendicular to each other and the filament of the light is not visible. (c) If a solution of sugar, which is optically active, is placed in the polarimeter, the plane of the light polarized by the immovable polarizer at the bottom is rotated and the filament of the light becomes visible. (d) Rotation of the dial 16° clockwise results in the rotatable polarizer being perpendicular to the plane of the light as rotated by the sugar solution, and the filament is again invisible. The rotation of the sugar solution is +16°.

If a second polarizer, which is called the analyzer, is placed in the beam of plane-polarized light, the plane-polarized light passes through the analyzer when the analyzer is lined up parallel to the polarizer [Figure 22.2(b)]. When the analyzer is perpendicular to the polarizer, no light passes through [Figure 22.2(c)].

If a solution that rotates the plane of polarized light is placed between the polarizer and the analyzer, the analyzer must be rotated to let light pass through [Figure 22.2(d)]. With a polarimeter, the angle through which the analyzer must be rotated can be measured. ■Figure 22.3 shows a simple student polarimeter.

The angle of rotation depends on the identity of the solute, the number of molecules of the solute in the light path (that is on the thickness of the sample and on its concentration), on the temperature, and on the wavelength of the light. *Materials that rotate the plane of polarized light* are said to be **optically active.** Most organic compounds that are obtained from plants and animals are optically active.

Enantiomers

The same van't Hoff was the first person to write equilibrium constant expressions and discovered the osmotic pressure law.

In 1874 the Dutch chemist van't Hoff and the French chemist Le Bel independently proposed that the four bonds formed by carbon are directed toward the corners of a tetrahedron. Soon afterward they realized that a tetrahedral carbon atom would explain the existence of pairs of isomers whose properties differ only in the direction of rotation of plane-polarized light. All the pairs of compounds that differ only in the direction of rotation of plane-polarized light that were known in 1874 contain a carbon atom with four different groups attached to it, as lactic acid does:

$$CH_3 - \overset{\overset{\displaystyle H}{|}}{\underset{\underset{\displaystyle OH}{|}}{C}} - COOH$$

This carbon has four different groups attached to it.

If four different atoms or groups are attached to a carbon atom in a species, two forms of the species exist that differ only in the arrangement in three-dimensional space of the attached atoms or groups. Using lactic acid as an example, the two

forms are

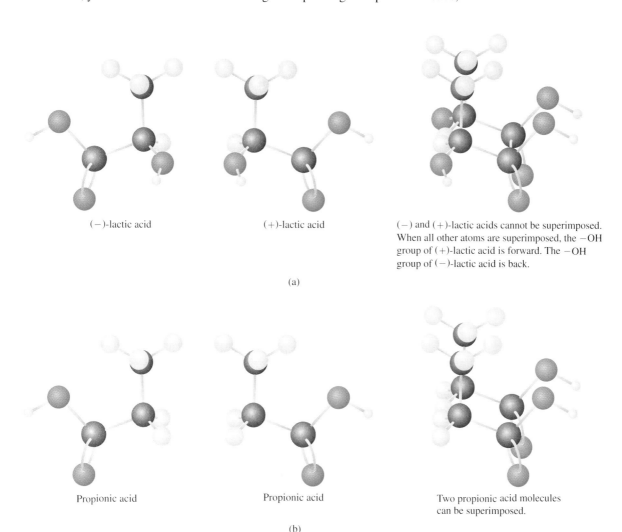

(−)-lactic acid   and   (+)-lactic acid

The symbol ($-$) indicates that the plane of polarized light is rotated counterclockwise, while ($+$) indicates that the plane of polarized light is rotated clockwise. X-ray crystallography is used to determine which isomer is represented by which stereochemical formula or model.

Models of ($-$)-lactic acid and ($+$)-lactic acid cannot be superimposed [see ►Figure 22.4(a)]. Therefore, the two forms of lactic acid are different compounds. (*You should use models to check the statements in this section.* If you do not have a model set, you can make models from small gumdrops using toothpicks for bonds;

(−)-lactic acid   (+)-lactic acid   (−) and (+)-lactic acids cannot be superimposed. When all other atoms are superimposed, the −OH group of (+)-lactic acid is forward. The −OH group of (−)-lactic acid is back.

(a)

Propionic acid   Propionic acid   Two propionic acid molecules can be superimposed.

(b)

► FIGURE 22.4   (a) Models of ($-$)- and ($+$)-lactic acid cannot be superimposed. There are two compounds $CH_3CH(OH)COOH$. (b) Models of propionic acid can be superimposed. There is only one compound CH_3CH_2COOH.

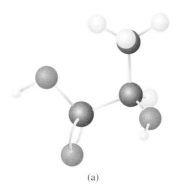

(a)

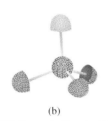

(b)

■ FIGURE 22.5 (a) Ball-and-stick model of (−)-lactic acid. (b) Gumdrop model. In the gumdrop model, the black gumdrop represents the carbon atom that has four different groups attached. Different colored gumdrops represent the four different groups. The green gumdrop represents the —COOH group, the yellow gumdrop the —CH₃ group, the purple gumdrop the H atom, and the red gumdrop the —OH group.

◐ Remember that in the stereochemical formulas used to show three-dimensional molecules on flat paper, ordinary lines show bonds that lie in the plane of the paper. Solid wedges show bonds that stick out toward the viewer and dashed lines show bonds directed away from the viewer behind the plane of the paper (Section 10.1).

Older terms for a stereocenter are chiral center *or* asymmetric atom.

See Mislow, K.; Siegel, J. J. Am. Chem. Soc. **1984,** 106, 3319–3328 *for the arguments in favor of using the term* stereocenter.

■Figure 22.5 shows models of (−)-lactic acid made from a commercial model set and from gumdrops and toothpicks.)

If all four atoms or groups attached to carbon are *not* different, only one form of the species exists. For example, if the —OH group in lactic acid is replaced by a hydrogen atom to give propionic acid, CH_3CH_2COOH,

$$HOOC-\overset{\overset{\displaystyle CH_3}{|}}{\underset{\underset{\displaystyle H}{|}}{C}}\cdots H \quad \text{and} \quad H\cdots\overset{\overset{\displaystyle H_3C}{|}}{\underset{\underset{\displaystyle H}{|}}{C}}-COOH$$

the two models can be superimposed [see Figure 22.4(b)]. There is only one compound CH_3CH_2COOH.

An atom with four different groups attached to it is called a **stereocenter.** Stereocenters are often marked with an asterisk

$$CH_3\overset{*}{C}H(OH)COOH$$

Sample Problem 22.1 shows how to identify a stereocenter.

SAMPLE PROBLEM

22.1 If a stereocenter is present, mark it with an asterisk.

(a) $CH_3CH_2\underset{\underset{\displaystyle CH_3}{|}}{C}HCH{=}CH_2$ (b) $CH_3\underset{\underset{\displaystyle CH_3}{|}}{C}HCH_2CH{=}CH_2$

Solution (a) A stereocenter must have four different groups attached to it. The carbon marked with an asterisk

$$CH_3CH_2\overset{*}{\underset{\underset{\displaystyle CH_3}{|}}{C}}HCH{=}CH_2$$

has four different groups, CH_3CH_2—, CH_3—, H—, and —CH=CH₂, attached to it. Note that the next carbon to the right of the stereocenter is not a stereocenter. The other ends of the bonds of a double or triple bond are attached to the same atom. Therefore, a doubly or triply bonded carbon can't be a stereocenter.
(b) No stereocenter is present in

$$CH_3\underset{\underset{\displaystyle CH_3}{|}}{C}HCH_2CH{=}CH_2.$$

No carbon has four different groups attached to it.

PRACTICE PROBLEM

22.2 If a stereocenter is present, mark it with an asterisk.

(a) $CH_3CH_2\underset{\underset{\displaystyle Cl}{|}}{C}HCH{=}CH_2$ (b) $CH_3\underset{\underset{\displaystyle Cl}{|}}{C}HCH_2CH{=}CH_2$

(c) $ClCH_2CH_2CH_2CH{=}CH_2$

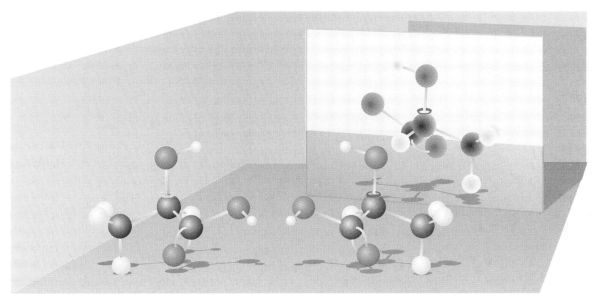

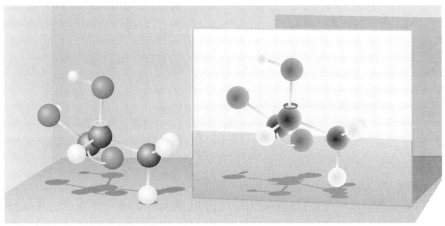

FIGURE 22.6 *Top:* Models of (−)-lactic acid (with yellow collar) and (+)-lactic acid (with blue collar) in front of mirror. *Bottom:* The model of (−)-lactic acid is identical with the image of (+)-lactic acid in the mirror. The computer-generated models in this book show H attached to O as being smaller than H attached to C because O is more electronegative than C. As a result, the electron density around H attached to O is less than the electron density of H attached to C.

Nature usually prefers one enantiomer over the other; for example, your liver is on your right side, and your stomach is on your left side. Most people are either right- or left-handed; few are ambidextrous. Seashells, plants, and bacteria display the property of handedness.

One isomer of lactic acid is the mirror image of the other isomer, just as your left hand is the mirror image of your right hand. ◾Figure 22.6 shows what is meant by saying that one isomer of lactic acid is the mirror image of the other isomer.

Mirror-image isomers are called **enantiomers** (optical isomers) and *objects that can exist as enantiomers* are said to be **chiral** (pronounced kī′ral). Some examples of chiral objects from everyday life are hands, feet, scissors, and screws. *Objects that are not chiral* are called **achiral;** balls, forks, and nails are achiral (see ◾Figure 22.7). The word *chiral* comes from the Greek word for "hand."

To convert one enantiomer to the other, bonds must be broken. Exchanging any pair of groups converts one enantiomer into the other:

$$HOOC-\underset{OH}{\overset{CH_3}{\underset{|}{\overset{|}{C}}}}\cdots H \longrightarrow HOOC-\underset{H}{\overset{CH_3}{\underset{|}{\overset{|}{C}}}}\cdots OH = H\cdots \underset{HO}{\overset{H_3C}{\underset{|}{\overset{|}{C}}}}-COOH$$

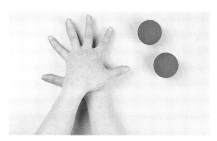

FIGURE 22.7 Hands are chiral. A left hand is a mirror image of a right hand, but left hands and right hands cannot be superimposed. The balls are achiral. The mirror image of one ball can be superimposed on the original ball.

An *arrangement of the parts of a molecule in space that is fixed by chemical bonding and can't be changed without breaking chemical bonds* is called a **configuration.**

A chiral probe must be used to distinguish between enantiomers. Plane-polarized light was the first chiral probe to be used and is still the one used most in chemistry laboratories. Your nose is a chiral probe and can distinguish between the enantiomers of carvone:

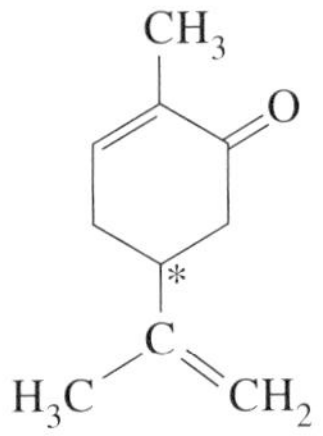

One enantiomer of carvone smells like caraway seeds, the other, like spearmint. Most biochemical reactions are chiral probes; biochemical reactions usually take place with only one of the two enantiomers. For example, Pasteur,* who discovered the existence of enantiomers in 1848, noted that microorganisms grow on one enantiomer of tartaric acid

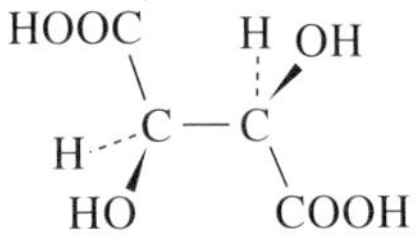

(+)-tartaric acid 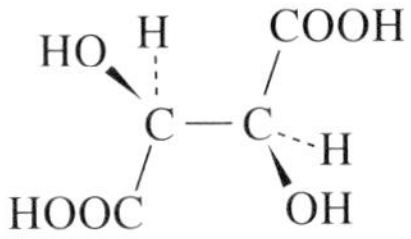   (−)-tartaric acid

but not on the other. The (−) enantiomer of dopa,

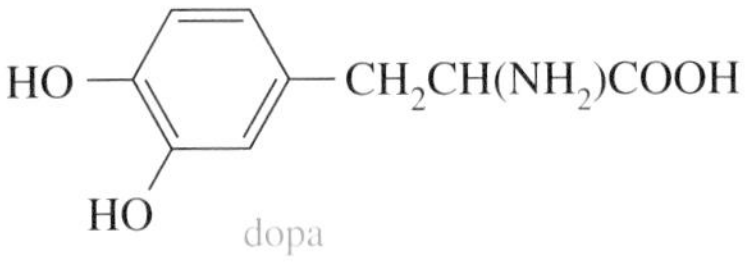

HO —⟨ ⟩— $CH_2CH(NH_2)COOH$

HO

dopa

which is commonly called L-dopa, is more effective for the treatment of Parkinson's disease than the (+) enantiomer. Parkinson's disease causes tremors and muscle rigidity in many elderly people. The (−) enantiomer of dopa is also less toxic than the (+) enantiomer.

The active sites of the biological catalysts called enzymes (Section 18.12) are chiral and can distinguish between enantiomers. Usually only one enantiomer fits into the chiral active site of an enzyme. The child's puzzle in ▪Figure 22.8 is a crude but simple analogy to the way one enantiomer fits into a chiral site but the other does not.

In mixtures of equal numbers of molecules or ions of each enantiomer, optical rotations cancel and no net rotation is observed. *Mixtures of equal numbers of molecules or ions of each enantiomer* are called **racemic mixtures.** As long as the environment in which a reaction takes place is not chiral, the probability of forming one enantiomer is the same as the probability of forming the other enantiomer. As a result, the products of ordinary laboratory and industrial syntheses are usually

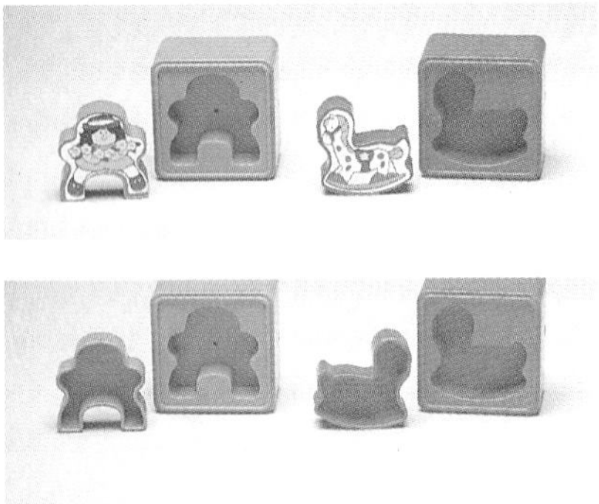

▪ FIGURE 22.8 The achiral figure on the left fits into its space even after it has been turned over. The rocking horse is chiral. The turned-over rocking horse, which is a mirror image of the right-side-up rocking horse, does not fit into the space in the orange block.

*Louis Pasteur was a French chemist and microbiologist. After graduating in chemistry with the equivalent of a C average, he discovered the existence of enantiomers when he was only 26 years old. Pasteur also proved that microorganisms (organisms too small to be seen except through a microscope) cause fermentation and disease, originated the process known as pasteurization (destruction of harmful microorganisms by heat), and originated and was the first to use vaccines for rabies, anthrax, and chicken cholera.

racic mixtures. On the other hand, products of biochemical reactions typically consist of a single enantiomer because enzymes are chiral. For example, all samples of lactic acid isolated from blood and muscle fluid are (+)-lactic acid.

The lactic acid in sour milk is a racemic mixture.

The use of racemic mixtures as drugs can cause problems. At best, the enantiomer of the pharmacologically active agent dilutes the active agent. At worst, the enantiomer may be harmful.

Diastereomers

Enantiomers are **stereoisomers**—that is, enantiomers are isomers that *differ only in the arrangement of atoms or groups in three-dimensional space. The order in which the atoms are connected to each other is the same.* However, not all stereoisomers are enantiomers. *Stereoisomers that are not enantiomers* are called **diastereomers.** The two kinds of 2-butene

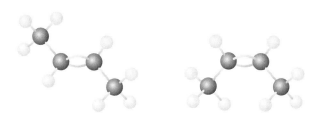

boiling point, ℃	0.9	3.7
melting point, ℃	−105.5	−138.9
density, g/mL	0.604	0.621

are diastereomers. Note that, unlike enantiomers, diastereomers have different physical properties; they also have different chemical properties.

The isomer on the left, which has the two methyl groups on opposite sides of the C=C bond, is called a ***trans*-isomer.** The isomer on the right, which has the two methyl groups on the same side of the C=C bond is referred to as a ***cis*-isomer.** *Cis-, trans*-isomers are often referred to as geometric isomers.

The stereoisomerism of 2-butene results from restricted rotation (Section 10.4). Rotation about single bonds is free but rotation about a double bond cannot take place unless the double bond is broken. Rings also restrict rotation so that stereoisomers such as the following exist:

▶ Remember that this type of stereoisomerism is the basis of the chemistry of vision (Section 10.4).

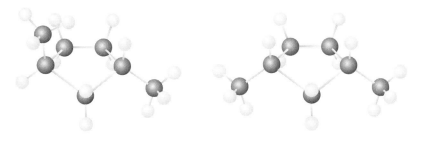

▶ You can tell that the two models represent different compounds because they are not superimposable.

PRACTICE PROBLEMS

22.3   Which of the two five-membered ring compounds is the *trans*-isomer?

22.4   Draw the structures of the two stereoisomers of BrCH=CHBr. Label the *cis*-isomer.

22.5   Do *cis*- and *trans*-isomers of $(CH_3)_2C=CH_2$ exist? Explain your answer.

IIIA	IVA	VA	VIA	VIIA
B–B 331	C–C 347	N–N 159	O–O 142	F–F 158
	Si–Si 192	P–P 197	S–S 268	Cl–Cl 244
	Ge–Ge 159	As–As 167	Se–Se 159	Br–Br 193
	Sn–Sn 151	Sb–Sb 142		I–I 151

■ FIGURE 22.9 Bond energies in kJ/mol. Data are from Jolly, W. L. *Modern Inorganic Chemistry,* 2nd ed.; McGraw-Hill: New York, 1991, pp 58 and 62.

◉ Remember that bond energy is the average amount of energy needed to break a covalent bond in a polyatomic molecule.

Thermodynamics predicts that people should react with the oxygen in the air! Fortunately for all of us, the activation energy is too high for reaction to take place under ordinary conditions.

A mill fire. Combustion of organic compounds is both spontaneous and rapid at high temperature.

WHY THERE ARE SO MANY CARBON COMPOUNDS

The enormous number of possible carbon compounds is one reason so many carbon compounds are known. Carbon's ability to form chains is the basis for the existence of a huge number of carbon compounds. However, other elements near carbon in the periodic table also catenate and compounds with B—B, Si—Si, Ge—Ge, Sn—Sn, N—N, P—P, O—O, S—S, Se—Se, and Te—Te bonds are known. Why aren't there millions of compounds of all these elements? These compounds are unstable and very reactive and do not last long whereas many compounds with C—C bonds exist for centuries under ordinary conditions.

The unusual stability of compounds with chains of carbon atoms is a result of the great strength of the C—C bond. The C—C bond is the second strongest single bond between like atoms after the H—H bond. ■Figure 22.9 shows bond energies (Section 9.12) for carbon and its neighbors in the periodic table. As you can see, the C—C bond is much stronger than the Si—Si, N—N, P—P, and most other bonds between like atoms.

Bond energy decreases from top to bottom of Group IVA. Bond formation requires good orbital overlap between adjacent atoms. Going down a group in the periodic table, the charge clouds become larger. The atoms cannot get as close together and bonds become longer and weaker. Silanes, hydrides of silicon, decompose at room temperature to silicon and hydrogen:

$$Si_4H_{10}(l) \longrightarrow 4Si(s) + 5H_2(g)$$

The only exception is SiH_4. The longest continuous chain silanes and germanes that have been made have fewer than ten atoms in the chain. The only catenated hydride of tin is Sn_2H_6 and no catenated hydride of lead is known.

The only stable catenated hydride of nitrogen is N_2H_4. A series of phosphanes, P_nH_{n+2}, up to $n = 6$ is known. However, the stability of the phosphanes decreases rapidly with increasing values of n. A number of boranes with B—B bonds are known but boranes are very reactive. Aluminum does not catenate. Thus, carbon is the only element that forms a very large number of stable, relatively unreactive hydrides.

The lack of reactivity of organic compounds is kinetic. For example, reaction of most organic compounds with oxygen in air is thermodynamically spontaneous under ordinary conditions, that is, ΔG is negative. The standard free energies of combustion of methane, ethyl alcohol, and acetic acid at 25 °C are typical:

$$CH_4(g) + 2O_2(g) \longrightarrow CO_2(g) + 2H_2O(l) \qquad \Delta G° = -817.90 \text{ kJ} \cdot \text{mol}^{-1}$$
$$CH_3CH_2OH(l) + 3O_2(g) \longrightarrow 2CO_2(g) + 3H_2O(l) \qquad \Delta G° = -1325.33 \text{ kJ} \cdot \text{mol}^{-1}$$
$$CH_3COOH(l) + 2O_2(g) \longrightarrow 2CO_2(g) + 2H_2O(l) \qquad \Delta G° = -873.1 \text{ kJ} \cdot \text{mol}^{-1}$$

Nevertheless, samples of methane, ethyl alcohol, and acetic acid can stand around for years without being oxidized by the oxygen in air.

The fact that combustion of organic compounds, although spontaneous, is usually slow at room temperature shows that activation energy is high. *No low-energy pathway for reaction of most carbon compounds exists.* Because carbon is in the second row of the periodic table with valence electrons in the second shell, a carbon atom can't have more than eight electrons around it. Only eight electrons can occupy the valence shell of second period elements. But carbon in organic compounds already has an octet of valence electrons. As a result, a new bond to carbon cannot form until an old bond breaks. Energy must be supplied to break the old bond before the new bond can form.

Carbon's neighbors in the periodic table can form new bonds without first breaking an old bond. Nitrogen and oxygen and the lower members of their groups all have unshared electron pairs in compounds. A new bond involving an unshared pair can form and provide energy to break an old bond. Silicon and the lower members of Group IVA can have more than eight electrons around them so that a new bond can form before an old bond breaks. Boron in boron hydrides is electron deficient; a new bond can form to boron providing energy to break an old bond. Activation energies for the reactions of catenated compounds of elements other than carbon are low, and the catenated compounds of other elements are very reactive. For example, silanes such as $SiH_3SiH_2SiH_2SiH_3$ (Si_4H_{10}) catch fire spontaneously or even explode in air:

$$2Si_4H_{10}(l) + 13O_2(g) \longrightarrow 8SiO_2(s) + 10H_2O(l)$$

Another factor that contributes to the lack of reactivity of carbon compounds is that the units of most carbon compounds are molecules. Molecules do not attract each other the way oppositely charged ions do. Moreover, many ions are spherical or approximately spherical; any collision results in reaction. But covalent bonds have definite length and direction in space. Collision must take place at the right spot for reaction to occur—just any old collision won't do. For example, for sodium metal to react with ethyl alcohol to form hydrogen gas,

$$2Na(s) + 2CH_3CH_2OH(l) \longrightarrow 2CH_3CH_2ONa(s) + H_2(g)$$

a sodium atom must hit an alcohol molecule on the end with the oxygen. Collision of sodium with the —CH_3 end of the alcohol molecule does not result in reaction:

Collision that can produce reaction.   Collision that can't produce reaction.

Besides being responsible for the existence of an enormous number of carbon compounds, the lack of reactivity of carbon compounds means that more time is needed to carry out experiments in organic chemistry laboratories than in general chemistry laboratories even though thermal energy and catalysts are often used to speed up reactions. Because they are predominantly covalent, many organic compounds have relatively low boiling points. Organic compounds are also usually flammable. Therefore, more elaborate equipment than a test tube is needed to contain them during a long reaction time. ▬Figure 22.10 shows a typical setup for carrying out an organic reaction.

In addition, organic molecules often undergo more than one reaction with a given set of reagents. Reactions other than the desired one occur leading to by-products. Time, materials, and equipment are required to remove the by-products and yield a pure product. Distillation, crystallization, extraction, and chromatography are often used to purify the crude products obtained from organic reactions.*

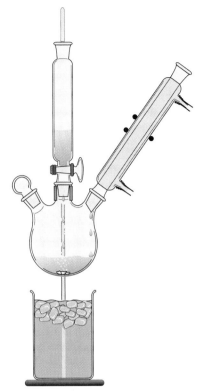

▬ FIGURE 22.10 Apparatus for carrying out an organic reaction. The reaction taking place in the flask releases enough thermal energy to boil the liquid in the flask. The condenser (right) condenses the vapors from the boiling liquid and returns the condensate to the flask. If the liquid starts to boil too vigorously, the ice bath can be raised to cool the contents of the flask. An addition funnel (center) is often used to add reagents gradually.

*Elaborate equipment and procedures are also needed to carry out many inorganic reactions, especially if air and moisture must be kept out. For example, as noted previously, silanes catch fire spontaneously or explode in air. Any reactions (except combustion) that involve silanes must be done in the absence of air. The inorganic reactions usually done in general chemistry labs are chosen because they are easy to carry out safely in simple equipment.

ALKANES

Because there are so many organic compounds, organic compounds are usually divided into families of compounds that have similar structures and similar properties for discussion. The alkanes make up the simplest family of organic compounds. **Alkanes** are *compounds of carbon and hydrogen that have no multiple bonds.* Methane, CH_4, ethane, CH_3CH_3, and isobutane, $(CH_3)_3CH$, are examples:

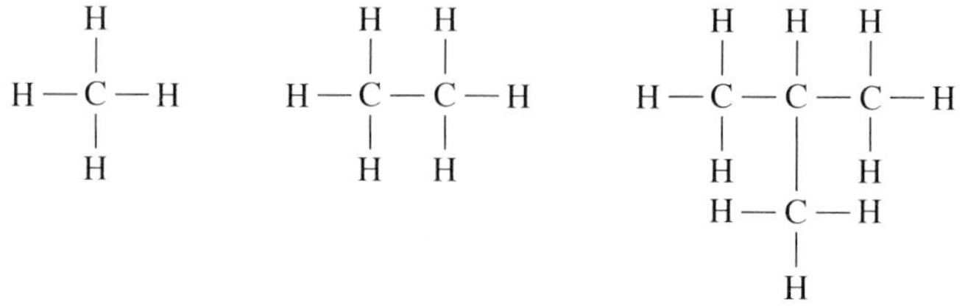

methane   ethane   isobutane (methylpropane)

Alkanes occur naturally in the plant and animal world. For example, cabbage leaves contain the alkane $CH_3(CH_2)_{27}CH_3$, and beeswax contains the alkane $CH_3(CH_2)_{29}CH_3$. However, natural gas and petroleum are by far the most important sources of alkanes. The formulas, names, and physical properties of the first ten continuous chain alkanes are given in Table 22.3. The molecular formulas of the alkanes can be represented by the general formula C_nH_{2n+2}, where n is any whole number. For example, $n = 1$ in methane CH_4 and 2 in ethane CH_3CH_3 (C_2H_6).

Alkanes are called **hydrocarbons** because they are compounds of carbon and hydrogen. *Hydrocarbons do not contain any kinds of atoms other than carbon and hydrogen.* All alkanes are hydrocarbons but not all hydrocarbons are alkanes. Other types of hydrocarbons exist that have multiple bonds. Ethylene, $H_2C{=}CH_2$, acetylene, $HC{\equiv}CH$, and benzene, ⬡, are examples. *Hydrocarbons that have rings but no multiple bonds* such as cyclohexane, ⬡, are known as **cycloalkanes.**

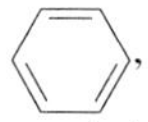

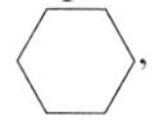

*Hydrocarbons that contain a double bond such as $H_2C{=}CH_2$ and $CH_3CH{=}CH_2$ are called **alkenes,** and hydrocarbons that contain a triple bond such as $HC{\equiv}CH$ are called **alkynes.***

PRACTICE PROBLEMS

22.6   Of the continuous chain alkanes in Table 22.3, which are liquids under ordinary conditions?

22.7   Of the following compounds

$$CH_3CH_2OCH_2CH_3 \qquad (CH_3)_2C{=}CH_2 \qquad \text{⬠}$$

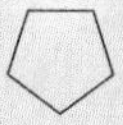

$$(CH_3)_2CHCH_2C(CH_3)_3 \qquad CH_3CH_2OH \qquad CH_2Cl_2 \qquad CH_3C{\equiv}CCH_3$$

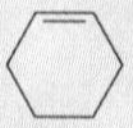

$$\text{⬡} \qquad CH_3CH_2CCH_3 \qquad CH_3OCH{=}CH_2 \qquad \text{⬡}{-}NH_2$$
$$\qquad\qquad\qquad \underset{O}{\overset{\|}{}}$$

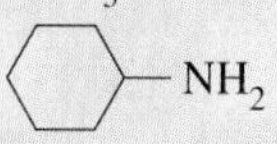

(a) which are hydrocarbons? (b) Which is an alkane? (c) Which is a cycloalkane?

22.8   Write the general formula for the cycloalkanes.

The continuous chain alkanes are an example of a **homologous series.** *Each member of a homologous series differs from the compound before and the compound*

Formula	Name	Melting Point, °C	Normal Boiling Point, °C	Density at 20 °C, g/mL
CH_4	Methane	-182	-164	0.000 667
CH_3CH_3	Ethane	-183	-89	0.001 25
$CH_3CH_2CH_3$	Propane	-190	-42	0.001 83
$CH_3(CH_2)_2CH_3$	Butane	-138	-0.5	0.002 42
$CH_3(CH_2)_3CH_3$	Pentane	-130	36	0.626
$CH_3(CH_2)_4CH_3$	Hexane	-95	69	0.660
$CH_3(CH_2)_5CH_3$	Heptane	-91	98	0.684
$CH_3(CH_2)_6CH_3$	Octane	-57	126	0.702
$CH_3(CH_2)_7CH_3$	Nonane	-51	151	0.718
$CH_3(CH_2)_8CH_3$	Decane	-30	174	0.730

TABLE 22.3 Names and Physical Properties of Continuous Chain Alkanes

after by $-CH_2-$. The physical properties of the members of a homologous series change in a predictable way as the number of carbon atoms increases. For example, as you can see from the data in Table 22.3, the boiling points and densities of the continuous chain alkanes all increase as the number of carbons in the alkanes increases.

The chemical properties of the members of a homologous series are similar. Alkanes are not very reactive. The only types of reactions that alkanes commonly undergo are combustion, pyrolysis (decomposition by thermal energy in the absence of air), and halogenation (substitution of one of the halogens—F, Cl, or Br—for H).

$$2CH_3CH_2CH_2CH_3(g) + 9O_2(g) \xrightarrow{\text{thermal energy}}$$

$$8CO(g) + 10H_2O(l) \qquad \text{incomplete combustion*}$$
excess

$$2CH_3CH_2CH_2CH_3(g) + 13O_2(g) \xrightarrow{\text{thermal energy}}$$

$$8CO_2(g) + 10H_2O(l) \qquad \text{complete combustion}$$
excess

$$CH_3CH_2CH_2CH_3(g) \xrightarrow{\text{thermal energy–no } O_2}$$

$$CH_4(g) + CH_3CH_3(g) + H_2C{=}CH_2(g) + CH_3CH{=}CH_2(g)^{\dagger} \qquad \text{pyrolysis}$$

$$CH_3CH_2CH_2CH_3(g) + Cl_2(g) \xrightarrow{\text{thermal energy and/or light}}$$
excess†

$$CH_3CH_2CH_2CH_2Cl(l) + CH_3CH_2CH(Cl)CH_3(l) + HCl(g)^{\dagger} \text{ halogenation (chlorination)}$$

---

*Incomplete combustion is accompanied by complete combustion and pyrolysis.

†*Not* an equation, merely a statement of what the reactants and the products are. The products result from more than one reaction. An equation must be written for *each* reaction; for example, one of the reactions involved in pyrolysis is

$$CH_3CH_2CH_2CH_3(g) \xrightarrow{\text{thermal energy–no } O_2} CH_3CH_3(g) + H_2C{=}CH_2(g)$$

‡An excess of $CH_3CH_2CH_2CH_3(g)$ must be used to prevent substitution of more than one hydrogen per molecule and formation of products such as $CH_3CH_2CH_2CHCl_2$, $CH_3CH_2CH(Cl)CH_2Cl$, and $CH_3CH_2CH_2CCl_3$.

■ FIGURE 22.11   "Cat cracker" in an oil refinery.

PRACTICE PROBLEMS

22.9   Why must oxygen be absent in order for pyrolysis to occur?

22.10   Write equations for each of the two reactions that take place when an excess of $CH_3CH_2CH_2CH_3(g)$ is chlorinated.

The electronegativity of carbon is 2.5, and the electronegativity of hydrogen is 2.2 (Figure 9.3); carbon and hydrogen have similar electronegativities. As a result, C—H bonds are only slightly polar, and alkanes do not act as Brønsted–Lowry acids. Alkanes do not act as Lewis acids either because neither the carbons nor the hydrogens in alkanes can accept any more electron pairs. Neither the carbons nor the hydrogens in alkanes have any unshared electron pairs and alkanes do not act as bases.

Both the C—C and the C—H bonds are strong. The average bond energies are $347\,kJ\cdot mol^{-1}$ and $414\,kJ\cdot mol^{-1}$, respectively, compared to $192\,kJ\cdot mol^{-1}$ for Si—Si and $318\,kJ\cdot mol^{-1}$ for Si—H. Neither C—C nor C—H bonds break unless alkanes are heated to high temperatures. When alkanes are heated to high temperatures in the absence of oxygen, pyrolysis takes place. Both C—C and C—H bonds break and a mixture of alkanes and alkenes results.

Decomposition of alkanes in petroleum by thermal energy is called **cracking.** Cracking of high molecular mass alkanes from petroleum increases the amount of gasoline that can be obtained from petroleum. A catalyst is usually used to lower the temperature required for cracking and to direct the cracking process toward specific products. ■Figure 22.11 shows a "cat cracker" in a refinery.

Alkanes are not usually affected by common oxidizing and reducing agents at room temperature. High-temperature oxidation by oxygen in air (combustion) is the most common reaction for which alkanes are used. Thermal energy from the combustion of alkanes in oil and natural gas is used to generate electricity and heat buildings. At the present time, combustion is people's main source of energy; natural gas and oil supply about two-thirds of our energy requirements in the United States. Natural gas, which is mostly methane, is also burned to heat clothes dryers, food on gas stoves, and the contents of test tubes in laboratories. Thermal energy from the combustion of gasoline is used to move cars.

Halogenation of alkanes is an important industrial process. An example is the chlorination of methane:

$$CH_4(g)\ +\ Cl_2(g)\ \xrightarrow[\text{and/or light}]{\text{thermal energy}}\ CH_3Cl(g)\ +\ CH_2Cl_2(l)\ +\ CHCl_3(l)\ +\ CCl_4(l)\ +\ HCl(g)$$

The common names, chloroform for $CHCl_3$ and carbon tetrachloride for CCl_4, are usually used. Chloroform was used in cough syrups and carbon tetrachloride for dry cleaning until their carcinogenicity was discovered.

Again, this is *not* an equation but simply a statement of what reacts and what forms. The mixture of products is formed by a series of reactions. Balanced equations must be written for *each* of them, for example,

$$CH_4(g)\ +\ Cl_2(g)\ \xrightarrow{\text{thermal energy and/or light}}\ CH_3Cl(g)\ +\ HCl(g)$$

$$CH_3Cl(g)\ +\ Cl_2(g)\ \xrightarrow{\text{thermal energy and/or light}}\ CH_2Cl_2(l)\ +\ HCl(g)$$

If the ratio of moles of CH_4 to moles of Cl_2 is 50:1, the product is mostly CH_3Cl. If the ratio of moles of CH_4 to moles of Cl_2 is 1:50, the product is mostly CCl_4. Ratios in between these two extremes yield mixtures of varying proportions of products.

Fluorine is very reactive and fluorination must be carried out at low temperatures or explosions are apt to result. Chorination and bromination do not take place in the absence of thermal energy or light. Bromine reacts at a practical rate but is rarely used in industry because it is much more expensive than chlorine. Iodine does not react.

Halogenation can be classified either as substitution (of Cl for H) or as oxidation. Chlorine is reduced from oxidation number 0 in Cl_2 to oxidation number -1 in HCl. Since chlorine is reduced, the alkane must be oxidized. Classification of changes in organic compounds as oxidation or reduction is *not* simple because oxidation numbers cannot be assigned to carbons in a chain of carbon atoms.

The mixture of CH_3Cl, CH_2Cl_2, $CHCl_3$, and CCl_4 formed by chlorination of methane can be separated by fractional distillation. About 10^9 lb of CH_3Cl, CH_2Cl_2, $CHCl_3$, and CCl_4 are produced each year. All are used as solvents and as starting materials for the manufacture of other organic compounds. Chloromethane is also used as a refrigerant and as a local anesthetic.

Even if a high enough ratio of alkane to chlorine is used so that only one hydrogen is replaced by halogen, halogenation of alkanes that contain three or more carbons usually yields mixtures. For example, both 1-chlorobutane and 2-chlorobutane are produced by chlorination of butane:

$$CH_3CH_2CH_2CH_3(g) + Cl_2(g) \xrightarrow{\text{thermal energy and/or light}}$$
large excess

$$CH_3CH_2CH_2CH_2Cl(g) + CH_3CH_2CH(Cl)CH_3(g) + HCl(g)$$

Mixtures of chlorinated alkanes are often used as solvents in industry. Only if a high enough ratio of halogen to alkane is used so that all hydrogens are replaced by halogen is a single product obtained by the halogenation of butane

$$CH_3CH_2CH_2CH_3(g) + 10Cl_2(g) \xrightarrow[\text{and/or light}]{\text{thermal energy}} CCl_3CCl_2CCl_2CCl_3(l) + 10HCl(g)$$
large excess

22.6 NOMENCLATURE OF ORGANIC COMPOUNDS

Because there are so many compounds of carbon, a system is needed for naming them. The system was developed by the International Union of Pure and Applied Chemistry (IUPAC) and is based on the continuous chain alkanes. You should memorize the names of the first ten continuous chain alkanes given in Table 22.3.

To name a branched chain alkane, *first find the longest continuous chain* in the alkane. The *longest continuous chain* is usually referred to as the **main chain** or **parent chain.** For example, the main chain in the compound

is five carbons long. There are two choices for five carbon chains in this compound:

and

Because the parent chain in the compound contains five carbons, the compound is named as a derivative of pentane, the five carbon continuous chain alkane.

In this case, the two choices of five carbon chains are equivalent—that is, either one leads to the same name for the compound so it does not matter which one you use for the second step. *The second step is to number the longest continuous chain starting from the end that gives the lower number to the carbon where the branch is attached:*

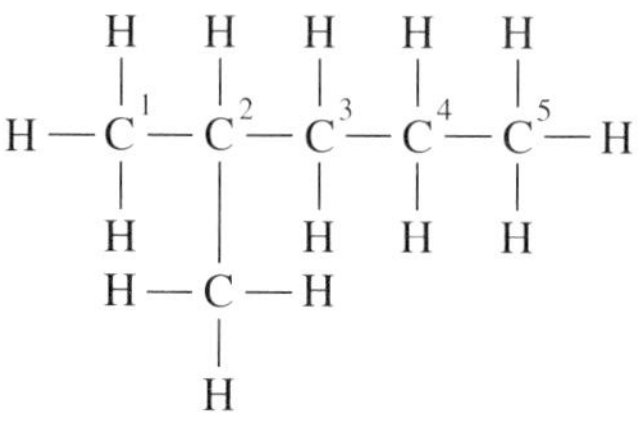

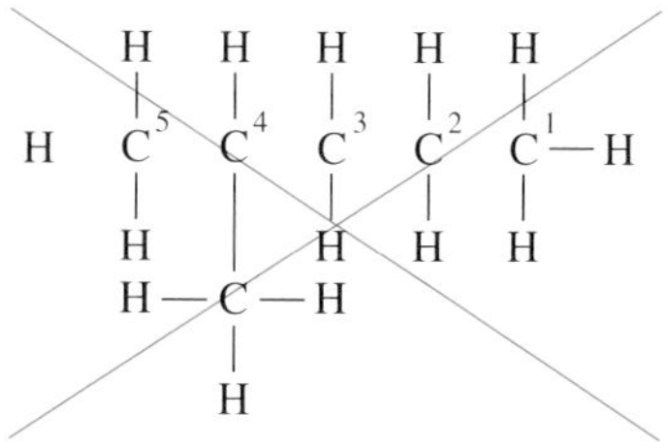

The branch is called a **side chain.** *The name of the side chain is obtained from the name of the straight chain alkane with the same number of carbons by changing the ending -ane of the alkane to -yl.* The name of the side chain —CH_3 is methyl, the name of the side chain —CH_2CH_3 is ethyl, and so forth. In general, side chains that are derived from alkanes are known as **alkyl groups.**

The position of attachment of the side chain and the number of carbons in it are shown as a prefix to the name of the longest continuous chain. The complete name of the compound is 2-methylpentane because the methyl group is attached to carbon number two. Note the spelling and punctuation carefully. The number is separated from the name of the side chain by a hyphen; methylpentane is a single word.

If the side chain is branched, the structure of the side chain must be shown in the name. Naming branched side chains is beyond the scope of a course in general chemistry; you will learn how to name branched side chains if you take organic chemistry. Fortunately, few common compounds have branched side chains.

PRACTICE PROBLEM

22.11   What is the name of the side chain —$CH_2CH_2CH_3$?

*If two or more identical groups are attached to the parent chain, the positions of attachment of the side chains are shown by numbers **and** prefixes, di-, tri-, and so forth (Table 1.4), are used to show how many of the groups there are.* For example, the compound $(CH_3)_3CCH_2CH_2CH_3$

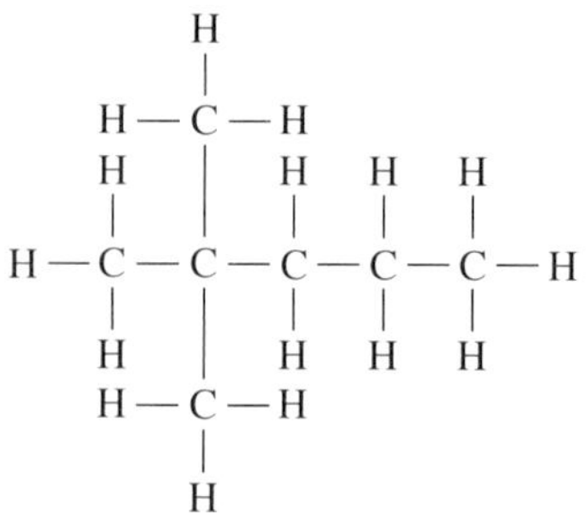

is called 2,2-dimethylpentane (the numbers must be separated by a comma so that

2,2- can be distinguished from 22):

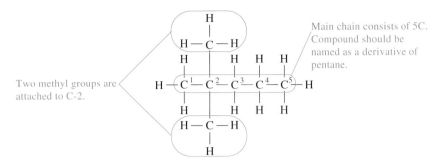

Two methyl groups are attached to C-2.

Main chain consists of 5C. Compound should be named as a derivative of pentane.

PRACTICE PROBLEM

22.12 In 3,3-dimethylpentane, does it matter which end of the main chain you number from? Explain your answer.

If the groups attached to the parent chain are different, the position of each is shown by a number, and the side chains are arranged in alphabetical order according to the name of the alkyl group. For example, the name of the compound $(CH_3CH_2)_2CHCH(CH_3)_2$ is 3-ethyl-2-methylpentane. This example also illustrates another rule: *If there are two or more longest chains containing the same number of carbons, the main chain is the one with the greater number of groups attached to it* (this rule makes the side chains smaller and simpler to name):

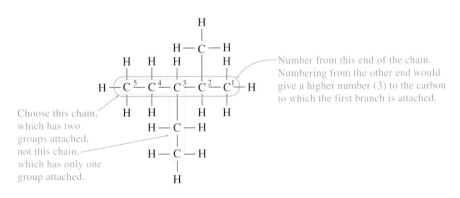

Choose this chain, which has two groups attached, not this chain, which has only one group attached.

Number from this end of the chain. Numbering from the other end would give a higher number (3) to the carbon to which the first branch is attached.

Rules for Naming Alkanes

1. Find the longest continuous chain and name it.
2. If there are two or more chains containing the same number of carbons, choose the one with the greater number of groups attached to it for the main chain.
3. Number the main chain starting from the end that gives the lower number to the carbon where the first branch is attached.
4. Show the position of attachment of the side chain and its structure as a prefix to the name of the main chain. If two or more identical groups are attached to the main chain, use a prefix to show how many of the groups there are.
5. If the groups attached to the main chain are different, the side chains are arranged in alphabetical order according to the name of the alkyl group.

All these rules may seem like a lot of picky detail but if you have to find a compound in an index that contains some 13 million compounds, it is important to look in the right place. With these five simple rules, you can name all alkanes up to and including the octane isomers (C_8H_{18}) as well as most of the common isomers of alkanes with more than eight carbon atoms.

Naming structures is the easiest and surest way to find out if two structures are the same or different. If the names are the same, the structures are the same. Sample Problem 22.2 shows how useful names are for detecting identical structures. You also need to be able to write complete structural formulas from names. Sample Problem 22.3 shows how.

For more information about the nomenclature of organic compounds see Fessenden, R. J. and Fessenden, J. S. *Organic Chemistry*, 4th ed.; Brooks/Cole, Pacific Grove, CA, 1990; pp 1033–1055 or *Nomenclature of Organic Chemistry*, 1979 ed.; Pergamon, New York.

SAMPLE PROBLEMS

22.2 Write complete structural formulas for all the isomers of C_7H_{16}.

Solution A system is necessary to find all the isomers of a given molecular formula. Begin by writing the formula of the continuous chain alkane with the proper number of carbons:

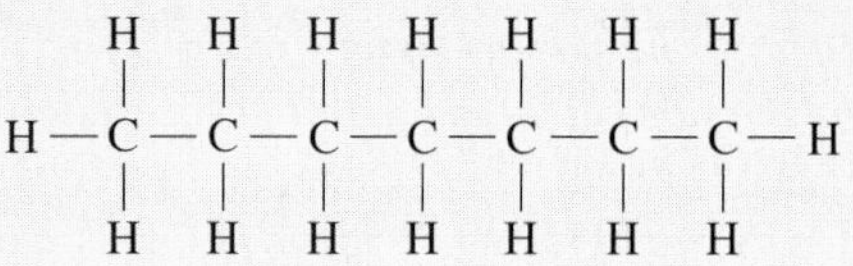

$$
\begin{array}{ccccccc}
 & H & H & H & H & H & H & H \\
 & | & | & | & | & | & | & | \\
H- & C- & C- & C- & C- & C- & C- & C-H \\
 & | & | & | & | & | & | & | \\
 & H & H & H & H & H & H & H
\end{array}
$$

To save time and focus attention on the carbon atoms, omit the hydrogens:

$$C—C—C—C—C—C—C$$

A structure like this, *which shows only the carbons,* is called a **carbon skeleton.**
   Then imagine that you can remove one carbon from the end of the chain and attach it to the next of the remaining carbons:

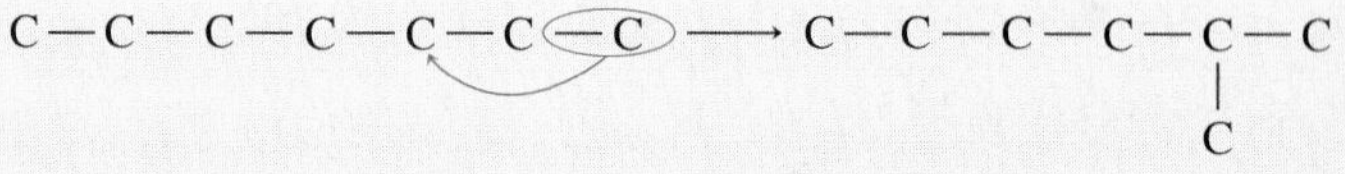

$$C—C—C—C—C—C\!=\!\!C \longrightarrow \begin{array}{c} C—C—C—C—C—C \\ | \\ C \end{array}$$

original compound (heptane)   isomer (2-methylhexane)

Note that C—C—C—C—C—C (with C branch) is just a conformation of C—C—C—C—C—C—C because rotation around single bonds is free. There are still seven carbons in a continuous chain in C—C—C—C—C—C (with C branch).

Continue moving the one carbon down the chain until you have written all the possible isomers with the same number of carbons in the main chain. Stop when you begin writing isomers you have already found:

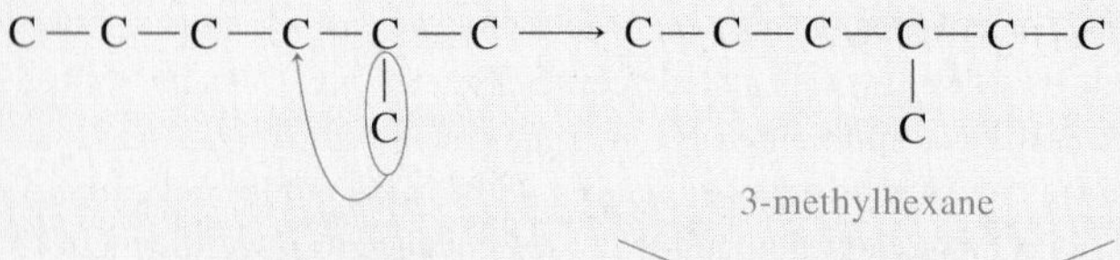

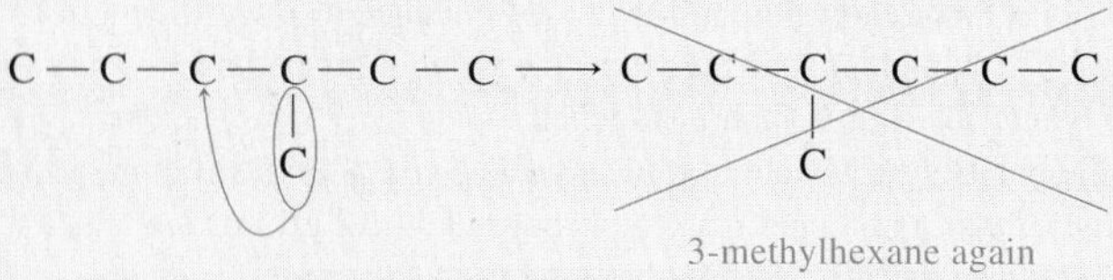

$$C—C—C—C—C—C \longrightarrow \begin{array}{c} C—C—C—C—C—C \\ | \\ C \end{array}$$

3-methylhexane

$$C—C—C—C—C—C \longrightarrow \begin{array}{c} C—C—C—C—C—C \\ | \\ C \end{array}$$

3-methylhexane again

With experience, you will be able to see that two structures are the same without naming them. In the meantime, naming them is good practice.

Next, remove two carbons from the end of the chain and attach them to the remaining carbons either as two methyl groups:

C — C — C — C — C — ⊂C — C⊃ → ⟶ C — C — C — C — C, C — C — C — C — C
                                           | | | |
                                           C   C   C   C

          2,3-dimethylpentane   2,4-dimethylpentane

               C   C
               | |
C — C — C — C, and C — C — C — C — C
      | |
      C   C

    2,2-dimethylpentane   3,3-dimethylpentane

or as one ethyl group. Attachment of the two carbons combined in the form of an ethyl group gives 3-methylhexane again and a new isomer, 3-ethylpentane:

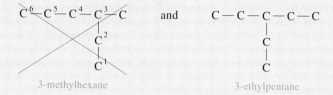

    3-methylhexane   3-ethylpentane

A common mistake is to call 3-methylhexane "2-ethylpentane." *Structures are not always written so that the longest continuous chain is obvious. To write a correct name, you must begin by finding the longest continuous chain.*

Now, remove three carbons from heptane and try attaching them both combined and broken into three methyl groups or into one ethyl and one methyl:

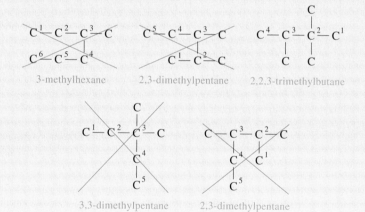

    3-methylhexane   2,3-dimethylpentane   2,2,3-trimethylbutane

    3,3-dimethylpentane   2,3-dimethylpentane

Of these five structures, only one, 2,2,3-trimethylbutane, is new. We have found all the isomers of C_7H_{14}. There are nine:

heptane	$CH_3(CH_2)_5CH_3$
2-methylhexane	$CH_3(CH_2)_3CH(CH_3)_2$
3-methylhexane	$CH_3(CH_2)_2CH(CH_3)CH_2CH_3$
2,3-dimethylpentane	$CH_3CH_2CH(CH_3)CH(CH_3)_2$
2,4-dimethylpentane	$CH_3CH(CH_3)CH_2CH(CH_3)_2$
2,2-dimethylpentane	$CH_3(CH_2)_2C(CH_3)_3$
3,3-dimethylpentane	$CH_3CH_2C(CH_3)_2CH_2CH_3$
3-ethylpentane	$(CH_3CH_2)_3CH$
2,2,3-trimethylbutane	$(CH_3)_2CHC(CH_3)_3$

● If you have any ideas for other possible structures, try writing and naming them to convince yourself that we have indeed found all the isomers. Check your structures to be sure they show four bonds to each C and one bond to each H.

If you have trouble naming the structures and writing condensed structural formulas from the carbon skeletons, add enough H's to each carbon skeleton to give every carbon four bonds. For example,

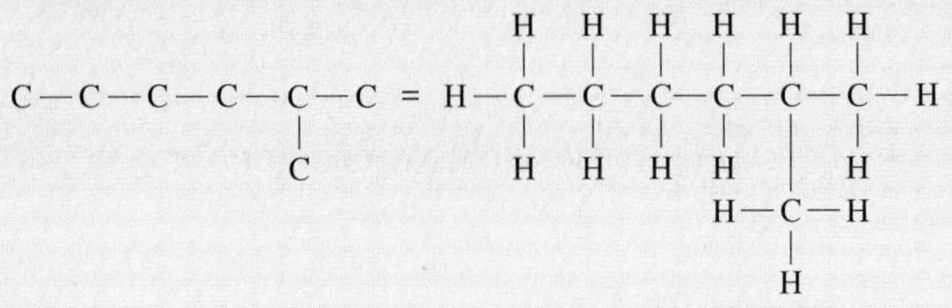

$$
C-C-C-C-C-C = H-C-C-C-C-C-C-H
$$

$$
= CH_3CH_2CH_2CH_2CH(CH_3)_2 \ [\text{or} \ CH_3(CH_2)_3CH(CH_3)_2]
$$

22.3 Write the complete structural formula for 6-ethyl-2,2-dimethyloctane. (Note that the prefix *di-* is *not* considered in alphabetizing.)

Solution Begin by writing and numbering the carbon skeleton of the parent chain, octane:

$$
C^1-C^2-C^3-C^4-C^5-C^6-C^7-C^8
$$

Attach the alkyl groups at the points specified in the name, that is, an ethyl group at C-6 and two methyl groups at C-2:

Finally, add enough H's to the carbon skeleton to give every carbon four bonds:

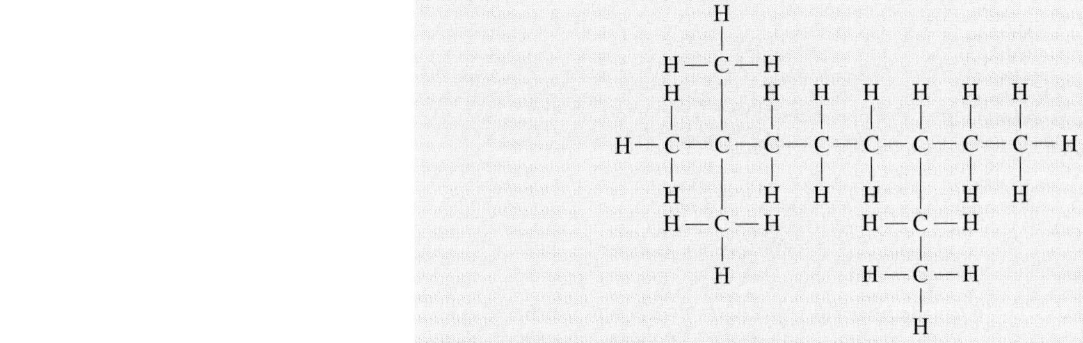

PRACTICE PROBLEMS

22.13 Which of the following names is the correct name for the compound

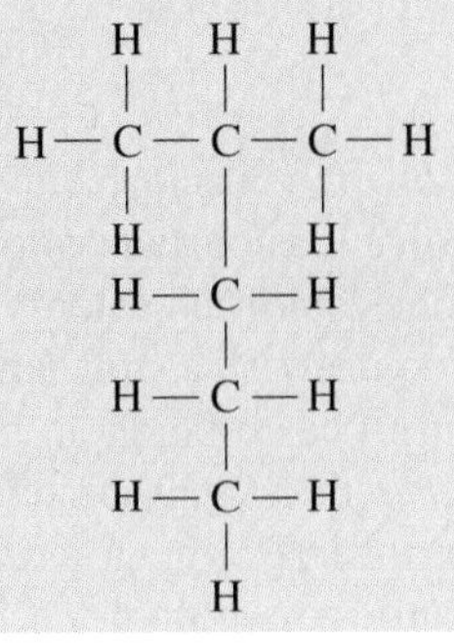

(a) 1,1-dimethylbutane (b) 2-methylpentane (c) 4-methylpentane
(d) 2-propylpropane. Explain why each of the incorrect answers is wrong.

22.14   Write carbon skeletons, complete structural formulas, condensed structural formulas, and names for all five isomers of C_6H_{14}.

22.15   Write structural formulas for each of the following compounds:
(a) 3-ethyl-2-methylhexane and (b) 3,3-dimethylheptane.

22.7   FUNCTIONAL GROUPS

Most organic compounds contain one or more functional groups. A **functional group** is *an atom or group of atoms in a molecule that accounts for the characteristic properties of the molecule.* All molecules that have the same functional group function or react similarly. The —COOH group in acetic acid (CH_3COOH), the —OH group in ethyl alcohol (CH_3CH_2OH), and the C=C in ethylene (H_2C=CH_2) are examples of functional groups.

In contrast to alkanes, most organic compounds that contain functional groups react readily with a variety of reagents. *It is usually the functional group that reacts. The hydrocarbon-like part of the molecule generally goes through the reaction unchanged providing that it does not contain any multiple bonds.* For example, like acetic acid, all compounds that contain the —COOH functional group react with bases to form salts and water:

$$CH_3COOH(aq) + NaOH(aq) \longrightarrow CH_3COONa(aq) + H_2O(l)$$
$$CH_3CH_2CH_2COOH(aq) + NaOH(aq) \longrightarrow CH_3CH_2CH_2COONa(aq) + H_2O(l)$$

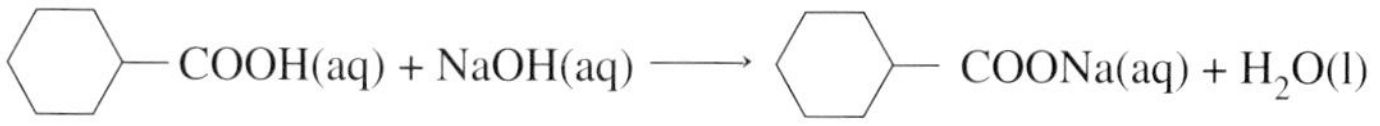

—COOH(aq) + NaOH(aq) $\longrightarrow$ — COONa(aq) + H$_2$O(l)

Like ammonia, amines (Section 15.5) react with acids to form salts:

$$CH_3NH_2(g) + HCl(g) \longrightarrow CH_3NH_3Cl(s) \qquad \text{(often written } CH_3NH_2 \cdot HCl)$$
$$(CH_3)_2NH(aq) + HCl(aq) \longrightarrow (CH_3)_2NH_2Cl(aq) \qquad [\text{or } (CH_3)_2NH \cdot HCl]$$

Alkenes can generally be catalytically hydrogenated (Section 18.12):

$$H_2C=CH_2(g) + H_2(g) \xrightarrow{\text{Pt, Pd, or Ni catalyst}} H_3C—CH_3(g)$$

Alcohols react with sodium (and other active metals) to form hydrogen gas as ethyl alcohol does:

$$2Na(s) + 2CH_3CH_2OH(l) \longrightarrow 2CH_3CH_2ONa(s) + H_2(g)$$

Thus, functional groups provide an important way of organizing the enormous amount of information that exists about organic compounds.

PRACTICE PROBLEM

22.16   Write the molecular equation for each of the following reactions:

(a) $CH_3CH_2COOH(aq) + NaOH(aq) \longrightarrow$ (b) $CH_3CH_2NH_2(aq) + HCl(aq) \longrightarrow$

(c) $CH_3CH=CH_2(g) + H_2(g) \xrightarrow{\text{catalyst}}$ (d) $CH_3OH(l) + Na(s) \longrightarrow$

Nomenclature of Compounds Containing Functional Groups

To name a compound that contains a functional group, you must first find the parent chain. *The parent chain of a compound that contains a functional group must either have the functional group in it or have the functional group attached to it.* For example, the main chain in the compound $CH_3CH_2CH(CH_2OH)CH_2CH_3$, which is circled in the complete structural formula

contains four carbons although the longest continuous chain contains five carbons.

The presence of most functional groups is shown by changing the *-e* at the end of the name of the parent alkane to an ending that shows the type of functional group. For example, the presence of an —OH group is shown by changing *-e* to *-ol*. The compound CH_3OH is called methanol, and CH_3CH_2OH is called ethanol.

The main chain is numbered starting from the end that gives the lower number to the carbon where the functional group is attached. The position of attachment of the functional group on the main chain is shown by a number that is placed in front of the name of the parent compound. For example, the compound $CH_3CH_2CH(CH_2OH)CH_2CH_3$ is called 2-ethyl-1-butanol:

Longest chain with functional group attached to it has 4 C's.

An ethyl group is attached to C-2.

Functional group —OH is attached to C-1.

Like alkyl groups, a few functional groups such as —F (fluoro-), —Cl (chloro-), —Br (bromo-), and —I (iodo-) are always named as prefixes. For example, the compound CCl_3F (Freon 12), which is used as a refrigerant, is called trichlorofluoromethane. Note that in the molecular formula for an organic compound, carbon is written first and hydrogen, if present, is written second. Other elements are then listed in alphabetical order. Prefixes such as *tri-* are ignored when the functional groups are alphabetized in the name.

22.4   What is the name of the compound

$$
\begin{array}{ccccccc}
 & H & H & H & H & H & H \\
 & | & | & | & | & | & | \\
H- & C & -C & -C & -C & -C & -C & -O \\
 & | & | & | & | & | & | \\
 & H & H & H & | & H & H & H \\
 & & & & H-C-H \\
 & & & & | \\
 & & & & H-C-H \\
 & & & & | \\
 & & & & H-C-H \\
 & & & & | \\
 & & & & H-C-H \\
 & & & & | \\
 & & & & H
\end{array}
$$

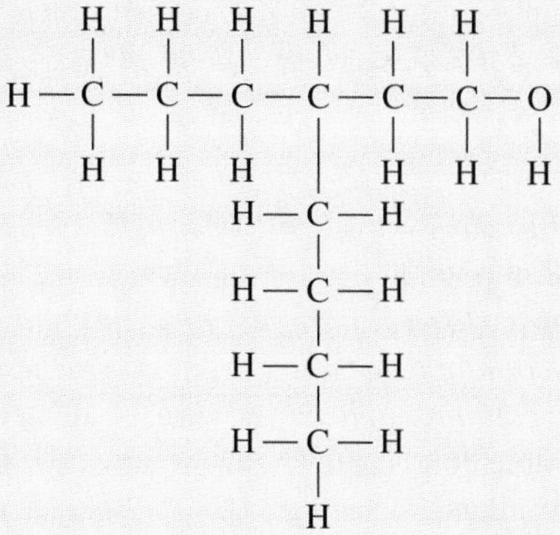

Solution   Find the longest chain that has the functional group attached. The main chain in the compound in this problem is seven carbons long:

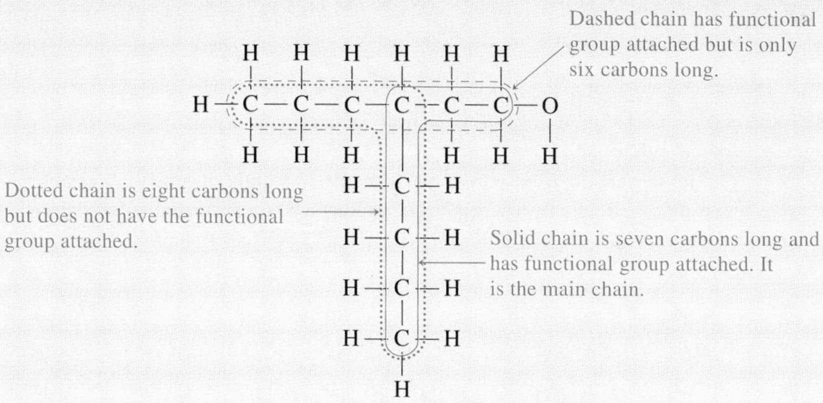

Dashed chain has functional group attached but is only six carbons long.

Dotted chain is eight carbons long but does not have the functional group attached.

Solid chain is seven carbons long and has functional group attached. It is the main chain.

The parent chain has a propyl group attached to C-3:

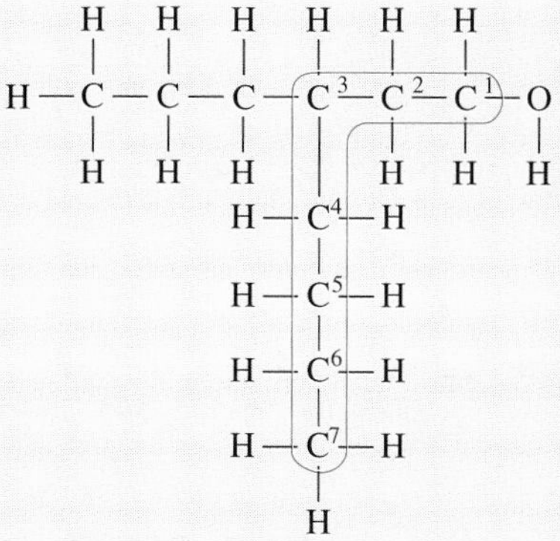

The name of the compound is 3-propyl-1-heptanol.

22.17 What is the name of $(CH_3)_2CHCH_2CH_2OH$? (*Hint:* Always write the carbon skeleton before trying to name a compound. Never try to name a compound from the condensed structural formula.)

22.18 Write the complete structural and condensed structural formulas for the following: (a) 2-methyl-2-propanol and (b) 1,2-dichloro-1,1,2,2-tetrafluoroethane.

Some common functional groups are listed in Table 22.4. Notice how five of the thirteen classes of compounds listed contain a $C{=}O$ group. The *group* $C{=}O$ is called a **carbonyl group.** *Observe carefully how the condensed structural formulas for the compounds that contain a carbonyl group look when written on a single line.*

Any compound that contains a benzene ring is an **aromatic compound.** The three carbon–carbon double bonds in the benzene ring do *not* behave like the carbon–carbon double bonds in alkenes (Section 10.7). Other aromatic structures besides the benzene ring exist. You will learn about them, about how to name aromatic compounds, and about the special properties of aromatic structures if you study organic chemistry.

You should be able to recognize the common functional groups listed in Table 22.4 in both complete and condensed structural formulas.

22.5 Circle the functional groups in the structural formula of glucose and give the classes of compounds to which glucose belongs. The condensed structural formula for glucose is $HOCH_2CH(OH)CH(OH)CH(OH)CH(OH)CHO$.

Solution

$$\boxed{HO}CH_2CH(\boxed{OH})CH(\boxed{OH})CH(\boxed{OH})CH(\boxed{OH})\boxed{CHO}$$

Glucose contains five —OH groups and one —CHO group. Glucose is both an alcohol and an aldehyde.

22.19 (a) Which of the compounds in Practice Problem 22.7 is an alkene? (b) A cycloalkene? (c) An alkyne?

22.20 Write complete structural formulas for all the compounds in Table 22.4 that contain a carbonyl group.

22.21 Circle the functional groups in the structural formulas of each of the following compounds and give the class (or classes) of compounds to which each belongs:

(a) $ClCH_2CH_2OH$ (b) ⬡—$CH{=}CH_2$ (c) $CH_3CH_2COOCH_2CH_3$

(d) CH_3CH_2Br (e) $HCON(CH_3)_2$

TABLE 22.4

TABLE 22.4   Common Functional Groups

Functional Group	Class of Compound	Example	IUPAC Name (Common Name)	Uses
$\diagdown C = C \diagup$	Alkene	$H_2C = CH_2$	Ethene (ethylene)	Intermediate, plant growth regulator
$- C \equiv C -$	Alkyne	$HC \equiv CH$	Ethyne (acetylene)	Intermediate, welding
$\overset{\displaystyle O}{\overset{\|}{-COH}}$	Carboxylic acid	CH_3COOH	Ethanoic acid (acetic acid)	Intermediate
$\overset{\displaystyle O}{\overset{\|}{-COC-}}$	Ester	$HCOOCH_2CH_3$	Ethyl methanoate (ethyl formate)	Flavor for lemonade
$\overset{\displaystyle O}{\overset{\|}{-CN\diagdown}}$	Amide	CH_3CONH_2	Ethanamide (acetamide)	Molten as solvent
$\overset{\displaystyle O}{\overset{\|}{-CH}}$	Aldehyde	$HCHO$	Methanal (formaldehyde)	Present in smoke produced for smoking ham and fish
$\overset{\displaystyle O}{\overset{\|}{-C-}}$	Ketone	CH_3COCH_3	Propanone (acetone)	Solvent (Keep away from plastics and rayon.)
$-OH$	Alcohol	CH_3CH_2OH	Ethanol (ethyl alcohol)	Alcoholic beverages, solvent
$-N-$	Amine	CH_3NH_2	Methanamine (methylamine)	Tanning leather
$-Cl$	Alkyl chloride	CH_3Cl	Chloromethane (methyl chloride)	Local anesthetic
$-O-$	Ether	$CH_3CH_2OCH_2CH_3$	Ethoxyethane (ethyl ether)	Solvent
(benzene ring)	Aromatic	(benzene ring)$-CH_3$	Methylbenzene (toluene)	Intermediate, solvent
$-OH$ directly attached to aromatic	Phenol	(benzene ring)$-OH$	Phenol (carbolic acid)	Disinfectant

22.8   MECHANISM

Reaction mechanisms provide another powerful way of organizing the vast amount of information available about the reactions of organic compounds. In this section, we will examine one very important reaction mechanism, the S_N2 mechanism, in some depth. The S_N2 reaction is used in the synthesis of a wide variety of organic compounds. For example, two S_N2 reactions are involved in the manufacture of phenobarbital

which is used to treat epilepsy. In addition, many biochemical reactions are S_N2 reactions, for example, the toxicity of chloromethyl methyl ether, $ClCH_2OCH_3$, which is believed to cause some types of cancer, is thought to result from the S_N2 reaction of the —Cl group with the —NH_2 groups of enzymes (biological catalysts). The usefulness of classifying reactions as S_N2 reactions lies in the fact that many reactions that do not appear to be similar are all S_N2 reactions. What is learned about one S_N2 reaction can be applied to all other S_N2 reactions.

The symbol, $\mathbf{S_N2}$, stands for *substitution nucleophilic bimolecular*. You already know what substitution means—one atom or group takes the place of another. In Section 18.11, you learned that bimolecular indicates that two species combine to form the activated complex. **Nucleophilic** means *nucleus loving;* a nucleophile is attracted by a positive charge. The reaction between chloromethane and hydroxide ion to form methyl alcohol and chloride ion

$$HO^- + CH_3Cl(aq) \longrightarrow CH_3OH(aq) + Cl^- \tag{22.1}$$

is a typical example of an S_N2 reaction. Because chlorine is more electronegative than carbon, the carbon–chlorine bond in chloromethane is polar with a partial positive charge on carbon. The negatively charged hydroxide ion is attracted by the partial positive charge on carbon. The hydroxide ion is the **nucleophile.** In the reaction in equation 22.1, chloromethane is called the **substrate,** a *species that undergoes reaction.* The functional group, —Cl, is called the **leaving group** because it is the group that *leaves the molecule and is replaced* by the —OH group.

The general equation for nucleophilic substitution is

$$\text{Nuc:}^- + RX \longrightarrow RNuc + \text{:}X^- \tag{22.2}$$

where Nuc:$^-$ stands for the nucleophile, which must have at least one unshared pair of electrons; R represents the alkyl group of the substrate RX, for example, in equation 22.1, R = CH_3— (R is not changed by the reaction); X is the leaving group.

Although the nucleophile is often an anion as shown in the general equation for nucleophilic substitution (equation 22.2), the nucleophile can be a molecule such as H_2O or NH_3. If the nucleophile is a molecule, the initial organic product of nucleophilic substitution is a cation:

$$H-\overset{\cdot\cdot}{\underset{|}{\overset{|}{O}}}: + CH_3Cl \longrightarrow H-\overset{+}{\underset{|}{\overset{\cdot\cdot}{O}}}:CH_3 + Cl^-$$
$$\quad\; H \qquad\qquad\qquad\quad H$$

If water is the solvent, the large excess of water present shifts the Brønsted–Lowry acid–base equilibrium

$$H_2\overset{+}{O}CH_3 + H_2O \rightleftarrows HOCH_3 + H_3O^+$$

to the right and the product that is actually obtained is $HOCH_3$ (usually written CH_3OH).

22.22   In the equation for the reaction

$$CH_3CH_2CH_2Br + CH_3O^- \longrightarrow CH_3CH_2CH_2OCH_3 + Br^-$$

(a) Circle and label the nucleophile. (b) Draw a rectangle around the substrate.
(c) Circle and label the leaving group. (d) What group would be represented by R
in the general equation for nucleophilic substitution?

Observations of S_N2 Reactions

The S_N2 mechanism was one of the first mechanisms to be studied. Let's look at
some of the observations that have been made about the S_N2 reaction.

Remember that the composition of the activated complex for a reaction with a
single slow step is given by the rate law for the reaction (Section 18.11). The rates
of S_N2 reactions are proportional to the first power of the concentrations of both
the substrate and the nucleophile:

$$\text{rate} = k[RX][\text{Nuc}:^-] \tag{22.3}$$

That is, the reaction is first order with respect to substrate and first order with respect
to nucleophile. From the rate law, we know that the activated complex for an S_N2
reaction consists of a combination of the substrate and the nucleophile.

The orientation of the substrate and the nucleophile in the activated complex can
be deduced from observations with stereoisomers. For example, *cis*-1-chloro-4-
methylcyclohexane yields *trans*-4-methyl-1-cyclohexanol:

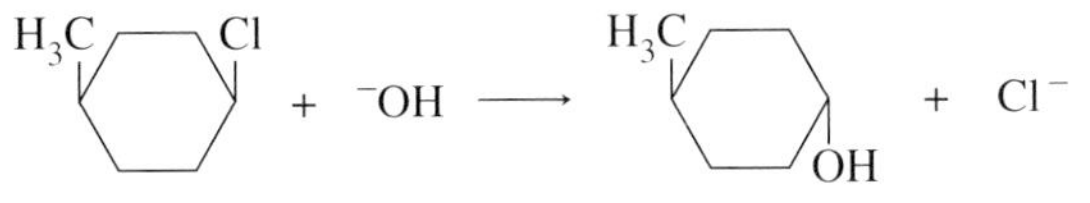

and *trans*-1-chloro-4-methylcyclohexane yields *cis*-4-methyl-1-cyclohexanol:

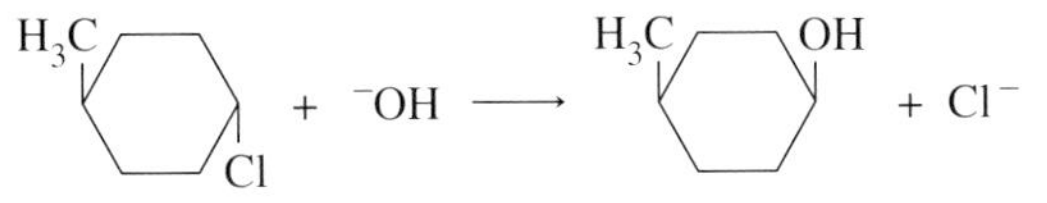

When the nucleophile, ⁻OH, takes the place of the leaving group, —Cl, the nucleophile becomes attached to the opposite or back side of the carbon to which the
leaving group was attached. The S_N2 reaction is said to take place with **inversion**
of configuration; the *configuration of the product is the opposite of the configuration
of the reactant.*

The S_N2 Mechanism

The following mechanism, which is shown in ▬Figure 22.12, was proposed to
account for these kinetic and stereochemical observations and is now generally
accepted.* The S_N2 reaction takes place in a single step. In this single step, the

---

*Remember that reaction mechanisms are theories. To be acceptable, a reaction mechanism must
explain all known observations about the reaction. A mechanism can be proved wrong but can
never be proved to be correct. The possibility always exists that a new observation will be made
that does not fit the mechanism. If this happens, the mechanism must be changed or replaced.

Remember that the curved arrows show how electrons move to form and break bonds.

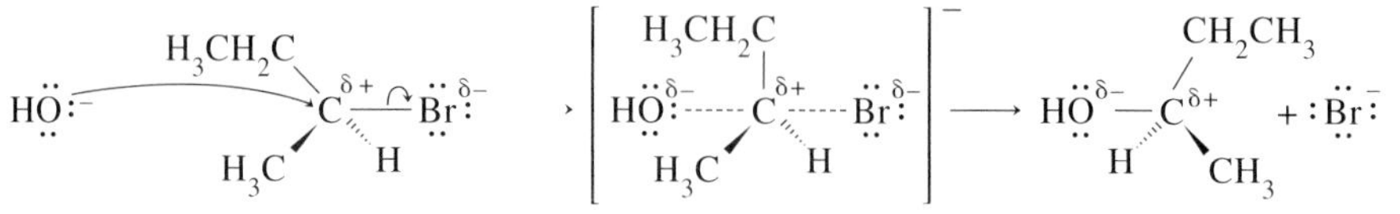

reactants — activated complex — products

The curved double-headed arrows show bond formation and bond breaking.

The bonds that are forming and breaking are shown by dashed lines.

The product has the opposite configuration from the reactant.

■ FIGURE 22.12 The S$_N$2 mechanism.

nucleophile moves toward the carbon to which the leaving group is attached because the negative charge on the nucleophile is attracted by the partial positive charge on the carbon.

To account for the observed inversion of configuration, the nucleophile must hit the back of the carbon to which the leaving group is attached. As the nucleophile begins to bond to this carbon, the orbital of the nucleophile that contains the unshared electron pair overlaps with an empty (antibonding) orbital of carbon. At the same time (because a carbon atom can't have more than eight valence electrons around it), the bond between carbon and the leaving group begins to break.

The reaction between one pure enantiomer of 2-bromobutane and hydroxide ion (Figure 22.12) is the simplest example of an S$_N$2 reaction in a noncyclic system in which inversion of configuration can be seen. In the reaction shown in Figure 22.12, inversion can be observed because substitution takes place at a stereocenter. All reactions that take place by the S$_N$2 mechanism are believed to take place with inversion of configuration even if inversion cannot be observed experimentally. The mechanism pictured in Figure 22.12 is in agreement with the experimental rate law (equation 22.3); the activated complex is a combination of nucleophile and substrate.

In the S$_N$2 mechanism, the nucleophile donates a pair of electrons to the substrate to form a covalent bond. The substrate accepts the pair of electrons. Thus, the nucleophile acts as a Lewis base, and the substrate acts as a Lewis acid.

The energy necessary for bond breaking is provided by collisions between the nucleophile and the substrate and by bond formation. Only a small fraction of collisions occurs with the necessary approach of the nucleophile to the substrate from the back. Of the collisions that take place with approach from the back, only a small fraction take place with enough energy for reaction to occur, that is, with energy greater than the activation energy. In the activated complex, there are five atoms or groups around carbon instead of the usual four. The activated complex is the highest energy arrangement of atoms along the reaction coordinate; only about 10^{-12} s are required for the atoms to pass through the transition state. ■Figure 22.13 is a sketch of the reaction profile for an S$_N$2 reaction.

PRACTICE PROBLEM

22.23 Write the mechanism for the S$_N$2 reaction of hydroxide ion, OH$^-$, with cis-1-chloro-3-methylcyclopentane.

Predicting the Product of an S$_N$2 Reaction

A reaction mechanism such as the S$_N$2 mechanism is a big help in organizing information about organic reactions so that a minimum of memorization is needed. Before organic chemists (and biochemists) had mechanisms to use, they had to memorize a great many reactions. Now the results of many reactions can be predicted.

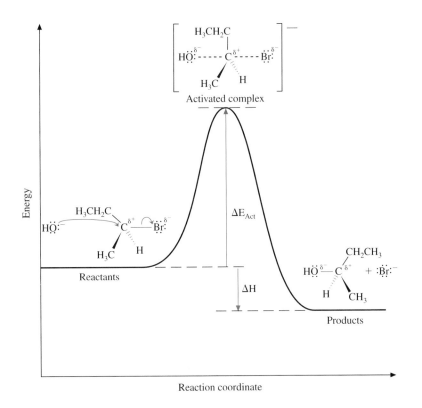

Energy

H₃CH₂C

$HO:^{\delta-}$ ------ $C^{\delta+}$ ----- $Br:^{\delta-}$

H₃C   H

Activated complex

ΔE_{Act}

H₃CH₂C

$HO:^-$ $C^{\delta+}$ $Br:^{\delta-}$

H₃C   H

Reactants

ΔH

CH₂CH₃

$HO:^{\delta-}$ — $C^{\delta+}$ $+ :Br:^-$

H   CH₃

Products

Reaction coordinate

■ FIGURE 22.13 Reaction profile for S_N2 reaction.

Sample Problem 22.6 shows how. The reaction in Sample Problem 22.6, as well as many others which at first glance appear quite different, all take place by the S_N2 mechanism.

SAMPLE PROBLEM

22.6 Predict the product of the following reaction, which takes place by an S_N2 mechanism:

$$CH_3CH_2O^- + CH_3CH_2CH_2Br \longrightarrow$$

Solution (a) Compare the reaction to the general reaction

$$Nuc:^- + RX \longrightarrow RNuc + :X^-$$

and identify the nucleophile, R, and the leaving group, X:

$$CH_3CH_2O^- + \underbrace{(CH_3CH_2CH_2)}_{R}\underbrace{(Br)}_{X} \longrightarrow$$

$$\underset{Nuc:^-}{\uparrow}$$

In an S_N2 reaction, the nucleophile, Nuc:⁻, takes the place of the leaving group, X:

$$CH_3CH_2O^- + \underbrace{(CH_3CH_2CH_2)}_{R}\underbrace{(Br)}_{X} \longrightarrow \underbrace{(CH_3CH_2CH_2)}_{R}\overset{Nuc:}{-}OCH_2CH_3 + \underbrace{(Br^-)}_{X^-}$$

$$\underset{Nuc:^-}{\uparrow}$$

PRACTICE PROBLEM

22.24   Predict the products of each of the following reactions:
(a) $HC \equiv C^- + CH_3CH_2CH_2CH_2Cl \longrightarrow$
(b) $CH_3COO^- + H_2C = CHCH_2Br \longrightarrow$

Predicting Whether or Not an S_N2 Reaction Will Take Place

Now that we know how to predict the products of S_N2 reactions, let's consider how to predict whether or not an S_N2 reaction will take place. In all the cases we have looked at in which an S_N2 reaction takes place, the nucleophile is a stronger base than the leaving group. The reaction in equation 22.4 is an example:

$$HO^- + CH_3Cl(aq) \longrightarrow CH_3OH(aq) + Cl^- \qquad (22.4)$$

The hydroxide ion, HO^-, is the conjugate base of water, which is a very weak acid. Therefore, the hydroxide ion is a strong base (Section 15.5). The chloride ion, Cl^-, is the conjugate base of HCl, which is a strong acid. Therefore, the chloride ion is a weak base.

The S_N2 reaction

$$Nuc:^- + RX \longrightarrow RNuc + X^-$$

is similar to a Brønsted–Lowry acid–base reaction

$$B:^- + HX \longrightarrow HB + X^-$$

In a Brønsted–Lowry acid–base reaction, the stronger acid and the stronger base react to form the weaker acid and the weaker base (Section 16.1); the stronger base displaces the weaker base from the acid. Similarly, *in S_N2 reactions, stronger bases displace weaker bases*. Reactions such as

$$HO^- + H_3C-H \longrightarrow H_3COH + H^-$$

Substitution of F, Cl, or Br for hydrogen in alkanes does not take place by an S_N2 mechanism.

in which a base displaces a much stronger base, do not take place. (Another reason that methane and hydroxide ion do not react to form methyl alcohol and hydride ion is that H is less electronegative than C. There is no partial positive charge on C in CH_4 to attract the nucleophile, HO^-.)

PRACTICE PROBLEM

22.25   Predict whether or not each of the following reactions will take place:
(a) $CN^- + CH_3CH_2CH_2Br \longrightarrow CH_3CH_2CH_2CN + Br^-$ (The negative charge on the cyanide ion is on the carbon, $:C :: N:$.)
(b) $Br^- + CH_3CH_2OH \longrightarrow CH_3CH_2Br + OH^-$

If the difference in base strength of a nucleophile and of a leaving group is not too large, the equilibrium can be shifted. For example, HI is a stronger acid than HBr (Section 16.2). As a result, Br^- is a stronger base than I^-. The equilibrium between iodide ion, 1-bromobutane, 1-iodobutane, and bromide ion

$$I^- + CH_3CH_2CH_2CH_2Br \rightleftharpoons CH_3CH_2CH_2CH_2I + Br^-$$

should lie to the left. Bromide ion should displace iodide ion from 1-iodobutane.

But if 1-bromobutane is treated with a solution of sodium iodide ion in acetone, 1-bromobutane is completely converted to 1-iodobutane. Reaction takes place to the right; iodide ion displaces bromide ion. Why? Sodium iodide is soluble in acetone, but sodium bromide is not. Precipitation of insoluble sodium bromide shifts the equilibrium to the right:

$$NaI(acetone) + CH_3CH_2CH_2CH_2Br(acetone) \longrightarrow CH_3CH_2CH_2CH_2I(acetone) + NaBr(s)$$

More Observations of S_N2 Reactions

As usual, the rate at which reaction takes place must be considered as well as the position of equilibrium. Comparisons of rates of reaction are in agreement with the S_N2 mechanism. For example, when $CH_3CH_2CH_2CH_2Br$ and NaI are mixed in acetone solution, a precipitate of NaBr forms rapidly. Reaction of $CH_3CH_2CH_2CH_2Cl$ with sodium iodide in acetone is slower. A C—Cl bond is shorter and stronger than a C—Br bond and more energy is required to break the stronger C—Cl bond. The activation energy for the reaction of $CH_3CH_2CH_2CH_2Cl$ is higher than that for the reaction of $CH_3CH_2CH_2CH_2Br$, and the reaction of $CH_3CH_2CH_2CH_2Cl$ is slower.

The S_N2 mechanism is also consistent with the effect of changes in the structure of the alkyl group, R, on rate of reaction. The rates of S_N2 reaction of CH_3Br, CH_3CH_2Br, $CH_3CH_2CH_2Br$, $(CH_3)_2CHCH_2Br$, and $(CH_3)_3CCH_2Br$ with chloride ion

$$RBr + Cl^- \longrightarrow RCl + Br^-$$

decrease in the order

$$CH_3Br > CH_3CH_2Br > CH_2CH_2CH_2Br > (CH_3)_2CHCH_2Br > (CH_3)_3CCH_2Br$$
relative rate   37   1.0   0.67   0.33   0.000 006

■Figure 22.14 shows space-filling models of these five compounds. You can see that a chloride ion would find it harder and harder to get at the back of the carbon to which the —Br is attached as H is replaced by larger and larger alkyl groups. The back of the carbon is said to be increasingly **hindered.**

The effect of solvents on the rate of substitution is also explained by the S_N2 mechanism. The rate of the reaction

$$CH_3Br + {}^-OH \longrightarrow CH_3OH + Br^-$$

is 10^{16} as fast in the gas phase as in aqueous solution and 10^6 times as fast in solution in dimethylformamide, $HCON(CH_3)_2$, as in alcohol or aqueous alcohol, $CH_3CH_2OH(aq)$. Both water and ethanol solvate the OH^- ion and keep it from reacting:

CH_3Br

CH_3CH_2Br

$CH_3CH_2CH_2Br$

$(CH_3)_2CHCH_2Br$

$(CH_3)_3CCH_2Br$

■ FIGURE 22.14 Approach to the back of the carbon to which Br is attached is increasingly blocked when —H is replaced by —CH₃.

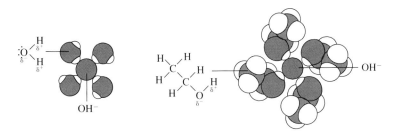

Dimethylformamide has no hydrogen attached to O or N; the hydrogens in dimethylformamide are attached to carbon and can't hydrogen bond to OH^-. As a result, OH^- in dimethylformamide solution is free to displace Br^-.

Petroleum currently provides not only about 42% of the energy used in the United States but also most of the organic chemicals that are used to make everything from adhesives, analgesics, and antifreeze to zip-lock bags and zippers. Petroleum refining is the largest manufacturing industry in the United States and provides about 10% of the gross domestic product. The recovery of oil and natural gas from continental shelves is by far the largest use of mineral resources from the oceans.

Crude petroleum is composed of thousands of different organic compounds. Continuous chain alkanes make up a larger fraction of most crudes than any other type of compound. The second most important fraction is cycloalkanes. Small amounts of branched chain alkanes, aromatics, sulfur-containing, oxygen-containing, and nitrogen-containing compounds, and salt (NaCl) are also present along with traces of metals such as nickel and vanadium.

Refining involves conversion and separation (■Figure 22.15). Fractional distillation is used to separate crude oil into various fractions as shown in Figure 13.10. The fractions obtained by the distillation of crude oil are converted to useful products by chemical reactions.

In the United States at the present time the largest market by far for petroleum products is for gasoline. To increase the proportion of gasoline obtainable from crude oil, large hydrocarbon molecules are broken down or cracked into smaller molecules by heating in the presence of appropriate catalysts. A typical reaction is

$$CH_3(CH_2)_{27}CH_3 \xrightarrow{\text{catalyst, thermal energy}}$$
$$\underbrace{CH_3(CH_2)_5CH_3 + CH_3(CH_2)_4CH{=}CH_2}_{\text{gasoline}}$$
$$+ \underbrace{CH_3(CH_2)_{12}CH{=}CH_2}_{\text{recycle}}$$

Cracking makes possible use of almost all of the crude.

For use as fuel for automobiles, appropriate volatility (boiling range and vapor pressure) and octane rating are important. Gasoline must be volatile enough to be vaporized. However, if gasoline is too volatile, vapor lock (formation of bubbles of vapor that block the fuel pump and fuel lines) takes place, and the engine stalls.

Octane rating is a measure of the antiknock properties of gasoline. Knocking occurs when the fuel burns too soon and wastes power; severe knocking may damage the engine. Continuous chain alkanes have very low octane ratings and knock badly when used as gasoline. Branched chain alkanes and aromatic hydrocarbons have high octane ratings and good antiknock properties. Therefore, cracking products are converted to branched chain alkanes and aromatic hydrocarbons. Small alkene molecules are linked to form larger, branched chain alkenes:

$$2(CH_3)_2C{=}CH_2 \xrightarrow[\text{pressure}]{\text{catalyst, thermal energy,}}$$
$$(CH_3)_3CCH_2C(CH_3){=}CH_2$$
$$+ (CH_3)_3CCH{=}C(CH_3)_2$$

Alkenes are linked to alkanes to form branched chain alkanes:

$$H_2C{=}CHCH_2CH_3 + (CH_3)_3CH \xrightarrow{\text{catalyst}}$$
$$(CH_3)_3CCH_2CH(CH_3)_2$$
$$+ (CH_3)_2CHC(CH_3)_2CH_2CH_3$$

Alkenes are linked to aromatics to form substituted aromatic hydrocarbons:

$$\text{(benzene ring)} + H_2C{=}CH_2 \xrightarrow{\text{catalyst}}$$

$$\text{(benzene ring)}{-}CH_2CH_3$$

Alkanes and cycloalkanes are reformed

22.9 SYNTHESIS

Synthesis is one of the major activities of organic chemists. Some organic compounds are obtained from natural sources, for example, sugar from sugarcane and ethyl alcohol by fermentation of plant materials such as grain and grapes. However, many compounds can be produced at lower cost and in larger quantities by synthesis. Most of the many organic compounds so widely used today are synthetic. *There is no difference between synthetic and natural compounds;* pure synthetic ethanol and pure ethanol from natural sources are identical. In the United States, most of the ethanol used commercially is synthetic because synthetic ethanol can be produced at a lower cost.

Organic chemists try to improve known syntheses so that a better quality product is obtained at lower cost. They attempt to find the best conditions for carrying out a particular reaction in the shortest time with the best yield. Organic chemists also look for new ways of making useful compounds that begin with less expensive starting materials or produce fewer or less harmful by-products.

Fermentation vats for wine. The alcohol in wine comes from sugar in the grapes.

to aromatics:

$$\text{C}_6\text{H}_{11}-\text{CH}_3 \xrightarrow[\text{and catalyst}]{\text{thermal energy}}$$

$$\text{C}_6\text{H}_5-\text{CH}_3 + 3\text{H}_2$$

$$\text{CH}_3(\text{CH}_2)_5\text{CH}_3 \xrightarrow[\text{and catalyst}]{\text{thermal energy}}$$

$$\text{C}_6\text{H}_5-\text{CH}_3 + 4\text{H}_2$$

Of the crude oil refined in the United States, about 47% is converted to gasoline and about 47% to other fuels such as heating oil, diesel fuel, and jet fuel. Only about 6% is used as a raw material for the production of organic chemicals and synthetic polymers although most organic chemicals and polymers are made from petroleum.

The first organic chemical to be made on an industrial scale from petroleum was isopropyl alcohol, $(\text{CH}_3)_2\text{CHOH}$, which was made from propylene by Standard Oil of New Jersey (now Exxon) in 1920:

$$\text{CH}_3\text{CH}{=}\text{CH}_2 + \text{H}_2\text{O} \xrightarrow{\text{H}_2\text{SO}_4}$$
$$\text{CH}_3\text{CH(OH)CH}_3$$

By 1925 production was about 1.6×10^5 lb per year, and the petrochemical industry was under way. From that time, the proportion of all organic chemicals that are produced from petroleum has increased until today over 80% of all organic chemicals are petrochemicals.

Ethylene is the principal raw material for the industrial synthesis of organic compounds. Of the 29 organic chemicals in the Top 50, seven are made from ethylene in one or more steps: ethylene dichloride, vinyl chloride, ethylbenzene, styrene, ethylene glycol, ethylene oxide, and vinyl acetate.

Use of petroleum products as fuels wastes resources by converting high-energy low-entropy materials to low-energy high-entropy carbon dioxide and water. The formation of oil fields took millions of years. The combustion of gasoline, heating oil, diesel fuel, and jet fuel takes but a moment. Fossil fuels are being used up 100 000 times faster than they are being formed. For the near future, efficient use is the best source of petroleum savings in quantities large enough to have a significant effect. Since 1975, conservation has yielded energy savings of 15–20%. For the long term, other sources of energy and of organic chemicals will have to be developed.

■ FIGURE 22.15   A petroleum refinery. The tall tower at the left is a cracking unit where large hydrocarbon molecules are broken down into smaller molecules.

Organic chemists are also continually trying to make new compounds that are better than known compounds, for example, detergents that do not form a scum in hard water as soap does or drugs that do not cause allergic reactions as natural penicillin does. These new compounds may be similar in structure to naturally occurring compounds or quite different. Compounds with properties not found in natural materials, such as nonstick Teflon linings for pans, can also be made. Many syntheses involve S_N2 reactions.

To plan a synthesis, organic chemists work backward from the structure of the product they want to make to the structures of possible starting materials. Sample Problem 22.7 illustrates planning a synthesis.

Drug resistance occurs more often with antibiotics than with their synthetic analogues.

SAMPLE PROBLEM

22.7   Write a structural formula for a substrate that could be used to make $\text{C}_6\text{H}_5\text{CH}_2\text{CN}$ by an S_N2 reaction and show the nucleophile that would be used.

PRACTICE PROBLEM

22.26 Write a structural formula for a substrate that could be used to make $H_2C{=}CHCH_2OH$ by an S_N2 reaction and show the nucleophile that would be used.

Although a synthesis may require more than one step, the synthesis is planned by working backward from the desired product to the starting materials one step at a time. The preparation of iodoethane, CH_3CH_2I, from ethane is a simple example of a synthesis that requires more than one step because iodine does not react with alkanes. However, iodoethane can be made from chloroethane by an S_N2 reaction:

$$\text{NaI(acetone)} + \text{CH}_3\text{CH}_2\text{Cl(acetone)} \longrightarrow \text{CH}_3\text{CH}_2\text{I(acetone)} + \text{NaCl(s)}$$

Chloroethane can be made from ethane by chlorination:

Why must an excess of CH₃CH₃ be used?

$$\text{CH}_3\text{CH}_3(g) + \text{Cl}_2(g) \xrightarrow{\text{thermal energy and/or light}} \text{CH}_3\text{CH}_2\text{Cl}(g) + \text{HCl}(g)$$
$$\text{excess}$$

SUMMARY

About 91% of the more than 13 million compounds that are known are organic compounds. Twenty-nine of the Top 50 chemicals are organic as are petroleum and natural gas. Biochemistry, the chemistry of living matter, is based on organic chemistry.

Carbon forms an enormous number of compounds with hydrogen. The key feature of the chemistry of carbon is **catenation,** the formation of chains of atoms of the same element in compounds. Carbon chains can be either **continuous (straight)** or **branched;** rings of carbon atoms also exist. Single, double, and triple bonds can occur, and carbon can be bonded to elements other than carbon and hydrogen. The number of possible isomers increases rapidly as the number of carbon atoms in the molecule increases.

The enormous number of possible carbon compounds is one reason so many carbon compounds are known. In addition, compounds with chains of carbon atoms are unusually stable as a result of the great strength of the carbon–carbon bond. They are also unusually unreactive because no low-energy pathway for reaction exists. Carbon in compounds has an octet of valence electrons, and a new bond can't form until an old bond breaks. Another factor in the lack of reactivity of carbon compounds is the fact that the units of most carbon compounds are uncharged molecules, not ions. Collisions between molecules must take place with the correct orientation for reaction to occur because covalent bonds have definite length and direction in space.

Two kinds of isomers exist: constitutional isomers and stereoisomers. In **constitutional isomers,** the atoms are connected in different ways. In **stereoisomers,** the atoms are connected in the same way; stereoisomers differ only in the arrangement of atoms or groups in space. The two types of stereoisomers are **geometric isomers** (*cis-, trans-*isomers) and **optical isomers** (enantiomers). **Enantiomers** are mirror-image isomers.

Objects that can exist as enantiomers are said to be **chiral.** Models of enantiomers are not superimposable. Many chiral molecules can be recognized by the presence of a **stereocenter,** an atom with four different atoms or groups attached. Objects that are not chiral are called **achiral.** Only a chiral probe such as a human nose or plane-polarized light can distinguish between enantiomers. In **plane-polarized light,** all waves vibrate in a single direction.

Products of biological reactions are usually pure enantiomers. Products of ordinary laboratory and industrial reactions are usually racemic mixtures. **Racemic mixtures** are mixtures that contain equal quantities of each enantiomer.

Because there are so many organic compounds, organic compounds are usually divided into families of compounds that have similar structures and properties. The alkanes make up the simplest family of organic compounds. **Alkanes** are **hydrocarbons** (compounds of carbon and hydrogen) that have no multiple bonds. The systematic names of organic compounds are based on the names of the continuous or straight chain alkanes. The **main chain** or **parent chain** of an alkane is the longest continuous chain in the molecule.

The continuous chain alkanes are members of a homologous series. In a **homologous series** each member differs from other members by multiples of —CH₂—. The physical properties of the members of a homologous series change in a predictable way as the number of carbon atoms increases. The chemical properties are similar. Alkanes are not very reactive and undergo only combustion, pyrolysis (decomposition by heat in the absence of oxygen), and halogenation. Pyrolysis of alkanes in petroleum is called **cracking.**

Most organic compounds contain one or more functional groups. A **functional group** is an atom or group of atoms in a molecule that accounts for the characteristic properties of the molecule. All molecules that have the same functional group function or react similarly. Functional groups are where reactions usually take place in organic molecules; the carbon chain generally does not change.

Reactions and synthesis and the study of reaction mechanisms are important areas of organic chemistry. Organic chemists are continually trying to make new compounds that are better for the purpose at hand than known compounds. Syntheses of organic compounds are planned by working backward from product to reactants.

Mechanisms are a powerful way of organizing the vast amount of information available about the reactions of organic compounds. The S_N2 mechanism (substitution nucleophilic bimolecular) is a very important type of mechanism. **Nucleophilic** means nucleus loving; a **nucleophile** is attracted by a positive charge. The species that undergoes reaction is called the **substrate.** The group that is replaced is called the **leaving group.** The S_N2 reaction takes place in a single step; configuration is **inverted**— that is, the nucleophile becomes attached to the back of the carbon to which the leaving group was attached. **Hindrance** by large groups slows the reaction. Nucleophilic substitution can be regarded as a Lewis acid–base reaction.

ADDITIONAL PRACTICE PROBLEMS

For information about the organization of Additional Practice Problems, Stop & Test Yourself, Putting Things Together, and Applications, see the beginnings of these sections in Chapter 1.

22.27 (a) What are essential elements? (b) Which two elements that are abundant on or near Earth's surface are not essential? (c) Why are these two elements probably not essential?

22.28 Define and give an example of each of the following terms: (a) catenation (b) continuous chain (c) branched chain (d) constitutional isomers

22.29 Do the structures in each pair represent the same compound, constitutional isomers, or different compounds that are *not* isomers?

(a) CH₃CH₂CH₂CH₂CH₃   CH₃CH₂CH₂
                                |
                            CH₂CH₃

(b) CH₃CH₂CHCH₂CH₃   CH₃CH(CH₃)CH₂CH₂CH₃
           |
           CH₃

(c)   H
        |
        C
H₃C ⁄   `CH₂CH₂CH₃   CH₃CH₂CH₂CH(CH₃)₂
   H₃C ⁄

(d) CH₃CH₂CHCH₃   CH₃CH₂CHCH₂CH₃
           | |
         CH₂CH₃   CH₃

(e) CH₃CH₂CH₂CH₂CH₃   H₂C—CH₂
                            H₂C   CH₂
                               CH₂

22.30 (a) Write a complete structural formula for [hexagon with N].

(b) Write a line formula for [ring structure]

22.31 Explain what is meant by each of the following terms: (a) plane-polarized light (b) optically active (c) stereocenter (d) chiral (e) achiral (f) racemic mixture

22.32 Explain what is meant by each of the following terms: (a) stereoisomers (b) enantiomers (c) diastereomers (d) geometric isomers (e) *trans*-isomer (f) *cis*-isomer

22.33 Of the following letters of the alphabet, which are chiral, and which are achiral (in the font shown)?

A, C, F, J, X

22.34 If a stereocenter is present, mark it with an asterisk. If no stereocenter is present, label the structure "achiral."

(a) $CH_3CH_2CH(OH)CH_3$

(b) $H_2C{=}C(CH_3)COOCH_3$

(c) Cl—⟨benzene⟩—CH—⟨benzene⟩—Cl with CHCl$_2$ on the central CH

(d) Cl—⟨benzene⟩—CH—⟨benzene⟩—CH$_3$ with CHCl$_2$ on the central CH

(e) H_3C—⟨cyclopentane ring⟩=O

22.35 How many stereocenters are there in each of the following compounds?

(a) $HOOCCH(OH)CH(OH)COOH$

(b) $HOCH_2CH(OH)CH(OH)CH(OH)CHO$

(c) $CHCl_2CHClCHF_2$

(d) HO—⟨benzene ring with two I substituents⟩—CH$_2$CH—⟨benzene⟩ with COOH

(e) CH_3CH_2—C—O—C with CH$_3$ above, CH$_2$CH$_3$ below, and =O and NH$_2$ on the right carbon

(f) H_3C—⟨cyclopentane ring⟩—CH$_3$

22.36 (a) Draw the structures of the two stereoisomers of $CH_3CH_2CH{=}CHCH_3$. Label the *trans*-isomer *"trans."*
(b) Draw the structures of the two stereoisomers of

⟨cyclobutane ring with CH$_3$ and CH$_3$⟩

Label the *trans*-isomer *"trans."*
(c) Draw the enantiomer of Br—C with CH$_2$CH$_3$, H, and CH$_3$

(d) Draw the structures of the two stereoisomers of $CH_3OCH(CH_3)CH_2CH_3$.

22.37 Draw the structures of the two stereoisomers of each of the following compounds if stereoisomers exist:

(a) $(CH_3CH_2)_2C{=}CHCH_3$

(b) $CH_3CH_2CH(CH_3)CH_2N(CH_3)_2$

(c) H_3C—⟨cyclohexane ring⟩—CH$_3$

(d) H_3C—⟨benzene ring⟩—CH$_3$

(e) $H_2C{=}CHC(Cl){=}CHCH_3$

22.38 (a) Explain briefly why so many more compounds of carbon exist than compounds of any other element.   (b) Give three reasons why experiments in the organic chemistry laboratory are likely to be more time-consuming than experiments in the general chemistry laboratory.

22.39 (a) Why is there a sudden difference in density between butane and pentane?   (b) From the data in Table 22.3, predict the density of $CH_3(CH_2)_9CH_3$ at 20 °C.

22.40 For each of the following, write an equation if a reaction will take place. If no reaction will occur, write "No reaction."

(a) $CH_3CH_2CH_3(g)\ +\ O_2(g)\ \xrightarrow[\text{excess}]{\text{thermal energy}}$

(b) $CH_3CH_2CH_3(g)\ +\ O_2(g)\ \xrightarrow[\text{excess}]{\text{thermal energy}}$

(c) $CH_3CH_2CH_3(g)\ +\ MnO_4^-\ +\ OH^-\ \xrightarrow{\text{thermal energy}}$

(d) $CH_3CH_2CH_3(g)\ \xrightarrow{\text{thermal energy—no } O_2}$

(e) $CH_3CH_2CH_3(g)\ +\ Na(s)\ \longrightarrow$

22.41 For each of the following, write an equation if a reaction will take place. If no reaction will occur, write "No reaction."

(a) $CH_3CH_2CH_3(g)\ +\ Cl_2(g)\ \xrightarrow[\text{excess}]{\text{thermal energy}}$

(b) $CH_3CH_2CH_3(g)\ +\ Cl_2(g)\ \xrightarrow[\text{excess}]{\text{thermal energy}}$

(c) $CH_3CH_2CH_3(g)\ +\ Cl_2(g)\ \xrightarrow[\text{excess}]{\text{room temperature in the dark}}$

(d) $CH_3CH_2CH_3(g)\ +\ HCl(g)\ \longrightarrow$

(e) $CH_3CH_2CH_3(g)\ +\ H_2SO_4(l)\ \longrightarrow$

22.42 For CH_3CH_3, (a) how many monochloro derivatives are there?   (b) How many dichloro?   (c) Trichloro?   (d) Tetrachloro?   (e) Pentachloro?   (f) Hexachloro? Write a structural formula for each of the chloro derivatives of ethane.

22.43 An alkane that has a molecular mass of 72 forms only 1 monochloro derivative. (a) What is the structure of the alkane?   (b) Write the equation for the monochlorination of the alkane.   (c) Should an excess of alkane or an excess of chlorine be used to stop at monosubstitution? Explain your answer.   (d) What is necessary to make reaction take place?

22.44 Name each of the following compounds:

(a) $(CH_3CH_2CH_2)_2CHCH_3$ (b) $CH_3(CH_2)_5CH(CH_3)_2$

(c)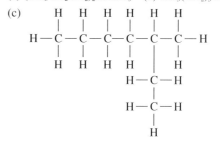

22.45 Write a structural formula for each of the following compounds: (a) 2,2-dimethylhexane (b) butylcyclopentane (c) 4-ethylheptane

22.46 To which class or classes of compounds does each of the following compounds belong?

(a) —CH_3 (b) $HCONH_2$

(c) $(CH_3)_2C{=}CH_2$ (d) $CH_3COOCH_2CH_3$
(e) $CH_3COCH_2CH_3$ (f) $H_3CC{\equiv}CCH_3$
(g) $(CH_3)_2CHCH_2CH_2CH_3$ (h) $CH_3CH_2NH_2$

22.47 To which class or classes of compounds does each of the following compounds belong?
(a) $(CH_3)_2CHCH_2OH$ (b) $CH_3OCH_2CH_3$
(c) $(CH_3)_2CHCl$ (d) CH_3CH_2I
(e) $CH_3CH_2CH_2COOH$ (f) CH_3CH_2CHO

(g) HO——CH_3 (h) —CH_2OH

22.48 Name each of the following compounds:
(a) $CH_3CH_2CH_2CH(Br)CH_3$
(b) $CH_3CH_2CH_2CH(CH_3)CH_2CH_2Cl$

22.49 Write a structural formula for each of the following compounds: (a) 1-chloro-2,3-dimethylbutane (b) 2,2-dimethyl-1-propanol (c) 1,2-dibromo-1,1-dichloroethane

22.50 The functional groups in each of the following pairs of classes of compounds are often confused. What do the members of each pair have in common and what is different about them? (a) an amine and an amide (b) an aldehyde and a ketone (c) a carboxylic acid and an ester (d) an ether and a ketone (e) an alcohol and a phenol

22.51 In the equation for the reaction

$$CH_3CH_2CH_2CH_2Br + NH_3 \longrightarrow$$
$$CH_3CH_2CH_2CH_2NH_3^+ + Br^-$$

(a) circle and label the nucleophile. (b) Draw a rectangle around the substrate. (c) Circle and label the leaving group. (d) What group would be represented by R in the general equation for nucleophilic substitution?

22.52 Predict the products of each of the following reactions, which take place by the S_N2 mechanism:
(a) $CH_3CH_2CH_2CH_2Cl + OH^- \longrightarrow$

(b) $CH_3CH_2CH_2CH_2Cl + OCH_3^- \longrightarrow$
(c) $CH_3CH_2CH_2CH_2Cl + SH^- \longrightarrow$
(d) $CH_3CH_2CH_2CH_2Cl + N_3^- \longrightarrow$
(e) $CH_3CH_2CH_2CH_2Cl + I^- \longrightarrow$

22.53 Predict the products of each of the following reactions, which take place by the S_N2 mechanism:
(a) $(CH_3)_2CHCH_2CH_2Cl + I^- \longrightarrow$
(b) $CH_3CH_2CH_2CH_2CH_2Cl + I^- \longrightarrow$
(c) $CH_3CH_2CH_2CH_2Br + OH^- \longrightarrow$
(d) $H_2C{=}CHCH_2Cl + SCH_3^- \longrightarrow$
(e) $CH_3CH_2CH_2Cl + OCH_2CH_3^- \longrightarrow$

22.54 Predict whether or not each of the following reactions will take place. Explain your answers.
(a) $CH_3COO^- + C_6H_5CH_2Br \longrightarrow$
$$C_6H_5CH_2OOCCH_3 + Br^-$$
(b) $OH^- + C_6H_5CH_2CH_3 \longrightarrow C_6H_5CH_2OH + CH_3^-$
(c) $Br^- + CH_3CH_2NH_2 \longrightarrow CH_3CH_2Br + NH_2^-$
(d) $CH_3S^- + CH_3CH_2OCH_2CH_3 \longrightarrow$
$$CH_3SCH_2CH_3 + OCH_2CH_3^-$$
(e) $SCN^- + CH_3I \longrightarrow CH_3SCN + I^-$

22.55 Write the mechanism for the S_N2 reaction of *trans*-1-chloro-3-methylcyclopentane with hydroxide ion.

22.56 Explain the following relative rates of S_N2 reaction with chloride ion in dimethylformamide solution:

$$CH_3Br > CH_3CH_2Br > (CH_3)_2CHBr > (CH_3)_3CBr$$

| relative rate | 37 | 1.0 | 0.02 | 0.0008 |

22.57 Write structural formulas for a substrate–nucleophile combination that could be used to make each of the following compounds: (a) $CH_3CH_2CH_2OH$
(b) $CH_3CH_2CH_2CH_2OCH_2CH_3$
(c) $CH_3CH_2CH_2CH_2CH_2SCH_2CH_3$
(d) $H_2C{=}CHCH_2OOCCH_3$ (e) $(CH_3)_4N^+I^-$

22.58 Outline a method for synthesizing iodocyclohexane (cyclohexyl iodide) from cyclohexane.

22.59 Write the molecular equation for each of the following reactions:
(a) $(CH_3)_3COH(l) + K(s) \longrightarrow$
(b) $(CH_3)_2NH(aq) + H_2SO_4(aq) \longrightarrow$
(c) $CH_3COOH(aq) + Ba(OH)_2(aq) \longrightarrow$

(d) $(l) + H_2(g) \xrightarrow{\text{catalyst}}$

(e) —$OH(aq) + NaOH(aq) \longrightarrow$

22.60 Write equations for each of the following reactions:
(a) $(CH_3)_3N(g) + H_2O(l)$ (b) $C_6H_5NH_2(l) + HCl(aq)$
(c) $CH_3CH_2COOH(l) + H_2O(l)$ (d) $C_6H_5COOH(s) + NaOH(aq)$ (e) $CH_3COOH(aq) + (CH_3)_2NH(aq)$.

1. Which compound is not organic?
   (a) $CoBr_2$ (b) CCl_4 (c) CH_3MgI (d) CH_3COOH
   (e) CH_3CH_2ONa

2. The IUPAC name of the following compound is

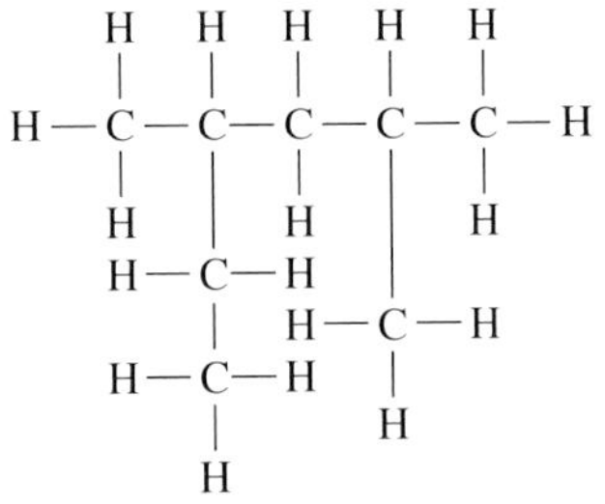

   (a) 1,1-dimethyl-3-ethylbutane
   (b) 2-methyl-4-ethylpentane
   (c) 4-ethyl-2-methylpentane
   (d) 3,5-dimethylhexane
   (e) 2,4-dimethylhexane

Use the following structures to answer questions 3 and 4.

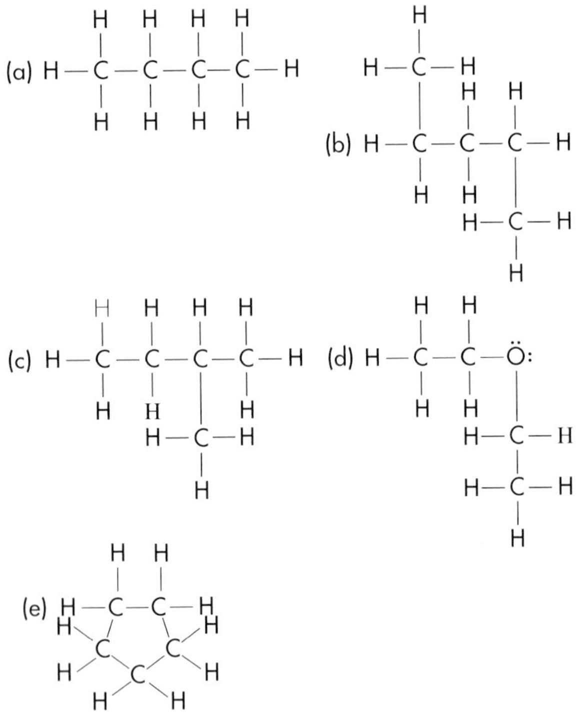

3. Which is the same as

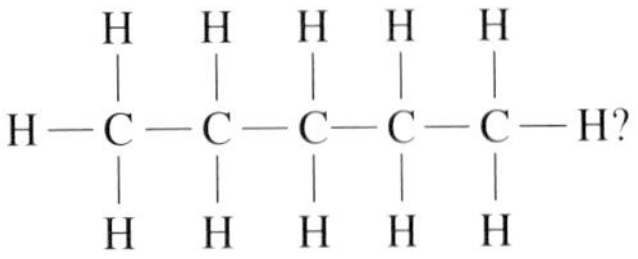

4. Which is an isomer of

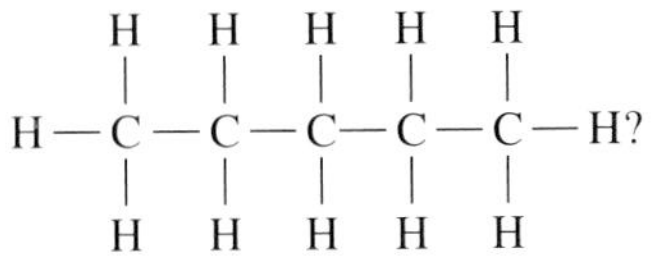

5. How many constitutional isomers of $C_4H_{10}O$ are there?
   (a) 3 (b) 4 (c) 5 (d) 6 (e) 7

6. Which of the following is *not* a reason for the existence of so many carbon compounds? (a) A large number of carbon compounds is possible. (b) Carbon–carbon bonds are unusually strong. (c) Carbon compounds are covalent. (d) Carbon can expand its octet. (e) Reactions of carbon compounds are usually slow at room temperature.

7. The compound $CH_3CH_2COCH_2CH_3$ is _____.
   (a) an alcohol (b) an aldehyde (c) an alkane (d) an ether
   (e) a ketone

8. Which of the following compounds do *not* have *cis-*, *trans-* isomers?
   (i) $CH_3CH_2CH{=}CHCH_3$ (ii) $CH_3CH_2C(CH_3){=}CH_2$
   (iii) $CH_3C{\equiv}CCH_3$

   (iv) 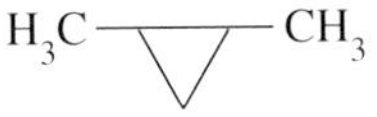 (v)

   (a) i and iv (b) ii and iii (c) ii and v (d) ii, iii, and v
   (e) iii, iv, and v

9. How many chiral carbons are there in the following?

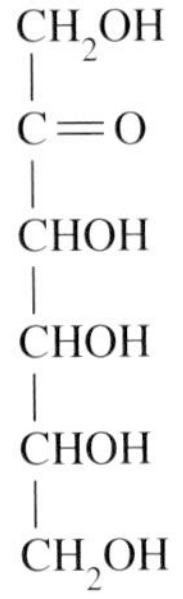

$$\begin{array}{c} CH_2OH \\ | \\ C{=}O \\ | \\ CHOH \\ | \\ CHOH \\ | \\ CHOH \\ | \\ CH_2OH \end{array}$$

   (a) 1 (b) 2 (c) 3 (d) 4 (e) 5

10. Arrange the compounds of the following set in order of increasing boiling point:

$$CH_3CH_2CH_2CH_2CH_2CH_3 \qquad CH_3CH_2CH_2CH_2CH_3$$
$$\text{A} \qquad\qquad \text{B}$$
$$CH_3CH_2CH(CH_3)_2 \qquad (CH_3)_4C$$
$$\text{C} \qquad\qquad \text{D}$$

   (a) D < C < B < A (b) A < B < C < D
   (c) B < C < D < A (d) A < D < C < B
   (e) D < A < B < C

11. Which of the following molecules is nonpolar?

   (a) 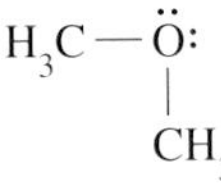 (b) 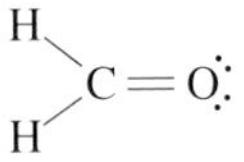 (c)

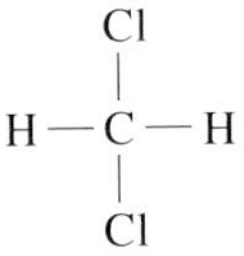

(d)
$$\underset{H}{\overset{H}{}}C=C\underset{H}{\overset{H}{}}$$

(e)
$$H-\overset{\overset{\displaystyle Cl}{|}}{\underset{\underset{\displaystyle Cl}{|}}{C}}-Cl$$

12. Which of the following combinations will result in reaction?

(i) $CH_3CH_2CH_2CH_2CH_3(l) \xrightarrow{\text{thermal energy, no } O_2}$

(ii) $CH_3CH_2CH_2CH_2CH_3(l) + Br_2 \xrightarrow{\text{light}}$

(iii) $CH_3CH_2CH_2CH_2CH_3(l) + Cr_2O_7^{2-} + H^+ \xrightarrow{\text{thermal energy}}$

(iv) $CH_3CH_2CH_2CH_2CH_3(l) + O_2(g) \xrightarrow{\text{thermal energy}}$

(v) $CH_3CH_2CH_2CH_2CH_3(l) + OH^- \xrightarrow{\text{thermal energy}}$

(a) all but i   (b) ii and iv   (c) iii and v   (d) i, ii, and iv
(e) All of them will result in reaction.

13. To make [cyclohexane ring with H_3C and OH substituents], use

(a) [ring with H_3C and H] and OH^- (b) [ring with H_3C and H] and O_2

(c) [ring with H_3C and H] and basic MnO_4^- (d) [ring with H_3C and Cl] and OH^-

(e) [ring with H_3C and Cl] and OH^-

PUTTING THINGS TOGETHER

22.61 (a) Discuss the reasons why there are so many more organic compounds than there are inorganic compounds.   (b) At the present time, what is the most important source of organic compounds?

22.62 What is the difference between *configurations* and *conformations?*

22.63 Write structural formulas for (a) all the aldehydes with molecular formula $C_5H_{10}O$ (b) all the alcohols with molecular formula $C_4H_{10}O$ (c) the continuous chain alkane with 26 H

22.64 (a) Write condensed structural formulas for the first five members of the homologous series of continuous chain alcohols.   (b) Write the general formula for these alcohols.

22.65 Predict the products of each of the following reactions. If no reaction will take place, write "No reaction."
(a) $CH_3CH_2CH_2Br + SH^- \longrightarrow$
(b) $CH_3CH_2CH_2OH + I^- \longrightarrow$
(c) $CH_3CH_2CH_2OOCCH_3 + Br^- \longrightarrow$

(d) $OCH_3^- +$ [benzene ring]$-CH_2Br \longrightarrow$

(e) $CH_3CH_2CH_2I + CN^- \longrightarrow$

(f) [cyclopentane ring with H_3C and CH_3] $+ OCH_3^- \longrightarrow$

(g) $\underset{H_3C}{\overset{CH_2CH_2CH_3}{}}\overset{|}{\underset{\overset{|}{Br}}{C}}{\cdots}H + CH_3CH_2S^- \longrightarrow$

22.66 Outline a method for bringing about each of the following conversions: (a) CH_4 to CH_3OCH_3 (b) CH_3CH_3 to CH_3CH_2SH (c) cyclopentane to cyclopentanol

22.67 A compound with molecular formula $C_3H_6Cl_2$ is a racemic mixture. (a) Write the complete and condensed structural formulas for this compound.   (b) What is the systematic name of the compound?   (c) How many other isomers of $C_3H_6Cl_2$ are possible? Give their complete structural formulas, a condensed structural formula for each, and the name of each.

22.68 A compound that has the molecular formula C_5H_{10} does not react with hydrogen gas in the presence of a catalyst. The compound reacts with chlorine gas when heated forming a single monochloro derivative, C_5H_9Cl. (a) Write a structural formula for the compound and give its name.   (b) Write the equation for the reaction of the compound with chlorine.

22.69 Write carbon skeletons, complete structural formulas, condensed structural formulas, and names for all the constitutional isomers of (a) C_4H_9Cl,   (b) $C_4H_8Cl_2$.

22.70 A gaseous compound is 85.63% C and 14.37% H. A 1.76-g sample occupies a volume of 764 mL at 22.3 °C and 758.2 mmHg. (a) What is the empirical formula?   (b) What is the molecular formula?   (c) Write structural formulas for the six isomers of this molecular formula.

22.71 All of the following solvents dissolve both substrates and nucleophiles. Which are good solvents for S_N2 reactions because they cannot solvate anions by hydrogen bonding?
(a) CH_3CN (b) CH_3COCH_3 (c) CH_3OH
(d) $(CH_3)_2S=O$ (e) $HCONH_2$

22.72 Write a resonance structure for CH_3CONH_2 that explains why this compound is not basic.

22.73 (a) Write the equation for the Brønsted–Lowry acid–base reaction between 1-propanol and hydrogen chloride.   (b) Explain why 1-propanol reacts with aqueous hydrogen chloride (hydrochloric acid) but not with aqueous sodium chloride to form 1-chloropropane.

22.74 An aqueous solution of a compound that has molecular formula $C_2H_4O_2$ is neutral to litmus. Write a structural formula for the compound.

22.75 How do single bonds compare with double bonds in length? In strength?

22.76 (a) Explain the following trend in normal boiling points:

$$CH_3CH_2CH_3 < CH_3OCH_3 < CH_3CH_2NH_2$$

bp, °C -42 -25 16.6

$$< CH_3CH_2OH < CH_3CH_2CH_2OH$$

               78.5   97.4

(b) Explain why the boiling point of 2-propanol (82.4 °C) is lower than the boiling point of 1-propanol.

22.77 Arrange the following species in order of increasing strength as Brønsted–Lowry bases: (a) CH_3^- (b) CH_3O^- (c) CH_3COO^- (d) $C_6H_5O^-$ (e) Cl^-. Explain your answer.

22.78 (a) What is the hybridization of the carbons in benzene? (b) Sketch the orbital model of the benzene molecule. (c) Explain why the benzene molecule is planar. (d) How does the structure of benzene compare with the structure of graphite?

22.79 (a) Draw stereochemical formulas for *cis-* and *trans-*1,2-dichloroethene. (b) One of the 1,2-dichloroethenes is polar and the other is nonpolar. Which is which? Explain your answer.

22.80 The reactivity of cyclohexane is similar to the reactivity of hexane. However, cyclopropane is much more reactive than propane. Explain why cyclopropane is less stable thermodynamically and more reactive kinetically than propane. (*Hint:* What is the hybridization of carbon in alkanes, and what is the normal angle between bonds to carbon for this hybridization?)

22.81 (a) Write the two important resonance structures for benzene.

(b) For aniline,   ⬡—ＮH₂   resonance structures such as   ⬡＝ＮH₂⁺   are important. How would you expect the basicity of aniline to compare with the basicity of ordinary amines such as CH_3NH_2? Explain your answer, then check it in Table 15.4. (c) Write the other three important resonance structures for aniline.

22.82 A 0.1436-g sample of an unknown organic acid isolated from lemon juice was titrated with 0.1000 M NaOH; 22.43 mL of 0.1000 M NaOH were required to neutralize the sample. A mixture of 0.0228 g of the unknown acid with 1.0000 g of camphor melted at 174.3 °C. How many —COOH groups are there in a molecule of the unknown acid?

22.83 A characteristic reaction of compounds with carbon–carbon double bonds is addition. An example is $H_2C=CH_2(g)$ + $HCl(g) \longrightarrow H_3CCH_2Cl$. The generally accepted mechanism of this reaction is

Step (1)

$$H_2C = CH_2 + H - \ddot{C}\ddot{l} \longrightarrow H_3\overset{+}{C} - CH_3 + :\ddot{C}\ddot{l}:^-$$

Step (2) $:\ddot{C}\ddot{l}:^- + H_2\overset{+}{C} - CH_3 \longrightarrow H_2C(Cl)CH_3$

(a) Which step is a Brønsted–Lowry acid–base reaction, and which step is a Lewis acid–base reaction? (b) In each step,

label the acid and the base. (c) In the overall reaction, does the entropy of the system increase or decrease? The enthalpy? The free energy? Explain your answers. (e) What other two reactions of carbon–carbon double bonds that have been discussed previously are addition reactions?

22.84 (a) Ethyl alcohol is converted to acetic acid by treatment with permanganate ion in basic solution. Manganese(IV) oxide precipitates. Write the net ionic equation for this reaction. (b) 2-propanol is converted to acetone by treatment with dichromate ion in acidic solution. Chromium(III) ion forms. Write the net ionic equation for this reaction. (c) Are the alcohols oxidized or reduced in these reactions? Give your reasoning. (d) Predict the product of the reaction of 1-propanol with basic permanganate. (e) Predict the product of the reaction of 2-butanol with acidic dichromate.

22.85 The boiling point of 1-butanol is 117 °C; the boiling point of diethyl ether is only 34.6 °C. However, the solubility in water of 1-butanol, 7.4 g/100 mL, is only slightly greater than the solubility of diethyl ether in water, 6.4 g/100 mL. Explain why the boiling points of 1-butanol and diethyl ether are very different but the water solubilities of the two compounds are similar.

22.86 Methyl iodide is made by an S_N2 reaction:

$$KI(aq) + CH_3OSO_2OCH_3(l) \longrightarrow$$
$$CH_3I(l) + KOSO_2OCH_3(aq)$$

(a) In this reaction, which substance is the substrate? Which species is the nucleophile? What is the leaving group? (b) Write the mechanism for the reaction. (c) If 800 (8.00×10^2) g KI and 473 mL $CH_3OSO_2OCH_3$ (density = 1.3283 g/mL) are used and 628 g of CH_3I are obtained, what is the percent yield?

22.87 For the reaction of

with KI in acetone, (a) write the mechanism. (b) Sketch the reaction profile. Potassium chloride precipitates; the reaction is exothermic. (c) Write the rate law.

22.88 Predict the product or products of each of the following reactions

(a) $CH_3CH=CH_2(g) + O_2(g) \xrightarrow[\text{excess}]{\text{thermal energy}}$

(b) $CH_3CH=CH_2(g) + O_2(g) \xrightarrow[\text{excess}]{\text{thermal energy}}$

(c) $CH_3CH=CH_2(g) + H_2(g) \xrightarrow{\text{catalyst}}$

22.89 What is the difference between (a) a continuous chain and a branched chain (b) constitutional isomers and stereoisomers (c) enantiomers and diastereomers (d) conformations and isomers (e) isomers and resonance structures.

22.90 A compound that has molecular formula C_3H_8O is inert toward metallic sodium. Write a structural formula for the compound.

22.91 An aqueous solution of an unknown compound was basic to litmus. A 0.023 06-g sample of the compound gave 0.032 68 g CO_2 and 0.033 45 g H_2O by combustion analysis, and a 0.825-g sample occupied a volume of 615 mL at 20.2 °C and 765 mmHg. Write the structural formula and give the name of the unknown compound.

22.92 Explain why 1-hexanol is soluble in hexane but only slightly soluble in water whereas glucose,

$$HOCH_2CH(OH)CH(OH)CH(OH)CH(OH)CHO,$$

is very soluble in water but insoluble in hexane.

22.93 Only one gram of procaine

$$H_2N-\bigcirc-COOCH_2CH_2N(CH_2CH_3)_2$$

will dissolve in 200 mL of water. However, procaine is very soluble in ethyl alcohol, diethyl ether, benzene, and hydrochloric acid. Explain the solubility of procaine.

22.94 (a) Indicate the type of hybrid orbital expected for each carbon in each of the following compounds and describe the geometry of the compound using stereochemical formulas. (i) CCl_4 (ii) $CH_3CH{=}CH_2$ (iii) $CH_3C{\equiv}CH$ (iv) CH_3CH_3 (v) CH_3OH (vi) $(CH_3)_2C{=}O$ (b) In part (a) (ii), how many sigma bonds are there? How many pi bonds?

22.95 Explain why cyclohexyne has not been prepared.

22.96 Write the Lewis formula for each of the following: (a) CH_3CN (b) CH_3NO_2 (c) C_6H_5COCl

22.97 (a) Calculate the pH of a 0.0150 M solution of phenol at 25 °C. (b) The pH of a 0.0150 M solution of cyclohexanol in pure carbon dioxide–free water at 25 °C is 7.0. Assuming that cyclohexanol is typical of alcohols, are alcohols acidic, basic, or neutral compared to water? (c) What can you conclude about the effect on the acidity of an —OH group of attaching an aromatic ring to the —OH group?

APPLICATIONS

22.98 Circle the functional groups in the structural formulas of each of the following compounds and give the class or classes of compounds to which each belongs:
(a) the low-calorie sweetener aspartame

$$\begin{array}{c}COOCH_3\\|\\H_2NCHCONHCHCH_2-\bigcirc\\|\\CH_2COOH\end{array}$$

(b) the analgesic and antipyretic phenacetin, which is listed as a carcinogen by the Environmental Protection Agency (EPA)

$$CH_3CH_2O-\bigcirc-NHCOCH_3$$

(c) the bronchodilator ephedrine, which is an ingredient of many cold medicines

$$\begin{array}{c}CH_3\\|\\HOCHCHNHCH_3\\|\\\bigcirc\end{array}$$

(d) the inhalation anesthetic methoxyflurane $CH_3OCF_2CHCl_2$

(e) the female hormone progesterone

(f) the flavoring agent vanillin

22.99 Today Neothyl, $CH_3OCH_2CH_2CH_3$, is preferred to diethyl ether as an anesthetic because it does not cause nausea and vomiting as diethyl ether does. Neothyl is prepared by an S_N2 reaction. Give two pairs of reactants that could be used to synthesize Neothyl.

22.100 Ethyl chloride, which is used as a refrigerant, solvent, and topical anesthetic, can be made by any one of three methods:

(i) $CH_3CH_3(g) + Cl_2(g) \xrightarrow{\text{thermal energy and/or light}}$

$$CH_3CH_2Cl(g) + HCl(g)$$

(ii) $H_2C{=}CH_2(g) + HCl(g) \longrightarrow CH_3CH_2Cl(g)$

(iii) $2CH_3CH_3(g) + 2HCl(g) + O_2(g) \xrightarrow{\text{catalyst}}$

$$2CH_3CH_2Cl(g) + 2H_2O(l)$$

(a) Which reaction is most atom economical? (The most *atom economical reaction* uses the most atoms of reactants to form the desired product and the fewest to form by-products. Atom economical reactions are examples of green chemistry.) (b) Which is least "green"? (*Hint:* Consider the percent by mass of reactants converted to products and the effect of by-products on the environment.)

22.101 Cholesterol, the name of the substance deposited on the walls of arteries, has become a household word. The line formula for cholesterol is

(a) To what class or classes of compounds does cholesterol belong? (b) Write the complete structural formula for cholesterol. (c) Write the line formula for the cycloalkane, cholestane, from which cholesterol is derived.

22.102 Citral, the principal component of oil of lemon grass, which is used in the synthesis of vitamin A, as a flavoring, and in perfumes, is a mixture of geranial and neral:

geranial

neral

(a) What is the relationship between geranial and neral? (b) Circle the functional groups in the structural formula of geranial and give the class or classes of compounds to which geranial belongs.

22.103 Hydroformylation (the "oxo" process)

$$RCH{=}CH_2 \xrightarrow{\text{CO, H}_2, \text{ catalyst}}$$

$$RCH_2CH_2CHO + RCH(CHO)CH_3$$

is the world's largest scale industrial homogeneous catalytic process. (a) What is a *homogeneous catalyst?* (b) Write condensed structural formulas for the products from $CH_3CH_2CH{=}CH_2$. (c) From propene, the yield of $CH_3CH_2CH_2CHO$ is 67%, and the yield of $(CH_3)_2CHCHO$ is 15%. How many pounds of each product will be obtained from one ton of propene?

22.104 Explain why naphthalene which is used to make mothballs, is quite soluble in benzene and ethyl alcohol but is insoluble in water.

22.105 Capacity for refining petroleum is usually stated in barrels (1 barrel = 42 U.S. gallons or 0.159 m^3). (a) How many gallons make up 100.0 barrels? (b) How many liters make up a barrel? (c) The current unit for petroleum reserves is metric tons (1 metric ton = 1000 kg). Why is interconversion of barrels and metric tons difficult?

22.106 The local anesthetic Novocaine is the salt of procaine (see Problem 22.93 for structural formula) and hydrochloric acid. The molecular formula for Novocaine is $C_{13}H_{21}ClN_2O_2$. Write the structural formula for Novocaine.

22.107 For $C_6H_{12}O_6(aq)$, $\Delta G_f^\circ = -911$ kJ/mol at 25 °C. (a) Calculate ΔG° for the photosynthesis of glucose in aqueous solution at 25 °C. (b) Calculate ΔG for the photosynthesis of glucose at 25 °C under current atmospheric conditions of $p_{O_2} = 0.21$ atm and $p_{CO_2} = 3.4 \times 10^{-4}$ atm if the concentration of glucose is 1.0×10^{-3} M. (c) Is photosynthesis spontaneous under current atmospheric conditions? (d) What is the function of sunlight in photosynthesis? (e) What is ΔG for the metabolism of glucose to $CO_2(g)$ and $H_2O(l)$ under current atmospheric conditions if $[C_6H_{12}O_6] = 1.0 \times 10^{-3}$?

22.108 How many milliliters of $H_2(g)$ at 23.0 °C and 754 mmHg are required to hydrogenate the $C{=}C$ bonds in a 0.4124-g sample of the typical vegetable oil?

$$CH_2OOC(CH_2)_{14}CH_3$$
$$|$$
$$CHOOC(CH_2)_7CH{=}CH(CH_2)_7CH_3$$
$$|$$
$$CH_2OOC(CH_2)_7CH{=}CHCH_2CH{=}CH(CH_2)_4CH_3$$

22.109 Pristane, 2,6,10,14-tetramethylpentadecane, which is obtained from shark liver oil, wool wax, and crude oil, is used as a lubricant. (a) Pentadecane is the name of the straight chain alkane with 15 carbons. Write the complete structural formula and a condensed structural formula for pristane. (b) To what class of compounds does pristane belong? (c) Which carbons are chiral?

An Organic Chemist in the Pharmaceutical Industry

TODD A. BLUMENKOPF

Department of Medicinal Chemistry
Central Research Division, Pfizer, Inc.

B.S. Chemistry
University of California, Los Angeles

Ph.D. Chemistry
University of California, Berkeley

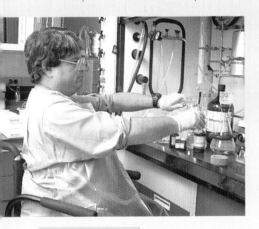

M y parents nurtured my early interests in science; they believed that scientists would be more likely to accept someone who was "different" than would professionals in the business world. They gave me a chemistry set when I was in fourth grade, but took it away shortly thereafter when I set fire to a table in my bedroom! I seriously started to consider chemistry as a career when my high school chemistry teacher challenged me; he wouldn't acknowledge a correct answer unless you could convince him (and yourself) that you understood why your answer was right. But I quickly became disillusioned with my freshman chemistry class in college. The course seemed designed more to "weed out the pre-meds" than to show students what chemistry was all about. Sometimes, I believe that I stuck with chemistry simply because I didn't find another subject that interested me, rather than because I

actually liked the classes. At the end of my junior year, I began a year-long independent research project with a professor. Finally, my interest in chemistry skyrocketed as I had an opportunity to interact with the professor and the graduate students in his group and experience a research project first-hand. We published the results in a major chemistry journal, and the experience resulted in my acceptance by a first-rate graduate program.

Graduate school made my undergraduate years look easy. The students were very dedicated and often worked in the laboratory late into the night. After the first year, there was no formal coursework or exams; instead, we attended seminars given by the faculty and by speakers from other colleges and universities. I also had to present two seminars to the department, one an overview of a subject from the chemical literature and one on my own research project. I made many good friends during those years, whom I see now when I travel to scientific conferences around the world.

I work as an organic chemist for a pharmaceutical company, synthesizing molecules that are tested to determine whether they can be used as drugs to fight or prevent disease. The biological properties of a chemical compound are determined by its structure and can be refined by varying the structures of the molecules synthesized. The chemist, therefore, is positioned at the very beginning of the drug development process, trying to unravel the relationship between chemical structure and biological activity, and to discover promising new compounds for further study. The branch of organic chemistry that deals with the discovery of therapeutic chemical agents is called medicinal chemistry.

A new molecule can be synthesized by a series of chemical reactions and, as a synthesis chemist, I attempt to predict which of a variety of possible routes is likely to be successful. I may rely on my understanding of reactions known in the chemical literature, or a new reaction may have to be invented. Every day on the job is different,

as almost every reaction I run is different from the one I ran before. Sometimes, the selected route to a compound fails due to unexpected reactions; indeed, observation of unexpected results often leads to the development of useful, new reactions. I must consider what I learned about chemical bonding and stoichiometry as I investigate new reactions. I must also prove that I have actually synthesized my desired target molecules. This is done by examining the physical properties of molecules, often by observing their behavior when exposed to light or magnetic fields. The spectra of molecules obtained in this manner provides clues to the structure of the compound. I then determine the structure of the molecule consistent with all the data.

One of the most exciting aspects of medicinal chemistry is the opportunity to interact with scientists from a wide variety of backgrounds. We work as a team to decide what desirable properties a drug must have. We also must consider biological properties to avoid, properties that might generate unacceptable or even toxic side effects. The drug discovery process requires frequent collaboration between chemists, biochemists, biologists, and physicians. Compounds synthesized by chemists are evaluated by collaborating biologists in a series of tests. Initial tests, often run on hundreds or even thousands of chemical compounds, provide data that are used to select the most promising compounds to undergo further testing. The chemist considers this initial biological data in selecting which new compounds to make next.

The real challenge for medicinal chemists is deciding what compounds to make. For example, to treat a bacterial or viral infection, a chemist may design a drug that selectively interferes with the biochemical processes of an invading organism. We may get ideas by examining the differences in metabolic pathways between a human and the organism and choose to synthesize a molecule that mimics an essential chemical component of the organism. Scientists may choose to exploit differences in properties of specific enzymes between the human host and a parasite. We can also use a variety of computer-based methods to examine the structures of enzymes determined by X-ray crystallography. Many different approaches may be used to examine a particular problem, and the methods utilized may vary depending on the disease or condition one is interested in treating. I am now working on a project whose outcome (hopefully!) will be a new drug to treat cancer.

Developing drugs is a very expensive and time-consuming process. More than $100 million is often spent to develop a single new drug, with ten years or more elapsing from the time a research program is started to the time the new drug reaches the marketplace. Thus, a successful medicinal chemist must be curious, creative, motivated, and patient!

I have been paraplegic since birth as a result of a congenital spinal cord injury known as spina bifida. I use a wheelchair.

I have found that educators and employers unacquainted with people with disabilities often focus on perceived limitations caused by my disability. Some have difficulty recognizing that I can accomplish my job safely and effectively, although I may accomplish it by slightly different means than my nondisabled colleagues. While my laboratory was modified somewhat to provide better access to my workspace (for example, the surface of my fume hood was lowered), often the only difference between my work and that of my colleagues is that I perform my experiments sitting down. Occasionally, I have had to challenge unnecessary restrictions imposed by individuals unfamiliar with the capabilities of persons with disabilities. Yet, I find this challenge to be consistent with successful research itself, as some of our most significant discoveries have been made by those who challenged conventional thinking. Like their able-bodied peers, people with disabilities must take positive action on their own behalf to obtain opportunities for a good education and for meaningful careers. I have found that even the most skeptical persons eventually come to accept my ability to work independently in the chemistry lab.

I am currently a member of the American Chemical Society's Committee on Chemists with Disabilities. Our goal is to make education and employment in the chemical sciences as accessible to disabled persons as they are to able-bodied people, while enhancing the general awareness of the abilities, skills, and accomplishments of scientists with disabilities. Chemistry is a rapidly expanding discipline, capable of accommodating people with a wide assortment of interests and abilities. Our committee's role is to show that chemistry can also accommodate people with a vast range of disabilities.

POLYMERS: SYNTHETIC AND NATURAL

23

The parrot's feathers are composed largely of the protein keratin, and the bamboo stalks of the polysaccharide cellulose. Both keratin and cellulose are biopolymers.

This chapter is about polymers. First, we will discuss the composition and uses of some common synthetic polymers and the reactions by which they are formed. Then, we will see how the properties of polymer molecules are affected by their large size. Finally, we will talk about the three major classes of **biopolymers** (*polymers made by living organisms*): proteins, polysaccharides, and nucleic acids. Natural rubber is also a biopolymer.

— *Why is a study of polymers a part of general chemistry?*

Compounds that contain 10^3 or more carbon atoms such as cellulose, starch, rubber, proteins, and nucleic acids occur widely in nature, and synthetic polymers such as PVC, nylon, polyester, and silicones are common in the modern world. The chair you are sitting on, the pages of this book, the pen you use to take notes, and the calculator you use to solve problems are all probably made of polymers. So are the clothes you are wearing, the tires on your car, your telephone and portable radio, the food you eat—the list could go on and on. More industrial chemists work in the field of polymer chemistry than in any other area of chemistry. Many engineers and all biologists and biochemists work with polymeric materials. Since 1976 synthetic polymers have been the most widely used materials in the United States. Many of the organic compounds in the Top 50 Chemicals list (Appendix E) are there because they are used to make polymers.

— What do you already know about polymers?

From your everyday experience, you know that most polymers are solid. You may know from disastrous personal experience that many solid polymers soften at a fairly low temperature and some burn. The other properties of polymers vary widely. Some are hard like those used to make telephones. Some are soft like erasers, and others are elastic like rubber bands. Some are transparent like large soft-drink bottles, and others are opaque like Clorox bottles. You have probably heard about proteins, the polysaccharides starch and cellulose, and the nucleic acids DNA and RNA, which carry the genetic information that directs all the activities of cells.

In Chapter 9, you learned that polymers are very large molecules made up of repeating units called monomers. The simplest example of a polymer is polyethylene. The monomer ethylene, $H_2C=CH_2$, is converted to polymer by heat and pressure in the presence of a catalyst:

$$x H_2C=CH_2 \xrightarrow[\text{100 °C, 7–20 atm}]{\text{catalyst}}$$

$$\sim CH_2-CH_2-CH_2-CH_2-CH_2-CH_2 \sim$$

part of polyethylene molecule

Different molecules of polyethylene have different numbers of CH_2-CH_2 units and thus different molecular masses. Like polyethylene, most synthetic polymers are made up of molecules with a range of different molecular masses. The polymers formed from unsymmetrically substituted ethylenes such as vinyl chloride, $H_2C=CHCl$, usually have head-to-tail structures.

$$\sim CH_2-\underset{\underset{Cl}{|}}{CH}-CH_2-\underset{\underset{Cl}{|}}{CH}-CH_2-\underset{\underset{Cl}{|}}{CH}-CH_2-\underset{\underset{Cl}{|}}{CH}-CH_2-\underset{\underset{Cl}{|}}{CH}-CH_2-\underset{\underset{Cl}{|}}{CH} \sim$$

Two or more different monomers can be polymerized together to form copolymers. By the proper choice of monomers and polymerization conditions and the use of additives, an almost infinite variety of materials having properties exactly suited for different purposes can be produced. Although polymerization results in a decrease in disorder or entropy, polymerization is spontaneous because it is exothermic.

Natural rubber latex being harvested from a rubber tree in Malaysia.

| 23.1 | ## SOME COMMON SYNTHETIC POLYMERS |

Because the monomer units in synthetic polymers (other than copolymers) are all the same, abbreviated formulas such as

$$+CH_2-CH_2+_n$$

for polyethylene and

$$+CH_2-\underset{\underset{Cl}{|}}{CH}+_n$$

for PVC, where n is the variable number of monomer units in the polymer molecules, can be used to show the structure of synthetic polymers. The formulas of monomers and polymers and the uses of some common synthetic polymers are shown in Table 23.1. Note that parentheses are placed around the name of the monomer if it consists of two words. For example, while the name of the polymer formed from ethylene is written simply "polyethylene," the name of the polymer formed from vinyl chloride is written "poly(vinyl chloride)." A polymer is only classified as a polyester

⏵ Note that polyethylene does not have any double bonds despite the -ene ending of its name. The alkanelike structure of polyethylene explains its lack of reactivity. A reactive polymer would not be very useful for making containers, one of the chief uses of polyethylene, because it would react with the contents.

TABLE 23.1   Some Common Synthetic Polymers

Name	Monomer	Formula	Recycling Symbol	Properties[a] and Uses
Polyethylene	$H_2C=CH_2$	$+CH_2-CH_2 \frac{}{}_n$	2 HDPE	Unreactive, flexible, impermeable to water vapor; packaging films, containers, toys, housewares
Polypropylene	$CH_3CH=CH_2$	$+CH-CH_2 \frac{}{}_n$ CH_3	5 PP	Lowest density of any plastic; indoor/outdoor carpeting, upholstery
Poly(vinyl chloride)(PVC)	$H_2C=CHCl$	$+CH_2-CH \frac{}{}_n$ Cl	3 V	Self-extinguishing to fire; pipe, siding, floor tile; when plasticized;[b] raincoats, shower curtains, imitation-leather upholstery
Polytetrafluoroethylene (Teflon)	$F_2C=CF_2$	$+CF_2-CF_2 \frac{}{}_n$	7 OTHER	Very unreactive, nonstick, relatively high softening point; liners for pots and pans, greaseless bearings, artificial joints
Polystyrene	$\bigcirc$-$CH=CH_2$	$+CH-CH_2 \frac{}{}_n$ $\bigcirc$	6 PS	Housings for large household appliances such as refrigerators, auto instruments and panels, clear cups and food containers, and foam cups and packing
Polyacrylonitrile (Orlon)	$H_2C=CHCN$	$+CH_2-CH \frac{}{}_n$ CN	7 OTHER	Carpets and knitwear
Poly(methyl methacrylate) (Lucite, Plexiglas)	$H_2C=CCOOCH_3$ CH_3	$+CH_2-C \frac{}{}_n$ $COOCH_3$ CH_3	7 OTHER	Substitute for glass
Polychloroprene (Neoprene)	$H_2C=CHC=CH_2$ Cl	$+CH_2CH=CCH_2 \frac{}{}_n$ Cl	7 OTHER	Rubberlike; more resistant to oil, gasoline, and other organic solvents than natural rubber
Polyamide (nylon 66)	$HOOC(CH_2)_4COOH$ and $H_2N(CH_2)_6NH_2$	$+C(CH_2)_4CNH(CH_2)_6NH \frac{}{}_n$ O O	7 OTHER	Strong, wear-resistant; hosiery; tire cord, door latches, gears
Polyester or polyethylene terephthalate (Dacron fiber and Mylar film)	CH_3OOC-$\bigcirc$-$COOCH_3$ and $HOCH_2CH_2OH$	$+OCH_2CH_2OC$-$\bigcirc$-$C \frac{}{}_n$ O O	1 PETE	Large soft-drink bottles, clothing, tire-cord, firehose, stuffing for pillows, sleeping bags, and comforters; magnetic recording tape

[a] Properties given are outstanding or unusual ones.
[b] A plasticizer is an organic compound that is added to a polymer to make the polymer softer and more flexible.

if the ester group is part of the polymer chain; thus, poly(methyl methacrylate)

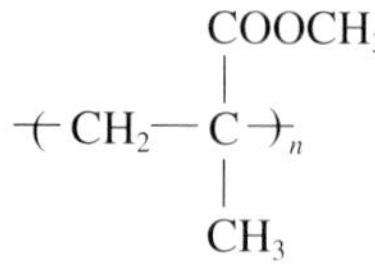

$$\begin{matrix} & COOCH_3 \\ & | \\ +CH_2-C \rightarrow_n \\ & | \\ & CH_3 \end{matrix}$$

is *not* a polyester, but polyethylene terephthalate

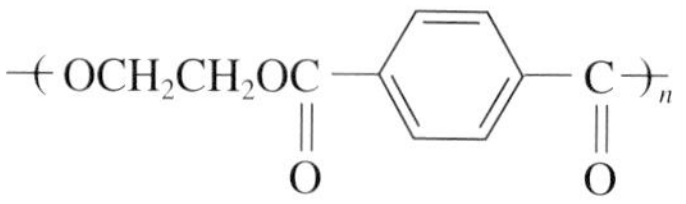

$$+OCH_2CH_2OC \!\!-\!\!\bigcirc\!\!-\!\! C \rightarrow_n$$
$$\quad\quad\quad\quad \| \quad\quad\quad\quad \|$$
$$\quad\quad\quad\quad O \quad\quad\quad\quad O$$

is a polyester.

Chain Polymerization

Synthetic polymers are formed either by a chain reaction or by stepwise polymerization. In a **chain reaction,** *an intermediate that is used up in one step is formed again in a later step so that the steps can take place again and again.* The intermediate in the polymerization of many alkenes is a radical. Production of poly(vinyl chloride) is an example:

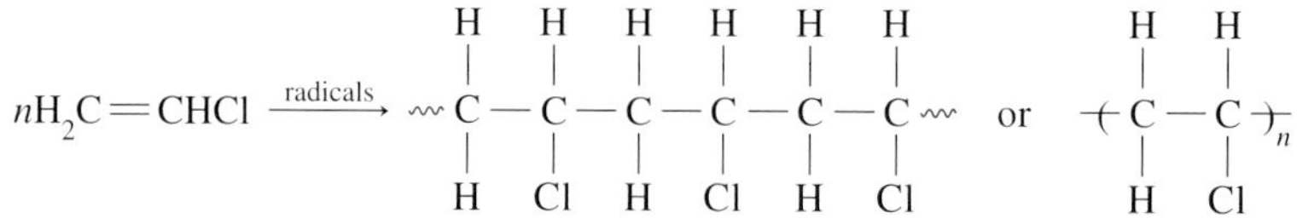

$$nH_2C=CHCl \xrightarrow{\text{radicals}} \begin{matrix} H & H & H & H & H & H \\ | & | & | & | & | & | \\ \sim C-C-C-C-C-C\sim \\ | & | & | & | & | & | \\ H & Cl & H & Cl & H & Cl \end{matrix} \quad or \quad \begin{matrix} H & H \\ | & | \\ +C-C\rightarrow_n \\ | & | \\ H & Cl \end{matrix}$$

poly(vinyl chloride)

The *first step in a radical-chain polymerization* is called a **chain-initiating** step. In the chain-initiating step of a radical-chain polymerization, a radical, Rad·, adds to a carbon–carbon double bond and supplies the reactive intermediate needed to start the chain:

Chain initiation $\quad\quad Rad\cdot + CH_2 \colon\!\colon CHCl \longrightarrow Rad\cdot\!\cdot CH_2\cdot\!\cdot\dot{C}HCl$

The source of the radical that starts the chain is called an **initiator.** The initiator is not a catalyst because it is used up in the reaction.

The radical formed in the chain-initiating step adds to another molecule of monomer and forms another radical. This step is part of the chain and is called a **chain-propagating** step:

chain propagation $\quad\left\{\begin{matrix} Rad\cdot\!\cdot CH_2\cdot\!\cdot\dot{C}HCl + CH_2\colon\!\colon CHCl \longrightarrow \\ \quad\quad\quad Rad\cdot\!\cdot CH_2\cdot\!\cdot CHCl\cdot\!\cdot CH_2\cdot\!\cdot\dot{C}HCl \text{ and then} \\ Rad\cdot\!\cdot CH_2\cdot\!\cdot CHCl\cdot\!\cdot CH_2\cdot\!\cdot\dot{C}HCl + CH_2\colon\!\colon CHCl \longrightarrow \\ \quad\quad\quad Rad\cdot\!\cdot CH_2\cdot CHCl\cdot\!\cdot CH_2\cdot\!\cdot CHCl\cdot\!\cdot CH_2\cdot\!\cdot\dot{C}HCl, \text{ etc.} \end{matrix}\right.$

The chain-propagating step takes place over and over again until the chain is ended by a **chain-terminating** step, the *combination of two radicals.* Some possible chain-terminating steps are

chain termination $\quad\quad Rad\cdot\!\cdot CH_2\cdot\!\cdot\dot{C}HCl + Rad\cdot \longrightarrow Rad\cdot\!\cdot CH_2\cdot\!\cdot CHCl\cdot\!\cdot Rad \quad\quad or$

$Rad\cdot\!\cdot CH_2\cdot\!\cdot\dot{C}HCl + \dot{C}HCl\cdot\!\cdot CH_2\cdot\!\cdot Rad \longrightarrow$

$\quad\quad\quad\quad Rad\cdot\!\cdot CH_2\cdot\!\cdot CHCl\cdot\!\cdot CHCl\cdot\!\cdot CH_2\cdot\!\cdot Rad$

Poly(vinyl chloride) used for upholstery in cars is plasticized (softened) with the ester dibutyl phthalate

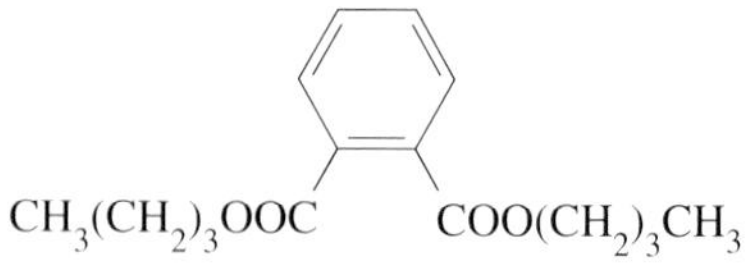

$$\begin{matrix} & \bigcirc \\ CH_3(CH_2)_3OOC & \quad COO(CH_2)_3CH_3 \end{matrix}$$

Gradual evaporation of the plasticizer results in cracking of the upholstery and fogging of the windows.

◗ Remember that a radical is a species that has one or more unpaired electrons. Radicals are usually very reactive (Section 9.10).

◗ A single-headed arrow indicates the movement of one electron.

The controlled nuclear fission that takes place in a critical-sized sample of uranium in a nuclear power plant is a nuclear chain reaction.

Chain-propagating steps take place thousands of times for each time a chain-terminating step occurs because the concentration of radicals is low compared to the concentration of the reactant, $H_2C{=}CHCl$. The chances of a radical bumping into a molecule of $H_2C{=}CHCl$ are much greater than the chances of one radical bumping into another radical.

In a chain polymerization, there is never a detectable number of middle-sized molecules in the reaction mixture. The reaction mixture consists of monomer and high molecular mass polymer. The only change with time is an increase in the number of polymer molecules. Polymers that contain 4000–400 000 carbon atoms are formed by radical-chain polymerization. The polymer chain is so long that the groups at the end (Rad), which come from the initiator, are an insignificant part of the whole molecule. Therefore, these groups are ignored when the structure is abbreviated.

■ FIGURE 23.1 Drawing nylon from the interface between adipoyl chloride and hexamethylenediamine solutions.

PRACTICE PROBLEMS

23.1 For the polymer of $CH_2{=}CHNO_2$, write the complete structural formula for a piece of the polymer chain that contains three monomer units.

23.2 (a) Write a structural formula for the monomer of the polymer

$$
\begin{array}{cccccccc}
\text{H} & \text{Cl} & \text{H} & \text{Cl} & \text{H} & \text{Cl} & \text{H} & \text{Cl} \\
| & | & | & | & | & | & | & | \\
{\sim}\text{C} & - \text{C} & - \text{C} & - \text{C} & - \text{C} & - \text{C} & - \text{C} & - \text{C}{\sim} \\
| & | & | & | & | & | & | & | \\
\text{H} & \text{Cl} & \text{H} & \text{Cl} & \text{H} & \text{Cl} & \text{H} & \text{Cl}
\end{array}
$$

(b) Write the abbreviated formula for the structure of the polymer.

23.3 Write the mechanism for the radical-chain polymerization of $H_2C{=}CHOCH_3$.

23.4 For the polymer of

$$
\begin{array}{c}
H_2C{=}CHC{=}CH_2 \\
| \\
CH_3
\end{array}
$$

(a) write the abbreviated formula. (b) Write the complete structural formula for a piece of the polymer chain that contains three monomer units.

Step Polymerization

The preparation of nylon is a typical example of a step polymerization. In a step polymerization, polymer size and amount of polymer increase with increasing time. As a lecture demonstration, nylon is often prepared by the reaction of adipoyl chloride and hexamethylenediamine (see ■Figure 23.1):

$$
n\text{ClC(CH}_2)_4\text{CCl} + n\text{H}_2\text{N(CH}_2)_6\text{NH}_2 \longrightarrow {-\!\!\left(\text{C(CH}_2)_4\text{CNH(CH}_2)_6\text{NH}\right)\!\!}_n + 2\text{HCl}
$$
$$
\overset{\|}{\text{O}} \quad \overset{\|}{\text{O}} \qquad\qquad\qquad\qquad \overset{\|}{\text{O}} \quad \overset{\|}{\text{O}}
$$

adipoyl chloride   hexamethylenediamine   nylon 66

The 66 in the name "nylon 66" indicates that there are six carbons in each of the two monomers from which the nylon is made. Nylon 66 was the first nylon to be produced and is still produced in larger quantity than any other nylon.

The reaction can be regarded as nucleophilic substitution of :N~ for Cl although the mechanism is more complicated than the S_N2 mechanism. The —COCl group of one molecule reacts with the —NH$_2$ group of another molecule:

$$
\text{ClC(CH}_2)_4\text{CCl} + \text{H}_2\text{N(CH}_2)_6\text{NH}_2 \longrightarrow \text{ClC(CH}_2)_4\text{CNH(CH}_2)_6\text{NH}_2 + \text{HCl}
$$
$$
\overset{\|}{\text{O}} \quad \overset{\|}{\text{O}} \qquad\qquad\qquad\qquad \overset{\|}{\text{O}} \quad \overset{\|}{\text{O}}
$$

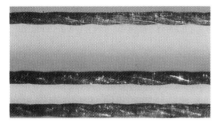

Top: *Nylon fibers enlarged 20 times.*
Bottom: *Silk fibers enlarged 20 times. What difference do you observe between nylon and silk fibers?*

Because both reactants have two functional groups, the product of the first step contains both a —COCl and an —NH_2 group. One end can react with another molecule of $H_2N(CH_2)_6NH_2$ and the other end with a molecule of $ClOC(CH_2)_4COCl$. More polymerization steps can take place.

Industrially, nylon 66 is produced by heating adipic acid and hexamethylenediamine at 280–300 °C under vacuum. The first step is

$$HOOC(CH_2)_4COOH + H_2N(CH_2)_6NH_2 \xrightarrow{\text{280 °C, vacuum}}$$
$$HOOC(CH_2)_4CONH(CH_2)_6NH_2 + H_2O$$

The structure of the polymer (showing three repeats) is

$$\sim\!\!\!\sim C(CH_2)_4 \underset{\overset{\|}{O}}{C} NH(CH_2)_6 NH \underset{\overset{\|}{O}}{C} (CH_2)_4 \underset{\overset{\|}{O}}{C} NH(CH_2)_6 NH \underset{\overset{\|}{O}}{C} (CH_2)_4 \underset{\overset{\|}{O}}{C} NH(CH_2)_6 NH \!\!\!\sim\!\!\!\sim$$

and the abbreviated formula is

$$\left(\!\!\!\; C(CH_2)_4 \underset{\overset{\|}{O}}{C} NH(CH_2)_6 NH \;\!\!\!\right)_{\!n}$$

The high temperature is needed because the —OH group is not as good a leaving group as —Cl.

Nylon produced industrially is composed of about 50 repeating units. About 2.4 billion pounds of nylon were produced in the United States in 1986. Most of this nylon was used for tire cord, carpet, and clothing. Nylon is similar to silk in many of its properties but can be produced at a lower cost. The similarity to silk is due to the fact that silk is also a polyamide, that is, a polymer with amide linkages, —CONH—, between the monomer units.

PRACTICE PROBLEM

23.5 (a) Write the equation for the first step in the polymerization of

$$HOOC-\!\!\!\left\langle \bigcirc \right\rangle\!\!\!-COOH \quad \text{and} \quad H_2N-\!\!\!\left\langle \bigcirc \right\rangle\!\!\!-NH_2$$

(b) Write a structural formula for the polymer that shows two pairs of monomer units (four benzene rings). (c) Write the abbreviated form of the structural formula for the polymer.

Silicones

Not all synthetic polymers are organic. The *synthetic polymers* called silicones *contain alternating silicon and oxygen atoms* like silica and the silicates (Section 21.8). An example of a **silicone** is

$$\sim\!\!\!\sim \underset{\underset{CH_3}{|}}{\overset{\overset{CH_3}{|}}{Si}}\!-\!O\!-\!\underset{\underset{CH_3}{|}}{\overset{\overset{CH_3}{|}}{Si}}\!-\!O\!-\!\underset{\underset{CH_3}{|}}{\overset{\overset{CH_3}{|}}{Si}}\!-\!O\!-\!\underset{\underset{CH_3}{|}}{\overset{\overset{CH_3}{|}}{Si}}\!-\!O\!-\!\underset{\underset{CH_3}{|}}{\overset{\overset{CH_3}{|}}{Si}}\!-\!O\!-\!\underset{\underset{CH_3}{|}}{\overset{\overset{CH_3}{|}}{Si}}\!-\!O\!-\!\underset{\underset{CH_3}{|}}{\overset{\overset{CH_3}{|}}{Si}}\!-\!O\!-\!\underset{\underset{CH_3}{|}}{\overset{\overset{CH_3}{|}}{Si}}\!-\!O\!-\!\underset{\underset{CH_3}{|}}{\overset{\overset{CH_3}{|}}{Si}}\!-\!O\!-\!\!\!\sim\!\!\!\sim$$

The methyl groups (or other alkyl groups) make silicones water repellant. Depending on their molecular mass and on what the alkyl group is, silicones may be oils, greases, or rubbery solids. Because both Si—O and Si—C bonds are strong (see Table 9.4), silicones are stable when heated. Silicones are very unreactive. As a result of their lack of reactivity, silicones were believed to be nontoxic and were used increasingly in medicine, for example, for artificial hearts and cosmetic surgery, such as breast implants, until 1992 when their safety began to be questioned. In dentistry, silicones are used to make impressions for inlays and dentures.

The viscosity of liquid silicones changes very little as temperature is raised or lowered. Therefore, silicone oils and greases still lubricate at very low and very high temperatures. Silicone rubbers remain elastic at low temperatures and are used as gaskets for refrigerator doors and airplane windows.

Silicones are used in a wide variety of consumer products from the antistick material for peel-off labels to lipsticks and suntan lotions. Silly Putty and superballs are made of silicones. Polishes for cars and furniture, cooking oils, and waterproofing treatments for fabrics, leather, and paper all contain silicones.

23.2 PHYSICAL AND CHEMICAL PROPERTIES OF POLYMERS

What difference does the extremely large size of polymer molecules make in physical and chemical properties? *Except when the carbon chain is very short, the reactivity of functional groups is the same regardless of the size of the molecule to which the functional groups are attached.* However, size has a large effect on physical properties, especially strength. Ordinary organic compounds form molecular crystals that are soft and have low melting points (Section 12.8). Intermolecular attractive forces consist of relatively weak London forces and dipole–dipole attractions (including hydrogen bonding). Intermolecular attractions between small molecules are weak. The atoms in polymer molecules are held together by covalent bonds, which are relatively strong; the breaking of a polymer chain is a chemical reaction. In addition, as a result of the large size of polymer molecules, intermolecular attractions are very strong too. Moreover, the long molecules can get tangled up in each other like strands of cooked spaghetti. *Compared with ordinary organic solids, polymeric organic solids are extremely strong.* Indeed, some polymers are strong even when compared with materials like steel.

In general, polymers do not exist in crystalline form. Remember that, as liquids are cooled, viscosity usually increases (Section 12.3). At lower temperatures, the molecules have less kinetic energy to overcome the attractive forces between them. Long polymer molecules get tangled up in each other and have difficulty arranging themselves in the regular pattern required for the formation of crystals. Polymers usually form solids made up of regions of order in amorphous material.

For each polymer there is a molecular mass below which the solid has low mechanical strength. Above this molecular mass, the mechanical strength of the solid increases with increasing molecular mass. The increase is rapid at first and then slower. The viscosity of solutions of a polymer or of molten polymer also increases with increasing molecular mass. A viscosity that is too high makes molding, rolling into sheets, and drawing into fibers difficult. As a result, for each polymer there is a best molecular mass at which the mechanical strength of the solid is adequate, but the solution or melt is still fluid enough to be handled easily.

Molecular mass distributions of polymers are usually determined by gel-permeation chromatography. In this technique, a column is packed with a gel that has pores of closely controlled size that act as sieves. The larger molecules cannot pass through the pores and travel down the column by the shortest possible route. Thus, the larger molecules pass through the column rapidly. The smaller molecules, which can pass through the pores, move down the column by a longer route and pass through slowly. As a result, large molecules are separated from small molecules.

Fibers

If intermolecular forces are strong and the molecules have a regular shape so that they pack well in the solid, a polymer is suitable for making fibers. The polymer chains are lined up by stretching. Once lined up, the strong intermolecular attractions

Nylon paintbrushes are recommended for use with latex paints.

hold them in position. Enthalpy favors the orderly arrangement. Polar groups or groups that can form hydrogen bonds lead to especially strong intermolecular forces. For example, nylons have amide groups that can form hydrogen bonds as shown in ▪Figure 23.2.

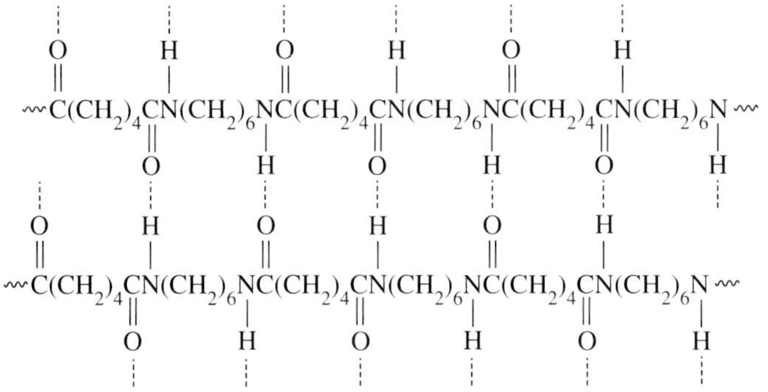

$$\cdots C(CH_2)_4 C N(CH_2)_6 N C(CH_2)_4 C N(CH_2)_6 N C(CH_2)_4 C N(CH_2)_6 N \cdots$$

▪ **FIGURE 23.2** Hydrogen bonding in nylon.

Elastomers

If intermolecular forces are weak and the molecules have an irregular shape so that they do not pack well in the solid, the polymer is elastic. Natural rubber is the classic example. Polymers with weak intermolecular attractions and irregular shapes do not stay lined up when a stretching force is removed. Entropy favors disorderly arrangement. When stretched, rubber gives off thermal energy. As rubber contracts, it absorbs thermal energy. Because rubber absorbs thermal energy when it contracts, heating rubber (carefully so as not to melt it or set it on fire) causes it to contract. Most substances expand when heated. The structure of natural rubber with the polymer chains stretched out is shown below:

You can test these statements by holding a wide rubber band across your lips and suddenly stretching it or by stretching the rubber band, placing it across your lips, and then suddenly letting it contract.

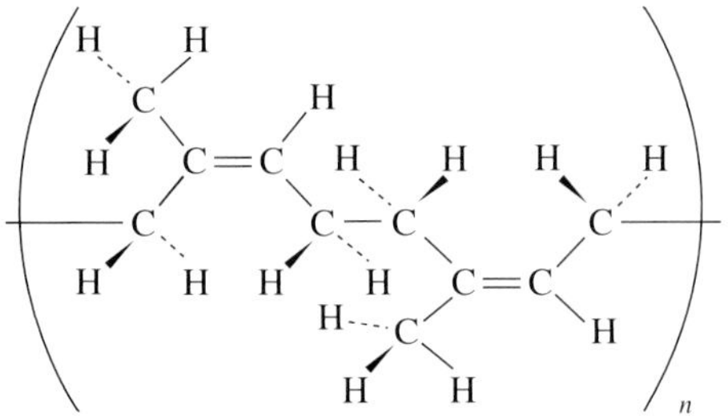

To be strong and elastic, natural rubber must be vulcanized, that is, heated with sulfur. Sulfur connects the chains of natural rubber with sulfur cross-links (see ▪Figure 23.3). The reactions involved in vulcanization are not fully understood. Rotation around single bonds is free. When rubber is unstretched, the polymer chains tend to curl and fold; when stretched, the polymer chains are straightened out. The sulfur cross-links pull the chains back into their original shape when the rubber is no longer stretched.

Vulcanization was discovered by the American inventor Charles Goodyear in 1839 when he accidentally dropped rubber mixed with sulfur on a hot stove. Although Goodyear's invention made millions of dollars for others who infringed his patent, he died in debt.

Plastics

The polymers used to make typical plastics are in between the two extremes of fibrous polymers with their strong intermolecular forces and efficient packing and elastomers with their weak intermolecular forces and irregularly shaped molecules. The strength of the intermolecular forces between polymer molecules and the shape of the molecules depend on the monomer used to make the polymer.

Some plastics are **thermoplastic,** that is, they *soften when heated.* Others are **thermoset** and *do not soften when heated.* The former are usually linear polymers such as polyethylene; the latter are highly cross-linked in three dimensions.

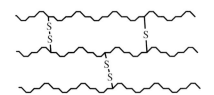

▪ **FIGURE 23.3** Schematic drawing showing cross-linking of vulcanized rubber.

The characteristic ability of thermoplastics to soften but not melt when heated results from the large size of the polymer molecules. When heated, the polymer molecules move far enough apart to slide over one another but not so far that the strong intermolecular attractions are weakened very much. The polymer molecules are able to move to new positions and become fixed in the new positions when the plastic is cooled. Thus, plastics can be molded into all sorts of shapes, rolled into sheets, or drawn into fibers. The small molecules of ordinary organic substances also move apart when heated. But because intermolecular attractions between small molecules are weak, ordinary organic substances melt.

23.3   PROTEINS

Proteins are the most plentiful organic substances in cells; about half of the dry mass of cells is composed of proteins. Like nylon, proteins are polyamides, but the building blocks of proteins are not dibasic acids such as $HOOC(CH_2)_4COOH$ and diamines such as $H_2N(CH_2)_6NH_2$. The building blocks of proteins are alpha-amino acids. An **alpha-amino acid** is *a carboxylic acid that has an amino group, —NH₂, attached to the same carbon as the carboxyl group, —COOH.* The carbon to which both —NH₂ and —COOH groups are attached is called the **alpha carbon.** (Alpha is the first letter in the Greek alphabet.) The simplest alpha-amino acid is glycine, H_2NCH_2COOH. Note that by agreement among scientists, the formulas for amino acids are written with the —NH₂ group on the left and the —COOH group on the right.

Amino acids have both an acidic group, —COOH, and a basic group, —NH₂, in the same molecule. The acidic group and the basic group neutralize each other; pure amino acids actually exist as inner salts. For example, the formula

explains the properties of glycine much better than the formula H_2NCH_2COOH. Glycine is a solid with a relatively high melting point (262 °C), not a liquid like $CH_3CH_2CH_2CH_2NH_2$ or CH_3CH_2COOH. Glycine has $K_a = 1.6 \times 10^{-10}$ and $K_b = 2.5 \times 10^{-12}$, whereas most carboxylic acids have K_a around 10^{-5} and most amines have K_b around 10^{-4}. In other words, glycine is less acidic than most carboxylic acids and less basic than most amines. The acidic group in glycine is —NH₃⁺, and the basic group in glycine is —COO⁻. A K_a of 1.6×10^{-10} is reasonable for a cation —NH₃⁺; a K_b of 2.5×10^{-12} is reasonable for an anion, —COO⁻.

All proteins, from the proteins of bacteria to the proteins of the human body, are synthesized from the same 20 alpha-amino acids. All 20 alpha-amino acids are related to glycine by substitution of a carbon-containing group for one of the hydrogens on the alpha carbon. In other words, the alpha-amino acids of which proteins are made can be represented by the general formula

$$H_2NCH(R)COOH$$

or better,

$$H_3\overset{+}{N}CH(R)COO^-$$

The *group R* is referred to as the **side chain** of the amino acid. The names, condensed structural formulas, and abbreviations of some of the simplest of the alpha-amino acids that occur in proteins are given in Table 23.2.

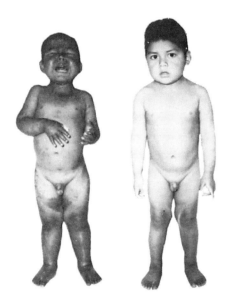

Left: *Little boy with kwashiorkor.* Right: *Same child after treatment. Kwashiorkor is the combination of symptoms that results from lack of protein, the most important and widespread nutritional problem among young children. Kwashiorkor is the outcome of a diet that provides enough calories but too little protein. It is usually caused by weaning a baby from breast milk to a starchy, protein-poor diet when another baby is born.*

TABLE 23.2	Some Simple Alpha-Amino Acids Found in Proteins	
Common Name	Condensed Structural Formula	Abbreviation
Glycine	$H_2NCH(H)COOH$	Gly or G
Alanine	$H_2NCH(CH_3)COOH$	Ala or A
Valine[a]	$H_2NCH[CH(CH_3)_2]COOH$	Val or V
Phenylalanine[a]	$H_2NCH(CH_2C_6H_5)COOH$	Phe or F
Serine	$H_2NCH(CH_2OH)COOH$	Ser or S
Cysteine	$H_2NCH(CH_2SH)COOH$	Cys or C
Lysine[a]	$H_2NCH(CH_2CH_2CH_2CH_2NH_2)COOH$	Lys or K
Glutamic acid	$H_2NCH(CH_2CH_2COOH)COOH$	Glu or E

[a]Essential amino acid.

Although people can synthesize most of the amino acids that are the building blocks of proteins in their bodies, a few cannot be synthesized in the human body and are known as **essential amino acids.** *Essential amino acids must be obtained from foods or dietary supplements.* Enzyme-catalyzed hydrolysis of proteins in the digestive system breaks down the proteins eaten into alpha-amino acids. The alpha-amino acids are reassembled in cells to form the proteins needed by the body. All of the essential amino acids must be eaten each day or a limiting reactant problem arises. Foods of animal origin—meat, poultry, seafood, milk, cheese, and eggs—contain all the essential amino acids. Many plant materials—such as beans, rice, and corn—are also good sources of protein. However, each plant material lacks at least one of the essential amino acids and combinations of plant foods—for example, beans and rice—must be eaten to provide all the essential amino acids.

The simple amino acids listed in Table 23.2 are typical of the amino acids found in proteins. As you can see from the table, some amino acids (alanine, for example) have nonpolar side chains. Others, such as serine, have polar side chains. Some have a second basic group in the side chain, like lysine, and others a second acidic group, like glutamic acid.

In all the alpha-amino acids from which proteins are built, except glycine, the alpha carbon is a stereocenter. *Most of the alpha-amino acids in proteins have the same configuration about the alpha carbon.* ▪Figure 23.4 shows stereochemical formulas of some of the simple amino acids found in proteins. In general, the enantiomers of the natural amino acids are not nutritious, and they may even be toxic.

▪ FIGURE 23.4   Stereoformulas of some simple amino acids. Almost all of the amino acids found in proteins have the same configuration about the alpha carbon.

PRACTICE PROBLEMS

23.6   For alanine, write the inner salt structure.

23.7   Of the following amino acids that have been found in proteins, (a) which has a nonpolar side chain? (b) Which has a second basic group in the side chain?

(c) Which has a second acidic group in the side chain? (d) Which have a second stereocenter in addition to the alpha carbon?

Asparagine, $H_2NCH(CH_2CONH_2)COOH$

Aspartic acid, $H_2NCH(CH_2COOH)COOH$

Hydroxylysine, $H_2NCH[CH_2CH_2CH(OH)CH_2NH_2]COOH$

Leucine, $H_2NCH[CH_2CH(CH_3)_2]COOH$

Threonine, $H_2NCH[CH(OH)CH_3]COOH$

23.8　Draw a stereochemical formula for the enantiomer of cysteine that is found in proteins.

In proteins, the —NH_2 group of one amino acid is linked to the —COOH group of another amino acid by *an amide group, —CO(NH)—*. The linkage is referred to as a **peptide link:**

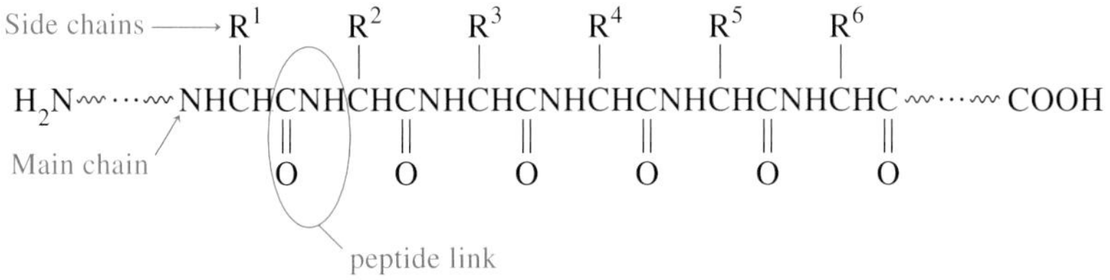

Molecules that have 1 to 50 peptide links are called **peptides.** Even small peptides are often active in living organisms; for example, the tripeptide glutathione

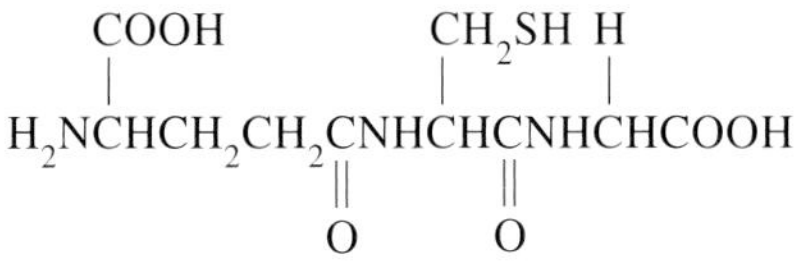

glutathione

is found in all cells. Glutathione protects red blood cells against hydrogen peroxide, H_2O_2, a toxic by-product of many metabolic reactions. Hydrogen peroxide is reduced to water, and the —SH group in glutathione is oxidized to a disulfide bridge, —S—S—:

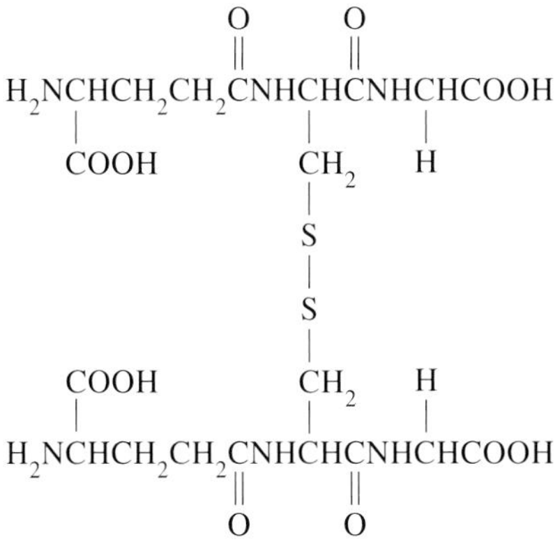

Disulfide bridges often connect two protein or peptide chains. Permanent waving of hair, which is a protein, involves breaking disulfide bridges and forming new ones.

Glutathione is classified as a tripeptide because it is made up of three amino acid units: one glutamic acid unit, one cysteine unit, and one glycine unit. **Proteins,** by definition, *have more than 50 peptide linkages.* Most proteins have over 100 peptide linkages; some have more than 8000 per molecule.

Primary Structure

In contrast to synthetic polymers and many other biopolymers, a given protein or peptide molecule always contains the same number of atoms. A protein or a peptide has a specific molecular mass, not a range of masses. In a given protein or peptide molecule, the order of the amino acids is also always the same; *the order in which the amino acids are arranged* is called the **primary structure** of the protein or peptide. *The primary structure determines the properties of the protein or peptide.**

The abbreviations for the various amino acids given in Table 23.2 are often used to show the primary structures of proteins and peptides. For example,

γ-Glu-Cys-Gly   or   γ-ECG

represents glutathione. The symbol γ (gamma) is used before the abbreviation for glutamic acid to show that the —COOH group on the third carbon from the one with the —NH₂ group is the one involved in the peptide link. (Gamma is the third letter in the Greek alphabet.) Usually the —COOH group attached to the same carbon as the —NH₂ group (the alpha carbon) is involved in peptide links. Reading from left to right, the order of the abbreviations shows the order of the amino acid units in the peptide or protein:

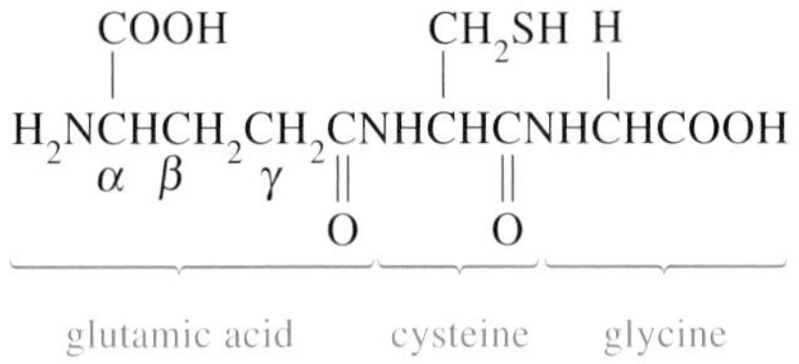

$$COOH \qquad CH_2SH \quad H$$
$$H_2NCHCH_2CH_2CNHCHCNHCHCOOH$$
$$\alpha \ \ \beta \ \ \gamma \quad \parallel \qquad \parallel$$
$$O \qquad O$$

glutamic acid   cysteine   glycine

An enormous number of different proteins is possible. For example, five other arrangements of the three amino acids of glutathione are possible:

Glu-Gly-Cys   Cys-Glu-Gly   Cys-Gly-Glu   Gly-Glu-Cys   Gly-Cys-Glu

Four different amino acid units can be arranged in 24 ways and eight, in 40 320 ways. Twenty different amino acids in a chain 100 units long can be arranged in more than 10^{100} ways! The human body contains about 10^5 different proteins, a large number but small compared with the number of proteins that are possible.

PRACTICE PROBLEM

23.9 For the dipeptide Val-Phe, write (a) the one-letter abbreviation, (b) a structural formula, and (c) the three-letter abbreviation and a structural formula for the other dipeptide that can be formed from valine and phenylalanine.

---

*Replacement of a single amino acid, glutamic acid, by another amino acid, valine, in a chain of 146 amino acids in hemoglobin results in the often fatal genetic disease sickle-cell anemia. Glutamic acid has a carboxylic acid group in its side chain, whereas valine has a nonpolar side chain (see Table 23.2). This information has been used to design drugs to treat sickle-cell anemia.

Peptides and proteins carry out a variety of jobs in the body as a result of the many different structures that are possible for them. Some proteins, such as collagen in tendons, are structural. Other proteins are enzymes and act as catalysts. Yet other proteins are regulators; for example, the hormone insulin, which is a protein, regulates glucose metabolism. Still other proteins are antibodies. The muscles we use to move are largely proteins, and other proteins carry small molecules and ions through the body. For example, hemoglobin transports oxygen in blood. The job that a peptide or protein does depends on which amino acids it is made from and the order in which the amino acids are joined, that is, on the primary structure of the protein.

The amino acid composition is obtained by complete hydrolysis. Complete hydrolysis means breaking down the protein into the individual amino acids by boiling for 24 hours with 6 M hydrochloric acid. Which amino acids are present and their relative proportions are then determined by chromatography. Together with a molecular mass determination, the amino acid composition tells how many of each kind of amino acid unit there are in a molecule of a given protein, just as percent composition and molecular mass give the molecular formula for an ordinary compound (Section 3.9). For example, complete hydrolysis of one mole of the peptide

Ala-Gly-Gly-Phe-Val

gives 1 mol of alanine, 2 mol of glycine, 1 mol of phenylalanine, and 1 mol of valine. Thus, the "molecular formula" of the peptide is $Ala_1Gly_2Phe_1Val_1$.

The order in which the amino acids are joined is found by partial hydrolysis, that is, by breaking the protein into pieces composed of two or more amino acid units by using milder conditions (more dilute acid or enzymes). From overlaps of the pieces, the structure of the whole protein can be deduced. To take a very simple example, partial hydrolysis of glutathione gives the two dipeptides Glu-Cys and Cys-Gly. Both pieces contain cysteine:

Glu-Cys
Cys-Gly

Therefore, cysteine must be in the middle of glutathione. The amino acid at the left end of the chain must be glutamic acid, and the amino acid at the right end of the chain must be glycine. Of the six possible arrangements of cysteine, glutamic acid, and glycine, only Glu-Cys-Gly could give these two dipeptides:

Cleavage of this bond gives Glu-Cys (and Gly).
↓
Glu-Cys-Gly
↑
Cleavage of this bond gives Cys-Gly (and Glu).

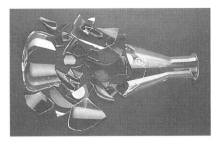

The primary structures of peptides and proteins can be deduced from the pieces formed by hydrolysis just as your laboratory instructor can figure out whether a crash resulted from the breaking of a beaker or a watch glass by looking at the pieces. Breakage of what object gave the pieces in the photograph?

PRACTICE PROBLEMS

23.10 What two dipeptides would be obtained by partial hydrolysis of Cys-Gly-Glu?

23.11 A tripeptide gives the dipeptides Glu-Cys and Gly-Glu on partial hydrolysis. Write the structure of the tripeptide (use three-letter abbreviations).

23.12 (a) Complete hydrolysis of one mole of a peptide gives 2 mol of alanine, 1 mol of glycine, and 1 mol of serine. What is the "molecular formula" of the peptide? (b) Fill in the blank. The peptide in part (a) is a _____ peptide. (c) Partial hydrolysis of the peptide yields Gly-Ser, Ala-Gly, and Ser-Ala. What is the structure of the peptide?

Secondary, Tertiary, and Quaternary Structure

On paper, showing the main chain of a protein stretched out in a straight line makes it easier to see the primary structure of the protein. However, real protein molecules are not stretched out but are coiled or folded. *The way the chains of amino acids are coiled or folded* is called the **secondary structure** of the protein. X-ray crystallography, together with a knowledge of the primary structure, is used to determine the secondary structures of proteins.*

Two ordered arrangements called an **alpha helix** and a **pleated sheet** are common. ➡Figures 23.5 and 23.6 show models of an alpha helix and a pleated sheet.

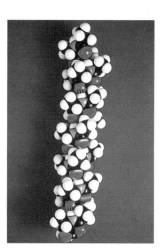

Carbon
Nitrogen
Oxygen
R group
Hydrogen

➡ FIGURE 23.5 *Top:* Ball-and-stick model of an alpha helix, a common secondary structure. Notice that the hydrogen bonds are between different units of the same chain and that the side chains are on the outside of the helix where there is room for large groups. *Bottom:* In the photograph of the space-filling model of the alpha helix, all the side chains are methyl.

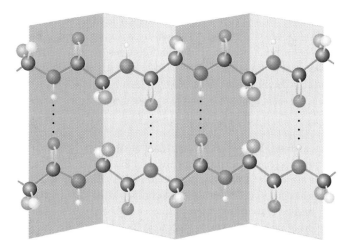

➡ FIGURE 23.6 *Top:* Ball-and-stick model of a pleated sheet, another common secondary structure. In pleated sheets, hydrogen bonds are between adjacent chains. *Bottom:* In the photograph of the space-filling model of the pleated sheet, all the side chains are methyl. Notice how the pairs of methyl groups touch on top of the pleated sheet. There is not room for large groups.

*The peptide linkage is planar with some double-bond character between carbon and nitrogen:

All bonds shown in the resonance structures and the atoms attached to them must lie in a plane. As a result, the peptide chain can bend only around the alpha carbon.

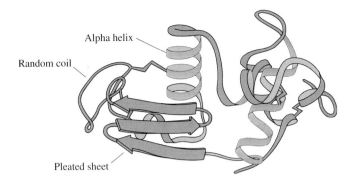

FIGURE 23.7 The tertiary structure of the enzyme egg lysozyme.

An alpha helix is a right-handed helix; the term *right-handed* refers to the direction of the turns in a helix. A left-handed helix is the mirror image of a right-handed helix. *Hydrogen bonding between —NH and C=O in the same chain holds protein molecules in an alpha helix, and hydrogen bonding between —NH and C=O in different chains holds protein molecules in a pleated sheet.*

Disorderly arrangements of proteins are called **random coils.** Most proteins have some regions in each of these three arrangements. The secondary structure of a protein depends on the primary structure.

Proteins also have tertiary structures, and some have quaternary structures. *The way alpha helices, folded sheets, and random coils fold and coil* is the **tertiary structure** (■Figure 23.7). To have a quaternary structure, a protein must be composed of more than one peptide chain, for example, hemoglobin has four. **Quaternary structure** is the way the peptide chains pack together (■Figure 23.8).

The natural arrangement of a protein is of lower energy than other possible arrangements. When proteins are made in the laboratory, they spontaneously coil or fold into the same arrangement as the natural material and have the same biochemical activity. Any factor, such as an increase in temperature, that disrupts the normal conformation of a protein **denatures** the protein, that is, *destroys its biological activity.* Disulfide bonds, hydrogen bonds, London forces, and attractions between charged groups all help to hold a protein in its characteristic tertiary structure. Interactions between the protein and the solvent are also important. The maximum possible number of nonpolar groups are on the inside away from the water in the cell, and the maximum possible number of polar groups are on the outside close to the water in the cell. One reason that tertiary structure is important is because it

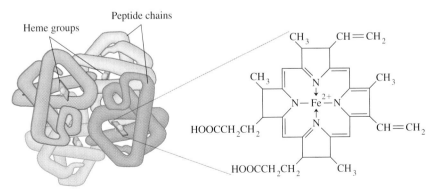

FIGURE 23.8 *Left:* The quaternary structure of hemoglobin, which is composed of four peptide chains and four heme groups. *Right:* Structural formula of a heme group. Oxygen binds to hemoglobin at the Fe^{2+} ions in the heme groups.

results in pockets and grooves that form the active sites of enzymes (remember Figure 18.20).*

Peptide Synthesis

The primary structure of a number of peptides and a few small proteins has been confirmed by synthesis. If a peptide or protein is synthesized by known reactions and the synthetic material has the same biological effect as the naturally occurring material, the structure of the peptide or protein is proved. Chemists and biochemists also synthesize peptides and proteins to obtain large quantities economically and to study the effect of changes in structure on behavior.

Synthesis of a peptide or protein in the laboratory is not easy. However, cells synthesize proteins with ease. In a cell, protein synthesis must be coupled with a reaction that releases energy because formation of a protein by itself is not a spontaneous process. The energy released by the formation of bonds is not enough to make up for the large decrease in entropy that takes place when amino acid molecules are arranged in order in a protein molecule.

23.4 POLYSACCHARIDES

Polysaccharides are *carbohydrates composed of hundreds or even thousands of simple sugar units* of which glucose is the most common. The condensed structural formula for glucose is

$$HOCH_2CH(OH)CH(OH)CH(OH)CH(OH)CHO$$

Glucose exists principally in one of two cyclic forms:

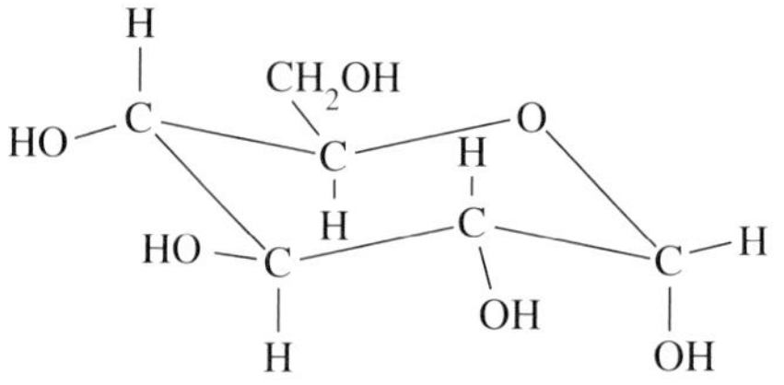

α-D-Glucose 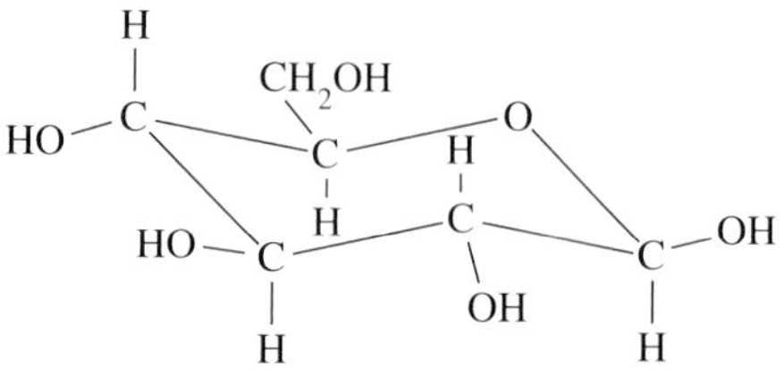   β-D-Glucose

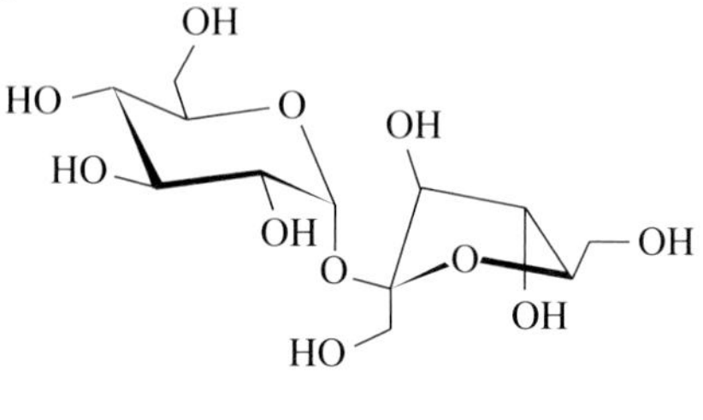

The D shows that the bond to the —CH$_2$OH group points up in the conventional way of drawing the ring shown above. ▪Figure 23.9 shows a glucose molecule in the proper conformation to form a ring. This conformation is the most stable (lowest energy) conformation for the straight-chain molecule. The —OH group that attacks the carbonyl group to form the ring can attack either from above yielding α-D-glucose or from below yielding β-D-glucose. Even in aqueous solutions containing glucose, such as blood, most glucose molecules are in a cyclic form, although a few straight-chain molecules are in equilibrium with the cyclic forms.

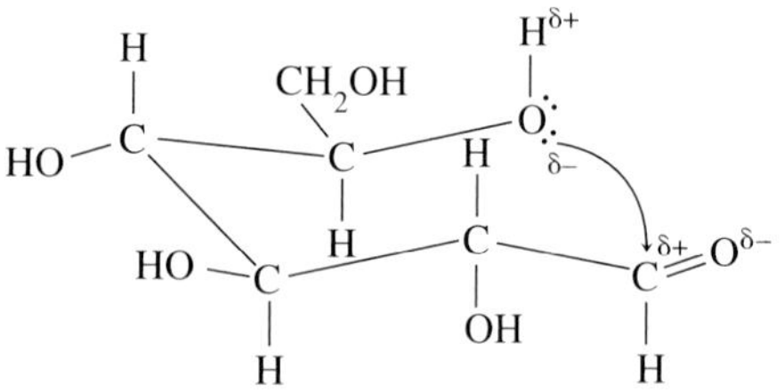

Common table sugar, $C_{12}H_{22}O_{11}$, (sucrose) is a disaccharide that contains one α-D-glucose unit.

▪ FIGURE 23.9   Formation of cyclic forms of glucose.

*Another reason that tertiary structure is important is because it accounts for the great strength of the collagen in skin, bone, tendons, cartilage, blood vessels, and teeth. Collagen consists of three helices wound into a super helix. A load of at least 10 kg is needed to break a fiber of collagen 1 mm in diameter.

Cellulose and starch are by far the most important polysaccharides. Cellulose has the simplest structure of the polysaccharides; it is composed of chains of β-D-glucose units. Part of a chain of glucose units from a molecule of cellulose is shown in ►Figure 23.10. One water molecule is split out each time two glucose units are joined. The molecular formula of a glucose unit is $C_6H_{12}O_6$; the empirical formula of cellulose is $C_6H_{10}O_5$. Each molecule of cellulose is made of more than 1500 glucose units.

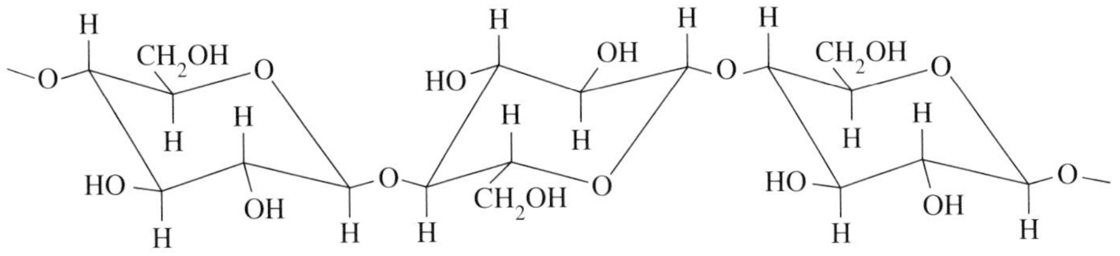

■ FIGURE 23.10 Three glucose units linked as they are in cellulose. For simplicity, the Cs in the rings are not shown. Notice that every other glucose unit is rotated by 180°.

Starch is a mixture of about 80% amylopectin and 20% amylose. Amylose is composed of continuous chains of α-D-glucose units; amylopectin is composed of branched chains of α-D-glucose units. Each molecule of amylose has 1000–4000 monomer units; a molecule of amylopectin may have as many as 1 million monomer units. ►Figure 23.11 shows pieces of amylose and amylopectin molecules.

Both cellulose and starch are formed in plants from carbon dioxide and water by photosynthesis. Cellulose is the chief structural material of plants and is probably

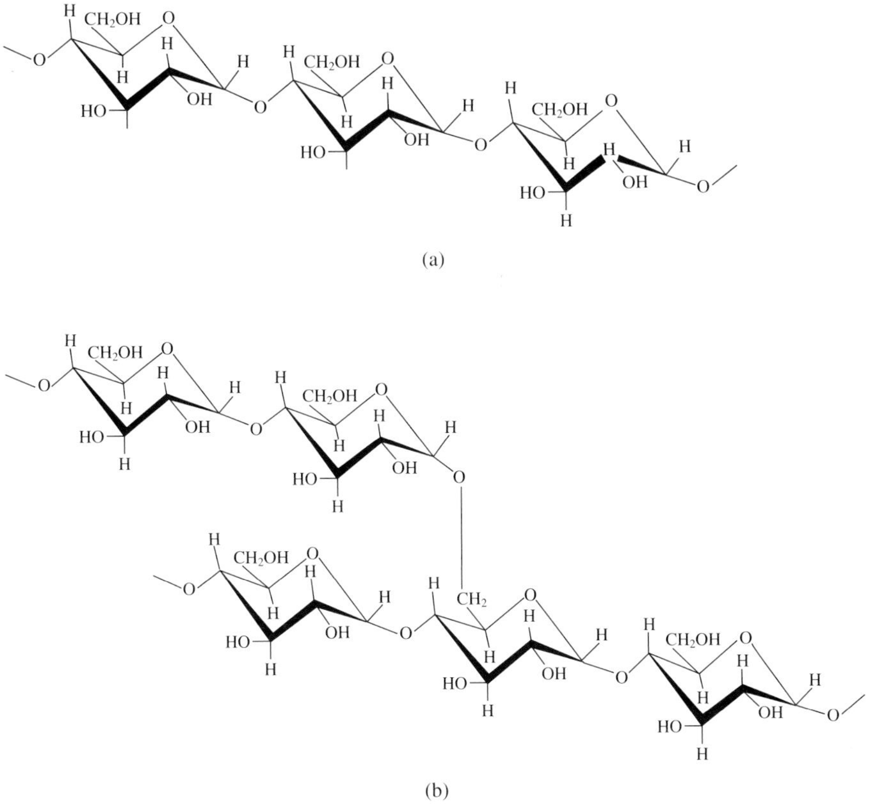

(a)

(b)

■ FIGURE 23.11 Forms of starch, a polymer of α-D-glucose. (a) Amylose and (b) amylopectin.

Glycoproteins

Glycoproteins are oligosaccharides that are covalently bonded to proteins. Oligosaccharides are sugar polymers composed of only a few sugar units. However, more different oligosaccharides can be formed from a given number of sugar units than peptides from the same number of amino acid units because sugar units can be joined to each other through any of the —OH groups, not just through one amino group and one carboxylic acid group. Fibrinogen, which is important in blood clotting, is a glycoprotein. Immunoglobulins, which are involved in the development of immunity to disease, are also glycoproteins. Blood groups are determined by glycoproteins on the surface of red blood cells.

The outer surface of many human cells is coated with glycoproteins. The sugar groups are always located on the outside of the cells because the —OH groups in

the sugars are hydrophilic. The proteins are anchored in the cell membrane (■Figure 23.12). Glycoproteins help cells recognize one another. The recognition of one cell surface by another is involved in the combination of different cells to form tissue and in the detection of foreign cells by the immune systems of higher organisms.

The oligosaccharides on the surfaces of cells play an important part in diseases ranging from the common cold to cancer. Bacteria, viruses, and toxins must bind to carbohydrates on the cell surface to cause infection. For example, flu virus binds to red blood cells through a carbohydrate called sialic acid. The binding of bacteria called *Helicobacter pylori (H. pylori)* to a carbohydrate on the surface of cells in the stomach lining is thought to cause many cases of gastritis, peptic ulcers, and duodenal ulcers.

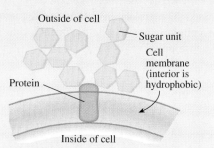

■ FIGURE 23.12 Diagram showing glycoprotein in a cell membrane.

Pharmaceutical researchers are trying to develop carbohydrate molecules that will prevent disease by keeping disease-causing microbes from adhering to the surfaces of cells. For example, an anti-adhesive carbohydrate drug to prevent binding of *H. pylori* may soon undergo clinical trials.

more widespread than any other organic material. People use cellulose as wood for houses, cotton for clothing, and paper for printing on and writing on. Acetate rayon, viscose rayon, and cellophane are modified forms of cellulose. Starch makes up the reserve food supply of plants; it occurs chiefly in seeds. Starch is more soluble in water than cellulose and can be digested by humans, whereas cellulose cannot be digested. Starch in potatoes, corn, rice, and so forth is used as food.

PRACTICE PROBLEMS

23.13 (a) When a glucose molecule cyclizes, what functional groups react? (b) How does the number of stereocenters in the ring form of glucose compare with the number of stereocenters in the straight-chain form? (c) Explain how glucose can show the characteristic reactions of *both* forms.

23.14 (a) What are polysaccharides? (b) What are the two most important polysaccharides? Briefly discuss their uses.

23.5 NUCLEIC ACIDS

Nucleic acids are *acidic substances present in the nuclei of cells.* Nucleic acids *store genetic information and direct the biological synthesis of proteins.* In many

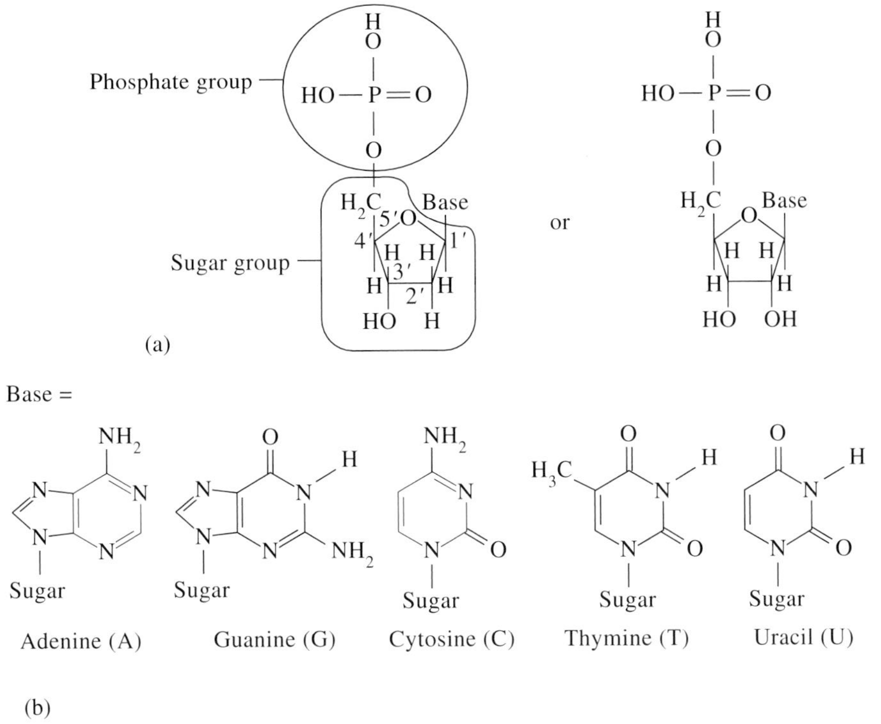

(a)

Base =

Adenine (A)   Guanine (G)   Cytosine (C)   Thymine (T)   Uracil (U)

(b)

■ FIGURE 23.13 Nucleotides and heterocyclic bases. (a) Nucleotides. Primed numbers must be used for the ring of the sugar group because unprimed numbers are used for the heterocyclic bases. What is the difference between the two nucleotides shown? (b) Heterocyclic bases showing point of attachment to the sugar group. In the free base, there is an H where "sugar" is shown.

ways, nucleic acids are similar to proteins. Nucleic acids are biopolymers and have both primary and secondary structures. However, nucleic acids are even larger and more complex than proteins. The molecular mass of collagen, which is one of the largest proteins, is about 1.3×10^5 u; the molecular mass of the polyoma virus, which is one of the smallest nucleic acids, is about 3.3×10^6 u. In addition, the repeating units of nucleic acids are more complicated than the alpha-amino acids that are the repeating units of proteins.

The *repeating units of nucleic acids* are **nucleotides,** *molecules made up of one unit each of phosphate, sugar, and one of five heterocyclic bases* (■Figure 23.13). **Heterocyclic compounds** are *cyclic organic compounds in which one or more of the ring atoms is not carbon.*

There are two types of nucleic acids, deoxyribonucleic acid (DNA) and ribonucleic acid (RNA). DNA stores genetic information. RNA is involved in the synthesis of proteins; some RNA molecules act as enzymes. The structures of DNA and RNA are shown in ■Figure 23.14. Notice how DNA is missing the oxygen at the 2′ position that is present in RNA. The prefix **deoxy-** means *with oxygen removed.* The backbone of DNA and RNA is alternating sugar and phosphate units. In Figure 23.14 the chains are shown stretched out so that you can see them easily; Figure 23.14 does *not* show either the correct bond angles or the relative bond lengths.

A typical strand of human DNA contains about 10^8 nucleotides and has a molecular mass of about 3.5×10^{10} u. Nevertheless, each time an organism synthesizes a particular nucleic acid molecule, the molecule has the same number and sequence of nucleotides. Molecules of DNA are very long but have an extremely small diameter. Although a stretched-out molecule of DNA would be several centimeters long, cells too small to see with the naked eye contain many molecules of coiled-up DNA.

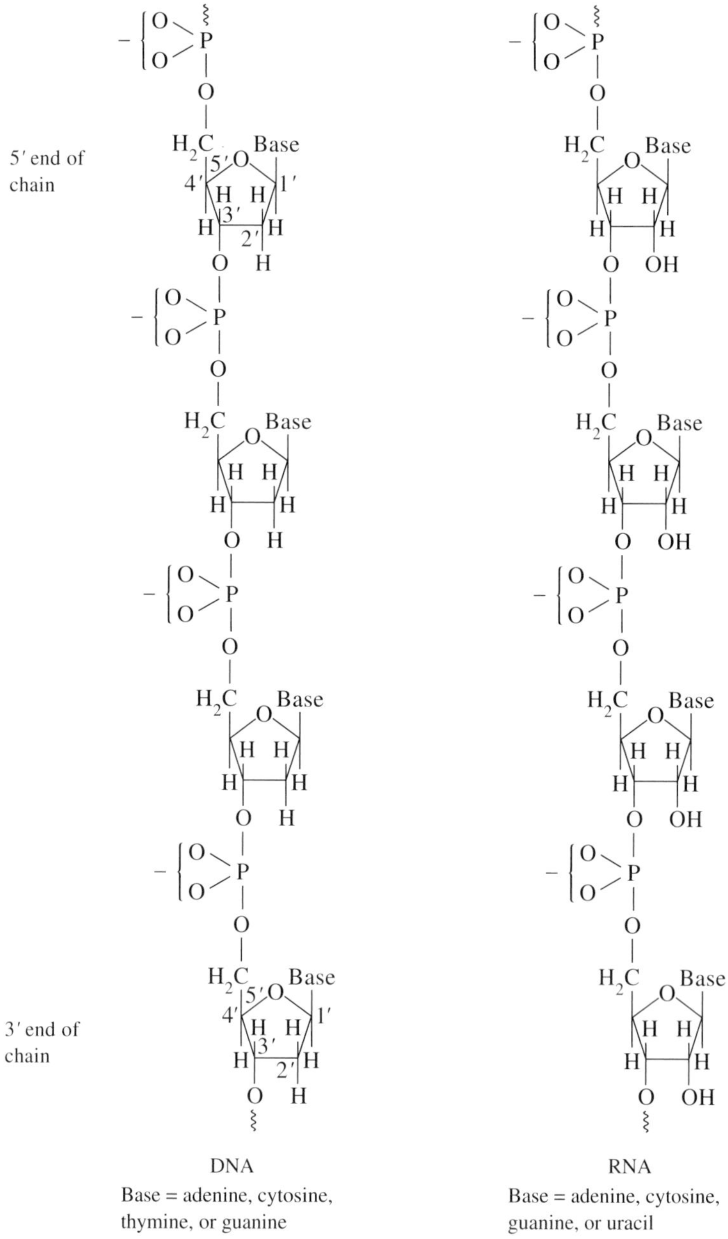
5′ end of chain
3′ end of chain
DNA
Base = adenine, cytosine, thymine, or guanine
RNA
Base = adenine, cytosine, guanine, or uracil

FIGURE 23.14 DNA and RNA.

The entire base sequence of an organism capable of living by itself (the bacterium Hemophilus influenzae, *which causes ear infections) was determined for the first time in 1995. This sequence is 1 830 121 nucleotides long!*

The order of the bases in DNA is called the primary structure of the nucleic acid and carries the information necessary for duplication of the cell and for the synthesis of proteins. Different DNAs have different orders of bases, just as different proteins have different orders of amino acids. The order of the bases can be determined by cleavage either with enzymes or with ordinary chemicals. In the laboratory, nucleic acids can be synthesized by a process similar to the process used to synthesize proteins. The laboratory synthesis of nucleic acids is even more difficult than that of proteins as a result of the greater size of nucleic acid molecules and the more complicated structure of the units.

Secondary Structure

The accepted secondary structure for nucleic acids was first proposed by Watson and Crick in 1953.* According to the Watson–Crick model, two nucleic acid chains are held together by hydrogen bonds on base pairs on opposite strands. In DNA, adenine always pairs with thymine and guanine with cytosine [■Figure 23.15(a)].

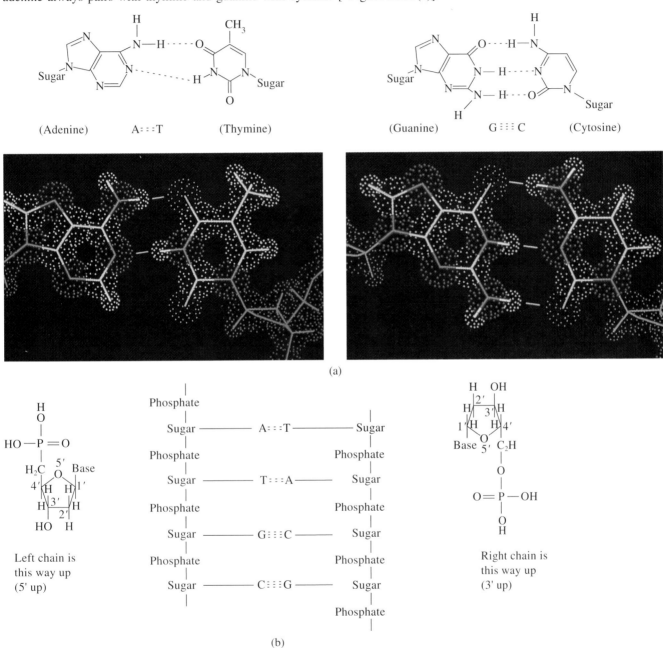

(Adenine)　　A∷∷T　　(Thymine)　　　　　(Guanine)　　G∷∷∷C　　(Cytosine)

(a)

Left chain is
this way up
(5' up)

Right chain is
this way up
(3' up)

(b)

■ FIGURE 23.15 Secondary structure of DNA. (a) Detail of base pairing. *Top:* Structural formulas. *Bottom:* Computer-generated diagrams showing bonds and charge clouds. (b) Schematic diagram of piece of DNA.

---

*James Watson is an American geneticist and biophysicist, and Francis Crick is a British biophysicist. Together with the British biophysicist Maurice Wilkins, who did some of the X-ray crystallography, Watson and Crick received the Nobel Prize for medicine and physiology in 1962. Watson described the work in a very entertaining book, *The Double Helix* (Atheneum, 1968). Rosalind Franklin, a British physical chemist, also provided X-ray crystallographic data but could not be included in the Nobel Prize nomination because she died of cancer in 1958 (age 38). Her story is told in *Rosalind Franklin and DNA* by A. Sayre, Norton, New York, 1975.

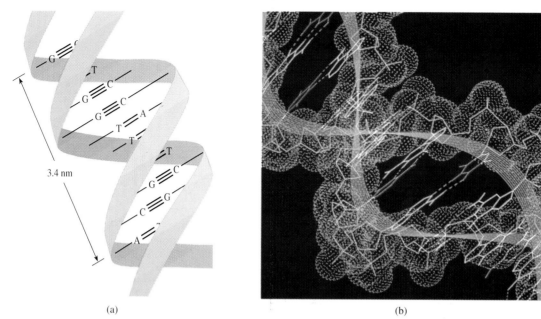

(a)   (b)

- FIGURE 23.16   The double helix. (a) Schematic drawing. (b) Computer-generated diagram showing bonds and charge cloud. Both helices are right-handed. There are ten base pairs in 3.4 nm.

Pairing of adenine with thymine and guanine with cytosine gives two pieces of about the same length and the best fit for hydrogen bonding. Adenine and thymine are said to be **complementary;** guanine and cytosine are also complementary. (In RNA, adenine pairs with uracil.) Figure 23.15(b) is a schematic diagram of a piece of DNA. In Figure 23.15(b), notice how the two strands run in opposite directions. The sequence in the left-hand strand in Figure 23.15(b) is ATGC. The complementary sequence in the right-hand strand is TACG. The double chain is wound into a helix with the base pairs on the inside and the sugar–phosphate backbone on the outside as shown in -Figure 23.16.

PRACTICE PROBLEMS

23.15   What sequence is complementary to the sequence AAG?

23.16   Structurally, what is the difference between DNA and RNA?

23.17   (a) How many hydrogen bonds are there between adenine and thymine? (b) How many hydrogen bonds are there between guanine and cytosine? (c) Which pair is more strongly bonded, adenine and thymine or guanine and cytosine?

23.18   Draw an adenine–uracil base pair.

Replication

Base pairing provides a way for DNA to copy itself when a cell divides. The two strands of a double helix begin to separate at one end like a zipper, and two new strands are made. Each of the new strands is complementary to one of the old strands. The *process by which DNA makes copies of itself* is called **replication.** -Figure 23.17 shows replication schematically.

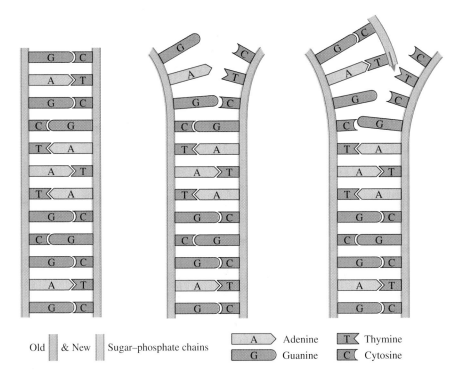

■ FIGURE 23.17   Replication of DNA. One new strand grows along one old strand in the 5′ to 3′ direction (downward) as the "zipper" opens. The next base on the old strand determines which base adds to the new strand. A second new strand forms in pieces in the 5′ to 3′ direction (upward) along the other old strand. The pieces then join to form a second new strand.

Old | & New | Sugar–phosphate chains

A ⟩ Adenine | T ⟨ Thymine
G ⟩ Guanine | C ⟨ Cytosine

DNA and the Primary Structure of Proteins

How does DNA determine the order of amino acids in proteins? *Pieces of DNA called* **genes** *contain sequences of nucleotides that are a code for the amino acid sequence of a particular protein.* There are only 16 possible combinations of four bases taken two at a time, not enough to code for 20 amino acids. Therefore, combinations of at least three bases are required to code for the amino acids used in protein synthesis. There are 64 possible combinations of four bases taken three at a time, more than enough to code for 20 amino acids. Three combinations are used as punctuation and tell where to start and where to stop. In addition, more than one combination often codes for the same amino acid. For example, the sequences AAA and AAG in DNA both indicate phenylalanine. In a complex series of enzyme-catalyzed reactions, the nucleic acid sequence of a gene on a DNA molecule is converted to a complementary sequence of RNA. The protein is then synthesized, one amino acid at a time, according to the code on the RNA:

$$DNA \longrightarrow RNA \longrightarrow protein$$

Details of how DNA copies itself and how DNA and RNA direct the synthesis of proteins are part of the field of molecular biology.

The commercial production of the hormone insulin for the treatment of diabetes is an example of the application of knowledge about biochemical protein synthesis. Insulin is composed of 51 amino acid units and is too large to be synthesized commercially. Isolation of insulin from the pancreas of animals such as cattle and pigs is very expensive. Recombinant DNA technology is used to introduce the gene that codes for insulin into the DNA of the common bacterium *Escherichia coli.* Cultures of the bacterium then produce insulin in large quantities at relatively low cost.

If you wish to cover lipids also, the Instructor's Resource Manual contains a complete text section available for reproduction and distribution as a handout.

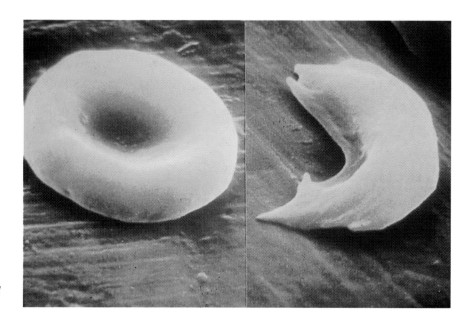

Left: *Normal red blood cell.* Right: *Sickled red blood cell (magnified 7 × 10³ times).*

DNA and Disease

Because of the part DNA plays in controlling reproduction and life processes, any change in the structure of DNA—a missing base, an extra base, or substitution of one base for another—may cause serious problems. For example, there is some evidence that chemical damage to DNA can interfere with hydrogen bonding and start the uncontrolled cell division called cancer. One type of chemical damage is caused by alkylating agents, compounds such as dimethyl sulfate, $CH_3OSO_2OCH_3$, that have good leaving groups and react with nucleophiles. Some antitumor agents replace adenine and thymine in DNA and slow cell growth. These antitumor agents affect fast-growing cancer cells more than normal cells.

Another change in a single amino acid may be responsible for allergic reactions (see Science *1994, 264, 1533).*

The hereditary disease sickle-cell anemia is caused by the change of a single base in DNA. The wrong base causes substitution of valine for glutamic acid at one point in the protein hemoglobin. Red blood cells that contain hemoglobin with this error become crescent-shaped and clog capillaries. Children born with sickle-cell anemia often die before they are two years old.

PRACTICE PROBLEM

23.19   The following DNA sequences code for glycine: CCA, CCG, CCT, and CCC. Write the RNA sequences that are complementary to each of these DNA sequences.

SUMMARY

Except when the carbon chain is very short, the reactivity of functional groups is the same regardless of the size of the molecule to which the functional groups are attached. However, molecular size has a large effect on physical properties; polymers are usually very strong compared to compounds composed of smaller molecules. The strength of the intermolecular forces be- tween polymer molecules and the shape of the molecules depend on the monomer used to make the polymer. If intermolecular forces are strong and the molecules have a regular shape so that they pack well in the solid, a polymer is suitable for making fibers. If intermolecular forces are weak and the molecules have an irregular shape so that they do not pack well in the solid, the polymer is elastic. Polymers used to make typical plastics are in between the two extremes of fibrous polymers with their strong intermolecular forces and efficient packing and elastomers with their weak intermolecular forces and irregularly shaped molecules. **Thermoplastics** soften when heated; **thermoset plastics** do not soften when heated.

Synthetic polymers are formed either by chain reactions or by stepwise reactions. Although most polymers are organic, some—for example, silicones—are inorganic.

Most synthetic polymers consist of molecules with a range of molecular masses. The molecular mass distributions of polymers are usually determined by **gel-permeation chromatography.**

Natural polymers, such as rubber, made by living organisms are called **biopolymers.** The most important types of biopolymers are proteins, polysaccharides, and nucleic acids. Proteins are polyamides; the units of proteins are **alpha-amino acids,** carboxylic acids that have an amino group, —NH_2, and a carboxyl group, —COOH, attached to the same carbon, which is called the **alpha** carbon. Almost all of the amino acids that are the building blocks of proteins are chiral.

Essential amino acids are amino acids that humans need for protein synthesis but cannot synthesize in their bodies. They must be obtained from foods or dietary supplements.

An amide bond between amino acids is referred to as a **peptide link.** Molecules that have 1–50 peptide links are called **peptides. Proteins** have more than 50 peptide links. The order in which the amino acids are arranged is called the **primary structure** of the peptide or protein. The primary structure determines the properties of a protein or peptide. The **secondary structure** is the way the chains of amino acids are coiled or folded. Some common secondary structures are **alpha helices, pleated sheets,** and **random coils.** The **tertiary structure** is the way secondary structures fold and coil. **Quaternary structure** is the way the peptides pack together in proteins that have more than one peptide chain.

Polysaccharides are compounds made up of hundreds or even thousands of sugar units. Cellulose and starch are the most important polysaccharides; starch can be digested by humans whereas cellulose cannot.

Nucleic acids are composed of **nucleotides,** molecules made up of one unit each of phosphate, sugar, and one of four heterocyclic bases. **Heterocyclic compounds** are cyclic organic compounds in which one or more of the ring atoms is not carbon. Deoxyribonucleic acid, DNA, stores genetic information. The prefix **deoxy-** means "with oxygen removed." In DNA two strands of nucleic acid are coiled in a double helix. The two strands are held together by hydrogen bonds between base pairs on opposite strands. Bases which pair are said to be **complementary.** Base pairing provides a way for DNA to copy itself. The process by which DNA copies itself is called **replication. Genes** are pieces of DNA containing sequences of nucleotides that are a code for the amino acid sequence of a particular protein. Ribonucleic acid, RNA, is involved in the synthesis of proteins; some RNA molecules act as enzymes.

ADDITIONAL PRACTICE PROBLEMS

For information about the organization of Additional Practice Problems, Putting Things Together, and Applications, see the beginnings of these sections in Chapter 1.

23.20 What is the difference between (a) a polymer and a monomer? (b) A polymer and a copolymer? (c) A chain polymer and a step polymer?

23.21 (a) Write structural formulas showing two repeats and abbreviated formulas for the polymers of
(a) $CH_3CH_2CH{=}CH_2$ (b) $H_2C{=}CHCH{=}CH_2$.
(c) $HOOC(CH_2)_{10}COOH$ and $H_2N(CH_2)_6NH_2$
(d)

$$HOOC-\text{⬡}-COOH \quad \text{and} \quad HOCH_2CH_2CH_2CH_2OH$$

23.22 Which of the following is a polyester?

(a) $\text{─}(CH_2CH_2O)_n$ (b) $\text{─}(CH_2CH)_n$
$\qquad\qquad\qquad\qquad\qquad\quad\ |$
$\qquad\qquad\qquad\qquad\qquad COO(CH_2)_3CH_3$

(c) $\text{─}(O(CH_2)_4OOC(CH_2)_4CO)_n$

23.23 Write the mechanism for the radical-chain polymerization of $H_2C{=}CHCN$.

23.24 Write equations for the first and second steps in the polymerization of

$$HOOCCH_2CH_2CH_2CH_2COOH \quad \text{and} \quad H_2N(CH_2)_9NH_2$$

23.25 (a) What functional group must be present in the monomers for chain polymerization? (b) What structural feature must be present in the monomers for step polymerization?

23.26 (a) What is a silicone? (b) Give two properties of silicones.

23.27 (a) The abbreviated formula for poly(methylphenylsiloxane) is $\text{─}(Si(CH_3)(C_6H_5)O)_n$. Write a structural formula for a piece of the polymer chain that contains three monomer units. (b) Write the abbreviated formula for the silicone shown in the text discussion and give its name. (c) Which silicone is more water repellant? Explain your answer.

23.28 In your own words, answer the question "What difference does the extremely large size of polymer molecules make in physical and chemical properties?"

23.29 (a) Briefly discuss the differences between fibers, elastomers, and plastics. (b) What is the difference between thermoplastic and thermosetting plastics?

23.30 What is meant by the term *essential amino acid*?

23.31 (a) Give the experimental evidence that amino acids exist as inner salts.   (b) Explain *why* amino acids exist as inner salts.   (c) What is the acidic group in alanine? The basic group?

23.32 Write structural formulas for the amino acids shown by the following models:

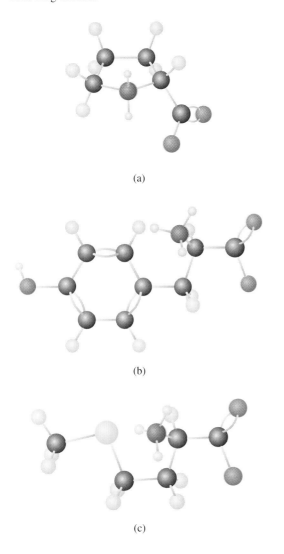

(a)

(b)

(c)

23.33 For the essential amino acid glutamine,

$$CH_2CH_2CONH_2$$
$$H---\overset{|}{\underset{H_2N}{C}}-COOH$$

(a) write the condensed structural formula as it would be shown in Table 23.2.   (b) Write the stereochemical formula of the enantiomer of the natural stereoisomer that is shown above.   (c) What functional group is present in the side chain?

23.34 What is the difference between a peptide and a protein?

23.35 For the tripeptide Ala-Lys-Ser, write (a) the structural formula   (b) the one-letter abbreviation   (c) one-letter abbreviations for all 26 other tripeptides that can be formed from these three amino acids.

23.36 A pentapeptide gives the dipeptides Glu-Phe, Lys-Glu, Phe-Val, and Val-Ser on partial hydrolysis. Write the structure of the pentapeptide.

23.37 (a) What is the difference between the primary, secondary, tertiary, and quaternary structures of proteins?   (b) How are primary structures determined?   (c) How are secondary structures determined?   (d) What are two common secondary structures?   (e) What is a random coil?

23.38 Give four jobs carried out by proteins in the body.

23.39 (a) From your everyday life, give an example of the denaturation of a protein.   (b) Explain what happens on a molecular level when a protein is denatured.

23.40 (a) Structurally, what is the difference between cellulose and starch?   (b) How are they formed in plants?   (c) What different jobs do they do in plants?   (d) What do people use them for?

23.41 What are (a) nucleic acids?   (b) Nucleotides?   (c) Heterocyclic compounds?

23.42 Draw a complete structural formula for the DNA nucleotide in which the base is cytosine.

23.43 (a) What is meant by the primary structure of DNA? The secondary structure?   (b) What is the function of DNA?

23.44 Which is stronger, guanine–cytosine or adenine–uracil bonding? Explain your answer.

23.45 If the sequence GGATCC appears in one strand of DNA, what sequence appears opposite it in the other strand?

23.46 What is replication?

PUTTING THINGS TOGETHER

23.47 Of the adipic acid made in the United States in 1993, 88% was used to make nylon 66. How many pounds of $H_2N(CH_2)_6NH_2$ are required to react with this much adipic acid?

23.48 Formaldehyde, which is a gas under ordinary conditions, is usually sold in the form of a 37% aqueous solution or as the solid polymer $-(CH_2O)_n$. Dry formaldehyde gas can be obtained by heating the solid polymer.   (a) Use bond energies to estimate ΔH for the polymerization of formaldehyde. Is the estimated value positive or negative?   (b) Is ΔS for the polymerization of formaldehyde positive or negative? Explain your answer.   (c) How do you account for the fact that formaldehyde polymerizes spontaneously?   (d) Formaldehyde also spontaneously forms a cyclic trimer that is a solid (mp 64 °C) that can be sublimed unchanged. (i) Write a structural formula for the trimer. (ii) What reaction conditions would you use to favor formation of the polymer over the trimer? (e) Although other low molecular mass aldehydes form trimers, no other aldehyde polymerizes; ketones neither cyclize nor polymerize. Write an abbreviated formula for poly(methyl vinyl ketone), $CH_3COCH=CH_2$.

23.49 Esters are made by the acid-catalyzed reaction of an acid with an alcohol:

$$RCOOH + R'OH \xrightarrow{H^+} RCOOR' + H_2O$$

(a) Write the equation for the reaction of CH_3COOH with $CH_3CH_2CH_2CH_2OH$.
(b) Write the equation for the first step in the polymerization of $HOOCCH_2CH_2COOH$ and $HOCH_2CH_2OH$.   (c) Write the equation for the first step in the polymerization of lactic acid, $CH_3CH(OH)COOH$.   (d) Write a structural formula that shows three repeats for the polymer in part (c).   (e) Write the abbreviated formula for the polymer in part (c).

23.50 (a) Write equations for the following reactions:

(i) $\left(CH_2C\right)_n$ with OH and COOH $+ \ nNaOH \longrightarrow$

(ii) $\left(CH_2C\right)_n$ with H and CH_2NH_2 $+ \ nHCl \longrightarrow$

(b) Why is polyethylene unreactive?

23.51 In the copolymer of styrene and maleic anhydride

styrene   maleic anhydride

the monomer units alternate.   (a) Show two repeats of the copolymer.   (b) For a regular polymer like this to form, equimolar quantities of the monomers must be present. What must also be true of the reactions rates of the monomers with each other compared to the reaction rate of each with itself? Explain your answer.   (c) If a random copolymer, such as the copolymer of $H_2C=CHCl$ and $H_2C=CCl_2$ (Section 9.11) forms, how must the relative rates compare?   (d) Must the monomers be present in equimolar amounts?

23.52 Explain why polystyrene can be made more rigid by copolymerizing divinylbenzene with the styrene.

$H_2C=CH$ —⟨ ⟩— $CH=CH_2$

divinylbenzene

23.53 Racemic mixtures are usually much cheaper than pure enantiomers. However, for about half the essential amino acids, the price of one pure enantiomer is substantially lower than the price of the racemic mixture. Explain.

23.54 The following titration curve is that of glycine hydrochloride, which behaves as a typical diprotic acid:

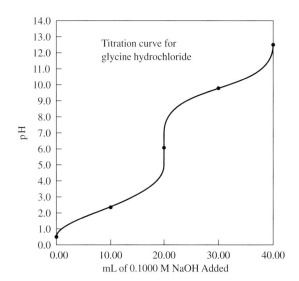

(a) Label the following points on the titration curve: (i) $H_3\overset{+}{N}CH_2COO^-$ is the predominant species. (ii) $H_3\overset{+}{N}CH_2COOH$ is the predominant species. (iii) $H_2NCH_2COO^-$ is the predominant species. (iv) $[H_3\overset{+}{N}CH_2COOH] = [H_3\overset{+}{N}CH_2COO^-]$. (v) $[H_3\overset{+}{N}CH_2COO^-] = [H_2NCH_2COO^-]$.   (b) What is the value of K_{a1}? K_{a2}?   (c) At what two pHs is the solution buffered? (d) What was the mass of the sample?

23.55 Because the chains of amino acids in protein molecules are so long, the side chains of different amino acids in the chain may interact with each other to hold the chain in its most stable conformation. What type of interaction would occur between each of the following pairs of groups?   (a) $—NH_3^+$ and $—COO^-$ (b) $—OH$ and $—OH$ (c) $—CH(CH_3)_2$ and $—CH_2CH(CH_3)_2$

23.56 In what ways are proteins and nucleic acids similar, and in what ways are they different?

23.57 How many chiral carbons are in each sugar group in (a) DNA? (b) RNA?

APPLICATIONS

23.58 Poly(vinyl acetate), which is the film-forming component of water-based (latex) paints, is made by polymerizing vinyl acetate, $CH_3COOCH=CH_2$. (a) Show the structure of poly(vinyl acetate).   (b) Polymerization of vinyl acetate is catalyzed by peroxides, ROOR, which decompose to form radicals, RO·. Write the mechanism for the polymerization of vinyl acetate.

23.59 (a) Pantyhose are made of nylon. Explain why a drop of sulfuric acid produces a hole. (b) The 66 in the name nylon 66 indicates that there are six carbons in each of the two monomers from which nylon 66 is made. A new nylon formed from 1,4-diaminobutane and adipic acid, HOOC(CH$_2$)$_4$COOH, which is tougher and has a higher melting point than nylon 66, has recently been introduced. Show the structure of this new nylon. (c) What do you think the new nylon in part (b) is called?

23.60 Laboratory wash bottles are made of polyethylene. A wash bottle has a mass (empty) of 61.17 g. (a) How many moles of ethylene were used to make the bottle? (b) If the average molecular mass of the polyethylene in the bottle is 1.0×10^6 u, how many moles of polyethylene are in the bottle? (c) Why can only an average molecular mass be given for polyethylene? (d) The type of polyethylene used to make wash bottles is produced by radical chain polymerization using oxygen gas as the initiator. To obtain polyethylene with a high average molecular mass, should the quantity of oxygen used be large or small? Explain your answer.

23.61 Kevlar

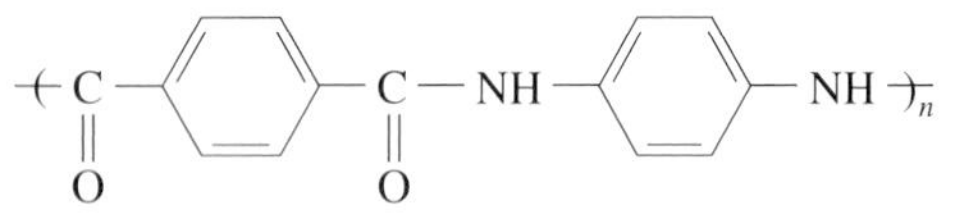

is very strong. For example, Kevlar ropes have 20 times the strength of steel ropes in seawater and are used on offshore oil-drilling platforms. (a) Write formulas for the monomers used to make Kevlar. (b) Draw two chains with two repeats each and show the hydrogen bonds between them with dashed lines. (c) How many of the atoms in Kevlar lie in the plane of the benzene rings? (d) The average molecular mass of the polymer chains is 10^5 u. How many repeats are in an average chain? (e) Besides strength, what advantage does Kevlar have over steel in saltwater?

23.62 Poly(vinyl alcohol), $-$(CH$_2$CH(OH)$)_n$, is used in textile sizing, adhesives, emulsifiers, and paper coatings. If a 1.35-g sample in 100.0 mL of aqueous solution has an osmotic pressure of 5.02 mmHg at 22.0 °C, what is the molecular mass?

23.63 Which is more cross-linked, the rubber in a tire or the rubber in a balloon?

23.64 Ethyl silicones are used in electrical insulation and in protective and decorative coatings. A piece of cross-linked ethyl silicone is shown below:

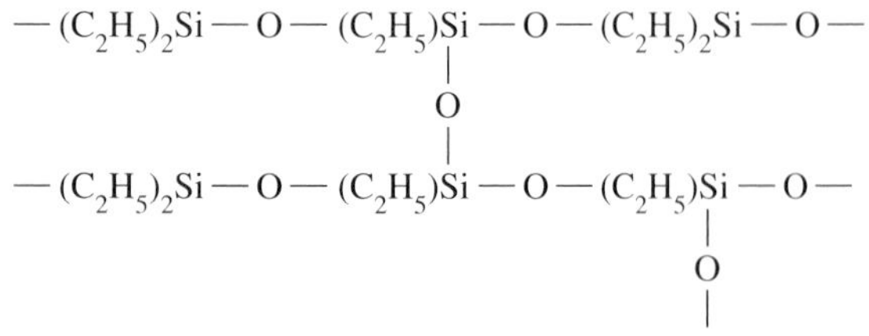

Below a C$_2$H$_5$:Si ratio of 0.5:1, the silicone is a brittle solid; above 1.5:1, the liquid is difficult to solidify. (a) Modify the drawing so that it shows a 1:1 ratio. (b) Explain the information given in the sentence that precedes part (a).

23.65 Besides silicones, another type of synthetic inorganic polymer is the polyphosphazenes $-$(X$_2$P$=$N$)_n$ where X can be any of a variety of atoms or groups, for example, $-$OCH$_2$CF$_3$ or $-$N(CH$_3$)$_2$. (a) Draw a general structural formula showing three repeats. (b) How would you expect the P—N bond lengths to compare with the length of P—N single bonds? (c) One of the two examples is water-soluble. Which one? Explain your answer.

23.66 Polymers can be made by bulk, solution, suspension, and emulsion processes. In bulk polymerization, only monomers and a small quantity of catalyst are present. (a) Give one advantage and one disadvantage of bulk polymerization over polymerization in solution. (b) In suspension polymerization, monomers and catalyst are suspended as tiny drops in a homogeneous medium, such as water, by continuous stirring. In emulsion polymerization, an emulsifying agent, such as soap, is used to form micelles where polymerization takes place. What is an emulsion? What are micelles?

23.67 What kind of polymer (elastomer, fiber, or plastic) would you use to make (a) rope? (b) A fork? (c) Sewing thread? (d) A bathing cap? (e) A telephone?

23.68 The artificial sweetener Nutrasweet is aspartame:

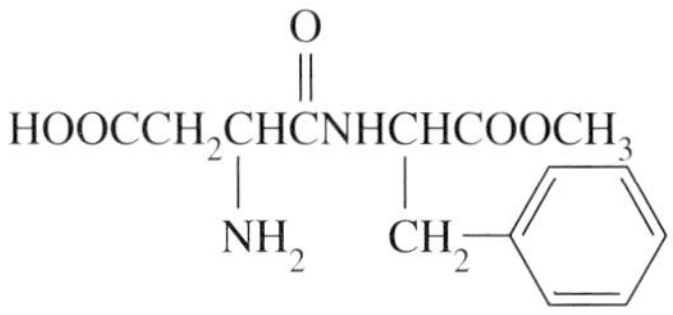

(a) To what classes of compounds does aspartame belong? (b) Circle the peptide link. (c) Mark the stereocenters with an asterisk.

23.69 One sequence of amino acids repeats for long distances in silk. Complete hydrolysis of one mole of a fragment with this sequence gives 2 mol alanine, 3 mol glycine, and 1 mol serine. Partial hydrolysis yields Ala-Gly-Ala, Gly-Ala-Gly, Gly-Ser-Gly, and Ser-Gly-Ala. What is the structure of the fragment?

23.70 Cytochromes are electron-transporting proteins generally regarded as universal catalysts of respiration. Cytochrome c occurs in the cells of all aerobic organisms. It consists of a single polypeptide chain of 104 or more amino acids and is 0.43% iron. (a) What is the minimum molecular mass of cytochrome c? (b) What is an *aerobic organism*?

23.71 Most enzymes are globular proteins like the one shown in Figure 23.7. (a) In an enzyme-catalyzed reaction, what is the reactant called? The pocket or groove that reaction takes place in? (b) Why do enzymes catalyze only a specific reaction, not a variety of reactions? Explain why the rate of enzyme-catalyzed reactions (c) increases when enzyme concentration is increased (d) at first increases and then remains constant as reactant concentration is increased (e) first increases and then decreases as temperature is increased (f) first increases and then decreases as pH is increased. (g) Of the amino acids in Table 23.2, which is most likely to be found in the hydrophobic interior of a globular protein such as egg lysozyme? Explain your answer.

23.72 Cotton towels are very good at soaking up water. Cotton–polyester blends wear better than pure cotton. Which would you buy—100% cotton towels or towels made of a cotton–polyester blend? Explain your answer.

23.73 In wood, long chains of cellulose lie side-by-side in bundles. The bundles are twisted together to form ropelike structures that are grouped to form the visible fibers. Explain why wood is strong.

23.74 Explain why a sweet taste is observed if bread or plain crackers are chewed for a long time without swallowing.

23.75 Starch is more easily synthesized in living organisms than are fats and oils, which are esters of glycerol, $(HOCH_2)_2CHOH$, such as

$$[CH_3(CH_2)_{16}COOCH_2]_2CHOOC(CH_2)_{16}CH_3$$

On the other hand, fats and oils provide more energy (40 kJ/g) than starch (17 kJ/g). Why do plants store energy in the form of starch whereas animals store energy in the form of fats?

23.76 Of the bases in human DNA, 19% are guanine and 31% are adenine. What percentages of cytosine and thymine does human DNA contain?

23.77 Human insulin consists of two polypeptide chains, one containing 21 and the other containing 30 amino acids. How many nucleotides must be present in DNA to code for each chain?

A Brewer Uses Chemistry

WILLIAM R. JENKINS

Package Production Manager
Pike Place Brewery

B.A. Chemistry
University of Southern Maine

Working in a microbrewery, I am constantly reminded that chemistry is an integral and inseparable part of the brewing day. From the inspection and analysis of the barley and hops when they arrive at the brewery to the finished beer, chemistry and the chemical method must be ever-present to ensure that the beer is as good as it can be. Whereas a larger brewery can generally compensate for undesirable qualities in its product by using flavor additives or stabilizers, microbreweries must rely instead on quality raw ingredients and a solid understanding of chemistry to ensure that their beer can compete in the marketplace.

Water, of course, is the main ingredient in all beers, yet its chemistry is often overlooked as an influence on the finished product. Different styles of beer benefit from the different balance of minerals present in the water from which they are brewed. The pale ales of England, for example, could never have achieved their style or their popularity if the water in the areas surrounding London had not had an extremely complex mineral composition: calcium, though having little direct effect on flavor, nevertheless is important in that it precipitates proteins, clarifying the beer, and also helps maintain the proper pH balance; magnesium is a vital cofactor in several enzymatic reactions; and the sulfate complement to these cations imparts a dryness to the beer and accentuates the bitterness of the hops. In contrast to this very hard water is the water from which the classic pilsner style, native to central Europe, is brewed. This beer, with its delicate bitterness and pronounced malt character, benefits from water having a relative absence of dissolved minerals.

Many municipal water supplies, however, must be chemically treated to meet quality standards of drinking water, and the results are certainly felt in the brewing industry: heavily chlorinated water will produce medicinal flavors and solventlike aromas, and highly alkaline water may inhibit essential steps in the fermentation cycle. The Seattle area owes a large part of its reputation for quality beers to the fact that its geography and climate minimize the need for most treatment methods. Thus, the water chemistry of a region affects the character of the beer even before the brewing process begins. In addition, many breweries must adjust the mineral composition of their water in order to brew a variety of beer styles, and a solid foundation in inorganic chemistry is certainly useful in this process.

A thorough understanding of both biochemistry and organic chemistry is also important to the brewer. The production of beer using only barley, hops, water, and yeast is completely dependent on microbial and enzymatic reactions, procedures far more complex than the simple dissolution of mineral salts in water. Once the brewery receives the malted barley, the mashing procedure begins, which converts the starches in the grain to fermentable sugars enzymatically. The starch in the barley kernels is a polymer composed of repeating glucose units. One form of starch, amylose, is simply a straight chain of glucose units like a string of beads. The other common form, called amylopectin, also has a beadlike backbone, but is characterized by the presence of numerous branching side chains.

Two principal diastatic, or starch-converting, enzymes are found in the barley malt: alpha-amylase and beta-amylase. Alpha-amylase attacks the bonds between the glucose units in both amylose and amylopectin at random, producing a mixture of sugars of different sizes. Beta-amylase, on the other hand, only attacks glucose-glucose bonds at the end of a chain and always produces a two-glucose unit, maltose. The two enzymes also have different optimum temperatures and pH ranges in which they operate, so the conditions of the mash greatly influence the character of the beer. Lower mash temperatures, around 140 °F (60 °C), and a pH near 5.0 favor beta-amylase activity, which results in a more fementable wort (the term given to beer prior to fermentation) and a drier taste, with correspondingly less body and a decreased malt character. Higher temperatures in the mash kettle, around 158 °F (70 °C), and a pH closer to 5.7 favor alpha-amylase activity (and also destroy the beta-enzyme), which produces a wort with more complex sugars that are harder for the yeast to ferment. The result is a heavier-bodied beer with a sweeter, more malty taste on the palate. Thus, by understanding the chemical processes going on in the mash tun, the brewer is able to custom craft the beer through the simple application of chemical principles.

The next step, boiling the wort, also benefits from an understanding of chemistry. Boiling denatures many organic molecules, especially proteins, and causes them to precipitate out, clarifying the beer. More importantly, the hops, the other principal flavoring ingredient, are added during the boil. Because the essential hop resins are only partially soluble in water, a vigorous and lengthy boiling period is required to extract and utilize the full flavor of the hops. The heat and agitation produced in the kettle chemically alter the bittering oils, allowing them to be assimilated into the beer. Successful fermentation is also vitally connected to chemistry. Yeast cells need certain narrow ranges of temperature, pH, and nutrient levels to produce the qualities we expect in a finely brewed beer. The presence of amino acids, lipids, and phosphates aids in the growth cycle of the yeast, and in the early stages of fermentation, the wort must also contain a sufficient level of dissolved oxygen.

Fermentation involves more than the simple conversion of sugars to alcohol and carbon dioxide. Other by-products, though produced in much smaller amounts, greatly influence the flavor of the finished beer; hence, the brewer can carefully craft the beer by manipulating the fermentation conditions. One of the most notorious of these by-products is 2,3-butanedione, or diacetyl, which is produced during the early aerobic phase of fermentation and normally is reduced to ethanol later in the cycle. Diacetyl imparts a strong butterscotch flavor and has a very low flavor threshold, below 0.1 ppm, so its presence can be detected even in very small amounts. Warmer temperatures toward the end of the fermentation cycle aid in the reduction of diacetyl, giving brewers a simple means of controlling its influences.

In addition, fusol oils, alcohols of higher molecular weight, are produced during fermentation. Fusol oils can result in off-flavors or may combine with various organic acids to form esters, which are responsible for unmistakable fruity aromas. Ester formation, however, like many of the reactions producing off-flavors or aromas, can easily be reduced or eliminated by changing conditions under which they are normally produced. In the case of unwanted esters, simply ensuring that the wort contains a sufficient level of dissolved oxygen prior to fermentation greatly inhibits their production.

Thus, aided by an appreciation of various chemical principles and an understanding of how they relate to the various steps in the brewing process, the brewer is able to fine-tune the beer-making procedure. The brewer's work serves as a reminder that the world of chemistry is not limited to the laboratory or lecture hall—it even has a place beside you at the next barstool.

24

A CLOSER LOOK AT INORGANIC CHEMISTRY
Transition Metals and Complexes

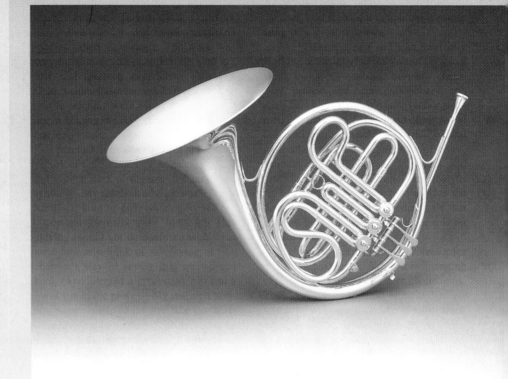

French horns are usually made of brass, an alloy of copper and zinc.

The major uses of transition metals are as metals. The properties of transition metals vary enormously; for example, iron rusts rapidly in moist air, but gold remains shining and beautiful for hundreds and even thousands of years (▬Figure 24.1). In addition, the properties of metals can be changed to suit different uses by alloying with other metals or with nonmetals.

In this chapter, the physical and chemical properties of the transition metals are related to atomic structure and position in the periodic table. Structure determina-

tion, nomenclature, a model used to explain the properties of complexes, and some applications of complexes are discussed. The formation of complexes is a characteristic property of transition-metal ions.

▬ *Why is a study of transition metals and complexes a part of general chemistry?*

A number of transition metals are of major importance. Iron is the cheapest and most used metal. Steel is used in the

construction of high-rise buildings and bridges, in the manufacture of automobiles and other vehicles, and to make machinery and household equipment, from large appliances like washing machines to flatware and pans. Copper and gold are used in the electrical industry; zinc is used to galvanize iron; and copper, zinc, and iron are used in alloys. Gold has been by far the most important means of payment between nations.

Many biologically important molecules such as hemoglobin, chlorophyll, and vitamin B_{12} are complexes. Indeed,

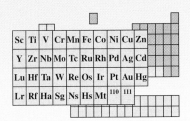

■ FIGURE 24.1 *Left:* The pre-Columbian Inca gold figure still looks like new after over 500 years. *Right:* The steel parts of the car are very rusty after less than 20 years.

the biological roles of the transition elements usually depend on the formation of complexes. A number of enzymes are complexes. In addition, complexes are important as catalysts in many industrial processes and are used in the separation of metals from their ores, in refining metals, and in electroplating. Complexes are also useful in water purification, as preservatives for vegetable oils and rubber, and in pharmacy.

— What do you already know about transition metals and complexes?

From your everyday life, you are familiar with gold, silver, copper, iron, nickel, and probably more transition metals. The reactivity of transition metals does not change in a regular way across a period of the periodic table. However, reactivity decreases from top to bottom of a group (Section 4.5).

In the simplest model of metallic crystals, the units are metal ions. The crystals are held together by the attraction between the positive charges of the metal ions and the negative charges on the mobile valence electrons. Metals are good conductors of thermal energy and electricity in both solid and liquid states because the valence electrons are free to move. When one layer of a metal crystal is moved past another, the positive charges on the metal ions are shielded from each other by the valence electrons and metals can be drawn into wires and hammered or rolled into sheets (Section 12.8).

Deterioration of metals by oxidation is called corrosion (Section 19.6). Electroplating of metals is an important method of preventing corrosion (Section 19.9). Electrolysis must be used to produce very reactive metals and is also used to purify less active metals.

In Section 16.4, you learned that complexes are species in which a central atom is surrounded by a set of Lewis bases that are covalently bonded to the central atom. Complexes can be charged like the $Al(H_2O)_6^{3+}$ and $Al(OH)_4^-$ ions or be molecules such as $Al(H_2O)_3(OH)_3$. The Lewis bases in complexes are usually referred to as ligands. Formation of complex ions other than hydrated ions in aqueous solution always involves substitution of other ligands for water. Substitution of other ligands takes place in steps and is reversible. The value of the equilibrium constant for each step is smaller than the value of the equilibrium constant for the preceding step. However, the differences between equilibrium constants for successive steps are small. Only when the equilibrium constant for the overall reaction is high and an excess of the ligand is present is calculation of the concentrations of the predominant species present in solution simple. Replacement of water in hydrated metal ions of the first transition series by other ligands usually takes place very rapidly, and equilibrium is reached almost immediately. Complexes are used to dissolve slightly soluble salts and hydroxides (Sections 16.6 and 16.7) and in electroplating (Section 19.9).

The green color of emeralds is caused by the presence of chromium.

Transition metals are responsible for the color of many gemstones and of colored glass. Ruby glass contains colloidal gold, and cobalt glass, which is blue, contains cobalt(II) phosphate.

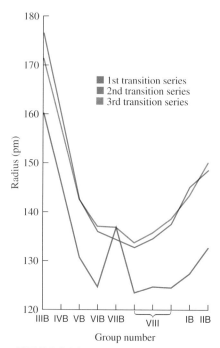

- FIGURE 24.2 Radii of transition metals.

TRANSITION METALS

Transition metals are enough less reactive than the metals of Groups IA and IIA and aluminum that their major uses are as metals. The only compound of a transition metal in the Top 50 is titanium(IV) oxide, which is the most important white pigment and is used in the manufacture of paint, plastics, and paper (Section 8.7).

Transition metals are *d*-block elements. All the transition metals except palladium have either one or two *s* electrons in the outermost shell (Figure 8.4). The elements in a given transition series differ from each other in the number of *d* electrons in the next-to-the-outermost ($n - 1$) shell. For example, the first transition series is in the fourth row of the periodic table. The elements in the first transition series have their outer or valence electrons in the fourth shell; they differ in the number of *d* electrons in the third shell.

The transition metals have some properties in common. Most of the transition metals have high melting and boiling points and high densities and are hard and strong. Most of the transition metals have more than one oxidation number in their common compounds; the different oxidation states often differ by one as iron(II) and iron(III) do. For most transition metals, the compounds in at least one oxidation state are colored. Most compounds of elements in Groups IA, IIA, and IIIA are colorless or white except for salts that have colored anions. Colored anions usually contain a transition metal; the permanganate ion, MnO_4^-, which is purple, and the dichromate ion, $Cr_2O_7^{2-}$, which is orange-red, are examples. For most transition metals, the compounds in at least one oxidation state have one or more unpaired electrons. Most compounds of elements in Groups IA, IIA, and IIIA have all electrons paired. Most of the transition metals form complexes with Lewis bases (ligands) such as $AgCl_2^-$, $Ni(CN)_4^{2-}$, and $Co(NH_3)_6^{3+}$.

Changes in Properties Down Groups

In general, the reactivity of transition metals *decreases* from top to bottom of a column in the periodic table. Although the transition elements have many similarities, each differs from the others. The elements in a group (vertical column) are usually most alike. With only a few exceptions, the elements in groups have similar outer electron configurations. For example, the electron configurations of titanium, zirconium, and hafnium are

Ti $[Ar]3d^24s^2$

Zr $[Kr]4d^25s^2$

Hf $[Xe]4f^{14}5d^26s^2$

These elements all have two *d* and two *s* valence electrons. There is little difference in atomic radius between second and third transition-series elements in the same group (see ▪Figure 24.2). As a result, second and third transition-series elements in the same group are usually more like each other than they are like the element in the first transition series. For example, zirconium and hafnium are so much alike that they always occur together in nature; however, titanium does not occur with zirconium and hafnium.

Changes in Properties Across Periods

The way the properties of transition elements change across a period is similar from period to period. For the most part, the chemistry of the elements in the first transition series is simpler than the chemistry of the elements in the second and third transition series. In addition, there are more familiar and important elements in the first transition series than there are in the second and third transition series. Therefore,

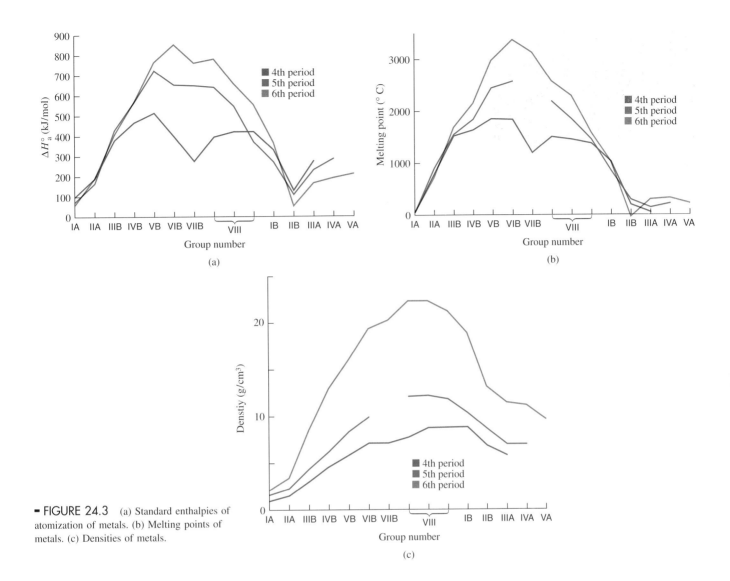

FIGURE 24.3 (a) Standard enthalpies of atomization of metals. (b) Melting points of metals. (c) Densities of metals.

elements of the first transition series will generally be used as examples of transition elements in this book.

Atomic radii of transition elements first decrease and then increase across a period, although there are a few irregularities (see Figure 24.2). As you can see, the minimum atomic radius in each series comes just after the center of the series. Both first ionization energies (Figures 8.10 and 8.11) and electronegativities (Figures 9.3 and 9.4) increase (with a few irregularities) across periods. Electron affinities of some transition elements are not known; those that are known appear to vary irregularly (Figures 8.13 and 8.14).

Atoms of elements in the third transition series are about the same size as atoms of elements in the second transition series because the lanthanides come between barium (atomic number 56) and lutetium (atomic number 71). Proceeding across the lanthanide series, nuclear charge and number of 4*f* electrons increase. However, 4*f* electrons do not shield 5*d* and 6*s* electrons very well, and atomic radius decreases across the lanthanide series. (Remember that this decrease in size is known as the lanthanide contraction.)

Figure 24.3(a) is a graph of standard enthalpies of atomization of the metals of periods 4, 5, and 6 as a function of group number. **Standard enthalpy of**

The standard enthalpy of atomization is the same as the standard enthalpy of vaporization if the element is monatomic in the gaseous state.

IIIB	IVB	VB	VIB	VIIB	VIII			IB	IIB
21 Sc +3	22 Ti +4 +3 +2	23 V +5 +4 +3 +2	24 Cr +6 +3 +2	25 Mn +7 +6 +4 +3 +2	26 Fe +3 +2	27 Co +3 +2	28 Ni +2	29 Cu +2 +1	30 Zn +2
39 Y +3	40 Zr +4 +3	41 Nb +5 +4 +2	42 Mo +6 +5 +4 +3	43 Tc +7 +5 +4	44 Ru +8 +5 +4 +3	45 Rh +4 +3	46 Pd +4 +2	47 Ag +1	48 Cd +2
*71 Lu +3	72 Hf +4 +3	73 Ta +5 +4 +3	74 W +6 +5 +4	75 Re +7 +5 +4	76 Os +8 +6 +4 +3 +2	77 Ir +4 +3 +1	78 Pt +4 +2	79 Au +3 +1	80 Hg +2 +1

■ FIGURE 24.4 Important oxidation numbers of the transition elements in their compounds. (Information is from Bard, A. J.; Parsons, R.; Jordan, J. *Standard Potentials in Aqueous Solution;* Dekker: Basel, 1985.)

atomization, ΔH_a°, or **atomization energy** is the enthalpy change for the conversion of an element in its usual form at 25 °C and one atmosphere to individual gaseous atoms. For example, the standard enthalpy of atomization for iron, 416 kJ/mol, is the enthalpy change for the process Fe(s) $\longrightarrow$ Fe(g) at 25 °C and 1 atm. Figure 24.3(b) shows melting points as a function of group number, and Figure 24.3(c) shows density as a function of group number. As you can see, going across the fourth, fifth, and sixth rows of the periodic table, enthalpies of atomization, melting points, and densities all reach maximums near the middle of the transition series. Boiling points are also highest near the middle of the transition series. Maximums in enthalpies of atomization, melting point, boiling point, and density also occur around the middle of the second and third rows. Why? High enthalpies of atomization, melting points, boiling points, and densities all show that attractions between atoms are strong. (High nuclear masses are also a factor in high densities.) Attractions between atoms are strongest when the valence orbitals are half-filled. Overlap of an orbital on one atom that contains one electron with an orbital on another atom that contains one electron results in a covalent bond between the two atoms. In diamond, every carbon atom has four sp^3 orbitals, each containing one electron. Each carbon in the interior of a diamond is bonded to four other carbons by four covalent bonds, and a crystal of diamond is a huge polymeric molecule. In transition-metal atoms with half-filled valence orbitals, considerable covalent bonding takes place; atoms are strongly bonded to each other, and the metals are hard, strong, and high-melting. Overlap of empty orbitals does not result in bond formation nor does overlap of orbitals that already contain two electrons. There is little or no covalent bonding in crystals of alkali and alkaline earth metals because there are too few valence electrons. There is little or no covalent bonding in crystals of elements such as zinc, cadmium, and mercury, which are at the end of the transition series, because there are too many valence electrons.

The number of oxidation states commonly observed for a transition element is also at a maximum near the middle of each transition series (see ■Figure 24.4). With the exception of mercury, the first elements and the last elements in each series have only a single important oxidation state; elements near the middle have as many as five. Many oxidation states vary by only one unit in contrast to the oxidation states for nonmetals, which usually differ by two units. From titanium to manganese, the highest oxidation number is equal to the group number, that is, to the total number of $(n - 1)d$ and ns electrons in the atom. The highest oxidation number is observed only in compounds of the metal with fluorine, oxygen, and chlorine. The stability of the highest oxidation state decreases from Ti(IV) to Mn(VII), and after manganese, high oxidation states are not important. The stability and importance of higher oxidation states increase down groups; for example, $+2$ is the most stable, characteristic oxidation state for manganese, whereas the $+5$ oxidation state is the most important oxidation state for Tc and Re and the $+2$ state is unimportant for these elements. The TcO_4^- and ReO_4^- ions are not nearly as strong oxidizing agents as the MnO_4^- ion.

Oxidation–Reduction Reactions

Because elements near the middle of each transition series exist in a number of oxidation states, much of the chemistry of these elements involves redox reactions. For example, ▪Figure 24.5 shows the steps in the reduction of a solution of pervanadyl sulfate, $(VO_2)_2SO_4$, in dilute sulfuric acid by an excess of zinc metal. The yellow pervanadyl ion, VO_2^+, is reduced to blue vanadyl ion, VO^{2+}, which is reduced first to green vanadium(III) ion, V^{3+}, and finally to violet vanadium(II) ion, V^{2+}.

The existence of a number of oxidation states for transition elements is explained by the use of both ns and $(n - 1)d$ electrons for bonding. The energies of ns and $(n - 1)d$ electrons are very similar. At the left side of a transition series, both ns and $(n - 1)d$ electrons are used for bonding; at the right side, only the s electrons are used.

Magnetic Properties

Many of the transition metals and their compounds have interesting and useful magnetic properties. Iron, and to a lesser extent, cobalt and nickel, are ferromagnetic, that is, they can be permanently magnetized.* For most transition metals, the compounds in at least one oxidation state are paramagnetic. Paramagnetic materials are weakly magnetized when brought near a magnet and are attracted into the magnetic field (Section 8.1). The magnetism of a paramagnetic material disappears when the magnet is removed. Diamagnetic materials are pushed out of a magnetic field.

Paramagnetic species have unpaired electrons, and the magnitude of the paramagnetism depends on the number of unpaired electrons. Thus, the number of unpaired electrons in a species can be found by measuring the force produced on the sample by a magnetic field. A sensitive analytical balance and a powerful magnet must be used. For example, from the electron configurations of its neighbors in the periodic table, an electron configuration of $[Ar]3d^4 4s^2$ would be expected for chromium, atomic number 24:

$$Cr \quad [Ar] \quad \boxed{\uparrow}\boxed{\uparrow}\boxed{\uparrow}\boxed{\uparrow}\boxed{} \quad \boxed{\uparrow\downarrow}$$
$$\qquad\qquad\qquad\qquad 3d \qquad\qquad 4s$$

In this electron configuration, the four $3d$ electrons are not paired. But experimental measurements show that chromium has six unpaired electrons; therefore, the electron

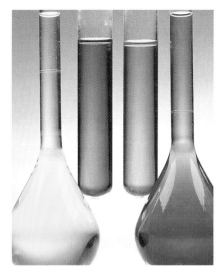

▪ **FIGURE 24.5** *Left to right:* Aqueous solutions containing vanadium(V), vanadium(IV), vanadium(III), and vanadium(II).

Pervanadyl sulfate is formed by dissolving vanadium(V) oxide, V_2O_5, in aqueous sulfuric acid:

$V_2O_5(s) + H_2SO_4(aq) \longrightarrow$
$\qquad\qquad (VO_2)_2SO_4(aq) + H_2O(l)$

◖ Remember Figure 8.3.

---

*Ferromagnetic materials are attracted into a magnetic field about a million times as strongly as paramagnetic materials. In a ferromagnetic material, the spacing of the units of the crystal is such that the electron spins in many particles, which are lined up when the material is placed in a magnetic field, remain lined up when the material is removed from the magnetic field. Permanent magnets can be demagnetized by heating or by hammering. Some paramagnetic elements that are not ferromagnetic, such as manganese, can be made ferromagnetic by alloying with elements that change the spacing of the units in the crystals.

(a)

(b)   (c)

■ FIGURE 24.6   The color of transition-metal complexes depends on the identity of the central atom, the oxidation state of the central atom, and the ligand. (a) A solution that contains $Co(H_2O)_6^{2+}$ is pink; a solution that contains $Cu(H_2O)_6^{2+}$ is blue. (b) A solution that contains $Fe(H_2O)_6^{2+}$ is pale green; a solution that contains $Fe(H_2O)_6^{3+}$ is pale purple. (c) A solution that contains $Cu(H_2O)_6^{2+}$ is greenish-blue; a solution that contains $Cu(H_2O)_2(NH_3)_4^{2+}$ is deep blue.

configuration of chromium must be $[Ar]3d^54s^1$:

$$Cr \quad [Ar] \quad \boxed{\uparrow}\,\boxed{\uparrow}\,\boxed{\uparrow}\,\boxed{\uparrow}\,\boxed{\uparrow} \quad \boxed{\uparrow}$$
$$\qquad\qquad\qquad\qquad 3d \qquad\qquad\quad 4s$$

In diamagnetic species, all electrons are paired. From the electron configurations of its neighbors in the periodic table, an electron configuration of $[Kr]4d^95s^1$ with two unpaired electrons would be expected for palladium, atomic number 46:

$$Pd \quad [Kr] \quad \boxed{\uparrow\downarrow}\,\boxed{\uparrow\downarrow}\,\boxed{\uparrow\downarrow}\,\boxed{\uparrow\downarrow}\,\boxed{\uparrow} \quad \boxed{\uparrow}$$
$$\qquad\qquad\qquad\qquad 4d \qquad\qquad\qquad 5s$$

However, palladium is diamagnetic, and the electron configuration of palladium must be $[Kr]4d^{10}$:

$$Pd \quad [Kr] \quad \boxed{\uparrow\downarrow}\,\boxed{\uparrow\downarrow}\,\boxed{\uparrow\downarrow}\,\boxed{\uparrow\downarrow}\,\boxed{\uparrow\downarrow} \quad \square$$
$$\qquad\qquad\qquad\qquad 4d \qquad\qquad\qquad 5s$$

PRACTICE PROBLEMS

24.1   How many unpaired electrons are there in the following? (a) an atom of iron (b) an Fe^{3+} ion (Remember that $4s$ electrons are lost before $3d$ electrons.)

24.2   Technetium does not occur in nature because all isotopes of technetium are radioactive with half-lives shorter than the age of Earth. Which element would you expect technetium to resemble most closely? Explain your answer.

24.2   OBSERVATIONS OF COMPLEXES AND COORDINATION COMPOUNDS

As a result of their bright colors, complexes are often used to detect trace quantities of transition elements.

Complexes are called complexes because, in the early days of chemistry, they seemed unusual and hard to understand.

Formation of complexes is a characteristic property of transition metals. The most easily observed property of transition-metal complexes is color. The color of transition-metal complexes depends on the identity of the central atom, the oxidation state of the central atom, and the ligand as shown in ■Figure 24.6.

Compounds that include one or more complexes are called **coordination compounds** because the bonds between the metal ions and the ligands in complexes are coordinate covalent bonds. Swiss chemist Alfred Werner, who won the 1913 Nobel Prize for chemistry for his research into the structure of coordination compounds, proposed the theory that is basic to modern views. A number of coordination compounds of cobalt(III), platinum(II), and platinum(IV) had been prepared before Werner first came up with his theory in 1893. The complexes in these compounds were unreactive enough to be studied by the methods available at that time.

The coordination compounds were formed by the combination of compounds. For example, 1 mol of the compound platinum(IV) chloride, $PtCl_4$, combined with

2, 3, 4, 5, or 6 mol of ammonia. The formulas for the five coordination compounds formed were written $PtCl_4 \cdot 2NH_3$, $PtCl_4 \cdot 3NH_3$, $PtCl_4 \cdot 4NH_3$, $PtCl_4 \cdot 5NH_3$, and $PtCl_4 \cdot 6NH_3$ to show the proportions of $PtCl_4$ and NH_3 that combined.

Reactivities of Chlorides

Werner observed that the reactivities of the chlorines in the five coordination compounds differed. For example, addition of silver nitrate solution to a solution of $PtCl_4 \cdot 2NH_3$ did not give any precipitate, but addition of silver nitrate to a solution of $PtCl_4 \cdot 3NH_3$ gave 1 mol of silver chloride per mole of coordination compound:

$$PtCl_4 \cdot 2NH_3(aq) + \text{excess } Ag^+ \longrightarrow \text{no precipitate}$$
$$PtCl_4 \cdot 3NH_3(aq) + \text{excess } Ag^+ \longrightarrow 1 \text{ AgCl(s)}$$

Formation of a precipitate with silver nitrate is a test for chloride ion, Cl^-. The coordination compound, $PtCl_4 \cdot 2NH_3$, must not contain any chloride ions; all four Cl must be firmly bonded to platinum. The coordination compound $PtCl_4 \cdot 3NH_3$ must contain one chloride ion per formula unit since 1 mol AgCl formed from one mole of coordination compound. The other three of the four Cl in $PtCl_4 \cdot 3NH_3$ must be firmly bonded to platinum.

Total Number of Ions

Werner also determined the total number of ions formed from a formula unit of the coordination compound when it was dissolved in water by comparing the conductivity of the solution with the conductivities of solutions of simple salts. For example, conductivity measurements showed that 1 mol $PtCl_4 \cdot 3NH_3$ gives 2 mol of ions just as 1 mol NaCl gives 2 mol of ions. Data for the five coordination compounds of $PtCl_4$ and NH_3 that were known to Werner are summarized in Table 24.1. Werner noticed that *in every case the number of firmly bound nonionic Cl's plus the number of NH_3's is equal to 6.* For example, $PtCl_4 \cdot 6NH_3$ has no firmly bound Cl and six NH_3; $PtCl_4 \cdot 5NH_3$ has one firmly bound Cl and five NH_3, and so forth. To explain this fact and the number of ions produced, Werner proposed that the platinum(IV) ion has a property that he called coordination number. He defined **coordination number** as *the number of atoms or groups that are firmly bound to the central atom;* the coordination number of Pt(IV) is 6. He then wrote the formulas for the five compounds in Table 24.1 with the coordinated atoms and groups inside square brackets and the free ions outside:

⬤ Remember that in connection with the arrangement of units in crystals (Section 12.9), coordination number was defined as the number of nearest neighbors that an atom has.

⬤ The square brackets are usually omitted when equilibrium constant expressions for complex formation are written. In equilibrium constant expressions, square brackets mean concentration in molarity.

$$PtCl_4 \cdot 6NH_3 = [Pt(NH_3)_6]Cl_4 \quad \text{or} \quad [Pt(NH_3)_6]^{4+} + 4Cl^-$$
$$PtCl_4 \cdot 5NH_3 = [PtCl(NH_3)_5]Cl_3 \quad \text{or} \quad [PtCl(NH_3)_5]^{3+} + 3Cl^-$$
$$PtCl_4 \cdot 4NH_3 = [PtCl_2(NH_3)_4]Cl_2 \quad \text{or} \quad [PtCl_2(NH_3)_4]^{2+} + 2Cl^-$$
$$PtCl_4 \cdot 3NH_3 = [PtCl_3(NH_3)_3]Cl \quad \text{or} \quad [PtCl_3(NH_3)_3]^+ + Cl^-$$
$$PtCl_4 \cdot 2NH_3 = [PtCl_4(NH_3)_2] \quad \text{or} \quad [PtCl_4(NH_3)_2] + 0Cl^-$$

TABLE 24.1	Observations of $PtCl_4$–NH_3 Coordination Compounds		
Empirical Formula	Number of Ions	Number of Cl^- Ions	Number of Nonionic Cl
$PtCl_4 \cdot 6NH_3$	5	4	0
$PtCl_4 \cdot 5NH_3$	4	3	1
$PtCl_4 \cdot 4NH_3$	3	2	2
$PtCl_4 \cdot 3NH_3$	2	1	3
$PtCl_4 \cdot 2NH_3$	0	0	4

In the *formulas*, notice that *the symbol for the central atom is placed first followed by the ionic and neutral ligands in that order.* The formula for the whole complex is enclosed in square brackets. *The charge of a complex is the sum of the charges of the species that must undergo a Lewis acid–base reaction to form the complex.* For example, the complex ion $[PtCl(NH_3)_5]^{3+}$ is formed from one platinum(IV) ion Pt^{4+} with a charge of $4+$, one chloride ion Cl^- with a charge of $1-$, and five ammonia molecules. The ammonia molecules have zero charge; therefore, the charge on the ion is

$$(4+) + 1(1-) + 5(0) = 3+$$

Notice how the number of atoms or groups in addition to Pt inside the square brackets totals 6 in each case and the number of ions formed and the number of Cl^- ions agree with the number found by experiment. Two more members of the series,

$$PtCl_5 \cdot 1NH_3^- = [PtCl_5(NH_3)_1]^- \text{ or } [PtCl_5(NH_3)]^-$$

$$\text{and } (PtCl_6 \cdot 0NH_3)^{2-} = [PtCl_6]^{2-}$$

might be expected to exist. Compounds containing the two anions, $K[PtCl_5(NH_3)]$ and $K_2[PtCl_6]$, have both been prepared.

SAMPLE PROBLEM

24.1 In the compound $[CoCl(NH_3)_5]Cl_2$, (a) which atoms or groups are part of the complex? (b) Which atom is the central atom of the complex? (c) What is the coordination number of the central atom? (d) What is the oxidation number of the central atom?

Solution (a) The atoms and groups inside the square brackets—Co, one Cl, and all five NH_3—are part of the complex.

(b) In the formulas of complexes (except structural formulas), the symbol for the central atom is placed first. The first symbol inside the square brackets is Co; therefore, Co (cobalt) is the central atom.

(c) One Cl and five NH_3 are also inside the square brackets and are the ligands. The coordination number of cobalt is 6.

(d) Because the charge on the complex ion is balanced by two chloride ions, Cl^-, outside the square brackets, the charge on the complex must be $2+$. The complex contains five NH_3 molecules that have 0 charge and one chloride ion with a $1-$ charge. For the charge on the complex to be $2+$, the charge on the cobalt ion must be $3+$. The oxidation number of cobalt is $+3$.

An aqueous solution of $[CoCl(NH_3)_5]Cl_2$.

PRACTICE PROBLEMS

24.3 What is the coordination number of silver in the complex ion $[Ag(NH_3)_2]^+$?

24.4 For the complex of gold(III) in which four fluoride ions are the ligands, (a) what is the formula (including charge)? (b) What is the formula of the potassium salt of this ion?

24.5 What is the oxidation number of the central metal atom in each of the following complexes? (a) $Fe(CN)_6^{3-}$ (b) $Co(NH_3)_6^{2+}$ (c) PtF_6^- (d) $Ni(CO)_4$

24.6 Given the following data:

Observations of PtCl₂–NH₃ Coordination Compounds			
Empirical Formula	Number of Ions	Number of Cl⁻ Ions	Number of Nonionic Cl
$PtCl_2 \cdot 4NH_3$	3	2	0
$PtCl_2 \cdot 3NH_3$	2	1	1
$PtCl_2 \cdot 2NH_3$	0	0	2

(a) What is the coordination number of platinum(II)? (b) Write the formulas of each of the three compounds.

STEREOISOMERISM IN COMPLEXES

Werner deduced the arrangement in space of the coordinated groups around the central atom from isomer numbers (the number of isomers formed). For example, there are two compounds of formula $[PtCl_2(NH_3)_2]$. If the arrangement of the four groups attached to platinum were tetrahedral as predicted by VSEPR, there could only be one compound of formula $[PtCl_2(NH_3)_2]$. *Valence shell electron pair repulsion cannot be used to predict the geometry of transition metal compounds.*

If the arrangement were rectangular planar (or rectangular pyramidal with platinum above or below the plane of the other four atoms), there would be three compounds:

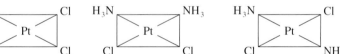

Because there are only two compounds of formula $[PtCl_2(NH_3)_2]$, these compounds must be square planar (or square pyramidal):

cis-$[PtCl_2(NH_3)_2]$ *trans*-$[PtCl_2(NH_3)_2]$

The two isomers of $[PtCl_2(NH_3)_2]$ are geometric isomers.

The *trans*-isomer of $[PtCl_2(NH_3)_2]$ has no dipole moment. Therefore, the *trans*-isomer must be square planar as shown. If the *trans*-isomer were square pyramidal, it would have a dipole moment. Only if all five atoms lie in one plane so that individual bond moments cancel can the *trans*-isomer have zero dipole moment. Although Werner deduced the shapes of complexes from isomer number, modern instrumental methods such as X-ray diffraction show that his conclusions are correct.

Some complexes exist in enantiomeric forms. Two of the simplest complexes that have been separated into pure enantiomers are the complexes of cobalt(III)

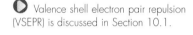

▶ Valence shell electron pair repulsion (VSEPR) is discussed in Section 10.1.

▶ Geometric isomerism is discussed in Section 22.3.

▶ Enantiomers are discussed in Section 22.3.

- FIGURE 24.7 *Left:* Ball-and-stick models of the enantiomers of $[Co(H_2NCH_2CH_2NH_2)_3]^{3+}$. *Right:* Simplifying the models by omitting the Hs and putting the two blue planes in the same direction, we can see that the two models are not superimposable.

with ethylenediamine, $H_2NCH_2CH_2NH_2$, and with oxalate ion, $^-O_2CCO_2{}^-$:

$$\left[Co \left(\begin{matrix} NH_2CH_2 \\ | \\ NH_2CH_2 \end{matrix} \right)_3 \right]^{3+} \quad \text{and} \quad \left[Co \left(\begin{matrix} O \\ \| \\ ^-O—C \\ | \\ ^-O—C \\ \| \\ O \end{matrix} \right)_3 \right]^{3-}$$

⬤ Racemic mixtures are discussed in Section 22.3.

■Figure 24.7 shows models of the complex of cobalt(III) with ethylenediamine. Although $[Co(C_2O_4)_3]^{3-}$ can be separated into pure enantiomers, $[Al(C_2O_4)_3]^{3-}$ and $[Fe(C_2O_4)_3]^{3-}$ cannot. Substitution of ligands attached to Co is slow, but substitution of ligands bonded to Al and Fe is very fast. One enantiomer is rapidly converted into the other, and a racemic mixture results.

⬤ You should work through this problem using models to convince yourself that the answer given is correct.

SAMPLE PROBLEM

24.2 (a) How many geometric isomers of $[CoCl_3(NH_3)_3]$ are there? (b) Draw stereochemical formulas for each of the geometric isomers of $[CoCl_3(NH_3)_3]$. (c) Do any of the geometric isomers have an enantiomer?

Solution Either commercial models or gumdrop–toothpick models (Section 22.3) should be used to answer questions about stereochemistry. Remember that *if two models are superimposable, the two models represent a single compound. If two models are not superimposable, the two models represent different compounds.* Draw stereochemical formulas from models.

(a) and (b) There are two geometric isomers of $[CoCl_3(NH_3)_3]$:

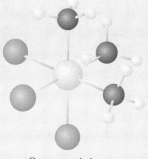

One geometic isomer

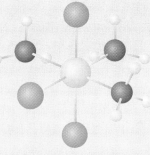

Other geometric isomer

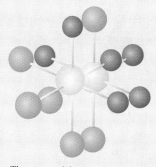

The two models are not superimposable. They represent different compounds and are geometric isomers. (Models have been simplified by omitting hydrogens.)

(c) Make a model of the mirror image of each geometric isomer and see if the two models that are mirror images of each other are superimposable. If they are, the two models represent a single compound. Neither of the two geometric isomers of [CoCl_3(NH_3)_3] has an enantiomer. There are just two stereoisomers of [CoCl_3(NH_3)_3].

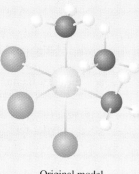

Original model

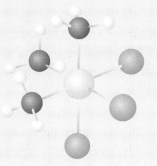

Mirror image

Models are superimposable; both represent the same compound.

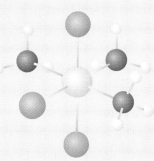

Original model

Mirror image

Models are superimposable; both represent the same compound.

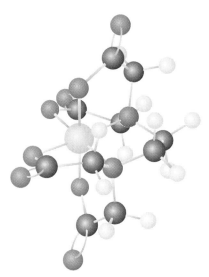

■ **FIGURE 24.8** Ball-and-stick model of a metal ion complexed with the hexadentate ligand EDTA.

PRACTICE PROBLEM

24.7 (a) How many geometric isomers of [CoCl$_2$(NH$_3$)$_4$] are there? (b) Draw stereochemical formulas for each of the geometric isomers of [CoCl$_2$(NH$_3$)$_4$]. (c) Do any of the geometric isomers have an enantiomer?

24.4 POLYDENTATE LIGANDS AND CHELATE COMPLEXES

The ligands H$_2$NCH$_2$CH$_2$NH$_2$ and $^-$O$_2$CCO$_2^-$ are called **didentate ligands** (or **bidentate ligands**) because they have two nitrogen atoms and two oxygen atoms, respectively, that can bond to Co^{3+}. Ammonia is a **monodentate ligand** because it has only one nitrogen atom that can bond to Co^{3+}. All ligands that we have discussed before this are monodentate ligands. All known monodentate ligands are too easily substituted for enantiomers to be isolated; one enantiomer is rapidly converted into the other enantiomer.

Ligands may have more than two atoms that can bond to a metal ion. For example, ethylenediamminetetraacetate

$$^-\text{O}_2\text{CCH}_2\diagdown\atop{^-\text{O}_2\text{CCH}_2}\diagup\text{NCH}_2\text{CH}_2\text{N}\diagup^{\text{CH}_2\text{CO}_2^-}\atop\diagdown_{\text{CH}_2\text{CO}_2^-}$$

EDTA is used to remove lead and other toxic metals from the body in the treatment of poisoning.

which is usually called EDTA for short, is a hexadentate ligand. Four O and two N from EDTA bond to a central metal ion as shown in ■Figure 24.8. In Figure 24.8, notice that the six atoms that are bonded to the central metal ion are located at the corners of an octahedron. Complexes in which the coordination number is 6 are almost always octahedral.

Ligands that attach themselves to metal ions by bonds from two or more donor atoms, such as ethylenediamine and EDTA, are called **polydentate ligands.** *Complexes in which polydentate ligands are bonded to a single central atom* are called **chelate complexes.** The atoms that bond to the metal ion must be far enough apart so that the rings formed have five or six members. Equilibria between complexes with monodentate ligands and chelate complexes such as the equilibrium

The word chelate comes from the Greek word chelé which means claw. In a chelate complex the metal ion is held in a claw.

$$\left[\text{Ni(NH}_3)_6\right]^{2+} + 3\text{H}_2\text{NCH}_2\text{CH}_2\text{NH}_2 \rightleftharpoons \left[\text{Ni}\left(\begin{matrix}\text{NH}_2\text{CH}_2\\|\\\text{NH}_2\text{CH}_2\end{matrix}\right)_3\right]^{2+} + 6\text{NH}_3$$

favor the chelate complex because there are more product particles than reactant particles and the entropy of the system increases from left to right. The enthalpy change is small because six Ni—N bonds are broken in the reactant and formed in the product. As a result, the entropy change determines the position of equilibrium. Once one end of a polydentate ligand has become attached to a metal ion, the rest of the polydentate ligand is held close to the metal ion. The probability that the other atoms that can bond to the central ion will do so is large.

In the scientific literature, abbreviations are often used for polydentate ligands, especially in formulas. For example, the abbreviation for ethylenediamine is en and the abbreviated formula for the complex ion [Co(H$_2$NCH$_2$CH$_2$NH$_2$)$_3$]$^{3+}$ is [Co(en)$_3$]$^{3+}$.

24.8 In the complex

$$\left[Co \left(\begin{array}{c} O \\ \parallel \\ {}^{-}O-C \\ | \\ {}^{-}O-C \\ \parallel \\ O \end{array} \right)_{3} \right]^{3-}$$

(a) what is the coordination number of cobalt? (b) What is the oxidation number of cobalt?

24.9 In a complex, one nickel(II) ion is bound to two oxalate ions $^{-}O_2CCO_2^{-}$ and two water molecules. Write a formula for this complex similar to the formula in Practice Problem 24.8.

24.5 CONSTITUTIONAL ISOMERISM IN COMPLEXES

Mirror-image isomers and geometric isomers are stereoisomers, that is, isomers that differ only in the arrangement of atoms or groups in three-dimensional space (Section 22.3). Constitutional isomers, isomers in which the atoms are connected in different ways, of complexes also occur. Constitutional isomers of complexes can be divided into two main types. In one type, linkage isomers, the composition of the complex remains the same. In the other type, the composition of the complex varies.

Linkage Isomers

Some ligands can be attached to the central metal ion by either of two different atoms. The classical example is the nitrite ion, NO_2^{-}, which can be attached either by the nitrogen or by one of the oxygens:

$$\left[\begin{array}{c} NH_3 \\ H_3N \cdots \overset{|}{\underset{\underset{O}{\overset{\overset{}{\underset{}{}}}{N}}{Co}} \cdots NH_3 \\ H_3N \qquad NH_3 \\ \end{array} \right]^{2+} \qquad \left[\begin{array}{c} NH_3 \\ H_3N \cdots \overset{|}{\underset{\underset{N}{\overset{\overset{}{\underset{O}{}}}{O}}{Co}} \cdots NH_3 \\ H_3N \qquad NH_3 \\ \end{array} \right]^{2+}$$

yellow   red

In ordinary formulas, the difference in structure is shown by the way the formula of the ligand is written. The atom of the ligand that is attached to the central atom is written first. For example, the formula for the yellow isomer is $[Co(NO_2)(NH_3)_5]^{2+}$; the formula for the red isomer is $[Co(ONO)(NH_3)_5]^{2+}$. The complex ion $[Co(NO_2)(NH_3)_5]^{2+}$ is called pentaamminenitrocobalt(III) ion; $[Co(ONO)(NH_3)_5]^{2+}$ is called pentaamminenitritocobalt(III) ion.

The red isomer is formed faster, but the yellow isomer is more stable. The red isomer changes slowly to the yellow isomer on standing at room temperature; conversion takes place more rapidly at higher temperatures. Because many inorganic reactions take place very rapidly, examples of kinetic control like this one are not as common in inorganic chemistry as in organic chemistry.

The yellow solution contains pentaamminenitrocobalt(III) ion; the red solution contains pentaamminenitritocobalt(III) ion.

The violet solution contains $[Cr(H_2O)_6]Cl_3$ and the green solution contains $[CrCl_2(H_2O)_4]Cl$.

Isomers in Which the Composition of the Complex Varies

This type of isomerism can arise in several different ways. *Different quantities of the solvent can be present in the complex.* For example, there are three compounds that have the empirical formula $CrCl_3(H_2O)_6$. The formulas for the three isomers of $CrCl_3(H_2O)_6$ are $[Cr(H_2O)_6]Cl_3$ (violet), $[CrCl(H_2O)_5]Cl_2 \cdot H_2O$ (pale blue-green), and $[CrCl_2(H_2O)_4]Cl \cdot 2H_2O$ (dark green). In the violet isomer, all six coordination positions are occupied by water molecules. In the pale blue-green isomer, only five of the six coordination positions are occupied by water, and the sixth is occupied by chlorine. In the dark green isomer, four of the six coordination positions are occupied by water, and the other two are occupied by chlorine. The equations for the reactions of solutions of the three compounds with excess silver ion are

$$[Cr(H_2O)_6]Cl_3(aq) + 3Ag^+ \longrightarrow 3AgCl(s) + [Cr(H_2O)_6]^{3+}$$
violet

$$[CrCl(H_2O)_5]Cl_2(aq) + 2Ag^+ \longrightarrow 2AgCl(s) + [CrCl(H_2O)_5]^{2+}$$
pale blue-green

$$[CrCl_2(H_2O)_4]Cl(aq) + 1Ag^+ \longrightarrow 1AgCl(s) + [CrCl_2(H_2O)_4]^+$$
dark green

Another possibility is for different ligands to be present in the complex. For example, two compounds that have the composition $CoBr(NH_3)_5SO_4$ exist. One is violet and gives a precipitate with Ba^{2+} but not with Ag^+. The other is red and gives a precipitate with Ag^+ but not with Ba^{2+}. Formation of a precipitate with Ba^{2+} is a test for sulfate ion, SO_4^{2-}; formation of a precipitate with Ag^+ is a test for bromide ion, Br^- (as well as for Cl^-). The two isomers are $[CoBr(NH_3)_5]SO_4$ and $[CoSO_4(NH_3)_5]Br$.

PRACTICE PROBLEM

24.10 Write equations for the reactions that will take place. If no reaction will occur, write "No reaction."
(a) $[CoBr(NH_3)_5]SO_4(aq) + Ba^{2+}$ (b) $[CoBr(NH_3)_5]SO_4(aq) + Ag^+$
(c) $[CoSO_4(NH_3)_5]Br(aq) + Ba^{2+}$ (d) $[CoSO_4(NH_3)_5]Br(aq) + Ag^+$

In compounds composed of both a complex cation and a complex anion, the distribution of ligands between complexes can vary. For example, two compounds with formula $CuPtCl_4(NH_3)_4$ exist. One is violet and the other is green. Their structures are $[Cu(NH_3)_4][PtCl_4]$ (violet) and $[Pt(NH_3)_4][CuCl_4]$ (green). In the compound $[Cu(NH_3)_4][PtCl_4]$, ammonia is bound to copper and chloride ion to platinum. In the compound $[Pt(NH_3)_4][CuCl_4]$, ammonia is bound to platinum and chloride ion to copper.

24.6 NOMENCLATURE OF COMPLEXES

Because thousands of complexes are known, a system is needed for naming complexes. Only the nomenclature of the simplest complexes will be covered here.*

---

*For information about naming more complicated complexes, see current inorganic textbooks such as Jolly, W. L. *Modern Inorganic Chemistry,* 2nd ed.; McGraw-Hill: New York, 1991; pp 602–605 and Huheey, J. E.; Keiter, E. A.; Keiter, R. L. *Inorganic Chemistry: Principles of Structure and Reactivity,* 4th ed.; HarperCollins: New York, 1993; pp A-63–A-77. Major changes were made in the rules for naming complexes in 1971. Papers written before 1971 used somewhat different names.

In the *name* of a complex, *the ligands are named first in alphabetical order.* Table 24.2 gives the names of some common ligands. Notice that the names of anionic ligands end in -o and that the name of ammonia as a ligand, ammine, has two m's.

Greek prefixes such as penta- are used to show how many of each kind of ligand there are in a complex. Greek prefixes are *not* considered in alphabetizing. The oxidation number of the central atom is shown by a Roman numeral except that an oxidation number of 0 is indicated by (0). For example, the complex ion $[CoCl_2(NH_3)_4]^+$ is called tetraamminedichlorocobalt(III) ion:

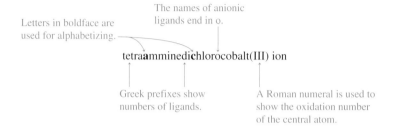

The names of anionic ligands end in o.

Letters in boldface are used for alphabetizing.

tetraamminedichlorocobalt(III) ion

Greek prefixes show numbers of ligands.

A Roman numeral is used to show the oxidation number of the central atom.

The name of the complex is all one word. There is a space between the name of the complex and the word ion.

The ending -ate is used to show that a complex is an anion. If the name of the metal ends in -ium or -um , the ending is changed to -ate. For example, the complex anion $[TiF_6]^{2-}$ is called the hexafluorotitanate(IV) ion. If the name of the metal does not end in -ium or -um, there does not seem to be any simple rule for naming an anion. To further complicate the matter, Latin names are used for some metals in anionic complexes. Table 24.3 gives the names for all the types of anions with unsystematic names that you are likely to meet (in alphabetical order by the name of the metal). As an example, the name of the complex ion $[Fe(CN)_5CO]^{3-}$ is carbonylpentacyanoferrate(II) ion.

When the formula for a compound that contains a complex is written, the cation is written on the left and the anion on the right as usual. Also as usual, the name of the cation comes first in the name of the compound. The compound $Na_3[Fe(CN)_5CO]$ is called sodium carbonylpentacyanoferrate(II). Note that there is a space between the names of the cation and anion. Sample Problem 24.3 shows how to write the name of a complex from the formula, and Sample Problem 24.4 shows how to write the formula from the name.

SAMPLE PROBLEMS

24.3   Name the compound $[CoCl(NH_3)_5]Cl_2$.

Solution   The central atom is Co, and the ligands are Cl^- and NH_3. (All these are *inside* the square brackets.) From Table 24.2, the ligand Cl^- is called chloro and the ligand NH_3 is called ammine. There are 5 NH_3s and 1 Cl^-. The prefix penta- should be used for ammine. As usual, the prefix mono- is omitted; no prefix is needed for Cl^-.

From Sample Problem 24.1, the oxidation number of Co in $[CoCl(NH_3)_5]Cl_2$ is +3. The name of the complex ion is pentaamminechlorocobalt(III) ion. Note that pentaammine- is spelled with two a's and two m's.

In the name of the compound, the cation is named first. The name of the compound $[CoCl(NH_3)_5]Cl_2$ is pentaamminechlorocobalt(III) chloride.

24.4   Write the formula for potassium tetracyanocadmate(II).

TABLE 24.2

Names of Some Common Ligands

Ligand	Name
F^-	Fluoro
Cl^-	Chloro
Br^-	Bromo
I^-	Iodo
OH^-	Hydroxo
CN^-	Cyano
NO_2^-	Nitro[a]
H_2O	Aqua
NH_3	Ammine
CO	Carbonyl

[a] If attached by N. Nitrito if attached by O.

⏵ Greek prefixes are listed in Table 1.4.

TABLE 24.3

Names for Anionic Complexes

Metal	Symbol	Name
Cobalt	Co	Cobaltate
Copper	Cu	Cuprate
Gold	Au	Aurate
Iron	Fe	Ferrate
Lead	Pb	Plumbate
Manganese	Mn	Manganate
Mercury	Hg	Mercurate
Molybdenum	Mo	Molybdate
Nickel	Ni	Nickelate
Silver	Ag	Argentate
Tin	Sn	Stannate
Tungsten	W	Tungstate
Zinc	Zn	Zincate

Solution The tetracyanocadmate(II) ion must be an anion because the name ends in -ate. The central atom is cadmium. In the formula of a complex, the central atom is written first. Then the ligand, CN^-, is written. Tetra- means 4. So far, the formula for the complex is

$$[Cd(CN)_4]^{n-}$$

A cadmium(II) ion has a charge of $2+$. The four CN^- ions in the complex have a total charge of $4-$. The complex must have a $2-$ charge. The formula for the complex is

$$[Cd(CN)_4]^{2-}$$

Because each potassium ion, K^+, has a $1+$ charge, two potassium ions are needed to balance the $2-$ charge on the complex. The formula for potassium tetracyanocadmate(II) is $K_2[Cd(CN)_4]$.

PRACTICE PROBLEMS

24.11 Name the following: (a) $[Al(OH)(H_2O)_5]^{2-}$, (b) $K_4[Fe(CN)_6]$, and (c) $V(CO)_6$.

24.12 Write formulas for the following compounds: (a) hexaaquachromium(III) chloride and (b) tetraamminecopper(II) tetrachloroplatinate(II).

If stereoisomerism is possible, information about stereochemistry, if available, should be included in the name of a complex. For example,

The cis- isomer of diamminedichloroplatinum(II) is called cisplatin and is one of the best current treatments for some types of cancer. The trans- isomer has no effect on cancers.

is called *cis*-diamminedichloroplatinum(II) and

is called *trans*-diamminedichloroplatinum(II).

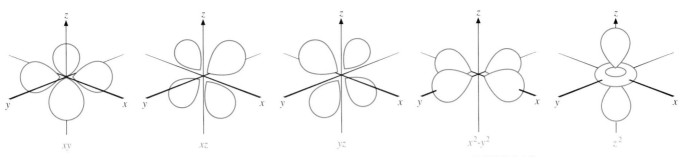

- FIGURE 24.9 Boundary surface diagrams for d orbitals.

24.7 THE d ORBITALS

Bonding in transition-metal complexes involves d orbitals, which, like p orbitals, are directed in space. Before we discuss the transition metals further, we need to consider the shapes and positions in space of the d orbitals. ▪Figure 24.9 shows boundary surface diagrams for a set of d orbitals. Remember that boundary surface diagrams show surfaces chosen so that there is a high probability of finding the electron inside the surface. They do *not* represent real surfaces.

There are five d orbitals; each diagram represents *one* orbital. The xy orbital consists of four lobes that lie between the x- and y-axes in the xy plane. The xz orbital consists of four lobes that lie between the x- and z-axes in the xz plane and the yz orbital consists of four lobes that lie between the y- and z-axes in the yz plane. The $x^2\text{-}y^2$ orbital consists of four lobes that lie along the x- and y-axes; the z^2 orbital consists of two lobes that lie along the z-axis and a doughnut (called a torus) in the xy plane. ▪Figure 24.10 shows a complete set of d orbitals. The size of the d orbitals depends on the principal quantum number, n. The higher the principal quantum number, the larger the d orbitals.

24.8 BONDING IN COMPLEXES

A completely satisfactory theory of bonding in complexes must explain all the observed properties of complexes: number of unpaired electrons, color, shape, stability, and reactivity. Most complexes contain a large number of electrons; for example, the complex ion $[\text{Fe(CN)}_6]^{4-}$ contains a total of 108 electrons of which 66 are valence electrons. For species that contain as many electrons as most complexes do, all theories of chemical bonding involve approximations. Different theories of bonding may be used; which theory is best depends on the purpose and skills of the user.

Crystal Field Theory

The crystal field theory explains both the colors of complexes and their magnetic properties. It was developed by physicists to account for the colors and magnetic properties of hydrated salts of transition metals. Originally, the crystal field theory was applied to the electrostatic effects of the surrounding anions and dipoles on the properties of metal ions in crystalline solids. That is why the theory is called crystal field theory. Chemists began to apply the crystal field theory to transition-metal complexes in the 1950s, and since that time inorganic chemistry has again become a very active field of research.

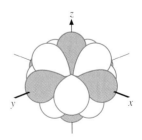

- FIGURE 24.10 A complete set of d orbitals. The shaded orbitals lie along the axes. The unshaded orbitals lie between axes in the plane of two axes. Note how spherical the complete set of d orbitals appears.

According to the **crystal field theory,** *electrostatic attraction between the positive charge on the central metal ion and the negative charges on the ligand bonds the ligands to the metal ion.* Both the central metal ion and the ligands are assumed to be point charges. Point charges are electric charges that exist at a single point; a point charge has neither area nor volume.

Octahedral Complexes

In an isolated gaseous metal ion, the five *d* orbitals are degenerate—that is, they have the same energy. Let's first consider the formation of an octahedral complex, the most common shape. The ligands (L), which are either anions such as Cl^- or polar molecules such as H_2O, are attracted by the + charge on the central metal ion. Imagine that the six ligands approach the central metal ion along the *x*-, *y*-, and *z*-axes:

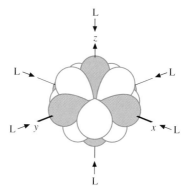

If the ligand is a polar molecule, the polar molecule approaches the central metal ion with its negative end in because of the attraction of the positive charge on the central metal ion for the negative charge on the ligand. Repulsion between the negative charges on the approaching ligands and the negative charges on the electrons present in the *d* orbitals raises the energy of all the *d* orbitals. However, the orbitals that lie along the axes (the x^2-y^2 and z^2 orbitals) are closer to the incoming ligands than the orbitals that lie between the axes (the *xy*, *xz*, and *yz* orbitals). The energies of the two orbitals that lie along the axes are raised more than the energies of the three orbitals that lie between the axes:

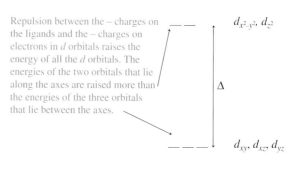

Repulsion between the – charges on the ligands and the – charges on electrons in *d* orbitals raises the energy of all the *d* orbitals. The energies of the two orbitals that lie along the axes are raised more than the energies of the three orbitals that lie between the axes.

$d_{x^2-y^2}, d_{z^2}$

Δ

d_{xy}, d_{xz}, d_{yz}

In an isolated gaseous metal ion, the five *d* orbitals have the same energy.

The *d* orbitals are split into a set of two high-energy orbitals and a set of three low-energy orbitals. The *difference in energy between the two sets of orbitals* is called the **crystal field splitting** and is represented by the symbol Δ (or Δ_O where the subscript O stands for octahedral).

Now consider what happens if electrons are placed in the split orbitals following the usual rules. After one electron has been placed in each of the three low-energy orbitals, the fourth electron must either go into a low-energy orbital and form a pair or go into one of the high-energy orbitals. Electrons repel each other. If the difference in energy between the two sets of orbitals is small, less energy will be required to place the fourth electron in one of the high-energy orbitals than to put a second electron into one of the low-energy orbitals and form a pair. The electron configuration will be

$$\frac{\uparrow}{\uparrow} \quad \frac{}{}$$
$$\frac{\uparrow}{} \ \frac{\uparrow}{} \ \frac{\uparrow}{}$$

with four unpaired electrons.

If the difference in energy between the two sets of orbitals is large, less energy will be required to place the fourth electron in one of the low-energy orbitals and form a pair than to put it in one of the high-energy orbitals. The electron configuration will be

$$\frac{}{} \quad \frac{}{}$$
$$\frac{\uparrow\downarrow}{} \ \frac{\uparrow}{} \ \frac{\uparrow}{}$$

with only two unpaired electrons. The former electron configuration is referred to as **weak field** or **high spin** and the latter as **strong field** or **low spin.**

The number of unpaired electrons can be found by experimental measurement of paramagnetism. From the number of unpaired electrons, you can tell whether a complex is low spin or high spin. Low-spin electron configurations with few unpaired electrons are the result of strong crystal fields, which lead to large splitting. High-spin electron configurations with many unpaired electrons are the result of weak crystal fields, which lead to small splitting.

SAMPLE PROBLEM

24.5   The complex ion $[CoF_6]^{3-}$ is paramagnetic. Sketch the d orbital splitting diagram.

Solution   The complex ion $[CoF_6]^{3-}$ is probably octahedral because there are six ligands around the central atom.

The electron configuration of the cobalt atom is $[Ar]3d^7 4s^2$; the electron configuration of the Co^{3+} ion is $[Ar]3d^6$. The fact that the $[CoF_6]^{3-}$ ion is paramagnetic tells us that this ion has at least one unpaired electron. Therefore, the ion is probably high spin, weak field and the appropriate orbital diagram is

$$\frac{}{} \quad \frac{}{}$$
$$\frac{}{} \ \frac{}{} \ \frac{}{}$$

with little difference in energy between the two sets of d orbitals. Placing six electrons in the orbital diagram gives

$$\frac{\uparrow}{} \ \frac{\uparrow}{}$$
$$\frac{\uparrow\downarrow}{} \ \frac{\uparrow}{} \ \frac{\uparrow}{}$$

with four unpaired electrons for the ground-state electron configuration. If the $[CoF_6]^{3-}$ ion were low spin, strong field, the orbital diagram would be

$$\frac{}{} \quad \frac{}{}$$
$$\frac{\uparrow\downarrow}{} \ \frac{\uparrow\downarrow}{} \ \frac{\uparrow\downarrow}{}$$

with no unpaired electrons.

The difference in energy between the higher energy and lower energy d orbitals, Δ, is of the same order of magnitude as the energy of chemical bonds and the energy of photons of visible light, 160–280 kJ/mol (Section 7.6). Photons of visible light have just the right energy to promote electrons from the lower energy d orbitals to the higher energy d orbitals. If some wavelengths are absorbed from white light, the light that is left is colored (remember Figure 7.17). Promotion of electrons from lower to higher energy d orbitals results in the colors of transition-metal complexes. Thus, the crystal field theory explains why transition-metal complexes are colored.

The magnitude of the crystal field splitting, Δ, can be determined by measuring the wavelength of light absorbed by a complex. The simplest example is the $Ti(H_2O)_6^{3+}$ ion, which has only one d electron (d^1). The $Ti(H_2O)_6^{3+}$ ion is octahedral and, in the ground state, has the electron configuration.

⚫ Remember that white light is a mixture of different wavelengths of light (Section 7.6).

⚫ Sample Problem 7.2 shows how to calculate the energy that corresponds to a given wavelength of electromagnetic radiation.

The $Ti(H_2O)_6^{3+}$ ion has a single absorption band centered at 493 nm (see ▬Figure 24.11). [Remember that polyatomic ions have band spectra instead of line spectra like atoms and monatomic ions have (Section 10.9).] Electromagnetic radiation with a 493-nm wavelength is blue-green. Absorption of blue-green light leaves yellow, orange, red, and violet light, and an aqueous solution of $Ti(H_2O)_6^{3+}$ looks reddish violet (see Figure 24.11). A wavelength of 493 nm corresponds to an energy of 243 kJ/mol; that is, 243 kJ are required to raise the electrons in 1 mol $Ti(H_2O)_6^{3+}$ from the lower to the higher energy level. In the first excited state, the $Ti(H_2O)_6^{3+}$ ion has the electron configuration

$$\uparrow \quad \quad \quad \updownarrow \ \Delta = 243 \text{ kJ/mol}$$

If the central metal ion in a complex has no d electrons like Sc^{3+} or full d orbitals like Zn^{2+} and Ag^{+}, the complex is colorless or white. The complex does not absorb visible light; it transmits or reflects all wavelengths (remember Figure 8.16).

The magnitude of the crystal field splitting depends on the identity and the oxidation state of the metal and on the ligand. The greater the charge on the central atom, the greater the attraction of the central atom for the ligands and the closer the ligands come to the central atom. The closer the ligands come to the central atom, the stronger the repulsion between the negative charge on the ligands and the electrons in the d orbitals of the central atom and the greater the splitting. Thus, for the same central atom and ligands, the higher the charge on the central atom, the greater the splitting. The complexes of cobalt show the effect of oxidation state on splitting. Most octahedral complexes of Co(II) are high spin, weak field; most octahedral complexes of Co(III) are low spin, strong field.

Nuclear charge increases going down a group because the number of protons in the nucleus increases; therefore, splitting is greater for second transition-series elements than for first transition-series elements. Splitting is sufficiently greater for second transition-series elements that high-spin complexes are rare and most complexes of second transition-series elements are low spin. For elements in the

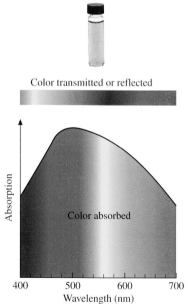

Color transmitted or reflected

Color absorbed

▬ FIGURE 24.11   *Top:* A solution containing $Ti(H_2O)_6^{3+}$. *Middle:* Color transmitted or reflected. *Bottom:* The absorption spectrum of $Ti(H_2O)_6^{3+}$.

third transition series, splitting is even larger than it is for elements in the second transition series.

Spectrochemical Series

If the identity and oxidation state of the central atom are kept constant, the effect of the ligand on the magnitude of the crystal field splitting can be observed. For example, measurements of paramagnetism show that the hexaamminechromium(II) ion, $[Cr(NH_3)_6]^{2+}$, has four unpaired electrons whereas the hexacyanochromate(II) ion, $[Cr(CN)_6]^{4-}$, has only two unpaired electrons:

$[Cr(NH_3)_6]^{2+}$ is a weak-field, high-spin complex.　　　$[Cr(CN)_6]^{4-}$ is a strong-field, low-spin complex.

Cyanide ions split the crystal field of the chromium(II) ion more than ammonia molecules split the crystal field of the chromium(II) ion.

In general the magnitude of the crystal field splitting increases along the series

$$I^- < Br^- < Cl^- < F^- < ONO^- < OH^- < C_2O_4^{2-}$$
$$< H_2O < NH_3 < en < NO_2^- < CN^- < CO$$

which is called the **spectrochemical series.** The spectrochemical series summarizes the results of many observations on complexes of a variety of central metal atoms. The wavelength of light absorbed shifts toward the blue side of the visible spectrum from the left side to the right side of the spectrochemical series. As a result, the color transmitted or reflected shifts toward the red. For example, $[CuBr_4]^{2-}$ is violet, $[Cu(H_2O)_4]^{2+}$ is blue, and $[Cu(en)_2]^{2+}$ is green:

$[CuBr_4]^{2-}$　　　$[Cu(H_2O)_4]^{2+}$　　　$[Cu(en)_2]^{2+}$
violet　　　　　　　blue　　　　　　　　green

The color observed along this series shifts toward the red (see Figure 24.11). Therefore, the wavelength absorbed has shifted toward the blue or high-energy side of the visible spectrum.*

Notice how ligands that have the same atom as the electron-pair donor are close together in the spectrochemical series; for example, the donor atom in OH^-, $C_2O_4^{2-}$, and H_2O is oxygen. Complexes containing these ligands have similar colors.

One might think that negative ions should produce larger splittings than polar molecules because they have a higher concentration of negative charge. However, this is obviously not correct. For example, the water molecule is to the right of the OH^- ion in the spectrochemical series, and the carbon monoxide molecule, which is almost nonpolar, is at the far right. You must refer to the spectrochemical series, which was determined by experiment, to find out whether a species produces much or little splitting.

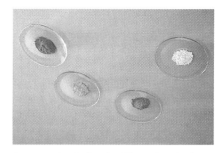

A spectrochemical series. The cations in these complex compounds are $[Co(NH_3)_5Br]^{2+}$ (dark red), $[Co(NH_3)_5Cl]^{2+}$ (rose-red), $[Co(NH_3)_5H_2O]^{3+}$ (orange-red), $[Co(NH_3)_6]^{3+}$ (yellow). The color reflected goes from red to yellow; the color absorbed shifts toward the blue side (see Figure 24.11).

Notice that the position of the nitrite ion in the spectrochemical series depends on whether O or N is attached to the central atom.

---

*When an oxygen molecule is released from oxyhemoglobin (see "The Oxygen–Hemoglobin Equilibrium" in Chapter 14), a water molecule takes its place. Oxygen lies to the far right in the spectrochemical series, and the oxygen-containing complex is strong field, low spin, whereas the water-containing complex is weak field, high spin. The wavelength of light absorbed by the complex decreases when oxygen is replaced by water, and the color observed changes from the bright red of arterial blood to the bluish color of venous blood.

24.16 The complex $[Co(NH_3)_6]^{3+}$ is yellow. If a water molecule is substituted for one of the ammonia molecules to form the complex $[Co(NH_3)_5H_2O]^{3+}$ (a) will the splitting be larger or smaller? (b) Will the wavelength of the light that is absorbed increase or decrease? (c) Will the color of the complex be shifted toward the blue end or to the red end of the visible spectrum?

The way the *d* orbitals are split depends on the number of ligands and on how the ligands are arranged around the central atom as well as on the identity and oxidation state of the central atom and the identity of the ligand. To see the effect of number and arrangement of ligands, let's consider first a tetrahedral complex and then a square planar complex.

Tetrahedral Complexes

Let's begin by considering eight ligands approaching the central metal ion from the corners of a cube:

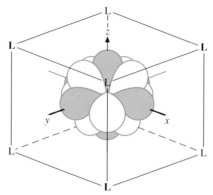

The shaded orbitals (x^2-y^2 and z^2), the two orbitals that lie along axes, are directed toward the centers of the faces of the cube. The energies of all the orbitals are raised. However, the energies are not raised as much as when the ligands approached from the corners of an octahedron because the ligands do not approach any of the orbitals directly. Although the ligands do not approach any of the orbitals directly, they come closer to the three orbitals that lie between the axes (*xy*, *xz*, and *yz*) than they do to the two orbitals that lie along the axes. The energies of the three orbitals are raised more than the energies of the two orbitals:

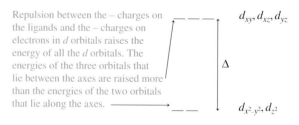

Repulsion between the – charges on the ligands and the – charges on electrons in *d* orbitals raises the energy of all the *d* orbitals. The energies of the three orbitals that lie between the axes are raised more than the energies of the two orbitals that lie along the axes. $\longrightarrow$

d_{xy}, d_{xz}, d_{yz}

Δ

$d_{x^2-y^2}, d_{z^2}$

In an isolated gaseous metal ion, the five *d* orbitals have the same energy.

The energy level pattern is the reverse of the energy level pattern for octahedral ligands. There are three orbitals in the high-energy set and only two orbitals in the low-energy set.

Now, if the four ligands that are in boldface in the cube are removed (one doesn't show because it is behind the model), the remaining four ligands form a tetrahedron around the central metal ion:

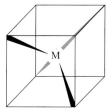

(In the picture, the cube has been rotated counterclockwise 45° so that you can see the bonds to the four ligands better.) The energy level pattern for four ligands at the corners of a tetrahedron is similar to the energy level pattern for eight ligands at the corners of a cube.

In a tetrahedral complex, not only do the ligands not approach any of the orbitals directly, but also only four ligands instead of the six of an octahedral complex split the *d* orbitals. Splitting in tetrahedral complexes is much smaller than splitting in octahedral complexes. The two sets of *d* orbitals are close in energy, and when the third, fourth, and fifth electrons are added, they never pair. Instead, the third, fourth, and fifth electrons go into the upper set of three orbitals:

$$\frac{\uparrow\quad\uparrow\quad\uparrow}{\underline{\uparrow}\quad\underline{\uparrow}}$$

Tetrahedral complexes are always weak field, high spin.

Square Planar Complexes

Splitting in square planar complexes is more complicated than splitting in octahedral and tetrahedral complexes. Imagine that the four ligands of a square planar complex approach the central metal ion along the *x*- and *y*-axes:

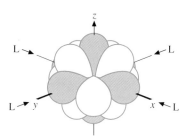

The energy of the x^2-y^2 orbital will be raised most because it lies along the *x*- and *y*-axes. The increase in energy of the *xy* orbital, which lies between the *x*- and *y*-axes, will be greater than the increase in energy of the *xz* and *yz* orbitals. However, because the doughnut part of the z^2 orbital lies in the *xy* plane, one cannot tell qualitatively where the z^2 orbital will come. Energies must be calculated. From calculations, the following orbital energy diagram is obtained for square planar complexes:

$$\underline{\quad} \; d_{x^2-y^2}$$

$$\underline{\quad} \; d_{xy}$$

$$\underline{\quad} \; d_{z^2}$$

$$\underline{\quad}\;\underline{\quad} \; d_{xz}\; d_{yz}$$

24.17 Sketch the *d* orbital splitting diagram for $[Ni(NH_3)_4]^{2+}$, which is tetrahedral.

24.18 Sketch the *d* orbital splitting diagram for $[Pt(NH_3)_4]^{2+}$, which is square planar and diamagnetic.

Because it is based on electrostatic attractions between point charges, the crystal field theory is not realistic. Molecular ligands are far from point charges; even anionic ligands are not point charges. In addition, the properties of complexes suggest that covalent bonding is important. For example, if bonding is electrostatic, how can the almost nonpolar CO molecule be such a good ligand? However, the crystal field theory is simple and surprisingly useful.

Molecular Orbital Theory

The molecular orbital theory takes both ionic and covalent character in the bonds of complexes into account and is the best explanation of the properties of complexes. But molecular orbital calculations are difficult, and the molecular orbital theory is not as easy to visualize as the crystal field theory. However, with increasingly accurate wave functions and increased availability of high-speed computers, the molecular orbital theory will probably become even more important in the future than it is at present.

24.9 STABILITY AND LABILITY OF COMPLEXES

The equilibrium constant for the equilibrium

$$Cu(H_2O)_4{}^{2+} + 4NH_3(aq) \rightleftharpoons Cu(NH_3)_4{}^{2+} + 4H_2O(l)$$

$$K_f = \frac{[Cu(NH_3)_4{}^{2+}]}{[Cu(H_2O)_4{}^{2+}][NH_3]^4} = 5.6 \times 10^{11} \text{ at } 25\ °C \quad \text{(Table 16.4)}$$

is large. The standard free-energy change for the reaction

$$\underset{\text{pale blue}}{Cu(H_2O)_4{}^{2+}} + 4NH_3(aq) \longrightarrow \underset{\text{deep blue}}{Cu(NH_3)_4{}^{2+}} + 4H_2O(l)$$

is negative (-72.4 kJ/mol), and the complex ion $[Cu(NH_3)_4]^{2+}$ is **stable.** *Stable refers to thermodynamics.*

However, if the deep blue solution of tetraamminecopper(II) ion is acidified, the color immediately changes to the pale blue of the tetraaquacopper(II) ion (■Figure 24.12):

$$\underset{\text{deep blue}}{Cu(NH_3)_4{}^{2+}} + 4H_3O^+ \longrightarrow \underset{\text{pale blue}}{Cu(H_2O)_4{}^{2+}} + 4NH_4{}^+$$

The complex ion $[Cu(NH_3)_4]^{2+}$ is labile. A **labile complex** *rapidly exchanges ligands.* A complex is classified as labile if equilibrium is reached in less than a minute; most complexes are labile.

A *complex that does not exchange ligands rapidly* is called an **inert complex.** The complex ion $[Co(NH_3)_6]^{3+}$ is an example of an inert complex. Although the equilibrium constant for the equilibrium

$$[Co(NH_3)_6]^{3+} + 6H_3O^+ \rightleftharpoons [Co(H_2O)_6]^{3+} + 6NH_4{}^+$$

■ FIGURE 24.12 The tetraamminecopper(II) ion is labile. If a solution that contains $[Cu(NH_3)_4]^{2+}$, which is deep blue, is acidified, pale blue $[Cu(H_2O)_4]^{2+}$ [and a precipitate of $Cu(OH)_2$] form immediately.

is extremely large (K_{eq} = ca. 10^{25} at 25 °C), over a month is required for this system to reach equilibrium under ordinary conditions. The activation energy, E_a (Section 18.9), is high. *Labile and inert refer to kinetics.* As usual, there is no relationship between the position of equilibrium and the time required to reach equilibrium.

The only common inert complexes are those of Co^{3+}, Cr^{3+}, Pt^{4+}, and Pt^{2+}. Most of the early observations of complexes described in Section 24.2 were made of complexes of these ions.

24.10 USES OF TRANSITION-METAL COMPLEXES

In most cases, the biological activity of transition metals depends on the formation of complexes. Nine transition metals are known to be essential to biological systems; these metals and their biological functions are shown in Table 24.4.

Nitrogen-fixing bacteria use enzymes that contain both iron and molybdenum. The use of cisplatin to treat cancer and of ethylenediamminetetraacetic acid (EDTA) to remove toxic heavy-metal ions from the body have already been mentioned.

Natural complexes are responsible for the color of many biological systems. For example, heme, the red pigment of blood, is an iron-containing complex compound. Some synthetic dyes and pigments are transition-metal complexes. Prussian blue, $Fe_4[Fe(CN)_6]_3 \cdot xH_2O$, has been used as a pigment by artists since the beginning of the eighteenth century.

Transition-metal complexes are used as catalysts for many important industrial processes. For example, $Co_2(CO)_8$ catalyzes the addition of hydrogen and carbon monoxide to alkenes to form aldehydes (the oxo process):

$$RCH{=\!=}CH_2 + H_2(g) + CO(g) \xrightarrow{\text{Co}_2\text{(CO)}_8} RCH_2CH_2CHO$$

Use of complexes as catalysts for the fixation of atmospheric nitrogen is being investigated. Bacteria and blue-green algae fix nitrogen without using high temperature and pressure.

This structure for Prussian blue is from Cotton, F. A.; Wilkinson, G. Advanced Inorganic Chemistry, 5th ed.; Wiley: New York, 1988; p 721.

TABLE 24.4	Essential Transition Metals	
Atomic Number	Symbol	Biological Function
23	V	When located inside a cell, V(V) inhibits the $Na^+ - K^+$ pump.
24	Cr	Chromium is a part of the glucose tolerance factor that, together with insulin, controls the removal of glucose from blood.
25	Mn	Manganese activates many enzymes.
26	Fe	Proteins containing iron take part in oxygen transport and electron transfer.
27	Co	Cobalt activates many enzymes. Vitamin B_{12}, which takes part in the development of red blood cells, contains cobalt.
28	Ni	The function of nickel has not been identified. The first enzyme to be crystallized (in 1920) contains nickel.
29	Cu	Redox enzymes and oxygen-transport pigments contain copper.
30	Zn	The digestive enzyme that hydrolyzes protein contains zinc.
42	Mo	The function of molybdenum has not been identified.

These horses atop St. Mark's Basilica in Venice are made of bronze.

Another important use of transition-metal complexes is extraction of gold (and silver if present) from ore. In the presence of oxygen from air, gold dissolves in dilute cyanide solution:

$$4Au(s) + 8NaCN(aq) + O_2(g) + 2H_2O(l) \longrightarrow 4Na[Au(CN)_2](aq) + 4NaOH(aq)$$

Any silver present also dissolves, forming $[Ag(CN)_2]^-$.

A complex is used to purify nickel. At low temperatures, nickel reacts with carbon monoxide to form tetracarbonylnickel(0):

$$Ni(s) + 4CO(g) \xrightarrow{80\ ^\circ C} Ni(CO)_4(g)$$

The $Ni(CO)_4$, which boils at 43 °C, is separated from impurities by distillation and decomposed to pure nickel:

$$Ni(CO)_4(g) \xrightarrow{230\ ^\circ C} Ni(s) + 4CO(g)$$

The nickel obtained is 99.95% pure; the carbon monoxide is recycled.

24.11 ALLOYS

Remember that an alloy is a material that has metallic properties but is not a pure metal (Section 11.9). You are undoubtedly familiar with a number of alloys from your everyday life. Table 24.5 lists some common alloys, the elements that compose them, and some uses of each alloy. Brass and bronze have been made for about 5000 years. Bronze is harder than pure copper and pure iron and was originally used to make weapons and tools. Iron took the place of bronze only because iron is more readily available. The relatively low melting point of bronze makes bronze easy to cast. The melt expands as it solidifies, filling every corner of the mold, and

Wood's metal bars melting over boiling water.

TABLE 24.5	Some Common Alloys	
Name	Composed of	Uses
Brass	Cu and Zn	Screws, window and door fittings
Bronze	Cu and Sn	Statues, machine parts
Dental amalgam	Ag, Hg, and Sn	Filling teeth
Gold	Au, Ag, and Cu	Jewelry and coins
Pewter	Sn, Cu, and Sb	Household objects such as pitchers
Sterling silver	92.5% Ag, 7.5% of another metal, usually Cu	Tableware and jewelry
Solder	Usually Sn and Pb	Electrical and plumbing industries to join metal surfaces without melting them (Solder has a low melting point.)
Steel	Iron, carbon, and other metals such as Cr, Ni, and W	Construction of high-rise buildings and bridges; autos and other vehicles; household products, from large appliances to tableware and cooking utensils
Wood's metal	50% Bi, 25% Pb, 12.5% Sn, and 12.5% Cd	Valves of automatic sprinkling systems (Wood's metal melts at 70 °C.)

Superconductors

Superconductors have no resistance to the movement of electric charge. When electricity moves from power plants to manufacturing plants, offices, and homes over power lines made of ordinary metallic conductors such as aluminum, a significant portion of the electrical energy is lost as thermal energy. Superconducting power lines would make possible movement of electricity over long distances without loss. Power plants could be located far from where people live, near the natural resources used to run them, and much energy would be saved. Superconducting chips would make possible smaller and more powerful computers because the chips would not get hot and need to be insulated from each other and cooled. Another potentially useful property of superconductors is their perfect diamagnetism. Weak magnets are repelled by superconductors and can be levitated (suspended in air).

Superconductivity was discovered by the Dutch physicist Kamerlingh Onnes in 1911 when he cooled mercury in liquid helium and found that its electrical resistance suddenly disappeared at about 4 K. Practical applications of superconductivity became possible in 1973 when a niobium–tin alloy was discovered that is superconducting up to 23.3 K and remains superconducting even when the current is large. Powerful electromagnets made of the niobium–tin alloy cooled in liquid helium are used in accelerators and in instruments for nuclear magnetic resonance spectroscopy (known in medicine as mag-

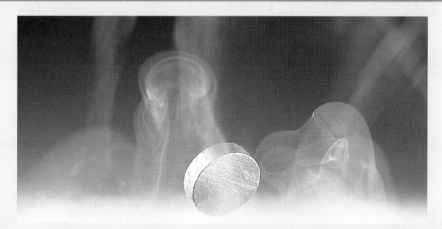

A small magnet levitated over a ceramic superconductor cooled by liquid nitrogen.

netic resonance imaging). However, liquid helium is expensive and inconvenient to use. Availability of higher temperature superconductors would make many more applications possible.

All of the early superconductors were metals or alloys. Then in 1986, K. Alex Müller and J. Georg Bednorz at IBM's labs in Switzerland discovered the first ceramic superconductor, which was composed of lanthanum, barium, copper, and oxygen and became superconducting at 35 K. Soon, several other ceramic superconductors were found that superconducted at temperatures greater than 77 K, the boiling point of liquid nitrogen. Liquid nitrogen is readily available and relatively inexpensive (cheaper than gasoline, milk, and beer). Researchers all over the world began looking for other high-temperature superconducting materials, trying to understand how ceramic superconductors

work, looking for possible applications of high-temperature superconductors, and trying to find ways to make ceramics into wires (or the equivalent).

The crystals of all high-temperature superconductors have planes of copper and oxygen sandwiched between layers of other elements. The copper acts as if it has an oxidation state between $+2$ and $+3$. As a result, electron-deficient holes are introduced into the conduction band; all but one of the known copper oxide superconductors are p-type conductors. All are substitutional solid solutions with some coppers in a $+2$ oxidation state and others in a $+3$ oxidation state. All are nonstoichiometric compounds with a slightly variable number of oxygens.

Discovery of a material that is superconducting at room temperature would make possible all sorts of applications. Maybe you will be the person to find one.

---

the solid then shrinks slightly as it cools making removal of bronze castings from a mold easy. Brass is more easily shaped by hammering or rolling than bronze.

Alloys are usually made by melting the mixture of elements and then allowing the melt to solidify. The alloying elements and their proportions can be changed. Thus, like the properties of synthetic polymers, the properties of alloys can be varied to suit a particular purpose. For example, addition of silver and copper to gold makes the gold harder (and cheaper) without noticeably changing the color. Pure gold, which is 24 carat, is very soft and dents easily. The gold used to make gold jewelry is usually 14 carat (58% by mass gold).

Alloys can be heterogeneous mixtures, solid solutions, or even intermetallic compounds. For example, Alnicos, alloys of iron with aluminum, nickel, and cobalt

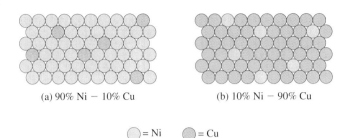

(a) 90% Ni − 10% Cu   (b) 10% Ni − 90% Cu

○ = Ni   ● = Cu

- FIGURE 24.13   Two alloys of nickel and copper that are substitutional solid solutions.

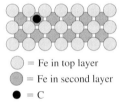

○ = Fe in top layer
● = Fe in second layer
● = C

- FIGURE 24.14   Steel, an interstitial solid solution. (Iron crystallizes in a body-centered cubic arrangement.)

that are used to make magnets up to 25 times as strong as ordinary magnets, are heterogeneous mixtures. Alnicos consist of two solid phases; fine crystals of one phase are distributed throughout another crystalline phase.

Two types of solid solution are possible. In one type, which is called **substitutional,** *atoms of one metal randomly take the place of atoms of another metal.* The two metals must have similar radii, crystal structures, and chemical properties. For example, nickel has an atomic radius of 125 pm and has cubic closest-packed crystals (Section 12.9). Copper has an atomic radius of 128 pm and also has cubic closest-packed crystals. Nickel and copper are completely soluble in each other in the solid state. Continuous variation in composition from 100% Ni–0% Cu to 0% Ni–100% Cu is possible. -Figure 24.13 shows one layer of two alloys of Ni and Cu that are substitutional solid solutions.

In the other type of solid solution, which is called **interstitial,** *small nonmetal atoms, such as H, B, C, and N, occupy holes in the crystal structures of the metal.* Steel is an example of an interstitial alloy. In steel, carbon atoms occupy holes in iron crystals as shown in -Figure 24.14.

Niobium–tin, Nb_3Sn, is an example of an alloy that is an intermetallic compound. Niobium–tin is superconducting (has zero resistance) below 18.4 K and is used to make superconducting magnets. The formulas for intermetallic compounds cannot usually be predicted by the ordinary rules of valence.

SUMMARY

The major uses of the transition metals are as metals. Many properties of metals are related to **standard enthalpy of atomization** or **atomization energy, ΔH_a°.** The standard enthalpy of atomization is the enthalpy change for the process by which an element in its usual form at 25 °C and 1 atm is converted to individual gaseous atoms. The enthalpy of atomization measures the strength with which atoms are held together. Attractions between atoms are strongest when valence shell orbitals are half-filled. In each row of the periodic table, enthalpies of atomization, melting points, boiling points, and densities are highest near the middle of the row. The number of oxidation states observed for a transition metal also reaches a maximum near the middle of each transition series.

Transition metals are *d*-block elements. All transition elements are metals and most have high melting points, boiling points, and densities and are strong. Most have more than one oxidation state in their compounds; the different oxidation states often vary by 1. Much of the chemistry of the transition metals involves redox reactions. Many transition-metal compounds are colored

and paramagnetic. Measurement of paramagnetism can be used to find the number of unpaired electrons in a species. Ferromagnetic substances are strongly and permanently magnetized.

Transition metals have a strong tendency to form complexes. The **coordination number** of a metal ion in a complex is the number of atoms or groups that are firmly bound to the metal ion, that is, the number of ligands. **Monodentate ligands** have only a single donor atom. **Di- or bidentate ligands** have two, and **polydentate,** two or more donor atoms. Complexes in which polydentate ligands are bonded to a single central atom are called **chelate complexes;** chelate complexes are very stable. Complexes have both stereoisomers and constitutional isomers. Compounds that include one or more complexes are called **coordination compounds.**

The crystal field theory is a simple way of explaining bonding in complexes that accounts for their magnetic properties and color. According to the **crystal field theory,** the bonds between ligands and the central atom in complexes are electrostatic. The ligands are held to the central atom by the attraction of the positive charge on the central atom for the negative charge of the ligands. The energies of all the d orbitals are increased by the ligands; however, the energies of some d orbitals are raised more than the energies of other d orbitals. The difference in energy between different d orbitals is called **crystal field splitting, Δ,** and is of the same order of magnitude as the energy of photons of visible light. Promotion of electrons from lower energy d orbitals to higher energy d orbitals results in the colors of transition-metal complexes. In **strong-field** complexes, the difference in energy between d orbitals is large. Electrons pair instead of going into higher energy orbitals, and **low-spin** complexes with few unpaired electrons result. In **weak-field** complexes, the difference in energy between d orbitals is small. Less energy is required for electrons to go into the higher energy d orbitals than to pair, and **high-spin** complexes with many unpaired electrons result.

The greater the effective nuclear charge of a metal ion, the greater the splitting. Within a group, Δ increases from top to bottom of the group. Most complexes of elements in the second and third transition series are strong-field, low-spin complexes.

The higher the oxidation state of a metal ion, the greater the splitting. The **spectrochemical series** shows the effect of ligands on splitting.

Equilibrium constants for the formation of **stable** complexes are large. **Labile** complexes rapidly exchange ligands with the solvent. A complex that does not exchange ligands rapidly is called an **inert complex.**

In most cases, the biological activity of transition metals depends on the formation of complexes. Complexes are widely used as catalysts in industry.

Atoms of elements in the second and third transition series are about the same size because of the lanthanide contraction, the decrease in atomic radius that takes place across the lanthanide series. Second and third transition-series elements that are in the same group are usually more like each other than they are like the element in the first transition series.

Alloys, as well as pure metals, have metallic properties, which can be varied to suit a particular purpose by changing the composition of the alloy. Alloys can be heterogeneous mixtures, solid solutions, or intermetallic compounds.

ADDITIONAL PRACTICE PROBLEMS

For information about the organization of Additional Practice Problems, Stop & Test Yourself, Putting Things Together, and Applications, see the beginnings of these sections in Chapter 1.

24.19 How many unpaired electrons are in (a) an atom of vanadium? (b) A V^{2+} ion?

24.20 (a) Which element resembles niobium most closely? (b) Which is more reactive, nickel or palladium? Explain your answers.

24.21 Refer to Figure 24.3(c) to answer the following questions. (a) How do the densities of metals change across a row in the periodic table? Account for the trends observed. (b) How does the change in density down a group for transition metals compare with the change in density down a group for the alkali and alkaline earth metals? Account for the difference.

24.22 (a) Define *enthalpy of atomization.* (b) Explain why the enthalpy of atomization of vanadium is much greater than the enthalpy of atomizaion of potassium. (c) Explain why the enthalpy of atomization of vanadium is much greater than the enthalpy of atomization of zinc. (d) What general relationship exists between the number of unpaired electrons in an atom of a transition element and the atomic radius of the element?

24.23 (a) What properties are characteristic of transition elements? (b) How do the melting points, boiling points, and densities of transition elements vary across rows in the periodic table? (c) Which element is most like molybdenum? (d) Why do most transition elements have compounds in which their oxidation number is $+2$? (e) Why are the transition elements used more than their compounds?

24.24 Given the complex ion $[CoBr(NH_3)_5]^{2+}$: (a) What are the ligands? (b) What is the charge on the central ion? (c) How many d electrons does the central metal ion have? (d) Write the formula of the nitrate of this complex ion. (e) What would be the charge on the complex ion if a sulfate ion replaced the bromide ion?

24.25 What is the coordination number of the central atom in each of the following coordination compounds? (a) $K[Au(CN)_2]$ (b) $[Cr(H_2O)_6]Cl_3$ (c) $Na_3[Ni(CN)_5]$ (d) $[Cu(NH_3)_4]SO_4$ (e) $Na_4[Fe(CN)_6]$

24.26 If one mole of each of the following compounds is dissolved in water, how many moles of ions will be formed? (Assume that dissociation is complete.) Show the formulas of the ions. (a) $[Cu(NH_3)_4]SO_4 \cdot H_2O$ (b) $[Co(NH_3)_6](NO_3)_3$ (c) $Na_2[Pt(CN)_4] \cdot 3H_2O$ (d) $K_4[Fe(CN)_6]$ (e) $[PtCl_2(NH_3)_2]$

24.27 Which of the following complexes have geometric isomers? If a complex has geometric isomers, sketch them. (a) $[PtCl(NH_3)_5]^{3+}$ (b) $[PtCl_3(NH_3)_3]^{+}$ (c) $[CuBr_2Cl_2]^{2-}$ (square planar) (d) $[CuBr_2I_2]^{2-}$ (tetrahedral)

24.28 (a) Sketch the three stereoisomers of

$$\left[Co\left(\begin{array}{c} NH_2CH_2 \\ | \\ NH_2CH_2 \end{array} \right)_2 Cl_2 \right]^{+}$$

(b) Classify each stereoisomer as *cis* or *trans*. (c) Identify the enantiomeric pair.

24.29 A, B, C, and D represent four different ligands. Why doesn't the tetrahedral complex ion $[ZnABCD]^{2+}$ display optical isomerism?

24.30 (a) How many compounds have the formula $[CoCl_3(NH_3)_3]$? (b) How many compounds have the formula $[CoCl_2(NH_3)_4]$? Sketch the compounds.

24.31 Classify each of the following ligands as monodentate, didentate, tridentate, tetradentate, pentadentate, or hexadentate:
(a) $(H_2NCH_2CH_2)_2NH$
(b) $CH_3CH(NH_2)COO^-$
(c) $H_2N(CH_2)_2NH(CH_2)_2NH(CH_2)_2NH_2$

(d) 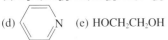   (e) $HOCH_2CH_2OH$

24.32 Name each of the following: (a) $[CuF_6]^{3-}$
(b) $[MoI(CO)_5]^-$ (c) $Na_4[Mn(CN)_6]$
(d) $[Ni(NH_3)_6](ClO_4)_2$ (e) $[CoCl_2(NH_3)_4]^+$

24.33 Write the formula for each of the following:
(a) hexacyanoferrate(II) ion   (b) potassium hexafluoronickelate(IV)   (c) tetrachlorocuprate(II) ion
(d) tetraamminechloronitrocobalt(III) sulfate
(e) tetraaquadihydroxoiron(III) ion

24.34 Write formulas for and name the complexes that contain (a) four cyanide ions bound to a gold(III) ion,   (b) two ammonia molecules bound to a silver(I) ion,   (c) four ammonia molecules and two bromide ions bound to a cobalt(II) ion,   (d) three bromide ions and three water molecules bound to a platinum(IV) ion,   (e) four chloride ions and two water molecules bound to an iron(III) ion.

24.35 Sketch boundary surface diagrams for each of the d orbitals.

24.36 (a) What are the two coordination numbers most common in transition-metal complexes?   (b) What shape or shapes are usually observed for complexes with each of these common coordination numbers?

24.37 According to the crystal field theory of bonding in transition-metal complexes, (a) what causes the splitting of the d orbitals of the metals into two sets in octahedral complexes?   (b) What two factors determine whether a complex will be high or low spin?   (c) Which d electron configurations can have both high- and low-spin octahedral complexes?

24.38 The complex ion $[Co(NH_3)_6]^{3+}$ is diamagnetic.   (a) Is the $[Co(NH_3)_6]^{3+}$ ion high or low spin?   (b) Is the $[Co(CN)_6]^{3-}$ ion high or low spin?   (c) The complex ion $[CoF_6]^{3-}$ is high spin. Is it diamagnetic or paramagnetic? If it is paramagnetic, how many unpaired electrons does it have?   (d) Is $[Co(NH_3)_6]^{2+}$ diamagnetic or paramagnetic? Explain your answers.

24.39 (a) Experimentally, how would you determine the number of unpaired electrons in a species?   (b) If you found that $[Fe(H_2O)_6]^{2+}$ is paramagnetic but $[Fe(CN)_6]^{4-}$ is diamagnetic, which ion is weak field, high spin and which is strong field, low spin?   (c) Is this result predicted by the spectrochemical series?

24.40 (a) Write the ground-state electron configuration of a cobalt atom.   (b) Write the ground-state electron configuration of an isolated cobalt(II) ion.   (c) Show the ground-state electron configuration of a high-spin octahedral cobalt(II) ion with an orbital diagram.   (d) Show the ground-state electron configuration of a low-spin octahedral cobalt(II) ion with an orbital diagram.   (e) For the octahedral cobalt(II) ion, the shift between high-spin and low-spin electron configuration comes between NH_3 and NO_2^- in the spectrochemical series. Is $[Co(CN)_6]^{4-}$ high spin or low spin?

24.41 The hydrated iron(II) ion $[Fe(H_2O)_6]^{2+}$ is light green.   (a) What does the fact that the color is light green, not dark green, indicate about the strength of the absorption band?   (b) Is the $[Fe(H_2O)_6]^{2+}$ ion paramagnetic or diamagnetic? Explain your reasoning.

24.42 Cobalt(II) ion forms both octahedral and tetrahedral complexes. In general, the octahedral complexes are pink, and the tetrahedral complexes are blue. Explain why.

24.43 Most complexes of iron(III) are octahedral and are high spin except for $[Fe(CN)_6]^{3-}$. Complexes of Ru(III) and Os(III) are also octahedral but are all low spin.   (a) Explain why $[Fe(CN)_6]^{3-}$ is low spin.   (b) Explain why most Fe(III) complexes are high spin but complexes of Ru(III) and Os(III) are all low spin.   (c) How many unpaired electrons are there in the low-spin octahedral Ru(III) and Os(III) complexes?

24.44 The strongest absorption band of *trans*-$[Co(NH_3)_4Cl_2]^{2+}$ is centered at 640 nm.   (a) What color light is absorbed?   (b) What color light is transmitted?   (c) What color is a solution of *trans*-$[Co(NH_3)_4Cl_2]Cl_2$?

24.45 Why is $[Cd(NH_3)_6]^{2+}$ colorless, whereas $[Co(NH_3)_6]^{2+}$ is golden brown?

24.46 Which member of each of the following pairs of complex ions absorbs at the longer wavelength? Explain your answers.
(a) $[Cu(H_2O)_6]^{2+}$ or $[Cu(NH_3)_4(H_2O)_2]^{2+}$ (b) $[Cr(H_2O)_6]^{3+}$ or $[Cr(H_2O)_6]^{2+}$ (c) $[Fe(H_2O)_6]^{3+}$ or $[Ru(H_2O)_6]^{3+}$.

24.47 Explain why $[CuBr_4]^{2-}$ is violet but $[Cu(H_2O)_4]^{2+}$ is blue and $[Cu(en)_2]^{2+}$ is green.

24.48 (a) How many unpaired electrons are there in a $[Ni(H_2O)_6]^{2+}$ ion?   (b) The $[Ni(H_2O)_6]^{2+}$ ion is green. Why are ammine complexes of nickel such as $[Ni(NH_3)_4(H_2O)_2]^{2+}$ blue-purple?   (c) The yellow $[Ni(CN)_4]^{2-}$ ion is diamagnetic. How are the cyanide ions arranged around the nickel ion in this complex? Explain your answer.

24.49 Which complex ion would absorb at a higher frequency, $[Cu(H_2O)_6]^{2+}$ or $[Cu(NH_3)_4(H_2O)_2]^{2+}$? Explain your answer.

24.50 (a) Why are octahedral complexes of elements in the second and third transition series usually low spin?   (b) Why are tetrahedral complexes always high spin?

24.51 The complex ion $[AuCl_4]^-$ is diamagnetic. Is this ion tetrahedral or square planar? Explain your answer.

24.52 What is the difference between an inert complex and a labile complex?

24.53 (a) What is an alloy?   (b) How are alloys usually made?   (c) What advantage do alloys have over pure metals?   (d) Give an example of an alloy that you are familiar with in your everyday life.

24.54 (a) What is the difference between a heterogeneous mixture and a solid solution?   (b) What two types of solid solution are possible? Describe each.   (c) What is an intermetallic compound?

1. The name of $[Co(CN)_5(OH)]^{3-}$ is
   (a) hydroxopentacyanocobaltate(III) ion
   (b) pentacyanohydroxocobaltate(II) ion
   (c) pentacyanohydroxocobalt(II) ion
   (d) pentacyanohydroxycobaltate(III) ion
   (e) pentacyanohydroxocobaltate(III) ion

2. For the coordination compound $[Co(NH_3)_5Cl]Cl_2$, the total number of ions formed in aqueous solution is _____, and the number of nonionic chlorines is _____.
   (a) 0, 3   (b) 2, 1   (c) 3, 1   (d) 3, 2   (e) 4, 2

3. Which of the following ligands is bidentate and can form a chelate ring?
   (a) NO_2^- (b) $N(CH_2COO^-)_3$ (c)

   (d)

   (e) $\begin{matrix} COO^- \\ COO^- \end{matrix}$

4. Which of the following is a d^6-type ion?
   (a) CrO_4^{2-} (b) Cr^{2+} (c) Mn^{2+} (d) Fe^{2+} (e) Co^{2+}

5. For which of the following types of ions is the number of unpaired electrons in octahedral complexes fixed at the same number as in the free ion no matter how weak or strong the crystal field?
   (a) d^3 (b) d^4 (c) d^5 (d) d^6 (e) d^7

6. Given the d-orbital splitting diagram

   $\uparrow \quad \underline{\ \ } \ \underline{\ \ }$
   $\underline{\uparrow} \ \underline{\uparrow} \ \underline{\uparrow}$

The complex is
   (a) tetrahedral with weak field and high spin.
   (b) octahedral with weak field and high spin.
   (c) octahedral with weak field and low spin.
   (d) octahedral with strong field and high spin.
   (e) octahedral with strong field and low spin.

7. Which of the following complex ions is colorless?
   (a) $[CoF_6]^{3-}$ (b) $[CuCl_4]^{2-}$ (c) $[Fe(SCN)_4]^-$ (d) $[NiCl_4]^{2-}$
   (e) $[Zn(CN)_4]^{2-}$

8. The complex ion $[CoCl_4]^{2-}$ absorbs strongly between 625 and 725 nm. A solution of $[CoCl_4]^{2-}$ is
   (a) blue-green   (b) colorless   (c) red   (d) violet   (e) yellow

9. All but one of the following ions are diamagnetic. Which one is paramagnetic?
   (a) $[Co(CN)_6]^{3-}$ (b) $[CoF_6]^{3-}$ (c) $[Co(H_2O)_6]^{3+}$
   (d) $[Co(NH_3)_6]^{3+}$ (e) $[Co(NO_2)_4(NH_3)_2]^-$

10. Which of the following complexes is most stable?

Complex	K_f
(a) $[CdBr_4]^{2-}$	10^4
(b) $[CdCl_4]^{2-}$	10^3
(c) $[ZnBr_4]^{2-}$	10^{-1}
(d) $[ZnCl_4]^{2-}$	1
(e) $[ZnI_4]^{2-}$	10^{-2}

PUTTING THINGS TOGETHER

24.55 Which transition metal forms an oxide, M_2O_5: Ti, V, or Cr?

24.56 Suggest an explanation for each of the following rules about the transition metals: (a) Across the first transition series, the stability of the highest oxidation state decreases from titanium(IV) to manganese(VII); after manganese(VII), oxidation states corresponding to the total number of d and s electrons in the atom are not commonly observed. (b) The highest oxidation state is usually observed only in oxo compounds, fluorides, and chlorides. (c) Considering elements in the same vertical column of the periodic table, high oxidation numbers are more stable for elements of the second and third transition series than for elements in the first transition series.

24.57 Consider the following data:

Observations of $CrCl_3$—$6H_2O$ Coordination Compounds			
Composition	Number of Ions	Number of Cl^- Ions	Number of Nonionic Cl
$CrCl_3 \cdot 6H_2O$	4	3	0
$CrCl_3 \cdot 6H_2O$	3	2	1
$CrCl_3 \cdot 6H_2O$	2	1	2

The coordination number of chromium in these compounds is 6. Write the formulas and names of the three compounds.

24.58 Addition of silver nitrate solution to a solution of a coordination compound with composition $CoCl_3 \cdot 4NH_3 \cdot H_2O$ gave 2 mol AgCl per mole of the coordination compound. According to conductivity measurements, the coordination compound was similar to $CaCl_2$. (a) Write the formula of the coordination compound. (b) Write the name of the coordination compound. (c) What is the coordination number of cobalt in this coordination compound?

24.59 (a) A green compound with composition $PtCl_2(NH_3)_2$ has a formula mass of 600 and consists of a complex cation and a complex anion. Write the formula of this compound and give its name. (b) The compound in part (a) is only slightly soluble in water and is made by a precipitation reaction. Write the net ionic equation for its preparation. (c) Give formulas and names for two compounds that might be used to make the compound in part (a) and write complete ionic and molecular equations for the reaction between them.

24.60 (a) If the tetraamminedichlorocobalt(III) ion is octahedral, how many geometric isomers can exist? (b) If the tetraamminedichlorocobalt(III) ion is planar hexagonal, how many geometric isomers can exist? (c) Can isomer number be used to choose between octahedral and planar hexagonal shapes? Explain your answers with drawings.

24.61 Two compounds with empirical formula PdF_3 and formula mass 327 are possible. What are their formulas and names?

24.62 (a) What is a chelate complex? (b) Which ion in the equilibrium

$$[Cd(NH_2CH_3)_4]^{2+} + 2H_2NCH_2CH_2NH_2 \rightleftharpoons$$
$$\left[Cd\left(\begin{matrix} NH_2CH_2 \\ | \\ NH_2CH_2 \end{matrix} \right)_2 \right]^{2+} + 4CH_3NH_2(aq)$$

is a chelate complex? (c) Show with a long arrow the side to which the equilibrium lies and explain your answer. (d) Assume that the product ion is tetrahedral and sketch it. (e) To what class of compounds does the product molecule belong?

24.63 Which member of each of the following pairs of complexes is more stable? (a) one with $K_f = 3 \times 10^5$ or one with $K_f = 5 \times 10^5$ (b) one with $K_f = 3 \times 10^5$ or one with $K_f = 3 \times 10^8$ (c) one with $K_f = 3 \times 10^{-5}$ or one with $K_f = 3 \times 10^{-7}$ (d) one with $pK_f = 5.2$ or one with $pK_f = 9.4$

24.64 Both titanium and germanium have four valence electrons. (a) Explain why the electrical conductivity of titanium is about a million times as great as the electrical conductivity of germanium at room temperature. (b) How do the electrical conductivities of titanium and germanium at high temperatures compare with their conductivities at low temperatures? Explain your answers. (c) How is the electrical conductivity of germanium increased?

24.65 Both silicon and titanium have four valence electrons. (a) Why is silicon a semiconductor whereas titanium is a metallic conductor? (b) What is the difference between a semiconductor and a metallic conductor?

24.66 The hydrated calcium ion is more stable than the hydrated Co^{2+} ion. The Zn^{2+}–Cl^- complex ion is more stable than the Co^{2+}–Cl^- complex ion. What will be observed if (a)

calcium ions are added to a pink solution containing the following equilibrium?

$$[CoCl_4]^{2-} + 6H_2O(l) \rightleftharpoons [Co(H_2O)_6]^{2+} + 4Cl^-$$
$$\text{blue} \qquad\qquad\qquad \text{pink}$$

(b) Zinc ions are added to a blue solution? Explain your answers.

24.67 What is the oxidation number of manganese in each of the following compounds? (a) $KMnO_4$ (b) MnO_2 (c) $MnSO_4$ (d) K_2MnO_4 (e) Mn_2O_3

24.68 Which is the stronger oxidizing agent in acidic aqueous solution, $Cr_2O_7^{2-}$ or MnO_4^-? Explain your answer.

24.69 Consider the following standard potentials:

$$O_2(g) + 4H^+(10^{-7}\,M) + 4e^- \longrightarrow$$
$$2H_2O(l) \qquad E° = \quad 0.82\text{ V}$$
$$Fe^{3+} + e^- \longrightarrow Fe^{2+} \qquad E° = \quad 0.77\text{ V}$$
$$2H_2O(l) + 2e^- \longrightarrow H_2(g) +$$
$$2OH^-(10^{-7}\,M) \qquad E° = \quad -0.42\text{ V}$$
$$Fe^{2+} + 2e^- \longrightarrow Fe(s) \qquad E° = \quad -0.44\text{ V}$$

Assuming that any spontaneous reactions will take place at a significant rate, (a) is Fe^{2+} stable in aqueous solution— that is, will it reduce water? Explain your answer. (b) Will Fe^{2+} disproportionate in aqueous solution under standard conditions? (c) Will Fe^{2+} be oxidized by $O_2(g)$ in water?

24.70 Although the metal ions of the first transition series shown in the table of standard reduction potentials (Table 19.1) are shown without any water of hydration, all first transition series ions are hydrated in aqueous solution. How does the formation of a complex with water affect the standard reduction potential for a metal ion? Explain your answer.

24.71 (a) Consider the following information:

$$Cu^{2+} + e^- \longrightarrow Cu^+ \qquad E° = \quad 0.159\text{ V}$$
$$Cu^+ + e^- \longrightarrow Cu(s) \qquad E° = \quad 0.52\text{ V}$$
$$Cu^{2+} + 2CN^- + e^- \longrightarrow [Cu(CN)_2]^- \qquad E° = \quad 1.12\text{ V}$$
$$[Cu(CN)_2]^- + e^- \longrightarrow$$
$$Cu(s) + 2CN^- \qquad E° = \quad -0.44\text{ V}$$

Show by calculating standard reduction potentials that in the presence of cyanide ion, copper(II) and copper(0) react spontaneously to form copper(I), whereas in the absence of cyanide ion, copper(I) spontaneously changes to Cu(0) and copper(II). (b) What is a reaction like the change of copper(I) to Cu(0) and copper(II) called?

24.72 Judging from the formation constants for common complex ions in Table 16.4, how does the stability of complexes change (a) when the oxidation number of the central ion increases? (b) Down a group for the central atom? (c) Down a group for the ligand? Give the information on which your conclusions are based and suggest an explanation for the trends observed.

24.73 Use formation constants from Table 16.4 to calculate the value of K for the equilibrium

$$[AgCl_2]^- + 2CN^- \rightleftharpoons [Ag(CN)_2]^- + 2Cl^-$$

part (a)?

24.75 With fluoride ion, aluminum forms the octahedral complex ion $[AlF_6]^{3-}$, but with chloride ion, aluminum forms only the tetrahedral complex ion $[AlCl_4]^-$. Suggest a reason for the difference.

24.76 The complex ion $[Co(NH_3)_5Cl]^{2+}$ is purple, and the complex ion $[Co(NH_3)_5H_2O]^{3+}$ is bright red. How can the rate of the reaction

$$[Co(NH_3)_5Cl]^{2+} + H_2O \longrightarrow [Co(NH_3)_5H_2O]^{3+} + Cl^-$$

be studied?

24.77 For nickel(II) compounds, oxidation–reduction reactions are unimportant, but the stereochemistry of complexes is of unusual interest. (a) Why are oxidation–reduction reactions unimportant for nickel(II) compounds? (b) Octahedral, tetrahedral, and square planar complexes of nickel(II) are all well known. For example, $[Ni(H_2O)_6]^{2+}$ and $[Ni(NH_3)_6]^{2+}$ are octahedral; $[NiCl_4]^{2-}$, $[NiBr_4]^{2-}$, and $[NiI_4]^{2-}$ are tetrahedral; and $[Ni(CN)_4]^{2-}$ is square planar. (i) Sketch a $[Ni(H_2O)_6]^{2+}$, a $[NiCl_4]^{2-}$, and a $[Ni(CN)_4]^{2-}$ ion. (ii) Draw the d orbital splitting diagram for each of the ions in part (i). (iii) Classify each of the ions in part (i) as paramagnetic or diamagnetic. If the ion is paramagnetic, give the number of unpaired electrons. (c) According to the valence bond theory, what is the hybridization of the central nickel ion in the octahedral and tetrahedral complex ions in part (b) (i)?

24.78 (a) Write the formula for a linkage isomer of $[Co(NCS)(NH_3)_5]Cl_2$. (b) Write the formula for an isomer of $[Co(NO_3)(NH_3)_5]SO_4$ in which the composition of the complex is different. (c) Draw the structures of all the stereoisomers of $[Co(NH_3)_3(NO_2)_3]$.

24.79 Why do the chemical properties of the transition elements vary from element to element whereas the chemical properties of all the lanthanide elements are similar?

24.80 Three stereoisomers of $[Co(en)_2F_2]^+$ exist. One is purplish red, and the other two are yellow. Sketch the three isomers and give the color of each.

24.81 A solution of 0.275 g of a compound with composition $CoBr_2Cl(NH_3)_3(H_2O)_2$ in 5.00 g of water had a freezing point of $-0.60\ °C$. Another solution of 0.275 g of the compound in 5.00 g of water yielded 0.1512 g of AgBr on treatment with excess silver nitrate solution. After standing over a drying agent in a desiccator until the mass became constant, a 2.750-g sample of the compound had a mass of 2.605 g. Write the formula and name of the compound.

24.82 What is the difference between diamagnetism, paramagnetism, and ferromagnetism?

24.83 Sketch the periodic table and label the areas where each of the following are located: (a) alkali metals (b) alkaline earth metals (c) Group IIIA metals (d) transition metals (e) first transition series (f) lanthanides (g) actinides

24.84 Absorption of blue light alone yields an orange solution. Absorption of what two colors of light will also yield an orange solution?

24.85 The strongest absorption band of $[Cr(H_2O)_6]^{3+}$ is centered at 5.22×10^{14} Hz. (a) The electronic transition that takes place in a hydrated chromium ion must involve an energy

change of how many joules? (b) What is the wavelength of this radiation in nanometers? (c) What color is a solution of $[Cr(H_2O)_6]^{3+}$?

24.86 The value of Δ for the $[Cu(H_2O)_6]^{2+}$ ion is 149 kJ/mol. (a) Calculate the wavelength of the light that is absorbed when an electron is raised from the low-energy d orbitals to the high-energy d orbitals. (b) From your calculation, what color is expected for aqueous solutions that contain copper(II) ions?

24.87 The positions of absorption bands are often given as frequencies per centimeter (cm^{-1}). The complex ion, $[Co(H_2O)_6]^{2+}$, has a strong band at $19\ 400\ cm^{-1}$ and a weak band at $16\ 000\ cm^{-1}$. (a) What are the wavelengths of these bands in nanometers? (b) What color is absorbed? (c) What color is a solution of $[Co(H_2O)_6](NO_3)_2$?

24.88 (a) The structures of the chlorides, bromides, and iodides of zinc and cadmium are close-packed arrays of halide ions. Why do zinc ions occupy tetrahedral holes whereas cadmium ions occupy octahedral holes? (b) The species present in aqueous solutions of zinc chloride are $[Zn(H_2O)_6]^{2+}$, $ZnCl^+$, $ZnCl_2$, and $[ZnCl_4(H_2O)_2]^{2-}$. Name each of these species.

24.89 The initial rate of the reaction

$$[Cr(NH_3)_6]^{3+} + H_2O(l) \longrightarrow [Cr(NH_3)_5(H_2O)]^{3+} + NH_3(aq)$$

at 25 °C was measured using different initial concentrations of $[Cr(NH_3)_6]^{3+}$. Results are summarized in the table below: (a) What is the order of the reaction with respect to $[Cr(NH_3)_6]^{3+}$? (b) What is the value of the rate constant k at 25 °C? (c) Why can't the order of the reaction with

Experiment Number	Conc. $[Cr(NH_3)_6]^{3+}$, M	Initial Rate, M/min
1	0.100	9.5×10^{-7}
2	0.050	4.8×10^{-7}

respect to water be measured in aqueous solution? (d) If the value of k is $8.0 \times 10^{-5}\ min^{-1}$ at 40 °C, what is the activation energy for this reaction?

24.90 All the iron(II) halides and iron(III) fluoride, iron(III) chloride, and iron(III) bromide are known, but iron(III) iodide can't be prepared. Explain why.

24.91 (a) The solubility product for zinc hydroxide is about 10^{-11}; however, the equilibrium constant for the equilibrium

$$Zn(OH)_2(s) \rightleftharpoons Zn(OH)_2(aq)$$

is about 10^{-6}. How does the solubility of zinc hydroxide in water compare with the solubility calculated from the solubility product constant? Explain your answer. (b) Write net ionic and complete ionic equations for the solution of zinc hydroxide in (i) aqueous sodium hydroxide (The predominant species in solution is aquatrihydroxozincate(II) ion.) (ii) hydrochloric acid (iii) an excess of concentrated ammonia (The tetraamminezinc(II) ion is formed.) (c)

What are substances that are soluble in both acid and base, such as zinc hydroxide, called?

24.92 (a) Titanium(IV) chloride is a yellow liquid that boils at 136 °C. Is titanium(IV) chloride predominantly ionic or predominantly covalent? Explain why. (b) Titanium(IV) chloride reacts with chloride ion to form the hexachlorotitanate(IV) ion. In this reaction, how is titanium(IV) chloride reacting? (c) Write an equation showing why aqueous solutions that contain hexaaquatitanium(III) ion are acidic. In this reaction, how is the hexaaquatitanium(III) ion reacting?

24.93 A blue coordination compound formed from cobalt(II) chloride and phosphine has the following mass percent composition: 29.9% Co, 35.8% Cl, 3.1% H, and 31.3% P. The molecular mass is 198. (a) What is the molecular formula? (b) This compound does not have any isomers. Sketch a molecule.

24.94 (a) Write equations for the reaction between magnesium oxide and titanium(IV) oxide to form (i) $MgTiO_3$ (ii) Mg_2TiO_4. (b) In the reactions in part (a), which metal oxide is acting as a Lewis base? As a Lewis acid?

24.95 What are the characteristic properties of metals?

24.96 (a) Mercury(I) chloride sublimes at 400 °C; mercury(II) chloride melts at 276 °C. What can you conclude about the strengths of the intermolecular attractions in the two chlorides and about the bonding? (b) Aqueous solutions of mercury(II) chloride do not conduct electricity very well. Suggest a reason. (c) Given the following data for saturated solutions of mercury(I) chloride,

[Mercury(I) ion]	[Cl$^-$]
1.5×10^{-16}	0.1
5.8×10^{-16}	0.05

show that the mercury(I) ion is Hg_2^{2+}, not Hg^+.

24.97 For the equilibrium

chelate complex   macrocylic ligand

macrocyclic complex   polydentate ligand

$K = 1.6 \times 10^5$ at 300 K. (a) Which complex is present in higher concentration at equilibrium? (b) Is ΔG for the forward reaction positive or negative? (c) Is ΔS for the forward reaction positive or negative? Suggest an explanation.

24.98 The complex ion in the compound $Cs[MnF_4(H_2O)_2]$ is octahedral with *trans*-water molecules and is high spin. (a) What is the name of the compound? (b) Sketch the complex ion. (c) What is the oxidation number of Mn in the compound? (d) Draw the d orbital splitting diagram.

24.99 The complex ion $[Cr(H_2O)_6]^{2+}$ has a d orbital electronic transition with energy of 169 kJ/mol. At what wavelength does the $[Cr(H_2O)_6]^{2+}$ ion absorb?

24.100 Explain the following observations. Write equations for all reactions that take place. (a) When copper wire is added to $AgNO_3$(aq), silvery crystals form rapidly, and the solution becomes blue-green. (b) When NH_3(aq) is added to $AgNO_3$(aq), a precipitate of Ag_2O forms and redissolves. When copper wire is added to the ammoniacal solution, silvery crystals form slowly, and the solution gradually becomes deep blue. (c) Hydrated ions such as $[Fe(H_2O)_6]^{3+}$ are very important for elements in the first transition series but are unimportant for elements in the second and third transition series.

24.101 Suggest a reagent that could be used to separate each of the following pairs of cations in aqueous solution:
(a) Ca^{2+} and Zn^{2+} (b) Ag^+ and Cu^{2+} (c) NH_4^+ and Ca^{2+}
(d) Al^{3+} and Mg^{2+} (e) Na^+ and Mg^{2+}.

24.102 Write an equation for each of the following: (a) solution of iron(II) sulfide in hydrochloric acid (b) hydrolysis of iron(III) chloride (c) replacement of copper(II) ion by iron (d) solution of iron metal by hydrochloric acid (e) reaction of iron metal with chlorine gas (f) reaction of iron metal with iodine vapor.

24.103 Define (a) complex, (b) coordination compound, (c) coordinate covalent bond, (d) coordination number, (e) total coordination number.

24.104 The thiocyanate ion SCN$^-$ forms the complex ion $[Zn(NCS)_4]^{2-}$ with zinc ions and the complex ion $[Cd(SCN)_4]^{2-}$ with cadmium ions. (a) What is the difference between the structures of these two complex ions? (b) Suggest a reason for the difference in structure between the two complex ions. (c) Write the Lewis formula (including formal charge) of the thiocyanate ion. (d) Describe the shape of the thiocyanate ion. (e) What is the hybridization of the carbon atom in the thiocyanate ion?

24.105 (a) Name the compound $Na_2[PtCl_6]$, which is used as a catalyst. (b) Write the formula for sodium dicyanoaurate(I), which is used in goldplating.

24.106 Potassium hexanitrocobaltate(III) is only slightly soluble, and formation of a precipitate with sodium hexanitrocobaltate(III) solution is used to confirm the presence of potassium ion. Write the net ionic equation for the precipitation of potassium hexanitrocobaltate(III).

24.107 Explain why oxalic acid removes rust stains.

24.108 The sodium salt of nitrilotriacetic acid, $N(CH_2COONa)_3$, which is biodegradable and does not cause algae to grow, was used instead of phosphates as a builder in synthetic detergents until it was found to increase the water solubility of poisonous heavy metal ions. How many dentate is the nitrilotriacetate ion?

24.109 Organic acids present in soils play an important part in the weathering of rocks. (a) High molecular mass acids are colloidal and absorb cations. How would this affect the solubility equilibria for calcium carbonate, iron(II) carbonate, and manganese(II) carbonate in aqueous carbon dioxide? (b) In addition, organic acids act as chelating agents for cations. How would this affect the equilibria in part (a)?

24.110 The catalyst for the oxo process used to make aldehydes is

$$
\begin{array}{c}
\text{OC} \quad \overset{\displaystyle CO}{\underset{\displaystyle \big|}{\big\uparrow}} \\
\text{OC} \diagup \overset{\displaystyle Co \!-\! H}{\underset{\displaystyle CO}{\big|}}
\end{array}
$$

(a) What is the name of this compound? (b) What is its shape called? (c) Why is CO attached to cobalt by C rather than by O?

24.111 The calcium disodium salt of ethylenediamine tetraacetic acid

$$
\begin{array}{c}
\text{NaOOCCH}_2 \qquad \text{CH}_2\text{CH}_2 \qquad \text{CH}_2\text{COONa} \\
\diagdown \; \text{N} \qquad\qquad\qquad \text{N} \; \diagup \\
\text{H}_2\text{C} \qquad\qquad \text{Ca}^{2+} \qquad\qquad \text{CH}_2 \\
\big| \qquad\qquad\qquad\qquad\qquad \big| \\
\text{C} \qquad\qquad\qquad\qquad\qquad \text{C} \\
\diagup \diagdown \qquad\qquad\qquad\qquad \diagup \diagdown \\
\text{O} \quad :\!\overset{-}{\text{O}}\!: \qquad\qquad -\!:\!\overset{}{\text{O}}\!: \quad \text{O}
\end{array}
$$

is used to treat metal poisoning because the simple sodium salt of ethylenediaminetetraacetic acid ties up serum calcium and causes muscle twitching, cramps, and convulsions. The calcium disodium salt exchanges its calcium for lead (and other heavy metal ions) forming stable water-soluble complexes that are excreted in the urine. (a) Show the structure of the soluble lead disodium chelate. (b) The formula for white lead, which used to be used as a pigment in paints, is $(PbCO_3)_2 \cdot Pb(OH)_2$. If a child ate chips of paint that contained 1.000 g of white lead, what is the minimum number of grams of the calcium disodium chelate of ethylenediaminetetraacetic acid necessary to remove the lead from the child?

24.112 (a) What experimental method do you think was used to determine the positions of the atoms in hemoglobin? (b) One molecule of oxygen is bound to each of the four atoms of iron in a molecule of oxyhemoglobin. How many milliliters of STP oxygen will 1.00 g of hemoglobin absorb? The molecular mass of hemoglobin is 64 500. (c) The oxygen-carrying capacity of a liter of blood is increased from 5 to 230 mL (STP) by the presence of hemoglobin. How many grams of hemoglobin does one liter of blood contain?

24.113 Chlorophyll absorbs light of wavelengths 400–500 nm and 600–700 nm strongly. What color is chlorophyll?

24.114 Manganese carbonyl is used as a catalyst. The empirical formula for manganese carbonyl is MnC_5O_5. (a) What is the oxidation number of manganese in manganese carbonyl? (b) How many electrons would a molecule with formula $Mn(CO)_5$ contain? (c) Suggest an explanation for the fact that the molecular mass of manganese carbonyl is 390. (d) Sketch a molecule of manganese carbonyl.

24.115 (a) For the manufacture of magnetic recording tapes, which type of material (diamagnetic, ferromagnetic, or paramagnetic) is needed? (b) Chromium(IV) oxide and Fe_3O_4 are used in the manufacture of magnetic recording tapes. What can you conclude about their magnetic properties? (c) Write the formula for chromium(IV) oxide. (d) What is the oxidation number of iron in Fe_3O_4?

24.116 Titration with standard potassium dichromate solution is used to determine the percent by mass of iron in iron ore. The ore is usually dissolved in hydrochloric acid. (a) Will potassium dichromate oxidize iron(II) ion under standard conditions? (b) Will potassium dichromate oxidize chloride ion under standard conditions? Explain your answer. (c) Write molecular, complete ionic, and net ionic equations for (i) solution of iron(II) carbonate in hydrochloric acid, (ii) reaction of potassium dichromate solution with aqueous iron(II) chloride in acidic solution. Chromium(III) ions are formed. (d) If titration of a 24.00-g sample of iron ore requires 22.63 mL of 0.016 670 M $K_2Cr_2O_7$, what is the percent by mass of iron in the ore?

24.117 The unit cell of a high-temperature ceramic superconductor is shown below. What is the empirical formula for this superconductor?

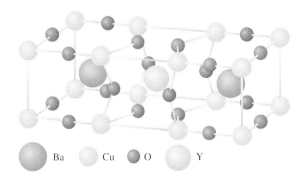

Ba   Cu   O   Y

24.118 Mercury(I) chloride (Hg_2Cl_2) is called calomel and is used in antiseptic salves. Mercury(II) chloride ($HgCl_2$), which is called corrosive sublimate and is a violent poison, is used as a fungicide in agriculture and as a catalyst. (a) Mercury(I) chloride sublimes at 400 °C and is insoluble in organic solvents, whereas mercury(II) chloride melts at 276 °C, boils at 302 °C, and is soluble in organic solvents. What can you conclude about the bonding in mercury(I) and mercury(II) chlorides? (b) At 500 °C and 1 atm, the density of the vapor formed from mercury(I) chloride is 3.72 g/L. What does the molecular formula of the vapor appear to be? Would a compound with this formula be diamagnetic or paramagnetic? (c) The vapor formed from mercury(I) chloride is diamagnetic and absorbs at 254 nm, which is characteristic of mercury vapor. Write the equation for the reaction that takes place when mercury(I) chloride is vaporized. (d) Write complete ionic and net ionic equations for a reaction that could be used to prepare mercury(I) chloride. (e) Suggest an explanation for the observation that the solubility of mercury(II) chloride in water is increased by the presence of chloride ion.

24.119 The compound $HSCH_2CH(SH)CH_2OH$ is known as dimercaprol or British anti-lewisite (BAL). This drug, which was developed to counteract the effects of the deadly war gas lewisite ($ClCH\!=\!CHAsCl_2$), is used to treat poisoning by metals and semimetals such as antimony, arsenic, cadmium, chromium, lead, and mercury that act by combining with —SH groups in cells. (a) Which of these elements are metals and which are semimetals? (b) BAL acts as a bidentate ligand. Sketch the most probable chelate ring formed by BAL^{2-} ions. Explain your answer. (c) Mark the chiral carbon in the formula for BAL with an asterisk. (d) BAL is dissolved in peanut oil for intramuscular injection because it is insoluble in water. Explain why BAL is insoluble although the corresponding oxygen-containing compound glycerol, $HOCH_2CH(OH)CH_2OH$, is miscible with water.

Nomenclature

Only the nomenclature of simple inorganic ions and compounds is discussed in this appendix. The nomenclature of complex ions and compounds is treated in Section 24.6 and the nomenclature of organic compounds in Sections 22.6 and 22.7.

The goal of any type of nomenclature is for each compound to have a distinctive name; each name should represent a single compound. In a systematic method of nomenclature, compounds with similar structures have similar names. Because a compound's name gives information about its structure, only a few rules, and not millions of individual names, need be memorized. A committee of the International Union of Pure and Applied Chemistry (IUPAC) is responsible for the official rules for naming elements and compounds, and these rules are revised from time to time. The most recent version of the rules was published in 1990.* The rules are a compromise, and chemists of each country use slightly different versions. More complete summaries of the United States version are given in Jolly, W. L. *Modern Inorganic Chemistry,* 2nd ed.; McGraw-Hill, New York, 1991; pp 598–607 and Huheey, J. E.; Keiter, E. A.; Keiter, R. L. *Inorganic Chemistry: Principles of Structure and Reactivity,* 4th ed.; HarperCollins, New York; 1993; pp A-46–77.

Cations

Metal cations are called by the name of the metal followed by ion, for example,

$$Na^+ \quad \text{sodium ion}$$
$$Mg^{2+} \quad \text{magnesium ion}$$
$$Al^{3+} \quad \text{aluminum ion.}$$

The Stock System (Roman numerals) is used to show which of more than one possible ion of a transition metal or metal following the transition metals in the periodic table is meant:

$$Fe^{2+} \quad \text{iron(II) ion}$$
$$Fe^{3+} \quad \text{iron(III) ion}$$
$$Hg_2^{2+} \quad \text{mercury(I) ion}$$
$$Hg^{2+} \quad \text{mercury(II) ion}$$
$$Sn^{2+} \quad \text{tin(II) ion}$$
$$Sn^{4+} \quad \text{tin(IV) ion}$$

Table A.1 lists Roman numerals and their number equivalents.

Systematic names are used in this book, but you need to be aware that common names such as ferrous ion for Fe^{2+}, ferric ion for Fe^{3+}, mercurous ion for Hg_2^{2+}, mercuric ion for Hg^{2+}, stannous ion for Sn^{2+}, and stannic ion for Sn^{4+} exist because you may come across these names elsewhere. If the symbol for the element comes from the Latin name, as do the symbols for iron, mercury, and tin, the common name usually comes from the Latin name, too. (Mercurous and mercuric ions are

TABLE A.1

Roman Numerals

Numeral	Number
I	1
II	2
III	3
IV	4
V	5
VI	6
VII	7
VIII	8

*See *Nomenclature of Inorganic Chemistry*; Blackwell Scientific Publications: Oxford, 1990.

TABLE A.2 Nonsystematic Names of Some Common Cations

Formula	Name	Formula	Name
Au^+	Aurous	Au^{3+}	Auric
Ce^{3+}	Cerous	Ce^{4+}	Ceric
Co^{2+}	Cobaltous	Co^{3+}	Cobaltic
Cr^{2+}	Chromous	Cr^{3+}	Chromic
Cu^+	Cuprous	Cu^{2+}	Cupric
Fe^{2+}	Ferrous	Fe^{3+}	Ferric
Hg_2^{2+}	Mercurous	Hg^{2+}	Mercuric
Mn^{2+}	Manganous	Mn^{3+}	Manganic
Pb^{2+}	Plumbous	Pb^{4+}	Plumbic
Sn^{2+}	Stannous	Sn^{4+}	Stannic

exceptions.) The old, nonsystematic names for a number of common cations are given in Table A.2.

The common names of cations do not provide for metals such as titanium, vanadium, chromium, and manganese that have more than two oxidation numbers in compounds. In addition, although the ending -ous always shows the lower oxidation number, you have to remember the ionic charges; for example, cuprous ion is Cu^+, ferrous ion is Fe^{2+}, and cerous ion is Ce^{3+}.

Roman numerals are not necessary for ions of Group IA and IIA metals and aluminum, nor for the transition-metal ions Ag^+, Cd^{2+}, Sc^{3+}, and Zn^{2+} because only one type of ion of each of these metals ordinarily exists. The ion Hg_2^{2+} is called mercury(I) ion because the average charge of *each* mercury atom in this ion is $1+$.

The only common nonmetal cations are

$$H^+ \quad \text{hydrogen ion}$$
$$H_3O^+ \quad \text{hydronium ion}$$
$$NH_4^+ \quad \text{ammonium ion}$$

Anions

Names of anions end in -ide, -ate, and -ite. Most ions with names ending in -ide are monatomic; the names of anions of oxo acids end in -ate and -ite. The names of a number of common anions are listed alphabetically in Table A.3.

Remembering that the rules of nomenclature are written so that similar compounds have similar names, you can use the names and formulas in Table A.3 to figure out the names and formulas of many other anions as shown in Sample Problem A.1.

SAMPLE PROBLEM

A.1 (a) What is the name of the HS^- ion? (b) What is the formula for the arsenate ion?

Solution (a) From Table A.3, the sulfide ion is S^{2-}. The hydrogen carbonate, hydrogen sulfate, and hydrogen sulfite ions all have one more hydrogen and one less negative charge than the carbonate, sulfate, and sulfite ions. The HS^- ion has one more hydrogen and one less negative charge than the sulfide ion, so it is called the hydrogen sulfide ion. (b) Arsenic is just below phosphorus in Group VA of the periodic table. From the table of common anions, the formula for the phosphate ion is PO_4^{3-}; therefore, the formula for the arsenate ion is AsO_4^{3-}.

Name	Formula	Name	Formula	Name	Formula
Acetate ion[a]	CH_3COO^- or $C_2H_3O_2^-$	Dihydrogen phosphate ion	$H_2PO_4^-$	Oxalate ion[a]	$^-OOCCOO^-$ or $C_2O_4^{2-}$
		Fluoride ion	F^-		
Amide ion	NH_2^-	Hydride ion	H^-	Oxide ion	O^{2-}
Azide ion	N_3^-	Hydrogen carbonate ion	HCO_3^-	Perchlorate ion	ClO_4^-
Borate ion	BO_3^{3-}	or bicarbonate ion[d]		Permanganate ion	MnO_4^-
Bromide ion	Br^-	Hydrogen phosphate ion	HPO_4^{2-}	Peroxide ion	O_2^{2-}
Carbide ion	$C_2^{2-a,b}$ or C^{4-c}	Hydrogen sulfate ion or bisulfate ion[d]	HSO_4^-	Phosphate ion	PO_4^{3-}
				Silicate ion	SiO_4^{4-}
Carbonate ion	CO_3^{2-}	Hydrogen sulfite ion or	HSO_3^-	Sulfate ion	SO_4^{2-}
Chlorate ion	ClO_3^-	bisulfite ion[d]		Sulfide ion	S^{2-}
Chloride ion	Cl^-	Hydroxide ion	OH^-	Sulfite ion	SO_3^{2-}
Chlorite ion	ClO_2^-	Hypochlorite ion	ClO^-	Thiocyanate ion[e]	SCN^-
Chromate ion	CrO_4^{2-}	Iodide ion	I^-	Thiosulfate ion[e]	$S_2O_3^{2-}$
Cyanate ion	OCN^-	Nitrate ion	NO_3^-	Triiodide ion	I_3^-
Cyanide ion	CN^-	Nitride ion	N^{3-}		
Dichromate ion	$Cr_2O_7^{2-}$	Nitrite ion	NO_2^-		

[a] A few common organic ions are included.
[b] Calcium carbide (CaC_2) is the most important example of this type of carbide.
[c] Silicon carbide (SiC), beryllium carbide (Be_2C), and (Al_4C_3) are examples.
[d] These common names are not used in this book but may be found elsewhere.
[e] Thio means that a sulfur has been substituted for an oxygen.

PRACTICE PROBLEM

A.1   (a) What is the name of the BrO^- ion? (b) What is the formula for the phosphide ion?

Compounds of Metals

Whether predominantly ionic or predominantly covalent, compounds of metals are named by giving the name of the metal followed by the name of an anion:

Na_2SO_4	sodium sulfate
Cu_2O	copper(I) oxide
CuO	copper(II) oxide
$AlCl_3$	aluminum chloride
$SnCl_4$	tin(IV) chloride

Binary Compounds of Nonmetals

Binary compounds of nonmetals are usually named by giving the name of the less electronegative nonmetal followed by the name of the anion of the more electronegative nonmetal:

HCl	hydrogen chloride
H_2S	hydrogen sulfide

However, if more than one compound of a nonmetal with another nonmetal exists, Greek prefixes, not Roman numerals, are used to show which compound is meant. Table A.4 lists Greek prefixes up to ten.

Common Greek Prefixes

Prefix	Number
Mono-	1
Di-	2
Tri-	3
Tetra-	4
Penta-	5
Hexa-	6
Hepta-	7
Octa-	8
Nona-	9
Deca-	10

The electronegativities of the elements are given in Figure 9.3.

Some examples of names of binary compounds of nonmetals are

CO	carbon monoxide
CO_2	carbon dioxide
PI_3	phosphorus triiodide
ICl	iodine monochloride
IF_5	iodine pentafluoride
N_2O_4	dinitrogen tetroxide

Spelling is important when names must be located in an index containing over 13 million compounds.

Note that mono- is usually used only before the name of the second element. The final -*o* of mono- and the final -*a* of any prefix that ends in -*a*, such as tetra-, are dropped before *o*-; however, the final -*i* of di- and tri- is kept before *i*-.

Common names, which must be memorized, are usually used for the following compounds:

H_2O	water
NH_3	ammonia
H_2NNH_2	hydrazine
NO	nitric oxide
N_2O	nitrous oxide
PH_3	phosphine (Hydrogen compounds of the other Group VA elements, arsenic and antimony, are called arsine and stibnine.)
CH_4	methane (The names of the hydrogen compounds of the other Group IVA elements and of boron also end in -ane: SiH_4 is called silane; GeH_4, germane; SnH_4, stannane; and B_2H_6, diborane.)

Acids

In inorganic chemistry, formulas of acids are usually written with the acidic hydrogens first. Some examples are HCl, H_2SO_4, H_3PO_4, and $HC_2H_3O_2$. Although pure HCl(g), which is a molecular compound, is called hydrogen chloride, aqueous solutions of hydrogen chloride, HCl(aq), which contain ions, are known as hydrochloric acid. Other hydrogen _____-ides are named similarly; for example, HCN(l) is called hydrogen cyanide, but HCN(aq) is known as hydrocyanic acid. Oxo acids such as sulfuric acid, H_2SO_4, have the same name both pure and in aqueous solution.

If two kinds of oxo acid are known, the more common (more stable under ordinary conditions) is named _____-ic acid; the acid with one less oxygen atom is called _____-ous acid. For example, H_2SO_4 is sulfuric acid and H_2SO_3 is sulfurous acid. Notice that -ic acids correspond to -ate ions and -ous acids correspond to -ite ions.

If more than two kinds of oxo acid are known, the one with one more oxygen atom than the _____-ic acid is called per-_____-ic acid, and the one with one less oxygen atom than the _____-ous acid is called hypo-_____-ous acid. The oxo acids of chlorine provide a good example:

Acid		Conjugate Base	
$HClO_4$	perchloric acid	ClO_4^-	perchlorate ion
$HClO_3$	chloric acid	ClO_3^-	chlorate ion
$HClO_2$	chlorous acid	ClO_2^-	chlorite ion
HClO	hypochlorous acid	ClO^-	hypochlorite ion

Ions with different numbers of hydrogens formed by stepwise neutralization of polyprotic acids are named to show the number of hydrogens:

$H_2PO_4^-$	dihydrogen phosphate ion
HPO_4^{2-}	hydrogen phosphate ion
PO_4^{3-}	phosphate ion

Hydrates

Compounds that contain water of crystallization can be named in one of two ways:

$CuSO_4 \cdot 5H_2O$	copper(II) sulfate pentahydrate
	copper(II) sulfate-water (1/5)
$2NaOCl \cdot 5H_2O$	sodium hypochlorite-water (2/5)

PRACTICE PROBLEM

A.2   Name each of the following: (a) Mn^{2+}, (b) $Hg(NO_3)_2$, (c) N_2O_3, (d) P_2O_5, (e) HIO_4, (f) KIO_3, (g) $LiMnO_4 \cdot 3H_2O$, (h) $HBr(g)$, (i) $HBr(aq)$, (j) $3CdSO_4 \cdot 8H_2O$, and (k) $MgSO_4 \cdot 7H_2O$ (two ways).

Mathematics Needed for General Chemistry

For general chemistry, you will need to know how to (1) use a handheld calculator, (2) use exponential notation, (3) use logarithms, (4) solve simple algebraic equations, (5) use percent, and (6) make and read graphs. This appendix reviews each of these skills. The geometric formulas needed to solve the problems in the text are given in the last section (B.7).

B.1   HANDHELD CALCULATOR

To learn how to operate your calculator, refer to its instruction manual.

When you use a calculator to carry out a series of calculations, you should round only once, after the last calculation (if possible). For example, to find the value of

$$\frac{3.82 \times 6.75}{4.91 \times 2.04}$$

Adopters of this text may request from the publisher a videotape produced for the text that advises students about the selection of a scientific calculator and teaches them how to use one efficiently.

you should multiply 3.82 by 6.75, divide the product first by 4.91 and then by 2.04, and then round off. Rounding after each step of a series can lead to a significant error.

Remember that a calculator does not display zeros at the end of a number if they are to the right of the decimal point. For example, according to many calculators,

$$7.0476/2.3492 = 3.$$

The correct answer is 3.0000.

A calculator does not round to the correct number of significant figures. For example, according to many calculators,

$$2.0/6.0 = 0.333333333$$

The correct answer, 0.33, has only two significant figures. It is your job to express each answer to the correct number of significant figures.

B.2   EXPONENTIAL NOTATION

Because many of the numbers used in science are either very large or very small, chemists and other scientists often use exponential notation. In exponential notation, a number is expressed as a number (usually a number between 1 and 10) times a power of 10; for example,

$$2425.3 = 2.4253 \times 10^3$$

where 10^3 means $10 \times 10 \times 10$. If no power of 10 is given, the power of 10 is assumed to be 0. Any number to the zeroth power equals 1:

$$10^0 = 1$$

therefore,

$$6.82 = 6.82 \times 10^0$$

Interconverting Ordinary Decimal Notation and Exponential Notation

When you go back-and-forth between ordinary decimal notation and exponential notation, you are simply writing the same number in a different way; therefore, the *value of the number must remain the same.* When converting a decimal to exponential notation, for each place the decimal point is moved to the left, the number must be multiplied by 10:

$$2425.3 = 2.4253 \times 10 \times 10 \times 10 = 2.4253 \times 10^3$$

If the number is made smaller by moving the decimal point to the left, the power of 10 must be made larger (more positive or less negative). For each place the decimal point is moved to the right, the number must be multiplied by 1/10:

$$0.0136 = 1.36 \times 0.1 \times 0.1 = 1.36 \times \frac{1}{10} \times \frac{1}{10} = 1.36 \times \frac{1}{10^2} = 1.36 \times 10^{-2}$$

If a number is made larger by moving the decimal point to the right, the power of 10 must be made smaller (more negative or less positive). Sample Problems B.1 and B.2 give more examples.

B.1 Write each of the following numbers in exponential notation:
(a) 7 653 192 and (b) 0.000 000 96.

Solution (a) To write 7 653 192 as a number between 1 and 10, the decimal point (which is not shown in the original number) must be moved six places to the left:

$$7\ 653\ 192 = 7\ 653\ 192. = 7.653\ 192 \times 10 \times 10 \times 10 \times 10 \times 10 \times 10$$
$$= 7.653\ 192 \times 10^6$$

Notice that the exponent of 10 is equal to the number of places the decimal point is moved. When a number ≥ 10 is written in exponential notation, the power of 10 is positive.

(b) To write 0.000 000 96 as a number between 1 and 10, the decimal point must be moved seven places to the right:

$$0.000\ 000\ 96 = 9.6 \times 0.1 \times 0.1 \times 0.1 \times 0.1 \times 0.1 \times 0.1 \times 0.1$$
$$= 9.6 \times 10^{-7}$$

Again, the exponent of 10 is equal to the number of places the decimal point is moved; however, when a number <1 is written in exponential notation, the power of 10 is negative.

B.2 Write each of the following numbers in ordinary decimal notation:
(a) 6.83×10^{-4} and (b) 5.4×10^5.

Solution (a) The power of 10, -4, means multiply by 0.1 four times,

$$6.83 \times 10^{-4} = 6.83 \times 0.1 \times 0.1 \times 0.1 \times 0.1 = 0.000\ 683$$

or, in other words, move the decimal point four places to the left:

$$6.83 \times 10^{-4} = 0.000\ 683$$

(b) The power of 10, 5, means multiply by 10 five times,

$$5.4 \times 10^5 = 5.4 \times 10 \times 10 \times 10 \times 10 \times 10 = 540\ 000,$$

or, in other words, move the decimal point five places to the right:

$$5.4 \times 10^5 = 540\ 000$$

Study these examples carefully so that you see the relationship between the number of places the decimal point is moved and the power of 10. Then do Practice Problems B.1 and B.2 both by hand and with your calculator. Answers to the problems in this appendix are given at the end of Appendix I.

B.1 Write each of the following numbers in exponential notation:
(a) 365.2, (b) 0.0007, (c) 0.0142, (d) 23 652, and (e) 7.5.

B.2 Write each of the following numbers in ordinary decimal notation:
(a) 7.4×10^{-7}, (b) 3.982×10^1, (c) 3.27×10^0, (d) 5.81×10^{-3}, and
(e) 7.456×10^8.

The number of significant figures in numbers such as 300 and 528 000 is uncertain. Exponential notation can be used to show the number of significant figures. To show that there are three significant figures in the number 300, write 3.00×10^2; to show two significant figures, write 3.0×10^2; to show one significant figure, write 3×10^2. The number 528 000 may have three, four, five, or six significant figures:

$$5.280\ 00 \times 10^5 \text{ has six significant figures}$$
$$5.2800 \times 10^5 \text{ has five significant figures}$$
$$5.280 \times 10^5 \text{ has four significant figures}$$
$$5.28 \times 10^5 \text{ has three significant figures}$$

If you have trouble telling how many significant figures there are in numbers like 0.002 50, write the number in exponential notation:

$$0.002\ 50 = 2.50 \times 10^{-3}$$

Thus, there are three significant figures in 0.002 50. The 0s on the left-hand side are not significant; they are necessary to show the magnitude of the number. The 0 on the right-hand side is significant; if it were not, it would be omitted and the number would be written 0.0025.

PRACTICE PROBLEMS

B.3   Rewrite 3200 to show the different numbers of significant figures that are possible.

B.4   How many significant figures are there in 0.060 700?

Multiplication of Exponential Numbers

To multiply exponential numbers, multiply the numbers between 1 and 10 and add the exponents to find the power of 10 in the product:

$$(2.3 \times 10^5)(3.42 \times 10^8)(6.1 \times 10^{-4}) = (2.3 \times 3.42 \times 6.1) \times \{10^{[5+8+(-4)]}\}$$
$$= 48 \times 10^9 = 4.8 \times 10^{10}$$

Notice that the product of the numbers, 47.9826, is rounded to two significant figures because there are only two significant figures in 2.3 and 6.1. In multiplication and division, the number of significant figures in the answer is limited by the number or numbers that have the smallest number of significant figures. The product is written 4.8×10^{10}—not 48×10^9—because, in exponential notation, the number should usually be between 1 and 10.

To convert a number from one power of 10 to another, remember that *the value of the number must not be changed when the number is rewritten*. If the number is made smaller, the power of 10 must be made larger by the same factor. If the number is made larger, the power of 10 must be made smaller by the same factor. For example, to convert 48 to 4.8, the number 48 must be multiplied by 0.1, or 10^{-1}; to keep the value of the number the same, the power of 10 must be multiplied by 10, or 10^{+1}. Any number to the zeroth power is equal to 1:

$$(10^{-1}) \times (10^{+1}) = 10^0 = 1$$

Multiplication by 1 does not change the value of a number.

B.3 The number 6.34×10^{-6} is equal to what number times 10^{-3}?

Solution To convert 10^{-6} to 10^{-3}, multiply by 10^{+3}; to keep the value of the number the same, multiply 6.34 by 10^{-3}:

$$6.34 \times 10^{-3} = 0.006\ 34$$

Therefore

$$6.34 \times 10^{-6} = 0.006\ 34 \times 10^{-3}$$

Check: Remember to check your answer.

$6.34 \times 10^{-6} = 0.000\ 006\ 34$ and $0.006\ 34 \times 10^{-3}$ is also equal to $0.000\ 006\ 34$.

◗ Always check your work.

B.5 The number 4.9×10^8 is equal to what number times 10^6?

B.6 What is the value of the following product?

$$(8.22 \times 10^7)(3.874 \times 10^{-9})(7.629 \times 10^3)$$

Division of Exponential Numbers

To divide exponential numbers, divide the numbers between 1 and 10, and subtract the sum of the powers of 10 in the denominator from the sum of the powers of 10 in the numerator to find the power of 10 in the quotient:

$$\frac{(4.5 \times 10^8)(6.32 \times 10^6)}{(8.41 \times 10^{10})(9.63 \times 10^{-4})} = \frac{(4.5 \times 6.32)}{(8.41 \times 9.63)} \times 10^{[(8+6)-[10+(-4)]]}$$

$$= 0.35 \times 10^8 = 3.5 \times 10^7$$

B.7 What is the value of each of the following quotients?

(a) $\dfrac{(9.6 \times 10^8)}{(3 \times 10^3)}$ (b) $\dfrac{(9.42 \times 10^7)}{(2.8 \times 10^{-6})}$ (c) $\dfrac{(8.6 \times 10^{-4})}{(2.3 \times 10^9)}$ (d) $\dfrac{(4.7 \times 10^{-6})}{(7.5 \times 10^{-10})}$

(e) $\dfrac{(8.76 \times 10^{14})(5.8 \times 10^9)}{(2.63 \times 10^{22})(4.54 \times 10^{12})}$

Addition and Subtraction of Exponential Numbers

Exponential numbers that are to be added or subtracted must first be written with the same power of 10. Sample Problem B.4 illustrates the addition of exponential numbers.

◗ To find the sum of one dollar and two cents, you would first either convert two cents to $0.02 and then add—$1.00 + $0.02 = $1.02—or you would convert the one dollar to 100 cents and add—100 cents + 2 cents = 102 cents. You would not simply add one dollar to two cents and get three dollars or three cents.

B.4   What is the sum of (3.2×10^8), (2.4×10^9), and (5.63×10^{10})?

Solution   Convert the other number or numbers to the same power of 10 as the largest number to be added or subtracted, 5.63×10^{10} in this example:

$$3.2 \times 10^8 = 0.032 \times 10^{10}$$
$$2.4 \times 10^9 = 0.24 \times 10^{10}$$
$$\underline{5.63 \times 10^{10}}$$
$$5.902 \times 10^{10} = 5.90 \times 10^{10}$$

Because the last two numbers have only two decimal places, the answer can have only two decimal places.

B.8   Write the answer to each of the following problems in exponential notation: (a) $(4.91 \times 10^7) + (3.62 \times 10^6)$, (b) $(5.14 \times 10^{23}) - (8.29 \times 10^{22})$, and (c) $(6.9 \times 10^{-13}) - (7.2 \times 10^{-14})$.

Powers and Roots

Raising an exponential number to a power is similar to multiplication: Find the power of the number between 1 and 10 and multiply the exponent of 10 by the power. For example,

$$(3.8 \times 10^7)^3 = (3.8)^3 \times 10^{7(3)}$$
$$= 55 \times 10^{21} = 5.5 \times 10^{22}$$

To find a root of an exponential number by hand, first rewrite the number so that the power of 10 is divisible by the root. For example, to take the square root of 4.9×10^9, first rewrite 4.9×10^9 so that the power of 10 is divisible by 2:

$$\sqrt{4.9 \times 10^9} = \sqrt{49 \times 10^8}$$

Then, take the square root of the number and divide the exponent by 2:

$$\sqrt{49 \times 10^8} = 7.0 \times 10^4$$

Be careful to show the correct number of significant figures in the root.

B.9   (a) What is the fourth power of 2.3×10^2?   (b) Find the square root of 8.10×10^{13}.   (c) Find the cube root of 6.4×10^{-11}.

B.3   LOGARITHMS

Logarithms are exponents. Two kinds of logarithms are used in general chemistry, the **common logarithm,** which is referred to as *log,* and the **natural logarithm,** which is referred to as *ln.* The logarithms used in pH are common logarithms, or logarithms to the base 10; that is, they are the power to which 10 must be raised to give a number. For example, $3 = \log (1000)$ because $10^3 = 1000$ and $-3 = \log (0.001)$ because $10^{-3} = 0.001$. The **antilogarithm** of a number x is the number that has x as its logarithm; that is, the antilog of 6 is 10^6, and the antilog of -6 is 10^{-6}. The antilogarithm is referred to as *antilog*.

PRACTICE PROBLEMS

B.10   Find the log of (a) 10^5 and (b) 10^{-4}.

B.11   Find the antilog of (a) -8 and (b) 15.

For most numbers, the power of 10 is not a whole number. Consult the instruction book for your calculator to learn how to find the logs of numbers such as 2762 and 0.002 762 and the antilogs of numbers such as 6.348 and -6.348 (log 2762 = 3.4412, log 0.002 762 = -2.5588, antilog 6.348 = 2.23×10^6, and antilog $-6.348 = 4.49 \times 10^{-7}$). The number of digits to the right of the decimal point in a logarithm should be equal to the number of significant figures as shown in the preceding examples.

PRACTICE PROBLEM

B.12   Use your calculator to find the value of each of the following expressions:
(a) log (3×10^{-12}), (b) log (8.42×10^{16}), (c) log 1, (d) antilog (-22.498), and
(e) antilog (5.7).

Except for the pH scale, this book uses only natural logarithms, or logarithms to the base e ($e = 2.7182 \ldots$):

The CRC Handbook gives the value of e to 50 decimal places!

$$\ln x = e^x$$

The number e occurs often in the mathematical relationships used in chemistry and physics. The Arrhenius equation (equation 18.8a) is an example. Consult the instruction book for your calculator to learn how to find the ln of numbers such as 2762 and 0.002 762 and the antiln of numbers such as 6.348 and -6.348 (ln 2762 = 7.9237, ln 0.002 762 = -5.8918, antiln 6.348 = 571, antiln -6.438 = 0.001 60). To convert a logarithm to the base 10 to a natural logarithm, multiply by 2.303:

$$\ln x = 2.303 \log x$$

PRACTICE PROBLEM

B.13   Use your calculator to find the value of each of the following expressions:
(a) ln (3×10^{-12}), (b) ln (8.42×10^{16}), (c) ln 1, (d) antiln (-22.498), and
(e) antiln (5.7).

Because natural logarithms are exponents, the same rules apply for multiplication and division:

$$\ln (a \times b) = \ln a + \ln b \qquad \text{and} \qquad \ln \frac{a}{b} = \ln a - \ln b$$

Since $\ln 1 = 0$,

$$\ln \frac{1}{a} = -\ln a$$

Also,

$$\ln a^b = b \ln a$$

These relationships apply to common logarithms as well.

B.4   SOLVING EQUATIONS

"Returning" students find a brief reminder about how to solve equations helpful.

Equations that contain only one unknown can be solved for the unknown by rearranging the equation so that the unknown is on one side and the knowns are on the other side. When an equation is rearranged, *whatever is done to one side of the equation must be done to the other side.* For example, to solve the equation

$$2.43 = \frac{x}{6.97}$$

multiply each side of the equation by 6.97 and simplify:

$$6.97 \times 2.43 = 6.97 \times \frac{x}{6.97}$$

or

$$6.97 \times 2.43 = x = 16.9$$

SAMPLE PROBLEM

B.5   Solve the following equations for x: (a) $3.54 = \dfrac{7.08}{x}$ and (b) $4.65 = x + 8.19$.

Solution   (a) To solve the equation $3.54 = \dfrac{7.08}{x}$ for x, multiply each side by x,

$$x \cdot 3.54 = x \cdot \frac{7.08}{x}$$

divide each side by 3.54, and simplify:

$$\frac{x \cdot 3.54}{3.54} = \frac{x \cdot 7.08}{x \cdot 3.54}$$

$$x = \frac{7.08}{3.54} = 2.00$$

(b) To solve the equation $4.65 = x + 8.19$ for x, subtract 8.19 from each side:

$$(4.65 - 8.19) = (x + 8.19) - 8.19$$

$$-3.54 = x$$

PRACTICE PROBLEM

B.14   Solve the following equations for x:
(a) $6.9 = \dfrac{2.0}{x}$, (b) $2.84 = \dfrac{x}{7.82}$, and (c) $5.82 = x - 6.43$.

Logarithms are useful for solving equations such as

$$3^x = 2 \qquad \text{and} \qquad x = 7^{0.25}$$

Sample Problems B.6 and B.7 show these types of calculations.

B.6 If $3^x = 2$, what is the value of x?

Solution To find the value of x in the equation $3^x = 2$, take the ln of each side:

$$x \cdot \ln 3 = \ln 2$$

Rearrangement gives

$$x = \frac{\ln 2}{\ln 3} = \frac{0.7}{1.1} = 0.6$$

B.7 If $x = 7^{0.25}$, what is the value of x? (Note that $x = 7^{0.25}$ means $x = 7^{1/4}$ or $\sqrt[4]{7}$.)

Solution Take the logarithm of each side:

$$\ln x = 0.25 \ln 7 = 0.25(1.9) = 0.48$$

Then solve for x by taking the antiln of each side:

$$x = \text{antiln} (0.48) = 1.6$$

B.15 (a) If $2.4^x = 8.6$, what is the value of x? (b) If $x = 6^{0.50}$, what is the value of x? (c) What is $\sqrt[8]{3.00}$?

Quadratic Equations

A *quadratic equation* is an equation in which the highest power of x is 2. Quadratic equations can be rearranged to the form

$$ax^2 + bx + c = 0$$

where a, b, and c are constants and $a \neq 0$, and then solved by means of the quadratic formula:

$$x = \frac{-b \pm \sqrt{b^2 - 4ac}}{2a}$$

For example, in Sample Problem 14.9 the following quadratic equation must be solved:

$$\frac{(2x)^2}{(0.0367 - x)} = 4.64 \times 10^{-3} \qquad (14.10)$$

where $x = $ mol N_2O_4 that have decomposed (in one liter) when the system has reached equilibrium. Equation 14.10 must first be put in the standard form for a quadratic equation; multiplication of each side by $(0.0367 - x)$ gives

$$(2x)^2 = (0.0367 - x)(4.64 \times 10^{-3})$$

or

$$(2x)^2 = (1.70 \times 10^{-4}) - (4.64 \times 10^{-3})x$$

Rearrangement gives

$$4x^2 + (4.64 \times 10^{-3})x - (1.70 \times 10^{-4}) = 0 \qquad (14.11)$$

Substitution of the coefficients from equation 14.11 in the quadratic formula gives

$$x = \frac{-(4.64 \times 10^{-3}) \pm \sqrt{(4.64 \times 10^{-3})^2 - 4(4)(-1.70 \times 10^{-4})}}{2(4)}$$

and the two solutions (or roots) are

$$x = 5.96 \times 10^{-3} \quad \text{and} \quad x = -7.12 \times 10^{-3}$$

The second solution, -7.12×10^{-3}, is physically impossible because $2x = [NO_2]_{eq}$, which can't be negative (less than zero). The answer is

$$x = 5.96 \times 10^{-3}$$

Check: Substitution of this value for x in the left-hand side of equation 14.10 gives 4.64×10^{-3}; the answer is correct.

○ Remember to stop and check your work.

○ Be especially careful not to round off until you reach the final answer or significant figures may be lost. If you have a programmable calculator, you can program it to solve quadratic equations by following the instructions in the user's manual.

PRACTICE PROBLEM

B.16 Use the quadratic formula to solve the following equation:

$$\frac{x^2}{(0.050 - x)} = 3.9 \times 10^{-2}$$

Higher Order Equations

Equations in which the highest power of x is greater than 2 are most easily solved by successive approximations (see Sample Problem 14.9). Whatever information is available should be used to make the best possible first approximation.

B.5 PERCENT

Many students don't really seem to understand percent and will turn in absurd answers such as 359% copper in a copper compound. They have particular difficulty finding the mass of the sample.

Percent is used often in general chemistry. **Percent** means parts per hundred; a general definition of percent is

$$\text{percent, \%} = \frac{\text{part}}{\text{whole}} \times 100$$

If all but one of the quantities in the definition of percent are known, the definition can be solved for the unknown variable as shown in Sample Problem B.8.

SAMPLE PROBLEM

B.8 If a sample of copper ore that is 2.76% copper contains 3.59 g of copper, what is the mass of the sample?

Solution The problem gives the percent copper in the ore (2.76%) and the part of the sample that is copper (3.59 g). The problem asks for the mass of the whole sample. The definition of percent

$$\text{percent, \%} = \frac{\text{part}}{\text{whole}} \times 100$$

must be solved for the mass of the whole sample:

$$\text{mass of } \textit{whole} \text{ sample} = \frac{\textit{part} \text{ of the sample that is copper}}{\textit{percent} \text{ of the sample that is copper}} \times 100$$

$$= \frac{3.59 \text{ g}}{2.76\%} \times 100 = 130 \text{ g}$$

Check: This answer is reasonable because the mass found for the whole sample is greater than the mass of the part of the sample that is copper:

$$\frac{3.59 \text{ g}}{130 \text{ g}} \times 100 = 2.76\%$$

> Always check to be sure that your work is reasonable.

PRACTICE PROBLEM

B.17 (a) If 37.97 g of a 52.40-g sample of an oxide of iron are iron, what is the percent iron in the sample? (b) If a 67.6-g sample of an iron ore is 9.31% iron, how many grams of iron are in the sample? (c) If a sample of iron ore that is 24.6% iron contains 7.18 g of iron, what is the mass of the sample?

B.6 GRAPHS

Graphs are scale drawings that present facts in picture form. **Line graphs** are the most commonly used graphs in chemistry because of the advantages they offer:

1. Line graphs show the relationships between variables (quantities that can have more than one numerical value).
2. Incorrect data are often obvious and can be ignored when drawing the line.
3. Predictions can be made from the graph about how the variables will behave at values for which data have not been recorded.
4. Values for quantities that are not easy to measure, such as slope, can be obtained from the graph.

Skill in making and using graphs is important not only in chemistry but in all sciences and in many other fields as well.

Although graphs are often drawn with computers or calculators, we feel that students still need to learn how to draw them because computers and calculators often choose scales that make graphs difficult to read.

Making a Line Graph

The first step in making a graph is to tabulate the data that are to be graphed. Imagine that a student has filled a container with warm water, placed a thermometer in the warm water, and recorded the temperature of the water at regular time intervals as it was cooled at a constant rate to obtain the data in Table B.1.

After the data have been tabulated, the next step in making a graph is to decide which is the independent and which is the dependent variable. Time is the variable that the student controlled by choosing when to record the temperature; therefore, time is the **independent variable.** The independent variable should be shown on the horizontal axis, or x-axis of the graph. The temperature observed by the student depended on the times at which temperature was observed. In this experiment, temperature is the **dependent variable** and should be placed on the vertical axis, or y-axis of the graph.

The third step in making a graph is to decide how many units the divisions along each axis are to represent. The scale should be chosen to make the points easy to

TABLE B.1

Time Required for Water to Cool

Time, s	Temperature Water, °C
0	46.1
100[a]	45.2
200	44.4
300	43.0
400	41.8
500	40.8

[a]Zeros are significant.

Adopters of this text may request from the publisher a videotape produced for this text that presents the principles of scientific graphing and guides students through several examples.

plot and the graph easy to read. In addition, the graph should fill more than half the graph paper in each direction.

To return to our example, let's first consider the horizontal axis, or *x*-axis, which is to be used for the independent variable, time. The difference between the starting time and the longest time at which the temperature was observed—500 s—must be shown on the *x*-axis. There are 38 divisions on the *x*-axis to the right of the *y*-axis in ▪Figure B.1. (Count them!) To find the minimum number of seconds that can be shown by each division on the *x*-axis, divide 500 s by 38 divisions:

$$\frac{500 \text{ s}}{38 \text{ divisions}} = 13 \text{ s/division}$$

● Avoid using scales that have three squares equal one unit or three squares equal ten units because points will be difficult to plot and graphs hard to read.

To make the data points easy to plot, use 20 s/division for the units on the *x*-axis.

The range of temperatures that must be shown on the *y*-axis is

$$(46.1 - 40.8) \text{ °C} = 5.3 \text{ °C}$$

and the number of divisions above the *x*-axis is 25. The minimum number of °C that can be shown by each division on the *y*-axis is

$$\frac{5.3 \text{ °C}}{25 \text{ divisions}} = 0.21 \text{ °C/division}$$

To make the points easy to plot, use 0.40 °C per division.

Look carefully at Figure B.1 and observe the following features of this graph:

1. Although an axis may start at 0, it does not have to do so as long as the starting point is clearly labeled. Beginning the vertical axis with 38.0 °C makes room for the graph to include all the experimental data; the extra space is left at the bottom because the temperature will continue to decrease as time passes. The temperature will never be greater than 46.1 °C.
2. Major divisions (heavier lines) are labeled with numbers ending in 0, such as 42.0 °C and 100 s. Numbers such as 46.1 °C and 103 s are *not* used for major divisions. This makes points easier to plot and the graph easier to read.
3. If possible, the graph paper should allow as many significant figures to be plotted as there are significant figures in the data. It may allow more. The number of digits used in labeling the major divisions should tell the number of significant figures in the data. Temperatures can be measured to ±0.1 °C with ordinary laboratory thermometers.

The ACS Style Guide parethesizes the units in graphs.

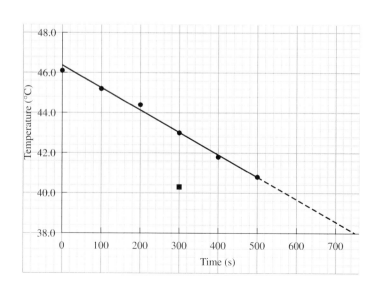

▪ FIGURE B.1 Time required for a sample of warm water to cool.

4. As well as being labeled with numbers, each axis must also each be labeled with the quantity that it represents and the unit used to measure the quantity. The units are written in parentheses.
5. Every graph must have a title that identifies the graph by giving a brief description of the data shown by the graph.

To plot a data point—for example, temperature = 45.2 °C and time = 100 s—find 45.2 °C on the y-axis and, with pencil, lightly draw a line perpendicular to the y-axis at 45.2 °C. Find 100 s on the x-axis and lightly draw a line perpendicular to the x-axis at 100 s. The intersection of the two lines is the data point; draw a dot at the data point with the data point at its center. Other data points are located and marked in a similar way.

Always be on the lookout for data points that are obviously wrong. For example, suppose that the student had interchanged 0 and 3 and recorded a temperature of 40.3 °C instead of a temperature of 43.0 °C at 300 s (the point shown by a square dot instead of a circular dot in Figure B.1). This point definitely does *not* lie on the line determined by the other five points. A mistake must have been made in recording the datum, and this point should be disregarded in drawing the line.

If the points fall on or near a straight line, as do the data points in Figure B.1, use a transparent straightedge to draw the straight line that comes closest to the data points. Usually you will find that, using experimental data, all the points do not lie exactly on a line. This is a result of uncertainty in the measurements and is illustrated by the first, third, and fifth points in Figure B.1. The lines should be drawn, as was the one in Figure B.1, so that the total number of divisions of points above the line is the same as the total number of divisions of points below the line. For example, in Figure B.1, the first point is about half a division below the line, and the fifth point is about a quarter of a division below the line, making a total of three-quarters of a division below the line. The third point is about three-quarters of a division above the line, and all the other points lie on the line. The total number of divisions of points to the left of the line should also be the same as the total number of divisions of points to the right of the line. Sometimes none of the points lie on the line. The important thing is that the total number of divisions that points lie above the line be the same as the total number of divisions that points lie below the line, and the total number of divisions that points lie to the left of the line should be the same as the total number of divisions that points lie to the right of the line. In the graph in ▪Figure B.2, the line is drawn incorrectly; the second, third, and fourth points are all above and to the right of the line while all the other points are

Computer programs (and programmable calculators) can be used to find the best straight line through data points.

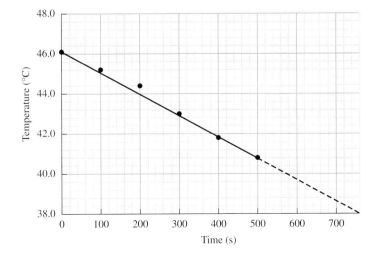

▪ FIGURE B.2   Time required for a sample of water to cool. In this graph, the line is not the best straight line through the data points because the three data points that are not on the line all lie above the line.

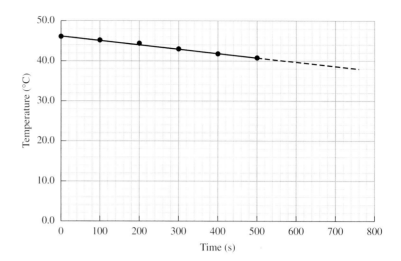

■ FIGURE B.3 Time required for a sample of warm water to cool. In this graph, the vertical scale is inappropriate. The graph occupies only a very small part of the space vertically; all the space below the line is wasted.

We prefer that our students learn to draw the best straight line by eye so that they understand the concept before they learn to use a least-squares treatment to calculate it (or a computer program to plot the whole graph).

on the line. The total number of divisions that points lie above and to the right of the line is *not* equal to the total number of divisions that points lie below and to the left of the line.

Another common graphing error is shown in ■Figure B.3. This graph occupies only a very small part of the height of the graph paper; most of the height of the graph paper is wasted. Look at Figure B.1 again to see a correct scale.

In Figure B.1, the vertical scale has been chosen so that the remaining space below the plotted data points may be used to **extrapolate** the graph to lower temperatures as shown by the broken line on the right-hand side. Extrapolations can be used to make predictions such as what the temperature will be after 600 s (39.7 °C). **Interpolations** between data points can be used to predict what the temperature will be after other periods of time between 0 and 500 s, such as 180 s (44.4 °C).

Slope, or steepness, is an important property of curves. The slope of a straight line is defined as the change in y, Δy, divided by the change in x, Δx; that is,

$$\text{slope of a straight line, } \lambda = \frac{\Delta y}{\Delta x} = \frac{y_2 - y_1}{x_2 - x_1}$$

The slope of a straight line is always constant. To find the slope of a straight line, first choose two points on the line that are near opposite ends and read their coordinates from the graph. For example, from Figure B.1 you might choose the points (0 s, 46.3 °C) and (500 s, 40.8 °C). Substitution of these values in the definition of slope gives

$$\lambda = \frac{(40.8 - 46.3)\,°C}{(500 - 0)\,s} = \frac{-5.5\,°C}{500\,s} = -0.011\,°C/s$$

Note that every slope has a sign. The straight line in Figure B.1 goes from the upper left to the lower right of the page; its slope is negative. A line going from the lower left to the upper right of a graph has a positive slope.

The general form of the mathematical equation for a straight line is

$$y = mx + b$$

where m is the slope of the line and b is the **intercept,** the value of y when $x = 0$. The equation for the line in Figure B.1 is

$$\text{temperature in °C} = -0.011\,\frac{°C}{s}\,(\text{time in s}) + 46.3\,°C$$

Not all data give a straight line. The graph showing the relation between pressure and volume of a sample of air at constant temperature in Figure 5.5, which is reproduced here for your convenience, is an example. When the data points do not lie on a straight line, use a French curve or flexible curve to draw the best smooth curve through the data points (▬Figure B.4). A thin plastic ruler can be used as a flexible curve if it is held perpendicular to the plane of the paper with one long edge on the paper.

The slope of a curved line changes continuously and is harder to determine than the slope of a straight line. To obtain the slope of a curved line at any point, draw a straight line tangent to the curve at the point. The slope of the straight line is the slope of the curve at the point. (See Sample Problem 18.2.)

Extrapolation and interpolation of a curved line are not as certain as extrapolation and interpolation of a straight line. Sometimes data that lead to a curved-line graph can be plotted in a different way so that a straight line results. For example, when the same data shown in Figure 5.5 are graphed as the reciprocal of pressure (1/pressure) against volume, the curve is a straight line as shown in ▬Figure B.5.

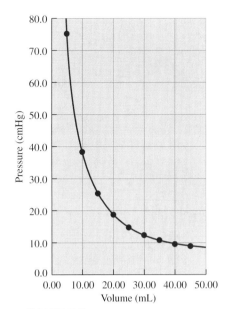

▬ FIGURE 5.5 Relation between pressure and volume of a sample of air at constant temperature.

Summary of Steps in Making a Graph

1. Tabulate the data to be graphed.
2. Identify the independent and dependent variables; use the x-axis for the independent variable and the y-axis for the dependent variable.
3. Decide how many units the divisions along each axis will represent. The scale should always make the data points easy to plot and the graph easy to read. In addition, a graph should always fill more than half the graph paper in each direction.
4. Label both axes; remember that one square should have the same value the whole length of an axis.
5. Plot the data points.
6. Draw the best curve through the data points, ignoring points that are obviously mistakes.
7. Give the graph a title.

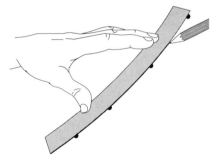

▬ FIGURE B.4 Use a flexible curve to draw the best smooth curve through the data points.

Reading a Line Graph

Sample Problem B.9 shows how to read a line graph.

SAMPLE PROBLEM

B.9 From Figure B.1, how long will it take for the sample of water to cool to 39.5 °C?

Solution Find 39.5 °C on the y-axis and lightly draw a line parallel to the x-axis from 39.5 °C on the y-axis to the curve. From the point where the pencil line intersects the curve, drop a perpendicular to the x-axis and read the time from the x-axis. The time is 610 s. It will take 610 s for the sample of water to cool to 39.5 °C.

PRACTICE PROBLEMS

B.18 Refer to Figure B.1. (a) How long will it take for the sample of water to cool to 43.6 °C? (b) What will be the temperature of the sample of water after 125 s have passed? (c) How many degrees will the water cool in 250 s?

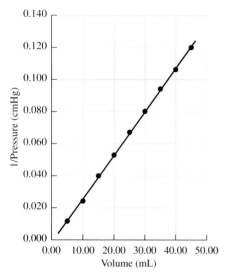

FIGURE B.5 A graph of the relationship between the pressure (P) and volume (V) of air at constant temperature is a straight line if $1/P$ instead of P is plotted against V.

B.19 In an experiment to study the density of aqueous solutions of ethylene glycol (antifreeze), a student prepared a series of solutions of ethylene glycol in water containing the concentrations of ethylene glycol shown in the following table. The student then measured the density of each solution at 20 °C.

Density of Aqueous Ethylene Glycol at 20 °C

Mass % Ethylene Glycol	Density, g/mL
10.0	1.011
28.0	1.035
44.0	1.057
60.0	1.077

(a) Graph the data. (b) What is the slope of the graph when the mass % ethylene glycol is 30.0? (c) What is the density of a 25.0 mass % ethylene glycol solution?

B.20 The following table shows the boiling point of ammonia at different pressures:

Boiling Point of Ammonia

Temperature, °C	Pressure, atm
−33.6	1.00
−18.7	2.00
+4.7	5.00
25.7	10.00
50.1	20.00
78.9	40.00
98.3	60.00

(a) Make a graph showing the data in the table. (b) What is the boiling point of ammonia at 30.00 atm?

B.7   GEOMETRIC FORMULAS

Triangle

$\text{area} = \dfrac{1}{2}(\text{base} \times \text{altitude})$

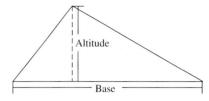

Rectangle

$\text{perimeter} = 2(\text{length} + \text{width})$

$\text{area} = \text{length} \times \text{width}$

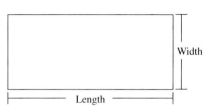

Rectangular Block

$\text{area} = 2(\text{length} \times \text{width}) + 2(\text{length} \times \text{height}) + 2(\text{width} \times \text{height})$

$\text{volume} = \text{length} \times \text{width} \times \text{height}$

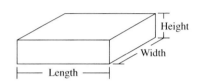

Circle

$\text{circumference} = 2\pi \times \text{radius} \quad \text{where } \pi = 3.1416$

$\text{area} = \pi(\text{radius})^2$

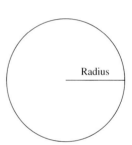

Cylinder

$\text{volume} = \text{area of base} \times \text{height} = \pi r^2 h$

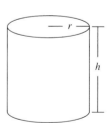

Sphere

$\text{area} = 4\pi r^2$

$\text{volume} = \dfrac{4}{3}\pi r^3$

Properties of Water

Normal Melting Point   0.0 °C

Normal Boiling Point   100.0 °C

Density in g/mL

Ice at 0 °C	0.917
Water at 0 °C	0.999 87
4 °C	1.000 00
15 °C	0.999 13
20 °C	0.998 23
25 °C	0.997 07
30 °C	0.995 67
100 °C	0.958 38
Steam at 100 °C	0.000 596

Vapor Pressure in mmHg

0 °C	4.6
10 °C	9.2
20 °C	17.5
30 °C	31.8
40 °C	55.3
50 °C	92.5
60 °C	149.4
70 °C	233.7
80 °C	355.1
90 °C	525.8
100 °C	760.0

Table 5.4 shows the vapor pressure of water at 0.2 °C intervals from 15.0 °C to 30.8 °C.

Thermodynamic Data for Selected Substances at 298 K[a]

Substance	ΔH°_f, kJ/mol	ΔG°_f, kJ/mol	S°, J/K·mol
Ag(s)	0	0	42.57
Ag$^+$(aq)	105.63	77.16	72.71
AgBr(s)	−100.42	−96.95	107.16
AgCl(s)	−127.13	−109.86	96.28
AgNO$_3$(aq)	−101.85	−34.24	219.35
Ag$_2$S(s)	−32.61	−40.69	144.08
Ag$_2$SO$_4$(s)	−716.22	−618.77	200.51
AlCl$_3$(s)	−705.6	−628.8	109.3
Al$_2$O$_3$(s)	−1675.7	−1582.3	50.9
Ar(g)	0	0	154.84
Br$^-$(aq)	−121.55	−103.97	82.4
C(s, graphite)	0	0	5.69
C(s, diamond)	1.895	2.832	2.37
CH$_4$(g)	−74.85	−50.79	186.2
HC≡CH(g)	226.75	209	200.82
H$_2$C=CH$_2$(g)	52.28	86.12	219.4
CH$_3$CH$_3$(g)	−84.67	−32.89	229.5
CH$_3$Br(g)	−36	−26	245.8
CH$_3$Cl(g)	−82	−59	234.2
CH$_3$I(g)	20.5	22	254.6
HCN(g)	131	120	201.8
HCN(l)	105	121	112.8
CN$^-$(aq)	151	166	118
CO(g)	−110.52	−137.27	197.91
CO$_2$(g)	−393.51	−394.38	213.64
CO$_2$(aq)	−412.9	−386.2	121
COCl$_2$(g)	−233.0	−210.4	289.2
CO(NH$_2$)$_2$(s)	−333.51	−197.33	104.60
HCHO(g)	−116	−110	218.6
CH$_3$OH(l)	−238.64	−166.3	127
CH$_3$OH(g)	−201.2	−161.9	238
CH$_3$CH$_2$OH(l)	−277.63	−174.8	161
CH$_3$CH$_2$OH(g)	−235.3	−168.6	282
HC$_2$H$_3$O$_2$(l)	−487.0	−392	160
HC$_2$H$_3$O$_2$(aq)	−488.45	−399.6	178.7
C$_2$H$_3$O$_2$$^-$(aq)	−488.87	−369.40	86.6
CaC$_2$(s)	−59.8	−64.9	69.96
CaCl$_2$(s)	−795.8	−748.1	104.6
CaCO$_3$(s, calcite)	−1206.92	−1128.84	92.9
CaF$_2$(s)	−1219.64	−1167.34	68.87
CaO(s)	−635.09	−604.04	39.75
Ca(OH)$_2$(s)	−986.09	−898.56	83.39

Substance	ΔH°_f, kJ/mol	ΔG°_f, kJ/mol	S°, J/K·mol
CaSO$_4$(s)	−1434.11	−1321.85	106.7
Cu^{2+}(aq)	−75.90	−77.61	−73.2
CdS(s)	−161.9	−156.5	64.9
Cl(g)	121.68	105.70	165.09
Cl$^-$(g)	−233.89	−240.12	153.25
Cl$^-$(aq)	−166.85	−131.06	56.70
Cl$^+$(g)	1383.23	1360.32	167.45
Cl$_2$(g)	0	0	222.96
Co(s)	0	0	28.5
CoO(s)	−211.3		43.9
Fe$_2$O$_3$(s, hematite)	−824.2	−742.2	87.40
Fe$_3$O$_4$(s, magnetite)	−1118.4	−1015.5	146.4
H$_2$(g)	0	0	130.68
H(g)	217.97	203.25	114.71
H$^+$(g)	1536.20		
H$^+$(aq)	0	0	0
H$^-$(g)	138.99		
HBr(g)	−36.40	−53.43	198.59
HCl(g)	−92.31	−95.30	186.80
HCl(aq)	−166.85	−131.06	56.70
HF(g)	−271.12	−273.22	173.67
HI(g)	25.94	1.30	206.33
H$_2$O(l)	−285.83	−237.18	69.91
H$_2$O(g)	−238.92	−228.59	188.72
H$_2$O$_2$(l)	−187.78	−120.41	109.6
He(g)	0	0	126.15
I$_2$(s)	0	0	116.73
I$_2$(g)	62.24	19.37	260.58
I(g)	106.62	70.15	180.68
I$^-$(aq)	−55.94	−51.67	111.3
K(s)	0	0	64.68
KCl(s)	−436.7	−408.8	82.6
Kr(g)	0	0	164.08
Li(s)	0	0	29.10
Mg^{2+}(aq)	−466.62	−454.6	−138
MgCO$_3$(s)	−1111	−1011.6	65.82
MgO(s)	−601.5	−569.2	27.0
Mg(OH)$_2$(s)	−924	−833.2	63.15
MgSO$_4$(s)	−1261.2	−1170.1	91.4
MnO$_2$(s)	−520.03	−465.14	53.05
N$_2$(g)	0	0	191.50
N(g)	472.70	340.9	153.2

Substance	ΔH°_f, kJ/mol	ΔG°_f, kJ/mol	S°, J/K·mol	Substance	ΔH°_f, kJ/mol	ΔG°_f, kJ/mol	S°, J/K·mol
$NH_3(g)$	−46.11	−16.5	192.3	$Na_2SiO_3(gl)^b$	−1506	−1426	
$NH_3(aq)$	−80.29	−26.6	111	$Ne(g)$	0	0	146.33
$NH_4^+(aq)$	−132.5	−79.37	113	$O(g)$	249.17	231.75	160.95
$N_2H_4(l)$	50.63	149.2	121.2	$O_2(g)$	0	0	205.03
$NH_4Cl(s)$	−314.43	−202.87	94.6	$O_2(aq)$	−11.7	16.3	110.9
$NH_4ClO_4(s)$	−295.3	−88.91	186.2	$O_3(g)$	142.7	163.2	238.82
$NO(g)$	90.25	86.57	210.65	$OH^-(aq)$	−229.99	−157.29	−10.75
$N_2O(g)$	82.05	104.2	291.7	$P(s,red)$	−17.56	−12.12	22.78
$NO_2(g)$	33.2	51.30	240.0	$P(s,white)$	0	0	41.05
$HNO_3(l)$	−174.1	−80.89	155.6	$PCl_3(g)$	−287.0	−267.8	311.78
$HNO_3(aq)$	−207.4	−111.3	146	$PCl_5(g)$	−374.5	−304.7	364.1
$NO_3^-(aq)$	−207.4	−111.3	146	$H_2S(g)$	−20.63	−33.56	205.69
$Na(s)$	0	0	51.45	$H_2S(aq)$	−39.7	−27.87	129
$Na(l)$	2.41	0.498	57.85	$HS^-(aq)$	−17.6	12.05	62.9
$Na(g)$	107.68	77.30	153.61	$S^{2-}(aq)$	33.05	86.31	−16.3
$Na^+(aq)$	−239.66	−261.87	60.25	$SO_2(g)$	−296.83	−300.19	248.12
$Na_2CO_3(s)$	−1130.9	−1048.1	138.8	$SO_3(g)$	−395.72	−371.08	256.72
$NaHCO_3(s)$	−947.7	−851.9	102.1	$SO_3(l)$	−441.0	−368.4	95.6
$NaCl(s)$	−411.12	−384.04	72.12	$SO_3(s)$	−454.5	−369.0	52.3
$NaCl(aq)$	−407.11	−393.04	115.48	$H_2SO_4(l)$	−813.99	−690.10	156.90
$NaClO_3(s)$	−358.69	−274.89	129.7	$H_2SO_4(aq)$	−909.27	−744.63	20.1
$NaOH(s)$	−427.77	−381.53	64.18	$HSO_4^-(aq)$	−887.34	−756.01	131.8
$NaOH(aq)$	−469.60	−419.1	49.79	$SO_4^{2-}(aq)$	−909.27	−744.63	20.1
$Na_2S(s)$	−372.4	−361.4	97.9	$TiO_2(s)$	−912.1	−888.4	50.25
$NaHS(s)$	−236.4	−213.4		$WO_3(s)$	−842.7	−763.87	75.94
$Na_2SO_4(s)$	−1381.1	−1265.7	157.9	$Xe(g)$	0	0	169.68
$Na_2SO_4(aq)$	−1386.8	−1265.7	137.7	$ZnO(s)$	−348.28	−318.32	43.64
$Na_2SiO_3(s)$	−1561.4	−1467.4	113.8	$ZnS(s)$	−206.0	−201.3	57.5

[a]Most data are from Bard, A. J.: Parsons, R.: Jordan, J. *Standard Potentials in Aqueous Solution*; Dekker: New York, 1985.
[b]glass

Top 50 Chemicals for 1993

TABLE E.1	Top 50 Chemicals for 1993[a]			

	Substance			
Rank	Name	Formula	Pounds	Uses[b]
1	Sulfuric acid	H_2SO_4	80.31×10^9	Manufacture of fertilizer
2	Nitrogen	N_2	65.29×10^9	Ammonia synthesis, inert atmosphere, enhanced oil recovery
3	Oxygen	O_2	46.52×10^9	Metallurgy
4	Ethylene	$H_2C{=}CH_2$	41.25×10^9	Manufacture of polymers
5	Lime	CaO	36.80×10^9	Metallurgy, pollution control
6	Ammonia	NH_3	34.50×10^9	Fertilizer
7	Sodium hydroxide	$NaOH$	25.71×10^9	Manufacture of organic and inorganic chemicals, pulp and paper
8	Chlorine	Cl_2	24.06×10^9	Manufacture of organic chemicals, bleaching pulp and paper
9	Methyl *tert*-butyl ether	$CH_3OC(CH_3)_3$	24.05×10^9	Octane-enhancing gasoline additive
10	Phosphoric acid	H_3PO_4	23.04×10^9	Manufacture of fertilizer
11	Propylene	$CH_3CH{=}CH_2$	22.40×10^9	Manufacture of polymers
12	Sodium carbonate	Na_2CO_3	19.80×10^9	Glass, exports, base in chemical manufacture
13	Ethylene dichloride	$ClCH_2CH_2Cl$	17.95×10^9	Manufacture of poly(vinyl chloride)
14	Nitric acid	HNO_3	17.07×10^9	Manufacture of ammonium nitrate
15	Ammonium nitrate	NH_4NO_3	16.79×10^9	Fertilizer and explosive
16	Urea	H_2NCONH_2	15.66×10^9	Fertilizer
17	Vinyl chloride	$H_2C{=}CHCl$	13.75×10^9	Manufacture of poly(vinyl chloride)
18	Benzene		12.32×10^9	Production of ethylbenzene, phenol, and acetone
19	Ethylbenzene	—CH_2CH_3	11.76×10^9	Production of styrene
20	Carbon dioxide	CO_2	10.69×10^9	Refrigeration, beverage carbonation
21	Methanol	CH_3OH	10.54×10^9	Manufacture of formaldehyde, methyl *tert*-butyl ether, and acetic acid
22	Styrene	—$CH{=}CH_2$	10.07×10^9	Manufacture of polystyrene
23	Terephthalic acid	$HOOC$—⬡—$COOH$	7.84×10^9	Production of polyester fiber, exports
24	Formaldehyde	$H_2C{=}O$	7.61×10^9	Manufacture of plastics and polyesters
25	Xylene	—[c]	6.84×10^9	Manufacture of *p*-xylene
26	Hydrochloric acid	HCl	6.45×10^9	Cleaning of metal surfaces, chemical and pharmaceutical manufacturing
27	Toluene	—CH_3	6.38×10^9	Gasoline, production of benzene, solvent
28	*p*-Xylene	CH_3—⬡—CH_3	5.76×10^9	Manufacture of polyester used in textile fibers, photographic film, and soft-drink bottles
29	Ethylene oxide	H_2C—CH_2 / O	5.68×10^9	Production of ethylene glycol

	Substance			
Rank	Name	Formula	Pounds	Uses[b]
30	Ethylene glycol	$HOCH_2CH_2OH$	5.23×10^9	Antifreeze, manufacture of polyester fiber
31	Ammonium sulfate	$(NH_4)_2SO_4$	4.80×10^9	Fertilizer
32	Cumene		4.49×10^9	Manufacture of phenol and acetone
33	Phenol		3.72×10^9	Manufacture of phenol–formaldehyde polymers used as the adhesive in plywood, and bisphenol A used in synthesis of polycarbonate and epoxy resins
34	Acetic acid	CH_3COOH	3.66×10^9	Synthesis of vinyl acetate and cellulose acetate, a polymer used mainly as a fiber in clothing
35	Potash	—[d]	3.31×10^{9e}	Fertilizer
36	Carbon black	C	3.22×10^9	Reinforcing agent for elastomers used in tires, black pigment
37	Butadiene	$H_2C{=}CHCH{=}CH_2$	3.09×10^9	Manufacture of synthetic rubber for automobile tires
38	Vinyl acetate	$CH_3COOCH{=}CH_2$	2.83×10^9	Manufacture of poly(vinyl acetate) for adhesives, coatings, and paints; exports
39	Propylene oxide		2.73×10^9	Manufacture of poly(propylene glycol), which is used to make polyurethane foams for automobile seats, furniture, bedding, and carpet
40	Titanium dioxide	TiO_2	2.56×10^9	White pigment for surface coatings, paper and paperboard, and plastics
41	Acrylonitrile	$H_2C{=}CHCN$	2.51×10^9	Exports, manufacture of acrylic fibers
42	Acetone	CH_3COCH_3	2.46×10^9	Synthesis of methyl methacrylate, which is used to make poly(methyl methacrylate), a glass substitute, and in solvents
43	Aluminum sulfate	$Al_2(SO_4)_3$	2.23×10^9	Pulp and paper, water purification
44	Cyclohexane		2.00×10^9	Synthesis of adipic acid and caprolactam
45	Sodium silicate	—[f]	1.97×10^9	Soaps and detergents, silica gel used for catalysis and column chromatography
46	Adipic acid	$HOOC(CH_2)_4COOH$	1.56×10^9	Manufacture of nylon 66 fibers
47	Sodium sulfate	Na_2SO_4	1.44×10^9	Detergents
48	Calcium chloride	$CaCl_2$	1.40×10^9	Road deicing and dust control
49	Caprolactam		1.36×10^9	Manufacture of nylon 6
50	n-Butyl alcohol	$CH_3CH_2CH_2CH_2OH$	1.33×10^9	Synthesis of butyl acrylate, methacrylate, glycol ethers

[a]Reprinted in part with permission from *Chem. Eng. News,* April 11, 1994, p 13, © 1994 American Chemical Society.

[b]Information is from Chenier, P. J. *Survey of Industrial Chemistry,* 2nd ed.; VCH: New York, 1992. Over half of the annual production is used for the purposes given, which are listed in order of decreasing use.

[c]Xylene is a mixture of

[d]Mainly KCl.

[e]As K_2O.

[f]"Sodium silicate" means compounds having the general formula $Na_2O \cdot nSiO_2$ such as $Na_2SiO_3(Na_2O \cdot SiO_2)$ and $Na_2Si_4O_9(Na_2O \cdot 4SiO_2)$.

Balancing Equations for Oxidation–Reduction Reactions

F.1   CHANGE-IN-OXIDATION-NUMBER METHOD

To balance an equation by the change-in-oxidation-number method, follow these steps:

Balancing Redox Equations by the Change-in-Oxidation-Number Method

1. Identify the reactants and products. The formation (or reaction) of water may be difficult to detect in aqueous solution.
2. Write symbols for elements (formulas for elements that exist as polyatomic molecules) and formulas for compounds.
3. Assign oxidation numbers to all atoms.
4. Identify the element oxidized and the element reduced. *If necessary, balance them by changing coefficients.*
5. Calculate the total increase and total decrease in oxidation number. Make the absolute values of the total increase and total decrease equal by changing coefficients.
6. Complete balancing by inspection. Charge can be balanced by adding H^+ if the solution is acidic or OH^- if the solution is basic. Water is present in all aqueous solutions and can be used to balance H and O. *Do **not** change the coefficients from step 5.*
7. Check to be sure there is the same number of each kind of atom and the same charge on both sides. If coefficients have a common denominator, simplify.
8. Add symbols showing whether substances are solids, liquids, gases, or in aqueous solution.

▶ The first two steps are the same as the first two steps for writing a chemical equation that you learned in Section 1.11.

Why are both methods of balancing equations for redox reactions treated in detail? Reviewers were about equally divided as to which method they preferred.

▶ The last two steps are also similar to the steps that you learned in Section 1.11.

Sample Problems F.1–F.4 illustrate the change-in-oxidation-number method and will introduce you to some common problems that you are likely to meet. The Sample Problems begin with step 3 so that you can focus your attention on the new material.

SAMPLE PROBLEMS

F.1   Write the molecular equation for the reaction between ethyl alcohol (CH_3CH_2OH, molecular formula C_2H_6O) and potassium dichromate ($K_2Cr_2O_7$) that takes place in aqueous solution in the presence of sulfuric acid (H_2SO_4). The products are acetic acid (CH_3COOH, molecular formula $C_2H_4O_2$), chromium(III) sulfate [$Cr_2(SO_4)_3$], and potassium sulfate (K_2SO_4):

$$C_2H_6O(aq) + K_2Cr_2O_7(aq) + H_2SO_4(aq) \longrightarrow$$
$$C_2H_4O_2(aq) + Cr_2(SO_4)_3(aq) + K_2SO_4(aq)$$

▶ Use of molecular formulas instead of structural formulas for organic compounds makes balancing equations easier.

Solution   Step 3. *Assign oxidation numbers to all atoms.*

$$\overset{-2\ +1\ -2}{C_2H_6O} + \overset{+1\ +6\ -2}{K_2Cr_2O_7} + \overset{+1+6-2}{H_2SO_4} \longrightarrow \overset{0\ +1\ -2}{C_2H_4O_2} + \overset{+3\ +6-2}{Cr_2(SO_4)_3} + \overset{+1+6-2}{K_2SO_4}$$

Step 4. *Identify the element oxidized and the element reduced.* **If necessary, balance them by changing coefficients.** The oxidation numbers of all elements, except carbon and chromium are the same before and after reaction. The oxidation number of carbon increases from -2 to 0; therefore, carbon is the element oxidized. The oxidation number of chromium decreases from $+6$ to $+3$; chromium is the element reduced. The expression in step 3 shows 2 C on each side; thus, the element oxidized is already balanced. The element reduced is also already balanced; the expression in step 3 shows 2 Cr on each side.

Step 5. *Calculate the total increase and the total decrease in oxidation number. Make the absolute values of the total increase and total decrease equal by changing coefficients.* The oxidation number of carbon increases by 2 (from -2 to 0). Two carbon atoms are oxidized, and the total change in oxidation number for carbon is 4. The oxidation number of chromium decreases by 3 (from $+6$ to $+3$). Two chromium atoms are reduced, and the total change in oxidation number for chromium is 6. To make the total change in oxidation numbers for carbon and chromium equal, we must multiply the change for carbon by 3 and the change for chromium by 2:

$$3(2) = 2(3)$$

In the equation, the coefficient for C_2H_6O should be 3, and the coefficient for $K_2Cr_2O_7$ should be 2. The coefficient for $C_2H_4O_2$ should be 3, and the coefficient for $Cr_2(SO_4)_3$ should be 2 so that there are the same number of C and the same number of Cr on both sides.

$$3C_2H_6O + 2K_2Cr_2O_7 + H_2SO_4 \longrightarrow 3C_2H_4O_2 + 2Cr_2(SO_4)_3 + K_2SO_4 \quad (F.1)$$

A diagram is helpful for carrying out step 5:

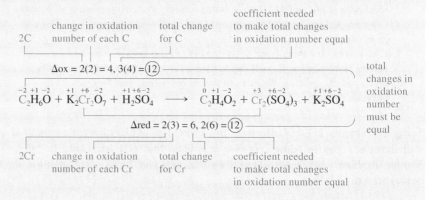

Step 6. *Complete balancing by inspection. Charge can be balanced by adding H^+ if the solution is acidic or OH^- if the solution is basic. Water is present in all aqueous solutions and can be used to balance H and O. Do* **not** *change the coefficients from step 5.* Expression F.1 shows 4 K on the left and only 2 K on the right. We must multiply K_2SO_4 by 2:

$$3C_2H_6O + 2K_2Cr_2O_7 + H_2SO_4 \longrightarrow 3C_2H_4O_2 + 2Cr_2(SO_4)_3 + 2K_2SO_4 \quad (F.2)$$

Expression F.2 shows 8 SO_4^{2-} on the right and only 1 SO_4^{2-} on the left. We must multiply H_2SO_4 by 8:

$$3C_2H_6O + 2K_2Cr_2O_7 + 8H_2SO_4 \longrightarrow 3C_2H_4O_2 + 2Cr_2(SO_4)_3 + 2K_2SO_4 \quad (F.3)$$

All elements except H and O now are balanced. There are 34 H and 49 O on the left and 12 H and 38 O on the right. We must add 22 H and 11 O or 11 H_2O to the right side of expression F.3:

$$3C_2H_6O + 2K_2Cr_2O_7 + 8H_2SO_4 \longrightarrow$$
$$3C_2H_4O_2 + 2Cr_2(SO_4)_3 + 2K_2SO_4 + 11H_2O \quad \text{(F.4)}$$

Step 7. Check to be sure there is the same number of each kind of atom and the same charge on both sides. If coefficients have a common denominator, simplify. There are 6 C, 34 H, 49 O, 4 K, 4 Cr, 8 S, and zero net charge on each side of equation F.4. The coefficients have no common denominator.

Step 8. Add symbols showing whether substances are solids, liquids, gases, or in aqueous solution. The molecular equation is

$$3C_2H_6O(aq) + 2K_2Cr_2O_7(aq) + 8H_2SO_4(aq) \longrightarrow$$
$$3C_2H_4O_2(aq) + 2Cr_2(SO_4)_3(aq) + 2K_2SO_4(aq) + 11H_2O(l)$$

F.2 Write the molecular equation for the following reaction:

$$Zn(s) + HNO_3(aq) \longrightarrow Zn(NO_3)_2(aq) + NH_4NO_3(aq)$$

Solution Step 3.

$$\overset{0}{Zn} + \overset{+1+5-2}{HNO_3} \longrightarrow \overset{+2\ +5-2}{Zn(NO_3)_2} + \overset{-3+1+5-2}{NH_4NO_3}$$

Step 4. The oxidation numbers of all elements except zinc and nitrogen are the same before and after reaction. The oxidation number of zinc increases from 0 to $+2$; therefore, zinc is the element oxidized. The oxidation number of nitrogen decreases from $+5$ in HNO_3 to -3 in the ammonium ion NH_4^+; nitrogen is the element reduced. However, the oxidation number of some N remains $+5$. Some nitrate ions are spectator ions. Let's write nitric acid twice so that we can treat the two types of nitrate ions separately:

$$\overset{0}{Zn} + \overset{+1+5-2}{HNO_3} + \overset{+1+5-2}{HNO_3} \longrightarrow \overset{+2\ +5-2}{Zn(NO_3)_2} + \overset{-3+1+5-2}{NH_4NO_3}$$

The expression shows 1 Zn on each side; the element that is oxidized is balanced. The expression shows 1 N that undergoes a change in oxidation number on each side; the element that is reduced is balanced too.

Step 5. The oxidation number of zinc increases by 2 (from 0 to $+2$). Only one zinc atom is oxidized; therefore, the total change in oxidation number for zinc is 2. The oxidation number of nitrogen decreases by 8 (from $+5$ to -3). Only one nitrogen atom is reduced; the total change in oxidation number for nitrogen is 8. To make the total change in oxidation numbers for zinc and nitrogen equal, we must multiply the change for zinc by 4:

$$4(2) = 8$$

The diagram is

$$\Delta ox = 2, 4(2) = 8$$
$$\overset{0}{Zn} + \overset{+1+5-2}{HNO_3} + \overset{+1+5-2}{HNO_3} \longrightarrow \overset{+2+5-2}{Zn(NO_3)_2} + \overset{-3+1+5-2}{NH_4NO_2}$$
$$\Delta red = 8$$

The coefficient for Zn in the equation should be 4 and the coefficient for the HNO_3 that is reduced should be 1. The coefficient for $Zn(NO_3)_2$ should also be 4, and the coefficient for NH_4NO_3 should be 1 so that there are the same numbers of Zn and the same number of N that undergo change on both sides:

$$4Zn + HNO_3 + HNO_3 \longrightarrow 4Zn(NO_3)_2 + NH_4NO_3 \quad \text{(F.5)}$$

◗ Don't forget to put (s), (l), (g), and (aq) back at the end.

◗ At first, you should follow along on the list of steps. With practice, you will remember what to do.

F.1 CHANGE-IN-OXIDATION-NUMBER METHOD

Step 6. Expression F.5 shows 9 NO_3^- on the right and only 1 (of the type that is a spectator ion) on the left. The second HNO_3 should be multiplied by 9:

$$4Zn + HNO_3 + 9HNO_3 \longrightarrow 4Zn(NO_3)_2 + NH_4NO_3$$

Both Zn and N are balanced. But there are 10 H and 30 O on the left and 4 H and 27 O on the right. We must add 6 H and 3 O or 3 H_2O to the right side. The HNO_3's should be combined:

$$4Zn + 10HNO_3 \longrightarrow 4Zn(NO_3)_2 + NH_4NO_3 + 3H_2O$$

Step 7. There are 4 Zn, 10 H, 10 N, 30 O, and a net charge of zero on each side.

Step 8. The molecular equation is

$$4Zn(s) + 10HNO_3(aq) \longrightarrow 4Zn(NO_3)_2(aq) + NH_4NO_3(aq) + 3H_2O(l)$$

F.3 Write the net ionic equation for the reaction

$$Cl^- + MnO_4^- \longrightarrow Mn^{2+} + Cl_2(g)$$

that takes place in acidic solution.

Solution Step 3.

$$\overset{-1}{Cl^-} + \overset{+7\ -2}{MnO_4^-} \longrightarrow \overset{+2}{Mn^{2+}} + \overset{0}{Cl_2(g)}$$

Step 4. The oxidation number of Cl increases from -1 to 0; Cl is the element oxidized. One Cl is shown on the left and 2 Cl on the right. A coefficient of 2 should be placed in front of Cl^-:

$$2\overset{-1}{Cl^-} + \overset{+7\ -2}{MnO_4^-} \longrightarrow \overset{+2}{Mn^{2+}} + \overset{0}{Cl_2}$$

The oxidation number of manganese decreases from $+7$ to $+2$; Mn is the element reduced. One manganese is shown on each side.

Step 5. The oxidation number of Cl increases by 1. However, 2 Cl are oxidized, and the total change in oxidation number for Cl is 2. The oxidation number of the Mn decreases by 5. Only one manganese atom is reduced and the total change in oxidation number for manganese is 5. To make the total change in oxidation numbers for chlorine and manganese equal, we must multiply the total change for chlorine by 5 and the total change for manganese by 2:

$$5(2) = 2(5)$$

The diagram is

$$
\begin{array}{c}
\overset{\Delta \text{ox} = 2,\ 5(2) = 10}{\overbrace{\qquad\qquad\qquad}} \\
2\overset{-1}{Cl^-} + \overset{+7\ -2}{MnO_4} \longrightarrow \overset{+2}{Mn^{2+}} + \overset{0}{Cl_2} \\
\underset{\Delta \text{red} = 5,\ 2(5) = 10}{\underbrace{\qquad\qquad\qquad}}
\end{array}
$$

The coefficient for Cl^- should be multiplied by 5 and the coefficient for MnO_4^- by 2. The coefficient for Mn^{2+} should also be 2; the coefficient for Cl_2 should be 5:

$$10Cl^- + 2MnO_4^- \longrightarrow 2Mn^{2+} + 5Cl_2 \qquad\qquad (F.6)$$

The elements oxidized and reduced are now balanced.

Step 6. The net charges on each side of expression F.6 are

left side: $10(1-) + 2(1-) = 12-$ right side: $2(2+) + 5(0) = 4+$

Thus, the difference between the net charge on the left and the net charge on the right is 16. Addition of 16 H^+ to the left side balances charge:

$$16H^+ + 10Cl^- + 2MnO_4^- \longrightarrow 2Mn^{2+} + 5Cl_2$$
$$16(1+) + 10(1-) + 2(1-) = 4+ = 2(2+)$$

Balance charge by adding H^+ in acidic solution.

Finally, add 8 H_2O to the right side to balance O's. Hydrogen should be balanced too. (If it is not, we have made a mistake and should check our work.)

$$16H^+ + 10Cl^- + 2MnO_4^- \longrightarrow 2Mn^{2+} + 5Cl_2 + 8H_2O$$

Step 7. There are 16 H, 10 Cl, 2 Mn, and 8 O on each side. Each side has a net charge of $4+$, and the coefficients have no common denominator.

Step 8. The net ionic equation is

$$16H^+ + 10Cl^- + 2MnO_4^- \longrightarrow 2Mn^{2+} + 5Cl_2(g) + 8H_2O(l)$$

F.4   Write the net ionic equation for the reaction

$$I^- + MnO_4^- \longrightarrow IO_3^- + MnO_2(s)$$

that takes place in basic solution.

Solution   Step 3.

$$\overset{-1}{I^-} + \overset{+7\ -2}{MnO_4^-} \longrightarrow \overset{+5-2}{IO_3^-} + \overset{+4\ -2}{MnO_2}$$

Step 4. Iodine is oxidized from -1 to $+5$, and Mn is reduced from $+7$ to $+4$. Both I and Mn are balanced.

Step 5. The increase in oxidation number is 6; the decrease in oxidation number is 3. To make the changes equal, we must multiply the decrease in oxidation number by 2. The diagram is

$$
\begin{array}{c}
\overbrace{}^{\Delta ox = 6} \\
\overset{-1}{I^-} + \overset{+7\ -2}{MnO_4^-} \longrightarrow \overset{+5-2}{IO_3^-} + \overset{+4\ -2}{MnO_2} \\
\underbrace{}_{\Delta red = 3,\ 2(3) = 6}
\end{array}
$$

The coefficient for I^- should be 1 and the coefficient for MnO_4^-, 2. The coefficient for IO_3^- should also be 1, and the coefficient for MnO_2, 2:

$$I^- + 2MnO_4^- \longrightarrow IO_3^- + 2MnO_2$$

Step 6. The net charge on the left is $3-$, and the net charge on the right is $1-$. Addition of 2 OH^- to the right will balance charge:

> Balance charge by adding OH^- in basic solution.

$$I^- + 2MnO_4^- \longrightarrow IO_3^- + 2MnO_2 + 2OH^-$$

Finally, addition of 1 H_2O to the left will balance O (and H):

> Finally, balance O by adding H_2O. Then H should also be balanced. If it is not, you should check your work.

$$I^- + 2MnO_4^- + H_2O \longrightarrow IO_3^- + 2MnO_2 + 2OH^- \tag{F.7}$$

Step 7. Equation F.7 shows 2 H, 9 O, 1 I, 2 Mn and a net charge of $3-$ on each side. The coefficients have no common denominator.

> Always check your work.

Step 8. The net ionic equation is

$$I^- + 2MnO_4^- + H_2O(l) \longrightarrow IO_3^- + 2MnO_2(s) + 2OH^-$$

PRACTICE PROBLEMS

F.1   Write molecular equations for each of the following reactions:
(a) $S(s) + HNO_3(aq) \longrightarrow SO_2(g) + NO(g)$
(b) $Cu_2O(s) + HNO_3(aq) \longrightarrow Cu(NO_3)_2(aq) + NO(g)$
(c) $Zn(s) + HNO_3(aq) \longrightarrow Zn(NO_3)_2(aq) + N_2(g)$ (Zinc undergoes a variety of reactions with nitric acid.)
(d) $Al(s) + NaOH(aq) \longrightarrow NaAl(OH)_4(aq) + H_2(g)$
(e) $C(s) + H_2SO_4(l) \longrightarrow CO_2(g) + SO_2(g)$

(f) $(CH_3)_2CHOH(aq) + CrO_3(aq) + H_2SO_4(aq) \longrightarrow$
$$(CH_3)_2CO(aq) + Cr_2(SO_4)_3(aq)$$
(g) $H_3AsO_4(aq) + Zn(s) + HNO_3(aq) \longrightarrow AsH_3(g) + Zn(NO_3)_2(aq)$
(h) $KI(aq) + KMnO_4(aq) \longrightarrow I_2(s) + MnO_2(s)$ (basic solution)

F.2   Write net ionic equations for each of the following reactions:
(a) $Cr_2O_7^{2-} + C_2O_4^{2-} \longrightarrow Cr^{3+} + CO_2(g)$ (acidic solution)
(b) $H_2O_2(aq) + MnO_4^- \longrightarrow O_2(g) + MnO_2(s)$ (basic solution)
(c) $AsH_3(g) + ClO_3^- \longrightarrow H_3AsO_4(aq) + Cl^-$ (acidic)
(d) $Cu(NH_3)_4^{2+} + S_2O_4^{2-} \longrightarrow Cu(s) + SO_3^{2-} + NH_3(aq)$ (basic)
(e) $TlOH(s) + NH_2OH(aq) \longrightarrow Tl(OH)_3(s) + N_2H_4(aq)$ (basic)
(f) $MnO_2(s) + O_2(g) \longrightarrow MnO_4^-$ (basic)
(g) $S_2O_3^{2-} + Cl_2(g) \longrightarrow HSO_4^- + Cl^-$ (acidic)

Why are both methods of balancing equations for redox reactions treated in detail? Reviewers were about equally divided as to which method they preferred.

F.2   HALF-REACTION METHOD

The steps for balancing equations for reactions in aqueous solution by the half-reaction method are summarized below.

Balancing Redox Equations by the Half-Reaction Method

1. Divide the reaction into an oxidation half-reaction and a reduction half-reaction and balance each half-reaction.
2. Balance elements other that oxygen and hydrogen first.
3. Use H_2O to balance O.
4. Use H^+ to balance H.
5. Use e^- to balance charge.
6. If reaction takes place in basic solution, add enough OH^- ions to each side to neutralize any H^+ ions shown. Combine H^+ and OH^- to form H_2O.
7. Combine the two half-reactions so that electrons cancel and simplify.
8. Check your work: Be sure there are the same number of each kind of atom and the same net charge on both sides. Add symbols showing whether substances are solids, liquids, gases, or an aqueous solution.

Hydroxide ions are added to the equations for half-reactions rather than to the equation for the overall reaction so that the half-reactions shown in the table of standard reduction potentials (Table 19.1) will be obtained.

Acidic Solution

The reaction

$$MnO_4^- + SCN^- \longrightarrow Mn^{2+} + SO_4^{2-} + CO_2(g) + N_2(g)$$

takes place in acidic solution. This reaction is an example of a reaction that is more easily balanced by the half-reaction method than by the change-in-oxidation-number method because assignment of oxidation numbers to sulfur and carbon in the thiocyanate ion is difficult. The thiocyanate ion SCN^- is composed of three atoms not covered by the rules for assigning oxidation numbers, and sulfur and carbon have similar electronegativities. Let's take this reaction as our first example of the half-reaction method.

Dimethylhydrazine, $(CH_3)_2NNH_2$, which was used as fuel for the Apollo Lunar Lander, is another example of a compound that cannot easily have oxidation numbers assigned to its atoms. There is no rule for assigning oxidation numbers to either C or N. Neither is there an easily predictable difference in electronegativity between the two N atoms.

Step 1. *Divide the overall reaction into half-reactions of oxidation and reduction.* Assigning oxidation numbers for the elements in species other than the thiocyanate ion is no problem:

$$\overset{+7\ -2}{MnO_4^-} + SCN^- \longrightarrow \overset{+2}{Mn^{2+}} + \overset{+6\ -2}{SO_4^{2-}} + \overset{+4\ -2}{CO_2} + \overset{0}{N_2}$$

The oxidation number of Mn in MnO_4^- is $+7$, and the oxidation number of Mn in Mn^{2+} is $+2$. Manganese is reduced, and the reduction half-reaction is

$$MnO_4^- \longrightarrow Mn^{2+}$$

The oxidation half-reaction must involve the SCN^- ion; therefore, the oxidation half-reaction must be

$$SCN^- \longrightarrow SO_4^{2-} + CO_2 + N_2$$

Thiocyanate ions are formed when cyanide ion, CN^-, is detoxified in living systems.

Let's begin by balancing the oxidation half-reaction.

Step 2. *Balance elements other than oxygen and hydrogen first.* Coefficients of 2 must be placed in front of the formulas for the thiocyanate ion, the sulfate ion, and the carbon dioxide molecule:

$$2SCN^- \longrightarrow 2SO_4^{2-} + 2CO_2 + N_2 \qquad \text{(F.8)}$$

Step 3. *Use H_2O to balance O.* The reaction takes place in aqueous solution, and water is readily available. Expression F.8 shows 12 O on the right and no O on the left. Addition of 12 H_2O to the left side makes 12 O on each side:

$$12H_2O + 2SCN^- \longrightarrow 2SO_4^{2-} + 2CO_2 + N_2$$

But now there are 24 H on the left and no H on the right.

Step 4. *Use H^+ to balance H.* In acidic solution, hydrogen ions are readily available. Add 24 H^+ to the right side to balance hydrogen:

$$12H_2O + 2SCN^- \longrightarrow 2SO_4^{2-} + 2CO_2 + N_2 + 24H^+$$

The half-reaction now shows the same number of each kind of atom on both sides. However, charge is not balanced. The net charge on the left side is $2-$, whereas the net charge on the right side is $20+$, a difference of 22.

Step 5. *Use e^- to balance charge.* Add 22 e^- to the right side to balance charge:

$$12H_2O + 2SCN^- \longrightarrow 2SO_4^{2-} + 2CO_2 + N_2 + 24H^+ + 22e^-$$

The charge is now $2-$ on both sides.

⊙ Remember that electrons appear on the product side of the equation for an oxidation half-reaction and on the reactant side of the equation for a reduction half-reaction.

Step 6. *If reaction takes place in basic solution, ...* This step is unnecessary because the reaction takes place in acidic solution.

The oxidation half-reaction is balanced. After one half-reaction has been balanced, the other half-reaction must be balanced. The same procedure is followed.

Step 2 is not necessary for the reduction half-reaction because 1 Mn is shown on each side:

$$MnO_4^- \longrightarrow Mn^{2+}$$

Step 3. There are 4 O on the left and no O on the right. Add 4 H_2O to the right side to balance oxygen:

$$MnO_4^- \longrightarrow Mn^{2+} + 4H_2O$$

Step 4. Now there are 8 H on the right and no H on the left. Add 8 H^+ to the left side to balance hydrogen:

$$8H^+ + MnO_4^- \longrightarrow Mn^{2+} + 4H_2O$$

Step 5. Five electrons must be added to the left side to balance charge:

$$5e^- + 8H^+ + MnO_4^- \longrightarrow Mn^{2+} + 4H_2O$$

The net charge on both sides is now $2+$.

Step 6 is not necessary since the reaction takes place in acidic solution.

Remember that electrons are matter and cannot be created or destroyed. Therefore, in the overall reaction, the same number of electrons released in the oxidation half-reaction must be used up in the reduction half-reaction.

Step 7. *Combine the equations for the two half-reactions so that electrons cancel and simplify.* We must multiply the oxidation half-reaction by 5 and the reduction half-reaction by 22 so that electrons will cancel when the half-reactions are added and add them:

$$5(12H_2O + 2SCN^- \longrightarrow 2SO_4^{2-} + 2CO_2 + N_2 + 24H^+ + 22e^-)$$
$$+ 22(5e^- + 8H^+ + MnO_4^- \longrightarrow Mn^{2+} + 4H_2O)$$

$$\overline{110e^- + \overset{56}{176}H^+ + 22MnO_4^- + 60H_2O + 10SCN^- \longrightarrow}$$
$$22Mn^{2+} + \overset{28}{88}H_2O + 10CO_2 + 10SO_4^{2-} + 5N_2 + 120H^+ + 110e^-$$

Step 8. *Check your work.* The equation shows 56 H, 22 Mn, 88 O, 10 S, 10 C, 10 N, and a net charge of 24+ on each side. The coefficients have no common divisor, and the net ionic equation is

$$56H^+ + 22MnO_4^- + 10SCN^- \longrightarrow$$
$$22Mn^{2+} + 28H_2O(l) + 10CO_2(g) + 10SO_4^{2-} + 5N_2(g)$$

SAMPLE PROBLEM

F.5 Use the half-reaction method to balance the equation for the reaction

$$Fe^{2+} + Cr_2O_7^{2-} \longrightarrow Fe^{3+} + Cr^{3+}$$

that takes place in acidic solution. Identify the element oxidized, the element reduced, the species that is acting as oxidizing agent, and the species that is acting as reducing agent.

Solution Step 1. Oxidation numbers can be assigned to all the elements involved in this reaction:

$$\overset{+2}{Fe^{2+}} + \overset{+6\,-2}{Cr_2O_7^{2-}} \longrightarrow \overset{+3}{Fe^{3+}} + \overset{+3}{Cr^{3+}}$$

The reaction

$$Fe^{2+} \longrightarrow Fe^{3+}$$

is the half-reaction of oxidation, and the reaction

$$Cr_2O_7^{2-} \longrightarrow Cr^{3+} \tag{F.9}$$

is the half-reaction of reduction.

Let's balance the half-reaction of oxidation first. This half-reaction is very easy to balance. The element other than oxygen or hydrogen, Fe, is already balanced. There is no oxygen or hydrogen on either side. All that's necessary to balance the half-reaction is to balance the charge. Addition of one electron to the right side results in a net charge of 2+ on each side:

$$Fe^{2+} \longrightarrow Fe^{3+} + 1e^- \tag{F.10}$$

Step 2. To balance the half-reaction of reduction, a coefficient of 2 must be placed in front of the formula for the chromium(III) ion:

$$Cr_2O_7^{2-} \longrightarrow 2Cr^{3+}$$

Step 3. Seven water molecules must be added to the right side to balance oxygen:

$$Cr_2O_7^{2-} \longrightarrow 2Cr^{3+} + 7H_2O$$

Step 4. Fourteen hydrogen ions must be added to the left side to balance hydrogen:

$$14H^+ + Cr_2O_7^{2-} \longrightarrow 2Cr^{3+} + 7H_2O \tag{F.11}$$

Step 5. Six electrons must be added to the left side of expression F.11 to balance charge:

$$6e^- + 14H^+ + Cr_2O_7^{2-} \longrightarrow 2Cr^{3+} + 7H_2O \qquad (F.12)$$

The half-reaction shown by equation F.12 is balanced both chemically and electrically and is the net ionic equation for the reduction half-reaction.

Step 7. The oxidation half-reaction must be multiplied by 6 before the half-reactions are added:

$$6(Fe^{2+} \longrightarrow Fe^{3+} + 1e^-) \qquad 6(F.10)$$
$$+ (6e^- + 14H^+ + Cr_2O_7^{2-} \longrightarrow 2Cr^{3+} + 7H_2O) \qquad +(F.12)$$
$$\overline{6Fe^{2+} + 6e^- + 14H^+ + Cr_2O_7^{2-} \longrightarrow}$$
$$6Fe^{3+} + 6e^- + 2Cr^{3+} + 7H_2O \qquad (F.13)$$

Step 8. Equation F.13 shows 6 Fe, 14 H, 2 Cr, and 7 O and a net charge of 24+ on each side; the coefficients have no common denominator.

Step 9. The net ionic equation is

$$6Fe^{2+} + 14H^+ + Cr_2O_7^{2-} \longrightarrow 6Fe^{3+} + 2Cr^{3+} + 7H_2O(l) \qquad (F.14)$$

In the reaction shown by equation F.14, Fe^{2+} reduces chromium and is the reducing agent; Fe^{2+} is oxidized to Fe^{3+}. Oxidation of Fe^{2+} is brought about by the dichromate ion, $Cr_2O_7^{2-}$, which is the oxidizing agent. Chromium is reduced from the +6 oxidation state (in $Cr_2O_7^{2-}$) to the +3 state (in Cr^{3+} ion).

▶ Don't forget to write the symbols (s), (l), (g), and (aq) in the final equation. The symbol (aq) after formulas for ions may be assumed (Section 4.4).

PRACTICE PROBLEM

F.3 Use the half-reaction method to balance the equations for the following reactions, which take place in acidic solution. For each reaction, identify the element oxidized, the element reduced, the species that is acting as oxidizing agent, and the species that is acting as reducing agent.
(a) $MnO_4^- + H_2C_2O_4(aq) \longrightarrow Mn^{2+} + CO_2(g)$
(b) $Zn(s) + NO_3^- \longrightarrow Zn^{2+} + N_2O(g)$
(c) $As_2O_3(s) + NO_3^- \longrightarrow H_3AsO_4(aq) + N_2O_3(aq)$
(d) $HIO_3(aq) + I^- \longrightarrow I_3^-$
(e) $ReCl_5(s) \longrightarrow HReO_4(aq) + ReO_2(s) + Cl^-$
(f) $Zn(s) + VO^{2+} \longrightarrow Zn^{2+} + V^{3+}$

Basic Solution

Oxidation–reduction reactions can be carried out in basic solution as well as in acidic solution. If a reaction is carried out in basic solution, hydroxide ions, OH^-, not hydrogen ions, should appear in the equation for the overall reaction. However, if OH^- and H_2O are used instead of H^+ and H_2O to balance the equations for the half-reactions, balancing the half-reactions is more difficult because addition of either OH^- or H_2O changes *both* the number of hydrogens *and* the number of oxygens. The simplest procedure is to balance the equations for the half-reactions with H^+ and H_2O and then add enough hydroxide ions *to each side* of each half-reaction to neutralize any hydrogen ions. Sample Problem F.6 shows how to balance the equation for a redox reaction that takes place in basic solution.

▶ Some hydrogen ions are present in all aqueous solutions, even basic ones (Section 15.2).

F.6   Write the net ionic equation for the reaction of glycerol, $HOCH_2CH(OH)CH_2OH$ ($C_3H_8O_3$), with permanganate ion, MnO_4^-, in basic solution:

$$C_3H_8O_3(aq) + MnO_4^- \longrightarrow CO_3^{2-} + MnO_4^{2-}$$

Solution   Step 1. For this reaction, one half-reaction involves carbon,

$$C_3H_8O_3 \longrightarrow CO_3^{2-} \tag{F.15}$$

and the other half-reaction involves manganese,

$$MnO_4^- \longrightarrow MnO_4^{2-} \tag{F.16}$$

Step 2. Let's balance the half-reaction shown by equation F.15 first. Balancing C gives

$$C_3H_8O_3 \longrightarrow 3CO_3^{2-}$$

Step 3. Six water molecules must be added to the left side:

$$6H_2O + C_3H_8O_3 \longrightarrow 3CO_3^{2-}$$

Step 4. Twenty H^+ must be added to the right side:

$$6H_2O + C_3H_8O_3 \longrightarrow 3CO_3^{2-} + 20H^+$$

Step 5. Fourteen electrons must be added to the right side:

$$6H_2O + C_3H_8O_3 \longrightarrow 3CO_3^{2-} + 20H^+ + 14e^- \tag{F.17}$$

Step 6. Because the reaction takes place in basic solution, add enough hydroxide ions to each side of the half-reaction to neutralize all hydrogen ions. Equation F.17 shows 20 H^+ on the right side; 20 OH^- are required to neutralize 20 H^+. Therefore, 20 OH^- must be added to *both* sides:

$$6H_2O + C_3H_8O_3 + 20OH^- \longrightarrow 3CO_3^{2-} + 20H^+ + 20OH^- + 14e^- \tag{F.18}$$

The $20H^+$ and $20OH^-$ on the right side of equation F.18 will immediately combine to form $20H_2O$. Equation F.18 becomes

$$6H_2O + C_3H_8O_3 + 20OH^- \longrightarrow 3CO_3^{2-} + 20H_2O + 14e^-$$

Simplification by subtraction of six molecules of H_2O from each side gives

$$C_3H_8O_3 + 20OH^- \longrightarrow 3CO_3^{2-} + 14H_2O + 14e^- \tag{F.19}$$

This half-reaction is now balanced both chemically and electrically and shows hydroxide ions, not hydrogen ions, as it should because the reaction takes place in basic solution.

The other half-reaction

$$MnO_4^- \longrightarrow MnO_4^{2-} \tag{F.16}$$

is balanced chemically. The net charge on the left is $1-$, and the net charge on the right is $2-$. One electron must be added to the left side:

$$1e^- + MnO_4^- \longrightarrow MnO_4^{2-} \tag{F.20}$$

Step 7. For electrons to cancel when the half-reactions are added, equation F.20 must be multiplied by 14:

$$(C_3H_8O_3 + 20OH^- \longrightarrow 3CO_3^{2-} + 14H_2O + 14e^-) \tag{F.19}$$
$$+14(1e^- + MnO_4^- \longrightarrow MnO_4^{2-}) \qquad +14(F.20)$$

$$\overline{14e^- + 14MnO_4^- + C_3H_8O_3 + 20OH^- \longrightarrow}$$
$$14MnO_4^{2-} + 3CO_3^{2-} + 14H_2O + 14e^-$$

Step 8. There are 14 Mn, 79 O, 28 H, and 3 C on each side, and the net charge on each side is $34-$. The net ionic equation is

$$14MnO_4^- + C_3H_8O_3(aq) + 20OH^- \longrightarrow 14MnO_4^{2-} + 3CO_3^{2-} + 14H_2O(l)$$

⬥ High concentrations of both H^+ and OH^- cannot exist in the same solution.

F.4   For the reaction in Sample Problem F.6, (a) which half-reaction is the half-reaction of oxidation? (b) Which half-reaction is the half-reaction of reduction? (c) Which element is oxidized? (d) Which element is reduced? (e) Which species is the oxidizing agent? (f) Which species is the reducing agent?

F.5   Use the half-reaction method to write balanced net ionic equations for the following reactions, which take place in basic solution. For each reaction, identify the element oxidized, the element reduced, the species that is acting as oxidizing agent, and the species that is acting as reducing agent.
(a) $Cr(OH)_6^{3-} + BrO^- \longrightarrow CrO_4^{2-} + Br^-$
(b) $H_2O_2(aq) + Cl_2O_7(aq) \longrightarrow ClO_2(aq) + O_2(g)$
(c) $CN^- + CrO_4^{2-} \longrightarrow CNO^- + Cr(OH)_4^-$

Writing Molecular Equations from Net Ionic Equations

Balancing equations for oxidation–reduction reactions by the half-reaction method gives net ionic equations. If an equation for a redox reaction is to be used to calculate the quantity of each reactant needed to carry out a reaction, a molecular equation is needed. The actual compounds to be used, not just the ions that take part in the reaction, must be known. For example, suppose the source of the Fe^{2+} ion in the reaction

$$Fe^{2+} + Cr_2O_7^{2-} \longrightarrow Fe^{3+} + Cr^{3+}$$

is $FeSO_4$, the source of the $Cr_2O_7^{2-}$ ion is $K_2Cr_2O_7$, and the acid used is H_2SO_4. The net ionic equation for the reaction that takes place in acidic solution (equation F.14) is

$$6Fe^{2+} + 14H^+ + Cr_2O_7^{2-} \longrightarrow 6Fe^{3+} + 2Cr^{3+} + 7H_2O(l)$$

One SO_4^{2-} must accompany each Fe^{2+}, and one SO_4^{2-} must accompany each two H^+; a total of

$$6Fe^{2+}\left(\frac{1\ SO_4^{2-}}{1\ Fe^{2+}}\right) + 14H^+\left(\frac{1SO_4^{2-}}{2H^+}\right) = 13SO_4^{2-}$$

should be shown on the reactant side of the complete ionic equation. Two K^+ must accompany each $Cr_2O_7^{2-}$, and two K^+ should be shown on the reactant side of the complete ionic equation. The SO_4^{2-} ions and the K^+ ions are spectator ions; the same number should be shown on the product side of the complete ionic equation. The complete ionic equation is

$$6Fe^{2+} + 14H^+ + 13SO_4^{2-} + 2K^+ + Cr_2O_7^{2-} \longrightarrow$$
$$6Fe^{3+} + 2Cr^{3+} + 13SO_4^{2-} + 2K^+ + 7H_2O(l)$$

The molecular equation is obtained from the complete ionic equation by combining ions in the correct proportions (so that the net charges on compounds are zero). The molecular equation is

$$6FeSO_4(aq) + 7H_2SO_4(aq) + K_2Cr_2O_7(aq) \longrightarrow$$
$$3Fe_2(SO_4)_3(aq) + Cr_2(SO_4)_3(aq) + K_2SO_4(aq) + 7H_2O(l)$$

F.6   In Sample Problem F.6, we found that the net ionic equation for the reaction of glycerol with permanganate in basic solution is

$$14MnO_4^- + C_3H_8O_3(aq) + 20OH^- \longrightarrow 14MnO_4^{2-} + 3CO_3^{2-} + 14H_2O(l)$$

If the source of the MnO_4^- ion is potassium permanganate, $KMnO_4$, and the source of the OH^- ion is potassium hydroxide, KOH, (a) write the complete ionic equation for the reaction. (b) write the molecular equation for the reaction.

F.7   Write the molecular equation for each of the following redox reactions:
(a) $C_6H_5OH(aq) + SO_3(g) \longrightarrow CO_2(g) + SO_2(g)$ (acidic solution)
(b) $FeSO_4(aq) + KMnO_4(aq) + H_2SO_4(aq) \longrightarrow MnSO_4(aq) + Fe_2(SO_4)_3(aq)$
(c) $HNO_3(aq) + Cu(s) \longrightarrow Cu(NO_3)_2(aq) + NO(g)$
(d) $NaCrO_2(aq) + NaClO(aq) + NaOH(aq) \longrightarrow Na_2CrO_4(aq) + NaCl(aq)$
(e) $Na_2SO_3(aq) + NaMnO_4(aq) \longrightarrow Na_2SO_4(aq) + MnO_2(s)$ (basic solution)

F.8   Write the molecular equation for the reaction of copper(II) sulfide with nitric acid. Sulfur and nitric oxide gas are formed.

APPENDIX G

Calculation of K_{sp}

Values of K_{sp} in tables such as the one in the *Handbook of Chemistry and Physics* are calculated from ΔG°. Solubility equilibria involve ions in solution. So far as calculations are concerned, the thermodynamic quantities ΔH_f°, ΔG_f°, and S° for ions are no different from thermodynamic quantities for solids, liquids, and gases. However, you need to understand how thermodynamic quantities for ions in solution are defined.

Because a solution cannot have a net electric charge, a solution must contain both positive and negative ions, and the properties of cations and anions cannot be measured separately. Therefore, by agreement among scientists, ΔH_f°, ΔG_f°, and S° for the hydrogen ion in aqueous solution $H^+(aq)$ are all set equal to zero. The thermodynamic properties of all other ions in aqueous solution are measured relative to those of the hydrogen ion. Determination of ΔH_f° for the chloride ion Cl^- in aqueous solution from the heat of solution of hydrogen chloride gas is an example. For the reaction

$$HCl(g) \longrightarrow H^+(aq) + Cl^-(aq)$$

$\Delta H_{rxn}^\circ = -74.54$ kJ. According to Hess's law (Section 6.9),

$$\Delta H_{rxn}^\circ = \Sigma\, n_p\, \Delta H_f^\circ(\text{product}) - \Sigma\, n_r\, \Delta H_f^\circ(\text{reactant}) \qquad (6.23)$$

where Σ means sum of, n_p is the number of moles of each product, $\Delta H_f^\circ(\text{product})$ is the standard enthalpy of formation of the product, n_r is the number of moles of each reactant, and $\Delta H_f^\circ(\text{reactant})$ is the standard enthalpy of formation of the reactant.

For the solution of hydrogen chloride gas in water

$$\Delta H_{rxn}^\circ = \{1 \text{ mol } H^+ \cdot \Delta H_f^\circ[H^+(aq)] + 1 \text{ mol } Cl^- \cdot \Delta H_f^\circ[Cl^-(aq)]\}$$
$$- 1 \text{ mol } HCl \cdot \Delta H_f^\circ[HCl(g)] \qquad (G.1)$$

From Appendix D, $\Delta H_f^\circ[\text{HCl}(g)] = -92.31$ kJ/mol. Solution of equation G.1 for $\Delta H_f^\circ[\text{Cl}^-(aq)]$ and substitution gives

$$\frac{-74.54 \text{ kJ} - 1 \text{ mol H}^+\left(0\frac{\text{kJ}}{\text{mol H}^+}\right) + 1 \text{ mol HCl}\left(-92.31\frac{\text{kJ}}{\text{mol HCl}}\right)}{1 \text{ mol Cl}^-}$$

● This calculation is similar to the one in Sample Problem 6.7.

$$= \Delta H_f^\circ[\text{Cl}^-(aq)] = -166.85\frac{\text{kJ}}{\text{mol Cl}^-}$$

In Chapter 17, we learned that, according to the third law of thermodynamics, the entropy of perfect crystalline substances is zero at the absolute zero of temperature, but it is impossible to reach absolute zero, and standard entropies are always positive. In Appendix D, some ions (Cu^{2+}, Mg^{2+}, OH^-, and S^{2-}) have negative entropies. How can this be? Remember that the entropy of ions in aqueous solution is relative to the entropy of $H^+(aq)$, which is arbitrarily set equal to zero. Hydration of ions that are either small or highly charged organizes the water molecules around the ion, and the entropy of the solution is lower than the entropy of an aqueous solution that contains hydrogen ions. ["Hydrogen ions" are *not* very small because they are really hydrated hydronium ions, $H_3O^+(aq)$.]

Sample Problem G.1 shows how to calculate the value of K_{sp} from ΔG°.

● This problem is similar to Sample Problem 17.10.

SAMPLE PROBLEM

G.1 Calculate the value of K_{sp} at 25.0 °C for AgBr from ΔG°.

Solution The relationship between ΔG° and the value of the equilibrium constant is given by equation 17.7:

$$\Delta G^\circ = -RT \ln K \qquad (17.7)$$

First, we must calculate ΔG° for the equilibrium using free energies of formation at 25 °C from Appendix D:

$$\text{AgBr}(s) \rightleftharpoons \text{Ag}^+(aq) + \text{Br}^-(aq)$$

ΔG_f°, kJ/mol -96.95 77.16 -103.97

For any reaction,

$$\Delta G_{rxn}^\circ = \Sigma\, n_p\, \Delta G_f^\circ(\text{product}) - \Sigma\, n_r\, \Delta G_f^\circ(\text{reactant}) \qquad (17.4)$$

where Σ means sum of, n_p is the number of moles of each product, ΔG_f°(product) is the standard free energy of formation of the product, n_r is the number of moles of each reactant, and ΔG_f°(reactant) is the standard free energy of formation of the reactant. For the forward reaction in this problem,

$$\Delta G^\circ = \left[1 \text{ mol Ag}^+\left(77.16\frac{\text{kJ}}{\text{mol Ag}^+}\right) + 1 \text{ mol Br}^-\left(-103.97\frac{\text{kJ}}{\text{mol Br}^-}\right)\right]$$
$$- 1 \text{ mol AgBr}\left(-96.95\frac{\text{kJ}}{\text{mol AgBr}}\right) = 70.14 \text{ kJ}$$

Solution of equation 17.7 for $\ln K$ and substitution gives

$$\frac{70.14 \text{ kJ}}{-\left(0.008\,315\frac{\text{kJ}}{\text{K}}\right)(298.2 \text{ K})} = \ln K_{sp} = -28.29$$

and

$$5.2 \times 10^{-13} = K_{sp}$$

Because we use molarities, not activities, in our calculations, we use 5×10^{-13} for the value of K_{sp} for AgBr.

PRACTICE PROBLEM

G.1 Calculate the value of K_{sp} at 25 °C for Ag_2SO_4 from ΔG°.

Qualitative Analysis for Metal Ions

Qualitative analysis is the determination of the identity of a pure substance or of the components of a mixture. The classical qualitative analysis scheme is designed to show the presence or absence of 20–25 cations. Unfortunately, reagents are not known that can be used to test for each ion in the presence of all the other ions. The other ions interfere with the identification. For example, the test for Pb^{2+} is formation of a yellow precipitate with potassium chromate solution. But Ba^{2+} also gives a yellow precipitate and Ag^+, a red precipitate. A yellow precipitate of lead chromate could not be seen in the presence of a red precipitate of silver chromate.

Therefore, an aqueous solution containing the metal ions is divided into groups by selective precipitation. After a group has been precipitated, the individual ions in the groups are separated and identified. The groups that are easiest to separate are precipitated first. For example, most chlorides are soluble except $AgCl$, Hg_2Cl_2, and $PbCl_2$. Therefore, addition of hydrochloric acid precipitates only Ag^+, Hg_2^{2+}, and Pb^{2+}. All the other ions are left in solution.

The ions that are precipitated must be removed as completely as possible so that they do not also precipitate when other groups are separated. In lab, directions must be followed carefully. For example, in the separation of the first group, an excess of chloride ion is desirable because it decreases the solubility of the chlorides by the common ion effect. However, if too large an excess is used, the precipitates dissolve as a result of complex ion formation such as

$$AgCl(s) + Cl^- \longrightarrow [AgCl_2]^-$$

The second group of cations to be separated is composed of ions that form very insoluble sulfides and the third, mainly of ions that form more soluble sulfides. *For salts with similar formulas,* salts with smaller values of K_{sp} are less soluble. For example, CdS, CuS, and SnS are much less soluble than FeS, MnS, and ZnS (see Table H.1).

As a result of the very low value of K_{a2} for H_2S, S^{2-} like O^{2-} is completely hydrolyzed in aqueous solution. The solubility equilibrium for a sulfide should be written

$$MS(s) + H_2O(l) \rightleftharpoons M^{2+} + HS^- + OH^-$$

not

$$\xcancel{MS(s) \rightleftharpoons M^{2+} + S^{2-}}$$

where M represents a metal. Therefore, $K_{sp} = [M^{2+}][HS^-][OH^-]$, not $[M^{2+}][S^{2-}]$. The less soluble sulfides are separated from the more soluble sulfides by adjusting the concentration of hydrogen ion. In acidic solution, the equilibrium

$$H_2S(aq) \rightleftharpoons H^+ + HS^-$$

is shifted to the left by H^+, and the concentration of HS^- is low. Only the very insoluble sulfides of group 2 precipitate. The pH is then increased to shift the $H_2S–HS^-$ equilibrium to the right so that the concentration of HS^- is high enough to precipitate the more soluble sulfides of group 3.

TABLE H.1

Solubility Product Constants for Sulfides at 25 °C

Compound	K_{sp}
CdS	1.2×10^{-30}
CuS	1.2×10^{-37}
SnS	5×10^{-31}
FeS	1.9×10^{-18}
MnS	4×10^{-15}
ZnS	9×10^{-23}

The values in Table H.1 are slightly different from the values in Table 16.5 for two reasons: the temperature is different and the values in Table H.1 are calculated for the

$$MS(s) + H_2O(l) \rightleftharpoons M^{2+} + HS^- + OH^-$$

equilibrium, whereas the values in Table 16.5 are for the $MS(s) \rightleftharpoons M^{2+} + S^{2-}$ *equilibrium.*

Group 4 consists of the ions of the alkaline earth metals, which are precipitated as carbonates, phosphates, or sulfates. The ions Na^+ and K^+, which form few insoluble compounds, are left in solution. Because sodium ions, and possibly some potassium ions as well, are added in the course of the separation, samples of the original solution are used to test for Na^+ and K^+. Sodium ion is identified by the yellow color that a solution containing sodium ion gives to a flame. This yellow color is invisible when the flame is viewed through blue glass. Potassium ion gives a purple–violet flame when the flame is observed through blue glass. The colors of the flames result from the emission of light of characteristic wavelengths. Flame tests are the predecessors of atomic absorption spectroscopy (see page 235). The separation scheme is summarized in ▬Figure H.1.

The separation and identification of the group 1 ions provides a simple example of the separation and identification of the ions in a group. Lead(II) chloride, which is more soluble than silver and mercury(I) chlorides, is separated by treating the precipitate with hot water. Only lead(II) chloride dissolves. Precipitate and solution

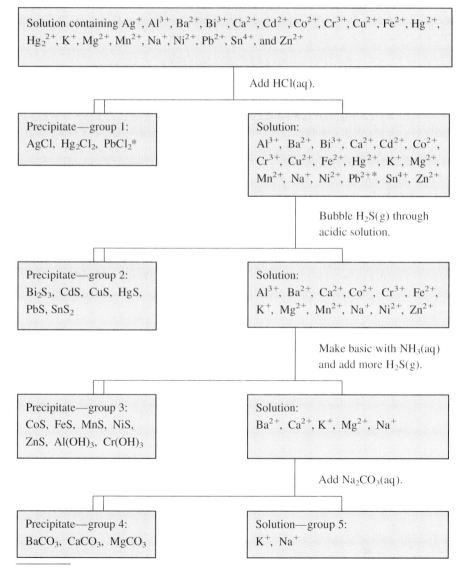

○ Note that the groups in the qualitative analysis scheme are not *the same as groups in the periodic table.*

▬ FIGURE H.1   Diagram of the separation of cations by selective precipitation.

*Lead appears in both the precipitate and the solution from the precipitation of group 1 because lead(II) chloride is slightly soluble.

are separated and the solution is treated with potassium chromate solution. If Pb^{2+} ion is present, a yellow precipitate of $PbCrO_4$ forms:

$$Pb^{2+} + CrO_4^{2-} \longrightarrow PbCrO_4(s)$$
$$\text{yellow}$$

The precipitate from which Pb^{2+} has been removed is treated with concentrated aqueous ammonia. If the precipitate contains Hg_2Cl_2, an oxidation–reduction reaction takes place and the precipitate turns black:

$$Hg_2Cl_2(s) + 2NH_3(aq) \longrightarrow Hg(l) + HgNH_2Cl(s) + NH_4^+ + Cl^-$$
$$\text{black} \qquad \text{white}$$

Thus, formation of a black precipitate proves that the unknown solution contains Hg_2^{2+}. Treatment of the precipitate with additional concentrated aqueous ammonia dissolves silver chloride, if present, by formation of diamminesilver ion:

$$AgCl(s) + 2NH_3(aq) \longrightarrow [Ag(NH_3)_2]^+ + Cl^-$$

Acidification of the ammoniacal solution with nitric acid results in the precipitation of silver chloride if Ag^+ is present in the unknown solution. Hydrogen ion from the acid reacts with ammonia, shifting the equilibrium

$$AgCl(s) + 2NH_3(aq) \rightleftharpoons [Ag(NH_3)_2]^+ + Cl^-$$

to the left. Silver chloride is not soluble in dilute nitric acid because it does not react with it. (Both possible products, $AgNO_3$ and HCl, are soluble, strong electrolytes.)

As you can see from this brief discussion, the classical qualitative analysis scheme illustrates much of the subject matter of general chemistry—for example, acid–base, solubility, and complex ion equilibria and oxidation–reduction reactions. In the real world, however, samples can contain other ions than the ones in the classical scheme; the identity of the ions in a sample is frequently not known. Samples are often very small and may not be replaceable. Therefore, few working chemists use the classical separation scheme any longer. Instead complex mixtures are usually separated by chromatography, and ions are identified by atomic absorption spectroscopy. Nevertheless, the principles on which the classical qualitative analysis scheme is based are still applied in quantitative analysis and in many industrial processes. General chemistry lab experiments using the classical scheme still provide good training in making observations, following directions, and thinking logically.

The combined text and laboratory manual Qualitative Inorganic Analysis *by William T. Scroggins introduces qualitative analysis techniques and their underlying principles and provides a range of experiments to guide students in analyzing solutions.* Inorganic Qualitative Analysis for IBM *by Trinity Software simulates the laboratory study of seven inorganic cations and four anions. Both are available from the publisher of this text.*

SAMPLE PROBLEM

H.1   A solution contains 0.10 M Ag^+ and 0.10 M Pb^{2+}. Can Ag^+ and Pb^{2+} be separated completely by selective precipitation with Cl^- if "completely" means that 99.99% of the Ag^+ can be precipitated without precipitating Pb^{2+}? (Remember that you can never reduce $[Ag^+]$ to zero because $[Ag^+][Cl^-]$ must equal 1.6×10^{-10}, the value of the solubility product constant for AgCl.)

Solution   If 99.99% of the Ag is precipitated, $(100.00 - 99.99) = 0.01\%$ must be left. The concentration of silver ion will be

$$\left(\frac{0.01}{100}\right)(0.10 \text{ M}) = 1 \times 10^{-5} \text{ M}$$

From Table 16.6, K_{sp} for AgCl is $1.6 \times 10^{-10} = [Ag^+][Cl^-]$. For $[Ag^+]$ to be 1×10^{-5},

$$[Cl^-] = \frac{1.6 \times 10^{-10}}{1 \times 10^{-5}} = 2 \times 10^{-5}$$

Will $PbCl_2$ precipitate if $[Cl^-] = 2 \times 10^{-5}$? From Table 16.6, the value of K_{sp} for $PbCl_2$ is $1.5 \times 10^{-5} = [Pb^{2+}][Cl^-]^2$, and

$$Q_{sp} = (0.10)(2 \times 10^{-5})^2 = 2 \times 10^{-11} < K_{sp}$$

No Pb^{2+} will precipitate. "Complete" separation of Ag^+ and Pb^{2+} can be achieved.

PRACTICE PROBLEMS

H.1 A solution contains 0.10 M Ca^{2+} and 0.10 M Sr^{2+}. Can Ca^{2+} and Sr^{2+} be separated completely by selective precipitation with SO_4^{2-} if "completely" means that 99.99% of the Sr^{2+} can be precipitated without precipitating Ca^{2+}?

H.2 (a) Make a diagram similar to Figure H.1 showing the separation and identification of the group 1 ions. (b) Describe exactly what will be observed if a solution containing Ag^+ and Hg_2^{2+} is analyzed.

H.3 How can you distinguish between the members of each of the following pairs?
(a) NaCl(s) and KCl(s)   (b) KCl(s) and $PbCl_2$(s)
(c) $Cu(NO_3)_2$(aq) and $Ca(NO_3)_2$(aq)

H.4 An unknown compound was soluble in water. No precipitate formed when HCl(aq) was added to an aqueous solution of the unknown or when H_2S was bubbled through the acidic solution. However, a precipitate did form after the solution was made basic, and more H_2S was bubbled through. Which one of the following was the precipitate?
(a) CoS   (b) CdS   (c) $CaCO_3$ (d) $Co(OH)_2$ (e) CuS

Answers to Text Problems

CHAPTER 1

1.1 The liquid in Figures 1.1(b) and (c) indicates that the solid in Figure 1.1(a) is melting. **1.2** *solid phases:* ice cubes and glass; *liquid phase:* cola; *gas phases:* "fizz" and air **1.3 (a)** physical **(b)** chemical **(c)** physical **(d)** physical **(e)** chemical **1.4** No. All samples of a substance have the same properties. **1.5 (a)** element **(b)** heterogeneous mixture **(c)** element **(d)** compound **(e)** solution **1.6 (a)** 13 **(b)** 6 **(c)** 10 **(d)** 47 **(e)** 92 **1.7 (a)** 13, 13+, 13 **(b)** 6, 6+, 6 **(c)** 10, 10+, 10 **(d)** 47, 47+, 47 **(e)** 92, 92+, 92 **1.8 (a)** Fe **(b)** Ar **(c)** P **(d)** Pb **(e)** U **1.9 (a)** zinc **(b)** selenium **(c)** plutonium **(d)** xenon **(e)** barium **1.10** very explosively **1.11 (a)** Ne **(b)** K **(c)** Ba **(d)** Cl **(e)** N **(f)** Cl **1.12** Na (sodium), Mg (magnesium), Al (aluminum), Si (silicon), P (phosphorus), S (sulfur), Cl (chlorine), Ar (argon) **1.13** seventh **1.14 (a)** calcium, molybdenum, uranium **(b)** antimony **(c)** argon, bromine **1.15 (a)** antimony, argon, bromine, calcium **(b)** molybdenum, uranium (inner transition element) **1.16 (a)** 1 atom C, 2 atoms O **(b)** 1 atom N, 3 atoms H **(c)** 2 atoms H, 1 atom S, 4 atoms O **(d)** 3 atoms H, 3 atoms Cl **1.17 (a)** N_2O **(b)** 4 atoms N **(c)** SO_2 **1.18 (a)** N_2 **(b)** Cl_2 **1.19 (a)** element **(b)** compound **(c)** compound **(d)** element **(e)** compound **1.20 (a)** ion **(b)** molecule **(c)** molecule **(d)** ion **1.21 (a)** cation **(b)** anion **(c)** cation **1.22 (a)** S^{2-} **(b)** Al^{3+} **1.23** Li_2O **1.24** AlF_3 **1.25 (a)** CaO **(b)** Al_2S_3 **1.26 (a)** ions **(b)** molecules **(c)** molecules **(d)** ions **(e)** ions **1.27** $HC_2O_4^-$ **1.28** ClO_2^- **1.29 (a)** $(NH_4)_2S$ **(b)** 2 N, 8 H, 1 S **1.30 (a)** organic **(b)** inorganic **(c)** inorganic **(d)** inorganic **(e)** organic **1.31 (a)** potassium iodide **(b)** magnesium bromide **1.32 (a)** VCl_2 **(b)** VO **(c)** V_2O_5 **1.33 (a)** chromium(II) sulfide **(b)** chromium(III) sulfide **(c)** titanium(II) oxide **(d)** titanium(IV) oxide **1.34 (a)** S_2Cl_2 **(b)** PBr_3 **1.35 (a)** dinitrogen monoxide **(b)** diphosphorus pentasulfide **1.36 (a)** $AgNO_3$, H_2S **(b)** Ag_2S, HNO_3 **(c)** 2 atoms Ag, 2 atoms N, 6 atoms O, 2 atoms H, 1 atom S **(d)** $AgNO_3$, HNO_3 **(e)** Ag_2S **1.37 (a)** It is the only one with the same number of atoms of each kind on both sides. **1.38 (a)** $2Mg(s) + O_2(g) \rightarrow 2MgO(s)$ **(b)** $N_2(g) + 3H_2(g) \rightarrow 2NH_3(g)$ **(c)** $H_2SO_4(aq) + 2NaOH(aq) \rightarrow Na_2SO_4(aq) + 2H_2O(l)$ **1.39** When a solution of sodium chloride in water is electrolyzed, hydrogen gas, chlorine gas, and an aqueous solution of sodium hydroxide are formed. **1.40** $2Cu(s) + O_2(g) \xrightarrow{heat} 2CuO(s)$ **1.41 (a)** $2LiNO_3(s) \xrightarrow{heat} 2LiNO_2(s) + O_2(g)$ **(b)** $2NaNO_3(s) \xrightarrow{heat} 2NaNO_2(s) + O_2(g)$ **1.83 (a)** Mn_2O_3 **(b)** Br_2O **(c)** $Sr(NO_3)_2$ **(d)** $NaClO_4$ **(e)** $(NH_4)_2CO_3$ **1.85 (a)** Yes. The original mixture of iron filings and sulfur powder has not chemically reacted. **(b)** When the mixture was heated, the iron filings and the sulfur powder reacted chemically to form a compound whose properties are different from those of the iron and sulfur. **1.87** Strontium metal would be expected to react rapidly with water at room temperature. **1.89** Each sulfide ion, S^{2-}, has two − charges, while each sodium ion, Na^+, has only one + charge. The compound sodium sulfide is electrically neutral. Therefore, each S^{2-} ion must have two Na^+ ions combined with it so that the net charge is 0.

STOP AND TEST YOURSELF

1. (b) *1.2*   **2.** (a) *1.3*   **3.** (d) *1.9*   **4.** (e) *1.8*   **5.** (a) *1.8* **6.** (d) *1.10*   **7.** (e) *1.9*   **8.** (c) *1.8, 1.9*   **9.** (e) *1.10* **10.** (e) *1.11*   **11.** (b) *1.11*   **12.** (b) *1.11*   **13.** (c) *1.9* **14.** (d) *1.9*   **15.** (b) *1.1*

1.91 (a) A law is a summary of many observations. An example is the law of constant composition. **(b)** A theory is an explanation of a law or a series of observations. An example is the atomic theory. **1.93 (a)** Na_2CO_3 **(b)** $NaHCO_3$ **(c)** KNO_2 **(d)** MgO **(e)** $CaCl_2$ **(f)** VS **(g)** $Fe_2(SO_4)_3$ **1.95 (a)** neither **(b)** neither **(c)** decomposition **(d)** neither **(e)** combination **1.97 (a)** an ionic compound **(b)** an element **(c)** an ionic compound **(d)** a molecular compound **(e)** a molecular organic compound **1.99 (a)** 17 electrons **(b)** 34 electrons **(c)** 18 electrons **(d)** 50 electrons **1.101 (a)** 50 **(b)** Group IVA **(c)** Period 5 **(d)** 50 **(e)** 50 **(f)** 48 **1.103** $CH_4(g) + 2O_2(g) \rightarrow CO_2(g) + 2H_2O(l)$ or (g) **1.105** ◯ sulfur **(a)** ◯ phosphorus **(b)** ◯ selenium **1.107 (a)** pp 8–9, 57, 218, 277 **(b)** pp 19–22, A1–A5 **(c)** pp 106–140, 460–489 **1.109** *Physical properties:* does not dissolve in water, bright color, soft and easily worked, can be drawn into a thin wire or beaten into a thin leaf. *Chemical properties:* forms few compounds, does not tarnish, is mostly found as the metal. **1.111** nitrogen dioxide, NO_2; chlorine dioxide, ClO_2; ozone, O_3 **1.113 (a)** Inorganic compounds are all the compounds that are *not* organic compounds; that is, they are compounds of all elements except carbon (except for a few simple carbon compounds such as calcium carbonate and carbon dioxide). **(b)** copper(I) sulfide **(c)** iron(III) oxide **(d)** iron(II) carbonate **(e)** aluminum oxide **(f)** molybdenum(IV) sulfide **(g)** silicon dioxide **(h)** calcium carbonate **(i)** Corundum is an ionic compound consisting of Al^{3+} and O^{2-} ions. These ions are held in place by the electrostatic attractions between opposite charges. Molecules do not exist in ionic compounds.

CHAPTER 2

2.1 (a) 6.39 kg **(b)** 0.000 015 K **(c)** 299 792 458 m/s **2.2 (a)** kilogram (kg) **(b)** (meter)^2 or m^2 **2.3 (a)** megasecond **(b)** picomole **(c)** centiampere

2.4 (a) mg **(b)** ns **(c)** kK **2.5 (a)** 10^6 **(b)** 10^6 or 1 000 000
2.6 (a) 10^3 or 1000 **(b)** millimeter **(c)** nanometer
(d) 10^9 or 1 000 000 000 **2.7** $\dfrac{10^{-3}\,\text{g}}{\text{mg}}$
2.8 (a) 6.4×10^{-9} or 0.000 000 0064 **(b)** 0.872
2.9 $\left(\dfrac{10^{-6}\,\text{g}}{\mu\text{g}}\right)$ and $\left(\dfrac{1\,\text{kg}}{10^3\,\text{g}}\right)$
2.10 (a) 4.9×10^5 **(b)** 2.648×10^{-6} **(c)** 3.41×10^{-4}
2.11 (a) 2.89×10^4 **(b)** 27.8 **2.12 (a)** 4.4, 4.5, or 4.6 cm
(b) 3.88 or 3.89 cm **(c)** 3.0 cm **(d)** 2.4, 2.5, or 2.6 cm
2.13 (a) ruler (b) **(b)**

2.14 (a) 3.21×10^5 **(b)** 3.210×10^5 **(c)** $3.210\,00 \times 10^5$
2.15 (a) 4 **(b)** 3 **(c)** 5 **(d)** 2 **(e)** 2 or 3 **(f)** 2
2.16 (a) exactly **(b)** some uncertainty **(c)** some uncertainty
2.17 (a) 62.4 cm **(b)** 62.1 cm **(c)** The median value. The
median is less affected by the one lower reading. **(d)** 62.5 m
2.18 Chances of measurements greater than the median = chances
of measurements less than the median. Normal distribution is
symmetrical about the mean. **2.19 (a)** 3.43 **(b)** 3.44 **(c)** 3.44
2.20 (a) 3 **(b)** 4 **2.21 (b)** wood, (d) rain water, (e) a dime
2.22 (a) extensive **(b)** intensive **(c)** extensive **(d)** intensive
(e) intensive **2.23 (a)** 11.3 g/cm³ **(b)** 1.13×10^4 kg/m³
2.24 61 cm³ **2.25** 34 g **2.26 (a)** 26 °C **(b)** 292.7 K
2.27 Celsius, 100 °C = 180 °F **2.28 (a)** 39 °C **(b)** 126 °F
2.29 (a) 86 400 or 8.6400×10^4 **(b)** 2.7×10^3 **(c)** 6.0×10^{10}
2.30 (a) 2 h, 50 min, 53 s **(b)** 1.0253×10^4 s **2.31 (a)** 50
(b) 68 **(c)** $^{118}_{50}$Sn **(d)** 48 **2.32 (a)** 17 **(b)** 18 **(c)** $^{35}_{17}$Cl **(d)** 18
(e) 34 **2.33** 50% bromine-79, 50% bromine-81 **2.34** 28.1 u
2.35 (a) The mass number is normally the nearest whole number to
the atomic mass of an isotope. **(b)** 14 **2.36** Naturally occurring
samples of carbon contain other isotopes that have higher atomic
masses than carbon-12 as well as carbon-12, causing the atomic
mass to be greater than 12.000 u. **2.37 (a)** 199.886 u
(b) 101.9613 u **(c)** 68.143 u **(d)** 153.822 u **2.38 (a)** 219.075 u
(b) hexahydrate **(c)** $CaCl_2 \cdot H_2O$ and $CaCl_2$ **2.39 (a)** 197.0 g
(b) 6.022×10^{23} atoms **(c)** 1.20×10^{24} atoms
(d) 3.01×10^{23} atoms **2.40 (a)** 123.9 g
(b) 6.022×10^{23} molecules **(c)** 2.409×10^{24} atoms
2.41 (a) 1.0000 mol **(b)** 6.0223×10^{23} atoms N,
1.8067×10^{24} atoms H **2.42** 1.2044×10^{24} ions
2.43 53 g **2.44** 0.2280 mol **2.45 (a)** 9.721×10^{23} atoms P
(b) 5.143×10^{-23} g **2.97** There are no neutrons in its nucleus.
2.99 1.2 g·mL⁻¹ **2.101** 5.49×10^{-4} g **2.103** Mg-24 = 79%,
Mg-25 = 10%, and Mg-26 = 11%

STOP AND TEST YOURSELF

1. (c) *2.1* **2.** (a) *2.2* **3.** (b) *2.10* **4.** (d) *2.2* **5.** (c) *2.2*
6. (e) *2.3* **7.** (c) *2.3* **8.** (e) *2.4* **9.** (d) *2.8* **10.** (d) *2.9*
11. (a) *2.11* **12.** (d) *2.11* **13.** (b) *2.12* **14.** (c) *2.13*
15. (c) *2.13*

2.105 No. The fact that volume can only be measured to three
significant figures limits the precision of density to three significant
figures. **2.107 (b)** CO_2, **(c)** H_2O, **(e)** CH_4. These are molecular

compounds. **2.109** Chemical properties of the elements are more
related to the number of electrons and protons than to the number
of neutrons. **2.111** More moles of ammonia were produced.
2.113 (a) volume **(b)** density **(c)** area **(d)** amount of substance
(e) mass **2.115 (a)** If two elements form more than one
compound, the masses of one element that combine with a fixed
mass of the other element are in ratios of small whole numbers.
(This generalization was first proposed by Dalton early in the
nineteenth century and is called the law of multiple proportions.)
(b) Possible pairs are CO and CO_2, H_2O and H_2O_2, $FeCl_2$ and
$FeCl_3$, SnO and SnO_2 **(c)** Yes because, according to Dalton's
atomic theory, compounds are combinations of indivisible units
called atoms. **2.117** 7×10^6–9×10^6 km² **2.119** 73 000 g
2.121 (a) 1.0, 1.0, 1.2, 1.3, 1.5, 1.6 **(b)** increases (beyond sulfur)
2.123 3×10^3 y **2.125** Empty, wash, and dry the can. Weigh. Fill
can with water. Be sure outside of can is dry and reweigh. Measure
temperature of water in can. From table of density of water at different temperatures in Appendix C, find density of water. Use mass
and density of water to calculate volume of empty can. Subtract
355 mL to get volume of empty space. **2.127** 38, 3.8×10^3, 0.48
2.129 (a) 1273 K, 671 mi/h **(b)** 2.1×10^{11}–4.2×10^{11} mol SO_2
(c) You need to know the world populations for these time periods.
2.131 (a) 6×10^3 molecules **(b)** 3×10^3 atoms

CHAPTER 3

3.1 (a) $4HNO_3(aq) \rightarrow 4NO_2(aq) + 2H_2O(l) + O_2(g)$
(b) $B_2O_3(s) + 3H_2O(l) \rightarrow 2H_3BO_3(aq)$
(c) $Cu(s) + 2AgNO_3(aq) \rightarrow 2Ag(s) + Cu(NO_3)_2(aq)$
(d) $4NH_3(g) + 5O_2(g) \rightarrow 4NO(g) + 6H_2O(l)$
(e) $C_3H_6O(l) + 4O_2(g) \rightarrow 3CO_2(g) + 3H_2O(l)$
3.2 (a) 2 mol **(b)** 0.3 mol **(c)** 9 mol **3.3 (a)** 28.5 g
(b) 92.0 g **3.4 (a)** 1.93×10^{-2} mol **(b)** 4.056 g
3.5 (a) 2.60×10^3 g **(b)** 199 g **(c)** 2.60×10^3 tons **(d)** 199 lb
3.6

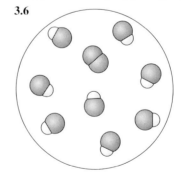

3.7 (a) H_2SO_4 **(b)** 73.8 g **(c)** 3.7 g $NH_3(g)$ **3.8 (a)** 9.59 g
(b) 87.4% **3.9 (a)** 92.61% **(b)** 7.39% **3.10 (a)** 64.80%
(b) 13.60% **(c)** 21.60% **3.11** C_2H_5 **3.12** CH_2
3.13 KH_2PO_4 **3.14** $Na_2Cr_2O_7$ **3.15** $C_3Cl_6O_3$
3.16 (a)

```
     H — N — N — H
         |   |
         H   H
```

(b)

```
      H   H   H   H   H
      |   |   |   |   |
  H — C — C — C — C — C — O
      |   |   |   |   |   |
      H   H   H   H   H   H
```

(c)

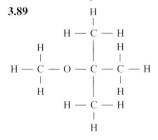

3.17 (a) C_6H_{14} **(b)** C_3H_7 **3.18 (a)** 12.6% **(b)** 87.4%
(c) No. The molecular formula is a whole number multiple of the empirical formula, and thus the ratio of N to H does not change.
3.19 Cr_2O_3 **3.53 (a)** $3A + B \rightarrow 2C$ **(b)** B is limiting
3.55 CCl_4 **3.57** 60.12% H_2O **3.59 (a)** 0.759 mol HCl
(b) 0.379 mol H_2 **(c)** 2.2×10^2 g Zn **(d)** 4.6×10^2 g $ZnCl_2$
3.61 The determination of the limiting reagent depends on the molar ratio of reactants in the balanced chemical equation as well as the actual number of moles of each reactant. **3.63** 14 g
3.65 86.0 tons **3.67 (a)** 81% **(b)** 73% **(c)** The overall yield decreases. **(d)** In order to maximize the overall yield, keep the number of steps as small as possible. **3.69 (a)** $(CH_3)_3CH$ or $(CH_3)CH(CH_3)_2$ **(b)** C_4H_{10} **(c)** C_2H_5

STOP AND TEST YOURSELF

1. (e) *3.2* **2.** (e) *3.2* **3.** (e) *3.3* **4.** (c) *3.4* **5.** (a) *3.4*
6. (e) *3.5* **7.** (d) *3.5* **8.** (a) *3.3* **9.** (d) *3.6* **10.** (c) *3.10*
11. (c) *3.10* **12.** (d) *3.8* **13.** (d) *3.9* **14.** (e) *3.9* **15.** (a) *3.7*

3.71 (a) magnesium **(b)** 3.0 g O_2 **(c)** 60.5 g **(d)** 97.7%
3.73 $n = 2$ **3.75** 97.2% Fe **3.77** 4 **3.79 (a)** C_5H_{12}
(b) C_5H_{12} **(c)** 0.016 32 g **3.81 (a)** 0.2332 g oxygen,
3.973 g Cu/g O **(b)** 0.2203 g oxygen, 7.943 g Cu/g O
(c) Yes. **(d)** CuO, copper(II) oxide and Cu_2O, copper(I) oxide
3.83 (a) 104 g **(b)** 48.8% **(c)** 29.3 g P_4
3.85 (a) $NH_4NO_3(l) \xrightarrow{heat} N_2O(g) + 2H_2O(g)$ **(b)** side reactions
(c) 0.021 613 mol **(d)** 84.30% **(e)** The chloride ion increases the reaction rate of the side reaction without being used up. This reduces the yield of the main product. The catalyst is shown above the arrow in an equation. **3.87** 0.6251% Ag

3.89

H
|
H — C — H
H | H
| | |
H — C — O — C — C — H
| | |
H H H
|
H — C — H
|
H

3.91 urea, CH_4N_2O **3.93** 4 atoms Fe **3.95** No. The % C and % H in the sample don't match the percent composition of cocaine.
3.97 (a) 657 g **(b)** 357 mL
3.99 (a) (i) $C_{15}H_{32}(l) + 23O_2(g) \rightarrow 15CO_2(g) + 16H_2O(l)$
(ii) $2C_{15}H_{32}(l) + 31O_2(g) \rightarrow 30CO(g) + 32H_2O(l)$
(iii) $C_{15}H_{32}(l) + 8O_2(g) \rightarrow 15C(s) + 16H_2O(l)$
(b) Formation of C requires the smallest number of oxygen per molecule of $C_{15}H_{32}$. When more oxygen is present, more of the

$C_{15}H_{32}$ can be converted to carbon monoxide and carbon dioxide; therefore, less smoke is formed. **3.101 (a)** 2 tablets
(b) Yes. Each 100 mg $CaCO_3$ contains 60 mg CO_3^{2-} but only 40 mg Ca^{2+}. In order to have more calcium than carbonate in your diet, you will need another source of calcium that does not contain carbonate. The composition of a compound is always the same.
3.103 0.250 kg NaOH, 0.221 kg Cl_2, 0.006 30 kg H_2
3.105 (a) $N_2(g) + O_2(g) \xrightarrow{heat} 2NO(g)$
(b) $2NO(g) + O_2(g) \rightarrow 2NO_2(g)$
(c) $2CO(g) + O_2(g) \xrightarrow{catalyst} 2CO_2(g)$
(d) $2C_8H_{18}(l) + 25O_2(g) \xrightarrow{catalyst} 16CO_2(g) + 18H_2O(l)$
(e) $2SO_2(g) + O_2(g) \xrightarrow{catalyst} 2SO_3(g)$

CHAPTER 4

4.1 (a) water **(b)** sugar **4.2 (a)** alcohol **(b)** water
4.3 (a) more soluble **(b)** decreases **4.4 (a)** saturated
(b) unsaturated **(c)** supersaturated **(d)** unsaturated
4.5 precipitation **4.6 (a)** saturated **(b)** dilute
4.7 (a) 3 **(b)** 1 **(c)** 2 **(d)** 0 **4.8** hydrobromic acid
4.9 (a) nitric acid **(b)** magnesium perchlorate
(c) potassium hydrogen carbonate **(d)** phosphoric acid
4.10 (a) CaS **(b)** H_2SO_4 **(c)** H_2CO_3 **(d)** $Al_2(SO_4)_3$
4.11 (a) KOH, (c) SrO **4.12 (a)** KI, (d) NH_4NO_3, (e) Ag_2SO_4
4.13 (a) $KClO_4$—potassium perchlorate **(b)** $Ca(NO_3)_2$—calcium nitrate **(c)** $KHCO_3$—potassium hydrogen carbonate and K_2CO_3—potassium carbonate **(d)** $Ca(HCO_3)_2$—calcium hydrogen carbonate and $CaCO_3$—calcium carbonate
(e) $AlCl_3$—aluminum chloride
4.14 (a) $H_2SO_4(aq) + 2NaOH(aq) \rightarrow Na_2SO_4(aq) + 2H_2O(l)$
(b) $2HNO_3(aq) + Ba(OH)_2(aq) \rightarrow Ba(NO_3)_2(aq) + 2H_2O(l)$
4.15 (a) weak electrolyte **(b)** strong electrolyte **(c)** strong electrolyte **(d)** nonelectrolyte **(e)** strong electrolyte
4.16 (a) slightly soluble **(b)** insoluble **(c)** soluble
(d) soluble **(e)** insoluble
4.17 (a) Yes. $2NaOH(aq) + MgCl_2(aq) \rightarrow$
$$Mg(OH)_2(s) + 2NaCl(aq)0$$
(b) Yes. $K_2SO_3(aq) + 2HCl(aq) \rightarrow SO_2(g) + H_2O(l) + 2KCl(aq)$
(c) no reaction
(d) Yes. $NaC_2H_3O_2(aq) + HBr(aq) \rightarrow HC_2H_3O_2(aq) + NaBr(aq)$
(e) Yes. $Na_2CO_3(aq) + CaCl_2(aq) \rightarrow CaCO_3(s) + 2NaCl(aq)$
4.18 (a) $CaCO_3(s)$ **(b)** $H^+ + NO_3^-$ or $H^+(aq) + NO_3^-(aq)$
(c) $HC_2H_3O_2(aq)$ **(d)** $Mg^{2+} + 2Cl^-$ or $Mg^{2+}(aq) + 2Cl^-(aq)$
(e) $Fe(OH)_2(s)$
4.19 (a) $H_2S(g) + Zn^{2+} + 2NO_3^- \rightarrow ZnS(s) + 2H^+ + 2NO_3^-$
(b) NO_3^- **(c)** $H_2S(g) + Zn^{2+} \rightarrow ZnS(s) + 2H^+$
(d) Yes. Zinc(II) chloride
4.20 (a) Complete ionic equation:
$2Na^+ + 2OH^- + Mg^{2+} + 2Cl^- \rightarrow Mg(OH)_2(s) + 2Na^+ + 2Cl^-$
Net ionic equation: $Mg^{2+} + 2OH^- \rightarrow Mg(OH)_2(s)$
(b) Complete ionic equation:
$2K^+ + SO_3^{2-} + 2H^+ + 2Cl^- \rightarrow SO_2(g) + H_2O(l) + 2K^+ + 2Cl^-$
Net ionic equation: $2H^+ + SO_3^{2-} \rightarrow SO_2(g) + H_2O(l)$
(c) Complete ionic equation:
$NH_4^+ + I^- + Zn^{2+} + 2Cl^- \rightarrow NH_4^+ + I^- + Zn^{2+} + 2Cl^-$
Net ionic equation: No net ionic equation (no reaction)

(d) Complete ionic equation:

$Na^+ + C_2H_3O_2^- + H^+ + Br^- \rightarrow HC_2H_3O_2(aq) + Na^+ + Br^-$

Net ionic equation: $H^+ + C_2H_3O_2^- \rightarrow HC_2H_3O_2(aq)$

(e) Complete ionic equation:

$2Na^+ + CO_3^{2-} + Ca^{2+} + 2Cl^- \rightarrow CaCO_3(s) + 2Na^+ + 2Cl^-$

Net ionic equation: $Ca^{2+} + CO_3^{2-} \rightarrow CaCO_3(s)$

4.21 (a) no reaction **(b)** Yes. Molecular equation:

$Sr(s) + 2H_2O(1) \rightarrow Sr(OH)_2(aq) + H_2(g)$

Complete and net ionic equations:

$Sr(s) + 2H_2O(1) \rightarrow Sr^{2+} + 2OH^- + H_2(g)$

(c) Yes. Molecular equation:

$2Al(s) + 6HCl(aq) \rightarrow 2AlCl_3(aq) + 3H_2(g)$

Complete ionic equation:

$2Al(s) + 6H^+ + 6Cl^- \rightarrow 2Al^{3+} + 6Cl^- + 3H_2(g)$

Net ionic equation: $2Al(s) + 6H^+ \rightarrow 2Al^{3+} + 3H_2(g)$

(d) no reaction **(e)** Yes. Molecular equation:

$2Cr(s) + 3Cu(NO_3)_2(aq) \rightarrow 2Cr(NO_3)_3(aq) + 3Cu(s)$

Complete ionic equation:

$2Cr(s) + 3Cu^{2+} + 6NO_3^- \rightarrow 2Cr^{3+} + 6NO_3^- + 3Cu(s)$

Net ionic equation: $2Cr(s) + 3Cu^{2+} \rightarrow 2Cr^{3+} + 3Cu(s)$

(f) no reaction

4.22 Molecular equation:

$Cl_2(aq) + 2NaBr(aq) \rightarrow 2NaCl(aq) + Br_2(aq)$

Complete ionic equation:

$Cl_2(aq) + 2Na^+ + 2Br^- \rightarrow 2Na^+ + 2Cl^- + Br_2(aq)$

Net ionic equation: $Cl_2(aq) + 2Br^- \rightarrow 2Cl^- + Br_2(aq)$

4.23 Yes. Molecular equation:

$Br_2(aq) + 2KI(aq) \rightarrow 2KBr(aq) + I_2(s)$

Complete ionic equation:

$Br_2(aq) + 2K^+ + 2I^- \rightarrow 2K^+ + 2Br^- + I_2(s)$

Net ionic equation: $Br_2(aq) + 2I^- \rightarrow 2Br^- + I_2(s)$

4.24 (a) No. Both H_2SO_4 and $BaCl_2$ are soluble, strong electrolytes.
(b) Yes. Weak electrolytes CO_2 and H_2O will be formed.

4.25 (a) $MgCl_2(aq) + 2NaOH(aq) \rightarrow Mg(OH)_2(s) + 2NaCl(aq)$

(b) $Mg(s) + 2HCl(aq) \rightarrow MgCl_2(aq) + H_2(g)$. Filter and evaporate.
  excess

(c) $KOH(aq) + HI(aq) \rightarrow KI(aq) + H_2O(1)$. Use stoichiometric amounts. Evaporate.

4.26 Measure 60.0 mL of ethyl alcohol into a graduated cylinder and add sufficient water to bring the total volume of the mixture to 500 (5.00×10^2) mL. Mix. **4.27** Measure 6.2 g $KI(s)$ into a suitable container and dissolve it in 69 mL H_2O.

4.28 330 ppb (volume)

4.29 parts per trillion (volume) $= \dfrac{\text{volume solute}}{\text{volume solution}} \times 10^{12}$

4.30 0.6965 M **4.31 (a)** 6.91 g

(b) 1. Measure 6.91 g $NaNO_3(s)$ using a balance.

2. Dissolve the $NaNO_3$ in a minimal amount of deionized water.

3. When the $NaNO_3$ is fully dissolved, transfer the solution to a 250-mL volumetric flask using a glass stirring rod and a funnel to reduce the risk of losing part of the solution.

4. Use a stream of water from a wash bottle containing deionized water to complete the transfer. If the solution does not fill two thirds to three quarters of the bulb of the flask, add more deioniz. d water.

5. Swirl the mixture and let it stand long enough to come to room temperature.

6. Dilute the solution up to the mark on the neck of the flask using a dropper to add the last few drops.

7. Mix thoroughly by repeatedly inverting the stoppered flask.

4.32 231 mL **4.33** Using a graduated cylinder, measure 154 mL concentrated solution (0.325 M) into about 500 mL of deionized water in a 1000-mL volumetric flask. Swirl until mixed. Fill to the mark with deionized water and mix thoroughly to make 1.000 L of dilute solution (0.0500 M). **4.34** 11.6 M **4.35** 0.273 M

4.36 $\boxed{\text{Vol HCl}} \xrightarrow{M_{HCl}} \boxed{\text{Mol HCl}} \xrightarrow[\text{from equation}]{\text{conversion factor}} \boxed{\text{Mol NaOH}} \xrightarrow{M_{NaOH}} \boxed{\text{Vol NaOH}}$

4.37 26.5 mL **4.38** 3.97 g H_2 formed, 0.29 mol HCl left
4.39 (a) 1.03 M **(b)** 0.302 g **(c)** 9.47%

4.75 Because the student has not labeled the numbers completely, he or she has not taken into account the fact that the solution is 6.00% by mass $BaCl_2$ nor that the density is in fact the density of the solution, not the density of $BaCl_2$. The correct solution is

$$\left(\frac{6.00 \text{ g BaCl}_2}{100.0 \text{ g soln}}\right)\left(\frac{1 \text{ mol BaCl}_2}{208.2 \text{ g BaCl}_2}\right)\left(\frac{1.063 \text{ g soln}}{\text{mL soln}}\right)\left(\frac{1000 \text{ mL soln}}{\text{L soln}}\right)$$
$$= 0.306 \text{ M}$$

4.77 (a) less than 0.200 M because part of what was used was water rather than LiBr. **(b)** less than 0.200 M because less LiBr is in solution than was intended. **(c)** less than 0.200 M because as the solution warms up, volume will increase.
(d) 0.200 M because once it is made, every part of the solution has the same concentration. **(e)** less than 0.200 M because the water on the inside of the bottle dilutes the solution.
4.79 Zn^{2+} because zinc sulfide is insoluble.

STOP AND TEST YOURSELF

1. (d) *4.1* **2.** (e) *4.8* **3.** (b) *4.8* **4.** (d) *4.8* **5.** (a) *4.8*
6. (c) *4.9* **7.** (b) *4.9* **8.** (d) *4.9* **9.** (d) *4.2* **10.** (e) *4.3*
11. (a) *4.3* **12.** (b) *4.3* **13.** (c) *4.4* **14.** (d) *4.6* **15.** (e) *4.5*

4.81 (a) Add $AgNO_3(aq)$ to the liquid. If a white precipitate of AgCl forms, the liquid is HCl(aq). **(b)** Add $AgNO_3(aq)$ to an aqueous solution of the white solid. If a white precipitate of AgCl forms, the white solid is NaCl. **(c)** Add $AgNO_3(aq)$ to an aqueous solution of the white solid. If a white precipitate of AgCl forms, the white solid is $CaCl_2$. **(d)** Add a concentrated aqueous solution of NaOH to the white solid. If the mixture has the odor of ammonia, the solid is NH_4Cl. **(e)** Attempt to dissolve the white solid in water. If it dissolves, it is Na_2CO_3.
4.83 (a) CuS **(b)** $FePO_4 \cdot 2H_2O$ **(c)** $MgBr_2 \cdot 6H_2O$
(d) $Sn(C_2H_3O_2)_2$ **(e)** $Ba(H_2PO_4)_2$ **4.85** 71 ppt
4.87 (a) 6.0 M **(b)** 4.8 M H^+ **4.89** 1.953×10^{-1} M
4.91 4.593×10^{-2} M NaOH **4.93** Yes. At the endpoint, the ions in solution have essentially been removed. **4.95** Six moles of HCl contain 3.613×10^{24} molecules of HCl. 6 M HCl refers to a solution containing 6 mol of HCl in each liter of solution.
4.97 (a) ammonia (NH_3), water (H_2O) **(b)** hydroxide ion (OH^-)
(c) ammonium ion (NH_4^+) **4.99** 6.02×10^{16}
4.101 (a) Molecular equation:

$Na_2CO_3(aq) + 2HC_2H_3O_2(aq) \rightarrow$
$\qquad\qquad 2NaC_2H_3O_2(aq) + CO_2(g) + H_2O(1)$

Complete ionic equation:

$2Na^+ + CO_3^{2-} + 2HC_2H_3O_2(aq) \rightarrow$
$\qquad\qquad 2Na^+ + 2C_2H_3O_2^- + CO_2(g) + H_2O(1)$

Net ionic equation:

$CO_3^{2-} + 2HC_2H_3O_2(aq) \rightarrow 2C_2H_3O_2^- + CO_2(g) + H_2O(1)$

(b) sodium acetate, carbon dioxide, and water
4.103 (a) $Cu(s) + 2AgNO_3(aq) \rightarrow Cu(NO_3)_2(aq) + 2Ag(s)$ is one example.   **(b)** $CH_4(g) + 2O_2(g) \rightarrow CO_2(g) + 2H_2O(g)$ is one example.   **(c)** All the oxidation-reduction reactions involve an element.   **4.105** 1.65 ppm   **4.107 (a)** 460 ppt and 40 ppt
(b) 23.2 ppm   **4.109** Tapwater in the bathtub, swimming-pool water, lake water, and seawater contain dissolved electrolytes that make the water conduct. Only pure water is a nonelectrolyte.
4.111 $Ca^{2+} + SO_4^{2-} \rightarrow CaSO_4(s)$; $Ba^{2+} + SO_4^{2-} \rightarrow BaSO_4(s)$; $Sr^{2+} + SO_4^{2-} \rightarrow SrSO_4(s)$
4.113 (a) $Mg(s) + Ag_2S(s) \rightarrow MgS(aq) + 2Ag(s)$
(b) The statement is inaccurate in that "tarnish" (Ag_2S) is not transferred to the magnesium bar. Rather the S^{2-} ion in tarnish is brought into solution by the reaction with $Mg(s)$. The bar of $Mg(s)$ does not "tarnish."   **(c)** No. The bar will be consumed during repeated uses when sufficient Ag_2S has been encountered to react with all of it.   **4.115** As carbonated beverages warm up, the solubility of CO_2 in them decreases, causing CO_2 to escape from solution. This causes the drink to taste "flat."
4.117 (a) In a 1.00-L volumetric flask, dissolve 56 g $C_6H_{12}O_6$ in about 500 mL deionized water. Add enough deionized water to this solution to fill the flask to the mark. Mix.   **(b)** 0.16 M NaCl
4.119 All of the major components of seawater are soluble ionic compounds that are strong electrolytes. Therefore, they all exist in a fully ionized state in water.   **4.121 (a)** 0.526 M Cl^- **(b)** 1.84%
4.123 (a) $CO(g)$ and $ZrO_2(s)$ **(b)** carbon monoxide and zirconium(IV) oxide
4.125 $Fe(s) + Cu^{2+} \rightarrow Fe^{2+} + Cu(s)$ or
$2Fe(s) + 3Cu^{2+} \rightarrow 2Fe^{3+} + 3Cu(s)$
4.127 (a) Barium ion is toxic. Chloride ion cannot be toxic because it is present in common table salt.   **(b)** Barium sulfate is not soluble.   **4.129 (a)** $Zn(s) + 2H^+ \rightarrow Zn^{2+} + H_2(g)$ **(b)** The "boiling" is bubbles of hydrogen gas. Hydrogen gas forms an explosive mixture with air. Smoking should be prohibited in the area, and nearby electric motors should be spark-proofed or relocated.
4.131 Dilute 60.00 mL of stock solution to 100.0 mL.

CHAPTER 5

5.1 (a) 1 Pa   **(b)** larger   **5.2 (a)** 473.1 torr
(b) 6.149×10^{-1} atm   **(c)** 3.67×10^3 mmHg   **(d)** 6.307×10^4 Pa
(e) 2.3×10^5 Pa   **5.3** volume increases   **5.4** less than 5.0 mL; data in Table 5.1(a) show that volume decreases as pressure increases.   **5.5** about 7 mL   **5.6** It is reduced to 1/2 its original volume.   **5.7** It is reduced to 1/3 its original pressure.
5.8 638 mL   **5.9 (a)** 714 Pa   **(b)** 714 Pa   **5.10 (a)** 273.2 K
(b) 313.2 K   **(c)** 235.8 K   **(d)** 92 °C   **(e)** −68 °C   **5.11** It will decrease to 1/2 the original volume.
5.12 It will increase to fourfold the original temperature.
5.13 3.2 L   **5.14** 158 °C   **5.15** 393 mL   **5.16** 302 mL
5.17 (a) 50.6 L   **(b)** 101.2 L (or 101 L)
5.18 (a) 1.748 L (or 1.75 L)   **(b)** 1.748 L H_2O (or 1.75 L), 0.874 L CO_2 **5.19 (a)** 10.8 L   **(b)** 21.6 L   **5.20** 68 L
5.21 0.23 mol   **5.22** 1.520 L   **5.23** 9.51 g/L   **5.24** Chlorine gas has a higher density owing to its higher formula mass.
5.25 28.1 u   **5.26** 0.641 atm   **5.27 (a)** $p_{N_2} = 61$ atm, $p_{H_2} = 76$ atm   **(b)** 136 atm   **5.28** 141 atm

5.29 (a) 13.634 mmHg   **(b)** 17.212 mmHg   **5.30** 0.550 L
5.31 $CO(g)$; it has a lower molecular mass.
5.32 Charles's law states that at constant pressure the volume of a gas is directly proportional to the Kelvin temperature. According to the kinetic-molecular theory, as the Kelvin temperature is increased, the average kinetic energy of the molecules is increased. This causes the molecules to have higher speeds and to have more frequent and forceful collisions with the walls of the container. If the pressure is held constant, the volume will increase. If the volume is held constant, the pressure of the gas will increase.
5.33 (a) intermolecular attraction; compressibility factor < 1.0.
(b) molecular volume; compressibility factor > 1.0.
5.34 low temperatures   **5.67** $p_{Ne} = 1.2 \times 10^2$ mmHg, $p_{Kr} = 1.9 \times 10^2$ mmHg, $P_{total} = 3.0 \times 10^2$ mmHg
5.69 The water in the glass tube will be pushed down 6 in., but no farther.   **5.71 (a)** near zero atmospheres   **(b)** 800 atm
(c) As pressure increases, the compressibility factors of real gases decrease due to stronger attractions between molecules. They also increase with increasing pressure due to the actual volume of the molecules becoming a significant part of the total gas volume. At this particular pressure, the two effects exactly offset each other.
5.73 The same. By increasing the diameter of the tube, the force exerted downward by the increased amount of mercury would increase. However, the area over which this force would be exerted would also increase, meaning that the pressure (force/area) would not change. In order for the height of the mercury to change, the pressure must change.
5.75 775.3 mmHg   **5.77** 2.60×10^2 mL
5.79 (a) The volume will decrease by a factor of 373 K/473 K.
(b) The volume will decrease by a factor of 323 K/373 K.

STOP AND TEST YOURSELF

1. (b) *5.1*   **2.** (d) *5.2*   **3.** (d) *5.3*   **4.** (a) *5.4*   **5.** (d) *5.5*
6. (a) *5.5*   **7.** (c) *5.7*   **8.** (d) *5.7*   **9.** (b) *5.8*   **10.** (c) *5.10*
11. (c) *5.10*   **12.** Curve (a) *5.10*   **13.** (c) *5.11*   **14.** (c) *5.11*
15. (c) *5.4*

5.81 0.232 atm   **5.83 (a)** $Cl_2(g)$ at 35 °C; molecules move faster at higher temperatures. (b) through a vacuum; it would not be slowed down by collisions with N_2 and O_2 molecules in air.
5.85 3.00×10^2 mL   **5.87 (a)** 17   **(b)** 0   **(c)** 6.90×10^2
5.89 (a) 1.06×10^{22} **(b)** 2.12×10^{22} **5.91** C_2H_6
5.93 (a) 2.37 L   **(b)** No. HCl is quite soluble in water.
5.95 8.9 L   **5.97** (b), (c), and (e). (b) and (e) are compounds of nonmetals and are molecular; (c) is a triatomic molecule of a nonmetal. (a) is an ionic compound and is a solid. (d) is also a solid under ordinary conditions (see Figure 2.16).   **5.99** Yes. Under ordinary conditions, all gaseous mixtures are solutions (homogeneous mixtures). Gases are mostly empty space. In the gaseous mixture, water molecules are in the space between the pentane molecules, and pentane molecules are in the space between water molecules.   **5.101 (a)** 16.0 kPa, 11 kPa (or 16 kPa, 1×10^1 kPa if zeros in data are not significant)   **(b)** 6.0×10^{-3} atm
(c) 5×10^4 mmHg   **5.103 (a)** The likely reason for having difficulty in opening the can is that the pressure of the air above the liquid in the can is lower than the surrounding atmospheric pressure. By warming the can, the pressure of the air in the can,

which is confined in a constant volume, will be increased, reducing the difference in pressure. **(b)** No. The reason the screw cap won't open is related to friction of the cap to the bottle rather than pressure differences. **5.105 (a)** The reduced pressure on the outside of the tennis ball causes the air inside the ball, which is at nearly twice the atmospheric pressure, to slowly leak out. **(b)** 1. Coat the inside of the ball with a flexible but impervious coating to reduce the air leakage. 2. Use a gas that effuses slower than air, such as carbon dioxide. 3. Place a higher pressure of air in the ball to allow it to remain at an increased level for a longer time. **(c)** $SF_6(g)$ has a larger formula mass (146 u) than air ($\sim$29 u) and therefore diffuses out of the tennis ball more slowly.
5.107 (a) ammonia, NH_3; argon, Ar; boron trichloride, BCl_3; carbon tetrafluoride, CF_4; chlorine, Cl_2; hydrogen bromide, HBr; hydrogen iodide, HI; hydrogen selenide, H_2Se; nitric oxide, NO; nitrogen dioxide, NO_2; nitrous oxide, N_2O; phosphorus pentafluoride, PF_5; silicon tetrafluoride, $SiCl_4$; sulfur hexafluoride, SF_6 **(b)** H_2Se, NO_2, N_2O **(c)** All are nonmetals.
5.109 2.46×10^{19} molecules/cm^3 **5.111** 2.6×10^{21} molecules
5.113 The units are "inches of Hg" or in. of Hg.
5.115 9.9×10^4 atm **5.117** 12.4 L **5.119** 7×10^1 °C
5.121 (a) -460 °F **(b)** 0.023 65 psi·ft^3·mol^{-1}·R^{-1}
5.123 (a) 19.7 L **(b)** 39.4 L **5.125** CO_2, 0.32 atm; N_2, 0.16 atm; He, 2.5 atm **5.127** 1.1 g/L

CHAPTER 6

6.1 (a) spontaneous **(b)** nonspontaneous **(c)** spontaneous **(d)** spontaneous **6.2 (a)** endothermic **(b)** exothermic **(c)** endothermic **6.3** No. The reverse of a spontaneous change is nonspontaneous. **6.4 (a)** 98.3 **(b)** 0.642 **(c)** 3.26×10^3 **(d)** 3.86×10^3 **6.5** 1.31×10^3 J must be added.
6.6 2.53×10^4 J must be removed.
6.7 (a) -53 kJ/mol **(b)** molar heat of neutralization HCl = $\frac{1}{2}$(molar heat of neutralization H_2SO_4)
6.8 1.77×10^4 kJ
6.9 (a) endothermic **(b)** exothermic **(c)** exothermic
6.10 $CaCO_3(s) + 178.3$ kJ → $CaO(s) + CO_2(g)$, $NH_3(g) + HCl(g) → NH_4Cl(s) + 176.01$ kJ
6.11 (a) hot; reaction is exothermic **(b)** -184.62 kJ
6.12 (a) $2HgO(s) → 2Hg(l) + O_2(g)$ $\Delta H_{rxn} = +181.66$ kJ
(b) $Hg(l) + \frac{1}{2}O_2(g) → HgO(s)$ $\Delta H_{rxn} = -90.83$ kJ or -90.830 kJ
6.13 -349.8 kJ
6.14 (a) $Na(s) + \frac{1}{2}Cl_2(g) + \frac{3}{2}O_2(g) → NaClO_3(s)$
(b) $\frac{1}{2}H_2(g) + \frac{1}{2}N_2(g) + \frac{3}{2}O_2(g) → HNO_3(l)$
(c) $2C(graphite) + 3H_2(g) + \frac{1}{2}O_2(g) → CH_3CH_2OH(l)$
6.15 (a) -104.86 kJ **(b)** -71.7 kJ **(c)** -1409.2 kJ
6.16 -487.0 kJ/mol
6.17 (a) neither formation nor combustion **(b)** formation **(c)** combustion **(d)** neither formation nor combustion (It is decomposition.) **(e)** formation and combustion **(f)** formation and combustion **6.18** greater; the piston would fall, adding energy to the system **6.57 (a)** neither **(b)** both **(c)** both **(d)** combustion **(e)** neither **6.59 (a)** (i) Changing a liquid to a gas is an endothermic process. **(b)** more negative
6.61 (a) $N_2(g) + 3H_2(g) → 2NH_3(g)$; $2SO_2(g) + O_2(g) → 2SO_3(g)$; $2HgO(s) → 2Hg(l) + O_2(g)$

(b) The difference is usually small compared with the magnitude of ΔH. **6.63 (a)** -1214.66 kJ **(b)** 252 kJ **(c)** -1169.54 kJ **(d)** 945.40 kJ **(e)** 126.4 kJ **6.65** -454 kJ

STOP AND TEST YOURSELF

1. (b) 6.5 **2.** (c) 6.5 **3.** (e) 6.9 **4.** (a) 6.9 **5.** (d) 6.9 **6.** (a) 6.8 **7.** (a) 6.2 **8.** (d) 6.9 **9.** (a) 6.9 **10.** (c) 6.8 **11.** (d) 6.9 **12.** (d) 6.6 **13.** (a) 6.8 **14.** (b) 6.9

6.67 (a)   **(b)**

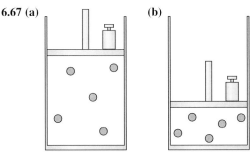

6.69 Real gases have intermolecular attractions. These intermolecular attractions must be overcome in order for the gas to expand. This requires energy from the surroundings, which results in a temperature decrease. **6.71** -24.8 kJ
6.73 $\frac{3}{2}NO_2(g) + \frac{1}{2}H_2O(l) → HNO_3(l) + \frac{1}{2}NO(g) + 35.0$ kJ or $\frac{3}{2}NO_2(g) + \frac{1}{2}H_2O(l) → HNO_3(l) + \frac{1}{2}NO(g)$ $\Delta H = -35.0$ kJ
6.75 -177 kJ **6.77** -53 kJ **6.79** 49 kJ **6.81** As water freezes, it gives off thermal energy to the surroundings.
6.83 27.4 g NH_4NO_3 **6.85 (a)** -930 kJ **(b)** -465 kJ **(c)** exothermic **6.87 (a)** 7×10^1 kcal (one significant figure) **(b)** 70 calories **(c)** 3×10^2 kJ **6.89 (a)** 2×10^3 lb **(b)** 1×10^4 kcal (using rounded value from part **a**)
6.91 5.7 °C **6.93 (a)** 141.79 kJ **(b)** 12.1 L **(c)** $\Delta H°_{combustion} = -5465.5$ kJ/mol, -47.846 kJ **(d)** 1.45 mL **(e)** Hydrogen gas takes up a lot of space! **6.95 (a)** dimethylhydrazine **(b)** dimethylhydrazine; it releases more energy per gram of fuel. **6.97 (a)** 6.36×10^2 kJ, 485 L **(b)** 485 L

CHAPTER 7

7.1 (a) wavelength of 1 m **(b)** wavelength of 10^{-10} m
7.2

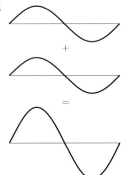

The amplitude of the new wave will be twice the amplitude of the original waves.

7.3 neither **7.4** 10^3 s^{-1} **7.5** 4.91×10^4 m **7.6** 54 MHz **7.7 (a)** yellow light **(b)** microwaves **7.8** violet or purple **7.9** 246 kJ/mol **7.10 (a)** 1.48×10^5 kJ/mol **(b)** X-rays **7.11** 6.947×10^{-8} m **7.12** photons of gamma rays

7.13 486.2 nm

7.14 (a) 292 kJ/mol **(b)**

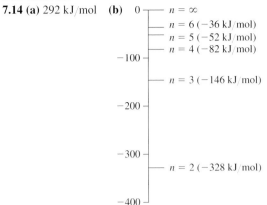

$$0 \quad — \quad n = \infty$$
$$— \quad n = 6\,(-36\ \text{kJ/mol})$$
$$— \quad n = 5\,(-52\ \text{kJ/mol})$$
$$— \quad n = 4\,(-82\ \text{kJ/mol})$$
$$-100$$
$$— \quad n = 3\,(-146\ \text{kJ/mol})$$
$$-200$$
$$-300$$
$$— \quad n = 2\,(-328\ \text{kJ/mol})$$
$$-400$$

7.15 (a)

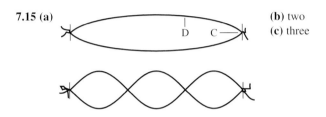

D C

(b) two **(c)** three

7.16

Principal Quantum Number, n (Shell)	Angular Momentum Quantum Number, l (Subshell)	Subshell Label	Magnetic Quantum Number, m_l	Number of Orbitals in Subshell
5	0	5s	0	1
	1	5p	−1, 0, +1	3
	2	5d	−2, −1, 0, +1, +2	5
	3	5f	−3, −2, −1, 0, +1, +2, +3	7
	4	5g	−4, −3, −2, −1, 0, +1, +2, +3, +4	9

7.17 (a) 3 **(b)** spherical **7.18 (a)** 10 **(b)** 18
7.55 1.8×10^{27} photons **7.57 (a)** green **(b)** green, blue, violet
(c) yellow, orange, red **(d)** 3.68×10^{-19} J
7.59 Ordinary uncertainty in measurements can be reduced by using more precise instruments. The uncertainty in measuring the position and speed of submicroscopic particles is unavoidable.
7.61 (a) No. Charges are not whole-number multiples of the smallest charge. **(b)** 1.55×10^{-19} C **(c)** 3

STOP AND TEST YOURSELF

1. (d) 7.6 **2.** (e) 7.6 **3.** (b) 7.6 **4.** (d) 7.6 **5.** (b) 7.6
6. (c) 7.6 **7.** (c) 7.6 **8.** (e) 7.6 **9.** (e) 7.10 **10.** (c) 7.7
11. (a) 7.11 **12.** (c) 7.12 **13.** (d) 7.12 **14.** (a) 7.10

7.63 (a) *Wavelength* is the distance between two similar points of a wave such as wave crests or wave troughs. The *frequency* is the number of cycles that pass a given point in one second.
(b) *Interference* occurs when two waves, which can be either in phase or out of phase with one another, are added together. *Diffraction* is the bending of waves around the edge of any object in their path that is about the same size or larger than the wavelength. **(c)** *Traveling waves* move from one site to another,

whereas *standing waves* are confined within a certain region of space. **(d)** When atoms are in the lowest energy state they are said to be in their *ground state*. Any state of the atom that has a higher energy than the ground state is said to be an *excited state*.
(e) An *orbit* is a defined pathway along which an object (such as a planet) moves around another object (such as the Sun). An *orbital* is a region of space around an atom's nucleus where a relatively high probability of finding an electron exists.
7.65 1. What is the charge on the nucleus of an atom?
2. What keeps the electrons from falling into the nucleus?
7.67 (a) six **(b)** $n = 4$ to $n = 1$ **(c)** $n = 4$ to $n = 3$
7.69 (a) 984 kJ/mol **(b)** 1.22×10^{-7} m **(c)** 2.47×10^{15} s^{-1}
(d) ultraviolet; radiation with frequencies between 10^{15} and 10^{16} s^{-1} is in the ultraviolet. **7.71** 1312 kJ/mol
7.73 According to Heisenberg's uncertainty principle, the speed and the position of a subatomic particle cannot both be determined simultaneously with any degree of certainty. This is not true of large objects, such as baseballs, cars, and people. This difference is due to the fact that light is used to determine the position and speed of objects. If the wavelength of the light used is small compared with the object, the uncertainty in measurement is very small. However, if the wavelength of light is about the same order of magnitude or larger than the object, either the speed and/or the position of the object is uncertain. **7.75** The negative charge and the relatively low mass of the electrons would have caused most of the electrons to be reflected back rather than to penetrate the negatively charged electron clouds of the metal atoms. If an electron did have enough kinetic energy to penetrate, it would have been captured by the positively charged nuclei of the atoms.
7.77 (a) decreases **(b)** increases **(c)** 5.86×10^{-19} J
(d) ultraviolet **7.79 (a)** 3×10^{15} g/cm^3 **(b)** 9.577×10^7 C/kg
(c) zero **7.81** Because of the wave properties of an electron, the electron microscope is possible. **7.83** 461 m
7.85 (b) < (c) < (d) < (e) < (a)
7.87 (a, b)

$\leftarrow$ wavelength $\rightarrow$

(c) The intensity of light is proportional to the amplitude of the wave (half the distance from the top of the wave to the bottom of the wave). Since each photon is in phase with the others and moving parallel to the others, significant constructive interference occurs, causing high-intensity light. **7.89 (a)** (i) Both X-rays and visible light move with the same speed in a vacuum.
(ii) The wavelength of an X-ray is significantly shorter than the wavelength of visible light. (iii) The frequency of an X-ray is significantly higher. (iv) The energy per photon of an X-ray is significantly higher. (v) The X-ray has a higher ability to penetrate flesh due to its significantly higher energy. **(b)** Lead is a very dense material that can absorb electromagnetic radiation of X-ray frequency, thus transmitting little radiation. **7.91** 12.9 kJ
7.93 from short radio waves and microwaves to X-rays

7.95 1×10^3 photon·s^{-1} **7.97 (a)** 1.661×10^{-19} J   **(b)** The energy per photon of visible light (2.8×10^{-19} J) is higher than the minimum required to darken silver bromide, whereas the energy per photon of the signal from a TV station (10^{-25} J) is not.
7.99 (a) ultraviolet   **(b)** 1.099×10^{-18} J
7.101 (a) hydrogen   **(b)** helium   **7.103** 1.88×10^{27} photons
7.105 (a) 3.843×10^5 km   **(b)** 11 significant figures
7.107 3.86×10^{-7} m   **7.109 (a)** green   **(b)** 6.14×10^{14} s^{-1} and 5.83×10^{14} s^{-1} **(c)** 18 protons, 17 electrons   **(d)** 1.000 mmHg, 1.316×10^{-3} atm, 133.3 Pa

CHAPTER 8

8.1 There are two electrons in the s subshell of the third main shell or principal energy level.   **8.2** $3s^1$ or $4s^1$, etc.
8.3 (a)

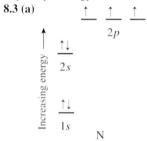

N

(b) $1s^2 2s^2 2p^3$ or $1s^2 2s^2 2p_x^1 2p_y^1 2p_z^1$
8.4 (a)

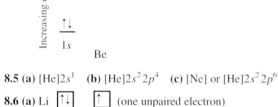

Be

    (b) $1s^2 2s^1 2p^1$ or $1s^2 2s^1 2p_x^1$

8.5 (a) [He]$2s^1$ **(b)** [He]$2s^2 2p^4$ **(c)** [Ne] or [He]$2s^2 2p^6$
8.6 (a) Li [↑↓]$_{1s}$ [↑]$_{2s}$ (one unpaired electron)

(b) O [↑↓]$_{1s}$ [↑↓]$_{2s}$ [↑↓ ↑ ↑]$_{2p_x\ 2p_y\ 2p_z}$ (two unpaired electrons)

(c) Ne [↑↓]$_{1s}$ [↑↓]$_{2s}$ [↑↓ ↑↓ ↑↓]$_{2p_x\ 2p_y\ 2p_z}$ (zero unpaired electrons)

8.7 C [He] [↑ ↑ ↑ ↓]$_{2s\ 2p_x\ 2p_y\ 2p_z}$

8.8 (a) [Ar]$4s^2 3d^9$ **(b)** [Ar]$4s^1 3d^{10}$ **8.9 (a)** 118   **(b)** 121
8.10 (i) (a) $1s^2 2s^2 2p^6$ **(b)** [Ne] or [He]$2s^2 2p^6$

(c) [Ne] or [He] [↑↓ ↑↓ ↑↓ ↑↓]$_{2s\ \ \ 2p}$ **(d)** zero

(ii) (a) $1s^2 2s^2 2p^6 3s^2 3p^5$ **(b)** [Ne]$3s^2 3p^5$

(c) [Ne] [↑↓ ↑↓ ↑↓ ↑]$_{3s\ \ \ 3p}$ **(d)** one

(iii) (a) $1s^2 2s^2 2p^6 3s^2 3p^6 4s^2 3d^{10} 4p^6 5s^2 4d^2$ **(b)** [Kr]$5s^2 4d^2$

(c) [Kr] [↑↓]$_{5s}$ [↑ ↑ □ □ □]$_{4d}$ **(d)** two

(iv) (a) $1s^2 2s^2 2p^6 3s^2 3p^6 4s^2 3d^{10} 4p^6 5s^2 4d^{10} 5p^6 6s^2 4f^5$
(b) [Xe]$6s^2 4f^5$

(c) [Xe] [↑↓]$_{6s}$ [↑ ↑ ↑ ↑ ↑ □ □]$_{4f}$ **(d)** five

8.11 (a) Cl$^-$ $1s^2 2s^2 2p^6 3s^2 3p^6$

(b) Cl$^-$ [Ne] [↑↓]$_{3s}$ [↑↓ ↑↓ ↑↓]$_{3p}$ or [Ar]

8.12 (a) Cu$^+$ $1s^2 2s^2 2p^6 3s^2 3p^6 3d^{10}$
Cu^{2+} $1s^2 2s^2 2p^6 3s^2 3p^6 3d^9$

(b) Cu$^+$ [Ar] [↑↓ ↑↓ ↑↓ ↑↓ ↑↓]$_{3d}$

Cu^{2+} [Ar] [↑↓ ↑↓ ↑↓ ↑↓ ↑]$_{3d}$

8.13 (a) Lithium and beryllium are both in the second period of the periodic table. This means that their outer shell electrons are being added to the same principal energy level (main shell). Since beryllium has one more proton in the nucleus than lithium, the outer shell electrons in beryllium feel a stronger force of attraction and are pulled in toward the nucleus.   **(b)** Chlorine and fluorine are in the same main group (Group VIIA), but they are in different periods, chlorine being in the third period and fluorine in the second period. In a chlorine atom, the outer electrons are in the third shell; in a fluorine atom the outer electrons are in the second shell. Electrons in the third shell are farther from the nucleus on average and are shielded from the positive charge of the nucleus by an additional layer of electrons. Therefore, a chlorine atom is larger than a fluorine atom in spite of the chlorine atom's higher nuclear charge.
(c) Between Ag and Au, the $4f$ orbitals of the lanthanides have been filled and many additional protons have been added to the nuclei. Thus, the higher energy level effects on size (5th level vs. 4th level) are offset by the substantial increase in nuclear charge, causing Au to be about the same size as Ag. This effect is referred to as the lanthanide contraction.
8.14 (a) K   **(b)** K   **8.15** B < Be < Mg $\xrightarrow{\text{increasing radius}}$

8.16 (a) When a sodium atom is ionized to a Na$^+$ ion, the one outer shell electron ($3s^1$) is removed. Thus, in effect, the outer shell is removed. In addition, the remaining electrons are held even more tightly by the nucleus owing to the excess positive charge.
(b) When a chlorine atom forms a chloride ion, Cl$^-$, the extra electron is added to a shell that already contains 7 electrons. Thus, electronic repulsion is increased, and the size increases.
8.17 (a) Al^{3+} **(b)** F$^-$ **(c)** Na$^+$ **8.18** K$^+$, Cl$^-$
8.19 (a) The $4s$ electron in K is in the 4th principal energy level and is farther from the nucleus than the $3s$ electron in Na. The $4s$ electron is also better shielded from the nucleus. Thus, the $4s$ electron in K is more easily removed than the $3s$ electron in Na.
(b) F has one more proton than O while having outer electrons in the same shell. Thus, F's outer shell electrons are held more tightly by the nuclear charge, giving F a higher first ionization energy.

(c) Oxygen has 2 paired electrons in the $2p_x$ orbital, while N has no paired electrons in the $2p$ sublevel. The pairing of electrons in O produces electron repulsion, making it easier to remove an electron. (d) Au has a much higher effective nuclear charge than does Ag while being about the same size. The higher effective nuclear charge is due to an increase of 32 protons in the nucleus from Ag to Au. This higher effective nuclear charge results in a higher first ionization energy. **8.20 (a)** He **(b)** Na **(c)** Hg

8.21 The y-axis on which the ionization energies are plotted would be much too long to fit on the page. **8.22 (a)** The second electron must be removed from a positively charged species, Be^+. **(b)** The 3rd ionization energy represents the energy required to remove a core electron from an electron configuration electronically identical to that of the noble gas He. This electron configuration is very stable. **8.23** 21 kJ/mol **8.24** When an electron is added to an oxygen atom to form O^-, that electron must go into a relatively small energy level (shell) that already contains six other electrons. By contrast, the 3rd, 4th, 5th, and 6th energy levels in S, Se, Te, and Po, respectively, are much bigger (they have a higher volume), and electronic repulsion in these elements is much less. Also, S, Se, Te, and Po have higher effective nuclear charges than O, causing more energy to be liberated when an electron is added to them. **8.25** For magnesium to add an electron, the electron must go into a higher energy sublevel due to the fact that magnesium's s sublevel is filled. Magnesium does not have a large enough effective nuclear charge for the addition of another electron to be an exothermic or energetically favorable change.

8.26 (a) Exothermic. Electron affinity of Na is positive. **(b)** Endothermic. Electron affinity of Mg is negative. **8.27 (a)** Exothermic. Electron attracted by $2+$ charge. **(b)** Endothermic. Electron repelled by $1-$ charge. **8.28 (a)** NaOH **(b)** Na_2CO_3 **(c)** Na_2SO_4 **(d)** SO_2 **(e)** HCl **8.29 (a)** sodium hydride **(b)** xenon difluoride **(c)** potassium hydroxide **(d)** carbon dioxide **(e)** magnesium nitride **8.30** Mg < Na < K **8.31 (a)** $Ca(s) + H_2(g) \rightarrow CaH_2(s)$ **(b)** $2K(s) + 2H_2O(l) \rightarrow 2KOH(aq) + H_2(g)$ **(c)** $Mg(s) + 2H_2O(g) \rightarrow Mg(OH)_2(s) + H_2(g)$ **(d)** $3Ca(s) + N_2(g) \rightarrow Ca_3N_2(s)$ **(e)** no reaction **(f)** $2K(s) + Cl_2(g) \rightarrow 2KCl(s)$ **(g)** $Ca(s) + Cl_2(g) \rightarrow CaCl_2(s)$ **(h)** $Cl_2(g \text{ or } aq) + 2NaBr(aq) \rightarrow Br_2(l \text{ or } aq) + 2NaCl(aq)$ **(i)** no reaction **(j)** $CH_3CH_3(g) + 2H_2O(g) \rightarrow 5H_2(g) + 2CO(g)$ **8.75 (a)** The ground state of hydrogen has one s electron. Similarly, the alkali metals in Group IA each have a single s electron in their outermost occupied shell. However, by gaining one additional electron, a hydrogen atom in its ground state can get the electron configuration of the next noble gas. In this respect, hydrogen is like the halogens in Group VIIA, each of which, by adding one electron, has the electron configuration of a noble gas. **(b)** Both of them have a half-filled outer shell. **8.77 (a)** nitrogen **(b)** lanthanum **(c)** actinium **8.79 (a)** Across a row or period of the periodic table, the nuclear charge increases but the principal electron energy level stays the same. Thus, a contraction in size occurs. For the transition metals, inner orbitals are being filled that serve to shield the outer electrons from the attraction of the nucleus. **(b)** Initially, the increased attraction of the nucleus for the electrons causes contraction. However, as more electrons are added to the d orbitals, electron-electron repulsion becomes important and a "swelling" of the atom

takes place. **(c)** This phenomenon is known as the lanthanide contraction. Between the second and third transition series a large number of protons are added to the nucleus, which causes significant contraction of the electron clouds. **8.81** Group VA **8.83 (a)** Mn^{3+} **(b)** Ag^+ **(c)** I^- **8.85** Metals are shiny when their surfaces are clean. With the exception of mercury, metals are solids under ordinary conditions. Metals conduct thermal energy and electricity. They can be rolled or hammered into sheets and pulled into wire. Their chemical properties vary from unreactive (e.g., gold and platinum) to very reactive (e.g., rubidium, which burns when exposed to air). **8.87 (a)** No. They are similar in reactivity to magnesium, which is not found free in nature. **(b)** Lanthanide ores are mixtures containing all of the lanthanides because the lanthanides have very similar chemical properties. **8.89** 5140 kJ **8.91** (b), (c), and (e)

STOP AND TEST YOURSELF

1. (b) 8.3 **2.** (c) 8.3 **3.** (c) 8.3 **4.** (d) 8.3 **5.** (d) 8.3 **6.** (a) 8.1 **7.** (c) 8.3 **8.** (d) 8.4 **9.** (c) 8.4 **10.** (e) 8.5 **11.** (a) 8.5 **12.** (c) 8.6 **13.** (d) 8.7 **14.** (d) 8.7 **15.** (c) 8.2

8.93 (a) 496 kJ/mol **(b)** 1447 kJ/mol **8.95** The properties of the elements are periodic functions of their atomic number. **8.97** (c) Cl **8.99 (a)** $1s^2 2s^2 2p^6 3s^2 3p^6 4s^2 \rightarrow 1s^2 2s^2 2p^6 3s^2 3p^6 4s^1 \rightarrow s^2 2s^2 2p^6 3s^2 3p^6 \rightarrow 1s^2 2s^2 2p^6 3s^2 3p^5$ **(b)** The second electron must be removed from a positively charged ion. Since it has a positive charge, the Ca^+ ion holds more tightly to the electrons that remain. **(c)** The first two electrons are removed from the fourth principal energy level, which is farther away from the nucleus and less shielded than the third energy level. The third electron not only must be removed from an energy level closer to the nucleus and from an electric field with two extra positive charges, but it also must be removed from a very stable electronic configuration identical to that of the noble gas argon. **(d)** For aluminum, the biggest increase in ionization energy is between the third and fourth ionization energies because the fourth electron must be removed from the core. **8.101** Magnesium has a higher first ionization energy than Na because Mg has a smaller radius and higher nuclear charge than Na. Magnesium's first ionization energy is also higher than that of Al because the first electron must be removed from an s energy level rather than from a p energy level. **8.103 (a)** Atoms with high first ionization potentials have a relatively large attraction for electrons. Thus, they also tend to have high electron affinities. Normally, by adding electrons, these atoms can attain a noble gas electron configuration. **(b)** The noble gases have very high ionization energies because of their very stable electron configuration. For the same reason, they have no affinity for any additional electrons. **8.105 (a)** Cl_2, H_2, and Ne **(b)** none **(c)** Ca and Na **8.107** The $2p$ orbital is degenerate with the $2s$ orbital, and the $3p$ and $3d$ orbitals are degenerate with the $3s$ orbital. **8.109 (a)** K^+, potassium ion; Ca^{2+}, calcium ion **(b)** Cl^-, chloride ion; S^{2-}, sulfide ion **(c)** H^+, hydrogen ion; transition metal ions such as Fe^{3+} **8.111 (a)** solid **(b)** 117, $[Rn]7s^2 5f^{14} 6d^{10} 7p^5$ or $[Rn]5f^{14} 6d^{10} 7s^2 7p^5$ **8.113 (a)** no reaction **(b)** $Ba(s) + 2H_2O(l) \xrightarrow{\text{room temp.}} Ba(OH)_2(aq) + H_2(g)$ **(c)** $2K(s) + Br_2(g) \rightarrow 2KBr(s)$ **(d)** no reaction **(e)** no reaction

8.115 (a) 24 cm³/mol   **(b)** No. At the same temperature and pressure, one mole of any gas (assuming ideal behavior) has the same volume as one mole of any other gas.   **8.117 (a)** 2.05 g/cm³ **(b)** 12 °C, liquid   **(c)** 285 pm   **(d)** Fr^+ **(e)** 185 pm **(f)** 320 kJ/mol   **(g)** greater than that of Cs; FrOH and $H_2(g)$ should be formed.   **(h)** FrCl   **(i)** soluble **8.119 (a)** Potassium nitrate must be the least soluble of the four compounds.   **(b)** concentrated **(c)** Molecular equation: $NaNO_3(aq) + KCl(aq) \rightarrow KNO_3(s) + NaCl(aq)$ Complete ionic equation: $Na^+ + NO_3^- + K^+ + Cl^- \rightarrow KNO_3(s) + Na^+ + Cl^-$ Net ionic equation: $NO_3^- + K^+ \rightarrow KNO_3(s)$ **8.121** 13.51 M   **8.123** Strontium is an alkaline earth metal, as is calcium; thus, they have very similar properties. Strontium has an identical outer shell electron configuration, ns^2, and is more reactive, in general, than calcium. It also has nearly the same atomic and ionic radii (slightly larger) and the same ionic charge (2+). Since strontium is more reactive, it can displace magnesium from its compounds.   **8.125 (a)** $1s^2 2s^2 2p^6 3s^0 3p^1$ or $[Ne]3s^0 3p^1$ **(b)** 3.37×10^{-19} J   **8.127 (a)** AlI_3, $BaBr_2$, CsI, $CoCl_2$, $ScCl_3$ **(b)** 5, the same number as the price. $2729.93   **(c)** 0.01% **(d)** to keep them out of contact with air, which usually contains water vapor   **8.129 (a)** 349 kJ/mol   **(b)** 349 kJ/mol **(c)** 3.43×10^{-7} m, ultraviolet   **8.131 (a)** 0.59 kg   **(b)** 34% **(c)** The aluminum part is lighter, which is especially important in transportation because less energy will be required to move the aluminum part.   **8.133 (a)** $2Mg(s) + O_2(g) \rightarrow 2MgO(s)$ **(b)** $Mg(s) + 2H_2O(l) \xrightarrow{heat} Mg(OH)_2(s) + H_2(g)$, $2H_2(g) + O_2(g) \xrightarrow{heat} 2H_2O(g)$ **(c)** -116 kJ. The reaction is exothermic. Therefore, the fire will burn more vigorously if a carbon dioxide extinguisher is used.

CHAPTER 9

9.1 (a)

(i) $[Ar]4s^1 + [Kr]5s^2 4d^{10} 5p^5 \rightarrow [Ar] + [Kr]5s^2 4d^{10} 5p^6$ (or $[Xe]$
  K + I $\rightarrow$ K^+ + I^-

(ii) $[Ar]4s^2 + \begin{array}{l}[Ar]4s^2 3d^{10} 4p^5 \\ [Ar]4s^2 3d^{10} 4p^5\end{array} \rightarrow [Ar] + \begin{array}{l}[Ar]4s^2 3d^{10} 4p^6 \text{ (or } [Kr]) \\ [Ar]4s^2 3d^{10} 4p^6 \text{ (or } [Kr])\end{array}$
  Ca + 2Br $\rightarrow$ Ca^{2+} + $2Br^-$

(iii) $\begin{array}{l}[Ne]3s^1 \\ [Ne]3s^1\end{array} + [Ne]3s^2 3p^4 \rightarrow \begin{array}{l}[Ne] \\ [Ne]\end{array} + [Ne]3s^2 3p^6$ (or $[Ar]$)
  2Na + S $\rightarrow$ $2Na^+$ + S^{2-}

(b) KI, $CaBr_2$, Na_2S
9.2 (a) two   **(b)** four   **(c)** $[Ar]3d^8$ **9.3** Both electron configurations have four unpaired electrons.   **9.4** two Fe^{3+} and three O^{2-}
9.5 (a) Al^{3+}, S^{2-} **(b)** Bi^{3+}, Cu^+, Hg^{2+}
9.6 (a) $MgCl_2$. Mg^{2+} is smaller.   **(b)** CaO. Ca^{2+} has higher charge.   **(c)** Fe_2O_3. Fe^{3+} is smaller and has higher charge. **(d)** NaBr. Br^- is smaller.

9.7 Na· Mg· ·Al· ·Si· ·P̈· ·S̈· :C̈l· :Är:

9.8 (a) 3   **(b)** 2   **9.9**

H:C:O:H (with H above and below C)

9.10 (a) H:F̈:   **(b)** H:P̈:H   **(c)** H:S̈:H   **(d)** H:C:H **(e)** :F̈:F̈:

9.11 (a) H—F̈:   **(b)** H—P̈—H   **(c)** H—S̈—H **(d)** H—C—H (with H above and below)   **(e)** :F̈—F̈:

9.12 (a) $\overset{\delta^+}{H}—\overset{\delta^-}{O}$ or H—O $\xrightarrow{}$ **(b)** $\overset{\delta^+}{H}—\overset{\delta^-}{N}$ or H—N $\xrightarrow{}$ **(c)** $\overset{\delta^+}{C}—\overset{\delta^-}{Cl}$ or C—Cl $\xrightarrow{}$ **(d)** $\overset{\delta^+}{Mg}—\overset{\delta^-}{C}$ or Mg—C $\xrightarrow{}$
9.13 (a) O—O   **(b)** Ba—O **(c)** O—O < C—O < Be—O < Ba—O   **9.14 (a)** ionic **(b)** ionic   **(c)** molecular   **(d)** ionic   **(e)** ionic   **(f)** molecular
9.15 $\left[H{:}\ddot{O}{:}\right]^-$, $\left[H—\ddot{O}{:}\right]^-$ **9.16 (a)** 12   **(b)** 10   **(c)** 24   **(d)** 32

9.17 (a) H:C::C:H or H—C=C—H (with H's)   **(b)** H:C⫶C:H or H—C≡C—H **(c)** H:C:C:Cl: or H—C—C—Cl: (with H's)

(d) $\left[\begin{array}{c}H\\H{:}C{:}\ddot{O}{:}\\\end{array}\right]^-$ or $\left[H—C—\ddot{O}{:}\right]^-$ (with H above and below)   **(e)** :F̈:Ö:F̈: or :F̈—Ö—F̈:

9.18 (a) the N—N single bond   **(b)** the triple bond **(c)** the double bond
9.19 Formal charge on H = $1 - \left[\frac{1}{2}(2) + 0\right] = 0$
Formal charge on N = $5 - \left[\frac{1}{2}(8) + 0\right] = +1$
Sum of formal charges = $4(0) + 1(+1) = +1$

9.20 oxygen   **9.21** $\left[\overset{(0)}{:}\overset{(0)}{O{::}C{::}}\overset{(-1)}{N{:}}\right]^-$

9.22 $\overset{(-2)}{:}\overset{(+2)}{C{::}O{::}}\overset{(0)}{\ddot{O}{:}}$ $\overset{(-1)}{:}\overset{(+2)}{C{::}O{:}}\overset{(-1)}{\ddot{O}{:}}$

$\overset{(0)}{:}\overset{(0)}{O{::}C{::}}\overset{(0)}{O{:}}$

The top structures have higher formal charges than the bottom structure. Thus, :C::O::Ö: is the best Lewis structure.

9.23 (a) $\left[:C≡N—\ddot{O}{:}\right]^-$ The central atom is N instead of C.

(b)

$$\left[:\overset{-1}{\ddot{N}}=\overset{0}{C}=\overset{0}{\ddot{O}}: \right]^{-}$$ has lower formal charge than $$\left[:\overset{-2}{\ddot{N}}-\overset{0}{C}\equiv\overset{+1}{O}: \right]^{-}$$

9.24 (a) $$\left[H_3C-C-\ddot{\underset{\displaystyle .\ddot{O}.}{O}}: \right]^{-}$$ **(b)** 1.5

9.25 (a)

$$\left[:\overset{..}{O}=C-\underset{\displaystyle :\overset{..}{O}:}{\ddot{O}}: \right]^{2-} \longleftrightarrow \left[:\overset{..}{\ddot{O}}-C=\underset{\displaystyle :\overset{..}{O}:}{\overset{..}{O}}: \right]^{2-} \longleftrightarrow \left[:\overset{..}{\ddot{O}}-C-\underset{\displaystyle .\overset{..}{O}.}{\overset{..}{O}}: \right]^{2-}$$

(b) $:\overset{..}{O}=\overset{..}{O}-\overset{..}{\ddot{O}}: \longleftrightarrow :\overset{..}{\ddot{O}}-\overset{..}{O}=\overset{..}{O}:$

9.26 (a) F As structure   **(b)** Xe F structure

9.27 (a) $H:\overset{\displaystyle :\ddot{O}:}{\underset{\displaystyle :\ddot{O}:}{O}}:S:\ddot{O}:H$ or $H-\overset{\displaystyle :\ddot{O}:}{\underset{\displaystyle :\ddot{O}:}{O}}-S-\ddot{O}-H$

(b) $H:\overset{\displaystyle .\ddot{O}.}{\underset{\displaystyle .\ddot{O}.}{O}}:S:\ddot{O}:H$ or $H-\overset{\displaystyle .\ddot{O}.}{\underset{\displaystyle .\ddot{O}.}{O}}-S-\ddot{O}-H$

(c) $H-\ddot{O}-\overset{\displaystyle .\ddot{O}.}{\underset{\displaystyle .\ddot{O}.}{S}}-\ddot{O}-H$ is better   **9.28** ClO_2

9.29 ～～ CH—CH₂—CH—CH₂—CH—CH₂～～ with CH₃ groups

9.30 Any of the following is correct: $CFCl=CF_2$, $ClFC=CF_2$, $FCCl=CF_2$, $CClF=CF_2$, $FClC=CF_2$, $ClCF=CF_2$

9.31 (a) $\Delta H_{rxn} = -8.0 \times 10^1$ kJ   **(b)** 12 kJ   **(c)** 13%

9.32 (a) Single bonds are weaker (compare C—O to C=O and C—C to C=C).
(b) decreases (H—F > H—Cl > H—Br > H—I)

9.69 Some of the common properties of ionic compounds are hardness, high melting and boiling points, ability to conduct electricity when melted but not when in the solid state, and ability to dissolve in water to form solutions that conduct electricity. Lattice energy (the energy required to separate the ions of an ionic compound) is a measure of the force of attraction between ions. The fact that the lattice energies of ionic compounds are high accounts for the hardness and high melting and boiling points of ionic compounds. In the solid state, the ions are fixed in position; therefore, ionic solids do not conduct electricity. In the liquid state and in aqueous solution, the ions are free to move and conduct electricity.
9.71 (a) BaF_2 and **(b)** CaO. Ba^{2+}, F^-, Ca^{2+}, and O^{2-} ions all have noble gas electron configurations and are combined so that net charge is zero.

9.73 (a)

$$H-\overset{+1}{\ddot{N}}=N=\overset{-1}{\ddot{N}}: \longleftrightarrow H-\overset{-1}{\ddot{N}}-N\equiv\overset{+1}{N}:$$

(b) lefthand bond, 1.5; righthand bond, 2.5   **9.75** Group VA

STOP AND TEST YOURSELF

1. (a) 9.2   **2.** (c) 9.2   **3.** (b) 9.10   **4.** (b) 9.5   **5.** (c) 9.6
6. (c) 9.2   **7.** (c) 9.10   **8.** (c) 9.12   **9.** (c) 9.6   **10.** (d) 9.9
11. (d) 9.8   **12.** (c) 9.8   **13.** (e) 9.7   **14.** (d) 9.11

9.77 (a) An ionic bond is the electrostatic attraction between positively and negatively charged ions.   **(b)** Covalent bonds consist of pairs of electrons shared between two atoms.

9.79 (a) I $H-C=C-C\equiv N:$ with H, H below

II $\longleftrightarrow H-\overset{+1}{C}-C=C=\overset{-1}{\ddot{N}}:$ with H, H below

III $\longleftrightarrow H-\overset{-1}{\ddot{C}}-C=C=\overset{+1}{N}:$ with H, H below

Lewis structure I has no atom with a formal charge.
(b) resonance structures   **(c)** $\longleftrightarrow$ **(d)** structure I
(e) structure III. In structure I, all atoms have zero formal charge. In structure III, the most electronegative atom has a positive formal charge.   **9.81** An anion with a 2− charge.
9.83 This results in both atoms having formal charges of zero.
9.85 *Ionic compounds* have high melting points, conduct electricity in the molten state, and conduct electricity in aqueous solutions. *Molecular compounds* have moderately low melting points and do not conduct electricity in the molten state.   **9.87** Acetylene is a molecular compound, so the molecular formula C_2H_2 is appropriate. Salt is an ionic compound consisting of positive Na^+ ions and negative Cl^- ions held together by the attractive forces between oppositely charged ions in a three-dimensional arrangement (remember Figure 1.18). No molecules of NaCl exist. Therefore, a molecular formula such as Na_2Cl_2 is inappropriate, and the empirical formula, NaCl, should be used for salt.
9.89 $[:C\equiv N:]^-$ $:C\equiv O:$ $:N\equiv N:$ $[:N\equiv O:]^+$ **(a)** similar Lewis formulas   **(b)** isoelectronic   **(c)** Yes. $HC\equiv CH$ also has the same number of valence electrons and a similar Lewis formula.
9.91 (a) BH_4^-. B has an octet.   **(b)** CaO. CaF needs another F^- ion.   **(c)** KCl. KCl_2 has one too many Cl^- ions.   **(d)** PF_5. P can expand its octet.   **(e)** NO_2^-. NO_2 has an odd electron.

9.93 (a)   **(b)**

A55

(c)

(d)

(e)

9.95

9.97

(a) $[Xe]6s^1 + [He]2s^2 2p^5 \rightarrow [Xe] + [He]2s^2 2p^6$ or $[Ne]$
  Cs + F $\rightarrow$ Cs$^+$ + F$^-$
  $2Cs(s) + F_2(g) \rightarrow 2CsF(s)$ cesium fluoride

(b) $[Xe]6s^2 + [He]2s^2 2p^4 \rightarrow [Xe] + [He]2s^2 2p^6$ or $[Ne]$
  Ba + O $\rightarrow$ Ba^{2+} + O^{2-}
  $2Ba(s) + O_2(g) \rightarrow 2BaO(s)$ barium oxide

(c) $[Xe]6s^2 + \begin{matrix}[Ne]3s^2 3p^5 \\ [Ne]3s^2 3p^5\end{matrix} \rightarrow [Xe] + \begin{matrix}[Ne]3s^2 3p^6 \text{ or } [Ar] \\ [Ne]3s^2 3p^6 \text{ or } [Ar]\end{matrix}$

  Ba + 2Cl $\rightarrow$ Ba^{2+} + 2Cl$^-$
  $Ba(s) + Cl_2(g) \rightarrow BaCl_2(s)$ barium chloride

(d) $[Ar]4s^2$ $[Ar]$

  $[Ar]4s^2 + \begin{matrix}[He]2s^2 2p^3 \\ [He]2s^2 2p^3\end{matrix} \rightarrow [Ar] + \begin{matrix}[He]2s^2 2p^6 \text{ or } [Ne] \\ [He]2s^2 2p^6 \text{ or } [Ne]\end{matrix}$

  $[Ar]4s^2$ $[Ar]$
  3Ca + 2N $\rightarrow$ 3Ca^{2+} + 2N^{3-}
  $3Ca(s) + N_2(g) \rightarrow Ca_3N_2(s)$ calcium nitride

(e) $[Ar]4s^2 + \begin{matrix}1s^1 \\ 1s^1\end{matrix} \rightarrow [Ar] + \begin{matrix}1s^2 \text{ or } [He] \\ 1s^2 \text{ or } [He]\end{matrix}$

  Ca + 2H $\rightarrow$ Ca^{2+} + 2H$^-$
  $Ca(s) + H_2(g) \rightarrow CaH_2(s)$ calcium hydride

9.99 (a) Molecular equation:
$PCl_3(l) + 3H_2O(l) \rightarrow H_3PO_3(aq) + 3HCl(aq)$
Ionic equations (complete and net):
$PCl_3(l) + 3H_2O(l) \rightarrow H_3PO_3(aq) + 3H^+ + 3Cl^-$
Molecular equation:
$PCl_5(s) + 4H_2O(l) \rightarrow H_3PO_4(aq) + 5HCl(aq)$

Ionic equations (complete and net):
$PCl_5(s) + 4H_2O(l) \rightarrow H_3PO_4(aq) + 5H^+ + 5Cl^-$

(b)

C has octet. C is in second period and cannot expand its octet. A bond must be broken before a new bond can form.

The unshared pair of electrons on P can form a bond to another atom.

P can expand its octet further so that a bond to P can form.

(c) A reaction will take place if $SiCl_4$ is added to water because Si is in the third period and can have more than eight electrons around it. (This prediction is correct.)   **(d)** Density of CCl_4 is greater than the density of water.   **9.101 (a)** -100 kJ   **(b)** -99 kJ   **(c)** The bond energy of a given bond depends on the character of the entire molecule.   **9.103** (a) C_3H_4, (c) NO_2^+, (d) SCN^-

9.105 The best Lewis formula that can be written for the cyanate ion, OCN^-, has the minimum formal charge of -1 on the most electronegative atom, N. The best Lewis formula that can be written for the fulminate ion, CNO^-, has more formal charges and has a positive formal charge on N. Therefore, it is not surprising that fulminates are unstable.

9.107

9.109 (a, b)

(c) The original Lewis formula contributes most to the hybrid because it has the smaller formal charges.

9.111 (a) H : As : H   **(b)** $2AsH_3(g) \xrightarrow{250-300\,°C} 2As(s) + 3H_2(g)$
        H

9.113 (a)

(b) The bonds to H are shortest. Of the bonds between second period elements, the C=O bond is shortest. The C—C bonds in the benzene ring are intermediate between the C=C bond length of 134 pm and the C—C bond length of 154 pm. The C—N and C—O bonds are slightly shorter than C—C single bonds.

9.115 322 kJ/mol   **9.117** $\left[H : \overset{\displaystyle H}{\underset{\displaystyle H}{N}} : H \right]^+$ $\left[\overset{\cdot\cdot}{O} :: Cl :: \overset{\cdot\cdot}{O} \right]^-$ with $\overset{\cdot\cdot}{\underset{\cdot\cdot}{O}}$

9.119 $H_2C\!=\!CH(OOCCH_3)$ or $CH_3COOCH\!=\!CH_2$

CHAPTER 10

10.1 (a)   **(b)** octahedral   **(c)** $SnCl_6$ structure with Cl atoms

10.2 linear   **10.3 (a)** SiH_4 structure   **(b)** AsF_5 structure

10.4 (a) (i) 4   (ii) 3   (iii) 6   (iv) 5   (v) 4   **(b)** (i) 0 unshared, 4 shared   (ii) 1 unshared, 2 shared   (iii) 2 unshared, 4 shared   (iv) 1 unshared, 4 shared   (v) 2 unshared, 2 shared

(c) (i) tetrahedral   (ii) bent   (iii) square planar   (iv) seesaw   (v) bent

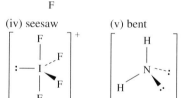

10.5 linear   **10.6 (a)** $BeCl_2$, BF_3, CH_4, PCl_5, SF_6 **(b)** GeF_2, SF_4, ClF_3, IF_5 **10.7** Sigma bonds. Ends of sp^3 orbitals overlap 1s orbitals of hydrogens.   **10.8 (a)** 1   **(b)** sp^3d **(c)** 5

10.9 (a)

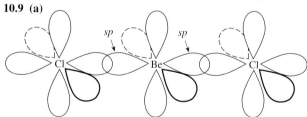

(b)

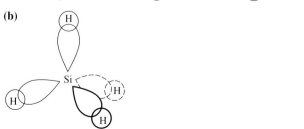

10.10 (a) 9   **(b)** 1   **(c)** 8

10.11 (a) H 1s overlaps with C sp to form a sigma bond; C sp overlaps with one N sp orbital to form a sigma bond; two C p orbitals overlap with two N p orbitals to form two pi bonds.   **(b)** The lefthand carbon has four sp^3 orbitals, three of which overlap with three H 1s orbitals to form sigma bonds and one of which overlaps with an sp^2 hybrid orbital of the middle carbon to form a sigma bond. The middle carbon also overlaps one sp^2 orbital with a hydrogen 1s to form a sigma bond, and another sp^2 orbital with an sp^2 from the righthand carbon to form a sigma bond. The righthand carbon forms a sigma bond with each of two hydrogen 1s orbitals, using its remaining two sp^2 orbitals. The remaining parallel p orbitals on the middle and righthand carbons overlap to form a pi bond.   **10.12 (a)** about 120°   **(b)** about 120°   **10.13** No. The three hydrogens on the lefthand C cannot all be in the same plane.

10.14 (a)

$(\sigma_{1s})^2(\sigma_{1s}^*)^2(\sigma_{2s})^2(\sigma_{2s}^*)^2(\sigma_{2p_z})^2$

$(\pi_{2p_x})^2(\pi_{2p_y})^2(\pi_{2p_x}^*)^2(\pi_{2p_y}^*)^2$

MO diagram (increasing energy):
$\sigma_{2p_z}^*$
$\pi_{2p_x}^*$ ↑↓ $\pi_{2p_y}^*$ ↑↓
π_{2p_x} ↑↓ π_{2p_y} ↑↓
σ_{2p_z} ↑↓
σ_{2s}^* ↑↓
σ_{2s} ↑↓
σ_{1s}^* ↑↓
σ_{1s} ↑↓

(b) zero
(c) one
(d) single bond, bond order = 1

10.15 (a)

$(\sigma_{1s})^2(\sigma_{1s}^*)^2(\sigma_{2s})^2(\sigma_{2s}^*)^2$

$(\pi_{2p_x})^1(\pi_{2p_y})^1$

MO diagram (increasing energy):
$\sigma_{2p_z}^*$
$\pi_{2p_x}^*$ $\pi_{2p_y}^*$
σ_{2p_z}
π_{2p_x} ↑ π_{2p_y} ↑
σ_{2s}^* ↑↓
σ_{2s} ↑↓
σ_{1s}^* ↑↓
σ_{1s} ↑↓

(b) bond order = 1
(c) paramagnetic, two

10.16 (a)

Energy level diagram for NO:

$$\sigma^*_{2p_z}$$

$$\pi^*_{2p_x} \uparrow \qquad \pi^*_{2p_y}$$

$$\pi_{2p_x} \uparrow\downarrow \qquad \uparrow\downarrow \ \pi_{2p_y}$$

$$\uparrow\downarrow \ \sigma_{2p_z}$$

$$\uparrow\downarrow \ \sigma^*_{2s}$$

$$\uparrow\downarrow \ \sigma_{2s}$$

$$\uparrow\downarrow \ \sigma^*_{1s}$$

$$\uparrow\downarrow \ \sigma_{1s}$$

NO (increasing energy ↑)

(b) $(\sigma_{1s})^2(\sigma^*_{1s})^2(\sigma_{2s})^2(\sigma^*_{2s})^2(\sigma_{2p_z})^2(\pi_{2p_x})^2(\pi_{2p_y})^2(\pi^*_{2p_x})^1$
(c) bond order = 2.5 **(d)** No. **(e)** bond energies and bond lengths
10.17 (a) $H_2C=CHCH=CHCH_3$. The p orbitals of the four carbons in the left formula overlap sidewise. **(b)** four
10.57 No. The p orbitals on C1 and C3 are perpendicular to each other.
10.59

	Li$_2$	Be$_2{}^+$	Be$_2$
σ^*_{2s}	—	↑	↑↓
σ_{2s}	↑↓	↑↓	↑↓
σ^*_{1s}	↑↓	↑↓	↑↓
σ_{1s}	↑↓	↑↓	↑↓

Bond order: Li$_2$ = 1, Be$_2{}^+$ = 0.5, Be$_2$ = 0
Stability: Li$_2$ > Be$_2{}^+$ > Be$_2$

10.61 (a) $sp \to sp^3$ **(b)** $sp^3 \to sp^2$ **(c)** no change in hybridization (sp^3) **(d)** $sp^3 \to sp^3d$ **(e)** $sp^3d \to sp^3d^2$
10.63 (a) i, iii, and iv; ii and v **(b)** i, iii, and iv; ii and v
(c) i and iv; ii, iii, and v

STOP AND TEST YOURSELF

1. (a) *10.1* **2.** (b) *10.1* **3.** (e) *10.2* **4.** (e) *10.4* **5.** (d) *10.4*
6. (c) *10.4* **7.** (c) *10.4* **8.** (c) *10.4* **9.** (d) *10.4* **10.** (a) *10.4*
11. (a) *10.6* **12.** (c) *10.6* **13.** (a) *10.7* **14.** (a) *10.8*

10.65 (a) $\left[:\ddot{I}:\ddot{I}:\ddot{I}:\right]^-$ **(b)** Br and I are able to have more than eight electrons around them. F atoms are unable to have more than eight electrons around them. **(c)** linear **10.67 (a)** superoxide ion, $O_2{}^-$; peroxide ion, $O_2{}^{2-}$; dioxygenyl ion, $O_2{}^+$

(b)

	$O_2{}^-$	$O_2{}^{2-}$	$O_2{}^+$
$\pi^*_{2p_x}\ \pi^*_{2p_y}$	↑↓ ↑	↑↓ ↑↓	↑ —
$\pi_{2p_x}\ \pi_{2p_y}$	↑↓ ↑↓	↑↓ ↑↓	↑↓ ↑↓
σ_{2p_z}	↑↓	↑↓	↑↓
σ^*_{2s}	↑↓	↑↓	↑↓
σ_{2s}	↑↓	↑↓	↑↓
σ^*_{1s}	↑↓	↑↓	↑↓
σ_{1s}	↑↓	↑↓	↑↓

(d)

Bond order	1.5	1	2.5

(e)

Magnetism	paramagnetic	diamagnetic	paramagnetic
# of unpaired electrons	1	0	1

(c) $O_2{}^-$:
$(\sigma_{1s})^2(\sigma^*_{1s})^2(\sigma_{2s})^2(\sigma^*_{2s})^2(\sigma_{2p_z})^2(\pi_{2p_x})^2(\pi_{2p_y})^2(\pi^*_{2p_x})^2(\pi^*_{2p_y})^1$
$O_2{}^{2-}$:
$(\sigma_{1s})^2(\sigma^*_{1s})^2(\sigma_{2s})^2(\sigma^*_{2s})^2(\sigma_{2p_z})^2(\pi_{2p_x})^2(\pi_{2p_y})^2(\pi^*_{2p_x})^2(\pi^*_{2p_y})^2$
$O_2{}^+$: $(\sigma_{1s})^2(\sigma^*_{1s})^2(\sigma_{2s})^2(\sigma^*_{2s})^2(\sigma_{2p_z})^2(\pi_{2p_x})^2(\pi_{2p_y})^2(\pi^*_{2p_x})^1$

10.69

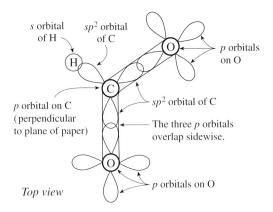

Top view

10.71 268 kJ/mol **10.73** A *resonance hybrid* is a molecule or polyatomic ion for which more than one Lewis formula must be written. *Hybrid orbitals* are formed by combination of two or more atomic orbitals of the same atom. **10.75** :N≡N:, :C≡O:, H:C≡C:H **10.77** The valence bond method predicts that oxygen is diamagnetic. The molecular orbital method predicts that oxygen is paramagnetic, in agreement with experiment.
10.79 (a) three **(b)** three **(c)** maximum number is two for both **(d)** Hybrid orbitals are orbitals formed by combination of atomic orbitals of the same atom; molecular orbitals are orbitals formed by combination of atomic orbitals from different atoms.
10.81 The valence bond method uses resonance to describe the bonding in ozone. Ozone is described as a hybrid of two structures.

The molecular orbital method assumes that the central oxygen atom is sp^2 hybridized, that sigma bonds are formed by endwise overlap of sp^2 hybrid orbitals of the central oxygen with p orbitals of the other two oxygens, and that the p orbital of the central oxygen overlaps sidewise with p orbitals on the other two oxygens, forming pi bonds. The electrons of the pi bonds are delocalized over all three oxygens atoms. Only one picture is needed.

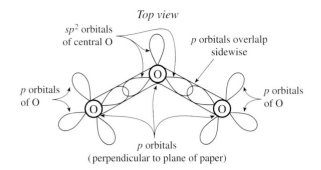

Top view

sp^2 orbitals of central O

p orbitals overlalp sidewise

p orbitals of O

p orbitals of O

p orbitals (perpendicular to plane of paper)

10.83 $O_2^{2-} < O_2^- < O_2 < O_2^+$

10.85 (a) $\Delta H^\circ = 248$ kJ   **(b)** 165 kJ

(c)

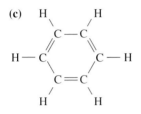

(d)

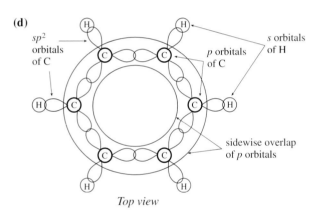

sp^2 orbitals of C

p orbitals of C

s orbitals of H

sidewise overlap of p orbitals

Top view

10.87 N_2: one sigma $sp-sp$, two pi $p-p$; NH_3: three sigma sp^3-s; HNO_3: H—O bond is sigma $s-sp^3$, O—N bond is sigma sp^3-sp^2, N=O bond is composed of one sigma sp^2-p and one pi $p-p$, N—O bond is sigma sp^2-p.

10.89
(a)

	O_2^+	N_2^+
$\pi_{2p_x}^*$ $\pi_{2p_y}^*$	↑ —	— —
π_{2p_x} π_{2p_y}	↑↓ ↑↓	↑ — σ_{2p_z}
σ_{2p_z}	↑↓	↑↓ ↑↓ π_{2p_x} π_{2p_y}
σ_{2s}^*	↑↓	↑↓
σ_{2s}	↑↓	↑↓
σ_{1s}^*	↑↓	↑↓
σ_{1s}	↑↓	↑↓

(c)   Bond order   2.5   2.5

(d) Magnetism   paramagnetic   paramagnetic
# of unpaired electrons   1   1

(b) electron configuration:
O_2^+: $(\sigma_{1s})^2(\sigma_{1s}^*)^2(\sigma_{2s})^2(\sigma_{2s}^*)^2(\sigma_{2p_z})^2(\pi_{2p_x})^2(\pi_{2p_y})^2(\pi_{2p_x}^*)^1$
N_2^+: $(\sigma_{1s})^2(\sigma_{1s}^*)^2(\sigma_{2s})^2(\sigma_{2s}^*)^2(\pi_{2p_x})^2(\pi_{2p_y})^2(\sigma_{2p_z})^1$

10.91 (a)   **(b)**

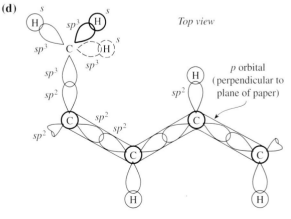

10.93 (a) $\ddot{O}=C=C=C=\ddot{O}$ **(b)** sp **(c)** 180°

(d) A polymer is a substance whose molecules are composed of thousands of atoms and which has repeating units.   **(e)** a monomer

10.95 (a)   [structure of SF_6]   **(b)** 1×10^{-6} ppb, 4×10^6 molecules SF_6

(c) 108 kg F_2 **(d)** sulfur hexafluoride   **(e)** It must not react with seawater or any of the solutes in seawater (such as Na^+ and Cl^-).
10.97 (a) H_2O, $CH_3CH_2CH_2OH$, and $CH_3(CH_2)_{14}COOCH_3$.
(b) Water has an unusually high specific heat.

(c) $H-\overset{\displaystyle H}{\underset{\displaystyle H}{C}}-\overset{\displaystyle H}{\underset{\displaystyle H}{C}}-\overset{\displaystyle H}{\underset{\displaystyle H}{C}}-\ddot{O}-H$

10.99 (a) 11-*trans*-retinal   **(b)** vitamin A: 10 atoms; β-carotene: 22 atoms   **(c)** 22 atoms

(d)   *Top view*

sp^3 orbitals of C, sp^2, p orbital (perpendicular to plane of paper)

(e) 3.61×10^{-17} J

11.1 (a) 0   **(b)** N = +5, O = −2   **(c)** N = −3, H = +1
(d) Mg = +2, N = −3   **(e)** H = +1, N = −3, O = −2
(f) Fe = +2, S = +6, O = −2   **(g)** Cr = +3, S = +6, O = −2
11.2 (a) P = +3, Cl = −1   **(b)** C = +2, N = −3
11.3 +5 and −3   **11.4** HNO_3. Nitrogen is in a higher oxidation
state in HNO_3.   **11.5 (a)** chromium(VI) oxide   **(b)** calcium
chlorite   **(c)** hypochlorite ion   **(d)** potassium chlorate   **(e)** periodic
acid   **11.6 (a)** $LiClO_2$ **(b)** $Ca(ClO)_2$ **(c)** Na_2SO_3 **(d)** $MnCl_2$
(e) Fe_2O_3 **11.7 (a)** Zn(s) is oxidized, H^+ is reduced.   **(b)** Na(s)
is oxidized, $Cl_2(g)$ is reduced.   **(c)** O^{2-} in HgO is oxidized and
Hg^{2+} is reduced.   **11.8 (a)** not redox   **(b)** redox Iron is
oxidized; chlorine is reduced.
Complete ionic equation:
$6Fe^{2+} + 18Cl^- + 6H^+ + Na^+ + ClO_3^- \rightarrow$
$$6Fe^{3+} + 19Cl^- + Na^+ + 3H_2O(l)$$
Net ionic equation:
$6Fe^{2+} + ClO_3^- + 6H^+ \rightarrow 6Fe^{3+} + Cl^- + 3H_2O(l)$
ClO_3^- is the oxidizing agent, and Fe^{2+} is the reducing agent.
(c) redox   Sulfur is oxidized; iodine is reduced.
Complete ionic equation:
$2Na^+ + S^{2-} + 4I_2(s) + 4H_2O(l) \rightarrow 8H^+ + 8I^- + 2Na^+ + SO_4^{2-}$
Net ionic equation:
$S^{2-} + 4I_2(s) + 4H_2O(l) \rightarrow 8H^+ + 8I^- + SO_4^{2-}$
S^{2-} is the reducing agent, and $I_2(s)$ is the oxidizing agent.
11.9 (a) $2Sb(s) + 3Cl_2(g) \rightarrow 2SbCl_3(s)$
(b) $2HBr(aq) + H_2SO_4(aq) \rightarrow Br_2(l) + SO_2(g) + 2H_2O(l)$

11.10 (a) $\overset{0}{2Sb}(s) + \overset{0}{3Cl_2}(g) \rightarrow \overset{+3\ -1}{2SbCl_3}(s)$

The oxidation number of two Sbs changes from 0 to +3, an
increase of 2(3) = 6; the oxidation number of six Cls changes from
0 to −1, a decrease of 6(1) = 6.

(b) $\overset{+1\ -1}{} \quad \overset{+1\ +6\ -2}{} \quad \overset{0}{} \quad \overset{+4\ -2}{} \quad \overset{+1\ -2}{}$

The oxidation number of two Brs changes from −1 to 0, an
increase of 2(1) = 2; the oxidation number of one S changes from
+6 to +4, a decrease of 1(2) = 2.
11.11 (a) $Zn(s) \rightarrow Zn^{2+} + 2e^-$ **(b)** $2H^+ + 2e^- \rightarrow H_2(g)$
11.12 (a) $2I^- + Br_2(aq) \rightarrow I_2(s) + 2Br^-$ **(b)** $2I^- \rightarrow I_2(s) + 2e^-$
11.13 (a) Ca   **(b)** F_2 **(c)** Fe^{2+}, SO_2, and Cu^+. Ca is in lowest
oxidation state, F_2 is in highest, and Fe^{2+}, S in SO_2, and Cu^+ are in
intermediate oxidation states.
11.14 (a) $2HS_2O_4^- + H_2O(l) \rightarrow S_2O_3^{2-} + 2H_2SO_3(aq)$ or
$2HS_2O_4^- + H_2O(l) \rightarrow S_2O_3^{2-} + 2H_2O(l) + 2SO_2(g)$ or
$2HS_2O_4^- \rightarrow S_2O_3^{2-} + H_2O(l) + 2SO_2(g)$
(b) sulfur
11.15 Complete ionic equation:
$3Br_2(aq) + 6Na^+ + 6OH^- \rightarrow 6Na^+ + BrO_3^- + 5Br^- + 3H_2O(l)$
Molecular equation:
$3Br_2(aq) + 6NaOH(aq) \rightarrow NaBrO_3(aq) + 5NaBr(aq) + 3H_2O(l)$
11.16 0.446 M   **11.17** (a), (c), and (d)
11.18 (a) $2Ca(s) + O_2(g) \overset{heat}{\rightarrow} 2CaO(s)$
(b) $Si(s) + O_2(g) \overset{heat}{\rightarrow} SiO_2(s)$
(c) $2C_6H_6(l) + 15O_2(g) \overset{heat}{\rightarrow} 12CO_2(g) + 6H_2O(l)$
(d) no reaction   **(e)** $4Li(s) + O_2(g) \overset{heat}{\rightarrow} 2Li_2O(s)$

(f) $2CH_3CH{=}CH_2(g) + 9O_2(g) \overset{heat}{\rightarrow} 6CO_2(g) + 6H_2O(l)$
(g) $2CH_3CH{=}CH_2(g) + O_2(g) \overset{catalyst}{\rightarrow} 2CH_3CH{-}CH_2(g)$
$$\diagdown \underset{O}{} \diagup$$

11.19 Like gold, iridium is not very reactive.
11.20 (a) $MoO_3(s) + 3H_2(g) \overset{heat}{\rightarrow} Mo(s) + 3H_2O(g)$
(b) $PbO(s) + C(s) \overset{heat}{\rightarrow} Pb(s) + CO(g)$
11.49 (a) Al = +3, N = −3   **(b)** Na = +1, O = −2, C = +4,
N = −3   **(c)** N = −3, H = +1, O = −2, Mo = +6
(d) S = 0   **(e)** Ti = +4, O = −2
11.51 $MnSO_4 < Mn_2O_3 < MnO_2 < MnO_4^-$
11.53 Yes.

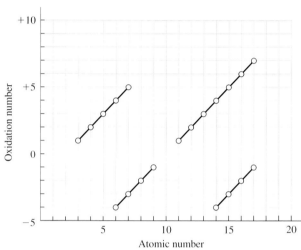

Oxidation number as a function of atomic number

11.55 (a) Limestone, $CaCO_3$, decomposes to form carbon dioxide
gas and the basic oxide CaO. The CaO reacts with acidic oxide
impurities in the iron ore, such as SiO_2, to form a less dense liquid
that floats on the surface of the iron.   **(b)** The incoming air reacts
with carbon to form carbon monoxide and to liberate considerable
amounts of thermal energy in the process. The carbon monoxide
acts as a reducing agent for the iron oxides. Thus, the oxygen in the
air blast generates the reducing agent from the coke and in so doing
generates enough thermal energy for the rest of the purification
process.   **(c)** The coke acts as the fuel and source of the reducing
agent CO. The reaction of carbon and limited quantities of oxygen
is highly exothermic; thus, the coke acts as a fuel by releasing
thermal energy.

STOP AND TEST YOURSELF

1. (d) *11.1*   **2.** (c) *11.1*   **3.** (d) *11.2*   **4.** (e) *11.4*
5. (a) and (c) *11.6*   **6.** (b) *11.6*   **7.** (b) *11.2*   **8.** (c) *11.3*
9. (a) *11.3*   **10.** (d) *11.5*   **11.** (c) *11.4*   **12.** (c) *11.4*
13. (a) *11.8*   **14.** (c) *11.7*   **15.** (a) *11.9*

11.57 (a) $3SO_3^{2-} + 2MnO_4^- + H_2O(l) \rightarrow$
$$3SO_4^{2-} + 2MnO_2(s) + 2OH^-$$
(b) Mn is reduced. S is oxidized.   **(c)** MnO_4^- is the oxidizing
agent. SO_3^{2-} is the reducing agent.   **11.59** 11.7 g CuS
11.61 (a) Complete ionic equation:
$14K^+ + 14MnO_4^- + 20K^+ + 20OH^- + C_3H_8O_3(aq) \rightarrow$
$$28K^+ + 14MnO_4^{2-} + 3CO_3^{2-} + 6K^+ + 14H_2O(l)$$

(b) Molecular equation:
$$14KMnO_4(aq) + 20KOH(aq) + C_3H_8O_3(aq) \rightarrow$$
$$14K_2MnO_4(aq) + 3K_2CO_3(aq) + 14H_2O(l)$$
11.63 (a) Cl_2O, dichlorine monoxide　**(b)** ClO_2, chlorine dioxide
(c) Cl_2O_6, dichlorine hexoxide
11.65 $HClO_4 > HClO_3 > HClO_2 > HClO$　　**11.67** In compounds,
$+4$ and $+2$. (The oxidation number of the free element would, of
course, be 0.)　　**11.69 (a)** -1　**(b)** Yes. H_2O　**(c)** Yes. O_2
11.71 (b) and (c)　　**11.73** The sulfur is oxidized. Copper, which is
in its highest oxidation state, and oxygen are reduced.
11.75 Two separate reactions are taking place simultaneously:
$HNO_3 \rightarrow HNO_2$ and $HNO_3 \rightarrow NO$. Two separate equations must be
written, not one combined equation.　　**11.77 (a)** $+8/3$　**(b)** $+2$
(c) $Mo = +4$, $S = -2$　**(d)** $+1$　**(e)** -3　　**11.79 (a)** copper(I)
iodide　**(b)** iron(II) phosphate　**(c)** iron(II) sulfate　**(d)** sulfuric
acid　**(e)** potassium hydrogen sulfite　**(f)** potassium sulfate
(g) sodium sulfite　**(h)** sulfur dioxide
11.81 (a) $O_2(g) + 2H_2O(l) + 4e^- \rightarrow 4OH^-$
(b) $4Fe^{2+} + O_2(g) + 2H_2O(l) \rightarrow 4Fe^{3+} + 4OH^-$　　**11.83** 61 ppm
11.85 (a) Molecular equation:
$$C_6H_{12}O_6(aq) + 6O_2(aq) \rightarrow 6CO_2(aq) + 6H_2O(l)$$
$$\Delta H = -2.87 \times 10^3 \text{ kJ}$$
(b) ΔH, the energy released or absorbed when a change takes place
at constant pressure, is a thermodynamic state function whose value
does not depend on the steps involved but only depends on the
initial and final states.
(c) Net ionic equations for half-reactions of reduction:
$$10e^- + 12H^+ + 2NO_3^- \rightarrow N_2(g) + 6H_2O(l)$$
$$8e^- + 8H^+ + SO_4^{2-} \rightarrow S^{2-} + 4H_2O(l)$$
11.87 (a)

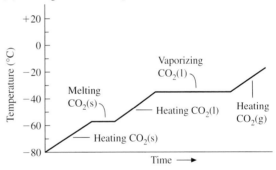

(b) $-1/3$　**(c)** No. No atoms undergo any change in oxidation
number.　**(d)** Lead azide is probably insoluble in water.
(e) Complete ionic equation:
$$Pb^{2+} + 2NO_3^- + 2Na^+ + 2N_3^- \rightarrow Pb(N_3)_2(s) + 2Na^+ + 2NO_3^-$$
Net ionic equation: $Pb^{2+} + 2N_3^- \rightarrow Pb(N_3)_2(s)$
(f) $Pb(N_3)_2(s) \rightarrow Pb(s) + 3N_2(g)$. Pb^{2+} is reduced. N is oxidized.
11.89 (a) All are redox.　**(b)** Step 3 is disproportionation.
(c) 15.6 lb HNO_3　**(d)** 14.8 lb HNO_3　**(e)** Volumes of gases will
be smaller. Therefore, either the plant can be smaller or else the
same size plant will yield a greater output.

CHAPTER 12

12.1 (a) CH_3CH_2OH and (d) NH_3. Both have H bonded to small,
electronegative atom (O, N).　　**12.2 (a)** CH_3CH_2OH,
(b) CH_3OCH_3, (d) NH_3
12.3

12.4 (a) NO. It is polar.　**(b)** NH_3. It is hydrogen bonded.
12.5 (a) argon, atoms larger　**(b)** Argon must only be cooled to
87.5 K; argon has stronger London forces.
12.6 (a) $CH_3CH_2CH_2CH_2Cl$, molecules longer
(b) $CH_3CH_2CH_2CH_2Cl$ condenses at 78 °C because it has stronger
London forces.
12.7 (a) $CH_3CH_2CH_2CH_2OH$, molecules linear [Molecules of
$(CH_3)_3COH$ are spherical.]　**(b)** $CH_3CH_2CH_2CH_2OH$ condenses at
117.2 °C because it has stronger London forces.　　**12.8 (a)** HCl;
Cl is more electronegative than H, so HCl is polar.　**(b)** HCl must
be cooled to 188 K. HCl has dipole-dipole attractions in addition to
London forces, which are similar for HCl and F_2.
12.9 (a) no change　**(b)** no shift in equilibrium　　**12.10** 78 °C
12.11 83 °C　　**12.12** approximately 220 mmHg
12.13 $CH_3(CH_2)_6CH_3$; vapor pressure is lower
12.14 decreases; solids are more orderly
12.15 24.2 kJ　　**12.16** gas　　**12.17 (a)** The temperature of the
solid CO_2 will increase until it reaches approximately -57 °C.
Then the temperature will remain constant until all of the $CO_2(s)$
has melted. The temperature will then again rise until it reaches
approximately -35 °C. The temperature will once again remain
constant until all of the $CO_2(l)$ has vaporized. When all of the
$CO_2(l)$ has vaporized, the temperature will rise to 0 °C.
(b) Heating curve for CO_2 at 13 atm:

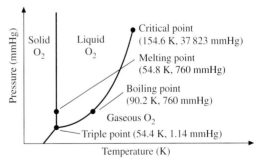

12.18 Phase diagram for O_2:

12.19 (a) molecular　**(b)** molecular　**(c)** metallic　**(d)** ionic
12.20 (a) London forces　**(b)** dipole-dipole attractions, London
forces　**(c)** metallic bonding　**(d)** electrostatic attractions
12.21 (a) soft, low melting, nonconducting solid　**(b)** soft, low
melting, nonconducting solid　**(c)** shiny, hard, high melting,
malleable, ductile, metallic solid with electrical and thermal
conductivity　**(d)** hard, brittle, high melting solid
12.22 molecular　　**12.23** two　　**12.24** 124 pm　　**12.25 (a)** *n*-type
semiconductor　**(b)** *p*-type semiconductor　　**12.77 (a)** Decrease.

When a glass crystallizes, the arrangement of the units becomes ordered. **(b)** Decreases. Thermal energy is released when a glass (or a liquid) crystallizes. **(c)** Increase. A liquid that crystallizes rapidly when scratched with a glass stirring rod must be supercooled, that is, cooled below its normal melting point. Thermal energy is released when the liquid crystallizes and the temperature rises. **12.79** 1.18 kJ **12.81 (a)** Temperature must be raised until the vapor pressure of the water equals atmospheric pressure. To convert liquid to vapor at the boiling point, thermal energy is needed to overcome intermolecular attractions. **(b)** Normally, the partial pressure of H_2O in the atmosphere is less than the vapor pressure of H_2O at room temperature. Water molecules will escape from the liquid faster than water molecules will condense from the vapor. If the container is open, all of the water will evaporate eventually. Because evaporation of the water takes place slowly, enough thermal energy is supplied by the surroundings so that the temperature does not change. **(c)** The vacuum pump reduces the atmospheric pressure to a pressure lower than the vapor pressure of the water. Thermal energy is required to vaporize the water, and this must be provided by the surroundings. If the pressure is very low, boiling will be rapid so that enough thermal energy is removed from the remaining water to cause it to freeze. **12.83 (a)** 12.1 cm^3 **(b)** 4.15 cm^3 **(c)** 66%
12.85 (a) Vapor pressure of Cl_2 as a function of temperature:

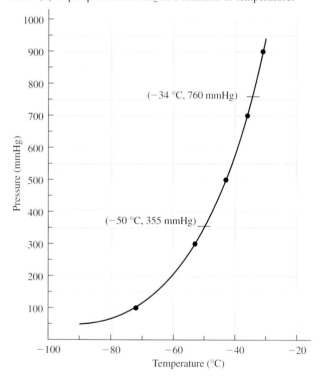

(b) −34 °C **(c)** 355 mmHg **12.87** The quantity of ethene may be determined by gauge pressure; the quantity of ethane must be determined by weighing. The critical temperature of ethene, 9.9 °C, is below room temperature. The contents of the ethene cylinder will be gaseous no matter how high the pressure. Part of the contents of the ethane container will be liquid at pressures above the critical pressure. The gauge will give the vapor pressure of the liquid. **12.89** Phase diagram for water, with slope of ice-liquid

water line exaggerated:

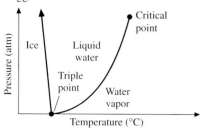

12.91 Because the sugar molecules have so many —OH groups, hydrogen bonding is strong and liquid sugars are very viscous. The molecules can't move into an orderly arrangement.
12.93 0.75 kJ (estimate) **12.95 (a)** 1.34 L **(b)** Some liquid will remain. Pressure will be the vapor pressure of water at 20.0 °C, 17.5 mmHg. **(c)** All water will evaporate and pressure will be 5, 5.1, or 5.07 mmHg. **12.97** 22.1 °C **12.99** 77.73% Fe
12.101 (a) Surface tension of water at different temperatures:

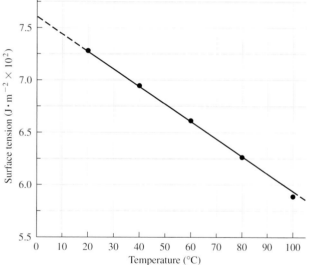

(b) a straight line
(c) surface tension = $(-1.74 \times 10^{-4} \text{ J/m}^2 \cdot °C)t_c + 0.0763 \text{ J/m}^2$
12.103 5.32×10^{-3} g **12.105** Heating curve for CH_3CH_2OH:

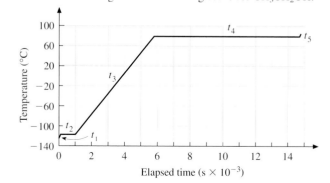

12.107 (b), (c), (e)   **12.109** The height of the liquid will be higher on the moon than on Earth because the force of gravity pulling down on the liquid will be less. Adhesive forces between liquid and glass and surface tension will remain the same.
12.111 (a) (i) N_2, (ii) Cl_2, (iii) NH_3 **(b)** (i) H_2NNH_2, (ii) Br_2, (iii) NH_3 **12.113** The surface tension of molten Pb is high enough that, while falling through air, the atoms move to the lowest energy shape, spherical. When the drops hit the water, cooling is so rapid that the drops stay round.   **12.115** Wetting agents decrease the surface tension of water so that the water is attracted more to solid surfaces than normally and spreads out.
12.117 The surface tension of molten or softened glass is high, so that the surface area tends to decrease to a minimum, which causes the rounding.   **12.119 (a)** The liquid is under pressure in the container so that boiling is prevented.   **(b)** The critical temperature must be above room temperature.   **12.121** 47 mmHg
12.123 TiO_2 **12.125** molecular crystals   **12.127 (a)** C_3N_4
(b) tricarbon tetranitride   **(c)** covalent network crystals
12.129 (a) 70.9%   **(b)** Assuming no further evaporation, the equilibrium vapor pressure will increase while partial pressure remains the same. Therefore, the relative humidity will decrease with increasing temperature.   **12.131 (a)** o-dichlorobenzene
(b) p-dichlorobenzene   **(c)** It must be at a pressure greater than 1 atm.   **12.133** 183 g NH_3 **12.135 (a)** nitrogen   **(b)** boron
12.137 The water in the squirrels' blood must supercool.

CHAPTER 13

13.1 (a) not miscible   **(b)** miscible   **(c)** miscible   **(d)** miscible Pentane is nonpolar. Water is polar; heptane and CCl_4 are nonpolar. Although 1-propanol is polar, it has a nonpolar part; because CH_3CH_2OH is soluble in $CH_3(CH_2)_4CH_3$, $CH_3CH_2CH_2OH$ is probably soluble in $CH_3(CH_2)_3CH_3$.   **13.2** $C_{12}H_{22}O_{11}$ can form hydrogen bonds with water. When sugar crystals and water are mixed, fast-moving water molecules hit vigorously vibrating sugar molecules at the surface of crystals and knock them out of the crystal. In solution, the sugar molecules are quickly surrounded by spherical shells of water molecules.
13.3 A lump of $CuSO_4 \cdot 5H_2O$ could be dissolved more quickly by (a) breaking it into smaller pieces (increasing the surface area), (b) raising the temperature, and (c) stirring the mixture.
13.4 (a) Mg^{2+}, smaller   **(b)** Ba^{2+}, higher charge   **(c)** Na^+, smaller   **13.5** (a) KI(s), (d) $CH_3OH(l)$. Potassium compounds are soluble; CH_3OH has only one carbon and can hydrogen bond with water. [Ti(s) is a metal, CCl_4 is nonpolar, and C (diamond) is a covalent network solid.]   **13.6** $CaSO_4$
13.7 about 25 g $KClO_3/100$ g H_2O **13.8** unsaturated
13.9 about 25 g H_2O **13.10** lower the temperature of the water
13.11 1.4×10^{-3} g $N_2/100$ g H_2O **13.12 (a)** 0.25 m
(b) 1.03×10^{-1} m **13.13** methyl alcohol—0.21, ethyl alcohol—0.26, water—0.53   **13.14** NaCl, 1.86×10^{-3}; H_2O, 0.9982
13.15 $p_{N_2} = 717$ mmHg, $p_{H_2O} = 35.4$ mmHg   **13.16** 98.7 m
13.17 14.9 M   **13.18** 1.09×10^{-1} **13.19 (a)** 335.5 mmHg
(b) It increases. Compare the mixture in this problem with the mixture in Sample Problem 13.6.   **13.20 (a)** 23.114 mmHg
(b) 0.362 mmHg   **13.21 (a)** 79.9 °C   **(b)** -120.0 °C
13.22 1.9 m **13.23 (a)** 0.1 m **(b)** 0.2 m **(c)** 0.3 m **(d)** 0.1 m
13.24 (a) $fp_{0.1\,m\,CaCl_2} < fp_{0.1\,m\,NaCl} < fp_{0.1\,m\,C_{12}H_{22}O_{11}}$

(b) $fp_{0.1\,m\,HCl} < fp_{0.1\,m\,HC_2H_3O_2} < fp_{0.05\,m\,HCl}$
13.25 -0.0558 °C   **13.26** 9.69 mmHg   **13.27** 3.45×10^4 u
13.77 32 g pentane, 78 g hexane, 115 g heptane
13.79 (a) For dilute solutions (0.1 M and lower), molality and molarity are equal for all practical purposes.   **(b)** No. The molarity and molality of dilute aqueous solutions are equal because the density of water equals 1.00 g/mL under ordinary conditions. Few other common solvents have a density of 1.00 g/mL.
13.81 (a) lyophobic—solvent hating; lyophilic—solvent loving
(b) hydrophobic—water hating; hydrophilic—water loving
(c) sol—a solid colloid in a liquid medium; gel—a semisolid sol
(d) foam—gas in a liquid or solid; emulsion—liquid in a liquid or solid   **(e)** aerosol—liquid or solid in a gas; foam—gas in a liquid or solid   **13.83 (a)** 1.3×10^2 u   **(b)** The melting point of camphor is above room temperature, making it easier to measure precisely and accurately. The K_{fp} for camphor is almost eight times that of benzene, which means for a given concentration of solute, a much larger Δfp is observed. This, too, should improve accuracy and precision.
13.85

STOP AND TEST YOURSELF

1. (d) *13.1, 13.2*   **2.** (b) *13.3, 13.4*   **3.** (b) *13.1*   **4.** (c) *13.7*
5. (e) *13.7*   **6.** (b) *13.5*   **7.** (c) *13.3*   **8.** (e) *13.7*   **9.** (d) *13.7*
10. (b) *13.6*   **11.** (d) *13.7*   **12.** (d) *13.5*   **13.** (a) *13.5*
14. (b) *13.8*   **15.** (a) *13.8*

13.87 (a) energy required to separate solute particles and to separate solvent molecules, energy released by solvation, entropy
(b) Small size and high charge both increase interionic attractions and result in strongly hydrated ions; entropy may increase or decrease.   **13.89 (a)** 13.8 mmHg   **(b)** 2.3 mmHg   **(c)** -17.1 °C
(d) 104.70 °C   **13.91** $Ba(OH)_2$ **13.93 (a)** right—increasing pressure favors solution of gases   **(b)** right—the solubility of most gases in water increases as temperature decreases.
13.95 (a) $C_6H_{12}O_6(s) \rightleftharpoons C_6H_{12}O_6(aq)$ **(b)** Some of the $C_6H_{12}O_6(s)$ will disappear.   **13.97 (a)** 239 u   **(b)** It is about double the value of 122.123 calculated for C_6H_5COOH.
(c) The molecule must form a dimer:

13.99 (a) $\dfrac{45 \text{ g NH}_3}{100 \text{ g H}_2\text{O}}$ **(b)** Since gases are usually not very soluble in water, the $NH_3(g)$ must have reacted with the water. **(c)** < 25 °C. Gases are more soluble in water at lower temperature. **13.101 (a)** 1.96 m **(b)** 1.82 M **(c)** $X_{NH_4Cl} = 0.0341$, $X_{H_2O} = 0.9659$ **(d)** 9.49% NH_4Cl, 90.51% H_2O **13.103 (a)** benzene; structure is similar to six-membered rings in buckyballs. **(b)** —NH_2; group capable of hydrogen bonding with water. **(c)** sodium benzoate; it is an ionic compound, and all sodium salts are water soluble. **13.105** The pressure of gas above the liquid must be maintained to keep the CO_2 dissolved in the desired concentration. **13.107** Slowly freeze the seawater. While some liquid remains, separate the ice, which is free of dissolved materials. **13.109** Osmosis occurs, removing water from the plant cells. **13.111** The water in the plant cells contains dissolved material, which lowers the freezing point. **13.113** The dissolved electrolytes in seawater precipitate the colloidal dispersion of mud in river water.

13.115 $\dfrac{0.580 \text{ g CO}_2}{100 \text{ g H}_2\text{O}}$ **13.117** 582 u **13.119 (a)** 29.5 atm **(b)** A pressure greater than 29.5 atm is needed. **(c)** As salt-free water is pushed out, M increases. This, in turn, increases the amount of pressure needed to continue the reverse osmosis process. **(d)** 5.68 L **13.121 (a)** 1.6×10^2 g urea, 1.0×10^2 g $CaCl_2$, 7.8×10^1 g NaCl **(b)** \$0.70 (urea), \$0.73 (NaCl), \$0.75 ($CaCl_2$) **(c)** A larger mass of urea is required to obtain the same amount of protection from freezing. Also, too much N in runoff is not good for the water supply. **13.123 (a)** Freezing-point depression calculations are based on colligative properties of solutions in which the solute is assumed to be nonvolatile. **(b)** 0.802 L methyl alcohol **(c)** \$1.30 per liter (methyl alcohol-water), \$2.71 per liter (ethylene glycol-water) **(d)** The boiling points of methyl alcohol-water mixtures are probably between the boiling points of CH_3OH (65.0 °C) and water (100 °C). Thus, the methyl alcohol solutions will evaporate much faster than ethylene glycol-water solutions and may even boil over. **13.125 (a)** water: carbon tetrachloride; air: heptane. Compounds with low values of H tend to accumulate in water; compounds with high values of H tend to accumulate in air. **(b)** smaller; solubility of gases in water increases with decreasing temperature. **13.127 (a)** The ability to use a wide variety of solvents for freezing-point depression is desirable because the method only works for soluble solutes. The wider the variety of solvents, the larger the number of substances whose molecular masses can be obtained. **(b)** -37.7 °C/m **13.129 (a)** 0.8 **(b)** orange-red **(c)** $4.74 \times 10^{14} \text{ s}^{-1}$ **(d)** 3.14×10^{-19} J

CHAPTER 14

14.1 (a) [HI] = 0.0200 M, $[H_2]$ = $[I_2]$ = 0.0000 M **(b)** 0.0175 M **(c)** 0.0025 mol **(d)** 0.0013 mol H_2, 0.0013 mol I_2 **(e)** $[H_2]$ = 0.0013 M, $[I_2]$ = 0.0013 M

14.2 (a) $K_c = \dfrac{[SO_2]^2[O_2]}{[SO_3]^2}$ **(b)** $K_c = \dfrac{[Sn^{2+}][Fe^{3+}]^2}{}$ **(c)** $K_c = \dfrac{[H^+][NO_2^-]}{[HNO_2]}$ **(d)** $K_c = \dfrac{[NH_3]^2}{[N_2][H_2]^3}$ **(e)** $K_c = \dfrac{[O_3]^2}{[O_2]^3}$

14.3 (a) homogeneous **(b)** heterogeneous **(c)** heterogeneous

(d) homogeneous **14.4 (a)** $K_c = [NH_3]^2[CO_2][H_2O]$ **(b)** $K_c = [Ag^+][Cl^-]$ **(c)** $K_c = \dfrac{1}{[O_2]}$ **(d)** $K_c = \dfrac{[OH^-]^2}{[O_2][SH^-]^2}$

14.5 No. In order for equilibrium to be reached, the products must be in contact with the reactants. In an open container, the $CO_2(g)$ is lost continuously from the reaction mixture, making it unavailable for the reverse reaction. **14.6 (a)** (iv) **(b)** (i) **(c)** (ii) and (iii) **14.7** 2.11×10^{-1} **14.8 (a)** 0.030 96 M **(b)** 0.212 or 2.12×10^{-1} **14.9** 1.38×10^5

14.10 (a) $K_p = \dfrac{p_{H_2}s^2}{p_{H_2}^2 p_{S_2}}$ **(b)** $K_p = p_{CO_2}$

14.11 (a) 2.43 **(b)** 2.70×10^{-2} **14.12 (a)** 9.050×10^{-8} **(b)** 3.324×10^3 **14.13 (a)** not at equilibrium, reverse direction **(b)** at equilibrium **14.14** $[COCl_2]_{eq} = 0.074$, $[CO]_{eq} = [Cl_2]_{eq} = 4.0 \times 10^{-6}$ **14.15** $[PCl_5]_{eq} = 0.0236$, $[PCl_3]_{eq} = [Cl_2]_{eq} = 0.0169$ **14.16** $p_{N_2O_4} = 2.5$ atm, $p_{NO_2} = 0.53$ atm **14.17** $[H_2] = 4.199 \times 10^{-3}$, $[I_2] = 3.76 \times 10^{-4}$, [HI] = 9.28×10^{-3} **14.18 (a)** shift to right **(b)** shift to right **(c)** shift to left **(d)** shift to left **14.19 (a)** shift to left **(b)** no shift **(c)** shift to right **14.20 (a)** shift to left **(b)** no shift **(c)** shift to right **(d)** no shift **14.21 (a)** (c) **(b)** increase **14.59** 1.24×10^{-5} **14.61 (a)** $F_2(g) \rightleftharpoons 2F(g)$ **(b)** $4NH_3(g) + 5O_2(g) \rightleftharpoons 4NO(g) + 6H_2O(g)$ **(c)** $C_6H_5COOH(aq) \rightleftharpoons C_6H_5COO^- + H^+$ **(d)** $2H_2O(l) \rightleftharpoons 2H_2(g) + O_2(g)$ **14.63 (a)** $[CO]_{eq} = 3.8 \times 10^{-6}$, $[COCl_2]_{eq} = 6.6 \times 10^{-2}$ **(b)** 6.6×10^{-2} M **14.65 (a)** low temperature **(b)** low pressure **(c)** Add a catalyst. **14.67 (a)** 1.6×10^{-21} **(b)** 3.0×10^{-2} **14.69 (a)** $[N_2O_4]_{eq} = 4.9 \times 10^{-2}$, $[NO_2]_{eq} = 1.02 \times 10^{-1}$ **(b)** $[N_2O_4] = 0.075$ **(c)** $[NO_2] = 0.080$ **(d)** Change in $[N_2O_4]$ and $[NO_2]$ as a function of time:

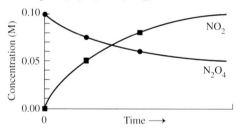

14.71 $[N_2O_4]_{eq} = 0.009\,48$, $[NO_2]_{eq} = 0.006\,64$ **14.73** $p_{PCl_3} = p_{Cl_2} = 0.691$ atm, $p_{PCl_5} = 0.96$ atm **14.75** 5.3×10^{-5}

14.77 (a) 0.403 **(b)** Either the reaction must be endothermic so that K_c is much smaller at room temperature or the rate of decomposition must be very low at room temperature. **14.79** $[SO_3] = 6.87 \times 10^{-3}$, $[SO_2] = 2.44 \times 10^{-3}$,

$[O_2] = 1.22 \times 10^{-3}$ **14.81 (a)** 0.20   **(b)** Decreasing the volume by a factor of two by increasing the pressure will result in a doubling of concentrations for both NO_2 and N_2O_4.   **(c)** No. The reaction will proceed from right to left to reestablish equilibrium.   **(d)** $[N_2O_4] = 0.211$, $[NO_2] = 0.206$
14.83 (a) $H_2O(l) \rightleftharpoons H_2O(g)$ **(b)** $K_p = p_{H_2O(g)}$
(c) 29.354 mmHg   **(d)** 1.560×10^{-3} **(e)** 1.560×10^{-3} M
14.85 1.6×10^{-15} **14.87 (a)** 0.75   **(b)** $X_{CO} = X_{H_2O} = 0.23$, $X_{CO_2} = 0.20$, $X_{H_2} = 0.35$ **(c)** Values are the same.   **(d)** No. In the first equilibrium, there are the same number of molecules of products as there are molecules of reactants. This is not true of the second equilibrium.   **14.89 (a)** 0.450 g   **(b)** 0.389 g HI
(c) Increase the concentration of H_2 or remove HI as it is formed.
14.91 27.1 M   **14.93 (a)** 5.64×10^{-2} **(b)** 12.5%
14.95 Ammonia will decompose.   **14.97** 5.1×10^{-4}
14.99 (a) The presence of water and carbon dioxide allows equilibrium to be established. When the water moves, removal of the ions causes the equilibrium to shift to the right, and $CaCO_3(s)$ dissolves.   **(b)** Heating shifts the equilibrium to the left, causing $CaCO_3(s)$ to precipitate.   **(c)** As water evaporates, the equilibrium is shifted to the left. This results in precipitation of $CaCO_3(s)$.
14.101 Don't print the story. Since catalysts speed up the rates of *both* the forward and reverse reactions, they do not change the position of the equilibrium.
14.103 (a) 203.25 kJ   **(b)** High temperature, low pressure, and a high concentration of $H_2O(g)$ would favor the forward reaction. The opposite is true for the reverse reaction.
14.105 (a) $p_{H_2O} = 2.51 \times 10^{-2}$ atm for $Na_2SO_4 \cdot 10H_2O$, $p_{H_2O} = 6.088 \times 10^{-8}$ atm for $CaCl_2 \cdot 6H_2O$ **(b)** $CaCl_2 \cdot 6H_2O$
14.107 (a) $[CO] = [H_2] = 0.3$, $[H_2O] = 0.5$ **(b)** endothermic
(c) The pressure should be low. One mole of gas is converted to two moles of gas by the reaction.   **(d)** At *low pressures*, gases take up a lot of room. The plant would have to be enormous. In addition, at low pressures the concentrations are low and reactions take place slowly compared with their rates at high pressures. *High pressures* require special strong pressure-containment vessels.
14.109 (a) 490   **(b)** $p_{CO_2} = 12.4$ atm, $p_{CO} = 2.53 \times 10^{-2}$ atm
(c) redox reactions   **14.111 (a)** 0.470   **(b)** 1.31   **(c)** 35.9%
14.113 (a) 14   **(b)** to the left   **(c)** $p_{CO} = 10$ atm, $p_{H_2} = 20$ atm

CHAPTER 15

15.1

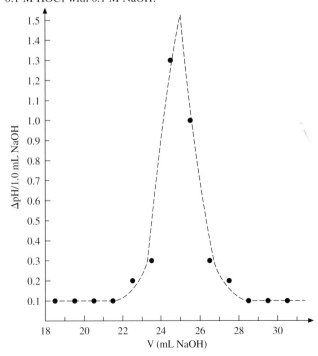

15.2 (a) $H_2PO_4^-(aq) + H_2O(l) \rightleftharpoons HPO_4^{2-}(aq) + H_3O^+(aq)$
   acid₁   base₂   base₁   acid₂
(b) $H_2PO_4^-$ **15.3 (a)** PO_4^{3-} **(b)** NH_2^- **(c)** $Cu(H_2O)_3(OH)^+$
15.4 (a) HNO_2 **(b)** $H_3C\ddot{O}:H^+$ **(c)** $Al(H_2O)_6^{3+}$
15.5 (a) autoionization   **(b)** amphoteric substances

15.6 (a) 1.7×10^{-6} **(b)** basic   **15.7 (a)** 1.3×10^{-11} **(b)** basic
15.8 (a) 5.00   **(b)** 3.08   **15.9 (a)** 2×10^{-9} **(b)** 2.3×10^{-6}
15.10 (a) 4.00   **(b)** 8.21   **(c)** 1.0×10^{-11} **(d)** 2.8×10^{-9}
(e) 6.01   **15.11** 3.1×10^{-4}
15.12 (a) $HCN(aq) \rightleftharpoons H^+(aq) + CN^-(aq)$ or

$HCN(aq) \rightleftharpoons H^+ + CN^-$ **(b)** $K_a = \dfrac{[H^+][CN^-]}{[HCN]}$

15.13 1.8×10^{-4} **15.14** 4.2×10^{-3}
15.15 so that the graph fits on the page   **15.16** 0.050 M
15.17 (a) $CH_3NH_2(aq) + H_2O(l) \rightleftharpoons CH_3NH_3^+ + OH^-$
(b) $K_b = \dfrac{[CH_3NH_3^+][OH^-]}{[CH_3NH_2]}$

15.18 (a) 6.2×10^{-6} M   **(b)** Yes. $[OH^-] > 1.0 \times 10^{-6}$
(c) 4.3×10^{-10} **15.19 (a)** 5.7×10^{-6} M   **(b)** 5.7×10^{-6} M
(c) 0.075 M   **15.20 (a)** chloroacetic acid; K_a for chloroacetic acid is larger.   **(b)** acetate ion; the conjugate of the weaker acid is the stronger base.   **15.21 (a)** 7.7×10^{-5} **(b)** stronger; K_b for the phenoxide ion is larger.   **15.22 (a)** 2.4×10^{-3} M   **(b)** more basic; phenol is a weaker acid than acetic acid
15.23 (a) acidic   **(b)** basic   **(c)** acidic   **(d)** neutral
15.24 1.8×10^{-4} **15.25** 3.9×10^{-4}
15.26 (a) and (b); (c) does not have enough capacity
15.27 (a) $[H^+] = 1.9 \times 10^{-4}$, pH = 3.72
(b) $[H^+] = 2.2 \times 10^{-4}$, pH = 3.66
15.28 (a) methylamine–methylammonium ion   **(b)** 0.70
15.29 3.37   **15.30 (a)** yellow (pH $\gg$ 2.8 and $<$ 8.0)   **(b)** 2
15.31 1.954   **15.32** 11.959   **15.33 (a)** 3.68   **(b)** 8.72
(c) 12.3632   **15.34** thymol blue
15.35 (a) $[H^+] = [HC_6H_6O_6^-] = 2 \times 10^{-3}$ **(b)** 3×10^{-12}
15.91 (a) 1.62×10^{-7} M   **(b)** 38.3%   **15.93** $[H^+] > 1$
15.95 (a) 2.30×10^{-7} M   **(b)** basic   **15.97** 6.5×10^{-2} M
15.99 0.1059 M   **15.101** Differential plot of titration of 0.1 M HOCl with 0.1 M NaOH:

The differential plot emphasizes the *change* in pH per mL that occurs at the equivalence point in a titration. It is easier to determine the equivalence point because there is an obvious change in direction on the graph. In the ordinary titration curve for this titration (Figure 15.13), the equivalence point is not obvious because the steep portion of the curve is short and far from vertical.

15.103 (a) ⠀⠀⠀⠀⠀**(b)**

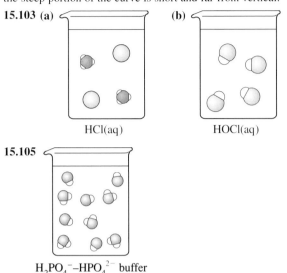

HCl(aq)⠀⠀⠀⠀⠀⠀HOCl(aq)

15.105

$H_2PO_4^- - HPO_4^{2-}$ buffer

STOP AND TEST YOURSELF

1. (c) *15.1*⠀⠀**2.** (d) *15.1*⠀⠀**3.** (e) *15.1*⠀⠀**4.** (a) *15.2*⠀⠀**5.** (e) *15.3*
6. (c) *15.4*⠀⠀**7.** (a) *15.4*⠀⠀**8.** (e) *15.4*⠀⠀**9.** (a) *15.4*
10. (b) *15.6*⠀⠀**11.** (e) *15.5*⠀⠀**12.** (b) *15.8*⠀⠀**13.** (b) *15.8*
14. (e) *15.10*⠀⠀**15.** (d) *15.11*

15.107 (a) $1.8 \times 10^{-7}\%$⠀**(b)** 6.0×10^{16}⠀**15.109** 7.0
15.111 (a) 1.60⠀**(b)** 1.90⠀⠀**15.113 (a)** 1.46⠀**(b)** 11.10⠀**(c)** 3.01
(d) 5.1⠀**(e)** 2.57⠀⠀**15.115 (a)** SO_3^{2-}, sulfite ion⠀**(b)** I^-, iodide
ion⠀**(c)** NH_3, ammonia⠀**(d)** O^{2-}, oxide ion⠀**(e)** CH_3COO^-,
acetate ion⠀⠀**15.117** 1.22⠀⠀**15.119 (a)** Hydrochloric acid is an
aqueous solution of a strong electrolyte and contains H^+ ions and
Cl^- ions. Hydrogen chloride gas is a molecular compound.
(b) The charge is carried by the movement of ions through the solution between the electrodes.⠀**(c)** Hydrogen chloride is not ionized
in benzene solution. Benzene is a nonpolar solvent and cannot
solvate ions. Thus, the H—Cl bond is not broken.
15.121 Ammonia and acetic acid are weak electrolytes. However, when their solutions are mixed, a Brønsted-Lowry acid-base
reaction takes place and ions are formed. The ionic equation for
this reaction is $NH_3(aq) + CH_3COOH(aq) \rightarrow NH_4^+ + CH_3COO^-$.
15.123 (a) Molecular equation:
$NaH_2PO_4(aq) + HCl(aq) \rightarrow H_3PO_4(aq) + NaCl(aq)$
Complete ionic equation:
$Na^+ + H_2PO_4^- + H^+ + Cl^- \rightarrow H_3PO_4(aq) + Na^+ + Cl^-$
Net ionic equation: $H_2PO_4^- + H^+ \rightarrow H_3PO_4(aq)$
(b) Molecular equation:
$NaH_2PO_4(aq) + NaOH(aq) \rightarrow Na_2HPO_4(aq) + H_2O(l)$
Complete ionic equation:
$Na^+ + H_2PO_4^- + Na^+ + OH^- \rightarrow 2Na^+ + HPO_4^{2-} + H_2O(l)$
Net ionic equation: $H_2PO_4^- + OH^- \rightarrow HPO_4^{2-} + H_2O(l)$
15.125 2.58⠀⠀**15.127** Either the sample contained a mixture of two

monoprotic acids with K_a's differing by 10^{-5} or so, or else the
sample was a diprotic acid with $K_{a2} \ll K_{a1}$.⠀⠀**15.129** 10.12
15.131 (a) C_6H_5⠀**(b)** $CH_3CH_2NH_2$, ethylamine; $(CH_3CH_2)_2NH$,
diethylamine; $(CH_3CH_2)_3N$, triethylamine
(c)

$$H-\overset{\overset{\textstyle H}{|}}{\underset{\underset{\textstyle H}{|}}{C}}-\ddot{N}\!: \;+\; \overset{\frown}{H}-\ddot{\underset{..}{\overset{..}{Cl}}}: \longrightarrow$$

$$H-\overset{\overset{\textstyle H}{|}}{\underset{\underset{\textstyle H}{|}}{C}}-\overset{+}{\underset{\underset{\textstyle H}{|}}{N}}\!-\!H \;+\; :\ddot{\underset{..}{Cl}}:^-$$

(d) $CH_3NH_2 + H^+ \rightarrow CH_3NH_3^+$
15.133 $Al(H_2O)_6^{3+} \rightleftharpoons Al(H_2O)_5(OH)^{2+} + H^+$
$Fe(H_2O)_6^{3+} \rightleftharpoons Fe(H_2O)_5(OH)^{2+} + H^+$⠀⠀**15.135** acidotic, basic
15.137 The basic group —NH_2 and the acidic group —COOH
must react to give a saltlike structure: $\overset{+}{H_3N}CH_2\overset{-}{COO}$
15.139 3×10^{-1} or 0.3

15.141 (a)

(b) No. The K_a value is for the formation of H^+ from the compound as given.⠀**(c)** The acidic H is the one from the —OH
attached to the benzene ring

The K_a value given (1.4×10^{-10}) is similar to the value for phenol
itself (1.3×10^{-10}).⠀⠀**15.143 (a)** 1.7×10^{-5}⠀**(b)** 5.37
15.145 $K_b = 2.5 \times 10^{-6}$⠀⠀**15.147 (a)** 0.01 M⠀**(b)** 0.4 g
(c) increases⠀**(d)** Classic Coke has a pH of 2.5 (Table 15.2).
Drinking colas replaces the acid and water lost by vomiting.
15.149 (a) 9×10^{-6} M⠀**(b)** CO_2 reacts with water producing
HCO_3^-, which buffers the solution so that the end point is not
sharp. CO_2-free water can be prepared by boiling the water and
cooling it in a CO_2-free atmosphere.⠀⠀**15.151** Ammonium nitrate
is the salt of a strong acid (HNO_3) with a weak base (NH_3) and is
acidic. It lowers the pH of soil. (Soil must be moist or plants will
not grow.)

CHAPTER 16

16.1 (a) to the right⠀**(b)** to the left⠀**(c)** to the left
16.2 (a) 5.6×10^2⠀**(b)** 5.1×10^{-2}⠀⠀**16.3** structure (a) has one
extra oxygen like phosphoric acid; monoprotic⠀⠀**16.4 (a)** CH_3SH;
S—H bond is longer and weaker than O—H bond⠀**(b)** H_2S;
bonds same length but S more electronegative than P⠀**(c)** H_2SO_3;
same number of extra oxygens, S more electronegative than P
(d) H_2SO_4; S has one more extra O in H_2SO_4⠀**(e)** CH_3COOH; C
in CH_3COOH has an extra O⠀⠀**16.5** hypobromous acid
16.6 (c); Ba^{2+} is larger than Sr^{2+}⠀⠀**16.7** CH_3COO^-; conjugate of

weaker acid **16.8 (a)** basic; oxide of active metal **(b)** acidic; nonmetal oxide **(c)** amphoteric; zinc is near metal-nonmetal border
16.9 (a) Net ionic equation: $O^{2-} + H_2O(l) \rightarrow 2OH^-$
Complete ionic equation: $2K^+ + O^{2-} + H_2O(l) \rightarrow 2K^+ + 2OH^-$
Molecular equation: $K_2O(s) + H_2O(l) \rightarrow 2KOH(aq)$
(b) Net ionic equation: $Fe_2O_3(s) + 6H^+ \rightarrow 2Fe^{3+} + 3H_2O(l)$
Complete ionic equation:
$Fe_2O_3(s) + 6H^+ + 6Cl^- \rightarrow 2Fe^{3+} + 6Cl^- + 3H_2O(l)$
Molecular equation:
$Fe_2O_3(s) + 6HCl(aq) \rightarrow 2FeCl_3(aq) + 3H_2O(l)$
(c) Net and complete ionic equations:
$N_2O_5(s) + H_2O(l) \rightarrow 2H^+ + 2NO_3^-$
Molecular equation: $N_2O_5(s) + H_2O(l) \rightarrow 2HNO_3(aq)$
16.10 (a) Cr^{3+}, $BeCl_2$ **(b)** CH_3NH_2 **(c)** CH_4, Na^+
16.11 (a) Lewis acid—$Al(OH)_3$, Lewis base—OH^-
(b) Lewis acid—Ag^+, Lewis base—NH_3
(c) Lewis acid—BF_3, Lewis base—$(CH_3CH_2)_2O$
16.12

16.13 1.1×10^{-12} M
16.14 (a) $AgCl(s) \rightleftharpoons Ag^+ + Cl^-$ $K_{sp} = [Ag^+][Cl^-]$
(b) $Fe(OH)_2(s) \rightleftharpoons Fe^{2+} + 2OH^-$ $K_{sp} = [Fe^{2+}][OH^-]^2$
(c) $PbBr_2(s) \rightleftharpoons Pb^{2+} + 2Br^-$ $K_{sp} = [Pb^{2+}][Br^-]^2$
(d) $Li_2CO_3(s) \rightleftharpoons 2Li^+ + CO_3^{2-}$ $K_{sp} = [Li^+]^2[CO_3^{2-}]$
(e) $Ag_2SO_4(s) \rightleftharpoons 2Ag^+ + SO_4^{2-}$ $K_{sp} = [Ag^+]^2[SO_4^{2-}]$
16.15 1.8×10^{-4} g AgCl/100 cm³ solution
16.16 3×10^{-8} **16.17 (a)** $MgCO_3$, **(b)** $Ca(OH)_2$, **(d)** ZnS
16.18 (a) Net ionic equation:
$Mg(OH)_2(s) + 2H^+ \rightarrow Mg^{2+} + 2H_2O(l)$
Complete ionic equation:
$Mg(OH)_2(s) + 2H^+ + 2Cl^- \rightarrow Mg^{2+} + 2Cl^- + 2H_2O(l)$
Molecular equation:
$Mg(OH)_2(s) + 2HCl(aq) \rightarrow MgCl_2(aq) + 2H_2O(l)$
(b) Net and complete ionic equations:
$AgCl(s) + 2NH_3(aq) \rightarrow Ag(NH_3)_2^+ + Cl^-$
Molecular equation: $AgCl(s) + 2NH_3(aq) \rightarrow Ag(NH_3)_2Cl(aq)$
(c) Net ionic equation: $AgCl(s) + Cl^- \rightarrow AgCl_2^-$
Complete ionic equation: $AgCl(s) + H^+ + Cl^- \rightarrow AgCl_2^- + H^+$
Molecular equation: $AgCl(s) + HCl(aq) \rightarrow HAgCl_2(aq)$
16.19 The solubility product constant, K_{sp}, for sulfides assumes equilibrium between undissolved solid and metal and sulfide ions (for example, $ZnS \rightleftharpoons Zn^{2+} + S^{2-}$). The solubilities of sulfides are much higher than predicted because S^{2-} is the conjugate base of a very weak acid, HS^-. This causes hydrolysis to take place when S^{2-} ions are present in aqueous solution (for example, $ZnS(s) + H_2O(l) \rightleftharpoons Zn^{2+} + SH^- + OH^-$). This type of equilibrium increases the solubility of sulfides in water.
16.20 8×10^{-4} **16.21** No.   **16.22 (a)** 9×10^{-9} M
(b) 9×10^{-9} M   **(c)** 2×10^{-5} M   **(d)** 4×10^{-15} M
16.73 (a)

(b)

(c)

(d)

16.75 (a) AgI   **(b)** 8×10^{-9}, 8×10^{-5}%

STOP AND TEST YOURSELF

1. (b) *16.1*   **2.** (d) *16.1*   **3.** (e) *16.2*   **4.** (e) *16.2*   **5.** (c) *16.2*
6. (b) *16.2*   **7.** (a) *16.2*   **8.** (a) *16.3*   **9.** (c) *16.4*
10. (b) *16.7*   **11.** (d) *16.7*   **12.** (e) *16.7*   **13.** (b) *16.6*
14. (c) or (d) (because Q_{sp} must be $1000 \times K_{sp}$ for the precipitate to be visible) *16.7*

16.77 (a), (c); $Zn(OH)_2$ and $Al(OH)_3$ are amphoteric
16.79 2.6×10^{-6} M
16.81 (a) $NaH(s) + H_2O(l) \rightarrow NaOH(aq) + H_2(g)$
(b) $CaH_2(s) + 2H_2O(l) \rightarrow Ca(OH)_2(s) + 2H_2(g)$
(c) $NaNH_2(s) + H_2O(l) \rightarrow NaOH(aq) + NH_3(aq)$
(d) $Na_2O(s) + H_2O(l) \rightarrow 2NaOH(aq)$
(e) $SO_2(g) + H_2O(l) \rightleftharpoons H^+ + HSO_3^-$ **(f)** no reaction
16.83 (a) $Al(OH)_3(s) + 3HCl(aq) \rightarrow AlCl_3(aq) + 3H_2O(l)$
(b) $Al(OH)_3(s) + NaOH(aq) \rightarrow NaAl(OH)_4(aq)$
(c) $NaHCO_3(s) + H_2O(l) \rightleftharpoons NaOH(aq) + H_2O(l) + CO_2(aq)$
(d) no reaction
(e) $NH_4Cl(s) + H_2O(l) \rightleftharpoons H_3O^+ + NH_3(aq) + Cl^-$
(f) $CaO(s) + SO_3(g) \rightarrow CaSO_4(s)$
16.85 (a)

(b)

(c)

(d)

Lewis base   Lewis acid

(e)

Note: $AlCl_3$ behaves as both a Lewis acid and a Lewis base.

16.87 Both Lewis and Brønsted-Lowry acid-base definitions can be applied to reactions in the gas phase and in solvents other than water as well as to reactions in aqueous solutions. The same substances are viewed as bases according to both definitions. However, Brønsted-Lowry acids must contain hydrogen, whereas Lewis acids do not need to contain hydrogen. Reactions such as the reaction between CaO and CO_2 to form $CaCO_3$ are acid-base reactions according to the Lewis definitions but not according to the Brønsted-Lowry definitions. On the other hand, typical hydrogen-containing acids such as HCl and CH_3COOH are more easily classified as acids using the Brønsted-Lowry definitions. Brønsted-Lowry acid-base reactions can easily be treated quantitatively; Lewis acid-base reactions cannot.

16.89 $Al^{3+} + 6H_2O \rightleftarrows Al(H_2O)_6^{3+}$
     Lewis   Lewis
     acid   base

$Al(H_2O)_6^{3+} + H_2O \rightleftarrows Al(H_2O)_5(OH)^{2+} + H_3O^+$
Brønsted-Lowry   Brønsted-Lowry
   acid   base

16.91 No.   **16.93** $[Ag^+] = 3.0 \times 10^{-6}$, $[Cl^-] = 5.4 \times 10^{-5}$
16.95 (a) 27 mL   **(b)** 23 mL   **(c)** 66 mg   **16.97** In the buffer solution, the $[CO_3^{2-}]$ is high enough to make $Q_{sp} > K_{sp}$ for $CaCO_3$ but not for $MgCO_3$.   **16.99 (a)** $KClO_2$ **(b)** $Ca(ClO)_2$
(c) NH_4ClO_4 **(d)** $NaClO_3$ **(e)** $Al(BrO_3)_3 \cdot 9H_2O$
16.101 (a) periodic acid, H_5IO_6 iodic acid, HIO_3

(b) Iodic acid is stronger because it has more extra oxygens (2) than does periodic acid (1).   **(c)** Periodic acid—octahedral; iodic

acid—trigonal pyramidal   **16.103** The ionization of weak acids at infinite dilution is limited by the ionization of water. The H^+ from the water shifts the equilibrium between the weak acid and its ions in solution to the left, thus limiting the percent ionization. Water does not produce any ions that participate in the equilibrium involving the solubility of silver chloride. Thus, the calculated solubility closely matches the observed solubility.
16.105 (a) Molecular equation:
$Ni(OH)_2(s) + 2HCl(aq) \rightarrow NiCl_2(aq) + 2H_2O(l)$
Complete ionic equation:
$Ni(OH)_2(s) + 2H^+ + 2Cl^- \rightarrow Ni^{2+} + 2Cl^- + 2H_2O(l)$
Net ionic equation:
$Ni(OH)_2(s) + 2H^+ \rightarrow Ni^{2+} + 2H_2O(l)$
(b) Molecular equation:
$FeCO_3(s) + 2HCl(aq) \rightarrow H_2O(l) + CO_2(g) + FeCl_2(aq)$
Complete ionic equation:
$FeCO_3(s) + 2H^+ + 2Cl^- \rightarrow H_2O(l) + CO_2(g) + Fe^{2+} + 2Cl^-$
Net ionic equation: $FeCO_3(s) + 2H^+ \rightarrow H_2O(l) + CO_2(g) + Fe^{2+}$
(c) Molecular equation:
$ZnS(s) + 2HCl(aq) \rightarrow ZnCl_2(aq) + H_2S(g)$
Complete ionic equation:
$ZnS(s) + 2H^+ + 2Cl^- \rightarrow Zn^{2+} + 2Cl^- + H_2S(g)$
Net ionic equation: $ZnS(s) + 2H^+ \rightarrow Zn^{2+} + H_2S(g)$
(d) Molecular equation:
$CoCl_3(aq) + 6NH_3(aq) \rightarrow Co(NH_3)_6Cl_3(aq)$
Complete ionic equation:
$Co^{3+} + 3Cl^- + 6NH_3(aq) \rightarrow Co(NH_3)_6^{3+} + 3Cl^-$
Net ionic equation: $Co^{3+} + 6NH_3(aq) \rightarrow Co(NH_3)_6^{3+}$
(e) Molecular equation:
$Al(OH)_3(s) + NaOH(aq) \rightarrow NaAl(OH)_4(aq)$
Complete ionic equation:
$Al(OH)_3(s) + Na^+ + OH^- \rightarrow Na^+ + Al(OH)_4^-$
Net ionic equation: $Al(OH)_3(s) + OH^- \rightarrow Al(OH)_4^-$
(f) Molecular equation:
$2AgNO_3(aq) + K_2CrO_4(aq) \rightarrow Ag_2CrO_4(s) + 2KNO_3(aq)$
Complete ionic equation:
$2Ag^+ + 2NO_3^- + 2K^+ + CrO_4^{2-} \rightarrow Ag_2CrO_4(s) + 2K^+ + 2NO_3^-$
Net ionic equation: $2Ag^+ + CrO_4^{2-} \rightarrow Ag_2CrO_4(s)$
16.107 (a) 1.9×10^{12} ions   **(b)** 3×10^3 ions   **(c)** No. Bi_2S_3, which has smaller K_{sp}, is more soluble.   **16.109 (a)** The *solubility product* is K_c for a solubility equilibrium; the concentrations in a solubility product are equilibrium concentrations. The *ion product* is Q for a solubility equilibrium; the concentrations may be any concentrations. **(b)** *Activity* is the "effective" concentration of species in solution; it includes effects due to interactions of ionic species. For dilute solutions, activity $\approx$ *concentration* (in molarity).
(c) A *saturated solution* is in equilibrium with undissolved solute (or has the same concentration as a solution that is in equilibrium with undissolved solute). A *supersaturated solution* contains more solute than is permitted by equilibrium considerations. **(d)** The *common ion effect* is a decrease in solubility of an ionic solute in a solution containing one of the ions of the solute. The *salt effect* is an increase in the solubility of an ionic solute in a solution containing ions that are not common to the added salt.
(e) A *Brønsted-Lowry acid* is a proton (H^+) donor. A *Lewis acid* is an electron pair acceptor.
16.111 (a) Lewis acid-base reaction   **(b)** -65.17 kJ/mol
(c) $Ca(OH)_2(s) + SO_2(g) \rightarrow CaSO_3(s) + H_2O(l)$

16.113 $Ca(OH)_2(s) + CO_2(g) \rightarrow CaCO_3(s) + H_2O(l)$

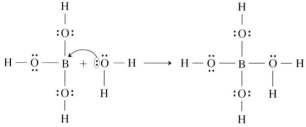

Ca^{2+} is a spectator ion.

16.115 9×10^5 **16.117 (a)** probably acidic; oxidation number of Cr is higher   **(b)** Cr_2O_3, chromium(III) oxide; CrO_3, chromium(VI) oxide   **(c)** CrO_2 **16.119 (a)** They are equal.   **(b)** 1.0×10^{-6} M   **(c)** 6.5×10^{-3} M   **16.121** 1.6×10^{-6}
16.123 (a)

(b) 6×10^{-10} **(c)** 5.5   **(d)** 9.2   **16.125 (a)** aragonite; K_{sp} for aragonite is larger.   **(b)** 0.7   **(c)** Hydrochloric acid will make the real carbonate coral fizz. Bubbles of $CO_2(g)$ will be given off.
16.127 (a) 0.30 g $Na_2S_2O_3$ **(b)** 0.47 g $Na_2S_2O_3 \cdot 5H_2O$
16.129 (a) An *aerosol* is a colloidal dispersion of a liquid or a solid in a gas.   **(b)** carbon dioxide (CO_2), sulfuric acid (H_2SO_4), nitric acid (HNO_3), nitrous acid (HNO_2), and sulfur dioxide (SO_2)
(c) Only nitric and sulfuric acids are strong acids. [H^+] from these two acids shift the weak acid equilibria toward molecules.
(d) oxidation-reduction (specifically, disproportionation)
16.131 (a) The solubility of gases increases with increasing pressure. Since the pressure at deep vents is greater than the pressure at shallow vents, gases are more soluble at deep vents and there are no bubbles.   **(b)** The fluids are acidic. Calcium carbonate is soluble in acidic solutions.

CHAPTER 17

17.1 (a) small and unpredictable; no change in moles gas
(b) negative; moles gas decrease, volume decreases   **(c)** negative; temperature decreases, volume decreases   **(d)** negative; pressure increases, volume decreases   **(e)** negative; moles gas decrease, volume decreases   **17.2 (a)** decreases; solid more ordered
(b) increases; volume increases   **(c)** increases; volume increases
17.3 (a) Ca(s); Ca below Mg in periodic table
(b) $H_2S(g)$; S below O in periodic table   **(c)** $PCl_5(g)$; more atoms
(d) $Cl_2(g)$; Cl below F in periodic table   **(e)** unpredictable; solid more ordered but I below Br in periodic table   **17.4 (a)** -5.2 J/K
(b) -242.8 J/K   **(c)** -88.95 J/K   **17.5** -281.07 J/K
17.6 (c) spontaneous at low temperatures but nonspontaneous at high temperatures   **17.7** -794.9 kJ   **17.8** -800.77 kJ

17.9 -793.8 kJ   **17.10** 77 °C   **17.11** 88.9 J/K·mol
17.12 -770.5 kJ/mol   **17.13** 1×10^{14} **17.14** -25.3 kJ/mol
17.41 No.   **17.43 (a)** 68.69 kJ   **(b)** No.   **(c)** No.   **(d)** No.
17.45 (a) left; larger volume   **(b)** left; larger volume   **(c)** left; more molecules in same volume

17.47 -46.3 kJ   **17.49** methyl alcohol
17.51 4.0×10^{-4} atm or 0.31 mmHg   **17.53 (a)** positive
(b) positive   **(c)** zero   **(d)** positive   **(e)** negative
17.55 -233.54 kJ   **17.57** reaction is nonspontaneous at all temperatures ($\Delta H > 0$, $\Delta S < 0$)   **17.59** 1.7×10^6
17.61 (a) -1.7 kJ   **(b)** 1.5   **(c)** $p_{PCl_3} = p_{Cl_2} = 0.68$ atm, $p_{PCl_5} = 0.31$ atm   **(d)** to the right   **(e)** At equilibrium, $\Delta G = 0$. For the forward (to the right) reaction to occur spontaneously, ΔG must be negative.   **17.63** ΔH = zero; no enthalpy change when an ideal solution forms; ΔS = positive; volume of the solution is greater than the volume of either component; ΔG = negative; solution forms spontaneously   **17.65** When small molecules are joined to form large ones, the entropy of the system decreases. However, the energy released as a result of bond formation increases the entropy of the surroundings more, so that the entropy of the universe increases as required by the second law of thermodynamics.   **17.67** No. The reaction is nonspontaneous at all temperatures ($\Delta H° > 0$, $\Delta S° < 0$).   **17.69 (a)** 3.06 g
(b) Reaction will be spontaneous because it is exothermic (ΔH is negative) and entropy increases.   **17.71** positive
17.73 (a) $\Delta H° = 180.50$ kJ for $N_2(g) + O_2(g) \rightarrow 2NO(g)$

$\Delta H° = -114.1$ kJ for $2NO(g) + O_2(g) \rightarrow 2NO_2(g)$

$\Delta H° = 306.2$ kJ for $NO_2(g) \xrightarrow{light} NO(g) + O(g)$

$\Delta H° = -106.5$ kJ for $O(g) + O_2(g) \rightarrow O_3(g)$

(b) The sign can be predicted for each reaction except $N_2(g) + O_2(g) \rightarrow 2NO(g)$.   **(c)** 24.77 J/K
(d) $N_2(g) + O_2(g) \rightarrow 2NO(g)$: nonspontaneous at low temperatures, spontaneous at high temperatures

   $2NO(g) + O_2(g) \rightarrow 2NO_2(g)$: spontaneous at low temperatures, nonspontaneous at high temperatures

   $NO_2(g) \xrightarrow{light} NO(g) + O(g)$: nonspontaneous at low temperatures, spontaneous at high temperatures

   $O(g) + O_2(g) \rightarrow O_3(g)$: spontaneous at low temperatures, nonspontaneous at high temperatures

(e) O_3:

NO:

NO_2:

(f) Both NO and NO_2 have an unpaired electron. Species with unpaired electrons (free radicals) are very reactive.
17.75 (a) -918 kJ/mol   **(b)** -2453.85 kJ   **17.77** spontaneous, because $\Delta G°$ is negative

18.1 (a) 0.000 30 mL/s   **(b)** Yes.   **18.2** 0.0013 mL/s
18.3 (a) -7.1×10^{-5} M/s   **(b)** 3.6×10^{-5} M/s
(c) It is about two times the initial rate of formation of O_2.

18.4 (a) rate $= -\dfrac{1}{5}\dfrac{\Delta[Br^-]}{\Delta t} = -\dfrac{\Delta[BrO_3^-]}{\Delta t} = -\dfrac{1}{6}\dfrac{\Delta[H^+]}{\Delta t}$

$= \dfrac{1}{3}\dfrac{\Delta[Br_2]}{\Delta t} = \dfrac{1}{3}\dfrac{\Delta[H_2O]}{\Delta t}$ **(b)** rate $= -\dfrac{\Delta[H_2O_2]}{\Delta t} = -\dfrac{1}{3}\dfrac{\Delta[I^-]}{\Delta t}$

$= -\dfrac{1}{2}\dfrac{\Delta[H^+]}{\Delta t} = \dfrac{\Delta[I_3^-]}{\Delta t} = \dfrac{1}{2}\dfrac{\Delta[H_2O]}{\Delta t}$

18.5 (a) Br_2 is formed three times as fast as BrO_3^- disappears.
(b) Br_2 is formed 3/5 as fast as Br^- disappears.   **(c)** No, because
water is the solvent for the reaction. Its concentration is very high
relative to the solute concentrations. This means that the changes
in the concentration of water are negligible.
18.6 1.28×10^{-4} M/min   **18.7 (a)** No. All of the reactions
except reaction (3) illustrate this point.   **(b)** No. In some cases,
such as reactions (4), (6), and (9), one or more of the reactants are
not even included in the rate law.   **18.8** (1) first order with
respect to (wrt) N_2O_5 (2) first order wrt NO_2 and F_2 (3) second
order wrt NO, first order wrt O_2 (4) zero order wrt NO, first order
wrt N_2O_5 (5) second order wrt NO   **18.9 (a)** extremely rare
(b) About 25% of reactions are zero order wrt a species.
18.10 rate $= k[I^-][H_3AsO_4][H^+]$ **18.11 (a)** first order
(b) second order   **(c)** rate $= k[HgCl_2][C_2O_4^{2-}]^2$
(d) $7.8 \times 10^{-3}\,s^{-1}\cdot M^{-2}$ **(e)** Initial rate will increase.
(f) 1.4×10^{-4} M/s   **18.12 (a)** increase fourfold   **(b)** no change
(c) increase threefold   **18.13** increase eightfold
18.14 (a) zero order   **(b)** 6.15×10^{-7} M/s   **18.15** 0.102 M
18.16 304.8 min   **18.17 (a)** and **(b)** reaction with $t_{1/2} = 273$ s
18.18 2×10^{-2} M   **18.19 (a)** 1.0×10^2 kJ/mol
(b) 1.1×10^{-3} **18.20** 102.0 kJ/mol
18.21

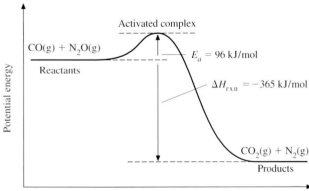

18.22 (3) $2NO(g) + O_2(g) \rightarrow 2NO_2(g)$
18.23 (1) N_2O_5 (3) N_2O_4 **18.24 (a)** $2NO_2 + O_3 \rightarrow N_2O_5 + O_2$
(b) step (1): rate $= k_1[NO_2][O_3]$, step (2): rate $= k_2[NO_3][NO_2]$
(c) overall reaction: rate $= k_1[NO_2][O_3]$ **(d)** step (1): NO_5,
step (2): N_2O_5, overall reaction: NO_5 **(e)** NO_3
18.25 No. The rate law would have to be rate $= k[NO][O_2]$,
whereas the experimentally determined rate law is
rate $= k[NO]^2[O_2]$.

18.26

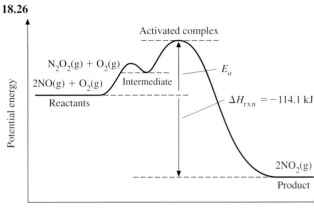

18.27 (a) 1 and 3   **(b)** 3
18.28 (a) $2Ce^{4+} + Tl^+ \rightarrow 2Ce^{3+} + Tl^{3+}$
(b) rate $= k_1[Ce^{4+}][Mn^{2+}]$ **(c)** Mn^{2+} **(d)** Mn^{3+} and Mn^{4+}
18.29 77 kJ/mol   **18.77** 0.020 M   **18.79 (a)** For a zero-order
reaction, the second half-life is shorter than the first half-life.
(b) For a second-order reaction, the second half-life is longer than
the first half-life.   **18.81** For zero-order reactions, rate $= k$; for
first-order reactions, rate $= k[A]$. The rates of zero-order reactions
are independent of concentration. Zero-order reactions do not slow
down as the reactant is used up. However, because the rates of first-
order reactions depend on concentration, reaction slows down as
the reactant is used up. If the values of k are similar, the time
required for A to react will be less if the reaction is zero order than
if it is first order.

18.83 (a) $M^{-2}\,s^{-1}$ **(b)** $\dfrac{1}{[A]_t^2}$

18.85 (a)

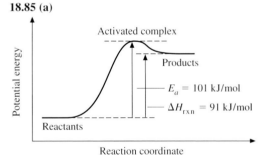

(b) The minimum activation energy for an endothermic reaction
is ΔH_{rxn}.   **(c)** No. A very endothermic reaction must have a very
high activation energy.
18.87 (a)

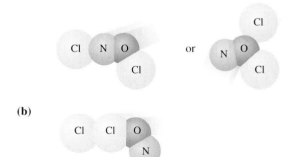

(b)

1. (a) *18.1*   **2.** (c) *18.4*   **3.** (b) *18.1*   **4.** (c) *18.1*   **5.** (d) *18.2*
6. (d) *18.3*   **7.** (b) *18.3*   **8.** (c) *18.3*   **9.** (a) *18.3*
10. (b) *18.4*   **11.** (a) *18.10*   **12.** (a) *18.11*   **13.** (c) *18.10*
14. (c) *18.12*

18.89 (a) $Cl\cdot + CH_4 \rightarrow HCl + \cdot CH_3$; chlorine atom can strike any one of four hydrogens, whereas with $CHCl_3$, there is only one hydrogen.   **(b)** (i) and (iii). In (i), no bonds need to be broken. In (iii), opposite charges attract, and also no bond needs to be broken.
18.91 (a) rate $= k[Cl_2][H_2S]$, rate $= k[Cl_2]^2$, rate $= k[H_2S]^2$
(b) See what happens to the initial rate if $[Cl]_0$ is kept constant and $[H_2S]_0$ is doubled and see what happens to the initial rate if $[H_2S]_0$ is kept constant and $[Cl]_0$ is doubled.
18.93 (a) The fraction of molecules having energy greater than the activation energy increases rapidly as temperature increases.
(b) If the reaction involves an intermediate that is the product of an exothermic equilibrium, a temperature increase will decrease the concentration of the intermediate (Le Châtelier's principle) and slow the reaction.
18.95 (a) The rate of disappearance of the red-brown color of $NO_2(g)$ could be measured with a spectrophotometer.   **(b)** The rate of formation of $O_2(g)$ could be measured by measuring the volume of the gas at different times.   **(c, d)** The volume of $CO_2(g)$ or the mass of $Hg_2Cl_2(s)$ could be measured. Also, small samples could be removed from the reaction mixture as the reaction progressed in order to determine (1) the decrease in $C_2O_4{}^{2-}$ concentration vs. time or (2) the increase in concentration of Cl^- vs. time.
(e) Titration of samples from the mixture would show the rate of formation of lactic acid or pH measurements vs. time might be used to determine the increase in acid concentration.
18.97 Reaction (a) fails to take place for kinetic reasons ($\Delta G° < 0$), and reaction (b) fails to occur for thermodynamic reasons ($\Delta G° > 0$).
18.99 (a) bromine is limiting   **(b)** The concentration of acetone does not change significantly because it is so much higher than the concentration of bromine. The concentration of HCl does not change because HCl is the catalyst.   **(c)** zero order; if the reaction rate had depended on the initial concentration of bromine, the time for experiment #2 would have been different.   **(d)** first order for both acetone and HCl; compare initial concentration of acetone vs. time required for experiments #1 and 3. Similarly, for HCl, compare these values for experiments #1 and 4.
(e) rate $= k[\text{acetone}][HCl]$ **(f)** 1.5×10^2 s   **(g)** 7.0×10^1 s
18.101 (a) rate $= k[(CH_3)_3CBr]$ **(b)** The rate would double.
(c) no effect   **(d)** Water is a much more polar solvent than alcohol and solvates the ions better.
18.103 (a) $2NO(g) + Cl_2(g) \rightarrow 2ONCl(g)$
(b) $2NO(g) + F_2(g) \rightarrow 2ONF(g)$ **(c)** The rate law differences indicate clearly that even though a similar overall reaction occurs, the mechanistic route to product in (a) differs from that in (b).
18.105 (a) The solution process was exothermic and, at the microscopic level, the acetone and water molecules occupied less volume when mixed than the sum of the volumes of the pure solvents. The attractive forces among the different molecular species was great enough to cause a shrinkage in total volume and a release of thermal energy.   **(b)** 29.5 M   **(c)** No.

18.107 (a) *Allotropes* are different forms of the same element that exist in the same physical state under the same conditions of temperature and pressure but differ in bonding.   **(b)** red
(c)

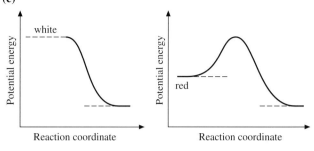

The energy of activation for the reaction of oxygen with white phosphorus is very small. The energy of activation for the reaction of red phosphorus with oxygen must be at least 17.56 kJ/mol higher; energy must be supplied for red phosphorus to begin to burn.   **(d)** tetraphosphorus decoxide, $P_4(g) + 5O_2(g) \rightarrow P_4O_{10}(s)$
18.109 *Advantages:* (1) It is easier to mix reactants. (2) It is easier to control concentrations of reactants without losses due to leakage of gases. (3) Rates can be controlled by adjusting concentrations. (4) Volatile organic solvents or water boil at convenient temperatures. The temperature of the solution cannot exceed the boiling point. (5) Flammable reactants can be made less flammable by dilution with nonflammable solvent. *Disadvantages:* (1) Solvents cost money. (2) Volatile organic solvents contribute to air pollution. (3) The solvent might participate in the reaction and/or change the mechanism of the reaction.
18.111 At constant volume and temperature, seven times the normal amount of gas resulted in seven times the normal pressure. Even more important, the increased concentration would increase the reaction rate. The reaction must be spontaneous and therefore is probably exothermic. Thermal energy released by greater quantities of reactant will further increase the reaction rate.   **18.113** 1/64
18.115 Assuming that physical activity is based on chemical reactions, the rates of these reactions will decrease with decreasing temperature.   **18.117 (a)** rate $= k[NO]^2[O_2]$
(b) $7.0 \times 10^3 \text{ M}^{-2}\text{s}^{-1}$ **(c)** $9.5 \times 10^{-5} \text{ M} \cdot \text{s}^{-1}$
18.119 about 90%   **18.121 (a)** ≈ 70 °C, ≈ 158 °F
(b) The ln time vs. $1/T$ plot would be linear, and thus extrapolation would be more accurate.
18.123 (a) No. Sunlight is not a catalyst because it is not a substance, but rather is a form of energy.   **(b)** Cl acts as a catalyst. It is present at the beginning and then is re-formed. ClO acts as an intermediate. It is formed and used up in the reaction. The overall reaction is $O_3 + O \rightarrow 2O_2$.
18.125 (a) about 6 h   **(b)** about 21 h   **(c)** about 27 h   **(d)** No, because some of the first dose remains.   **(e)** increasing the size of the first dose   **(f)** By taking two capsules to start, time required to reach minimum effective relative concentration is reduced. After this level is reached, a smaller dose is all that is needed to maintain it.   **(g)** first order
18.127 6×10^1 kJ·mol^{-1}
18.129 (a) 63 kJ·mol^{-1} **(b)** 10^{13} **(c)** Yes. The rate law has exponents equal to the coefficients in the equation.
(d) 7.9×10^{-16} M·s^{-1} at 25 °C, 3.0×10^{-14} M·s^{-1} at 75 °C
(e) about 38 times as fast   **(f)** -199.8 kJ

(g)

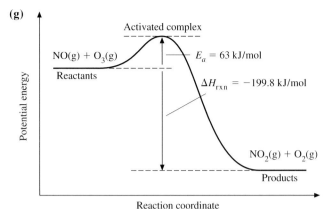

(h) 263 kJ·mol⁻¹ **(i)** No.

18.131 (a) $CH_3CH_2OH(g) + 3O_2(g) \rightarrow 2CO_2(g) + 3H_2O(l \text{ or } g)$
(b) The mechanism probably takes place by a series of steps. A single-step mechanism based on the molecularity of the equation would require instantaneous collision of four particles. This is highly improbable. **18.133** a steady state

18.135 (a) transfer of an electron from NO to O⁺; formation of a new N—O bond would have caused the NO⁺ to become enriched in O-18. **(b)** 8 protons, 10 neutrons, and 7 electrons

CHAPTER 19

19.1 (a)

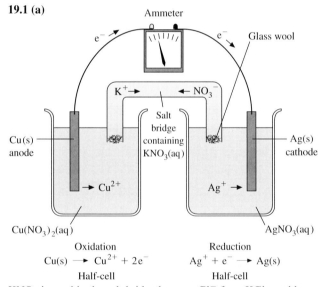

Oxidation
$Cu(s) \longrightarrow Cu^{2+} + 2e^-$
Half-cell

Reduction
$Ag^+ + e^- \longrightarrow Ag(s)$
Half-cell

KNO_3 is used in the salt bridge because Cl^- from KCl would form a precipitate with Ag^+.
(b) Net ionic equation: $2Ag^+ + Cu(s) \rightarrow Cu^{2+} + 2Ag(s)$
Molecular equation:
$2AgNO_3(aq) + Cu(s) \rightarrow Cu(NO_3)_2(aq) + 2Ag(s)$
19.2 $Cu(s)\,|\,Cu^{2+}(aq)\,\|\,Ag^+(aq)\,|\,Ag(s)$
19.3 (a) oxidizing agent **(b)** oxidizing agent **(c)** reducing agent
(d) reducing agent **(e)** oxidizing agent **(f)** both
19.4 (a) $Fe^{2+} < Ag^+ < NO_3^- < Br_2(l)$ **(b)** $Fe^{2+} < I^- < Al(s)$
19.5 Yes. H^+ has a higher standard reduction potential than Al^{3+}.
19.6 (a) +0.04 V, spontaneous **(b)** +0.46 V, spontaneous
19.7 −0.14 V

19.8 (1) $F_2(g) + 2Cl^- \rightarrow 2F^- + Cl_2(g)$
(2) $F_2(g) + 2Br^- \rightarrow 2F^- + Br_2(l)$
(3) $F_2(g) + 2I^- \rightarrow 2F^- + I_2(s)$
(4) $Cl_2(g) + 2Br^- \rightarrow 2Cl^- + Br_2(l)$
(5) $Cl_2(g) + 2I^- \rightarrow 2Cl^- + I_2(s)$
(6) $Br_2(l) + 2I^- \rightarrow 2Br^- + I_2(s)$
19.9 Ni^{2+} will oxidize Zn to Zn^{2+}.
E_{cell}° for $Zn(s) + Ni^{2+} \rightarrow Zn^{2+} + Ni(s)$ is +.
19.10 +0.37 V **19.11** -2.12×10^5 J or −212 kJ

19.12 +0.162 V **19.13** $K = \dfrac{[Au^{3+}]}{[Au^+]^3} = 10^{16}$

19.14 Eventually the gutter will fall from the house as holes form in the aluminum. In contact with the iron nails, the aluminum in the gutter is like zinc on galvanized iron or a sacrificial anode.
19.15 (a) Anode: $2Br^- \rightarrow Br_2(l) + 2e^-$
Cathode: $2H_2O(l) + 2e^- \rightarrow H_2(g) + 2OH^- (10^{-7}$ M)
Overall: $2Br^- + 2H_2O(l) \rightarrow Br_2(l) + H_2(g) + 2OH^- (10^{-7}$ M)
(b) Anode: $2H_2O(l) \rightarrow O_2(g) + 4H^+(10^{-7}$ M) + 4e⁻
Cathode: $Ag^+ + e^- \rightarrow Ag(s)$
Overall: $4Ag^+ + 2H_2O(l) \rightarrow 4Ag(s) + O_2(g) + 4H^+(10^{-7}$ M)
(c) Anode: $2H_2O(l) \rightarrow O_2(g) + 4H^+(10^{-7}$ M) + 4e⁻
Cathode: $2H_2O(l) + 2e^- \rightarrow H_2(g) + 2OH^- (10^{-7}$ M)
Overall: $2H_2O(l) \rightarrow O_2(g) + 2H_2(g)$
(d) Anode: $Ni(s) \rightarrow Ni^{2+} + 2e^-$
Cathode: $Ni^{2+} + 2e^- \rightarrow Ni(s)$
No net reaction occurs; nickel is transferred from anode to cathode.
19.16 0.771 g **19.17** 0.966 A **19.55** 1.08 g Cu, 2.24 g Au
19.57 (a) stable **(b)** stable **(c)** redox reaction
19.59 neodymium (Nd) **19.61 (1)** $B^+ + e^- \rightarrow B$
(2) $A^+ + e^- \rightarrow A$ **(3)** $C^+ + e^- \rightarrow C$

STOP AND TEST YOURSELF

1. (d) *19.2* **2.** (d) *19.2* **3.** (c) *19.2* **4.** (e) *19.3* **5.** (b) *19.3*
6. (b) *19.4* **7.** (c) *19.4* **8.** (a) *19.7* **9.** (c) *19.7* **10.** (c) *19.8*

19.63 (a)

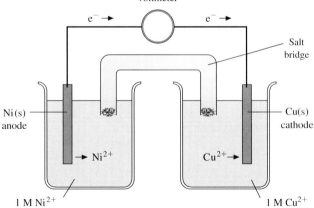

Anode: $Ni(s) \rightarrow Ni^{2+} + 2e^-$ Cathode: $Cu^{2+} + 2e^- \rightarrow Cu(s)$
(b) $Ni(s) + Cu^{2+} \rightarrow Ni^{2+} + Cu(s)$
19.65 (a) -1.5×10^4 J ($Cu^{2+} \rightarrow Cu^+$), -5.0×10^4 J ($Cu^+ \rightarrow Cu$)
(b) 0.34 V **(c)** Take the weighted average of the two half-cell

potentials E_1° and E_2°: $E^{\circ} = \dfrac{n_1 E_1^{\circ} + n_2 E_2^{\circ}}{n_1 + n_2}$ **19.67** −48 kJ

19.69 (a) 0.00 V (b) No. No reaction will take place.
(c) remove $Cl_2(g)$ as formed or increase $[H^+]$
19.71 (a) +0.828 V (b) E_{cell} will be lower. (c) +0.688 V
19.73 $1.602\ 177 \times 10^{-19}$ C/e^-
19.75 (a) 1.23 V (b) 1.0 atm (c) 1.22 V
(d) $CH_3COOH(aq) \rightleftharpoons CH_3COO^- + H^+$
$\quad H_2O(l) + H^+ \rightleftharpoons H_3O^+$
(e) $\qquad\qquad$ *Table 15.3* $\qquad\qquad$ *Table 19.1*
Top left: $\quad$ most easily ionized acids $\;$ most easily reduced species
Bottom right: $\quad$ strongest bases $\qquad$ strongest reducing agents
Reactions: $\qquad$ acid-base involve $\qquad$ redox involve transfer
$\qquad\qquad\qquad$ transfer of H^+ $\qquad\qquad\qquad$ of e^-
A similar diagonal relationship can be used to make predictions
from both tables.
19.77 (a) $E° = (T\Delta S° - \Delta H°)/nF$ (b) small value of $\Delta S°$
19.79 (a) 7×10^{-7} (b) 5×10^{-13}
(c)

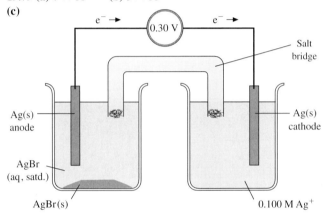

19.81 (a) $\Delta G° = -3164.6$ kJ for $4Al(s) + 3O_2(g) \rightarrow 2Al_2O_3(s)$,
$\Delta G° = -636.64$ kJ for $2Zn(s) + O_2(g) \rightarrow 2ZnO(s)$
(b) 1.33×10^4 kJ/lb Al, 2.21×10^3 kJ/lb Zn
(c) $E° = 2.73$ V for the Al battery, $E° = 1.65$ V for the Zn battery
19.83 (a) Bubbles are formed on the surface of both copper and
magnesium. Solution remain colorless (b) Conclusions: a reaction
is taking place (formation of bubbles) but copper is not reacting
(solution not blue). (c) Copper simply conducts electrons from
magnesium to hydrogen ions in water
19.85 The iron will act as a sacrificial anode for the copper, which
will cause the iron pipe to spring a leak. **19.87** 2.8×10^6 C
19.89 (a) $Cd(s) \rightarrow Cd(OH)_2(s)$ (anode),
$NiO_2(s) \rightarrow Ni(OH)_2(s)$ (cathode)
(b) $NiO_2(s) + 2H_2O(l) + 2e^- \rightarrow Ni(OH)_2(s) + 2OH^-$
$\qquad\qquad\qquad\qquad\qquad\qquad E° = -0.49$ V
$\quad Cd(s) + 2OH^- \rightarrow Cd(OH)_2(s) + 2e^-$
$\qquad\qquad\qquad\qquad\qquad\qquad E° = 0.82$ V
$\overline{Cd(s) + NiO_2(s) + 2H_2O(l) \rightarrow Cd(OH)_2(s) + Ni(OH)_2(s)}$
(c) +0.33 V
19.91 (a) 11.4 s (b) 35.4 g C (all to CO_2), 70.8 g C (all to CO)
(c) 67.9% (d) 98.7% **19.93** (a) 1.01×10^4 A (b) 9.8 kg/h
(c) converted to thermal energy **19.95** (a) 0.76 V (b) 12.50 mL
(c) 0.77 V (d) Solutions of reducing agents are oxidized by O_2 in
air, and thus their concentrations change. **19.97** (a) ultraviolet
region (b) $O_3(g) + 3I^- + 2H^+ \rightarrow O_2(g) + I_3^- + H_2O(l)$
(c) 1.54 V (d) 1.14 V

(e) $H_2O_2(aq) + 3I^- + 2H^+ \rightarrow 2H_2O(l) + I_3^-$
$\qquad Cl_2(g) + 3I^- \rightarrow 2Cl^- + I_3^-$

CHAPTER 20

20.1 $^{222}_{86}Rn \rightarrow ^{218}_{84}Po + ^{4}_{2}He$
20.2 (a) beta decay (b) positron emission (c) spontaneous
fission (d) alpha decay (e) electron capture (f) beta decay
20.3 (a) $^{139}_{57}La$ (b) $^{23}_{11}Na$ (c) $^{176}_{77}Ir$
20.4 (a) $^{245}_{100}Fm \rightarrow ^{241}_{98}Cf + ^{4}_{2}He$ (b) $^{251}_{100}Fm + ^{0}_{-1}e \rightarrow ^{251}_{99}Es$
(c) $^{18}_{7}N \rightarrow ^{18}_{8}O + ^{0}_{-1}e$ (d) $^{14}_{8}O \rightarrow ^{14}_{7}N + ^{0}_{+1}e$
(e) $^{258}_{102}No \rightarrow ^{120}_{47}Ag + ^{135}_{55}Cs + 3^{1}_{0}n$
20.5 target: $^{27}_{13}Al$, particle ejected: $^{1}_{0}n$ (neutron),
projectile: $^{4}_{2}He$ (alpha ray), product nuclide: $^{30}_{15}P$
20.6 (a) (i) $^{239}_{94}Pu + ^{2}_{1}H \rightarrow ^{240}_{95}Am + ^{1}_{0}n$
(iv) $^{139}_{58}Ce + ^{1}_{0}n \rightarrow ^{140}_{58}Ce + \gamma$ (v) $^{239}_{94}Pu + ^{4}_{2}He \rightarrow ^{242}_{96}Cm + ^{1}_{0}n$
(b) (iv) $^{139}_{58}Ce + ^{1}_{0}n \rightarrow ^{140}_{58}Ce + \gamma$ **20.7** (a) 0.17 g (b) 0.22 g
20.8 (a) 93 events/s (b) 1.4×10^{16} atoms (c) 4.4 μg
20.9 (a) probably stable (b) probably radioactive, beta emission
(c) probably radioactive, electron capture (d) probably radioactive, beta decay or alpha decay (e) probably radioactive,
alpha decay **20.10** Fluorine-17. It is further from band of stability than fluorine-18. **20.11** (a) $-0.000\ 167$ u (b) Yes. ΔE is
negative. (c) -1.50×10^7 kJ/mol **20.12** (a) $-0.528\ 46$ u
(b) -7.8868×10^{-11} J (c) -1.4084×10^{-12} J/nucleon
20.13 2.50×10^3 y **20.59** $^{9}_{4}Be + ^{4}_{2}He \rightarrow ^{12}_{6}C + ^{1}_{0}n$

20.61 $[\ddot{\text{O}}\!-\!\ddot{\text{S}}\!-\!\ddot{\text{O}}\!-\!\ddot{\text{S}}\!-\!\ddot{\text{O}}\!:]^{2-}$; S from this structure would
appear in both sulfur and SO_2

STOP AND TEST YOURSELF

1. (d) *20.1, 20.6* **2.** (e) *20.1* **3.** (d) *20.4* **4.** (c) *20.4*
5. (c) *20.2* **6.** (e) *20.3* **7.** (d) *20.3* **8.** (b) *20.3* **9.** (b) *20.4*
10. (c) *20.6* **11.** (b) *20.6* **12.** (d) *20.8* **13.** (e) *20.9*
14. (c) *20.10* **15.** (c) *20.11*

20.63 Nuclear reactions involve protons and neutrons within the
nucleus, whereas chemical reactions involve electrons outside the
nucleus. In nuclear reactions, elements are transmuted into other
elements, unlike chemical reactions, which have the same number
of each kind of atom as both reactants and products. Different
isotopes of an element will undergo different nuclear reactions, but
behave similarly in chemical reactions. Unlike chemical reactions,
nuclear reactions are independent of state of chemical combination.
The energy changes of nuclear reactions are of the order of
10^8–10^9 kJ/mol; for chemical reactions, however, the energies are
much smaller (10–10^3 kJ/mol). Changes in mass are detectable
in nuclear reactions but undetectable in chemical reactions.
20.65 (a) In electron capture, one of the electrons, usually one
from the first shell, becomes part of the nucleus. This leaves a
hole in the first energy shell into which an electron from an outer
shell falls. The transition from an outer energy level to $n = 1$
corresponds to radiation in the X-ray region of the spectrum.
(b) Nuclear reactions involve the release of energy from an
unstable nucleus. The emission of γ rays is one way a nucleus
with excess energy can relax to its ground state.

20.67 (a) $^{14}_{7}N + ^{1}_{0}n \rightarrow ^{3}_{1}H + ^{12}_{6}C$ **(b)** 17.7 y

(c) $^{3}_{1}H \rightarrow ^{3}_{2}He + ^{0}_{-1}e$ **(d)** 9.80×10^{13} disintegrations

(e) 2.65×10^3 Ci **(f)** 1.20×10^{-4} disintegrations

20.69 2.88×10^9 y old

20.71 NO-30 : NO-31 : NO-32 : NO-33 = 99.39 : 0.41 : 0.20 : 0.00

20.73 (a) metal **(b)** IIB **(c)** mercury **(d)** highest state: +2; lowest state: +1 **20.75** Prepare saturated solutions of labeled and unlabeled barium sulfate by stirring 1 mg of barium sulfate with 100 mL of water overnight. From the labeled solution, remove the undissolved solid by filtration. Add the filtrate to the saturated solution of unlabeled barium sulfate and stir the mixture overnight. Then separate the undissolved solid by filtration, wash thoroughly to remove the mother liquor, and measure the radioactivity of the solid. If the solubility equilibrium is indeed dynamic, the solid will be radioactive because it will have exchanged some of its unlabeled sulfate ions for radioactive sulfate ions from the labeled solution.

20.77 (a) A *nuclide* is any atom defined by a specific atomic and mass number. A *nucleon* is a general term for the building blocks of the nucleus—neutrons and protons. **(b)** *Gamma rays* are electromagnetic radiation of very high energy, typically released in nuclear reactions as a means for the nucleus to stabilize. *X-rays* are electromagnetic radiation of slightly less energy than gamma rays, typically released when electrons make transitions from higher to lower energy levels. **(c)** *Atomic number* represents the number of protons in a nucleus. *Mass number* represents the total number of nucleons. **(d)** *Fission* is the splitting of one heavy nucleus into two nuclei with smaller mass numbers. *Fusion* is the combining of two light nuclei to form a heavier, more stable nucleus. **(e)** A *beta particle* is an electron and is negatively charged. A *positron* is a particle with the same mass as a beta particle but opposite in charge. **(f)** A *rad* is a measure of the amount of energy deposited in tissue as radiation passes. A *rem* is a rad times a correction factor that takes into account the effectiveness of the radiation in causing biological damage. **(g)** *Slow neutrons* have average kinetic energies similar to those that they would have if they existed as a gas at room temperature. *Fast neutrons* have higher kinetic energies. **20.79** reduction of IO_4^-; IO_3^- formed by oxidation of $^{128}I^-$ would contain I-128.

20.81 (a) 9.57×10^6 y **(b)** $\approx 0\%$; none of the originally formed Al is present **20.83 (a)** 7.7×10^{-13} % **(b)** 1.3×10^{14} y

20.85 (a) Positron emission: P-28, P-29, P-30; beta emission: P-32, P-33, P-34 **(b)** P-32 and P-33 **20.87** Gamma radiation is detected because it penetrates through more material than alpha or beta radiation can. Thus, the prospector can detect uranium buried deeply in the ground by detecting gamma radiation.

20.89 12 μg B-12 **20.91** Fuel for fusion is plentiful and large quantities of radioactive wastes are not produced. However, release of useful amounts of energy by controlled fusion has not yet been achieved because of the extremely high temperatures required to initiate fusion. The major problem with fission as a source of energy is the radioactive wastes produced. Neither fission nor fusion produce harmful oxides of nitrogen and sulfur or large quantities of carbon dioxide as combustion does. Another problem with combustion is limited fuel supplies. However, fossil fuels are convenient and equipment for using them is readily available. Substitution of plant materials for fossil fuels would use up carbon dioxide produced by combustion in photosynthesis and provide a renewable energy source. **20.93 (a)** lower

(b) positron emission: $^{40}_{19}K \rightarrow ^{40}_{18}Ar + ^{0}_{+1}e$

electron capture: $^{40}_{19}K + ^{0}_{-1}e \rightarrow ^{40}_{18}Ar$

(c) beta decay: $^{40}_{19}K \rightarrow ^{40}_{20}Ca + ^{0}_{-1}e$

(d) K-40 has 21 (odd number) neutrons and 19 (odd number) protons. K-39 and K-41 both have even numbers of neutrons.

(e) All three types yield stable nuclides.

20.95 (a) $^{57}_{27}Co + ^{0}_{-1}e \rightarrow ^{57}_{26}Fe$, $^{60}_{27}Co \rightarrow ^{60}_{28}Ni + ^{0}_{-1}e$; Co-57 has too few neutrons, and Co-60 has too many

(b) Co-57; shorter half-life, lower energy gamma rays

(c) Co-60; higher energy gamma rays

20.97 (a) 4.00 mCi **(b)** 98.9% remains **(c)** 5.25×10^3 mL

(d) 1.43×10^{-9} g NaTcO$_4$ **(e)** Prepare a solution of 1.00×10^{-5} g NaTcO$_4$ in 1.000 L. Dilute 0.143 mL of this solution to 1.00 mL.

20.99 (a) as a catalyst **(b)** $^{13}_{7}N$, $^{13}_{6}C$, $^{14}_{7}N$, $^{15}_{8}O$, $^{15}_{7}N$

(c) The energy released should be the same as that released when four H-1 nuclei fuse by the proton-proton chain, since the overall equation is the same. The difference in energy between reactants and products is independent of the path from reactants to products.

20.101 (a) $^{99}_{43}Tc^* \rightarrow ^{99}_{43}Tc + ^{0}_{0}\gamma$, $^{99}_{43}Tc \rightarrow ^{99}_{44}Ru + ^{0}_{-1}e$

(b) 0.480 events/s **(c) (i)** 1.36×10^7 kJ/mol

(ii) 8.79×10^{-12} m **(iii)** 1.51×10^{-4} g

CHAPTER 21

21.1 $CH_3CH_2CH_3(g) + 3H_2O(g) \xrightarrow[heat]{catalyst} 7H_2(g) + 3CO(g)$

21.2 (a) $[Ar]4s^2 + \begin{matrix} 1s^1 \\ 1s^1 \end{matrix} \rightarrow [Ar] + \begin{matrix} 1s^2 \text{ or } [He] \\ 1s^2 \text{ or } [He] \end{matrix}$

$Ca + 2H \rightarrow Ca^{2+} + 2H^-$

(b) $Ca(s) + H_2(g) \rightarrow CaH_2(s)$

(c) Net ionic: $H^- + H_2O(l) \rightarrow H_2(g) + OH^-$

Complete ionic:
$Ca^{2+} + 2H^- + 2H_2O(l) \rightarrow 2H_2(g) + Ca^{2+} + 2OH^-$

Molecular: $CaH_2(s) + 2H_2O(l) \rightarrow 2H_2(g) + Ca(OH)_2(s)$

21.3 (a) H_2O, proton donor **(b)** H^-, proton acceptor

21.4 (a) hydrogen in H^- **(b)** one hydrogen in $H_2O(l)$

(c) H_2O **(d)** H^- **21.5**

Lewis base Lewis acid

21.6 (a) $CH_3CH{=}CH_2(g) + H_2(g) \xrightarrow{catalyst} CH_3CH_2CH_3(g)$

(b) $MoO_3(s) + 3H_2(g) \rightarrow Mo(s) + 3H_2O(g)$

21.7

21.8 37%

21.9 (a) $CaO(s) + H_2O(l) \rightarrow Ca(OH)_2(s)$

(b) $HNO_3(l) + H_2O(l) \rightarrow H_3O^+ + NO_3^-$

(c) $2Na(s) + 2H_2O(l) \rightarrow 2NaOH(aq) + H_2(g)$

(d) $CH_3COOH(l) + H_2O(l) \rightleftarrows H_3O^+ + CH_3COO^-$

21.10 (a)

(b) Polar.

21.11 (a) $Pb^{2+} + H_2S(g) \rightarrow PbS(s) + 2H^+$

(b) $ZnS(s) + 2HCl(aq) \rightarrow ZnCl_2(aq) + H_2S(g)$

21.12 (a) potassium hydrogen sulfate **(b)** iron(II) sulfide

(c) sulfite ion **21.13 (a)** $MgSO_4 \cdot 7H_2O$ **(b)** $Ca(HSO_3)_2$

(c) K_2SO_4 **21.14 (a)** $2Al(s) + N_2(g) \xrightarrow{heat} 2AlN(s)$

(b) $NH_4Cl(s) + NaOH(aq) \rightarrow NaCl(aq) + NH_3(g) + H_2O(l)$

(c) $NH_3(g) + HCl(g) \rightarrow NH_4Cl(s)$

(d) $NH_4Cl(s) \xrightarrow{heat} NH_3(g) + HCl(g)$

(e) $Fe(s) + 4HNO_3(6\ M) \rightarrow Fe(NO_3)_3(aq) + NO(g) + 2H_2O(l)$

(f) $CaCO_3(s) + 2HNO_3(6\ M) \rightarrow Ca(NO_3)_2(aq) + CO_2(g) + H_2O(l)$

21.15 (a) $+5$ **(b)** $+5$ **(c)** $+4$ **(d)** $+3$ **(e)** -3

21.16 (a) $2H_3PO_4(aq) + K_2CO_3(aq) \xrightarrow{excess}$

$$2KH_2PO_4(aq) + CO_2(g) + H_2O(l)$$

(b) $H_3PO_4(aq) + \underset{excess}{K_2CO_3(aq)} \rightarrow K_2HPO_4(aq) + CO_2(g) + H_2O(l)$

(c) $H_3PO_4(aq) + \underset{excess}{3KOH(aq)} \rightarrow K_3PO_4(aq) + 3H_2O(l)$

(d) $3Zn(s) + 2H_3PO_4(aq) \rightarrow Zn_3(PO_4)_2(s) + 3H_2(g)$

21.17 (a) $C(graphite) + 2S(g) \xrightarrow{heat} CS_2(g)$

(b) $MnO_2(s) + C(graphite) \xrightarrow{heat} Mn(l) + CO_2(g)$

(c) $CH_3COCH_3(l) + \underset{excess}{4O_2(g)} \xrightarrow{spark} 3CO_2(g) + 3H_2O(l)$

(d) $Na_2CO_3(s) + 2HNO_3(aq) \rightarrow 2NaNO_3(aq) + CO_2(g) + H_2O(l)$

21.19 The density of $H_2(g)$ is very low. Therefore, hydrogen gas was not held by Earth's gravitational field and escaped into space.

21.23 (a, b) $2Na(s) + 2D_2O(l) \rightarrow 2NaOD(soln) + D_2(g)$

(c) $Cl_2(g) + D_2(g) \rightarrow 2DCl(g)$ **21.27 (a)** The *greenhouse effect* is caused by substances that are transparent to visible radiation but absorb infrared radiation. They let the sun warm Earth during the day but prevent loss of thermal energy during the night. **(b)** water, H_2O; carbon dioxide, CO_2, methane, CH_4 **21.31 (a)** Nitric acid is colorless but turns yellow as the result of the formation of $NO_2(g)$. **(b)** Copper and silver are soluble in nitric acid because the nitrate ion in acidic solution is a powerful oxidizing agent.

21.35 The HPO_4^{2-} ion can behave as an acid or as a base: addition of H^+: $HPO_4^{2-} + H^+ \rightleftarrows H_2PO_4^-$
addition of OH^-: $HPO_4^{2-} + OH^- \rightleftarrows PO_4^{3-} + H_2O$

21.39 two **21.43 (a)** *Allotropes* are different forms of the same element that differ in bonding. **(b)** O, S, P, C **(c)** O: $O_2(g)$, $O_3(g)$; S: rhombic, monoclinic; P: white, red; C: diamond, graphite, buckyballs, $\sim\!\!\sim\! C\!\equiv\!C\!-\!C\!\equiv\!C\!-\!C\!\equiv\!C \sim\!\!\sim$

(d) Silicon does not readily form double bonds.

21.47 (a)

Lewis base Lewis acid

(b) $NH_2^- + H_2O \rightarrow NH_3 + OH^-$
  B-L   B-L
  base   acid

21.51 (a) covalent bonds **(b)** electrostatic attraction
(c) London forces **(d)** hydrogen bonds **(e)** London forces
21.55 (a) one **(b)** sodium hypophosphite monohydrate **(c)** $+1$
21.59 (a) $T = 937.5\ K$ for $SnO_2(s) + 2C(s) \rightarrow Sn(s) + 2CO(g)$,
$T = 2706\ K$ for $CaO(s) + C(s) \rightarrow Ca(s) + CO(g)$
(b) The first reaction requires a temperature of 937.5 K, which can be reached relatively easily. The second reaction requires a temperature of 2706 K, which is very high and probably not reachable in ancient times. Thus, the first reaction was probably the one used in ancient times.

21.63 (a)

$Cl + H_2 \rightarrow HCl + H$
$+ (H + Cl_2 \rightarrow HCl + Cl)$
———————————————
$H_2 + Cl_2 \rightarrow 2HCl$

(b) Cl and H are radicals.

Step (1) $Cl_2 \xrightarrow{\text{light (or thermal energy)}} 2Cl\cdot$
Step (2) $Cl\cdot + H_2 \rightarrow HCl + H\cdot$
Step (3) $H\cdot + Cl_2 \rightarrow HCl + Cl\cdot$

(c) one, one **(d)** $Cl\cdot + Cl\cdot \rightarrow Cl_2$ or $H\cdot + Cl\cdot \rightarrow HCl$
(e) In steps (2) and (3), $H_2 + Cl_2$ are reactants; their concentration is high, at least until reaction approaches completion. $H\cdot$ and $Cl\cdot$, on the other hand, are reactive intermediates, and their concentration is low. **(f)** -92.31 kJ/mol **(g)** Once the reaction begins, much thermal energy is released. Thermal energy, as well as light, can cause chains to start. The reaction gets faster and faster, yielding an explosion. **21.67** 30.3 g $H_2(g)$ **21.71 (a)** NO
(b) SO_2 **(c)** O_3 and NO **(d)** hydrocarbons containing double bonds such as isoprene $[H_2C\!=\!C(CH_3)CH\!=\!CH_2]$
21.75 3.11×10^3 kJ
21.79 (a) Yes. H_2O_2 appears in both left and right columns of a table of standard reduction potentials. From the diagonal relationship in the table, the disproportionation reaction is spontaneous. **(b)** 1.078 V
(c) $3H_2O_2(aq) + 2Cr^{3+} + H_2O(l) \rightarrow Cr_2O_7^{2-} + 8H^+$; hydrogen peroxide is acting as an oxidizing agent.
(d) $Cr_2O_7^{2-} + 8H^+ + 3H_2O_2(aq) \rightarrow 2Cr^{3+} + 7H_2O(l) + 3O_2(g)$; hydrogen peroxide is acting as a reducing agent.
(e) HO_2^- **21.83 (a)** 33% P_2O_5 **(b)** 14% P **21.87** $KAlSi_3O_8$
21.91 (a) 5.85 atm **(b)** 96.5% He, 3.5% O_2 **(c)** 0.222 g/L
(d) 1.32 g/L **21.95** Ice is a thermal insulator. Once the ice has formed, it is a poor conductor of thermal energy and does not transfer thermal energy from the plant to the outside air. Also, freezing of the water releases thermal energy that warms the trees and fruit.
21.99 (a, b) $H_2O(s) \rightarrow H_2O(l)$: melting or fusion; reverse phase change is freezing or crystallization.
$CO_2(s) \rightarrow CO_2(g)$: sublimation; reverse phase change is deposition.
$N_2(l) \rightarrow N_2(g)$: boiling or evaporation or vaporization; reverse phase change is condensation.
21.103 (a) 5×10^9 kg **(b)** 3×10^{11} L **21.107 (a)** Mass of $CaCl_2$ is greater, probably making transportation and labor costs greater. **(b)** $CaCl_2$ costs almost twice as much as NaCl.
21.111 (a) Water extinguishes burning paper by lowering the temperature of the paper and by excluding oxygen, a necessary reactant for combustion. **(b)** Water can't be used to extinguish a

sodium fire because sodium reacts violently with water. Furthermore, one of the products of this reaction is the highly flammable gas H_2: $2Na(s) + 2H_2O(l) \rightarrow 2NaOH(aq) + H_2(g)$.
(c) Water can't be used to extinguish burning oil because oil is less dense than water. As a result, the oil floats on the surface of the water and the fire is spread out over a larger area.   **(d)** If you lie down and roll, the supply of oxygen will be cut off and the fire will be extinguished. Running to a faucet will replenish the oxygen consumed by the fire, and the fire will burn more vigorously.
(e) Ammonium bromide is water soluble and washes out when the fabric is laundered.   **21.115 (a)** 1×10^3 ppm   **(b)** 49 g CO
(c) 8.6×10^{-3} M   **(d)** 4×10^{-5} M   **(e)** 2×10^2 **(f)** 1
(g) 50% of the hemoglobin is tied up as Hb(CO).

CHAPTER 22

22.1 (a)

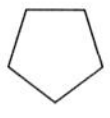

(b)

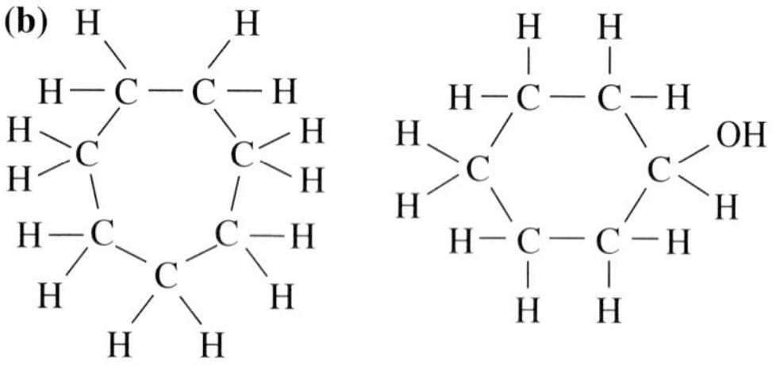

22.2 (a)   **(b)**

(c) No stereocenter is present.   **22.3**

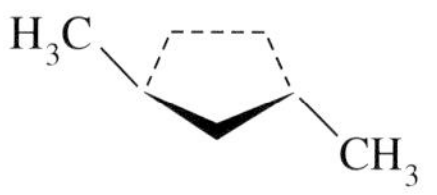

22.4

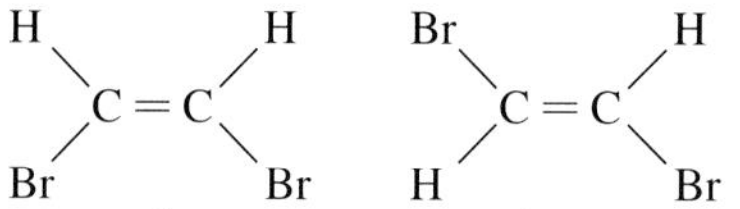

22.5 No. The two CH_3— groups are identical and interchangeable.
22.6 pentane, hexane, heptane, octane, nonane, and decane
22.7 (a) $(CH_3)_2C{=}CH_2$, 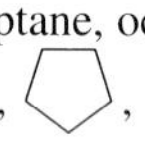 , $(CH_3)_2CHCH_2C(CH_3)_3$,

$CH_3C{\equiv}CCH_3$, 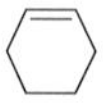   **(b)** $(CH_3)_2CHCH_2C(CH_3)_3$ **(c)**

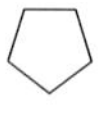

22.8 C_nH_{2n} **22.9** If O_2 is present, complete or incomplete combustion will be the predominant reaction.

22.10 $CH_3CH_2CH_2CH_3(g) + Cl_2(g)$ $\xrightarrow[\text{excess}]{\substack{\text{light}\\\text{and/or}\\\text{thermal}\\\text{energy}}}$ $CH_3CH_2CH_2CH_2Cl(l) + HCl(g)$

$CH_3CH_2CH_2CH_3(g) + Cl_2(g)$ $\xrightarrow[\text{excess}]{\substack{\text{light}\\\text{and/or}\\\text{thermal}\\\text{energy}}}$ $CH_3CH_2CH(Cl)CH_3(l) + HCl(g)$

22.11 propyl   **22.12** No. The two methyl groups will be located on the third carbon regardless of which end carbon is number 1.
22.13 (b) 2-methylpentane; **(a)** and **(d)** are wrong because the longest chain is 5 Cs long. **(c)** is wrong because 4 > 2.

22.14
(i) 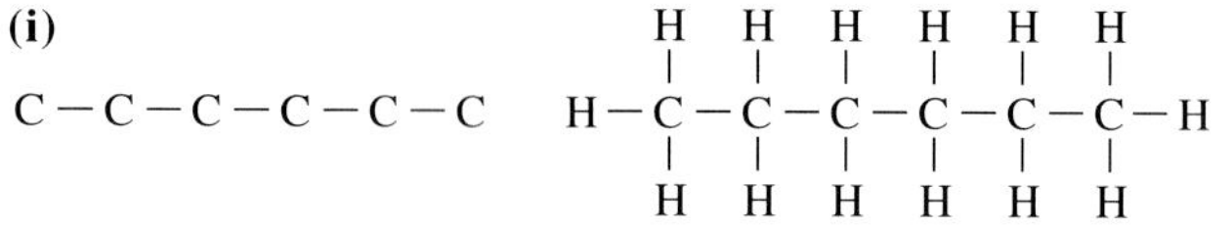
$CH_3CH_2CH_2CH_2CH_2CH_3$, hexane
(ii) 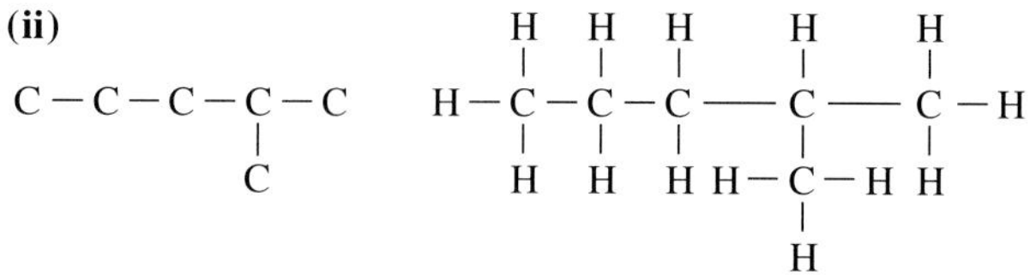
$CH_3CH_2CH_2CH(CH_3)_2$, 2-methylpentane
(iii) 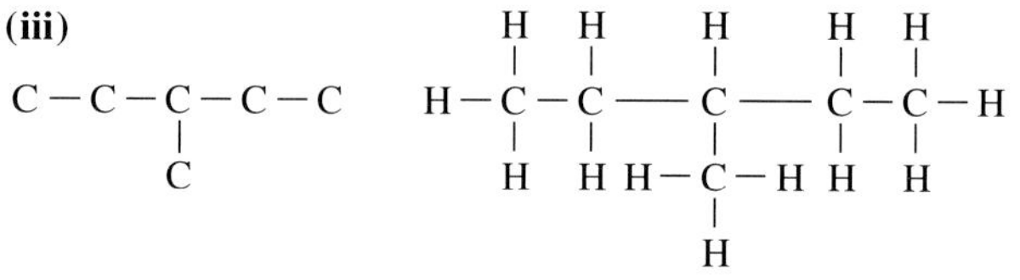
$CH_3CH_2CH(CH_3)CH_2CH_3$, 3-methylpentane
(iv) 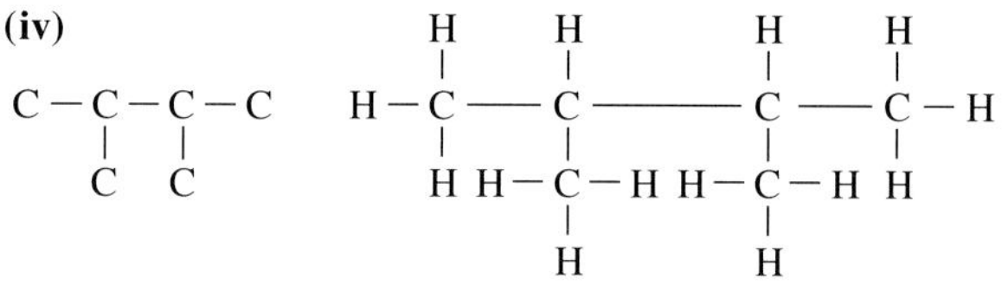
$(CH_3)_2CHCH(CH_3)_2$, 2,3-dimethylbutane
(v) 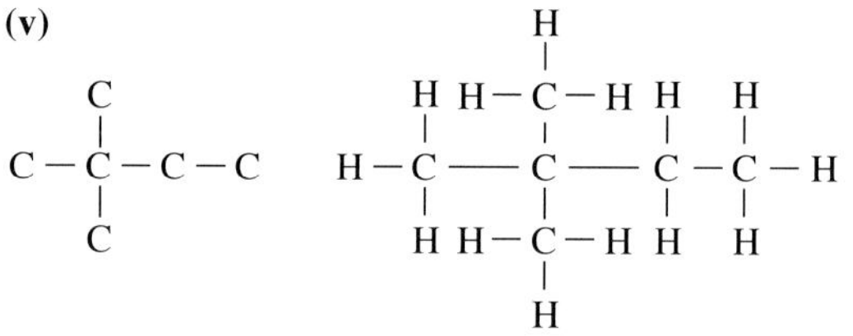
$(CH_3)_3CCH_2CH_3$, 2,2-dimethylbutane

22.15 (a)

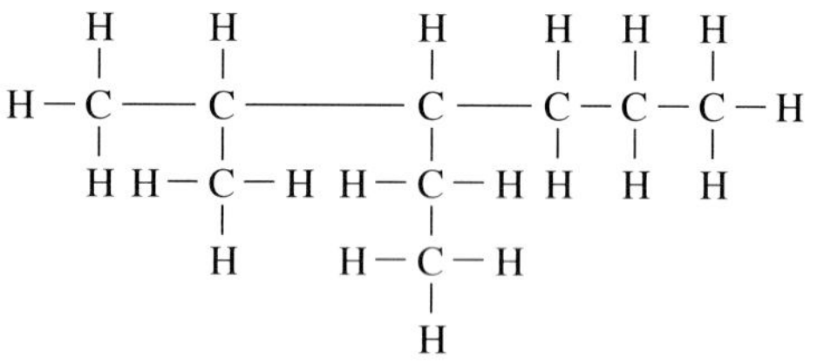

(b)

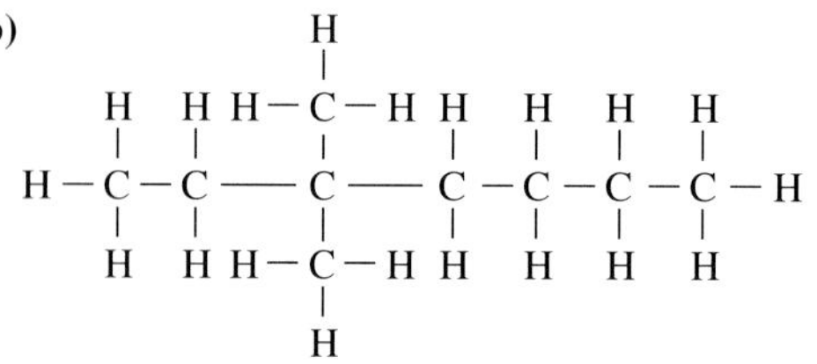

22.16 (a) $CH_3CH_2COOH(aq) + NaOH(aq) \rightarrow$
$CH_3CH_2COONa(aq) + H_2O(l)$
(b) $CH_3CH_2NH_2(aq) + HCl(aq) \rightarrow CH_3CH_2NH_3Cl(aq)$
(c) $CH_3CH{=}CH_2(g) + H_2(g) \xrightarrow{\text{catalyst}} CH_3CH_2CH_3(g)$
(d) $2CH_3OH(l) + 2Na(s) \rightarrow 2CH_3ONa(\text{soln}) + H_2(g)$

22.17 3-methyl-1-butanol

22.18 (a)

H–C(H)(H)–C(H)(C H₂H...)

Structure: $(CH_3)_3COH$ or $(CH_3)_2C(OH)CH_3$

(b)

$F–C(Cl)(F)–C(F)(Cl)–F$, $CClF_2CClF_2$

22.19 (a) $(CH_3)_2C=CH_2$ **(b)** ⬡ **(c)** $CH_3C\equiv CCH_3$

22.20

$H–C(H)(H)–C(=O)–O–H$

$H–C(=O)–O–C(H)(H)–C(H)(H)(H)$

$H–C(H)(H)–C(=O)–N(H)(H)$

$H–C(=O)–H$

$H–C(H)(H)–C(=O)–C(H)(H)(H)$

22.21 (a) Cl⟩CH_2CH_2⟨OH   alkyl chloride and alcohol

(b) ⬡–⟨$CH=CH_2$⟩   aromatic and alkene

(c) CH_3CH_2⟨COO⟩CH_2CH_3 ester

(d) CH_3CH_2⟨Br⟩   alkyl bromide   **(e)** H⟨CON⟩$(CH_3)_2$ amide

22.22 (a)–(c)

leaving group ⟶

$\boxed{CH_3CH_2CH_2}$⟨Br⟩ + ⟨CH_3O^-⟩ ⟶ $CH_3CH_2CH_2OCH_3$ + Br^-

substrate   nucleophile

(d) $CH_3CH_2CH_2$— (propyl group)

22.23

HO^- (nucleophile) ... :Cl: ... substrate ... activated complex ...

⟶ product + :Cl:⁻ (leaving group)

22.24 (a) $CH_3CH_2CH_2CH_2C\equiv CH + Cl^-$
(b) $H_2C=CHCH_2OOCCH_3$ (or $H_2C=CHCH_2OCCH_3$, with =O) + Br^-

22.25 (a) Reaction will occur; CN^- is a stronger base than Br^-.
(b) Reaction will not occur; OH^- is a stronger base than Br^-.

22.26 $H_2C=CHCH_2Br$ + :Ö–H

substrate   nucleophile

22.27 (a) Elements termed "essential" are elements other than O, C, H, N, Ca, P, Cl, K, S, and Na that are needed for life.
(b) aluminum and titanium   **(c)** At the pH of biological systems, the oxides of aluminum and titanium are insoluble.
22.31 (a) *plane polarized light:* a light beam with waves vibrating in only one direction or plane   **(b)** *optically active:* a material that rotates the plane of polarized light passed through it
(c) *stereocenter:* an atom with four different groups attached
(d) *chiral:* objects that can exist as enantiomers
(e) *achiral:* objects that are not chiral
(f) *racemic mixture:* a mixture of exactly 50% of one form of an optically active material with 50% of its enantiomer.
22.35 (a) two **(b)** three   **(c)** one **(d)** one   **(e)** zero   **(f)** two
22.39 (a) Butane is a gas under ordinary conditions; pentane is a liquid.   **(b)** 0.740 g/mL
22.43 (a) $(CH_3)_4C$

(b) $(CH_3)_4C(g) + Cl_2(g) \xrightarrow[\text{or light}]{\text{thermal energy}} (CH_3)_3CCH_2Cl(l) + HCl(g)$ (excess)

(c) An excess of the alkane is required. At a lower ratio of alkane to chlorine, disubstitution and higher substitutions occur.
(d) Thermal energy or light is required to initiate the reaction.
22.47 (a) alcohol **(b)** ether   **(c)** alkyl chloride   **(d)** alkyl iodide
(e) carboxylic acid   **(f)** aldehyde   **(g)** phenol   **(h)** aromatic alcohol
22.51 (a)–(c)

leaving group ⟶

$\boxed{CH_3CH_2CH_2CH_2}$⟨Br⟩ + ⟨NH_3⟩ ⟶ $CH_3CH_2CH_2CH_2NH_3^+ + Br^-$

substrate   nucleophile

(d) $CH_3CH_2CH_2CH_2—$
22.55

HO^- (nucleophile) ... substrate ... activated complex ...

⟶ product + :Cl:⁻ (leaving group)

22.59 (a) $2(CH_3)_3COH(l) + 2K(s) \rightarrow H_2(g) + 2(CH_3)_3COK(soln)$
(b) $2(CH_3)_2NH(aq) + H_2SO_4(aq) \rightarrow [(CH_3)_2NH_2]_2SO_4(aq)$ or
$(CH_3)_2NH(aq) + H_2SO_4(aq) \rightarrow [(CH_3)_2NH_2]HSO_4(aq)$
(c) $2CH_3COOH(aq) + Ba(OH)_2(aq) \rightarrow (CH_3COO)_2Ba(aq) + 2H_2O(l)$

(d) ⬡(l) + H_2(g) $\xrightarrow{\text{catalyst}}$ ⬡(l)

(e) ⬡–OH(aq) + NaOH(aq) → ⬡–ONa(aq) + H_2O(l)

1. (a) *1.9*   **2.** (e) *22.6*   **3.** (b) *22.6*   **4.** (c) *3.10*   **5.** (e) *22.7*
6. (d) *22.4*   **7.** (e) *22.7*   **8.** (d) *22.3*   **9.** (c) *22.3*
10. (a) *12.2*   **11.** (d) *9.12*   **12.** (d) *22.5*   **13.** (e) *22.8, 22.9*

22.63 (a) ① $CH_3CH_2CH_2CH_2CHO$
　　② $CH_3CH_2CH(CH_3)CHO$
　　③ $(CH_3)_2CHCH_2CHO$
　　④ $(CH_3)_3CCHO$

(b) ① $CH_3CH_2CH_2CH_2OH$
　　② $CH_3CH_2CH(OH)CH_3$
　　③ $(CH_3)_2CHCH_2OH$
　　④ $(CH_3)_3COH$

(c) $CH_3(CH_2)_{10}CH_3$

22.67 (a) H—C—C*—C—Cl, $CH_3CH(Cl)CH_2Cl$

(b) 1,2-dichloropropane
(c) Three others are possible:

① H—C—C—C—Cl, $CH_3CH_2CHCl_2$, 1,1-dichloropropane

② Cl—C—C—C—Cl, $CH_2ClCH_2CH_2Cl$, 1,3-dichloropropane

③ H—C—C—C—H, $CH_3CCl_2CH_3$, 2,2-dichloropropane

22.71 (a) CH_3CN, **(b)** CH_3COCH_3, **(d)** $(CH_3)_2S{=}O$
22.75 Double bonds are shorter and stronger than single bonds.
22.79 (a)

　　　cis　　　　　　*trans*

(b) *trans*-1,2-dichloroethene is nonpolar as a result of its symmetry.
22.83 (a) Step (1) is Brønsted-Lowry; Step (2) is Lewis.
(b)

$H_2C{=}CH_2 + H—\ddot{\underset{\cdot\cdot}{Cl}}:$ ⟶ $H_2\overset{+}{C}—CH_3 + :\ddot{\underset{\cdot\cdot}{Cl}}:^-$
BL base　　　BL acid

$:\ddot{\underset{\cdot\cdot}{Cl}}:^- + H_2\overset{+}{C}—CH_3$ ⟶ $H_2C(Cl)CH_3$
Lewis　　　Lewis
base　　　acid

(c) Entropy of system decreases; two molecules of gas combine to form one. Enthalpy decreases; thermal energy is released, and the reaction is exothermic. Since the reaction takes place spontaneously, free energy must decrease.
(d) $H_2C{=}CH_2 + H_2 \xrightarrow{\text{catalyst}} H_3CCH_3$ (catalytic hydrogenation)

◯ + Br_2 ⟶ ◯ (Section 10.7)
　　　　　　　Br　Br

22.87 (a)

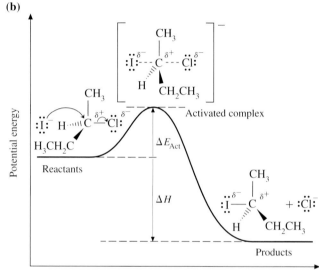

reactants　　　　　　activated complex

products

(b)

Potential energy / Reaction coordinate diagram:
Reactants, Activated complex, ΔE_{Act}, ΔH, Products

(c) rate $= k[I^-][CH_3CH_2CHClCH_3]$
22.91 CH_3NH_2, methylamine　**22.95** The two atoms on either side of a triple bond lie on a straight line with the triply bonded atoms C—C≡C—C. This leaves only two more carbon atoms to complete the 6-membered ring. They can't reach.
22.99 $CH_3I + {}^-OCH_2CH_2CH_3 \rightarrow CH_3OCH_2CH_2CH_3 + I^-$
$CH_3CH_2CH_2Br + {}^-OCH_3 \rightarrow CH_3CH_2CH_2OCH_3 + Br^-$

22.103 (a) A *homogeneous catalyst* is in the same phase as the reactants.
(b) $CH_3CH_2CH_2CH_2CHO$, $CH_3CH_2CH(CHO)CH_3$
(c) 2.3×10^3 lb $CH_3CH_2CH_2CHO$, 5.1×10^2 lb $(CH_3)_2CHCHO$
22.107 (a) 2878 kJ　**(b)** 2956 kJ　**(c)** No.　**(d)** Sunlight provides the energy required to make the nonspontaneous process take place.
(e) -2956 kJ

CHAPTER 23

23.1

23.2 (a) $CH_2{=}CCl_2$ **(b)**

$$\left(\!-CH_2\underset{\underset{Cl}{|}}{\overset{\overset{Cl}{|}}{C}}-\!\right)_n$$

23.3 Chain initiation:

Rad$\cdot$ + $CH_2{::}\dot{C}HOCH_3 \longrightarrow$ Rad$\cdots CH_2\cdots\dot{C}HOCH_3$

Chain propagation:

Rad$\cdots CH_2\cdots\dot{C}HOCH_3$ + $CH_2{::}\dot{C}HOCH_3 \rightarrow$

  Rad$\cdots CH_2\cdots CH(OCH_3)\cdots CH_2\cdots\dot{C}HOCH_3$

and then

Rad$\cdots CH_2\cdots CH(OCH_3)\cdots CH_2\cdots\dot{C}HOCH_3$ +

   $CH_2{::}\dot{C}HOCH_3 \rightarrow$

Rad$: CH_2: CH(OCH_3): CH_2: CH(OCH_3): CH_2: \dot{C}HOCH_3$,

etc.

Chain termination:

Rad$\cdot$ + $\dot{C}H_3O\dot{C}H\cdots CH_2\cdots CH(OCH_3)\cdots CH_2\cdots$Rad or

Rad$\cdots CH_2\cdots\dot{C}HOCH_3$ + $\dot{C}H_3O\dot{C}H\cdots CH_2\cdots$Rad or

Rad$\cdot$ + $\cdot$Rad or any other step in which two radicals combine.

23.4 (a) $-(CH_2CH{=}CCH_2)_n$
          with CH_3 substituent

(b)

$$\sim\!\!\!\sim\!\! \underset{H}{\overset{H}{C}}-\underset{CH_3}{\overset{H}{C}}{=}\underset{H}{\overset{}{C}}-\underset{H}{\overset{H}{C}}-\underset{H}{\overset{H}{C}}-\underset{CH_3}{\overset{H}{C}}{=}\underset{H}{\overset{}{C}}-\underset{H}{\overset{H}{C}}-\underset{H}{\overset{H}{C}}-\underset{CH_3}{\overset{H}{C}}{=}\underset{H}{\overset{}{C}}-\underset{H}{\overset{H}{C}}\!\!\sim\!\!\!\sim$$

23.5 (a) $HOC{-}\langle\text{ring}\rangle{-}COH$ + $H_2N{-}\langle\text{ring}\rangle{-}NH_2 \longrightarrow$
(each C with =O)

$HOC{-}\langle\text{ring}\rangle{-}CNH{-}\langle\text{ring}\rangle{-}NH_2$ + H_2O
(each C with =O)

(b)

$\sim\!\!\!\sim C{-}\langle\text{ring}\rangle{-}CNH{-}\langle\text{ring}\rangle{-}NHC{-}\langle\text{ring}\rangle{-}CNH{-}\langle\text{ring}\rangle{-}NH\!\sim\!\!\!\sim$
(C's with =O)

(c)

$$\left(\!-\underset{O}{\overset{}{C}}{-}\langle\text{ring}\rangle{-}\underset{O}{\overset{}{C}}NH{-}\langle\text{ring}\rangle{-}NH-\!\right)_n$$
(C's with =O)

23.6 $H_3\overset{+}{N}CH(CH_3)COO^-$ **23.7 (a)** leucine **(b)** hydroxylysine
(c) aspartic acid **(d)** hydroxylysine and threonine

23.8

$$\underset{H_2N}{\overset{CH_2SH}{\underset{\,}{\overset{|}{H{-}\!-\!-\!C{-}COOH}}}}$$

23.9 (a) VF **(b)**

$$H_2N{-}CH{-}\underset{O}{\overset{O}{C}}{-}NH{-}CH{-}COOH$$
with CH (H_3C, CH_3) and CH_2 (phenyl)

(c) Phe-Val

$$H_2N{-}CH{-}\underset{}{\overset{O}{C}}{-}NH{-}CH{-}COOH$$
with CH_2 (phenyl) and CH (H_3C, CH_3)

23.10 Cys-Gly and Gly-Glu   **23.11** Gly-Glu-Cys
23.12 (a) $Ala_2Gly_1Ser_1$ **(b)** tetra- **(c)** Ala-Gly-Ser-Ala
23.13 (a) During cyclization, the second-to-the-last —OH group reacts with the carbonyl (aldehyde) group. **(b)** The ring form has one more stereocenter than the straight-chain form. **(c)** In aqueous solution, the cyclic forms are in equilibrium with the straight-chain form. **23.14 (a)** Polysaccharides are macromolecules made up of hundreds or thousands of simple sugar units. **(b)** The two most important polysaccharides are cellulose and starch. Cellulose is the chief structural material of plants. It is used as wood for building, cotton for clothing, and paper. Acetate rayon, viscose rayon, and cellophane are modified forms of cellulose. Starch functions as a reserve food supply in plants. Starch in potatoes, corn, and rice is used as food. **23.15** TTC **23.16** DNA is missing the oxygen at the 2′ position that is present in RNA. DNA contains thymine; RNA contains uracil. **23.17 (a)** two **(b)** three **(c)** guanine and cytosine

23.18

23.19 The complementary triplets of RNA to the glycine-coding CCA, CCG, CCT, and CCC sequences are GGU, GGC, GGA, and GGG, respectively.

23.21 (a) $\sim\!\!\!\sim CH{-}CH_2{-}CH{-}CH_2\!\sim\!\!\!\sim$
            with CH_2CH_3 substituents
$-(CH(CH_2CH_3)CH_2)_n$
(b) $\sim\!\!\!\sim CH_2CH{=}CHCH_2{-}CH_2CH{=}CHCH_2\!\sim\!\!\!\sim$
$-(CH_2CH{=}CHCH_2)_n$
(c) $\sim\!\!\!\sim OC(CH_2)_{10}CONH(CH_2)_6NHOC(CH_2)_{10}CONH(CH_2)_6NH\!\sim\!\!\!\sim$
$-(OC(CH_2)_{10}CONH(CH_2)_6NH)_n$
(d)
$\sim\!\!\!\sim OCH_2(CH_2)_2CH_2OC{-}\langle\text{ring}\rangle{-}COCH_2(CH_2)_2CH_2OC{-}\langle\text{ring}\rangle{-}C\!\sim\!\!\!\sim$
(C's with =O)
$-(OCH_2(CH_2)_2CH_2OC{-}\langle\text{ring}\rangle{-}C)_n$
(C's with =O)

23.25 (a) a C=C double bond **(b)** The monomers must have two functional groups (either two different molecules each with two identical functional groups or one molecule with two different groups).

23.29 (a) Molecules of *fibers* are regularly shaped and have strong intermolecular forces. Molecules of *elastomers* are irregularly shaped and have weak intermolecular forces. Molecules of *plastics* are intermediate between those of fibers and those of elastomers. **(b)** *Thermoplastic plastics* soften when heated, whereas *thermosetting plastics* do not soften when heated.

23.33 (a) $H_2NCH(CH_2CH_2CONH_2)COOH$

(b) $H_2NOCCH_2CH_2$ **(c)** amide

23.37 (a) The *primary structure* is the order in which the amino acids are arranged in a protein or peptide molecule. The *secondary structure* is the way that the chains of amino acids are coiled or folded. The *tertiary structure* is the way that the secondary structures fold and coil up (see Figure 23.7). *Quaternary structure* only exists in proteins containing more than one peptide chain and is the way that peptide chains pack together (see Figure 23.8). **(b)** hydrolysis **(c)** X-ray crystallography and a knowledge of the primary structure **(d)** alpha helix and pleated sheet **(e)** a disorderly arrangement (see Figure 23.7)

23.41 (a) *Nucleic acids* are acidic substances present in the nuclei of cells that store genetic information and direct the biological synthesis of proteins. **(b)** *Nucleotides* are the repeating units of nucleic acids and are composed of one unit each of phosphate, sugar, and one of five heterocyclic bases. **(c)** *Heterocyclic compounds* are cyclic organic compounds in which one or more of the ring atoms is not carbon.

23.45 CCTAGG

23.49 (a) $CH_3COOH + CH_3CH_2CH_2CH_2OH \xrightarrow{H^+}$
$CH_3COOCH_2CH_2CH_2CH_3 + H_2O$

(b) $HOOCCH_2CH_2COOH + HOCH_2CH_2OH \xrightarrow{H^+}$
$HOOCCH_2CH_2COOCH_2CH_2OH + H_2O$

(c) $2CH_3CH(OH)COOH \xrightarrow{H^+}$
$CH_3CH(OH)COOCH(CH_3)COOH + H_2O$

(d) ∿ $OCH(CH_3)COCH(CH_3)COCH(CH_3)C$ ∿
with $\parallel$ O below each

(e) $\pm OCH(CH_3)C \pm_n$ with $\parallel$ O below

23.53 One pure enantiomer of an essential amino acid occurs naturally. A racemic mixture obtained by synthesis does *not* have to be separated to obtain this enantiomer.

23.57 (a) three **(b)** four

23.61 (a) $HOOC-\langle\rangle-COOH$ $H_2N-\langle\rangle-NH_2$

(b)

(c) all of them **(d)** $400\ (4 \times 10^2)$ **(e)** Kevlar does not corrode.

23.65 (a)

(b) P-N bond lengths in the polymer are shorter than P-N single bonds. (This prediction is correct.) **(c)** $-N(CH_3)_2$; its structure is similar to NH_3 **23.69** Gly-Ser-Gly-Ala-Gly-Ala **23.73** Wood is strong because London forces between the long chains of cellulose are very large. There is also considerable hydrogen bonding between chains. The twisting together of the bundles of cellulose chains adds to wood's strength. **23.77** For the 21 amino acid chain, 69 and for the 30 amino acid chain, 96.

CHAPTER 24

24.1 (a) four **(b)** five **24.2** Rhenium would be most likely to resemble Tc because of their similarities in size and electron configuration. **24.3** two **24.4 (a)** $[AuF_4]^-$ **(b)** $K[AuF_4]$
24.5 (a) +3 **(b)** +2 **(c)** +5 **(d)** 0 **24.6 (a)** four
(b) $[Pt(NH_3)_4]Cl_2$; $[PtCl(NH_3)_3]Cl$; $[PtCl_2(NH_3)_2]$
24.7 (a) two **(b)**   **(c)** No.

24.8 (a) six **(b)** +3 **24.9**

24.10 (a) $[CoBr(NH_3)_5]SO_4(aq) + Ba^{2+} \rightarrow$
$BaSO_4(s) + [CoBr(NH_3)_5]^{2+}$
(b) no reaction **(c)** no reaction
(d) $[CoSO_4(NH_3)_5]Br(aq) + Ag^+ \rightarrow AgBr(s) + [CoSO_4(NH_3)_5]^+$
24.11 (a) pentaaquahydroxoaluminum ion **(b)** potassium hexacyanoferrate(II) **(c)** hexacarbonylvanadium
24.12 (a) $[Cr(H_2O)_6]Cl_3$ **(b)** $[Cu(NH_3)_4][PtCl_4]$

24.13 ↑↓ ↑↓ ↑↓

24.14 No. Only three electrons are available.

24.15 ↑↓ ↑↓ ↑↓

24.16 (a) smaller **(b)** increase **(c)** blue end

24.17 ↑↓ ↑ ↑   **24.18** —
↑↓ ↑↓
↑↓
↑↓
↑↓ ↑↓

24.19 (a) three **(b)** three **24.23 (a)** Characteristic properties of transition elements: metallic, high melting and boiling points, high densities, hard and strong. Most transition elements have more than one oxidation number in their common compounds; the oxidation numbers differ by one. Compounds in at least one oxidation state are usually colored. Compounds in at least one oxidation state are usually paramagnetic. Most transition metal ions form complexes readily. **(b)** Melting points, boiling points, and densities increase to a maximum toward the middle of a row in the periodic table and then decrease. **(c)** tungsten **(d)** Outer electron configuration is ns^2 for most transition elements. **(e)** Transition metals are sufficiently unreactive for the free elements to be used and have many desirable properties. For example, some, such as iron, are very strong and/or can be magnetized. Some, such as copper, are good conductors of electricity and thermal energy. Some, such as gold, are unusually unreactive and do not corrode.

24.27 (a) no geometric isomers
(b)

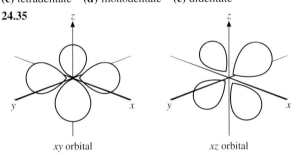

(c)

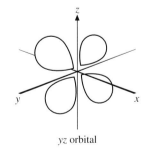

(d) no geometric isomers **24.31 (a)** tridentate **(b)** didentate
(c) tetradentate **(d)** monodentate **(e)** didentate

24.35

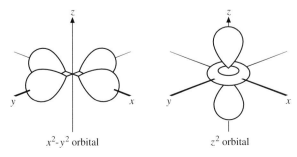

xy orbital   xz orbital

yz orbital

x^2-y^2 orbital z^2 orbital

24.39 (a) The number of unpaired electrons is determined by measuring with a sensitive balance the effect of a strong magnetic field on a sample of the material suspended in the magnetic field. (Remember Figure 8.3.) The more unpaired electrons present, the stronger the attraction of the sample to the magnet.
(b) weak field, high spin = $[Fe(H_2O)_6]^{2+}$;
strong field, low spin = $[Fe(CN)_6]^{4-}$ **(c)** Yes.
24.43 (a) Cyanide ion is at the far right of the spectrochemical series. As a result, the crystal field splitting is large and $[Fe(CN)_6]^{3-}$ is low spin. **(b)** Ruthenium and osmium have higher nuclear charges than iron. The ligands are more strongly attracted to Ru^{3+} and Os^{3+} than to Fe^{3+} and approach the nucleus more closely, producing greater splitting. As a result, no weak field, high spin complexes are known for Ru(III) and Os(III), although many weak field, high spin complexes of Fe(III) are known. **(c)** one
24.47 The ligands vary in strength in the order $Br^- <$ $H_2O <$ en. Thus, the energy absorbed increases from left to right. The wavelength of the light absorbed decreases. Thus, the wavelength of light tranmitted or reflected (the "color" of the compound) increases from violet to blue to green. **24.51** The ion is square planar. Putting 8 e^- in the splitting diagram for a square planar complex results in zero unpaired electrons, whereas putting 8 e^- in the splitting diagram for a tetrahedral complex results in two unpaired electrons.

STOP AND TEST YOURSELF

1. (e) *24.6* **2.** (c) *24.2* **3.** (e) *24.4* **4.** (d) *24.8* **5.** (a) *24.8*
6. (b) *24.8* **7.** (e) *24.8* **8.** (a) *24.8* **9.** (b) *24.8*
10. (a) *24.9*

24.55 Vanadium forms the oxide V_2O_5.
24.59 (a) $[Pt(NH_3)_4][PtCl_4]$: tetraammineplatinum(II) tetrachloroplatinate(II)
(b) $[Pt(NH_3)_4]^{2+} + [PtCl_4]^{2-} \rightarrow [Pt(NH_3)_4][PtCl_4](s)$
(c) $[Pt(NH_3)_4]Cl_2$: tetraammineplatinum(II) chloride
$Na_2[PtCl_4]$: sodium tetrachloroplatinate(II)
24.63 (a) one with $K_f = 5 \times 10^5$ **(b)** one with $K_f = 3 \times 10^8$
(c) one with $K_f = 3 \times 10^{-5}$ **(d)** one with $pK_f = 5.2$
24.67 (a) +7 **(b)** +4 **(c)** +2 **(d)** +6 **(e)** +3
24.71 (a) In the presence of cyanide ion, the spontaneous reaction is $Cu^{2+} + Cu(s) + 4CN^- \rightarrow 2[Cu(CN)_2]^-$. For this cell, $E^\circ_{cell} = +1.12\,V - (-0.44\,V) = +1.56\,V$. In absence of cyanide ion, spontaneous reaction is $2Cu^+ \rightarrow Cu^{2+} + Cu(s)$. For this cell, $E^\circ_{cell} = +0.52\,V - (+0.159\,V) = +0.36\,V$. **(b)** disproportionation
24.75 The aluminum ion is small. Six small fluoride ions can approach the aluminum ion to form an octahedral complex. Chloride ions are larger that fluoride ions, and only four chloride ions can get close enough to form a complex, which is therefore tetrahedral.
24.79 The chemical properties of the transition elements vary from element to element because the configuration of d electrons varies and the sizes of the atoms vary across a period. The lanthanides are similar in chemical properties because the valence electron configuration of the elements is the same, $6s^2$, and they all form a 3+ ion with configuration $4f^n 5d^0 6s^0$. In addition, the lanthanides all have similar radii.

24.83

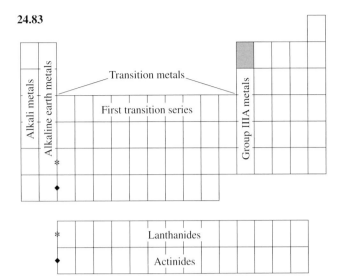

24.87 (a) 515 nm and 625 nm, respectively **(b)** green **(c)** red to red-blue **24.91 (a)** The actual solubility of zinc hydroxide will only be about 1% greater than the solubility calculated from K_{sp}. **(b)** (i) Net ionic equation:
$$Zn(OH)_2(s) + OH^- + H_2O(l) \rightarrow [Zn(OH)_3H_2O]^-$$
Complete ionic equation:
$$Zn(OH)_2(s) + OH^- + Na^+ + H_2O(l) \rightarrow [Zn(OH)_3H_2O]^- + Na^+$$
(ii) Net ionic equation:
$$Zn(OH)_2(s) + 2H^+ \rightarrow Zn^{2+} + 2H_2O(l)$$
Complete ionic equation:
$$Zn(OH)_2(s) + 2H^+ + 2Cl^- \rightarrow Zn^{2+} + 2Cl^- + 2H_2O(l)$$
(iii) Net and complete ionic equation:
$$Zn(OH)_2(s) + 4NH_3(aq) \rightarrow [Zn(NH_3)_4]^{2+} + 2OH^-$$
(c) amphoteric
24.95 Properties of metals: ductile, malleable, good conductors of thermal energy and electricity (electrical conductivity decreases with increasing temperature), lustrous, usually have high melting and boiling points, have relatively low ionization energies.
24.99 708 nm **24.103 (a)** A *complex* is a species in which a central atom is surrounded by a set of Lewis bases that are covalently bonded to the central atom. **(b)** *Coordination compounds* include one or more complexes. **(c)** A *coordinate covalent bond* is a shared electron pair of which both electrons originally belonged to the same atom. **(d)** The *coordination number* is the number of atoms or groups firmly bound to the central atom. **(e)** The *total coordination number* is the total number of atoms and sets of unshared electron pairs around a central atom. **24.107** Oxalic acid removes rust stains because the oxalate ion, $^-OOCCOO^-$, is a bidentate ligand that forms stable chelate complexes with iron(III) ions.

24.111 (a)

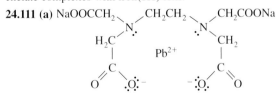

(b) 1.448 g **24.115 (a)** ferromagnetic **(b)** Chromium(IV) oxide and Fe_3O_4 are ferromagnetic. **(c)** CrO_2 **(d)** +8/3
24.119 (a) metals: Cd, Cr, Pb, Hg; semimetals: As, Sb

(b) H_2C — $CHCH_2OH$

The hydrogens on sulfur are more acidic than those on oxygen, so the negative charges in BAL^{2-} are on sulfur.

(c) $HSCH_2\overset{*}{C}H(SH)CH_2OH$ **(d)** Hydrogens bonded to sulfur do not form effective hydrogen bonds, so BAL can only form one-third as many hydrogen bonds as glycerol. Since BAL has a larger molecular mass than glycerol, it is not water soluble.

APPENDIXES

Appendix A

A.1 (a) hypobromite ion **(b)** P^{3-}
A.2 (a) manganese(II) ion **(b)** mercury(II) nitrate
(c) dinitrogen trioxide **(d)** diphosphorus pentoxide **(e)** periodic acid **(f)** potassium iodate **(g)** lithium permanganate trihydrate **(h)** hydrogen bromide **(i)** hydrobromic acid **(j)** cadmium sulfate-water (3/8) **(k)** magnesium sulfate heptahydrate, magnesium sulfate-water (1/7)

Appendix B

B.1 (a) 3.652×10^2 **(b)** 7×10^{-4} **(c)** 1.42×10^{-2}
(d) 2.3652×10^4 **(e)** 7.5×10^0 **B.2 (a)** 0.000 000 74
(b) 39.82 **(c)** 3.27 **(d)** 0.005 81 **(e)** 745 600 000
B.3 Two, three, or four significant figures are possible (3.2×10^3, 3.20×10^3, or 3.200×10^3). **B.4** five significant figures
B.5 490 **B.6** 2.43×10^3 **B.7 (a)** 3×10^5 **(b)** 3.4×10^{13}
(c) 3.7×10^{-13} **(d)** 6.3×10^3 **(e)** 4.3×10^{-11}
B.8 (a) 5.27×10^7 **(b)** 4.31×10^{23} **(c)** 6.2×10^{-13}
B.9 (a) 2.8×10^9 **(b)** 9.00×10^6 **(c)** 4.0×10^{-4}
B.10 (a) 5 **(b)** -4 **B.11 (a)** 10^{-8} **(b)** 10^{15}
B.12 (a) -11.5 **(b)** 16.925 **(c)** 0.0 **(d)** 3.18×10^{-23}
(e) 5×10^5 **B.13 (a)** -26.5 **(b)** 38.972 **(c)** 0.0
d) 1.70×10^{-10} **(e)** 3×10^2 **B.14 (a)** 0.29 **(b)** 22.2
(c) 12.25 **B.15 (a)** 2.5 **(b)** 2.4 **(c)** 1.15
B.16 $x = 2.9 \times 10^{-2}$ and $x = -6.8 \times 10^{-2}$
B.17 (a) 72.46% **(b)** 6.29 g Fe **(c)** 29.2 g
B.18 (a) 250 s **(b)** 45.0 °C **(c)** 2.7 °C
B.19 (a) Density of aqueous ethylene glycol at 20 °C:

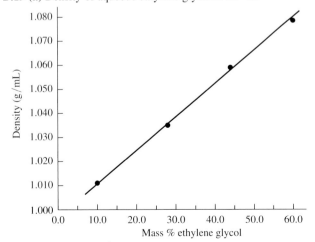

(b) $0.0014 \dfrac{g/mL}{mass\ \%\ ethylene\ glycol}$ **(c)** 1.030 g/mL

B.20 (a) Boiling point of ammonia at different pressures:

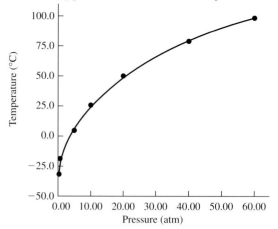

(b) 66.0 °C

Appendix F

F.1 (a) $3S(s) + 4HNO_3(aq) \rightarrow 3SO_2(g) + 4NO(g) + 2H_2O(l)$
(b) $3Cu_2O(s) + 14HNO_3(aq) \rightarrow$
$$6Cu(NO_3)_2(aq) + 2NO(g) + 7H_2O(l)$$
(c) $5Zn(s) + 12HNO_3(aq) \rightarrow 5Zn(NO_3)_2(aq) + N_2(g) + 6H_2O(l)$
(d) $2Al(s) + 2NaOH(aq) + 6H_2O(l) \rightarrow$
$$2NaAl(OH)_4(aq) + 3H_2(g)$$
(e) $C(s) + 2H_2SO_4(l) \rightarrow CO_2(g) + 2SO_2(g) + 2H_2O(l)$
(f) $3C_3H_8O(aq) + 2CrO_3(aq) + 3H_2SO_4(aq) \rightarrow$
$$3C_3H_6O(aq) + Cr_2(SO_4)_3(aq) + 6H_2O(l)$$
(g) $H_3AsO_4(aq) + 4Zn(s) + 8HNO_3(aq) \rightarrow$
$$AsH_3(g) + 4Zn(NO_3)_2(aq) + 4H_2O(l)$$
(h) $4H_2O(l) + 6KI(aq) + 2KMnO_4(aq) \rightarrow$
$$3I_2(s) + 2MnO_2(s) + 8KOH(aq)$$
F.2 (a) $Cr_2O_7^{2-} + 3C_2O_4^{2-} + 14H^+ \rightarrow$
$$2Cr^{3+} + 6CO_2(g) + 7H_2O(l)$$
(b) $3H_2O_2(aq) + 2MnO_4^- \rightarrow$
$$3O_2(g) + 2MnO_2(s) + 2OH^- + 2H_2O(l)$$
(c) $3AsH_3(g) + 4ClO_3^- \rightarrow 3H_3AsO_4(aq) + 4Cl^-$
(d) $Cu(NH_3)_4^{2+} + S_2O_4^{2-} + 4OH^- \rightarrow$
$$Cu(s) + 2SO_3^{2-} + 4NH_3(aq) + 2H_2O(l)$$
(e) $TlOH(s) + 2NH_2OH(aq) \rightarrow Tl(OH)_3(s) + N_2H_4(aq)$
(f) $4MnO_2(s) + 3O_2(g) + 4OH^- \rightarrow 4MnO_4^- + 2H_2O(l)$
(g) $S_2O_3^{2-} + 4Cl_2(g) + 5H_2O(l) \rightarrow 2HSO_4^- + 8Cl^- + 8H^+$
F.3 (a) $2MnO_4^- + 5H_2C_2O_4(aq) + 6H^+ \rightarrow$
$$2Mn^{2+} + 10CO_2(g) + 8H_2O(l)$$
The element oxidized is carbon, and $H_2C_2O_4$ is the reducing agent.
The element reduced is Mn, and MnO_4^- is the oxidizing agent.
(b) $4Zn + 10H^+ + 2NO_3^- \rightarrow 4Zn^{2+} + N_2O(g) + 5H_2O(l)$
The element oxidized is Zn, and Zn is the reducing agent.
The element reduced is N, and NO_3^- is the oxidizing agent.
(c) $As_2O_3(s) + 2NO_3^- + 2H_2O(l) + 2H^+ \rightarrow$
$$2H_3AsO_4(aq) + N_2O_3(aq)$$
The element oxidized is As, and As_2O_3 is the reducing agent.
The element reduced is N, and NO_3^- is the oxidizing agent.
(d) $HIO_3(aq) + 8I^- + 5H^+ \rightarrow 3I_3^- + 3H_2O(l)$
The element oxidized is I, and I^- is the reducing agent.
The element reduced is I, and HIO_3 is the oxidizing agent.
(e) $3ReCl_5(s) + 8H_2O(l) \rightarrow$
$$HReO_4(aq) + 2ReO_2(s) + 15Cl^- + 15H^+$$

The element oxidized is Re, and $ReCl_5$ is the reducing agent.
The element reduced is Re, and $ReCl_5$ is the oxidizing agent.
(f) $Zn(s) + 2VO^{2+} + 4H^+ \rightarrow Zn^{2+} + 2V^{3+} + 2H_2O(l)$
The element oxidized is Zn, and Zn is the reducing agent.
The element reduced is V, and VO^{2+} is the oxidizing agent.
F.4 (a) $C_3H_8O_3 + 20OH^- \rightarrow 3CO_3^{2-} + 14H_2O + 14e^-$
(b) $MnO_4^- + 1e^- \rightarrow MnO_4^{2-}$ **(c)** C is oxidized.
(d) Mn is reduced. **(e)** MnO_4^- is the oxidizing agent.
(f) $C_3H_8O_3$ is the reducing agent.
F.5 (a) $2Cr(OH)_6^{3-} + 3BrO^- \rightarrow$
$$2CrO_4^{2-} + 3Br^- + 5H_2O(l) + 2OH^-$$
The element oxidized is Cr, and $Cr(OH)_6^{3-}$ is the reducing agent.
The element reduced is Br, and BrO^- is the oxidizing agent.
(b) $3H_2O_2(aq) + Cl_2O_7(aq) \rightarrow 2ClO_2(aq) + 3O_2(g) + 3H_2O(l)$
The element oxidized is O, and H_2O_2 is the reducing agent.
The element reduced is Cl, and Cl_2O_7 is the oxidizing agent.
(c) $3CN^- + 2CrO_4^{2-} + 5H_2O(l) \rightarrow$
$$3CNO^- + 2Cr(OH)_4^- + 2OH^-$$
The elements oxidized are C and N, and CN^- is the reducing agent.
The element reduced is Cr, and CrO_4^{2-} is the oxidizing agent.
F.6 (a) $34K^+ + 14MnO_4^- + 20OH^- + C_3H_8O_3(aq) \rightarrow$
$$34K^+ + 14MnO_4^{2-} + 3CO_3^{2-} + 14H_2O(l)$$
(b) $14KMnO_4(aq) + 20KOH(aq) + C_3H_8O_3(aq) \rightarrow$
$$14K_2MnO_4(aq) + 3K_2CO_3(aq) + 14H_2O(l)$$
F.7 (a) $C_6H_5OH(aq) + 14SO_3(g) \rightarrow$
$$6CO_2(g) + 14SO_2(g) + 3H_2O(l)$$
(b) $10FeSO_4(aq) + 2KMnO_4(aq) + 8H_2SO_4(aq) \rightarrow$
$$2MnSO_4(aq) + 5Fe_2(SO_4)_3(aq) + K_2SO_4(aq) + 8H_2O(l)$$
(c) $8HNO_3(aq) + 3Cu(s) \rightarrow 3Cu(NO_3)_2(aq) + 2NO(g) + 4H_2O(l)$
(d) $2NaCrO_2(aq) + 3NaClO(aq) + 2NaOH(aq) \rightarrow$
$$2Na_2CrO_4(aq) + 3NaCl(aq) + H_2O(l)$$
(e) $H_2O(l) + 3Na_2SO_3(aq) + 2NaMnO_4(aq) \rightarrow$
$$3Na_2SO_4(aq) + 2MnO_2(s) + 2NaOH(aq)$$
F.8 $3CuS(s) + 8HNO_3(aq) \rightarrow$
$$3S(s) + 3Cu(NO_3)_2(aq) + 2NO(g) + 4H_2O(l)$$

Appendix G

G.1 1.0×10^{-5}

Appendix H

H.1 No.
H.2 (a)

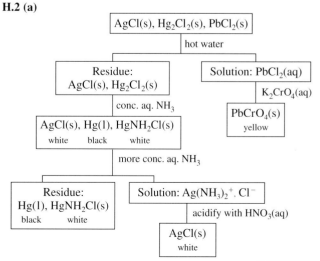

(b) No yellow precipitate will be observed when potassium chromate is added to solution from hot-water extraction. Residue will turn black when treated with concentrated aqueous ammonia. Solution from treatment of black residue with more concentrated aqueous ammonia will give a white precipitate of silver chloride on acidification with nitric acid.

H.3 (a) Use flame test. Sodium ion gives yellow color that is invisible when flame is viewed through blue glass; potassium ion gives a purple-violet flame when the flame is viewed through blue glass.   (b) Treat with room temperature water. If solid dissolves, it is KCl. If it does not dissolve, it is $PbCl_2$.   (c) If solution is blue, it is $Cu(NO_3)_2(aq)$; if it is colorless, it is $Ca(NO_3)_2$.   H.4 (a) CoS

Boldface number indicates a definition and "n" indicates a footnote.

Pa **150**
Paints, water-based (latex) 931
Paleontology 806
Palladium 259, 283, 942
  catalyst 687, 698
  cathode, and cold
    fusion 788n
PAN 377, 713
Paper chromatography 494,
  495
Paper production and
  processing 333
Paracelsus 7
Parallel-connected cells 736
Paramagnetism **257**, 956
  of oxygen 359, 364, 823
  of transition elements 941
Parent, of radioactive decay
  series **776**
Parent chain **875**–76, 882
Parentheses 18
Parkinson's disease 868
Partial charge 308, 344
Partial pressure **168**–70
  calculation from mole
    fraction 475
  of carbonated beverages 841
  Dalton's law of 168–70
  equilibrium constant (K_p)
    518–19, 526–28
  standard cell potential 720
Particle(s)
  colloidal **490**
  precipitation from smoke 496
  properties of electromagnetic
    radiation 227–30
  size, and reaction rates 686
  subatomic, symbols for 765
Particle–particle reactions **765**
Parts per billion (ppb) **126**
Parts per million (ppm) **126**
Pascal, Blaise 150n
Pascal (Pa) **150**
Pasteur, Louis 868, 868n
Patina 757
Pauli, Wolfgang 244
Pauli exclusion principle **244**,
  337, 348, 359, 360
Pauling, Linus 308n
PCBs 711
Pencil "lead" 5, 437, 808
Penetrate **243**, 256
Penicillin G 840n
Pentaamminenitritocobalt(III)
  ion 949
Pentaamminenitrocobalt(III)
  ion 949
Pentadecane 902

Pentane 462, 860, 873
Pepsin 489
Peptide link **915**–16, 918n
Peptides **915**
  functions 917
  primary structure **916**–17
  synthesis 920
per- prefix 387
Percent **A14**–A15
  composition 124–25
    empirical formula from 91
    from formulas 96
  concentration 124–25
  converting to molarity
    132–33
  hydrolysis 624
  mass 88–89, 125
  volume 125
  yield **87**
Percent ionization 559–60, 565
  and acid concentration 563
  and acid ionization constant
    559–60, 563
  and acid strength 563
  and base ionization constant
    565–66
Perchloric acid 387, 607
Perfect wave 223
Perfluorodecalin 823
Perfluorotripropylamine 823
Period, in periodic table
  12, 254
Periodic chart *See* Periodic
  table
Periodic law 276
Periodic repeating systems 661
Periodic table **10**–13
  and acidity 605–12
  alternative group numbering
    system 285–86
  and atomic radii 266–68
  and basicity 610–12
  bond energies 870
  and chemical properties
    279–85
  and electron affinity **274**–78
  and electron configuration
    258–62
  and electronegativity 308
  and electronic structure
    253–88
  expanded form of 260
  and first ionization energies
    269–72
  and Mendeleev 276–77
  metal order of reactivity
    119–20
  and Meyer 292

and molecular orbital energy
  level diagrams 362
  nonmetal reactivity 121
  and number of valence
    electrons 299
  and oxidation numbers
    385–86
  and predicting chemical
    changes 25
  and silicon 847
  and standard entropies
    644–45
  and transition metal
    properties 937, 938–41
Permanganate 283, 938
  absorption spectrum 370
  as oxidizing agent 723
  reaction with chloride ion
    725–26
  titration 393–95
  ultraviolet-visible
    spectrum 370
Peroxides, oxygen-oxygen
  bond 311
Peroxyacetylnitrate (PAN)
  377, 713
Perrin, Jean-Baptiste 174
Perutz, Max 446
Pervanadyl sulfate 941
Pesticides 37
PET 760
Petrochemical industry 893
Petroleum 892–93
  barrels unit 902
  combustion of 660
    environmental effects 208
  consumption 208
  cracking 687, 874, 892
  refining 892–93
  uses 892, 893
Petrology 488n
Pewter 962
pH **556**–58
  of blood 575, 590, 597, 598
  and buffer solutions
    575–78
  of common liquids 557
  at equivalence point
    strong acid–strong base
      titration 583
    weak acid–strong base
      titration 584, 585
  and hemoglobin oxygen-
    carrying capacity 575
  Henderson-Hasselbalch
    equation 578
  of lakes 617
  of rain 616

range of indicators
  579–80, 581
  of soft drinks 838
  of stomach acid 598
  of strong acids 558–59
  titration curve **580**
  and titration of strong acid
    with strong base 582–84
pH meter 557, 586
  electrodes 732
  and voltaic cells 732
pH test paper 557, 580
Pharmaceutical industry 903–4
Phase **3**
  boundaries, and reaction
    rates 686
  diagrams **431**–34
  heterogeneous mixtures 6
  symbols for reactions 24
Phenacetin 901
Phenobarbital 886
Phenol 108, 638
  functional group 885
Phenolphthalein 138, 580,
  738, 739
Phenylketonuria (PKU) 334
Phosgene 333
Phosphanes 870
Phosphate 836
Phosphate rock 836, 852–53
Phosphine 831
Phosphoric acid 387, 549,
  607–8, 808, 837–838
  in body 590
  in carbonated beverages 841
  ionization 587
  manufacture of 839
  salts of 838
Phosphorous acid 607–8
Phosphorus 13, 807, 836–39
  content in phosphate rock
    852–53
  in matches 837n
  occurrence 836
  radioisotopes 803
  reactivity 13
  red 837
  sources of 836
  uses 836, 837n
  white 836–37
Phosphorus-30 765
Phosphorus cycle 838–39
Phosphorus dichloride 334
Phosphorus pentafluoride 321
Phosphorus pentoxide 837, 838
Phosphorus trichloride
  313, 334
Photochemical reactions 251

PHOTO CREDITS

Education Development Center, Inc., Newton, MA; **7.29** CDC/RG/Peter Arnold, Inc.; **p. 245** Carl Anderson/Science Photo Library/Photo Researchers, Inc.

CHAPTER 8

p. 253 Mark Harwood/Tony Stone; **8.15** University of Pennsylvania; **p. 280** Mark C. Burnett/Stock Boston; **p. 283, top** Barry L. Runk/Grant Heilman Photography; **8.16** E. R. Degginger; **p. 284** Richard Megna/Fundamental Photographs; **p. 291, left and right** Joel Gordon

CHAPTER 9

p. 297 Courtesy of Bufftech; **p. 298, top** Northern Natural Gas Company; **p. 298, bottom** Visuals Unlimited; **p. 299** E. R. Degginger; **9.1** E. R. Degginger; **p. 304** Nimtallah/Art Resource, NY. Cellini, The Salt Cellar, Vienna, Kunsthistorisches Museum; **9.8** E. R. Degginger; **9.10** Richard Megna/ Fundamental Photographs; **p. 325, top** Fred Bavendam/Peter Arnold, Inc.; **p. 325, bottom** Richard Megna/Fundamental Photographs; **p. 326** Linda Bartlett/Photo Researchers, Inc.; **p. 328** Willard Clay/ Tony Stone

CHAPTER 10

p. 336 Dr. Jeremy Burgess/Science Photo Library/Photo Researchers, Inc.; **10.1** Phil Degginger; **p. 345, top and bottom** Joel Gordon; **p. 347** Richard Megna/Fundamental Photographs; **10.29** Richard Megna/Fundamental Photographs; **10.33** E. R. Degginger; **10.39b** E. R. Degginger; **p. 371** Milton Roy Company, Rochester, NY

CHAPTER 11

p. 380 John Neubauer/Photo Edit; **p. 381** John D. Cunningham/Visuals Unlimited; **p. 386** E. R. Degginger; **p. 388** E. R. Degginger; **p. 391** E. R. Degginger; **11.2a–b** Richard Megna/Fundamental Photographs; **p. 394** E. R. Degginger; **p. 396** Firth Photobank; **p. 397** AP Wide World; **11.5b** Brownie Harris/The Stock Market; **11.5c** Courtesy of Bethlehem

Steel; **11.6, bottom** H. P. Merten/The Stock Market

CHAPTER 12

p. 410 James Randklev/Tony Stone; **p. 411, top right** John Shaw/Bruce Coleman Inc.; **p. 411, bottom right** Steve McCutcheon/Visuals Unlimited; **12.1a–c** Richard Megna/Fundamental Photographs; **12.3a** Tom Harm/Quest Photographics; **12.3b** Richard Megna/Fundamental Photographs; **12.4** Phil Degginger; **12.6a** Paul Silverman/Fundamental Photographs; **12.6b** E. R. Degginger; **12.6c** Fundamental Photographs; **12.6d** Arthur Hill/Visuals Unlimited; **12.6e** Grant Heilman Photography; **12.15** E. R. Degginger; **12.16a–b** E. R. Degginger; **12.16c–d** Phil Degginger; **12.17** E. R. Degginger; **12.20a** E. R. Degginger; **12.20b** Phil Degginger; **12.21** Richard Megna/Fundamental Photographs; **12.23** Richard Megna/Fundamental Photographs; **12.25** Richard Megna/Fundamental Photographs; **12.26** Barry Iverson, Time Magazine; **p. 430** Richard Megna/Fundamental Photographs; **12.31b** E. R. Degginger; **p. 434, top and bottom** E. R. Degginger; **p. 435, top** Richard C. Walters/Visuals Unlimited; **p. 435, center and bottom** E. R. Degginger; **12.34a** Cabisco/Visuals Unlimited; **12.35a** John Cancalosi/Peter Arnold Inc.; **p. 437** Richard Megna/Fundamental Photographs; **12.37b** Courtesy of Bethlehem Steel Company; **12.39a** John D. Cunningham/Visuals Unlimited; **12.39b** Wendell Metzen/Bruce Coleman Inc.; **p. 440, lower left** Photo courtesy of Y. Z. Li, M. Chandler, J. C. Patrin, J. H. Weaver, Department of Materials Science and Chemical Engineering, University of Minnesota. From *Science,* 25 July 1991; **p. 443, top and bottom** NASA; **12.46b** E. R. Degginger; **12.46d** Richard C. Walters/Visuals Unlimited; **12.47a** Kodak RL/Visuals Unlimited; **12.47b** Professor Haikon Hope, University of California at Davis; **p. 446** Photo Courtesy Siemens Energy and Automation, Inc., Analytical Instrumentation; **p. 447** Richard Megna/Fundamental Photographs; **p. 448** Grant Heilman Photography; **12.48** Photo courtesy of Digital Instruments, Inc.; **12.50** Photo courtesy of IBM Corporation, Research Division, Almaden

Research Center; **p. 451** Charles Falco/Photo Researchers, Inc.

CHAPTER 13

p. 460 Peter Aprahamian/Science Photo Library/Photo Researchers, Inc.; **13.1a–c** Richard Megna/Fundamental Photographs; **13.2a–c** Richard Megna/Fundamental Photographs; **p. 466** Richard Megna/Fundamental Photographs; **p. 469** Richard Megna/Fundamental Photographs; **p. 472** E. R. Degginger; **13.9** E. R. Degginger; **13.10a–b** E. R. Degginger; **13.12** Richard Megna/Fundamental Photographs; **p. 485** LINK/Visuals Unlimited; **13.15a–b** E. R. Degginger; **13.17** Richard Megna/Fundamental Photographs; **p. 493, top and bottom** E. R. Degginger; **13.22a–d** Richard Megna/Fundamental Photographs; **13.23** E. R. Degginger; **13.24** E. R. Degginger; **p. 502** Richard Megna, Fundamental Photographs, from Greenberg, F. H., *J. Chem. Educ.* **1992,** *69,* 654; **p. 503** Richard Megna, Fundamental Photographs, from Greenberg, F. H., *J. Chem. Educ.* **1992,** *69,* 654

CHAPTER 14

p. 505 Gary Ladd/Photo Researchers, Inc.; **14.6** Richard Megna/Fundamental Photographs; **14.7** Richard Megna/Fundamental Photographs; **p. 537** Mike Schmitt, Soil Scientist, University of Minnesota; **p. 538** Galen Rowell/Peter Arnold, Inc.

CHAPTER 15

p. 549 Tony Stone; **p. 550** Leonard Lessin/Peter Arnold Inc.; **15.1** Richard Megna/Fundamental Photographs; **p. 552** Richard Megna/Fundamental Photographs; **p. 554** E. R. Degginger; **15.2** Richard Megna/Fundamental Photographs; **15.3** Richard Megna/Fundamental Photographs; **p. 565** Leonard Lessin/Peter Arnold Inc.; **15.7** E. R. Degginger; **15.8a–c** Richard Megna/Fundamental Photographs; **p. 575, bottom** Peter Arnold, Inc.; **15.9** E. R. Degginger; **15.10** Courtesy of Micro Essential Laboratories; **15.11** E. R. Degginger; **15.14** Courtesy of Brinkmann Instruments; **p. 590** CNRI/Science Photo Library/Photo Researchers Inc.

Frequently Used Formula Masses

Name	Formula	Formula Mass	Name	Formula	Formula Mass
Acetic acid	CH_3COOH	60.053	Hydrogen peroxide	H_2O_2	34.0147
Ammonia	NH_3	17.030 56	Hydrogen sulfide	H_2S	34.082
Ammonium chloride	NH_4Cl	53.4912	Iodine	I_2	253.808 94
Ammonium nitrate	NH_4NO_3	80.0434	Iron(III) oxide	Fe_2O_3	159.692
Ammonium sulfate	$(NH_4)_2SO_4$	132.141	Magnesium oxide	MgO	40.3044
Benzoic acid	C_6H_5COOH	122.123	Methane	CH_4	16.043
Bromine	Br_2	159.808	Methanol	CH_3OH	32.042
Butane	$CH_3CH_2CH_2CH_3$	58.123	2-Methylpropane	$(CH_3)_3CH$	58.123
Calcium carbonate	$CaCO_3$	100.087	Nitric oxide	NO	30.0061
Calcium chloride	$CaCl_2$	110.983	Nitrogen	N_2	28.013 48
Calcium hydroxide	$Ca(OH)_2$	74.093	Oxygen	O_2	31.9988
Calcium oxide	CaO	56.077	Pentane	$CH_3(CH_2)_3CH_3$	72.150
Carbon dioxide	CO_2	44.010	Phosphorus	P_4	123.895 048
Carbon monoxide	CO	28.010	Potassium chlorate	$KClO_3$	122.5492
Chlorine	Cl_2	70.9054	Potassium hydrogen phthalate	$KHC_8H_4O_4$	204.224
Chloromethane	CH_3Cl	50.488	Potassium iodide	KI	166.0028
Copper(I) sulfide	Cu_2S	159.158	Potassium nitrate	KNO_3	101.1032
Copper(II) sulfate	$CuSO_4 \cdot 5H_2O$	249.686	Propane	$CH_3CH_2CH_3$	44.097
Ethanol	CH_3CH_2OH	46.069	Sodium chloride	$NaCl$	58.4425
Fluorine	F_2	37.996 8064	Sodium hydroxide	$NaOH$	39.9971
Glucose	$C_6H_{12}O_6$	180.158	Sulfur dioxide	SO_2	64.065
Heptane	$CH_3(CH_2)_5CH_3$	100.204	Sulfur trioxide	SO_3	80.064
Hexane	$CH_3(CH_2)_4CH_3$	86.177	Sulfuric acid	H_2SO_4	98.079
Hydrogen	H_2	2.015 88	Urea	H_2NCONH_2	60.056
Hydrogen chloride	HCl	36.4606	Water	H_2O	18.0153

Where to find it:

Rules for Using SI Section 2.1
List of Weak and Strong Electrolytes Table 4.3
Solubility Rules for Inorganic Compounds Table 4.4
Vapor Pressure of Water Vapor Table 5.4 and Appendix C
Distribution of Molecular Speeds Figure 5.16
Electromagnetic Spectrum Figure 7.15
Electron Configurations Figure 8.4
Atomic Radii Figure 8.7
Ionic Radii Figure 8.9
First Ionization Energies Figure 8.10
Successive Ionization Energies Table 8.1
Electron Affinities Figure 8.13
Electronegativities Figure 9.3
Bond Energies Table 9.4
Shapes of Molecules Based on VSEPR Table 10.1
Rules for Assigning Oxidation Numbers Section 11.1
Important Oxidation Numbers of the Elements in Their Compounds Figure 11.1

Acid Ionization Constants, K_a: Monoprotic Acids Table 15.3
Acid Ionization Constants, K_a: Polyprotic Acids Table 15.7
Acid Ionization Constants, K_a: Hydrated Cations Table 16.3
Base Ionization Constants, K_b: Table 15.4
Acid-base Indicators Table 15.6
Formation Constants for Complex Ions Table 16.4
Solubility Product Constants, K_{sp} Table 16.6
Rate Laws for Common Reaction Orders Table 18.3
Distribution of Kinetic Energies Figure 18.10
Standard Reduction Potentials Table 19.1
Geometric Formulas Appendix B.7
Nomenclature Rules: Simple Inorganic Appendix A
Nomenclature Rules: Complexes Section 24.6
Nomenclature Rules: Organic Sections 22.6 and 22.7
Thermodynamic Data Appendix D
Top 50 Chemicals Appendix E

SI Prefixes

a	f	p	n	μ	m	c	d	da	h	k	M	G	T	P	E
atto	femto	pico	nano	micro	milli	centi	deci	deka	hecto	kilo	mega	giga	tera	peta	exa
10^{-18}	10^{-15}	10^{-12}	10^{-9}	10^{-6}	10^{-3}	10^{-2}	10^{-1}	10^{1}	10^{2}	10^{3}	10^{6}	10^{9}	10^{12}	10^{15}	10^{18}